Transistor Applications from RF to Microwave Frequencies

Transistor Applications from RF to Microwave Frequencies

Theory of SiGe HBTs, pHEMT and CMOS and Practical Circuits up to Sub-terahertz Frequencies

Ulrich L. Rohde

Professor of Radio Frequency and Microwave Techniques
Brandenburg University of Technology Cottbus-Senftenberg
Cottbus, Germany

Professor of Microwave Systems, Faculty of Technical Informatics
Universität der Bundeswehr (Federal University of the Armed Forces)
Munich, Germany

Visiting Scientist, MIT Microsystems Technology Laboratories, MA, USA

Ajay K. Poddar

Chief Scientist, Synergy Microwave Corp., Paterson, NJ, USA

Matthias Rudolph

Chair of Radio Frequency and Microwave Techniques
Brandenburg University of Technology Cottbus-Senftenberg
Cottbus, Germany

WILEY

Published by John Wiley & Sons, Inc., Hoboken, New Jersey.

Library of Congress Cataloging-in-Publication Data has been applied for:

Print ISBN: 9781394162000
ePDF ISBN: 9781394162017
ePUB ISBN: 9781394162024
oBook ISBN: 9781394162031

Cover Design: Wiley
Cover Image: Courtesy of Ulrich L. Rohde

Set in 9.5/12.5pt STIXTwoText by Lumina Datamatics, Madurai, India

SKY10169031_092126

Contents

Foreword by Rüdiger Quay

Advanced microwave transistors have always been the powerhouses behind progress in radio frequency (RF) and microwave applications. Since the first commercial products were created in the 1960s, their performance has been a critical factor that has dominated system performance in communications, sensing, and analog design. Ever since that time, engineers have exploited state-of-the-art performance on the circuit, subsystem, and system levels using the latest combination of devices, integrated circuits, and approaches on the basis of heterogeneous integration.

It is the genuine nature of RF and microwave applications that the different functions of a transceiver structure are critically determined through the "right" mix of technologies along with individual performance at the highest level. This book addresses this specific feature and is thus devoted to questions of design and advancement in microwave and mm-wave technologies, including a deep focus on applications involving computer-aided design and related techniques. For each application, an individually optimized "mix" of technology and specific component features is needed. Thus, apart from learning about the transistors and devices, the book is devoted to the latest combinations of advanced design techniques for their application.

The three authors from very different backgrounds in academia, industry, and research have again created an innovative overview of relevant and diligently chosen examples for the whole spectrum of RF and microwave to terahertz (THz) frequencies, as relevant in the third decade of the twenty-first century.

The book specifically includes aspects of conventional passive elements and resonators with their evolution, and the status of microwave transistors as lumped versus distributed elements. This is mirrored from the perspective of typical system specifications, which are summarized with great completeness for their application. Another chapter gives a good overview of all the relevant high-performance transistor technologies and their modeling. For specific topologies, quality-factor–enhanced passive microwave resonators are addressed with new technological approaches included. Quality-factor–enhanced planar resonators with active functions are also given, again for low phase-noise oscillator circuits. From a different point of view, low-loss micro electro mechanical systems (MEMS) switches with metamaterial contacts are discussed for the specific applications in switched-band approaches. The important topic of amplifier designs up to the sub-THz range is discussed, including wideband and power amplifiers with a great variety of examples for various output power levels and frequencies. Mixers and their different topologies are given, especially with respect to their noise performance. Great effort is spent to explain complex RF and microwave oscillators and their predictive realization, again in combination with the latest low-noise (hetero) approaches. It is a key feature of the book to also explain modern simulation program with integrated circuit emphasis (SPICE)-based nonlinear computer aided design (CAD) tools, operating in the frequency and time domain ("harmonic balance method"), and to verify their usage for the various examples given. In order to build trust in the planar as well as the 2.5D simulators, a full 3D simulation validation (HFSS program) was performed for resonance structures up to 1 THz. This assurance of correct simulation is important for new users as well as for experts.

I am confident that motivated readers and designers all over the globe will make good use of this handy set of examples to learn about the latest advances in design on the circuit and subsystem level in preparation for a new era of hetero-integration of the latest technologies. This will enable the most advanced systems in the coming decades, allowing continuous further development based on even more sophisticated combinations of the techniques described in this book.

Freiburg im Breisgau, Germany
Prof. Dr. Rüdiger Quay
Fritz-Hüttinger Professor for
Energy-Efficient High-Frequency Electronics
University of Freiburg
Fraunhofer Institute for Applied Solid-State Physics

Foreword by Peter Russer

The transistor is probably one of the most influential inventions of the last century. It enabled the realization of huge networks of telecommunications systems, computers and the internet and the rapid development of the information age [1, 2]. As a basic and key technology, semiconductor electronics is of fundamental importance for large areas of industrial production and value creation. It is a future-oriented and pace-setting technology with extensive significance for technological development in many areas of industrial production. In view of the importance that electronics will have in the future in the areas of communication, information processing, automatization, energy efficiency, mobility, medical technology, and environmental protection, its importance as a key technology can hardly be overestimated.

Let us take a look back at the development of semiconductor technology, which began over 150 years ago, in 1874 with the discovery of unipolar conduction at metal–semiconductor contacts by Ferdinand Braun in the physics laboratory of the St. Thomas School in Leipzig [3, 4]. It is worth mentioning that the St. Thomas School is the school where Johann Sebastian Bach worked as Thomas Cantor of Church Music in Leipzig from 1723 until his death in 1750 [5]. He was the head of the leading cantorate in Protestant Germany and is regarded as one of the greatest composers in the history of Western music. The memory of these important creators in art and science, Johann Sebastian Bach, and Ferdinand Braun, is still kept alive today in the St. Thomas School in Leipzig.

We can trace the beginning of semiconductor electronics to Ferdinand Braun's experiments with copper–galena junctions [6–8]. The first to realize a semiconductor rectifier for the detection of electromagnetic waves was Jagadish Chandra Bose, Professor of Physics at the Presidency College in Kolkata, India. He used galena crystals contacted by a metal wire to detect millimeter waves. He reported this invention in the April 27, 1899, meeting of the Royal Society, London [9], and filed a U.S. patent in 1901 for the point-contact semiconductor rectifier for detecting radio signals [10]. Applying this detector and the coupled resonant circuits introduced by Ferdinand Braun, Guglielmo Marconi succeeded in the first transatlantic radio transmission in 1901 [4, 11, 12].

It took more than 50 years from Ferdinand Braun's discovery until the work of Arnold Sommerfeld and Felix Bloch in 1928 led to an understanding of the electronic properties of solids based on quantum theory [13, 14]. Based on this, Walter Schottky developed a theory of junction and peak rectifiers, which is fundamental to the understanding of all modern semiconductor components [15]. The breakthrough in solid-state electronics came with the transistor. Julius Edgar Lilienfeld developed concepts of field-effect transistors at the University of Leipzig in the mid-nineteen twenties. For those, he was granted a patent in the USA; however, he never realized them [16–18]. At Bell Labs, John Bardeen, William Shockley, and Walter Brattain developed and realized the first bipolar transistor and applied for a patent on June 17, 1948 [19]. Independently, Heinrich Welker and Herbert Franz Mataré developed a bipolar transistor in France and applied for a patent on August 13, 1948 [20]. The first bipolar transistors were point-contact transistors [7]. These were replaced in the next development step in 1951 by the junction transistor, developed by the group around Gordon Teal, Morgan Sparks, and William Shockley at Bell Labs [21, 22]. In 1955, after seven decades of the dawn of semiconductor technology, Texas Instruments launched the first transistor-based product, the Regency TR-1 transistor radio [23]. The Regency TR-1 was a success in many ways: 100,000 units were sold in its first year and it contributed to the boom in the transistor market. The word "transistor" now become a social term. The transistor made radios portable and transformed radio listening into a mobile phenomenon, embraced by a new youth culture [24].

Initially, the application areas of bipolar transistors were severely limited due to their low transit frequencies [25]. The small size, low energy consumption, and increasing transit frequencies of transistors due to technological progress meant that from the 1960s onwards, electron tubes were replaced by transistors in almost all electronic systems [26–28]. In 1959, Martin M. Atalla and Dawon Kahng at Bell Labs developed the first metal–oxide–semiconductor field-effect transistor

(MOSFET), which forms the basis of CMOS technology [29]. The MOSFET enables the development of highly integrated circuits thanks to its scalability, low power consumption, and high feature density.

The SiGe heterobipolar transistor (HBT) with an emitter with a larger energy gap than the base has enabled the breakthrough of Si-based monolithic microwave integrated circuit (MMIC) technology [30]. The principle of a transistor with a higher emitter band gap was already mentioned in William Shockley's original patent [31]. The HBT was first proposed by Alfons Hähnlein [32]. In July 1954, Herbert Kroemer formulated the concept of the broadband emitter transistor [33, 34]. The practical implementation failed due to problems of lattice mismatch. To solve this problem, Erich Kasper and Peter Russer proposed an HBT based on a monocrystalline silicon–germanium mixed-crystal system in 1977 [35]. Based on this principle, the first SiGe HBT was realized by IBM researchers in 1987 [36, 37]. Later, SiGe HBTs with transit frequencies and maximum oscillation frequencies above 300 GHz and monolithically integrated millimeter-wave circuits based on them were developed by several groups [30, 38, 39].

The higher charge carrier density compared to vacuum electronics enabled miniaturization, which then experienced rapid development through the monolithic integration of circuits. The planar semiconductor technologies developed in the 1960s enabled the rapid development of the integration of semiconductor components. The first patent for an integrated circuit was filed by Werner Jacobi and was submitted by Siemens & Halske on April 15, 1949 [40]. Patent claim 1 of this invention reads: "Semiconductor amplifier, characterized in that several electrode systems acting in different switching and amplifier stages are placed on the semiconductor." Jacobi's patent remained largely unknown and was not used commercially. The first integrated circuit with only two transistors was realized in 1958 by Jack S. Kilby at Texas Instruments [41, 42]. Kilby's invention was not a monolithic integrated circuit (IC), but a hybrid integrated circuit, since the components were still connected with external wires. Robert Noyce invented the first real monolithic IC chip six months after Kilby [43]. His chip was made of silicon, while Kilby's chip was made of germanium. The most important substrate material for integrated circuits today is silicon. Very large-scale integration (VLSI) technology makes it possible to accommodate computing power on a single chip that exceeds that of mainframe systems of the 1960s by several orders of magnitude. Over the last 60 years, the development of microelectronics has followed Moore's Law, an empirical law formulated by Intel founder Gordon Moore in 1965 that predicts that the component density and performance of monolithic silicon integrated circuits doubles approximately every 18 months [44–47]. In the period from 1968 to 2022, the feature sizes of MOS transistors decreased from 20 μm to about 3 nm. The component density is inversely proportional to the square of the feature size. The period for doubling the component density in this case is 25.5 months. As the size of the components has decreased, it has become possible to accommodate increasingly complex circuits with an ever-increasing number of circuit elements on the chip, despite the chip area remaining constant. Modern integrated circuits such as microprocessors can contain several billion transistors and other circuit elements.

In 1865, James Clerk Maxwell with his publication "A Dynamical Theory of the Electromagnetic Field" established the analytic and conceptual framework of electromagnetism, and in 1888, Heinrich Hertz generated, transmitted and received electromagnetic waves with wavelengths of 0.5 and 5 m [48–56]. Based on Maxwell's theory, extensive and powerful mathematical tools have been developed for the technical application of electromagnetic waves [57–69]. The development of microwave technology from Hertz's time to the second third of the twentieth century encompassed applications such as communications, radar, heating, and medical applications [70]. Later, large areas of electronics evolved to ever higher frequencies. The clock frequencies of digital systems have now advanced into the millimeter-wave range. In November 1971, the world's first single-chip microprocessor, the Intel 4004, was released. It featured 2300 transistors and operated at a clock frequency of up to 740 kHz. The computing power of microprocessors increased exponentially over the next four decades. Today's microprocessor chips, with billions of transistors on a single silicon chip, run at clock frequencies in the gigahertz range and offer several million times the performance of the original 4004 [71].

An important prerequisite for the design and implementation of complex electronic circuits is the availability of powerful tools for computer-aided circuit design [72, 73]. Georg Simon Ohm (1789–1854) and Gustav Robert Kirchhoff (1824–1887) can be regarded as the founders of topological circuit theory [74–76]. The resistor constitutive equation was formulated by Georg Simon Ohm [77]. In his first work, while still a student, Kirchhoff established the rules later named after him, which are still the basis of network analysis today [78, 79]. Network analysis involves determining the voltages and currents of all network components. It is assumed that the state of the network is completely described by the currents through its branches and the voltages at the branches. Only the topological structure of the network connecting the circuit elements is important, not its geometry. A variety of methods have been developed to calculate the current and voltage values [80–84]. From the mid-1950s onwards, the breadboard design of electronic circuits was gradually replaced by computer-aided circuit design [85–88]. The Electronic Circuit Analysis Program (ECAP) from IBM and SPICE

(Simulation Program with Integrated Circuit Emphasis) were early circuit design programs, developed in the 1960s, that already used computer models for the bipolar transistor and the field effect transistor [89–96]. Stephen W. Director and Ronald A. Rohrer developed a frequency-domain network design program that calculated the sensitivity of any associated network function with respect to all pertinent network parameters in terms of analysis of its topologically equivalent adjoint network [97]. Les Besser developed in 1970 a fast computer routine to design high-frequency circuits, based on two-port manipulations [98], followed by the program COMPACT (an acronym for "Computer Optimization of Microwave Passive and Active CircuiTs") based on a nodal or n-port analysis with the capability of handling virtually unlimited circuit connections [99–101].

During the 1950s, there was a transition from point-to-point wiring of electronic circuits on a chassis to circuit construction on printed circuit boards, first developed by Paul Eisler in 1936. [54, 102]. This transition was facilitated by the replacement of vacuum tubes by much smaller transistors. Computer-aided simulation and optimization greatly speeds up the development of broadband analog and digital monolithic integrated circuits. The era of computational RF and microwave engineering began with sophisticated analytically based models of microwave components and simple computational techniques for connecting linear models of transistors and other components to determine the properties of linear microwave circuits. Development has progressed rapidly as today's computational microwave techniques can model and optimize large circuits with complex geometries.

The microwave circuit board has all the advantages of lumped element circuit boards, such as light weight, low cost, miniaturization, and the ability to operate at frequencies up to the millimeter-wave range. The basis for this technique is the planar or stripline circuitry for which an adequate theoretical analysis was developed subsequently. Its basis is the strip lines and planar circuit elements [103–105]. A broad methodological basis has been developed for the theoretical analysis of planar microwave and millimeter-wave structures [106–112]. Planar transmission line devices, including striplines, microstrips, slot lines, coplanar waveguides, and strips, phase shifters planar ferrimagnetic devices, such as phase shifters, isolators, and circulators, are treated in [113].

Noise is the reason why reliable representation and transmission of information requires a certain amount of energy to ensure a sufficient signal-to-noise ratio. Therefore, an important aspect of analog RF circuit design is optimization for low noise [114–116]. In 1919, Walter Schottky investigated the insurmountable limits for amplification with hot cathode and gas discharge tubes [117]. He concluded, "As a result of the thermal movement, continuous oscillations of small amplitude occur in electrical oscillating circuits, which should become noticeable with sufficiently high amplification and would mask very weak signals. The energy of these thermal oscillations is k_{BT} in all systems that are approximately in thermal equilibrium, i.e., it is independent of the material and the size of the electrical elementary charge." The electromotive force due to thermal agitation in conductors was calculated by Harry Nyquist in 1928 [118]. To describe the noise of linear active two-port devices, the noise figure F was introduced independently by Kurt Fränz and Harald T. Friis, which indicates the quotient of the signal-to-noise ratios at the input and output of the two-port device [119–122]. Initially, noise sources were represented by nonrandom sinusoidal sources and the noise analysis was carried out using an existing AC analysis program. However, this method is limited to cases in which the noise sources involved are either uncorrelated or fully correlated [123–125]. Gaussian noise signals can be described by power spectra [126]. The statistical dependence of noise signals yields non-vanishing cross correlation spectra [127, 128]. Rothe and Dahlke introduced the correlation admittance to describe the partial correlation of the two equivalent noise sources in noisy two ports [129]. Based on the spectral representation of noise signals in correlation matrices the fundamentals of the theory by Hermann A. Haus and Richard B. Adler [127] and by H. Bosma [130]. A method for correlation matrix-based noise analysis of linear circuits was presented by Herbert Hillbrand and Peter Russer [131, 132] and extended to multiport circuits in [133–140]. Pucel et al. developed a general automated noise measurement system with a noise de-embedding procedure based on the noise correlation technique used for data acquisition [141]. Peter Heymann and Matthias Rudolph presented a guide to noise computation and measurement in microwave devices and circuits [142].

In addition to thermal noise, we also have to deal with other electromagnetic interference. These are electromagnetic phenomena that can affect the performance of devices. Electromagnetic interference can be thermal or terrestrial electromagnetic noise or any unwanted signal. Time domain EMI measurement systems are based on a broadband analog-to-digital conversion and subsequent digital real-time processing of the EMI measurement signals, allowing the fast and accurate detection of electromagnetic interference within a wide frequency range [143–145]. Near-field scanning of the EMI radiated from electronic components or circuit boards and characterization of the radiated EMI via two-point correlation measurements facilitates an accurate modeling of the expected EMI spectral energy distribution inside a piece of electronic equipment under design and is a requirement for modern computer-aided design and fabrication of electronic systems [146–148]. Cyclostationary stochastic electromagnetic fields originating from digital circuits also affect the operation of a circuit [149]. As the on-chip data transmission rate increases to several tens of Gbit/s, the spatiotemporal inter-symbol interference within the multiwire interconnect systems begins to have an impact on device performance [150]. Shielding cables leads to an increase in system volume, weight, and cost. To minimize the effects of parasitic phenomena due to Intersymbol interference caused by cable coupling, spatial coding techniques are proposed [151].

Nonlinear RF circuits, such as mixers, oscillators, and amplifiers, require nonlinear design methods. A variety of methods for analyzing nonlinear circuits, both in the frequency domain and in the time domain, on the basis of the large-signal models of active components with the various intrinsic noise sources of the active and passive components, have been developed. The frequency domain methods mainly are based on harmonic balance and Volterra series techniques [152–158]. The time domain methods are based on the integration of nonlinear systems of ordinary differential equations using numerical methods [159–162]. A method for calculating the steady state of oscillators with lumped nonlinear active circuit elements and linear distributed elements combines the advantages of both time domain and frequency domain simulation. The oscillator circuit is split into a linear and a nonlinear circuit, with the linear circuit part being solved in the frequency domain and the nonlinear circuit part in the time domain [163]. While linear circuit design is based on measured two-port parameters of the active components, the design of nonlinear circuits requires nonlinear transistor models, which are created on the basis of nonlinear physical transistor models by adapting the parameters to measured data [164–166].

The application of linear and nonlinear techniques to analysis of microwave circuit design is discussed in [167]. Cai et al. developed a nonlinear model for accurate dc, small-signal, and noise characterization of AlGaAs–GaAs heterojunction bipolar transistors (HBTs), based on a method for accurate HBT model parameter extraction by fitting the dc, multibias S-parameter, and noise measurements simultaneously [168]. The design of CMOS RF integrated circuits is treated in [169].

Ulrich L. Rohde's work now spans more than six decades. Over 60 years ago while at the University of Darmstadt, he wrote his first book on transistors at the highest frequencies [170]. Ulrich L. Rohde is an enthusiastic radio amateur and 40 years ago while at RCA, in charge of the military radio division, together with his team he was the first to propose and demonstrate the software defined radio (SDR), where today signal processing is mainly carried out in programmable devices such as Field Programmable Logic Arrays (FPGA) and digital signal processors (DSP) and has now become the dominant technology in radio communications [171]. Over the decades, numerous books on microwave technology followed [172–178], but also on the mathematical principles [179–180], culminating in the encyclopedic work *Fundamentals of RF and Microwave Techniques and Technologies*, edited two years ago together with Hans L. Hartnagel, Rüdiger Quay and Matthias Rudolph [178].

This brings me to a few personal remarks about the beginning of my contact with Ulrich L. Rohde and Matthias Rudolph. In the mid-1970s, at the AEG-Telefunken Research Institute in Ulm, I worked with Herbert Hillbrand to develop methods for computer-aided optimization of the signal and noise properties of integrated analog circuits and applied these to silicon RF IC design [133, 137, 138]. Les Besser incorporated some of these methods into the circuit analysis program COMPACT, which was in 1973 the only commercially available microwave circuit computer aided design and optimization software. From 1980 onwards, this PC-based CAD software was successfully further developed by Ulrich L. Rohde by introducing the capability of simulating noise correctly in nonlinear circuits and other needed enhancements [181]. Since the 1980s, there have been many technical discussions in the field of noise in electronic circuits between Ulrich L. Rohde and me, which, when they took place at my institute at the Technical University of Munich, usually ended with the colleagues involved in the no longer existing Chinese restaurant Canton in the Theresienstrasse. I have also known Matthias Rudolph for more than three decades. I remember him very well as a student in my course on quantum electronics at the TU Berlin and many encounters since then.

This book covers the history from RF up to the sub Terahertz frequency range, also covers the development and application of RF and microwave transistors, microwave, and RF circuit design based on the required system properties, discussion of CAD tools for linear and nonlinear circuit design and optimization of the signal and noise properties of the circuits, treatment of distributed circuit elements and finally examples of circuit concepts. I am convinced that this remarkable book, written for RF and microwave engineers and scientists working in the field, will be a valuable basis and reference for their work.

Peter Russer
TUM School of Computation
Information and Technology
TUM Senior Excellence Faculty
Computational Photonics TUEICPH
Technical University Munich

References

1 Drews, J., Gomez, U., Gutheit, T. et al. (2022). *Hidden Electronics III*. Frankfurt am Main: VDE.

2 Russer, P. (2024). Neue physikalische Technologien für Bauelemente und System-Ideen, *ITG News*, Sonderheft "70 Jahre ITG," pp. 8–16.

3 Braun, F. (1877). Über Abweichungen vom Ohm'schen Gesetz in metallisch leitenden Körpern, *Annalen der Physik* 237 (5): 95–110.

4 Russer, P. (2012). Ferdinand Braun—A pioneer in wireless technology and electronics. In *A Wireless World, ser. Contributions to the history of the Royal Swedish Academy of Sciences*, (ed. K. Grandin et al.), vol. 42, 228–247, Firenze University Press.

5 Gardiner, J.E. (2014). *Music in the Castle of Heaven: A Portrait of Johann Sebastian Bach*. London, England: Penguin Books.

6 Pearson, G.L. and Brattain, W.H. (1955). History of semiconductor research. *Proceedings of the IRE* 43 (12): 1794–1806.

7 Spenke, E. (1965). *Elektronische Halbleiter*. Berlin: Springer.

8 Riordan, M. and Hoddeson, L. (1999). *Crystal Fire: The Invention of the Transistor and the Birth of the Information Age*. Thames & Hudson.

9 Bose, J.C. (1899). On a self-recovering coherer and the study of the cohering action of different metals. *Proceedings of the Royal Society of London* 65: 166–172.

10 Bose, J.C. (1904). Detector for electrical disturbances. US Patent No. 755840, Application filed 30 September 1901.

11 Braun, K.F. (1967). Electrical oscillations and wireless telegraphy. In *Nobel Lectures, Physics 1901–1921*. 226–245, Amsterdam: Elsevier Publishing Company.

12 Bondyopadhyay, P. (1998). Sir J.C. Bose diode detector received Marconi's first transatlantic wireless signal of December 1901 (the "Italian navy coherer" scandal revisited). *Proceedings of the IEEE* 86 (1): 259–285.

13 Sommerfeld, A. (1928). Zur Elektronentheorie der Metalle. *Naturwissenschaften* 16 (21): 374–381.

14 Bloch, F. (1929). Uber die Quantenmechanik der Elektronen in Kristallgittern. *Zeitschrift für Physik* 52 (7): 555–600.

15 Schottky, W. (1939). Zur Halbleitertheorie der Sperrschicht-und Spitzengleichrichter. *Zeitschrift für Physik A* 113 (5): 367–414.

16 Lilienfeld, J. (1926). Method and apparatus for controlling electric currents. US Patent 1,745,175, 8 October 1926.

17 Crawford, B.E. (1991). The invention of the transistor. Ph.D. dissertation. University of Vermont, Burlington, VT.

18 Hillmer, H. (2008). Die vielen V¨ater des Transistors. Meilensteine in der Geschichte des Festk¨orperverst¨arkers. *Physik in unserer Zeit* 39 (6): 296–300.

19 Bardeen, J., Brattain, W., and Shockley, W. (1950). Three-electrode circuit element utilizing semiconductive materials. US Patent 2524035, filed 17 June 1948, granted 3 October 1950.

20 Mataré, H.F. and Welker, H. (1952). Nouveau système cristallin à plusieurs électrodes réalisant des effects de relais électroniques. Fr. Pat. FR1010427, demand´e 13 août 1948, published 11 juin.

21 Sparks, M. and Teal, G.K. (1950). Method of making P-N junctions in semiconductor materials. US Patent US2, 631,356, angemeldet am 15. Juni 1950.

22 Sparks, M. and Teal, G.K. (1951). p-n junction transistors. *Physical Review* 83 (1): 151–162.

23 Marsh, A. (2024). Past forward: the incredible story behind the first transistor radio: the Regency TR-1 helped Texas Instruments launch its semiconductor business. *IEEE Spectrum* 1–4. https://doi.org/10.1109/MSPEC.2024.10746251.

24 Gopinath, S. and Stanyek, J. (2014). *The Oxford Handbook of Mobile Music Studies*, Vol. 1. OUP USA.

25 Mende, H.G. (1959). *Leitfaden der Transistortechnik*. Franzis-Verlag.

26 Sze, S.M. (1981). *Physics of Semiconductor Devices*. Wiley.

27 Sze, S.M. (1990). *High-speed Semiconductor Devices*. Wiley.

28 Luy, J.-F. Russer, P. (1994). *Silicon-Based Millimeter-Wave Devices*. Springer Series in Electronics and Photonics, vol. 32. Berlin Heidelberg New York: Springer.

29 Kahng, D. (1976). A historical perspective on the development of MOS transistors and related devices. *IEEE Transactions on Electron Devices* 23 (7): 655–657.

30 Kasper, E., Kissinger, D., Russer, P. et al. (2009). High speeds in a single chip. *IEEE Microwave Magazine* 10 (7): 28–33.

31 Shockley, W. (1951). Circuit element utilizing semiconductive material. US Patent 2,569,347, September 1951.

32 Hähnlein A. (1958). Halbleiter-Kristallode der Schichtenbauart. German Patent DE 1 021 488, filed 19 February 1954, issued 10 July 1958.

33 Kroemer, H. (1954). Zur Theorie des Diffusions-und Drifttransistors -III Dimensionierungsfragen. *Archiv für Elektrischen Übertragung* 8: 363–369.

34 Kroemer, H. (1957). Theory of a wide-gap emitter for transistors. *Proceedings of the IRE* 45 (11): 1535–1537.

35 Kasper, E. and Russer, P. (1977). Verfahren zur Herstellung von bipolaren Hochfrequenztransistoren. German Patent DE 2719464, filed 30 April 1977.

36 Iyer, S.S., Patton, G.L., Delage, S.S. et al. (1987). Silicon-germanium base heterojunction bipolar transistors by molecular beam epitaxy. *IEDM Technical Digest*, 874–875.

37 Cressler, J.D. and Niu, G. (2003). *Silicon-Germanium Heterojunction Bipolar Transistors*. Norwood, Ma: Artech House.

38 Wanner, R., Lachner, R., Olbrich, G.R. et al. (2007). A SiGe monolithically integrated 278 GHz push-push oscillator. *2007 IEEE MTT-S International Microwave Symposium Digest* (June 4–8). Honolulu, USA, 333–336.

39 Russer, P. and Fichtner, N. (2010). Nanoelectronics in radio-frequency technology. *IEEE Microwave Magazine* 11 (3): 119–135.

40 Jacobi, W. (1952). Halbleiterverstaerker. German Patent DE 833366B, filed on 15 April 1949.

41 Kilby, J. (1976). Invention of the integrated circuit. *IEEE Transactions on Electron Devices* 23 (7): 648–654.

42 Reid, T. (2001). *The Chip*. New York: Random House.

43 Noyce, R. (1961). Semiconductor device-and-lead structure. US Patent 2981877, filed 30 July 1959.

44 Moore, G. (1965). Cramming more components onto integrated circuits. *Electronics* 19 (3): 114–117.

45 Thompson, S.E. and Parthasarathy, S. (2006). Moore's law: the future of Si microelectronics. *Materials Today* 9 (6): 20–25.

46 Schaller, R. (1997). Moore's law: past, present and future. *IEEE Spectrum* 34 (6): 52–59.

47 Moore's law (2024). Moore's law. Page Version ID: 1245754212. https://en.wikipedia.org/wiki/Moore's law (accessed July 2026)

48 Maxwell, J.C. (1998). *A Treatise on Electricity and Magnetism*. Vol. 1. New York: Oxford University Press.

49 Maxwell, J.C. (1998). *A Treatise on Electricity and Magnetism*. Vol. 2. New York: Oxford University Press.

50 Hertz, H. (1889). Die Kräfte electrischer Schwingungen behandelt nach der Maxwell'schen Theorie. *Annalen der Physik und Chemie* 3 (1): 1–22.

51 Hertz, H. (1894). *Gesammelte Werke, Untersuchungen über die Ausbreitung der elektrischen Kraft*. Gesammelte Werke. Vol. 2. Leipzig: Johann Ambrosius Barth.

52 Schelkunoff, S.A. (1970). Forty years ago: Maxwell's theory invades engineering -and grows with it. *IEEE Transactions on Antennas and Propagation* 18 (3): 309–322.

53 Rautio, J.C. (2005). Maxwell's legacy. *IEEE Microwave Magazine* 6 (2): 46–53.

54 Sarkar, T.K., Mailloux, R., Oliner, A.A. et al. (2006). *History of Wireless*. John Wiley & Sons.

55 Hertz, J. and Hertz, H. (1977). *Heinrich Hertz: Erinnerungen, Briefe, Tagebücher*, (ed. M. Hertz and C. Susskind). Weinheim, Germany and San Francisco, Cal.: Physik Verlag and San Francisco Press.

56 Gerhard-Multhaupt, R., Ihlenburg, L., von Reden, W. et al. (1988). *Heinrich Hertz: Festschrift anlässlich der Erforschung der elektromagnetischen Wellen vor 100 Jahren*. Heinrich-Hertz-Institut.

57 Sommerfeld, A. (1947). *Elektrodynamik*. Leipzig: Akademische Verlagsgesellschaft Geest & Portig.

58 Sommerfeld, A. (1947). *Partielle Differentialgleichungen der Physik*. Leipzig: Akademische Verlagsgesellschaft Geest & Portig.

59 Stratton, J.A. (1941). *Electromagnetic Theory*. New York, London: John Wiley.

60 Harrington, R.F. (1961). *Time Harmonic Electromagnetic Fields*. New York: McGraw-Hill.

61 Ramo, S., Whinnery, J. R., and van Duzer, T. (1965). *Fields and Waves in Communication Electronics*. New York: John Wiley & Sons.

62 Felsen, L. and Marcuvitz, N. (1972). *Radiation and Scattering of Waves*. Englewood Cliffs, NJ: Prentice Hall.

63 Balanis, C.A. (1989). Advanced Engineering Electromagnetics. New York: John Wiley & Sons.

64 Pozar, D.M. (1990). *Microwave Engineering*. Reading: Addison-Wesley.

65 Collin, R. (1991). *Field Theory of Guided Waves*, 2e. New York: IEEE Press, Inc.

66 Collin, R.E. (1992). *Foundations of Microwave Engineering*. New York: McGraw-Hill.

67 Lindell, I.V. (1992). *Methods for Electromagnetic Field Analysis*. Oxford: Clarendon Press.

68 Lindell, I.V. (2004). *Differential Forms in Electromagnetics*. New York: IEEE Press.

69 Russer, P. (2006). *Electromagnetics, Microwave Circuit and Antenna Design for Communications Engineering*. Boston: Artech House.

70 Sobol, H. and Tomiyasu, K. (2002). Milestones of microwaves. *IEEE Transactions on Microwave Theory and Techniques* 50 (3): 594–611.

71 Danowitz, A., Kelley, K., Mao, J. et al. (2012). CPU DB: recording microprocessor history: with this open database, you can mine microprocessor trends over the past 40 years. *Queue* 10 (4): 10–27.

72 Steer, M., Bandler, J., and Snowden, C. (2002). Computer-aided design of RF and microwave circuits and systems. *IEEE Transactions on Microwave Theory and Techniques* 50 (3): 996–1005.

73 Russer, P. (2024). Network and circuit models in electromagnetic field computation. *IEEE International Symposium on Antennas and Propagation*, Florence, Italy, 1771–1772.

74 Guarnieri, M. (2021). A glance at circuit theory development. *IEEE Industrial Electronics Magazine* 15 (1): 79–83.

75 Mathis, W. (2023). Gustav Robert Kirchhoff. *IEEE Circuits and Systems Magazine* 23 (3): 68–69.

76 Mathis, W. (2024). Georg Simon Ohm. *IEEE Circuits and Systems Magazine* 24 (2): 102–104.

77 Ohm, G.S. (1827). Versuch einer Theorie der durch galvanische Kräfte hcrvorgebrachten elektroskopischen Erscheinungen. *Annalen der Physik* 83 (5): 45–54.

78 Kirchhoff, G. (1845). Ueber den Durchgang eines elektrischen Stromes durch eine Ebene, insbesondere durch eine kreisf¨ormige. *Annalen der Physik* 64 (4): 497–514.

79 Kirchhoff, G. (1958). On the solution of the equations obtained from the investigation of the linear distribution of galvanic currents. *IRE Transactions on Circuit Theory* 5 (1): 4–7. (Translated by J.B. O'Toole).

80 Cauer, W. (1954). *Theorie der linearen Wechselstromschaltungen*. Berlin: Akademie-Verlag Berlin.

81 Carlin, H. and Giordano, A. (1964). *Network Theory*. Englewood Cliffs, NJ: Prentice-Hall.

82 Chua, L.O., Desoer, C.A., and Kuh, E.S. (1987). *Linear and Nonlinear Circuits*. New York: Mc Graw Hill.

83 Belevitch, V. (1968). *Classical Network Theory*. San Francisco: Holden-Day.

84 Mathis, W. (1987). *Theorie nichtlinearer Netzwerke*. Berlin: Springer.

85 Kuo, F. (1966). Network analysis by digital computer. *Proceedings of the IEEE* 54 (6): 820–829.

86 Branin, F. (1967). Computer methods of network analysis. *Proceedings of the IEEE* 55 (11): 1787–1801.

87 Chua, L.O. and Lin, P.Y. (1975). *Computer–aided Analysis of Electronic Circuits*. Englewood Cliffs, NJ: Prentice-Hall.

88 Maas, S. (2012). *Circuit Simulation in the Dark Ages. IEEE Microwave Magazine* 13 (4): 89–99.

89 Branin, F., Hogsett, G., Lunde, R. et al. (1971). ECAP II-a new electronic circuit analysis program. *IEEE Journal of Solid-State Circuits* 6 (4): 146–166.

90 Nagel, L.W. (1975). SPICE2: a computer program to simulate semiconductor circuits. *College of Engineering, University of California, Berkeley*. https://cir.nii.ac.jp/crid/1573105974314515840

91 Banzhaf, W. (1992). *Computer-Aided Circuit Analysis Using PSPICE*, 2e. Englewood Cliffs, NJ: Regents/Prentice Hall, Inc.

92 Sandler, S.M. and Hymowitz, C.E. (2006). *SPICE Circuit Handbook*. New York: McGraw-Hill.

93 Arsov, G.L. (2012). The 40-th anniversary of the simulation program with integrated circuit emphasis SPICE. *IX Symposium Industrial Electronics INDEL 2012*, Banja Luka, Macedonia (01–03 November 2012), pp. 16–21.

94 Jensen, R. (1966). Charge control transistor model for the IBM electronic circuit analysis program. *IEEE Transactions on Circuit Theory* 13 (4): 428–437.

95 Roberts, B. and Harbourt, C. (1967). Computer models of the field-effect transistor. *Proceedings of the IEEE* 55 (11): 1921–1929.

96 Dowell, R.I. (2011). Hijacked by SPICE. *IEEE Solid-State Circuits Magazine* 3 (2): 14–16.

97 Director, S. and Rohrer, R. (1969). Automated network design-the frequency-domain case. *IEEE Transactions on Circuit Theory* 16 (3): 330–337.

98 Besser, L. (1970). A fast computer routine to design high frequency circuits. *IEEE ICC Conference*, San Francisco, California.

99 Bodharamik, P., Besser, L., and Newcomb, R. (1971). Two scattering matrix programs for active circuit analysis. *IEEE Transactions on Circuit Theory* 18 (6): 610–619.

100 Besser, L., Ghoul, F., and Hsieh, C. (1973). Computerized optimization of transistor amplifiers and oscillators using COMPACT. *1973 3rd European Microwave Conference*, vol. 1. IEEE, pp. 1–4. https://ieeexplore.ieee.org/abstract/document/4130197/.

101 Besser, L. (2008). Compact-microwave circuit optimization through commercial time sharing. *2008 IEEE MTT-S International Microwave Symposium Digest*. IEEE, pp. 711–714.

102 Eisler, P. (1948). Manufacture of electric circuit components. US Patent 2,441,960, May 1948.

103 Barrett, R. (1955). Microwave printed circuits -a historical survey. *IRE Transactions on Microwave Theory and Techniques* 3 (2): 1–9.

104 Oliner, A. (1955). Equivalent circuits for discontinuities in balanced strip transmission line. *IRE Transactions on Microwave Theory and Techniques* 3 (2): 134–143.

105 Robertson, I.D., Karacaoglu, U., and Sanchez-Hernandez, D. (1996). Cad techniques for microwave circuits. *Electronics & Communication Engineering Journal* 8 (6): 245–256.

106 Harrington, R. (1967). Matrix methods for field problems. *Proceedings of the IEEE* 55 (2): 136–149.

107 Harrington, R.F. (1968). *Field Computation by Moment Methods*. San Francisco: IEEE Press.

108 Gupta, K. (1998). Emerging trends in millimeter-wave CAD. *IEEE Transactions on Microwave Theory and Techniques* 46 (6): 747–755.

109 Bahl, I.J. and Bhartia, P. (1988). *Microwave Solid State Circuit Design*. New York: Wiley.

110 Bhat, B. and Koul, S.K. (1989). *Stripline-like Transmission Lines for Microwave Integrated Circuits*. New Age International.

111 Felsen, L.B., Mongiardo, M., Russer, P. et al. (1997). Waveguide component analysis by a generalized network approach. *1997 27th European Microwave Conference*, vol. 2, pp. 949–954.

112 Felsen, L.B., Mongiardo, M., and Russer, P. (2009). *Electromagnetic Field Computation by Network Methods*. Berlin, Heidelberg, New York: Springer-Verlag.

113 Di Paolo, F. (2018). *Networks and Devices Using Planar Transmissions Lines*. Boca Raton: CRC Press.

114 der Ziel, A.V. (1954). *Noise*. London: Chapman and Hall.

115 der Ziel, A.V. (1986). *Noise in Solid State Devices and Circuits*. New York: John Wiley.

116 L¨ocherer, K.-H. (1987). Bemerkungen zum Rauschen von Vierpolen. *Frequenz* 41 (1-2): 18–21.

117 Schottky, W. (1918). Über spontane Stromschwankungen in verschiedenen Elektrizit¨atsleitern. *Annalen der Physik* 362 (23): 541–567. WILEY-VCH Verlag Leipzig.

118 Nyquist, H. (1928). Thermal agitation of electric charge in conductors. *Physical Review* 32 (1): 110–113.

119 Fr¨anz, K. (1939). Uber die Empfindlichkeitsgrenze beim Empfang elektrischer Wellen und ihre Erreichbarkeit. *Elektrische Nachrichten-Technik* 16: 92–96.

120 Fr¨anz, K. (1942). Messung der Empf¨angerempfindlichkeit bei kurzen elektrischen Wellen. *Zeitschrift für Hochfrequenztechnik* 59: 105–112 and 143–144.

121 Friis, H. (1944). Noise figures of radio receivers. *Proceedings of the IRE* 32 (7): 419–422.

122 Goldberg, H. (1948). *Some notes on noise figures. Proceedings of the IRE* 36 (10): 1205–1214.

123 Baechtold, W., Kotyczka, W., and Strutt, M. (1969). Computerized calculation of small signal and noise properties of microwave transistors. *IEEE Transactions on Microwave Theory and Techniques* 17 (8): 614–619.

124 Rohrer, R., Nagel, L., Meyer, R. et al. (1971). Computationally efficient electronic-circuit noise calculations. *IEEE Journal of Solid-State Circuits* 6 (4): 204–213.

125 Hartmann, K., Kotyczka, W., and Strutt, M. (1973). Computerized determination of electrical network noise due to correlated and uncorrelated noise sources. *IEEE Transactions on Circuit Theory* 20 (3): 321–322.

126 Davenport, W.B. and Root, W.L. (1958). *Random Signals and Noise*. New York: McGraw-Hill.

127 Haus, H.A. and Adler, R.B. (1959). *Circuit Theory of Linear Noisy Networks*. Technology Press of Massachusetts Institute of Technology.

128 König, H.W. and Pötzl, H. (1959). Schwankungsvorgänge in Elektronenstrahlen. In: *Fortschritte der Hochfrequenztechnik*, vol. 4, pp. 195–239. Frankfurt, Germany: Akademische Verlagsgesellschaft.

129 Rothe, H. and Dahlke, W. (1956). Theory of noisy four poles. *Proceedings of the IRE* 44 (6): 811–818.

130 Bosma, H. (1967). On the theory of linear noisy systems. Ph.D. dissertation. Technische Hogeschool Eindhoven.

131 Hillbrand, H. and Russer, P. (1976). An efficient method for computer aided noise analysis of linear amplifier networks. *IEEE Transactions on Circuits and Systems* 23 (4): 235–238.

132 Hillbrand, H. and Russer, P. (1976). Correction to 'An efficient method for computer aided noise analysis of linear amplifier networks'. *IEEE Transactions on Circuits and Systems* 23 (11): 691.

133 Russer, P. and Hillbrand, H. (1976). Rauschanalyse von linearen Netzwerken. *Wissenschaftliche Berichte AEG Telefunken* 49: 127–138.

134 Rizzoli, V. and Lipparini, A. (1985). Computer-aided noise analysis of linear multiport networks of arbitrary topology. *IEEE Transactions on Microwave Theory and Techniques* 33 (12): 1507–1512.

135 Dobrowolsky, J.A. (1989). A CAD-oriented method for noise figure computation of two-ports with any internal topology. *IEEE Transactions on Microwave Theory and Techniques* 37: 15–20.

136 Rizzoli, V., Mastri, F., and Cecchetti, C. (1989). Computer-aided noise analysis of MESFET and HEMT mixers. *IEEE Transactions on Microwave Theory and Techniques* 37 (9): 1401–1410.

137 Russer, P. and Müller, S. (1990). Noise analysis of linear microwave circuits. *International Journal of Numerical Modelling, Electronic Networks, Devices and Fields* 3: 287–316.

138 Russer, P. (1996). Noise analysis of linear microwave circuits with general topology. In: *The Review of Radio Science 1993–1996*, 887–890, England: Oxford.

139 Maas, S.A. (2005). *Noise in Linear and Nonlinear Circuits*. Boston: Artech House.

140 Dobrowolski, J.A. (2016). *Scattering Parameters in RF and Microwave Circuit Analysis and Design*. Boston: Artech House.

141 Pucel, R., Struble, W., Hallgren, R. et al. (1992). A general noise de-embedding procedure for packaged two-port linear active devices. *IEEE Transactions on Microwave Theory and Techniques* 40 (11): 2013–2024.

142 Heymann, P. and Rudolph, M. (2021). *A Guide to Noise in Microwave Circuits: Devices, Circuits and Measurement*. John Wiley & Sons.

143 Krug, F. and Russer, P. (2003). The time-domain electromagnetic interference measurement system. *IEEE Transactions on Electromagnetic Compatibility* 45 (2): 330–338.

144 Krug, F., Mueller, D., and Russer, P. (2004). Signal processing strategies with the TDEMI measurement system. *IEEE Transactions on Instrumentation and Measurement* 53 (5): 1402–1408.

145 Russer, P. (2011). EMC measurements in the time-domain. *General Assembly and Scientific Symposium, 2011 XXXth URSI*, pp. 1–35. IEEE.

146 Russer, J.A. and Russer, P. (2015). Modeling of noisy EM field propagation using correlation information. *IEEE Transactions on Microwave Theory and Techniques* 63 (1): 76–89.

147 Russer, J.A., Uddin, N., Awny, A.S. et al. (2015). Near-field measurement of stochastic electromagnetic fields. *IEEE Electromagnetic Compatibility Magazine* 4 (3): 79–85.

148 Gradoni, G., Russer, J., Baharuddin, M.H. et al. (2018). Stochastic electromagnetic field propagation—measurement and modelling. *Philosophical Transactions of the Royal Society A: Mathematical, Physical and Engineering Sciences* 376 (2134): 20170455.

149 Kuznetsov, Y.V., Baev, A.B., Konovalyuk, M.A. et al. (2020). Cyclostationary characterization of radiated emissions in digital electronic devices. *IEEE Electromagnetic Compatibility Magazine* 9 (4): 63–76.

150 Yordanov, H., Ivrlać, M., Nossek, J. et al. (2007). Field modelling of a multiconductor digital bus. *European Microwave Integrated Circuit Conference*, pp. 579–582.

151 Ivrlac, M.T. and Nossek, J.A. (2006). Challenges in coding for quantized MIMO systems. *2006 IEEE International Symposium on Information Theory*, pp. 2114–2118.

152 Nakhla, M. and Vlach, J. (1976). A piecewise harmonic balance technique for determination of periodic response of nonlinear systems. *IEEE Transactions on Circuits and Systems* 23 (2): 85–91.

153 Rizzoli, V. and Neri, A. (1988). State of the art and present trends in nonlinear microwave CAD techniques. *IEEE Transactions on Microwave Theory and Techniques* 36 (2): 343–365.

154 Gilmore, R.J. and Steer, M.B. (1991). Nonlinear circuit analysis using the method of harmonic balance—A review of the art. Part I. Introductory concepts. *International Journal of Microwave and Millimeter-Wave Computer-Aided Engineering* 1 (1): 22–37.

155 Rizzoli, V., Lipparini, A., Costanzo, A. et al. (1992). State-of-the-art harmonic-balance simulation of forced nonlinear microwave circuits by the piecewise technique. *IEEE Transactions on Microwave Theory and Techniques* 40 (1): 12–28.

156 Anzill, W. and Russer, P. (1993). A general method to simulate noise in oscillators based on frequency domain techniques. *IEEE Transactions on Microwave Theory and Techniques* 41 (12): 2256–2263.

157 Rohde, U., Change, C.-R., and Gerber, J. (1994). Design and optimization of low-noise oscillators using nonlinear CAD tools. *Proceedings of IEEE 48th Annual Symposium on Frequency Control*, pp. 548–554.

158 Maas, S.A. (2003). *Nonlinear Microwave and RF Circuits*. Boston: Artech House.

159 Mathis, W. and Russer, P. (2005). Oscillator design. In*: Encyclopedia of RF and Microwave Engineering*. Wiley.

160 Colon, F. and Trick, T. (1973). Fast periodic steady-state analysis for large-signal electronic circuits. *IEEE Journal of Solid-State Circuits* 8 (4): 260–269.

161 Egami, S. (1974). Nonlinear, linear analysis and computer-aided design of resistive mixers. *IEEE Transactions on Microwave Theory and Techniques* 22 (3): 270–275.

162 Kaertner, F.X. (1989). Determination of the correlation spectrum of oscillators with low noise. *IEEE Transactions on Microwave Theory and Techniques* 37 (1): 90–101.

163 Schwab, M.H. (1991). Determination of the steady state of an oscillator by a combined time-frequency method. *IEEE Transactions on Microwave Theory and Techniques* 39 (8): 1391–1402.

164 Rohde, U.L., Chang, C.-R., Gerber, J. et al. (1993). Parameter extraction for large signal noise models and simulation of noise in large signal circuits like mixers and oscillators. *1993 23rd European Microwave Conference*, pp. 465–470.

165 Sinnesbichler, F., Fischer, M., and Olbrich, G. (1998). Accurate extraction method for 1/f-noise parameters used in Gummel-Poon type bipolar junction transistor models. *1998 IEEE MTT-S International Microwave Symposium Digest*, vol. 3, pp. 1345–1348.

166 Sinnesbichler, F. and Olbrich, G. (2000). Accurate large-signal modeling of SiGe HBTs. *2000 IEEE MTT-S International Microwave Symposium Digest*, vol. 2, pp. 749–752.

167 Vendelin, G.D., Pavio, A.M., Rohde, U.L. et al. (2021). *Microwave Circuit Design Using Linear and Nonlinear Techniques*, 3e. John Wiley & Sons.

168 Cai, Q., Gerber, J., Rohde, U. et al. (1997). HBT high-frequency modeling and integrated parameter extraction. *IEEE Transactions on Microwave Theory and Techniques* 45 (12): 2493–2502.

169 Caverly, R. (2007). *CMOS RFIC Design Principles*. Boston: Artech House.

170 Rohde, U.L. (1965). *Transistoren bei höchsten Frequenzen*. Verlag für Radio-Foto-Kinotechnik.

171 Rohde, U. (1984). Digital hf radio: A sampling of techniques. *Third International Conference on HF Communication Systems and Techniques*, London.

172 Rohde, U.L. (1983). *Digital PLL Frequency Synthesizers*. Englewood Cliffs, NJ: Prentice-Hall.

173 Rohde, U.L. and Bucher, T.T.N. (1988). *Communications Receivers – Principles & Design*. McGraw-Hill.

174 Vendelin, G.D., Pavio, A.M., and Rohde, U.L. (1990). *Microwave Circuit Design Using Linear and Nonlinear Techniques*. John Wiley & Sons.

175 Rohde, U.L., and Rudolph, M. (2012). *RF/Microwave Circuit Design for Wireless Applications*. John Wiley & Sons.

176 Rohde, U.L., Poddar, A.K., and Böck, G. (2005). *The Design of Modern Microwave Oscillators for Wireless Applications: Theory and Optimization*. John Wiley & Sons.

177 Rohde, U.L., Rubiola, E. and Whitaker, J.C. *Microwave and Wireless Synthesizers: Theory and Design*. John Wiley & Sons.

178 Hartnagel, H.L., Quay, R., Rohde, U.L. (eds.) et al. (2023). *Fundamentals of RF and Microwave Techniques and Technologies*. Cham: Springer International Publishing.

179 Rohde, U.L., Jain, G.C., Poddar, A.K. et al. (2012). *Introduction to Differential Calculus: Systematic Studies with Engineering Applications for Beginners*. John Wiley & Sons.

180 Rohde, U.L., Jain, G.C., Poddar, A.K. et al. (2012). *Introduction to Integral Calculus: Systematic Studies with Engineering Applications for Beginners*. John Wiley & Sons.

181 Compact Software (2024). Page Version ID: 1219677937. https://en.wikipedia.org/w/index.php?title=CompactSoftware &oldid=1219677937 (accessed date July 2026)

Foreword by Tomás Palacios

Today's RF and microwave electronics underpin a vast range of critical technologies, from 5G and emerging 6G communications to advanced radar, sensing, and medical systems. At the same time, they are enabling entirely new paradigms, including dielectric waveguides for ultra-low-loss interconnects in data centers and novel approaches to wireless energy transmission and conversion. Few individuals have contributed as profoundly to this evolution as Prof. Ulrich L. Rohde, whose work over more than half a century has helped shape both the theoretical foundations and the practical realization of high-frequency systems. Beginning with his seminal 1965 book, *Transistoren bei höchsten Frequenzen,* he has consistently anticipated and advanced the frontiers of transistor technology, circuit design, and system-level innovation, pushing performance into ever higher frequency regimes. This new book on RF and microwave electronics continues that legacy, arriving at a moment when the field he helped define is undergoing a remarkable transformation.

We are now witnessing a rare convergence of technological advances that makes this an exceptionally fertile period for innovation in RF and microwave engineering. Breakthroughs in semiconductor materials, particularly GaN and other wide-bandgap and III–V technologies, are delivering unprecedented capabilities in terms of electric field strength, power density, and carrier transport. In parallel, heterogeneous integration has matured, enabling the seamless co-design of high-performance RF front-ends with silicon-based control and digital electronics. This progress is further amplified by advances in passive components, driven by silicon back-end-of-the-line processes and chiplets, as well as modern packaging and interposer technologies. Together, these developments are unlocking system architectures and performance levels that were previously out of reach, a breadth of opportunity that is reflected throughout this book.

Against this backdrop, Prof. Rohde's textbook stands out as both timely and impactful. It is comprehensive in scope, covering the essential building blocks of modern RF and microwave electronics with rigor and clarity. Equally important, it is firmly rooted in applications and guided by a system-level perspective, ensuring its relevance to both practicing engineers and researchers. The presentation reflects Prof. Rohde's distinctive pedagogical style—precise, insightful, and shaped by decades of experience across academia and industry. As microwave engineering enters a new era defined by advanced materials, sophisticated integration, and expanding application domains, this book is well positioned to inspire innovation, educate the next generation of engineers, and serve as a valuable reference for seasoned professionals alike.

On a personal note, it is especially gratifying to see this book emerge during Prof. Rohde's association with the Microsystems Technology Laboratories at MIT, the laboratory I have the privilege to direct. His engagement with our community has been both inspiring and impactful, characterized by tireless energy, intellectual curiosity, and a profound commitment to the education and mentorship of our students. His presence has not only enriched our community but also exemplified a deep dedication to advancing both knowledge and people, precisely the mission that underlies this book. I hope the reader will find it as rewarding and inspiring as we have.

Tomás Palacios
Director, Microsystems Technology Laboratories
Massachusetts Institute of Technology

The Authors

Ulrich L. Rohde, Prof. Dr.-Ing. habil., received his Dr.-Ing. degree from the Technical University, Berlin, and his Dr.-Ing. habil degree from the Brandenburg University of Technology, Cottbus, Germany. He is a member of the Faculty of Technical Informatics as the Professor of Microwave Systems at the Universität der Bundeswehr (Federal University of the Armed Forces), Munich, Germany. He also holds appointments at other universities worldwide. In April 2025, the MIT Microsystems Technology Laboratories welcomed him as visiting scientist.

Dr. Rohde is a Partner of Rohde and Schwarz, Munich, Germany, as well as Chairman of the Board of Synergy Microwave Corporation. Formerly Professor of Electrical Engineering at George Washington University and the University of Florida, Gainesville, Dr. Rohde has consulted on numerous communication projects in industry and government. He has authored over 350 papers and 6 books, and holds 35 patents.

In 1983, with his team at the RCA Radio Business Unit, he designed and proved the concept of a system that today is referred to as the software-defined radio (SDR). He later acquired Compact Software and introduced the ability to accurately simulate the noise contribution from active devices on signals in nonlinear systems and its influence and distortion in oscillators, frequency doublers, mixers, and amplifiers.

Dr. Rohde has received numerous prestigious awards, including honorary PhD degrees. He is an honorary member of the Bavarian Academy of Science and Humanity, a Life Fellow of the IEEE, a Life Fellow of the IEIT, and a Fellow of the Indian National Academy of Engineering (INAE).

https://orcid.org/0009-0009-2271-4438.

Ajay K. Poddar, Dr.-Ing. habil., is an IEEE Fellow, working as a Chief Scientist at Synergy Microwave, NJ, USA, where he is responsible for the design and development of signal generation and signal processing electronics for industrial, medical, and space applications. He also serves as a visiting professor at the University of Oradea, Romania, Indian Institute of Technology Jammu, India, and as a guest lecturer at Technical University Munich, Germany. He has published over 400 papers in technical conferences and journals, and co-authored half a dozen books and over six dozen patents in his credit.

https://orcid.org/0000-0002-2733-8517.

Matthias Rudolph, Prof. Dr.-Ing., holds the Ulrich-L.-Rohde Chair for radio-frequency (RF) and Microwave Techniques and represents the RF and Microwave activities at the Brandenburgische Technische Universität Cottbus-Senftenberg (BTU).

The focus of his research is on microwave electronics, specifically on monolithically integrated III–V circuits for advanced mobile communications. Additionally, he addresses the related topics of transistor modeling, microwave measurement techniques, and electromagnetic field simulation.

The department is in close cooperation with the Ferdinand-Braun-Institut, Leibniz-Insitut für Höchstfrequenztechnik in the framework for the Joint Lab BTU-CS—FBH Microwaves.

https://orcid.org/0000-0002-0331-8394.

Introduction

This book was triggered by revisiting my German book, see below, which I published during the 1965 time frame as a student at the Technical University of Darmstadt, Germany, at a time when tubes were still most relevant and silicon-based radio frequency (RF) transistors were just evolving into useful devices.

Teaching RF transistor technology and techniques had just begun at this time, and I was told that if you write a peer-reviewed book on the topic, and if it was accepted and goes into print then, you really conquered the topic. So, I put together the needed mathematics and technology and circuitry to describe how to use transistors at very high frequencies, actually low frequencies compared with today's sub-terahertz technology.

In addition, the transition from coaxial circuit design to microstrip happened, and there were no useful microwave CAD tools available that could handle both planar circuit design, the time-domain nonlinear signal and noise behavior. Today, these tools are available and reliable.

For me, it all began with the: MIMIC Phase 1 teams Phase 1, Raytheon and Texas Instruments, Aerojet, Airtron, Compact Software, Consilium, General Dynamics, Magnavox, Norden, and Teledyne.

The acronym "MIMIC" stands for microwave/millimeter-wave monolithic integrated circuits. The DARPA program manager was Dr. Eliot Cohen. In 2012, Eliot Cohen wrote an article on the MIMIC Program for IEEE MTT-S Microwave Magazine, which is a good resource if you want to learn about this historic program. Here are the details: (E. D. Cohen, "The MIMIC Program—A Retrospective," in *IEEE Microwave Magazine*, vol. 13, no. 4, pp. 77–88, May–June 2012, https://doi.org/10.1109/MMM.2012.2189989). The important tasks for the software company were not only speed and accuracy but also "seamless" integration from a schematic with correct models for all the microwave components to the transfer to a printed circuit board, assembly, and the correct prediction of the interaction of the components, simply called crosstalk.

Designing circuits at microwave and mm-wave frequencies is not possible without thorough validation of models based on curve-fitting numerical methods, and we received a big library of measurements and test structures. Example: Reimut K. Hoffmann, *Integrierte Mikrowellen Schaltungen*, Springer-Verlag, 1983.

Figure I.1a–e is an example of what modern 3D CAD can do.

The meshing generator reduces the problem of too many small segments as shown in Figure I.1a,b. Figure I.1b shows the meshing applied to the structure, which is generated by the software, to calculate the 3D performance of the filter.

The actual simulation of the field distribution is shown in Figure I.1c.

Finally, the frequency response is shown in Figure I.1d.

Throughout this book, we have used mostly the Serenade program from my Compact Software time or its successor the Ansys Electronics Desktop 2022 R2 and also very successful ADS from Keysight and Microwave Office from AWR (Cadence), available through academic support programs. These programs offer good layout and other important simulation capabilities like transient analysis and a large variety of time domain semiconductor models. We performed some of the transistor parameter extraction ourselves at the time, as these data were not available from the manufacturer. Today, some commercial semiconductor houses and/or research companies provide them.

Figure I.1a The meshing generator reduces the problem of too many small segments.

Figure I.1b The meshing applied to the structure generated by the software to calculate the 3D performance of the filter.

Figure I.1c The simulation of the field distribution.

Despite being slower in computational time, this Serenade program on purpose used practically no mathematical shortcuts. An example is shown Figure I.1e, and the simulation results in Figure I.1f.

The software was developed jointly with Professor Vittorio Rizzoli and team at the University of Bologna, UNIBO, "Guglielmo Marconi" Department of Electrical, Electronic and Information Engineering DEI, Fred J. Rosenbaum, Professor of Electrical Engineering at Washington University, St. Louis, and with us at Compact Software, where Rowen Gilmore, Jason Gerber, and many more engineers, and I incorporated a fast multidimensional sparse matrix solver in double precision and many necessary microwave related capabilities. Later, Krylov subspaces which are used in algorithms for finding approximate solutions to high-dimensional linear algebra problems, and then added the noise correlation part for the active devices and their complex noise contribution. Modern iterative methods can be used for finding one (or few) eigenvalues of large sparse matrices or solving large systems of linear equations. They try to avoid matrix–matrix operations, but rather multiply vectors by the matrix and work with the resulting vectors. All algorithms that work this way are referred to as Krylov subspace methods.

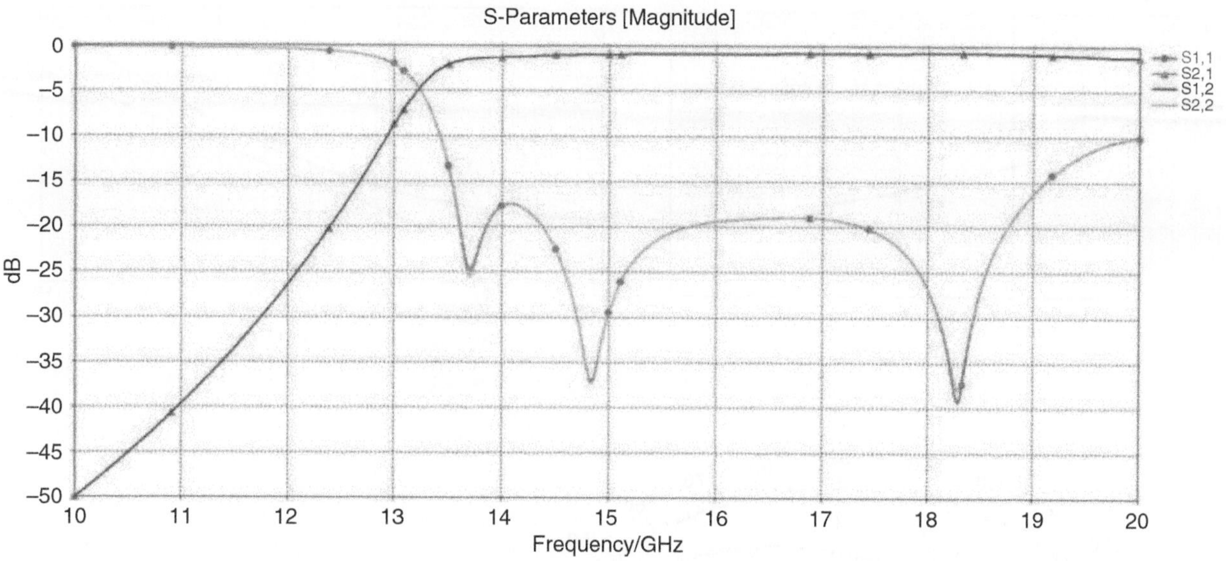

Figure I.1d Frequency response plot of the filter (Figure I.1a). Courtesy of Rohde & Schwarz Munich, Germany.

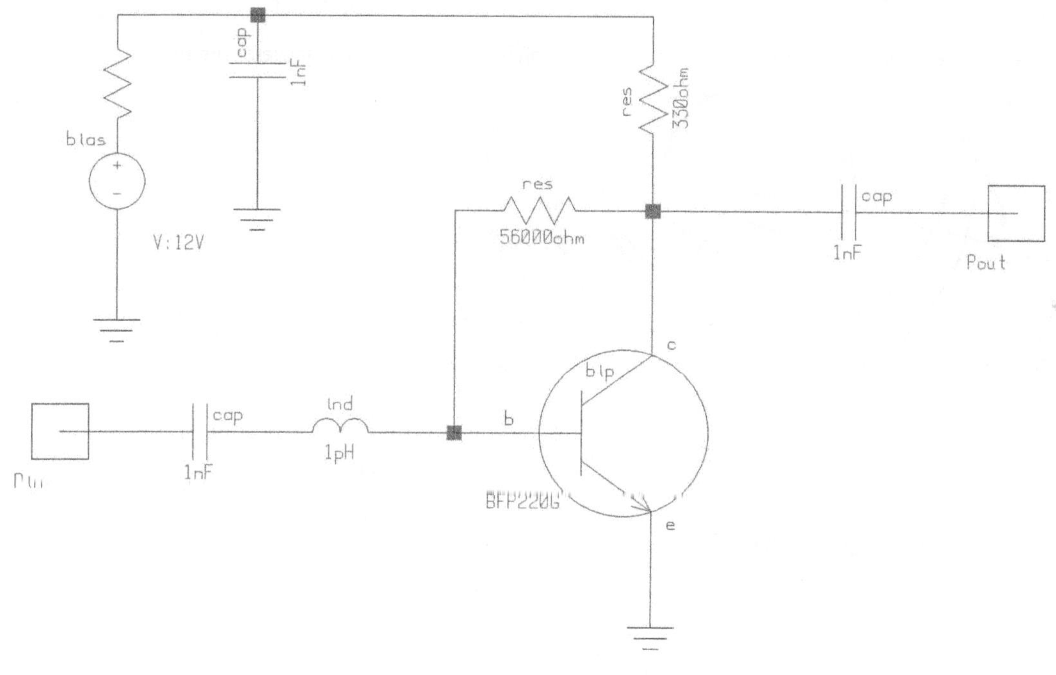

Figure I.1e Simulation using the Serenade program.

The noise correlation part was based on: H. Hillbrand and P. Russer, "An efficient method for computer aided noise analysis of linear amplifier networks," in *IEEE Transactions on Circuits and Systems*, vol. 23, no. 4, pp. 235–238, April 1976, https://doi.org/10.1109/TCS.1976.1084200. Other very advanced mathematical subroutines were also incorporated. Their work was probably based on: H. Statz, H. A. Haus and R. A. Pucel, "Noise Characteristics of Gallium Arsenide Field-effect Transistors," in *IEEE Transactions on Electron Devices*, vol. 21, no. 9, pp. 549–562, Sept. 1974, https://doi.org/10.1109/T-ED.1974.17966.

For the Harmonic-Balance Algorithms, refer to (V. Rizzoli, D. Masotti, F. Mastri, and E. Montanari, "System-Oriented Harmonic-Balance Algorithms for Circuit-Level Simulation," in *IEEE Transactions on Computer-Aided Design of Integrated Circuits and Systems*, vol. 30, no. 2, pp. 256–269, Feb. 2011, https://doi.org/10.1109/TCAD.2010.2092250), and also see "Further Reading" and "References" throughout this book.

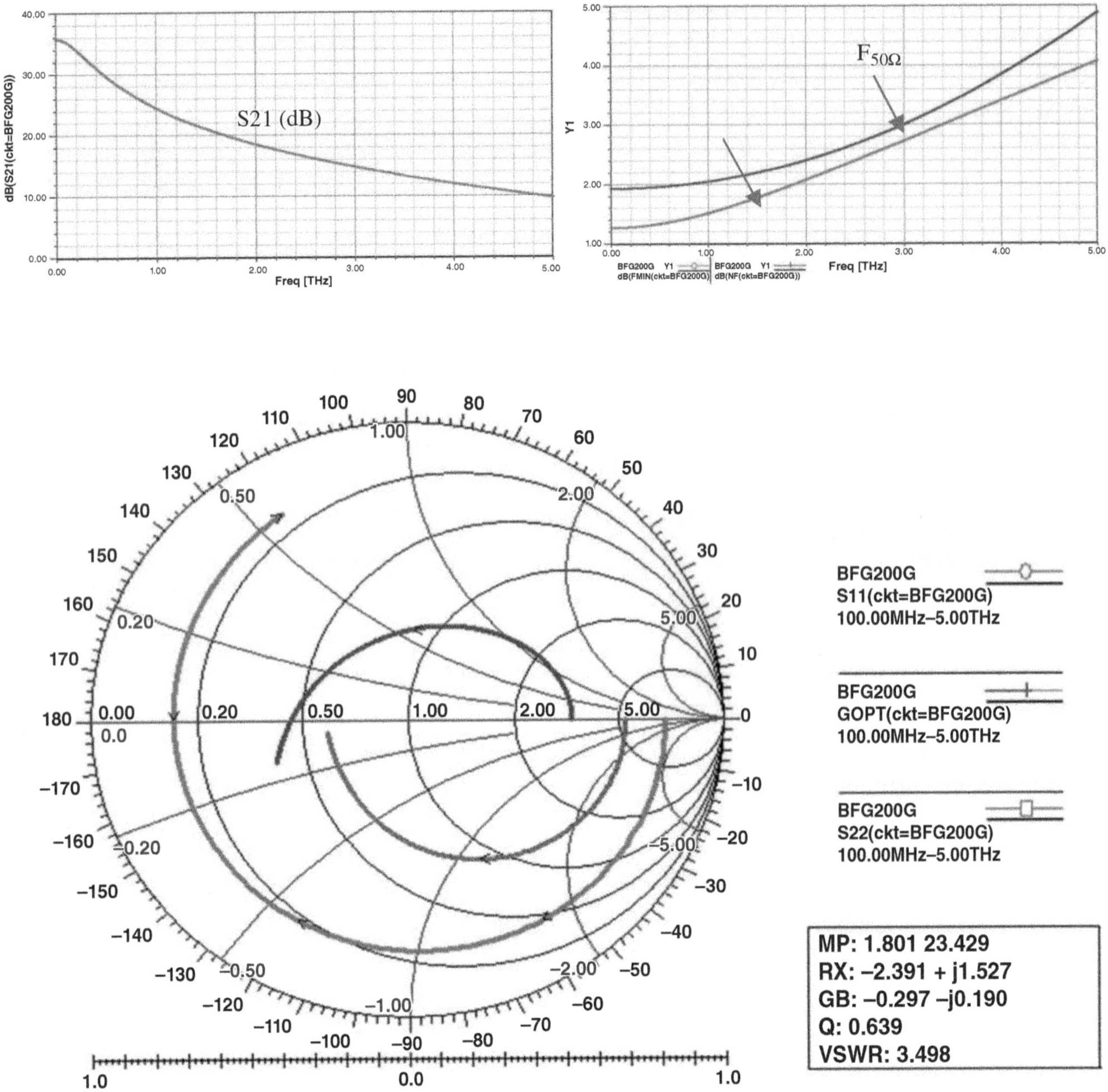

Figure I.1f Simulation results using the Serenade program.

Regarding numerical accuracy and further enhancements, I am very grateful for the work done by Jacob White and team from MIT for their contribution.

In 2014, my effort to have a really powerful general-purpose simulator was rewarded by receiving the IEEE C. B. Sawyer Memorial Award, with the citation as follows: "*For the development of PC software now allowing nonlinear noise analysis of RF circuit and the founding of Synergy Microwave, including the design and manufacture of ultra-low noise sources, RF components and subsystems exhibiting state-of-the-art performance.*" For more details see link (https://ieee-uffc.org/contact/ulrich-l-rohde).

The award is: "To recognize entrepreneurship or leadership in the frequency control community; or outstanding contributions in the development, production, or characterization of resonator materials or structure."

The original idea to develop the program came from Robert Pucel, a good friend of mine, from Raytheon who heard about the work of the Rizzoli group.

While the Compact Software company was the very first to offer a fully functional software solution for the analysis of nonlinear noise effects in oscillators, mixers, and frequency doublers Compact Software was never given official credit for this important contribution.

A very useful feature of the Serenade program is the fact that we found and implemented a reliable mathematical approach to calculate from the SPICE parameters both the spot noise figure as well as the large signal performance, including multitone nonlinear performance. During the cable TV time special highly linear transistors were developed, that were called CATV transistors. Unfortunately, most emphasis is now on noise figure and highest f_T and less on dynamic range. However, we will address this in the book. A perfect example of these highly linear transistors is the BFG540, as shown in Figure I.2.

Next is the BFP 840 low noise transistor from Infineon.

This datasheet from Infineon's BFP 840, Figure I.3 shows a somewhat unusual display for the large signal behavior. Its f_T is 85 GHz.

Figure I.4 shows the third-order intercept point at output O/P_3 (dBm) $= f(I_c,\ V_{\mathrm{CE}})$, $Z_S = Z_L = 50\ \Omega$, $f = 5.5$ GHz.

There are two problems with this transistor: One is that the package severely limits the practical usable upper frequency of operation and the other is the f_T curve vs. I_{ce}, shown in Figure I.5, which is highly nonlinear, with a peak at 25 mA.

While f_T of 85 GHz sounds attractive, we found no useful way to use the transistor in an oscillator above 20 GHz. This topic will be addressed further and the HICUM model will help.

While there are today voluminous textbooks/reference books on microwave circuit design available, this book is focusing on how to get started on designing amplifiers, mixers, and oscillators successfully and systematically using both available

NXP Semiconductors — Product specification

NPN 9 GHz wideband transistor — **BFG540; BFG540/X; BFG540/XR**

Three tone tests for – 3 dBm output

$d_{\mathrm{im}} = -60$ dB (DIN 45004B); $I_c = 40$ mA; $V_{CE} = 8$ V; $Z_L = Z_S = 75\ \Omega$; $T_{\mathrm{amb}} = 25^\circ$C;

$V_p = V_O$; $V_q = V_O - 6$ dB; $V_r = V_O - 6$ dB

$F_p = 795.25$ MHz; $f_q = 803.25$ MHz, $f_T = 805.25$ MHz;

Measured at $f_{(p+q-r)} = 793.25$ MHz

Figure I.2 Datasheet from Infineon's BFP 840 shows a somewhat unusual display for the large signal behavior.

BFP840ESD

Robust very low noise SiGe:C Bipolar RF Transistor in easy to use industry standard package

Features

- Robust very low noise amplifier based on Infineon's reliable, high volume SiGe:C technology
- Unique combination of high end RF performance and robustness: high maximum RF input power, 1.5 kV HBM ESD hardness
- Very high transition frequency $f_T = 80$ GHz enables very low noise figure at high frequencies:

  $NF_{\min} = 0.85$ dB at 5.5 GHz, 1.8 V, 6 mA

- High gain $|S_{21}|^2 = 18.5$ dB @ 5.5 GHz, 10 mA, 1.8 V
- $OIP_3 = 23$ dBm at 5.5 GHz, 1.5 V, 6 mA
- Ideal for low voltage applications e.g. $V_{CC} = 1.2$ V and 1.8 V (2.85 V, 3.3 V, 3.6 V requires corresponding collector resistor)
- Low power consumption, ideal for mobile applications
- Easy to use Pb- and halogen free (RoHS compliant) industry standard package with visible leads

Figure I.3 Part of the datasheet for the BFP840 transistor.

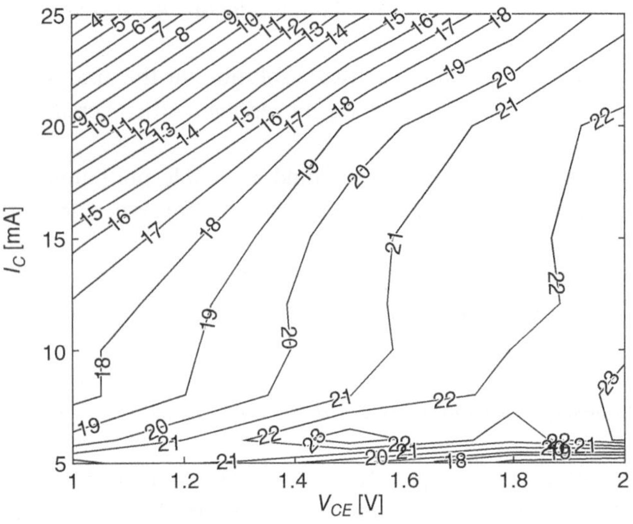

Figure I.4 Third-order intercept point at output O/P_3 [dBm] $= f(I_c, V_{\text{CE}})$, $Z_S = Z_L = 50\ \Omega$, $f = 5.5$ GHz.

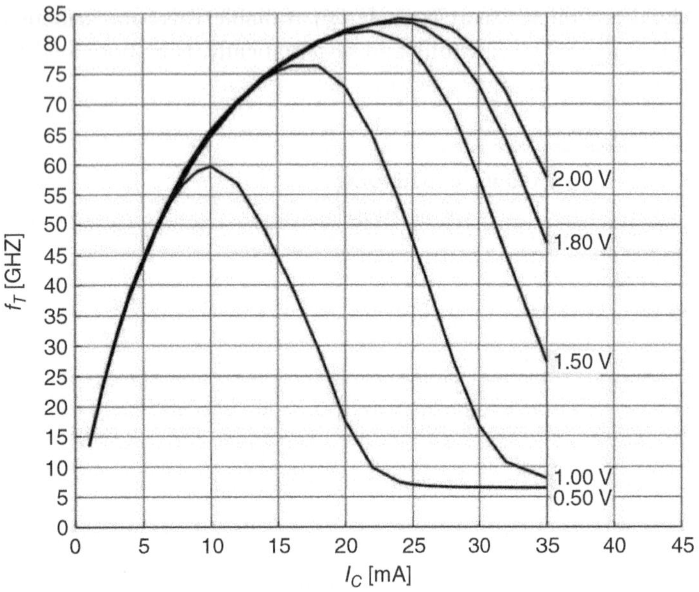

Figure I.5 f_T vs. I_{ce} curve.

and advanced transistors. In my book published in 1965, I predicted the development of GaAs heterojunction and FET transistors; it took until 1980 that they became available. So, I am curious which of my new predictions will come through.

In this book, each design will be unique and optimized for a particular purpose, we will show some useful examples with the proper choice of the semiconductors, and will discuss what to look for, we will apply the necessary mathematics and show effective use of CAD tools and what to expect at the end. Today's modern heterojunction bipolar transistors (HBT's) and MESFETs and CMOS will be covered and for essentially passive circuits, Hot Carrier diode applications as frequency multipliers with critical structures will be the starter. The mm-wave oscillators will be designed to be tunable by Si or GaAs tuning diodes so a brief introduction to them will be found in the diode chapter.

Both lumped and distributed designs will be addressed, and besides single circuit designs, more complex designs will be shown at the end.

For a more traditional and more demanding education a book is available: *Microwave Circuit Design Using Linear and Nonlinear Techniques*, 3rd Edition, Wiley 2021, by George Vendelin, Anthony M. Pavio, Ulrich L. Rohde and Matthias

Rudolph, and *Fundamentals of RF and Microwave Techniques and Technologies*, 1st Edition, by Hans Ludwig Hartnagel (Editor), Rüdiger Quay (Editor), Ulrich L. Rohde (Editor), Matthias Rudolph (Editor) ISBN-10, 3030940985, Springer 20223.

Also, *RF/Microwave Circuit Design for Wireless Applications*, John Wiley and Sons, 2013, By Ulrich L. Rohde, and Matthias Rudolph.

A book dedicated to wide band amplifier design is, *Fundamentals of RF and Microwave Transistor Amplifier*, 2009 by Inder J. Bahl, John Wiley and Sons, ISBN 978-0-4-470-130199-2

Another book totally dedicated to oscillators is: *The Design of Modern Microwave Oscillators for Wireless Applications: Theory and Optimization*, Ulrich L. Rohde, Ajay K. Poddar, Georg Böck.

Most of the publications on microwave circuits show the use of foundries and the foundry manual. All foundries request a non-disclosure agreement, and really do not give the user a detailed insight into the semiconductor used. Also, coming from the university, to get started, it is useful to work from discrete transistors up into the use of foundries. This explains why this book does not address foundry details but rather provides details about the information a foundry manual provides. The disadvantages of highly integrated circuits are that they work for one and one purpose only, while discrete circuits can be modified and yet the majority of microwave circuits are now built with complex integrated circuits. So, the purpose of this book is to enable the reader an easier transition from RF to sub-terahertz designs. This is necessary to understand how these circuits work.

Now many decades later, I have decided to "refresh" the 1965 book with the help of my colleagues:

Dr. Ing. habil. Ajay Poddar, Chief Scientist of my company **Synergy Microwave Corporation**, Paterson, USA, who in 2015 received the Walter G. Cady Award, "To recognize technical contributions in the area of piezoelectric frequency control devices," with the citation: "For the analysis, design, and development of a host of frequency control products exhibiting state-of-the-art performance, including the development of extremely low noise crystal oscillator circuitry."

And,

Professor Dr.-Ing. Matthias Rudolph, Faculty 1, Chair of Radio Frequency and Microwave Techniques – Ulrich L. Rohde endowed Professorship, who also cooperates with the Ferdinand Braun Institute, Leibniz-Institute für Höchstfrequenztechnik, which is a member of the Gottfried Wilhelm Leibniz Scientific Community, Berlin, a research institute.

Because of his affiliation, we had access to some rare and very useful data from the dissertation, "Signal Generation for Millimeter Wave and THz Applications in InP-DHBT and InP-on-BiCMOS Technologies, Muhammad Maruf Hossain, M.Sc., Eng., Rajshahi, Bangla, Fakultät IV - Elektrotechnik und Informatik der Technischen Universität Berlin."

In Chapter 4, Tools and Technology Explained, Sections 4.1 to 4.5 Professor Rudolph gives a detailed insight into the transistor technologies for microwave applications. The remaining chapter gives insights into the software tools used.

Key was to keep this book both thorough and easy to follow and yet complete and make the reader curious enough to possibly dive deeper into this topic. While supervising various PhD dissertations, I decided to sponsor a more detailed work to be performed by Trusha Kared of Synergy Microwave as a topic for a dissertation and look into the modeling and designing of microwave mixers. Both passive and active mixers are covered in a part of her work.

One more important fact: While at frequencies above 10 GHz a CAD tool is really necessary, the user may lose some insight of parts of the circuit. To minimize this, we have used at times a somewhat ideal transistor, its response is then limited by adding the parasitics and the response explained. Our goal is also to provide insight in the physics-based parameters like gain, noise and distortion, and how to deal with them optimally. As a reference and for scaling, we used the BFP 620 and the BFP 840 ($f_T = 80$ GHz). After scaling the BFP 840 to $f_T = 750$ GHz transistor, we called it BFP 750 in reference to the f_T.

During the last few years, Dr. Poddar and I have been involved in developing "Printed Multi Coupled Resonators, Slow wave Evanescent–mode Metamaterial based resonators," which will allow us the transition into the frequencies above 1000 GHz = 1 THz. In slow-wave propagation, the electromagnetic wave is transmitted in the guided-wave media with a slower phase velocity, namely, a shorter guided wavelength, at a specified operating frequency. This is achieved by modifying electric and magnetic energy storage in the guided-wave media. SWRs are attractive due to several reasons: the slow-wave effect makes them very compact and can support evanescent mode coupling, enabling the Q-multiplier effect at resonant frequency, thus providing a low phase-noise signal source solution. This new topic will be presented here, and we will introduce the potential of using these metamaterials-based high-Q resonators and the ability to change the frequency with magnetically coupled Möbius Loops and show its verification by using these resonators in practical oscillator applications. Here is a layout example, as shown in Figure I.6, and Figure I.7 shows its frequency response.

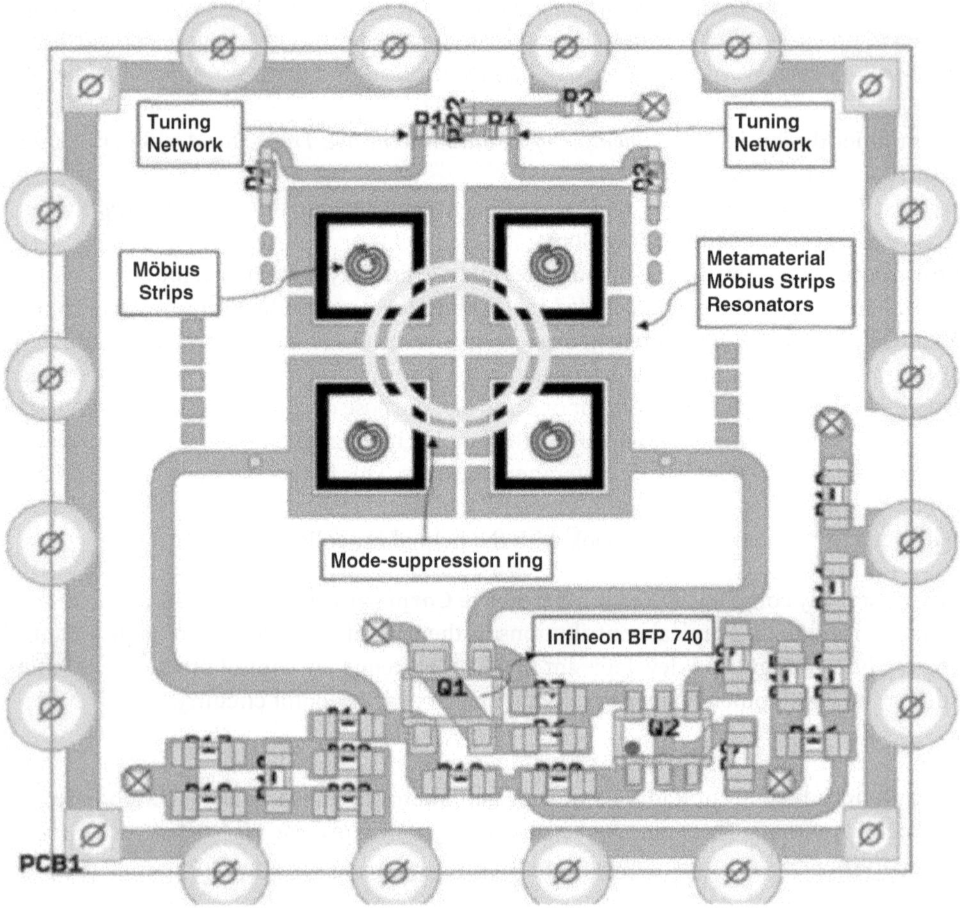

Figure I.6 Printed multi-coupled resonators, slow wave evanescent–mode metamaterial-based resonators [1].

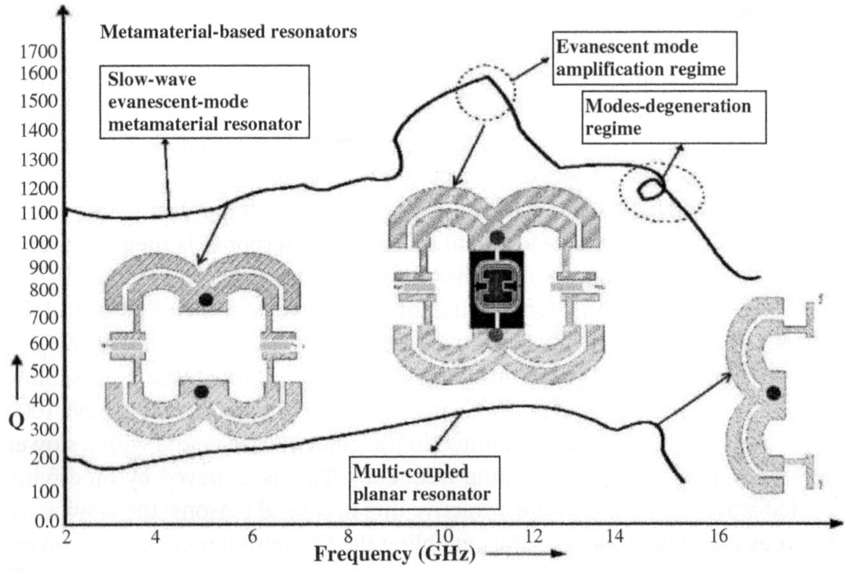

Figure I.7 Metamaterial Resonator. *Source*: Adapted from [2].

In 1965, microwave transistors were mostly used in single-stage circuits and comparatively simple modulations were used. Today's waveforms contain both important amplitude and gain information; also, phase and gain linearity are important. New parameters like differential gain and phase have been introduced.

Therefore, a special chapter on "Fundamentals of Linearity and Noise" is needed.

While the semiconductor development is going at a very fast pace, there are limits given by physics. There are two papers on the topic that stand out.

Implementation Challenges and Opportunities in Beyond-5G and 6G, Part III Sub-Millimeterwave/THz communication; Semiconductor Technology Capabilities, page 88 and SiGe HBTs and BiCMOS Technology for Present and Future Millimeter-Wave Systems, page 288; all in the *IEEE Journal of Microwave*, January 2021, inaugural Issue Volume 1 IJMEMX (ISSN 2692-8388). I am sure the next issues of the journal will likewise be very educational.

Also, Dr. Frank Schwierz, from the Technical University Ilmenau, Germany contributed a very educational summary of the evolution of the microwave transistors.

Finally, here is a very good introduction to what is coming:

https://spectrum.ieee.org/indium-phosphide-ics-unleash-the-high-frequency-spectrum

The news of the day, *Microwave Journal, April 10, 2025,* "Revolutionizing Cellphone and Wi-Fi RF Front-Ends with GaN-on-Si, Hal Emmer, Michael Guyonnet and Vincent Johnson, Finwave Semiconductor, Waltham, Mass.

The real highlight of the book project was the invitation from Tomás Palacios
Clarence J. LeBel Professor, EECS Director,

Microsystems Technology Laboratories

Massachusetts Institute of Technology

to join his team and the ability to contribute to the research at MIT.
https://www.mtl.mit.edu/news/mit-microsystems-technology-laboratories-welcomes-visiting-scientist-ulrich-l-rohde-expert

I hope this book will be useful for the microwave community, and I must thank all my friends and colleagues who contributed to this new adventure, not to forget Anisha M. Apte, who spent endless hours carefully and successfully optimizing this manuscript, and part of her PhD work is included here.

References

1 Poddar, A. K. (2014). Slow wave resonator based tunable multi-band multi-mode injection-locked oscillators. Dr.-Ing.-habil thesis, Chapter 6, BTU Cottbus, Germany.

2 Lei, H.-W., Wang, Y., Chiong, C.-C. et al. (2018). A 2.5-31-GHz high gain LNA in 0.15-µm GaAs pHEMT for radio astronomical application. In *2018 Asia-Pacific Microwave Conference (APMC)*, Kyoto, Japan, (228–230). https://doi.org/10.23919/APMC.2018.8617300.

Ulrich L. Rohde, Marco Island, Florida, USA

Technical University of Munich

TUM Distinguished Affiliated Professor

School of Computation, Information and Technology

Department of Electrical Engineering

1

Microwave Frequencies, Conventional Passive Elements, and Resonators

1.1 Introduction

Maxwell's equations govern all types of electromagnetic behavior at all frequencies. At lower frequencies (below about 500 MHz) where lumped components are less than $\lambda/8$ in length, it is convenient to define the component as a frequency-independent R, L, or C. As the frequency increases, the component will have distributed effects or added phase shift, which must be accounted for in the analysis. These effects will be discussed in this chapter [1–5].

1.2 Transition from Radio Frequency to Microwave Circuits

As electrical engineers, we first learn to analyze lumped-element circuits consisting of R, L, and C components, which are independent of frequency; the impedance or admittance of the components is linearly dependent on frequency. Then we learn that at high frequencies, additional parasitic effects must be included in the model of the component. For lumped components at low frequencies, signals travel at essentially the speed of light instantaneously through points in space where the component is located. At microwave frequencies (defined as 300 MHz–300 GHz, with frequencies above 30 GHz also called millimeter waves), the electrical components used to generate and process signals have a size similar to one-eighth of a wavelength. Then we must return to Maxwell's equations that are subject to the appropriate boundary conditions.

The radio frequency (RF) range is generally defined as 30–300 MHz, but it should be a frequency where all of the lumped components are shorter than about $\lambda/8$ in length. In the 100–500 MHz range, we are in a gray area where the circuit might be treated as either lumped or distributed, provided the lumped equivalent circuit is accurate. The distributed transmission lines are very long in this frequency range and thus are not used very often. Even at 850 MHz, lumped components are successfully used (e.g., cell phones). The distributed transmission line circuits would be very large for an 850 MHz cell phone but much more realistic at 5.8 GHz, a future band for cell phones and other nonlicensed products [2].

To illustrate this point, we calculate the $\lambda/4$ length on FR-4, a low-cost substrate at 850 MHz using an effective dielectric constant of 3:

$$\frac{\lambda}{4} = \frac{c}{4f\sqrt{\varepsilon_{\text{eff}}}} = \frac{7.5}{1.732 \cdot 0.85} = 5.09 \text{ cm} = 2.00 \text{ in} \tag{1.1a}$$

which is extremely large for a hand-held cellular phone circuit. A higher dielectric constant will reduce this dimension.

For transmission line circuits, we use transverse electromagnetic (TEM) or quasi-TEM microstripline for the passive circuits, which became available in 1965 [6]. The impedance and phase velocity were given by Wheeler in this chapter. It was immediately applied to the design of an X-band solid-state radar at Texas Instruments [7] using hybrid technology based upon alumina substrates and millimeter-wave monolithic circuits using gallium arsenide (GaAs) substrates at 94 GHz [8].

To understand this transition from RF to microwave or distributed elements, a plot of wavelength versus frequency is given in Figure 1.1 for various dielectric constants, where dielectrics effectively slow the velocity as ε_r increases.

It is important for the loss tangent to be low, which is true for alumina and GaAs but not for silicon. With GaAs substrates, $\rho > 10^6$ Ω-cm is easily achieved and maintained throughout the processing cycle. With Si substrates of high resistivity (10^3 Ω-cm), the n-type may become p-type with processing, and the resistivity is likely to lower, which is unacceptable for monolithic circuits. One of the most challenging problems today is to achieve a high-Q inductor on silicon, where a Q above 10 at 2 GHz is about the best that can be achieved at the time of this writing, as shown later in this chapter.

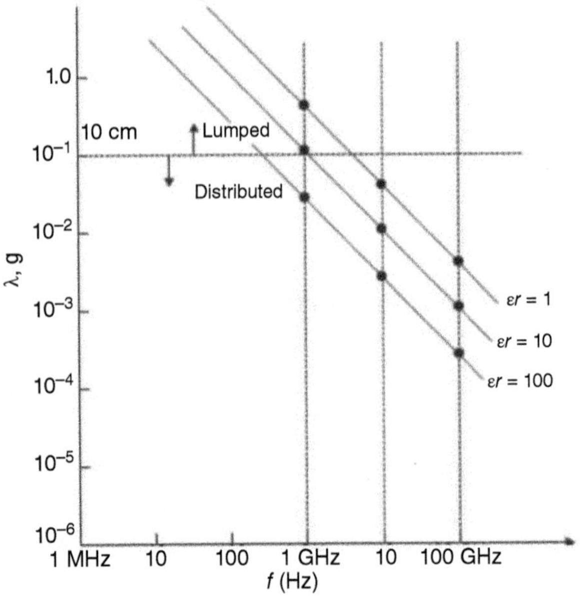

Figure 1.1 Guide wavelength versus frequency with dielectric constant as a variable. [2] / John Wiley & Sons.

Referring to Figure 1.1, the dielectric constant of most interesting dielectrics for monolithic integrated circuits (MICs) is of the order of 10, so the border between lumped and distributed may be estimated at $\lambda = 0.1$ m, or 10 cm, which coincides with the previous division at about 0.5–1 GHz.

The calculation of the guide wavelength for an effective dielectric constant of 10 follows at 1 GHz:

$$\lambda g = \frac{c}{f\sqrt{\varepsilon_{ff}}} = \frac{3 \times 10^{10}}{10 \times 10^{9}} = 10 \text{ cm} = 0.1 \text{ m} = 4 \text{ in}; \ \frac{\lambda g}{8} = 0.5 \text{ in} \tag{1.1b}$$

In other words, at frequencies above 1 GHz, the circuit must be treated as distributed since a length of more than 0.5 in will have distributed or phase properties that cannot be modeled by a simple R, L, C component [2].

At low frequency, we use lumped components with transistors and diodes as needed, that is, R, L, and C. When the components become about $\lambda/8$ long (approximately 500 MHz–1 GHz), we may add transmission line components (usually microstripline) in addition to lumped components. When the free-space wavelength becomes less than 1 mm (millimeter wave), designers are usually forced to use distributed transmission line elements where possible. Other forms of transmission are also used due to the limitations of TEM stripline/microstripline transmission lines [9], such as waveguides, surface modes, slotlines, coplanar waveguides, inverted microstriplines, and suspended microstriplines [10]. The geometry for these forms of TEM, transverse-electric (TE), and transverse-magnetic (TM) lines is given in Figure 1.2. It is useful to keep in mind that two wires (or conductors) are needed for TEM and only one conductor is required for TE and TM waves, which are generally at higher frequencies [2, Chapter 1].

Transmission lines may be modeled as in Figure 1.3, leading to the telegrapher equations, which are a time-domain description of the line:

$$\frac{\partial v(z,t)}{\partial z} = -RI(z,t) - L\frac{\partial I(z,t)}{\partial t} \tag{1.2}$$

$$\frac{\partial I(z,t)}{\partial z} = -Gv(z,t) - C\frac{\partial v(z,t)}{\partial t} \tag{1.3}$$

For sinusoidal steady-state conditions, this may be simplified to

$$\frac{dV(z)}{dz} = -(R + j\omega L)I(z) \tag{1.4}$$

$$\frac{dI(z)}{dz} = -(G + j\omega C)V(z) \tag{1.5}$$

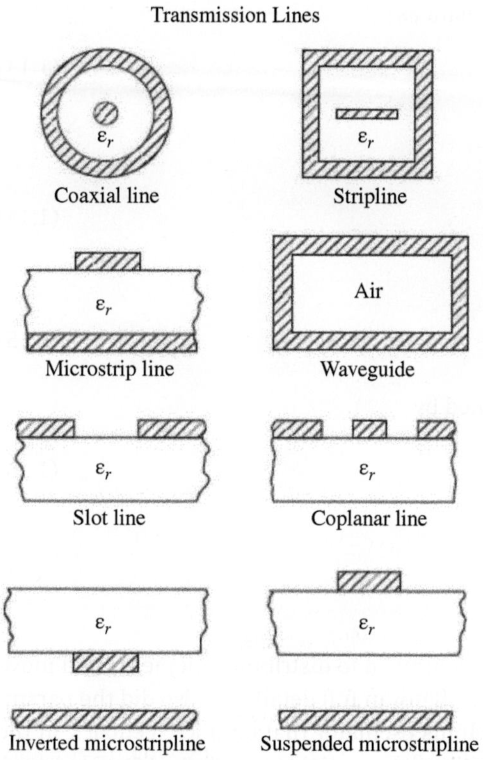

Figure 1.2 Geometry for microwave transmission. [2] / John Wiley & Sons.

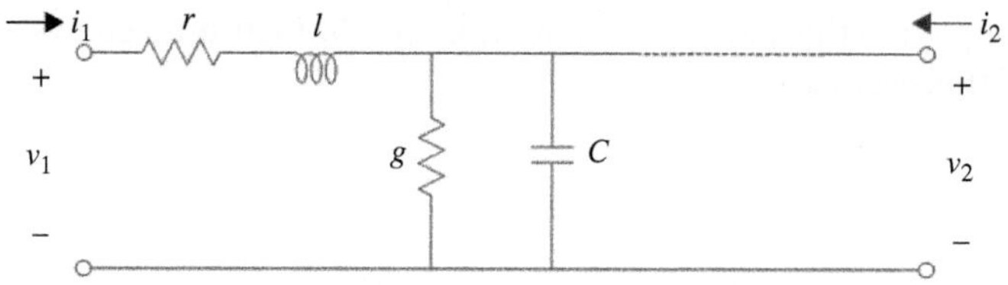

Figure 1.3 Lumped-element equivalent circuit of transmission line. [2] / John Wiley & Sons.

which is noted to be very similar to Maxwell's curl equations:

$$\nabla \times E = -j\omega\mu H \tag{1.6}$$

$$\nabla \times H = j\omega\varepsilon E \tag{1.7}$$

Combining the telegrapher equations leads to

$$\frac{d^2V(z)}{dz^2} = -\gamma^2 V(z) = 0 \tag{1.8}$$

$$\frac{d^2I(z)}{dz^2} - \gamma^2 I(z) = 0 \tag{1.9}$$

where $\gamma = \alpha + j\beta = \sqrt{(R + j\omega L)(G + j\omega C)}$ (1.10)

is the complex propagation constant, which is a function of frequency. Traveling-wave solutions can be found as

$$V(z) = V_0^+ e^{-\gamma \cdot z} + V_0^- e^{\gamma \cdot z} \tag{1.11}$$

$$I(z) = I_0^+ e^{-\gamma \cdot z} + I_0^- e^{\gamma \cdot z} \tag{1.12}$$

With a few more steps, we may obtain the voltage waveform in the time domain as

$$V(z,t) = \left|V_0^+\right| \cos(\omega t - \beta z + \varphi^+)e^{-\alpha z} + \left|V_0^-\right| \cos(\omega t + \beta z + \varphi^-)e^{-\alpha z} \tag{1.13}$$

where $\varphi^{\pm}$ is the phase angle of the complex voltage $V_0^{\pm}$.

The wavelength on the line is

$$\lambda = \frac{2\pi}{\beta} \tag{1.14}$$

and the phase velocity is

$$v_p = \frac{\omega}{\beta} = \lambda f \tag{1.15}$$

where the actual time delay must be calculated using the group velocity, defined by

$$v_g = \frac{d\omega}{d\beta} \tag{1.16}$$

For a TEM wave, these two velocities are the same.

1.2.1 Example: 800 MHz Voltage-controlled Oscillator

In order to demonstrate a transition from RF to microwave and ultimately from lumped to distributed elements and move toward microwave frequencies, we analyzed an 800 MHz voltage-controlled oscillator in full detail. We also did the parameter extraction for the Motorola transistor. Figure 1.4 shows the circuit, a Colpitts oscillator that uses RF feedback in the form of a 15 Ω resistor between the emitter and the base feedback. The capacitive voltage divider consists of 1 pF between the base of bipolar junction transistor (BJT) and the feedback resistor and 1 pF between the feedback resistor and common. The difference between the conventional analysis and this case is that the parasitic parts of the lumped elements were taken into consideration. On purpose, the tuned circuit is loosely coupled to this part of the transistor circuit. Figure 1.5 shows a comparison between predicted and measured phase noise for this oscillator.

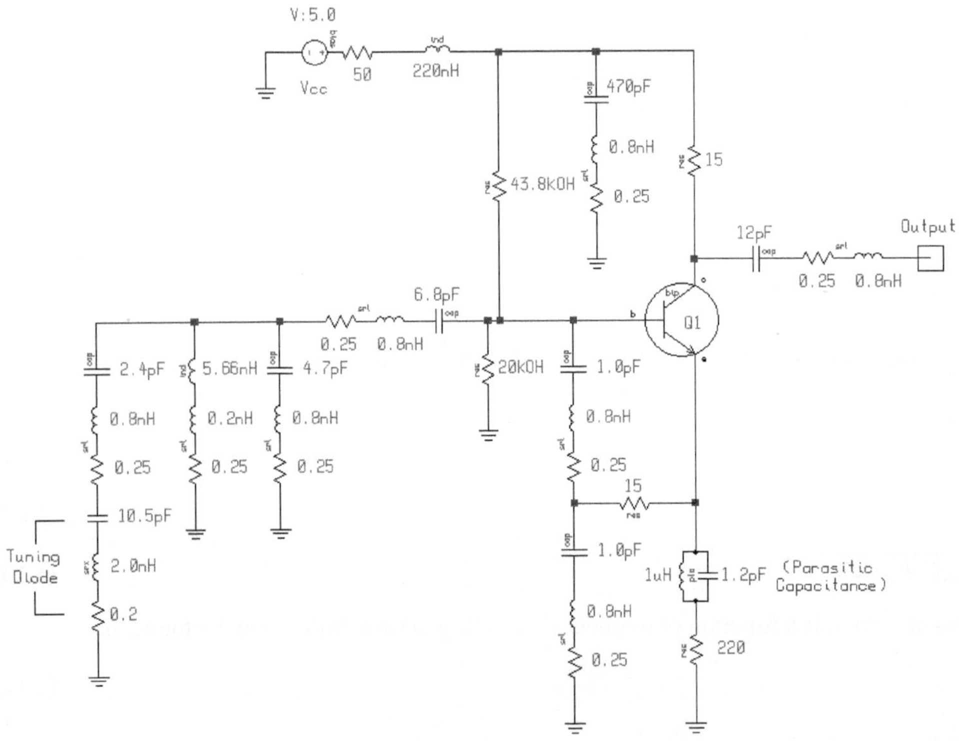

Figure 1.4 Colpitts oscillator that uses RF negative feedback between the emitter and capacitive voltage divider.

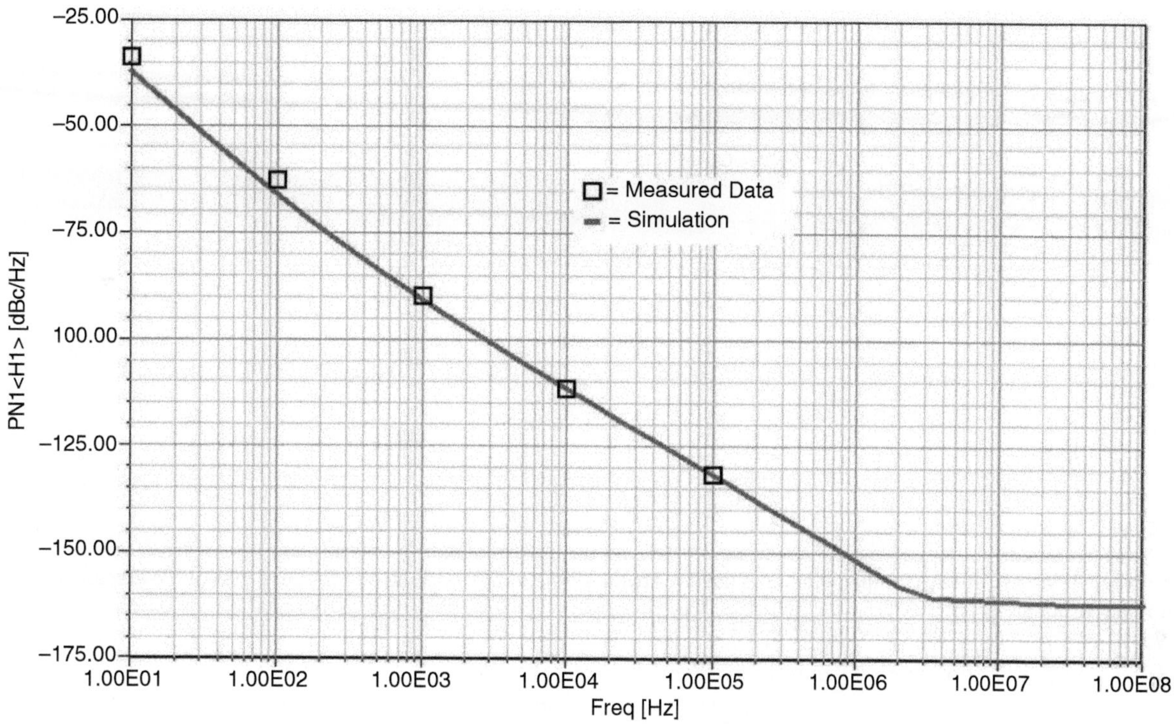

Figure 1.5 Comparison between predicted and measured phase noise for the oscillator shown in Figure 1.4.

To be realistic, we have also used real components rather than ideal ones. The suppliers for the capacitors and inductors provide some typical values for the parasitics. The major changes are 0.8 nH and 0.25 Ω in series with the 1.0 pF capacitors. The same thing applies for the main inductance, which has a parasitic connection inductance of 0.2 nH in series with a 0.25 Ω resistance. These types of parasitics are valid for a fairly large range of components assembled in surface-mount applications. Most engineers model the circuit only by assuming non-lossy/perfect devices, not adding these important parasitics. One of the side effects we have noticed is that the output power is more realistic and, needless to say, the simulated phase noise now agrees quite well with measured data. This circuit can also serve as an example for modeling amplifiers and mixers using surface-mount components. Good computer aided design (CAD) tools provide a library of passive elements in the form of measured S-parameters where all the parasitics are considered.

Figure 1.5 shows the simulated phase noise of the oscillator and some of the measured points. Following this example, we now apply the same logic to an amplifier.

1.3 RF Versus Microwave Modeling

Let us now take a look at a simple amplifier that shows a textbook-type approach to an RF circuit.

The circuit shown in Figure 1.6 is too simplified, as it assumes ideal components that are not realistic. For easy understanding, the transistor bias circuit is not shown [1].

In order to decouple the transistor from the biasing, we use RF chokes and DC decoupling (bypass) capacitors. As frequency increases, these inductors are either not manufacturable or have such a low Q that their use becomes questionable. This is the point where one should introduce the so-called distributed elements.

Figure 1.7 shows the very same circuit but resorting to distributed rather than lumped elements. The amplifier shown earlier, on purpose, illustrates a somewhat exaggerated case as we show interdigital capacitors, radial stubs, and transmission lines. Not all of them are necessary at the same time, but the two T-junctions are really important for interconnection. Most designers do not go to this level of detail, including showing vias for grounding. In summary, this circuit diagram may be a little overboard, but it indicates how complicated things can become [1].

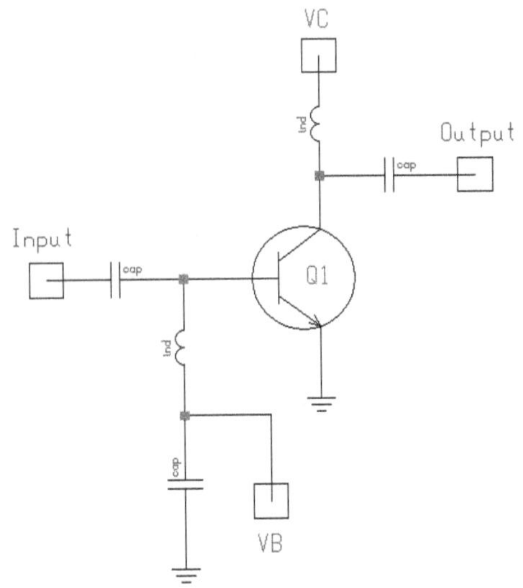

Figure 1.6 Simple BJT RF amplifier with lumped elements used for bypassing and DC blocking and for base bias and collector supply feeds. [1] / John Wiley & Sons.

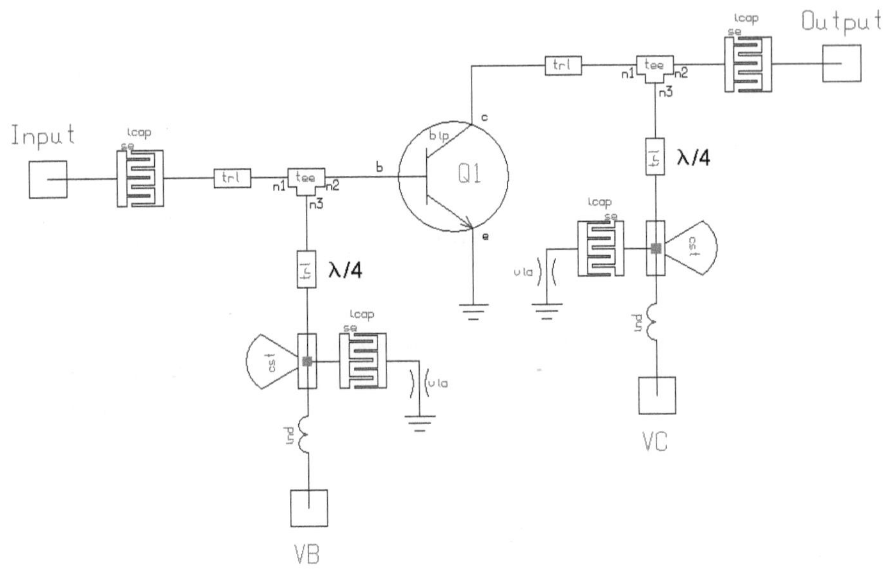

Figure 1.7 Simple BJT RF amplifier with distributed elements used for bypassing and DC blocking and for base bias and collector supply feeds. [1] / John Wiley & Sons.

The elements we have introduced now (which are part of any good, up-to-date CAD tool; see the element library for details on how to use their physical element descriptions) are listed as follows:

The transmission line is one of the most important interconnections at microwave frequencies.

Any printed connection between two points on a circuit board is a transmission line (Figure 1.8). Its characteristic electrical impedance is a function of the square root of the dielectric constant (ε_r), the width, metallization thickness, height above substrate of the line, and the loss tangent of the substrate. Since lines frequently have to be laid out in the form of curved connections or have a bend in their direction, we have to add elements capable of describing the high-frequency consequences of such connections such as mitered and radial bend elements that perform this function [1].

Now to the printed circuit board (PCB) carrier materials, the following is a list of carrier materials useful from very high frequency (VHF) to super high frequency (SHF) range.

Dielectric and material properties of common RF and microwave substrates and base materials (hard and soft substrates) (based on Rogers's corporation).

Material	Dielectric permittivity at 10 GHz	Loss tangent at 10 GHz (10e4 tan)	Surface roughness (μm)	Thermal conductivity (W/mK)	Dielectric strength (kV/cm)	Soft or hard substrate
Alumina (Al_2O_3)	9.9	1–2	0.05–0.25	30–37	4000	Hard
Aluminum nitride (AlN)	8.9	3–5	0.05–0.6	150–170	150	Hard
Gallium arsenide (GaAs)	12.85	6	0.025	30	350	Hard
Gallium nitride (GaN)	9.5/10.4	–	0.025	66–225	4000	Hard
Indium phosphide (InP)	12.4	10	0.025	40	350	Hard
Silicon (Si)	11.9	10–100	<0.001	100–150	300	Hard
Silicon carbide (SiC)	10.8	20	<0.001	120–200	2.5×10^3	Hard
Duroid 5870	2.33	12	–	0.26	–	Soft
Duroid 5880	2.2	12	–	0.26	–	Soft
Duroid 6002	2.94	12	–	0.44	–	Soft
Duroid 6010	10.2–10.8	27	–	0.48	–	Soft
Duroid RO3003	3	13	–	0.50	–	Soft
Duroid RO3006	6.15	25	–	0.61	–	Soft
Duroid RO3010	10.2	35	–	0.66	–	Soft
Duroid RO4003	3.38	27	–	0.64	–	Soft
Duroid RO4350B	3.48	40	–	0.62	–	Soft
TMM 3	3.27	20	–	0.7	–	Soft
TMM 4	4.5	20	–	0.7	–	Soft
TMM 6	6	23	–	0.72	–	Soft
TMM 10	9.2	22	–	0.76	–	Soft
TMM 10i	9.8	20	–	0.76	–	Soft
CuFlon	3.05	30	–	0.25	–	Soft
CLTE	2.94	0.0025	–	0.5	–	Soft
FR-4	4.5–4.8	220	~6	0.16	–	Soft
Norclad	3.8	11	–	0.23	–	Soft
Polyimide	2.3–13.4	50	–	0.04–1.7	–	Soft
PTFE (Teflon)	3.05	30	–	0.25	–	Soft

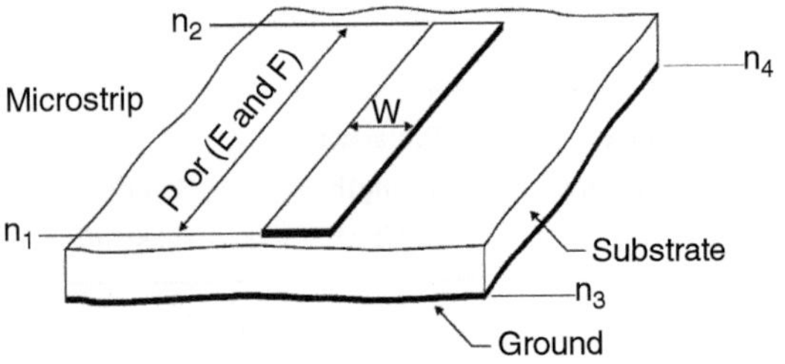

Figure 1.8 Transmission line in microstrip. [1] / John Wiley & Sons.

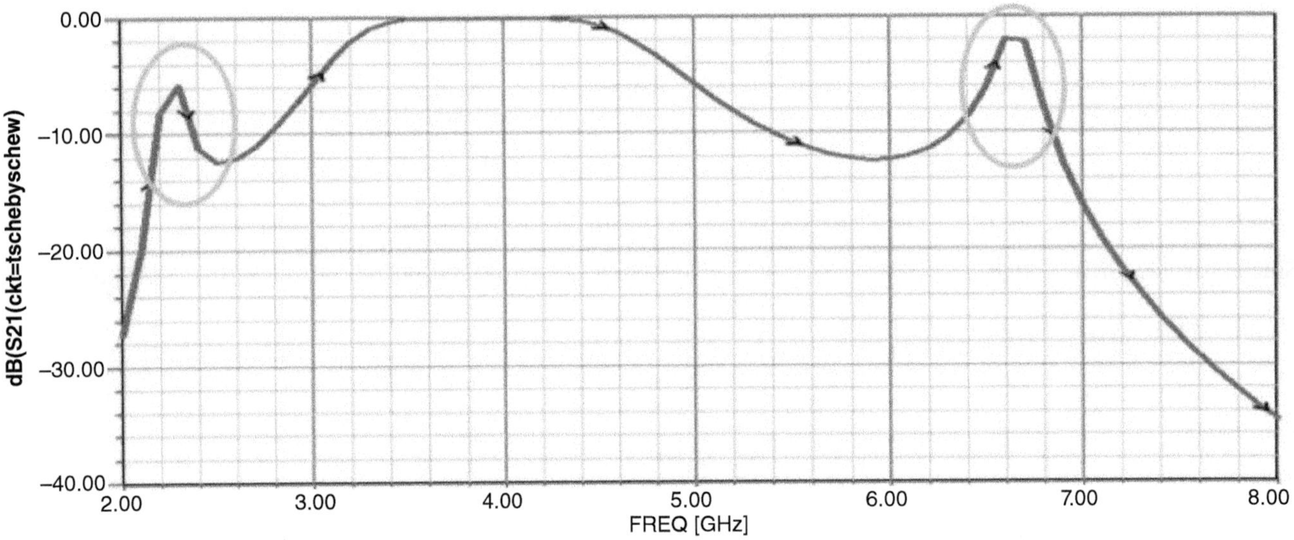

Figure 1.9 The frequency response of the circuit.

1.4 Application of Passive Devices in Filters

An important application of passive devices, such as lumped and distributed elements, is their use in filters. The following are some practical applications.

1.4.1 Example: 4 GHz Filter

The first example is a 4 GHz filter circuit diagram published in 1977 at the European Microwave Conference. In essence, it has two parallel-tuned circuits, with two series resonant circuits.

Figure 1.9 shows some interesting frequency responses outside the pass band, as the ears on the frequency–response curve. Figure 1.10 shows the simulated return loss in dB of the 4 GHz filter. The −50 dB return loss is based on numerical accuracy, not on actual losses. In reality, the return loss would be more like −30 dB.

The filter in Figure 1.11 is a fifth-order 4 GHz Chebyshev filter, calculated for lumped elements.

Using a modern filter-synthesis program with internal impedance transformation, we get the schematic and CAD simulated results for a higher-order 4 GHz filter shown in Figure 1.12, with much better behavior.

Essentially, the filter in Figure 1.12 is also a fifth-order filter, which looks more complicated because of the impedance transformation, but is now much better behaved, as seen in Figure 1.13.

Frequency response of the synthesized filter using realistic components (Q = 50), resulting in an insertion loss of 1.4 dB. The filter has no spurious resonances.

The return loss of about −20 dB is quite reasonable (Figure 1.14).

At frequencies above 500 MHz, lumped elements are typically replaced by transmission line elements.

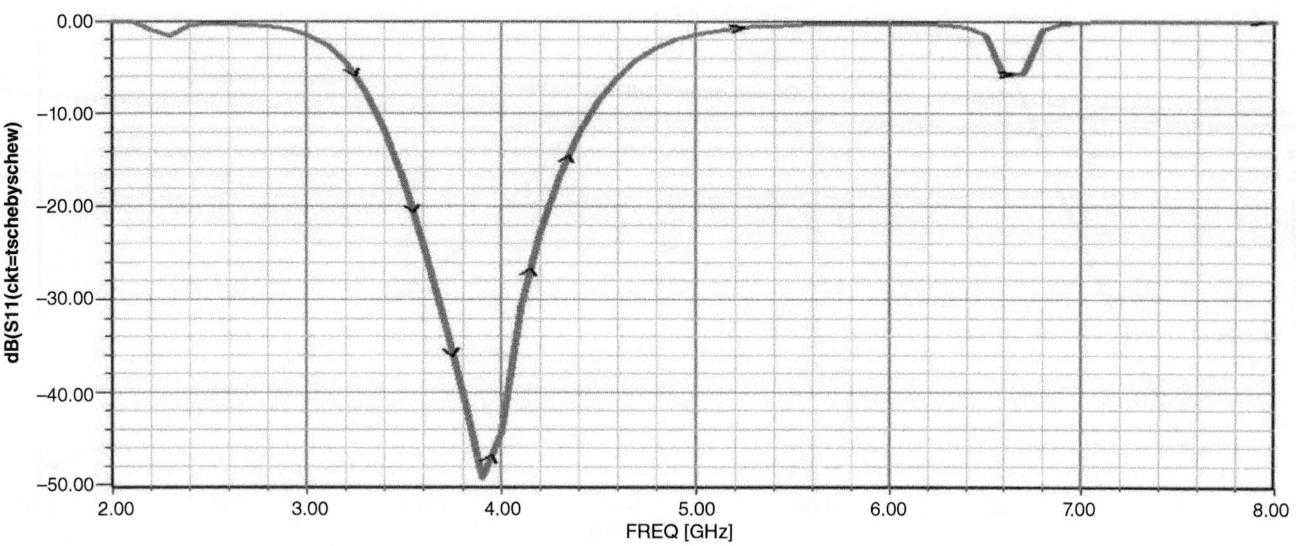

Figure 1.10 Return loss in dB of the 4 GHz filter.

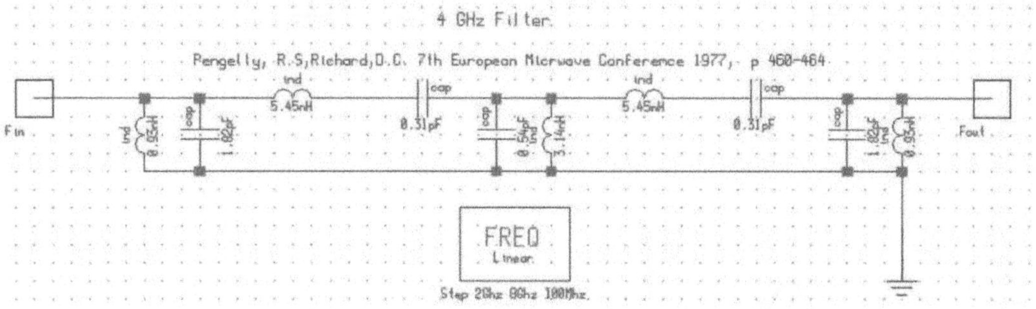

Figure 1.11 4 GHz center-frequency bandpass filter using lumped elements.

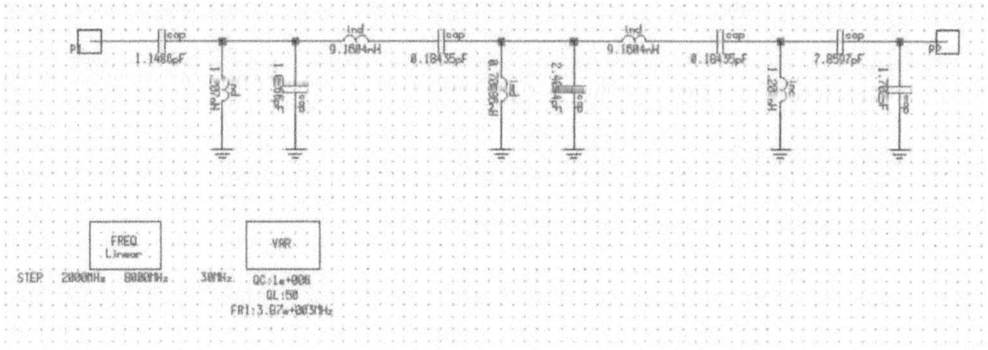

Figure 1.12 New 4 GHz synthesized filter. *Note:* The impedance transformation at the input and output of the filter.

1.5 Richards Transformation

Most classical transmission line-based circuits including filters are designed on the basis of one of the fundamental properties of the transmission line, as they cannot be realized in the form of lumped elements. Consider a length of transmission line with characteristic impedance Z_t and electrical length θ, short-circuited at one end. The impedance at the input will be

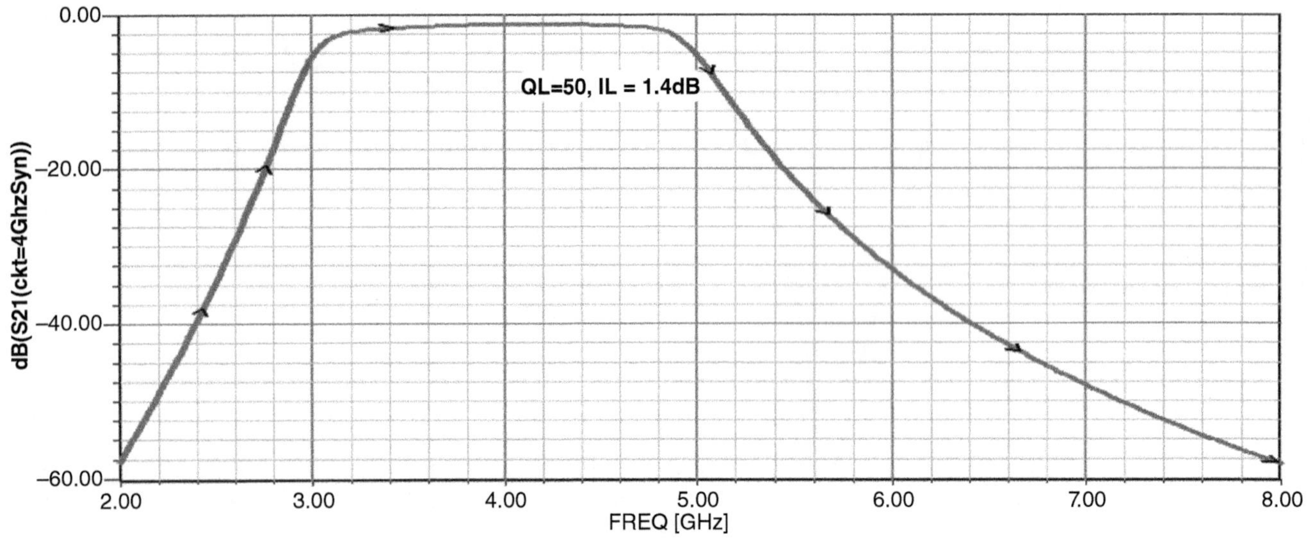

Figure 1.13 Frequency response of the synthesized filter using realistic components (Q = 50).

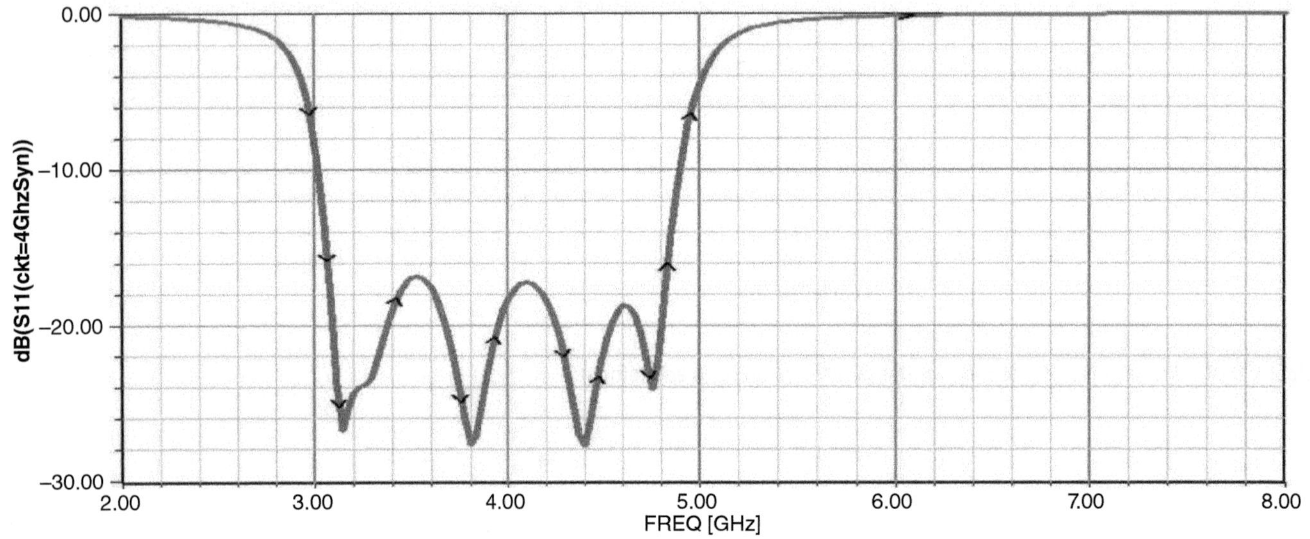

Figure 1.14 Return loss of the 4 GHz filter shown earlier.

$$Z_{in} = jZ_t \tan(\theta) \tag{1.17}$$

Define a new frequency variable $\lambda = \Sigma + j\Omega = \tan h(sT)$, where $s = \sigma + j\omega$ is the standard complex frequency variable and T is the time delay through the transmission line. Along the imaginary frequency axis, $\lambda = j\Omega = \tan(\theta)$. Now, we can express the impedance of the shorted stub in the new frequency variable as

$$Z_{in} = j\Omega Z_t \tag{1.18}$$

In this new frequency variable, the shorted stub takes on the appearance of an inductor. By a similar process, an open-circuited stub is a capacitor. The transformation used to establish λ is known as the Richards transformation, after its author [4]. The new complex frequency variable, λ, is often known as the Richards variable. Circuits designed to take advantage of this analogy consist of assemblages of shorted stubs (inductors), open stubs (capacitors), and cascaded transmission lines (unit elements), all having the same phase length. Formally, these are *commensurate transmission line networks* and form the basis of much transmission line theory.

We can use Richards transformations to replace the inductors and capacitors of a lumped-element filter design. In fact, for low-pass filter design, the frequency ωc is the filter's cutoff frequency. Using these stubs to replace inductors and capacitors will result in a filter response similar to that of the lumped-element design—a low-pass filter with a cutoff frequency ωc. However, the behavior of the filter in the stopband will be very different from the lumped-element design. For example, at (high) frequencies where the stub length becomes a multiple of $\lambda/2$, the filter response will be that of $\omega = 0$—near-perfect transmission!

$j\omega L = jZ_0 \tan(\pi/4) = jZ_0$, short-circuited stub as $\tan(\pi/4)$, and $-j/\omega C = jZ_0 \cot(\pi/4) = -jZ_0$; open-circuited stub.

This is correct as for $ag << 1$ $\tan(ag) = ag$ and $\cot(ag) = 1/ag$ and the electrical length is one-eighth of the wavelength, meaning that the available variable is the line impedance [5, Chapter 6].

1.6 Kuroda's Identities

We find that Kuroda's identities can be very useful in making the implementation of Richards transformations more practicable. Kuroda's identities essentially provide a list of equivalent two-port networks. By equivalent, we mean that they have precisely the same scattering/impedance/admittance/transmission matrices. In other words, we can replace one two-port network with its equivalent in a circuit, and the behavior and characteristics (e.g., its scattering matrix) of the circuit will not change! We want to do this because one of the equivalents may be more practical to implement! For example, we can use Kuroda's identities to (1) physically separate transmission line stubs; (2) transform series stubs into shunt stubs; and (3) change impractical characteristic impedances into more realizable ones.

Kuroda's identities (aka Kuroda's transforms) are used to convert a section of transmission line with an open-parallel stub into an electrically equivalent section of transmission line with a shorted series stub. As a result, an identical S-parameter matrix is produced, which performs the same function. The technique is handy when designing distributed element circuits where one configuration is possible and the other is not. Filters are a good example, because in the physical layout, open parallel stubs are difficult (or impossible) to realize, whereas series shorted series stubs are not [5, pp 318].

Regarding transformation, a simple modeling may be used to find parameter values for an equivalent network. This example demonstrates the technique by optimizing a transmission line to perform like a lumped inductor. The example uses a transmission line with 1 pF in parallel to behave like a parallel LC circuit at a specific frequency (Figure 1.15a–c).

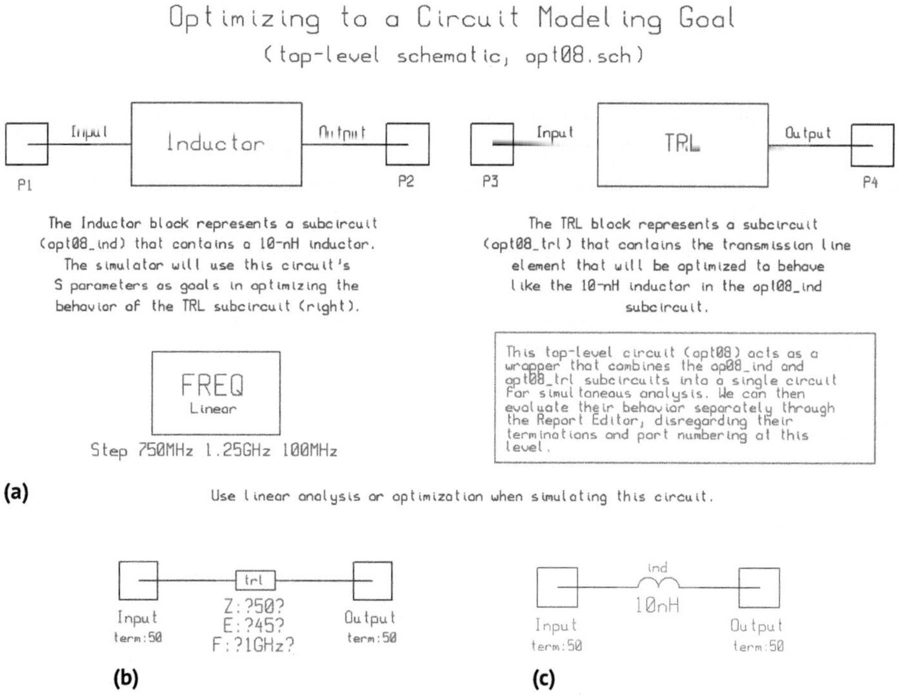

Figure 1.15 (a) Optimizing a transmission line to be used as an inductor. (b) Optimizable transmission line. (c) Reference inductor.

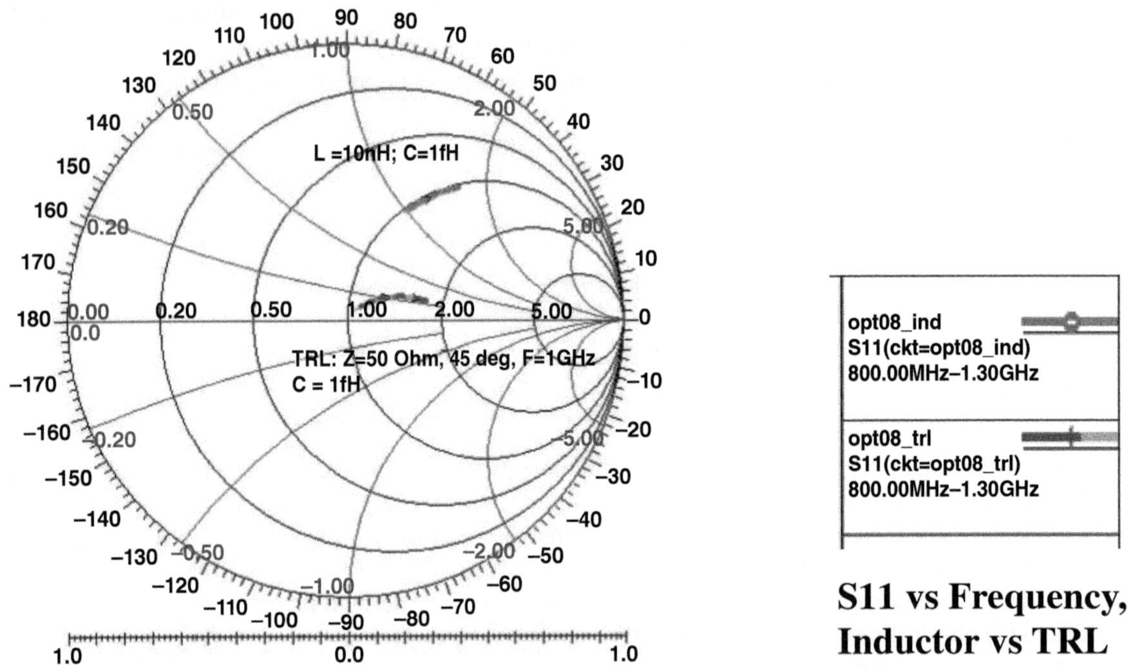

Figure 1.16 Result of optimization of the 750–1250 MHz.

The optimizer will adjust the transmission line to behave like a 10 nH inductor from 750 MHz to 1.25 GHz. This can be seen in the plot of the Smith diagram in Figure 1.16, showing S11 versus frequency.

1.7 10.2 GHz Test Circuit

The example in Figure 1.17a is the synthesis of a lumped-element filter for 10.2 GHz bandpass filter (ham RF). The synthesis program provides results for L and C's, which are not practical.

Figure 1.17b shows S21 and S11 in dB as a function of frequency. Insertion loss is about 5 dB.

Figure 1.18a shows a bandpass filter using transmission lines and the cross element.

Frequency response of the filter using lossless elements See Figure 1.18b shows a slight dip; the center frequency of the filter is about 16 GHz. By introducing losses, a more realistic frequency response can be obtained (see Figure 1.18c).

This type of transformation introduces a notch, which is at about 13.4 GHz.

Now let us apply electromagnetic (EM) coupling. The following is a Richards transformation of a double EM-coupled tuned circuit. Figure 1.19a shows the circuit arrangement. It is actually quite a complex filter, with typical distributed elements such as crosses and coupled lines, including the end effects of the transmission line (radiation).

Figure 1.19b shows the frequency response of the bandpass filter at 16.3 GHz, having an insertion loss of about 2.61 dB.

1.7.1 Example: Broadside-coupled Filter at 10.2 GHz (Ham RF)

The following circuit (Figure 1.20a) is a perfect symmetrical filter at 10.4 GHz and is the result of a filter synthesis using five coupled elements. It is designed on a material with $\varepsilon = 10$. Figure 1.20b shows the S11 and S21 of the synthesized filter at 10.4 GHz.

The next example, Figure 1.21a shows a four-stage advanced filter with EM coupling, a result of a synthesis program, while the results of a synthesis program shows the S11 and S21 in dB of the 10.2 GHz synthesized filter in Figure 1.21b.

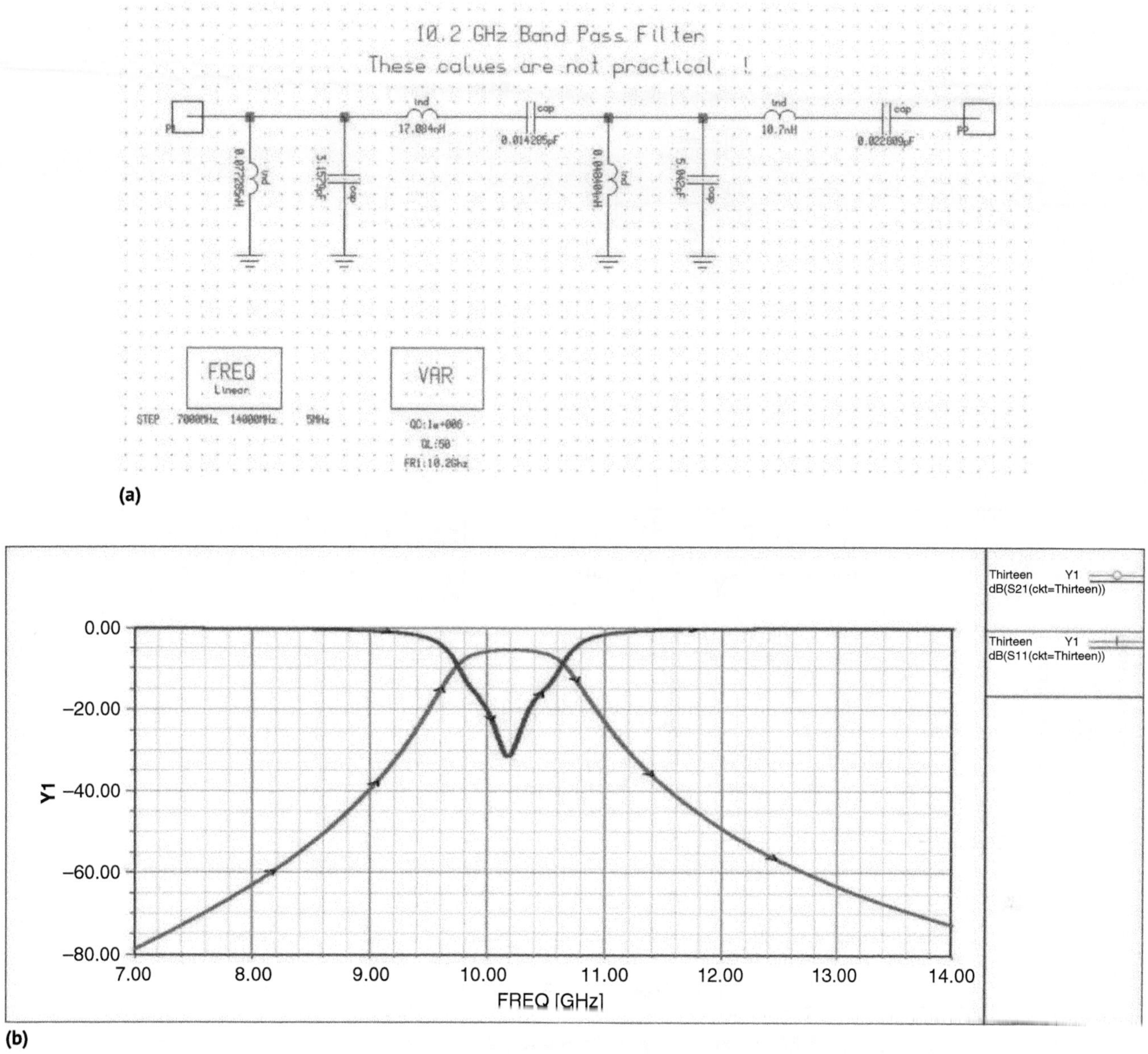

Figure 1.17 (a) 10.2 GHz bandpass filter (values not practical). (b) S21 and S11 in dB as a function of frequency.

1.7.2 Example: Two-stage Matching with a Two-section Series Transformers

Another interesting problem that can be solved using transmission lines is impedance matching. Let us assume that we have to match a 25 Ω source to a 75 Ω load. A typical answer would be to use a transmission line of the impedance sqrt(75 × 25) = 43 Ω. Now let us try the optimizer; but using two transmission lines complicates the situation. The desired center frequency is 10 GHz and we will try to get the maximum bandwidth possible while maintaining a return loss of better than −30 dB.

After optimization, in the schematic shown in Figure 1.22a, the characteristic impedance of TRL1 needs to be ≈ 33.5 Ω and the second TRL2 ≈ 56.3 Ω. This is a very elegant method to generate a transmission line-based wideband matching. The specified frequency at the transmission line needs to be the highest; in this case, it is 12 GHz. The worst case S21 in dB (no losses) is −0.04 dB (see Figure 1.22b).

In the previous simulations of filters, we used ideal transmission lines. For the transition to real transmission lines, meaning losses from skin effect and radiation, and loss tangent of the material, we now use a transmission line calculator/synthesizer.

The following calculations, Figure 1.23, are based on Duroid-based material.

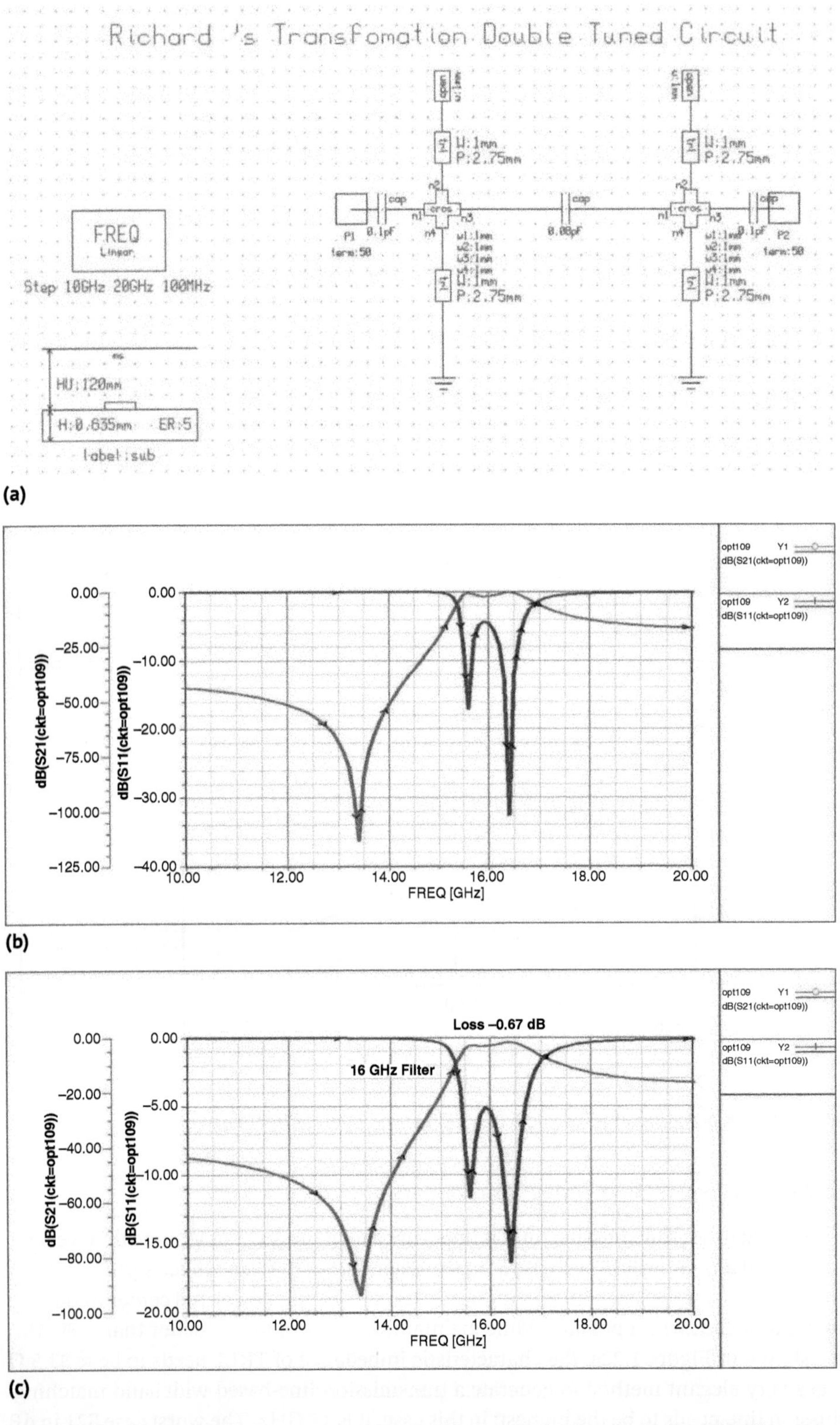

Figure 1.18 (a) The Richards transformation double-tuned circuit schematic of the filter using the ideal case. (b) Frequency response of the filter using lossless elements. (c) Frequency response of the filter using lossless elements—with loss.

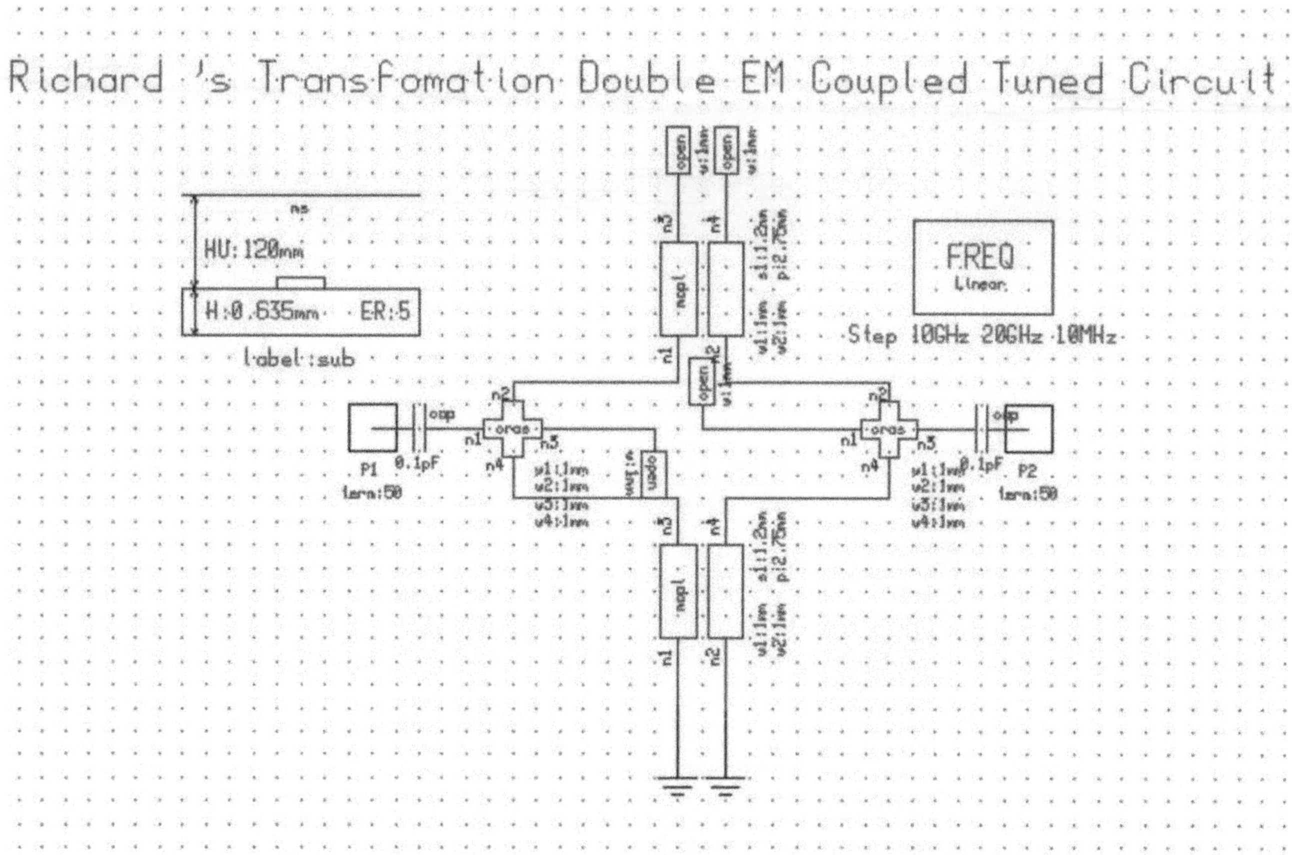

(a)

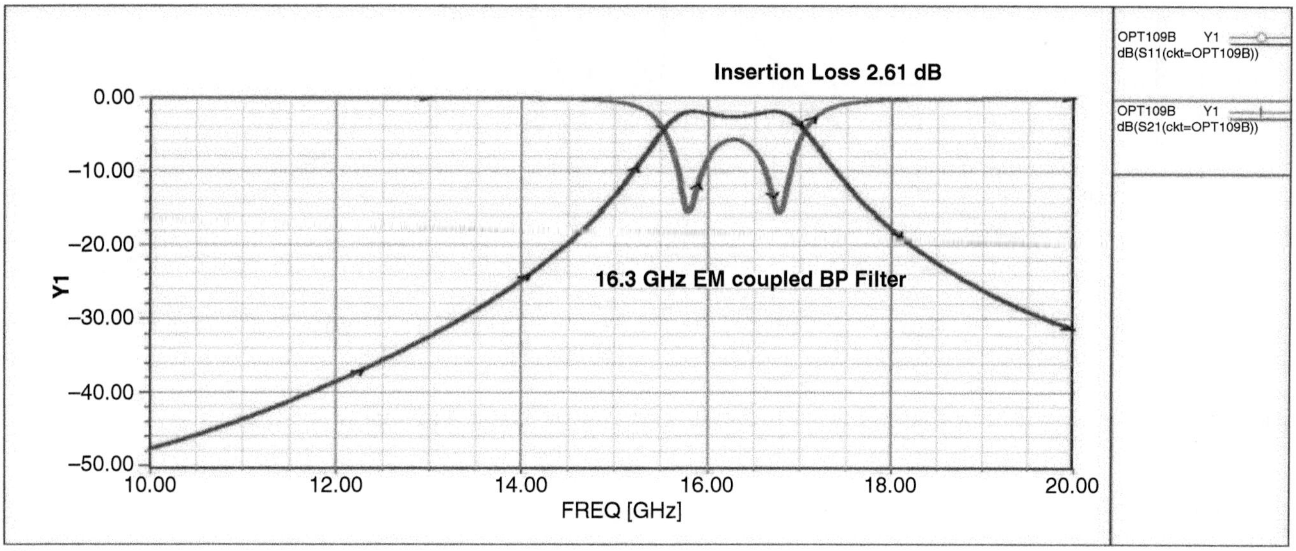

(b)

Figure 1.19 (a) Richards transformation double EM-coupled tuned circuit. (b) Frequency response of the bandpass filter at 16.3 GHz, having an insertion loss of about 2.61 dB.

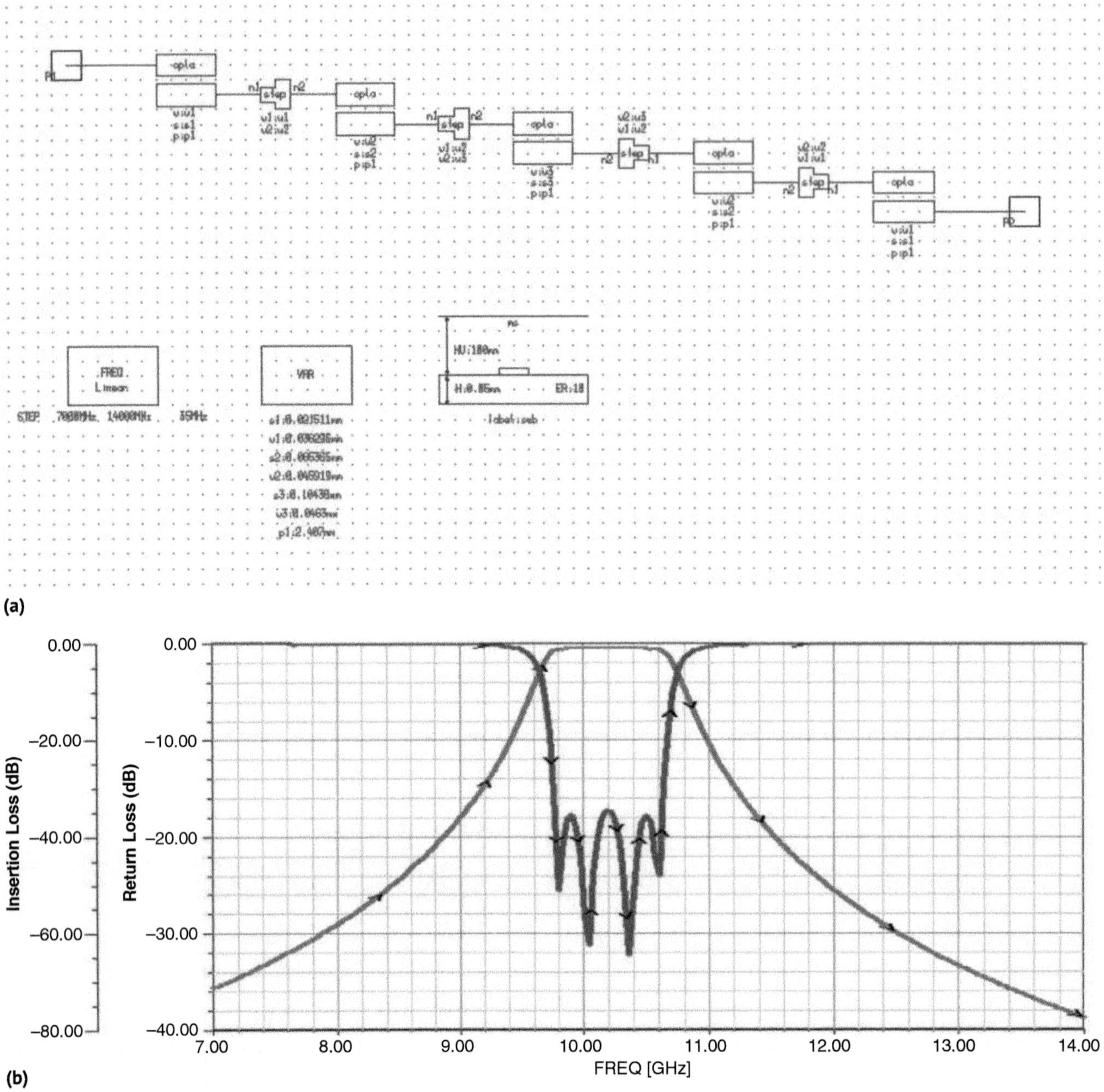

Figure 1.20 (a) A perfect symmetrical filter at 10.4 GHz. (b) S11 and S21 of the synthesized filter at 10.4 GHz.

1.8 Validation

As we are going to use this synthesis program in the range from 10 GHz to 1 THz, we will build test structures and compare measurements with simulations. For critical applications at 1 THz, we decided to use high frequency structure simulation software (HFSS), an Ansys product, which is considered as the gold standard, in cases where very accurate predictions are necessary.

The measurements were done with the R&S ZVA network analyzer, as shown in Figure 1.24. The basic model covers the frequency range from 10 MHz to 67 GHz. With external sources, the frequency range can be currently extended up to 200 GHz (approximately).

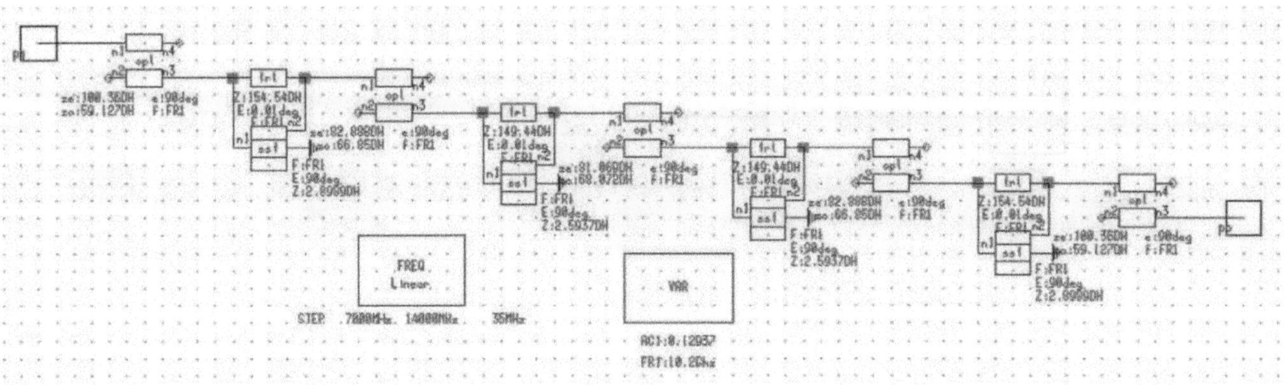

(a)

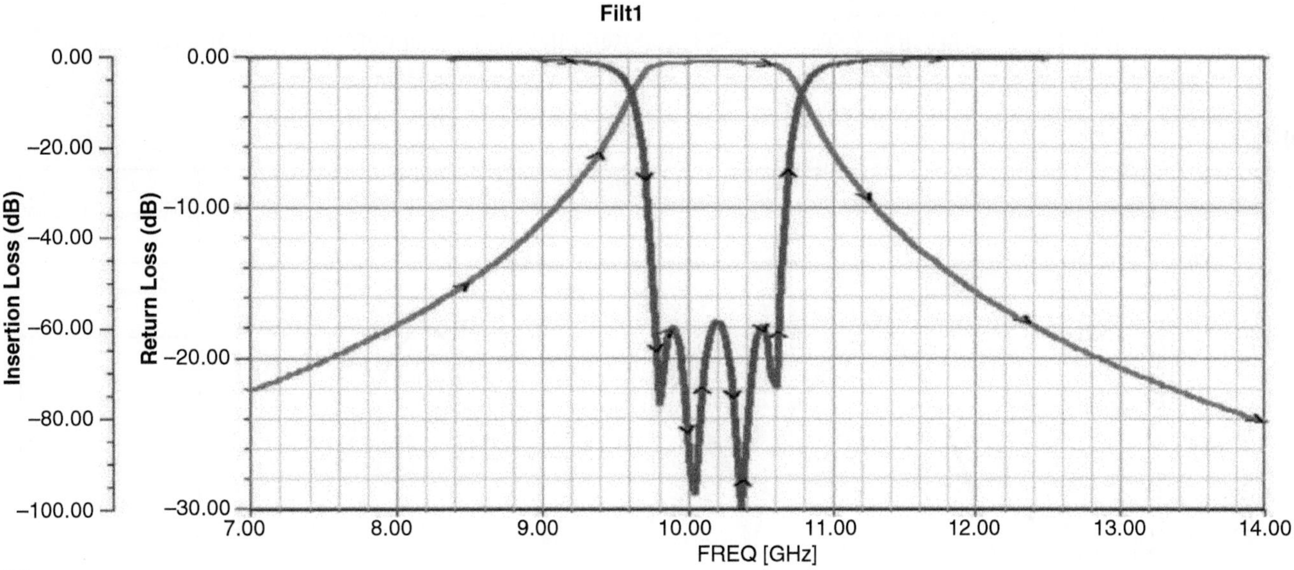

(b)

Figure 1.21 (a) Four-stage advanced filter with EM coupling, a result of a synthesis program. (b) S11 and S21 in dB of the 10.2 GHz synthesized filter.

The vector network analyzer shown above in Figure 1.24b, model ZNA, is the successor of the model seen in Figure 1.24a.

Next, we will validate some structures, beginning with a stub to be resonant at 10 GHz. Figure 1.25a shows the layout of the actual structure. The following stub uses dimensions as calculated by the synthesis program, to be resonant at 10 GHz (Figure 1.25b).

It turns out that the resonant frequency is 9.64 GHz; because the transmission line calculator does not add the end effect (for transmission line, it needs to be 220 mil), the calculated frequency is 1.38% too high.

Figure 1.26a is a simulation in the Serenade program.

The simulator itself calculates the resonant frequency at 9.77 GHz (calculation error is about 1%), but the synthesis is 1.38%. In Figure 1.26b, the left curve was adjusted to reflect the measured value, by adjusting the physical length of the stub. The right curve shows the calculated resonant frequency f or the actual physical length. What did we learn from this? The EM simulation in Serenade and the synthesis program are accurate within 1.4%, while the difference between the measurement and the simulation is 1%.

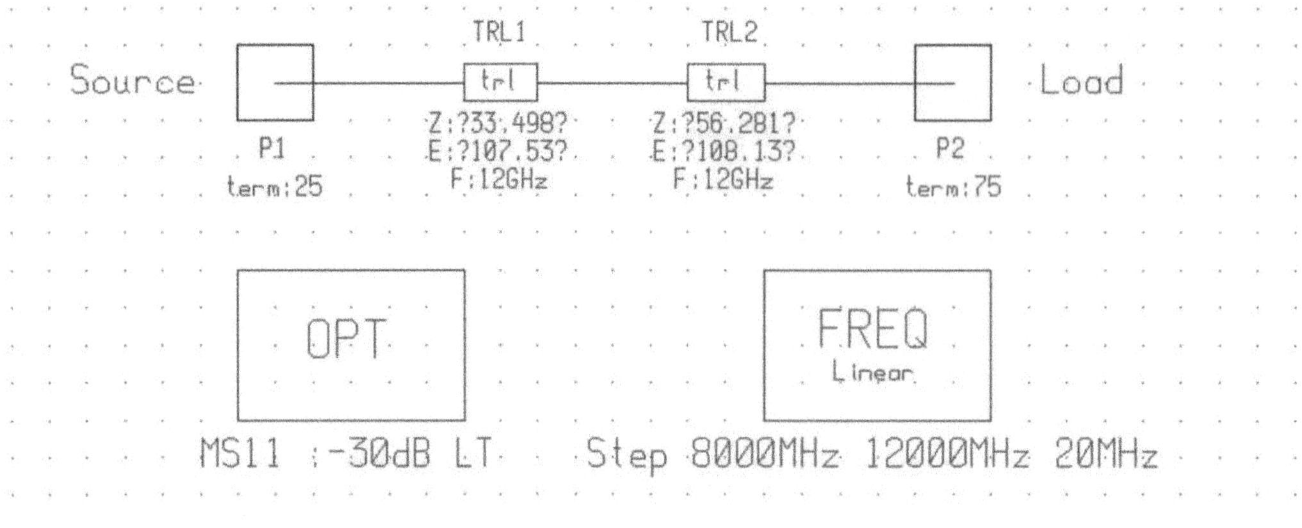

(a)

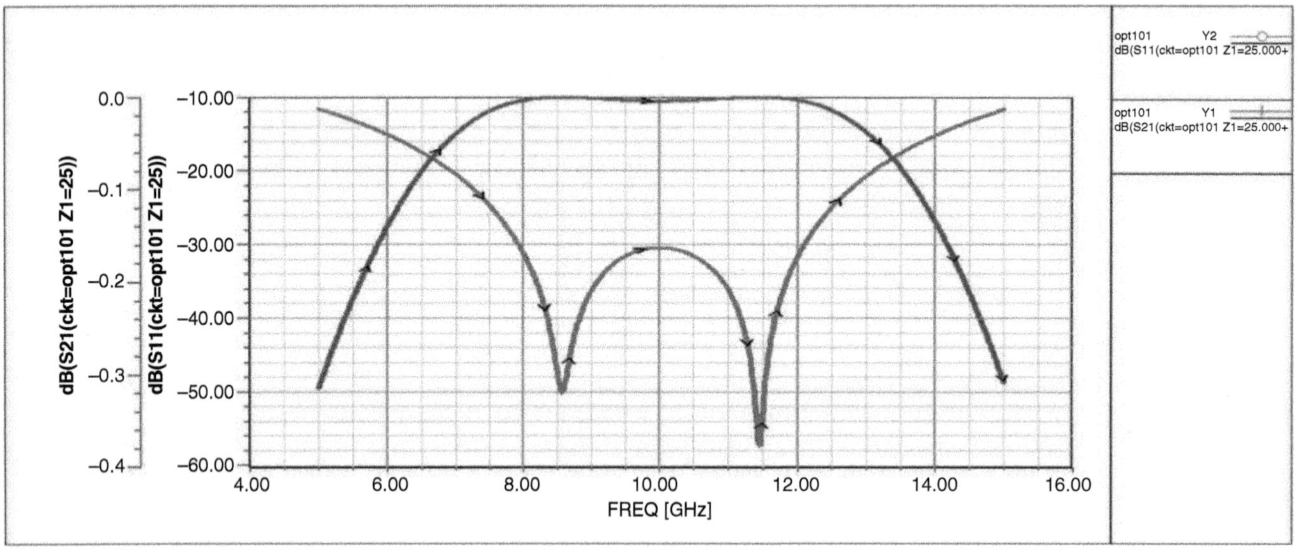

(b)

Figure 1.22 (a) Two-stage matching with a two-section series transformers. (b) S21 (gain) in dB and S11 return loss (worst case is −30 dB) of the filter circuit of Figure 1.22a.

1.8.1 100 GHz Test Circuit

The next test case, in Figure 1.27a is 100 GHz. To obtain the physical dimensions, the same Serenade transmission line synthesis program was used to find the necessary values. Again we use the synthesis program to give us the dimensions of the structure, length = 389 μm, width = 34 μm, and height = 20 μm. The simulation in Serenade is shown in Figure 1.27b, while ADS simulation is shown in Figure 1.27c.

The error of the synthesis program relative to the ADS simulation is about 1%. The same test in Serenade practically shows no difference in the resonance frequency, between the synthesized and simulated frequencies.

Now is the final test for 1 THz. The synthesis program provides the following dimensions: length = 39.5 μm, width = 8.4 μm, and height = 5 μm. We have used the Serenade program and ADS to perform the same analysis.

Based on the analysis of Figure 1.28a using the two programs, the error is 1.02 for Serenade and 1.05% for ADS, as shown in Figure 1.28b,c.

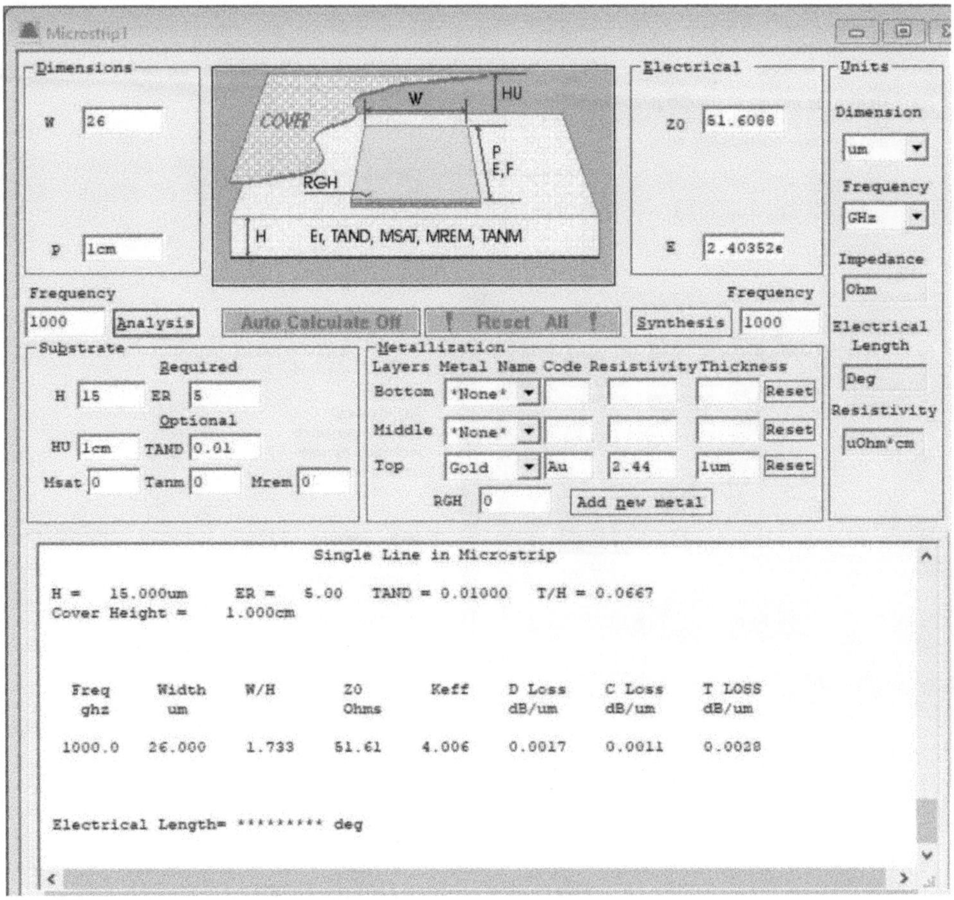

Figure 1.23 A picture of the transmission line synthesis user interface.

This exercise was necessary to calibrate the simulator. Both simulators give a high-resolution solution that is better than the actual result obtainable in production.

1.9 More EM Simulation

Having validated a transmission line accuracy, we will show some filter examples.

Figure 1.29a is a four-stage filter with real values using EM coupling rather than a curve-fitting model.

The simulation of the synthesized filter, in Figure 1.29b, shows that the synthesis program is well-behaved and accurate.

After this example, let us take a look at more filters.

1.9.1 Edge-coupled Filter Using Models and Electromagnetic Results

Serenade has a built-in spectral Domain simulator, quasi, 2.5-dimensional simulators, which can handle up to 12 different parallel transmission lines of unequal spacing and widths. Figure 1.30a shows the schematic of a proposed 110 GHz filter with different TL widths, while Figure 1.30b shows the layout (not to scale) of the circuit simulated.

The values for S11 (−12 dB worst case) are not too impressive (see Figure 1.30c).

ADS simulation is shown next, in Figure 1.30d.

The simulation result in Figure 1.30e shows a slight frequency shift of about 2.5 GHz or 2.2%.

The previous simulation results were of the simulation of the circuit schematic built with the substrate, and the plots in Figure 1.30f are the simulation results of the layout.

The filter shown in Figure 1.31a uses ideal elements and a coupled filter model. While this filter cannot be built like this, a simulation was done to validate the mathematical stability of the simulator.

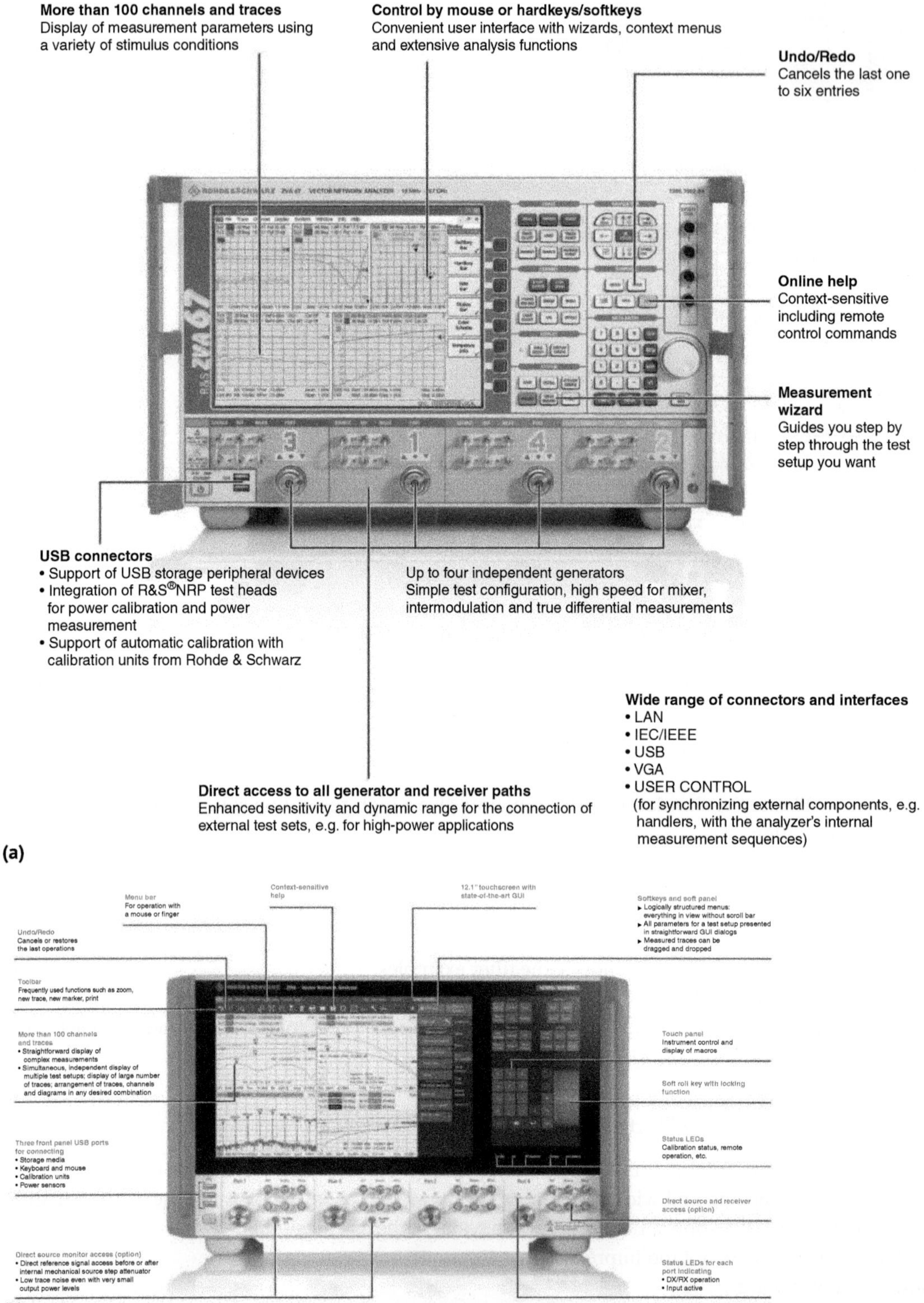

Figure 1.24 (a) R&S ZVA67 vector network analyzer, 10 MHz–67 GHz. *Source*: Courtesy of Dr. Ulrich L. Rohde. (b) R&S ZVA vector network analyzer, 10 MHz–41.5 GHz. *Source*: Courtesy of Dr. Ulrich L. Rohde.

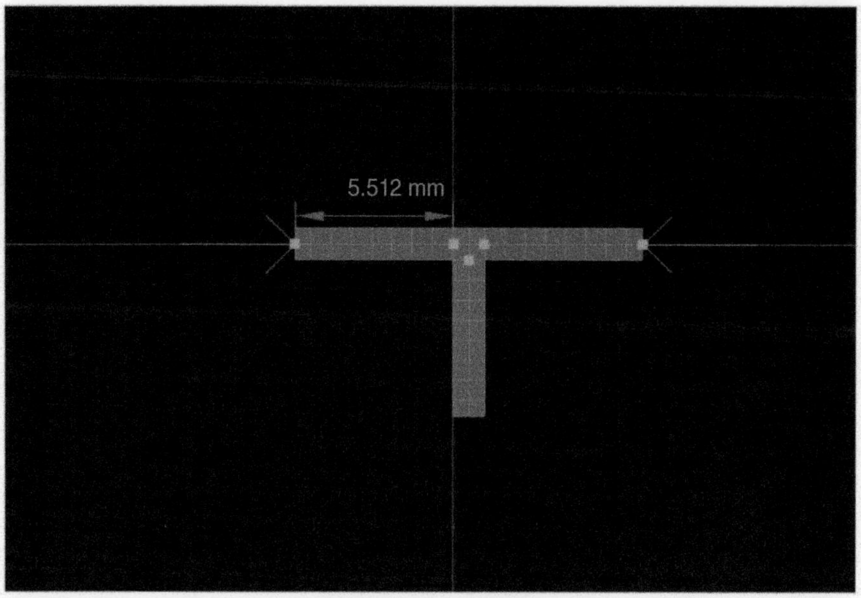

(a) Stub test@10GHz

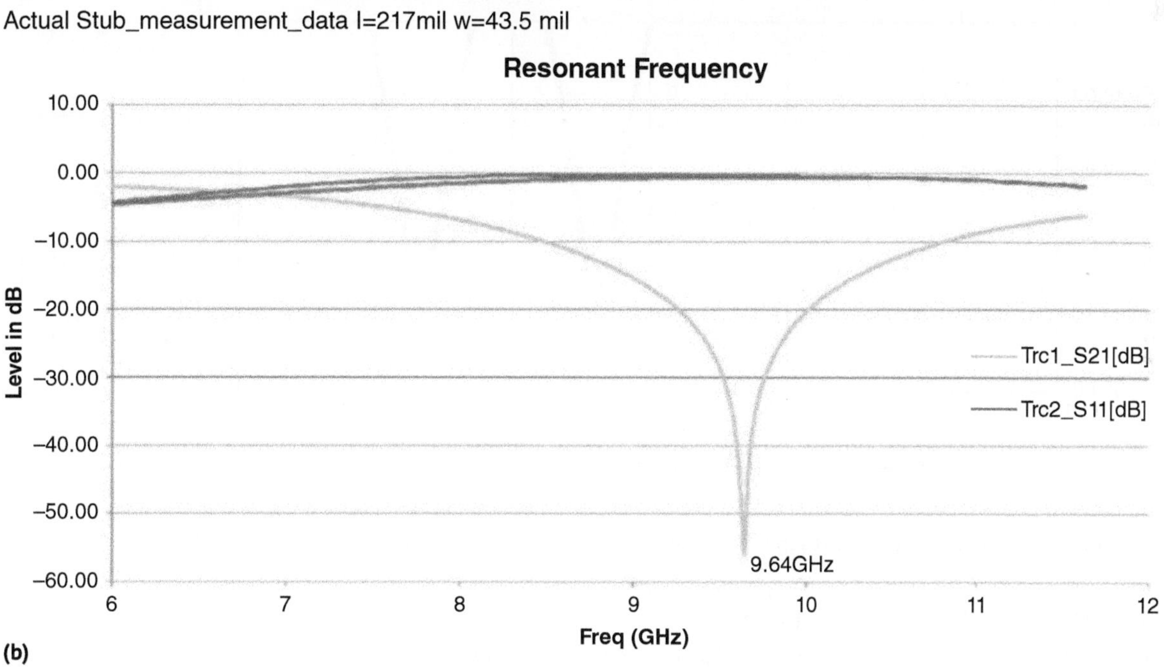

(b)

Figure 1.25 (a) T-stub resonant at 10 GHz. (b) The resonant frequency at 9.64 GHz.

The results of the simulation in Figure 1.31b shows the worst-case return loss of about −17 dB, and the ripple is less than 1 dB.

1.10 Transmission Line Test

Figure 1.32a shows a simulation up to 2 THz.

Figure 1.32b shows the return loss (S11), the curve indicates about −22 dB at 1 THz, with the loss being about 0.15 dB.

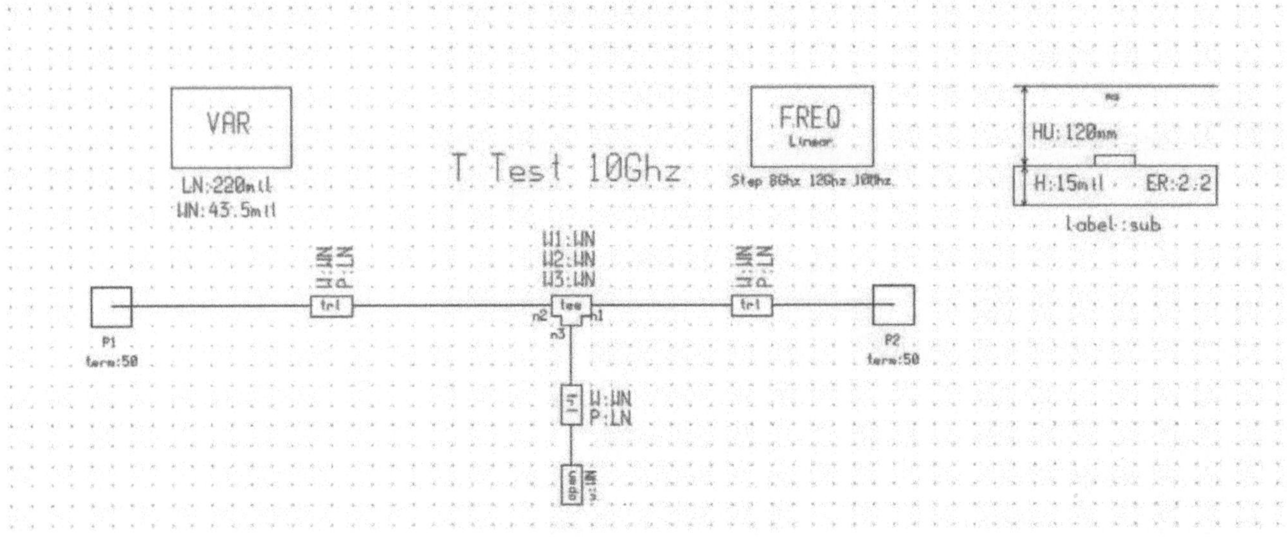

(a)

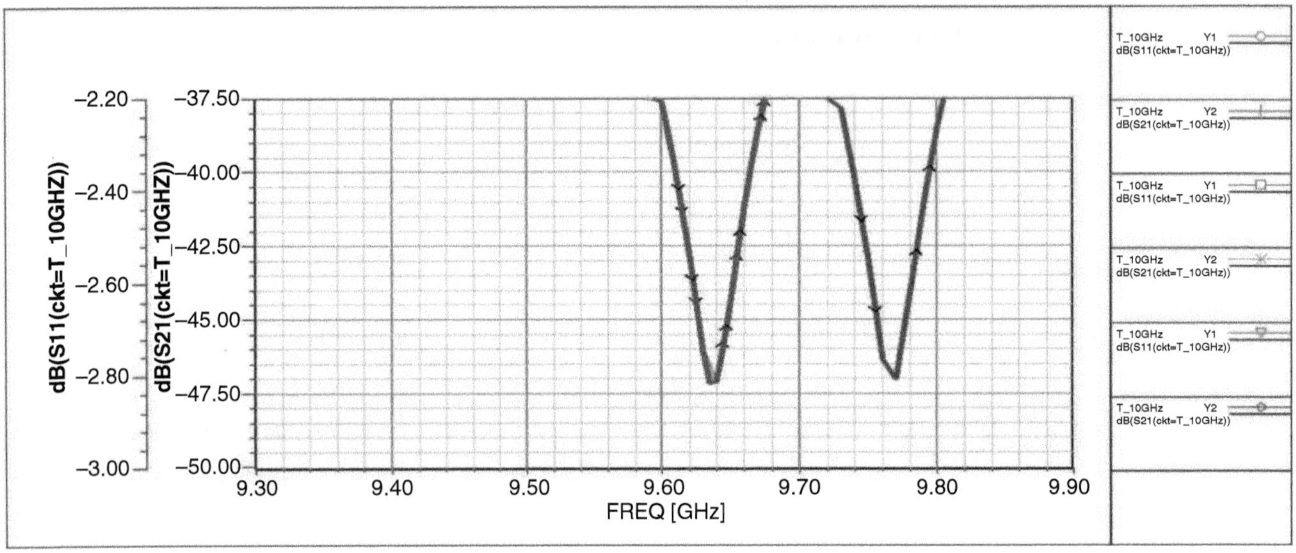

(b)

Figure 1.26 (a) 10 GHz test circuit (TRL). (b) The resonant frequency: measured and calculated for the 10 GHz stub.

Figure 1.32c is an HFSS-based simulation up to 2.2 THz.

This simulation predicts insertion loss of about 0.4 dB at 1 THz and the higher-order modes are much more complicated.

1.11 Branch Line and the Rat-race Coupler

An example of a branch line coupler is shown in Figure 1.33a.

Figure 1.33b shows the asymmetrical behavior based on the limitation of the models.

Figure 1.33c shows the S11 and S21 in dB of the branch line coupler simulation, which is symmetrical but slightly off in the resonant frequency.

A variation of this is the rat-race coupler, which can also be modeled as a coupler with four bends, as shown in Figure 1.34a, and its S21, S31, and S41 in dB of the variation are shown in Figure 1.34b.

And, finally, Figure 1.35a shows the well-published rat-race coupler.

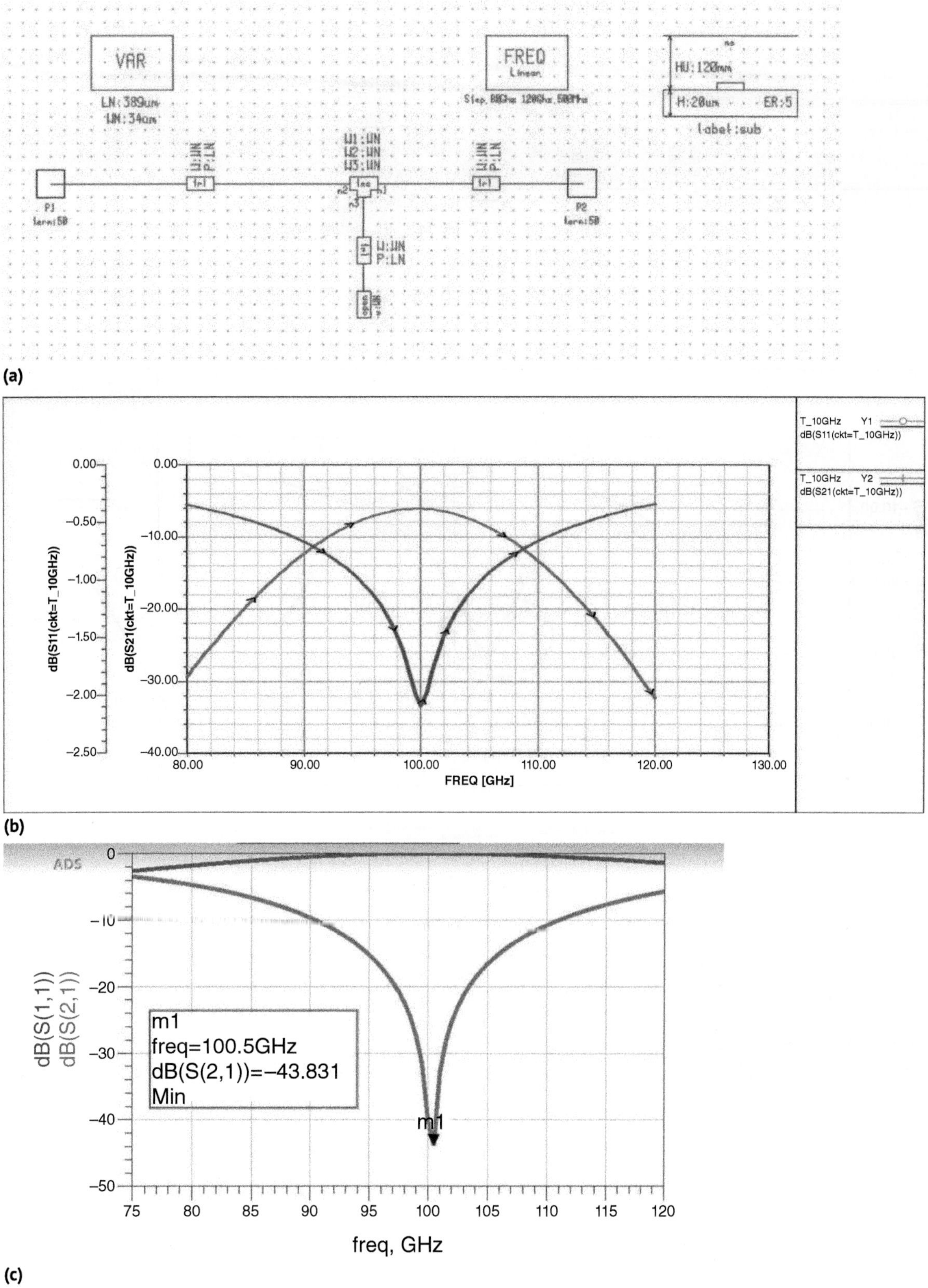

Figure 1.27 (a) Simulation of the 100 GHz test circuit. (b) S21 and S11 of the 100 GHz test circuit. (c) Simulation of the resonant frequency in ADS.

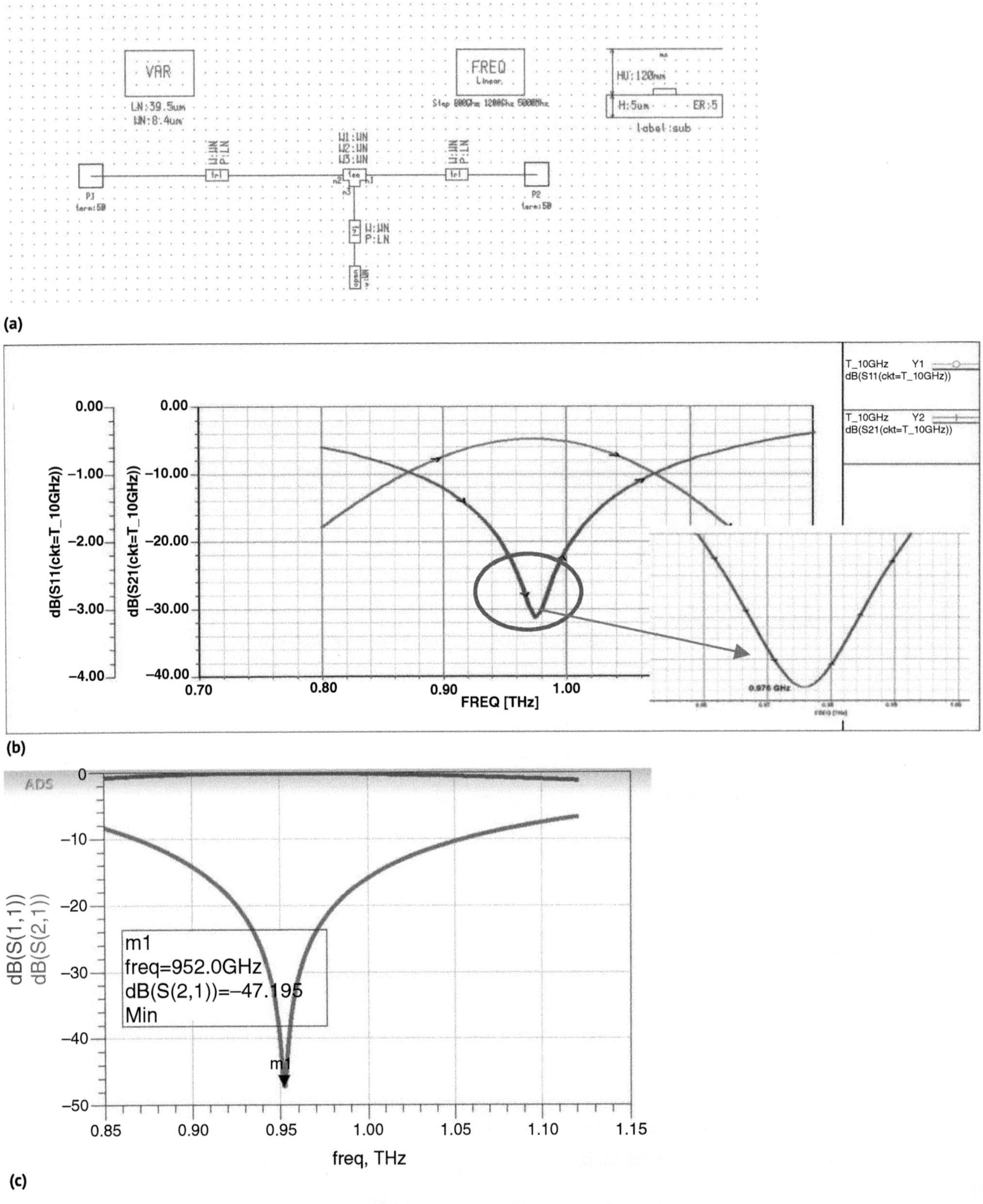

Figure 1.28 (a) Schematic of the test structure for 1 THz. (b) S21 and S11 in dB of the 1 THz resonator test structure, a closer look at the S21 of the test structure for 1 THz. (c) Simulation of the resonant frequency using ADS shows 952 GHz.

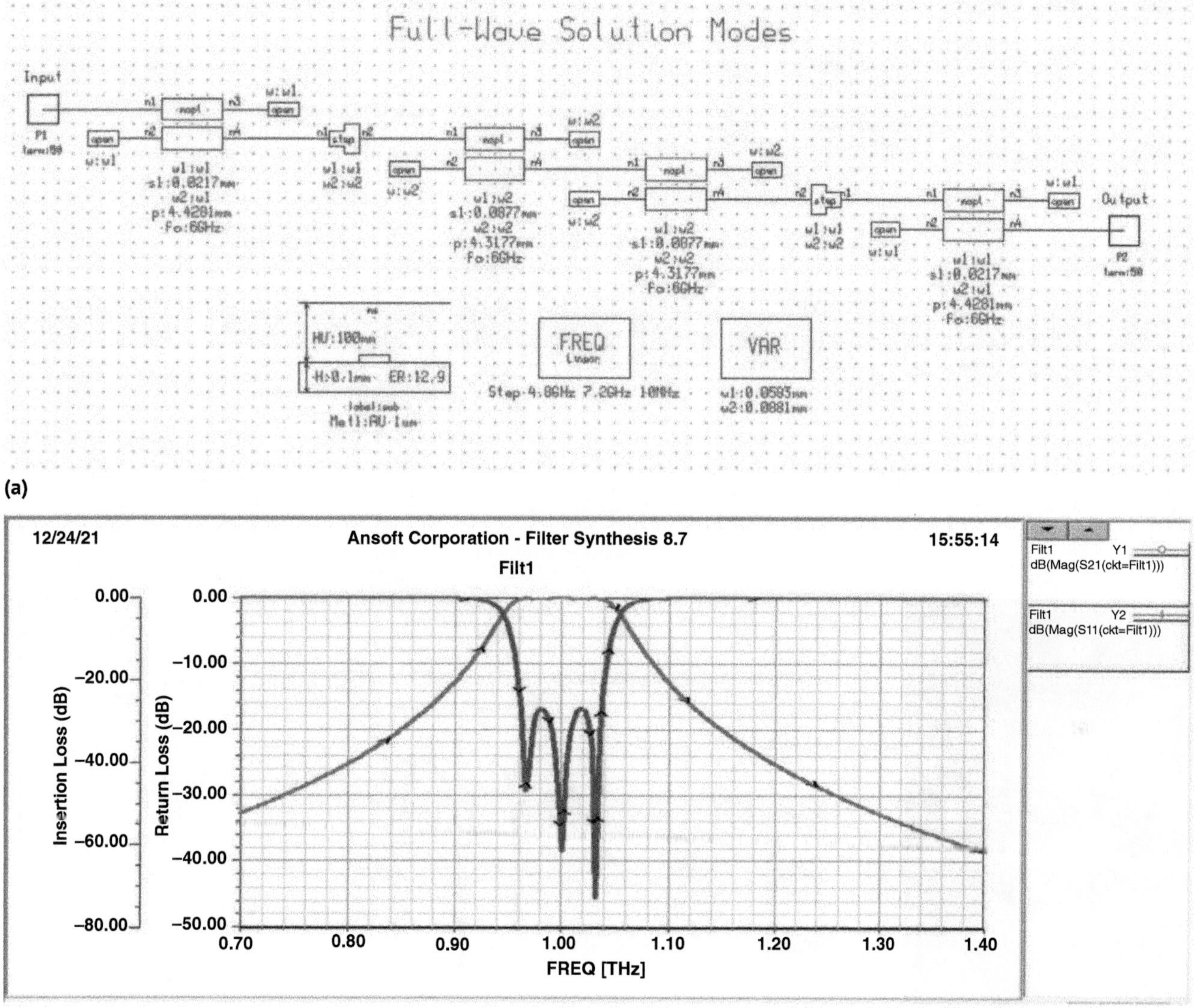

Figure 1.29 (a) The four-stage filter with EM coupling. (b) Simulation of the four-stage filter with EM coupling.

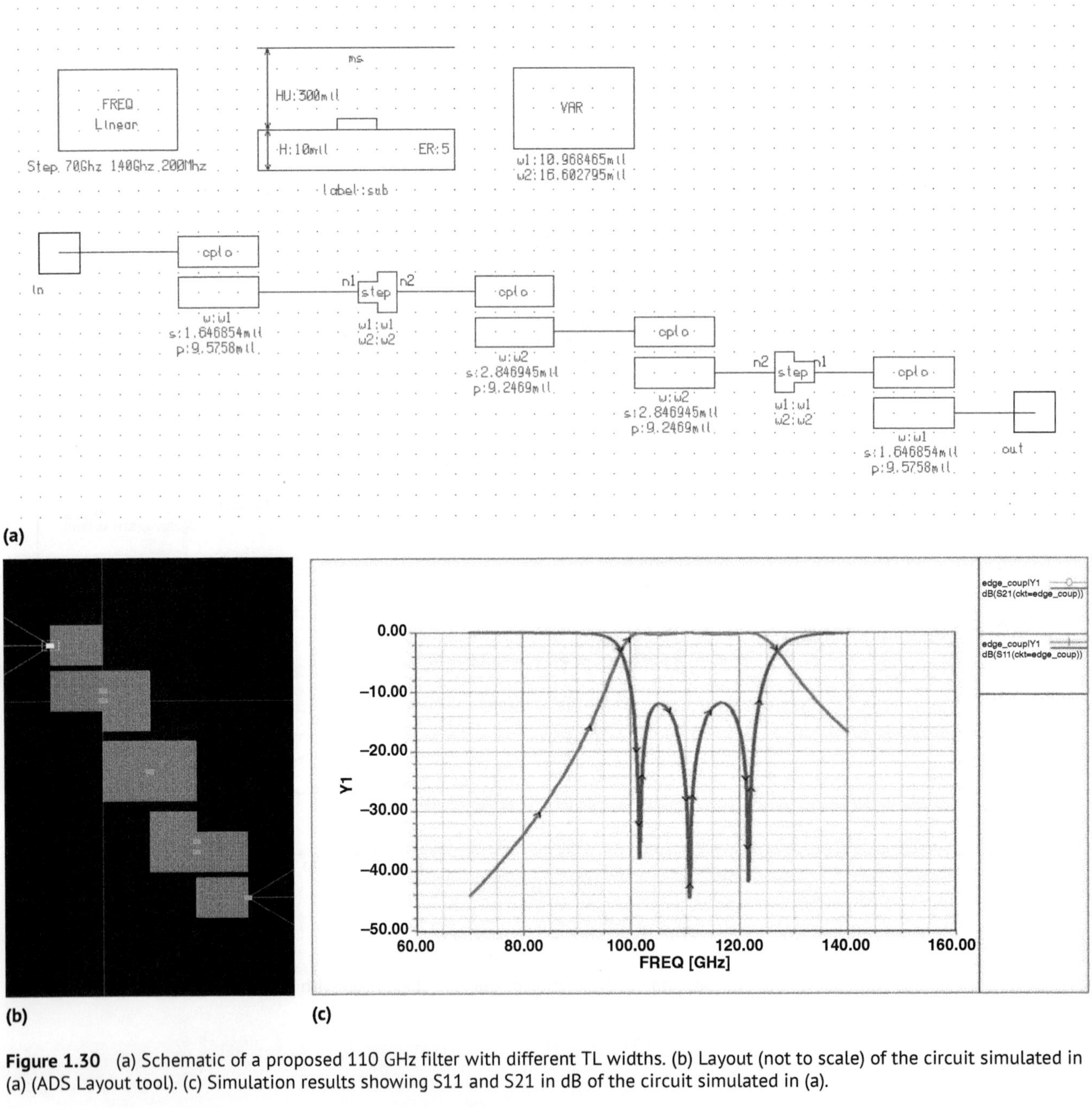

Figure 1.30 (a) Schematic of a proposed 110 GHz filter with different TL widths. (b) Layout (not to scale) of the circuit simulated in (a) (ADS Layout tool). (c) Simulation results showing S11 and S21 in dB of the circuit simulated in (a).

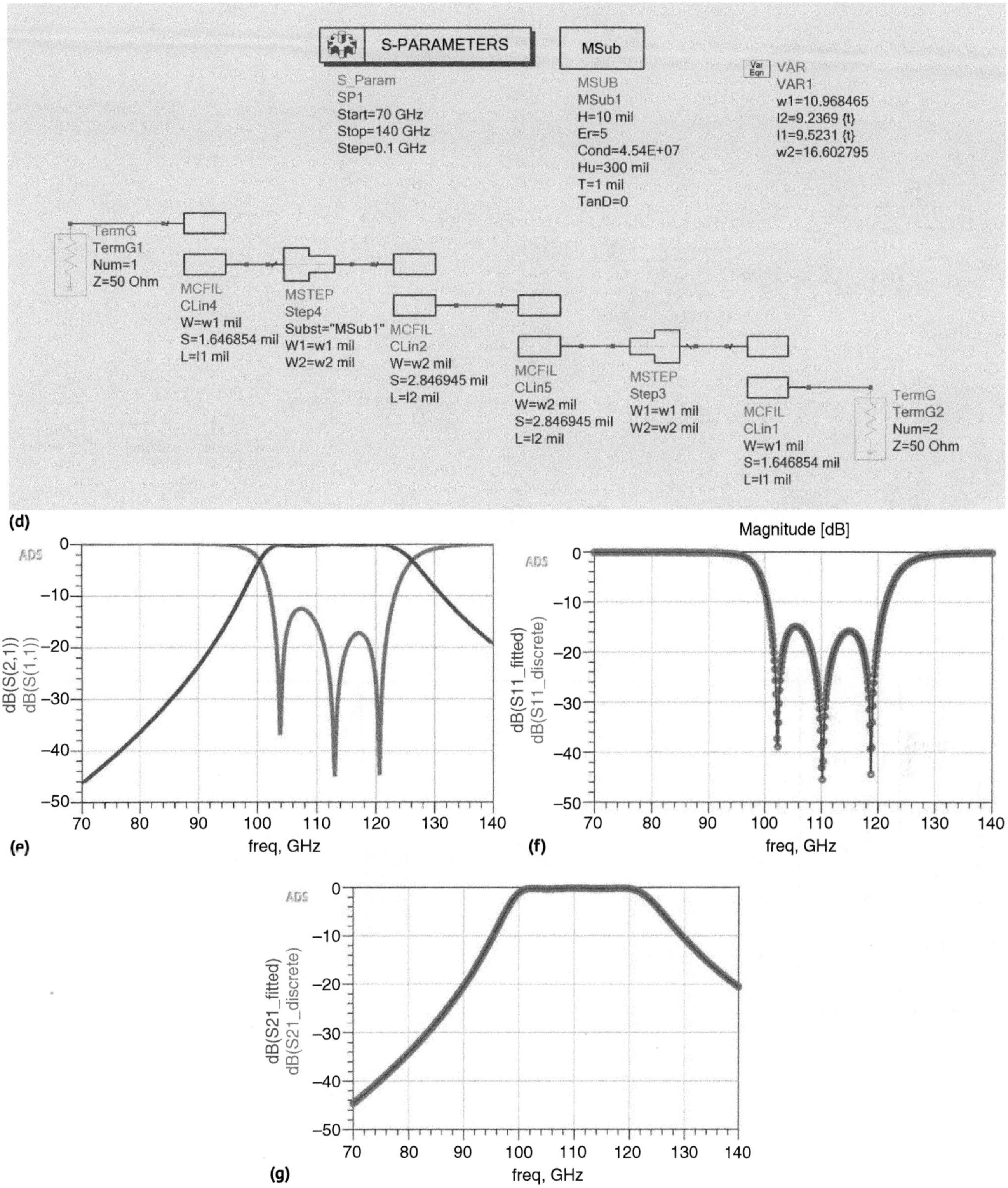

Figure 1.30 (Cont'd) (d) ADS simulation of the four-stage filter at 110 GHz. (e) S21 and S11 in dB of the 110 GHz filter. (f) Return loss (S11) of the circuit simulated in the layout (a). (g) Gain (S21) of the circuit simulated in the layout (a). *Note*: In Figure 1.30g, the thin and the thick traces indicate two different resolution settings used in the simulator: synthesis of a filter designed with a center frequency of 320 GHz and a bandwidth of about 40 GHz.

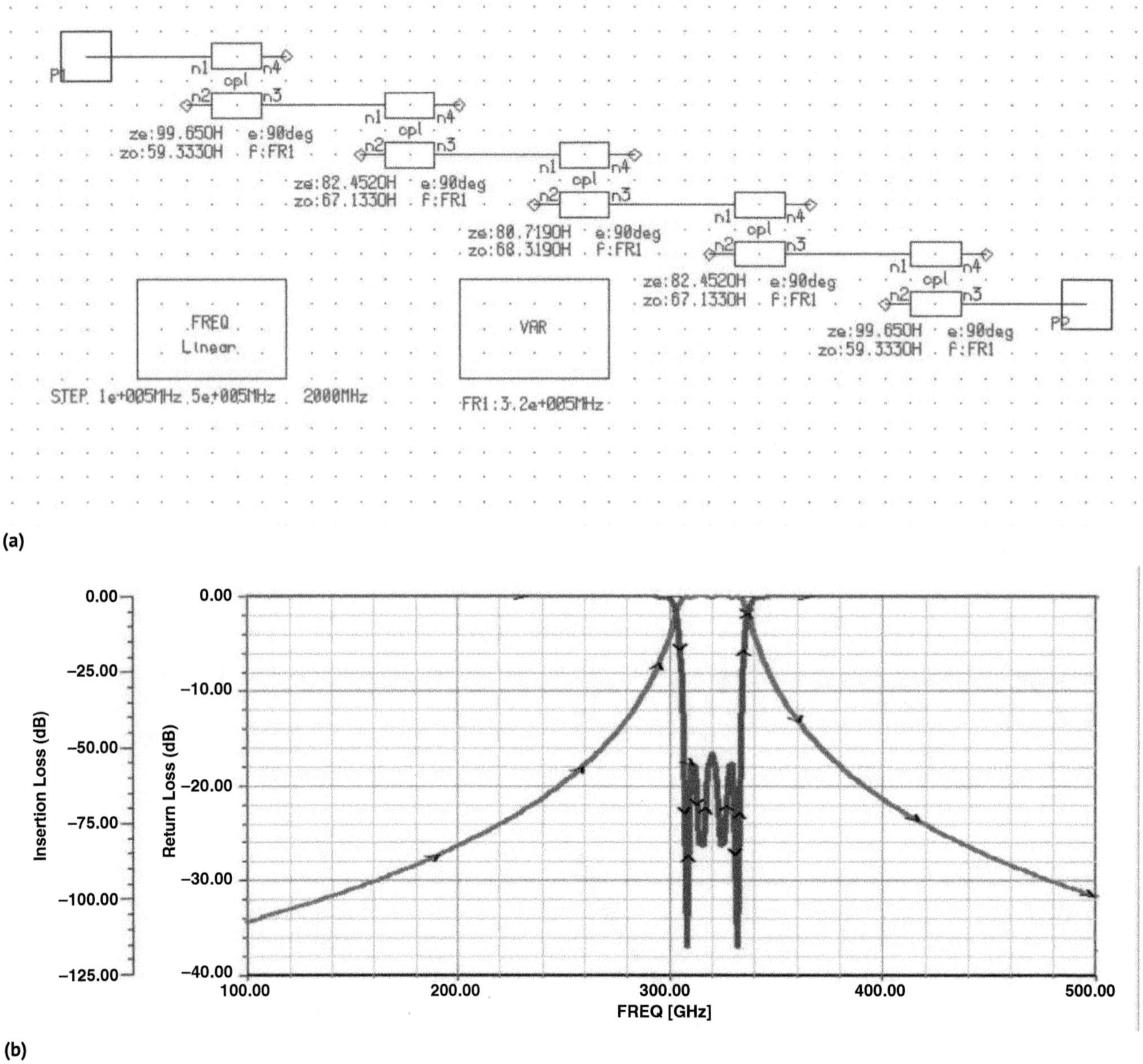

Figure 1.31 (a) Schematic of the 320 GHz filter with four poles. (b) Simulated result showing S11 and S21 in dB of the 320 GHz filter with four poles.

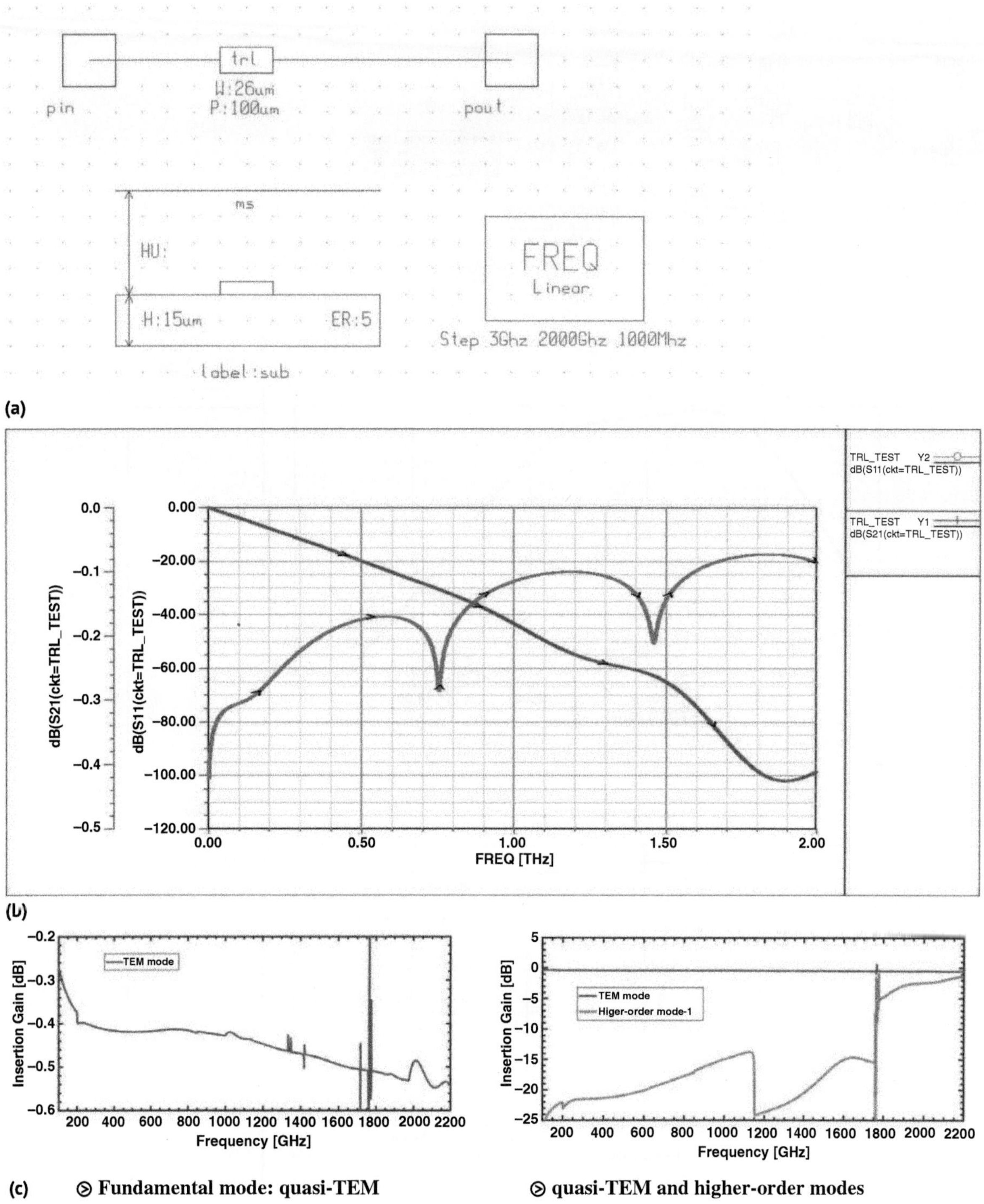

Figure 1.32 (a) Simulation of transmission line behavior up to 2 THz. (b) Simulation results of a transmission line behavior up to 2 THz. (c) HFSS-simulated forward transmission (W = 25.15 μm, h = 15 μm, ε_r = 5, t = 0).

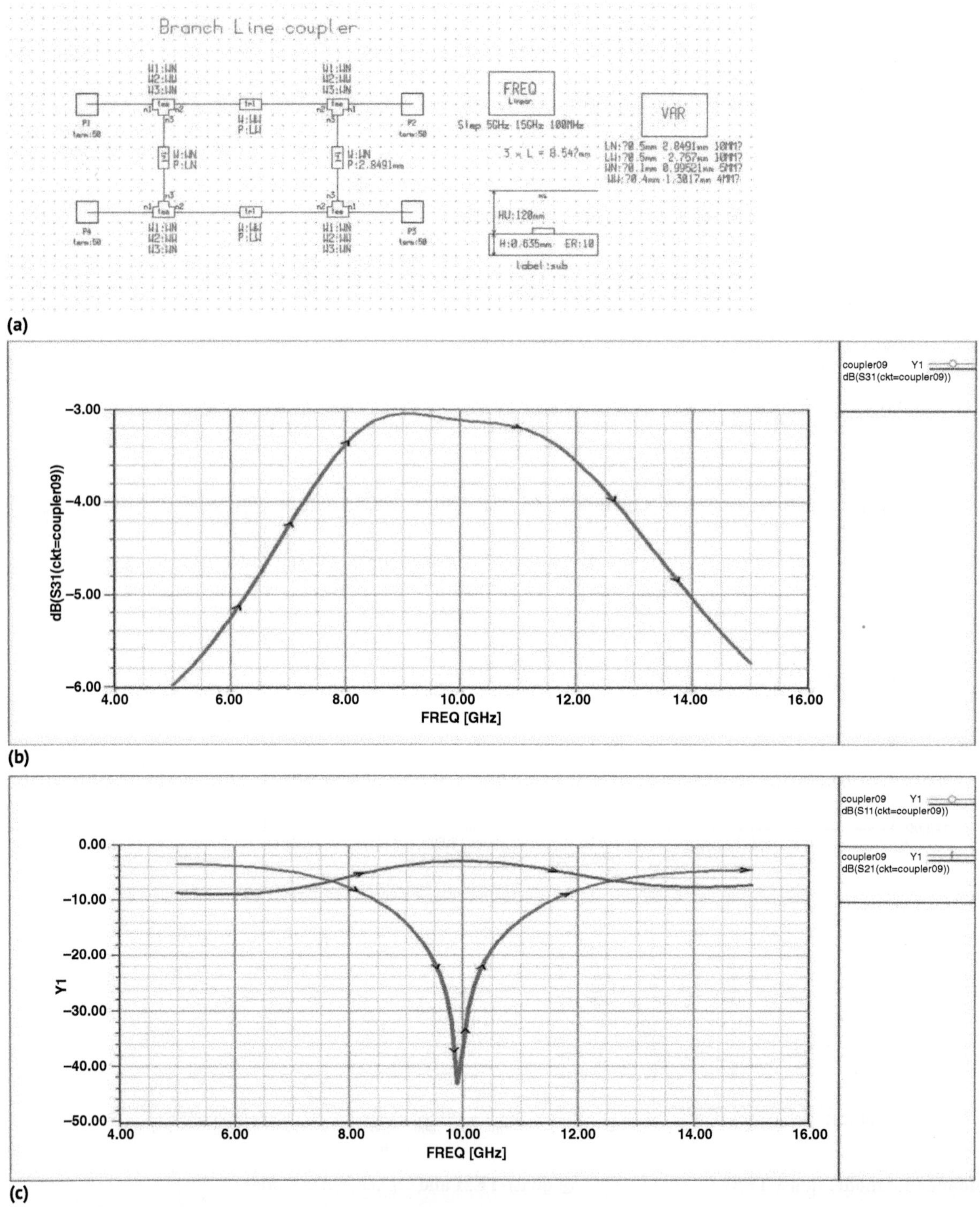

Figure 1.33 (a) Example of a branch line coupler simulation. (b) S31 in dB of the branch line coupler simulation. (c) S11 and S21 in dB of the branch line coupler simulation.

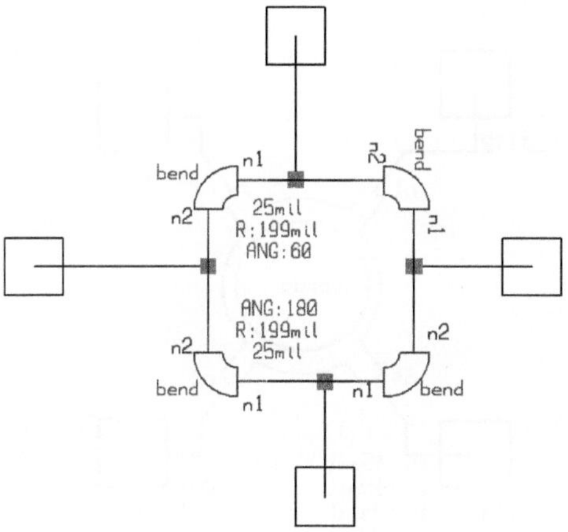

(a)

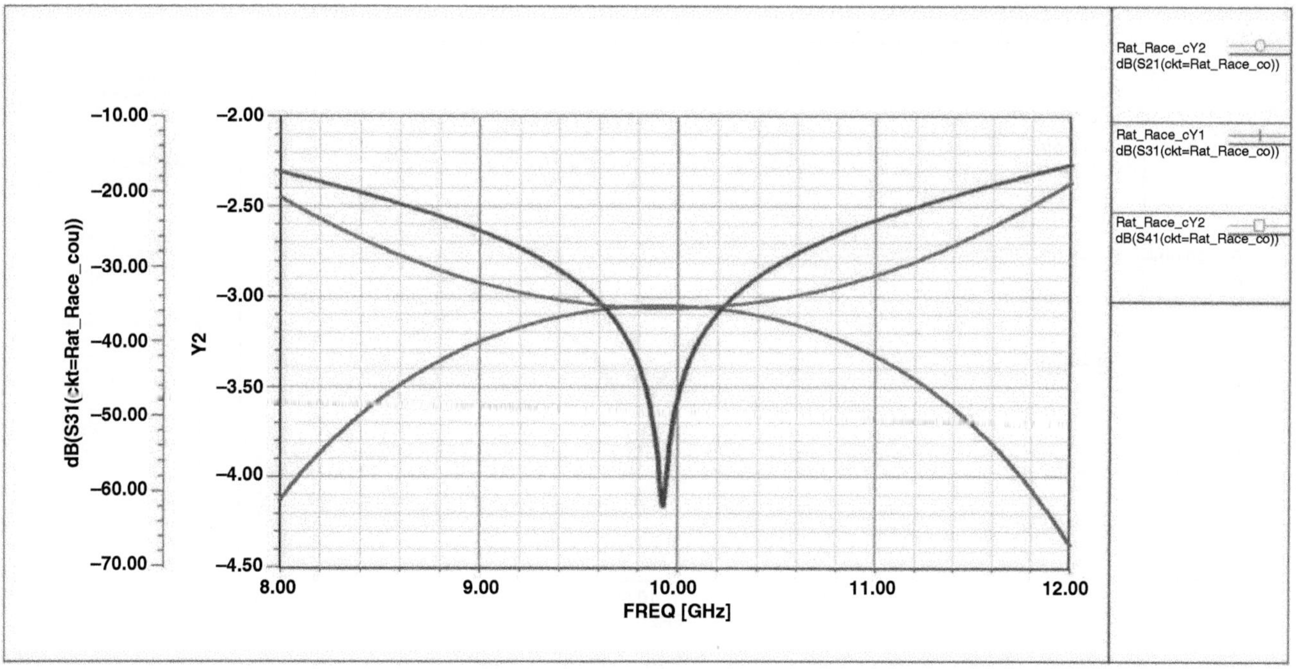

(b)

Figure 1.34 (a) Variation of the rat-race coupler. (b) S21, S31, and S41 in dB of the variation of the rat-race coupler.

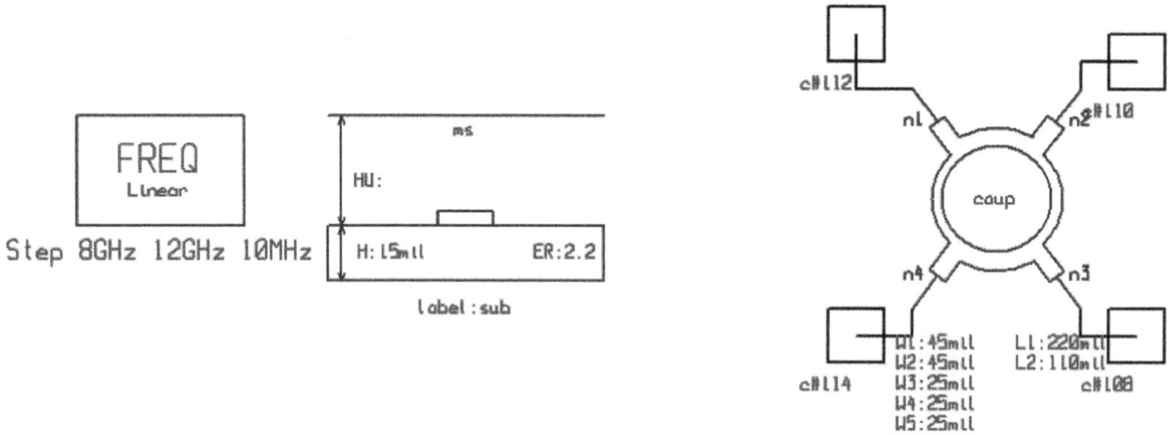

(a)

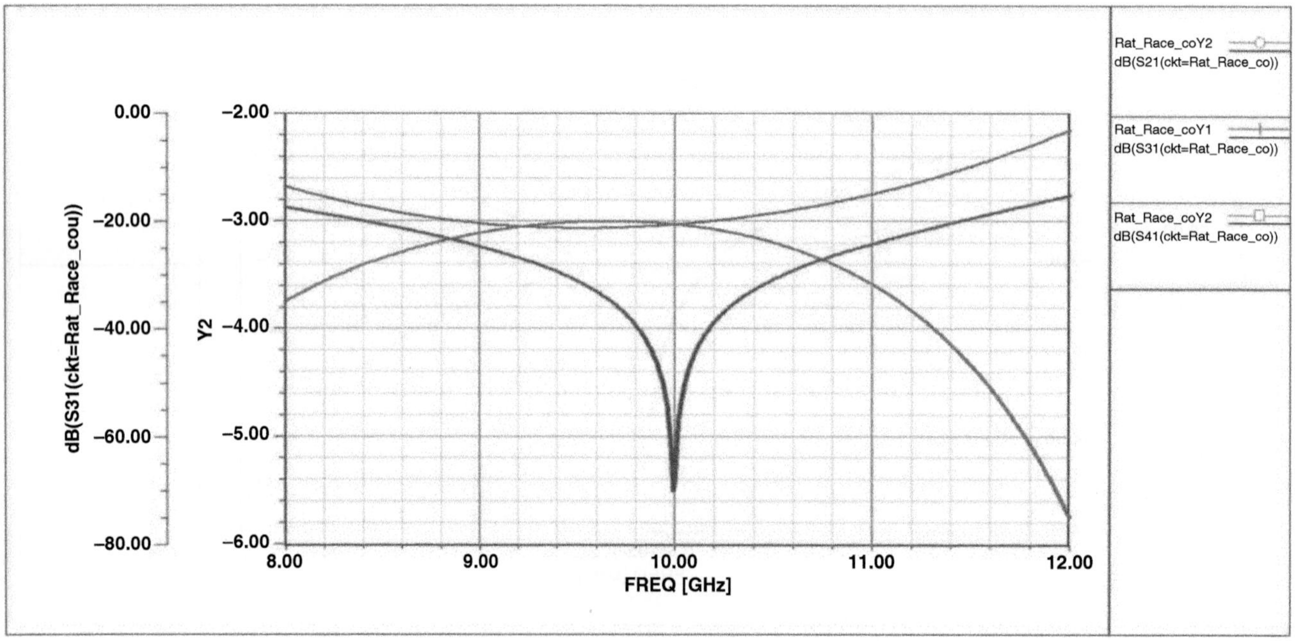

(b)

Figure 1.35 (a) A rat-race coupler. (b) S21, S31, and S41 in dB. *Source*: Rat-race coupler [11].

The three ports are quarter-wavelength and the fourth is $3/4^{th}$ λ.

All these couplers (Figures 1.33–1.35) are related to each other. The branch line coupler and its cousin using bent elements all give the same results.

The next example is the Wilkinson coupler.

1.12 Wilkinson Coupler

Figure 1.36a shows the schematic for the Wilkinson coupler.

The Wilkinson coupler is used to combine two signal sources into one output with up to 45 dB isolation between the ports. A typical example is a power divider, meaning the coupler is used backward. The input is symmetrically fed to two outputs. If the termination has a mismatch, then the 220 Ω resistor absorbs the reflected energy (see Figure 1.36a,b). Thus, this coupler can be used either as a power adder or divider. There are also asymmetrical cases in the literature.

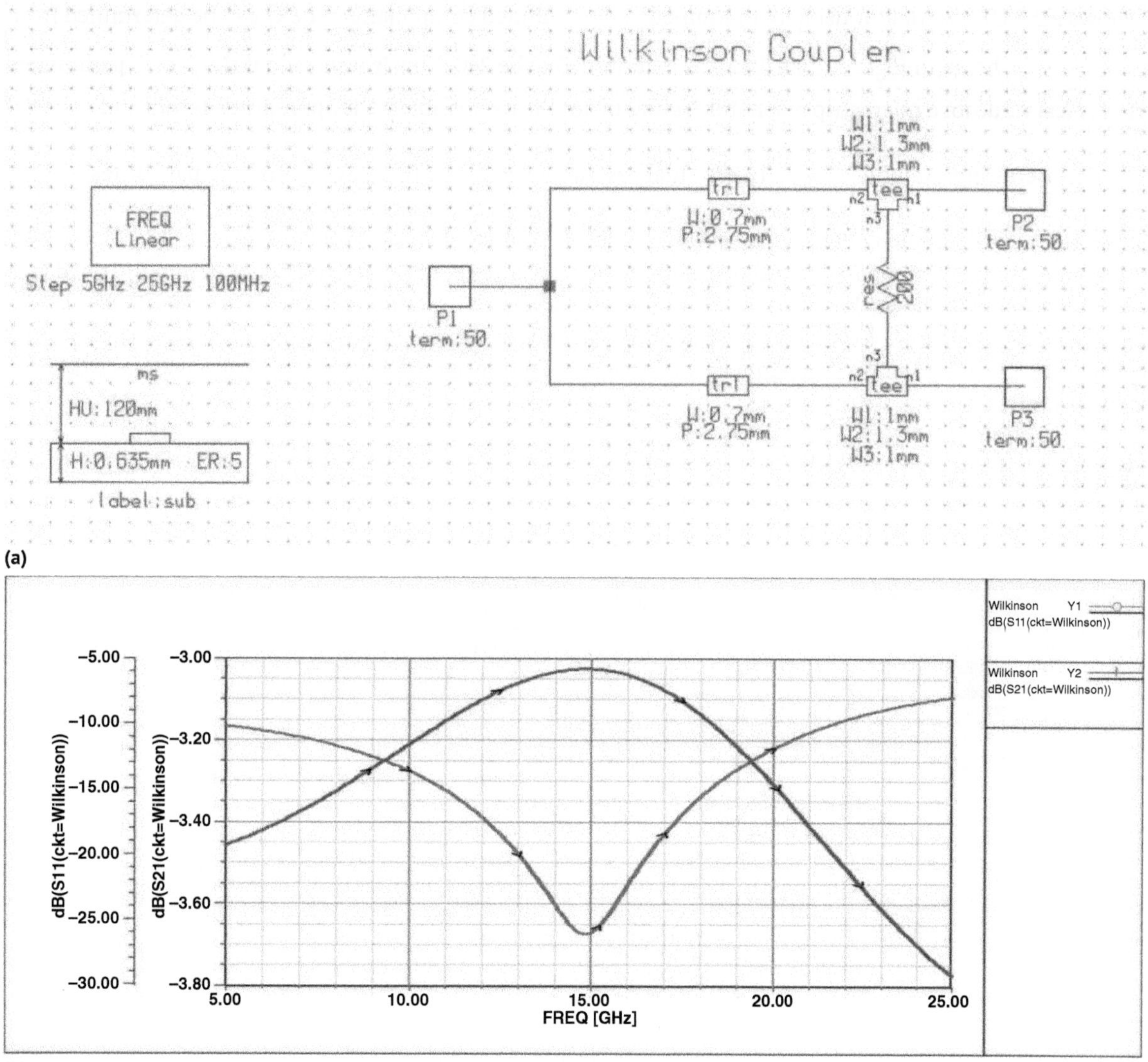

Figure 1.36 (a) Wilkinson coupler. (b) Simulated results for Wilkinson coupler showing S21 and S11 in dB.

1.13 Microstrip Resonators

1.13.1 Element Library

The following list of symbols represents some of the distributed elements in use for PC or other material-based circuit designs. By connecting them properly, resonant circuits can be created/configured. As an example, some elements such as dielectric resonators or Yttrium Iron Garnet (YIG) resonators are already part of the library.

An interesting question is the validity or boundary condition of these models as well as how to model some more exotic structure as we will see.

Coupling (a form of crosstalk) can be used for many purposes. For the calculation of the performance of this element, a three-dimensional (3D) simulator may be useful, but a special spectral-domain simulator may be sufficiently accurate but faster by several orders of magnitude.

The same parallel-tuned circuit can be generated by using printed transmission line type elements instead of the lumped inductor while maintaining the same capacitance (Figure 1.37). Since the transmission line has losses due to the material, they need to be considered. It is not practical to calculate these by hand, but rather to use a CAD program that does this accurately [3, 12, 13]. These references describe how to get the Q factor from S11 measurements.

Figure 1.38a shows a 2.4 GHz resonator using both lumped and distributed components. The Q can be determined from the 3 dB bandwidth $\left(\frac{\Delta f}{f_0}\right)$ shown in Figure 1.38b and was determined to be 240. This is also valid if the Y or Z parameters are used. This is a typical value for a microstrip resonator. Values up to 300 are possible if the appropriate layout and material are used.

Figure 1.39a is an interesting example of a 400 GHz dual-resonator bandpass filter. Again, this is part of testing the simulator.

The 0.01 pF capacitor is really implemented by capacitive coupling of the resonators.

S21 in dB of the dual-resonator bandpass filter shows a spurious resonance at 612 GHz and is not optimally tuned (see Figure 1.39b), while Figure 1.39c shows the return loss (S11) in dB of the dual-resonator circuit.

The next example shown in Figure 1.40 is a 4.5 GHz filter built on Duroid substrate.

Shown in Figure 1.41a is the simulation schematic of the filter provided earlier. It uses multiple coupled lines (EM coupling) for the simulation.

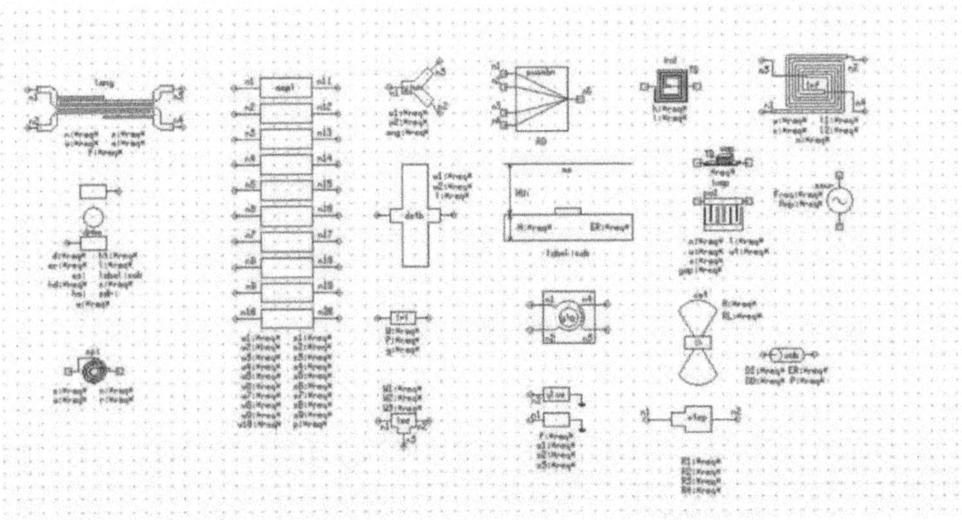

Figure 1.37 Schematics of the transmission line elements typically found in a good simulator library.

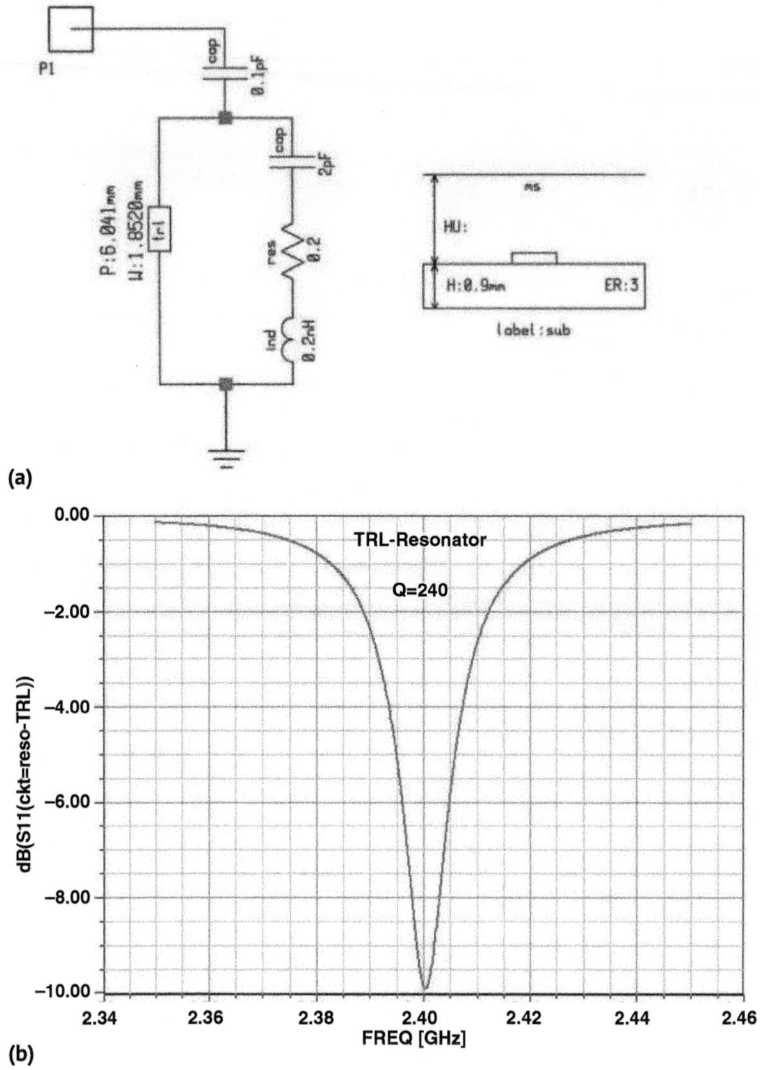

Figure 1.38 (a) 2.4 GHz resonator using both lumped and distributed components (physical dimensions are given). *Source*: [3] / John Wiley & Sons. (b) Simulated reflection coefficient S11 to determine the operating Q. Since this material has fairly high losses, an operating Q of only 240 was achieved. *Source*: [3] / John Wiley & Sons.

The resonant frequency is at 4.5 GHz, with about 4 dB losses, and has an image reject notch at about 5.06 GHz (see Figure 1.41b).

1.14 Integrated Resonators

The same circuit can be generated not only using printed circuit board material but also in GaAs or in silicon. Figure 1.42 shows the schematic of a parallel-tuned circuit using a rectangular inductor and an interdigital capacitor. The ground connection is achieved using vias. At 2.4 GHz, the number of turns and the size of the inductor would be significant. The same applies to the capacitor. This arrangement should be reserved for much higher frequencies, above 5 GHz. The inductor losses, both in GaAs and silicon, are substantial, and this case is shown only for completeness. Where possible, an external resonator should be used.

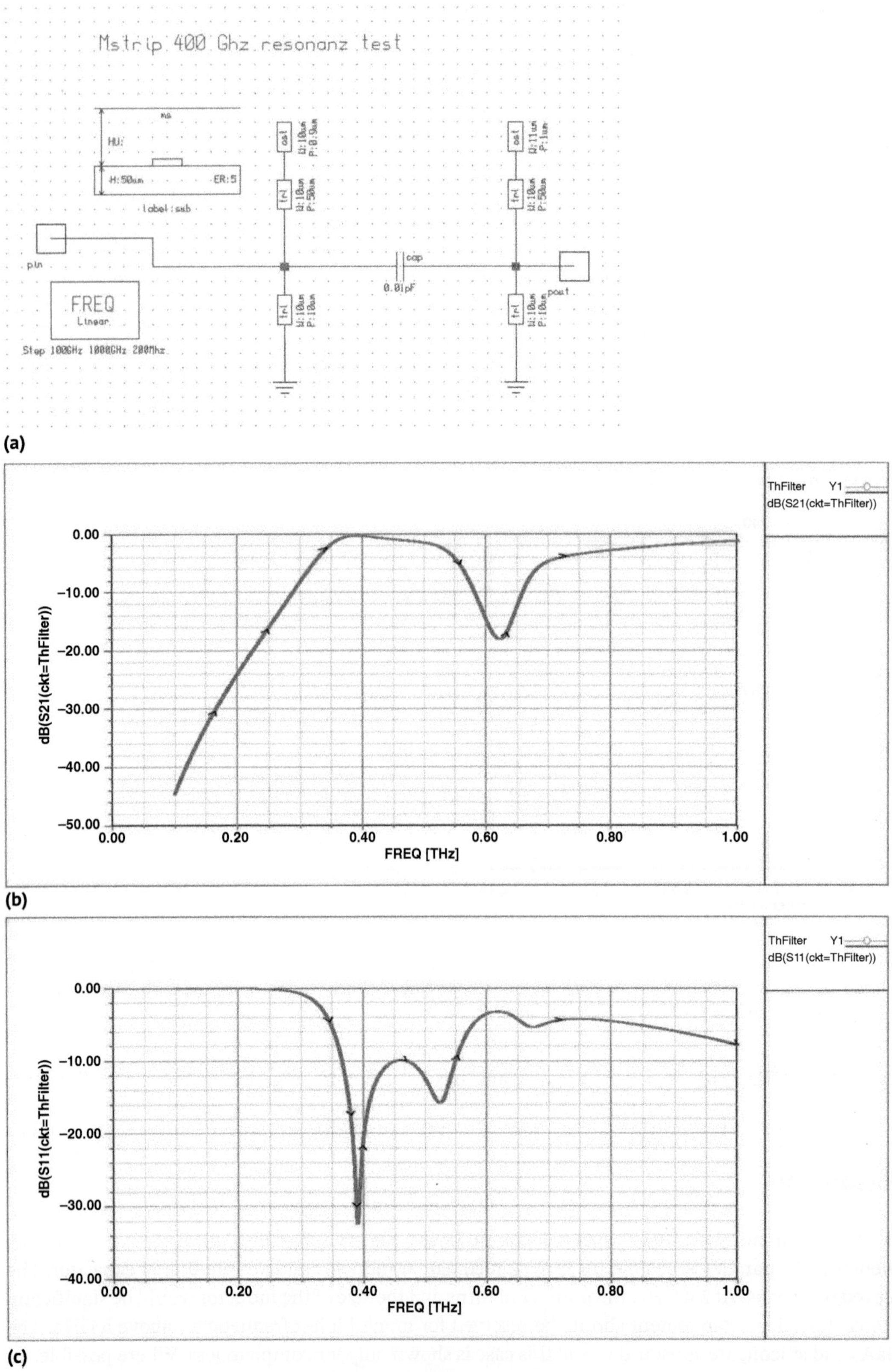

Figure 1.39 (a) Schematic of 400 GHz dual microstrip resonator bandpass filter. (b) S21 in dB of the dual-resonator bandpass filter. (c) Return loss (S11) in dB of the dual-resonator circuit.

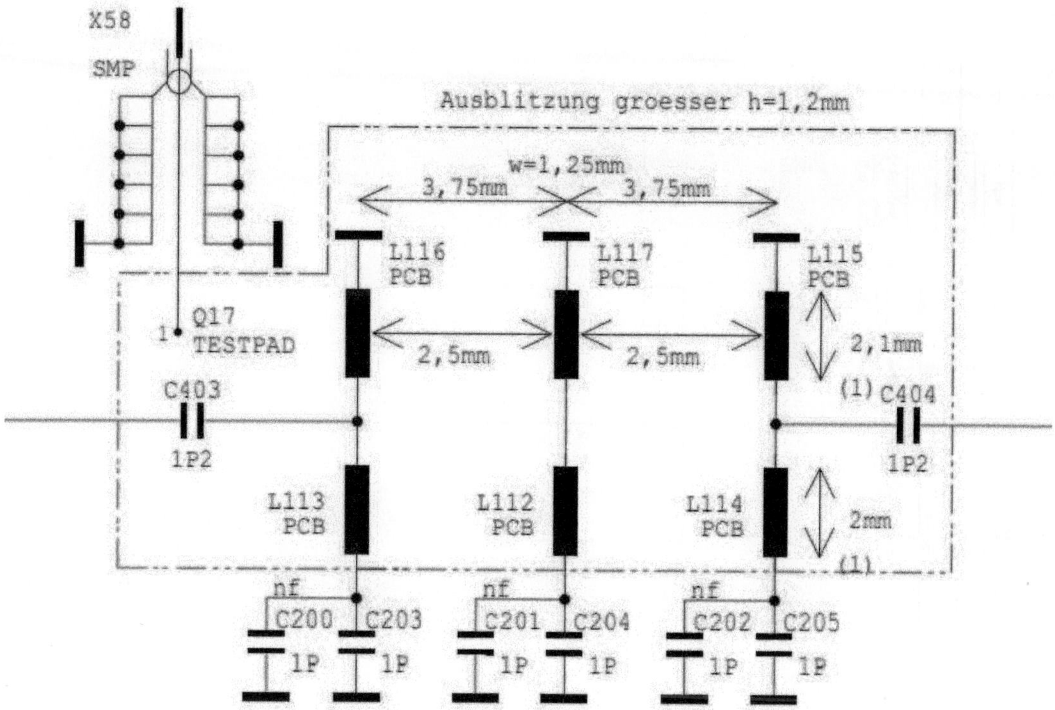

Figure 1.40 Schematic of a 4.5 GHz bandpass filter built on Duroid substrate. *Source*: Courtesy of Rohde & Schwarz.

Referring to integrated resonators, a high-Q resonator consisting of two coupled inductors has been developed by us. Figure 1.43 shows a 3D array in which the two coupled resonators are easily identifiable. One side of the resonator is connected to ground through vias.

Throughout the book, we are referring to coupled resonators for this arrangement, but purists could argue rightfully that these are coupled inductors. In reality, we are dealing with coupled microstrips that behave differently from coupled inductors. We have validated this experimentally, and the exact explanation for this behavior would require a lengthy treatment (Figure 5.107 from [1] gives a detailed insight into this). The reason it is frequently called coupled transmission lines or inductors is the fact that the elements themselves are not resonant and are much shorter than quarter-wavelength. This is applicable to two or more coupled transmission lines. The resulting Q can be as much as 10 times as the Q of a single-tuned circuit. The rate of change of phase versus frequency in this configuration is much faster and is responsible for the higher Q.

The 3D drawing can be reduced to a two-dimensional drawing (layout), as shown in Figure 1.44, which gives further details about the resonator. The resonator analysis was performed using Ansoft Designer, specifically the 2.5-D simulator Ensemble.

Taking the *S*-parameters obtained from the structure, a more conventional resonator analysis can be performed. Figure 1.45a shows the electrical equivalent circuit of the coupled microstrip line resonator.

Finally, the S11 resonant curve is analyzed. The curve seen in Figure 1.45b shows a coupled microstripline resonator response, and the resulting Q is determined to be 560. This structure and application is covered by the US patent applications 60/564173 and 60/501371.

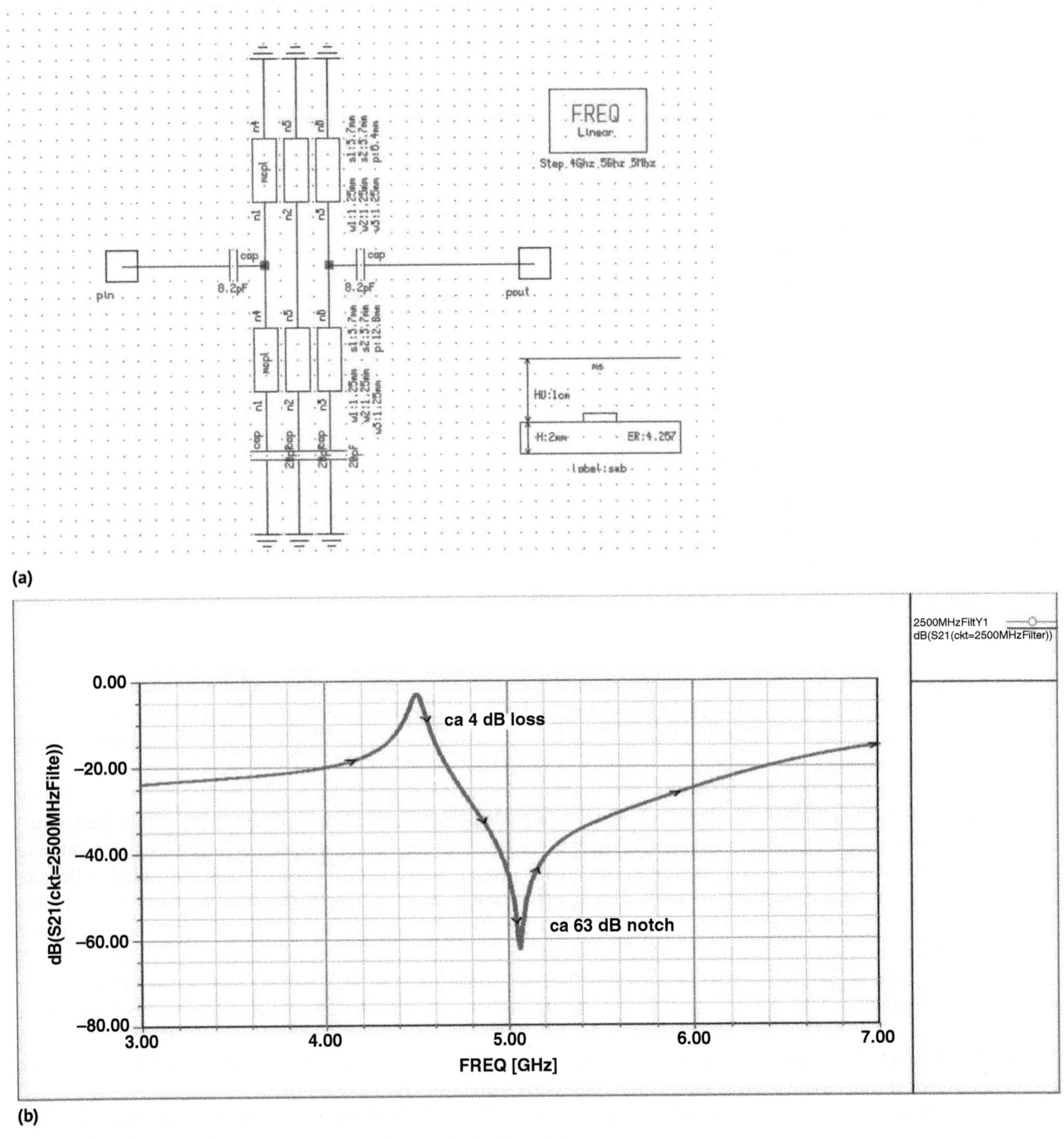

Figure 1.41 (a) Simulation schematic of the aforementioned filter. (b) Simulation result of the aforementioned filter.

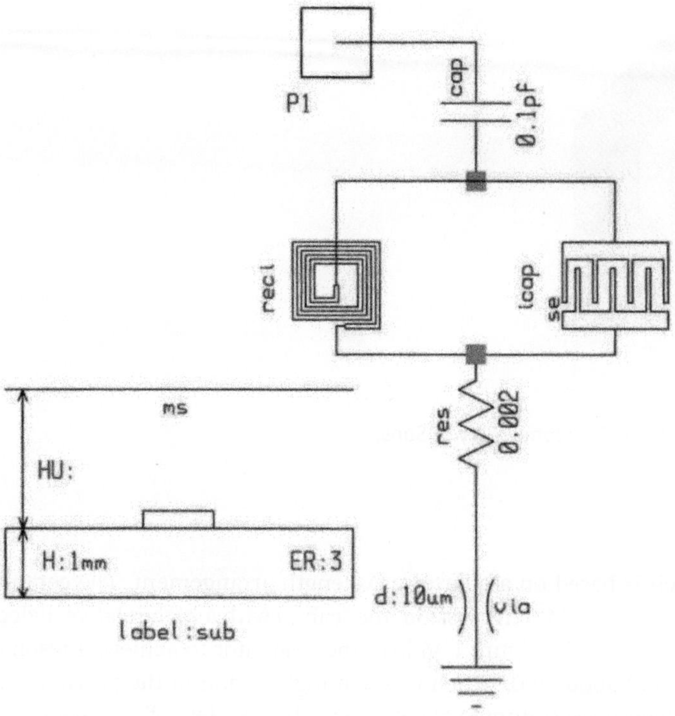

Figure 1.42 Parallel-tuned circuit using a rectangular inductor (could also use a spiral) and an interdigital capacitor. If implemented on GaAs or silicon, it exhibits low Q. *Source*: [3] / John Wiley & Sons.

Figure 1.43 3D view of the coupled resonator (patent pending). *Source*: [3] / John Wiley & Sons.

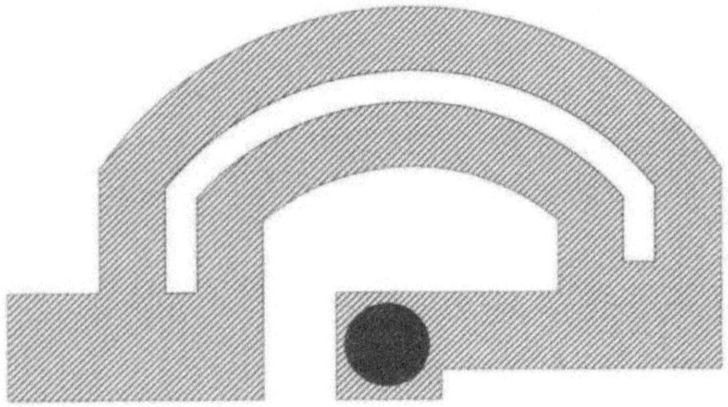

Figure 1.44 A 2D view of the coupled microstrip resonator. *Source*: [3] / John Wiley & Sons.

1.15 Ceramic Resonators

A very popular resonator is the ceramic resonator, which is based on a quarter-wavelength arrangement. The ceramic is silver-plated and can be modeled as a quarter-wave section of a TEM transmission line (cable) with a high relative dielectric constant ranging from 33 to 88. The two 1 pF capacitors shown in Figure 1.46 load the resonator to achieve a resonance frequency of 2.4 GHz. This resonator has an operating Q of about 400, which is much higher than in the previous cases. By varying the capacitors, the frequency can be altered. These are commercial parts made by a variety of companies. Most high-performance (800 MHz–2.4 GHz) oscillators use these resonators.

Figure 1.47 shows a coaxial resonator, while another important resonator, useful up to 40 GHz, is the dielectric resonator (Figures 1.48 and 1.49).

Limitations:

$$1 \le \varepsilon_s \le \varepsilon_r \text{ of substrate}$$

$$1 \le \varepsilon_r < 100$$

$$2\,\text{mm} < D < 50\,\text{mm}$$

$$0.015\,\text{cm} < H_d < 10\,\text{cm}$$

$$0.1 < D/H_d < 10$$

$$H_t \ge 0.0$$

$$0.5\,D < L < 0.1\,\text{m}$$

$$0.0 \le S \le 5\,\text{cm}$$

$$0.0 \le H_S < H_d$$

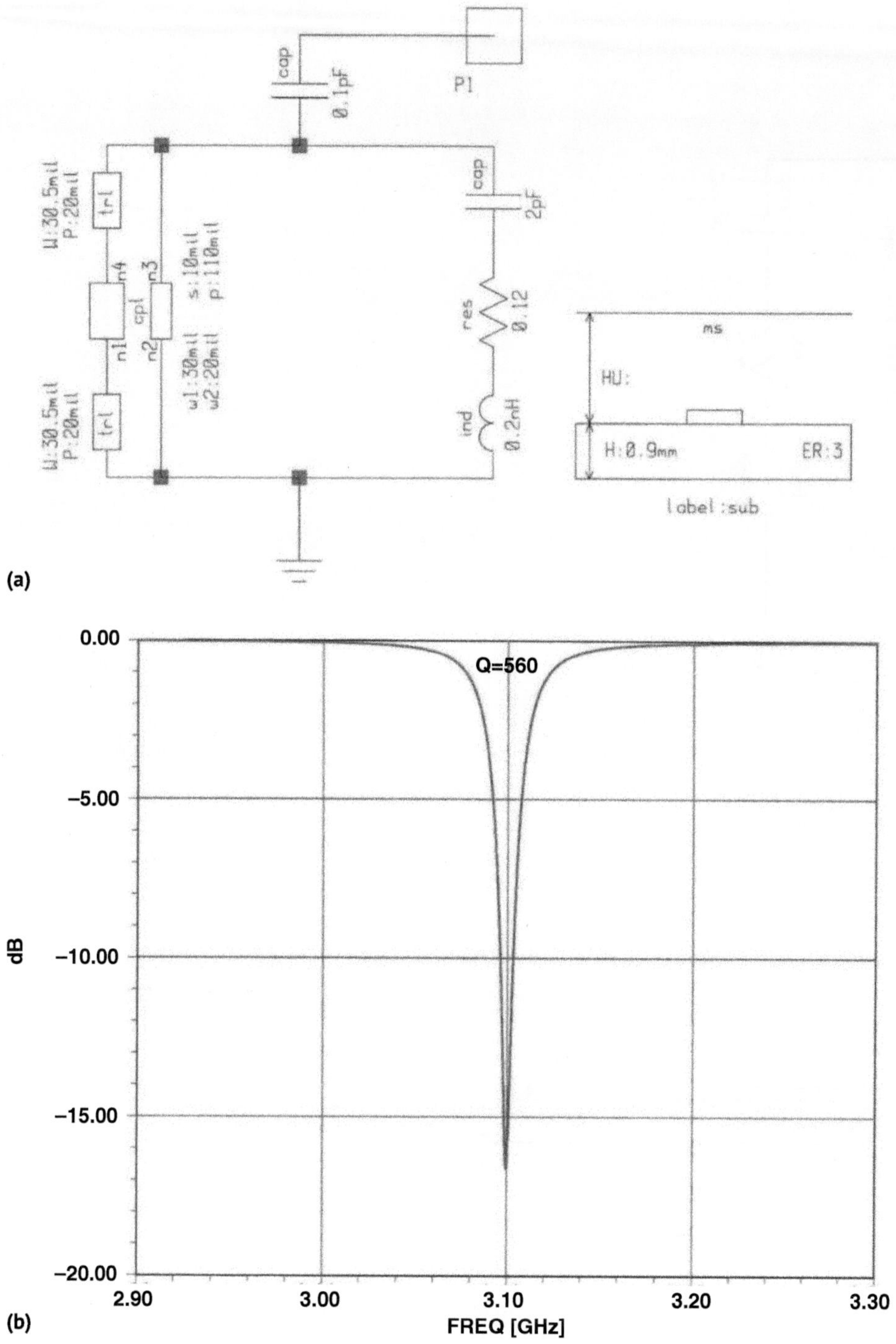

Figure 1.45 (a) Electrical equivalent of the coupled microstrip line resonator. *Source*: [3] / John Wiley & Sons. (b) Frequency response of the coupled microstrip line resonator. *Source*: [3] / John Wiley & Sons.

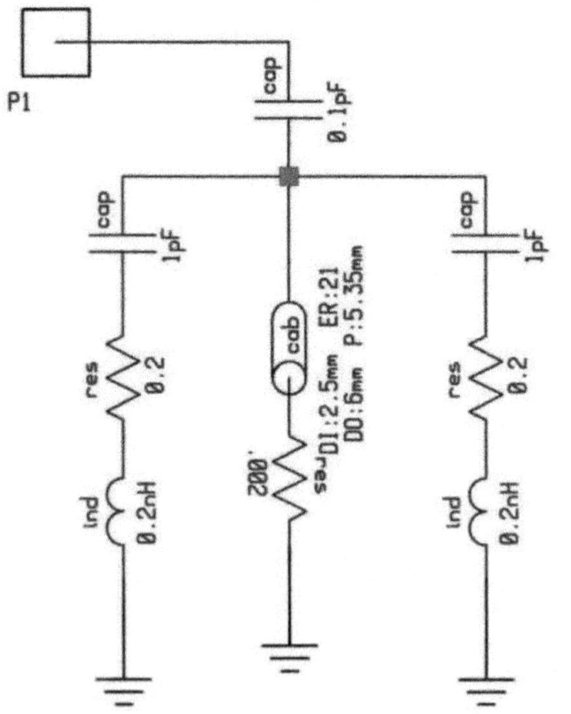

Figure 1.46 A high-Q resonator based on a quarter-wavelength ceramic resonator (CRO). Operating Q values of up to 500 are easily achievable. *Source*: [3] / John Wiley & Sons.

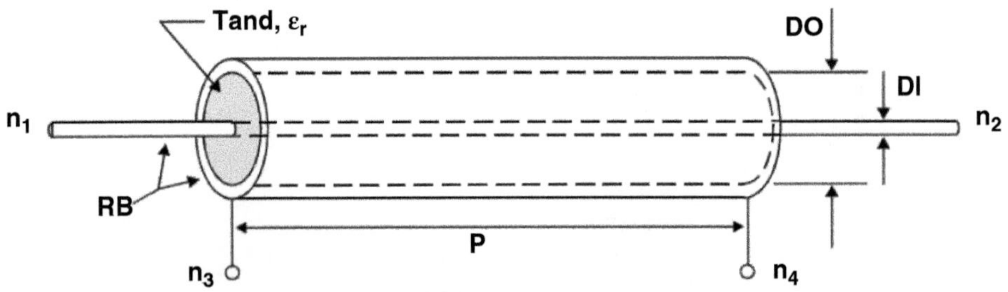

Figure 1.47 Coaxial resonator [14, pp. 2–51].

DRMS—Dielectric Resonator and Microstrip Coupling

The Dielectric Resonator and Microstrip Coupling has two options:

- Option 1: Specify keyword BSF for BandStop structure
- Option 2: Specify keyword BPF for BandPass structure

DRMS BandStop structure

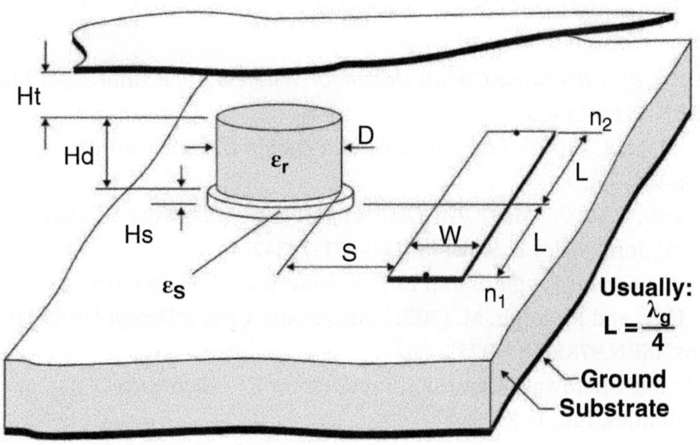

Figure 1.48 Dielectric resonator and microstrip coupling. *Source*: [5] / John Wiley & Sons.

DRMS BandPass structure

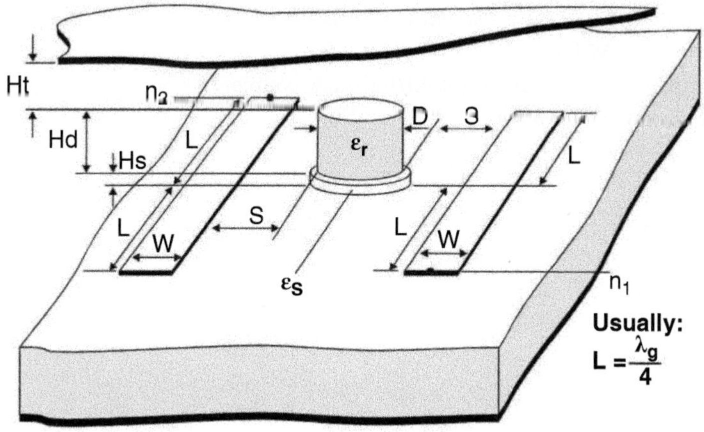

Figure 1.49 Dielectric resonator and microstrip coupling. *Source*: [5] / John Wiley & Sons.

1.16 Conclusion

Designing circuits at low frequencies is mostly trivial. Already at RF frequencies above 400 MHz, complications start as coupling and parasitics begin to play a larger role. In general, the lossy components are the inductors, while the capacitors maintain a high Q up to higher frequencies. The general rule "make the inductor about four times as long as its diameter to have a high Q" loses its meaning for small inductors. Practical experience shows that above 7–10 GHz, only truly dedicated distributed elements are useful. Hopefully, this chapter has provided some inside information and advice about this topic.

References

1 Rohde, U.L. and Rudolph, M. (2013). *RF/Microwave Circuit Design for Wireless Applications*, 2e, 745–746. New York: John Wiley & Sons. ISBN 1118431405.

2 Vendelin, G., Pavio, A.M., and Rohde, U.L. (2005). *Microwave Circuit Design Using Linear and Nonlinear Techniques*, 2e. John Wiley & Sons. ISBN 10 0-471-41479-4.

3 Rohde, U.L., Poddar, A.K., and Boeck, G. (2005, May). *The Design of Modern Microwave Oscillators for Wireless Applications: Theory and Optimization*. New York: John Wiley & Sons. ISBN 0-471-72342-8.

4 Temes, G.C. and LaPatra, J.W. (1977). *Circuit Synthesis and Design*. New York: McGraw Hill.

5 Vendelin, G. Pavio, A.M., Rohde, U.L., and Rudolph, M. (2021). *Microwave Circuit Design Using Linear and Nonlinear Techniques*, 3e. John Wiley & Sons. ISBN 9781118449752.

6 Application Notes: Temperature Stable Microwave Ceramics. *Products for RF/Microwave*. Adamstown, MD, USA: Application, Transactions on Techniques, Inc., 6–85.

7 Kajfez, D. (1995, October). Q factor measurement with a scalar network analyzer. *IEE Proceedings on Microwave Antennas Propagation* 142: 389–372.

8 Wheeler, H.A. (1965, March). Transmission-Line Properties of Parallel Strips Separated by a Dielectric Sheet. *IEEE Transactions on Microwave Theory and Techniques* 13 (2): 172–185. https://doi.org/10.1109/TMTT.1965.1125951.

9 1999 Pioneer Award - IEEE Xplore [PDF], *IEEE Transactions on Aerospace and Electronic Systems*, vol. 36, no. 3, July 2000, pp. 1013, Texas Instruments MERA (Microwave Electronics Radar Applications) Program with WPAFB, 1965–1970; development of a solid-state X-band radar.

10 Mao, S., Jones, S., and Vendelin, G.D. (1968, July). Millimeter-Wave Integrated Circuits. *IEEE Transactions on Electron Devices* 15 (7): 517–523.

11 Vendelin, G.D. (1970, May). Limitations on stripline Q. *Microwave Journal* 13 (5): 63–69.

12 Pozar, D. (1998). *Microwave Engineering*, 2e. New York: Wiley.

13 Rat-race coupler. (2023, April 16). Wikipedia. https://en.wikipedia.org/wiki/Rat-race_coupler.

14 Ansoft Corporation (2001). Serenade Elements Library, Version 8.7.

2

The History, Evolution, and Current Status of Microwave Transistors

Frank Schwierz

Department of Micro- and Nanoelectronic Systems, Technische Universität Ilmenau, Germany

2.1 Introduction

The term *microwave frequencies* is quite common and frequently used in electrical engineering, yet there is no unambiguous definition for it. In the following, we simply designate frequencies in the range between 1 GHz (10^9 Hz) and 1 THz (10^3 GHz = 10^{12} Hz) as microwave frequencies, and correspondingly, in our terminology, microwave transistors are transistors capable of operating in this frequency range.

For many years, microwave electronics has been a dynamic and fast-growing branch of semiconductor electronics, and this holds until the present day. In particular, the explosion of wireless communications since the 1990s and the increasingly popular use of mobile internet access during the last 15 years created huge consumer markets for microwave products. About 50 years ago, the situation was very different. At that time, microwave electronics was somewhat mysterious, used mainly in defense and aerospace applications, and the funding for its development came mostly from government agencies. In the first half of the 1980s, satellite television using low-noise transistors operating at 12 GHz in the receiver front-ends and the first-generation (1G) mobile communication systems emerged as the first consumer applications for microwave transistors with market volumes worth mentioning. Since then, subsequent new generations of mobile communication systems, each operating at higher frequencies and having more users than its predecessor, and new applications such as the global positioning system, wireless local area networks, or automotive radar systems were introduced. In 2020, the first 5G systems went into operation and beyond-5G systems were under development.

On the hardware side, the backbone of all the microwave systems mentioned earlier consists of microwave transistors capable of operating at GHz frequencies. In the following sections of this chapter, we introduce important microwave transistor figures of merit (FOMs) and establish a set of very basic design rules to be obeyed to achieve a good microwave transistor. Moreover, we provide an overview of the different phases in the evolution of microwave transistors, thereby highlighting major achievements and also mentioning research efforts that led to a dead end, and discuss the state of the art of microwave transistors in early 2025.

Part of the material presented in this chapter has been discussed in more detail in a review paper on GHz and THz transistor published in September 2024 [1]. Moreover, we refer the interested reader to several excellent review papers highlighting the state of the art of several types of microwave transistors in the early 2020s [2–6] and to textbooks dealing with the basics and various aspects of the operation and design of microwave transistors and integrated circuits (ICs) [7–10].

2.2 Microwave Transistor FOMs

A view at the technical literature and textbooks on microwave electronics reveals that a variety of different FOMs are used to assess and evaluate the performance of microwave transistors. To keep the discussion within reasonable limits, here in Chapter 2, we focus on a limited set of just four transistor FOMs, namely the characteristic frequencies f_T and f_{max}, the noise figure NF, and the microwave output power P_{out} [1, 7–10]. In addition, we introduce the dynamic range DR as an important parameter. Strictly speaking, *DR* is not a microwave transistor FOM, but a metric of microwave circuits, and it is closely related to the behavior of the transistors the circuit is composed of.

In microwave electronics, the ability of transistors to deliver gain and to amplify signals is exploited. It is a general trend that transistor performance is getting worse with increasing operating frequency, and this is particularly true for the transistor's amplification capabilities and for the gain. At a certain frequency, the transistor completely loses its ability to provide gain, its output signal will be equal to the one at the input, and the transistor can no longer be used as an amplifier [11].

The cutoff frequency f_T is defined as the frequency at which the magnitude of the small-signal current gain of a transistor, $|h_{21}|$, has decreased to unity or, if the gain is given in units of dB (decibel), to 0 dB. The maximum frequency of oscillation f_{max}, on the other hand, is the frequency at which the unilateral power gain of the transistor, U, becomes unity, i.e., 0 dB. Figure 2.1 shows the measured h_{21} and U of a microwave field-effect transistor (FET). As can be seen, both h_{21} and U roll off with a characteristic slope of −20 dB/dec in a wide frequency range. For the specific transistor from Figure 2.1, a GaAs metal-semiconductor FET (MESFET), this is the case for frequencies above 3 GHz, while at lower frequencies, the slope of the gains is flatter. It can also be seen that f_T and f_{max} are estimated by extrapolating the gains measured up to a certain frequency (in Figure 2.1 up to 25 GHz) down to the intersection with the frequency axis using the −20 dB/dec slope mentioned earlier. This practice is convenient, and in some cases even imperative, because of the limitations of the measurement equipment. In a nice review on high-speed GaAs transistors published in 1984, P. Greiling, at that time head of the GaAs IC design and analysis group at Hughes Research Laboratories, stated: "For those of us associated with this technology, this measurement problem always seems to exist. We are in a 'catch 22' situation in which we are developing circuits for instruments that are needed to measure the circuits we are developing" [12].

It should be noted that at high frequencies, the slopes of $|h_{21}|$ and U may again be different from −20 dB/dec. Moreover, particularly U is sensitive to noise. To get the correct values for f_T and f_{max}, extrapolation should be carried out from a frequency range in which the slope of the gains is actually −20 dB/dec and measurement errors are negligible or least tolerable.

If the cutoff frequency of a microwave transistor is known, using the −20 dB/dec slope, its small-signal current gain at the operating frequency f_{op} can be calculated by

$$|h_{21}|[\text{dB}] = 20 \log \frac{f_T}{f_{op}} \tag{2.1}$$

Conversely, if $|h_{21}|$ is known at an operating frequency somewhere in the range where $|h_{21}|$ follows the −20 dB/dec slope, f_T is given by

$$f_T = f_{op} 10^{|h_{21}|\,[\text{dB}]/20} \tag{2.2}$$

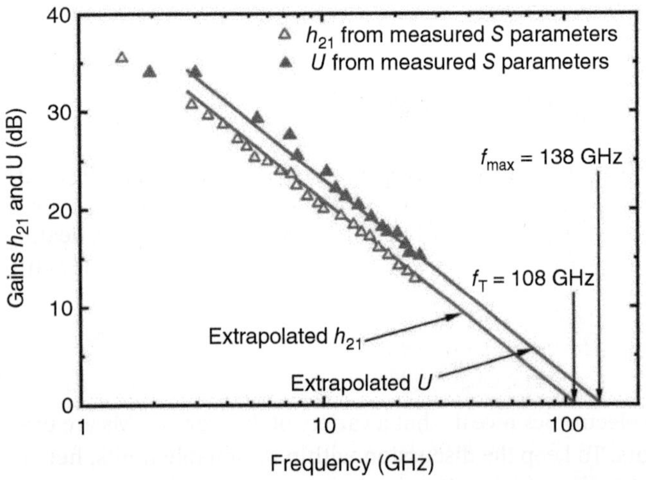

Figure 2.1 Current gain, unilateral power gain, and extrapolated f_T and f_{max} of a GaAs MESFET. *Source*: Data taken from [13].

The fit based on the measured h_{21} data in Figure 2.1 and on the −20 dB/dec slope results in a small-signal current gain of 15.3 dB at 18.5 GHz, and from Eq. (2.2), we obtain an f_T around 108 GHz. Expressions corresponding to Eqs. (2.1) and (2.2) are easily obtained also for U and f_{max}.

We mention that for the vast majority of microwave applications, transistors showing simultaneously high f_T and f_{max} are desirable, but that for most applications, a high power gain and thus a high f_{max} are more important than a high current gain and a high f_T.

At its input terminals, a microwave transistor receives the signals it is supposed to amplify, superimposed by noise. As the transistor is unable to distinguish between signal and noise, both the signal and the noise will be amplified and thus one expects the amplified signal and the amplified noise at the transistor's output. In reality, however, a second noise component is found at the output, which is the noise generated in the transistor. For front-end amplifiers where the signal-to-noise ratio is small, this noise generated in the transistor must be kept as small as possible. A FOM describing the noise generated in a microwave transistor is the noise figure NF [14], which is usually given in units of dB and is defined as

$$NF[\mathrm{dB}] = 10 \log \frac{P_{\text{S-in}}/P_{\text{N-in}}}{P_{\text{S-out}}/P_{\text{N-out}}} \tag{2.3}$$

In Eq. (2.3), P_{Si} and P_{So} are the signal powers at the input and the output of the transistor and P_{Ni} and P_{No} are the noise powers at its input and output, respectively. An ideal transistor internally not generating any noise would show a noise figure of 0 dB. The noise figure increases with increasing frequency and depends on the matching and bias conditions. At a given frequency, NF reaches a minimum value NF_{min}, the minimum noise figure, under optimal matching and bias conditions.

Particularly for microwave power transistors, the output power P_{out} is an important FOM. It represents the amount of microwave power that can be fed from the transistor to a load. For a given transistor, P_{out} decreases with increasing frequency and depends on both the bias conditions and the type of the amplifier circuit, e.g., class-A amplifier or class-B amplifier.

As mentioned earlier, the dynamic range DR is a FOM of microwave circuits. Let us consider a microwave amplifier as an example. Ideally, this amplifier could handle input signals of any signal level, and at the amplifier output, we would observe the amplified signal with exactly the shape of the input signal, i.e., without any distortion. In other words, an ideal sine-shaped input signal would appear as an amplified (increased amplitude) ideal sine signal at the output.

In reality, the situation is different. Here, the input signal the amplifier receives from the signal source is superimposed by noise, and this noise sets the lower limit of the input signal power the amplifier can handle. A resistive signal source generates a noise power P_N given by

$$P_N = k_B\, T\, \Delta f \tag{2.4}$$

where k_B is the Boltzmann constant, T is the temperature, and Δf is the bandwidth. At room temperature, Eq. (2.4) yields a noise power of 4.143×10^{-21} Ws $\times\, \Delta f$. Note that in microwave electronics, the power is frequently given in units of dBm and not in W. The power in dBm is obtained from the power in W using

$$P[\mathrm{dBm}] = 10 \log P[\mathrm{mW}] \tag{2.5}$$

Note that at the right-hand side, the power in mW = 10^{-3} W has to be used. Applying (2.5), the room-temperature noise power given earlier converts to −174 dBm $\times\, \Delta f/\mathrm{Hz}$.

The upper limit of the input signal level is set by distortions. Even if we had an ideal sine input signal, we would see an amplified but distorted sine signal at the output. The origin of such distortions lies in the nonlinearities of the transistor characteristics. These nonlinearities lead to a situation in which the output signal contains components with frequencies different from the frequency of the input signal called fundamental frequency or, for short, the fundamental. Of these additional frequency components, particularly the third-order intermodulation products are critical. A useful measure for the strength of the third-order intermodulation product is the third-order intercept point IP_3 with the corresponding input and output powers IIP_3 and OIP_3. For more information on intermodulation issues and the definition of intercept points, see [10].

As shown in Figure 2.2, the fundamental output power increases on a 1:1 (dB) basis with the input power (before it goes into saturation at higher input powers), while the output powers of the second- and third-order intermodulation products increase on a 2:1 and 3:1 basis, respectively.

We see in Figure 2.2 that the minimum input signal power $P_{\text{in-min}}$ our amplifier can handle, i.e., the minimum detectable signal input power, is defined by the noise floor (i.e., the input noise power) and that the maximum input signal power

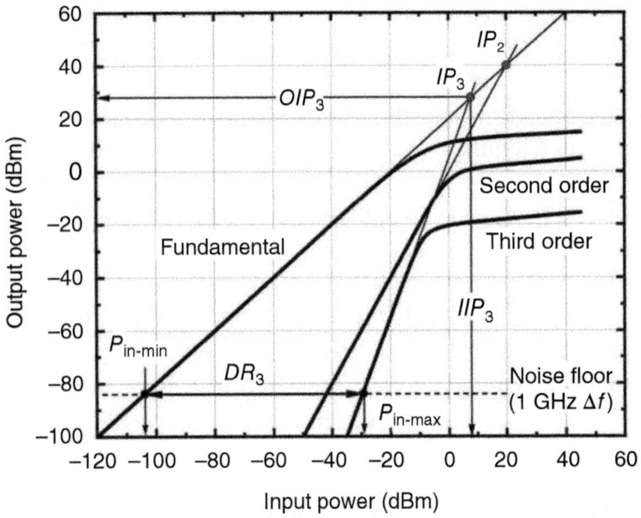

Figure 2.2 First-order (fundamental), second-order, and third-order output powers versus input power of a microwave amplifier. Also shown are the second-order and third-order intercept points, the noise power delivered from the signal source for a bandwidth of 1 GHz (here designated as noise floor), the dynamic range DR_3 related to the third-order intermodulation product, and the lower and upper limits of the input power $P_{\text{in-min}}$ and $P_{\text{in-max}}$.

$P_{\text{in-max}}$ is related to the intercept point of the intermodulation product of interest. From these limits, we get the dynamic range DR_n related to the nth order intermodulation product as

$$DR_n = P_{\text{in-max}} - P_{\text{in-min}} = \frac{(n-1)}{n}(IIP_n - P_{\text{in-min}}) \tag{2.6}$$

where DR_n is the dynamic range in dB and IIP_n and $P_{\text{in-min}}$ are the nth order input intercept power and the minimum detectable signal input power, respectively, in dBm. For the dynamic range related to the critical third-order intermodulation product, we arrive at

$$DR_3 = \frac{2}{3}(IIP_3 - P_{\text{in-min}}) \tag{2.7}$$

For the example from Figure 2.2, we obtain a dynamic range DR_3 of around 74 dB. Note that for an increasing bandwidth, the dynamic range decreases.

2.3 How to Design a Good Microwave Transistor?

In general, we can distinguish between small-signal low-noise microwave transistors and microwave power transistors. A good small-signal low-noise transistor combines a low noise figure NF with a high operating frequency, i.e., a high cut-off frequency f_T and a high maximum frequency of oscillation f_{max}, whereas power transistors should offer high output powers at high operating frequencies. Moreover, regarding the operation principle, we distinguish between bipolar transistors and FETs. On the bipolar side, conventional bipolar junction transistors (BJTs) and heterojunction bipolar transistors (HBTs) are used in microwave circuits, while metal-oxide-semiconductor FETs (MOSFETs), high electron mobility transistors (HEMTs), and also MESFETs are popular types of microwave FETs.

Designing a good microwave transistor, whether it is a FET or a bipolar transistor, is a challenging and complex task. A first aspect of the problem is that microwave transistors are used in a wide variety of different circuit types and applications, each with its own specific requirements regarding transistor performance. Furthermore, when designing a transistor, not only device physics but also the limitations of the available processing technology, questions of reliability and yield, and economic constraints have to be considered. Notwithstanding the complexity of the design task, four general basic rules for the design of microwave transistors have been established [1].

i. The output current of a microwave transistor should be comprised of fast carriers. This rule consists of two parts, the first of which relates to the type of carriers (i.e., electrons or holes) and the second to the choice of the semiconductor material for the active region of the transistor. In most semiconductors, electrons travel much faster than holes (i.e., the mobility of electrons μ_n is higher than the mobility of holes μ_p). For example, in undoped bulk Si at room temperature, μ_n is around 1400 cm^2/Vs compared to $\mu_p \approx 460$ cm^2/Vs. Therefore, n-channel FETs and npn bipolar transistors are preferable to *p*-channel FETs and pnp bipolar transistors. Furthermore, the carrier mobilities as well as the maximum carrier drift velocities vary considerably between different semiconductors. The electron mobility in undoped bulk GaAs, for instance, amounts to about 8600 cm^2/Vs, i.e., more than six times the electron mobility in Si (note that the hole mobility in GaAs is only around 380 cm^2/Vs), and the maximum stationary electron drift velocity in GaAs is about 2×10^7 cm/s compared to 1×10^7 cm/s in Si. Therefore, from the speed point of view, the channel of a microwave FET and the base of a microwave bipolar transistor should consist of a semiconductor showing both a high electron mobility and high maximum electron drift velocity.

ii. The length of the active region of the transistor should be short in the direction of carrier flow. In other words, the channel of a fast microwave FET needs to be short, and the base of a fast bipolar microwave transistor needs to be thin.

iii. Parasitic effects occurring in the transistor, i.e., all effects that do not directly contribute to signal amplification, must be minimized.

iv. For small-signal microwave FETs, the channel material should have a minimum bandgap somewhere between the gap of InAs (0.35 eV) and that of $In_{0.53}Ga_{0.47}As$ (0.74 eV), and for microwave high-power FETs, a wide bandgap of the channel material is beneficial. In microwave bipolar transistors, the space-charge region of the reverse-biased collector–base junction should be located in a region with a sufficiently wide bandgap.

If ***ALL*** requirements mentioned earlier are met simultaneously, we will automatically achieve a fast microwave transistor. This claim is supported by the results of a great many theoretical and experimental studies on the design of microwave transistors. Unfortunately, the situation is different for the linearity performance of microwave transistors. Though there is only a limited number of publications on the relation of the transistor FOMs f_T, f_{max}, and NF to the linearity parameters, there are strong indications that a good $f_T - f_{max} - NF$ performance is not necessarily connected to a good linearity behavior. In [15, 16], for example, it has been shown that gate length scaling of Si microwave MOSFETs leads, as expected, to higher f_T and f_{max} and to lower noise figures, but simultaneously causes degraded linearity parameters of the scaled transistors. This indicates that optimizing the transistor design for good linearity performance while keeping the other FOMs sufficiently high (gains, f_T, f_{max}) or low (NF) represents a formidable challenge.

2.4 The Evolution of Microwave Transistors

2.4.1 The Very Early Years: Ge BJTs Appear as the First Microwave Transistors

Since the demonstration of the first transistor, a Ge BJT, in 1947, device physicists and engineers devoted a lot of effort to increase the speed and the operating frequencies of transistors. As a result of these endeavors, in the early 1950s, Ge BJTs showing maximum frequencies of oscillation in the 100 ... 300 MHz range were reported by Bell Laboratories and RCA [17, 18], and in 1959, the first transistors capable of amplifying signals at frequencies around 1 GHz, again Ge BJTs from Bell Laboratories, were demonstrated [19].

2.4.2 The Period Between 1960 and 1980: Si BJTs and GaAs MESFETs Become the Workhorses of Microwave Electronics

After 1960, the dominance of Ge BJTs gradually faded away, Si BJTs moved into the center of attention, and in the early 1970s, almost all microwave transistors were planar Si BJTs [20]. We note that the reason for the transition from Ge to Si was not the fact that the electronic properties of Si are better than those of Ge. Instead, the replacement of Ge was related to fabrication technology issues and to the excellent quality of the native oxide (SiO_2) that can be formed on Si. Around 1970, state-of-the-art Si BJTs showed cutoff frequencies of 15 GHz [21], maximum frequencies of oscillation in excess of 25 GHz [21], and output powers of 100, 20, and 5 W at frequencies of 1.2, 2, and 4 GHz, respectively [20].

In 1966, C. Mead from the California Institute of Technology presented a new type of FET, the GaAs MESFET, thereby laying the foundation for a revolution in microwave electronics [22]. Only one year later, a GaAs MESFET with an f_{max} of 3 GHz was reported [23]; in 1970, a GaAs MESFET with a record f_{max} around 30 GHz was demonstrated [24], which clearly exceeded the performance of any other transistor type at that time; and in 1973, the 100 GHz f_{max} mark was reached [25].

Moreover, the noise performance of GaAs MESFETs could be significantly improved over the years, as depicted in Figure 2.3. In 1967, the best MESFETs showed a minimum noise figure of 2.8 dB at 0.5 GHz, while comparable noise figures were achieved by state-of-the-art MESFETs at 12 GHz in 1976 and around 20 GHz in 1980. Both low-noise and power GaAs MESFETs became commercially available in the mid-1970s and enjoyed increasing popularity until the end of the decade. GaAs is a III-V compound semiconductor and is attractive for microwave transistors, particularly because its electron mobility is much higher compared to Si; see our discussion on design rule (i) in Section 2.3. Also shown in Figure 2.3 is the noise performance of Si BJTs.

As a side note for readers interested in the history of microwave electronics, we mention here that in 1965, the first two textbooks on microwave transistors and circuits were published [28, 29]. In the book by U.L. Rohde [28], a co-author of the present book, the physics of bipolar microwave transistors and the operation of different types of microwave circuits were discussed, and the frequency range considered was several tens of MHz up to 4.6 GHz. On the other hand, the book by A. Lytel [29] was exclusively devoted to microwave circuits operating at frequencies up to 0.5 GHz.

Si BJTs and GaAs MESFETs were the only two microwave transistor types available in the late 1970s. Si BJTs were commonly used at frequencies below 4 GHz, whereas in the frequency range between 4 and 18 GHz, the GaAs MESFET was the device of choice. The performance enhancement during the 1960–1980 period had mainly been achieved by shrinking the gate length of GaAs MESFETs and by decreasing the base thickness of Si BJTs, i.e., by following design rule (ii) defined in Section 2.3.

As we will see later, III–V semiconductors still play an extremely important role in microwave electronics today. Therefore, we state that the invention of the GaAs MESFET and the introduction of III–V compounds to microwave devices were true revolutionary achievements.

2.4.3 The 1980–2000 Period: The Triumph of the Heterostructures

2.4.3.1 The HEMT Enters the Stage

In the late 1970s, researchers at Bell Laboratories observed the formation of two-dimensional electron gases (2DEG) with high electron mobility in epitaxially grown heterostructures consisting of undoped GaAs layers and n-doped AlGaAs barriers [30]. Due to the almost identical lattice constants of GaAs and AlGaAs, the resulting AlGaAs/GaAs heterostructure is lattice-matched. The underlying physics of the 2DEG formation in an AlGaAs/GaAs heterojunction is shown in Figure 2.4. Caused by the combined effect of the conduction band offset ΔE_C and the doping gradient at the interface, electrons transfer from the conduction band of the doped AlGaAs to the energetically lower lying conduction band of the undoped GaAs. Here, the transferred electrons are confined in a narrow potential well and form a 2DEG channel that is spatially separated from the donor ions in the AlGaAs barrier, which originally delivered the electrons. Thus, ionized impurity scattering is largely suppressed, leading to a high mobility of the 2DEG electrons.

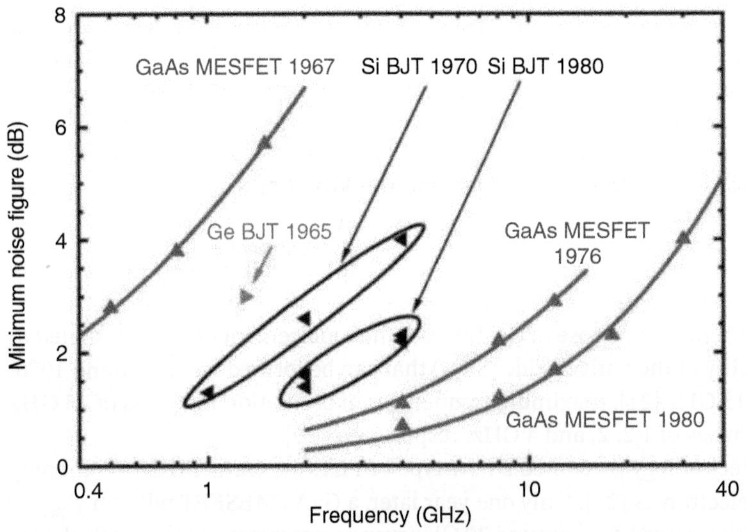

Figure 2.3 Evolution of the noise performance of Si BJTs and GaAs MESFETs. *Source:* Data taken from [7, 26, 27].

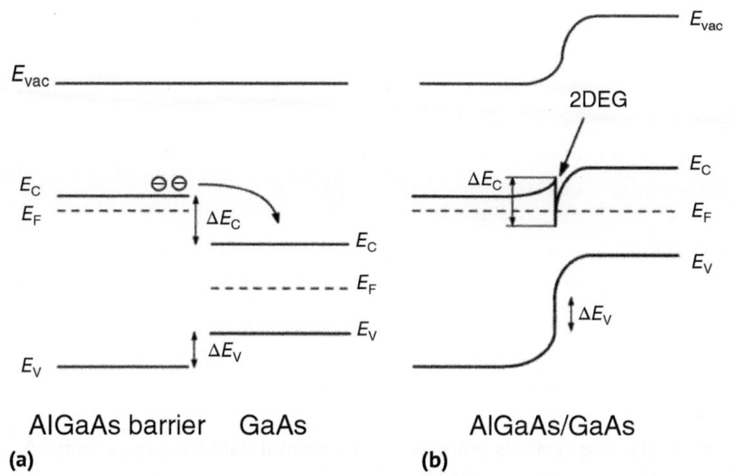

Figure 2.4 Band diagram of an AlGaAs/GaAs heterostructure as used in HEMTs. (a) The AlGaAs and GaAs layers are not in contact with each other but are considered separately. Here, the vacuum level E_{vac} is used as the reference energy, E_C and E_V are the conduction and valence band edges, E_F is the Fermi level, and ΔE_C and ΔE_V are the conduction band and valence band offsets. (b) AlGaAs and GaAs now in contact with each other.

It was obvious to elaborate a FET concept taking advantage of the high 2DEG mobilities. The basic idea for such an AlGaAs/GaAs FET with a 2DEG channel and a gate on top of the barrier came again from Bell Laboratories [31], but the first experimental transistors of this type were reported by researchers from Fujitsu Laboratories in 1980 [32]. The Fujitsu group gave its device the name HEMT, whereas the researchers at Bell Laboratories called their transistor selectively doped heterojunction transistor (SDHT) and published the first experimental data for such transistors in 1982 [33]. The naming of this transistor type became even more confusing because other groups working on transistors based on the same principle called their devices modulation-doped FET (MODFET, University of Illinois, Rockwell) and two-dimensional electron gas FET (TEGFET, Thomson-CSF, France). After a period where all these designations were used in parallel, eventually the name HEMT prevailed and is widely used in the microwave community ever since.

AlGaAs/GaAs HEMTs showed promising performance and competed successfully with GaAs MESFETs having comparable gate lengths, particularly in terms of noise figure and output power. However, the improvements of f_T and f_{max} were less than anticipated. It was found that by introducing a thin InGaAs channel layer between the AlGaAs barrier and the GaAs substrate, a larger conduction band offset ΔE_C is obtained, resulting in an enhanced 2DEG sheet density in the InGaAs layer. Moreover, it was assumed that the 2DEG mobility in the InGaAs channel of AlGaAs/InGaAs/GaAs structures would be higher than that in AlGaAs/GaAs structures, an expectation that was only partially fulfilled (see Figure 3.15 in [7]). In 1985, the first HEMTs using AlGaAs/InGaAs/GaAs heterostructures were reported [34]. Since ΔE_C increases and the 2DEG mobility was expected to improve with an increasing amount of In in the channel, $In_xGa_{1-x}As$ layers with high In content x would be desirable. However, because the lattice constant of InGaAs is larger than that of AlGaAs and GaAs, InGaAs layers epitaxially grown on GaAs are strained and the In content x is limited to about 0.2. The strained InGaAs layers are said to be pseudomorphic and therefore AlGaAs/InGaAs/GaAs HEMTs are called pHEMTs (pseudomorphic HEMT).

GaAs pHEMTs were able to outperform GaAs MESFETs and AlGaAs/GaAs HEMTs in terms of f_T, f_{max}, and noise figure, became commercially available in the early 1990s, and were (and still are) in widespread use in both low-noise and power amplifiers. Figure 2.5 shows the microwave performance of the best GaAs pHEMTs reported until 2000 in terms of f_T and f_{max}.

From the empirical upper-limit trend curves in Figure 2.5, we can learn an important lesson that improves our understanding not only of the speed limits of GaAs pHEMTs but also of the speed limits of microwave FETs in general. For longer gate lengths ($L \geq 0.3$ μm for f_T and $L \geq 0.6$ μm for f_{max} in the case of GaAs pHEMTs), the trend curves are linear in the log frequency versus log gate length plot, and shortening the gate length leads to considerably improved f_T and f_{max}; see design rule (ii). For shorter gates, however, the trend curves flatten and the beneficial effects of a gate length reduction diminish. The reasons for this are (i) that in FETs with short gates, the devices' parasitics, such as fringing capacitances, contact and access resistances, etc., do not scale as much as the active part of the transistor and (ii) that undesirable effects such as the degradation of the drain current saturation become stronger at short gate length levels. Thus, when shrinking

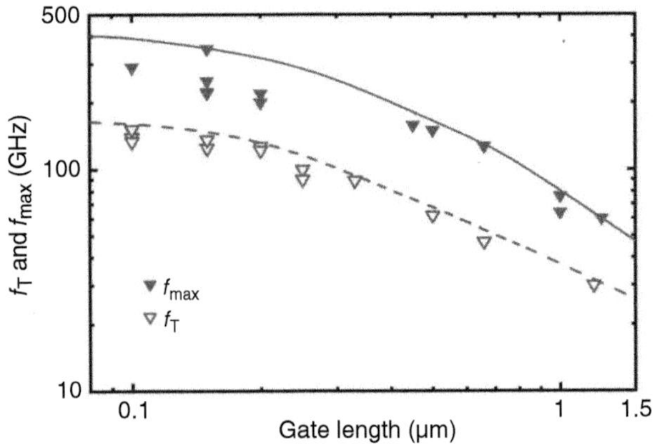

Figure 2.5 Reported f_T and f_{max} versus gate length for GaAs pHEMTs. The symbols represent experimental data from the literature, and the full and dashed lines indicate the empirical upper limit f_T and f_{max} trend curves [1, 7, 26–27].

the gate length, adverse effects of the parasitics are getting more effective and seriously degrade transistor performance at very short gate length levels.

During the 1980–2000 time span, three further III–V HEMT types were introduced and quickly gained popularity. The first of these is the InP HEMT. Here, an $In_yAl_{1-y}As/In_xGa_{1-x}As$ heterostructure is grown on an InP substrate. For $y = 0.52$ and $x = 0.53$, the InAlAs and InGaAs layers are lattice-matched to InP. The advantage of this system compared to the pseudomorphic AlGaAs/InGaAs structures used in GaAs pHEMTs is twofold. First, the conduction band offset at the $In_{0.52}Al_{0.48}As/In_{0.53}Ga_{0.47}As$ heterojunction is actually larger than that in GaAs pHEMTs, leading to a higher 2DEG sheet concentration, and second, due to the higher In content in the channel, the 2DEG mobility is significantly enhanced. These advantages can even be increased by using a strained, i.e., pseudomorphic, InGaAs channel with an In content above 0.53. The work on InP HEMTs started in the early 1980s [35, 36]. As can be seen from Table 2.1, summarizing the $f_T - f_{max}$ performance of the best microwave HEMTs reported until late 2000, InP HEMTs behave much better than GaAs pHEMTs. Their downside is, however, the higher InP wafer and device processing costs.

This brings us to the GaAs mHEMT, where the "m" stands for metamorphic. The motivation for the metamorphic approach is to realize transistors with InP-HEMT-like performance on large-diameter, cost-effective GaAs substrates. The key feature of a GaAs mHEMT is an $In_{0.52}Al_{0.48}As/In_xGa_{1-x}As$ heterostucture with an In content $x \geq 0.53$ as used in InP HEMTs, but grown on a GaAs substrate. The large lattice mismatch between InGaAs with high In content and GaAs prohibits growing the InGaAs channel layer with high In content directly onto the GaAs substrate. Instead, first a thick graded buffer is grown on the substrate, leading to a gradual transformation of the lattice constant from that of the GaAs substrate to that of the $In_xGa_{1-x}As$ channel layer with $x \geq 0.53$. Research on GaAs mHEMTs began in the early 1990s [37] and led, as can be seen from Table 2.1, to transistors with microwave parameters better than those of GaAs pHEMTs, although until the year 2000 not yet as good as those of InP HEMTs.

Finally, the emerging GaN HEMTs whose development began in the early 1990s are worth mentioning. The heart of these transistors is a heterostructure consisting of an AlGaN barrier and a GaN channel layer. This heterostructure has two characteristic features. First, both AlGaN and GaN possess a hexagonal lattice structure (this is in contrast to the other III–V HEMTs discussed earlier, in which all layers consisted of cubic materials) and a wide bandgap above 3 eV compared to 1.1 eV for Si and 1.4 eV for GaAs. Second, due to strong polarization effects, a 2DEG with very high sheet concentration ($>10^{13}$ cm^{-2} and thus much higher than in the other heterostructures discussed earlier) forms at the GaN side of the AlGaN/GaN junction, even if the barrier is undoped [38]. The combination of the wide bandgap, the high 2DEG sheet concentration, and a reasonably high 2DEG mobility (of the order of 1500 cm^2/Vs) made the GaN HEMT a highly promising candidate for microwave power amplification. The first GaN HEMT was reported in 1993 [39], and already two years later, a transistor of this type showing an f_T in excess of 20 GHz and an f_{max} of 70 GHz was demonstrated [40]. At the end of the 1990s, the best GaN HEMTs featured f_T and f_{max} above 100 GHz (see Table 2.1), microwave output powers > 10 W at 8.2 GHz, and output power densities approaching 10 W/mm [41]. Note that the latter is about a factor of 5 higher than the best power densities reported for GaAs and InP HEMTs.

Table 2.1 State of the art of different types of microwave MESFETs and HEMTs in terms of f_T and f_{max}, status late 2000. L is the gate length.

FET type	f_T (GHz)	L (nm)	References	f_{max} (GHz)	L (nm)	References
GaAs MESFET	168	60	[42]	177	120	[43]
AlGaAs/GaAs HEMT	113	100	[44]	151	240	[45]
GaAs pHEMT	152	100	[46]	350	150	[47]
GaAs mHEMT	204	180	[48]	400	100	[49]
InP HEMT	362	50	[50]	600	100	[51]
GaN HEMT	110	50	[52]	140	50	[52]

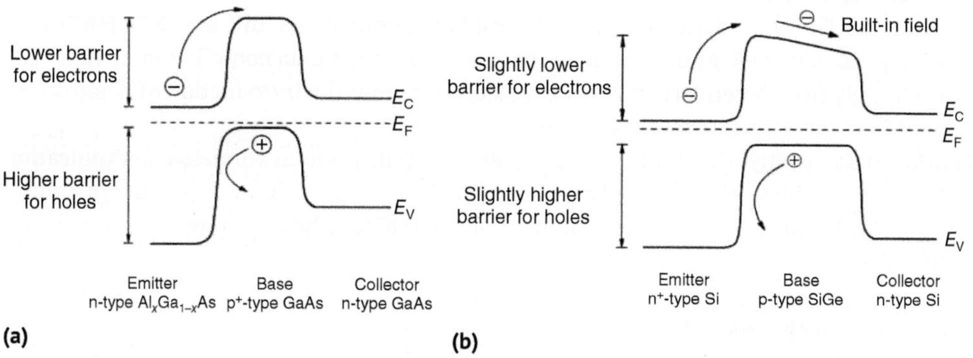

Figure 2.6 Band diagram of HBTs under equilibrium conditions (no voltages applied). *Source:* [1] / AIP Publishing / CC BY 4.0. (a) True HBT, here for a GaAs HBT with graded AlGaAs emitter. (b) Graded-base SiGe HBT.

2.4.3.2 The Development of HBTs

The idea of the HBT was elaborated by W. Shockley in a patent specification in 1948 [53], i.e., only one year after the demonstration of the first bipolar transistor. However, it took until the late 1970s before the epitaxial growth technologies needed to realize high-quality heterostructures for HBTs became available and well-operating HBTs could be demonstrated.

The basic concept of an HBT is shown in Figure 2.6a. The heart of this transistor is the emitter–base heterojunction, where the emitter consists of a semiconductor with a wider bandgap compared to that of the base material. Because of the bandgap difference at the emitter–base junction, electrons moving from the emitter to the base encounter a lower energy barrier to be surmounted compared to the barrier faced by holes when moving from the base to the emitter. Thus, the undesirable hole injection into the emitter that limits the common emitter current gain in conventional bipolar transistors is suppressed very effectively. This allows for the realization of very thin and heavily doped base layers, thereby leading to short base transit times and low base resistances. The transistor design from Figure 2.6a is frequently called a true HBT and represents the common design used for GaAs and InP HBTs. GaAs HBTs consist of an n-type $Al_xGa_{1-x}As$ (Al content x around 0.3) or $In_{0.49}Ga_{0.51}P$ (for this composition, InGaP is lattice-matched to GaAs) emitter, a p-type GaAs base, and an n-type GaAs collector. InP HBTs, on the other hand, have an n-type InP emitter, a p-type InGaAs base, and either an n-type InGaAs/InP or an n-type InP collector.

There is a second HBT type called graded base HBT, which is a very popular design for the Si-based SiGe HBTs (see Figure 2.6b). Here, the bandgap of the $Si_{1-x}Ge_x$ base is wider close to the emitter–base junction (almost as wide as in the Si emitter, since the Ge content x at the emitter side of the base is small) and is gradually becoming narrower toward the collector–base junction (due to an increasing Ge content). This means that the bandgap difference at the emitter–base junction is smaller than in a true HBT and, as a consequence, the suppression of the hole current from base to emitter is less pronounced; however, this is more than outweighed by the built-in field in the base, which causes the electrons to move very quickly through the base and leads to a small base transit time.

Table 2.2 State-of-the-art $f_T - f_{max}$ performance of bipolar microwave transistors, status late 2000.

Transistor type	f_T (GHz)	References	f_{max} (GHz)	References
Si BJT	100	[58]	84	[58]
GaAs HBT	156	[59]	275	[60]
InP HBT	305	[61]	300	[61]
SiGe HBT	156	[62]	180	[63]

In his famous paper from 1982, H. Kroemer expressed his expectation that the HBT would become an important and successful microwave transistor type [54]. This turned out to be true, and in the second half of the 1980s, GaAs, InP, and SiGe HBTs operating at GHz frequencies were reported [55–57]. Ever since, the advances in the development of HBT transistors have been enormous. Table 2.2 showing the $f_T - f_{max}$ performance of the best Si BJTs and GaAs, InP, and SiGe HBTs as of late 2000 clearly indicates the rapid progress achieved. Moreover, as we have seen from the data compilation in Table 2.1, microwave FETs benefited similarly strongly from heterostructures. So, we state here that the introduction of heterostructures was a revolutionary achievement in microwave transistor research.

In the mid-1990s, commercial GaAs HBTs, mostly used for microwave power amplification in wireless communication systems, appeared on the market, and in 1998, the world's first industrial fab line for SiGe HBT ICs was put into operation at IBM [64]. Significant progress could also be observed in the development of InP HBTs, although these devices were not yet commercially available in the late 1990s.

2.4.3.3 The Si MOSFET Becomes a Hot Topic in Microwave Electronics

In addition to the successful introduction of heterostructures in high-performance microwave transistors, the efforts undertaken during the 1990s to explore the microwave potential of Si MOSFETs should definitely be included in our foray through the history of microwave transistors. A look back at the 1970s and 1980s reveals that at that time Si MOSFETs were not considered seriously for microwave applications due to their limited speed. However, continuous MOSFET scaling, a key trend in digital Si CMOS, and the increasing maturity of Si CMOS technology changed the picture over the years, and the Si MOSFET became a serious and recognized candidate for microwave circuits. In fact, the topic of microwave CMOS was discussed at all major device conferences around the world in the second half of the 1990s. At that time, Si laterally diffused MOSFETs (LDMOSFETs) for high-power amplification up to 2.5 GHz and small-signal microwave MOSFETs became commercially available.

Highlights in the field of Si microwave MOSFETs during the 1990s were the demonstration of an LDMOSFET delivering 120 W output power at 2 GHz [65] and of small-signal silicon-on-insulator MOSFETs showing an f_T of 150 GHz and an f_{max} of 66 GHz [66, 67].

2.4.4 The Period 2001–2025: Evolutionary Advancements and Revolutionary Dead Ends

After the turn of the millennium, the dynamic evolution of microwave transistors continued. By optimizing the design and improving the processing technology, continuous performance enhancements of microwave transistors could be achieved. We start with considering the $f_T - f_{max}$ performance of these transistors. A particularly impressive highlight here was overcoming the 1 THz f_{max} barrier by InP HEMTs and InP HBTs in 2011 [68, 69]. Furthermore, the fastest GaN HEMTs today reach both f_T and f_{max} in excess of 450 GHz [70], and for the best SiGe HBTs, $f_T > 500$ GHz and $f_{max} > 700$ GHz were reported [71].

The evolution of small-signal microwave Si MOSFETs between 2001 and 2025 was exciting on the one hand, but it also revealed some aspects that gave pause for thought. During the early 2000s, Si MOSFETs broke performance records in terms of f_T and f_{max} year by year, and transistors with f_T up to 395 GHz and f_{max} up to 420 GHz were demonstrated [72, 73], but this trend came to an abrupt end in late 2008. It took 10 years until a new f_{max} record of 450 GHz for Si MOSFETs could be reported in 2018 [74], while the f_T record of 395 GHz set in 2008 still stood in early 2025. The record-f_{max} transistor from 2018 was a FinFET having a fin-shaped channel around which the gate was arranged from three sides (at the two vertical sidewalls and at the top). The FinFET architecture was introduced by Intel in its commercial digital CMOS circuits in 2011 [75], quickly followed by other leading chip manufacturers such as Samsung, TSMC, and GlobalFoundries. During the 2010s, the FinFET became the standard MOSFET architecture for the most advanced digital CMOS technologies. A highlight of the early work on microwave FinFETs was the demonstration of 60 nm gate laboratory-stage Si FinFETs showing both f_T and f_{max} around 100 GHz in 2006 [76]. In order to assess this achievement properly, it should be mentioned that at that time, the fastest Si microwave MOSFETs with the conventional planar single-gate architecture and comparable gate lengths showed twice the f_T and f_{max} of these FinFETs. However, by improving FinFET technology and optimizing FinFET design, Si microwave FinFETs became more and more competitive to planar single-gate MOSFETs over time.

Next, let us take a look at the performance of Si *p*-channel microwave MOSFETs. In basic rule (i) for the design of microwave transistors established in Section 2.3, we stated that n-channel FETs are to be preferred because of the higher electron mobility compared to holes. This holds in particular for III–V semiconductors, where the hole mobilities are much lower than the electron mobilities. According to the mobility numbers we provided in Section 2.3, the μ_n/μ_p ratio in undoped GaAs is above 20 compared to about 3 in undoped Si. Moreover, in compressively strained channels of Si pMOSFETs, the hole mobility can be enhanced significantly, resulting in a μ_n/μ_p ratio close to unity. Because of this, Si microwave *p*-channel MOSFETs can achieve competitive $f_T - f_{max}$ performance as shown in Table 2.3.

Figure 2.7 summarizes the state of the art of microwave transistor performance in terms of f_T and f_{max} (status early 2025).

Figure 2.7b shows the cut off frequency in GHz versus the maximum frequency of oscillation for some microwave bipolar transistors

Now we turn to the noise performance of microwave transistors. We focus here on FETs since for microwave low-noise amplifiers (LNAs), HEMTs and MOSFETs are more popular than bipolar transistors. Figure 2.8 shows the minimum noise figures of different types of microwave FETs as a function of frequency. As can be seen, at high frequencies, InP HEMTs and GaAs mHEMTs show by far the best noise behavior, followed by GaAs pHEMTs. Not as good as the HEMTs, but still quite impressive, is the noise behavior of Si MOSFETs, with reported noise figures of 0.6 dB at 26 GHz and below 2 dB at 40 GHz.

Table 2.3 Comparison of the $f_T - f_{max}$ performance of n-channel ($f_{T\text{-}n}$ and $f_{max\text{-}n}$) and *p*-channel ($f_{T\text{-}p}$ and $f_{max\text{-}p}$) Si microwave MOSFETs.

MOSFET type	L (nm)	$f_{T\text{-}n}$ (GHz)	$f_{T\text{-}p}$ (GHz)	$f_{max\text{-}n}$ (GHz)	$f_{max\text{-}p}$ (GHz)	References
Planar single-gate	28	395	300	410	325	[73]
FinFET	32	230	238	284	242	[77]
FinFET	32	306	300	452	300	[74]

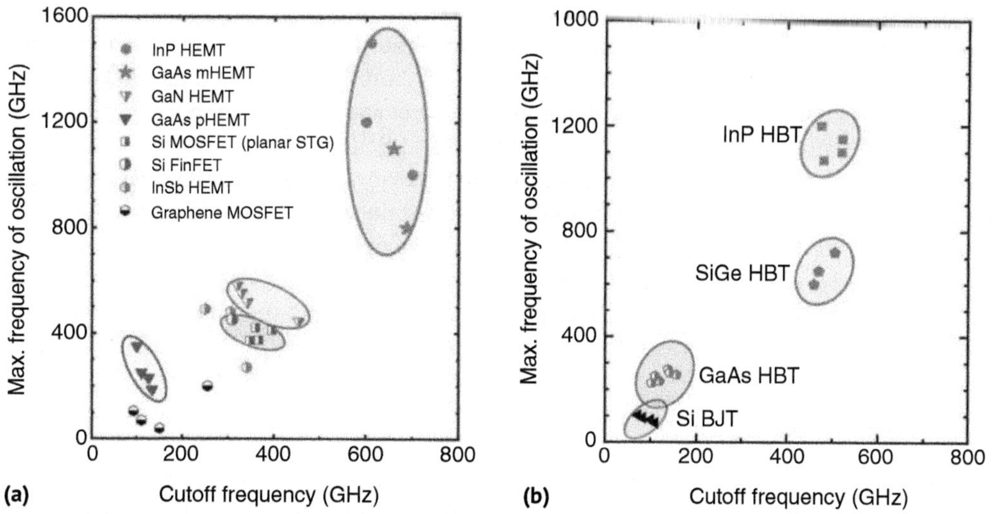

Figure 2.7 Compiled $f_T - f_{max}$ data of (a) microwave FETs and (b) microwave bipolar transistors [1, 7, 26, 27, 78–80]. STG: single-top-gate.

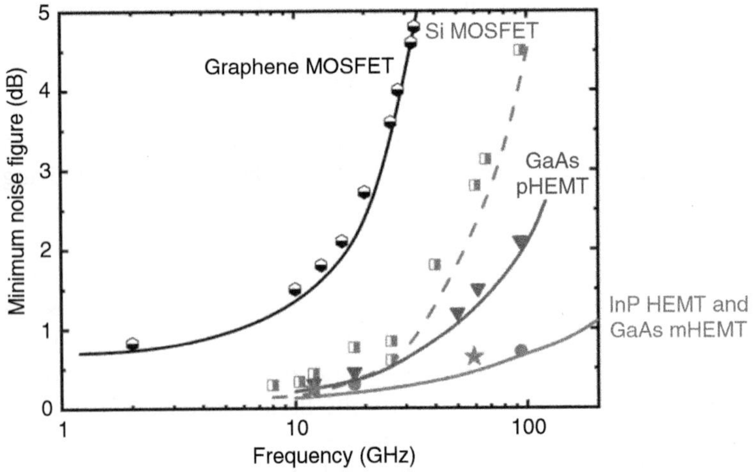

Figure 2.8 Reported minimum noise figures of microwave FETs versus frequency. The lines are a guide for the eye and indicate the empirical lower limit of the minimum noise figure for each transistor type [1, 7, 26, 27, 78]. Circles: InP HEMT. Downward-pointing triangles: GaAs mHEMT.

As we have seen from Figures 2.7a and 2.8, among the microwave FETs discussed so far, the transistors with the highest channel mobilities (InP HEMTs and GaAs mHEMTs with their high-In-content channels) show the highest f_T and f_{max} and the lowest noise figures. This is not a surprise and is in accordance with design rule (i) from Section 2.3. The close connection between mobility and FET performance was the motivation to explore novel alternative channel materials with ultra-high mobilities for microwave FETs during the 2000s and 2010s.

The first of these high-mobility channel materials we discuss here is graphene. Graphene is a so-called two-dimensional material and consists of one single layer of carbon atoms arranged in a hexagonal lattice. Already in their pioneering work on the exfoliation of graphene published in 2004, the Novoselov–Geim group from Manchester University reported carrier mobilities around 10,000 cm^2/Vs in this material [81], and shortly afterward, an upper limit of 200,000 cm^2/Vs for the room-temperature carrier mobility in graphene was suggested [82]. These high mobilities raised expectations that MOSFETs with graphene channels could operate extremely fast and be capable of outperforming the traditional microwave transistors [83]. Generous programs for graphene research with huge funding volumes were established during the 2000s and 2010s, and numerous groups around the world started working on graphene microwave MOSFETs. The first graphene MOSFET was demonstrated in 2007 [84], and since 2008, microwave graphene MOSFETs have been reported and microwave performance data for such MOSFETs have been published [85–87].

Had it been possible to transfer the super-high mobilities mentioned earlier into graphene MOSFET channels and had such graphene MOSFETs followed the mobility–performance trend known from the III–V HEMTs, another revolution with consequences similarly far-reaching as those of the introduction of heterostructures would indeed have taken place. However, things turned out differently. It is important to recognize that graphene is not a semiconductor but a material with zero bandgap. What was overlooked in the exuberance over the first successes of graphene transistor research was the fact that the missing bandgap of the channel material has devastating effects on the behavior of graphene MOSFETs. Relevant for microwave MOSFETs is that a gapless channel leads to a poor saturation of the drain current caused by a change of the conductivity type from n-type to p-type (or from p-type to n-type) at some point in the channel. This, in turn, particularly deteriorates the transistor's power gain and f_{max} [88, 89]. The best reported $f_T - f_{max}$ data of graphene MOSFETs included in Figure 2.7a shows that these transistors, although capable of GHz operation, fall far behind the traditional microwave FETs, and the data in Figure 2.8 indicates that the noise performance of graphene MOSFETs does not benefit from the high mobility either.

A second effort we deal with was focused on exploring InSb for microwave FET channels. InSb also features very high electron mobilities (around 80,000 cm^2/Vs predicted and exceeding 30,000 cm^2/Vs measured at 300 K in AlInSb/InSb heterostructures [90, 91]). Unlike graphene, InSb is a semiconductor, but its bandgap is quite narrow (170 meV). Intensive work on InSb HEMTs was conducted during the 2000s, and transistors showing cutoff frequencies in the 250–340 GHz range and maximum frequencies of oscillation up to almost 500 GHz were reported [91–93]. While these numbers look impressive at first glance, Figure 2.7a indicates that the InSb FETs also fall significantly behind the corresponding data for

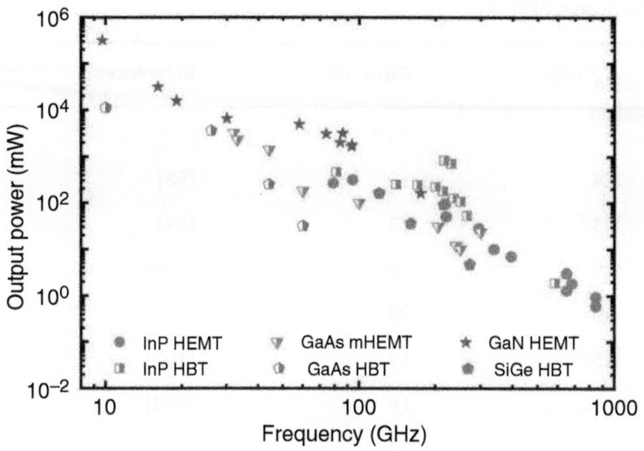

Figure 2.9 Reported output power of MMIC power amplifiers versus frequency. *Source*: Data taken from the compilations published in [1, 7, 26–27].

InP HEMTs and GaAs mHEMTs. We note that the reason for this lag is not a change of the conductivity type in the channel, as discussed earlier for graphene MOSFETs, but rather the consequence of excessive impact ionization caused by the narrow bandgap of InSb. The performance data for graphene MOSFETs and InSb HEMTs presented earlier tells us that high mobility alone is not sufficient for achieving outstanding microwave FET performance. Instead, it provides evidence for the fact that all requirements for a good microwave FET we established in the basic design rules (i) to (iv) in Section 2.3, including the need for a sizable bandgap of the channel, must be met simultaneously.

The aforementioned discussion leads us to the conclusion that the research efforts devoted to alternative high-mobility materials did not lead to another revolution in microwave electronics but rather to dead ends. Instead, the progress in the field of microwave transistors achieved since 2001 has come from evolutionary achievements.

In the early stages of their development, transistors of a new type are fabricated and characterized as discrete devices. When, however, their use in real-word applications comes to the agenda, these transistors must be integrated into ICs. In microwave electronics, such ICs are frequently called monolithic microwave integrated circuits (MMICs), and typical examples for MMICs are amplifiers, oscillators, mixers, and transceivers. We close our discussion on the current status of microwave transistors by exemplarily reviewing the performance of power amplifier and LNA MMICs. Figure 2.9 shows the output power of MMICs based on various types of microwave transistors as a function of frequency. As can be seen, up to 100 GHz GaN HEMT MMICs show the highest output powers, while between 100 and 250 GHz, InP HBT MMICs perform best. Finally, at 300 GHz and beyond, the InP HEMT MMICs are currently unchallenged. Moreover, Figure 2.9 reveals that the good old GaAs HBT delivers reasonably high output powers up to 50 GHz and that between 30 and 300 GHz, power amplifier MMICs based on GaAs mHEMTs show competitive performance. Finally, up to 250 GHz MMICs based on SiGe HBTs represent a reasonable alternative for applications where only moderate output powers are needed.

Examples for the performance of LNA MMICs are compiled in Table 2.4. We note that LNA MMICs have also been realized using other types of microwave transistors not included in the table. For example, a GaN HEMT LNA MMIC showing noise figures of 1.9 and 4.1 dB and gains of 11 and 12.6 dB at frequencies of 20 and 50 GHz, respectively, has been reported in [94]. Moreover, noise figures/gains of 1.7 dB/16 dB at 16 GHz [95] and 7.6 dB/17.2 dB at 183 GHz [96] have been achieved with SiGe HBT LNA MMICs.

2.5 Summary and Outlook

During the almost 70 years of microwave transistor development, the operating frequencies, the frequency limits, and the microwave performance of these transistors have been enhanced continuously. This trend is illustrated in Figure 2.10, which shows the evolution of the best reported data for the cutoff frequency f_T and the maximum frequency of oscillation f_{max}. While f_T and f_{max} of the fastest transistors in 1960 were slightly above 1 GHz, the record performance as of early 2025 was defined by InP HEMTs and HBTs, showing maximum frequencies of oscillation in excess of 1 THz and cutoff frequencies of 700 GHz and more.

Table 2.4 Performance of LNA MMICs based on different types of microwave FETs.

FET type used for the LNA	f (GHz)	NF (dB)	$Gain$ (dB)	References
Si MOSFET (planar single-gate)	60	3	18.5	[97]
	89	6.4	10	[98]
	135	8.3	17	[99]
Si MOSFET (FinFET)	5.7	1.9	30.5	[100]
	75	4.0	20	[101]
	80	4.5	17	[102]
GaAs pHEMT	5	0.76	16	[103]
	28	1.4	13	[104]
	40	3.7	23	[105]
GaAs mHEMT	90	3.3	17.5	[106]
	243	5.6	30.6	[107]
	600	15	14.1	[108]
InP HEMT	90	2	29	[109]
	660	10	28	[110]
	850	11.2	13.6	[111]

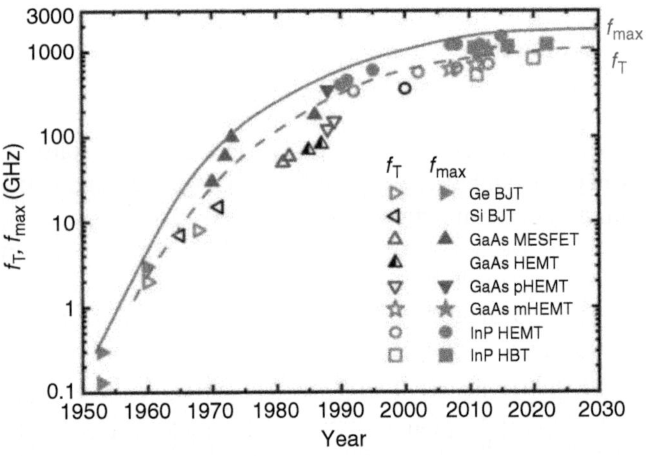

Figure 2.10 Evolution of the best reported $f_{max}$and f_T data of microwave transistors from the 1950s until 2025. Note that for Si BJTs and GaAs HEMTs (conventional AlGaAs/GaAs HEMTs), no f_{max} data is included since these two transistor types were never the fastest in terms of f_{max}. *Source*: Data taken from the compilations published in [1, 7, 26, 27].

This became possible by shrinking the critical device dimensions, by introducing heterostructures, by exploiting the high electron mobilities of medium-sized bandgap III–V semiconductors, and by minimizing the transistor parasitics by optimizing the transistor designs. Figure 2.11, showing some important milestones, summarizes the development of microwave transistors.

To date, InP heterostructure transistors, i.e., InP HEMTs and HBTs, show the highest cutoff frequencies and the highest maximum frequencies of oscillation, and thus the highest operating frequencies of all transistor types. GaAs mHEMTs perform very competitively and show only slightly higher noise figures and slightly lower cutoff frequencies than InP HEMTs.

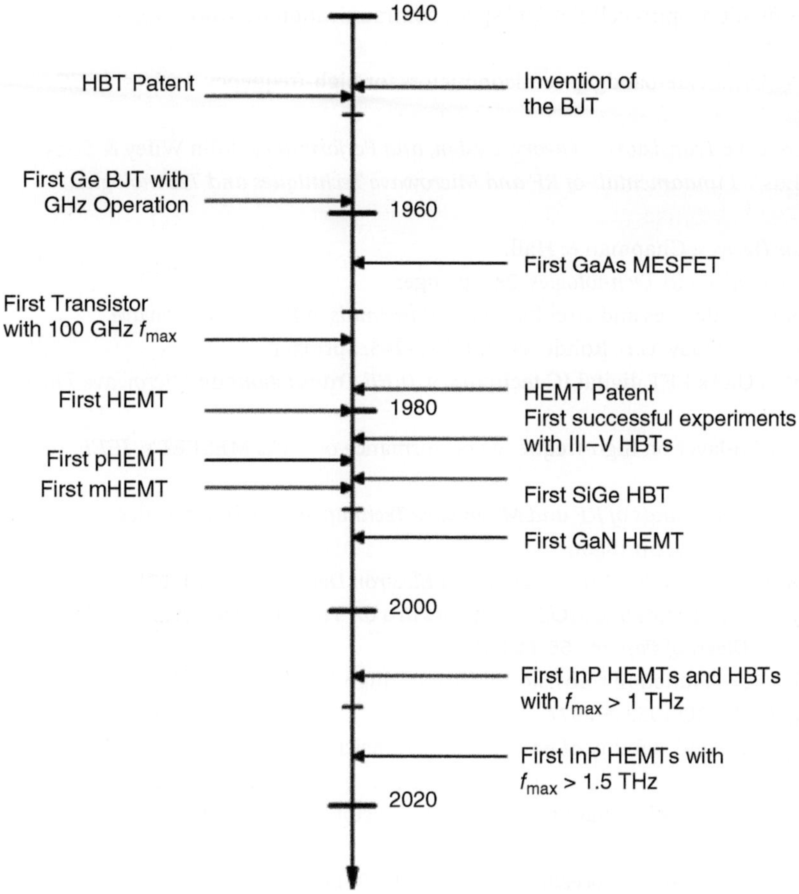

Figure 2.11 Important milestones of the evolution of microwave transistors.

Moreover, GaN HEMTs are very useful for power amplifiers up to more than 100 GHz and can also be used for robust LNAs. Furthermore, the Si-based microwave transistors, i.e., SiGe HBTs and Si MOSFETs, are very successful types of microwave transistors and can be used at frequencies far beyond 100 GHz. Thus, we see that several competing types of transistors are available for a variety of different microwave applications.

We conclude our journey through the history of microwave transistors with two simple statements. First, in the foreseeable future, the dynamic growth of microwave electronics will continue and new applications will emerge. Second, further improvements in the performance of microwave transistors and MMICs can be expected, although increasing the speed (i.e., enhancing f_T and f_{max}) of microwave transistors beyond the current state of the art will be anything but easy. Since power dissipation is an important one of Prof. Rohde's student is currently evaluating FETs on diamonds because of the very good thermal conductivity

References

1 Liou, J.J., Ziegler, M., and Schwierz, F. (2024). Gigahertz and terahertz transistors for 5G, 6G, and beyond communication systems. *Applied Physics Reviews* 11: 031318.

2 Maiwald, T., Li, T., Hotopan, G.-R. et al. (2023). A review of integrated systems and components for 6G wireless communication in the D-band. *Proceedings of the IEEE* 111: 220–256.

3 Medjdoub, F. et al. (2024). Emerging GaN technologies for next-generation millimeter-wave applications. *IEEE Microwave Magazine* 25 (10): 18–37.

4 Markman, B. and Schröter, M. (2024). Indium-phosphide transistors: a review of current state and suitability for commercial > 100-GHz wireless communication systems. *IEEE Microwave Magazine* 25 (10): 38–53.

5 Chevalier, P. and Pallotta, A. (2024). Silicon germanium spins cellular and space communication networks. *IEEE Microwave Magazine* 25 (10): 54–67.

6 Pavlidis, S. Medwig, G., and Thomas, M. (2024). Ultrawide-bandgap semiconductors for high-frequency devices. *IEEE Microwave Magazine* 25 (10): 68–79.

7 Schwierz, F. and Liou, J.J. (2003). *Modern Microwave Transistors – Theory, Design, and Performance*. John Wiley & Sons.

8 Hartnagel, H.L., Quay, R., Rohde, U.L. et al. (Eds.). *Fundamentals of RF and Microwave Techniques and Technologies*. Springer, (2023).

9 Beneking, H. (1994). *High Speed Semiconductor Devices*. Chapman & Hall.

10 Ellinger, F. (2008). *Radio Frequency Integrated Circuits and Technologies*, 2e. Springer.

11 Quay, R. (2023). Semiconductors and semiconductor devices and circuits. In *Fundamentals of RF and Microwave Techniques and Technologies*, (ed. H.L. Hartnagel, R. Quay, U.L. Rohde, et al.), 551–745. Springer.

12 Greiling, P. (1984). The historical development of GaAs FET digital IC technology. *IEEE Transactions on Microwave Theory and Techniques* 32: 1144–1156.

13 Onodera, K. et al. (1991). Effects of neutral buried p-layer on high-frequency performance of GaAs MESFET's. *IEEE Transactions on Electron Devices* 38: 429–436.

14 Rudolph, M. (2023). Interference and noise. In *Fundamentals of RF and Microwave Techniques and Technologies*, (ed. H.L. Hartnagel, R. Quay, U.L. Rohde, et al.), 747–791. Springer.

15 Woerlee, P.H. et al. (2001). RF-CMOS performance trends. *IEEE Transactions on Electron Devices* 48: 1776–1782.

16 Niu, G. et al. (2005). Intermodulation linearity characteristics of CMOS transistors in a 0.13 μm process. In *2005 IEEE Radio Frequency integrated Circuits (RFIC) Symposium - Digest of Papers*, (65–68). IEEE.

17 Wallace, R.L., Schimpf, L.G., and Dickten, E. (1952). A junction transistor tetrode for high-frequency use. *Proceedings of the IRE* 40 (11): 1395–1400. https://doi.org/10.1109/JRPROC.1952.273972.

18 Slade, B.N. (1952). The control of frequency response and stability of point-contact transistors. *Proceedings of the IRE* 14: 1382–1384.

19 Davis, R.E., Bittmann, C.A., and Gnaedinger, R.J. (1959). Microwave germanium transistor. In *1959 International Electron Devices Meeting*, 46. IEEE.

20 Cooke, H.F. (1971). Microwave transistors: theory and design. *Proceedings of the IEEE* 59: 1163–1181.

21 Archer, J. (1971). Design and performance of C-band transistors. In *1971 International Electron Devices Meeting*, 150. IEEE.

22 Mead, C.A. (1966). Schottky barrier gate field effect transistor. *Proceedings of the IEEE* 54: 307–308.

23 Hooper, W.W. and Lehrer, W.I. (1967). An epitaxial GaAs field-effect transistor. *Proceedings of the IEEE* 55: 1237–1238.

24 Drangeid, K.E., Sommerhalder, R., and Walter, W. (1970). High-speed gallium-arsenide Schottky-barrier field-effect transistors. *Electronics Letters* 6: 228–229.

25 Baechtold, W. et al. (1973). Si and GaAs 0.5 μm gate Schottky-barrier field-effect transistors. *Electronics Letters* 9: 232–234.

26 Liou, J.J. and Schwierz, F. (2002). Evolution, current status and future trend of RF transistors. In *Proceedings of the IEEE Hong Kong Electron Devices Meeting*, 5–10. IEEE.

27 Schwierz, F. (2025). Microwave transistors: state of the art and evolution during the 1950s to the 2020s. A compilation of 2000 top references, unpublished.

28 Rohde, U.L. (1965). *Transistoren bei höchsten Frequenzen: Theorie und Schaltungspraxis von Diffusionstransistoren im VHF- und UHF-Bereich*, Verlag für Radio-Foto-Kinotechnik.

29 Lytel, A. (1965). *Transistor AF and RF Circuits*. Howard W. Sams & Co.

30 Dingle, R. et al. (1978). Electron mobilities in modulation-doped semiconductor heterojunction superlattices. *Applied Physics Letters* 33: 665–667.

31 Dingle, R. et al. (1979). High mobility multilayered heterojunction devices employing modulated doping, United States patent 4,163,237.

32 Mimura, T. et al. (1980). A new field-effect transistor with selectively doped GaAs/n-$Al_xGa_{1-x}As$ heterojunctions. *Japanese Journal of Applied Physics* 19: L225–L227.

33 DiLorenzo, J.V. et al. (1982). Material and device considerations for selectively doped heterojunction transistors. In *1982 International Electron Devices Meeting*, (578–581). IEEE.

34 Rosenberg, J.J. et al. (1985). An $In_{0.15}Ga_{0.85}As$/GaAs pseudomorphic single quantum well HEMT. *IEEE Electron Device Letters* 10: 491–493.

35 Ohno, H. et al. (1980). Double heterostructure $Ga_{0.47}In_{0.53}As$ MESFET by MBE. *IEEE Electron Device Letters* 1: 154–155.

36 Eastman, L.F. (1986). III-V heterojunction field effect transistors using indium alloys. In *1986 International Electron Devices Meeting*, (456–459). IEEE.

37 Win, P. et al. (1992). Metamorphic $In_{0.27}Ga_{0.73}As/In_{0.25}Al_{0.75}As$ HEMT on GaAs: MBE growth and device performance. *Microelectronic Engineering* 19: 317–320.

38 Ambacher, O. et al. (1999). Two-dimensional electron gases induced by spontaneous and piezoelectric polarization charges in N- and Ga-face AlGaN/GaN heterostructures. *Journal of Applied Physics* 85: 3222–3233.

39 Asif Khan, M. et al. (1993). High electron mobility transistor based on a GaN-$Al_xGa_{1-x}N$ heterojunction, *Applied Physics Letters* 63: 1214–1215.

40 Asif Khan, M. et al. (1995). Temperature activated conductance in GaN/AlGaN heterostructure field effect transistors operating at temperatures up to 300 °C, *Applied Physics Letters* 69: 1083–1085.

41 Wu, Y.-F. et al. (1999). High AI-content AIGaN/GaN HEMTs on SiC substrates with very-high power performance. In *International Electron Devices Meeting 1999. Technical Digest (Cat. No.99CH36318)*, (925–927). IEEE.

42 Tokumitsu, M. et al. (1996). A 0.1 - μm self-aligned-gate GaAs MESFET with multilayer interconnection structure for ultra-high-speed ICs. In *International Electron Devices Meeting. Technical Digest*, (211–214). IEEE.

43 Caruth, D.C. et al. (2000). Low-cost 38 and 77 GHz CPW MMICs using ion-implanted GaAs MESFETs. in *2000 IEEE MTT-S International Microwave Symposium Digest (Cat. No.00CH37017)*, (995–998). IEEE.

44 Lepore, A.N. et al. (1988). 0.1 μm gate length MODFETs with unity current gain cutoff frequency above 110 GHz. *Electronics Letters* 24: 364–366.

45 Hanyu, I. et al. (1988). Super low-noise HEMTs with a T-shaped WSix gate. *Electronics Letters* 24: 1327–1328.

46 Nguyen, L.D. et al. (1989). Characterization of ultra-high-speed pseudomorphic AlGaAs/InGaAs (on GaAs) MOSFET's. *IEEE Transactions on Electron Devices* 36: 2243–2248.

47 Lester, L.F. et al. (1988). 0.15 μm gate-length double recess pseudomorphic HEMT with Fmax of 350 GHz. In *Technical Digest., International Electron Devices Meeting*, (172–175). IEEE.

48 Dumka, D.C. et al. (1999). Metamorphic $In_{0.52}Al_{0.48}As/In_{0.53}Ga_{0.47}$GaAs HEMTs on GaAs substrate with f_T over 200 GHz. In *Technical Digest, International Electron Devices Meeting*, (783–786). IEEE.

49 Zaknoune, M. et al. (1999). 0.1 μm high performance metamorphic $In_{0.32}Al_{0.68}As/In_{0.33}Ga_{0.67}As$ HEMT on GaAs using inverse step InAlAs buffer. *Electronics Letters* 35: 1670–1671.

50 Endoh, A. et al. (2000). High f_T 50-nm-gate lattice-matched InAlAs/InGaAs HEMTs. In *Conference Proceedings. 2000 International Conference on Indium Phosphide and Related Materials (Cat. No.00CH37107)*, (87–90). IEEE.

51 Smith, P.M. et al. (1995). W-band high efficiency InP-based power HEMT with 600 GHz f_{max}. *IEEE Microwave and Guided Wave Letters* 5: 230–232.

52 Micovic, M. et al. (2000). GaN/AlGaN high electron mobility transistors with f_T of 110 GHz. *Electronics Letters* 36: 358–359.

53 Shockley, W. Circuit element utilizing semiconductive material. U.S. Patent 2,569,347, Filed Jun. 26, 1948, Issued Sep. 25, 1951.

54 Kroemer, H. (1982). Heterostructure bipolar transistors and integrated circuits. *Proceedings of the IEEE* 70: 13–25.

55 Ito, H. et al. (1984). High-frequency characteristics of AIGaAs/GaAs heterojunction bipolar transistors. *IEEE Electron Device Letters* 5: 214–216.

56 Mishra, U.K. et al. (1988). 48 GHz AlInAs/GaInAs heterojunction bipolar transistors, Tech. Dig. IEDM, 873–875.

57 Fischer, S. et al. (1989). A 45 GHz strained-layer SiGe heterojunction bipolar transistor fabricated with low temperature epitaxy. In *International Technical Digest on Electron Devices Meeting*, (890–892). IEEE.

58 Kiyota, Y. et al. (1996). Lamp-heated rapid vapor-phase doping technology for 100-GHz Si Bipolar Transistors. In *Proceedings of the 1996 BIPOLAR/BiCMOS Circuits and Technology Meeting*, (173–176). IEEE.

59 Oka, T. et al. (1998). Advanced performance of small-scaled InGaP/GaAs HBT's with f_T over 150 GHz and f_{max} over 250 GHz. In *International Electron Devices Meeting 1998. Technical Digest (Cat. No.98CH36217)*, (653–656). IEEE.

60 Oka, T. et al. (1997). InGaP/GaAs HBT's with high-speed and low-current operation fabricated using WSi/Ti as the base electrode and burying SiO_2 in the extrinsic collector. In *International Electron Devices Meeting. IEDM Technical Digest*, (739–742). IEEE.

61 Dvorak, M.W. et al. (2000). Abrupt junction InP/GaAsSb/InP double heterojunction bipolar transistors with F_T as high as 250 GHz and BV_{CEO}>6V. In *International Electron Devices Meeting 2000. Technical Digest. IEDM (Cat. No.00CH37138)*, (178–181). IEEE.

62 Gruhle, A. et al. (1999). SiGe heterojunction bipolar transistors with 156 GHz transit frequency. In *Proceedings of the State-of-the-Art Program on Compound Semiconductors*, (198–200).

63 Washio, K. et al. (2000). A 0.2 - μm 180 - GHz - f_{max} 6.7-ps-ECL SOI/HRS self-aligned SEG SiGe HBT/CMOS technology for microwave and high-speed digital applications. In *International Electron Devices Meeting 2000. Technical Digest. IEDM (Cat. No.00CH37138)*, (741–744). IEEE.

64 Telford, M. (2004). SiGe slips into main fabs. *III-Vs Review* 17 (2): 32–34.

65 Wood, A. et al. (1998). 120 Watt, 2GHz, Si LDMOS RF power transistor for PCS base station applications. In *1998 IEEE MTT-S International Microwave Symposium Digest (Cat. No.98CH36192)*, (707–710). IEEE.

66 Wann, C. et al. (1997). High-performance 0.07 - μm CMOS with 9.5-ps gate delay and 150 GHz f_T. *IEEE Electron Device Letters* 18: 625–627.

67 Johnson, R.A. et al. (1998). Advanced thin-film silicon-on-sapphire technology: microwave circuit applications. *IEEE Transactions on Electron Devices* 45: 1047–1054.

68 Deal, W. et al. (2011). THz monolithic integrated circuits using InP high electron mobility transistors. *IEEE Transactions on Terahertz Science and Technology* 1: 25–32.

69 Urteaga, M. et al. (2011). 130nm InP DHBTs with $f_T > 0.52$ THz and $f_{max} >1.1$ THz. In *69th Device Research Conference*, (281–282). IEEE.

70 Shinohara, K. et al. (2013). Scaling of GaN HEMTs and Schottky diodes for submillimeter-wave MMIC applications. *IEEE Transactions on Electron Devices* 60: 2982–2996.

71 Heinemann, B. et al. (2016). SiGe HBT with f_T/f_{max} of 505 GHz/720 GHz. In *2016 IEEE International Electron Devices Meeting (IEDM)*, (51–54). IEEE.

72 Post, I. et al. (2006). A 65nm CMOS SOC technology featuring strained silicon transistors for RF applications. In *2006 International Electron Devices Meeting*, (1–3). IEEE.

73 Jan, C.-H. et al. (2008). A 45nm low power system-on-chip technology with dual gate (logic and I/O) high-k/metal gate strained silicon transistors. In *2008 IEEE International Electron Devices Meeting*, (1–4). IEEE.

74 Lee, H.-J. et al. (2018). Intel 22nm FinFET (22FFL) process technology for RF and mm wave applications and circuit design optimization for FinFET technology. In *2018 IEEE International Electron Devices Meeting (IEDM)*, (316–319). IEEE.

75 Damaraju, S. et al. (2012). A 22nm IA multi-CPU and GPU system-on-chip. In *2012 IEEE International Solid-State Circuits Conference*, (56–57). IEEE.

76 Lederer, D. et al. (2006). Dependence of FinFET RF performance on fin width. In *Digest of Papers. 2006 Topical Meeting on Silicon Monolithic Integrated Circuits in RF Systems*, (8–110). IEEE.

77 Sell, B. et al. (2017). 22FFL: A high performance and ultra low power FinFET technology for mobile and RF applications. In *2017 IEEE International Electron Devices Meeting (IEDM)*, (685–688). IEEE.

78 Schwierz, F. (2015). Performance of graphene and beyond graphene 2D semiconductor devices. *ECS Transactions* 69: 231–240.

79 Schwierz, F. (2008). The frequency limits of field-effect transistors: MOSFET versus HEMT. In *9th International Conference on Solid-State and Integrated-Circuit Technology*, (1433–1436). IEEE.

80 Schwierz, F. (2004). SiGe HBTs for ultra-high speed applications. In *Proceedings of 7th International Conference on Solid-State and Integrated Circuits Technology*, Beijing, China (2114–2119). IEEE.

81 Novoselov, K.S. et al. (2004). Electric field effect in atomically thin carbon films. *Science* 306: 666–669.

82 Morozov, S.V. et al. (2008). Giant intrinsic carrier mobilities in graphene and its bilayer. *Physical Review Letters* 100: 016602.

83 Geim, A.K. and Novoselov, K.S. (2007). The rise of graphene. *Nature Materials* 6: 183–191.

84 Lemme, M.C. et al. (2007). A graphene field-effect device. *IEEE Electron Device Letters* 28: 282–284.

85 Meric, I. et al. (2008). RF performance of top-gated, zero-bandgap graphene field-effect transistors. In *2008 IEEE International Electron Devices Meeting*, San Francisco, CA (1–4). IEEE.

86 Wu, Y.Q. et al. (2011). Record high RF performance for epitaxial graphene transistors. In *2011 International Electron Devices Meeting*, (528–530). IEEE.

87 Cheng, R. et al. (2012). High-frequency self-aligned graphene transistors with transferred gate stacks, *Proceedings of the National Academy of Sciences* 119: 11588–11592.

88 Schwierz, F. (2011). Graphene transistors 2011. In *Proceedings of 2011 International Symposium on VLSI Technology, Systems and Applications*, (16–17). IEEE.

89 Schwierz, F. (2013). Graphene transistors: status, prospects, and problems. *Proceedings of the IEEE* 101: 1567–1584.

90 Rodilla, H. et al. (2009). High-mobility heterostructures based on InAs and InSb: a Monte Carlo study. *Journal of Applied Physics* 105: 113705.

91 Ashley, T. et al. (2009). High-performance InSb based quantum well field effect transistors for low-power dissipation applications. In *2009 IEEE International Electron Devices Meeting (IEDM)*, (849–852). IEEE.

92 Ashley, T. et al. (2006). Indium antimonide based technology for RF applications. In *2006 IEEE Compound Semiconductor Integrated Circuit Symposium*, (121–124). IEEE.

93 Ashley, T. et al. (2010). High performance InSb QWFETs for low power dissipation millimetre wave applications, *Proceedings of the European Microwave Integrated Circuits Conference*, (158–161). IEEE.

94 Kobayashi, K.W. et al. (2021). A baseband-65GHz high linearity-bandwidth GaN LNA using a 1.7 A/mm high current density ScAlN based GaN HEMT Technology, In *IEEE Microwave Theory and Technology Society (MTT-S)*, (772–775). IEEE.

95 Kanar, T. et al. (2014). X- and K-band SiGe HBT LNAs with 1.2- and 2.2-dB mean noise figures. *IEEE Transactions on Microwave Theory and Techniques* 62: 2381–2389.

96 Coen, C.T. et al. (2016). Design and on-wafer characterization of G-band SiGe HBT low-noise amplifiers. *IEEE Transactions on Microwave Theory and Techniques* 64: 3631–3642.

97 Plouchart, J.-O. et al. (2015). A 18mW, 3.3dB *NF*, 60GHz LNA in 32 nm SOI CMOS technology with autonomic *NF* calibration. In *2015 IEEE Radio Frequency Integrated Circuits Symposium (RFIC)*, (319–322). IEEE.

98 Yeh, P.-C. and Kuo, C.-N. (2019). A 94 GHz 10.8 mW low-noise amplifier with inductive gain boosting in 40 nm digital CMOS technology. *Proceedings of the 2019 IEEE Asia-Pacific Microwave Conference (APMC)*, (1357–1359). IEEE.

99 Hamani, A. et al. (2020). A 125.5-157 GHz 8 dB *NF* and 16 dB of gain D-band low noise amplifier in CMOS SOI 45 nm. In *2020 IEEE/MTT-S International Microwave Symposium (IMS)*, (197–200). IEEE.

100 Yeh, Y.-S. and Lee, H.-J. (2020). A 5-6 GHz low-noise amplifier with > 65-dB variable-gain control in 22nm FinFET CMOS technology. In *2020 IEEE Radio Frequency Integrated Circuits Symposium (RFIC)*, (371–374). IEEE.

101 Shin, W. et al. (2018). A compact 75 GHz LNA with 20 dB gain and 4 dB noise figure in 22nm FinFET CMOS technology. In *2018 IEEE Radio Frequency Integrated Circuits Symposium (RFIC)*, (284–287). IEEE.

102 Landsberg, N. et al. (2022). High linearity 76-81 GHz LNA using a 16 nm FinFET technology for phased array radar applications. In *Proceedings of the 2022 17th European Microwave Integrated Circuits Conference (EuMIC)*, (123–126). IEEE.

103 Choi, B.G. et al. (2000). A low noise on-chip matched MMIC LNA of 0.76 dB noise figure at 5 GHz for high speed wireless LAN applications. In *GaAs IC Symposium. IEEE Gallium Arsenide Integrated Circuits Symposium. 22nd Annual Technical Digest 2000. (Cat. No.00CH37084)*, (143–146). IEEE.

104 Lee, L.-Y. et al. (2021). A 25-31 GHz LNA in GaAs 0.15-μm pHEMT. In *2021 IEEE International Symposium on Radio-Frequency Integration Technology (RFIT)*, (1–4). IEEE.

105 Chou, Y.-T. et al. (2016). A Q-band LNA with 55.7% bandwidth for radio astronomy applications in 0.15-μm GaAs pHEMT process. In *2016 IEEE International Symposium on Radio-Frequency Integration Technology (RFIT)*, (1–4). IEEE.

106 Tessmann, A. et al. (2019). 20-nm $In_{0.8}Ga_{0.2}As$ MOSHEMT MMIC technology on silicon. *IEEE Journal of Solid-State Circuits* 54: 2411–2418.

107 Tessmann, A. et al. (2013). A 243 GHz low-noise amplifier module for use in next-generation direct detection radiometers. In *2013 European Microwave Integrated Circuit Conference*, (220–223). IEEE.

108 Tessmann, A. et al. (2014). A 600 GHz low-noise amplifier module. In *2014 IEEE MTT-S International Microwave Symposium (IMS2014)*, (1–4). IEEE.

109 Farkas, D. et al. (2014). A W-band 100 nm InP HEMT ultra low noise amplifier. In *2014 Asia-Pacific Microwave Conference*, (229–231). IEEE.

110 Deal, W.R. et al. (2016). A low-power 670-GHz InP HEMT receiver. *IEEE Transactions on Terahertz Science and Technology* 6: 862–864.

111 Leong, K.M.K.H. et al. (2017). 850 GHz receiver and transmitter front-ends using InP HEMT. *IEEE Transactions on Terahertz Science and Technology* 7: 466–475.

3

System Specifications

3.1 System Specifications and Their Relationship to Circuit Design

Wireless communication involves a large range of signal powers—from levels on the order of 10^{-18} W (at the receiver input) to 10^{2} W (at a base-station transmitter output). A receiver must be able to demodulate signals that have been attenuated billions of times through propagation; a transmitter must be able to produce a properly modulated signal at a frequency suitable for propagation, at a level high enough to overcome worst-case propagation losses, and provide a useful signal at the receiver. Gain is therefore an essential attribute of wireless systems. Because no single active device can provide all the gain required for transmission or reception, we must distribute the gain among multiple stages, designing each for optimum performance across the power span it bridges.

Two inescapable realities impose limits on the gain and absolute power output we may achieve with a given circuit: All real electrical and electronic networks generate noise to some degree, and all real electrical and electronic networks distort the signals applied to them to some degree. A signal weaker than a circuit's inherent noise cannot be amplified by that circuit because it remains indistinguishable from the noise. A signal that exceeds the power-handling capability of the circuit to which it is applied may be degraded, even rendered useless, by the resulting distortion. The building blocks of the following sections are parts of systems, so we will first examine in detail the issues that are particular to system noise and linearity performance as needed with modern complex waveforms [1].

3.1.1 System Noise and Noise Floor

Assuming that a system's gain is sufficient, the weakest signal that may be processed satisfactorily, a figure of merit referred to as the *noise floor* or (in receivers) *minimum detectable signal (MDS)*, is limited by thermal noise, assumed to be equal to the noise power available from a resistor at 290 K (about 17 °C or 62 °F), an arbitrary reference value near standard room temperature.

$$\text{This noise power is equal to } P_n = kTB \tag{3.1}$$

where P_n is the noise power, k is Boltzmann's constant (1.38×10^{-23} W/K), T is the temperature in kelvins, and B is the bandwidth (in Hertz) in which the noise appears.

For $T = 290$, P_n is therefore 4.00×10^{-18} W, or −174 dBm in a 1 Hz bandwidth.

Increasing B to a value suitable for digital communications, such as 160 kHz for a GSM system, admits more noise to the network, raising the minimum noise power to −122 dBm, against which an incoming signal must compete.

If the noise figure and bandwidth are known, the system noise floor can be calculated using the equation

$$\text{Noise floor} = -174\ \text{dBm} + NF + 10 \log B \tag{3.2}$$

The trouble with this equation is that the "integrated" bandwidth depends so much on the selectivity shape factor, which is not always known. Figure 3.1 shows the translation of the bandwidth of a single-tuned circuit with its Gaussian shape into its rectangular equivalent. The transformation is done by sizing the rectangle such that A′ = A and B′ = B; when this is true, the area of the rectangle equals the area under the curve.

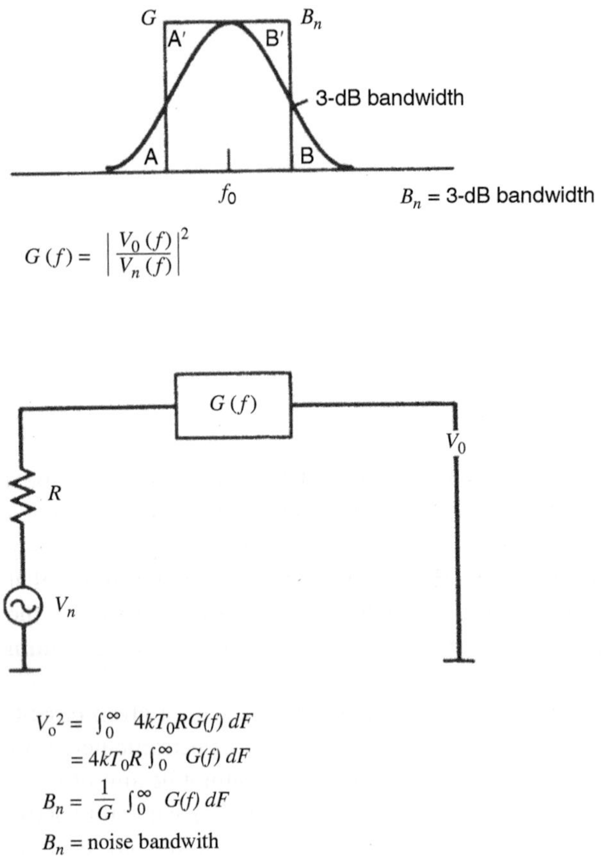

Figure 3.1 Graphical and mathematical explanation of the noise bandwidth from a comparison of the Gaussian-shaped bandwidth to the rectangular filter response. *Source*: [1] / John Wiley & Sons.

3.2 Signal-to-noise Ratio and Sensitivity

Successful radio communication depends on the achievement of a specified minimum ratio of signal power to noise power, expressed in decibels (dB), at the output of the receiver. The input voltage, expressed in absolute units or decibels relative to a microvolt (dBμV), necessary to achieve a particular signal-to-noise ratio (S/N, SNR) in a particular bandwidth, may be specified as a figure of merit called *sensitivity*. Because techniques used to measure S/N actually measure the ratio of signal-*plus-noise* to noise, specifications may refer to (or imply) (S+N)/N rather than S/N. The difference between (S+N)/N and S/N becomes negligible at high ratios of signal to noise; even at an SNR of 10 dB—a common value—the difference is only 0.46 dB [1, 2].

Most receivers are designed to operate optimally when connected to an antenna system of a specified impedance (commonly 50 Ω), but relatively few receivers exhibit this design load impedance at their input terminals; that is, they are not designed for a conjugate input match. It is therefore customary to specify sensitivity in terms of "open circuit" voltage—the signal voltage that, with the receiver's antenna input terminated in its design antenna impedance, results in the desired ratio of signal to noise. If, as is commonly the case, the input voltage for a given SNR is determined using instrumentation calibrated in terms of *closed*-circuit voltage—that is, in terms of voltage across a load resistance equal to the instrument's source resistance—the voltage indicated will be 1/2 the open-circuit value for the SNR specified [3]. By convention, the open-circuit measurement condition is indicated by a sensitivity specification in volts of electromotive force (EMF). Specifying sensitivity in terms of available signal power (usually decibels relative to 1 mW, or dBm) eliminates this open/closed-circuit confusion.

3.2.1 SINAD Ratio

Extending the measurement of signal-plus-noise to noise to include distortion results in a figure of merit called *SINAD* (signal-plus-noise-and-distortion), commonly applied to FM receivers:

$$\text{SINAD} = 10\log_{10}\frac{S+N+D}{N+D} \tag{3.3}$$

where SINAD is in decibels, S is signal power, N is noise power, and D is distortion power. At a SINAD ratio of 12 dB—a common specification—the noise-and-distortion power is 25% that of the desired signal. As is true of (S+N)/N, SINAD closely approximates S/N at high ratios of signal to noise [1].

3.3 Bit Error Rate and Noise

For digital systems, SNR and bit error rate (BER) are related. As introduced earlier, depending on the waveform, coding, and filtering, different BERs are related to particular SNRs (Figure 3.2). The SNR also depends on the interference from the synthesizer, as will be seen in the section on adjacent-channel power ratio (ACPR). Since a digital transmitter is always involved, in the on-the-air testing, the laboratory measurement of SINAD cannot quite be correlated. At the intermediate frequency (IF) level, we find the handover point between the transceiver's analog front end and its digital portion, and a standard noise-figure test setup, such as the one by Hewlett–Packard, is still a valuable instrument for evaluating noise figure and SNR.

Here is an example for a digital measurement: assume that we use a test generator like the SMIQ to evaluate a front end having an 8.8 dB noise figure. A 384 kbps $\pi/4$ DQPSK digital modulation signal (-5 dBm) is applied to the transceiver front end. Testing at 2.47 GHz, we find that the error vector magnitude (EVM; defined later in this section) is about 3.89% in transmit and about 1.37% in receive. To determine the receive sensitivity of the RF front end, the signal from the digitally modulated generator (at -95 dBm) is applied to the RF front end in receive mode. The measured EVM at 2.47 GHz is now 20%. By using the equation

$$\text{S/N} = -20\log(\text{EVM}) \tag{3.4}$$

we see that the ratio that corresponds to an EVM of 20% is 14 dB.

Since

$$\text{S/N} = \frac{(E_s R_S)}{N_0 B} \tag{3.5}$$

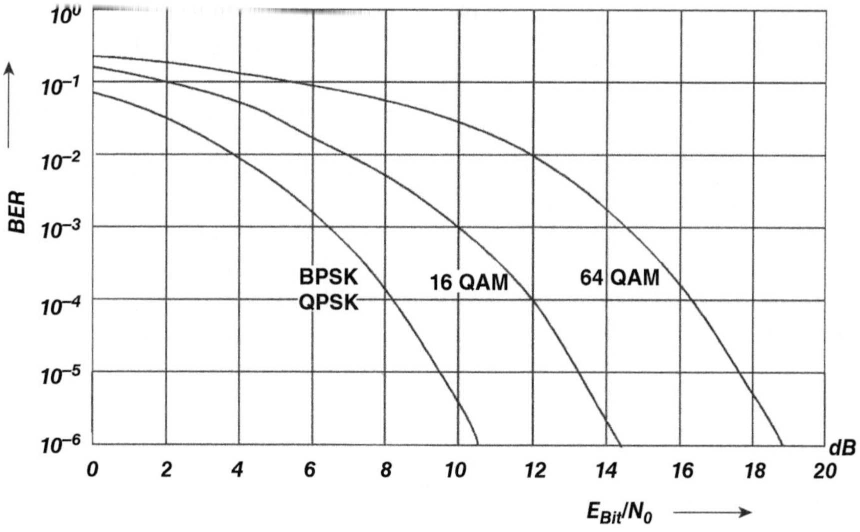

Figure 3.2 BER versus Ebit/N0 for BPSK/QPSK, 16-QAM, and 64-QAM, showing the significantly greater SNR necessary for a given BER as the number of signal states is increased. *Source*: [1] / John Wiley & Sons.

where E_S is the energy of a symbol, R_S is the symbol rate, N_0 is the noise power density, and B is the bandwidth. Assuming a $\pi/4$ DQPSK signal and ratio of 14 dB, it can be determined that

$$E_s/N_0 = S/N\ (B/R_S) = 17 \text{ dB} \tag{3.6}$$

Assuming that the BER is 1E–5, the required E_S/N_0 is 10 dB. Hence, from the definition of the sensitivity:

$$\text{Sensitivity} = -144 + NF + R_S + (E_s/N_0) = 144 + 8.8 + 10\log\left(\frac{384}{2}\right) + 10 = -102 \text{ dBm} \tag{3.7}$$

The value −144 is derived from kT_0 by taking the bandwidth (1 kHz) into consideration.

The result obtained in (3.7) is consistent with the obtained E_S/N_0 of 17 dB from the measurement at −95 dBm input power (since −102 + [17–10] = −95). Hence, it was determined that the receiver sensitivity of the RF front end is approximately −102 dBm. We can then determine the dynamic range (DR) from

$$\text{DR} = P_{-1\text{ dB}} - \text{Sensitivity} = -21 - 102 = 81 \text{ dB} \tag{3.8}$$

3.4 Noise Factor and Noise Figure

The noise floor predicted by (3.1) cannot be achieved or maintained in any real network or system of networks because all real networks generate noise. Determining how closely the SNR achieved at a given input level approaches the achievable SNR at that input level in a noiseless network is therefore of high interest to the circuit and system designer. The degree to which a network's noise contribution degrades the noise floor predicted by (3.1) is evaluated by its *noise factor (F)*, which is expressed as the ratio

$$F = \frac{N_{\text{in}} + N_{\text{added}}}{N_{\text{in}}} \tag{3.9}$$

where F is the noise factor, N_{in} is the noise power available from the source, and N_{added} is the noise power added by the network, with both powers determined over the same bandwidth. Expressing this ratio in decibels $(10\log_{10} F)$ returns *noise figure (NF)*, a bandwidth-independent figure of merit of great value in evaluating the noise performance of networks and communication systems. We can also express NF as the ratio of the network's input SNR to its output SNR:

$$NF = 10\log_{10}\left[\frac{(S_{\text{in}}/N_{\text{in}})}{(S_{\text{out}}/N_{\text{out}})}\right] \tag{3.10}$$

where NF is the noise figure in decibels, S is the signal power, and N is the noise power, with the input and output values of these quantities signified by the subscripts and all powers determined in the same bandwidth. The NF of an ideal noiseless network is 0 dB; for all real, noisy networks, NF is positive. The NF of a lossy passive device is equal to its insertion loss.

Noise figure can also be defined for antennas and antenna systems:

$$NF_{\text{ant}} = 10\log_{10}\left[\frac{N_t + N_{\text{ant}}}{N_t}\right] \tag{3.11}$$

where NF_{ant} is the antenna system noise figure in decibels, N_t is the antenna system thermal noise power, as defined in (3.1), and N_{ant} is the total noise power picked up by the antenna system.

From the lower end of the radio spectrum up to approximately 400 MHz, noise intercepted by an antenna system from atmospheric, man-made, and galactic sources will dominate NF_{ant} (Figure 3.3) and N_t can be considered as equivalent to the noise power of a resistor at 290 K. Atmospheric noise subsides above 40 MHz; from this region to perhaps 700 MHz, N_t is still largely negligible, with noise from man-made and/or sky sources largely determining an antenna's noise figure. At these frequencies, an antenna's directivity and orientation relative to noise sources play a critical role in determining its noise figure; a strongly directive antenna located in a city suburb, for instance, exhibits a significantly higher noise figure when pointed toward the city center than when pointed toward a more sparsely populated area. At frequencies above about 700 MHz, the RF environment is generally quieter, allowing receiver *NF*, not antenna *NF*, to more routinely limit a wireless receiver's sensitivity. At these frequencies, the concept of *noise temperature* is commonly used to evaluate the quietness of a receiver, its antenna system, and its RF environment. Noise temperature is particularly useful in designing systems for space communication, for which an antenna may be cooled, through radiation of a portion of its own noise into the RF-cold sky, to a noise temperature an order of magnitude below 290 K [1].

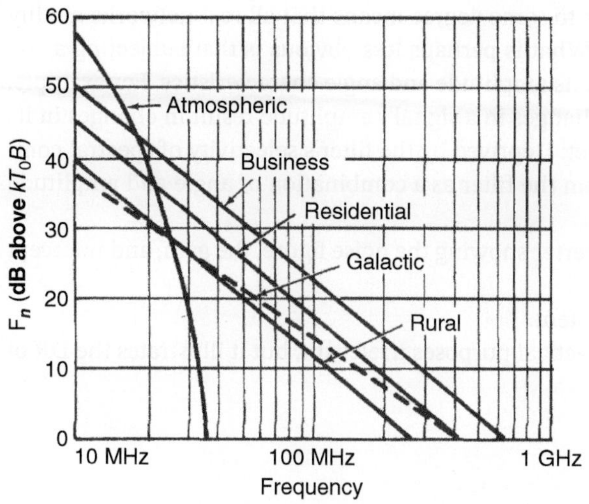

Figure 3.3 The higher the frequency, the quieter the RF environment becomes, although the noise profile of specific sources may contradict this general rule. *Source*: [1] / John Wiley & Sons.

3.4.1 Noise Figure of Cascaded Networks

The noise figure of two networks in cascade may be determined from

$$NF_{\text{total}} = 10 \log_{10}\left(F_1 + \frac{F_2 - 1}{G_1}\right) \tag{3.12}$$

where NF is noise figure in dB, F_1 is the noise factor of the first network, F_2 is the noise factor of the second network, and G_1 is the gain (as a numerical ratio, *not* in dB) [4]. The noise figure of a system with more than two stages can be evaluated through repeated iterations of (3.12). Note that (3.12) assumes two conditions: (1) that F_1 and F_2 are determined in the same bandwidth and (2) that the network input and output terminations are resistive—a condition that is commonly *not* true of RF amplifiers optimized for lowest noise. Eq. (3.12) can be expanded to account for bandwidth, but accounting for complex terminations requires the use of *noise correlation matrix* techniques as described in [2]. See also the important Ref. [5].

In the case that one of the parts of a system chain is a passive/active mixer, the noise contribution of the image frequency needs to be considered. See Ref. [6] and the mixer chapter in this book.

3.5 System Amplitude and Phase Behavior

If we could build electronic systems that were absolutely amplitude- and phase-linear, radio communication system design would be greatly simplified. An amplifier designed for a power gain of 10 dB, for example, would merely increase the magnitude of all signals at its input by a factor of 10, regardless of their frequencies, while perfectly maintaining their relative phases. But all real electrical and electronic networks, even those designed (or supposed) to be amplitude- and phase-linear, exhibit amplitude and phase nonlinearity to some degree, just as they all generate noise to some degree.

The effects of amplitude nonlinearity, generically referred to as *nonlinear distortion*, include the generation, through *harmonic distortion* and/or *intermodulation distortion* (IMD), of output signals at frequencies not present at a system's input. Nonlinear distortion also results in *gain compression*—changes in system gain with changes in input signal level. By convention, when workers in electronics refer to or consider a network's "linearity," they usually mean its *amplitude* linearity; likewise, by "distortion," they usually mean *nonlinear* distortion.

The effects of phase or frequency nonlinearity are generically referred to as *linear distortion* because they occur independently of signal amplitude and polarity. We often intentionally apply linear distortion through *filtering*, which modifies the amplitude relationships among existing spectral components of a signal without creating any new frequencies. Another linear distortion effect, *phase* or *delay distortion*, results in the delay of signals of differing frequencies by differing amounts of time. In a system where signal phase conveys information, as is true of most wireless links, phase distortion can seriously degrade communication.

The fact that all real networks are amplitude- and angle-nonlinear to some degree means that all real networks modify the amplitude and angle characteristics of the signals they handle. What is perhaps less obvious is that subjecting a signal to amplitude and angle nonlinearities causes "crosstalk" between its amplitude and angle characteristics. For example, through a nonlinear distortion effect called AM-to-PM conversion, changes in a signal's amplitude result in changes in its phase. Filtering an angle-modulated signal produces the reverse effect: Deprived by the filter's selectivity of spectral components necessary to maintain its envelope constancy, it emerges from the filter as a combination of angle and amplitude modulation [1].

As an example, Figure 3.4 shows a block diagram of a 2.4 GHz converter showing the noise figure, the gain, and intercept point of the system.

Figure 3.5 shows the frequency response and noise figure of the system.

Figure 3.6 shows the close-in mixing products, which are for all practical purposes irrelevant, but it illustrates the DR of the simulator.

For clarity, the detailed calculation is given in Figure 3.7.

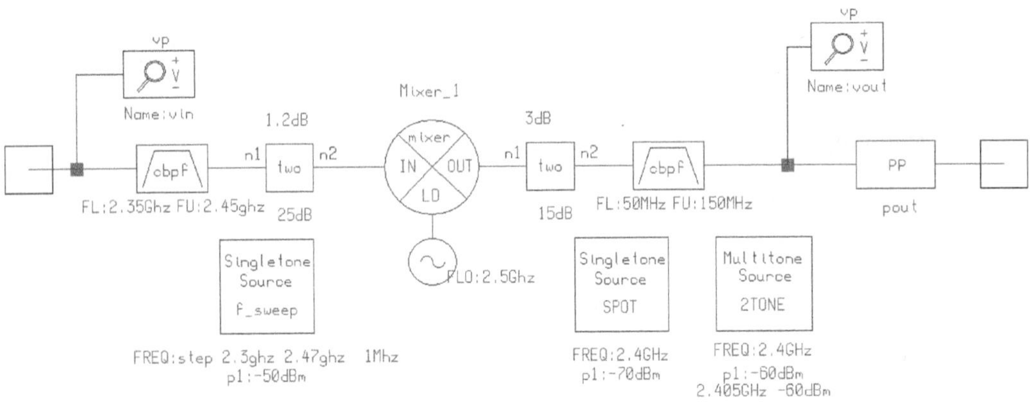

Figure 3.4 2.4 GHz converter block diagram with performance metrics.

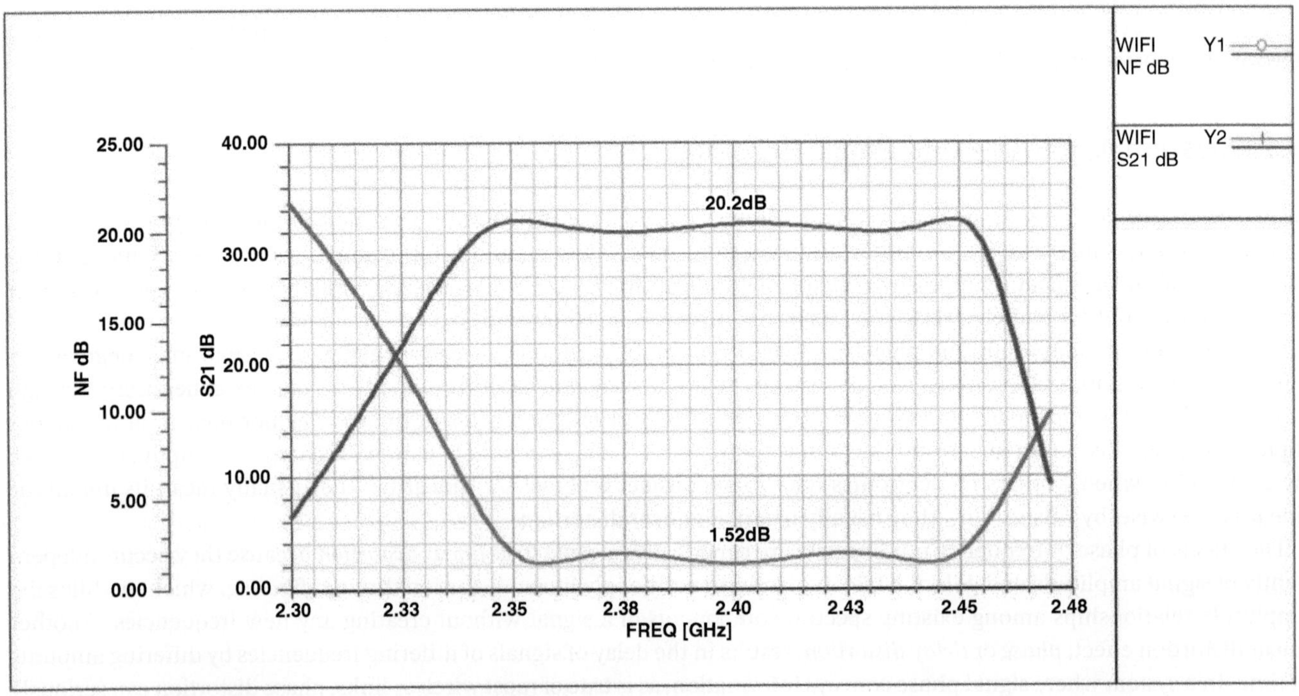

Figure 3.5 Frequency response and noise figure of the example system.

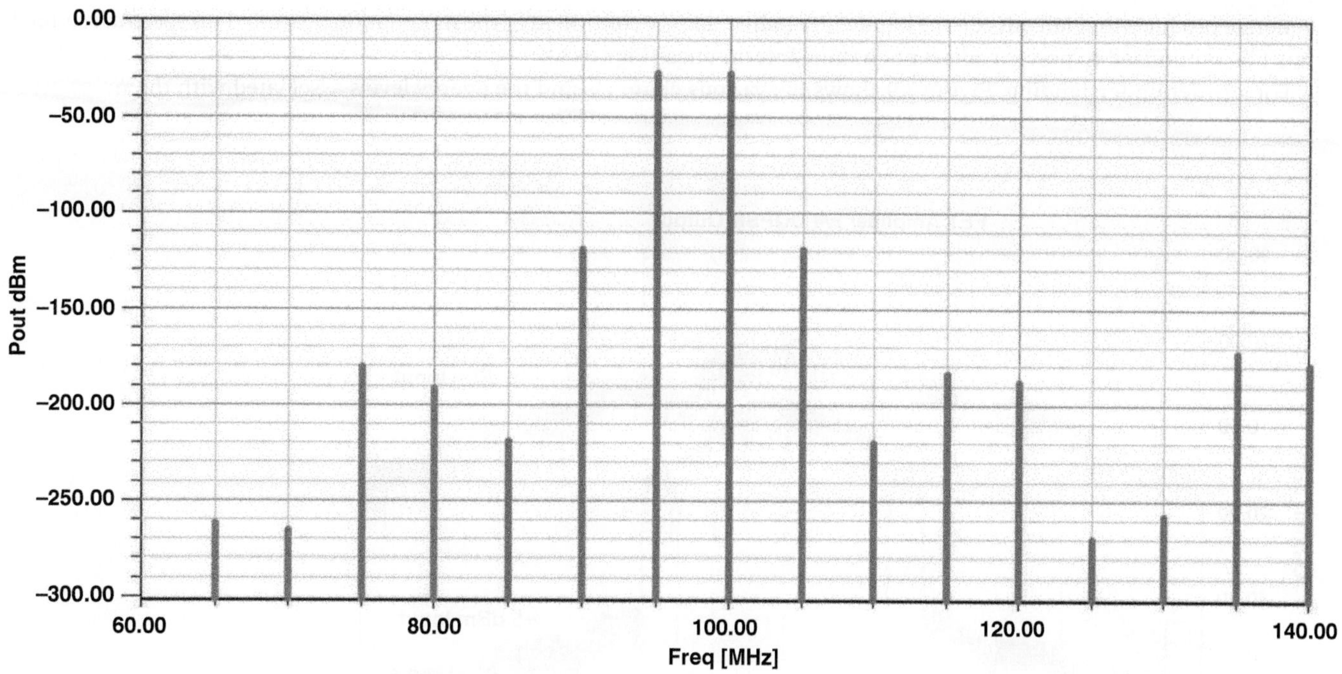

Figure 3.6 Close-in mixing products of the example system.

	Names	S21 dB 50ohms	S21 dB cum	S21 dB delta	NF dB cum	IP3 dB cum	IP3 dB delta	NF dB delta
1	CBPF	-0.13	-0.43	-0.43	0.13	Inf	0.00	0.13
2	TWO	25.00	24.76	25.19	1.33	-10.00	-Inf	1.20
3	MIXER_1	-8.00	17.65	-7.11	1.39	-17.11	-7.11	0.06
4	TWO	15.00	32.77	15.12	1.47	-10.60	6.51	0.08
5	CBPF	-0.14	32.72	-0.05	1.47	-10.95	-0.35	0.00

Figure 3.7 Detailed calculation of the example system parameters.

3.6 Spectral Considerations of Analog and Digitally Modulated Signals

Many wireless modulation schemes exist, and even more are proposed; all use angle (usually phase) modulation or a combination of phase and amplitude modulation using an emission format engineered to achieve multiple goals of information throughput, robustness, spectral and hardware efficiency, and reproducibility. Digital modulation is standard in mainstream wireless applications because it allows increased channel capacity, and immunity to noise and distortion, compared to analog systems. At the circuit design level, the particulars of whether, or to what degree, a modulation scheme is AM, PM, analog, or digital matter less than the actual angle and amplitude characteristics of the signal(s) involved and the tolerances within which these characteristics can be expected and allowed to vary.

A system's amplitude linearity is of major concern because of its relation to energy efficiency, receiver DR, and transmitter spectral purity. Energy efficiency, important because many wireless applications use battery power, generally decreases inversely with amplitude linearity. Yet, sufficient amplitude linearity must be guaranteed in the front-end circuitry of wireless receivers subjected to multiple strong signals, and in all wireless receiver and transmitter stages handling variable-envelope signals.

Some digital modulation schemes, designed to be distortion-tolerant, produce constant-envelope signals; Gaussian minimum shift keying (GMSK) and Feher's quadrature phase-shift keying (FQPSK) are examples. Such signals can be processed in highly nonlinear circuitry—for example, an RF power amplifier operated in saturation to maximize efficiency—without degrading their spectral composition. Other digital modulation schemes result in the emission of—ideally—a single PM sideband, with the carrier and all other sideband components suppressed. Such a signal exhibits, of necessity, phase and amplitude variations; the IS-54 cellular standard $\pi/4$ differential quadrature phase-shift keying ($\pi/4$ DQPSK) scheme, which results in envelope fluctuations of 3–6 dB, exemplifies this [7]. IMD among such a signal's spectral components can generate products that fall outside the bandwidth painstakingly controlled in the modulation process. This *spectral regrowth* must be minimized to prevent adjacent-channel interference (Figure 3.8).

For a better understanding, Figure 3.9 shows the various channels and the energy levels associated with them.

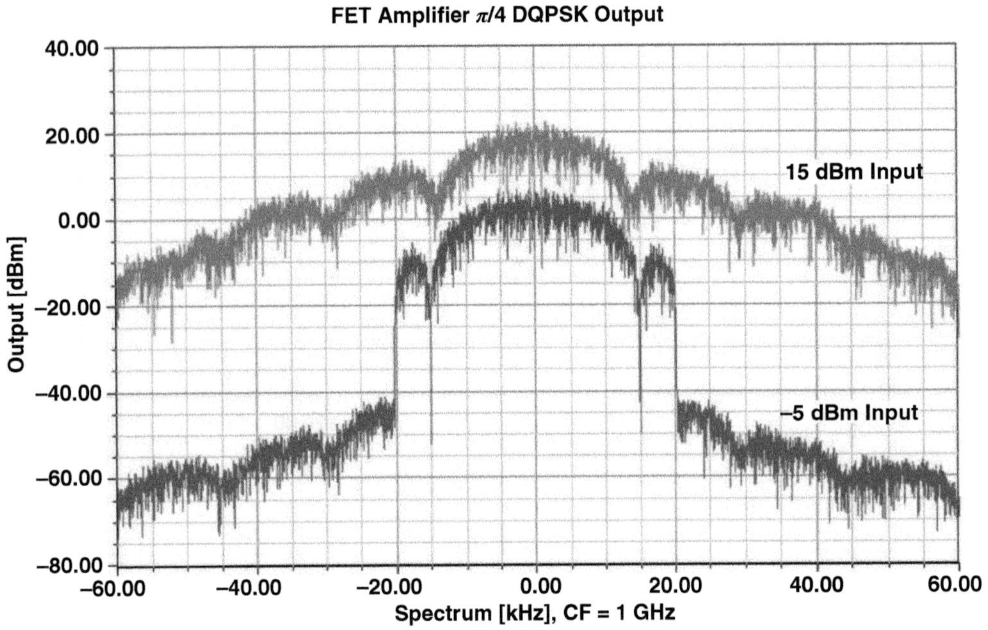

Figure 3.8 *Spectral regrowth* results when a variable-envelope emission, $\pi/4$ DQPSK in this case, is subjected to significant nonlinear distortion. This graph shows the simulated performance of a MESFET power amplifier operating at 1 GHz; the amplifier is 6 dB into compression when driven at 15 dBm. *Source*: [1] / John Wiley & Sons.

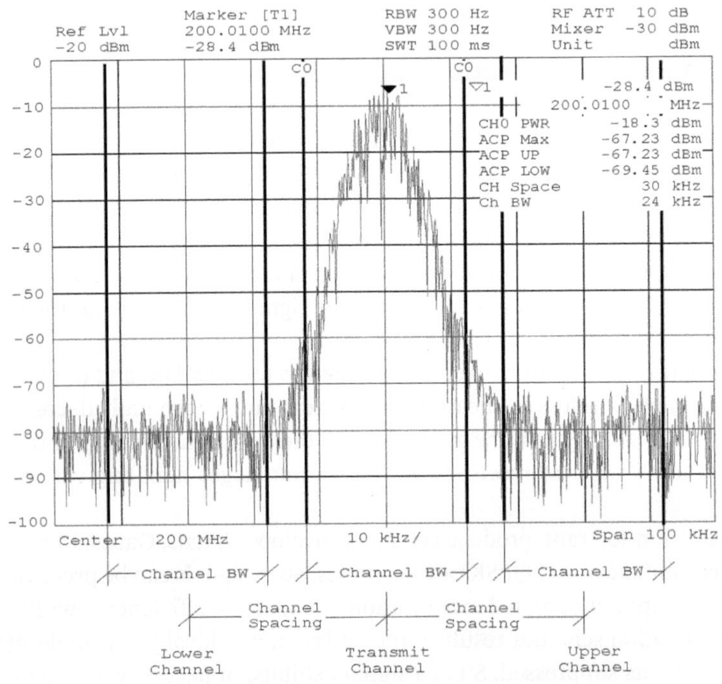

Figure 3.9 NADC signal and parameters, including channel spacing and channel bandwidth. *Source*: [1] / John Wiley & Sons.

3.7 Amplitude Linearity Issues and Figures of Merit

A network's amplitude nonlinearity can be characterized by the expansion

$$y = k_1 f(x) + k_2[f(x)]^2 + k_3[f(x)]^3 + \text{higher-order terms} \tag{3.13}$$

where y represents the output, the coefficients k_n represent complex quantities whose values can be determined by an analysis of the output waveforms, and $f(x)$ represents the input. Even though all practical networks exhibit amplitude nonlinearity, we can (and often do) refer to many networks as "linear." We say this of networks that are *sufficiently amplitude-linear for our purposes*—for example, weakly nonlinear networks in which small-signal operation is assumed even though the signal levels involved are sufficient to cause slight distortion. For many practical purposes, the first three terms of (3.13) adequately describe such a network's nonlinearity:

$$y = k_1 f(x) + k_2[f(x)]^2 + k_3[f(x)]^3 \tag{3.14}$$

In adopting this simplification, we assume also that the nonlinearity is frequency-independent—that is, that the network has sufficient bandwidth to allow all of the products predicted by (3.14) to appear at its output terminals unperturbed [8].

When multiple signals are present in a network, even weak nonlinearity can result in profound consequences. To illustrate this, we'll let $f(x)$ consist of two sinusoidal signals:

$$f(x) = A_1 \cos \omega_1 t + A_2 \cos \omega_2 t \tag{3.15}$$

We will assume that ω_1 and ω_2 are close enough so that the coefficients k_i can be considered equal for both signals. We will also assume for simplicity that all of the k_i are real. If (3.14) describes the network's response to an input $f(x)$, the response will be

$$\begin{aligned}
y &= k_1(A_1 \cos \omega_1 t + A_2 \cos \omega_2 t) + k_2(A_1 \cos \omega_1 t + A_2 \cos \omega_2 t)^2 + k_3(A_1 \cos \omega_1 t + A_2 \cos \omega_2 t)^3 \\
&= k_1(A_1 \cos \omega_1 t + A_2 \cos \omega_2 t) \\
&\quad + k_2\left[A_1^2 \frac{1 + \cos 2\omega_1 t}{2} + A_2^2 \frac{1 + \cos 2\omega_2 t}{2} + A_1 A_2 \frac{\cos(\omega_1 + \omega_2)t + \cos(\omega_1 - \omega_2)t}{2}\right] \\
&\quad + k_3\left\{\begin{array}{l}
\left[A_1^3\left(\dfrac{\cos \omega_1 t}{2} + \dfrac{\cos \omega_1 t}{4} + \dfrac{\cos 3\omega_1 t}{4}\right) + A_2^3\left(\dfrac{3 \cos \omega_2 t}{4} + \dfrac{\cos 3\omega_2 t}{4}\right)\right] \\
+ A_1^2 A_2\left[\dfrac{3}{2} \cos \omega_2 t + \dfrac{3}{4} \cos(2\omega_1 - \omega_2)t + \dfrac{3}{4} \cos(2\omega_1 - \omega_2)t\right] \\
+ A_2^2 A_1\left[\dfrac{3}{2} \cos \omega_1 t + \dfrac{3}{4} \cos(2\omega_2 + \omega_1)t + \dfrac{3}{4} \cos(2\omega_2 - \omega_1)t\right]
\end{array}\right\}
\end{aligned} \tag{3.16}$$

The k_1 term of Eq. (3.16) represents the results of amplitude-linear behavior. No new frequency components have appeared; the two sine waves have merely been "rescaled" by k_1.

The second- and third-order terms of (3.16) represent the effects of harmonic distortion and IMD. Second-order effects include second-harmonic distortion (the production of new signals at $2\omega_1$ and $2\omega_2$) and IMD (the production of new signals at $\omega_1 + \omega_2$ and $\omega_1 - \omega_2$). Third-order effects include gain compression, third-harmonic distortion (the production of new signals at $3\omega_1$ and $3\omega_2$), and IMD (the production of new signals at ($2\omega_1 \pm \omega_2$ and $2\omega_2 \pm \omega_1$).

3.8 Gain Compression

Gain compression occurs when a network cannot increase its output amplitude in linear proportion to an amplitude increase at its input; gain *saturation* occurs when a network's output amplitude stops increasing (in practice, it may actually decrease) with increases in input amplitude.

We can deduce from (3.16) that the amplitude of the $\cos \omega_1 t$ signal has become

$$A_1' = k_1 A_1 + k_3\left(\frac{3}{4}A_1^3 + \frac{3}{2}A_1 A_2^2\right) \tag{3.17}$$

Because k_3 will normally be negative, a large signal $A_2 \cos \omega_2 t$ can effectively mask a smaller signal $A_1 \cos \omega_1 t$ by reducing the network's gain. This third-order effect, known as *blocking* or *desensitization* when it occurs in a receiver, is a special case of gain compression. The presence of additional signals results in a greater reduction in gain; the gain reduction for each signal is a function of the relative levels of all signals present. A receiver's blocking behavior may be characterized in terms of the level of off-channel signal necessary to reduce the strength of an in-passband signal by a specified value, typically 1 dB; alternatively, the decibel ratio of the off-channel signal power to the receiver noise floor power may be cited as *blocking DR*. Desensitization may be also characterized in terms of the off-channel signal power necessary to degrade a system's SNR by a specified value.

Multiple signals need not be present for gain compression to occur. If only one signal is present, the ratio of gain with distortion to the network's idealized (linear) gain is

$$A_1' = \frac{k_1 + k_3\left(\frac{3}{4}A_1^2\right)}{k_1} \tag{3.18}$$

This is referred to as the *single-tone gain-compression factor*. Figure 3.10 shows how the k_3 term causes a network's gain to deviate from the ideal. The point at which a network's power gain is down 1 dB from the ideal for a single signal is a figure of merit known as the *1 dB compression point* ($P_{-1\,\text{dB}}$). Many networks (including many receiving and low-level transmitting circuits, such as low-noise amplifiers, mixers, and IF amplifiers) are usually operated under small-signal conditions—at levels sufficiently below $P_{-1\,\text{dB}}$ to maintain high linearity. As we will see, however, some networks (including power amplifiers for wireless systems) may be operated under large-signal conditions—near or in compression—to achieve optimum efficiency at some specified level of linearity. Figure 3.11 shows what happens when a digital emission that uses amplitude to convey information is subjected to amplitude compression [1].

The 1 dB compression point can be expressed relative to input power $\left(P_{-1\,\text{dB,in}}\right)$ or output power $\left(P_{-1\,\text{dB,out}}\right)$. For the amplifier simulated here, Figure 3.10 shows the $P_{-1\,\text{dB,in}} \approx -14.5$ dBm and $P_{-1\,\text{dB,out}} \approx -1.3$ dBm [1]. Figure 3.11 shows the influence of differential amplitude error (compression) on a QAM constellation.

3.9 Intermodulation

The new signals produced through IMD can profoundly affect performance, even of systems operated far below gain compression (Figure 3.12). IMD products of significant power can appear at frequencies remote from, in, and/or near the system passband, resulting in demodulation errors (in reception) and interference to other communications (in transmission). Where an IMD product appears relative to the passband depends on the passband width and center frequency, the frequencies of the signals present at the system input, and the order of the nonlinearity involved. These factors also determine the strength of an IMD product relative to the desired signal [1].

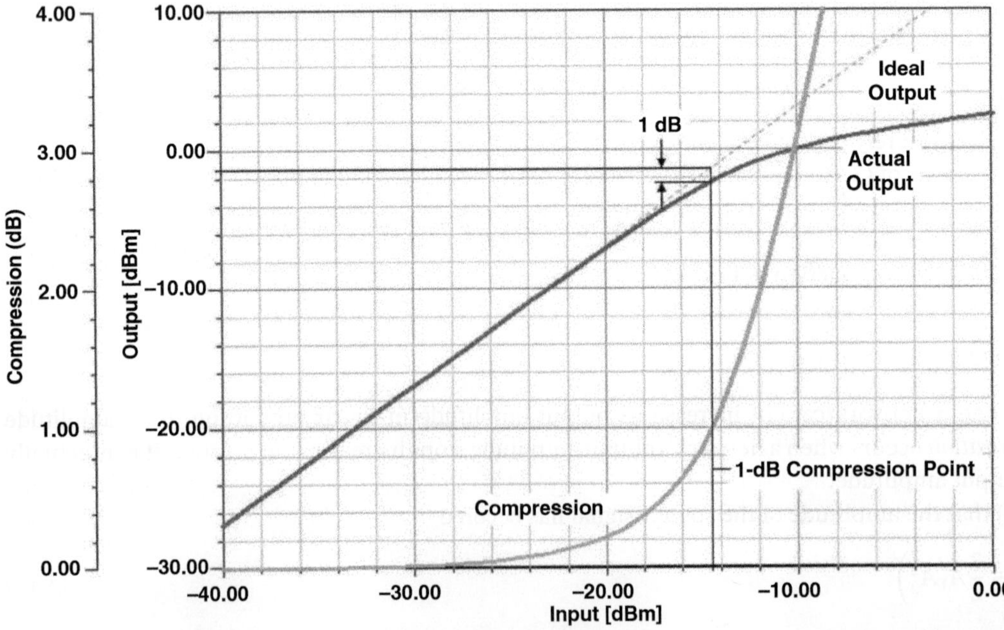

Figure 3.10 The power level at which a network's power output is down 1 dB relative to that of its ideally linear equivalent is a figure of merit known as the *1 dB compression point* ($P_{-1\,\text{dB}}$).

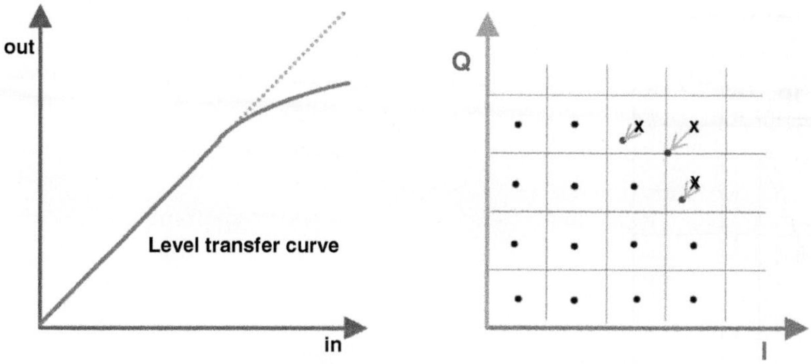

Figure 3.11 Influence of differential amplitude error (compression) on a QAM constellation. *Source*: [1] / John Wiley & Sons.

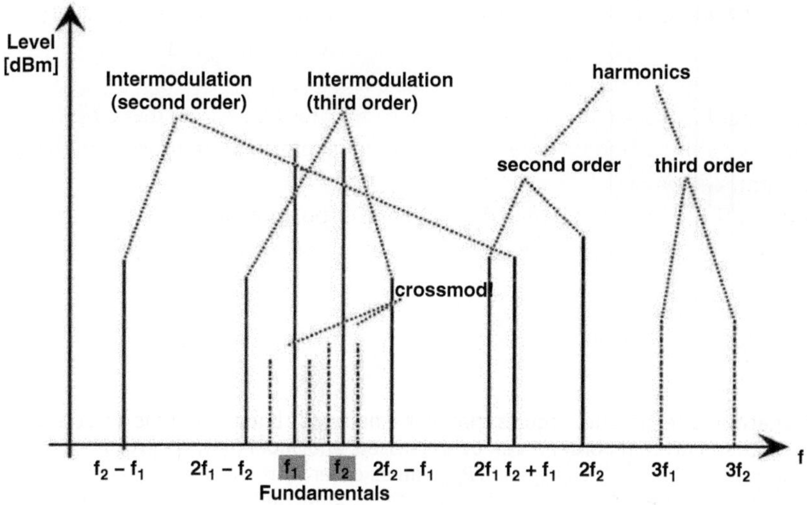

Figure 3.12 Relationships between fundamental and spurious signals, including harmonics and products of intermodulation. *Source*: [1] / John Wiley & Sons.

Second-order IMD (IM_2) results, for an input consisting of two signals ω_1 and ω_2, in the production of new signals at $\omega_1 + \omega_2$ and $\omega_1 - \omega_2$; third-order IMD (IM_3) results, for an input consisting of two signals ω_1 and ω_2, in the production of new signals at $2\omega_1 \pm \omega_2$ and $2\omega_2 \pm \omega_1$.

Under small-signal conditions—that is, at levels well below compression—the power of an IM_2 product varies by 2 dB, and the power of an IM_3 product varies by 3 dB, per decibel change in input power level. This allows us to derive a network figure of merit, the *intermodulation intercept point* (*IP*), for a given IM order by extrapolating a network's linear and IM responses to their point of intersection (Figure 3.13)—the point at which their powers would be equal if compression did not occur. Because of the system noise and/or IMD products, there is a minimum discernible signal (MDS) that limits the DR at the lower end. Theoretically, Figure 3.13 should show a noise floor or IMD-spur floor for a given input signal that represents a lower limit below which signals cannot be detected. The intercept point for a given IM order n can be expressed, and should always be characterized, relative to input power $(IP_{n,\text{in}})$ or output power $(IP_{n,\text{out}})$; the IP_{in} and IP_{out} values differ by the network's linear gain. For equal-level test tones, $IP_{n,\text{in}}$ can be determined by:

$$IP_{n,\text{in}} = \frac{nP_A - P_{IM_n}}{n-1} \tag{3.19}$$

where n is the order, P_A is the input power (of one tone), P_{IM_n} is the power of the IM product, and IP is the intercept point. The intercept point for cascaded networks can be determined from

$$IP_{2,\text{in}} = \frac{1}{\left(\frac{1}{\sqrt{IP_1}} + \frac{G}{\sqrt{IP_2}}\right)^2} \tag{3.20}$$

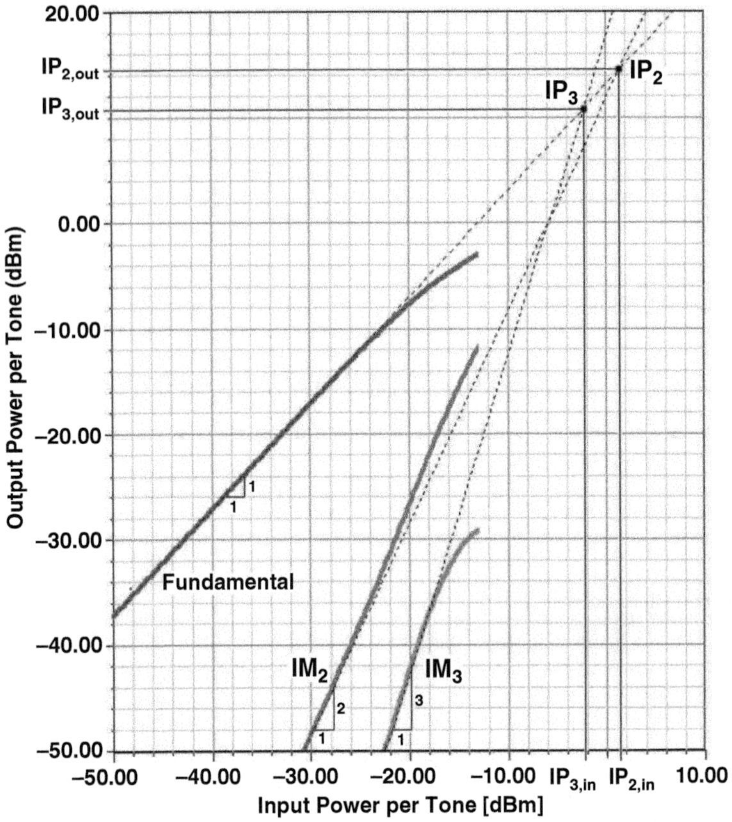

Figure 3.13 The level at which the power of one of a network's IM products equals that of the network's linear output is a figure of merit known as the *intermodulation* IP. The intercept point for a given IM order n can be expressed, and should always be characterized, relative to input power ($IP_{n,\text{in}}$) or output power ($IP_{n,\text{out}}$); the IP_{in} and IP_{out} values differ by the network's linear gain. For the amplifier simulated here, $IP_{2,\text{in}} \approx 1.5$ dBm, $IP_{2,\text{out}} \approx 14.5$ dBm, $IP_{3,\text{in}} \approx -2.3$ dBm, and $IP_{3,\text{out}} \approx 10.7$ dBm. Each curve depicts the power in one tone of the response evaluated. *Source*: [1] / John Wiley & Sons.

for IP_2 and from

$$IP_{3,\text{in}} = \frac{1}{\left(\frac{1}{IP_1} + \frac{G}{IP_2}\right)} \tag{3.21}$$

for IP_3. In both equations, IP_1 is the input intercept of Stage 1 in watts, IP_2 is the input intercept of Stage 2 in watts, and G is the gain of Stage 1 (as a numerical ratio, *not* in decibels). Both equations assume the worst-case condition, in which the distortion products of both stages add in-phase.

The ratio of the signal power to the IM-product power, the *distortion ratio*, can be expressed as

$$R_{dn} = (n-1)\lfloor IP_{n(\text{in})} - P_{(\text{in})} \rfloor \tag{3.22}$$

where n is the order, R_{dn} is the distortion ratio, $IP_{n(\text{in})}$ is the input intercept point, and $P_{(\text{in})}$ is the input power of one tone.

Discussions of IMD have traditionally downplayed the importance of IM_2 because the incidental distributed filtering contributed by the tuned circuitry once common in radiocommunication systems was usually enough to render out-of-passband IM_2 products caused by in-passband signals, and in-passband IM_2 products caused by out-of-passband signals, vanishingly weak compared to fundamental and IM_3 signals. In broadband systems that operate at bandwidths of an octave or more, however, in-passband signals may produce significantly strong in-passband IM_2 and second-harmonic products. In such applications, balanced circuit structures (such as push–pull amplifiers and balanced mixers) can be used to minimize IM_2 and other even-order nonlinear products.

As with IM_2, IM_3 products depend on the spacing of the signals involved and the relative width of the system passband. If ω_1 and ω_2 are of approximately the same frequency, the additive products $2\omega_1 + \omega_2$ and $2\omega_2 + \omega_1$ will be outside the passband of a narrowband system. The subtractive products $2\omega_1 - \omega_2$ and $2\omega_2 - \omega_1$, however, will likely appear near or within the system passband. The IM_3 performance of any network subjected to multiple signals is therefore of critical importance, and an array of IM_3-related, sometimes application-specific, figures of merit has evolved as a result [1].

3.10 Dynamic Range

As we have seen, thermal noise sets the lower limit of the power span over which a network can operate. Distortion—that is, degradation by distortion of the signal's ability to convey information—sets the upper limit of a network's power span. Because the power level at which distortion becomes intolerable varies with signal type and application, a generic definition has evolved: The upper limit of a network's power span is the level at which the power of one IM product of a specified order is equal in power to the network's noise floor. The ratio of the noise floor power to the upper-limit signal power is referred to as the network's DR, often more carefully characterized as *two-tone IMD DR*, which, when evaluated with equal-power test tones, is a figure of merit commonly used to characterize receivers. The MDS relative to the input, as already defined, is

$$\mathrm{MDS_{in}} = kTB + 3\ \mathrm{dB} + \mathrm{NF}$$

When $\mathrm{IP}_{(n)\mathrm{in}}$ and MDS are known, IMD DR can be determined from

$$\mathrm{DR}_n = \frac{(n-1)\left[\mathrm{IP}_{n(\mathrm{in})} - \mathrm{MDS_{in}}\right]}{n} \tag{3.23}$$

where DR is the dynamic range in decibels, n is the order, $\mathrm{IP}_{(\mathrm{in})}$ is the input intercept power in dBm, and MDS is the minimum detectable signal power in dBm. The so-called *spurious-free dynamic range* (SFDR or $\mathrm{DR_{SF}}$) is calculated from

$$\mathrm{DR_{SF}} = \frac{2}{3}(\mathrm{IP_3} - 174\ \mathrm{dBm} + \mathrm{NF} + 3\ \mathrm{dB}) \tag{3.24}$$

This equation allows us to determine how to measure the SFDR. This is done by applying the two-tone signals (in the case of $\mathrm{IP_3}$) and increasing the two signals to the point where the SNR deteriorates by 3 dB or, if the measurement is done relative to MDS, where the noise floor rises by 3 dB. The factor 2/3 is derived from the fact that the levels of IM_3 outputs increase 3 dB for 1 dB of input increase. This definition of DR now is referenced to a noise figure rather than a minimum level in dBm and is therefore independent of bandwidth. (By choosing smaller bandwidths [1 kHz instead of 10 kHz], a DR measurement can be made to look better. Basing the specification on noise figure directly avoids this problem [1].)

3.11 Triple-beat Distortion and Cross-modulation

$\mathrm{P_{-1\,dB}}$ is a single-tone figure of merit; blocking, intercept point, and DR evaluate two-tone behavior. For networks that must handle AM and composite (AM and angle modulation) signals very linearly, a three-tone figure of merit called *triple-beat distortion* has gained acceptance. Signals at ω_1 and ω_2 (closely spaced) and ω_3 (positioned far away from ω_1 and ω_2) are applied to the network under test, at levels, frequencies, and spacings that vary with the application. One triple-beat distortion figure of merit is the ratio, expressed in decibels, of the IM product at $\omega_3 + (\omega_2 - \omega_1)$ to one of the network's linear outputs at a specified output level. Alternatively, the triple-beat figure of merit may express the network output level at which a specified triple-beat ratio occurs.

Triple-beat distortion is the mechanism underlying cross-modulation, a form of intermodulation in which one or more AM signals present in a network amplitude-modulate all signals present in the network [9]. Angle modulation-based wireless systems are largely immune to such effects.

Figures 3.14 and 3.15 graph the results of gain compression, two-tone intermodulation, cross-modulation, and triple-beat testing on a wideband (5–1000 MHz) amplifier.

3.12 Noise Power Ratio

Triple-beat testing is one way of improving on two-tone testing as a means of evaluating a network's intermodulation behavior in the presence of multiple signals. Another figure of merit, *noise power ratio (NPR)*, uses thermal noise as a test signal. The test measures the introduction, by IM, of noise into a quiet slot created by the insertion of a band-stop filter, equal in stopband width to the width of the measurement channel, between the noise generator and the network under test (Figure 3.16).

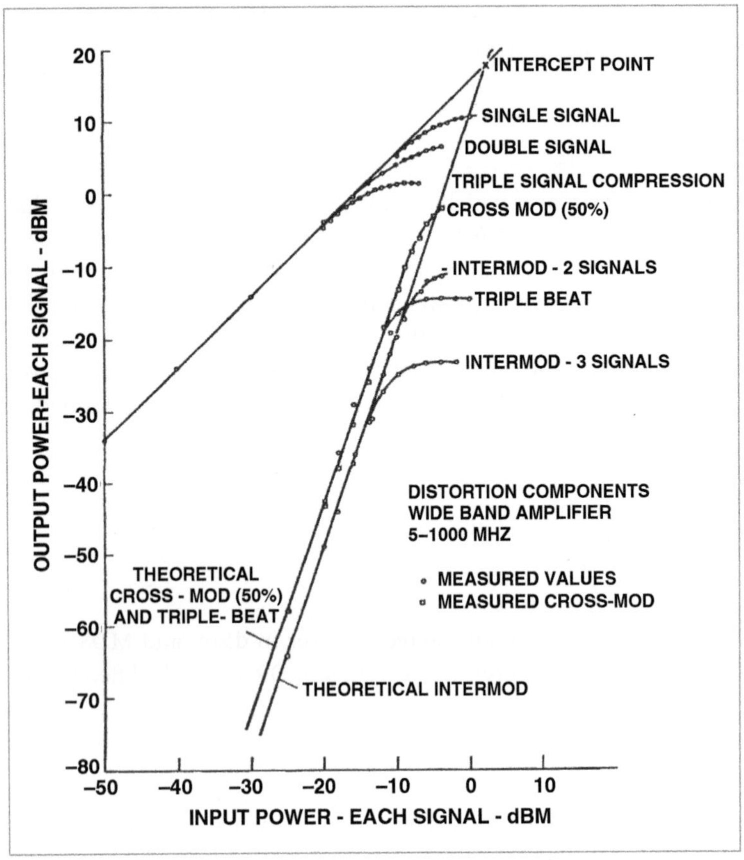

Figure 3.14 Measured distortion components in a wideband (5–1000 MHz) amplifier. Figure 3.15 shows a magnified view of the gain-compression region. *Source*: Adapted from [10].

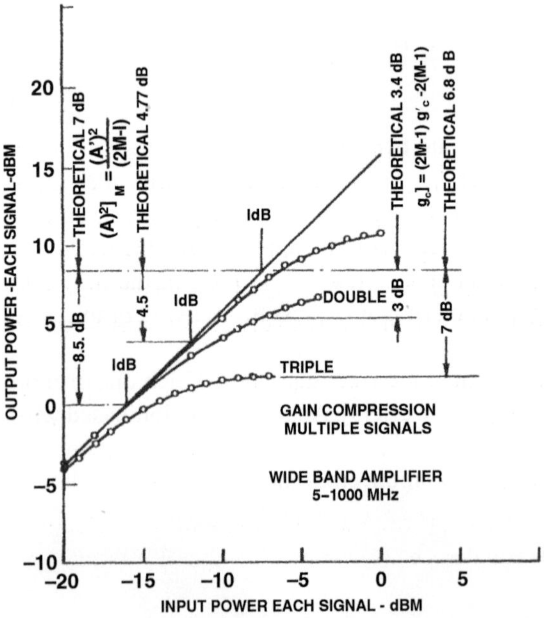

Figure 3.15 Measured multiple-signal gain compression of the 5–1000 MHz amplifier. *Source*: Adapted from [10].

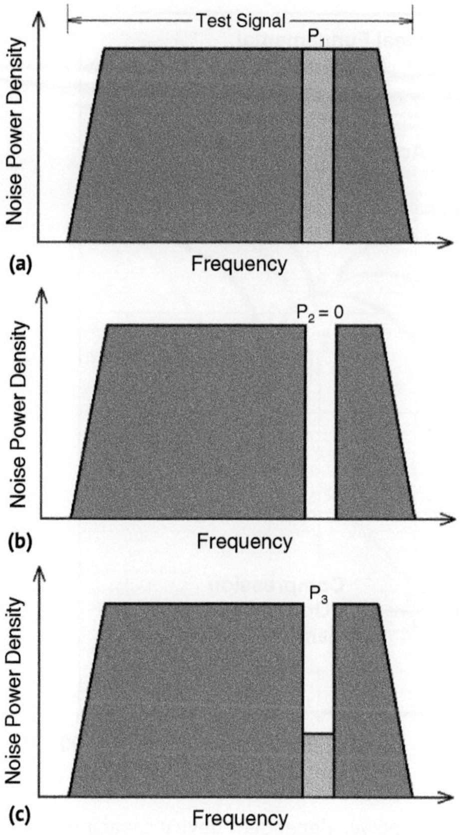

Figure 3.16 Determining a network's *NPR* involves the application of a test signal consisting of thermal noise [10]. The reference measurement-channel noise power, P_1, is then measured (a). Next, a stop-band filter is placed between the noise generator and network under test to keep the test signal out of the measurement channel (b). Assuming sufficient filter attenuation, if the network were absolutely noiseless and linear, the ideal noise power in the measurement channel, P_2, would then be zero. In practice, the network's own thermal noise and intermodulation between noise components outside the measurement channel result in an actual measurement-channel noise power (P_3) greater than zero (c). The NPR equals P_1/P_3. *Source*: [1] / John Wiley & Sons.

3.13 Large-signal Effects

Except for $P_{-1\,dB}$, the figures of merit discussed so far evaluate amplitude nonlinearity under small signal conditions. At input powers that drive a network into gain compression and saturation, IM products of odd orders higher than 3 become significant, and curves, dips, and nulls appear in their characteristics (Figure 3.17). Phase shifts related to nonlinear (primarily voltage-dependent) capacitances in solid-state devices are one cause of these effects. Under such conditions, a network may exhibit hysteresis, with its behavior at any given instant depending not only on the voltage or current applied to it but also on its recent history [11].

As a network is driven into compression, IM products at odd orders higher than 3 become significant, and phase shifts in power-dependent device capacitances cause curves and dips in the IM characteristics.

The onset of these departures from IM-response linearity occurs at generally lower input power levels for higher IM orders; their severity, and their position on the IM curves, differs among the various products of a given order and varies with network topology and tone spacing. Figures of merit based on straight-line IM responses fail to usefully predict nonlinear network behavior under these conditions. Figure 3.18 shows the simulated performance of a single-BJT broadband amplifier driven by two equal-amplitude tones at 10 and 11 MHz [1].

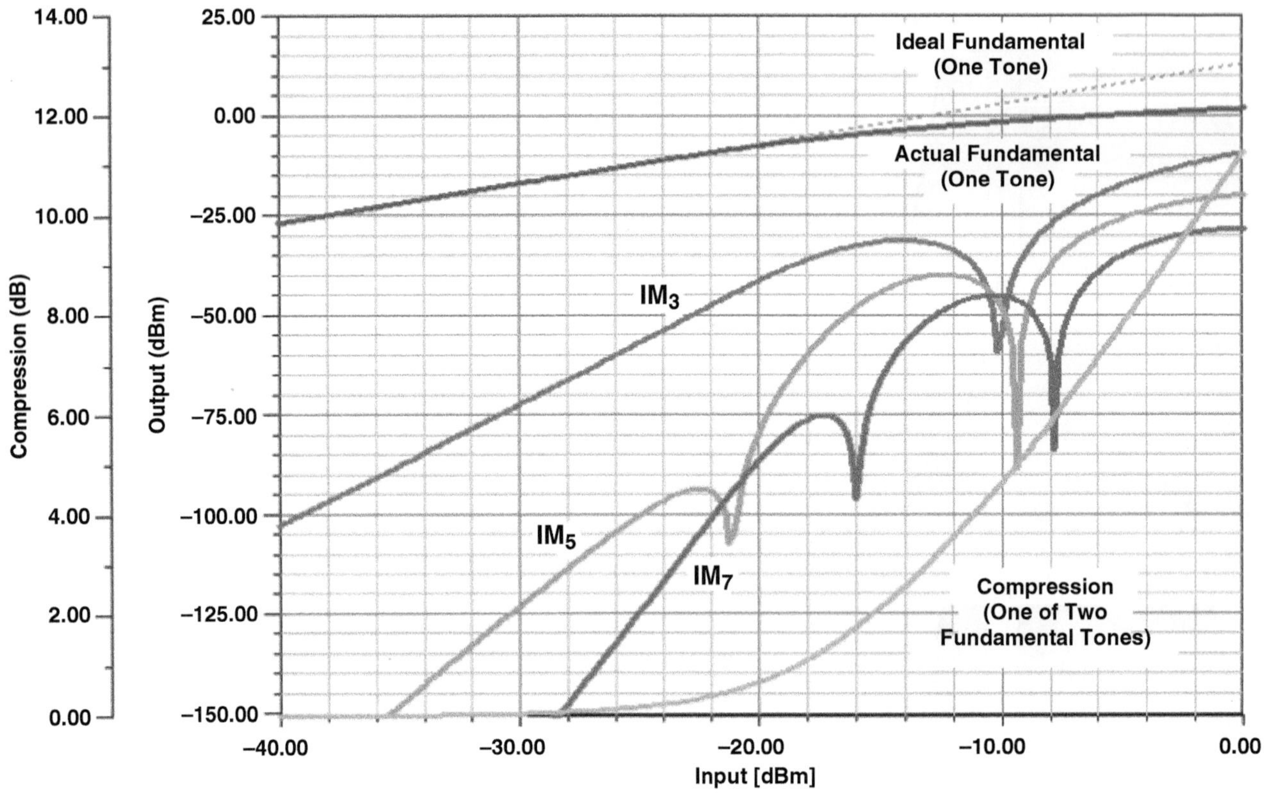

Figure 3.17 IM products at odd orders higher than 3 become significant, and phase shifts in power-dependent device capacitances cause curves and dips in the IM characteristics. *Source*: [1] / John Wiley & Sons.

3.14 AM-to-PM Conversion

The nonlinear distortion effects we have discussed so far can be termed *AM-to-AM distortion*—distortion that, to a degree that depends on the amplitude of the signal(s) applied to the network, results in changes in the network's gain and/or production of signals at new frequencies. AM-to-PM distortion can also occur. As a network nears saturation, part of its driving signal goes into shifting the bias point(s) of its active device(s), changing their drive-dependent reactances and shifting the phase of the output signal relative to its value at input levels below compression (Figures 3.19 and 3.20). This effect, *AM-to-PM conversion*, can cause incidental phase modulation that degrades the performance of digital communication systems [1].

The difference is 0.2 dB in gain change (6.45–6.25) and 0.2 dB in phase shift over the range of −10 dBm to 5 dBm input power.

3.15 Spectral Regrowth and Adjacent-channel Power Ratio

Spectral regrowth occurs largely as a result of third-, fifth-, and seventh-order IMD in power amplifiers operated near or in compression—at power levels where hysteretic IM effects result in poor agreement between measured behavior and predictions based on small-signal IM figures of merit [12]. We therefore evaluate the impact of spectral regrowth more directly, using a figure of merit called ACPR. ACPR measurement techniques that incorporate memory can be used to increase ACPR predictions for networks that exhibit saturation hysteresis [13]. Figure 3.21 shows the critical relationship between compression, power-added efficiency (PAE), and ACPR in a MESFET power amplifier.

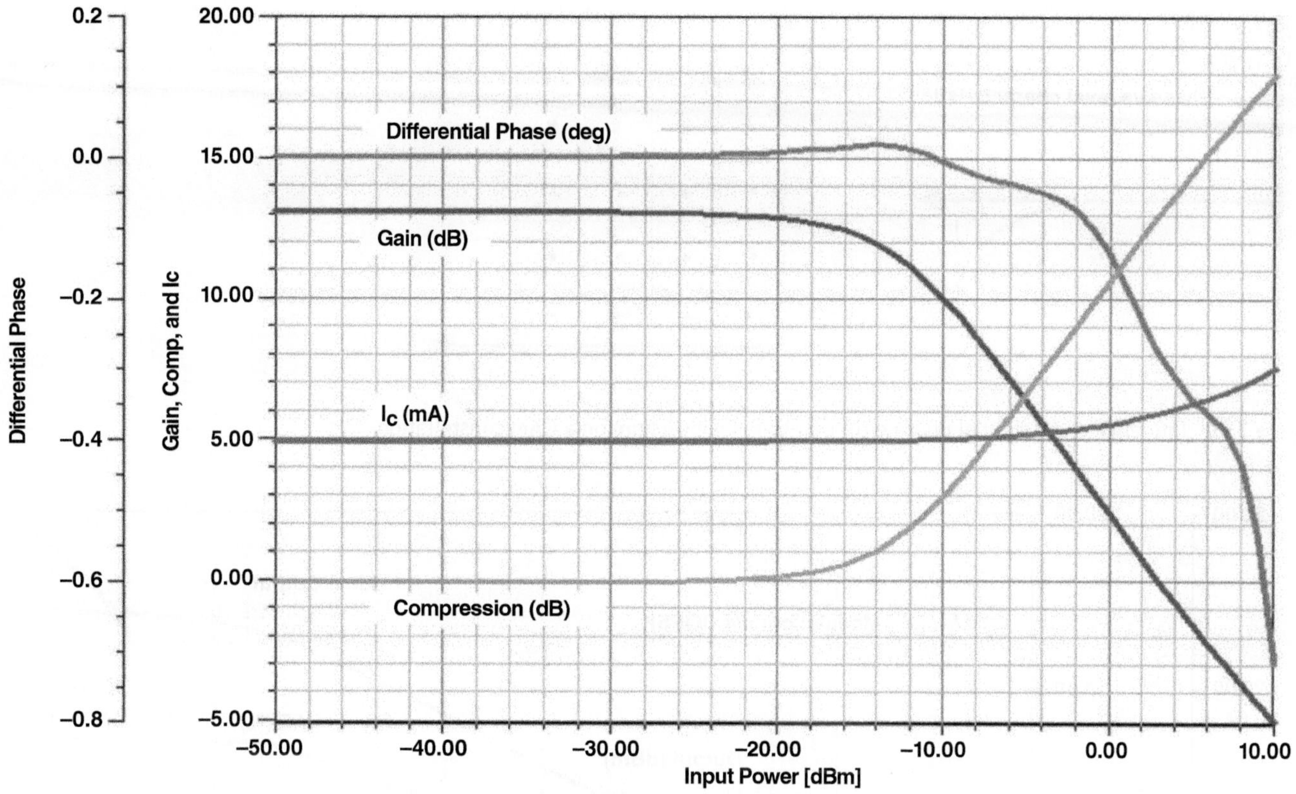

Figure 3.18 Driving a network into compression and saturation shifts the bias point(s) of its active device(s), changing their drive-dependent reactances and shifting the phase of the output signal relative to its value at input levels below compression. This graph shows the simulated performance of a single-BJT broadband amplifier driven by a single tone at 10 MHz. *Source*: [1] / John Wiley & Sons.

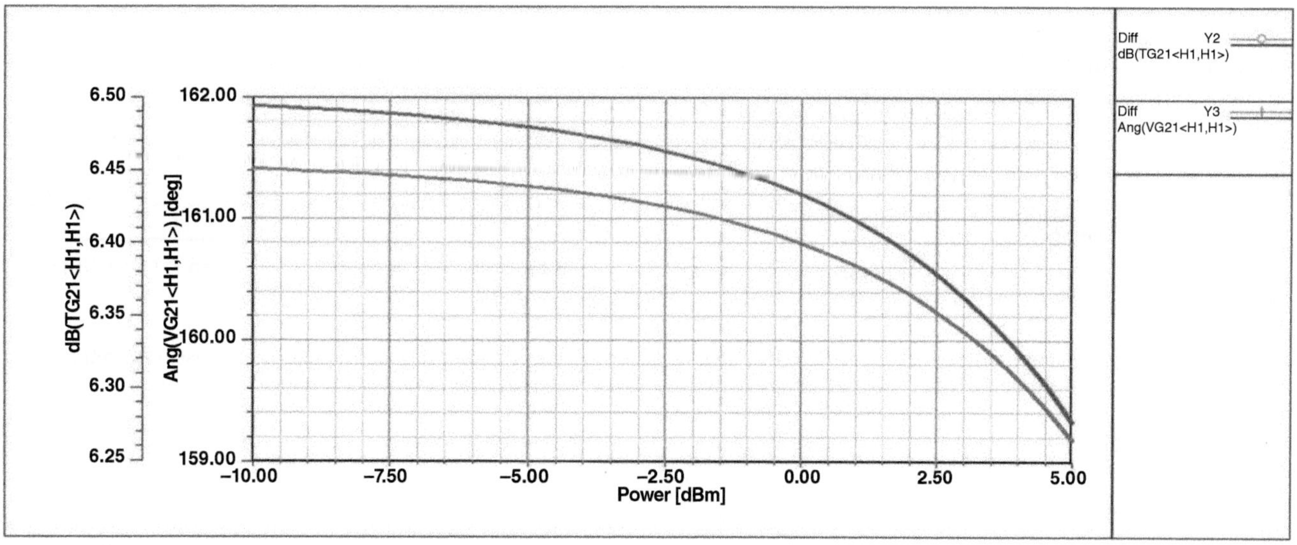

Figure 3.19 Differential gain and phase as a function of input power with high resolution.

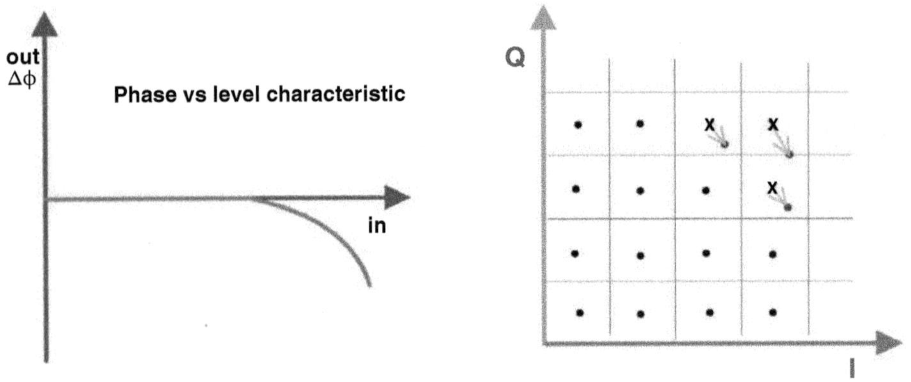

Figure 3.20 Influence of differential phase error (AM-to-PM conversion) on a QAM constellation.

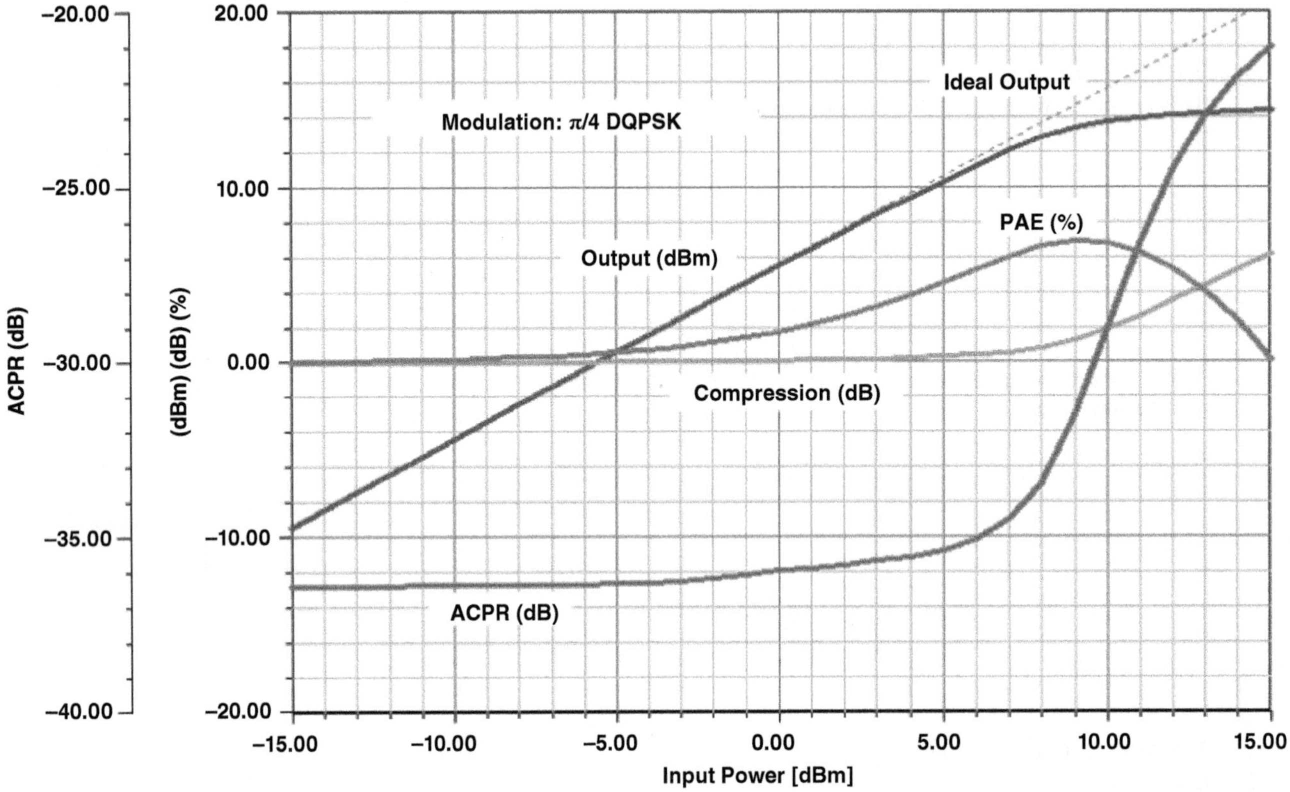

Figure 3.21 Keeping adjacent-channel interference under control can involve a critical trade-off between a wireless transmitter's PAE and ACPR, as shown in this graph of the simulated behavior of the 1 GHz MESFET power amplifier introduced in Figure 3.10. As the amplifier is driven into compression, a peak occurs in the PAE response—in this case, at $P_{-1\,dB}$—and ACPR rises sharply.
Source: [1] / John Wiley & Sons.

3.16 Phase Response Issues and Figures of Merit

We have already seen how large-signal nonlinear distortion can result in amplitude-dependent phase shifts through AM-to-PM conversion. Because phase linearity is critical at *all* signal levels in PM systems, especially those using digital modulation, we must also consider linear distortion in evaluating networks used in wireless systems [1].

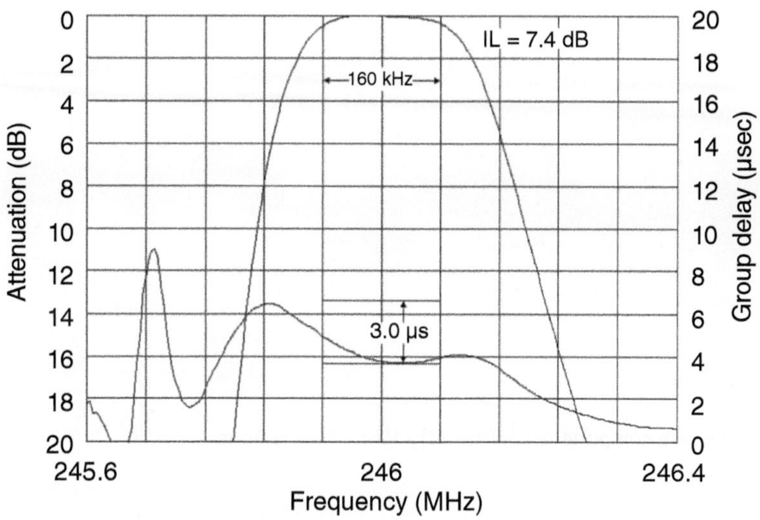

Figure 3.22 Close-in amplitude and group delay responses for a 246 MHz SAW filter designed for GSM applications [14]. This filter is well within its 3.0 μs differential group delay specification across its passband (160 kHz at −3 dB); the peaks just outside the passband limits are characteristic of a network's transition-band phase response.

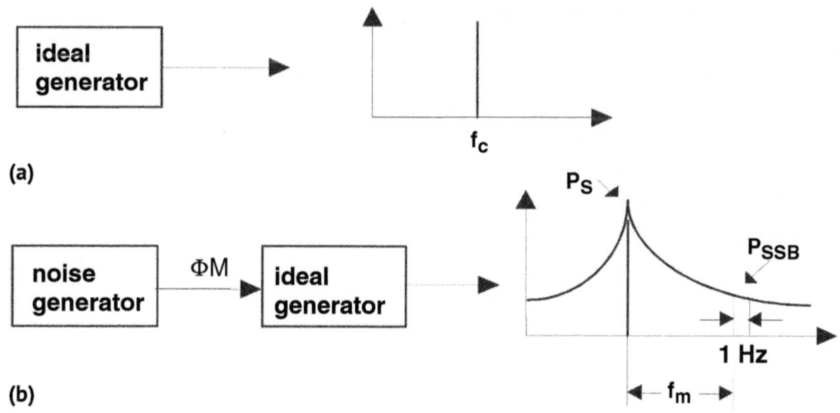

Figure 3.23 SSB phase noise. An ideal signal generator (a) would produce an absolutely pure carrier. A real signal generator (b) acts like an ideal generator driven by a noise generator, producing a noise-modulated carrier. *Source*: [1] / John Wiley & Sons.

3.17 Differential Group Delay

Every frequency-selective network subjects signals passing through it to some degree of time delay. Ideally, this delay, also known as *group* or *envelope delay*, does not vary with frequency; that is, the network's phase shift versus frequency response is monotonic and linear. In practice, a network's time delay varies across its passband, transition bands and stop-bands, exhibiting curvature, ripple, and transition-band peaks (Figure 3.22). The network's *differential* group delay—its group delay spread—is therefore of considerable importance. This is especially so in digitally modulated systems, where the resulting phase distortion can cause errors in modulation and demodulation [1].

3.18 Effects of Phase Noise

The phase of an oscillator's output signal is subject to random phase variations (Figures 3.23 and 3.24) ([1], Chapter 5 discusses this in detail). Called *phase noise*, this effect is often quantified as the decibel ratio of the phase noise power in a single (the upper or lower) phase-noise sideband, in a 1 Hz bandwidth centered at a specified frequency offset from the

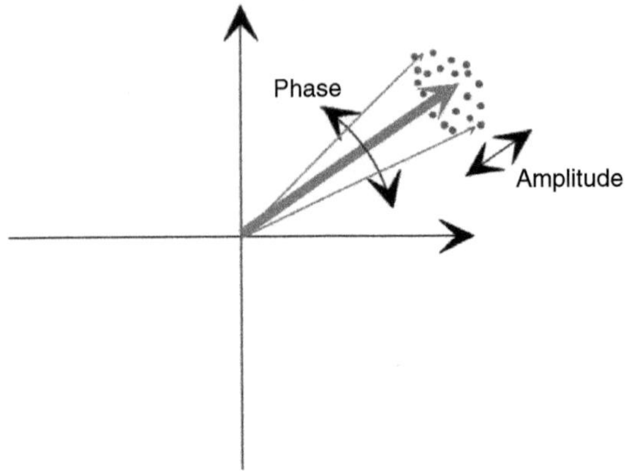

Figure 3.24 Oscillator noise can be split into amplitude and phase components. *Source*: [1] / John Wiley & Sons.

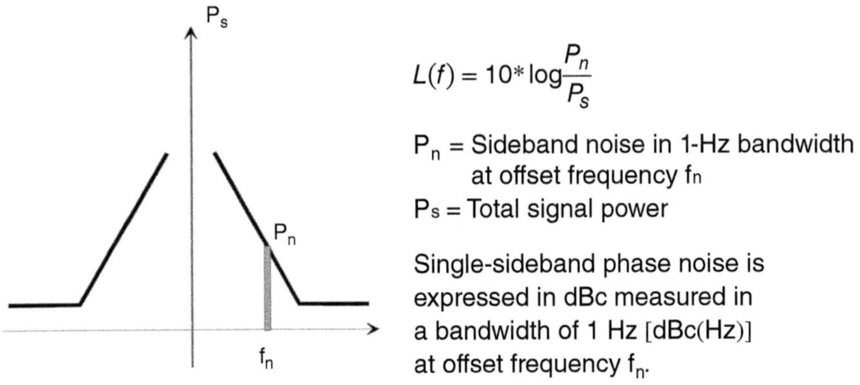

Figure 3.25 Phase noise calculation.

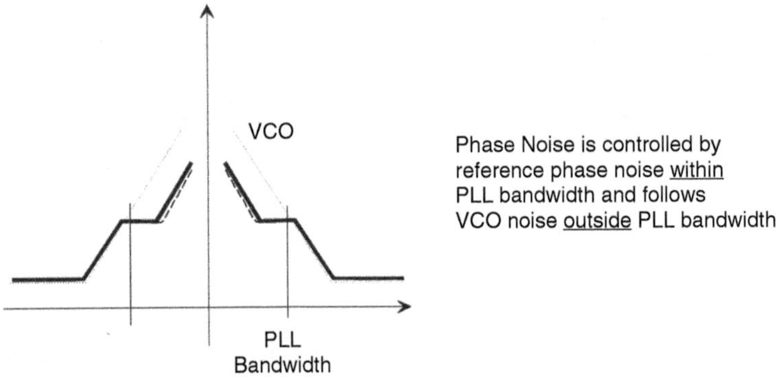

Figure 3.26 Phase noise of an oscillator controlled by a phase-locked loop. *Source*: [1] / John Wiley & Sons.

oscillator carrier, to the carrier power (Figure 3.25); alternatively, it may be specified in degrees rms. A microwave voltage-controlled oscillator, for instance, might exhibit an single side band (SSB) phase noise of −95 dBc/Hz at 10 kHz. Oscillator phase noise may manifest itself, through a mechanism known as *reciprocal mixing*, as the emission of unacceptably strong noise outside a transmitter's occupied bandwidth or as an increase in receiver noise floor. Phase noise may also directly introduce phase errors that result in modulation and demodulation errors [1].

Because the oscillators used for frequency translation in wireless systems are usually embedded in phase-locked loops, their phase-noise characteristics differ from those of "bare" oscillators as shown in Figures 3.26 and 3.27. Figures 3.28

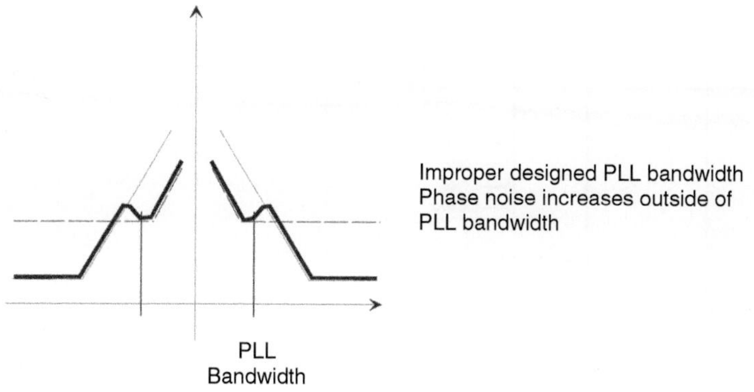

Figure 3.27 Effect of improper loop-filter design. *Source*: [1] / John Wiley & Sons.

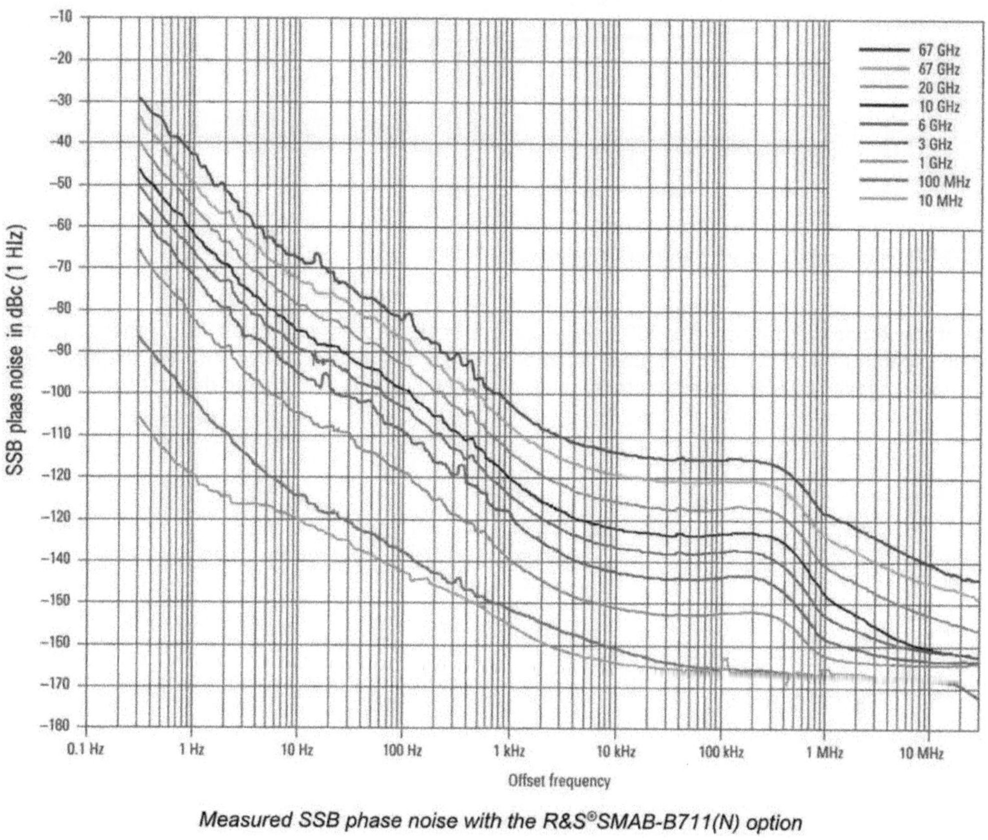

Figure 3.28 Measured phase noise of the Rohde & Schwarz SMA 100 B signal generator at various frequencies. This signal generator has no provision for digital modulation and therefore shows the best possible phase noise in its class.

and 3.29 show the measured phase noise of the Rohde & Schwarz SMA 100B and SMW 200A signal generators. The SMY is a best-in-class analog source; the SMW 200A is a very high-performance signal generator capable of being programmed for all digital modulations; therefore, their PLL systems exhibit different phase noise versus frequency responses as the measured results show.

3.19 Reciprocal Mixing

In reciprocal mixing, incoming signals mix with LO-sideband energy to produce IF output (Figure 3.30). Because one of the two signals is usually noise, the resulting IF output is usually noise. (Reciprocal mixing effects are not limited to noise; discrete-frequency oscillator sideband components, such as those resulting from crosstalk to or reference energy on a VCO's

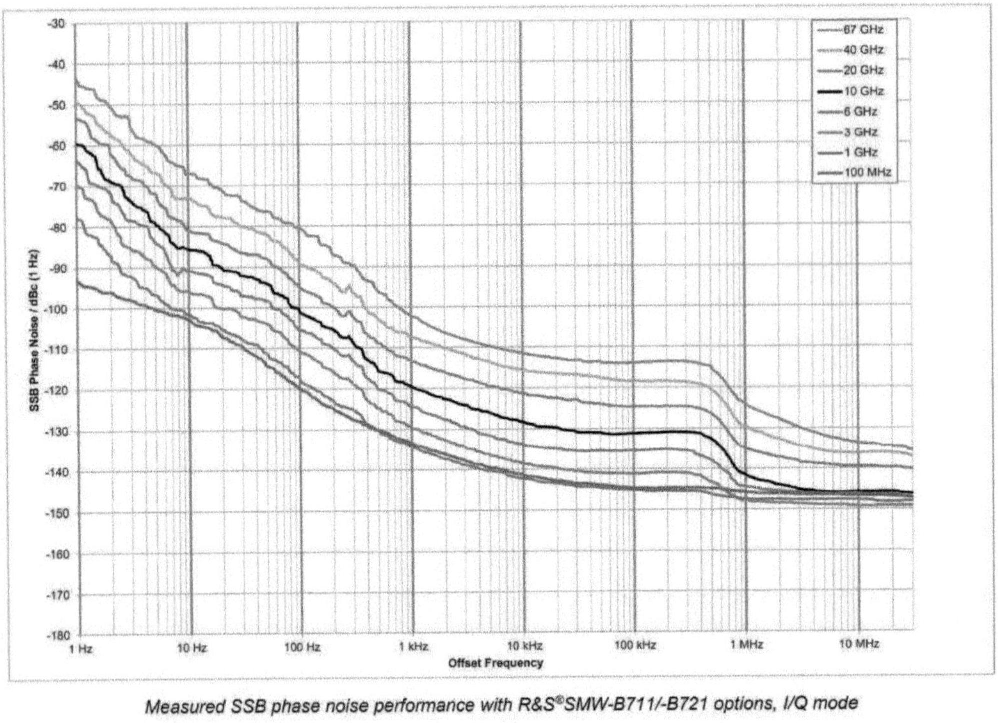

Figure 3.29 Measured phase noise of the Rohde & Schwarz SMW 200A signal generator at various frequencies. This signal generator is optimized for all digital modulation capabilities and can be configured via appropriate programming. Above 300 kHz, the influence of the wideband loop becomes noticeable; above 700 kHz, the resonator Q takes over. The noise floor is limited due to the many modulation states.

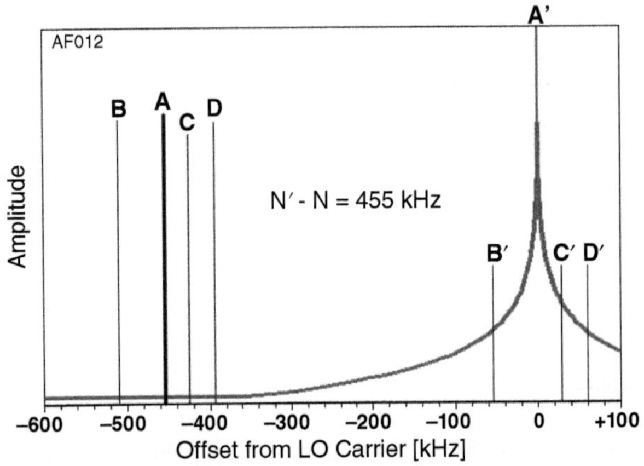

Figure 3.30 Reciprocal mixing occurs when incoming signals mix energy from an oscillator's sidebands to the IF. In this example, the oscillator is tuned so that its carrier, at A′, heterodynes the desired signal, A, to the 455 kHz as intended; at the same time, the undesired signals B, C, and D mix the oscillator noise sideband energy at B′, C′, and D′, respectively, to the IF. Depending on the levels of the interfering signals and the noise sideband energy, the result may be a significant rise in the receiver noise floor. *Source*: [1] / John Wiley & Sons.

control line, or the discrete-frequency spurious signals endemic to direct digital synthesis, can also mix incoming signals to IF.) In practice, the resulting noise floor increase can compromise the receiver's ability to detect weak signals and achieve a high IMD DR; on the test bench, noise from reciprocal mixing may invalidate desensitization, cross-modulation, and IM testing by obscuring the weak signals that must be measured in making these tests [1].

Figure 3.31 shows a typical arrangement of a dual-conversion receiver with local oscillators. The signal coming from the antenna is filtered by an arrangement of tuned circuits referred to providing as *input selectivity*. For a minimum attenuation in the passband, an operating bandwidth is given by

$$B = \frac{f}{\sqrt{2} \cdot Q_L} \tag{3.25}$$

This approximation formula is valid for the insertion loss of about 1 dB due to loaded Q.

The filter in the first IF is typically either a SAW filter (in the frequency range from 500 MHz to 1 GHz) or a crystal filter (45 to 120 MHz). Typical insertion loss is 6 dB. Since these resonators have significantly higher Q than LC circuits, the bandwidth for the first IF will vary from ±5 kHz to ±500 kHz. It is now obvious that the first RF filter does not protect the first IF because of its wider bandwidth. For typical communication receivers, IF bandwidths from 150 Hz to 1 MHz are found; for digital modulation, the bandwidth varies roughly from 30 kHz to 1 MHz. Therefore, the IF filter in the second IF has to accommodate these bandwidths; otherwise, the second mixer easily gets into trouble from overloading. This also means that both synthesizer paths must be of low-noise and low-spurious design. The second IF of this arrangement (Figure 3.31) can be either analog or digital, or even zero-IF. In practical terms, there are good reasons for using IFs like 50/3 kHz, as in hi-fi audio equipment, with DSP processing at this frequency (50/3 kHz) using the low-cost modules found in mass-market consumer products.

Figures 3.32 and 3.33 show the principle of selectivity measurement both for analog and digital signals. The main difference is that the occupied bandwidth for the digital system can be significantly wider, and yet both signals can be interfered with by either a noise synthesizer/first LO or a synthesizer that has unwanted spurious frequencies. Such a spurious

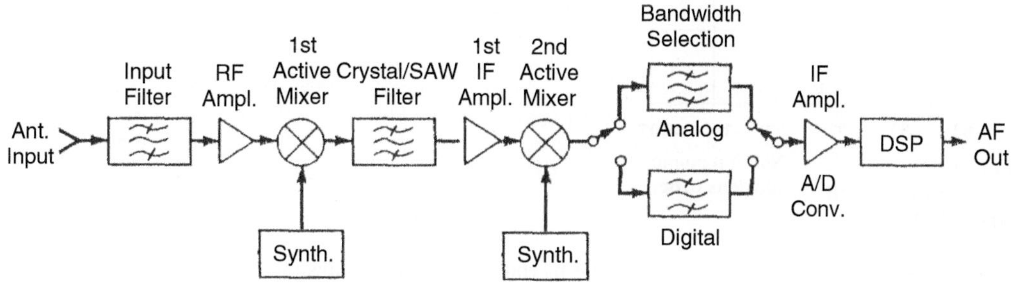

Figure 3.31 Block diagram of an analog/digital receiver showing the signal path from antenna to audio output. No AGC or other auxiliary circuits are shown. This receiver principle can be used for all types of modulation, since the demodulation is done in the DSP block. *Source*: [1] / John Wiley & Sons.

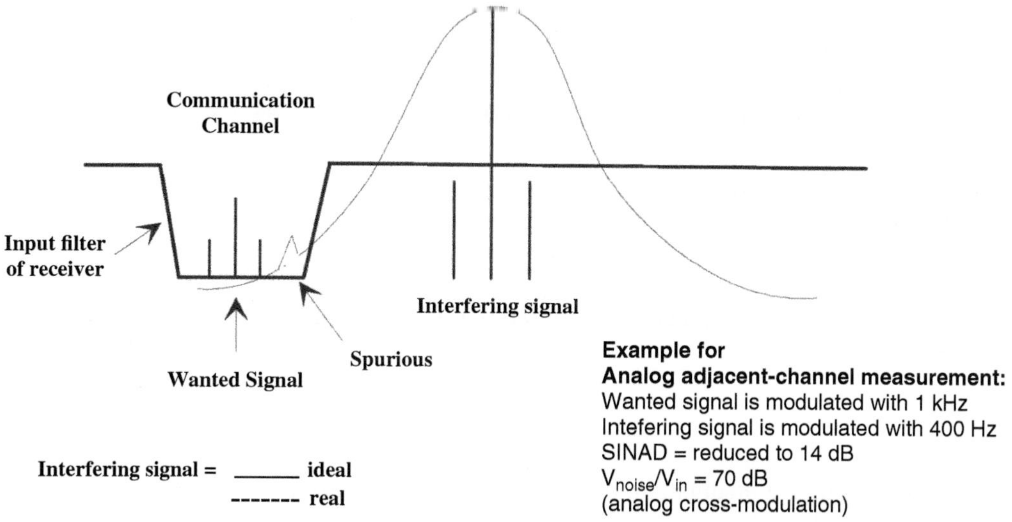

Figure 3.32 Principle of selectivity measurement for analog receivers. *Source*: [1] / John Wiley & Sons.

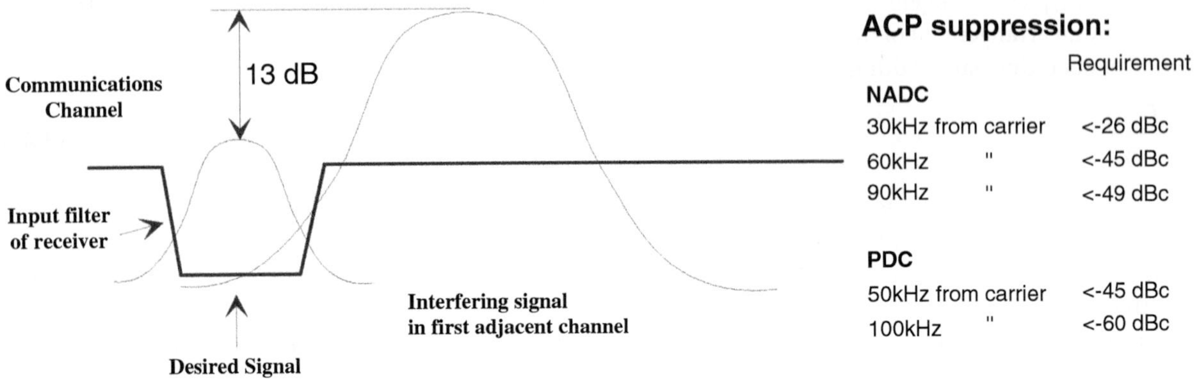

Figure 3.33 Principle of selectivity measurement for digital receivers. *Source*: [1] / John Wiley & Sons.

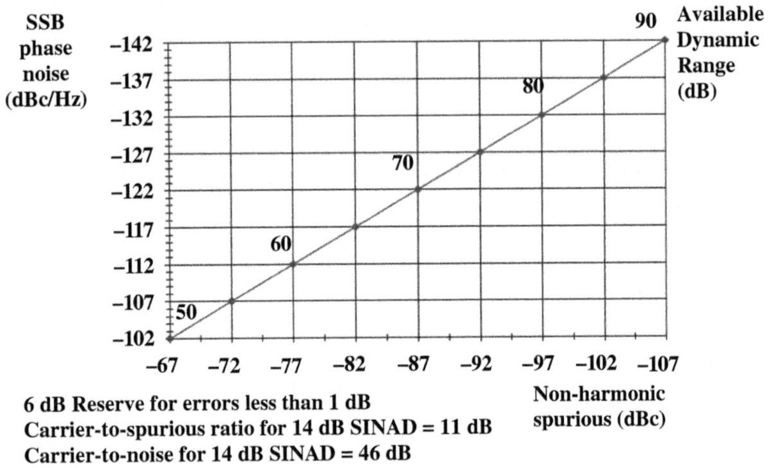

Figure 3.34 This graph shows the available DR, which is determined either by the masking of the unwanted signal by phase noise or by discrete spuria. As far as the culprit synthesizer is concerned, it can be either the local oscillator or the effect of a strong adjacent-channel signal that takes over the function of the local oscillator. *Source*: [1] / John Wiley & Sons.

signal is shown in Figure 3.32. In this case, the analog adjacent-channel measurement has some of the characteristics of cross-modulation and intermodulation, while in the digital system the problem with the adjacent-channel power suppression in modern terms is more obvious. Rarely has the concept of adjacent-channel power (ACP) been used with analog systems. Also, to meet the standards, signal generators have to be used that are better, with some headroom, than the required dynamic measurement. We have therefore included in Figure 3.33 the achievable performance for a practical signal generator—in this case, the Rohde & Schwarz SMA100B.

Although additional RF selectivity can reduce the number of signals that contribute to the noise, improving the LO spectral purity is the only effective way to reduce reciprocal mixing noise from *all* signals present at a mixer's RF port.

Factoring in the effect of discrete spurious signals with that of oscillator phase noise can give us the useful DR of which an instrument or receiver is capable (Figure 3.34).

In evaluating the performance of digital wireless systems, we are interested in determining a receiver's resistance to adjacent-channel signals.

3.20 Phase Errors

In PM systems, especially those employing digital modulation, oscillator phase noise contributes directly to modulation and demodulation errors by introducing random phase variations that cannot be corrected by phase-locking techniques (Figure 3.35). The greater the number of phase states a modulation scheme entails, the greater its sensitivity to phase noise [1].

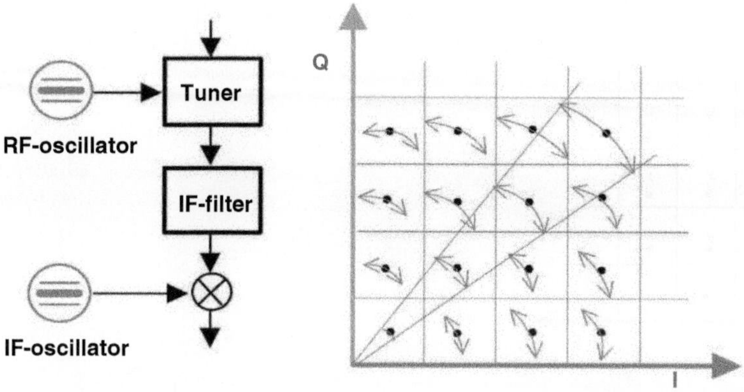

Figure 3.35 Phase noise is critical to digitally modulated communication systems because of the modulation errors it can introduce. Intersymbol interference (ISI), accompanied by a rise in BER, results when state values become so badly error-blurred that they fall into the regions of adjacent states. This drawing depicts only the results of phase errors introduced by phase noise; in actual systems, thermal noise, AM-to-PM conversion, differential group delay, propagation, and other factors may also contribute to the spreading of state amplitude and phase values. *Source*: [1] / John Wiley & Sons.

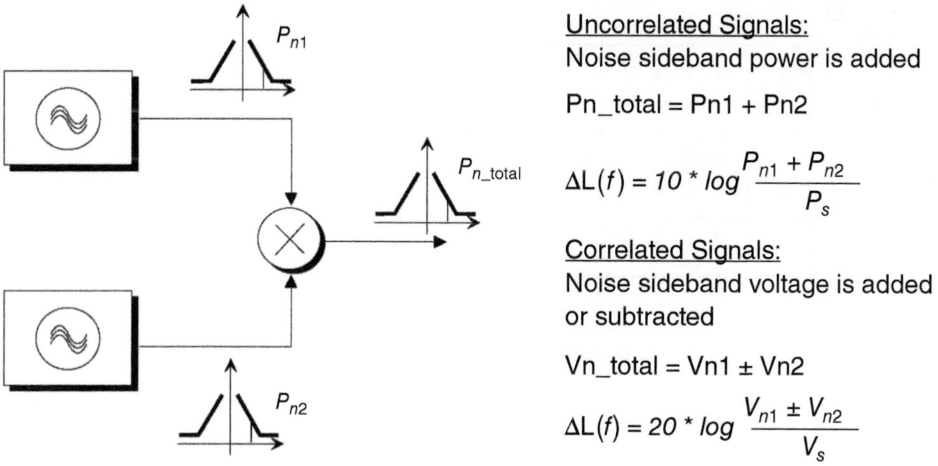

Figure 3.36 The noise sideband power of a signal that results from mixing two signals depends on whether the signals are correlated—referenced to the same system clock—or uncorrelated. *Source*: [1] / John Wiley & Sons.

When an output signal is produced by mixing two signals, the resulting phase noise depends on whether the input signals are *correlated*—all referred to the same system clock—or uncorrelated, as shown in Figure 3.36.

3.21 Error Vector Magnitude

Several earlier figures (Figures 3.12, 3.20, and 3.35) have shown how particular sources of amplitude and/or phase error can shift the values of a digital emission's data states toward decision boundaries, resulting in increased BER due to ISI. Figures 3.37–3.39 show three additional sources of such errors [15].

A figure of merit known as *error vector magnitude* (EVM) has been developed as a sensitive indicator of the presence and severity of such errors. An emission's error vector magnitude is the magnitude of the phasor difference as a function of time between an ideal reference signal and the measured transmitted signal after its timing, amplitude, frequency, phase, and DC offset have been modified by circuitry and/or propagation. Figure 3.40 illustrates the EVM concept.

Test equipment capabilities required for the validation of circuit parameters described earlier are listed in Table 3.1.

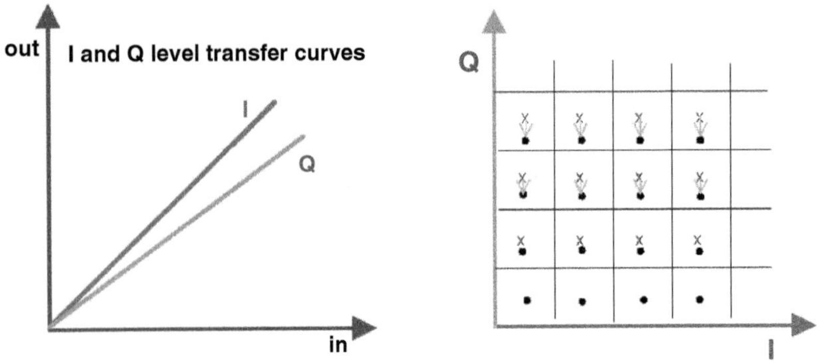

Figure 3.37 Effect of gain imbalance between I and Q channels on a data signal's phase constellation. *Source*: [1] / John Wiley & Sons.

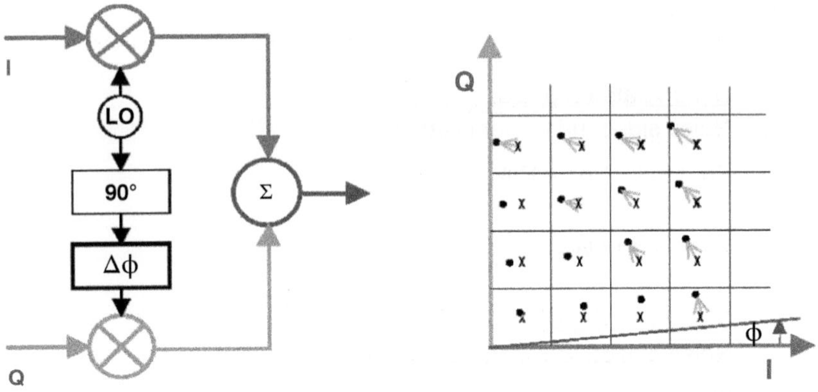

Figure 3.38 Effect of quadrature offset on a data signal's phase constellation. *Source*: [1] / John Wiley & Sons.

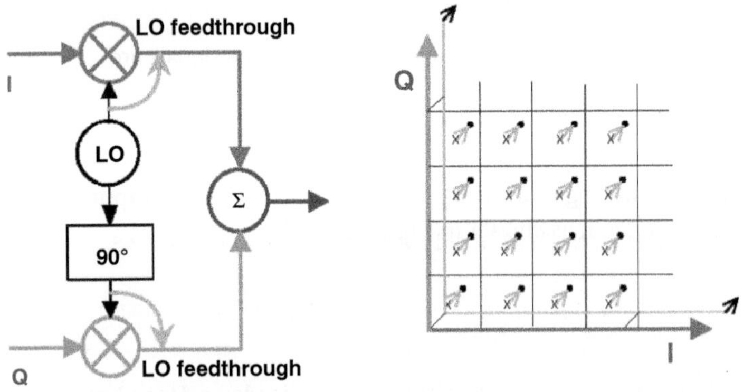

Figure 3.39 Effect of LO-feedthrough-based IQ offset on a data signal's phase constellation. *Source*: [1] / John Wiley & Sons.

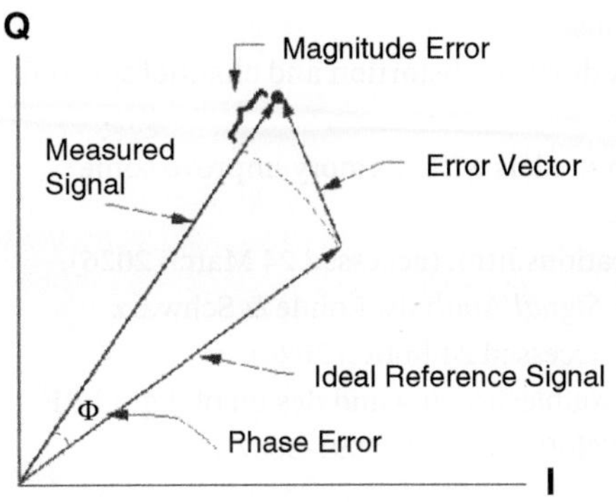

Figure 3.40 Representing errors in a digitally modulated signal's in-phase (I) and quadrature (Q) values relative to the ideal as a single error vector allows us to derive the resulting *EVM*, a sensitive indicator of transmission quality. *Source*: [16] / Agilent Technologies, Inc.

Table 3.1 Desired test equipment capabilities.

High-end network analyzer (such as the R&S ZNA)	• Compression point measurements • Intermodulation and noise figure measurements on amplifiers and mixers • Pulsed measurements • Spectrum analysis with multichannel view • Time domain analysis • Signal integrity measurements • Millimeter wave measurements
High-end signal generator (such as the R&S SMW 200A and SMA 100B)	Arbitrary multitone signal testing of various waveforms (see the aforementioned examples)
High-end analysis of signal sources (such as R&S FSWP phase noise analyzer and VCO tester)	• High sensitivity for phase noise measurements, including pulsed signals through cross-correlation with high performance close to carrier • Frequency range from 1 MHz up to 50 GHz (up to 500 GHz with external mixers) • Internal source for measuring residual phase noise, also on pulsed signals • Simultaneous measurement of phase noise and amplitude noise

The following tutorial is recommended in Ref. [6].

References

1 Rohde, U.L. and Rudolph, M. (2013). *RF/Microwave Circuit Design for Wireless Applications*, 2e, 745–746. New York: John Wiley & Sons. ISBN 1118431405.

2 Rohde, U.L. and Whitaker, J.C. (2017). *Communications Receivers*, 4e. McGraw Hill Professional. ISBN 0071843345, 9780071843348.

3 Sabin, W.E. and Schoenike, E.O. (1995). *Single Sideband Systems and Circuits*, 2e. New York, NY, USA: McGraw Hill.

4 Friis, H.T. (1944, July). Noise figures of radio receivers. *Proceedings of IRE* 32 (7): 419–422.

5 McCorkle, J.W. Noise figure one and two, Friis and IEEE. *Microwaves* 101, Feb. 2019. [Online]. Available: https://www.microwaves101.com/encyclopedias/noise-figure-one-and-two-friis-and-ieee.

6 Razzell, C. (2013). System noise-figure analysis for modern radio receivers. Maxim Integrated (a division of analog devices). https://www.analog.com/media/en/technical-documentation/tech-articles/system-noisefigure-analysis-for-modern-radio-receivers.pdf (accessed 24 March 2026).

7 Williams, M., Bonn, F., Gong, C. et al. (1994, July). GaAs RF ICs target 2.4-GHz frequency band. *Microwaves & RF* 33 (7): 111.

8 Smith, J. (1997). *Modern Communication Circuits*, 2e, 93–99. New York: McGraw-Hill.

9 Perlow, S.M. (1976, June). Third-order distortion in amplifiers and mixers. *RCA Review* 37 (6): 234–265.

10 Jordan, E.C. (ed.) (1985) *Reference Data for Engineers*, 7e, 12-33–12-35. Indianapolis, IN: Howard W. Sams & Co, Inc.

11 Maas, S.A. (1988). *Nonlinear Microwave Circuit*, 18–169. Norwood, MA: Artech House.
12 Staudinger, J. (1997, July/August). Specifying power amplifier linearity via intermodulation distortion and channel spectral regrowth. *Applied Microwaves and Wireless* 9 (4): 62–70.
13 Heutmaker, M.S., Wu, E., and Welch, J.R. (1997, July/August). Envelope distortion models with memory improve some amplifier spectral regrowth predictions. *Applied Microwaves and Wireless* 9: 72–78.
14 SAW Filter Applications. Toko [Online]. Available: http://tokoam.com/saw/applications.html (accessed 24 March 2026).
15 Daniel, R. (2022). *RF Fundamentals Seminar Part 5: Digital Modulation and Vector Signal Analysis*. Rohde & Schwarz. Columbia, MD, USA, 2022. [Online]. Available: https://www.rohde-schwarz.com (accessed 24 March 2026).
16 Hewlett-Packard Co. (1994). Using vector modulation analysis in the integration, troubleshooting and design of digital RF communications systems. Product Note HP89400-8. http://www.tmo.hp.com/tmo/Notes/English/5091-8687E.html (accessed 24 March 2026).

4

The Tools and Technology Explained

Huge advances have been made in CAD tools and, in parallel, in the development of semiconductor technology from the 1970s, when SPICE was created, to today. Likewise, in the area of modeling passive structures like transmission lines and other distributed elements, the approaches have been refined.

To begin with, we want to bring an overview of the software tools and provide a high-end application example and information about the inner working of the harmonic balance method-based implementation [1, 2]. Then we want to walk the reader through the basic physics of bipolar transistors of different designs and likewise for microwave field-effect transistors (FETs), with emphasis on discrete devices and hot topics.

The first step in using CAD was the introduction of the SPICE program,[1] with the simulation capabilities of nonlinear DC analysis, small-signal analysis, and nonlinear transient analysis in a simulation program. Particular emphasis was placed upon the circuit models for the bipolar junction transistor (BJT) and the FET, which are implemented in SPICE. SPICE stands for *Simulation Program with Integrated Circuit Emphasis*. Researchers at the University of California, Berkeley, developed this computer program during the mid-70s. Analog devices offer such an up-to-date program free of charge called LTspice®. It is a high-performance SPICE simulation software, schematic capture, and waveform viewer with enhancements and models for easing the simulation of analog circuits. Included in the download of LTspice are macromodels for a majority of analog devices, switching regulators, and amplifiers, as well as a library of devices for general circuit simulation.[2]

As a test case, we will compare the results for an oscillator design with the LTspice and the Ansys Designer next, using the published parameters for the BFP 620.

The following are the SPICE parameters from Infineon for this microwave transistor without package parasitics:

```
.Model bfp620 Bip( rc2 = 2.2 re1 = 0.4 NPN is = 354e-18 ise = 2.978pA
+isc = 12.23fA bf = 557 br = 100 nf = 1.02 ne = 3.355
+ nc = 2.18 va = 1000 vb = 1.2 ikr = 6.31e-3 ikf = 2.26 rbm = 2.5
+ rb = 2.6 irb = 18E-6 tf = 1.3pS itf = 2.44 fcc = .82
+ vje = 0.89 mje = 0.315 cjc = 225fF cje = 371fF xcjc = 1
+ vjc = .74 mjc = 0.3926 vcmx = 4 ibmx = 1mA area = 1
+ tnom = 298 xtb = -.9 eg = 1.114 kf = 1e-9 af = 2)
```

The SPICE program is lacking all the needed distributed elements for microwave circuit analysis, so Less Besser & Company developed a high-performance but linear microwave tool by the name of Compact Software.[3]

Today, the introduction of modern nonlinear microwave programs like Ansys 2023R2 Electronics Desktop or Cadence Microwave Office® and Keysight PathWave Advanced Design System (ADS) or others based on the harmonic balance method (see below) allows a design engineer to produce simulations of amplifier, mixer, and oscillator designs to achieve given specifications. These programs operate under the Windows PC environment. In this book, we will consider passive devices and address some of the behavior of the distributed elements including resonators and the very advanced application of metamaterial structures.

1 http://bwrcs.eecs.berkeley.edu/Classes/IcBook/SPICE/
2 https://www.analog.com/en/design-center/design-tools-and-calculators/ltspice-simulator.html
3 https://en.wikipedia.org/wiki/Compact_Software

4.1 Transistor Classification

The microwave circuit designer can choose from a broad range of transistor technologies. Bipolar and field-effect devices are available, also many different material systems, ranging from traditional GaAs to silicon, GaN, or InP. Choosing a transistor type means choosing the technology that fits the needs and constraints of the intended application in an optimal way.

The first criterion could be to select the material system that shows the best potential for microwave performance. A good measure is Johnson's figure of merit (JFoM) [3], a figure that is proportional to the square of the maximum carrier saturation velocity multiplied by the maximum breakdown electrical field strength. These two material properties relate to cutoff frequencies and supply voltages of bipolar and FETs. Even though it is a rough estimate, it is a good method to judge material systems relatively to each other. Other than the criterion of switching speed that follows Moore's law by aggressive downscaling, JFoM also considers the maximum power that a device can handle. In terms of potential microwave performance, III–V materials are superior to silicon, and a wide bandgap boosts the figure. With JFoM being normalized to unity for silicon, it reaches 7 for gallium arsenide and 760 for gallium nitride. These numbers explain why the semiconductor industry is still willing to invest significantly in emerging technologies even though the well-established silicon technologies already extended their frequency range into the microwave and millimeter wave field.

Another criterion in selecting is cost, which usually directs us toward CMOS. The direct cost per chip is determined by the cost per wafer, by the wafer diameter, and by the processing costs. The advantage for silicon is the generally lower cost per wafer as compared to the less mature lower volume production III–V wafers. But also the wafer diameter differs, ranging between 100 and 150 mm for III–Vs compared to up to 300 mm for silicon wafers. For consumer products serving a mass market, silicon technologies will therefore be the best choice if the achievable performance matches the requirements. But microwave and millimeter wave applications don't always target mass markets. Space applications or radio astronomy often require the production of just a few circuits, and even infrastructure applications like amplifiers for overseas optical cables often might satisfy global needs with a few thousand circuits, if at all. For small-scale production, the cost calculation looks different, as the development costs, especially for the lithographic mask set, are a dominant factor. Producing just a few wafers in a III–V technology might be a more cost-efficient solution in small scale, even if the wafers are shared in multi-project runs offered by commercial foundries. Total cost, however, is not only the cost of the microwave circuit alone. Packaging and integration of different technologies contribute significantly. Therefore, the possibility to integrate digital baseband data processing on the same chip is a clear advantage of silicon technologies.

But many applications demand highest performance while cost is important but comes second. In these cases, the technology of choice depends on the specific application.

For power amplifiers, it is desirable that the device is capable to operate at the desired power level and can also be easily matched. Technologies that operate at higher voltages have an advantage in this respect, as they allow achieving a certain power at lower currents. Higher operation voltage at lower current means a higher DC load resistance, but also a smaller device size and thus lower output capacitance. Depending on frequency and absolute power level, HBTs provide better performance than HEMTs in GaAs and InP, and GaN devices provide higher power levels than comparable GaAs technologies. In terms of absolute currents, HBTs use wafer space more efficiently since the current flows vertically through the area defined by the emitter contact. The current in HEMT devices, on the other hand, flows horizontally in the 2D electron gas at the hetero interface. Due to the lower current density per chip area, HEMTs will consume more space and provide higher output capacitance than HBTs. However, GaN provides superior properties in terms of breakdown voltages over all other commercially available technologies, as well as the highest power densities available today.

Frequencies beyond 100 GHz were, for some time only, reached by HEMTs on GaAs or InP featuring gate lengths of 100 nm or less. But today, HBTs in SiGe or InP technologies provide cutoff frequencies of 500 GHz and beyond. And especially if some power is required, HBTs are the devices of choice. A special application is, e.g., laser modulator drivers for fiber optic links, requiring fast switching of rather high voltages.

For low-noise amplification, HEMTs are again superior. The thermal noise generated in the 2D electron gas channel is very low, gain is high, and access resistances are minimized in modern processes. Since the noise sources are of thermal nature, cooling of FETs improves the low-noise properties even further. HBTs, on the other hand, are limited by the shot noise at the pn junctions, and by the thermal noise of the rather high intrinsic base resistance.

If nonlinear noise is relevant, e.g., in case of low phase noise oscillators, we get another picture. Due to the nonlinearities, the low-frequency $1/f$ noise is upconverted to the carrier frequency and dominates at the relevant range of frequencies close to the carrier. As one origin of $1/f$ noise is generation and recombination, the noise level strongly depends on the quality of the crystal. Epitaxy and process technology are quite mature and provide very good crystal quality today, but in terms of $1/f$ noise it is advisable to keep the current away from hetero interfaces or surfaces that were exposed during device processing. Bipolars are in advantage for these applications as the current flows vertically in the bulk through the hetero interface and only a minor fraction of the current flows at the edges, while the 2D channel of HEMTs lies along the hetero interface. SiGe technologies are usually superior to III–V HBTs.

This chapter addresses the relevant transistor technologies, bipolars first, followed by field-effect devices. From the viewpoint of circuit design, the relevant device physics will be explained and relevant compact transistor modeling concepts. Contemporary transistor models, however, can be quite involved, requiring 200 parameters or more, and are still being improved. Documentation of model mathematics is commonly provided with the models. Thus, this chapter aims to explain the physics that are relevant in contemporary microwave transistors, and how these can be translated into an equivalent circuit-based model, i.e., to provide the basis that is frequently omitted in model documentation.

Before the chapter introduces the different types of transistors individually, the transistor selection criteria as discussed earlier shall be illustrated by a few more detailed application examples: low-noise and power amplification as well as oscillator design. In the examples, we also discuss the steps to be taken prior to circuit design: choosing a technology, and then determining the optimal operation conditions based on information as provided by a data sheet or process development kit (PDK) documentation.

4.1.1 Example: Selecting Transistor and Bias for Low-noise Amplification

The low-noise amplifier (LNA) is usually the first active component in a receiver system. The idea is to take advantage of Friis' formula that allows for calculation of the linear noise factor of cascaded two ports:

$$F_{1,2} = F_1 + \frac{F_2 - 1}{G_1} \tag{4.1}$$

with the noise factors of the first, second and cascaded two ports. F_1, F_2, and $F_{1,2}$, respectively, and the linear gain of the first two-port, G_1. Obviously, if the gain of the LNA is very high, one could neglect the second term on the right-hand side, and only the noise of the LNA itself determines the noise figure of the receiver system.

We need the transistor that provides minimum noise figure and maximum small-signal gain at the same time. As mentioned before, HEMT devices outperform bipolars in terms of white noise. While the channel of HEMTs consists of an electron gas that exhibits very low thermal noise, bipolars suffer from the shot noise source at the base-emitter junction and from the thermal noise of the intrinsic base resistance. These two noise sources in bipolars are, unfortunately, located at the transistor's input terminal, and dominate the noise characteristics of these devices. A comparison of different transistor technologies is shown in Figure 4.1. In general, the noise as a function of frequency is characterized by the DC limit, which is lower for the HEMT devices if comparably complex technologies are compared. The slope, on the other hand, is determined by the technology's cutoff frequency. Let's assume a careful selection leads to the choice of an Avago ATF34143 low-noise pseudomorphic HEMT packaged in a surface-mount package. The device is chosen for the example partly for its performance and partly for the quality of its data sheet.

The first look should consider the maximum ratings of the device. The data sheet tells us that V_{DS} should not exceed 5.5 V, and for HEMT devices, the maximum I_{DS} is reached for $V_{GS} = 0$ V, a current that is usually called I_{DSS}. According to the data

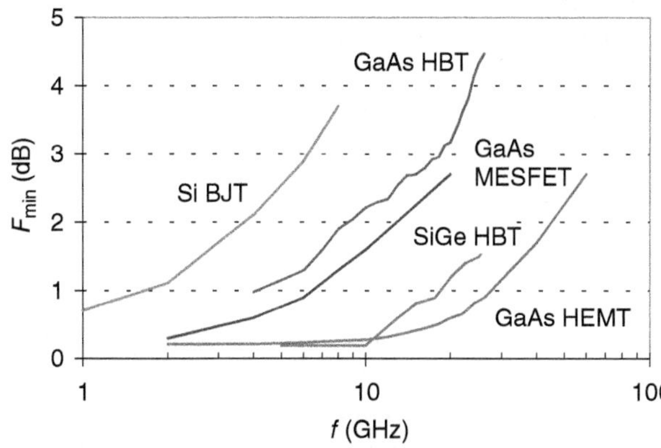

Figure 4.1 Typical microwave noise properties of different transistor technologies.

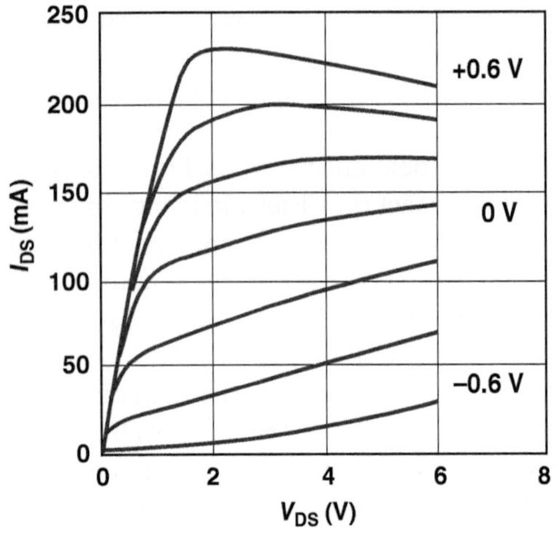

Figure 4.2 Output I–V curves of packaged Avago ATF34143 low-noise pseudomorphic HEMT (from data sheet).

sheet, we should therefore stay within $-5\text{ V} < V_{GS} < 0\text{ V}$ and $0\text{ V} < V_{DS} < 5.5\text{ V}$. The data sheet gives the measured output I–V curves (Figure 4.2), even exceeding the safe operating area. The safe operating area doesn't mean that the device will face immediate destruction if one of the parameters is outside. But it can be assumed that the overall lifetime of the device is reduced, thus it is better to stay within the limits. This step is crucial, and it is important to memorize the safe operating area's boundaries throughout the design process: The circuit simulator will almost certainly not warn you if you leave it. Also in experiment, it might seem like optimum performance is achieved toward or beyond the safe operation area, with the device working well, at least for the duration of the measurement.

The output I–V curve is a good starting point for the bias selection. We need the device to operate in the saturation area, not too close to the linear area. If the linear area is in the range $V_{DS} < 2$ V in our case, we could consider operating the HEMT at $V_{DS} = 3$ V or higher. Which current to choose? A general rule of thumb suggests to set the current I_{DS} to 10% of I_{DSS}. In our example, we would end up with about 12 mA, which needs to be verified. The rationale behind this first suggestion of the bias point is finding an optimum. We need the transistor to provide low noise and high gain. It needs to be operated in the saturation region drawing a significant drain current. But on the other hand, our LNA should not consume too much power. As it is a small-signal amplifier, consider all power $P = I_{DS} \cdot V_{DS}$ to be dissipated, and minimum power consumption is imperative in order to minimize the power requirements of the entire system.

In order to determine the exact bias point, we need more than the output I–V plot. Noise and gain in the chosen bias area need to be inspected. The data sheet provides us with a number of measurement curves, and I pick the values determined for 2 GHz (Figure 4.3a). For the reasons discussed already, we are only shown the noise in the range between $V_{DS} = 3$ V and 4 V. The first observation is that the noise at 3 V seems to be slightly better than for 4 V. This is reasonable, as the power dissipation at 4 V is higher, and a higher channel temperature directly translates into noise figure. However, it is a slight difference. Before we start optimizing on this level, we should ask for device parameter spread. Luckily, the data sheet provides the information, e.g., for the bias point $V_{DS} = 4$ V, $I_{DS} = 60$ mA (Figure 4.3b). The graph shows the statistic distribution of the noise figures that were measured at a total of 450 samples from 9 wafers. Given the device spread, the advantage of 3 V over 4 V looks less pronounced.

Concerning I_{DS}, Figure 4.3a shows the typical behavior of a HEMT. It starts with a steep slope from $I_{DS} = 0$, which is not really surprising as the device would block the input signal at $I_{DS} = 0$ while still exhibiting thermal noise to the output, and the noise figure would approach infinity. With increasing current, the HEMT starts operating and the noise figure drops. Toward high I_{DS}, a slight slope back toward higher noise figures is observed. Again, more power dissipation results in higher noise. From this graph, it seems wise to choose I_{DS} in the valley of noise figures, between 30 and 60 mA, which is more than the 10% I_{DSS} that are suggested by the rule of thumb.

What remains to be considered is the gain that we could expect from our device when it is matched for minimum noise. The graph in the data sheet shows that gain and noise are somehow correlated, Figure 4.4, but gain is less affected by temperature than the noise figure. Considering gain, it would be safe to choose 40 mA or beyond.

At this stage, the transistor and a bias point are chosen. The device alone would provide about $NF = 0.5$ dB and $G_{assoc} = 18$ dB. What performance could be expected from the LNA in the best case? From the data sheet, we learn that the production test board that Avago used for the measurement provided an input matching providing an optimum between

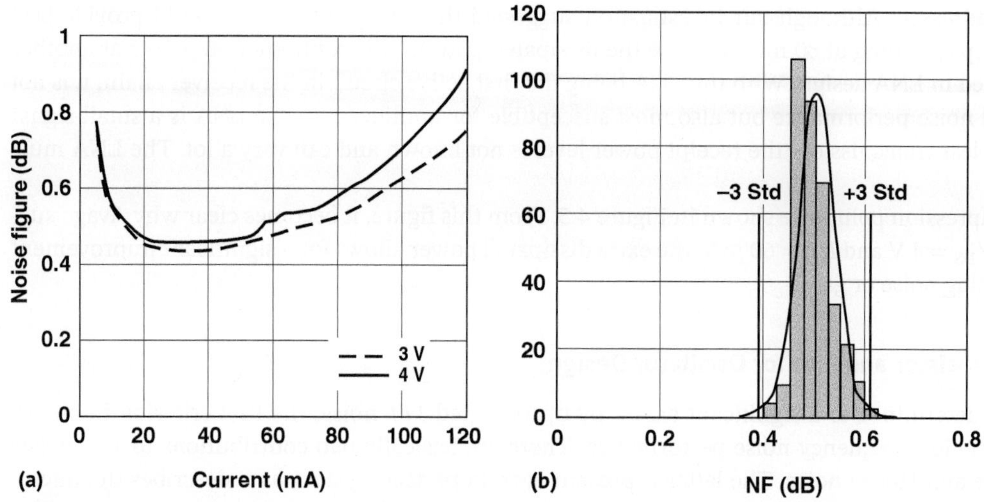

Figure 4.3 Noise figure of packaged Avago ATF34143 low-noise pseudomorphic HEMT at 2 GHz (a), and statistical distribution (b) of noise figure measured at $V_{DS} = 4$ V, $I_{DS} = 60$ mA at 450 samples from 9 wafers (from data sheet).

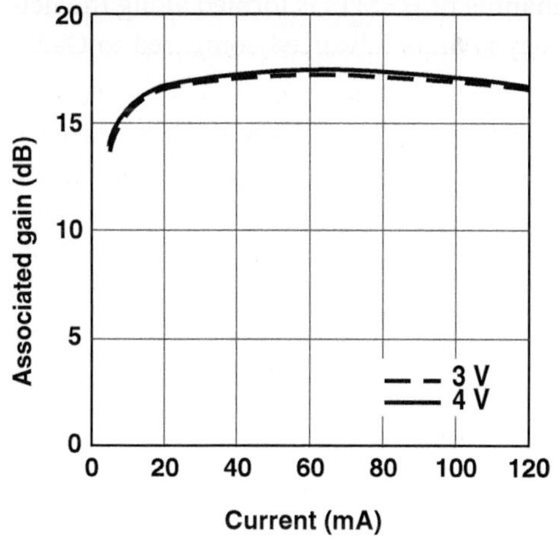

Figure 4.4 Associated gain of packaged Avago ATF34143 low-noise pseudomorphic HEMT at 2 GHz (from data sheet).

power and noise match, and additionally lines with bias supplies. Losses at the input of the device add up to 0.9 dB in this circuit. Probably an optimized circuit could be a bit better, but let's consider 0.9 dB loss for the moment. Since the input circuit is lossy and passive, a gain of −0.9 dB directly translates into a noise figure of 0.9 dB, since the signal is attenuated, but the thermal noise available at the output is determined only by Boltzmann constant and temperature. If source and input circuit are assumed to be at 290 K, the signal is attenuated by 0.9 dB, but the noise is unchanged, which reduces the SNR exactly by the attenuation. The linear gain would be $G_1 = 0.81$, the linear noise factor $F_1 = 1.23$. The linear noise factor of the matched transistor is $F_2 = 1.12$. Applying Friis' formula yields

$$F_{1,2} = 1.23 + \frac{1.12 - 1}{0.81} = 1.38 \tag{4.2}$$

or $NF = 1.4$ dB, which shows that 0.9 dB in terms of gain and noise figure would be lost due to the matching. Optimizing the noise matching circuit is imperative, since every tenth of a dB is visible in the total noise figure of the LNA. On the other hand, one should consider the question of noise matching already upon selecting the device. When dealing with packaged devices as in our case, it is highly likely that the company already did its best to produce transistors that are relatively easy to match. But in an MMIC process, the designer can often influence device size, like the number of gate fingers, or total gate width. A careful search for the optimum device considering noise, gain, dissipated power, and prospective effort in noise and gain matching will be required in order to obtain an optimum LNA performance.

One last issue needs to be addressed. Although our investigation suggested that an I_{DS} of 30 mA would provide best performance, the data sheet keeps pointing at 60 mA, at twice the dissipated power. This contradiction points at another constraint that is to be considered in LNA design. With the LNA being the first active circuit in the receiver chain, it is not only responsible for the overall noise performance but also most susceptible for nonlinearity. The LNA is a small-signal amplifier, but especially in wireless transmission, the receipt power level is not known and can vary a lot. The LNA must be absolutely linear.

The output IP3 and 1-dB compression points are shown in Figure 4.5. From this figure, it becomes clear why Avago suggests to select the bias point at $V_{DS} = 4$ V and $I_{DS} = 60$ mA: the extra dissipated power allows for a significant improvement in linearity, without compromising noise or gain.

4.1.2 Example: Selecting Transistor and Bias for Oscillator Design

As the noise of an oscillator is determined to a significant extent by upconverted $1/f$ noise, the first criterion in choosing a transistor technology is the low-frequency noise performance. There are basically two contributions to $1/f$ noise: generation-recombination noise and Hooge noise. The latter is present even in perfect crystals, as it describes the fluctuation in carrier mobility. It is usually shadowed by the much higher generation-recombination noise that is promoted by crystal imperfections that are always present at surfaces and interfaces, but their density is reduced as technology improves. Figure 4.6 compares typical low-frequency noise performances of different technologies. Bipolar transistors are superior to HEMTs, since in the HBT current flows through the bulk crystal while the channel of HEMTs is formed along the hetero interface. SiGe HBTs are superior to GaAs HBTs, since the silicon technology is more advanced compared to GaAs.

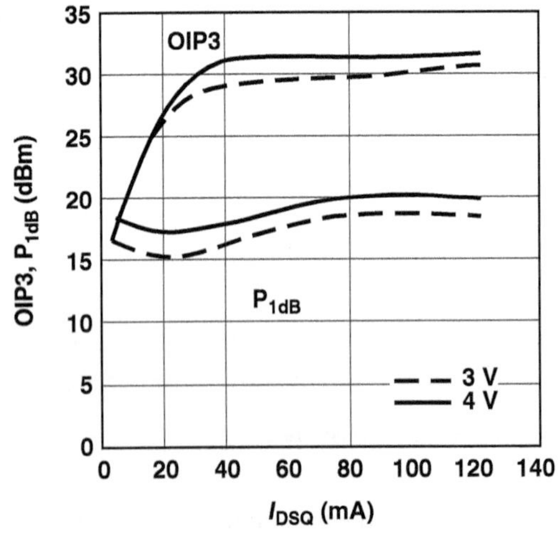

Figure 4.5 Output IP3 and P_{1dB} of packaged Avago ATF34143 low-noise pseudomorphic HEMT that is matched for best noise performance at 2 GHz (from data sheet).

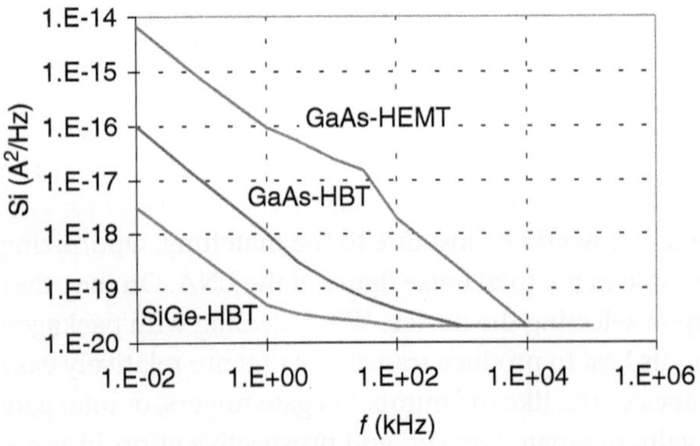

Figure 4.6 Typical $1/f$ noise properties of different transistor technologies.

The differences in the noise corner frequencies, i.e., the frequency where the noise reaches the white noise level, can be significant, as seen in the figure. For high-performance oscillators, we would therefore select a SiGe HBT device.

But let's consider an InGaP/GaAs HBT in this example, since we have good data available [4].

The importance of the transistor noise properties can be demonstrated considering Leeson's formula describing oscillator phase noise. It is observed that the 1-Hz bandwidth f_m noise sideband relative to the carrier power in dBc, $\mathscr{L}(f_m)$, is proportional to:

$$\mathscr{L}(f_m) \propto 10 \log\left\{\left(1 + \frac{f_c}{f_m}\right)\frac{FkT}{2P_{\text{sav}}}\right\} \tag{4.3}$$

with the offset frequency f_m, the device's $1/f$ noise corner frequency f_c, the linear noise factor under large-signal excitation F, the oscillator output power P_{sav}, Boltzmann constant k and temperature T.

The first thing one would do now is, of course, try to minimize f_c/f_m. As the application commonly dictates a certain f_m, the only freedom lies in the choice of a transistor with low $1/f$ noise corner frequency.

The next step would be to select a suitable bias point. As described for the case of the LNA in the previous section, the first consideration is to determine the bias range that provides good gain and is sufficiently far away from the maximum ratings as outlined in the data sheet.

The Leeson formula suggests that the phase noise is proportional to the oscillator power. Since the transistor in the oscillator runs in a large-signal amplifier mode, it is to be expected that increasing collector current provides more headroom for the current swing leading to higher output power and lower phase noise. This is indeed the case, as shown in the measurement given in Figure 4.7. This measurement was taken from a 19-GHz InGaP/GaAs HBT oscillator varying the bias point. The advantage of this circuit is the fact, that the resonator quality factors achievable on chip in the GHz range are very low, around Q $\approx$ 20. The oscillator works, but the resonator is not able to limit the upconverted transistor noise since its filter curve is too broad. Which is good for us, as we intend to see the impact of the transistor.

Increasing the drain current from 50 to 75 mA provided an improvement of about 15 dBc in phase noise. One important effect is indeed that the white noise floor, defined by the noise factor F, is assumed to be more or less independent of the increase in power. It is a noise power that adds to the output spectrum. Increasing signal power under these conditions enhances the signal-to-noise ratio. A second look at the graphic reveals that this is probably not the only effect, since the two curves show different frequency dependence. While the curve measured at 50 mA follows a 30 dB/decade slope almost up to an offset frequency of 10 MHz, one can see a transition between the 30 dB/decade slope to a less steep 20 dB/decade slope in case of the 75 mA curve around $\Delta f = 200$ kHz. The interpretation of the different slopes is the following: the white noise and the upconverted $1/f$ noise of the transistor are filtered by the 20 dB/decade roll-off of the resonator filter curve. An upconverted $1/f$ noise is transferred into the 30 dB/decade slope and the white noise into the 20 dB/decade slope.

The noise is not only lower for 75 mA, but the mechanisms governing the generation of the phase noise have changed, too. And obviously, upconversion of the low-frequency noise is less pronounced, which points to the assumption that the device is also less nonlinear when operated at 75 mA.

An important factor of nonlinearity in terms of phase noise is how the phase of S_{21} changes with changes in current. It could be expected that the phase noise is somehow proportional to the current derivative of S_{21} and should show a minimum if the transistor is biased at

$$\frac{\partial \angle S_{21}}{\partial I_c} = 0 \tag{4.4}$$

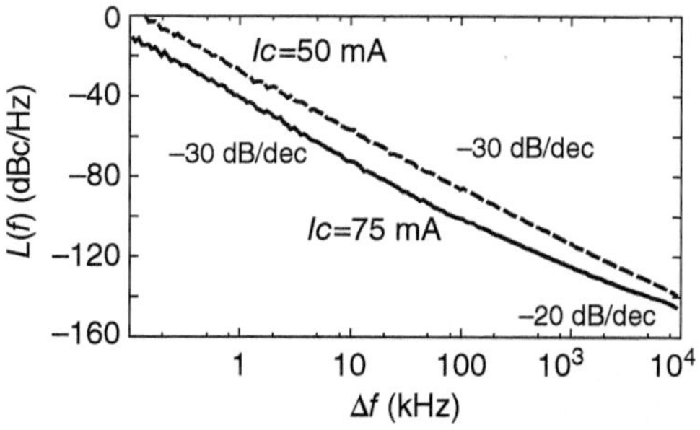

Figure 4.7 Phase noise of a 19-GHz InGaP/GaAs HBT MMIC oscillator as a function of collector current. *Source:* [4]/IEEE.

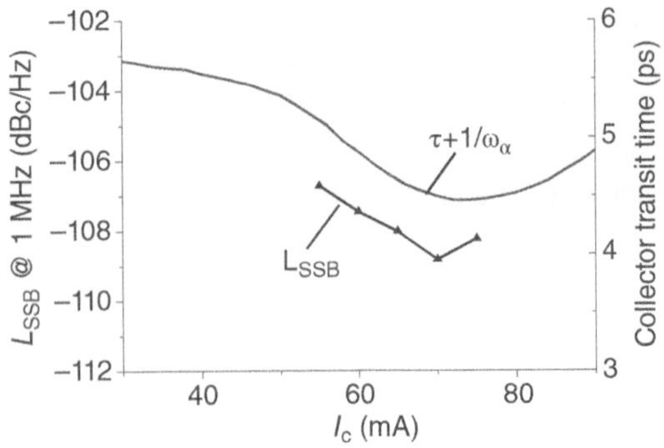

Figure 4.8 Phase noise of a 19-GHz InGaP/GaAs HBT MMIC oscillator as a function of collector current (symbols). The line shows the current dependence of the HBT time delay τ, which is correlated to $\partial\angle S_{21}/\partial I_c$. *Source*: [4]/IEEE.

The measurement was therefore repeated for a number of collector currents; the measured phase noise at 100 kHz offset is shown in Figure 4.8. Although the values scatter a bit, a clear minimum around 75 mA is clearly visible. The line in the graph indicates an estimated shape of the phase noise curve, assuming the changes in $\partial\angle S_{21}/\partial I_c$ and output power.

The phase of S_{21} of an HBT is linked to its internal transit time, which is also inversely proportional to the transit frequency of the device. The condition $\partial\angle S_{21}/\partial I_c = 0$ translates thereby directly into the condition $\partial f_t/\partial I_c = 0$. Graphs of f_t as a function of bias are commonly given in data sheets, and regions with a flat current dependence are easily identified.

4.1.3 Example: Selecting Transistor and Bias for Power Amplification

In a power amplifier, one needs to draw the maximum RF power from a transistor. The peaks of current and voltage should be the maximum that is possible, and the swing should reach its minimum at zero current or a minimum offset voltage, respectively. Independent of the type of transistor one chooses, the supply voltage needs to be set with a sufficient safety margin toward the device's breakdown voltage. The idealized class-A amplifier would show a sinusoidal voltage between zero voltage and the double of the bias supply voltage. For real-life applications, significant headroom needs to be planned to ensure that a singular voltage peak doesn't destroy the transistor. Real-life PA designs also frequently show peak voltages exceeding 3.6 times the supply voltage depending on the class of operation and harmonics termination. Determining the supply voltage therefore needs some consideration. In practical cases, the bias is often determined by system supply voltages, and the question would be how a certain transistor type could be operated safely.

Figure 4.9 shows a sketch of an output I–V curve of a transistor (dotted lines), together with an ideal class-A loadline (dashed line) and a possible loadline extending beyond the double supply voltage V_{DC} (solid line), in case that the load is purely resistive. A first estimate of the output impedance of the power transistor can be derived from the loadline, as the ratio of voltage and current amplitudes:

$$R_{out} = \frac{V_{max} - V_{off}}{I_{max}} \approx \frac{V_{DC}^2}{2P_{out}} \tag{4.5}$$

with the output power estimated by $P_{out} = 0.5 \cdot \hat{V} \cdot \hat{I} \approx 0.5 \cdot V_{DC} \cdot I_{DC}$, assuming class-A regime and neglecting V_{off}. Thus, in the best case, one can not expect a higher output resistance than $R_{out} = 4.5\ \Omega$ for a 1 W transistor operated from 3 V supply. 10 W would result in 0.45 Ω, which is impossible to match to a 50 Ω load without significant losses in the matching circuit. However, using a transistor operating at 28 V, $R_{out} = 39\ \Omega$ could theoretically be reached. This simplified example highlights the difficulties that result from constraints in supply voltage, and why semiconductor manufacturers are spending huge efforts in wide-bandgap technologies providing high breakdown voltages.

Increasing output power by increasing the transistor size also increases the output capacitances that shunt the theoretical output resistance R_{out} discussed so far. Bipolar transistors benefit also in this respect, since the current density per transistor area can be much higher than for HEMT devices, simply because the current flows vertically through the bulk instead of horizontally in a channel. The advantage of HBTs over FETs in the same technology and power level is therefore reduced space requirement, easier matching, and higher cutoff frequency. Comparing different material systems, GaN allows not only for higher operating voltages but also for higher current densities than other commercial technologies.

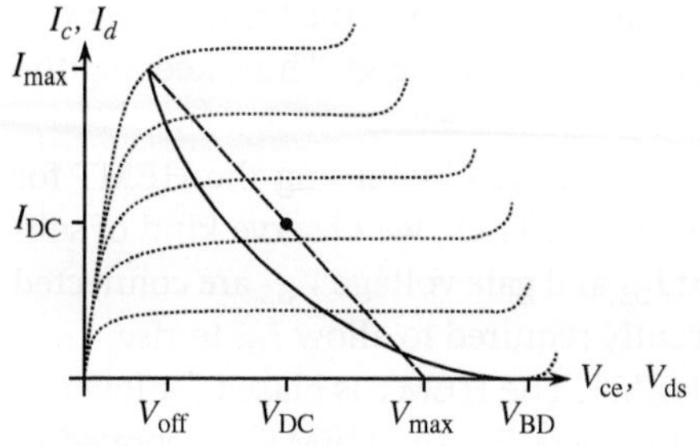

Figure 4.9 Sketch of an output $I-V$ curve of a transistor (dotted lines), together with an ideal loadline for class-A operation (dashed line) with DC bias point (dot), and typical realistic loadline showing higher peak voltage (solid line).

Regarding the optimum load impedance, we can conclude that it is determined by large-signal quantities like maximum voltage and current and by the effective output capacitance that is observed in the large-signal operation. The output capacitance will certainly be in the same order of magnitude for the large-signal and small-signal regimes, but small-signal match to S_{22}^* is no more than a rough estimate of the optimum load for the large-signal operation.

Self-heating is significant in power amplifiers. At low absolute power levels, a transistor might just heat up, which is to be considered in circuit simulation. But how much heat gets dissipated, and if the device requires heat sinking, or if it heats up other parts of the electronics, requires consideration. Again, GaN provides superior thermal conductivity, especially if it is realized on SiC wafers, while GaAs-based devices suffer from the poor thermal conductivity of the substrates.

The maximum thermal load is commonly translated into a maximum dissipated power level in the data sheets. If the gain of the amplifier is sufficiently high so that the input power can be neglected with respect to the output power P_{out}, the dissipated power P_{diss} can be determined by

$$P_{diss} = P_{DC} - P_{out} = (1 - \eta) \cdot P_{DC} \tag{4.6}$$

with the DC power P_{DC} and the amplifier efficiency η. Considering the linear class-A amplifier again, what do we expect in terms of η, or, with other words, which relation do we expect for P_{diss} and P_{DC}?

The classical textbook number for η in class-A operation is a maximum of 0.5, which is reached when the amplitudes of the RF voltage and current, $\hat{V}$ and $\hat{I}$, equal the DC voltage and current, V_{DC} and I_{DC}. In this case, we obtain $P_{DC} = V_{DC} \cdot I_{DC}$, and $P_{out} = 0.5 \cdot \hat{V} \cdot \hat{I}$. Therefore, 50% of the power drawn from the bias supply is converted to heat, while the other half is converted into RF output power. It is important to keep in mind that these 50% are only reached for maximum output power for the linear class-A power amplifier. But in the real world, the amplifier is expected to work for output powers between zero and maximum power. Now, if the DC bias point is kept constant, η can change between $\eta = 0$ for no output power up to $\eta = 0.5$ for maximum output power. In worst case, the whole power drawn from the bias supply is converted into heat.

With other words, the transistor at fixed bias point would become cooler with increasing output power, and one would need to reduce the P_{DC} proportionally to the RF power in order not to burn the transistor. Commonly, high-power transistors need to be operated at rather high efficiencies in order to provide the maximum output power level while staying below the maximum operating temperature.

Quiescent bias for high-power transistors is therefore usually at a rather low collector or drain current, and the biasing scheme allows the transistor to draw more DC current as output power increases.

4.1.3.1 Biasing HEMTs

The HEMT requires two bias voltages, at the gate and at the drain, if source is grounded. In normally-on transistors, one always needs to set the gate bias first in order not to blow the device or any other part of the circuit when switching on. Drain bias is usually provided through a choke inductor that presents low resistance to the DC current but blocks RF in combination with a large capacitance providing a short to the RF behind the choke. On the gate side, we only need to provide the DC voltage, no current. An easy solution is to feed the DC voltage over a high-ohmic resistor that decouples the RF path from the DC network.

In power amplifiers, one needs to consider dissipated power, which depends mainly on the DC power drawn from the bias supply minus the RF power delivered to the load. For example, if a HEMT is properly biased to deliver 100 W of output power to the load with an efficiency of 50%, it might heat up internally to 120 °C, which would be within the specifications. The dissipated power in this case is also 100 W, and the DC bias provides a total of 200 W to the transistor. If the ambient temperature is about 20 °C, our device shows self-heating by 1 °C/W. If the RF power is switched off and the DC bias remains constant, all 200 W are dissipated in the device that now heats up to 220 °C and probably is damaged. Thus, keeping the same bias independent of output power is not always a good idea.

A simple way to obtain a biasing that dynamically adjusts to output power can be achieved by biasing the HEMT for low drain current. With increasing input power, output power and drain current increase. Thus, we observe kind of self-biasing of the HEMT device despite the fixed gate supply voltage. Since drain current I_{DS} and gate voltage V_{GS} are connected through the transconductance G_m by $I_{DS} = G_m \cdot V_{GS}$, something nonlinear is apparently required to allow I_{DS} to rise.

Figure 4.10 shows a sketch explaining the underlying operation principle of the HEMT. The HEMT is biased for low I_{DS}, by setting the gate bias voltage V_{GS} close to the threshold voltage V_{th}. At low input powers P_{in}, the HEMT is operated in class-A, as indicated by [1]. The DC drain current I_{DS} and the RF drain current i_{ds} are given by $I_{DS} = G_m \cdot V_{GS}$, indicated by the dotted lines, and by $i_{ds} = g_m \cdot v_{gs}$, following the sinusoidal gate voltage. As long as the gate voltage swing is moderate, output power P_{out} will increase while I_{DS} remains constant.

At higher input powers, indicated by [2] in the figure, the negative voltage swing eventually exceeds the threshold voltage V_{th}, so that i_{ds} approximately equals $g_m \cdot v_{gs}$ for $v_{gs} > V_{th}$, and remains at $i_{ds} = 0$ otherwise. The waveform of i_{ds} thus resembles a cropped sinusoidal shape. Cropping the negative currents effectively shifts the average toward higher current values, which is observed as a rising DC drain current I_{DS}, indicated by the dashed line. The HEMT is, therefore, biased in class-AB to class-B, and output power and DC drain current eventually saturate once the loadline reaches into the linear operation area that limits maximum i_{ds} as indicated by condition [3] in the graph.

4.1.3.2 Biasing HBTs

When defining the biasing circuitry of an HBT, one needs to consider its implication on breakdown and thermal stability on top of the self-biasing capability. Especially in GaAs devices that generally show a higher thermal resistance than Si-based devices, the wrong biasing scheme can lead to thermal runaway and destruction of the transistor. The reason is that the base voltage V_{be} decreases for a fixed base current I_b and increasing temperature. Biasing with a constant voltage V_{be}, similar to the common bias scheme for HEMTs, would allow the current to rise with temperature. If self-heating is significant, thermal runaway may be observed, where an increase of current leads to an increase in temperature and vice versa. Therefore, output I–V curves are commonly measured by setting collector-emitter voltage and forcing a base current.

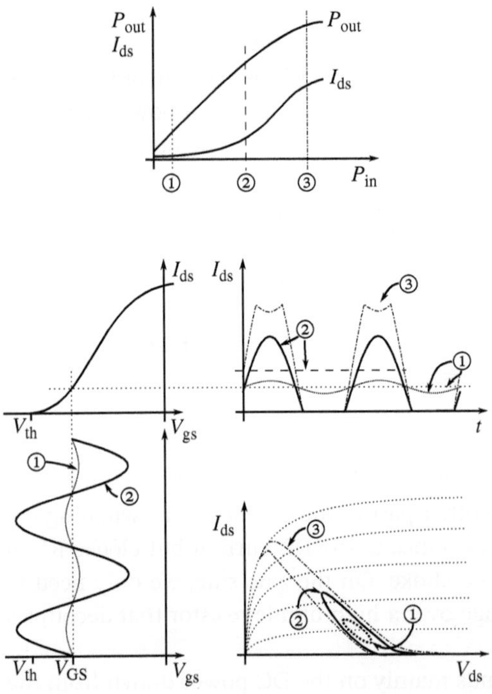

Figure 4.10 Self-biasing of a HEMT device biased for low power dissipation in case of low output powers.

Biasing the base by a current source, on the other hand, prevents the device from shifting its bias point with output power. In the forward-active operation region, the collector current I_c depends on the base current I_b by the current gain β, that ideally should, for the moment, be considered a constant value. Thereby, $I_c = \beta \cdot I_b$, and if one fixes the DC base current I_B to a certain value, also the DC collector current is fixed. The RF currents will deviate from the perfect sine shape, but the average value will stay constant at base and collector terminals. One can add a degree of freedom to the transistor by biasing with the potentially dangerous constant voltage bias, allowing the HBT to show a self-biasing behavior equivalent to that of the HEMT.

Figure 4.11 illustrates the impact of the choice of a DC source on the power behavior of an HBT [5]. Let's first only discuss the measured and modeled data up to 15 dBm of input power. The solid lines and bullets denote four different measurements with different quiescent bias points. Figure 4.11b,d shows the collector and base currents, which are constant over the power range up to 15 dBm. Output power increases slowly with input power (Figure 4.11a). The base voltage, shown in Figure 4.11c, decreases, since the voltage swing increases with input power, but the diode characteristics of the base-emitter junction prevent the voltage to exceed the built-in voltage of about 1.4 V. The higher voltage swings at the base therefore lead to higher negative voltage peaks, resulting in a shift of the average voltage toward negative voltages.

The dashed lines in the figure and the triangle symbols show simulation and measurement with constant voltage bias, V_{be} stays constant in Figure 4.11c. The device shows a typical self-biasing behavior, as increasing RF input voltage swings result in higher base-current peaks and in turn, yield a higher DC component of base and collector currents (Figure 4.11d,b). The output power follows this trend and increases fast, with a much steeper slope than in the case of the fixed base current.

For the range of input powers beyond 15 dBm, a kind of self-biasing is observed that contradicts the theory presented so far. The collector current definitely shows self-biasing behavior together with the characteristic increase in output power. The base current, on the other hand, still remains constant while the negative trend in the base voltage with input powers

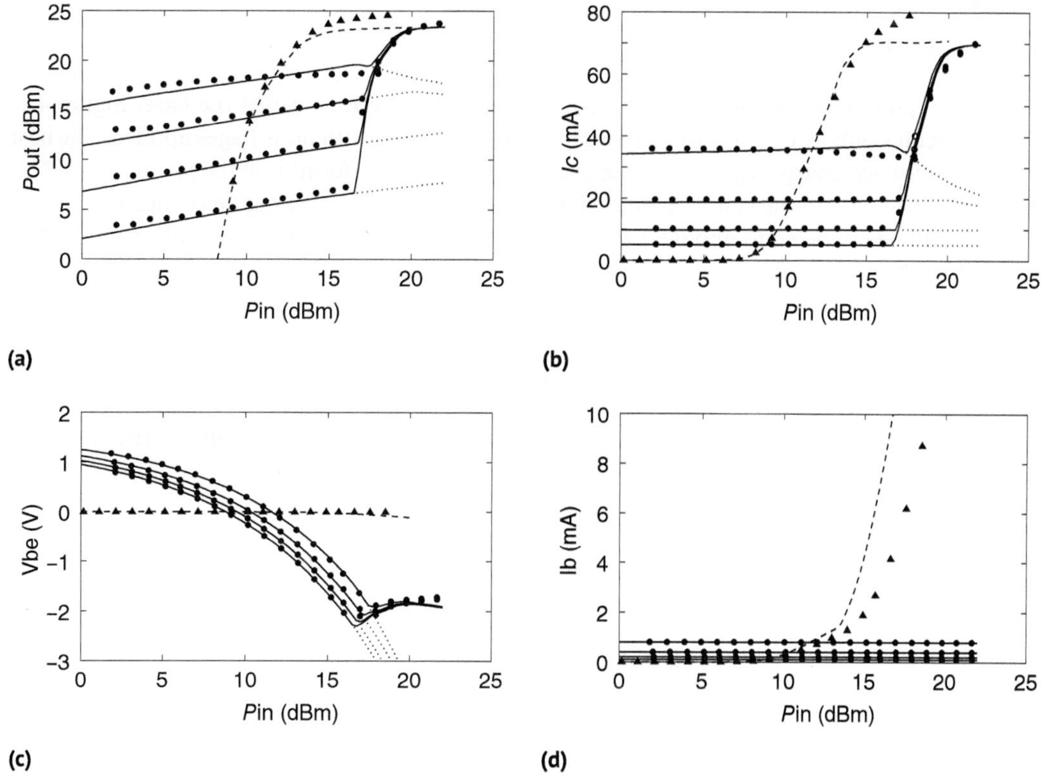

Figure 4.11 Bias and output power of unit cell HBT (emitter area $3 \times 30\ \mu\text{m}^2$) at 2 GHz, $V_{\text{CE}} = 5$ V, measured (symbols) and simulated (lines) as function of available input power. Constant V_{BE} (- -), constant I_B (—), constant I_B simulated without base-emitter breakdown (···). *Source:* [5]/IEEE.

is stopped. Especially the decoupling of base and collector current is unexpected, and also the sharp kink in the curves indicates a qualitative change in the transistor behavior. In fact, what happens in this power ranges is that the negative voltage swing at the base reaches the base-emitter breakdown voltage. The breakdown causes a negative base current allowing for higher forward currents while maintaining the DC current.

Driving the HBT into breakdown should be avoided, therefore it is not advisable to provide the base current through a perfect current source if the transistor is driven into saturation as in our example. On the other hand, a perfect voltage source is also not advisable since thermal runaway could occur. Providing DC through a source with a limited output impedance solves the issue, as it allows the base current to shift with power and thereby allows for self-heating. The voltage feedback over the source impedance, on the other hand, yields negative feedback if the base current increases due to self-heating and can prevent thermal runaway.

The question of the source impedance is not limited to the DC path. It will be shown later, that bipolar transistors provide two different kinds of collector breakdown voltages, with BV_{CE0} being much lower than BV_{BC0}. While BV_{BC0} corresponds to the condition that the base-emitter junction is short-circuited, BV_{BC0} is determined for an open-circuit at the base terminal. These two extremes again correspond to the two extremes of negligible source resistance and infinite source resistance. Selecting a rather low impedance RF source therefore allows the HBT to operate at voltages well beyond BV_{CE0} observed in the traditional output I–V curves.

The source impedances presented to the input terminal of the HBT should therefore be chosen carefully, considering self-biasing on the DC path. Also the thermal stabilization through the feedback of the source impedance concerns the DC path rather than the RF path since self-heating is determined by rather long time constants. Providing the RF signal by a low-impedance source, on the other hand, can improve breakdown voltages on the collector side.

4.2 Heterojunction Bipolar Transistors

4.2.1 A Brief Review of BJT and HBT Physics

Bipolar transistors are basically two pn diodes back-to-back, sharing one common electrode called the base. Figure 4.12 shows the band diagram and the circuit symbol of a bipolar transistor in equilibrium, i.e., with no voltages applied. The base region is supposed to be very thin, and space-charge regions (denoted SCR in the figure) are forming at the pn junctions. This device is an npn transistor, as a p-doped base in the center is sandwiched between two n-doped regions called the emitter and the collector. This derivation will be restricted to npn devices, as pnp devices are operating in an analog fashion. The figure also shows the relevant voltages and currents. We will consider the emitter potential as reference potential (ground), as it simplifies the discussion slightly. Operation cases, where the emitter is not grounded or when the emitter current I_e is of interest, can be derived applying Kirchhoff's current law.

If a base-emitter voltage below the diode's diffusion voltage V_D is applied, it is expected to be in reverse operation so that no current flows between the base and emitter. If a positive collector-emitter voltage V_{CE} is applied, the base-collector junction is forced into reverse operation as well, since $V_{BC} = V_{BE} - V_{CE} \leq V_D$ (see Figure 4.13). In this condition, the transistor

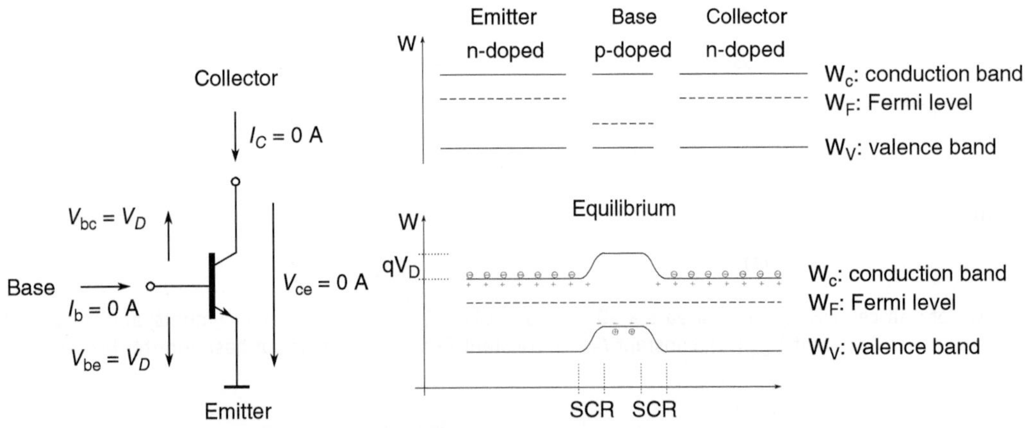

Figure 4.12 Band diagram and circuit symbol of an npn transistor in equilibrium.

acts as a closed switch, where the base potential controls the collector current, forcing it to zero. The maximum collector voltage is limited by breakdown, which might be lower than the breakdown voltage expected for the base-collector diode alone. Breakdown depends on external circuitry, especially on the impedance seen from the base terminal, which will be discussed in the respective section. The equivalent circuit of the transistor in off state in the ideal case reduces to the junction capacitances.

If both pn junctions are biased in forward direction, with $V_{\mathrm{be}} \geq V_D$ and $V_{\mathrm{bc}} \geq V_D$, a base current I_b flows into the device. This state is depicted in Figure 4.14, and it can be seen that electrons are able to diffuse into the base from collector and emitter, since the barrier at the pn junctions is reduced due to the applied voltage. For a fully symmetrical case, with $V_{\mathrm{be}} = V_{\mathrm{bc}}$, $V_{\mathrm{ce}} = 0$, the base current will split and leave through the collector and emitter terminals, so that $I_b = I_c + I_e$ holds. Of higher technical relevance is the case, where both junctions are still forward biased, but a low collector-emitter voltage $V_{\mathrm{ce}} > 0\,\mathrm{V}$ is applied. Now, a higher number of electrons enter the base from the emitter than from the collector, as $V_{\mathrm{be}} > V_{\mathrm{bc}}$ holds. The majority of the electrons directly diffuses toward the collector terminal, so that the collector-emitter current I_{ce} is a multiple of the base current I_b. In this state, the transistor can be represented by its low-ohmic pn-junction resistances and can be understood as an open switch. The switch is controlled through the voltage $V_{\mathrm{be}} \geq V_D$, which can also be ensured forcing $I_b > 0$ if the emitter is set to ground as in our case. The transistor leaves the saturation region once V_{ce} gets too high, so that $V_{\mathrm{bc}} = V_{\mathrm{be}} - V_{\mathrm{ce}} < V_D$. This operation is therefore theoretically limited to very low values of V_{ce}, as expected from a perfect switch. In reality, however, the main resistance limiting I_c in the saturation region is the extrinsic collector resistance R_C, so that resistive current–voltage characteristics are observed, as expected from a real switch. R_C describes all resistive contributions of semiconductor and metal between the ideal intrinsic transistor and the externally accessible collector terminal.

If the base-emitter junction is kept with forward bias $V_{\mathrm{be}} > V_D$, while the base-collector junction is reverse-biased, the device operates in the so-called forward-active mode. As depicted in Figure 4.15, electrons now diffuse from emitter into the base, while holes diffuse into opposite direction. Since the base-collector junction is reversely biased, diffusion of majority carriers out of the base into the collector and vice versa is not possible. As indicated in the figure, a significant fraction

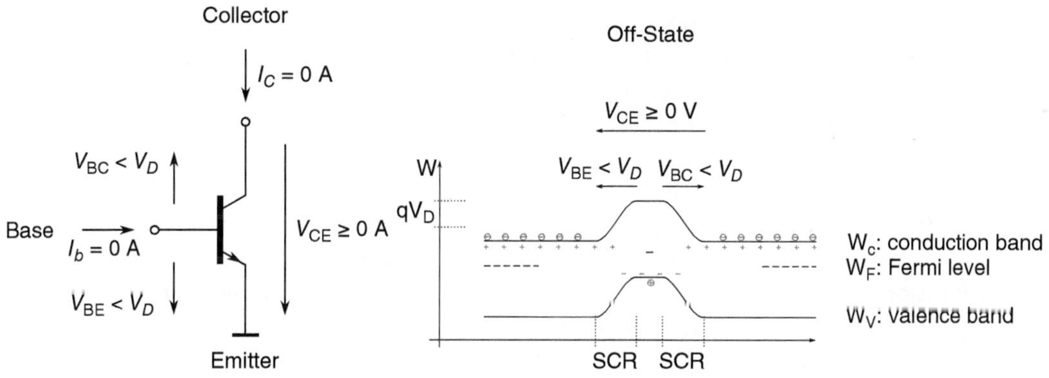

Figure 4.13 Band diagram and circuit symbol of an npn transistor in off state.

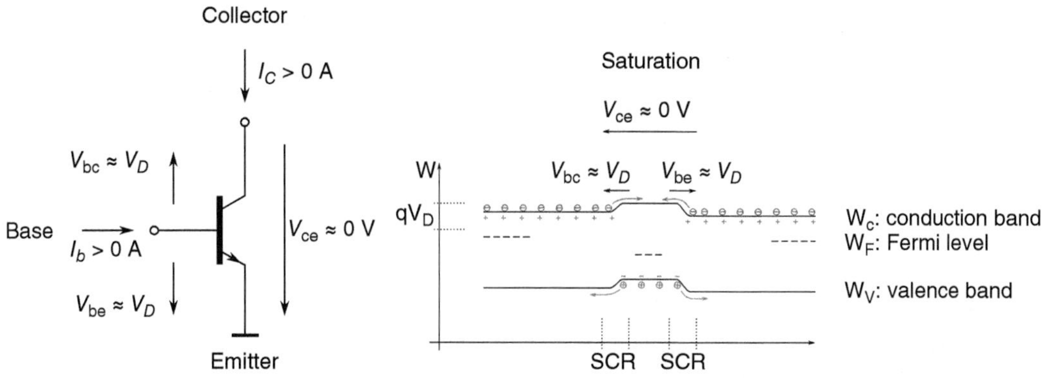

Figure 4.14 Band diagram and circuit symbol of an npn transistor in saturation.

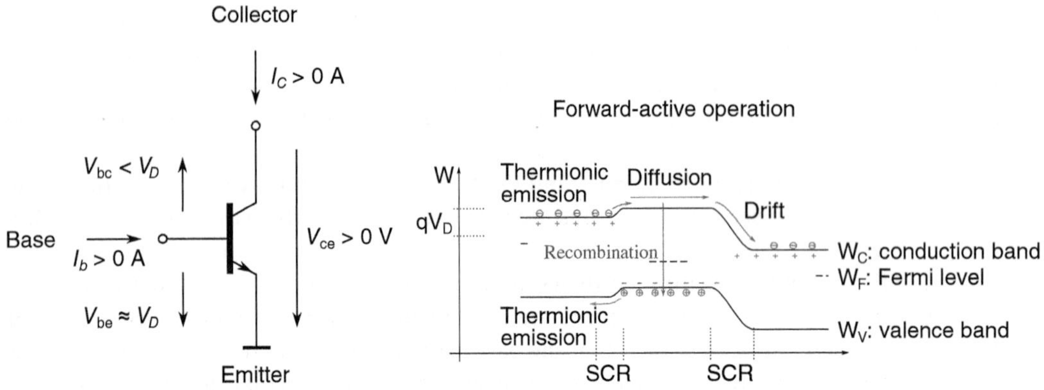

Figure 4.15 Band diagram and circuit symbol of an npn transistor in forward-active operation.

of the electrons injected from the emitter into the base diffuses through it and reaches the base-collector SCR. Once an electron enters the SCR, it is accelerated by the electric field toward the collector, so that these electrons add up to a collector current I_c. Assuming that the electrons diffuse through the base, the drift-diffusion approach can be used to derive the base transport factor α_T from the electron current density at the collector side of the base $J_n(W_B)$ in relation to the electron current at the emitter side of the base, $J_n(0)$. It can be approximated as:

$$\alpha_T = \frac{J_n(W_B)}{J_n(0)} \approx \frac{1}{\cosh(W_B/L_n)} \tag{4.7}$$

Here, W_B denotes the length of the neutral base between the two SCRs, and L_n denotes the diffusion length for electrons in the base. It is therefore imperative to employ a thin base of highest crystal quality, so that L_n is as high as possible and W_B assumes a low value, so that α_T approaches unity.

In the forward-active region, the collector current is thereby controlled by the base-emitter voltage, or base current if the emitter is set to ground. The collector-emitter voltage, on the other hand, is decoupled from the current, so that the transistor operates as a voltage- or current-controlled current source. The transistor can now be used to amplify a signal applied to the base-emitter terminals, to a higher power signal between either the base-collector or collector-emitter terminals, depending on the application.

Concerning this amplification, it is important to note that the current of holes injected into the emitter is unwanted. It is not taking part in the amplification process, but causes unnecessary power dissipation. Therefore, another important quantity is the emitter efficiency γ, which is defined as the ratio of the electron current at the base-emitter junction to the total current. It depends on the diffusion constants D_p, D_n and diffusion lengths L_p, L_n for holes and electrons, respectively, and on the emitter and base doping concentration N_D, N_A, respectively, that control the electron and hole densities in the respective regions:

$$\gamma = \frac{J_n}{J_n + J_p} = \frac{1}{1 + \frac{D_p}{D_n}\frac{L_n}{L_p}\frac{N_A}{N_D}} \tag{4.8}$$

The current gain I_c/I_e, which is also denoted as α and equals the product of emitter efficiency and base transport factor

$$\alpha = \frac{I_c}{I_e} = \gamma \cdot \alpha_T \leq 1 \tag{4.9}$$

is therefore smaller than unity. If the current gain is given as the relation of collector to base currents, $\beta = I_c/I_b$, quite high numbers can be achieved if α approaches unity, since:

$$\beta = \frac{I_c}{I_b} = \frac{\alpha I_e}{I_e - \alpha I_e} = \frac{\alpha}{1-\alpha} \gg 1 \tag{4.10}$$

Before we discuss the DC performance of the bipolar transistor further, limitations of the traditional BJT will be discussed, which led to the development of heterojunction bipolar transistors (HBTs).

4.2.1.1 Limitations of Bipolar Junction Transistor

Since the doping is the only quantity that can be directly controlled, BJTs feature a high emitter doping, while the base is very thin to boost α_T and weakly doped to enhance γ. A thin, lightly doped base, on the other side, has three drawbacks: first, it is of a high resistance, second, the SCR on the side toward the collector can consume a significant portion of the total base width and third, high-current effects set in early. The base layer resistance is of practical importance, as the base layer is used to contact the intrinsic base from the side. The bipolar transistor, as an amplifying element, thereby comes with a high series resistor at the input, which increases noise figure through thermal noise and reduces the maximum frequency of oscillation $f_{\max}$, as will be discussed below when it comes to the RF behavior. The second drawback is the dependence of the neutral base width on the base-collector voltage. As indicated in Figure 4.16, the total base width W_B can be understood to be effectively reduced by a length ΔW_B corresponding to the length of the SCR within the base. This effect causes a feedback of the base-collector voltage on the amplification, as it modulates the distance through which the electrons diffuse, which alters the transport factor: The higher the collector voltage, the shorter the neutral base and the lower the recombination. Consequently, the current gain increases. This behavior is called the Early effect. The third issue with a lightly doped base results from the fact that the ideal transistor behavior requires that minority carriers are in strict sense minority carriers. This condition is easily violated, if a strong electron current is injected into a weakly doped base. With increasing current density, the charge of the electrons adds to the negative charge of the base doping, so that the base can only stay electrically neutral if the number of holes increases by the same amount. As a consequence, emitter efficiency degrades due to the higher density of holes in the base (see Figure 4.16).

These three issues result from the limited degrees of freedom when it comes to the choice of doping and thickness of the base layer. Therefore, the potential of the BJT to be scaled to the microwave range and beyond is rather limited.

4.2.1.2 The Heterojunction Bipolar Transistor

Optimization of bipolar transistors requires the combination of a thin, but highly doped base, with a high emitter efficiency. Ideally, the electrons would not diffuse through the base, which is a rather slow process, but drift. This all comes into reach if the electrons and holes are treated separately through careful engineering of conduction and valence bands. Controlling bands independently is not possible within a single semiconductor, where the bandgap is fixed, but modern epitaxy allows to combine different semiconductor materials within a single crystal.

III–V HBTs (based on GaAs or InP) use an emitter built from a higher bandgap material than the material of the base. Choosing the right materials results in a step ΔE_v in the valence band that equals the difference in the bandgaps (see Figure 4.17). This step has the power to decouple the emitter efficiency γ, which is the ratio of electrons injected into the base J_n to the total junction current $J_n + J_p$, from the ratio of the emitter doping N_D to base doping N_A:

$$\gamma = \frac{J_n}{J_n + J_p} = \frac{1}{1 + \left(\frac{D_p}{D_n}\frac{L_n}{L_p}\frac{N_A}{N_D}\right) e^{-\Delta E_v}} \tag{4.11}$$

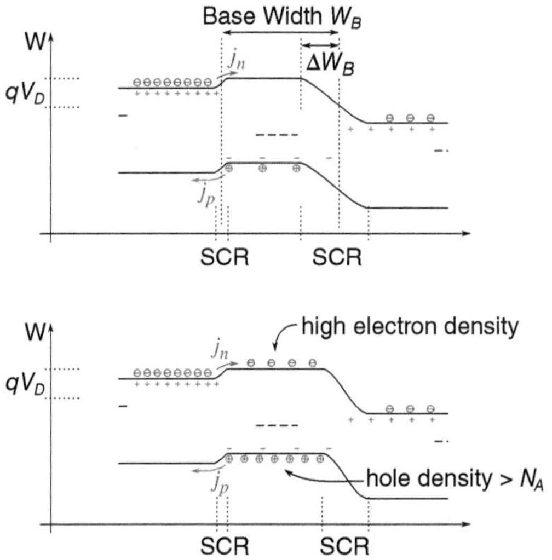

Figure 4.16 Band diagrams illustrating the Early and Webster effects.

with the diffusion constants for electrons D_n and holes D_p, and the respective diffusion length L_n and L_p. Since ΔE_v stands in the exponent, it dominates the denominator, so that the ratio N_A/N_D can practically be chosen freely. The band diagram for the emitter-base junction is depicted in Figure 4.17. Ideally, the bandgap difference between emitter and base materials leads to a high step in the valence band ΔW_V, while the conduction band stays continuous or with a negligible ΔW_C. As shown in the graph, the step in the valence band leads to blocking of holes injection into the emitter, while a possible step in the conduction band results in a spike under forward bias. Such a spike can be crossed by the electrons by tunneling, but should rather be suppressed by proper bandgap engineering.

This heterojunction allows to thin down the base and thereby reducing base transit time. A low base resistance is maintained by increasing the base doping to very high levels. The operation principle was patented by W. Shockley in the 1950s, and thoroughly derived by H. Kroemer [6] in 1957. But it took until the mid-1980s for the epitaxial technology to become able to actually grow these layers.

A special type of HBTs are double heterojunction bipolar transistors (DHBTs). These devices also realize the collector in a wider bandgap material. The benefit is the higher breakdown voltage that is achieved even for thinner collectors, allowing to speed up the device without compromising output powers. On the negative side, the base-collector heterojunction can block the electrons at high emitter currents and low V_{CE}.

HBTs on silicon substrate benefit from the much more advanced features of silicon technology as compared to III–V technologies. Often, these bipolars are integrated with CMOS processes, in so-called BiCMOS technologies. Thereby, CMOS digital power and integration density combine with bipolar RF power on a single chip.

Silicon-based HBTs are realized according to a slightly different concept than their III–V counterparts. Instead of using a higher bandgap material for the emitter, Germanium is introduced into the base in order to lower the bandgap. A typical band diagram of a SiGe HBT is shown in Figure 4.18, where the dotted line denotes the conduction band of a standard Si

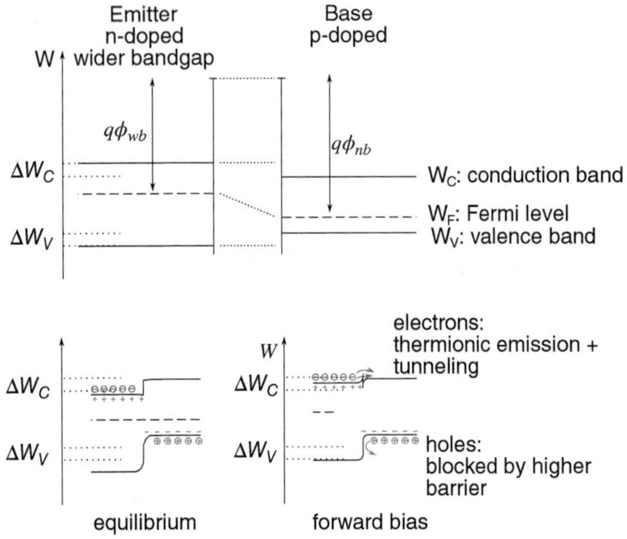

Figure 4.17 Typical band diagram of the emitter-base heterojunction of a III–V HBT.

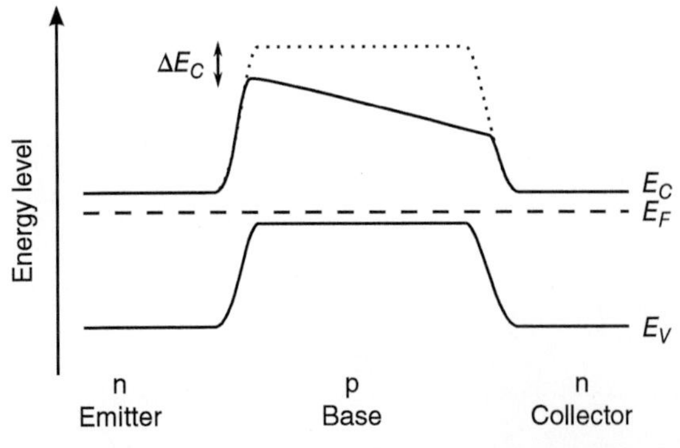

Figure 4.18 Typical band diagram of a SiGe HBT.

BJT. The effect of the narrower bandgap is twofold. First, it reduces the barrier of the conduction band, thereby promoting the injection of electrons into the base, increasing the emitter efficiency and decoupling it from the ratio of emitter and base doping. The second effect is that a variation of the Germanium content realizes a slope in the conduction band. The resulting built-in drift field increases the electron speed and thereby the cutoff frequency. In order to optimize the current gain β, one now has two degrees of freedom. The first one is the step in the valence band at the base-emitter junction $\Delta E_{g,0}$, which enhances emitter efficiency γ_e. In contrast to the III–V HBTs discussed before, lowering the bandgap in the base promotes the injection of electrons into the base, while the step in the valence band remains unaffected. The effect on γ_e, however, is the same in both cases. The second effect is the slope in the conduction band, which can be expressed by the difference of the conduction band offsets in the SiGe base with respect to Si emitter, $\Delta E_{g,0}$, and Si base $\Delta E_{g,W_B}$:

$$\beta_{\mathrm{SiGe}} \propto e^{\Delta E_{g,0}/(kT)} \frac{\Delta E_{g,W_B} - \Delta E_{g,0}}{1 - e^{-\frac{\Delta E_{g,W_B} - \Delta E_{g,0}}{kT}}} \tag{4.12}$$

A schematic doping profile of a SiGe HBT is shown in Figure 4.19. Typical dopants for n-emitter and collector are As and P, while the n-base is highly B-doped. Some processes, such as the IHP BiCMOS process, add carbon to the base layer, which has been proven to prevent the B dopants from diffusing out of the base, which would lead to degradation of the devices. This process is therefore also called a SiGe:C process.

4.2.2 Small-signal Model and Cutoff Frequencies

The small-signal model is derived from the large-signal model by linearizing it for a given bias point, thus replacing current sources by differential resistances and charges by capacitances. A typical bipolar equivalent circuit for the forward-active state in common-emitter configuration is shown in Figure 4.20. It consists of an intrinsic part, which describes the transistor

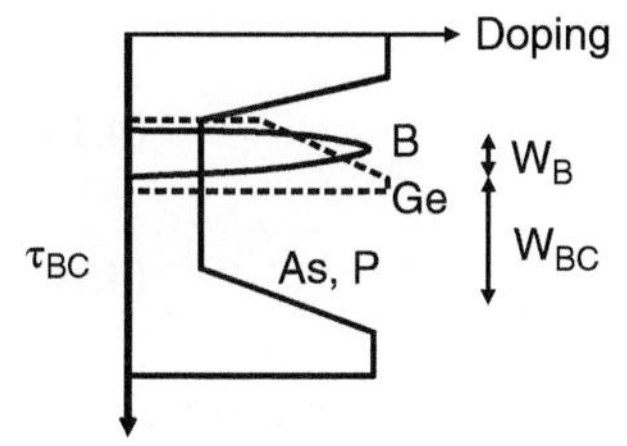

Figure 4.19 Typical doping profile and Ge content distribution of a SiGe HBT. *Source*: Courtesy of HP Microelectronics.

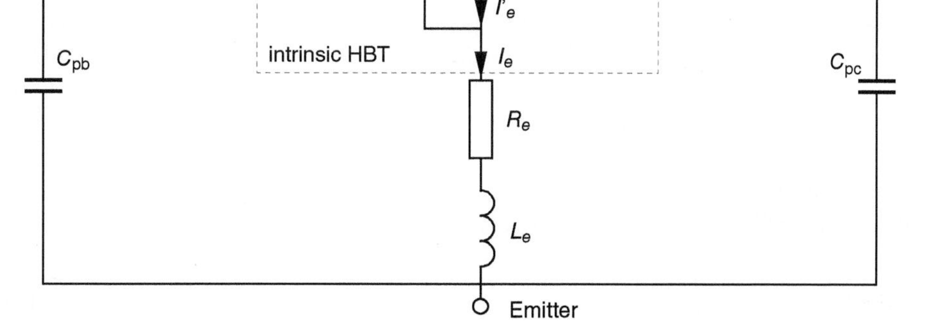

Figure 4.20 Small-signal equivalent circuit of GaAs HBT.

itself, and an extrinsic part, which describes the contact resistances and feed lines from the bond pads to the transistor itself. Since the small-signal model is valid only for a certain bias point, it is not symmetric as the large-signal model.

The small-signal model shown in the figure is a T-topology, thus derived from an injection type model. It would, however, also be possible to turn this circuit into its equivalent π-topology that matches the linearized Gummel–Poon model. The pn-junctions are represented by $R||C$ branches, and no distinction is made between diffusion and depletion capacitances. In case of the base-collector junction, the resistor is omitted since it is very large. This might not hold close to the saturation region. It is also seen that the base-collector capacitance is split into C_{bc} and C_{ex}, and that the base resistance is split into R_b and R_{b2}. This split should account for a topology where the base-collector junction is wider than the active area of the transistor. This equivalent circuit is commonly valid for modern HBTs, too.

The transit frequency f_t is defined as the frequency at which the two-port parameter $h_{21} = i_2/i_1|_{u_2=0}$ reaches unity, usually written in the form of $|h_{21}|^2 = 1$ for $f = f_t$. Translated into the equivalent circuit world, it means that the common-emitter current gain $\beta_F(f)$ becomes equal to unity. For the derivation of f_t, we will consider the injection-type model, as it is closer linked to physics. From Eq. (4.10), we see that $\beta_F(f_t) = 1$ means for the common-base current gain $\alpha_F(f_t) = 1/2$. The transit frequency is the 3-dB cutoff frequency of α_F.

If, as a first-order approximation, the base-collector capacitances C_{bc} and C_{ex} are neglected (the same holds for the much smaller pad capacitances C_{pb} and C_{pc}), we see from Figure 4.20 that f_t is fully described by the low-pass behavior of the current source α:

$$\alpha = \frac{\alpha_0\, e^{-j\omega\tau_\alpha}}{1 + j\omega/\omega_\alpha} \tag{4.13}$$

Two time constants are present in α. The first one is the RC time constant of the base-emitter pn-junction, which is represented in this formula by its equivalent angular frequency ω_α. The other time constant is, as a first approximation, an ideal time delay due to the transit time through the neutral base and the collector SCR. Linearizing this formula through the first term of the respective Taylor series yields:

$$\alpha \approx \frac{\alpha_0}{1 + j2\pi f\,(\tau_\alpha + 1/\omega_\alpha)} \tag{4.14}$$

$$\rightsquigarrow f_t \approx \frac{1}{2\pi(\tau_\alpha + 1/\omega_\alpha)} \tag{4.15}$$

While this approximation is frequently given in textbooks, it dates back to times where the transit time through the thick base layers limited the transistor's RF performance. This assumption is not wise when modern high-speed HBTs are considered featuring base widths below 100 nm, as transit times might be in the same order of magnitude as the RC time constant of the rest of the equivalent circuit. Following the method of open-circuit time constants, we define an integral time constant from the sum of the relevant resistances multiplied with the sum of the relevant capacitances. This second-order approximation of f_t reads:

$$f_t \approx \frac{1}{2\pi\left[\tau + 1/\omega_\alpha + (R_c + R_e + R_{be}) \cdot (C_{\text{bc}} + C_{\text{ex}} + C_{\text{pc}})\right]} \tag{4.16}$$

This formula will be used in the following when discussing simulated values of f_t as compared to the values that were extracted from two-port measurements directly. It also shows the limitations of transistor downscaling, since access resistances R_c and R_e increase as the area is shrunk, and pad capacitances remain constant.

The maximum frequency of oscillation is defined as the frequency, where the maximum available power gain (MAG) reaches unity. Once the MAG is lower than or equal to unity, the device can't possibly be brought to oscillation by feeding the output back to the input. Besides of regarding MAG, also the unilateral gain (U) can be considered. The value of U is always higher than MAG, since the definition of the unilateral gain assumes that, somehow, the internal feedback of the device is compensated. However, the internal feedback is irrelevant when the gain becomes unity, and input and output powers are identical.

The traditional textbook formula for $f_{\max}$ assumes a simplified equivalent circuit, where the base resistances and base-collector capacitances are lumped together: $C_{\text{BC}} = C_{\text{bc}} + C_{\text{ex}}$, $R_B = R_b + R_{b2}$. It reads:

$$f_{\max} \approx \sqrt{\frac{f_t}{8\pi R_B C_{\text{BC}}}} \tag{4.17}$$

The ladder structure of the base-collector capacitance and resistance, however, plays an important role at higher frequencies. Simply adding up the component values will not yield a good estimate of f_{max} for modern HBTs, as the values obtained will be too pessimistic. Neglecting the extrinsic parameters C_{ex} and R_b, on the other hand, will be too optimistic.

In [7], MAG and U were calculated for an equivalent circuit as in Figure 4.20, and the respective f_{max} were estimated taking into account; the definition of f_{max} assumes the frequency slope of a single-pole low-pass. As a result, approximate effective time constants $(R_B C_{BC})_{eff}$ are given:

$$(R_B C_{BC})^U_{eff} = R_{b2}C_{bc} + R_b(C_{ex} + C_{bc}) \quad \text{for Magon's gain } U \tag{4.18}$$

$$(R_B C_{BC})^{MAG}_{eff} = R_{b2}(C_{bc} + C_{ex}/2) + R_b(C_{ex} + C_{bc}) \quad \text{for MAG} \tag{4.19}$$

As stated earlier, $U = \text{MAG} = 0$ dB holds necessarily at a certain frequency. But the definition of the maximum frequency of oscillation assumes a two-port that shows a frequency response like a single-pole low pass, i.e., a slope of −20 dB/decade. Since a transistor's frequency response is a bit more complex than that, neither MAG nor U provides a slope of exactly −20 dB/decade over the whole frequency range. To make it short: U yields the higher value from the extrapolation, and these are the values that are commonly cited.

Increasing cutoff frequencies in bipolar technologies requires conceptionally to move from standard homojunction bipolar transistor to HBT structures, and then to shrink the devices geometrically. Electron traveling times through the bipolar structure need to be minimized, but also parasitic capacitances and resistances are to be reduced. As an example for technological challenges, an approach to reduce the depleted collector width W_C in a SiGe HBT technology shall be given. In the process shown in Figure 4.21, a selectively implanted collector (SIC) is realized in order to minimize W_C. In this region, the collector doping is locally increased in order to restrain the SCR to the thin layer between the base and SIC. The SIC layer itself forms a low-impedance contact. The drawback of a narrow collector depletion region is a reduced breakdown voltage. Breakdown is observed in the output I–V curves shown in Figure 4.22 beyond $V_{CE} = 1.7$ V. While the maximum

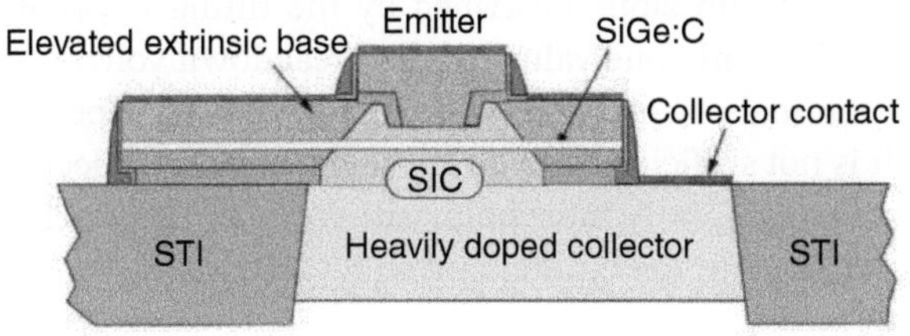

Figure 4.21 Schematic cross-section of a SiGe HBT (STI: shallow trench isolation, SIC: selectively implanted collector). *Source*: Courtesy of IHP Microelectronics.

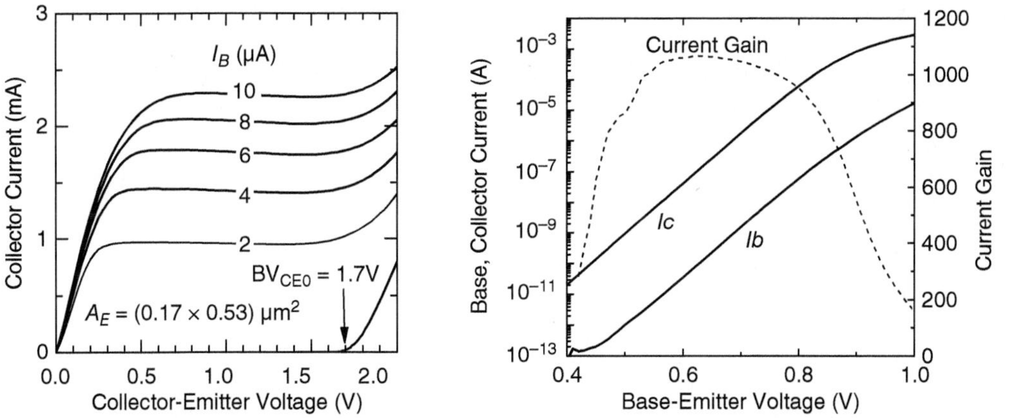

Figure 4.22 DC characteristics of a 0.17×0.53 μm² SiGe HBT from the IHP technology. *Source*: Courtesy of IHP Microelectronics.

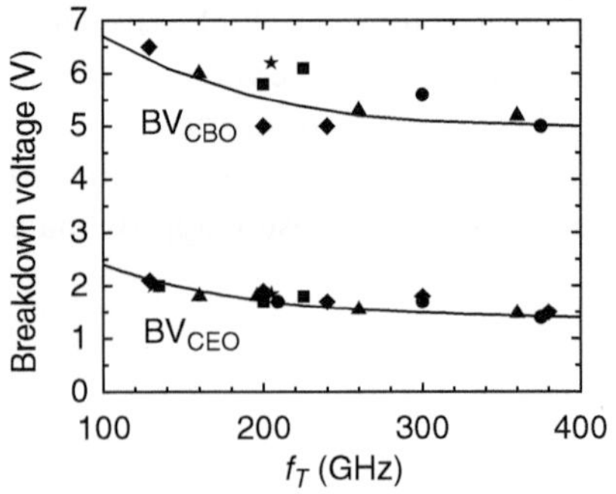

Figure 4.23 Published breakdown voltages BV_{CE0} and BV_{BE0} as a function of maximum transit frequency for different technologies. *Source*: Courtesy of IHP Microelectronics.

voltage ratings are generally lower for faster HBTs, this is not a linear relation. As shown in Figure 4.23 for many different SiGe HBT technologies, the slope of the breakdown voltages with cutoff frequencies flattens out. Before discussing the means to enhance cutoff frequencies without compromising the breakdown behavior, the two breakdown voltages shown in the figure, BV_{CE0} and BV_{CB0}, should be discussed. Since HBTs are intended for higher-power operation, a breakdown slightly below 2 V according to BV_{CE0} would be quite a disadvantage. However, getting roughly 5 V as suggested by BV_{CB0} would significantly enhance the power handling capability of the devices. In fact, the breakdown depends on the avalanche multiplication factor, M, with

$$M = \frac{1}{1 - (V_{CB}/BV_{CB0})^n} \tag{4.20}$$

The collector-base reverse current would be multiplied by M with a certain slope described by the fitting factor n, until the junction breaks down at $V_{CB} = BV_{CB0}$. The reason why there is not only one value for the breakdown voltage is that the transistor is a three-terminal device. Breakdown between two terminals, in our case collector and emitter, depends on the third terminal, which is the base. If the base current is forced, it is not sufficient only to consider the base-collector junction. If the base current is set to zero, as required by the definition of BV_{CE0}, $I_C = I_E$ must hold, with $I_C = M \cdot (I_{C0} + \alpha I_E)$ in the presence of an avalanche. I_{C0} denotes the base-emitter reverse current. For the emitter current, it follows:

$$I_E = \frac{MI_{C0}}{1 - \alpha M} \tag{4.21}$$

and the collector-emitter breakdown occurs already when M equals $1/\alpha$, at a voltage $BV_{CE0} = (1-\alpha)^{1/n} BV_{CB0}$. This lower breakdown voltage is observed in the output I–V curves that are measured with forced I_B. In circuit design, however, this value marks the lower bound for the breakdown, for the case that the base is connected to a high-impedance source. If the source impedance connected to the base is reduced, the breakdown behavior is improved and would reach BV_{CB0} if the base were short-circuited.

Typical S-parameters of a high-speed InP HBT ($f_t > 300$ GHz) and a GaAs HBT for mobile communications ($f_t \approx 30$ GHz) are shown in Figure 4.24. Within the frequency range of interest for circuit design, the S-parameters seem to follow semi-circles of similar shape.

Regarding S_{11}, the input resistance corresponding to the two intersections with the real axis can be estimated to be:

$$R_{\text{in}}(f \to 0) = \frac{R_e + R_{\text{be}}}{1 - \alpha_0} + R_{b2} + R_b \tag{4.22}$$

$$R_{\text{in}}(f_{\text{res}}) = R_b + R_{b2} \frac{\frac{\tau_\alpha}{C_{\text{ex}}+C_{\text{bc}}} + \frac{C_{\text{bc}}}{C_{\text{ex}}+C_{\text{bc}}}(Z_e + Z_c)}{\frac{\tau_\alpha}{C_{\text{ex}}+C_{\text{bc}}} + Z_e + Z_c} + \frac{Z_c Z_e}{\frac{\tau_\alpha}{C_{\text{ex}}+C_{\text{bc}}} + Z_e + Z_c} \tag{4.23}$$

with $Z_e = R_e + R_{\text{be}}$ and $Z_c = R_c + 50\ \Omega$. From the formula for the intersection with the real axis at higher frequencies, it is clear that a proper distinction between intrinsic and extrinsic base resistance and base-collector capacitance is required to model the high-frequency behavior of S_{11}.

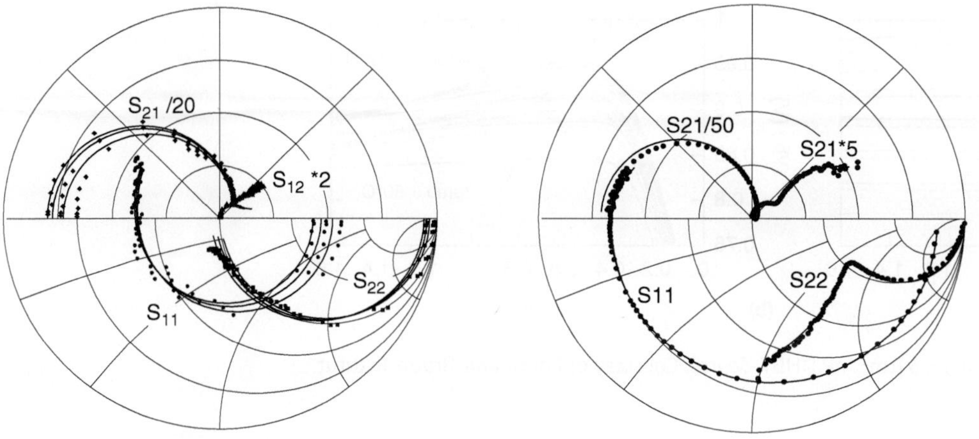

Figure 4.24 Left: Measured (symbols) and modeled by FBH-HBT model (lines) S-parameters of a $0.8 \times 5\ \mu m^2$ TS DHBT, $f = 50$ MHz – 110 GHz, $V_{ce} = 2$ V, $I_c = 15, 20, 25$ mA. Right: $3 \times 30\ \mu m^2$ InGaP/GaAs HBT $V_{ce} = 3$ V, $I_c = 25$ mA, $f = 0.05 - 50$ GHz. *Source*: Courtesy of Ferdinand-Braun-Institut.

These formulae assume that the extrinsic inductances and pad capacitances are already subtracted. The points on the curve of S_{11} will be transformed in the Smith Chart due to the series inductance L_b. This inductance is responsible for S_{11} reaching into the inductive half of the Smith Chart. L_e, on the other hand, performs a transformation of the intrinsic S_{11} as expected from a parallel inductance.

The output reflection coefficient starts from the open circuit for the GaAs HBT, which could be expected from the equivalent circuit (Figure 4.20). The extrapolation $f \to 0$ for the InP DHBT also yields a high output resistance, but the respective conductance is not zero. The reason can be, for example, Early effect. S_{22} shows a kink at a certain frequency. As discussed in [8] for the analog case of an FET equivalent circuit, one can approximate the transistor's output impedance by a series $R - C$ branch for low frequencies and by a parallel $R||C$ branch for high frequencies. The crossover frequency, of course, depends on the values of the equivalent circuit elements. It is in the lower GHz region for the GaAs HBT and seems to be around 100 GHz for the InP DHBT. Accordingly, the limit $f \to \infty$ for S_{22} will be the short circuit.

The feedback parameter S_{12} starts at zero, and its shape reflects that feedback happens through the base-collector capacitances.

The limit for $f \to 0$ for the forward gain S_{21} is given by:

$$S_{21}(f \to 0) = -\frac{\alpha_0(R_c + 50\,\Omega)}{R_e + R_{be} + (1 - \alpha_0)(R_{b2} + R_b)} \cdot (1 + S_{11}) \tag{4.24}$$

and it approaches zero for $f \to \infty$. However, rather than following a semi-circle, the locus resembles a spiral due to τ_α.

4.2.3 Temperature and Self-heating

The electrical properties of semiconductors in general depend on temperature. With respect to circuit design, the range of temperatures is fortunately limited, e.g., to typically −40 °C–120 °C ambient temperature for commercial applications. Since time constants of degradation mechanisms in transistors decrease exponentially with increasing temperature, the safe internal transistor temperature—the junction temperature T_j in bipolars, or the channel temperature T_{ch} in FETs—is limited.

The influence of temperature on HBT DC performance can be seen in the measurement presented in Figure 4.25a. It shows output I–V curves measured at 17 °C and 50 °C ambient temperature. Regarding collector current, it is observed that increasing temperature yields reduced current. The respective base-emitter voltages are shown in Figure 4.25b. Since I_b is forced in these measurements, V_{be} drops with increasing temperature. We not only observe a shift in the curves measured at different ambient temperatures, but also a negative slope for increasing V_{ce}. This slope is caused by self-heating of the device, which depends on the dissipated power. It is interesting to note, that we would expect to see a similar behavior in I_{ce}, Figure 4.25a. It seems that the Early effect in this type of device compensates for the self-heating in the I–V measurement. The voltage V_{be} is therefore a much better indicator for the investigation of self-heating than the current.

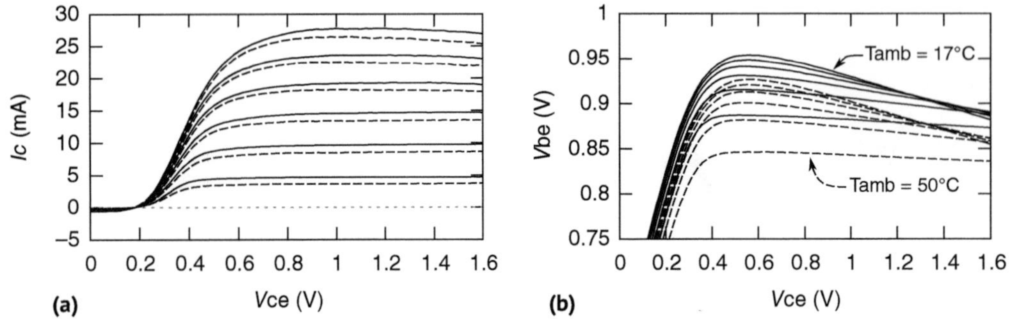

Figure 4.25 Output I–V curves of $0.8 \times 6\ \mu m^2$ InP DHBT. *Source*: Courtesy of Ferdinand-Braun-Institut.

Since only a rather narrow range of temperatures is considered, transistor models usually linearize the temperature dependence of model parameters around the temperature at which the model was extracted, T_{nom}. This temperature is not necessarily identical to the ambient temperature T_{amb} for which the simulation is carried out.

If for a parameter X, a temperature coefficient TX is given, the simulator will calculate the value of $X(T_j)$ by:

$$X(T_j) = X + TX \cdot \Delta T \tag{4.25}$$

with $\Delta T = T_j - T_{nom}$. Diode currents, however, depend exponentially on temperature, which cannot be linearized. The exponential term in the current–voltage dependence already contains the temperature, but also the saturation currents, e.g., $I_S(T_j)$, change with temperature according to an exponential law:

$$I_S(T_j) = I_S \left(\frac{T_j}{T_{nom}}\right)^{XTI} e^{\left(\frac{VG}{V_{th0}} - \frac{VG}{V_{th}}\right)} \tag{4.26}$$

with the parameters XTI and VG, with VG expressing the diode's activation energy in terms of a voltage. The thermal voltages are given by $V_{th} = kT_j/q$ and $V_{th0} = kT_{nom}/q$.

It depends on the model and how detailed the temperature dependence is modeled. Usually, only first-order parameters are considered. As long as it doesn't compromise accuracy, reducing the number of temperature-dependent parameters simplifies model extraction and enhances numerical stability of the model if dynamic self-heating is considered.

One should not overestimate the capability of a model to extrapolate the transistor performance toward extreme temperatures. Especially, if cryogenic or high-temperature electronics are to be designed, it cannot be expected that a model extracted only around room temperature still predicts the transistor performance well.

Early models, such as the Gummel–Poon model, only accounted for temperature scaling. This allows to modify the ambient temperature T_{amb}, which is a simulator variable and thereby constant during one simulation and equal for all circuit components. But it ignores that a device might heat up due to the power it dissipates, that different devices might operate at different temperatures, and that some components can exchange heat. Dynamic self-heating is a first-order effect, especially in power amplifiers. Depending on the mode of operation, the transistors in the final stage dissipate more or less power, heat up accordingly, and also increase the ambient temperature in their vicinity.

Fortunately, the transport of heat follows similar rules that apply for electrical quantities. It is straightforward to trick the circuit simulator to simulate an electro-thermal model using the following equivalences:

Thermal quantity	**Electrical equivalent**
Temperature [K]	Voltage [V]
Dissipated power P_{diss} [W]	Current [A]
Heat [J]	Charge [C]
Thermal resistance R_{th} [K/W]	Resistance [Ω]
Thermal capacitance C_{th} [J/K]	Capacitance [F]

The self-heating of a transistor can therefore be simulated through a thermal subcircuit as shown in Figure 4.26. The power the device dissipates is calculated in the time domain by $P_{\mathrm{diss}}(t) = \sum |V(t) \cdot I(t)|$, where the powers dissipated at the various branches of the transistor equivalent circuit are added up. Feeding the equivalent current to the $R_{\mathrm{th}}||C_{\mathrm{th}}$ network results in a voltage that is equivalent to the rise in temperature. Since the temperature typically needs about a microsecond to settle, the junction temperature will not follow instantaneous dissipated powers at frequencies beyond the lower MHz range. The thermal capacitance C_{th} is thus often not determined accurately, but rather with the assumption that a low-pass that prevents RF power from being considered in the calculation of self-heating might be enough. While this approximation holds for narrowband circuits, it is problematic for broadband amplifiers, where the distortion behavior can be impacted by third-order nonlinearities causing downconversion to the base band and upconversion back to RF. In this case, the thermal frequency response might be of importance.

The model shown in the figure is the simplest implementation. It is basically valid only if the transistor chip backside is held at T_{amb}. In reality, packaging will impact the cooling of the transistor, e.g., when the chip is mounted into a package, which is again soldered on a heat sink. For cases like these, a ladder structure of $R_{\mathrm{th}}||C_{\mathrm{th}}$ elements can be used, with one $R_{\mathrm{th}}||C_{\mathrm{th}}$ element describing one layer like chip substrate, glue, package base, soldering, or heat sink. Especially thin layers of low thermal conductivity should not be ignored, as they can easily dominate the total thermal resistance.

4.2.4 DC Characteristics and the Gummel–Poon Charge Control Model

The DC characteristics of (hetero) bipolar transistors can best be explained using an equivalent circuit. Figure 4.27 shows two representations. The left one reproduces physics: The current through the emitter-base junction I_{Ei} is represented by a diode, while the transport of the electrons toward the collector according to $\alpha_f \cdot I_{\mathrm{Ei}}$ is represented by a controlled current source between base and collector. Since the device is considered to be somehow symmetrical, a corresponding diode and current source are added to complete the equivalent circuit. This injection representation is best suited to explain transistor physics.

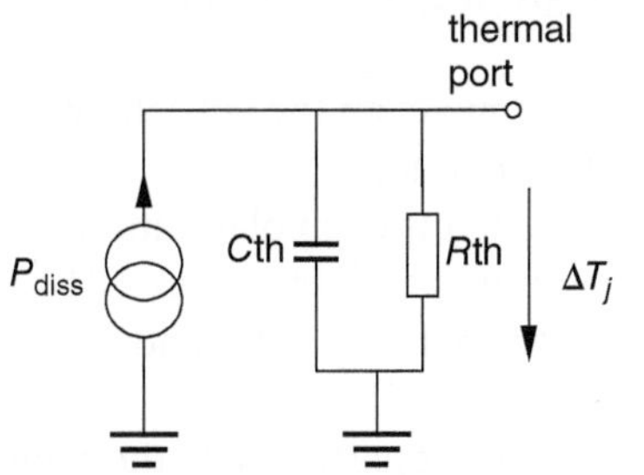

Figure 4.26 Thermal equivalent circuit.

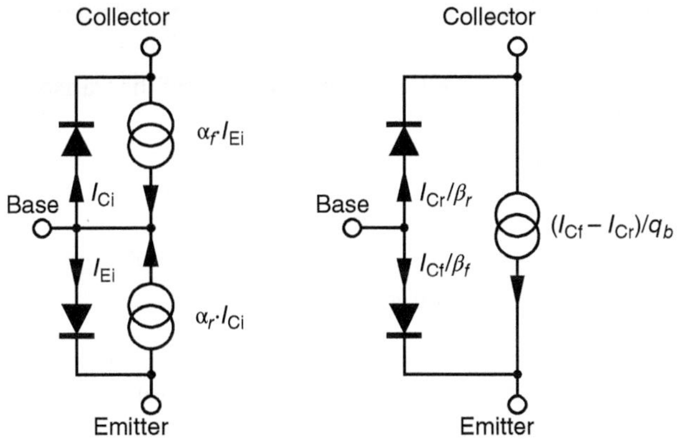

Figure 4.27 DC equivalent circuits of the intrinsic (hetero) bipolar transistor.

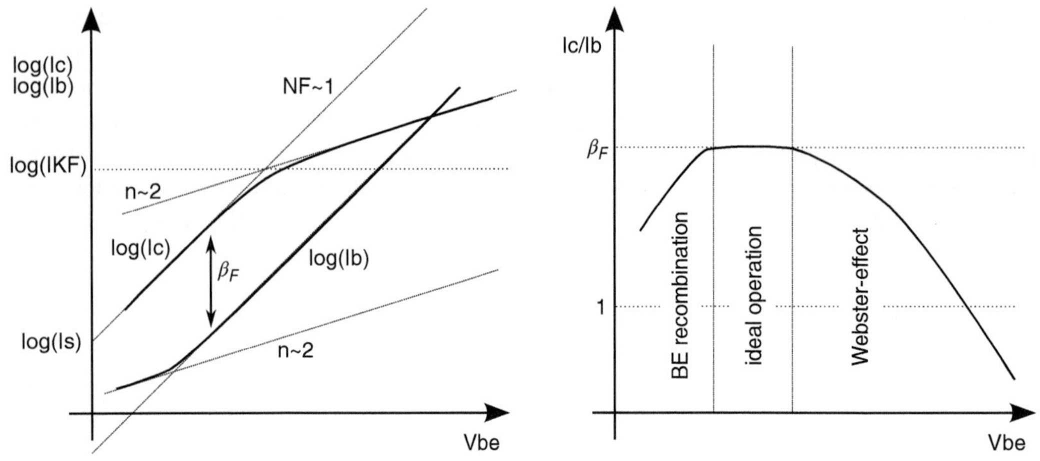

Figure 4.28 Ideal Gummel plot of npn transistor (a), and current gain obtained from Gummel plot (b).

For circuit application, the transport representation is commonly preferred, as shown on the right side in Figure 4.27. If we consider ideal forward-active operation [$I_{\mathrm{Cr}} \rightarrow 0$, $q_b = 1$], the collector current $I_C \approx I_{\mathrm{Cf}}$ is taken as the reference, and the base current is derived from it as $I_B = I_C/\beta_f$.

The advantage of the transport model is its relation to measurement. Figure 4.28 shows a so-called Gummel plot measurement, where $V_{\mathrm{BC}} = 0$ V ensures that the base-collector junction is reverse-biased. Varying V_{BE} yields currents as expected from pn-diode characteristics for I_B and I_C, which ideally should be two straight lines in log scale, separated by the current gain β_f. Already in the ideal Gummel plot, it is apparent that β_f is constant only for a certain range of voltages. For this ideal range, the model parameters can be directly determined based on the dashed lines in the figure: I_{Cf} is characterized by its saturation current equal to the intersection with the y-axis, and by the ideality factor NF given by the slope of collector current. Current gain β_f follows from the offset between I_C and I_B.

Toward lower current densities, a higher base current is observed, which apparently has no impact on the collector current. This additional current is understood to be caused by recombination at the base-emitter interface.

In the early 1970s, Gummel and Poon introduced the so-called charge-control theory to the modeling of bipolar transistors [9–11]. This sophisticated approach yields a very elegant model that describes the most prominent effects in the BJT of that time in a unified way. This is the reason why the model was dominant in simulator software for decades and still is generally supported.

The basis of the model derivation is the finding that the current gain $\beta = I_C/I_B$ is inversely proportional to the total majority charge of the base, Q_B. The reason to investigate the *majority* carriers while I_C is the current of *minority* carriers through the base is justified as the base is neutral. Thus changes in the number of minority carriers will also affect the number of majority carriers. The advantage of investigating majority carriers is that they are much easier to calculate, at least under certain assumptions.

The impact of the second-order effects in the transistor on the current gain is described through a variation of the transport current I_{CT}:

$$I_{\mathrm{CT}} = \frac{I_{\mathrm{CC}} - I_{\mathrm{EC}}}{q_b} \tag{4.27}$$

where q_b is the normalized unified base charge.

The Gummel–Poon model accounts for the following effects through the unified charge formula:

- Early effect, which is base width modulation due to the variation of the base-collector SCR. Changing V_{BC} results in variations of the base-collector's pn-diode depletion region. Increasing the reverse base-collector voltage, or increasing V_{CE}, yields a wider SCR that extends more into the base, and in consequence, results in an effective reduction of base width. As the effective base width decreases, current gain increases as less minority carriers recombine there, and the base transport factor (number of electrons injected into the base/number of electrons reaching the collector) increases. Figure 4.29 shows typical output I–V curves in the presence of Early effect. The Gummel–Poon model determines q_B by integration over the effective base. Reducing the effective base width reduces q_B accordingly and $\beta \propto 1/q_B$ increases.

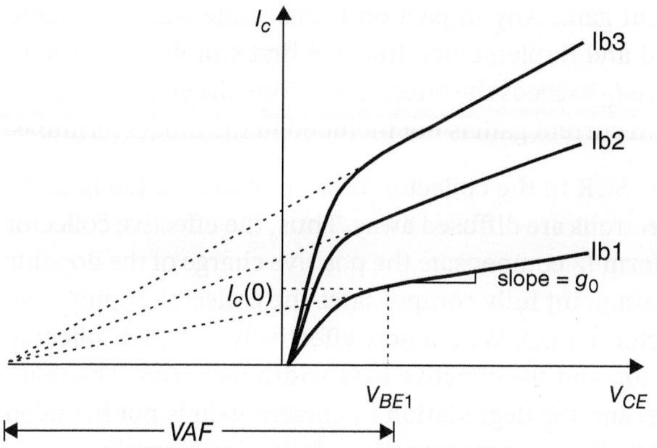

Figure 4.29 $I-V$ curves showing forward Early voltage VAF of npn transistor.

In the Gummel–Poon model, the Early effect is implemented through a charge q_1:

$$q_1 = \frac{1}{1 - \frac{V_{\text{be}}}{\text{VAR}} - \frac{V_{\text{bc}}}{\text{VAF}}} \tag{4.28}$$

The two model parameters VAF and VAR denote the so-called Early voltages, an ideal point on the negative axis, where the tangents of the I–V curves would meet, as shown in Figure 4.29. If the bipolar transistor is somehow symmetrical and operated in reverse mode, the Early effect takes place at the base-emitter interface. This is sometimes called Late effect—just for the fun of it.

- Webster effect, which means high-current injection into the base. If the number of minority carriers in the base gets into the same order of magnitude as the doping, it becomes necessary to also consider the free majority carriers in the base. For an npn transistor, for example, the base is neutral, and the number of electrons injected into it is controlled by V_{BE}. Increasing the number of electrons requires an equal number of holes to be present in the base in order to maintain neutrality, where the sum of base doping and holes can be understood as effective base doping. The total base charge q_B would therefore increase with increasing current. The current gain is compromised by high-current injection, since the emitter efficiency (electrons injected into the base/holes injected into the emitter) depends on the ratio of the effective doping in base and emitter.

  In the Gummel–Poon model, a charge q_2 describes the Webster effect:

$$q_2 = \frac{I_{\text{CC}}}{\text{IKF}} + \frac{I_{\text{EC}}}{\text{IKR}} \tag{4.29}$$

  where IKF and IKR denote the critical current for the onset of Webster effect in forward- and reverse operation.

The total normalized charge q_b is then determined by:

$$q_b = \frac{q_1}{2}\left(1 + \sqrt{1 + 4\,q_2}\right) \tag{4.30}$$

A so-called Gummel plot is shown in Figure 4.28. This measurement is performed by setting the base-collector voltage to zero volts while sweeping base-emitter voltage. This condition ensures that the BJT is in forward-active operation, while base-width modulation or power dissipation due to high collector voltages is minimized. Plotting the collector current in logarithmic scale yields I_S from an extrapolation toward $V_{\text{BE}} = 0$, and NF from its slope. The current gain β_F is determined from the difference of the I_C and I_B curves. This measurement therefore directly gives the parameters of the transport model. The Gummel plot also allows to determine the onset of Webster effect directly, as it is the transition current IKF where the emission coefficient approaches 2, since the emitter efficiency degrades under high-current injection into the base. For lower currents, recombination at the base-emitter junction can significantly increase the base current. Since these electrons do not contribute to the collector current, only I_B is affected. The emission coefficient for this current is also ideally 2. In the model, additional base-emitter diodes are added to account for these parasitic currents. Note that V_{BE} on the x-axis needs to be the internal voltage at the junction. The impact of parasitic resistances would stretch the axis with increasing current and flatten the curves.

Base modulation and high-current injection primarily affect current gain. Any impact on transit time was considered a third-order effect at the time the Gummel–Poon model was derived and implemented into the first simulator codes. In contrast, base push-out will result in a collapse of transit frequency once I_C exceeds the onset. Therefore, the effect on transit time is considered in the Gummel–Poon model, and any degradation in current gain is not included in the model formulas.

- Base push-out is also known as Kirk effect [12, 13]. Consider the SCR in the collector at the interface to the base. In an npn transistor, only the positive dopants remain, while the electrons are diffused away. Thus, the effective collector doping is proportional to the positive charge. The electrons that form I_C compensate the positive charge of the dopants and reduce the effective collector doping. Increasing I_C at a certain point fully compensates the collector doping, and even higher currents result in an effective inversion of the collector doping. With a now effectively p-doped collector, the base seems to be effectively being pushed out into the collector, and the effective base width increases. The main consequence of this effect is the degradation of base transit time, and the degradation in current gain is not included in the model formulas. The low-current forward transit time TF is altered according to the following formula:

$$T_{\mathrm{FF}} = \mathrm{TF}\left[1 + \mathrm{XTF}\left(\frac{I_{\mathrm{CC}}}{I_{\mathrm{CC}} + \mathrm{ITF}}\right)^2 \cdot e^{V_{\mathrm{bc}}/(1.44\,\mathrm{VTF})}\right] \tag{4.31}$$

where XTF denotes the slope of T_{FF} under base push-out, and ITF and VTF are critical current and voltage, since base push-out sets in earlier for low collector voltages.

4.2.5 Large-signal Modeling

The Gummel–Poon model was a major breakthrough in BJT modeling, and it is still the basic bipolar model available in circuit simulators. The full model accounts, in addition to the DC part discussed in the previous section, also for the diffusion and depletion capacitances of the two pn-junctions. The base-emitter diffusion capacitance is expressed in terms of a constant forward time constant τ_F, that also takes over the description of the base and collector transit times.

The Gummel–Poon model is a good basis for the understanding how bipolar large-signal models work. It is, however, designed with the transistors in mind that were the standard at the time: homojunction devices with rather large dimensions. Also, computing power in circuit simulation was very limited and, with few exceptions, not existing in measurement and model parameter extraction.

Modern HBT models are much more involved. SiGe models, such as HICUM [14] or MEXTRAM followed a similar basic approach as Gummel and Poon in their derivation of the model, as they are derived bottom up, i.e., from physics toward the compact model formulation. Models for GaAs and InP HBTs, such as the AHBT [15] and FBH HBT [16] model were derived the other way around—as the physics are much more involved and the technology at least historically repeatable much less predictable, analytical models were preferred. The starting point of these models is a basic understanding of device physics, and empirical measurement data. From there, a mathematical description is derived, and questions of nonlinear RF accuracy, numerical convergence and reliable parameter extraction possibilities are the main focus.

While the Webster effect is not relevant anymore due to the highly doped base in HBTs, and the Early effect is reduced, the Gummel–Poon model lacks quite a number of features that are considered in modern HBT models:

- Self-heating
- Different properties of base-emitter and base-collector junction
- Extrinsic base-collector diode (in GaAs and InP Meas-type transistors), and interaction with the substrate (in SiGe HBTs)
- Parasitic base-emitter diode currents
- Soft-knee effect, if present in the device
- Bias dependence of the collector transit time
- Collector breakdown

In addition, a model accounting for breakdown of the emitter-base junction can be advantageous. In GaAs HBTs, the collector breakdown can be much higher than the breakdown of the base-emitter junction, e.g., 25 V compared to 7 V. Driving the HBT with high input powers while biasing through a high-impedance source can drive the emitter into breakdown [5].

A typical large-signal equivalent circuit for an III–V HBT large-signal model is shown in Figure 4.30. It features the thermal subcircuit that allows to calculate dynamic self- and mutual heating. It also accounts for the distributed nature of

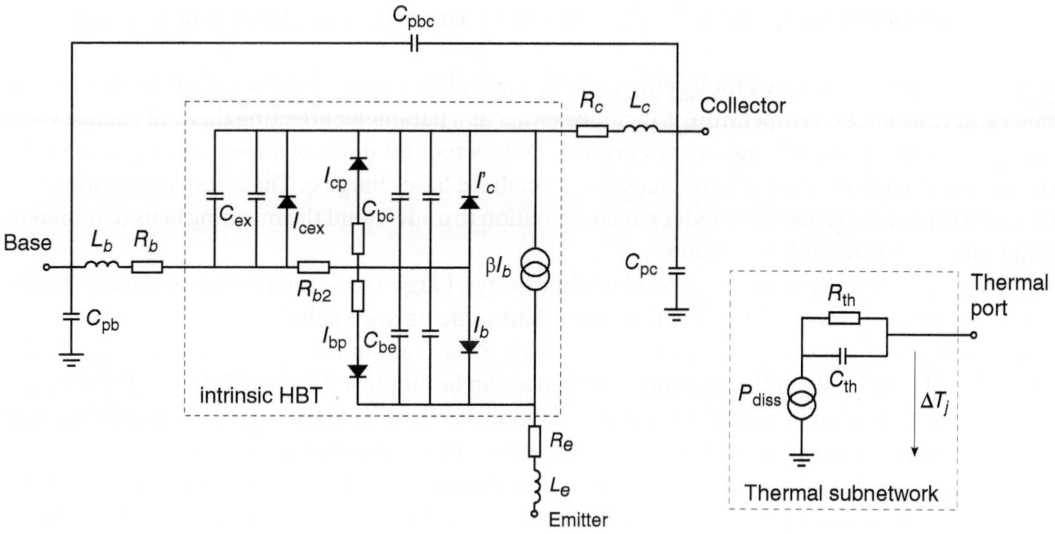

Figure 4.30 Large-signal equivalent circuit of FBH HBT model.

the base by splitting the base resistance into the extrinsic R_b and the intrinsic R_{b2}, and describes the extrinsic base-collector diode by the current I_{cex} and the respective capacitances such as C_{ex}. A breakdown model is also present, but not reflected in the equivalent circuit. The topology of SiGe models is more or less identical, but additionally features a subcircuit that accounts for possible interaction with a conductive substrate. Parasitic base-emitter or base-collector currents are modeled through diode currents I_{bp} and I_{cp}. The respective currents originate from recombination at the heterointerface, so that the electrons are not even reaching the base. They are commonly only important at very low current densities and often saturate, which is modeled by series resistances.

The soft-knee effect affects the transition from saturation to forward-active operation. One would expect a sharp transition of the output I–V curve from saturation, where I_{ce} is linearly increasing according to the parasitic resistances to forward-active, where the current is constant since the base-collector junction is closed. In case of a soft knee region, a smooth transition is observed, as seen in Figure 4.34. Possible reasons are:

Conduction band spike at base-collector junction: InP DHBTs feature a higher bandgap material also in the collector, in order to increase breakdown voltage while keeping the layer very thin. The drawback is that an unwanted spike in the conduction band can prevent electrons from leaving the base, if the collector-base voltage is not sufficiently high. This effect is suppressed in advanced InP devices by separating the base-collector pn-junction from the heterojunction. If the spike is situated deeper within the high-field drift zone, its effect on current gets suppressed.

Quasi-saturation: In SiGe HBTs featuring a rather long collector region, the SCR can extend only through a part of the collector at low voltages. The remaining part of the collector is only weakly doped and presents an additional parasitic collector resistance. This resistance is bias-dependent, as an increasing collector-base voltage reduces it to zero, while increasing current, through the voltage drop over the undepleted part, increases the effect. Modern SiGe HBTs, however, usually feature very short collectors so that the effect should be suppressed.

Base push-out: Once the collector current density is high enough, the charge of the electrons compensates the charge of the few acceptors in the lightly doped collector. At the point where the collector becomes an electrically neutral region, it appears to be an extension of the base. The pn-junction is thereby shifted to the collector–subcollector interface. In consequence, the effective base length on which the electrons diffuse and recombine is extended, and β drops. This effect was described first for silicon BJTs, and also observed in early GaAs HBTs. Considering the high crystal quality of the epitaxial layer stack today, base push-out is hardly seen in DC parameters anymore, but remains visible in a reduction of transit frequency.

Most models follow the basic concept of the Gummel–Poon approach, which means that base and collector currents are linked through a parameter `BetaF` representing the ideal (forward) current gain β. If I_b and I_c show different ideality factors, "parasitic" diodes are added to the netlist in parallel to the junction's "ideal" diode. Another concept is to decouple the two currents. I_c and I_b are modeled separately, and $\beta = I_c/I_b$ is still observed for the HBT, but it is no longer a model parameter.

Most effects observed in the HBT can be accounted for by either adding equivalent-circuit elements of a familiar type like diodes, depletion capacitances, or resistances. Temperature is included either as a parameter affecting thermal voltage $V_{\mathrm{th}} = q/(kT)$ or by scaling model parameters like diode quiescent currents. Others require more complex variations of model parameters, similar to how the Gummel–Poon model introduces the normalized base charge q. These implementations are quite involved, and it is advisable to read the respective model's documentation to understand the meaning of its parameters, the physical background and mathematical implementation.

Concept-wise interesting is the implementation of bias-dependent time delays. Large-signal models need to be compatible with any kind of solver, and time domain solvers like SPICE, or even harmonic balance solvers, basically consider only the current and previous time step of the calculation. Even a constant time delay cannot be implemented directly in this setting. In the following, it will be shown how a collector transcapacitance can be employed instead [16–18]. The example is a InGaP/GaAs HBT, but the concept is used by virtually all advanced models. Some models disguise it slightly through offering partitioning factors, which allow for a shift of charge from base-emitter to base-collector junction.

For GaAs HBTs, it was shown quite early, and also derived from physics, that not only the collector transit time, but also the base-collector capacitance C_{bc} is changed by velocity modulation, i.e., a modification of the electron velocity in the collector according to bias [19]. The bias dependence of the total transit time τ_{eff} and C_{bc} is shown in Figure 4.31.

In fact, the collector charge Q_C, i.e., the charge stored in the base-collector SCR, depends on base-collector voltage, as any pn-junction and also on I_C, since the charge of the electrons cannot be neglected. Thus, $Q_C = f(V_{\mathrm{CE}}, I_C)$, and two derivatives exist:

$$\frac{\partial Q_C}{\partial V_{\mathrm{BC}}} = C_{\mathrm{bc}} \tag{4.32}$$

$$\frac{\partial Q_C}{\partial I_C} = \tau_C \tag{4.33}$$

The dashed lines in Figure 4.31 were extracted from simulations based on the FBH HBT model, which provides an appropriate description of Q_C. This model starts with the description of C_{bc}. It is assumed that C_{bc} basically follows the known voltage dependence of a pn-junction capacitance, which would be $C_{\mathrm{bc}} = C_{\mathrm{jc}}(1 - V_{\mathrm{bc}}/V_{\mathrm{jc}})^{(-M_{\mathrm{jc}})}$, with the zero-voltage capacitance C_{jc}, the built-in potential V_{jc}, and the grading exponent M_{jc}. However, there is a lower bound to C_{bc}, that is when the SCR fills the whole collector, and C_{bc} will not fall below C_{min}. Therefore, the formula for C_{bc} is modified to be $C_{\mathrm{bc}} = (C_{\mathrm{jc}} - C_{\mathrm{min}})(1 - V_{\mathrm{bc}}/V_{\mathrm{jc}})^{(-M_{\mathrm{jc}})} + C_{\mathrm{min}}$. This way, C_{jc} keeps its meaning as zero-voltage capacitance. Finally, the current dependence is introduced, which is more or less linear. A tanh-based function is applied, and a partition factor X_{j0} is introduced that allows to distinguish for C_{min} between voltage- and current dependence. The full formula reads:

$$C_{\mathrm{bc}} = [1 - \tanh(J_c/J_0)] \cdot \left(\frac{C_{\mathrm{jc}} - C_{\mathrm{min}}}{(1 - V_{\mathrm{bc}}/V_{\mathrm{jc}})^{M_{\mathrm{jc}}}} + (1 - X_{j0}) C_{\mathrm{min}} \right) + X_{j0}\, C_{\mathrm{min}} \tag{4.34}$$

with the collector current density J_c.

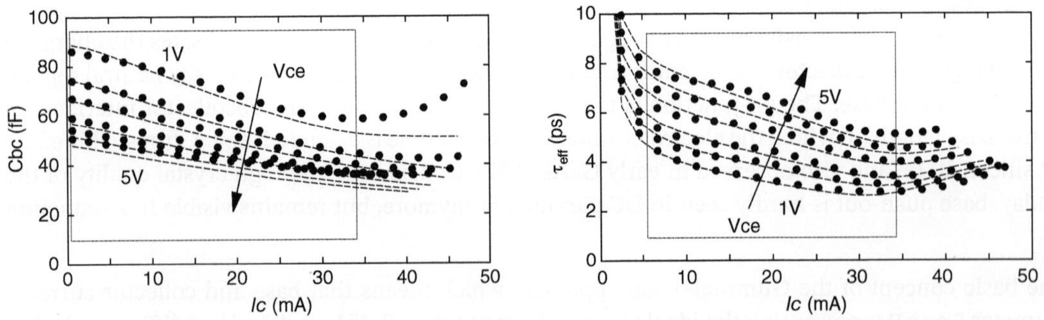

Figure 4.31 Base-collector capacitance C_{bc} and effective total transit time τ_{eff} of an InGaP/GaAs HBT, emitter size $3 \times 15\ \mu\mathrm{m}^2$.

The collector charge is then obtained by integration

$$Q_C = \int C_{\mathrm{bc}} dV_{\mathrm{bc}} \tag{4.35}$$

which, in turn, automatically yields a collector time delay

$$\tau_c = \frac{\partial Q_C}{\partial J_c} = \frac{1}{J_0 \cosh^2(J_c/J_0)} \cdot \left((C_{\mathrm{jc}} - C_{\mathrm{min}}) V_{\mathrm{jc}} \frac{\left(1 - V_{\mathrm{bc}}/V_{\mathrm{jc}}\right)^{1-M_{\mathrm{jc}}}}{1 - M_{\mathrm{jc}}} \right) \tag{4.36}$$

which did prove to be the portion of the total transit time present in the GaAs HBT that was missing in previous models [17]. A formulation for Q_C is therefore the core of III–V models that attempt to describe velocity modulation. But it is not required to start with the bias dependence of C_{bc}. The AgilentHBT model, for example, gives a formula for τ_C in order to determine Q_C. C_{bc} then follows automatically.

While this derivation seems to be straightforward, how is the time delay realized in the large-signal model? Figure 4.32 shows a large-signal model topology of the intrinsic HBT in forward-active operation. It consists of the forward-biased base-emitter diode, the current-controlled current source that describes the injection of current into the collector, and two charge sources that describe the diffusion and depletion charges of the pn junctions. Q_C, in addition, takes the collector current charge into account.

On the right-hand side, the respective small-signal equivalent circuit is shown. The diode is replaced by its differential resistance, the injected current source remains as it is assumed to be constant, and the charges show up as their derivatives, which are capacitances. At the base-collector junction, a transcapacitance exists, its value is $C_{\mathrm{tr}} = \partial Q_C / \partial V_{\mathrm{be}}$. The current through the transcapacitance is $i_{\mathrm{tr}} = j\omega C_{\mathrm{tr}} v_{\mathrm{be}} = j\omega C_{\mathrm{tr}} R_{\mathrm{be}} i'_e$. The total collector current that is driven by the emitter current i'_e thus reads $i'_c = (\alpha_0 - j\omega C_{\mathrm{tr}} R_{\mathrm{be}}) i'_e$. The current gain $\alpha' = i_c / i_e$ assumes the following form:

$$\begin{aligned} \alpha' &= \frac{\alpha_0 - j\omega C_{\mathrm{tr}} R_{\mathrm{be}}}{1 + j\omega C_{\mathrm{be}} R_{\mathrm{be}}} = \frac{\alpha_0 - j\omega \tau_C}{1 + j\omega \tau_{\mathrm{be}}} \approx \left(\frac{\alpha_0^{-j\omega\tau_C}}{1 + j\omega \tau_{\mathrm{be}}} \right) \\ &\approx \frac{\alpha_0}{1 + j\omega(\tau_{\mathrm{be}} + \tau_C)} \\ &= \frac{\alpha_0}{1 + j\omega \tau_{\mathrm{eff}}} \end{aligned} \tag{4.37}$$

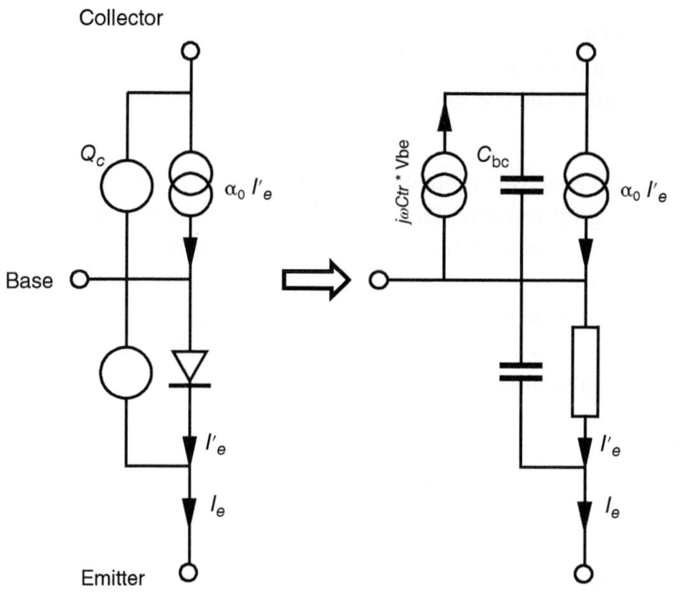

Figure 4.32 The current dependence of Q_C yields a transcapacitance that can be interpreted as collector time delay.

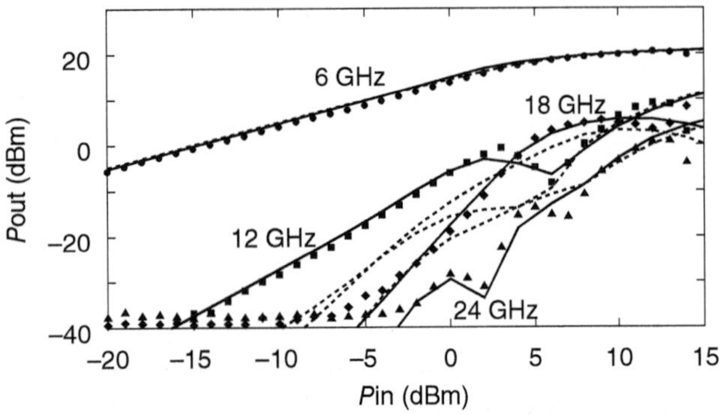

Figure 4.33 Power-spectral measurement at 6 GHz (symbols) and simulation (lines) of a $3 \times 30\ \mu m^2$ InGaP/GaAs HBT, for $V_{CE} = 6$ V, $I_C = 20$ mA. *Source*: [20]/IEEE.

Employing the transcapacitance thereby is a very elegant way to introduce the collector time delay in a consistent way. First, it solves the issue of bias-dependent f_t and at the same time the current dependence of C_{bc}. Second, it effectively realizes a time delay although circuit simulators in general don't allow time delays in large-signal models. And if they do, which prohibits the use of the model in SPICE, only constant delays are possible. Therefore, time delays anyways need to be approximated through *RC* time constants. Finally, the approach doesn't modify the current source, it only yields a transcapacitance. It is therefore applicable with no impact on model performance also in transport-type large-signal models.

The importance of the velocity modulation on the accuracy of the large-signal model should not be underestimated, as it is the prime cause of weak nonlinearities [20, 21]. Figure 4.33 shows a power-spectral measurement at 6 GHz in a 50-Ω system. Input power is swept, and the first four harmonics are measured at the collector. Two models are compared to the measurement. One model neglects the velocity modulation effects, but the parameter T_F is chosen so that the total transit time is correct for the bias point considered. The result is shown as dashed lines in the plot. Only the model featuring velocity modulation (solid lines) is capable to predict the weak nonlinearity, i.e., the region where the higher harmonics grow out of the noise floor.

4.3 Field-effect Transistor Basics

The current in FETs consists purely on majority carriers accelerated by an electric field. In case of an n-channel, electrons flow from a source contact toward a drain contact, while holes in a p-channel FET travel in the opposite direction. A gate contact which is insulated from the channel controls the conditions of the current.

FETs are commonly used in two operation conditions:

As a switch: The gate contact is used to fully deplete the channel through an applied electric field vertical to it. The output resistance between drain and source contacts thus becomes highly resistive. Or, in the other case, the gate voltage is set to a value that allows for a maximum of carriers in the channel, so that the output resistance becomes very low. By selecting the gate voltage, we are therefore able to control whether an open or a closed switch is seen at the other two electrodes. Switching behavior is technically used not only in digital applications. FET switches are also used in mobile communication devices, e.g., as antenna switches. Mixers commonly translate frequencies by chopping the signal by means of a transistor switch. DC-DC converters in power electronics also require switches, and so do switching-type RF power amplifiers such as classes D, E, and S.

As a voltage-controlled current source: If the channel is conducting, we might apply a voltage at the drain. which increases the gate-to-channel voltage locally at the drain to a value that would be expected to fully deplete the channel. Since the conditions in the rest of the channel still demand that a drain-source current flows, the FET gains an important property: The drain voltage is decoupled from the amount of current flowing. In this so-called saturation region, the FET behaves like a voltage-controlled current source. The gate voltage determines the amount of current flowing between drain and source, while providing a high output resistance. This operation condition is instrumental for any kind of analog or RF amplification.

FETs are built on different material systems and different channel types, although the operation principle remains the same. The vast majority of FETs are MOSFETs, silicon devices with a gate that is separated through an isolator. These devices are basically easy to manufacture. It would mean to dope the silicon wafer, metallize source and drain contacts, and oxidize the surface in between them so that a silicon oxide isolator is formed on which the gate contact metal can be deposited. Optimization of these devices in terms of speed requires, as is commonly known, downscaling, i.e., the reduction of their dimensions. Today's MOSFETs, therefore, are much more complicated in their technology. The gate insulator is replaced by a material that can be deposited as a very thin layer with a high breakdown field in order to allow for gate voltages high enough to provide the required electrical field in the semiconductor. For the same reason, the channel might be a ridge allowing for the gate to be wrapped around it, which leads to a better control of the channel by the gate. These devices are known under the name FinFET and mainly used in digital electronics. In digital electronics, downscaling is advantageous in almost any respect except, maybe, for the setup cost implied by the very small dimensions. Smaller dimensions enable faster switching, denser packaging, and lower power dissipation per switching event. What is an advantage in the digital domain can be a disadvantage in the RF domain: power handling is instrumental in communication systems. For RF applications, the optimization direction was therefore rather to reduce parasitic capacitive effects by building the MOSFET on an isolation layer (SOI MOSFET). In case of really high powers, e.g., up to the 100 W range, drain-source breakdown is improved by extending the gate drain reverse-biased depletion region. At least in the early days, the doping in this region was achieved through lateral diffusion, hence the name LDMOS FETs.

Digital and mass market domains are clearly governed by MOSFETs, but the high-performance and high-speed domain takes advantage of the higher electron mobility and higher breakdown voltages of III–V material systems. These devices use Schottky contacts for the gate, which are isolated from the channel through the SCR that forms as long as the gate-to-channel voltage is low enough. Some devices feature a gate isolated through a thin dielectric (MISFETs), but traditional MOS-type FETs cannot be processed as it was not possible to oxidize the semiconductor surface reliably as part of the transistor process. The oldest type of III–V FET is the GaAs MESFET, where the gate-induced SCR depletes a resistive channel. These devices are in principle even easier to process than the traditional MOSFET: the surface of the semiconductor has to be n-doped, as the substrate by itself is semi-isolating. This channel region is contacted by ohmic drain and source contacts, and the gate is formed by a Schottky contact in between. MESFETs are still in use in the SiC material system for power applications. Concerning RF applications, huge steps in performance were possible when complex material layer stacks became possible through epitaxial growth methods like molecular beam epitaxy or metal-organic vapor-phase epitaxy. Devices for highest frequencies and lowest noise (on InP) or highest power (based on GaN), or traditionally on GaAs substrates, are therefore commonly HEMT devices, where the channel is formed as a two-dimensional electron gas (2DEG) that exists at the conduction band step of a hetero-interface. The advantage of a 2DEG channel is that the electron density is not controlled by the channel doping, and that the electron mobility is higher than in the bulk semiconductor. A disadvantage of III–V devices is that they all feature an n-channel, so that complementary logic (CMOS) cannot be realized. Integration with highly integrated logic functions is therefore not possible on the same chip.

In this section, we will first introduce the basic operation principle of the classic MESFET and MOSFET, which should then help to understand the operation principle of HEMT devices. A first-order approximation will be given to enable easier understanding, followed by a discussion of second-order effects. The section concludes with an example of a modern large-signal model.

4.3.1 Ideal FET Model

A MESFET and a MOSFET in equilibrium are shown in Figure 4.34. The MESFET consists of a layer of n-doped GaAs on top of a semi-insulating not intentionally doped GaAs substrate. Source and drain form ohmic contacts, while the gate forms a Schottky contact, under which an SCR partly depletes the channel. Controlling the channel charge is therefore done by controlling the undepleted channel cross-section. The MOSFET cross-section looks quite different. Source and drain contact n-doped regions, while the region underneath the gate, where one would expect the channel, is p-doped. The gate is isolated from the semiconductor by means of a thin insulator, which historically could be SiO_x.

4.3.1.1 Off-state

In order to suppress the flow of current through the channel, the SCR needs to extend through the full channel width in case of the MESFET (see Figure 4.35a). Its height h thus needs to be equal or higher than the channel thickness a:

$$h = \sqrt{\frac{2\epsilon(V_{\text{bi}} - V_{\text{gs}})}{qN_D}} \geq a \tag{4.38}$$

with the permittivity ϵ, the diode's built-in voltage V_{bi}, doping N_D and the electron charge q. Therefore, a threshold voltage $V_{\text{gs}} = V_{\text{th}}$ can be defined according to

$$V_{\text{th}} = V_{\text{bi}} - \frac{qN_D a^2}{2\epsilon} \tag{4.39}$$

below which the channel is closed. We can also define the so-called pinch-off voltage $V_p = V_{\text{bi}} - V_{\text{th}}$, which denotes the corresponding internal voltage. For $V_{\text{gs}} \geq V_{\text{th}}$, no current can flow even if a drain-to-source voltage V_{ds} is applied, obviously. FETs of this type, where a current flows unless a negative V_{gs} is applied, are also called commonly-on FETs.[4]

The MOSFET, on the other hand, is a normally-off FET. If no gate voltage is applied, we see a np-pn diode sequence between drain and source (see Figure 4.35b). In equilibrium, the two SCRs suppress any current. If, for example, a positive drain voltage is applied, while source and gate remain grounded, the reversely biased diode between gate and drain suppresses any current.

4.3.1.2 Linear Region—The Channel

If the drain-source voltage is still quite low so that we can approximately neglect it when determining the channel charge, the FET behaves like a controllable resistor. The channel resistance R_{ch} is given by:

$$R_{\text{ch}} = \frac{L_g}{W_g d\kappa} \tag{4.40}$$

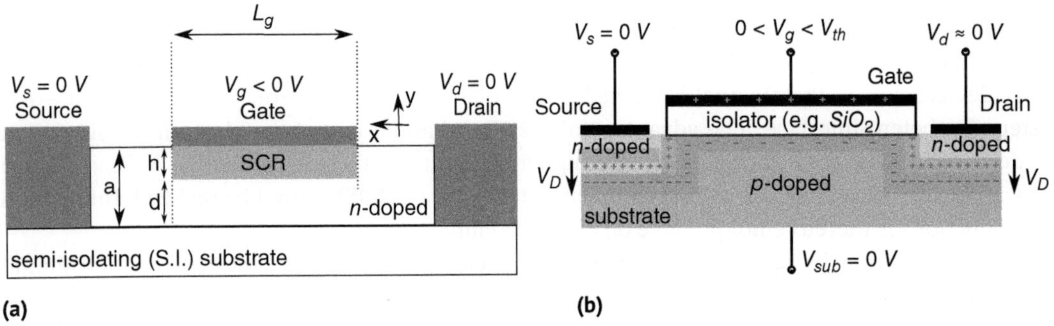

Figure 4.34 Schematic cross-section of (a) a MESFET and (b) a MOSFET.

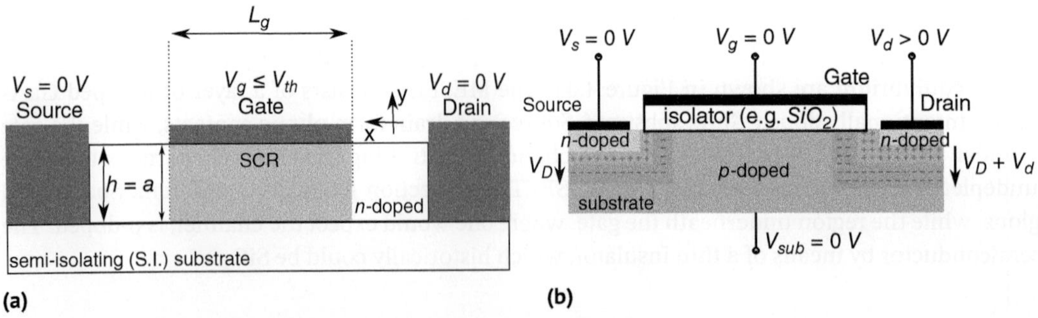

Figure 4.35 Schematic cross-section of (a) a MESFET and (b) a MOSFET in pinch-off.

4 The sequence in applying bias voltage to this type of FET therefore requires to apply a negative gate voltage first, before applying V_{ds}.

with $\varkappa = qN_D\mu_n$, with the electron mobility μ_n that is considered to be constant, L_g denotes the gate length (in direction of the current), and W_g the gate width (lateral to the gate current direction). The channel thickness d is given by, cf. Figure 4.35a:

$$d = a - h = a - \sqrt{\frac{2\epsilon(V_{\text{bi}} - V_{\text{gs}})}{qN_D}} \tag{4.41}$$

In case of the MOSFET, a positive charge needs to be applied to the gate in order for a conductive channel to form. Figure 4.36 depicts the principle of how the so-called inversion channel is formed. The sketch shows a vertical cross-section below the gate. What we see here is a metal–insulator–semiconductor (MIS) structure. Considering that it resembles a parallel plate capacitor, we can assume that a negative charge in the semiconductor must exist, if a positive voltage charges the gate positive. At first, the field drives the majority carriers, holes, from the interface, so that a depletion region forms. Minority carriers, electrons, on the other hand, are accelerated toward the interface and accumulate accordingly. In case of low voltages, recombination might prevent the cumulation of electrons. If a voltage V_{th} is crossed, however, electrons accumulate at the interface and form a so-called inversion channel. Since the gate metal extends over the pn-interfaces at the drain and source side, the n-regions can supply electrons to the inversion channel.

The channel charge Q_{ch} can therefore be calculated considering the gate oxide capacitance C_{ox} and the threshold voltage V_{th} according to:

$$Q_{\text{ch}} = W_g \cdot C_{\text{ox}} \cdot (V_{\text{gs}} - V_{\text{th}}) \tag{4.42}$$

The channel resistance in case of the MOSFET therefore can be given by:

$$R_{\text{ch}} = \frac{L_g}{Q_{\text{ch}}\mu_e} \tag{4.43}$$

with the electron mobility in the inversion channel μ_e.

The nature of the inversion channel can be better understood by inspection of the band diagram of a MIS structure, which would be a vertical cut through the FET, as shown in Figure 4.37. In the idealized case, there is no accumulation of charges at the semiconductor–isolator interface, and the valence and conduction bands are flat, Figure 4.37a. The channel in this example is p-doped, and therefore the Fermi (W_F) level is shifted from the center between conduction and valence bands (W_i) toward the valence band (W_v). If a negative voltage is applied, the gate electrode gets negatively charged, which is compensated on the side of the semiconductor by the positive charges of holes that accumulate at the interface (Figure 4.37b). This condition is called accumulation. Although the number of free carriers is enhanced, no current flows in the channel, as it is still p-doped and the pn-junctions at drain and gate prevent it. On the other hand, if a positive voltage is applied, Figure 4.37c holes are driven out of the interface, and the bands are bent toward lower energies. As highlighted through the circle in the graphic, W_i might cross the Fermi level at a certain gate voltage, so that the material now in fact behaves rather like an n-doped semiconductor than as p-doped. Electrons can now flow in from the source, as this very thin area beneath the surface now practically looks like an n-region connecting drain to source. This so-called inversion channel is confined to a very thin film at the surface.

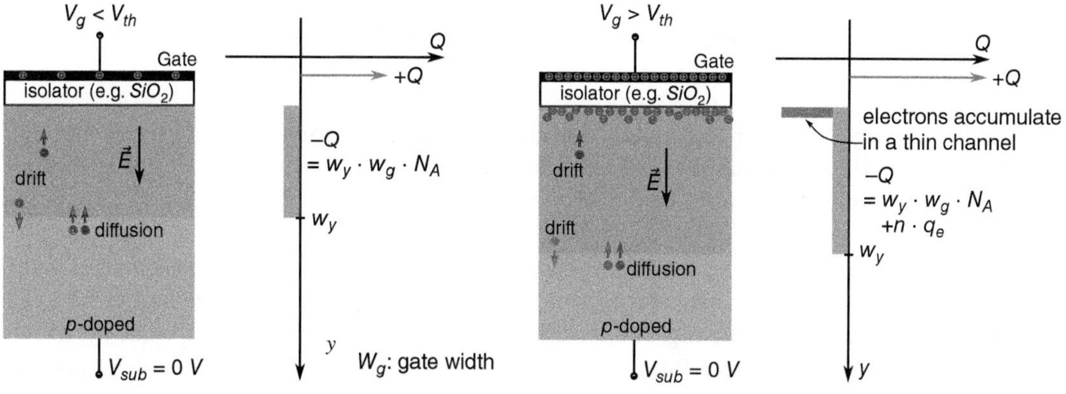

Figure 4.36 Schematic cross-section underneath the gate, depicting the formation of an inversion channel.

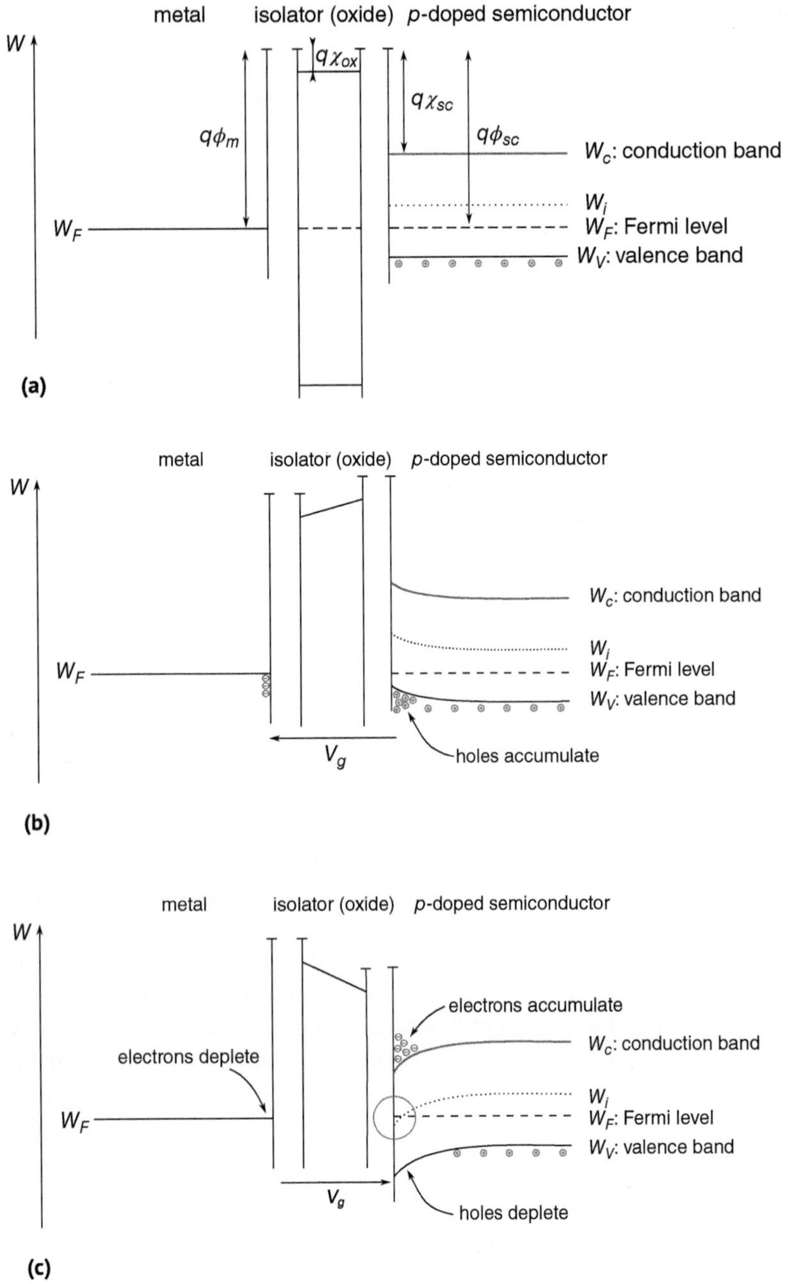

Figure 4.37 Idealized band diagrams of a MIS structure (vertical cut through MOSFET gate) in (a) equilibrium, (b) accumulation, and (c) inversion.

In reality, surface states and fixed charges inside the insulator bend the bands in equilibrium toward accumulation or inversion, see Figure 4.38. Accordingly, V_{th} is shifted, so that even normally-on MOSFETs can be produced. A positive or negative "flatband voltage" has therefore to be applied to the gate in order to achieve the flat bands as shown in the beginning, Figure 4.37a.

Finally, a third type of channel should be discussed. It is the 2DEG channel of high electron-mobility transistors (HEMTs), that are realized in III–V material systems, employing heterostructures. The structure of a HEMT resembles the MESFET structure in a way that the channel is by default conducting, and directly connected by source and drain, while the gate is isolated through a reverse-biased Schottky diode contact. But the channel works similar to a MOSFET channel, by appropriately bending the conduction band.

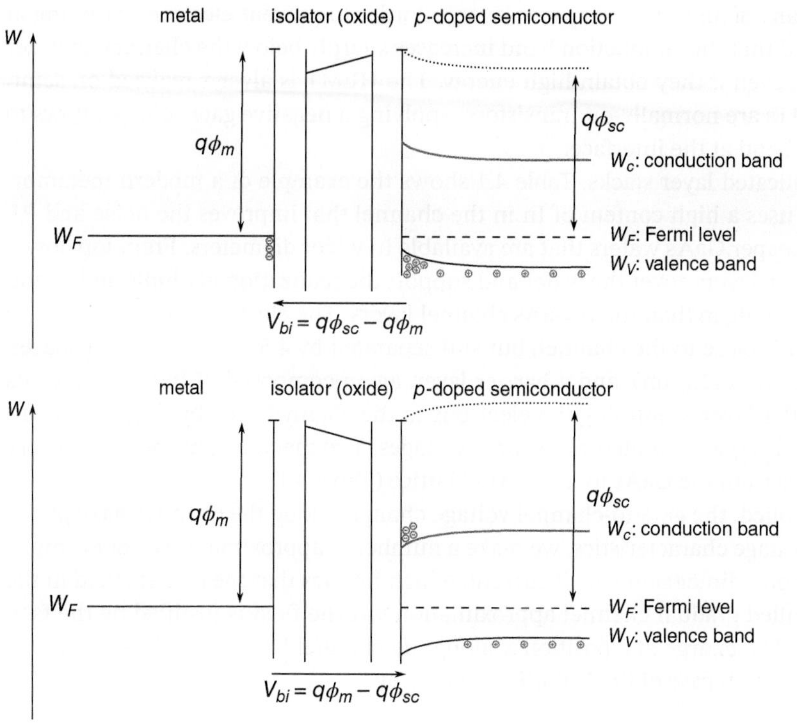

Figure 4.38 Band diagram of a MIS structure considering semiconductor surface states and oxide charges, resulting in a V_{th} shift toward positive (a) and negative (b) voltages.

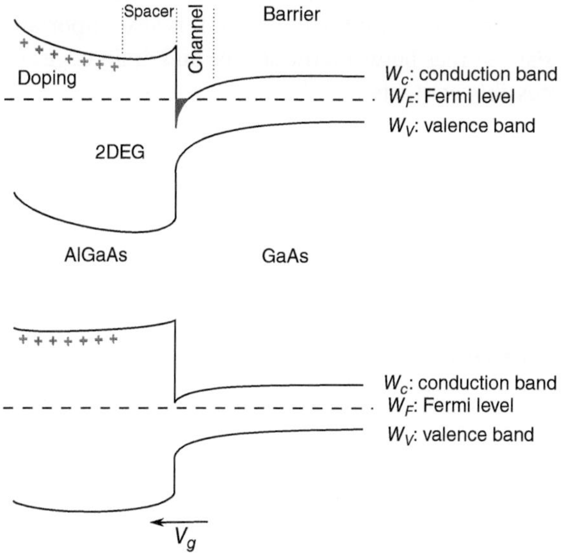

Figure 4.39 Sketch of the band diagram of a AlGas/GaAs HEMT.

A band diagram of a basic AlGaAs/GaAs HEMT is sketched in Figure 4.39. The gate contact would be situated on the left-hand side, and it would form a Schottky contact to a material like AlGaAs, which provides a higher bandgap. This top layer is n-doped, in order to supply electrons for the channel. The channel itself consists of a material like GaAs, which provides a lower bandgap, which causes a step in the conduction band. The electrons accumulate here, at the interface, at the lowest energy level. The step in the bandgap prevents the electrons to return to the location of the dopants, and their negative charge results in a bending of the conduction band to below the Fermi level. This type of channel therefore differs from the inversion channel; it is rather a quantum film at the hetero-interface. Two important details: The doped cap region is separated from the channel by means of an undoped spacer layer, in order to prevent electron scattering at the doping. The second is that it has to be ensured that the conduction band increases sharply below the channel, in order to confine the electrons within the thin channel even if they obtain high energy. The HEMT is always realized on semi-isolating substrates. As the figure suggests, HEMTs are normally-on transistors; applying a negative gate voltage allows to deplete the channel by lifting up the conduction band at the interface.

Real HEMT devices feature much more complicated layer stacks. Table 4.1 shows the example of a modern metamorphic HEMT (mHEMT). The mHEMT approach uses a high content of In in the channel that improves the noise and RF performance, but still can be processed on the cheaper GaAs wafers that are available in wider diameters. From top down, we see two cap and a Schottky contact layers. These layers cover the wafer and support the realization of ohmic and Schottky contacts. The InAlAs layers provide a higher bandgap than the InGaAs channel layers. The electrons are supplied by a very thin n-doped layer (so-called δ-doping) that is close to the channel, but still separated by 4.5 nm of undoped spacer. The narrower bandgap channel layer is also very thin (15 nm), and a barrier layer, again composed of InAlAs, prevents electrons from being scattered away into the buffer layer. Confining the electrons to the channel thereby suppresses parasitic substrate current, that could otherwise be observed at high drain-source voltages. The metamorphic buffer is a very elaborate crystal structure, that gradually changes from the GaAs to the InAlAs lattice (Table 4.1).

Once a voltage between drain and source is applied, the gate-to-channel voltage changes along the channel, as depicted in Figure 4.40. In order to estimate the current–voltage characteristics, we make a number of approximations. For example, it is assumed that the current in the channel is a one-dimensional drift current, which implies that the electric field in the channel only points into its direction. This so-called gradual channel approximation wrt the field is justified by the fact, that the channel is much longer than it is thick. The charge at a position x along the channel has then to be determined considering that the gate-to-channel voltage $V_{\text{SCR}}(x)$ in case of the MESFET is given by

$$V_{\text{SCR}}(x) = V_D - V_{\text{gs}} + V(x) \tag{4.44}$$

with the channel potential $V(x)$, gate-source voltage V_{gs}, and the Schottky diode's diffusion voltage V_D. This voltage is oriented vertically, the respective field is orthogonal to gate contact surface and channel. We also approximate the electron mobility μ_n to be constant, neglecting any velocity saturation or other high-field effect for the moment. It is also important to note, that the following discussion ignores the voltage drop in the resistive areas between the source and drain edges of the channel (marked as dotted lines in the figure) and the respective accessible contacts.

The drain-source current is with these assumptions given by:

$$I_{\text{ds}} = q \cdot \mu_n \cdot N_D(a - h(x)) \cdot W_g \cdot E(x) \tag{4.45}$$

Table 4.1 Typical layer structure of an mHEMT.

Function	Material	Doping (cm^{-3})	Thickness (nm)
Cap	$\text{In}_{0.43}\text{Ga}_{0.57}\text{As}$	n, $5 \cdot 10^{18}$	20
Cap	$\text{In}_{0.43}\text{Ga}_{0.57}\text{As}$	n, $3 \cdot 10^{17}$	20
Schottky contact layer	$\text{In}_{0.42}\text{Al}_{0.58}\text{As}$	n.i.d	15
δ-doping	$\text{In}_{0.42}\text{Al}_{0.58}\text{As}$	n, $5 \cdot 10^{12}$	
Spacer	$\text{In}_{0.42}\text{Al}_{0.58}\text{As}$	n.i.d	4.5
Channel	$\text{In}_{0.43}\text{Ga}_{0.57}\text{As}$	n.i.d	15
Barrier	$\text{In}_{0.44}\text{Al}_{0.56}\text{As}$	n.i.d	300
Metamorphic buffer			1500
s.i. substrate	GaAs	n.i.d.	

Source: Adapted from [22].

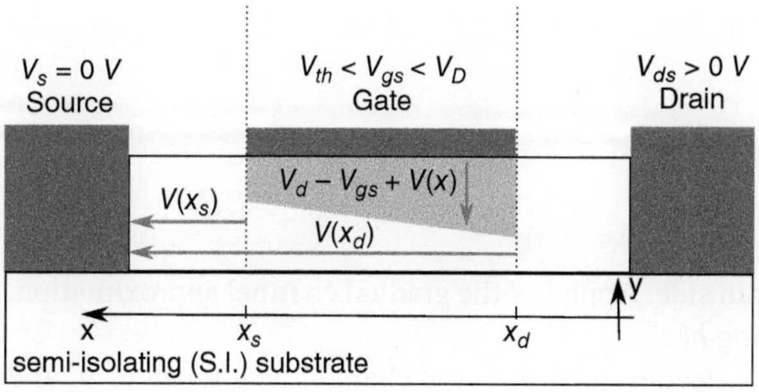

(a)

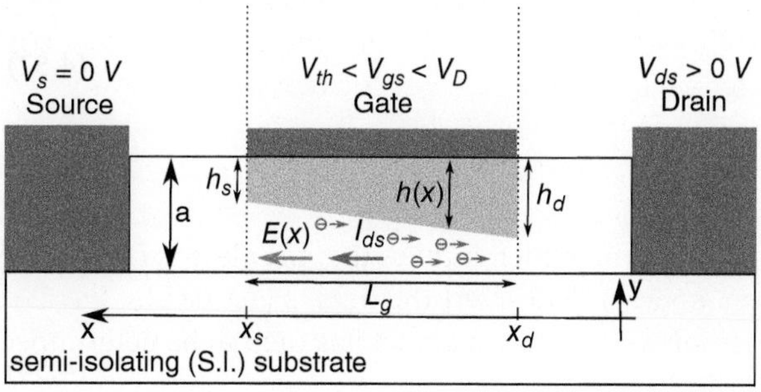

(b)

Figure 4.40 (a) Voltage and (b) current definitions in the FET channel.

with the electron charge q, channel doping N_D, the height of the SCR at a position x and the respective field in the channel $E(x)$. Note that I_{ds} is not considered to be depending on x, since we assume that the electrons flow from source to drain, and generation and recombination are negligible. The height of the SCR and the electric field need to be known in order to solve the equation. However, these quantities are linked, since the channel potential $V(x)$ controls the height and depends on the channel's electrical field $E(x)$. The width of the SCR is given by:

$$d(x) = a - h(x) = a - \sqrt{\frac{2\epsilon\left(V_D - V_{gs} + V(x)\right)}{qN_D}} \tag{4.46}$$

And for the electrical field, we have

$$E(x) = \frac{V(x)}{dx} = \frac{V(x)}{dh} \cdot \frac{dh}{dx} \tag{4.47}$$

the term dh/dx can be derived from eq. 4.46, so that we obtain

$$E(x) = \frac{h(x) \cdot q \cdot N_D}{\epsilon} \cdot \frac{dh}{dx} \tag{4.48}$$

Thereby, the equation for I_{ds} becomes:

$$I_{ds} = \frac{q^2 \cdot \mu_n \cdot N_D^2 \cdot h(x)}{\epsilon} \cdot (a - h(x)) \cdot W_g \cdot \frac{dh}{dx} \tag{4.49}$$

which can be solved by multiplying the equation on both sides with dx, and then integrating I_{ds} along the channel, and the right-hand term over the SCR height h.

$$\int_{x=x_d}^{x_s} I_{ds}\, dx = I_{ds} \cdot L_g \tag{4.50}$$

with the gate length L_g. The right-hand side yields:

$$\frac{q^2 \cdot \mu_n \cdot N_D^2 \cdot W_g}{\epsilon} \cdot \int_{h=h_s}^{h_d} \left(ah(x) - h(x)^2\right) \cdot dh$$
$$= \frac{q^2 \cdot \mu_n \cdot N_D^2 \cdot W_g}{\epsilon} \cdot \left(\frac{a}{2}(h_d^2 - h_s^2) - \frac{1}{3}(h_d^3 - h_s^2)\right) \tag{4.51}$$

since the SCR height varies from h_s at the source side to h_d at the drain side. Applying the gradual channel approximation, we also assume that the SCR height changes linearly when integrating $h(x)$.

As a final step to the derivation of the I–V characteristics, we now need to substitute the channel height values h_s and h_d by terms that depend on voltages as independent variables. Since it leads to a nicer formula, also the maximum channel height a will be related to voltages:

$$h = \sqrt{\frac{2\epsilon(V_D - V_{gs} + V(x))}{qN_D}} \tag{4.52}$$

$$a = \sqrt{\frac{2\epsilon(V_p)}{qN_D}} \tag{4.53}$$

which leads to, for source and drain side, respectively:

$$\frac{h_s}{a} = \sqrt{\frac{V_D - V_{gs}}{V_p}} \text{ since } V(x_s) = 0 \tag{4.54}$$

$$\frac{h_d}{a} = \sqrt{\frac{V_D - V_{gs} + V_{ds}}{V_p}} \text{ since } V(x_d) = V_{ds} \tag{4.55}$$

The equation for I_{ds} therefore becomes:

$$I_{ds}(V_{gs}, V_{ds}) = I_{DSS} \cdot \left(3\frac{V_{ds}}{V_p} - 2\frac{(V_D - V_{gs} + V_{ds})^{3/2} - (V_D - V_{gs})^{3/2}}{V_p^{3/2}}\right) \tag{4.56}$$

with

$$I_{DSS} = \frac{q^2 \cdot \mu_n \cdot N_D^2 \cdot W_g \cdot a^3}{6\epsilon L_g} \tag{4.57}$$

A similar estimation of the I–V characteristics can be done for the MOSFET, considering that the channel charge Q_{ch} is understood from the capacitor formed by the MIS structure:

$$Q_{ch}(x) = C_{ox}[V_{gs} - V(x) - V_t] \tag{4.58}$$

with C_{ox} denoting the MIS capacitance. The current now can be derived, by integration, analogous to the case mentioned earlier:

$$\int_0^{L_g} I_{ds}\,\mathrm{d}x = \int_0^{V_{Ds}} W_g \mu_n C_{ox}(V_{gs} - V(x) - V_t)\mathrm{d}V \tag{4.59}$$

which results in

$$I_{ds} = \frac{k'}{2}\frac{W_g}{L_g}[2(V_{gs} - V_t)V_{ds} - V_{ds}^2] \tag{4.60}$$

where

$$k' = \mu_n C_{ox} = \frac{\mu_n \epsilon_{ox}}{t_{ox}} \tag{4.61}$$

with the thickness t_{ox} and dielectric constant ϵ_{ox} of the gate oxide separating contact from semiconductor.

In both cases, I_{ds} increases approximately linearly with V_{ds}, for low values of V_{ds}, while V_{gs} controls the slope. At a certain drain-source voltage, the equations reach a maximum, where the slope naturally reaches zero. The drain-to-source resistance therefore starts with a constant value, which increases for higher values of V_{ds}.

At this point, it is required to determine the maximum I_{ds}, which is reached for the case that the channel is completely depleted at the drain side. In case of the MESFET, the condition is reached once the SCR at the drain side fills the whole channel, i.e., $h_d = a$. It reads:

$$I_{\mathrm{ds,sat}}(V_{gs}) = I_{\mathrm{DSS}} \cdot \left(1 - 3\frac{V_D - V_{\mathrm{gs}}}{V_p} + 2\left(\frac{V_D - V_{\mathrm{gs}}}{V_p}\right)^{3/2}\right) \tag{4.62}$$

In case of the MOSFET, it is reached if $Q_{\mathrm{ch}}(d) = C_{\mathrm{ox}}[V_{gs} - V_{ds} - V_t] = 0$:

$$I_{\mathrm{ds,sat}}(V_{gs}) = \frac{k'}{2}\frac{W_g}{L_g}(V_{\mathrm{gs}} - V_t)^2 \tag{4.63}$$

Figure 4.41 shows a plot of I_{ds} as a function of V_{ds} with V_{gs} as a parameter. This type of graph is called output I–V plot. The area labeled *linear region* has been treated so far, and we can see that for very low values of V_{ds}, the current shows a linear slope, which can be controlled by the gate voltage. The gray line denotes the boundary, where the current reaches its maximum, and the channel gets fully depleted at the drain side. The validity of the theory discussed so far ends at this condition—the so-called *saturation region* will be discussed in the following section.

4.3.1.3 Saturation Region

The derivation of the FET's current–voltage characteristics so far treated the case where gate and drain voltages changed the channel charge. This approach is mathematically only valid until the channel becomes fully depleted on the drain side. When this condition is met, the condition in the FET seems to be contradictory at first: on one hand, there is depletion on the drain end of the channel. This means that the channel ceased to exist and we would not expect any current to flow. On the other hand, in the rest of the channel, carriers are still accelerated according to the lateral field and the current is flowing. Our first-order model may be too simple to explain what is happening now, but it can be said that the current keeps flowing. Otherwise, we would see the accumulation of carriers on the drain side of the channel, which eventually would change the field as well.

In MESFETs and HEMTs, it can be assumed that the channel is not fully depleted due to the charge of the electrons that are accelerated toward the drain. In MOSFETs, the electrons are accelerated toward a SCR of a pn junction, and since they enter the SCR on the p-side, the field inside the SCR accelerates them away toward the n-doped drain contact area.

Under depletion condition, the voltage between drain end of the open channel to the source end, V'_{ds} stays constant, as it equals the value of $V_{\mathrm{ds,sat}} = V_p - V_D + V_{\mathrm{gs}}$ required to pinch-off the channel at the drain end, for the example of the MESFET, or in case of the MOSFET: $V_{\mathrm{ds,sat}} = V_{\mathrm{gs}} - V_{\mathrm{th}}$. Under the assumption that the pinch-off takes place exactly at the drain end, so that the channel length is equal to the gate length L_g, we can assume that I_{ds} stays constant and equal to $I_{\mathrm{ds,sat}}$, which now only depends on V_{gs}.

Rather long and highly doped channels reduce the impact of increasing drain voltages to negligible channel length reduction. In MESFETs and HEMTs, the SCR also extends to a great part into the channel region between gate and drain contact. The electric field in the channel peaks at the drain side of the gate electrode, and the acceleration of the electrons toward the drain certainly delays the extension of the SCR toward the source.

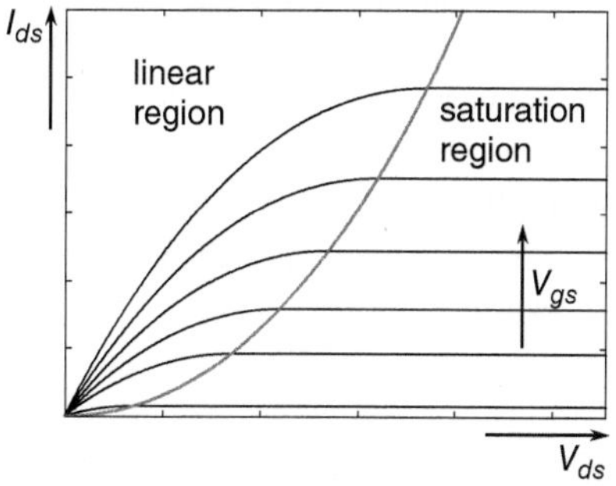

Figure 4.41 MESFET output I–V plot calculated according to Eq. (4.56).

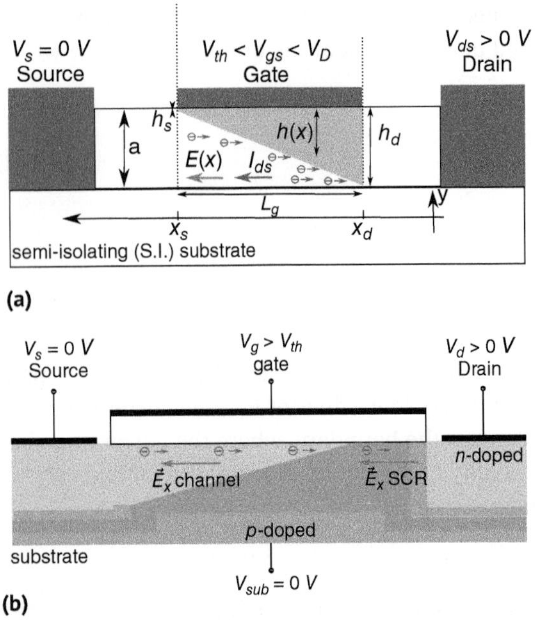

Figure 4.42 Sketch of MESFET (a) and MOSFET (b) in saturation.

As a first-order approximation, we can therefore assume that in the saturation region, the FET behaves like a voltage-controlled current source. As stated earlier, the voltage between drain end and source end of the channel underneath the gate, V'_{ds}, remains constant and equal to $V_{ds,sat}$, so that I_{ds} stays constant at $I_{ds,sat}$ (Figure 4.42).

4.3.1.4 Limitations of the Simple Theory and Short-channel Effects

While the theory presented earlier followed the classic derivation by Shockley [23] and allows for a simple derivation of the transistor I–V curve, it obviously serves a good understanding of the transistor operation at the cost of neglecting effects that are important in real devices.

First of all, the derivation stopped at the saturation voltage, claiming that the extension of the SCR underneath the gate, and therefore the reduction of effective channel length, could be neglected. In reality, this is often not the case, as modern gates are very short, and even a tiny bit of channel modulation will result in an approximately linear increase of I_{ds} in the saturation region. The effective gate length was hidden in the derivation of the drain-source currents in the saturation current $I_{DSS} \propto 1/L_g$.

MOSFETs are realized on top of a conducting substrate, from which they are isolated either through a pn-SCR or through a buried oxide layer. It is therefore possible to apply a voltage to the substrate, which shifts the threshold voltage. This is called body-effect or back-gate effect.

An other effect to consider is that electron (or hole) mobility is by no means constant in modern FETs. The channels are too short and the voltages too high, so that the electrons are accelerated toward velocity saturation. More elaborate analytical models therefore divide the channel into a region of constant mobility and a region of constant velocity [24]. This approach is more precise, but still not catching everything, as the gate length today is very short and fields are very high. Therefore, electrons are accelerated in a way that their energy can be considered to be equivalent to a temperature that is a lot higher than the lattice temperature: so-called hot electrons. On one side, ballistic transport might happen on the very short distances. On the other side, these electrons can be scattered out of the channel, e.g., toward the surface or toward the buffer below the channel. Proper design of the layer structure of HEMTs, for example, prevents electrons from leaving the channel through a back-barrier, as given in the example of the mHEMT layer structure above. This prevents the formation of a second, parasitic, channel that reacts differently to the applied voltages and is not as well controlled by the gate voltage. Even if there is no parasitic channel, it often has a negative effect on FET performance. The electrons reach regions of the FET, where no electrons should be, above or below the channel. If they accumulate in these nonconducting regions, or compensate trap charges, they need to be taken into account when calculating the channel charge as a function of surface and gate contact charges. Therefore, until the scattered electrons recombine, a reversible temporal change in the FET properties is observed, a so-called memory effect that will be addressed in a later section.

Another effect is the so-called drain-induced barrier lowering (DIBL), which may occur in weak inversion in MOSFETs as well as in HEMTs in the regime where low I_{ds} is expected. Under this condition, the field between drain and gate can affect the conduction band also in the channel region, and thereby support conduction. As a result, a kink in the I_{ds}/V_{ds} curve is observed, from where a very low constant I_{ds} starts to increase approximately linearly as if a parallel resistive current would be added.

Another important effect in transistors is self-heating. It occurs if significant bias is applied to the FET in terms of DC, with I_{DS} and V_{DS} being the main quantities in normal operation. The respective power $P = I_{DS} \cdot V_{DS}$ is partially transformed into RF output power. The rest is transformed into thermal energy inside the transistor and leads to a dynamic increase in temperature. Temperature changes rather slowly, so that device temperature after switch-on often stays constant, but it shows a number of negative effects: the drain current is reduced, and device lifetime can be affected if specified maximum channel temperatures are exceeded.

High voltages in the device lead to high electric fields, which may support impact ionization. Breakdown due to impact ionization causes the device to melt. But also below this threshold, an increase in drain current can be observed in certain devices.

Measured output I–V curves of a GaAs power pHEMT are shown in Figure 4.43. A pHEMT is comparable to the mHEMT discussed earlier, but features a lower In content in the channel. The figure compares a standard measurement (dotted line) to a curve determined in pulsed conditions (solid line). The pulsed measurement applies a quiescent bias point that usually is at low drain and gate voltages. From there, the output I–V curve is measured during very short V_{gs}, V_{ds} pulses. The sub-microsecond bias pulses are too short for self-heating, memory or avalanche effects to occur.

The I–V curve in the figure differs from the ideal model in various ways. A few effects are highlighted in the figure, as they become visible in the comparison between static and pulsed measurement. Trapping, a memory effect associated with scattering electrons to deep traps at surface or in the buffer, affects the knee voltage, while self-heating reduces I_{ds} with increasing dissipated powers, and impact ionization is observed at high voltages. It is in addition observed that the current increases linearly even in the saturation region, which is a sign for gate length modulation. At a drain voltage of about 4 V, the DIBL effect leads to an increase of I_{ds} that was fully suppressed by the gate voltage at lower drain voltages.

4.3.2 FET Capacitances

Applying voltages to the FET affects the charges stored in the region under the gate. In case of the MESFET taken as an example, the SCR depends on V_{GS} and V_{DS}, reducing the volume that contains electrons. In the MOSFET, the voltages control the charge density in the inversion layer. Figure 4.44 depicts the charges.

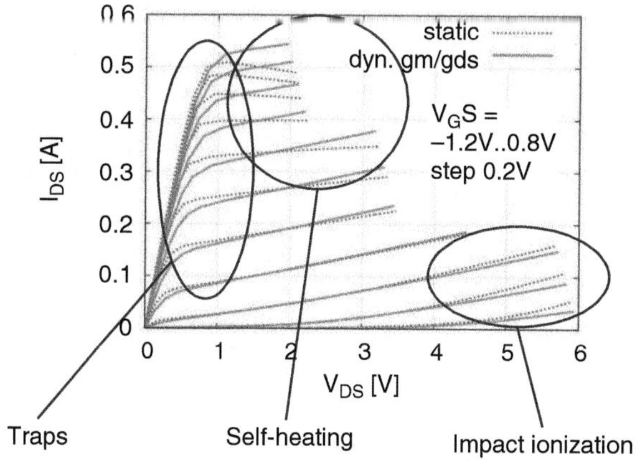

Figure 4.43 Static and dynamic I–V curves of a 8×100 μm GaAs power pHEMT [25]. *Source:* Courtesy of I. Kallfass.

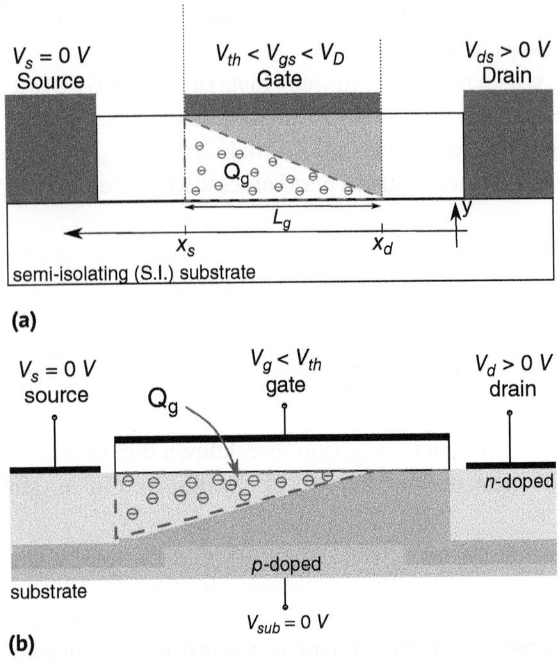

Figure 4.44 Sketch of (a) MESFET and (b) MOSFET charges.

In the small-signal regime, the respective depletion capacitances will be visible. In general, three capacitances are visible in the intrinsic FET between the three terminals, the gate-source capacitance C_{gs}, gate-drain capacitance C_{gd}, and a smaller capacitance between gate and drain C_{ds}.

These three capacitances describe one bulk three-terminal charge Q_g under the gate. Therefore special care has to be taken when defining the nonlinear capacitances for the FET model [26, 27].

First of all, the three-terminal charge Q_g needs to be split into two charge sources Q_{gs} in the gate-source branch, and Q_{ds} in the drain-source branch. The topology of the equivalent circuits is given in Figure 4.45. Only the capacitive elements are shown for sake of clarity.

These charges contribute to gate and drain currents by their time derivative:

$$I_g = \frac{dQ_{gs}(V_{gs}, V_{ds})}{dt} \tag{4.64}$$

$$I_d = \frac{dQ_{ds}(V_{gs}, V_{ds})}{dt} + I_{ds}(V_{gs}, V_{ds}) \tag{4.65}$$

which becomes in the small-signal case

$$i_g = \frac{\partial Q_{gs}}{\partial V_{GS}} j\omega v_{gs} + \frac{\partial Q_{gs}}{\partial V_{DS}} j\omega v_{ds} \tag{4.66}$$

$$= j\omega C_{11} v_{gs} + j\omega C_{12} v_{ds} \tag{4.67}$$

$$i_d = \frac{\partial Q_{ds}}{\partial V_{GS}} j\omega v_{gs} + \frac{\partial Q_{ds}}{\partial V_{DS}} j\omega v_{ds} + \frac{\partial I_{ds}}{\partial V_{GS}} v_{gs} + \frac{\partial I_{ds}}{\partial V_{DS}} v_{ds} \tag{4.68}$$

$$= j\omega C_{21} v_{gs} + j\omega C_{22} v_{ds} + g_m v_{gs} + g_{ds} v_{ds} \tag{4.69}$$

Since the charges depend on both terminal voltages, the elements C_{12} and C_{21} show up in the small-signal case, as shown in Figure 4.45b. These are called transcapacitances in analogy to the transconductance g_m. The transcapacitances are current sources controlled by the derivative of a voltage. If not properly implemented, these transcapacitance elements will drive currents that are quite different from the currents that flow due to the storage of charges. What is intended to be described through the two functions Q_{gs} and Q_{ds} is the bulk charge underneath the gate, which is energy conserving. This is not automatically the case with transcapacitances present.

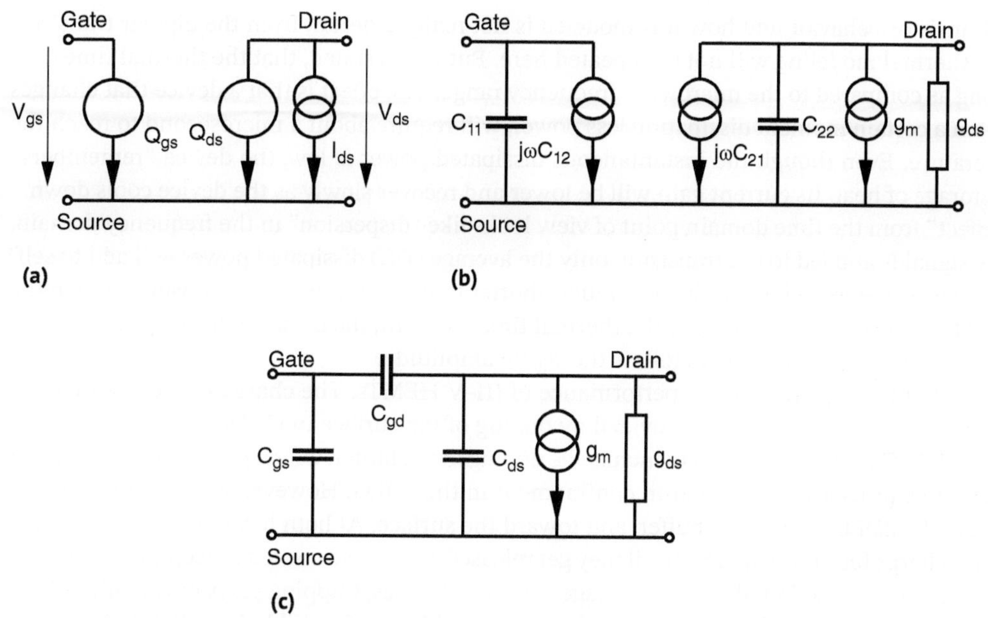

Figure 4.45 Interpretation of transcapacitances as capacitance using the example of a FET (only the capacitive elements of the model are shown). (a) Gate and drain charge sources, (b) corresponding small-signal elements, and (c) common capacitance setup.

In order to solve the issue, these small-signal currents can be compared to the characteristics of the common small-signal model. The common π-type equivalent circuit of the FET shown in Figure 4.45c, and it yields:

$$i_g = j\omega(C_{gs} + C_{gd})\, v_g - j\omega C_{gd}\, v_d \tag{4.70}$$

$$i_d = -j\omega C_{gd}\, v_g + j\omega(C_{ds} + C_{gd})v_d + g_m v_{gs} + g_{ds}v_{ds} \tag{4.71}$$

Enforcing that the two small-signal descriptions become equivalent ensures that the large-signal model observes energy conservation in the nonlinear capacitances. This is possible if the condition $C_{12} \stackrel{!}{=} C_{21}$, or $\partial Q_{gs}/\partial V_d \stackrel{!}{=} \partial Q_{ds}/\partial V_g$, is valid, and it follows:

$$C_{gs} + C_{gd} = C_{11} \tag{4.72}$$

$$C_{gd} = -C_{12} = -C_{21} \tag{4.73}$$

$$C_{ds} + C_{gd} = C_{22} \tag{4.74}$$

The issue of energy conservation or even charge conservation is a big issue in the formulation of FET models. It also strictly limits the degrees of freedom in the formulation of the different charges, and often a difference remains between the bias dependence of measurement-extracted capacitances and the functions that are available to describe it.

Without going into detail, it should just be mentioned that the issue of energy conservation is not better solved by defining nonlinear capacitances instead of charges. While this is a quick solution to describe the small-signal behavior, such a model can behave unexpectedly in the general large-signal simulation case. Even if the numerical inaccuracy might be tolerable, it is often observed that the convergence of the nonlinear simulator is compromised.

4.3.3 Dispersion or Memory Effects, and Self-heating

The term "memory effects" denotes the property of a transistor to "remember" previous operation conditions in a way that it changes its actual characteristics. One effect that falls into this category, although it is usually not called a memory effect, is self-heating.

The way self-heating affects transistor behavior and how it is modeled is not really different from the bipolar transistor models, therefore the section on thermal modeling will not be repeated here. But it was stated, that the thermal time constant is to be considered very long as compared to the microwave frequency range. The effect is that a device that changes its operation conditions, e.g., from a certain moment dissipating less power, will require about a microsecond to reach the respective lower channel temperature. Even though the instantaneous dissipated power is low, the device "remembers" the previous state through the storage of heat. Its current gain will be lower and recover slowly as the device cools down.

What looks like a "memory effect" from the time domain point of view looks like "dispersion" in the frequency domain. If a single-frequency microwave signal is applied to the transistor, only the average (DC) dissipated power will add to self-heating. This will be the case as long as the period of the signal is much shorter than the thermal time constant. When the frequency is reduced further and the period is in the range of the thermal time constant, the device will show a frequency-dispersive behavior, once the channel temperature starts to follow the signal amplitude.

In a similar way, surface and bulk traps compromise the performance of III–V HEMTs. The charge in the channel of a HEMT depends on the opposite positive charge, which is provided by doping of the surface in GaAs HEMTs or through piezo-electric charge in case of GaN. In GaN, it is also usual to include Fe doping in the buffer, which pushes the conduction band toward higher energies and thus provides better electron confinement in the 2DEG. However, at high voltages, hot electrons might be scattered out of the 2DEG toward the buffer, and toward the surface. At both locations, they might get trapped and compensate a positive charge for a limited time, until they get released again. As a consequence, channel charge is reduced, which has a similar effect as a reduction of the gate voltage. In GaAs devices, trapping plays only a minor role. As shown in Figure 4.43, it only gradual changes the transistor performance and is mostly visible in a slightly increased small-signal output conductance g_{ds} and transconductance g_m.

In GaN devices, the effect is more pronounced; see Figure 4.46 as an example. In these devices, a shift in threshold voltage is observed due to trapping in the buffer layer within the intrinsic transistor region, underneath the gate. This is an effect that is also called back-gating, as the trapped charge acts equivalent to a virtual gate electrode below the channel. The second effect is electrons getting trapped at the surface, in the access region between gate and drain contact. The effect on this extrinsic region is an increase of the drain access resistance R_d, as the 2DEG charge density is reduced. This therefore affects much more the knee region, the transition of linear to saturation regions, and is called knee walkout. This dispersion in GaN devices is also known to be linked to leakage currents in the buffer, or (surface) leakage current of the gate contact, which supports the electrons to the traps.

In MOSFETs, dispersion might be observed due to electrons scattered into the gate insulator, or into the backside insulator in case of SOI transistors.

The first models for trapping that made it into widespread transistor models were aiming at closing the gap between the values of g_m and g_{ds} as obtained from small-signal extraction and the model predicted $G_m = \partial I_{DS}/\partial V_{GS}$ and $G_{ds} = \partial I_{DS}/\partial V_{DS}$. Lowering the output impedance of a transistor model is easily achieved by adding an appropriate resistor in parallel to the drain-source current source. In order to prevent this resistor to degrade DC performance, a DC blocking capacitor is connected in series to this resistor. This approach was introduced in the mid-1980s for GaAs transistors [28]. The transconductance can be treated similarly. Since the RF g_m is lower than $\partial I_{DS}/\partial V_{GS}$, it is possible to employ an *RC* low-pass voltage divider that reduces V_{gs} at high frequencies.

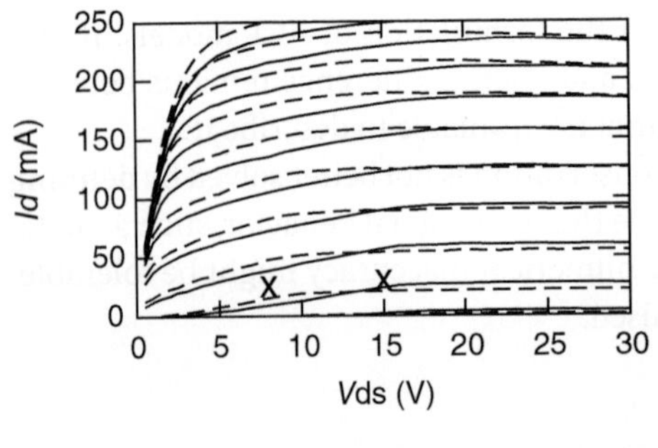

Figure 4.46 Dispersion in GaN HEMTs: output *I*–*V* curve, measured in pulsed condition with quiescent biases marked by X of 15 V, 30 mA (solid lines), and 8 V, 25 mA (dashed lines). *Source*: Courtesy of Ferdinand-Braun-Institut.

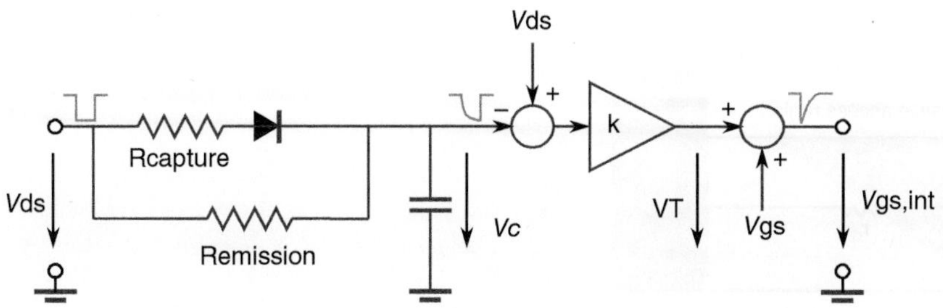

Figure 4.47 A subcircuit describing the effect of drain lag on the internal gate voltage. *Source*: [29]/IEEE.

The advantage of these approaches is their simplicity. Inherently, the DC behavior will be unaffected, since the dispersion model is purely additive. But the scope of these implementations is clearly limited, since g_m and g_{ds} are just lowered by adding a bias-independent current or voltage divider for g_{ds} or g_m, respectively. Therefore, these models are not really helpful if the main issue is far off the quiescent bias point. For example, the power amplification properties of GaN HEMTs can be compromised by knee walk out, i.e., a shift of the high-current–low-voltage region toward higher V_{ds} and lower I_{ds}.

A recent compact approach to introduce trapping effects in GaN models attempts to describe the physical effect of charge accumulation through a subcircuit, a similar approach that was already successfully applied for self-heating. The respective equivalent circuit is shown in Figure 4.47 [29]. This circuit generates V_s, a delayed replica of the drain-source voltage V_{ds}. The difference between these two voltages represents a quantity that is proportional to the internal potential maintained by the traps. The *RC* time constants are different depending on the polarity of the change in V_{ds}, which is realized by the diode. Through this approach, the model can realize short time constants for electron capture and long time constants for electron release. Finally, the voltage is appropriately scaled in order to adjust it to the drain lag observed in measurement and added to the applied voltage V_{gs} in order to obtain the effective internal voltage $V_{gs,int}$. The necessity of relying on two different time constants can be motivated by the measurement shown in Figure 4.46a. This I–V plot shows almost identical drain currents for the region $V_{ds} > 15$ V, which is the region where the drain voltage during the pulses is higher than the quiescent voltage. This is an indication, that the capture of electrons happens almost instantaneously, within the pulse duration, so that the conditions for both measurements are the same. For the range $V_{ds} < 15$ V, the curves differ, and knee walk-out is observed. The release time constant is long compared to the duration of the pulses, and with increasing V_{ds}, memory effects become more pronounced.

This approach is very elegant, as it accounts for the rather involved drain lag in a consistent way, requiring a minimum of parameters, i.e., the trapping and detrapping time constants and the scaling factor k. If gate lag is also to be considered, it becomes necessary to enhance the model accordingly. Also, if more than one trap time constant should be modeled, it would be required to use multiples of this model and add the results. The major advantage of the model is that it is not just fixing a mismatch in g_{ds}, but it aims at modeling the physical mechanism, the trapping. Thereby, the validity of the model is not restricted to a small range in bias, but should establish that a generally accurate model stays applicable even in presence of drain or gate lag.

4.3.4 Small-signal Model and Cutoff Frequencies

Figure 4.48 shows the small-signal equivalent circuit commonly used for FETs. In order to motivate the topology on the background of device physics, the equivalent circuit elements were already drawn inside the cross-section of a MESFET in Figure 4.48a. The elements g_m and g_{ds} describe small-signal transconductance and drain-source conductance. The capacitances C_{gs} and C_{gd} account for the charge stored underneath the gate and are usually much larger than the third charge C_{ds}. The resistances are access resistances, representing the metal resistance, the metal-semiconductor contact, and the access resistance inside the semiconductor.

Typical S-parameters measured on-wafer are shown in Figure 4.49. Two 0.25 μm GaN HEMT devices were measured, a small device of 5 × 50 μm gate width, and a large one of 8 × 250 μm gate width. The S-parameters look quite similar to those of HBTs discussed earlier. Even the kink in S_{22} is observed at a certain frequency [8]. The main difference is that the FET gate is isolated, so that the shape of S_{11} looks like a *RC* series connection, starting at the open circuit and following a semi-circle. The access inductances L_g and L_s transform this curve as discussed earlier for the HBT. If the device is not

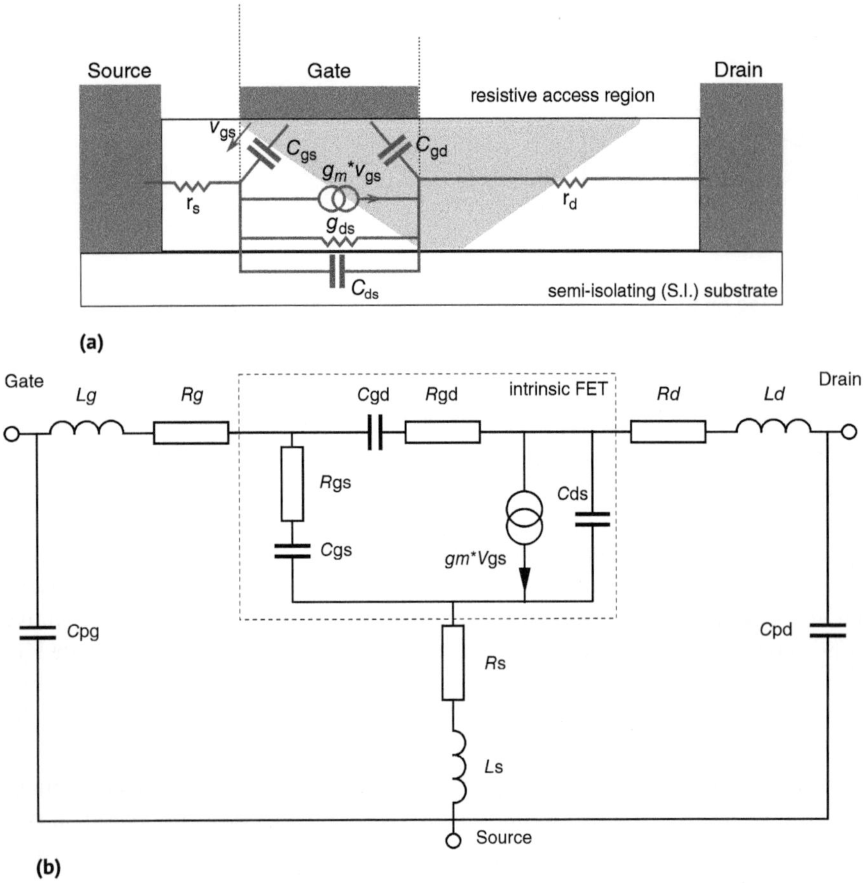

Figure 4.48 Small-signal equivalent circuit of a FET. (a) Drawn into the MESFET cross-section and (b) complete model.

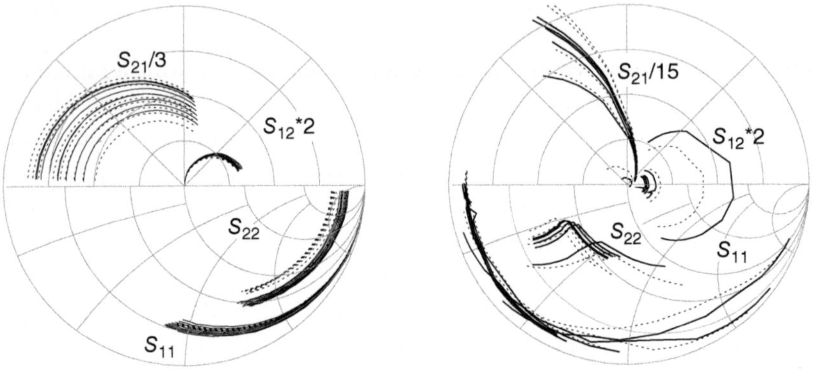

Figure 4.49 Left: On-wafer measured (dashed lines) and modeled (solid lines) S-parameters of 2 × 50 μm GaN HEMT, at $V_{ds} = 20$ V, $f = 0.1...20$ GHz, $I_d = 10...100$ mA. Right: 8 × 250 μm GaN power cell, at $V_{ds} = 26$ V, $f = 0.1...10$ GHz, $I_d = 20...600$ mA [30]. *Source*: Courtesy of Ferdinand-Braun-Institut.

saturating ideally, and $g_{ds} \neq 0$, the locus of S_{22} will not start at the open circuit, but rather at a high resistance according to the drain-source conductance.

The transit frequency f_t was introduced for HBTs above as the frequency where $|h_{21}|^2$ in common source configuration equals 0 dB. For the sake of simplicity, C_{gd} and extrinsic elements are usually neglected in the derivation, which yields for h_{21}:

$$|h_{21}| = \frac{g_m}{\omega C_{gs}} \tag{4.75}$$

and the resulting transit frequency reads

$$f_t = \frac{g_m}{2\pi C_{gs}} \tag{4.76}$$

The maximum frequency of oscillation can also be calculated based on the simplified intrinsic equivalent circuit. It is derived from the maximum available gain, which reads

$$|\mathrm{MAG}| = \frac{|Y_{21}|^2}{4\mathrm{Re}(Y_{11}) \cdot \mathrm{Re}(Y_{22})} \tag{4.77}$$

which yields the following formula for $f_{\max}$:

$$f_{\max} = \frac{g_m}{4\pi C_{gs}}\sqrt{\frac{R_{ds}}{R_{gs}}} = \frac{f_t}{2}\sqrt{\frac{R_{ds}}{R_{gs}}} \tag{4.78}$$

These formulas are commonly given for FETs, as they allow for a good guess on how scaling and technological changes affecting the channel will impact the cutoff frequencies.

However, with aggressive downscaling, it becomes necessary to also consider the extrinsic elements and their contribution to the FET's low-pass behavior. A more sophisticated derivation yields for the maximum frequency of oscillation:

$$f_{\max} = \frac{1}{4\pi\sqrt{R_s C_{gd}\tau + (C_{gd}/g_m)R_g C_{gs} + (C_{gd}/g_m)^2 g_{ds}(R_g + R_S + 1/g_{ds})}} \tag{4.79}$$

4.3.5 Large-signal Modeling

Numerous models are available for FET devices at different levels of abstraction. The most abstract type relies on artificial neural networks (ANN) to describe the relation of port currents and charges to port voltages. ANN-based models have the advantage to be very precise even if the underlying physics are not understood in detail. On the other hand, training of the ANN requires a lot of measurement data. On the other side, physics-based models describe the equivalent circuit element of a FET through parameterized formulas, which are derived from device physics. The advantage of these models is their relatively close connection to physics, which enables, e.g., modeling of the technological spread of parameters and scaling the model with device size. Also, the number of measurements to determine the model is much reduced. The drawback of this approach is that it requires in-depth knowledge of transistor physics, so that the development time of new model features can take up to several years. Fully featured models like the BSIM4 for MOSFETs also take so many effects into account that the number of parameters to be determined might exceed 200. In between, you find analytical models, which are mathematical models that are derived on a basic understanding of device physics, but starting from what is actually visible in measurement. The underlying mathematics are therefore much simpler than in case of the physics based models, and simulation time and numerical stability should be better. Traditionally, the domain of III–V devices was modeled by analytical models, as physics were more complex and technology less mature than in case of MOSFETs on silicon. However, today the modeling pushes toward physics-based models for III–V devices as well, while analytical models are being tested for MOSFETs in order to simplify model parameter extraction and to speed up simulation time. In the following, large-signal modeling will be explained with the example of analytical models for GaAs, InP, or GaN HEMTs.

The GaAs MESFET was the best microwave transistor available at the time when simulators became available, and the models grew in number and matured together with computer power and simulator codes. First-generation models like Statz or Curtice Cubic had to satisfy two objectives: describe the device while adding a very light-weight numerical load to the limited computer power of those times. Practically, this meant to approximate the transfer functions by cubic functions and to keep capacitances constant. These approaches enabled simulation in the early years, but restricted the validity of the model to a narrow range of bias, showing nonphysical behavior for high voltages [31]. Especially, a current approximated by a cubic function increases according to V_{ds}^3 at larger V_{ds}, which is not physical and possibly harmful to numerical stability.

Modern transistor models, such as the EEHEMT and the Angelov (Chalmers) model, are no longer limited by maximum computer power. They strongly rely on tanh functions that are numerically well behaved and for quite a while no longer really an issue concerning computer power. Additionally, these models are formulated in order to describe even the weakly nonlinear behavior of the transistors, with a number of parameters that allow to fine-tune not only the transfer characteristics but also their third-order derivative g_3.

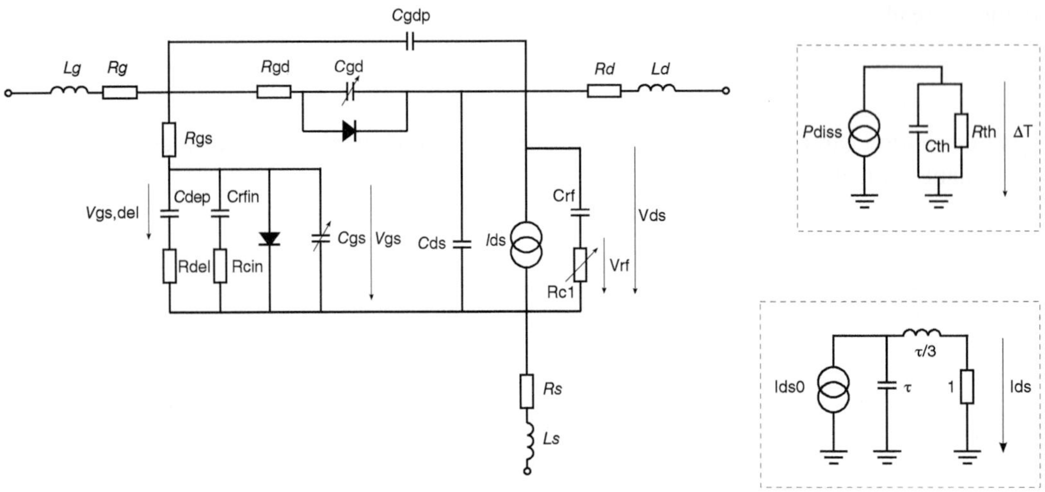

Figure 4.50 Equivalent circuit of the Angelov (Chalmers) HEMT model.

As an example, the Angelov (Chalmers) model will be treated here. Its equivalent circuit is shown in Figure 4.50. It consists of the model for I_{ds}, which is time-delayed using a Bessel-type low-pass filter shown in the lower right corner, providing a time delay of τ. The main capacitances are C_{gs} and C_{gd}, which are modeled through bias-dependent charge sources. The drain-source capacitance C_{ds} is assumed to be constant. The two diodes in the equivalent circuit describe the gate Schottky contact. In order to account for dispersion, the two RC branches at gate-source and one RC branch at drain-source are added. At the gate, these describe a variation of C_{gs} with frequency and are also used to obtain a delayed voltage $V_{gs,del}$. At the output, the bias-dependent resistor R_{c1} is used to describe the reduction of the output impedance at RF frequencies in contrast to the DC value. The voltage V_{rf} across this resistance is used to model the backgating effect. The thermal subcircuit on the right side allows to simulate the actual channel temperature of the device. The model accounts for the feed resistances at gate, drain, and source, R_g, R_d, and R_s, respectively, and also for the feed inductances L_g, L_d, and L_s.

This model provides a number of different functions for the drain current, which are based on the first current model for I_{ds}:

$$I_{ds} = I_{pkoT} \cdot (1 + \tanh\psi) \cdot \tanh(\alpha V_{ds}) \cdot \left(1 + \underbrace{\Lambda \cdot V_{ds}}_{\text{slope in saturation}} + \underbrace{L_{sboT} \cdot e^{(V_{dg} - V_{tr})}}_{\text{breakdown}} \right) \tag{4.80}$$

In general, this model relies on two tanh functions to describe I_{ds} as a function of V_{gs} (through $(1 + \tanh\psi)$) and on V_{ds} (through $\tanh(\alpha V_{ds})$). The parameter I_{pko} is used to scale the function. $I_{ds} = I_{pko}$ is reached at the point of maximum g_m for $V_{ds} = V_{pko}$. The parameter V_{pko} will be discussed later. The index T is used to indicate that this parameter depends on temperature, this nomenclature will also be used for the other temperature-dependent parameters. The tanh function used for the V_{ds} dependence describes the linear region, but would yield perfectly constant currents in saturation. The parameter Λ allows to add a slope to the curves, and breakdown is described by the parameters L_{sbo} and V_{tr}.

The dependence $I_{ds}(V_{gs})$ is mainly governed by the function ψ, that reads:

$$\psi = P_{1m}T_1 + P_2T_1^2 + P_3T_1^3 \tag{4.81}$$

with the model parameters P_2 and P_3. Parameter P_1 is turned into a function P_{1m} that alters its value for small values of V_{ds}:

$$P_{1m} = P_{1T}(1 + B_1/T_0^2) \tag{4.82}$$

$$T_0 = \cosh(B_2 V_{ds}) \tag{4.83}$$

For $V_{ds} \to 0$, P_{m1} becomes equal to $P_{1T}(1 + B_1)$, and for $V_{ds} \to \infty$, P_{m1} approaches P_{1T}. B_1 and B_2 therefore introduce kind of a cross-coupling of the V_{gs} and V_{ds} dependencies in the linear region.

The dependence of ψ on V_{gs} is described through the adjusted gate-source voltage function T_1.

$$T_1 = V_{\mathrm{gs,del}} - V_{\mathrm{pkm}} \tag{4.84}$$

$$V_{\mathrm{pkm}} = V_{\mathrm{pks}} + D_{\mathrm{vpks}}(\tanh(\alpha_s V_{ds}) - 1) - \underbrace{V_{\mathrm{bg}}}_{\text{dispersion}} - \underbrace{V_{\mathrm{sb2}}(V_{\mathrm{dg}} - V_{\mathrm{tr}})^2}_{\text{breakdown}} \tag{4.85}$$

T_1 is the delayed gate voltage shifted by V_{pkm}. For $V_{\mathrm{gs,del}} = V_{\mathrm{pkm}}$ it follows $\psi = 0$, and tanh ψ is at the point of its steepest slope. Thus g_m is maximum for this gate voltage. The parameter D_{vpks} together with the parameter α_s allows to describe the shift in V_{pkm} when V_{ds} is varied.

Dispersion is included through the backgating voltage V_{bg}. It introduces a dynamic feedback from V_{ds} to V_{gs}, using the parameter K_{bgate}:

$$V_{\mathrm{bg}} = K_{\mathrm{bgate}} V_{\mathrm{rf}} \tag{4.86}$$

The transition of I_{ds} from the linear to the saturation range with increasing V_{ds} is controlled by α, according to

$$\alpha = \alpha_r + \alpha_s(1 + \tanh\psi) \tag{4.87}$$

For low values of V_{gs}, at low currents, $\alpha \approx \alpha_r$, while at high currents at low currents, $\alpha \approx \alpha_r + 2\alpha_s$. Thereby, the slope of the I_{ds} function can be adjusted for high- and low-current regimes independently.

While the basic idea behind the I_{ds} function is quite clear, with

- α describing the slope $I_{\mathrm{ds}}/V_{\mathrm{ds}}$ in the linear range,
- Λ describing the slope $I_{\mathrm{ds}}/V_{\mathrm{ds}}$ in the saturation range,
- V_{pkm} describing V_{gs} at the peak of g_m, and
- I_{pk0} describing I_{ds} at $V_{\mathrm{gs}} = V_{\mathrm{pkm}}$,

it is rather complicated to determine the full set of model parameters by hand. Especially since the model's strength lies in the description of even weakest nonlinearities that are not visible without looking at the derivatives of I_{ds}, it is advisable to numerically fine-tune the model parameter set at least.

The capacitance model is less complicated, since the basic cause of nonlinearity was found to be nonlinearity of the current I_{ds}. The capacitances are given by:

$$C_{\mathrm{gs}} = C_{\mathrm{gspi}} + C_{\mathrm{gs0T}}(1 + \tanh\psi_1)(1 + \tanh\psi_2) \tag{4.88}$$

$$C_{\mathrm{gd}} = C_{\mathrm{gdpi}} + C_{\mathrm{gd0T}}((1 + \tanh\psi_3 - P_{111})(1 + \tanh\psi_4) + 2P_{111}) \tag{4.89}$$

$$C_{\mathrm{ds}} = C_{\mathrm{ds0}} \tag{4.90}$$

with the parameters

$$\psi_1 = P_{10} + P_{11} V_{\mathrm{gs}} + P_{111} V_{\mathrm{ds}} \tag{4.91}$$

$$\psi_2 = P_{20} + P_{21} V_{\mathrm{ds}} \tag{4.92}$$

$$\psi_3 = P_{30} - P_{31} V_{\mathrm{ds}} \tag{4.93}$$

$$\psi_4 = P_{40} + P_{41} V_{\mathrm{gd}} - P_{111} V_{\mathrm{ds}} \tag{4.94}$$

The capacitances are again described through tanh-functions, with a minimum value of C_{gspi} or C_{gdpi}, respectively, at pinch off. The maximum values of the capacitances are $C_{\mathrm{gspi}} + C_{\mathrm{gs0T}}$ and $C_{\mathrm{gdpi}} + C_{\mathrm{gd0T}}$.

4.4 An Improved EM Simulation Procedure to Extract Extrinsic Elements of Terahertz InP DHBTs[5]

For accurate small-signal modeling of DHBT, electromagnetic (EM) simulations of the outer shell of the DHBT are used for de-embedding the external parasitic parameters. However, under different simulation setup conditions, the parasitic elements can be overestimated or underestimated. This paper presents a systematic approach to set up an EM simulation and obtain accurate parasitic elements of DHBT devices using Ansys high-frequency structure simulator (HFSS). An innovative simulation method is also introduced in order to calculate parasitic base capacitance (C_{pb}) and base-collector capacitance (C_{pbc}), which cannot be extracted accurately from "off-state" measurements. Finally, the EM simulation de-embedded small-signal model is found to provide a good fit to the measured data from 50 MHz to 150 GHz [32].

The terahertz regime has immense potential to enhance the quality of life for humans within a wide range of applications. Due to the fact that the penetrative power and irradiation effects of terahertz radiation are less harmful than X-rays, it is finding use in many clinical applications such as cancer imaging and sensing disorders [33]. Its imaging capabilities can also be used for nondestructive testing or inspection of specific materials in industries [34]. Furthermore, the terahertz regime is quite attractive for high data rates in future space communication [35].

The commercial potential of the terahertz applications can only be realized through ongoing research on InP HBT and high electron mobility transistor (HEMT) technologies [36]. The InP HBT technology has demonstrated maximum frequency of oscillation (f_{max}) of more than 1 THz [36]. The transferred-substrate InP DHBTs (used in this work) have shown a higher breakdown voltage with power handling capacity of over 200 mW at W-band [37].

For implementing terahertz technologies, there is a need for developing highly accurate small-signal models of DHBTs. To achieve this, a more precise de-embedding technique is required as it becomes progressively more critical from millimeter-wave to terahertz. This is because, as the size of the transistors becomes smaller and smaller, the values of extrinsic elements (bias-independent) become more comparable to intrinsic elements (bias-dependent). Traditionally, this was achieved by using measuring techniques like open collector (for extracting extrinsic inductance and resistance) and off-state (for extracting extrinsic capacitances) [16]. The transistor is operated in saturation mode while using open-collector measurements. To obtain accurate results, a high base current is required which may damage the device. Furthermore, it is difficult to separate the outer shell resistances from internal resistances of the device. In the off-state measurement, it is difficult to separate the extrinsic capacitances from bias-dependent junction capacitances. Only the collector capacitance can be accurately extracted as it is not influenced by junction capacitances. A way to graphically extrapolate the extrinsic capacitance values by using the reverse bias-dependent formula of junction capacitance is presented in Rudolph et al. [38]. Unfortunately, it is not possible to obtain accurate results as the junctions of the InP DHBTs are fully depleted even at low reverse bias.

To avoid potential errors from measurement data and to obtain reliable extrinsic parameters, this paper presents an improved and systematic approach to setup the EM simulation using Ansys HFSS. Although, the EM simulation procedure is used in refs. [39, 40], this paper introduces an innovative EM setup procedure to accurately calculate the extrinsic capacitances, parasitic base-emitter capacitance (C_{pb}) and parasitic base-collector capacitance (C_{pbc}), which cannot be calculated from "off-state" measurements. Using this EM simulation de-embedding procedure, the small-signal model is in good agreement with the measurement results from 50 MHz to 150 GHz. The extracted EM simulated values can potentially be used to de-embed the parasitic elements of DHBTs up to 0.4 THz or beyond.

4.4.1 Technology and Calibration Procedure

4.4.1.1 Technology

The device considered in this work has been fabricated in a transferred-substrate InP DHBT technology at the Ferdinand-Braun-Institute (FBH). In the transferred-substrate technology, the InP DHBTs are first processed up to the emitter as in a conventional triple-mesa process. Subsequently, the semi-processed InP wafer is transferred emitter down onto a host substrate (e.g., Si or AlN) by adhesive wafer bonding. Following the wafer bonding step, the original InP substrate is removed by wet-chemical etching allowing the definition of isolated collector mesa regions. The main advantage of the transferred-substrate technology over a conventional triple-mesa technology is that the extrinsic base-collector capacitance is greatly

5 This section was originally published as: S. V. Pawan et al., "An Improved EM-Simulation Procedure to Extract Extrinsic Elements of Terahertz InP DHBTs," in 2020 German Microwave Conference (GeMiC), Mar. 2020, pp. 240–243. Reprinted by permission.

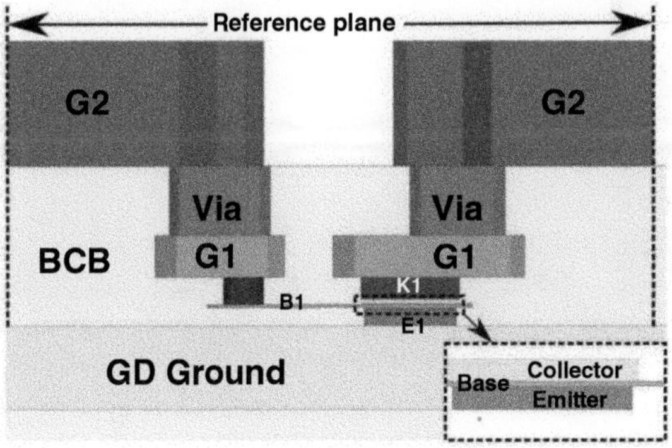

Figure 4.51 Detailed cross-section of outer shell and active region of InP DHBT in the FBH transferred-substrate process.

reduced due to the possibility to contact the collector mesa from the top. This results in an increase of the maximum frequency of oscillation (f_{max}). The process contains three gold-based interconnect layers, GD, G1, and G2, separated by BCB dielectric layers (Figure 4.51) [41]. There is an ongoing activity to go from 800 nm emitter width devices achieving $f_T/f_{max} \sim 350$ GHz to devices with $f_{max} > 1$ THz. This will be achieved by lateral scaling to 100–200 nm emitter widths, optimization of the epitaxial structure, and improved contacts. As an intermediate step, laterally downscaled InP DHBT devices with 500 nm emitter width have been fabricated. By compacting the layout cell but keeping the epitaxial structure and contacts unchanged, these devices achieve $f_T \sim 360$ GHz and $f_{max} \sim 490$ GHz. The InP DHBTs with 500 nm emitter width will be used for the modeling presented in the following.

4.4.1.2 Calibration Procedure

For the characterization and modeling of THz InP DHBTs, setting a well-defined reference plane is mandatory. Our approach is based on the multiline TRL calibration technique [42] using on-wafer thin-film microstrip structures. The S-parameters are subsequently renormalized to 50 Ω by estimating the line capacitance per unit length from resistive loads embedded into a line structure identical to that as used for the multiline TRL calibration technique [43]. After the aforementioned multiline TRL calibration and re-normalization steps, additional 80 μm of uniform thin-film microstrip line still needs to be de-embedded to reach the reference planes coinciding with the layout cell of the InP DHBT (see Figure 4.51). For the de-embedding of the additional uniform line, the complex propagation constant $\gamma = \alpha + j\beta$ and characteristic impedance Z_0 as found from the multiline TRL calibration procedure are applied.

4.4.2 EM Simulation Setup

The EM simulation setup procedure used in this work is explained in three steps using Ansys HFSS software. In the first step, a modal type solution is selected, which provides the S-parameter solution in terms of incident and reflected power of the network. The modal solution allows us to define an incident field pattern at the ports using an integration line. This solution type can be used for lumped port excitation or wave port excitation.

In the second step, the mesh settings of the 3D model are selected. A proper meshing is significant to obtain precise results. The HFSS uses tetrahedral elements for meshing the 3D object on the surface and inside the volume to generate a field solution. For the field solution generation inside metal lines, it is important to include the conductor losses. That is why the bulk conductivity of all the different metals is included in the HFSS simulation. To ensure these fields are generated inside the volume of metal objects, “solve inside” option is selected.

Inside selection option of “length-based mesh operations” in HFSS is used to generate a fine mesh of 3D objects as shown in Figure 4.52. In that option, we can limit the maximum number of tetrahedral elements to a certain number, which will be sufficient to get accurate results and have reasonable simulation time. In this work we have found that having 6000 maximum elements has optimum results.

In the third step, the ports for excitation, the radiation boundary, and solution setup are performed. Lumped ports are used to set up the port excitation because the ports are employed inside the solution space. The inductance introduced

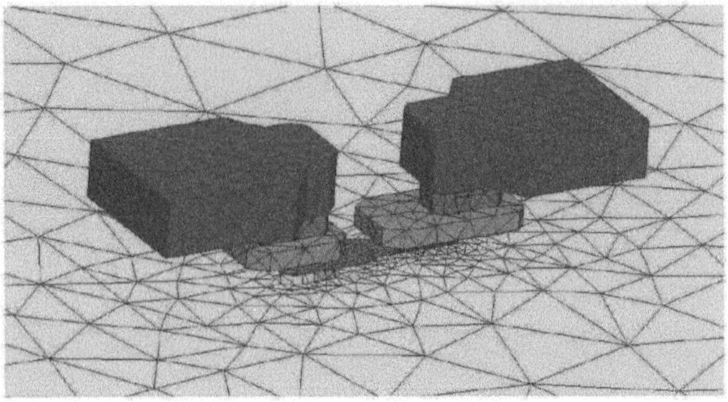

Figure 4.52 Meshing of solid objects with 6000 elements with a length-based (inside selection) meshing operation.

by the lumped port's current is removed by selecting the "de-embed" option in the port options. In the solution setup, a solution frequency sweep is selected (0–400 GHz). The initial mesh setup of lambda refinement is set to a value of 0.1 or lower. For frequency sweep, either interpolating or discrete sweeps are used.

4.4.3 EM Simulation Results

To find the parasitic elements of the DHBT device, an open and a short test structure of Figure 4.51 is used. The active region of the device is replaced with highly conductive metal to make the device short. The equivalent network of the short device can be described as "T" network. The parasitic inductances and resistances can be calculated from this network similar to [39]. The extracted values are shown in Table 4.2. The parasitic capacitance extraction is explained in the next section.

4.4.3.1 Parasitic Capacitance Extraction

Generally, to calculate the parasitic capacitance, the active region of the device is replaced with material having properties of BCB or InP, as in refs. [39, 40], respectively. This setup does not represent an actual device condition during measurement setup. The electric fields generated around the active semiconductor by different metal layers are shielded by the depletion regions. This reduces the effective capacitance of the device.

In order to improve the simulation results, the active region in the EM simulation can be replaced by material having relative permittivity (ϵ_r) of zero. If $\epsilon_r = 0$, the capacitance experienced in that region is zero, as the capacitance is directly proportional to permittivity. In HFSS for EM simulation, the ϵ_r is set to a low value as it is not possible to select zero. Figure 4.53 shows the magnitude of the E-field of the transistor in logarithmic scale (viewed from the left side). The maximum E-field intensity can be seen in and around the active region of DHBT. This can be explained by boundary conditions. The normal D-field is continuous from BCB to the active region. However, the normal E-field is given by $E_{\text{active-region}} = \epsilon_{\text{BCB}} \cdot E_{\text{BCB}} / (\epsilon_{\text{active-region}} \approx 0)$, which makes the $E_{\text{active-region}}$ have a very high value.

The open test structure setup can be described by a Π network having parasitic base capacitance C_{pb}, parasitic base-collector C_{pbc} and parasitic collector capacitance C_{pc}. The extracted capacitance values are shown in Figure 4.54. It is interesting to note that the C_{pc} of "off-state" measurement and EM simulation fits with good agreement. This further validates the EM simulation setup results. The values of C_{pb} and C_{pbc} measurements are higher because of the influence of junction capacitances. As there is no influence of the junction capacitance in the EM simulation, the extracted values of C_{pb} and C_{pbc} should be closer to actual parasitic capacitance values.

4.4.4 EM Simulation-assisted Small-signal Model Comparison with Measurements

The EM simulated values in Table 4.2 are used to de-embed the DHBT. The intrinsic parameters calculated by both the T-topology [44] and the hybrid-Π topology [39] are found to improve the small-signal model. Figure 4.55 shows S-parameters of the model which are in good agreement with the measured data from 50 MHz to 150 GHz for bias points $V_{\text{ce}} = 1.8$ V and $I_c = 5.2$ mA and 15.3 mA.

Table 4.2 Parasitic elements extracted from EM simulation.

	EM simulation test structure results	
Open	$C_{\text{pc}} = 3.2$ fF	$C_{\text{pb}} = 5.36$ fF
	$C_{\text{pbc}} = 0.89$ fF	
Short	$L_{\text{pc}} = 4.4$ pH	$L_{\text{pb}} = 2.9$ pH
	$L_{\text{pe}} = 0.17$ pH	
	$R_{\text{pc}} = 1.61\ \Omega$	$R_{\text{pb}} = 0.33\ \Omega$
	$R_{\text{pe}} = 0.013\ \Omega$	
	$R_{\text{pc,ac}} = 1.4^{-6}\ \Omega/\sqrt{\text{Hz}}$	$R_{\text{pb,ac}} = 7.6^{-7}\ \Omega/\sqrt{\text{Hz}}$
	$R_{\text{pe,ac}} = 9^{-8}\ \Omega/\sqrt{\text{Hz}}$	

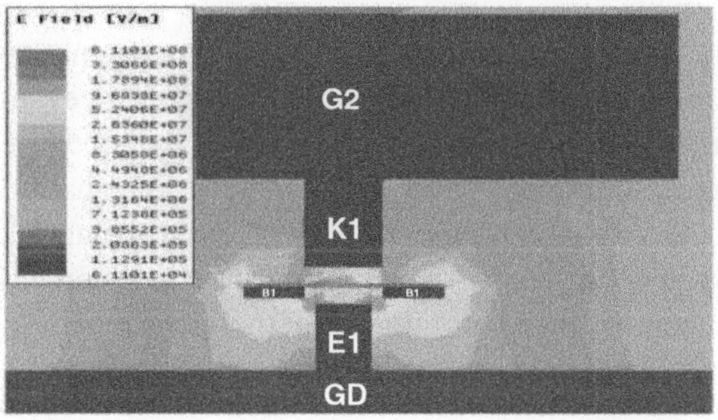

Figure 4.53 Magnitude of electric field around the active region of DHBT in logarithmic scale.

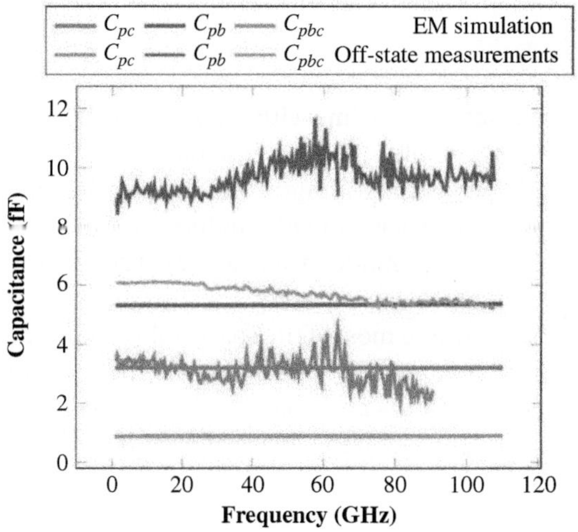

Figure 4.54 Comparison of EM simulation of open test structure with "off-state" measurements.

4.5 Conclusion to the Semiconductor Introduction

The actual application of transistors falls in chip transistors/housed transistors or ready applications like gain blocks. Depending on the assembly, they are used effectively up to 10 GHz and a bit higher when bonded. For the very latest bipolar transistor or CMOS designs, these are accomplished together with a foundry and a capable CAD tool.

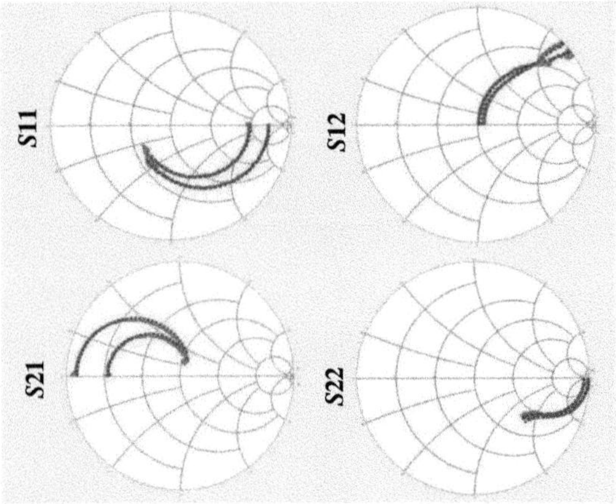

Figure 4.55 Comparison between measured (solid line) and calculated (dotted line) parameters over the frequency range from 50 MHz to 150 GHz. The bias points are $V_{ce} = 1.8$ V and $I_c = 5.2$ mA and 15.3 mA.

In designing a microwave circuit, which will be built on a material such as GaAs, one needs to go to a GaAs alternative silicon or silicon germanium foundry. The foundries seem to be very secretive about their information, and the first experience with a foundry is a shock because one needs to sign a nondisclosure agreement and pay up to US$5000 to obtain a foundry manual. These foundry manuals are somewhat unique to each foundry and vary. The foundry supplies information about passive and active structures as well as interconnect information. The following is an example of a foundry manual, which was put together from information from various foundries. It does not apply to a particular foundry but is generic in its contents. In the area of passive components, typically a variety of inductor cells are recommended: the same applies to spiral inductors/rectangular inductors and interconnect components such as bends, tees, crosses, and other components as outlined.

4.6 Example: GaAs MESFET Foundry Design Manual

4.6.1 Example for the Use of the Foundry

The examples in Figures 4.56 and 4.57 are simple cases for foundry use and were done at the time with the TriQuint foundry service. Figure 4.56 shows the layout of a simple LO medium-power amplifier. The input is 50 Ω; the output will drive the double-balanced mixer. The input of the amplifier is on the top. The center pad is the RF input, and the pads left and right are at ground. The signal travels via a capacitor to the gate, which is grounded via a rectangular inductor and a small resistor. On the drain side, we have a similar inductor as an RF choke. The output is grounded via a larger capacitor. The drain voltage is applied to the square pad. The output from the drain is also available via a capacitor and a similar output terminal with two grounds. It is noteworthy to look at the sizes. The inductors determine most of the space followed by one large capacitor and the transistor. Since the cost is determined by the surface area, higher frequencies would mean lower cost. As outlined earlier, there are much more complex foundry models and interconnects available.

And now, because of the complexity of the circuits we need to resort to CAD. For low frequencies and most transient analysis (time domain) Spice is a good choice.

Once distributed microwave elements come in to play, we will need a Harmonic Balanced bases CAD tool that also, based on the noise correlation mathematics can correctly predict noise in non linear circuits. Now let's start with a mixer

Besides amplifiers and mixers, the oscillator is a fairly complex circuit as it is both an amplifier and a mixer that converts different noise voltages and currents to the final output signal. And SPICE cannot handle both the single-sideband (SSB) noise and the distributed elements.

4.6.2 Example of a 3.6 GHz Microwave Oscillator Design

Oscillator design is an autonomous problem in which there is no external driving source. This complicates the design process considerably, as the frequency of an oscillator is then an additional degree of freedom in the circuit that must be

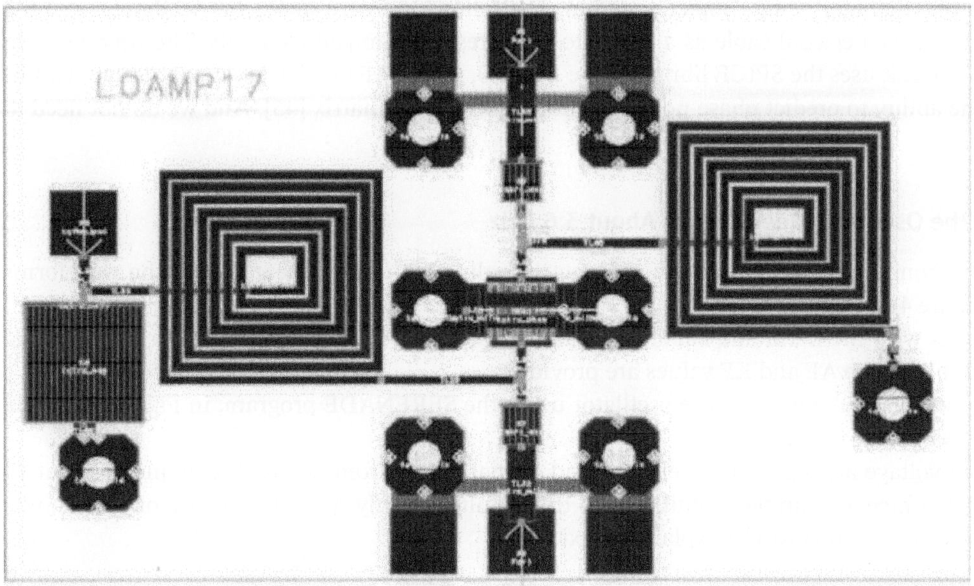

Figure 4.56 Layout of Star Mixer with four FETs switching to ground. The left lower corner contains the local oscillator (LO) amplifier. The RF input occurs in the middle of the transformer on the left, and because of a very low IF frequency, a huge transformer, as shown on the right, is needed. It may be a better decision to use an external transformer, but this was an experiment, again built around the TriQuint foundry, to evaluate the mixer performance.

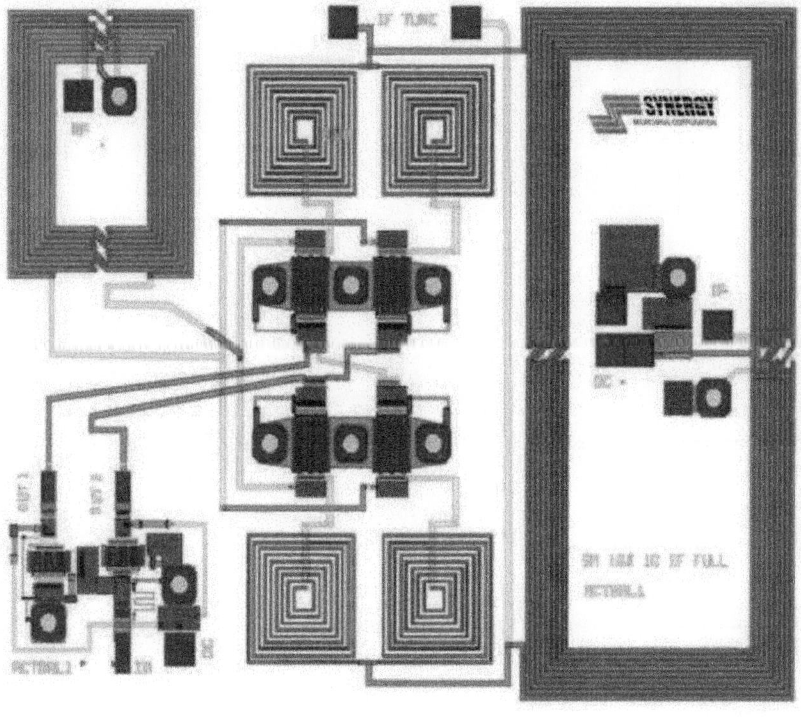

Figure 4.57 MMIC for Star Mixer using TriQuint foundry.

accounted for. First, we analyze the oscillator using LTspice program. The following figure shows the schematic and the simulation results. The simulation uses a coaxial cable as a resonator. The rest are lumped elements. The time domain displays are self-explanatory. The circuit uses the SPICE library of the BFP620, where AF and KF are incorporated. Unfortunately, SPICE does not have the ability to predict phase noise. See noise-correlation matrix [45]. And we do not need to use distributed elements.

4.6.3 The LTspice Version of the Oscillator Operating at About 3.6 GHz

Above are the various waveforms from the SPICE simulation. As we are dealing with a nonlinear circuit, the waveforms are distorted, as expected (see Figure 4.58).

Now we use the same oscillators with the harmonic balance simulator Serenade. To be fair, we need to use identical elements. In addition, in the NOI block, the AF and KF values are provided.

Figure 4.59 shows the SSB phase noise simulation of the oscillator using the SERENADE program, in Figure 4.60. The flicker corner frequency for this example is about 1 kHz.

Figure 4.61 shows the expected voltage and current waveforms, and output power from this oscillator and is about 18 dB harmonic suppression. Overall a nice and simple circuit. To get this result was only possible because of a harmonic balance-based simulator, the principle of which will be explained next.

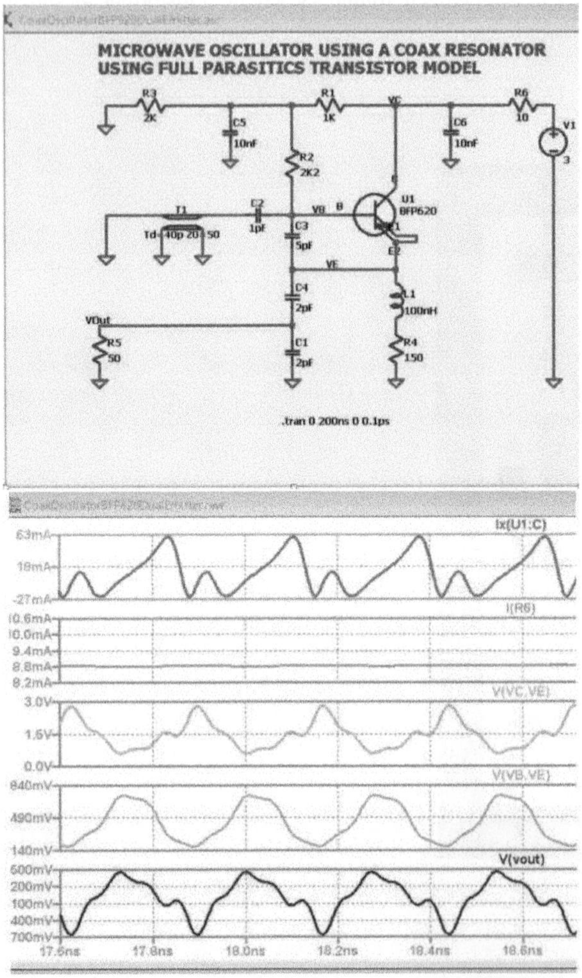

Figure 4.58 Microwave oscillator using a COAX resonator using full parasitics transistor model.

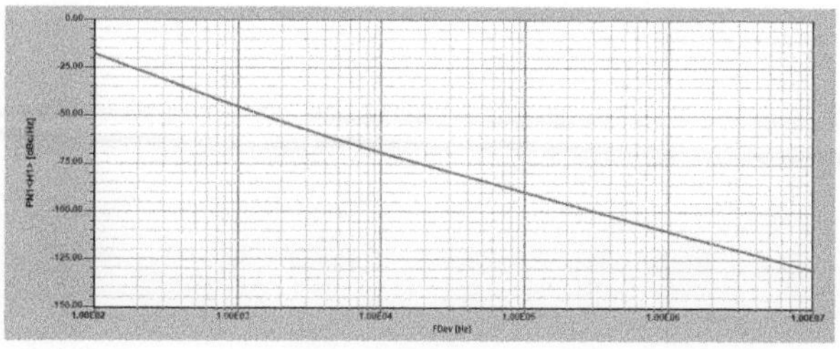

Figure 4.59 Microwave oscillator using a COAX resonator CAD simulation circuit—simulated phase noise.

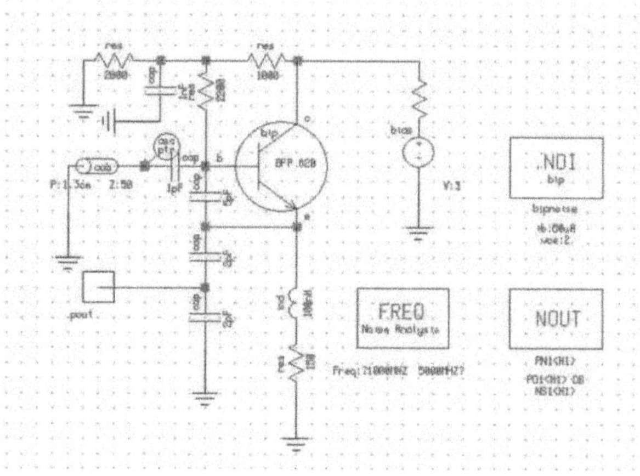

Figure 4.60 Microwave oscillator using a COAX resonator CAD simulation circuit.

4.7 Introduction of the Harmonic Balance Method and Nonlinear Noise Calculation

There are now nonlinear Microwave CAD packages using the harmonic balance technique. Because of its nonlinear optimization capabilities, it is uniquely suited to the task of oscillator design [46–48]. The harmonic balance method seeks a solution to a steady-state nonlinear design problem by iteratively solving for a set of variables, referred to as state variables. The state variables can typically be chosen as the voltages at the linear–nonlinear interface in a circuit partitioned into linear and nonlinear segments. They are expressed as the phasor components, and their harmonics, of a sinusoidal excitation frequency. The state variables are usually found iteratively by a gradient-based technique which seeks a simultaneous solution for Kirchhoff's equations applied to the linear and nonlinear sides of the network separately [49–53].

For the nonautonomous, or driven, circuit, the driving frequency is known a priori and Kirchhoff's equations are a well-determined system of equations in which the phase and amplitude of the excitation appear on one side of the equations as forcing terms. In an autonomous circuit, the only excitation terms appearing in Kirchhoff's equations are DC sources. One stable solution to these equations for a circuit with no applied RF drive will always be the DC (or degenerate) solution, as all phasor terms at an arbitrarily chosen frequency can be set to zero and still satisfy the RF driving conditions (of zero excitation). For an autonomous circuit to have a solution to Kirchhoff's equations at nonzero frequencies, at least one additional degree of freedom is required, as there is one additional unknown in the equations—the oscillation frequency.

For oscillator analysis purposes (i.e., for a fixed circuit topology), the additional free parameter is just the unknown frequency of oscillation. For oscillator design purposes, the frequency is fixed as a design goal, and the required degree of freedom must be represented by a free circuit variable, such as a bias voltage or other tuning element. In this way, a solution can be found to the conditions for oscillation (which are just Kirchhoff's equations) by varying some circuit parameter. This parameter must be adjusted so that the equations can be satisfied at the oscillation frequency, with some set of (solution) state variables, which are determined at the same time.

The harmonic balance problem must then allow for the simultaneous solution of both the state variables and the circuit elements to satisfy Kirchhoff's equations under the chosen conditions (i.e., at the design frequency). In fact, these programs allow the user to set additional circuit parameters to be variables in order to optimize for other circuit responses, such as output power, efficiency, spectral purity, or distortion, while simultaneously satisfying Kirchhoff's equations. Consequently,

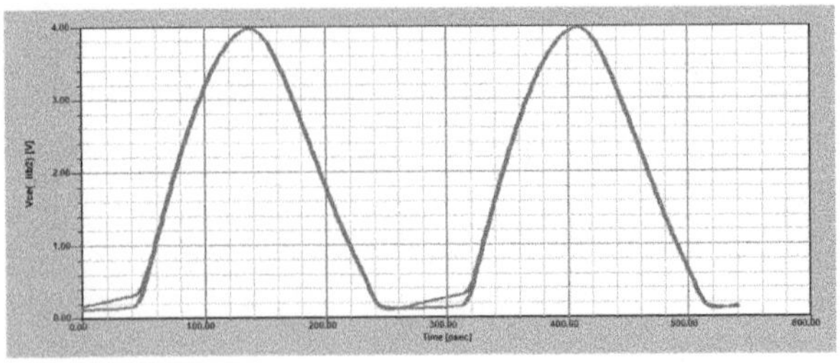

(a)

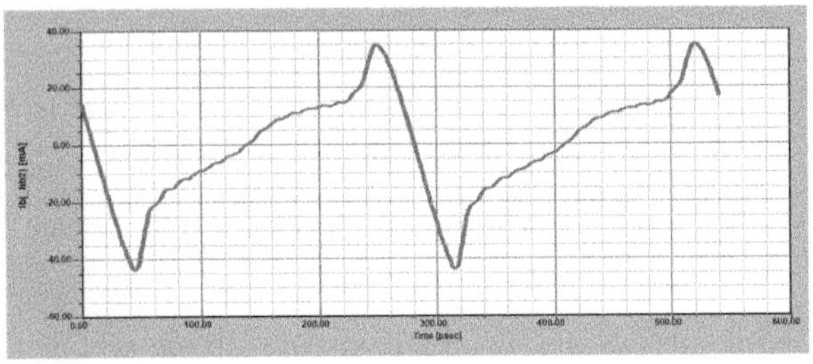

(b)

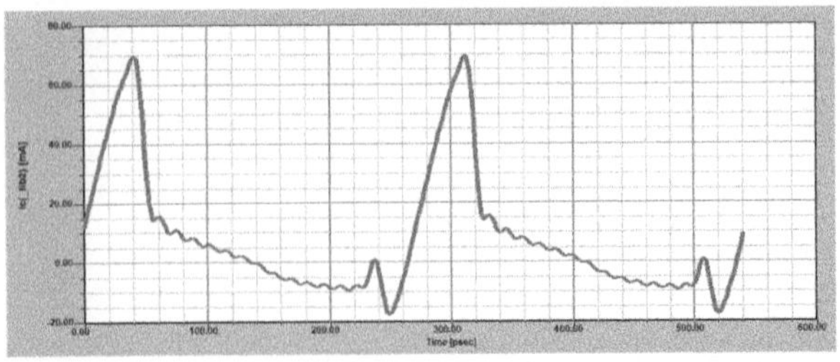

(c)

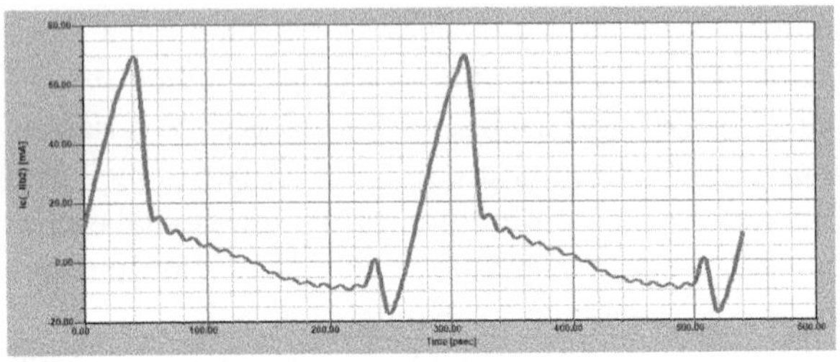

(d)

Figure 4.61 Microwave oscillator using a COAX resonator CAD simulation circuit. (a) Simulated V_{ce}. (b) Simulated I_b. (c) Simulated I_c. (d) Simulated P_{out} (dBm).

the methodology of introducing additional degrees of freedom into the harmonic balance problem allows not only for the solution of autonomous designs but also for the optimization of all types of circuits for desired response. The example shown in Figure 4.62 is a 6.6 GHz dielectric resonator-based oscillator.

4.7.1 6.6 GHz Dielectric Resonator Based Oscillator

As an example, shown in Figure 4.62, we show a DRO design using a bipolar transistor embedded in a simple microstrip circuit. The specified optimization goals were 20 mW output power, at a frequency of 6.6 GHz. In the design, the additional degrees of freedom for the optimization problem and the oscillator synthesis were provided by the resonator diameter and the microstrip lengths together with the FET bias voltages.

Prior to circuit optimization, the program automatically optimizes the DRO dimensions to resonate at the $TE_{01}\delta$ mode at 6.6 GHz. The pre-optimization is needed to ensure that the initial starting point for the design equations is within the operating regime. A sequence of optimizations was then performed by increasing the lower bounds on the output power. The maximum power available was found to be 6 dBm with an efficiency of 3.28% but optimized for phase noise −112 dBc/Hz indicates a high Q and low phase noise design. The bias point was $V_D = 5.9$ V, $I_b = 0.138$ mA, and $I_c = 13.6$ mA. Output harmonics were about 24 dB below the fundamental.

The circuit can also be tuned by moving the metal tuning plate over the resonator. To tune to a frequency of 6.05 GHz, the circuit must be re-optimized with the plate distance s as the only tuning variable. This change to the single circuit parameter is needed to allow there to be a solution to Kirchhoff's equations at the new oscillation frequency. The values of the state variables are, of course, also different at the new point, and the output power is reduced correspondingly (Figure 4.63).

As oscillator analysis can be performed by repetitively reoptimizing the tuning parameter "s" at a series of frequencies in the program and constructing a tuning curve of oscillation frequency versus s. For an oscillator of unknown frequency, the actual oscillation frequency can be determined by entering the tuning curve at the known value of s. Note that in the case of an oscillator analysis only a single tuning variable is adjusted at each frequency point, so that a tuning curve can

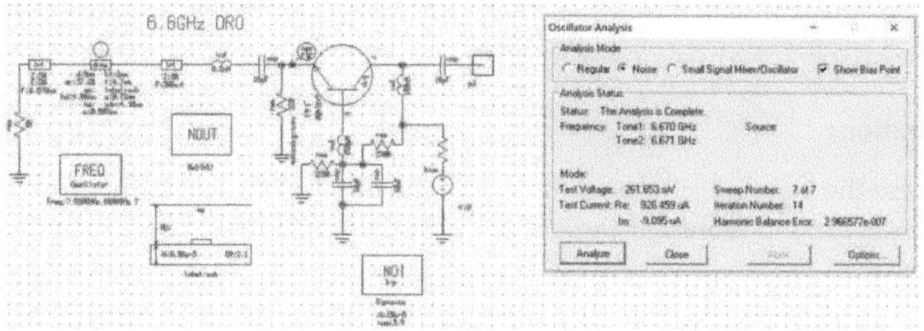

Figure 4.62 DRO design example [52, 54].

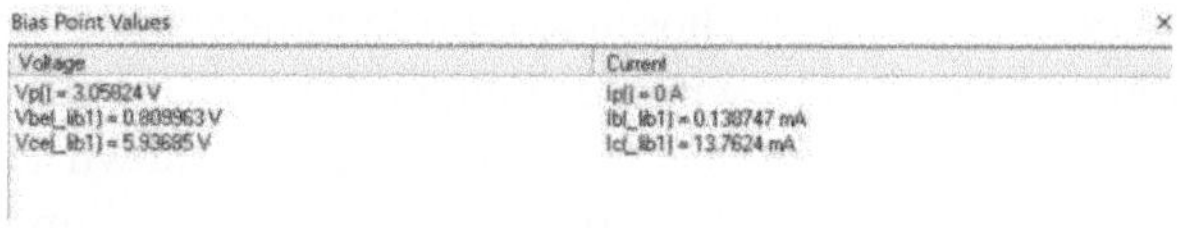

(a)

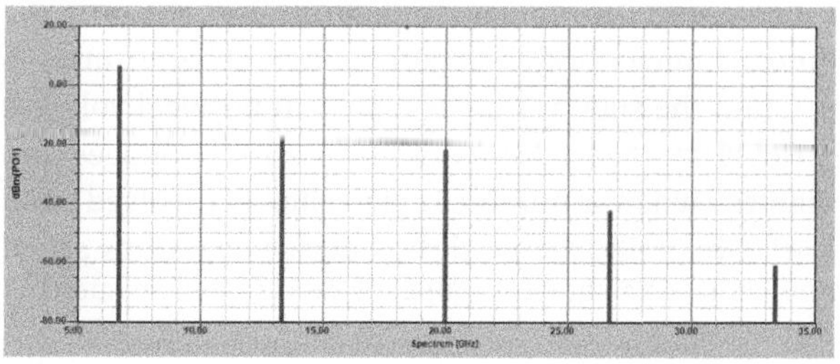

(b)

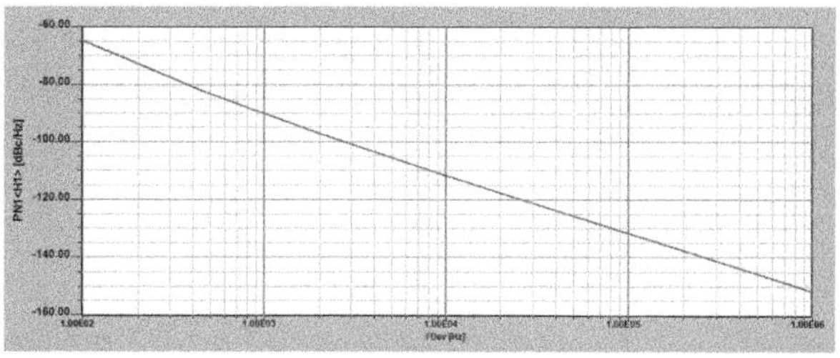

(c)

Figure 4.63 6.6 GHz DRO design example from Figure 4.62 [52, 54]. (a) Bias point values. (b) Simulated P_{out} (dBm). (c) Simulated phase noise spectrum.

be constructed to give a single, unique relationship between the design frequency and the actual circuit parameter value in the circuit being analyzed.

Although the harmonic balance approach ensures that a steady-state solution exists, the buildup of oscillations and the stability of the steady-state operating point also need to be addressed. One way to provide a numerical solution to this problem is to use the principles of bifurcation theory [52, 55]. Using the frequency of oscillation as a continuation parameter [56] and choosing the output power as a parameter to describe the circuit state, a solution path may be built by stepping the frequency through a prescribed range.

In summary, the advent of nonlinear CAD programs now allows the designer to verify many aspects of circuit operation not previously obtainable through linear programs alone. In this way, the levels of output power, harmonic content, DC-to-RF efficiency, device currents, and load pushing and pulling can be examined before the oscillator is constructed.

Next some more explanations about the nonlinear program will follow.

4.8 Regular Analysis with the Harmonic Balance Program

Harmonic balance analysis consists of determining the periodic steady-state response of a fixed circuit given a predetermined set of fundamental tones [1, 2].

4.8.1 Formation of the Harmonic Balance Equations

The analysis is limited to periodic responses because the basis set, chosen to represent the physical signals in the circuit, is sinusoids, which are periodic. The Fourier series is used to represent these signals. In the single-tone case, a signal is given by

$$x(t) = \sum_{k=-NH}^{NH} X_k e^{jk\omega_0 t} \tag{4.95}$$

where $X_k = X^*_{-k}$, ω_0 is the fundamental frequency and NH is the number of harmonics chosen to represent the signal.

In harmonic balance, the circuit is usually divided into two subcircuits connected by wires forming multiport. One subcircuit contains the linear components of the circuit and the other contains the nonlinear device models as shown in Figure 4.64. The linear subcircuit response is calculated in the frequency domain at each harmonic component $(k \cdot \omega_0)$ and is represented by a multiport Y matrix. This is the function performed by linear analysis.

The nonlinear subcircuit contains the active devices whose models compute the voltages and currents at the intrinsic ports of the device (parasitic elements are linear and absorbed by the linear subnetwork). The port voltages (v) and currents (i) are analytic or numeric functions of the device state variables (x). Often the state variables represent physical voltages such as diode junction voltage or FET gate voltage, but are not restricted to physical quantities. The port voltages and currents are often functions of the time derivatives of the state variables (when a nonlinear capacitor is involved) and of time-delayed state variables (such as a time-delayed current source). Generally, the nonlinear device equations are of the form:

$$v(t) = \Phi\left[x(t), \frac{dx}{dt}, \ldots, \frac{d^n x}{dt^n}, x(t-\tau)\right] \tag{4.96}$$

$$i(t) = \Psi\left[x(t), \frac{dx}{dt}, \ldots, \frac{d^n x}{dt^n}, x(t-\tau)\right] \tag{4.97}$$

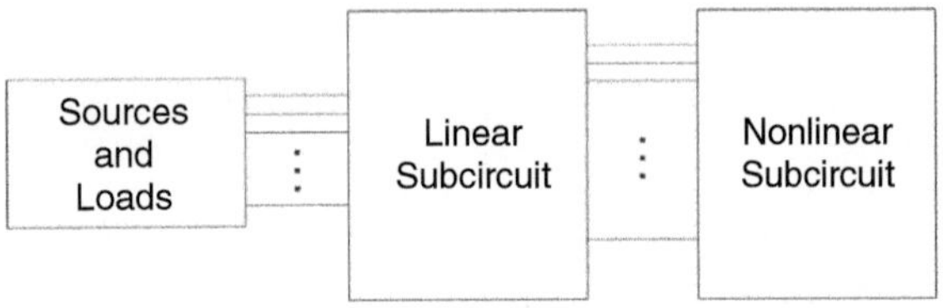

Figure 4.64 Separation of the linear and nonlinear subcircuits.

The device state variables, port voltages, and currents are transformed to the frequency domain using the discrete Fourier transforms as $\mathbf{X}$, $\mathbf{V}_k(\mathbf{X})$ and $\mathbf{I}_k(\mathbf{X})$, respectively. Kirchhoff's current law is applied to the interface between the subcircuits at each harmonic frequency:

$$\mathbf{Y}_k\mathbf{V}_k(\mathbf{X}) + \mathbf{I}_k(\mathbf{X}) + \mathbf{J}_k = 0 \tag{4.98}$$

where $\mathbf{J}_k$ are the Norton equivalents of the applied generators. This constitutes the harmonic balance equations at each harmonic frequency. The object of the analysis is to find the set of state variables, $\mathbf{X}$, to satisfy Eq. (4.98).

When the analysis begins, the state variables are typically set to zero and the left side of Eq. (4.98) is nonzero. We can write an error vector

$$\mathbf{E}_k(\mathbf{X}) = \mathbf{Y}_k\mathbf{V}_k(\mathbf{X}) + \mathbf{I}_k(\mathbf{X}) + \mathbf{J}_k \tag{4.99}$$

whose Euclidean norm $\mathbf{E}^t\mathbf{E} = ||\mathbf{E}||$ is called the harmonic balance error (HBE). If the HBE is reduced below a tolerance, we say that Eq. (4.98) is satisfied and a solution has been obtained.

4.8.2 Solving Methods

The process of solving the harmonic balance equations is an iterative one. An estimate of $\mathbf{X}$ is inserted into Eq. (4.99); $\mathbf{E}$ is calculated and if it is not below the tolerance, then a new value of $\mathbf{X}$ must be determined and tried. Each such loop is termed "iteration." There have been several methods used in the past to determine new values of $\mathbf{X}$ and two that have proven to be the most general and efficient are discussed here.

The state variables $\mathbf{X}$ and harmonic balance residuals $\mathbf{E}$ are complex-valued. In practice, these are decomposed into their real and imaginary parts so that the number of real unknowns in $\mathbf{X}$ is $ND \cdot (2 \cdot N_t + 1)$ where ND is the total number of nonlinear device ports and N_t is the number of frequency components ($= NH$ for single-tone analysis). Now we can write $\mathbf{E}(\mathbf{X}) = \mathbf{0}$ as a Taylor series expansion truncated after the first derivative term:

$$\mathbf{E}(\mathbf{X}) = 0 \approx \mathbf{E}(\mathbf{X}^{(n)}) + \mathbf{J}(\mathbf{X}^{(n)})(\mathbf{X}) - \mathbf{X}^{(n)}) \tag{4.100}$$

where $\mathbf{J}$, the Jacobian matrix, is the first derivative matrix of $\mathbf{E}$ with respect to $\mathbf{X}$ and superscript "n" indicates the current iteration. Solving for $\mathbf{X}$ and using this for the next trial:

$$\mathbf{X}^{(n+1)} = \mathbf{X}^{(n)}\mathbf{J}^{-1}(\mathbf{X}^{(n)}) \cdot \mathbf{E}(\mathbf{X}^{(n)}) \tag{4.101}$$

This is the Newton–Raphson update method where the last right-hand term is the update. This method works in one iteration, if the set of equations is linear, but will take an unknown number of iterations if nonlinear. Often the update is reduced by a factor called the Newton damping factor, so the method takes smaller steps for each of the next iterations. Convergence to a solution is not guaranteed and the results of the iterations may diverge if not controlled. Harmonica uses enhanced versions of the Newton–Raphson method to improve convergence and speed.

Harmonica contains an algorithm that dynamically changes the Newton damping factor during solving, based on the rate of convergence. If the solver has trouble converging, the factor will be made smaller to improve convergence. If it has been reduced by more than a predetermined factor, the solver will stop and an error will be reported.

An important aspect to note is the size of the Jacobian. If $\mathbf{X}$ contains $ND \cdot (2 \cdot N_t + 1)$ elements, then $\mathbf{J}$ contains this number squared. As a practical example, if $ND = 10$ (5 FETs) and $N_t = 4$, then there are 8100 entries in $\mathbf{J}$, which takes 63 kB. This is relatively small, but N_t becomes much larger when multitone excitation is considered.

4.8.3 HBE Tolerance—HBTOL x

Here x is the tolerance per device port per frequency component. The absolute HBE allowed is scaled by the number of device ports and number of frequency components so that large circuits with many frequency components meet similar HBE criteria as small circuits. The default for HBTOL is $1.0 \cdot 10^{-6}$. For the case of two-tone intermodulation analysis and three-tone analysis, the allowed harmonic balance is also scaled by the relative currents of the circuit. This reduces the allowed error (effectively reducing HBTOL) to provide better accuracy of the intermodulation products.

4.8.4 Multitone Analysis

The aforementioned discussion was based on single-tone analysis for conceptual simplicity. Multitone analysis is simply an extension of single-tone [46–48]. In the single-tone case, a circuit is excited with an RF source and harmonics of that source are produced by the nonlinearities of the circuit. The set of harmonics, the frequency of excitation and DC are called the *spectrum* of the analysis. The single-tone spectrum is defined as:

$$S_1 = k \cdot f_0 k = 0, 1 \ldots NH \tag{4.102}$$

where f_0 is the fundamental frequency. In multitone analysis the spectrum is modified to include the harmonic products of each fundamental tone. The harmonic products are just integer functions of the fundamental frequencies and indicate the allowed "bins" for power conversion within a circuit. The rest of the harmonic balance analysis is exactly the same.

The conversion between time domain waveforms and Fourier coefficients is accomplished by the discrete Fourier transform in single-tone analysis. For each additional fundamental tone, a dimension is added in the transform. This allows efficient computation between domains, but becomes CPU-intensive when more than three dimensions are encountered.

4.8.5 Two-tone Intermodulation Case

The two-tone analysis spectrum is defined as:

$$S_2 = |p \cdot f_1 + q \cdot f_2| |p| = 0, 1, \ldots, m_1, \quad |q| = 0, 1, \ldots, m_2 \tag{4.103}$$

where f_1 and f_2 are the first and second fundamentals, and m_1 and m_2 define the bounds of the spectrum. When the powers of the fundamentals are of similar magnitude, the bounds are chosen such that the harmonics and intermodulation products can be accurately assessed. The bounds are chosen as:

$$m_1 = m_2 = M \tag{4.104}$$

$$|p| + |q| \leq M \tag{4.105}$$

M is called the intermodulation order since it determines the order of intermodulation products that will be included in the spectrum. This selection results in a spectrum as shown in Figure 4.65, for $M = 3$. The intermodulation products and harmonics are clearly seen. The total number of spectral components for this selection algorithm is:

$$N_t = M \cdot (M + 1) \tag{4.106}$$

4.8.6 Two-tone Mixer Case

When the power contained in one of the fundamentals is much greater than the other, as in a mixer case, then a spectrum is selected such that the harmonics of the strong signal (LO) and the sidebands of the LO and weak signal (RF) can be assessed. The bounds are then chosen as:

$$m_1 = M_1 = m_2 = M_2 0 \leq |p| \leq M_1 \text{ and } 0 \leq |q| \leq M_2 \tag{4.107}$$

where M_1 is the number of LO harmonics and M_2 is the number of sidebands on each side of the LO. This selection results in a spectrum as shown in Figure 4.66 for $M_1 = 4$, $M_2 = 2$. The LO harmonics, sidebands, and difference frequency are clearly seen. The total number of spectral components for this selection algorithm is:

$$N_t = M_1 \cdot (2 \cdot M_2 + 1) + M_2 \tag{4.108}$$

4.8.7 Three-tone Intermodulation Case

The three-tone analysis spectra are very similar to two-tone and are defined by:

$$S_3 = |p \cdot f_1 + q \cdot f_2 + r \cdot f_3| |p| = 0, \ldots, m_1, \quad |q| = 0, \ldots, m_2, \quad |r| = 0, \ldots, m_3 \tag{4.109}$$

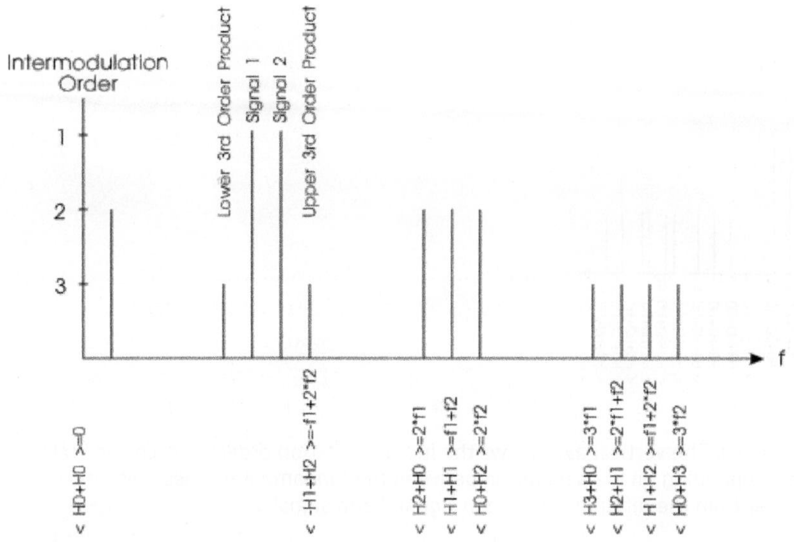

Figure 4.65 Two-tone intermodulation spectrum for $M = 3$. The vertical axis shows the intermodulation order for each spectral component.

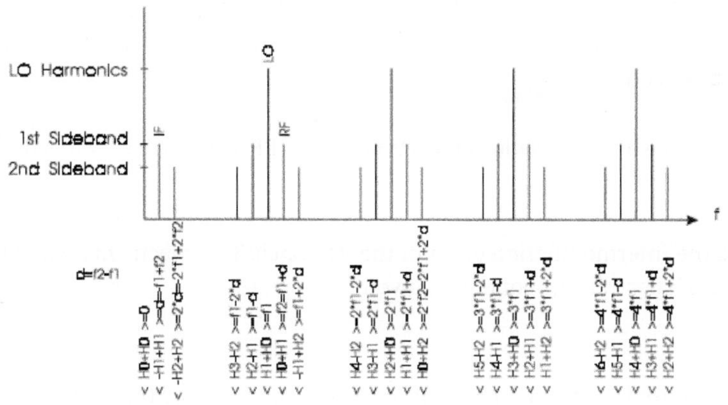

Figure 4.66 Two-tone spectrum for $M_1 = 4$, $M_2 = 2$. The vertical axis delineates the LO harmonics and sideband products.

where f_3 is the frequency of the third fundamental tone. The spectrum for the intermodulation case, where the power in the three tones is of the same order, and is

$$m_1 = m_2 = m_3 = M0 \leq |p| + |q| + |r| \leq M \tag{4.110}$$

This selection results in a spectrum as shown in Figure 4.67, for $M = 3$. Notice the much larger number of tones in the spectrum as compared with the two-tone case for $M = 3$. The total number of spectral components is given by:

$$N_t = \frac{1}{3} \cdot M \cdot (M + 1) \cdot (2M + 1) + 1 \tag{4.111}$$

4.8.8 Three-tone Mixer Case

For the mixer case, the spectrum is chosen assuming that one LO and two RF signals are present. The spectrum is selected by recognizing the intermodulation products of the RF signals as the quantities of interest. The intermodulation spectrum of the two isolated RF signals is then repeated on each side of the LO harmonics to form the three-tone mixer spectrum. This is given by:

$$m_1 = M_1, m_2 = m_3 = M_2, 0 \leq |p + q + r| \leq M_1, \quad 0 \leq |q| + |r| \leq M_2 \tag{4.112}$$

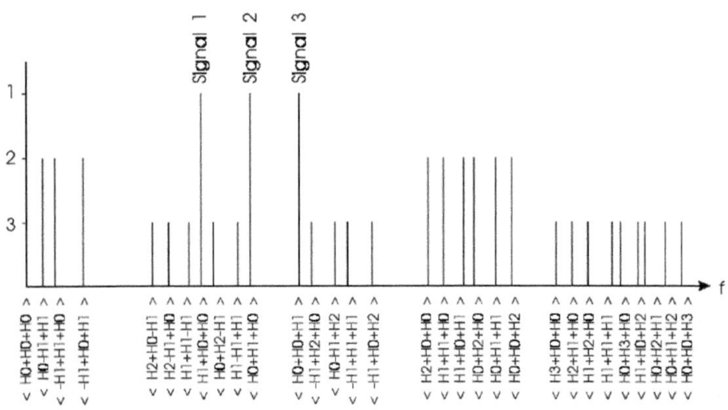

Figure 4.67 Three-tone intermodulation spectrum for $M = 3$. The vertical axis shows the intermodulation order for each spectral component. Note that the products will change position depending on the separation between the fundamental tones. Here the difference between signal 3 and signal 2 is slightly greater than the difference between signal 2 and signal 1.

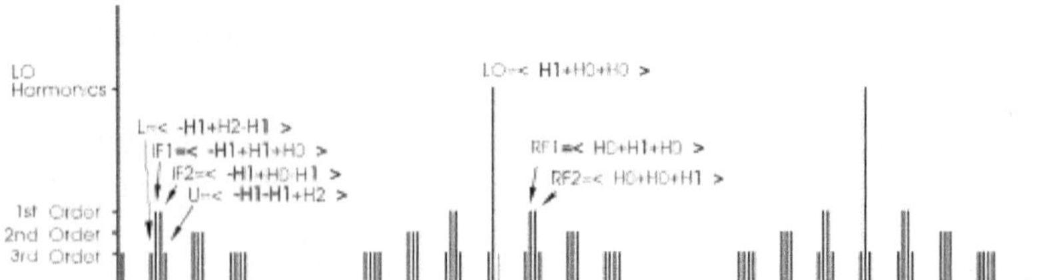

Figure 4.68 Three-tone mixer spectrum for $M_1 = 2$, $M_2 = 3$. The intermodulation products of the two RF signals are shown to third order. Important spectral components are labeled.

where M_1 is the number of LO harmonics and M_2 is the intermodulation order of the RF tones. The spectrum is shown in Figure 4.68 for $M_1 = 2$ and $M_2 = 3$. The total number of spectral components is given by

$$N_t = M_1 \cdot (2 \cdot M_2^2 + 2 \cdot M_2 + 1) + M_2^2 + M_2 \tag{4.113}$$

Generally, the number of frequency components required for three-tone analysis grows much faster than for two-tone analysis. This restricts the nonlinearity that can be computed. If four or more tones were considered, the number of frequency components becomes prohibitive unless smaller selection schemes are used.

4.8.9 Local Oscillator Spectrum Initialization of Mixer Circuits

For mixer analysis cases where the primary interest is the conversion gain and the RF signal powers are small compared to the LO, the circuit can be analyzed using the LO signal only and the conversion gain is determined using small-signal (linear) frequency-conversion methods.

This is performed using the small-signal mixer analysis option (see the next section in this chapter).

For cases where the RF signal power is not insignificant compared to the LO, a full mixer spectrum must be used. Compression of the conversion gain due to high RF powers can then be analyzed. Here, the mixer problem can be divided into two parts to help speed the analysis.

Firstly, the LO signal is analyzed using single-tone analysis; the RF signal is turned off. Single-tone analysis is usually very fast compared with a full two-tone analysis. Once the LO signal spectrum is found, the results are used to initialize the full mixer spectrum and the RF signal is turned back on. The full spectrum is then analyzed.

This method is most useful for three-tone mixer problems, due to the large number of spectral components used in the analysis. The primary use of the three-tone mixer analysis is to determine the intermodulation products of the IF products. This precludes the use of small- signal mixer analysis (since the intermodulation products cannot be determined using linear frequency conversion methods), but the RF signals are generally small compared to the LO. By solving the LO problem first, which is the primary nonlinear problem, and then introducing the RF signals, the analysis time can be reduced by a factor of about three. The actual time reduction depends on the circuit, the RF power levels, and the conversion gain.

4.8.10 Sparse Jacobian Techniques

The Jacobian matrix, when properly arranged, can be treated as a sparse matrix by pre-setting some entries to zero [49].

The physical reason for doing this is that most of the power transfer takes place between the harmonic frequencies of the fundamentals and much less takes place between the other frequencies in the spectrum. We can therefore set these derivatives to zero within the Jacobian. When this criterion is not met, the band of nonzero entries is widened to include cross-harmonic terms.

As the Jacobian structure is properly arranged, sparse matrix techniques are efficiently employed. General-purpose sparse matrix (Figure 4.69: Sparse Jacobian block structure) solvers that analyze the sparsity structure are avoided and specialized solvers can be used that are much more efficient. Harmonica automatically sets the bandwidth of the sparse tridiagonal matrix and dynamically alters it if the nonlinearity of the circuit is too great for the sparse assumptions. In this way the simulator achieves convergence using the minimal amount of computation time and memory that is possible for a given problem. For circuits with many devices under multitone operation, the CPU time may be decreased by a factor of 40.

The sparsity parameter will be dynamically altered during execution if needed. If $\boldsymbol{n}$ is greater than $N_t/3$ (N_t is the number of frequencies), then the program will use the full Jacobian. If only the full Jacobian is desired, then set n to a large number.

NOTE Using a sparse Jacobian does not affect the final values or accuracy of the results. It will only affect the convergence properties of the particular problem.

4.8.11 Iterative Newton Method

One of the shortcomings of harmonic balance methods is the large memory requirements when a circuit has many nonlinear devices and/or multitone analysis is needed. The Jacobian system matrix grows large and must be stored and factored. Sparse methods may not be enough to keep the problem within the memory bounds and acceptable computational resources of desktop computers. Harmonica uses a technique that efficiently solves large systems of equations without direct factorization of the system matrix. In this way, there is no simplification or approximation made to the problem and the full accuracy of the conventional harmonic balance method is completely maintained. The convergence and power-handling capabilities of conventional harmonic balance analysis are also fully maintained. The method is completely automatic and does not require any user intervention. An internal software switch detects when the new method should be used and automatically invokes it.

A brief summary of the method and its advantages is given: Conventional harmonic balance computes and stores the Jacobian matrix. The iterative solution of the harmonic balance equations requires factorization of the Jacobian to obtain updates of the circuit voltages. As the number of nonlinear devices in the circuit increases and the number of spectral components used to analyze the circuit increases, the Jacobian matrix can become very large, requiring tens or hundreds

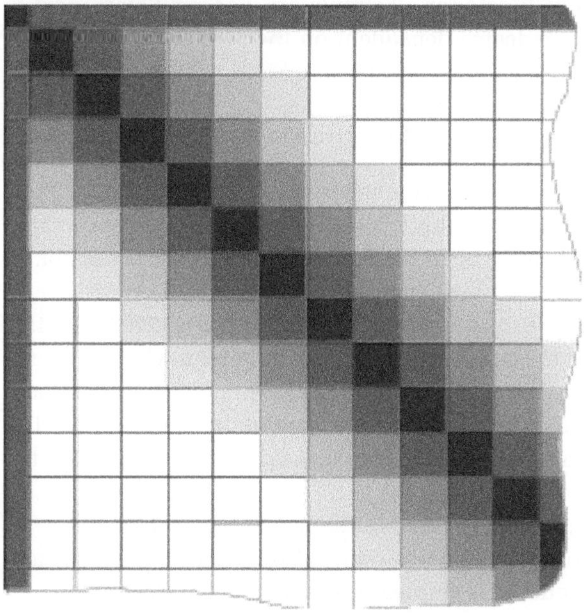

Figure 4.69 Sparse Jacobian block structure.

of megabytes of storage and much of CPU time to factor it. The calculation and factorization of the Jacobian typically occur several times during a single harmonic balance solution.

A very efficient method, based on an iterative approach, known as *Krylov Subspace Methods*, avoids direct storage and factorization of the Jacobian. Rather, a series of matrix-vector operations replaces the full storage and factorization steps while retaining full numerical accuracy. A good overview on the topic is provided by the summary papers [57, 58], and the probably best presentation of these techniques is found in "Steady-State Methods for Simulating Analog and Microwave Circuits" by Kundert, White, and Sangiovanni-Vincentelli [59]. Also see [60] as follows:

As a widely adopted frequency-domain method, harmonic balance (HB) provides efficient steady-state circuit analysis for analog and RF circuits. The conventional matrix-implicit Krylov subspace technique with the block-diagonal (BD) preconditioner has made it possible to compute the steady-state responses of large-scale circuits. However, not all HB problems, particularly strongly nonlinear circuit problems, can be solved reliably or efficiently using the standard BD-preconditioning technique. In this paper, hierarchical HB methods are proposed wherein robust preconditioning is provided via solution of a set of approximate linearized HB problems of progressively smaller size across multiple levels of the problem hierarchy. These subproblems are constructed using the same matrix-implicit formulation to retain the memory efficiency of Krylov subspace methods. Moreover, the number of allocated Krylov subspace matrix solvers, hence the memory usage, is significantly reduced via a recently introduced solver-sharing technique. The efficiency of our hierarchical preconditioning technique is further improved by adopting a one-step correction to the standard BD preconditioner and a multigrid-motivated iterative scheme. It has been shown that the proposed approaches can achieve up to 10 runtime speedup over the popular BD preconditioner and robust convergence even for strongly nonlinear circuits for which the BD preconditioner fails to converge.

The observed speed-up factors depend on the number of nonlinear devices in the circuit and the number of spectral components used in the analysis as well as the convergence properties of the harmonic balance algorithm. Speed improvements over conventional harmonic balance analysis from 2× to 10× for circuits consisting of a few transistors under two- and three-tone excitation have routinely been observed. A circuit containing 20 FETs under three-tone analysis exhibited a speed improvement factor of 30×. Memory requirements have also been tremendously reduced. The 20 FET circuit originally required >200 MB and now will analyze with 64 MB. As the circuit becomes more "complex," the new methods provide better speed and memory improvements. The frequency-doubler example mentioned in Rohde and Whitaker [61] will be demonstrated later.

The key quality of the core of the simulator is its numerical accuracy. To get some feel for this, we took a circuit from our textbook [62].

As can be seen from the various publications, including the first ever good explanation by Leeson, why all oscillators are noisy is the fact that the oscillator has both conversion noise and modulation noise contributions, similar to a mixer.

Here is the test circuit diagram; it's a patented circuit with a wideband DC control loop that improves the noise behavior, and this circuit has only lumped elements and uses the identical SPICE data for all simulators. For simulators, next to accuracy speed is an important parameter. The simulation time is to a degree dependent on the maximum numbers of harmonics considered.

4.8.12 CAD Simulations Using Three Different Circuit Simulators

The three simulations differ very little in phase noise and output power at the fundamental frequency. Far out the ADS correctly predicts the influence of the DC loop bandwidth; the AWR program shows this to a lesser degree. The first and second harmonic output power varies between the simulations. This depends on the different default settings of the simulator.

The circuit shown in Figure 4.70 is a Colpitts oscillator, not quite; as the tuned circuit consisting of a 12 kΩ resistor, in parallel with a 5 nH inductor and 4.7 pF capacitor, $1/\omega C$ is approximately 8 Ω, at 1 GHz, so the operating Q is 12000/80 = 150. For the oscillator to operate, the loosely coupled parallel circuit has to act like an inductor. The reason why the energy is taken out at the emitter is because this is part of the feedback loop. The collector circuit monitors the DC current and has the same temperature coefficient as the oscillator transistor, so the overall circuit is temperature-compensated. The 82 Ω resistor in the collector samples the voltage across the DC feedback circuit and maintains a constant current. The top transistor acts like a reference diode.

Based on analytical formulae (for details see the oscillator section), we obtain results as shown in Figure 4.71.

This formula, containing nonlinear equations, which required a complicated derivation, has been tested very often and next to having a simulator gives fairly accurate results. The flicker frequency is defined by the AF and KF factors.

And now the phase noise calculation of the simulator shows the performance of the DC feedback well.

The simulation of the circuit using the Serenade harmonic balance simulator indicates an output power of 14.35 dBm (Figure 4.72a) and the predicted phase noise as shown in Figure 4.72b. The phase noise calculation and simulation are fairly close.

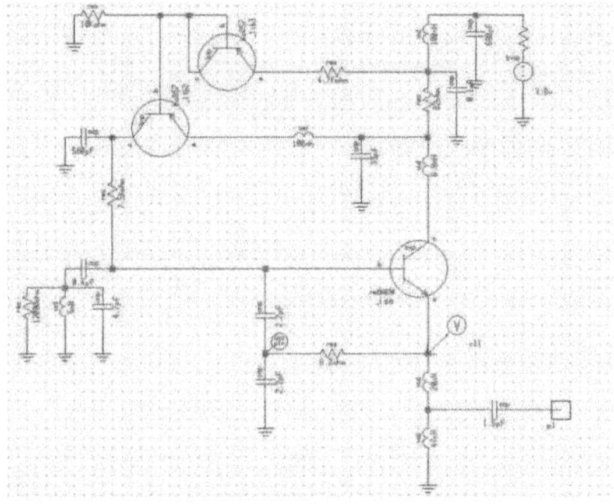

Figure 4.70 Colpitts oscillator.

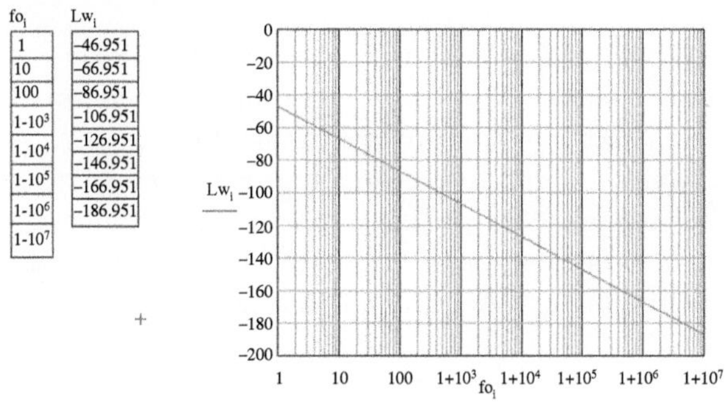

Figure 4.71 Calculated phase noise plot of the Colpitts oscillator in Figure 4.70.

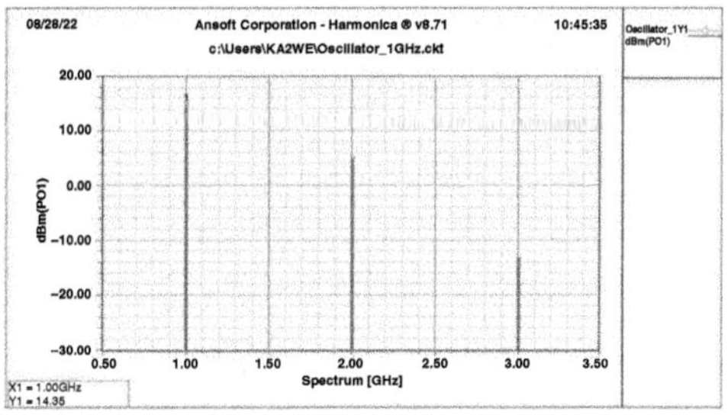

(a)

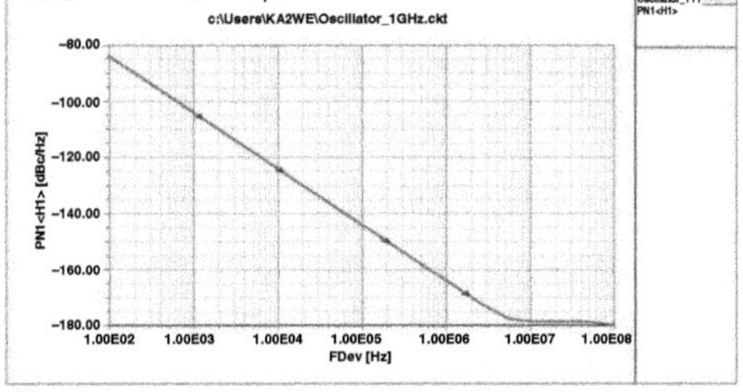

(b)

Figure 4.72 Simulations of the circuit in Figure 4.60 using Serenade. (a) Simulated power output in dBm. (b) Simulated phase noise.

The identical circuit using the identical SPICE parameters in the Keysight program, ADS gives the results as shown in Figures 4.73 and 4.74.

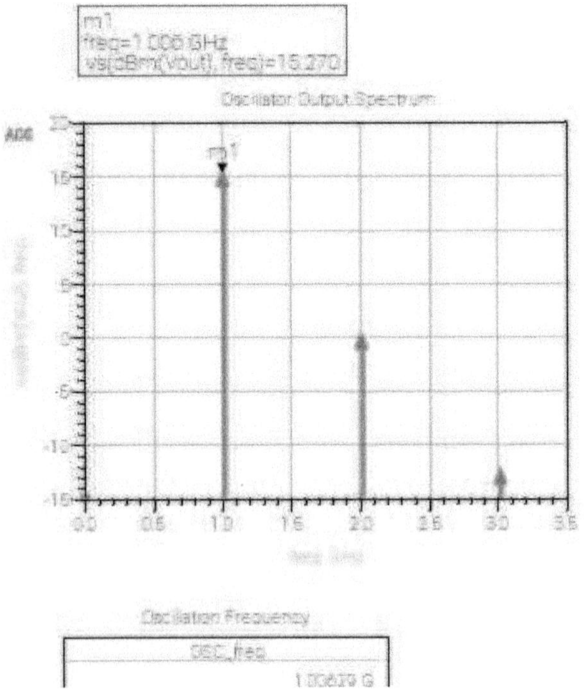

(a)

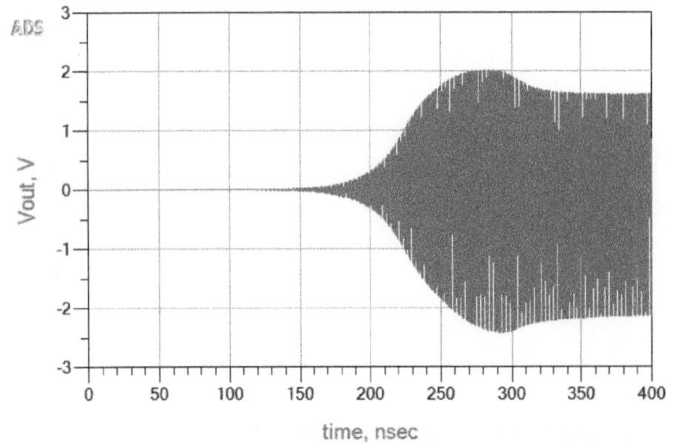

(b)

Figure 4.73 Simulations of the circuit in Figure 4.60 using Keysight ADS. (a) Simulated power output in dBm. (b) Simulated output voltage response in the time domain.

For further validation, the same circuit was analyzed in the CADENCE AWR circuit simulator.

Finally, Figure 4.75 shows the third simulation.

This exercise shows that the modern harmonic balance simulators are well-matured.

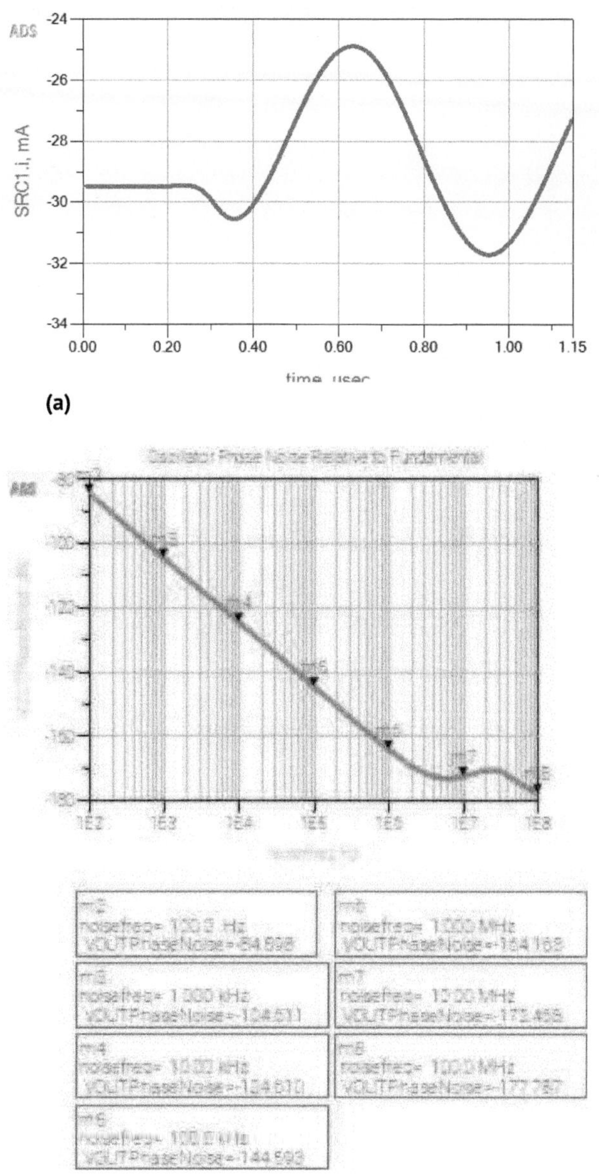

(a)

(b)

Figure 4.74 Simulations of the circuit in Figure 4.60 using Keysight ADS. (a) Collector current time domain response and (b) phase noise.

4.9 Coupled-line Resonator Oscillator (EM Simulation)

The following oscillator is based on loosely coupled transmission lines and operates at 6.37 GHz. It's a good test to validate the two simulators to give essentially the same result but we can also demonstrate the frequency shift due to the layout coupling.

The simulation of the output power of the circuit in Figure 4.76 shows that the harmonic suppression is about 13 dB, see Figure 4.77a.

The simulated phase noise of the circuit in Figure 4.77b shows a phase noise 107 dBc/Hz at 10 kHz offset for the operating frequency of 6.37 GHz and is therefore noisier than Figure 4.75. The flicker corner frequency appears to be around 1 kHz. Next we repeat the simulation with the Keysight's ADS and Cadence AWR circuit simulators. The results are shown in the following figures. The results among all the three simulators agree well, as can be seen in Figures 4.75, 4.78, and 4.79.

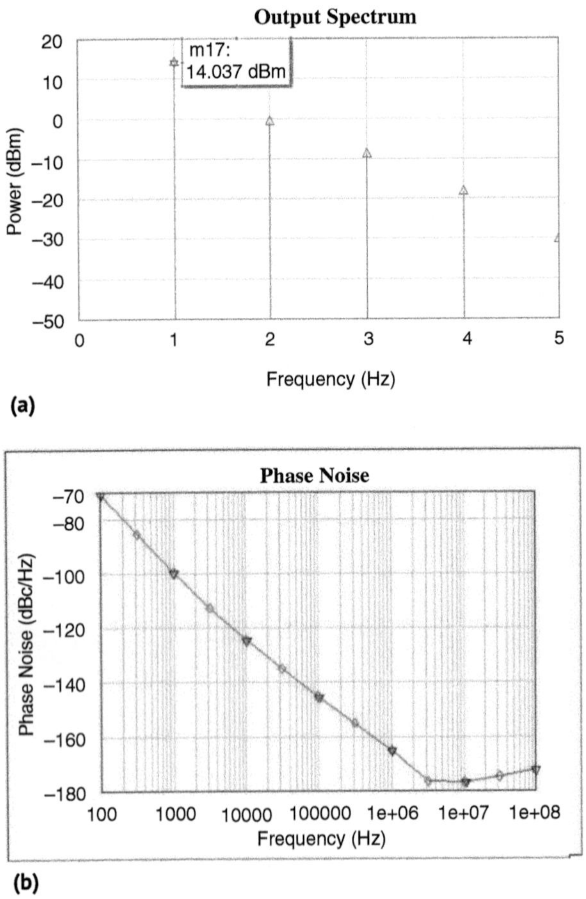

Figure 4.75 Simulations of the circuit in Figure 4.60 using Cadence AWR. (a) Output power spectrum and (b) phase noise.

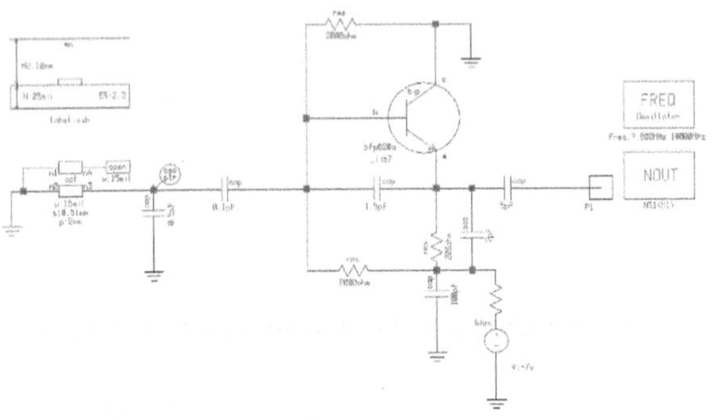

Figure 4.76 6.37 GHz coupled resonator oscillator.

When the circuit was transferred into a layout, shown in Figure 4.79, it resulted in shift of frequency and phase noise (little) as was expected (Figure 4.80).

Simulation based on the influence of the printed circuit shows a shift in the oscillating frequency (Figure 4.81).

These two examples were necessary to have enough confidence in the simulators.

Under large-signal conditions, standard S-parameters are no longer valid, and oscillator operation as well as mixers and doublers are operating in large-signal conditions. The generation of the large-signal S-parameters is an important capability.

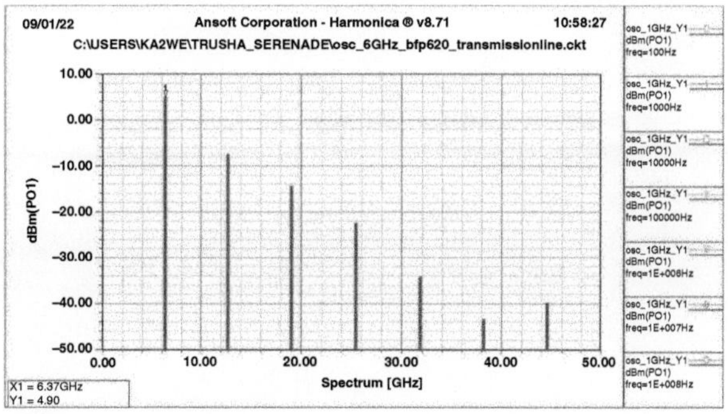

(a)

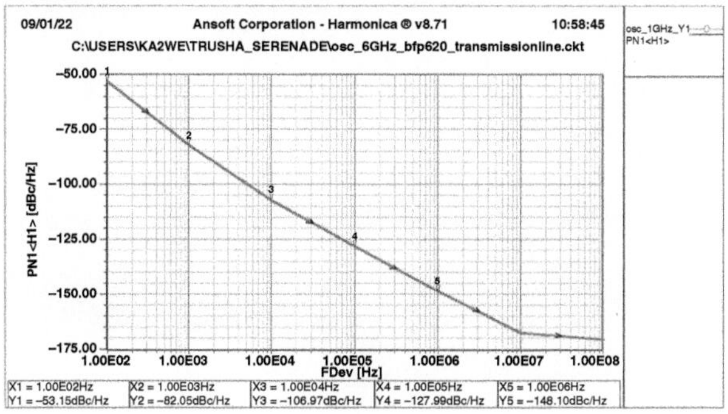

(b)

Figure 4.77 Simulations of the circuit in Figure 4.76. (a) Output power spectrum and (b) phase noise.

4.10 Generating Large-signal S-parameters

The calculation of S-parameters in the large-signal regime is not as straightforward as it is in the linear, small-signal regime. The "large-signal S-parameters" are dependent on the power of the excitation sources at each external circuit port as well as the circuit bias and terminations.

Guidelines will be given here on using Harmonica to generate large-signal S-parameters, but the proper use of these S-parameters in circuit design is up to you.

Consider a two-port circuit whose large-signal S-parameters are desired. If we apply a source at port 1 with port 2 terminated, we could measure the reflected and transmitted waves, and conversely for a source applied to port 2 [50]

However, this assumes that when the device under test is actually used, it will be terminated in the same impedance as it was tested. This is rarely the case. Typically the device is embedded in some matching network which presents complex impedance to the device. Therefore, the operating regime of the device will change and its large-signal S-parameters will be altered.

We could then hypothesize that a source can be placed at each port, and the traveling waves could be measured at each port. The problem here is that it is not possible to distinguish between the reflected wave at a port and the transmitted wave due to the source at the other port because the sources are the same frequency. If we perturb the frequency of one of the sources, then the reflected and transmitted waves due to each source can be resolved. This, however, requires a two-tone analysis. The situation is illustrated in Figure 4.82 for the two-port device under test.

The difference in frequency between the two sources can be made small, on the order of 0.001%. This is recommended for circuits of large Q. Typically, the difference used is about 0.1% because the S-parameters of the device under test do not change rapidly with frequency. To measure the traveling waves, we use the return loss and transducer gain output responses in the *nout* block.

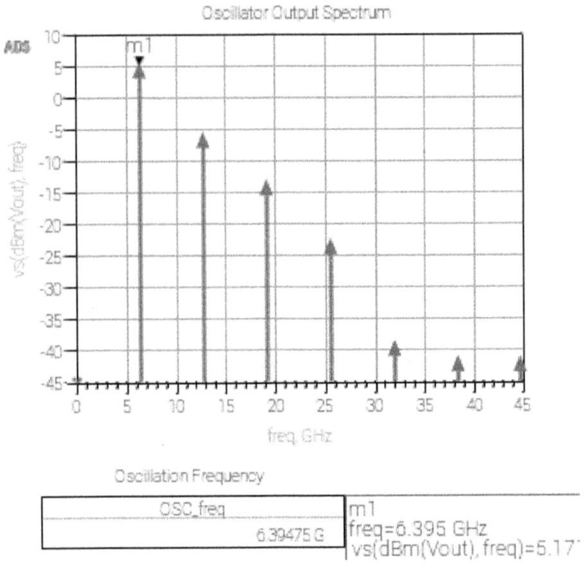

(a)

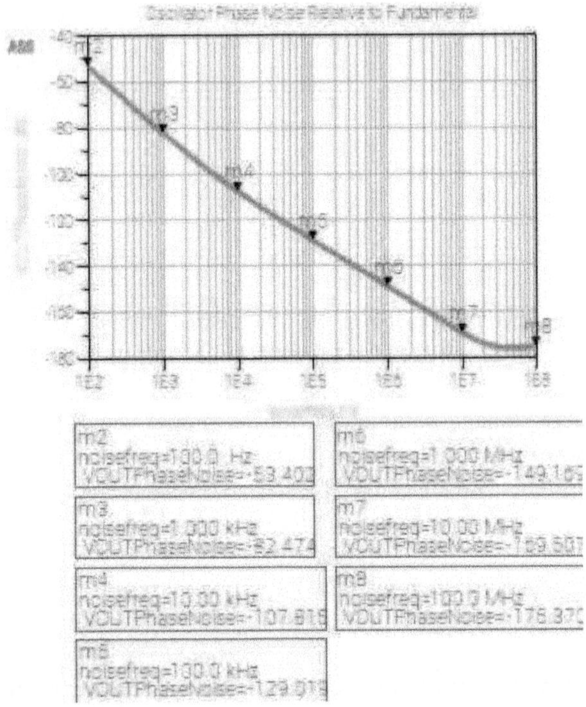

(b)

Figure 4.78 Simulations of the circuit in Figure 4.76 with Keysight ADS. (a) Output power spectrum and (b) phase noise.

4.11 Nonlinear Noise Analysis—An Overview

Noise performance of microwave circuits is one of the major concerns for circuit design engineers. It is an important determining factor of receiver system sensitivity and dynamic range. Linear simulators are ideal to predict the noise performance of circuits using arbitrary combinations of active and passive elements. However, most microwave circuits need to be analyzed using nonlinear analysis techniques since the applied signals may cause the circuit to operate closely to a saturated mode and result in nonlinear operation. This section illustrates how the noise performance of general mixer or multiplier/divider circuits is simulated using the techniques implemented in Harmonica [63, 64].

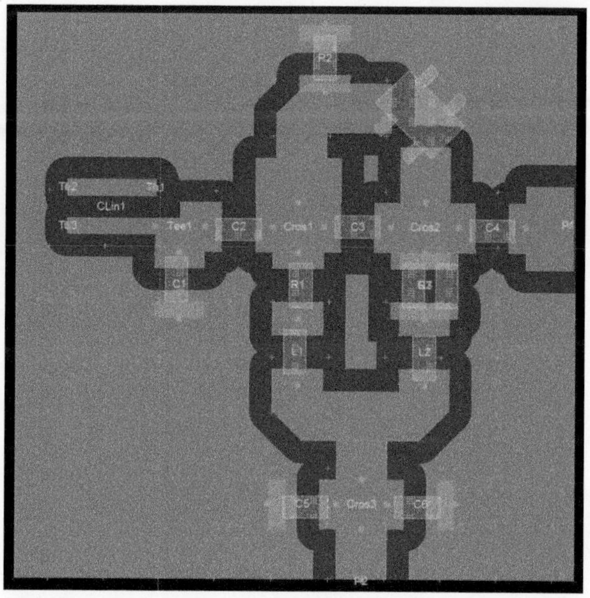

Figure 4.79 Printed circuit board layout for the coupled resonator oscillator.

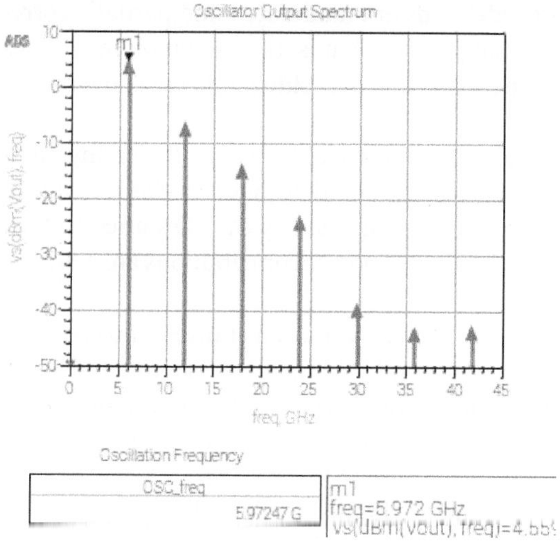

Figure 4.80 Simulated output power of the oscillator on the printed circuit board.

The goal of the noise analysis is to find the spectral distribution of the noise power actually delivered to the load. Three different noise analysis capabilities have been implemented:

- For mixer noise analysis, the noise figure from the RF signal input to the IF load can be determined [63, 64].
- For single-tone circuits with a large-signal input, such as a frequency multiplier or divider, the noise analysis can determine the noise power spectral distribution and amplitude and phase noise near a specified harmonic [65, 66].
- For oscillator circuits, the phase and amplitude noise power delivered to a load can be determined.

The first two topics are discussed in this section. For a discussion on oscillator noise analysis, please refer to the *Oscillator Analysis* chapter in this volume.

4.11.1 Basic Information About Noise Calculations

A circuit biased at DC without RF excitation generates noise due to thermal noise and the bias current in the active devices (flicker noise and shot noise). The noise is dependent on the bias point of the active devices and is uncorrelated. When

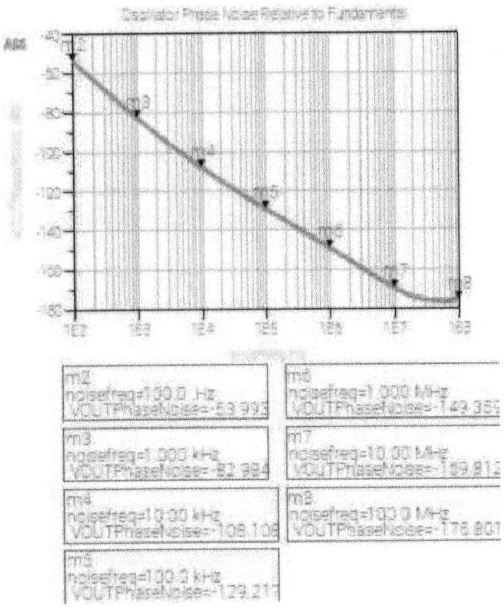

Figure 4.81 Simulated phase noise of the oscillator on the printed circuit board (Figure 4.79) (using ADS).

the device is pumped by an RF excitation (e.g., the LO), the nonlinear noise sidebands are modulated and partially correlated because each sideband is a combination of the original uncorrelated DC sidebands. That is, the excitation modulates the instantaneous operating point of the device, and due to the device nonlinearity, the noise sidebands are converted in frequency to other harmonic sidebands.

During the frequency conversion, each sideband generates correlated components in the vicinity of the IF (or any other spectral frequency). Noise figure can be determined from these noise components that are dissipated in the IF load. In addition, contributions to the noise power at the IF load are also made by both the noise injected by the LO source and the thermal noise generated by the linear network. Because they are uncorrelated, their noise power contributions are additive. Figure 4.83 illustrates the general concept of the noise calculation.

Figure 4.83 can be viewed in three pieces. The nonlinear devices are modulated by the LO (or other source), and the noise due to the devices is contributed to the IF load through frequency conversion. The second piece is the noise from the linear subnetwork, which, through frequency conversion, makes it to the IF load. Finally, noise present in the LO can make it to the IF load through frequency conversion. The summation of noise powers indicates that noise from the three sources is uncorrelated.

In the following, various noise sources considered in the noise analysis of nonlinear circuits are described.

4.11.2 Thermal Noise Sources in a Passive Subnetwork

In a linear passive network, noise arises only from the losses in the network. The noise correlation matrix for a linear passive subnetwork is directly related to losses represented in the admittance matrix:

$$C_n = 2K_B T\,\Delta f(Y + Y^*) \tag{4.114}$$

where Y is the admittance matrix of the passive network, Y^* indicates the conjugate transpose, K_B is Boltzmann's constant, T is the ambient temperature, and Δf is the noise bandwidth. The linear analysis of the circuit automatically computes the noise matrix. No special keywords are required.

4.11.3 Noise Models of Active Devices

The noise models of the active devices vary from device to device. All have a shot noise component, which is proportional to the instantaneous current through the active region of the device. The models also have a flicker noise ($1/f$) component, which is important for low-frequency mixer IF and near-carrier noise in oscillators.

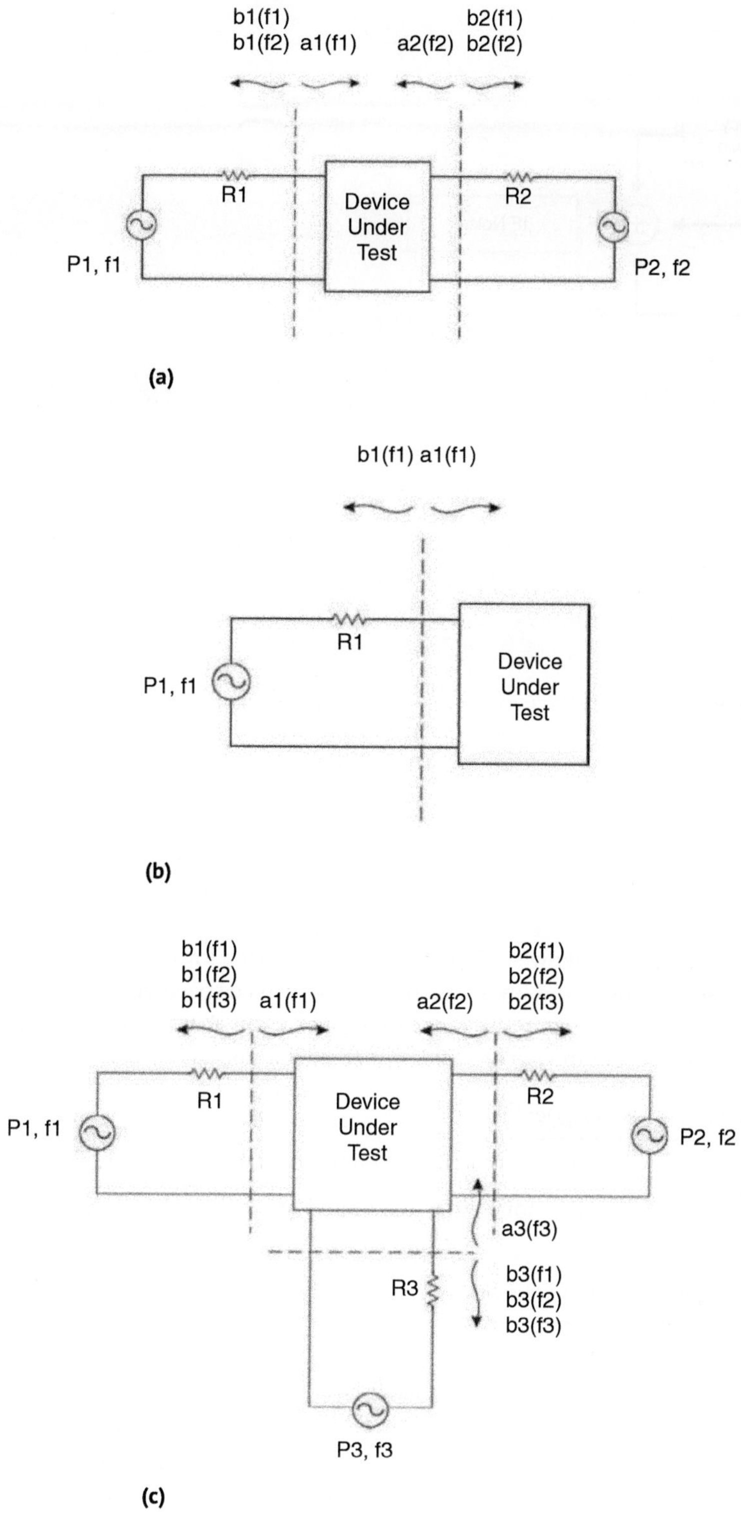

Figure 4.82 Test setup for determining large-signal S-parameters. (a) Two-port setup. (b) One-port setup. (c) Three-port setup.

4.11.4 Statistical Noise of an Exciting Source

The specification of the noise associated with the excitation is shown graphically in Figure 4.84. FDEVi is the frequency deviation of the ith noise sideband. RFUPi is normalized upper sideband of the ith noise power component in units of dBc/Hz.

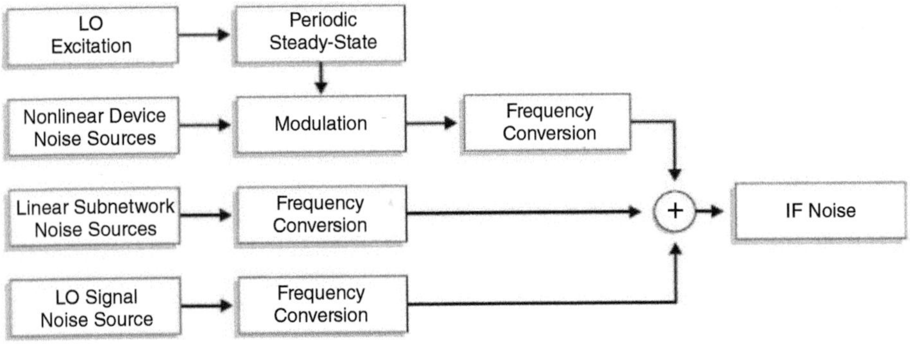

Figure 4.83 IF noise in a general nonlinear mixer circuit.

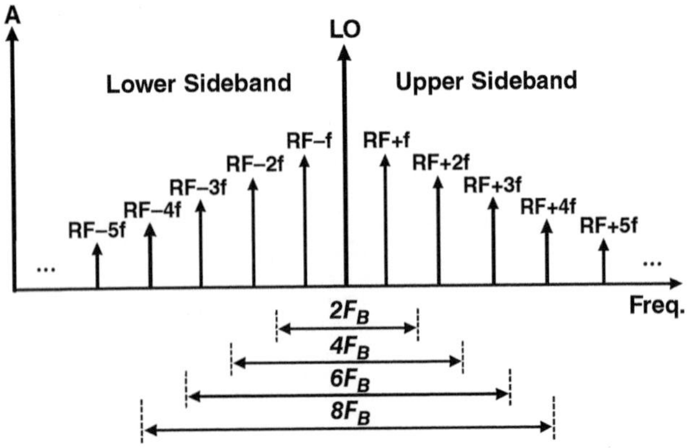

Figure 4.84 FDEV*i* is the frequency deviation of the *i*th noise sideband.

It is specified in decibels relative to the carrier (LO in the figure). Likewise, RFLO*i* is the normalized lower sideband. For a complete statistical description, the correlation between the upper and lower sidebands can also be specified.

The mean-square value of the amplitude fluctuations (normalized amplitude noise) is given from the upper and lower complex phasors of the noise components in a 1 Hz band located at the lower and *R* upper sidebands:

$$\left\langle |\delta A|^2(f_d) \right\rangle = 2\frac{\left\langle |V_1(f_d)|^2 \right\rangle + \left\langle |V_u(f_d)|^2 \right\rangle + 2Re[\langle V_1^*(f_d)V_u^*(f_d)\rangle e^{2j\Phi_0}]}{|V_0|^2} \tag{4.115}$$

The mean-square value of the phase fluctuations (normalized phase noise) in a 1 Hz band at a frequency deviation f_d from the carrier (PM noise) is given by:

$$\left\langle |\delta \Phi|^2(f_d) \right\rangle = 2\frac{\left\langle |V_1(f_d)|^2 \right\rangle + \left\langle |V_u(f_d)|^2 \right\rangle - 2Re[\langle V_1^*(f_d)V_u^*(f_d)\rangle e^{2j\Phi_0}]}{|V_0|^2} \tag{4.116}$$

A complete statistical description of the noisy source requires that the two real quantities, $\langle V_1(f_d)V_1^*(f_d)\rangle$ and $\langle V_u(f_d)V_u^*(f_d)\rangle$, and the complex quantity, $\langle V_1^*(f_d)V_u^*(f_d)\rangle$, be assigned as a function of the frequency deviation from the carrier. However, in many practical cases, a full statistical description of the noisy source is not available, because the information on the source noise is produced by measurements that only provide the RF spectrum and the AM noise (or equivalently the FM and AM noise). In this case, one of the alternatives of the reduced data set can be used.

It is worth noting that in practice the normalized amplitude noise is often several tens of dBs lower than the phase noise (or, equivalently, than the RF spectrum), especially at low-frequency deviations. In such cases, the noise analysis may become ill-conditioned from the viewpoint of numerical accuracy. Relatively large values of the numbers of sampling points may then be required to produce correct results.

4.11.5 Noise Figure and Noise Spectrum Calculations

A mixer noise figure calculation determines the noise added by the circuit under investigation relative to the noise injected into the circuit by the source. It is given by:

$$\mathrm{NF} = \frac{\mathrm{SNR}_{\mathrm{in}}^{\mathrm{RF}}}{\mathrm{SNR}_{\mathrm{out}}^{\mathrm{IF}}} = \frac{N_{\mathrm{out}}^{\mathrm{IF}}}{N_{\mathrm{in}}^{\mathrm{RF}}} \frac{1}{G_c} \tag{4.117}$$

where N_p^f is the noise power at port p, frequency f. G_c is the conversion gain of the mixer.

Within the simulation, the RF source and LO source are considered noisy and the IF termination is considered noiseless. These sources contribute noise at all frequencies and the double-sideband noise figure is calculated.

The calculation of noise spectrum for single-tone analysis (including amplifiers, oscillators, multipliers, etc.) determines the noise power relative to the specified harmonic, the absolute noise power absorbed by the load at the specified harmonic is computed and normalized (in dBc) to that harmonic power. In this case, both the source termination and the load termination are considered noisy.

4.11.6 Conclusion

While this was a major effort to go through all the details as discussed earlier, at least now we have the knowledge to see how all these things fit together.

References

1 Rizzoli, V., Lipparini, A., and Marazzi, E. (1983). A general-purpose program for nonlinear microwave circuit design. *IEEE Transactions on Microwave Theory and Techniques* 31 (9): 762–770.

2 Rizzoli, V., Lipparini, A., Costanzo, A. et al. (1992). State-of-the-art harmonic-balance simulation of forced nonlinear microwave circuits by the piecewise technique. *IEEE Transactions on Microwave Theory and Techniques* 40 (1): 12–28.

3 Johnson, A. (1965). Physical limitations on frequency and power parameters of transistors. *RCA Review* 26: 163–177.

4 Rudolph, M. and Heymann, P. (2004). The influence of microwave two-port noise on residual phase noise in GaAs-HBTs. *Proceedings of the European Microwave Conference*, vol. 2, 945–948.

5 Rudolph, M., Schnieder, F., and Heinrich, W. (2003). Modeling emitter breakdown in GaAs-based HBTs. *IEEE MTT-S International Microwave Symposium Digest*, 651–654.

6 Kroemer, H. (1957). Theory of a wide-gap emitter for transistors. *Proceedings of the IRE* 45: 1535–1537.

7 Vaidyanathan, M. and Pulfrey, D.L. (1999). Extrapolated $f_{\max}$ of heterojunction bipolar transistors. *IEEE Transactions on Electron Devices* 46 (2): 301–309.

8 Lu, S.-S., Meng, C., Chen, T.-W., et al. (2001). The origin of the kink phenomenon of transistor scattering parameter S_{22}. *IEEE Transactions on Microwave Theory and Techniques* 49 (2): 333–340.

9 Gummel, H.K. and Poon, H.C. (1970). An integral charge control model of bipolar transistors. *Bell System Technical Journal* 49 (5): 827–852.

10 Gummel, H.K. (1970). A charge control relation for bipolar transistors. *Bell System Technical Journal* 49 (1): 115–120.

11 Getreu, I. (1976). *Modeling the Bipolar Transistor*. Beaverton, OR: Tektronix, Inc.

12 Kirk, C.T. Jr. (1962). A theory of transistor cutoff frequency (f_t) falloff at high current densities. *IRE Transactions on Electron Devices* 9 (2): 164–174.

13 Whittier, R.J. and Tremere, D.A. (1969). Current gain and cutoff frequency falloff at high currents. *IEEE Transactions on Electron Devices* ED-16 (1): 39–57.

14 The HICUM model page at TU Dresden. https://www.iee.et.tu-dresden.de/iee/eb/hicineq-810new/hic_intro.html

15 Agilent Technologies (2004). AgilentHBT model (Agilent heterojunction bipolar transistor model). *Nonlinear Devices Manual for Advanced Design System*, 2-4–2-42.

16 Rudolph, M. (2006). *Introduction to* Modeling *HBTs*. Artech House Microwave Library. Artech House Publishers. ISBN: 10:1-58053-144-x.

17 Rudolph, M., Doerner, R., and Beilenhoff, K. et al. (2002). Unified model for collector charge in heterojunction bipolar transistors. *IEEE Transactions on Microwave Theory and Techniques* 50: 1747–1751.

18 Rudolph, M. and Doerner, R. (2005). Consistent modeling of capacitances and transit times of GaAs-based HBTs. *IEEE Transactions on Electron Devices* 52: 1969–1975.

19 Camnitz, L.H. and Moll, N. (1993). An analysis of the behaviour of microwave heterostructure bipolar transistors. In: *Compound Semiconductor Transistors — Physics and Technology* (ed. S. Tiwari), 21–45. New York: IEEE Press.

20 Rudolph, M. and Doerner, R. (2004). Large-signal HBT model requirements to predict nonlinear behaviour. *IEEE MTT-S International Microwave Symposium Digest*, 43–46.

21 Iwamoto, M., Asbeck, P.M., Low, T.S. et al. (2000). Linearity characteristics of GaAs HBTs and the influence of collector design. *IEEE Transactions on Microwave Theory and Techniques* 48: 2377–2386.

22 Huang, C.-C., Chen, T.-Y., Hsu, C.-S. et al. (2011). Comprehensive temperature-dependent studies of metamorphic high electron mobility transistors with double and single-doped structures. *IEEE Transactions on Electron Devices* 58 (12): 4276–4282.

23 Shockley, W. (1932). A unipolar field-effect transistor. *Proceedings of the IEEE* 20: 1365–1376.

24 Pucel, R.A., Haus, H.A., and Statz, H. (1975). Signal and noise properties of gallium arsenide microwave field-effect transistors. *Advances in Electronics and Electron Physics* 38: 195–265.

25 Kallfass, I. (2006). Comprehensive nonlinear modelling of dispersive heterostructure field effect transistors and their MMIC applications. Dissertation, University of Ulm.

26 Maas, S.A. (2003). *Nonlinear Microwave and RF Circuits*. Norwood, MA: Artech House, Chapters 1 and 2.

27 Kallio, A.I. and Valtonen, M.E. (2004). A new rule for MESFET gate charge division based on the energy conservation principle. *International Journal of Circuit Theory and Applications* 32: 139–165.

28 Camacho-Peñalosa, C. and Aitchison, C. (1985). Modelling frequency dependence of output impedance of a microwave MESFET at low frequencies. *Electronics Letters* 21 (12): 528–529.

29 Jardel, O., De Groote, F., Reveyrand, T. et al. (2007). An electrothermal model for AlGaN/GaN power HEMTs including trapping effects to improve large-signal simulation results on high VSWR. *IEEE Transactions on Microwave Theory and Techniques* 55: 2660–2669.

30 Rudolph, M. (2010). Modeling GaN power transistors. In: *11th IEEE Wireless and Microwave Technology Conference (WAMICON)*, 1–4.

31 Maas, S. (2002). Fixing the Curtice FET model. *Microwave Journal* 45 (3): 68–75.

32 Sriperumbuduri, V.P., Johansen, T.K., and Erkelenz, K. et al. (2020). An improved EM-simulation procedure to extract extrinsic elements of terahertz InP DHBTs. *German Microwave Conference (GeMiC)*, 240–243.

33 Son, J.-H., Oh, S.J., and Cheon, H. (2019). Potential clinical applications of terahertz radiation. *Journal of Applied Physics* 125 (19): 190901.

34 Kawase, K., Shibuya, T., Hayashi, S., et al. (2010). THz imaging techniques for nondestructive inspections. *Comptes Rendus Physique* 11: 510–518.

35 Deal, W.R., Foster, T., Wong, M.B. et al. (2017). A 666 GHz demonstration crosslink with 9.5 Gbps data rate. *IEEE MTT-S International Microwave Symposium (IMS)*.

36 Urteaga, M., Griffith, Z., Seo, M. et al. (2017). InP HBT technologies for THz integrated circuits. *Proceedings of the IEEE* 105 (6): 1051–1067.

37 Al-Sawaf, T., Nosaeva, K., Weimann, N.G. et al. (2016). A 200 mW InP DHBT W-band power amplifier in transferred substrate technology with integrated diamond heat spreader. *IEEE MTT-S International Microwave Symposium*.

38 Rudolph, M., Fager, C., and Root, D.E. (2011). *Nonlinear Transistor Model Parameter Extraction Techniques*. The Cambridge RF and Microwave Engineering Series. Cambridge University Press. ISBN: 970-0-521-76210-6.

39 Johansen, T.K., Doerner, R., Weimann, N. et al. (2018). EM simulation assisted parameter extraction for transferred-substrate InP HBT modeling. *EuMW Access* 10 (5-6): 700–708.

40 Chen, Y., Zhang, Y., Xu, Y. et al. (2018). Investigation of terahertz 3D EM simulation on device modeling and a new InP HBT dispersive inter-electrode impedance extraction method. *IEEE Access* 6 (10): 45772–45781.

41 Weimann, N.G., Stoppel, D., Schukfeh, M.I. et al. (2016). SciFab - a wafer-level heterointegrated InP DHBT/SiGe BiCMOS foundry process for mm-wave applications. *Physica Status Solidi A* 4: 909–916.

42 Marks, R.B. (1991). A multiline method of network analyzer calibration. *IEEE Transactions on Microwave Theory and Techniques* 39 (7): 1205–1215.

43 Williams, D.F. and Marks, R.B. (1991). Transmission line capacitance measurement. *Proceedings of the IEEE* 1 (9): 243–245.

44 Lenk, F. and Rudolph, M. (2002). New extraction algorithm for GaAs-HBTs with low intrinsic base resistance. *IEEE MTT-S International Microwave Symposium Digest*, 725–728.

45 Hillbrand, H. and Russer, P. (1976). An efficient method for computer aided noise analysis of linear amplifier networks. *IEEE Transactions on Circuits and Systems* 23 (4): 235–238. https://doi.org/10.1109/TCS.1976.1084200

46 Rizzoli, V., Cecchetti, C., and Lipparini, A. (1987). A general-purpose program for the analysis of nonlinear microwave circuits under multitone excitation by multidimensional Fourier transform. *European Microwave Conference*, 635–640.
47 Rizzoli, V. and Neri, A. (1988). State-of-the-art and present trends in nonlinear microwave CAD techniques. *IEEE Transactions on Microwave Theory and Techniques* 36 (2): 343–365.
48 Rizzoli, V., Cecchetti, C., Lipparini, A. et al. (1988). General-purpose harmonic-balance analysis of nonlinear microwave circuits under multitone excitation. *IEEE Transactions on Microwave Theory and Techniques* 36 (12): 1650–1660.
49 Rizzoli, V., Mastri, F., Sgallari, F. et al. (1990). The exploitation of sparse-matrix techniques in conjunction with the piecewise harmonic-balance method for nonlinear microwave circuit analysis. In: *IEEE MTT-S International Microwave Symposium Digest*, 1295–1298.
50 Rizzoli, V., Lipparini, A., and Mastri, F. (1988). Computation of large-signal S-parameters by harmonic-balance techniques. *Electronics Letters* 24: 329–330.
51 Rizzoli, V., Mastri, F., and Sgallari, F. et al. (1996). Harmonic-balance simulation of strongly nonlinear very large-size microwave circuits by inexact Newton methods. *IEEE MTT-S*, 1357–1360.
52 Vendelin, G.D., Pavio, A.M., and Rohde, U.L. et al. (2021). *Microwave Circuit Design: Using Linear and Nonlinear Techniques*, 3e. Hoboken, NJ: Wiley.
53 Rizzoli, V., Neri, A., and Costanzo, A. (1988). Analysis and optimization of DROs using a general purpose CAD program. *Alta Frequenza* 57: 389–398.
54 Neri, A. and Rizzoli, V. (1987). Global stability analysis of microwave circuits by a frequency-domain approach. *IEEE MTT-S International Microwave Symposium Digest*, 689–692.
55 Wacker, H. (1978). *Continuation Methods*. New York: Academic Press.
56 Iooss, G. and Joseph, D.D. (1980). *Elementary Stability and Bifurcation Theory*. New York: Springer-Verlag.
57 Telichevesky, R., Kundert, K.S., and White, J.K. (1995). Efficient steady-state analysis based on matrix-free Krylov-subspace methods. *32nd Design Automation Conference*, San Francisco, CA, USA, 480–484. https://doi.org/10.1145/217474.217574
58 Feng, D., Phillips, J., Nabors, K. et al. (1999). Efficient computation of quasi-periodic circuit operating conditions via a mixed frequency/time approach. *Design Automation Conference*, New Orleans, LA, USA, 635–640. https://doi.org/10.1109/DAC.1999.782020
59 Kundert, K.S., White, J.K., and Sangiovanni-Vincentelli, A. (1990). *Steady-state Methods for Simulating Analog and Microwave Circuits*. New York: Springer. https://doi.org/10.1007/978-1-4757-2081-5
60 Dong, W. and Li, P. (2007). Hierarchical harmonic-balance methods for frequency-domain analog-circuit analysis. *IEEE Transactions on Computer-Aided Design of Integrated Circuits and Systems* 26 (12): 2089–2101. https://doi.org/10.1109/TCAD.2007.907018
61 Rohde, U.L. and Whitaker, J.C. (2017). *Communications Receivers*, 4e. New York, NY: McGraw-Hill Education.
62 Rohde, U.L., Poddar, A.K., and Böck, G. (2005). *The Design of Modern Microwave Oscillators for Wireless Applications: Theory and Optimization*, 396. John Wiley & Sons. https://doi.org/10.1002/0471727172
63 Rizzoli, V., Mastri, F., and Cecchetti, C. (1989). Computer-aided noise analysis of MESFET and HEMT mixers. *IEEE Transactions on Microwave Theory and Techniques* 37 (9): 1401–1410.
64 Chang, C.R. and Gerber, J. (1993). Nonlinear noise analysis for wireless communication systems. *RF Expo West*, 254–260.
65 Rizzoli, V., Mastri, F., and Masotti, D. (1992). General-purpose noise analysis of forced nonlinear microwave circuits. *Military Microwaves*, 293–298.
66 Rizzoli, V., Mastri, F., and Masotti, D. (1992). Advanced piecewise-harmonic-balance noise analysis of nonlinear microwave circuits with applications to Schottky-barrier diodes. *IEEE MTT-S International Microwave Symposium Digest*, 247–250.

Further Reading

Schwierz, F. and Liou, J.J. (2003). *Modern Microwave Transistors: Theory, Design, and Performance*. Wiley-Interscience.
Beleniotis, P., Zervos, C., Krause, S. et al. (2025). Traps in GaN HEMTs: From characterization to model development and device optimization. *2025 IEEE BiCMOS and Compound Semiconductor Integrated Circuits and Technology Symposium (BCICTS)*, 1–8. https://doi.org/10.1109/BCICTS63111.2025.11211504.

Beleniotis, P., Krause, S., Zervos, C. et al. (2024). A physics-based model for slow gate-induced electron trapping in GaN HEMTs. *IEEE Transactions on Electron Devices* 71 (7): 4058–4065. https://doi.org/10.1109/TED.2024.3397234.

Beleniotis, P., Zervos, C., Krause, S. et al. (2024). Investigation of traps impact on PAE and linearity of AlGaN/GaN HEMTs relying on a combined TCAD–compact model approach. *IEEE Transactions on Electron Devices* 71 (6): 3582–3589. https://doi.org/10.1109/TED.2024.3388386.

Marzall, L. (2026). Teaching GaAs and GaN MMIC design: An education project between WIN semiconductors and UC boulder. *IEEE Microwave Magazine* 27 (8): 71.

5

Q-Enhanced Passive Microwave Resonators

5.1 Introduction

A resonator is an element that can store frequency-dependent electric and magnetic energy [1]. At microwave frequencies, resonators can take various shapes and forms. The shape of the microwave structure affects the field distribution and, hence, the stored electric and magnetic energies. Potentially, any microwave structure should be capable of constructing a resonator whose resonant frequency is determined by the structure's physical characteristics and dimensions [2–6]. A simple example is a series or parallel combination of an inductor (L) and a capacitor (C), where the magnetic energy is stored in the inductance L, and the electric power is stored in the capacitance C [7]. The resonant frequency of a resonator is the frequency at which the energy stored in an electric field equals the energy stored in the magnetic field. As components, resonators allow a selective transmission or blocking of signals and serve as 1-port or 2-port frequency-determining elements for oscillator applications [1–60].

5.2 Linear Passive 1-Port

A circuit network with a single port is referred to as a 1-port network. Figure 5.1 illustrates the typical schematic and equivalent representation of a 1-port network. The signal at the port can be described by the complex amplitudes a and b of the incident and reflected, or by the voltage $\overline{V}$ and $\overline{I}$ waves. If the relation between $\overline{V}$ and $\overline{I}$ or a and b are linear, the 1-port network is defined as linear 1-port, and the complex power P flowing into the 1-port can be given in terms of active and reactive power as [7]

$$P = \frac{1}{2}\overline{V \times I^*} = P = \frac{1}{2}Z|\overline{I}|^2 = \frac{1}{2}Y^*|\overline{V}|^2 = P_{\text{active}} + jP_{\text{reactive}} \tag{5.1}$$

For the source-free 1-port network, complex power P can be described by

$$= P_{\text{active}} + 2j\omega(W_m - W_e) \tag{5.2}$$

where W_m and W_e are the average stored magnetic and electrical energy.

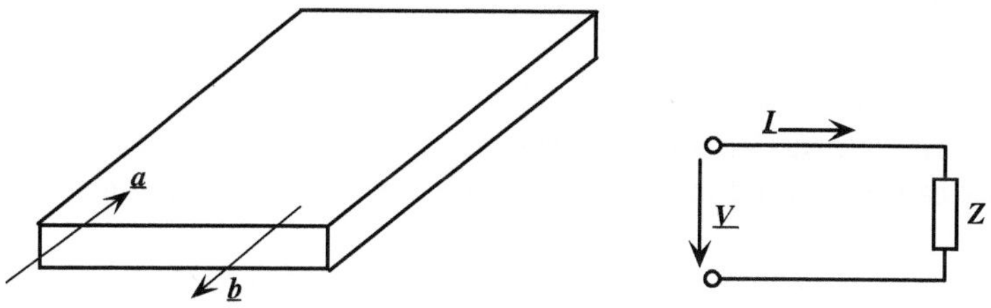

Figure 5.1 1-port: (a) a typical schematic drawing and (b) equivalent representation of 1-port network.

From (5.1) and (5.2), complex impedance [Z] and admittance [Y] can be characterized as

$$Z = \frac{P_{\text{active}} + 2j\omega(W_m - W_e)}{\frac{1}{2}|\bar{I}|^2} \Rightarrow X = \frac{4\omega(W_m - W_e)}{|\bar{I}|^2} \tag{5.3}$$

$$Y = \frac{P_{\text{active}} + 2j\omega(W_e - W_m)}{\frac{1}{2}|\bar{V}|^2} \Rightarrow B = \frac{4\omega(W_e - W_m)}{|\bar{V}|^2} \tag{5.4}$$

where X and B are the reactance and susceptance.

From (5.3) and (5.4)

$$\frac{dX}{d\omega}\left|\bar{I}\right|^2 = \frac{dB}{d\omega}\left|\bar{V}\right|^2 = 4(W_m - W_e) \tag{5.5}$$

$$W_e = \frac{1}{8}\left(\frac{dX}{d\omega} - \frac{X}{\omega}\right)\left|\bar{I}\right|^2 = \frac{1}{8}\left(\frac{dB}{d\omega} + \frac{B}{\omega}\right)\left|\bar{V}\right|^2,\ W_m = \frac{1}{8}\left(\frac{dX}{d\omega} + \frac{X}{\omega}\right)\left|\bar{I}\right|^2 = \frac{1}{8}\left(\frac{dB}{d\omega} - \frac{B}{\omega}\right)\left|\bar{V}\right|^2 \tag{5.6}$$

For a passive lossless 1-port network, $P_{\text{active}} = 0$, the port is also called a reactive 1-port network, and complex power P is given by

$$P = 2j\omega(W_m - W_e) \tag{5.7}$$

Since W_m and W_e are positive values for a passive lossless 1-port network, Foster [9] established the frequency dependence of a reactance as

$$\frac{dX}{d\omega} > \left|\frac{X}{\omega}\right| \text{ and } \frac{dB}{d\omega} > \left|\frac{X}{\omega}\right| \tag{5.8}$$

From (5.8), the instantaneous values of W_m and W_e oscillate with double the resonance frequency of oscillation, and transform their stored energy from W_m to W_e and vice versa within a quarter of the period of oscillation. However, for $W_m \neq W_e$ a periodical energy exchange can also occur with an external circuitry (active device that generates gain or negative resistance for stable oscillation), driving the resonator tank circuit. In this case, the reactive power flows through the port.

5.3 Resonator Networks

The resonator is the core component of the filter and oscillator circuits. The selection of a resonator for an oscillator or filter application involves several trade-offs: insertion loss, Q-factor, size, cost, power-handling requirements, etc. The main design considerations of resonators are the resonator size, unloaded Q, spurious performance, and power-handling capability. The unloaded Q represents the inherent losses in the resonator, and the higher the loss, the lower the Q value. It is therefore desirable to use resonators with high Q values since this reduces the insertion loss of the tuned resonator and improves its selectivity performance. Microwave resonator networks are grouped into three categories: lumped-element LC resonators, planar resonators (distributed printed transmission line resonators), and three-dimensional (3D) cavity-type resonators.

Figures 5.2 and 5.3 show the typical comparison between these resonators. Typically, lumped-element resonators are employed in low-frequency applications. Figure 5.4a shows a lumped-element resonator constructed by using a coil inductor and a chip capacitor. The lumped-element resonator can be printed on a dielectric substrate in the form of a spiral inductor and an inter-digital capacitor, as shown in Figure 5.4b.

As shown in Figure 5.4a, lumped element resonators are large but can be made small at microwave frequencies and offer a wide spurious-free window; however, they have a relatively low Q value. A typical Q value for lumped LC resonators is between 10 and 50 at 1 GHz.

Planar resonators can take the form of the length of a microstrip transmission line, terminated in a short circuit or open circuit, or can take the form of a meander line, folded line, ring resonator, patch resonator, or any other configuration [8–11]. Any printed structure effectively acts as a resonator whose resonant frequency is determined by the resonator's dimensions, substrate dielectric constant, and substrate height, which can be used for covering both narrowband and broadband tuning

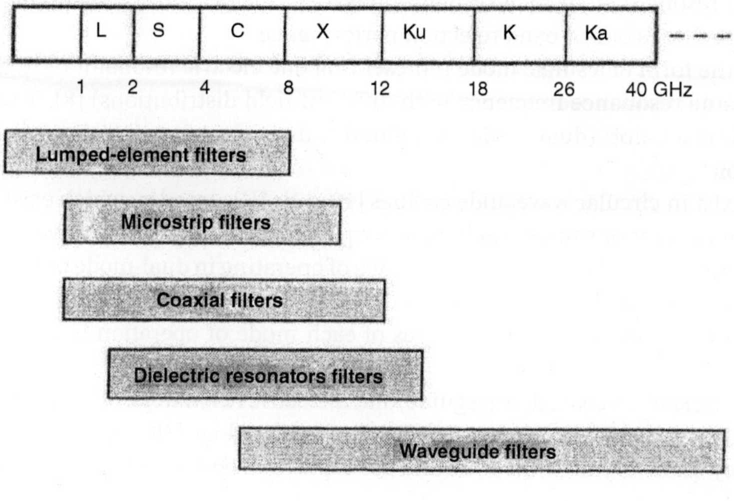

Figure 5.2 Application of the various resonator configurations. *Source*: [8]/John Wiley & Sons.

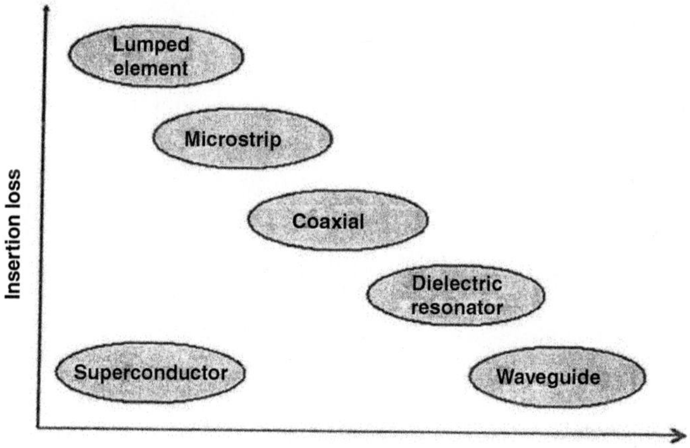

Figure 5.3 A typical relative size and insertion loss of various resonators. *Source*: [8]/John Wiley & Sons.

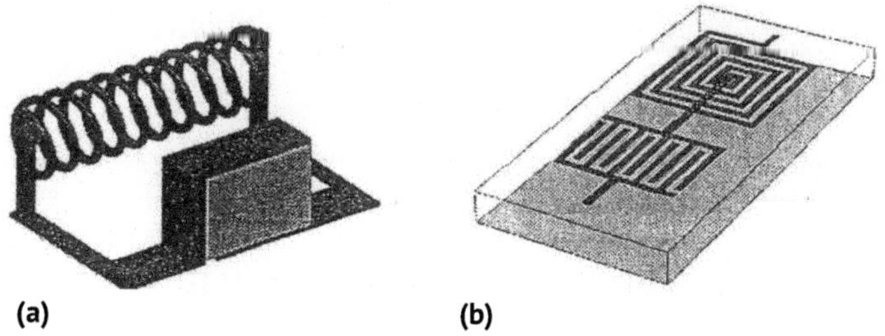

Figure 5.4 Lumped-element resonators realized by (a) coil inductor and chip capacitor, and (b) spiral inductor and inter-digital capacitor. *Source*: [8]/John Wiley & Sons.

characteristics of frequency source applications for current and later-generation communication systems [12]. Figure 5.5 shows typical planar resonators, usually employed in wideband, compact, and low-cost signal source and filter applications. The typical Q value for planar resonators is 50–300 at 1 GHz [13–17]. The state-of-the-art planar resonator circuit using superconductor techniques can exhibit typical Q values ranging from 20,000 to 50,000 at 1 GHz, but is not a cost-effective alternative because resonator circuitry needs to be cooled down to very low temperatures below 90 K [18].

In contrast to lumped LC resonators, which have only one resonant frequency, printed transmission line resonators can support infinite electromagnetic field configurations or resonant modes. The spurious performance of a resonator is determined by how close the neighboring resonant modes are to the operating mode. The neighboring resonant modes act as spurious modes, interfering with the fundamental resonant mode's performance. It is therefore desirable to increase the spurious-free window of the resonator to improve the filter's out-of-band rejection performance.

The resonant modes in planar resonators exist in the form of a single mode representing one electric resonator or in the form of degenerate modes (i.e., modes having the same resonance frequency with different field distributions) [8]. These degenerate modes allow the realization of two electric resonators (dual-mode resonators) or three electric resonators (triple-mode resonators) within the same physical resonator.

Examples of dual modes are TE_{11} modes, which exist in circular waveguide cavities [10–13]; HE_{11} modes, which exist in dielectric resonators [14, 15]; or TM_{11}, which exist in circular or square patch microstrip resonators [16]. Cubic waveguide cavities and cubic dielectric resonators can support triple modes [17]. The key advantage of operating in dual-mode or triple-mode configuration is size reduction. However, these modes do have an impact on the unloaded Q, spurious performance, and power-handling capability of the cavity resonator. A summary of the features of each mode of operation is given in Table 5.1.

Figure 5.6 shows that the typical 3D cavity resonators, such as coaxial, waveguide, and dielectric resonators, offer Q values ranging from 3000 to 30,000 at 1 GHz but are not amenable to integration into RFIC/MMIC technology [19, 20]. In addition to this, 3D resonators are bulky in size; however, they offer very high Q values and are capable of handling high RF power levels.

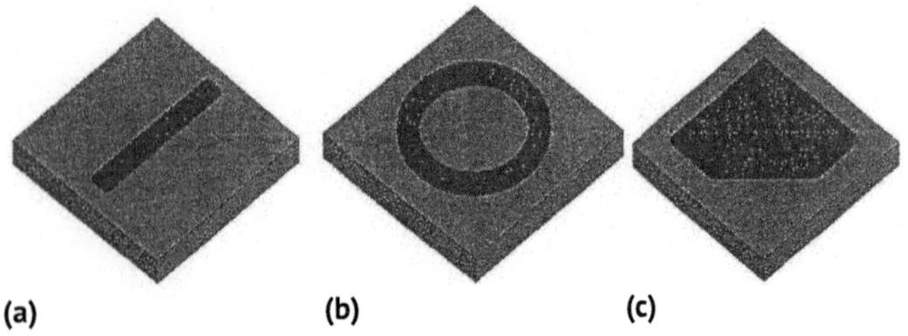

Figure 5.5 Examples of microstrip resonator configurations: (a) half-wavelength resonator, and (b) spiral inductor and interdigital capacitor. *Source*: [8]/John Wiley & Sons.

Table 5.1 Comparison between various modes of operation.

Parameter	Single Mode	Dual Mode	Triple Mode
Size	Large	Medium	Small
Spurious response	Good	Fair	Fair
Unloaded Q	High	Medium	Medium
Power-handling capability	Low	Medium	High
Design complexity	Low	Medium	High

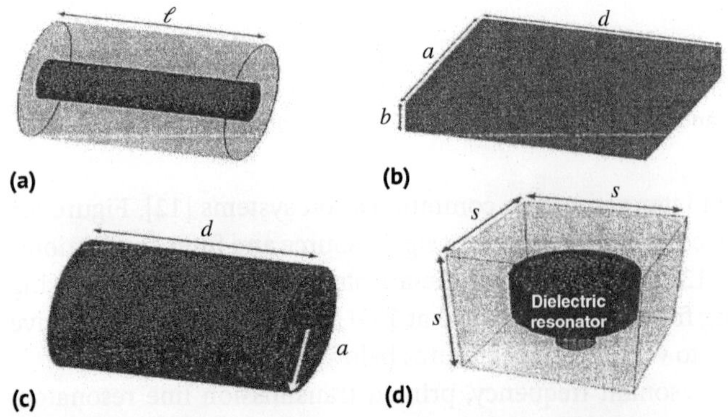

Figure 5.6 Examples of 3D cavity resonators: (a) coaxial resonator, (b) rectangular waveguide resonator, (c) circular waveguide resonator, and (d) dielectric resonator. *Source*: [8]/John Wiley & Sons.

5.4 Resonator Q-factor

Energy dumping in electromagnetic resonance defines the Q-factor for passive networks and is 2π times the ratio of the reactively stored energy to the energy dissipated in a unit cycle [16].

$$Q_{\text{loaded}} = 2\pi\left[\frac{\text{average reactive energy stored}}{\text{energy dissipated in unit cycle}}\right] = \omega_0\left[\frac{W_T}{P_T}\right] \Rightarrow \frac{1}{Q_{\text{loaded}}} = \left[\frac{1}{Q_{\text{unloaded}}} + \frac{1}{Q_{\text{external}}}\right] \tag{5.9a}$$

$$Q_{\text{loaded}} = Q_{\text{unloaded}} \text{ (loosely–coupled resonator)} \tag{5.9b}$$

$$Q_{\text{loaded}} = Q_{\text{external}} \text{ (tightly–coupled resonator)} \tag{5.9c}$$

where

$$Q_{\text{unloaded}} = \omega_0\left[\frac{W_T}{P_i}\right] \tag{5.9d}$$

$$Q_{\text{external}} = \omega_0\left[\frac{W_T}{P_e}\right] \tag{5.9e}$$

$$P_T = P_i + P_e \tag{5.9f}$$

where ω_0 is the resonant frequency, W_T is average stored energy, and P_T is the total power loss in watts, P_i and P_e are the dissipated energies due to the resonator's internal losses and the external loadings, respectively.

From (5.9b) and (5.9c), the loaded Q-factor is equal to the unloaded Q-factor ($Q_{\text{loaded}} = Q_{\text{unloaded}}$). For a loosely coupled resonator, the internal losses are more dominant as compared to external loss; conversely, in a tightly coupled resonator, the external loading is much more dominant, thus $Q_{\text{loaded}} = Q_{\text{external}}$ [17–20].

This definition (Eqs. 5.9a–e) is valid for circuits involving passive and active devices or energy sources until the open-loop gain approaches unity. In positive feedback filters and multipliers or active inductors, frequency selectivity can be as sharp as desired by increasing the open-loop gain, and if this exceeds the limit, oscillation takes place in the circuit. In such a state, the above definition of the Q-factor is no longer valid because it runs into infinity. Therefore, a unified definition of Q is necessary to evaluate the performance of active circuits even during the transient state of regenerative circuits (oscillators).

The quality factor (Q-factor) is the most important parameter of a resonant circuit for low-phase noise oscillator (signal sources) application. The Q-factor is a versatile index of resonator and oscillator performance, but its definition is not unified. Ohira [21] did a rigorous analysis of the characterization of a Q-factor for general RF components and circuits. Two common definitions of the Q-factor are discussed for the understanding of the dynamics of this parameter on autonomous circuits (oscillators). The first definition is based on field theory, which relates the Q-factor to frequency-selectivity, energy storage, and dissipation in resonance for stable passive circuits, whereas the second one relates to resonance for unstable active circuits (oscillators, regenerative circuits). The energy-based definition is ambiguous when a resonator has no energy storage elements, as in the case of rings or distributed oscillators. The two common definitions of Q-factor (i) for the passive resonant circuit and (ii) for the active resonant circuit (described in Sections 5.4.1 and 5.4.2) are explained in a step-by-step process by Ohira et al. [21–29].

a. ***Definition of Q-factor for Passive Resonant Circuit***
   i. Fractional 3-dB bandwidth
   ii. Phase-to-frequency slope
   iii. Stored-to-dissipated energy ratio
b. ***Definition of Q-factor for Active Resonant Circuit***
   i. Noise spectrum basis
   ii. Source-pull/push basis
   iii. Injection locking basis

5.4.1 Definition of Q-factor for Passive Resonant Circuit

i. Fractional 3-dB bandwidth
ii. Phase-to-frequency slope
iii. Stored-to-dissipated energy ratio

5.4.1.1 Fractional 3-dB Bandwidth

Figure 5.7 shows the typical 1-port resonator and its impedance versus frequency response function for the evaluation of fractional 3-dB bandwidth and quality factor, which is based on the frequency selectivity of resonator and filter networks.

Steps for Q-factor:

1. Get $|Z(\omega)|^2$ in frequency domain
2. Find ω_0 by solving $\frac{d}{d\omega}|Z(\omega_0)|^2 = 0$
3. Find ω_1 and ω_2 by solving $|Z(\omega_1)|^2 = |Z(\omega_2)|^2 = \frac{1}{2}|Z(\omega_0)|^2$
4. Q-factor from the definition of frequency selectivity: $Q = \frac{\omega_0}{\omega_2 - \omega_1} = \frac{f_0}{f_2 - f_1}$

5.4.1.2 Phase-to-frequency Slope

An alternative definition of Q-factor is based on the resonator's transfer function in the frequency domain. The transfer function is a complex function that governs the relationship between the input and output as

$$H(j\omega) = A(\omega)\ \exp\left[j\varnothing(\omega)\right] \tag{5.10}$$

where $A(\omega)$ and $\varnothing(\omega)$ are the amplitude and phase response of the resonator. The Q-factor in terms of the transfer function is given by [34]

$$Q = \frac{\omega_0}{2}\left|\sqrt{\left(\frac{\partial A(\omega)}{\partial \omega}\right)^2 + \left(\frac{\partial \varnothing(\omega)}{\partial \omega}\right)^2}\right|_{\omega=\omega_0} \tag{5.11}$$

$$Q \approx \frac{\omega_0}{2}\left|\frac{\partial \varnothing(\omega)}{\partial \omega}\right|_{\omega=\omega_0} \text{ for } \left|\frac{\partial A(\omega)}{\partial \omega}\right|_{\omega=\omega_0} \to 0 \tag{5.12}$$

From (5.12), the approximation $\left(\left|\frac{\partial A(\omega)}{\partial \omega}\right|_{\omega_0} = 0\right)$ is because phase variation is more predominant as compared to amplitude variation in well-designed oscillator circuits. From (5.12), the shift in frequency ($\Delta\omega$) from ω_0 can cause a large rate of change in phase if the Q-factor is high, forcing the frequency to return to ω_0.

Figure 5.8 shows the typical 1-port resonator and phase response function for the evaluation of the rate of change of phase and quality factor.

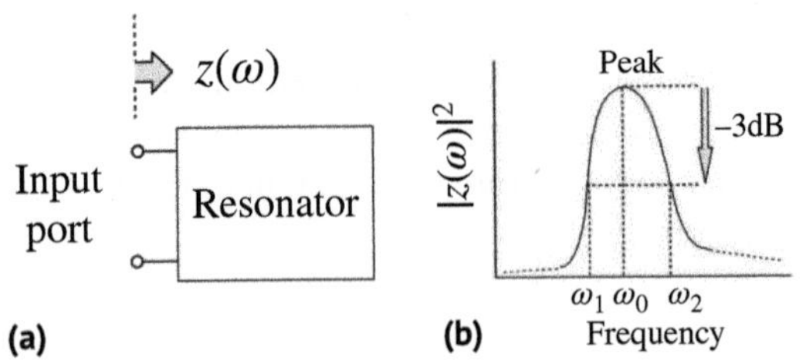

Figure 5.7 Examples of 1-port resonators (a) schematic of the 1-port resonator, and (b) impedance versus frequency response [27].

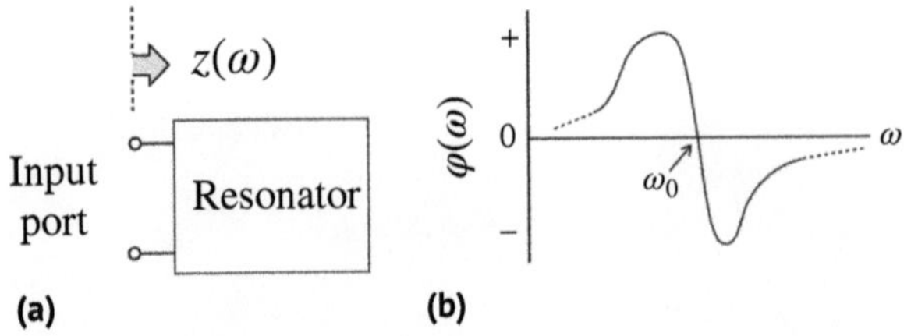

Figure 5.8 Examples of 1-port resonators (a) schematic of 1-port resonator, and (b) phase versus frequency response [27].

Steps for Q-factor (neglects amplitude slope):

1. Get $\varphi(\omega)$ in frequency domain $\varphi(\omega) = \angle[Z(\omega)]$

$$\varphi'(\omega) = \frac{d}{d\omega}\angle[Z(\omega)]$$

2. Find ω_0 by solving

$$\varphi(\omega_0) = 0$$

3. Calculate the Q-factor from

$$Q = \frac{\omega_0}{2}\left|\frac{\partial\varphi(\omega)}{\partial\omega}\right|_{\omega=\omega_0} = \frac{\omega_0}{2}|\varphi'(\omega_0)|$$

5.4.1.3 Stored-to-dissipated Energy Ratio

The electromagnetic field inside the ideal resonator network stores energy at the resonant frequency, where equal storage of electrical and magnetic energies occurs. However, in reality, part of the stored energy is dissipated due to losses across the loads, thereby reducing the resonator's frequency selectivity [23, 24].

Figure 5.9 shows the typical 1-port resonator for the evaluation of stored energy rate and quality factor.

Steps for Q-factor

1. Estimate input power and stored energy in each component

$$P(\omega) = \frac{1}{2}Re\{v_{ex}i_{ex}^*\}, \quad W_e(\omega) = \frac{1}{4}\sum_k^{\text{cap}} C_k|v_k|^2, \quad W_m(\omega) = \frac{1}{4}\sum_k^{\text{ind}} L_k|i_k|^2$$

2. Find ω_0 by solving $W_e(\omega_0) = W_m(\omega_0)$
3. Calculate Q-factor from $Q = \omega_0\dfrac{W_e(\omega_0) + W_m(\omega_0)}{P(\omega_0)}$

The above method for the evaluation of the Q-factor is valid for massive resonators; this definition is no longer valid when the 1-port circuit is unstable and oscillates (for example: regenerative active circuits) [26]. Therefore, a new definition of Q-factor for active circuits is required for both analysis and synthesis of the oscillator circuits.

5.4.2 Definition of Q-factor for Active Resonant Circuit: Sensitive to Oscillation

Table 5.2 describes the formulation of oscillation-stability conditions for active networks under various conditions.

i. Noise spectrum basis
ii. Source-pull/push basis
iii. Injection locking basis

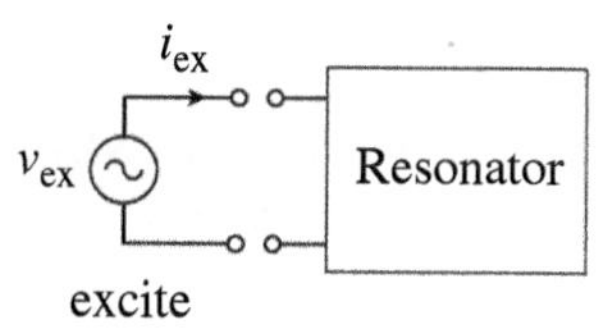

Figure 5.9 Schematic of the 1-port resonator (for the estimation of stored energy in each resonating component) [27].

Table 5.2 Types of unstable active networks [27].

<table>
<tr><th colspan="3">Unstable active network (closed-loop gain>= 1) Sensitive to oscillations ($\omega = \omega_0$)</th></tr>
<tr><th>Type</th><th>Scheme</th><th>Condition of oscillation-stability</th></tr>
<tr><td>NINO (No Input-No Output)</td><td></td><td>Power and energy equilibrium condition stability criterion $W_e(\omega_0) = W_m(\omega_0) \rightarrow \Delta W = 0$</td></tr>
<tr><td>NISO (No Input-Signal Output)</td><td>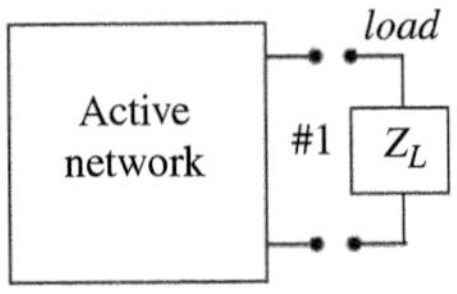</td><td>Sideband noise load pulling</td></tr>
<tr><td>SINO (Signal Input-No Output)</td><td>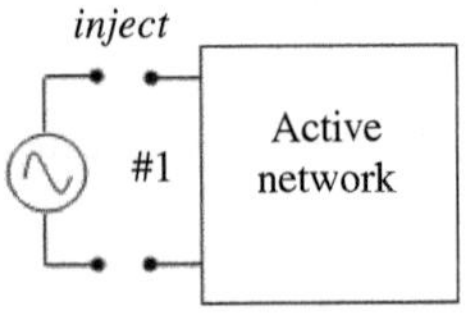</td><td>Injection locking</td></tr>
<tr><td>SISO (Signal Input-Signal Output)</td><td>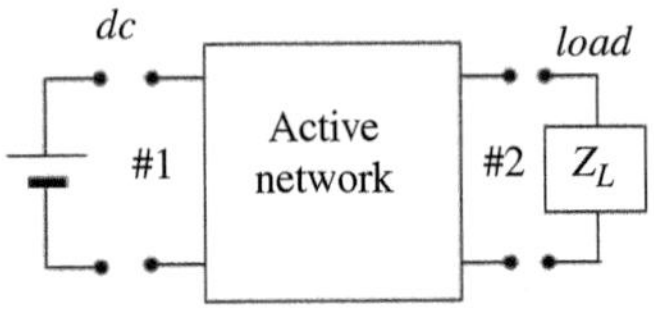</td><td>Source pushing</td></tr>
</table>

NINO (No Input-No Output) active network obeys the power and energy balance equilibrium condition. Thus, elements for NINO active network are confined to inside and transfer energy for equilibrium condition, the complex power $\Psi(\alpha, \omega)$ can be defined as [28]

$$\underbrace{\Psi(\alpha, \omega)}_{\text{complex power}} = \frac{1}{2}\left[v_1(\omega)i_1^*(\omega) + v_2(\omega)i_2^*(\omega) + v_3(\omega)i_3^*(\omega) + \ \dots + v_K(\omega)i_K^*(\omega)\right]$$
$$= P_a - P_r + 2j\omega W_e - 2j\omega W_m = \Delta P + 2j\omega\Delta W \tag{5.13}$$

From (5.13), the equilibrium state for the oscillation condition of the zero port (NINO) active circuit can be formulated by [29, 30]

$$\underbrace{[\Psi(\alpha, \omega)]_{\alpha=\alpha_0, \omega=\omega_0}}_{\text{complex power}} = 0 \rightarrow \text{Equlibrium state for oscillation condition} \tag{5.14}$$

From (5.14), once the equilibrium point is reached ($\alpha = \alpha_0, \omega = \omega_0$), the important thing is whether the zero port active circuit (NINO) remains in a steady state at that point ($\alpha = \alpha_0, \omega = \omega_0$) against the small perturbation ($\alpha = \alpha_0 + \delta\alpha,\ \omega = \omega_0 + \delta\omega$), while maintaining the complex power balance [$\Psi(\alpha = \alpha_0 + \delta\alpha, \omega_0 + \delta\omega) = 0$], i.e.,

$$\frac{\partial\Psi(\alpha, \omega)}{\partial\alpha}\delta\alpha + \frac{\partial\Psi(\alpha, \omega)}{\partial\omega}\delta\omega = 0 \Rightarrow \delta\omega = \left(\frac{d\phi}{dt} - j + \frac{d\alpha}{dt}\right) \tag{5.15}$$

From (5.15), the instantaneous frequency deviation ($\delta\omega$) depends not only on phase change in time but also on logarithmic amplitude change for the imaginary part, which can be described by [25]

$$\frac{\partial\Psi(\alpha,\omega)}{\partial\alpha}\delta\alpha + \left(\frac{d\phi}{dt} - j + \frac{d\alpha}{dt}\right)\frac{\partial\Psi(\alpha,\omega)}{\partial\omega} = 0 \tag{5.16a}$$

$$\text{Imag}\left[\frac{\partial\Psi(\alpha,\omega)}{\partial\alpha}\left(\frac{\partial\Psi(\alpha,\omega)}{\partial\omega}\right)^{-1}\right] = \frac{1}{\delta\alpha}\frac{d\alpha}{dt} \tag{5.16b}$$

stability criterion $\delta\alpha\dfrac{d\alpha}{dt} < 0\ at\ (\alpha_0,\ \omega_0)$

$$\underbrace{\text{Im}\left\{\frac{\partial\Psi}{\partial\omega}\frac{\partial\Psi^*}{\partial\alpha}\right\}_{\substack{\alpha=\alpha_0\\ \omega=\omega_0}}}_{\text{discriminant}} > 0 \text{ Or } \underbrace{\begin{vmatrix}\dfrac{\partial\Delta P}{\partial\alpha} & \dfrac{\partial\Delta P}{\partial\omega}\\ \dfrac{\partial\Delta U}{\partial\alpha} & \dfrac{\partial\Delta U}{\partial\omega}\end{vmatrix}_{\substack{\alpha=\alpha_0\\ \omega=\omega_0}}}_{\text{Jacobian discriminant}} > 0 \tag{5.17}$$

5.4.2.1 Active NISO Circuit Q-factor (Noise Spectrum Basis)

Figure 5.10 shows the typical 1-port oscillator circuit model, as depicted, both active and passive components are embedded in a black box, and represented by the $Z(\omega)$ without the load (R).

Assuming the noise current source is in parallel, this yields single-sideband (SSB) power on the spectrum. The power observed when the noise source is solely connected to load R is given by

$$P_n = \left[\frac{R}{2}\right]|n|^2 \tag{5.18}$$

The output sideband noise power $P_{\text{SSB}}(\omega)$ can be described by [29]

$$P_{\text{SSB}}(\omega) = \left[\frac{R}{2}\right]\left|\frac{z(\omega)n}{R + z(\omega)}\right|^2 = P_n\left|\frac{z(\omega)}{R + z(\omega)}\right|^2 \tag{5.19}$$

$$[z(\omega) + R]_{\omega\to\omega_0} = 0 \text{ (at the equilibrium state, the resonance frequency } \omega = \omega_0) \tag{5.20}$$

Frequency deviation ($\omega = \omega_0 + \delta\omega$) expands output impedance $z(\omega)$ as

$$z(\omega) = z(\omega_0) + \frac{dz(\omega_0)}{d\omega_0}\delta\omega + \ldots \quad \text{(Taylor expansion)} \tag{5.21}$$

$$|z(\omega)|^2 = \left|z(\omega_0) + \frac{dz(\omega_0)}{d\omega_0}\delta\omega\right|^2 \approx |z(\omega_0)|^2 \tag{5.22}$$

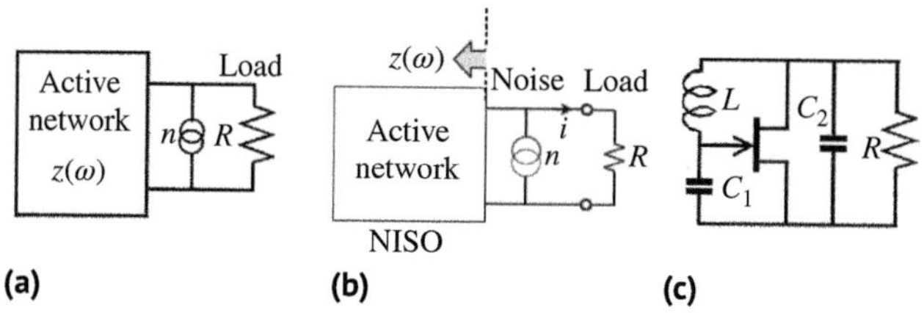

Figure 5.10 Shows the typical 1-port oscillator: (a) general model, (b) equivalent noise model of NISO, and (c) a colpitts example representing NISO model [27].

From (5.19) to (5.22)

$$P_{SSB}(\omega) = P_n\left|\frac{z(\omega)}{z(\omega) - z(\omega_0)}\right|^2 = P_n\left|\frac{z(\omega_0) + \frac{dz(\omega_0)}{d\omega_0}\delta\omega}{\frac{dz(\omega_0)}{d\omega_0}\delta\omega}\right|^2 = P_n\left|1 + \frac{z(\omega_0)}{\frac{dz(\omega_0)}{d\omega_0}\delta\omega}\right|^2 \approx P_n\left|\frac{z(\omega_0)}{z'(\omega_0)\delta\omega}\right|^2 \tag{5.23}$$

From [34]

$$P_{SSB}(\omega) = \left[\frac{\omega_0}{2Q\delta\omega}\right]^2 [P_n] \Rightarrow Q = \frac{\omega_0}{2}\left|\frac{z'(\omega_0)}{z(\omega_0)}\right| \tag{5.24}$$

From (5.24)

$$\left\{\begin{aligned} &Q(pz) &&= \frac{\omega_0}{2}\left|\frac{pz'(\omega_0)}{pz(\omega_0)}\right| = \frac{\omega_0}{2}\left|\frac{z'(\omega_0)}{z(\omega_0)}\right| = Q(z) \rightarrow \text{Scaling operation} \\ &Q(z^{-1}) = Q(y) = &&\frac{\omega_0}{2}\left|\frac{y'(\omega_0)}{y(\omega_0)}\right| = \frac{\omega_0}{2}\left|\frac{z'(\omega_0)}{z(\omega_0)}\right| = Q(z) \rightarrow \text{Inverse operation} \\ &Q(z^*) &&= \frac{\omega_0}{2}\left|\frac{z'(\omega_0)^*}{z(\omega_0)^*}\right| = \frac{\omega_0}{2}\left|\frac{z'(\omega_0)}{z(\omega_0)}\right| = Q(z) \rightarrow \text{Conjugate operation} \end{aligned}\right\} \tag{5.25}$$

$P_{SSB}(\omega)$ expressed in (5.25) represents the Q-factor of the oscillator that includes both active and passive devices embedded in the 1-port black box shown in Figure 5.10.

In other words, the above formula is valid for any active network, regardless of oscillator topology that is comprised of types of active devices for providing closed-loop gain ≥ 1 and compensating for the loss of resonator. From (5.25), Q is invariant against the three operations (scaling, inverse, and conjugate) [30].

5.4.2.2 Active NISO Circuit Q-factor (Reflection Coefficient $\Gamma(\omega)$ Basis)

From (5.25)

$$Q = \frac{\omega_0}{2}\left|\frac{z'(\omega_0)}{z(\omega_0)}\right|, \quad z'(\omega) = \left[\frac{dz(\omega)}{d\omega}\right] \tag{5.26}$$

The expression for reflection coefficient $\Gamma(\omega)$ for the 1-port network shown in Figure 5.11 can be given by [21]

$$\Gamma(\omega) = \left|\frac{z(\omega) - z_0}{z(\omega) + z_0}\right| \Rightarrow \Gamma'(\omega) = \left|\left\{\frac{[z(\omega) - z_0]\, z'(\omega) - [z(\omega) + z_0]z'(\omega)}{[z(\omega) + z_0]^2}\right\}\right|, \quad \Gamma'(\omega) = \left[\frac{d\Gamma(\omega)}{d\omega}\right] \tag{5.27}$$

From (5.26) and (5.27)

$$Q = \omega_0\left|\frac{\Gamma'(\omega)}{1 - \Gamma^2(\omega_0)}\right| \tag{5.28}$$

5.4.2.3 Active NISO Circuit Q-factor (Energy Basis)

Figure 5.12 represents the general model of circuits without ports for understanding the energy equilibrium dynamics, even though this kind of configuration does not have practical usefulness. Considering such a system when oscillation takes place, Figure 5.12a can be illustrated as a zero-port closed-circuit network, as shown in Figure 5.12b for the analysis of equilibrium and stability criteria.

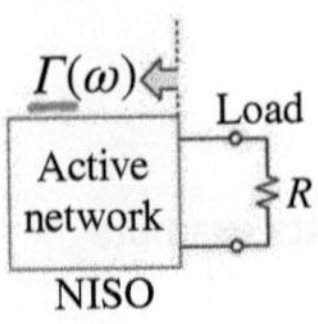

Figure 5.11 Shows the typical 1-port oscillator.

The total complex power of the circuit represented in Figure 5.13 can be described as

$$\psi(\omega) = \frac{1}{2}(+v_1 i_1^* + v_2 i_2^* + v_3 i_3^* \,\dots\, ..v_k i_k^*) \tag{5.29}$$

$$\psi(\omega) = \Delta P + j2\omega \Delta E \tag{5.30}$$

$$\Delta P = P_a - P_r \tag{5.31}$$

$$\Delta E = E_e - E_m \tag{5.32}$$

where v_k and i_k are sinusoidal voltage and current, respectively, at the kth branch in the system. An asterisk indicates a complex conjugate.

Active devices generate RF power P_a, dissipating this power in resistive elements P_r. Some energy is reactively stored in inductive (E_m) and capacitive (E_e) elements. $\Delta P = 0$, $\Delta E = 0$ (Energy interaction must be completed within the system since there is no transaction between inside and outside in the circuit shown in Figure 5.12). Therefore, both real and imaginary parts of $\psi(\omega)$ must vanish, which implies a need for balance in both power and energy.

For circuit designers who are familiar with impedance parameters rather than with terms of energy, it is worthwhile to translate the above criterion into the impedance domain. To carry this out, the zero-port oscillator is divided into two pieces as shown in Figure 5.12b. As illustrated in Figure 5.12, Parts I and II are interconnected in multiplex, and therefore they are each regarded as a multi-port network with impedance matrix $[Z_{\mathrm{I}}]$ or $[Z_{\mathrm{II}}]$, respectively.

From Kirchhoff's law:

$$v = z_I\, i = -z_{II}\, i,\quad Z = Z_I\ + Z_{II} \Rightarrow |Z| = 0 \text{ at the resonance for multi-connected active circuits} \tag{5.33}$$

From (5.29)

$$\psi(\omega) = \frac{1}{2}\left(v_I^T i^* - v_{II}^T i^*\right) = \frac{1}{2} i^T \left(Z_I^T + Z_{II}^T\right) i^* = \frac{1}{2} i^T Z^T i^* = \frac{1}{2} i^T \left(|Z| \widetilde{Z}^{-1}\right)^T i^* = \frac{1}{2} |Z| i^T \widetilde{Z}^{-T} i^* \tag{5.34}$$

where superscripts ^{T} and $^{-T}$ stand for transposition and inverse transposition, and tilde ~ designates cofactor matrix. Note that term $v_I^T i^*$ for Part II has negative polarity due to the opposite direction of currents.

A loaded 1-port oscillator with R can be considered as a subset of zero-port if the load R is in circuit, as shown in Figure 5.13, which is embedded in $Z(\omega)$ so that the resulting impedance is $Z = [Z(\omega) + R]$. From (5.34)

$$\psi(\omega) = \frac{1}{2} |Z| i^T \widetilde{Z}^{-T} i^* = \frac{1}{2}\{Z(\omega) + R\} |i|^2 = -\frac{Z(\omega)}{2} |i|^2 - P_{\text{out}}(\omega) \tag{5.35}$$

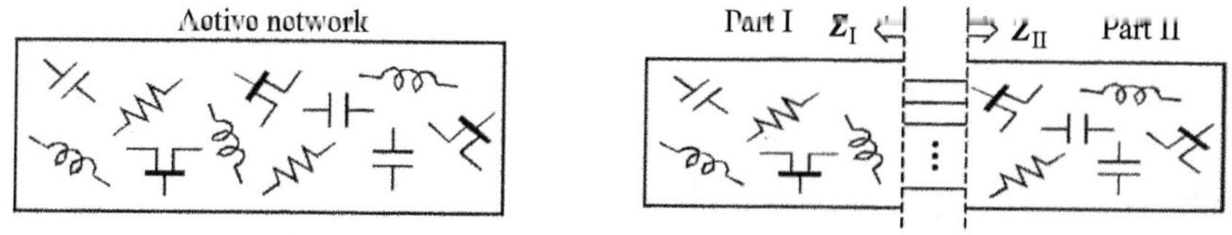

(a) (b)

Figure 5.12 A typical zero-port oscillator: (a) general model and (b) separated into two parts [27].

Figure 5.13 Shows the typical 1-port oscillator.

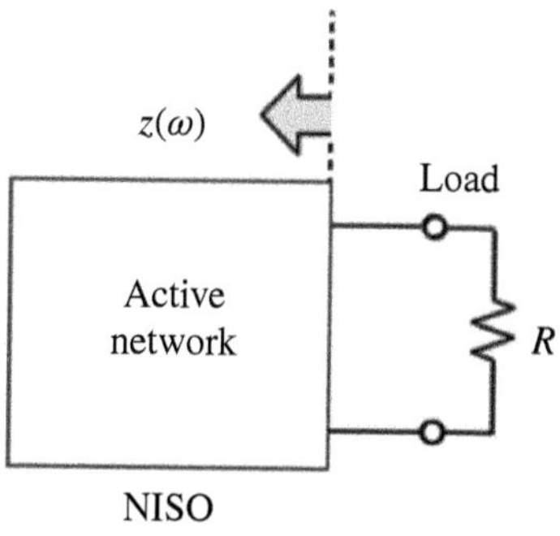

where $P_{out}(\omega)$ is power output to the load. At the equilibrium, oscillation condition forces $\Psi(\omega_0) = 0$. However, its frequency slope remains finite as $z(\omega_0)\Psi'(\omega_0) = -z'(\omega_0)\,P_{out}(\omega_0)$. From (5.29) to (5.35), the expression of Q is given by [29–32]

$$Q = \frac{\omega_0}{2}\left|\frac{\Psi'(\omega_0)}{P_{out}(\omega_0)}\right| = \frac{\omega_0}{P_{out}}\sqrt{\frac{1}{4}\left(\frac{\partial \Delta P}{\partial \omega}\right)^2 + \left(\frac{\omega_0 \partial \Delta W}{\partial \omega}\right)^2} \tag{5.36}$$

Considering that the power dissipation takes place only in the load and neglecting frequency slope of the active devices, i.e., $P_{out} = P_r$ and $\frac{\partial P_a}{\partial \omega} = 0$, (5.36) can be written as

$$Q = \frac{\omega_0}{P_r}\sqrt{\frac{1}{4}\left(\frac{\partial P_r}{\partial \omega}\right)^2 + \left(\frac{\omega_0 \partial \Delta W}{\partial \omega}\right)^2} \tag{5.37}$$

Equation (5.57) agrees with Eq. (8) in [10] and Eq. (2) in [11]. Assuming $\left|\frac{\partial P_r}{\partial \omega}\right| \ll \left|\frac{\omega_0 \partial \Delta U}{\partial \omega}\right|$, Q-factor can be given by [30]

$$Q \cong \frac{\omega_0^2}{P_r}\left|\frac{\partial W_e}{\partial \omega} - \frac{\partial W_m}{\partial \omega}\right| \tag{5.38}$$

This agrees with [3, eq. (31)]. If we further additionally assume special relation $(2\omega_0 - \omega)U_m = \omega U_e$ *or* $\omega U_m = (2\omega_0 - \omega)U_e$ at $\omega = \omega_0$. The Q equation ultimately reduces to

$$Q = \omega_0 \frac{(W_e + W_m)}{P_r} \tag{5.39}$$

Equation (5.39) concurs with the energy-dumping phenomena of Q-factor.

5.4.2.4 Active SISO Circuit Q-factor (Source-push and Load-pull Basis)

Figure 5.14 shows the typical 2-port oscillator model (SISO) for the derivation of Q-factor, taking into account the DC source pushing and RF load pulling effect. For high-performance oscillator circuit, Q-factor plays an important role in making the circuit insensitive to pushing and pulling [33].

The steady-state oscillation condition for oscillator circuit shown in Figure 5.14 can be expressed as

$$Z_{out}(\omega, V) + Z_L(\omega) = 0 \tag{5.40}$$

$$[R_{out}(\omega, V) + R_L(\omega)] + j[X_{out}(\omega, V) + X_L(\omega)] = 0 \tag{5.41}$$

From (5.41) $R_{out}(\omega, V)$ must exhibit a negative resistance while oscillation exists because R_L is passive load and intrinsically has positive resistance. The source pushing on the oscillation frequency can be formulated by perturbation techniques:

$$\omega \rightarrow \omega_0 + \Delta\omega \text{ and } V \rightarrow V_0 + \Delta V \tag{5.42}$$

where $|\Delta\omega| \ll \omega_0$ and $|\Delta V| \ll V_0$

With the help of Taylor series expansion, Eq. (5.42) can be expanded as [31]

$$Z_{out}(\omega_0, V_0) + Z_{out}^{\omega}(\omega_0, V_0)\Delta\omega + Z_{out}^{V}(\omega_0, V_0)\Delta V + \ldots + Z_L(\omega_0) = 0 \tag{5.43}$$

where

$$Z_{out}^{\omega}(\omega, V) = \frac{\partial Z_{out}(\omega, V)}{\partial \omega},\ Z_{out}^{V}(\omega, V) = \frac{\partial Z_{out}(\omega, V)}{\partial V} \tag{5.44}$$

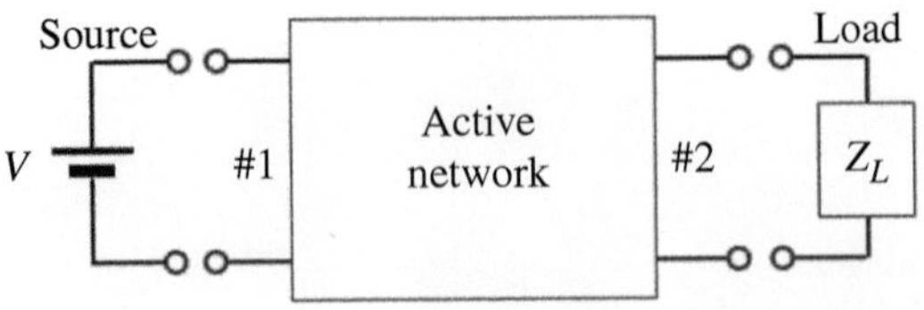

Figure 5.14 A typical 2-port oscillator model.

Neglecting higher-order partial differential term,

$$Z_{out}(\omega_0, V_0) + Z_L(\omega_0) = 0 \tag{5.45}$$

$$Z_{out}^{\omega}(\omega_0, V_0)\Delta\omega + Z_{out}^{V}(\omega_0, V_0)\Delta V = 0 \tag{5.46}$$

$$\Delta\omega = \left|\frac{Z_{out}^{V}(\omega_0, V_0)}{Z_{out}^{\omega}(\omega_0, V_0)}\right|\Delta V \tag{5.47}$$

Normalizing the frequency deviation ($\Delta\omega$) and voltage deviation (ΔV) on both sides of Eq. (5.48) by their original values (ω_0, V_0):

$$\frac{\Delta\omega}{\omega_0 V_0} = \frac{1}{\omega_0 V_0}\left|\frac{Z_{out}^{V}(\omega_0, V_0)}{Z_{out}^{\omega}(\omega_0, V_0)}\right|\Delta V \tag{5.48}$$

$$\left|\frac{\Delta\omega}{\omega_0}\right| = \frac{V_0}{\omega_0}\left|\frac{Z_{out}^{V}(\omega_0, V_0)}{Z_{out}^{\omega}(\omega_0, V_0)}\right|\left|\frac{\Delta V}{V_0}\right| = \frac{1}{Q}\left|\frac{\Delta V}{V_0}\right| \tag{5.49}$$

$$\left|\frac{\Delta\omega}{\omega_0}\right|_{SSB} = \frac{1}{Q}\left|\frac{\Delta V}{V_0}\right| \tag{5.50}$$

$$\left|\frac{2\Delta\omega}{\omega_0}\right|_{DSB} = \frac{1}{Q}\left|\frac{\Delta V}{V_0}\right| \tag{5.51}$$

On the left-hand side of (5.51), frequency deviation is doubled to measure double sideband (upper and lower sideband frequencies from the carrier). The coefficient shown in (5.51) can be expressed as $1/Q$ because the frequency deviation ($\Delta\omega$) must be inversely proportional to a certain figure of merit (FOM) for the sustained oscillation condition. From (5.47) and (5.51), the constant Q can be defined as a dimension-free positive scalar quantity [33]

$$Q_{source-pull} = \frac{\omega_0}{2V_0}\left|\frac{Z_{out}^{\omega}(\omega, V)}{Z_{out}^{V}(\omega, V)}\right|_{(\omega=\omega_0, V=V_0)} \tag{5.52}$$

From the duality and superposition theorem,

$$Q_{source-pull} = \frac{\omega_0}{2V_0}\left|\frac{Y_{out}^{V}(\omega, V)}{Y_{out}^{\omega}(\omega, V)}\right|_{(\omega=\omega_0, V=V_0)} \tag{5.53}$$

Equations (5.52) and (5.53) are referred to as a source-pull Q-factor of oscillator circuit.

Similarly, load-pull Q-factor can be formulated to estimate the frequency stability due to load variation.

From (5.40), the condition for stable oscillation for a constant voltage V can be expressed as

$$Z_{out}(\omega, g_m) + Z_L(\omega) = 0 \tag{5.54}$$

$$[R_{out}(\omega, g_m) + R_L(\omega)] + j[X_{out}(\omega, g_m) + X_L(\omega)] = 0 \tag{5.55}$$

From (5.55), $R_{out}(\omega, g_m)$ must exhibit a negative resistance while oscillation exists because R_L is passive load with intrinsically positive resistance. The load-pulling on oscillation frequency can be formulated by perturbation of frequency and transconductance of the active device [32]. For stable oscillation, the transconductance of the active device must be adjusted to maintain the gain-to-loss equilibrium condition due to the variation in the load impedance $Z_L(\omega)$. These phenomena can be defined as [34]

$$Z_L(\omega) \to Z_{Lo}(\omega) + \Delta Z_L,\ \ \omega \to \omega_0 + \Delta\omega,\ \ g_m \to g_{mo} + \Delta g \tag{5.56}$$

where $|\Delta\omega| \ll \omega_0$ and $|\Delta g| \ll g_{mo}$

From (5.55) and (5.56), Taylor series expansion:

$$Z_{out}(\omega_0, g_{mo}) + Z_{out}^{\omega}(\omega_0, g_{mo})\Delta\omega + Z_{out}^{g}(\omega_0, g_{mo})\Delta g + \dots + Z_{Lo}(\omega_0) + \Delta Z_L = 0 \tag{5.57}$$

where

$$Z_{\text{out}}^{\omega}(\omega, g_m) = \frac{\partial Z_{\text{out}}(\omega, g_m)}{\partial \omega},\ Z_{\text{out}}^{g}(\omega, g_m) = \frac{\partial Z_{\text{out}}(\omega, g_m)}{\partial g} \tag{5.58}$$

Neglecting higher-order partial differential term,

$$Z_{\text{out}}(\omega_0, g_{mo}) + Z_{Lo}(\omega_0) = 0 \tag{5.59}$$

$$Z_{\text{out}}^{\omega}(\omega_0, g_{mo})\Delta\omega + Z_{\text{out}}^{g}(\omega_0, g_{mo})\Delta g + \Delta Z_L = 0 \tag{5.60}$$

From (5.60), for a given perturbation in load ΔZ_L, there are two unknown variants ($\Delta\omega$, Δg), the solution of (5.60) can be formulated by applying the complex conjugate function (*)

$$Z_{\text{out}}^{\omega\ *}(\omega_0, g_{mo})\Delta\omega + Z_{\text{out}}^{g\ *}(\omega_0, g_{mo})\Delta g + \Delta Z_L^{\ *} = 0 \tag{5.61}$$

From (5.60) and (5.61)

$$\begin{bmatrix} \Delta\omega \\ \Delta g \end{bmatrix} = -\begin{bmatrix} Z_{\text{out}}^{\omega}(\omega_0, g_{mo}) & Z_{\text{out}}^{g}(\omega_0, g_{mo}) \\ Z_{\text{out}}^{\omega\ *}(\omega_0, g_{mo}) & Z_{\text{out}}^{g\ *}(\omega_0, g_{mo}) \end{bmatrix}^{-1} \begin{bmatrix} \Delta Z_L \\ \Delta Z_L^{\ *} \end{bmatrix} \tag{5.62}$$

$$\begin{bmatrix} \Delta\omega \\ \Delta g \end{bmatrix} = -\frac{1}{J}\begin{bmatrix} Z_{\text{out}}^{g\ *}(\omega_0, g_{mo}) & -Z_{\text{out}}^{g}(\omega_0, g_{mo}) \\ -Z_{\text{out}}^{\omega\ *}(\omega_0, g_{mo}) & Z_{\text{out}}^{\omega}(\omega_0, g_{mo}) \end{bmatrix}^{-1} \begin{bmatrix} \Delta Z_L \\ \Delta Z_L^{\ *} \end{bmatrix} \tag{5.63}$$

where $[J]$ denotes the Jacobian determinant [29]

$$J = \begin{vmatrix} Z_{\text{out}}^{\omega}(\omega_0, g_{mo}) & Z_{\text{out}}^{g}(\omega_0, g_{mo}) \\ Z_{\text{out}}^{\omega\ *}(\omega_0, g_{mo}) & Z_{\text{out}}^{g\ *}(\omega_0, g_{mo}) \end{vmatrix} = 2j\gamma \tag{5.64}$$

$$\gamma = Im\left[Z_{\text{out}}^{\omega}(\omega_0, g_{mo}) Z_{\text{out}}^{g\ *}(\omega_0, g_{mo})\right] \tag{5.65}$$

$$\Delta\omega = -\frac{1}{J}\left[Z_{\text{out}}^{g\ *}(\omega_0, g_{mo})\Delta Z_L - Z_{\text{out}}^{g}(\omega_0, g_{mo})\Delta Z_L^{\ *}\right] = \frac{1}{\gamma} Im\left[Z_{\text{out}}^{g}(\omega_0, g_{mo})\Delta Z_L^{\ *}\right] \tag{5.66}$$

$$\Delta\omega = \frac{1}{\gamma}\left|Z_{\text{out}}^{g}(\omega_0, g_{mo})\Delta Z_L\right| \sin|(\emptyset - \theta)| \tag{5.67}$$

where

$$Z_{\text{out}}^{g}(\omega_0, g_{mo}) = \left|Z_{\text{out}}^{g}(\omega_0, g_{mo}) e^{j\emptyset}\right| \tag{5.68}$$

$$\Delta Z_L = |\Delta Z_L| e^{j\theta} \tag{5.69}$$

Figure 5.15 shows the load impedance locus on a complex plane. From (5.57), frequency shift $\Delta\omega$ depends on the phase of the load deviation ΔZ_L. Revolving θ from zero to 2π as shown on a dashed circumference in Figure 5.15, the frequency shift reaches its maximum, given by [33]

$$|\Delta\omega|_{\max} = \left|\left\{\frac{\Delta Z_L}{\gamma}\right\} Z_{\text{out}}^{g}(\omega_0, g_{mo}) \sin\frac{\pi}{2}\right|_{(\emptyset - \theta) = \frac{\pi}{2}} \tag{5.70}$$

Normalizing the deviations in Eq. (5.70) on both sides by their original values as

$$\left|\frac{2|\Delta\omega|_{\max}}{\omega_0}\right|_{DSB} = \frac{1}{Q}\left|\frac{\Delta Z_L}{Z_{Lo}}\right| \tag{5.71}$$

On the left-hand side of (5.71), the frequency deviation is doubled to measure the double sideband (upper and lower sideband frequencies from the carrier). The coefficient shown in Eq. (5.71) can be expressed as $1/Q$ because the frequency deviation ($\Delta\omega$) must be inversely proportional to a certain FOM for sustained oscillation conditions.

From (5.48) and (5.52), the constant Q can be defined as a dimension-free positive scalar quantity [33]

$$Q = \frac{\omega_0}{2}\left|\frac{\gamma(\omega_0, g_{mo})}{Z_{Lo}Z^{g}_{\text{out}}(\omega_0, g_{mo})}\right| \tag{5.72}$$

From (5.59) and (5.72)

$$Q = \frac{\omega_0}{2}\left|\frac{\gamma(\omega_0, g_{mo})}{Z^{\omega}_{\text{out}}(\omega_0, g_{mo})Z^{g}_{\text{out}}(\omega_0, g_{mo})}\right| = \frac{\omega_0}{2}\left|\frac{Im\left[Z^{\omega}_{\text{out}}(\omega_0, g_{mo})Z^{g\,*}_{\text{out}}(\omega_0, g_{mo})\right]}{Z^{\omega}_{\text{out}}(\omega_0, g_{mo})Z^{g}_{\text{out}}(\omega_0, g_{mo})}\right| \tag{5.73}$$

$$Q_{\text{load-pull}} = \frac{\omega_0}{2}\left|\frac{Im\left[Z^{\omega}_{\text{out}}(\omega, g_m)Z^{g\,*}_{\text{out}}(\omega, g_m)\right]}{Z^{\omega}_{\text{out}}(\omega, g_m)Z^{g}_{\text{out}}(\omega, g_m)}\right|_{(\omega=\omega_0, g_m=g_{mo})} \tag{5.74}$$

$$Q_{\text{load-pull}} = \frac{\omega_0}{2}\left|\frac{Im\left[Y^{\omega}_{\text{out}}(\omega, g_m)Y^{g\,*}_{\text{out}}(\omega, g_m)\right]}{Y^{\omega}_{\text{out}}(\omega, g_m)Y^{g}_{\text{out}}(\omega, g_m)}\right|_{(\omega=\omega_0, g_m=g_{mo})} \qquad (Y \equiv Z^{-1}) \tag{5.75}$$

Equations (5.52) and (5.75) give insight into pushing and pulling characteristics of an oscillator for the optimization of the oscillator design for stable operation against those variations.

5.4.2.5 Active SIBO Circuit Q-factor (Injection Locking Basis)

Figure 5.16 shows the typical injection-locked oscillator employing 2-port active device: (a) general model, (b) typical Colpitts oscillator, and (c) equivalent representation of (a).

The impedance matrix of Figure 5.17 can be given by [28, 33]

$$Z(\omega) = \begin{bmatrix} Z_{11}(\omega) & Z_{12}(\omega) & Z_{13}(\omega) \\ Z_{21}(\omega) & Z_{22}(\omega) & Z_{23}(\omega) \\ Z_{31}(\omega & Z_{32}(\omega) & Z_{33}(\omega) \end{bmatrix} = \begin{bmatrix} R_{11}(\omega) & R_{12}(\omega) & R_{13}(\omega) \\ R_{21}(\omega) & R_{22}(\omega) & R_{23}(\omega) \\ X_{31}(\omega & X_{32}(\omega) & X_{33}(\omega) \end{bmatrix} + \begin{bmatrix} X_{11}(\omega) & X_{12}(\omega) & X_{13}(\omega) \\ X_{21}(\omega) & X_{22}(\omega) & X_{23}(\omega) \\ X_{31}(\omega & X_{32}(\omega) & X_{33}(\omega) \end{bmatrix} \tag{5.76}$$

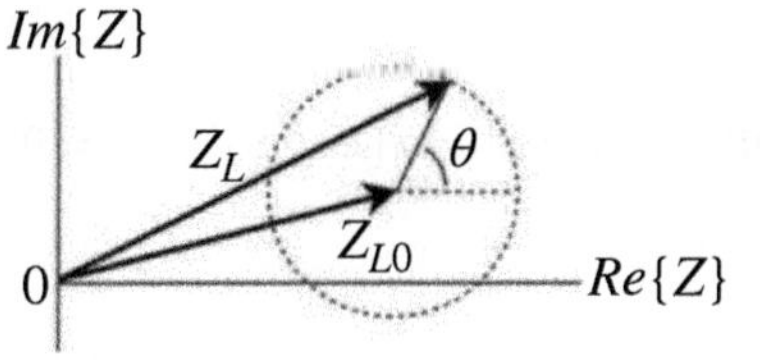

Figure 5.15 Deviated load impedance locus on a complex plane.

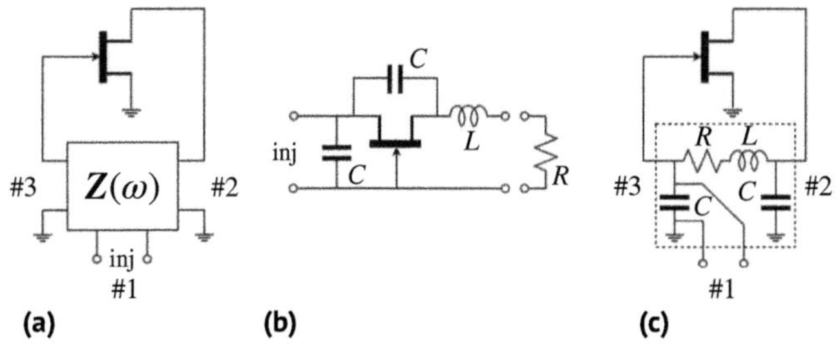

Figure 5.16 Injection-locked oscillator employing 2-port active device: (a) general model, (b) typical Colpitts oscillator, and (c) equivalent representation of (a) [27].

Case I (no injected current at port 1), $i_{inj} = i_1(\omega) = 0$

From (5.76)

$$v_3(\omega) = R_{32}(\omega)i_2(\omega) + jX_{32}(\omega)i_2(\omega) = [R_{32}(\omega) + jX_{32}(\omega)]i_2(\omega) \tag{5.77}$$

$$i_2(\omega) = -g_m(\omega)v_3(\omega) \tag{5.78}$$

For stable oscillation to take place at $\omega \to \omega_0$, neither v_3 nor i_2 should vanish ($v_3 \neq 0$ and $i_2 \neq 0$).

From (5.77) and (5.78)

$$[1 + g_m(\omega)R_{32}(\omega)]_{\omega\to\omega_0} = 0,\ [X_{32}(\omega)]_{\omega\to\omega_0} = 0 \tag{5.79}$$

where ω_0 = free running oscillation frequency

Case II [injection-locked state: $i_{inj} = i_1(\omega) \neq 0$**]**

$$i_{inj} = i_1(\omega) = I_1e^{j\theta_1} \text{ (Current represented in exponent form)} \tag{5.80}$$

$$i_2(\omega) = -g_m(\omega)v_3(\omega) = I_2e^{j\theta_2} \text{ (Current represented in exponent form)} \tag{5.81}$$

From (5.76)

$$v_3(\omega) = [R_{31}(\omega) + jX_{31}(\omega)]I_1e^{j\theta_1} + [R_{32}(\omega) + jX_{32}(\omega)]I_2e^{j\theta_2} \tag{5.82}$$

From (5.80) to (5.82)

$$[R_{31}(\omega)\sin\Delta\theta + jX_{31}(\omega)\cos\Delta\theta]I_1 + X_{32}(\omega)I_2 = 0 \tag{5.83}$$

where $\Delta\theta = \theta_1 - \theta_2$ is the phase difference between the currents at two ports.

The injection-locked state oscillation frequency ω is slightly offset ($\Delta\omega$) from the free-running frequency ω_0, the impedance matrix (8) can be expanded with the help of the Taylor series.

$$R_{31}(\omega) \cong R_{31}(\omega_0) + \Delta\omega\frac{d}{d\omega}[R_{31}(\omega)]_{\omega\to\omega_0} = R_{31}(\omega_0) + \Delta\omega R'_{31}(\omega_0) \tag{5.84}$$

$$X_{31}(\omega) \cong X_{31}(\omega_0) + \Delta\omega\frac{d}{d\omega}[X_{31}(\omega)]_{\omega\to\omega_0} = X_{31}(\omega_0) + \Delta\omega X'_{31}(\omega_0) \tag{5.85}$$

$$X_{33}(\omega) \cong X_{33}(\omega_0) + \Delta\omega\frac{d}{d\omega}[X_{33}(\omega)]_{\omega\to\omega_0} = X_{33}(\omega_0) + \Delta\omega X'_{33}(\omega_0) \tag{5.86}$$

$$X_{32}(\omega) \cong X_{32}(\omega_0) + \Delta\omega\frac{d}{d\omega}[X_{32}(\omega)]_{\omega\to\omega_0} = X_{32}(\omega_0) + \Delta\omega X'_{32}(\omega_0) \tag{5.87}$$

With the help of Taylor series expansion, based on $\omega \to \omega_0 + \Delta\omega$ and free-running oscillation condition $[X_{32}(\omega)]_{\omega\to\omega_0} = 0$; $|\Delta\omega| \ll \omega_0$

From (5.76) to (5.87)

$$\Delta\omega \cong -\left[\frac{R_{31}(\omega_0)\sin\Delta\theta + X_{31}(\omega_0)\cos\Delta\theta}{X'_{32}(\omega_0)}\right]\frac{I_1}{I_2} \tag{5.88}$$

$$\Delta\omega \cong -\left[\frac{\sqrt{R_{31}^2(\omega_0) + X_{31}^2(\omega_0)}}{X'_{32}(\omega_0)}\sin\varnothing\right]\frac{I_1}{I_2} = -\frac{|Z_{31}(\omega_0)|\sin\varnothing}{X'_{32}(\omega_0)}\frac{I_1}{I_2} \tag{5.89}$$

where $\varnothing = \Delta\theta + \tan^{-1}\left[\frac{X_{31}(\omega_0)}{R_{31}(\omega_0)}\right]$, assuming phase difference $\Delta\theta$ is not constrained i.e., $(0 \le \Delta\theta < 2\pi) \to 0 \le |\sin\varnothing| \le 1$, from (5.89)

$$0 \le |\Delta\omega| \le \left|\frac{Z_{31}(\omega_0)}{X'_{32}(\omega_0)}\right|\frac{I_1}{I_2} \tag{5.90}$$

Equation (5.90) is identical to Adler's injection-lock range Ref. [31] and Eq. (10.90) Ref. [33]

$$0 \leq |\Delta\omega| \leq \left[\frac{\omega_0}{2Q}\right]\left[\frac{I_1}{I_2}\right] \tag{5.91}$$

From (5.90) and (5.91)

$$Q = \frac{\omega_0}{2}\left|\frac{X'_{32}(\omega_0)}{Z_{31}(\omega_0)}\right|, \text{ where } X'_{32}(\omega_0) = \frac{d}{d\omega}[X_{32}(\omega)]_{\omega\to\omega_0} \tag{5.92}$$

The circuit shown in Figure 5.16c, Q-factor can be given by

$$Q = \frac{\omega_0}{2}\left|\frac{X'_{32}(\omega_0)}{Z_{31}(\omega_0)}\right| \tag{5.93}$$

where

$$Z_{31}(\omega_0) = \left\{\frac{1}{j\omega C + \frac{1}{R + j\omega L + \frac{1}{j\omega C}}}\right\}, \text{ and } X_{32}(\omega) = \frac{1}{\omega C}\left\{\frac{LC\omega^2 - 2}{\omega^2C^2R^2 + (LC\omega^2 - 2)^2}\right\} \tag{5.94}$$

From (5.79), (5.93), and (5.94)

$$X'_{32}(\omega_0) = \frac{d}{d\omega}[X_{32}(\omega)]_{\omega\to\omega_0} = \frac{d}{d\omega}\left[\frac{1}{\omega C}\left\{\frac{LC\omega^2 - 2}{\omega^2C^2R^2 + (LC\omega^2 - 2)^2}\right\}\right]_{\omega=\omega_0} \tag{5.95}$$

$$Q = \frac{\omega_0}{2}\left|\frac{X'_{32}(\omega_0)}{Z_{31}(\omega_0)}\right| = \frac{\omega_0 L}{R}\left(\frac{L}{L + 2CR^2}\right)^{1/2} \tag{5.96}$$

5.5 Resonator Design Criteria for Low Phase Noise Oscillator Applications

The Leeson phase noise equation is given by [35]

$$£(f_m) = 10\log\left\{\left[1 + \frac{{f_0}^2}{(2f_mQ_L)^2\left(1 - \frac{Q_L}{Q_0}\right)^2}\right]\left(1 + \frac{f_c}{f_m}\right)\frac{FkT}{2P_o}\right\} \tag{5.97}$$

where

$£(f_m)$ = ratio of sideband power in a 1Hz bandwidth at f_m to total power in dB
f_m = frequency offset from the carrier
f_0 = center frequency
f_c = flicker frequency
Q_L = loaded Q of the tuned circuit
Q_0 = unloaded Q of the tuned circuit
F = noise factor
$kT = 4.1 \times 10^{-21}$ at 300 K (room temperature)
P_o = average power at oscillator output

From (5.97), phase noise in oscillator is inversely proportional to the square of the resonator loaded Q-factor, indicating that insertion loss of the resonator is linked with FOM as [32–34]

$$[\text{Resonator}]_{\text{FOM}} = 10\log\left(\frac{I.L}{Q_L^2}\right) \tag{5.98}$$

From (5.98), for low phase noise, designer must minimize the resonator FOM.

At microwave frequencies, resonators can take various shapes and forms [35–44]. The shape of microwave structure affects the field distribution and influence the tuning range of the oscillator circuits. Usually, passive lumped LC and planar transmission line resonator networks are used in low-cost broadband oscillator circuits.

5.5.1 Passive Lumped LC Resonator

Figure 5.17 shows the typical doubly loaded shunt resonator. Lumped or quasi-lumped resonator will oscillate at $f_o = \dfrac{1}{2\pi\sqrt{LC}}$.

The S_{21} (transmission gain) can be given for doubly loaded shunt resonator (Figure 5.17) is given by [36]

$$S_{21}(\omega) = \frac{2}{\left(\frac{Z_O}{\frac{1}{R_r}+\frac{1}{Z_O}+\frac{1}{Z_O}}\right) + jZ_O\left(\omega C - \frac{1}{\omega L}\right)} \tag{5.99}$$

$$Q = \frac{\omega_0}{2}\left|\frac{\partial\varphi(\omega)}{\partial\omega}\right|_{\omega=\omega_0} \quad \text{where } (\varphi(\omega) = \angle S_{21}) \tag{5.100}$$

$$I.L = \left|\frac{1}{S_{21}}\right|^2_{\omega=\omega_0} = \left(1+\frac{Z_O}{2R_r}\right)^2 = \left(1+\frac{Q_e}{Q_u}\right)^2 \tag{5.101}$$

From (5.98),

$$[\text{Resonator}]_{\text{FOM}} = 10\log\left(\frac{I.L}{Q_L^2}\right) = 10\log\left[\left(1+\frac{Q_e}{Q_u}\right)^2\left(\frac{1}{Q_e}+\frac{1}{Q_u}\right)^2\right] \tag{5.102}$$

$$\frac{d[\text{Resonator}]_{FOM}}{dQ_e} = 0 \Rightarrow (Q_u = Q_e) \Rightarrow Q_{\text{loaded}} = \frac{Q_{\text{unloaded}}}{2} \tag{5.103}$$

5.5.2 Planar Transmission Line

Various forms of planar transmission lines have been developed. Some examples are strip line, microstrip line, slot line, and coplanar waveguide. The microstrip line is the most popular type of resonator used for oscillator and filter applications. Figure 5.18 shows the typical structure of a microstrip transmission line. A conducting strip (microstrip line) with a width w and a thickness t is on the top of a dielectric substrate that has a relative dielectric constant ε_r and a thickness h and the bottom of the substrate is a ground (conducting) plane.

The fields in the microstrip extend within two media—air above and dielectric below so that the structure is inhomogeneous. Due to this inhomogeneous nature, the microstrip does not support a pure TEM wave. This is because a pure TEM wave has only transverse components, and its propagation velocity depends only on the material properties, namely the permittivity ε and the permeability μ. However, with the presence of the two guided wave media (the dielectric substrate and the air), the waves in a microstrip line will have no vanished longitudinal components of electric and magnetic fields, and their propagation velocities will depend not only on the material properties, but also on the physical dimensions of

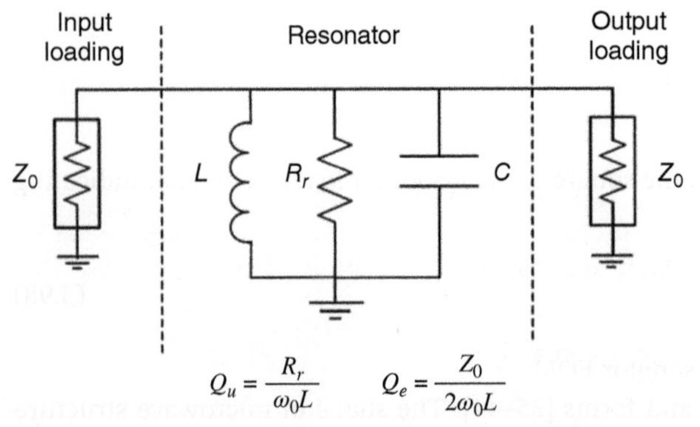

Figure 5.17 A typical doubly loaded shunt resonator.

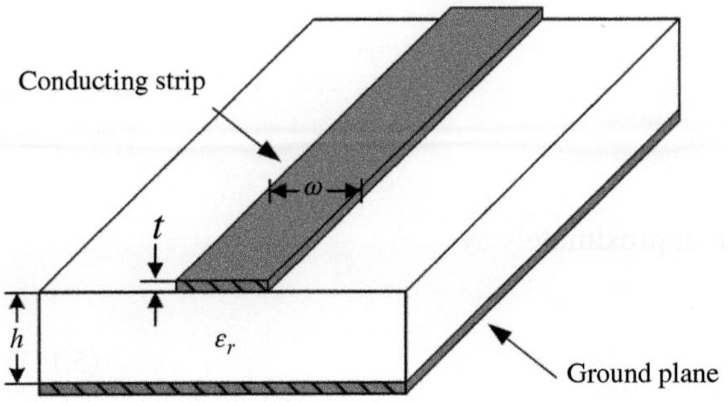

Figure 5.18 A typical structure of microstrip Line.

the microstrip. When the longitudinal components of the fields for the dominant mode of a microstrip line remain very much smaller than the transverse components, they may be neglected. In this case, the dominant mode then behaves like a TEM mode, and the TEM transmission line theory is applicable for the microstrip line as well. This is called the quasi-TEM approximation, and it is valid over most of the operating frequency ranges of microstrip.

5.5.2.1 Effective Dielectric Constant and Characteristic Impedance of a Microstrip Line

The microstrip transmission line is an inhomogeneous transmission line. The field between the strip and the ground plane is not contained entirely in the substrate but extends within two media, air and dielectric. Hence, the microstrip line cannot support a pure TEM wave. The mode of propagation is quasi-TEM. In the quasi-TEM approximation, a homogeneous dielectric material with an effective dielectric permittivity replaces the inhomogeneous dielectric-air media of microstrip. Transmission characteristics of microstrip lines are described by two parameters, namely the effective dielectric constant ε_{re} and characteristic impedance Z_c, which may then be obtained by quasi-static analysis. In quasi-static analysis, the fundamental mode of wave propagation in a microstrip transmission line is quasi-TEM; however, for simplification in analysis, it is assumed pure TEM. The above two parameters of microstrip transmission lines are then determined from the values of two capacitances as follows:

$$\varepsilon_{re} = \frac{C_d}{C_a} \tag{5.104}$$

$$Z_c = \frac{1}{c\sqrt{C_d C_a}} \tag{5.105}$$

where C_d is the capacitance per unit length with the dielectric substrate present, $C_a$is the capacitance per unit length with the dielectric substrate replaced by air, and c is the velocity of electromagnetic waves in free space $(c = 3.0 \times 10^8\ m/s)$. The phase velocity and propagation constant can be expressed as

$$v_p = \frac{\omega}{\beta} = \frac{c}{\sqrt{\varepsilon_{re}}} \tag{5.106}$$

$$\beta = k_0\sqrt{\varepsilon_e} \tag{5.107}$$

where c is the velocity of light in free space.

The electrical length θ for a given physical length l of the microstrip is defined by

$$\theta = \beta l \tag{5.108}$$

Therefore, $\theta = \pi/2$ when $l = \lambda_g/4$, and $\theta = \pi$ when $l = \lambda_g/2$. These so-called quarter-wavelength and half-wavelength microstrip lines are important for the design of microstrip resonators.

For very thin conductors (i.e., $t \to 0$), the closed-form expressions that provide accuracy better than one percent are given as follows [43, 57]:

For $\frac{w}{h} \le 1$: effective dielectric constant of a microstrip line is given approximately by

$$\varepsilon_{re} = \frac{\varepsilon_r + 1}{2} + \frac{\varepsilon_r - 1}{2}\left\{\left(1 + 12\frac{h}{w}\right)^{-1/2} + 0.04\left(1 - \frac{w}{h}\right)^2\right\} \tag{5.109}$$

$$Z_c = \frac{\eta}{2\pi\sqrt{\varepsilon_{re}}} \ln\left(\frac{8h}{W} + \frac{w}{4h}\right) \tag{5.110}$$

where $\eta = 120\pi$ (Ω) is the wave impedance in free space.

For $\left(\frac{w}{h} \geq 1\right)$: effective dielectric constant of a microstrip line is given approximately by

$$\varepsilon_{re} = \frac{\varepsilon_r + 1}{2} + \frac{\varepsilon_r - 1}{2}\left(1 + 12\frac{h}{w}\right)^{-1/2} \tag{5.111}$$

$$Z_c = \frac{\eta}{\sqrt{\varepsilon_{re}}}\left\{\frac{w}{h} + 1.393 + 0.6697 \ln\left(\frac{w}{h} + 1.444\right)\right\}^{-1} \tag{5.112}$$

The effective dielectric constant can be interpreted as the dielectric constant of a homogeneous medium that replaces the air and dielectric regions of the microstrip. Given the dimensions of the microstrip line, the characteristic impedance can be written as

$$Z_c = \begin{cases} \dfrac{60}{\sqrt{\varepsilon_e}} \ln\left(\dfrac{8h}{w} + \dfrac{w}{4h}\right) & \text{for } w/h \leq 1 \\ \dfrac{120\pi}{\sqrt{\varepsilon_{re}}\left[\frac{w}{h} + 1.393 + 0.6697\ \ln\left(\frac{w}{h} + 1.444\right)\right]} & \text{for } w/h \geq 1 \end{cases} \tag{5.113}$$

For given characteristic impedance Z_c and dielectric constant ε_{re} the w/h ratio can be found as

$$\frac{w}{h} = \begin{cases} \dfrac{8\ \exp(A)}{\exp(2A) - 2} & \text{for } w/h < 2 \\ \dfrac{2}{\pi}\left[B - 1 - \ln(2B - 1) + \dfrac{\varepsilon_{re} - 1}{2\varepsilon_{re}}\left[\ln(B - 1) + 0.39 - \dfrac{0.61}{\varepsilon_{re}}\right]\right] & \text{for } w/h > 2 \end{cases} \tag{5.114}$$

where

$$A = \frac{Z_c}{60}\sqrt{\frac{\varepsilon_{re} + 1}{2}} + \frac{\varepsilon_r - 1}{\varepsilon_r + 1}\left(0.23 + \frac{0.11}{\varepsilon_{re}}\right) \tag{5.115}$$

$$B = \frac{60\pi^2}{Z_c\sqrt{\varepsilon_r}} \tag{5.116}$$

These expressions also provide accuracy better than one percent. If values that are more accurate are needed, an iterative or optimization process can be employed.

In general, there is dispersion in microstrip; its phase velocity is not a constant but depends on frequency. The effective dielectric constant ε_{re} is a function of frequency and can be defined as the frequency-dependent effective dielectric constant ε_{re} (f) [58].

$$\varepsilon_{re}(f) = \varepsilon_r - \frac{\varepsilon_r - \varepsilon_{re}}{1 + (f/f_{50})^m} \tag{5.117}$$

where

$$f_{50} = \frac{f_{TM_0}}{0.75 + (0.75 - 0.332\varepsilon_r^{-1.73})w/h} \tag{5.118}$$

$$f_{TM_0} = \frac{c}{2\pi h\sqrt{\varepsilon_r - \varepsilon_{re}}} \tan^{-1}\frac{\sqrt{\varepsilon_r - 1}}{\varepsilon_r - \varepsilon_{re}} \tag{5.119}$$

$$m = m_0 m_c \leq 2.32 \tag{5.120}$$

$$m_0 = 1 + \frac{1}{1 + \sqrt{w/h}} + 0.32\left(\frac{1}{1 + \sqrt{w/h}}\right)^3 \tag{5.121}$$

$$m_c = \begin{cases} 1 + \dfrac{1.4}{1 + w/h}\left\{0.15 - 0.235 \exp\left(\dfrac{-0.45f}{f_{50}}\right)\right\} & \text{for } \dfrac{w}{h} \leq 0.7 \\ 1 & \text{for } \dfrac{w}{h} \geq 0.7 \end{cases} \tag{5.122}$$

where, c is the velocity of light in free space, and whenever the product $m_0 m_c$ is greater than 2.32, the parameter m is chosen equal to 2.32

The dispersion model shows that the $\varepsilon_{re}(f)$ increases with frequency and $\varepsilon_{re}(f) \to \varepsilon_r, as\ f \to \infty$. The accuracy is estimated to be within 0.6% for $0.1 \leq w/h \leq 10$, for $1 \leq \varepsilon_r \leq 128$ and for any value of $\left(\dfrac{h}{\lambda_0}\right)$.

The effect of dispersion on the characteristic impedance may be estimated by [44]

$$Z_c(f) = Z_c \frac{\varepsilon_{re}(f) - 1}{\varepsilon_{re} - 1} \sqrt{\frac{\varepsilon_{re}}{\varepsilon_{re}(f)}} \tag{5.123}$$

5.5.2.2 Planar Transmission Line Bends

The compact printed resonator is formed by bending the transmission line in a closed form, such as ring resonator, square loop resonator, hairpin resonator, etc. Right-angle bend and mitered bend of microstrip transmission lines may be modeled by an equivalent T-network, as shown in Figure 5.19, and its closed-form expressions for evaluation of capacitance and inductance are given by [43]:

$$C = 0.001h\left[(10.35\varepsilon_r + 2.5)\left(\frac{w}{h}\right)^2 + (2.6\varepsilon_r + 5.64)\left(\frac{w}{h}\right)\right] pF \tag{5.124}$$

$$L = 0.22h\left\{1 - 1.35 \exp\left[-0.18\left(\frac{w}{h}\right)^{1.39}\right]\right\} nH \tag{5.125}$$

For the microstrip mitered bend, and as

$$C = 0.001h\left[(3.93\varepsilon_r + 0.62)\left(\frac{w}{h}\right)^2 + (7.6\varepsilon_r + 3.8)\left(\frac{w}{h}\right)\right] pF \tag{5.126}$$

$$L = 0.44h\left\{1 - 1.062 \exp\left[-0.177\left(\frac{w}{h}\right)^{0.947}\right]\right\} nH \tag{5.127}$$

5.5.3 Planar Transmission Line Resonator

Planar transmission line resonators are formed by using microstrip lines of various wavelengths ($\lambda/4$, $\lambda/2$, λ), where λ is the guided wavelength at the fundamental resonant frequency f_0. The quarter-wavelength resonator $\lambda/4$ long resonates at the fundamental frequency f_0 and at its multiple frequencies of $f = (2n - 1)f_0$ for $n = 2, 3, 4, 5$ ...

The half-wavelength resonator $\lambda/2$ long resonates at the fundamental frequency f_0 and at its multiple frequencies of $f = n f_0$ for $n = 2, 3$ ... etc. This type of resonator can also be shaped into an open-loop resonator. The full wavelength

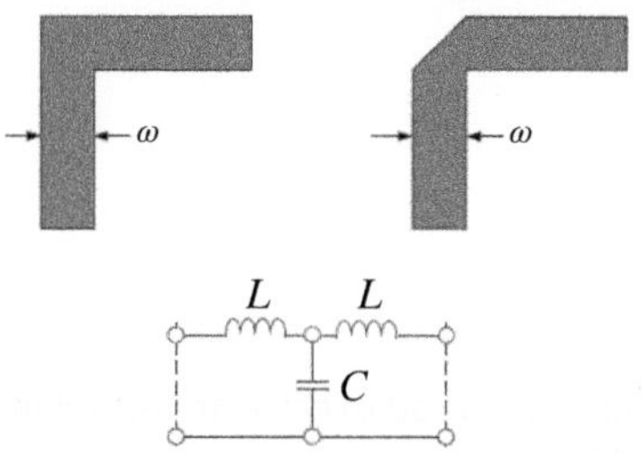

Figure 5.19 Equivalent lumped LC representation of right-angle bend, mitered bend, and model [57, 58].

resonator, λ long resonates at the fundamental frequency f_o and at other frequencies of $f = nf_o$ for $n = 2, 3, \ldots$ etc. This type of resonator is commonly found in the form of ring or closed-loop resonators with a median circumference $2\pi r = \lambda$, where r is the radius of the ring. Because of its symmetrical geometry, a resonance can occur in either of 2 orthogonal coordinates. This type of transmission line resonator has a distinct feature; it can support a pair of degenerate modes that have the same resonant frequencies but orthogonal field distributions. This feature can be utilized to design dual-mode filters.

5.5.3.1 Microstrip Resonator

A microstrip resonator is any structure that is able to contain at least one oscillating EM field. There are many forms of microstrip resonators; however, their large physical size can present a drawback. Hence, there is a strong interest in miniaturizing such resonators, particularly for oscillator applications. Miniaturization of microstrip resonators can be achieved by using either high dielectric constant substrates or by meandering the lines to create a folded microstrip resonator [37].

5.5.3.2 Folded Open Loop Microstrip Resonator

Figure 5.20 shows the typical square open-loop resonator, which can be obtained by folding a straight open resonator (as illustrated in Figure 5.20a). Due to the corners and the fringing capacitance between the open ends, a rigorous calculation of the electromagnetic fields in the square resonator is impractical. However, it is possible to study the main characteristics of the resonant modes of the square open-loop resonator by analogy to those of the straight resonator. This qualitative analysis can shed some light on the behavior of the resonator with minimum effort. The conclusions drawn using this approach can then be compared for validation against the actual distribution of the electromagnetic fields obtained with the aid of full-wave simulators.

The resonant frequency of the straight transmission line (Figure 5.21a) can be obtained by looking at the input admittance from any point within its length. Figure 5.21b shows an equivalent circuit that can be used to calculate this admittance as [38]

$$Y_{\text{in}} = jY_O(\tan\theta_1 + \tan\theta_2) = jY_O\left(\frac{\sin\theta_T}{\cos\theta_1\cos\theta_2}\right) \tag{5.128}$$

$$[Y_{\text{in}}]_{\left(\theta_1=\theta_1=\frac{\pi}{2}\right)} = \infty \Longrightarrow [Z_{\text{in}}]_{\left(\theta_1=\theta_1=\frac{\pi}{2}\right)} = 0 \tag{5.129}$$

where $\theta_T = \theta_1 + \theta_2$, is the total electrical length of the resonator. A standing wave can be maintained in the resonator whenever $Y_{\text{in}} = 0$. This yields infinite resonant frequencies at: $\theta_T = n\pi$ or $l = n\frac{\lambda}{2}$.

Figure 5.22 shows the voltage distribution at the first two resonant frequencies ($n = 1, 2$). Since the open ends of the resonator force the current to be zero there, the voltage attains a maximum, and the modes shown are the only ones allowed at those frequencies. If the loop were closed, this boundary condition would not apply, and two orthogonal modes would exist at each frequency [37–39].

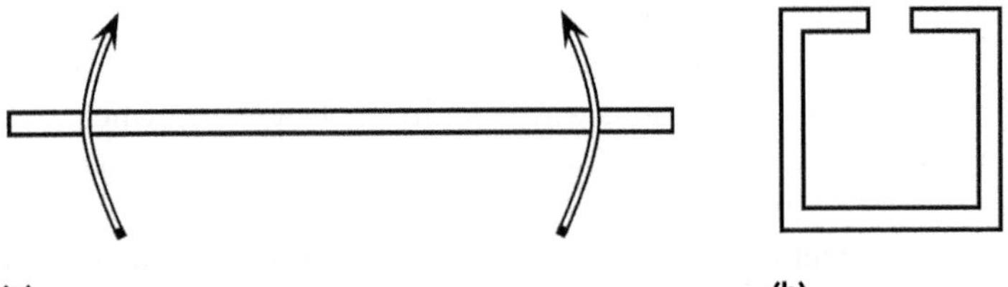

(a) (b)

Figure 5.20 The square open-loop resonator can be obtained by folding a straight open resonator: (a) straight open resonator, and (b) square open-loop resonator. *Source*: [37]/John Wiley & Sons.

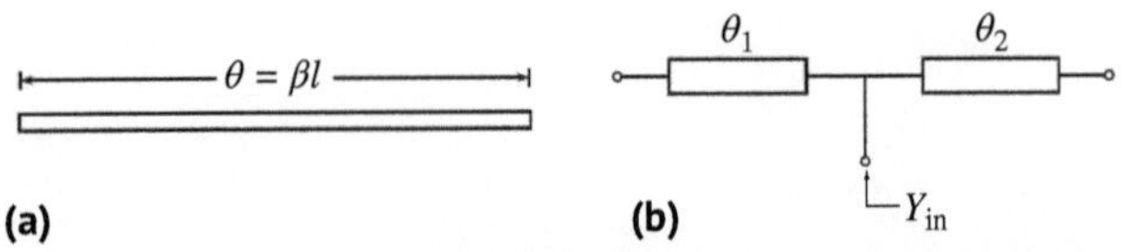

(a) (b)

Figure 5.21 Microstrip open resonator: (a) top view of a microstrip straight resonator, (b) equivalent circuit used to calculate the input admittance from an arbitrary point within the length of the resonator. *Source*: [37]/John Wiley & Sons.

Figure 5.22 shows the positions of the voltage nulls in the mode diagram. At the first resonant frequency, there is only one such null at $\theta_1 = \theta_2 = \pi/2$, while at the second resonance, there are two of them at $\theta_1 = \pi/2$ and $\theta_2 = 3\pi/2$. From (5.129), the location of voltage nulls is important because the resonator cannot be excited there. However, by choosing the feeding point of the resonator, it is possible to excite only the odd or even modes of the resonator. As an interesting consequence, the fundamental resonance ($n = 1$) or any other odd mode resonance cannot be excited at the center of the resonator. This phenomenon translates to the square open-loop resonator as is shown in Figure 5.23.

The Q_{ex} (external quality factor) of the square open-loop resonator obtained by tapping into the resonator depends on the voltage level at the tapping point at resonance. If the tapping point coincides with a voltage null, then no coupling is achieved between the resonator and the external circuit, and the resulting quality factor is very large. Similarly, if the voltage at the tapping point is high, then the external quality factor will be low. Referring to the voltage distribution of the conventional and the miniaturized resonator, it is possible to predict that in the case of the conventional resonator, the external Q will decrease rapidly as the tapping point is moved away from the voltage null at the center of the resonator.

For comparative analysis, Figure 5.24 shows the external quality factor of both conventional and miniaturized resonator with the same resonant frequency (1 GHz) [39]. As illustrated in Figure 5.24, Q_{ex} diminishes relatively fast for unloaded resonator. But in the case of the miniaturized resonator loaded with 1 pF capacitor, the change in the Q_{ex} is slower. Therefore, the tapping distance from the null point necessary to obtain a given Q_{ex} is larger in the case of the miniaturized resonator. The tapping point is selected on the side of the miniaturized loaded resonator; this is because the quality factors obtained by tapping closer to the null point produce large values of Q_{ex} that are difficult to estimate accurately.

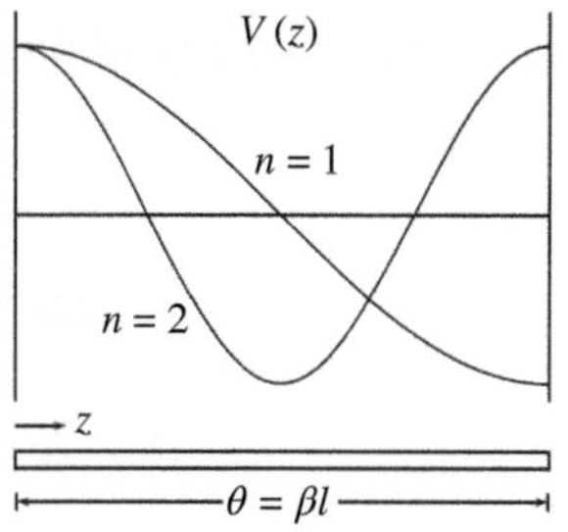

Figure 5.22 Voltage distribution in a straight open resonator. *Source*: Adapted from [39].

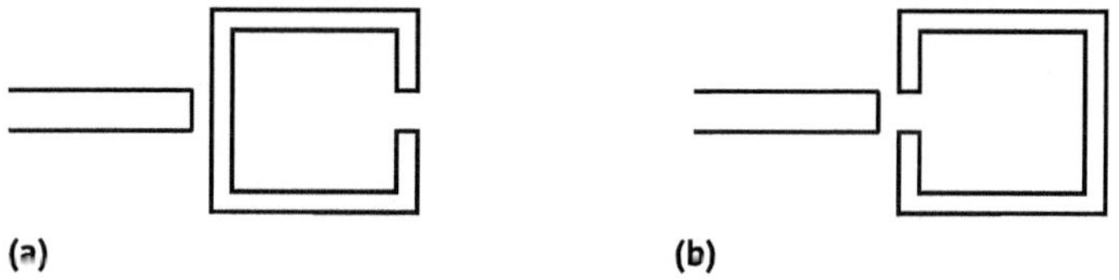

Figure 5.23 Two ways of exciting only the even modes of the square open-loop resonator: (a) excitation of the resonator in a null of the fundamental mode, and (b) excitation of the resonator symmetrically with respect to both open ends. *Source*: Adapted from [39].

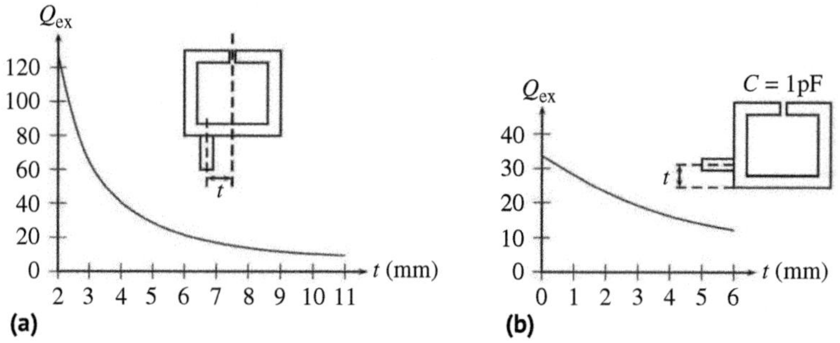

Figure 5.24 External quality factor of an: (a) unloaded resonator, and (b) a resonator loaded with $C = 1$ pF. *Source*: Adapted from [39].

5.5.3.3 Folded Hairpin Resonator

Figure 5.25 shows the basic layout of the typical stepped impedance hairpin resonator, which consists of a single transmission line l_s and coupled lines with a length of l_c; Z_s is the characteristic impedance of the single transmission line l_s.

As shown in Figure 5.25, Z_{oe} and Z_{oo} are the even- and odd-mode impedances of the symmetric capacitance-load parallel-coupled lines with a length of l_c. The drawback of the stepped impedance hairpin resonator is its large size, which can be made smaller by proper selection of the value of Z_s $\left(Z_s > \sqrt{Z_{oe}Z_{oo}}\right)$.

The effect of the loading capacitance shifts the spurious resonant frequencies of the resonator from integer multiples of the fundamental resonant frequency, thereby reducing interferences from highorder harmonics [40]. For the analysis purpose, Figure 5.26 shows the typical configuration of stepped impedance hairpin resonator built from open-end transmission line.

As shown in Figure 5.26a, the single transmission line is modeled as an equivalent LC π-network. Assuming the lossless single transmission line with a length of l_s, the ABCD matrix is given by

$$\begin{bmatrix} A & B \\ C & D \end{bmatrix} = \begin{bmatrix} \cos(\beta_s l_s) & jZ_s \sin(\beta_s l_s) \\ jY_s \sin(\beta_s l_s) & \cos(\beta_s l_s) \end{bmatrix} \tag{5.130}$$

$$\begin{bmatrix} A & B \\ C & D \end{bmatrix} = \begin{bmatrix} 1 + Z_L Y_c & Z_L \\ Y_c(2 + Z_L Y_c) & 1 + Z_L Y_c \end{bmatrix} \tag{5.131}$$

where $Z_L = j\omega L_s$, $Y_c = j\omega C_s$, ω is the angular frequency, L_s and C_s are the equivalent inductance and capacitance of the single transmission line.

From (5.130) with (5.131), the equivalent L_s and C_s can be described by

$$L_s = \frac{Z_s \sin(\beta_s l_s)}{\omega} H \tag{5.132a}$$

$$C_s = \frac{1 - \cos(\beta_s l_s)}{\omega Z_s \sin(\beta_s l_s)} F. \tag{5.132b}$$

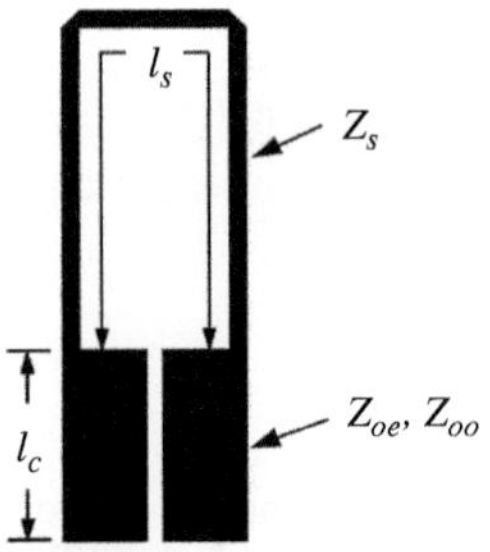

Figure 5.25 A typical layout of the stepped impedance hairpin resonator. *Source*: [40]/L-Hwa Hsieh.

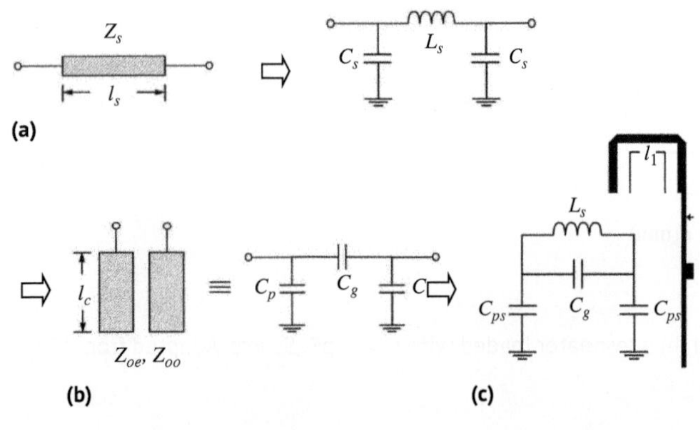

Figure 5.26 Equivalent circuit of (a) single transmission line, (b) symmetric coupled lines, and (c) stepped impedance hairpin resonator. *Source*: [41]/IEEE.

In case of the symmetric parallel-coupled lines, the electrical equivalent circuit is capacitive π-network. The ABCD matrix of the lossless parallel-coupled lines is expressed as [41]

$$\begin{bmatrix} A & B \\ C & D \end{bmatrix} = \begin{bmatrix} \dfrac{Z_{oe}+Z_{oo}}{Z_{oe}-Z_{oo}} & \dfrac{-j2Z_{oe}Z_{oo}\cot(\beta_c l_c)}{(Z_{oe}-Z_{oo})} \\ \dfrac{j2}{(Z_{oe}-Z_{oo})\cot(\beta_c l_c)} & \dfrac{Z_{oe}+Z_{oo}}{Z_{oe}-Z_{oo}} \end{bmatrix} \tag{5.133}$$

$$\begin{bmatrix} A & B \\ C & D \end{bmatrix} = \begin{bmatrix} 1+Z_gY_p & Z_g \\ Y_p(2+Z_gY_p) & 1+Z_gY_p \end{bmatrix} \tag{5.134}$$

where β_c is the phase constant of the coupled lines, $Z_g = 1/j\omega C_g$, $Y_p = j\omega C_p$.

From (5.133) and (5.134), the equivalent capacitances of the π-network can be given by

$$C_g = \left[\frac{(Z_{oe}-Z_{oo})}{2\omega Z_{oe}Z_{oo}\cot(\beta_c l_c)}\right]F,\ C_p = \left[\frac{1}{\omega Z_{oe}\cot(\beta_c l_c)}\right]F \tag{5.135}$$

The realization of stepped impedance hairpin resonator can be formulated by combining the equivalent circuits of the single transmission line and coupled lines, as shown in Figures 5.25a and 5.26b, the equivalent circuit of the stepped impedance hairpin resonator in terms of lumped elements L and C is shown in Figure 5.25c, where $C_{ps} = (C_P + C_s + \Delta C)$ is the sum of the capacitances of the single transmission line, coupled lines, and the junction discontinuity (ΔC) [42] between the single transmission line and the coupled lines. The widths of the single transmission line and coupled lines resonator can be obtained by selecting the impedances that satisfy the condition $Z_s > \sqrt{Z_{oe}Z_{oo}}$. The lengths of the single transmission line and coupled lines of the filter are transformed from (5.132) and (5.135):

$$l_s = \frac{\sin^{-1}(\omega_c L_{st}/Z_s)}{\beta_s} \text{ and } l_c = \frac{\tan^{-1}[\omega_c Z_{oe}(C_{pst} - C_s - \Delta C)]}{\beta_c} \tag{5.136}$$

where ω_c is the 3 dB cutoff angular frequency, L_{st} and C_{pst} are the inductance and capacitance chosen from the available LC tables.

Figure 5.27 shows the typical configuration of hairpin resonator.

5.5.3.4 Ring Resonator

A most common closed-loop resonator type is the ring resonator, which consists of a transmission line of a full wavelength λ long, formed in a circular closed loop, and resonates when the mean circumference of the ring resonator is equal to an integral multiple of a guided wavelength, can be described by [44]

$$2\pi r = n\lambda_g, \text{ for } n = 1, 2, 3, \ldots \tag{5.137}$$

$$f = \frac{nc}{2\pi\sqrt{\epsilon_{eff}}} = nf_0 \tag{5.138}$$

$$\lambda_g = \frac{c}{f\sqrt{\epsilon_{eff}}} \tag{5.139}$$

where r is the mean radius of the ring that equals the average of the outer and inner radii, λ_g is the guided wavelength and n is the mode number.

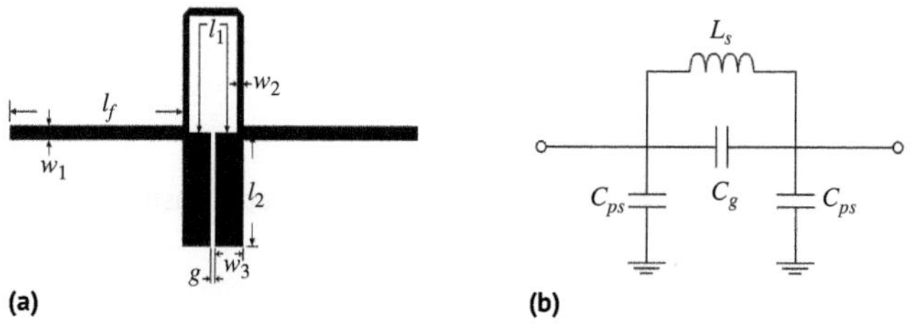

Figure 5.27 A typical low-pass filter using one hairpin resonator: (a) layout and (b) equivalent circuit. *Source:* [43]/John Wiley & Sons.

Figure 5.28 shows the typical ring resonator, which consists of feed lines, coupling gaps, and the circular closed-loop printed transmission line resonator. As shown in Figure 5.28, power is coupled into and out of the resonator through feed lines and coupling gaps. For the first mode, the maxima of fields occur at the coupling gap locations, and nulls occur at 90° from the coupling gap locations, valid only for the weakly coupled case, as it does not account for loading effects from the ports [44].

Coupling is said to be weak or "loosely coupled" if the separation between the feed lines and the resonator is large such that the resonant frequency of the ring is unaffected [45–53]. However, if the separation is reduced, the gap capacitance increases, thereby resonator loading will occur and may cause the resonant frequency of the circuit to deviate from the inherent resonant frequencies of the ring.

The ring resonator can be fed using only one feed line; this configuration is used in dielectric constant, Q-measurements, and ring-stabilized oscillations. As shown in Figure 5.29, for the first mode, the maximum field occurs at the coupling gap; however, a minimum occurs at the opposite side 180° from the coupling gap. Thus, when using a single feed, the ring behaves as a half-wavelength resonator. Resonance occurs when the ring circumference equals half of the guide wavelength:

$$2\pi r = n\frac{\lambda_g}{2}, \text{ for } n = 1, 2, 3, \ldots \tag{5.140}$$

$$f = \frac{nc}{4\pi r\sqrt{\epsilon_{\text{eff}}}} \tag{5.141}$$

5.5.3.5 Annular Ring Resonator Model

Figure 5.30 shows the 2-port lumped electrical equivalent model of the ring resonator, which can be reduced to a 1-port circuit by terminating one of the ports with an arbitrary impedance that corresponds with the feed impedance (usually 50 Ω) [45].

The electrical model exhibits symmetry of the circuit; therefore, input impedance can be calculated by simplifying parallel and series combinations of the equivalent electrical model, as shown in Figure 5.30.

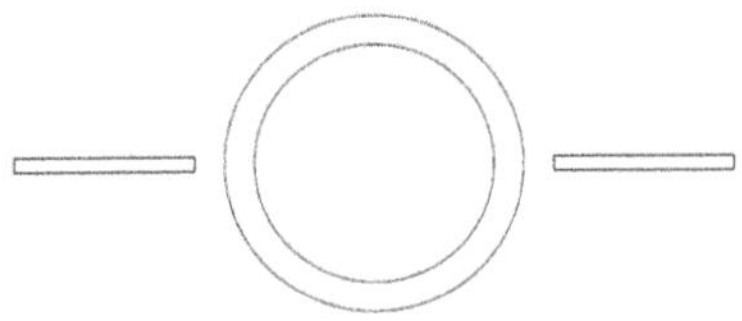

Figure 5.28 A typical ring resonator consists of feed lines, coupling gaps, and the circular closed-loop printed transmission line resonator. *Source*: Adapted from [44].

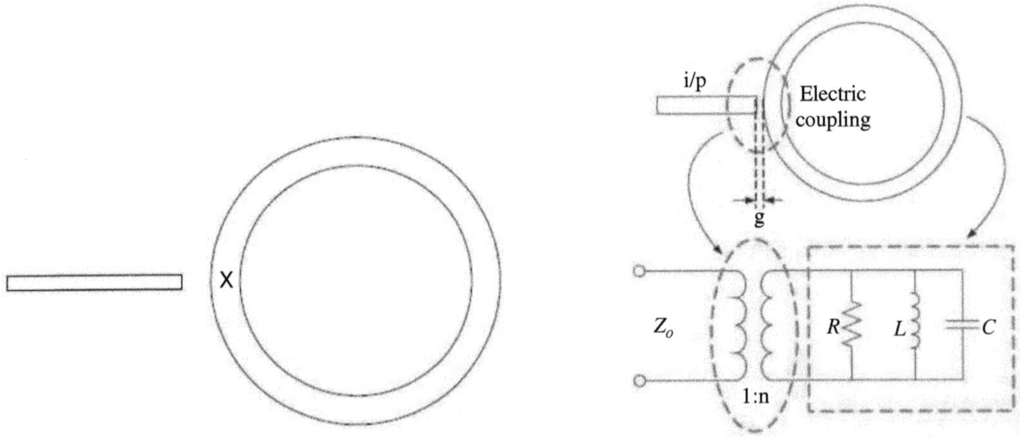

Figure 5.29 A typical ring resonator consists of single feed lines, coupling gaps, and the circular closed-loop printed transmission line resonator. *Source*: Adapted from [44].

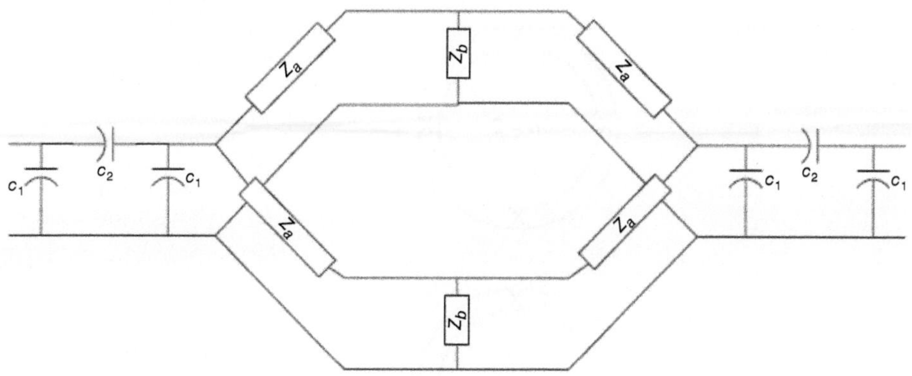

Figure 5.30 Simplified 2-port lumped electrical equivalent model of the ring resonator. *Source:* [43]/John Wiley & Sons.

From Figure 5.30, the input impedance is expressed as [45]:

$$R_{\text{in}} = \frac{C(C_1 + C_2)[(C_1 + C_2) - \omega D(C_1^2 + 2C_1C_2)]}{[(C_1 + C_2) - \omega D(C_1^2 + 2C_1C_2)]^2 + [\omega C(C_1^2 + 2C_1C_2)]^2} + \frac{[D(C_1 + C_2) - \omega^{-1}][\omega C(C_1^2 + 2C_1C_2)]}{[(C_1 + C_2) - \omega D(C_1^2 + 2C_1C_2)]^2 + [\omega C(C_1^2 + 2C_1C_2)]^2} \tag{5.142a}$$

$$X_{\text{in}} = \frac{[D(C_1 + C_2) - \omega^{-1}][(C_1 + C_2) - \omega D(C_1^2 + 2C_1C_2)]}{[(C_1 + C_2) - \omega D(C_1^2 + 2C_1C_2)]^2 + [\omega C(C_1^2 + 2C_1C_2)]^2} + \frac{[D(C_1 + C_2) - \omega C^2(C_1 + C_2)(C_1{}^2 + 2C_1C_2)]}{[(C_1 + C_2) - \omega D(C_1^2 + 2C_1C_2)]^2 + [\omega C(C_1^2 + 2C_1C_2)]^2} \tag{5.142b}$$

$$C = \frac{AZ_b^2}{(2A)^2 + (Z_a - 2B - Z_b)^2} \tag{5.143}$$

$$D = \frac{1}{2}\left[(Z_a - Z_b) - \frac{Z_b^2(Z_a - 2B - Z_b)}{(2A)^2 + (Z_a - 2B - Z_b)^2}\right] \tag{5.144}$$

$$A = \frac{RC_2^2}{(C_1 + C_2)^2 + [\omega R(C_1{}^2 + 2C_1C_2)]^2} \tag{5.145}$$

$$B = \frac{(C_1 + C_2) + \omega^2R^2(C_1{}^2 + 2C_1C_2)(C_1 + C_2)}{\omega(C_1 + C_2)^2 + \omega[\omega R(C_1{}^2 + 2C_1C_2)]^2} \tag{5.146}$$

where R is the terminated load, and the input impedance is ($Z_{\text{in}} = R_{\text{in}} + jX_{\text{in}}$), resonance occurs when $X_{\text{in}} = 0$.

5.5.3.6 Ring Resonator Modes

The ring resonator exhibits different modes depending upon excitation and the perturbation, broadly categorized into regular and forced modes.

a. **Regular Resonant Modes:**

The regular resonant mode is realized by applying symmetric input and output feed lines on the annular ring resonator element, where resonant wavelengths of the regular mode are determined by $2\pi r = n\lambda g$ [44].

For simplicity, the annular ring is analyzed as two half-wavelength linear resonators connected in parallel, assuming the parallel connection suppresses radiation from open ends, resulting a higher Q-factor compared to straight open transmission line resonator shown in Figure 5.21a. The resonant condition is enforced when standing waves are setup in the annular ring when the circumference is an integer multiple of guided wavelength.

As shown in Figure 5.31, in the absence of gaps or other discontinuities, maximum field occurs at the position where the feed line excites the resonator. The number of maximum field points increases with the mode order as illustrated in Figure 5.31.

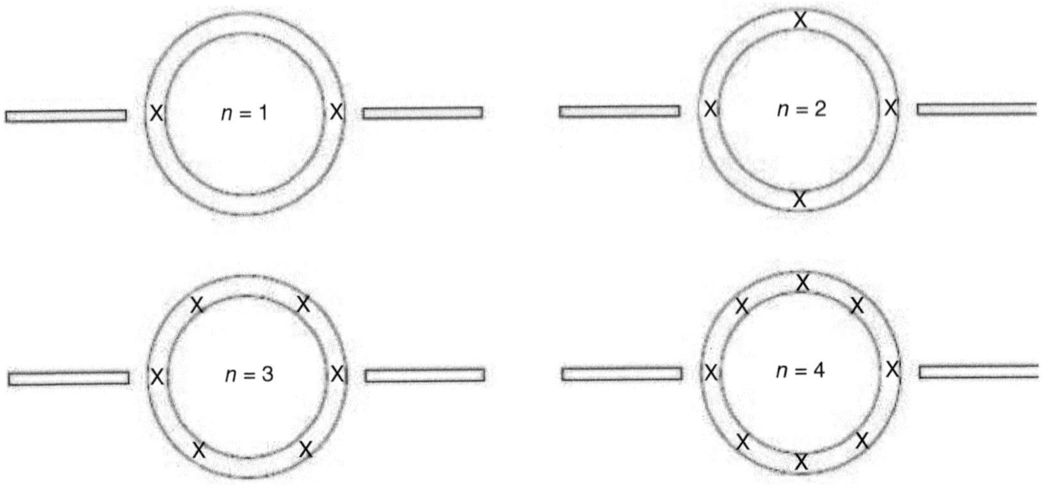

Figure 5.31 Simplified representation of maximum field points on annular ring for different resonant modes. *Source*: Adapted from [44].

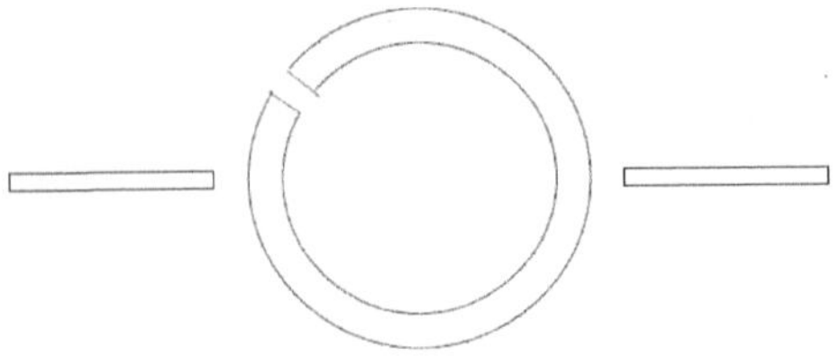

Figure 5.32 Shows a typical open annular ring with slit for satisfying open boundary condition. *Source*: Adapted from [45].

b. Forced Resonant Modes:
Usually, forced resonant modes are excited by forced boundary conditions on a microstrip annular ring element (ring can be open or shorted to ground). In principle, the boundary condition can either be open, as shown in Figure 5.32, or short depending upon the geometry of annular ring. For example, the open boundary condition is realized by cutting slits on the annular ring element, whereas the shorted boundary condition is realized by inserting vias to ground inside substrate, which forces minima of electric field to occur on both sides of the shorted plane. After analyzing the boundary conditions, the standing wave pattern and maximum field points inside the ring can be determined for the evaluation of the forced resonant modes.

5.6 Printed Coupled Slow-wave Resonator

A printed coupled resonator network realized by slow-wave dynamics is attractive due to its compact size, providing a wide spurious-free band [61–97]. In addition to this, the physical layout of slow-wave resonators (SWRs) enables the implementation of optimum EM (electromagnetic) couplings [62]. Printed resonator-based tunable oscillator circuits are large, especially at lower operating frequencies. The physical dimension of the conventional printed transmission line resonator can be reduced by incorporating slow-wave propagation characteristics, thereby reducing the size of the circuits [63]. Typically, periodic shunt loading of the transmission lines can exhibit a simplified slow-wave structure. This periodic shunt loading reduces the phase velocity, thereby increasing the effective electric length of the line [64]. As a result, resonators are not only compact in size but also exhibit high-frequency selectivity and wider stop-band due to the slow-wave effect and dispersion phenomena. In order to facilitate a broad yet precise description of the VCO topology, the layout of the SWR structure is selected in such a way that it minimizes the effect caused by temperature and mechanical stresses, and supports uniform negative resistance over a wide tuning range [68–75].

In slow-wave propagation, the electromagnetic wave is transmitted in the guided-wave media with a slower phase velocity, namely, shorter guided wavelength, at a specified operating frequency. This is achieved by modifying electric and magnetic energy storage in the guided-wave media. SWRs are attractive due to several reasons: the slow-wave effect makes them very compact and can support evanescent mode coupling, enabling Q-multiplier effect at resonant frequency, thus providing a low phase noise signal source solution [68, 69]. Additionally, the geometrical configuration of SWR resonators makes possible the implementation of compact layout and is amenable for RFIC/MMIC realizations [61–70]. In conventional transmission lines, the phase velocity v_p is controlled only by the dielectric material and can be expressed as [63]

$$v_P = f \times \lambda = \frac{c_0}{\sqrt{\mu_r \varepsilon_{\text{reff}}}} \tag{5.147}$$

where c_0 is the velocity of light, μ_r is the effective relative *permeability*, and $\varepsilon_{\text{reff}}$ is the effective relative permittivity. From (5.147), increasing the effective relative permittivity of a dielectric material at a given operating frequency decelerates the propagation of electromagnetic (EM) waves in a guided medium. The Slow-Wave Effect can be realized by adding periodical shields that decelerate the propagation of electromagnetic (EM) waves in a guided medium. From (5.147), an equivalent inductance $L_{\text{eq},}$ equivalent capacitance C_{eq}, and propagation constant β can be described as [62]:

$$\beta = \omega\sqrt{L_{\text{eq}}C_{\text{eq}}} \Rightarrow V_P = \frac{1}{\sqrt{L_{\text{eq}}C_{\text{eq}}}} \tag{5.148}$$

From (5.148), both distributed inductance (L) and capacitance (C) along the transmission line in the guided medium can be increased for the realization of slow-wave effects, owing to an increase in the propagation constant β, consequently, reducing the loss of the resonator network [2]. Figure 6.1 shows the simplified topology of a single-coupled SWR, which consists of a SWR loaded at its near and far ends with series capacitors C_S.

The resonator itself is made up of a transmission line of characteristic impedance Z_0 and electrical length "θ," periodically loaded by shunt capacitors C_p. The electrical length θ of the unloaded transmission line is defined at frequency f as [61, 64]

$$\theta = \beta d = \frac{2\pi f\sqrt{\varepsilon_{\text{reff}}}}{C} d \Rightarrow \theta|_{f=f_0} = \theta_0 \tag{5.149}$$

where $\varepsilon_{\text{reff}}$ is the effective relative permittivity, d is the physical length of the unloaded line, and C is the free space light velocity. At the center frequency f_0, the electrical length is θ_0.

By using ABCD (cascade) matrices, the equivalent characteristic impedance Z_{eq} of a coupled SWR can be easily extracted.

As shown in Figure 5.33, the real estate of SWR layout is quite large, therefore not suitable for the application where size is a constraint. For brief insights about the minimization of the size of slow-wave structure, the characteristics of the Lossless transmission and capacitive-loaded transmission line (CTL) are discussed.

5.6.1 Lossless Transmission Line

Figure 5.34 shows the typical lossless transmission line circuit, which can be characterized by Z_0 (characteristic impedance) and V_{p}(phase velocity) as

$$Z_0 = \sqrt{\frac{L}{C}} \tag{5.150}$$

$$v_p = \frac{1}{\sqrt{\mu\varepsilon_e}} = \frac{1}{\sqrt{LC}} \Rightarrow LC = \mu\varepsilon_e = \text{k (constant depending on medium)} \tag{5.151}$$

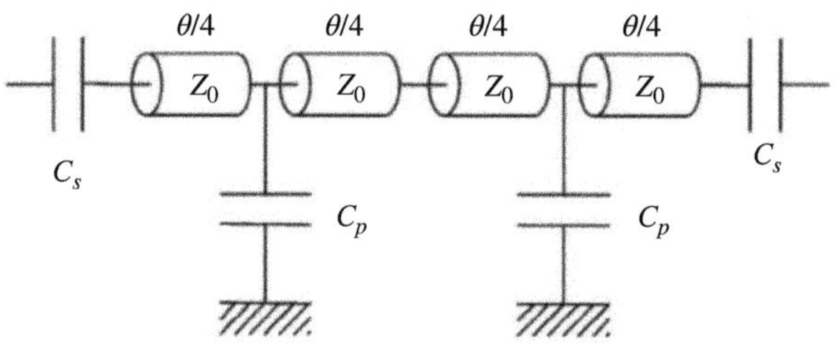

Figure 5.33 A typical topology of a single-coupled SWR loaded at the near and far ends with series capacitors C_s. *Source:* [61]/IEEE.

where

$$L = \frac{Z_O}{v_p},\ C = \frac{1}{v_p Z_O} \quad (5.152)$$

From (5.151), $LC = \mu\varepsilon_e$; therefore, for a given ε_e and μ, it is not possible to reduce V_p by increasing inductance or capacitance per unit length, because an increase in inductance L leads to a decrease in capacitance C, ($L \downarrow\uparrow \equiv C \uparrow\downarrow$). Hence, for a physically smooth transmission line, reduction in phase velocity v_p is only possible by increasing ε_r.

5.6.2 Capacitive-loaded Transmission Lines

By removing the constraint that the line should be physically smooth, an effective increase in the shunt capacitance per unit length C can be obtained without a reduction in inductance L. This can be realized by loading a printed transmission line with shunt capacitance C_p at periodic intervals d. Figure 5.35 shows the typical CTL line circuit formed by loading a printed microstrip line with open stubs that exhibit shunt capacitance at periodic intervals, which are shorter than the guide wavelength and cause the periodic structure to exhibit slow-wave characteristics [62–68].

The effective characteristic impedance and phase velocity of the CTL circuit shown in Figure 5.35 can be described by [62–64]

$$Z_{0\text{CTL}} = \left(\frac{L}{C + \frac{C_p}{d}}\right)^{1/2} \quad (5.153)$$

$$Z_{\text{in}} = Z_0 \frac{Z_L + jZ_0 \tan(\beta l)}{Z_L + jZ_L \tan(\beta l)} \Rightarrow Z_\infty \text{ (open circuit stub)} = \frac{Z_{\text{stub}}}{j\tan[\beta l]} = \frac{1}{j\omega C_p} \quad (5.154)$$

$$v_{p\text{CTL}} = \frac{1}{\sqrt{L\left(C + \frac{C_p}{d}\right)}} = \left[\frac{Z_0}{v_p}\left(\frac{1}{v_p Z_0} + \frac{C_p}{d}\right)\right]^{-1/2} \quad (5.155)$$

where $\frac{C_p}{d}$ = lumped capacitance per unit length, $Z_0 = \sqrt{\frac{L}{C}}$

For an m-section CTL, the equivalent electrical length in angle ($\varnothing_{\text{CTL}}$) and the loaded capacitance of a unit cell C_p can be given by

$$\varnothing_{\text{CTL}} = \frac{2\pi m d f_0}{v_{pCTL}} = m d\omega_0 \sqrt{L\left(C + \frac{C_p}{d}\right)} \text{ (rad)} \quad (5.156a)$$

$$C_p = \frac{\varnothing_{CTL}(Z_0^2 - Z_{CTL}^2)}{2\pi m f Z_0^2 Z_{CTL}} \text{ (F)} \quad (5.156b)$$

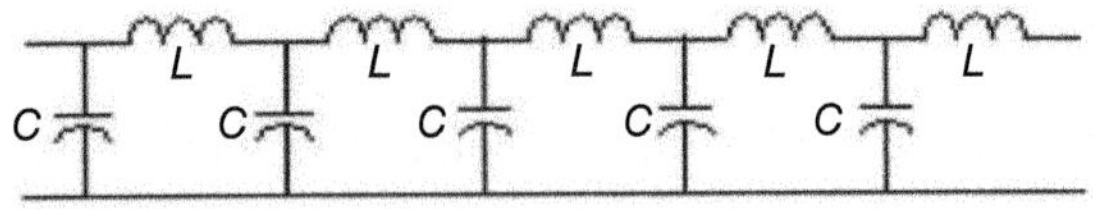

Figure 5.34 A typical lossless transmission line circuit. *Source:* [60]/Ajay K. Poddar.

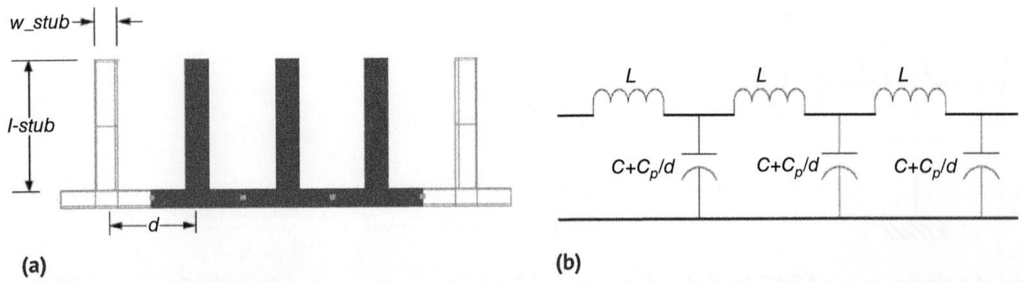

Figure 5.35 A typical capacitive-loaded transmission line circuit: (a) printed layout, and (b) lumped equivalent network. *Source:* [60]/Ajay K. Poddar.

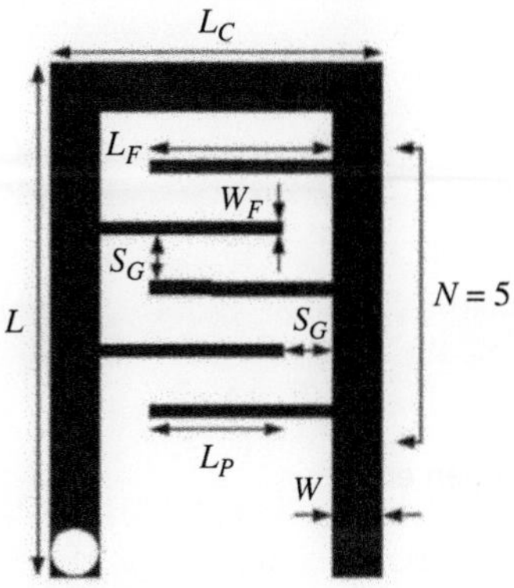

Figure 5.36 A typical slow-wave coupled resonator (no. of section, $m = 5$). *Source*: [62]/Asia-Pacific Microwave Conference.

where f_0 is the operating frequency.

From (5.154) to (5.156), the phase velocity v_{pCTL} can be slowed down either by one or a combination of the following: (i) Increase the characteristic impedance of the unloaded unit cell Z_0, achieved by reducing the microstrip line width w_{TL}, (ii) Reduce the distance between stubs d, and (iii) Increase the load capacitance C_p (increasing the stub electrical length $\to \frac{\pi}{2}$, reducing the stub characteristic impedance $\downarrow Z_{\text{stub}}$. This is achieved by increasing the width of the stub $\uparrow w_{\text{stub}}$).

Figure 5.36 shows the typical layout of the compact capacitive-coupled printed SWR. As shown in Figure 5.36, capacitive coupling between two arms of the resonator leads to the slow-wave dynamics.

The drawback of a periodically loaded line is dispersion in phase velocity; therefore, phase velocity of the capacitive-loaded transmission line is frequency dependent, whereas the unloaded transmission line does not exhibit dispersion in the phase velocity [69–84].

5.7 Slow-wave Resonator

SWR using stepped impedance hairpin printed transmission line structure can exhibit larger group delay in a compact size, resulting in improved Q-factor [85–97]. The hairpin resonator was introduced initially to reduce the size of the conventional parallel-coupled half-wavelength resonator, with subsequent improvements made to reduce its size [98].

Beyond the advantage of the compact size, the spurious frequencies of the stepped impedance hairpin resonator are shifted from the integer multiples of the fundamental resonant frequency due to the effect of the capacitance-load coupled lines, including the low-quality factor and high insertion loss [99].

As illustrated in Figure 5.37, the transmission line is periodically loaded with identical open stub elements for the realization of high Q-factor SWR structure. The input impedance Z_{in2} for a lossless line is given by [100]

$$Z_{\text{in2}}\big|_{\tan(\beta l)\to 0} = \frac{Z_0 Z_L}{Z_0} \Rightarrow Z_{\text{in2}} \propto Z_L \tag{5.157}$$

where L and C are the inductance and capacitance per unit length of the line, Z_0 and β are the characteristic impedance and phase constant of the open stub, respectively.

From (6.11), for infinitesimal value of $\tan(\beta l)$, the input impedance $Z_{\text{in2}}$is proportional to Z_L, therefore, $Z_{\text{in2}} \to \infty$ or 0 for corresponding $Z_L = \infty$ or 0. Under these cases, the slow-wave periodic structure loaded by Z_{in2} in Figure 6.5b provides the passband ($Z_{\text{in2}} \to \infty$) and stopband ($Z_{\text{in2}} \to 0$) characteristics.

As depicted in Figure 5.37, the layout size is comparatively large and not attractive for VCO applications used in small modern handheld portable and test equipment systems.

Figure 5.38 shows the modified version of SWR, where the transmission line is loaded by a square ring resonator with a line-to-ring coupling structure [98–100].

As shown in Figure 5.38b, Z_{in3} is the input impedance looking into the transmission line l_b toward the ring resonator with the line-to-ring coupling. The input impedance Z_{r1} looks into the line-to-ring coupling structure toward the ring resonator. The input impedance Z_{in3} is given by

$$Z_{\text{in3}} = Z_0 \frac{Z_{r1} + jZ_0 \tan(\beta l_b)}{Z_0 + jZ_{r1} \tan(\beta l_b)} \tag{5.158}$$

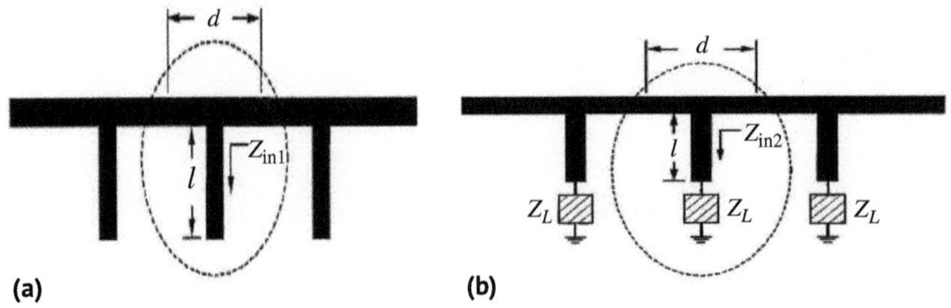

Figure 5.37 A typical slow-wave structure: (a) a periodic SW structure and (b) with loading Z_L at the open end. *Source*: Adapted from [100].

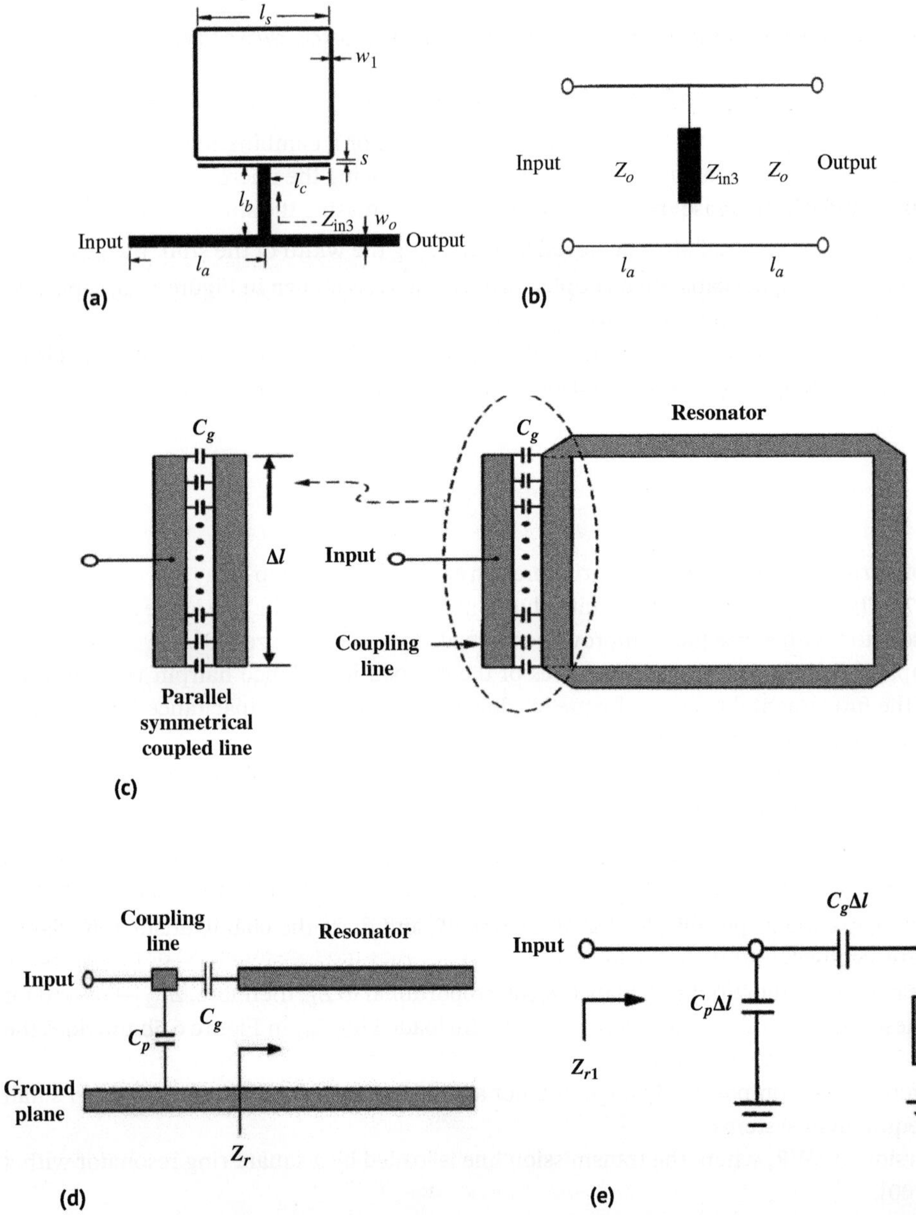

Figure 5.38 A typical slow-wave ring resonator with single edge coupling gap (slower the phase velocity, causing slow-wave effect): (a) printed layout of single edge coupled slow-wave ring resonator network, (b) equivalent representation of single edge coupled slow-wave ring resonator circuit, (c) top view of line-to-ring coupling structure, (d) side view of line-to-ring coupling structure, and (e) equivalent representation of circuit of line-to-ring coupling structure. *Source*: Adapted from [100].

where

$$Z_{r1} = \frac{(Z_r + Z_g)Z_p}{(Z_r + Z_g + Z_p)}, \quad Z_g = \frac{1}{j\omega C_g \Delta l}, \quad Z_p = \frac{1}{j\omega C_p \Delta l}, \quad \omega = 2\pi f \tag{5.159}$$

The parallel f_p and series f_s resonances of the slow-wave ring resonator, as shown in Figure 5.38 can be obtained by setting $|Y_{in3}| = 0$ and $|Z_{in3}| = 0$. The ABCD matrix of the ring circuit (Figure 6.6a) can be described by [100]

$$\begin{bmatrix} A & B \\ C & D \end{bmatrix} = \begin{bmatrix} \cos(\beta l_a) & jZ_o \sin(\beta l_a) \\ jY_o \sin(\beta l_a) & \cos(\beta l_a) \end{bmatrix} \begin{bmatrix} 1 & 0 \\ Y_{in3} & 1 \end{bmatrix} \begin{bmatrix} \cos(\beta l_a) & jZ_o \sin(\beta l_a) \\ jY_o \sin(\beta l_a) & \cos(\beta l_a) \end{bmatrix} =$$

$$\begin{bmatrix} 1 - 2\sin^2(\beta l_a) + jZ_o Y_{in3} \sin(\beta l_a)\cos(\beta l_a) & -Z_o^2 Y_{in3} \sin^2(\beta l_a) + j2Z_o \sin(\beta l_a)\cos(\beta l_a) \\ Y_{in3}\cos^2(\beta l_a) + j2Y_o \sin(\beta l_a)\cos(\beta l_a) & 1 - 2\sin^2(\beta l_a) + jZ_o Y_{in3} \sin(\beta l_a)\cos(\beta l_a) \end{bmatrix} \tag{5.160}$$

where $Y_o = 1/Z_o$. Using $Y_{in3}(f_p)$ and $Z_{in3}(f_s)$, the passband and stopband of the ring circuit can be obtained by calculating S_{11} and S_{21} from the ABCD matrix in (5.160).

5.7.1 Slow-wave Evanescent Mode (SWEM) Propagation

The slow-wave evanescent-mode can represent surface wave propagation in planar printed resonators or the lattice waves in waveguides. In slow-wave propagation, the electromagnetic wave is transmitted in the guided-wave media with a slower phase velocity, namely, a shorter guided wavelength, at a specified operating frequency [61–68]. This is achieved by modifying electric and magnetic storage energy by incorporating perturbation in the guided-wave media. SWRs are attractive for several reasons: the slow-wave effect makes them very compact and can support evanescent mode coupling for obtaining Q-multiplier effect at resonant condition, thus low phase noise signal source solutions. The wave propagation in SWEM structure can be described well using Maxwell's equations by partitioning into (i) transmission line (TL) equations in the direction of travel, and (ii) orthogonal transverse evanescent (TE_V) wave equations [64–66].

For the characterization of evanescent-mode propagation, we need four pairs of separate, partially coupled TL and TEυ wave equations for the four fields E, D, B, and H. The four slow-wave evanescent-mode equations for E, D, B, and H are all radial profiles of stored energy. The seven main processes in the formation of evanescent waves are: (i) the spreading function, (ii) the self-coupling function, (iii) (radial) standing wave function, (iv) time variation of phase of stored energy components, (v) root sum of the squares (RSS) combination of the four coupled equations giving dominance to the strongest field in the resonator, (vi) dissipation, radiation or absorption of energy in the resonator, (vii) exchange of energy between inter-coupled evanescent modes in SWEM resonator [67–73].

5.7.2 SWEM Resonator Modes and Noise Dynamics

In contrast to lumped LC resonators, which have only one resonant frequency, SWRs can support an infinite number of electromagnetic field configurations or resonant modes [71–76]. The resonant modes in SWR structure exist in the form of degenerate modes (i.e., modes having the same resonance frequency with different field distributions). These modes allow the realization of two electric resonators within the same physical resonator (dual-mode resonators) or three electric resonators within the same physical resonator (triple-mode resonators), and n-electric resonators within the same physical resonator (nth-mode resonators).

The single–mode resonator possesses two degrees of freedom, namely, the electric and magnetic fields (voltage and current standing waves), each storing a mean thermal energy of $kT/2$ (k: Boltzmann's constant; T: temperature), resulting in improved unloaded quality factor. The multi-mode resonator exhibits 2nd-degrees of freedom, causing neighboring resonant modes, which act as spurious modes interfering with the fundamental resonant mode's performance and exhibiting higher noise than a single-mode resonator oscillator. It is therefore desirable to increase the spurious-free window of the multi-mode resonator oscillator in order to improve the stability, phase noise, and mode-jumping problems. An experimental validation supports the convergence of degenerated modes in SWR structure, resulting in improved dynamic loaded Q-factor. It is interesting to note that slow-wave propagation in SWEM structure mode-locking dynamics under large-signal drive-level condition, which is opposite to the analysis, because of the fact that slow-wave evanescent modes are inter-coupled in phase for a given topology [77–86].

The inconsistency of multi-mode oscillator phase noise dynamics calls for a revisit to the phase noise analysis based on physics-based noise modeling in SWEM oscillator circuits. This can be physically understood in time domain: noise perturbation generated at any point $p(x, y, z)$ in SWEM resonator affects the oscillator's phase and timing dynamics only when the respective modes pass through the point $p(x, y, z)$. This notion makes sense as explained above, i.e., noise at any given point $p(x, y, z)$ in SWEM resonator structure has less chance to be involved in the phase-noise dynamics for an inter-coupled mode-locked transmission line, leading to lower phase noise than the single-mode resonator oscillator [6]. The Fourier-domain phase noise argument would predict wrong result, i.e., higher phase noise for SWEM oscillator that has a large number of harmonic modes than the single-mode oscillator circuits. Thus, the general phase noise theory is not applicable to the inter-coupled mode-locked resonator oscillator circuits.

It is to note that SWEM design parameters (coupling coefficient, multi-mode sensitivity, and Q-factor) can be obtained using full-wave EM simulation. It has been found that degenerate modes play an important role in improving the Q-factor; therefore, rigorous mathematical treatment is necessary for the minimization of the spurious and jitters in passband. The theoretical treatment and experimental validation suggest the possibility of low-cost, high-performance synthesizer using SWEM VCO for low jitter and low phase noise applications.

5.7.3 SWR Oscillator

Figure 5.39a–c shows the typical block diagram and layout of the VCO using mode-coupled SWR, which validates a novel SWR approach, using a SiGe Hetero-junction-bipolar-transistor (HBT) active device fabricated on low-loss RF dielectric substrate material with a dielectric constant of 3.38 and a thickness of 22 mils printed structure [84]. The SWR-inspired oscillator circuits are discussed in Chapter 6, which shows the variation of signal generation circuits based on SWR resonator tank.

The SWR structure is modeled using 3D EM (electromagnetic) CAD simulator and incorporated into optimized nonlinear oscillator circuit to enable configurable and low phase noise operation over the band. This enables SWR structure to set up optimum standing waves (within the resonator) and the noise impedance transfer function over the tuning range by controlling m_{opt} (by optimizing injection locking) and β_{opt} (by optimizing mode tuning) [77–84].

The nonlinear circuit contains the oscillator's active device, with S-parameters. This partitioning of the oscillator into its modeled component parts works quite well, and the combination of the S-parameters and the nonlinear circuit model agrees closely with the measured data in the circuits we built [60]. The S-parameters used for these transistors are large-signal S-parameters, which improve the optimization cycles using commercial CAD tools such as ADS (Agilent), AWR, CST, and Ansys [86–91].

5.7.3.1 SWR Coupling Characteristic and Q-factor

The Q (quality) factor of the printed transmission line resonator can be enhanced by introducing coupling mechanism related to the relative orientation of the neighboring resonators (electric/magnetic/hybrid). Figure 6.8 illustrates the layout of the typical electric, magnetic, and hybrid-coupling planar resonator networks and oscillator circuits for comparative analysis [92–97].

As described in Figure 5.40, the coupling dynamics can be characterized by proximity effect through the fringing fields, which exponentially decay outside the region; electric and magnetic field intensity tends to concentrate near the side having the maximum field distribution. The coupling coefficient "β_j" depends upon the geometry of the perturbation, given by

$$\beta_j = \left[\left(\frac{\int \varepsilon E_a . E_b dv}{\sqrt{\int \varepsilon E_a^2 dv \int \varepsilon E_{bd}^2 dv}} \right)_{\text{electrical-coupling}} + \left(\frac{\int \mu H_a . H_b dv}{\sqrt{\int \mu H_a^2 dv \int \mu H_b^2 dv}} \right)_{\text{magnetic-coupling}} \right] \tag{5.161}$$

where E_a and H_a are, respectively, the electric and magnetic fields produced by the square loop ring resonator, and E_b, H_b are the corresponding *fields* due to the perturbation ($d \neq 0$) or nearby adjacent resonator (second square loop resonator).

From (5.161), the first term represents the coupling due to the interaction between the electric fields of the resonators, and the second term represents the magnetic coupling between the resonators. Depending on the strength of interaction, multi-mode dynamics exist related to electrical, magnetic, and hybrid coupling. The configuration of Figure 5.40a produces an electric coupling since the electric field is maximum near the open ends, maximizing the numerator of the first term of Eq. (5.161) [60].

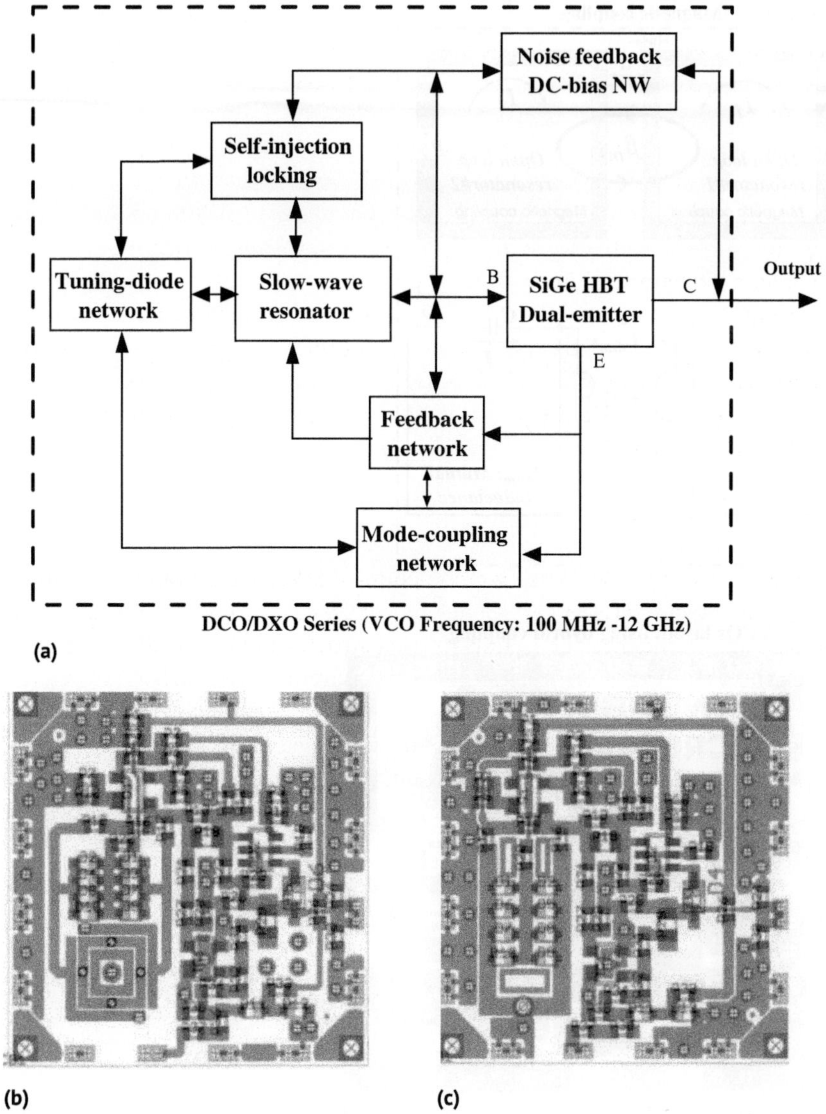

Figure 5.39 (a) A typical Block diagram of SWR VCO, (b) layout of SWR VCO, and (c) layout of mode-coupled SWR VCO. *Source*: [60]/Ajay K. Poddar.

As depicted in Figure 5.40c, when the resonators are operating near their first resonant frequency, the pair of resonators interacts mainly through their magnetic field, this is because the magnetic fields is maximum near the center of the resonator opposite to its open ends, maximizing the numerator of the second term of Eq. (5.161). The coupling produced by the two configuration (open-loop resonator # 1 and open-loop resonator #2), as shown in Figure 5.40f, is referred as mixed coupling or hybrid coupling because neither the electric fields nor the magnetic fields dominate the interaction between the resonators.

The definition of "β_j" given in (5.161) involves complex mathematical analysis and is not suited for practical calculation since it requires the knowledge of the electromagnetic fields everywhere. A useful alternative expression for "β_j" can be obtained from a well-known fact in physics: when multiple resonators are coupled to each other, they resonate together at different, distinct frequencies (f_{ee}, f_{em}, f_{eh}, f_{mh}) which are in general different from their original resonant frequency f_0.

Furthermore, these frequencies are associated with corresponding to their normal modes of oscillation of the coupled system (electric/magnetic/hybrid), and their difference increases as the coupling "β_j" (β_e: electric, β_m: magnetic and β_h: hybrid) between the resonators increases [91–97].

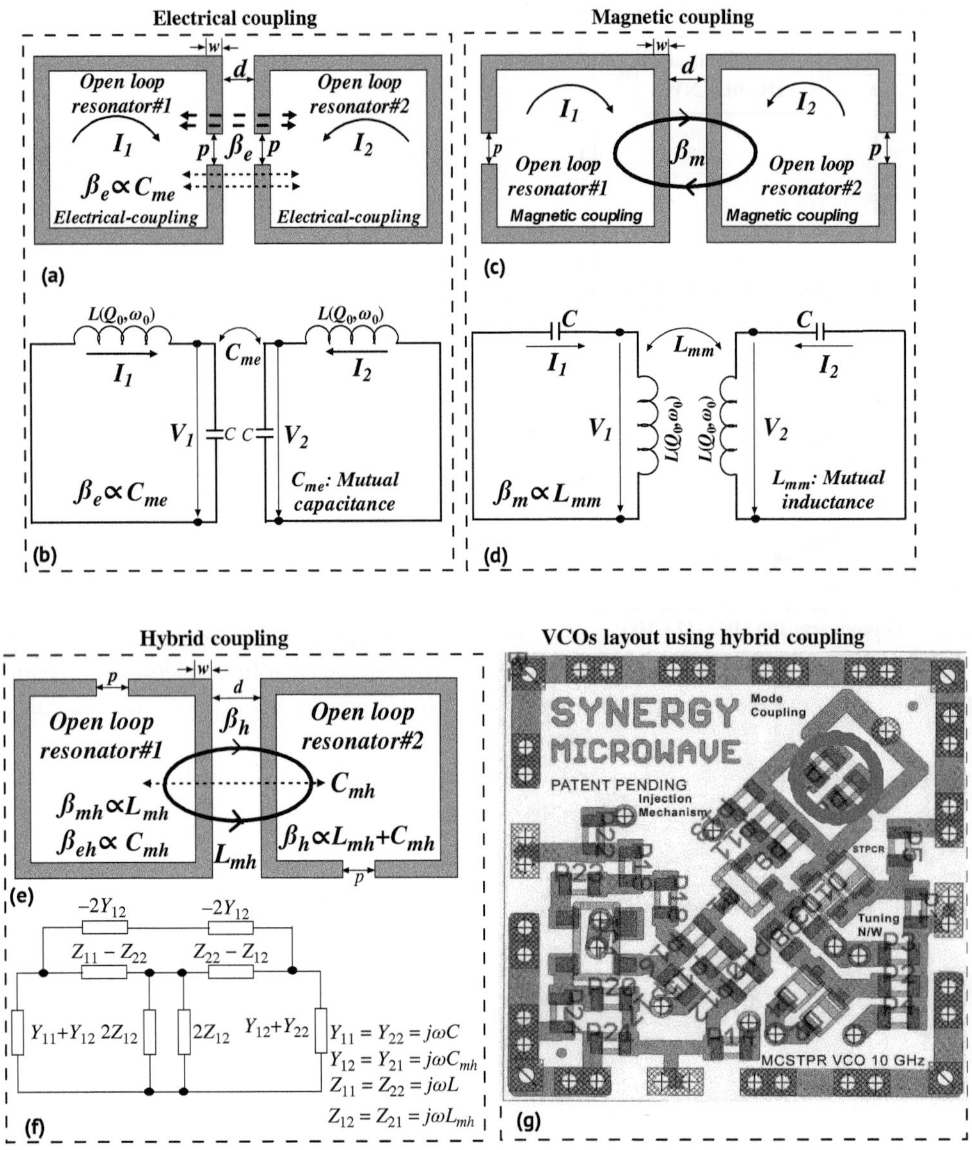

Figure 5.40 A typical simplified structure of open-loop microstrip line coupled resonator networks: (a) electrical coupling, (b) equivalent lumped model of electrical coupling, (c) magnetic coupling, (d) equivalent lumped model of magnetic coupling, (e) hybrid coupling, (f) equivalent lumped model of hybrid coupling, and (g) layout of VCO using electric and magnetic coupling. *Source*: [60]/Ajay K. Poddar.

The main interaction mechanism between resonators is proximity coupling and can be characterized by a coupling coefficient "β_j" that depends upon the ratio of coupled energy to stored energy, described by

$$\beta_e \cong \frac{\text{coupled} - \text{electrical} \quad \text{energy}}{\text{stored} - \text{energy of uncoupled} - \text{resonator}} \cong \frac{f_{me}^2 - f_{ee}^2}{f_{me}^2 + f_{ee}^2} \cong \frac{C_{me}}{C} \tag{5.162}$$

$$\beta_m \cong \frac{\text{coupled} - \text{magnetic} \quad \text{energy}}{\text{stored} - \text{energy of uncoupled} - \text{resonator}} \cong \frac{f_{em}^2 - f_{mm}^2}{f_{em}^2 + f_{mm}^2} \cong \frac{L_{mm}}{L} \tag{5.163}$$

$$\beta_h \cong \frac{\text{coupled} - \text{electro} - \text{magnetic} \quad \text{energy}}{\text{stored} - \text{energy of uncoupled} - \text{resonator}} \cong \frac{f_{eh}^2 - f_{mh}^2}{f_{eh}^2 + f_{mh}^2} \cong \frac{CL_{mh} + LC_{mh}}{LC + L_{mh}C_{mh}} \tag{5.164}$$

where

$$f_{ee} = \frac{1}{2\pi\sqrt{L(C + C_{me})}},\ f_{me} = \frac{1}{2\pi\sqrt{L(C - C_{me})}},\ C_{\text{me}}:\ \text{Mutual Capacitance} \tag{5.165}$$

$$f_{em} = \frac{1}{2\pi\sqrt{C(L - L_{mm})}},\ f_{mm} = \frac{1}{2\pi\sqrt{C(L + L_{mm})}},\ L_{\text{mm}}:\ \text{Mutual Inductance} \tag{5.166}$$

$$f_{eh} = \frac{1}{2\pi\sqrt{(L - L_{mh})(C - C_{mh})}},\ f_{mh} = \frac{1}{2\pi\sqrt{(L + L_{mh})(C + C_{mh})}},\ L_{\text{mh}}:\ \text{Hybrid Inductance} \tag{5.167}$$

$$f_0 = \frac{1}{2\pi\sqrt{LC}},\ f_0:\ \text{fundamental resonance frequency of uncoupled resonator} \tag{5.168}$$

The time average loaded Q-factor of SWR: $\overline{Q_{\text{swr}}(\omega)}_{\omega\to\omega_0}$ can be described by

$$\overline{Q_{\text{swr}}(\omega)}_{\omega\to\omega_0} = \left[\frac{\omega}{2(I_{\max} - I_{\min})}\int_{I_{\min}}^{I_{\max}} Q_{\text{swr}}(\omega.i)di\right]_{\omega\to\omega_0} \tag{5.169}$$

where $I_{\min}$ and $I_{\max}$ are the minimum and maximum resonator currents of the SWR network associated with the fundamental modes of the coupled resonator networks, the $Q_{\text{swr}}(\omega, i)$ is the instantaneous quality factor at frequency ω and current i provides an effective means to quantify the Q-multiplier effect of SWR when operated in an evanescent-mode coupling condition, especially in printed coupled resonator-based oscillator circuits.

From (5.169), the loaded quality factor Q_L of the coupled resonator network is given in terms of unloaded quality factor Q_o as

$$Q_L(\omega_0) = \frac{\omega_0}{2}\left[\frac{\partial\varphi}{\partial\omega}\right] \tag{5.170}$$

$$[Q_L(\omega_0)]_{\text{electrical-coupling}} \cong 2\left[\frac{Q_0}{(1 + \beta_e)}\right]_{\beta_e << 1} \cong 2Q_0 \tag{5.171}$$

$$[Q_L(\omega_0)]_{\text{magnetical-coupling}} \cong 2[Q_0(1 + \beta_m)]_{\beta_m \to 1} \cong 2Q_0 \tag{5.172}$$

$$[Q_L(\omega_0)]_{\text{hybrid-coupling}} \cong 2\left[Q_0\frac{(1 + \beta_{mh})}{(1 + \beta_{eh})}\right]_{\beta_e << 1,\ \beta_m \to 1} \cong 2Q_0 \tag{5.173}$$

where $\frac{\partial\phi}{\partial\omega}$ is the rate of change of the phase, and Q_0 is the unloaded Q-factor of the uncoupled single open-loop microstrip line resonator.

From (5.171) to (5.173), there is trade-off between improving the Q-factor and the permissible attenuation required (which is compensated by active device for oscillation build up).

The coupling mechanism described in Figure 5.40 shows improvement in quality factor in comparison to single uncoupled planar resonator, but the drawback is limited tuning range (less than 1%). By introducing tunable capacitor across the open end of uncoupled planar open-loop resonator, dynamic unloaded Q-factor can be improved, but with a limited tuning range (<25%). This is due to minimization of the radiation losses from the open ends of the resonator because of capacitor loading, causing dielectric polarization in the capacitor since most of the electric field resides inside it [98–102]. Therefore, a high Q capacitor could actually increase the unloaded quality factor of the whole resonator. This is analogous to the case of dielectric resonators, where the fields are constrained to a small volume dielectric with high permittivity and low loss tangent, resulting in a high overall Q [103–109]. For wideband tunability (>100% tuning), adjacent coupled open-loop resonator network is preferred, but at the cost of large real estate area.

In general, the miniaturization of the open-loop resonator reduces its capacity to couple to adjacent structures. This is because smaller resonator size represents a smaller volume of electromagnetic interactions between its coupled arms;

reason being that a smaller size represents a smaller volume of electromagnetic interaction between coupled resonators. The fact that the majority of the electric field that existed in the volume surrounding the open ends of a resonator is now confined to the interior of a capacitor, limiting its possibility to interact with a neighboring resonator. It can be seen from Figure 5.41 that the effect on the magnetic and mixed coupling is less severe than for the electric coupling, where the magnetic coupling coefficient is plotted against the separation between resonators for different loading capacitors [103–109].

The microstrip square open-loop resonator is one of the most used structures for multi-mode oscillator resonator applications due to its compact size ($a = \lambda/8$). For low-phase noise multi-octave band tunability, the loaded quality factor (Q_L) as described in (5.170)–(5.173) can be maximized by either lowering the value of mutual capacitance (C_m) and inductance (L_m) or maximizing the self-capacitance (C) and inductance (L). Therefore, the upper limit of the loaded Q-factor is dependent on the coupling "β_j" (β_e: electric, β_m: magnetic and β_h: hybrid) that can be optimized by controlling the width of the transmission line (w), gap of the open line resonator (p), and spacing between the two open line resonators (d) [110–115].

For a low-phase noise tunable oscillator, the coupling coefficient β_j should be dynamically tuned over the operating frequency band. However, dynamic controlling and tuning of the parameters (w, p, a, and d) as shown in Figure 5.40 at high frequency is a challenging task. The alternative tuning mechanism is capacitive loading by incorporating tuning diodes [98–104].

Figure 5.42 shows the typical tunable square open-loop resonator in a compact size ($\lambda/8$ by $\lambda/8$) for the applications in oscillator circuits.

As shown in Figure 5.42c, the goal is to minimize the real estate area by using meander line in the inner part of the resonator. To optimize the geometry of the coupled resonator, they are excited with a pair of loosely coupled feed lines to obtain a transmission parameter $S_{21}(\omega)$ from which the two resonant frequencies f_1 and f_2 can be obtained for a given geometry and values of d between resonators.

The resonator shown in Figure 5.42d offers a compact size and exhibits two independent modes (dual-modes), the coupling between them can be optimized by the geometry of the inner structure. This tunable dual-mode resonator can then function as two independent tunable resonators, providing an immediate size reduction of 50%.

For brief insights about the tuning capability of square open-loop resonator (Figure 5.43), a simple equivalent circuit model of the varactor diode-loaded resonator, as shown in Figure 5.43a–c is used to derive its multi-mode tuning dynamics and regime.

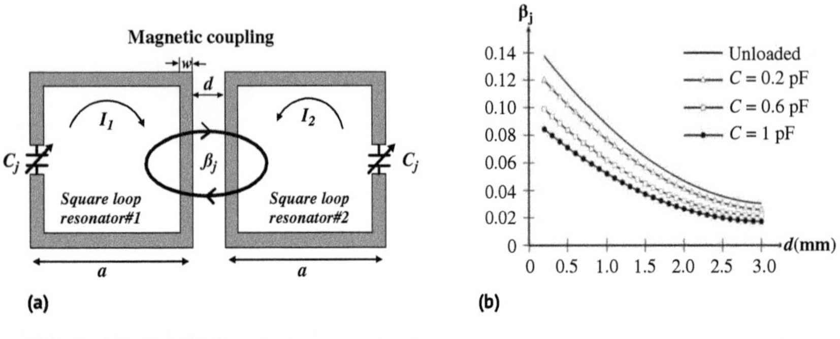

Figure 5.41 A typical capacitive-loaded magnetic-coupled square loop resonator characteristics: (a) printed layout, (b) plot of magnetic coupling "β_j" as a function of the distance between resonators for a given capacitive (C_j) loading with resonator's physical dimension $\omega = 2$ mm and $a = 26$ mm, fabricated using Roger RO4003c substrate with a dielectric constant of 3.55 and a thickness of 60 mil). *Source*: [103]/John Wiley & Sons.

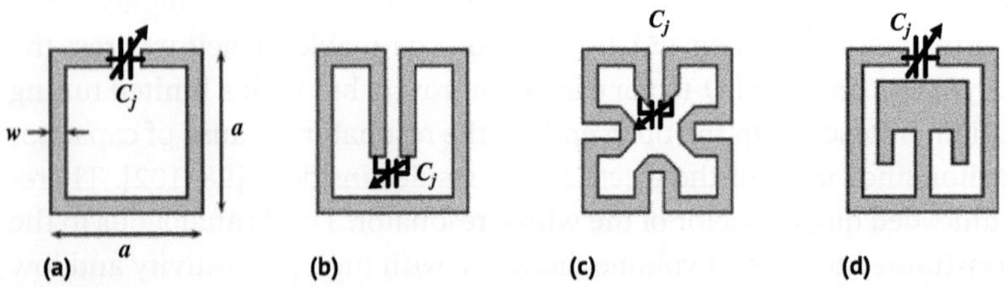

Figure 5.42 A typical layout of tunable square open-loop resonator: (a) conventional square open-loop hairpin resonator, (b) folded arms square open-loop resonator, (c) meander line square open-loop resonator, and (d) dual mode square open-loop resonator. *Source*: [43]/John Wiley & Sons.

The voltage and current distribution at resonance can be described by using transmission line theory and the shunt equivalent model of Figure 5.43c as:

$$V(\theta) = \left[\cos(\theta) - \frac{\sin(\theta)}{\tan(\theta_T/2)}\right] 0 < \theta < \theta_T \tag{5.174}$$

$$I(\theta) = \frac{j}{Z_0}\left[\frac{\cos(\theta)}{\tan(\theta_T/2)} + \sin(\theta)\right] 0 < \theta < \theta_T \tag{5.175}$$

where $\theta = \beta z$ is the electrical length measured from one open end of the resonator, and θ_T is the total electrical length of the resonator.

Figure 5.44 shows the plots of $V(\Theta)$ and $I(\Theta)$ for loaded and unloaded square open-loop resonator network.

As shown in Figure 5.44, the current never goes to zero but remains near its maximum value along the resonator and the voltage varies almost linearly between open ends (Assumptions: these typical plots are just to compare the distribution of the voltages and currents between the loaded and unloaded resonators, but not their amplitudes; they are normalized with respect to their respective maxima and they have either different frequencies or different resonator sizes) [116–144].

Using Miller's theorem, tunable square loop open resonator (SLOR) shown in Figure 5.43a can be equivalently represented as Figure 5.43b,c, if the following conditions are satisfied [60]:

$$Y_1(\omega) = Y(\omega)\left(1 - \frac{V_2(\omega)}{V_1(\omega)}\right) \rightarrow Y_1(\omega) = Y(\omega)[1 - A(\omega)] \tag{5.176}$$

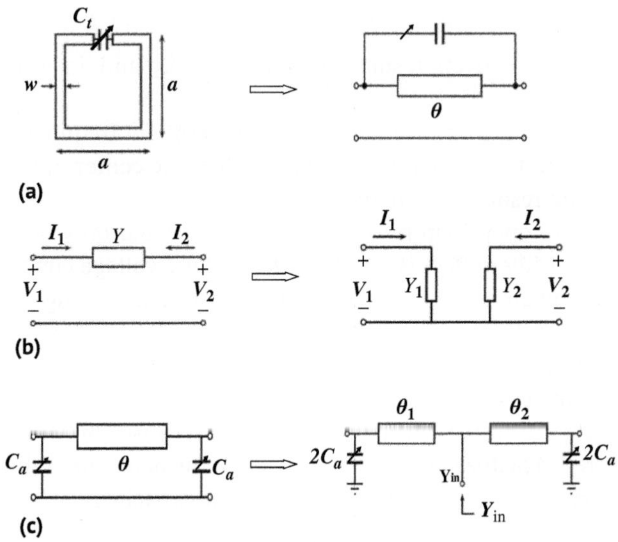

Figure 5.43 A square open-loop resonator loaded with a tunable series capacitor: (a) simplified equivalent circuit model for square open-loop resonator, (b) Miller-transformed equivalent circuit model, and (c) equivalent circuit of square open-loop resonator with two tunable shunt capacitors at open ends. *Source*: [103]/John Wiley & Sons.

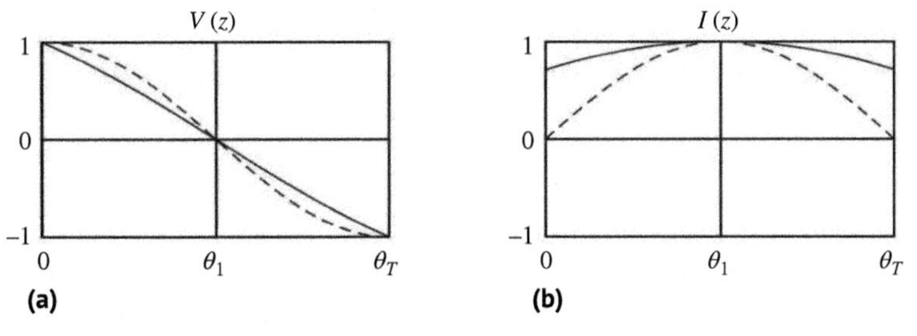

Figure 5.44 Voltage and current distribution at resonance of SLOR shown in Figure 5.42: (a) voltage distribution in a loaded and unloaded open-loop resonator and (b) current distribution in loaded and unloaded open-loop resonator (dashed lines are for the unloaded case). *Source*: [103]/John Wiley & Sons.

$$Y_1(\omega \to \omega_0)|_{A=-1} = 2Y(\omega_0) = j\omega_0 C_{t1} = j2\omega_0 C_t \tag{5.177}$$

$$Y_2(\omega) = Y(\omega)\left(1 - \frac{V_1(\omega)}{V_2(\omega)}\right) \to Y_2(\omega) = Y(\omega)\left(1 - \frac{1}{A(\omega)}\right) \tag{5.178}$$

$$Y_2(\omega \to \omega_0)|_{A=-1} = 2Y(\omega_0) = j\omega_0 C_{t2} = j2\omega_0 C_t \tag{5.179}$$

where $A = V_2/V_1$ is the voltage gain from node 1 to node 2, near resonance ($\omega \to \omega_0$), both voltages (V_1, V_2) are in opposite phase ($V_1 = -V_2$), gain $A = -1$ (odd-mode resonance).

From Figure 5.43c, the input admittance can be described by

$$Y_{\text{in}}(\omega) = jY_O\left[\frac{(2\omega C_t + Y_O \tan\theta_1)}{(Y_O - 2\omega C_t \tan\theta_1)} + \frac{(2\omega C_t + Y_O \tan\theta_2)}{(Y_{O-2\omega C_t \tan\theta_2})}\right] \tag{5.180}$$

From (5.180), at fundamental odd-mode resonance ($V_1 = -V_2$), $Y_{\text{in}}(\omega \to \omega_0) = 0$

$$Y_{\text{in}}(\omega \to \omega_0) = 0 \Longrightarrow jY_O\left[\frac{(2\omega C_t + Y_O \tan\theta_1)}{(Y_O - 2\omega C_t \tan\theta_1)} + \frac{(2\omega C_t + Y_O \tan\theta_2)}{(Y_O - 2\omega C_t \tan\theta_2)}\right] = 0 \Longrightarrow \tan(\theta_1 + \theta_2) = \frac{4Z_O\omega_0 C_t}{4Z_O^2{\omega_0}^2{C_t}^2} \tag{5.181}$$

$$\tan(\theta_1 + \theta_2) = \tan(\theta_T)|_{\theta_T=(\theta_1+\theta_2)} = \frac{4Z_O\omega_0 C_t}{4Z_O^2{\omega_0}^2{C_t}^2} = \frac{1}{Z_O\omega_0 C_t} \tag{5.182}$$

$$Y_{\text{in}}(\omega \to \omega_0) = \infty \Longrightarrow Z_{\text{in}} = 0 \Longrightarrow (Y_O - 2\omega C_t \tan\theta_1) \Longrightarrow \tan(\theta_1)|_{(\theta_1=\theta_2)} = \frac{1}{2Z_O\omega_0 C_t} \tag{5.183}$$

where $\theta_T = \theta_1 + \theta_2$ is the total length of the resonator.

From (5.182), the electrical length ($\theta_T = \theta_1 + \theta_2$) is calculated for a given resonance frequency ω_0 and loading capacitance C_t.

From (5.183), for $\theta_1 = \theta_2$, $Z_{\text{in}} = 0$, indicating the center of the resonator is a voltage null at the first resonant frequency. This implies that for $\theta_1 = \theta_2$; Eqs. (5.182) and (5.183) are equivalent, and there will be a voltage null at the center of the resonator whenever the resonance condition exists in square open-loop resonator structure.

From (5.176) to (5.183), the resonant condition of the loaded square open-loop resonator structure is valid only near an odd mode resonance ($V_1 = -V_2$), where the parameter "A" of the Miller effect is −1. Away from these voltage conditions ($V_1 = -V_2$) at odd-mode resonant frequencies, the voltage relation among the open ends changes, causing change in equivalent admittances Y_1 and Y_2 as shown in Figure 5.43b. This leads to difference between the behavior of the series and shunt-loaded resonators away from the odd resonant frequencies. The two most important differences are the performance near the even resonant mode and the existence of anti-resonance for the case of series loading [103–114].

For even-mode ($V_1 = V_2$), the capacitor (C_{t1} and C_{t2}) as shown in Figure 5.43c is virtually open circuited caused by Miller effect ($A = 1$). Therefore, the series capacitor (C_t) as shown in Figure 5.42a does not have any effect on the behavior of the resonator near even mode resonances; these resonant frequencies are unchanged by the presence of the capacitor.

5.7.3.2 Loaded Open Loop Printed Resonator Coupling and Mode-characteristics

Figure 5.45 shows the typical setup used for measurement of the $S_{21}(\omega)$ for deriving the coupling characteristics of the tunable open-loop resonator loaded with the varactor diode. The parameter of interest is the transmission coefficient $S_{21}(\omega)$, where the resonant frequencies are manifested as peaks of maximum transmission between ports.

Figure 5.46 shows a CAD-simulated (Ansoft Designer) plot of the varactor-loaded open square loop printed resonator with $\omega = 2$ mm and $a = 26$ mm, fabricated using Roger RO4003c substrate with a dielectric constant of 3.55 and a thickness of 60 mil (1.524 mm) [101–115]. As shown in Figure 5.46, the first resonant frequency is shifted down with different values of C, whereas the second resonance frequency remains at the exact location. Nevertheless, as varactor diode capacitance increases beyond a certain value (for example, 1.4 pF), a couple of frequencies where the transmission coefficient S_{21} is zero appear between the first and second resonant frequencies, which is observed in Figure 5.46 for $C = 2$ pF.

As shown in Figure 5.46, the physical size of the resonator ($\omega = 2$ mm and $a = 26$ mm) is kept constant while C is varied causing a shift in the first resonant frequency. This frequency shift can be capitalized into miniaturization if we let the size of the resonator vary and is minimized while keeping the fundamental resonance fixed.

Table 5.3 describes a summary of the results obtained by fixing the resonant frequency at 1 GHz using the same substrate as before [102–116].

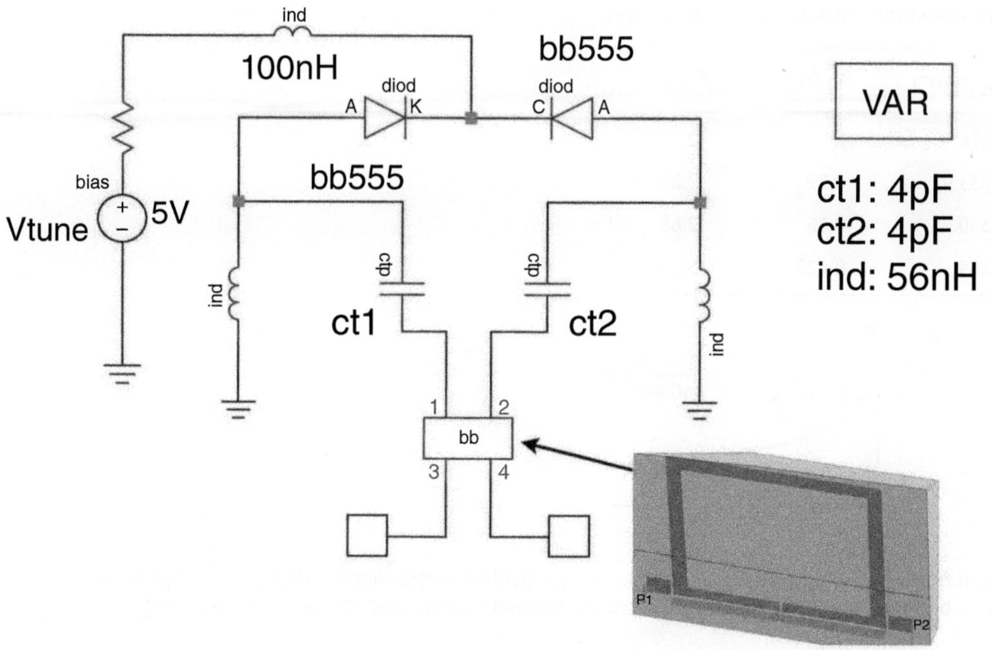

Figure 5.45 A typical setup for carrying out the measurement of the transmission coefficient $S_{21}(\omega)$ for analyzing the coupling characteristics of the varactor loaded tunable open-loop resonator [60, 104].

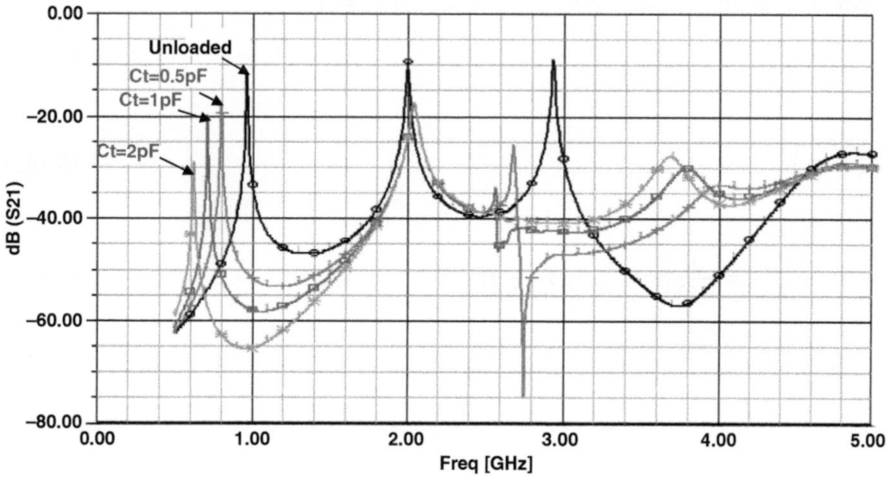

Figure 5.46 A CAD-simulated (Ansoft Designer) plot of $S_{21}(\omega)$ of a varactor-loaded SOLR for different capacitance values. *Source*: [60]/Ajay K. Poddar.

As shown in Table 5.3, the ratio of f_1/f_0 increases as the loading increases, when the loading capacitance is 1 pF, the last row of Table 6.1 indicates that the area of the miniaturized resonator is 36% that of the conventional resonator (shown in the fourth column of Table 5.3).

where

A_m: Area occupied by the varactor-loaded miniaturized resonator

A_c: Area of the conventional (unloaded) resonator

w = width (w = 2 mm in all resonators), kept constant for the simplification

a = length of the SLOR, varying for the miniaturization

f_0 = fundamental frequency is fixed at 1 GHz

f_1 = first spurious frequency

Table 5.3 Resonator characteristics as a function of loading capacitance.

C_t(pF)	a (mm)	A_m (mm^2)	A_m/A_c	f_1/f_0
0	26	676	1	2
0.2	23.2	535.92	0.79	2.23
0.6	19.5	380.25	0.56	2.66
1	17	289	0.427	3.06

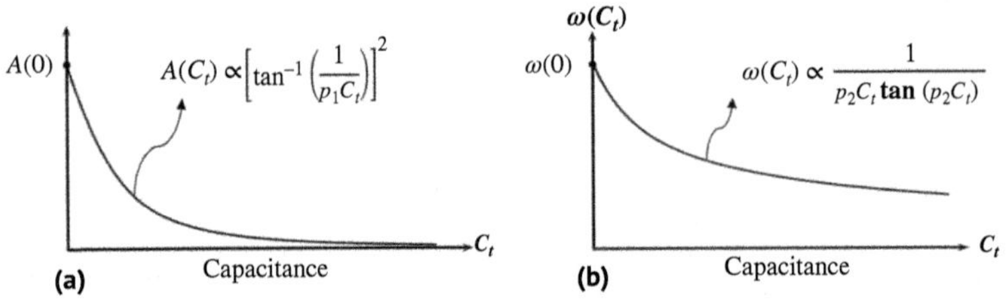

Figure 5.47 Plots of area $A(C_t)$ and frequency (C_t) versus capacitive loading provided by the varactor diode in printed square open-loop resonator: (a) relationship between the capacitance value and the resonator's area and (b) resonant frequency. *Source*: [103]/John Wiley & Sons.

Figure 5.47 shows the plot of area $A(C_t)$ and frequency $\omega(C_t)$ versus capacitive loading provided by the varactor diode in printed square open-loop resonator.

From (5.183), the total length of the square open-loop resonator is $\theta_T = \theta_1 + \theta_2$, for $\theta_1 = \theta_2$:

$$\tan(\theta_1)|_{\left(\theta_1 = \frac{\theta_T}{2}\right)} = \frac{1}{2Z_O\omega_0 C_t} \tag{5.184}$$

$$\theta_1 = \frac{\theta_T}{2} = \omega\left(\frac{l}{2v_p}\right) = \frac{\beta l}{2} = \tan^{-1}\left(\frac{1}{2Z_0\omega C_t}\right) \tag{5.185}$$

$$A\text{ (Square loop resonator area)} = \left(\frac{l}{4}\right)^2 = \frac{\theta_T{}^2}{16\beta^2} = \frac{1}{4\beta^2}\left[\tan^{-1}\left(\frac{1}{2Z_0\omega C_t}\right)\right]^2 \tag{5.186}$$

$$A(C_t) = k_1\left[\tan^{-1}\left(\frac{1}{p_1 C_t}\right)\right]^2 \Rightarrow A(C_t) \propto \left[\tan^{-1}\left(\frac{1}{p_1 C_t}\right)\right]^2 \tag{5.187}$$

From (6.39)

$$C_t = \frac{k_2}{p_2\omega\tan(p_2\omega)} \Rightarrow \omega(C_t) \propto \frac{1}{p_2 C_t \tan(p_2 C_t)} \tag{5.188}$$

Equations (5.187) and (5.188) are trigonometric functions, where "l" represents the total physical length (perimeter) of the resonator (neglecting the size of the gap where the capacitor is mounted, v_p is the phase velocity, k_1 and p_1, k_2 and p_2 are constant, and C_t is the capacitive loading provided by the varactor diode.

Since the constants k_1, k_2 and p_1, p_2 have the effect of scaling the x and y axes respectively, any choice of them gives a good representation of the general tendency. A plot of this curve, where, for simplicity $k_1 = p_1 = k_2 = p_2 = 1$, is shown in Figure 5.47.

Both curves shown in Figure 5.47 indicate that there is maximum variability for relatively small values of capacitance. It can be noticed from Figure 5.47 that as the capacitance increases, the rate of change decreases, which leads to the following [103]:

- Designs with large capacitance values are more robust (less sensitive to capacitor tolerances).
- For tunable operation, small values of capacitance are better.
- Increasing the capacitance after a certain value does not provide significant advantage.

Based on these observations, oscillator circuits are designed for validation purpose.

5.8 Printed Coupled Möbius Resonator Oscillators

The geometrical phenomenon of anholonomy depends on the failure of a quantity to recover its original value when the parameters on which it depends are varied around a closed circuit. Möbius strip provides one of the simplest examples of anholonomy, as the normal to the surface of the strip does not return to its original direction even though the radius vector does. The strip therefore deforms in such a way that its metrical properties are barely changed; some nanostructures have the same elastic properties.

A necessary and sufficient condition for a Möbius surface to be developable is that its Gaussian curvature must vanish everywhere. Given a curve with non-vanishing curvature, there exists a unique flat ruled surface (the so-called rectifying developable) on which this curve is a geodesic curve, illustrated in Figure 5.48, is described by [171]

$$\vec{x}(s,t) = \vec{r}(s) + t\left[\vec{b}(s) + \eta(s)\vec{t}(s)\right] \tag{5.189}$$

$$\tau(s) = \eta(s)k(s), \quad s = [0, L], \; t = [-w, \; w] \tag{5.190}$$

where $\vec{r}$ is a parameterization of a strip with r as the centerline and of length L and width $2w$, where $\vec{t}$ is the unit tangent vector, $\vec{b}$ the unit binormal, k the curvature, and τ the torsion of the centerline, the parameterized lines $s =$ const. are the generators, which make an angle $\beta = \mathrm{arc}[\mathrm{Tan}(1/\eta)]$ with the positive tangent direction.

The unique properties of Möbius strips, the shape that minimizes the deformation energy, which is entirely due to bending, can be described by

$$V = \frac{1}{2}D \int_{0}^{L}\int_{-w}^{w} k_1^2(s,t)\,dt\,ds \tag{5.191}$$

where $D = 2h^3E/\left[3(1 - \upsilon^2)\right]$, with $2h$ the thickness of the strip, and E and ν Young's modulus and Poisson's ratio of the material.

In this section, a planar Möbius-coupled resonator is described, and the method for miniaturization can be applied to tunable oscillator circuits and memory devices used for receivers, filters, antennas, and matching networks.

5.8.1 Planar Möbius Strip Resonator

A typical Möbius is a surface with only one side and only one boundary component, the mathematical property of being non-orientable (Figure 5.48). A unified system of differential-algebraic equations that describes models of this type was first reported by Starostin [171] together with its numerical solution and many technical applications.

The concept of the Möbius strip is based on the fact that a signal coupled to a strip shall not encounter any obstruction when travelling around the loop, the loop shall behave like an infinite transmission line, enabling compact high Q-factor resonators [172]. This characteristic enables many radio and microwave applications, including: (i) a compact resonator with the resonance frequency, which is half that of identically constructed linear coils [172], (ii) Tesla coil for global transmission of electricity without wires [173], and (iii) high temperature superconductors [174], to name a few.

Recent publication describes the Möbius strip resonator [172–177], but the surface is nonplanar 3D structure depicted in Figure 5.49, which is not suitable for integration and surface-mounted device (SMD) technology applications. Printed

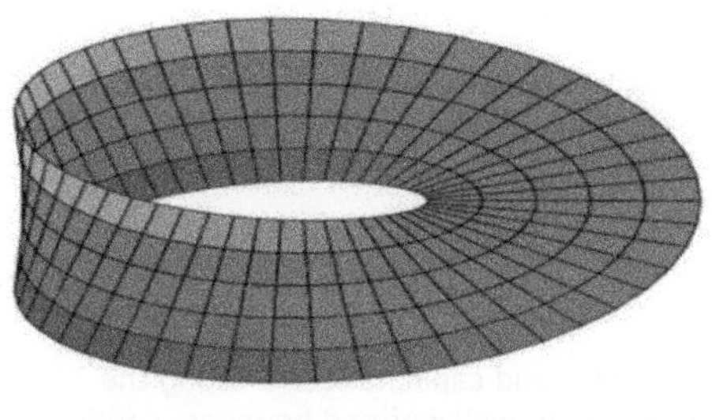

Figure 5.48 A typical Möbius strip, one can move along the length of the strip, return to its starting point having traversed the entire length of the strip without ever crossing an edge. *Source*: [184]/IEEE.

resonators are a special class of transmission lines of finite extent with well-defined boundary conditions. The particular interest here is a one-dimensional (1D) printed ring resonator that can be equivalently represented by the simple lumped LC network shown in Figure 5.50.

In solving, the electric currents on the resonator can be formulated by a periodic boundary condition of the form described by [178]

$$I_{j+k} = I_j \tag{5.192}$$

where I_k represents the electric current around the nth closed loop on the periodic ladder structure of k-elements.

The boundary condition of the general form shown in (5.192) governs that I_k is a conserved quantity that gives invariance of solutions under a 2π rotation with a definite handedness.

The solution for the ring resonator shown in Figure 5.50a with the periodic boundary condition given in (5.192) for the kth element can be given in the form

$$I_k = A_1 e^{\left(\frac{i2\pi jp}{k}\right)} + A_2 e^{-\left(\frac{i2\pi jp}{k}\right)} \tag{5.193}$$

where p is an integer specifying the normal mode and k is number of element structure.

From the circuit theory, Kirchhoff's voltage relation for the kth element is given by (assuming that the equivalent low-pass ladder network is driven across any equivalent lumped inductor L or capacitor C:

$$\left(\omega^2 - \frac{1}{LC}\right)I_k - \gamma\left(\omega^2 - \frac{1}{2\gamma LC}\right)(I_{k+1} + I_{k-1}) = 0 \tag{5.194}$$

where γ is mutual coupling coefficient (Mutual inductance "M" = 2γL) between nearest neighbor elements shown in Figure 5.50, and k is number of element structure.

Equation (5.193) is a 1D non-dissipative wave equation on the LC network (assuming that the zero resistive loss and electromagnetic wavelength are much larger than the linear dimensions of the network) with a discrete rather than

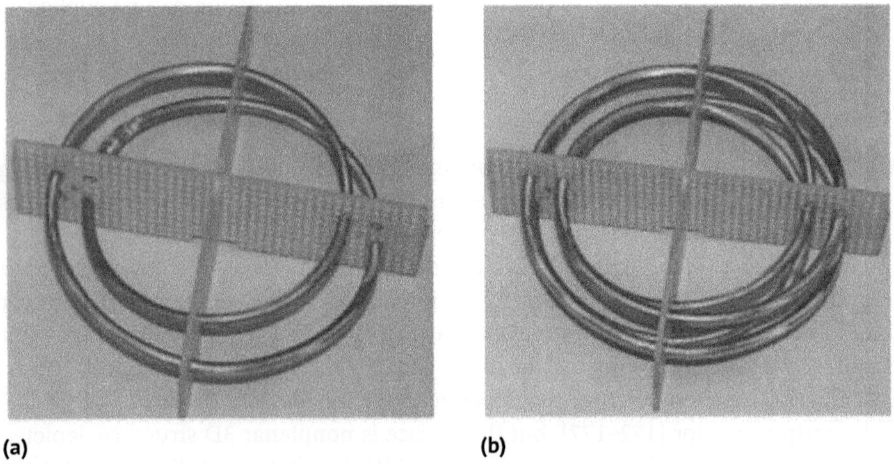

Figure 5.49 Photograph of the nonplanar 3D Möbius wire resonator: (a) dual mode and (b) quad mode. *Source*: [173]/IEEE.

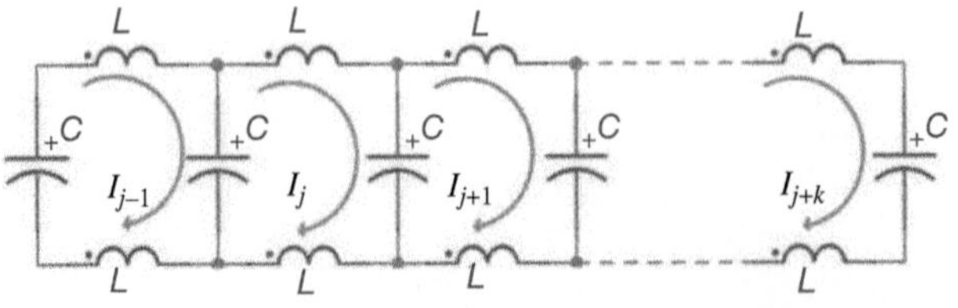

Figure 5.50 Shows the typical low-pass ladder network consisting of a series of inductances L and capacitances C along the transmission line (printed resonators are a special class of transmission lines of finite extent with well-defined boundary conditions; the 1D printed ring resonator can be equivalently represented by the simple lumped LC network). *Source*: Adapted from [184].

continuous spatial variable [178]. From (5.193) and (5.194), the dispersion relation for the allowed frequency spectrum can be described by [178]

$$\omega^2 = \left\{ \frac{2\sin^2\left(\frac{p\pi}{k}\right)}{LC\left(1 - 2\gamma\cos\frac{2p\pi}{k}\right)} \right\} \tag{5.195}$$

where p is an integer specifying the normal mode, γ is mutual coupling coefficient (Mutual inductance "M" = 2γL) between nearest neighbor elements shown in Figure 5.50, and k is number of element structure.

From (5.193) to (5.195) for even value of k, there are $k - 1$ eigenvalues, including $(k - 2)/2$ degenerate doublets and one singlet. A typical ring resonator, whose eigenfunction satisfies (5.192), defines a distinct inner and outer surface of the ring shown in Figure 5.51a. Figure 5.51b shows a topological transformation resulting in a Möbius strip resonator, whose current dynamics can be formulated by applying twisted boundary conditions as

$$I_{j+k} = -I_j \tag{5.196}$$

From (5.196), a simple topological transformation on the resonator ring, as depicted in Figure 5.5, results in a sign reversal of current (I_j) upon a 2π rotation of the solutions, and a 4π rotation is now required for invariance of the eigenfunctions [176–186].

Note that the eigenfunctions satisfying the condition for twisted boundary are of the same form as (5.193) provided that the mode indices are given half-integral values ($p = 1/2,\ 3/2,\ 5/2, \ldots (k - 1)/2$) relative to a ring consisting of identical components. The dispersion relation for Möbius ring is same as (5.195), however, the wave-vectors are shifted by

$$\Delta\lambda = -\left(\frac{\pi}{k}\right) \tag{5.197}$$

The two distinct topologies shown in Figure 5.51a,b can be considered as a complementary pair related by a single transformation.

From (5.192) and (5.196), it is evident that there is no additional structure associated with the Möbius ring resonator, since a second topological half-twist transformation on the Möbius resonator leads back to the boundary condition of (5.192). Their description divides into half-integral and integral normal mode indices. The eigenfunctions of the Möbius resonator form an orthogonal basis set; this presents an interesting possibility for the design of metamaterials for the application in tunable oscillators, antennas, and filter circuits [6]. The oscillator's loaded Q-factor Q_L is described by [185]

$$Q_L = \frac{\omega_0}{2}\left|\frac{d\varphi(\omega)}{d\omega}\right|_{\omega=\omega_0} = \frac{\omega_0}{2}\tau_d;\quad \tau_d = \left|\frac{d\varphi(\omega)}{d\omega}\right|_{\omega=\omega_0} \tag{5.198}$$

$$\tau_d = \left.\frac{d\varphi(\omega)}{d\omega}\right|_{\omega=\omega_0} = \frac{\varphi(\omega_0 + \Delta\omega) - \varphi(\omega_0 - \Delta\omega)}{2\Delta\omega} \tag{5.199}$$

where $\varphi(\omega)$ is the phase of the oscillator's open-loop transfer function at a steady state and τ_d is the group delay of the resonator.

From (5.198), Q_L is proportional to the group delay; therefore, for low phase noise application, the design goal is to maximize the group delay of Möbius strip resonator by incorporating phase-injection techniques. The eigenfunctions of the Möbius resonator form an orthogonal basis set; therefore, all eigenfunctions of the pair are mutually orthogonal; exhibits

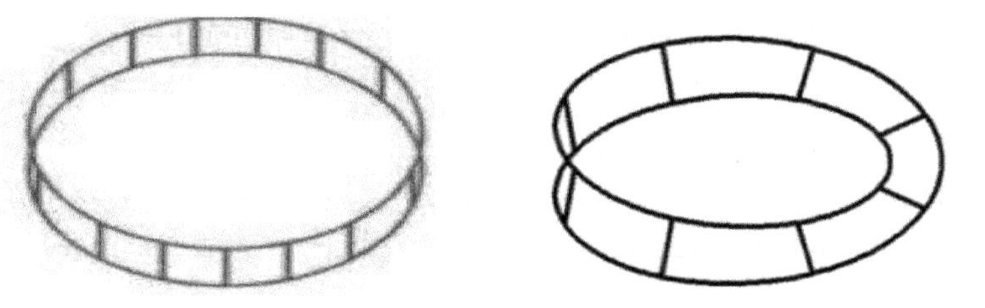

(a) (b)

Figure 5.51 Orientation of typical closed-loop resonator: (a) the ring resonator and (b) the Möbius strip resonator (a topological transformation of ring resonator into a Möbius strip resonator). *Source*: Adapted from [185].

minimal electromagnetic interactions [178]. This presents an interesting possibility for incorporating phase injection along the mutually coupled surfaces of the strips, resulting in higher quality factor for a given size of the printed transmission line resonator [182]. The Möbius strip resonator shows an interesting possibility, for example, if the arbitrary structure is composed of an array of typical Möbius strips as its basis, complementary strips could in principle be introduced without much disturbance to the coupling dynamics for microwave component (sensors, resonators, antennas, and filters) applications [183–185].

5.8.2 MCPR Oscillator: Inexpensive Alternative of DRO

Dielectric resonators (DRs), for example, exhibit a high Q-factor and have been used in high-spectral-purity signal sources at RF and microwave frequencies. However, the DR requires precise machining for fabrication, careful placement of the dielectric puck for optimal coupling, and involves manual mechanical tuning of the DR for the desired frequency of operation [177–186].

The printed coupled resonator-based oscillator can overcome problems associated with DRO, except for the poor phase noise performance due to inherent low Q-factor associated with printed transmission line resonator [187–192]. The Q-factor of the coupled planar resonator network can be enhanced by introducing slow-wave propagation dynamics as discussed in Chapter 6; however, large physical size and mode-jumping restrict the application, especially at lower frequencies. The loaded quality factor Q_L of the coupled resonator network is given in terms of unloaded quality factor Q_o as [60]

$$[Q_L(\omega_0)]_{\text{electrical-coupling}} \cong 2\left[\frac{Q_0}{(1+\beta_e)}\right]_{\beta_e<<1} \cong 2Q_0 \tag{5.200}$$

$$[Q_L(\omega_0)]_{\text{magnetical-coupling}} \cong 2[Q_0(1+\beta_m)]_{\beta\to 1} \cong 2Q_0 \tag{5.201}$$

$$[Q_L(\omega_0)]_{\text{hybrid-coupling}} \cong 2\left[Q_0\frac{(1+\beta_{mh})}{(1+\beta_{eh})}\right]_{\beta_e<<1,\,\beta_{mh}\to 1} \cong 4Q_0 \tag{5.202}$$

where coupling factor β (β_e for electric field, β_m for magnetic field, and β_h for hybrid field).

From (5.200) to (5.202), hybrid coupling (β_h) shown in Figure 5.40f acts as a Q-multiplier effect, but this phenomenon holds good for fixed frequency operation only, and very difficult to realize broadband Q-multiplier dynamics, which is necessary for wideband oscillators. In addition to this, mode jumping caused by undesirable higher-order mode generation in hybrid coupling makes this structure not suitable for tunable wideband VCO (voltage-controlled oscillator) applications. The alternative approach is to tune the coupling factor β_j over the desired operating frequency band in conjunction with higher-order mode-suppressing network. But broadband mode suppression is not an easy task in multimode resonator structure.

A simplified approach is to use injection-locking and phase-synchronization techniques for broadband applications. The Möbius strip resonator topology described in Figure 5.48 can offer promising solution for broadband oscillator circuits in compact size, amenable for integration in chip, and inexpensive alternative of expensive DROs [182].

Figure 5.52 shows the typical layout of 10 GHz Möbius resonator-based oscillator for building low phase noise signal source for RADAR applications. As shown in Figure 5.52, Möbius strips resonator is formed by connecting the outer and inner strips by via hole for the realization of planar structure. It can be seen that two concentric rings inside the Möbius loop minimize the spurious modes, thereby stable oscillation.

Figure 5.53 shows the measured plot of unloaded Q-factor of the printed resonator structures (electrically coupled, magnetically coupled, hybrid or mixed coupled, and Möbius coupled), giving brief insight into the improvement in quality factor in planar domain.

As shown in Figure 5.53, the Möbius coupled printed resonator (MCPR) structure and hybrid coupled act like a Q-multiplier, peaking at 8.6 GHz due to the convergence of evanescent modes. It can be seen in Figure 5.54 that Möbius resonator exhibits the undesired multi-mode resonant characteristics because of nonuniform coupling across the strips (ideally Möbius strip is 3D structure, depicted in Figure 5.48), and care must be taken to suppress these spurious modes for low jitter, especially in RADAR applications.

The drawback of this approach is that the multimode frequencies and mode-jumping (encircling the resonator characteristics) restrict the broadband operation. To overcome this problem, mode stabilization (injection of higher order modes into the resonant cavity with proper phase shift that improves the stability factor), a control circuit is incorporated [60].

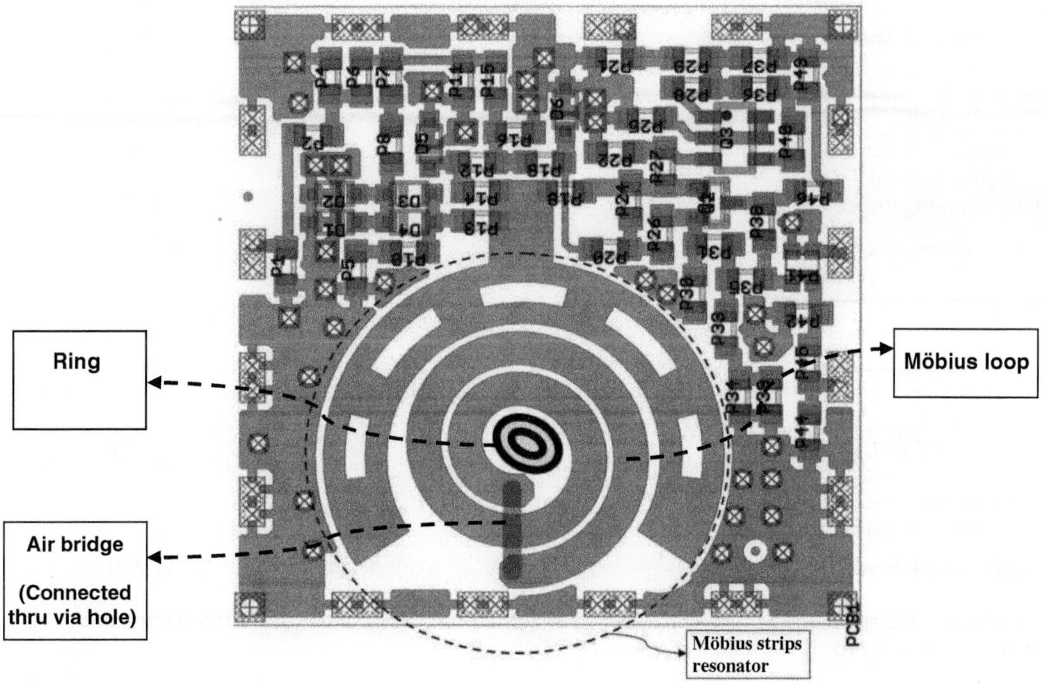

Figure 5.52 The layout of 10 GHz oscillator (depicts the Möbius strips resonator used for high Q-factor resonator; PCB layout is done with 22 mil substrate thickness with 3.38 dielectric constant, 0.5 × 0.5 × 0.18 inches).

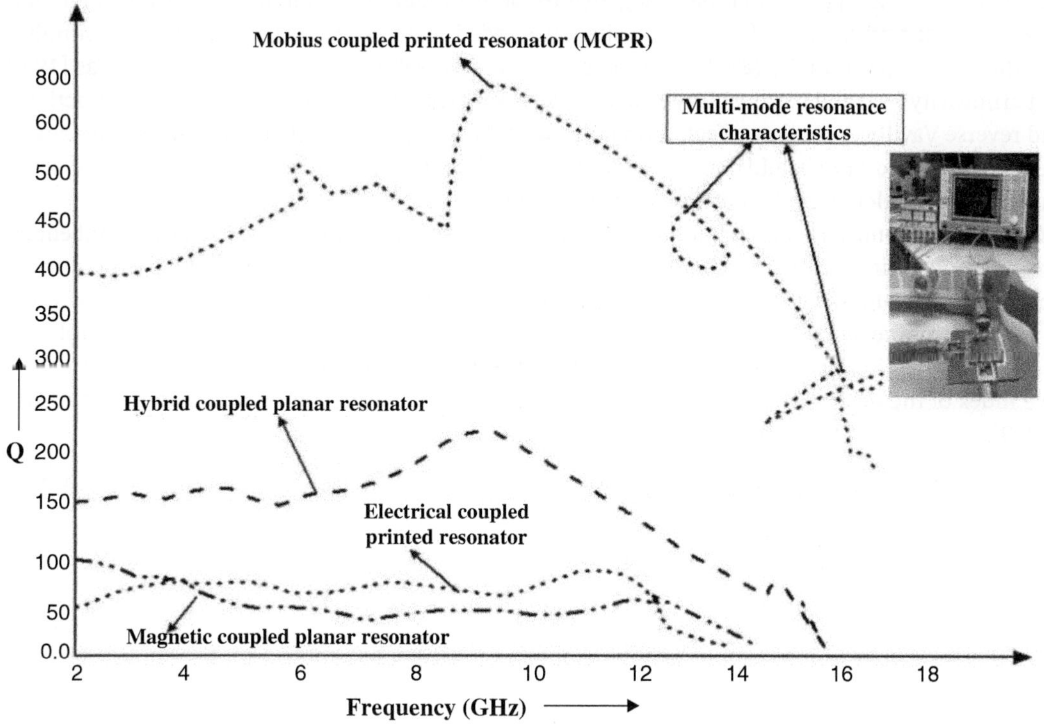

Figure 5.53 The measured plot of unloaded Q-factor of the printed resonator structures (electrical coupled, magnetic coupled, hybrid or mixed coupled, and Möbius coupled) for giving brief insight about the improvement in quality factor in planar domain (PCB layout for these resonators done with 20 mil substrate thickness with 2.22 dielectric constant and PCB size: 0.5 × 0.5 × 0.18 inches). *Source*: [60]/Brandenburg university of technology.

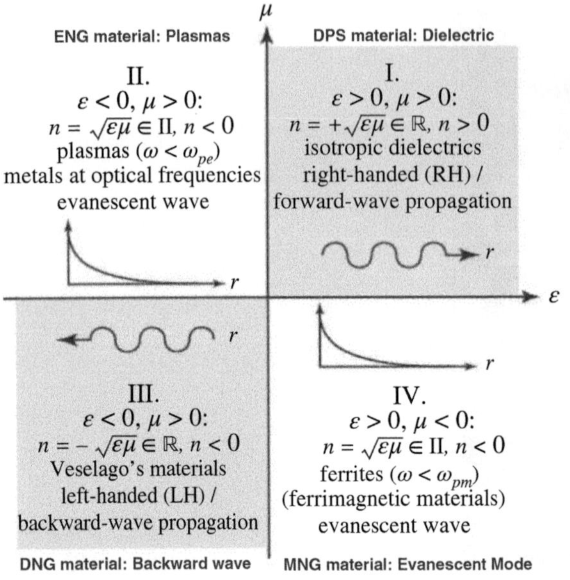

Figure 5.54 Permittivity-permeability diagram, R and I represents real and imaginary terms; ω_{pe} and ω_{pm} are the electric and magnetic plasma frequencies, respectively. *Source*: [220]/IEEE.

5.9 Printed Coupled Metamaterial Resonator

Metamaterials are engineered periodic composites that have negative refractive-index characteristics not available in natural materials [193]. In 1968, Veselago reported artificial composite left-handed material (LHM), which exhibits simultaneously negative values of the electric permittivity ($\varepsilon < 0$) and the magnetic permeability ($\mu < 0$), characterized as [194]: (i) Simultaneous negative permittivity ($-\varepsilon$) and permeability ($-\mu$), (ii) Reversal of Snell's Law (negative index of refraction), reverse Doppler effect, and reverse Vavilov-Cerenkov effect, diffraction-limit breaking imaging, (iii) electric field, magnetic field, and wave-vector of electromagnetic wave in a LHM form a left-handed triad, (iv) LHMs support backward-wave propagation: antiparallel group and phase velocity, and (v) artificial homogenous structure.

The characteristics of the artificial composite metamaterial depend on the properties of the host materials, embedded material, volume of the fraction, operating frequency, and the morphology of the composite material, such as the dimensions and shapes of the host structure and the guest structure [195–197]. Therefore, when all other variables are known, the important parameter is to control the dynamics of the morphology of the embedded structure that allows control over a change of the properties (permittivity, permeability, and the refractive index) in the desired frequency range.

In general, the refractive index of the medium can be characterized by four possible sign combinations in the pair ($\mu\epsilon$), can be described as [198–219]:

$$n = \sqrt{(+\epsilon)(+\mu)} = +\sqrt{\mu\epsilon} \rightarrow (\text{DPS} - \text{double positive material}) \tag{5.203}$$

$$n = \sqrt{(-\epsilon)(+\mu)} = j\sqrt{\mu\epsilon} \rightarrow (\text{ENG} - \text{epsilon negative material}) \tag{5.204}$$

$$n = \sqrt{(-\epsilon)(-\mu)} = -\sqrt{\mu\epsilon} \rightarrow (\text{DNG} - \text{double negative material}) \tag{5.205}$$

$$n = \sqrt{(+\epsilon)(-\mu)} = j\sqrt{\mu\epsilon} \rightarrow (\text{MNG} - \text{mu negative material}) \tag{5.206}$$

From (5.205), double negative (DNG) material is nomenclature as metamaterial, which is broadly defined as left-handed (LH) material with artificial homogeneous structures related to negative refractive index as

$$n = -\sqrt{\mu\epsilon} = -\sqrt{\mu_0\mu_r\varepsilon_o\varepsilon_r} \tag{5.207}$$

where ε_o is 8.85×10^{-12} F/m and $\mu_0 = 4\pi \times 10^{-7}$H/m, n is refractive index of the medium.

From (5.203) to (5.206), there are four possible classes of materials, as illustrated in Figure 5.54 [220]:

i. DPS ($\varepsilon > 0$ and $\mu > 0$)
These are conventional materials, also named as DPS or right-handed materials (RHM), and dielectrics are their examples. The propagation of electromagnetic waves is possible in such materials.

ii. ENG ($\varepsilon < 0$ and $\mu > 0$)
These are epsilon-negative materials (ENG), having a characteristic of negative value of permittivity, normally shown by many types of plasma in a particular frequency region. Metals like gold, silver, etc., demonstrate this negative permittivity in the infrared and visible frequency domains. The propagation of electromagnetic waves is not possible in ENG materials.

iii. MNG ($\varepsilon > 0$ and $\mu < 0$)
These materials are named as MNG, having a characteristic of negative value of permeability, shown by gyrotropic materials in certain frequency bands. The propagation of electromagnetic waves is not possible in such materials.

iv. DNG ($\varepsilon < 0$ and $\mu < 0$)
These materials are named as DNG or left-handed materials (LHM). These materials do not exist in nature but are physically realizable by making composite structures. The propagation of electromagnetic waves is possible in such a material.

From (5.201), the negative refractive index referred to as left-handed (LH) material is characterized as antiparallel phase and group velocity material that exhibits unusual properties. It can allow an electromagnetic wave to convey energy and group velocity in the opposite direction to its phase velocity, causing strong localization and enhancement of fields, thereby significant enhancement in the slew rate and group velocity of resonator, resulting in enhancement of quality factors of its resonant tank circuits [199].

These unusual characteristics of metamaterial open new degrees of freedom in resonator design for high-frequency oscillator, filter, phase-shifter, antenna, and other microwave applications in compact size. The artificial metamaterial components (negative value of permittivity and permeability) have been demonstrated over a large portion of the electromagnetic spectrum, from visible to radiofrequency [204].

The sign (positive or negative values) of permittivity and permeability is not restricted by any physical law, and can be positive or negative. In natural materials, negative permeability can occur during resonant condition [215]. The real part of permeability may take negative values near a resonance of sufficient strength, as for example in ferrites near the ferromagnetic resonance frequency.

For artificial materials, manipulation of a negative permeability response can be realized using electrically small resonant metal inclusions of special shapes, e.g., split rings. Among metamaterial components, a split-ring resonator (SRR) is used for microwave oscillator circuits and planar microwave devices [204–216].

5.9.1 The Electromagnetic Wave Propagation Dynamics of Metamaterial

The Maxwell's equations can be given by

$$\nabla \times \vec{\mathrm{E}} = -\frac{\partial \vec{\mathrm{B}}}{\partial t} - \overline{M_s} \text{ (Faraday's Law)} \tag{5.208}$$

$$\nabla \times \vec{\mathrm{H}} = \vec{J} + \frac{\partial \vec{D}}{\partial t} \text{ (Ampere's law)} \tag{5.209}$$

$$\nabla \cdot \vec{D} = \rho V \text{ (Electric Gauss's law)} \tag{5.210}$$

$$\nabla \cdot \vec{B} = \rho_m \text{ (Magnetic Gauss's law)} \tag{5.211}$$

where

$$\overline{B} = \mu \overline{H} = \mu_0 \mu_r \overline{H} \tag{5.212}$$

$$\overline{D} = \in \overline{E} = \varepsilon_o \varepsilon_r \overline{E} \tag{5.213}$$

$$\varepsilon_r = \varepsilon' - j\varepsilon'' = \varepsilon'[1 - j\tan(\delta_e)], \quad \tan(\delta_e) = \frac{\omega\varepsilon'' + \sigma_e}{\omega\varepsilon'} \tag{5.214}$$

$$\mu_r = \mu' - j\mu'' = \mu'(\delta_m), \quad \tan(\delta_m) = \frac{\omega\mu'' + \sigma_m}{\omega\mu'} \tag{5.215}$$

where $\vec{E}$ (V/m) is the electric field intensity, $\vec{H}$ (A/m) is the magnetic field intensity, $\vec{D}$ (Coulomb/m^2) is the electric flux density, $\vec{B}$ (W/m^2) is the magnetic flux density, M_s (Volt/m^2) is the fictitious magnetic current density, $\vec{J}$ (A/m^2) is the electric current density, ρ_V (Coulomb/m^3) is the electric charge density, ρ_m (Coulomb/m^3) is the magnetic charge density, ε_o is 8.85×10^{-12} (Farad/m), $\mu_0 = 4\pi \times 10^{-7}$ (Henry/m), ε_r is the relative permittivity, and μ_r is the relative permeability of the medium.

From (8.6) to (8.7), conventional medium follows right-hand rule, which means the electric field intensity vector $\vec{E}$, magnetic field intensity $\vec{H}$ and Poynting vector $\overline{k}$ inside medium form right-handed triplet.

The plane wave propagation along $+z$ direction can be described by Helmholtz equations [208]:

$$\nabla^2\vec{E} + k^2\vec{E} = 0 \text{ (Vector wave equation-Helmholtz equations)} \tag{5.216}$$

$$\frac{d^2E_x}{dz^2} + k^2E_x = 0 \text{ (Simplified scalar wave equation-Helmholtz equations)} \tag{5.217}$$

$$\vec{E}(z,t) = \hat{a}_x E_0^+ \cos(\omega t - kz) \text{ (Traveling wave E-field solution in time-domain)} \tag{5.218}$$

$$H(z,t) = \hat{a}_y \frac{E_0^+}{\eta} \cos(\omega t - kz) \text{ (Traveling wave H-field solution in time-domain)} \tag{5.219}$$

$$\eta = \frac{\overline{[E]}}{\overline{[H]}} \text{ (Wave impedance)} \tag{5.220}$$

For right-handed material ($\mu,\ \epsilon > 0$), triad $\left[\overline{E}, \overline{H}, \overline{k}\right]$ can be illustrated by [220]

$$\overline{k} \times \overline{E} = +\omega\mu\overline{H} \tag{5.221}$$

$$\overline{k} \times \overline{H} = -\omega\varepsilon\overline{E} \tag{5.222}$$

For left-handed material ($\mu,\ \epsilon < 0$), triad $[\overline{E}, \overline{H}, \overline{k}]$ can be illustrated by [220]

$$\overline{k} \times \overline{E} = -\omega|\mu|\overline{H} \tag{5.223}$$

$$\overline{k} \times \overline{H} = +\omega|\varepsilon|\overline{E} \tag{5.224}$$

Assuming the frequency is positive quantity, the phase velocity is given by

$$\overline{v}_p = \frac{\omega}{k}\hat{k}, \text{ where } \left(\hat{k} = \overline{k}/|k|\right) \tag{5.225}$$

$$\text{LH medium}: k < 0 \ (v_p < 0) \tag{5.226}$$

$$\text{RH medium}: k > 0 \ (v_p > 0) \tag{5.227}$$

$$\overline{k} \times \overline{E} = s\omega|\mu|\overline{H} \tag{5.228}$$

$$\overline{k} \times \overline{H} = -s\omega|\varepsilon|\overline{E} \tag{5.229}$$

$$s = \begin{pmatrix} +1, & RH \\ -1, & LH \end{pmatrix}, \text{ (where “s” is definition of handedness with respect to } k) \tag{5.230}$$

$$\overline{E}, \overline{H} \approx e^{+j(\omega t + |n|k_0 r)} \tag{5.231}$$

$$U_n = \frac{1}{4}\left\{\varepsilon|E|^2 + \mu|H|^2\right\} (U_n: \text{Energy}) \tag{5.232}$$

where $k = \dfrac{\omega}{u_p}$ (wave number or phase constant), $v_p = \dfrac{1}{\sqrt{\mu \in}}$ (Phase velocity), $k = \dfrac{2\pi}{\lambda}$ (phase constant in terms of wavelength).

Figure 5.55a,b illustrate the plane wave propagation characteristics in right-handed and left-handed orthogonal coordinate systems.

From (5.217) to (5.218), $[\overline{E}, \overline{H}, \overline{k}]$ form a right-handed orthogonal coordinate system for positive values of ε and μ, whereas from (5.222) to (5.223), $[\overline{E}, \overline{H}, \overline{k}]$ form a left-handed orthogonal coordinate system for negative values of ε and μ [145–170, 209, and 220].

Figure 5.55c depicts the 2-layer structure (left layer is a DPS material and right layer is a DNG material), the direction of Poynting vector $\vec{S}$ is parallel with the direction of phase velocity or wave vector $\overline{k}$ in DPS material, but these two directions are antiparallel in DNG material (right layer).

Figure 5.55d exhibits CAD HFSS simulation illustrating the plane wave propagation dynamics using an effective medium [220].

From (5.220) in LH (left-handed) medium phase velocity v_p propagates backward to the source in the opposite direction to that of power, related to group velocity v_g (Figure 5.55c). The backward propagation implies that the fields have time-space dependence. Backward-wave media are media in which the energy velocity direction is opposite to the phase velocity

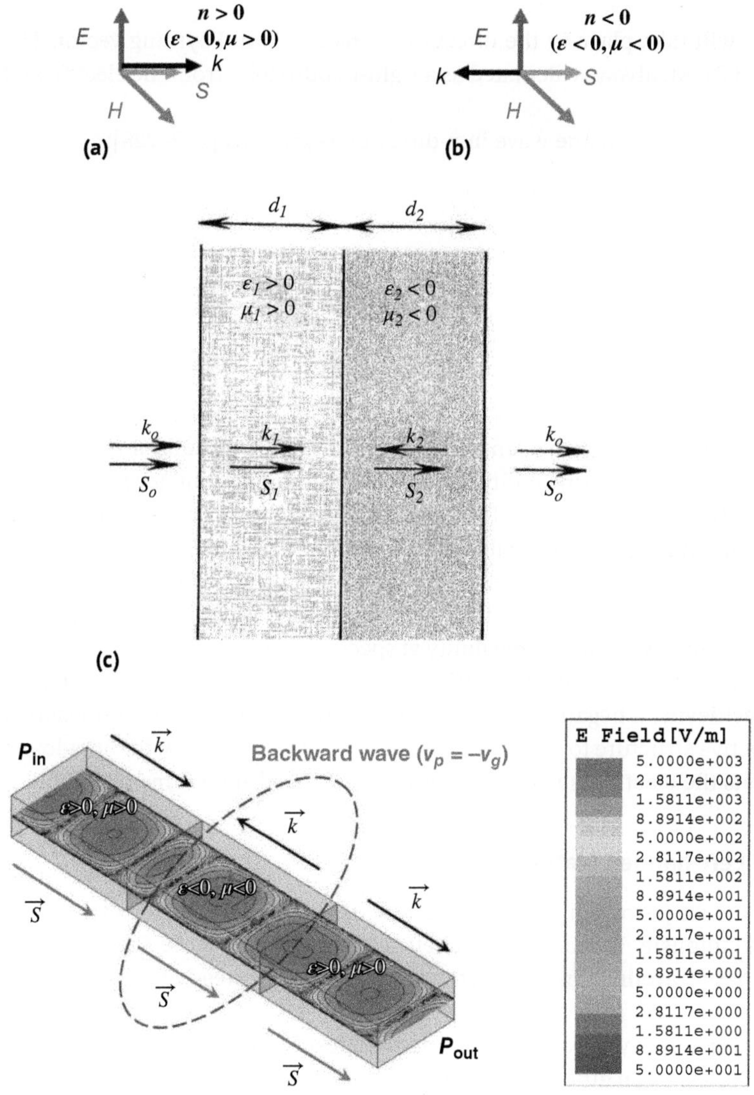

(d)

Figure 5.55 (a) Right-handed orthogonal coordinate system for materials with positive ε and μ, (b) left-handed orthogonal coordinate system for materials with negative ε and μ, (c) depicts the 2-layer structure (left layer is a DPS material and right layer is a DNG material), and (d) CAD HFSS simulation shows the plane wave propagation dynamics using effective medium. *Source*: [220]/IEEE.

direction (or the scalar product between the corresponding vectors is negative). In particular, this takes place in isotropic materials with negative permittivity and permeability (double negative media). Negative refraction takes place on interfaces between "usual" materials and backward-wave media. Surface waves (waves propagating along a surface but exponentially decaying in the directions orthogonal to the surface) can exist on interfaces between media having material parameters of the opposite sign. The anomalous refraction at the boundary of such media and the fact that for plane wave the direction of the Poynting vector is antiparallel with the direction of phase velocity, can be advantageous in designing compact high-quality factor resonator circuit.

From (5.212) to (5.213), the power flow as shown in Figure 5.55d is given by the Poynting theorem as [221]

$$\oint_s (\vec{\mathrm{E}} \times \vec{\mathrm{H}}) \cdot d\vec{s} = -\frac{\partial}{\partial t} \int \left(\frac{1}{2}\varepsilon \mathrm{E}^2 + \frac{1}{2}\mu \mathrm{H}^2\right) dv - \int \sigma \mathrm{E}^2 dv \tag{5.233}$$

$$\oint_s (\vec{\mathrm{E}} \times \vec{\mathrm{H}}) \cdot d\vec{s} = \oint_s \vec{S}.d\vec{s} \tag{5.234}$$

where, $\vec{S} = (\vec{\mathrm{E}} \times \vec{\mathrm{H}})$ is the Poynting vector.

From (5.212) to (5.213), the propagation of energy will take place in the direction extracted from Poynting vector. The direction of Poynting vector is same for both LHM and RHM, always calculated as a right-hand-triplet from the electric and magnetic field.

From (5.228), average Power density transmitted by a uniform plane wave in z-direction is given as [222–228]:

$$\vec{S}_{av}(z) = \hat{a}_z \frac{\mathrm{E}_0^2}{2|\eta_c|} e^{-2\alpha z} \cos\theta_\eta \text{ (Lossy medium)} \tag{5.235}$$

$$\vec{S}_{av}(z) = \hat{a}_z \frac{\mathrm{E}_0^2}{2\eta} \text{ (Lossless medium)} \tag{5.236}$$

$$\vec{S}_{av} = \frac{1}{2} R_e(\vec{\mathrm{E}} \times \mathrm{H}^*) \text{ (Phasor form)} \tag{5.237}$$

The realization of true DNG material (metamaterial) is questionable! In general, it is difficult to build a material or medium simultaneously with negative permittivity and negative permeability for a broad operating frequency ranges from a set of arbitrary passive structure unit cells arranged in predefined order. This may lead to violation of energy conservation principle at the intersecting plane between a RHM and LHM (left-handed material) media because of the generation of energy. However, an attempt to build the DNG material or medium is based on the fact that for a specific orientation and arrangement of the passive structure, the values of permittivity and permeability reduce as the frequency increases till these material or medium initiates demonstrating negative permittivity and permeability at specific narrowband frequencies and lacks the validity for broadband operation. It is, however, easy to demonstrate MNG (negative permeability) or ENG (negative permittivity) characteristics independently, but achieving a negative regime for permittivity and negative permeability could be a challenging task and needs complex and active structure to support the evidence of DNG material for broadband operation. For clear understanding, Figure 8.3 exhibits the electromagnetic wave propagation in different media [208].

5.9.2 Backward Wave Propagation Dynamics in Left-handed Material

A typical material that has both negative permittivity and permeability, nomenclature as left-handed (LH), is where the electric field intensity vector $\vec{\mathrm{E}}$ and the magnetic field intensity $\vec{\mathrm{H}}$ vector forms a left-handed triplet (Figure 5.55b).

From (5.224), a backward-wave implies that the phase traveling direction is toward the source instead of away from the source after the electromagnetic wave enters the Metamaterial, as shown in (Figure 5.56).

From (5.228) and (5.231), the energy flowing direction is always away from the source, irrespective of the type of medium the wave is traveling, as shown in Figures 5.55 and 5.56, the main difference is which side of the normal face the energy will be flowing to [217]. This intrinsic property of negative phase advance in left-handed material exhibits properties well-suited for perfect lenses or super lenses [217].

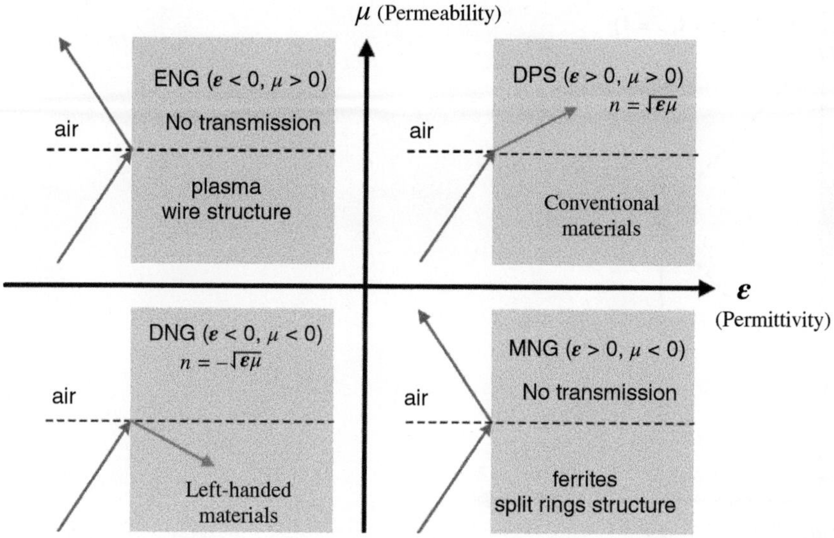

Figure 5.56 The electromagnetic wave propagation in different media. *Source*: Adapted from [208].

If the electromagnetic wave propagation is in the z axis, in order to transmit all the energy through the DNG material slab, it is required that we have a propagation constant k_z' [218]:

$$k_z' = -\sqrt{\frac{\omega^2}{c^2} - (k_x^2 + k_y^2)} \tag{5.238}$$

The overall transmission coefficient for the wave entering into the DNG slab ($\varepsilon = -1$ and $\mu = -1$) as shown in Figure 5.57 is given by [218]

$$T = tt' = e^{-j\left[\sqrt{\frac{\omega^2}{c^2} - (k_x^2 + k_y^2)}\right]d} \tag{5.239}$$

$$T = e^{j(k_z' d)} \tag{5.240}$$

where t and t' are the transmission coefficients of DPS and DNG media, and d is the distance the wave travels inside the DNG (metamaterial slab).

The choice of the propagation constant k_z' is made in order to maintain causality, and the required phase correction grants the DNG material capability of refocusing the image by canceling the phase of the transmitted wave as it propagates from its source [193]. This is an interesting phenomenon, which can be useful toward improving the phase-dispersion and group delay in a resonator for oscillator and filter applications.

5.9.3 Evanescent Wave Propagation Dynamics

Considering the field expression for TE wave propagation in vacuum (medium 1) is

$$\overline{E_1} = e^{j(\overline{z}k_z + \overline{x}k_x - \omega t)} \tag{5.241}$$

where k_z is a propagation constant,

$$k_z = j\sqrt{(k_x^2 + k_y^2) - \frac{\omega^2}{c^2}} \text{ with } \frac{\omega^2}{c^2} < (k_x^2 + k_y^2) \tag{5.242}$$

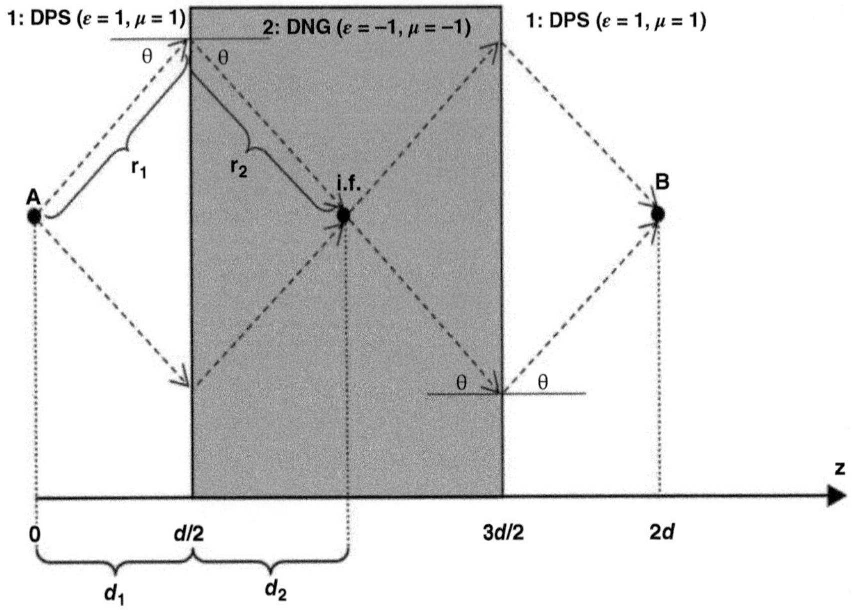

Figure 5.57 The wave propagation from free space (DPS) into metamaterial slab (DNG medium). *Source*: Adapted from [218].

In order to maintain the causality, the field must decay as it goes away from the interface; the transmitted $\left(\overline{E_{t2}}\right)$ and reflected field $\left(\overline{E_{r1}}\right)$ can be described by

$$\overline{E_{t2}} = Te^{j(\overline{z}k_z'+\overline{x}k_x-\omega t)} \tag{5.243}$$

$$\overline{E_{r1}} = Re^{j(-\overline{z}k_z'+\overline{x}k_x-\omega t)} \tag{5.244}$$

where

$$[k_z']_{(k_x^2+k_y^2)>\epsilon_2\mu_2\left(\frac{\omega^2}{c^2}\right)} = j\sqrt{(k_x^2+k_y^2)-\epsilon_2\mu_2\frac{\omega^2}{c^2}} \tag{5.245}$$

where ϵ_2 and μ_2 are the permittivity and permeability of medium 2 (slab) and $\frac{\omega^2}{c^2} < \frac{(k_x^2+k_y^2)}{\epsilon_2\mu_2}$. The transmission, T, and reflection coefficients, R, can be derived by matching the electromagnetic wave fields at the interface from medium # 1 and medium #2 [218]:

$$T = \frac{2\mu k_z}{\mu k_z + k_z'} \tag{5.246}$$

$$R = \frac{\mu k_z - k_z'}{\mu k_z + k_z'} \tag{5.247}$$

The transmission, T', and reflection coefficients, R', of the transition from slab (inside medium # 2) to medium # 1 are given by

$$T' = \frac{2k_z'}{\mu k_z + k_z'} \tag{5.248}$$

$$R' = \frac{k_z' - \mu k_z}{k_z' + \mu k_z} \tag{5.249}$$

The expression of the wave transmission coefficient (T_e) through both the interfaces can be derived by adding the multiple scattering dynamics, described by

$$T_e = \frac{\overline{E_{\text{transmitted}}}}{\overline{E_{incident}}} \tag{5.250a}$$

$$T_e = TT'e^{j(k'_z d)} + TT'R'^2e^{j(3k'_z d)} + TT'R'^3e^{j5(5k'_z d)} + TT'R'^4e^{j(7k'_z d)} + (\ldots) \tag{5.250b}$$

$$T_e = \frac{TT'e^{j(k'_z d)}}{1 - R'^2e^{j(2k'_z d)}} \tag{5.251}$$

For the metamaterial (DNG medium), $\varepsilon = -1$ and $\mu = -1$

$$[T_e]_{\varepsilon\to-1,\mu\to-1} = \left[\frac{\left(\frac{2\mu k_z}{\mu k_z + k'_z}\right)\left(\frac{2k'_z}{\mu k_z + k'_z}\right)e^{j(k'_z d)}}{1 - \left(\frac{k'_z - \mu k_z}{k'_z + \mu k_z}\right)^2 e^{j(2k'_z d)}}\right] \tag{5.252}$$

$$[T_e]_{\varepsilon\to-1,\mu\to-1} = \left[\frac{\left(\frac{-2k_z}{-k_z + k'_z}\right)\left(\frac{2k'_z}{-k_z + k'_z}\right)e^{j(k'_z d)}}{1 - \left(\frac{k'_z + k_z}{k'_z - k_z}\right)^2 e^{j(2k'_z d)}}\right] \cong e^{j(k'_z d)} \tag{5.253}$$

where

$$[k'_z]_{\varepsilon\to-1,\mu\to-1} = j\left(\sqrt{(k_x^2 + k_y^2) - \epsilon_2\mu_2\frac{\omega^2}{c^2}}\right) = j\left(\sqrt{(k_x^2 + k_y^2) - \frac{\omega^2}{c^2}}\right) = k_z \tag{5.254}$$

From (5.247)

$$[T_e]_{\varepsilon\to-1,\mu\to-1} = e^{j(k'_z d)} = e^{j(k_z d)} \tag{5.255}$$

The overall reflection coefficient can be given by (R_r) as [218]

$$[R_r] = \frac{\overline{E_{\text{reflected}}}}{\overline{E_{\text{incident}}}} = R\left[1 - \frac{(1 - R'^2)e^{j(2k'_z d)}}{1 - R'^2e^{j(2k'_z d)}}\right] = R\left[\frac{1 - e^{j(2k'_z d)}}{1 - R'^2e^{j(2k'_z d)}}\right] \tag{5.256}$$

Simplifying the Eqs. (5.241), (5.243), and (5.250)

$$[R_r]_{\varepsilon\to-1,\mu\to-1} \to 0 \tag{5.257}$$

From (5.251), the reflection coefficient R_r is zero in DNG medium. The consequence of having negative index ($\varepsilon \to -ve,\ \mu \to -ve$) refraction results in the wave of the form $e^{(-k_z d)}$ outside the slab (DNG) that couple to waves of the form $e^{(k_z d)}$ inside the slab (DNG). Therefore, even if the waves decay outside the slab, they are amplified on the inside of it, recovering an image on the opposite side of the slab, from the source, and all done by the transmission process [218]. Since the wave decay in amplitude, not in phase, as they get further from the source, the DNG slab focuses the image by amplifying these waves rather than correcting the phase.

Figure 5.58 illustrates the evanescent field variation in DNG slab.

These features can be utilized for improving the Q-factor of the evanescent mode resonator because both, the propagating and evanescent waves contribute to improving the group delay. This phenomenon does not violate energy conservation law because evanescent wave do not transport the energy [219].

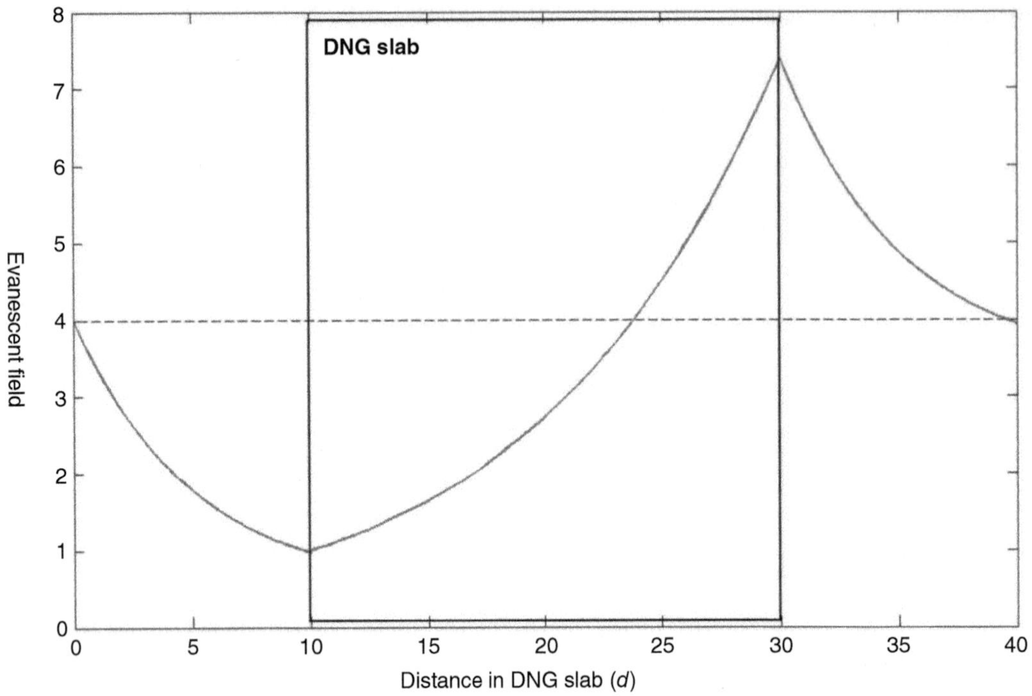

Figure 5.58 Evanescent field variation in metamaterial slab (DNG), shows amplification of evanescent wave inside the DNG slab. *Source*: Adapted from [218].

5.10 Phase Velocity, Group Velocity, Energy Density

5.10.1 Phase Velocity

From (5.219), the phase velocity of a wave is the rate at which the phase of the wave propagates in space or medium. The phase velocity is given in terms of the wave's frequency ω, wavenumber k, velocity of light c, and refractive index n, given by [210, 211]

$$v_p = \frac{\omega}{k} = \frac{\omega}{nk_o} \ (k = \omega\sqrt{\epsilon\mu} = \omega n\sqrt{\epsilon_o\mu_o} = nk_o) \tag{5.258}$$

$$v_p = \frac{c}{n} = \frac{c}{\sqrt{\epsilon\mu}} \tag{5.259}$$

5.10.2 Group Velocity

The group velocity is the rate at which the envelope of the waveform is propagating in space or medium, i.e., the rate of variation of the amplitude of the waveform. Assuming the waveform is not distorted significantly during propagation, it is the group velocity, which represents the rate at which information (and energy) is transmitted by the wave. However, if the wave is travelling through an absorptive and strongly dispersive medium, this condition no longer holds. For certain conditions, it is possible to reduce the group velocity to zero, stopping the wave, or have negative group velocity, making the wave appear to propagate backwards. In isotropic media, the group velocity is given as [210]

$$v_g(\omega_0) = \frac{d\omega}{dk} = \left(\frac{dk}{d\omega}\right)^{-1} \tag{5.260}$$

where v_g is the group velocity, ω is the wave's angular frequency, k is the wave number.

The function $\omega(k)$, which gives ω as a function of k, is known as the dispersion relation. If ω is directly proportional to k, then the group velocity is exactly equal to the phase velocity. Otherwise, the envelope of the wave will become distorted as it propagates.

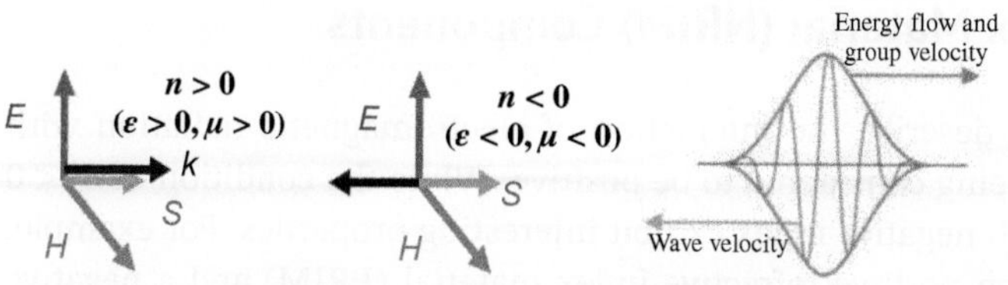

Figure 5.59 The typical representation of electromagnetic wave velocity.

Figure 5.59 shows the typical pictorial representation of the electromagnetic wave velocity.
From (5.252) and (5.253) [211],

$$v_g(\omega_0) = v_p + k\frac{d(v_p)}{dk} \tag{5.261}$$

$$v_g(\omega_0) = c\left[\frac{d(n\omega)}{d\omega}\right]^{-1} = \frac{c}{n + \omega_0\left[\frac{d(n)}{d\omega}\right]}, \text{ where } \frac{d(n)}{d\omega} > 0 \tag{5.262}$$

From (5.256), for positive refractive index n, the group velocity is always smaller than the phase velocity and smaller than the speed of light in vacuum.

From (5.256), for negative refractive index material ($n < 0$), the group velocity can take any value because $\frac{d(n)}{d\omega} > 0$ in a low-loss media, including materials with negative refractive index.

5.10.3 Energy Density

It is well-known that any physical media other than vacuum is dispersive [220]. The expression of energy density U for non-dispersive media can be described by [222]:

$$U = \frac{1}{4}\{\varepsilon|E|^2 + \mu|H|^2\} \to U = \frac{1}{4}\left\{\left[\frac{\partial(\omega\varepsilon)}{\partial\omega}\right]|E|^2 + \left[\frac{\partial(\omega\mu)}{\partial\omega}\right]|H|^2\right\} \tag{5.263}$$

From (5.257), the physical requirement of the existence of the positive energy density is given by

$$\frac{\partial(\omega\varepsilon)}{\partial\omega} > 0, \ \frac{\partial(\omega\mu)}{\partial\omega} > 0 \tag{5.264}$$

From (5.257), for metamaterial (μ, $\epsilon < 0$) the energy density results of negative value, which yields nonphysical result. Eq. (5.258) is compatible with (μ, $\epsilon < 0$), provided $\frac{\partial(\omega\varepsilon)}{\partial\omega} > \frac{|\varepsilon|}{\omega}$ and $\frac{\partial(\omega\mu)}{\partial\omega} > \frac{|\mu|}{\omega}$. Hence, physical left-handed media is highly dispersive (metamaterial resonator exhibits a negative value of ε and μ in the highly dispersive regions just above resonances). For Metamaterial, backward-wave propagation implies opposite signs between phase and group velocities [223], which can be expressed as [221]

$$\frac{\partial k^2}{\partial\omega} = 2k\left(\frac{\partial(k)}{\partial\omega}\right) = 2\left(\frac{\omega}{v_p v_g}\right) \tag{5.265}$$

$$\frac{\partial k^2}{\partial\omega} = \omega\varepsilon\left(\frac{\partial(\omega\mu)}{\partial\omega}\right) + \omega\mu\left(\frac{\partial(\omega\varepsilon)}{\partial\omega}\right) < 0 \tag{5.266}$$

$$v_p v_g < 0 \tag{5.267}$$

where $v_p = \frac{\omega}{k}$ and $v_g = \frac{d\omega}{dk}$ are the phase and group velocity, respectively, and $k = \omega\sqrt{\varepsilon\mu}$. Backward wave transmission line can form a nonresonant LHM (left-handed material) [220, 227].

5.11 Realization of Negative Refractive Index Material (NRIM) Components

The refractive index is the most fundamental parameter to describe the interaction of electromagnetic radiation with matter. It is a complex $n = (n' + jn'')$ where n' is generally being considered to be positive. While the condition of $n' < 0$ does not violate any fundamental physical law, materials with negative index exhibit interesting properties. For example, electromagnetic wave when refracted at an interface between positive refractive index material (PRIM) and a negative refractive index material (NRIM), is bent in the "wrong" way with respect to the normal, resulting in group and phase velocity are antiparallel.

A possible-but not only-approach to achieve an NRIM in a passive medium is to design a material in which the (isotropic) permittivity $\varepsilon = (\varepsilon' + j\varepsilon'')$ and the (isotropic) permeability $\mu = (\mu' + j\mu'')$ obey the inequality [227–233]

$$\varepsilon'|\mu| + \mu'|\varepsilon| < 0 \tag{5.268}$$

This leads to a negative real part of the refractive index $n = (n' + jn'') = \sqrt{\mu\varepsilon}$. Eq. (5.262) only holds good for passive media, and the equality condition is satisfied if $\varepsilon' < 0$ and $\mu' < 0$. But this is not a necessary condition; there may be magnetically active media ($\mu \neq 1$) with a positive real part μ' for which (5.262) is justified, and show a negative real part of the refractive index n'. The incorporating impedance term Z in the general relationship between the refractive index can lead to more general relationship [229]

$$2\,Re\left(\frac{1}{Z}\right)n' = \varepsilon' + \mu'\frac{|\varepsilon|}{|\mu|} \tag{5.269}$$

$$2n'n'' = \varepsilon'\mu'' + \mu'\varepsilon'' \tag{5.270}$$

where $Z = \sqrt{\left(\frac{\mu}{\varepsilon}\right)}$, the direction of the Poynting vector and wave vectors are governed by the sign of $Re\left(\frac{1}{Z}\right)$ and n' respectively.

Therefore, the sign of $\varepsilon' + \mu'\frac{|\varepsilon|}{|\mu|}$ should be negative for the NRIM. Eq. (5.264) requires $(\varepsilon'\mu'' + \mu'\varepsilon'')$ being negative for the absorbing effective medium and positive for an NRIM. The key to the design of the artificial NRIM resonator is to select novel geometric shape so that induced currents form loops with a relatively uniform distribution of the induced current, which produces a strong magnetic moment.

In general, three types of typical geometry are used for the realization of NIRM components: (i) split-ring resonators, (ii) coupled wire sets, (iii) dielectric resonators. Figure 5.60 depicts these resonators, which exhibit negative refractive index property described in (5.262) [230–244].

5.11.1 Split-ring Resonator

As shown in Figure 5.60a, SRR produces a negative permeability just above its resonant frequency. Typically, these structures are bi-anisotropic, exhibiting coupled electric and magnetic effects [243, 244]. A negative refractive index transmission passband occurs at the resonant frequency of the split ring when wires that provide a negative permittivity are aligned with their axis parallel to the plane of the loop [246–248]. By incorporating concentric split rings in conjunction with varactor diodes, as shown in Figure 5.61, resonant frequency can be switched and tuned for multi-band applications [209, 238, 249].

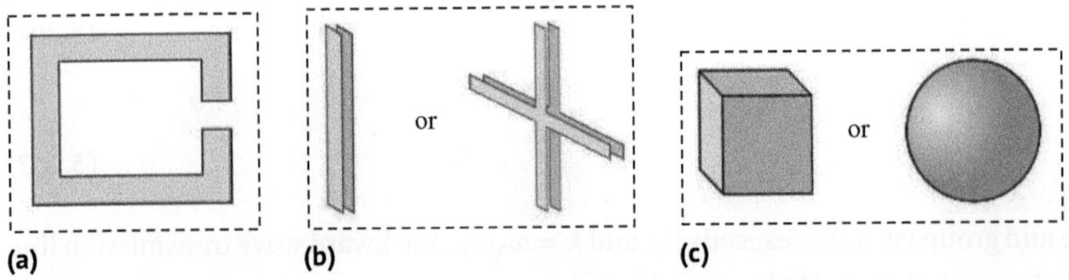

Figure 5.60 Typical resonator geometries that exhibit negative refractive index property: (a) split-ring resonator, (b) coupled wire sets, and (c) dielectric resonators [245].

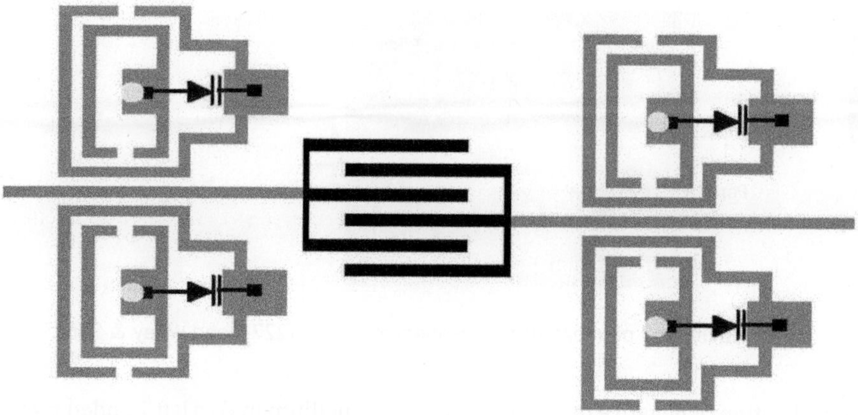

Figure 5.61 Typical geometries of metamaterial resonators for switchable band operation. *Source:* [60]/Ajay K. Poddar.

5.11.2 Coupled Wire Sets NRIM

Figure 5.60b shows a typical coupled wire constructed as negative refractive index material (NRIM) that uses pairs of conducting wire strips, or other shapes, which are spaced apart in the direction of wave propagation at a distance much smaller than the wire's width [245].

As shown in Figure 5.60b, the electric field interacts with linear elements of the structure, leading to negative permittivity, whereas the magnetic field interacts with the small gaps between conducting elements, exciting antiparallel currents in the open loop of the two conducting structures, leading to negative permeability [250–288]. While coupled wire structures can be built as a bulk Metamaterial, since the desired properties are achieved face-on, the structure is planar by design. Thus, the axis perpendicular to the material face has a different permittivity and permeability than do the axes parallel to the material face. Even though the working angle of the metamaterial is broad, the refraction angles of different polarizations will eventually decouple because the structure is uniaxial [244].

5.11.3 Dielectric Material Resonator NRIM

Dielectric material resonator NRIMs (Figure 5.60c) uses a nonconductive particle with high permittivity, compared to that of the surrounding material, to create negative permeability [58]. The shape of the inclusions determines the resonant frequencies as well as the level of anisotropy. The resonances are Mie resonances (named after Gustav Mie for his theory of the electromagnetic wave scattering of spheres) for cubes and spheres: the first resonance is magnetic; the second resonance is electric [251–264].

5.12 Metamaterial (NRIM) Model

There are different models to describe the metamaterial behavior (negative permittivity and negative permeability), but the resonator and transmission line models are most common, which use the analogy between natural medium and artificial metamaterial [213].

5.12.1 Resonator Model

Figure 5.62 shows the typical artificial metamaterial resonator model, which consists of dipole-shaped thin wire for creating negative permittivity and split ring for negative permeability medium. By adjusting the spacing and dimension of dipole-shaped thin wire and split ring, negative permittivity and permeability medium in microwave frequency region can be realized simultaneously, known as double negative medium (DNG).

Figure 8.9 shows the 3D SRR-metal wire, which exhibits negative refractive index properties ($\varepsilon < 0$ and $\mu < 0$) for the realization of metamaterial resonator, therefore restrict the uses in planar microwave circuit applications [214]. For the application in planar microwave devices, two-dimensional (2D) transmission-line networks can be periodically loaded with

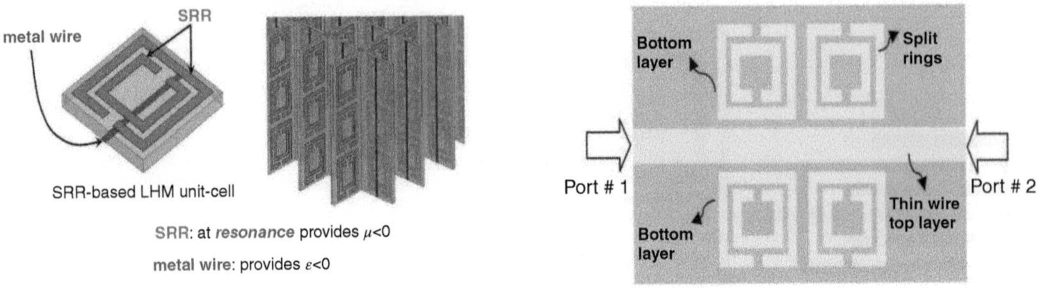

Figure 5.62 Typical coupled SRR, exhibits negative permittivity and permeability at resonance. *Source*: [227]/John Wiley & Sons.

series capacitors and shunt inductors for the realization of negative refractive index, such medium makes left-handed material and supports propagating backward waves exhibiting a negative refractive index over a wide bandwidth [216].

5.12.2 Transmission Line Model

A typical transmission line is a uniform and homogenous structure with invariant cross-section along the direction of propagation, which can transmit signals from zero to very high frequencies [221]. In general, TLM can be classified as right-handed (printed microstripline). The Left-handed (metamaterial) material is not available in nature and can be artificially created, known as CRLH (composite right/left-handed) material. Transmission line model (TLM) assumes metamaterial as an isotropic and homogeneous medium, contrary to resonator model. A TLM can be represented by distributed series impedance (Z) and shunt admittance (Y) in terms of electric permittivity (ε) and magnetic permeability (μ) [214] as

$$Z'(\omega) = \frac{Z(\omega)}{d} = j\omega\mu(\omega) \Rightarrow \mu(\omega) = \frac{Z(\omega)}{j\omega d};\ Z' \text{ is impedance per unit length} = Z(\omega)/d \tag{5.271}$$

$$Y'(\omega) = \frac{Y(\omega)}{d} = j\omega\varepsilon(\omega) \Rightarrow \varepsilon(\omega) = \frac{Y(\omega)}{j\omega d};\ Y' \text{ is admittance per unit length} = Y(\omega)/d \tag{5.272}$$

The effective propagation constant $\gamma(\omega)$ of right-handed (normal) material is given by

$$\gamma(\omega) = \sqrt{Z'(\omega)Y'(\omega)} \approx j\beta = j\omega\sqrt{\mu(\omega)\varepsilon(\omega)} \tag{5.273}$$

The series inductance and shunt capacitance of right-handed material are given by

$$L = \mu_O d,\ L' = \frac{L}{d} = \mu_O,\ C = \epsilon_O\epsilon_r d,\ C' = \frac{C}{d} = \epsilon_O\epsilon_r \tag{5.274}$$

$$\beta = \omega\sqrt{\mu(\omega)\varepsilon(\omega)} = \omega\sqrt{L'C'} \tag{5.275}$$

From (5.265) to (5.269), the effective permittivity is connected with series inductance, and permeability is connected with shunt capacitance for right-handed material.

For left-handed material (metamaterial) that exhibits negative permittivity and permeability, change in polarity is implemented for the realization of negative series impedance for (5.265) and negative admittance for (5.266). There exists a reverse symmetry between right-handed medium and left-handed medium; hence, series inductor and shunt capacitor in RHM are replaced by series capacitor and shunt inductor in the unit cell of the left-handed material (metamaterial) as shown in Figure 5.63 [213, 214].

$$Z'(\omega) = \frac{1/j\omega C}{d} = j\omega\mu(\omega) \Rightarrow \mu(\omega) = \frac{\frac{1}{j\omega C}}{j\omega d} = -\frac{1}{\omega^2 C'} \tag{5.276}$$

$$Y'(\omega) = \frac{1/j\omega L}{d} = j\omega\varepsilon(\omega) \Rightarrow \varepsilon(\omega) = \frac{\frac{1}{j\omega L}}{j\omega d} = -\frac{1}{\omega^2 L'} \tag{5.277}$$

From (5.270) to (5.271), the LHM distributed series capacitance $C' = C/d$ and shunt inductance $L' = L/d$.

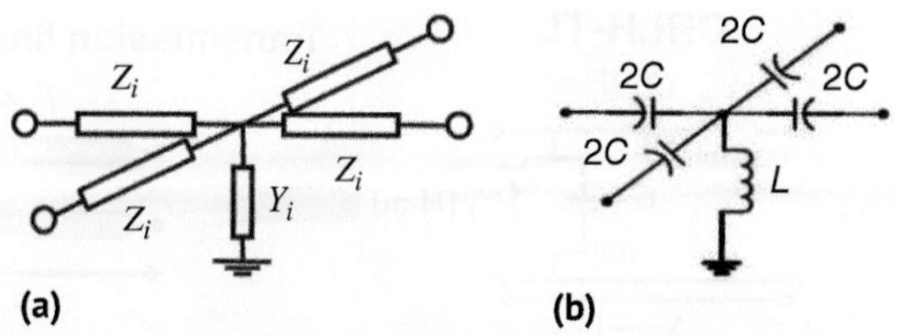

Figure 5.63 Equivalent circuit of the metamaterial unit cell: (a) general model, (b) LC model (using shunt inductors for affecting ε_r and series capacitors for affecting μ_r). *Source*: Modified from [60].

The effective propagation constant $\gamma(\omega)$ of left-handed (Metamaterial) is shown in Figure 5.63 and is given by [295–301]

$$\gamma(\omega) = \sqrt{Z'(\omega)Y'(\omega)} \approx j\beta = j\omega\sqrt{\mu(\omega)\varepsilon(\omega)} = j\omega\sqrt{\left(\frac{1/j\omega C}{d}\right)\times\left(\frac{1/j\omega L}{d}\right)} = \beta = -\frac{1}{\omega\sqrt{L'C'}} \tag{5.278}$$

From (5.272), the phase velocity and group velocity of left-handed material is given by

$$v_p(\omega) = \frac{\omega}{\gamma(\omega)} = -\omega^2\sqrt{L'C'} \tag{5.279}$$

$$v_g(\omega_0) = \frac{d\omega}{d\gamma} = \omega^2\sqrt{L'C'} \tag{5.280}$$

From (5.272) and (5.273), direction of group velocity is opposite to the phase velocity, and group velocity has positive value.

This implies that phase will propagate backward and the group velocity direction (Poynting vector) is forward as the direction of energy flowing away from the source. In an isotropic medium for RHM, the phase and group velocity has same direction, therefore no dispersion phenomenon. However, in LHM the phase velocity and group velocity are 180° out of phase (direction is antiparallel with group velocity direction), which implies medium is dispersive. From (5.273), the negative phase velocity in LHM medium leads to the negative refractive index, which can be given by

$$n = \frac{c}{v_p(\omega)} = -\frac{\sqrt{\mu(\omega)\varepsilon(\omega)}}{\sqrt{\mu_o\varepsilon_o}} = -\frac{1}{\omega^2\sqrt{L'C'\mu_o\varepsilon_o}} \tag{5.281}$$

$$c = \frac{1}{\sqrt{\mu_o\varepsilon_o}} \tag{5.282}$$

where c is the velocity of light in free space.

5.12.3 Physical Realization of Metamaterial Component for Oscillator Circuits

Cascaded composite right/left-handed structures (CRLH) with unit dimension much smaller than the guided wavelength behave as metamaterial, producing simultaneous negative permittivity (ε_r) and permeability (μ_r) properties. A CRLH-TL consists of a conventional transmission line (TL), or a right-handed (RH) TL, and its dual, consisting of series capacitors (C_L) and shunt inductor (L_L), termed a left-handed (LH) TL as shown in Figure 5.64. CRLH structures may be realized in many different forms, such as in microstrip TL, substrate integrated waveguide (SIW), and CPW [231]. These types of TLs can be designed to provide a nonlinear phase response with respect to frequency [328–346]. With proper coupling and termination (either open or short), CRLH resonators can be designed.

Comparison between the dispersion relations of the conventional resonators and the CRLH resonator under unbalanced condition is illustrated in Figure 5.65. Note that the spurious (higher-order) resonances are not at the harmonic frequencies, and the mode spacing may be manipulated. Balanced condition may also be achieved by overlapping the shunt and series resonant frequencies, thereby creating smooth transition from LH to RH regions. CRLH resonator exhibits a unique resonance $m = 0$, at the frequencies:

$$\omega_{\text{res}}^{\text{open}} = \omega_{sh} = \frac{1}{\sqrt{L_L C_R}} \tag{5.283}$$

$$\omega_{\text{res}}^{\text{short}} = \omega_{se} = \frac{1}{\sqrt{L_R C_L}} \tag{5.284}$$

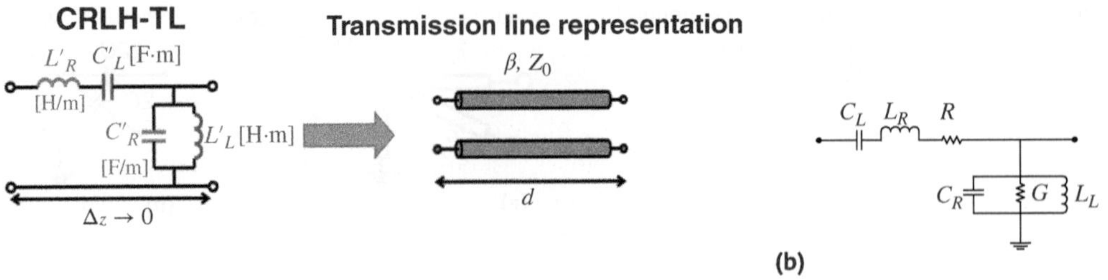

(a) (b)

Figure 5.64 A typical equivalent circuit model of CRLH unit-cell: (a) lossless ($R = 0$) transmission line representation of CRLH unit cell and (b) lossy circuit ($R > 0$) model of CRLH unit-cell. *Source*: [231]/John Wiley & Sons.

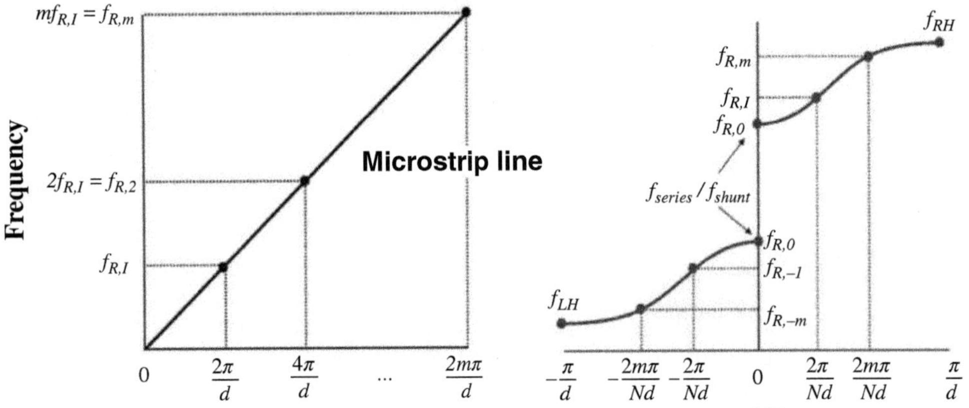

(a) (b)

Figure 5.65 Dispersion diagram of (a) conventional microstrip line and (b) CRLH transmission line. *Source*: [231]/John Wiley & Sons.

The unloaded quality factor Q_o is independent of the physical length of the structure and only depends on one of two present losses, either G or R,

$$Q_o^{\text{open}} = \frac{1}{G}\sqrt{\frac{C_R}{L_L}} \tag{5.285}$$

$$Q_o^{\text{short}} = R\sqrt{\frac{C_L}{L_R}} \tag{5.286}$$

For instance, short-circuited format is preferred in order to avoid conductor loss of the shunt inductance L_L. In addition, unlike the conventional resonator, spurious resonances can be engineered to an arbitrary frequency away from harmonic frequencies and Q-factor can be improved as compared to the conventional resonators [232].

When aforementioned zeroth-order resonators (ZOR) operate under balanced condition, the mode spacing and Q-factor become dependent. The phase slope of the resonant structure can be manipulated to either increased mode spacing/decreased Q-factor or decreased mode spacing/increased Q-factor. However, under unbalanced condition, the mode spacing and Q-factor become independent, and it has been demonstrated where first-order mode can be considerably increased (from $f_{\text{mode1}}/f_{\text{mode0}} = 3.5$ to 5.33), without sacrificing the Q-factor [231].

A high-Q resonator based on unbalanced CRLH resonator was built using lumped components with the first-order resonance occurring at around five times f_{mode0}. However, the performance was degraded due to undesired leakage when the resonator was built using distributed technology. To prevent the radiation, proper shielding may be added, or lower modes such as $n = -1, \ -2$, etc. may be used to prevent leakage of radiation. If the size is not the constraining factor, higher-order modes such as $n = +1, +2$, etc. can be explored.

Figure 5.66 shows the typical CRLH SIW structure, which is a good candidate for oscillator resonators because of high Q-factor. In addition, CRLH SIW structure can support zeroth order and negative order resonance and thereby miniaturize the circuit [235].

As shown in Figure 5.67, the one-unit-cell CRLH SIW structure and miniaturized CRLH SIW antennas based on negative order resonance. The radiation through the interdigital capacitor creates radiation loss, and it is difficult to use this as a guided-wave application, such as a cavity. Shielded CRLH SIW can eliminate the radiation by enclosing the top side with the metallic shield. The proposed structure has small size when operated in the left-handed frequency region and high quality factor due to the waveguide nature.

Figure 5.67 shows the proposed shielded CRLH SIW unit-cell and the dispersion relation of the structure. Even with the shield, this structure shows CRLH characteristics. Figure 5.68 shows possible candidates of shielded CRLH SIW resonators. It is noted that the resonators are directly excited by a strip-line type feeding structure. However, we could use microstrip line input using the transition of microstrip line to strip line.

Figure 5.69 shows the typical layout of the CRLH CPW resonator for low phase noise oscillator applications. CRLH-based ZOR using CPW technology may provide high-Q with spurious resonances far away from the harmonic resonances since this resonator dimension is not proportional to the resonant frequency as the conventional microstrip resonator. Also, when operated in zeroth-order mode, it can be miniaturized. CRLH-based ZOR in CPW technology has been used for miniaturized band-pass filter application [236].

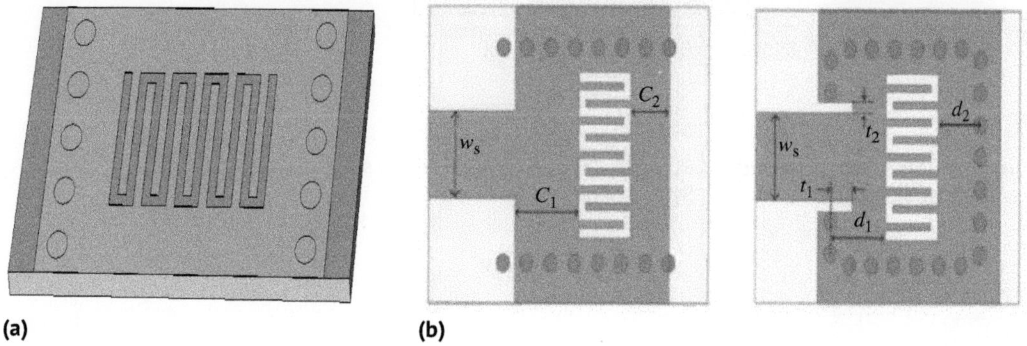

Figure 5.66 (a) The configuration of the CRLH SIW unit-cell and (b) its applications to resonator antennas. *Source:* [235]/IEEE.

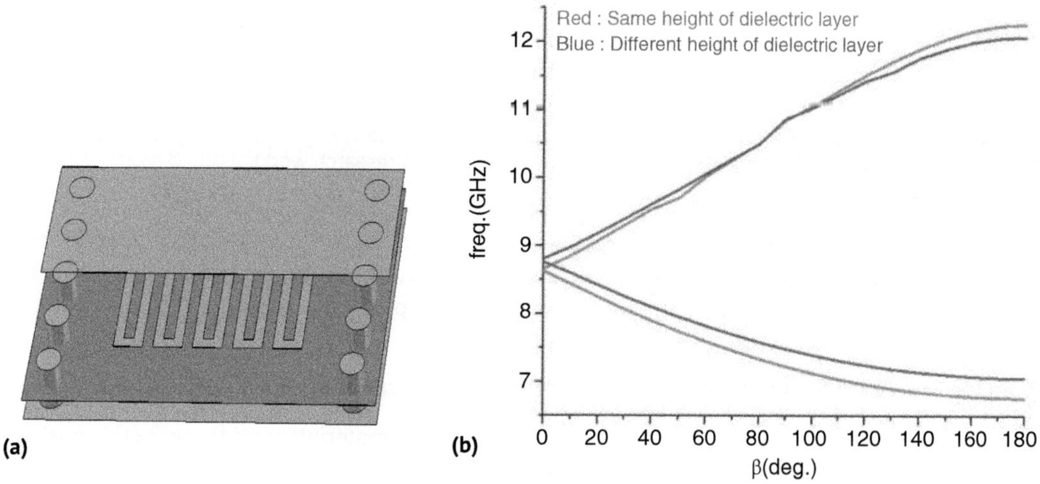

Figure 5.67 (a) The structure and (b) the dispersion relation of the shielded CRLH SIW unit-cell (a preliminary work carried out at UCLA as a joint collaboration research work with Synergy Microwave Corp).

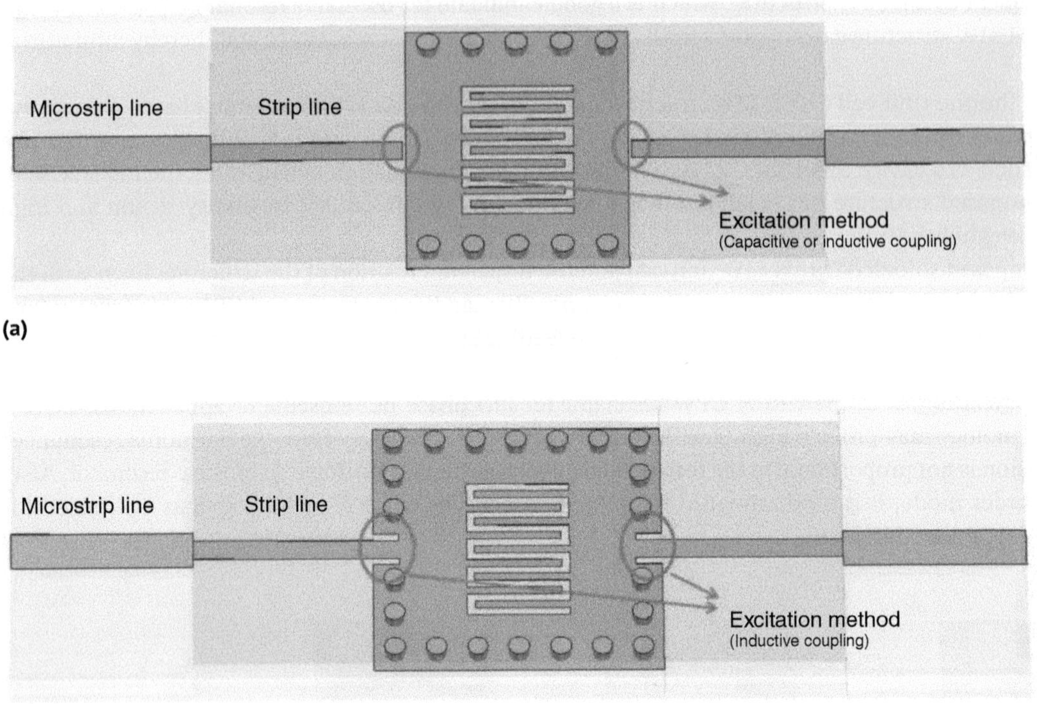

Figure 5.68 Possible resonator structures using the shielded CRLH SIW: (a) open ended and (b) short ended (a preliminary work carried out at UCLA as a joint collaboration research work with Synergy Microwave Corp).

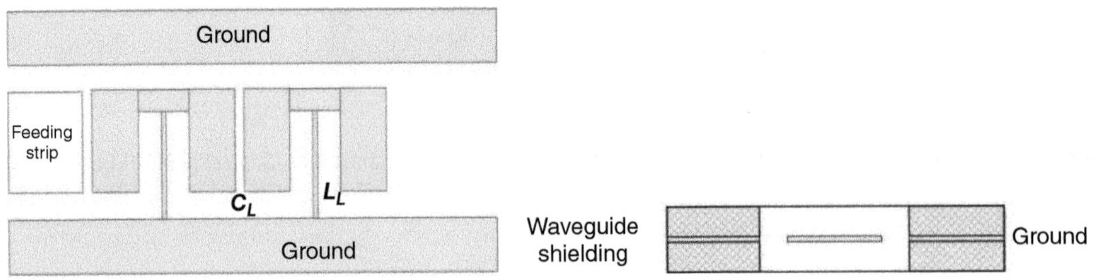

Figure 5.69 Possible resonator structure using the shielded CRLH CPW: (a) top view and (b) side view (a preliminary work carried out at UCLA as a joint collaboration research work with Synergy Microwave Corp. as a part of this research work).

Similar concept may be applied in designing resonator for an oscillator application. However, as previously mentioned, one major problem with the structure may be the undesired radiation leakage [41]. This may be alleviated by enclosing the structure inside the waveguide shield or using MIM capacitors rather than edge-coupled capacitors.

5.12.4 CSRR (Complementary Split-ring Resonator) for Oscillator Circuit Applications

Figure 5.70 shows the typical circular split-ring resonator (SRR) that is one kind of metamaterial, providing effective positive permittivity (ε) and negative permeability (μ). As shown in Figure 5.70a, SRRs can induce resonating currents in the loop and thus generate equivalent magnetic dipole moments. On the other hand, complementary split-ring resonators (CSRRs), a duality form of SRRs shown in Figure 5.70b, behave as electric dipoles that need an axial electric excitation and are able to exhibit negative permittivity [234].

Typically, waveguide structures are often utilized when it comes to the need of high Q-factor and high power-handling capability. When these waveguide structures are loaded with MTMs such as SRRs or CSRRs, they can operate below their original cutoff frequency and thus achieve size reduction. However, the main disadvantages of using conventional waveguides are high cost and the difficulty in fabrication.

In addition, it is also not easy to load MTMs into the conventional waveguides. Substrate-integrated waveguides synthesized in planar form by two rows of metallic via embedded in a dielectric substrate have served as a good candidate in realizing low-cost and highly integrated waveguide components, since SIWs can be easily fabricated using printed circuit board (PCB) technology.

Moreover, it is very convenient and efficient to create CSRRs on the surface of SIWs, which makes the structures more appealing and competitive (Figure 5.71).

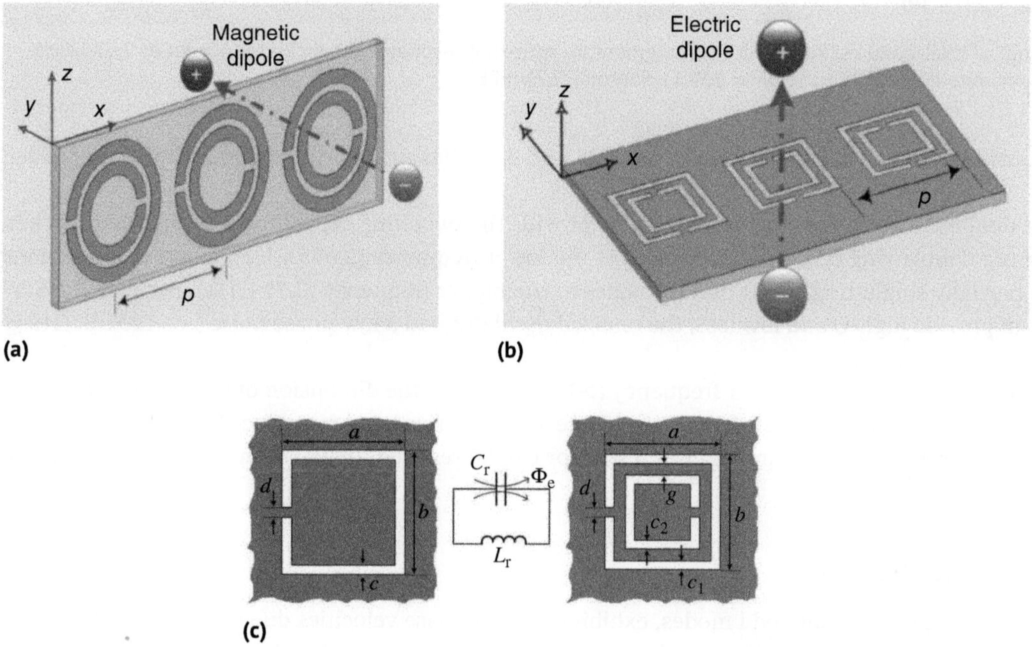

Figure 5.70 (a) SRRs, as equivalent magnetic dipoles, (b) CSRRs, as equivalent electric dipoles, and (c) equivalent LC model of CSRR. *Source*: [234]/IEEE.

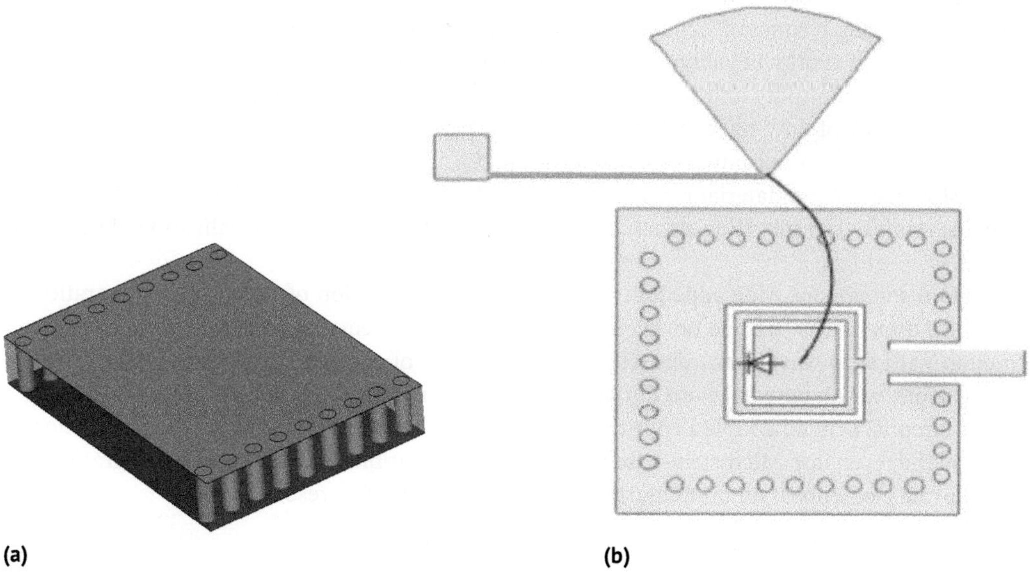

Figure 5.71 (a) A typical SIW cavity and (b) a typical layout of the SIW cavity loaded with CSRR and diode. *Source*: [237]/IEEE.

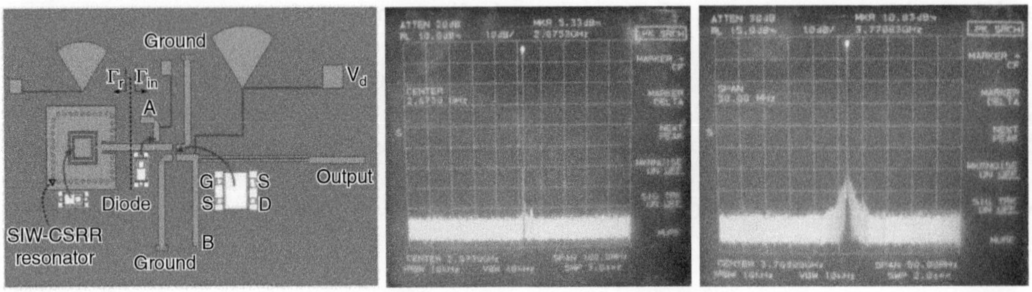

(a) (b) (c)

Figure 5.72 (a) A typical layout of dual-band oscillator with complementary split-ring resonator, (b) oscillator output at 2.675 GHz (FOM = 171.8), and (c) oscillator output at 3.77 GHz (FOM = 169.1). *Source*: [237]/IEEE.

Figure 5.72 shows the reconfigurable dual-band oscillator at 2.675 and 3.77 GHz using printed planar CSRR embedded in an SIW cavity [237].

As shown in Figure 5.71, dual-band oscillator contains a diode to provide the switching capability of the oscillator. When the diode is on, it behaves as a double-ring CSRR that resonates at the lower frequency (2.675 GHz), whereas if the diode is turned off, it becomes a partially single-ring CSRR that resonates at the higher frequency (3.77 GHz). Because the SIW cavity loaded with CSRR can provide high-Q and low loss, the realized oscillator had a low phase noise (~ −120 dBc/Hz @ 1 MHz offset).

This structure is promising to operate at even higher frequency (>15 GHz) since the dimension of the SIW can be easily scaled down. With loaded CSRR on the SIW cavity surface, it can create a very narrow band resonance below the waveguide cutoff frequency, and hence an electrically small and high-Q resonator can be realized. Installation of a shielding may be necessary.

5.12.5 Slow-wave Metamaterial Resonator

Conventional microstrip coupled lines, even and odd modes, exhibit different phase velocities due to the inhomogeneous dielectric medium [312–327]. Signals on a coupled line are the superposition of an even and odd mode, and a discrepancy between the phase velocities of the two modes is undesirable for a high-quality resonator for low-phase-noise oscillator applications [313]. Slow wave has been demonstrated in many systems by using very different methods, such as the electromagnetic induced transparency [210] and the coupled resonator structures [211, 212].

Slow waves on printed cascaded split-ring resonators or lattices of resonant elements exhibit interesting characteristics in terms of phase velocity, group velocity, energy velocity, radiation and attenuation characteristics, quality factor, and left-handed material features (negative permittivity and negative permeability) near the resonance condition [284–311]. Figure 5.73 shows the typical representation of slow-wave dynamics in series-coupled split-ring and nanoparticle resonators that exhibit left-handed and evanescent mode propagation characteristics [229].

There is significant interest in slow-wave metamaterial resonator (SWMR) for the design of low-phase-noise microwave signal sources [60]. Slow-wave metamaterial coupled line structure improves the group delay and equalization of the phase velocity.

The metamaterial-based high-Q resonators provide added advantages such as prevention of unintended generation of harmonic oscillation and structural dimensions that are not dictated by the resonant frequency. Metamaterials with their unique properties, in conjunction with slow-wave dynamics, allow for the design of compact tunable oscillators, filters, phase-shifters, and microwave sensors, which otherwise are difficult to achieve using conventional design approaches.

To design integrated oscillator circuits, printed coupled split-ring resonator structures enable readily the integration of active devices as an integral part of the resonator. Microstrip-based designs are compatible with wafer-level integration and lead to the integration of active device elements as an integral part of the coupled split-ring resonators.

5.12.6 Complementary Coupled Resonator using Substrate Integrated Waveguide Cavity

Figure 5.74 shows the geometry of a conventional microstrip coupled line resonator and the proposed complementary structure. The conventional coupled line resonator is excited at its differential mode, and its electric (E) field is plotted

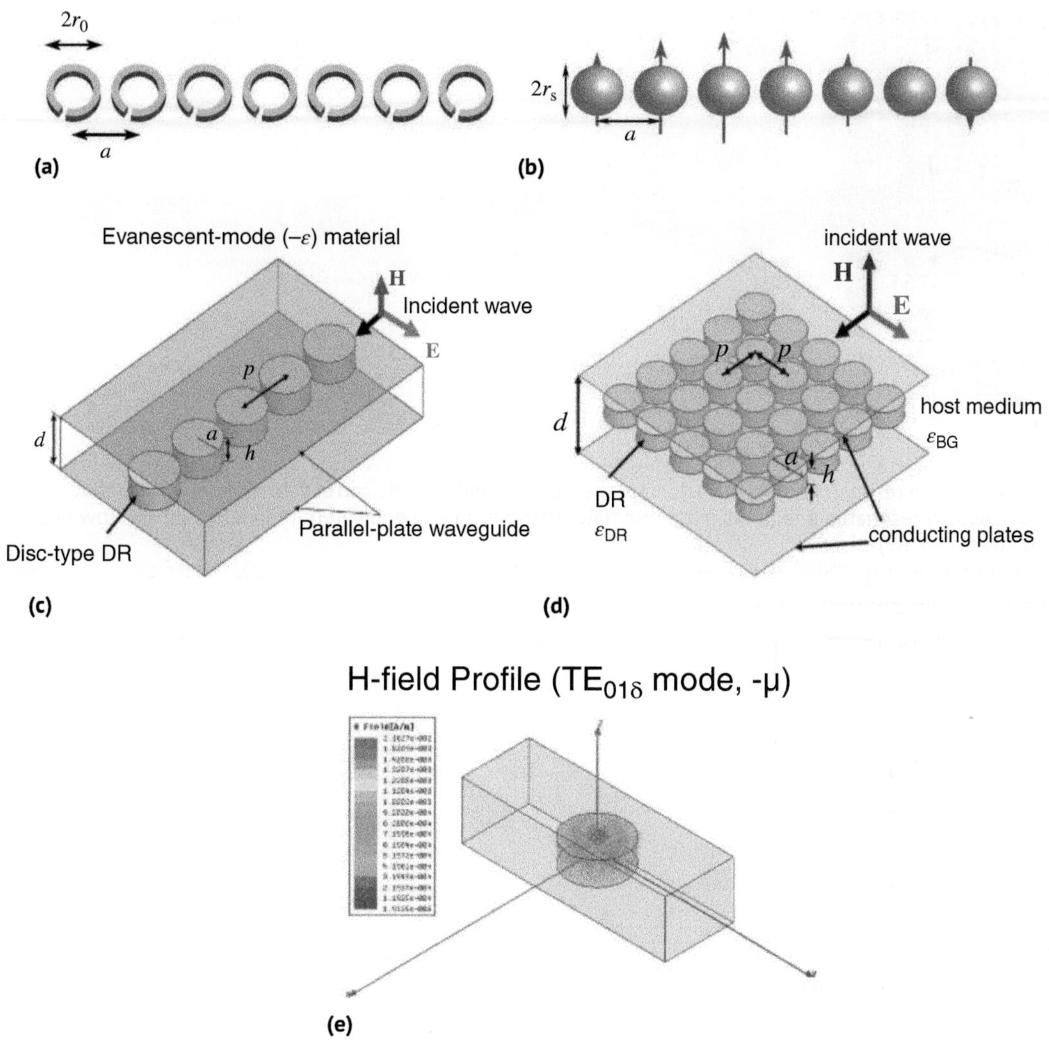

Figure 5.73 A typical representation of slow-wave dynamics in (a) series coupled split ring, (b) nanoparticle resonators that exhibit left-handed and evanescent mode propagation characteristics, (c) 1D LHM: cylindrical DRs in TE mode cutoff parallel plate waveguide (−ε), (d) 2D lattice structure for one dielectric-resonator scheme in the cutoff waveguide, and (e) 3D negative permeability material. *Source:* (a–c) Adapted from [36]; (e) adapted from [229].

in Figure 5.74b [69]. Using the principle of duality, the coupling mechanism for the proposed complementary coupled resonator is through magnetic (H) field as shown in Figure 5.74d, and has a similar distribution as E field in the conventional one [60, 69].

Unlike the conventional coupled resonator, where the differential mode needs to be excited by a pair of inputs with opposite polarity, the proposed complementary coupled resonator is excited at differential mode due to the nature of the fundamental mode in the waveguide. As a result, the equivalent magnetic currents on the slots of the complementary coupled resonator flow in opposite directions at the symmetrically opposite edges of the SIW resonator, and therefore, the radiation can be minimized, which generates a high Q-factor of the resonator.

From [28], the unloaded Q can be calculated to be about 270 at 5.35 GHz.

$$Q = \frac{\omega_0}{2}\left|\frac{Z_{11}(\omega_0)'}{Z_{11}(\omega_0)}\right| \tag{5.287}$$

where $Z_{11}(\omega_0)$ and $Z_{11}(\omega_0)'$signify the impedance and its derivative, respectively of the SIW resonator structure at a frequency of ω_0.

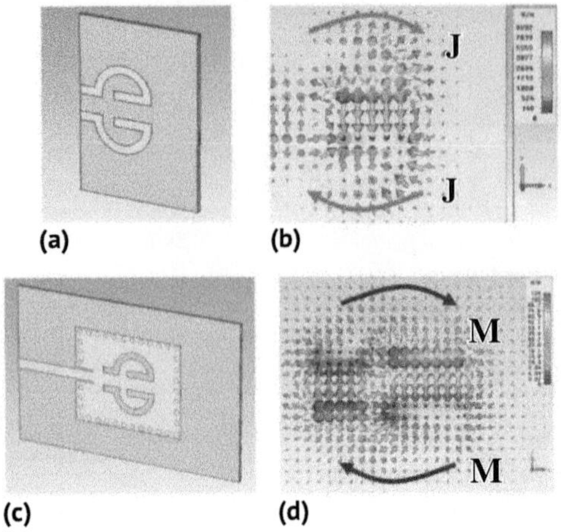

Figure 5.74 (a) A microstrip coupled line resonator, (b) its electric field distribution and electric current flow under differential excitation, (c) a complementary coupled resonator, and (d) its magnetic field distribution and equivalent magnetic current flow under the fundamental mode of SIW cavity [60, 69].

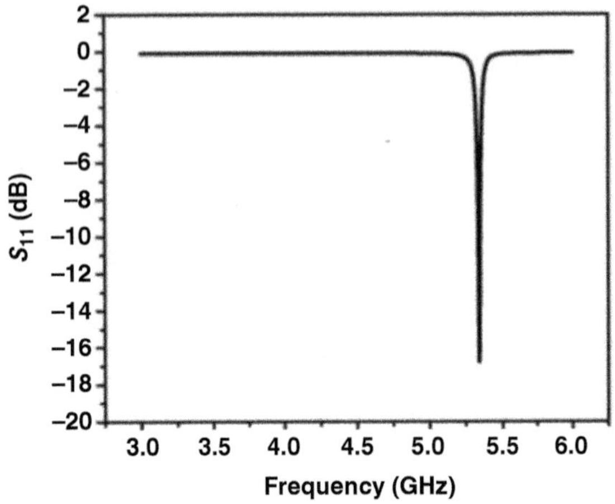

Figure 5.75 Simulated S_{11} of the 1-port resonator shown in Figure 5.74c.

Figure 5.75 shows CAD-simulated plots of S_{11} of the proposed 1-port complementary coupled resonator shown in Figure 5.74c. The resonant frequency for this fixed frequency resonator is 5.34 GHz; as its simulated unloaded Q is around 272, matches with the theoretical values within 2 dB variations.

The resonator can also be made into a 2-port structure, as shown in Figure 5.76a, in order to perform a band-pass filtering function. The simulated and measured results are plotted in Figure 5.76b, showing a good agreement. The measured insertion loss is less than 1 dB at 5.1 GHz; the measured unloaded Q of the two-port resonator can be calculated by Eq. (5.282), and is found out to be around 300 [28].

$$Q = \frac{\omega_0}{2}\left|\frac{Z_{21}(\omega_0)'}{Z_{21}(\omega_0)}\right| \tag{5.288}$$

In an attempt to realize a tunable resonator using the proposed complementary coupled resonator topology, two Varactors are exploited and placed in the location where the E field on the slot is maximum as shown in Figure 5.77a in order to achieve a wide tuning range of the resonant frequency.

Figure 5.77b shows the fabricated prototype of the tunable resonator, and Figure 5.77c plots the measured S_{11} with various reverse bias voltages of both Varactors from 0 to 20 V, indicating the resonant frequency can be tuned from 4.579 to 4.984 GHz, or 8.84% of relative bandwidth.

It is worth mentioning that the tuning range can be further increased by reducing the gap between the DC bias line and SIW cavity, which may also boost the tuning range of the oscillator. The Varactors we use here are MA/COM MA46H120.

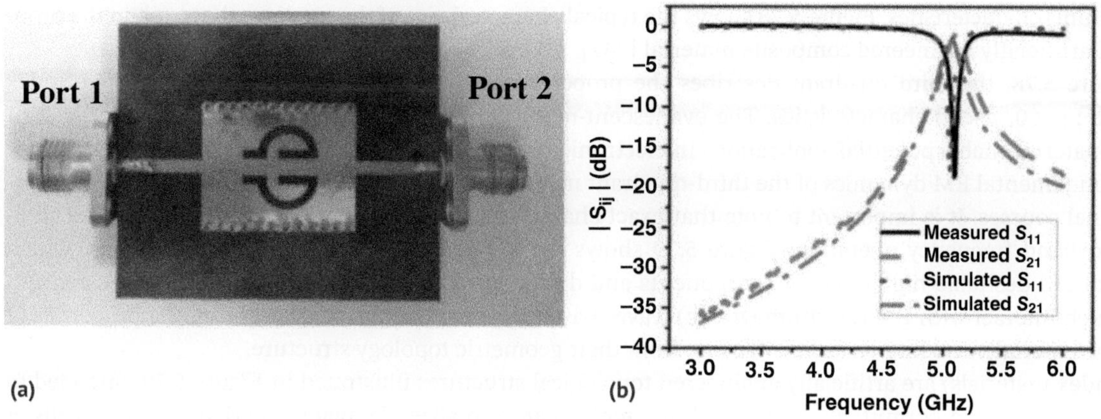

Figure 5.76 (a) Fabricated prototype of a 2-port complementary coupled resonator and (b) its S-parameters. *Source*: [69]/IEEE.

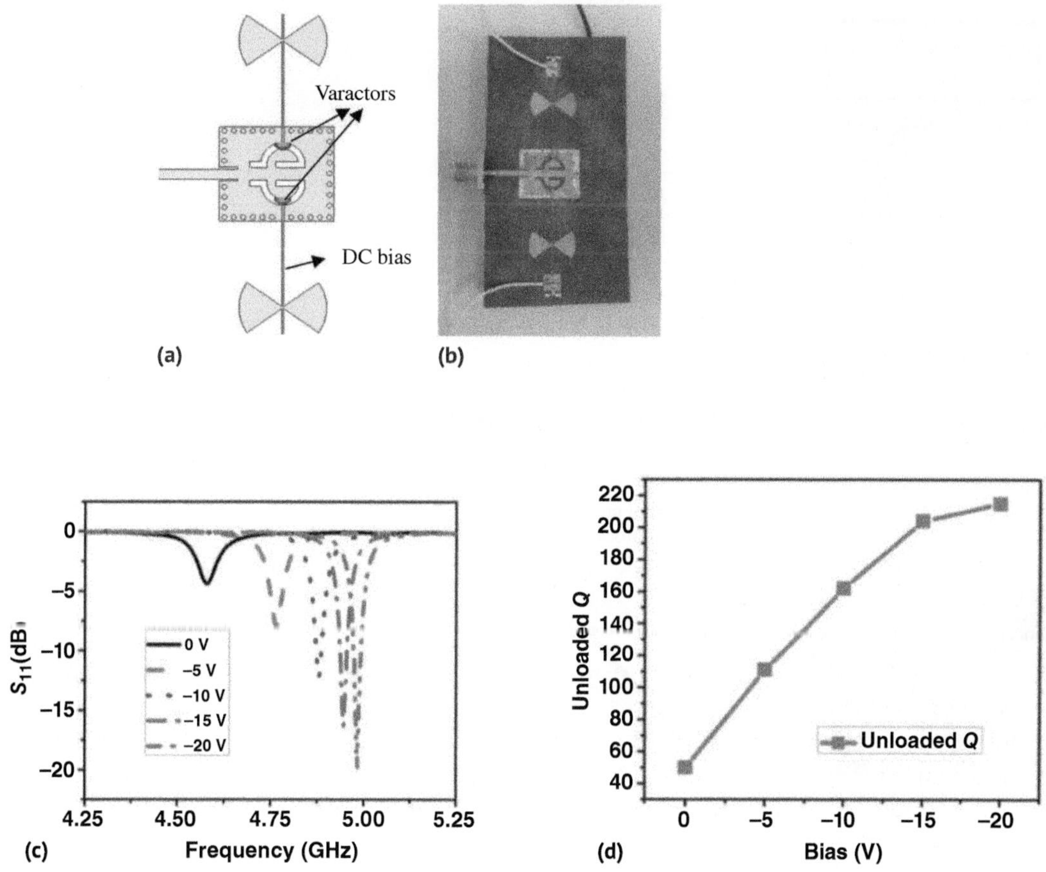

Figure 5.77 (a) Topology of the proposed resonator with the bias network, (b) fabricated prototype, (c) measured results of S_{11}, and (d) measured unloaded Q of the proposed tunable resonator with various reverse bias voltages [60, 69].

The measured unloaded Q, calculated from (5.282), of the resonator equipped with two Varactors varies from 50 to 220 as shown in Figure 5.77d.

5.12.7 Retrieval of Metamaterial Resonator's Effective Parameters

MTM (Metamaterial) is an artificial composite structure, nomenclature as NIM (negative index material), which exhibits negative index ($n = \sqrt{\mu\varepsilon}$; $\varepsilon < 0$, $\mu < 0$, where n is refractive index, ε is electric permittivity of medium, and μ is magnetic permeability of medium) characteristics. Figure 5.78 shows the typical characteristics of the medium that explain the properties of natural and artificially engineered composite material [347].

As shown in Figure 5.78, the third quadrant describes the properties of LHM that possesses NIM (negative index material: $n = -\sqrt{\varepsilon\mu}$; $\varepsilon < 0$, $\mu < 0$) characteristics. The evanescent-mode energy storing and resonant characteristics of the third-quadrant material enable potential applications in electronics, medical, space, and optics. Hence, it is important to understand the fundamental EM dynamics of the third-quadrant material ($\varepsilon < 0$, $\mu < 0$) for designing NIMs resonator-based electronic signal sources. It is important to note that exact characterization of ε and μ of NIM structure is a challenging task for wideband frequency operation. Figure 5.79 shows the MTM-inspired grapheme nano-ribbons, which possess topology-induced desirable high Q-factor components and desirable electromagnetic properties. Unlike conventional materials, which interact with EM (electromagnetic) waves based on their chemical composition, the properties of NIM (negative index material), also known as MTM, come from their geometric topology structure.

NIMs (negative index materials) are artificially engineered topological structures illustrated in Figure 5.79, intended to modify the bulk permeability and permittivity of the medium, which can be realized by placing periodically unit cells of NIM structures under the constraints of their size (typically much less than the wavelength of incident EM wave). It is

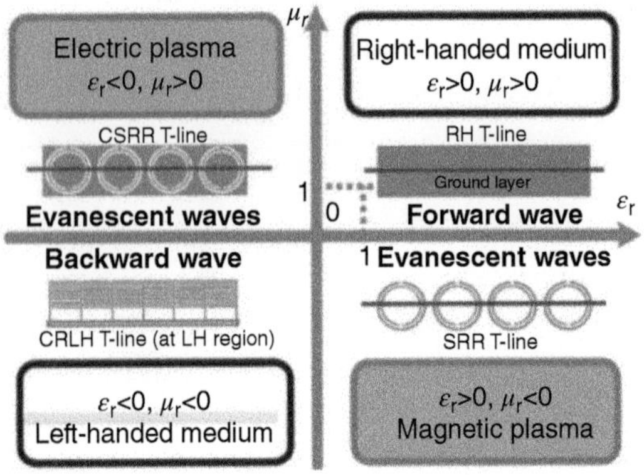

Figure 5.78 A typical representation of the medium: the third quadrant explains LHM, NIM (negative index) material properties. *Source*: Adapted from [347].

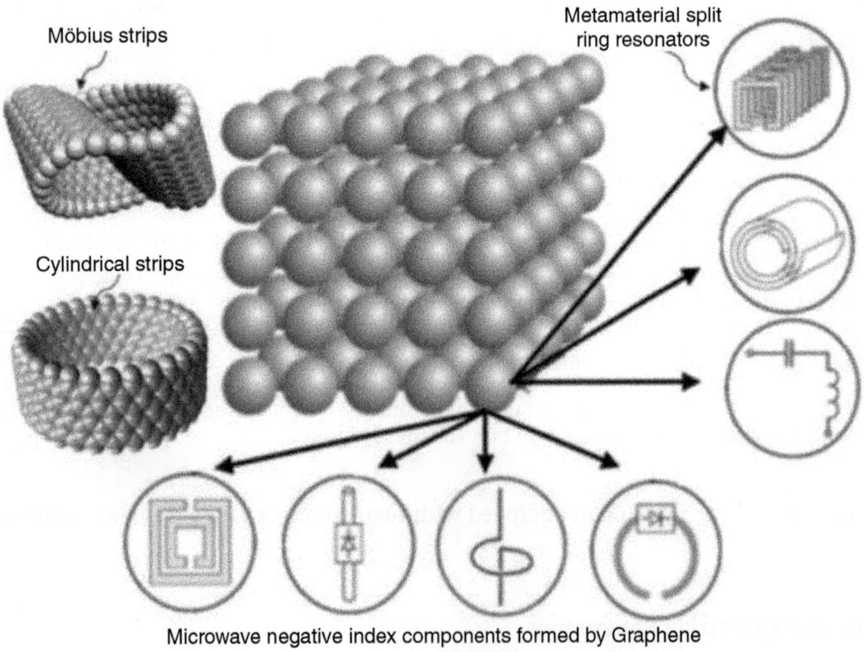

Figure 5.79 Möbius-strips structure formed by graphene nanoribbons exhibits negative index properties ($\varepsilon < 0$, $\mu < 0$), well-suited for high-quality factor microwave frequency components. *Source*: Adapted from [381].

interesting to note that by changing the position, orientation, excitation of NIM structures, important parameters (permittivity, permeability, refractive index, transmission, reflection, impedance, and coupling) can be tuned and optimized for desired applications. There are many ways to vary the properties of NIM, such as by using ferrites, liquid crystals, frequency-selective surfaces, and manipulating the lattice structure; but these methods are band-limited [347–385].

While it is possible to characterize the NIMs in terms of refractive index (n) and normalized impedance (z), these parameters do not provide meaningful information. The ongoing effort by research scientists is to categorize NIM-structure by effective constitutive parameters: $\varepsilon_r = n/z$, $\mu_r = n \times z$, which gives a direct and understandable interpretation of medium such as NIMs.

Figure 5.80 shows the typical circular and square split ring structures for the realization of NIM (negative index material); the size of SRs (split rings) and the distances between them are kept smaller than the wavelength (λ) [373–384]. Under this assumption, SR-NIM structure may present bulk properties and can be characterized by a macroscopic model, with effective index parameter (ε and μ). The expression for index (n) is given by

$$n = \pm\sqrt{(\pm\varepsilon)(\pm\mu)} = \pm\sqrt{\mu\varepsilon} \tag{5.289}$$

From (5.283), the positive sign is used for n when $\varepsilon, \mu > 0$, whereas the negative sign is used when $\varepsilon, \mu < 0$. The assumption of positive index ($n > 0$) implies that the index n is a scalar, and does not depend on frequency. But in reality, this is not always true. The energy of the field (W) would be negative ($W = \varepsilon E^2 + \mu H^2$) value when ε and μ are negative, and this is not possible, hence ε and μ bound to depend on frequency and fall into complex domain [2]. The unified expression of energy W can be described by

$$W = \frac{\partial(\omega\varepsilon)}{\partial\omega}E^2 + \frac{\partial(\omega\mu)}{\partial\omega}H^2 \tag{5.290}$$

where $\varepsilon(\omega) = \varepsilon'(\omega) + j\varepsilon''(\omega)$ and $\mu(\omega) = \mu'(\omega) + j\mu''(\omega)$

From (5.284), the electrodynamics of negative index materials ($n < 0$; $\varepsilon < 0$ and $\mu < 0$) would give positive value of W for a very broad class of dispersive material characterized by $\varepsilon(\omega)$ and $\mu(\omega)$, exhibit dependency on frequency.

Any material supporting single propagating mode at a known frequency usually exhibits well-defined index (n), despite the material is homogeneous/continuous or not. But it is not easy to assign normalized impedance (z) to a nonhomogeneous material, because z strongly depends on surface termination and overall size of the material [347]. The classical homogenization theories are typically valid when the unit-cell size is insignificant with respect to the wavelength (the zero frequency limits) and thus might be expected to result in a poor description of negative index materials. The key supposition is that if the wavelength of the incident wave is much larger than the size and spacing between the NIM unit cell element, then it is possible to perform homogenization, and to obtain an approved value of z within an acceptable limit of errors [348, 349].

Assuming the size of NIM structure is much smaller than $\lambda/10$, homogenization theory can be applied to topology-inspired Möbius strip structure, averaging of Maxwell's equations over small volumes (with respect to wavelength) can be performed to get analogous uniform continuous harmonized medium, described by effective constitutive parameters (ε and μ). It is possible to retrieve values for the complex refractive index (n) and wave impedance (z) of inhomogeneous periodic NIMS structure, by prior information about termination of the NIMS-unit cell structure, phases, and amplitudes of the waves transmitted/reflected from the NIMS sample. But the challenge is if the NIMS structures are not symmetric along

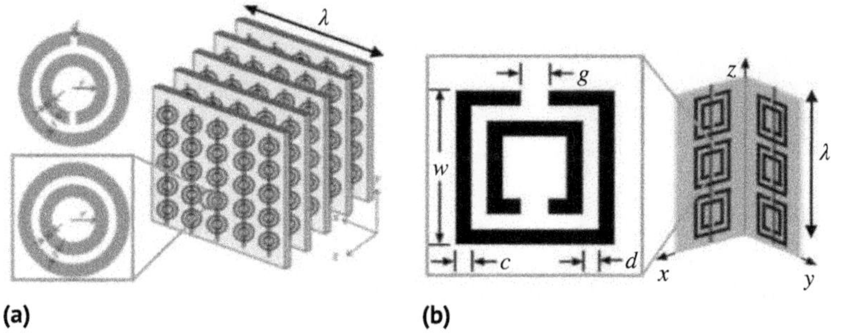

Figure 5.80 Typical SRRs and the construction of NIM with closed SRR wire structure: (a) circular split ring and (b) rectangular split ring [373–384].

the wave propagation direction, two different values of impedance (z) are retrieved corresponding to two incident directions of propagation. This ambiguity in the computation of impedance (z) leads to a fundamental ambiguity in the definitions of ε and μ for MIMS, which increases as the ratio of NIMS unit cell dimension to wavelength increases.

We can possibly determine index (n) and wave impedance (z) of nonhomogeneous periodic MTM structure but requires prior information about the termination of unit cell, phases, and amplitudes of the transmitted/reflected wave from the MTM structure. The problem is that if the structures are not symmetric along the wave propagation direction, lead to two different values of "z" corresponding to two incident directions of propagation. This creates uncertainty in the computation of "z," ε and μ, increases as the ratio of MTM unit cell size to wavelength increases [347–354]. The simplest approach to the theoretical formulation of SR-inspired NIM (negative index material) can be deduced from an equivalent circuit model that gives insight into the relationship between the physical properties and geometrical parameters of SRRs, as shown in Figures 5.80 and 5.81. SRR is considered as the simplest unit MTM cell.

Figure 5.80 shows the typical SRR unit cell structure for a theoretical formulation that gives brief insights into physical properties and geometrical parameters. The analytical values of constitutive parameters (ε_{zz}, μ_{yy}, ξ_0) is given by [351, 359, 363]

$$\varepsilon_z = 1 + \frac{1}{\varepsilon_0 a^3}\left\{\varepsilon_0 \frac{16}{3} r_{\text{ext}}^3 + 4d_{\text{eff}}^2 r_0^2 C_{\text{pull}}^2 L\left(\frac{\omega^2}{\omega_0^2 - \omega^2 - i\omega\gamma}\right)\right\} \tag{5.291}$$

$$\mu_y = 1 + \frac{\mu_0}{a^3}\left\{\frac{\pi^2 r_0}{L}\left(\frac{\omega^2}{\omega_0^2 - \omega^2 - i\omega\gamma}\right)\right\} \tag{5.292}$$

$$\xi_0 = -i\sqrt{\frac{\mu_0}{\varepsilon_0 a^3}}\left\{-2ir_0^3 d_{\text{eff}} C_{\text{pull}} \frac{\omega_0^2}{\omega}\left(\frac{\omega^2}{\omega_0^2 - \omega^2 - i\omega\gamma}\right)\right\} \tag{5.293}$$

where $d_{\text{eff}} = c + s$, $\gamma = \frac{R}{L}$, $r_0 = r + c + s/2$ and $r_{\text{ext}} = r + 2c + s$; L and C_{pull} are SRR total inductance and capacitance between the rings. As depicted in Figure 5.80, SRR acts as a resonant LC circuit, and the frequency of the resonance is given by

$$\omega_0^2 = \frac{1}{\left(\frac{1}{2}\pi r_0 C_{\text{pull}} + C_{\text{gap}}\right)L} \tag{5.294}$$

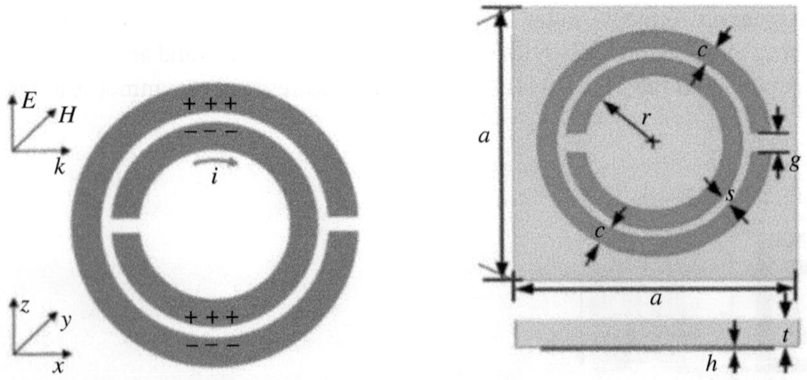

Figure 5.81 A typical SRR unit cell shows the asymmetric rings. *Source*: [366]/American Physical Society.

L and C_{pull} is given by analytical equations [350]. C_{gap} is the capacitance of the gap illustrated in Figure 5.80, for narrow gap, its capacitance can be theoretically estimated, as reported in [351]

$$C_{\text{gap}} = \frac{\varepsilon_0 ch}{g} + C_0 \text{ with } C_0 = \varepsilon_0 (h + c + g) \tag{5.295}$$

For example, the dimension of SRR is: $a = 3$ mm, $r = 0.74$ mm, $c = 0.2$ mm, $s = 0.1$ mm, $g = 0.2$ mm, $t = 0.3$ mm, $h = 35$ μm, $\varepsilon_0 = 8.8541 \times 10^{-12}$F/m.

From (5.283) to (5.287), L = 3 : 9287 × 10^{-9} H, $C_{\text{pull}} = 1 : 6806 \times 10^{-11}$ F, $C_{\text{gap}} = 4 : 155 \times 10^{-15}$ F and $R = 0 : 4901$ Ω at the frequency of resonance.

The numerical values obtained using Eqs. (347)–(350) are approximate, resulting in poor accuracy of analytical method, errors increase for thicker structures. The alternative technique is based on S-parameters extraction. There are three imperative methods generally used for the determination of effective constitutive parameters based on applications and acceptable limits of errors [349]. The first method is to numerically calculate the ratios of the EM (electromagnetic) field inside MTM's structure, but this approach is good for numerical simulations, and not for experimental measurement [352]. The second method calculates the effective constitutive parameters by using analytical models of MTM's structures, including numerical averaging of fields. But this technique is also not suitable for complex structures such as superconducting MTM. The third method is a retrieval technique based on the inversion of scattering data (S-parameters) of a finite slab, called the NRW (Nicolson–Ross–Weir) procedure [353–355].

The problem with NRW method arises in cases when topology-driven MTM's structure undergoes asymmetric reflections. Smith et al. [348] reported a modified approach to resolve asymmetric reflections concern with the help of averaged value of reflection coefficients. As we know that isotropic medium model cannot replace this property; hence, it is important to understand the EM dynamics of SR-inspired MTM medium that intrinsically possesses reflection asymmetric structures ($S_{11} \neq S_{22}$) [376].

5.12.8 NIMs Inspired Resonator for Oscillator Circuit

NIMs resonator can be classified into two types: resonant type and nonresonant type: examples are SRR, and left-handed (LH) transmission line (T-line). Resonant-type metamaterial exhibits larger dynamic range for material parameter ($\varepsilon_r < 0$, $\mu_r < 0$), while nonresonant type offers wide bandwidth and lower loss. Traditional metamaterial transmission lines (T-lines) can also be categorized into resonant type and nonresonant type, represented by (split-ring-resonator) SRR/(complementary split-ring-resonator) CSRR-loaded T-line, and composite right/left-handed (CRLH) T-line, with the corresponding layout shown in Figures 5.82 and 5.83, where C and L are per-unit length capacitance and inductance that

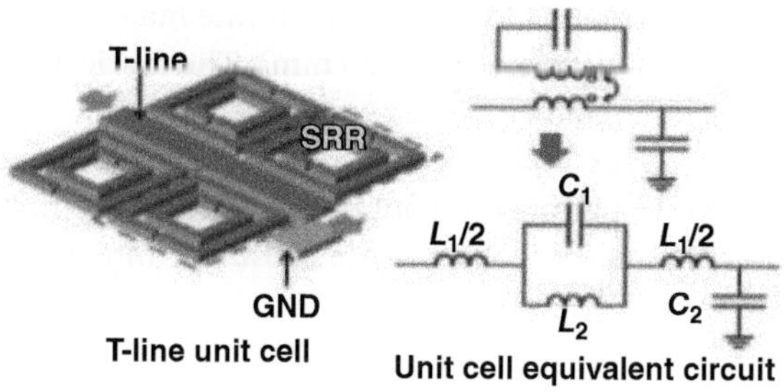

Figure 5.82 SRR loaded T-line unit cell and equivalent lumped LC model [344].

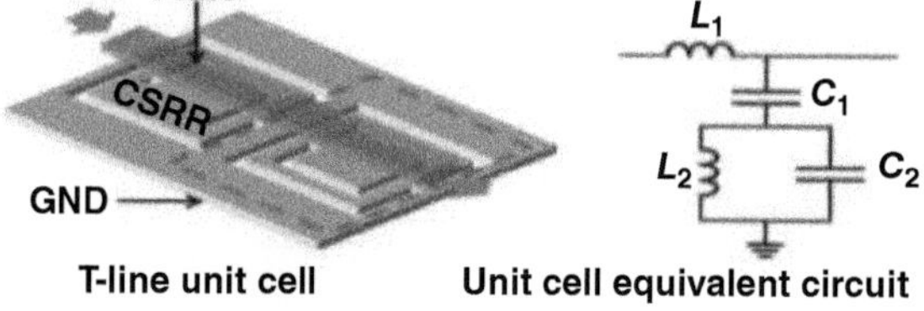

Figure 5.83 CSRR-loaded T-line unit cell and equivalent lumped LC model [344].

determine the metamaterial properties of T-lines. For T-line loaded with SRR (Figure 5.82), L_1 and C_2 linked with T-line, whereas C_1 and L_2 are linked to magnetic coupling. Figure 5.82 shows the typical split-ring resonator and the construction of NIM.

The physical properties of Figure 5.82 are given by:

$$\varepsilon = \frac{Y}{j\omega} = C_2 \text{ is} + \text{ve} \rightarrow \varepsilon > 0\,(+ve) \tag{5.296}$$

$$\mu = \frac{Z}{j\omega} = \frac{L_1 + L_2 - \omega^2 L_1 L_2 C_1}{1 - \omega^2 L_2 C_1} \tag{5.297}$$

$$\mu < 0\,(-\text{ve}) \text{ for } \left(\frac{1}{L_2 C_1}\right)^{1/2} < \omega < \left(\frac{L_1 + L_2}{L_1 L_2 C_1}\right)^{1/2} \tag{5.298}$$

From (5.290) and (5.291), Re (μ) is negative and Re (ε) is positive. For ($\varepsilon > 0$, $\mu < 0$), propagating waves become evanescent waves; therefore, energy cannot propagate through the resonator, reflected back to establish a standing-wave.

As a result, SRR-loaded T-line structure stores the energy and forms a high-Q resonator tank circuit for low phase noise oscillator application. Similarly, for T-line loaded with CSRR (Figure 5.82), L_1 associated with T-line, C_1, C_2, and L_2 account from electric coupling of CSRR on the ground. The physical properties of Figure 5.83 are given by

$$\mu = \frac{Y}{j\omega} = L_1 \text{ is } + \text{ve} \rightarrow \mu > 0\,(+\text{ve}) \tag{5.299}$$

$$\varepsilon = \frac{Y}{j\omega} = \frac{C_1(1 - \omega^2 L_2 C_2)}{1 - \omega^2 (C_1 + C_2) L_2} \tag{5.300}$$

$$\varepsilon < 0\,(-\text{ve}) \text{ for } \left(\frac{1}{(C_1 + C_2) L_2}\right)^{1/2} < \omega < \left(\frac{1}{L_2 C_2}\right)^{1/2} \tag{5.301}$$

From (5.293) and (5.295), Re (μ) is positive and Re (ε) is negative, in this case ($\varepsilon < 0$, $\mu > 0$), electric plasma is formed, where propagating waves become evanescent waves; hence, energy cannot propagate through the resonator either and is reflected back to establish a standing-wave. As a result, CSRR-loaded T-line can also be used to form a high-Q resonator.

Figure 5.84a shows the structure of the broadside-coupled split-ring resonator; there are two relative slips S_x and S_y between the two broadsides coupled rings. Figure 5.84b shows the corresponding equivalent circuit, where L, R are the equivalent inductance and resistance, respectively; C_0 is mutual capacitance between the two rings; C_s is the capacitance of the split. As shown in Figure 5.84a, the tuning can be realized by minor slips along and perpendicular to the gap direction of two broadside-coupled rings [368].

Figure 5.85 shows the (a) layout of the broad coupled split ring (dark region denotes etched on signal plane, and light region denotes broadsides coupled triangular SR etched on ground plane), and (b) equivalent lumped circuit.

For the validation of reported research work, resonators were fabricated on low-loss quartz and Borosilicate material. As shown in Figure 5.86, the metal structures are processed on top of quartz wafers (diameter $D = 150$ mm). The fabrication process uses a double-sided coating of the substrates using a metallization, sputtered metal gold (Au) is used.

Unlike SRR/CSRR as depicted in Figure 5.86, the geometry of Negative Index Möbius-Strips (NIMS) is conformal, continuous, and maps one-to-one onto itself. These unique properties of NIMS permit EM coupling in such a way that signal coupled to loop does not encounter any obstructions when travelling around the loop, emulate like an infinite transmission line, hence large group delay and high Q-factor.

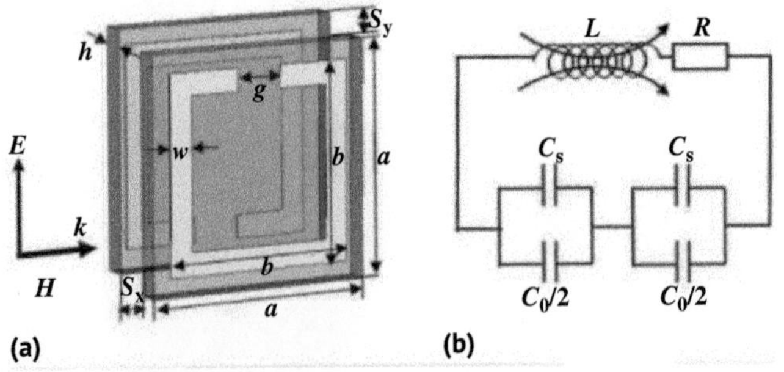

Figure 5.84 (a) A broadside-coupled split-ring resonator and (b) equivalent circuit. *Source*: [368]/The Electromagnetics Academy.

Figure 5.85 (a) Layout of the BCTSR (dark region denotes BCTSR etched on signal plane, and light region denotes broadsides coupled triangular SR etched on ground plane) and (b) equivalent lumped circuit. *Source*: [368]/The Electromagnetics Academy; [477]/ Microwave Journal.

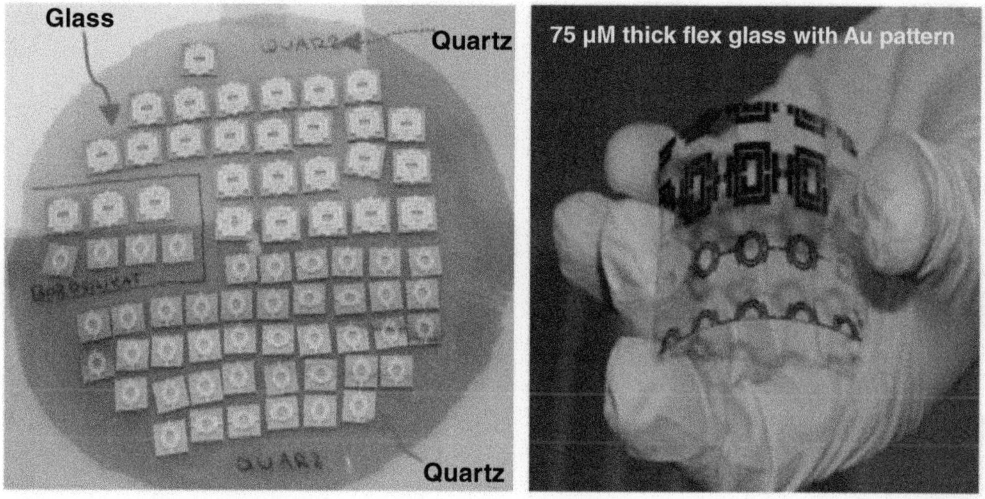

Figure 5.86 Photos of metamaterial resonator fabricated on glass/quartz.

5.12.9 Electromagnetic Bandgap Resonator

The limited spectral resources at microwave frequencies and the demand for high data capacity have led us to seek unexploited frequency resources in millimeter and sub-terahertz (THz) wave bands [385].

Enrico Lia et al. [386] reported millimeter wave EBG (electromagnetic bandgap resonator for the application in millimeter wave oscillators. The design of the silicon EBG resonator relies upon the creation of a surrounding region of the silicon exhibiting an EBG, and this is realized using a triangular lattice of air holes perforating the planar silicon wafer. The resonator is formed by the concatenation of unit cells with a small increase in horizontal width period from the central unit cell to the outer unit cells, as shown in Figure 5.87 to confine the resonant mode.

The periodic EBG structure shown in Figure 5.87 prohibits propagating modes within its bandgap, and the energy stored inside a localized mode cannot be transported through this region of the silicon wafer. For this type of resonator, the approach employs the idea of reducing the dielectric filling factor of the resonator to reduce overall losses. The filling factor (κ) is defined as the ratio of the electromagnetic energy of a resonator stored inside the silicon to the total stored electromagnetic energy. By lowering (κ) the achievable quality factor increases. To minimize out-of-plane losses, it is necessary to shape the electric and magnetic fields in the resonator to reduce wave-vector components inside the light cone.

A common approach is to create a Gaussian envelope of the field by linearly increasing reflection from the canter to outer unit cells. This is achieved by varying the horizontal period of the unit cells quadratically along the x-directions, as shown in Figure 5.87.

As shown in Figure 5.88, Enrico Lia et al. [386] reported design of low filling factor EBG resonator, formed by the concatenation of unit cells with a gradual increase in horizontal width from the central unit cell to the outer unit cells. The unit cells are engineered so that most of the energy is concentrated in an air gap rather than in the silicon. The simulated unloaded Q-factor obtained was 89,400, and measured 108,300.

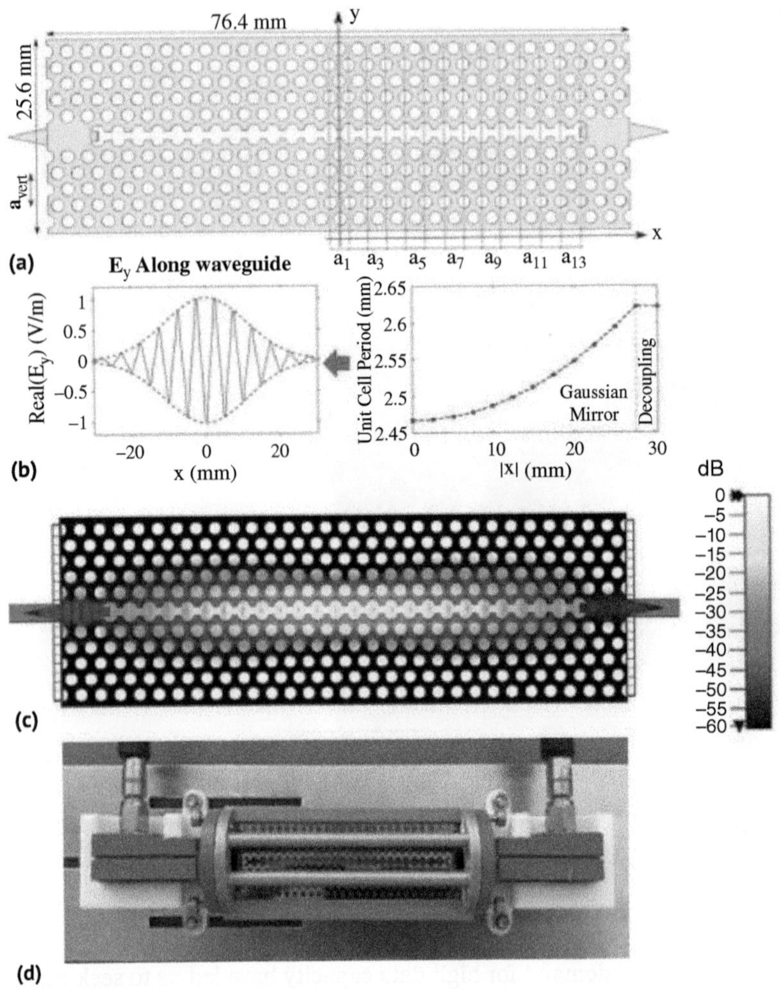

Figure 5.87 Typical EBG Resonator: (a) EBG resonator with its dimensions and variable periodicity (a_n) along the x-axis, (b) Gaussian envelope of the resonant mode achieved by linearly increasing the reflection from the center to the outer unit cells, (c) EM field distribution-E field strongly confined in the slot regions, and (d) EBG resonator in its test jig interfaced with the waveguide-to-coaxial adapters. *Source*: [386]/IEEE.

5.12.10 Terahertz Metamaterial Resonator

THz technology has garnered significant attention due to its broad spectrum and low latency, positioning it as a promising window for the frequency range of next-generation wireless communication networks [474, 475]. As a critical component in communication technology, resonators are one of the key areas of interest in research and technology development.

Ho-Jin Song and Namyoon Lee [385] reported the potential benefit of the huge spectral resources in the Sub-THz band (above 100 GHz) for the next-generation 6G and later-generation communication networks.

Figure 5.89 shows the demonstration of data rates reaching 100 Gbps or higher with several potential device technologies, including not only III–V heterojunction bipolar transistors (HBTs) and high-electron-mobility transistors (HEMTs) but also silicon complementary metal-oxide-semiconductor (CMOS) and photonic transmitters [387–436].

Photonic devices and technologies are very effective for high-frequency radio transmitters due to their broadband and low-loss characteristics. However, receivers that require reception devices, such as antennas and front-end functional blocks containing a local oscillator, amplifier, and mixer, are challenging and not cost-effective to realize by optical means. In general, wireless applications have been implemented with electronic devices that provide various functionalities for transmitters and receivers. In particular, compound semiconductor devices on GaAs or InP substrates have been attracting great attention for high-frequency applications due to their superior electron mobility and saturation velocities over silicon-based

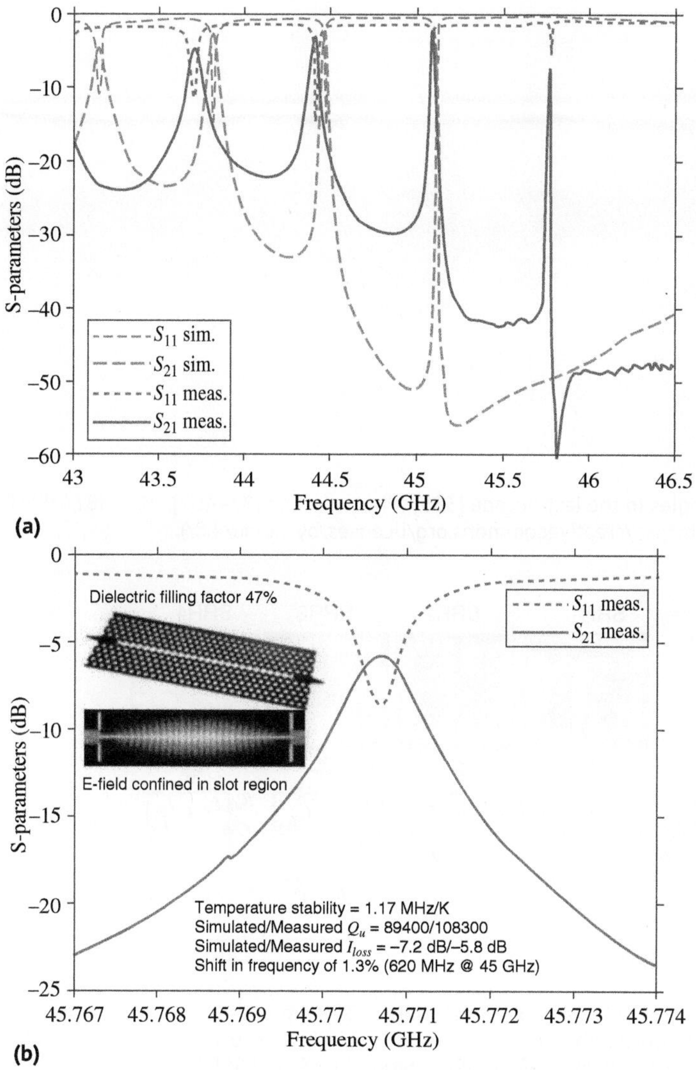

Figure 5.88 Plot of S-parameter of EBG resonator: (a) simulated and measured S-parameter for frequency 43–46.5 GHz, and (b) simulated and measured S-parameter for frequency 45.767–45.774 GHz. *Source*: [386]/IEEE.

devices [385]. The maximum oscillation frequency (f_{MAX}) of HEMTs and HBTs has been improved considerably, having already exceeded 1 THz around ten years ago [469–472].

Various functional integrated circuits, such as amplifiers, oscillators, phase-locked loops, and signal generators, have been demonstrated in HEMT or HBT technologies at THz frequencies [409, 437–467]. Thanks to the signal amplification with III–V devices, HBT and HEMT-based electronics modules capable to generate output power 10 mW plus at 300 GHz, which is approximately a hundred times larger than that of photonic transmitters [385].

Yangbin Yu and Yu-Sheng Lin [447] proposed metamaterial resonators tailored with single-, dual-, triple-, and quad-split ring electrical resonators, respectively, shown in Figures 5.90 and 5.91. The reported ESRR (electrical split-ring resonator) is made of Al materials of 300 nm in thickness on flexible polydimethylsiloxane (PDMS) substrate, with the relative permittivity of PDMS is 2.82.

In order to improve the adaptability of THz device working in various environments, there were reported many methods to fulfill the tunable mechanisms including ferroelectric material [448, 449], semiconductor diode [450–452], thermal tuning [453, 454], and reconfigurable tuning [455–458]. Among these methods, reconfigurable metamaterial becomes more practical than the others because the incident electromagnetic wave could be optimized by reconfiguring the dimension of metamaterial unit cell [459, 460] reshaping the structural element [461], rotating the unit cell [462, 463] or bending the substrate or lattice [464], which are other than the use of other methods are highly dependent on the nonlinear properties of

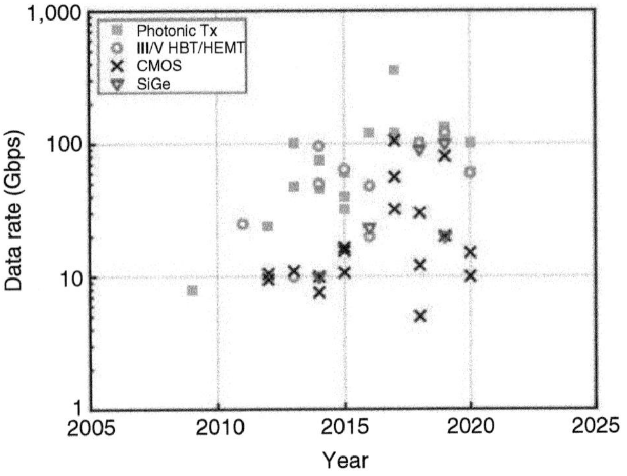

Figure 5.89 Data rates reported with different device technologies in the last decade [385] (Photonic TX [387–402], III/V HBT/HEMT [403–411], SiGe devices [412–416], CMOS [417–435]) (Credit: https://creativecommons.org/licenses/by-nc-nd/4.0/).

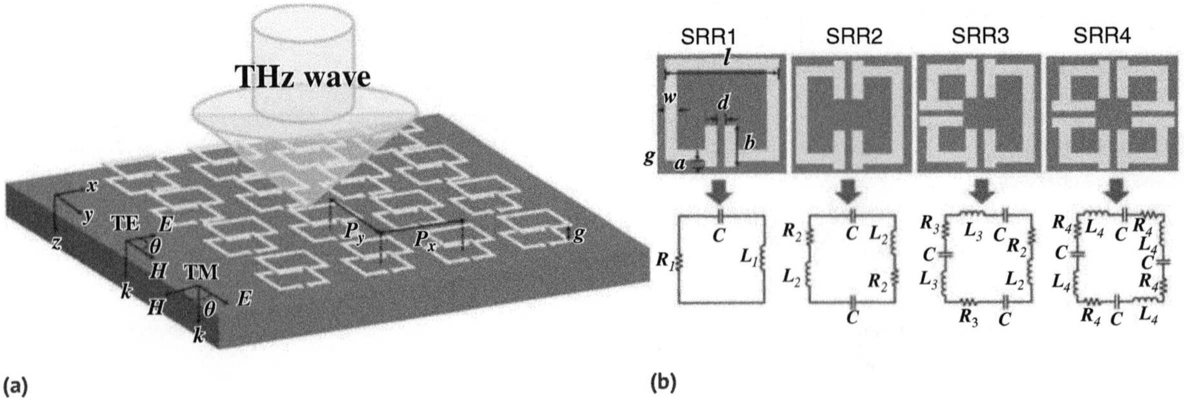

Figure 5.90 A typical metamaterial structure with a double layer of SRR made of Al materials of 300 nm in thickness on flexible PDMS substrate with relative permittivity 2.82: (a) geometrical representation of SRR1 with a double layer showing incident TE and TM waves, normal incidence along z-axis, and (b) equivalent electric circuit of SRRs (SRR1, SRR2, SRR3, and SRR4).
Source: [447]/Elsevier/CC BY ND 4.0.

natural materials. Moreover, the variations in geometry of metamaterial structure often result from mechanical movement or deformation of the metamaterials [465].

The resonant frequency of composite SRR can be expressed by Drude–Lorentz model [447]

$$\eta_{\text{em}} = \sqrt{\varepsilon(\omega)\mu(\omega)} = \sqrt{\left(1 - \frac{\kappa\omega_{pe}^2}{\omega^2 - \omega_{LCe}^2}\right)\left(1 - \frac{\kappa\omega_{pm}^2}{\omega^2 - \omega_{LCm}^2}\right)} \tag{5.302}$$

where ω_p is plasma frequency, ω_{LC} is resonant frequency, k is a dimensionless quantity, and subscripts e and m refer to electric and magnetic response [466].

As illustrated in Figure 5.91, SRRs (SRR1, SRR2, SRR3, and SRR4) can be modeled as an (LC) resonator, the metal frame acts as the resistance (R), ring as inductance (L), and split is the capacitance (C) respectively. The corresponding impedances of SRRs depicted in Figure 5.91 can be described by [447]

$$Z_{\text{SRR1}} = R_1 + j\omega L_1 + \frac{1}{j\omega C} \tag{5.303}$$

$$Z_{\text{SRR1}} = R_1 + j\omega L_1 + \frac{1}{j\omega C} \tag{5.304}$$

$$Z_{\text{SRR3}} = R_2 + 2R_3 + 2j\omega L_3 + \frac{3}{j\omega C} \tag{5.305}$$

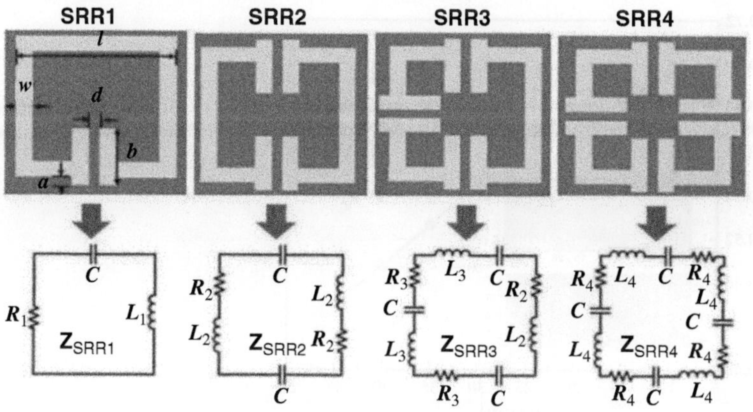

$$Z_{SRR1} = R_1 + j\omega L_1 + \frac{1}{j\omega C} \qquad Z_{SRR3} = R_2 + 2R_3 + 2j\omega L_3 + \frac{3}{j\omega C}$$

$$Z_{SRR2} = 2R_2 + 2\left(j\omega L_2 + \frac{1}{j\omega C}\right) \qquad Z_{SRR4} = 4R_4 + 4\left(j\omega L_4 + \frac{1}{j\omega C}\right)$$

$$L = \frac{\mu_0 l^2}{t} \qquad C = \frac{\varepsilon_0 \varepsilon_r wt}{d} \qquad f_0 = \frac{1}{2\pi\sqrt{LC}} = \left(\frac{c_0}{2\pi l\sqrt{\varepsilon_r}}\right)\sqrt{\frac{d}{w}}$$

Figure 5.91 SRRs (SRR1, SRR2, SRR3, and SRR4) which can be modeled as an (LC) resonator, the metal frame acts as the resistance (R), ring as inductance (L), and split is the capacitance (C), physical dimensions of SRR are: $l = 60$ μm, $w = 6$ μm, $b = 4$ μm, $a = 2$ μm, $d = 4$ μm, and $g = 10$ μm, respectively. *Source*: [447]/Elsevier/CC BY ND 4.0.

$$Z_{SRR4} = 4R_4 + 4\left(j\omega L_4 + \frac{1}{j\omega C}\right) \tag{5.306}$$

$$\text{where } L \text{ and } C \text{ are given as: } L = \frac{\mu_0 l^2}{t}, \quad C = \frac{\varepsilon_0 \varepsilon_r\, wt}{d} \tag{5.307}$$

ε_0 and ε_r are the free space permittivity and relative permittivity of the materials within split.

The SRR resonant frequency can be expressed by [468–473]

$$f_0 = \frac{1}{2\pi\sqrt{LC}} = \left(\frac{c_0}{2\pi l\sqrt{\varepsilon_r}}\right)\sqrt{\frac{d}{w}} \tag{5.308}$$

Figure 5.92 shows the E-field and H-field distributions and resonant frequency of SRR1 with $b = 20$ μm, resonant frequency f_{0TE} observed for TE mode is 0.73 THz and for TM mode, f_{0TM} is 0.62 THz: (a) SRR1 exhibits tuning range of 0.10 THz in TE mode from 0.76 to 0.66 THz by increasing b value from 10 to 45 μm, and (b) SRR1 exhibits tuning range of 0.22 THz in TM mode from 0.67 to 0.45 THz by increasing b value from 10 to 45 μm [447]. This implies that which means the resonant frequency of SRR1 is sensitive to geometrical dimensions at TM mode.

Figure 5.93 shows the E-field and H-field distributions and resonant frequency of SRR2 with $b = 20$ μm, resonant frequency f_{0TE} observed for TE mode is 0.90 THz and for TM mode, f_{0TM} is 0.59 THz: (a) SRR2 exhibits tuning range of 0.25 THz in TE mode from 1.05 to 0.80 THz by increasing b value from 10 to 30 μm, and (b) SRR2 exhibits tuning range of 0.22 THz in TM mode from 0.67 to 0.45 THz by increasing b value from 10 to 30 μm [447].

Figure 5.94. The E-field and H-field distributions and resonant frequency of SRR3 with $b = 15$ μm, resonant frequency f_{0TE} observed for TE mode is 0.91 THz and for TM mode, f_{0TM} is 0.46 and 0.83 THz respectively: (a) SRR3 exhibits tuning range of 0.22 THz in TE mode from 1.02 to 0.80 THz by increasing b value from 10 to 20 μm, and (b) SRR3 exhibits two resonances in TM mode, causing from the top and bottom splits of SRR3, tuning range of 0.06 THz from 0.50 to 0.44 THz and 0.15 THz from 0.88 to 0.73 THz by increasing b value from 10 to 20 μm [447]. Figure 5.95 Shows the E-field and H-field distributions and resonant frequency of SRR4 with $b = 15$ μm, resonant frequency f_{0TE} observed for TE mode is 0.87 THz and for TM mode, f_{0TM} is 0.87 THz: (a) SRR4 exhibits tuning range of 0.22 THz in TE mode from 1.00 to 0.78 THz by increasing b value from 10 to 20 μm, and (b) SRR4 exhibits tuning range of 0.22 THz in TM mode from 1.00 to 0.78 THz

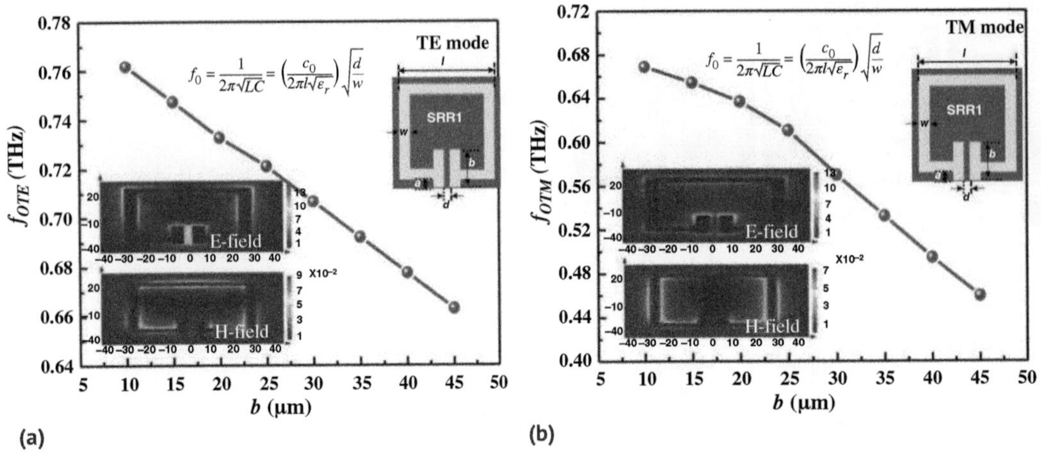

Figure 5.92 E-field and H-field distributions and resonant frequency of SRR1 with $b = 20$ μm, resonant frequency $f_{0\text{TE}}$ observed for TE mode is 0.73 THz and for TM mode, $f_{0\text{TM}}$ is 0.62 THz: (a) SRR1 exhibits tuning range of 0.10 THz in TE mode from 0.76 to 0.66 THz by increasing b value from 10 to 45 μm, and (b) SRR1 exhibits tuning range of 0.22 THz in TM mode from 0.67 to 0.45 THz by increasing b value from 10 to 45 μm. *Source*: [447]/Elsevier/CC BY ND 4.0.

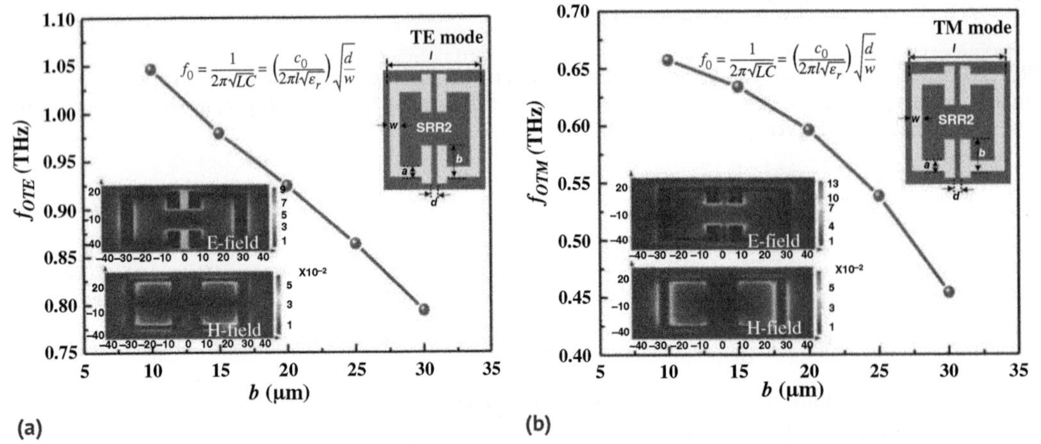

Figure 5.93 E-field and H-field distributions and resonant frequency of SRR2 with $b = 20$ μm, resonant frequency $f_{0\text{TE}}$ observed for TE mode is 0.90 THz and for TM mode, $f_{0\text{TM}}$ is 0.59 THz: (a) SRR2 exhibits tuning range of 0.25 THz in TE mode from 1.05 to 0.80 THz by increasing b value from 10 to 30 μm, and (b) SRR2 exhibits tuning range of 0.22 THz in TM mode from 0.67 to 0.45 THz by increasing b value from 10 to 30 μm. *Source*: [447]/Elsevier/CC BY ND 4.0.

by increasing b value from 10 to 20 μm [447]. Both TE and TM mode behaviors are identical because SRR4 is symmetrical structure.

Yangbin Yu and Yu-Sheng Lin Yu [447] reported four types of composite SRRs (SRR1, SRR2, SRR3, and SRR4) structures to investigate the resonant and tuning characteristics by varying the physical dimensions of SRR, described in Figures 5.90–5.95, offers potential use of metamaterial structures for numerous applications (tunable oscillator, sensors, filters, switching devices etc.).

Compact continuous-wave oscillators in the sub-THz and THz range are pivotal for various innovative applications, offering advantages such as low power consumption and ease of deployment. With advancements in semiconductor processing and manufacturing technologies, solid-state THz compact oscillators have successfully expanded coverage into the THz frequency range. One notable technology, resonant tunneling diodes (RTDs), emerges as a strong contender for fundamental THz oscillators. These devices operate at room temperature, consume minimal power, and are inherently compact, making them ideal for practical applications. In a significant contribution to the field, X. Yu et al. [476] experimentally demonstrated highly efficient compact RTD-based THz oscillators. Their work underscores the potential of RTDs in advancing THz applications across various domains, such as wireless communications, spectroscopy, and imaging.

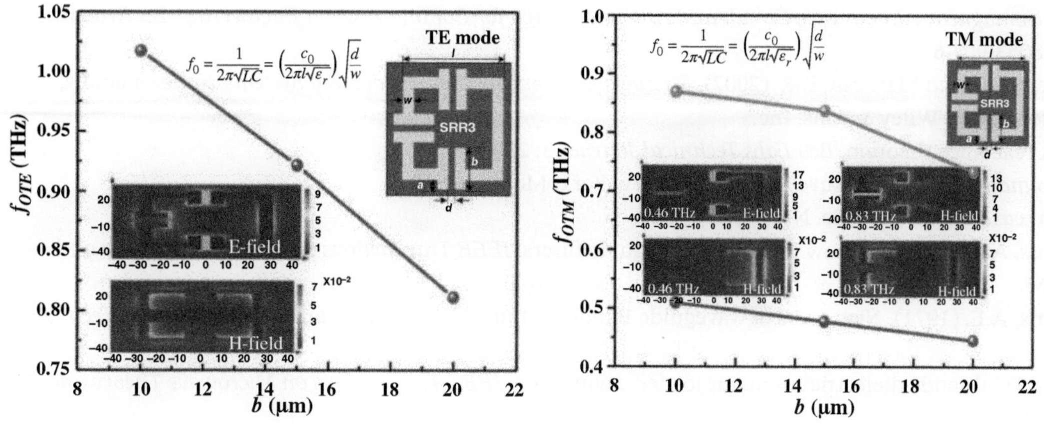

(a) (b)

Figure 5.94 The E-field and H-field distributions and resonant frequency of SRR3 with $b = 15$ μm, resonant frequency f_{0TE} observed for TE mode is 0.91 THz and for TM mode, f_{0TM} is 0.46 and 0.83 THz, respectively: (a) SRR3 exhibits tuning range of 0.22 THz in TE mode from 1.02 to 0.80 THz by increasing b value from 10 to 20 μm, and (b) SRR3 exhibits two resonances in TM mode, causing from the top and bottom splits of SRR3, tuning range of 0.06 THz from 0.50 to 0.44 THz and 0.15 THz from 0.88 to 0.73 THz by increasing b value from 10 to 20 μm. *Source*: [447]/Elsevier/CC BY ND 4.0.

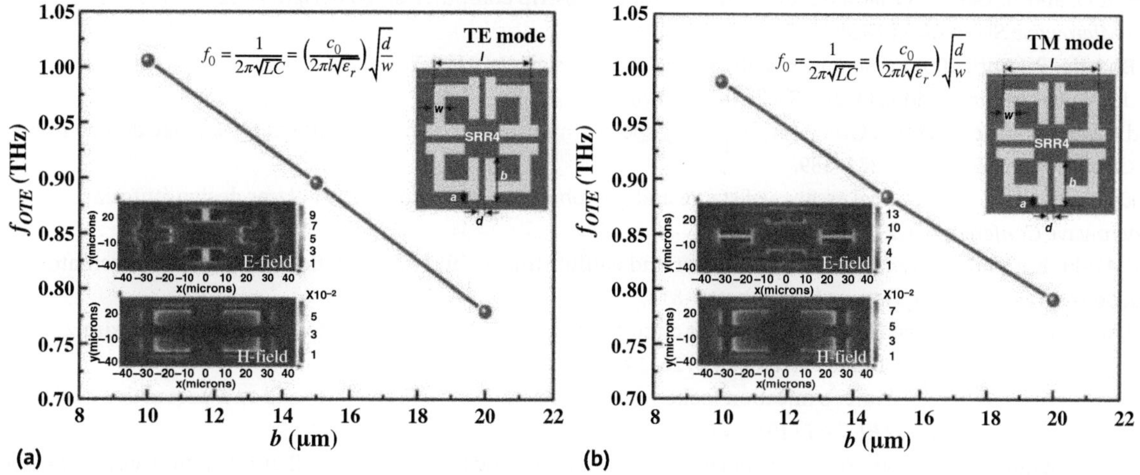

(a) (b)

Figure 5.95 E-field and H-field distributions and resonant frequency of SRR4 with $b = 15$ μm, resonant frequency f_{0TE} observed for TE mode is 0.87 THz and for TM mode, f_{0TM} is 0.87 THz: (a) SRR4 exhibits tuning range of 0.22 THz in TE mode from 1.00 to 0.78 THz by increasing b value from 10 to 20 μm, and (b) SRR4 exhibits tuning range of 0.22 THz in TM mode from 1.00 to 0.78 THz by increasing b value from 10 to 20 μm [101]. Both TE and TM mode behaviors are identical because SRR4 is symmetrical structures. *Source*: [447]/Elsevier/CC BY ND 4.0.

References

1 Vendelin, G.D., Pavio, A.M., and Rohde, U.L. (1990). *Microwave Circuit Design Using Linear and Nonlinear Techniques*. New York: John Wiley & Sons.

2 Kotzebue, K.L. (1984). A technique for the design of microwave transistor oscillators. *IEEE Transactions on Microwave Theory and Techniques* 32: 719–721, 1984.

3 Cohn, D. and Mitchel, R. (1973). Design of microstrip transistor oscillators. *International Journal of Electronics* 35: 385–395.

4 Zysman, G.I. and Johnson, A.K. (1969). Coupled transmission line networks in an inhomogeneous dielectric medium. *IEEE Transactions on Microwave Theory and Techniques*, 17: 753–759.

5 Gupta, K.C., Garg, R., and Chadha, R. (1981). *Computer-Aided Design of Microwave Circuits*. Dedham: Artech House.

6 Howe, H. (1974). *Stripline Circuit Design*. Dedham: Artech House.

7 Russer, P. (2006). *Electromagnetics, Microwave Circuit and Antenna Design for Communication Engineering*, 2e. Artech House, Inc., ISBN 1-58053-907-6

8 Cameron, R.J., Kudsia, C.M., and Mansour, R.R. (2007). *Microwave Filters for Communication Systems*. Fundamentals, Design, and Applications, John Wiley & Sons, Inc.

9 Foster, R.M. (1924). A reactance theorem. *Bell Labs Technical Journal* 3: 259–267.

10 Collin, R.E. (1966). *Foundation for Microwave Engineering*. New York: McGraw-Hill.

11 Pozar, D. (1998). *Microwave Engineering*, 2e. New York: Wiley.

12 Atia, A.E. and Williams, A.E. (1972). Narrow Bandpass waveguide filters. *IEEE Transactions on Microwave Theory and Techniques* 20: 238–265.

13 Atia, A.E. and Williams, A.E. (1971). New types of waveguide Bandpass filters for satellite transponders. *Comsat Technical Review* 1 (1): 21–43.

14 Fiedziusko, S.J. (1982). Dual-mode dielectric resonator loaded cavity filter. *IEEE Transactions on Microwave Theory and Techniques* 30: 1311–1316.

15 Kajfez, D. and Guilon, P. (1986). *Dielectric Resonators*. Norwood, MA: Artech House.

16 Mansour, R.R. (1994). Design of superconductive multiplexes using single-mode and dual-mode filters. *IEEE Transactions on Microwave Theory and Techniques* 42: 1411–1418.

17 Walkar, V. and Hunter, I.C. (June 2002). Design of triple mode TE01 resonator transmission filters. *IEEE Microwave Wireless Component Letters* 12: 215–217.

18 Mansour, R.R. (1972). Microwave superconductivity. *IEEE Transactions on Microwave Theory and Techniques* 50: 750–759.

19 Hammerstad, E.O. and Jensen, O. (1980). Accurate models for microstrip computer-aided design. *IEEE MTT-S International Microwave Symposium Digest*, 407–409.

20 Bahl, I. and Bhartia, P. (April 2003). *Microwave Solid State Circuit Design*, 2e. New York: Wiley.

21 Ohira, T. (2016). What in the world is Q? *IEEE Microwave Magazine*, 42–49, 2016.

22 Ohira, T. and Araki, K. (August 2006). Oscillator frequency spectrum as viewed from resonant energy storage and complex Q factor. *IEICE Electronics Express* 3: 385–389.

23 Ohira, T. (2006, September). Theoretical essence of stable and low phase noise microwave oscillator design (tutorial). *European Microwave Conference SC2-1*. Manchester.

24 Ohira, T. and Araki, K. (2007, September). Active Q factor and equilibrium stability formulation for sinusoidal oscillators. *IEEE Transactions on Circuits and Systems II* 54 (9): 810–814.

25 Ohira, T. and Wuren, T. (2008, November). Pseudolinear circuit theory for sinusoidal oscillator performance maximization (invited). *IEICE Transactions on Electronics* 91-C (11): 1726–1737.

26 Ohira, T. (2010, October). Extended Adler's injection locked Q factor formula for general one- and two-port active device oscillators. *IEICE Electronics Express ELEX* 7 (19): 1486–1492.

27 Ohira, T. (2012). Emerging Technology and Technological Challenges in Low Phase Noise Oscillator Circuit Designs: Seven different aspects on what role Q factor plays in resonators and oscillators. Toyohashi University of Technology IMS Workshop Tutorial Lecture 2012 (IMS 2012 workshop organizer: Kenji Itoh, Ajay Poddar).

28 Ohira, T. (2005, December). Rigorous Q factor formulation for one- and two-port passive linear networks from an oscillator noise spectrum viewpoint. *IEEE Transactions Circuits Systems II* 52 (12): 846–850.

29 Tellegen, D.B.H. (1953, November). A general network theorem with applications. *Proceedings IRE Australia*, 265–270.

30 Kurokawa, K. (1969, July). Some basic characteristics of broadband negative resistance oscillator circuits. *Bell System Technical Journal* 48 (6): 1937–1955.

31 Adler, R. (1946, June). A study of locking phenomena in oscillators. *Proceedings IRE* 34: 351–357.

32 Ohira, T. (2012). Dedicated Q factor formulas stemming from oscillation frequency stability against source and load deviations. *IEICE Electronics Express* 9: 616–621.

33 Rohde, U.L., Poddar, A.K., and Boeck, G. (2005). *The Design of Modern Microwave Oscillators for Wireless Applications: Theory and Optimization*. John Wiley & Sons Inc.

34 Leeson, D.B. (1966, February). A simple model of feedback oscillator noise spectrum. *Proceedings of the IEEE* 54 (2): 329–330.

35 Razavi, B. (1996, March). A study of phase-noise in CMOS oscillators. *IEEE Journal Solid-State Circuits* 31 (3): 331–343.

36 Bal S. Virdee and Grassopoulosf', C. (2003, June). Folded microstrip resonator. *Microwave Symposium Digest, 2003 IEEE MTT-S International* 3: 8–13.

37 Chang, K. and Hsieh, L.-H. (2004). *Microwave Ring Circuits and Related Structures*. Wiley-IEEE.

38 Ledezma, L.M. (2011). A study on the miniaturization of microstrip square open loop resonators. USF, MS thesis.

39 Yabuki, H., Sagawa, M., and Makimoto, M. (1991). Voltage controlled push-push oscillators using miniaturized hairpin resonators. *IEEE Microwave Theory and Technology Society Digest*, 1175–1178.

40 Hsieh, L.-H. (2004). Analysis, modeling and simulation of ring resonators and their applications to filters and oscillators. PhD dissertation. Texas A&M University.

41 Sagawa, M., Takahashi, K., and Makimoto, M. (1989, December). Miniaturized hairpin resonator filters and their application to receiver front-end MIC. *IEEE Transactions on Microwave Theory and Techniques* 37: 1991–1997.

42 Wadell, B.C. (1991). *Transmission Line Design Handbook*, 321. MA: Artech House.

43 Chang, K. (1996). *Microwave Ring Circuits and Antennas*. Wiley-Inter Science.

44 Yu, C.C., Chang, K. (1997, November). Transmission-line analysis of a capacitively coupled microstrip-ring resonator. *IEEE Transactions on Microwave Theory and Techniques* 45 (11): 2018–2024.

45 Hoffmann, K. and Skvor, Z. (2001, July). Active resonator. *International Conferences Trends, Communications, EUROCON'2001* 1: 164–166.

46 Alinikula, P., Kaunisto, R., and Stadius, K. (1994, May). Monolithic active resonators for wireless application. *IEEE MTT-S Digest* 2: 1151–1154.

47 Cenac, C., Jarry, B., and Guillon, P. (1996, April). X-band filters with half-wave or ring resonators and variable gain and phase monolithic circuits. *Microwave and Optical Technology Letters* 11 (5): 254–257.

48 Chang, C.-Y. and Itoh, T. (1990, December). Microwave active filters based on coupled negative resistance method. *IEEE Transactions on MTT* 38 (12): 1879–1884.

49 Hopf, B.P., Wolff, I., and Giglielmi, M. (1994, December). Coplanar MMIC active bandpass filters using negative resistance circuits. *IEEE Transactions on MTT* 42 (12): 2598–2602.

50 Karacaoglu, U. and Robertson, I.D. (1995, December). MMIC Active bandpass filters using varactor-tuned negative resistance elements. *IEEE Transactions on Microwave Theory and Techniques* 43(12): 2926–2932.

51 Ishikawa, Y., Yamashita, S., and Hidaka, S. (1991, December). Noise design of active feedback resonator BEF. *IEEE Transactions on Microwave Theory and Techniques* 41 (12): 2133–2138.

52 Ezzedine, H., Billonnet, L., Jarry, B. et al. (1998, December). Optimization of noise performance for various topologies of planar microwave active filters using noise wave techniques. *IEEE Transactions on Microwave Theory and Techniques* 46 (12): 2484–2492.

53 Nick, M. (2011). New Q-enhanced planar resonators for low phase-noise radio frequency oscillators. PhD dissertation. Electrical Engineering, University of Michigan.

54 Gardner, P. and Paul, D.K. (1997, May). Optimum noise measure configurations for transistor negative resistance amplifiers. *IEEE Transactions on Microwave Theory and Techniques* 45 (5): 580–586.

55 Grebennikov, A. (2007). *RF and Microwave Transistor Oscillator Design*. New York: John Wiley & Sons.

56 Hauss, H.A. and Adler, R.B. (1959). *Circuit Theory of Linear Noisy Networks*. New York: Wiley.

57 Hong, J.-S. and Lancaster, M.J. (2001). *Microwave Filter for RF/Microwave Application*. New York: Wiley.

58 Rohde, U.L. and Rudolph, M. (2012, December). *RF/Microwave Circuit Design for Wireless Applications*. Wiley & Sons, ISBN-13: 978-1118431399.

59 Rohde, U.L. and Poddar, A.K. (2007, January). *Active Planar Coupled Resonators Replace Traditional High Q Resonators in Low Phase Noise Oscillators/VCOs. 2007 IEEE Radio and Wireless Symposium*, 39–42, CA, USA.

60 Ajay Poddar, K. (2013). Slow wave resonator based tunable multi-band multi-mode injection-locked oscillators. Research report, Germany: Brandenburg University of Technology Cottbus-Senftenberg.

61 Pistono, E., Robert, M., Duvillaret, L. et al. (2006, June). Compact fixed and tune-all bandpass filters based on coupled slow-wave resonators. *IEEE Transactions on Microwave Theory and Techniques* 54 (6): 2790–2798.

62 Chang, W.-S., Liang, C.-H., and Chang, C.-Y. (2010). Compact microstrip bandpass filters using miniaturized slow-wave quarter-wavelength resonators. *Proceedings of Asia-Pacific Microwave Conference*, 1958–1961.

63 Hsieh, L.-H. and Chang, K. (2002, July). Slow-wave bandpass filters using ring or stepped-impedance hairpin resonators. *IEEE Transactions on Microwave Theory and Techniques* 50 (7): 1795–1800.

64 Cho, H.-Y., Yeh, T.-J., Liu, S. et al. (2009, August). High-performance slow-wave transmission lines with optimized slot-type floating shields. *IEEE Transactions on Electron Devices* 56 (8): 1705–1711.

65 Masuda, T., Shiramizu, N., Nakamura, T. et al. (2008). Characterization and modeling of microstrip transmission lines with slow-wave effect. *2008 IEEE Topical Meeting on Silicon Monolithic Integrated Circuits in RF Systems*, 155–158.

66 Wang, H., Zhanget, L., Yang, D. et al. (2011). A 60 GHz wideband injection-locked frequency divider with adaptive-phase-enhancing technique. *2011 IEEE Radio Frequency Integrated Circuits Symposium*, 75–78.

67 Poddar, A.K., Koul, S.K., and Bharthi Bhat (1997, July). Millimeter wave evanescent mode Gunn diode oscillator in suspended Stripline configuration. *IR &MM Wave, 22nd International Conference*, 265–266.

68 Poddar, A.K. and Rohde, U.L. (2014, May). Metamaterial Möbius Strips (MMS): Application in resonators for oscillators and synthesizers. *IEEE UFFC Symposia*, 1–9.

69 Wu, C.-T.M., Poddar, A.K., Rohde, U.L. et al. (2014). A C-band tunable Oscillator based on complementary coupled resonator using substrate integrated waveguide cavity. *European Microwave Conference*, 715–718.

70 Poddar, A.K. and Rohde, U.L. (2012, May). Slow-wave evanescent-mode coupled resonator oscillator. *IEEE Frequency Control Symposium, Proceedings*, 01–07.

71 Poddar, A.K. and Rohde, U.L. (2011, September). Adaptive mode-coupled harmonically tuned high-Q ultra low phase noise sources. *Microwave Journal* 54: 106–120.

72 Rohde, U.L. and Poddar, A.K. (2011, May). Slow-wave structures slice VCO phase noise. *Microwave & RF* 50 (5): 136.

73 Poddar, A.K. and Rohde, U.L. (2011, May). Slow wave resonator based tunable oscillators. *Proceedings of 2011 Joint Conference of the IEEE International Frequency Control Symposium & European Frequency and Time Forum*, 549–558.

74 Rohde, U.L. and Poddar, A.K. (2011, May). A digital frequency synthesizer using adaptive mode-coupled resonator mechanism for low noise and low jitter applications. *IEEE (ISCAS)*, 414–417, Brazil.

75 Chang, W.-S. and Chang, C.-Y. (2012, November). A high slow-wave factor microstrip structure with simple design formulas and its application to microwave circuit design. *IEEE Trans. on Microwave Theory and Techniques* 60 (11): 3376–3383.

76 Gupta, V.K., Kumar, P.P., and Rao, C.N. (2013). A novel miniaturized, wideband, high directivity microstrip coupler on FR4. *International Conference on Microwave and Photonics (ICMAP)*, 1–4.

77 Rohde, U.L. and Poddar, A.K. (2008, April). *Large-Signal Approach Yields Low-Noise VHF/UHF Oscillators*. Part 1, 62–76.

78 Rohde, U.L. and Poddar, A.K. (2008, May). *Large-Signal Approach Yields Low-Noise VHF/UHF Oscillators*. Part 2, 84–96.

79 Rohde, U.L. and Poddar, A.K. (2008, June). Planar resonators arm tunable oscillators. *Microwave & RF*, Part 1, 92–96.

80 Rohde, U.L. and Poddar, A.K. (2008, July). Planar resonators arm tunable oscillators. *Microwave & RF*, Part 2.

81 Rohde, U.L. and Poddar, A.K. (2008, December). Miniature VCOs shrink wideband synthesizers. *Microwave & RF*, 87–92.

82 Rohde, U.L. and Poddar, A.K. (2008, June). STPCR offers integrable alternatives of DRO. *Microwave Symposium Digest*, 2008 *IEEE MTT-S*, 233–236, Atlanta, USA.

83 Rohde, U.L. and Poddar, A.K. (2008, April). Mode-coupled stubs-tuned planar resonator offers promising and integrable alternatives of DRO (Dielectric Resonator Oscillator). *IEEE Sarnoff Symposium*, 1–7, Princeton, NJ, USA.

84 Rohde, U.L. and Poddar, A.K. (2008, January). Injection-tuned coupled oscillators. *IEEE Radio Wireless Symposium 2008*, 367–370, Orlando, FL, USA.

85 Rohde, U.L. and Poddar, A.K. (2008, January). Reconfigurable concurrent oscillator (RCO). *IEEE Radio Wireless Symposium 2008*, 371–374, Orlando, FL, USA.

86 Rohde, U.L. and Poddar, A.K. (2007, September). Stub-tuned planar coupled resonator based spectral pure signal source for wireless communication systems. *IEEE/PIMRC 2007*, 01–05, Greece.

87 Rohde, U.L. and Poddar, A.K. (2007, September). Reconfigurable concurrent oscillator (RCO). *IEEE 18th International Personal, Indoor and Mobile Radio Communications, IEEE/PIMRC 2007*, 01–05, Greece.

88 Rohde, U.L. and Poddar, A.K. (2007, August). Self-injection locked compact coupled planar resonator based cost-effective ultra low phase noise VCOs for wireless systems. *IEEE/ECCTD 2007*, 998–1001, Spain.

89 Rohde, U.L. and Poddar, A.K. (2007, August). Noise cancellation and noise minimization techniques for low cost compact size configurable RF oscillators/VCOs. *European Conference on Circuit Theory and Design*, 986–989, ECCTD 2007, Spain.

90 Rohde, U.L. and Poddar, A.K. (2007, October). Concurrent, reconfigurable and concurrent oscillators for multi-band multi-mode communication systems. *European Microwave Conference, EUMC 2007*, 1050–1053, Munich.

91 Rohde, U.L. and Poddar, A.K. (2007, October). Concurrent, reconfigurable and concurrent oscillators for multi-band multi-mode communication systems. *European Conference on Wireless Technology, ECWT 2007*, 292–295, Munich.

92 Rohde, U.L. and Poddar, A.K. (2007, October). Low cost configurable RF signal source for wireless applications. *European Microwave Conference, EuMiC 2007*, 508–511, Munich.

93 Rohde, U.L. and Poddar, A.K. (2007, September). Noise reduction techniques for configurable RF signal source. *6th International Conference on Antenna Theory and Techniques, IEEE/ICATT 2007*, 163–166, Ukraine.

94 Rohde, U.L. and Poddar, A.K. (2007, September). Minimization of phase noise in scaled device coupled mode oscillators. *2007 6th International Conference on Antenna Theory and Techniques, IEEE/ICATT 2007*, 51–56, Ukraine.

95 Rohde, U.L. and Poddar, A.K. (2007, September). Low cost signal source for multi-band multi-mode wireless systems. *2007 6th International Conference on Antenna Theory and Techniques, IEEE/ICATT 2007*, 167–169, Ukraine.

96 Shen, F., Wei, Y., Yin, H. et al. (2012, February). A novel V-shaped microstrip meander-line slow-wave structure for W-band MMPM. *IEEE Transactions on Plasma Science* 40 (2): 463–469.

97 Rohde, U.L. and Poddar, A.K. (2007, December). STPCR VCO replaces expensive DRO. *IEEE/APMC 2007*, 1–4, Bangkok, Thailand.

98 Hong, J.-S. and Lancaster, M.J. (1997, December). Theory and experiment of novel microstrip slow-wave open-loop resonator filters. *IEEE Transactions on Microwave Theory and Techniques* 45: 2358–2365.

99 Rusdiyanto, D., Zakaria, Z., and Wibisono, G. (2026). Integrated hairpin bandpass filter and open-loop resonator sensor operating at distinct frequencies for solid material detection. *2026 Symposium on Microwave, Antenna, and Propagation (SMAP)*, Jakarta, Indonesia, (115–118). https://doi.org/10.1109/SMAP67528.2026.11439103.

100 Rouchaud, F., Madrangeas, V., Aubourg, M. et al. (1998, June). New classes of microstrip resonators for HTS microwave filters applications. *IEEE MTT-S International Microwave Symposium Digest* 2: 1023–1026.

101 Kuo, J.T., Maa, M.J., and Lu, P.H. (2000, March). A microstrip elliptic function filter with compact miniaturized hairpin resonators. *IEEE Microwave and Guided Wave Letters* 10: 94–95.

102 Chen, J.-X., Zhu, P.-X., Tang, S.-C. et al. (2026, April). Dual-wideband high-selectivity BPF based on fundamental and high-order modes of ridge WG SSPPs. *IEEE Microwave and Wireless Technology Letters* 36 (4): 521–524. https://doi.org/10.1109/LMWT.2025.3645766.

103 Cameron, R.J., Kudsia, C.M., and Mansour, R.R. (2018). *Microwave Filters for Communication Systems: Fundamentals, Design, and Applications*. John Wiley & Sons, Inc.

104 Poddar, A.K., Rohde, U.L., Apte, A.M. et al. (2023). Möbius-coupled SRR (split ring resonator) oscillator extends high-performance signal generations into K-band. *2023 IEEE Microwaves, Antennas, and Propagation Conference (MAPCON)*, Ahmedabad, India (1–6). https://doi.org/10.1109/MAPCON58678.2023.10463818.

105 Hong, J. and Lancaster, M. (2001). *Microstrip Filters for RF/Microwave Applications*. John Wiley and Sons.

106 Sinnesbichler, F.X., Hautz, B., and Olbrich, G.R. (2000, April). A Si/SiGe HBT dielectric resonator push-push oscillator at 58 GHz. *IEEE Microwave and Guided Wave Letters* 10, 145–147.

107 Kumar, A., Kumari, S., and Mahto, M. (2026, April). Investigation of bi-frequency microwave generation using a sectioned metamaterial slow wave structure. *IEEE Transactions on Plasma Science* 54 (4): 1474–1479. https://doi.org/10.1109/TPS.2026.3666795.

108 Yun, T.Y. and Chang, K. (2002, January). Analysis and optimization of a phase shifter controlled by a piezoelectric transducer. *IEEE Transactions on MTT* 50: 105–111.

109 Poplavko, Y.M., Prokopenko, Y.V., Molchanov, V.I. et al. (2001, June). Frequency-tunable microwave dielectric resonator. *IEEE Transactions on MTT* 49: 1020–1026.

110 Shi, H., Wang, Z., Ma, K. et al. (2026). An ultra-wideband 125-MHz-to-8.4-GHz fractional-N PLL featuring an isolated multi-core VCO and an active-feedback doubler. *IEEE Microwave and Wireless Technology Letters*. https://doi.org/10.1109/LMWT.2025.3645772.

111 Hong, J.-S. and Lancaster, M.J. (1998, January). Cross-coupled microstrip hairpin-resonator filters. *IEEE Transactions on Microwave Theory and Techniques* 46 (1): 118–122.

112 Ameri, A., Chien, J.-C., and Niknejad, A.M. (2025). Theoretical studies of sub-THz active split-ring resonators for near-field imaging. *IEEE Transactions on Circuits and Systems I: Regular Papers*. https://doi.org/10.1109/TCSI.2026.3685973.

113 Kirschning, M., Jansen, R.H., and Koster, N.H.L. (1983). Measurement and computer aided modeling of microstrip discontinuities by an improved resonator method. *IEEE MTT-S International Microwave Symposium Digest*, 495–497.

114 Lee, S.-Y. and Tsai, C.-M. (2000, December). New cross-coupled filter design using improved hairpin resonators. *IEEE Transactions on Microwave Theory and Techniques* 48: 2482–2490.

115 Tsai, C.M., Lee, S.Y., and Tsai, C.C. (2001). Hairpin filters with tunable transmission zeros. *IEEE MTT-S International Microwave Symposium Digest*, 2175–2178.

116 Gupta, K.C., Garg, R., Bahl, I. et al. (2013). *Microstrip Lines and Slotlines*, 2e. Boston: Artech House.

117 Kaajakari, V., Koskinen, J.K., and Mattila, T. (2005, December). Phase noise in capacitively coupled micromechanical oscillators. *IEEE Transactions on UFFC* 52 (12): 2322–2331.

118 Shakib, S.H., Niloy, N.H., and Ahammad, A. (2025). High Q-factor resonances based on bound states in the continuum in an all-dielectric metasurface with sensing application. *Heliyon* 11 (5): e43072.

119 Rohde, U.L. and Poddar, A.K. (2007, February). Wideband voltage controlled oscillators employing evanescent mode coupled resonators. US Patent No 7,180, 381.

120 Rohde, U.L. and Poddar, A.K. (2006, August). Integrated low noise microwave wideband push-push VCO. US Patent No. 7,088189.

121 Rohde, U.L. and Poddar, A.K. (2007, September). User-definable thermal drift voltage controlled oscillator. US Patent No. 7, 265,642 B2.

122 Rohde, U.L. and Poddar, A.K. (2007, August). Low thermal drift tunable frequency voltage controlled oscillatory. US Patent No. 7, 262,670 B2.

123 Rohde, U.L. and Poddar, A.K. (2007, November). Multi-octave band tunable coupled resonator oscillator. US Patent No. 7, 292,113.

124 Rohde, U.L. and Poddar, A.K. (2007, March). Tunable frequency low phase noise and low thermal drift oscillator. US Patent No. 7196591.

125 Rohde, U.L. and Poddar, A.K. (2008, April). Low noise, hybrid tuned wideband voltage controlled oscillator. US Patent No. 7,365,612 B2.

126 Rohde, U.L. and Poddar, A.K. (2009, June). Tunable frequency low phase noise and low thermal drift oscillator. US Patent No. 7, 545, 229.

127 Liu, L.Y., Poddar, A.K., Rohde, U.L. et al. (2025). A high-Q microwave antenna sensor based on multiple ring resonators for ice monitoring. *2025 IEEE International Symposium on Antennas and Propagation and North American Radio Science Meeting (AP-S/CNC-USNC-URSI)*, Ottawa, ON, Canada, (2606–2609). https://doi.org/10.1109/AP-S/CNC-USNC-URSI55537.2025.11266013.

128 Rohde, U.L. and Poddar, A.K. (2009, September). User-definable, low cost, low phase hit and spectral pure tunable oscillator. US Patent No. 7,586,381.

129 Rohde, U.L. and Poddar, A.K. (2009, October). User-definable, low cost, low noise, and phase hit insensitive multi-octave-band tunable oscillator, phase hit and spectral pure tunable oscillator. US Patent No. 7,605,670.

130 Rohde, M. and Poddar, A.K. (2009, November). Visually inspectable surface mount device pad. US Patent No. 7612296.

131 Rohde, U.L. and Poddar, A.K. (2009, December). Low noise and low phase hits tunable oscillator. US Patent No. 7,636, 021.

132 Rohde, U.L. and Poddar, A.K. (2009, July). Wideband voltage controlled oscillators employing evanescent mode coupled resonators. Canadian Patent No. 2,563, 174.

133 Rohde, U.L. and Poddar, A.K. (2009, April). User-definable thermal drift voltage controlled oscillator. Canadian Patent No. 2,548, 317.

134 Rohde, U.L. and Poddar, A.K. (2009, April). Integrated low noise microwave wideband push-push VCO. Canadian Patent No. 2,548, 311.

135 Rohde, U.L. and Poddar, A.K. (2006, June). Novel multi-coupled line resonators replace traditional ceramic resonators in oscillators/VCOs. *2006 IEEE International Frequency Control Symposium and Exposition*, 432–442, Florida, USA.

136 Voinigescu, S.P., Tomkins, A., Dacquay, E. et al. (2013). A study of SiGe HBT signal sources in 220-330-GHz range. *IEEE Journal of SSC* 48 (9): 1–11, 2013.

137 Tanaka, T., Takata, K., Aikawa, M. et al. (2012). A Ku band Oscillator array using positive feedback type push-push oscillators. *IEEE APMC Conference Proceedings*, 112–114.

138 Sirburanon, T., Sato, T., Musa, A. et al. (2012). A 20 GHz push-push voltage-controlled oscillator for a 60 GHz frequency synthesizer. *IEEE APMC Conference Proceedings*, 205–207.

139 Chu, S.-W. and Wang, C.-K. (2012). An 80 GHz wide tuning range push-push VCO with gm-boosted full-wave rectification technique in 90 nm CMOS. *IEEE Microwave and Wireless Components Letters* 22 (4): 203–205.

140 Yaoming, S. and Scheytt, C.J. (2012). A low phase noise 61 GHz push-push VCO with divider chain buffer in SiGe BiCMOS for 122 GHz ISM applications. *2012 IEEE RFIC Symposium*, 79–82.

141 Nakamura, T., Masuda, T., Washio, K. et al. (2012). A push-push VCO with 13.9 GHz wide tuning range using loop-ground transmission line for full band 60-GHz transceiver. *2102 IEEE Journal of SSC* 47 (6): 1267–1277.

142 Stuenkel, M. and Feng, M. (2012). A W-band tunable push-push oscillator with 128X division for frequency synthesis applications. *2013 IEEE MTT-S Digest*, 1–3.

143 Wang, T.-P. (2011). A fully integrated W-band push-push CMOS VCO with phase noise and wide tuning range. *2011 IEEE Transactions on UFFC* 58 (7): 1307–1319.

144 Poddar, A.K., Rohde, U.L., and Daryoush, A.S. (2013, February). Integrated production of self injection locked self phase loop locked Opto-electronic Oscillators. US Patent application No. 13/760767.

145 Poddar, A.K., Rohde, U.L., and Daryoush, A.S. (2012, December). Self injection locked phase locked looped optoelectronic oscillator. US Patent application No. 61/746, 919.

146 Yao, X.S. and Maleki, L. (1996). Optoelectronic oscillator for photonic systems. *IEEE Journal of Quantum Electronics* 32 (7): 1141–1149.

147 Moyer, H.P. and Daryoush, A.S. (2000). A unified analytical model and experimental validations of injection-locking processes. *IEEE Transactions on Microwave Theory and Techniques* 48 (4): 493–499.

148 McNeilage, C., Ivanov, E.N., Stockwell, P.R. et al. (1998, May). Review of feedback and feedforward noise reduction techniques. *Proceedings IEEE Frequency Control Symposium*, 146–155, Pasadena, CA.

149 Suarez, A. and Ramirez, F. (2005, September). Analysis of stabilization circuits for phase-noise reduction in microwave oscillators. *IEEE Transactions on Microwave Theory and Techniques* 53: (9): 2743–2751.

150 Chang, H.C. (2003, September). Phase noise in self-injection locked oscillators – Theory and Experiment. *IEEE Transactions on Microwave Theory Techniques* 51 (9): 1994–1999.

151 Lee, K.H. Kim, J.-Y., and Choi, W.-Y. (2007, December). A 30GHz self injection locked oscillator having a long Optical delay line for phase noise reduction. *IEEE Photonics Technology Letters* 19 (24): 1982–1984.

152 Pillet, G., Morvan, L., Brunel, M. et al. (2008, August). Dual-frequency laser at 1.5 μm for optical distribution and generation of high-purity microwave signals. *Journal of Lightwave Technology* 26 (15): 2764–2773.

153 Zhou, X.S., Zhang, X., and Daryoush, A.S. (1997, February). A new approach for a phase controlled self-oscillating mixer. *IEEE Transactions on MTT* 45 (2): 196–204.

154 Surzbecher, D., Zhou, X., Zhang, X.-S. et al. (1993, June/July). Optically controlled oscillators for millimeter-wave phased-array antennas. *IEEE Transactions on Microwave Theory and Techniques* 41 (6): 998–1004.

155 Chang, H.-C., Borgioli, A., Yeh, P. et al. (1999, August). Analysis of oscillators with external feedback loop for improved locking range and noise reduction. *IEEE Transactions on Microwave Theory and Techniques* 47 (8): 1535–1543.

156 Zhang, X., Zhou, X., and Daryoush, A. S. et al. (1992, May). A theoretical and experimental study of the noise behavior of subharmonically injection locked local oscillators. *IEEE Transactions on Microwave Theory and Techniques* 40 (5): 895–902.

157 Best, R.E. (1999). *Phase-locked Loops: Design, Simulation, and Applications*, 4e. 25–26. New York: McGraw-Hill.

158 Xu, F. and Wu, K. (2005, January). Guided-wave and leakage characteristics of substrate integrated waveguide. *IEEE Transactions on Microwave Theory Techniques* 53 (1): 66–73.

159 Dong, Y. and Itoh, T. (2012, June). A dual-band oscillator with reconfigurable cavity-backed complementary split-ring resonator. *IEEE MTT-S International Microwave Symposium Digest*, 1–3.

160 He, F.F., Wu, K., Hong, W. et al. (2010, December). A low phase-noise VCO using an electronically tunable substrate integrated waveguide resonator. *IEEE Transactions on Microwave Theory and Techniques* 58 (12): 3452–3458.

161 Sirci, S., Martinez, J.D., Taroncher, M. et al. (2011). Varactor loaded continuously tunable SIW resonator for reconfigurable filter design. *Proceedings 41st European Microwave Conference*, 436–439.

162 Michael Wu, C.-T., Poddar, A.K., Rohde, U.L. et al. (2014). Active complementary coupled resonator for low phase noise X-band oscillator. *European Frequency and Time Forum*, 356–359.

163 Prabha, R. and Varshney, P.K. A novel innovative bi-complementary resonator coupled advanced substrate integrated waveguide sensor for angular positioning measurement. *IEEE Sensors Journal*. https://doi.org/10.1109/JSEN.2026.3685514.

164 Chen, Z., Hong, W., and Chen, J.X. (2012, May). High-Q planar active resonator based on substrate integrated waveguide technique. *Electronics Letters* 48 (10): 575–577.

165 Maxwell, C.J. (1892). *A Treatise on Electricity and Magnetism*, 3e, vol. 2. Oxford: Clarendon, 68–73.

166 Nick, M. and Mortazawi, A. (2010, May). Low phase-noise planar oscillators based on low-noise active resonators. *IEEE Transactions on Microwave Theory and Techniques* 58 (5): 1133–1139.

167 Leong, K. (2000). Precise measurements of surface resistance of HTS thin films using a novel method of Q-factor computations for sapphire dielectric resonators in the transmission mode. PhD thesis. James Cook University.

168 Leong, K. and Mazierska, J. (2001). Accurate measurements of surface resistance of HTS films using a novel transmission mode Q-factor technique. *Journal of Superconductivity* 14 (1): 93–103.

169 Leong, K., Mazierska, J., Jacob, M.V. et al. (2002). Comparing unloaded Q-factor of a high-Q dielectric resonator measured using the transmission mode and reflection mode methods involving S-parameter circle fitting. *IEEE MTT-S International Microwave Symposium Digest*, 1665–1668. https://arxiv.org/ftp/arxiv/papers/1207/1207.5622.pdf.

170 Wu, C.-T.M. (2014, June). Active integrated metamaterial-based antennas and microwaves components. PhD thesis. University of California, Los Angeles: Department of Electrical Engineering.

171 Starostin E.L. and van der Heijden G.H.M. (2007). The shape of a Möbius strip. *Nature Materials* 6 (8): 563–567. https://doi.org/10.1038/nmat1929. PMID 17632519.

172 Pond, J.M. (2000, December). Möbius dual mode resonators and bandpass filter. *IEEE Transactions of MTT* 48 (12): 2465–2471.

173 Pond, J.M., Liu, S., and Newman, N. (2001, December). Bandpass filters using dual-mode and quad-mode Möbius resonators. *IEEE Transactions on MTT* 49: 2363–2368.

174 Borselli, M., Johnson, T.J., and Painter, O. (2005, March). Beyond the Rayleigh scattering limit in high-Q silicon microdisks: Theory and experiment. *Optics Express* 13 (5): 1515–1530. https://doi.org/10.1364/OPEX.13.001515.

175 Honkote V. (2009). Capacitive load balancing for Möbius implementation of standing wave oscillator. *52nd IEEE MWSCAS*, 232–235.

176 Hoffman, A.J. Alekseyev, L., Howard, S.S. et al. (2007). Negative refraction in semiconductor metamaterials. *Nature Materials* 6: 946–950.

177 Gravel, J.-F. and Wight, J.S. (2006, January). On the conception and analysis of a 12-GHz push-push phase locked DRO. *IEEE Transactions on Microwave Theory and Techniques* 54 (1): 153–159.

178 Ballon, D.J. and Voss, H.U. (2008). Classical Möbius-ring resonators exhibit fermion-boson rotational symmetry. *Physical Review Letters* 101: 247701. American Physical Society.

179 Tesla, N. (1894, January). Coil for electromagnet. US Patent 512340.

180 Perez-Enriquez, R. (2002). A Structural parameter for high Tc superconductivity from an Octahedral Möbius Strip in BaCuO: 123 type Perovskites. *Revista Mexicana de Fisica* 48 (suppl. 1): 262.

181 Watkins-Johnson Company (1974, March). *Development in FML and Miniature Millimeter Devices*, 1 (2). Watkins Johnson, Tech Note.

182 Rohde, U.L. and Poddar, A.K. (2013, February). DRO drops phase-noise levels. *Microwaves RF*, 80–84.

183 Rohde, U.L., Poddar, A.K., and Apte, A.M. (2013, April). Phase noise measurement and its limitations. *Microwave Journal* 56 (5): 22–46.

184 Poddar, A.K., Sundararajan, D., and Rohde, U.L. (2013, June). Real time signal retention device using co-planar waveguide (CPW) as Möbius strip. *2013 IEEE MTT-S Digest*, 1–3.

185 Poddar, K. and Rohde, U.L. (2013, July). A novel evanescent-mode Möbius-coupled resonator oscillators. *IEEE Joint UFFC Symposia with European Frequency and Time Forum (EFTF) and Piezo Response Force Microscopy.*

186 Anritsu Application Note (2013). Time domain reflectometry using vector network analyzers.

187 Agilent Technologies (2012, January). Agilent time domain analysis using a network analyzer. Application Note 1287-12.

188 Agilent Technologies (2012, May). Agilent time domain analysis using a network analyzer. Literature Number 5989-5723EN.

189 Rohde, U.L., Poddar, A.K., and Sundararajan, D. (2013, November). Printed resonators: Möbius strips theory and applications. *Microwave Journal* 56 (11): 24–54.

190 Poddar, A.K. and Rohde, U.L. (2014, May). The pursuit for low cost and low phase noise synthesized signal sources: Theory & optimization. *IEEE UFFC Symposia.*

191 Poddar, K., Rohde, U.L., and Itoh, T. (2014, June). Metamaterial Möbius Strips (MMS): Tunable oscillator circuits. *IEEE MTT-S Digest*, 1–4.

192 Sheleg, S., Sliusar, I. and Slyusar, V. (2025). Study of a metamaterial unit cell with a flat Möbius strip: frequency characteristics and DNG bands. *2025 IEEE 30th International Seminar/Workshop on Direct and Inverse Problems of Electromagnetic and Acoustic Wave Theory (DIPED)*, Tbilisi, Georgia (259–264). https://doi.org/10.1109/DIPED66951.2025.11194616.

193 Veselago, V.G. (1968). The electrodynamics of substance with simultaneously negative values of ε and μ. *Soviet Physics Uspekhi* 10: 509–514.

194 Pendry, J.B., Holden, A.J., Robbins, D.J. et al. (1999). Magnetism from conductors and enhanced nonlinear phenomena. *IEEE Transactions on Microwave Theory and Techniques* 47: 2075–2084.

195 Shelby, R.A., Smith, D.R., and Schultz, S. (2001, April). Experimental verification of a negative index of refraction. *Science* 292 (5514): 77–79.

196 Pendry, J.B. (2000). Negative refraction makes a perfect lens. *Physical Review Letters* 85 (18): 3966–3969.

197 Jacob, Z., Alekseyev, L.V., and Narimanov, E. (2006). Optical hyperlens: Far-field imaging beyond the diffraction limit. *Optics Express* 14: 8247–8256.

198 Liu, Z., Lee, H., Xiong, Y. et al. (2007). Optical hyperlens magnifying sub-diffraction-limited object. *Science* 315: 1686–1687.

199 Smolyaninov, I.I., Hung, Y.J., and Davis, C.C. (2007). Magnifying superlens in the visible frequency range. *Science* 315: 1699–1701.

200 Pendry, J.B., Schurig, D., and Smith, D.R. (2006). Controlling electromagnetic fields. *Science* 312: 1780–1782.

201 Schurig, D., Mock, J.J., Justice, B.J. et al. (2006). Metamaterial electromagnetic cloak at microwave frequencies. *Science* 314: 977–980.

202 Cai, W., Chettiar, U.K., Kildishev, A.V. et al. (2007). Optical cloaking with metamaterials. *Nature Photonics* 1: 224–227.

203 Pendry, J.B. (2006). Metamaterials in the sunshine. *Nature Materials* 5: 599–600.

204 Martin, F., Falcone, F., Bonache, J. et al. (2003). Miniaturized coplanar waveguide stop band filters based on multiple tuned split ring resonators. *IEEE Microwave and Wireless Components Letters* 13: 511–513.

205 Joan, G.G., Martin, F., Falcone, F. et al. (2005). Microwave filters with improved stopband based on sub wavelength resonators. *IEEE Transactions on Microwave Theory and Techniques* 53: 1997–2006.

206 Bonache, J., Gil, I., Joan, G.G. et al. (2005). Novel microstrip bandpass filters based on complementary split-ring resonators. *IEEE Transactions on Microwave Theory and Techniques* 54: 265–271.

207 Nanostructure Metamaterials (2010). Directorate-General for research unit G3-Vlaue-added materials. EUR 24409 EN.

208 Suresh, N.V., Sridhar, G., Prakash, K.B., et al. (2025). Advances in meta-materials. In: *New Frontiers in Materials Science: Interdisciplinary Approaches to Innovation and Technologies*, 17–36. Wiley. https://doi.org/10.1002/9781394314935.ch2.

209 Chen, H.-T., O'Hara, J.F., Taylor, A.J. et al. (2007). Complementary planar terahertz metamaterials. *Optical Society of America* 15 (3): 1084–1095.

210 Hau, L.V. (2008). Optical information processing in Bose-Einstein condensates. *Nature Photonics* 2: 451–453.

211 Baba, T. (2008). Slow light in photonic crystals. *Nature Photonics* 2: 465–473.

212 Notomi, M., Kuramochi, E., and Tanabe, T. (2008). Large-scale arrays of ultrahigh-Q coupled nanocavities. *Nature Photonics* 2: 741–747.

213 Wang, X. (2009). Simulation of simultaneously negative medium metamaterials. *Master of Science in Electrical Engineering*. Falls Church, Virginia.

214 Eleftheriades, G.V., Siddiqui, O., and Iyer, A.K. (2003, February). Transmission line models for negative refractive index media and associated implementations without excess resonators. *IEEE Microwave and Wireless Components Letters* 13 (2): 51–53.

215 Iyer, A.K. and Eleftheriades, G.V. (2002, June). Negative refractive index metamaterials supporting 2-D waves. *IEEE MTT-S Symposium Digest*, 2: 1067–1070. Seattle, WA.

216 Grbic, A. and Eleftheriades, G.V. (2002, July). Experimental verification of backward-wave radiation from a negative refractive index metamaterial. *Journal of Applied Physics* 92 (10): 5930–5935.

217 Wang, W.-C., Sharma, A., Mo, C.-T. et al. (2025). Reconfigurable tunable fishnet metamaterial for terahertz wave control. *2025 50th International Conference on Infrared, Millimeter, and Terahertz Waves (IRMMW-THz)*, Helsinki, Finland (1–2). https://doi.org/10.1109/IRMMW-THz61557.2025.11319634.

218 Manuel, J. and Alves, T. (2010, October). Metamaterials with negative permeability and permittivity: Analysis and application. MS thesis. Universidade Técnica De Lisboa, Instituto Superior Técnico.

219 Nawaz, M. (2009). Low impedance wheel resonators for low voltage and low power applications. Dr.-Ing dissertation. Technischen FakultÄat der UniversitÄat Erlangen-NÄurnberg.

220 Caloz, C. and Itoh, T. (2006). *Electromagnetic Metamaterials: Transmission Line Theory and Microwave Applications: The Engineering Approach*. Hoboken, NJ: IEEE Press.

221 Lai, A. and Itoh, T. (2004, September). Composite right/left-handed transmission line metamaterials. *IEEE Microwave Magazine* 5 (3): 34–50.

222 Pacheco, J., Grzegorczyk, T.M., Wu, B.-I. et al. (2002). Power propagation in homogeneous isotropic frequency dispersive left-handed media. *Physical Review Letters* 89: 257401.

223 Qu, L., Zheng, L., Liu, R. et al. (). Near net shape fabrication of tunable Ku-band metamaterial filters. *IEEE Transactions on Industrial Electronics*. https://doi.org/10.1109/TIE.2026.3666062.

224 Li, L., Xu, F., Wu, K. et al. (2007, November). Slow-wave line coupler with interdigital capacitor loading. *IEEE Transactions on Microwave Theory and Techniques* 55 (11): 2427–2433.

225 Sun, S. and Zhu, L. (2005, April). Guided-wave characteristics of periodically non uniform coupled microstrip lines-even and odd modes. *IEEE Transactions on Microwave Theory and Techniques* 53 (4): 1221–1227.

226 Liu, Z. and Weikle, R.M. (2005, December). A compact quadrature coupler based on coupled artificial transmission lines. *IEEE Microwave Wireless Component Letters* 15 (12): 889–891.

227 Caloz, C. and Itoh, T. (2005). *Electromagnetic Metamaterials: Transmission Line Theory and Microwave Applications*. Hoboken, NJ: Wiley and IEEE Press.

228 Ueda, T., Lai, A., and Itoh, T. (2007, June). Demonstration of negative refraction in a cutoff parallel-plate waveguide loaded with 2-D square lattice of dielectric resonators. *IEEE Transactions on Microwave Theory and Techniques* 55 (6): 1280–1287.

229 Zedler, M., Russer, P., and Caloz, C. (2007, June). Circuital and experimental demonstration of a 3D isotropic LH metamaterial based on the rotated TLM scheme. *IEEE-MTT Int'l Symposium*. Honolulu, HI.

230 Riaz, L. et al. (2026). Electronically tunable half-mode substrate-integrated waveguide coupler. *IEEE Access* 14: 9183–9192. https://doi.org/10.1109/ACCESS.2026.3654176.

231 Liu, N., Wang, B., Sheng, X. et al. (2026, January). A noncontact permittivity sensor based on EBG and metasurface. *IEEE Sensors Journal* 26 (1): 528–535. https://doi.org/10.1109/JSEN.2025.3629090.

232 Abielmona, S., Nguyen, H.V., and Caloz, C. (2006, December). CRLH zeroth order resonator (ZOR): Experimental demonstration of insensitivity to losses and to size. *Proceedings Asia-Pacific Microwave Conference*, 657–662.

233 Allen, C.A., Leong, K.M.K.H., and Itoh, T. (2006, July). Design of microstrip resonators using balanced and unbalanced composite right/left-handed transmission lines. *IEEE Transactions on Microwave Theory and Techniques* 54 (7): 3104–3112.

234 Baena, J.D., Bonache, J., Martin, F. et al. (2005, April). Equivalent circuit models for split ring resonators and complementary split ring resonators coupled to planar transmission lines. *IEEE Transactions on Microwave Theory and Techniques* 53 (4): 1451–1461.

235 Dong, Y. and Itoh, T. (2010, December). Miniaturized substrate integrated waveguide slot antennas based on negative order resonance. *IEEE Transactions on Antennas and Propagation* 58 (12): 3856–3864.

236 Li, C., Liu, K.Y., and Li, F. (2007, May). Composite right/left-handed coplanar waveguide bandpass filter using capacitively-coupled zeroth-order resonators. *Applied Physics A: Materials Science and Processing* 87 (2): 317–319.

237 Majidimosleh, M., Akhtari, S., Helaoui, M. et al. (2026). Concurrent dual-band quadrature phase shift frequency-selective receiver. *IEEE Transactions on Microwave Theory and Techniques*. https://doi.org/10.1109/TMTT.2026.3683009.

238 Choi, J. and Seo, C. (2007, November). Broadband and low phase noise VCO using tunable metamaterial transmission line based on varactor-loaded split-ring resonator. *Korea–Japan Microwave Conference*, 145–148.

239 Im, S., Seo, C., Kim, J. et al. (2002, June). Improvement of microstrip open loop resonator filter using aperture. *IEEE MTT-S International* 3: 1801–1804.

240 Park, E. and Seo, C. (2006, June). Low phase noise oscillator using microstrip square open loop resonator. *IEEE MTT-S International Microwave Symposium*.

241 Lee, Y.-T., Lim, J.-S., Park, J.-S. et al. (2002, February). A novel phase noise reduction technique In Oscillators using defected ground structure. *IEEE Microwave and Wireless Components Letters* 12 (2): 39–41.

242 Smith, D.R., Padilla, W.J., Vier, D.C. et al. (2000, May). Composite medium with simultaneously negative permeability and permittivity. *Physical Review Letters* 84 (18): 4184–4187.

243 Smith, D.R., Rye, P., Vier, D.C. et al. (2004, March). Design and measurement of anisotropic metamaterials that exhibit negative refraction. *IEICE Transactions on Electronics* E87-C (3): 359–370.

244 Marqu´es, R., Medina, F., and Rafii-El-Idrissi, R. (2002, April). Role of bianisotropy in negative permeability and left-handed metamaterial. *Physical Review B* 65 (14): 144440.

245 Cramer, J. (2011). Dielectric resonator fabrication and assembly methods for terahertz metamaterials. PhD thesis. University of Massachusetts Lowell.

246 Pendry, J.B., Holden, A.J., Stewart, W.J. et al. (1996, June). Extremely low frequency plasmons in metallic mesostructures. *Physical Review Letters* 76 (25): 4773–4776.

247 Pendry, J.B., Holden, A.J., Robbins, D.J. et al. (1998, June). Low frequency plasmons in thin-wire structures. *Journal of Physics: Condensed Matter* 10 (22): 4785–4809.

248 Rotman, W. (1962, January). Plasma simulation by artificial dielectrics and parallel-plate media. *IRE Transactions on Antennas and Propagation* 10 (1): 82–95.

249 Wiwatcharagoses, N., Park, K.Y., and Chahal, P. (2011). A new metamaterial unit cell for compact microstrip circuit designs. *Electronic Components and Technology Conference*, 169–172.

250 Zhou, J., Zhang, L., Tuttle, G. et al. (2006, January). Negative index materials using simple short wire pairs. *Physical Review B* 73 (4): 041101(R).

251 Drachev, V.P., Cai, W., Chettiar, U. et al. (2006, October). Experimental verification of an optical negative index material. *Laser Physics Letters* 3 (1): 49–55.

252 Kildishev, A.V., Cai, W., Chettiar, U.K. et al. (2006, March). Negative refractive index in optics of metal-dielectric composites. *Journal of the Optical Society of America B*, 23 (3): 423–433.

253 Zhang, S., Fan, W., Panoiu, N.C. et al. (2005, September). Experimental demonstration of near-infrared negative-index metamaterials. *Physical Review Letters* 95 (13): 137404.

254 Zhang, S., Fan, W., Malloy, K.J. et al. (2005, June). Near-infrared double negative metamaterials. *Optics Express* 13 (13): 4922–4930.

255 Dolling, G., Enkrich, C., Wegener, M. et al. (2006, June). Low-loss negative-index metamaterial at telecommunication wavelengths. *Optics Letters* 31 (12): 1800–1802.

256 Dolling, G., Enkrich, C., Wegener, M. et al. (2006, May). Simultaneous negative phase and group velocity of light in a metamaterial. *Science* 312 (5775): 892–894.

257 Dolling, G., Wegener, M., Soukoulis, C.M. et al. (2007, January). Negative-index metamaterial at 780 nm wavelength. *Optics Letters* 32 (1): 53–55.

258 Valentine, J., Zhang, S., Zentgraf, T. et al. (2008, September). Three-dimensional optical metamaterial with a negative refractive index. *Nature* 455 (7211): 376–379.

259 Mie, G. (1908). Beiträge zur Optik trüber Medien, speziell kolloidaler Metallösungen. *Annalen der Physik, Vierte Folge, Band* 25 (3): 377–445.

260 Mie, G. (1978). Contributions on the optics of turbid media, particularly colloidal metal solutions. Translation. Albuquerque, New Mexico: Sandia Laboratories. SAND78-6018. Chicago, IL: National Translation Center, Translation 79-21946.

261 Lewin, L. (1947). The electrical constants of a material loaded with Sphereical particles. *The Proceedings of the Institution of Electrical Engineers* 94: 65–68.

262 O'Brien, S. and Pendry, J.B. (2002, April). Photonic band-gap effects and magnetic activity in dielectric composites. *Journal of Physics: Condensed Matter* 14 (15): 4035–4044.

263 Holloway, C.L., Kuester, E.F., J. Baker-Jarvis et al. (2003, October). A Double Negative (DNG) composite medium composed of magneto dielectric spherical particles embedded in a matrix. *IEEE Transactions on Antennas and Propagation* 51 (10): 2596–2601.

264 Baker-Jarvis, J., Janezic, M.D., Love, D. et al. (2006, October). Phase velocity in resonant structures. *IEEE Transactions on Magnetics* 42 (10): 3344–3346.

265 Naqui, J., Duran-Sindreu, M., and Martin, F. (2013). Modeling split-ring resonator (SRR) and complementary split-ring resonator (CSRR) loaded transmission lines exhibiting cross-polarization effects. *IEEE Antennas and Wireless Propagation Letters* 12: 178–181. https://doi.org/10.1109/LAWP.2013.2245095.

266 Sisó, G., Gil, M., Aranda, M. et al. (2009, June). Miniaturization of planar microwave devices by means of complementary spiral resonators (CSRs): Design of quadrature phase shifters. *Radioengineering* 18 (2): 144–148.

267 Tiebout, M. (2002). A CMOS fully integrated 1 GHz and 2 GHz dual band VCO with a voltage controlled inductor. *Proceedings of the IEEE European Solid-State Conference*, 799–802.

268 Goel, A. and Hashrmi, H. (2007, March). Frequency switching in dual-resonance oscillators. *IEEE Journal of Solid-State Circuits* 42 (3): 571–582.

269 Liu, L.Y., Poddar, A.K., Rohde, U.L. et al. (2025). A CPW-based microwave sensor with integrated interdigitated electrode structure for permittivity characterization of solid materials. *2025 IEEE International Symposium on Antennas and Propagation and North American Radio Science Meeting (AP-S/CNC-USNC-URSI)*, Ottawa, ON, Canada, (2688–2691). https://doi.org/10.1109/AP-S/CNC-USNC-URSI55537.2025.11266803.

270 Berny, A.D., Niknejad, A.M., and Meyer, R.G. (2005). A 1.8 GHz LC VCO with 1.3 GHz tuning range and digital amplitude calibration. *IEEE Journal of Solid-State Circuits* 40 (4): 909–917.

271 Rohde, U.L. and Poddar, A.K. (2006, September). Low noise, low power consumption, configurable, and adaptable ultrawideband VCOs. *IEEE International Conference on Ultrawideband (ICUWB), ICUWB Conference proceedings*, 545–550.

272 Bavisi, A. (2006, May). Integrated multi-mode oscillators and filters for multi-band radios using liquid crystalline polymer based packaging technology. PhD thesis. Atlanta, GA: Georgia Institute of Technology.

273 Ravi, A., Carlton, B.R., Banerjee, G. et al. (2005). A 1.4-V, 2.4/5.2-GHz, 90-nm CMOS system in a package transceiver for next generation WLAN. *IEEE Symposium on VLSI Technology, Digest of Technical Papers*. New York: IEEE.

274 Lam, C. and Razavi, B. (2000, May). A 2.6/5.2 GHz frequency synthesizer in 0.4-{LC MU}m CMOS technology. *IEEE Journal of Solid-State Circuits* 35: 788–790.

275 Razavi, B. (1998). *RF Microelectronics*, 2e. Englewood Cliffs, NJ: Prentice-Hall.

276 Yim, M.-S. and O, K.K. (2006, January). Switched resonators and their applications in a dual band monolithic CMOS LC tuned VCO. *IEEE Transactions on Microwave Theory and Techniques* 54 (1): 74–81.

277 Klepser, B., Punzenberger, M., Ruhlicke, T. et al. (2003). 5 GHz and 2.4 GHz dual-band RF-transceiver for WLAN 802.11a/b/g applications. *IEEE RFIC Symposium Digest Papers*, 37–40.

278 Araki, K., Nakagawa, T., Kawashima, M. et al. (2004). Implementation and performance of a multi-band transceiver for software defined radio. *Proceedings of the 2004 IEEE Radio and Wireless Conference (RAWCON)* (207–210).

279 Wiggins, S. (1997). *Introduction to Applied Nonlinear Dynamics and Chaos*, 2e. New York: Springer-Verlag.

280 van der Pol, B. (1922). On oscillation hysteresis in a triode generator with two degree of freedom. *Philosophy Magazine* 43: 700–719.

281 Schaffner, J.S. (1954, June). Simultaneous oscillations in oscillators. *IRE Transactions on Circuit Theory* 1 (2): 2–8.

282 Endo, T. and Mori, S. (1976, February). Mode analysis of a multimode ladder oscillator. *IEEE Transactions on Circuits & Systems* CAS-23: 100–113.

283 Poddar, A.K. (2004, December). A novel approach for designing integrated ultra low noise microwave wideband voltage controlled oscillators. Dr.-Ing. Diss., TU- Berlin, Germany.

284 H. Liu, Y. Wu, K. Ma, S. Liang and S. LI, “Miniaturized Bandpass Filter Using MISL-Based Slow Wave Meander Line Resonators”, *IEEE Microwave and Wireless Letters (Early access)*, pp. 1-4, Feb 26, 2026, https://doi.org/10.1109/LMWT.2026.3666042.

285 Bruyland, I. (1968, January). Simultaneous oscillations at two frequencies in RLC circuits. *Proceedings of the IEEE* 1: 82–83.

286 Poddar, A.K., Bansal, J.K., and Pandey, K.N. (1998, June). Millimeter wave evanescent mode power combiner Gunn oscillator in suspended Stripline configuration. *Mathematical Methods in Electromagnetic Theory, IEEE/MMET*, 384–386.

287 Engheta, N. (2001, September). Compact cavity resonators using metamaterials with negative permittivity and permeability. *Proceedings International Conference Electromagnetics in Advanced Applications (ICEAA)*, 739–742. Torino, Italy.

288 Zhou, Y., Zhong, J., Sun, Y. et al. (2025). MTJ based temperature-adaptive VCO (TAVCO) for compensating CP-PLL frequency drift. *2025 IEEE International Symposium on Circuits and Systems (ISCAS)*, London, United Kingdom (1–5). https://doi.org/10.1109/ISCAS56072.2025.11043994.

289 Sanada, A. and Matsuda, I. (2005). Experimental verification of surface waves of metamaterials. IMS 2005, MTT-S Digest, 1737–1740.

290 Poddar, A.K., Koul, S.K., and Bhat, B. (1997). Millimeter Wave Evanescent Mode Gunn Diode Oscillator in Suspended Stripline Configuration. *Proceedings of the 22nd International Conference on Infrared and Millimeter Waves (IRMMW 1997)* (265–266).

291 Sotiriadis, P.P. (2010, September). Exact spectrum and time-domain output of flying-adder frequency synthesizers. *IEEE Transactions on Ultrasonics, Ferroelectrics, and Frequency Control* 57 (9): 1926–1935.

292 Wang, X. and Kennedy, M.P. (2026, May). Unified analysis of digital Δ-Σ modulators (DDSMs) for fractional-N frequency synthesis—introducing the PASS family of DDSMs featuring independent shaping of the probability density and spectral envelope. *IEEE Transactions on Circuits and Systems I: Regular Papers* 73 (5): 3245–3258. https://doi.org/10.1109/TCSI.2026.3673203.

293 Calbaza, D. and Savaria, Y. (2002, August). A direct digital periodic synthesis circuit. *IEEE Journal Solid-State Circuits* 37 (8): 1039–1045.

294 Rohde, U.L. and Poddar, A.K. (2011, October). A novel low cost high performance synthesizer. *Proceedings of the IEEE EuMic*, 200–203.

295 Rohde, U.L. and Poddar, A.K. (2011, May). A novel high frequency synthesizer using adaptive injection mode-coupled VCSO for low jitter and low noise applications. *Proceedings of 2011 Joint Conference of the IEEE International Frequency Control Symposium & European Frequency and Time Forum*, 452–457.

296 Lin, D.N., Poddar, A.K., Rohde, U.L. et al. (2024). A novel polarization-reconfigurable dielectric resonator antenna based on liquid material. *2024 Photonics & Electromagnetics Research Symposium (PIERS)*, Chengdu, China, (1–5). https://doi.org/10.1109/PIERS62282.2024.10617911.

297 Egan, W. (2000). *Frequency Synthesis by Phase Lock*. Hoboken, NJ: John Wiley & Sons, Inc.

298 Narasapura Ramesh, S., Singhal, K., and Semnani, A. (2026). An evanescent-mode cavity-based frequency-selective plasma limiter. *IEEE Transactions on Microwave Theory and Techniques*. https://doi.org/10.1109/TMTT.2026.3680125.

299 Barajas, B., Molavi, R., and Mirabbasi, S. (2024). A wideband 65-nm 60-GHz push-push LC VCO using a nonlinear varactor array. *2024 IEEE 67th International Midwest Symposium on Circuits and Systems (MWSCAS)*, Springfield, MA, USA (367–371). https://doi.org/10.1109/MWSCAS60917.2024.10658873.

300 Yu, J., Yu, Z., Liu, C. et al. (2025, October). A MEMS oscillator and TDC-based temperature sensing architecture with embedded calibration. *IEEE Transactions on Nuclear Science* 72 (10): 3442–3450. https://doi.org/10.1109/TNS.2025.3605918.

301 Zare, A., Jafari, A., Ali, A. et al. (2026). Passive thermal compensation of silicon resonators through modal engineering. *IEEE Open Journal of Ultrasonics, Ferroelectrics, and Frequency Control* 6: 112–121. https://doi.org/10.1109/OJUFFC.2026.3681525.

302 Rohde, U.L. and Poddar, A.K. (2007, November). Tunable oscillator. US Patent No. 7, 292,113.

303 Kang, T., Zhu, H., Liang, J. et al. (2025, May). A miniaturized 2.49-GHz low-phase-noise temperature-stable SAW oscillator. *IEEE Microwave and Wireless Technology Letters* 35 (5): 601–604. https://doi.org/10.1109/LMWT.2025.3546238.

304 Li, Z., Quan, X., Zhao, X. et al. (2024). A 6.4-to-16.1GHz hybrid parallel-series resonator mode-switching oscillator with 206.6dBc/Hz FoMT at 1MHz offset in 40nm CMOS. *2024 IEEE European Solid-State Electronics Research Conference (ESSERC)*, Bruges, Belgium (313–316). https://doi.org/10.1109/ESSERC62670.2024.10719465.

305 Lee, J., Adrian, V. and Chang, J.S. (2026). A novel quadrature-phase sampling technique for filterless, low-complexity, high-alias-suppression in VCO-ADCs. *IEEE Open Journal of Circuits and Systems* 7: 218–229. https://doi.org/10.1109/OJCAS.2025.3631878.

306 Psycharis, I.-D., Tsourtis, V., and Kalivas, G. (2026). A 30 GHz inductively coupled VCO with a FoMT of −191.8 dBc/Hz in 40-nm CMOS. *IEEE Microwave and Wireless Technology Letters*. https://doi.org/10.1109/LMWT.2026.3684874.

307 Wang, L. et al. (2026, May). A 4-GHz ring-VCO-based wideband double-sampling PLL with −88.2-dBc reference spur, 119-fsRMS jitter and −244.1-dB FOM. *IEEE Transactions on Circuits and Systems II: Express Briefs* 73 (5): 533–537. https://doi.org/10.1109/TCSII.2026.3680570.

308 Lu, T., Liu, Z., Yang, H. et al. (2026). A 0.5–2.5-GS/s resettable ring-VCO-based ADC eliminating quantization-noise shaping. *IEEE Journal of Solid-State Circuits*. https://doi.org/10.1109/JSSC.2026.3668599.

309 Wu, W., Zhang, X., Zhang, N. et al. (2026, May). A 16-to-24.9 GHz dual-core dual-mode noise-circulating VCO using single-phase switch with FOMT of 209.0 dBc/Hz. *IEEE Transactions on Circuits and Systems II: Express Briefs* 73 (5): 508–512. https://doi.org/0.1109/TCSII.2026.3669392.

310 Yin, Y., Han, Y., Lu, C. et al. (2026). A fractional-N PLL for low-power GNSS achieving 0.72-ppm/°C frequency stability and 0.3-dBc/Hz phase noise variation from −40 °C to 125 °C. *IEEE Transactions on Very Large Scale Integration (VLSI) Systems*. https://doi.org/10.1109/TVLSI.2026.3660636.

311 Breun, S., Schrotz, A.-M., Koch, M. et al. (2025). A 148–166 GHz Meissner push-push VCO with–168 dBc/Hz FOMT using a single transformer in SiGe BiCMOS technology. *2025 IEEE MTT-S Latin America Microwave Conference (LAMC)*, San Juan, PR, USA (35–37). https://doi.org/10.1109/LAMC63321.2025.10880535.

312 Low Noise VCOs product datasheet. Available at http://www.synergymwave.com/products.

313 Khanna, P.S. (Paul) (2006, April). Microwave oscillators: The state of the technology. *Microwave Journal* 49 (4): 22–42.

314 Chenakin, A. (2008). Building a microwave frequency synthesizer, Part 2: Component selection. *High Frequency Electronics* 17 (6): 18–26.

315 Gonzalez, G. (2007). *Foundations of Oscillator Circuit*. London: Ch-IV, Artech House, Inc.

316 Osadchuk, I., Osadchuk, O., Osadchuk, V. et al. (2020). Microwave oscillator on transistor structures with dielectric resonators. *2020 IEEE Ukrainian Microwave Week (UkrMW)*, Kharkiv, Ukraine (902–906). https://doi.org/10.1109/UkrMW49653.2020.9252748.

317 Everard, J.K.A. and Zhou, L. (2006, May). Non linear Effects in varactor tuned resonators. *IEEE Transactions on Ultrasonics, Ferroelectrics and Frequency Control* 53 (5): 853–861.

318 Ohira, T. and Araki, K. (2009). Active Q factor and equilibrium stability formulation for sinusoidal oscillators. *IEEE Transactions on Circuits and Systems Pt II* 54 (9): 810–814.

319 Nallatamby, J.-C., Prigent, M., Camiade, M. et al. (2003, March). Extension of the Leeson formula to phase noise calculation in transistor oscillators with Complex tanks. *IEEE Transactions on Microwave Theory and Techniques* 51 (3): 690–696.

320 Chae, M., Jang, S., Hwanq, C. et al. (2025). A 65 fs rms jitter 10.1 GHz fractional-N digital PLL with quantization-error compensation. *Proceedings of IEEE International Solid-State Circuits Conference (ISSCC)* (554–556).

321 Pamarti, S. and Delshadpour, S. (2008, July). A spur elimination technique for phase interpolation-based fractional-N PLLs. *IEEE Transactions Circuits Systems I, Regular Papers* 55 (6): 1639–1647.

322 Sun, Y., Yu, X., Rhee, W. et al. (2010, March). Dual-path LC VCO design with partitioned coarse-tuning control in 65 nm CMOS. *IEEE Microwave and Wireless Components Letters* 20 (3): 169–171.

323 Yu, X., Sun, Y., Rhee, W. et al. (2009, August). A fractional-N frequency synthesizer with customized noise shaping for WCDMA/HSDPA applications. *IEEE Journal of Solid-State Circuits* 44: 2193–2201.

324 Wu, T., Hanumolu, P., Mayaram, K. et al. (2007, September). A 4.2 GHz PLL frequency synthesizer with an adaptively tuned coarse loop. *Proceedings IEEE CICC*, 547–550.

325 Chenakin, A. (2007, May). Frequency synthesis: Current solutions and new trends. *Microwave Journal* 50: 256–266.

326 Browne, J. (2006, March). Frequency synthesizers tune communication systems. *Microwaves & RF* 45 (3): 22–30.

327 Rohde, U.L. and Poddar, A.K. (2012, May). Synthesizers shave size and noise. *Microwave & RF* 51 (5): 132–164.

328 Mao, S.-G., Wu, M.-S., Chueh, Y.-Z. et al. (2005, November). Modeling of symmetric composite right/left-handed coplanar waveguides with applications to compact bandpass-filters. *IEEE Transactions on Microwave Theory and Techniques* 53 (11): 3460–3466.

329 Poddar, A.K. and Rohde, U.L. (2014). A novel metamaterial resonator based signal sources for applications in current and later generation radio communication systems. *IEEE BenMAS*. Philadelphia: Drexel Univ.

330 Apte, A., Madhavan, V., Poddar, A. et al. (2014). A novel low phase noise X-band oscillator. *IEEE BenMAS*. Philadelphia: Drexel Univ.

331 Liu, J., Cheng, Y., Nie, Y. et al. (2013, December). Metamaterial extends microstrip antenna. *Microwaves & RF* 52 (12): 69–73.

332 S. Benmostefa et al. (2012, September). Metamaterials form miniature bandstop filters. *Microwaves & RF* 51 (9): 69–73.

333 Katko, A.R., Barrett, J.P., and Cummer, S.A. (2014). Time-varying transistor-based metamaterial for tunability, mixing, and efficient phase conjugation. *Journal of Applied Physics* 115 (14): 144501–144506.

334 Tupin, J.P., Wu, Q., Werner, D.H. et al. (2014). Near-zero-index metamaterial lens combined with AMC metasurface for high-directivity low-profile antennas. *IEEE Transactions on AP-S* 62 (4): 1928–1936.

335 Gupta, M. and Saxena, J. (2014). Superdirective patch antenna array using metamaterial. *2014 IEEE International Conference on Electronics, Computing and Communication Technologies (IEEE CONECCT)*, 1–6.

336 Engheta, N. (2002). An idea for thin subwavelength cavity resonators using metamaterials with negative permittivity and permeability. *IEEE Antenna and Wireless Propagation Letters* 1: 10–13.

337 Poddar, A.K. and Rohde, U.L. (2014, May). Metamaterial Möbius Strips (MMS): Application in resonators for oscillators and synthesizers. *IEEE UFFC Symposia*.

338 Chen, P.-C., Wang, W.-T., and Lee, T.-C. (2026). A fast-locking phase-locked loop with joint phase and frequency compensation technique. *IEEE Transactions on Very Large Scale Integration (VLSI) Systems*. https://doi.org/10.1109/TVLSI.2026.3684335.

339 Poddar, A.K. and Rohde, U.L. (2014, June). Evanescent-mode phase-injection mode-coupled (EMPIMC) metamaterial resonator based tunable signal sources. *IEEE Wamicon*.

340 Poddar, A.K., Rohde, U.L., and Itoh, T. (2014, June). Metamaterial Möbius strips (MMS): Tunable oscillator circuits. *IEEE MTT-S Digest*, 1–4.

341 Abdul Alim, S.K., Moznebi, A.-R., and Mirzavand, R. (2026, April). An SIW sensor with environmental compensation for permittivity measurement or high-sensitivity displacement sensing. *IEEE Sensors Journal* 26 (8): 11815–11826. https://doi.org/10.1109/JSEN.2026.3669593.

342 Zhang, L., Poddar, A.K., Rohde, U.L. et al. (2014, May). Oscillator phase noise reduction using self-injection locked and phase locked loop (SILPLL). *IEEE IFCS*.

343 Wu, C.-T.M., Poddar, A.K., Rohde, U.L. et al. (2014, June). Active complementary coupled resonator for low phase noise X-band oscillator. *2014 European Frequency and Time Forum*, Neuchâtel, Switzerland.

344 Fei, W., Yu, H., Shang, Y. et al. (2013, March). A 96-GHz oscillators by high Q-differential transmission line loaded with complementary split-ring resonator in 65-nm CMOS. *IEEE Transactions on Circuits and Systems-II, Express Briefs* 60 (3): 127–131.

345 Cano, S., Faura, M., Garcia, L. et al. (2026). Frequency-dependent impedance matching synthesis methodology for filters or matching networks with equiripple responses. *IEEE Transactions on Microwave Theory and Techniques*. https://doi.org/10.1109/TMTT.2026.3674572.

346 Seenivasan, M. and Paranthaman, M. (2026). Metamaterial inspired MIMO antenna for 5G applications. *2026 International Conference on Wireless Communications Signal Processing and Networking (WiSPNET)*, Chennai, India (1–6). https://doi.org/10.1109/WiSPNET69615.2026.11489458.

347 Smith D.R., Schultz, S., Markoš, P. et al. (2002). Determination of effective permittivity and permeability of metamaterials from reflection and transmission coefficients. *Physical Review B* 65 (19): 195104.

348 Angiulli, G., Versaci, M., Calcagno, S. et al. (2024). Analytic continuation, phase unwrapping, and retrieval of the refractive index of metamaterials from S-parameters. *Sensors* 24 (3): 912.

349 Smith, D.R., Vier, D.C., Koschny, Th. et al. (2005). Electromagnetic parameter retrieval from inhomogeneous metamaterials. *Physical Review E* 71 (3): 36617.

350 Hasar, U.C. and Barroso, J.J. (2011). Retrieval approach for determination of forward and backward wave impedances of bianisotropic metamaterials. *PIER* 112: 109–124.

351 Kong, H., Chen, X. and He, Y. (2026). Radiation improvement for chain-type FSS integrated conformal omnidirectional antenna using spoof surface plasmon polaritons. *IEEE Antennas and Wireless Propagation Letters*. https://doi.org/10.1109/LAWP.2026.3689052.

352 Nicolson, A.M. and Ross, G.F. (1970). Measurement of the intrinsic properties of materials by time-domain techniques. *IEEE Transactions on Instrumentation and Measurement* 19 (4): 377–382.

353 Weir, W.B. (1974). Automatic measurement of complex dielectric constant and permeability at microwave frequencies. *Proceedings IEEE* 62 (1): 33–36.

354 Milosevic, V., Jokanovic, B., Bojanic, R. (2013, August). Effective electromagnetic parameters of metamaterial transmission line loaded with asymmetric unit cells. *IEEE Transactions on MTT* 61 (8): 2761–2772.

355 Landau, L. (1984). *Electrodynamics of Continuous Media*, 2e. Oxford, UK: Pergamum.

356 Graham, J.W. (2012, May). Arbitrary oriented biaxial anisotropic media: Wave behaviors and microstrip antennas. PhD dissertation. Syracuse University.

357 Koschny, Th., Markoš, P., Economou, E.N. et al. (2005, June). Impact of inherent periodic structure on effective medium description of left-handed and related metamaterials. *Physical Review B* 71: 245105.

358 Alù, A. (2011, February). Restoring the physical meaning of metamaterial constitutive Parameters. *Physical Review B* 83: 081102.

359 Chen, L., Lei, Z.-Y., Yang, R. et al. (2013). Determining the effective electromagnetic parameters of bianisotropic metamaterials with periodic structures. *PIERM* 29: 79–93.

360 Chen, X., Grzegorczyk, T.M., Wu, B.-I. et al. (2004). Robust method to retrieve the constitutive effective parameters of Metamaterial. *Physical Review E* 70 (1): 16608.

361 Rohde, U.L. and Poddar, A.K. (2016, May). Moebius metamaterial inspired next generation circuits and systems. *Microwave Journal* 59 (5): 62–88.

362 Chen, X., Wu, B.-I., Kong, J.A. et al. (2005). Retrieval of the effective constitutive parameters of bianisotropic Metamaterial. *Physical Review E* 71: 046610.

363 Marques, F. and Sorolla, M. (2008). *Metamaterial with Negative Parameters: Theory, Design and Microwave Applications*. Wiley.

364 Sydoruk, O., Tatartschuk, E., Shamonina, E. et al. (2009). Analytical formulation for the resonant frequency of split rings. *Journal of Applied Physics* 105 (1): 14903.

365 Chen, X., Grzegorczyk, T.M., and Kong, J.A. (2006). Optimization approach to the retrieval of the constitutive parameters of a slab of general bianisotropic medium. *Progress in Electromagnetics Research, PIER* 60: 1–18.

366 Li, Z., Aydin, K., and Ozbay, E. (2009). . *Physical Review E* 79: 026610-1-7.

367 Yang, C., Yu, H., Shang, Y. et al. (2015). Characterization of CMOS metamaterial transmission line by compact fractional-order equivalent circuit model. *IEEE Transactions on Electron Devices* 62 (9): 3012–3018.

368 Wang, J., Qu, S., Zhang, J. et al. (2009). A tunable left-handed metamaterial based on modified broadside-coupling split-ring resonators. *Progress In Electromagnetics Research Letters* 6: 35–45.

369 Nakamura, T., Masuda, T., Shiramizu, N. et al. (2012, September). A 11.1-V regulator-stabilized 21.4-GHz VCO and a 11.5% frequency-range dynamic divider for K-band wireless communication. *IEEE Transactions on Microwave Theory and Techniques* 60 (9): 2823–2832.

370 Chen, Z., Hong, W., Chen, J. et al. (2013, January). Design of high-Q tunable SIW resonator and its application to low phase noise VCO. *IEEE Microwave and Wireless Components Letters* 23 (1): 43–45.

371 Collado, A., Mira, F., and Georgiadis, A. (2013, September). Mechanically tunable substrate integrated waveguide (SIW) cavity based oscillator. *IEEE Microwave and Wireless Components Letters* 23 (9): 489–491.

372 Chang, C.L., Tseng, C.H., and Chang, H.Y. (2014, February). A new monolithic Ka-band filter-based voltage controlled oscillator using 0.15 μm GaAs PHEMT Technology. *IEEE MWCL Letters* 24 (2): 111–113. https://doi.org/10.1109/LMWC.2013.2290224.

373 Chen, Y., Mouthaan, K., and Lin, F. (2010, August). Design of X-band and Ka-band Colpitts oscillators using a parasitic cancellation technique. *IEEE Transactions on Circuits and Systems-I* 57: 1817–1828.

374 Poddar, A., Rohde, U., and Sundararajan, D. (2013, November). Printed resonators: Möbius strips theory and applications. *Microwave Journal* 56: 24–54.

375 Wang, N., Wang, S., Zhang, Z.-Q. et al. (2018). Closed-form expressions for effective constitutive parameters: Electrostrictive and magnetostrictive tensors for bianisotropic metamaterials. *Physical Review B* 98 (4): 045426. https://doi.org/10.1103/PhysRevB.98.045426.

376 Hasar, U.C. (2010). Procedure for accurate and stable constitutive parameters extraction of materials at microwave frequencies. *PIERS* 109: 107–121.

377 Chen, L., Lei, Z., Yang, R. et al. (2013). Determining the effective electromagnetic parameters of bianisotropic metamaterials with periodic structures. *Progress in Electromagnetics Research M* 29: 79–93.

378 Graham, J.W. and Lee, J.K. (2013). Reflection and transmission from biaxially anisotropic–isotropic interfaces. *Progress In Electromagnetics Research (PIER)* 136: 681–702.

379 Li, B., Zhao, Z., Song, L. et al. (2025). Numerical simulation and electromagnetic parameter retrieval: performance evaluation of metamaterials under TE and TM polarization conditions. *Modelling and Simulation in Materials Science and Engineering* 33 (1): 015015.

380 Chen, X., Wu, B.-I., Kong, J.A. et al. (2005). Retrieval of the effective constitutive parameters of bianisotropic metamaterials. *Physical Review E* 71 (4): 046610. https://doi.org/10.1103/PhysRevE.71.046610.

381 Wang, X., Zheng, X., Ni, M. et al. (2010). Theoretical investigation of Möbius strips formed from graphene. *Applied Physical Letters* 97: 123103.

382 Shelby, R.A., Smith, D.R., Nemat-Nasser, S.C. et al. (2001, January). Microwave transmission through a two-dimensional, isotropic, left-handed metamaterial. *Applied Physics Letters* 78: 489–491.

383 Aydin, K., Guven, K., Kafesaki, M. et al. (2004). Experimental observation of true left-handed transmission peaks in metamaterials. *Optics Letters* 29: 2623–2625.

384 Lapine, M. and Tretyakov, S. (2007, March). Contemporary notes on metamaterials. *IET Microwave Antennas Propagation* 1 (1): 3–11.

385 Song, H.-J. and Lee, N. (2022, March). Terahertz communications: Challenges in the next decade. *IEEE Transactions on Terahertz Science and Technology* 12: 105–117.

386 Lia, E., Ghosh, I., Hanham, S.M. et al. (2023, June). Novel mm-wave oscillator based on an electromagnetic bandgap resonator. *IEEE MTT-S International Microwave Symposium (IMS 2023)*. San Diego, CA, USA.

387 Dan, I., Szriftgiser, P., Peytavit, E. et al. (2020, May). A 300-GHz wireless link employing a photonic transmitter and an active electronic receiver with a transmission bandwidth of 54 GHz. *IEEE Transactions on Terahertz Science and Technology* 10 (3): 271–281.

388 Ducournau, G., Engenhardt, K., Szriftgiser, P. et al. (2015). 32 Gbit/s QPSK transmission at 385 GHz using coherent fibre-optic technologies and THz double heterodyne detection. *Electronics Letters* 51 (12): 915–917.

389 Ducournau, G., Szriftgiser, P., Beck, A. et al. (2014, May). Ultrawide-bandwidth single-channel 0.4-THz wire-less link combining broadband quasi-optic photomixer and coherent detection. *IEEE Transactions on Terahertz Science and Technology* 4 (3): 328–337.

390 Hirata, A., Kosugi, T., Takahashi, H. et al. (2012, March). 120-GHz-band wireless link technologies for out- door 10-Gbit/s data transmission. *IEEE Transactions on Microwave Theory and Techniques* 60 (3): 881–895.

391 Kanno, A., Kuri, T., Morohashi, I. et al. (2015). Coherent terahertz wireless signal transmission using advanced optical fiber communication technology. *Journal of Infrared, Millimeter, and Terahertz Waves* 36 (2): 180–197.

392 Koenig, S., Lopez-Diaz, D., Anteset, J. et al. (2013). Wireless sub-THz communication system with high data rate. *Nature Photonics* 7 (12): 977.

393 Li, X., Yu, J., Wang, K. et al. (2019, January). 120 Gb/s wireless terahertz-wave signal delivery by 375 GHz-500 GHz multi-carrier in a 2 × 2 MIMO system. *Journal of Lightwave Technology* 37 (2): 606–611.

394 Li, X., Yu, J., Zhang, J. et al. (2013). A 400 G optical wireless integration delivery system. *Optics Express* 21 (16): 18812–18819.

395 Li, X., Yu, J., Zhao, L. et al. (2019). 132-Gb/s photonics-aided single-carrier wireless terahertz- wave signal transmission at 450 GHz enabled by 64 QAM modulation and probabilistic shaping. *Proceedings Optical Fiber Communications Conference and Exhibition*, 1–3.

396 Nagatsuma, T., Horiguchi, S., Minamikata, Y. et al. (2013). Terahertz wireless communications based on photonics technologies. *Optics Express* 21 (20): 23736–23747.

397 Puerta, R., Yu, J., Li, X. et al. (2017). Demonstration of 352 Gbit/s photonically-enabled D-band wireless delivery in one 2 × 2 MIMO system. *Proceedings Optical Fiber Communications Conference and Exhibition*, 1–3.

398 Shams, H., Fice, M.J., Balakier, K. et al. (2014). Photonic generation for multichannel THz wireless communication. *Optics Express* 22 (19): 23465–23472.

399 Song, H.-J., Ajito, K., Hirata, A. et al. (2009). 8 Gbit/s wireless data transmission at 250 GHz. *Electronics Letters* 45 (22): 1121–1122.

400 Song, H.-J., Ajito, K., Muramoto, Y. et al. (2012). 24 Gbit/s data transmission in 300 GHz band for future terahertz communications. *Electronics Letters* 48 (15): 953–954.

401 Yu, X., Asif, R., Piels, M. et al. (2015). 60 Gbit/s 400 GHz wireless transmission. *Proceedings International Conference on Photonics in Switching*, 4–6.

402 Zhou, W., Zhao, L., Zhang, J. et al. (2020, July). 135-GHz D-band 60-Gbps PAM-8 wireless transmission employing a joint DNN equalizer with BP and CMMA. *Journal of Lightwave Technology* 38 (14): 3592–3601.

403 Boes, F., Messinger, T., Antes, J. et al. (2014). Ultra-broadband MMIC-based wireless link at 240 GHz enabled by 64GS/s DAC. *Proceedings 39th International Conference on Infrared, Millimeter, and Terahertz Waves*, 1–2.

404 Carpenter, S., Nopchinda, D., Abbasi, M. et al. (2016, April). A D-band 48-Gbit/s 64-QAM/QPSK direct- conversion I/Q transceiver chipset. *IEEE Transactions on Microwave Theory and Techniques* 64 (4): 1285–1296.

405 Dan, I., Ducournau, G., Hisatake, S. et al. (2020, January). A terahertz wireless communication link using a super-heterodyne approach. *IEEE Transactions on Terahertz Science and Technology* 10 (1): 32–43.

406 Hamada, H., Fujimura, T., Abdo, I. et al. (2018). 300-GHz. 100-Gb/s InP-HEMT wireless transceiver using a 300-GHz fundamental mixer. *Proceedings IEEE/MTT-S International Microwave Symposium*, 1480–1483.

407 Kallfass, I., Antes, J., Schneider, T. et al. (2011, November). All active MMIC-based wireless communication at 220 GHz. *IEEE Transactions on Terahertz Science and Technology* 1 (2): 477–487.

408 Kallfass, I., Boes, F., Messinger, T. et al. (2015). 64 Gbit/s transmission over 850 m fixed wireless link at 240 GHz carrier frequency. *Journal of Infrared, Millimeter, and Terahertz Waves* 36 (2): 221–233.

409 Song, H.-J., Kim, J.-Y., Ajito, K., et al. (2014, March). 50-Gb/s direct conversion QPSK modulator and demodulator MMICs for tera-hertz communications at 300 GHz. *IEEE Transactions on Microwave Theory and Techniques* 62 (3): 600–609.

410 Song, H.-J., Kosugi, T., Hamada, H. et al. (2016). Demonstration of 20-Gbps wireless data transmission at 300 GHz for KIOSK instant data downloading applications with InP MMICs. *Proceedings IEEE MTT-S International Microwave Symposium*, 1–4.

411 Takahashi, H., Kosugi, T., Hirata, A. et al. (2013, December). 120-GHz-band fully integrated wireless link using QSPK for real time 10-Gbit/s transmission. *IEEE Transactions on Microwave Theory and Techniques* 61 (12): 4745–4753.

412 Foulon, S., Pruvost, S., Pache, D. et al. (2014). A 140 GHz multi-gigabits self-heterodyne transceiver for chip-to-chip communications. *Proceedings 44th European Microwave Conference*, 901–904.

413 Rodríguez-Vázquez, P., Grzyb, J., Heinemann, B. et al. (2019, April). A 16-QAM 100-Gb/s 1-M wireless link with an EVM of 17% at 230 GHz in an SiGe technology. *IEEE Microwave and Wireless Components Letters* 29 (4): 297–299.

414 Rodríguez-Vázquez, P., Grzyb, J., Sarmah, N. et al. (2018). Towards 100 Gbps: A fully electronic 90 Gbps one meter wireless link at 230 GHz. *Proceedings 48th European Microwave Conference*, 1389–1392.

415 Sarmah, N., Vazquez, P.R., Grzyb, J. et al. (2016). A wideband fully integrated SiGe chipset for high data rate communication at 240 GHz. *Proceedings 11th European Microwave Integrated Circuits Conference*, 181–184.

416 Wang, H., Mohammadnezhad, H., and Heydari, P. (2019, November). Analysis and design of high-order QAM direct-modulation transmitter for high-speed point- to-point mm-wave wireless links. *IEEE Journal Solid-State Circuits* 54 (11): 3161–3179.

417 Hara, S., Katayama, K., Takano, K. et al. (2017). A 32 Gbit/s 16 QAM CMOS receiver in 300 GHz band. *Proceedings IEEE MTT-S International Microwave Symposium*, 1703–1706.

418 Kang, S., Thyagarajan, S.V., and Niknejad, A.M. (2015, October). A 240 GHz fully integrated wideband QPSK transmitter in 65 nm CMOS. *IEEE Journal Solid- State Circuits* 50 (10): 2256–2267.

419 Katayama, K., Motoyoshi, M., Takano, K. et al. (2013). 209 mW 11 Gbps 130 GHz CMOS transceiver for indoor wireless communication. *Proceedings IEEE Asian Solid-State Circuits Conference*, 409–412.

420 Katayama, K., Takano, K., Amakawa, S. et al. (2016, December). A 300 GHz CMOS transmitter with 32-QAM 17.5 Gb/s/ch capability over six channels. *IEEE Journal Solid-State Circuits* 51 (12): 3037–3048.

421 Kim, Y., Huet, B., Du, Y. et al. (2019, June). A millimeter-wave CMOS transceiver with digitally Pre- distorted PAM-4 modulation for contactless communications. *IEEE Journal Solid-State Circuits* 54 (6): 1600–1612.

422 Lee, S., Hara, S., Yoshida, T. et al. (2019, December). An 80-Gb/s 300-GHz-band single-chip CMOS transceiver. *IEEE Journal Solid-State Circuits* 54 (12): 3577–3588.

423 Moghadami, S., Hajilou, F., Agrawal, P. et al. (2015, September). A 210 GHz fully-integrated OOK transceiver for short-range wireless chip-to-chip communication in 40 nm CMOS technology. *IEEE Transactions on Terahertz Science and Technology* 5 (5): 737–741.

424 Momson, I., Dong, S., Yelleswarapu, P. et al. (2020). 315-GHz self-synchronizing minimum shift keying receiver in 65-nm CMOS. *Proceedings IEEE Symposium VLSI Circuits*, 1–2.

425 Nakajima, K., Maruyama, A., Kohtani, M. et al. (2014). 23 Gbps 9.4 pJ/bit 80/100 GHz band CMOS transceiver with on-board antenna for short-range communication. *Proceedings IEEE Asian Solid-State Circuits Conference*, 173–176.

426 Nazari, P., Jafarlou, S., and Heydari, P. (2020, February). A CMOS two-element 170-GHz fundamental-frequency transmitter with direct RF-8PSK modulation. *IEEE Journal Solid-State Circuits* 55 (2): 282–297.

427 Ono, N., Motoyoshi, M., Takano, K. et al. (2012). 135 GHz 98 mW 10 Gbps ASK transmitter and receiver chipset in 40 nm CMOS. *Proceedings Symposium VLSI Circuits*, 50–51.

428 Park, J.-D., Kang, S., Thyagarajan, S.V. et al. (2012). A 260 GHz fully integrated CMOS transceiver for wireless chip-to-chip communication. *Proceedings Symposium VLSI Circuits*, 48–49.

429 Standaert, A. and Reynaert, P. (2018). A 410 GHz OOK transmitter in 28 nm CMOS for short distance chip-to-chip communications. *Proceedings IEEE Radio Frequency Integrated Circuits Symposium*, 240–243.

430 Takano, K., Amakawa, S., Katayama, K. et al. (2017). 17.9 A 105 Gb/s 300 GHz CMOS transmitter. *Proceedings IEEE International Solid-State Circuits Conference*, 308–309.

431 Takano, K., Katayama, K., Amakawa, S. et al. (2017). 56-Gbit/s 16-QAM wireless link with 300-GHz-band CMOS transmitter. *Proceedings IEEE MTT-S International Microwave Symposium*, 793–796.

432 Thyagarajan, S.V., Kang, S., and Niknejad, A.M. (2015, October). A 240 GHz fully integrated wideband QPSK receiver in 65 nm CMOS. *IEEE Journal Solid- State Circuits* 50 (10): 2268–2280.

433 Wang, Z., Chiang, P.-Y., Nazari, P. et al. (2014, March). A CMOS 210-GHz fundamental transceiver with OOK modulation. *IEEE Journal Solid-State Circuits* 49 (3): 564–580.

434 Yu, B., Ye, Y., Ding, X. et al. (2018). Sub-THz interconnect for planar chip-to-chip communications. *Proceedings IEEE 18th Topical Meeting Silicon Monolithic Integrated Circuits RF Systems*, 54–56.

435 Zhong, Q., Chen, Z., Sharma, N. et al. (2018). 300-GHz CMOS QPSK transmitter for 30-Gbps dielectric waveguide communication. *Proceedings IEEE Custom Integrated Circuits Conference*, 1–4.

436 Song, H.-J., Kim, J.-Y., Ajito, K. et al. (2014, March). 50-Gb/s direct conversion QPSK modulator and demodulator MMICs for terahertz communications at 300 GHz. *IEEE Transactions on Microwave Theory and Techniques* 62 (3): 600–609.

437 Deal, W., Mei, X.B., Radisic, V. et al. (2010, May). Demonstration of a 0.48 THz amplifier module using InP HEMT transistors. *IEEE Microwave and Wireless Components Letters* 20 (5): 289–291.

438 Urteaga, M., Seo, M., Hacker, J. et al. (2010). InP HBT integrated circuit technology for terahertz frequencies. *Proceedings IEEE Compound Semiconductor Integrated Circuit Symposium*, 1–4.

439 Seo, M., Urteaga, M., Rodwell, M. et al. (2011). A 300 GHz PLL in an InP HBT technology. *Proceedings IEEE MTT-S International Microwave Symposium Digest*, 1–4.

440 Radisic, V., Leong, K.M.K.H., Mei, X. et al. (2012, March). Power amplification at 0.65 THz using InP HEMTs. *IEEE Transactions on Microwave Theory and Techniques* 60 (3): 724–729.

441 Kim, J.-Y., Song, H.-J., Ajito, K. et al. (2013, August). A 325 GHz quadrature voltage controlled oscillator with superharmonic-coupling. *IEEE Microwave and Wireless Components Letters* 23 (8): 430–432.

442 Mei, X., Yoshida, W., Lange, M. et al. (2015, April). First demonstration of amplification at 1 THz using 25-nm InP high electron mobility transistor process. *IEEE Electron Device Letters* 36 (4): 327–329.

443 Samoska, L.A. (2011, September). An overview of solid-state integrated circuit amplifiers in the submillimeter-wave and THz regime. *IEEE Transactions on Terahertz Science and Technology* 1 (1): 9–24.

444 Urteaga, M., Hacker, J., Griffith, Z. et al. (2016). A 130 nm InP HBT integrated circuit technology for THz electronics. *Proceedings IEEE International Electron Devices Meeting*, 29.2.1–29.2.4.

445 Leuther, A., Tessmann, A., Dammann, M. et al. (2013). 35 nm mHEMT technology for THz and ultra-low noise applications. *Proceedings International Conference Indium Phosphide Related Materials*, 1–2.

446 Tessmann, A., Leuther, A., Wagner, S. et al. (2017). A 300 GHz low-noise amplifier S-MMIC for use in next-generation imaging and communication applications. *Proceedings IEEE MTT-S International Microwave Symposium*, 760–763.

447 Yu, Y. and Lin, Y.-S. (2019). Multi-functional terahertz metamaterial using symmetrical and asymmetrical electric split-ring resonators. *Results in Physics* 13: 102321. ISSN 2211-3797. https://doi.org/10.1016/j.rinp.2019.102321.

448 Lee, S.H., Choi, M., Kim, T.T. et al. (2012). Switching terahertz waves with gate-controlled active graphene metamaterials. *Nature Materials* 11: 936–941.

449 Jiang, R., Wu, Z.R., Han, Z.Y. et al. (2016). HfO 2-based ferroelectric modulator of terahertz waves with graphene metamaterial. *Chinese Physics B* 25: 106803.

450 Qi, L., Li, C., and Fang, G. (2014). Tunable terahertz metamaterial absorbers using active diodes. *International Journal of Electromagnetics and Applications* 4: 57–60.

451 Zhu, B., Feng, Y.J., Zhao, J.M. et al. (2010). Switchable metamaterial reflector/ absorber for different polarized electromagnetic waves. *Applied Physics Letters* 97: 051906.

452 Zhao, J., Cheng, Q., Chen, J. et al. (2013). A tunable metamaterial absorber using varactor diodes. *New Journal of Physics* 15: 043049.

453 Han, J. and Lakhtakia, A. (2009). Semiconductor split-ring resonators for thermally tunable terahertz metamaterials. *Journal of Modern Optics* 56: 554–557.

454 Wen, Q.Y., Zhang, H.W., Yang, Q.H. et al. (2010). Terahertz metamaterials with VO_2 cut-wires for thermal tunability. *Applied Physics Letters* 97: 021111.

455 Chen, H.T., O'Hara, J.F., Azad, A.K. et al. (2008). Experimental demonstration of frequency-agile terahertz metamaterials. *Nature Photonics* 2: 295–298.

456 Coldren, L.A. (2000). Monolithic tunable diode lasers. *IEEE Journal of Selected Topics in Quantum Electronics* 6: 988–999.

457 Chen, C.Y., Pan, C.L., Hsieh, C.F. et al. (2006). Liquid-crystal-based terahertz tunable Lyot filter. *Applied Physics Letters* 88: 101107.

458 Cong, L.Q., Pitchappa, P., Wu, Y. et al. (2017). Active multifunctional microelectromechanical system metadevices: Applications in polarization control, wave front deflection, and holograms. *Advanced Optical Materials* 5: 1600716.

459 Tao, H., Strikwerda, A.C., Fan, K. et al. (2009). Reconfigurable terahertz metamaterials. *Physical Review Letters* 103: 147401.

460 Unlu, M., Hashemi, M.R., Berry, C.W. et al. (2014). Switchable scattering meta-surfaces for broadband terahertz modulation. *Scientific Reports* 4: 5708.

461 Zhu, W.M., Liu, A.Q., Zhang, X.M. et al. (2011). Switchable magnetic metamaterials using micromachining processes. *Advanced Materials* 23: 1792–1796.

462 Liu, H., Liu, Y.M., Li, T. et al. (2009). Coupled magnetic plasmon in metamaterials. *Physica Status Solidi B* 246: 1397–1406.

463 Hesmer, F., Tatartschuk, E., Zhuromskyy, O. et al. (2007). Coupling mechanisms for split ring resonators: Theory and experiment. *Physica Status Solidi B* 244: 1170–1175.

464 Khodasevych, I.E., Shah, C.M., Sriram, S. et al. (2012). Elastomeric silicone substrates for terahertz fishnet metamaterials. *Applied Physics Letters* 100: 061101.

465 Chen, C.C., Hsiao, C.T., Sun, S. et al. (2012). Fabrication of three dimensional split ring resonators by stress-driven assembly method. *Optics Express* 20: 9415–9420.

466 Soukoulis, C.M., Kafesaki, M., and Economou, E.N. (2006). Negative-index materials: New frontiers in optics. *Advanced Materials* 18: 1941–1952.

467 Linden, S., Enkrich, C., Wegener, M. et al. (2004). Magnetic response of metamaterials at 100 terahertz. *Science* 306: 1351–1353.

468 Lin, Y.S., Yan, K.H., Yao, D.Y. et al. (2019). Investigation of electromagnetic response of terahertz metamaterial by using split-disk resonator. *Optics & Laser Technology* 111: 509–514.

469 Seo, M., Urteaga, M., Young, A. et al. (2012). A single-chip 630 GHz transmitter with 210 GHz subharmonic PLL local oscillator in 130 nm InP HBT. *Proceedings IEEE/MTT-S International Microwave Symposium Digest*, 1–3.

470 Lai, R., Deal, W.R., Mei, X.B. et al. (2008). Fabrication of InP HEMT devices with extremely high Fmax. *Proceedings 20th International Conference on Indium Phosphide and Related Materials*, 1–3.

471 Leuther, A., Tessmann, A., Doria, P. et al. (2014). 20 nm Metamorphic HEMT technology for terahertz monolithic integrated circuits. *Proceedings of the 14th European Microwave Integrated Circuits Conference*, 84–87.

472 Urteaga, M., Pierson, R., Rowell, P. et al. (2011). 130nm InP DHBTs with ft >0.52 THz and fmax >1.1 THz. *Proceedings of 69th Device Research Conference*, 281–282.

473 Marques, R., Medina, F., and Rafii-El-Idrissi, R. et al. (2002). Role of bianisotropic in negative permeability and left-handed metamaterials. *Physical Review B* 65: 144440-1-6.

474 Asgari, S. and Fabritius, T. (2025). Multi-purpose graphene-based terahertz metamaterial and its equivalent circuit model. *IEEE Access* 13: 56808-56819. https://doi.org/10.1109/ACCESS.2025.3554530.

475 Chouhan, B.S., Barman, B., Vaswani, L.K. et al. (2024, May). Broadening of resonance with the bending of strongly coupled strip resonators in a planar terahertz metamaterial. *Journal of Lightwave Technology* 42 (9): 3283–3289. https://doi.org/10.1109/JLT.2024.3353330.

476 Yu, X., Suzuki, Y., Van Ta, M. et al. (2021, July). Highly efficient resonant tunneling diode terahertz oscillator with a split ring resonator. *IEEE Electron Device Letters* 42 (7): 982–985. https://doi.org/10.1109/LED.2021.3082577.

450 Qi, L., Li, C., and Fang, G. (2014). Tunable terahertz metamaterial absorbers using active diodes. International Journal of Electromagnetics and Applications 4: 57–60.
451 Zhu, B., Feng, Y.J., Zhao, J.M. et al. (2010). Switchable metamaterial reflector/absorber for different polarized electromagnetic waves. Applied Physics Letters 97: 051906.
452 Zhao, J., Cheng, Q., Chen, J. et al. (2013). A tunable metamaterial absorber using varactor diodes. New Journal of Physics 15: 043049.
453 Han, J. and Lakhtakia, A. (2009). Semiconductor split-ring resonators for thermally tunable terahertz metamaterials. Journal of Modern Optics 56: 554–557.
454 Wen, Q.Y., Zhang, H.W., Yang, Q.H. et al. (2010). Terahertz metamaterials with VO2 cut-wires for thermal tunability. Applied Physics Letters 97: 021111.
455 Chen, H.T., O'Hara, J.F., Azad, A.K. et al. (2008). Experimental demonstration of frequency-agile terahertz metamaterials. Nature Photonics 2: 295–298.
456 Coldren, L.A. (2000). Monolithic tunable diode lasers. IEEE Journal of Selected Topics in Quantum Electronics 6: 988–999.
457 Chen, C.Y., Pan, C.L., Hsieh, C.F. et al. (2006). Liquid-crystal-based terahertz tunable Lyot filter. Applied Physics Letters 88: 101107.
458 Cong, L.Q., Pitchappa, P., Wu, Y. et al. (2017). Active multifunctional microelectromechanical system metadevices: Applications in polarization control, wavefront deflection, and holograms. Advanced Optical Materials 5: 1600716.
459 Tao, H., Strikwerda, A.C., Fan, K. et al. (2009). Reconfigurable terahertz metamaterials. Physical Review Letters 103: 147401.
460 Unlu, M., Hashemi, M.R., Berry, C.W. et al. (2014). Switchable scattering meta-surfaces for broadband terahertz modulation. Scientific Reports 4: 5708.
461 Zhu, W.M., Liu, A.Q., Zhang, X.M. et al. (2011). Switchable magnetic metamaterials using micromachining processes. Advanced Materials 23: 1792–1796.
462 Liu, H., Liu, Y.M., Li, T. et al. (2009). Coupled magnetic plasmons in metamaterials. Physica Status Solidi B 246: 1397–1406.
463 Hesmer, F., Tatartschuk, E., Zhuromskyy, O. et al. (2007). Coupling mechanisms for split ring resonators: Theory and experiment. Physica Status Solidi B 244: 1170–1175.
464 Khodasevych, I.E., Shah, C.M., Sriram, S. et al. (2012). Elastomeric silicone substrates for terahertz fishnet metamaterials. Applied Physics Letters 100: 061101.
465 Chen, C.C., Hsiao, C.T., Sun, S. et al. (2012). Fabrication of three dimensional split ring resonators by stress-driven assembly method. Optics Express 20: 9415–9420.
466 Soukoulis, C.M., Kafesaki, M., and Economou, E.N. (2006). Negative-index materials: New frontiers in optics. Advanced Materials 18: 1941–1952.
467 Linden, S., Enkrich, C., Wegener, M. et al. (2004). Magnetic response of metamaterials at 100 terahertz. Science 306: 1351–1353.
468 Lin, Y.S., Yan, K., Liao, D.Y. et al. (2019). Investigation of electromagnetic response of terahertz metamaterial by using split-disk resonator. Optics and Laser Technology 111: 509–514.
469 Seo, M., Urteaga, M., Young, A. et al. (2012). A single-chip 630 GHz transmitter with 210 GHz sub-harmonic PLL local oscillator in 130 nm InP HBT. Proceedings IEEE/MTT-S International Microwave Symposium Digest, 1–3.
470 Lai, R., Deal, W.R., Mei, X.B. et al. (2008). Fabrication of InP HEMT devices with extremely high Fmax. Proceedings 2008 International Conference on Indium Phosphide and Related Materials, 1–3.
471 Leuther, A., Tessmann, A., Doria, P. et al. (2014). 20 nm Metamorphic HEMT technology for terahertz monolithic integrated circuits. Proceedings of the 44th European Microwave Integrated Circuits Conference, 84–87.
472 Urteaga, M., Pierson, R., Rowell, P. et al. (2011). 130nm InP DHBTs with ft>0.52THz and fmax>1.1THz. Proceedings of 69th Device Research Conference, 281–282.
473 Marqués, R., Medina, F., and Rafii-El-Idrissi, R. et al. (2002). Role of bianisotropy in negative permeability and left-handed metamaterials. Physical Review B 65: 144440-1-6.
474 Asgari, S. and Fabritius, T. (2023). Multi-purpose graphene-based terahertz metamaterial and its equivalent circuit model. IEEE Access 11: 56808–56819. https://doi.org/10.1109/ACCESS.2023.3284336.
475 Chouhan, B.S., Barman, B., Vaswani, L. et al. (2024, May). Broadening of resonance with the bandgap of strongly coupled strip resonators in a planar terahertz metamaterial. Journal of Lightwave Technology 42(9): 3283–3289. https://doi.org/10.1109/JLT.2024.3353520.
476 Yu, X., Suzuki, Y., Van Ta, M. et al. (2021, July). Highly efficient resonant tunneling diode terahertz oscillator with a split ring resonator. IEEE Electron Device Letters 42(7): 982–985. https://doi.org/10.1109/LED.2021.3082577.

6

Q-Enhanced Planar Resonators for Low Phase-noise Oscillators

In this chapter [1], a comprehensive overview of active microwave resonators and their role in high-frequency system design is provided. The discussion begins with the fundamental performance parameters that characterize resonators, including the unloaded and loaded quality factors, figure of merit, frequency selectivity, and energy storage efficiency [1]. These characteristics form the foundation for evaluating resonator suitability in practical applications. Attention is then directed to resonator design criteria, where factors such as material properties, fabrication tolerances, coupling configurations, and geometrical constraints are critically examined, as they directly influence device performance in terms of frequency stability, insertion loss, and tunability. Building upon these foundations, the chapter also explores oscillator design methodologies that incorporate active microwave resonators as frequency-determining elements. This includes an examination of theoretical models, stability conditions, and phase-noise considerations, as well as design trade-offs associated with power consumption, integration, and tuning mechanisms. Collectively, the material presented establishes the link between resonator properties, oscillator design strategies, and the overarching goals of modern microwave engineering, namely achieving compact, efficient, and high-performance frequency sources for applications in communication systems, radar, and advanced sensing technologies [1].

6.1 Active Resonator

A resonator constitutes one of the most critical elements in oscillators and tunable voltage-controlled oscillators (VCOs) circuits, as its performance strongly dictates the overall frequency stability, spectral purity, and efficiency of the system. The effectiveness of a resonator is governed by multiple factors, including its physical geometry, fabrication cost, achievable quality factor (Q-factor), ease of manufacturability, and suitability for seamless integration into larger circuit architectures [1–8]. Since most modern integrated circuits employ planar topologies, substantial research has been devoted to replacing conventional discrete high-Q resonators—such as ceramic, surface acoustic wave (SAW), cavity, dielectric, substrate-integrated waveguide (SIW), bulk acoustic wave (BAW), optoelectronic (OE), and yttrium iron garnet (YIG) structures—with compact, integrable alternatives [8, 9]. Within this context, planar resonators based on microstrip and stripline technologies have gained considerable attention. These devices are realized by patterning conductive traces on dielectric substrates, enabling cost-effective fabrication and compatibility with standard integrated circuit processes. Despite these advantages, they present notable trade-offs: planar implementations typically occupy larger areas and exhibit reduced Q-factors compared with their bulk or cavity-based counterparts. Moreover, a well-established trend in resonator physics indicates that the Q-factor decreases as the operating frequency increases, a limitation primarily associated with the reduced skin depth and the corresponding rise in conductor losses at higher frequencies [7].

$$Q = [f]^{-1/2} \tag{6.1}$$

In Eq. (6.1), Q is the quality factor.

Equation (6.1) [7] defines the quality factor (Q) as a constant that represents the energy storage ability of a resonator compared to its energy loss. In passive resonators (PRs), energy is mainly lost due to internal friction and radiation, which both lower the resonator tank's Q-factor. To maintain a stable oscillation signal, the resonator is connected to an external circuit with active devices and supporting components that offset these energy losses [10]. Although loose coupling reduces the loading effect on the resonator, it also increases the attenuation.

In reality, the unloaded quality factor of a resonator, labeled "Q_u," is limited and will degrade when connected to the external oscillator circuit [10]. As a result, even a perfect resonator with minimal internal losses (i.e., $Q_u \rightarrow \infty$) will only have a finite loaded quality factor, " Q_l," once external loading is added. The unloaded quality factor of a PR is mathematically expressed as follows [7, 10]:

$$Q_{u(\mathrm{PR})} = \frac{2\pi f_0 E}{W} = \frac{2\pi f_0 C}{G} \tag{6.2}$$

In Eq. (6.2), f_0 denotes the resonant frequency, and E represents the energy stored in the PR, and W corresponds to the energy dissipated within one oscillation period [10]. The parameters C and G describe the capacitance and conductance, respectively, in the parallel-equivalent circuit representation of the PR tank circuit [7, 8].

Although PRs play a central role in oscillator design, their performance is inherently constrained by energy dissipation, which limits the achievable quality factor (Q). To overcome these limitations, various active resonator (AR) topologies have been reported in the literature, offering a promising pathway toward the realization of highly spectrally pure signal sources [8]. Unlike conventional passive implementations, the AR approach enhances both the unloaded (Q_u) and loaded (Q_l) quality factors by introducing a gain element that compensates for the intrinsic energy losses of the resonator. In practice, this is accomplished by coupling an active circuit (gain block) with the PR network, thereby achieving effective loss cancellation. Such configurations provide a viable alternative to bulky and costly high-Q PRs, while maintaining compactness and integrability in modern oscillator systems.

Figure 6.1 illustrates a representative example of an AR circuit, in which a PR is coupled to a transmission line [10]. As shown, the active circuit introduces a negative conductance, denoted as $|-G_n|$, which counteracts the intrinsic positive conductance (G) of the resonator. As a result, the effective conductance of the AR can be expressed as [8]:

$$G_a = -|G_n| + G \tag{6.3}$$

From (5.148), the unloaded quality factor of the AR can be expressed as [8]:

$$Q_{u(\mathrm{PR})} = \frac{2\pi f_0 C}{G_a} Q_{u(\mathrm{AR})} > Q_{u(\mathrm{CPR})} |G_a| < |G| \tag{6.4}$$

where f_0 is the resonant frequency, C is the equivalent capacitance, G is the conductance of the PR, and G_a is the effective conductance including the contribution of the active circuit [11].

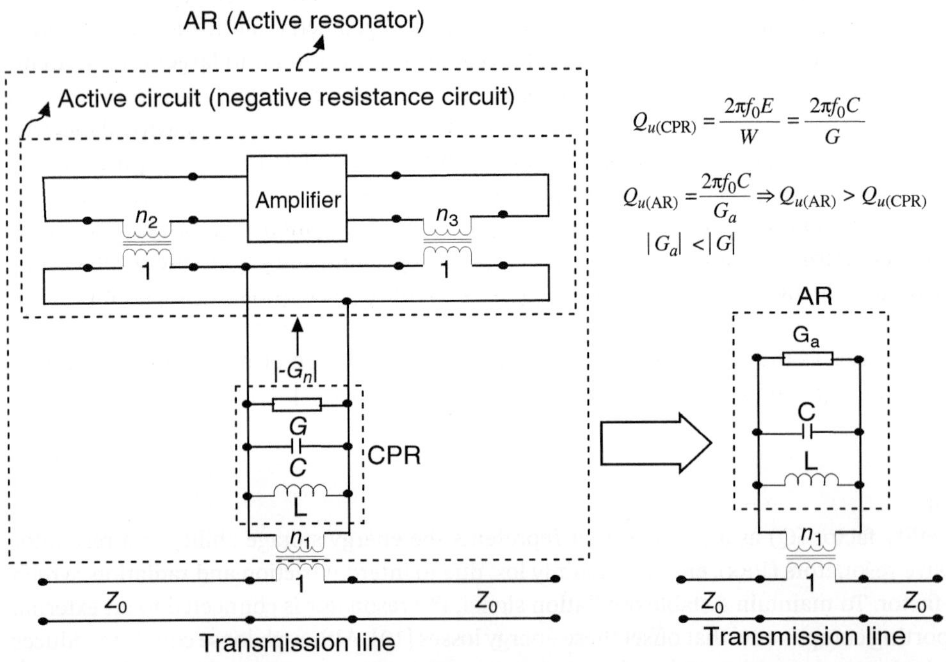

Figure 6.1 A typical AR circuit with a feedback configuration. *Source*: Adapted from [8, 10].

Based on Eqs. (6.3) and (6.4), an AR achieves a higher quality factor compared to a PR, which is limited by resistive and radiative losses. In practice, the gain Ga must remain positive to avoid spurious oscillations while maintaining the enhanced loaded quality factor (Q).

From Eqs. (6.3) and (6.4), it is evident that an AR can achieve a considerably higher quality factor than a PR. Passive structures inherently suffer from performance degradation due to dissipative mechanisms, including ohmic losses in the conductive paths and radiation leakage from the resonant cavity, both of which limit the attainable Q-factor. By contrast, an AR introduces a controlled gain mechanism that compensates for these intrinsic losses, effectively restoring or even enhancing the stored energy within the system. For stable oscillator operation, it is crucial that the small-signal gain parameter, GaG_aGa, remains greater than zero. This condition ensures that the gain compensates only for unavoidable losses without exceeding the stability threshold that would otherwise lead to unwanted spurious oscillations or mode competition. At the same time, maintaining a positive "G_a" allows the loaded quality factor, "Q_l," to increase, which directly enhances frequency stability, narrows the spectral linewidth, and reduces phase noise. Consequently, the careful adjustment of G_a not only prevents instability but also enables ARs to achieve superior performance compared to their passive counterparts, making them highly attractive for advanced oscillators and filter applications in modern microwave systems.

From (6.3) and (6.4), it follows that the AR can achieve a significantly higher quality factor compared to the PR [1], particularly in scenarios where the PR suffers from losses caused by resistive dissipation and radiation [8]. In practical implementations, however, G_a must remain greater than zero to suppress spurious oscillations, while still enabling an enhanced loaded quality factor, Q-factor "Q_l."

From Eqs. (6.3) and (6.4), it can be deduced that an AR is capable of achieving a substantially higher quality factor than its passive counterpart (PR), particularly in cases where the PR performance is limited by resistive dissipation and radiation losses [8]. In practical implementations, however, the small-signal gain parameter, G_a, must be maintained at a value greater than zero. This condition ensures that spurious oscillations are effectively suppressed, while at the same time enabling the enhancement of the loaded quality factor, "Q_l," which directly contributes to improved frequency stability and reduced phase noise.

6.2 AR Topology

In the AR topology [10], the PR is interfaced with an auxiliary network designed to provide negative resistance, typically implemented using active devices such as transistors or operational amplifiers [8]. This configuration allows the AR to emulate the behavior of a conventional oscillator [9], but with a key distinction: whereas a standard oscillator requires both amplitude and phase conditions to be satisfied for sustained oscillations at the resonant frequency, f_0, an AR requires only the phase condition. The negative-resistance network compensates for the intrinsic losses of the resonator—such as resistive and radiation losses—maintaining a stable oscillation amplitude without the need for an explicit amplitude constraint [10, 11]. Schematic representations of the AR topology [10] typically illustrate the PR coupled to the active network, forming a feedback loop that reinforces oscillations [11]. The negative-resistance element effectively cancels the energy dissipation in the PR, while the feedback path ensures that the phase condition is met precisely at f0f_0f0. This design approach not only enhances the loaded quality factor, Q_l, of the resonator but also provides greater frequency stability, reduced phase noise, and improved overall performance compared to purely passive structures. By decoupling the amplitude control from the oscillation condition, the AR topology simplifies oscillator design and enables integration into compact, high-performance microwave circuits.

As illustrated in Figure 6.1 [8, 10], the active amplifier operates in the small-signal linear regime, providing just enough gain to partially or fully offset losses without inducing instability. Careful design is necessary to prevent uncontrolled oscillation buildup in the AR circuit [10]. Typically, the amplifier's gain compensates for the intrinsic losses of the AR, complete loss compensation ($-|G_n| + G = 0$) [10] would theoretically result in an infinite unloaded Q-factor and zero bandwidth. When this AR is subsequently loaded with the oscillator circuitry to generate the desired output, the resonator bandwidth increases, enabling its use in tunable signal sources with enhanced phase-noise performance [12].

Although ARs using the negative-resistance approach [10] provide substantially higher Q-factors, they come with specific challenges. Their circuit design is inherently more intricate, necessitating a feedback component and matching networks to realize the negative conductance $|-G_n|$. Such designs are particularly prone to spurious oscillations if the start-up condition is accidentally satisfied. Unlike conventional oscillators, which require both amplitude and phase conditions for sustained oscillation, an AR needs only the phase condition to oscillate [9]. Consequently, oscillations do not spontaneously buildup across the AR, allowing the module to operate in the small-signal regime. The added negative resistance reduces intrinsic losses of the PR, yielding a high-Q element. However, this advantage comes with the caveat that ARs remain susceptible to unwanted oscillation modes under certain conditions.

Printed PRs—such as hairpin, spiral, ring, and interdigital types—are commonly used in tunable oscillator circuits [10]. Their quality factor, however, is inherently limited by dielectric, conductor, and radiation losses, which restrict the oscillator's phase-noise performance [9, 10]. AR topologies provide a means to overcome these limitations by significantly enhancing the Q-factor. Yet, if not carefully optimized for the specific oscillator design and resonator figure of merit (FOM), ARs can introduce additional noise [10].

AR designs typically employ either active feedback loops [10] or integrate negative-resistance elements with PRs to compensate for inherent losses [13]. A key challenge in this approach is the additional noise [10] introduced by active components, such as transistors, requiring precise design strategies to ensure low phase-noise performance in oscillators utilizing AR networks [10]. Figure 6.2 presents a 3.2 GHz voltage-controlled oscillator (VCO) [10] that incorporates an AR network, serving as a benchmark for phase-noise performance comparison [10].

PRs—such as hairpin, spiral, ring, and interdigital configurations—are widely used in tunable oscillator circuits because of their compact size and straightforward fabrication. However, their quality factor (Q) is inherently constrained by dielectric, conductor, and radiation losses, which in turn directly impact the oscillator's phase-noise performance [9].

AR architectures provide an effective means to overcome the limitations of PRs by enhancing the resonator's quality factor (Q). AR designs commonly utilize active feedback loops or negative-resistance elements combined with PRs to compensate for inherent losses [10]. Nonetheless, these designs pose challenges: active devices, such as transistors, can introduce additional noise that may degrade phase-noise performance if the circuit is not carefully optimized [14]. Achieving low phase noise in AR-based oscillators, therefore, requires meticulous consideration of both the resonator FOM and the oscillator topology [10]. Figure 6.2 illustrates a 3.2 GHz VCO [10] employing an AR network, providing a benchmark for assessing phase-noise improvements enabled by AR integration [10].

As shown in Figure 6.3 [10], simulations indicate that the Q-factor of the AR is approximately three times higher than that of an equivalent PR. This improvement comes with a power consumption penalty of 60 mW. The typical RF output power

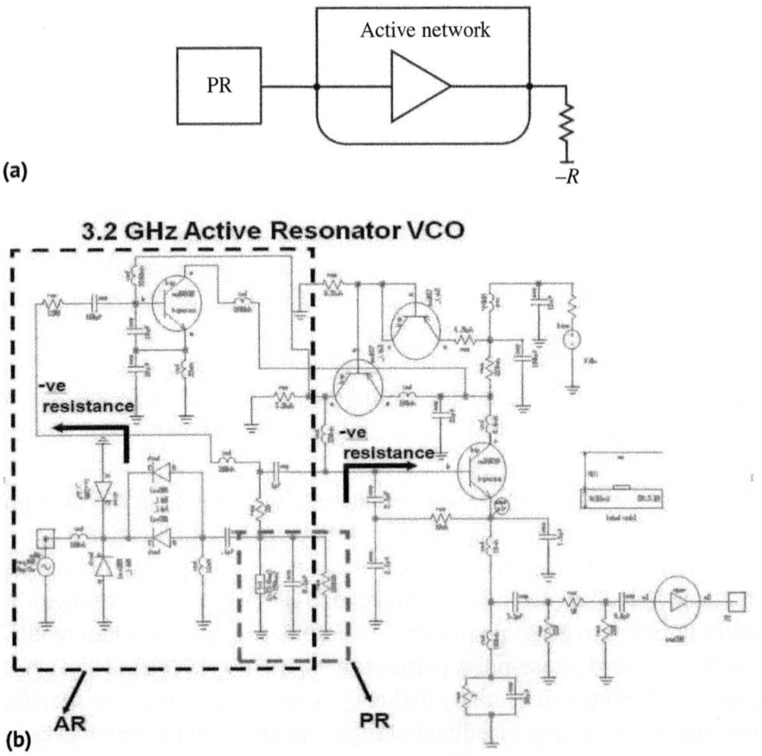

Figure 6.2 (a) Simplified schematic of an oscillator employing an AR network, illustrating the coupling between the resonator and the negative-resistance network for sustaining oscillations, and (b) typical 3.2 GHz VCO based on an AR network, with a power consumption of 150 mW [10].

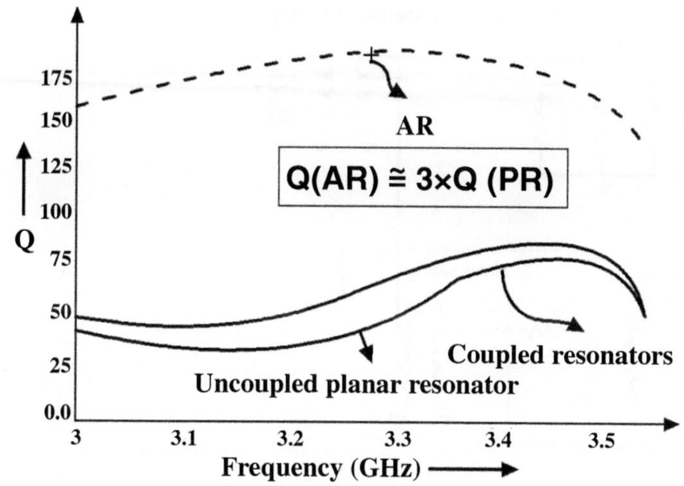

Figure 6.3 Simulated quality factor (Q) comparison for PRs-both uncoupled and coupled planar types, and an AR employing a negative-resistance configuration (see Figure 6.2) [10].

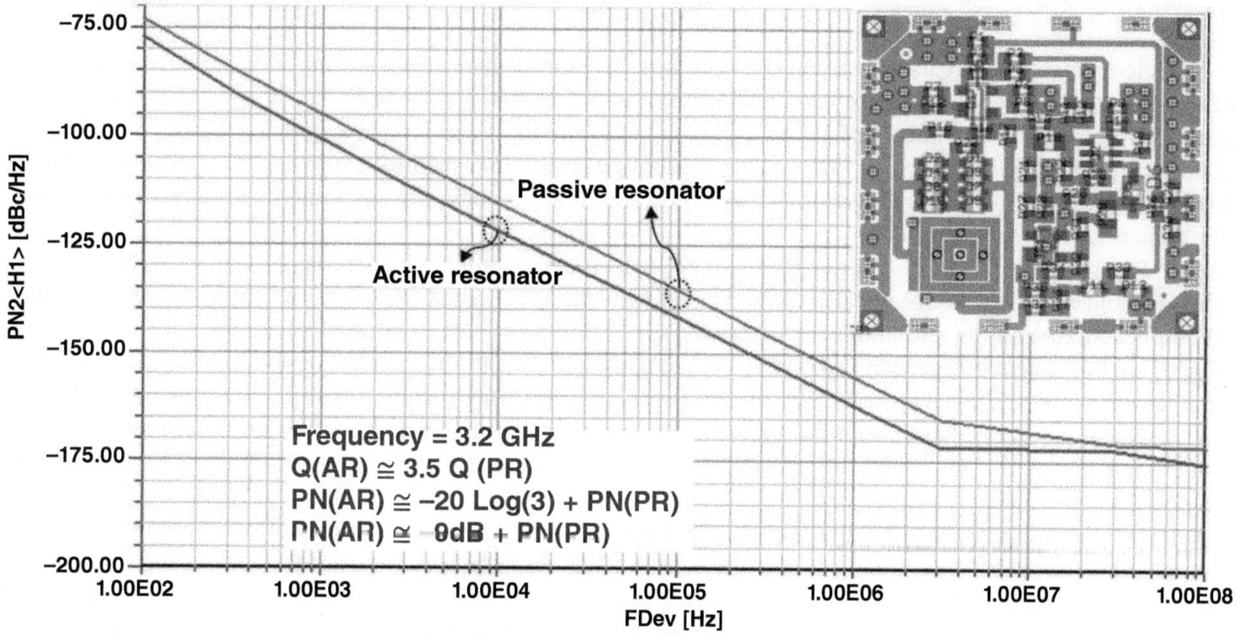

Figure 6.4 CAD-simulated phase-noise characteristics of a 3.2 GHz VCO [10] incorporating an AR network with a 200 MHz tuning range (FOM = −210.3, output power = 5 dBm, DC-to-RF efficiency = 2.1%) [15].

of the oscillator is +5 dBm, with a DC-to-RF conversion efficiency of 2.1%, yielding a FOM of −210.3, and an operating DC power consumption of 150 mW (5 V, 30 mA).

Figure 6.4 [10] presents the CAD-simulated phase-noise (PN) profile, showing a reduction of approximately 9 dB in PN, which corresponds to a threefold enhancement in the resonator Q-factor relative to a PR network. Various AR implementations are discussed to provide an overview of their relative advantages and limitations compared to traditional PR-based designs. However, it is important to note that VCOs employing AR technology typically incur trade-offs, including increased printed circuit board (PCB) area and higher power consumption.

6.2.1 AR Using Gain Feedback Loop

As noted by Nick [1], "an active resonator can be realized using a gain feedback loop," a concept central to his thesis. Figure 6.5 illustrates a typical AR circuit topology that implements such a feedback loop [16, 17]. In this configuration, the

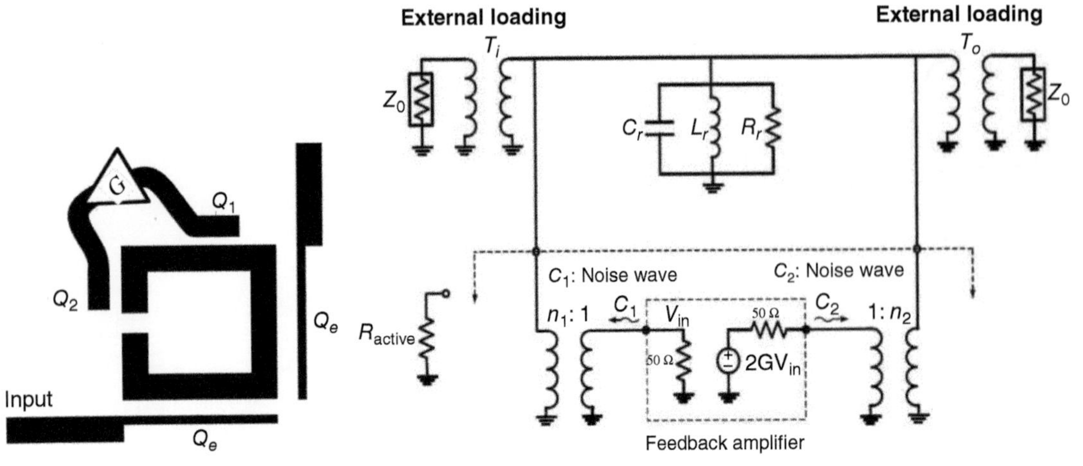

(a) (b)

Figure 6.5 Square open-loop resonator featuring an active feedback loop to offset losses, doubly coupled through input (I/P) and output (O/P) lines with external quality factor Q_e: (a) layout view and (b) simplified circuit diagram showing the amplifier matched to 50 Ω at both I/P and O/P [1, 17].

active feedback loop injects energy into the resonator to compensate for intrinsic losses arising from conductor, dielectric, and radiation effects. By offsetting these losses, the effective quality factor (Q) of the resonator is significantly enhanced, which directly improves the phase-noise performance and frequency stability of oscillators employing this topology. Careful design of the loop gain and phase is critical, as excessive feedback can lead to instability, while insufficient feedback limits the achievable Q-factor improvement. This approach demonstrates how AR networks can outperform conventional PRs in high-performance oscillator applications.

As shown in Figure 6.5a [1], the external quality factor, Q_e, is a crucial parameter for assessing the loading effects at the resonator's input and output ports. Specifically, the external quality factors Q_1 and Q_2 describe the impact of the amplifier's input and output coupling on the resonator's performance [1].

The voltage gain of the active stage, denoted by G, works in conjunction with these couplings to determine the degree to which the resonator's intrinsic losses are compensated. For stable and sustained oscillations, the phase of the active feedback loop must satisfy the condition of an integer multiple of 360°, ensuring that the injected signal reinforces the oscillations rather than diminishing them. The careful design of Q_1, Q_2, and G is therefore critical to optimizing both amplitude stability and phase-noise performance.

Figure 6.5b [18] presents a simplified equivalent circuit of the AR, which provides a framework for analyzing its noise characteristics and performance limitations. In this model, the PR is represented as a series RLC network, capturing its intrinsic resonant behavior and losses [9]. Coupling to the active amplifier is realized through transformers with turn ratios n_1 and n_2 [1], which control the effective energy transfer between the resonator and the gain stage. The input and output external couplings of the resonator are implemented via transformers T_i and T_o, respectively, allowing precise adjustment of loading effects and impedance matching.

The combined influence of the external quality factors, transformer ratios, and amplifier gain directly affects the resonator's effective Q-factor and, consequently, the oscillator's phase-noise performance. A higher effective Q reduces the linewidth of the oscillation, leading to lower phase noise, while mismatched couplings or improper gain can increase noise and potentially destabilize the oscillation. This detailed equivalent-circuit approach provides critical insights into the design trade-offs inherent in AR networks, highlighting the interplay between energy compensation, loading, and noise management in high-performance VCOs.

For analytical simplicity, the amplifier is assumed to be bilaterally conjugate-matched at both its input and output terminals. Conjugate matching ensures maximum power transfer between the amplifier and resonator while minimizing reflections, which is essential for sustaining stable oscillations. Under this condition, the resonator achieves full loss compensation when the negative resistance introduced by the amplifier precisely cancels the intrinsic positive resistance of the PR, Rr. This balance is attained through careful selection of the amplifier gain [1, 16]. By properly tuning the gain, the AR not only compensates for losses but also enhances the effective quality factor (Q), thereby improving both the oscillator's phase-noise performance and frequency stability.

$$G = \frac{\sqrt{n_1 n_2}}{2}\left[\frac{1}{Q_1} + \frac{1}{Q_2} + \frac{1}{Q_3}\right] \tag{6.5}$$

$$\text{where} \quad Q_1 = \frac{50n_1^2}{\omega_0 L_r} \tag{6.6}$$

$$Q_2 = \frac{50n_2^2}{\omega_0 L_r} \tag{6.7}$$

$$Q_u = \frac{R_r}{\omega_0 L_r} \tag{6.8}$$

The active feedback loop shown in Figure 6.5a can be represented equivalently as a negative resistance as [1]

$$R_{\text{active}} = -\left(\frac{(50n_2^2) \times (50n_1^2)}{\left(2G\frac{n_2}{n_1} - 1\right)\left\{\left(\frac{-50n_2^2}{2G\frac{n_2}{n_1} - 1}\right) + 50n_1^2\right\}}\right) \tag{6.9}$$

In the feedback loop [1], the amplifier's inherent noise generates two noise waves, denoted as C_1 and C_2 (Figure 6.5b), which contribute to the degradation of the noise performance of the doubly loaded resonator. Minimizing the impact of this additional noise is critical, particularly when the AR is used in low-noise oscillator designs [1].

Prior studies on the noise behavior of ARs [16, 17] have demonstrated that the effect of the noise wave C_1 (shown in Figure 6.5) on the resonator's overall noise figure can be effectively eliminated by appropriately designing the active feedback loop according to the following condition [1]:

$$Q_2 = G^2 Q_1 \Rightarrow \frac{Q_2}{G^2 Q_1} = 1 \tag{6.10}$$

Based on Eqs. (6.6)–(6.10), the optimal design parameters for the feedback loop of a lossless, high-Q resonator that minimizes noise degradation can be expressed as [1]:

$$Q_1 = \frac{(G^2 - 1)}{G^2} Q_u \tag{6.11}$$

$$Q_2 = (G^2 - 1) Q_u \tag{6.12}$$

$$F_{\min} = \left[1 + \left(\frac{2Q_e}{Q_u}\right) + m\left(\frac{2Q_e}{Q_u}\right)\right] = \left[F_{\text{passive-resonator}} + m\left(\frac{2Q_e}{Q_u}\right)\right] \tag{6.13}$$

Here $F_{\text{passive-resonator}}$ represents the noise figure of the PR, while m and F_A correspond to the noise measure and noise figure of the amplifier, respectively, with m defined as [1]:

$$m = \frac{F_A - 1}{1 - G^{-2}} \tag{6.14}$$

The noise figure of the PR is given by [10]:

$$F_{\text{passive-resonator}} = 1 + \left(\frac{2Q_e}{Q_u}\right) \tag{6.15}$$

Figure 6.6 presents the CAD-simulated S_{21} responses of both passive and ARs with gain feedback, providing a preliminary view of their frequency characteristics.

Figure 6.7 illustrates the CAD-simulated noise figures of passive and ARs for different amplifier noise-measure values ($m = 0.103, 0.425, 0.778$) with $Q_u = 200$ and $Q_e = 200$. Here, m represents the noise measure of the amplifier employed in the active feedback loop [1]. The results emphasize the impact of amplifier noise on resonator performance, demonstrating how variations in the noise-measure parameter "m" affect the phase-noise characteristics of oscillators employing AR networks.

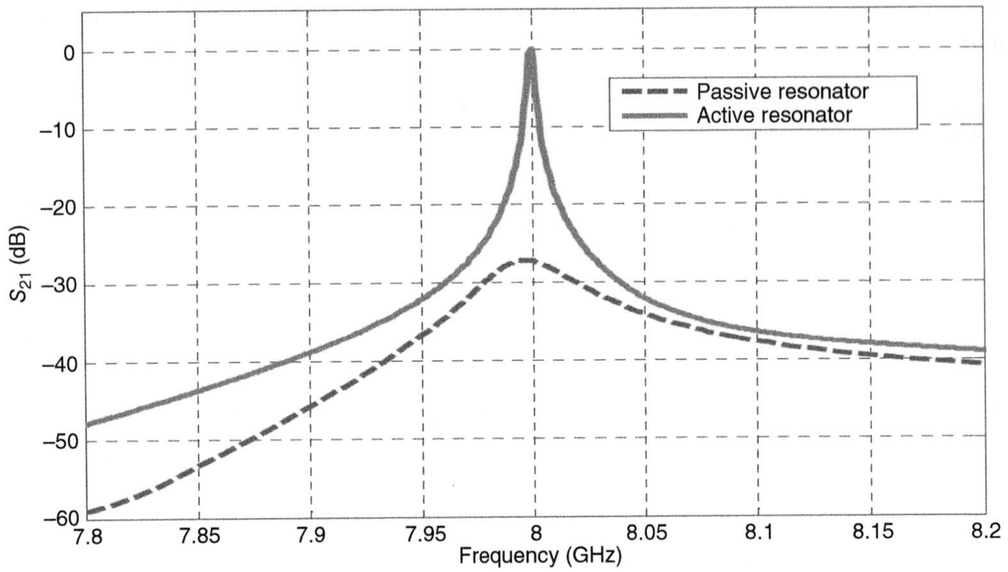

Figure 6.6 CAD-simulated frequency-response (S_{21}) of the passive and ARs obtained using CAD analysis ($Q_u = 200$ and $Q_e = 200$) [1].

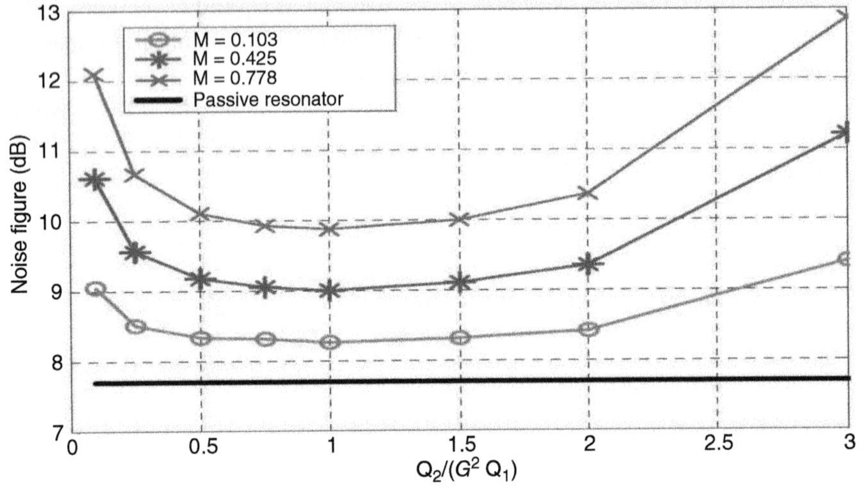

Figure 6.7 CAD-simulated noise figures of passive and ARs for different amplifier noise measures (m = 0.103, 0.425, and 0.778), $Q_u = 200$ and $Q_e = 200$. Here, m denotes the noise measure of the amplifier employed in the active feedback loop [1].

As illustrated in Figure 6.7 [17], the noise figure is evaluated at the resonant frequency (ω_o) with each marker corresponding to an AR configured with distinct feedback-loop parameters [1]. From (6.10), the minimum noise figure is achieved when $\frac{Q_2}{G^2Q_1} = 1$, as also observed in Figure 6.7, the noise degradation for the AR is about 0.5 dB as compared to the (PR). However, if the design parameter is not optimized $\left(\frac{Q_2}{G^2Q_1} \neq 1\right)$, significant noise-figure degradation could occur.

6.2.2 AR Using Negative Resistance Devices

Figure 6.8 shows a square-loop resonator combined with a negative-resistance circuit intended to offset the intrinsic losses of the PR [1]. The external quality factor, QeQ_eQe, quantifies the loading effects at the input and output ports, whereas QrQ_rQr represents the effective quality factor arising from coupling to the negative-resistance network. Together, these parameters govern the energy transfer efficiency and the active network's capability to enhance the resonator's effective Q-factor.

The square-loop AR can be represented by an equivalent shunt RLC circuit, modeling the PR coupled to the amplifier via input and output transformers with a turn ratio n_e [1]. As depicted in Figure 6.9 [1], this model incorporates a current source

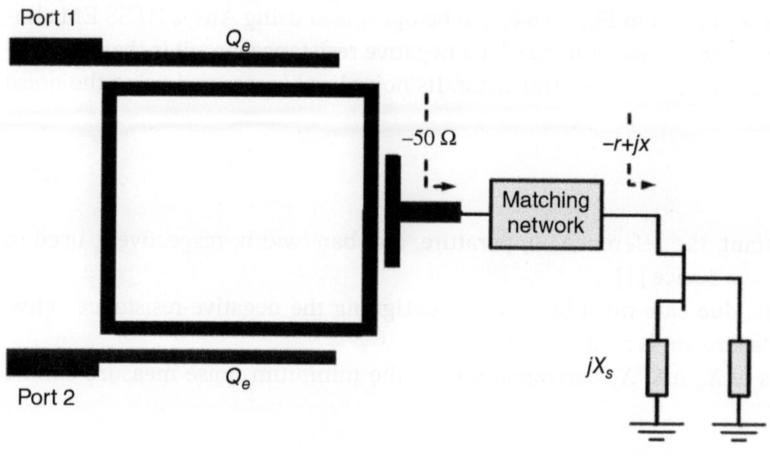

Figure 6.8 Printed square-loop resonator integrated with a negative-resistance circuit for compensating losses [1, 17].

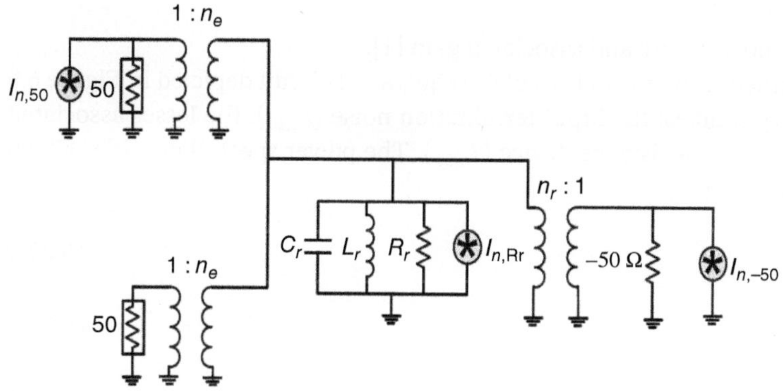

Figure 6.9 Simplified circuit model of an AR, illustrating a microstrip square-loop resonator coupled with a negative-resistance network for loss compensation [1, 17].

I_n, to account for noise contributions from the active device. Coupling to the negative-resistance network is implemented through a transformer with a turn ratio n_r which determines the level of negative resistance applied to the resonator and directly influences the compensation of its intrinsic losses. This equivalent representation provides a clear framework for analyzing both the resonator's energy dynamics and the noise performance of the active network.

The transformer ratios n_e and n_r play a critical role in determining oscillator performance. A higher n_e increases the effective coupling between the PR and the active stage, enhancing energy transfer but potentially increasing the noise contribution from the amplifier. Conversely, n_r controls the amount of negative resistance applied; optimizing n_rr ensures that losses are fully compensated without driving the circuit into instability. Proper selection of these ratios is therefore essential to maximize the effective Q-factor, minimize phase noise, and maintain stable oscillations in AR-based oscillator designs. This framework provides a comprehensive methodology for analyzing and designing high-performance ARs with predictable noise and stability characteristics.

Theoretically, complete full loss compensation in the resonator is obtained when the negative resistance introduced into RLC network precisely cancels the intrinsic positive resistance of the PR, denoted as R_r [1, 19, 20]:

$$R_r = 50n_2^2 \Rightarrow Q_u = Q_{-r} \tag{6.16}$$

$$Q_u = \frac{R_r}{\omega_0 L_r} \tag{6.17}$$

$$Q_e = \frac{50n_e^2}{\omega_0 L_r} \tag{6.18}$$

$$Q_{-r} = \frac{50n_r^2}{\omega_0 L_r} \tag{6.19}$$

From (6.16), the negative-resistance coupling gap, as shown in Figure 6.8, can be optimized using Ansys HFSS EM simulation under the specified condition $Q_u = Q_{-r}$. However, a disadvantage of the negative resistance circuit is that it generates excess noise, which is associated with the active device, such as a transistor. Its noise level is assessed using the noise measurement equation described in [21].

$$m = \frac{P_e}{KT_0B} \tag{6.20}$$

where K, T_0, and B represent the Boltzmann constant, the reference temperature, and bandwidth, respectively, used to evaluate the exchangeable noise power Pe of the active device [1].

For the application of low-phase-noise oscillators, due care must be taken in designing the negative-resistance active device such that the minimum achievable noise measure, m, is attained [1].

As illustrated in Figure 6.8, the optimum values of X_g and X_s corresponding to the minimum noise measure can be expressed as [10]:

$$m_{\min} = \frac{(F_{min} - 1)}{\left(1 - \frac{1}{G_a}\right)} \tag{6.21}$$

Here, $F_{\min}$ and G_a denote the transistor's minimum noise figure and associated gain [1].

The minimum noise figure of the AR can be determined through analysis of the equivalent circuit depicted in Figure 6.9 [1]. The overall output noise arises from three primary sources: the input termination noise $(I_{n,50})$, the losses associated with the PR $(I_{n,Rr})$, and the noise generated by the negative-resistance device $(I_{n,50})$. The power spectral densities (PSD) corresponding to these noise contributions are expressed [1]:

$$\underline{i^2_{n,50}} = \frac{4kT_0B}{50} \tag{6.22a}$$

$$\underline{i^2_{n,R_r}} = \frac{4kT_0B}{R_r} \tag{6.22b}$$

$$\underline{i^2_{n,-50}} = \frac{4kT_0B}{50} m_{\min} \tag{6.23}$$

$$\begin{aligned} F_{\min}(\text{active} - \text{resonator}) &= \left[1 + \left(\frac{2Q_e}{Q_u}\right) + m_{\min}\left(\frac{2Q_e}{Q_u}\right)\right] \\ &= \left[F_{\text{passive-resonator}} + m_{\min}\left(\frac{2Q_e}{Q_u}\right)\right] \end{aligned} \tag{6.24}$$

$$\text{where} \quad F_{\text{passive-resonator}} = 1 + \left(\frac{2Q_e}{Q_u}\right) \tag{6.25}$$

From (6.24), F_{active} and F_{passive} denote the noise figures of the active and PRs, respectively, capturing the influence of the active device on the overall resonator noise performance [1]. This relationship highlights the importance of selecting a transistor with a low noise measure for AR implementations.

As illustrated in Figure 6.9, the negative-resistance device supplies RF power to counteract the intrinsic losses of the PR. If the input power at port #1 is P_{in}, the voltage developed across the LC resonator can be expressed as [1]:

$$V_{\text{resonator}} = \sqrt{2\omega_o L_r Q_e P_{\text{in}}} \tag{6.26}$$

The additional power contributed by the negative-resistance device (P_{added}) is equivalent to the power dissipated by the tank's positive resistance, and can be described by [1]:

$$P_{\text{added}} = \frac{V^2_{\text{resonator}}}{2R_r} = \left(\frac{2Q_e}{Q_u}\right)P_{\text{in}} \tag{6.27}$$

6.2.3 Oscillator Design Methodology Using AR

As outlined in Sections 6.2.1 and 6.2.2 [1], ARs can be implemented either through active feedback loops or by employing negative-resistance devices, both approaches enabling partial or complete loss compensation with minimal degradation in noise figure. In principle, these two compensation techniques offer comparable performance in terms of noise characteristics and power efficiency; however, the selection of a suitable method depends on the specific implementation requirements and the desired FOM [22]. It is important to note that, in the case of ARs, the loaded quality factor (Q_l) and the external quality factor (Q_e) become identical, since the unloaded quality factor (Q_u) approaches infinity under conditions of complete loss compensation [1]. The relationships derived in (6.24)–(6.27) provide the governing expressions for this behavior [1, 17].

$$Q_{L,\text{active}} = Q_e \tag{6.28}$$

$$F_{\text{passive}} = 1 + \frac{2Q_e}{Q_u} \tag{6.29}$$

$$F_{\text{active}} = F_{\text{passive}} = M\left(\frac{2Q_{L,\text{active}}}{Q_u}\right) \tag{6.30}$$

$$\frac{P_{\text{added}}}{P_{\text{in}}} = \frac{2Q_{L,\text{active}}}{Q_u} \tag{6.31}$$

Here, F_{passive} and F_{active} denote the noise figures of the passive and ARs, respectively [1]. The terms $Q_{L,\text{active}}$, Q_e, and Q_u correspond to the loaded quality factor of the AR, the external quality factor, and the unloaded quality factor of the PR, respectively. P_{in} represents the input power supplied to the resonator, while P_{added} refers to the compensating power contributed by the loss-recovery compensation circuit network.

Based on the observations from Eqs. (6.28) to (6.31), it can be concluded that resonators with elevated quality factors are particularly suitable for the development of low phase-noise oscillators [1]. This suitability hinges on the availability of adequate power resources, as ARs inherently necessitate a DC power supply to function effectively. Such resonators, when properly powered, offer the potential for enhanced frequency stability and minimized phase noise, making them invaluable in precision applications. Furthermore, the adverse effect of excess noise introduced by the active device can be alleviated through noise-filtering or cancellation techniques, or by optimizing the resonator design for a minimum noise measure [3–8, 16, 17, 19–21, 23–26]. In other words, oscillators that utilize a solitary AR as the key element for frequency stabilization face an intrinsic compromise. While enhancements in phase-noise performance can be attained, this improvement comes at the cost of diminished efficiency when converting direct current (DC) to radio frequency (RF) power.

6.3 Example Printed Resonator Oscillators

Microstrip, stripline, and suspended-stripline resonators are commonly employed in planar oscillator designs owing to their low fabrication cost, compact form factor, and seamless integration with active circuits, including monolithic microwave integrated circuits (MMICs) and radio-frequency integrated circuits (RFICs) [1]. However, a significant limitation of these printed planar resonators is their relatively large physical size and low quality factor, which pose challenges in realizing compact, low phase-noise oscillator designs. Various methods have been explored [1–8, 16, 17, 19–21, 23–39] to improve the Q-factor and achieve miniaturization; these are categorized into two groups: (i) passive planar resonator oscillator, and (ii) active planar resonator oscillator [1].

6.3.1 Oscillators Based on Passive Planar Resonator

Figure 6.10 [1] illustrates an 8 GHz oscillator incorporating an amplifier gain block to ensure adequate loop gain, a 4-pole elliptical bandpass filter for frequency stabilization, and a parallel feedback network for phase matching [1, 28]. As depicted in Figure 6.10, the active device is a packaged SiGe heterojunction bipolar transistor (HBT) (NEC NESG2030M04) biased at $V_{CE} = 2\,\text{V}$, $I_c = 11\,\text{mA}$ [1].

Figures 6.11–6.13 illustrate the BPF layout, plots of printed BPF characteristics for the X-band oscillator, and the phase-noise plot (simulated and measured) [1].

6.3.2 Oscillator Based on Active Planar Resonator

A planar AR with a high quality factor (Q) can be realized by embedding a suitably designed gain element that compensates for the intrinsic dissipative losses of the PR. In principle, such a configuration can facilitate the development of oscillators exhibiting exceptionally low phase noise. However, achieving this performance in practice is not straightforward [13, 31]. The introduction of an active gain block inherently contributes excess noise, which increases the effective noise within the

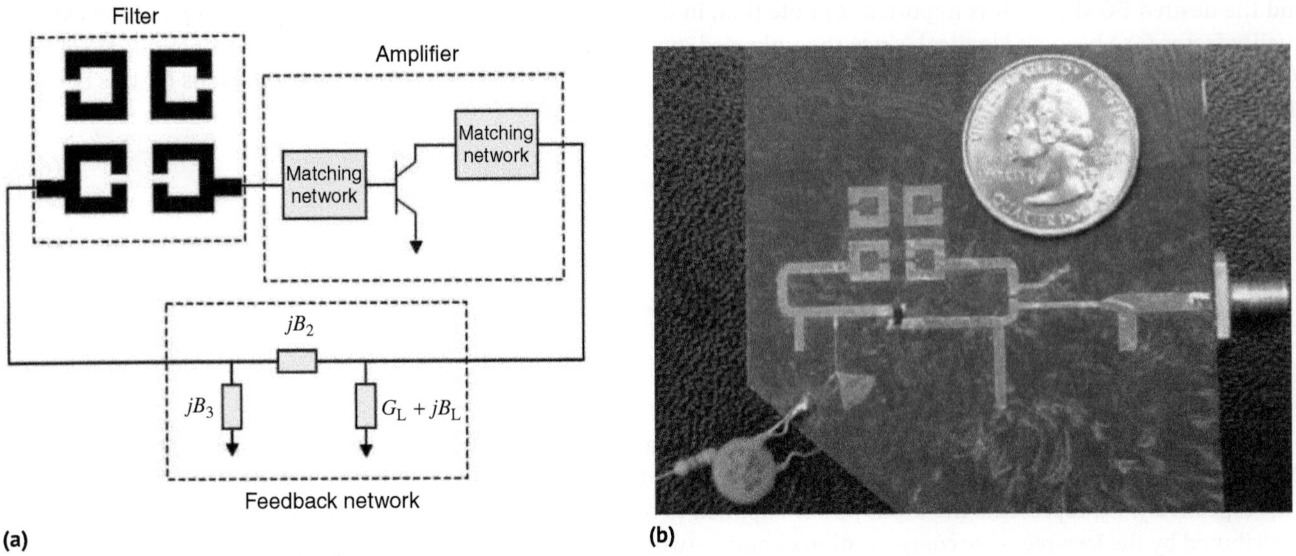

Figure 6.10 The typical printed resonator oscillator [10]: (a) schematic of printed resonator oscillator, and (b) layout of printed resonator oscillator (Rogers RT/Duroid 5880 substrate). Frequency = 8.05 GHz, power output = 3.5 dBm, DC power consumption = 22 mW, DC-RF conversion efficiency = 10%, oscillator FOM = −207 dBc/Hz [1, 17].

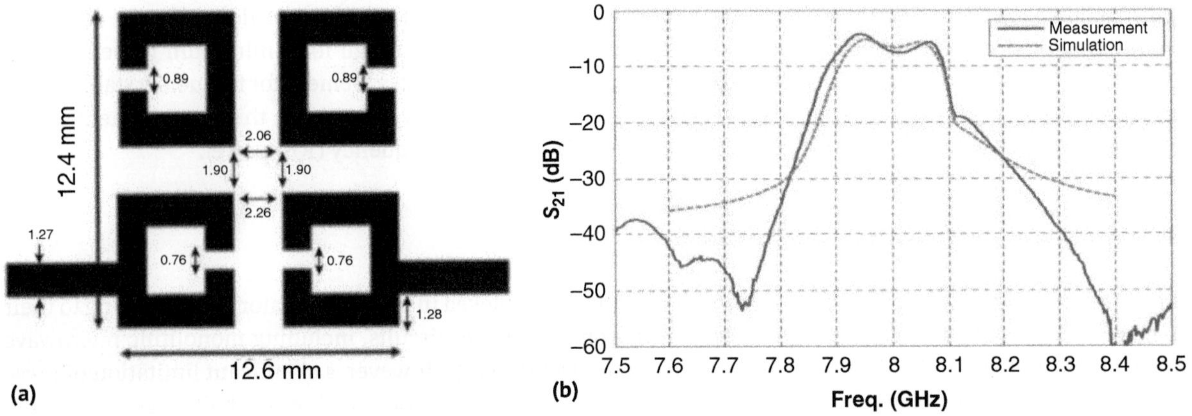

Figure 6.11 (a) Layout of printed resonator (Roger 5880) BPF and (b) insertion loss of the 4-pole elliptical BPF [1].

oscillator feedback loop and can ultimately deteriorate phase-noise characteristics. To address this challenge, it is essential to carefully engineer the active device and its biasing network to minimize both thermal and flicker noise contributions [32]. Furthermore, the noise reduction benefits afforded by AR architectures are typically accompanied by higher power dissipation, since the gain stage must continuously supply energy to offset resonator losses. To evaluate the trade-off between noise performance and power consumption, the FOM for an AR, denoted as $[\mathrm{FOM}]_{\mathrm{AR}}$ is expressed as [1]:

$$[\mathrm{FOM}]_{\mathrm{AR}} = 10\log\left(\frac{F_{\mathrm{AR}}}{Q_{L-\mathrm{AR}}^2}\right) \tag{6.32}$$

$$[\mathrm{FOM}]_{\mathrm{AR}} = 10\log\left(\frac{1+(m+1)\frac{2Q_{L-\mathrm{AR}}}{Q_u}}{Q_{L-\mathrm{AR}}^2}\right) \tag{6.33}$$

$$\text{where}\quad F_{\mathrm{AR}} = \left[1+\left(\frac{2Q_{L-\mathrm{AR}}}{Q_u}\right)+m\left(\frac{2Q_{L-\mathrm{AR}}}{Q_u}\right)\right] \tag{6.34}$$

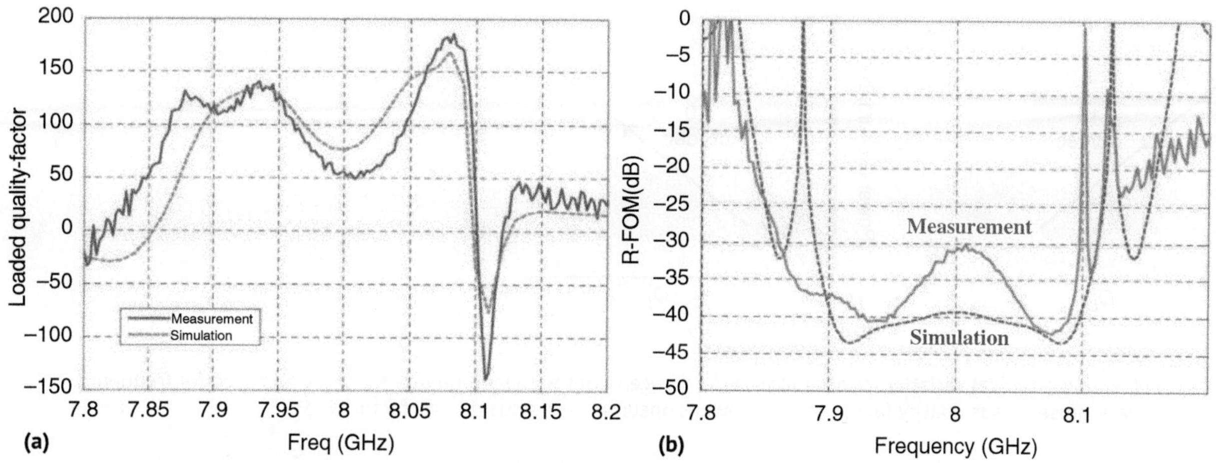

Figure 6.12 CAD-simulated and measured results for the 4-pole bandpass filter: (a) quality factor and (b) resonator FOM [1, 40].

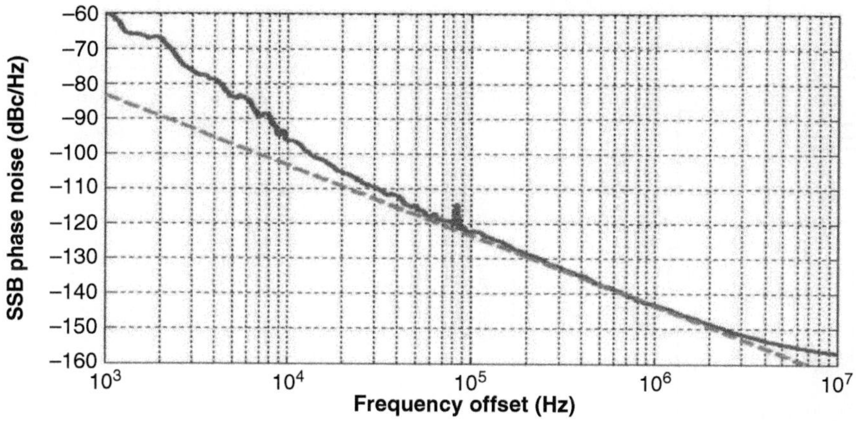

Figure 6.13 CAD-simulated (dashed-line) and measured (solid-line) phase-noise performance of the oscillator shown in Figure 6.10 [10].

In an AR configuration, the loss compensation mechanism relies on an embedded active device that converts supplied direct current (DC) power into radio frequency (RF) energy [1]. This generated RF energy is injected back into the resonant structure to offset the inherent dissipative losses of the PR elements, thereby sustaining resonance with minimal attenuation. The contribution of the active device to the overall power balance of the system can be mathematically represented as follows [33–35]:

$$P_{\text{added}} = \frac{V^2_{\text{resonator}}}{2R_r} = \left(\frac{2Q_e}{Q_u}\right)P_{\text{in}} \Rightarrow \frac{P_{\text{added}}}{P_{\text{in}}} = \frac{2Q_{L-\text{AR}}}{Q_u} \tag{6.35}$$

Equation (6.35) illustrates how the power contributed by the active device depends on the loaded quality factor of the resonator [1, 10]. Increasing the loaded Q of the AR reduces phase noise but simultaneously results in higher power consumption; scaling at approximately 10 dB per decade, as depicted in Figure 6.14 [3].

Figure 6.15 illustrates the standard circuit schematic of a printed active-resonator oscillator, which plays a crucial role in various electronic applications, including frequency generation and signal processing. The circuit comprises three main components: the amplifier, the resonator, and a π-feedback network [1]. In this design, the amplifier is responsible for boosting the signal, ensuring that the oscillations sustain and maintain stability. The resonator, which is typically designed to operate at a specific frequency, interacts with the amplifier to produce a sinusoidal output. This component is usually characterized by its resonant frequency and quality factor (Q), which indicates how selective the resonator is in terms of its frequency response. The π-feedback network is used to create the necessary phase shift for oscillation. Its configuration allows for precise control over the feedback signals, which ultimately determine the stability and amplitude of the oscillations produced by the circuit. One key aspect of this configuration is that the loaded quality factor of the AR can be fine-tuned by adjusting the input and output coupling gaps (g). By manipulating these gaps, designers can optimize the

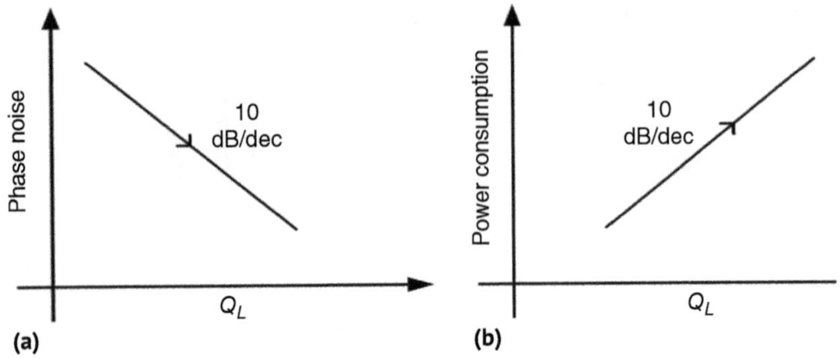

Figure 6.14 Graph shows a typical variation of oscillator performance with the loaded quality factor of ARs used for frequency stabilization: (a) phase noise versus quality factor, and (b) power-consumption versus quality factor [1, 3].

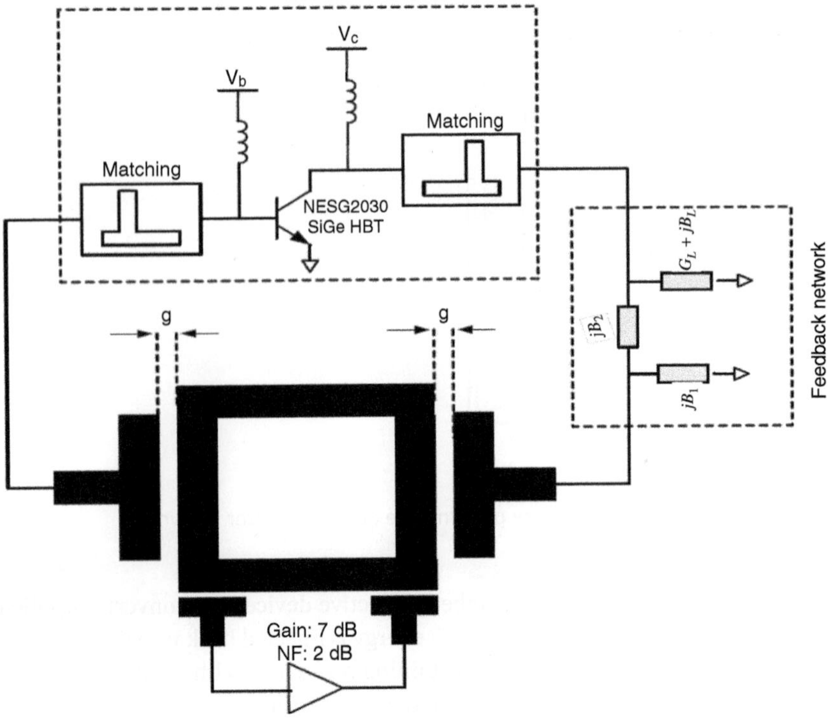

Figure 6.15 A schematic of a C-band oscillator operating at 8 GHz, incorporating a square-loop AR, which generates 7-dBm O/P (output) power [1].

energy exchange between the resonator and the surrounding components, which impacts the overall performance of the oscillator in terms of bandwidth and efficiency [40]. These adjustments are critical for tailoring the oscillator's response to meet specific application requirements, ensuring that it operates effectively within its intended frequency range. [1, 17].

Figure 6.16 illustrates the CAD-simulated phase-noise plot as a function of the loaded quality factor of the AR [1]. As illustrated in Figure 6.16a [10], the CAD-simulated phase-noise plot decreases monotonically with an increase in the resonator loaded Q (achieved by increasing g), exhibiting an enhancement rate of approximately 10 dB/decade, consistent with the theoretical prediction in Figure 6.14 (Eq. 6.35). The loaded quality factor of the active resonator is controlled by adjusting the input/output coupling gaps (g). ARs with different loaded quality factors were designed, and their simulated noise figures are shown in Figure 6.16b.

Figure 6.17a demonstrates that an AR with a loaded quality factor of 2000 improves phase-noise performance by 25 dB (−150 dBc/Hz at 1 MHz offset) relative to an oscillator employing a passive square-loop resonator [1]. However, this enhancement comes at the cost of significantly higher power consumption.

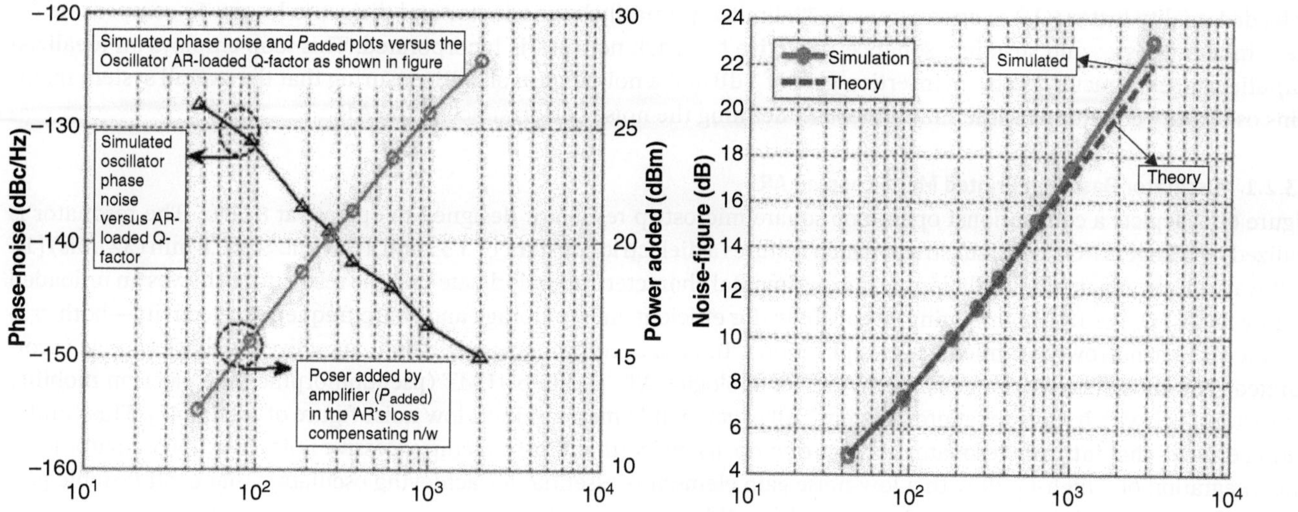

Figure 6.16 (a) CAD-based simulation of oscillator phase noise at a 1 MHz frequency offset from the 8 GHz carrier, plotted against the loaded quality factor of the AR, and (b) corresponding CAD-simulated noise figure [1].

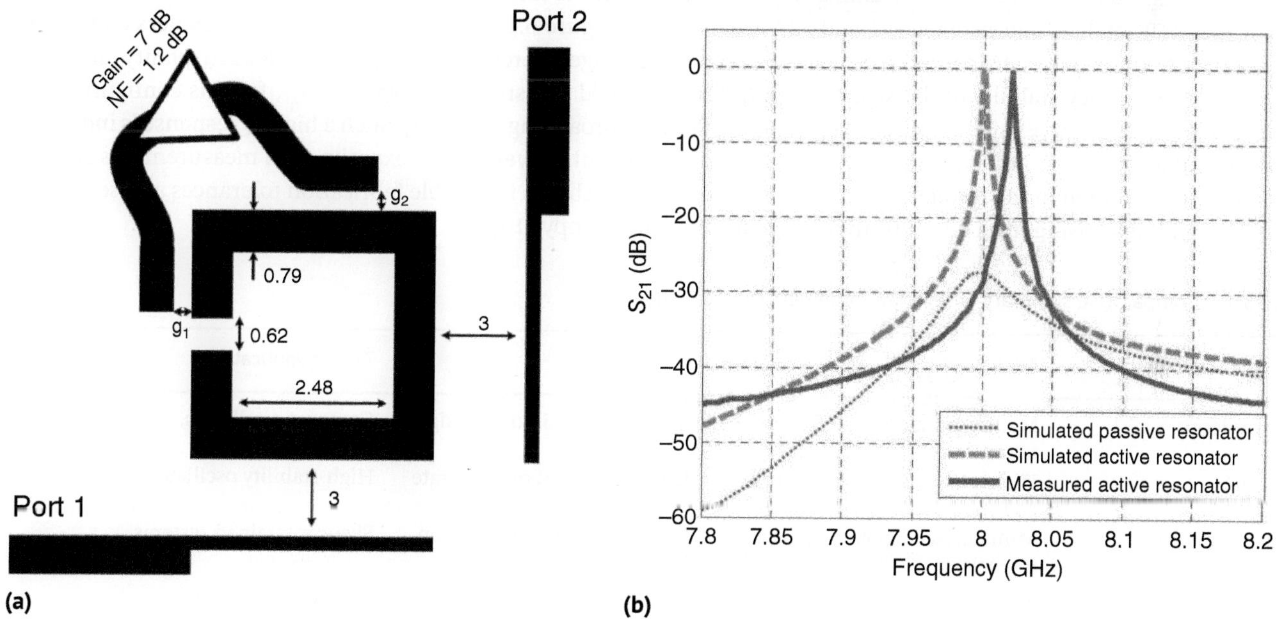

Figure 6.17 (a) Open-square printed resonator incorporating an active feedback loop for loss compensation, and (b) comparison of measured and simulated frequency responses of the loosely coupled AR with those of an equivalent PR [1].

As depicted in Figure 6.16a [1], with a loaded Q of 2000, the amplifier within the resonator's loss-compensation network is required to generate approximately 28 dBm of RF power—around 125 times greater than the output power of the oscillator [1]. As a result, the DC-to-RF power conversion efficiency is remarkably low (<0.5%) [1], making this method impractical for handheld applications [15]. This limitation can be mitigated by employing a multi-coupled resonator-based elliptical filter, which achieves higher loaded quality factors compared to single resonators [1]. The enhanced frequency selectivity, arising from closely spaced transmission zeros near the passband, enables improved phase-noise performance without excessive power consumption [13, 28–38].

As illustrated in Figure 6.15 [1, 10], the reference structure consists of a passive microstrip square-loop resonator fabricated on a 32-mil RO4003C dielectric substrate [1]. Electromagnetic simulations indicate that this resonator exhibits an unloaded quality factor (Q_u) of approximately 72, limited primarily by conductor and dielectric losses. To compensate for these inherent losses and sustain resonance, an active feedback network is incorporated. This loop employs an idealized amplifier stage characterized by an insertion gain of 7 dB and a noise figure of 2 dB, ensuring that the overall system maintains oscillatory conditions while simultaneously defining the noise performance of the AR.

6.3.2.1 Oscillator Based on Printed Multi-coupled AR

Figure 6.17 depicts a conventional open-loop square microstrip resonator designed to operate at 8 GHz. The resonator is realized on a Rogers RO3035™ substrate, which features a dielectric constant of 3.55 and a thickness of 0.8 mm (32 mils) [1]. Full-wave electromagnetic simulations and experimental characterization indicate that the resonator achieves an unloaded quality factor (Q_u) of 180, highlighting its capability for efficient energy storage and sharp frequency selectivity—both critical metrics for narrowband microwave applications. To enable continuous oscillation, an active feedback loop is incorporated. The chosen active device is the Avago Technologies ATF-33143 pHEMT (pseudomorphic high electron mobility transistor) [40], which provides approximately 7 dB of gain while maintaining a low noise figure of 1.2 dB [41]. This combination ensures that intrinsic resonator losses are compensated without introducing excessive noise into the oscillator loop. The integration of a high-Q PR with a low-noise gain element is essential for achieving oscillators that exhibit stable performance and favorable phase-noise characteristics at X-band frequencies.

From a broader perspective, the square-loop geometry used here offers a balance between compact layout and high Q-factor, making it suitable for planar circuit integration. However, alternative resonator topologies and substrate selections present different trade-offs in performance, described in Tables 6.1 and 6.2.

The optimization of the feedback-loop configuration in AR systems is fundamentally governed by the choice of coupling gaps at the input and output, denoted ($g1$ and $g2$) as depicted in Figure 6.17a [1]. These coupling parameters control the balance between external energy exchange and internal energy storage, thereby dictating both the loaded quality factor and the resonant frequency stability of the structure [40]. The simulated transmission response, S_{21}, for this configuration is presented in Figure 6.17b, exhibiting an unloaded quality factor approaching 4000 [13]. Such a high-Q response is indicative of minimal internal losses and efficient electromagnetic confinement. Nevertheless, experimental measurements reveal a slight upward deviation in the resonant frequency, an effect attributable to inevitable fabrication tolerances and secondary parasitic couplings arising from discontinuities and substrate anisotropy [10].

Table 6.1 Comparative summary of resonator geometries trade-offs.

Resonator geometry	Advantages	Limitations	Typical application
Square-loop	Moderate size, good Q, simple fabrication	Q lower than circular designs	Compact oscillators, filters
Circular	Higher Q (less edge loss), strong confinement	Larger footprint on substrate	High-stability oscillators
Hairpin/folded	Miniaturized layout, compact integration	Reduced Q, higher insertion loss	Size-constrained systems
DGS/metamaterial	Multi-band, tunability, bandgap engineering	Complex fabrication, higher sensitivity	Reconfigurable RF front-ends

Table 6.2 Comparative summary of substrate trade-offs.

Substrate type	Dielectric constant (ε_r)	Advantages	Trade-offs
RO3035™	3.55	Balance of loss, size, and cost	Moderate Q compared to ceramics
RO4350B™	3.48	Lower loss tangent, higher Q	More expensive than RO3035
RO5880™	2.2	Very low loss, wide bandwidth	Larger resonator size
Alumina/ceramic	~9.8	Very high Q, excellent thermal stability	High cost, challenging fabrication

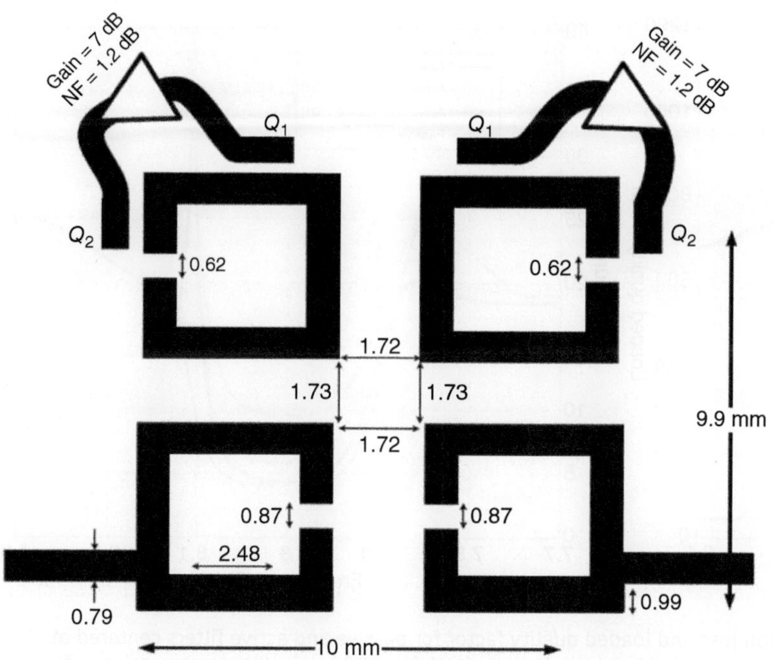

Figure 6.18 Typical configuration of an active four-pole bandpass filter employing an elliptic response characteristic [1].

Building upon the resonator's fundamental properties, the elliptic filter topology shown in Figure 6.18 demonstrates the utility of engineered cross-coupling between nonadjacent resonators. This deliberate coupling pathway introduces transmission zeros at finite frequencies, thereby providing enhanced selectivity and steeper skirt characteristics. The filter network is realized on a Rogers RO3035™ dielectric substrate [1], chosen for its favorable permittivity consistency, low dielectric loss tangent, and compatibility with precision fabrication at X-band frequencies. Unlike conventional designs that employ auxiliary amplification blocks to mitigate insertion losses, this filter excludes the integration of the 7 dB gain/1.2 dB NF amplifier. This decision is based on the observation that the loaded quality factors of the resonators are predominantly determined by external loading effects rather than intrinsic resonator dissipation, rendering the gain block unnecessary [1].

The electromagnetic simulation results of the filter, illustrated in Figure 6.19, quantify the influence of the active architecture. The analysis reveals that the loaded quality factor reaches a peak value of approximately 720 near the passband edge. This represents nearly a sixfold increase over the equivalent passive implementation, confirming the active filter's ability to significantly sharpen frequency selectivity. Such an enhancement is particularly advantageous in oscillator design, where low phase-noise performance is critically dependent on the effective loaded Q of the frequency determining network.

The insertion loss characteristics, as presented in Figure 6.20, demonstrate a residual attenuation of approximately 2.2 dB. The origin of this loss is primarily linked to uncompensated dissipation within the input and output resonator loops. While this residual insertion loss is not fully eliminated in the absence of a gain stage, the overall improvement in loaded Q substantially outweighs this drawback, ensuring superior phase-noise suppression capability. These results confirm that the integration of optimized coupling parameters with an elliptic filter topology on a low-loss substrate provides a viable pathway toward compact, high-performance active filters suitable for next-generation low phase-noise oscillators [28].

Figure 6.20 illustrates the experimentally measured insertion loss characteristics of the active filter configuration previously introduced in Figure 6.18 [1]. The data reveal a slight upward displacement of the resonant frequency, a deviation that can be reasonably attributed to variations arising from fabrication tolerances during the manufacturing process. To quantitatively evaluate the resonator's performance, the FOM is determined in accordance with relation (6–11). In this evaluation, the quality factor (Q), which is extracted directly from experimental measurements, is combined with the noise figure obtained through simulation. The resulting formulation for the FOM, incorporating both empirical and modeled parameters, can be expressed as follows [1]:

$$[\mathrm{FOM}]_{\mathrm{AR}} = 10\log\left(\frac{F_{\mathrm{AR}}}{Q_{L-\mathrm{AR}}^2}\right) = [\mathrm{FOM}]_{\mathrm{AR}} = 10\log\left(\frac{1+(m+1)^{\frac{2Q_{L-\mathrm{AR}}}{Q_u}}}{Q_{L-\mathrm{AR}}^2}\right) \tag{6.36}$$

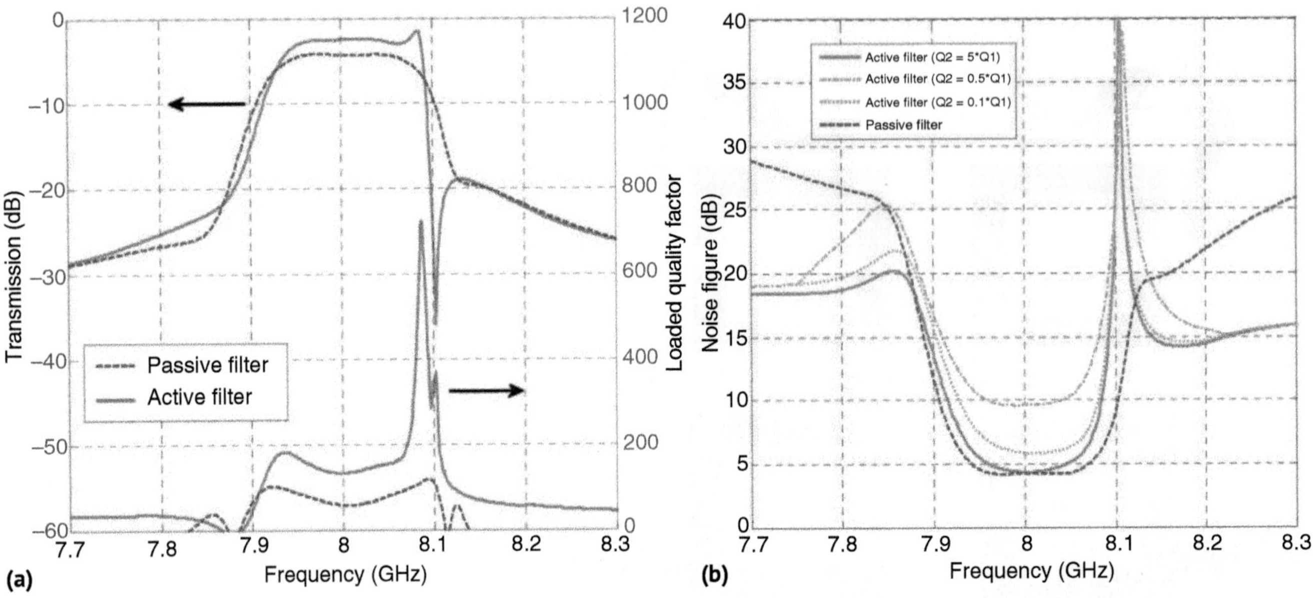

Figure 6.19 CAD-simulated performance [1]: (a) insertion loss and loaded quality factor for passive and active filters centered at 8 GHz with 130 MHz bandwidth, 10 dB return loss, and a normalized transmission zero at 1.2, and (b) noise figure comparison of passive and active filters, including simulations of additional active designs with varied feedback-loop parameters (Q_1 and Q_2) [1].

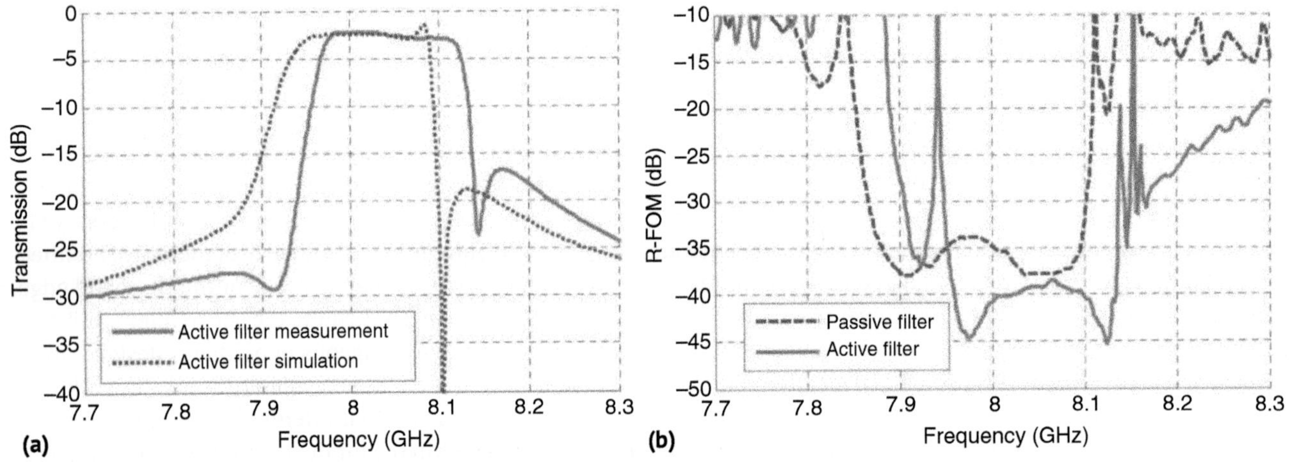

Figure 6.20 Frequency response of the active filter: (a) comparison of measured and simulated insertion loss from the structure in Figure 6.19 [27, 28], and (b) resonator figure-of-merit (R-FOM) for both active and passive filters without Q_1 and Q_2, where lower R-FOM values are favorable for low phase-noise oscillator design [1].

As shown in Figure 6.20b, the incorporation of the active filtering topology [14] results in a substantial improvement in the resonator's performance, with the FOM [14] exhibiting an enhancement of approximately 8 dB when compared to a passive filter of equivalent structure. This improvement highlights the effectiveness of the active feedback mechanism in reducing losses and improving overall efficiency.

Figure 6.21 provides a detailed view of the fabricated oscillator prototype. In this implementation, the feedback path is realized using a HBT, specifically the NEC NESG2030M04 device, which delivers a small-signal gain of about 8 dB [40]. The inclusion of this transistor within the loop architecture plays a crucial role in sustaining oscillations while simultaneously compensating for resonator losses, thereby ensuring stable operation and improved signal integrity [1, 17, 28].

The experimental results are consolidated in Figure 6.22a,b, which respectively depict the phase-noise characteristics and the output power spectrum of the fabricated oscillator [42]. At a frequency offset of 1 MHz from the carrier, the oscillator

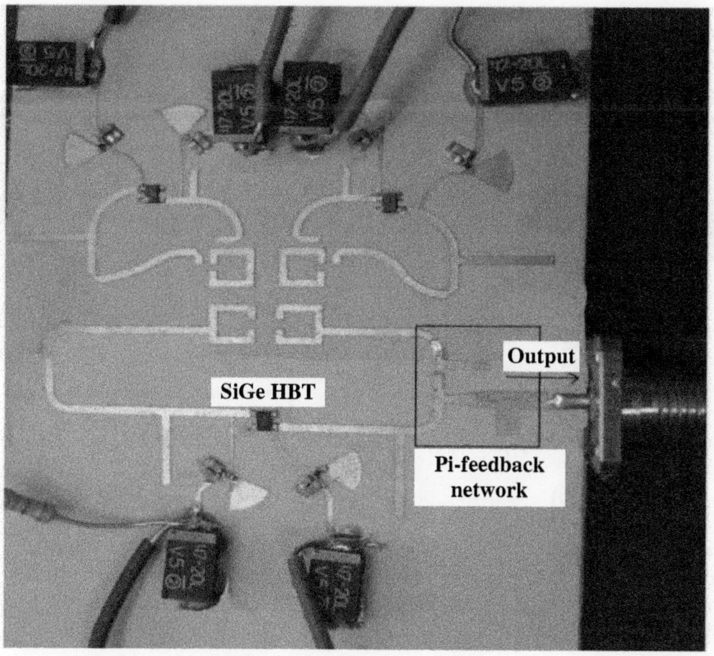

Figure 6.21 Illustrates the configuration of an 8 GHz oscillator that employs a multi-coupled active filter [1]. The active component in the oscillator circuit is an NESG2030M04 HBT SiGe transistor, which operates at a collector voltage of 2 V and a quiescent current of 10 mA. The active filter consists of ATF-33143 pHEMT transistors, while the π-network is made up of a series capacitor and two shunt microstrip open stubs [1, 42].

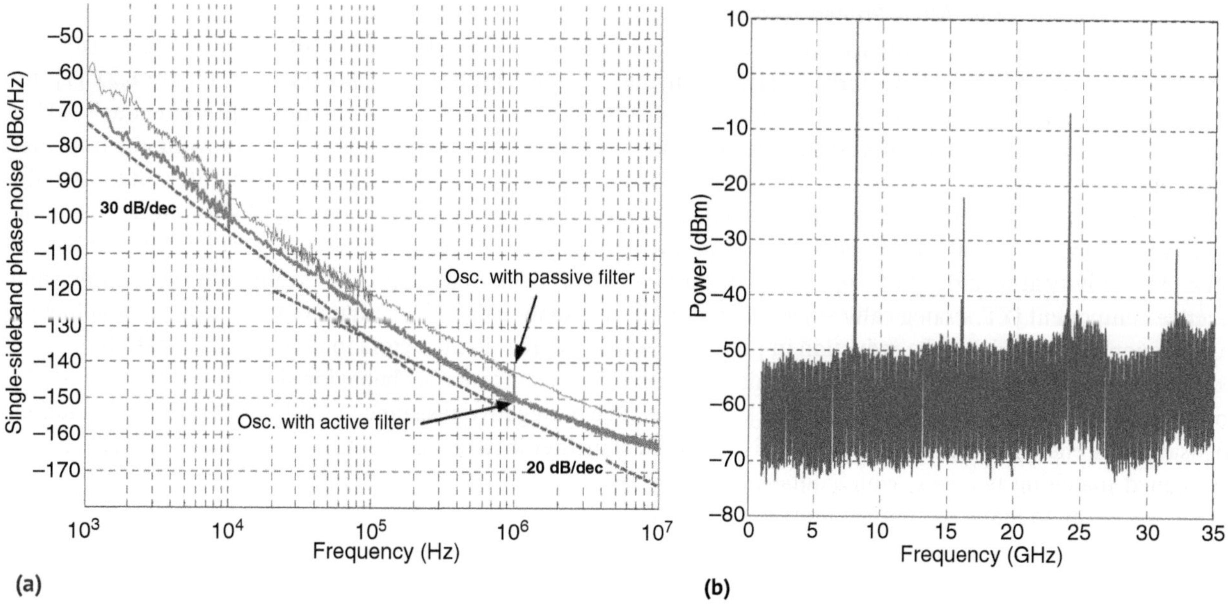

Figure 6.22 Measured results of the 8 GHz oscillator shown in Figure 6.21 [10] using the multi-coupled active filter of Figure 6.18: (a) phase noise of approximately −150 dBc/Hz at 1 MHz offset from the carrier, and (b) corresponding output spectrum [1].

that incorporates the active filter demonstrates a phase-noise level of approximately −150 dBc/Hz [1]. This represents a performance gain of nearly 7 dB when compared with a counterpart oscillator employing a purely PR structure. Further operational details are illustrated in Figure 6.23b, where the oscillator is observed to sustain oscillations at 7.990 GHz with an output power of 10 dBm [1]. From a power perspective, the complete circuit consumes 200 mW of DC input power. This

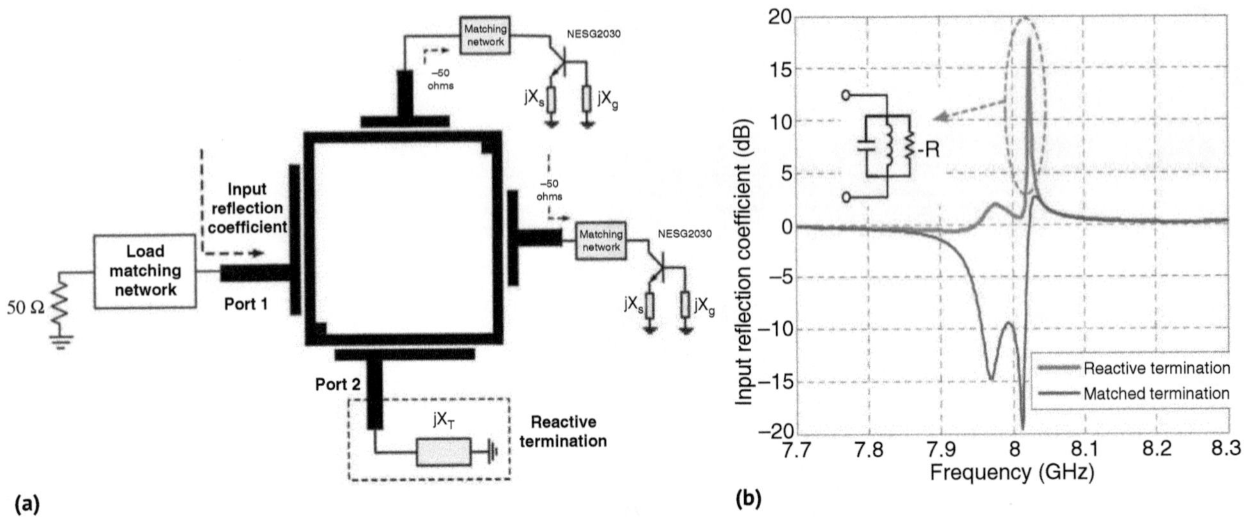

Figure 6.23 (a) Dual-mode oscillator setup incorporating a reactive termination (XT) and load-matching network derived from the dual-mode active elliptic filter, and (b) input reflection coefficient of the active filter. The reactive termination produces a prominent peak, around which the filter operates as a high-Q shunt LC resonator with negative resistance [1].

corresponds to an overall DC-to-RF conversion efficiency of about 5%. However, when the AR contribution is excluded from the calculation, the intrinsic oscillator efficiency improves significantly, reaching approximately 31%. These figures underscore the ability of the active filter approach to simultaneously enhance spectral purity and maintain reasonable power efficiency in high-frequency oscillator implementations.

6.3.2.2 Compact Oscillator Using Active Printed Resonator

Figure 6.23 presents the detailed circuit schematic of a compact dual-mode oscillator that utilizes a series feedback network to enhance its performance characteristics [1]. Specifically, in the configuration illustrated in Figure 6.21 [1], the series feedback is achieved by connecting the second port of the filter to a reactive termination represented as (jXT) in Figure 6.23a. This connection is crucial as it significantly increases the filter's instability, thereby allowing for improved oscillation conditions. Figure 6.23b showcases the simulated input reflection coefficient of the active filter under two different conditions [1]: matched termination and reactive termination. The simulation results highlight that the introduction of reactive termination creates a pronounced peak in the input reflection coefficient, which is indicative of enhanced reflection and selectivity at specific frequencies [1]. The position of this peak frequency is determined by the specific value of the reactive component jXT, strategically aligned at the filter's passband edge. This alignment is essential for achieving maximum frequency selectivity and a reduction in phase noise, which are critical factors in oscillator design [38].

At frequencies near this peak, the reactively terminated active filter demonstrates behavior akin to that of a high-Q LC resonator imbued with shunt negative resistance [1]. The presence of negative resistance facilitates sustained oscillations by compensating for energy losses in the system. This resonator is further interfaced with an external load through a meticulously designed matching network, which adheres to the principles delineated in established device-line theory. This design approach is key to maximizing the output power of the oscillation, ultimately enhancing the overall efficiency of the oscillator [1, 3]. In Figure 6.24, a photograph of the compact prototype oscillator is displayed, revealing the intricate layout that includes a high-Q dual-mode active filter, a reactive termination component, and an external load matching network [1]. The physical dimensions of the resonator are notably compact, measuring just 7×7 mm^2, which underscores the effectiveness of the design in achieving miniaturization without compromising performance [1, 13, 17].

Figure 6.24 illustrates the oscillator configuration in which two NEC NESG2030M04 SiGe HBTs are integrated into the feedback path. Each device provides a gain of approximately 8 dB with a noise figure around 2 dB, enabling the generation of effective negative resistance to counteract intrinsic resonator losses [17, 40]. Frequency agility is realized through the use of hyper-abrupt junction varactor diodes (MA46H120), whose capacitance varies between 0.2 and 1 pF. These varactors are coupled to the dual-mode resonator structure depicted in Figure 6.25 [1], allowing fine control over the oscillation frequency [1, 17]. Experimental measurements confirm that the oscillator exhibits a tuning range extending from 8.150

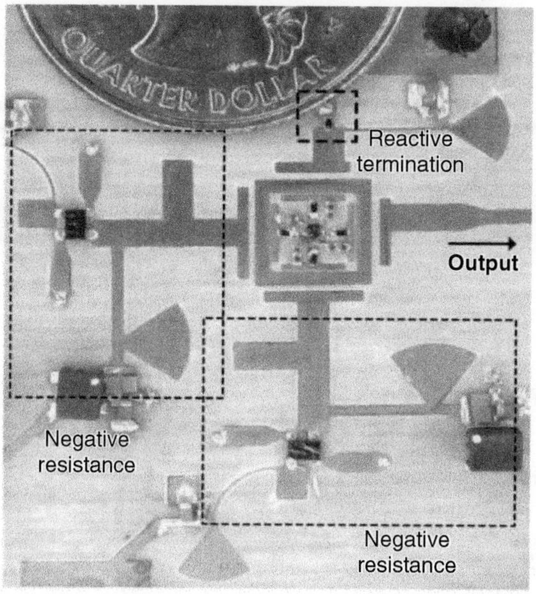

Figure 6.24 The photo displays the tunable oscillator operating between 8.15 and 8.25 GHz [1].

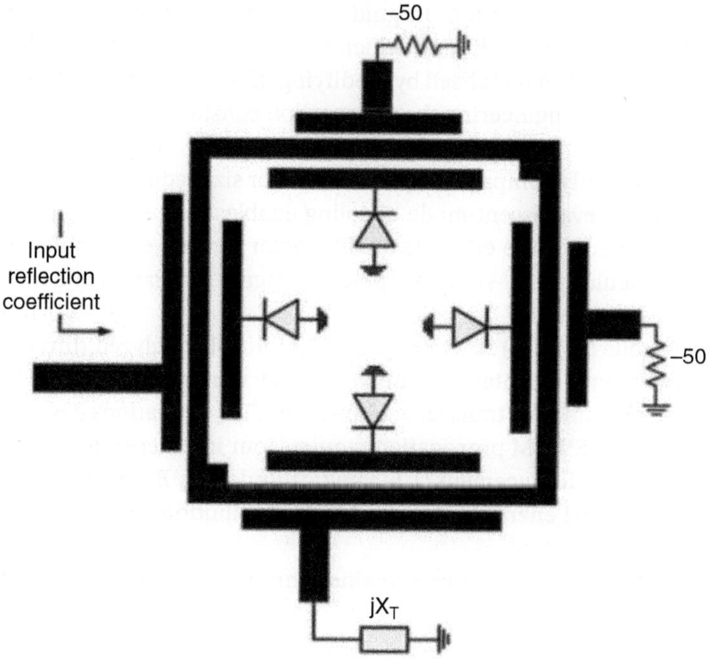

Figure 6.25 Schematic of the resonator with varactor diodes incorporated for frequency tuning [1].

to 8.255 GHz, while consistently producing an output power near 7 dBm. The corresponding DC-to-RF power conversion efficiency is approximately 12.5%. The phase-noise performance, presented in Figure 6.26, shows suppression to better than −149 dBc/Hz at a 1 MHz offset from the carrier [39]. The flicker noise corner occurs at about 90 kHz, beyond which the noise floor stabilizes.

At a nominal oscillation frequency of 8.2 GHz, the VCO achieves a continuous tuning span of roughly 105 MHz, sustaining a 7 dBm output level with 12.5% efficiency. Based on these parameters, the FOM is calculated to be −211.7 dBc/Hz, underscoring the high spectral purity and efficiency of the active-resonator design [13, 17].

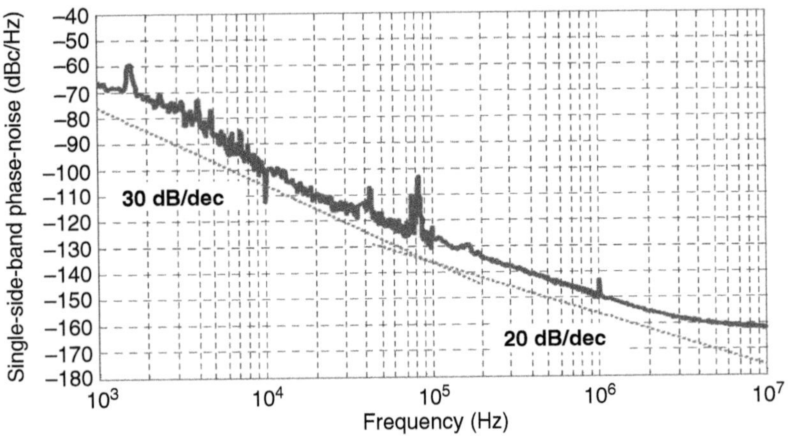

Figure 6.26 Measured phase-noise performance of the oscillator using a dual-mode resonator at 8.2 GHz [1, 10].

6.4 Printed Coupled Slow-wave Resonator Oscillators

As reported by Poddar in his doctoral research [10], a printed coupled slow-wave resonator (SWR) oscillator was introduced, demonstrating a compact design that achieved wide frequency tuning capability while maintaining excellent phase-noise performance. The concept of slow-wave propagation underpins this innovation. In guided-wave media, electromagnetic signals propagate with a reduced phase velocity compared to free-space conditions, which effectively shortens the guided wavelength at a specified operating frequency [12, 22]. This phenomenon is realized by modifying the balance of stored electric and magnetic field energy within the guiding structure, thereby engineering the propagation constant of the medium [43–50].

SWRs offer distinct advantages in oscillator design. Their inherently compact footprint allows for size reduction without sacrificing resonant characteristics, while their ability to support evanescent-mode coupling enables the realization of a Q-multiplier effect under resonance conditions. This enhancement of the effective quality factor directly contributes to superior phase-noise suppression, making SWR oscillators particularly well-suited for precision signal generation and low-noise source applications in advanced communication and radar systems.

The electromagnetic behavior in a slow-wave evanescent-mode (SWEM) resonator can be rigorously analyzed through Maxwell's equations, which may be decomposed into two complementary formulations: (i) transmission-line (TL) equations governing wave propagation along the longitudinal axis, and (ii) transverse evanescent (TE_V) equations describing the orthogonal field components. A complete representation of SWEM propagation requires four interdependent pairs of TL and TE_V relations corresponding to the electric-field E, electric displacement D, magnetic flux density B, and magnetic field intensity H. Together, these coupled equations define the radial energy storage and field distribution characteristics of the resonator [12, 46–48].

The emergence and evolution of evanescent waves in SWEM structures can be explained through a sequence of seven fundamental processes [43–50]:

- Spreading function—the spatial expansion of fields away from the guiding path,
- Self-coupling function—interaction of the resonator's own fields within the structure,
- Radial standing-wave formation—development of localized oscillatory field patterns in the radial direction,
- Temporal phase variation of stored energy—modulation of electric and magnetic energy with respect to time,
- Root-sum-square (RSS) field superposition—vector combination of the coupled TL–TE_V equations, dominated by the strongest resonant field component,
- Dissipation, radiation, or absorption of energy—mechanisms contributing to intrinsic losses within the resonator,
- Inter-modal energy exchange—transfer of stored energy between coupled evanescent modes.

These processes collectively determine the field dynamics, stored energy distribution, and resonance properties of SWEM resonators, directly influencing their quality factor and suitability for compact, low phase-noise oscillator applications.

Figure 6.27a–c depicts both the system-level block diagram and the fabricated layout of a VCO employing a mode-coupled SWR topology [51]. In this design, a SiGe HBT serves as the active device, integrated with the SWR to realize a compact,

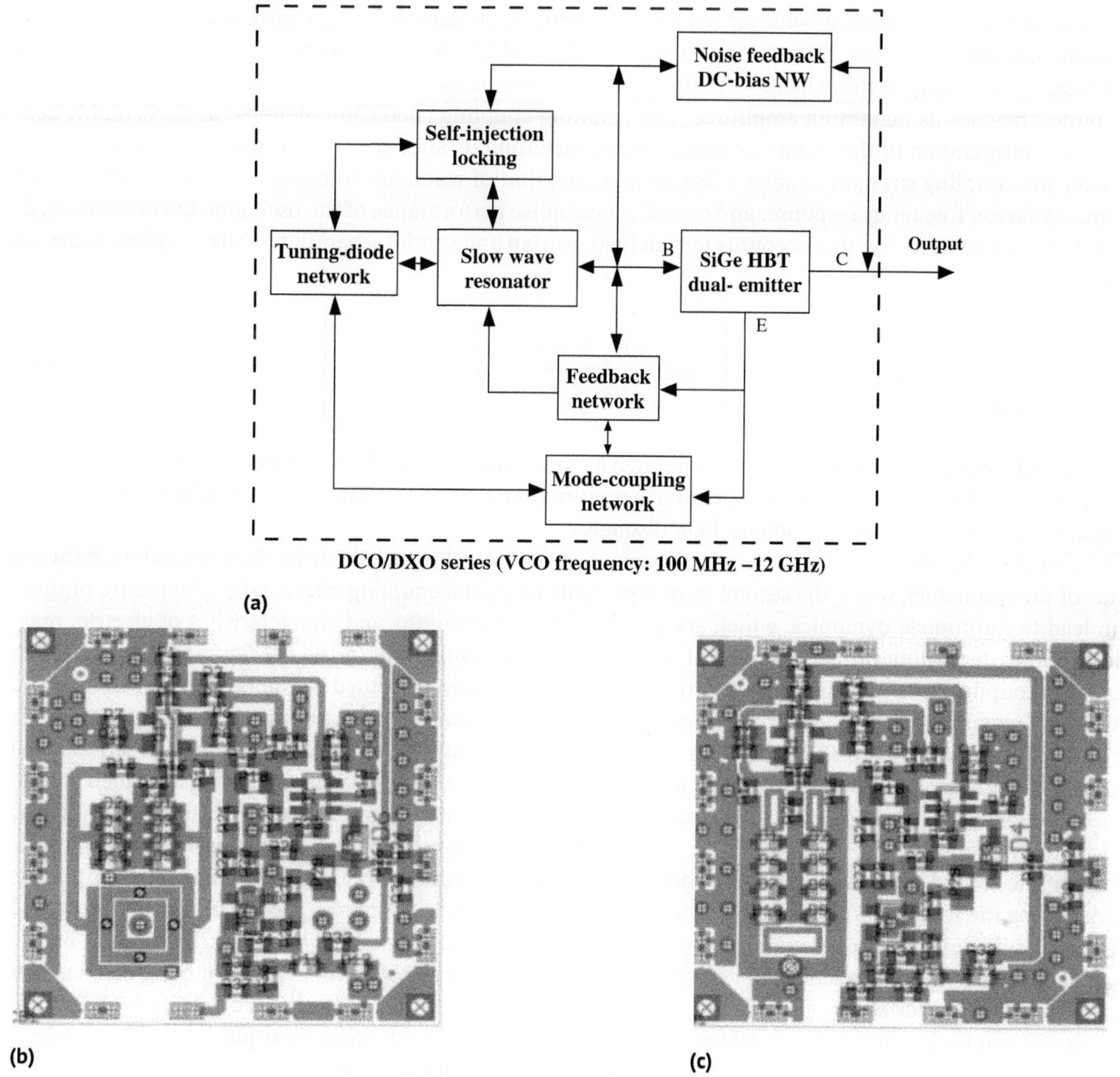

Figure 6.27 (a) Block diagram of a standard SWR-based VCO, (b) physical layout of the SWR VCO, and (c) layout of a mode-coupled SWR VCO [10, 51].

tunable oscillator. The circuit is implemented on a low-loss RF dielectric substrate characterized by a relative permittivity of $\varepsilon_r = 3.38$ and a thickness of 22 mils [10, 52], providing low dissipation and stable high-frequency performance.

To accurately predict and optimize the oscillator's behavior, the SWR geometry is first analyzed with a full-wave three-dimensional electromagnetic (3D EM) simulator. The resonator model is then incorporated into a nonlinear oscillator design environment, where biasing and feedback conditions are adjusted for low phase-noise operation across the target frequency range [51]. The SWR establishes well-defined standing-wave distributions and simultaneously regulates the noise impedance transfer function throughout the tuning band. This is achieved through dual-parameter optimization: m_{opt}, obtained via injection-locking adjustment, and β_{opt}, determined through mode-tuning optimization [15, 52–58].

The effective quality factor (Q) of the printed transmission-line resonator can be enhanced by exploiting inter-resonator coupling mechanisms. Depending on the spatial orientation and field overlap, these mechanisms may be predominantly electric, magnetic, or a hybrid combination of both [51]. For comparative analysis, Figure 6.18 illustrates several planar resonator network layouts and corresponding oscillator topologies based on electric, magnetic, and hybrid coupling schemes [10, 51, 59–64]. These variations enable evaluation of coupling-dependent Q-enhancement and their impact on oscillator stability, phase-noise reduction, and frequency agility.

As depicted in Figure 6.27, the coupling behavior between adjacent resonators is strongly governed by the proximity effect. This phenomenon arises from the interaction of fringing fields, which decay exponentially outside the resonator boundaries. The localized intensity of these fields—whether electric or magnetic—tends to peak along the regions where the modal distribution reaches its maximum amplitude. The resulting coupling coefficient, denoted as β_j, is highly sensitive to the physical configuration of the resonator perturbation, including its size, shape, and relative orientation. This precise control over the coupling strength enables effective manipulation of resonator interactions, which is critical for optimizing the quality factor, frequency response, and overall phase-noise performance of the oscillator. Quantitatively, β_j can be expressed through a formulation that accounts for the field overlap and spatial separation of the coupled modes, as detailed in [10].

$$\beta_j = \left[\left(\frac{\int \varepsilon E_a . E_b dv}{\sqrt{\int \varepsilon E_a^2 dv \int \varepsilon E_{bd}^2 dv}} \right)_{\text{electrical-coupling}} + \left(\frac{\int \mu H_a . H_b dv}{\sqrt{\int \mu H_a^2 dv \int \mu H_b^2 dv}} \right)_{\text{magnetic-coupling}} \right] \tag{6.37}$$

where (E_a, H_a) refer to the electric and magnetic fields generated by the primary square-loop resonator. In contrast, (E_b, H_b) denote the corresponding fields that arise either as a result of a perturbation (where $d \neq 0$) or from the influence of a nearby adjacent resonator, specifically the secondary square-loop resonator.

From Eq. (6.37), we can identify that the first term corresponds to the coupling arising from the interactions between the electric fields of the resonators, while the second term represents magnetic coupling effects [10]. The nature of these interactions can lead to multimode dynamics, which are governed by the strengths and characteristics of electric, magnetic, or hybrid coupling depending on the operational conditions of the resonators. In the configuration illustrated in Figure 6.28a, electric coupling is the dominant interaction. This dominance is attributed to the behavior of the electric field, which attains its maximum intensity near the open ends of the resonators. This positioning effectively maximizes the numerator of the first term in Eq. (6.37), enhancing the overall electric coupling strength and leading to pronounced interaction effects [10]. Conversely, Figure 6.28b depicts a scenario in which the resonators are operating near their fundamental resonance frequency. Under these conditions, magnetic coupling becomes more pronounced. The intensity of the magnetic field is notably stronger at the midpoint of the resonator—situated opposite to the open ends—thereby optimizing the numerator of the second term in Eq. (6.37) [10, 40]. This shift in dominance from electric to magnetic coupling is critical in understanding the resonators' dynamic behavior at different frequencies (Figure 6.28c–e). In the case presented in Figure 6.28f, the coupling interaction between resonator #1 and resonator #2 is classified as hybrid. This hybrid nature arises because neither electric-field interactions nor magnetic-field interactions solely dominate the coupling mechanism. Instead, the interaction is characterized by a balanced contribution from both electric and magnetic fields, which complicates the dynamics and can lead to rich multimodal behaviors (6.28g). Such nuanced interactions play a significant role in tuning the resonators' responses and optimizing their performance for various applications.

The formal definition of β_j in (6.37) requires detailed electromagnetic analysis, which is often complex and impractical for routine calculations. A practical alternative is provided by coupled-resonator theory, where the resonators oscillate together at shifted normal-mode frequencies (f_{ee}, f_{em}, f_{eh}, f_{mh}) that differ from the uncoupled resonant frequency f_0 [10]. These mode frequencies are associated with the respective electric, magnetic, and hybrid coupling mechanisms, and their separation increases with the strength of the coupling coefficient β_j (β_e for electric, β_m for magnetic, β_h for hybrid) [10, 18, 59–64].

The primary mechanism governing energy exchange between adjacent resonators is proximity coupling, which arises from the overlap of their fringing electromagnetic fields. This interaction is quantitatively described by the coupling coefficient, β_j, defined as the ratio of the energy transferred between resonators to the total energy stored within the resonant structure [10]:

$$\beta_e \cong \frac{\text{coupled} - \text{electrical energy}}{\text{stored} - \text{energy of uncoupled} - \text{resonator}} \cong \frac{f_{me}^2 - f_{ee}^2}{f_{me}^2 + f_{ee}^2} \cong \frac{C_{me}}{C} \tag{6.38}$$

$$\beta_m \cong \frac{\text{coupled} - \text{magnetic energy}}{\text{stored} - \text{energy of uncoupled} - \text{resonator}} \cong \frac{f_{em}^2 - f_{mm}^2}{f_{em}^2 + f_{mm}^2} \cong \frac{L_{mm}}{L} \tag{6.39}$$

$$\beta_h \cong \frac{\text{coupled} - \text{electro} - \text{magnetic energy}}{\text{stored} - \text{energy of uncoupled} - \text{resonator}} \cong \frac{f_{eh}^2 - f_{mh}^2}{f_{eh}^2 + f_{mh}^2} \cong \frac{CL_{mh} + LC_{mh}}{LC + L_{mh}C_{mh}} \tag{6.40}$$

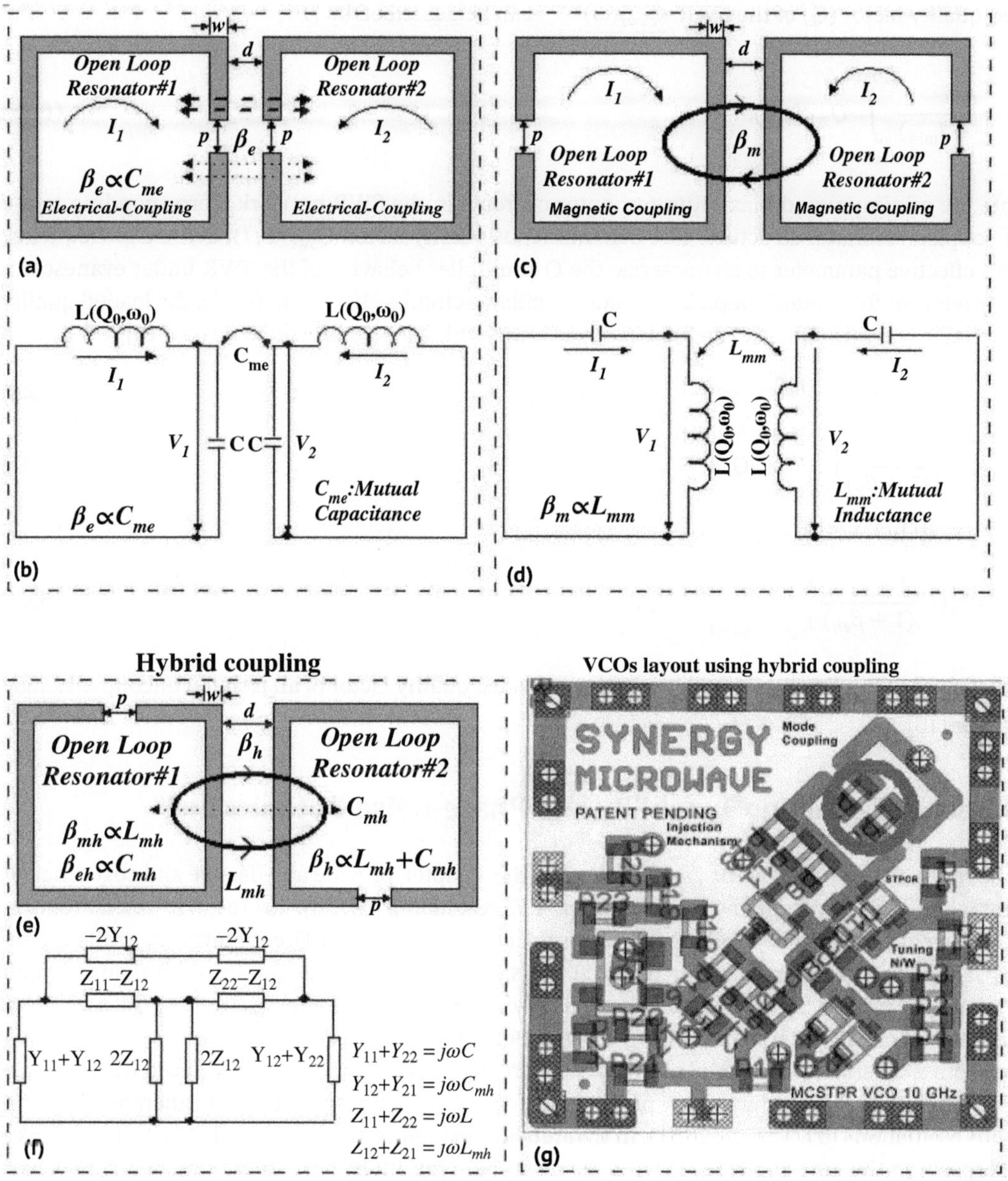

Figure 6.28 Simplified views of open-loop microstrip line coupled-resonator networks: (a) electric coupling, (b) lumped-element representation of electric coupling, (c) magnetic coupling, (d) lumped-element model of magnetic coupling, (e) hybrid coupling, (f) lumped-element model of hybrid coupling, and (g) layout of a VCO employing both electric and magnetic coupling [10].

where

$$f_{ee} = \frac{1}{2\pi\sqrt{L(C + C_{me})}}, f_{me} = \frac{1}{2\pi\sqrt{L(C - C_{me})}}, \quad C_{me} : \text{Mutual capacitance} \tag{6.41}$$

$$f_{em} = \frac{1}{2\pi\sqrt{C(L - L_{mm})}}, f_{mm} = \frac{1}{2\pi\sqrt{C(L + L_{mm})}}, \quad L_{mm} : \text{Mutual inductance} \tag{6.42}$$

$$f_{eh} = \frac{1}{2\pi\sqrt{(L - L_{mh})(C - C_{mh})}}, f_{mh} = \frac{1}{2\pi\sqrt{(L + L_{mh})(C + C_{mh})}}, \quad L_{mh} : \text{Hybrid inductance} \tag{6.43}$$

$$f_0 = \frac{1}{2\pi\sqrt{LC}}, f_0 : \text{fundamental resonance frequency of uncoupled resonator} \tag{6.44}$$

The time-averaged loaded quality factor (Q) of the SWR: $\underline{Q_{swr}(\omega)}_{\omega\to\omega_0}$ can be described by [10]

$$\underline{Q_{swr}(\omega)}_{\omega\to\omega_0} = \left[\frac{\omega}{2(I_{max}-I_{min})}\int_{I_{min}}^{I_{max}} Q_{swr}(\omega.i)di\right]_{\omega\to\omega_0} \tag{6.45}$$

Here, I_{min} and I_{max} denote the minimum and maximum resonator currents in the SWR network, corresponding to the fundamental modes of the coupled-resonator structure. The instantaneous quality factor, $Q_{swr}(\omega, i)$, defined at frequency ω and current i, serves as an effective parameter to characterize the Q-multiplier behavior of the SWR under evanescent-mode coupling—particularly relevant for printed coupled-resonator oscillator circuits [10]. From (6.45), the loaded quality factor Q_L of the coupled-resonator network can be expressed in terms of the unloaded quality factor Q_o as [10]:

$$Q_L(\omega_0) = \frac{\omega_0}{2}\left[\frac{\partial\phi}{\partial\omega}\right] \tag{6.46}$$

$$[Q_L(\omega_0)]_{electrical-coupling} \cong 2\left[\frac{Q_0}{(1+\beta_e)}\right]_{\beta_e \ll 1} \cong 2Q_0 \tag{6.47}$$

$$[Q_L(\omega_0)]_{magnetical-coupling} \cong 2[Q_0(1+\beta_m)]_{\beta_m\to 1} \cong 2Q_0 \tag{6.48}$$

$$[Q_L(\omega_0)]_{hybrid-coupling} \cong 2\left[Q_0\frac{(1+\beta_{mh})}{(1+\beta_{eh})}\right]_{\beta_e\ll 1,\, \beta_m\to 1} \cong 2Q_0 \tag{6.49}$$

where $\frac{\partial\phi}{\partial\omega}$ represents the rate of phase variation, and Q_0 denotes the unloaded quality factor of an isolated uncoupled single open-loop microstrip resonator [10].

6.5 Oscillator Architectures Enabling Tunability and Phase-noise Optimization

Tunable oscillator circuits based on SWR [10] networks, as examined in this chapter, provide an effective and cost-efficient alternative to conventional high-Q oscillators employing dielectric or YIG resonators [49, 50]. These SWR-based architectures are capable of delivering frequency stability and tunability comparable to their high-Q counterparts, while simultaneously offering reduced fabrication complexity, lower material costs, and simplified integration with planar circuit technologies. The compact nature of printed SWR networks, combined with their ability to support evanescent-mode coupling, enables efficient miniaturization without compromising spectral performance.

Figure 6.29a depicts a conceptual schematic of a typical oscillator comprising a resonator module interfaced with external circuitry. The active component within this configuration provides the necessary gain to counteract inherent resonator losses and sustain continuous oscillations [39, 43–50, 52–54]. In theoretical treatments, the resonator is often modeled as a linear system, with the superposition theorem applied to analyze its response to multiple excitation frequencies. However, in real-world implementations, high-Q resonators—whether they are quartz, ceramic, dielectric, or MEMS-based—exhibit nonlinear behavior that is strongly dependent on the amplitude of the driving signal [10, 65–70].

These nonlinearities can lead to the undesired folding of low-frequency noise components into the oscillator's carrier sidebands, a phenomenon commonly referred to as noise aliasing, as illustrated in Figure 6.29b. In some cases, this aliased noise may exceed the thermal noise floor of high-Q piezoelectric quartz resonators [10], degrading spectral purity and phase-noise performance. To mitigate these effects, resonator linearization techniques are often employed, ensuring that precision frequency sources maintain optimal stability, minimal sideband noise, and high signal integrity across the operational bandwidth [65–69]. The combination of SWR-based miniaturization, tunable operation, and enhanced phase-noise control establishes these designs as a practical solution for modern communication systems, radar modules, and high-performance signal generators, where compactness and low-cost implementation are critical.

The tunable oscillators analyzed in this chapter are implemented using readily available discrete silicon (Si) and silicon-germanium (SiGe) HBTs in combination with abrupt and hyper-abrupt junction varactor diodes, enabling wideband frequency tuning [10]. Although transistor S-parameters supplied by manufacturers provide reliable characterization under small-signal conditions, oscillator circuits inherently operate in the large-signal regime, where device nonlinearities play a dominant role. Direct extraction of large-signal parameters typically requires advanced instrumentation and expensive measurement setups; as a practical alternative, synthesis-driven modeling and estimation techniques are often employed

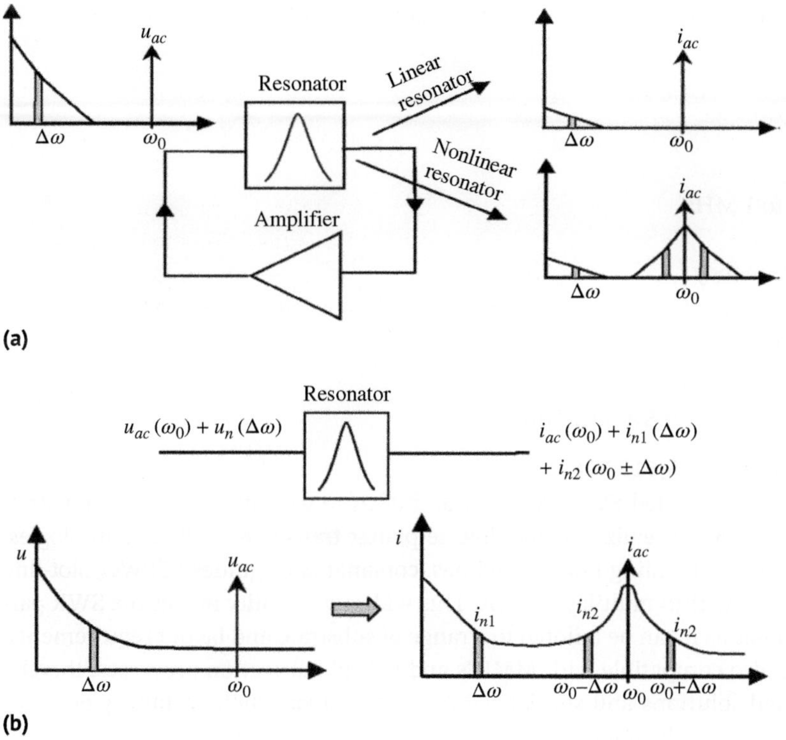

Figure 6.29 Noise aliasing in nonlinear resonator-based oscillators: (a) schematic of an oscillator incorporating a nonlinear resonator, and (b) low-frequency noise $u_n(\Delta\omega)$ at the filter input being aliased to carrier sidebands $(\omega_0 + \Delta\omega)$ due to nonlinear mixing within the resonator network [10, 65].

to approximate these nonlinear characteristics with sufficient accuracy for circuit design. The oscillator prototypes presented in this study exhibit notable enhancements in FOM, balancing phase-noise suppression, broad tuning capability, and optimized power consumption [10]. However, direct comparison of phase-noise performance across different oscillator topologies is only meaningful when evaluated under identical conditions—specifically, the same carrier frequency, frequency offset, tuning range, and output power [10]. To enable fair benchmarking across diverse oscillator architectures and operating regimes, a unified FOM expression is therefore adopted as a standardized performance metric [22].

The FOM (in dBc/Hz) and the power–frequency tuning–normalized metric (PFTN, in dB) provide standardized measures for evaluating oscillator performance across different designs and operating conditions. These metrics are mathematically defined as [71]:

$$\text{FOM}\Big|_{f_{\text{offset}}} = \left[\pounds(f_{\text{offset}}) - 20\left(\frac{f_0}{f_{\text{offset}}}\right) + 10\left(\frac{P_{\text{DC}}}{1\,\text{mW}}\right)\right]\left(\frac{\text{dBc}}{\text{Hz}}\right) \tag{6.50a}$$

$$\text{PFTN} = -\left[\pounds(f_{\text{offset}}) - 20\left(\frac{\Delta f}{f_{\text{offset}}}\right) - 10\left(\frac{P_{\text{DC}}}{kT}\right)\right](\text{dB}) \tag{6.50b}$$

where f_0 is the operating oscillator frequency, $\pounds(f_{\text{offset}})$ is the phase-noise at the offset frequency f_{offset}, k is the Boltzmann constant, $\Delta f = f_{\max} - f_{\min}$ is the tuning range, T is the temperature in Kelvin, and P_{DC} is the total DC power consumption in milliwatts [41]. Larger absolute values of $|\text{FOM}|\left(\frac{dBc}{\text{Hz}}\right)$ and PFTN (dB) correspond to superior oscillator performance, reflecting lower phase noise, higher efficiency, and broader tuning capability.

Additionally, from (6.50a) and (6.50b), the integrated FOM over the range 1 kHz–1 MHz can be expressed as $\text{FOM}|_{\text{Integrated(1 kHz–1 MHz)}}$ [22]:

$$\text{FOM}|_{\text{Integrated (1 kHz–1 MHz)}} = 10\log(P^2(\varnothing)) + 10\log\left(\frac{P_{\text{RF}}}{P_{\text{DC}}}\right) - 20\log\left(\frac{2\Delta f}{f_{\max} + f_{\min}}\right) \tag{6.51}$$

where

$$\Delta f = f_{\max} - f_{\min}$$

f_{max} = maximum oscillation frequency

f_{min} = minimum oscillation frequency

$P^2(\emptyset)$ = integrated phase noise from 1 kHz to 1 MHz

P_{RF} = signal output power averaged over frequency

P_{DC} = DC power consumption of the oscillator

6.5.1 Practical Examples: Tunable Oscillators Based on SWR Circuits

6.5.1.1 Tunable VCO (2000–3200 MHz) (US Patent No. 7,365,612 B2)

As reported by Poddar [10, 72], an innovative printed coupled SWR oscillator architecture was introduced, demonstrating that a compact and efficient resonator design could be realized using diverse planar transmission-line technologies. The SWR concept supports multiple implementations, including microstrip lines, coplanar waveguides (CPWs), slot-line structures, as well as embedded resonators integrated within multilayer PCBs. This wide adaptability makes the SWR particularly attractive for practical oscillator applications, as it can be tailored to a range of substrate and layout requirements.

Beyond PCB-based realizations, the topology is also compatible with MMICs and hybrid fabrication processes, thereby extending its utility to both high-volume integrated solutions and specialized discrete designs. Such versatility provides designers with the freedom to optimize performance while balancing fabrication complexity and cost.

A representative example of this methodology is illustrated in Figure 6.30, which presents a tunable oscillator that employs a dynamically tuned microstrip-line coupled resonator functioning as a slow-wave element. In this configuration, the resonator is electronically synchronized with the VCO tuning port, ensuring that the resonant response tracks the oscillator's frequency variation. This electronic synchronization plays a pivotal role in minimizing phase deviations, thereby stabilizing the oscillation condition and significantly reducing phase noise. The resulting oscillator structure offers high spectral purity and excellent frequency stability, making it a compelling solution for advanced communication systems and precision frequency source applications [10, 72].

A key advantage of the distributed resonator–based configuration lies in its inherent capability to achieve ultra-low phase-noise performance without relying on expensive high-Q dielectric resonators or YIG resonator networks, which are traditionally used in precision oscillators. This design approach circumvents the need for specialized and often bulky tuning mechanisms while still providing wide frequency coverage, making it both technically efficient and economically practical. The implementation leverages the integration of an open-strip element at the electrical midpoint of the coupled-line resonator, which serves two critical functions: (i) it enables the realization of the distributed resonator effect, and (ii) it preserves a wide tuning range while simultaneously enhancing the effective quality factor (Q) of the overall resonant structure. This dual optimization ensures that the oscillator maintains high frequency stability and spectral purity across its entire tuning band.

Compared to conventional oscillator designs that employ high-Q dielectric or YIG resonators, this approach offers several distinct benefits: reduced system complexity, improved frequency agility, and enhanced manufacturability using standard planar fabrication processes. These advantages make the architecture particularly well-suited for modern communication and radar systems requiring compact, tunable, and low-noise frequency sources. The practical realization of this methodology is presented in Figure 6.31a,b [10], which illustrates the corresponding circuit schematic and the physical layout, respectively, demonstrating the effectiveness of the distributed resonator configuration in practical oscillator design.

The oscillator layout illustrated in Figure 6.31b [10] demonstrates highly stable functionality across the extended temperature range of –40 °C to +85 °C. This robustness provides ample margin to counteract performance degradations arising from thermal frequency drift, parasitic effects introduced by the package, and inherent tolerances of the discrete components. The measured phase-noise characteristics, presented in Figure 6.32, highlight the oscillator's capability to maintain superior spectral purity, achieving values consistently better than –105 dBc/Hz at a 10 kHz frequency offset throughout its wide tuning span of 1900–3200 MHz [9]. Furthermore, the oscillator sustains an output power level exceeding +3 dBm while operating with a total power consumption of 240 mW under the bias condition of $V_{CC} = 8$ V and $I_c = 30$ mA [10].

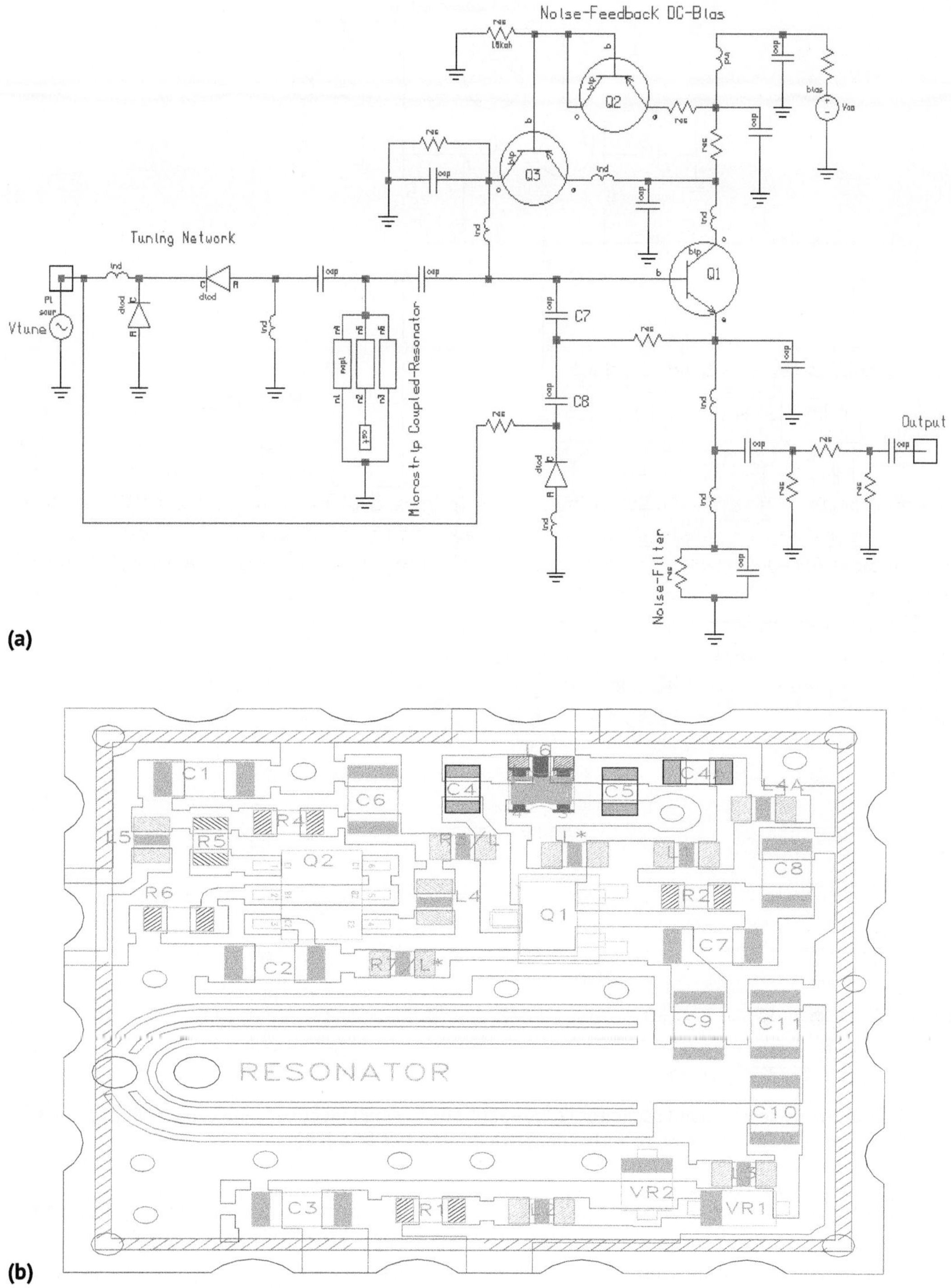

Figure 6.30 (a) Schematic of the 2000–3200 MHz oscillator, and (b) layout implemented on a 32-mil substrate with a dielectric constant of 3.38 [40, 72].

Quantitative assessment of oscillator performance is further emphasized by the calculated FOM, obtained from Eq. (6.50), which yields −194.49 dBc/Hz [10, 40]. And, the PFTN metric, derived from Eq. (6.51), is determined as 15.08 dB [10]. These benchmarks provide valuable performance indicators, with FOM serving as a widely recognized reference standard in evaluating oscillator phase-noise efficiency relative to operating frequency and power dissipation.

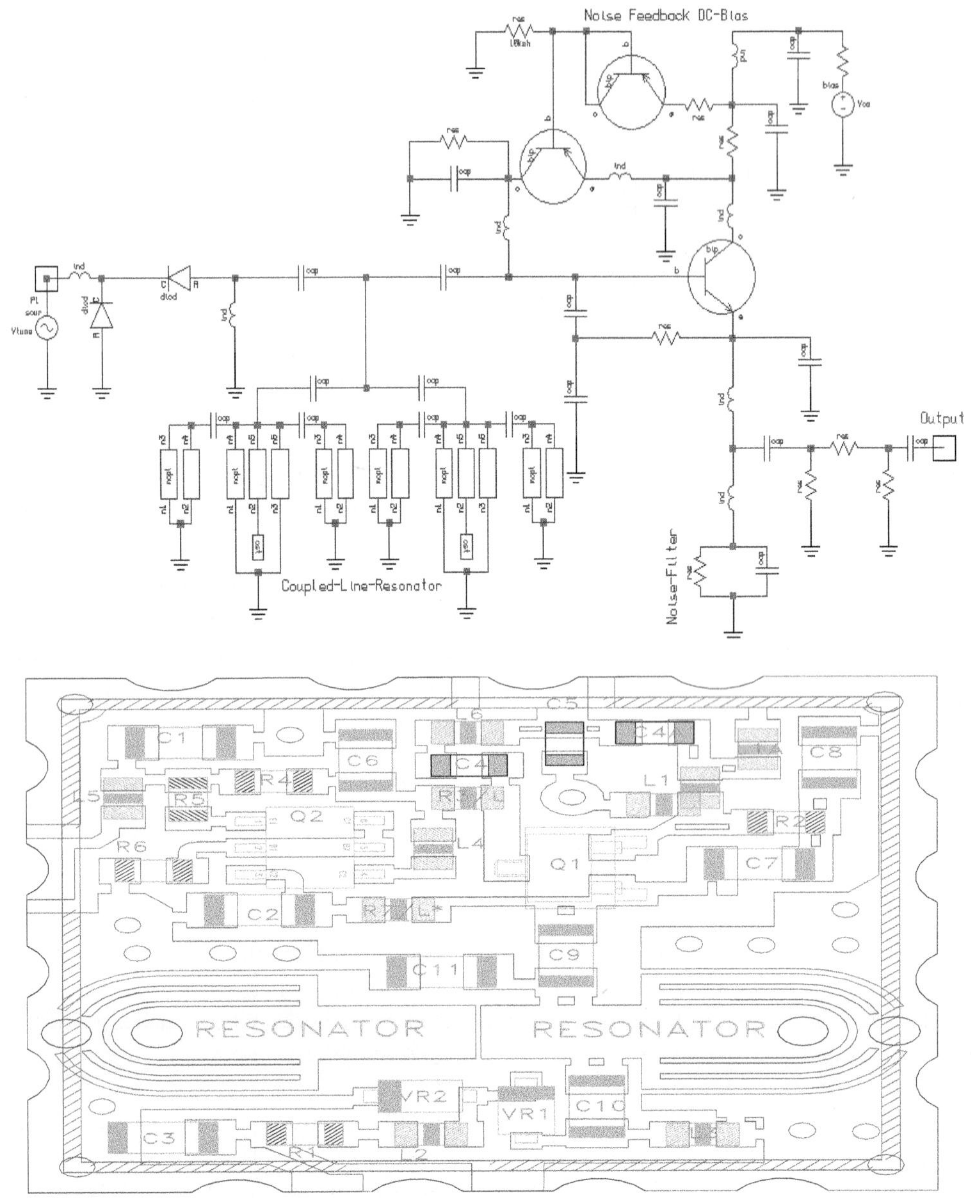

(a)

(b)

Figure 6.31 (a) Schematic of an ultra-low-noise wideband oscillator (2000–3200 MHz) using inductively and capacitively coupled microstrip-line resonators, and (b) layout on a 32-mil substrate with a dielectric constant of 3.38 [72].

Complementary designs are depicted in Figure 6.33a,b [10], which illustrate the schematic and physical layout, respectively, of a wideband, power-efficient slow-wave resonator oscillator (SWRO) covering the frequency range of 2000–3000 MHz [41]. The implementation adopts a current-efficient circuit architecture based on a hybrid configuration incorporating both common-collector and common-emitter stages. This cascaded arrangement generates the required negative resistance, with both active devices sharing a unified bias current to optimize efficiency. When biased at 10 V with a supply current of 15 mA, the circuit described in Figure 6.33a achieves remarkable performance, with phase-noise measurements demonstrating typical values better than –156 dBc/Hz at a 1 MHz frequency offset, as reported in Figure 6.34. The oscillator simultaneously provides an output power level above –3 dBm, affirming its suitability as a compact, power-conscious, wideband solution for advanced RF and microwave systems.

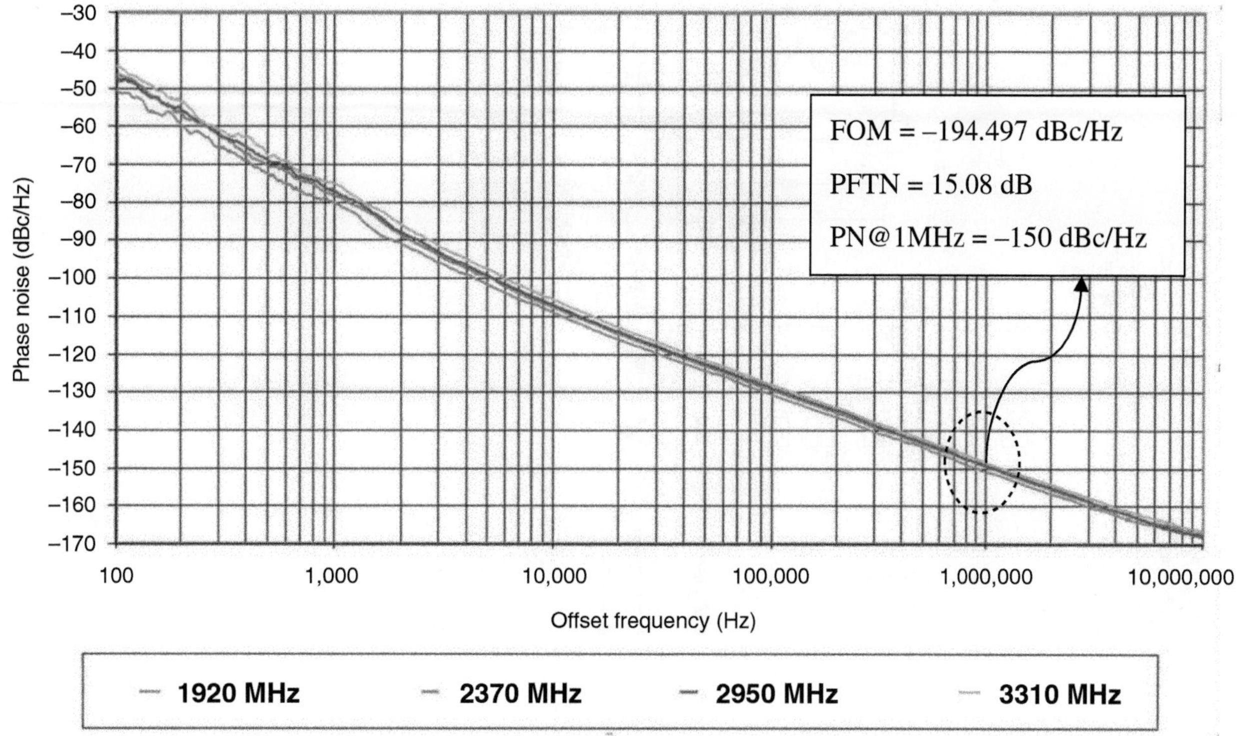

Figure 6.32 Measured phase noise of the oscillator circuit depicted in Figure 6.22a, with a FOM (Eq.6.50) of −194.49 dBc/Hz and a PFTN (Eq.6.51) value of 15.08 dB [10, 72].

6.5.1.2 Multi-octave Band SWRO Circuit (US Patent No. 7,605,670)

Poddar [10, 76] introduced a novel multi-coupled slow-wave resonator (MCSWR) VCO architecture that is inherently planar and wideband, thereby offering significant advantages for cost-effective monolithic microwave integrated circuit (MMIC) implementation. Unlike conventional oscillator designs that often rely on bulky and expensive discrete resonators such as YIG spheres, the proposed MCSWR-based approach enables the realization of compact, lithographically defined resonators that are fully compatible with integrated circuit fabrication processes. This architecture not only supports wide operational bandwidths but also incorporates high-quality-factor planar resonators that play a critical role in achieving low phase-noise performance [11]. Furthermore, because the fabrication is compatible with standard IC and MMIC technologies, the design lends itself to scalability and mass production [76–81]. Owing to these advantages, MCSWR VCOs demonstrate strong potential for deployment in diverse broadband communication platforms, including advanced microwave communication systems, precision test and measurement instrumentation, radar systems, local multipoint distribution systems (LMDS), and multichannel multipoint distribution systems (MMDS).

Figure 6.35 [10] illustrates the block diagram of a multi-octave-band oscillator that employs printed multi-mode, progressive-wave, and slow-wave coupled resonators to achieve ultra-low phase noise and broadband frequency coverage [76]. The design exploits the Q-multiplier effect generated by evanescent-mode progressive delay [10], which effectively enhances the time-averaged loaded quality factor (Q) of the planar resonator throughout the operational frequency range. To ensure that the oscillator maintains a stable and uniform negative resistance across its wide tuning span, a varactor-tuned coupled resonator is strategically connected between the base and collector terminals of the active device [10]. This configuration compensates for inherent resonator losses by providing dynamically adjustable negative resistance, thereby preserving high-Q performance. Additionally, the phase shift introduced by the negative-resistance network is continuously tuned across the frequency band to satisfy the resonance condition, ensuring reliable multi-octave operation with superior spectral purity [40].

In the configuration shown in Figure 6.35, the variable coupling capacitor (C_c) is carefully optimized to ensure appropriate loading of the coupled-resonator network across the active device and is dynamically adjusted to achieve peak oscillator performance [10]. The oscillator employs a three-terminal active device, such as the NEC 68830 bipolar transistor, in conjunction with both a noise-filtering network and a noise-cancellation network connected in parallel between the base and

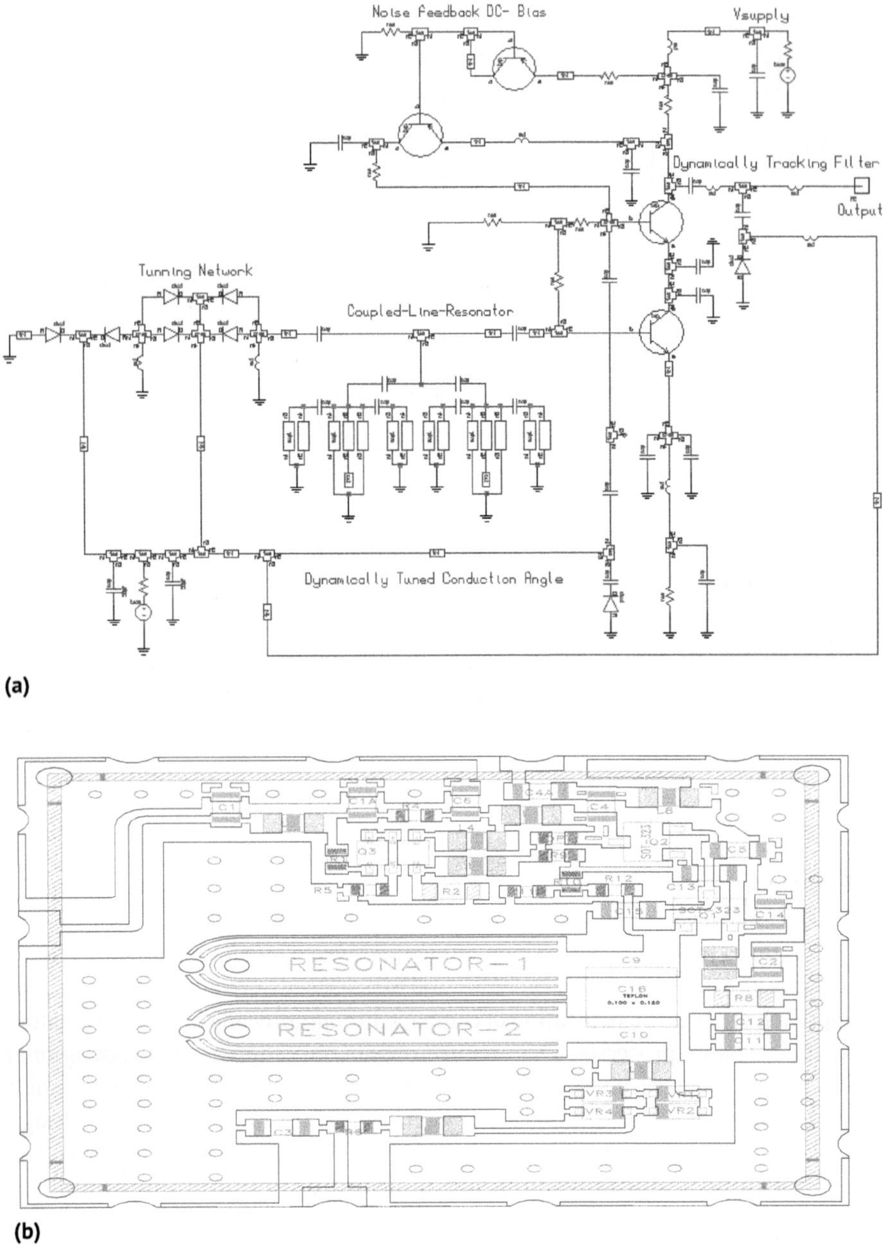

(a)

(b)

Figure 6.33 Power-efficient wideband SWRO circuit: (a) cascode configuration schematic, and (b) layout on a 32-mil substrate with dielectric constant 3.38 [10, 72, 73].

collector terminals. To further enhance stability, an active feedback network is incorporated, which includes a regulated DC bias supply designed to compensate for variations in bias voltage arising from temperature fluctuations in the operating environment [85].

In the MCSWR oscillator, the three-terminal active device generates an effective impedance that exhibits a negative real component accompanied by a frequency-dependent imaginary component. The negative real part provides the necessary energy to compensate for intrinsic losses within the MCSWR network, thereby sustaining oscillation. Its magnitude, however, is intrinsically linked to the imaginary component, which is carefully tuned to coincide with the maximum-slope inflection point of the oscillator's phase response [10]. This design choice ensures optimal group delay characteristics, directly enhancing phase stability and reducing spectral noise. In essence, the interplay between the real and imaginary components of the active impedance governs both the oscillation condition and the efficiency with which resonator losses are offset.

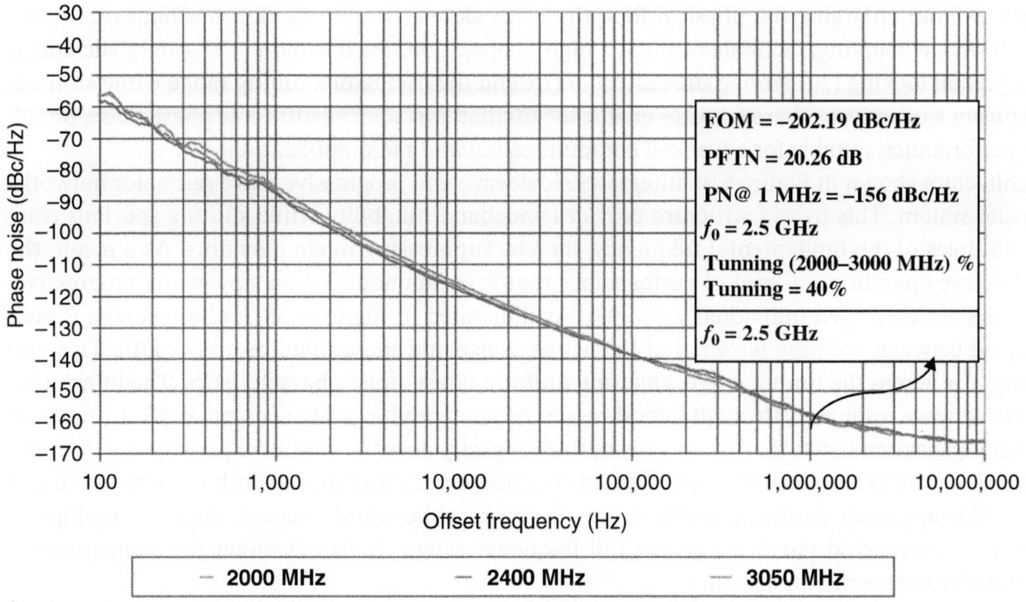

Figure 6.34 Measured phase-noise of a power-efficient wideband SWRO (2–3 GHz). Poddar [74] reports that *"the figure-of-merit (FOM, Eq. 6.50) is −202.19 dBc/Hz, and the power-frequency tuning-normalized metric (PFTN, Eq. 6.51) is 20.26 dB."* The circuit exhibits efficient operation, consuming 150 mW of power (V_{CC} = 10 V, I_c = 15 mA) and delivering an output power of 0 dBm [10, 72, 75].

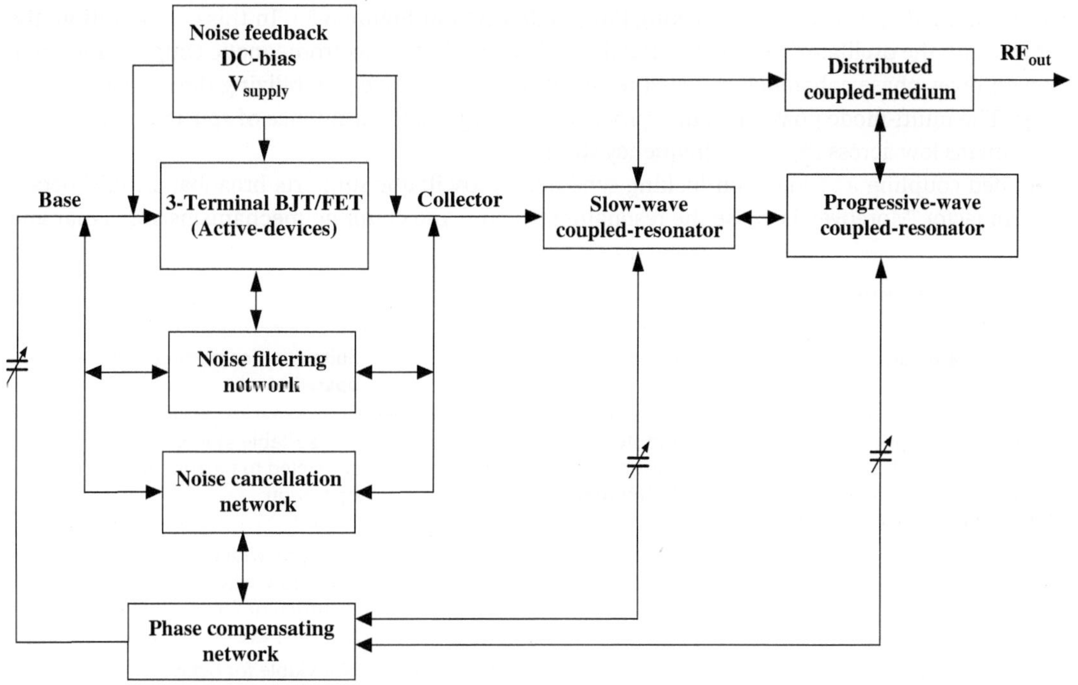

Figure 6.35 Functional block diagram of an oscillator using multi-mode, progressive-wave, and slow-wave coupled resonators for ultra-low phase noise across multiple octaves, operating with a DC bias of 5 V and 40 mA [10].

Figure 6.36 [10] presents the PCB layout of the VCO corresponding to the block diagram in Figure 6.35, highlighting its multi-octave tuning capability. The resonator network is composed of planar multi-mode, progressive-wave, and slow-wave coupled resonators that are electromagnetically coupled to one another. The SWR employs planar projections that interleave with complementary openings—such as meander-line structures—to slow down the guided wave, thereby increasing the effective electrical length without enlarging the physical footprint. This slow-wave propagation mechanism, in conjunction with controlled multi-mode coupling, facilitates compact high-Q operation. Furthermore, the synergistic effects of mode coupling and self-injection locking [10] among the resonators extend the oscillator's tuning range while simultaneously improving spectral purity. Collectively, these features enable the oscillator to achieve ultra-low phase noise, broadband tunability, and robust performance suitable for advanced communication and radar applications [1].

The unified resonator architecture shown in Figure 6.36 integrates slow-wave and progressive-wave resonator networks into a single high-Q composite system. This hybrid structure provides wideband tunability while shifting spurious resonances away from integer multiples of the fundamental frequency, thereby suppressing mode crowding. As a result, the system enables stable multi-octave operation—a level of performance that is unattainable when slow-wave, progressive-wave, or multi-mode resonators are employed individually [77–93]. As illustrated in Figure 6.35, a phase-compensation network is capacitively coupled between the base terminal of the active device and the unified resonator [10]. This network equalizes the phase response across the tuning range, ensuring uniform phase-noise characteristics. The interaction of the slow-wave and progressive-wave resonators through hybrid-mode coupling produces a convergence effect, which, in conjunction with the phase-compensation network, dynamically optimizes group delay to minimize phase noise throughout the operating band. The RF output is extracted via a distributed coupling medium spanning both the slow-wave and progressive-wave resonators. This approach ensures consistent output power while simultaneously suppressing higher-order harmonics, thereby improving spectral purity across the full frequency range. Table 6.3 shows the comparison of resonator architectures for multi-octave oscillator design.

The mode-coupling methodology employed in the oscillator integrates a dynamic coupling approach that continuously adapts to operating conditions, thereby enhancing the loaded quality factor (Q) of the resonator system [10]. This mechanism not only increases spectral purity by suppressing phase hits but also minimizes microphonic sensitivity, leading to improved stability under environmental perturbations. As a result, the oscillator achieves a substantial reduction in phase noise while simultaneously maintaining a broadband, linear tuning response. A central innovation within this methodology is the incorporation of multi-mode phase injection locking [10], as depicted in Figure 6.36. In this configuration, the reference signal (P1) interacts with the oscillator's RF signal (P2) through a distributed electromagnetic coupling medium. This controlled interaction enforces phase coherence between resonant modes, effectively stabilizing the oscillator while broadening its tuning range. The multi-mode phase injection process also mitigates the influence of spurious resonances, ensuring that phase noise remains low across the entire frequency spectrum.

By leveraging this distributed coupling and injection-locking scheme, the oscillator supports broadband multi-octave tuning within a compact form factor. Moreover, because the resonator structure and coupling mechanisms are planar and

Table 6.3 Comparison of resonator architectures.

Resonator type	Key characteristics	Limitations	Suitability for multi-octave operation
Slow-wave resonator	Compact footprint, supports high-Q through extended effective electrical length (e.g., meander-line structures).	Limited tuning range; prone to spurious responses near harmonics.	Not suitable alone; restricted to narrow-band operation.
Progressive-wave resonator	Provides distributed phase shift; enhances bandwidth when cascaded.	Low-Q when used alone; higher insertion losses.	Supports wider bandwidth than slow-wave but insufficient for multi-octave operation.
Multi-mode resonator	Supports multiple resonant frequencies within same structure.	Spurious modes appear near harmonic frequencies; mode crowding degrades spectral purity.	Not viable for octave-spanning oscillators.
Unified hybrid resonator (slow-wave + progressive-wave)	Combines advantages of both resonators; achieves high-Q, wide tunability, and shifted spurious modes.	Increased design complexity; requires precise hybrid-mode coupling and phase compensation.	Enables stable multi-octave operation with ultra-low phase noise.

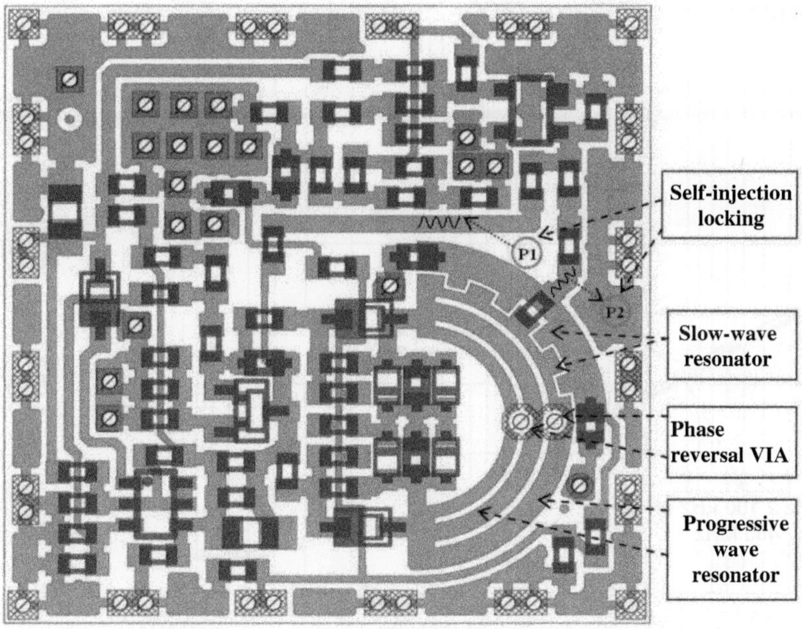

Figure 6.36 Printed layout of the oscillator corresponding to the block diagram in Figure 6.35, implemented using MMIC technology on a 32-mil substrate with dielectric constant 3.38, dimensions 0.75 × 0.75 × 0.18 inches [10].

lithographically defined, the design is inherently compatible with conventional integrated circuit (I_c) and MMIC fabrication processes. This makes the architecture particularly well-suited for scalable, cost-effective implementation in modern broadband communication and radar systems.

The oscillator topology further suppresses phase noise by dynamically optimizing both the impedance transfer function and the coupling factor throughout the planar multi-coupled resonator network. This adaptive mechanism ensures that the resonator system maintains high loaded-Q characteristics across the entire tuning range. Figure 6.37 presents the measured phase-noise performance of the proposed oscillator, demonstrating typical values of −157 dBc/Hz at a 1 MHz offset over a multi-octave tuning span of 500–2500 MHz [94]. The oscillator operates under a DC supply of 5 V with a bias current of 40 mA [40]. Across the full tuning range, phase-noise variation remains within 3 dB, indicating robust consistency in noise suppression. The RF output power lies between 3 and 5.2 dBm, making the design suitable for integration into high-performance frequency synthesizers and precision signal sources.

The measured FOM (Eq. 6.50) is −197.5 dBc/Hz, highlighting the oscillator's exceptional phase-noise efficiency relative to its operating frequency and power consumption. Additionally, the PFTN (Eq. 6.51) reaches 26.02 dB, underscoring the design's superior trade-off between tuning range and power efficiency. The oscillator's total power consumption is approximately 200 mW ($V_{CC} = 5$ V, $I_c = 40$ mA), while delivering an average output power of +3.2 dBm. The tuning range covers 136% (500–2500 MHz), with additional margin demonstrated at both extremes of operation—achieving a minimum frequency of 450 MHz at ($V_t = 0.2$ V) and a maximum frequency of 2600 MHz at ($V_t = 26$ V). Collectively, these results confirm that the oscillator topology not only achieves ultra-low phase noise across multiple octaves but also maintains high output stability and efficiency, establishing it as a strong candidate for advanced broadband communication and radar systems.

6.5.1.3 Broadband Tunable Push–Push SWRO Topology (US Patent No. 7, 292,113)

High-frequency signal generation can be accomplished using either fundamental-frequency oscillators or harmonic oscillator architectures [10, 73]. Conventional fundamental-frequency oscillators, however, face intrinsic challenges at elevated frequencies, including reduced quality factor (Q), diminished device gain, and increased phase noise. To address these limitations, harmonic oscillators are widely adopted in two primary configurations:

1. Cascade structures, which generate second-harmonic signals via frequency doubling, and
2. Parallel (N-push) structures, such as push–push or N-push topologies, which employ coupled oscillators to generate higher-order harmonics.

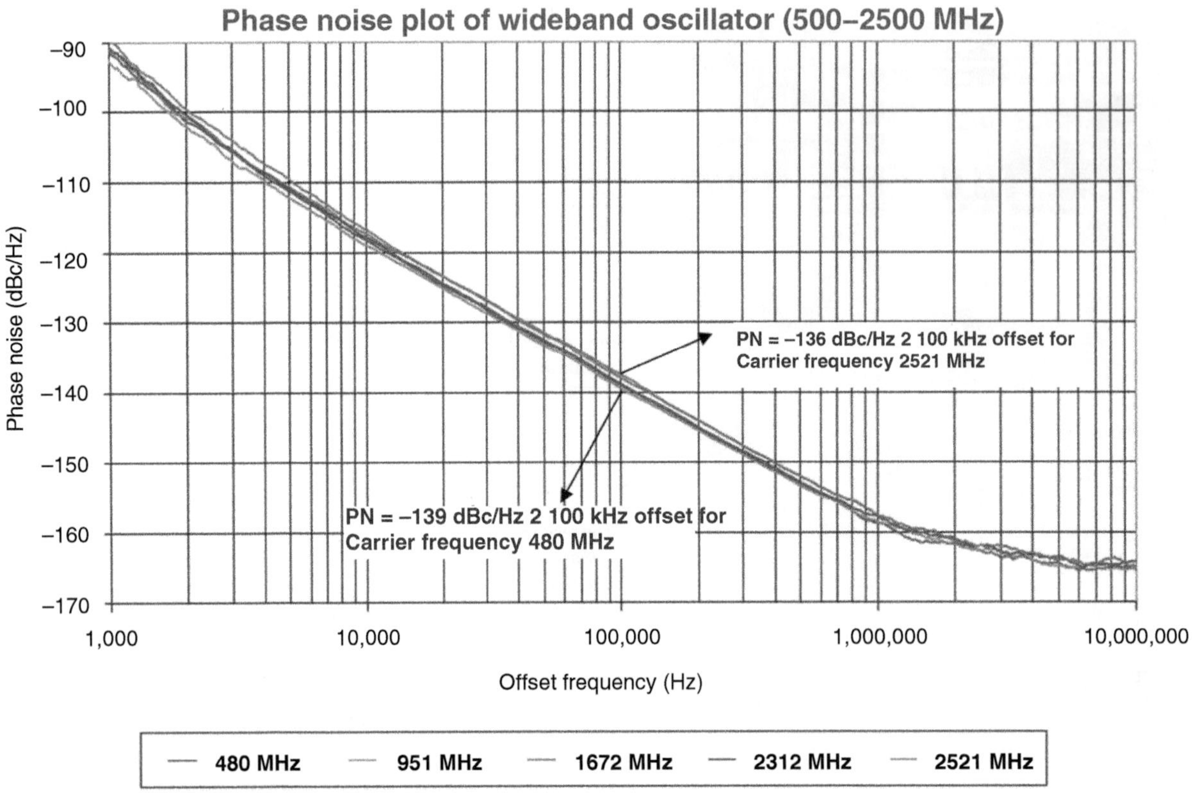

Figure 6.37 Measured phase noise of the oscillator, exhibiting typical values below −136 dBc/Hz at a 100 kHz offset over the 500–2500 MHz range [10]. Poddar [74] reports that *"the figure-of-merit (FoM, Eq. 6.50) is −197.5 dBc/Hz, and the power-frequency tuning-normalized metric (PFTN, Eq. 6.51) is 26.02 dB."* The circuit demonstrates efficient performance, consuming 200 mW ($V_{CC} = 5$ V, $I_c = 40$ mA) while delivering an output power of +3.2 dBm [10].

Although frequency doublers and related up-conversion methods provide a direct means to achieve high-frequency signals from lower-frequency sources, they often introduce signal distortion and degrade phase-noise performance. These trade-offs have spurred intensive research in microwave and millimeter-wave engineering to develop cost-effective, low-phase-noise high-frequency sources. Among these, push–push oscillators offer notable advantages over single-ended designs, including superior phase-noise characteristics, an extended usable frequency range beyond transistor fundamental limits, and the ability to use larger transistors operating at reduced current densities—thereby mitigating $1/f$ noise. Furthermore, coupled-oscillator N-push configurations can further suppress phase noise and support operation beyond the cutoff frequencies of available active devices [95]. The state-of-the-art oscillator topology [10] incorporates several advanced features to achieve exceptional performance over broad frequency ranges:

- Uniform negative resistance: The design sustains a nearly constant negative resistance across an entire octave band, ensuring stable oscillation and minimal distortion over wide frequency spans.
- Wideband tuning mechanism: A novel tuning scheme enables wide frequency agility without degrading the loaded Q of the resonator network [10]—a common shortcoming of conventional designs.
- Coupled-resonator architecture: Multiple resonators are coupled to sustain resonance across multi-octave bandwidths, enhancing spectral purity and extending operational versatility for applications ranging from communication systems to high-resolution sensing.
- Optimized device sizing: Device parameters (e.g., $I_{c_{max}}/I_{d_{ss}}$) are carefully optimized to improve linearity and noise handling, thereby reducing phase noise and improving stability.
- Enhanced phase-noise suppression: Leveraging N-push or coupled-oscillator methods inherently improves phase-noise performance and enables frequencies beyond single-ended limits.
- Dynamically tuned phase-coupling network: An adaptive phase-coupling network is dynamically adjusted to preserve oscillator stability and maximize bandwidth under varying operating conditions.

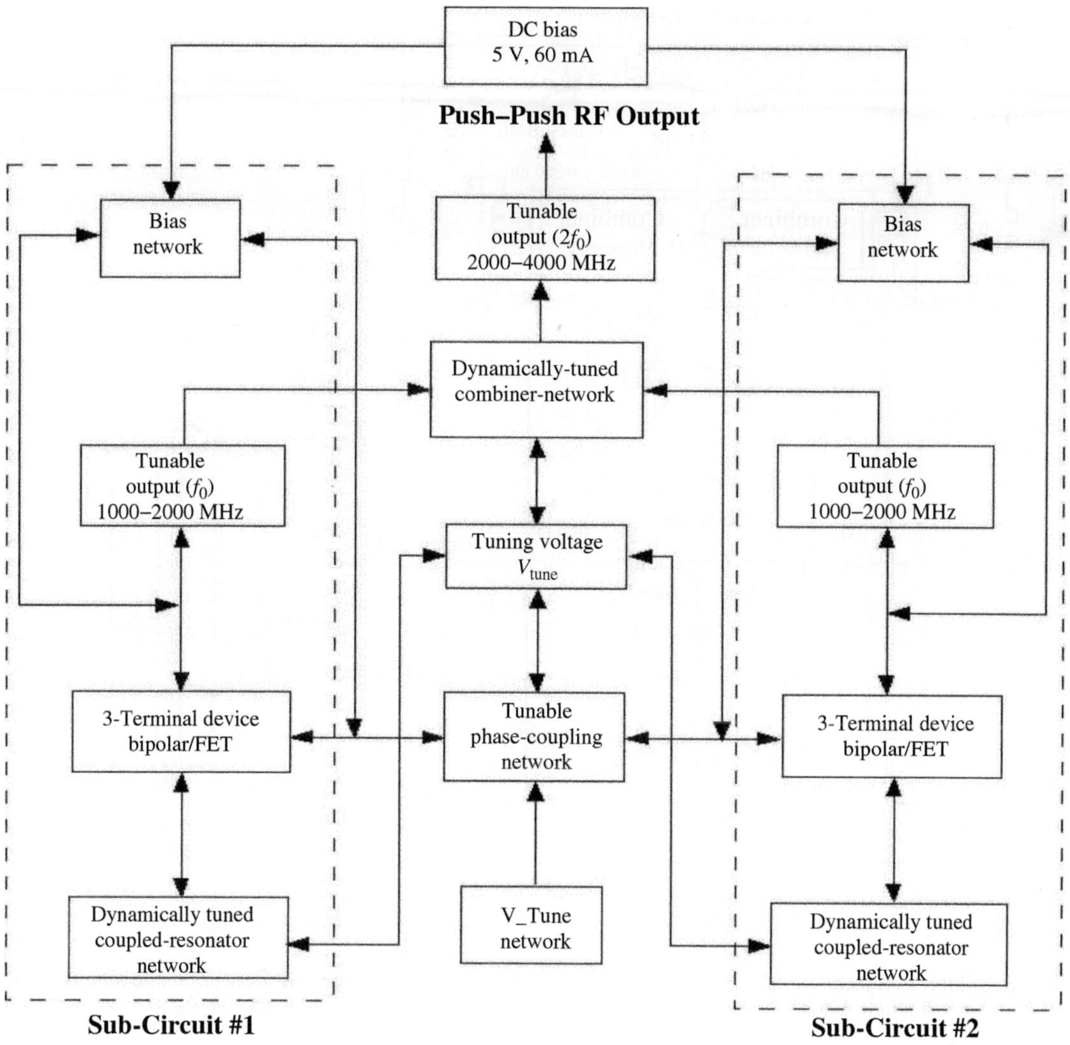

Figure 6.38 Block diagram showing the main modules of a low-noise octave-band VCO covering 1000–2000 MHz and 2000–4000 MHz frequency ranges (US Patent No. 7,292,113) [10, 95, 96].

This combination of design strategies enables robust, low-noise signal generation at frequencies approaching and exceeding device cutoff limits, making the topology particularly well-suited for next-generation microwave, millimeter-wave, and terahertz systems. The oscillator architecture described represents a major advancement in wideband signal source design, providing effective solutions to long-standing problems of frequency stability, spectral purity, and efficient use of bandwidth. Figure 6.38 illustrates the functional organization of a broadband VCO capable of operating in two wide ranges: 1000–2000 MHz and 2000–4000 MHz. The design incorporates three dynamically reconfigurable building blocks: a coupled-resonator network, a phase-control network, and a signal-combining network. Together, these modules enable an octave-band push–push mode of operation. In this mode, two identical oscillator cores run 180° out of phase at the fundamental frequency, and their outputs are combined in such a way that the fundamental components cancel while the second harmonic adds coherently, producing a strong and spectrally clean output across the tuning range.

Unlike traditional push–push or N-push oscillators, which are typically restricted by narrow tuning limits or fixed-frequency operation, this design leverages active tuning of both resonant and phase-coupling elements. The coordinated control of these networks provides wideband frequency agility and stable harmonic generation across the full octave range. This ensures that the oscillator maintains low phase noise and reliable harmonic suppression while extending the tuning capability well beyond that of earlier implementations [77–84]. Figures 6.39 and 6.40 further present the detailed circuit schematic and PCB layout, respectively, confirming the feasibility of octave-spanning tuning in practical implementations [95, 96].

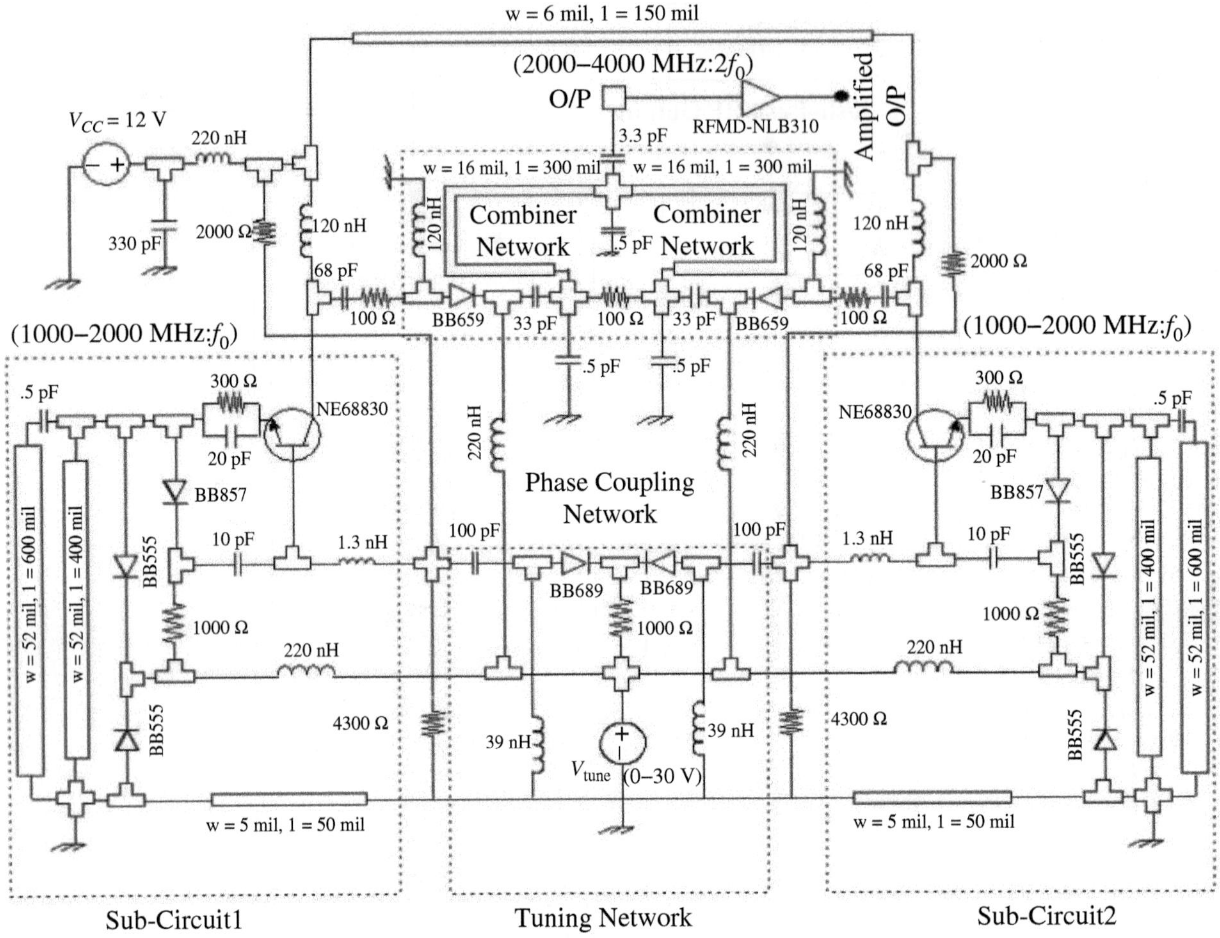

Figure 6.39 Schematic of the oscillator integrating a printed coupled-resonator network, a phase-coupling network, and a combiner network to achieve octave-band push–push operation, corresponding to the block diagram in Figure 6.38 [10, 95, 96].

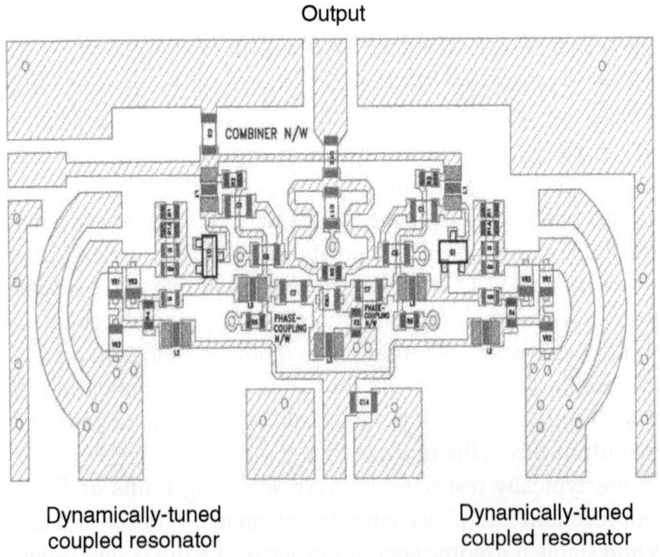

Figure 6.40 The layout of the schematic of the push–push oscillator circuit shown in Figure 6.39 (multilayer 32-mil substrate, Dielectric constant 3.38), 1.75 × 1.75 × 0.18 inches ROGERS PCB (US Copyright Registration No. Vau-603-982) [10, 95, 96].

The block-level elements in Figure 6.38 are engineered to support miniaturized integration and are compatible with monolithic chip fabrication techniques. As documented in US Patents 7,292,113 and 7,088,189, the circuit's modular approach allows each oscillator cell to generate signals at half of the intended output frequency (f_0). When these are combined, the even-order harmonic ($2f_0$) is enhanced, while the fundamental frequency is inherently suppressed. This eliminates the need for external frequency-selective filters, thereby reducing circuit size and complexity. Crucially, the dynamically tuned phase-coupling network maintains the strict 180° phase separation between the two oscillator cores across the tuning span, ensuring proper push–push operation and robust harmonic control [96].

Figure 6.39 illustrates how the dynamically adjustable coupled-resonator network is directly interfaced with the emitter terminal of the NE68830 transistor. This configuration is selected to preserve a stable loaded quality factor (Q) across the full tuning band, thereby supporting low-noise and reliable oscillation. The corresponding physical implementation is presented in the PCB layout of Figure 6.40.

The circuit is fabricated on a Rogers laminate with a thickness of 32 mils, a relative dielectric constant of 3.38, and a loss tangent of 2.7×10^{-4}. The finished board has compact dimensions of $1.75 \times 1.75 \times 0.18$ inches, making it well-suited for high-frequency applications requiring reduced form factor. Furthermore, the approach is compatible with integrated chip-scale realization, consistent with the fabrication strategies outlined in US Patents 7,292,113 and 7,088,189.

As shown in Figure 6.41, the circuit employs a configuration designed to compensate for phase errors that arise from unequal path lengths between the two oscillator branches during frequency tuning. While this approach reduces phase imbalance, tuning irregularities still appear at certain frequencies as a result of package parasitics and tolerances associated with discrete components [10]. These discontinuities can be mitigated by incorporating a phase detector, which enforces synchronization between the oscillator branches and enables continuous tuning across multiple octaves while simultaneously improving phase-noise characteristics.

The central design objective is to identify the mechanisms that limit wideband tunability and to establish an oscillator topology that minimizes unwanted phase deviations without sacrificing low-noise performance. Compared with conventional push–push or N-push designs, the integration of phase-synchronization circuitry based on a phase detector (PD) provides a practical pathway toward realizing broadband, dynamically reconfigurable coupled oscillators that achieve both extended frequency agility and stable spectral behavior.

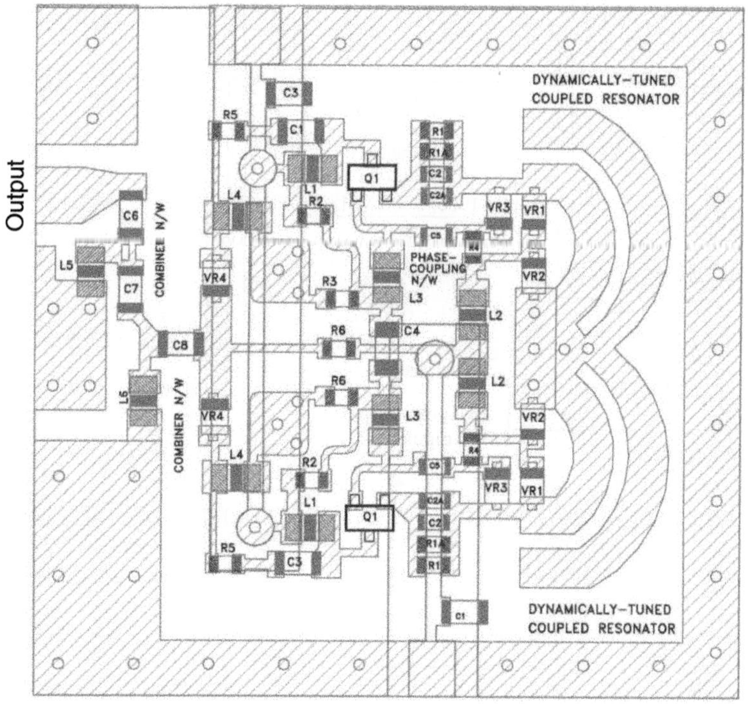

Figure 6.41 The compact layout of the integrated structure (coupled horse-shoe microstrip resonator) of the dynamically tuned coupled resonator of both the sub-circuits shown in Figure 6.40. The layout is built using a multilayer with a 64-mil substrate and a 3.38 dielectric constant of $0.75 \times 0.75 \times 0.18$ inches (US Copyright Registration No. Vau-603-982) [10, 95, 96].

In the coupled-oscillator arrangement shown in Figure 6.38 [10], the two oscillators inherently operate at slightly different free-running frequencies due to unavoidable variations in circuit parameters and component tolerances. This mismatch, however, is resolved through the phenomenon of injection locking, which forces both oscillators to synchronize and oscillate coherently at a single common frequency. The extent of this locking range is determined by the external quality factor (Q) of the oscillators: systems with lower external Q values exhibit broader locking ranges, allowing synchronization even when the natural free-running frequencies differ considerably. Accurate evaluation of such systems requires large-signal analysis, since small-signal models cannot adequately represent the nonlinear interactions that govern odd-mode operation in coupled oscillators. Moreover, the design of the biasing network must carefully address two distinct critical frequency points, ensuring that stable oscillation and high performance are maintained across the entire intended frequency band.

Figure 6.42 illustrates the architecture of a broadband, low-phase-noise push–push VCO capable of operating across two octave-spanning frequency ranges: 1–5 GHz and 5–10 GHz.

The circuit incorporates a dynamically tunable resonator network, a phase-control section, a signal-combining network, and a phase detector. All modules are designed with compatibility for MMIC processes, making the topology suitable for integrated implementations. The inclusion of a phase detector, together with varactor-based tuning elements, allows active

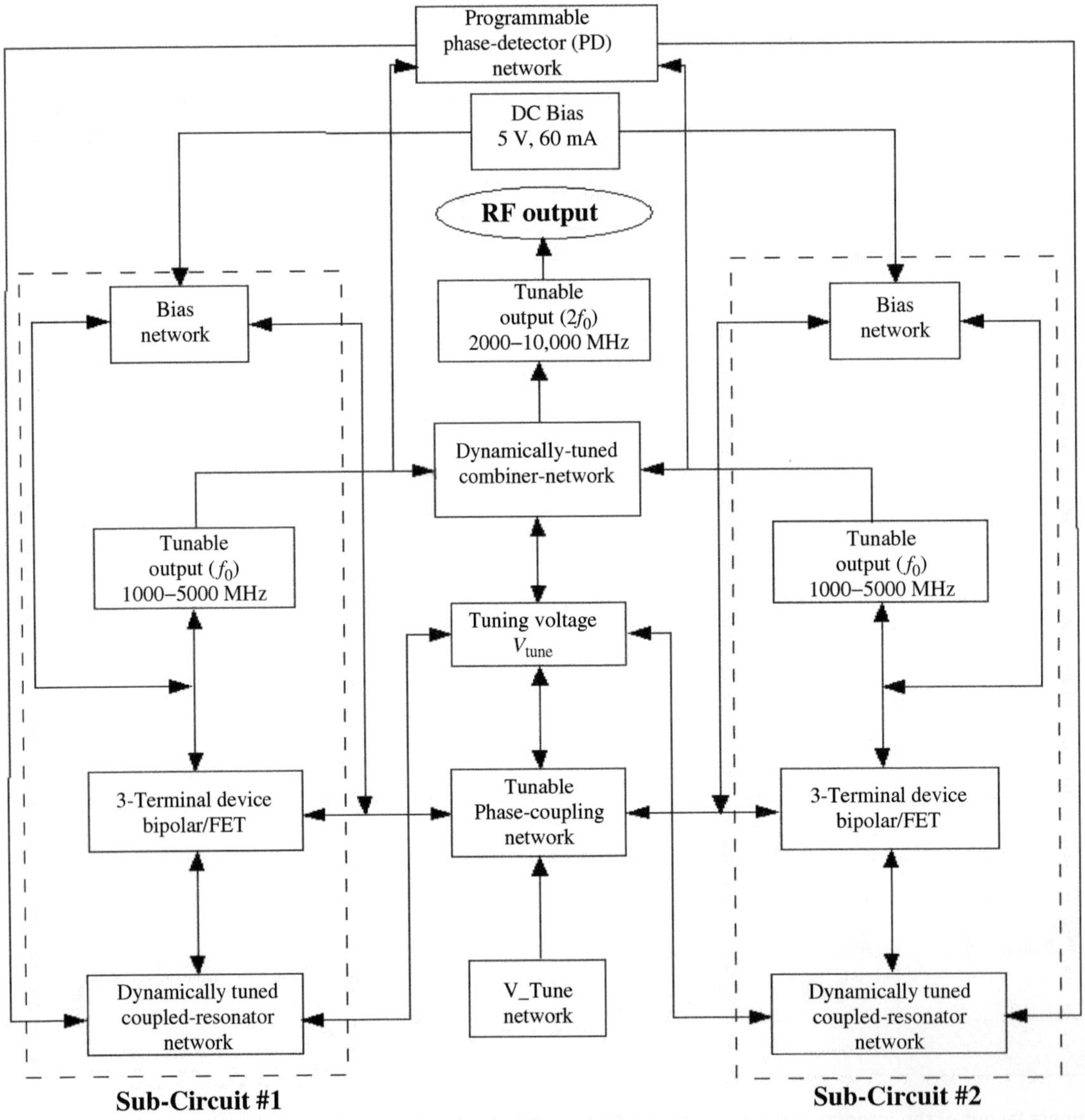

Figure 6.42 Block diagram of a multi-octave push–push VCO covering 1–5 GHz and 5–10 GHz [10]. The design includes dynamically tuned coupled-resonator, phase-coupling, combiner, and phase-detector networks, suitable for MMIC implementation [10, 73].

correction of phase mismatches that typically restrict wideband operation. By continuously monitoring random fluctuations in the relative frequencies of the oscillator cores, the detector generates phase-error signals that are used to control both the combiner network—stabilizing the buffered output—and the tunable resonators, thereby maintaining coherent oscillation across the entire tuning span of 1–5 GHz and 5–10 GHz.

Figures 6.43 and 6.44 present both simulated and experimental phase-noise results for this multi-octave configuration. In the 1–5 GHz band, the measured data align closely with CAD predictions, showing only a 3–5 dB discrepancy.

At higher frequencies, however, performance divergence becomes more pronounced: in the 5–10 GHz range, measured phase noise is degraded by approximately 5–8 dB relative to simulation results. This degradation is attributed primarily

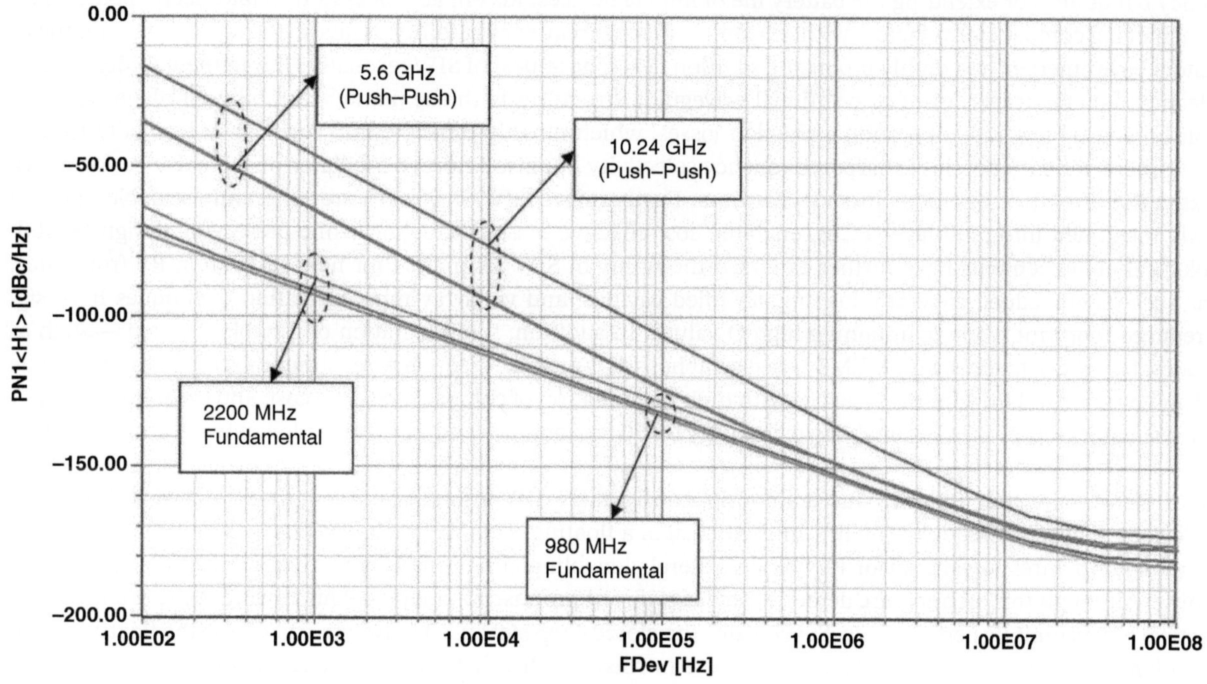

Figure 6.43 CAD-simulated phase-noise performance of the multi-octave VCO spanning [10] 1–5 GHz and 5–10 GHz, implemented in the push–push configuration shown in Figure 6.42 [10].

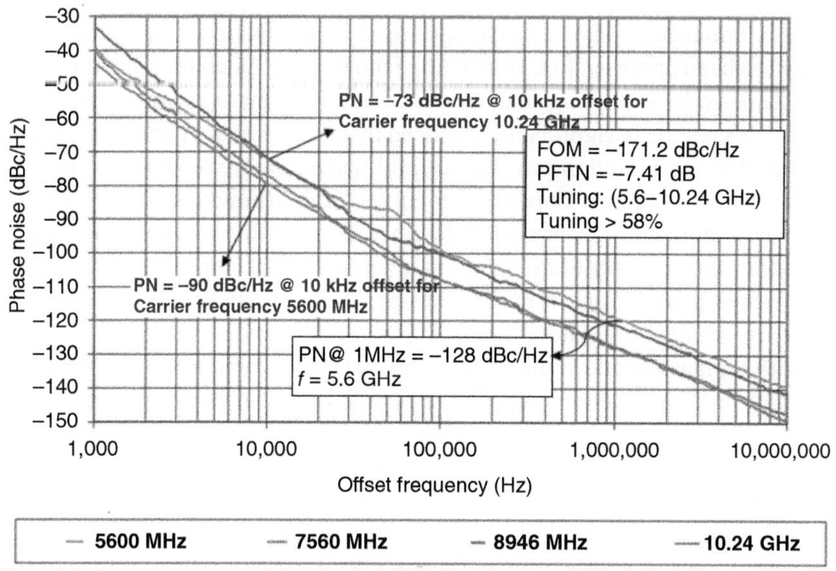

Figure 6.44 Measured phase-noise of the multi-octave VCO (1–5 GHz/5–10 GHz) implemented in a push–push configuration (see Figure 6.28b) [10]. Oddar [10] reported that *"the voltage-controlled oscillator (VCO) achieves a figure-of-merit (FOM, Eq. 6.50) of* $-171.2\ dBc/Hz$ *and a normalized power-frequency tuning (PFTN, Eq. 6.51) of −7.41 dB. The circuit operates with a* $5\ V$ *supply and a* $60\ mA$ *current, corresponding to a total power consumption of* $300\ mW$*, and delivers an output power of* $-3.0\ dBm$*"* [10].

to practical limitations, including manufacturing tolerances in discrete components and the increased phase dispersion effects that emerge at higher operating frequencies.

6.6 SIW Resonator Oscillators

In state-of-the-art portable and wireless communication systems, the demand for both high DC-to-RF power conversion efficiency and excellent phase-noise performance continues to intensify due to stringent constraints on energy consumption, linearity, and spectral purity. Meeting these requirements is vital for ensuring reliable operation in densely occupied frequency bands and for extending the battery life of mobile devices. Recent advances in resonator technologies have focused on achieving compact form factors without compromising frequency stability or quality factor (Q). Among these, SIW resonators have emerged as a highly promising solution [97]. The appeal of SIW resonators lies in their ability to combine the electromagnetic properties of conventional waveguide structures with the manufacturability of planar circuits. They exhibit inherently low radiation and conduction losses, which directly translate into superior unloaded Q factors. These high-Q characteristics not only sharpen frequency selectivity but also improve oscillator phase-noise performance by minimizing the impact of resonator loss mechanisms. Furthermore, SIW structures are inherently scalable and can be seamlessly integrated into multilayer PCBs, enabling cost-effective mass production while preserving design flexibility. Technological enhancements have further extended the utility of SIW resonators for next-generation RF front-ends. Innovations such as embedded dielectric loading, air-filled cavities, and via-array miniaturization techniques have significantly reduced footprint while maintaining high Q values. In addition, the integration of tunable elements—such as varactor diodes, ferroelectric films, and MEMS-based switches—has enabled frequency reconfigurability, supporting agile multi-band and multi-standard wireless systems. With their combination of compactness, low cost, and exceptional performance, SIW resonators represent a key enabling technology for 5G, beyond-5G, and future ultra-low-power communication platforms.

Wu, Poddar, Rohde, and Itoh [98] introduced a novel reconfigurable SIW resonator concept that leverages active semiconductor devices to achieve frequency agility. In their design, a floating metallic patch was placed above the SIW cavity and coupled via jump wires to varactor or PIN diodes, thereby enabling either continuous frequency tuning or discrete state switching. Although this architecture demonstrated significant promise for adaptable microwave front-end modules, several practical issues limit its performance. Specifically, the closed-loop slots used to isolate and define the floating patch generate unintended radiation leakage, which reduces the intrinsic quality factor (Q) of the resonator and compromises frequency selectivity [99–101]. Furthermore, the reliance on jump wires for biasing introduces fabrication complexity and additional parasitic inductances and capacitances, both of which adversely affect high-frequency response and consistency across wideband applications.

To address these shortcomings, a technologically enhanced resonator topology based on complementary coupled resonators (CCRs) integrated within SIW technology has been proposed [98, 102]. Unlike the floating-patch method, the CCR-based approach inherently suppresses radiation leakage by exploiting magnetic-field coupling mechanisms rather than relying on geometrically isolated metallic patches. This results in significantly reduced insertion loss and preserved high-Q operation across the tuning range. In addition, the CCR configuration provides a simplified and more reliable biasing network, eliminating the need for wire-bonded connections and thereby enhancing manufacturability and long-term stability.

Advanced implementations of CCR-based SIW resonators have further expanded their applicability through the integration of voltage-controlled varactors, ferroelectric thin films, and MEMS actuators, offering both continuous and discrete tuning capabilities. These improvements enable broader tuning ranges while retaining high selectivity and low phase-noise performance, making them highly attractive for use in agile oscillators, tunable filters, and compact RF front-end systems. The compatibility of CCR-SIW structures with standard planar printed circuit board (PCB) processes further supports low-cost, scalable production for next-generation communication systems, including 5G, beyond-5G, and adaptive radar applications.

The CCR can be viewed as the dual counterpart of a conventional microstrip coupled-line resonator, as described in [94, pp. 358] [102]. When adapted to a SIW environment, the CCR primarily supports differential-mode operation [102]. In this mode, the equivalent magnetic currents along the surrounding slot apertures are oriented in opposite directions. This counter-propagating current distribution leads to a partial cancellation of the associated radiated fields, which in turn minimizes unwanted radiation leakage and improves the overall energy confinement within the resonant structure. As a

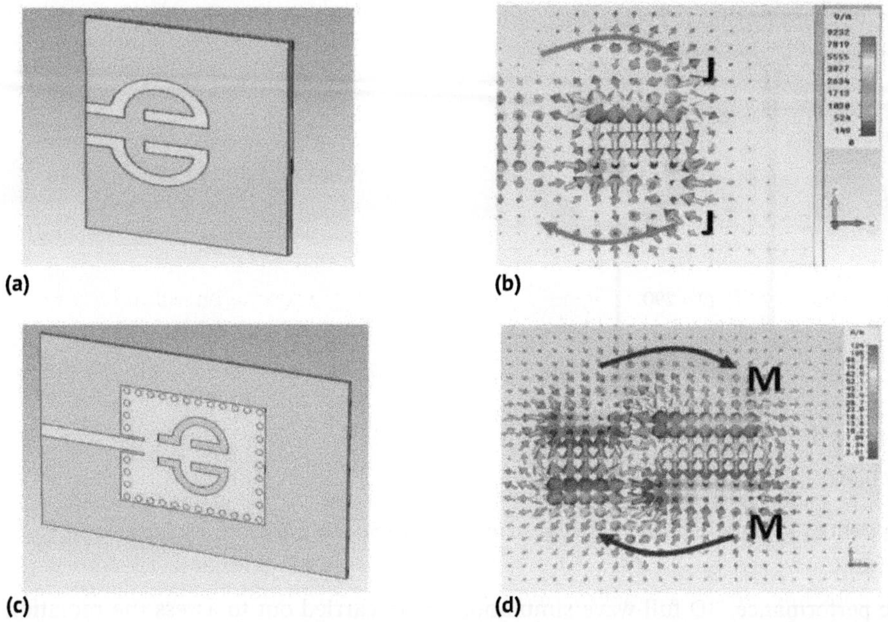

Figure 6.45 Geometry of coupled planar resonators [10]: (a) microstrip coupled-line resonator, (b) corresponding electric-field distribution and current flow under differential excitation, (c) CCR, and (d) magnetic-field distribution with equivalent magnetic current in the fundamental mode of an SIW cavity [94, 98, 102, 103].

direct consequence, the CCR achieves a substantially higher unloaded quality factor (Q) compared with its conventional counterpart.

The electromagnetic behavior of this configuration is illustrated in Figure 6.45. Figure 6.45a presents the structure of a traditional microstrip coupled-line resonator, while Figure 6.45b shows the electric-field (E-field) distribution of the proposed complementary design operating near 5 GHz [9]. The comparison highlights how the differential-mode field interactions in the CCR facilitate radiation suppression and contribute to enhanced frequency selectivity, stability, and efficiency. This makes CCR-based SIW resonators particularly attractive for high-performance RF and microwave front-end modules where low loss and compact integration are essential.

By applying the principle of electromagnetic duality, the CCR employs magnetic-field (H-field) coupling rather than the conventional electric-field coupling mechanism, as depicted in Figure 6.45d. The resulting H-field distribution is the dual of the E-field pattern typically observed in standard coupled-line resonators. Unlike traditional microstrip coupled-line structures, which require two out-of-phase input signals to drive the differential mode, the CCR benefits from the intrinsic modal characteristics of the SIW cavity. Specifically, the fundamental cavity mode naturally excites the differential response without the need for additional input circuitry.

This inherent excitation mechanism ensures that the equivalent magnetic currents along the slots flow in opposite directions at the symmetrically placed edges of the resonator [10]. Such counter-propagating current flow leads to effective cancellation of radiated fields, thereby minimizing radiation leakage and enhancing electromagnetic confinement within the structure. As a result, the CCR demonstrates improved energy efficiency and a significantly higher unloaded quality factor (Q). The unloaded Q, denoted as $Q(\omega_0)$, can be analytically estimated using the formulation provided in [102].

$$Q(\omega_0) = \frac{\omega_0}{2}\left|\frac{Z_{11}(\omega_0)'}{Z_{11}(\omega_0)}\right|_{f_0=5.34\ \mathrm{GHz}} \approx 287 \tag{6.52}$$

where $Z_{11}(\omega_0)$ represents the input impedance of the SIW resonator at the frequency ω_0, while $Z_{11}(\omega_0)'$ denotes its derivative with respect to frequency.

Figure 6.46 shows the simulated reflection coefficient $[S_{11}]$ response of the 1-port CCR illustrated in Figure 6.45c, configured as a single-port element within a Colpitts oscillator architecture. The simulation results indicate a well-defined resonance at 5.34 GHz. Based on the fractional 3-dB bandwidth of the $[S_{11}]$ response, the unloaded quality factor $Q(\omega_0)$ is determined to be approximately 290, which is in close agreement with the analytical estimate obtained from Eq. (6.52).

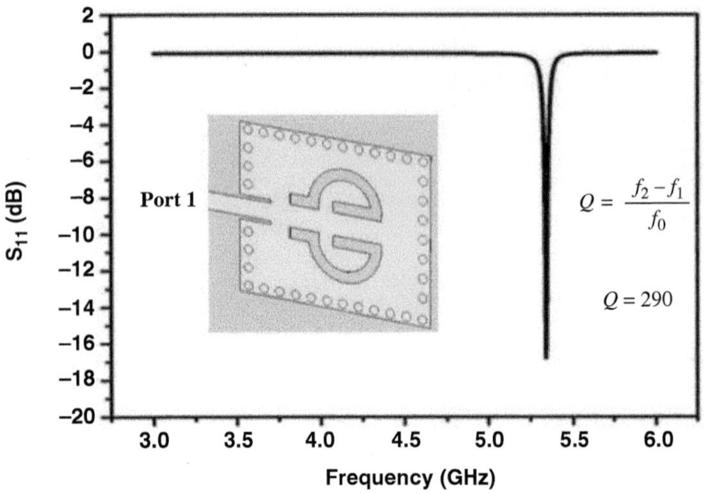

Figure 6.46 CAD-simulated reflection coefficient (S_{11}) of the 1-port CCR depicted in Figure 6.45c [98, 102].

To further evaluate the electromagnetic performance, 3D full-wave simulations were carried out to assess the radiation behavior of the resonator, as shown in Figure 6.27c [40]. These analyses reveal a radiation efficiency of 6.96% and a corresponding radiation gain of −4.64 dB [10]. The results verify that the resonator maintains strong field confinement and exhibits only minimal radiation leakage, thereby confirming the suitability of the CCR structure for low-loss, high-Q oscillator applications [98].

Figure 6.47a depicts a representative 2-port CCR implemented as a dual-port filtering element within a feedback oscillator topology. This configuration leverages the CCR's differential-mode operation to provide sharp frequency selectivity while maintaining minimal radiation loss. Figure 6.47b presents a comparison between CAD-simulated and experimentally measured [S_{11}] responses for the 2-port CCR. The results exhibit strong agreement, confirming the accuracy of the simulation model. The measured insertion loss is observed to be below 1 dB at the resonant frequency of 5.1 GHz, indicating highly efficient energy transfer through the resonator. Using the methodology outlined in [102], the unloaded quality factor (Q) of the 2-port CCR can be calculated from the fractional bandwidth, providing a reliable metric of its resonant performance and suitability for integration in low-loss, high-Q oscillator circuits.

$$Q(\omega_0) = \frac{\omega_0}{2}\left|\frac{Z_{21}(\omega_0)'}{Z_{21}(\omega_0)}\right|_{f_0=5.1\text{ GHz}} \approx 312 \tag{6.53}$$

As demonstrated in Figure 6.47a,b, the SIWCCRs naturally operate at fixed resonant frequencies determined by their geometrical and electromagnetic properties. Broadband frequency tunability is achieved by strategically incorporating varactor diodes at locations corresponding to the maximum electric-field intensity along the resonator slots, as highlighted in Figures 6.45b and 6.47a.

This approach allows continuous adjustment of the resonant frequency while preserving the high quality factor and low radiation losses inherent to the CCR design [94, 98]. Figure 6.48 illustrates a physical prototype of a 1-port complementary coupled tunable resonator (CCTR) realized with MA/COM MA46H120 varactor diodes. The integrated varactors enable voltage-controlled frequency tuning, converting the fixed-frequency SIW CCR into a compact, broadband, and reconfigurable resonator suitable for agile oscillator and filter applications in advanced RF and microwave systems.

Figure 6.49a presents the measured reflection coefficient [S_{11}] responses of the varactor-loaded CCTR under reverse-bias voltages ranging from 0 to 20 V. The results demonstrate a continuous tuning of the resonant frequency from 4.579 to 4.984 GHz, corresponding to a relative bandwidth of approximately 8.84%. The frequency tuning range can be further expanded by reducing the separation between the DC bias line and the SIW cavity, introducing a coupling capacitor between the bias line and the resonator, or employing hyper-abrupt varactor diodes. Such enhancements make the CCTR highly suitable for broadband filtering and VCO applications in modern RF and microwave systems [10].

Figure 6.49b summarizes the measured unloaded quality factor (Q) across the tuning range, showing values between 50 and 220. The variation in (Q) reflects the trade-off between tuning sensitivity and energy confinement, with higher bias voltages generally reducing effective Q due to increased diode losses. These results confirm that the varactor-loaded CCR

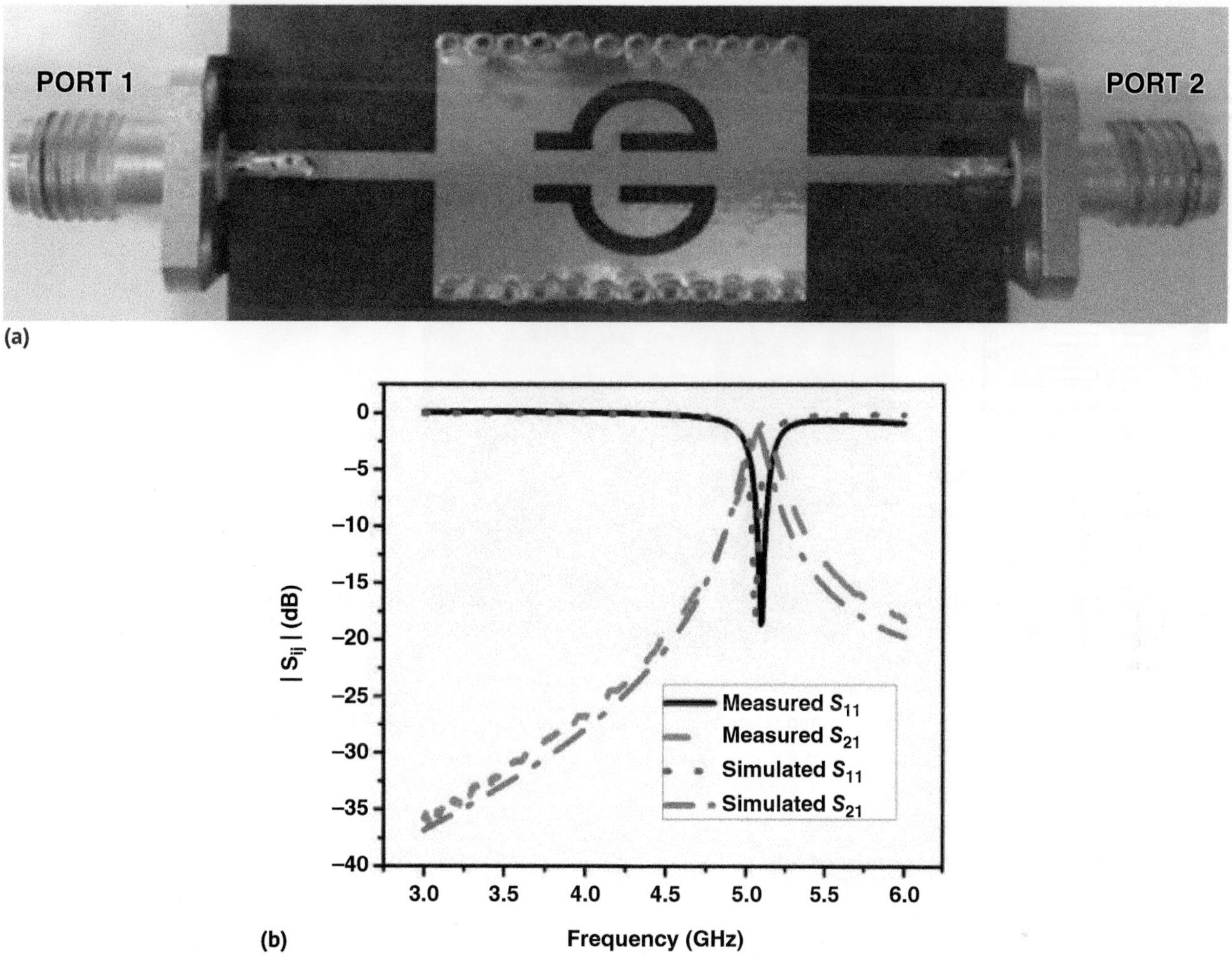

Figure 6.47 2-port CCR: (a) fabricated prototype and (b) measured and simulated S-parameter responses [10, 98].

maintains both tunability and reasonably high-Q performance, supporting its integration in compact, agile RF front-end modules.

Figure 6.50 shows a photograph of the fabricated 2-port CCTR. The resonator design incorporates a coupling capacitor between the DC bias line and the SIW cavity, which substantially extends the frequency tuning range. With this modification, the resonant frequency can be adjusted from the original 4.579–4.984 GHz (8.84% bandwidth) up to a broader span of 2.73–4.535 GHz, corresponding to an effective tuning range of 66%.

Figure 6.51a illustrates the measured tuning characteristics of the 2-port CCTR [10, 104]. Poddar [74] reports that *"as the reverse-bias voltage is varied from 0 to −15 V, the resonant frequency shifts smoothly from 2.73 GHz (marker m16) to 4.535 GHz (marker m15)."* [10]. This demonstrates the capability of the enhanced CCTR to achieve wideband tunability while maintaining the inherent high-Q performance of the CCR structure, making it highly suitable for broadband filters and agile oscillator implementations in advanced RF systems.

The significant extension of the tuning range introduces a notable trade-off in resonator performance. As illustrated in Figure 6.51b, the unloaded quality factor (Q) of the 2-port CCTR varies between 10 and 130 across the 2.73–4.535 GHz tuning span. This reduction in Q is primarily due to the additional losses and reactive loading introduced by the coupling capacitors used to extend the frequency range [10]. While the use of low-loss multilayer capacitors can partially mitigate this degradation, maintaining a multi-octave tuning capability without substantial reduction in Q remains a complex design challenge. Achieving both wideband tunability and high energy confinement requires careful optimization of resonator geometry, coupling networks, and biasing schemes to balance bandwidth, efficiency, and spectral purity in compact RF and microwave systems.

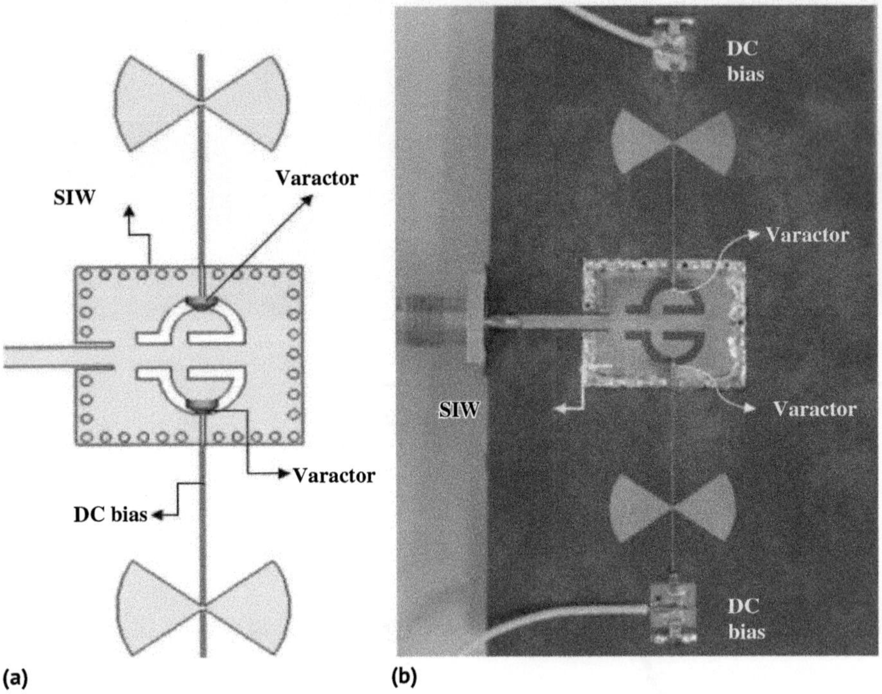

Figure 6.48 Tunable 1-port CCR: (a) varactor-tuned schematic, and (b) fabricated prototype on RT/Duroid 5880 substrate (0.508 mm thick, dielectric constant 2.2) [10, 98].

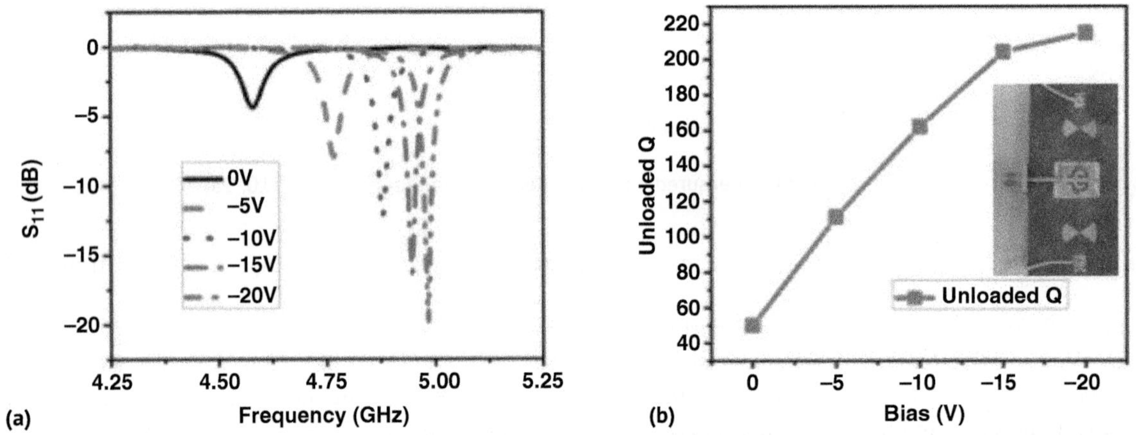

Figure 6.49 Measured results for the tunable 1-port resonator: (a) $[S_{11}]$ responses under different reverse bias voltages, and (b) corresponding unloaded Q-factor variations [98].

Various techniques have been investigated to simultaneously maximize the unloaded quality factor (Q) and tuning range of resonators for high-performance signal sources [14]. A particularly effective approach involves compensating intrinsic resonator losses by employing the negative resistance characteristics of active devices [10], thereby enhancing the effective Q and improving overall oscillator performance [105–107].

Poddar [74] describes that *"for the complementary coupled tunable resonator (CCTR) shown in Figure 6.50, an active heterojunction field-effect transistor (HJ-FET NE3210S01, NEC) [10] is integrated directly into the resonator circuit,"* highlighting its role in enhancing the resonator's tunability and performance. The device, supplied with a controlled DC bias, injects negative resistance that counteracts the resonator's inherent losses across the entire tuning range. This strategy allows the realization of a compact, high-Q resonator structure, which is crucial for designing low-phase-noise X-band oscillators. The

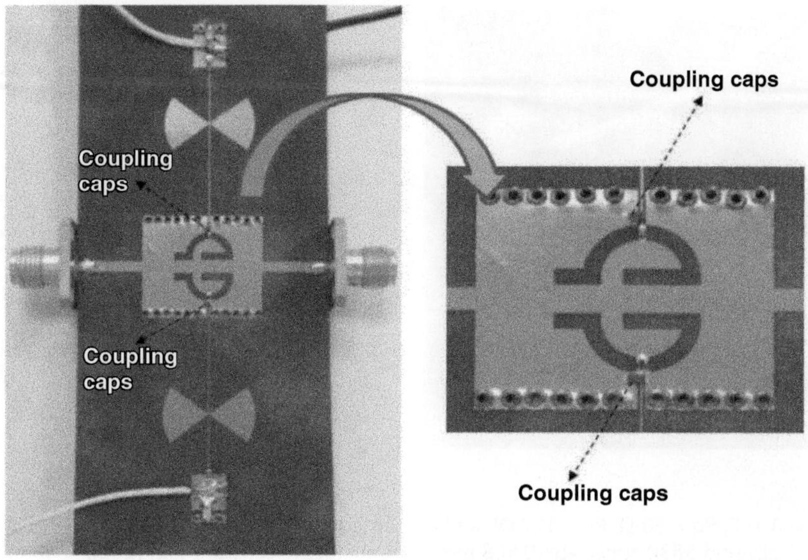

(a) (b)

Figure 6.50 Photograph of the fabricated 2-port CCTR prototype: (a) showing an additional capacitor between the bias line and resonator, and (b) close-up view [98, 102].

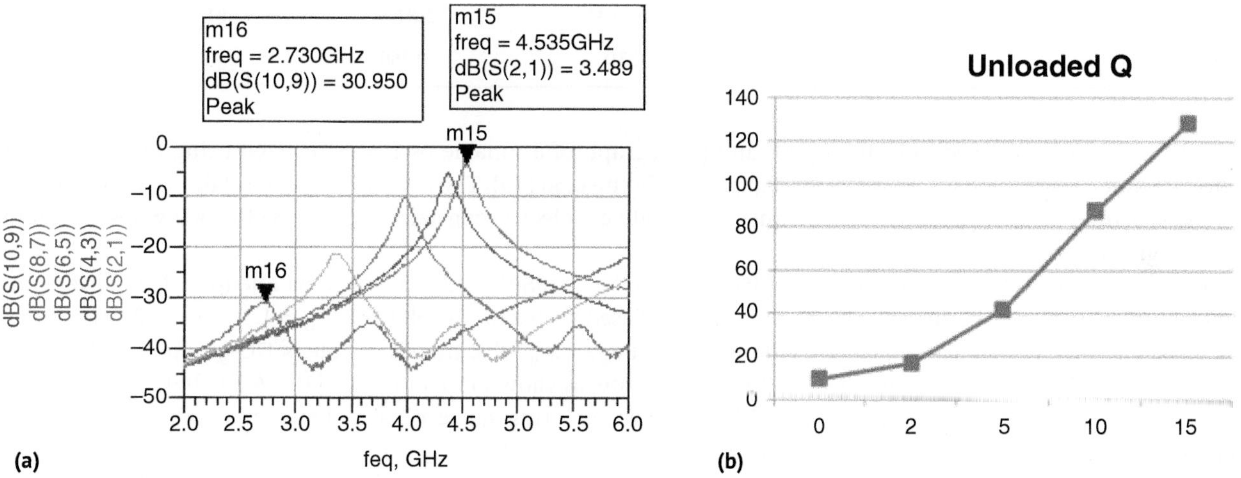

(a) (b)

Figure 6.51 Measured results: (a) S_{21} of the tunable resonator from Figure 6.34, and (b) unloaded quality factor (Q) as a function of reverse-bias voltage [10, 14].

resulting system is particularly suitable for radar and other advanced microwave applications, where spectral purity and energy efficiency are critical.

Figure 6.52a,b depict, respectively [10], the schematic and physical layout of the substrate-integrated waveguide-based active complementary coupled tunable resonator (SIW-ACCTR). The resonator is fabricated on an RT/Duroid 5880 substrate with a thickness of 0.508 mm and a relative dielectric constant of 2.2, providing a low-loss platform suitable for high-frequency applications [10, 108, 109]. Experimental characterization, summarized in Table 6.4, indicates that the unloaded quality factor (Q) reaches a peak value of 21,172 when the active transistor is biased at 1.8 V. These results confirm that appropriate DC biasing of the active device dramatically enhances the resonator's Q, demonstrating that the integration of negative-resistance elements is an effective strategy for achieving high-Q performance in tunable SIW resonators [1, 98, 108]. This high-Q operation is particularly beneficial for low-phase-noise oscillator design in advanced RF and microwave systems.

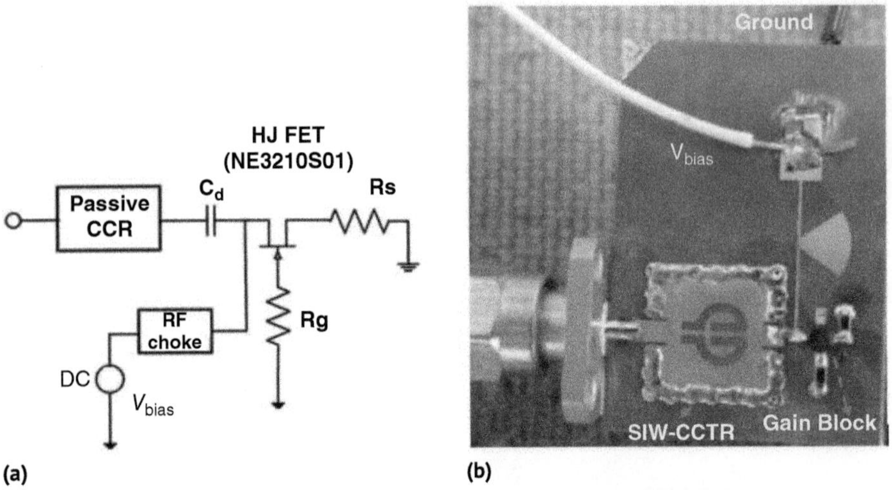

Figure 6.52 (a) Circuit layout of the active CCR (Cd = 0.3 pF, Rg = 50 Ω, Rs = 127 Ω), and (b) photograph of the fabricated SIW-ACCTR (substrate-integrated waveguide active CCTR) on an RT/Duroid 5880 substrate (0.508 mm thick, dielectric constant 2.2) [102].

Table 6.4 Measured unloaded quality factor (Q).

V_{Bias}	**0.0 V**	**0.652 V**	**1.01 V**	**1.21 V**
Unloaded Q-factor	131	101	715	1915
V_{Bias}	**1.41 V**	**1.62 V**	**1.81 V**	**2.01 V**
Unloaded Q-factor	3402	7310	21260	7089

Figure 6.53a,b illustrate, respectively, the layout and photograph of a tunable oscillator realized using the SIW-CCTR (substrate-integrated waveguide CCTR) network depicted in Figure 6.50 [10]. The oscillator circuit is fabricated on an RT/-Duroid 5880 substrate with a thickness of 0.508 mm and a relative dielectric constant of 2.2, providing a low-loss medium suitable for high-frequency operation [109].

For active loss compensation, a heterojunction FET (HJ-FET NE3210S01, NEC) is employed due to its low power consumption and high-frequency performance. The transistor is biased with a drain voltage of $V_d = 1.5\text{ V}$, $I_{ds} = 22\text{ mA}$, while the varactor diodes are reverse-biased at a common tuning voltage V_{tune}. The transistor gate is DC-grounded $(V_g = 0\text{ V})$ through an inductive load integrated into the SIW-CCTR structure, as shown in Figure 6.32 [10]. A DC-blocking capacitor is included in the output path to prevent unwanted DC leakage, ensuring stable RF signal extraction while maintaining high-Q performance [10].

To ensure stable oscillation, $\boldsymbol{\Gamma}_{\text{in}}$ must exceed unity [102]. This condition is achieved by connecting two short-circuited stubs with series inductors to the transistor's source terminals, increasing instability within the desired frequency range. The stub lengths are optimized so that $R_{\text{res}} = -\frac{R_{\text{in}}}{3}$, $X_{\text{res}} = -X_{\text{in}}$ satisfying the oscillation criteria [98]. The measured oscillation frequency and output power, calibrated for cable loss versus bias voltage, are shown in Figures 6.38–6.41.

Figure 6.54 illustrates the continuous frequency tuning of the SIW-CCTR oscillator from 4.850 to 5.100 GHz, achieved by varying the varactor reverse-bias voltage from 0 to 20 V. This corresponds to a total tuning span of 5.16%. Within this operating range, the measured output power remains relatively stable, ranging from 6 to 8 dBm, demonstrating consistent RF performance across the tuning band.

Figure 6.55 shows the measured output spectrum of the oscillator at a nominal frequency of 5.0933 GHz. The phase-noise measurement indicates a value of −115.2 dBc/Hz at a 1 MHz offset, highlighting the oscillator's low-noise performance [40]. These results confirm that the combination of the SIW-CCTR structure and active loss compensation provides both wide tunability and high spectral purity, making it suitable for applications requiring agile, low-phase-noise microwave sources [1].

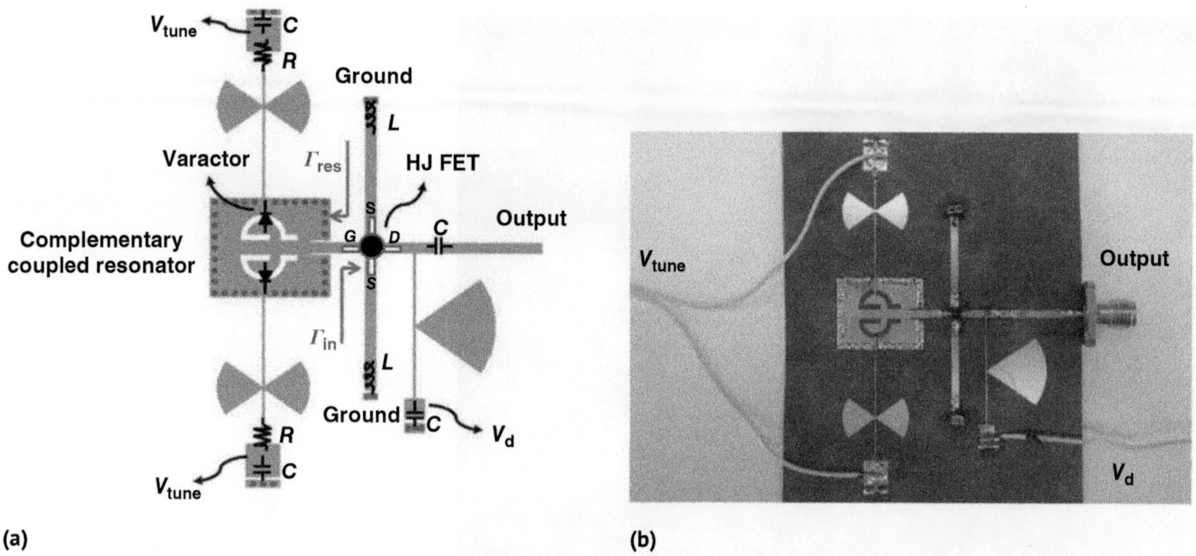

Figure 6.53 (a) Layout of a tunable oscillator utilizing the SIW-CCTR (substrate-integrated waveguide CCTR) network from Figure 6.32; (b) photograph of the fabricated oscillator implementing the same SIW-CCTR configuration [10, 98, 102].

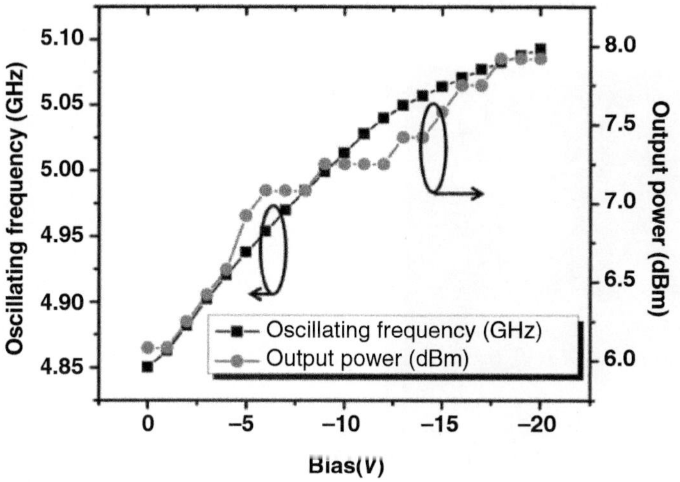

Figure 6.54 The measured tuning characteristics and output power changes of the tunable VCO circuit from Figure 6.46, under different bias voltages [98].

The FOM is calculated using Eq. (6.50) [10].

$$\mathrm{FOM}\big|_{f_{\mathrm{offset}}} = \left[\pounds(f_{\mathrm{offset}}) - 20\left(\frac{f_0}{f_{\mathrm{offset}}}\right) + 10\left(\frac{P_{\mathrm{DC}}}{1\,\mathrm{mW}}\right)\right]\left(\frac{\mathrm{dBc}}{\mathrm{Hz}}\right) \tag{6.54}$$

$$\mathrm{FOM}\big|_{f_{\mathrm{offset}=1\,\mathrm{MHz}}} = -174.1\left(\frac{\mathrm{dBc}}{\mathrm{Hz}}\right) \tag{6.55}$$

where £(offset) denotes the phase noise at offset frequency f_{offset}, f_0 is the oscillation frequency, f_{offset} is expressed in MHz, and P_{DC} represents the total DC power consumption in milliwatts.

Figure 6.56a,b depict the layout and photograph of a tunable X-band oscillator based on an AR (as shown in Figure 6.52), fabricated on an RT/Duroid 5880 substrate with a thickness of 0.508 mm and a dielectric constant of 2.2. In Figure 6.56a, a heterojunction FET (HJ-FET NE3210S01, NEC) is employed due to its low power consumption and high DC-to-RF conversion efficiency [11]. To promote oscillation at the desired frequency, a shunt stub is connected to the

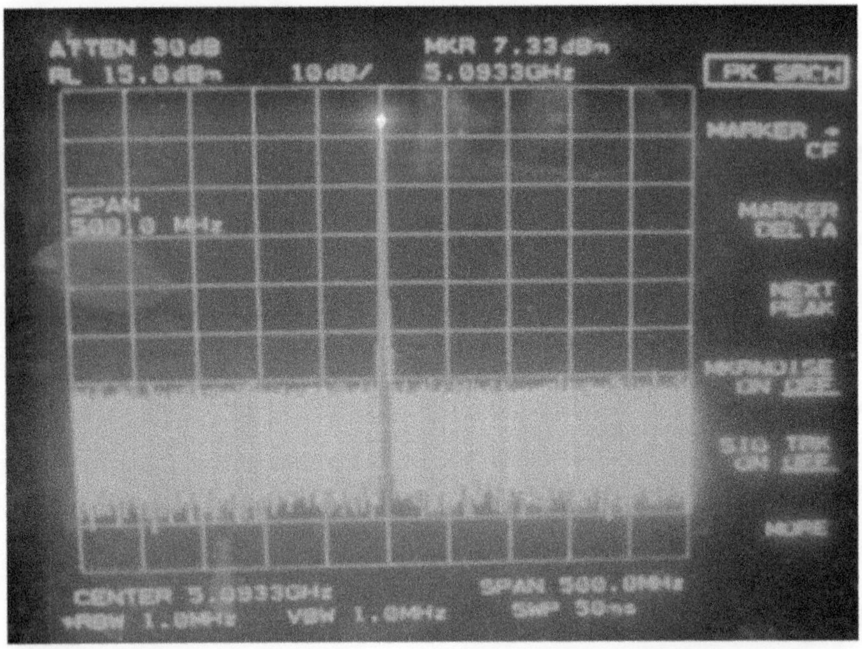

Figure 6.55 Measured output spectrum of the oscillator shown in Figure 6.46 at 5.0933 GHz [98].

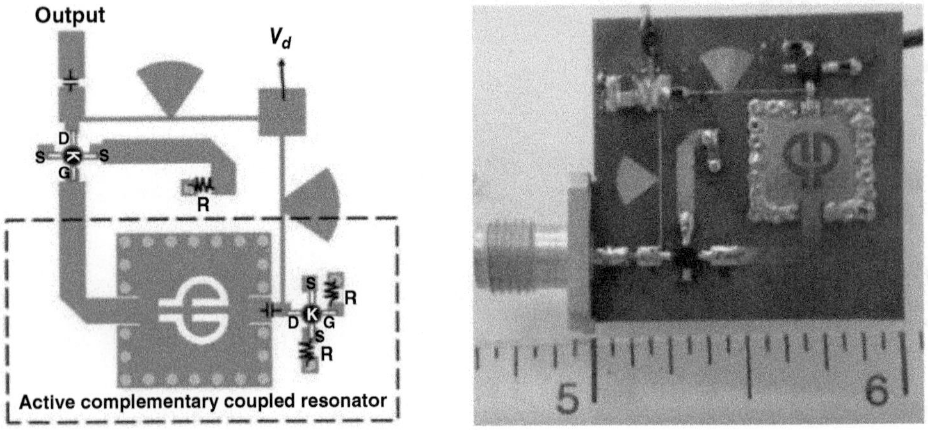

(a) (b)

Figure 6.56 (a) Layout of a tunable X-band oscillator using the AR from Figure 6.34 [10], and (b) photograph of the fabricated oscillator on an RT/Duroid 5880 substrate (0.508 mm thick, dielectric constant 2.2) [102].

transistor source, increasing negative resistance and enhancing instability within the target band. For experimental characterization, the oscillator is biased at a drain voltage of $V_d = 2.501$ V and $I_d = 6.01$ mA, optimizing both DC-to-RF efficiency and the FOM. Importantly, the entire oscillator operates with a single bias line (V_d), simplifying the circuitry and enabling a compact, high-performance implementation suitable for X-band applications.

The oscillator design employs the negative-resistance method [94]. The stub length for the primary oscillating transistor is selected such that $R_{res} = -\frac{R_{in}}{3}$, $X_{res} = -X_{in}$, satisfying the oscillation startup conditions, where R_{res} and R_{in} are the real parts Z_{res} and Z_{in}, and X_{res} and X_{in} are their respective imaginary parts. Appropriate DC biasing is achieved by using lumped resistors to self-bias the two transistors, while two capacitors decouple DC from the RF paths.

Figure 6.57a shows the measured RF output spectrum of the X-band oscillator based on the AR illustrated in Figure 6.56. The fundamental oscillation occurs at 9.930 GHz with an output power of 2.361 dBm [10]. The corresponding second harmonic is observed at 19.860 GHz with a significantly lower output power of –18.331 dBm, as presented in

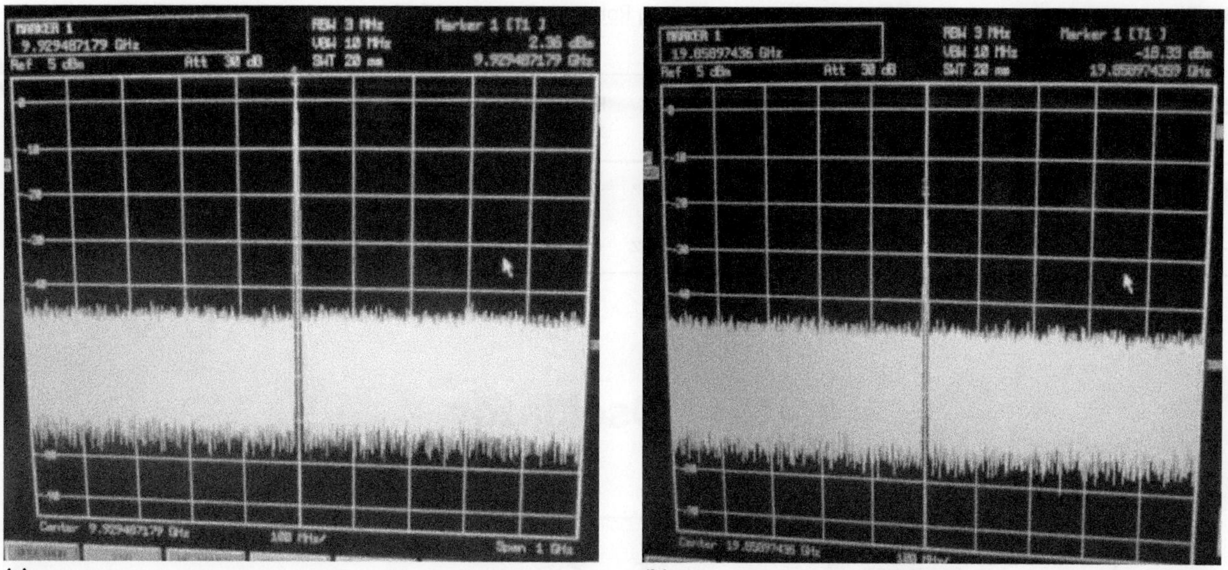

(a) (b)

Figure 6.57 (a) Measured RF output spectrum of the oscillator from Figure 6.56, and (b) second harmonic at −18.33 dBm observed at 19.8589 GHz for the same oscillator [10, 102].

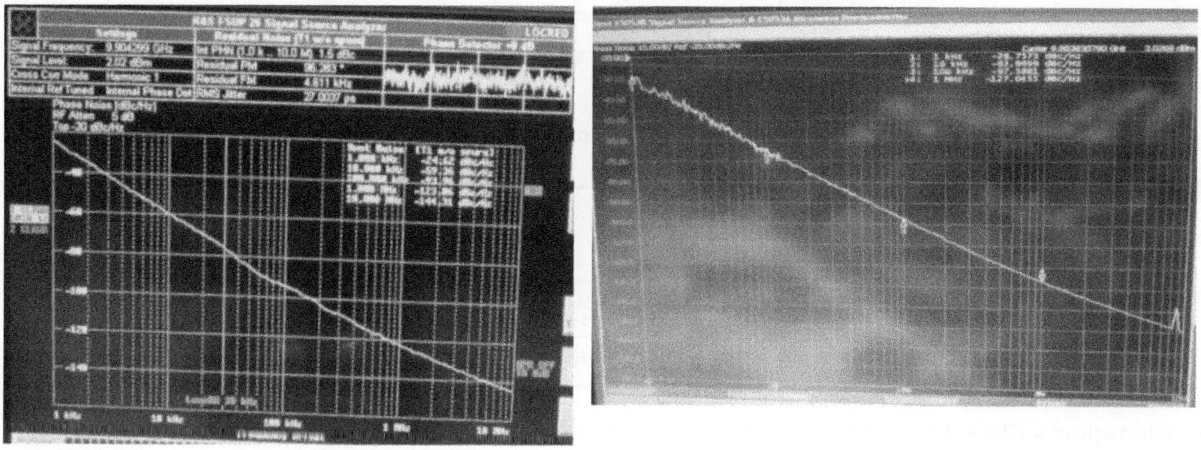

(a) (b)

Figure 6.58 Phase-noise measurements performed with two setups: (a) R&S FSUP26 Signal Source Analyzer, and (b) Agilent-5052B combined with E5053A Microwave Down-Converter [102].

Figure 6.57b [10]. These results highlight the oscillator's strong fundamental signal and effective suppression of higher-order harmonics, confirming the suitability of the AR design for low-phase-noise, high-purity X-band applications. Phase-noise performance of the X-band oscillator was characterized by using two independent measurement setups. In the first setup, a Rohde & Schwarz FSUP26 Signal Source Analyzer measured phase-noise values of −93.961 dBc/Hz at a 100 kHz offset and −123.861 dBc/Hz at a 1 MHz offset, as shown in Figure 6.58a. The second measurement employed an Agilent −5052B Signal Source Analyzer in conjunction with an E5053A Microwave Down-Converter [10]. This configuration recorded phase noise of −97.651 dBc/Hz at 100 kHz offset and −127.02 dBc/Hz at 1 MHz offset, illustrated in Figure 6.58b.

A consolidated summary of these phase-noise measurements is provided in Table 6.5 [102]. These results confirm the oscillator's low-phase-noise performance across multiple measurement platforms, demonstrating the effectiveness of the SIW-based AR design for high-purity X-band signal generation.

As illustrated in Figure 6.56, the oscillator operates at 10 GHz and incorporates a SiGe HBT active device supplied by NEC [10]. The circuit is implemented on a Rogers substrate characterized by a dielectric constant of 3.38 and a thickness of 32 mils, employing microstrip transmission lines as the guiding medium [1]. The choice of active device plays a critical role in

Table 6.5 Phase-noise measurements of the X-band oscillator using Rohde & Schwartz FSUP and Keysight Agilent E5052B measurement instruments [102].

	Oscillating Frequency	Output Power	Phase Noise@ 100 kHz	Phase Noise@ 1 MHz	DC Power Consumption	FOM@ 100 kHz	FOM@ 1 MHz
Rohde & Schwarz FSUP	9.9042 GHz	2.01 dBm	−93.962 dBc/Hz	−123.87 dBc/Hz	15.01 mW	−182.210	−192.12
Agilent-5052B	9.9294 GHz	2.36 dBm	−97.651 dBc/Hz	−127.02 dBc/Hz	15.01 mW	−185.890	−195.26

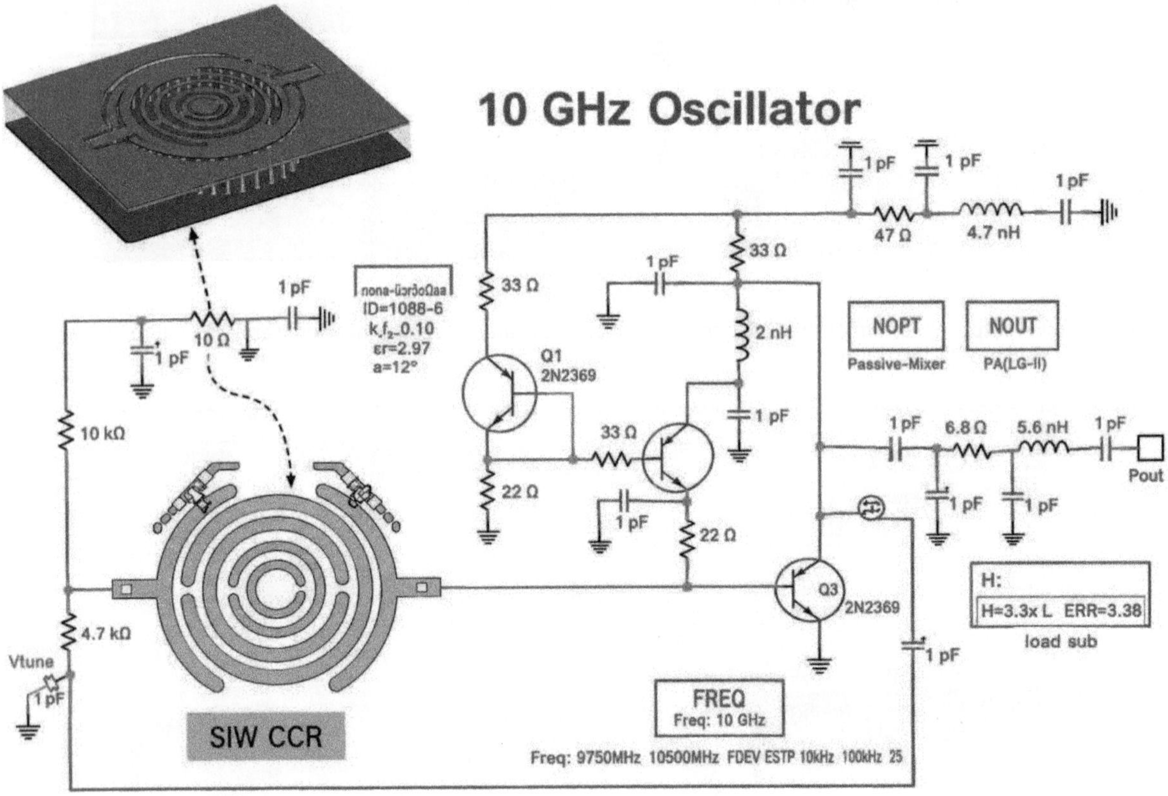

Figure 6.59 Schematic of a 10.0 GHz oscillator employing a printed resonator on a 32 mil Rogers R4003C substrate ($\varepsilon_r = 3.380$, board size 0.50 × 0.50 inches) with DC bias 5.0 V and 12.0 mA [10].

determining the oscillator's spectral purity. Conventional heterojunction field-effect transistors (HJ-FETs) exhibit elevated phase noise at low-frequency offsets (<100 kHz) due to the dominant contribution of $1/f$ (flicker) noise, which modulates the carrier through upconversion mechanisms. By substituting the HJ-FET with a SiGe HBT, the flicker noise contribution is substantially suppressed. This improvement stems from the superior noise performance and higher transconductance efficiency of SiGe HBTs, which effectively reduce the close-in phase noise and enhance overall oscillator stability [14, 94].

A novel oscillator configuration, depicted in Figures 6.59 and 6.60 [102], employs Infineon BFP540 SiGe HBT devices and demonstrates a remarkable phase-noise performance enhancement. Specifically, the circuit achieves more than 40 dB reduction in phase noise at a 10 kHz offset from the 10 GHz carrier when compared with conventional oscillator implementations, though at the cost of increased DC power dissipation [12, 50, 51]. Figures 6.59 and 6.60 present the oscillator schematic and the corresponding phase-noise measurement results, respectively [10, 102].

A key innovation in this design is the incorporation of a bias-stabilization network, which maintains a highly stable operating point across a broad temperature range (−40 °C to +85 °C). This feature minimizes performance drift under environmental variations and contributes to the robustness of the oscillator. Experimental results confirm that the oscillator delivers a consistent output power of 4.75 dBm while operating with a 5 V supply and a bias current of 12 mA. The substrate-integrated waveguide complementary coupled-resonator (SIW-CCR) [10] oscillator architectures shown in

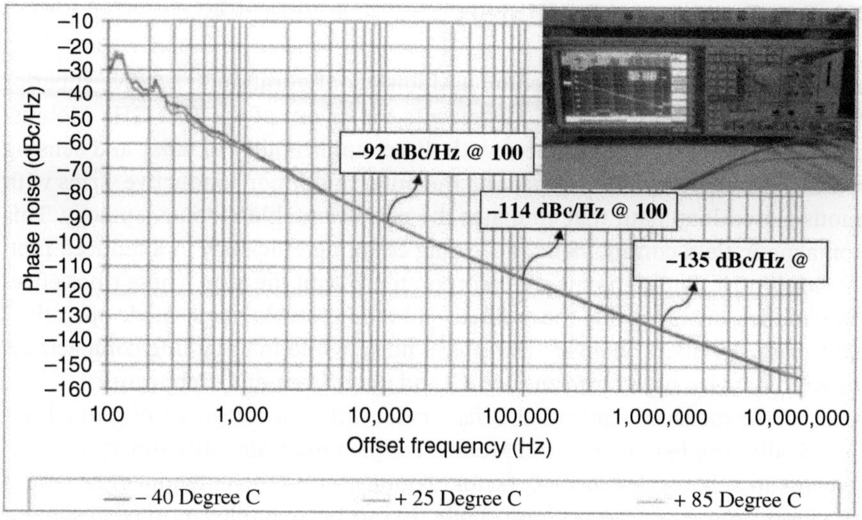

Figure 6.60 Measured phase noise of the 10.0 GHz oscillator from Figure 6.59 [10], showing stable performance over −40 °C to +85 °C, thanks to the novel bias-stabilization circuit that optimizes phase-noise behavior [10].

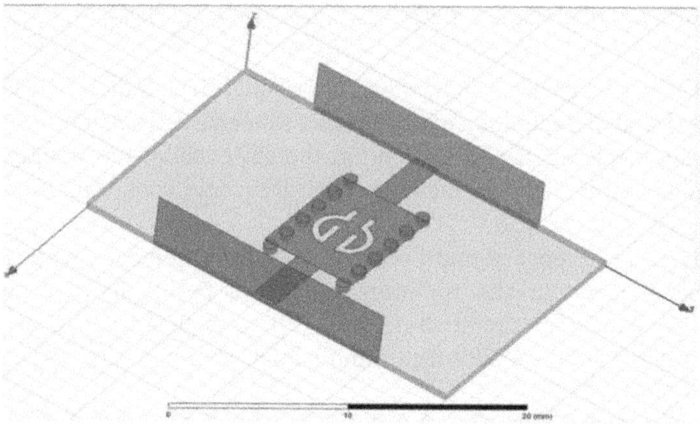

Total X size	15.2 mm
X_res	7.2 mm
Y_res	6 mm
Zsub	.508 mm
Via diameter	0.8 mm

Figure 6.61 CAD schematic of a typical Ku-band SIW-CCR [10].

Figures 6.56, 6.59, and 6.60 [10], originally demonstrated in the C- and X-band frequency ranges, exhibit excellent scalability. Owing to the inherent field-confinement and low-loss characteristics of SIW structures, these topologies can be straightforwardly extended to higher frequency regimes, including Ku- and Ka-bands, where stringent requirements on phase-noise and power stability are critical.

Figure 6.61 illustrates the CAD-simulated layout of a Ku-band substrate-integrated waveguide complementary coupled resonator (SIW-CCR) [10]. The structure occupies a compact footprint of 7.2 mm × 6 mm and is designed to resonate at a center frequency of approximately 16 GHz. Full-wave electromagnetic simulations performed using Ansys HFSS predict an unloaded quality-factor (Q) of about 350 for this configuration, highlighting its suitability for low-loss oscillator applications in the Ku-band.

The same CCR topology can be readily scaled for higher-frequency operation. For example, when miniaturized to a footprint of 3.6 mm × 3 mm, the structure supports a Ka-band resonance near 32 GHz, with a simulated unloaded Q of approximately 330. This demonstrates the scalability and versatility of the SIW-CCR architecture, enabling efficient oscillator design across multiple millimeter-wave frequency bands. In addition to SIW-CCR designs, the chapter explores Möbius-inspired resonators derived from slow-wave structures. These resonators introduce a unique geometry that enables high-Q performance within a compact form factor while also supporting wide tuning capability. When implemented in oscillator circuits, Möbius SWRs have been shown to achieve low phase noise and strong frequency agility, underscoring their potential as a promising alternative for next-generation compact, high-performance RF and millimeter-wave systems.

6.7 Tunable Möbius Coupled Printed Resonator Oscillators

Poddar [10] introduced a novel printed coupled-resonator topology based on a Möbius configuration to achieve superior performance in a compact form factor operating at 10.0 GHz. Figure 6.62 presents the layout of the 10.0 GHz Möbius-resonator-based oscillator [9], designed as a low-phase-noise signal source with direct applicability to radar and sensing systems. The planar Möbius resonator [10] is implemented by interconnecting the outer and inner conductive strips with a metallized via, thereby creating a continuous closed-loop path that emulates the non-orientable Möbius topology. This unique structure provides enhanced electromagnetic field confinement and strong coupling efficiency in a reduced footprint. To further improve spectral purity, two concentric ring elements are incorporated within the loop. These rings effectively suppress parasitic and spurious resonant modes, ensuring that the oscillator maintains stable single-mode operation. Through this approach, the Möbius-resonator-based oscillator delivers low phase noise while maintaining compactness and design simplicity, making it a promising candidate for high-performance RF and radar front-end architectures.

Figure 6.63 presents a comparative summary of the measured unloaded Q-factors for different categories of printed resonators, including electrically coupled, magnetically coupled, hybrid, and Möbius-coupled resonator structures [10].

The results highlight significant improvements in achievable planar Q-factor, demonstrating the potential of advanced coupling mechanisms to enhance resonator performance. Among these, both the Möbius-coupled printed resonator (MCPR) and the hybrid resonator act as effective Q-multipliers [10]. A notable peak Q is observed near 9.7 GHz, attributed to mode convergence effects that enhance energy storage within the resonant structure. While this property contributes to superior spectral purity, the Möbius resonator exhibits an additional design challenge—its tendency toward multi-mode behavior. This arises from non-uniform coupling along the resonator strips, a limitation imposed by approximating the inherently 3D Möbius geometry in a planar implementation.

To address this issue, meticulous design refinement and electromagnetic optimization are essential. Techniques such as mode-selective loading, careful adjustment of strip geometry, and incorporation of suppressive elements can help mitigate unwanted resonances. These measures are critical to ensuring stable single-mode operation, thereby enabling the Möbius-based oscillators to deliver the low-jitter performance required for demanding radar and high-frequency communication applications.

A key limitation of Möbius-resonator-based oscillators is the occurrence of multi-mode frequencies and mode-jumping within the resonant structure, which constrain broadband operation and compromise spectral purity. To mitigate these effects, a mode-stabilization circuit is introduced. This auxiliary circuit strategically injects higher-order modes with precisely controlled phase shifts, thereby enhancing the overall stability factor and suppressing undesired mode

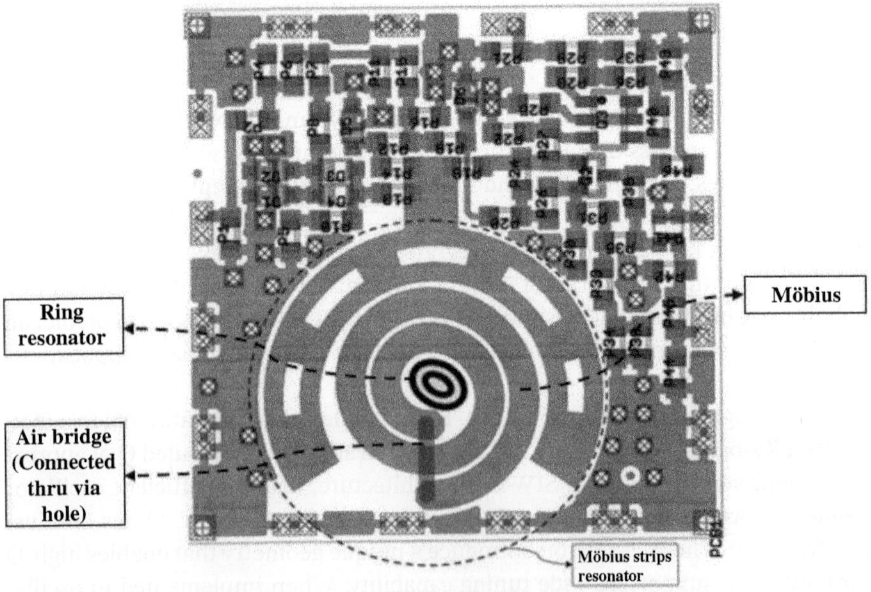

Figure 6.62 Layout of the 10.0 GHz oscillator using a Möbius strip resonator for high-Q performance. PCB implemented on a 220-mil substrate (ε_r = 3.380, size 0.50 × 0.50 × 0.18 inches) [10].

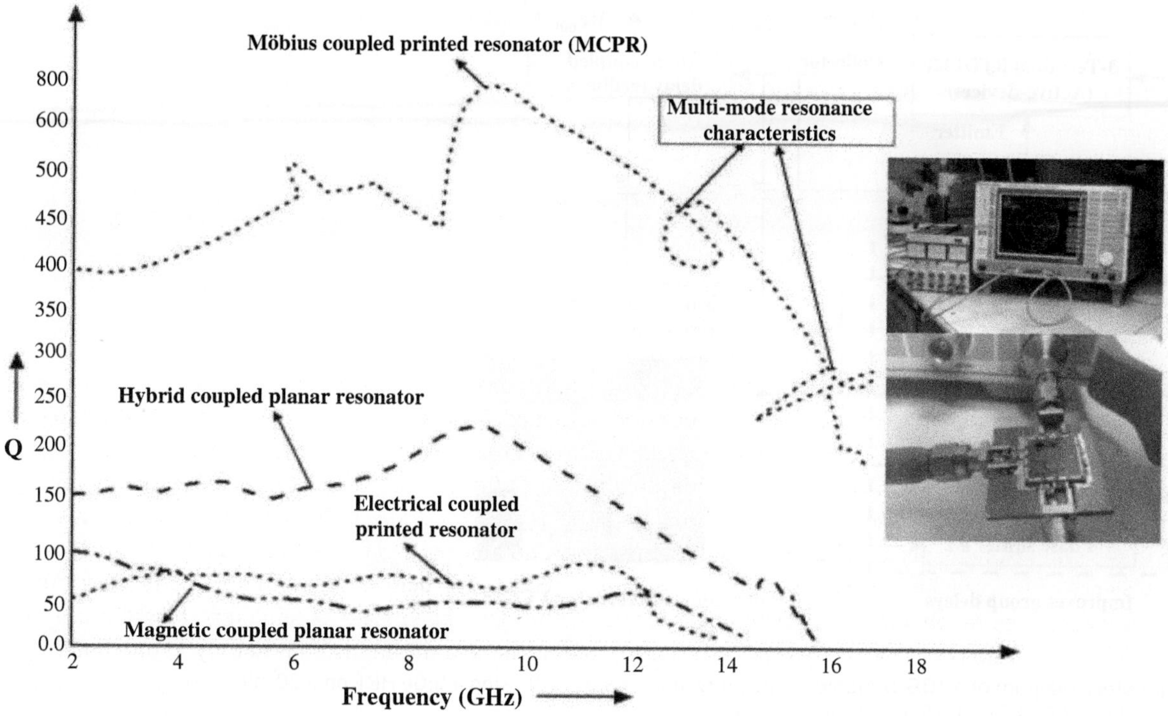

Figure 6.63 Measured unloaded Q-factor of various printed resonator structures—electrically coupled, magnetically coupled, hybrid, and Möbius-coupled—highlighting Q-factor improvements in the planar domain. PCB implemented on a 20.0 mil substrate ($\varepsilon_r = 2.22$, size 0.50 × 0.50 × 0.18 inches) [10].

transitions. The incorporation of this stabilization mechanism, however, introduces certain trade-offs. Specifically, the unloaded Q-factor is reduced by approximately 10–25%, and the phase noise degrades by 2.5–6.5 dB relative to the uncompensated design.

Despite these penalties, the suppression of mode-jumping ensures consistent single-mode oscillation, enabling robust and predictable performance. This design approach provides a practical and cost-effective solution when compared to dielectric resonator oscillators (DROs), particularly for X- and Ka-band radar systems where low phase noise, compactness, and affordability are critical. By balancing stability with acceptable performance compromises, the stabilized Möbius-resonator oscillator offers a viable alternative for next-generation radar and high-frequency communication applications.

Figure 6.64 illustrates the block diagram of a VCO [40] based on a Möbius coupled resonator topology, designed to operate at 10.000236 GHz. The circuit employs a SiGe HBT as the active device, chosen for its high transconductance efficiency and superior close-in noise performance. The oscillator is implemented on a 20 mil-thick substrate with a relative dielectric constant ($\varepsilon_r = 2.20$), providing a suitable balance between low dielectric loss and compact physical dimensions, thereby validating the feasibility of the proposed planar architecture.

Figure 6.65 presents the measured phase-noise characteristics of the prototype. At a 10 kHz frequency offset from the 10 GHz carrier, the oscillator achieves a phase noise of −112 dBc/Hz. This performance underscores the effectiveness of the Möbius-resonator topology in achieving high-Q resonance while maintaining a compact footprint. The results confirm that the proposed design offers a promising low-phase-noise alternative for radar, communication, and sensing applications, where both miniaturization and spectral purity are critical [22, 41, 42, 110].

As illustrated in Figure 6.65, the measured oscillator FOM, calculated using Eq. (6.50), highlights the comparative performance of different resonator topologies. The hybrid-coupled resonator achieves an FOM of −196 dBc/Hz, while the Möbius-coupled resonator demonstrates improved noise efficiency with an FOM of −203 dBc/Hz. The mode-locked MCPR [10] network exhibits the best performance, reaching −205 dBc/Hz under identical operating conditions.

All three oscillator configurations were characterized with the same DC power consumption of 160.1 mW ($V_{CC} = 5.0$ V, $I_c = 32.0$ mA). The corresponding DC-to-RF conversion efficiencies are 1.20% for the hybrid-coupled design, 1.30% for the Möbius-coupled version [10], and 1.70% for the mode-locked MCPR structure [10]. These results underscore the advantage of the Möbius-based architectures, particularly the mode-locked MCPR, in achieving higher spectral efficiency

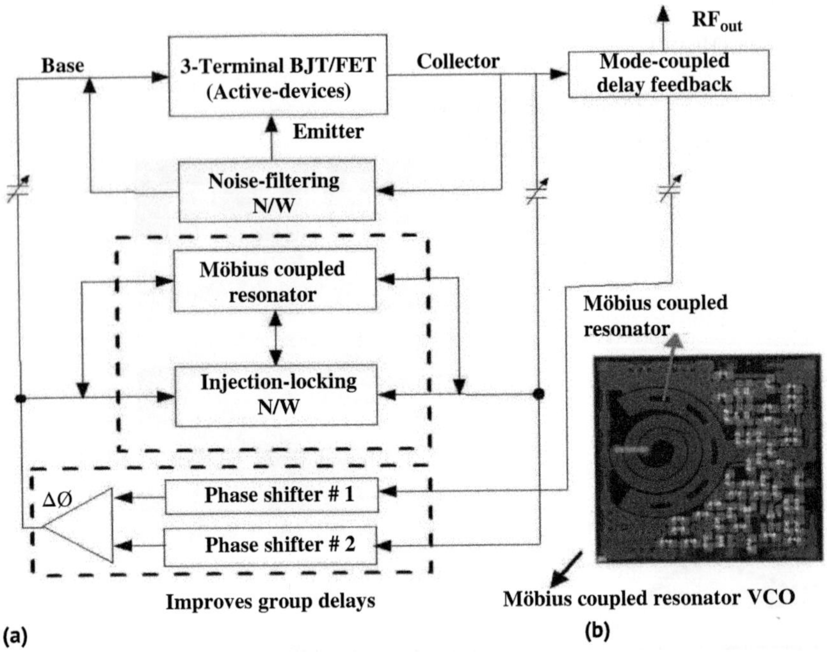

Figure 6.64 (a) Block diagram of a 10 GHz Möbius-coupled resonator VCO [10] using a SiGe HBT on a 20 mil substrate ($\varepsilon_r = 2.20$), and (b) the corresponding PCB (0.75 × 0.75 × 0.18 inches) [10].

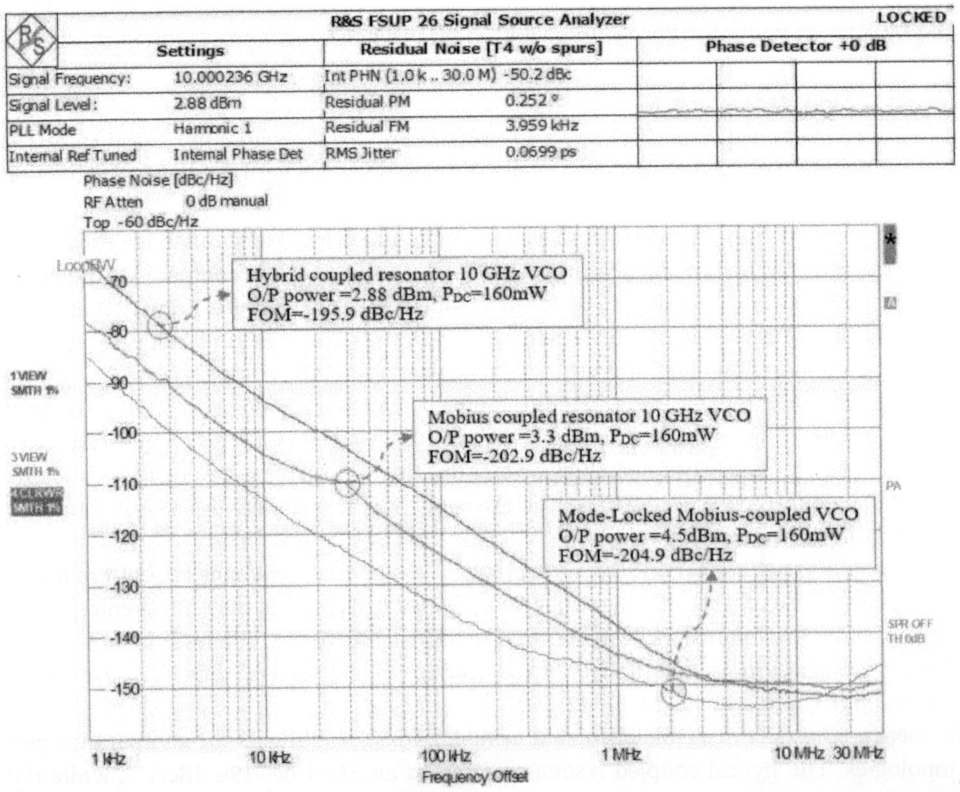

Figure 6.65 Measured phase noise of the 10.000236 GHz oscillator using hybrid-coupled, Möbius-coupled, and mode-locked Möbius-coupled printed resonator networks. The corresponding FOMs (Eq. 6.50) are −196, −203, and −205 dBc/Hz, respectively, with a DC power consumption of 160 mW ($V_{CC} = 5.0$ V, $I_c = 32.0$ mA) [9, 10, 14].

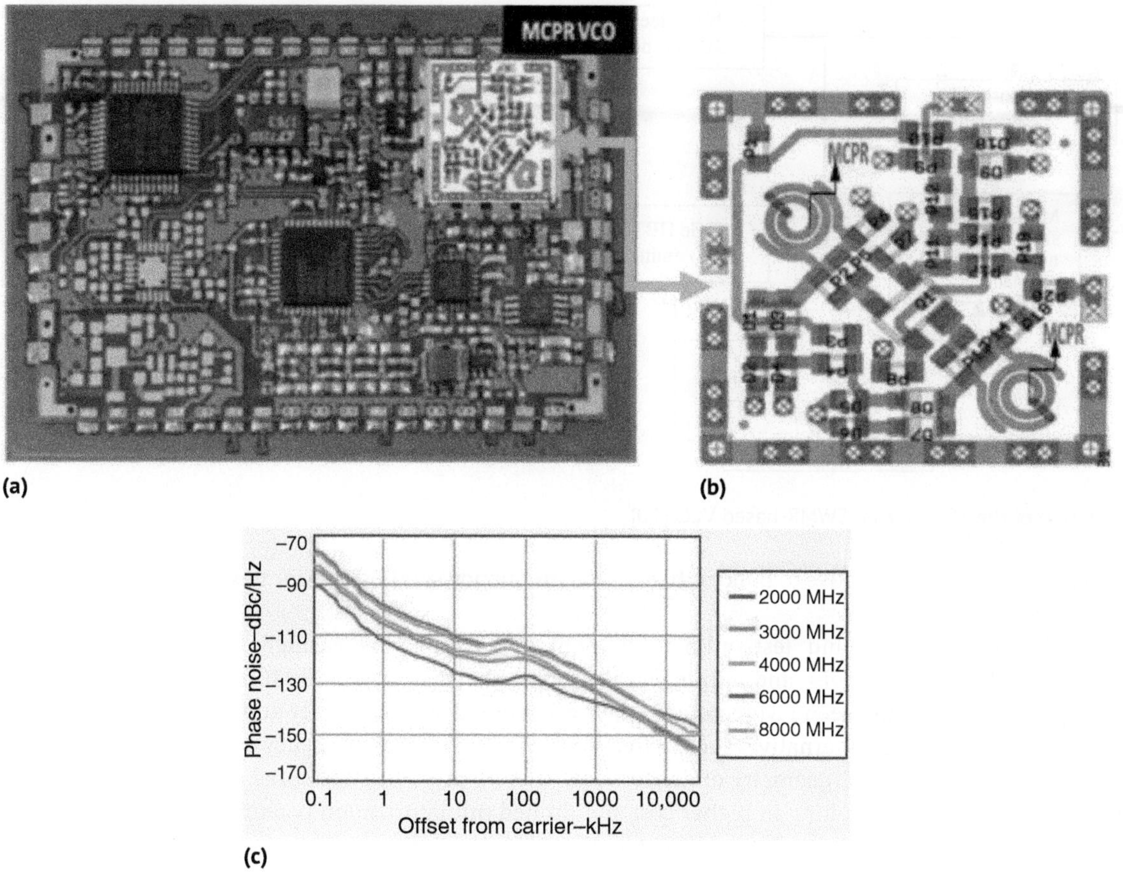

Figure 6.66 High-performance wideband signal sources for modern communication systems [1]: (a) PCB layout of the 1.9–8.1 GHz configurable synthesizer module; (b) layout of the 1.8–8.5 GHz MCPR VCO; (c) measured phase noise of the synthesizer from Figure 6.65 [10].

and superior phase-noise performance without increasing power dissipation. Collectively, they demonstrate that optimized resonator topologies can provide compact, low-noise, and energy-efficient solutions for advanced RF and radar oscillator applications [10].

Figure 6.66a presents the layout of a broadband frequency synthesizer covering the 1.9–8.1 GHz range. The design is realized using a wideband tunable MCPR VCO, which itself achieves a tuning span from 1.9 to 8.5 GHz. The corresponding VCO layout is shown in Figure 6.64b [11]. This MCPR-based approach offers a practical and cost-efficient alternative to YIG resonator oscillators, which, while traditionally used for broadband synthesis, suffer from high cost, large physical size, and sensitivity to environmental factors [11]. In contrast, the MCPR VCO provides comparable wide tuning capability in a compact planar implementation, while demonstrating significantly reduced susceptibility to thermal drift, mechanical vibration, and microphonic disturbances. These advantages make the MCPR-based synthesizer particularly attractive for portable, low-cost, and robust RF systems requiring broadband frequency agility.

The broadband synthesizer operates from a 5.0 V supply and draws approximately 200 mA, integrating a multi-band, multi-mode MCPR VCO that itself consumes 32.0 mA at 5.0 V. This configuration results in a cost-effective, power-efficient, and reconfigurable design suitable for wideband RF applications. Figure 6.66b shows the physical layout of the 2–8 GHz MCPR VCO [10], implemented on a 22-mil substrate ($\varepsilon_r = 2.20$) with compact dimensions of 0.30 × 0.30 × 0.18 inches.

The measured phase-noise characteristics of the mode-stabilized synthesizer, based on the VCO in Figure 6.65a, are presented in Figure 6.66c. Across the 2.1–7.9 GHz tuning range, the synthesizer achieves phase noise better than −108 dBc/Hz at a 10 kHz offset from the carrier. In addition, the design demonstrates strong harmonic suppression (>−20 dBc), fast frequency settling time (<1 ms), and low sideband spurious levels (<−60 dBc) [9, 40]. With overall dimensions of just 1.0 × 1.0 × 0.2 inches, the synthesizer offers a compact and low-power solution while leveraging the benefits of MCPR technology. The combination of wide tuning capability, excellent spectral purity, and robust suppression of unwanted signals

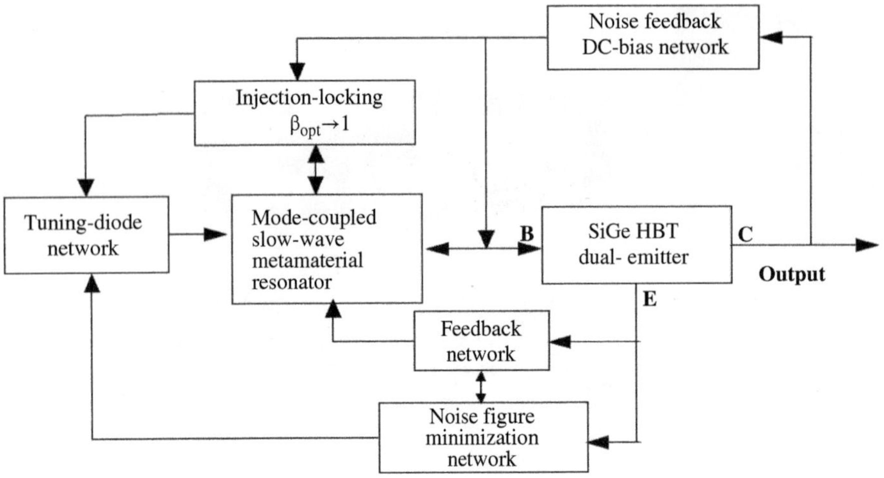

Figure 6.67 Block diagram of the 10.2 GHz MCSWMR-based VCO [10].

establishes this design as a practical alternative to bulkier, higher-cost frequency synthesizers for next-generation communication and radar systems.

Modern military, industrial, medical, and test-and-measurement applications increasingly require highly stable frequency sources that exhibit low phase noise and minimal sensitivity to thermal variations [10]. Within the 3–30 GHz frequency range, DROs have traditionally been favored due to their exceptional spectral purity and ultra-low phase-noise performance, significantly outperforming alternatives such as frequency-multiplied low-frequency oscillators. Despite their performance advantages, the inherently 3D geometry of dielectric resonators presents substantial challenges for integration into planar or monolithic circuits, resulting in higher cost and limited miniaturization potential. To overcome these constraints, a novel mode-coupled self-injection-locked VCO has been developed, providing a cost-effective and compact alternative to conventional DROs for high-performance reference signal generation in modern RF systems.

This design exploits the mode-coupled slow-wave metamaterial resonator (MCSWMR) topology [10], which offers a high unloaded Q-factor, low phase noise, and wideband tunability while remaining fully compatible with planar fabrication techniques. The MCSWMR approach enables seamless integration into compact RFIC or PCB platforms, delivering DRO-level performance in a significantly reduced footprint. Consequently, the MCSWMR-based VCO provides a practical, scalable, and manufacturable solution for applications that have traditionally depended on bulky and costly DRO architectures.

The proposed design methodology for tunable oscillators facilitates the development of compact VCOs with outstanding phase-noise performance, fully compatible with contemporary RFIC and MMIC fabrication processes. By leveraging this approach, the resulting VCOs deliver performance comparable to traditional high-cost, large-footprint, and power-intensive DROs and YIG oscillators, while providing substantial advantages in size reduction, integration potential, and manufacturability [11].

Figures 6.67–6.69 [10] illustrate the block diagram, physical layout, and measured phase-noise characteristics of a 10.212 GHz VCO, respectively [14, 111]. These oscillators utilize a SiGe HBT as the active device and are fabricated on a 20-mil Rogers substrate with a dielectric constant (ε_r) of 2.2, employing microstripline and stripline techniques for planar implementation. The combination of high-performance active-device selection, optimized resonator topology, and precise planar fabrication yields VCOs that achieve low phase noise, compact size, and cost efficiency. This makes them highly suitable for modern RF applications, where miniaturization, manufacturability, and spectral purity are critical, providing a practical and scalable alternative to conventional bulky resonator-based solutions.

The oscillator circuit depicted in Figure 6.67 is designed to operate over a supply voltage range of +5 to +10 V, incorporating internal voltage regulation to ensure strong immunity against power-supply fluctuations and noise [10]. Under typical operating conditions, the circuit draws approximately 20 mA of supply current and is capable of functioning reliably over a wide temperature range from −40 °C to +85 °C, making it suitable for both commercial and harsh-environment applications. For the specific layout shown in Figure 6.68, applying a DC bias of 10.0 V with a current of 30.0 mA results in an output power exceeding +5.0 dBm. This demonstrates that the oscillator can deliver robust RF performance while maintaining compact size and low power consumption, consistent with the design objectives for high-performance planar VCOs in modern RF systems.

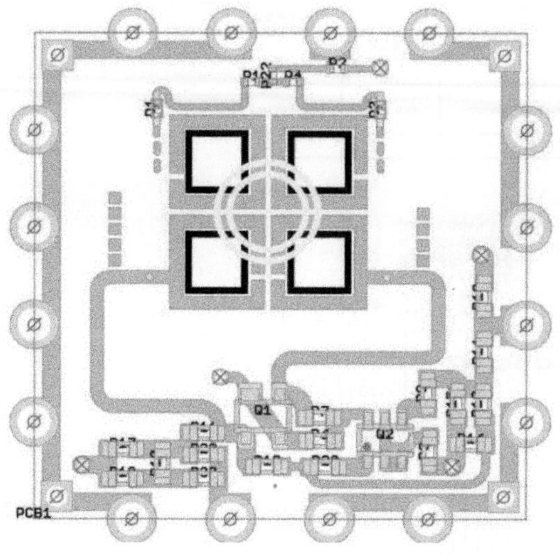

Figure 6.68 Layout of the 10.2 GHz MCSWMR-based VCO on 20 mil Rogers's substrate ($\varepsilon_r = 2.2$) [10, 42].

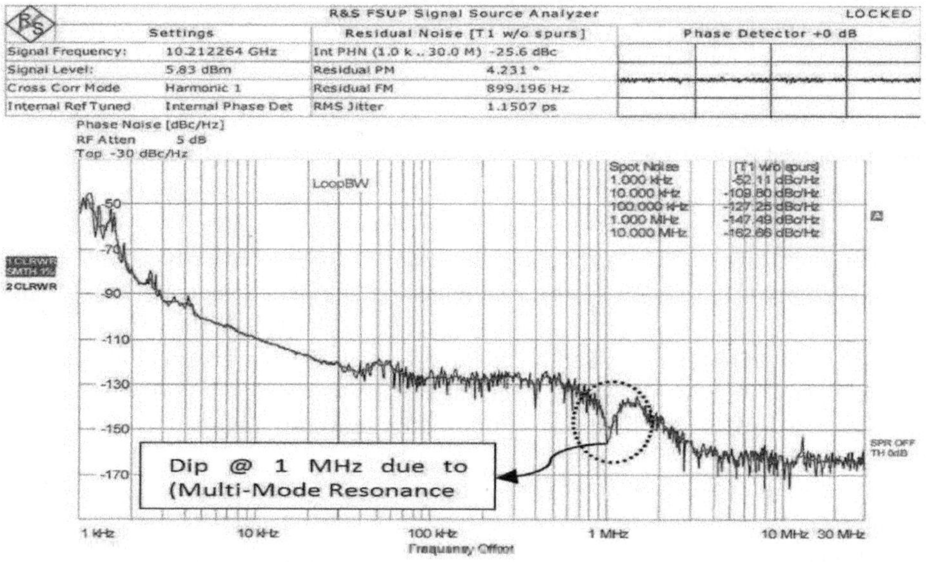

Figure 6.69 Measured phase-noise plot of the MCSWMR-based VCO [10].

As observed in Figure 6.69, the measured phase-noise profile exhibits a pronounced hump between 40 kHz and 1 MHz, followed by a dip near the 1 MHz offset [10]. These features are attributed to multi-mode resonances and mode-jumping phenomena within the resonator, which can degrade oscillator stability and spectral purity. To mitigate these effects, a multi-mode injection-locking technique is employed. By selectively reinforcing specific resonator modes with controlled phase and amplitude, this approach effectively suppresses undesired mode transitions, thereby stabilizing the oscillation frequency and improving overall phase-noise performance. The result is a VCO that maintains low phase noise across the operating offset range while ensuring robust single-mode operation, critical for high-performance RF and radar applications.

Figure 6.70 presents the measured phase-noise spectrum of the oscillator, demonstrating a clean and stable output across the evaluated offset range [10]. The oscillator achieves a FOM (Eq. 6.50) of −203 dBc/Hz for a PFTN (Eq. 6.51) value of −14.1 dB, highlighting its high spectral efficiency. The circuit operates with a DC supply of 10 V and a bias current of 30 mA, resulting in a total power consumption of 300 mW, and delivers an output power of +10.05 dBm [40]. With compact physical dimensions of 0.75 × 0.75 × 0.18 inches, the oscillator maintains stable operation across a wide temperature

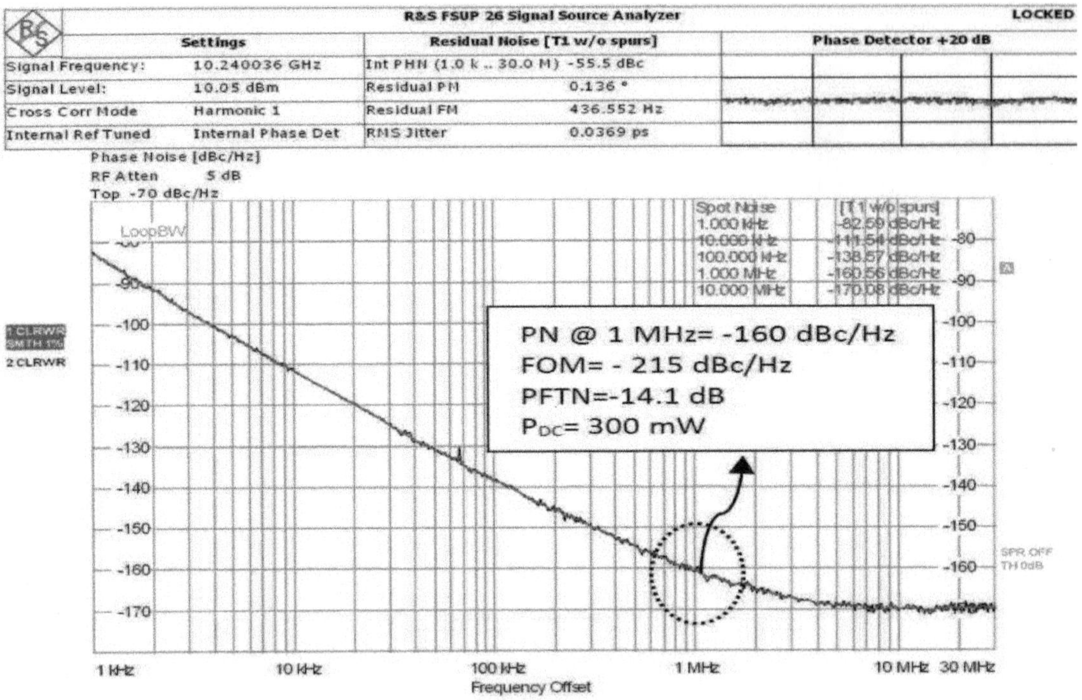

Figure 6.70 Measured phase noise of the mode-locked 10.24 GHz oscillator, showing suppression of the hump/dip observed in Figure 6.69 [10].

range of –40°C to +85°C. This thermal robustness provides ample margin to accommodate frequency drift caused by temperature variations, package parasitics, and component tolerances, ensuring reliable performance in demanding RF and radar system applications.

For compact VCO applications, a series of printed resonators based on the mode-coupled slow-wave metamaterial resonator (MCSWMR) topology have been developed [10]. Figure 6.71a–f depicts the surface-mount-device (SMD) implementations, each packaged in a compact 0.50 × 0.50 × 0.18-inch footprint, optimized for low-cost, high-performance signal sources.

Figure 6.72 shows the CAD-simulated phase-noise characteristics of these designs, highlighting performance trends as a function of operating frequency. As the frequency increases from 3.2 to 18 GHz, a gradual degradation in phase-noise performance is observed, influenced by factors such as the planar resonator topology, substrate properties, and limited layout area. These simulations underscore the trade-offs inherent in miniaturized planar resonator designs, where higher frequencies demand careful optimization of coupling, Q-factor, and layout to maintain low phase-noise performance in compact VCO implementations.

With recent advancements in MMIC fabrication, resonator networks inspired by artificial composite negative-index metamaterials have emerged as a highly promising approach for high-frequency signal generation and processing in modern electronic systems. These metamaterial-based resonators leverage engineered sub-wavelength structures to realize tailored electromagnetic responses, offering enhanced quality factors, compact physical footprints, and frequency-selective characteristics that are difficult to achieve using conventional resonator designs.

Figure 6.73 presents CAD-simulated S_{11} and S_{21} transmission characteristics for three resonator configurations: a conventional metamaterial split-ring resonator, a Möbius-strip resonator, and a metamaterial Möbius-strip resonator. All designs are implemented on a planar substrate with a dielectric constant (ε_r) of 2.2 and a thickness of 12 mils [42]. The simulation results reveal distinct resonance behaviors and frequency-selective properties for each topology, highlighting how metamaterial-inspired geometries can provide enhanced spectral control. These findings underscore the potential of such resonator networks for integration into high-frequency MMIC circuits, where compactness, tunability, and low insertion loss are critical for next-generation RF and microwave applications.

As depicted in Figure 6.73, the metamaterial Möbius strip (MMS) resonators demonstrate superior S_{21} transmission characteristics compared to both conventional metamaterial split-ring and standard Möbius-strip resonators. This enhanced

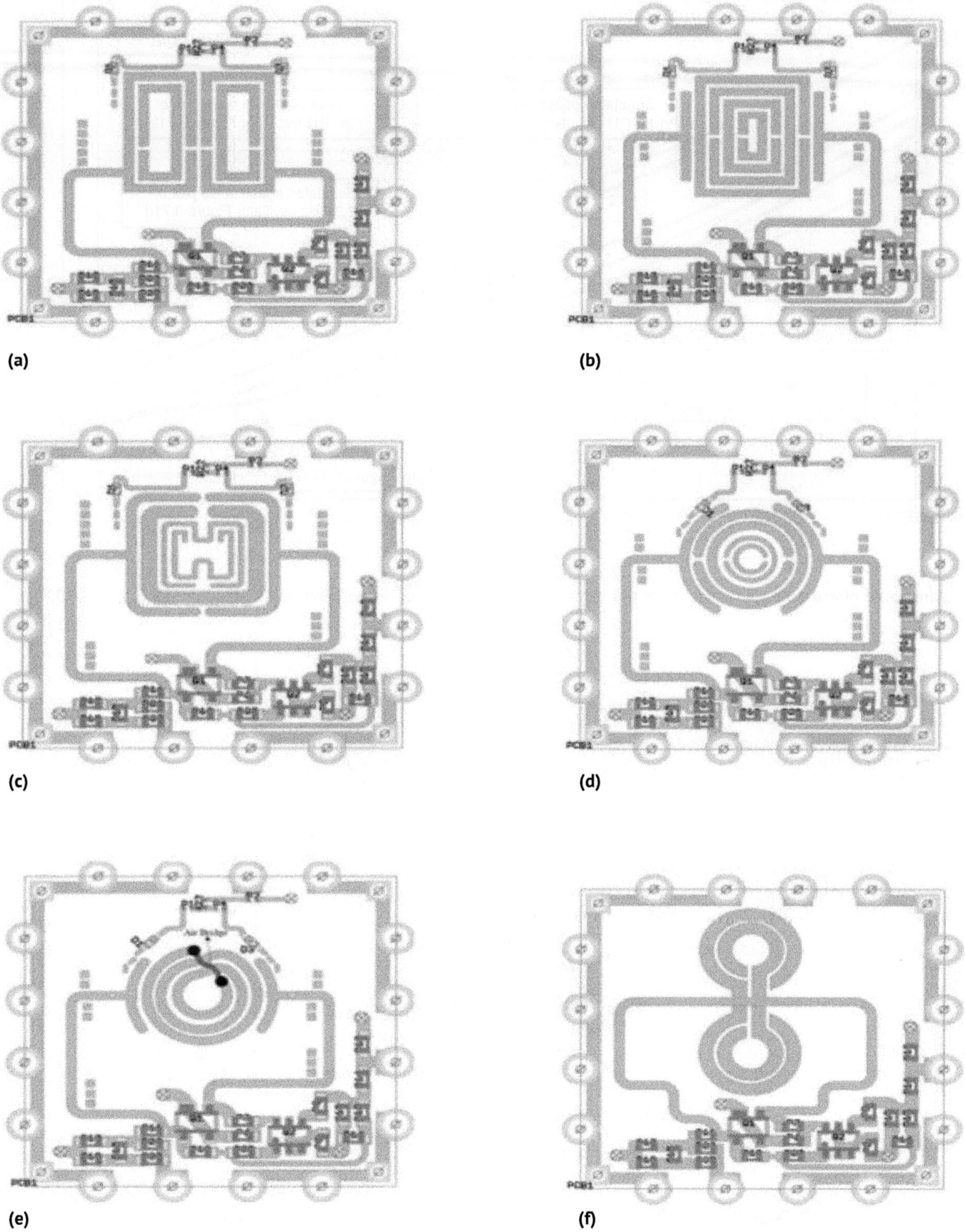

Figure 6.71 Layouts of SMD (surface-mounted device) packages (0.50 × 0.50 × 0.18 inches) for low-cost signal sources [10]: (a) 3.20 GHz, (b) 4.70 GHz, (c) 8.20 GHz, (d) 12.20 GHz, (e) 14.20 GHz, and (f) 18.0 GHz oscillators, all on a 22 mil substrate ($\varepsilon_r = 2.22$) [42].

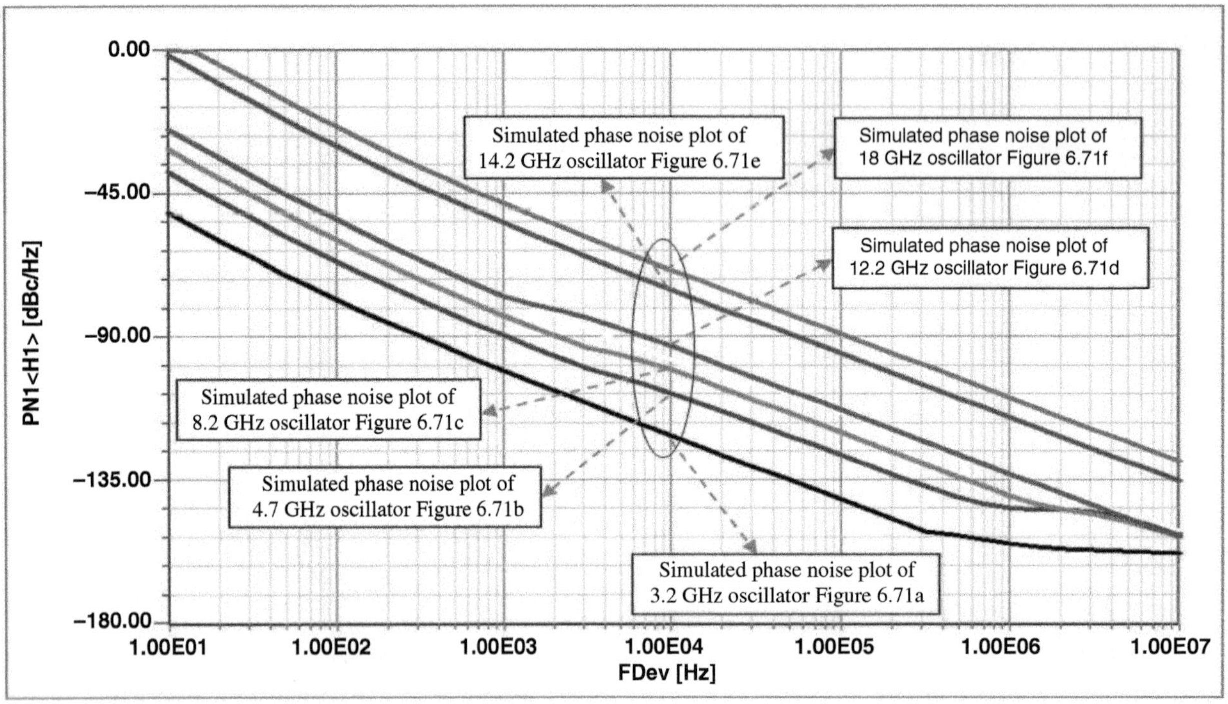

Figure 6.72 Simulated phase-noise performance of the metamaterial resonator oscillator circuits depicted in Figure 6.71 [10].

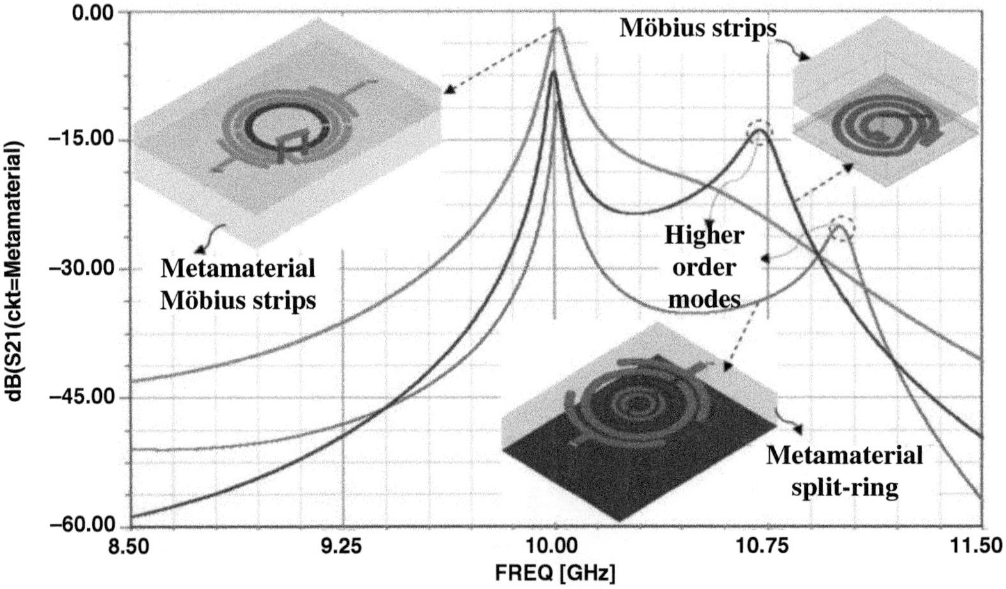

Figure 6.73 Simulated phase-noise of the metamaterial resonator oscillator circuits shown in Figure 6.71 [10].

transmission behavior corresponds to a higher unloaded quality factor (Q), enabling more efficient energy storage and improved spectral purity. In addition, the MMS topology effectively suppresses spurious resonances, which are evident in the conventional designs—specifically, undesired second-order modes appear in the standard Möbius strips and metamaterial split-ring resonators illustrated in Figure 6.73. By mitigating these parasitic modes, the MMS resonators provide more stable broadband operation, making them particularly suitable for high-performance RF and microwave systems where robust single-mode behavior and high-Q resonance are critical.

Oscillator designs based on conventional split-ring resonators, such as those illustrated in Figure 6.71, are inherently prone to mode-jumping, which can degrade frequency stability and compromise overall reliability. To address this issue, a mode-stabilization technique is employed, leveraging precise control of the phase velocity within the resonator using a mode-suppression ring, as shown in Figure 6.74. This structure enables controlled multi-mode self-injection into the Möbius cavity [10], effectively reinforcing the desired oscillation mode while suppressing spurious and parasitic resonances. By selectively shaping the resonator's electromagnetic response, the mode-suppression ring improves single-mode stability, reduces phase noise, and enhances the robustness of the oscillator across variations in temperature, supply voltage, and fabrication tolerances. This technique is particularly advantageous for high-frequency RF and microwave systems where reliable, low-noise operation is essential.

Figure 6.75 presents the measured phase-noise characteristics of the mode-stabilized oscillator, highlighting its high-performance operation. Sivaraj et al. [112] and Cao et al. [113] reported that *"the figure-of-merit (FOM) is a key metric for evaluating oscillator performance."* As shown in Figure 6.64, the oscillator topology achieves an exceptional FOM (Eq. 6.50) of −216.5 dBc/Hz at a 1 MHz offset, demonstrating superior spectral purity and noise efficiency.

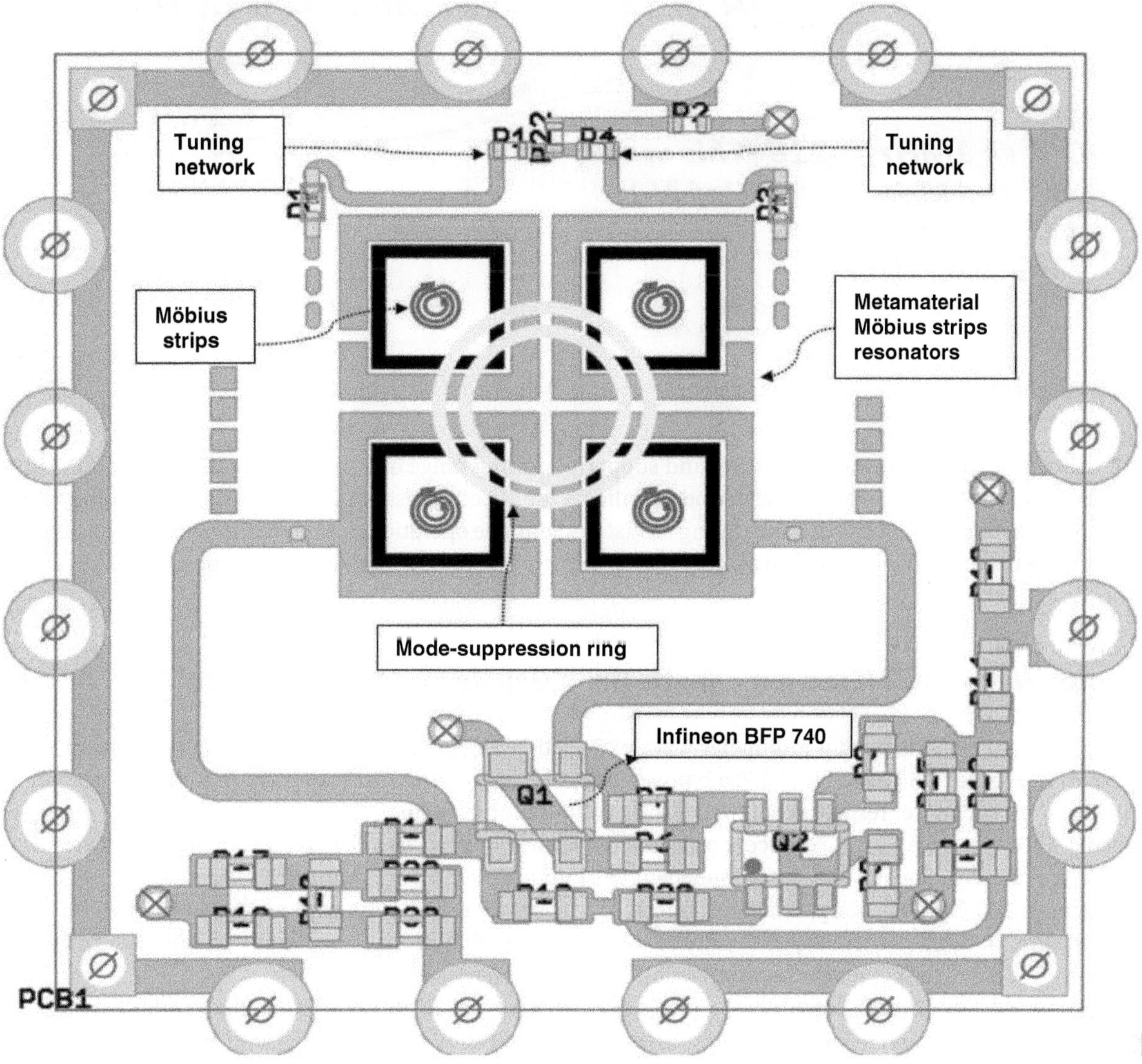

Figure 6.74 A typical layout of X-Band (10.24 GHz) metamaterial Möbius VCO (0.9 × 0.9 inches) [10].

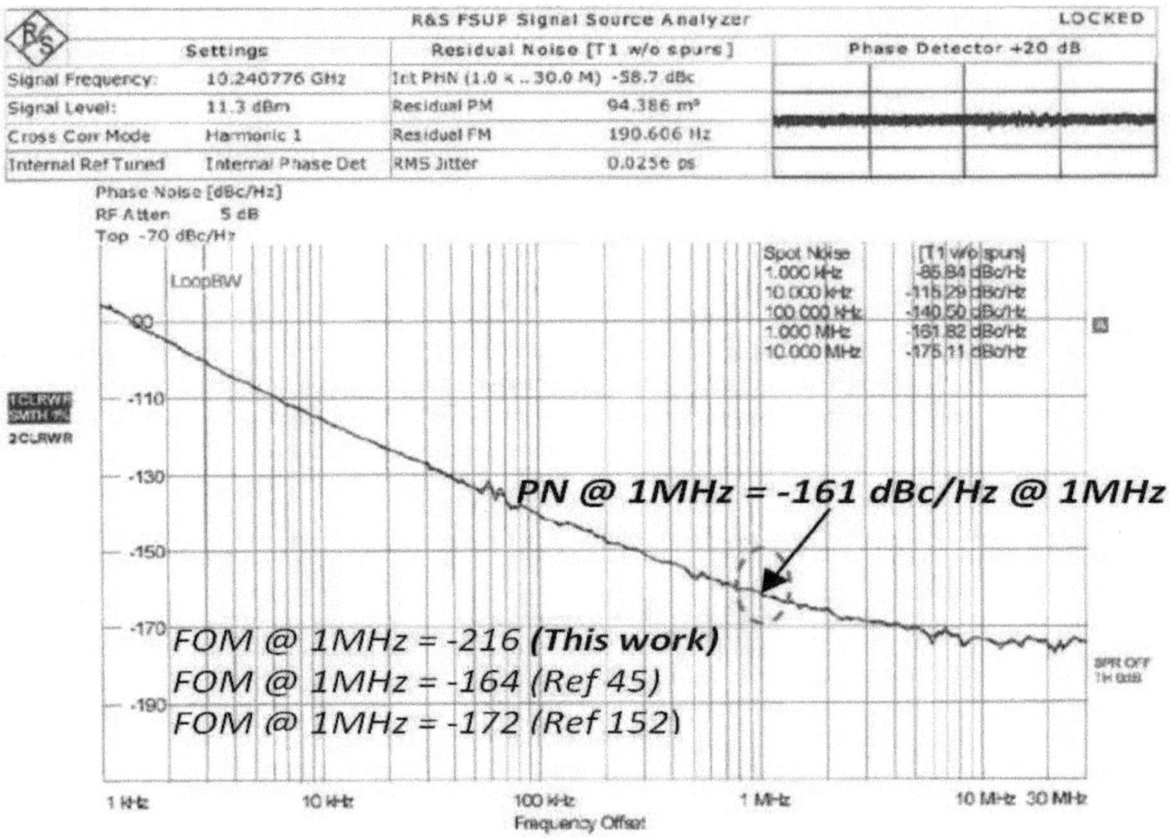

Figure 6.75 Measured phase-noise of the X-band metamaterial Möbius VCO shown in Figure 6.61 [10].

The oscillator delivers an output power of 11.5 dBm and supports a tuning range of 0.5 GHz, operating under a DC bias of 5.0 V with a bias current of 80.0 mA. These measurements validate the effectiveness of the MMS resonator topology combined with mode-stabilization techniques, demonstrating robust single-mode operation, low phase noise, and stable broadband performance. The results underscore the suitability of this architecture for high-frequency RF applications requiring compact, low-noise, and reliable signal sources.

6.8 Millimeter Wave EBGR (Electromagnetic Band Gap Resonator) Oscillator

Enrico Lia et al. [114, 115] demonstrated a millimeter-wave extended-bandgap resonator (EBGR) oscillator operating at 45.8 GHz, as shown in Figure 6.76 [10]. Figure 6.76a,b presents the corresponding block diagram and the physical oscillator assembly within its enclosure, respectively [10]. In the configuration depicted in Figure 6.76a, the active circuit is implemented using a 130-nm SiGe BiCMOS process provided by innovations for high performance (IHP) Microelectronics. This advanced technology supports a cutoff frequency (f_t) of up to 250 GHz and a maximum oscillation frequency (f_{max}) of approximately 340 GHz, enabling robust millimeter-wave operation with high gain and wide bandwidth. Such characteristics make this platform particularly well-suited for the EBGR oscillator, ensuring stable oscillation and efficient performance at 45.8 GHz for high-frequency communication and sensing applications.

The millimeter-wave EBGR oscillator employs a loop-feedback architecture featuring a two-stage amplification chain that integrates both single-ended and balanced amplifier stages. The chain is followed by a 10 dB directional coupler and an electronically controlled phase shifter to maintain stable oscillation. This design achieves an overall loop gain of 12.5 dB, ensuring reliable startup and sustained oscillation. To selectively excite a TE-like mode within the feed waveguide, the input and output ports are connected via bond wires to microstrip transition probes. These probes interface with a waveguide bandpass filter, fine and coarse phase shifters, and the EBGR through triangular taper structures, providing precise mode

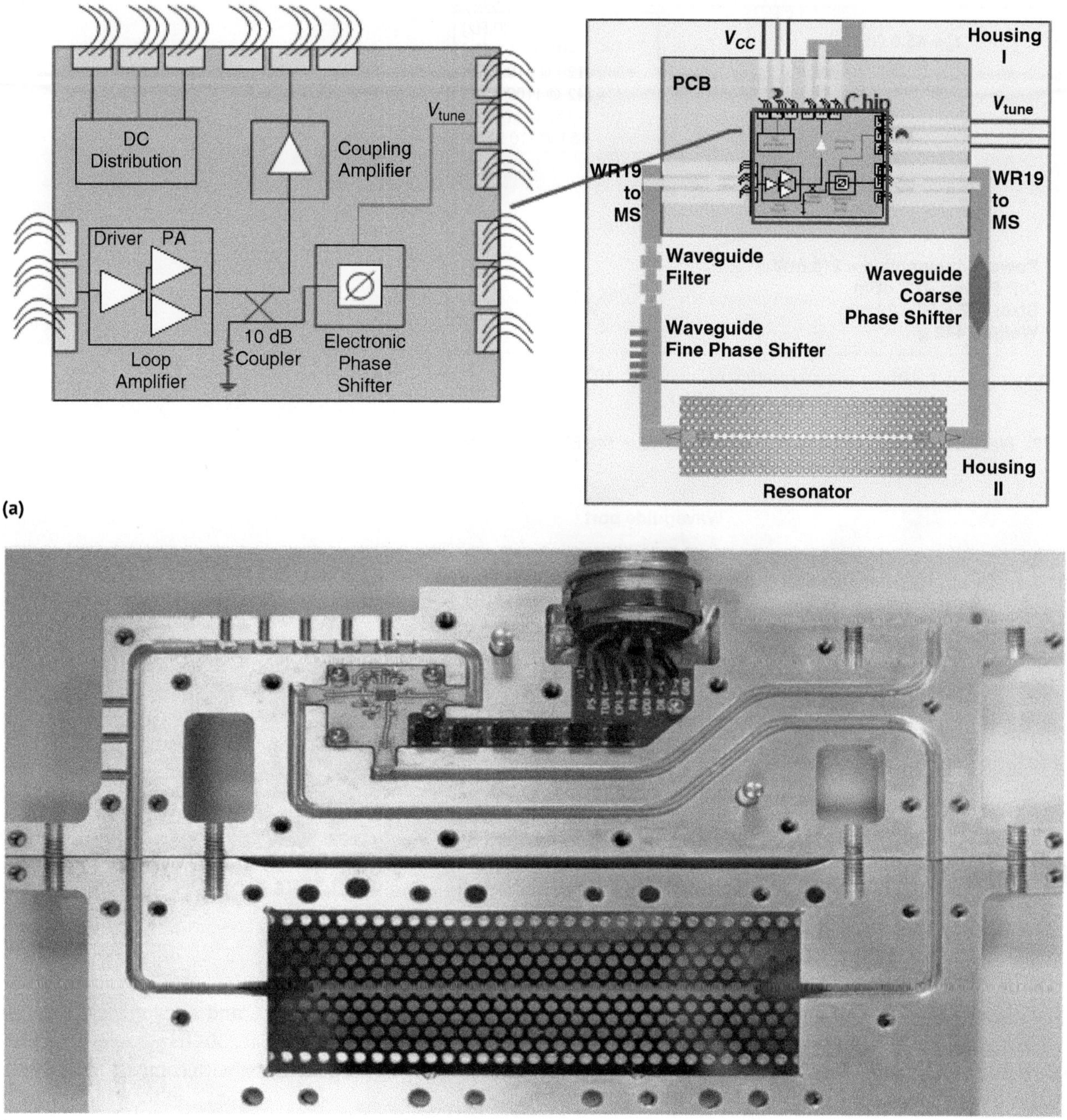

(a)

(b)

Figure 6.76 EBGR oscillator: (a) block diagram and (b) MMIC VCO assembly with the housing top removed [114, 115].

control and frequency selection. The RF output is extracted using a coupling amplifier with 7.8 dB gain and a waveguide-to-coaxial adapter, enabling efficient signal delivery for measurement or system integration. The complete oscillator assembly consumes 215 mW of DC power and occupies a volume of 157 × 88 × 29 mm, with a total mass of 442 g, including connectors [115]. Figure 6.77 presents the measured phase-noise characteristics of this millimeter-wave EBGR oscillator at 45.8 GHz, demonstrating excellent low-phase-noise performance, confirming the effectiveness of the loop-feedback architecture and precision mode control for high-frequency RF applications [115].

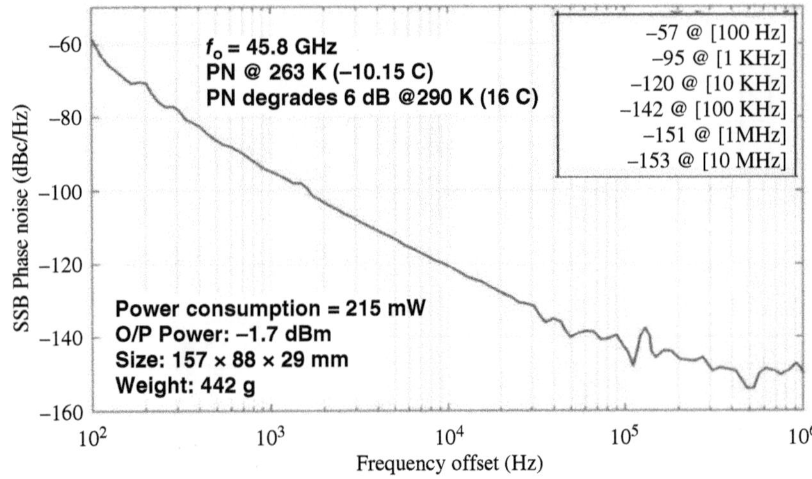

Figure 6.77 Measured single-sideband (SSB) phase-noise of the EBGR oscillator [114].

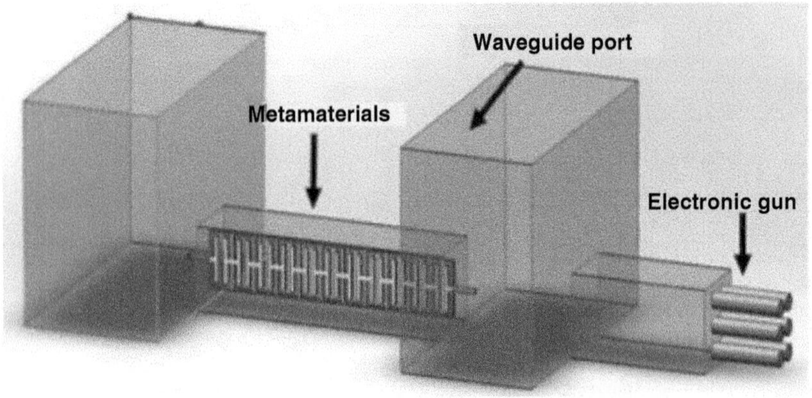

Figure 6.78 Schematic of a typical terahertz signal source employing a metamaterial resonator [116].

6.9 Terahertz Metamaterial Resonator Oscillator

The terahertz (THz) frequency spectrum, extending from 100 GHz to 10 THz, corresponds to quantum energies in the range of 0.4 meV to 0.4 eV and represents a frontier for high-speed communication, imaging, and sensing technologies. Within this regime, Feng, Yuming, et al. [116, 117] reported a terahertz signal source operating at 230 GHz that leverages an artificial composite metamaterial resonator to achieve efficient high-frequency signal generation with compact form factor and enhanced performance metrics.

The resonator design, illustrated in Figure 6.78, integrates sub-wavelength engineered elements to create a tailored electromagnetic response, resulting in a high unloaded Q-factor and precise mode selectivity. This metamaterial-based approach facilitates tunable frequency operation while maintaining strong confinement of electromagnetic energy, minimizing radiative losses, and suppressing spurious modes. The integration of the resonator into the THz source exemplifies how metamaterial-inspired structures can overcome conventional resonator limitations, enabling compact, planar, and high-performance oscillators suitable for next-generation terahertz applications, including spectroscopy, high-resolution imaging, and ultra-fast wireless communication systems.

Feng [116] utilized the Drude model in combination with full-wave electromagnetic simulations in Ansys HFSS to characterize the effective permittivity and permeability of the artificial composite metamaterial resonator [40]. The extracted parameters, presented in Figure 6.79, reveal that the resonator exhibits simultaneously negative effective permittivity and permeability over the 220–300 GHz frequency range. This behavior confirms the resonator's capability to support backward-wave propagation and establishes its suitability for sub-terahertz signal generation in high-performance applications, including medical imaging, high-resolution spectroscopy, and precision sensing. Figure 6.80 illustrates the transient

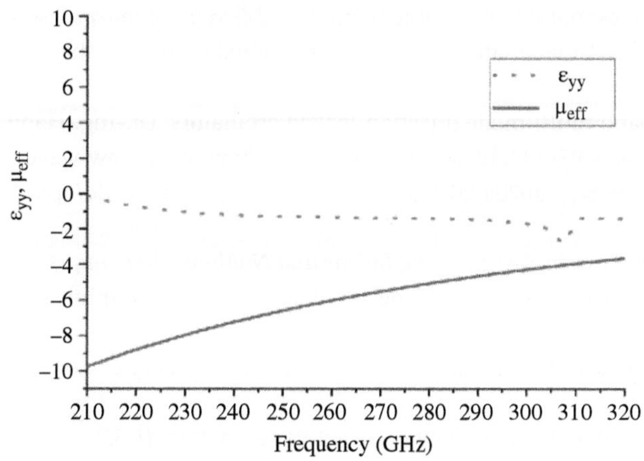

Figure 6.79 Effective permittivity and permeability [116].

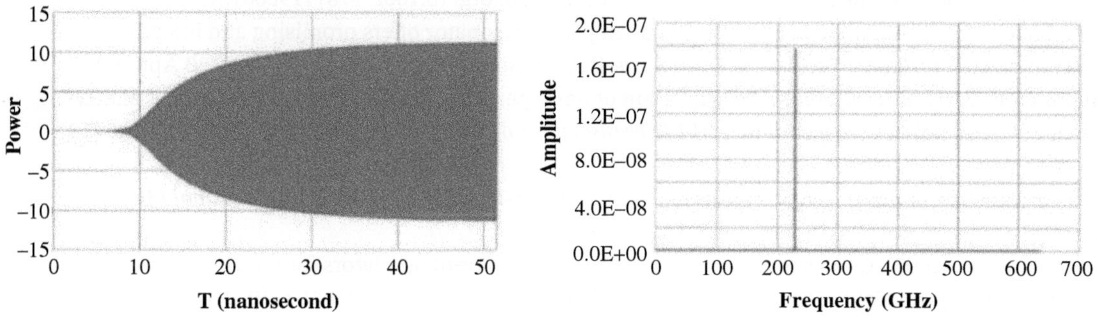

Figure 6.80 The transient analysis and output spectrum of the R-band BWO described in Figure 6.78 [116].

response and output spectrum of an R-band backward-wave oscillator (BWO) incorporating the metamaterial resonator described in Figure 6.78 [116]. The results demonstrate stable high-frequency oscillation with a sharply defined spectral profile and minimal spurious content. These findings highlight the effectiveness of metamaterial-inspired resonator designs in supporting compact, tunable, and high-Q terahertz oscillators, providing a pathway toward miniaturized, high-performance THz sources for advanced imaging, communication, and sensing systems.

References

1 Nick, M. (2011). New Q-enhanced planar resonators for low phase-noise radio frequency Oscillators. PhD dissertation. Electrical Engineering, University of Michigan. https://hdl.handle.net/2027.42/89682, https://deepblue.lib.umich.edu/bitstream/handle/2027.42/89682/mornick_1.pdf?sequence=1.

2 Gardner, P. and Paul, D.K. (1997). Optimum noise measure configurations for transistor negative resistance amplifiers. *IEEE Transactions on Microwave Theory and Techniques* 45 (5): 580–586.

3 Grebennikov, A. (2007). *RF and Microwave Transistor Oscillator Design*. New York: John Wiley & Sons.

4 Hauss, H.A. and Adler, R.B. (1959). *Circuit Theory of Linear Noisy Networks*. New York: Wiley.

5 Hong, J.-S. and Lancaster, M.J. (2001). *Microstrip Filter for RF/Microwave Application*. Wiley-Interscience. ISBN-10: 9780471388777.

6 Rohde, U.L. and Rudolph, M. (2012). *RF/Microwave Circuit Design for Wireless Applications*. Wiley & Sons. ISBN-13: 978-1118431399.

7 Hoffmann, K. and Skvor, Z. (2001). Active resonator. *EUROCON 2001 Proceeding*, Bratislava. Vol. 1, 164–166. IEEE.

8 Rohde, U.L. and Poddar, A.K. (2007). Active planar coupled resonators replace traditional high Q resonators in low phase noise oscillators/VCOs. *IEEE Radio and Wireless Symposium*, CA, USA (09–11 January 2007), 39–42.

9 Rohde, U.L. and Poddar, A.K. (2015). Metamaterial Möbius strip resonators for tunable oscillators. *Microwave Journal* 58 (1): 64–73. https://www.microwavejournal.com/articles/print/23633-metamaterial-mbius-strip-resonators-for-tunable-oscillators

10 Poddar, A.K. (2014). Slow wave resonator based tunable multi-band multi-mode injection-locked oscillators. Dr.-Ing.-Habil Thesis. Chapter 6, BTU Cottbus, Germany. https://d-nb.info/1114664936/34, https://pdfsecret.com/download/slow-wave-resonator-based-tunable-multi-band-multi-mode-injection_59f70036d64ab20a7510a819_pdf, https://opus4.kobv.de/opus4-btu/frontdoor/index/index/docId/3193.

11 Vandelin, G.D., Pavio, A.M., Rohde, U.L. et al. (2021). *Microwave Circuits Design Using Linear and Nonlinear Techniques*, 3e. John Wiley & Sons. ISBN: 978-1-119-74170-1. https://epdf.pub/microwave-circuit-design-using-linear-and-nonlinear-techniques.html.

12 Poddar, A.K. and Rohde, U.L. (2012). Slow-wave evanescent-mode coupled resonator oscillator. *IEEE Frequency Control Symposium Proceedings*, 1–7.

13 Nick, M. and Mortazawi, A. (2010). Low phase-noise planar resonators based on low noise active resonators. *IEEE Transactions on Microwave Theory and Techniques* 55 (5): 1133–1139.

14 Rohde, U.L., Rubiola, E., and Whitaker, J. (2021*). Microwave and Wireless Synthesizers: Theory and Design*, 2e. John Wiley & Sons. Print ISBN:9781119666004, Online ISBN:9781119666127. https://doi.org/10.1002/9781119666127.

15 Rohde, U.L. and Poddar, A.K. (2008). Mode-coupled stubs-tuned planar resonator offers promising and integrable alternatives of DRO (dielectric resonator oscillator). *IEEE Sarnoff Symposium*, Princeton, NJ, USA (28–30 April), 1–7.

16 Ezzedine, H., Billonnet, L., Jarry, B. et al. (1998). Optimization of noise performance for various topologies of planar microwave active filters using noise wave techniques. *IEEE Transactions on Microwave Theory and Techniques* 46 (12): 2484–2492.

17 Nick, M. (2011). A very low phase-noise voltage-controlled oscillator at X-band. *IEEE MTT-S International Microwave Symposium*, 1–4.

18 Rohde, U.L. and Poddar, A.K. (2007). Concurrent, reconfigurable and concurrent oscillators for multi-band multi-mode communication systems. *European Conference on Wireless Technology, ECWT 2007*, Munich (08–10 October), 292–295.

19 Paul, D.K. and Gardner, P. (1995). Microwave oscillators and filters based on microstrip ring resonators. *Proceedings of 1995 IEEE MTT-S International Microwave Symposium*, Orlando, FL, USA. Vol. 1, 83–86. https://doi.org/10.1109/MWSYM.1995.406092.

20 Grebennikov, A. (2019). *Radio Frequency and Microwave Power Amplifiers: Efficiency and Linearity Enhancement Techniques (Materials, Circuits and Devices)*. IET (Institution of Engineering and Technology). ISBN-10:1839530383.

21 Mirzaei, A., Heidari, M.E., and Abidi, A.A. (2006). Analysis of oscillators locked by large injection signals: generalized Adler's equation and geometrical interpretation. *IEEE Custom Integrated Circuits Conference 2006*, San Jose, CA, USA, 737–740. https://doi.org/10.1109/CICC.2006.320928.

22 Poddar, A.K. and Rohde, U.L. (2014). The pursuit for low cost and low phase noise synthesized signal sources: theory & optimization. *IEEE UFFC Symposia* (19–22 May).

23 Hong, J.-S. (2011). *Microwave Filter for RF/Microwave Application*, 2e. Wiley-Interscience. ISBN-13: 978-0470408773.

24 Hartnagel, H., Quay, R., Rohde, U.L. et al. (2023). *Fundamentals of RF and Microwave Techniques and Technologies*, 1e. Springer. ISBN-10 3030940985.

25 Hoffmann. K. and Skvor, Z. (2001). Active microwave and millimeter wave resonator. *2001 31st European Microwave Conference*, London, UK, 1–4. https://doi.org/10.1109/EUMA.2001.339030.

26 Zhang, S., Deng, W., Jia, H. et al. (2025). A series-parallel-resonance oscillator with a 191.5dBc/Hz FoM. *IEEE Solid-State Circuits Letters*. https://doi.org/10.1109/LSSC.2025.3564312.

27 Sun, M., Lu, D., Zhu, B. et al. (2025). Low-phase-noise negative-resistance oscillator design based on phase-noise synthesis technique. *IEEE Transactions on Microwave Theory and Techniques*. https://doi.org/10.1109/TMTT.2025.3552322.

28 Tsai, C.M. and Lee, H.-M. (2006). The effect of component Q distribution on microwave filters. *IEEE Transactions on Microwave Theory and Techniques* 54 (4): 1545–1553.

29 Hsieh, L. and Chang, K. (2003). High-efficiency piezoelectric-transducer tuned feedback microstrip ring-resonator oscillators operating at high resonant frequencies. *IEEE Transactions on Microwave Theory and Techniques* 51: 1141–1145.

30 Dussopt, L., Guillois, D., and Rebeiz, G.M. (2002). A low phase noise silicon 9 GHz VCO and 18 GHz push-push oscillator. *IEEE MTT-S International Microwave Symposium Digest* 2: 695–698.

31 Lee, Y.-T., Lee, J., and Nam, S. (2004). High-Q active resonators using amplifiers and their applications to low-phase noise free-running and voltage-controlled oscillators. *IEEE Transactions on Microwave Theory and Techniques* 52: 2621–2626.

32 Choi, J. and Mortazawi, A. (2007). A new X-band low phase-noise multiple-device oscillator based on the extended-resonance technique. *IEEE Transactions on Microwave Theory and Techniques* 55 (8): 1642–1648.

33 Lee, J., Lee, Y.-T., and Nam, S. (2002). A phase noise reduction technique in microwave oscillator using high-Q active filter. *IEEE Microwave and Wireless Components Letters* 12 (11): 426–428.

34 Choi, J., Nick, M., and Mortazawi, A. (2009). Low-phase-noise planar oscillators employing elliptic-response bandpass filters. *IEEE Transactions on Microwave Theory and Techniques* 57 (8): 1959–1965.

35 Rael, J.J. and Abidi, A.A. (2000). Physical processes of phase-noise in differential LC oscillators. *Proceedings of the Custom Integrated Circuits Conference*, 569–572.

36 Krizhanovski, V., Rassokhina, Y., Kryzhanovskyi, V. et al. (2023). Mutually synchronized stabilized class E oscillators. *IEEE International Conference on Information and Telecommunication Technologies and Radio Electronics (UkrMiCo)*, Kyiv, Ukraine, 29–33, https://doi.org/10.1109/UkrMiCo61577.2023.10380334.

37 Tiebout, M. (2001). Low-power low-phase-noise differentially tuned quadrature VCO design in standard CMOS. *IEEE Journal of Solid-State Circuits* 36 (7): 1018–1024.

38 Khanna, A.P.S., Topacio, E., Gane, E. et al. (2001). Low jitter silicon bipolar based VCOs for applications in high speed optical communication systems. *IEEE MTT-S International Microwave Symposium Digest*, 1567–1570.

39 Victor, A. and Steer, M.B. (2007). Reflection coefficient shaping of a 5-GHz voltage tuned oscillator for improved tuning. *IEEE Transactions on Microwave Theory and Techniques* 55 (12): 2488–2494.

40 Prodromakis, T. and Papavassiliou, C. (2010). An experimental technique for characterizing slow-wave characteristics of MIS-like transmission lines using aqueous dielectrics. *IEEE Transactions on Microwave Theory and Techniques* 58 (4): 985–993. https://doi.org/10.1109/TMTT.2010.2042521.

41 Rohde, U.L., Poddar, A.K., and Sundararajan, D. (2013). Printed resonators: Möbius strips theory and applications. *Microwave Journal*, pp. 24–54. https://ar.scribd.com/document/192482078/Mwjournalmwj201311-Dl.

42 Poddar, A.K., Rohde, U.L., and Itoh, T. (2014). Metamaterial Möbius Strips (MMS): tunable oscillator circuits. *IEEE MTT-S Digest*, 1–4.

43 Pistono, E., Robert, M., Duvillaret, L. et al. (2006). Compact fixed and tune-all bandpass filters based on coupled slow-wave resonators. *IEEE Transactions on Microwave Theory and Techniques* 54 (6): 2790–2798.

44 Chang, W.S., Liang, C.H., and Chang, C.Y. (2010). Compact microstrip bandpass filters using miniaturized slow-wave quarter-wavelength resonators. *Proceedings of Asia-Pacific Microwave Conference*, 1958–1961.

45 Hsieh, L.H. and Chang, K. (2002). Slow-wave bandpass filters using ring or stepped-impedance hairpin resonators. *IEEE Transactions on Microwave Theory & Techniques* 50 (7): 1795–1800.

46 Cho, H.Y., Yeh, T.J., Liu, S. et al. (2009). High-performance slow-wave transmission lines with optimized slot-type floating shields. *IEEE Transactions on Electron Devices* 56 (8): 1705–1711.

47 Masuda, T., Shiramizu, N., Nakamura, T. et al. (2008). Characterization and modeling of microstrip transmission lines with slow-wave effect. *IEEE Topical Meeting on Silicon Monolithic Integrated Circuits in RF Systems*, 55–158.

48 Wang, H., Zhang, L, Yang, D. et al. (2011). A 60 GHz wideband injection-locked frequency divider with adaptive-phase-enhancing technique. *Radio Frequency Integrated Circuits Symposium*, 75–78.

49 Poddar, A.K., Koul, S.K., and Bhat, B. (1997). Millimeter wave evanescent mode Gunn diode oscillator in suspended stripline configuration. *IR &MM Wave, 22nd International Conference*, 265–266.

50 Poddar, A.K. and Rohde, U.L. (2014). Metamaterial Möbius Strips (MMS): application in resonators for oscillators and synthesizers. *IEEE UFFC Symposia*, 19–22.

51 Poddar, A.K. and Rohde, U.L. (2011). Slow wave resonator based tunable oscillators. *Proceedings of 2011 Joint Conference of the IEEE International Frequency Control Symposium & European Frequency and Time Forum*, 549–558.

52 Rohde, U.L. and Poddar, A.K. (2008). Injection-tuned coupled oscillators. *IEEE Radio Wireless Symposium*, Orlando, FL, USA (22–24 January), 367–370.

53 Rohde, U.L. and Poddar, A.K. (2008). Large-signal approach yields low-noise VHF/UHF oscillators, part 1. *Microwaves & RF* (April), pp. 62–76.

54 Rohde, U.L. and Poddar, A.K. (2008). Large-signal approach yields low-noise VHF/UHF oscillators, part 2. *Microwaves & RF* (May), pp. 84–96.

55 Rohde, U.L. and Poddar, A.K. (2008). Planar resonators arm tunable oscillators, part 1. *Microwaves & RF* (June), pp. 92–96.

56 Rohde, U.L. and Poddar, A.K. (2008). Planar resonators arm tunable oscillators, part 2. *Microwaves & RF* (July).

57 Rohde, U.L. and Poddar, A.K. (2008). Miniature VCOs shrink wideband synthesizers. *Microwaves & RF* (Dec), pp. 87–92.

58 Rohde, U.L. and Poddar, A.K. (2008). STPCR offers integrable alternatives of DRO. *2008 IEEE MTT-S Microwave Symposium Digest*, Atlanta, USA (15–20 June), 233–236.

59 Rohde, U.L. and Poddar, A.K. (2007). Low cost configurable RF signal source for wireless applications. *European Microwave Conference, EuMiC*, Munich (09–12 October), 508–511.

60 Rohde, U.L. and Poddar, A.K. (2007). Noise reduction techniques for configurable RF signal source. *6th International Conference on Antenna Theory and Techniques, IEEE/ICATT 2007*, Ukraine (17–21 September), 163–166.

61 Rohde, U.L. and Poddar, A.K. (2007). Minimization of phase noise in scaled device coupled mode oscillators. *2007 6th International Conference on Antenna Theory and Techniques, IEEE/ICATT 2007*, Ukraine (17–21 September), 51–56.

62 Rohde, U.L. and Poddar, A.K. (2007). Low cost signal source for multi-band multi-mode wireless systems. *6th International Conference on Antenna Theory and Techniques, IEEE/ICATT 2007*, Ukraine (17–21 September), 167–169.

63 Shen, F., Wei, Y., Yin, H. et al. (2012). A novel V-shaped microstrip meander-line slow-wave structure for W-band MMPM. *IEEE Transactions on Plasma Science* 40 (2): 463–469.

64 Rohde, U.L. and Poddar, A.K. (2007). STPCR VCO replaces expensive DRO. *IEEE/APMC 2007*, Bangkok, Thailand (11–14 December), 1–4.

65 Kaajakari, V., Koskinen, J.K., and Mattila, T. (2005). Phase noise in capacitively coupled micromechanical oscillators. *IEEE Transactions on Ultrasonics, Ferroelectrics, and Frequency Control* 52 (12): 2322–2331.

66 Kirschning, M., Jansen, R.H., and Koster, N.H.L. (1983). Measurement and computer aided modeling of microstrip discontinuities by an improved resonator method. *IEEE MTT-S International Microwave Symposium Digest*, 495–497.

67 Lee, S.-Y. and Tsai, C.-M. (2000). New cross-coupled filter design using improved hairpin resonators. *IEEE Transactions on Microwave Theory and Techniques* 48: 2482–2490.

68 Tsai, C.M., Lee, S.Y., and Tsai, C.C. (2001). Hairpin filters with tunable transmission zeros. *IEEE MTT-S International Microwave Symposium Digest*, 2175–2178.

69 Gupta, K.C., Garg, R., Bahl, I. et al. (1996). *Microstrip Lines and Slotlines*, 2e. Boston: Artech House.

70 Sinnesbichler, F.X., Hautz, B., and Olbrich, G.R. (2000). A Si/SiGe HBT dielectric resonator push-push oscillator at 58 GHz. *IEEE Microwave and Guided Wave Letters* 10: 145–147.

71 Hamand, D. and Hajimiri, A. (2001). Concepts and methods in optimization of integrated LC VCOs. *IEEE Journal of Solid-State Circuits* 36 (6): 896–909.

72 Rohde, U.L. and Poddar, A.K. (2008). Low noise, hybrid tuned wideband voltage controlled oscillator. US Patent No. 7,365,612 B2, April 29. http://www.freepatentsonline.com/7605670.html.

73 Rohde, U.L. and Poddar, A.K. (2007). Multi-octave band tunable coupled resonator oscillator. US Patent No. 7, 292,113, November 6. http://www.freepatentsonline.com/7605670.html.

74 Poddar, A.K. (2009). User-definable, low cost, low phase hit and spectral pure tunable oscillator. US Patent No. 7,586,381, September 8, 2009. https://patents.google.com/patent/US7586381B2/en.

75 Choi, J. and Seo, C. (2008). Broadband VCO using electronically controlled metamaterial transmission line based on varactor-loaded split-ring resonator. *Microwave and Optical Technology Letters*. https://doi.org/10.1002/mop.23305.

76 Rohde, U.L. and Poddar, A.K. (2009). User-definable, low cost, low noise, and phase hit insensitive multi-octave-band tunable oscillator. Phase Hit and Spectral Pure Tunable Oscillator, US Patent No. 7,605,670, October 20. http://www.freepatentsonline.com/7605670.html.

77 Voinigescu, S.P., Tomkins, A., Dacquay, E. et al. (2013). A study of SiGe HBT signal sources in 220-330-GHz range. *IEEE Journal of Solid-State Circuits* 48 (9): 1–11.

78 Tanaka, T., Takata, K., Aikawa, M. et al. (2012). A Ku band Oscillator array using positive feedback type push-push oscillators. *IEEE APMC Conference Proceedings*, 112–114.

79 Sirburanon, T., Sato, T., Musa, A. et al. (2012). A 20 GHz push-push voltage-controlled oscillator for a 60 GHz frequency synthesizer. *IEEE APMC Conference Proceedings*, 205–207.

80 Chu, S.W. and Wang, C.K. (2012). An 80 GHz wide tuning range push-push VCO with gm-boosted full-wave rectification technique in 90 nm CMOS. *IEEE Microwave and Wireless Components Letters* 22 (4): 203–205.

81 Yaoming, S. and Scheytt, C.J. (2012). A low phase noise 61 GHz push-push VCO with divider chain buffer in SiGe BiCMOS for 122 GHz ISM applications. *2012 IEEE RFIC Symposium*, 79–82.

82 Nakamura, T., Masuda, T., Washio, K. et al. (2012). A push-push VCO with 13.9 GHz wide tuning range using loop-ground transmission line for full band 60-GHz transceiver. *IEEE Journal of SSC* 47 (6): 1267–1277.

83 Stuenkel, M. and Feng, M. (2012). A W-band tunable push-push oscillator with 128X division for frequency synthesis applications. *2013 IEEE MTT-S Digest*, 1–3.

84 Wang, T.P. (2011). A fully integrated W-band push-push CMOS VCO with phase noise and wide tuning range. *2011 IEEE Transactions on Ultrasonics, Ferroelectrics, and Frequency Control* 58 (7): 1307–1319.

85 Rohde, U.L. and Poddar, A.K. (2007). Tunable frequency low phase noise and low thermal drift oscillator. US Patent No. 7196591, March.

86 Poddar, A.K., Rohde, U.L., and Daryoush, A.S. (2013). Integrated production of self injection locked self phase loop locked opto-electronic oscillators. US Patent application No. 13/760767 (filed on Feb 06, 2013). http://www.freepatentsonline.com/7605670.html.

87 Poddar, A.K., Rohde, U.L., and Daryoush, A.S. (2012). Self injection locked phase locked looped optoelectronic oscillator. US Patent application No. 61/746, 919 (filed on Dec 28, 2012).

88 Yao, X.S. and Maleki, L. (1996). Optoelectronic oscillator for photonic systems. *IEEE Journal of Quantum Electronics* 32 (7): 1141–1149.

89 Moyer, H.P. and Daryoush, A.S. (2000). A unified analytical model and experimental validations of injection-locking processes. *IEEE Transactions on Microwave Theory and Techniques.* 48 (4): 493–499.

90 McNeilage, C., Ivanov, E.N., Stockwell, P.R. et al. (1998). Review of feedback and feedforward noise reduction techniques. *Proceedings of the 1998 IEEE International Frequency Control Symposium*, Pasadena, CA, 146–155.

91 Suarez, A. and Ramirez, F. (2005). Analysis of stabilization circuits for phase-noise reduction in microwave oscillators. *IEEE Transactions on Microwave Theory and Techniques* 53 (9): 2743–2751.

92 Chang, H.C. (2003). Phase noise in self-injection locked oscillators – theory and experiment. *IEEE Transactions on Microwave Theory and Techniques* 51 (9): 1994–1999.

93 Lee, K.H., Kim, J.Y., and Choi, W.Y. (2007). A 30GHz self-injection locked oscillator having a long optical delay line for phase noise reduction. *Photonics Technology Letters* 19 (24): 1982–1984.

94 Rohde, U.L., Poddar, A.K., and Boeck, G. (2005). *The Design of Modern Microwave Oscillators for Wireless Applications: Theory and Optimization.* New York: Wiley.

95 Rohde, U.L. and Poddar, A.K. (2009). Integrated low noise microwave wideband push-push VCO. Canadian Patent No. 2,548, 311, April 14. http://www.freepatentsonline.com/7605670.html.

96 Rohde, U.L. and Poddar, A.K. (2006). Integrated low noise microwave wideband push-push VCO. US Patent No. 7,088189, August. http://www.freepatentsonline.com/7605670.html.

97 Xu, F. and Wu, K. (2005). Guided-wave and leakage characteristics of substrate integrated waveguide. *IEEE Transactions on Microwave Theory and Techniques* 53 (1): 66–73.

98 Wu, C.-T.M., Poddar, A.K., Rohde, U.L. et al. (2014). A C-band tunable oscillator based on complementary coupled resonator using substrate integrated waveguide cavity. *European Microwave Conference.*

99 Dong, Y. and Itoh, T. (2012). A dual-band oscillator with reconfigurable cavity-backed complementary split-ring resonator. *IEEE MTT-S International Microwave Symposium Digest* (17–22 June), 1–3.

100 He, F.F., Wu, K., Hong, W. et al. (2010). A low phase-noise VCO using an electronically tunable substrate integrated waveguide resonator. *IEEE Transactions on Microwave Theory and Techniques* 58 (12): 3452–3458.

101 Sirci, S., Martinez, J.D., Taroncher, M. et al. (2011). Varactor loaded continuously tunable SIW resonator for reconfigurable filter design. *Proceedings of the 41st European Microwave Conference*, 436–439.

102 Wu, C.-T.M., Poddar, A.K., Rohde, U.L. et al. (2014). Active complementary coupled resonator for low phase noise X-band oscillator. *European Frequency and Time Forum.* https://doi.org/10.1109/EFTF.2014.7331509.

103 Zhang, T., Xu, R., and Wu, C.-T.M. (2017). Unconditionally stable non-foster element using active transversal-filter-based negative group delay circuit. *IEEE Microwave and Wireless Components Letters* 27 (10): 921–923. https://doi.org/10.1109/LMWC.2017.2745487.

104 Zhang, N. (2012). Investigation and design of wideband microwave VCO. Thesis. Linköping University Electronic Press. http://diva-portal.org/smash/get/diva2:556643/FULLTEXT01.pdf.

105 Chen, Z., Hong, W., and Chen, J.X. (2012). High-Q planar active resonator based on substrate integrated waveguide technique. *Electronics Letters* 48 (10): 575–577.

106 Maxwell, J.C. (1892). *A Treatise on Electricity and Magnetism*, 3e, Vol. 2. Oxford: Clarendon, 68–73.

107 Nick, M. and Mortazawi, A. (2010). Oscillator phase-noise reduction using low noise high-Q resonators. *2010 IEEE MTT-S International Microwave Symposium*, 276–279.

108 Wu, C.-T.M. (2014). Active integrated metamaterial-based antennas and microwaves components. PhD Thesis. Department of Electrical Engineering, University of California, Los Angeles.

109 Yan, F., Huang, Y.M., Huang, T. et al. (2019). Transversely compact single-ended and balanced bandpass filters with source–load-coupled spurlines. *Electronics* 8: 416. https://doi.org/10.3390/electronics8040416, https://www2.mdpi.com/2079-9292/8/4/416.

110 Poddar, A.K., Rohde, U.L., and Sundararajan, D. (2013). Real time signal retention device using co-planar waveguide (CPW) as Möbius strip. *2013 IEEE MTT-S Digest*, 1–3.

111 Poddar, A.K., Rohde, U.L., and Sundararajan, D. (2013). A novel Möbius-coupled printed resonator based signal sources. *2013 IEEE MTT-S International Microwave Symposium Digest (MTT)*, Seattle, WA, USA, 1–4. https://doi.org/10.1109/MWSYM.2013.6697691.

112 Sivaraaj, N.R. and Abdul Majeed, K.K. (2025). High-performance LC-VCO achieving −201.7 dBc/Hz FoMT and 29.23% tuning range for Ku-band. *AEU - International Journal of Electronics and Communications* 200: 155918. ISSN 1434-8411. https://doi.org/10.1016/j.aeue.2025.155918, https://www.sciencedirect.com/science/article/pii/S1434841125002596.

113 Cao, Y., Leroux, P., and Steyart, M. (2025). *Radiation-Tolerant Delta-Sigma Time-to-Digital Converter*. Analog Circuits and Signal Processing. Springer. https://doi.org/10.1007/978-3-319-11842-0.

114 Lia, E., Ghosh, I., Hanham, S.M. et al. (2023). Novel mm-wave oscillator based on an electromagnetic bandgap resonator. *IEEE MTT-S International Microwave Symposium (IMS 2023)*, San Diego, CA, USA (11–16 June).

115 Lia, E., Ghosh, I., Hanham, S.M. et al. (2023). Novel mm-wave oscillator based on an electromagnetic bandgap resonator. *IEEE Microwave and Wireless Technology Letters* 33 (6): 863–866. https://doi.org/10.1109/LMWT.2023.

116 Feng, Y., Zhang, K., Tang, L. et al. (2023). R-band backward wave oscillator based on metamaterial structure and multi-electron beams. *24th International Vacuum Electronics Conference (IVEC)*, 1–2.

117 Yu, K., Wang, C., Wen, Y. et al. (2024). Nonlinear metamaterial for efficient terahertz second-harmonic generation from anharmonic oscillation of 2DEG. *2024 Conference on Lasers and Electro-Optics (CLEO)*, Charlotte, NC, USA, 1–2.

7

MEMS Switches with Metamaterial Contacts: Applications in Switched Band Oscillators

7.1 A MEMS Switch with Metamaterial Contacts: Theory, Concepts, and Technology

Microelectromechanical system (MEMS) switch technology has become a pivotal enabler in high-frequency and microwave electronics, offering distinct advantages over traditional electromechanical counterparts [1–51]. These advantages include a substantially reduced footprint, ultralow power consumption, and minimized insertion loss, making MEMS switches indispensable for compact, energy-efficient, and high-performance systems. As a result, they are increasingly adopted across a wide spectrum of advanced applications, such as multiband tunable oscillators, reconfigurable signal-routing networks in transmit/receive (T/R) front ends, switched-line phase shifters [22] for phased-array antennas [23], wideband frequency-agile tuning circuits [26], and precision instrumentation subsystems [24]. From both structural and functional perspectives, MEMS switches are broadly categorized by their contact mechanism [1–15]:

- Capacitive-contact (metal–insulator–metal [MIM]) switches [7]: Optimized for high-frequency operation, these devices exhibit excellent isolation characteristics, superior linearity, and low intermodulation distortion, making them particularly well suited for radio frequency (RF) and millimeter-wave applications [22].
- Resistive-contact (metal–metal) switches: Distinguished by their low ON-state resistance, they enable efficient broadband signal conduction and are widely employed in signal-routing architectures that demand minimal insertion loss across wide operating bandwidths [7].

The choice between capacitive and resistive-contact mechanisms is often dictated by the intended application, performance trade-offs, and integration requirements within the broader RF/microwave system [8–13].

Recent innovations in MEMS switch design and fabrication are progressively extending the usable frequency range of these devices well beyond conventional limits, with demonstrated reliability into the sub-terahertz domain (>100 GHz). This advancement has been driven by several key design strategies, including (i) the implementation of defected ground structures (DGS) [27] to enhance electromagnetic performance and suppress unwanted parasitic modes, (ii) the integration of secondary switching elements to improve port-to-port isolation in complex RF networks, and (iii) the adoption of metamaterial-inspired capacitive contacts, which simultaneously reduce insertion loss and broaden operational bandwidth. Collectively, these techniques represent a significant leap over earlier MEMS switch generations, enabling high-efficiency operation in regimes once considered impractical.

Despite these gains, reliability remains one of the most persistent challenges in MEMS switch deployment. Issues such as stiction, surface adhesion, and long-term mechanical instability limit device endurance, particularly under continuous or high-power cycling. To address these constraints, emerging design methodologies are exploring the controlled utilization of Casimir forces—quantum-induced attractive interactions that arise at nanoscale separations. By carefully engineering switch geometries and actuation mechanisms to harness, rather than mitigate, these forces, researchers have demonstrated improvements in both device lifetime and switching repeatability, effectively transforming a traditional failure mechanism into a functional design advantage [10].

The integration of metamaterial concepts into MEMS switch architectures further enhances their potential by addressing long-standing limitations associated with contact degradation. Conventional MEMS switches—whether capacitive (MIM) or resistive (metal–metal)—often suffer from adhesion forces, surface asperities, and electromechanical fatigue, particularly in RF and millimeter-wave environments. In contrast, metamaterial-engineered surfaces provide tailored electromagnetic and mechanical properties that improve field confinement, reduce effective contact area, and mitigate failure-inducing stress concentrations. These artificial structures therefore enable superior contact reliability, reduced stiction, and improved high-frequency performance.

Taken together, these advances establish MEMS switches as key building blocks for next-generation communication and sensing systems, supporting reconfigurable, energy-efficient architectures for both signal generation and signal processing across microwave, millimeter-wave, and sub-terahertz bands [1–47].

By exploiting engineered surface textures, nano-patterned coatings, and materials with tailored permittivity, permeability, and surface energy, metamaterial-based contacts can suppress van der Waals and Casimir attraction, thereby enabling repulsive Casimir interactions that mitigate stiction. Additionally, metamaterial-inspired electrodes can enhance isolation, lower insertion loss, and improve broadband performance by shaping the local electromagnetic field distribution at the contact region. From a technological perspective, the concept of metamaterial contacts involves [1, 10]:

- Surface engineering: Incorporation of subwavelength periodic structures or patterned films at the contact interface to reduce effective adhesion and minimize asperity welding.
- Electromagnetic tuning: Embedding metamaterial elements within the ground planes or signal lines to achieve desired impedance characteristics, suppression of parasitic modes, and enhanced ON/OFF capacitance ratio.
- Reliability enhancement: Leveraging metamaterial coatings and textures to reduce stiction, prevent micro-welding, and increase the lifetime of metal-to-metal (M2M) contacts under repeated actuation cycles.
- Scalability for millimeter-wave/THz applications: Designing metamaterial surfaces compatible with miniaturized MEMS geometries, ensuring low loss and high linearity in next-generation communication and sensing systems.

The fusion of MEMS switch technology with metamaterial engineering provides a hybrid platform capable of addressing the limitations of conventional MEMS switches. Such devices hold strong potential for applications in reconfigurable RF front ends, adaptive filters, phased-array antennas, and 5G/6G millimeter-wave systems, where high isolation, low power consumption, and reliability are critical [7–15].

7.1.1 Innovative Concepts in MEMS Switches

MEMS switches are predominantly realized using silicon-based integrated circuit fabrication technologies, in which movable mechanical elements are released by selectively etching sacrificial silicon dioxide (SiO_2) layers [1]. The typical MEMS switch structure consists of a coplanar waveguide (CPW) [19] or microstrip signal line with defined input and output ports positioned between upper and lower ground planes [22]. A suspended microbeam or membrane acts as the primary actuation element, enabling or interrupting the signal path when subjected to electrostatic, thermal, or piezoelectric actuation forces. In advanced switch topologies, architectural variations incorporate DGS patterned in one or both ground planes. Over each DGS, a secondary deflectable beam may be positioned to control localized electromagnetic coupling or to provide additional reconfigurability. Such structural modifications enable enhanced frequency-selective behavior and broaden the functionality of MEMS-based reconfigurable devices. One critical challenge in MEMS switch operation is stiction, which arises when contacting surfaces adhere due to intermolecular forces, including van der Waals and Casimir interactions [9]. To mitigate this, engineered artificial structures, specifically electromagnetic metamaterials, have been proposed to tailor and induce repulsive Casimir forces, thereby reducing adhesion and extending device reliability [1–6].

Relative to conventional solid-state switching technologies, such as PIN diodes and field-effect transistors, MEMS switches demonstrate significant performance advantages. These include ultralow power consumption (due to near-zero static current), excellent signal isolation in the OFF state [7], low insertion loss in the ON state, and superior linearity, making them highly attractive for RF and microwave system applications. Additionally, MEMS switches can be manufactured at a lower cost for certain applications due to compatibility with batch-fabricated silicon processes. Despite these benefits, MEMS switches face several intrinsic drawbacks [6]. First, relatively high actuation voltages [7]—often tens of volts—are required to deflect the beam, limiting integration with low-voltage complementary metal-oxide-semiconductor (CMOS) circuits. Second, they may exhibit increased insertion loss and poor return loss under specific design or fabrication constraints. Third, electromechanical reliability remains a major limitation: repeated switching cycles, particularly under "hot switching" conditions (i.e., when the RF signal is present during actuation), can lead to failure modes such as contact degradation, dielectric charging, or mechanical fracture.

Among these, stiction is a dominant reliability concern. During operation, the movable beam may be pulled into contact with a signal line or electrode. Once engaged, surface adhesion forces combined with static friction can prevent the beam from releasing, effectively causing the switch to remain stuck in the closed state. In severe cases, a high external voltage is required to overcome the stiction force [5]. At low operating voltages, the movable element may remain permanently "welded" [9] to the contact surface, rendering the switch nonfunctional and severely limiting its operational lifetime [1–12].

MEMS switches are generally classified into two primary categories based on their contact mechanism [20]: capacitive (MIM) and resistive (M2M) configurations [21]. Capacitive MEMS switches operate by introducing a thin dielectric layer between two conductive electrodes. When the movable electrode is actuated, it approaches the fixed electrode but remains electrically isolated by the dielectric film, thereby preventing direct M2M contact. This structural feature significantly reduces stiction-related reliability concerns, since adhesion forces between contacting metallic surfaces are effectively eliminated. However, capacitive switches inherently exhibit frequency-dependent performance. The dielectric-separated capacitor behaves as a high-pass filter, allowing efficient RF signal transmission only above a certain cutoff frequency, where the capacitive reactance becomes sufficiently low to ensure reasonable insertion loss. Furthermore, the isolation bandwidth of capacitive switches is constrained by the ratio of the ON-state capacitance to the OFF-state capacitance [24]. A large capacitance ratio is desired for broadband isolation, but is often limited by practical constraints in fabrication and dielectric layer thickness. Resistive (M2M) MEMS switches, in contrast, rely on direct mechanical contact between metallic electrodes. In the actuated (ON) state, the electrical continuity is established with inherently low contact resistance, yielding excellent insertion loss performance over a very wide frequency range, extending from DC through RF and even into millimeter-wave bands. In the open (OFF) state, isolation is defined by the parasitic coupling capacitance between the separated electrodes [7, 26, 27].

The physics of M2M contact is governed by surface roughness [20]. Due to the presence of surface asperities, the *true* contact area is much smaller than the *apparent* or geometric area [25]. For low contact resistance, some degree of plastic deformation of these asperities is required, which increases the true contact area under higher pull-down force. Consequently, contact resistance decreases with greater actuation force. However, this comes at the cost of increased adhesion: adhesion forces scale with both the effective contact area and the local temperature rise associated with plastic deformation [21]. Thus, a fundamental trade-off exists between minimizing contact resistance and mitigating stiction. In RF applications, additional reliability concerns emerge. When high RF power passes through the switch—particularly under impedance mismatch conditions—localized Joule heating can occur at the contact interface. This heating may lead to partial micromelting of asperities, which not only alters the contact morphology but can also result in the formation of permanent electrical shorts. Such degradation severely impacts the lifetime and reliability of M2M MEMS switches, posing a barrier to their deployment in high-frequency applications such as 5G millimeter-wave communication systems [13, 20].

To address these challenges, various mitigation techniques have been investigated. These include [1]:

- Material engineering, such as selecting contact metals with inherently low adhesion properties (e.g., noble metals, alloys).
- Surface modification, through chemical treatments or nanoscale coatings that reduce surface energy and adhesion.
- Contamination control, using plasma cleaning or in-situ passivation to minimize adsorbates that exacerbate stiction and contact degradation.

Ultimately, while capacitive switches are more robust against stiction but limited by bandwidth and frequency dependence, resistive switches offer superior broadband performance at the expense of significant reliability challenges associated with contact adhesion and thermal degradation. Continued progress in contact engineering and surface science is essential before large-scale integration of M2M MEMS switches into next-generation millimeter-wave systems becomes feasible [27].

Here's a detailed comparative table highlighting the differences between capacitive and M2M MEMS switches, including their performance trade-offs and reliability aspects (Table 7.1).

Figure 7.1 [1] illustrates a representative example of a cantilever-type out-of-plane MEMS switch. The figure is composed of three views for clarity [1, 7, 22]:

- Figure 7.1a [1] shows a cross-sectional view of the device along the x-axis.
- Figure 7.1b [1] provides a top view of the switch layout.
- Figure 7.1c [1] presents a cross-sectional view along the y-axis.

The switch is fabricated on a CPW structure [1], denoted as 101, in which a central signal line 110 is flanked by two ground planes, 102 and 104, patterned on a dielectric substrate 105 (e.g., high-resistivity silicon, quartz, or alumina). The signal line 110 is terminated with an input port (112) and an output port (114) at opposite ends of the substrate, providing the RF signal path [1, 7].

Table 7.1 Comparison of capacitive versus M2M MEMS switches.

Parameter	Capacitive MIM [7]	Resistive M2M [7]
Operating principle	Actuation brings a movable electrode close to a fixed electrode, separated by a thin dielectric layer.	Actuation forces direct physical contact between metallic electrodes to establish conduction.
Frequency range	RF to millimeter-wave; not suitable for DC due to dielectric isolation.	DC to RF/millimeter-wave; broadband performance across a wide frequency spectrum.
Insertion loss (ON state)	Low at high frequencies (above cutoff frequency); increases at lower frequencies.	Very low across entire spectrum (DC–millimeter-wave) due to direct conduction.
Isolation (OFF state)	Determined by capacitance ratio (C_{ON}/C_{OFF}); typically narrowband.	Defined by parasitic coupling capacitance between open contacts; broadband but may degrade.
Linearity	Excellent; dielectric layer prevents nonlinearity from direct contact.	Excellent; low distortion due to ohmic conduction.
Power handling	Limited by dielectric breakdown and charging effects.	Higher power handling, but limited by contact heating and potential micro-melting.
Stiction susceptibility	Very low (no direct metal contact).	High; adhesion forces (van der Waals, Casimir, micro-welding) cause reliability issues.
Actuation voltage	Typically 20–80 V (depends on beam stiffness and dielectric thickness).	Typically 20–80 V; may require higher force to ensure low contact resistance.
Reliability issues	Dielectric charging, dielectric breakdown at high fields.	Contact degradation (wear, pitting, micro-welding), stiction, and contamination.
Lifetime (switching cycles)	Up to 10^{11} cycles (with optimized dielectric).	Typically 10^7–10^9 cycles; limited by contact wear and adhesion.
Fabrication complexity	Requires precise dielectric deposition and thickness control.	Requires surface roughness and material engineering to minimize adhesion.
Suitability for 5G/millimeter-wave	Attractive due to low loss at millimeter-wave frequencies, but bandwidth is limited.	Potentially ideal due to broadband/DC–millimeter-wave capability, but stiction and reliability remain challenges.

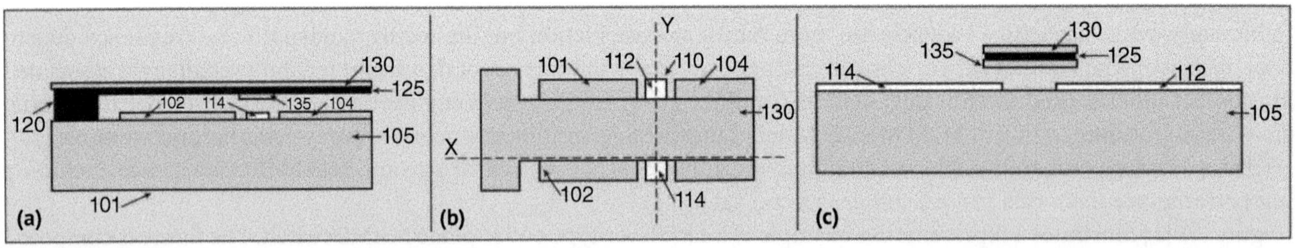

Figure 7.1 Representative MEMS switch configuration: (a) cross-section along the x-axis, (b) planar top view, and (c) cross-section along the y-axis [1].

The MEMS cantilever is anchored to the substrate through a fixed support post or anchor (120), which mechanically secures the beam at one end. The cantilever extends outward in a direction perpendicular to the signal line. Structurally, the cantilever is composed of a dielectric base layer (125)—typically silicon dioxide, silicon nitride, or another insulating silicate—and a top conductive layer (130), such as gold, aluminum, or a refractory metal, which provides both structural reinforcement and electrical functionality.

A contact dimple (135) [1] is formed beneath the cantilever, typically at its free end, and aligned precisely above the CPW signal line 110, in line with the input and output ports (112 and 114). During actuation, when the cantilever is deflected downward, the dimple 135 establishes contact with the signal line, thereby completing the circuit and providing a conductive path between the input and output ports.

Actuation is achieved via an electrostatic actuator [7] (not explicitly shown in the figure). By applying a DC bias voltage between the conductive layer of the cantilever and the adjacent ground planes (102, 104) of the CPW [1], an electrostatic force is generated. This force pulls the cantilever downward toward the substrate (the "ON" state), establishing M2M or metal-to-dielectric contact depending on the switch type. Removal or reduction of the bias voltage allows the cantilever's restoring spring force to lift it back to its original position (the "OFF" state). Thus, the cantilever oscillates vertically (out-of-plane), moving toward or away from the signal line in response to the applied actuation voltage [27].

Alternative MEMS switch configurations may employ lateral motion mechanisms, in which the movable beam or plate slides parallel to the substrate surface to make or break contact. Furthermore, the nature of the contacting surfaces distinguishes between resistive (M2M) switches—where conduction occurs through direct metallic contact—and capacitive (MIM) switches, in which a dielectric layer separates the electrodes and the signal is coupled capacitively at RF frequencies.

7.1.2 Next-generation MEMS Switch Employing Capacitive MIM Contact

Figures 7.2–7.6 illustrate the design, mechanical behavior, and RF performance of various capacitive-contact shunt MEMS switches fabricated on CPWs. These switches employ suspended metallic beams with engineered perforations and geometries to optimize actuation voltage, mechanical compliance, and RF characteristics [27].

7.1.2.1 Doubly Clamped Beam Switch

Figures 7.2 and 7.3 depict a shunt MEMS switch featuring a doubly supported cantilever beam 210 suspended over a CPW fabricated on substrate 201. As illustrated in Figures 7.2 and 7.3, describe the side and top views, respectively, of a capacitive-contact shunt MEMS switch incorporating a doubly supported cantilever beam 210, which is suspended above a CPW fabricated on substrate 201 [1].

As depicted in Figure 7.2, the beam 210 [1] is anchored at its first end 212 and second end 214 by the corresponding ground planes 202 and 204 of the CPW structure. The central portion of the beam 210 is suspended over the CPW signal line 220 [1], allowing for capacitive coupling between the beam and the underlying conductive path. The beam 210 is mechanically coupled to an external actuator (not shown), which applies a controlled DC bias voltage between the beam 210 and the ground planes 202 and 204. Application of the DC bias induces electrostatic attraction, causing the beam 210 to deflect downward toward the signal line 220, thereby modulating the RF signal path. The signal line 220 [7] itself consists of a conductive layer 222 overlaid with a thin dielectric layer 224—commonly silicon nitride—to provide electrical insulation while maintaining effective capacitive interaction with the actuated beam [27].

The dielectric layer 224 [1] is selected to have a thickness of approximately 0.2 μm, providing electrical insulation between the CPW signal line 220 and the actuated beam 210 [7]. When a DC bias voltage is applied, the electrostatic force causes the beam 210 to deflect downward toward the signal line 220. This deflection increases the shunt capacitance, which effectively blocks the propagation of RF signals along the CPW, defining the ON state of the switch. Upon removal of the

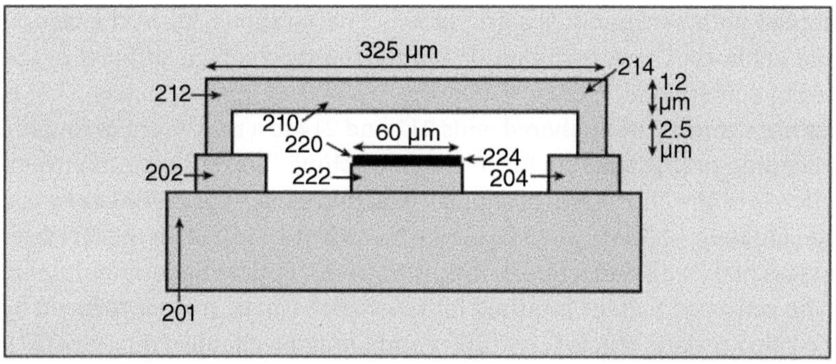

Figure 7.2 A side-view illustration of the MEMS switch employing a doubly anchored cantilever beam structure fabricated above a CPW [1].

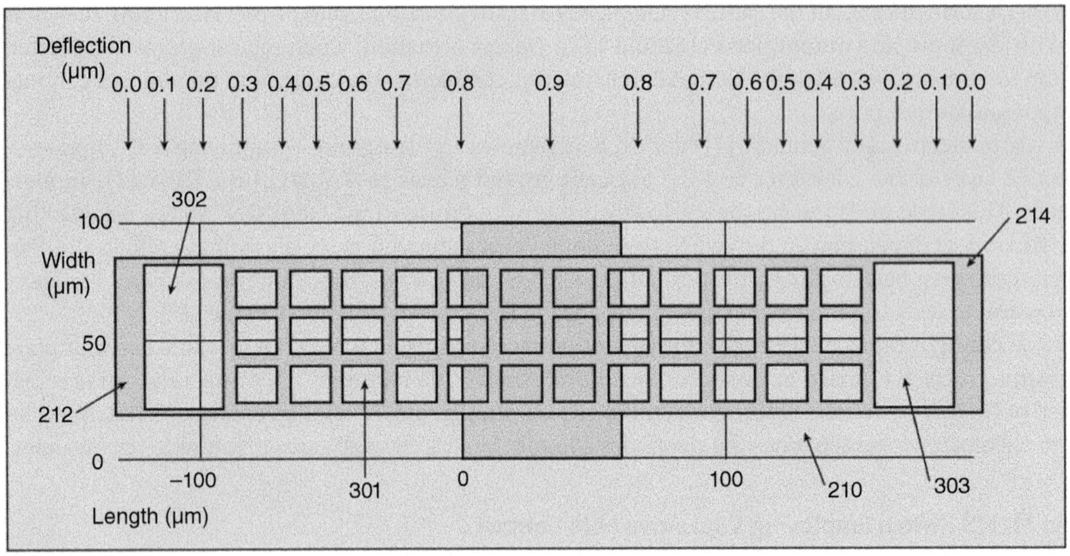

Figure 7.3 A plan-view representation of the MEMS switch featuring a doubly anchored cantilever beam positioned over a CPW [1].

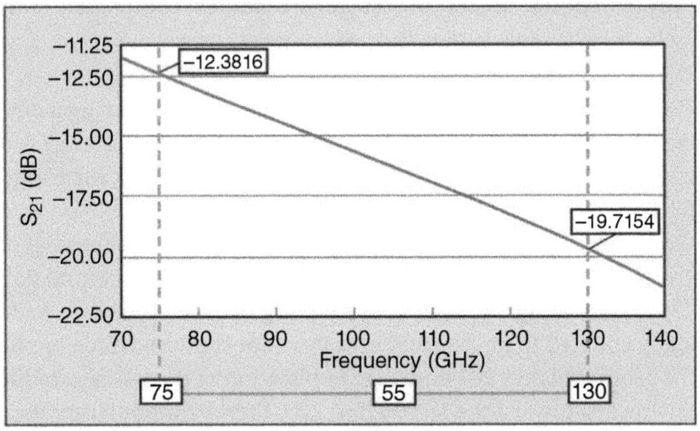

Figure 7.4 Isolation characteristics of the MEMS switch corresponding to the configurations shown in Figures 7.2 and 7.3 [1].

DC bias, the beam 210 returns to its original suspended position, reducing the shunt capacitance and allowing RF signals to propagate unimpeded, corresponding to the OFF state. In the illustrated design (Figure 7.2), the beam 210 [27] is fabricated from molybdenum with a length of approximately 325 μm, a width of 60 μm, and a thickness of 1.2 μm. The CPW signal line 220 extends continuously beneath the beam, with a width of about 60 μm, and is separated from the beam by an air gap of approximately 2.5 μm. The thin dielectric layer of 0.2 μm ensures adequate insulation while maintaining strong capacitive coupling during actuation.

As shown in Figure 7.3 [1], the beam 210 is patterned with perforations: a grid of small perforations 301 in the central region and larger perforations 302 and 303 near the anchors. These perforations reduce the mechanical stiffness of the beam, enhancing downward deflection and improving actuation efficiency. When the DC bias of approximately 37 V is applied, the vertical displacement of the beam varies from zero at the anchored ends 212 and 214 to a maximum of roughly 0.91 μm at the center, demonstrating a controlled bending profile suitable for reliable switching. The RF performance of the switch is illustrated in Figure 7.4. When the switch is in the OFF state (beam up), isolation across the millimeter-wave frequency band of 75–130 GHz ranges [7] from approximately −12.4 dB at 75 GHz to −19.7 dB at 130 GHz. In the ON state (beam down), the switch exhibits a typical insertion loss of 0.74 dB and a return loss of 10.04 dB, indicating minimal signal attenuation and acceptable impedance matching. The actuation voltage required for the switch can be further reduced by optimizing the perforation layout, allowing enhanced electrostatic efficiency while maintaining mechanical integrity [27].

7.1.2.2 Perforated and Serpentine Beam Switch

Figures 7.5 and 7.6 illustrate an improved MEMS switch design using a gold rectangular beam 510 with an engineered perforated and serpentine structure. In the example illustrated in Figure 7.5 [1], the MEMS switch employs a rectangular beam 510 fabricated from gold with a perforated structure to enhance mechanical compliance. The central portion 516 of the beam forms a perforated grid or lattice, while each corner of this lattice extends outward in a serpentine pattern toward the anchored first and second ends 512 and 514 of the beam [1]. The serpentine patterns at both ends are interconnected, forming continuous first and second serpentine structures that anchor the beam. This serpentine design reduces the effective stiffness of the beam, enabling downward deflection with a significantly lower actuation voltage compared to conventional doubly clamped beams. The overall dimensions of the MEMS switch in Figure 7.5 are comparable to those of Figure 7.3 [1], with minor adjustments: the beam length is approximately 345 μm, and the width is about 65 μm. Under a DC bias of only 17 V, the beam achieves a maximum downward deflection of approximately 0.9 μm, demonstrating improved actuation efficiency relative to the prior design [7].

The RF performance of the switch is shown in Figure 7.6. When the switch is in the OFF state (beam up), the isolation across the 75–130 GHz millimeter-wave band ranges from approximately −22 dB at 75 GHz to −14.7 dB at 130 GHz [27], with a minimum isolation of around −24.8 dB observed near 86 GHz. In the ON state (beam down), the insertion loss is measured to be only 0.6 dB, and the return loss is approximately 15.15 dB, indicating excellent signal transmission and impedance matching [1]. The combination of the perforated lattice and serpentine end structures provides both enhanced isolation and reduced actuation voltage, highlighting the design's superior electromechanical and RF performance characteristics.

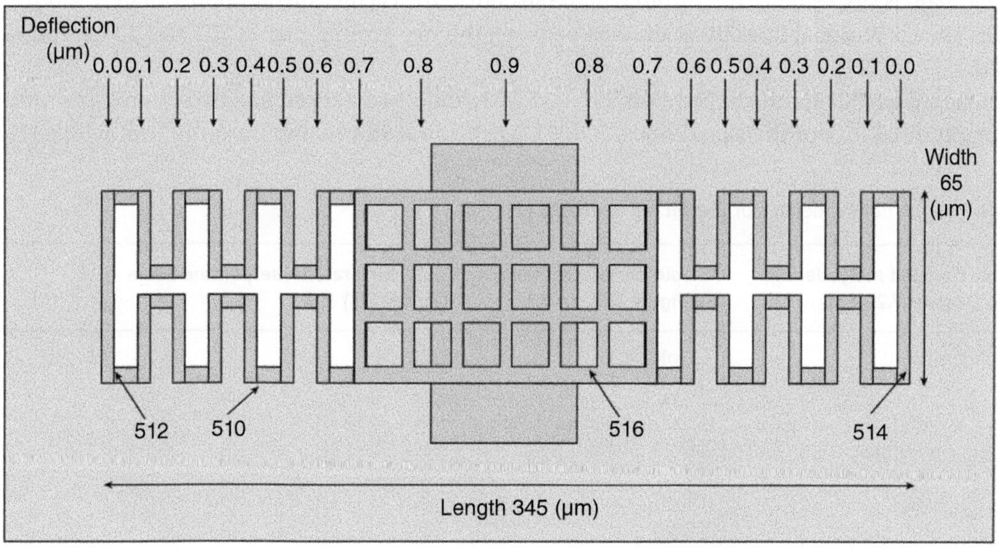

Figure 7.5 A top-view layout of a capacitive shunt MEMS switch incorporating a serpentine cantilever beam structure [1].

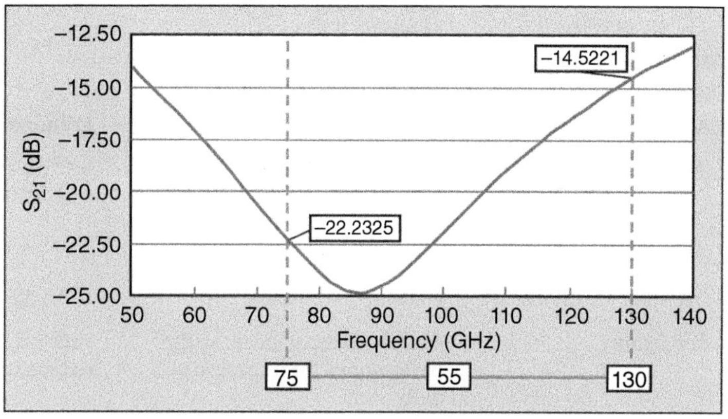

Figure 7.6 Isolation response of the MEMS capacitive shunt switch illustrated in Figure 7.5 [1].

By comparing the designs, it is evident that introducing perforations and serpentine geometries in MEMS switch beams enhances mechanical compliance, lowers actuation voltage requirements, and improves RF performance metrics such as insertion loss, return loss, and isolation. The choices of beam material, perforation pattern, and air gap dimensions are critical in balancing mechanical robustness with high-frequency electrical performance, making these MEMS switches suitable for millimeter-wave applications spanning 75–130 GHz. Table 7.2 summarizes the key parameters and performance metrics [7] of the MEMS switches.

7.1.3 MEMS Switch Featuring Defected Ground Structure Capacitive Contacts

The incorporation of a two-dimensional DGS [1] into the MEMS switch design substantially improves the isolation characteristics of capacitive shunt switches [7], particularly within the millimeter-wave spectrum spanning 75–130 GHz [27]. By strategically introducing periodic or patterned defects in the ground plane beneath the signal line, the DGS perturbs the local electromagnetic fields, suppresses parasitic modes, and increases the effective impedance between the signal and ground. This engineered modification not only reduces undesired signal leakage and cross talk but also enhances the overall RF performance of the switch, enabling reliable operation at high frequencies while maintaining low insertion loss and minimal intermodulation distortion. The DGS introduces localized discontinuities in the ground planes 702 and 704, which perturb the current distribution and effectively act as a band-stop filter. This modifies the equivalent capacitance and inductance seen by the CPW signal line 720, thereby increasing the shunt impedance in the OFF state and suppressing RF signal propagation.

Figure 7.7 depicts a MEMS switch with a beam 710 that retains the same perforated and serpentine arrangement as the gold beam 510 in Figure 7.5. The beam is suspended over a ground plane structure 701 [1] measuring approximately 320 μm in length and 400 μm in width. The CPW signal line 720 is situated between the two ground planes 702 and 704, forming the standard switch configuration [7, 22, 27].

In this example, the DGS is realized as four spiral-shaped slots 731, 732, 733, and 734 arranged in a two-by-two grid with mirror symmetry along the longitudinal axis of the signal line 720 [27]. Each spiral slot maintains a uniform width [1],

Table 7.2 The key parameters and performance metrics of the MEMS switches [1].

Parameter	Doubly clamped molybdenum beam (Figures 7.2–7.3)	Gold perforated beam (Figure 7.5)	Gold perforated + serpentine beam (Figure 7.5)
Beam material	Molybdenum	Gold	Gold
Beam length (μm)	325	325	345
Beam width (μm)	60	60	65
Beam thickness (μm)	1.2	1.2	1.2
Air gap (μm)	2.5	2.5	2.5
Dielectric layer thickness (μm)	0.2	0.2	0.2
Perforation	Central small grid, large end holes	Central grid	Central grid + serpentine ends
Maximum deflection (μm)	0.91	0.9	0.9
Actuation voltage (V)	37	37 (approx.)	17
OFF-state isolation	−12.4 dB at 75 GHz −19.7 dB at 130 GHz	−22 dB at 75 GHz −14.7 dB at 130 GHz	−22 dB at 75 GHz −14.7 dB at 130 GHz Minimum −24.8 dB at 86 GHz
ON-state insertion loss	0.74 dB	0.6 dB	0.6 dB
ON-state return loss	10.04 dB	15.15 dB	15.15 dB
Key feature	Doubly clamped beam with perforations for improved deflection	Perforated beam for reduced stiffness	Perforated + serpentine beam for lower actuation voltage and improved isolation

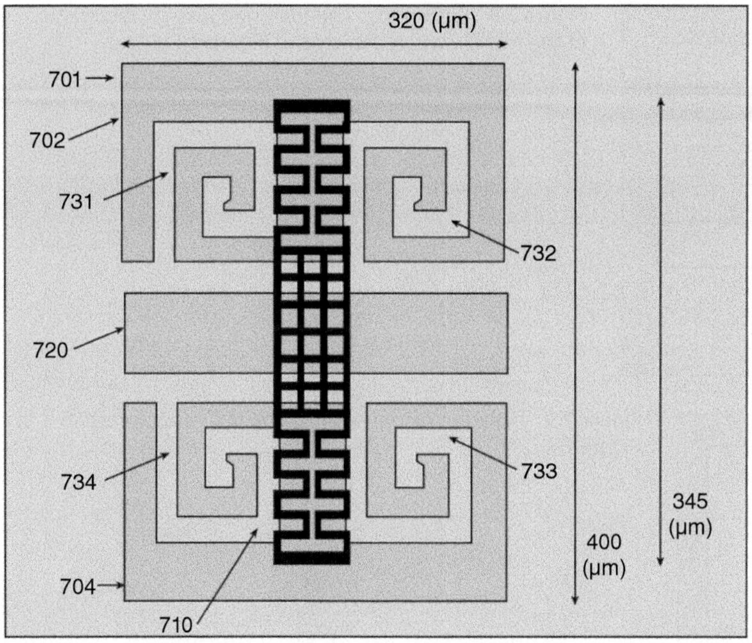

Figure 7.7 A top-view illustration of the MEMS switch incorporating a DGS [27].

ensuring consistent perturbation of the ground plane current distribution and predictable modification of the switch's RF transmission characteristics. The incorporation of such a two-dimensional DGS enhances the effective shunt impedance in the OFF state, leading to improved isolation across the millimeter-wave band without significantly affecting the insertion loss in the ON state.

The electromagnetic behavior of the DGS can be approximated using a lumped-element model [8], where each spiral slot introduces an equivalent series inductance L_{DGS} and parallel capacitance C_{DGS} to the CPW structure. The resulting resonant frequency f_r of the DGS is given by [1]:

$$f_r = \frac{1}{2\pi\sqrt{L_{\text{DGS}}C_{\text{DGS}}}} \tag{7.1}$$

At frequencies near f_r, the DGS presents a high-impedance path, effectively increasing the shunt capacitance between the beam 710 and the ground planes, which enhances the OFF-state isolation. This modification reduces leakage of RF signals in the OFF state without substantially increasing the ON-state insertion loss, thus preserving signal integrity when the switch is closed.

By carefully designing the geometry of the spiral slots [6]—including the number of turns, width, and spacing—the resonant frequency of the DGS [1] can be tuned to align with specific millimeter-wave bands, thereby optimizing isolation across 75–130 GHz. This combination of perforated and serpentine beam geometry with a two-dimensional DGS provides a high-performance MEMS shunt switch suitable for next-generation millimeter-wave applications.

Figure 7.8 presents the insertion and return loss characteristics of the MEMS switch depicted in Figure 7.7 [27] when the switch is in the ON state (beam closed). The insertion loss varies from approximately 2.2 dB at 75 GHz to 10.4 dB at 130 GHz, with a peak value reaching 16.6 dB near 105 GHz. Correspondingly, the return loss ranges from about 24 dB at 75 GHz to 11.2 dB at 130 GHz, with a minimum of approximately 9.5 dB around 105 GHz. Figure 7.9 [1] illustrates the isolation performance of the same DGS-enhanced MEMS switch in the OFF state (beam open). Isolation measures approximately −17.1 dB at 75 GHz and −11.5 dB at 130 GHz, with a pronounced minimum of about −32.5 dB observed near 82 GHz [7, 27].

While the DGS integration in Figure 7.7 improves isolation in the millimeter-wave band, the increased insertion loss in the ON state limits the practical applicability of the switch. To address this limitation, modified DGS geometries are implemented, as shown in Figure 7.10. The MEMS switch in Figure 7.10 [7] retains the same general structure as the Figure 7.7 design, featuring two ground planes 1102 and 1104 separated by a CPW signal line 1120. Four DGS structures

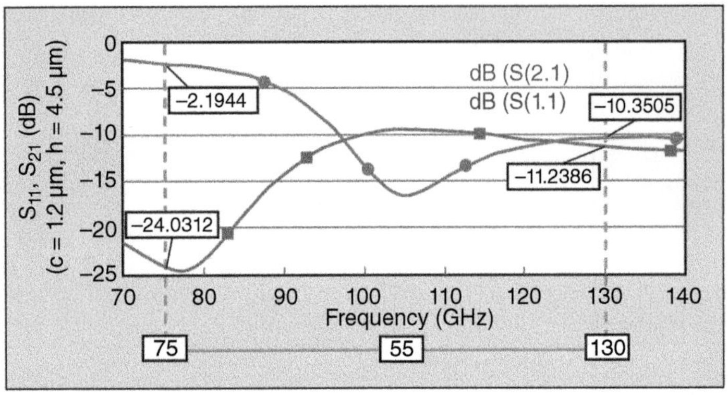

Figure 7.8 Return loss and insertion loss characteristics of the MEMS switch depicted in Figure 7.7 [27].

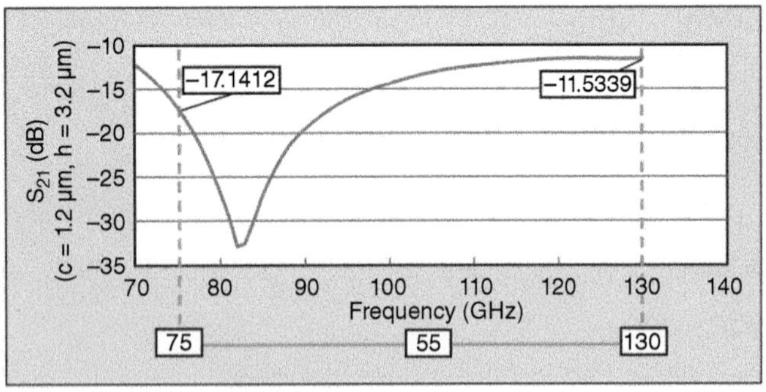

Figure 7.9 Isolation response of the MEMS switch presented in Figure 7.7 [27].

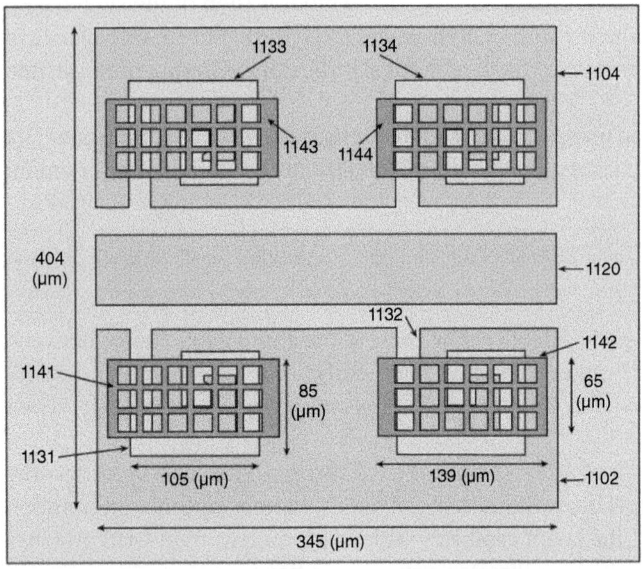

Figure 7.10 Top view of the MEMS switch, illustrating secondary switches positioned above the DGS [27].

1131, 1132, 1133, and 1134 [27] are incorporated into the ground planes [1], with geometry adjustments intended to maintain high isolation while reducing the ON-state insertion loss. These modifications optimize the electromagnetic response of the ground plane, mitigating the excessive energy dissipation introduced by the original DGS configuration.

In the MEMS switch configuration shown in Figure 7.10 [1], the CPW ground planes 1102 and 1104 and the signal line 1120 have a length of approximately 340 μm, while the overall lateral width of the switch is about 404 μm. This design extends the concept introduced in Figure 7.7 by integrating secondary MEMS switches 1141, 1142, 1143, and 1144 above each DGS 1131–1134. While both the DGS structures and the secondary switches may assume a rectangular geometry, the secondary switches are typically longer along the longitudinal axis of the signal line, whereas the DGS structures are comparatively wider in the transverse direction [7].

In the example shown, each DGS structure [1] comprises a perforated lattice measuring approximately 105 μm in length and 85 μm in width. Each secondary MEMS switch is overlaid directly above the corresponding DGS, with a beam length of approximately 139 μm and a width of 65 μm. The structures are fabricated on substrate 1101, which supports a ground plane 1102 with a thickness of about 2 μm. The DGS slots, such as those in 1131, are formed entirely through the ground plane, extending to its full thickness, thereby perturbing the current distribution in the CPW and affecting RF transmission.

The secondary switch 1141 includes a beam 1151 [7] supported by two anchoring feet 1162 and 1164, each approximately 1 μm in height. Consequently, the beam 1151 is suspended about 1 μm above the underlying DGS structure and ground plane, forming a precise air gap [1]. The beam itself has a thickness of roughly 1.2 μm. An actuator (not shown) applies a DC bias voltage between the beam 1151 and the ground plane 1102 through the supporting feet, generating an electrostatic force that deflects the beam downward toward the ground plane [1]. This actuation modulates the local capacitive characteristics of the underlying DGS, thereby influencing the overall RF behavior of the main MEMS switch [27].

The applied bias voltage to the secondary switch can be continuously varied, enabling fine-tuning of the DGS capacitance and, in turn, adjustment of the transmission and reflection characteristics of the primary switch [20]. This arrangement effectively imparts metamaterial-like properties to the CPW structure: by comparing the transmission and reflection phase of the signal line with and without the DGS-secondary switch configuration, one observes engineered modifications of the electromagnetic response. The combination of tunable capacitive elements above the DGS with the primary MEMS switch introduces a reconfigurable metamaterial behavior, enhancing isolation, insertion loss control, and overall RF performance across millimeter-wave frequency bands [22].

The transmission and reflection phase characteristics [24] of a CPW without any DGS structures were first analyzed to establish a baseline for comparison. Figure 7.11 presents the results over the millimeter-wave frequency range of 50–140 GHz. As observed, the transmission phase of a signal propagating along the CPW closely follows the reflection phase, with any phase shift in transmission being accompanied by an approximately equal shift in reflection—typically within ±20°—indicating a conventional CPW response without significant perturbation.

Figure 7.12 [1] illustrates the transmission and reflection phase behavior for the same CPW incorporating the two-dimensional DGS structure, with the secondary MEMS switch suspended above the DGS at a height of 2.2 μm. This distance corresponds to the separation from the top surface of the substrate (i.e., the base of the DGS slots) [1] to the bottom surface of the secondary switch positioned above the DGS. Unlike the baseline case, the transmission and reflection phases no longer shift in unison across the frequency band. Instead, the phases exhibit nonuniform and even opposing behavior, crossing one another at approximately 85 GHz and returning to cross again near 96 GHz [27]. This nonreciprocal phase interaction demonstrates that the DGS-secondary switch combination introduces a frequency-dependent perturbation to the CPW, modifying both the effective capacitance and inductance along the transmission path. These engineered phase shifts are indicative of metamaterial-like behavior, where the local electromagnetic response is controlled by the geometry and positioning of the DGS and the overlying secondary switch, enabling tunable RF characteristics for millimeter-wave applications.

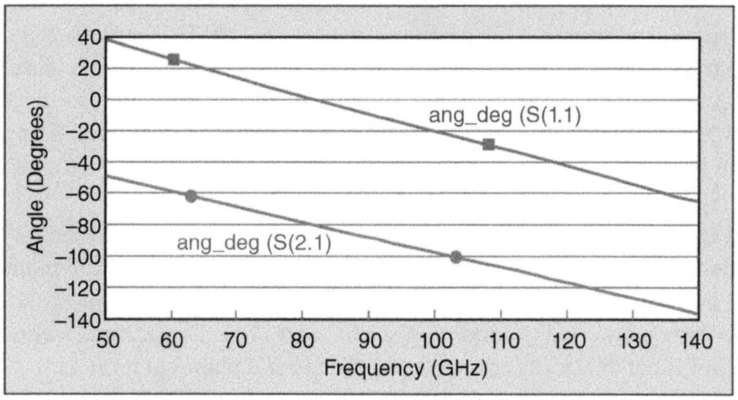

Figure 7.11 Transmission and reflection phase characteristics of the coplanar line in a MEMS switch without a DGS [27].

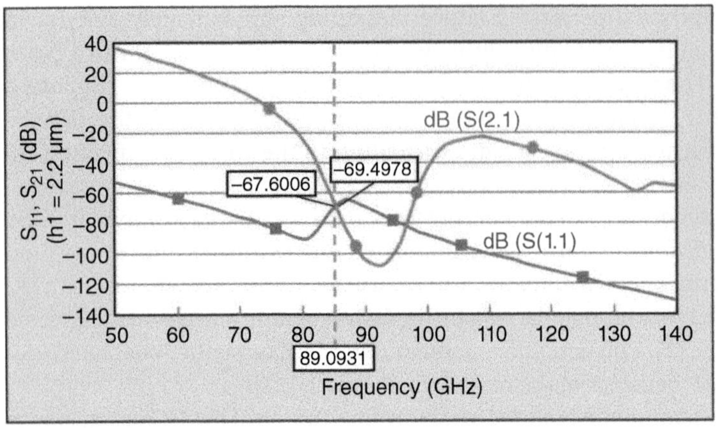

Figure 7.12 Transmission and reflection phase plots for the CPW incorporating a DGS, as shown in Figure 7.10, measured at a height of 2.2 μm [1].

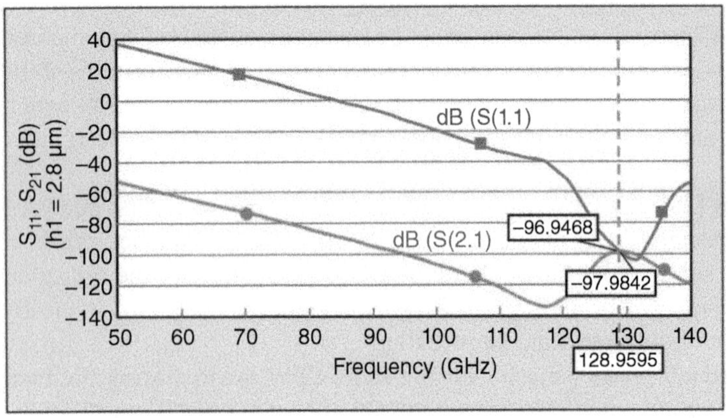

Figure 7.13 Transmission and reflection phase plots for the CPW with DGS, as depicted in Figure 7.10, at a height of 2.8 μm [1].

Figure 7.13 [1] presents the transmission and reflection phase characteristics of the same CPW incorporating the DGS structure, with the secondary MEMS switch positioned at a height of 2.8 μm above the substrate. At this increased separation, the transmission and reflection phases remain approximately equal up to around 110 GHz, indicating conventional CPW behavior. However, at frequencies above 115 GHz, the phases begin to diverge, even crossing one another near 128 GHz [8]. This behavior demonstrates that the effective resonant frequency of the DGS structure is strongly dependent on the air gap between the ground plane and the overlying beam. By varying this separation, the electromagnetic response of the CPW—including the relative phase of transmitted and reflected signals—can be actively tuned, providing a mechanism for dynamic control of the switch's RF characteristics and further emphasizing the metamaterial-like properties of the DGS-secondary switch combination [9].

Figure 7.14 [1] illustrates the isolation performance of five secondary MEMS switches [7] positioned over DGS structures at varying air gap heights above the ground plane. The results indicate that the resonant frequency of the DGS structure systematically shifts toward higher frequencies as the separation between the ground plane and the overlying beam increases [27]. This trend demonstrates that the air gap serves as a tunable parameter, allowing precise control over the resonance and, consequently, the isolation characteristics of the MEMS switch. By adjusting the vertical position of the secondary switches, the electromagnetic response of the CPW can be actively modified, enabling dynamic tuning of the switch's performance across the millimeter-wave frequency band [8]

Figure 7.15 [1] presents a complete MEMS shunt switch configuration incorporating DGS and overlaid secondary switches. The CPW comprises a signal line 1720 positioned between a first ground plane 1702 and a second ground plane 1704, with the signal line separated from each ground plane by gaps 1703 and 1705, respectively. A primary shunt MEMS switch 1710 is suspended above the CPW, bridging the first and second ground planes 1702 and 1704. The primary switch 1710 is oriented perpendicular to the signal line 1720 and is mechanically connected to the ground planes at its ends.

When a DC bias voltage is applied to the primary switch 1710, electrostatic forces induce downward deflection of the beam toward the underlying signal line 1720, increasing the shunt capacitance and effectively blocking RF signal propagation along the CPW (ON state). Upon removal of the bias voltage, the beam returns to its initial suspended position, restoring the normal propagation path of the signal line (OFF state). This primary shunt switch configuration provides the

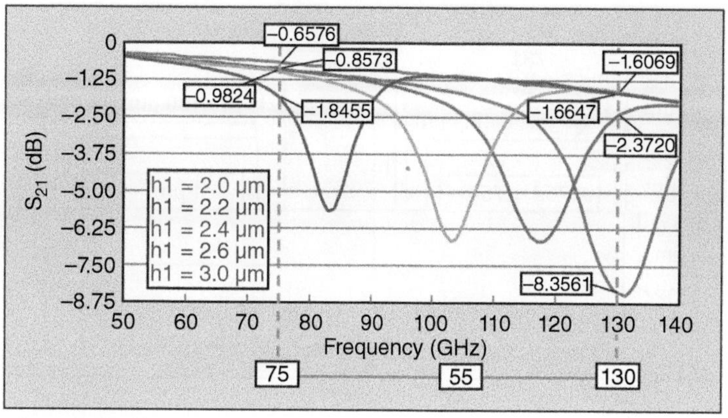

Figure 7.14 Isolation characteristics of five secondary MEMS switches arranged over DGS structures at different heights.

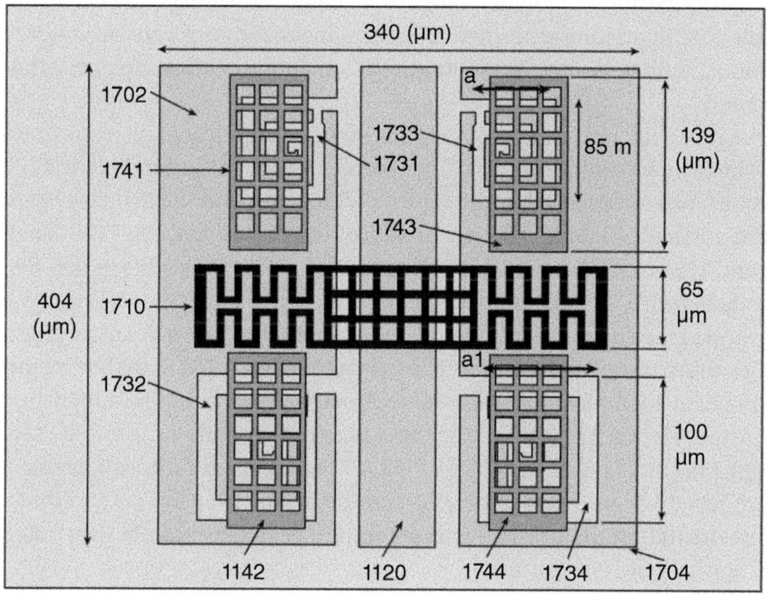

Figure 7.15 A top-view illustration of a metamaterial-inspired MEMS switch featuring a DGS and secondary switches, highlighting an aspect of the novel design [1].

fundamental actuation mechanism, while the inclusion of DGS structures and secondary MEMS switches above the ground planes enables further tuning of the switch's RF response and isolation characteristics across millimeter-wave frequencies.

In the MEMS switch configuration shown in Figure 7.15 [1], four DGS are incorporated to enhance isolation and tunability. The first DGS structure 1731 and the second DGS structure 1732 are formed within the first ground plane 1702 [7], while the third DGS structure 1733 and the fourth DGS structure 1734 [1] are formed within the second ground plane 1704 [27]. The first and third DGS structures, 1731 and 1733, exhibit mirror symmetry along the longitudinal x-axis of the primary shunt switch 1710 and share a similar geometry. Similarly, the second and fourth DGS structures, 1732 and 1734, are mirror-symmetric along the same axis and are identical in shape [1].

In the example illustrated, the dimensions of the DGS structures vary: the second slots of the first and third DGS structures 1731 and 1733 are approximately 85 μm in length, whereas the second slots of the second and fourth DGS structures, 1732 and 1734, are approximately 100 μm in length. The DGS structures are separated by air gaps to allow proper capacitive interaction with overlaid MEMS elements.

Rectangular secondary shunt switches 1741, 1742, 1743, and 1744 are positioned above each DGS structure. These secondary switches are oriented parallel to the CPW signal line 1720 and perpendicular to the primary shunt switch 1710 [1]. The secondary switches above the first and third DGS structures (1741 and 1743) are mirror-symmetric to those above the second and fourth DGS structures (1742 and 1744) along the longitudinal x-axis of the primary switch. Additionally, the secondary switches above the first and second DGS structures (1741 and 1742) [1] are mirror-symmetric to those above the third and fourth DGS structures (1743 and 1744) along the transverse y-axis of the signal line 1720 [1].

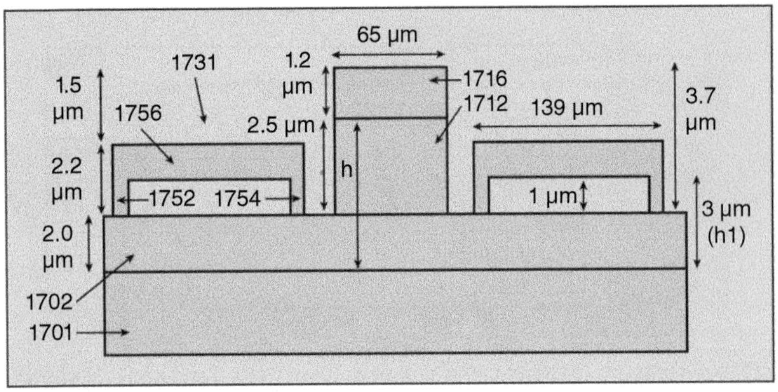

Figure 7.16 A side-view illustration of the MEMS switch depicted in Figure 7.15 [7].

Each secondary [7] switch incorporates a grid-like lattice perforation, which reduces mechanical stiffness, lowers actuation voltage, and enhances downward deflection. This combination of patterned DGS structures and perforated secondary switches enables precise tuning of the CPW's electromagnetic response, improving isolation and providing metamaterial-like behavior across the millimeter-wave frequency band.

Figure 7.16 [1] presents a side view of the MEMS switch depicted in Figure 7.15. The switch is fabricated on a substrate 1701, over which a ground plane 1702 is deposited. The primary shunt switch 1710 is formed above the ground plane 1702 and is supported at its ends by two feet 1712 (the second foot is obscured in the figure), which suspend the primary beam 1716 [1]. Two secondary switches, 1731 and 1732, are positioned laterally on either side of the primary switch 1710. Each secondary switch includes two supporting feet 1752 and 1754 that hold a secondary beam 1756. DGS, not visible in this side view, are located within the ground plane 1702 directly beneath the secondary switches [27].

In the illustrated example [1], the substrate and ground planes have lateral dimensions of approximately 404 μm in length and 340 μm in width, with the ground planes having a thickness of about 2 μm. The primary switch 1710 spans the length of the substrate, with each foot 1712 and the primary beam 1716 having a width of approximately 65 μm. The supporting feet 1712 have a height of about 2.5 μm, while the primary beam 1716 has a thickness of approximately 1.2 μm. The secondary switches 1731 and 1732 have a length of about 139 μm [1], with their feet 1752, 1754 and beam 1756 also having a width of approximately 65 μm. The secondary switch feet 1752 are 1 μm in height, and the beams 1756 are 1.2 μm thick. Consequently, the complete MEMS switch structure—including the primary and secondary switches—can be fabricated on the substrate 1701 within a total vertical space of approximately 5.7 μm [7].

7.1.4 MEMS Switch Featuring Metamaterial-inspired Architecture

MEMS switches have emerged as a promising technology for reconfigurable [7] RF and microwave circuits due to their low insertion loss, high isolation, and excellent linearity [7]. Traditional MEMS switches are typically classified as either capacitive (MIM) or resistive (M2M) types [1], each with distinct advantages and limitations. Capacitive switches rely on a thin dielectric layer to prevent direct metal contact, thereby reducing stiction but limiting bandwidth and frequency performance. In contrast, M2M switches offer broadband operation from DC to millimeter-wave frequencies with low insertion loss but are inherently susceptible to stiction, adhesion, and micro-welding at the contact interface, particularly under high RF power or repeated actuation cycles [13].

To overcome these limitations, metamaterial-based contacts [2–5] have been introduced as an innovative approach to enhance the performance and reliability of MEMS switches. Metamaterials [17] are engineered structures with properties not found in natural materials, allowing precise control over electromagnetic response, surface energy, and mechanical interactions at the nanoscale. By integrating metamaterial features into the contact regions of MEMS switches, adhesion forces such as van der Waals and Casimir interactions can be tailored, even producing repulsive forces that mitigate stiction. This capability significantly improves the lifetime, switching reliability, and operational consistency of M2M MEMS switches under both hot and cold switching conditions. The integration of metamaterial contacts offers multiple performance improvements:

- Reduced stiction and adhesion: Engineered surface structures modulate Casimir and van der Waals forces [11], significantly lowering the probability of micro-welding or permanent sticking.

- Enhanced RF performance: Localized electromagnetic field shaping at the contact interface improves ON/OFF capacitance ratio, increases isolation, and reduces parasitic losses, particularly at millimeter-wave frequencies.
- Broadband operation: M2M metamaterial contacts retain the intrinsic broadband capabilities of resistive switches, enabling DC–millimeter-wave operation without the frequency-dependent limitations of dielectric-separated capacitive switches.
- Improved reliability: By mitigating adhesion and controlling mechanical stress distribution, metamaterial contacts extend the switch lifetime, often by several orders of magnitude relative to conventional M2M switches.

The fabrication of metamaterial-contact MEMS switches builds upon standard MEMS processing techniques while integrating nano-patterning methods, such as electron-beam lithography, nanoimprint lithography [19], or focused ion beam milling. Critical steps include:

- Deposition of multilayer cantilevers (dielectric + conductive layers)
- Formation of subwavelength metamaterial structures on contact surfaces
- Release of movable elements through sacrificial layer etching
- Surface treatment or passivation to minimize contamination and optimize surface energy

Material selection is also crucial. Noble metals such as gold or platinum are often used for the contact layer to reduce oxidation, while dielectric coatings or ultrathin insulating films can be incorporated to further suppress stiction without compromising RF conduction. Table 7.3 shows the comparison with conventional MEMS switch.

Metamaterial-contact MEMS switches are highly suitable for reconfigurable RF front ends, phased-array antennas, tunable filters, and 5G/6G millimeter-wave systems, where performance, reliability, and low power operation are critical. Future research is likely to focus on the following:

- Hybrid designs combining lateral and out-of-plane actuation
- Advanced metamaterial geometries for multifunctional surfaces
- Integration with CMOS circuits for compact, low-voltage operation
- High-power RF testing to validate thermal and mechanical robustness

By merging MEMS switch technology with metamaterial engineering, a new class of high-performance, reliable, and scalable RF-MEMS devices is emerging, capable of meeting the stringent demands of next-generation communication and sensing systems.

Figure 7.17 [1] illustrates an example layout of a metamaterial-inspired MEMS [11] shunt switch, highlighting the interconnections between the primary switch 1910, secondary switches 1941–1944, and their respective actuators 1962 [1] and 1964 [7]. The first actuator 1962 [1] is electrically connected to the primary switch 1910 [1] and is configured to supply a controllable DC bias voltage, enabling actuation of the primary beam. The second actuator 1964 [1] is connected to all secondary switches 1941–1944 [1] and provides a bias voltage to control their deflection independently of the primary switch [7, 13, 15].

During operation, the primary switch 1910 can be set to an ON or OFF state [1]. In the ON state, the bias voltage from actuator 1962 causes the primary switch beam to deflect downward toward the signal line 1920, generating a large shunt capacitance that effectively blocks RF signal propagation along the CPW. When the primary switch is in the OFF state, the

Table 7.3 Comparison with conventional MEMS switches.

Parameter	Conventional M2M	Metamaterial contact MEMS
Contact resistance	Low, but dependent on asperity deformation	Low, with reduced adhesion forces
Stiction/adhesion	High, prone to micro-welding	Low, engineered repulsive interactions
Lifetime (switching cycles)	10^7–10^9	10^9–10^{11} or higher
RF bandwidth	DC to millimeter-wave	DC to millimeter-wave, with optimized field distribution
Actuation voltage	20–80 V	20–80 V, similar to conventional MEMS
Fabrication complexity	Standard MEMS processes	Requires nano-patterning and surface engineering

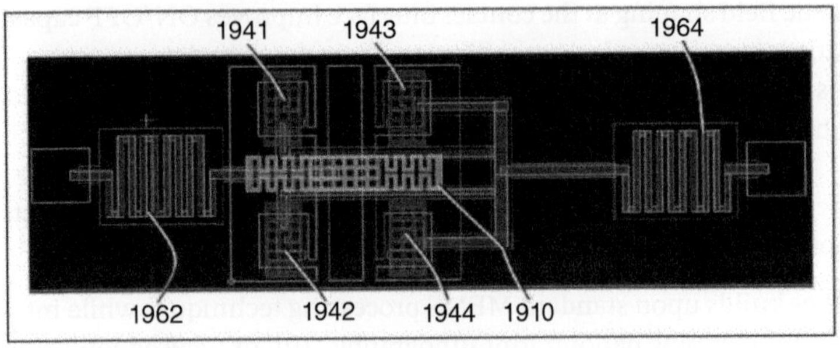

Figure 7.17 Representative layout of the metamaterial-based MEMS switch [1, 7].

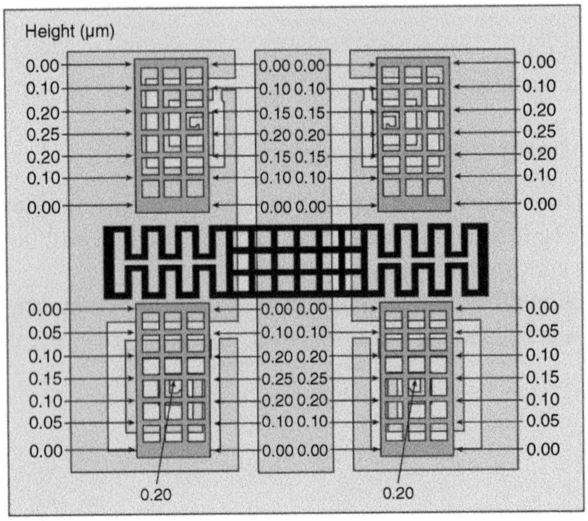

Figure 7.18 Downward deflection of multiple points on the secondary MEMS switches (in μm) during actuation [7].

beam returns to its rest position, the shunt capacitance decreases, and RF signals propagate unimpeded along the signal line [7].

When the primary switch 1910 [1] is OFF, the secondary switches 1941–1944 can be actuated independently via the second actuator 1964. Application of a bias voltage to these secondary switches causes their beams to deflect downward toward the underlying DGS structures, creating a localized shunt capacitance that counteracts the capacitive effects of the DGS. This mechanism reduces the impact of the DGS on the ON-state insertion loss and return loss of the primary switch, effectively providing tunable compensation and enabling dynamic control over the overall RF performance of the metamaterial-inspired MEMS switch.

Figure 7.18 [1] presents the measured downward deflection profiles of the secondary switches, expressed in micrometers (μm), under actuation. The displacement data correspond to multiple points along the span of the switch beams, thereby characterizing the out-of-plane motion induced by the applied electrostatic force [3].

Figure 7.19 [1] illustrates the RF performance of the MEMS switch configuration described in Figure 7.17, specifically the return loss and insertion loss responses, under the condition in which the primary switch (1910) is held in the OFF state while the secondary switches (1941–1944) are simultaneously actuated. At 75 GHz, the insertion loss is approximately 0.6 dB, while the return loss achieves a relatively high value of about 21.1 dB, indicating excellent impedance matching and low signal attenuation. At 130 GHz, the insertion loss remains acceptably low at around 1.5 dB, with the corresponding return loss of 14.5 dB still demonstrating effective performance at higher frequencies [7].

Referring again to the switching architecture shown in Figure 7.17, when the primary switch (1910) is turned ON, the secondary switches (1941–1944) can be deactivated in order to leverage the DGS elements for enhanced isolation. In this state, the second actuator does not supply any bias voltage to the secondary switches, leaving them suspended above the DGS regions with an air gap, thus minimizing unwanted electromagnetic coupling [1, 15].

Figure 7.20 depicts the measured downward deflection profiles of the primary switch under actuation, also recorded in micrometers. The results show that the displacement is spatially uniform across the width of the switch beam for any given longitudinal position along its length, confirming consistent mechanical behavior and stable actuation characteristics [10–16].

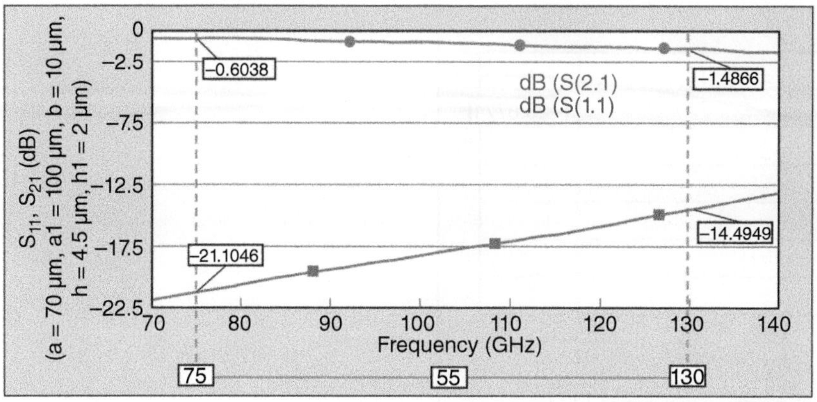

Figure 7.19 Return loss and insertion loss characteristics of the MEMS switch shown in Figure 7.17 with secondary switches activated [7].

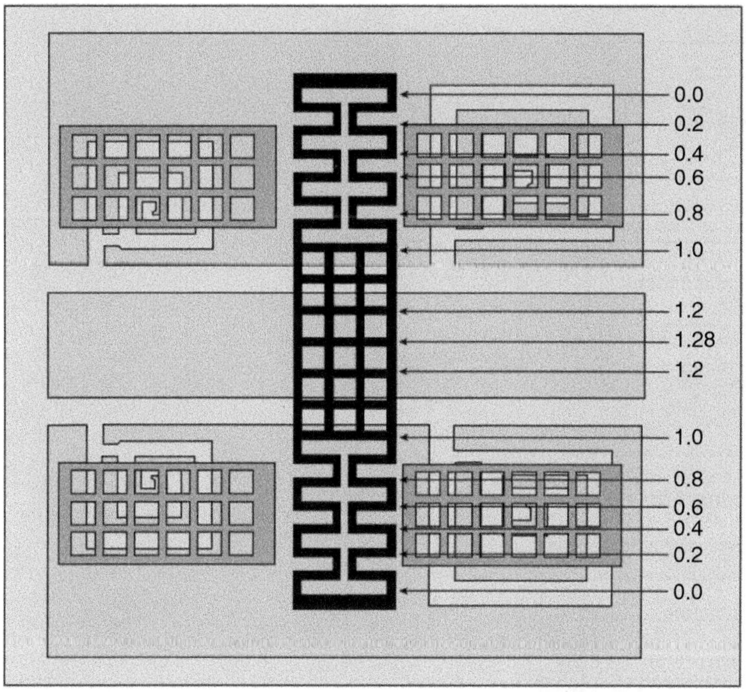

Figure 7.20 Downward deflection measured at multiple cross-sections of the primary MEMS switches (in μm) [1, 13].

Figures 7.21 and 7.22 [1] present the isolation performance of the MEMS switch architecture illustrated in Figure 7.17 under the operating condition where the primary switch (1910) is actuated (ON) while the secondary switches (1941–1944) remain deactivated (OFF) [1]. In Figure 7.21, the isolation response is obtained using a conventional DGS. This configuration enhances isolation but only within a relatively narrow frequency window, less than approximately 10 GHz, centered around 90–100 GHz. Specifically, the isolation magnitude is measured to be about −23.1 dB at 75 GHz and −23.9 dB at 130 GHz, while at 95 GHz, a pronounced improvement is observed, with isolation reaching approximately −52 dB [27].

Figures 7.22 and 7.23 [1] extend this analysis by comparing the isolation and insertion loss characteristics of MEMS switches implemented with alternative resonant structures [27]. In particular, the adoption of metamaterial-inspired geometries yields superior isolation performance over a much broader frequency range compared to the conventional DGS-based design. This broadband enhancement highlights the role of engineered metamaterial resonators in suppressing electromagnetic coupling across multiple frequency bands, thereby providing improved scalability and robustness for high-frequency and millimeter-wave applications [7].

The configuration depicted in Figure 7.15 [1] demonstrates significant isolation enhancement at specific frequency points [23–25]. In particular, isolation reaches approximately −51 dB near 84 GHz and improves further to about −59 dB near

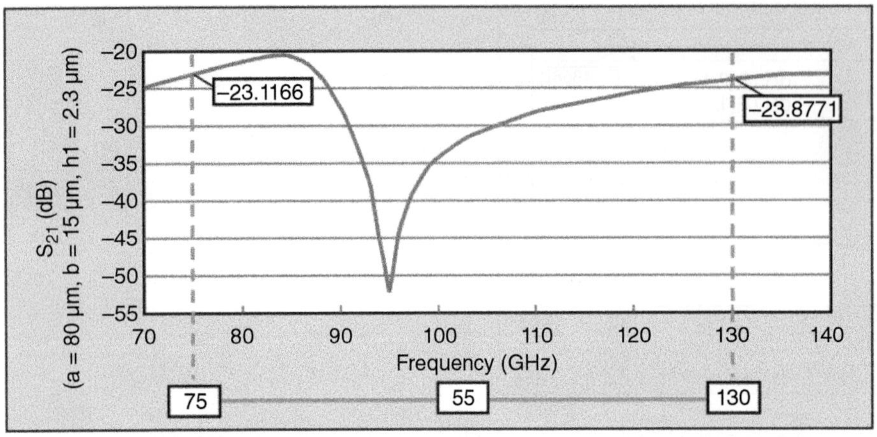

Figure 7.21 Isolation performance of the MEMS switch shown in Figure 7.17 at different air gap heights [1, 22].

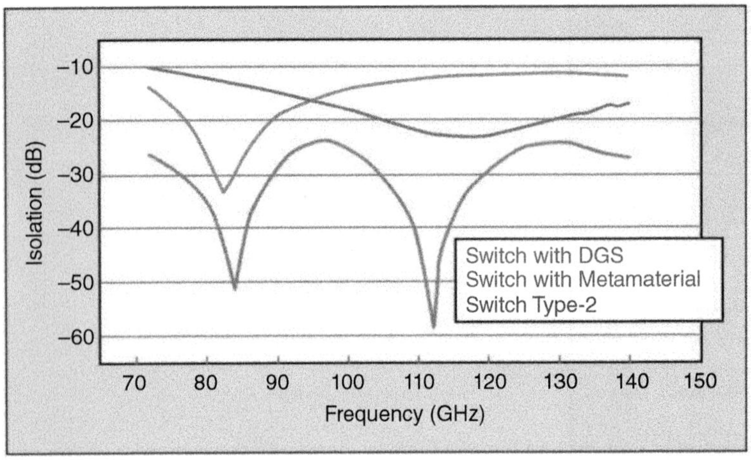

Figure 7.22 Solation performance of MEMS switches featuring different topological configurations [1, 22].

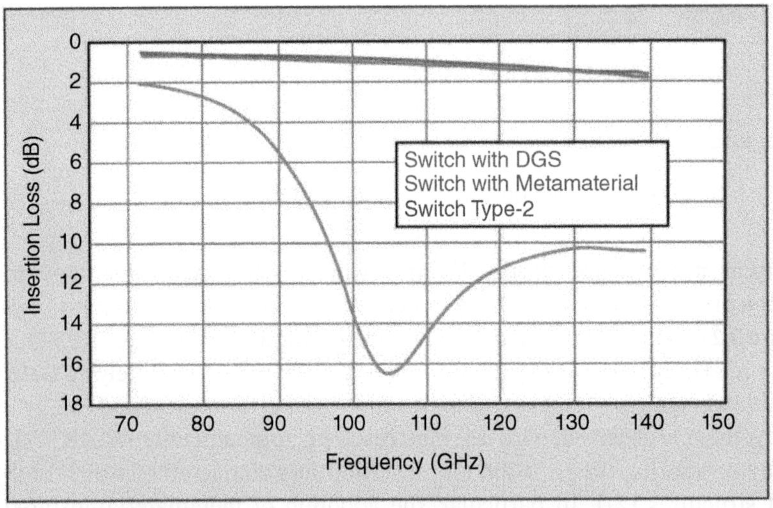

Figure 7.23 Insertion loss performance of MEMS switches with varying topological configurations [1, 23].

112 GHz, while maintaining no worse than –24 dB across the broader 75–130 GHz range. However, the insertion loss performance of this DGS-based switch [7] is inferior when compared to both the baseline (regular) switch and the metamaterial-inspired designs, highlighting a key trade-off in its implementation [27].

Switch type 2 [7] represents the initial design iteration of the proposed MEMS switch employing capacitive-contact elements [25]. As evident from the attenuation profiles shown in Figures 7.19 and 7.21, integrating DGS with capacitive MEMS shunt switches located directly above them provides a balanced design strategy. This architecture effectively leverages the isolation benefits of DGS elements when blocking RF signals, while simultaneously mitigating the adverse impact that the DGS would otherwise impose on insertion loss and return loss during signal transmission. Thus, the hybrid use of DGS structures in conjunction with shunt MEMS elements constitutes a substantive advancement in RF-MEMS switch design and operation.

Table 7.4 consolidates the key performance parameters—actuation voltage [1], isolation levels, and insertion loss values [1]—for the various switch designs described earlier, all fabricated with nominal air gaps (cantilever beam heights) of approximately 2.5 μm [1].

Further insight is provided by Figure 7.24 [1, 25], which plots isolation characteristics for multiple MEMS switch configurations featuring different DGS and secondary switch arrangements, with both primary and secondary switches actuated [5]. The results reveal that secondary switch activation leads to isolation enhancement over a relatively narrow frequency band. Moreover, the frequency at which peak isolation occurs is strongly dependent on the air gap spacing between the suspended beams and the underlying DGS features [3]. For instance, with an air gap of 2.2 μm, isolation [7] reaches approximately −52 dB near 85 GHz, whereas with a larger air gap of 3.0 μm, peak isolation shifts upward to approximately 122 GHz, with a corresponding isolation level of about −44 dB. These results highlight the tunability and design flexibility offered by the combined use of DGS elements and secondary switches, enabling optimized isolation across a wide frequency range within the millimeter-wave spectrum [2].

Section 7.1 [1] demonstrates that the integration of DGS together with secondary MEMS switches provides simultaneous enhancement of both insertion loss and isolation characteristics. This dual improvement directly addresses the limitations of conventional approaches, where the use of a single shunt switch typically results in low insertion loss but insufficient

Table 7.4 Comparison of MEMS switches performance.

Parameters	Shunt switch (Figures 7.2–7.3)	Shunt switch (Figure 7.5)	Shunt switch + DGS without switches (Figure 7.7)	Shunt switch DGS with switches (Figure 7.17)
Actuation voltage	37 V	17 V	17 V	17 V
Isolation (75–130 GHz)	−12 to −19 dB	−15 to 24 dB	−11 to −32 dB	−24 to −59 dB
Insertion loss	0.74 dB	0.6 dB	−2 to −11 dB	0.6 dB
Material	Molybdenum	Gold	Gold	Gold
Cantilever height	2.5 μm	2.5 μm	2.5 μm	2.5 μm

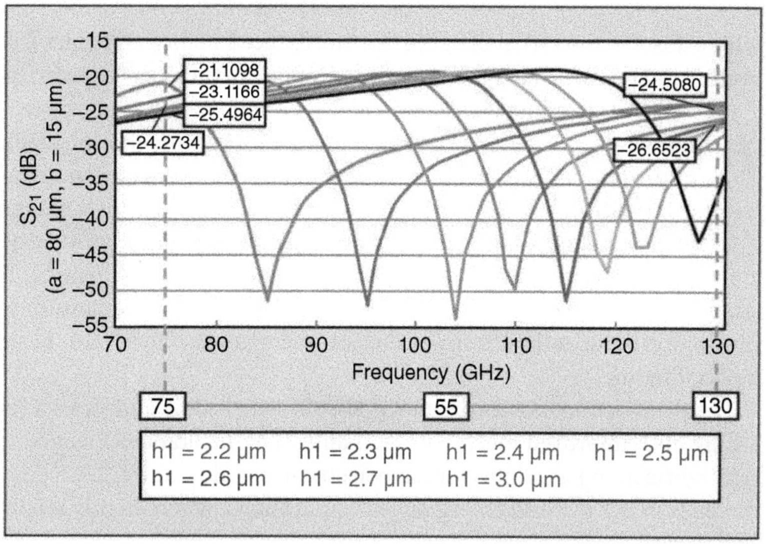

Figure 7.24 Isolation performance of different MEMS switch designs [1, 25].

isolation, while reliance solely on DGS elements achieves high isolation at the expense of elevated insertion loss. By combining a primary shunt switch [27] with DGS [1] features and secondary shunt switches, the proposed architecture exhibits electromagnetic behavior analogous to that of a metamaterial. The periodic arrangement and coupling of these elements enable a tailored dispersion response that supports both low-loss transmission and strong suppression of undesired signal leakage.

Beyond the electromagnetic performance benefits, the design also mitigates a critical reliability concern for RF-MEMS switches—stiction. The incorporation of metamaterial-inspired layers into the contact interface reduces the effective contact area and modifies the local surface energy distribution, thereby minimizing adhesion forces that typically lead to permanent switch failure. These structural innovations not only enhance switching reliability but also extend device lifetime under repeated actuation cycles. The detailed mechanisms by which metamaterial layers contribute to stiction resistance are analyzed comprehensively in Sections 7.2 and 7.3.

7.2 Casimir Force: MEMS Switch Design Leveraging the Casimir Effect

MEMS switches have emerged as highly promising candidates for compact, low-cost, and high-performance components in RF and millimeter-wave applications [9]. Their intrinsic advantages—such as low insertion loss, high isolation, and minimal power consumption—make them ideal for integration into reconfigurable signal-routing networks, phased-array antennas, and tunable oscillators [32, 33]. Despite these benefits, the broad adoption of MEMS switches has been constrained by reliability challenges, with stiction at the contact interfaces representing the most critical limitation [1–33].

Stiction arises primarily from adhesion forces that dominate at micro- and nanoscale separations, where surface interactions such as van der Waals forces [40] and Casimir forces become significant. In conventional MEMS switches, the attractive Casimir force between closely spaced metallic electrodes—typically separated by nanometer- to micrometer-scale gaps—can prevent the movable beam or cantilever from fully returning to its initial unactuated position. This phenomenon can lead to permanent contact adhesion, mechanical fatigue, or even complete device failure [41].

The impact of stiction is particularly severe in high-frequency systems, where the switch must reliably cycle over millions of actuation events while maintaining precise electrical characteristics. Uncontrolled adhesion at the contact interface compromises long-term performance, manifesting as increased insertion loss, reduced isolation, and inconsistent switching response. Consequently, understanding and mitigating Casimir- and van der Waals-induced stiction is essential for ensuring the durability and repeatability of MEMS switches in demanding RF and millimeter-wave applications [8–13].

In this work [22–31], a new MEMS switch configuration incorporates artificial composite (metamaterial) layers into the contact interface [5]. The effective electromagnetic response of the metamaterial is engineered to produce a *repulsive Casimir force*, counteracting the attractive interactions responsible for stiction [1–12]. The Casimir force between two planar surfaces separated by a medium can be described by the Lifshitz formulation [14, 39, 41, 51]:

$$F(d) = \frac{\hbar}{2\pi^2} \int_0^\infty d\xi \int_0^\infty k_\perp dk_\perp \sum_{p=\mathrm{TE,TM}} \ln\left[1 - r_{1p}(i\xi, k_\perp) r_{2p}(i\xi, k_\perp) e^{-2k_\perp d}\right] \tag{7.2}$$

where d is the separation distance, $\hbar$ is the reduced Planck constant, ξ is the imaginary frequency, $k\perp$ is the transverse wave vector, and r_{jp} represents the Fresnel reflection coefficients for polarization p at interface j.

For conventional metallic boundaries, r_{1p} and r_{2p} are both positive, resulting in an attractive force. However, repulsion can occur when the dielectric permittivity $\varepsilon(i\,\xi)$ of the intervening medium (or engineered metamaterial layer) lies between those of the bounding materials across the relevant frequency spectrum [2–6]. Under these conditions, the product $r_{1p}r_{2p}$ becomes negative, leading to a logarithmic term that produces a net *repulsive* Casimir interaction [3, 4].

By integrating metamaterial layers with tailored effective permittivity and permeability into the MEMS switch design, the contact interface can be engineered to achieve this repulsive regime. High-conductivity metamaterial structures not only support low-loss RF operation but also enable a controlled quantum vacuum response that mitigates stiction forces. This innovation reduces the effective adhesion energy, ensuring reliable beam release after actuation and significantly improving device lifetime.

The proposed metamaterial-based MEMS switch thus addresses a fundamental limitation of conventional designs, providing both improved mechanical reliability and favorable RF characteristics. Beyond switches, this approach introduces new fabrication paradigms for nanoelectromechanical and MEMS devices, offering a foundation for next-generation technologies in 5G, 6G, and emerging IoT applications, where highly reliable, low-loss, and broadband switching elements are indispensable.

7.2.1 Casimir Interaction Effect

The Casimir force [11] arises from the interaction [12] between material boundaries and the zero-point fluctuations [3] of the surrounding electromagnetic vacuum [4]. Unlike classical forces, it originates from the quantum electrodynamic (QED) nature of the vacuum, in which even in the absence of real photons, the electromagnetic field exhibits fluctuating energy modes. When two closely spaced surfaces are introduced, these vacuum fluctuations are modified by the boundary conditions imposed by the materials, resulting in a net attractive or, in some configurations, repulsive force between the surfaces.

The magnitude and nature of the Casimir interaction are determined by several factors. Key among these is the dielectric response of the interacting materials, which governs how the surfaces interact with the electromagnetic field at different frequencies. Additionally, the electromagnetic properties of the intervening medium, such as permittivity and permeability, influence the spectral distribution of vacuum fluctuations and consequently the strength and direction of the Casimir force.

From a QED perspective [34], Casimir interactions represent the macroscopic manifestation of fundamental quantum phenomena also observed at the atomic and molecular scales. They are closely related to van der Waals forces, which similarly arise from correlations in zero-point energy fluctuations, but operate predominantly at short (nanometer-scale) separations. Whereas van der Waals forces describe interatomic or intermolecular interactions, the Casimir effect extends these concepts to macroscopic bodies, producing measurable forces between micrometer- and nanometer-scale structures.

In practical terms, the Casimir effect [33] becomes particularly significant in micro- and nanoscale devices, such as MEMS and nanoelectromechanical systems (NEMS) switches, where electrode separations approach the sub-micrometer regime. Here, the Casimir force can rival or even exceed electrostatic actuation forces, influencing the mechanical behavior of movable elements, affecting contact reliability, and becoming a critical design consideration for stable device operation.

In the canonical case of two parallel, uncharged, perfectly conducting plates placed in close proximity within a vacuum, the modification of the allowed electromagnetic modes between the plates results in a net attractive force that pulls the plates together [2]. This effect has been experimentally observed and is illustrated schematically in Figure 7.25. The magnitude of the force per unit area, F/A, exhibits a strong inverse dependence on the plate separation distance d and is given in the idealized case by [2]:

$$F = \frac{\pi^2 \hbar c}{240} \frac{A}{d^4} \tag{7.3}$$

where $\hbar$ is the reduced Planck constant, c is the speed of light, A is the plate area, and d is the separation distance. This $1/d^4$ scaling highlights the rapid increase in the Casimir force as the gap size decreases, which becomes particularly problematic in nanoscale and microscale electromechanical systems.

More generally, the magnitude of the Casimir force is material-dependent rather than universal, influenced by the effective permittivity and permeability of the interacting surfaces. For metallic interfaces, higher effective permittivity tends to amplify the attractive force [2, 3]. This observation has practical implications: by designing materials or surface composites with reduced effective permittivity, the net Casimir attraction can be mitigated. Such strategies are particularly relevant for MEMS switches, where excessive Casimir-induced adhesion at nanoscale gaps is a primary contributor to stiction, preventing the movable beam or cantilever from fully returning to its unactuated state and compromising device reliability [1].

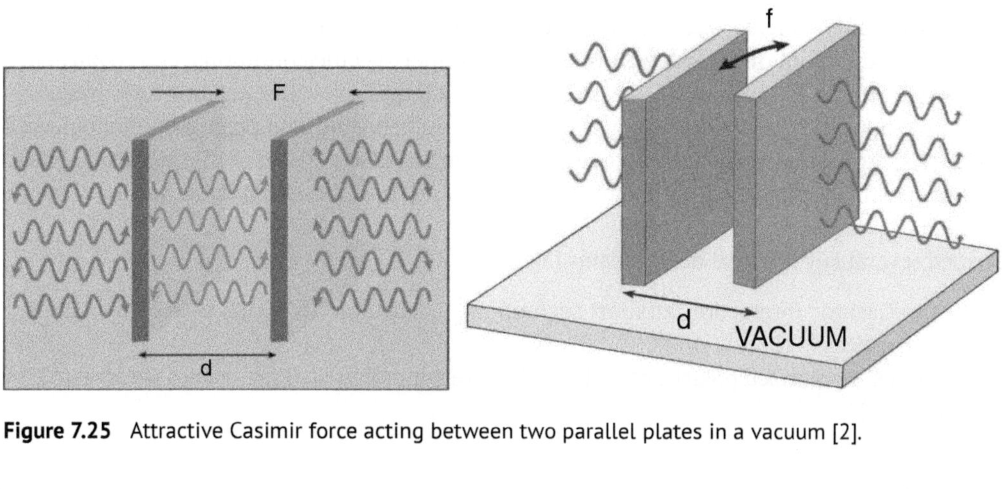

Figure 7.25 Attractive Casimir force acting between two parallel plates in a vacuum [2].

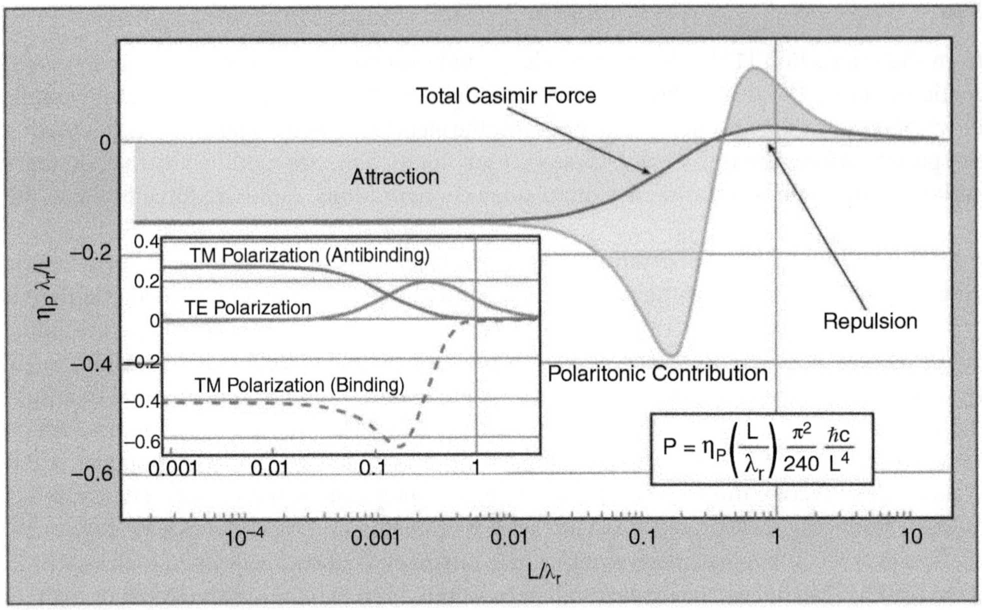

(a)

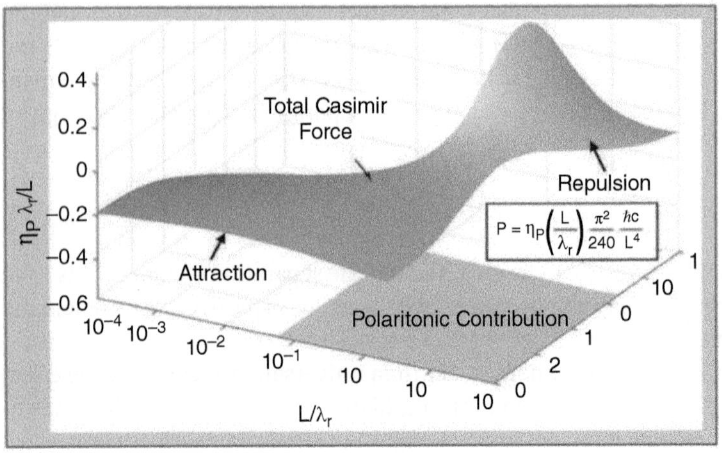

(b)

Figure 7.26 (a) Two-dimensional plots showing the regions in which the Casimir force exhibits attractive and repulsive behavior [3]. (b) Three-dimensional plots depicting the regions where the Casimir force is either attractive or repulsive [3].

Recent advances in artificial electromagnetic media, such as metamaterials and polaritonic structures, have shown that Casimir interactions can be tailored to exhibit not only reduced attraction but, under specific conditions, even *repulsion*.

Intravaia and Henkel [3] demonstrated that in metamaterial-based mirror configurations, “the Casimir force is significantly modified and may reverse its sign” as a result of surface plasmon–polariton contributions arising from engineered polaritonic dispersion. Their analysis shows that artificially structured metamaterials can reshape the electromagnetic mode density at subwavelength scales, thereby altering the balance of vacuum fluctuations responsible for the Casimir interaction, illustrated in Figure 7.26.

Subsequent studies have clarified several key physical mechanisms [52]:

- Surface polaritons dominate the Casimir interaction at short separations, particularly through discrete evanescent modes localized near material interfaces [52].
- In magnetodielectric configurations, *s*-polarized polaritons—absent in purely metallic cavities—play a decisive role in producing repulsive interactions [52].

- The interplay between binding and antibinding p-polariton modes partially cancels attractive forces, thereby amplifying the relative contribution of repulsive s-polariton modes [52].
- Bulk continuum modes within the magnetodielectric band, especially for s-polarization, contribute appreciably to the Casimir force—a feature not observed in conventional metallic cavities [52].

Taken together, these insights establish that metamaterials with engineered polaritonic dispersion provide a powerful route to precisely tune the Casimir interaction, enabling controlled transitions between attractive and repulsive regimes, as illustrated in Figure 7.26. Incorporating such metamaterial layers into MEMS switch architectures allows designers to exploit engineered repulsive Casimir forces to mitigate stiction—a critical failure mechanism in conventional MEMS devices. This approach promises to enhance switching stability, extend device lifetime, and substantially improve overall operational reliability.

The probability of stiction in MEMS switches can be reduced through several design and operational strategies. One straightforward method is to increase the applied bias voltage, as discussed previously in Section 7.1, thereby enhancing the electrostatic actuation force relative to the adhesion forces at the contact interface [5]. An alternative approach involves increasing the effective electric field strength by modifying the electrode geometry. For instance, the top electrode may be displaced further from the ground plane by embedding the conductive layer (e.g., gold) between two dielectric films [1], such as silicon oxynitride [3], thereby reshaping the electric field distribution and improving separation upon release [5].

A second class of solutions targets the mechanical properties of the movable beam to maximize its restoring force [41] without requiring a higher bias voltage [42]. Parameters that directly influence restoring force include the lateral dimensions of the electrode (with larger plate areas increasing stiffness), the beam length (shorter beams providing greater elastic restoring force), and the dielectric thickness (which modifies both compliance and capacitance). By carefully optimizing these structural parameters, the restoring force can be increased sufficiently to overcome stiction while preserving desirable actuation characteristics.

Beyond geometric and material modifications, another promising strategy is to address the fundamental origin of adhesion forces. Specifically, this work proposes the weakening—or in some cases, reversal—of quantum fluctuation-induced forces (e.g., van der Waals and Casimir interactions) that arise due to the close proximity of the switch contacts [35–39]. By engineering the electromagnetic environment at the contact interface, for example, through the incorporation of metamaterial or layered dielectric structures, it is possible to suppress attractive interactions or induce repulsive Casimir forces, thereby fundamentally reducing the likelihood of stiction.

7.2.2 Exploiting the Casimir Effect for Enhanced MEMS Switch Performance

Reducing the effective permittivity between interacting surfaces can induce a repulsive Casimir force [52] if the material properties are engineered appropriately, as in the case of metamaterials [4, 11]. By carefully designing the electromagnetic response of these structures, the normally attractive Casimir interaction can be reversed or significantly weakened, thereby decreasing the likelihood of stiction-induced adhesion between MEMS switch contacts. This repulsive Casimir force provides an additional mechanism to prevent the contacts from becoming "welded" together during operation, enhancing device reliability [37, 41, 42].

Figures 7.27 and 7.28 [27] illustrate the concept of Casimir repulsion in a MEMS-scale experimental setup. In Figure 7.27, a metal plane (2910) is positioned parallel to a metamaterial layer (2920) with a separation distance d. The interaction forces between the two surfaces are represented by arrows (2930), indicating the attractive tendency of the Casimir force under conventional conditions [1].

However, under the engineered conditions shown in Figure 7.28 [1], the system exhibits a repulsive force, FFF, acting to separate the metal and metamaterial planes. This repulsive interaction arises from the reduced effective permittivity of the metamaterial layer, which modifies the boundary conditions of the electromagnetic vacuum fluctuations between the surfaces [33]. As a result, the net force experienced by the metal and metamaterial planes is directed away from each other, counteracting adhesion and mitigating stiction. These observations demonstrate that incorporating metamaterial layers into MEMS switch contacts provides a practical method to exploit Casimir repulsion, enabling enhanced mechanical reliability while preserving high-frequency RF performance [1–12].

Figure 7.28 [1] illustrates a representative levitating mirror system [4], demonstrating the effect of metamaterial-induced Casimir repulsion. In this configuration, the repulsive Casimir force [41] generated by the metamaterial can counterbalance the weight of one of the mirrors, enabling it to levitate above the opposing surface due to zero-point electromagnetic fluctuations [41, 42].

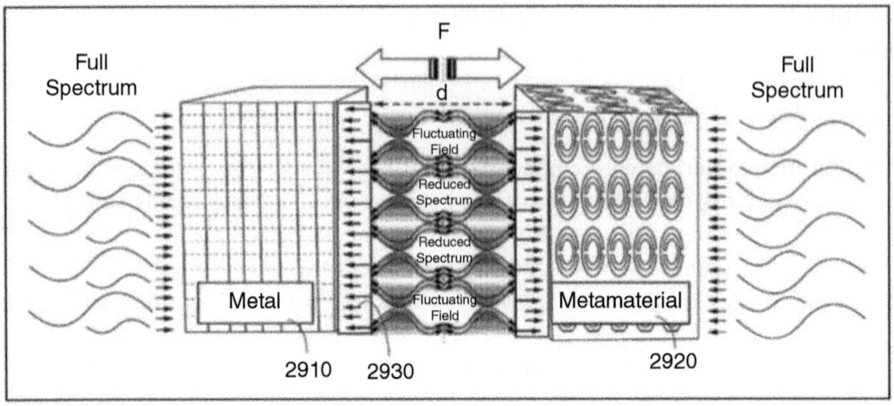

Figure 7.27 Illustration of the Casimir force acting within a metamaterial structure [1–4].

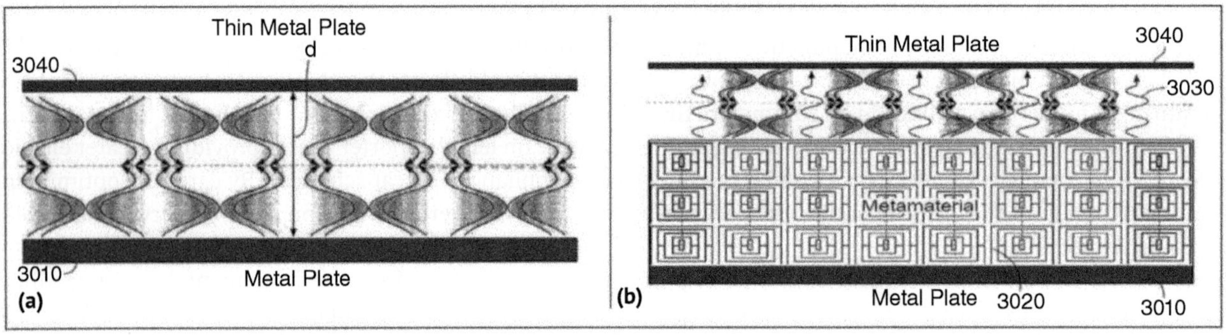

Figure 7.28 Illustration of the Casimir force between (a) conventional metal plates and (b) a metamaterial structure [1–4].

In Figure 7.28a, a first metal plate (3010) [1] is positioned parallel to a second metal plate (3040) with a separation distance *d*. These plates can be interpreted as analogous to opposing contacts in a MEMS switch, which are susceptible to permanent adhesion when the separation distance is sufficiently small, leading to stiction [34].

By contrast, Figure 7.28b [1] shows a thin metamaterial layer (3020) affixed to the surface of the first metal plate (3010) [7] and interposed between the two metal plates (3010 and 3040). The metamaterial modifies the electromagnetic boundary conditions at the interface, producing a repulsive Casimir force (3030) acting on the second metal plate (3040). This repulsive interaction increases the effective separation between the plates, and under appropriate conditions, it can overcome gravitational forces, causing the second plate to levitate. Metamaterials for this application can be realized using structures such as split-ring resonators fabricated from high-conductivity layers, for example, gold foil [3–11].

In the context of MEMS switch design, a typical structure comprises a deflectable beam with a shorting bar mounted on its surface and aligned with the signal line contact. The shorting bar is typically metallic, for instance, a thin layer of gold foil [4]. Upon actuation, when the shorting bar makes contact with the signal line, strong M2M adhesion may occur, giving rise to stiction. This undesired adhesion can electrically short the switch and may require significant force to disengage the contacts [44].

Conventionally, MEMS switches rely on the elastic restoring force generated by stresses accumulated in the deflectable beam during actuation. This restoring force serves to overcome adhesive interactions and return the beam to its equilibrium or at-rest position. However, in many practical scenarios, the mechanical restoring force alone is insufficient to counteract the adhesive forces between the metallic contacts, making stiction a persistent reliability concern [1].

By integrating a metamaterial layer between the shorting bar and the signal line, the repulsive Casimir force generated at close proximity can supplement the beam's restoring force. As the shorting bar approaches the signal line, the engineered metamaterial modifies the local electromagnetic boundary conditions, producing a repulsive quantum force that counteracts adhesion. This dual mechanism—combining mechanical restoring forces with Casimir-induced repulsion—effectively mitigates stiction, enhances contact reliability, and ensures consistent electrical performance under repeated actuation cycles [46].

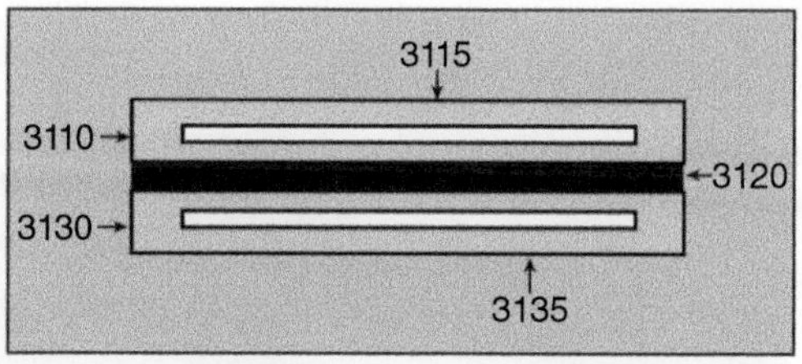

Figure 7.29 Metamaterial unit cell comprising a three-layer composite: first layer 3110 with permittivity ε_1, second layer 3120 with permittivity ε_2, and third layer 3130 with permittivity ε_3 [13].

The magnitude and direction of the Casimir force [48] can be actively controlled by engineering a permittivity gradient across the contact interface of the deflectable beam. This gradient can be realized by stacking three distinct material layers with systematically varying permittivity, arranged in either ascending or descending order. As illustrated in Figure 7.29, a representative three-layer configuration is considered. The first layer (3110) [1] possesses permittivity ε_1, the second layer (3120) [1] has permittivity ε_2, and the third layer (3130) exhibits permittivity ε_3. The first and third layers may be metallic, while the intermediate layer is typically a dielectric. The layers can be arranged such that either $\varepsilon_1 < \varepsilon_2 < \varepsilon_3$ or $\varepsilon_1 > \varepsilon_2 > \varepsilon_3$, allowing precise tailoring of the Casimir interaction [40].

Such a gradient may be implemented [50], for example, by combining metal layers with positive and negative effective permittivity. In one embodiment, the first layer is a metallic material such as gold, exhibiting a very high or nearly infinite permittivity [46]. The second layer may be a conventional dielectric, such as silicon mononitrate (SiN), with a modest but positive permittivity. The third layer can incorporate a metamaterial unit cell (3135) [7], engineered to achieve near-zero or negative permittivity [47]. In alternative configurations, the first layer itself may also include a metamaterial structure (3115) [1] to further adjust the effective permittivity and achieve the desired gradient. By establishing this permittivity gradient, the electromagnetic boundary conditions across the interface are modified, enabling fine control over the Casimir force. This approach allows for the generation of a repulsive, reduced, or tailored Casimir interaction, which can supplement the mechanical restoring forces of the beam, mitigate stiction, and enhance the reliability and repeatability of MEMS switch operation [41].

7.2.3 Example: Repulsive Casimir Force-inspired Resistive-contact MEMS Switch

Figure 7.30 [1] illustrates a MEMS switch design incorporating metamaterial unit cells to generate a repulsive Casimir force between the movable switch contacts, thereby mitigating stiction. The switch is implemented within a CPW structure (3201) consisting of two ground planes (3202 and 3204) fabricated above a substrate (3205). A longitudinal channel separates the ground planes, and a central signal line (3210) extends along the length of this channel. The signal line is interfaced with an input port (3212), through which the RF signal is injected, and an output port, through which the signal is transmitted [1].

The MEMS switch features a cantilevered beam that deflects out of plane to make or break contact with the signal line, enabling switching action. The beam is composed of multiple functional layers to achieve both mechanical resilience and electromagnetic control. From top to bottom, these layers include a dielectric top layer (3420), a first metallic layer (3210), an intermediate dielectric layer (3220), and a second metallic layer (3230). Both the first and second metallic layers can incorporate embedded metamaterial unit cells (3215 and 3235, respectively), as depicted in the cross-sectional view in Figure 7.30c [27].

The integration of metamaterial elements within the metallic layers enables the engineering of local permittivity and permeability, producing a repulsive Casimir force at the interface with opposing contacts [41]. This quantum-induced repulsion supplements the elastic restoring forces of the cantilevered beam, reducing the probability of stiction and ensuring reliable contact disengagement [42]. Furthermore, the layered configuration maintains the RF performance of the CPW structure while simultaneously enabling the active control of adhesion forces at the microscale, offering a practical pathway for high-frequency, low-loss, and robust MEMS switch operation [37].

The top metallic layer (3210) [1] and the first dielectric layer (3220) [1] of the cantilevered beam may extend along the full length of the beam to provide structural integrity and uniform electromagnetic properties. In contrast, the length of the sandwiched dielectric layer (3220) and the second metallic layer (3230) can be restricted to the region directly above the signal line, concentrating the functional elements where contact occurs. Alternatively, the dielectric layer may span the entire beam length, while the second metallic layer remains limited to the contact region above the signal line (3210), depending on design requirements [27].

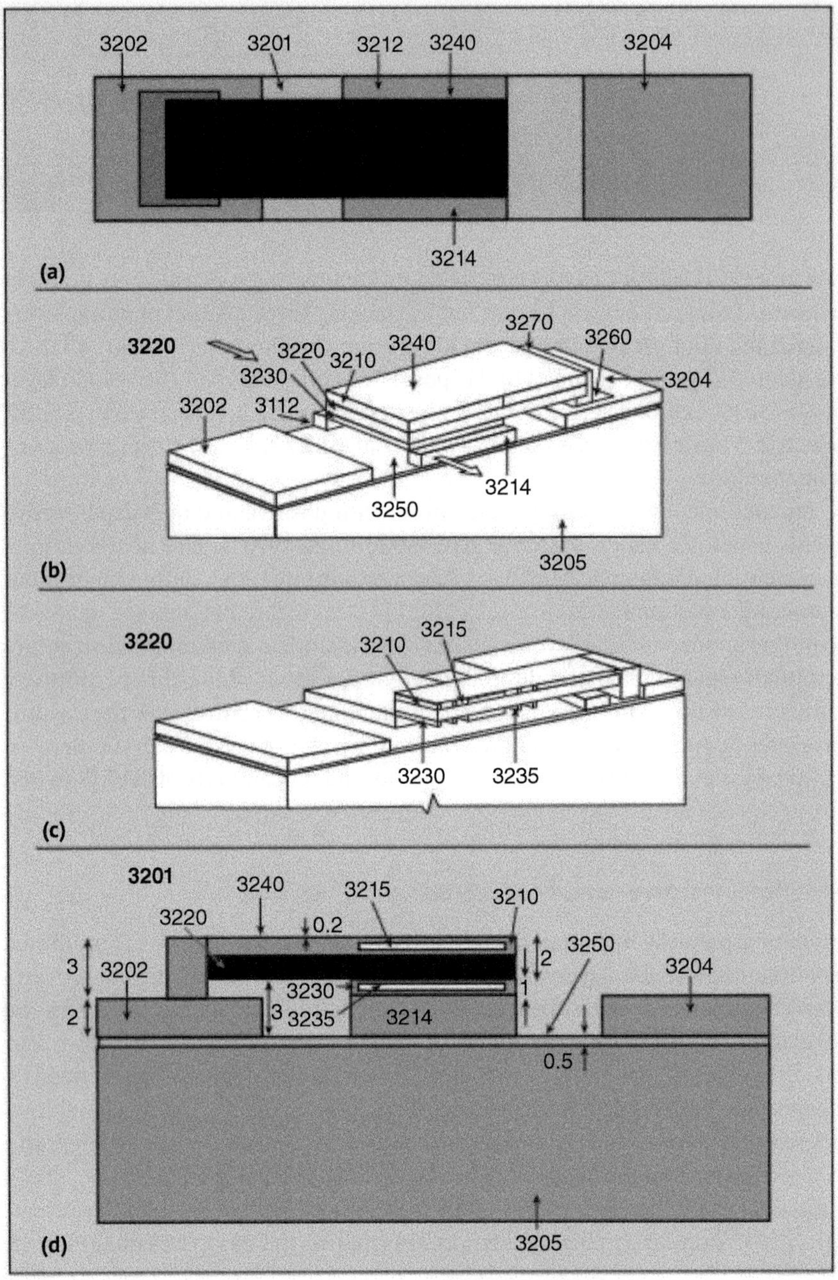

Figure 7.30 Example of a MEMS switch integrating metamaterial cells to induce a repulsive Casimir force between the switch contacts: (a) top view, (b) perspective view, (c) cross-sectional perspective, and (d) side view in the closed state [1].

The CPW ground planes (3202 and 3204) and the signal line ports (3212 and 3214) are isolated from the substrate (3205) by a thin dielectric layer (3250), such as silicon nitride (SiN) or silicon dioxide (SiO_2), to provide electrical insulation and maintain the desired RF characteristics. Actuation of the MEMS switch is achieved by manipulating the anchor (3270) to which the cantilevered beam is attached. The anchor is capable of moving in and out of the CPW plane, thereby controlling the engagement of the switch contacts. In this configuration, the ground plane (3202) includes an access hole (3260) through which the beam anchor (3270) passes. Vertical motion of the anchor moves the beam up or down, causing the contacts to either separate or make electrical contact.

Figure 7.30d [1] illustrates the switch in the closed state, where the contacts are in electrical engagement. For the example depicted in Figure 7.30, the CPW section shown has a length of approximately 100 μm, while the beam width is about

75 μm, dimensions that are typical for high-frequency MEMS switch applications. This configuration ensures precise contact alignment, mechanical robustness, and integration of metamaterial layers to leverage Casimir repulsion for stiction mitigation [52].

The anchor supporting the cantilevered beam may have dimensions of approximately 11.25 μm in the longitudinal direction (along the beam length) and 75 μm in width. The opening in which the beam is anchored may be larger, for example, around 80 μm × 30 μm, to allow for secure attachment and controlled deflection. The overall length of the CPW section along the beam axis may be about 330 μm, with the ground planes and signal line each having widths of approximately 75 μm and separated by 38 μm channels. The beam itself may extend roughly 140 μm, excluding the anchor portion. When in the closed position, the total beam height relative to the dielectric surface supporting the ground planes and signal line is approximately 5 μm. The ground planes and signal line are typically 2 μm thick, with the beam contributing an additional 3 μm to the switch height. The beam is constructed from multiple layers: the first and second metallic layers (3210 and 3230) are each about 1 μm thick, separated by a 1 μm dielectric layer, while a top dielectric layer (3440) may add approximately 0.2 μm to the overall height.

Metamaterial unit cells, often realized as split-ring resonators [1–6], can be embedded within the second metal layer (3230) and optionally in the first metal layer (3210) to engineer the Casimir interaction. Figure 7.31 provides an example of such a structure: a metallic layer (3310) contains a first split ring (3322) of width L and a second concentric split ring (3324) formed within the same layer. The rings may be fabricated by patterning the metal layer, removing material to create the ring structures. Each ring maintains a uniform width W, and the splits (3330) in the rings are positioned on opposite sides of the layer to achieve the desired electromagnetic response. The rings are separated by a uniform spacing S (3332). By adjusting parameters such as ring width, gap size, separation, and orientation, various engineered metamaterial unit cells can be realized to exhibit targeted electromagnetic characteristics across the relevant frequency bands for MEMS switch operation [1–2].

Figures 7.32 through 7.34 present experimental results for the transmission and reflection characteristics of distinct metamaterial unit cell structures integrated into MEMS switch components [1]. In Figure 7.32, a metamaterial unit cell, similar to that illustrated in Figure 7.8, is embedded within a metallic layer spanning the full width of the cantilevered beam (3402). The measured response of this unit cell demonstrates frequency-dependent behavior characteristic of metamaterials. Specifically, the unit cell exhibits high transmission at lower frequency bands, notably near 300 GHz and again around 470 GHz. Conversely, at approximately 150 GHz and again near 300 GHz, the unit cell shows a reflection-dominated response, where the attenuation of transmitted signals exceeds that of reflected signals. This alternating transmission–reflection behavior confirms the presence of engineered metamaterial properties within the composite structure [39].

Figure 7.33 depicts a second composite structure (3501) [1] incorporated within a metallic layer whose length corresponds to the width of the underlying signal line. This layer is mounted on a cantilevered beam (3502) that is narrower than the metallic layer, allowing for localized electromagnetic interactions. Testing of this configuration reveals that the unit

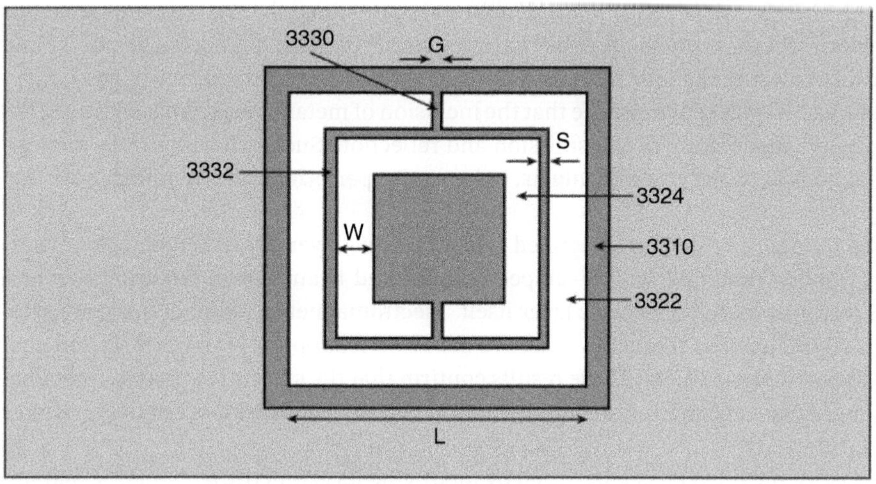

Figure 7.31 Configuration of a split-ring resonator [1].

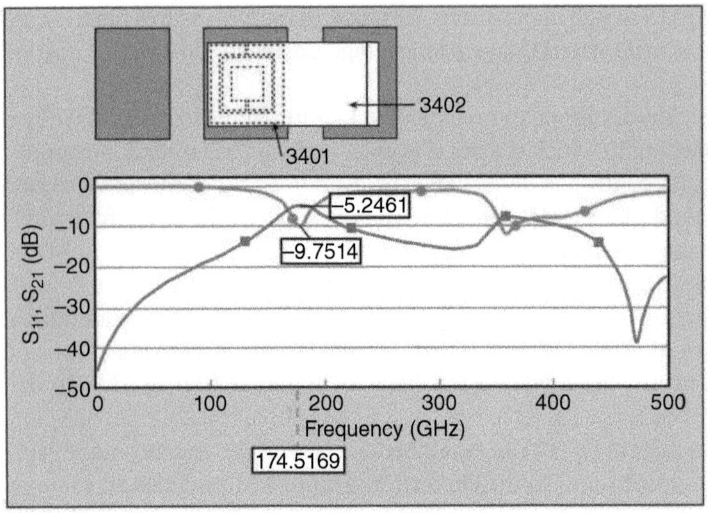

Figure 7.32 The transmission and reflection responses of the artificial composite structure example #1 [1].

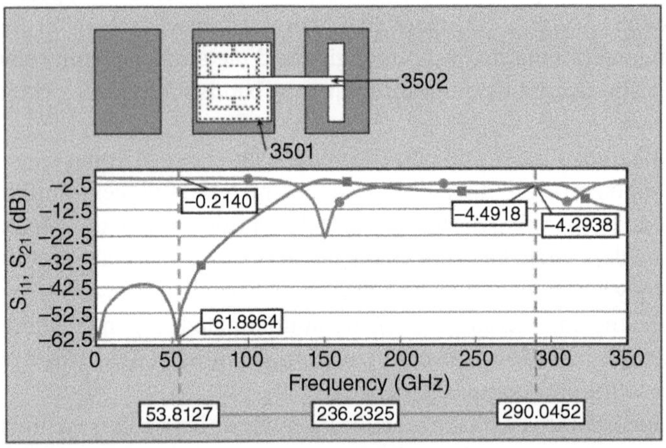

Figure 7.33 The transmission and reflection behavior of the artificial composite structure example #2 [1].

cell exhibits strong transmission at around 54 GHz, while reflection dominates at approximately 150 GHz. These results similarly confirm that the composite structure possesses metamaterial characteristics, with the engineered geometry and layer composition enabling selective control of electromagnetic wave propagation within specific frequency bands [38].

Collectively, the results illustrated in Figures 7.32–7.34 demonstrate that the inclusion of metamaterial unit cells in MEMS switch components enables frequency-selective modulation of transmission and reflection. Such behavior can be strategically leveraged to tailor the Casimir interactions between switch contacts, enhance RF performance, and mitigate stiction in high-frequency MEMS applications.

The composite structure (3601) depicted in Figure 7.10 [7] is integrated into a metallic layer whose length corresponds to the width of the underlying signal line and is mounted on a U-shaped cantilevered beam (3602) featuring two narrow branches, each considerably smaller in width than the metallic layer itself. Electromagnetic testing of this configuration demonstrates that the metamaterial unit cell exhibits frequency-selective behavior: it supports high transmission near 80 GHz, while reflection dominates at approximately 163 GHz. These results confirm that the structure possesses metamaterial characteristics, with the observed frequency-dependent transmission and reflection confined to a relatively narrow bandwidth of roughly 80 GHz.

Furthermore, the electromagnetic response of the unit cell can be tuned by modifying its geometric and material parameters [7]. Variations in dimensions, spacing, or the composition of the metamaterial inclusions allow designers to shift the

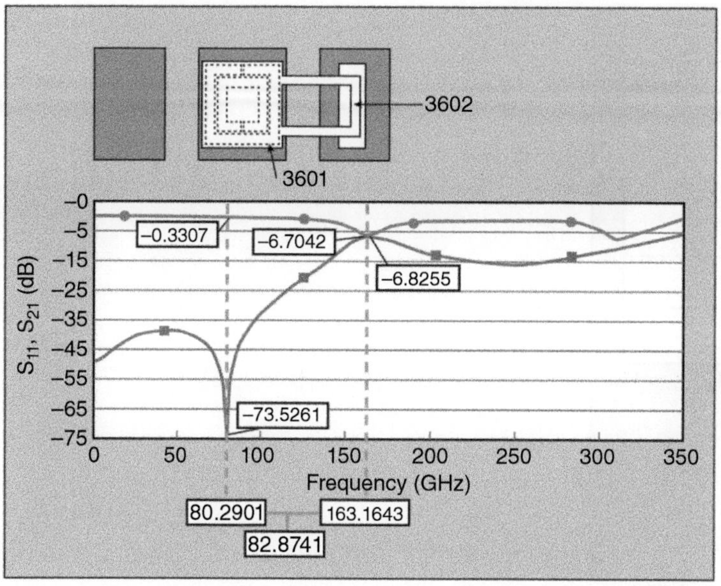

Figure 7.34 The transmission and reflection behavior of the artificial composite structure example #3 [1].

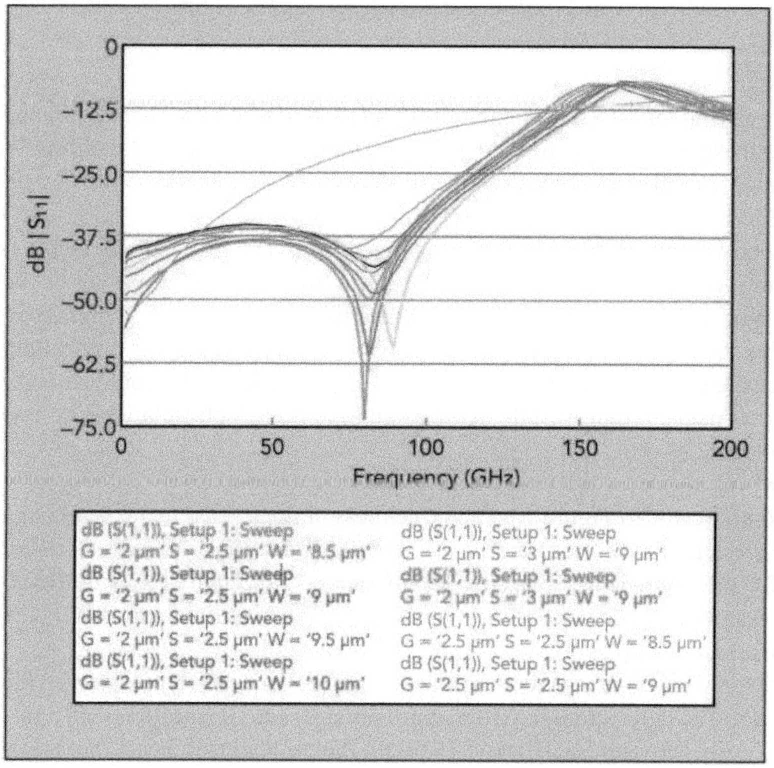

Figure 7.35 Reflection characteristics of a metamaterial cell as a function of the parameters G, S, and W (related to Figure 7.31) [1].

transmission and reflection peaks, thereby tailoring the frequency range over which metamaterial behavior is exhibited. This tunability provides a versatile tool for engineering the RF characteristics of MEMS switch components, enabling precise control over signal propagation, isolation, and interactions at the microscale.

Figure 7.35 [1] presents a representative graph illustrating the reflection characteristics of a metamaterial unit cell, highlighting the effects of varying key geometric parameters: the split width G, the ring separation S, and the ring width W, as defined in Figure 7.31.

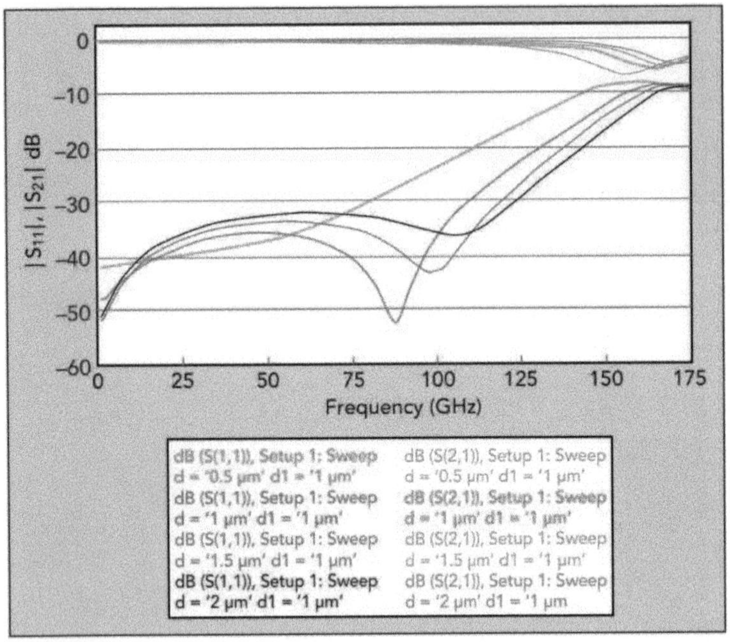

Figure 7.36 Transmission and reflection characteristics of the switch, as the thickness of the second metal layer "d" (e.g., 3230 in Figure 7.30) varies from 0.5 μm to 2 μm [1].

The results demonstrate that the frequency at which the reflection is maximally attenuated can be tuned across a range from approximately 80 GHz to 90 GHz by adjusting these parameters [27]. For example, when G is 2 μm, S is 3 μm, and W is 9 μm, the insertion loss reaches a minimum of about 74 dB at 80 GHz. In contrast, alternative combinations of G, S, and W result in reflection minima of approximately −60 dB near 90 GHz.

These observations underscore the critical role of geometric tuning in engineering the electromagnetic response of metamaterial unit cells [37]. Beyond variations in the unit cell geometry itself, the dimensions and material properties of the metallic layers in the MEMS switch can also be adjusted independently [35]. By carefully selecting layer thicknesses, widths, and lengths, designers can further optimize the RF performance of the switch while simultaneously controlling the metamaterial-induced Casimir forces that influence stiction and contact reliability [36]. Such design flexibility enables tailored performance over specific frequency bands, ensuring that MEMS switches achieve both high isolation and low insertion loss while mitigating adhesion-related failures [37–42].

Figures 7.36 through 7.38 illustrate the effects of varying layer thicknesses on the transmission and reflection properties of MEMS switches incorporating metamaterial unit cells. In Figure 7.36, the thickness of the second metal layer (e.g., 3230 in Figure 7.30) is varied between 0.5 μm and 2 μm, and the corresponding transmission and reflection characteristics are plotted. Figure 7.37 shows analogous results for the sandwiched dielectric layer (e.g., 3220 of Figure 7.6) with thicknesses ranging from 1.5 μm to 5 μm. Figure 7.38 examines the impact of varying the first metal layer thickness d1d_1d1 (e.g., 3210 in Figure 7.30) between 0.5 μm and 2 μm.

Across all three scenarios, the overall transmission properties of the switches remain largely consistent [52]. However, the frequency at which transmission attenuation occurs shifts slightly, ranging from approximately 160 GHz to 180 GHz, while the reflection characteristics exhibit variability primarily between 60 GHz and 150 GHz. These results indicate that while layer thickness has a moderate effect on frequency-dependent transmission, it plays a more pronounced role in shaping the reflection response of the composite structure.

Using the measured transmission and reflection data [1], the effective permittivity and permeability of the metamaterial unit cells can be extracted through standard parameter retrieval techniques [12]. Figure 7.39 presents an example of such an extraction, revealing that the composite structure exhibits near-zero permittivity and permeability over a frequency band centered on 85 GHz. This behavior suggests that the structure is capable of generating a repulsive Casimir force within the frequency range of approximately 60–130 GHz. Such tunability provides a pathway to engineer MEMS switch contacts with reduced adhesion forces, enabling stiction mitigation while maintaining high-frequency RF performance [44].

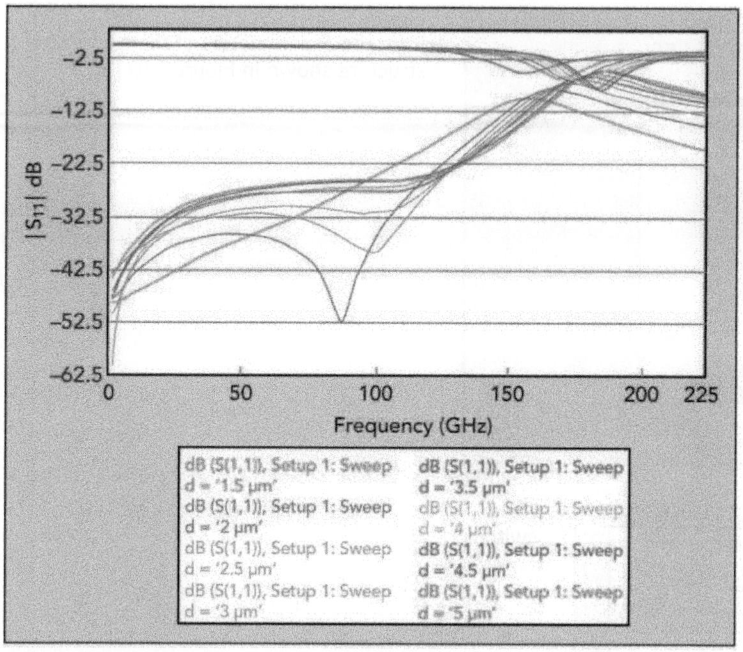

Figure 7.37 Transmission and reflection behavior of the switch, as the thickness of the second metal layer "d" (e.g., 3230 in Figure 7.30) is varied from 1.5 μm to 5 μm [1].

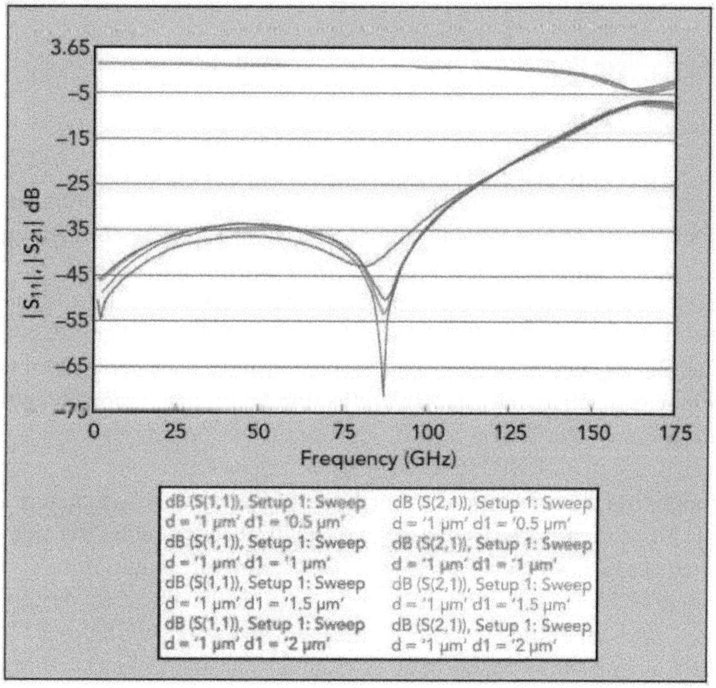

Figure 7.38 Transmission and reflection characteristics of the switch, as the thickness of the first metal layer "d_1" (e.g., 3210 in Figure 7.30) varies from 0.5 μm to 2 μm [1].

Figures 7.40 and 7.41 illustrate the comprehensive RF response of the MEMS switch in its OFF and ON states [1], respectively. In Figure 7.40, the switch is in the OFF state, meaning that the transmitted signal between the input and output ports is effectively blocked. Under these conditions, the reflection characteristics remain slightly below 0 dB across the examined

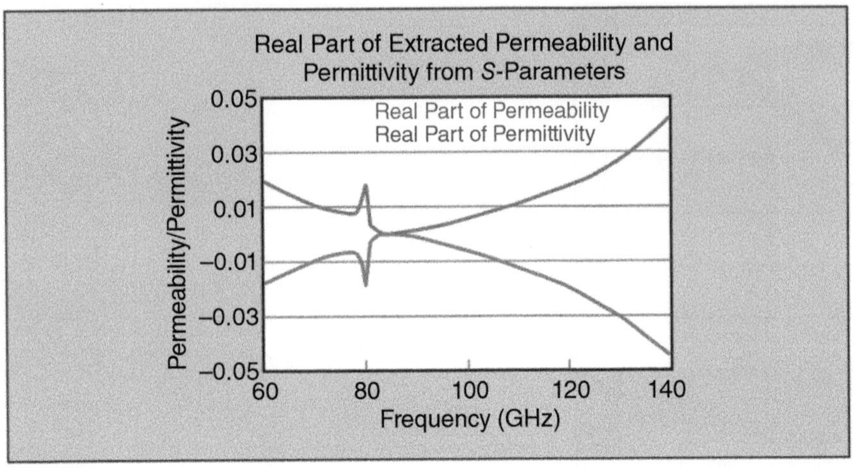

Figure 7.39 Effective permittivity derived from the S-parameters of the composite structure shown in Figure 7.30 [1].

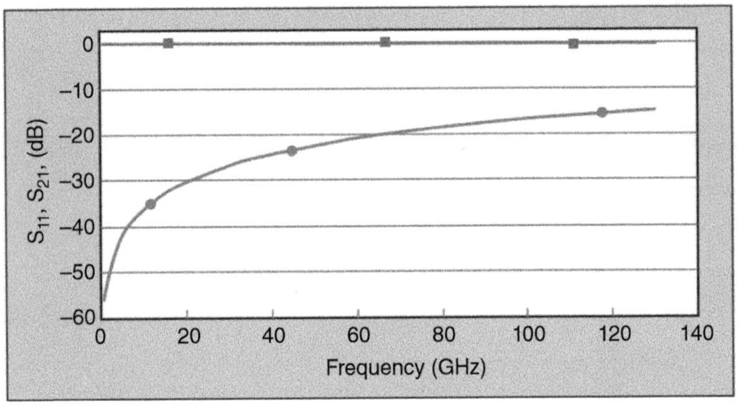

Figure 7.40 Dynamic response of the MEMS switch [1].

frequency range, up to 130 GHz, indicating minimal signal absorption or loss. The transmission characteristics are measured between approximately −20 dB and −15 dB across the operating band from roughly 60 GHz to 130 GHz, confirming effective signal isolation when the switch is disengaged.

Figure 7.41 shows the switch in the ON state, where the cantilevered beam makes contact with the signal line, allowing transmission between the input and output ports. In this configuration, the reflection characteristics are significantly reduced, reaching values as low as −73.5 dB at 80 GHz, while the transmission characteristics achieve high performance, with insertion loss as low as −0.33 dB [7]. At higher frequencies, such as 163 GHz, the reflection and transmission characteristics are measured at approximately −6.75 dB, indicating some degradation but still acceptable performance within the high-frequency band.

Collectively, the results presented in Figures 7.30 through 7.41 demonstrate the feasibility of integrating metamaterial structures into a high-frequency, resistive MEMS switch [1]. The incorporation of metamaterial unit cells enables the engineering of Casimir-induced repulsive forces between switch contacts, effectively mitigating stiction while maintaining low insertion loss and high isolation over a broad frequency range [1, 2]. These findings highlight a practical pathway for designing reliable, high-performance MEMS switches suitable for next-generation RF and millimeter-wave applications.

Section 7.2 demonstrates the application of repulsive Casimir forces to mitigate stiction in resistive-contact MEMS switches. This same principle can be extended to capacitive-contact MEMS switches to enhance their reliability. In both cases, a multilayered "sandwich" structure comprising metallic and dielectric layers is employed to engineer the desired permittivity profile at the contact interface. Typically, this arrangement consists of a metallic layer—such as gold—with a very high or effectively infinite permittivity, a dielectric layer with a modest positive permittivity, and a metamaterial layer engineered to exhibit near-zero or even negative effective permittivity.

In contrast to the resistive-contact switch examples discussed earlier, in capacitive MEMS switches, the metamaterial layer can be incorporated directly into the signal line contact rather than the cantilevered beam contact. This configuration

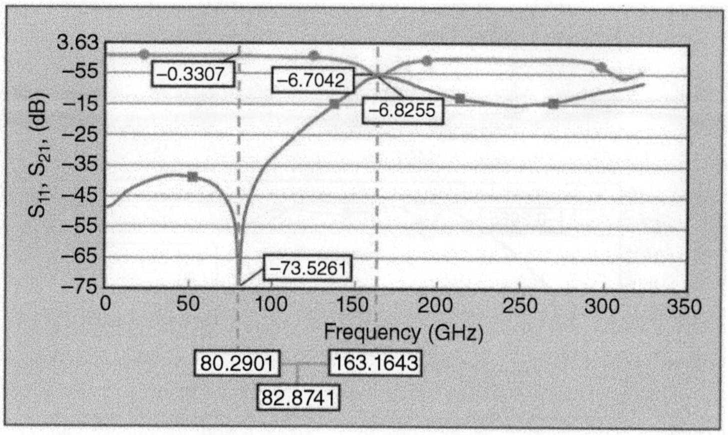

Figure 7.41 Performance response of the MEMS switch [1].

allows the Casimir-induced repulsive force to act at the interface of the signal line, thereby reducing adhesion while maintaining the capacitive coupling necessary for switch operation. The detailed design, integration, and performance characteristics of such capacitive switches are further discussed in Section 7.3.

7.3 Casimir Force-inspired MEMS Switch

Section 7.2 presented the theoretical foundations of repulsive Casimir forces and demonstrated their application in mitigating stiction in resistive-contact MEMS switches. Building upon this [33–52], Section 7.3 introduces a novel approach to reduce stiction in capacitive-contact (MIM) MEMS switches, which are widely employed in low signal-to-noise ratio (SNR) signal routing for advanced electronic systems, including next-generation 5G communication networks [1–17].

This section details both the design methodology and experimental validation of capacitive MEMS switches incorporating metamaterial-inspired structures in conjunction with DGS. Experimental results are provided to corroborate the simulated performance metrics presented earlier, highlighting the efficacy of the metamaterial-assisted DGS approach in enhancing isolation, minimizing insertion loss, and mitigating stiction effects at high operating frequencies [42].

7.3.1 Capacitive MEMS Switch Leveraging Casimir Repulsive Force

The metamaterial unit cell described in reference [13], realized using a carefully engineered composite structure, exhibits highly desirable electromagnetic characteristics within the operational frequency band relevant for MEMS switch applications, specifically between 60 GHz and 130 GHz. Figures 7.42 through 7.44 present the measured transmission and reflection responses for several representative unit cell configurations [1].

In Figure 7.42, the composite structure (4401) is embedded within a metallic layer of the signal line contact and interfaces with a cantilevered beam (4402). In this configuration, the beam is thinner than the metamaterial structure and is supported by a single anchor extending from one of the adjacent ground planes. The unit cell demonstrates transmission-dominated behavior at approximately 34 GHz, with reflection measured at −83.75 dB and transmission at −0.29 dB, while at around 120 GHz, it exhibits reflection-dominated behavior. These results confirm that the composite structure possesses frequency-dependent metamaterial characteristics, making it suitable for implementing a stiction-resistant capacitive (MIM) contact MEMS switch [1].

Similarly, Figure 7.43 depicts a composite structure (4501) incorporated into the metallic layer of a signal line contact and interfacing with a beam (4502) [1]. In this example, the beam is again thinner than the metamaterial layer but is supported at both ends by posts flanking the signal line, providing enhanced mechanical stability. The unit cell exhibits strong transmission at approximately 40 GHz, with reflection characteristics of −54 dB and transmission at −0.5 dB, while reflection-dominated behavior is observed near 140 GHz. This confirms that the structure also demonstrates metamaterial properties, allowing its integration into stiction-free capacitive MEMS switch designs.

Collectively, these examples illustrate that by embedding engineered metamaterial unit cells within signal line contacts, capacitive MEMS switches can achieve selective frequency-dependent transmission and reflection, while the repulsive Casimir interactions associated with the metamaterial layers significantly reduce stiction at the contact interface [37–40].

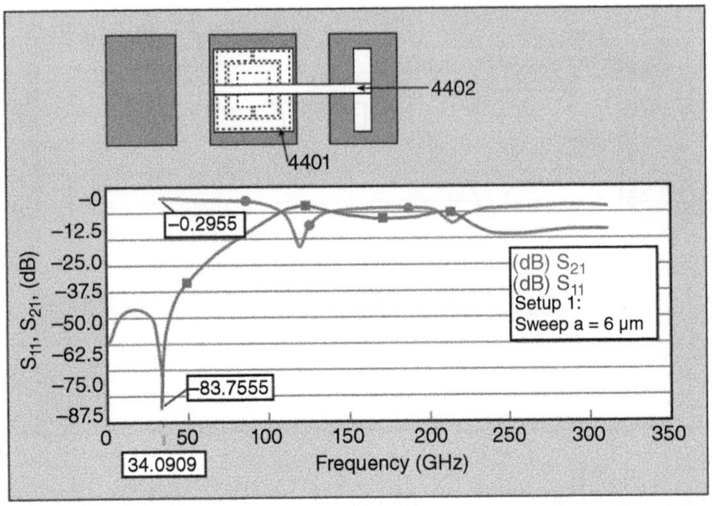

Figure 7.42 Transmission and reflection responses of the composite structure represented by 4401 and 4402 [1].

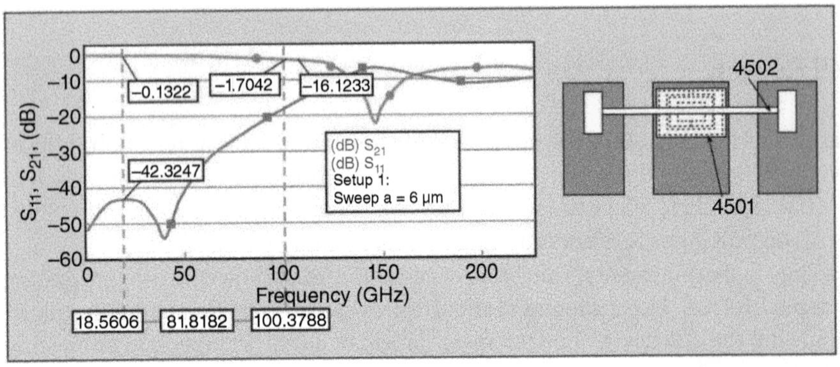

Figure 7.43 Transmission and reflection responses of the composite structure corresponding to 4501 and 4502 [1].

Figure 7.44 depicts a configuration in which two composite metamaterial structures, 4601 and 4603, are positioned at the input and output sides of the signal line, respectively [1]. Each composite structure is embedded within a metallic layer corresponding to the signal line contact. Above each metamaterial layer, doubly supported cantilevered beams, 4602 and 4604, are arranged, with beam thicknesses smaller than those of the underlying metamaterial structures, consistent with the design illustrated in Figure 7.43.

Electromagnetic characterization of the unit cells demonstrates frequency-selective behavior: at approximately 8 GHz, the structures are transmission-dominated, exhibiting reflection of −60 dB and transmission of −0.01 dB, while at around 160 GHz, the unit cells exhibit reflection-dominated behavior. These results confirm that the composite structures maintain metamaterial properties, enabling selective control of transmission and reflection across a wide frequency range. Consequently, this configuration is well suited for implementing stiction-free capacitive (MIM) contact MEMS switches, where the repulsive Casimir forces induced by the metamaterial layers help prevent undesirable adhesion at the signal line contacts while maintaining optimal electrical performance [47].

Figure 7.45 presents another example of a capacitive (MIM) MEMS switch [27]. The switch is formed over a signal line that includes an input port (4712) and an output port (4714), through which RF signals are received and transmitted, respectively. Embedded within the signal line contact is a metamaterial structure comprising an outer split ring (4722) and an inner split ring (4724), positioned between the input and output ports to interact with the propagating signal.

The ground planes (4702 and 4704) and the signal line are constructed from a high-conductivity material, such as gold, and are fabricated atop a dielectric layer (4740), for example, silicon nitride (Si_3N_4), which in turn rests on a substrate (4705). Ground plane 4702 includes a post (4770) that extends downward through the dielectric layer and serves as an anchor for a cantilevered beam (4780). The beam extends laterally from the post in the direction of the signal line, with its tip precisely aligned beneath the metamaterial split-ring structures (4722 and 4724) of the input and output ports [1].

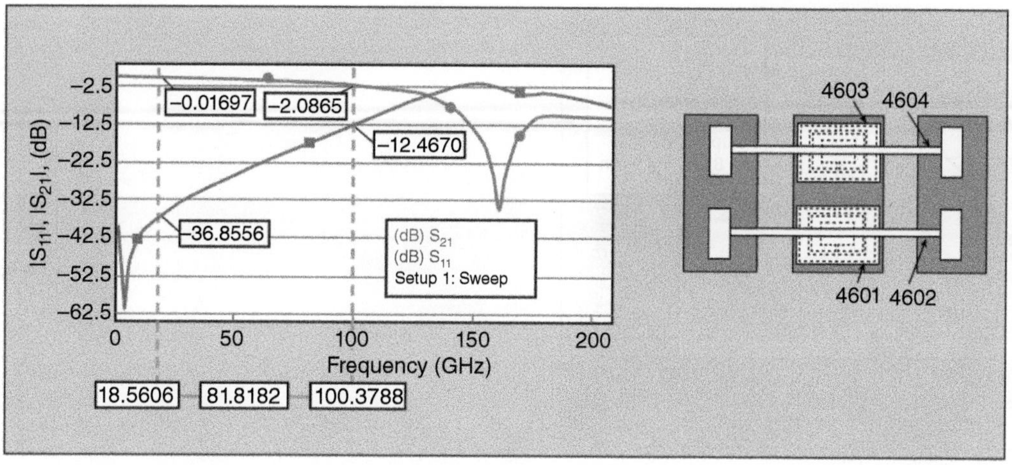

Figure 7.44 Transmission and reflection responses of the composite structure represented by 4601 and 4602 [1].

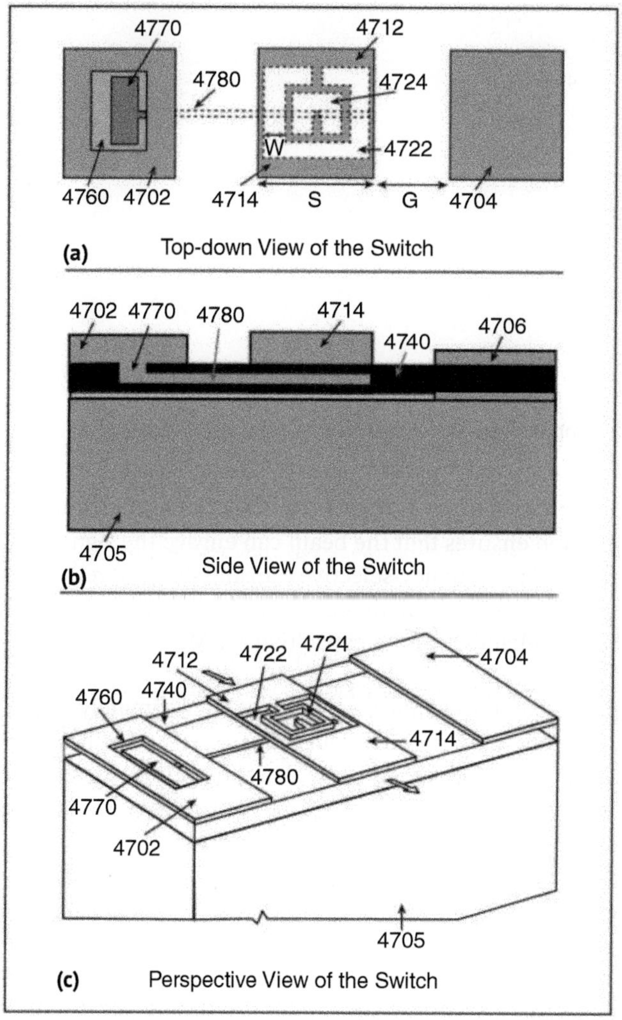

Figure 7.45 Schematic of a capacitive MEMS switch with integrated metamaterial cells generating a repulsive Casimir force between the contacts: (a) top view, (b) side view, and (c) perspective view [1].

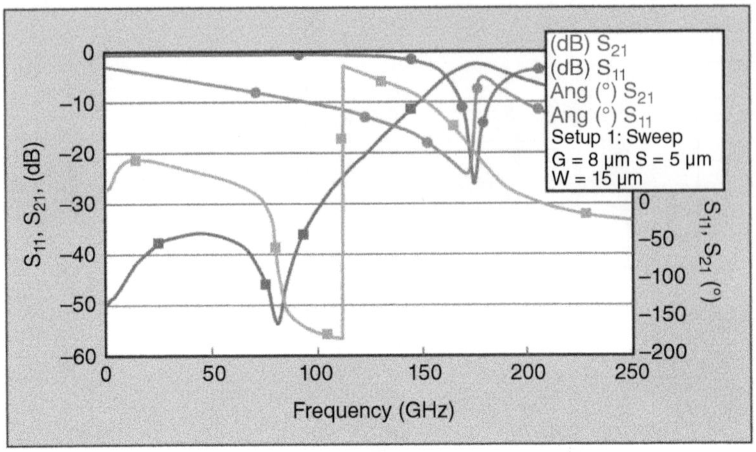

Figure 7.46 Transmission and reflection behavior of the capacitive MEMS switch shown in Figure 7.45 [1].

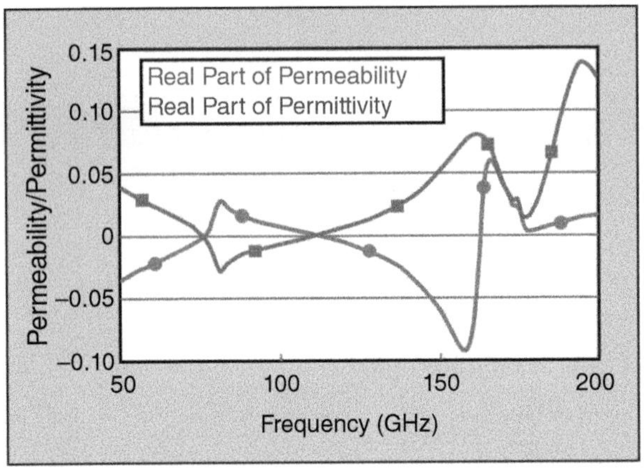

Figure 7.47 Effective permittivity derived from the S-parameters of the composite structure depicted in Figure 7.45 [1].

In Figure 7.45a,c, the post (4770) [7] is visible through an opening (4760) in the ground plane (4702), illustrating how the beam is supported while allowing vertical deflection. This configuration ensures that the beam can engage the signal line contact in a controlled manner, with the metamaterial structures providing frequency-selective transmission and reflection properties. Additionally, the integration of the metamaterial layers facilitates the generation of repulsive Casimir forces at the contact interface, which helps mitigate stiction and enhances the reliability of the capacitive MEMS switch in high-frequency applications [39–46].

In the configuration shown in Figure 7.45, the ground planes and signal line each have a width of approximately 73 μm along the longitudinal axis of the cantilevered beam (4780), while the beam itself extends about 168 μm in length. The metamaterial structure integrated into the signal line contact is composed of split-ring resonators with a ring width W of approximately 15 μm, a split width G of approximately 8 μm, and an inter-ring spacing S of approximately 5 μm. These geometric parameters are critical in determining the frequency-dependent transmission and reflection behavior of the switch [48].

Figure 7.46 [1] presents the measured transmission and reflection characteristics of the switch across a broad frequency spectrum. The metamaterial exhibits its maximum reflective behavior near 175 GHz, while it demonstrates peak transmission around 80 GHz, illustrating the frequency-selective properties of the embedded split-ring structures. Based on these experimental results, standard material parameter extraction techniques [14] can be employed to determine the effective permittivity and permeability of the metamaterial layer. The extracted frequency-dependent permittivity and permeability profiles are shown in Figure 7.47, confirming the metamaterial's engineered electromagnetic response and its suitability for stiction mitigation in capacitive (MIM) MEMS switch applications [41].

As illustrated in Figure 7.47 [1, 7], the metamaterial structure exhibits near-zero effective permittivity and permeability over a frequency range of approximately 50–150 GHz. This behavior confirms that the design presented in Figure 7.45 is

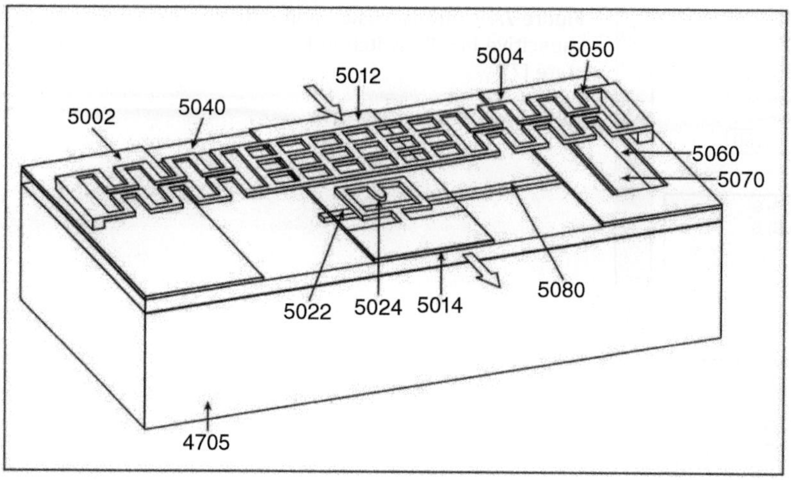

Figure 7.48 Perspective view of a capacitive shunt MEMS switch featuring a metamaterial signal line contact designed to minimize stiction [1].

well suited to leverage repulsive Casimir forces for stiction mitigation within the targeted operational frequency band. By generating a repulsive interaction between the cantilevered beam and the signal line contact, the metamaterial reduces the likelihood of adhesive contact, thereby enhancing switch reliability.

Figure 7.48 [1, 7] presents a perspective view of a capacitive shunt MEMS switch incorporating a metamaterial-enabled signal line contact specifically designed to minimize stiction. Many features of this switch are analogous to those described in Figure 7.45. The device includes a deflectable beam (5050), which is comparable to the rectangular, perforated, serpentine-patterned gold beam 510 illustrated in Figure 7.5. Beam 5050 is anchored to the substrate via a pair of posts formed atop the ground planes (5002 and 5004), providing mechanical support and controlled flexibility. When a bias voltage is applied, the beam deflects downward toward the signal line, enabling capacitive coupling while maintaining minimal adhesion due to the repulsive Casimir force induced by the metamaterial contact layer [41].

This configuration demonstrates how metamaterial integration in capacitive MEMS switches can simultaneously achieve high-frequency signal-routing performance and effective stiction reduction, offering a practical design pathway for advanced RF and millimeter-wave applications [48].

During operation [1], the application of a bias voltage causes the midpoint of the cantilevered beam (5050) to deflect downward toward the signal line contact. This deflection establishes capacitive coupling, effectively switching the signal line either OFF—or in alternative configurations, ON. Upon removal of the bias voltage, the midpoint of the beam retracts upward toward its equilibrium position. Because the beam's midpoint is aligned with the metamaterial structures (5022 and 5024) integrated into the signal line contact, the resulting Casimir force at the interface is significantly reduced and can even become repulsive. This repulsive interaction mitigates the risk of stiction between the beam (5050) and the signal line, enhancing the switch's reliability during repeated actuation cycles.

Although not explicitly illustrated in Figure 7.48 [1], the signal line contact may also include a dielectric layer above the metallic layer containing the metamaterial structures. This dielectric layer serves as an isolation medium, establishing a controlled permittivity gradient across the contact interface. Following the design principle discussed in Section 7.2 (Figure 7.5), the cantilevered beam can be considered as having an effectively infinite permittivity, the isolation layer as having a smaller positive permittivity, and the metallic layer—including the embedded metamaterial structures—as having near-zero, zero, or even negative permittivity [39]. This arrangement satisfies the condition $\varepsilon_1 < \varepsilon_2 < \varepsilon_3$ (or the reverse ordering), which is critical for optimizing the repulsive Casimir force effect [51].

The high-frequency performance of the capacitive MEMS switch illustrated in Figure 7.48 is shown in Figures 7.49 and 7.50. These plots present the reflection and transmission characteristics over a wide range of RF frequencies, confirming that the metamaterial-assisted contact design not only reduces stiction but also preserves efficient signal routing in high-frequency applications.

Figure 7.49 [1, 7] illustrates the capacitive MEMS switch operating in the ON state, where signals are actively transmitted between the input and output ports, while Figure 7.50 shows the switch in the OFF state, effectively blocking signal transmission. In the ON state depicted in Figure 7.49, the switch exhibits excellent high-frequency performance. Specifically, at

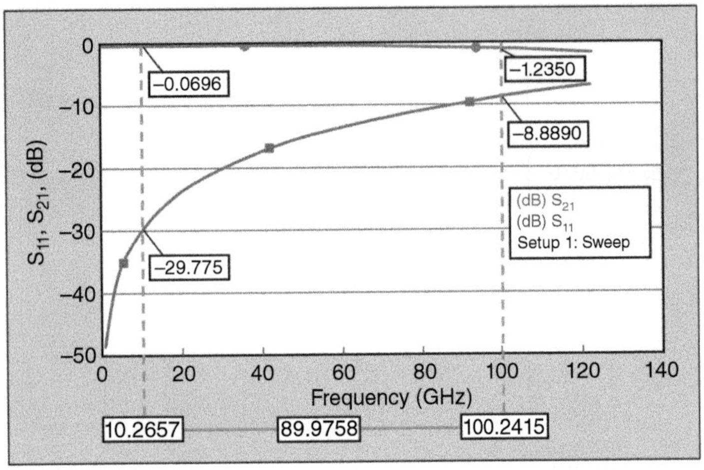

Figure 7.49 Transmission and reflection responses of the capacitive MEMS switch in Figure 7.48 when in the ON state [7].

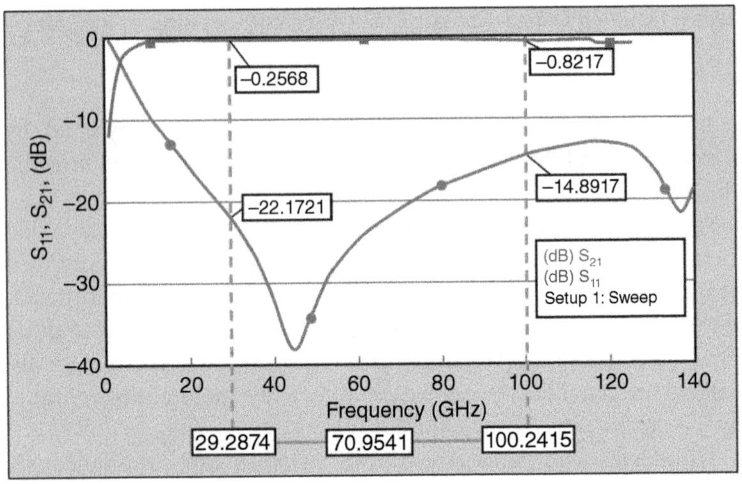

Figure 7.50 Transmission and reflection responses of the capacitive MEMS switch in Figure 7.48 when in the OFF state [1].

10.3 GHz, the return loss reaches approximately 29.8 dB, indicating minimal signal reflection, while the insertion loss is as low as 0.07 dB, signifying near-ideal signal transmission. Even at 100.2 GHz, the switch maintains good performance, with a return loss of about 8.9 dB and an insertion loss of approximately 1.23 dB. These results confirm that the capacitive MEMS switch delivers reliable signal routing across a broad frequency range from 10 GHz to 100 GHz, demonstrating the effectiveness of the metamaterial-assisted design in maintaining low-loss transmission while mitigating stiction effects at the contact interface.

In Figure 7.50, the capacitive MEMS switch is in the OFF state, operating in a reflective mode rather than a transmissive mode [1]. At 29.3 GHz, the insertion loss reaches approximately 22.2 dB, while the return loss is as low as 0.26 dB, indicating that the majority of the signal is reflected rather than transmitted. Even at 100.2 GHz, the switch maintains its reflective functionality, with an insertion loss of approximately 14.9 dB and a return loss of about 0.82 dB. These results demonstrate that the switch provides reliable operation in the OFF state across a wide frequency range of approximately 20–100 GHz.

Overall, the MEMS switch exhibits strong insertion and return loss performance in both ON and OFF states over the 30–100 GHz range [27]. This broadband, high-frequency performance makes the described switch an excellent candidate for advanced RF applications requiring low-loss, high-isolation switching. Potential applications include 5G communication systems, high-frequency switching networks, electronically scanned phased-array antennas (phase shifters), and emerging Internet of Things (IoT) devices [42]. The integration of metamaterial-based contacts further enhances switch reliability by mitigating stiction, ensuring consistent performance in demanding high-frequency environments.

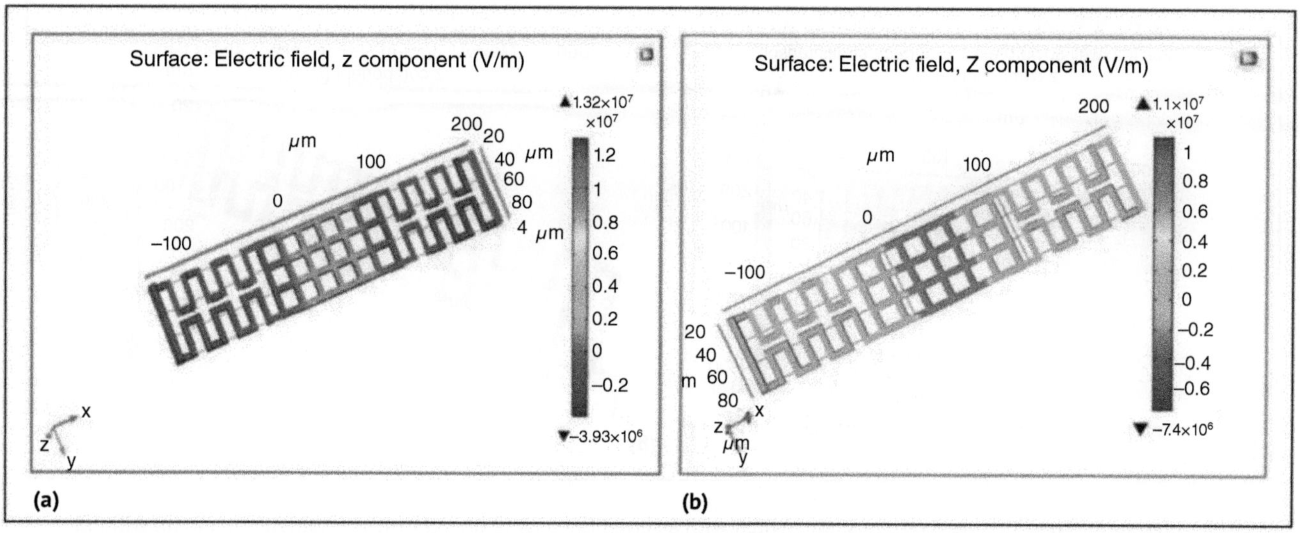

Figure 7.51 Electric field distribution of a shunt switch with a serpentine signal line: (a) incorporating metamaterial and (b) without metamaterial.

7.3.2 Investigation and Study of the Casimir Force

Section 7.1 [22–31] demonstrated that the integration of a primary shunt switch with DGS and secondary shunt switches produces a behavior analogous to that of a metamaterial, effectively combining frequency-selective isolation with low insertion loss. Building upon this, Section 7.2 examined the enhancement of stiction resistance in MEMS switches by incorporating metamaterial layers within the resistive switch contact design. Extending this concept, Section 7.3.2 employed metamaterial layers as part of the signal line contacts in capacitive (MIM) MEMS switches, thereby improving both stiction mitigation and device reliability [37].

To quantitatively assess the repulsive Casimir forces generated in these metamaterial-integrated structures, a comprehensive numerical study was conducted on a shunt switch featuring a serpentine signal line [52]. Simulations were performed with and without metamaterial layers positioned beneath the signal line using COMSOL Multiphysics® software, enabling evaluation of the influence of metamaterial-induced permittivity gradients on the Casimir interactions and the resulting stiction reduction.

Figures 7.51 through 7.53 [1, 7] illustrate the simulated electric field distribution, surface charge distribution, and electric surface stress tensor of a shunt MEMS switch featuring a serpentine signal line, both with and without an underlying metamaterial layer. Analysis of these distributions enables the estimation of the Casimir forces acting between the switch contacts. The estimated Casimir force for the serpentine shunt switch, comparing configurations with and without the metamaterial layer, is presented in Figure 7.54. Table 7.5 summarizes key geometric and material parameters of the serpentine structure and highlights the effect of incorporating the metamaterial layer, demonstrating that the inclusion of the metamaterial induces a repulsive Casimir force.

Table 7.5 shows the key parameters for serpentine structure with and without metamaterial in the switch.

This repulsive interaction contributes to reduced stiction, enhancing the reliability and operational stability of the MEMS switch, particularly under repeated actuation cycles and high-frequency operation.

As observed from the simulations, the MEMS switch structure incorporating a metamaterial layer underneath the signal line exhibits a repulsive Casimir force, effectively mitigating stiction [41]. To validate these simulation results, sample structures were fabricated using the PolyMUMPs process. For instance, the metamaterial-inspired MEMS shunt switch illustrated in Figure 7.15 (Section 7.1) was successfully fabricated and characterized. SEM and optical microscope images of the fabricated device are shown in Figure 7.53, confirming the fidelity of the fabrication process.

Both the primary and secondary switches were experimentally characterized, including eigenfrequency measurements, capacitance–voltage (C–V) characteristics, profiling, and laser Doppler vibrometry [44]. The experimental data closely matched the simulation results, confirming the accuracy of the numerical models. Figures 7.54 and 7.55 present the

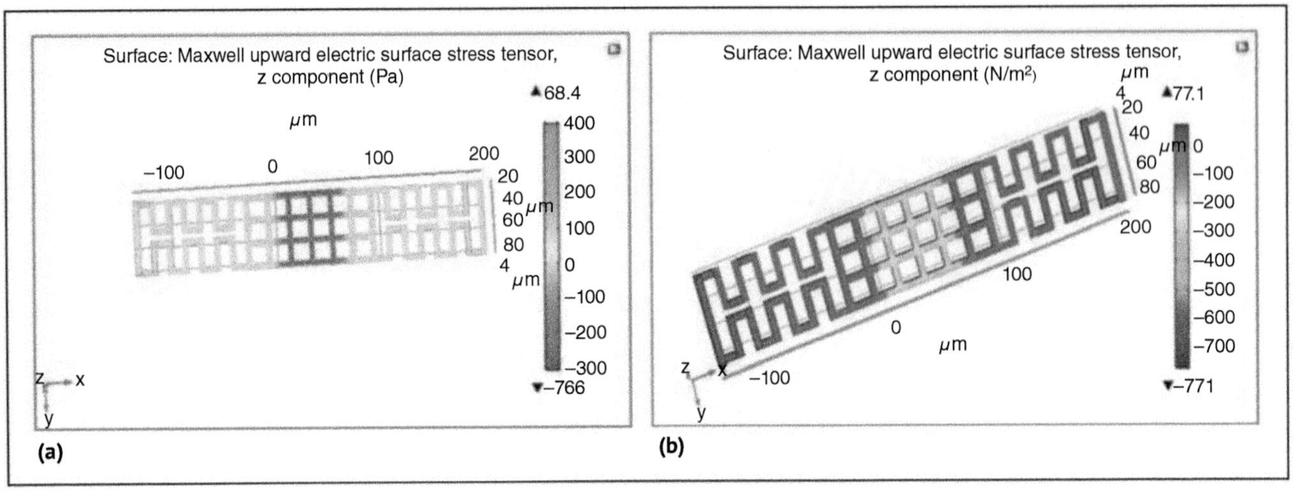

Figure 7.52 Electric surface stress tensor of a shunt switch with a serpentine signal line: (a) with metamaterial and (b) without metamaterial.

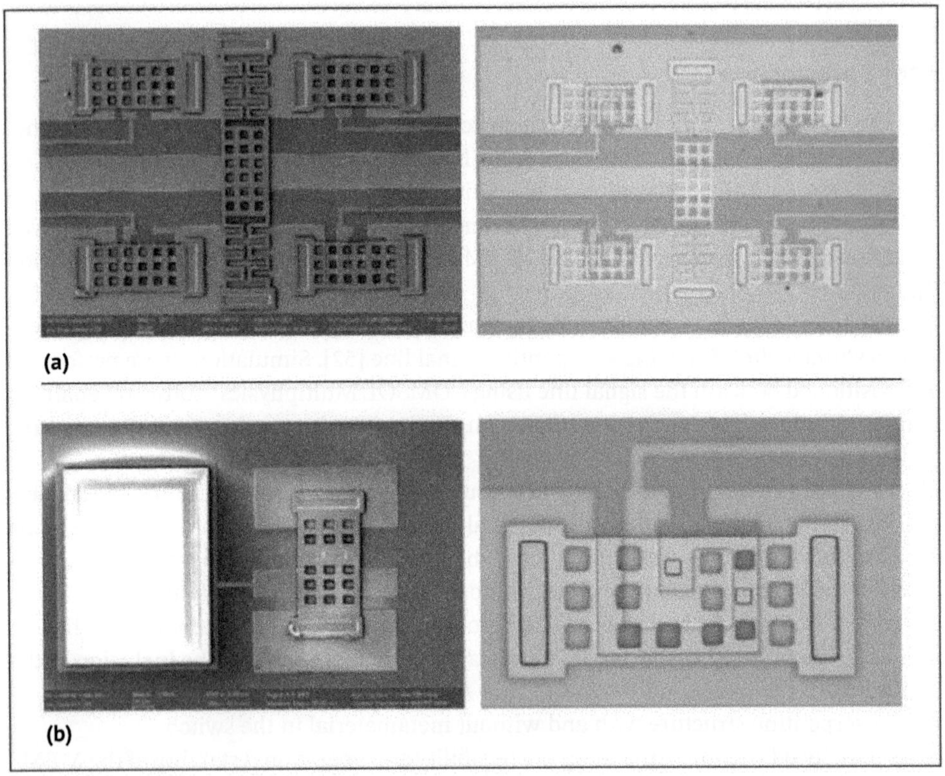

Figure 7.53 Scanning electron microscope (SEM) and optical microscope images of (a) the primary shunt switch and (b) the secondary shunt switch.

measured performance results obtained from the bare die. It was observed that implementing a serpentine beam design reduced the pull-in voltage by approximately 9 V, demonstrating improved actuation efficiency [45].

Additionally, work on packaging techniques suitable for the present MEMS switch design was conducted separately [18, 19]. Measurements carried out post-packaging showed no significant deviations in switch performance, confirming the robustness of the metamaterial-assisted MEMS switch design under practical operating conditions [48].

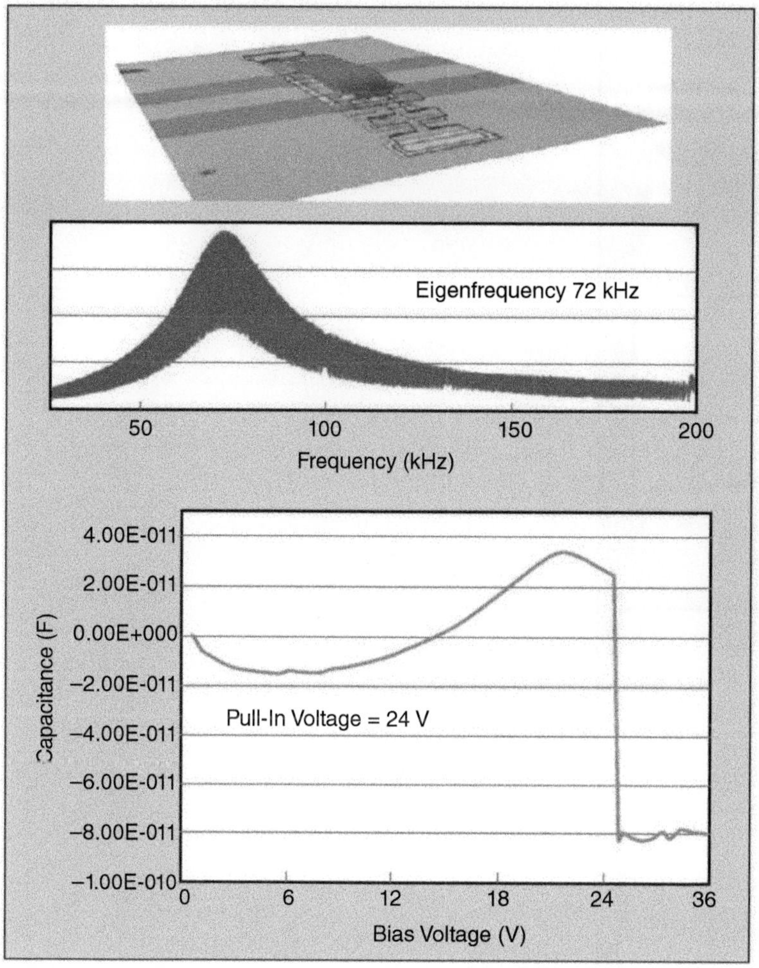

Figure 7.54 Measured plots of eigenfrequencies and pull-in voltage for the primary shunt switches [1].

Table 7.5 Key parameters for serpentine structure with and without metamaterial in the switch.

No.	Parameter	Serpentine structure with metamaterial underneath	Serpentine structure without metamaterial underneath
1	Electric field (V/m)	1.32×10^7	1.1×10^7
2	Electric charge (C)	-9.65×10^{-5}	-9.98×10^{-5}
3	Electric surface Tensor (Pa)	−766	−771
4	Calculated force (10^{-5} N)	−18.5	1,980

7.3.3 Metamaterial MEMS Switches Applications

RF-MEMS switches offer promising solutions for next-generation 5G and IoT applications, where reconfigurable wideband and frequency-agile components—such as high-order switching elements, tunable filters, multistate attenuators, and phase shifters—are essential to support millimeter-wave communications, small-cell networks, and advanced beamforming architectures [20, 21].

As discussed in Sections 7.1 and 7.2, key limitations preventing widespread commercial adoption of MEMS devices have historically included concerns related to reliability, packaging, and seamless integration with standard semiconductor

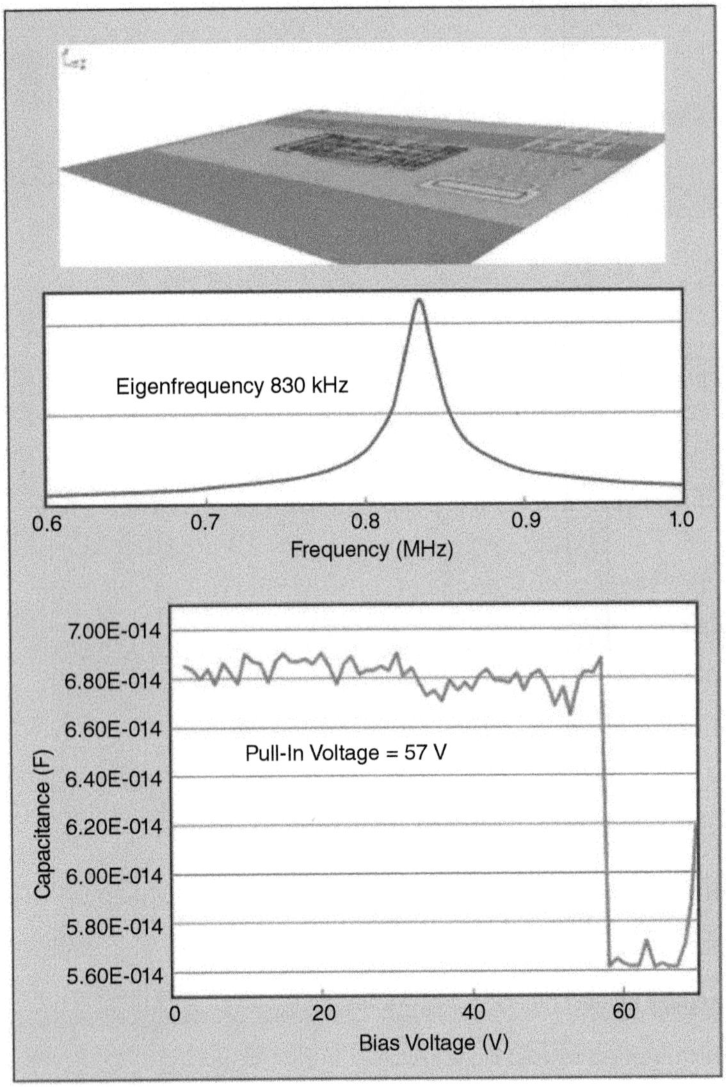

Figure 7.55 Measured plots of eigenfrequencies and pull-in voltage for the secondary shunt switches [1].

technologies [22–25]. Section 7.3 detailed research efforts addressing these challenges by leveraging the repulsive Casimir effect through metamaterial-inspired MEMS switch designs. This approach has yielded significant improvements in operational stability, stiction mitigation, and overall switch performance, demonstrating the potential of these advanced MEMS devices to meet the demanding requirements of 5G and IoT systems [26].

As discussed in Sections 7.1 through 7.3, the metamaterial structures employed in the MEMS switch designs are primarily split-ring resonators. Nevertheless, alternative metamaterial topologies can be utilized, provided they exhibit similar permittivity and permeability characteristics across the target frequency range [50]. For example, a topology-inspired Möbius transformation metamaterial (MTM) structure—characterized by a continuous closed path that maps onto itself—may offer particular advantages for generating repulsive Casimir forces [13].

It is important to note that the MEMS switch configurations described in Sections 7.1–7.3 represent specific embodiments developed to illustrate the design principles and functional applications explored in this research. These examples are not limiting; a wide variety of modifications and alternative designs can be implemented while still leveraging the same fundamental concepts demonstrated herein. Such flexibility allows the application of these principles to diverse MEMS switch architectures and frequency-selective devices for high-performance RF and millimeter-wave systems [1].

7.4 Metamaterial-inspired Tunable Switch-band Oscillator

Metamaterial-integrated MEMS (MetaMEMS) resonator oscillators represent a transformative advancement in high-performance device technology, merging the precise mechanical stability of MEMS resonators with the engineered electromagnetic responses of metamaterials [10, 26, 27]. MEMS resonators are well known for their ability to achieve high-frequency, stable oscillations while maintaining low power consumption, providing a reliable and compact mechanical platform. By integrating metamaterials—whose electromagnetic, acoustic, or mechanical responses can be precisely tailored—these oscillators achieve functionalities that surpass conventional resonator designs [39–51].

This synergistic combination enables tunability across multiple operational parameters, including frequency, amplitude, and phase, allowing the oscillators to maintain optimal performance under varying environmental conditions or external perturbations. Such dynamic adaptability is particularly advantageous for real-time, high-precision applications. Additionally, the inherent compactness of MEMS technology, coupled with the lightweight and highly configurable design of metamaterials, facilitates significant miniaturization. This makes MetaMEMS oscillators ideally suited for deployment in constrained environments, including wearable electronics, implantable medical devices, and other space-limited high-frequency systems [50].

The benefits of MetaMEMS resonator oscillators extend beyond their compact form factor and operational tunability. These devices also demonstrate outstanding energy efficiency, minimizing power consumption while maintaining high-performance operation. Such efficiency aligns with the growing emphasis on sustainable and environmentally responsible technologies. Furthermore, the ability to engineer the electromagnetic properties of the metamaterial component allows for precise tailoring of the device to specific application requirements. This includes functions such as advanced noise suppression in communication systems, improved signal fidelity in radar and sensing applications, and enhanced sensitivity in environmental monitoring and diagnostic platforms [50, 51].

The ongoing drive to miniaturize wireless communication hardware has spurred the development of compact radio architectures featuring tunable oscillators, such as voltage-controlled oscillators (VCOs), capable of covering multiple frequency bands [26, 28–32]. Conventional approaches to designing multiband VCOs with low phase noise, however, often result in bulky, high-power circuits that are inefficient for modern portable and integrated applications [32].

A variety of techniques have been explored to implement compact, multiband signal sources, including switching among separate VCOs, employing intermodal multiple-frequency circuits, or toggling between multiple resonators [31, 32]. Despite these efforts, most solutions remain constrained by large footprint, elevated power consumption, or compromised phase-noise performance. For instance, switched-resonator designs inherently suffer from resistive and capacitive parasitics introduced by the switching elements, which can degrade oscillator performance. Mode-switching strategies, which select among multiple oscillation modes within a single resonant circuit, mitigate the adverse effects of switch-induced losses and nonlinearities on steady-state operation and phase noise. Nonetheless, these methods do not support the simultaneous generation of multiple signals and remain limited in reducing overall circuit size, power requirements, and design cycle time.

Conventional multiband signal sources typically rely on arrays of resonators, VCOs, or other tunable oscillator architectures. However, any implementation of a band-selection switch in these designs inherently introduces performance degradation, often manifesting as increased insertion loss, phase noise, or reduced signal fidelity. Figure 7.56 illustrates representative geometries of metamaterial-inspired slow-wave resonators, which are engineered to enable switchable band operation while mitigating some of the limitations associated with traditional band-selection approaches [26, 27].

This emerging class of oscillators integrates a metamaterial-based slow-wave resonator network with a MEMS switch that exploits the repulsive Casimir effect to minimize stiction and enhance device reliability. By leveraging the tunable properties of both the metamaterial resonators and the MEMS switching mechanism, this architecture achieves the functionality of multiple independent oscillators while significantly reducing overall circuit size and power consumption. Such a design is particularly well suited for multiband, multimode wireless communication systems, where compact, energy-efficient, and reliable oscillators are critical [26, 27].

Figure 7.57 provides a representative layout of a multiband MEMS-switched oscillator incorporating the Möbius metamaterial resonator network depicted in Figure 7.56. This layout illustrates how the metamaterial network and MEMS switch are integrated to enable selective band operation and maintain high-performance characteristics across multiple frequency bands [26, 27].

In modern telecommunications, metamaterial-enabled MEMS resonator oscillators—often referred to as MetaMEMS oscillators—offer an innovative solution to the persistent challenges of frequency agility, bandwidth constraints, and signal integrity. By integrating the mechanical precision of MEMS resonators with the tailored electromagnetic responses of metamaterials, these devices enable the creation of reconfigurable systems capable of operating seamlessly across multiple frequency bands without compromising performance. This capability is particularly critical for emerging high-frequency technologies, such as 5G, 6G, and beyond, where requirements for ultralow latency, high throughput, and uninterrupted connectivity demand advanced, adaptive oscillator architectures.

Beyond communications, MetaMEMS oscillators have transformative implications in sensing applications. The metamaterial elements can manipulate electromagnetic fields to enhance signal focus or amplification, thereby improving detection sensitivity, spatial resolution, and overall operational range. For example, these oscillators can increase the accuracy of radar or environmental monitoring systems, enabling reliable detection of weak or distant signals that conventional resonators might miss [50].

The development of MetaMEMS resonator oscillators represents a significant leap in both design methodology and functional capability [49]. They open avenues for applications in high-performance computing, where they can serve as precision timing elements in quantum computing platforms or as core components in neuromorphic computing architectures that emulate neural networks. Furthermore, their compactness, tunability, and robustness render them highly suitable for aerospace and defense applications, where resilience, adaptive performance, and operational reliability are paramount.

As ongoing research continues to refine the integration of metamaterials with MEMS technology, it is expected that these oscillators will achieve even greater performance benchmarks. Advances in material engineering, resonator topology, and system-level integration are likely to expand their operational bandwidth, reduce energy consumption, and enhance device miniaturization. Consequently, MetaMEMS oscillators are poised to become foundational components in next-generation electronic and electromechanical systems, bridging the gap between high-performance functionality and scalable, low-power design [50, 51].

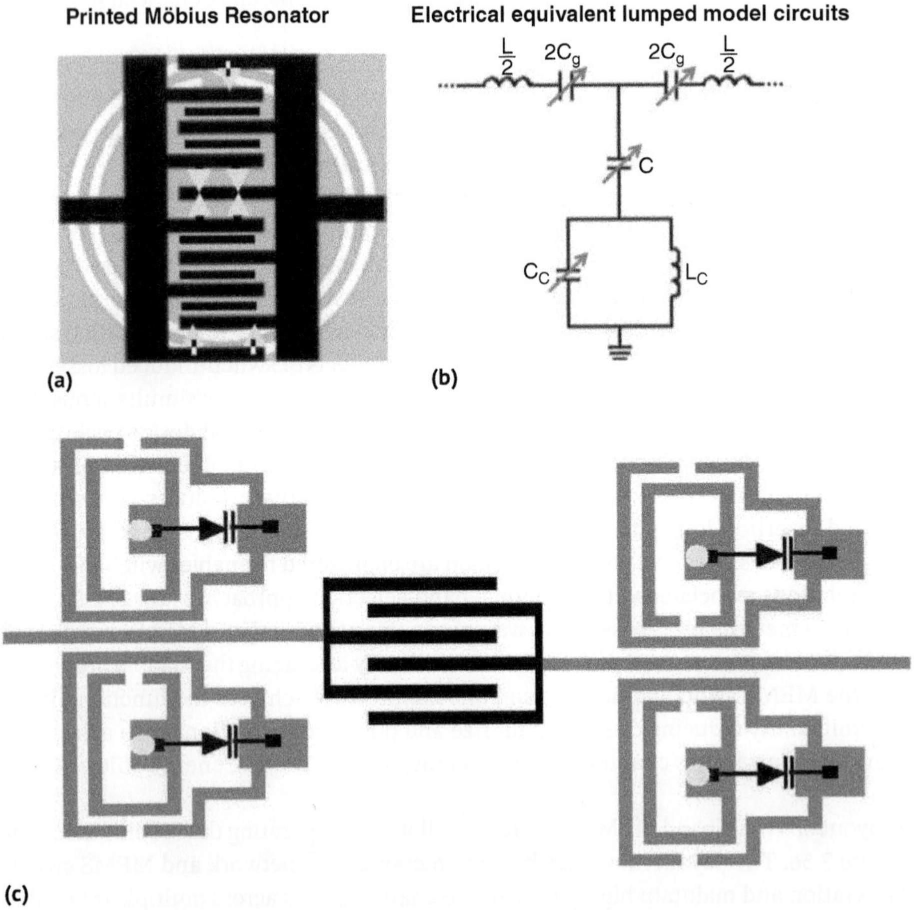

Figure 7.56 A typical configuration of metamaterial resonators for switchable band operation [26, 27]. (a) A typical layout of metamaterial Möbius resonator. (b) An electrical equivalent lumped model circuit. (c) Typical geometries of metamaterial resonators enabling switchable band operation.

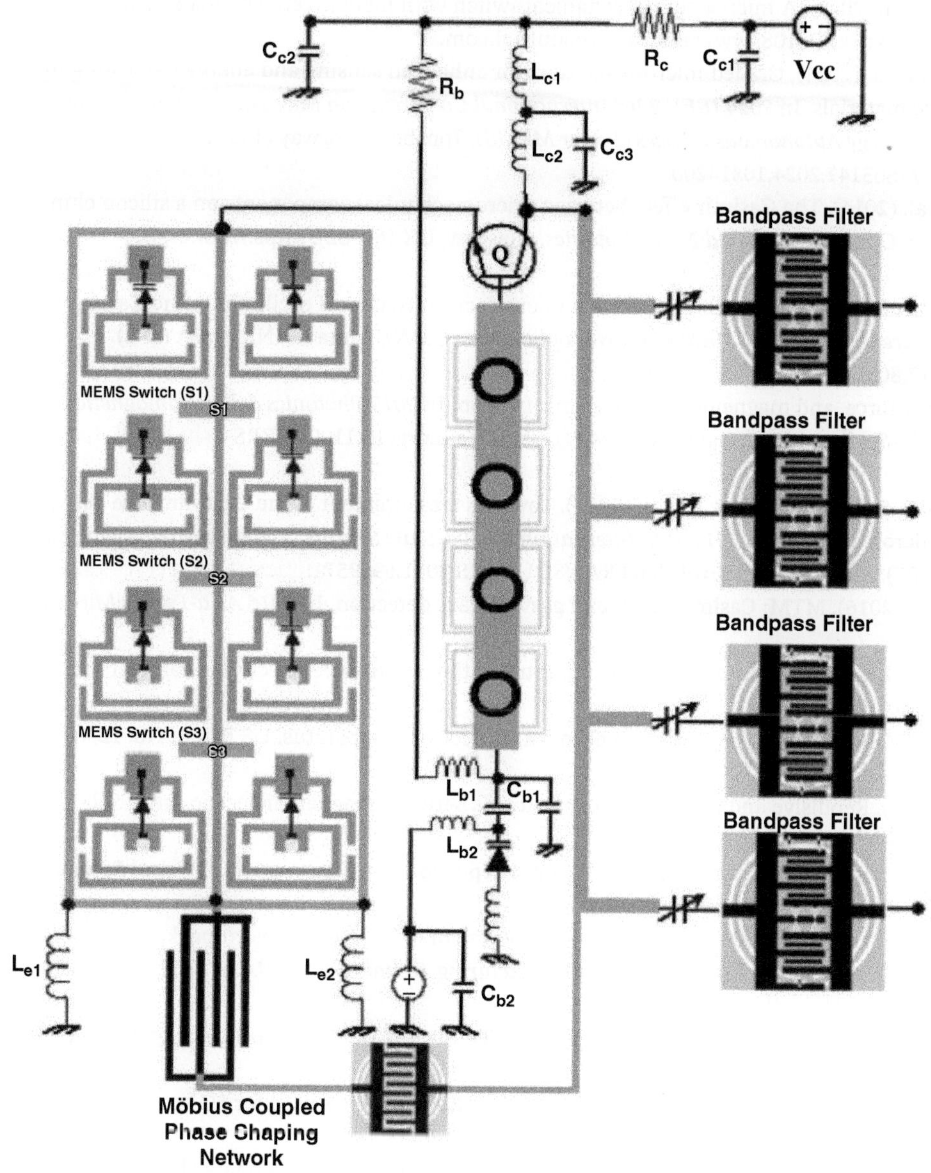

Figure 7.57 Layout of a multiband MEMS-switched oscillator employing a Möbius metamaterial resonator network, as depicted in Figure 7.56 [27].

References

1 Koul, S., Poddar, A., and Rohde, U. (2018). Microelectromechanical switch with metamaterial contacts. US Patent Pub. No. 10,784, 066, 22 September 2020. patents.google.com.

2 Lambrecht, A. (2002). The Casimir effect: a force from nothing. *Physics World* 15 (9): 29.

3 Intravaia, F. and Henkel, C. (2009). *New Frontiers of Casimir Force Control*, 27–29. Santa Fe, New Mexico, USA.

4 Leonhardt, U. and Philbin, T.G. (2007). Quantum levitation by left-handed metamaterials. *New Journal of Physics* 9: 1367–2630.

5 Mercado, L.L., Koo, S.M., Lee, T.Y.T. et al. (2003). A mechanical approach to overcome RF MEMS switch stiction problem. In *Proceedings of 53rd Electronic Components and Technology Conference*, (377–384). IEEE.

6 Mahajan, C. (2018). Design and analysis of RF MEMS switches (70 GHz to 130 GHz). M. Tech dissertation. CARE, IIT Delhi.

7 Koul, S., Mahajan, C., Poddar, A. et al. (2020). A microelectromechanical switch with metamaterial contacts, part I: concepts and technology. *Microwave Journal* 63 (5): 82–108. www.microwave.journal.com.

8 De Ponti, J.M., Iorio, L., Rosso, M. et al. (2024). Graded micro-resonators for enhanced sensing and energy harvesting in MEMS with lead-free piezoelectric materials. In *2024 IEEE 23rd International Conference on Micro and Miniature Power Systems, Self-Powered Sensors and Energy Autonomous Devices (Power MEMS)*, Tonsberg, Norway (155–158). https://doi.org/10.1109/PowerMEMS63147.2024.10814200.

9 Chan, H.B., Zou, J., Marcet, Z. et al. (2014). The Casimir effect between micromechanical components on a silicon chip. In *2014 International Conference on Optical MEMS and Nanophotonics*, Glasgow, UK (61–62). https://doi.org/10.1109/OMN.2014.6924543.

10 Decca, R.S. (2017). Fourier reconstruction of the force signal using a microelectromechanical oscillator in the Casimir regime. In *2017 International Conference on Optical MEMS and Nanophotonics (OMN)*, Santa Fe, NM, USA (1–2). https://doi.org/10.1109/OMN.2017.8051491.

11 Brevik, I. (2019). Repulsive Casimir force, and magnetoelectric Casimir friction. In *2019 Photonics & Electromagnetics Research Symposium-Spring (PIERS-Spring)*, Rome, Italy (2454–2459). https://doi.org/10.1109/PIERS-Spring46901.2019.9017679.

12 Nawazuddin, M.B.S., Lammerink, T.S.J., Wiegerink, R.J. et al. (2011). Towards measurement of the Casimir force between parallel plates separated at sub-micron distance. In *2011 16th International Solid-State Sensors, Actuators and Microsystems Conference*, Beijing, China (434–437). https://doi.org/10.1109/TRANSDUCERS.2011.5969571.

13 Rohde, U., Poddar, A., and Koul, S. (2016). MTM: Casimir effect and gravity wave detection. In *2016 Asia-Pacific Microwave Conference (APMC)*, (1–4). IEEE.

14 Chen, X., Grzegorczyk, T.M., Wu, B.I. et al. (2004). Robust method to retrieve the constitutive effective parameters of metamaterials. *Physical Review E* 70: 016608.

15 Shi, Y., Zhou, C., and Li, Y. (2026). MEMS switch, preparation method, and electronic apparatus. US Patent No. 12,466, 723.patents.google.com.

16 Sun, J., Hua, X.K., and Gao, L. (2012). Repulsive and attractive Casimir forces between magnetodielectric slabs. *Solid State Communication* 152 (17): 1666–1669.

17 Yannopapas, V. and Vitanov, N.V. (2009). First-principles study of Casimir repulsion in metamaterials. *Physical Review Letters* 103 (12): 120401.

18 Chaturvedi, S., Bhalke, S., Bhat, M. et al. (2010). EM simulation and development of wafer level micro-packaging technique for GaAs-based RFMEMS switches. In *Compound Semiconductor Manufacturing Techniques (CSMANTECH) Conference*, (17–20), Portland, U.S.A (303–305).

19 Chaturvedi, S. (2012). Modelling of packages for MMICs and MEMS. Ph.D. Thesis. IIT Delhi under the supervision of Shiban K Koul. CARE, IIT Delhi.

20 Ma, L.-Y., Soin, N., Daut, M.H.M. et al. (2019). Comprehensive study on RF-MEMS switches used for 5G scenario. *IEEE Access* 7: 107506–107522.

21 Iannacci, J., Resta, G., Bagolini, A. et al. (2020). RF-MEMS monolithic K and Ka-band multi-state phase shifters as building blocks for 5G and internet of things (IoT) applications. *Sensors* 20 (9): 2612. MDPI Journal. www.mdpi.com.

22 Dey, S., Koul, S.K., Poddar, A.K. et al. (2020). RF MEMS switches, switching networks and phase shifters for microwave to millimeter wave applications. *ISSS Journal of Micro and Smart Systems* 9: 33–47.

23 Karthikeya, G.S., Koul, S.K., Poddar, A.K. et al. (2020). Ultra-compact orthogonal pattern diversity antenna module for 5G smartphones. *Microwave and Optical Technology Letters*, 63: 2003–2012.

24 Dey, S., Koul, S.K., Poddar, A.K. et al. (2019). Compact, broadband, and reliable lateral MEMS switching networks for 5G communications. *Progress in Electromagnetics Research M* 86: 163–171.

25 Dey, S., Koul, S.K., Poddar, A.K. et al. (2018). Reliable and compact 3- and 4-bit phase shifters using MEMS SP4T and SP8T switches. *Journal of Micromechanical Systems* 27 (1): 113–124.

26 Poddar, A.K. (2014). Slow wave resonator based tunable multi-band multi-mode injection-locked oscillator. Dr.-Ing.-Habil Thesis. BTU Cottbus, Germany.

27 Koul, S.K., Mahajan, C., Poddar, A.K. et al. (2020). Innovative Concepts in MEMS Switches with Metamaterial Contacts. e-Book, *Microwave Journal* 63 (5): 1–24. www.microwave.journal.com.

28 Rohde, U.L. and Poddar, A.K. (2009). Low noise and low phase hits tunable oscillator. U.S. Patent No. 7,636, 021.patents. google.com.

29 Rohde, U.L. and Poddar, A.K. (2009). Wideband voltage-controlled oscillators employing evanescent mode coupled resonators. Canadian Patent No. 2,563, 174.patents.google.com.

30 Rohde, U.L. and Poddar, A.K. (2009). User-definable thermal drift voltage controlled oscillator. Canadian Patent No. 2,548, 317.patents.google.com.

31 Rohde, U.L. and Poddar, A.K. (2009). Integrated low noise microwave wideband push-push VCO. Canadian Patent No. 2,548, 311.patents.google.com.

32 Sun, Y., Yu, X., Rhee, W. et al. (2010). Dual-path LC VCO design with partitioned coarse-tuning control in 65nm CMOS. *IEEE Microwave and Wireless Components Letters* 20 (3): 169–171.

33 Rodrigues, J.R., Rosa, F.S.S. and Almeida, V.R. (2016). Casimir and optical forces acting on a silicon NOEMS device based on slot-waveguide structure. *IEEE Photonics Technology Letters* 28 (5): 589–592. https://doi.org/10.1109/LPT.2015.2508808.

34 Chan, H.B., Bao, Y., and Zou, J. (2012) The Casimir effect on silicon micromechanical systems: forces due to virtual photons. In *2012 International Conference on Optical MEMS and Nanophotonics*, Banff, AB, Canada (29–30). https://doi.org/10.1109/OMEMS.2012.6318786.

35 Lia, X.-W., Huang, Z.-X., and Wu, X.-L. (2013). The restoring Casimir force between doped silicon slab and metamaterials. In *2013 Proceedings of the International Symposium on Antennas & Propagation*, Nanjing, China (1264–1267).

36 Shimofuri, M., Banerjee, A., Hirai, Y. et al. (2022). Observation of pull-in by Casimir force in MEMS-controlled nanogap fabricated by silicon cleavage. In *2022 IEEE 35th International Conference on Micro Electro Mechanical Systems Conference (MEMS)*, Tokyo, Japan (511–514). https://doi.org/10.1109/MEMS51670.2022.9699645.

37 Mukherjee, B. and Sen, S. (2015). Generalized closed form approximations for pull-in characteristics of electrostatic nanotweezers under the influences of van der Waals and Casimir forces. In *2015 IEEE 15th International Conference on Nanotechnology (IEEE-NANO)*, Rome, Italy (842–845). https://doi.org/10.1109/NANO.2015.7388744.

38 Maleki, F.M. and Tajik, F. (2024). Influence of piezoelectric actuator on the stability of micromechanical device via Casimir force. In *2024 6th Iranian International Conference on Microelectronics (IICM)*, Tabriz, Iran, Islamic Republic of (1–5). https://doi.org/10.1109/IICM65053.2024.10824340.

39 Xia, T., Atkins, P.R., Sha, W.E.I. et al. (2021). Casimir force: vacuum fluctuation, zero-point energy, and computational electromagnetics. *IEEE Antennas and Propagation Magazine* 63 (5): 14–24. https://doi.org/10.1109/MAP.2021.3073118.

40 Salazar-Lazaro, C. H., Chew, W.C., and Stone, M. (2019). Casimir energy for perfect electric conductors revisited. In *2019 IEEE MTT-S International Conference on Numerical Electromagnetic and Multiphysics Modeling and Optimization (NEMO)*, Boston, MA, USA (1–4). https://doi.org/10.1109/NEMO.2019.8853712.

41 Henkel, C. and Joulain, K. (2005). Fundamental limits for Casimir repulsion between metamaterials. In *CLEO/Europe. 2005 Conference on Lasers and Electro-Optics Europe, 2005*, Munich, Germany (609–609). https://doi.org/10.1109/CLEOE.2005.1568385.

42 Elsaka, B., Kästner, P., Hamarneh, Y. et al. (2024). Casimir forces leading to 3D self-assembled paired metallic microshutters. In *2024 International Conference on Optical MEMS and Nanophotonics (OMN)*, San Sebastian, Spain (1–2). https://doi.org/10.1109/OMN61224.2024.10685251.

43 Cui, S. and Soh, Y.C. (2010). An accurate separation estimation algorithm for the Casimir oscillator. *Journal of Microelectromechanical Systems* 19 (5): 1153–1161. https://doi.org/10.1109/JMEMS.2010.2067433.

44 Carter, E.L., Ward, M., and Anthony, C. (2009). Design and fabrication of novel devices using the Casimir force for non-contact actuation. *SENSORS,* 2009 *IEEE*, Christchurch, New Zealand (229–233). https://doi.org/10.1109/ICSENS.2009.5398181.

45 Ardito, R., de Masi, B., Frangi, A. et al. (2010). An on-chip experimental assessment of Casimir force effect in micro-electromechanical systems. In *2010 11th International Thermal, Mechanical & Multi-Physics Simulation, and Experiments in Microelectronics and Microsystems (EuroSimE)*, Bordeaux, France (1–8). https://doi.org/10.1109/ESIME.2010.5464555.

46 Chew, W.C., Xiong, J. L., Qian, Z.G. et al. (2009). Computational electromagnetics: Casimir force, multiscale calculations, coordinate stretching. In *2009 Computational Electromagnetics International Workshop*, Izmir, Turkey (50–55). https://doi.org/10.1109/CEM.2009.5228105.

47 Calis, M. and Desmulliez, M.P.Y. (2008). Haptic sensing for MEMS with application for cantilever and Casimir effect. In *2008 Symposium on Design, Test, Integration and Packaging of MEMS/MOEMS*, Nice, France (80–84). https://doi.org/10.1109/DTIP.2008.4752957.

48 Ardito, R., Comi, C., Zega, V. et al. (2023). Metamaterials and MEMS (MetaMEMS): a promising trend in Microsystems technology. In *2023 24th International Conference on Thermal, Mechanical and Multi-Physics Simulation and Experiments in Microelectronics and Microsystems (EuroSimE)*, Graz, Austria (1–4). https://doi.org/10.1109/EuroSimE56861.2023.10100824.

49 Ekmekci, E., Topalli, K., Akin, T. et al. (2009). A tunable multi-band metamaterial design using micro-split SRR structures. *Optics Express* 17: 16046–16058.

50 Xu, R.-J. and Lin, Y.-S. (2022). Actively MEMS-based tunable metamaterials for advanced and emerging applications. *Electronics* 11: 243. https://doi.org/10.3390/electronics11020243.

51 Chang, Y., Wei, J., and Lee, C. (2020). Metamaterials – from fundamentals and MEMS tuning mechanisms to applications. *Nanophotonics* 9 (10): 3049–3070. https://doi.org/10.1515/nanoph-2020-0045.

52 Haakh, H.R. and Intravaia, F. (2013). Mode structure and polaritonic contributions to the Casimir effect in a magnetodielectric cavity. *Physical Review A* 88 (5): 052503. https://doi.org/10.1103/PhysRevA.88.052503.

8

Amplifier Design with BJTs and FETs up to the Sub THz Range

8.1 Properties of Amplifiers

The properties of amplifiers are typically obtained by measurements. A good choice is the R&S ZNA (Figure 8.1).

The goal of this chapter is to move from the details on the semiconductor devices themselves to our first practical application. One of the more interesting features of amplifiers is that they can easily become what is covered in Chapter 10 about *oscillators*. This is due to the high gain of the devices, unaccounted for parasitic elements, and other design flaws. Part of the material in this chapter is taken from our joint book [1].

In general, the amplifiers we need will fall into three categories:

- Low Noise Amplifiers
- High-gain Amplifiers
- Medium- to High-power Amplifiers

See Figure 8.2. The low-noise amplifier always operates in Class A, typically at 15–20% of its maximum useful current. The high-gain amplifier can operate in Class A, as well as B (mostly push–pull). The higher dc current for the same device results in a higher noise figure and more gain, and ultimately more output power. Class C operating mode is really reserved for either FM transmissions or constant-carrier modes like CW. Some of the modern digital modulation types are sensitive to phase distortion rather than amplitude changes and because of the resulting output spectrum designers have stayed away from Class C operation [1].

8.1.1 Some Examples of Practical Amplifiers

Some examples of these practical amplifiers, such as bipolar versions, can be dc-coupled with very few difficulties; others, like those in the FET families, will cause more headaches. An interesting example, although much too high in frequency for the purpose of this book, is shown in Figures 8.3–8.5. It shows a three-stage amplifier that has noise feedback for the first two stages and resistive feedback for the output stage. These are the three types of amplifiers, regardless of technology, we will be evaluating. The achievement of accurate noise analysis of this circuit, reported in [1, 2], marked for the first time a complete linear noise model with essentially no frequency limitation having been developed. Its accuracy depends solely on the accuracy of the element values of the linear equivalent circuit, which can be obtained from measured S parameters by an optimization process. Needless to say, this technique is also applicable at lower frequencies, such as 1.5–3 GHz, and its deviation from measured values has never been worse than 0.2 dB or, relative to the noise figure, 10% expressed in dB.

The results of Figure 8.5a were obtained by using the linear FET model as supplied by Texas Instruments. It becomes somewhat obvious that a match for S_{11} of −60 dB is not likely. We then replaced the linear transistor with the SPICE-type nonlinear model, resulting in the responses shown in Figure 8.5b. S_{11} now looks much more realistic in the center of the range, but at the same time exhibits a tendency toward instability at frequencies around 8 GHz. The actual circuit, when measured, did not exhibit this potential instability; on the other hand, the noise figure and gain agreed quite well. The discrepancies between the results of the linear and nonlinear models emphasize the importance of accurate device modeling.

Figure 8.6 shows a wideband amplifier, specifically a distributed amplifier that covers 1–20 GHz. Figure 8.7 shows its simulated frequency-dependent gain, matching, and noise performance. This example is given here as the upper frequency for wireless systems is rapidly expanding and the technology to accommodate this is already in place. The other reason is to show that the modern CAD tools can handle such high frequencies accurately, which means one can trust them at the

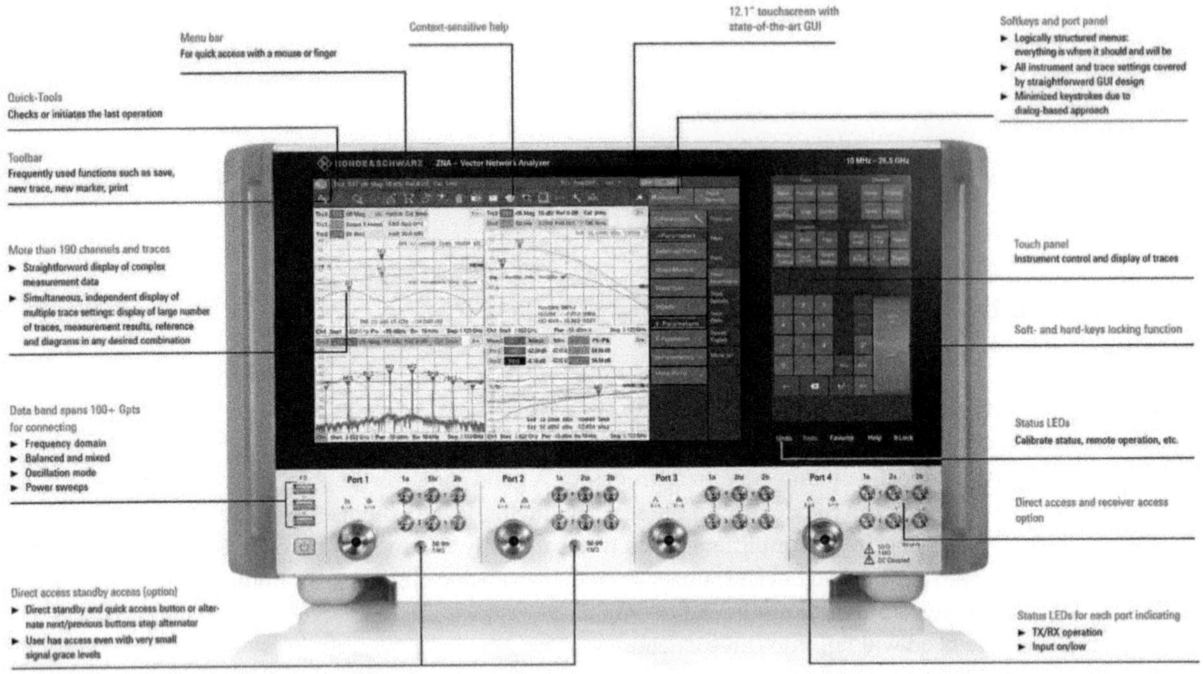

Figure 8.1 Rohde & Schwarz vector network analyzer (ZNA67) up to 67 GHz.

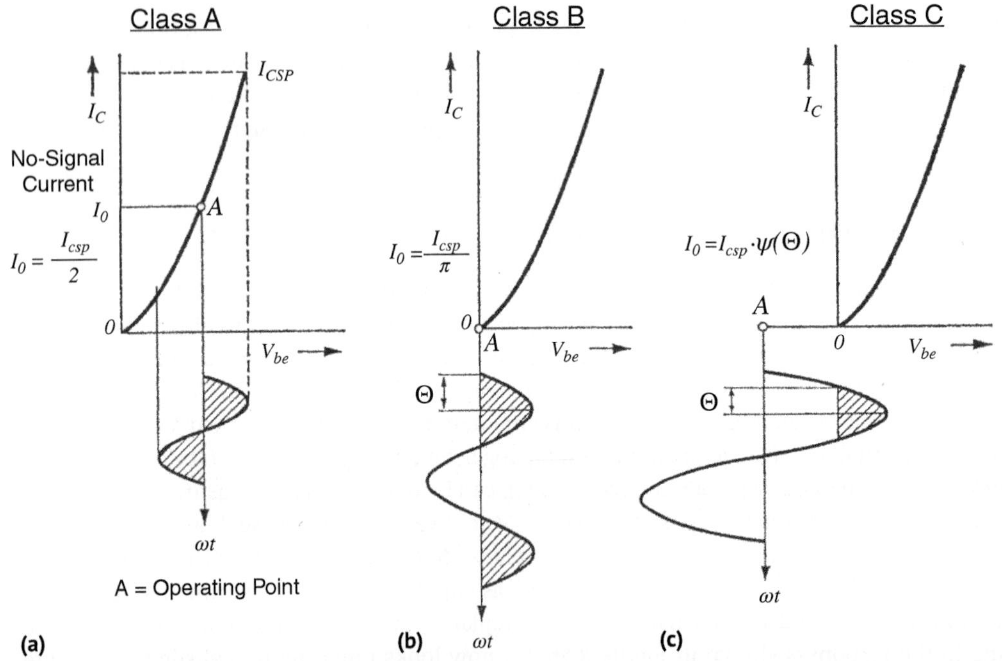

Figure 8.2 Definition of Class A, B, and C operation and resulting bias, including conduction angle (Θ, to be discussed in Section 8.23. The transfer characteristic can be either quadratic or exponential. This results in different distortions, but does not change the basic operating mode. *Source*: [1]/John Wiley & Sons.

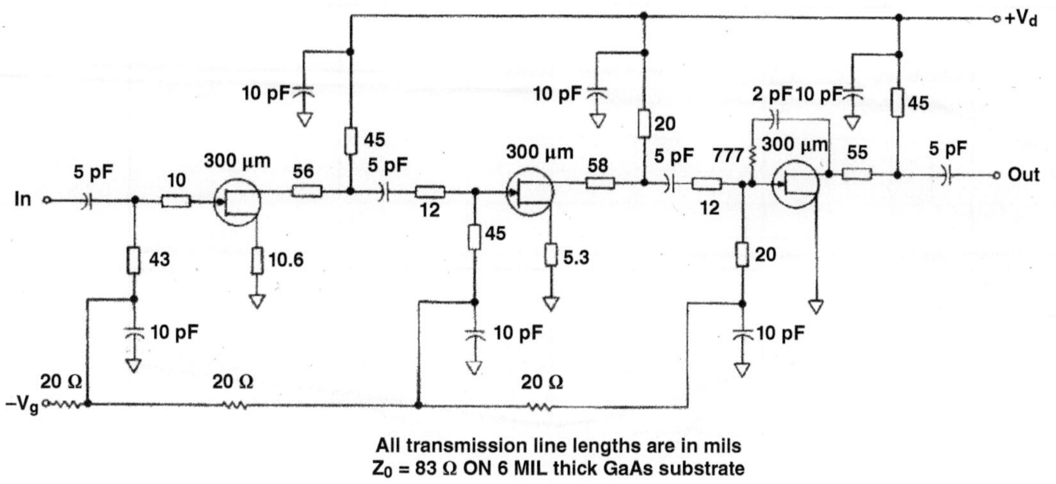

Figure 8.3 Schematic of the X-band GaAs monolithic low-noise amplifier (Texas Instruments EG8021). For reasons of linearity, all three stages operate in Class A. *Source*: [1]/John Wiley & Sons.

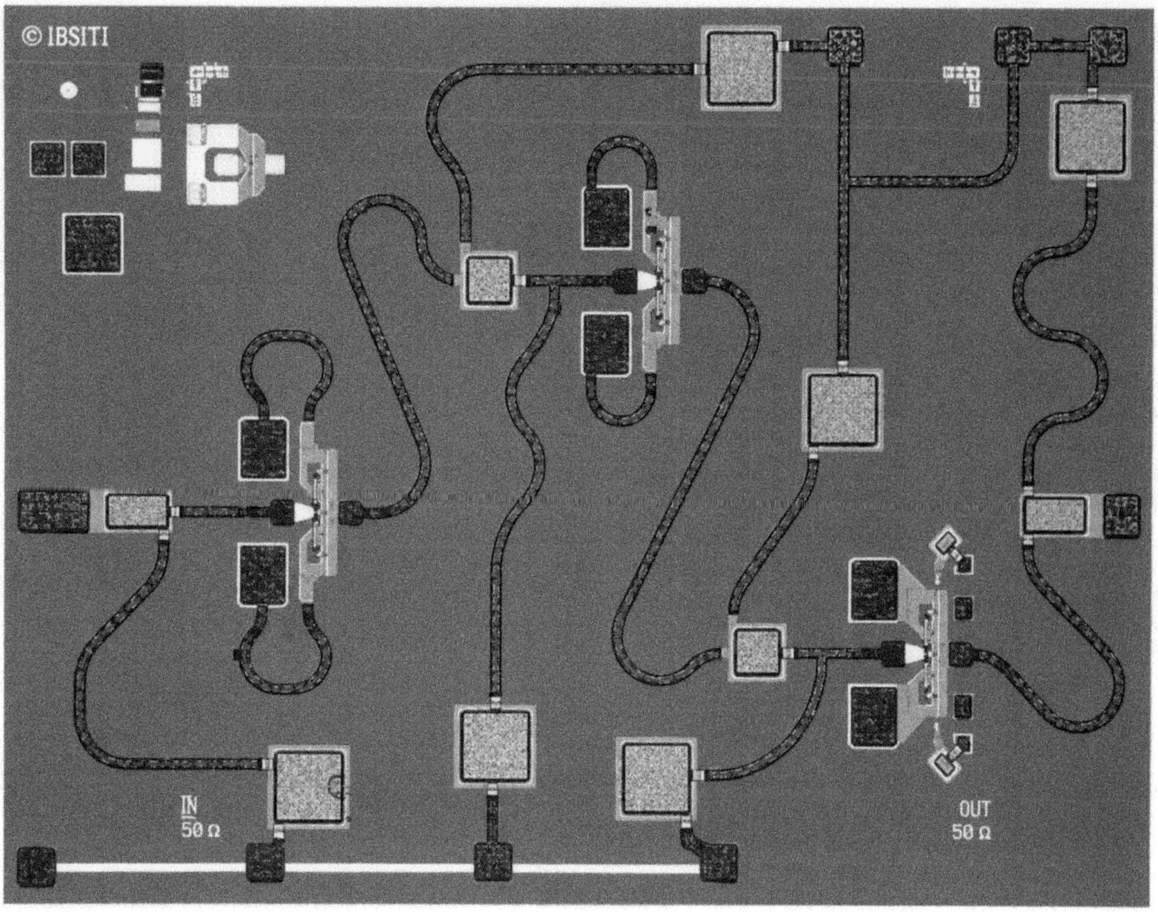

Figure 8.4 Photograph of the EG8021 monolithic amplifier chip. The area pictured is 0.09 inch by 0.12 inch in size. *Source*: [1]/John Wiley & Sons.

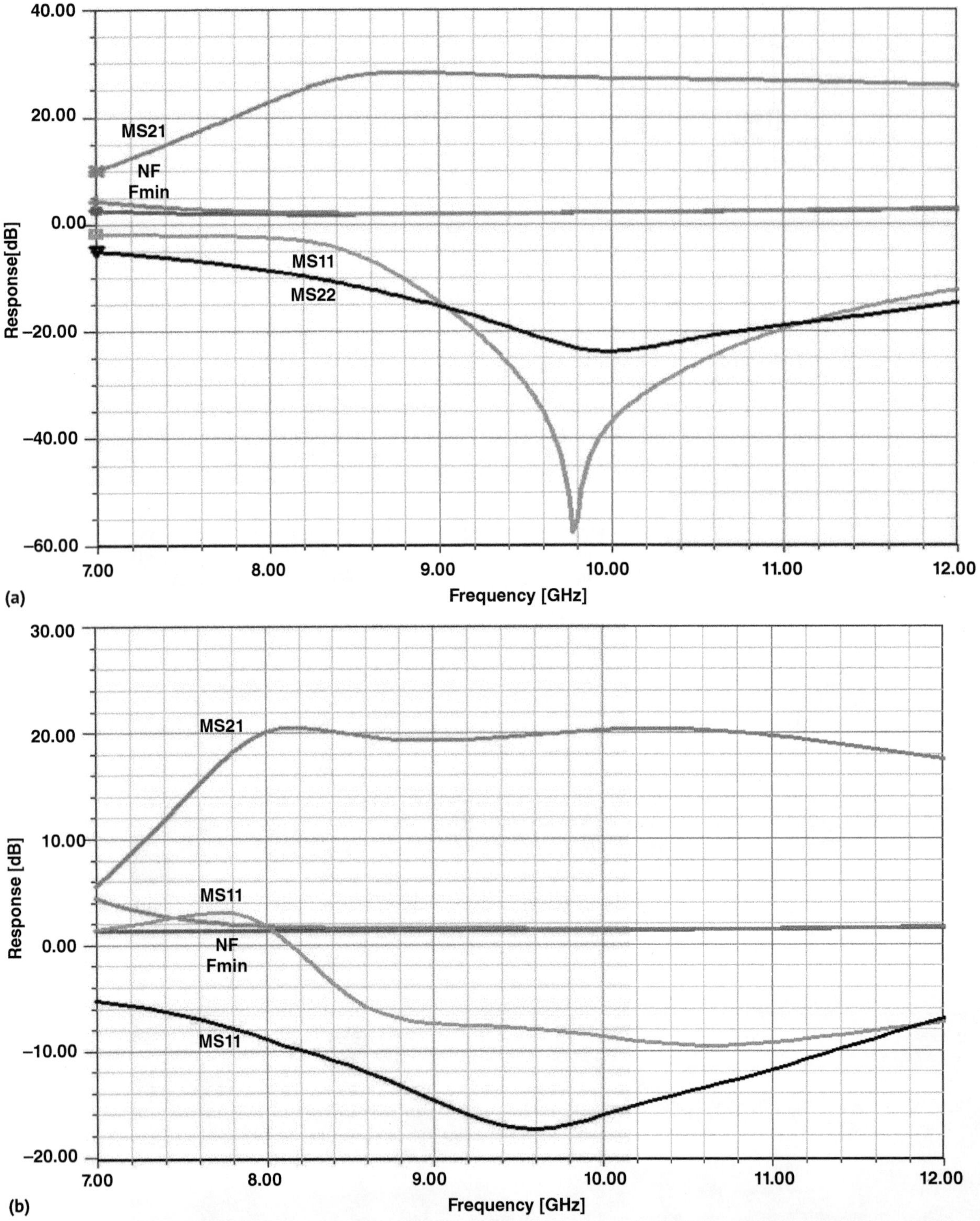

Figure 8.5 (a) Simulated F_{min}, noise figure, S_{11}, S_{21}, and S_{22} responses for the three-stage GaAsFET amplifier using the TI linear FET model. The values at 10 GHz are: F_{min}, 2.20 dB; NF, 2.21 dB; S_{11}, −36.8 dB; S_{21}, 27.2 dB; and S_{22}, −23.7 dB [1]. (b) Simulated F_{min}, noise figure, S_{11}, S_{21}, and S_{22} responses for the three-stage GaAsFET amplifier using the nonlinear FET model. The values at 10 GHz are: F_{min}, 1.53 dB; NF, 1.65 dB; S_{11}, −8.5 dB; S_{21}, 20.3 dB; and S_{22}, −15.9 dB. *Source*: [1]/John Wiley & Sons.

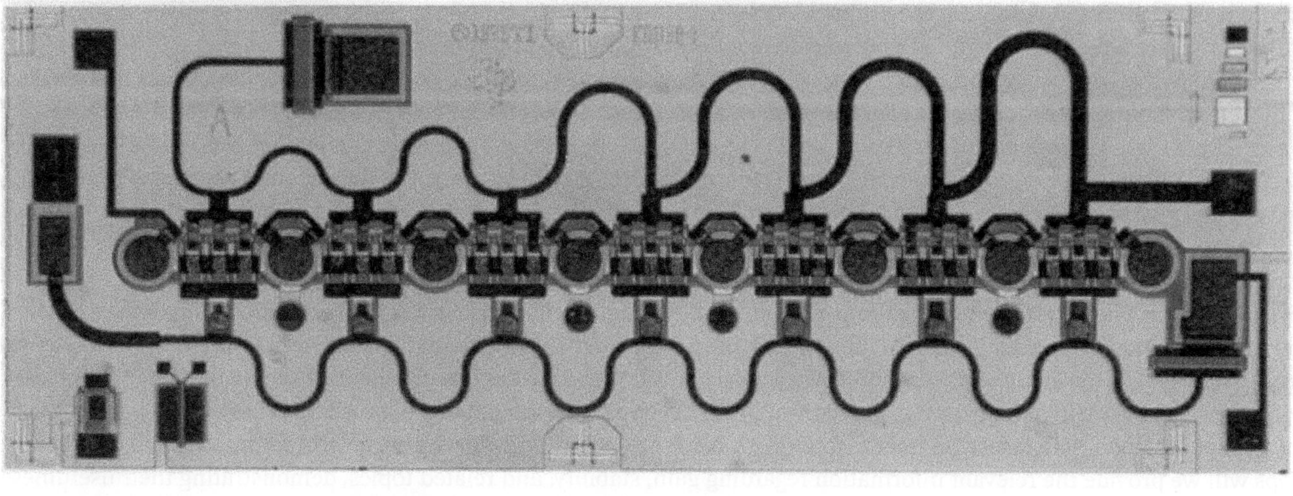

Figure 8.6 Photo of the 1–20 GHz distributed amplifier. *Source*: [1]/John Wiley & Sons.

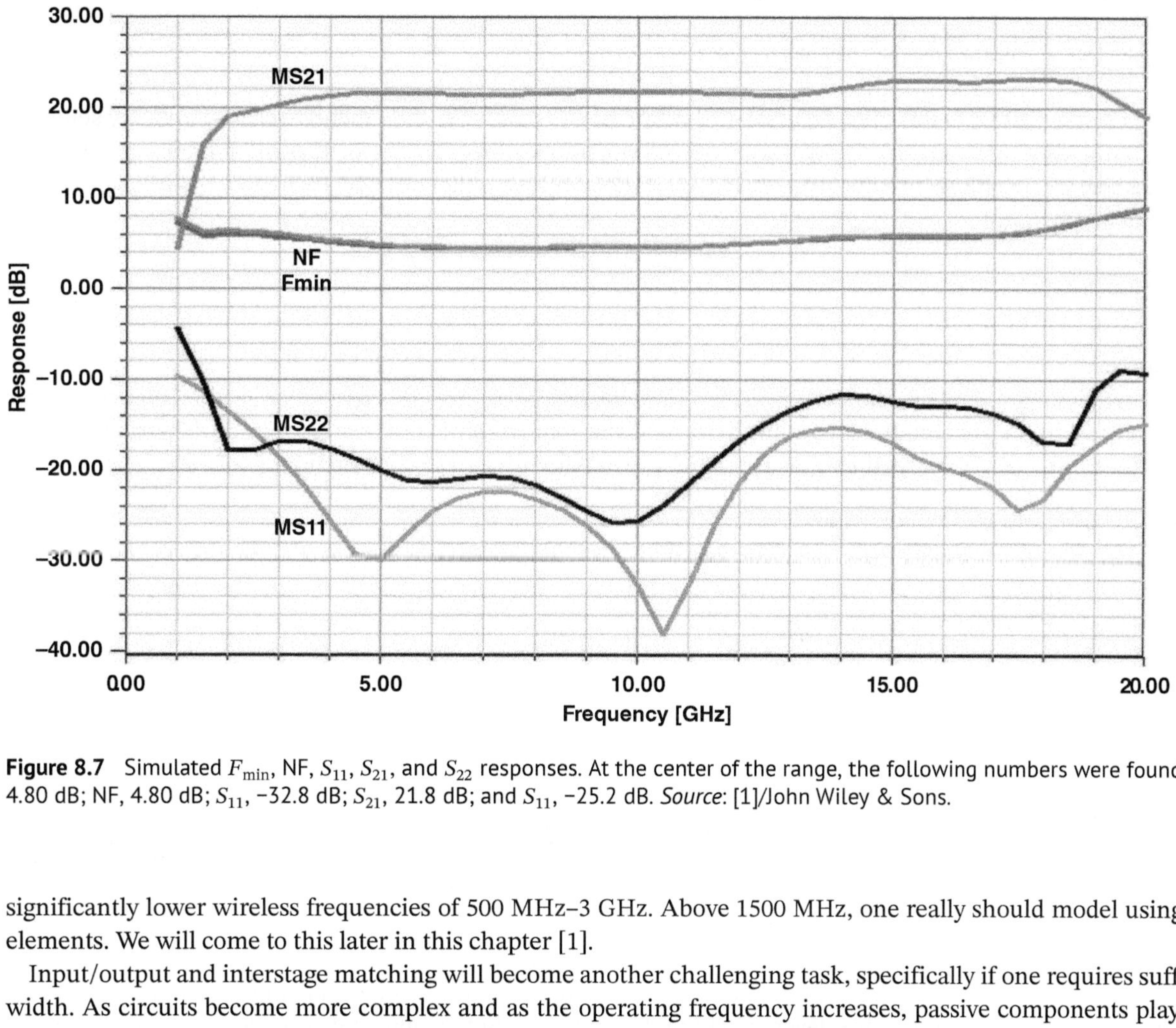

Figure 8.7 Simulated F_{min}, NF, S_{11}, S_{21}, and S_{22} responses. At the center of the range, the following numbers were found: F_{min}, 4.80 dB; NF, 4.80 dB; S_{11}, −32.8 dB; S_{21}, 21.8 dB; and S_{11}, −25.2 dB. *Source*: [1]/John Wiley & Sons.

significantly lower wireless frequencies of 500 MHz–3 GHz. Above 1500 MHz, one really should model using distributed elements. We will come to this later in this chapter [1].

Input/output and interstage matching will become another challenging task, specifically if one requires sufficient bandwidth. As circuits become more complex and as the operating frequency increases, passive components play an increasingly important role; the use of distributed elements is practically a necessity above 1500 MHz. For this reason, we begin this chapter by covering the application of transistors as amplifiers, following this with a discussion of the surrounding external circuitry, such as transmission lines, inductors, capacitors, and resistors. Monolithic ICs on silicon are a totally different ball game, and neither of the authors has sufficient expertise in this area. For this reason, we refer readers who will concentrate on MMIC design to the literature mentioned at the end of this chapter.

The goal of the applications presented here is to address, where possible, high-performance wireless stages that are less subject to constraints of space, power consumption, and cost than portable/mobile applications. This is consistent with our preface, in which we have clearly stated our goal. The designers of handsets typically cannot consider the full variety of circuits we will present here. These types of stages (low noise, high gain, and power) will frequently be used with filtering, and we will give some examples of input filters, including a tracking high-performance design.

A key topic besides the frequently mentioned gain is the whole noise issue. The following subchapter gives the reader a complete introduction into this important area, followed by a similar introduction into the gain and matching of transistors. A caution: The majority of applications published go through the exercise of using S parameters manually to design low-noise and small-signal amplifiers. While we have been tempted to repeat the information spread around in several textbooks, we feel that the engineers today will have access to CAD tools one way or another. It is highly unlikely that readers will undertake the inefficient process of going through all the design steps manually (a procedure that can take days); hopefully, engineers today will resort to a linear/nonlinear CAD tool to do so. Only for the purpose of summarizing the necessary steps will we provide the relevant information regarding gain, stability, and related topics, demonstrating their usefulness or possible misuse with a modern CAD tool—in our case, the Serenade Design Environment product from Ansoft. As this manuscript nears completion, several additional companies besides Ansoft and Hewlett–Packard have emerged, but it is dangerous to list all of them as they come and go [1].

8.1.2 Gain Calculations

In practical terms, when we talk about gain, we mean the output power relative to the input power. The industry standard now uses 50 Ω; however, at the point where we have to match high-impedance devices such as monolithic crystal filters, where the impedance jumps, things are not so obvious. The transformation equation, which allows us to transform from one impedance to another, is:

$$m = \sqrt{\frac{R_2}{R_1}} \tag{8.1}$$

This correction factor, m, is required to do the transformation from R_1 (50 Ω) to R_2 (1.2 kΩ, for example). It needs to be understood that while the power gain remains the same, the voltage gain requires this correction factor. Since the power gain is expressed as 10 log A, A being the loop gain, 20 log A is the voltage gain, both expressed in dB. The reason for 10 versus 20 is that the power gain is always the voltage squared, which accounts for the 2. As a reminder of these relationships:

Distinction: power ratios – voltage ratios $P = V^2/Z$

$$\text{ndB} = 10\log_{10}\left(\frac{P2}{P1}\right) ======\Longrightarrow \text{n} = 10\log_{10}\frac{\frac{V_2^2}{Z_2}}{\frac{V_1^2}{Z_1}}\text{dB}$$

$$Z_2 = Z_1 \rightarrow \text{ndB} = 10\log_{10}\frac{V_2^2}{V_1^2} = 10\log_{10}\left(\frac{V_2}{V_1}\right)^2 ===\Longrightarrow \text{ndB} = 20\log_{10}\left(\frac{V_2}{V_1}\right)\text{dB}$$

The gain is typically defined under the condition of real resistive source and load terminations, which means that all the available power is being supplied to the amplifier, and likewise, through proper matching at the output, all the available power is fed to the termination. This gain figure is typically measured at a single spot frequency, but since most amplifiers are not totally flat, one needs to specify the gain tolerances or gain flatness. A good number is x dB ± 0.5 dB gain variation. In the case of a narrowband amplifier, the gain is typically shaped like a tuned circuit or like a critically coupled bandpass filter frequency response, while in the case of a real wideband application the desired gain has some kind of ripple, which is due to the various compensation components in the circuit. An extreme case of a wideband amplifier is a 6- to 18-GHz single-stage amplifier which uses some very clever matching. This is shown in the following example. Figure 8.8 shows the circuit diagram, and Figure 8.9 shows its frequency-dependent performance [1].

While this is a cute approach of generating a single-stage amplifier over such a wide bandwidth, in real life it is not acceptable. The reason for this is that neither S_{11} nor S_{22} meet any useful specifications over the entire range. Typically,

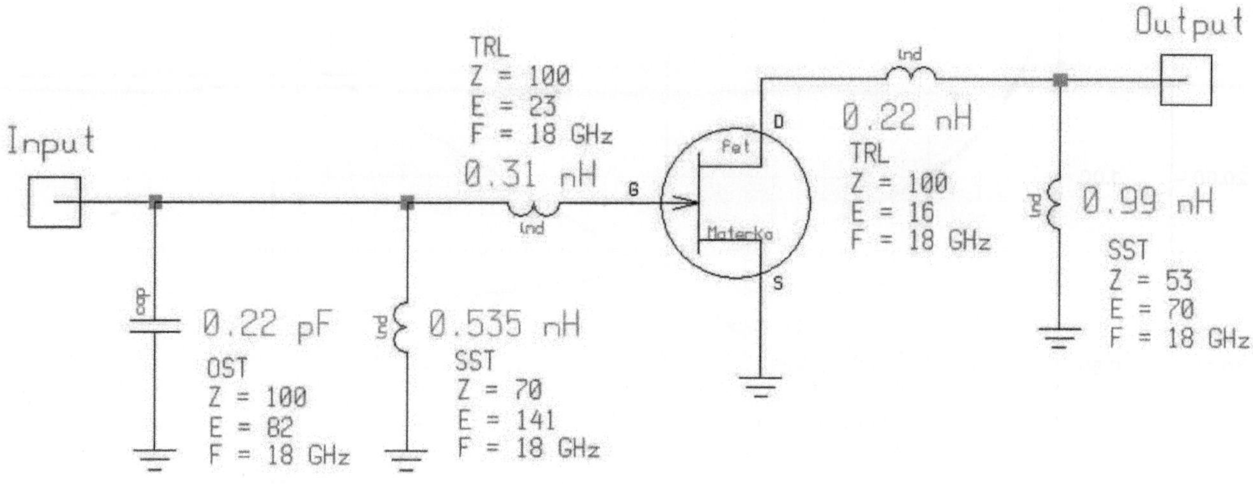

Figure 8.8 A 6–18 GHz FET amplifier. *Source*: [1]/John Wiley & Sons.

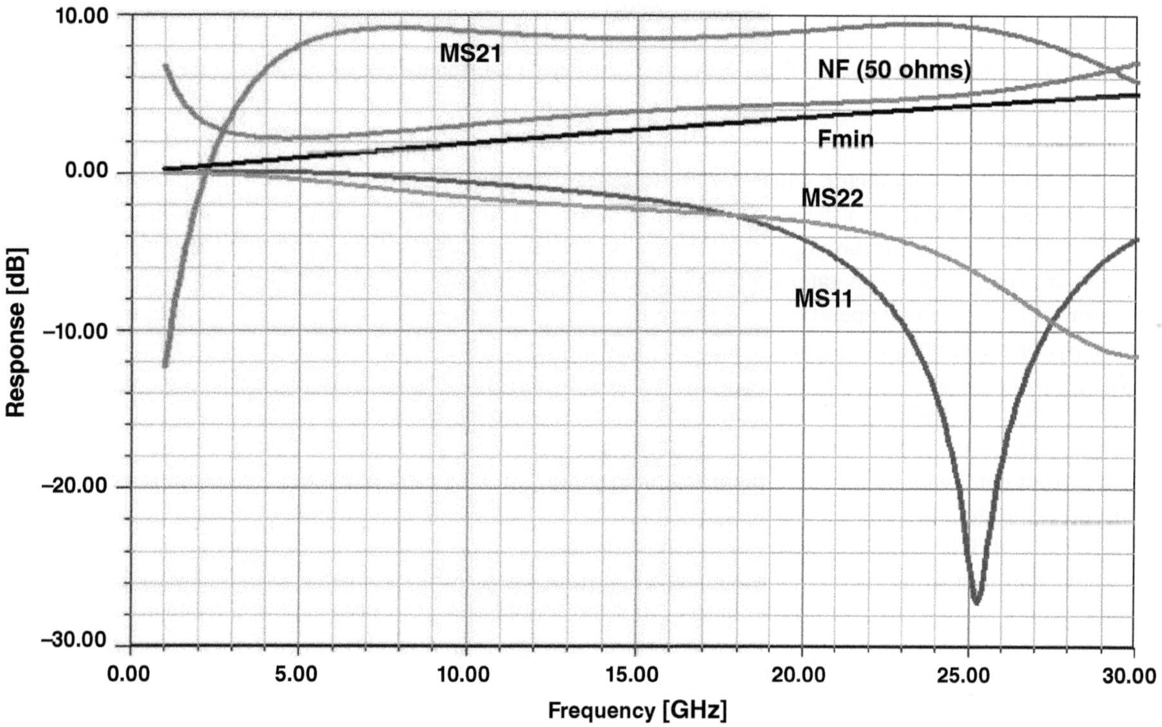

Figure 8.9 Frequency-dependent gain, matching, and noise performance of the Figure 8.8 amplifier. *Source*: [1]/John Wiley & Sons.

S_{11}/S_{22} values are only a few dB, while in reality, more than 10 dB is required. We have chosen this example to show that gain and matching, depending upon the circuit topology, are connected. We get similar results for the noise figure whereby the best achievable noise figure (F_{min}) and the actual 50 Ω noise figure is only a few dB apart; however, its frequency dependence in a wideband stage is certainly not acceptable.

Another parameter associated with gain is the stability factor K of the circuit. A value of $K > 1$ is mandatory to guarantee stability, together with the magnitude of S_{11} and S_{22} being less than 1.

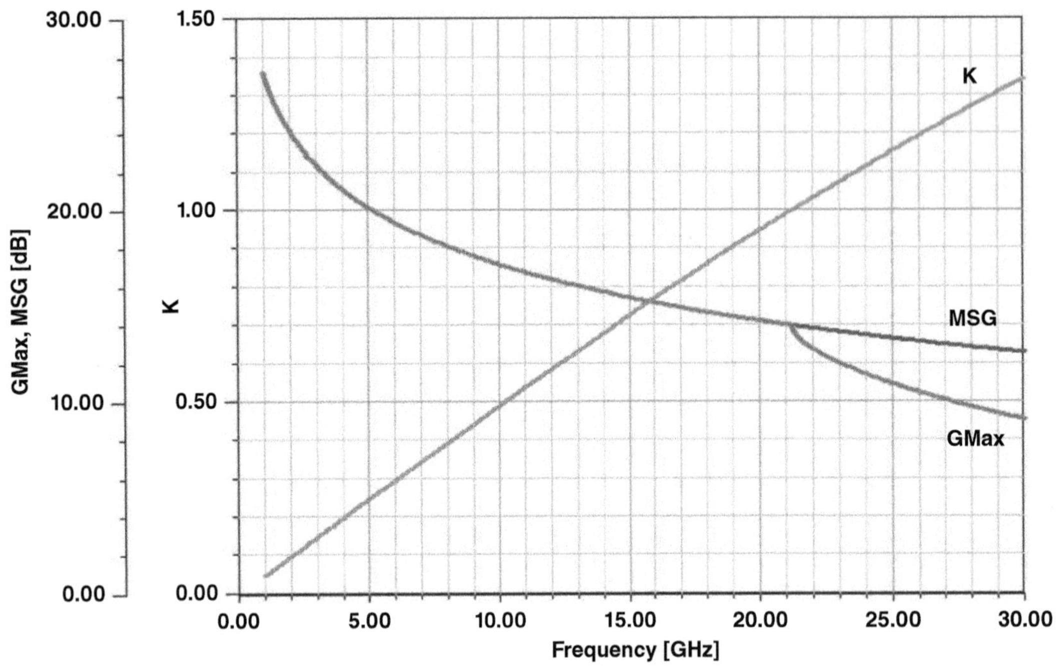

Figure 8.10 G_{max}, maximum stable gain (MSG), and K performance of the Figure 8.8 amplifier. *Source*: [1]/John Wiley & Sons.

Figure 8.10 plots both the K stability factor, which for absolute stability must be > 1, and maximum stable gain (MSG) as well as G_{max}. It turns out that MSG is a highly artificial definition of stability that implies that input and output are terminated with pure resistances and then conjugate matched. There are four ways of achieving this resistive loading, either in the emitter (as a shunt resistor between gate/drain or base/collector), or finally in parallel to the input and output to ground. Needless to say, this is not a real operating mode because it deteriorates both noise and gain. In our case, the entity G_{max} is a function of S_{12} of the circuit, not just the transistor, and depends highly on S_{12}. If we bring the system close to oscillation, then G_{max} will increase at the expense of bandwidth. Unfortunately, some of these microwave definitions of gain (see Table 8.1) have become widely used in industry without always being useful. Evaluating these equations for S_{12} is not legitimate because other quantities like K become infinite, and the entire system of equations falls apart. The power gain with input and output conjugate matched, or the transducer power gain for arbitrary source and load matching, still gives more insight into the system [3].

Definition for G_{max}: If $k > 1$, it is called G_{ma}, and if $k < 1$, it is called G_{ms}.

In the above example, we specifically stated that the input and output VSWR was totally unacceptable, and this was due to a one-stage design. Figure 8.11 shows an amplifier chip built around two transistors with standard feedback techniques applied; Figures 8.12 and 8.13 show its frequency-dependent gain, matching, stability, and noise performance. This amplifier was one of the early examples of GaAsFET wideband amplifiers, but it is interesting because while it has wild (and wide) gain swings, it exhibits a fairly constant input and output impedance or reflection coefficient. A further evaluation of the circuit shows that not only is the circuit unconditionally stable, but that the maximum stable gain has an extremely high value based on essentially minimal S_{12}, the K factor is always significantly higher than 1, and G_{max} follows, to some degree, the actual gain response of the amplifier. (Ultimately, designers gave up on simple *RCL* feedback amplifiers for such a wide bandwidth, replacing them with distributed amplifiers previously called traveling wave amplifiers.) The problem of gain and stability starts at audio-frequency amplifiers, where the motorboating phenomenon has been known for years, and the problem moves up in frequency with more modern technology, and is largely dependent not only on the device and the circuit, but also on the layout. The motorboating effect has to do with ground loops and the fact that if the printed circuit traces from the power supply connection to the output power stage present too high a resistance, then at high currents the voltage will drop to a point below useful operation, and as the volume is lowered, become stable again. In this case, the amplitude of the signal through the amplifier controls the stability of the output stage. Needless to say, this most

Table 8.1 Nine power gains.

Transducer power gain in 50 Ω system	$G_T = \|S_{21}\|^2$
Transducer power gain for arbitrary Γ_G and Γ_L	$G_T = \dfrac{(1-\|\Gamma_G\|^2)\|S_{21}\|^2(1-\|\Gamma_L\|^2)}{\|(1-S_{11}\Gamma_G)(1-S_{22}\Gamma_L)-S_{12}S_{21}\Gamma_G\Gamma_L\|^2}$
Unilateral transducer power gain	$G_{TU} = \dfrac{\|S_{21}\|^2(1-\|\Gamma_G\|^2)(1-\|\Gamma_L\|^2)}{\|1-S_{11}\Gamma_G\|^2\|1-S_{22}\Gamma_L\|^2}$
Power gain with input conjugate matched	$G = \dfrac{\|S_{21}\|^2(1-\|\Gamma_L\|^2)}{\|1-S_{22}\Gamma_L\|^2(1-\|S'_{11}\|^2)} = \dfrac{\|S_{21}\|^2}{1-\|S_{11}\|^2}$ (for $\Gamma_L = 0$)
Available power gain with output conjugate matched	$G_A = \dfrac{\|S_{21}\|^2(1-\|\Gamma_G\|^2)}{\|1-S_{11}\Gamma_G\|^2(1-\|S'_{22}\|^2)} = \dfrac{\|S_{21}\|^2}{1-\|S_{22}\|^2}$ (for $\Gamma_G = 0$)
Maximum available power gain	$G_{ma} = \left\|\dfrac{S_{21}}{S_{12}}\right\|\left(k-\sqrt{k^2-1}\right)$
Maximum unilateral transducer power gain	$G_{TUmax} = \dfrac{\|S_{21}\|^2}{(1-\|S_{11}\|^2)(1-\|S_{22}\|^2)}$
Maximum stable power gain	$G_{ms} = \dfrac{\|S_{21}\|}{\|S_{12}\|}$
Unilateral power gain	$U = \dfrac{1/2\|S_{21}/S_{12}-1\|^2}{k\|S_{21}/S_{12}\| - \mathrm{Re}(S_{21}/S_{12})}$

Source: [1]/John Wiley & Sons.

unpleasant effect applies to high-power stages for all output transistors, where we use Class B or C operation, meaning that the dc operating current as a function of output does not remain constant.

In terms of gain, there is also an important gain characteristic called *differential gain* and *differential phase*. These expressions were first used in television circuits, where linearity in both areas is an absolute must. The amplitude of the signal determines the black and white contrast, and the phase determines the color. As an example, transmitting an image of one of our presidents playing golf in California to the New York area is done by many microwave hops. Any multipath reception, such as that caused by reflection from airplanes, may have little effect on the signal amplitude, but the resulting phase shift can transform a green lawn to a red lawn if a total shift of 180° occurs. Given a particular operating point, if the 1 dB compression point of multiple amplifiers in the chain is approached or exceeded, their differential gain, which is defined by the IEEE as "the difference between (a) the ratio of the output of a small high-frequency sine-wave signal at two different levels of a low-frequency signal on which it is superimposed, and (b) unity," may increase. This mode of saturation charges all the capacitances of the amplifier transistors, causing huge phase shifts that result in the known problem of changing colors. The same applies, of course, in test equipment amplifiers, in which such changes cannot be tolerated. A somewhat similar, but not quite the same, problem is spectral regrowth (evaluated in terms of adjacent channel power ratio [ACPR]), which is also due to nonlinearities. Needless to say, any overdrive condition also changes an amplifier's input and output SWR [1].

8.1.3 Noise Figure (NF)

8.1.3.1 Introduction

In this section, we look at the best way to describe noise in active devices (black boxes). Even when a 2-port is linear, the output waveform will differ from the input because of the failure to transmit all spectral components with equal gain (or attenuation) and delay. By careful design of the 2-port, or by limitation of the bandwidth of the input waveform, such distortions can largely be avoided. However, noise generated within the 2-port can still change the waveform of the output signal. In a linear passive 2-port, noise arises only from the losses in the 2-port; thermodynamic considerations indicate that

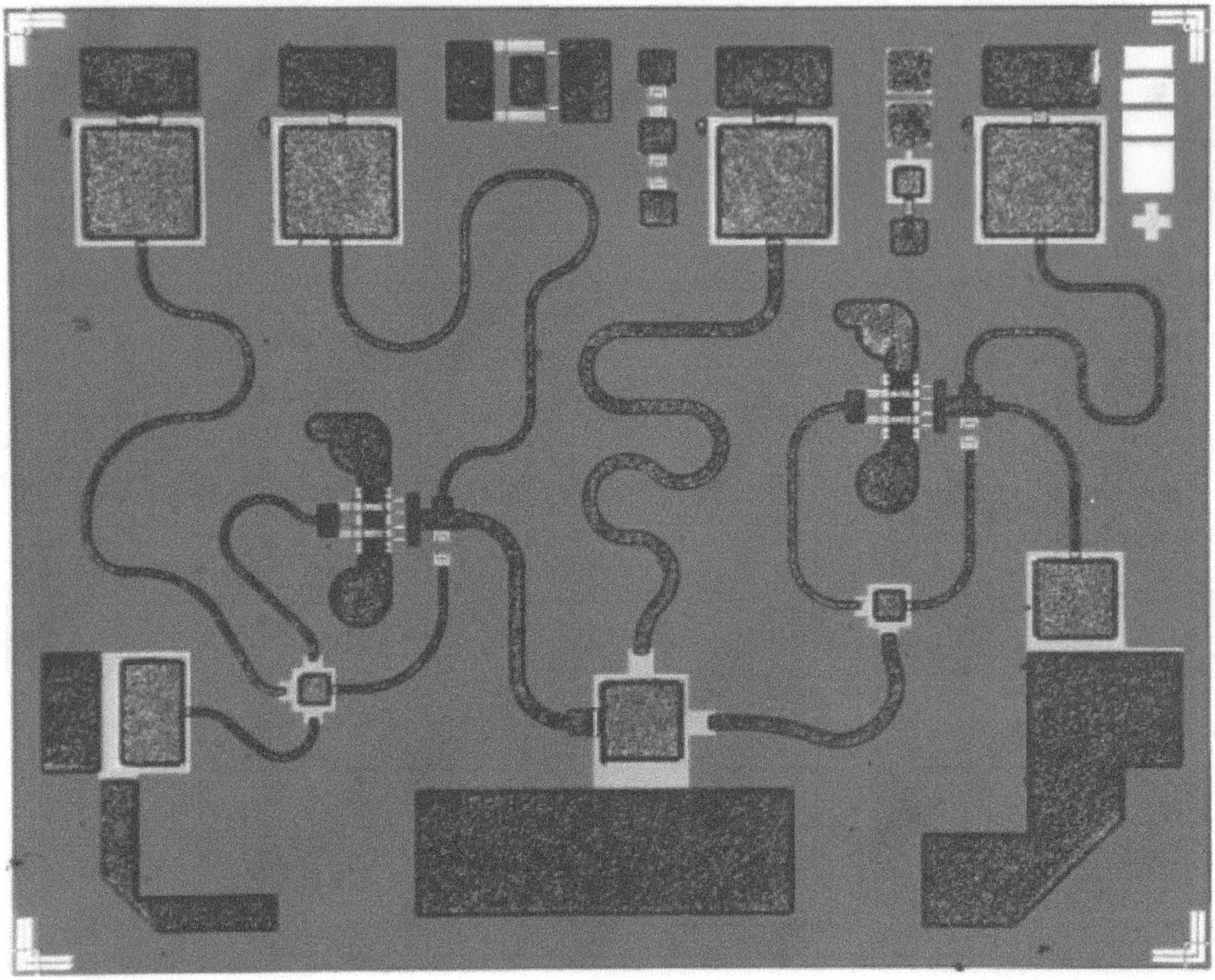

Figure 8.11 A two-stage, wideband feedback amplifier. *Source*: [1]/John Wiley & Sons.

such losses result in the random changes that we call noise. When the 2-port contains active devices, such as transistors, there are other noise mechanisms that are present. A very important consideration in a system is the amount of noise that it adds to the transmitted signal. This is often judged by the ratio of the output signal power to the output noise power (S/N). The ratio of signal plus noise power to noise power [$(S+N)/N$] is generally easier to measure, and approaches S/N when the signal is large.

In the evaluation of a 2-port, it is important to know the amount of noise added to a signal passing through it. An important parameter for expressing this characteristic is the noise factor. The signal energy coming from a generator or antenna is amplified or attenuated in passing from the input to the output of a 2-port, as is the noise that accompanies the input signal energy. A system generally includes a cascade of 2-port networks that constitute one overall 2-port that amplifies the signal to a high-enough power level for its intended use. The *noise factor* of a system is defined as the ratio of the signal-to-noise ratios available at the input and output:

$$F = \frac{(S/N)_{\text{input}}}{(S/N)_{\text{output}}} \geq 1 \tag{8.2}$$

The noise figure (or factor) of a receiver is an easily measured quantity that describes the signal-to-noise ratio reduction of that receiver.

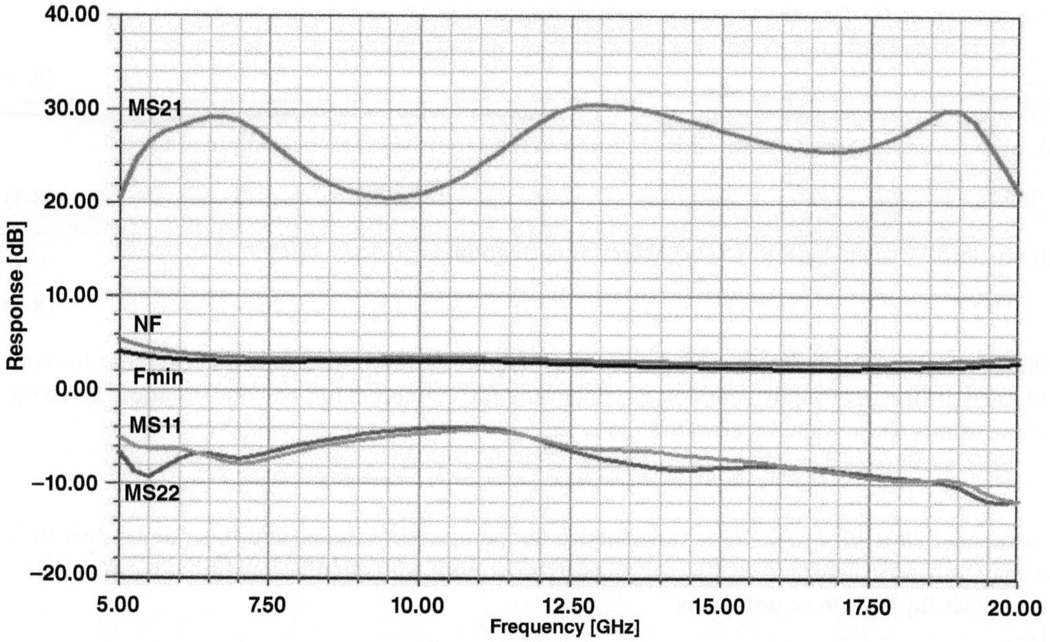

Figure 8.12 Frequency-dependent gain, matching, and noise performance for the Figure 8.11 amplifier. *Source*: [1]/John Wiley & Sons.

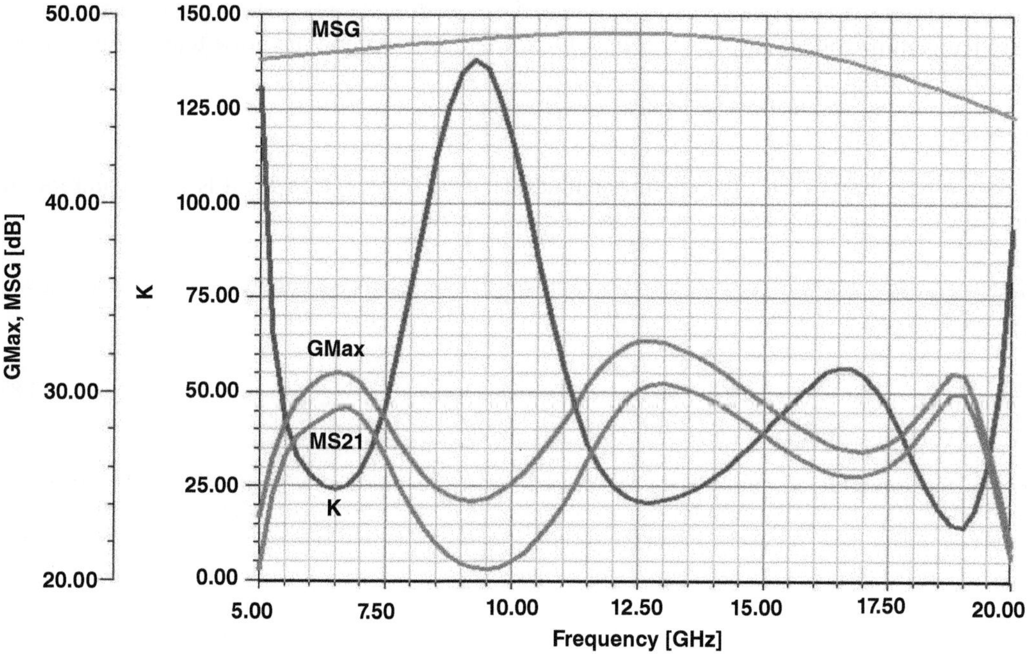

Figure 8.13 Frequency-dependent $G_{\max}$, K, maximum stable gain and gain performance for the Figure 8.11 amplifier. *Source*: [1]/John Wiley & Sons.

When this ratio of powers is converted to decibels, it is generally referred to as the *noise figure* rather than noise factor. Various conventions are used to distinguish the symbols used for noise factor and noise figure. Here we use F to represent the noise factor and NF to represent the noise figure, although the terms are usually used interchangeably.

For an amplifier with the power gain G, the noise factor can be rearranged as

$$F = \frac{S_i/N_i}{GS_i/G(N_i + N_a)} \tag{8.3}$$

when N_a is the additional noise power added by the amplifier referred to the input. This can be computed to be

$$F = 1 + N_a/N_i \tag{8.4}$$

The noise factor is often replaced by the noise figure (NF), which is defined in decibels as

$$\text{NF} = 10 \log_{10} F \tag{8.5}$$

In applications like satellite receivers, the noise factor becomes such a small number that it is inconvenient to work with. Many people have adopted the use of an equivalent noise temperature for a circuit to remedy this situation. Since the thermal noise power available from a resistor at temperature T_e is

$$N = kT_eB \tag{8.6}$$

where k is Boltzmann's constant (1.38×10^{-23} J/K), T_e is the effective temperature in kelvins, and B is the bandwidth in hertz. The equation above may be used to associate an effective noise temperature with circuits containing more than just thermal noise sources. This allows Eq. (8.4) to be written as

$$F = 1 + \frac{kT_eB}{kT_0B} = 1 + \frac{T_e}{T_0} \tag{8.7}$$

where T_e is the effective noise temperature of the circuit and T_0 is the temperature of the generator resistor in kelvins. The noise temperature T_e now characterizes our circuit noise contribution and can be directly related to the noise factor.

Assuming a reference noise temperature of 290 K ($-273 + 290 = 17$ °C), let us determine the noise temperature of the system with a noise factor of 2.6 (4.15 dB):

$$T_e = (2.6 - 1)(290) = 464 \text{ K}$$

This temperature T_e should not be confused with the environmental operating temperature T_0. It is quite common to operate low-noise amplifiers with T_e below 100 K at an ambient temperature of 290 K [1].

Signal-to-noise Ratio: Let us consider the signal-to-noise ratio of power delivered from a generator to a load as shown in Figure 8.14. The signal power delivered to the input is given by

$$S_{in} = P_{in} = \frac{E_g^2 \,\text{Re}(Z_{in})}{\left|Z_g + Z_{in}\right|^2} \tag{8.8}$$

where E_g is the rms voltage of the input signal supplied to the system, and the noise power supplied to the input is expressed by

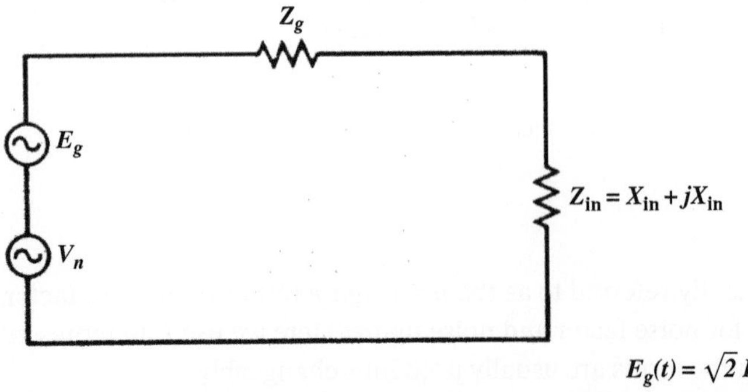

Figure 8.14 Combination of signal and noise voltages is supplied to a complex termination. *Source:* [1]/John Wiley & Sons.

$$N_{\text{in}} = \frac{\overline{v_n^2}\,\text{Re}(Z_{\text{in}})}{\left|Z_g + Z_{\text{in}}\right|^2} \tag{8.9}$$

where the noise power at the input is provided by the noise energy of the real part of Z_g. The input impedance Z of the system, in the form $Z = R_{\text{in}} + jX_{\text{in}}$ is assumed to be complex.

The Johnson noise of a resistor [here $\text{Re}(Z_g)$], given by the mean-square voltage

$$\overline{v_n^2} = 4kTRB \tag{8.10}$$

where k (Boltzmann's constant) = 1.38×10^{-23} J/K, T is the absolute temperature of the resistor, and B is the bandwidth, is sufficiently small that the resistive component of impedance does not change. The available signal power from the generator has a lower limit, even if the signal is attenuated by the highest possible attenuation. The generator resistor acts as a Johnson noise generator, with its power being

$$P_A = \frac{4kTRB}{4R} = kTB \tag{8.11}$$

where k is Boltzmann's constant, T is the absolute temperature, and B is the bandwidth. This power is the maximum available output power.

For an ambient temperature of 290 K, $kT = 4 \times 10^{-21}$ W/Hz. This expression is also given as $kT = -204$ dBW/Hz = -174 dBm/Hz = -114 dBm/MHz. We can combine Eqs. (8.8)–(8.10) to obtain

$$\left(\frac{S}{N}\right)_{\text{in}} = \frac{E_g^2}{4kT\,\text{Re}(Z_g)B} \tag{8.12}$$

This is the value of S/N contributed by the generator, which does not include the noise generated by the load, in this case, $\text{Re}(Z_{\text{in}})$, which would need to be included in the measurement of the total S/N across the input impedance.

A critical parameter is the noise bandwidth, B_n, which is defined as the equivalent bandwidth, as shown in Figure 8.15. For reasons of group delay correction, most practical filters have round rather than sharp corners. The noise figure measurements shown later can be used to determine the "integrated" bandwidth, which is B_n.

An active system, such as a combination of amplifiers and mixers, will add noise to the input signals, and the noise factor that describes this is defined as the S/N ratio at the input to the S/N ratio at the output, which is always greater than unity [4]. In practice, a certain minimum signal-to-noise ratio is required for operation. For example, in a communication system, a minimum is required for intelligible transmission, either voice or data. For high-performance TV reception, to provide a noise-free picture to the eye, a typical requirement is for a 60 dB S/N. In the case of a TV system, a large dynamic range is required, as well as a very large bandwidth to reproduce all colors truthfully and all shades from high-intensity white to black. Good systems will have a bandwidth of 8 MHz or more [1].

Noise Figure Measurements: Some of the noise equations are based on mathematical models and physics. To understand some of these expressions, it is useful to look at a practical case of a system with amplifiers that has to be evaluated.

Let us look at Figure 8.16, which consists of a signal generator, the system or device under test (DUT), and a selective receiver with a built-in root-mean-square (rms) voltmeter to determine the signal and the noise voltage. It is necessary that the system have enough gain so that the noise voltage supplied by the generator will be indicated [1, 5].

If we assume that our selective receiver is a video noise meter calibrated in rms voltage levels, we can perform two measurements. With an input termination connected to the TV system (typically, 75 Ω for cable TV, 50 Ω for satellite TV), the noise receiver/meter will read a value for proper termination that can easily be calculated. Since one-half of the mean-square noise voltage appears across the input,

$$v_{\text{in}} = \frac{v_n}{2} = \frac{\sqrt{4kTRB}}{2} \tag{8.13}$$

With $B = 10$ MHz, $T = 290$ K (T is always expressed in absolute temperature; $T_0 = -273$ °C), and $k = 1.38 \times 10^{-23}$ J/K, then for $R = 75$ Ω,

$$v_{\text{in}} = \frac{v_n}{2} = 1.73\ \mu\text{V} \tag{8.14}$$

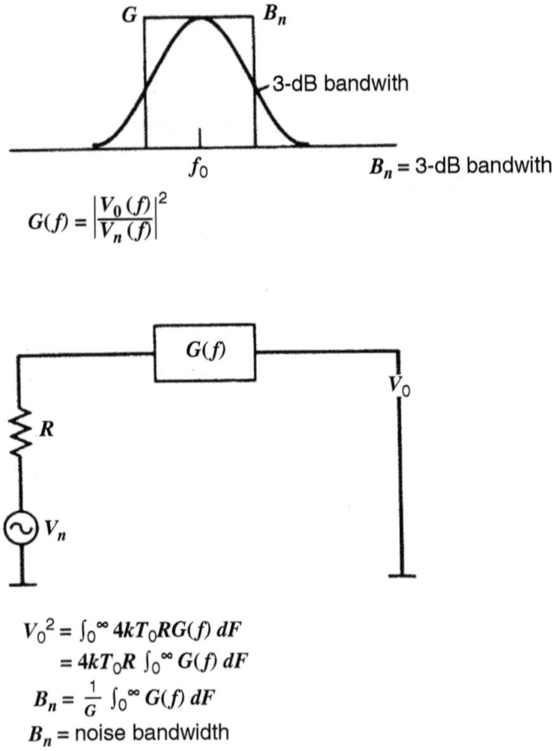

Figure 8.15 Graphical and mathematical explanation of the noise bandwidth from a comparison of the Gaussian-shaped bandwidth to the rectangular filter response. *Source*: [1]/John Wiley & Sons.

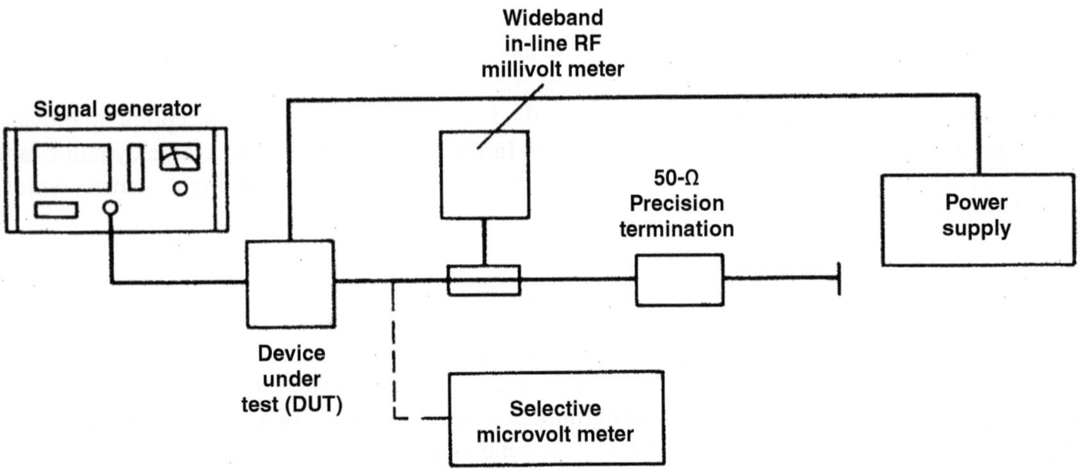

Figure 8.16 Test setup to measure signal-to-noise ratio. *Source*: [1]/John Wiley & Sons.

where the rms noise voltage has been referred to the input port. We can verify this with our first measurement.

Now we increase the input voltage of the signal generator to a value that indicates a 60 dB S/N ratio at the output port. This should be about

$$E_g = \frac{v_n}{2}\sqrt{F} \times 1000 = 1.73\sqrt{F}\ \text{mV}$$

where F is the noise factor of the receiver. For a receiver noise factor of 10, we would obtain $E_g = 5.48$ mV (rms value). If the noise energy equivalent to a noise factor of F is assumed, we need $\sqrt{F}$ times more voltage. For a 60 dB ratio, this means that $E_g = 1000 \times (v_n/2) \times \sqrt{F}$.

As they are done here, over a power range of 60 dB, the measurements can be performed over such a wide range only if special equipment is available. In cases where the internal detector of a piece of communications equipment is used, the signal-to-noise ratio measurements are performed over much smaller power ranges.

Let us assume that for the above-mentioned case ($F = 10$), we find a S/N ratio of 10 dB at the output for an input signal of 5.47 μV. By rewriting Eq. (8.7) as

$$E_g = \frac{v_n}{2}\sqrt{F} = \sqrt{kTRBF} \tag{8.15}$$

with F being the noise factor, we can solve for F with

$$F = \frac{P_s}{P_n} = \frac{E_g^2/R}{kTB} \tag{8.16}$$

While the input power from the thermal energy of the input termination resistor was $kTB = 4 \times 10^{-4}$ W, the input power required for the 10 dB S/N ratio was

$$P_s = \frac{(5.47 \times 10^{-6})^2}{75} = 3.98 \times 10^{-13}\ \text{W} \tag{8.17}$$

The noise factor is defined as the ratio P_s/P_n:

$$F = \frac{3.98 \times 10^{-13}}{4 \times 10^{-14}} = 10 \tag{8.18}$$

which is the proof.

This method is used more frequently at the 3 dB point, or double the input power if the dynamic range of the detector is small or only a linear indicator is available. Because of hum and other pickup, this is not an easy measurement. Using a signal generator is very expensive because, in a laboratory or production environment, a wide frequency range requires several generators.

Another method is the use of a wideband noise generator. Modern gas discharge diodes or avalanche diodes are available that provide essentially white noise energy over a large frequency range. These microwave diodes typically have an output of 30 dB above kT when switched on and kT when switched off. To provide good matching at microwave frequencies, a 15 dB attenuator is cascaded. This means that the noise power of the source in the ON condition is about 15 dB above kT.

In the early 1960s, low-cost noise figure test equipment was built around vacuum diodes whose operating range was limited to 1200 MHz due to the resonance effects of the structure. Today, the automatic noise gain analyzer offered currently by Hewlett–Packard and Eaton/AIL uses calibrated solid-state noise sources up to 26.5 GHz. It appears that the upper frequency limit has to do with matching, and the lower-frequency limit with $1/f$ noise [1].

Noisy 2-port Description: Based on the convention by Rothe and Dalke [6], any linear 2-port can be in the general form shown in Figure 8.17a. This general case of a noisy 2-port can be redrawn showing noise sources at the input and at the output. Figure 8.17b shows this in admittance form, and Figure 8.17c in impedance form. The internal noise sources are assumed to produce very small currents and voltages, and we assume that linear 2-port equations are valid. The internal noise contributions have been expressed by using external noise sources:

$$\begin{aligned} I_1 &= y_{11}V_1 + y_{12}V_2 + I_{K1} \\ I_2 &= y_{21}V_1 + y_{22}V_2 + I_{K2} \end{aligned} \tag{8.19}$$

$$\begin{aligned} V_1 &= z_{11}I_1 + z_{12}I_2 + V_{L1} \\ V_2 &= z_{21}I_1 + z_{22}I_2 + V_{L2} \end{aligned} \tag{8.20}$$

where the external noise sources are I_{K1}, I_{K2}, V_{L1}, and V_{L2}.

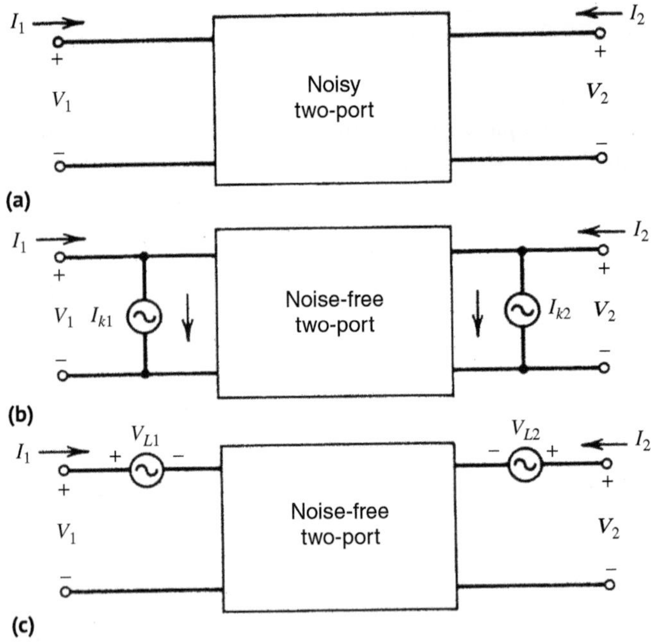

Figure 8.17 Noise linear 2-ports: (a) general form, (b) admittance form, and (c) impedance form Rohde and Rudolph. *Source*: [1]/John Wiley & Sons.

Since we want to describe our noisy circuit in terms of the noise figure, the *ABCD*-matrix description will be more convenient since it refers both noise sources to the input of the 2-port [7]. This representation is given below (note the change in direction of I_2):

$$\begin{aligned} V_1 &= AV_2 + BI_2 + V_A \\ I_1 &= CV_2 + DI_2 + I_A \end{aligned} \tag{8.21}$$

where V_A and I_A are the external noise sources.

It is important to remember that all of these matrix representations are interrelated. For example, the noise sources for the *ABCD*-matrix description can be obtained from the *z*-matrix representation shown in Eq. (8.20). This transformation is

$$V_A = -\frac{I_{K2}}{y_{21}} = V_{L1} - \frac{V_{L2}z_{11}}{z_{21}} \tag{8.22}$$

$$I_A = I_{K1} - \frac{I_{K2}y_{11}}{y_{21}} = -\frac{V_{L2}}{z_{21}} \tag{8.23}$$

The *ABCD* representation is particularly useful based on the fact that it allows us to define a noise temperature for the 2-port referenced to the input. The 2-port itself (shown in Figure 8.18) is assumed to be noise-free.

In the past, *z* and *y* parameters have been used, but in microwave applications, it has become common to use *S*-parameter definitions. This is shown in Figure 8.19. The previous equations can be rewritten in their new form using *S* parameters:

$$\begin{bmatrix} b_1 \\ b_2 \end{bmatrix} = \begin{bmatrix} S_{11} & S_{12} \\ S_{21} & S_{22} \end{bmatrix} \begin{bmatrix} a_1 \\ a_2 \end{bmatrix} + \begin{bmatrix} b_{n1} \\ b_{n2} \end{bmatrix} \tag{8.24}$$

There are different physical origins for the various sources of noise. Typically, thermal noise is generated by resistances and losses in the circuit or transistor, whereas shot noise is generated by current flowing through semiconductor junctions and vacuum tubes. Since these many sources of noise are represented by only two noise sources at the device input, the two equivalent noise sources are often a complicated combination of the circuit's internal noise sources. Often, some fraction

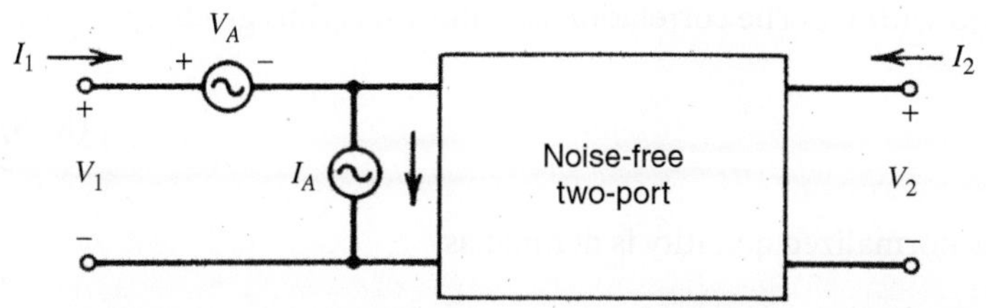

Figure 8.18 Chain-matrix form of linear noisy 2-ports. *Source*: [1]/John Wiley & Sons.

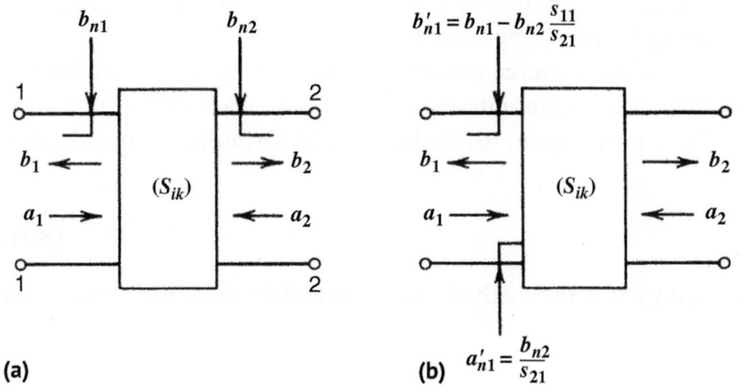

Figure 8.19 *S*-parameter form of linear noisy 2-ports. *Source*: [1]/John Wiley & Sons.

of V_A and I_A is related to the same noise source. This means that V_A and I_A are not independent in general. Before we can use V_A and I_A to calculate the noise figure of the 2-port, we must calculate the correlation between the V_A and I_A shown in Figure 8.18.

The noise source V_A represents all the device noise referred to the input when the generator impedance is zero; that is, when the input is short-circuited. The noise source I_A represents all the device noise referred to the input when the generator admittance is zero; that is, the input is open-circuited.

The correlation of these two noise sources considerably complicates the analysis. By defining a correlation admittance, we can simplify the mathematics and get some physical intuition for the relationship between noise figure and generator admittance. Since some fraction of I_A will be correlated with V_A, we split I_A into correlated and uncorrelated parts as follows:

$$I_A = I_n + I_u \tag{8.25}$$

I_u is a part of I_A uncorrelated with V_A. Since I_n is correlated with V_A, we can say that I_n is proportional to V_A and the constant of proportionality is the correlation admittance.

$$I_n = Y_{cor} V_A \tag{8.26}$$

This leads us to

$$I_A = Y_{cor} V_A + I_u \tag{8.27}$$

The following derivation of the noise figure will use the correlation admittance. Y_{cor} is not a physical component located somewhere in the circuit. Y_{cor} is a complex number derived by correlating the random variables I_A and V_A. To calculate Y_{cor}, we multiply each side of Eq. (8.27) by V_A^* and average the result. This gives

$$\overline{V_A^* I_A} = Y_{cor} \overline{V_A^2} \tag{8.28}$$

where the I_u term averaged to zero since it was uncorrelated with V_A. The correlation admittance is thus given by

$$Y_{cor} = \frac{\overline{V_A^* I_A}}{\overline{V_A^2}} \tag{8.29}$$

Often, people use the term "correlation coefficient." This normalized quantity is defined as

$$c = \frac{\overline{V_A^* I_A}}{\sqrt{\overline{V_A^2}\,\overline{I_A^2}}} = Y_{cor}\sqrt{\frac{\overline{V_A^2}}{\overline{I_A^2}}} \tag{8.30}$$

Note that the dual of this admittance description is the impedance description. Thus, the impedance representation has the same equations as above with Y replaced by Z, I replaced by V, and V replaced by I.

V_A and I_A represent internal noise sources in the form of a voltage source acting in series with the input voltage and a source of current flowing in parallel with the input current. This representation conveniently leads to the four noise parameters needed to describe the noise performance of the 2-port. Again, using the Nyquist formula, the open-circuit voltage of a resistor at the temperature T is

$$\overline{V_A^2} = 4kTRB \tag{8.31}$$

This voltage is a mean-square fluctuation (or spectral density). It is the method used to calculate the noise identity. We could also define a noise equivalent resistance for a noise voltage as

$$R_n = \frac{\overline{V_A^2}}{4kTB} \tag{8.32}$$

The resistor R_n is not a physical resistor but can be used to simulate different portions of the noise equivalent circuit.

In a similar manner, a mean-square current fluctuation can be represented in terms of an equivalent noise conductance G_n, which is defined by

$$G_n = \frac{\overline{I_A^2}}{4kTB} \tag{8.33}$$

and

$$G_u = \frac{\overline{I_u^2}}{4kTB} \tag{8.34}$$

for the case of the uncorrelated noise component. The input generator to the 2-port has a similar contribution.

$$G_G = \frac{I_G^2}{4kTB} \tag{8.35}$$

with Y_G being the generator admittance and G_G being the real part. With the definition of F above, we can write

$$F = 1 + \left|\frac{I_A + Y_G V_A}{I_G}\right|^2 \tag{8.36}$$

The use of the voltage V_A and the current I_A has allowed us to combine all the effects of the internal noise sources.

We can use the previously defined Eq. (8.29) correlation admittance, $Y_{cor} = G_{cor} + jB_{cor}$, to simplify Eq. (8.36). First, we determine the total noise current:

$$\overline{I_A^2} = 4kT\left(|Y_{cor}|^2 R_n + G_u\right)B \tag{8.37}$$

where R_n and G_u are defined in Eqs. (8.32) and (8.34). The noise factor can now be determined.

$$F = 1 + \frac{G_u}{G_g} + \frac{R_n}{G_g}\left[(G_G + G_{cor})^2 + (B_G + B_{cor})^2\right] \tag{8.38}$$

$$F = 1 + \frac{R_u}{R_g} + \frac{G_n}{R_g}\left[(R_G + R_{cor})^2 + (X_G + X_{cor})^2\right] \tag{8.39}$$

The noise factor is a function of various elements, and the optimum impedance for the best noise figure can be determined by minimizing F with respect to generator reactance and resistance. This gives

$$R_{0n} = \sqrt{\frac{R_u}{G_n} + R_{cor}^2} \tag{8.40}$$

$$X_{0n} = -X_{cor} \tag{8.41}$$

and

$$F_{min} = 1 + 2G_nR_{cor} + 2\sqrt{R_uG_n + (G_nR_{cor})^2} \tag{8.42}$$

(To distinguish between optimum noise and optimum power, we have introduced the convention $0n$ instead of the more familiar abbreviation opt.) At this point, we see that the optimum condition for minimum noise figure is not a conjugate power match at the input port. We can explain this by recognizing that the noise source V_A and I_A represent all the 2-port noise, not just the thermal noise of the input port. We should observe that the optimum generator susceptance, $-X_{corr}$, will minimize the noise contribution of the two noise generators.

In rearranging for conversion to S parameters, we write

$$F = F_{min} + \frac{g_n}{R_G}|Z_G - Z_{0n}|^2 \tag{8.43}$$

$$F = F_{min} + \frac{R_n}{G_G}|Y_G - Y_{0n}|^2 \tag{8.44}$$

From the definition of the reflection coefficient,

$$\Gamma_G = \frac{Y_0 - Y_G}{Y_0 + Y_G} \tag{8.45}$$

and with

$$g_G = \frac{G_G}{Y_0} \tag{8.46}$$

$$r_n = \frac{R_n}{Z_0} \tag{8.47}$$

the normalized equivalent noise resistance

$$F = F_{min} + \frac{4r_n|\Gamma_G - \Gamma_{0n}|^2}{g_G(1 - |\Gamma|^2)|1 + \Gamma_{0n}|^2} \tag{8.48}$$

$$r_n = (F_{50} - F_{min})\frac{|1 + \Gamma_{0n}|^2}{4|\Gamma_{0n}|^2} \tag{8.49}$$

$$\Gamma_{0n} = \frac{Z_{0n} - Z_0}{Z_{0n} + Z_0} \tag{8.50}$$

The noise performance of any linear 2-port can now be determined if the values of the four noise parameters, F_{min}, $r_n = R_n/50$, and Γ_{0n} are known.

Figure 8.20 shows the noise factor of a high-frequency transistor as a function of B_g for G_g = constant and as a function of G_g for $B_g = B_{opt}$ [1].

Noise Figure of Cascaded Networks: In a system with many circuits connected in cascade (Figure 8.21), we must consider the contributions of the various circuits. In considering the equivalent noise resistor R_n in series with the input circuit [1].

$$F = \frac{R_G + R_n}{R_G} \tag{8.51}$$

$$F = 1 + \frac{R_n}{R_G} \tag{8.52}$$

The excess noise added by the circuit is R_n/R_G.

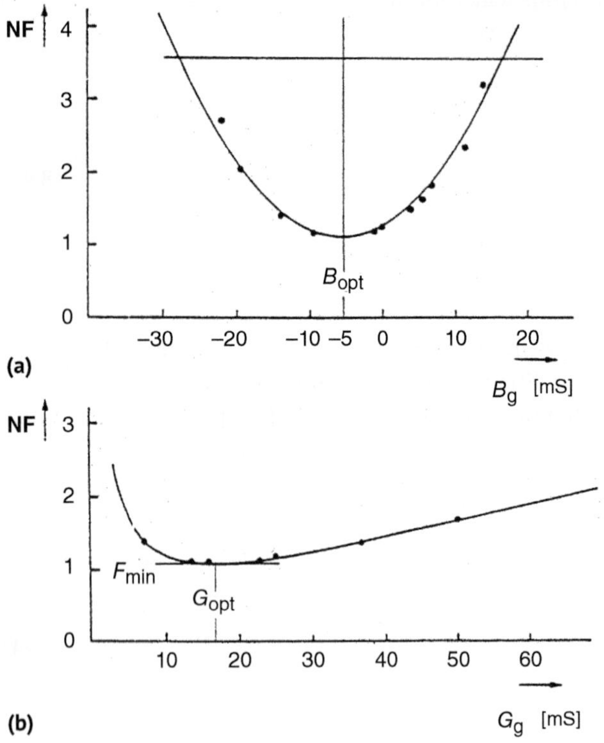

Figure 8.20 Noise factor in high-frequency BJTs for $f = 600$ MHz: (a) as a function of B_g for $G_g =$ constant, (b) as a function of G_g for $B_g = B_{opt}$. *Source*: [1]/John Wiley & Sons.

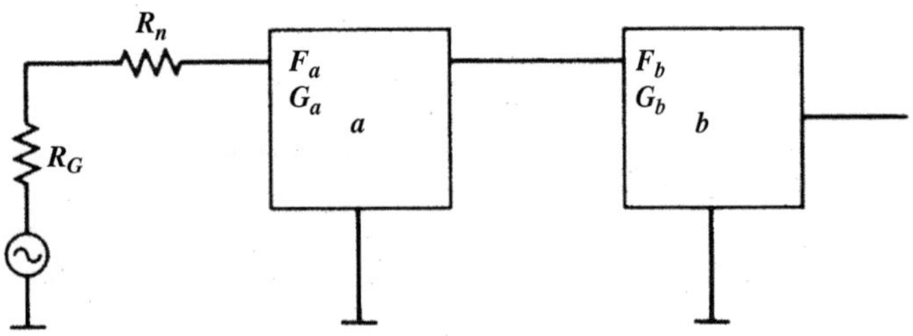

Figure 8.21 Cascaded noisy 2-ports with the noise figures F_a and F_b and the gain figure G_a and G_b. *Source*: [1]/John Wiley & Sons.

In considering two cascaded circuits, a and b, by definition, the available noise at the output of b is

$$N_{ab} = F_{ab}G_{ab}kTB \tag{8.53}$$

with B the equivalent noise bandwidth in which the noise is measured. The total available gain G is the product of the individual available gains, so

$$N_{ab} = F_{ab}G_aG_bkTB \tag{8.54}$$

The available noise from network a at the output of network b is

$$N_{a/b} = N_aG_b = F_aG_aG_bkTB \tag{8.55}$$

The available noise added by network b (its excess noise) is

$$N_{b/b} = (F_b - 1)G_bkTB \tag{8.56}$$

The total available noise N_{ab} is the sum of the available noise contributed by the two networks:

$$N_{ab} = N_{a/b} + N_{b/b} = F_a G_a G_b kTB + (F_b - 1)G_b kTB$$
$$= \left(F_a + \frac{F_b - 1}{G_a}\right) G_a G_b kTB \tag{8.57}$$

$$F_{ab} = F_a + \frac{F_b - 1}{G_a} \tag{8.58}$$

For any number of circuits, this can be extended to be

$$F = F_1 + \frac{F_2 - 1}{G_1} + \frac{F_3 - 1}{G_1 G_2} + \frac{F_4 - 1}{G_1 G_2 G_3} + \cdots \tag{8.59}$$

When considering a long chain of cascaded amplifiers, there will be a minimum noise figure achievable for this chain. This is a figure of merit and was proposed by Haus and Adler [8]. It is calculated by rearranging Eq. (8.59).

$$(F_{tot})_{min} = (F_{min} - 1) + \frac{F_{min} - 1}{G_A} + \frac{F_{min} - 1}{G_A^2} + \cdots + 1 \tag{8.60}$$

where F_{min} is the minimum noise figure for each stage and G_A is the available power gain of the identical stages. Using

$$\frac{1}{1 - X} = 1 + X + X^2 + \cdots \tag{8.61}$$

we find a quantity $(F_{tot} - 1)$, which is defined as the noise measure M. The minimum noise measure

$$(F_{tot})_{min} - 1 = \frac{F_{min} - 1}{1 - 1/G_A} = M_{min} \tag{8.62}$$

refers to the noise of an infinite chain of optimally tuned, low-noise stages, so it represents a lower limit on the noise of an amplifier.

The minimum noise measure M_{min} is an invariant parameter and is not affected by feedback. It is somewhat similar to a gain-bandwidth product in its use as a system invariant. The minimum noise measure is achieved when the amplifier is tuned for the available power gain and $\Gamma_G = \Gamma_{0n}$, given by Eq. (8.50) [1].

Influence of External Parasitic Elements: Mounting an active 2-port, such as a transistor, usually adds stray capacitance and lead inductance to the device, as shown in Figure 8.22. These external components, consisting of transmission lines and parasitic reactances, modify the noise parameters and the gain. Some researchers have published the results of these parasitic effects and have made manual computations or used some limited computer programs.

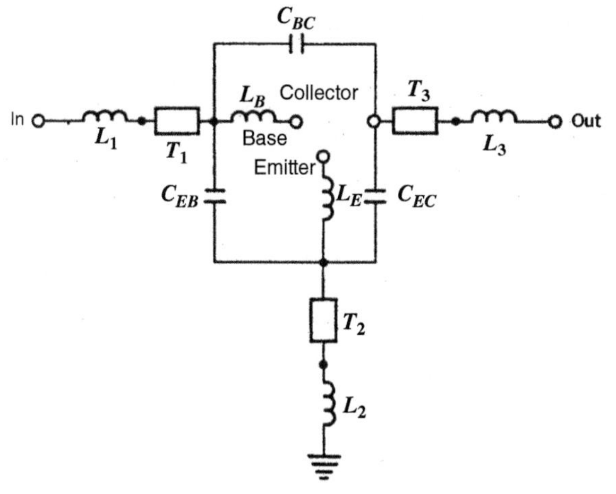

Figure 8.22 Equivalent circuit of the transistor package. *Source:* [1]/John Wiley & Sons.

In a paper by Fukui [9], an attempt was made to determine the necessary equations, but the formulas are too involved even for pocket calculators. A more generic study by Iversen [10] is also very involved because of the various matrix manipulations, and is more suitable for a computer. Besser's paper in the IEEE MTT-S in 1975 [11] and Vendelin's paper [12] in the same issue have shown for the first time some practical results using computers and even optimization methods, using an early version of COMPACT. The intention of these investigations was to find feedback that modifies the device noise and scattering parameters such that a noise match could also provide a low-input VSWR. It can be seen from these discussions that some feedback, besides resulting in some gain reduction, may improve the noise matching at the input for a limited frequency range.

A more recent paper by Suter [13] based on a report by Hartmann and Strutt [14] has given a simple transformation starting from the S parameters and the noise parameters from common-source (or common-emitter) measurements. The noise parameters for the "packaged" device are calculated. This means that the parameters for the "new" device, including the common-gate (or common-base) case, are calculated. The equations are device-dependent. They are valid for any active 2-port.

A transformation matrix, n, may be used to combine the noise sources of the various circuit configurations. The transformation matrix parameters are given in Table 8.2 for series feedback, shunt feedback, and the common-gate (base) case, which will be important for oscillator analysis. The transformation matrix gives the new four noise parameters as follows:

$$R_n' = R_n|n_{11} + n_{12}Y_{cor}|^2 + G_n|n_{12}|^2 \tag{8.63}$$

Table 8.2 Transformation of Matrix Parameters.

Series Feedback

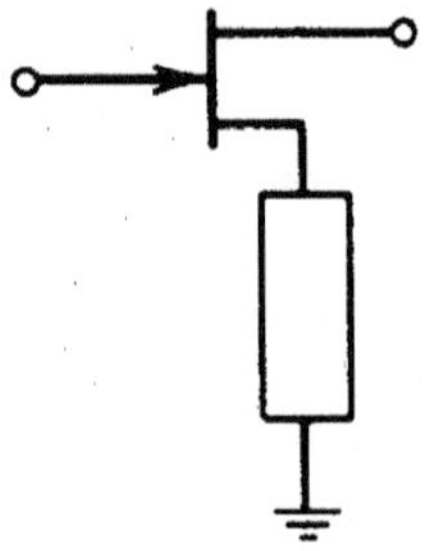

$Z_s = R_s + jX_s = Z_s$

$$[n] = \begin{bmatrix} n_{11} = 1 & n_{12} = Z_0 \dfrac{S_{21}M - S_{21}'N}{S_{21}C_1' + S_{21}'C_1}\Omega \\ n_{21} = 0 & n_{22} = \dfrac{S_{21}C_1'}{S_{21}C_1' + S_{21}'C_1} \end{bmatrix}$$

where

$$S_{11}' = S_{22}' = \frac{-1}{1 + 2Z_S}$$

$$S_{12}' = S_{21}' = \frac{2Z_S}{1 + 2Z_S}$$

$$M = (1 + S_{11}')(1 - S_{22}') + S_{12}'S_{21}'$$

$$N = (1 + S_{11})(1 - S_{22}) + S_{12}S_{21}$$

$$C_1 = (1 - S_{11})(1 - S_{22}) - S_{12}S_{21}$$

$$C_1' = (1 + S_{11}')(1 - S_{22}') - S_{12}'S_{21}'$$

$$[S]_{DEVICE} = \begin{bmatrix} S_{11} & S_{12} \\ S_{21} & S_{22} \end{bmatrix}$$

Shunt Feedback

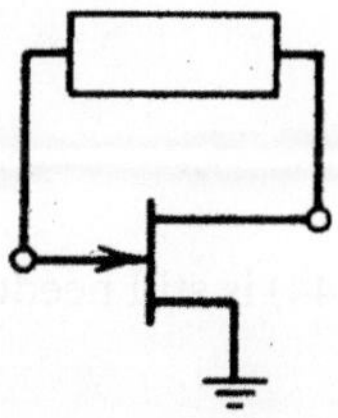

$$Z_P = R_P + jX_P = Z_P$$

$$[n] = \begin{bmatrix} n_{11} = \dfrac{S_{21}C_2'}{S_{21}C_2' + S_{21}'C_2} & n_{12} = 0\ \Omega \\ n_{21} = \dfrac{1}{Z_0}\dfrac{S_{21}P - S_{21}'Q}{S_{21}C_2' + S_{21}'C_2}S & n_{22} = 1 \end{bmatrix}$$

where

$$S_{11}' = S_{22}' = \frac{Z_P}{2 + Z_P}$$

$$S_{12}' = S_{21}' = \frac{2}{2 + Z_P}$$

$$P = (1 - S_{11}')(1 + S_{22}') + S_{12}'S_{21}'$$

$$Q = (1 - S_{11})(1 + S_{22}) + S_{12}S_{21}$$

$$C_2 = (1 + S_{11})(1 + S_{22}) - S_{12}S_{21}$$

$$C_2' = (1 + S_{11}')(1 + S_{22}') - S_{12}'S_{21}'$$

$$[S]_{\text{DEVICE}} = \begin{bmatrix} S_{11} & S_{12} \\ S_{21} & S_{22} \end{bmatrix}$$

Common Gate

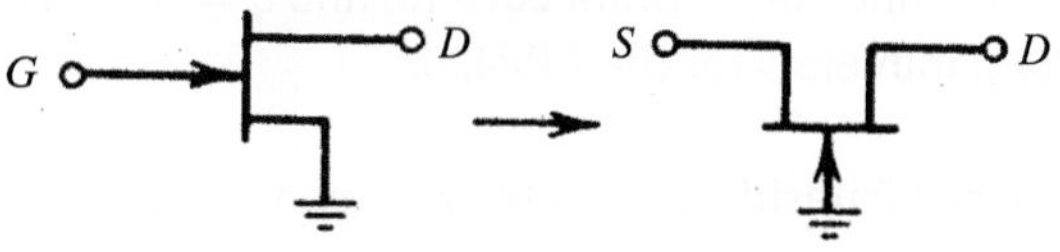

$$[n] = \begin{bmatrix} n_{11} = \dfrac{2S_{21}}{-2S_{21} + C_4} & n_{12} = 0\ \Omega \\ n_{21} = \dfrac{1}{Z_0}\dfrac{C_3C_4 - 4S_{12}S_{21}}{V(-2S_{21} + C_4)}S & n_{22} = -1 \end{bmatrix}$$

where

$$V = (1 + S_{11})(1 + S_{22}) - S_{12}S_{21}$$

$$C_3 = (1 - S_{11})(1 + S_{22}) + S_{12}S_{21}$$

$$C_4 = (1 + S_{11})(1 - S_{22}) + S_{12}S_{21}$$

$$[S]_{\text{DEVICE}} = \begin{bmatrix} S_{11} & S_{12} \\ S_{21} & S_{22} \end{bmatrix} = \text{common} - \text{source } S \text{ parameters}$$

Source: [1]/John Wiley & Sons.

$$G'_n = \frac{G_n R_n}{R'_n} |n_{11}n_{22} - n_{12}n_{21}|^2 \tag{8.64}$$

$$Y'_{cor} = \frac{R_n}{R'_n}(n_{21} + n_{22}Y_{cor})(n^*_{11} + n^*_{12}Y^*_{cor}) + \frac{G_n}{R'_n} n_{22}n^*_{12} \tag{8.65}$$

A final transformation to the more common noise-parameter format given by Eq. (8.44) is still needed [12]:

$$F_{min} = 1 + 2R'_n(G'_{cor} + G'_{0n}) \tag{8.66}$$

$$R_n = R'_n \tag{8.67}$$

$$G_{0n} = \sqrt{\frac{G'_n}{R'_n} + G'^2_{cor}} \tag{8.68}$$

$$B_{0n} = -B'_{cor} \tag{8.69}$$

Figure 8.23 shows the noise figure as a function of external feedback for a low-noise microwave bipolar transistor, the AT-41435.

Bias-dependent Noise Parameters: Since we have elaborated on the effect of parasitics, it is now useful to point out that noise is also a strong function of bias point and frequency. Figure 8.24 shows the noise figure expressed in dB as a function of different dc currents and frequencies for a BFP420 BJT by Siemens. A change in the collector voltage will also affect the noise parameters, not quite as strongly as shown here, but still significantly. The reason for the turning point of the noise figure has to do with the fact that for lower currents the f_t cutoff frequency has not reached its peak yet, and therefore, the noise increases. The increase in the noise at higher currents is due to the Schottky noise calculated from the equation [1]

$$P_n = 2Iq \tag{8.70}$$

where P_n is the noise power, I is the saturation, collector, or drain current, and q = charge of an electron (1.60×10^{-19} C).

Table 8.3 shows the common emitter noise parameters for the BFP420.

Noise Circles: From the "Influence of External Parasitic Elements" section, we see that the noise factor is a strong function of the generator admittance (or admittance) presented to the input terminals of the noisy 2-port. Noise tuning is the method to change the values of the input admittance to obtain the best noise performance. There is a range of values of input reflection coefficient over which the noise figure is constant. In plotting these points of constant noise figure, we obtain the so-called noise circles, which can be drawn on the Smith chart Γ_G plane [15]. Starting with the noise equation [see Eq. (8.44)] for a 50-Ω generator impedance, we find that

$$F_{50} = F_{min} + 4r_n \frac{|\Gamma_{0n}|^2}{|1 + \Gamma_{0n}|^2} \tag{8.71}$$

We want to find the position of the reflection coefficient on the Smith chart, as in the case of the gain circles, for which F is constant. First, we rearrange Eq. (8.67) to read

$$r_n = (F_{50} - F_{min}) \frac{|1 + \Gamma_{0n}|^2}{4|\Gamma_{0n}|^2} \tag{8.72}$$

By introducing

$$N_i = \frac{F_i - F_{min}}{4r_n} |1 + \Gamma_{0n}|^2 \tag{8.73}$$

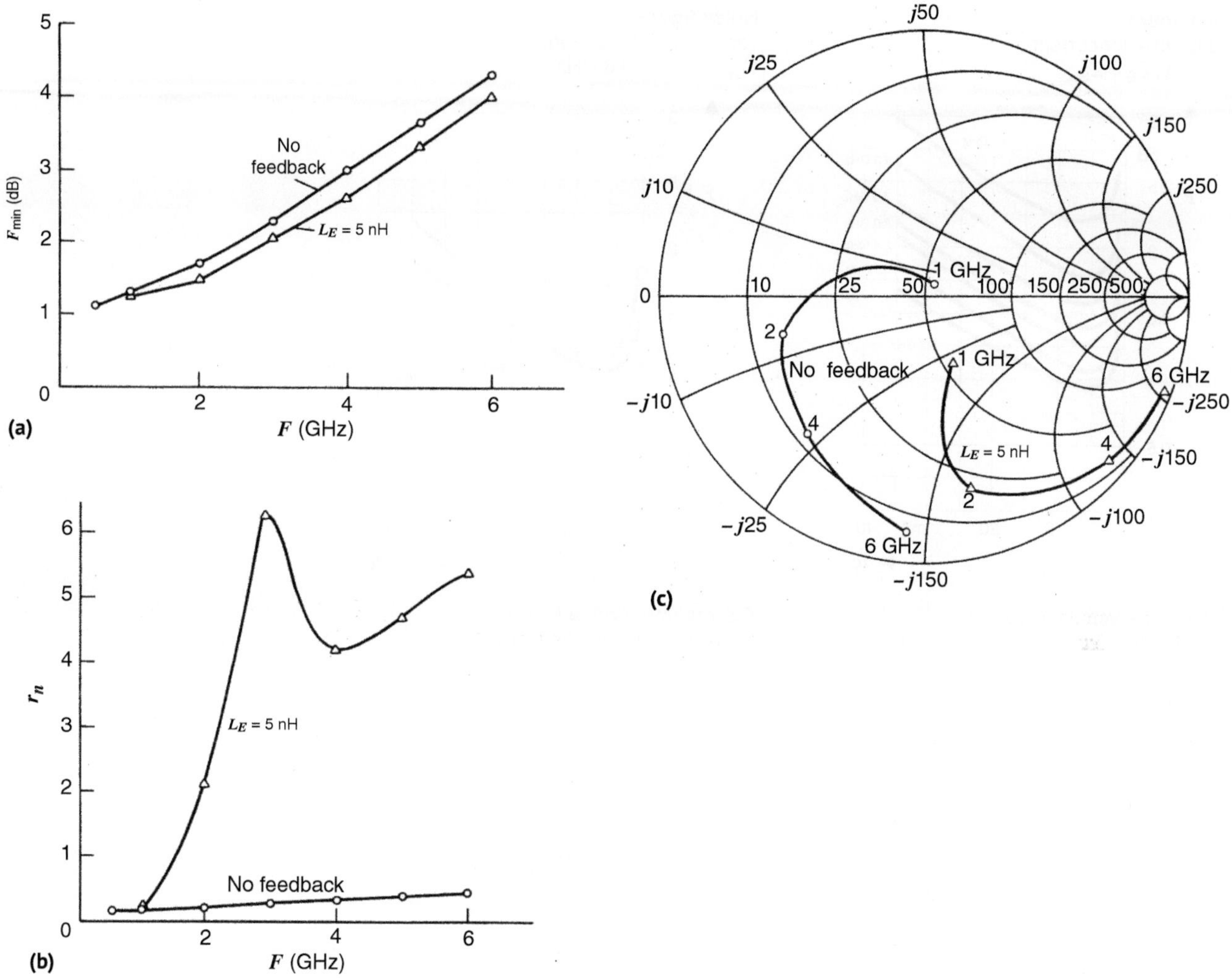

Figure 8.23 Noise parameters versus feedback for the AT-41435 silicon bipolar transistor: (a) $F_{\min}$ versus frequency and feedback, (b) r_n versus frequency and feedback, (c) Γ_{0n} for AT-41435 versus frequency and feedback. *Source*: [1]/John Wiley & Sons.

we can find an expression for a circle of constant noise figure as introduced by Rothe and Dahlke [6, 15]. The center of the noise circle is

$$C_i = \frac{|\Gamma_{0n}|}{1 + N_i} \tag{8.74}$$

and the radius

$$r_i = \frac{\sqrt{N_i^2 + N_i\left(1 - |\Gamma_{0n}|^2\right)}}{1 + N_i} \tag{8.75}$$

with the definition of N used previously. However, if we consider only the minimum noise figure for a given device, we will not obtain the minimum noise figure for the multistage amplifier system. This was explained when the noise measure was introduced (see Figure 8.25.). Therefore, a better way to design the amplifier would be to use circles at constant

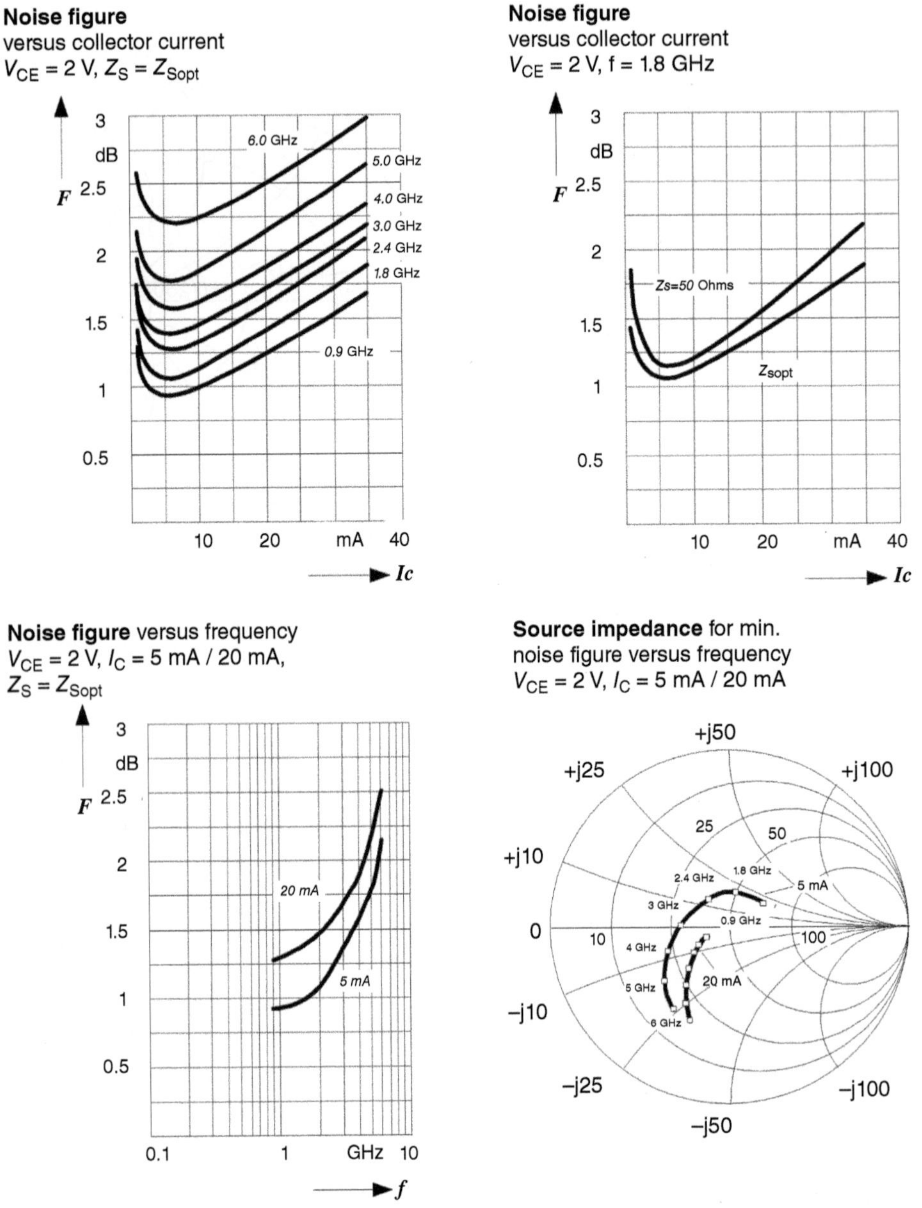

Figure 8.24 Noise figure versus collector current, noise figure versus collector current, and source impedance for minimum noise figure versus frequency for the Siemens BFP420 transistor. *Source*: [1]/John Wiley & Sons.

noise measure instead of circles of constant noise figure. This was recently done by Poole and Paul [16]. They derived the expressions for the noise measure circles as a function of S parameters, noise parameters, and Γ_G, using

$$G_A = \frac{|S_{21}|^2\left(1 - |\Gamma_G|^2\right)}{\left(1 - |S_{22}|^2\right) + |\Gamma_G|^2\left(|S_{11}|^2 - |\Delta|^2\right) - 2\,\mathrm{Re}(\Gamma_G C_1)} \tag{8.76}$$

where

$$C_1 = S_{11} - S_{22}^*\Delta$$

$$\Delta = S_{11}S_{22} - S_{12}S_{11}$$

Table 8.3 BFP420 common-emitter noise parameters.

f	F_{min}[a]	G_a[a]	Γ_{opt}		R_N	r_n	$F_{50\,\Omega}$[b]	$\lvert S_{21}\rvert^2$[b]
GHz	dB	dB	MAG	ANG	Ω	-	dB	dB
$V_{CE} = 2$ V, $I_C = 5$ mA								
0.9	0.90	20.5	0.28	41.0	8.7	0.17	1.02	20.3
1.8	1.05	15.2	0.20	82.0	6.7	0.13	1.11	15.8
2.4	1.25	13.0	0.20	124.0	5.5	0.11	1.32	13.5
3.0	1.38	12.1	0.22	−175.0	5.0	0.10	1.48	11.6
4.0	1.55	10.3	0.33	−157.0	5.5	0.11	1.83	9.1
5.0	1.75	8.6	0.45	−142.0	5.0	0.10	2.20	7.0
6.0	2.20	6.4	0.53	−123.0	15.0	0.30	3.30	5.3

a. Input matched for minimum noise figure, output for maximum gain.
b. $Z_s = Z_L = 50\ \Omega$.
Source: [1]/John Wiley & Sons.

In terms of the generator reflection coefficient, the noise measure can be expressed as

$$M = \left[(F_{\min} - 1)\left(1 - |\Gamma_G|^2\right)|S_{21}|^2|1 + \Gamma_{0n}|^2 + 4r_n|S_{21}|^2|\Gamma_G - \Gamma_{0n}|^2\right] \\ \times \left\{|1 + \Gamma_{0n}|^2\left[|S_{21}|^2\left(1 - |\Gamma_G|^2\right) - \left(1 - |S_{22}|^2\right) - |\Gamma_G|^2\left(|S_{11}|^2 - |\Delta|^2\right) + 2\text{Re}(\Gamma_G C_1)\right]\right\}^{-1} \tag{8.77}$$

Equation (8.77) can be shown to represent circles in the source reflection coefficient plane described by the following:

$$|\Gamma_G|^2 + |\Gamma_m|^2 - \Gamma_G\Gamma_m^* - \Gamma_G^*\Gamma_m = r_m^2 \tag{8.78}$$

The centers and radii of the constant-noise-measure circles are given by

$$C_m = \frac{M|1 + \Gamma_{0n}|^2 C_1^* + 4r_n|S_{21}|^2\Gamma_{0n}}{M|1 + \Gamma_{0n}|^2 P + |S_{21}|^2(4r_n - W)} \tag{8.79}$$

$$r_m = \frac{\sqrt{M^2 M_a + M M_b + M_c}}{M|1 + \Gamma_{0n}|^2 P + |S_{21}|^2(4r_n - W)} \tag{8.80}$$

where

$$P = |S_{21}|^2 + |S_{11}|^2 - |\Delta|^2$$

$$Q = |S_{21}|^2 + |S_{22}|^2 - 1$$

$$W = |1 + \Gamma_{0n}|^2(F_{\min} - 1)$$

$$M_a = |1 + \Gamma_{0n}|^4\left(PQ + |C_1|^2\right)$$

$$M_b = |1 + \Gamma_{0n}|^2|S_{21}|^2 \left[-\left(4r_n|\Gamma_{0n}|^2 + W\right)P - (W - 4r_n)Q\right]$$

$$M_c = |S_{21}|^4|1 + \Gamma_{0n}|^2(F_{\min} - 1) \left[W - 4r_n\left(1 - |\Gamma_{0n}|^2\right)\right]$$

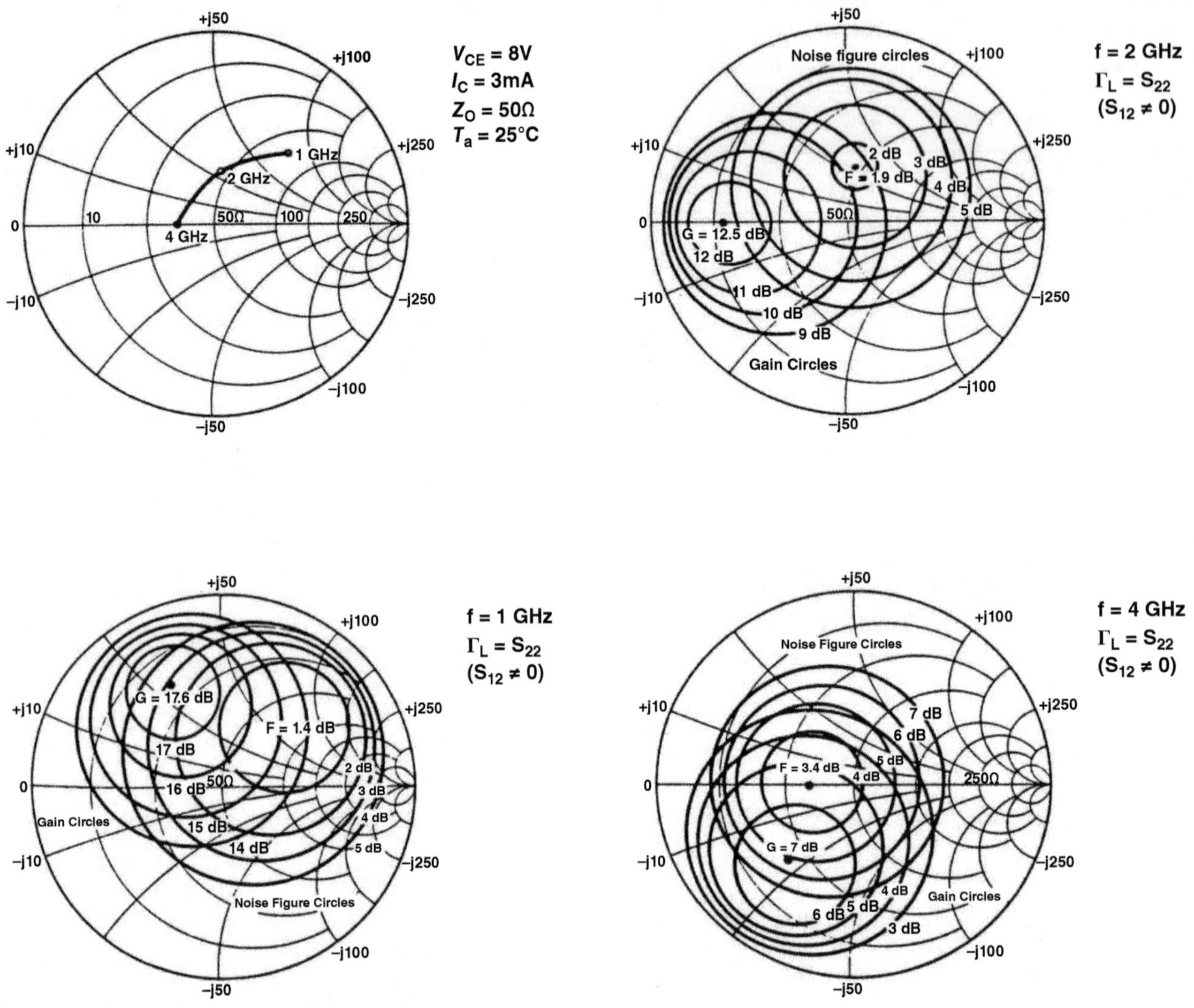

Figure 8.25 Typical noise figure circles and gain circles. *Source*: [1]/John Wiley & Sons.

The value of the minimum noise measure can be found by considering the noise measure circle of zero radius; that is, set r_m equal to zero in Eq. (8.80). This results in

$$M_{\text{min}} = \frac{-M_{\text{b}} \pm \sqrt{M_{\text{b}}^2 - 4M_{\text{a}}M_c}}{2M_{\text{a}}} \tag{8.81}$$

Equation (8.81) yields the same value of M_{min} as would have been obtained by using the immittance parameter equation given by Fukui [15] and can, therefore, be considered as the reflection coefficient plane analog of Fukui's expression. The elimination of the need for the parameter R_{eg}, however, results in a considerable simplification as compared with the earlier approach [15].

The value of the minimum noise measure is taken as the smallest nonnegative value of M_{min} given by Eq. (8.81). The source reflection coefficient that results in the minimum noise measure can now be obtained by employing Eq. (8.79):

$$\Gamma_{\text{Om}} = \frac{M_{\text{min}}|1 + \Gamma_{\text{On}}|^2 C_1^* + 4r_{\text{n}}|S_{21}|^2\Gamma_{\text{On}}}{M_{\text{min}}|1 + \Gamma_{\text{On}}|^2 P + |S_{21}|^2(4r_{\text{n}} - W)} \tag{8.82}$$

The output reflection coefficients of the device, when Γ_{0m} is presented to the input port, and is given by

$$S'_{22} = \frac{S_{22} - \Delta\Gamma_{0m}}{1 - S_{11}\Gamma_{0m}} \tag{8.83}$$

Noise Correlation in Linear 2-ports Using Correlation Matrices: In the introduction to 2-port noise theory, it was indicated that noise correlation matrices form a general technique for calculating noise in n-port networks. Haus and Adler have described the theory behind this technique [8]. In 1976, Hillbrand and Russer published equations and transformations that aid in supplying this method to 2-port CAD [17].

This method is useful because it forms a base from which we can rigorously calculate the noise of linear 2-ports combined in arbitrary ways. For many representations, the method of combining the noise parameters is as simple as that for combining the circuit element matrices. In addition, noise correlation matrices can be used to calculate the noise in linear frequency conversion circuits. The following is an introduction to this subject.

Linear, noisy 2-ports can be modeled as a noise-free 2-port with two additional noise sources. These noise sources must be chosen so that they add directly to the resulting vector of the representation, as shown in Eqs. (8.84) and (8.85) and Figure 8.17.

$$\begin{bmatrix} I_1 \\ I_2 \end{bmatrix} = \begin{bmatrix} y_{11} & y_{12} \\ y_{21} & y_{22} \end{bmatrix} \begin{bmatrix} V_1 \\ V_2 \end{bmatrix} + \begin{bmatrix} i_1 \\ i_2 \end{bmatrix} \tag{8.84}$$

$$\begin{bmatrix} V_1 \\ V_2 \end{bmatrix} = \begin{bmatrix} z_{11} & z_{12} \\ z_{21} & z_{22} \end{bmatrix} \begin{bmatrix} I_1 \\ I_2 \end{bmatrix} + \begin{bmatrix} v_1 \\ v_2 \end{bmatrix} \tag{8.85}$$

where the i and v vectors indicate noise sources for the y and z representations, respectively. This 2-port example can be extended to n-ports in a straightforward, obvious way.

Since the noise vector for any representation is a random variable, it is much more convenient to work with the noise correlation matrix. The correlation matrix gives us deterministic numbers to calculate with. The correlation matrix is formed by taking the mean value of the outer product of the noise vector. This is equivalent to multiplying the noise vector by its adjoint (complex conjugate transpose) and averaging the result:

$$\left\langle \overline{ii}^{+} \right\rangle = \begin{bmatrix} i_1 \\ i_2 \end{bmatrix} \begin{bmatrix} i_1^* & i_2^* \end{bmatrix} = \begin{bmatrix} \langle i_1 i_1^* \rangle \langle i_1 i_2^* \rangle \\ \langle i_1^* i_2 \rangle \langle i_2 i_2^* \rangle \end{bmatrix} = [C_y] \tag{8.86}$$

where the angular brackets denote the average value.

Note that the diagonal terms are the "power" spectrum of each noise source, and the off-diagonal terms are complex conjugates of each other and represent the cross "power" spectrums of the noise sources. "Power" is used because these magnitude-squared quantities are proportional to power.

To use these correlation matrices in circuit analysis, we must know how to combine them and how to convert them between various representations. An example using y matrices will illustrate the method for combining 2-ports and their correlation matrices. Given two matrices y and y', when we parallel them, we have the same port voltages, and the terminal currents add (Figure 8.26):

$$\begin{aligned} I_1 &= y_{11}V_1 + y_{12}V_2 + y'_{11}V_1 + y'_{12}V_2 + i_1 + i'_1 \\ I_2 &= y_{21}V_1 + y_{22}V_2 + y'_{21}V_1 + y'_{22}V_2 + i_2 + i'_2 \end{aligned} \tag{8.87}$$

or

$$\begin{bmatrix} I_1 \\ I_2 \end{bmatrix} = \begin{bmatrix} y_{11} + y'_{11} & y_{12} + y'_{12} \\ y_{21} + y'_{21} & y_{22} + y'_{22} \end{bmatrix} \begin{bmatrix} V_1 \\ V_2 \end{bmatrix} + \begin{bmatrix} i_1 + i'_1 \\ i_2 + i'_2 \end{bmatrix} \tag{8.88}$$

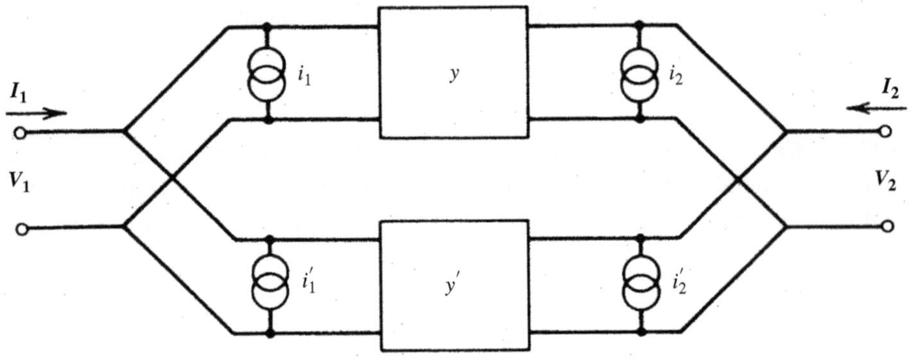

Figure 8.26 Parallel combination of 2-ports using y parameters. *Source*: [1]/John Wiley & Sons.

Here we can see that the noise current vectors add just as the y parameters add. Converting the new noise vector to a correlation matrix yields

$$\left\langle \bar{i}_{\mathrm{new}}\,\bar{i}^{+}_{\mathrm{new}} \right\rangle = \left\langle \begin{bmatrix} i_1 + i_1' \\ i_2 + i_2' \end{bmatrix} \left[i_1^* + i_1'^* i_2 i_2'^* \right] \right\rangle \tag{8.89}$$

$$= \begin{bmatrix} \langle i_1 i_1^* \rangle + \langle i_1' i_1'^* \rangle & \langle i_1 i_2^* \rangle + \langle i_1' i_2'^* \rangle \\ \langle i_2 i_1^* \rangle + \langle i_2' i_1'^* \rangle & \langle i_2 i_2^* \rangle + \langle i_2' i_2'^* \rangle \end{bmatrix} \tag{8.90}$$

The noise sources from different 2-ports must be uncorrelated, so there are no cross-products of different 2-ports. By inspection, Eq. (8.90) is just the addition of the correlation matrices for the individual 2-ports, so

$$[C_{\mathrm{ynew}}] = [C_{\mathrm{y}}] + [C_{\mathrm{y}}'] \tag{8.91}$$

The same holds true for g, h, and z parameters, but $ABCD$ parameters have the more complicated form shown below. If

$$[\mathrm{A}_{\mathrm{new}}] = [\mathrm{A}][\mathrm{A}'] \tag{8.92}$$

then

$$[C_{\mathrm{A\,new}}] = [C_{\mathrm{A}}] + [\mathrm{A}][C_{\mathrm{A}'}][\mathrm{A}]^{+} \tag{8.93}$$

The transformation of one representation to another is best illustrated by an example. Let us transform the correlation matrix for a Y representation to a Z representation. Starting with

$$\begin{bmatrix} I_1 \\ I_2 \end{bmatrix} = [Y] \begin{bmatrix} V_1 \\ V_2 \end{bmatrix} + \begin{bmatrix} i_1 \\ i_2 \end{bmatrix} \tag{8.94}$$

we can move the noise vector to the left side and invert y:

$$\begin{bmatrix} V_1 \\ V_2 \end{bmatrix} = [Y^{-1}] \begin{bmatrix} I_1 - i_1 \\ I_2 - i_2 \end{bmatrix} = [Y^{-1}] \begin{bmatrix} I_1 \\ I_2 \end{bmatrix} + [Y^{-1}] \begin{bmatrix} -i_1 \\ -i_2 \end{bmatrix} \tag{8.95}$$

Since $(Y)^{-1} = (Z)$, we have

$$\begin{bmatrix} V_1 \\ V_2 \end{bmatrix} = [Z] \begin{bmatrix} I_1 \\ I_2 \end{bmatrix} + [Z] \begin{bmatrix} -i_1 \\ -i_2 \end{bmatrix} \tag{8.96}$$

so

$$= [Z]\begin{bmatrix} -i_1 \\ -i_2 \end{bmatrix} = [T_{yz}]\begin{bmatrix} -i_1 \\ -i_2 \end{bmatrix} \tag{8.97}$$

where the signs of i_1 and i_2 are superfluous since they will cancel when the correlation matrix is formed. Here, the transformation of the Y noise current vector to the Z noise voltage vector is done simply by multiplying by (Z). Other transformations are shown in Table 8.4.

To form the noise correlation matrix, we gain from the mean of the outer product:

$$\langle vv^+\rangle = \begin{bmatrix} \langle v_1 v_1^*\rangle\langle v_1 v_2^*\rangle \\ \langle v_1^* v_2\rangle\langle v_2 v_2^*\rangle \end{bmatrix} = [Z]\left\langle \begin{bmatrix} i_1 \\ i_2 \end{bmatrix}\begin{bmatrix} i_1^* & i_2^* \end{bmatrix}\right\rangle [Z]^+ \tag{8.98}$$

or

$$[C_z] = [Z][C_y][Z]^+ \tag{8.99}$$

where

$$v^+ = \begin{bmatrix} i_1^* & i_2^* \end{bmatrix}[Z]^+ \tag{8.100}$$

This is called a congruence transformation. The key to all of these derivations is the construction of a correlation matrix from the noise vector, as shown in Eq. (8.90).

These correlation matrices may easily be derived from the circuit matrices of passive circuits with only thermal noise sources. For example,

$$[C_z] = 2kT\Delta f\ \mathrm{Re}([Z]) \text{ and} \tag{8.101}$$

$$[C_y] = 2kT\Delta f\ \mathrm{Re}([Y]) \tag{8.102}$$

The $2kT$ factor comes from the double-sided spectrum of thermal noise. The correlation matrix from the *ABCD* matrix may be related to the noise figure, as shown by Hillbrand and Russer [17]. We have

$$F = 1 + \frac{\overline{Y}[C_a]\overline{Y}^+}{2kT\ \mathrm{Re}(Y_G)} \tag{8.103}$$

Table 8.4 Noise matrix $T_{\alpha\beta}$ transformation.

		Original form (α form)		
		Y	**Z**	**A**
Resulting form (β form)	Y	1 0	y_{11} y_{12}	$-y_{11}$ 1
		0 1	y_{21} y_{22}	$-y_{21}$ 0
	Z	z_{11} z_{12}	1 0	1 $-z_{11}$
		z_{21} z_{22}	0 1	0 $-z_{21}$
	A	0 A_{12}	0 $-A_{11}$	1 0
		1 A_{22}	1 $-A_{21}$	0 1

Source: [1]/John Wiley & Sons.

where

$$\overline{Y} = \begin{bmatrix} Y_G \\ 1 \end{bmatrix}$$

Expressing the noise factor in terms of the correlation matrix, here is a complete formula:

$$F = 1 + \frac{C_{22}^{A}(f) + 2\,\mathrm{Re}\{Y_g(f)C_{12}^{A}(f)\} + |Y_g(f)|^2 C_{11}^{A}(f)}{2kT_0\,\mathrm{Re}\{Y_g(f)\}} \tag{8.104}$$

Once we transform this into the Y parameter form, we obtain the following equation:

$$F(f) = F_{\min}(f) + \frac{R_n(f)|Y_{\mathrm{opt}}(f) - Y_g(f)|^2}{\mathrm{Re}\{Y_g(f)\}} \tag{8.105}$$

It should be noted that all these values are frequency-dependent as expressed in this equation.
The $ABCD$ correlation matrix can be written in terms of the noise-figure parameters as (double-sided spectrum)

$$[C_a] = 2kT\begin{bmatrix} R_n & \frac{F_0 - 1}{2} - R_n Y_{0n}^* \\ \frac{F_0 - 1}{2} - R_n Y_{0n} & R_n |Y_{0n}|^2 \end{bmatrix} \tag{8.106}$$

The noise correlation matrix method forms an easy and rigorous technique for handling noise in networks. This technique allows us to calculate the total noise for complicated networks by combining the noise matrices of subcircuits. It should be remembered that although noise correlation matrices apply to n-port networks, noise-figure calculations apply only to pairs of ports. The parameters of the C_a matrix can be used to give the noise parameters:

$$Y_{0n} = \sqrt{\frac{C_{ii^*}}{C_{uu^*}} - \left[\mathrm{Im}\left(\frac{C_{ui^*}}{C_{uu^*}}\right)\right]^2} + j\,\mathrm{Im}\left(\frac{C_{ui^*}}{C_{uu^*}}\right) \tag{8.107}$$

$$F_0 = 1 + \frac{C_{ui^*} + C_{uu^*} Y_{0n}^*}{kT} \tag{8.108}$$

$$R_n = C_{uu^*} \tag{8.109}$$

Noise Figure Test Equipment [1]: Figure 8.27 shows the block diagram of a noise test setup. It includes the noise source and the other components. The metering unit has a special detector that is linear and over a certain dynamic range measures linear power. The tunable receiver covers a wide frequency range (for example, 10–1800 MHz) and controls the noise source. The receiver is a double-conversion superheterodyne configuration with sufficient image rejection to avoid double-sideband noise measurements that would give the wrong results.

These receivers are microprocessor-controlled, and the measurement is a two-step procedure. The first is a calibration step that measures the noise figure of the receiver system and a reference power level. Then the device under test (DUT) is inserted, and the system noise figure and total output power are measured. The noise factor is calculated by:

$$F_1 = F_{\mathrm{system}} - \frac{F_2 - 1}{G_1} \tag{8.110}$$

and the gain is given by the change in output power from the reference level [18]. The noise of the system is calculated by measuring the total noise power with the noise source on and off. With the ENR (excess noise ratio) known [18],

$$F_{\mathrm{system}} = \frac{\mathrm{ENR}}{Y - 1} \tag{8.111}$$

The noise bandwidth is usually set by the bandwidth of the receiver, which is assumed to be constant over the linear range. The ENR of the noise source is given by

$$\mathrm{ENR} = \frac{T_{\mathrm{hot}}}{T_{\mathrm{cold}}} - 1 \tag{8.112}$$

where T_{cold} is usually room temperature (290 K). This ENR number is about 15 dB for noise sources with a 15-dB pad and 5 dB for noise sources with a 25-dB pad. Since both gain and noise were stored in the initial calibration, a noise/gain sweep can be performed.

For frequencies above 1800 MHz, we can extend the range with the help of the external signal generators, as shown in Figure 8.28. As shown, a filter ahead of the external mixer reduces the noise energy in the image band. If the DUT has a very broad frequency range and has flat gain and noise over that range, a double-sideband (DSB) measurement is possible, with the image-rejection filter removed. However, a single-sideband (SSB) measurement is always more accurate [19].

How To Determine the Noise Parameters [1]: The noise figure of a linear 2-port network as a function of source admittance may be represented by

$$F = F_{\min} + \frac{R_{\mathrm{n}}}{G_G}\left[(G_{0\mathrm{n}} - G_G)^2 + (B_{0\mathrm{n}} - B_G)^2\right] \tag{8.113}$$

where $G_G + jB_G$ = generator admittance presented to the input of the 2-port
$G_{0n} + jB_{0n}$ = generator admittance at which the optimum noise figure occurs
R_n = empirical constant relating to the sensitivity of the noise figure to generator admittance, with dimensions of resistance.

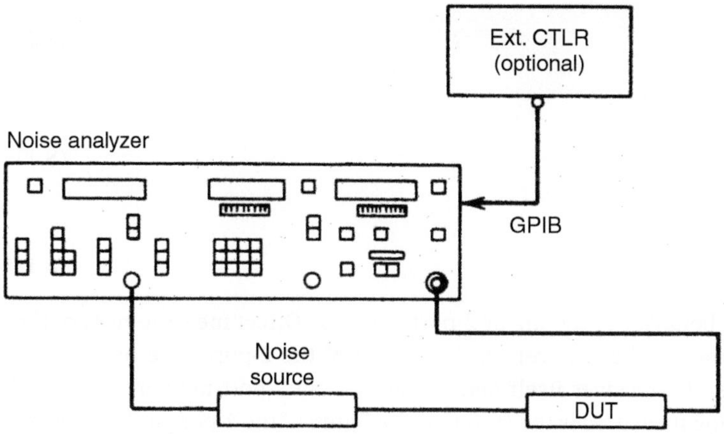

Figure 8.27 Noise figure measurement. *Source*: [1]/John Wiley & Sons.

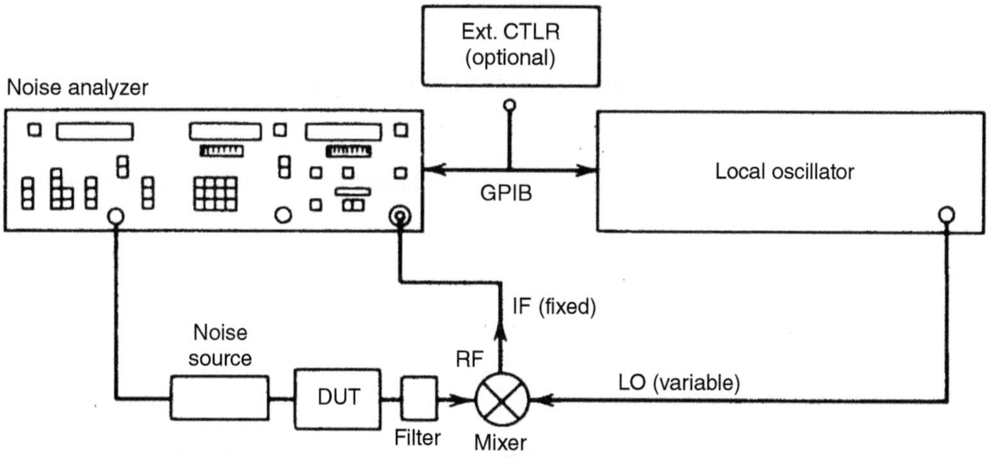

Figure 8.28 Single-sideband (SSB) noise figure measurements using an external mixer. *Source*: [1]/John Wiley & Sons.

It may be noted that for an arbitrary noise-figure measurement with a known generator admittance, Eq. (8.113) has four unknowns, $F_{\min}$, R_n, G_{0n}, and B_{0n}. By choosing four known values of generator admittance, a set of four linear equations are formed, and the solution of the four unknowns can be found [20, 21]. Equation (8.113) may be transformed to

$$F = F_{\min} + \frac{R_n|Y_{0n}|^2}{G_G} - 2R_nG_{0n} + \frac{R_n|Y_G|^2}{G_G} - 2R_nB_{0n}\frac{B_G}{G_G} \quad (8.114)$$

$$F = F_{\min} + \frac{R_n}{G_G}|Y_G - Y_{0n}|^2 \quad (8.115)$$

Let

$$X_1 = F_{\min} - 2R_nG_{0n}$$

$$X_2 = R_n|Y_{0n}|^2$$

$$X_3 = R_n$$

$$X_4 = R_nB_{0n}$$

Then the generalized equation may be written as

$$F_i = X_1 + \frac{1}{G_{si}}X_2 + \frac{|Y_{si}|^2}{G_{si}}X_3 - 2\frac{G_{si}}{B_{si}}X_4 \quad (8.116)$$

or, in matrix form,

$$[F] = [A][X] \quad (8.117)$$

and the solution becomes

$$[X] = [A]^{-1}[F] \quad (8.118)$$

These parameters completely characterize the noise behavior of the linear 2-port network. Direct measurement of these noise parameters by this method would be possible only if the receiver on the output of the 2-port were noiseless and insensitive to its input admittance. In actual practice, the receiver itself behaves as a noisy 2-port network and can be characterized in the same manner. What is actually being measured is the system noise figure of the 2-port and the receiver. Thus, it becomes apparent that to do a complete 2-port noise characterization, the gain of the 2-port must be measured [22]. In addition, any losses in the input-matching networks must be carefully accounted for, because they add directly to the measured noise figure reading [23].

8.1.4 Linearity

Consistent with the elaborate discussion of linearity questions above, here is a quick refresher on what must be considered in amplifiers [1].

Dynamic Range, Compression, and IMD: An amplifier's dynamic range is related to the minimum discernible signal and the 1-dB compression point according to the following equation

$$DR_n = \frac{(n-1)[IP_{n(in)} - MDS]}{n} \quad (8.119)$$

where DR is the dynamic range in decibels, n is the order, $IP_{(in)}$ is the input intercept power in dBm, and MDS is the minimum detectable signal power in dBm. Once the 1-dB compression point is known, it is a fair assumption that the 3rd-order intercept point, expressed in dBm, is 10 dB above this.

The intermodulation distortion gives the quality of an amplifier to withstand multiple signals without generating large intermodulation products or when using amplitude-modulated signals, cross-modulation. Figure 8.29 provides a quick reminder of IMD issues.

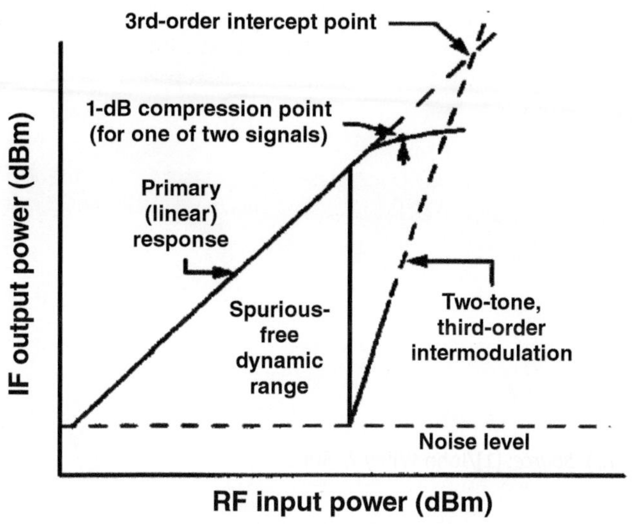

Figure 8.29 Amplifier linearity evaluation, including compression and two-tone IMD dynamic range. P_{-1dB} for a single-tone cannot be read directly from this graph because the values shown are the result of two equal-power tones. *Source*: [1]/John Wiley & Sons.

8.2 Single-stage Feedback Amplifiers

8.2.1 Wideband Feedback Amplifier [1, pg. 437]

The amplifier shown in Figure 8.30 is a standard minimal configuration amplifier with a reduction of current distortion because of the emitter feedback. The voltage gain is:

$$VG = \frac{R_L}{\frac{26\text{ mV}}{I_C} + R_e} \tag{8.120}$$

where R_L is the total load seen at the collector.

The feedback amplifier in Figure 8.31 uses voltage feedback to "correct" the input and output impedance to be close to 50 Ω, but does not improve the current distortion. The voltage gain of this circuit is:

$$VG = \frac{R_F}{R_g} \tag{8.121}$$

$$\begin{aligned} R_{IN} &= R_{OUT} \\ &= R_F R_D \text{ with } R_D = \frac{26\text{ mV}}{I_C} \end{aligned} \tag{8.122}$$

Example for $I_C = 10$ mA → $R_D = 2.6$ Ω. Assuming $R_F = 200$ Ω, the input resistance $R_{IN} =$ 22.63 Ω. The gain is calculated by getting the inverse of $R_D = g_m =$ 0.39 A/V. Once this is multiplied by the output impedance of 22.63 Ω, the resulting gain becomes 10 log (8.6) or 9.3 dB. The gain can be increased by either using a larger collector current or a larger feedback resistor. It should be remembered that this circuit does not improve the distortion caused by the base-emitter junction.

To reduce the current distortion as well, we use the circuit shown in Figure 8.32, which contains both types of feedback. In reality, the collector-base resistor is frequently split into an unbypassed section responsible for the feedback and a larger resistor responsible for the appropriate biasing. The example circuit in Figure 8.32 shows this.

In this book, we concentrate on using high-frequency transistors, and parasitic elements are going to play an important role. The very moment we design our own transistors, and then we use a foundry, things get complex. To explain how complex things would be, here is an example (Figure 8.32):

Some Useful Examples: The following examples are a potpourri of useful circuits, either a single or dual stage, or with bipolar or field-effect transistors.

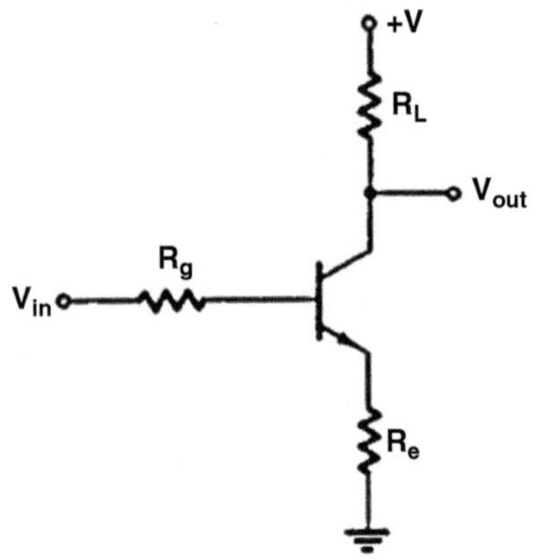

Figure 8.30 Single-stage amplifier with current feedback (provided by R_e). *Source*: [1]/John Wiley & Sons.

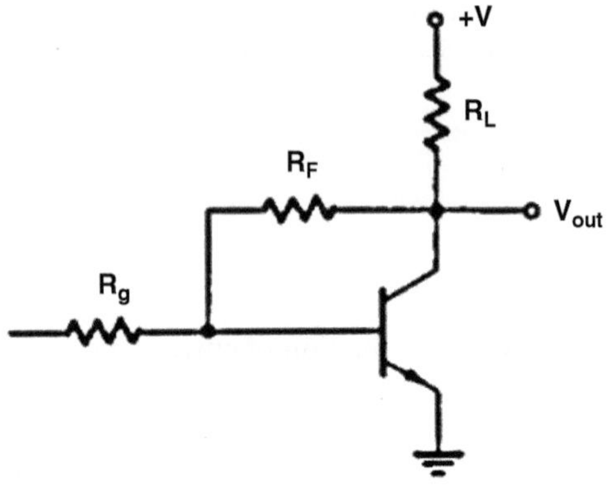

Figure 8.31 Single-stage amplifier with voltage feedback (provided by R_F). *Source*: [1]/John Wiley & Sons.

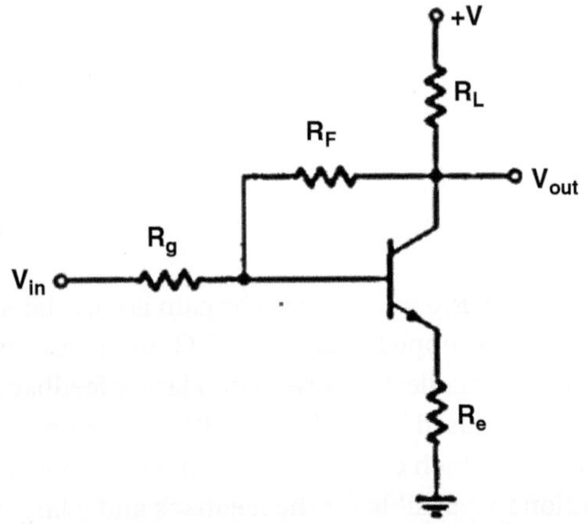

Figure 8.32 Single-stage amplifier with voltage and current feedback. *Source*: [1]/John Wiley & Sons.

8.3 Example: 6... 18 GHz Single Ended Amplifier Using a Foundry FET

The frequency range is a popular range for many applications, and so we chose this example.

This example came from the work done as a member of the Raytheon-TI joint venture DARPA program, and the particular FET was supplied by Texas Instruments. This foundry no longer exists.

Following the negative voltage supply is a matching circuit which provides a wideband 50 Ω transformation to the gate. Likewise, in the collector, transmission lines transform the output conductance (typically 1/400 Ω) into the 50 Ω load (Figure 8.33).

The following netlist describes an actual circuit, assuming a linear FET model, as frequently supplied by the foundries. It includes the SPICE parameters of the transistor.

```
blk
  fet  1 2 0 g=0.045 cgs=.3pf f=1e19 t=3ps gds=0.0029 cdg=0.02pf
+ cds=0.085pf  ri=4.111 rd=3.8 rs=3.5 lg=0.01127nh  ls=.02nh ld=.02nh
  ti:2por 1 2
end
*
blk
   ind  1  5   l=1nh
   res  5  0   r=100
   ost  1  0   z=?100? e=?64? f=18ghz
   sst  1  0   z=70    e=131   f=18ghz
   trl  1  2   z=100   e=23  f=18ghz
   ti   2  3
*  TI fet
   trl 3 4 z=100 e=16 f=18ghz
   sst  4  0   z=?40 51 100? e=70 f=18ghz
  amp:2por 1  4
end
freq
  step 4ghz 22ghz.5ghz
end
.MODEL FET
+ idss=.45449e-01  vp0=-2.0843  gama=-.10738  e=1.5585
+ ke=-.41829  sl=.90179e 01  kg=.000  t= 15881e-13
+ ss=.19998e-02  ig0=2.26e-12 afag=13.5  ib0=129e-6
+ afab=.389  vbc=9.9  r10=1.395  kr=.35605
+ c10=.22836e-12 k1=1.9646 c1s=.1e-14 cf0=.62559e-14
+ kf=.25384  rg=.12  rd=1.89 rs=1.8955
+ lg=.15e-10 ld=.27725e-10 ls=.20563e-10
+ cds=0.0 cdsd=.1e-8  rdsd=331.98
+ cge=.14329e-14 cde=.13695e-14 vdmx=6)
```

In Figure 8.33c, the first curve is S_{21} in dB, the second curve is the noise figure in dB, the third curve is the input match (S_{11} in dB), and the fourth curve is S_{22} in dB. These results are prior to optimization.

In Figure 8.33d, the first curve is S_{21} in dB, the second curve is the noise figure in dB, the third curve is the input match (S_{11} in dB), and the fourth curve is S_{22} in dB. These results are after optimization.

The gain optimization was done at the cost of increased noise figure, as seen from Figure 8.33d. It shows the difference between S_{11} complex conjugate and Γ_{opt}, the best noise figure. The S_{11} increases the noise figure to about 4 dB from the 2 dB prior to the gain optimization.

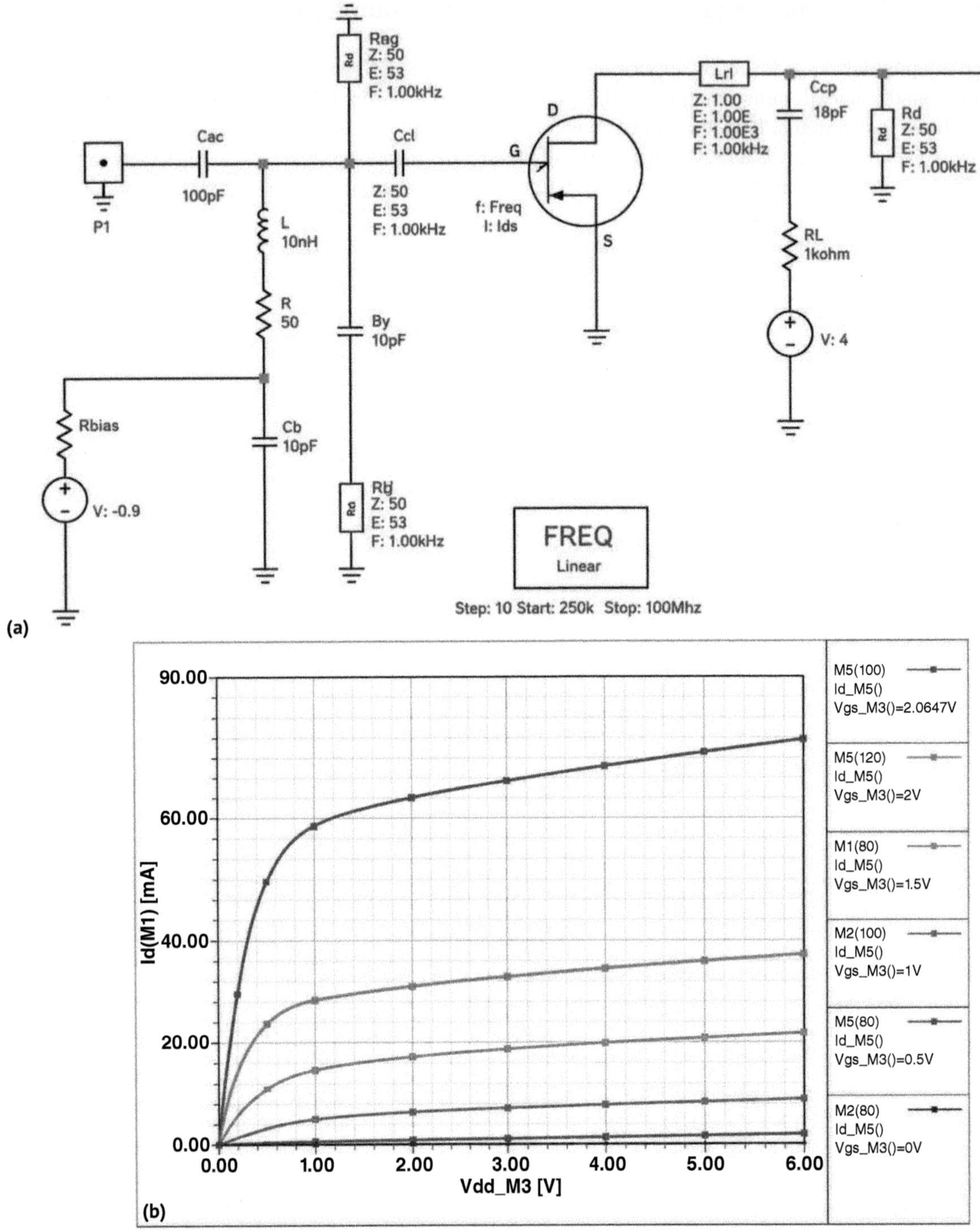

Figure 8.33 (a) 6… 18 GHz single-ended amplifier using the former TI-Foundry FET. (b) DC-IV curve of the FET used for the circuit of (a).

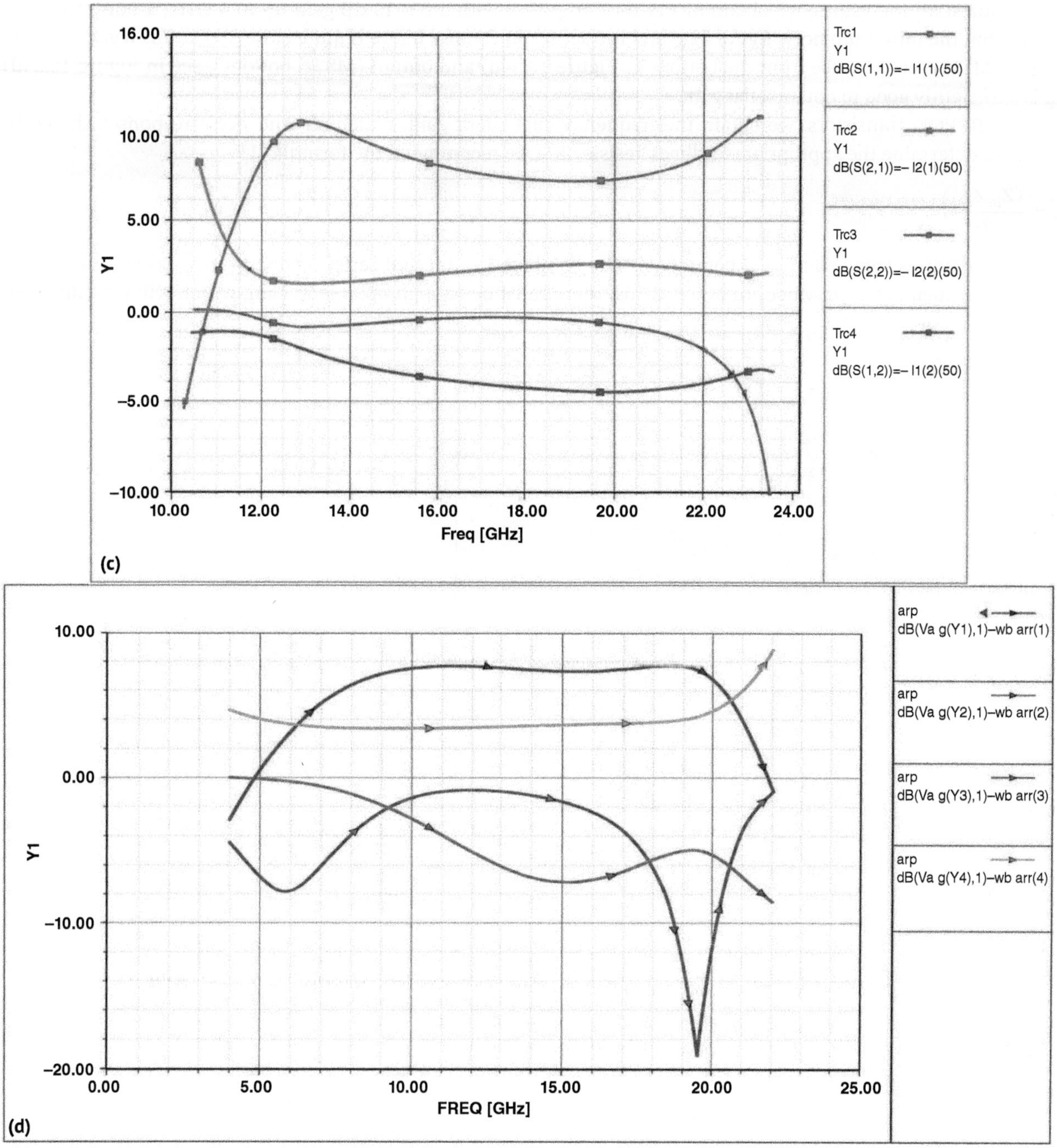

Figure 8.33 (Cont'd) (c) Initial frequency response of the amplifier circuit shown in (a). (d) Frequency response of the optimized FET amplifier—from 5 to 20 GHz.

8.4 Example: Single-stage Amplifier with Voltage and Current Feedback

Figure 8.34a is a practical amplifier circuit showing voltage and current feedback, and illustrating how reactances (in this case, 500-pF bypass capacitors) and multiple resistors can be used to provide different ac and dc resistances for the paths from collector to base and emitter to ground.

Having optimized all the values, we obtain a very nice amplifier with a flat 10 dB gain up to 4 GHz, a noise figure of 4 dB. The fact that the minimum noise figure F_{min} and the actual 50 Ω noise figure is quite close, indicating a good design. Finally, input (MS11, second trace from the bottom in Figure 10.34d) and output (MS22, bottom trace in Figure 10.34d) matching is sufficiently good to obtain a flat gain.

By using the BFP620 transistors, we obtain the frequency-dependent gain, matching, and noise responses shown in Figure 8.34d. To determine the appropriate feedback resistor R_F, we recommend the formula [1]

$$R_F = \frac{(Z_{in}Z_{out})}{R_E} \tag{8.123}$$

This equation requires some additional comments. The combined shunt and series feedback only works as long as a sufficient transconductance is provided. In other words, there must be enough open-loop gain, which will be reduced to

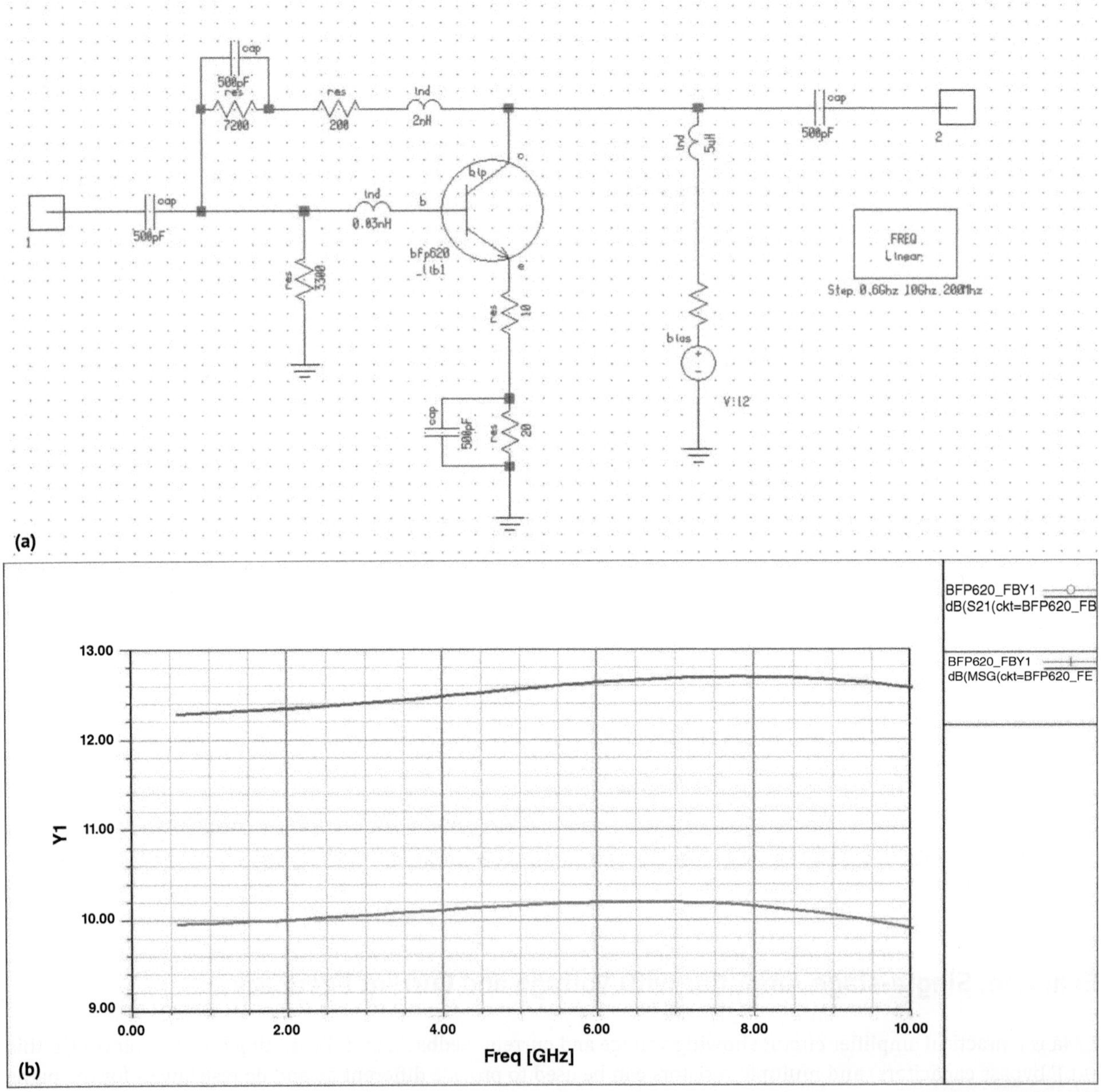

Figure 8.34 (a) CAD simulation of the single-stage amplifier with voltage and current feedback. (b) Simulated maximum stable gain (top curve) and the practical available gain.

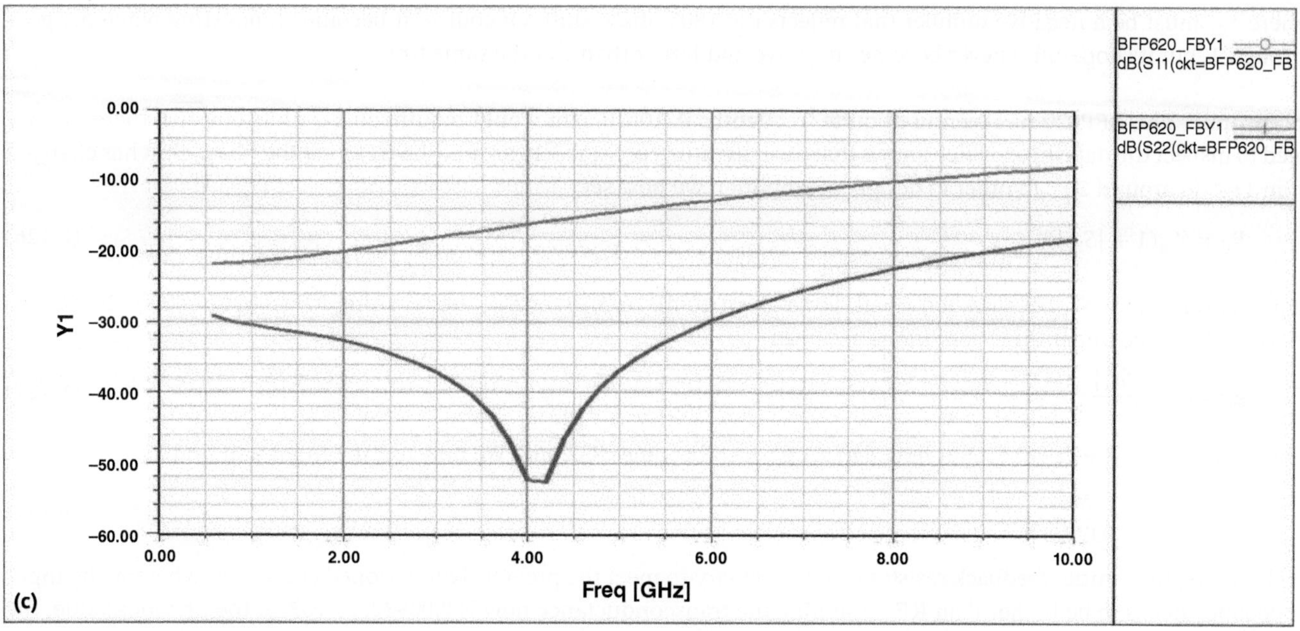

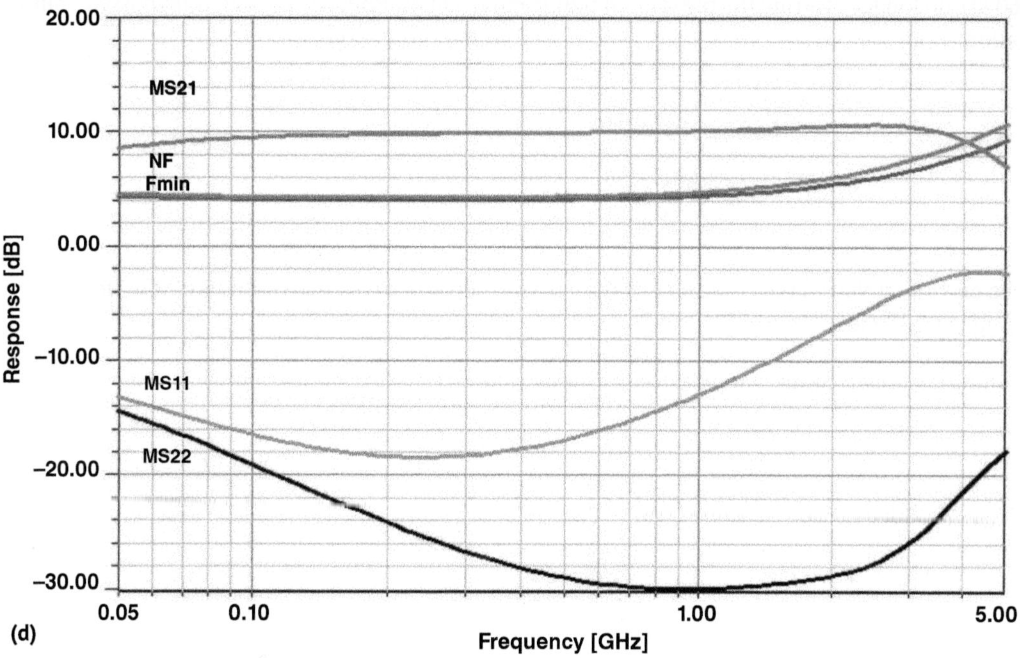

Figure 8.34 (Cont'd) (c) Simulated input (S_{11}) and output (S_{22}) matching. (d) Frequency-dependent gain, matching, and noise performance of the circuit shown in (a). *Source*: [1]/John Wiley & Sons.

obtain more bandwidth. A necessary condition for this approach, which really comes from the frequency range of less than 30 MHz, is that

$$Z_{\text{in}} = \frac{26\ \text{mV}}{I_C}\beta >> R_{\text{F}} \tag{8.124}$$

This requires a minimum transconductance of

$$g_{\text{m}} = \frac{1 + S_{21}}{Z_{\text{in}}} \tag{8.125}$$

where S_{21} must be a negative number that reflects the 180° phase shift. Of course, it becomes interesting when S_{21} goes through 1, but it (hopefully) never becomes negative and larger than 1 at the same time.

Example: The BFP620 has been used quite frequently. A 10-dB gain would require an S_{21} of $3.16(\sqrt{10})$; therefore, we need to inspect the datasheet, which shows that the magnitude of S_{21} is 3.16 around 4.5 GHz, yet the phase shift has changed from 180° to around 20°. In order to design an amplifier, we now set

$$R_F = Z_{in}(1 + |S_{21}|) \tag{8.126}$$

which is valid until S_{21} becomes too small. For a given 10 dB or 3.16 for S_{21}, the resulting RF is 4.16 × 50, or 208Ω. The transconductance under these conditions becomes

$$g_m = \frac{1 + |S_{21}|}{Z_0} \tag{8.127}$$

where $Z_0 = 50\ \Omega$ (the desired input impedance). g_m now becomes 83.2 mS. We now test the input and find

$$\frac{\beta}{g_m} = Z_{in} = \frac{50}{0.0832} = 600\ \Omega \tag{8.128}$$

Therefore, the emitter feedback resistor is not necessary to meet the previously mentioned condition, whereby the input impedance needs to be higher than RF. Assuming the transconductance now is 0.00832, or 10% of the previous value, the input impedance would be 60 Ω), no longer meeting the prescribed requirement. Therefore, we need an unbypassed emitter resistor that fulfills the equation

$$R_E = \frac{Z_0^2}{R_F} - \frac{1}{g_{m_{dc}}} \tag{8.129}$$

where

$$g_{m_{dc}} = \frac{1 - S_{21_{max}}}{Z_0} \tag{8.130}$$

Again, the reader should be cautioned that in the common-emitter circuit, there is 180° phase shift, and therefore $S_{21_{max}}$ is a negative quantity, resulting in reality in

$$g_{m_{dc}} = \frac{1 + S_{21_{max}}}{Z_0}$$

In our case,

$$g_{m_{dc}} = \frac{-S_{21}}{2Z_0} = \frac{38}{2 \times 50} = \frac{38}{100} = 0.38$$

so

$$R_E = \frac{50^2}{208} - \frac{1}{0.38} = 12 - 2.6 = 9.4\ \Omega \tag{8.131}$$

In practice, the designer would use a 10-Ω resistor.

Since the gain of the transistor continues to fall as a function of frequency, one can compensate to a degree of the cutoff frequency by putting an inductance in series that follows the equation

$$L_F = \frac{R_F}{2\pi f_{3\,dB}} \tag{8.132}$$

Assuming $f_{1\,db} =$ 26 GHz, the resulting series inductance $L_F =$ 2nH. Basically, the same applies to circuits where the second transistor is used as an emitter follower to feed the signal back from the collector to the base of the first transistor.

In summary (see Figures 8.35 and 8.36):

- Resistive negative feedback at lower frequencies reduces open-loop gain and simultaneously matches input and output.
- Reactive elements set gain and match at high frequencies. Feedback is reduced at high frequencies.
- Emitter bypass capacitors are not used, as they would destroy the impedance match.

This is equally valid for bipolar transistors, as we will show.

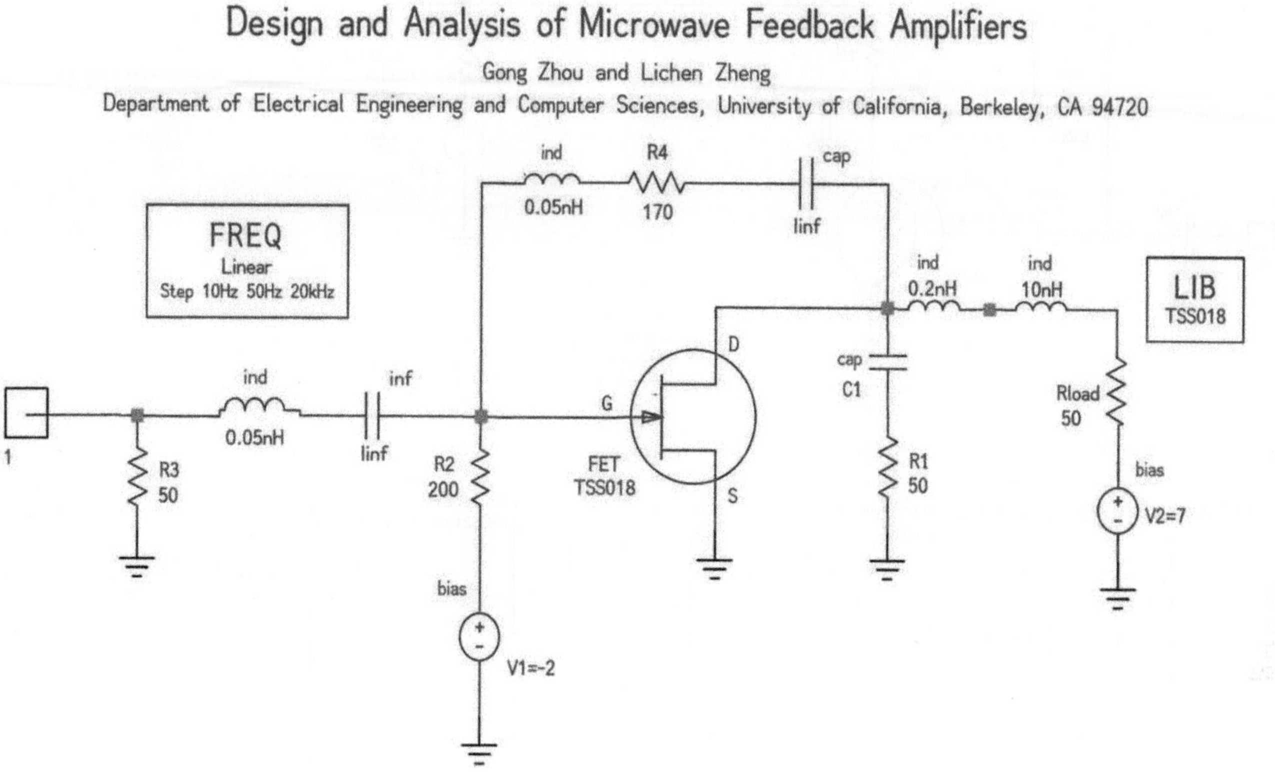

Figure 8.35 Microwave feedback amplifier.

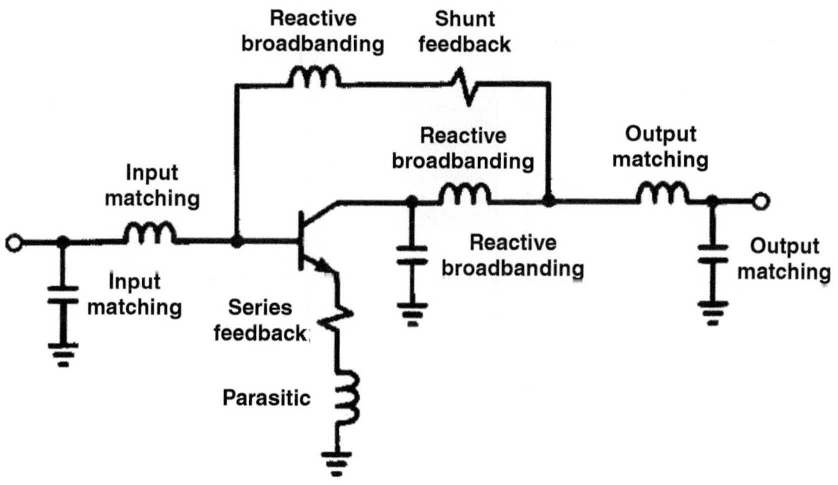

Figure 8.36 Optimized wideband feedback amplifier. *Source:* [1]/John Wiley & Sons.

8.5 Example: Numerical Example of a Dual Feedback Amplifier Using a BFP840 Transistor

Here is an example of designing a simple dual-feedback amplifier. Figure 8.37a shows the collector-base part as the voltage-feedback, which controls the input and output impedance, while the emitter resistor (10 Ω) is the current feedback. We all need to note that the current feedback reduces the distortion, while the collector feedback improves matching.

Figure 8.37b shows the frequency-dependent S_{21} in dB Flat Gain response up to 2 GHz (in dB).

Figure 8.37c shows the frequency-dependent S_{11} and S_{22} in dB.

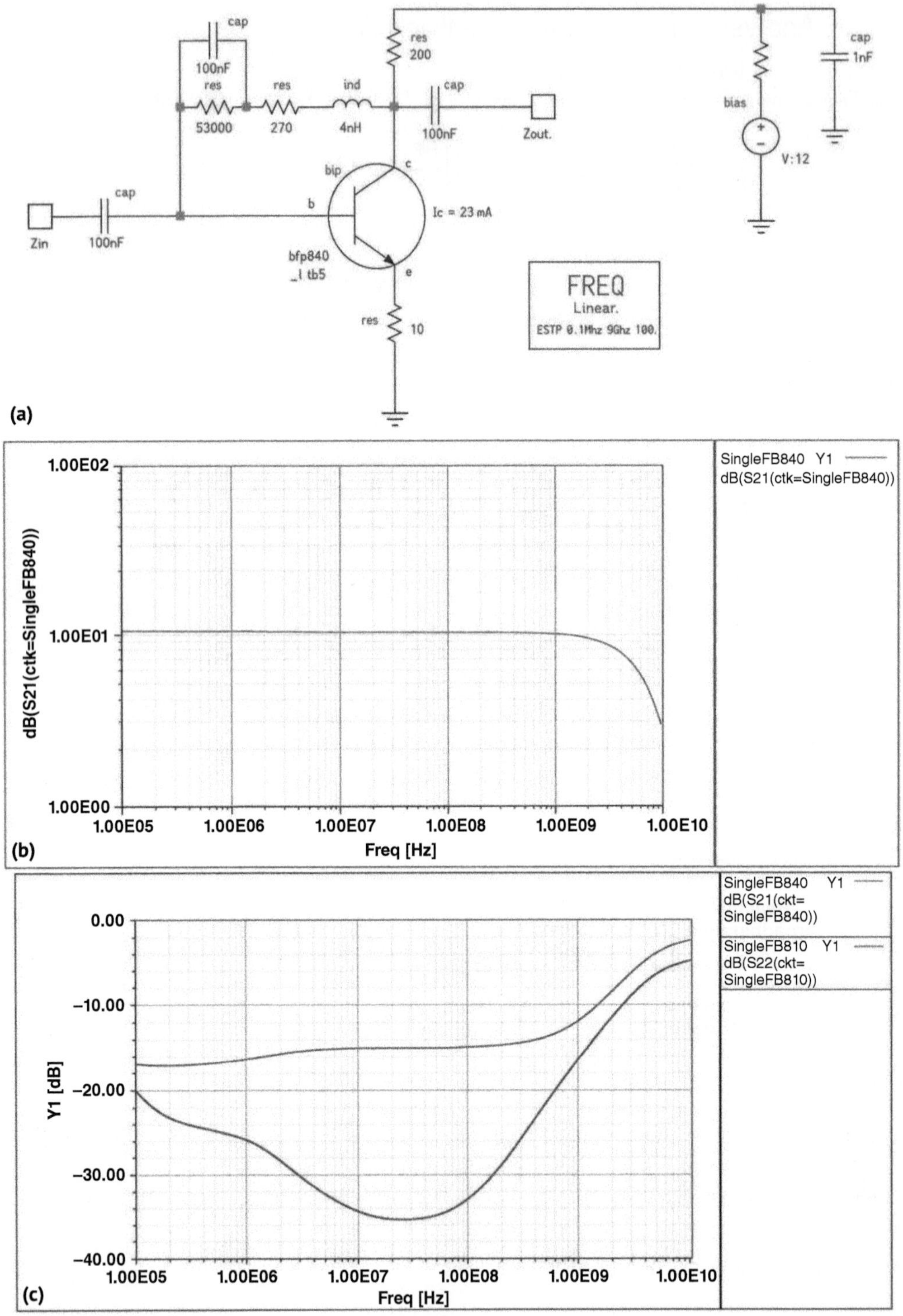

Figure 8.37 (a) Test circuit for a dual-feedback amplifier. (b) Flat gain response up to 2 GHz (in dB) of the amplifier circuit shown in (a). (c) Simulated input and output matching of the amplifier circuit in (a).

Figure 8.37d shows the F_{min} and F50 ohms as a function of frequency, of about 3.3 dB, which means the circuit is perfectly matched.

Figure 8.37e shows the reverse isolation S_{12} in dB, which varies very little, between −14.5 and −16 dB.

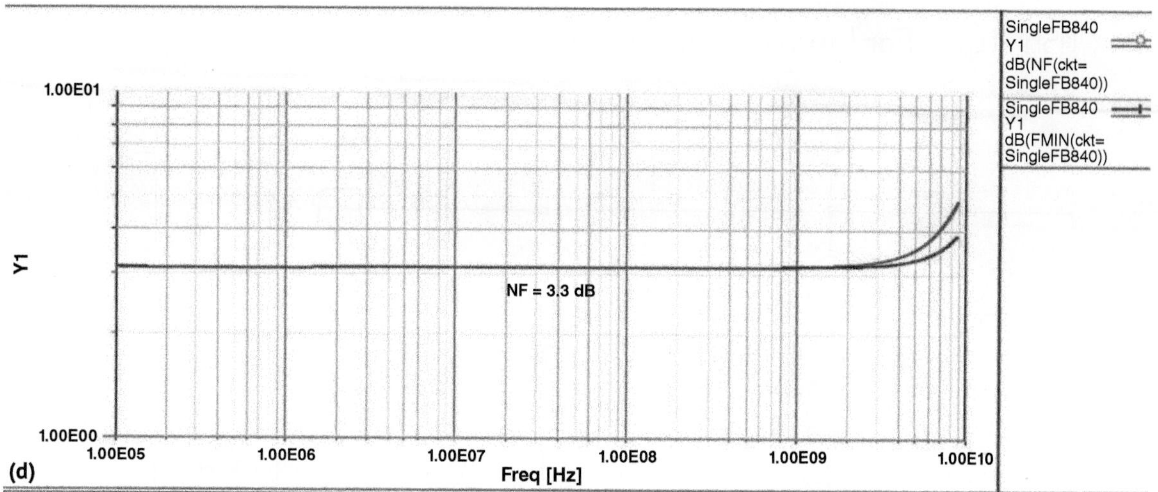

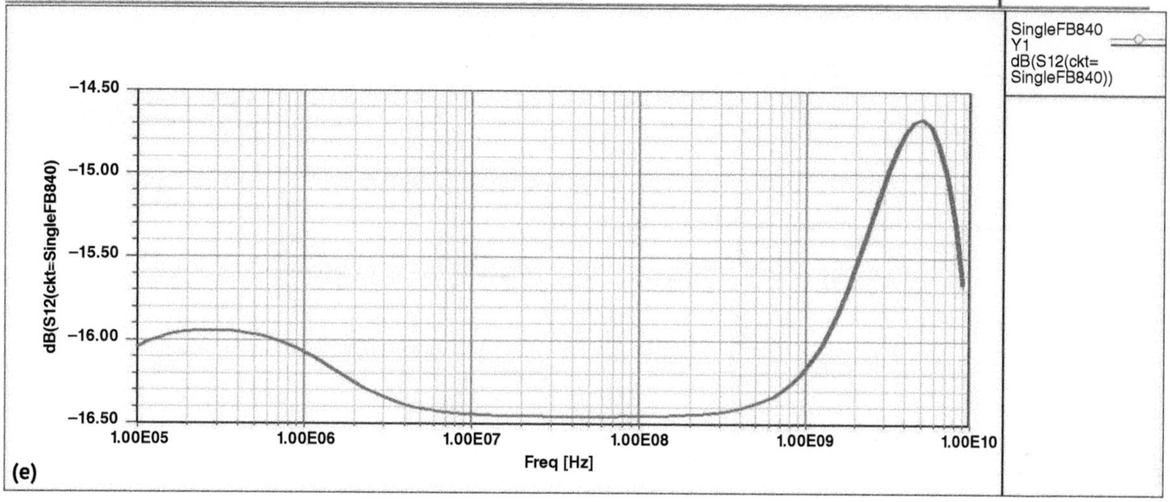

Figure 8.37 (Cont'd) (d) Simulated minimum and 50 Ω noise figure as a function of frequency of the amplifier circuit in (a). (e) Simulated reverse isolation S_{12} (in dB) of the amplifier circuit in (a).

8.6 Multi-stage Amplifier

There is a need for wideband integrated amplifiers with minimal external components. A good example is a dual-Darlington amplifier. The first stage is an actual emitter follower that drives the output stage. Both stages have a common-collector output impedance, in our case 200 Ω, but the RF termination is 50 Ω. So the gain of this amplifier is (50 Ω||200 Ω)/10 Ω (Emitter resistor of the last transistor).

The manufacturer referred to this device as a multi-stage amplifier (MSA) device. The DC current of the second transistor (26 mA) is significantly larger than that of the first transistor (4.8 mA). For simulation purposes, we use the BFP840 and BFP640, seen in Figure 8.38a.

The difference between the two transistors can be seen easily in Figure 8.38b.

Figure 8.38c shows the DC operating points of the two transistors, while Figure 8.38d shows the simulated Gain (S_{21} in dB) and Stability factor K (dimensionless).

Figure 8.38e shows the simulated minimum (F_{min}) and a 50 Ω Noise Figure of about 4 dB.

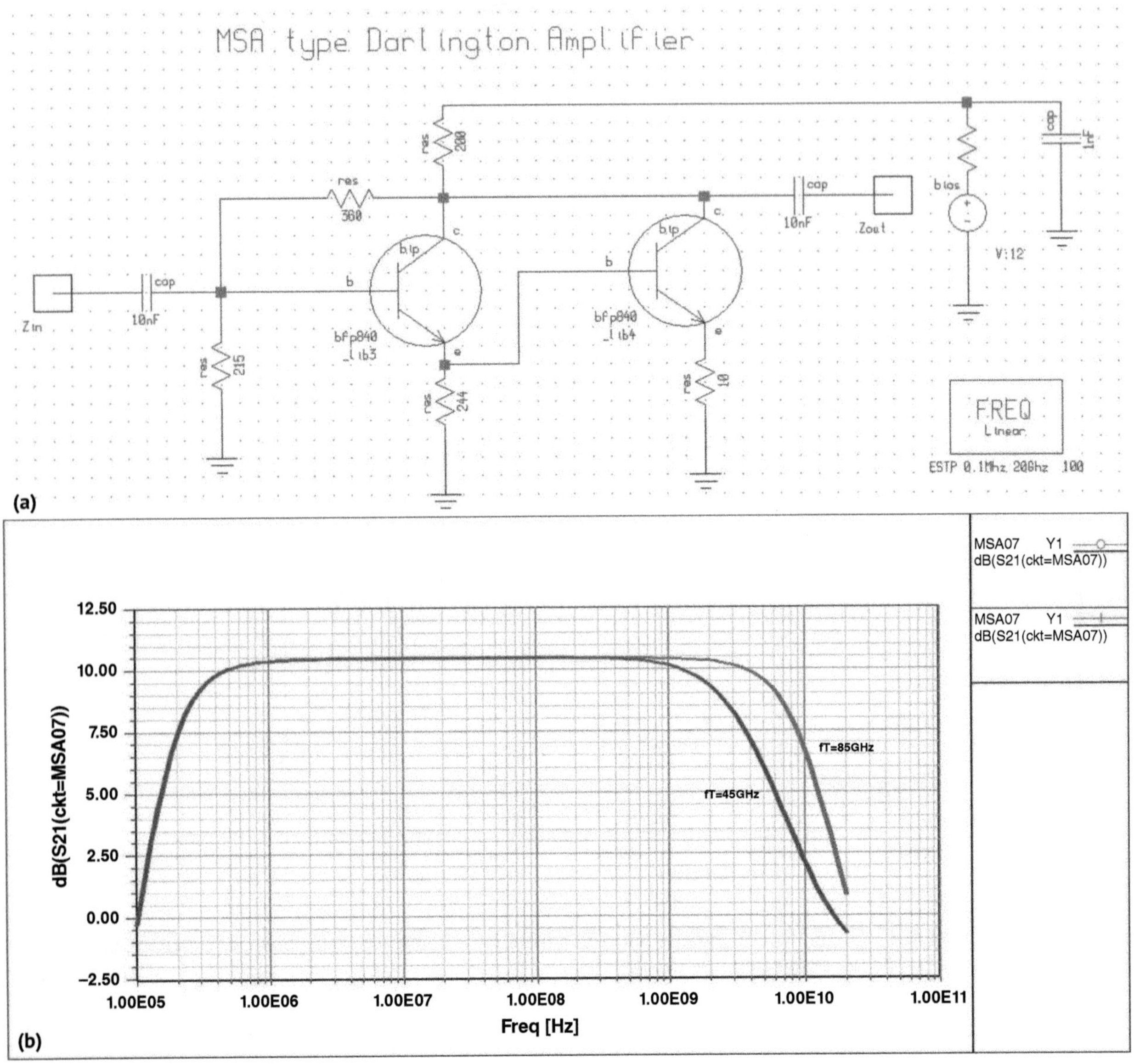

Bias Point Values

Voltage	Current
Vbe(_lib3) = 0.827194 V	Ib(_lib3) = 15.7152 uA
Vce(_lib3) = 3.21045 V	Ic(_lib3) = 4.86383 mA
Vbe(_lib4) = 0.885899 V	Ib(_lib4) = 0.149484 mA
Vce(_lib4) = 4.09634 V	Ic(_lib4) = 26.674 mA

(c)

Figure 8.38 (a) Multi-stage type Darlington amplifier. (b) Simulated frequency response of the MSA device. (c) DC operating points of the two transistors.

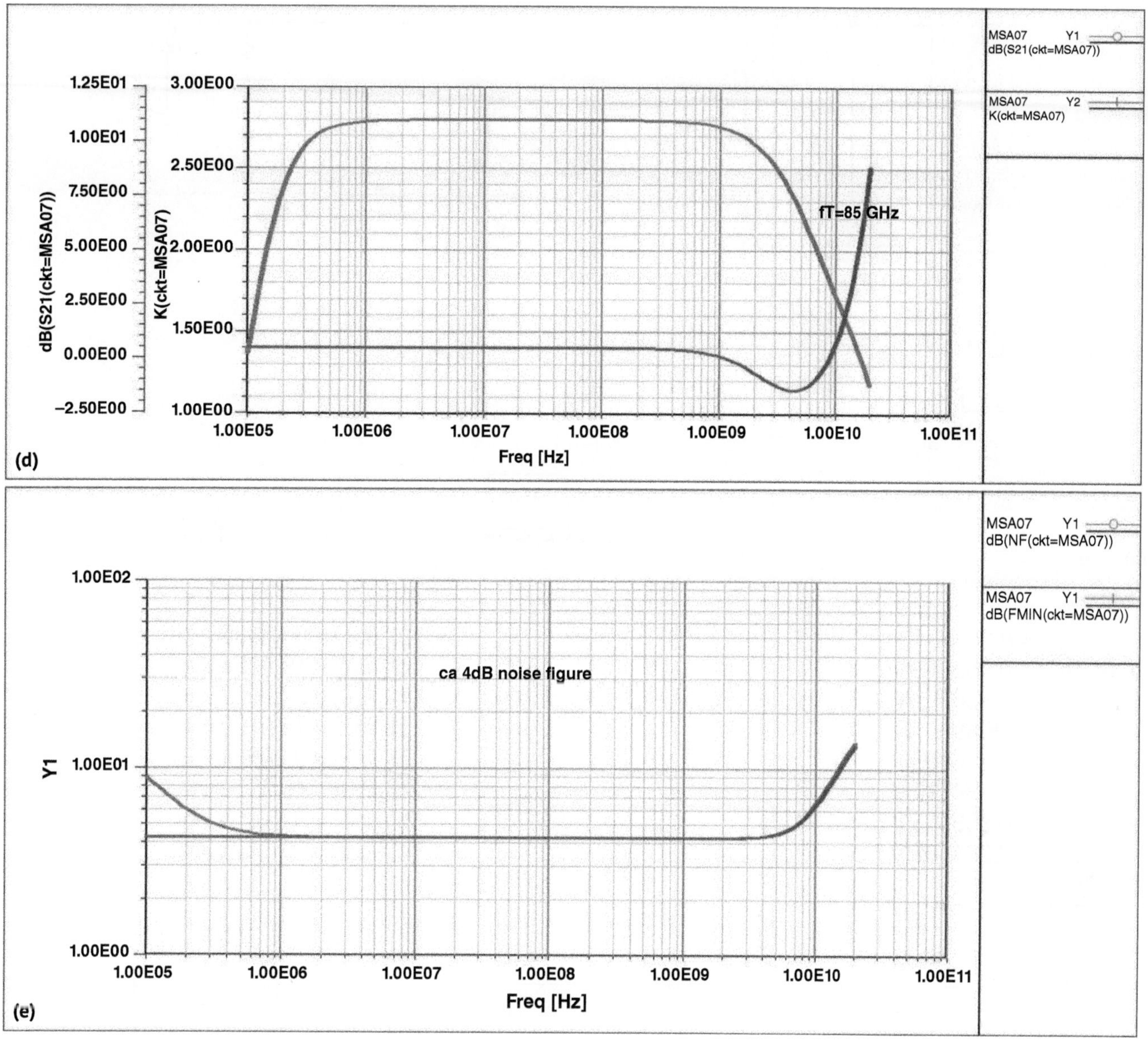

Figure 8.38 (Cont'd) (d) Simulated gain (S_{21} in dB) and stability factor K (dimensionless). (e) Simulated minimum ($F_{\min}$) and 50 Ω noise figure.

8.7 Example: Monolithic Selective Amplifier

In the introduction statement, we have pointed out that we are somewhat limited in using information about the foundry data. Figure 8.39 is a simulated foundry amplifier, which normally would not be built in a discrete form, shown here. The main RF portion is a cascode amplifier, with a low-Q tuned circuit in the collector, a matching circuit to an emitter-follower, which provides the output. Therefore, this amplifier should be referred to as a selective amplifier.

Based on the feedback, the resulting NF is about 4 dB.

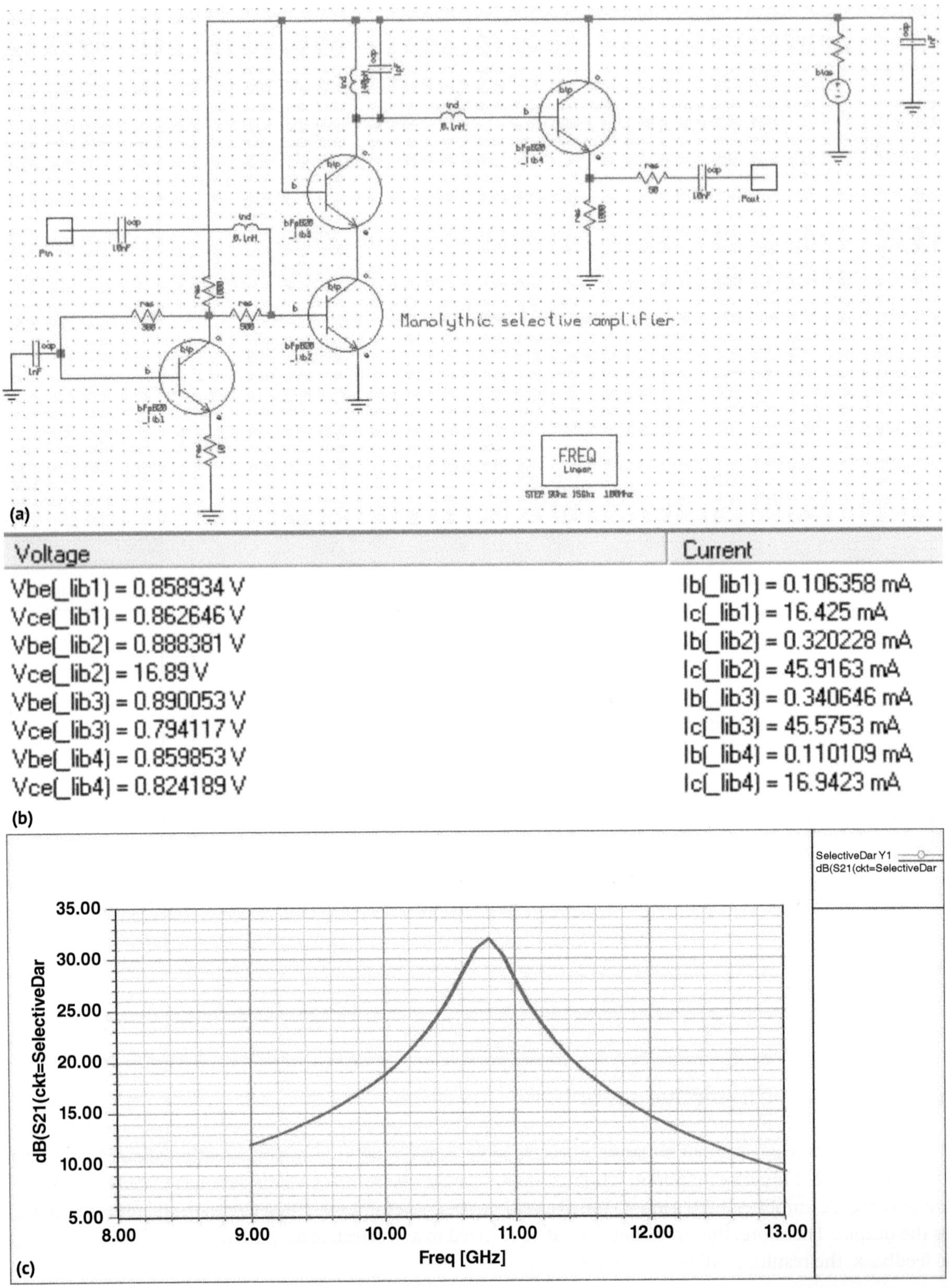

Figure 8.39 (a) Monolithic selective amplifier. (b) DC-operating points of the various transistors used in the circuit of (a). (c) Simulated gain of the circuit of (a) peaking at 10.8 GHz.

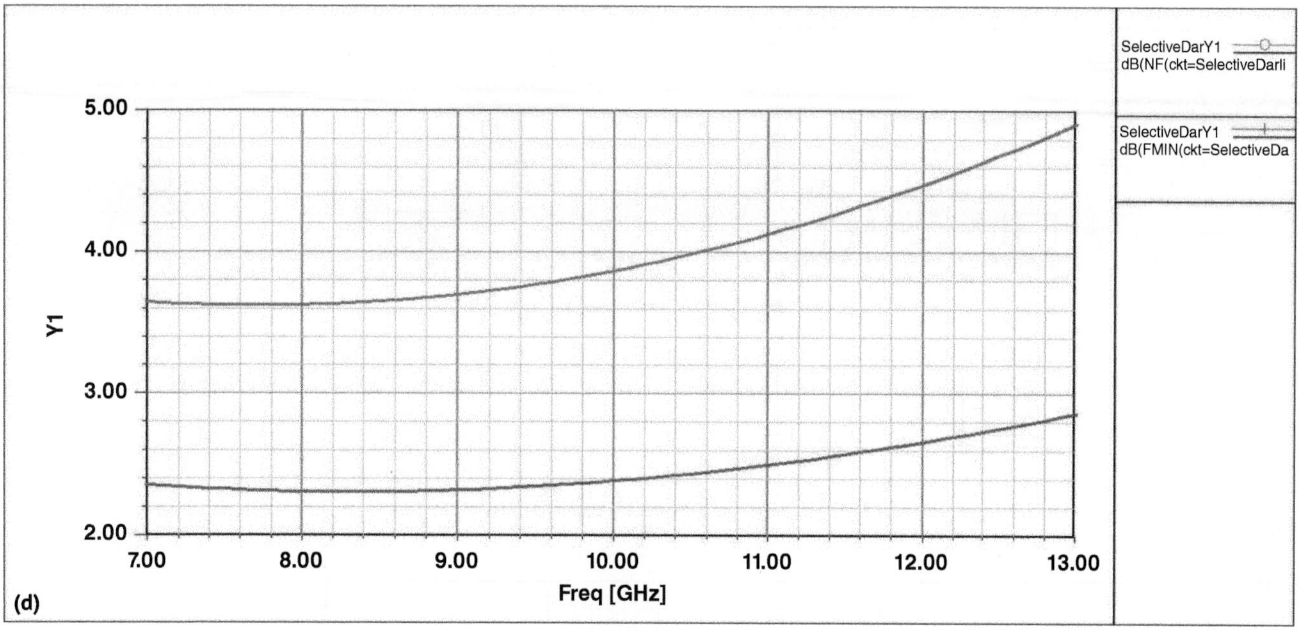

Figure 8.39 (Cont'd) (d) Simulated minimum NF and 50 Ω NF of the circuit of (a).

8.8 25 GHz C-SiGe Transistor Amplifier with Gain Peaking

We already looked at the multi-stage amplifier MSA above, and now we introduce the concept of gain peaking. The inductor in series with the first transistor (0.3 nH) and the matching inductor in the first and the second transistor will provide gain peaking, recommended by Yuan [24] and other references [25, 26] (Figure 8.40).

8.9 Example: 2N4417 Cascode Amplifier

The Si-junction FETs (JFETs) fell out of grace a long time ago, as the production showed too many tolerances of Y21, almost 3:1. The 2N4417 is a microwave packaged version of a familiar 2N4416. The following circuit shows a cascode, with a matching inductor between the two transistors and a matching circuit at the input and output (Figure 8.41).

The radio amateurs among us recognize this as the 70 cm band.

For 432 MHz, the noise figure is about 1.2 dB as seen in Figure 8.41d.

The maximum stable gain, maximum available gain, minimum noise figure, and available gain up to 1 GHz are seen in Figure 8.41e.

8.10 Single Stage Amplifier with CFY99A GaAs FET

Previously, we have been using a Silicon JFET, and now we switch to a more exotic GaAs FET, which will operate at 90 GHz. The circuit seen in Figure 8.42a is optimized for a low noise figure. This amplifier shows input/output matching better than 30 dB, and a simulated gain of about 12 dB, as shown in Figure 8.42b. The simulated Noise Figure was less than 4 dB.

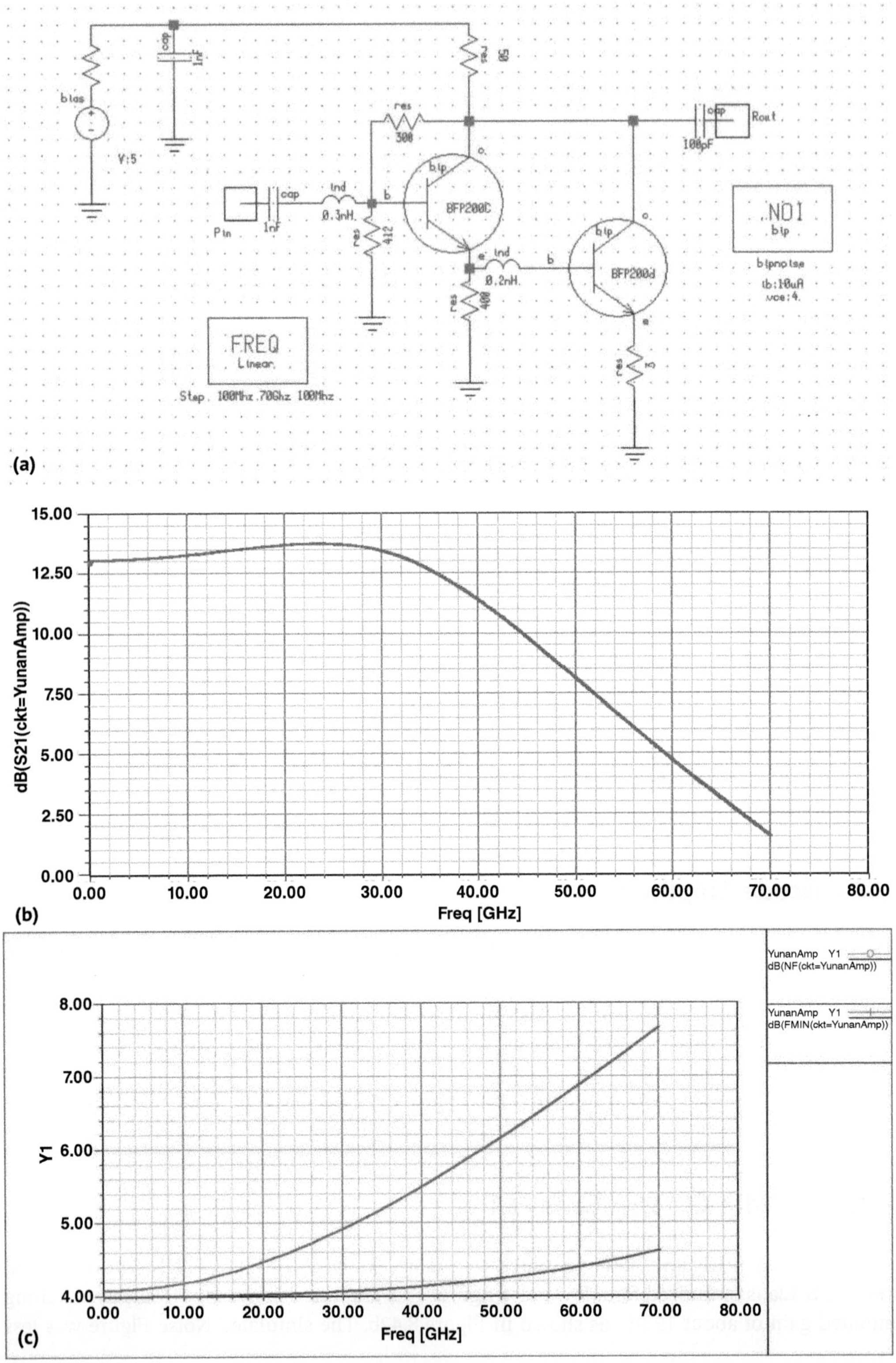

Figure 8.40 (a) 200 GHz C-SiGe transistor amplifier with gain peaking. (b) Simulated gain (S_{21} in dB) peaking at about 25 GHz. (c) Simulated $F_{\min}$ and a 50 Ω noise figure.

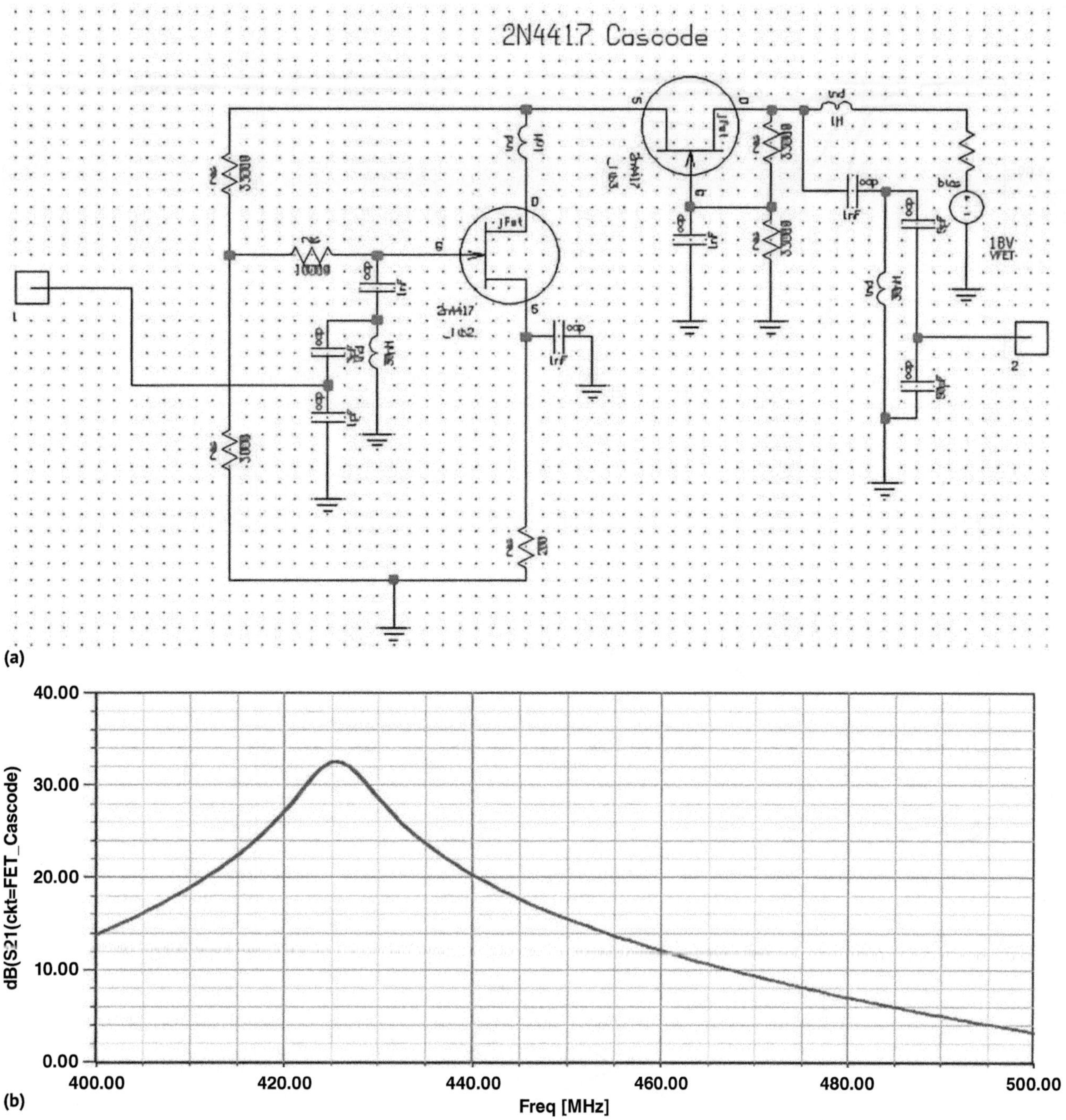

Figure 8.41 (a) 2N4417 Cascode amplifier. (b) The gain simulation indicates that this amplifier was designed for 432 MHz.

(Continued)

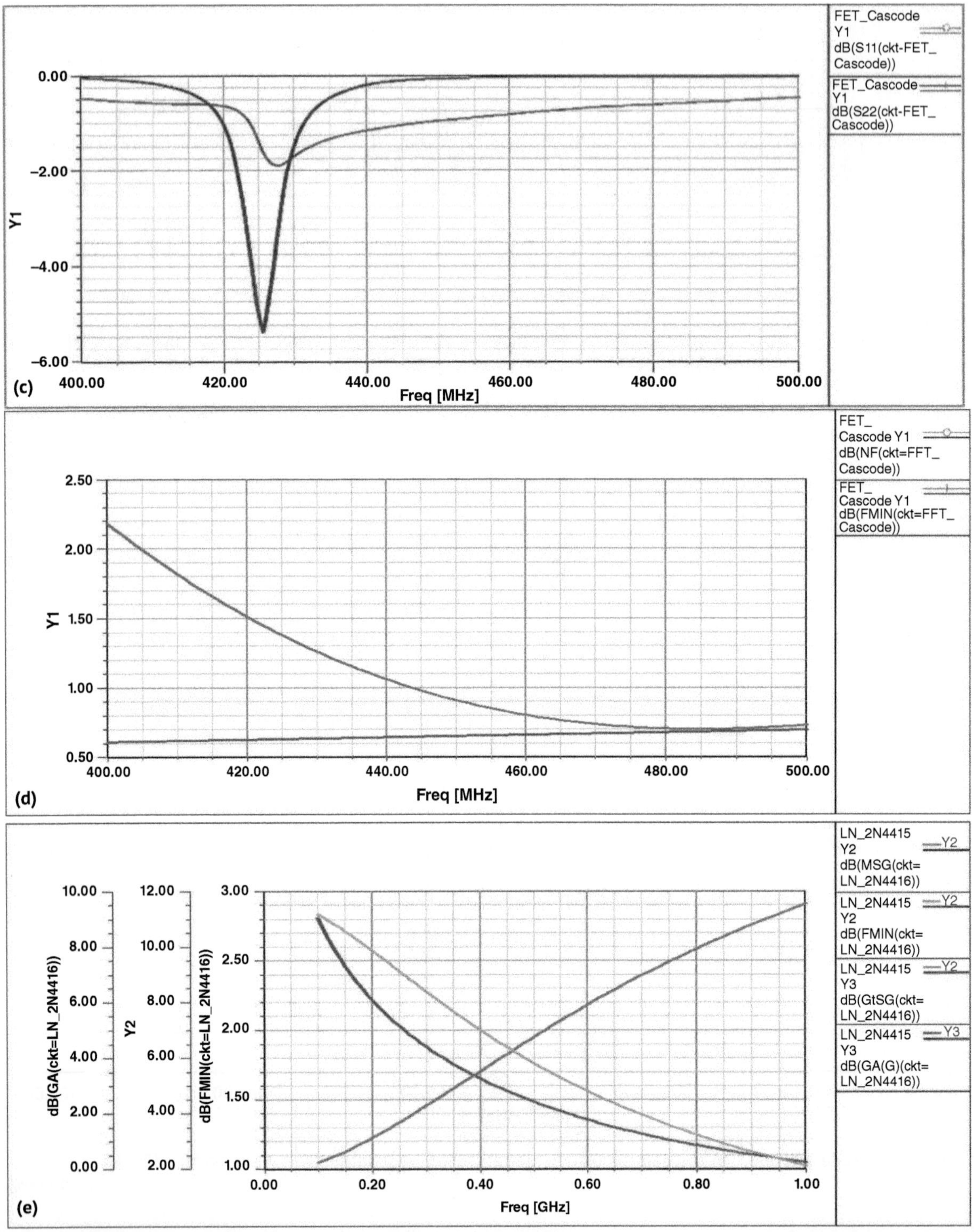

Figure 8.41 (Cont'd) (c) Simulated input and output matching of the circuit in (a). (d) Simulated min noise figure (FMIN the lower trace) and 50 Ω noise figure (NF the upper curved trace). (e) Max stable gain, maximum available gain, minimum noise figure, and available gain up to 1 GHz.

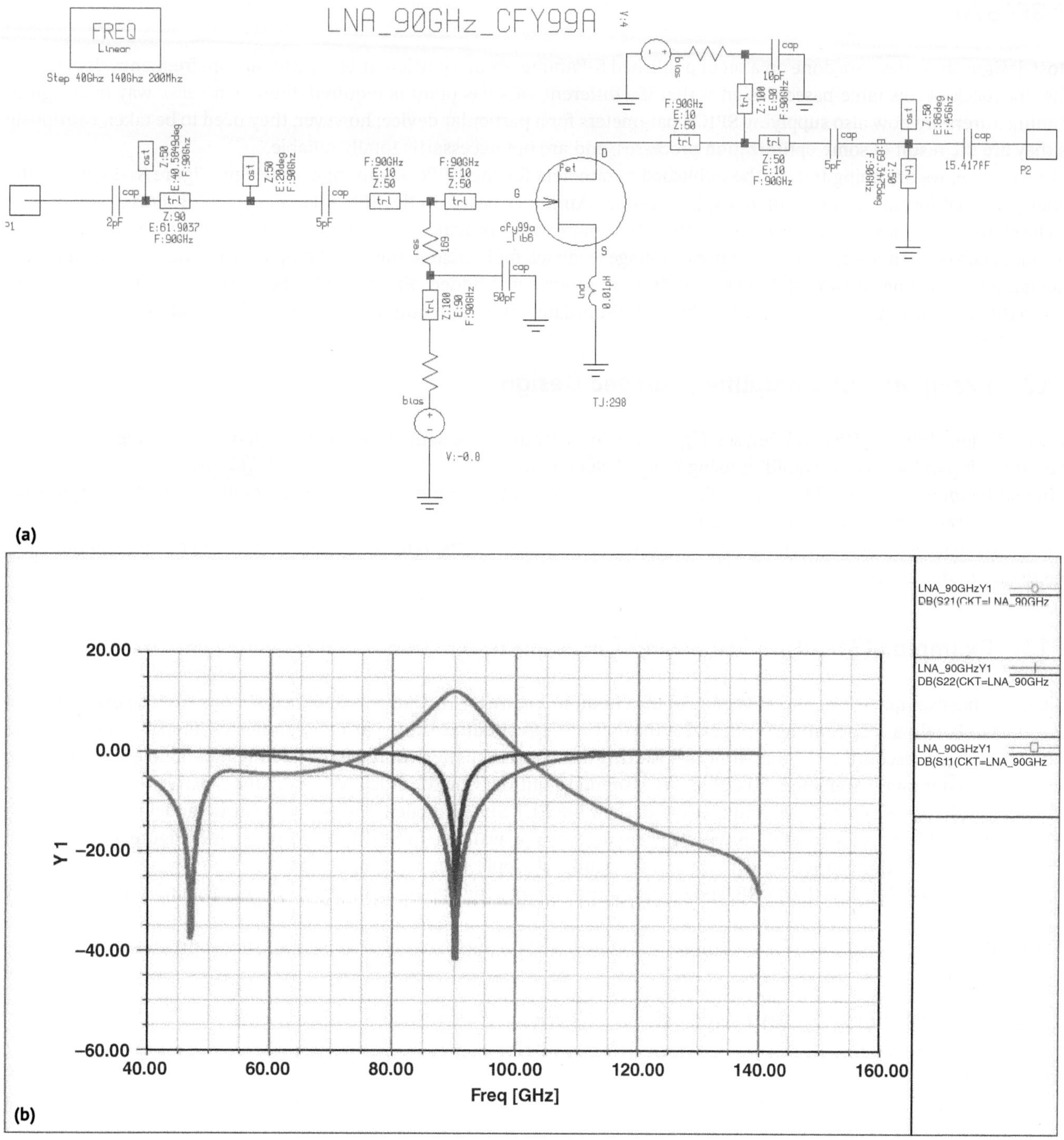

Figure 8.42 (a) Schematic for a 90 GHz GaAs FET-based low noise amplifier. (b) Simulated gain and I/P and O/P matching for the 90 GHz GaAs FET-based low noise amplifier.

8.11 Example: Optimization of a Single-stage Amplifier Based on the SPICE Parameters of BFP620

Most designs these days are done on a set of published S-Parameters at a particular bias point and are frequency dependent. The drawback of this table-based design is that if a different DC-Bias point is required, there is no easy way to design it. Manufacturers are now also supplying SPICE parameters for a particular device; however, they need to be taken cautiously as they are the result of some optimization processes, and are not necessarily totally reliable.

In this case, we are going to take the published parameters for the BFP620 and make a design. Figure 8.43a shows the schematic used for the optimization of the Single-stage Amplifier based on the SPICE Parameters of BFP620.

The simulated performance was based on the SPICE parameters provided by the manufacturer and agrees very well with the published performances. The circuit uses a voltage feedback for increased input and output matching. To avoid spurious oscillations, a loading resistor of 200 Ω from drain to ground was added. Figure 8.43b shows the simulated power gain (S_{21} in dB) of the circuit, and Figure 8.43c shows the simulated input and output matching of this amplifier.

8.12 Example: 7 GHz Amplifier, Lumped Design

Having designed the 7 GHz amplifier, see Figure 8.44, now let us do the same using lumped elements. See Figure 8.44a for the simulation of the 7 GHz Amplifier using lumped elements.

In reality, some of the small inductor values, such as the 3 nH at the input, are really traces on the printed circuit boards. It is probably correct to call it a hybrid circuit.

The gain peaks at about 6.5 GHz, the noise figure is about 2 dB, and the return loss at the input and output is about 8 dB. See Figure 8.44b.

8.13 Example: 430 MHz JFET 2N4417 Preamplifier

The following example is a modified version of the circuit in Figure 8.41a. Compared to the previous cascode example, this circuit has a band-pass filter at the input and at the output. This circuit is shown just for demonstration purposes, since the 2N4417 has been discontinued a long time ago. During the 1990s time there were no useful SPICE parameters available, so we had to do our own parameter extraction. We also found out that the Modified Materka Model extraction gives a more accurate result, between predicted and measured, than the FET SPICE model. The Materka GaAs model is similar to the JFET model and is therefore useful. Initially, this model was not available in SPICE-based programs (Figure 8.45).

The filter is not optimized yet, which explains the bandpass ripple, seen in Figure 8.45b.

The flat line, which is less than 1 dB, shows the minimum noise figure, and the curved trace is the input tuned, as seen in Figure 8.45c.

The following is a set of interesting examples partially based on publications with incomplete results shown.

8.14 Example: Design and Analysis of Microwave Feedback Amplifiers

The first example:

This is a case of a wideband amplifier from 6 to 18 GHz, a single-stage amplifier that uses some very clever matching (Figure 8.46).

The first thing that is obvious is that in the published circuit, a DC-block capacitor between the gate and drain is missing.

The voltage feedback reduces the output impedance and not the gain, but in order to maintain a reasonable gain, there is a 0.2 nH inductance between the drain and the output.

This circuit is derived from a publication, Zhou and Zheng [27].

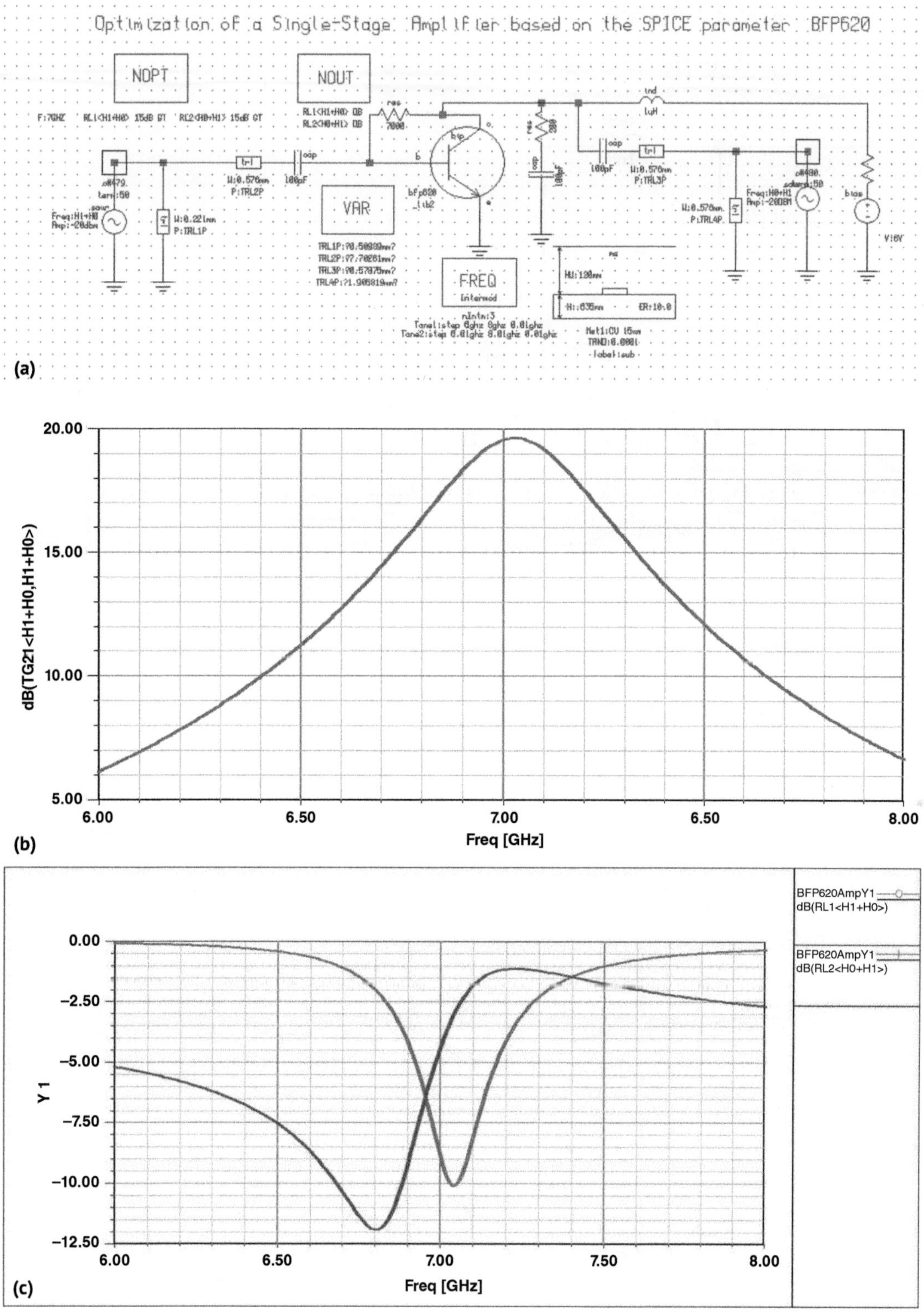

Figure 8.43 (a) Optimization of the Single-stage Amplifier based on the SPICE Parameters of BFP620. (b) Simulated Power Gain (S_{21} in dB) of the circuit in (a). (c) Simulated input and output matching of the amplifier in (a).

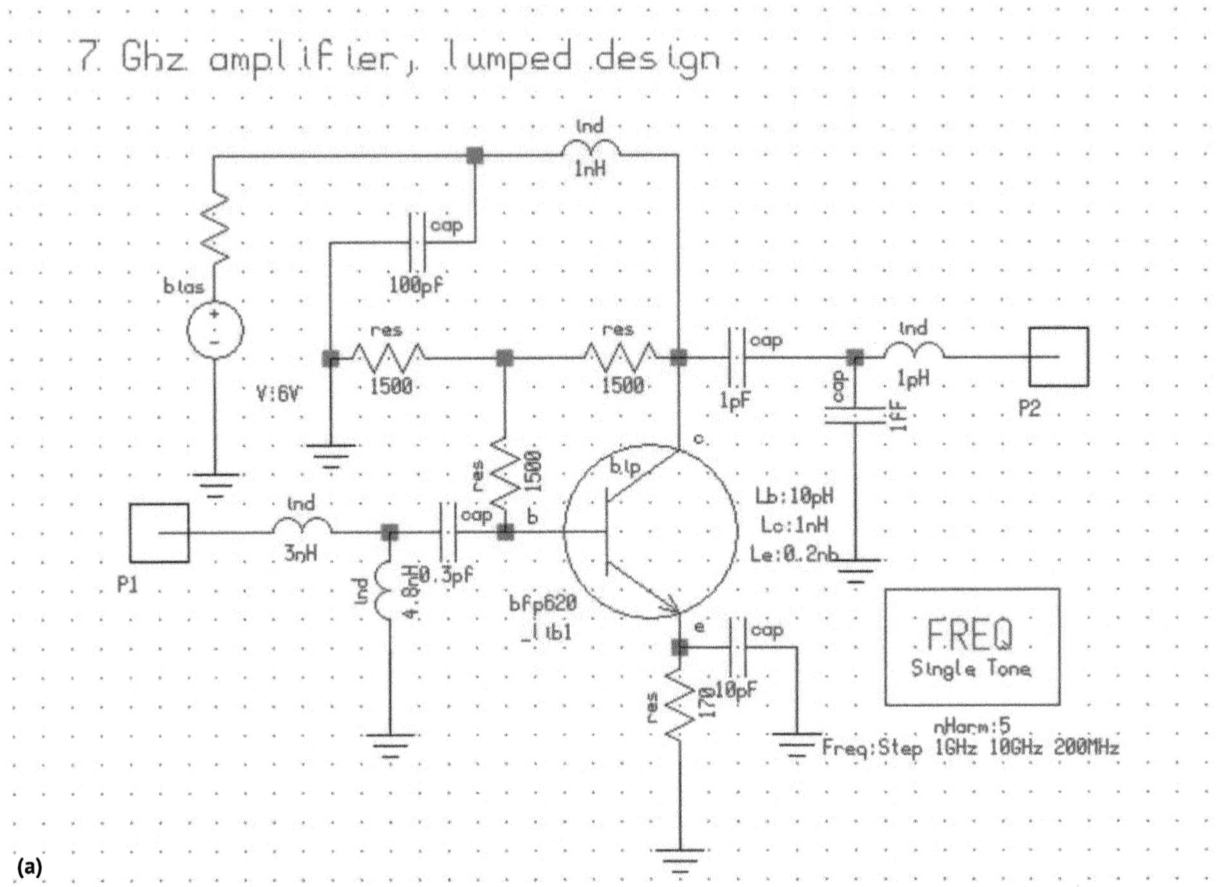

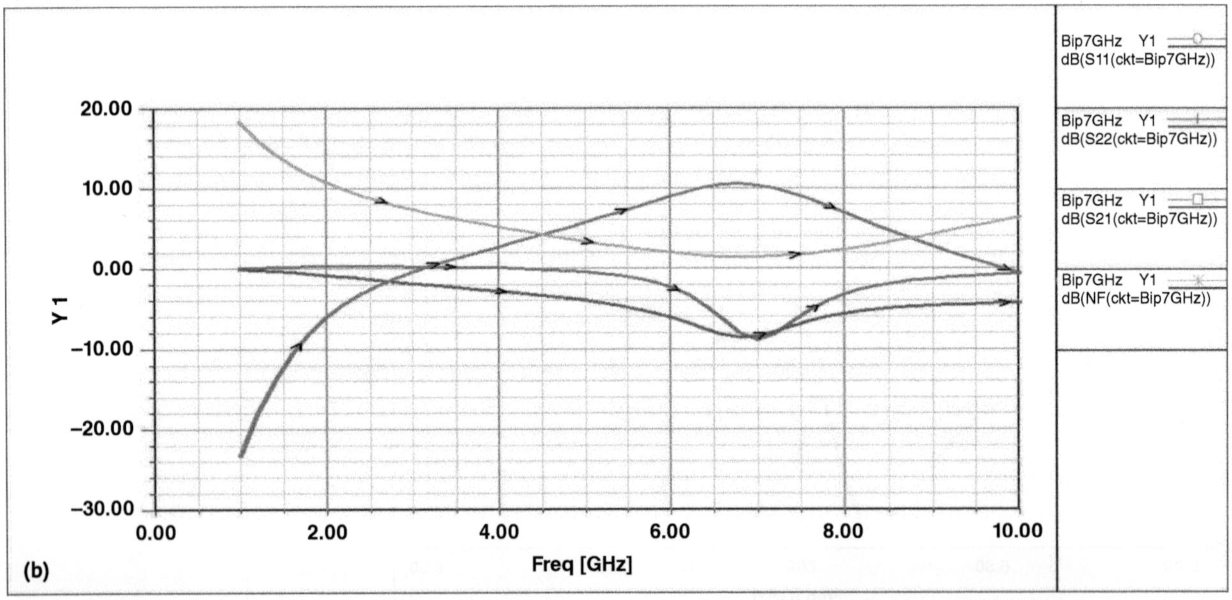

Figure 8.44 (a) Circuit diagram of the 7 GHz Amplifier using lumped elements. (b) Simulated performance of the 7 GHz Amplifier circuit of (a).

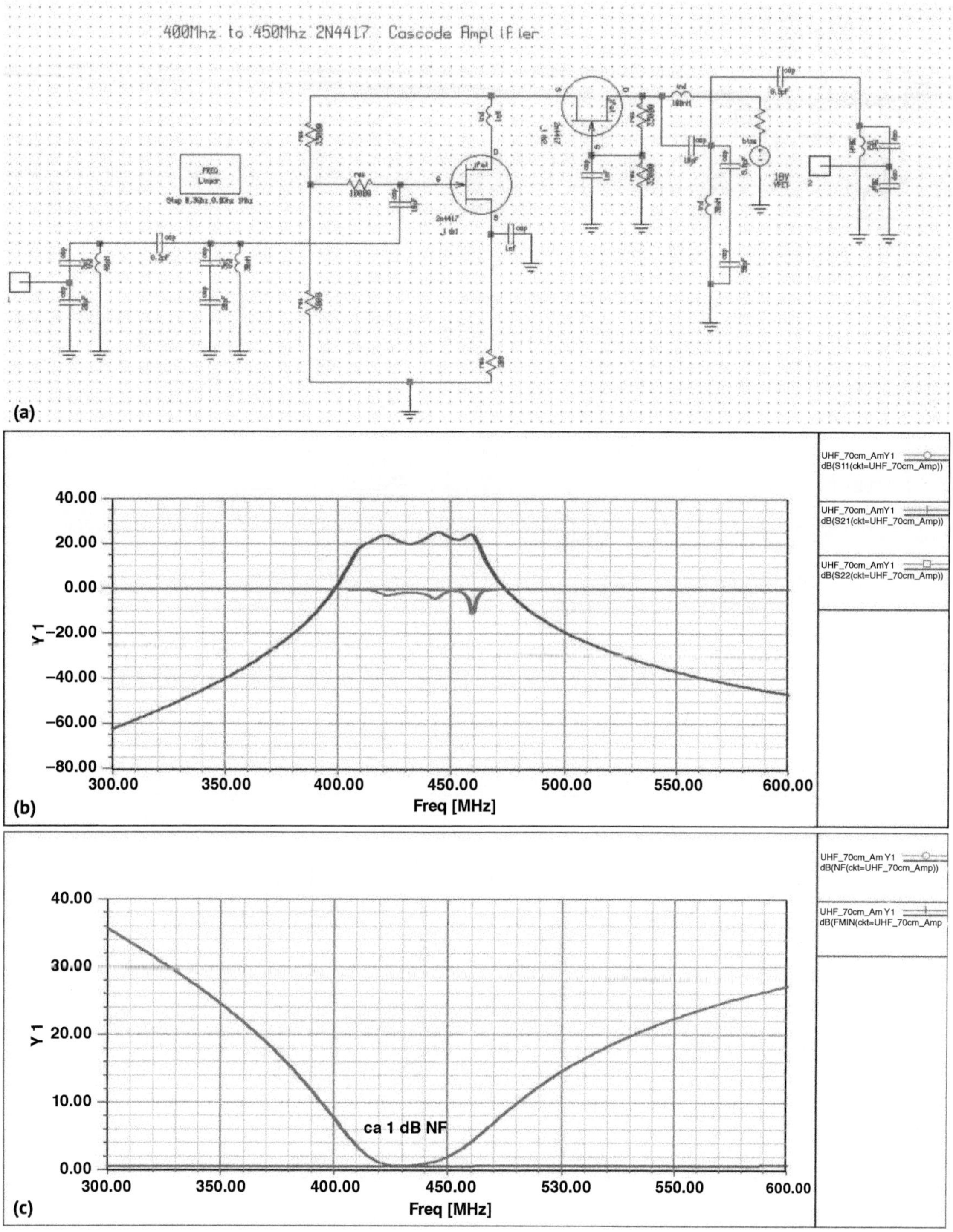

Figure 8.45 (a) 430 MHz JFET 2N4417 Cascode-based preamplifier using input and output bandpass filters. (b) Simulated gain and input and output matching of the circuit in (a). (c) Simulated $F_{\min}$ (min noise figure) and the 50 Ω noise figure of the circuit in (a).

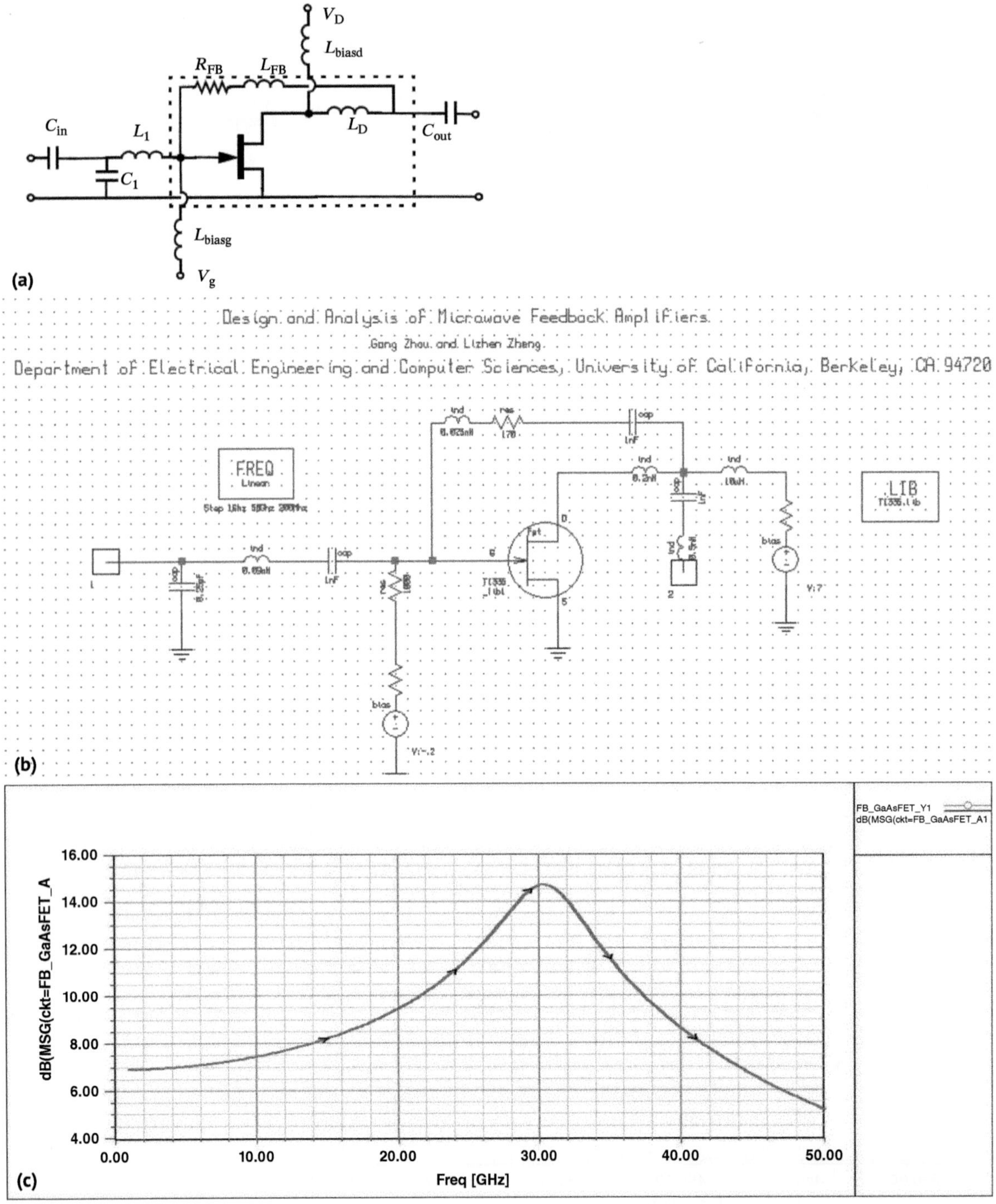

Figure 8.46 (a) A single-stage feedback amplifier with input matching network and biasing RF chokes $R_{FB} = 170\ \Omega$, $L_D = 0.10$ nH, $L_{FB} = 0.73$ nH, $L_1 = 0.76$ nH, $C_1 = 0.25$ pF. *Source*: [27]/University of California, Berkeley. (b) Simulation circuit for the amplifier design and analysis of microwave feedback amplifiers. *Source*: [27]/University of California, Berkeley. (c) Simulated gain of the GaAs FET circuit from Figure 8.47a.

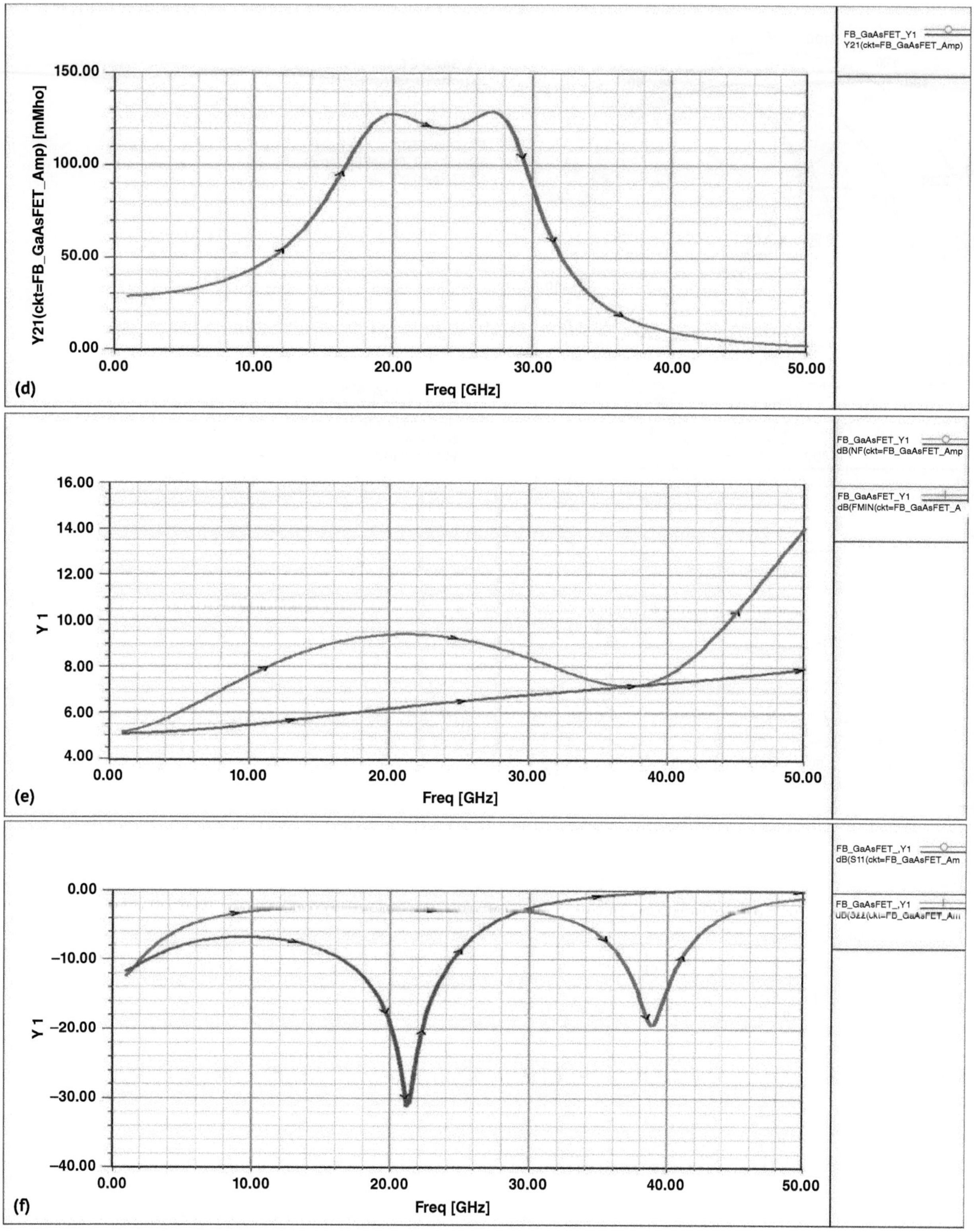

Figure 8.46 (Cont'd) (d) Bandpass characteristics of the amplifier (S_{21} not in dB). (e) Simulated F_{min} and 50 Ω noise figure. (f) Input (S_{11} in dB) and Output (S_{22} in dB) matching.

(Continued)

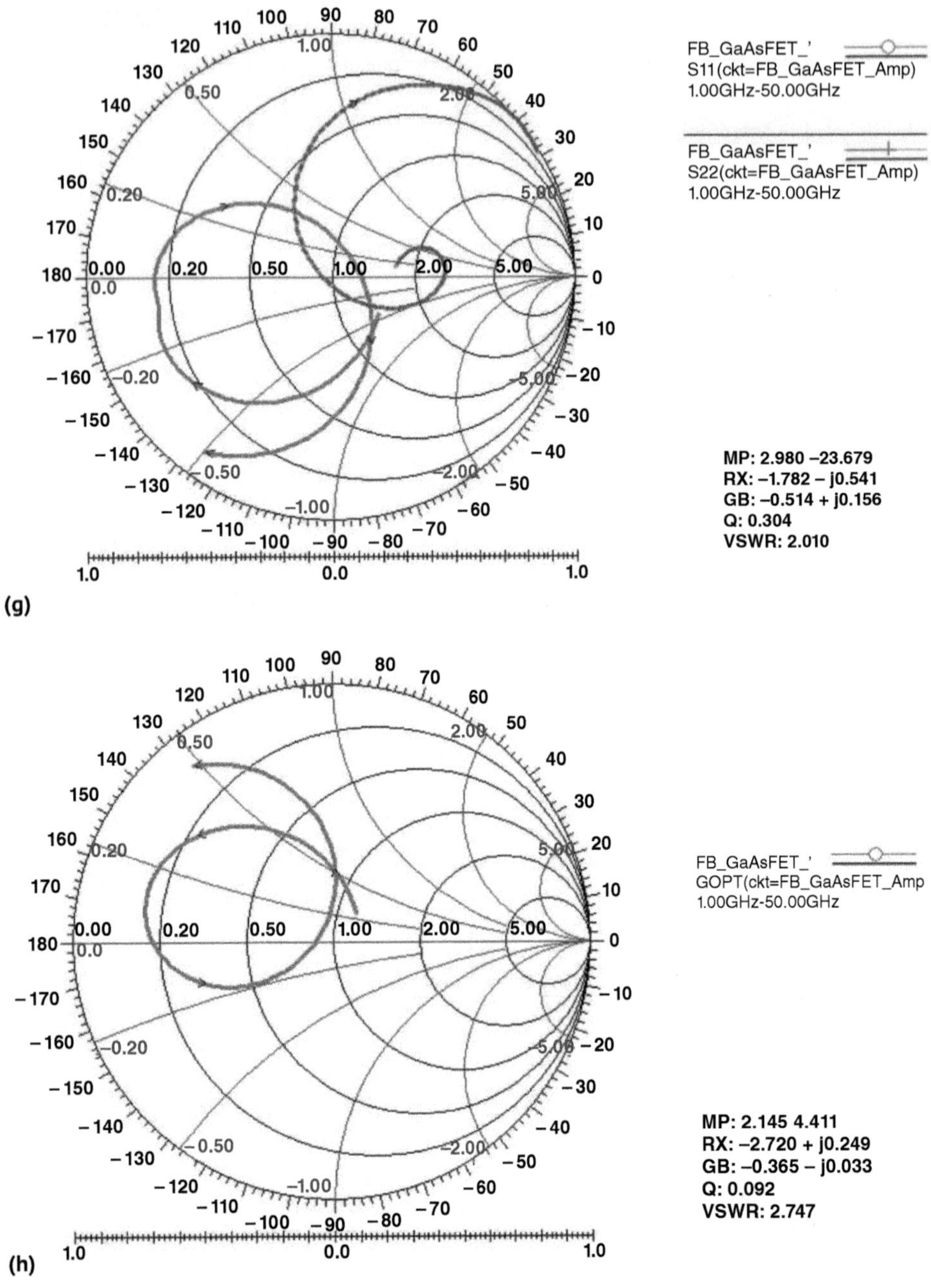

Figure 8.46 (Cont'd) (g) Smith diagram representation of the input (S_{11}) matching and output (S_{22}) matching. (h) Smith diagram of Γ_{opt} (noise matching).

8.15 Example of an 80–100 GHz Amplifier

Talking about exotic devices, shown here is a two-stage amplifier from 80 to 100 GHz. Its schematic is shown in Figure 8.47a, and its Gain vs. frequency plots at the set junction temperature of 22 K are shown in Figure 8.47b, and the predicted noise figure is between 3.12 and 3.54 dB as seen in Figure 8.47c.

The CFY99a GaAs FET. Needless to say, this is a very tricky circuit, and in practical life, it can only be built using a foundry. However, for better understanding, we are showing here all the details and simulating access to the foundry.

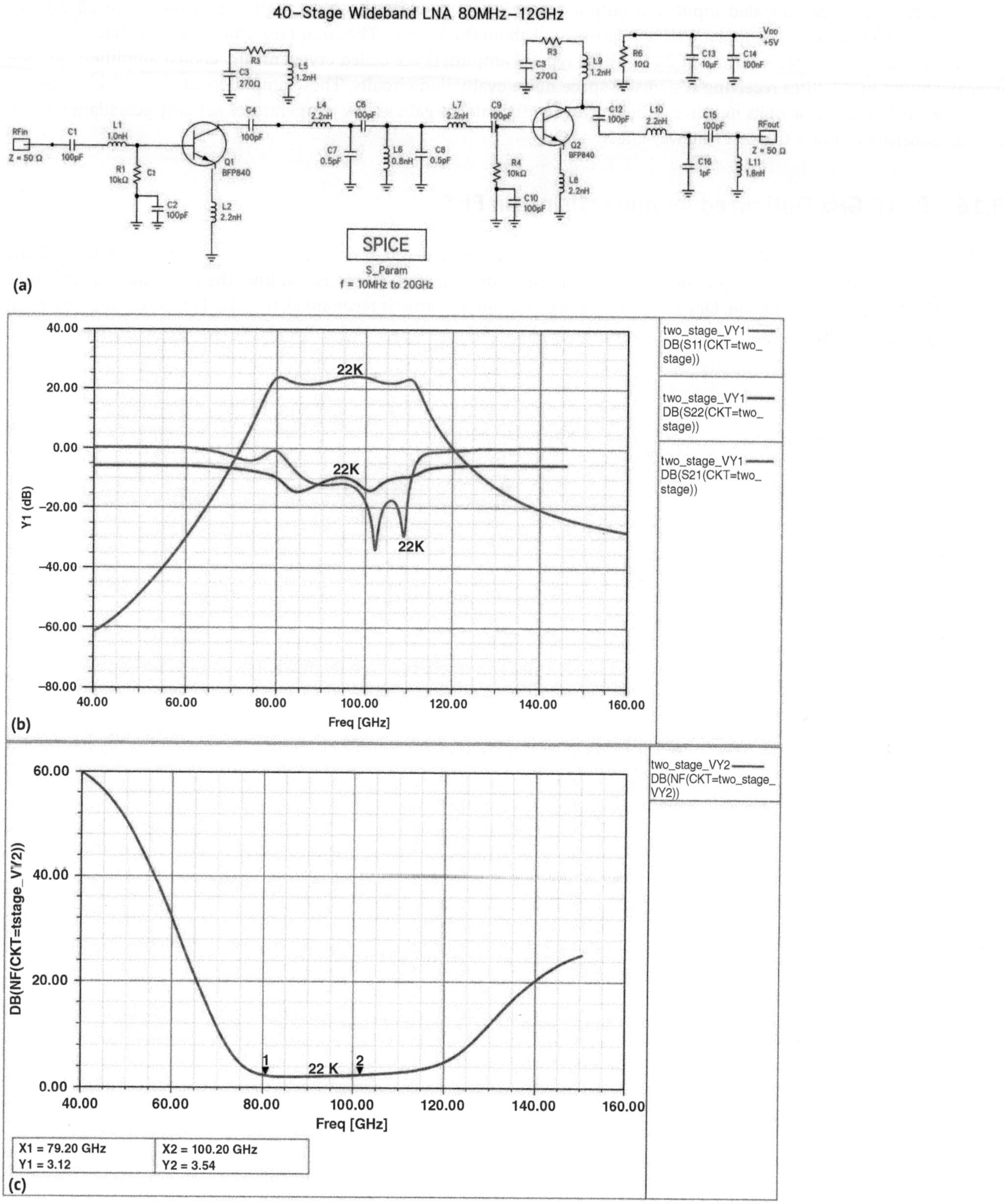

Figure 8.47 (a) Two stage wideband low-noise amplifier 80–100 GHz. (b) Gain vs. frequency plots at the set junction temperature of 22 K. (c) The predicted noise figure is between 3.12 and 3.54 dB.

This circuit has a complicated input and output matching, and while the average gain is a bit above 20 dB, see Figure 8.47b, it has quite some ripple. What else is unusual about this circuit? The circuit uses the modified Materka model, and has set the junction temperature at 22 K. These type of amplifiers are called cryogenically cooled amplifiers and are typically found in satellites receiving systems or space noise evaluation circuits. These amplifiers are typically connected with large dishes. Because of its negative temperature coefficient, the gain at low temperatures actually gets higher, while bipolar transistors show a thermal runaway effect.

8.16 8–18 GHz Optimized Amplifier Using an FET

With this design, we try to show a wideband amplifier with a nice flat gain. This is typically tricky to obtain, and the matching circuit, therefore, is very critical. After the input transformative transmission line, the open and shorted stubs provide the necessary termination. One is resonant at 8 GHz, and the other is resonant at 18 GHz. This forces the wideband matching (Figure 8.48).

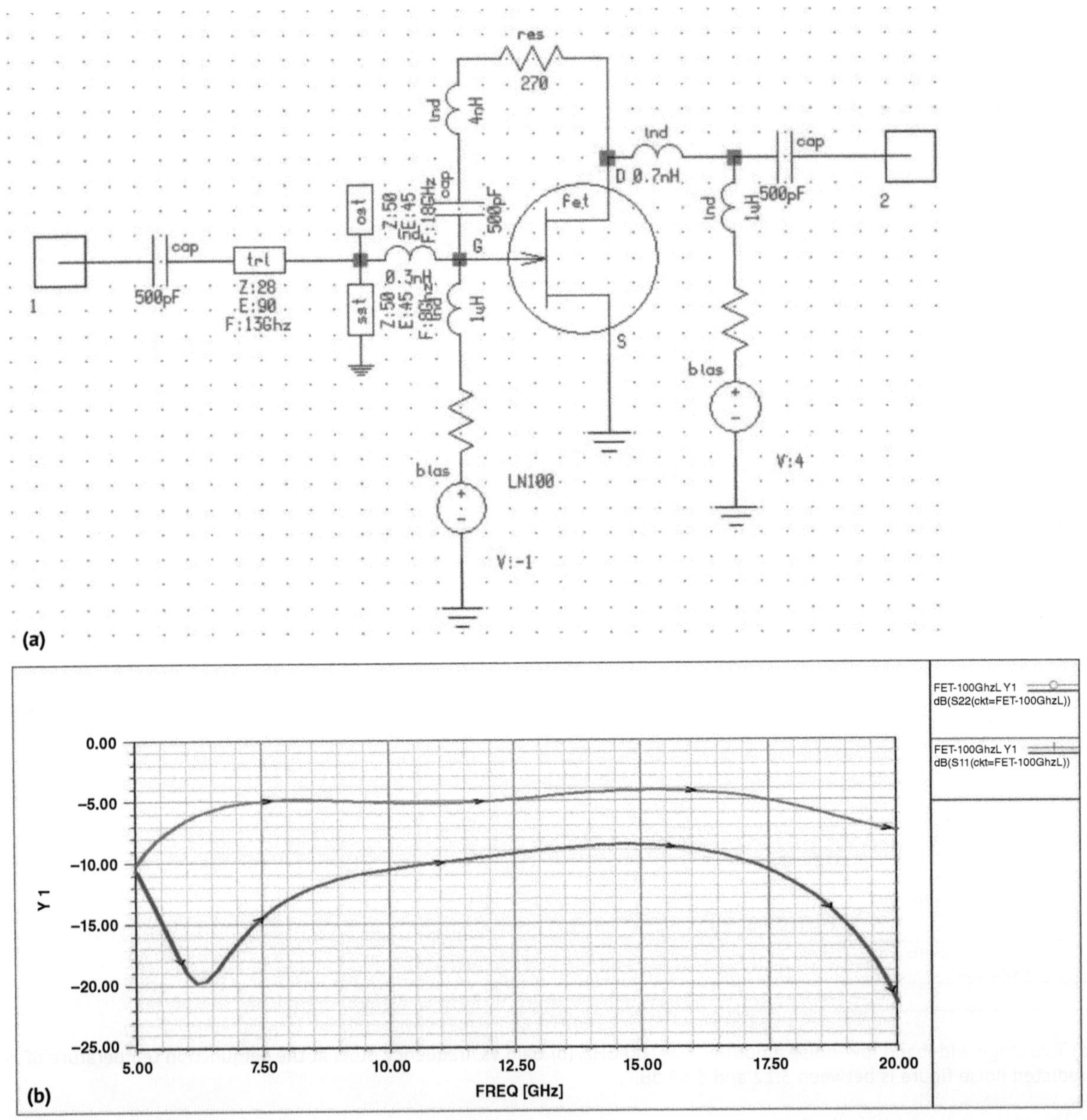

Figure 8.48 (a) Circuit diagram of the 8–18 GHz optimized amplifier. (b) Input and output matching of the amplifier in (a).

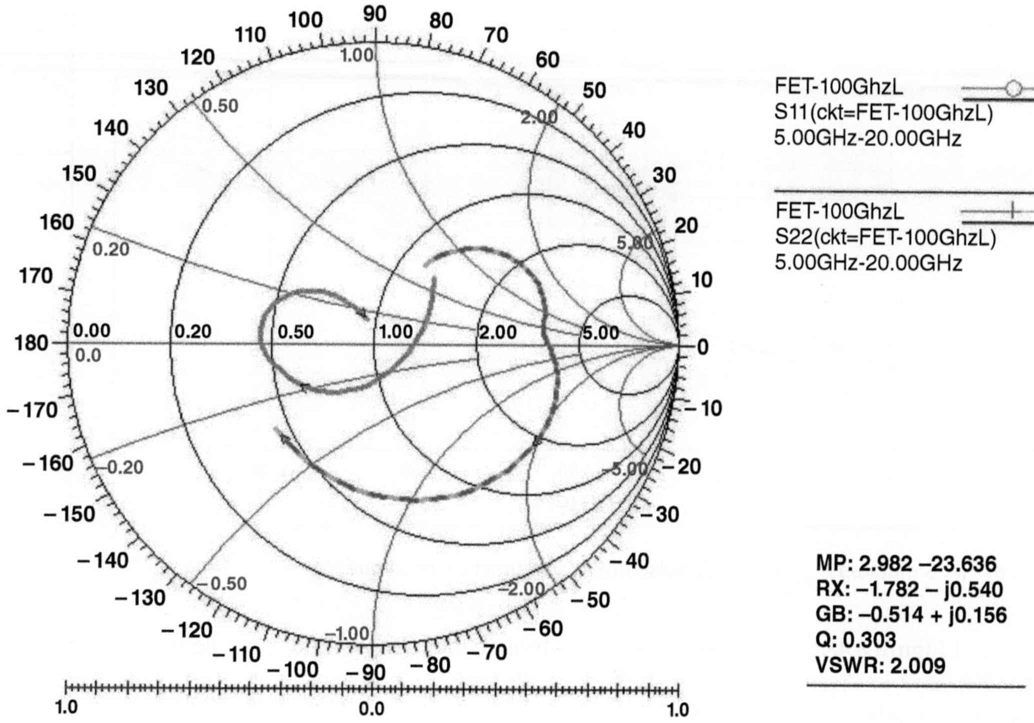

(c)

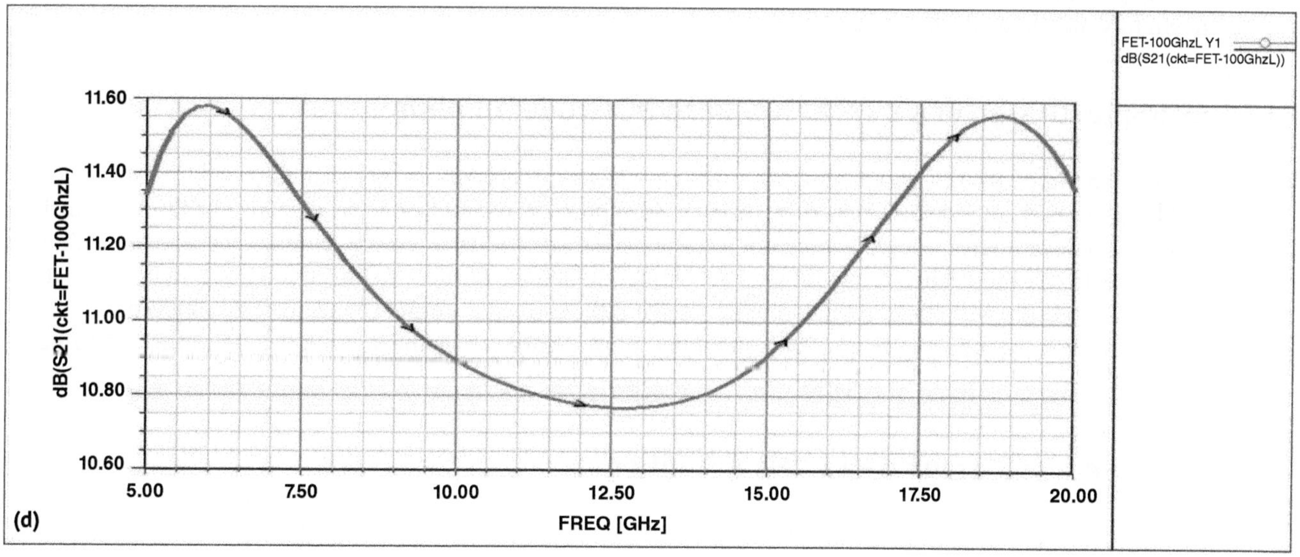

(d)

Figure 8.48 (Cont'd) (c) Smith diagram representation of the S_{11} and S_{22} (input and output matching). (d) High resolution gain (S_{21}) plot with a variation of about 1 dB.

(Continued)

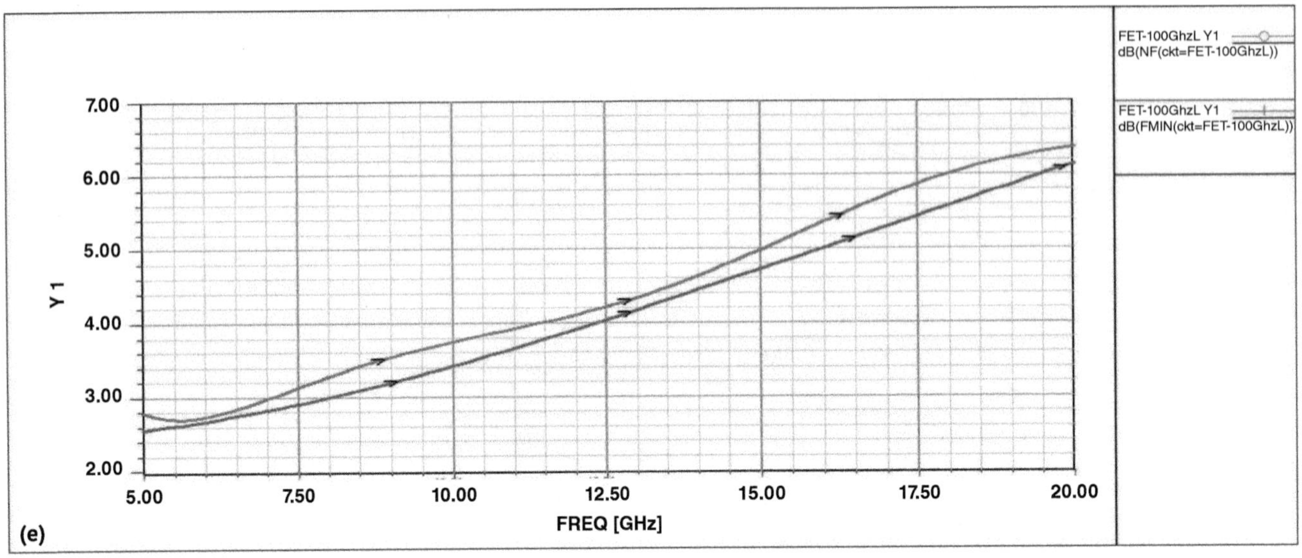

Figure 8.48 (Cont'd) (e) Simulated $F_{\min}$ and 50 Ω noise figure as a function of frequency.

In this amplifier of Figure 8.48a, the output matching is broadband, about (−5 dB), while the input matching is always better than (−10 dB), seen in Figure 8.48b.

This feedback circuit made a very flat gain possible.

This was only possible because of the use of a very clever circuit design. The basic principle on which this example is based is shown in Vendelin et al. [28, pg. 285].

8.17 Example: Design of a 4 GHz MOS Based Cascode Amplifiers

The circuit shown in Figure 8.49a is based on foundry data, as discrete transistors are not available.

The circuit below uses N-MOS devices to build a Cascode Amplifier, similar to the one we did with the bipolar transistor [29–37]. It is designed for 4 GHz. The left transistor is responsible for the biasing of the circuit. The input matching is achieved using the 18 nH inductance at the input. The source inductance of the 1 nH does the noise matching. At the output, a resonant condition with the drain inductor of 6 nH and the transistor output capacitance is achieved.

The simulated N-MOS cascode 4 GHz amplifier simulated noise figure shows the simulated gain (8 dB), noise figure (1 dB), input (−10 dB), and Output (−1 dB, poor) matching, Figure 8.49b.

8.18 Example of a 2-stage 6–12 GHz Amplifier

The circuit diagram of the two-stage amplifier with the various matching circuits is self-explanatory (Figure 8.50).

This amplifier has a fairly wide and flat bandwidth.

The matching circuits result in a NF of 4.4 dB, and the minimum NF would be 3.5 dB at 7.5 GHz.

Now we introduce a completely different type of amplifiers, called distributed amplifiers. At the lowest operating frequency, these three transistors are essentially in parallel, while at higher frequencies, the phase from one transistor to another changes based on the delay line.

The input capacitance of the transistors together with the inductances between them form a low-pass filter resonant at the highest frequency of operation. The output transmission line has a similar function.

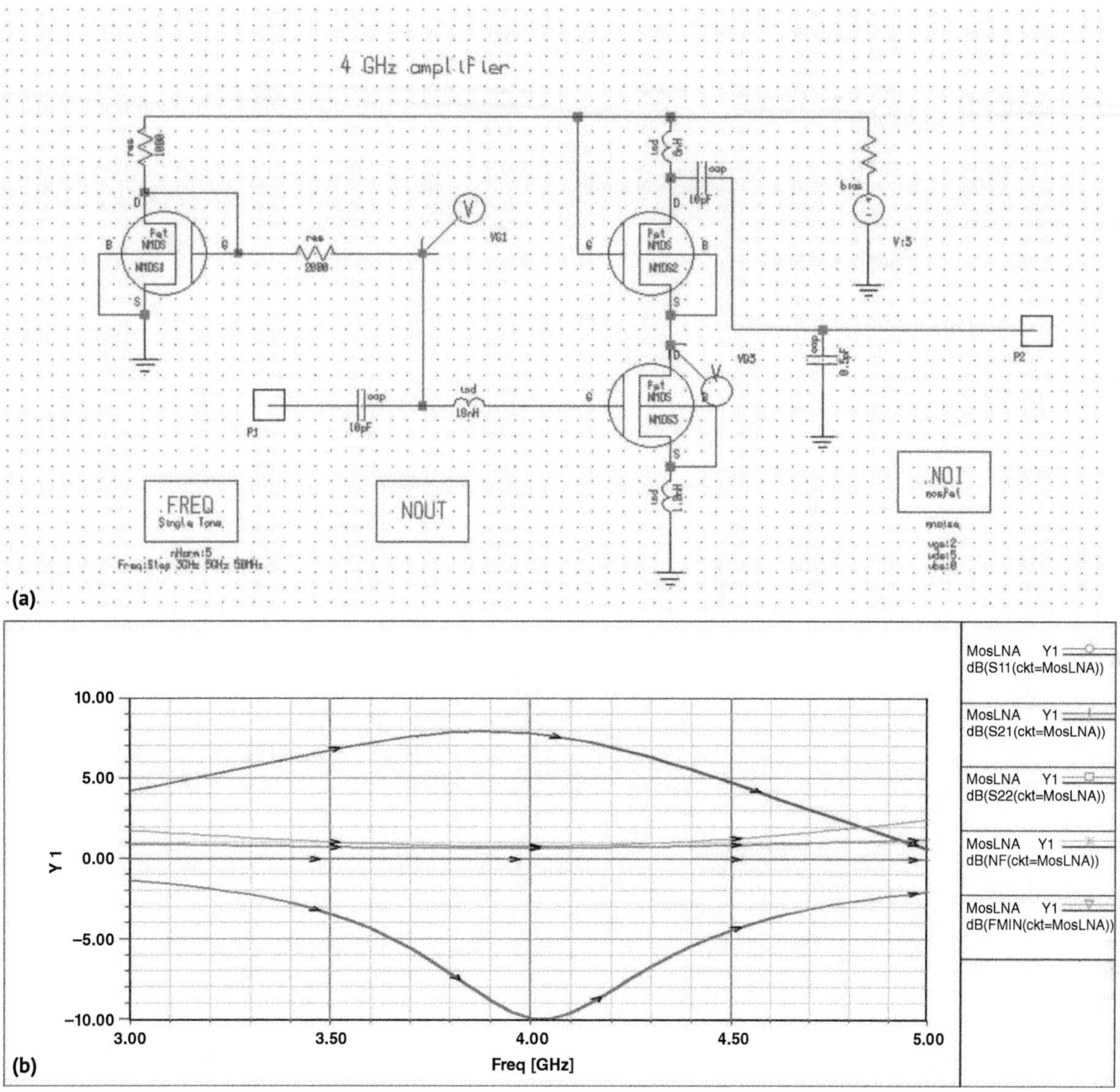

Figure 8.49 (a) Circuit diagram of the N-MOS Cascode amplifier. (b) Simulated gain (8 dB), noise figure (1 dB), input (−10 dB), and output (−1 dB, poor) matching.

8.19 Simulation of a 3 Stage Distributed FET Amplifier

The example shown below is a modified example, based on the [38, Nonlinear Examples] (Figure 8.51).

The input cutoff frequency is about 2 GHz, and the output cutoff frequency is 22 GHz (peaking).

The simulation, Figure 8.51a, shows a flat gain from 2 GHz to 20 GHz (Figure 8.51b). The input matching curve varies between 8 dB S_{11} or better and 10 dB S_{22}. The gain goes into compression at higher input levels. (+10 dBm) is about the maximum that the amplifier can tolerate.

The simulated noise figure and Fmin remain pretty constant at about 5 dB. See Figure 8.51d for the simulated noise figure ($F_{\min}$ and 50 Ω).

Figure 8.51e: Large signal simulation of the amplifier shows the gain compression at different input levels.

The next section considers examples of the large signal behavior of amplifiers.

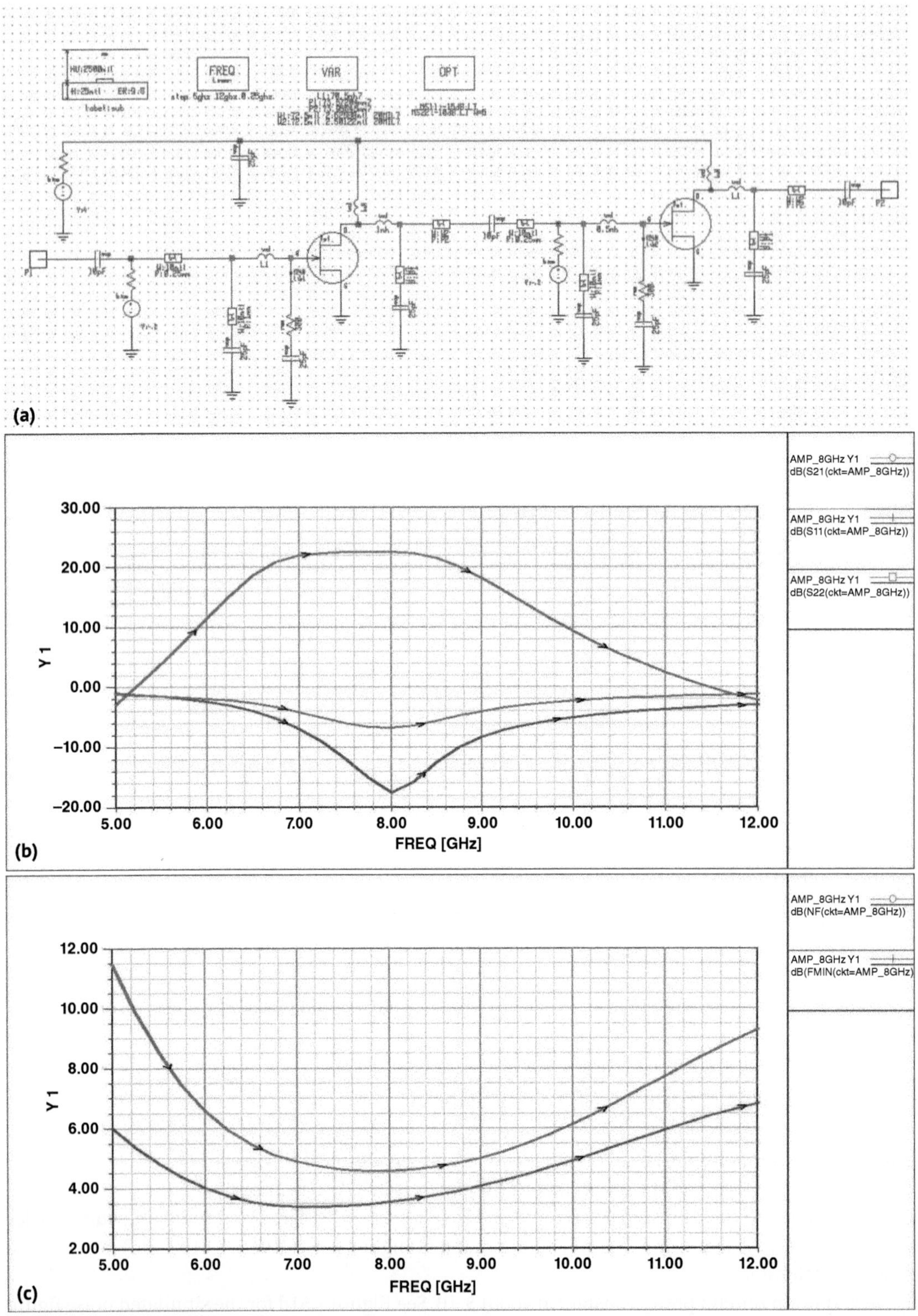

Figure 8.50 (a) Circuit diagram of a 2-stage 6–12 GHz amplifier. (b) Simulation of the gain and matching of the 2-stage 6–12 GHz amplifier (a) optimized for 8 GHz. (c) Simulation of the F_{min} and 50 Ω noise figure of the 2-stage 6–12 GHz amplifier (a) optimized for 8 GHz.

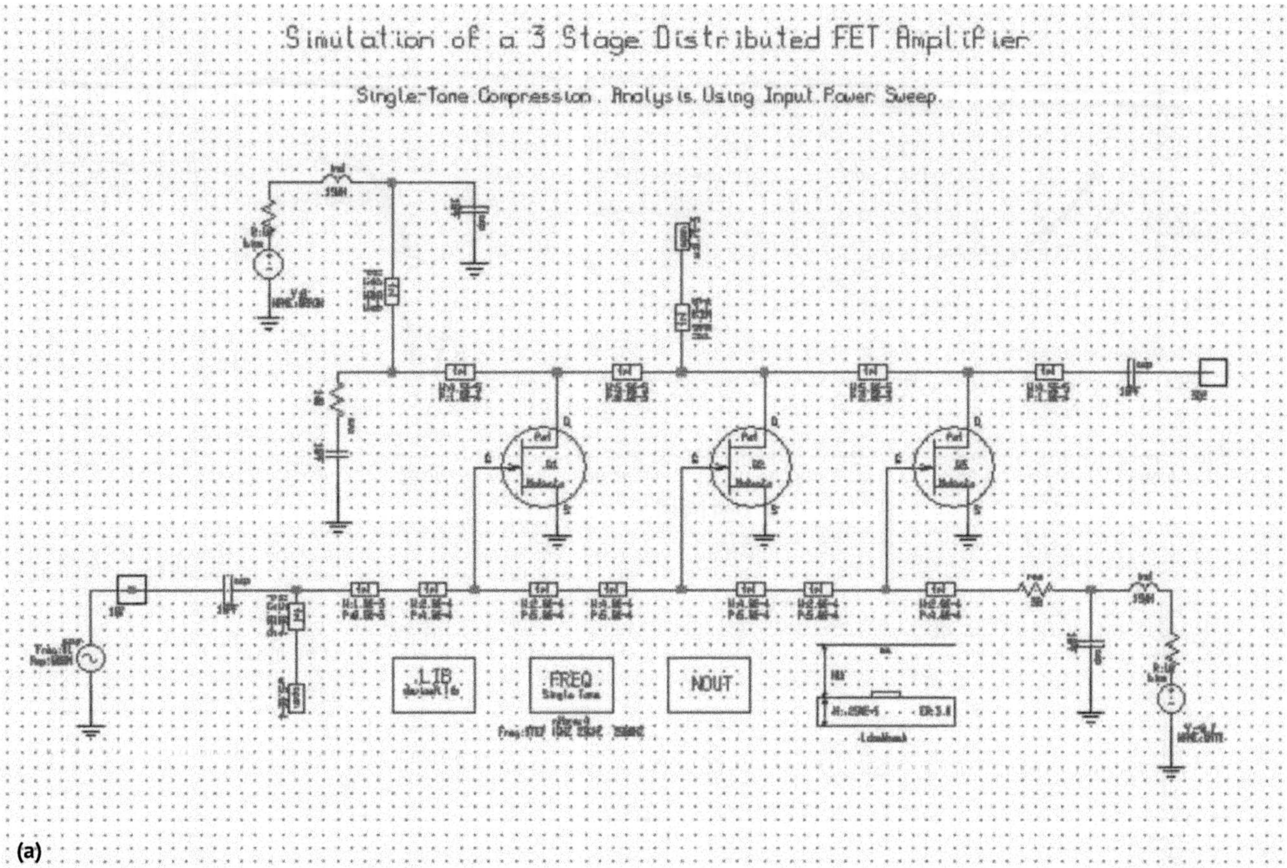

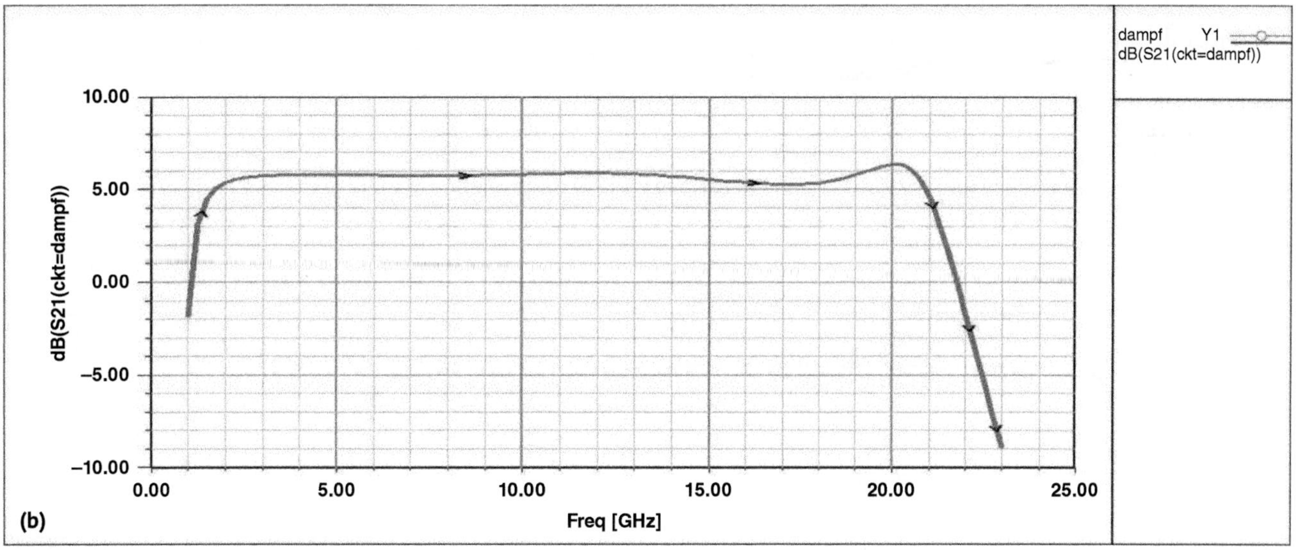

Figure 8.51 (a) Simulation of a 3 stage distributed FET amplifier [38, nonlinear examples]. (b) Gain curve of the distributed amplifier (a) [38, nonlinear examples].

(Continued)

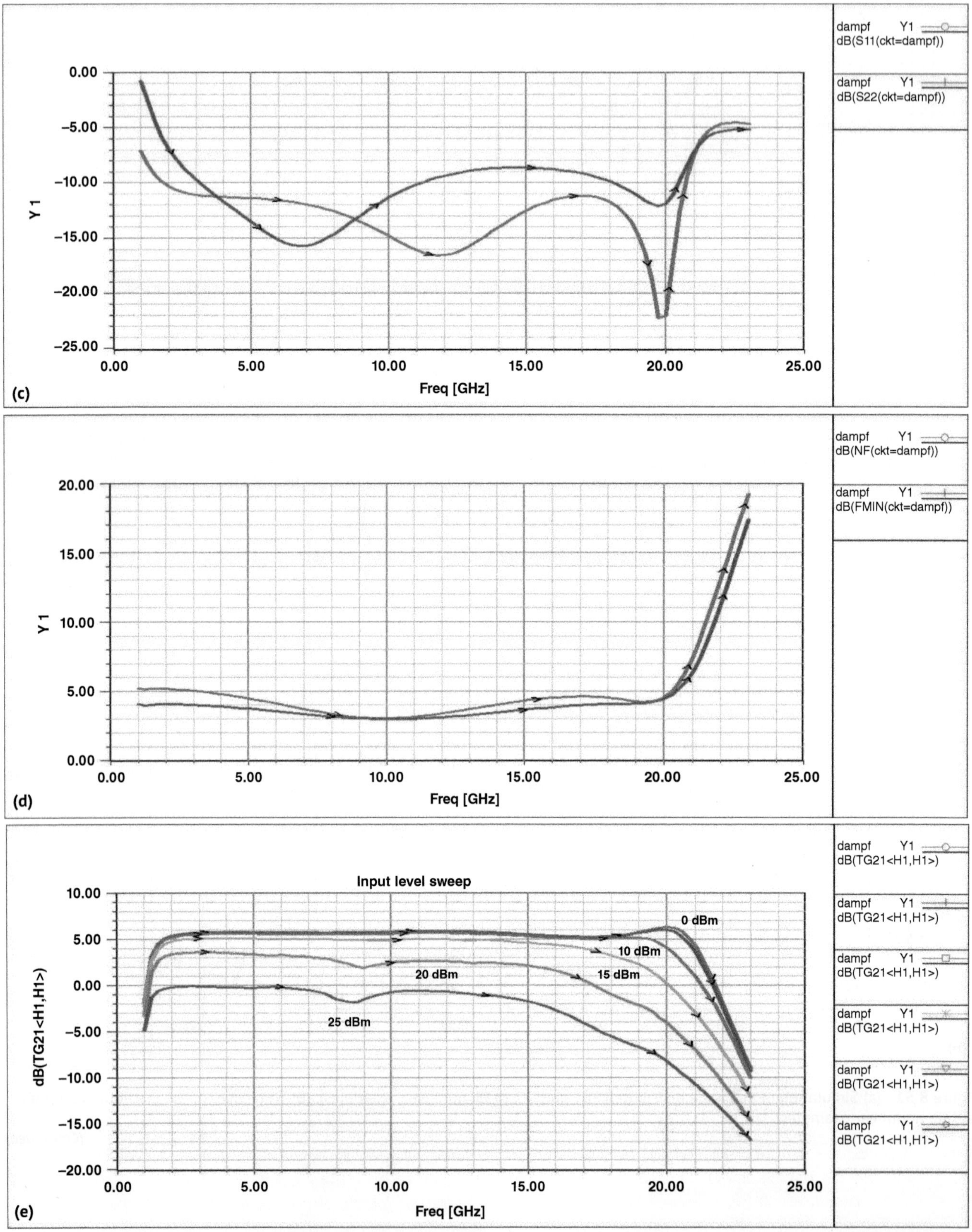

Figure 8.51 (Cont'd) (c) Simulated input and output matching (S_{11} and S_{22} in dB) [38, nonlinear examples]. (d) Simulated noise figure (F_{min} and 50 Ω) [38, nonlinear examples]. (e) Large signal simulation of the amplifier [38, nonlinear examples].

8.20 Example: Two Tone Test of a FET Amplifier at 5 GHz

So far, we have only looked at low-level noise and the gain of the amplifier. In real life, there is more than one signal in a system. Therefore, we have to look at the nonlinearity, specifically intermodulation. Therefore, we connect to sources at the input with sufficient spacing and isolation and look at the output spectrum. The example shown in Figure 8.52 is based on the examples from [38, nonlinear Examples].

The top curve shows the output power as a function of the input power. The 1 dB compression point occurs at about 1 dBm. We typically differentiate between the input and output intercept points. The difference is the gain of the amplifier. It is obvious that the lower curve, which is the input intercept point, shows a significant distortion around 7 dBm input. In the literature, one typically refers to the 3 dB/dB distortion products, also called IP3. However, close to the compression point, the transistor is totally overdriven, as can be seen by the lower curve. This is due to the various non-linearities in the transistor, which the standard textbooks do not show. This behavior depends to a large degree on a particular transistor.

8.21 Power Amplifiers

Power amplifiers are devices that are tasked to transform as much dc power into RF/microwave power as possible. Depending upon the technology, devices are available that can produce at least 100 W up to 2 GHz. Since the transconductance of FETs at conducting angles of less than 180° becomes very small, the actual power gain sharply decreases compared to Class A operation, while the power-added efficiency is optimized to values of 55% (Class B) or so. The dissipation of these devices is, of course, quite high, while in Class B operation, this is even more of an issue. Power devices in Class C are dying to have big heatsinks to survive. One of the nice things about FETs is a much reduced thermal runaway effect compared to their bipolar brothers [1].

Power amplifier classes are defined in terms of the device conduction angle and/or the type of device operation:

Class A	"Linear" operation; 180° conduction angle
Class AB	"Linear" operation; conduction angle between 90° and 180°
Class B	"Linear" operation; 90° conduction angle
Class C	Fixed drive; less than 90° conduction angle
Class D	Switched operation, conduction range may vary with time from 0° to 180° or may be fixed. There are also Classes E and F, which are currently under evaluation and showing some promising results.

Figure 8.53 shows Class C operation. The output of the amplifier is a current pulse train, which frequently leads to misunderstanding as to the calculation of its load. Given the fact that the amplifier is a current generator, it is silly to assume one has to do a conjugate matching looking backward into the amplifier. This is the condition for Class A operation. Unfortunately, this is still not understood by some engineers, but we hope we can clear this up today. The theoretical efficiency is 89.7% at a conducting angle of less than 180°; in practice, efficiencies more than 85% are not obtainable. (All of these efficiencies are those of the device itself; losses in matching networks are not taken into account in these efficiency values.) Considering optimal gain and optimal output power, the conducting angle actually settles somewhere between 100 and 140°. Again, the power gain in Class C operation is less than in Class A and Class B operation [1].

8.21.1 Example 1: 7-W Class C BJT Amplifier for 1.6 GHz [1]

The following demonstrates how to design a 1.6-GHz, 7-W amplifier operating in Class C. We need to point out again that all modulation forms that have the information in both amplitude and phase require more linearity than Class C affords. Needless to say, this affects the efficiency; therefore, at a later point, we will also show a Class AB power amplifier. This is a tutorial, and we are going to take a UHF example, and we will walk the reader through all the necessary problems arising from the design. The selection of the transistor is done by first looking at the f_T, which in our case should be at least 4–5 GHz. The first order of business is to define the point of operation for the output amplifier. To do this, we need to maximum voltage and current values supplied by the manufacturer. We have chosen a device that operates at 28 V and can sustain a peak collector current of 1.5 A. Figure 8.53 lets us determine the conducting angle Θ from the I_C/V_{BE} plane.

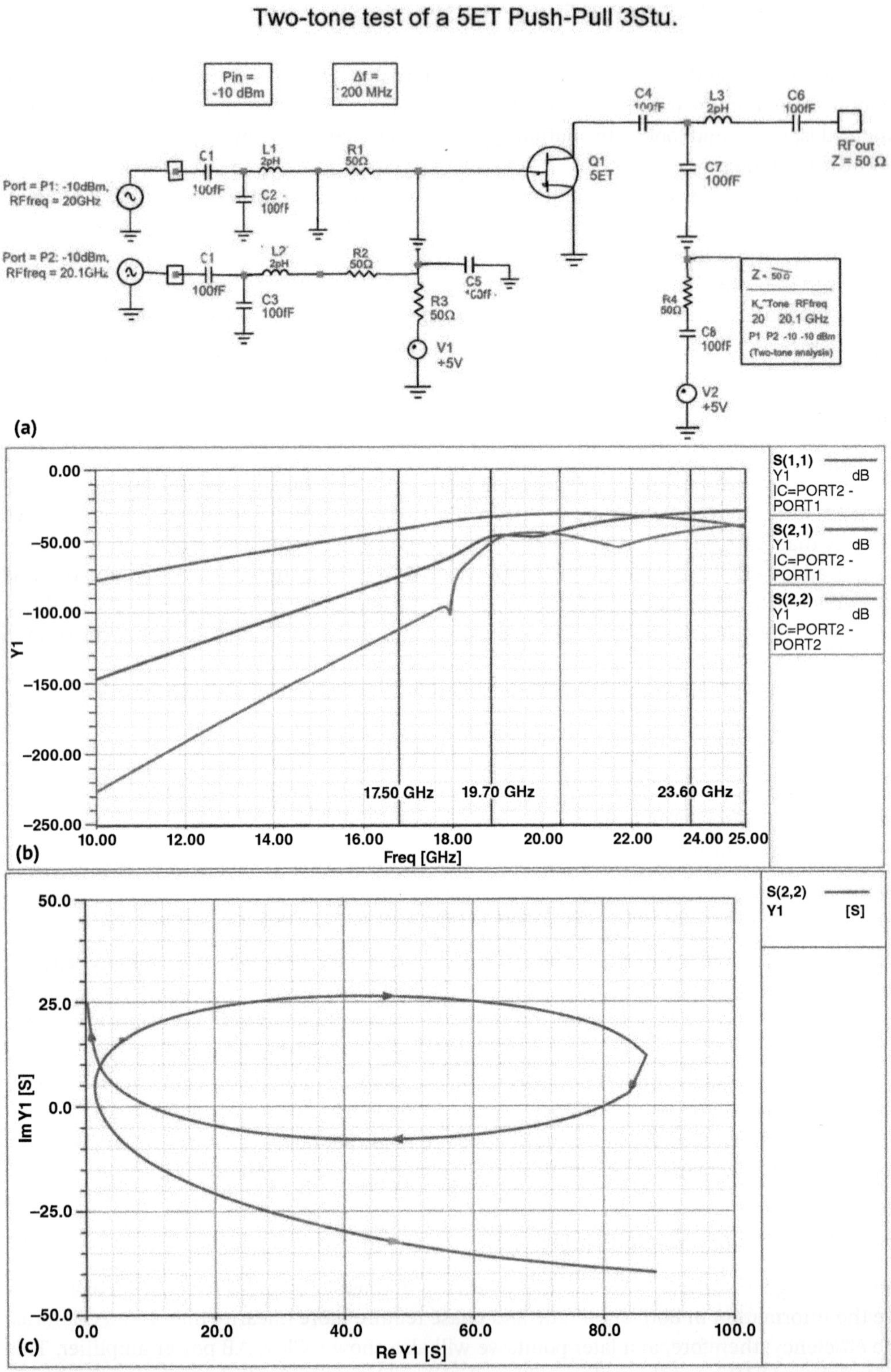

Figure 8.52 (a) Two tone test circuit of a FET amplifier at 5 GHz [38, nonlinear examples]. (b) Simulation of the third-order intermodulation distortion as a function of input power and output compression [38, nonlinear examples]. (c) Plot of the real and imaginary input impedance of the circuit, indicating a fairly wide swing in the reactive part [38, nonlinear Examples].

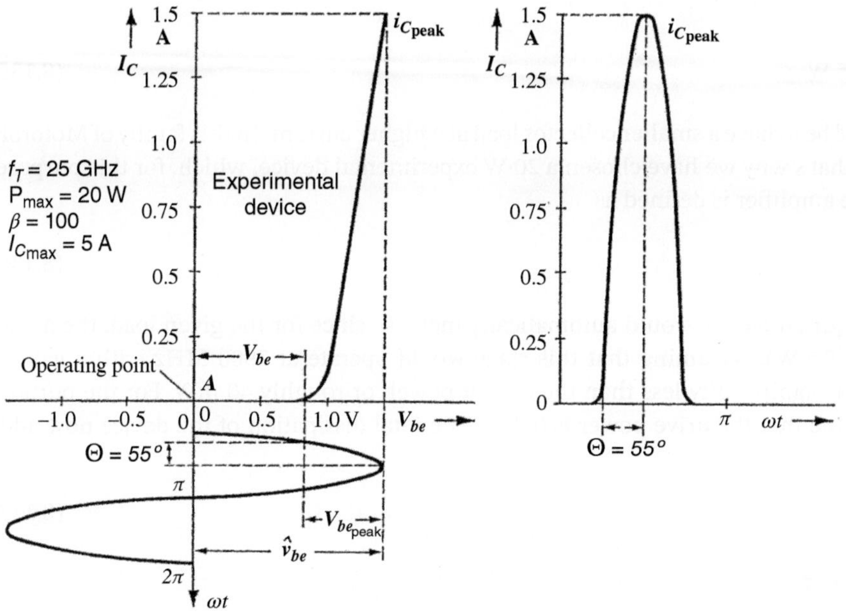

Figure 8.53 Output current as a function of drive voltage, taking into consideration the knee voltage of the base-emitter p–n junction. The output is a series of current pulses with a duty cycle determined by the conduction angle. *Source*: [1]/John Wiley & Sons.

This figure shows on the left side the I_C/V_{BE} plane and the necessary voltage to drive the amplifier.

Initially, we assume that the base-emitter junction is operated without any bias ($V_{BE} = 0$). The dc I–V curve shows that the collector current starts at about 650 mV. The amplifier itself, therefore, is already in Class C operation. For full output power, the drive voltage V_{BEpeak} has to be 1.14 V. Therefore, the conducting angle can be determined by:

$$\cos\Theta = \frac{V_{be}}{\hat{v}_{be}} = \frac{0.65}{1.14} = 0.57 \tag{8.133}$$

$$\cos 0.57 = 55°$$

V_{be} is the dc voltage at which the transistor starts conducting. From Figure 8.65 and Eqs (8.160)–(8.162), the fundamental current i_{c1} can be determined from

$$i_{c1} = i_{peak} \cdot f_1(\Theta)$$

or

$$i_{c1} = 1.5\text{ A} \times 0.36 = 0.54\text{ A}$$

The average collector dc current can be determined from

$$i_{c2} = i_{peak} \cdot f_2(\Theta) \tag{8.134}$$

or

$$i_{c2} = 1.5 \times 0.21 = 0.315\text{ A}$$

The collector saturation voltage $V_{ce(sat)}$, unfortunately is around 2–3 V; for the purpose of the calculation, we assume that the saturation voltage is 3 V, and therefore we can assume that the maximum collector ac voltage $\hat{v}_{ce} \approx (V_B - 3) = 25$.

The appropriate collector load has to be

$$R_c = \frac{\hat{v}_{ce}}{i_{c1}} = \frac{25}{0.54} = 46.3\ \Omega$$

The maximum output power is

$$P_{out} = \frac{\left(V_{ce} - V_{ce(sat)}\right)^2}{2R_c} = \frac{625}{92.6} = 6.75\ \text{W} \tag{8.135}$$

The only way to increase the power would be to have a smaller collector load at a higher current. In this family of Motorola transistors, there was no 10-W transistor; that's why we have chosen a 20-W experimental device, which, for this purpose, is overkill. The theoretical efficiency of the amplifier is defined as

$$\eta = \frac{P_{out}}{P_{in}} = \frac{6.75\ \text{W}}{0.315 \times 28} = 85\% \tag{8.136}$$

If the saturation voltage would be less, our efficiency would automatically increase since for the given load, the actual output power would increase (6.75 W → 7.5 W). Assuming that this stage would operate at 1900 MHz with a gain of roughly 22 dB, the required drive would be significantly less than the output power, or roughly 40 mW. For the purpose of the following calculations, we will assume that the drive power is 0.7 W. The total dissipation of the device now adds up to be

$$P_T = P_{IN(dc)} - P_{OUT} + P_{DRIVE} \tag{8.137}$$

$$P_T = (28 \times 0.315) - 6.75 + 0.7 = 2.77$$

Power-added efficiency is derived from this. Knowing the power dissipation, one can calculate the thermal resistance R_{therm} of the heatsink and the transistor:

$$R_{therm} = \frac{T_j - T_{amb}}{P_T} \tag{8.138}$$

Where T_j is the device junction temperature and T_{amb} is the ambient temperature. For a maximum ambient temperature of 60 °C and a maximum junction temperature of 150 °C, the thermal resistance

$$R_{therm} = \frac{120\text{–}60}{2.77} = 21.7\ °\text{C/W} \tag{8.139}$$

R_{therm} is the sum of the thermal resistance of the transistor and its package, the thermal resistance between the transistor package and the heat sink, and the thermal resistance between the heat sink and the ambient air. Therefore

$$R_{therm} = R_{th(case)} + R_{th(case/heatsink)} + R_{th(heatsink)} \tag{8.140}$$

The required thermal resistance of the heatsink, which is used to determine its size or surface area, is calculated from

$$R_{th(heatsink)} = 21.7 - 10 - 1.4 = 15.6\ °\text{C/W}$$

Assuming an aluminum heat sink (a 2-mm-thick plate), the required heatsink surface area A (cm^2) is

$$\text{A} = \frac{1}{\alpha R_{th(heatsink)}} : \alpha \approx 1.5 \frac{\text{mW}}{°\text{Ccm}^2} \tag{8.141}$$

Therefore, the required area is

$$\text{A} = \frac{1}{1.5 \times 0.0156} = 42.7\ \text{cm}^2$$

The next task is to match both the input and output to the (most likely) 50-Ω source and termination. The following section will guide us through how to do this. The input impedance Z_{11} (after some lengthy search) was found to be $1 + j10\ \Omega$. This means that the input consists of the base spreading resistance (approximately 1 Ω) and an input capacitance, which, because of phase shift, now turns out to be inductive. Since the popular method for input matching these days is a pi configuration, we are going to use a simple pi filter both for input and output matching. There are several equations in the literature that give the values for the two capacitors and inductors; most of them result in an incorrect resonance frequency. Since we have access to software tools, we decided to take three approximate values and use the optimizer to give us the appropriate match. The only remaining variable in such an approach is the bandwidth. If we specify $S_{21} = 1$ and $S_{11} = 0$ (100% energy transfer and perfect return loss), we will realize that this is possible only at one frequency. We therefore need to identify a frequency band or bandwidth over which an acceptable match is needed. As the bandwidth increases, the match for such a simple circuit will get poorer; there are not enough elements of freedom available. In this case, we took the center frequency plus/minus 100 MHz. Figures 8.54 and 8.55 show the actual circuit with the element values and the resulting input match. Likewise, we did the same thing at the output, but the actual resulting dynamic output capacitance is not known *a priori*. For low frequencies, the integral equation gives a value of $2C_{CE}$; however, practical tests in frequencies above 500 MHz show that the actual increase is only 1.3 times the dc output capacitance. Therefore, we decided to design the filter with no output capacitance assumed, and the input capacitance of about 5 pF needs to be reduced for proper reactive matching. Mathematically, this could result in a negative number, specifically if the resulting output capacitance

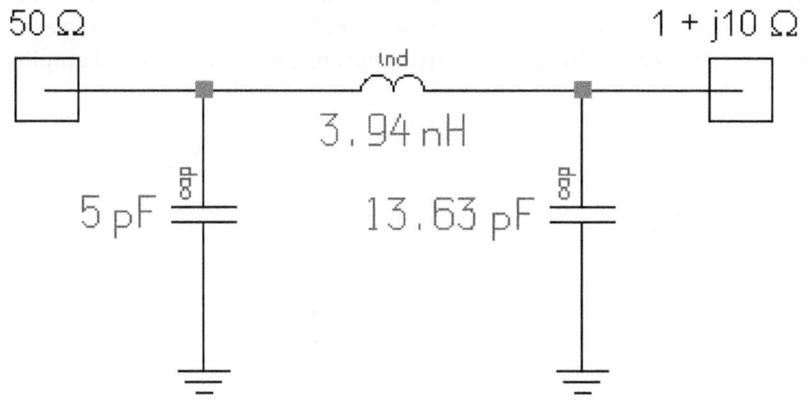

Figure 8.54 Input matching network for the amplifier used in this tutorial. *Source*: [1]/John Wiley & Sons.

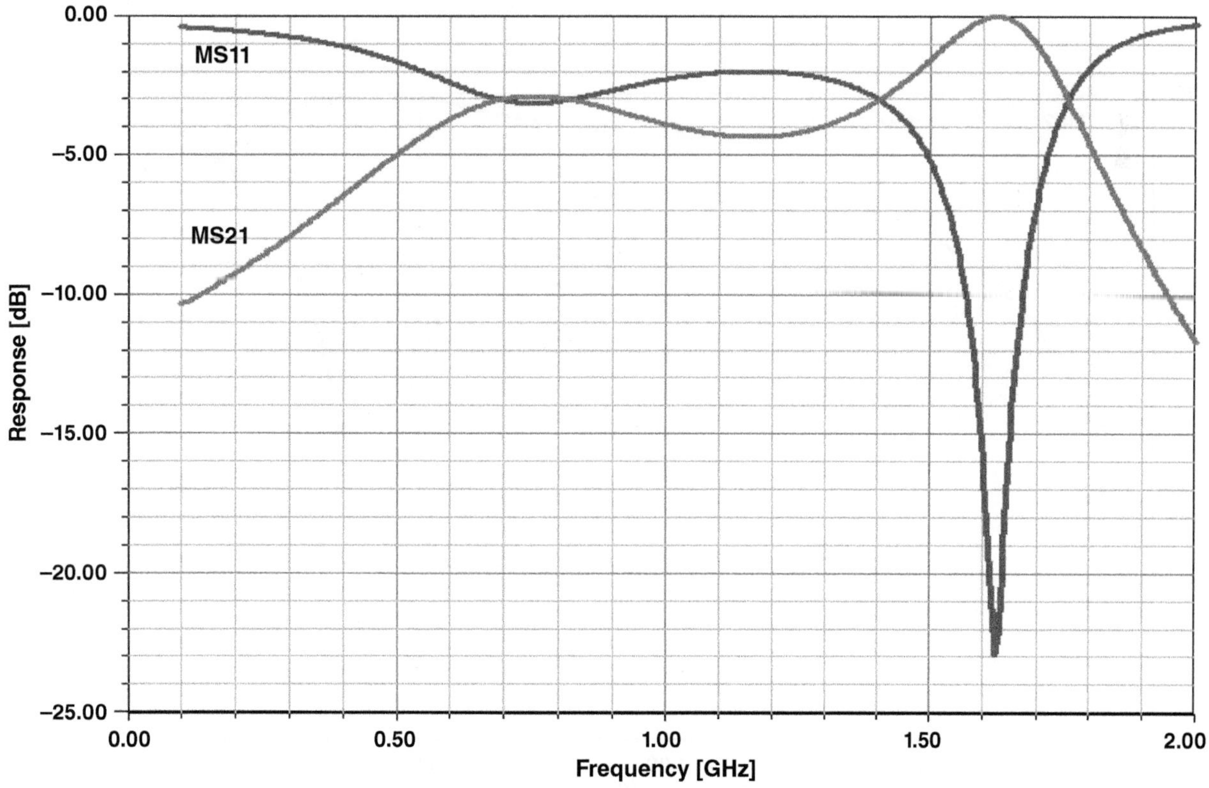

Figure 8.55 Frequency response of the input matching network. *Source*: [1]/John Wiley & Sons.

exceeds the 5 pF value. To compensate for this, one needs to use an RF choke that tunes out the resulting dynamic output capacitance, and then one can revert to the original 5 pF as calculated [1].

The selection of the actual component is another issue, because these capacitors need to handle several amperes of RF current. It is not always easy to find capacitors with the appropriate value that can handle the current. The inductors can be printed. The paragraph following this section, therefore, will deal with the issue of how to translate the lumped values into distributed values, which will make the current handling much easier (we hope).

For this, we did the same optimizing approach by using a 50-Ω source and a 46.3-Ω load. Figures 8.56 and 8.57 show the frequency-dependent responses and the circuit with the appropriate values.

The next step is to connect the input and output matching network to an "actual" transistor. This is a point where difficulties become very high, since very few, if any, companies provide SPICE-type parameters for transistors operating at currents more than 200 mA. Figures 8.58 and 8.59 show the overall schematic and frequency response of the actual amplifier

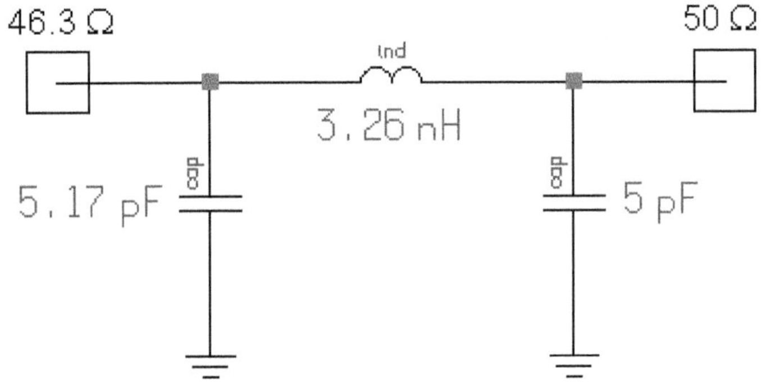

Figure 8.56 Output matching network for the amplifier. *Source*: [1]/John Wiley & Sons.

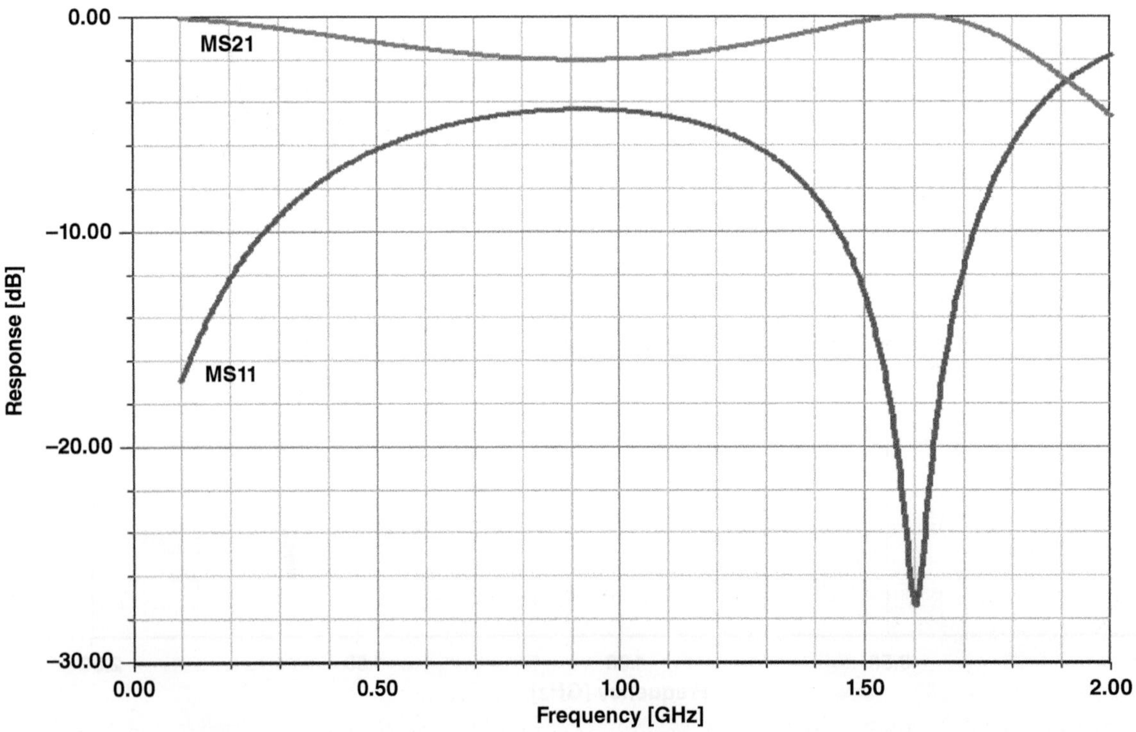

Figure 8.57 Frequency response of the output matching network. *Source*: [1]/John Wiley & Sons.

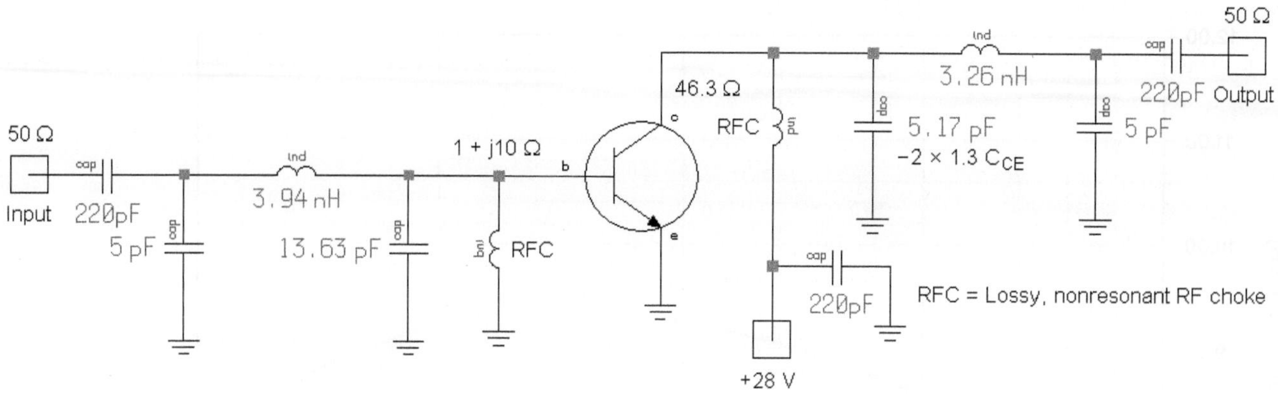

Figure 8.58 Actual schematic of our experimental 1.6-GHz amplifier. The RF chokes have enough RF losses not to be resonant and yet don't load the circuit. This is frequently accomplished either with ferrite beads or resistors in parallel with the chokes. *Source*: [1]/John Wiley & Sons.

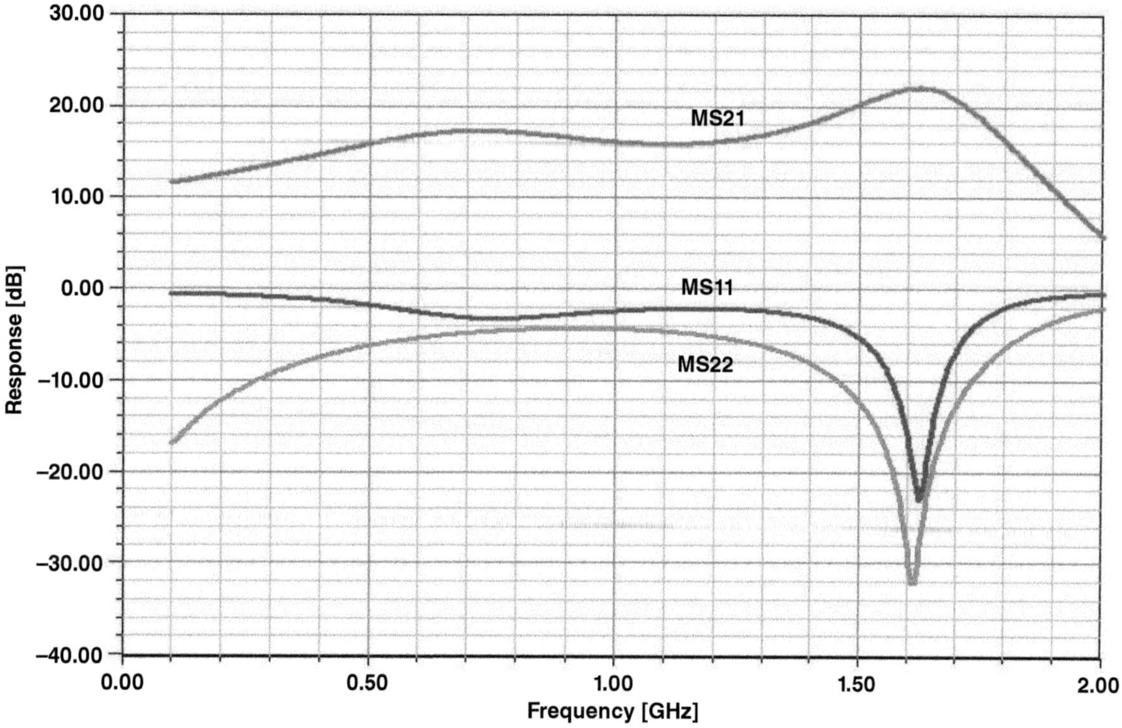

Figure 8.59 Overall frequency response of our experimental amplifier. Input and output matching, and gain, are shown as a function of frequency. *Source*: [1]/John Wiley & Sons.

with all filters connected. Figure 8.60 details the amplifier's peak gain response. If more bandwidth is needed, the matching circuit must be increased in complexity, as will be shown in the following section.

Now is a good time to look at the final results of our simulation, which is based on the various assumptions as outlined earlier. Figure 8.61 shows the dc I–V curve with a straight load line and no reactances. This was accomplished by using a wideband transformer instead of an output matching circuit. Figure 8.62, however, shows the "load line" that results when

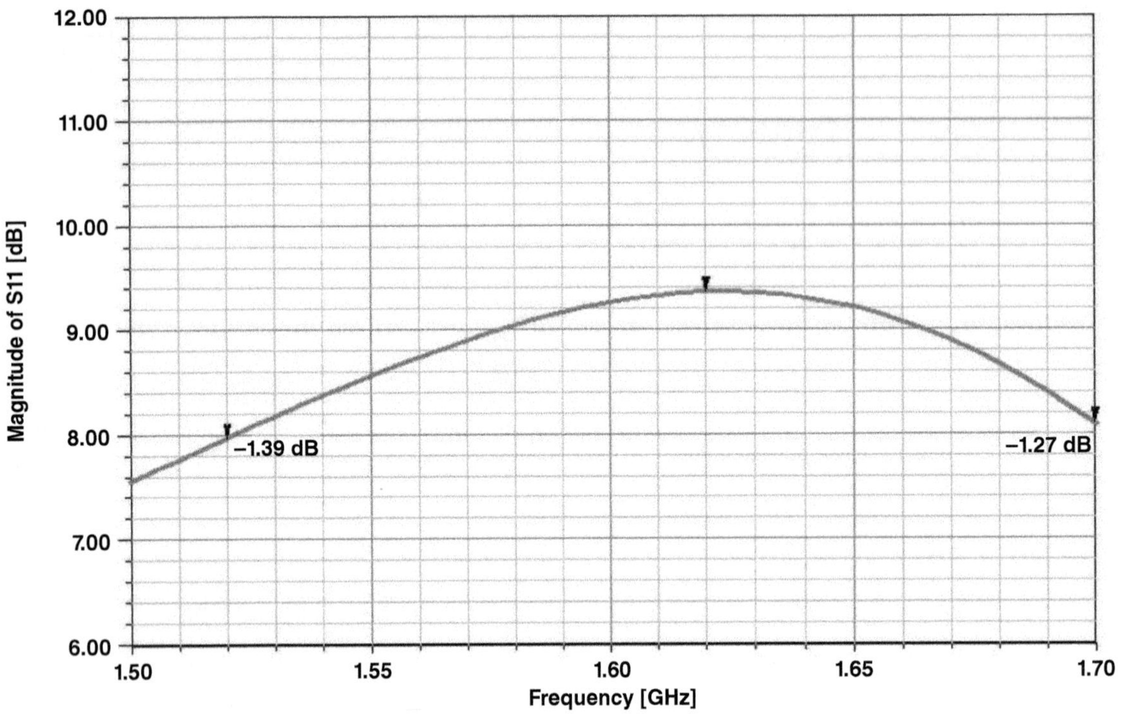

Figure 8.60 Close-in display of the frequency-dependent gain. *Source*: [1]/John Wiley & Sons.

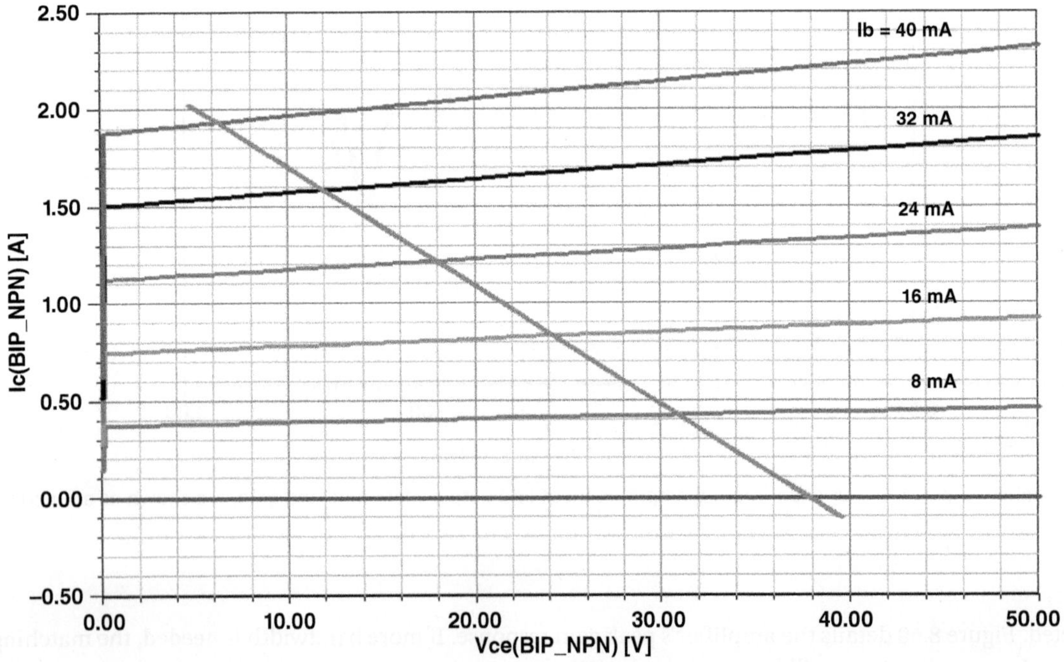

Figure 8.61 Dc $I-V$ curves assuming a perfect resistive load. No RF saturation voltage or other RF saturation effects are evident. *Source*: [1]/John Wiley & Sons.

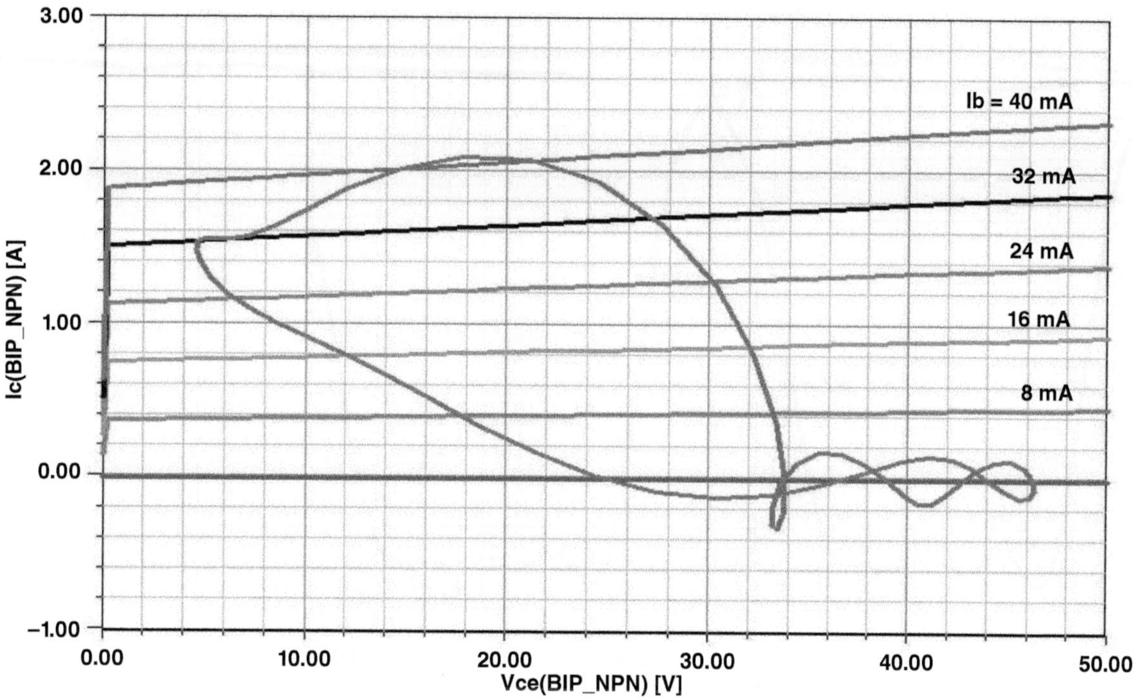

Figure 8.62 Load line with the output transformer replaced with a real matching circuit. *Source:* [1]/John Wiley & Sons.

the transformer is replaced with a real matching network. Based on the harmonic contents and their improper termination, one can see how the output load line opens and shows ringing in the low-current area. If we expect to see the saturation voltage, we will be disappointed because the saturation voltage is frequency-dependent, increases with frequency, and this figure would give an erroneous impression. Hence, we show dc I–V curves. To display the conduction angle, which we showed previously, we are going to look at the output current as a function of time at a drive level of +16.5 dBm (Figure 8.63). During the negative half, the transistor is not conducting as predicted, so this validates our assumptions. Next, we are interested in determining the power gain. Therefore, we look at the output spectrum for a drive level of +16.5 dBm (Figure 8.64). It shows an output level of 38.6 dBm. If we subtract the drive level (16.5 dBm) from this, we see that the power gain is 22.1 dB at this drive level. Since we are used to thinking in watts, Figure 8.65 shows the absolute output power and the harmonics prior to filtering. Figure 8.66 shows the output spectrum after replacing the transformer with a simple pi output network [1].

During the conduction period, there will be a voltage across the transistor 180° out of phase, as shown in Figure 8.67. Finally, to show the lack of linearity of a Class C amplifier, Figure 8.68 shows the amplifier's output power versus drive, and Figure 8.69 shows the amplifier's gain versus drive. When the transistor turns on is clearly evident.

A particular difficulty arises in the case of FETs, where the charge energy makes for a difference between dc and pulsed measurements. Figure 8.70 shows this. Some of the more modern transistors, such as SiGe types, may need some pulsing also, and as can be seen in Figure 8.70, even depending on the pulsing, one may get different dc I–V curves. This area is still under investigation, and even the "experts" do not agree on all aspects of this. In real life, it turns out that a lot of experimentation is the answer. Anybody who thinks that a pure CAD approach will give the right design answers immediately will encounter many surprises! This is not a fault of the CAD tools—assuming that they are properly engineered—but rather the availability of appropriately "tweaked" models for high-power applications, and the willingness of device suppliers to standardize on certain models and to support them [1].

Because of the difficulties involved in cooling transistors under test and simulating RF pulse and CW conditions, depending on the application, manufacturers have so far stayed away from investing money in this area. The designer must therefore resort to application notes published by device manufacturers. Because even slight layer changes can have a large impact on the frequency response of a given device, application-note-based designs can run into difficulty if a device's fabrication has significantly evolved since the publication of its application notes.

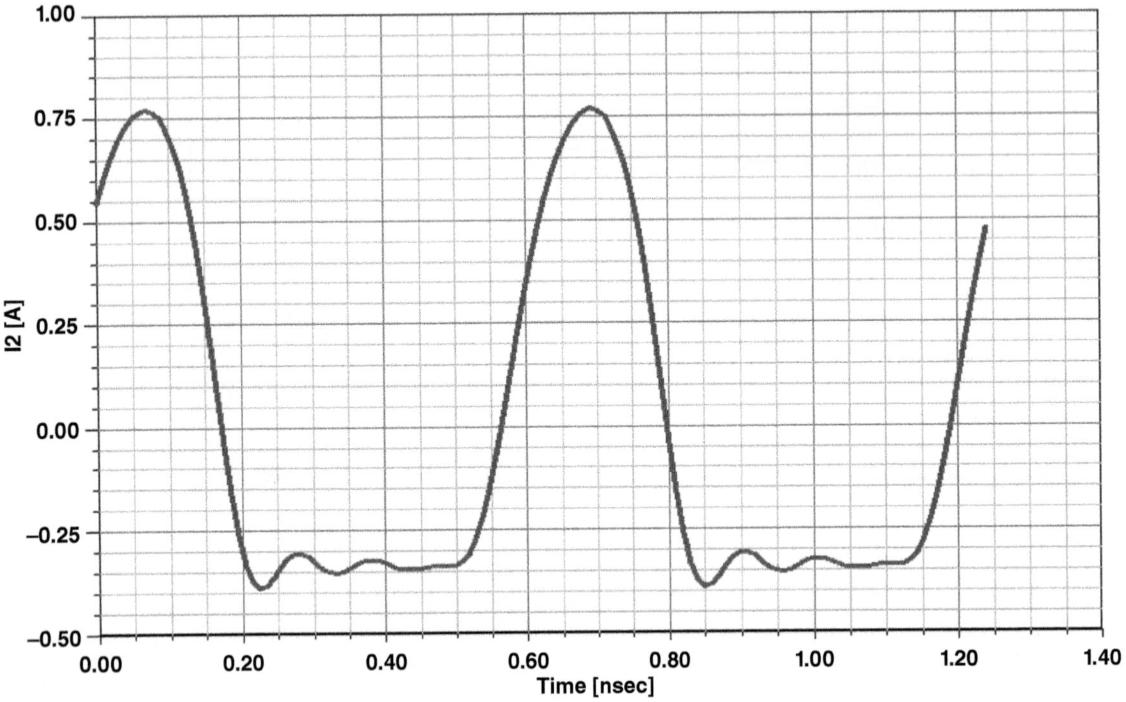

Figure 8.63 Output current as a function of time. The duty cycle is determined by the device conduction angle. *Source*: [1]/John Wiley & Sons.

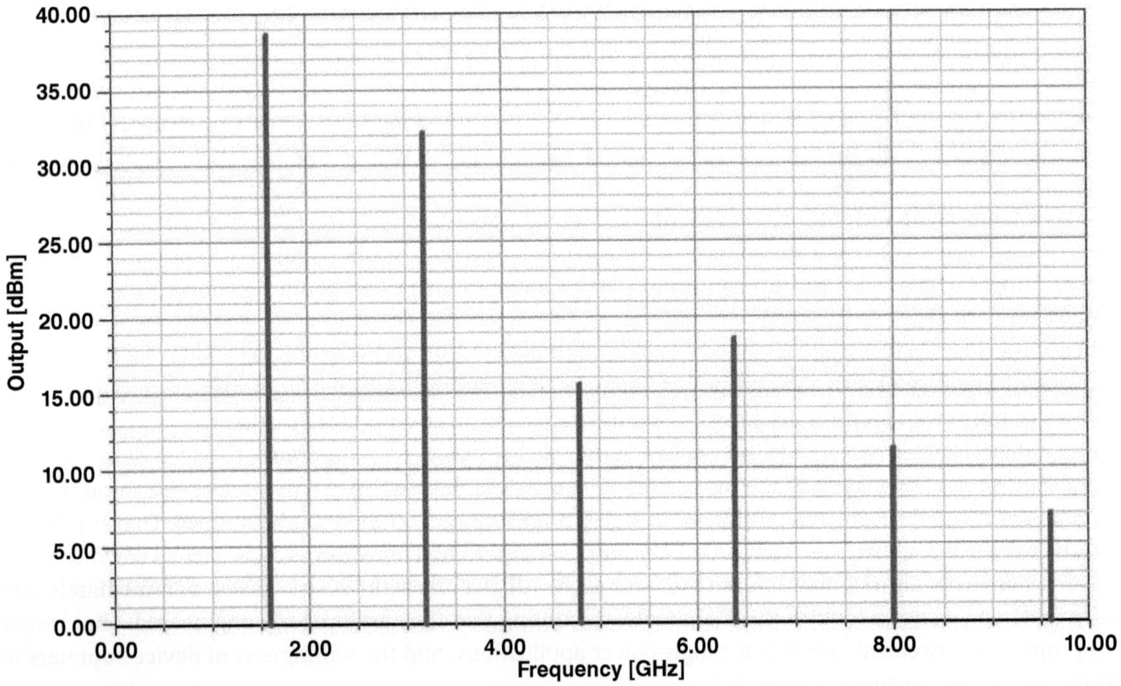

Figure 8.64 Power output at the collector prior to any filtering, assuming a real output termination. *Source*: [1]/John Wiley & Sons.

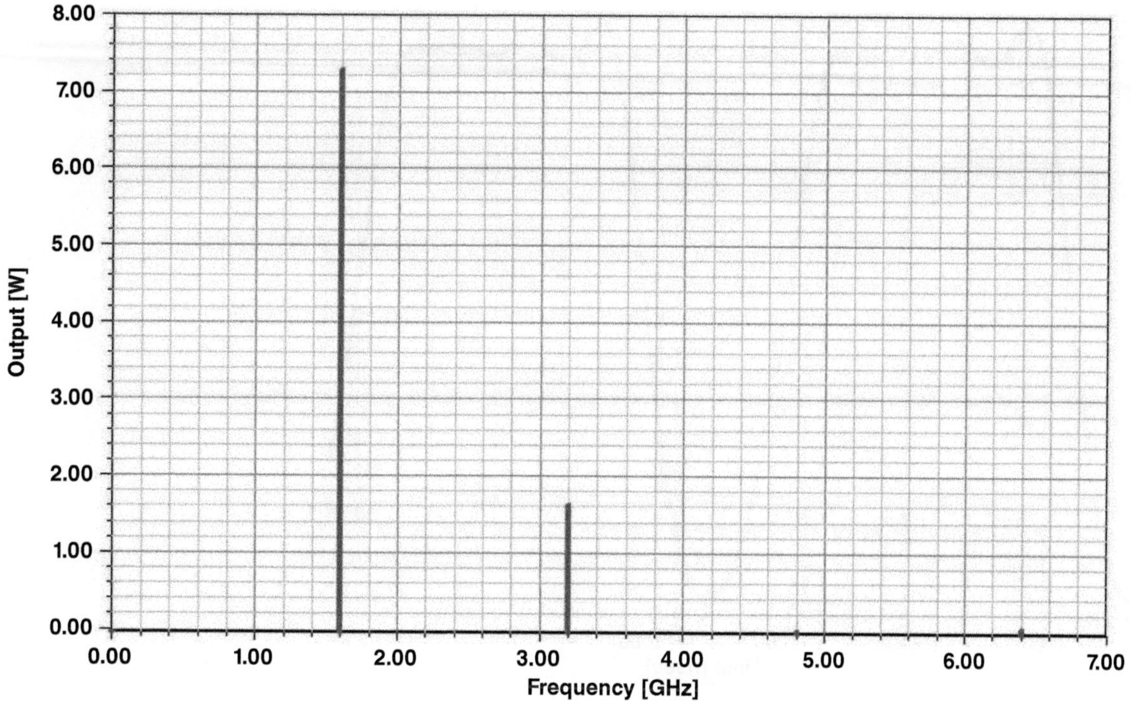

Figure 8.65 Available output power in watts, prior to any filtering, assuming a real output termination. *Source*: [1]/John Wiley & Sons.

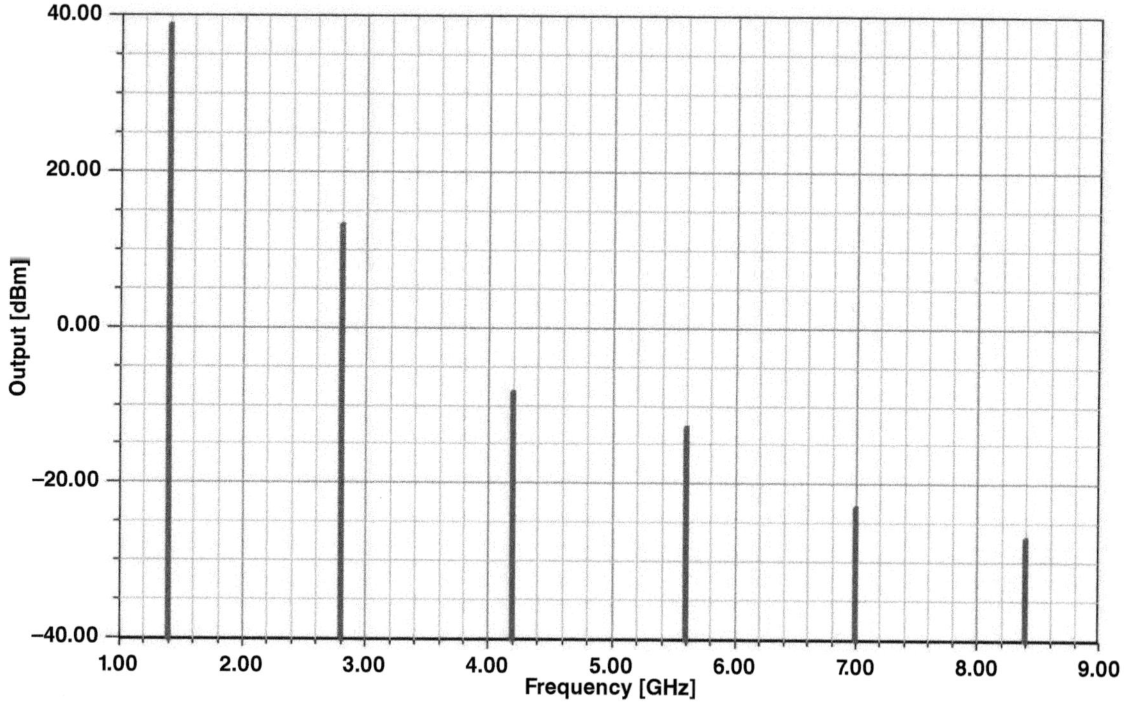

Figure 8.66 The output spectrum after replacing the transformer with a simple pi output network. *Source*: [1]/John Wiley & Sons.

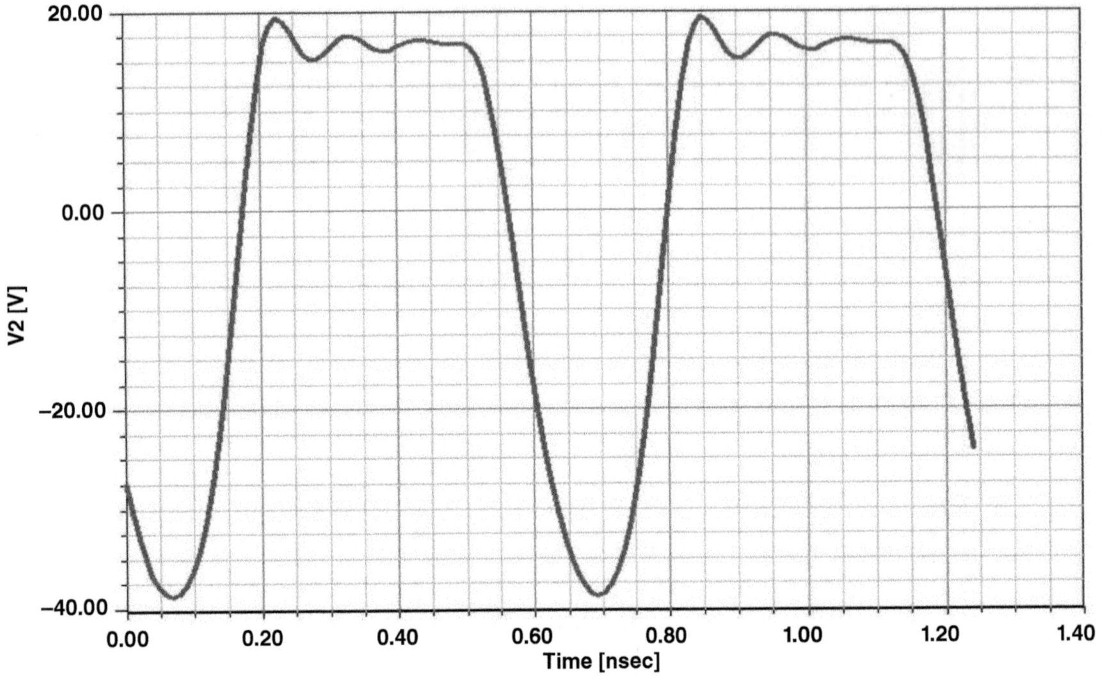

Figure 8.67 Output voltage at the collector. The voltage is asymmetrical compared to 0; this is important in the selection of a device capable of handling these voltages. *Source*: [1]/John Wiley & Sons.

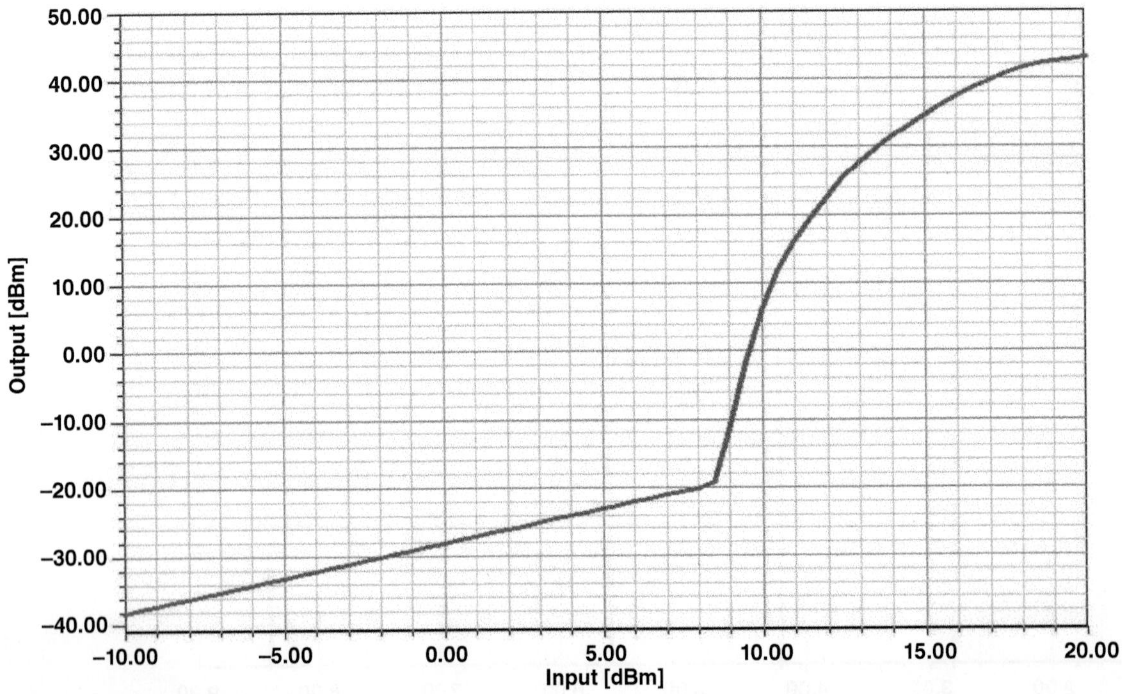

Figure 8.68 Output power versus input power for the Class C amplifier. Until the transistor turns on (at a drive level of about 8.5 dBm), only fed-through power appears at the output. *Source*: [1]/John Wiley & Sons.

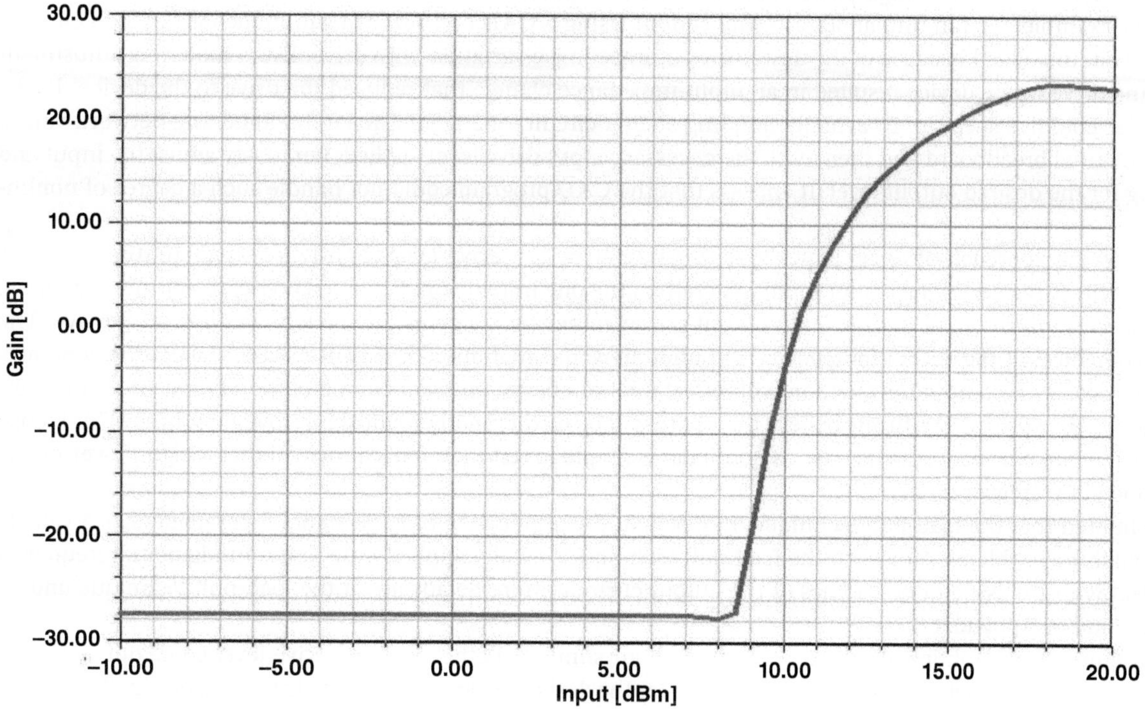

Figure 8.69 Amplifier gain as a function of drive. It is evident that one needs to cross the 0.65 V point for the transistor to become conductive and show gain. The initial amplification as a function of drive shows a very steep curve that then levels off and droops as the amplifier moves into saturation. *Source*: [1]/John Wiley & Sons.

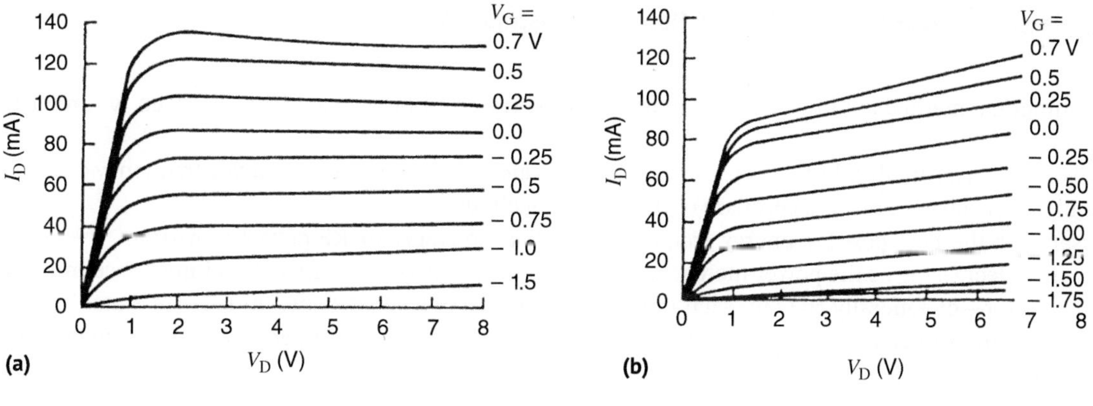

Figure 8.70 Comparison of FET drain current characteristics obtained from (a) pulsed *I*–*V*, (b) rectified sine-wave evaluation. The rectified sine-wave curves can be considered equivalent to curves obtained at dc. *Source*: [1]/John Wiley & Sons.

Given the multitude of models outlined in Chapter 2, the CAD user is sometimes left hanging as to which is the best model to use. The latest indications are that the MEXTRAM (Most Exquisite Transistor Model) model, developed by the University of Delft, Holland, with Philips of Holland, shows the most promising result, specifically in the area of IMD products. While other late models have up to seven internal nodes, MEXTRAM fortunately uses only five. (The MEXTRAM model actually consists of three transistors, which makes modeling fun.) Power-added efficiency simulation is not available, since the MEXTRAM model was developed for small-signal applications, meaning circuits that operate at less than 100 mW of RF output power.

As if the absence of trustworthy high-power device models is not a challenge enough, in developing this 1.6-GHz Class C amplifier example, we ran into another complication: Using a CAD optimizer to simultaneously match circuit's input and output circuits failed because of the device's input impedance variation with drive. Every matching adjustment that improved the drive to the device resulted in an input-impedance change that reduced the drive to the device! In the end, the optimizer couldn't keep up. In actually building such a circuit, one would make the matching-network values adjustable on the initial breadboard and then, with the necessary input drive level applied, iteratively adjust the input and output matching for the desired output level. It appears that the CAD program could not handle such a degree of nonlinearity.

We then went to the approach of using large-signal S parameters at a drive level slightly less than 16 dBm, but this detuned the input filter, resulting in a poorer input match than the one initially predicted. Likewise, the output matching, which in a power stage must result in maximum output power rather than a conjugate match, was less impressive. The large-signal S parameters at the output for a power amplifier are somewhat dubious because Class C operation generates so many harmonics that the output termination will reflect a lot of energy, and, depending on the output filter topology, different results can be expected. We firmly believe that there will be further mathematical improvement in the CAD tools, and we very much invite our readers to use such a nonlinear example to test their tools, provided that a high-power model becomes available in this power range.

It has been rumored that the load-pull technique, rather than the application of large-signal S parameters, is a vehicle to solve such matching problems. However, the load-pull technique is really applied only to the fundamental frequency and does not deal with the issue of harmonics of the output current. More details about the load-pull technique and its capabilities are found in Appendix B.

Actual measurements of amplifiers of this type show that our simulated efficiency and drive level come quite close to reality, but at a frequency somewhat offset from that chosen for the design. As an example, the input and output filters were tuned at 1.4 GHz instead of 1.6 GHz using the large-signal S parameters. Again, an iterative process, similar to manually trimming a prototype, will give a similar result. This is consistent with our earlier statement that most final designs are really hand-tweaked, and that CAD can only bring us close, but not all the way, to a good solution. For passive problems, the CAD does a significantly better job; the nonlinear cases are the ones that still cause headaches [1].

8.21.2 Impedance Matching Networks Applied to RF Power Transistors[1] [39]

8.21.2.1 Introduction

Some graphical and numerical methods of impedance matching will be reviewed here. The examples given will refer to high-frequency power amplifiers.

Although matching networks normally take the form of filters and therefore are also useful to provide frequency discrimination, this aspect will be considered only as a corollary of the matching circuit.

Matching is necessary for the best possible energy transfer from stage to stage. In RF-power transistors, the input impedance is of low value, decreasing as the power increases or as the chip size becomes larger. This impedance must be matched either to a generator—of generally 50 Ω internal impedance—or to a preceding stage. Impedance matching has to be made between two complex impedances, which makes the design still more difficult, especially if matching must be accomplished over a wide frequency band.

8.21.2.1.1 Input Impedance

The general shape of the input impedance of RF-power transistors is as shown in Figure 8.71. It is a large-signal parameter, expressed here by the parallel combination of a resistance R_p and a reactance X_p [40].

The equivalent circuit shown in Figure 8.72 accounts for the behavior illustrated in Figure 8.71. With the presently used stripline or flange packaging, most of the power devices for VHF low band will have their R_p and X_p values below the series resonant power f_s. The input impedance will be essentially capacitive.

Most of the VHF high-band transistors will have the series-resonant frequency within their operating range; that is, they will be purely resistive at one single frequency f_s, whereas the parallel-resonant frequency f_p will be outside.

Parameters for 1- or 2-GHz transistors will be beyond f_s and approach f_p. They show a high value of R_p and X_p with an inductive character.

1 Based on Motorola application note AN-721 ("Impedance Matching Networks Applied to RF Power Transistors"). Copyright of Motorola; used by permission.

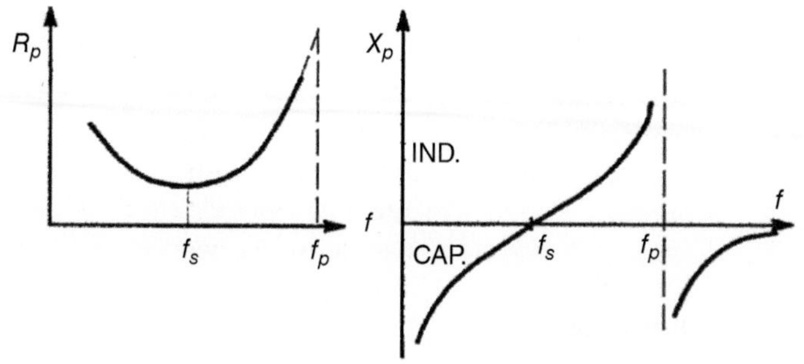

Figure 8.71 Input impedance of RF power transistors as a function of frequency. *Source*: [39]/Motorola.

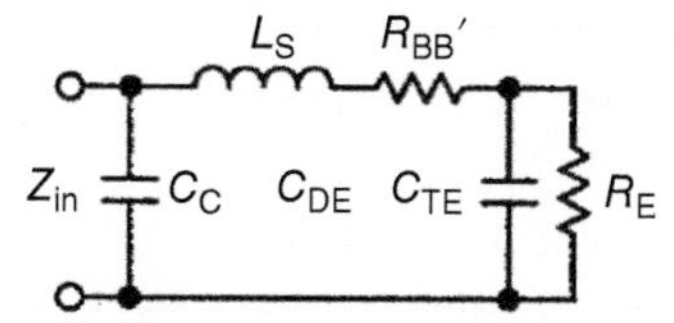

Figure 8.72 Equivalent circuit for the input impedance. R_E = emitter diffusion resistance; C_{DE}, C_{TE} = diffusion and transition capacitances, respectively, of the emitter junction; $R_{BB'}$ = base spreading resistance; C_C = package capacitance; L_S = base lead inductance.

A parameter that is very often used to judge the broadband capabilities of a device is the input Q or Q_{IN}, defined simply as the ratio R_p/X_p. Practically, Q_{IN} ranges around 1 or less for VHF devices and around 5 or more for microwave transistors.

Q_{IN} is an important parameter to consider for broadband matching. Matching networks normally are low-pass or pseudo-low-pass filters. If Q_{IN} is high, it can be necessary to use band-pass filter types matching networks and to allow insertion losses. But broadband matching is still possible. This will be discussed later.

8.21.2.1.2 Output Impedance

The output impedance of RF power transistors, as given by all manufacturers' datasheets, generally consists of only a capacitance C_{OUT}. The internal resistance of the transistor is supposed to be much higher than the load and is normally neglected. In the case of a relatively low internal resistance, the efficiency of the device would decrease by the factor

$$1 + R_L/R_T \tag{8.142}$$

where R_L is the load resistance, seen at the collector-emitter terminals, and R_T the internal transistor resistance is equal to

$$\frac{1}{\omega_T(C_{TC} + C_{DC})} \tag{8.143}$$

defined as a single parameter, where:

ω_T = transit angular frequency
C_{TC} and C_{DC} = transition and diffusion capacitances, respectively, at the collector junction

The output capacitance, C_{OUT}, which is a large-signal parameter, is related to the small-signal parameter C_{CB}, the collector-base transition capacitance.

Since a junction capacitance varies with the applied voltage, C_{OUT} differs from C_{CB} in that it has to be averaged over the total voltage swing. For an abrupt junction and assuming certain simplifications, $C_{OUT} = 2C_{CB}$.

Figure 8.73 shows the variation of C_{OUT} with frequency. C_{OUT} decreases partly due to the presence of the collector lead inductance, but mainly because of the fact that the base-emitter diode does not shut off anymore when the operating frequency approaches the transit frequency, f_T.

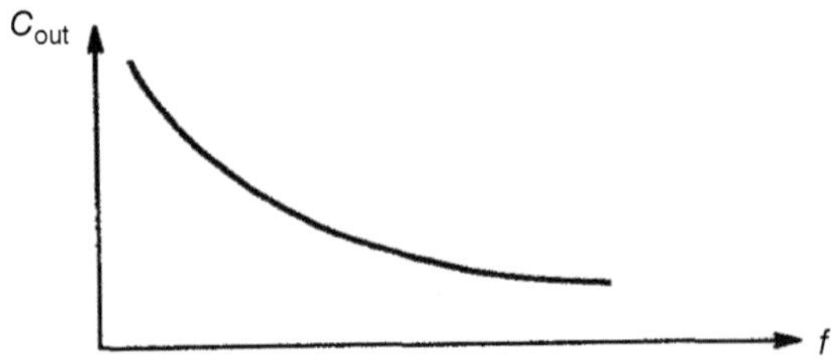

Figure 8.73 Output capacitance C_{OUT} as a function of frequency. *Source*: [39]/Motorola.

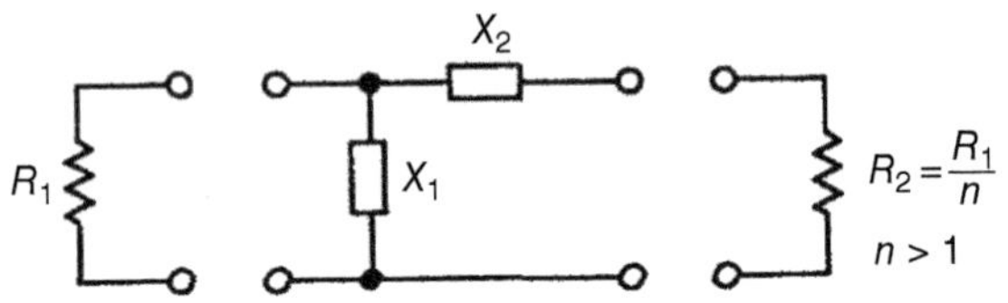

Figure 8.74 Two-reactance matching network. *Source*: [39]/Motorola.

Output Load: In the absence of a more precise indication, the output load R_L is taken equal to:

$$R_L = \frac{[V_{CC} - V_{CE(sat)}]^2}{2P_{OUT}} \tag{8.144}$$

with $V_{CE(sat)}$ equal to 2 or 3 volts, increasing with frequency.

The above equation just expresses a well-known relation, but also shows that the load, in first approximation, is not related to the device, except for $V_{CE(sat)}$. The load value is primarily dictated by the required output power and the peak voltage; it is not matched to the output impedance of the device.

At high frequencies, this approximation becomes less exact, and for microwave devices, the load that must be presented to the device is indicated on the datasheet.

Strictly speaking, impedance matching is accomplished only at the input. Interstage and load matching are more impedance transformations of the device input impedance and of the load into a value R_L (sometimes with an additional reactive component) that depends essentially on the power demanded and the supply voltage.

Matching Networks: In the following, matching networks will be described by the order of complexity. These are ladder-type reactance networks.

The different reactance values will be calculated and determined graphically. Increasing the number of reactances broadens the bandwidth. However, matching networks consisting of more than four reactances are rare. Above four reactances, the improvement is small.

8.21.2.2 Numerical Design

8.21.2.2.1 Two-resistance Networks

Resistance terminations will first be considered. Figure 8.74 shows the reactive L section and the terminations to be matched.

Matching or exact transformation from R_2 into R_1 occurs at a single frequency f_0. At f_0, X_1, and X_2 are equal to:

$$X_1 = \pm R_1\sqrt{\frac{R_2}{R_1 - R_2}} = R_1\frac{1}{\sqrt{n-1}} \tag{8.145}$$

$$X_2 = \mp\sqrt{R_2(R_1 - R_2)} = R_1\frac{\sqrt{n-1}}{1} \tag{8.146}$$

At f_0, $X_1X_2 = R_1R_2$. X_1 and X_2 must be of opposite sign. The shunt reactance is in parallel with the larger resistance.

The frequency response of the L section is shown in Figure 8.75, where the normalized current is plotted as a function of the normalized frequency.

If X_1 is capacitive and consequently X_2 is inductive, then:

$$X_1 = -\frac{f_o}{f}R_1\sqrt{\frac{R_2}{R_1 - R_2}} = -\frac{f_o}{f}R_1\frac{1}{\sqrt{n-1}} \tag{8.147}$$

and

$$X_2 = \frac{f}{f_o}\sqrt{R_2(R_1 - R_2)} = \frac{f}{f_o}R_1\frac{\sqrt{n-1}}{1} \tag{8.148}$$

The normalized current absolute value is equal to:

$$\left|\frac{I_2}{I_o}\right| = \frac{2\sqrt{n}}{\sqrt{(n-1)^2 \cdot \left(\frac{f}{f_o}\right)^4 - 2\left(\frac{f}{f_o}\right)^2 + (n+1)^2}} \tag{8.149}$$

where $I_o = \frac{\sqrt{n}E}{2 \cdot R_1}$, and is plotted in Figure 8.75 [41].

If X_1 is inductive and consequently X_2 is capacitive, the only change required is a replacement of f by f_0 and vice versa. The L section has low-pass form in the first case and high-pass form in the second case.

The Q of the circuit at f_0 is equal to:

$$Q_0 = \frac{X_2}{R_2} = \frac{R_1}{X_1} = \sqrt{n-1} \tag{8.150}$$

For a given transformation ratio n, there is only one possible value of Q. On the other hand, there are two symmetrical solutions for the network that can be either a low-pass filter or a high-pass filter.

The frequency f_0 does not need to be the center frequency, $(f_1 + f_2)/2$, of the desired band limited by f_1 and f_2. In fact, as can be seen from the low-pass configuration of Figure 8.75, it may be interesting to shift f_0 toward the high band edge frequency f_2 to obtain a larger bandwidth w, where

$$w = \frac{2(f_1 + f_2)}{f_2 - f_1} \tag{8.151}$$

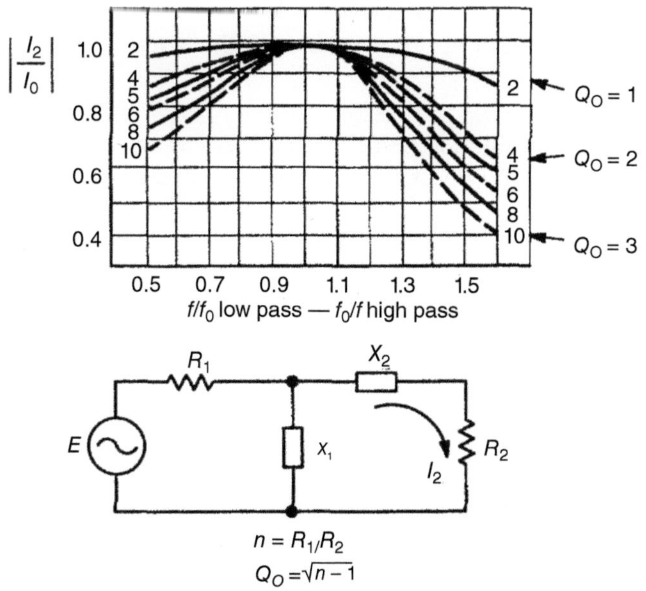

Figure 8.75 Normalized frequency response for the L section in low-pass or high-pass form. *Source:* [39]/Motorola.

This will, however, be at the expense of poor harmonic rejection.

Example: For a transformation ratio n = 4, it can be determined from the above relations:

Bandwidth w	0.1	0.3
Max insertion losses	0.025	0.2
X_1/R_1	1.730	1.712

If the terminations R_1 and R_2 have a reactive component X, the latter may be taken as part of the external resistance, as shown in Figure 8.76. This compensation is applicable as long as

$$Q_{INT} = \frac{X_{INT}}{R_2} \text{ or } \frac{R_1}{X_{INT}} < n - 1 \tag{8.152}$$

Tables giving reactance values can be found in [42, 43].

8.21.2.2.2 Use of Transmission Lines and Inductors

In the preceding section, the inductance was expected to be realized by a lumped element. A transmission line can be used instead (Figure 8.77).

As can be seen from the computed selectivity curves (Figure 8.78) for the two configurations, transmission lines result in a larger bandwidth. The gain is important for a transmission line having a length $L = \lambda/4$ ($\Theta = 90°$) and a characteristic impedance $Z_0 = \sqrt{R_1 R_2}$. It is not significant for lines short with respect to $\lambda/4$. One will notice that there is an infinity of solutions, one for each value of C, when using transmission lines.

8.21.2.2.3 Three-reactance Matching Networks

The networks that will be investigated are shown in Figure 8.79. They are made of three reactances alternately connected in series and shunt.

A three-reactance configuration allows the designer to make the quality factor (Q) of the circuit and the transformation ratio $n = R_2/R_1$ independent of each other and consequently to choose the selectivity between certain limits.

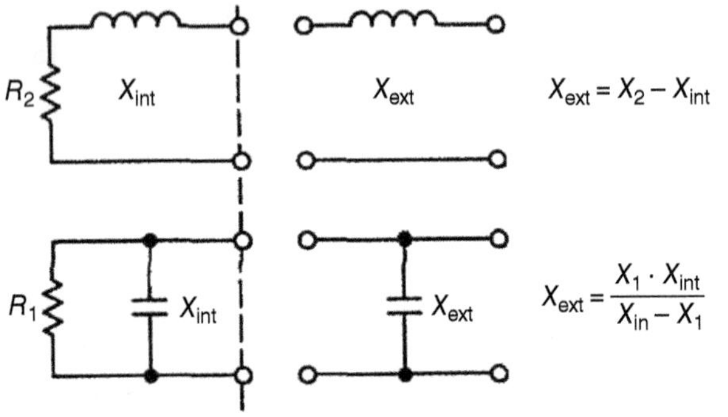

Figure 8.76 Termination reactance compensation. *Source*: [39]/Motorola.

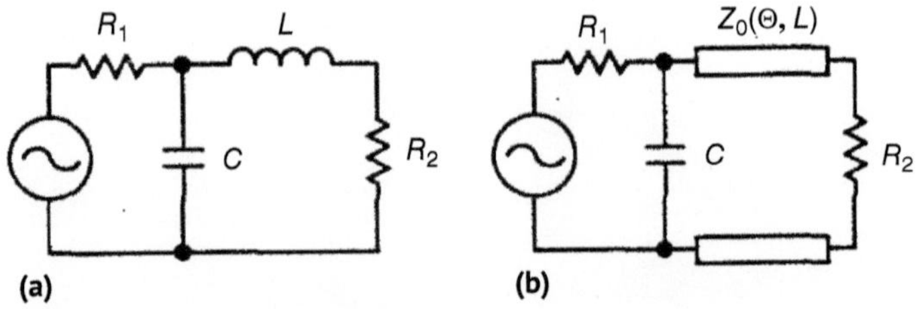

Figure 8.77 Use of a transmission line in the L section. *Source*: [39]/Motorola.

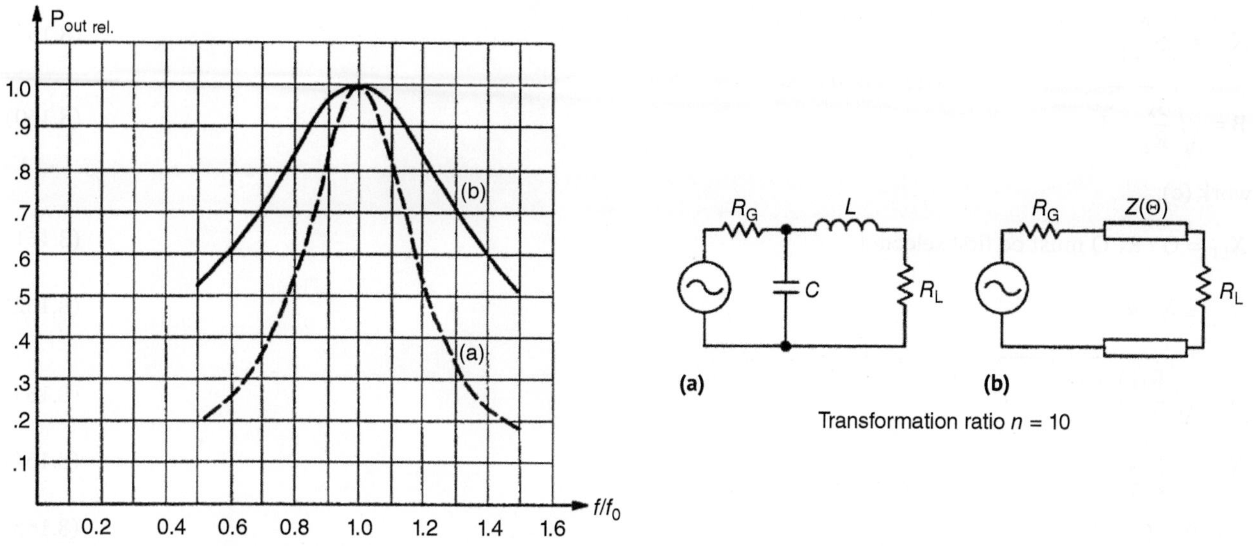

Figure 8.78 Bandwidth of the L section for $n = 10$ (a) with lumped constants and (b) with transmission line ($\lambda/4$). *Source*: [39]/Motorola.

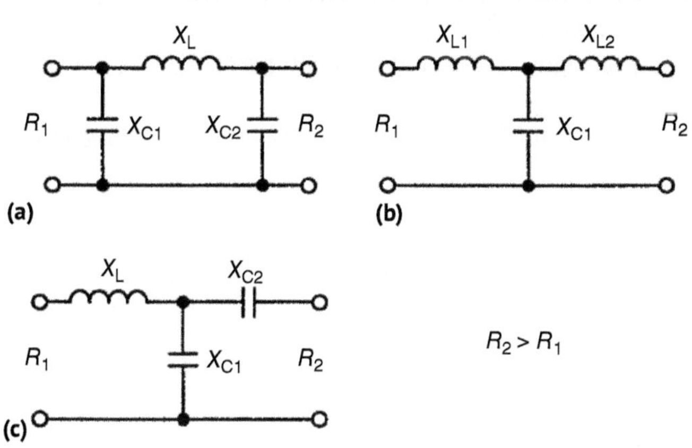

Figure 8.79 Three-reactance matching networks. *Source*: [39]/Motorola.

For narrowband designs, one can use the following formulas (see [44]), where tables are given:
Network (a):

$$X_{C1} = R_1/Q \text{ Q must be first selected} \tag{8.153}$$

$$X_{C2} = R_2\sqrt{\frac{R_1R_2}{(Q^2+1) - \frac{R_1}{R_2}}} \tag{8.154}$$

$$X_L = \frac{QR_1 + (R_1R_2/X_{C2})}{Q^2+1} \tag{8.155}$$

Network (b):

$$X_{L1} = R_1Q \text{ Q must be first selected} \tag{8.156}$$

$$X_{L2} = R_2 \cdot B \tag{8.157}$$

$$A = R_1(1+Q^2) \tag{8.158}$$

$$X_{C1} = \frac{A}{Q+B} \tag{8.159}$$

$$B = \sqrt{\frac{A}{R_2} - 1} \tag{8.160}$$

Network (c):

$$X_{L1} = Q \cdot R_1 \text{ Q must be first selected} \tag{8.161}$$

$$X_{C2} = A \cdot R_2 \tag{8.162}$$

$$A = \sqrt{\frac{R_1(1+Q^2)}{R_2} - 1} \tag{8.163}$$

$$X_{C1} = \frac{B}{Q-A} \tag{8.164}$$

$$B = R_1 \cdot (1+Q^2) \tag{8.165}$$

The network that yields the most practical component values should be selected for a given application.

The three-reactance networks can be thought of as being formed of an L section (two reactances) and of a compensation reactance. The L section essentially performs the impedance transformation, while the additional reactance compensates for the reactive part of the transformed impedance over a certain frequency band.

Figure 8.80 shows a representation in the Z plane of the circuit of Figure 8.79a split into two parts: $R_1 - C_1 - L_1$ and $C_2 - R_2$.

Exact transformation from R_1 into R_2 occurs at the points of intersection M and N. Impedances are then conjugate, or $Z' = R' + jX'$ and $Z'' = R'' + jX''$ with $R' = R''$ and $X' = -X''$.

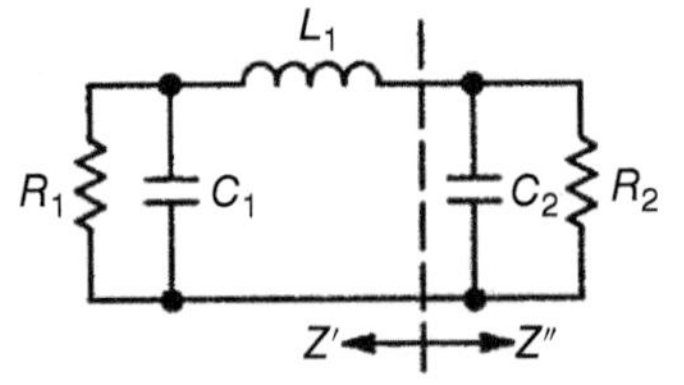

$$Z' = \frac{R_1}{1+\omega^2 R_1^2 C_1^2} + j\omega\left[L1 - \frac{R_1^2 C_1}{1+\omega^2 R_1^2 C_1^2}\right] = R' + jX'$$

$$Z'' = \frac{R_2}{1+\omega^2 C_2^2 R_2^2} - \frac{j\omega^2 C_2 R_2^2}{1+\omega^2 C_2^2 R_2^2} = R'' + jX''$$

$$|X'| = \left(\frac{L_1}{C_1 R_1^2} - \frac{R'}{R_1}\right)\sqrt{\frac{R_1}{R'} - 1}$$

$$|X''| = \sqrt{R' R_2 - R'^2}$$

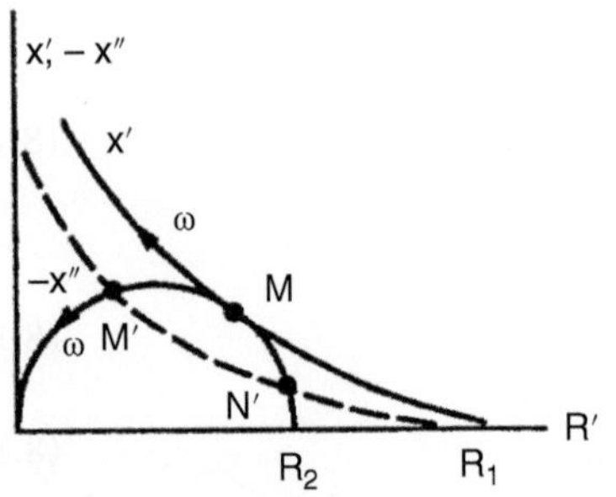

Figure 8.80 Z-plane representation of the circuit of Figure 8.79a. *Source:* [39]/Motorola.

The only possible solution is obtained when X' and $-X''$ are tangential to each other. For the dashed curve, representing another value of L_1 or C_1, a wider frequency band could be expected at the expense of some ripple inside the band. However, this can only be reached with four reactances, as will be shown below.

With a three-reactance configuration, there are not enough degrees of freedom to permit $X' = -X''$ and simultaneously obtain the same variation of frequency on both curves from point M' to point N'. Exact transformation can, therefore, be obtained at only one frequency.

The values of the three reactances can be calculated by making

$$X' = -X'', R' = R'' \text{ and } \frac{dX'}{dR'} = -\frac{dX''}{dR''} \tag{8.166}$$

The general solution of these equations leads to complicated calculations. Therefore, computed tables should be used.

One will note from Figure 8.80 that the compensation reactance contributes somewhat to impedance transformation; that is, R' varies when going from M to R_2.

The circuit of Figure 8.79b is dual with respect to the first one and gives exactly the same results in a Y-plane representation. The circuit of Figure 8.79c is somewhat different since only one intersection M exists as shown in Figure 8.81. Narrower frequency bands must be expected from this configuration. The widest band is obtained for $C_1 = \infty$.

Again, if one of the terminations has a reactive component, the latter can be taken as a part of the matching network, provided that it is not too large (see Figure 8.78).

8.21.2.2.4 Four-reactance Networks

Four-reactance networks are used essentially for broadband matching. The networks that will be considered in the following consist of two two-reactance sections in cascade. Some networks have pseudo-low-pass-filter character; others, bandpass-filter character. In principle, the former show a narrower bandwidth since they extend the impedance transformation to very low frequencies unnecessarily, while the latter insure good matching over a wide frequency band around the center frequency only (see Figure 8.84).

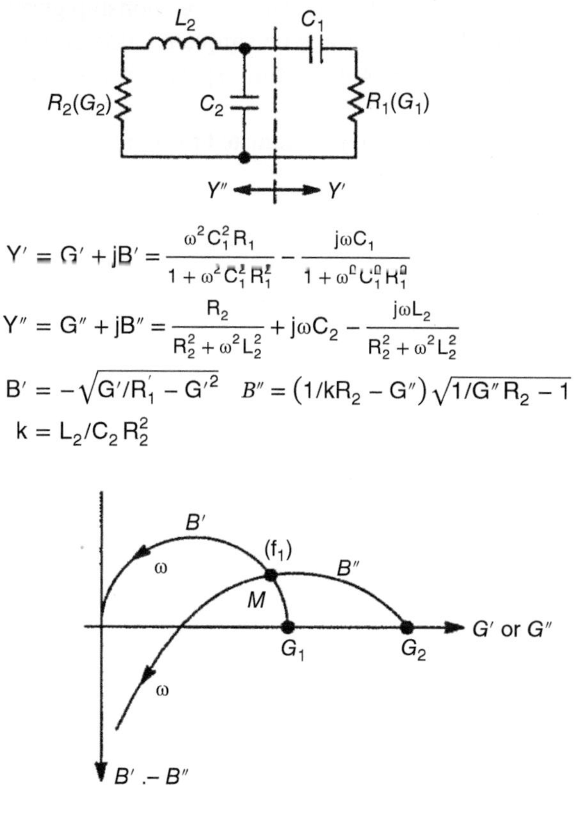

Figure 8.81 Y-plane representation of the circuit of Figure 8.79c. *Source:* [39]/Motorola.

The two-reactance sections used in the networks in Figure 8.82 have either transformation properties or compensation properties. Impedance transformation is obtained with one series reactance and one shunt reactance. Compensation is made with both reactances in series or in shunt. If two cascaded transformation networks are used, transformation is accomplished partly by each one [39].

With four-reactance networks, there are two frequencies, f_1 and f_2, at which the transformation from R_1 into R_2 is exact. These frequencies may also coincide. For the network in Figure 8.82b, for instance, at point M, R_1 or R_2 is transformed into $\sqrt{R_1 R_2}$ when both frequencies fall together. At all points (M), Z_1 and Z_2 are conjugate if the transformation is exact.

In the case of Figure 8.82b, the reactances are easily calculated for equal frequencies:

$$X_1 = \frac{R_1}{\sqrt{\sqrt{n} - 1}} \tag{8.167}$$

$$X_2 = R_1 \sqrt{\frac{\sqrt{n} - 1}{n}} \tag{8.168}$$

$$X_1 \cdot X_4 = R_1 \cdot R_2 = X_2 \cdot X_3 \tag{8.169}$$

$$X_3 = \frac{R_1}{\sqrt{n\left(\sqrt{n} - 1\right)}} \tag{8.170}$$

$$X_4 = \frac{R_1}{n} \sqrt{\sqrt{n} - 1} \tag{8.171}$$

For the network in Figure 8.82a, normally, at point (M) Z_1 and Z_2 are complex. This pseudo-low-pass filter has been computed elsewhere [42]. Many tables can be found in the literature for networks of four or more reactances having Chebyshev character or maximally flat response [42, 43, 45].

Figure 8.83 shows the transformation path from R_1 to R_2 for networks (a) and (b) in Figure 8.82 on a Smith Chart.

Case (a) of Figure 8.83 has been calculated using tables in [43]. Case (b) has been obtained from the relationship given above for $X_1 \cdots X_4$. Both apply for a transformation ratio equal to 10 and for $R_1 = 1$. There is no simple relationship for $X'1 \cdots X'4$ of network (b) if f_1 is made different from f_2 for a larger bandwidth. Figure 8.84 shows the respective bandwidths of network (a) and (b) for the circuits shown in Figure 8.83.

If the terminations include a reactive component, the computed values for X_1 or X_4 may be adjusted to compensate for this.

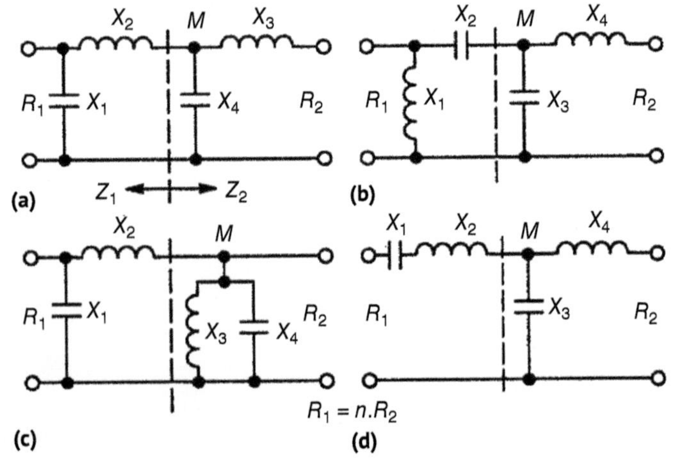

Figure 8.82 Four-reactance networks. *Source*: [39]/Motorola.

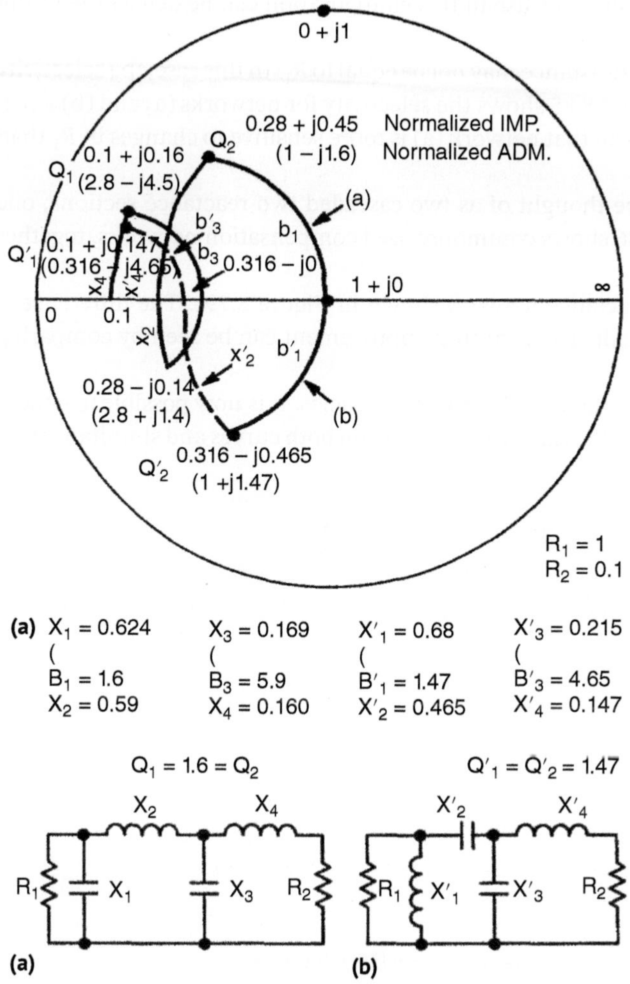

Figure 8.83 Transformation paths for networks (a) and (b). *Source:* [39]/Motorola.

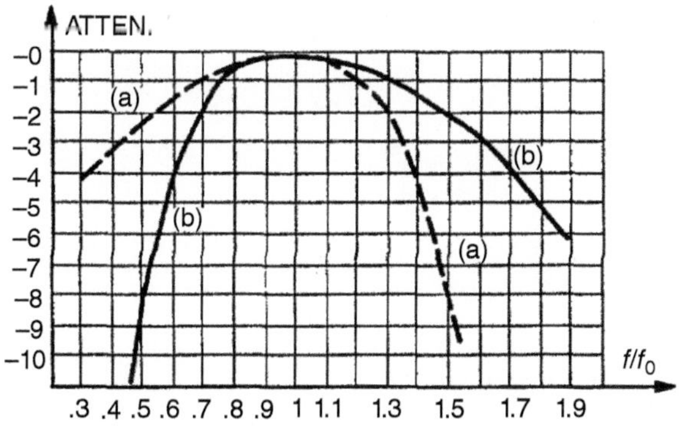

Figure 8.84 Selectivity curves for networks (a) and (b) of Figure 8.83. *Source:* [39]/Motorola.

For configuration (a), it can be seen from Figure 8.83 that, in the considered case, the *Q*s are equal to 1.6. For configuration (b), Q'_1, which is equal to Q'_2, is fixed for each transformation ratio

n	2	4	8	10	16
$Q'_1 = Q'_2$	0.65	1	1.35	1.46	1.73

$$Q' = \sqrt{\sqrt{n} - 1} \tag{8.172}$$

The maximum value of reactance that the terminations may have for use in this configuration can be determined from the above values of Q'.

If R_1 is the load resistance of a transistor the internal transistor resistance may not be equal to R_1. In this case, the selectivity curve will be different from the curves given in Figure 8.84. Figure 8.85 shows the selectivity for networks (a) and (b) when the source resistance R_1 is infinite. From Figure 8.85, it can be seen that network (a) is more sensitive to changes in R_1 than network (b).

As mentioned earlier, the four-reactance network can also be thought of as two cascaded two-reactance sections; one used for transformation, the other for compensation. Figure 8.86 shows commonly used compensation networks, together with the associated L section.

The circuit of Figure 8.86a can be compared to the three-reactance network shown in Figure 8.79c. The difference is that capacitor C_2 if that circuit has been replaced by an LC circuit. The resulting improvement can be seen by comparing Figure 8.87 with Figure 8.81.

By adding one reactance, an exact impedance transformation is achieved at two frequencies. It is now possible to choose component values such that the point of intersection M' occurs at the same frequency f_1 on both curves and simultaneously

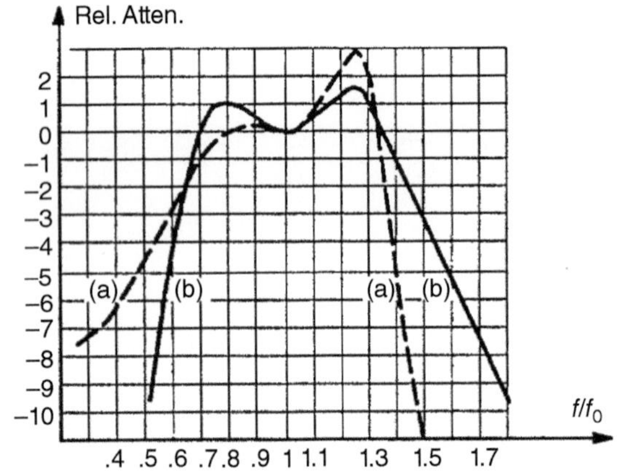

Figure 8.85 Selectivity curves for networks (a) and (b) of Figure 8.83 with infinite R_1. *Source*: [39]/Motorola.

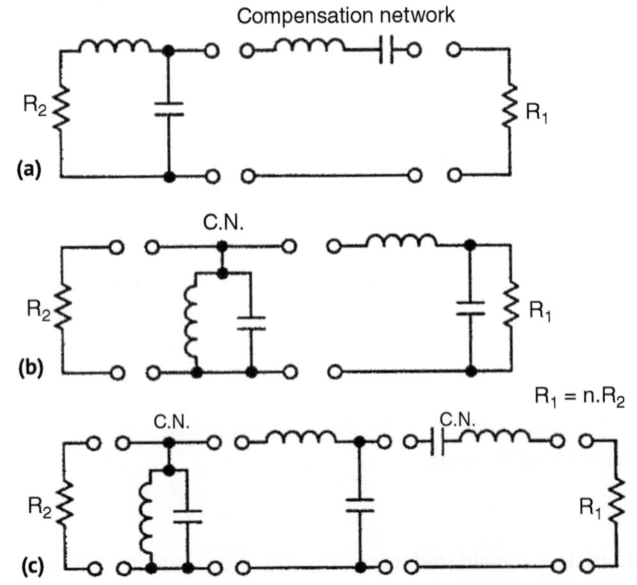

Figure 8.86 Compensation networks used with an L section. *Source*: [39]/Motorola.

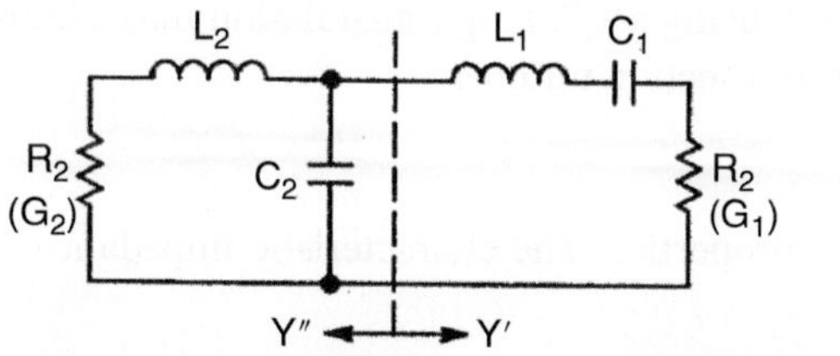

$$Y' = G' + jB' = \frac{R_1}{R_1^2 + (\omega L_1 - 1/\omega c_1)^2} - \frac{j(\omega L_1 - 1/\omega c_1)}{R_1^2 + (\omega L_1 - 1/\omega c_1)^2}$$

$$Y'' = G'' + jB'' = \frac{R_2}{R_2^2 + \omega^2 L_2^2} + j\omega\left(C_2 - \frac{L_2}{R_2^2 + \omega^2 L_2^2}\right)$$

$$B' = \pm\sqrt{\frac{G'}{R_1} - G'^2}$$

$$B'' = (1/kR_2 - G'')\sqrt{\frac{1}{G''R_2} - 1} \quad k = \frac{L_2}{C_2 R_2^2}$$

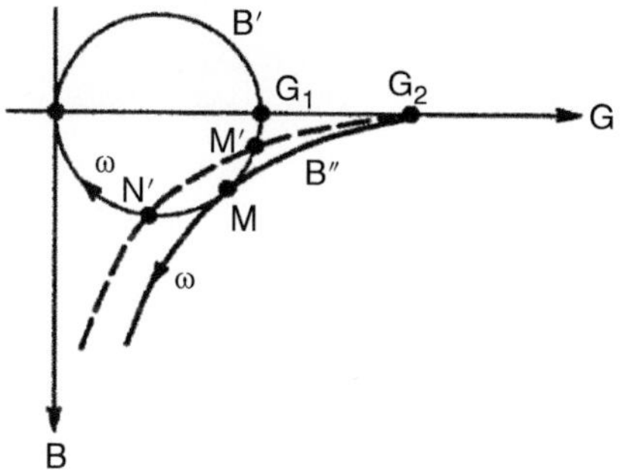

Figure 8.87 Y-plane representation of the circuit of Figure 8.86a. *Source:* [39]/Motorola.

that N' occurs at the same frequency f_2 on both curves. Among the infinite number of possible intersections, only one allows the achievement of this.

When M' and N' coincide in M, the new $\mathrm{d}X'/\mathrm{d}f = \mathrm{d}X''/\mathrm{d}f$ condition can be added to the condition $X' = -X''$ (for three-reactance networks) and similarly $R' = R''$ and $\mathrm{d}R'/\mathrm{d}f = \mathrm{d}R''/\mathrm{d}f$.

Again, a general solution of the above equations leads to still more complicated calculations than in the case of three-reactance networks. Therefore, tables are preferable [42, 43, 45].

The circuit of Figure 8.86b is the dual of the circuit of Figure 8.84a and does not need to be treated separately. It gives exactly the same results in the Z plane. Figure 8.86c shows a higher-order compensation requiring six reactive elements.

The above-discussed matching networks employing compensation circuits result in narrower bandwidths than the former solutions (see Subsection 8.21.2.2.4) using two transformation sections. A matching arrangement with higher-order compensation, such as in Figure 8.86c, is not recommended. Better use can be made of the large number of reactive elements by using them all for transformation.

When the above configurations are realized using short portions of transmission lines, the equations or the usual tables no longer apply. The calculations must be carried out on a computer because of their complexity. However, a graphical method can be used (see the next section) that will consist essentially of tracing a transformation path on the Z-Y chart using the computed lumped element values and replacing it with the closest path obtained with distributed constants. The bandwidth change is not significant as long as short portions of lines are used.

8.21.2.2.5 Matching Networks Using Quarter-wave Transformers

At sufficiently high frequencies, where $\lambda/4$-long lines of practical size can be realized, broadband transformations can easily be accomplished by the use of one or more $\lambda/4$ sections. Figure 8.88 summarizes the main relations for (a) one-section and

(b) two-section transformation. A compensation network can be realized using a $\lambda/2$-long transmission line. Figures 8.89 and 8.90 show the selectivity curves for different transformation ratios and section numbers.

8.21.2.2.6 Exponential Lines

Exponential lines have largely frequency-independent transformation properties. The characteristic impedance of such lines varies exponentially with their length l:

$$Z = Z_0 \cdot e^{kl} \tag{8.173}$$

where k is a constant, but these properties are preserved only if k is small.

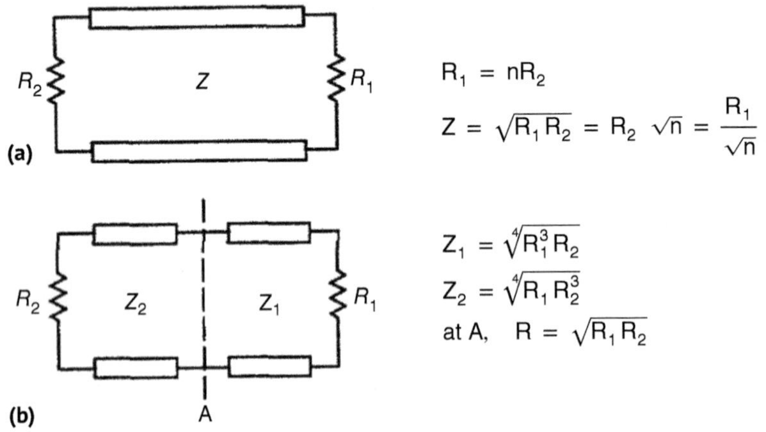

Figure 8.88 Transformation networks using $\lambda/4$-long transmission lines. *Source*: [39]/Motorola.

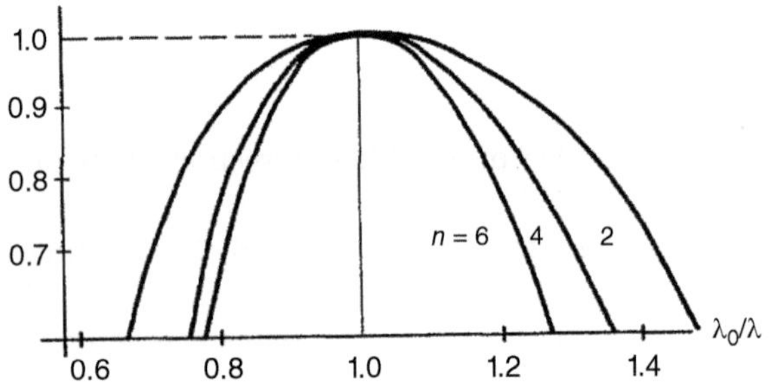

Figure 8.89 Selectivity curves for two $\lambda/4$-section networks at different transformation ratios. *Source*: [39]/Motorola.

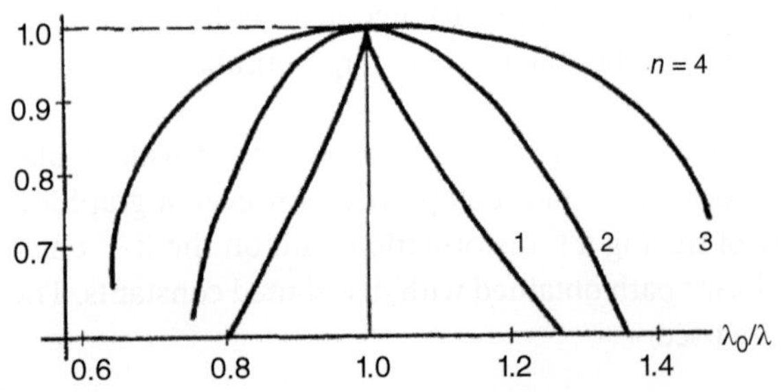

Figure 8.90 Selectivity curves for one, two, and three $\lambda/4$ sections. *Source*: [39]/Motorola.

8.21.2.2.7 Broadband Matching Using Bandpass-filter Networks–High-Q Case

The above circuits are applicable to devices having low input or output Q, if broadband matching is required. Generally, if the impedances to be matched can be represented, for instance, by a resistor R in series with an inductor L (sometimes a capacitor C) within the band of interest, and if L is sufficiently low, the latter can be incorporated into the first inductor in the matching network. This is also valid if the representation consists of a shunt combination of a resistor and a reactance [39].

Practically, this is feasible for Qs around 1 or 2. For higher Qs or for input Impedances consisting of a series or parallel resonant circuit (see Figure 8.72), as it appears to be for large bandwidths, a different treatment must be followed.

Let us first recall that, as shown by Bode and Fano [46, 47], limitations exist on the impedance matching of a complex load. In the example of Figure 8.91, the load to be matched consists of a capacitor C and a resistor R in shunt.

The reflection coefficient between the transformed load and generator is equal to:

$$\Gamma = \frac{Z_{IN} - R_g}{Z_{IN} + R_g} \tag{8.174}$$

When $\Gamma = 0$, there is a perfect matching; when $\Gamma = 1$, there is total reflection. The ratio of reflected to incident power is:

$$\frac{Pr}{Pi} = |\Gamma|^2 \tag{8.175}$$

The fundamental limitation on the matching takes the form:

$$\int_{\omega=0}^{\infty} \ln\left(\frac{1}{|\Gamma|}\right) d\omega \le \frac{\pi}{RC} \text{ Bode equation} \tag{8.176}$$

and is represented in Figure 8.92.

The meaning of the Bode equation is that the area S under the curve cannot be greater than π/RC, and therefore, if matching is required over a certain bandwidth, this can only be done at the expense of less power transfer within the band. Thus, power transfer and bandwidth appear as interchangeable quantities.

It is evident that the best utilization of the area S is obtained when $|\Gamma|$ is kept constant over the desired band ω_c and made equal to 1 over the rest of the spectrum. Then

$$|\Gamma| = e - \frac{\pi}{\omega_C RC} \tag{8.177}$$

within the band and no power transfer happens outside. A network fulfilling this requirement cannot be obtained in practice as an infinite number of reactive elements would be necessary.

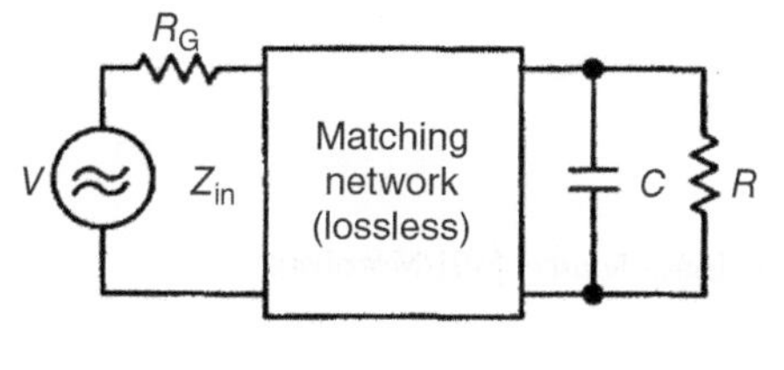

Figure 8.91 General matching conditions. *Source*: [39]/Motorola.

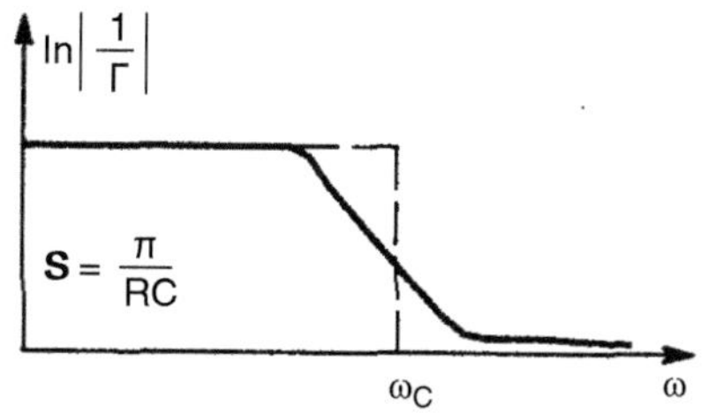

Figure 8.92 Representation of the Bode equation. *Source*: [39]/Motorola.

If the attenuation is plotted versus frequency for practical cases, one may expect to have curves like the ones shown in Figure 8.93 for a low-pass filter having Chebyshev character.

For a given complex load, an extension of the bandwidth from ω_1 to ω_2 is possible only with a simultaneous increase of the attenuation a. This is especially noticeable for Qs exceeding 1 or 2. (see Figure 8.94).

Thus, devices having relatively high input Qs are usable for broadband operation, provided the consequent higher attenuation or reflection introduced is acceptable.

The general shape of the average insertion losses or attenuation a (neglecting the ripple), the attenuation decreases if the number n of the network element increases. But above n = 4, the improvement is small. For a given attenuation a and bandwidth, the larger the n, the smaller the ripple; for a given attenuation, the larger the n, the larger the bandwidth.

Computations show that for $Q < 1$ and $n \leq 3$, the attenuation is below 0.1 dB (approximately). The impedance transformation ratio is not free here. The network is a true low-pass filter. For a given load, the optimum generator impedance will result from the computation.

Before impedance transformation is introduced, a conversion of the low-pass prototype into a bandpass filter network must be made. Figure 8.95 summarizes the main relations for this conversion.

r is the conversion factor. For the bandpass filter, $Q_{\text{IN-max}}$, or the maximum possible input Q of the device to be matched, has been increased by the factor r (from Figure 8.95, $Q'_{\text{IN-max}} = rQ_{\text{IN-max}}$).

Impedance inverters will be used for impedance transformation. These networks are suitable for insertion into a bandpass filter without affecting the transmission characteristics.

Figure 8.96 shows four impedance inverters. It will be noticed that one of the reactances is negative and must be combined in the bandpass network with a reactance of at least an equal positive value. Insertion of the inverter can be made at any convenient place [41, 48].

When using the bandpass filter for matching the input impedance of a transistor, reactances $L'_1C'_1$ should be made to resonate at ω_0 by the addition of a convenient series reactance.

As stated above, the series combination of R_0, L'_1 and C'_1 normally constitutes the equivalent input network of a transistor when considered over a large bandwidth. This is a good approximation up to about 500 MHz.

In practice, the normal procedure for using a bandpass filter matching network will be the following:

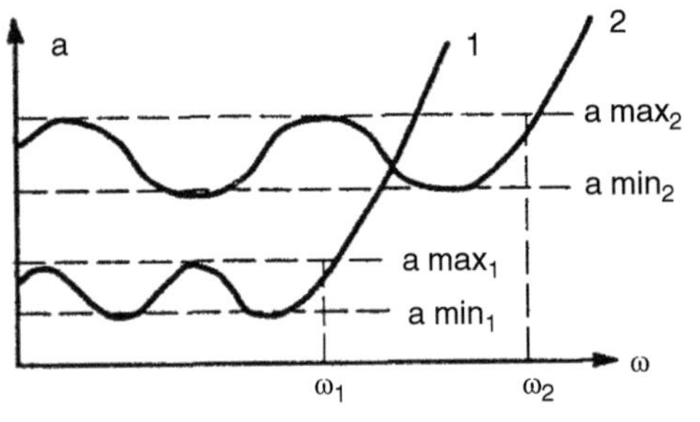

Figure 8.93 Attenuation versus angular frequency for different bandwidths with the same load. *Source*: [39]/Motorola.

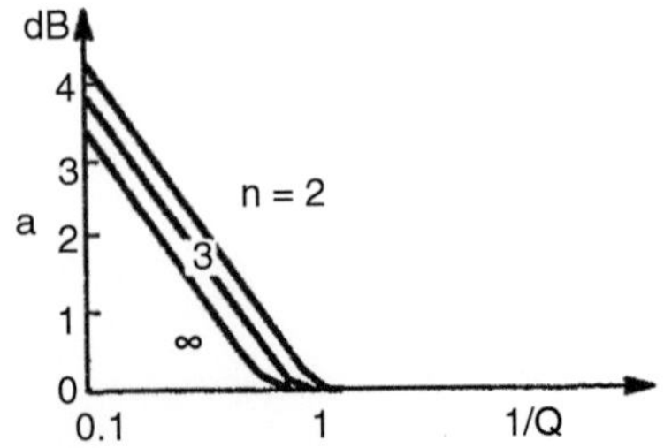

Figure 8.94 Insertion losses as a function of $1/Q$. *Source*: [39]/Motorola.

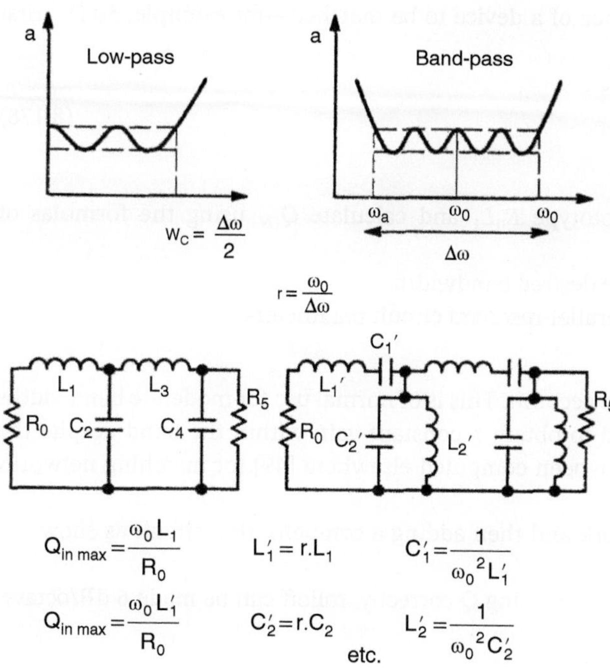

Figure 8.95 Conversion from a low-pass filter into a bandpass filter. *Source:* [39]/Motorola.

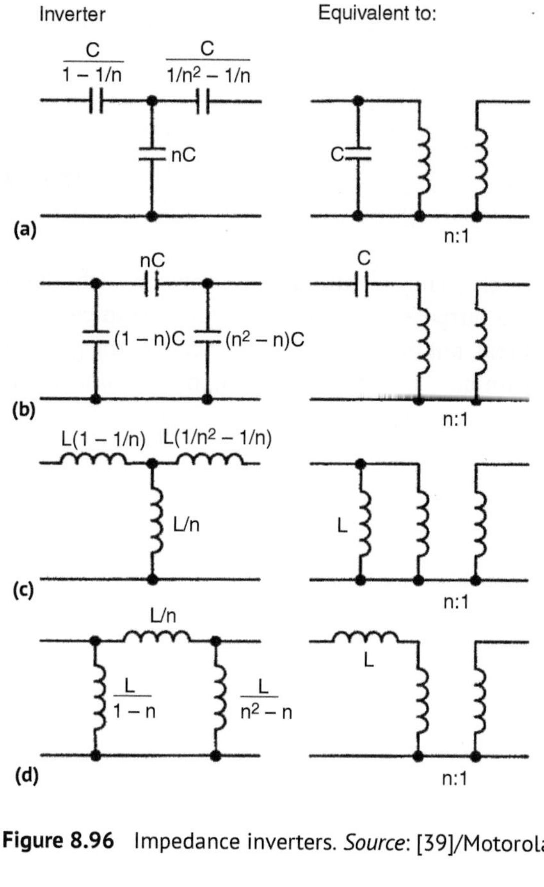

Figure 8.96 Impedance inverters. *Source*: [39]/Motorola.

1. For a given bandwidth center frequency and input impedance of a device to be matched—for example, 50 Ω—first determine Q'_{IN} from the data sheet as

$$\frac{\omega_0 L'1}{R_0} \tag{8.178}$$

after having eventually added a series reactor for centering.
2. Convert the equivalent circuit $R_0L'_1C'_1$ into a low-pass prototype R_0L_1 and calculate Q_{IN} using the formulas of Figure 8.95.
3. Determine the other reactance values from tables [42] for the desired bandwidth.
4. Convert the element values found by Step 3 into series- or parallel-resonant circuit parameters.
5. Insert the impedance inverter in any convenient place.

In the above discussions, the gain rolloff has not been taken into account. This is of normal use for moderate bandwidths (for example, 30%). However, several methods can be employed to obtain a constant gain within the band despite the intrinsic gain decrease of a transistor with frequency. Tables have been computed elsewhere [49] for matching networks approximating 6 dB/octave attenuation versus frequency.

Another method consists of using the above-mentioned network and then adding a compensation circuit as shown in Figure 8.97.

Resonance ω_b is placed at the high edge of the frequency band. Choosing Q correctly, rolloff can be made 6 dB/octave. The response of the circuit of Figure 8.97 is expressed by:

$$\frac{1}{1+Q^2\left(\frac{\omega}{\omega_b}-\frac{\omega_b}{\omega}\right)^2} \tag{8.179}$$

where $\omega < \omega_b$. This must be equal to ω/ω_b for 6 dB/octave attenuation.

At the other band edge a, an exact compensation can be obtained if:

$$Q = \frac{\left(\frac{\omega_b}{\omega_a}\right)^2 - 1}{\frac{\omega_a}{\omega_b} - \left(\frac{\omega_b}{\omega_a}\right)^2} \tag{8.180}$$

There are several methods available for the synthesis of matching networks. They should not be confused with filters because they can be either high-pass, low-pass, or bandpass, and their main purpose is to match one complex impedance to another. In many cases, one of the terminations happens to be 50 Ω and real, but this is just a "lucky" case of the general approach. An excellent paper on this [50] refers to the "real frequency" technique. Another equally important publication is the one by Cuthbert [51], followed by his book [52]. Two more good references are [53, 54].

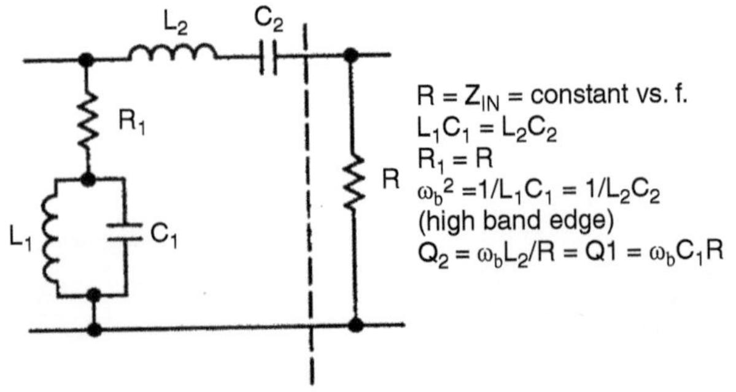

Figure 8.97 Roll-off compensation network. *Source:* [39]/Motorola.

8.21.3 Example 2: Low-noise Amplifier Using Distributed Elements

In our previous examples, the amplifiers were built around lumped elements. Taking the fact that a transmission line length of less than $\lambda/4$ is an inductor, and $3/4\ \lambda$ is a capacitor. The whole amplifier can be built around microstrip technology [1].

If anybody wonders why we are going to show a low-noise amplifier in the middle of the power section, the reason is that we will select a low-power Siemens transistor, which, being operated in Class A, has an intercept point of +30 dBm and can easily achieve a noise figure of less than 2 dB. Such an amplifier is attractive when feeding an input stage from an antenna that sees many signals and curtails the bandwidth with an input filter of good selectivity that needs a good output termination, which in many cases works against the achievement of a good noise figure because when designing for the best noise figure, the input impedance of the transistor stage is far from 50 Ω.

First, we will start with the textbook approach and use the Smith diagram to obtain the best noise figure the device is capable of, and see what happens. Figure 8.98 shows the actual approach to input and output matching.

We start our design by querying the Smith tool about the minimum noise figure and the gain that comes with it. We mark the point of the lowest noise figure (labeled Γ_{opt}). The small noise circle yields the location for Γ_{opt} and tells us the minimum noise figure is going to be 1.91 dB. (By the way, this assumes no further losses, even in soldering the transistor to the PC board.) Then we construct the input gain circle (G1), which touches Γ_{opt} and turns out to represent 13.47 dB of available gain.

The next step is to find a matching network between Γ_{opt} and the 50-Ω source impedance. Immediately, there is a conflict: If we look at S_{11}, which is the input reflection coefficient shown in the Smith diagram, it is on the other side of the complex plane and sufficiently far away, indicating that there is going to be quite a difference between noise matching and power-gain matching. We have two options for matching the input: Either we start from the origin and move toward the position of Γ_{opt}, or we start with Γ_{opt} and go to 50 Ω. Depending on the direction in which we do this, this can result in different topologies, such as a high-pass or low-pass filter. The high-pass filter is obtained with starting with Γ_{opt}; the low-pass topology is

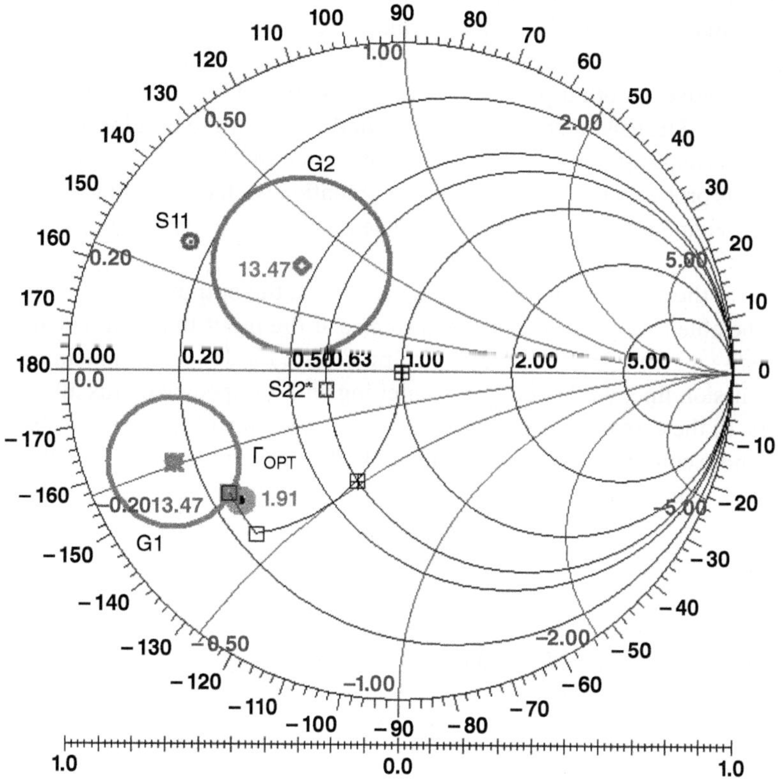

Figure 8.98 Input and output matches for 13.47 dB of gain and NF = 1.91 dB as displayed by the interactive Smith tool. *Source*: [1]/John Wiley & Sons.

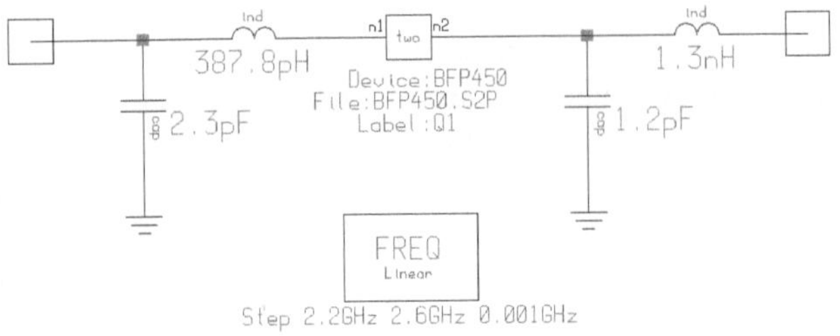

Figure 8.99 A 2.4-GHz amplifier using a BFP450 transistor and lumped-element input and output matching. *Source*: [1]/John Wiley & Sons.

obtained when starting at the 50-Ω point. The Smith Tool then provides us with the corresponding values of capacitors and inductors after the transformation has been accomplished.

The final step is to do the same type of matching at the output. We start with S_{22}, and either select $S_{22}*$ (the complex conjugate for the highest gain) or S_{22load}. As a rule, the second one is always a lower impedance (in terms of the real part) than the conjugate match. This graphic procedure is somewhat of a trial and error. As to the topology, a high-pass match should be preferred at the input and a low-pass match at the output. The reason for this is that the high-pass filter will tend to protect the stage from the enormous number of signals present below 1 GHz; using a low-pass filter at the output suppresses unwanted harmonics that the amplifier may generate. Figure 8.99 shows one of the approaches we like best simply because it gives a good input termination, and we were trying for a simultaneous match for best noise figure and power gain [1].

The way we were able to achieve this was to take advantage of the existing feedback in the transistor and select a less-than-optimal output termination, but gain at the input. The way this works is, taking advantage of the Miller effect, we rotate the phase angle of the feedback with the load to move the optimum noise and gain matching points closer. The danger of this is that this condition is now highly sensitive to both input and output termination. The output *must* see 50 Ω to maintain this good input match, but since the second transistor stage can now be optimized for the best input match, a two-stage combination can be configured to meet this requirement. Also, let us not forget that the device is running 50 mA and as far as bipolar transistors are concerned, this is probably the best amplifier combination we have seen for getting the highest dynamic range and lowest noise figure, providing a high dynamic range. The resulting gain, matching, and noise figure characteristics are shown in Figure 8.100.

Now we turn to using distributed elements and, at the same time, use a shunt feedback at the output; to be specific, we will load it with 150 Ω. The reason for this is to be able to get a still good compromise at the input that will depend less on the output, and at the same time improve the S_{22} matching at the output. The procedure for doing the matching now offers only certain degrees of freedom as a transmission line, while doing the matching with lumped elements allows the choice of series or parallel inductors or capacitors. In the case of transmission line, we have open stubs, shorted stubs, and transmission lines. This may result in somewhat more complex matching schemes. On the positive side, they can be easily trimmed, are reproducible, and certainly cost less than exotic capacitors, which need to have small tolerances and be able to tolerate high RF currents.

The resulting input and output matching is shown in Figure 8.101, and its frequency response in Figure 8.102. We were able to maintain a good input and output match with an insignificant deterioration of the noise figure.

Since this is a high-dynamic-range amplifier, we also show its predicted two-tone response in Figure 8.103. The manufacturer specifies an $IP_{3,in}$ of 29 dB for a bias point of $V_{CE} = 3$ and $I_C = 50$ mA with $Z_S = Z_{Sopt}$ and $Z_L = Z_{Lopt}$ at 1.8 GHz. Significantly, however, the manufacturer-supplied nonlinear BFP450 library we used exhibits considerably different characteristics at $V_{CE} = 4$ and $I_C = 50$ than those reflected in the corresponding S parameters supplied by the same manufacturer. In our circuit, the S parameters resulted in about 13.5 dB of gain (based on a compromise between matching for optimum noise and maximum gain) versus about 9.5 dB of gain for the nonlinear library (match optimized for maximum gain). The difference between the 2.4 GHz F_{min} values returned for the two modeling approaches are similarly striking: 1.90 dB for the bare S parameter set and 7.76 dB for the bare nonlinear library biased to $V_{CE} = 4$ and $I_C = 50$ mA. Clearly, the nonlinear modeling of the BFP450 needs work!

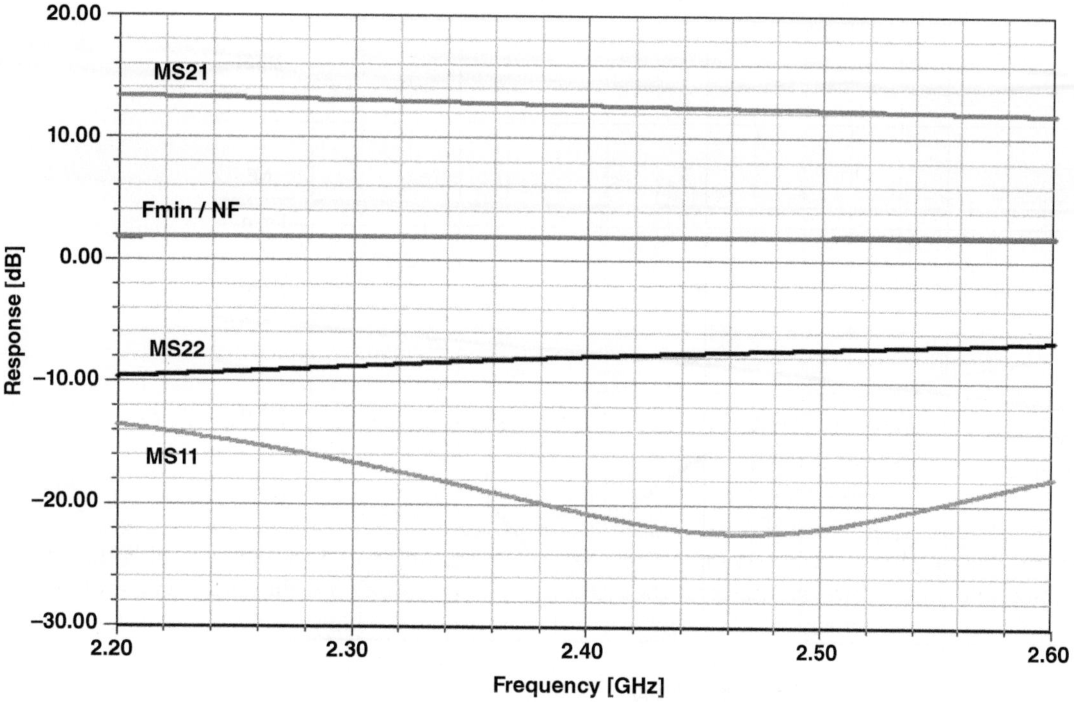

Figure 8.100 Frequency-dependent gain, matching, and noise characteristics of the matched high-dynamic-range BFP450 2.4 GHz amplifier. *Source*: [1]/John Wiley & Sons.

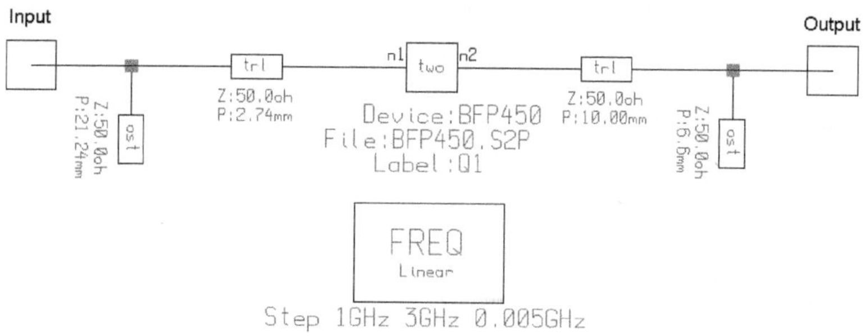

Figure 8.101 Distributed-element input and output matching for the 2.4 GHz BFP450 amplifier. *Source*: [1]/John Wiley & Sons.

Having fully evaluated the amplifier at 2.4 GHz as a working example, we note that this frequency is used by radio amateurs for some applications. We will therefore supply examples for 432 and 1296 MHz, popular spot frequencies in the amateur 70- and 23-cm bands, respectively. Figures 8.104 and 8.105 show the 432-MHz version's schematic and frequency-dependent responses. Figures 8.106 and 8.107 show the schematic and frequency-dependent responses for the 1296-MHz version.

Using a bipolar transistor that operates at much less current, the noise figure can be improved at the expense of the dynamic range/intercept point. It is possible to reduce the noise figure by about 1 dB or so, but the reduction in intercept point will be much more dramatic. Another choice is the use of GaAsFETs for this application, but their matching in comparable circuits has been described as a nightmare simply because the devices have too much gain and very high input impedances (essentially 0.2 pF in series with 3 Ω) combined with reasonable output impedances in the vicinity of 240 Ω. Here is a wide field for engineers to experiment with and look at price/performance and performance/real estate issues. GaAsFETs also tend to require more than 30–40 mA to be alive, and even the low operating voltage of 4–5 V makes the device take a lot of dc power [55].

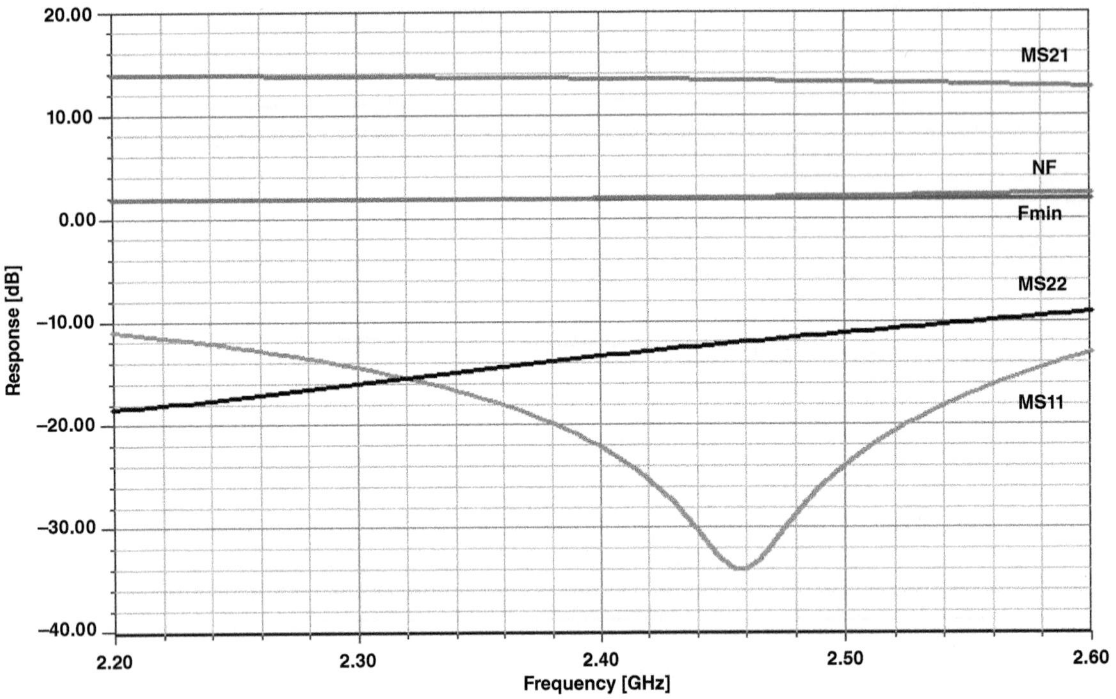

Figure 8.102 Predicted frequency-dependent gain, matching, and noise performance for the BFP450 amplifier with distributed-element matching. *Source*: [1]/John Wiley & Sons.

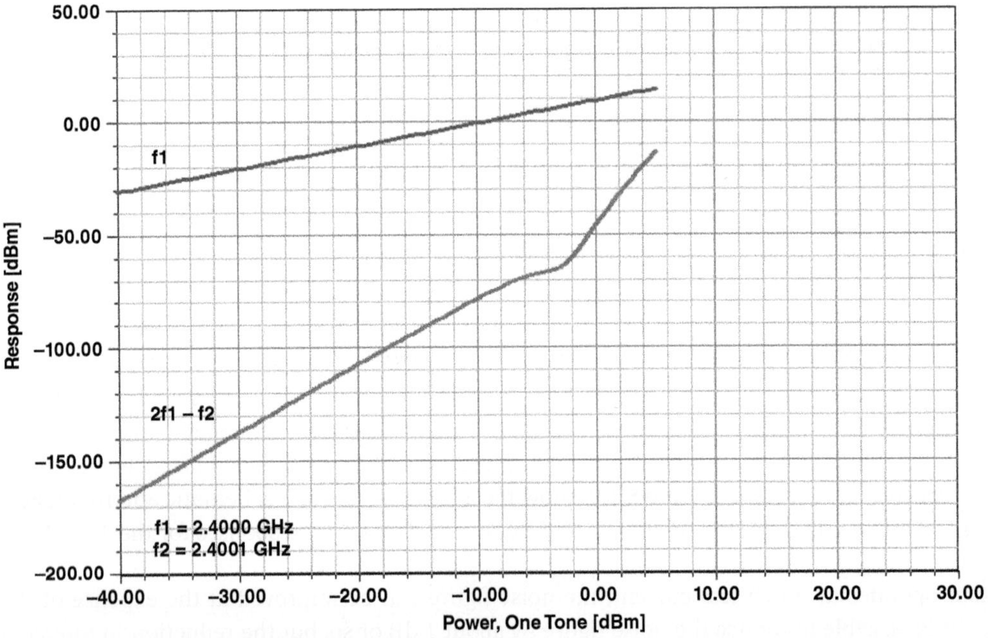

Figure 8.103 Predicted two-tone response of the BFP450 amplifier with the circuit adjusted for maximum gain (9.5 dB): $IP_{3,in} \approx 28.5$ dBm, and $IP_{3,out} \approx 38$ dBm. That this version's gain is significantly lower than that of the S-parameter-based equivalent illustrates the need for better accuracy in parameter extraction. *Source*: [1]/John Wiley & Sons.

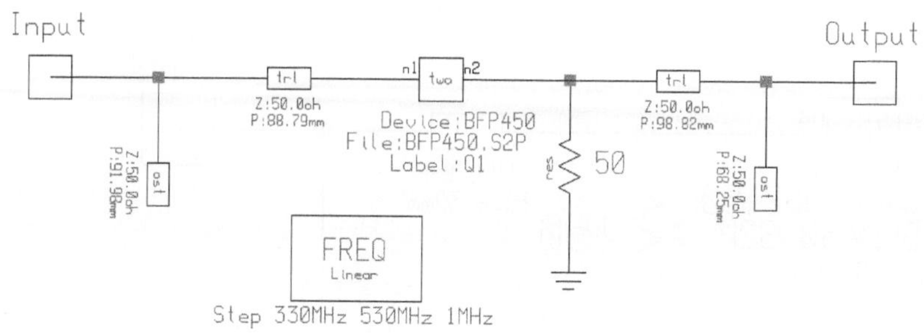

Figure 8.104 Schematic of BPF450 amplifier for 432 MHz. Because the BFP450's $f_T = 25$ GHz, at this lower frequency, we must shunt the device output with a 50-Ω resistor to maintain stability. *Source*: [1]/John Wiley & Sons.

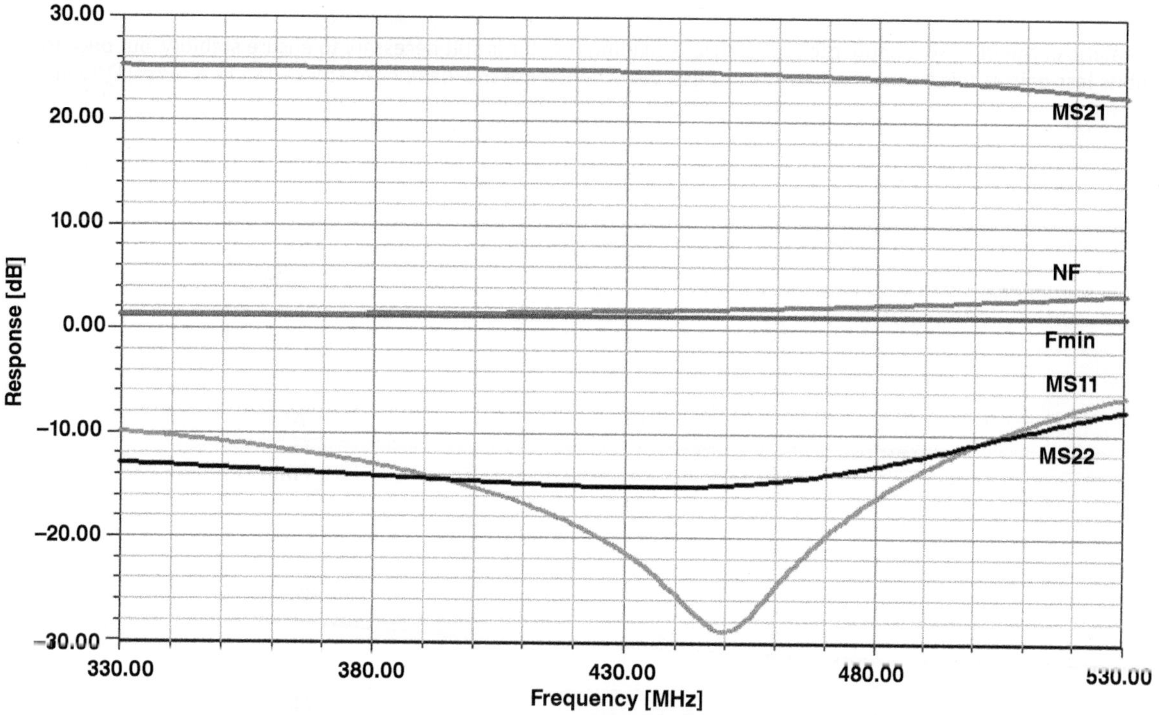

Figure 8.105 Frequency-dependent gain, matching, and noise for the BPF450 amplifier at 432 MHz. As with the circuit in Figure 8.105, the input and output network values have been chosen to achieve a low NF and good input matching at the expense of the output match. *Source*: [1]/John Wiley & Sons.

8.21.4 Example 3: 1-W Amplifier Using the CLY15 [1]

The following is an application showing a fairly wideband amplifier with +30 dBm (1 W) output. We need to point out that, in accordance with the various digital modulation modes used with this amplifier, this is not continuous operation, but a much less than 1:1 duty cycle operation. The device would probably not survive operating at full output under CW conditions. This is subject to a discussion with the manufacturer, Siemens, who has not released sufficient data for us to make a judgment as to the CLY15's continuous power capability. Since there is no useful large-signal model available at the time of this writing, we will design this Class A amplifier using large dc bias *S* parameters. In the actual design, there will be some cut-and-try, but we find that in this type of application, this is still the best way of handling this.

The CLY15 (gate width 16 mm) is the largest of a series of Siemens GaAsFETs; it was already mentioned in Chapter 2, where its datasheet is reproduced. Its device family also includes the CLY2 (gate width 2 mm), CLY5 (4 mm),

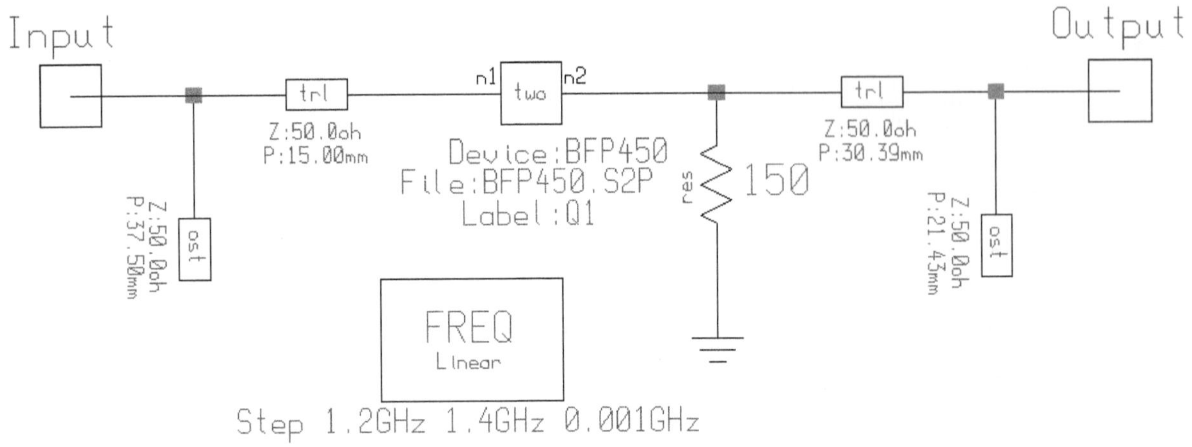

Figure 8.106 Schematic of the BPF450 amplifier for 1296 MHz. A loading resistor is still necessary to ensure stability, but because the transistor exhibits less gain at this frequency than at 432 MHz, we can increase the loading resistor to 150 Ω. *Source*: [1]/John Wiley & Sons.

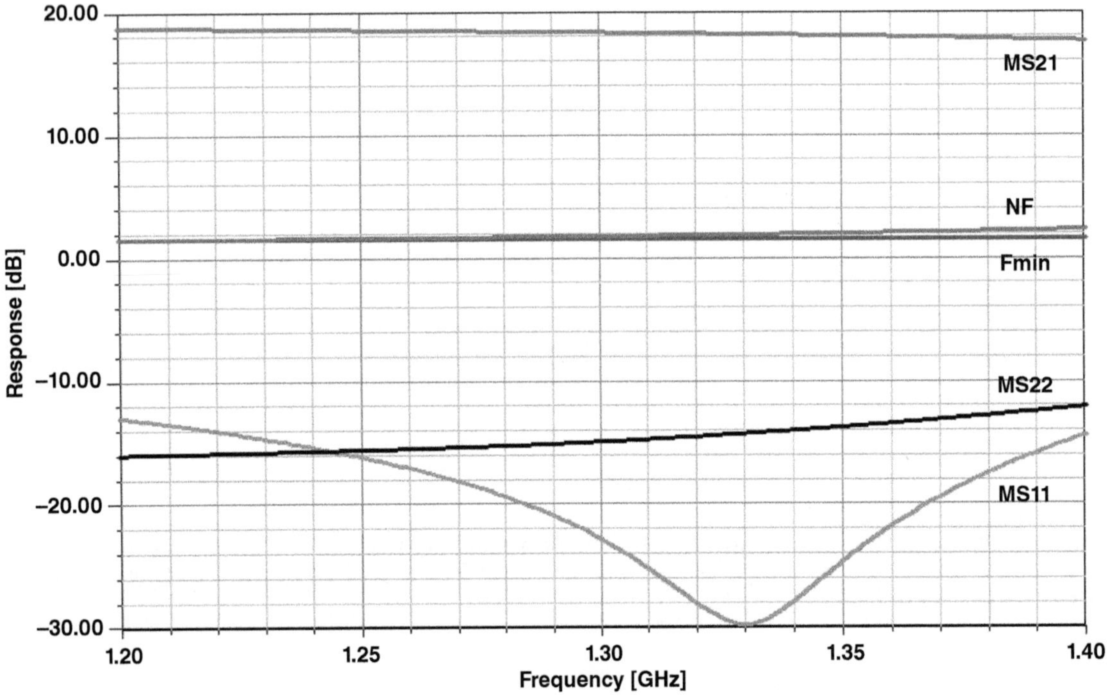

Figure 8.107 Frequency-dependent gain, matching, and noise responses of the BFP450 amplifier at 1296 MHz. As with the two previous versions, the input and output network values have been chosen to achieve a low NF and good input matching at the expense of the output match. *Source*: [1]/John Wiley & Sons.

and CLY10 (8 mm). It is possible that bigger devices will be offered in the future. One of the most critical parameters will be the saturation voltage, from the datasheet of the CLY15, we can see that at 1.5 A the saturation voltage is approximately 1 V. This drop of 1 V includes losses that result from other parasitic elements; the dc I–V curve (Figure 8.108) indicates a saturation voltage of 0.75. Since we aim for 1 W output power, the resistive portion of the load will be $(5 - 1)^2/(1 \times 2) = 8\ \Omega$. This assumes a supply voltage of 5.

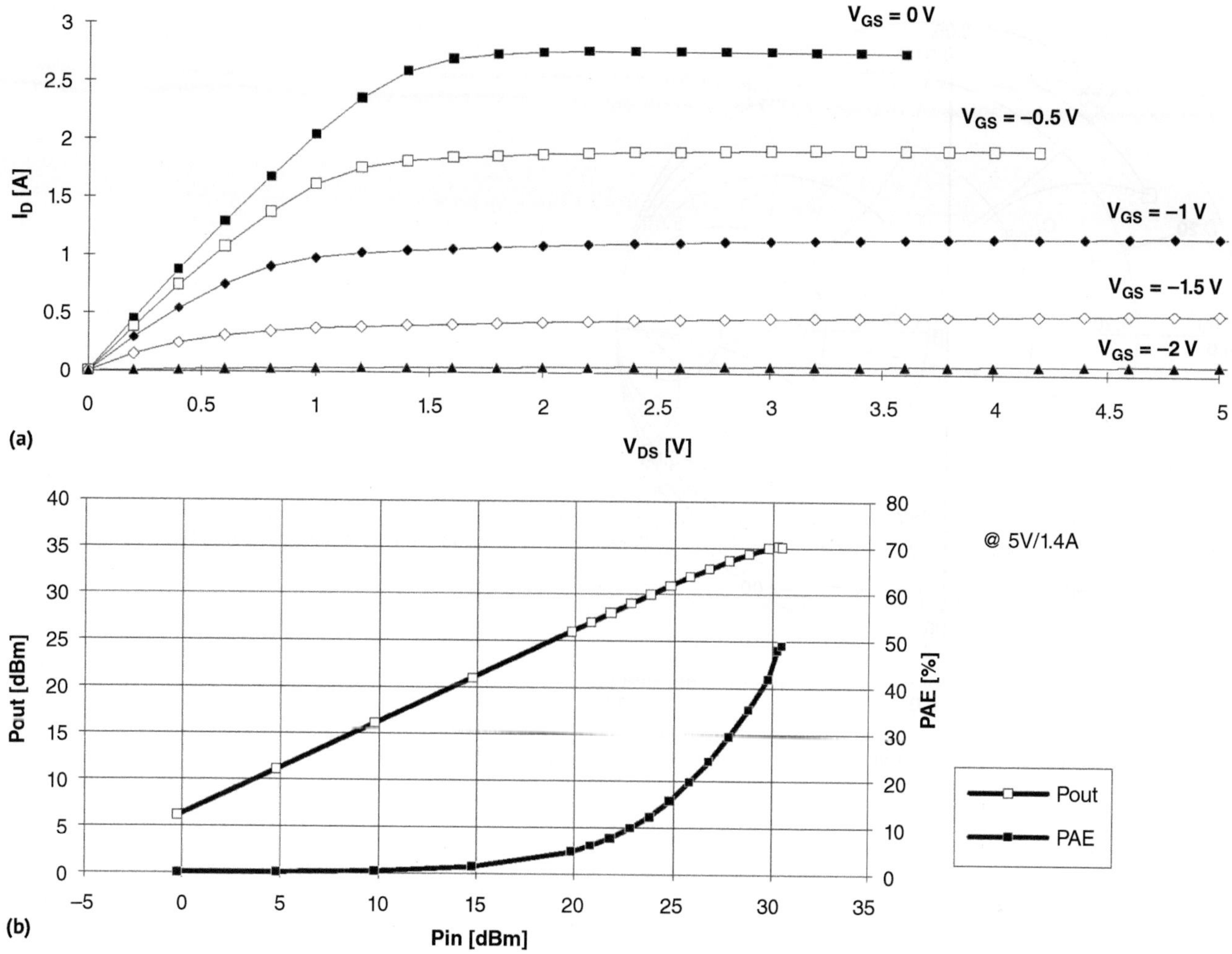

Figure 8.108 Output and power characteristics for the CLY15 medium-power GaAsFET. *Source*: [1]/John Wiley & Sons.

The next step is to design the input and output matching network. Previewing the S parameters, it will be noticed that they are essentially inductive, which means that the input match, using lumped elements, will result in a combination of capacitors both at the input and the output. Figures 8.109 and 8.110 show the way the input and output matching was done on the Smith diagram, resulting in lumped matching elements. By inspecting the lumped input matching, here is what we find. We start with S_{11}, which is in the lower part of the Smith diagram, and determine its conjugate match, $S_{11}*$. We already pointed out that this device is inductive, which means that in order to match it from the conjugate point, we need two capacitors. The first one is a series capacitor, as identified in the Smith diagram, and then finally a shunt capacitor to transform the impedance to 50 Ω. A close inspection of the Smith diagram shows that the resulting point is not quite on S_{11}; it is an iterative process to exactly find the crossover point, yet the input matching of better than 20 dB is more than acceptable.

As far as the output is concerned, we use the same technique by moving from S_{22} to $S_{22}*$. S_{22} has been determined from the required output power; as we showed above, the real portion is 8 Ω. Using the lumped elements, the matching technique is identical; we first need to use a series capacitor and then a shunt capacitor for the transformation. The arc for the series capacitor this time is much longer, resulting in a series capacitance of 11.2 pF instead of the 2.1 pF capacitance used at the input. The actual schematic is shown in Figure 8.111. The drawback of the lumped elements is that it is not practical to buy components that can handle the heavy RF currents that flow in the input and output networks. Taking the values obtained from the lumped input and output matching approach, Figure 8.112 shows the resulting frequency response.

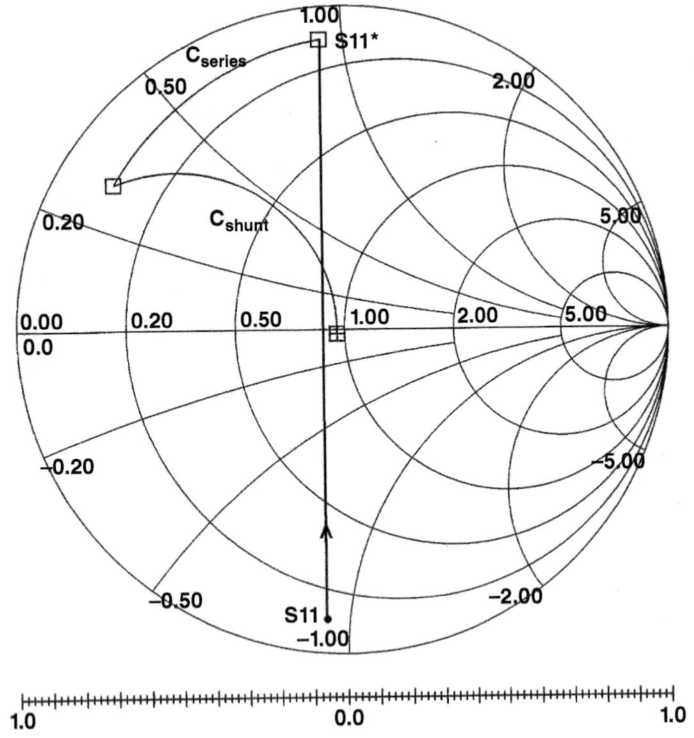

Figure 8.109 Input match for S_{11} * using lumped elements. *Source*: [1]/John Wiley & Sons.

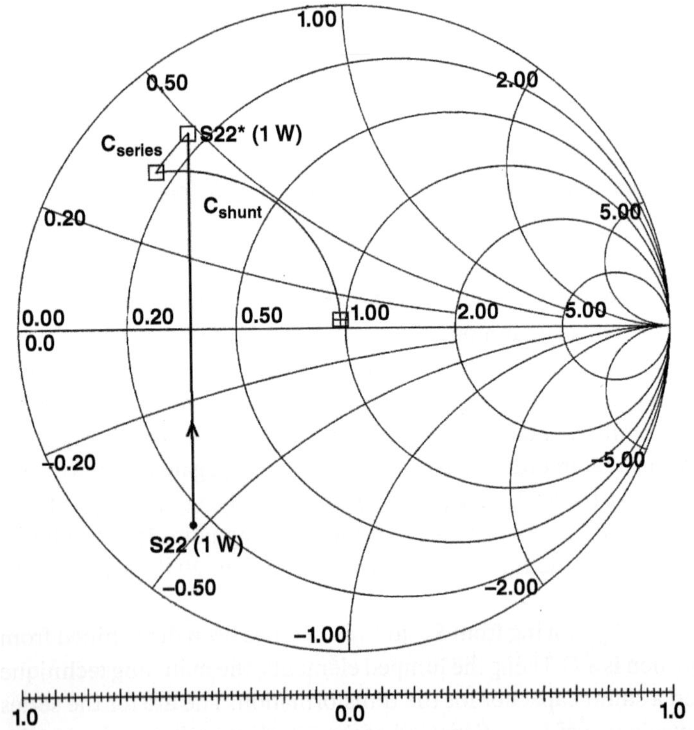

Figure 8.110 Output match for S_{22} * (1 W) using ideal lumped elements. *Source*: [1]/John Wiley & Sons.

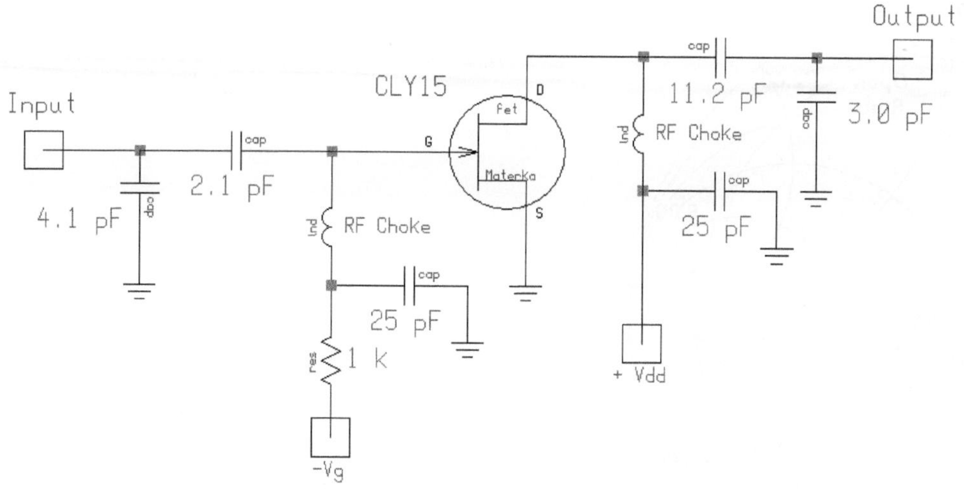

Figure 8.111 The complete amplifier using the CLY15 medium-power GaAsFET, with lumped-element input and output matching. *Source*: [1]/John Wiley & Sons.

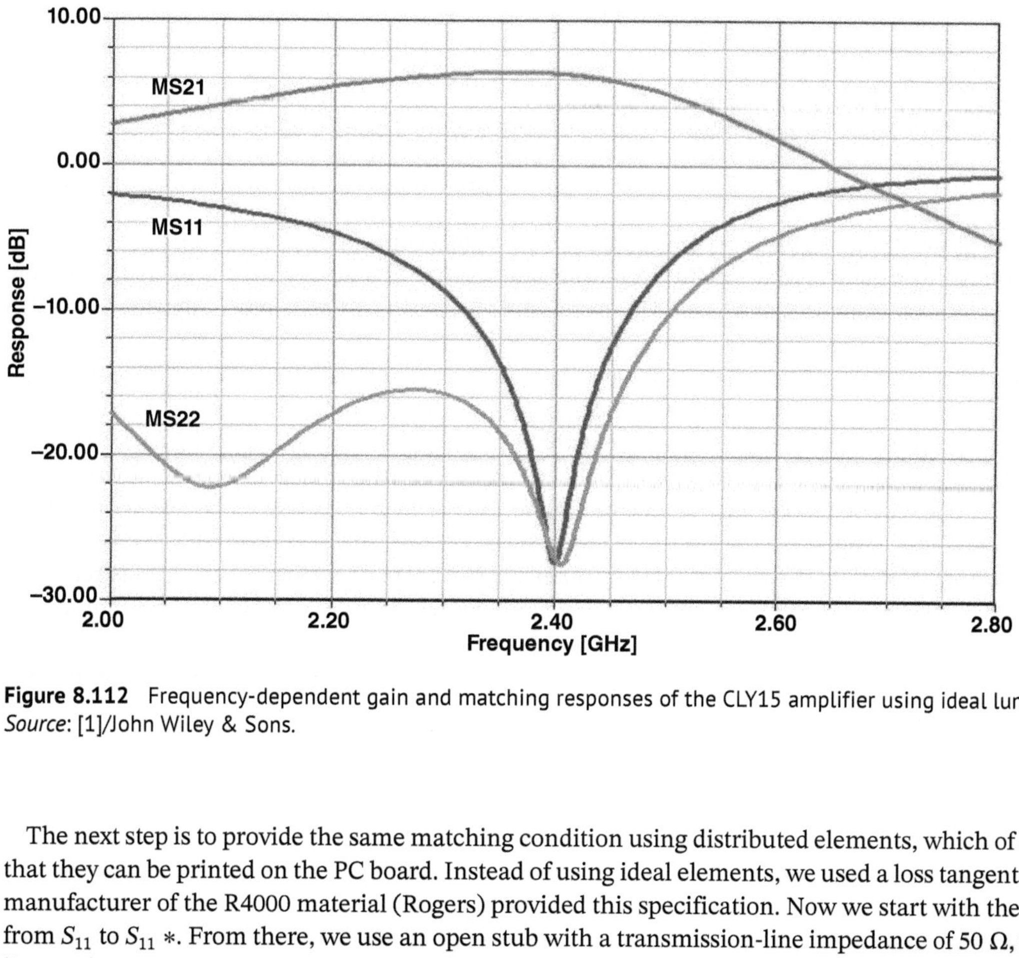

Figure 8.112 Frequency-dependent gain and matching responses of the CLY15 amplifier using ideal lumped elements. *Source*: [1]/John Wiley & Sons.

The next step is to provide the same matching condition using distributed elements, which of course, have the advantage that they can be printed on the PC board. Instead of using ideal elements, we used a loss tangent of 2E−3, or a Q of 500. The manufacturer of the R4000 material (Rogers) provided this specification. Now we start with the same approach by moving from S_{11} to $S_{11}*$. From there, we use an open stub with a transmission-line impedance of 50 Ω, followed by a transmission line to the point in the lower left corner, and finally, a shorted transmission-line stub allows the 50-Ω matching. For this amplifier, we are not looking for noise matching but for input and output power matching. Figure 8.113 shows this match on the Smith chart.

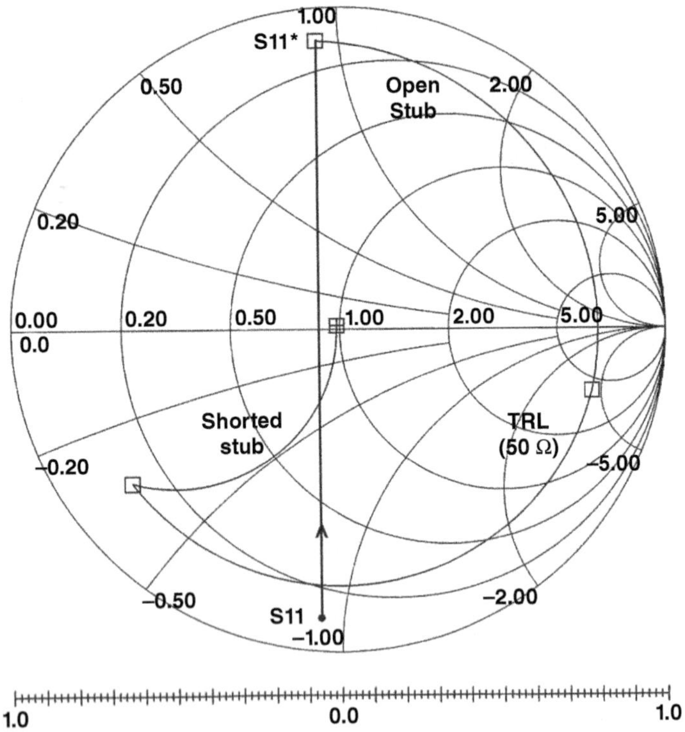

Figure 8.113 Input match for S_{11} * (1 W) using distribution elements. *Source*: [1]/John Wiley & Sons.

At the output, the procedure again is similar: We move from the S_{11} for 1 W to a conjugate value, then use an open stub that acts like a capacitor until we cross the real axis; finally, a $\lambda/4$ transmission line is used to match from approximately 75 – 50 Ω. Figure 8.114 shows this match on the Smith chart. This concludes the distributed matching design for the CLY15 amplifier.

Finally, the complete circuit using distributed elements is shown in Figure 8.115. From a dc point of view, we took advantage of the incoming shorted stub by using a transmission line of the appropriate length and a bypass capacitor to ground. This allows us to feed the gate with the required negative voltage without interfering with the frequency response of the amplifier. When evaluating the total frequency response (Figure 8.116), we noticed that this circuit seems to have a higher Q than the lumped equivalent. Again, to obtain a wider bandwidth, more transformation stages will be needed. We leave this task to the interested reader.

The 1-dB bandwidth of the amplifier using distributed, lossy elements is 120 MHz. If a wider bandwidth is required, matching networks with more elements are required. Because of the gain-bandwidth product, the peak gain for such a wider-bandwidth arrangement will be less than that of the narrowband example.

Having used a lumped to distributed transformation, we next present a power amplifier that uses coaxial lines, baluns, and other transformation steps capable of handling the increased power requirement.

8.21.5 Example 4: 90-W Push–Pull BJT Amplifier at 430 MHz [1]

The last complete application we want to take a look at is a 90-W-output push–pull amplifier using two TRW transistors (that are no longer made—an all-too-frequent occurrence after a design is complete, resulting in a scramble for replacements). Figure 8.117 shows the circuit diagram of this UHF transistor amplifier, and Figure 8.118 shows its mechanical design. The amplifier requires 28 V dc and will draw approximately 6 A, requiring 15 W of drive at 438 MHz. This amplifier is a Class AB-type amplifier, which means its application is not limited to FM and other angle-modulated signals, but accommodates even SSB (single sideband) operation.

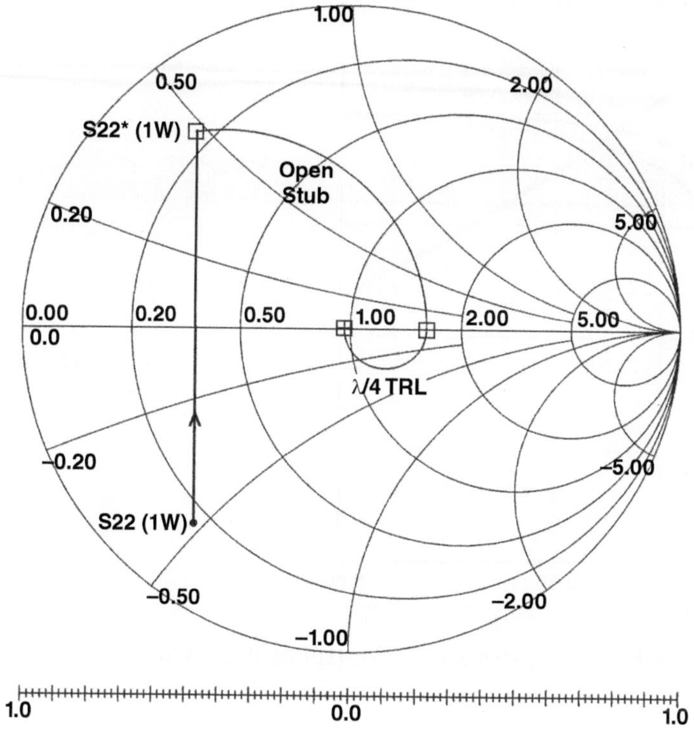

Figure 8.114 Output match for $S_{22}*$ (1 W) using distributed elements. *Source*: [1]/John Wiley & Sons.

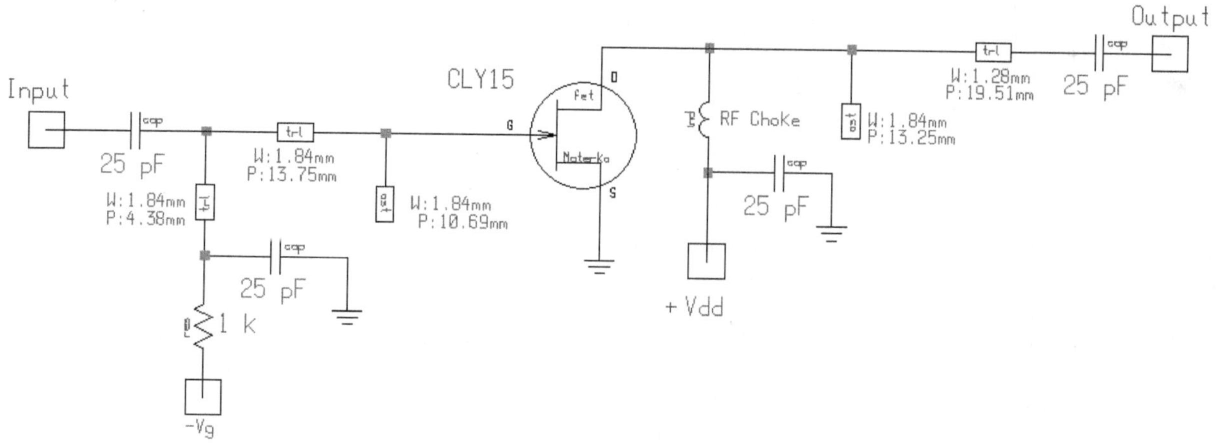

Figure 8.115 The complete amplifier using the CLY15 medium-power GaAsFET, with distributed-element input and output matching. *Source*: [1]/John Wiley & Sons.

The drive power coming from the source connected to the input N connector (15 W) is fed through a short piece of coaxial cable to the PC board sitting on top of the heat sink. To change from single-ended to push–pull, a 50-Ω coaxial cable is coiled up with sufficient electrical length and then feeds two symmetrical paths through the amplifier. The *RC* section of 10 Ω in parallel with a 220-pF capacitor stabilizes the amplifier at lower frequencies, while at higher frequencies the 10-Ω resistor has no contribution. There are two independent 4:1 transformers, one for each section; they are built using 25-Ω coaxial cable with a set of two binocular ferrite cores to force the transformation. At this point, the dc voltage is also

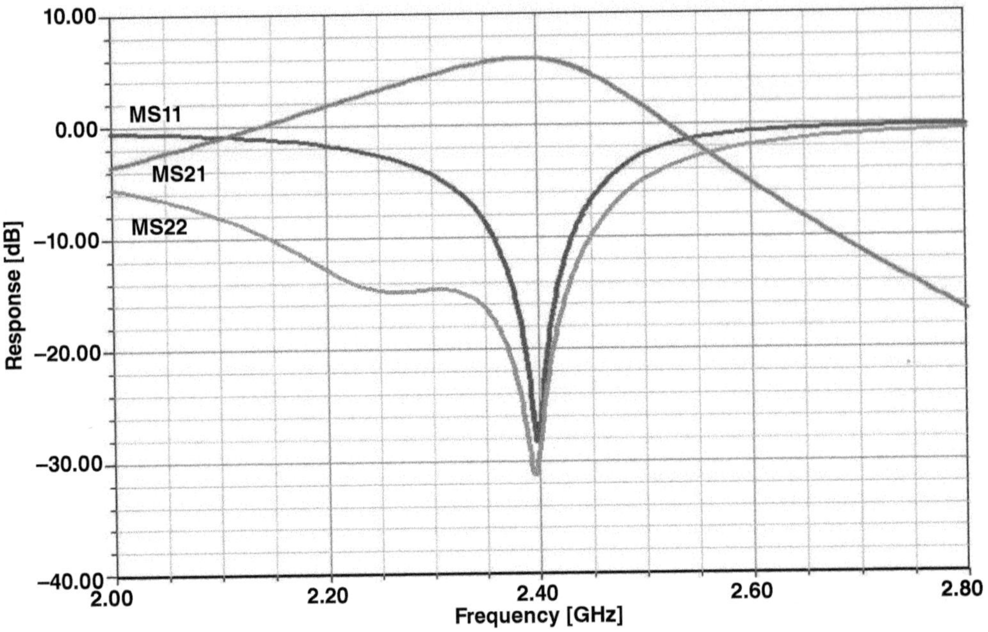

Figure 8.116 Frequency-dependent gain and matching using distributed, lossy elements. *Source*: [1]/John Wiley & Sons.

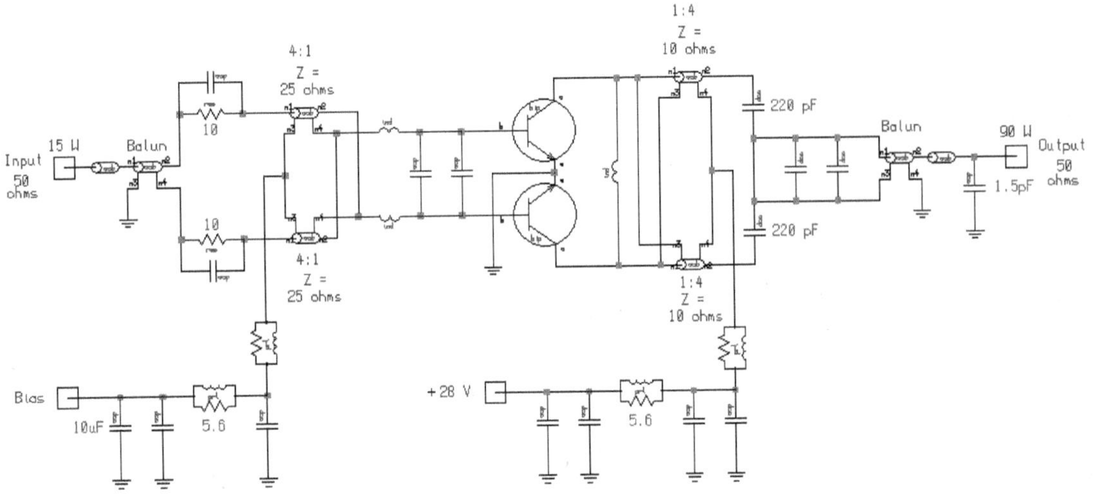

Figure 8.117 Circuit diagram of the 90-W, push–pull, 438-MHz linear amplifier. *Source*: [1]/John Wiley & Sons.

fed to the transistors. The stabilization circuit consists of a temperature monitor using the *p*-n junction of a transistor, side-mounted on the heat sink, and a combination of operational amplifiers and adjustments. This guarantees proper, temperature-independent biasing. The 4:1 input baluns result in 12.5 Ω impedance, or 3.125 Ω per transistor. Since the input impedance at this drive level is somewhere in the vicinity of $[(1 \cdots 2) + j(3 \cdots 5)]$ Ω. The final matching is accomplished by an *RC* section whereby the inductance is actually printed and the shunt capacitor across is soldered from base to base as close to the transistor bodies as possible. A ceramic-dielectric variable capacitor in parallel with this allows fine tuning.

The output consists of an inductor from collector to collector that tunes out the capacitive reactance and makes the transistor output impedance real. We then follow the same pattern by taking a 1:4 transformer, but this time built using 10-Ω rigid line instead of 25-Ω rigid line. Together with the output capacitance, prior to the output Balun, this is a medium-bandwidth matching arrangement that transforms up to 50 Ω. The collector impedance per collector had to be $(28-2)^2/2 \times 90$, resulting in roughly 3.8 Ω—not that different from the input match at the balun point. This way, the matching

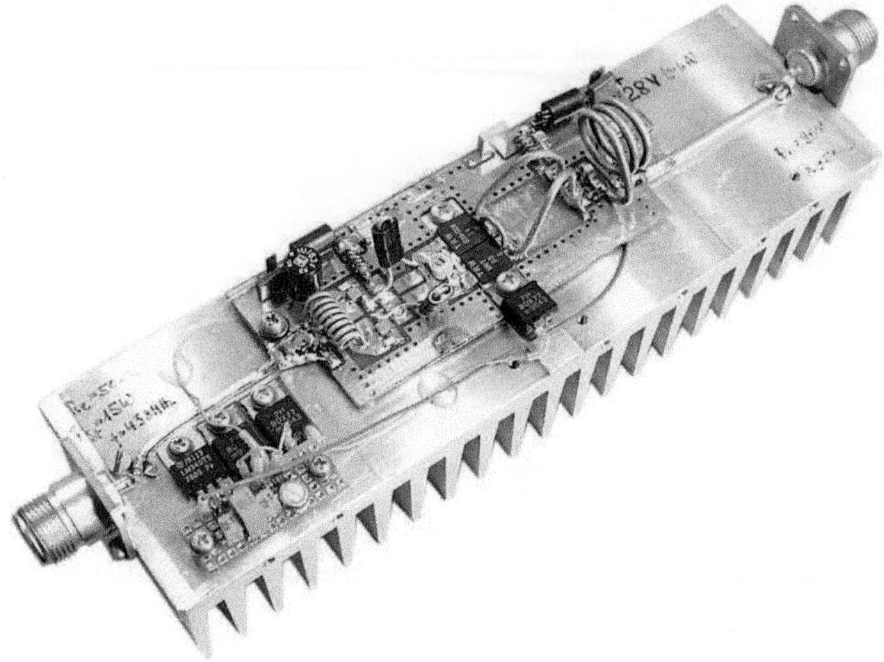

Figure 8.118 Mechanical assembly of the 90-W amplifier shown in Figure 8.117. *Source*: [1]/John Wiley & Sons.

networks at the input and output can be identical. At the output of the matching network, another coiled 50-Ω coaxial cable serves as a balun; because it must operate at the 90 W level, a larger-diameter line was used than in the input balun. Finally, another section of rigid line is used to connect the output balun on the PC board to the output N connector. The output capacitor again helps in the transformation.

Both the base and collector voltage supplies are heavily filtered using either ferrite-bead-based RF chokes or heavy wire inductors wound on low-impedance resistors. A fairly large number of these amplifiers were actually built and are still in use [1].

8.21.6 Quasiparallel Transistors for Improved Linearity [1]

Some applications require really low intermodulation distortion. One of the authors (Rohde), during his tenure at RCA Government Systems Division, and his team developed a parallel-device topology that is somewhat similar to the conventional parallel approach but shows vastly increased linearity. In creating higher-power devices, manufacturers parallel several transistors in one package, connecting them in a clever way so the overall drive, dissipation, and output power are equally divided between all the devices. Special nickel–chromium ballasting resistors in the many fingers of the emitter connections of the multiple internal devices equalize any differences among the transistors while increasing the saturation voltage only by a small amount. In our high-linearity design, we used two transistors in parallel, one being biased somewhere between Classes A and B, and the second one somewhere between B and C (Figure 8.119). The exact bias voltages and operating points had to be found through experiment. In practice, Q1, the "initially on" transistor, is driven to a point close to that at which its IMD products would increase. Q2 is biased such that at this same drive level, it adds to the output power.

We frequently referred to this configuration as "turbocharged." The IMD products at full output power were down 40–45 dB, compared to 26–30 dB for a single transistor. While such an arrangement shows superior IMD performance, because of the somewhat abrupt kick-in of the second transistor, the transconductance response is semi-smooth, a characteristic that might make alert readers think that IMD will occur at this point. On the other hand, the transconductance for

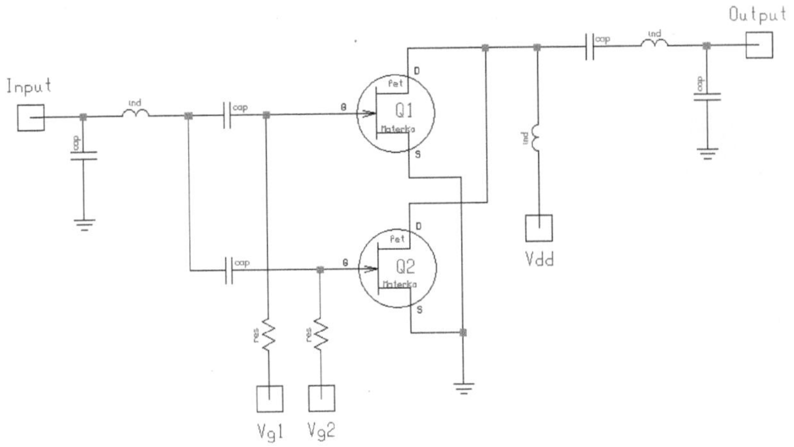

Figure 8.119 The quasiparallel amplifier topology looks like a standard parallel amplifier. Biasing the devices to different operating points nets an improvement in IMD performance. *Source*: [1]/John Wiley & Sons.

Q1 in effect saturates or stays constant, so when the transconductance responses of both transistors are overlaid, the result is a smooth response overall.

This technique has been validated up to 100 W and up to at least 50 MHz. Its drawbacks include the need for two transistors with their associated cost, additional real estate, and higher input and output capacitances. Nonetheless, considering that MOS transistors capable of operation to 1 GHz were unavailable when this technique was developed, we believe that it should be revisited [29–37].

The Class D and higher operating modes require a large number of additional components and will make the efficiency higher (selective in frequency), but not necessarily reduce the IMD products. A further point of consideration is that it is incorrect to assume that reducing the input level to a given Class A will necessarily reduce its IMD products. As with tubes, the closer one gets to the knee, even in Class A operation, the worse the IMD. (This phenomenon was particularly troublesome in systems using 1-kW amplifiers up to 150 MHz; underdrive with the intent of getting better IMD performance produced just the opposite result.) If a wide power range is considered, we recommend using a bias scheme that includes sufficient intelligence to find the best operating point.

Another hot amplifier issue is that of IMD as a function of load impedance. If an amplifier is used at output power considerably below its design value, its load impedance will be too low and, depending on the circuit configuration, the result may be additional IMD [1].

8.21.7 Distribution Amplifiers [1]

It is not uncommon that a signal from a particular source, like an antenna, has to be distributed to several places. A condominium is a typical place where, from a main signal source, the energy has to be split to provide a larger number of users—enough voltage or power to produce a noise-free picture. An amplifier capable of doing this is a *distribution* amplifier, not to be confused with a *distributed* (or traveling-wave) amplifier, such as that shown earlier in this chapter at Figure 8.6. An interesting means of achieving this is to use multiple Wilkinson couplers following an amplifier. Figure 8.120 shows an arrangement that makes heavy use of multiple power dividers to provide multiple outputs. Theoretically, the reverse is also possible: If we want to collect energy from a number of sources and combine them, as on a cable, we can use Wilkinson couplers at the *input* of a gain block. Note, however, that depending on the level and number of the combined signals, this may put enormous strain on the linearity of the amplifier. What we want to avoid is a lot of distortion, which, in the case of television, will show up as color confetti in the noise and signal.

8.21.8 Stability Analysis of a Power Amplifier

One of the hidden pitfalls in narrowband circuit design is that a circuit may be unstable outside the frequency range of interest. Circuits containing structures built from distributed elements, which exhibit multiple resonances, can be particularly susceptible to such instabilities. In this example, we analyze the stability of a simple FET power amplifier designed using small-signal concepts and the interactive Smith Tool utility in Ansoft's Serenade Design Environment. The evaluation shows how K factor analysis alone is not sufficient to check for the stability of a design.

Small-signal Ac Analysis. We begin by taking the Figure 8.121 amplifier, which was designed for about 16 dB of gain at 2.2 GHz. Due to the very low, capacitive input impedance of the FET and the proximity of the source-plane gain circles and stability circles to the edge of the Smith chart, the matching network attempts to transform 50 Ω to an impedance of approximately (2.5 + j15) Ω. The impedance lies close to the stability circles at 2 GHz, but remains in a stable region. The bias point selected (−3.5 V) is chosen for Class AB operation. Higher linear gain can be achieved by reducing the gate voltage to −2.0 V; at this bias point, the amplifier gain will increase to about 19 dB.

First, we will do a small-signal ac analysis and examine the S parameters, K, and B_1 factors from 1 to 6 GHz—well beyond the bandwidth of the design. Figure 8.122 shows the circuit's frequency-dependent gain and matching characteristics.

If we examine B_1 and the stability factor K, we see that the amplifier is not unconditionally stable in the 2.2-GHz passband as shown in Figure 8.123. K dips slightly below 1, although B_1 remains above zero. In practice, a slight change in design would be required to bring K above 1 and remove this potential instability. However, for this example, this is of no consequence, as we shall see below.

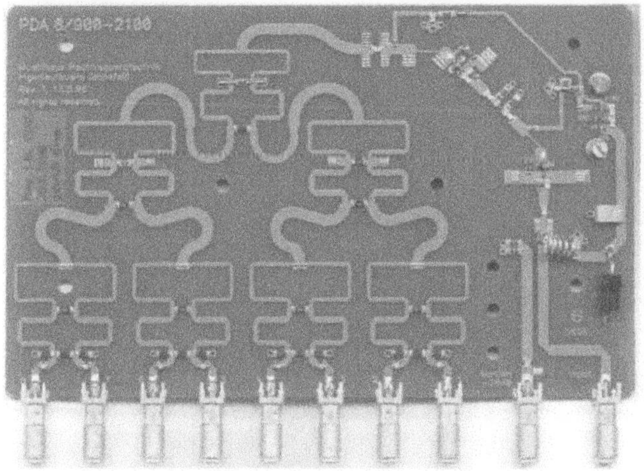

Figure 8.120 Distribution amplifier showing the use of multiple Wilkinson power dividers. *Source*: [1]/John Wiley & Sons.

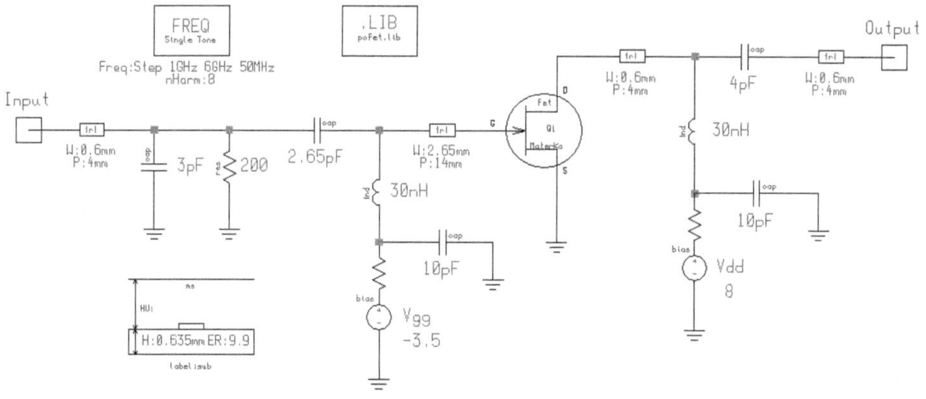

Figure 8.121 2.2-GHz amplifier design for stability analysis. *Source*: [1]/John Wiley & Sons.

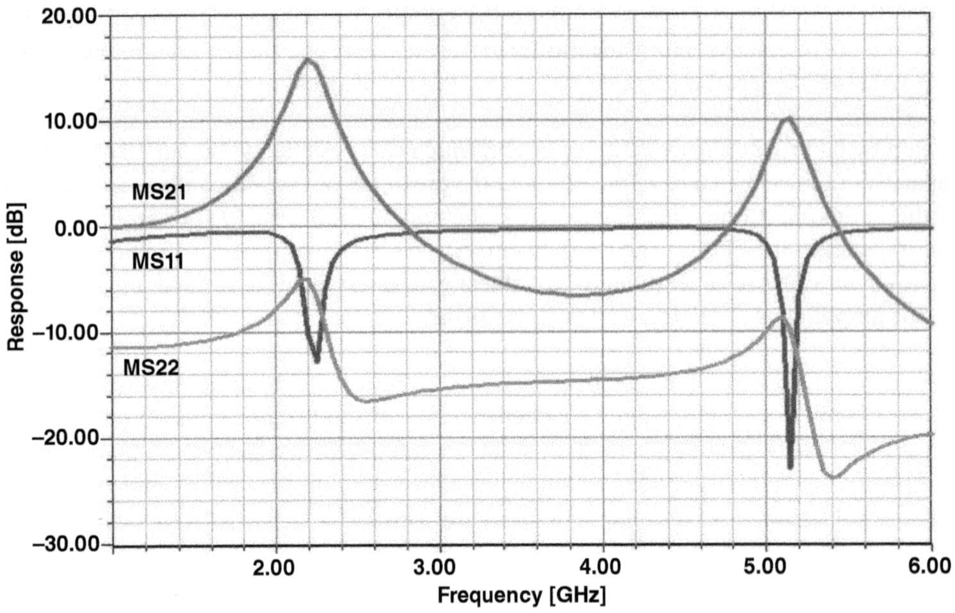

Figure 8.122 Frequency-dependent gain and matching characteristics of the amplifier. Although the circuit is designed for 2.2-GHz operation, the periodic responses of its distributed-element-based input and output matching networks result in a second gain peak near 5.1 GHz. As we shall see, however, instabilities may result at frequencies that cannot be intuited from this graph. *Source*: [1]/John Wiley & Sons.

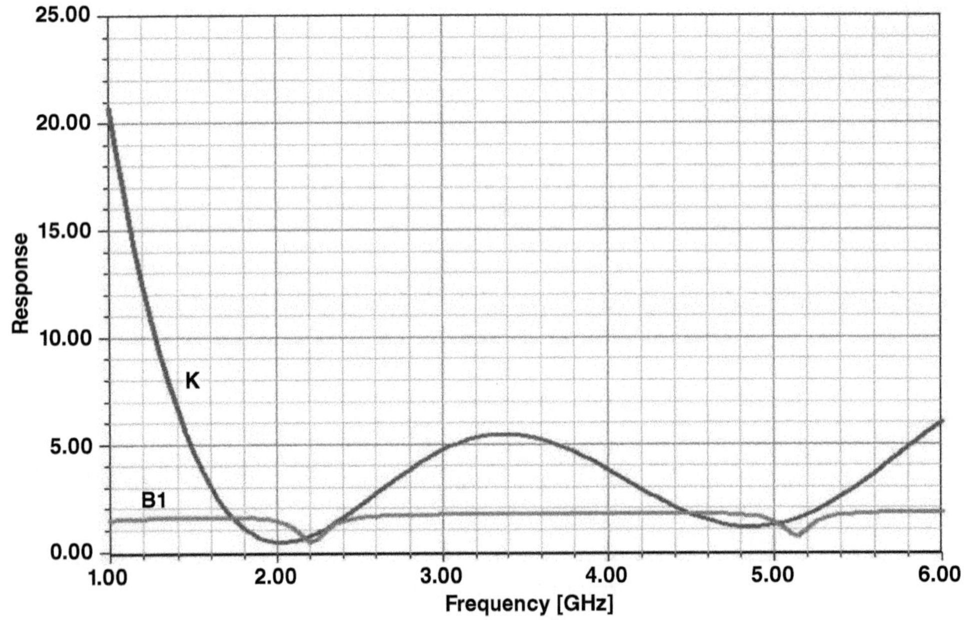

Figure 8.123 K and B_1 for the 2.2-GHz amplifier. *Source*: [1]/John Wiley & Sons.

Nyquist Stability Analysis: Now we will utilize a Nyquist stability analysis to check the stability of the amplifier. To fully realize the advantages of a Nyquist stability analysis, we must sweep the frequency over a very wide range. We want to be able to pick up instabilities at *any* frequency, not just near the design frequency. In this example, we sweep from 1 kHz to 20 GHz. A start frequency of 1 kHz is chosen so we can determine the starting point of the Nyquist plot, and we ensure we pick up any low-frequency resonances. The stop frequency is chosen past the $f_{\max}$ of the transistor (where we know it cannot oscillate).

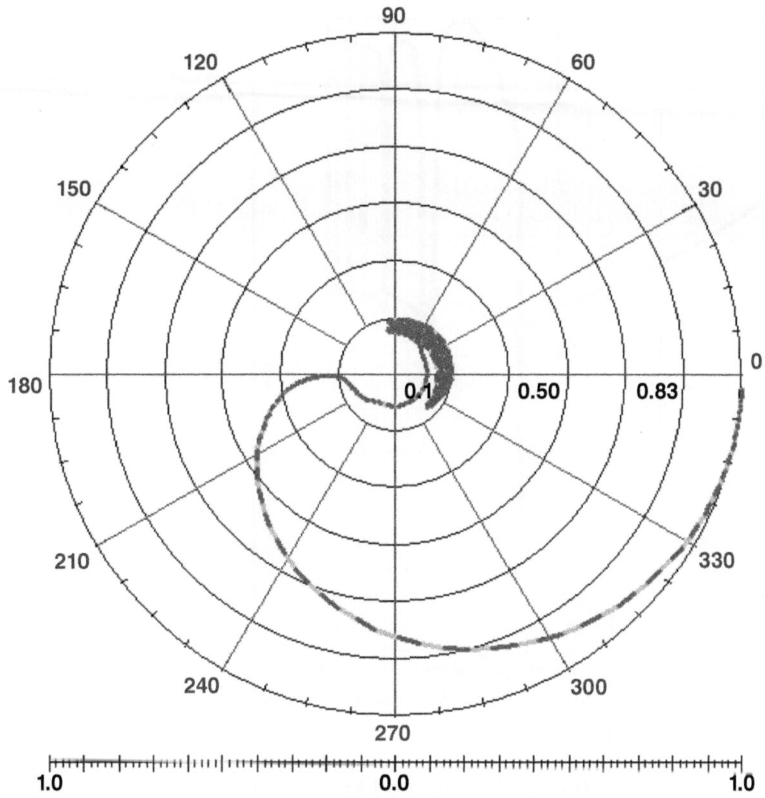

Figure 8.124 Polar Nyquist plot for the 2.2-GHz amplifier covering 1 kHz–20 GHz. *Source*: [1]/John Wiley & Sons.

The analysis results are shown in polar plot form in Figure 8.124 and in magnitude-angle form in Figure 8.125. The polar graph shows the real and imaginary parts of the system determinant as frequency is swept. The magnitude-angle graph shows the magnitude and cumulative angle (no cut at ±180°) of the system determinant as frequency is swept. The key to interpreting these graphs lies in the following statement drawn from the Nyquist criterion:

The circuit will be unstable if the Nyquist plot encircles the origin in a clockwise direction.

The term "unstable" refers to the existence of natural frequencies lying in the right half of the complex frequency plane (RHP). These natural frequencies on the RHP will cause oscillations. To encircle the origin in a clockwise direction means that the Nyquist plot will travel in a clockwise path, cross the negative real axis, and continue to completely surround the origin. If the path appears to encircle the origin but then "unravels" itself in a counter-clockwise direction, the origin is not actually encircled. Figure 8.124 shows that the origin is not encircled and therefore this circuit is stable.

The magnitude-angle plot can assist you to determine if a clockwise encirclement has actually been made by recognizing the following:

- To travel in a clockwise direction on a polar chart means to decrease the phase angle. On a magnitude-angle plot, the phase will become more negative as frequency is increased (although it may not be monotonic).
- If the path crosses the negative real axis on the polar chart, it will cross −180° on the magnitude-angle plot.
- Encirclement will occur if the phase angle continues to decrease to, say, −360°. If "unraveling" occurs, the phase angle will not stay below −360°. Figure 8.125 clearly shows that the phase angle decreases to about −180°, but does not decrease anymore. Rather, the angle returns to a positive value and oscillates between about −40° and +100°. Therefore, an encirclement did not occur, and the circuit is stable.

An Unstable Case: Let us now change the gate bias to −2 V—a value that increases the amplifier gain enough to cause instability. Figures 8.126 and 8.127 show the results. From Figure 8.126, we can see that the origin is encircled in a clockwise direction. Figure 8.127 confirms this since the phase angle goes through −180° and continues without unraveling.

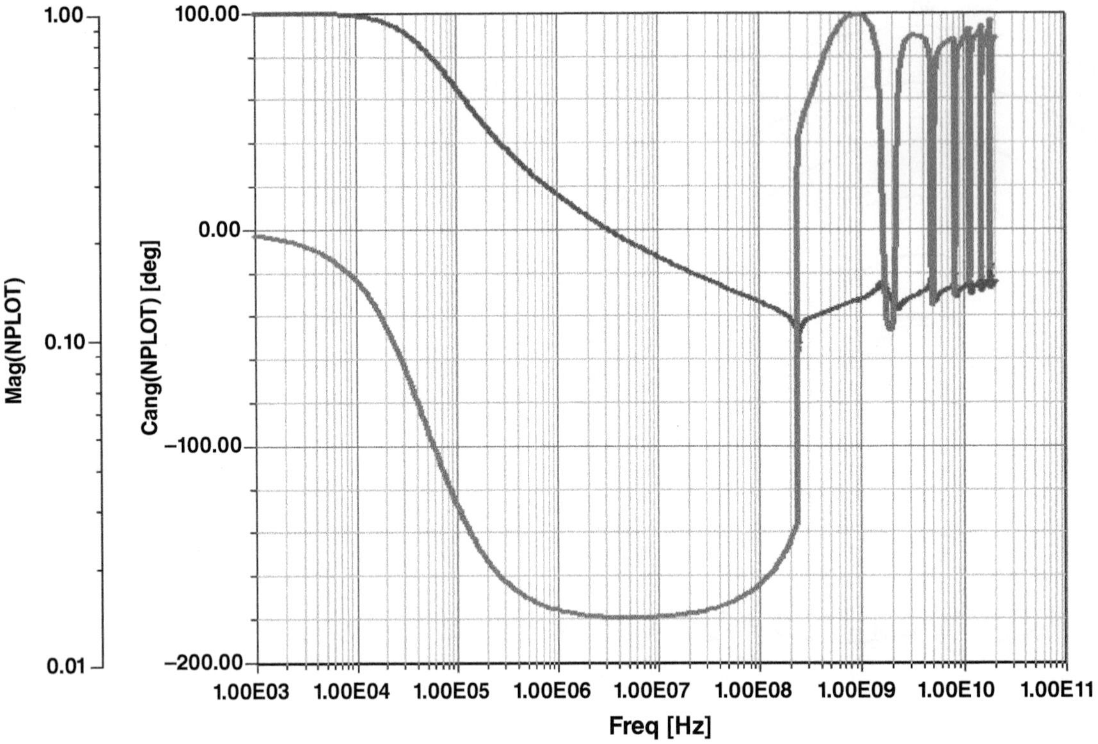

Figure 8.125 Magnitude and cumulative angle Nyquist plot for the 2.2-GHz amplifier covering 1 kHz–20 GHz. *Source*: [1]/John Wiley & Sons.

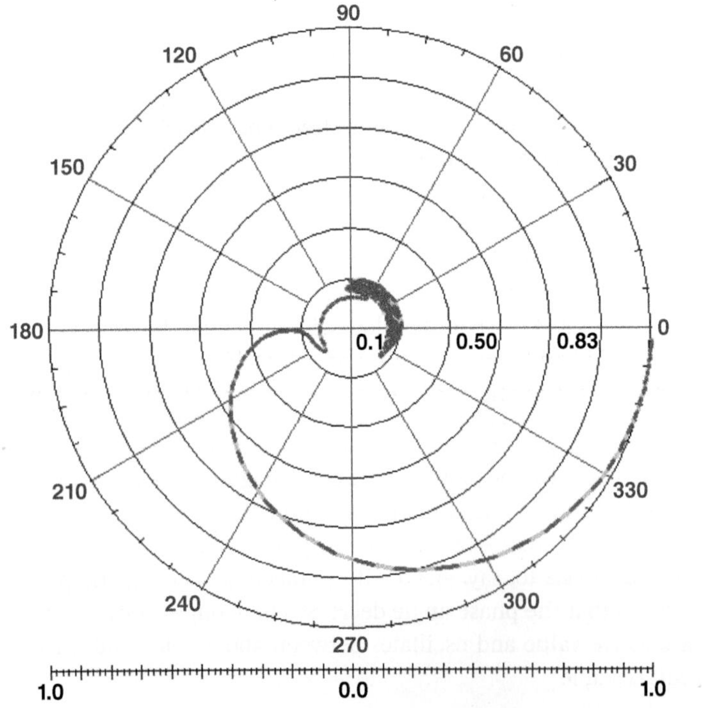

Figure 8.126 Polar Nyquist plot of the unstable amplifier. *Source*: [1]/John Wiley & Sons.

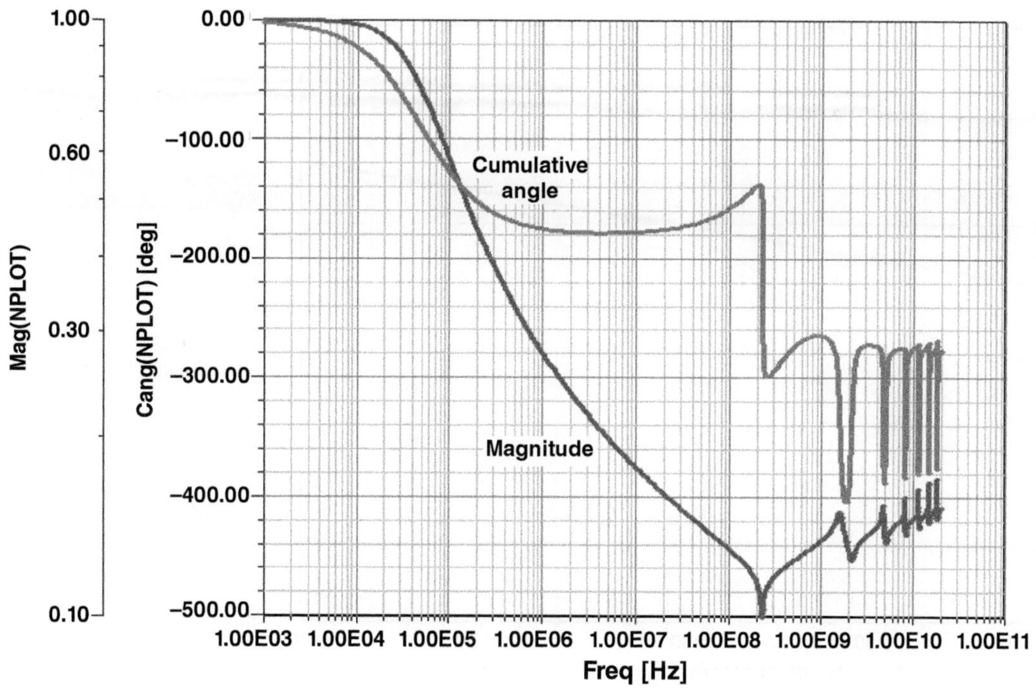

Figure 8.127 Magnitude-cumulative angle Nyquist plot of the unstable amplifier. *Source*: [1]/John Wiley & Sons.

Approximate Frequency of Oscillation: The frequency at which the Nyquist plot crosses −180° is often the approximate frequency of the unstable right-half plane zeros and provides an estimate of the potential oscillation. For this example, there is a sharp crossing at 200 MHz that is characteristic of a circuit resonance. Note that the frequency is not related to a resonant frequency in a rigorous mathematical fashion, but it has been our experience that this frequency approximates the resonant frequency. To confirm the circuit's readiness for oscillation in the vicinity of this frequency, we evaluate it with Serenade's Oscillator Design Aid from 10 MHz–1 GHz. The Design Aid finds an initial resonant frequency near 230.6 MHz—not too far from the −180° crossing frequency of 200 MHz of the Nyquist plot. Figure 8.128 shows this point.

Oscillator analysis finds a final oscillation frequency of 228.48 MHz. Figure 8.129 shows the resulting output spectrum at Port 2. At this point, we decided to do a quick jump and take advantage of the CAD capability to predict the resulting phase noise, which is shown in Figure 8.130. As we compare this with other measured results, the approximately 228-MHz oscillator has a poor phase noise since the same phase noise is achievable at 1 GHz and higher. The reader needs to be reminded that this is an unwanted effect, but all of us who have built amplifiers will have experienced either low-frequency oscillations (motorboating) or very high-frequency oscillations, sometimes caused by the components in the immediate vicinity of the transistor itself and found somewhere above 3 GHz. This case is really a demonstration case; a closer look reveals that the transconductance required for this condition is unrealistically high, but regardless, it is a good example to demonstrate the stability issue.

Where Does Oscillation Begin? An interesting question to solve now is, "What gate bias voltage is needed for oscillation to begin?" Ansoft's Serenade Design Environment includes oscillator synchronous stability analysis, which uses a modification of the harmonic-balance technique to determine critical points, such as the voltage needed to "turn on" an oscillator. Mathematically, this is called a *Hopf bifurcation* because the behavior of the circuit changes dramatically from a dc steady state to an oscillatory response. Figure 8.131 shows this effect by plotting output power versus gate bias. The turning point at −3.2 V (22 mW) is clearly visible, as is the bifurcation at −3.1 V (0 mW output).

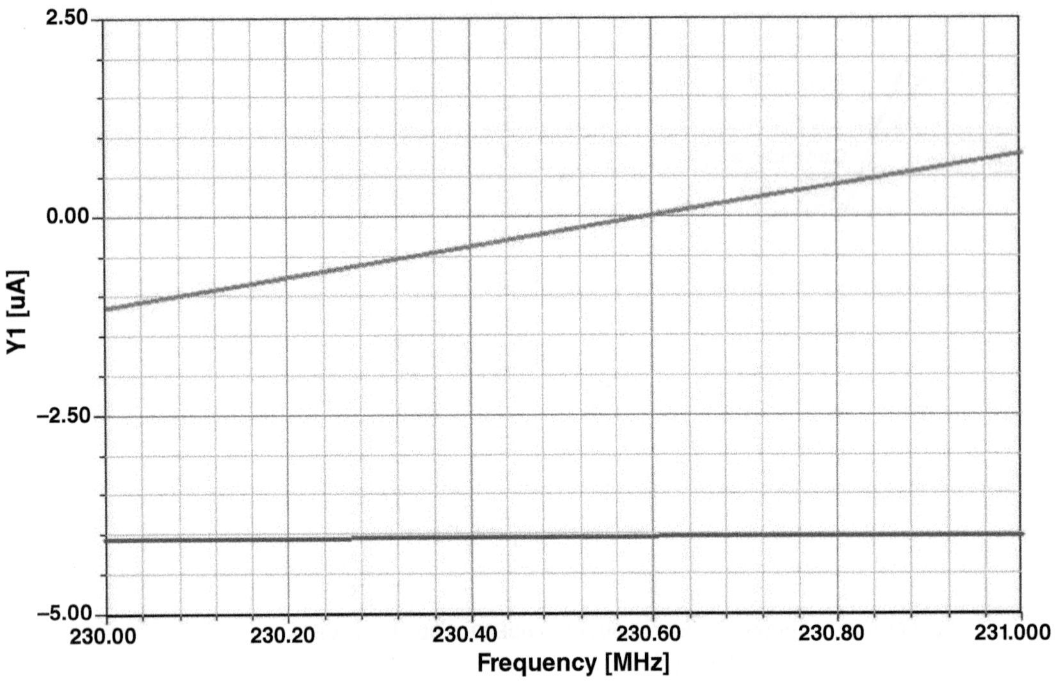

Figure 8.128 Graph from Serenade oscillator design aid analysis of the amplifier. A resonant frequency is indicated where the imaginary part of the test current equals zero, and the real part is negative. *Source*: [1]/John Wiley & Sons.

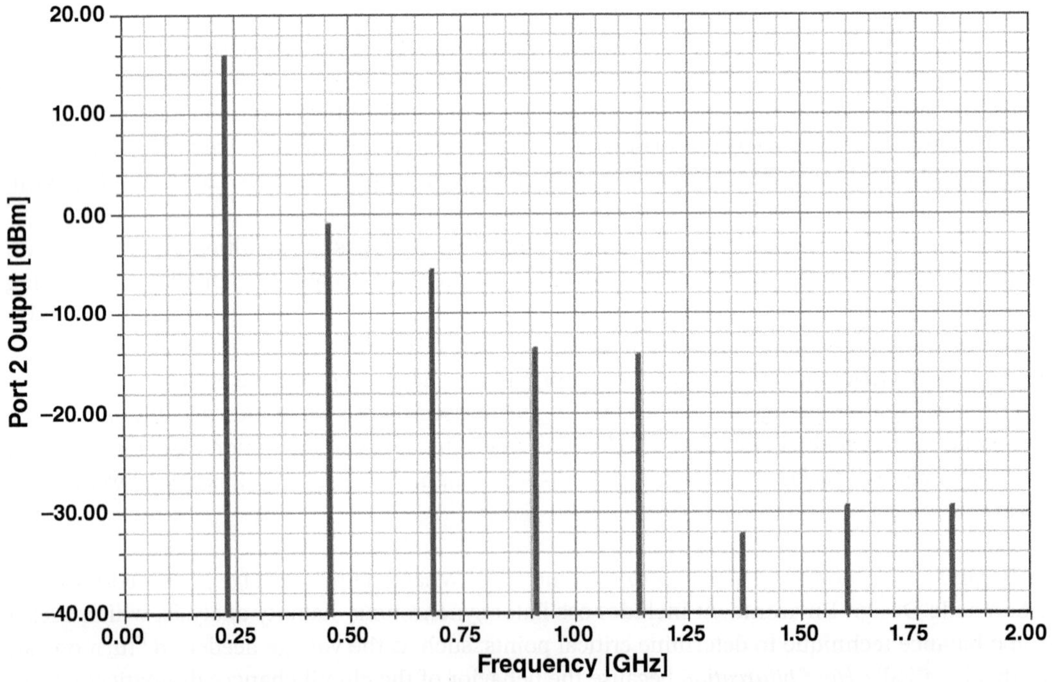

Figure 8.129 Predicted output spectrum of the unintended oscillator. *Source*: [1]/John Wiley & Sons.

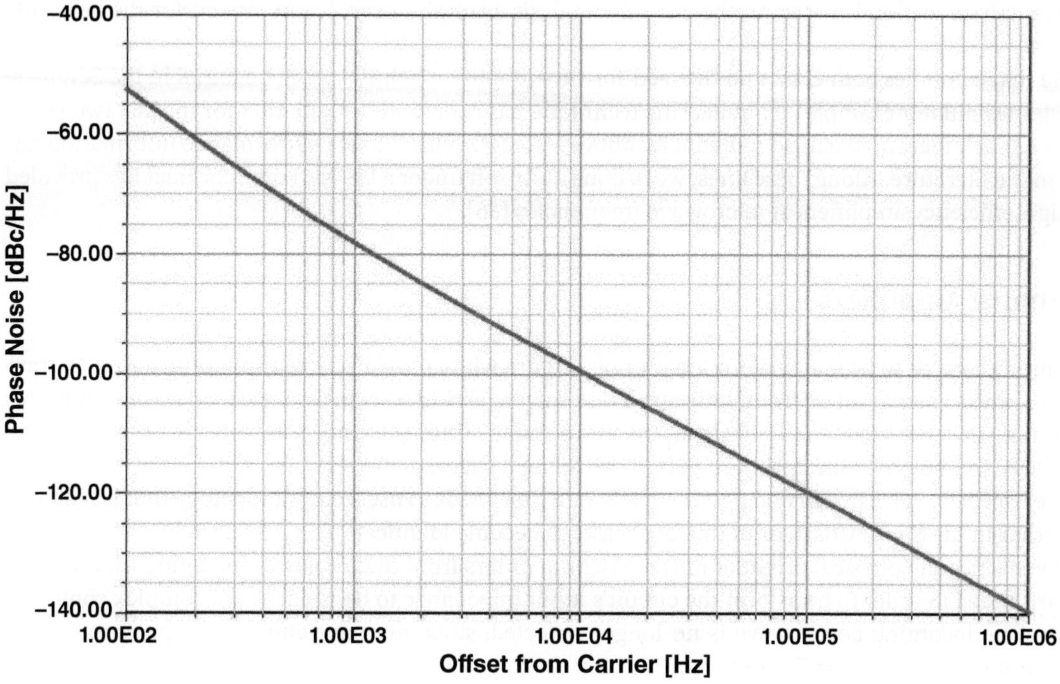

Figure 8.130 Predicted phase noise of the oscillating amplifier. *Source*: [1]/John Wiley & Sons.

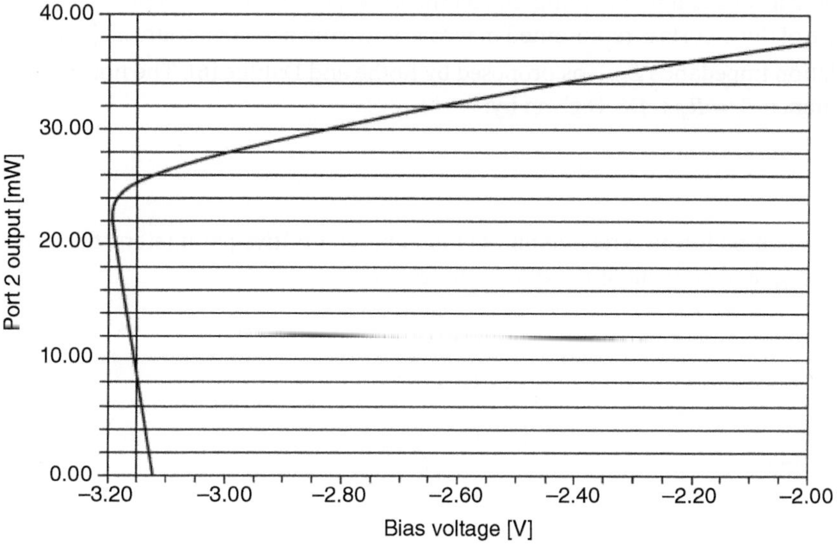

Figure 8.131 Output power versus gate bias for the unstable amplifier. *Source*: [1]/John Wiley & Sons.

Figure 8.131 can be interpreted as follows:

1. If the bias voltage is increased from, say, −4.0 V, the circuit will begin to oscillate at the bifurcation voltage, and the output power will jump to 26 mW.
2. If the circuit is oscillating and the voltage is decreased from −2.0 V, the circuit will cease oscillating at the turning voltage of −3.2 V.
3. The branch between the turning point and the bifurcation point is unstable and cannot be realized in practice. This forms a hysteresis loop around points (−3.1 V, 0 mW), (−3.1 V, 26 mW), (−3.2 V, 22 mW), and (−3.2 V, 0 mW).

This amplifier stability analysis section is based on the stability analysis examples provided by Ansoft for the Serenade Design Environment.

An interesting example that combines both CAD and the feed-forward amplifier technique can be found in the Serenade Design Environment system simulator examples manual. This technique has been under evaluation for the last two years, but our feeling is that not enough measured data at frequencies above 500 MHz with reliable performance information has yet been made available in the literature. Along these lines, work done at Ansoft under a US Air Force contract has provided very good insight into high-efficiency amplifiers at microwave frequencies [56].

8.22 CMOS Low-noise Amplifiers

An optimal low-noise amplifier needs to be matched for lowest noise figure while providing low input return loss. In transistors, however, optimum power match differs from optimum noise match, and MOS transistors are no exception in this respect. Just adding matching networks to a MOS transistor, therefore, means finding a compromise between noise figure and return loss. This strategy will not yield the optimum performance the device can provide.

With the aim to provide excellent noise match and power match, one first needs to use a circuit topology that transforms the two to get closer together in the Smith Chart, or in the ideal case, to become identical.

This goal can be achieved since the physical origin of noise matching differs from that of power matching. In the latter case, reactive elements are added in order to transform the circuit's input impedance to be equal to the complex conjugate of the source impedance. The incoming power wave is no longer reflected, since resonant circuits compensate for any mismatch in phase and amplitude.

Noise match, on the other hand, means to exploit the fact that the short-circuit noise current and the open-circuit noise voltage at the input of the device are partly correlated. Minimum noise figure is achieved by choosing a source impedance that forces the correlated part of the noise to cancel out.

In order to have a closer look at how noise-free feedback and matching impact the noise performance, a generic 2-port is considered, as shown in Figure 8.132. The noise of the 2-ports is represented in terms of two uncorrelated sources, $\langle |v_{\mathrm{n}}|^2 \rangle = 4kT_0BR_{\mathrm{u}}$ and $\langle |i_{\mathrm{n}}|^2 \rangle = 4kT_0BG_{\mathrm{n}}$, and a correlation impedance Z_{corr}, as proposed by Rothe and Dahlke [6]. The noise factor of the 2-port as a function of the source admittance $Z_s = R_s + jX_s$ is given by:

$$F = 1 + \frac{R_{\mathrm{u}} + G_{\mathrm{n}}|Z_s + Z_{\mathrm{corr}}|^2}{R_s} \tag{8.181}$$

This formulation is equivalent to the commonly used formula relying on the much more intuitive parameters, the minimum noise factor F_{min}, optimum noise match Z_{opt}, and equivalent noise admittance G_{n}.

$$F = F_{\mathrm{min}} + \frac{G_{\mathrm{n}}}{R_s}\left|Z_s - Z_{\mathrm{opt}}\right|^2 \tag{8.182}$$

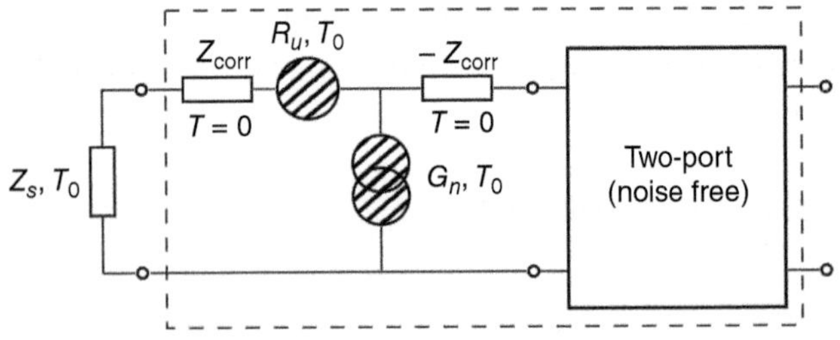

Figure 8.132 Representation of the noise of a 2-port through two noise sources and a correlation impedance.

The parameter G_n in fact is identical in both descriptions; the other parameters can be calculated by:

$$Z_{opt} = R_{opt} + jX_{opt}$$
$$X_{opt} = -X_{corr}$$
$$R_{opt} = \sqrt{\frac{R_u}{G_n} + R_{corr}^2} \tag{8.183}$$
$$F_{min} = 1 + 2G_n(R_{opt} + R_{corr})$$

The common approach to alter the noise matching is to introduce inductive series feedback at the source. Now, adding a lossless inductor at the source will not add any noise to the circuit. But feedback will impact noise and signal differently, and in general, alter F_{min}. However, due to the simplified equivalent circuit regarded here, F_{min} is not changed. For matching purposes, a series inductance is also connected to the gate, which should also be considered at this point. This configuration is shown in Figure 8.133. The generic 2-port is now replaced by a basic MOS transistor equivalent circuit. The MOS transistor's noise is described by the Rothe-Dahlke description, not by its internal noise model. One might expect to see the internal FET noise current sources describing channel noise and induced gate noise. But this description is preferred since we are interested in seeing how the external inductances alter the noise four parameters, while the physics governing the noise is not in focus.

The inductive feedback obviously has an impact on G_n and Z_{corr}. These parameters read [29]:

$$G_n = G_n^o \cdot \left| \frac{Z_{21}}{Z_{21} + j\omega L_s} \right|^2$$
$$Z_{corr} = j\omega L_g + j\omega L_s\left(1 - \frac{Z_{11}}{Z_{21}}\right) + Z_{corr}^o\left(1 + \frac{j\omega L_s}{Z_{21}}\right) \tag{8.184}$$

where the superscript o denotes the original parameter.

Assuming the MOS transistor can be approximated by the simple equivalent circuit shown in Figure 8.133, the formulae can be simplified as follows:

$$G_n = G_n^o \cdot \left| \frac{g_m}{g_m + j\omega^3 L_s C_{gs} C_{ds}} \right|^2 \approx G_n^o$$
$$Z_{corr} = j\omega L_g + j\omega L_s\left(1 - \frac{j\omega C_{ds}}{g_m}\right) + Z_{corr}^o\left(1 + \frac{j\omega^3 L_s C_{gs} C_{ds}}{g_m}\right) \approx j\omega(L_g + L_s) + Z_{corr}^o \tag{8.185}$$

The optimum noise match in terms of optimum source impedance Z_{opt} can therefore be approximated by:

$$Z_{opt} = Z_{opt}^o - j\omega(L_g + L_s) \tag{8.186}$$

The real part of Z_{opt} therefore remains unchanged, but its imaginary part is reduced. That's an important finding since it means that $Re(Z_{opt})$ is unchanged by the matching and feedback inductances. Optimizing this value to the needs of the circuit design, therefore, means choosing a suitable device and bias point.

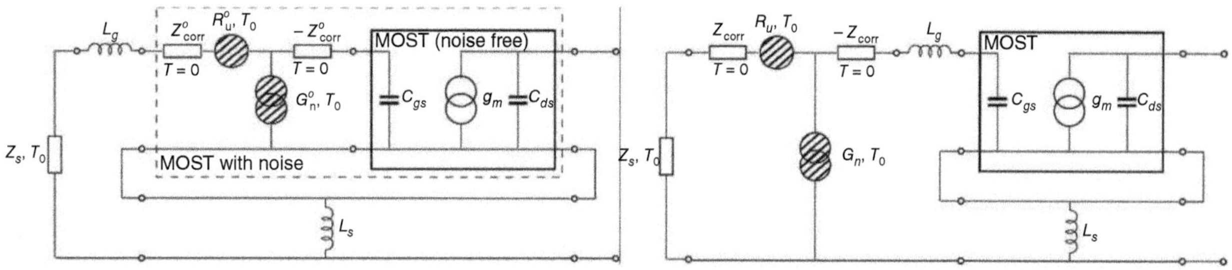

Figure 8.133 MOS transistor with source and gate inductances. (a) The noise of the transistor is described by the Rothe–Dahlke equivalent circuit. (b) The noise sources are transformed in order to describe the noise properties of the whole circuit.

Regarding the imaginary part of Z^o_{opt}, it is now required to know at least its sign, and its frequency dependence in order to see whether it can be compensated by the inductances or not. This calculation requires considering the physics-based noise model of the MOS transistor, and quite a bit of calculations, which, in all details, are found in the literature [30]. In the end, Eq. (8.186) can be rewritten by approximating $Im\left(Z^o_{\text{opt}}\right)$ by the following:

$$Z_{\text{opt}} = \text{Re}\left(Z^o_{\text{opt}}\right) - m\frac{1}{j\omega C_{\text{gs}}} - j\omega(L_{\text{g}} + L_{\text{s}}) \tag{8.187}$$

with the parameter m in the range of 0.6 for long channel devices, while it is expected to approach 1 as the gate width is scaled down. The imaginary part of the optimum noise match impedance is capacitive, which can very well be compensated by properly choosing the values of the inductances.

But noise match alone is not sufficient for a good LNA, it also requires good input matching. The input impedance for the device with gate and source inductances, as shown in Figure 8.133 reads:

$$Z_{\text{in}} = j\omega(L_{\text{s}} + L_{\text{s}}) + \frac{1}{j\omega C_{\text{gs}}} + \frac{g_m L_{\text{s}}}{C_{\text{gs}}} = j\omega(L_{\text{s}} + L_{\text{s}}) + \frac{1}{j\omega C_{\text{gs}}} + \omega_{\text{T}} L_{\text{s}} \tag{8.188}$$

with the transit frequency $\omega_T = 2\pi f_T = g_m/C_{\text{gs}}$.

The source inductance L_s introduces a real part in the input impedance, which is the main benefit of the inductive feedback. The formulae describing Z_{in} and Z_{opt} become similar in structure. With the designer's freedom to choose suitable values for inductance and device size, it becomes possible to bring these two impedances as close together as possible, allowing for simultaneous noise and power match. While $Re(Z_{\text{in}})$ is affected by ω_T, that depends on the transistor and how it is biased, the sum of the inductances $L_s + L_{\text{g}}$ can be used to compensate for the capacitive component of Z_{opt} and Z_{in}.

In conclusion, the design of a CMOS LNA follows the following steps:

1. Choose a device of appropriate size that satisfies $Re\left(Z_{\text{opt}}\right) = Re(Z_s)$.
2. Select source inductance L_s and bias point to satisfy $\omega_T L_s = Re(Z_s)$.
3. Determine the value of the gate inductance L_{g} so that $Im\left(Z_{\text{opt}}\right) = Im(Z_s)$ holds.

In certain cases, especially if a very low-power design is required, it is not possible to choose the device size freely. In this case, usual matching techniques have to be applied instead of just compensating C_{gs} and forcing $Re\left(Z_{\text{opt}}\right) = Re(Z_s)$. In this case, the inductive degeneration still has its benefit, since it allows, in principle, simultaneous power and noise matching, when $\omega_T L_s = \text{Re}\left(Z_{\text{opt}}\right)$ is enforced. However, this is not always possible in practice, Z_{opt} increases with decreasing device size, requiring large inductance values. But if L_s excites a certain limit, noise figure is compromised. On one hand, large inductances come with higher parasitic resistances and therefore add thermal noise. On the other hand, as the inductive feedback increases, the transistor's internal feedback will also gain importance and degrade the minimum noise factor.

If it is practically not possible to select a source inductance, often an extrinsic gate-source capacitance C_{ex} is added. Regarding Z_{in}, it simply adds to C_{gs}, with the negative impact on the external transit frequency that now becomes $\omega_T' = g_{\text{m}}/(C_{\text{gs}} + C_{\text{ex}})$. But considering noise matching, this approach has the power to lower $Re\left(Z_{\text{opt}}\right)$ [30]. In this case, $Re\left(Z_{\text{opt}}\right)$ reads:

$$Re\left(Z_{\text{opt}}\right) = \frac{\text{A} \cdot B}{\omega C_{\text{gs}}\left[\text{A}^2 \cdot B^2 + \left(\frac{C_{\text{gs}} + C_{\text{ex}}}{C_{\text{gs}}} + \text{A} \cdot |c|^2\right)^2\right]} \tag{8.189}$$

with $\text{A} = \alpha\sqrt{\delta/(5\gamma)}$, $B = \sqrt{1 - |c|^2}$. The parameters α, γ, δ refer to the noise sources of the FET [31]. The channel noise current $\left\langle |i_{\text{nd}}|^2 \right\rangle$ and the induced gate noise current $\left\langle |i_{\text{gd}}|^2 \right\rangle$ are given by:

$$\left\langle |i_{\text{ng}}|^2 \right\rangle = 4kTB\delta\frac{\omega^2 C_{\text{gs}}^2}{5g_{d0}} \tag{8.190}$$

$$\left\langle |i_{\text{nd}}|^2 \right\rangle = 4kTB\gamma g_{d0} \tag{8.191}$$

where g_{d0} is the drain-source conductance at zero drain-source voltage V_{ds}. The typical values for the noise parameters are the following. For long-channel devices, γ starts at unity for zero V_{DS}, and approaches 2/3 in saturation. However, it increases at high V_{GS} and V_{DS}. For short-channel devices, it can exceed 2. The parameter δ is constant in long channel devices, its value is around 3/4. It is higher for short channel devices. The value α is defined as the ratio of g_m to g_{d0}, which is unity for long-channel devices and lower in the case of a short-channel device. The correlation coefficient c can be given for long channel devices, its value is purely imaginary $c = j0.395$.

Equation (8.189) shows that $Re(Z_{opt})$ is reduced by adding C_{ex}. This therefore is an adequate measure for the case in question. Lowering $Re(Z_{opt})$ reduces the requirement of a high value of $\omega_T L_s$. It thereby allows to reduce feedback, and it enables the use of very small, low-power, transistors that inherently have high impedance levels.

Thus, for a given device size where $Re(Z_{opt})$ is too high for the original approach, the design procedure is as follows:

1. Select a source inductance L_s and capacitance C_{ex} to obtain matching of $Re(Z_s)$ and $Re(Z_{opt})$ without compromising effective transit frequency (and, thus, gain) and noise.
2. Determine the value of the gate inductance L_g so that $Im(Z_{opt}) = Im(Z_s)$ holds.

The whole approach up to now, however, has a drawback. Introducing source and gate inductances is also a viable starting point in oscillator design. Another point is the low reverse isolation of a single-transistor LNA, which should also be reduced because of the Miller effect. Therefore, two CMOS transistors are commonly employed in cascode configuration.

Another issue in CMOS design is power dissipation. Being energy efficient is imperative for battery-powered mobile systems, which first impacts the choice of the transistor size, as just discussed. But it also means that the supply voltage is constrained to a lower value [32].

The standard cascode is disadvantageous in this respect, as it requires the supply voltage to be twice the voltage V_{ds} required for a single transistor. The drain voltage of each of the transistors should be twice the overdrive voltage $V_{od} = (V_{gs} - V_t)$ in order to provide high gain. The supply voltage can therefore only be lowered if it is not applied to the full cascode stack, but instead each transistor is biased individually. This requires decoupling the RF path from the DC path, resulting in a topology known as folded cascode [33–35].

The folded cascode stage is shown in Figure 8.134. The supply voltage is fed in between the two transistors that are, in fact, DC-wise connected in parallel. CMOS offers the advantage that it provides NMOS and PMOS devices that require supply voltages of opposite sign. Thereby, the folded cascode transistors can be DC coupled. In the original approach based on bipolar technology, it was necessary to decouple the two transistors by means of capacitors and to bias them individually [33]. But in CMOS, the DC is supplied through an RF choke that provides an open to the RF signal. The RF therefore directly passes the two transistors in cascode configuration, without the need for any DC blocking capacitance in-between.

The folded cascode is operated at lower supply voltages compared to the original cascode, but it still dissipates the same power. An approach to lower voltage and current at the same time is the complementary current-reuse architecture, as shown in Figure 8.135 [36]. As in case of the folded cascode, DC and RF paths are decoupled in this architecture. The two transistors are in series concerning DC, thus the second transistor "reuses" the current of the first. But for the RF signal, the structure acts as a two-stage amplifier. The inductances are employed like inductors to ground for interstage matching, together with the coupling capacitance. Although it looks like a cascode at first glance, the complementary current-reuse stage works differently. Thus, it does not provide the benefits inherent to the cascode structure, like reduction of the Miller capacitance or improved stability. Its main advantage is that it provides the highest gain at the lowest supply voltage. Due to the complementary MOS architecture, the required supply voltage is reduced by one overdrive voltage V_{od}

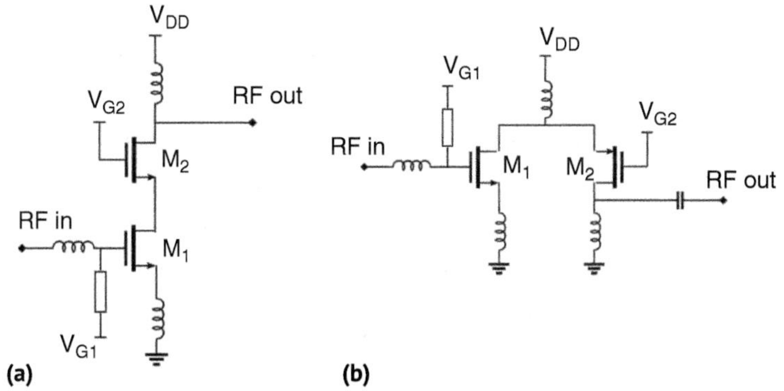

Figure 8.134 Cascode (a) and folded cascode (b) CMOS LNA circuit.

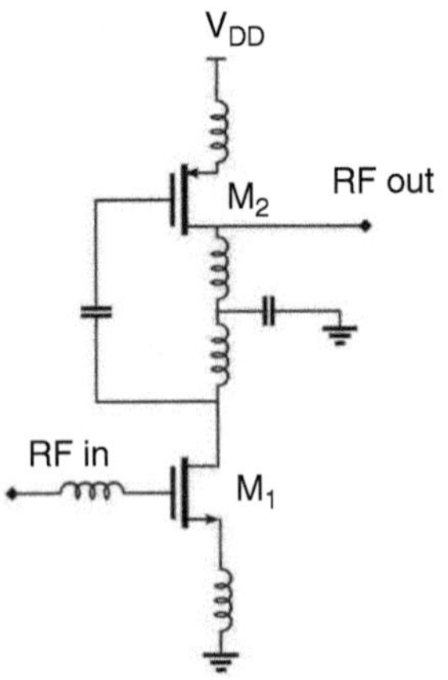

Figure 8.135 Complementary current-reuse CMOS LNA circuit.

as compared to the standard cascode architecture, without operating the transistors in weak inversion. Thereby, the highest gain for minimum dissipated power is achieved. The drawback of this architecture, however, is the need for quite a number of inductors, which are not easily realized with high Q on-wafer, and which also consume a lot of space.

8.22.1 Passive Elements Available in CMOS Technologies

Monolithic integrated CMOS processes use a larger number of interconnect layers that can be used to design passive elements such as capacitors, inductors, and transformers. Traditional MMIC processes only provide one metal layer, and transmission lines are either of the microstrip or coplanar waveguide (CPW) type. In either case, the metallization is high-conductivity electroplated gold, and the substrate is semi-insulating. Chips are often thinned down to 100 m in order to suppress higher modes of the microstrip lines. Gold metallization, semi-insulating substrate, and the appropriate dimensions of the lines allow for optimum propagation of the electromagnetic waves along the transmission lines. CMOS is different and a rather hostile environment for RF. The standard silicon wafers are of the low-resistivity type to save on cost. And aluminum and copper are used instead of gold for the transmission lines. Therefore, it is to be expected that the losses due to the finite conductivity of the transmission line metal will be higher than on similar lines on GaAs, GaN, or InP processes. In addition, it is advisable to keep the electric field out of the substrate, as the substrate losses are significant and, if neglected, may render any RF design impossible.

On the other hand, there are some advantages of CMOS technologies besides the aspect of low cost in high-volume production. The high number of metal layers enable the custom design of various capacitor and inductor structures, a feature that is not available in the traditional GaAs world.

The metallization layer stack of a typical RFCMOS process is shown in Figure 8.136 [37]. This process provides six metallization layers; the lower four are made of copper, and the upper ones are aluminum layers. Basically, the higher layers provide thicker metal. The layers are embedded in a low-loss dielectric, on top of the substrate. The present process uses two types of dielectric, first silicon oxide, and on top polyimide. The electric fields should be confined within these dielectric layers, as they are good isolators. Often, even a lower metallization layer is employed to shield the passive structures from the lossy substrate. Good RF transmission lines can be realized within the oxide if the RF ground is also realized within this metallization layer stack, forming a microstrip or stripline.

Capacitance elements are designed straightforwardly in the form of parallel-plate capacitors. Figure 8.137a shows an example of such an approach using multiple layers to enhance the capacitance per square. In the example, four layers are used. The interconnection between the plates belonging to the same electrode is realized through via holes. Figure 8.137b shows the three-dimensional layout of the capacitor. The via-holes are closely spaced and form a type of via fence in order to provide a good connection between the layers.

These structures, however, don't just behave like pure capacitances. The following effects are expected to be observed:

- The finite conductivity of the metal layers will lead to parasitic resistance.
- A parasitic capacitance between the metals and the substrate below will establish a lossy connection to ground.
- The structure has a certain dimension, giving rise to parasitic inductances at higher frequencies.

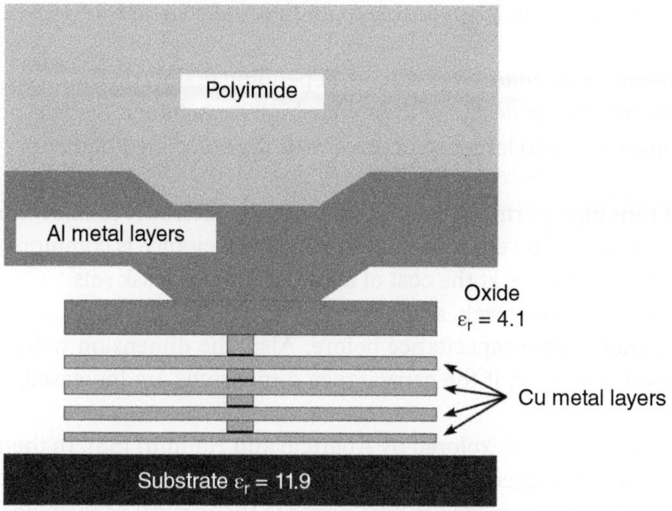

Figure 8.136 Metal-layer stack of a typical RFCMOS process. *Source*: [37]/Wojciech Andrzej Debski.

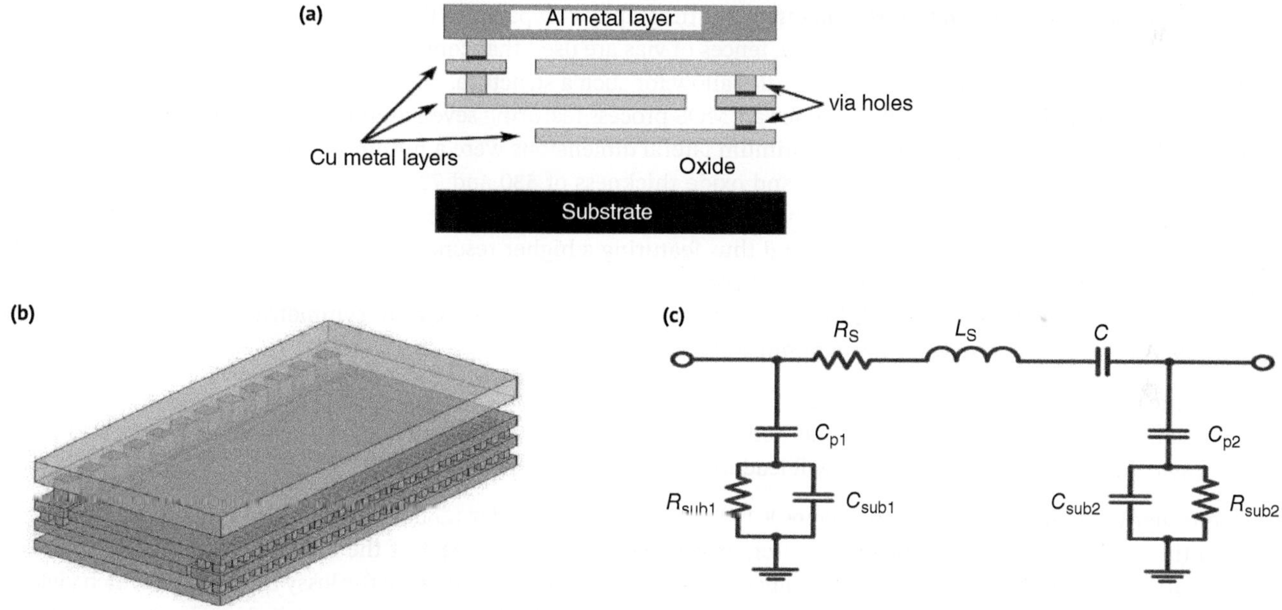

Figure 8.137 Parallel-plate capacitor realized in a typical RFCMOS process. *Source:* Adapted from [37, 57].

These effects are accounted for in the first-order equivalent circuit shown in Figure 8.137c. The capacitance C is the desired element, ballasted with the parasitic resistance R_s and inductance L_s. The impact of coupling to substrate and substrate losses is described through $C_{p1,2}$, $R_{\text{sub}1,2}$ and $L_{\text{sub}1,2}$. The equivalent circuit reveals that this structure is a resonance circuit, not only the desired capacitance. Thus, it shows capacitive behavior only up to a certain frequency. Beyond this resonance frequency, it becomes inductive. Since the basic geometrical parameters for a certain technology are fixed, it is observed that the resonance frequency of the structure decreases with increasing size of the structure. Higher capacitance values will come with lower usable maximum frequencies.

Regarding lumped equivalent models like the one in Figure 8.137c, it must be kept in mind that the model also has an upper usable frequency, not only the device. Even though the model consists of a number of elements, it only applies a first-order model to each of the physical effects. If the wavelength and structure size come into the same order of magnitude, distributed effects need to be considered. As a rule of thumb, such models can be expected to be reasonably accurate up to the resonance frequency, and definitely fail beyond. This rule of thumb is applicable to all first-order models of passive components, and even transistors.

The parallel-plate capacitances discussed so far provide a certain capacitance per square, depending mainly on the dielectric properties of the oxide isolation layers. These layers, however, are not designed to provide high capacitance. Staggering multiple layers as shown in the figure yields higher capacitance, but also larger structures with the associated higher parasitic effects.

A number of processes therefore provide an extra layer of thin high-permittivity, high-breakdown dielectric to allow for metal–insulator–metal (MIM) capacitances, and an extra metal for the top electrode. This type of capacitance is commonly available in III-V processes, and can be integrated into RFCMOS processes at the cost of a few additional mask sets.

Figure 8.138 shows a schematic of a MIM capacitance within the layer stack, and also the associated equivalent circuit. The equivalent circuit looks a bit simpler than that of the parallel plate capacitance before. Also, the dimension will be much smaller due to the thin high-permittivity dielectric used. However, if the capacitance dimensions are increased, it will also show some parasitic inductive effect.

Theoretical limits of capacitance values and resonance frequencies were explored by Aparicio and Hajimiri [58]. In their work, a vast number of configurations to design capacitances are discussed. A main result is shown in Table 8.5. A third configuration, in addition to the types of capacitance discussed so far, regarded in this paper is the vertical capacitance. The basic idea behind the vertical capacitance is simple: The achievable lateral dimensions are much smaller than the vertical dimensions, thus allowing for higher capacitance per size. While vertical dimensions are fixed by the dielectric layer thickness, the lateral dimensions depend on how narrow via holes and metallization can be placed. Also, tolerances are lower in the lateral dimension that is structured by lithography compared to the vertical layers formed by deposition. However, in order to get a vertical capacitor plate, fences of vias are used that connect many metal layer strips. It depends on the individual technology whether its design rules allow for such a structure.

The values shown in the table were obtained for a CMOS process featuring seven aluminum metal layers, four of which were used to fabricate the capacitances. The minimum lateral dimensions were a minimum 240 nm for distance between and for width of metal structures, and a metal and oxide thickness of 530 and 730 nm, respectively. Statistical data was obtained by measurements on different spots of two 8-inch wafers. As expected, the vertical capacitor shows less tolerances, and higher capacitance, consuming less space and thus featuring a higher resonance frequency. The values for the MIM capacitance are the typical values shown for comparison.

Inductances can be realized in monolithic processes through spiral lines. A fully symmetric structure is shown in Figure 8.139. Also, this structure will show a number of parasitic effects:

- The finite conductivity of the metal layers will lead to parasitic resistance.
- A parasitic capacitance between the metals and the substrate below will establish a lossy connection to ground.
- The capacitance between the windings will lead to a parasitic parallel capacitance.

The equivalent-circuit model shown in Figure 8.139b provides a first-order model for the spiral inductance. It is quite similar to the model of the parallel-plate capacitor, at least regarding the impact of the lossy substrate that is modeled through $C_{ox1,2}$, $R_{sub1,2}$, and $C_{sub1,2}$. The parasitic capacitance turns out to be in parallel to the lossy inductance, which yields a

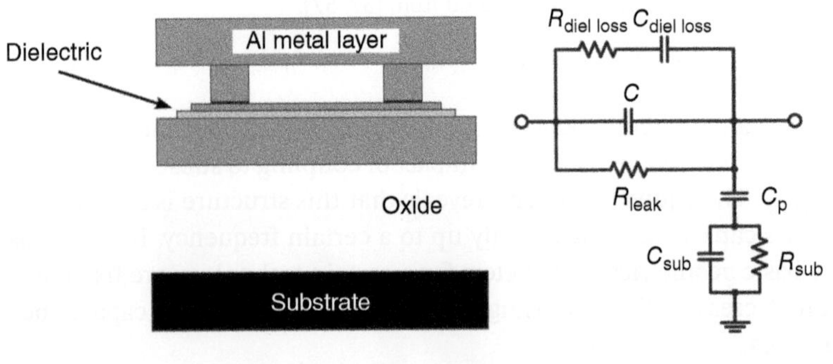

Figure 8.138 Metal–insulator–metal capacitor realized in a RFCMOS process. *Source:* [37]/Wojciech Andrzej Debski.

Table 8.5 Comparison of performance parameters for different types of CMOS capacitances.

Structure	Capacitance density [fF/m^2]	Average capacitance [pF]	Area [m^2]	Standard deviation σ [fF]	Resonance frequency [GHz]	Q at 1 GHz	Breakdown Voltage [V]
Vertical plate	1.51	1.01	669.9	5.06	>40	83.2	128
Horizontal plate	0.20	1.09	5378.2	26.11	21	63.8	500
MIM	1.1	1.05	960.9			11	

Source: [58]/IEEE.

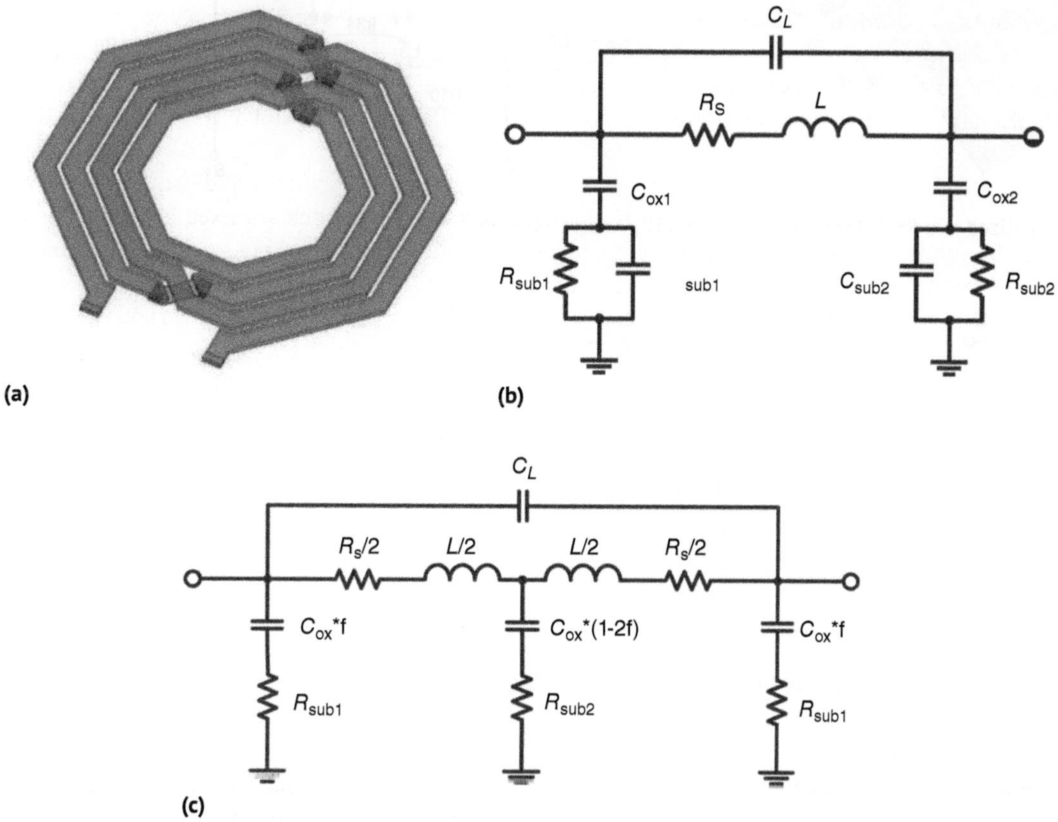

Figure 8.139 Spiral inductor realized in a RFCMOS process. *Source*: [37]/Wojciech Andrzej Debski.

parallel resonance circuit. There is, therefore, a maximum achievable inductance, depending on frequency and technology, since adding winding will increase the parasitic capacitance and loss resistance together with the inductance.

Variations and refinements of the equivalent circuit are possible, for example, the symmetrical structure shown in Figure 8.139c. It intends to enhance the validity of the equivalent circuit model somehow toward higher frequencies, by somehow distributing the coupling to the lossy substrate. However, even such a model fails close to or beyond the resonance frequency of the structure.

Transformers are important building blocks in many CMOS integrated circuits. These can be realized by designing concentric spiral inductors. An example is shown in Figure 8.140. This transformer consists of one turn on the primary side, and two turns on the secondary side.

An equivalent-circuit model of a transformer is slightly more complicated than that of a single inductor. For a structure providing a center connection, the basic model topology is shown in Figure 8.140b. It can be understood as four inductor models with additional coupling between the inductances. Although this model already consists of 30 elements, it still

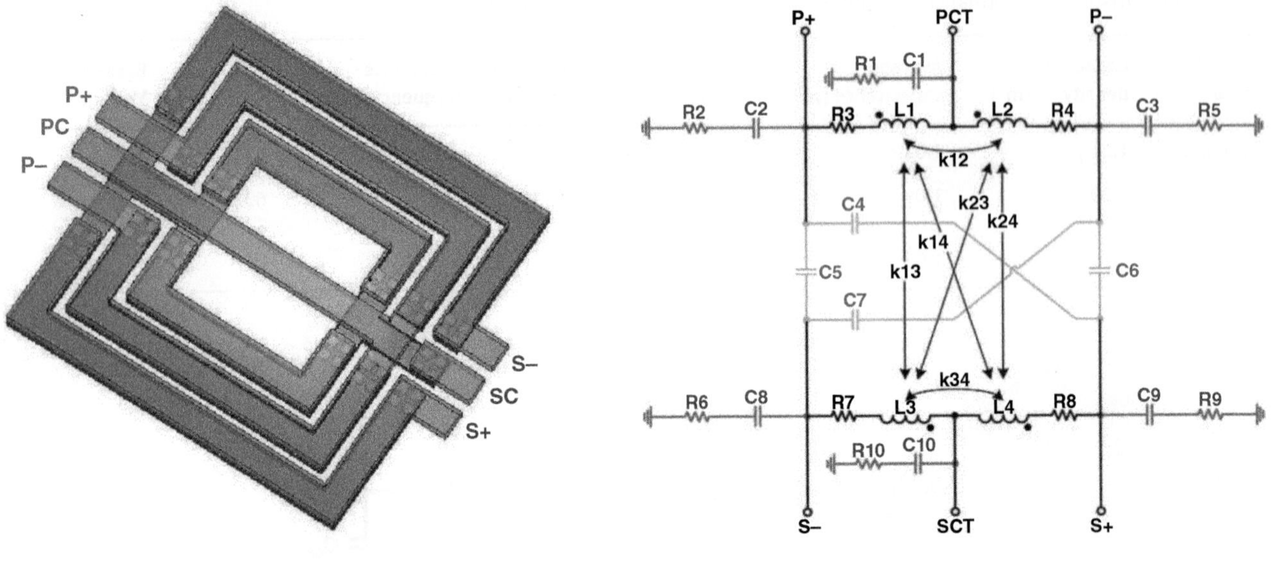

Figure 8.140 Transformer realized in a RFCMOS process. *Source*: [57]/Andriy Vasylyev.

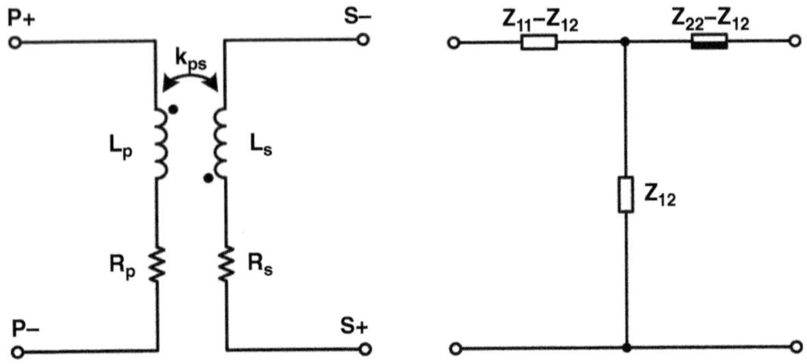

Figure 8.141 Simplified transformer equivalent-circuit model. *Source*: [57]/Andriy Vasylyev.

catches only first-order effects. Since this model is quite complicated, it can be reduced to the one shown in Figure 8.141. This model is, of course, very basic, as it only accounts for the self- and mutual inductance and line losses.

The accuracy of equivalent-circuit models is constrained to a certain frequency range, as already mentioned. Another issue is parameter extraction for such a model. It is quite tedious to determine a vast number of parameters, basically from 2-port measurements. In order to obtain physically significant model parameters, it is required to analytically develop a suitable algorithm. It also commonly requires investigating dedicated test structures. On the positive side, it is in principle possible to develop parameterized models for a class of elements. It can be accounted for geometrical variations if it is known how the model parameters depend on geometry, and an appropriate formula is used to determine specific parameters.

A faster and, at least at first sight, more accurate approach is to use measured S-Parameters of the passive models, or to rely on electromagnetic simulation. Both approaches promise to provide highly accurate descriptions of the device performance, since no assumptions restrict the validity of the model. The designer also saves the time required to determine the equivalent circuit model; simulation becomes both faster and more accurate.

However, there must be a reason why equivalent-circuit models are still in use. The first is that not all circuit simulation tools calculate in the frequency domain. S-parameters are frequency-domain data. Using an S-parameter model in a time-domain solver like SPICE requires applying an inverse Fourier transform in order to determine the time-domain response of the model. First of all, that requires a certain simulator kernel that allows for this so-called convolution algorithm.

Second, will the time-domain response be causal? This is definitely required for simulation, but ensuring it for just any set of S-parameters requires elaborate mathematical treatment of the data. Thus, simulating the same circuit in the time domain and in frequency domain will show slightly different results. A model based on a lumped-element equivalent circuit, on the other hand, works with any circuit simulation tool.

It is another issue that S-parameters are only known for the measured or simulated frequency range. The fundamental and the first harmonics are usually covered. But what about baseband and DC? Also, it might be desirable in nonlinear Harmonic-Balance simulations to use many harmonics. If one leaves it to the simulation tool to extrapolate the data easily lead to non-physical parameters. An equivalent-circuit model won't be accurate beyond a certain frequency as discussed above. But it will never become non-physical and predict, for example, active behavior for an inductor. DC and baseband performance are also well covered.

Finally, there is basically no way to derive a scaled S-parameter-based model. Only discrete measured or simulated structures are covered. Parameterized equivalent-circuit-based models have the capabilities to interpolate within certain geometrical limits. This allows us to define the models behind the well-known pick-and-place symbols in circuit simulators that describe an element like a spiral inductors on the basis of a few electrical and geometrical parameters.

Design Examples: Introducing the following two subsections on the 130 nm CMOS LNAs.

Throughout this book, we always address the issue that we have to separate the battery-operated handheld applications from the more stationary designs like those for base stations. In the case of the handheld or low-voltage applications, the dynamic range requirements are significantly less than the front-end of a base station.

8.22.2 130-nm CMOS Low-noise Amplifier

As an example, let's consider a LNA for 2.41 GHz applications that should be fabricated in a 130-nm CMOS technology. We choose devices with a gate width of 250 m and a bias of $V_{DD} = 1\,V$, $I_{DD} = 20$ mA. A generic model for such a device can be found, e.g., on the predictive technology model website [59]. The schematic of the circuit is given in Figure 8.142.

How simultaneous noise and power matching is achieved is shown in Figure 8.143. The starting point is the unmatched cascode, the respective values of S_{11}^* and Γ_{opt} at 2.4 GHz are indicated by the larger symbol. Obviously, the real part of the noise match impedance is higher than 50 Ω, which requires either using a larger device, or starting by adding an extrinsic gate-source capacitance C_{ex}. It is also obvious that the optimum noise and power match are quite far apart.

The impact of adding C_{ex} is shown by the bullet symbols. The capacitance values are increased by 100 fF from one symbol to the next, the highest value shown is $C_{ex} = 1.4$ pF, which is used in this example. Since noise and power match are still too far apart from each other, a source degeneration inductance L_s is added. In the example, L_s is increased in steps of 50 pH up to a maximum of 0.45 nH. The diamond symbols in the Smith charts show that indeed, S_{11} changes much faster

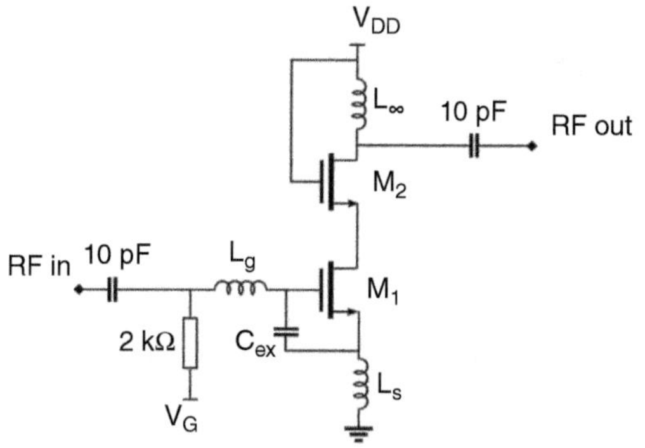

Figure 8.142 Schematic of the CMOS LNA. Both transistors are 130 nm NMOS, gate width 250 m, and 20 gate fingers. Model parameters are given in the table. The values of the other circuit elements are: $C_{ex} = 1.4$ pF, $L_s = 0.45$ nH, $L_g = 2.7$ nH. The quality factor of these inductances is set to $Q = 6$ at the target frequency of 2.41 GHz. The biasing network and output matching are omitted for simplicity, $V_{DD} = 2$ V, and $I_{DD} = 23$ mA.

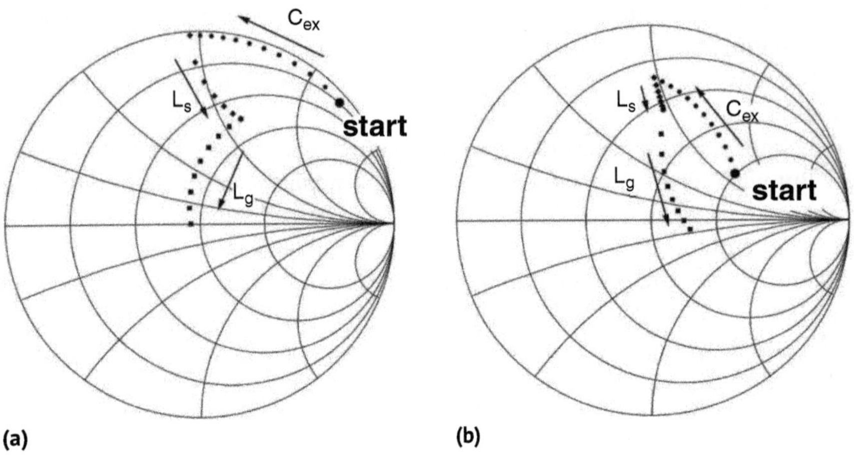

Figure 8.143 Impact of external gate-source capacitance C_p, source degradation inductance L_s and gate inductance L_g on (a) complex conjugate of S_{11} and (b) optimum noise match Γ_{opt}.

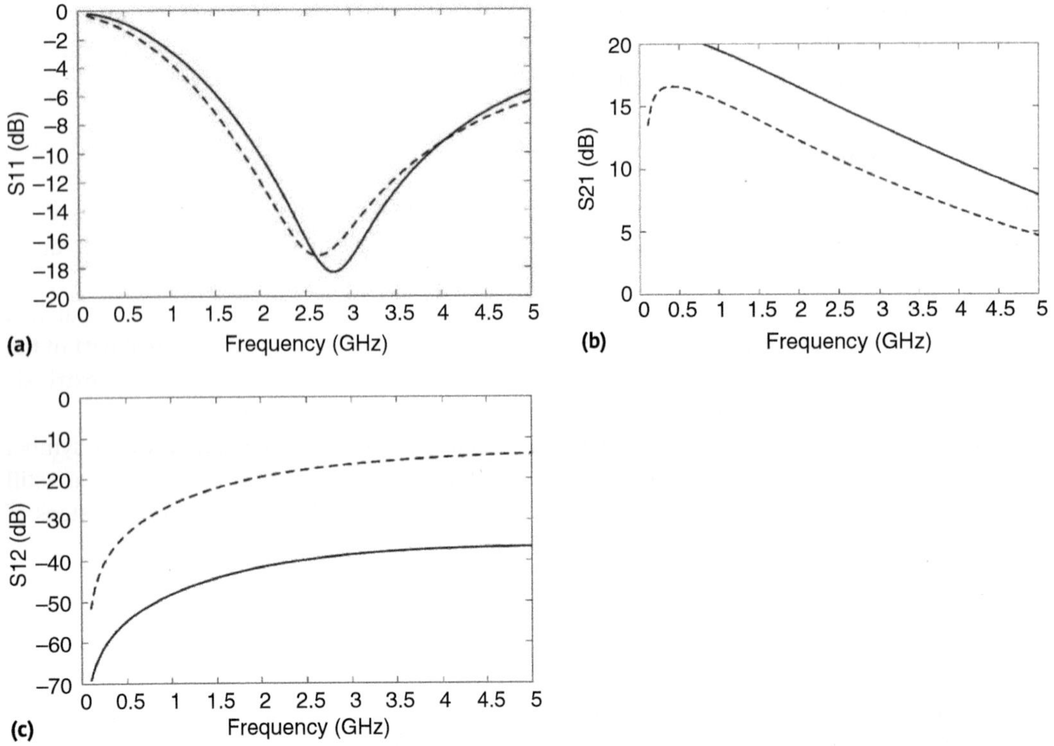

Figure 8.144 Simulated S-parameters of the circuit shown in Figure 8.142 (solid lines) compared to the results obtained when the cascode is replaced by a single transistor (dashed lines), (a) S_{11}, (b) S_{21}, and (c) S_{12}.

than Γ_{opt}. Finally, a gate inductance L_s is added to obtain input matching; it is increased in steps of 300 pH up to a total value of $L_g = 2.7$ nH. In this simulation, it is accounted for inductor losses by setting the inductors' quality factor to $Q = 6$.

This practical example shows some discrepancies with the theory presented earlier. For example, it doesn't seem to be possible to achieve perfect simultaneous power and noise matching. But despite the fact that the theory was based on many simplifications, it proves to be applicable for real-life circuit design.

Figure 8.144 presents the S-parameters of the circuit after input matching. Noise figure and stability factor are shown in Figure 8.145. In these figures, the performance of the cascode LNA (solid lines) is compared with a single-transistor

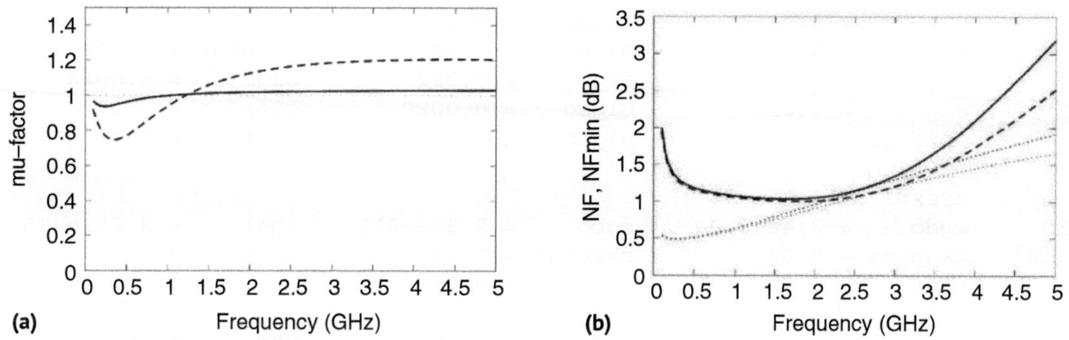

Figure 8.145 Simulated results of the circuit shown in Figure 8.142 (solid lines), compared to the results obtained when the cascode is replaced by a single transistor (dashed lines). (a) stability factor and (b) noise figures. The dotted lines denote the respective minimum noise figures.

LNA (dashed lines). The input matching for the single transistor LNA was achieved with similar component values, i.e., $C_{ex} = 1.4$ pF, $L_s = 0.64$ nH, $L_g = 2.2$ nH. The benefit of the cascode configuration is obviously its better isolation, as S_{12} is about 20 dB lower than in the single transistor case. The price is, however, a slightly higher achievable noise figure.

Other interesting design examples can be found online [60].

```
* Beta Version released on 2/22/06

* PTM 130nm NMOS

.model  nmos  nmos  level = 54

+version = 4.0        binunit = 1          paramchk= 1          mobmod  = 0
+capmod  = 2          igcmod  = 1          igbmod  = 1          geomod  = 1
+diomod  = 1          rdsmod  = 0          rbodymod= 1          rgatemod= 1
+permod  = 1          acnqsmod= 0          trnqsmod= 0

+tnom    = 27         toxe    = 2.25e-9    toxp    = 1.6e-9     toxm    = 2.25e-9
+dtox    = 0.65e-9    epsrox  = 3.9        wint    = 5e-009     lint    = 10.5e-009
+ll      = 0          wl      = 0          lln     = 1          wln     = 1
+lw      = 0          ww      = 0          lwn     = 1          wwn     = 1
+lwl     = 0          wwl     = 0          xpart   = 0          toxref  = 2.25e-9
+xl      = -60e-9
+vth0    = 0.3782     k1      = 0.4        k2      = 0.01       k3      = 0
+k3b     = 0          w0      = 2.5e-006   dvt0    = 1          dvt1    = 2
+dvt2    = -0.032     dvt0w   = 0          dvt1w   = 0          dvt2w   = 0
+dsub    = 0.1        minv    = 0.05       voffl   = 0          dvtp0   = 1.2e-010
+dvtp1   = 0.1        lpe0    = 0          lpeb    = 0          xj      = 3.92e-008
+ngate   = 2e+020     ndep    = 1.54e+018  nsd     = 2e+020     phin    = 0
+cdsc    = 0.0002     cdscb   = 0          cdscd   = 0          cit     = 0
+voff    = -0.13      nfactor = 1.5        eta0    = 0.0092     etab    = 0
+vfb     = -0.55      u0      = 0.05928    ua      = 6e-010     ub      = 1.2e-018
+uc      = 0          vsat    = 100370     a0      = 1          ags     = 1e-020
+a1      = 0          a2      = 1          b0      = 0          b1      = 0
+keta    = 0.04       dwg     = 0          dwb     = 0          pclm    = 0.06
+pdiblc1 = 0.001      pdiblc2 = 0.001      pdiblcb = -0.005     drout   = 0.5
+pvag    = 1e-020     delta   = 0.01       pscbe1  = 8.14e+008  pscbe2  = 1e-007
+fprout  = 0.2        pdits   = 0.08       pditsd  = 0.23       pditsl  = 2.3e+006
+rsh     = 5          rdsw    = 200        rsw     = 100        rdw     = 100
+rdswmin = 0          rdwmin  = 0          rswmin  = 0          prwg    = 0
+prwb    = 6.8e-011   wr      = 1          alpha0  = 0.074      alpha1  = 0.005
+beta0   = 30         agidl   = 0.0002     bgidl   = 2.1e+009   cgidl   = 0.0002
+egidl   = 0.8
```

```
+aigbacc  = 0.012        bigbacc  = 0.0028        cigbacc  = 0.002
+nigbacc  = 1            aigbinv  = 0.014         bigbinv  = 0.004         cigbinv  = 0.004
+eigbinv  = 1.1          nigbinv  = 3             aigc     = 0.012         bigc     = 0.0028
+cigc     = 0.002        aigsd    = 0.012         bigsd    = 0.0028        cigsd    = 0.002
+nigc     = 1            poxedge  = 1             pigcd    = 1             ntox     = 1

+xrcrg1   = 12           xrcrg2   = 5
+cgso     = 2.4e-010     cgdo     = 2.4e-010      cgbo     = 2.56e-011     cgdl     = 2.653e-10
+cgsl     = 2.653e-10    ckappas  = 0.03          ckappad  = 0.03          acde     = 1
+moin     = 15           noff     = 0.9           voffcv   = 0.02

+kt1      = -0.11        kt1l     = 0             kt2      = 0.022         ute      = -1.5
+ua1      = 4.31e-009    ub1      = 7.61e-018     uc1      = -5.6e-011     prt      = 0
+at       = 33000

+fnoimod  = 1            tnoimod  = 0

+jss      = 0.0001       jsws     = 1e-011        jswgs    = 1e-010        njs      = 1
+ijthsfwd= 0.01          ijthsrev= 0.001          bvs      = 10            xjbvs    = 1
+jsd      = 0.0001       jswd     = 1e-011        jswgd    = 1e-010        njd      = 1
+ijthdfwd= 0.01          ijthdrev= 0.001          bvd      = 10            xjbvd    = 1
+pbs      = 1            cjs      = 0.0005        mjs      = 0.5           pbsws    = 1
+cjsws    = 5e-010       mjsws    = 0.33          pbswgs   = 1             cjswgs   = 3e-010
+mjswgs   = 0.33         pbd      = 1             cjd      = 0.0005        mjd      = 0.5
+pbswd    = 1            cjswd    = 5e-010        mjswd    = 0.33          pbswgd   = 1
+cjswgd   = 5e-010       mjswgd   = 0.33          tpb      = 0.005         tcj      = 0.001
+tpbsw    = 0.005        tcjsw    = 0.001         tpbswg   = 0.005         tcjswg   = 0.001
+xtis     = 3            xtid     = 3

+dmcg     = 0e-006       dmci     = 0e-006        dmdg     = 0e-006        dmcgt    = 0e-007
+dwj      = 0.0e-008     xgw      = 0e-007        xgl      = 0e-008

+rshg     = 0.4          gbmin    = 1e-010        rbpb     = 5             rbpd     = 15
+rbps     = 15           rbdb     = 15            rbsb     = 15            ngcon    = 1
```

8.22.3 130-nm CMOS Transformer-coupled Low-noise Amplifier

An alternative approach to the inductively degenerated low-noise amplifier is the transformer-coupled LNA, a topology initially introduced as *zwischenbasis*-configuration by Cantz in 1953 [61]. The schematic is shown in Figure 8.146. The topology differs from the previously discussed LNA by the fact that the source and gate inductance are now mutually coupled. This feedback supports simultaneous noise and power matching.

In the present case, the transformer was modeled by assuming a fixed ratio of 4:1 for the primary and secondary inductances L_s and L_p, respectively. Also, the coupling factor was fixed to $k = 0.6$. In order to account for the transformer losses, it is assumed that each of the inductances has a parasitic resistance of 1.1 Ω/100 pH. The transformer, together with the extrinsic gate-source capacitance C_{ex}, was employed to bring noise and power match to the same impedance, with a real part of 50 Ω. The gate inductance L_g was finally used to transform the resulting input impedance to the 50 Ω source. This inductance is assumed to have a quality factor limited to $Q = 6$ at the target frequency of 2.41 GHz. The parameter values are: $L_s = 200$ pH and $L_p = 0.8$ nH, $C_{ex} = 1.5$ pF, and $L_g = 1.7$ nH. The transistor model parameters are given in Table 8.5. The original transformer-coupled LNA relies on a single transistor instead of a cascode. Therefore, simulation results obtained with the same topology except that the cascode is replaced by a single MOS transistor are included in the results shown in Figures 8.147 and 8.148. In the case of the single-transistor amplifier, matching differs slightly. The element values are: $L_s = 350$ pH, $L_p = 1.4$ nH, $C_{ex} = 1.2$ pF, and $L_g = 0.4$ nH.

The benefit of the transformer-coupled configuration over the traditional source degeneration topology is obviously the lower total inductance values that are required in order to achieve a good matching. In the present simulation, the main noise sources are not the transistors but the losses of the inductors.

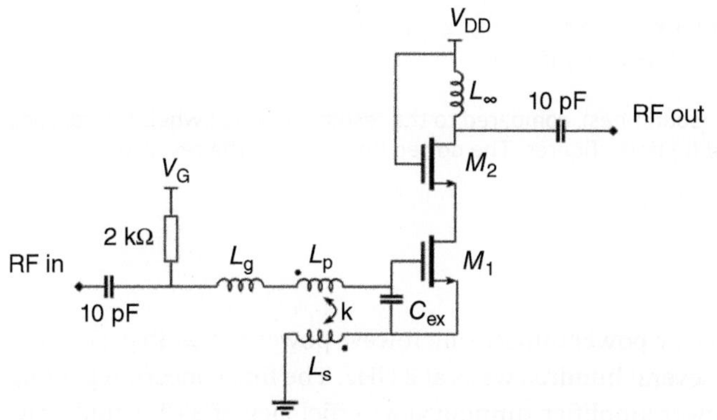

Figure 8.146 Schematic of the transformer-coupled CMOS LNA. Both transistors are 130nm NMOS, gate width 250 m, and 20 gate fingers. Model parameters are given in the table. The values of the other circuit elements are: $C_{ex} = 1.5$ pF, the transformer parameters are $L_p = 0.8$ nH, $L_s = 0.2$ nH, $k = 0.6$, and inductor losses are modeled by adding resistances of 1.1 Ω/0.1 pH to each of the inductances. $L_g - 1.7$ nH with a quality factor of $Q = 6$ at the target frequency of 2.41 GHz. The biasing network and output matching are omitted for simplicity, $V_{DD} = 2$ V, and $I_{DD} = 23$ mA.

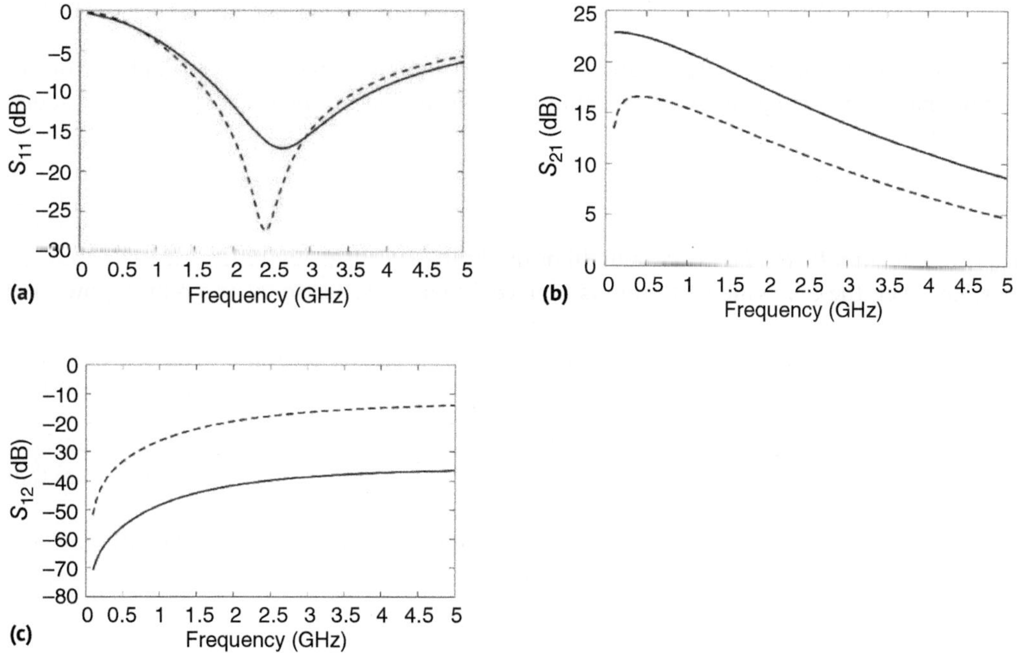

Figure 8.147 Simulated S-parameters of the circuit shown in Figure 8.146 (solid lines). Compared to the results obtained when the cascode is replaced by a single transistor (dashed lines) (a) S_{11}, (b) S_{21}, and (c) S_{12}.

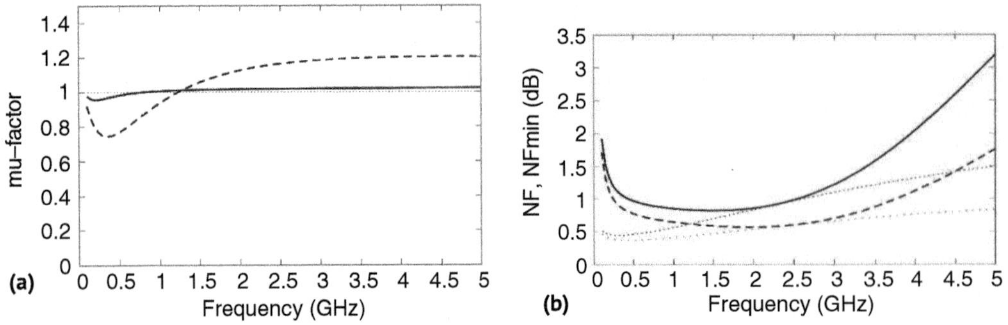

Figure 8.148 Simulated results of the circuit shown in Figure 8.146: (solid lines), compared to the results obtained when the cascode is replaced by a single transistor (dashed lines). (a) stability factor and (b) noise figures. The dotted lines denote the respective minimum noise figures.

8.23 Power Amplifiers—Theory

Power amplifiers are devices that are tasked to transform as much dc power into RF/microwave power as possible. Depending upon the technology, devices are available that can produce several hundred watts at 2 GHz. The first concern regarding power amplification is efficiency. As an example, the 100 W power amplifier running at an efficiency of 33% would draw 300 W from the grid and produce 200 W of heat. It is even worse than just that 2/3 of the energy is lost: the lifetime of semiconductor devices decreases exponentially with temperature. Appropriate cooling is therefore imperative for successful power amplifier design.

The definitions for the efficiency can be given as the ratio of the desired RF output power P_{out} to the power that is delivered to the amplifier. The first is the drain (or collector) efficiency η which is just the ratio of P_{out} to the DC power consumed by the amplifier

$$\eta = \frac{P_{\text{out}}}{P_{\text{DC}}} \tag{8.192}$$

This definition of efficiency neglects the input power P_{in} required to drive the amplifier. In the wireless and microwave range, power amplifiers might not have too much gain, and neglecting P_{in} yields too optimistic values. The power added efficiency (PAE) accounts also for P_{in}:

$$PAE = \frac{P_{\text{out}} - P_{\text{in}}}{P_{\text{DC}}} \tag{8.193}$$

This definition will yield negative values for PAE if $P_{\text{out}} < P_{\text{in}}$. But this is rarely an issue for an amplifier, and this definition is widely accepted. Sometimes, however, the following definition is preferred that is limited to 0% (no output power) to 100%:

$$PAE = \frac{P_{out}}{P_{DC} + P_{\text{in}}} \tag{8.194}$$

The efficiency issue can be highlighted regarding Figure 8.149. A generic power amplifier consists basically of a transistor that converts the DC power provided by the power supply into the output RF power. A filter is required to make sure that only power at the desired frequency is finally delivered to the load.

Figure 8.149a shows current and voltage at the collector of a bipolar transistor in 'class A' operation. The RF power is delivered toward the load, as current and voltage are in antiphase. The DC power, however, is delivered toward the transistor where it is dissipated and converted to heat. In other words, the dissipated power is generated when collector current I_c and collector-emitter voltage V_{ce} exist at the same time:

$$P_{\text{diss}}(t) = i_c(t) \cdot v_{ce}(t) \tag{8.195}$$

An effective means to reduce the dissipated power is therefore to bias the transistor in a way that it only provides current pulses instead of a full sine wave, as shown in Figure 8.149b. How the current waveforms are obtained through proper biasing is shown in Figure 8.151. In order to distinguish the different modes of operation, the current conduction angle Θ is defined in Figures 8.151 and 8.152.

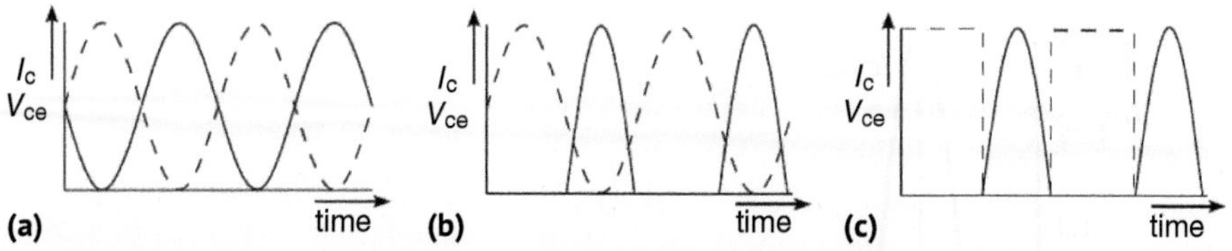

Figure 8.149 Sketch of collector current (solid lines) and collector voltage (dashed lines) of a power-amplifier transistor. (a) class A operation, (b) class C operation, (c) class F operation.

Table 8.6 Power amplifier classes A to C.

Class A	"Linear" operation; 180 conduction angle
Class AB	"Linear" operation; conduction angle between 90 and 180
Class B	"Linear" operation; 90 conduction angle
Class C	Fixed drive; less than 90 conduction angle

Power amplifier classes A to C are defined in terms of the device conduction angle and/or the type of device operation (Table 8.6):

Since the transconductance of FETs at conducting angles of less than 180 degrees becomes very small, the actual power gain sharply decreases compared to Class A operation, while the power-added efficiency is optimized to values of 55% (Class B) or so. The dissipation of these devices is, of course, quite high, while in Class B operation, this is even more of an issue. Power devices in Class C are dying to have big heatsinks to survive. One of the nice things about FETs is a much-reduced thermal runaway effect compared to their bipolar brothers.

Figure 8.150 shows Class C operation. The output of the amplifier is a current pulse train, which frequently leads to misunderstanding as to the calculation of its load. Given the fact that the amplifier is a current generator, it is silly to assume one has to do a conjugate matching looking backward into the amplifier. This is the condition for Class A operation. Unfortunately, this is still not understood by some engineers, but we hope we can clear this up today. The theoretical efficiency is 89.7% at a conducting angle of less than 180; in practice, efficiencies of more than 85% are not obtainable. (All of these efficiencies are those of the device itself; losses in matching networks are not taken into account in these efficiency values.) Considering optimal gain and optimal output power, the conducting angle actually settles somewhere between 100 and 140. Again, the power gain in Class C operation is less than in Class A and Class B operation.

While the equation $P_{\text{OUT}} = (V_{\text{bat}} - V_{\text{sat}})^2/(2R_L)$ is valid, it really applies only to power amplifiers in the sense that for Class A, high-gain operation R_L can be made significantly higher as the absolute power output is not the important item. While the supply voltage V_{bat} is always known, the saturation voltage V_{sat} has to be obtained by the load line dc analysis. Note: The RF saturation voltage is always higher than the dc one. When looking at Figure 8.151, it becomes obvious that the conducting angle is 180, which implies that the entire sine wave at the input is reproduced as an essentially undistorted sine wave at the output.

We can now calculate the components of the distorted current in separating them into the dc component, the fundamental frequency, and its harmonics:

$$
\begin{aligned}
i_{\text{cn}} = i_{\text{peak}} \cdot &\left[\frac{2\Theta}{2\pi[1-\cos(\Theta)]} \cdot \{\text{Si}[(\text{n}-1)\Theta] + \text{Si}[(\text{n}+1)\Theta]\} - \right. \\
&\left. - \frac{2\Theta\cos(\Theta)}{\pi[1-\cos(\pi/2)]} \cdot \text{Si}(\text{n}\Theta)\right]
\end{aligned}
\tag{8.196}
$$

With n being the index of the harmonic, starting with n = 0 at DC, and fundamental n = 1, the instantaneous current maximum i_{peak}, the function $\text{Si}(x) = \sin(x)/x$, and the conduction angle Θ. The equation is easily obtained from deriving the Fourier series for a clipped sine wave with DC offset, as shown in Figure 8.152. These harmonic components are a first-order approximation of the current waveform, but it is close enough to reality to explain the general concept. Figure 8.153 shows the relative amplitudes of the current harmonics as a function of conduction angle Θ. As expected, the harmonic

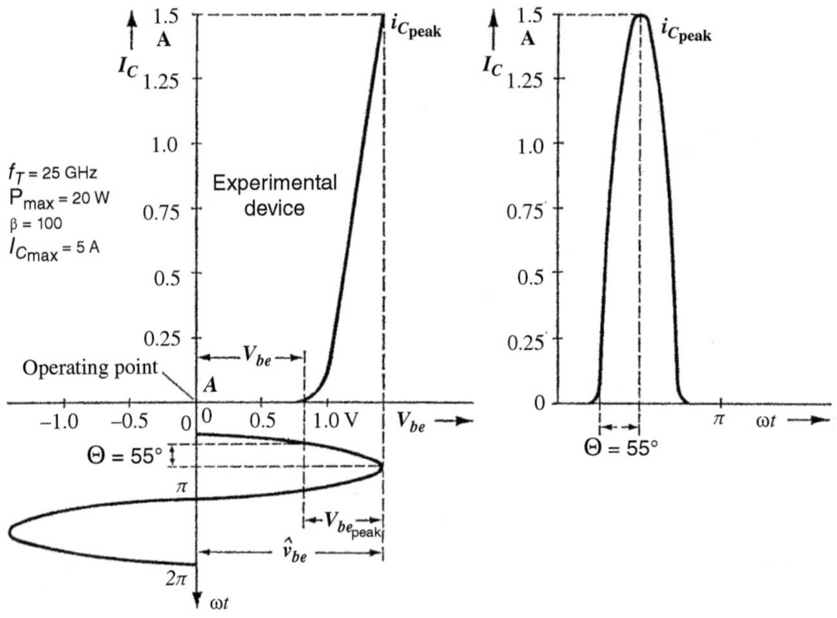

Figure 8.150 Output current as a function of drive voltage, taking into consideration the knee voltage of the base-emitter p–n junction. The output is a series of current pulses with a duty cycle determined by the conduction angle.

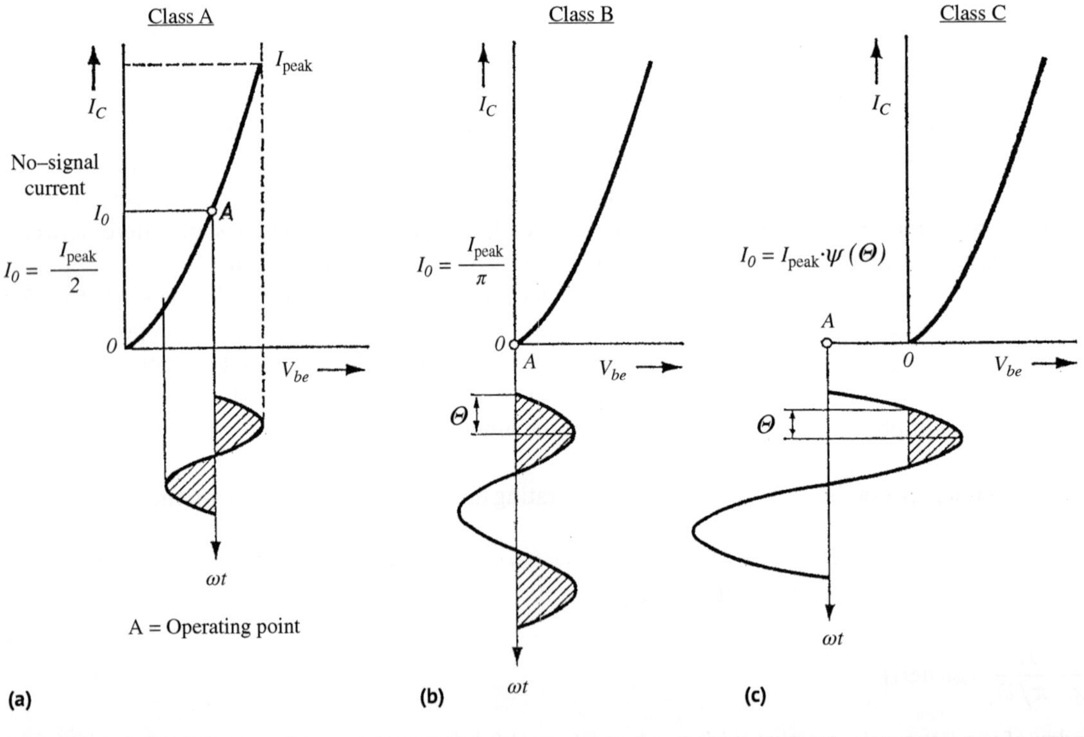

Figure 8.151 Definition of Classes A, B, and C operation as mentioned above, including conduction angle (Θ). Figure 8.152 shows the determination of Θ.

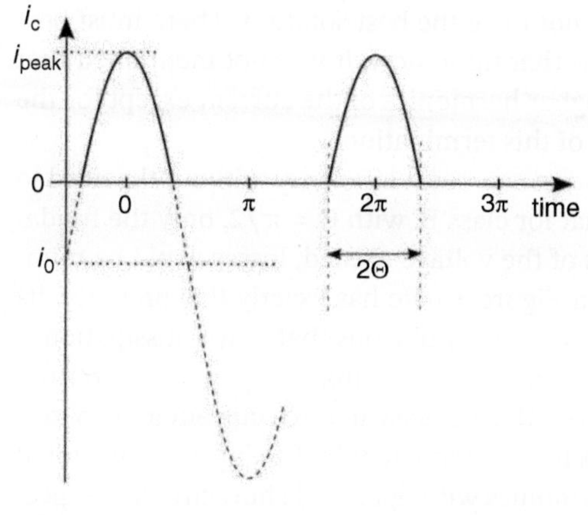

Figure 8.152 Definition of conduction angle Θ.

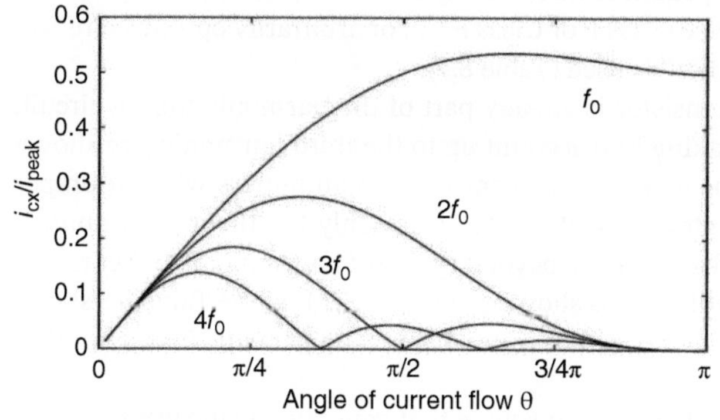

Figure 8.153 Function values for the fundamental, to the fourth harmonic (f_0, $2f_0$, $3f_0$, and $4f_0$, respectively) versus device conduction angle.

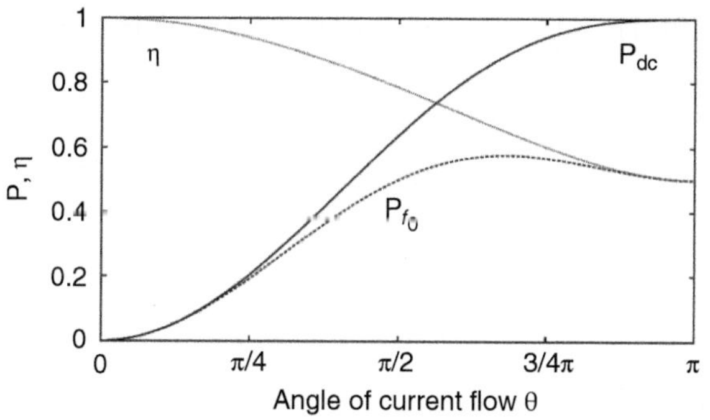

Figure 8.154 Values for the DC and fundamental frequency power, and resulting drain efficiency η versus device conduction angle.

content increases as the conduction angle decreases. Also, the output power approaches zero with Θ, but peaks somewhere in class AB. Figure 8.154 finally shows the emitter efficiency η calculated from Eq. (8.153), together with DC and fundamental power. It is obvious that increasing the efficiency by reducing the conduction angle comes at the expense of reduced output power. It finally leads to the trivial solution of 100% efficiency at no power.

It should be noted, however, that these amplifier classes are defined only under ideal conditions, for pure sinewaves of constant amplitude, and that the theoretical efficiencies are just drain/collector efficiencies, not taking into account any kind of required drive power or parasitic or matching losses. Therefore, realistic *PAE* values are significantly lower, and the question of which class an amplifier is becomes questionable without a constant drive signal.

Enhancing efficiency only by reducing conduction angle, therefore, seems not to be the best solution. There must be a way to obtain higher efficiency without degrading output power. The point is that up to now, it was not mentioned that the Classes A–C assume that the load is formed by a filter that shortens all higher harmonics of the current except for the fundamental. The purely sinusoidal shape of the voltage waveform is a result of this termination.

It has already been proposed quite early that appropriate harmonic termination can boost efficiency without the need to sacrifice fundamental power [62, 63]. A closer look at Figure 8.153 reveals that for class B, with $\Theta = \pi/2$, only the fundamental and even harmonics of the current exist at all. If only odd harmonics of the voltage existed, losses could be minimized. Fortunately, the square-wave shape of the voltage waveform shown in Figure 8.149c has exactly this property. Its spectrum is proportional to $\mathrm{Si}(n\pi/2)$. Looking at the time domain waveforms, it is pretty obvious that power dissipation is zero since current and voltage don't overlap at any time. In the frequency domain, this means that the power drawn from the voltage supply equals the RF power at the fundamental delivered to the load. It is not easy to force current and voltage exactly into these shapes just by tuning the load. But what can be achieved is to approximate it by forcing zero current at even harmonics and zero voltage at odd harmonics, by terminating higher harmonics with open and short circuits, respectively. This is the approach taken in Class F amplifiers [64, 65]. Of course, in reality, higher harmonics than the third are usually neglected. Also, there might be different optimum load impedances than exactly short or open. If the sequence of short and open circuits is reversed, for example, one speaks of inverse Class F or Class F^{-1}. For arbitrarily optimized reflective terminations, commonly the term "harmonically tuned amplifier" is used (Table 8.7).

It is necessary to consider that the output capacitance of the transistor is already part of the harmonic tuning circuit. Designs of optimized tuning circuits for ideal Class F operation, taking into account up to the third harmonic, are shown in Figures 8.155 and 8.156 [66]. The Class-F concept allows to boost efficiency compared to amplifiers without proper harmonic termination. But the ideal waveforms that provide an efficiency of $\eta = 100\%$ are only the theoretical limit. It would require to control all harmonics, while in practice, it is not feasible to go beyond the third harmonic. The theoretical efficiency limit for the case where only the lower harmonics are controlled is shown in Figure 8.151, while Table 8.8 shows the theoretical maximum efficiency of Class-F Pas. Practical PAE values will be even lower as no parasitic losses or other non-ideal transistor behavior is regarded so far.

The power amplifiers so far all relied on the transistor as a controlled current source that generates a sine wave or parts of it. Another concept relies on the transistor as switch, just as in a high-efficiency DC–DC converter. A switch is either on or off, short or open circuit, respectively. Thus, it ideally consumes no power. Probably the most famous concept relying on this principle is the Class-E amplifier.

This concept was proposed and patented by the Sokals in 1975 [67, 68]. The Class-E amplifier was derived to match the following conditions:

- account for device output capacitance
- minimize switching times
- voltage delay at switch turn-off
- zero voltage at switch turn-on
- zero voltage slope at switch turn-on
- flat top of voltage and current waveforms

These conditions are defined to make the concept useful in practical circuits. First, the condition of *zero voltage at switch turn-on* ensures that no capacitor is discharged when the transistor becomes a short circuit. Otherwise, all the stored energy would be lost. This, of course, requires *accounting for device output capacitance*, too. During the event of switching, the most power is lost. Thus, the output circuit is optimized to support *minimized switching times*. Finally, *voltage delay at switch turnoff* and *zero voltage slope at switch turn-on* ensure low voltage while the transistor is switching, thereby reducing

Table 8.7 Class F and Class F^{-1} For arbitrarily optimized reflective terminations.

Class F	90 conduction angle, harmonics tuned in order to force the voltage waveform into square pulses at the times where the current is zero
Class inverse F	basically Class F, but with voltage and current waveforms interchanged
Harmonically tuned	individually optimized current conduction angle and reflective terminations for higher harmonics

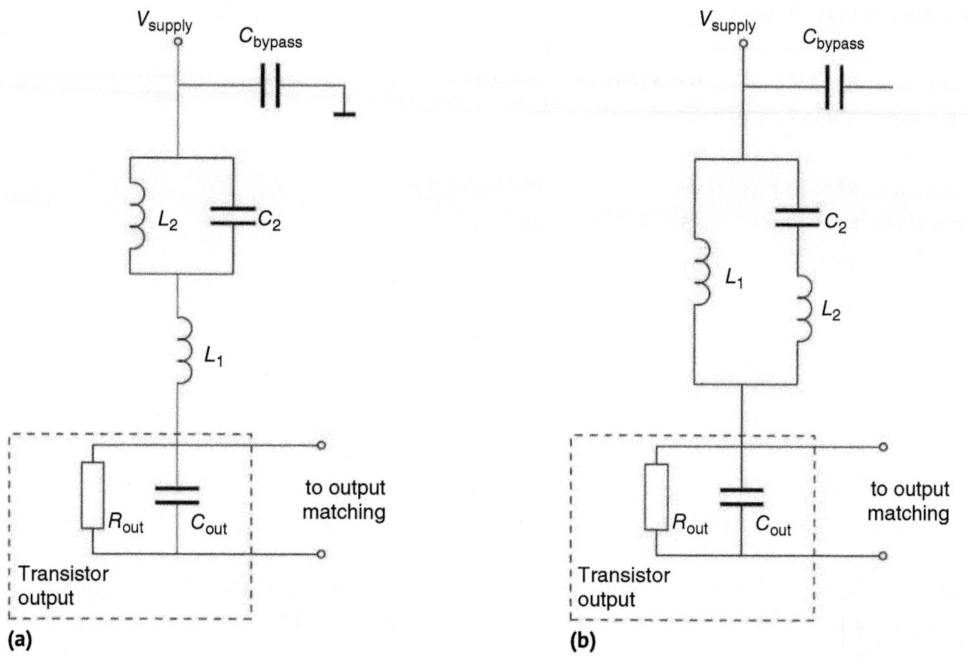

Figure 8.155 Class-F harmonic tuning circuits taking the transistor output capacitance into account, as proposed in. *Source*: Adapted from [66]. The dimensions of the elements are: (a) $L_1 = 1/(6\omega^2 C_{out})$, $L_2 = 5/3 \cdot L_1$, $C_2 = \frac{12}{5C_{out}}$; (b) $L_1 = 4/(9\omega^2 C_{out})$, $L_2 = 9/15 \cdot L_1$, $C_2 = 15/16 \cdot C_{out}$.

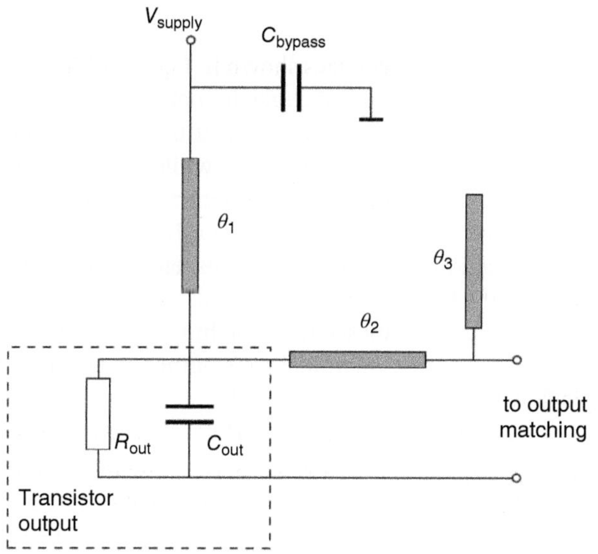

Figure 8.156 Class-F harmonic tuning circuits taking the transistor output capacitance into account, as proposed in. *Source*: Adapted from [66]. The lengths of the transmission lines are: $\theta_1 = \pi/2$, $\theta_2 = 1/3 \tan^{-1}[1/(3Z_0\omega_0 C_{out})]$, $\theta_3 = \pi/6$.

the dissipated power. *Flat top of voltage and current waveforms* has no impact on efficiency, but is required to protect the transistor from breakdown.

Figure 8.157 shows the idealized equivalent circuit of a first-order class-E amplifier. The transistor is acting as a switch. The transistor's output capacitance is not neglected but absorbed into the capacitance C_1, which is a significant advantage of the class-E concept.

Table 8.8 Theoretical maximum efficiency of Class-F Pas.

n harmonics of voltage	n = 1 (%)	n = 3 (%)	n = 5 (%)	n = ∞ (%)
m harmonics of current				
$m = 1$	50	58	60	64
$m = 2$	71	82	85	90
$m = 4$	75	87	91	96
$m = \infty$	79	91	95	100

Source: [64]/IEEE.

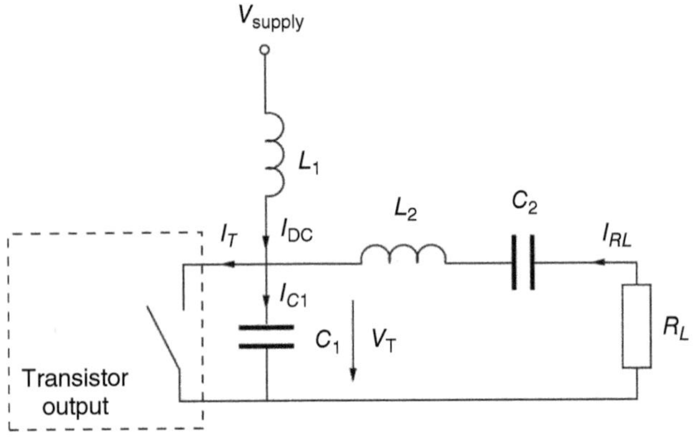

Figure 8.157 Circuit schematic of a first-order Class-E amplifier. The currents and voltages are shown in Figure 8.158.

In order to understand the principle of operation, we take a look at the currents and voltages shown in Figure 8.158. The currents in the collector (or drain) node are defined through the respective branches: The RF choke inductor L_1 forces I_{DC} to be a pure DC current. The series resonance of C_2 and L_2 guarantee that I_{RL} is a pure sinus signal at the fundamental frequency. The resulting current is a bias-shifted sine wave that is used to charge *and discharge* C_1 while the switch is open. Thus I_{C1} has a positive and a negative portion, and as a result, the voltage V_T builds up *and returns to zero* again. After the voltage became zero, the switch is closed, and for the remaining time of the period, the current I_T flows through the transistor while the voltage is ideally zero. Therefore, by appropriate selection of the element values, either current or voltage at the transistor will be zero, and no charged capacitance is shortened when switching. Thus, the class-E amplifier reaches the theoretical limit of $\eta = 100\%$. It is, however, usually not possible to actually see the switching of the current, even in circuit simulation. This is due to the fact that a major portion of C_1 might be the transistor's output capacitance. Hence, at the drain (or collector) node, the sum of I_T and part of I_{C1} is flowing and can't be separated.

It is, of course, necessary to choose the right circuit element values in order to achieve Class-E performance. Numerous works have been published that derive these values, e.g., [67–71]. Also, non-ideal behavior and current conduction angles other than the standard 50% duty cycle were investigated. Values for the class-E amplifier with a duty cycle of 50% are given in Table 8.9 [68]. With these relations between the different element values, an amplifier can be designed. It starts by choosing the DC voltage V_{CC}:

$$V_{\mathrm{CC}} = \frac{BV_{\mathrm{CEV}}}{3.65} SF \tag{8.197}$$

Where BV_{CEV} is the transistor's breakdown voltage, and $SF \leq 0.8$ is a security factor. Since the maximum voltage the transistor must withstand builds up when C_1 is charged, it is not limited to $2 \cdot V_{\mathrm{CC}}$ as for Class-A amplifiers. In Class-E, the voltage peaks at values in excess of $3 \cdot V_{\mathrm{CC}}$.

From a given V_{CC}, C_1, C_2, and R are determined from Table 8.9. This is an optimization procedure, as output power P should be maximized with realistic element values. Also, the transistor output capacitance constrains the minimum value

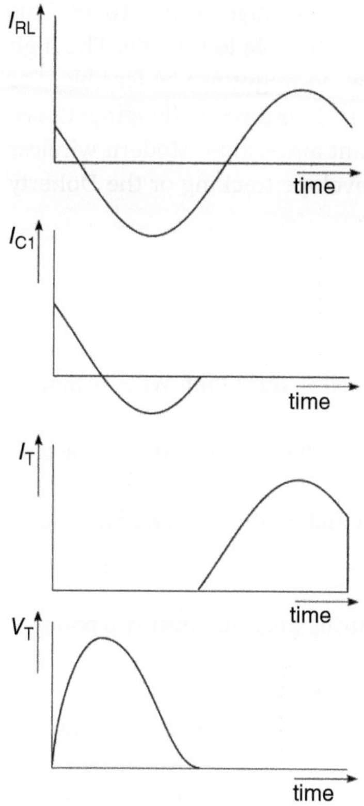

Figure 8.158 Currents and voltages of the Class-E amplifier shown in Figure 8.157.

Table 8.9 Dependence of Class-E circuit elements on loaded Q and output power.

Q_L	$\frac{P \cdot R}{(V_{CC} - V_0)^2}$	$\omega C_1 R$	$\omega C_2 R$
∞	0.576801	0.18360	0
20	0.56402	0.19111	0.05313
10	0.54974	0.19790	0.11375
5	0.51659	0.20907	0.26924
3	0.46453	0.21834	0.63467
2 .5	0.43550	0.22036	1.01219
2	0.38888	0.21994	3.05212
1 .7879	0.35969	0.21770	∞

Source: [68]/Australian National University.

for C_1. Table 8.9 also accounts for the saturation offset voltage V_0 that reduces the effective voltage swing. Finally, the value of L_2 is determined from the loaded Q of the series resonance circuit $Q_L = \omega L_2/R$.

It is the promise of the Class-E concept that an efficiency close to 100% can be achieved without any kind of transistor modeling. Provided that the transistor behaves like a decent switch, no load-pull measurement and individual power matching optimization is required. The transistor just becomes a switch, and the rest is determined through the other circuit elements, which are known a priori.

Regarding the different classes of amplifiers, it has to be noted that the definitions idealize the situation in a way that only a single frequency at a fixed amplitude is regarded. The issue of feedback, or how to drive the amplifier, is also not considered. But assuming a constant amplitude is much more severe. Modern communication standards use RF signals that don't fall into this category. It is rather common to have a bandwidth in excess of 10 MHz and a peak-to-average ratio of 10 dB. This means that the average output power is only 10% of the peak power that the amplifier must be able to provide. The high peak-to-average ratio means that, e.g., optimizing an amplifier to optimum Class-C operation might not be feasible. For tuned concepts like the Classes E and F approaches, the relative bandwidth of modern signals is also challenging. Class-E specifically is rather suited for narrowband PM/FM signals that inherently are of constant amplitude. Modern wireless power amplifiers, therefore, rely on linearization and efficiency boosting concepts like envelope tracking or the Doherty technique rather than on the pure single-transistor amplifier stage.

References

1 Rohde, U.L., Rudolph, M. (2013). *RF/Microwave Circuit Design for Wireless Applications*, 2e. New York: John Wiley & Sons., ISBN 1118431405.

2 Rohde, U.L., Pavio, A.M. and Pucel, R.A. (1988). Accurate noise simulation of microwave amplifiers using CAD. *Microwave Journal* 31 (12): 130–141.

3 Vendelin, G.D., Pavio, A.M., Rohde, U.L. et al. (1990). Microwave circuit design using linear and nonlinear techniques, 3e. New York: John Wiley & Sons, ISBN 0-471-58060-0.

4 Friis, H.T. (1944). Noise figures of radio receivers. *Proceedings of the IRE* 32: 419–422.

5 Haus, H.A., Atkinson, W.R., Branch, G.M. et al. (1960). IRE standards on methods of measuring noise in linear two ports. *Proceedings of the IRE* 48: 60–68.

6 Rothe, H. and Dhalke, W. (1956). Theory of noisy four poles. *Proceedings of the IRE* 44: 811–818.

7 Haus, H.A., Atkinson, W.R., Branch, G.M. et al. (1960). Representation of noise in linear two ports. *Proceedings of the IRE* 48 (1): 69–74.

8 Haus, H.A. and Adler, R.B. (1959). *Circuit Theory of Linear Noisy Networks*. Cambridge, MA: MIT Press; New York: Wiley.

9 Fukui, H. (1966). The noise performance of microwave transistors. *IEEE Transactions on Electron Devices* ED-13 (3): 329–341.

10 Iversen, S. (1975). The effect of feedback on noise figure. *Proceedings of the IEEE* 63 (3): 540–542.

11 Besser, L. (1975). Stability considerations of low-noise transistor amplifiers with simultaneous noise and power match. *IEEE MTT-S International Microwave Symposium Digest*, 327–329.

12 Vendelin, G.D. (1975). Feedback effects on the noise performance of GaAs MESFETs. *IEEE MTT-S International Microwave Symposium Digest*, 324–326.

13 Suter, W.A. (1983). Feedback and parasitic effects on noise. *Microwave Journal* 26 (2): 123–129.

14 Hartman, K. and Strutt, M.J.O. (1973). Changes in four noise parameters due to general changes in linear two-port circuits. *IEEE Transactions on Electron Devices* ED-20 (10): 874–877.

15 Fukui, H. (1966). Available power gain, noise figure, and noise measure of two-ports and their graphical representation. *IEEE Transactions on Circuit Theory* CT-13 (2): 137–142.

16 Poole, C.R. and Paul, D.K. (1985). Optimum noise measure terminations for microwave transistor amplifiers. *IEEE Transactions on Microwave Theory and Techniques* MTT-33 (11): 1254–1257.

17 Hillbrand, H. and Russer, P.H. (1976). An efficient method for computer aided noise analysis of linear amplifier networks. *IEEE Transactions on Circuits and Systems* CAS-23 (4): 235–238.

18 Pastori, W.E. (1983). Topics on noise. *Los Angeles: Eaton Electronics Instrumentation Division, (EID) Seminar Notes*, Los Angeles.

19 Pastori, W.E. (1983). Image and second stage corrections resolve noise figure measurement confusion. *Microwave Systems News*, 67–86.

20 Lane, R.Q. (1969). The determination of device noise parameters. *Proceedings of the IEEE* 57 (8): 1461–1462.

21 Lane, R.Q. (1978). A microwave noise and gain parameter test set. *IEEE International Solid-State Circuits Conference*, 172–173.

22 Kuhn, N. (1980). Accurate and automatic noise figure measurements with standard equipment. *IEEE MTT-S International Microwave Symposium Digest*, 425–427.

23 Strid, E.W. (1981). Measurement of losses in noise-matching networks. *IEEE Transactions on Microwave Theory and Techniques* MTT-29 (3): 247–252.

24 Yuan, J.S. (1999). *SiGe, GaAs, and InP Heterojunction Bipolar Transistors*, Wiley Series in Microwave and Optical Engineering. New York: Wiley.

25 Schmacher, H., Gruhle, A., Erben, U. et al. (1995). A 3V supply voltage, DC–18 GHz SiGe HBT wideband amplifier. *BCTM Technical Digest*, 190.

26 Sheng, N.H., Ho, W.J., Wang, N.L. et al. (1991). A 30 GHz bandwidth AlGaAs-GaAs HBT direct-coupled feedback amplifier. *IEEE Microwave and Guided Wave Letters* 8: 208.

27 Zhou, G. and Zheng, L. *Design and Analysis of Microwave Feedback Amplifiers*. Berkeley, CA: University of California.

28 Vendelin, G.D., Pavio, A.M. and Rohde, U.L. (2005). *Microwave Circuit Design Using Linear and Nonlinear Techniques*, 2e. Hoboken, NJ: John Wiley & Sons.

29 Dhalke, W. (1956). Transformation der Rauschkennwerte durch Netzwerke. In: Rothe, H. (ed.), *Die Telefunken-Röhre* vol. 33 30–45.

30 Nguyen, T.-K., Kim, C.-H., Ihm, G.-J. et al. (2004). CMOS low-noise amplifier design optimization techniques. *IEEE Transactions on Microwave Theory and Techniques* 52 (5): 1433–1442, and (2004). Comments on "CMOS low-noise amplifier design optimization techniques." *IEEE Transactions on Microwave Theory and Techniques* 52 (7): 3155.

31 van der Ziel, A. (1986). *Noise in Solid State Devices and Circuits*. New York: Wiley.

32 Lu, L.-H. and Chen, H.-S. (2010). Lower the voltage for CMOS RFIC. *IEEE Microwave Magazine*, 70–77.

33 Manku, T., Beck, G. and Shin, E.J. (1998). A low-voltage design technique for RF integrated circuits. *IEEE Transactions on Circuits and Systems II* 45 (10): 1408–1413.

34 Tsang, T.K.K. and El-Gamal, M.N. (2002). Gain and frequency controllable sub-1 V 5.8 GHz CMOS LNA. *IEEE International Circuits and Systems Symposium* 4: IV-795–IV-798.

35 Hsieh, H.-H., Wang, J.-H. and Lu, L.-H. (2008). Gain-enhancement techniques for CMOS folded cascode LNAs at low-voltage operation. *IEEE Transactions on Microwave Theory and Techniques* 56 (8): 1807–1816.

36 Hsieh, H.-H. and Lu, L.-H. (2007). Design of ultra-low-voltage RF front ends with complementary current-reused architectures. *IEEE Transactions on Microwave Theory and Techniques* 55 (7): 1445–1458.

37 Debski, W.A. (2007). Multi-band adaptive WLAN receivers in 0.13 μm CMOS. Ph.D. thesis. Brandenburg University of Technology, Cottbus, Germany, http://opus.kobv.de/btu/volltexte/2007/368/

38 Ansoft Corporation (2001). Serenade examples volume, version 8.7

39 Motorola application note AN-721 ("Impedance Matching Networks Applied to RF Power Transistors").

40 Motorola Application Note AN-282A, *Systemizing RF Power Amplifier Design*.

41 Everitt, W.E. and Anner, G.E. *Communication Engineering*. McGraw-Hill Book Company, Inc.

42 Matthaei, G.L., Young, L., and Jones, E.M.T. *Microwave Filters, Impedance-Matching Networks and Coupling Structures*. McGraw-Hill Book Company, Inc.

43 Matthaei, G.L. (1964). Tables of Chebyshev impedance-transforming networks of low-pass filter form. *Proceedings of the IEEE*.

44 Motorola Application Note AN-267, *Matching Network Designs with Computer Solutions*.

45 Cristal, E.G. (1965). Tables of maximally flat impedance-transforming networks of low-pass filter form. *IEEE Transactions on Microwave Theory and Techniques* MTT-3 (5): 693–695.

46 Bode, H.W. (1975). *Network Analysis and Feedback Amplifier Design*. New York: D. Van Nostrand Company.

47 Fano, R.M. (1950). Theoretical limitations on the broadband matching of arbitrary impedances. *Journal of the Franklin Institute* 249 (1): 57–83.

48 Horwitz, J.H. (1969). Design wideband UHF-power amplifiers. *Electronic Design* 11.

49 Pitzalis, O. and Gilson, R.A. (1971). Tables of impedance matching networks which approximate prescribed attenuation versus frequency slopes. *IEEE Transactions on Microwave Theory and Techniques* MTT-9 (4): 381–386.

50 Gerkis, A.N. (1998). Broadband impedance matching using the "real frequency" network synthesis technique. *Applied Microwave & Wireless*: 10: 26–36.

51 Cuthbert, T. (1994). Broadband impedance matching methods. *RF Design* 17 (8): 64–71.

52 Cuthbert, T. (1999). *Broadband Direct-Coupled and Matching RF Networks*.

53 Vendelin, G.D., Pavio, A.M. and Rohde, U.L. (1990). *Microwave Circuit Design Using Linear and Nonlinear Techniques*, 279–283. New York: John Wiley & Sons, ISBN 0-471-58060-0.

54 Potter, A. (1999). HP RF compiler automates schematic capture and extends capabilities of circuit synthesis. *Applied Microwave & Wireless* 11 (6): 106–118.

55 Statz, H., Haus, H. and Pucel, R.A. (1974). Noise characteristics of gallium arsenide field-effect transistors. *IEEE Transactions on Electron Devices* ED-21: 549–562.

56 Kain, A. Final report for an integrated CAD tool for high-power, high-efficiency RF and microwave amplifiers, STTR Phase II, contract no. F33615-95-C-174, Dept. of the Air Force, Small Business Technology Transfer (STTR) Program.

57 Vasylyev (2006). Integrated RF power amplifier design in silicon-based technologies. Ph.D. thesis. Cottbus: Brandenburg University of Technology. http://opus.kobv.de/btu/volltexte/2006/20/.

58 Aparicio, R. and Hajimiri, A. (2002). Capacity limits and matching properties of integrated capacitors. *IEEE Journal of Solid-State Circuits* 37 (3): 384–393.

59 The Predictive Technology Model (PTM). http://mec.umn.edu/ptm.

60 Silver, J.P. *RF, RFIC & Microwave Theory, Design*. http://www.odyseus.nildram.co.uk/. This website provides a number of highly interesting RFIC design tutorials and examples.

61 Cantz, R. (1953). Hochfrequenzverstärkung mit trioden. *Telefunken-Röhre* 30: 52–69(in German).

62 Tyler, V.J. (1958). A new high-efficiency high-power amplifier. *Marconi Review* 21: 96–109.

63 Snider, D.M. (1967). A theoretical analysis and experimental confirmation of the optimally loaded and overdriven RF power amplifier. *IEEE Transactions on Electron Devices* ED-14: 851–857.

64 Raab, F.H. (1997). Class-F power amplifiers with maximally flat waveforms. *IEEE Transactions on Microwave Theory and Techniques* 45: 2007–2012.

65 Gao, S. (2006). High-efficiency class-F RF/microwave amplifiers. *IEEE Microwave Magazine*, 40–48.

66 Grebennikov, A.V. (2000). Circuit design technique for high efficiency class-F amplifiers. *IEEE MTT-S International Microwave Symposium Digest*, Boston, MA, June 11–16, 2000, vol. 2 771–774.

67 Sokal, N.O. and Sokal, A.D. (1975). Class E – a new class of high-efficiency tuned single-ended switching power amplifiers. *IEEE Journal of Solid-State Circuits* SC-10: 168–176.

68 Sokal, N.O. (2001). Class-E RF power amplifiers. *QEX*: 9–20.

69 Raab, F.H. (1977). Idealized operation of the class E tuned power amplifier. *IEEE Transactions on Circuits and Systems* CAS-24: 725–735.

70 Kazimierczuk, M.K. and Puczko, K. (1987). Exact analysis of class E tuned power amplifier at any Q and switch duty cycle. *IEEE Transactions on Circuits and Systems* CAS-34: 149–159.

71 Kessler, D.J. and Kazimierczuk, M.K. (2004). Power losses and efficiency of class-E power amplifier at any duty ratio. *IEEE Transactions on Circuits and Systems I* 51 (9): 1675–1689. https://doi.org/10.1109/TCSI.2004.834505

Further Reading

Abrie, P.L.D. (1985). *The Design of Impedance-Matching Networks for Radio-Frequency and Microwave Amplifiers*. Norwood, MA: Artech House.

Abrie, P.L.D., Rademeyer, P. et al. (1985). A method for evaluating and the evaluation of the influence of the reverse transfer gain on the transducer power gain of Some Microwave Transistors. *IEEE Transactions of Microwave Theory and Techniques* MTT-33 (8): 711–713.

Ambrózy, A. (1982). *Electronic Noise*. New York: McGraw-Hill.

Baechtold, W. and Strutt, M.J.O. (1968). Noise in microwave transistors. *IEEE Transactions on Microwave Theory and Techniques* MTT-16 (9): 578–585.

Cappy, A. (1988). Noise modeling and measurement techniques. *IEEE Transactions on Microwave Theory and Techniques* MTT-36: 1–10.

Carson, R.S. (1975). *High Frequency Amplifiers*. New York: John Wiley & Sons.

Carioca, C.A.A. and de Carvalho, P.H.P. (1999). Computer-aided design of diode frequency multipliers. *Applied Microwave & Wireless* 11: 38–46.

Carlin, H.J. (1977). A new approach to gain-bandwidth problems. *IEEE Transactions on Circuits and Systems* CAS-23: 170–175.

Carlin, H. J. and Amstutz, P. (1977). On optimum broad-band matching. *IEEE Transactions on Circuits and Systems* CAS-23401–405.

Carlin, H.J. and Komiak, J.J. (1979). A new method of broad-band equalization applied to microwave amplifiers. *IEEE Transactions on Microwave Theory and Techniques* MTT-27: 93–99.

Carlin, H.J. and Amstutz, P. (1981). On optimum broadband matching. *IEEE-CAS* 28 (5): 401–405.

Carlin, H.J. and Yarman, B. (1982). A simplified real frequency technique applied to broad-band multistage amplifiers. *IEEE-MTT* 30 (12): 2216–2222.

Carlin, H.J. and Yarman, B.S. (1983). The double matching problem: analytic and real-frequency solutions. *IEEE Transactions on Circuits and Systems* CAS-30 (1): 15–28.

Caruso, G. and Sannino, M. (1978). Computer-aided determination of microwave two-port noise parameters. *IEEE Transactions on Microwave Theory and Techniques* MTT-26: 639–642.

Compact Software. (1988). *Complex Math User's Guide*. Paterson, NJ.

Cuthbert, T. (1983). *Circuit Design Using Personal Computers*. John Wiley & Sons.

Eisenberg, J.A. (1974). Designing amplifiers for optimum noise figure. *Microwaves* 13: 36–41.

Fano, R.M. (1960). Theoretical limitations on the broadband matching of arbitrary impedances. *Journal of the Franklin Institute* 249: 57–83, 139–155.

Fukui, H. (1981). *Low-Noise Microwave Transistors and Amplifiers*. New York: IEEE Press.

Gentili, C. (1987). *Microwave Amplifiers and Oscillators*. New York: McGraw-Hill.

Georgi, S., Rohde, U.L. (1997). A novel approach to voltage-controlled tuned filters using CAD validation. *Microwave Engineering Europe*, 35–42.

Gonzales, G. (1984). *Microwave Transistor Amplifier Analysis and Design*. Englewood Cliffs: Prentice-Hall.

Haus, H.A. and Adler, R.B. (1958). Optimum noise performance of linear amplifiers. *Proceedings of the IRE* 46: 1517–1533.

Kain, A. (1997). *Final Report for Bias Dependence Noise Modeling of Heterojunction Bipolar Transistors*, USAF SBIR Phase II (PIIN), F33615-95-C-1707, November 1997. Issued by USAF/AFMC/ASC, Wright Laboratory WL/AAKE BLD 7, 2530 C ST, Wright-Patterson AFB, OH 45433-7607.

Köppen, M.J. (1970). A single-stage power amplifier for 17 watts output in the 470-MHz band using the 266BLY, operating from 12.5 volts supply voltage. C. A. B. Communications, Eindhoven, Holland, Electronic Application Laboratory Report, August 8, 1970.

Kuhn, W.B. and Yanduru, N.K. (1999). Spiral inductor substrate loss modeling in silicon RFICs. *Microwave Journal*, 66–81.

Lange, J. (1967). Noise characterization of linear two-ports of invariant parameters. *IEEE Journal of Solid-State Circuits* SC-2: 37–40.

Liu, S.Y. and Chuang, H.R. (1998). A 2.4-GHz transceiver RF front-end for ISM-band digital wireless communications. *Applied Microwave & Wireless*: 32–48.

Lovelace, D., Camilleri, N., and Kannell, G. (1994). Silicon MMIC inductor modeling for high volume, low cost applications. *Microwave Journal* 60–71.

Lucek, J. and Damen, R. (1999). Designing an LNA for a CDMA front end. *RF Design* 20–30.

Maas, S.A. (1999). FET models for Volterra-series analysis. *Microwave Journal* 260–270.

Marzall Laila, "Teaching GaAs and GaN MMIC Design: An Education Project Between WIN Semiconductors and UC Boulder" *IEEE Microwave Magazine*, 27 (8), pp 71, August 2026.

McClure, D.A. (1997). Transmission-line transformers. *Communications Quarterly* 45–58.

Mellor, J. and Linvill, J.G. (1975). Synthesis of interstage networks and prescribed gain versus frequency slopes. *IEEE Transactions on Microwave Theory and Techniques* MTT-23 (12): 1013–1020.

Mercer, S. (1999). Exact simulation of LNAs reduces design cycle time. *Applied Microwave & Wireless*: 74–82.

Meys, R.P. (1978). A wave approach to the noise properties of linear microwave devices. *IEEE Transactions on Microwave Theory and Techniques* MTT-26: 34–37.

Miller, K.S., Daywitt, W.C., and Arthur, M.G. (1967). Noise standards, measurements and receiver noise definitions. *Proceedings of the IEEE* 55: 865–877.

Minnis, B.J. (1996). *Designing Microwave Circuits by Exact Synthesis*. Boston: Artech House, ISBN 0-89006-741-4.

Minton, R. (1964). Design of large signal VHF transistor power amplifiers. RCA Application Note SMA-36.

Minton, R. and Lee, H.C. (1965). Frequency multiplication using overlay transistors. RCA Application Note SMA-40.

Mitama, M. and Katoh, H. (1979). An improved computational method for noise parameter measurement. *IEEE Transactions on Microwave Theory and Techniques* MTT-27: 612–615.

Motorola Tutorial. (1999). The cascode amplifier: a useful RF building block. *Microwave Product Digest* 39, 59.

Mukaihata, T. (1968). Applications and analysis of noise generation in N-cascaded mismatched two-port networks. *IEEE Transactions on Microwave Theory and Techniques* MTT-16: 699–708.

Mumford, W.W. and Scheibe, E.H. (1968). *Noise Performance Factors in Communications Systems*. Dedham, MA: Horizon House.

Norton, E. (1976). High dynamic range transistor amplifier using lossless feedback. *Microwave Journal* 53–57.

Norton, D.E. Strong signal survivability design for lossless feedback amplifiers. *Proceedings of the European Microwave Conference.*

Okean, H.S. and Lombardo, P.O. (1973). Noise performance of M/W and MM-wave receivers. *Microwave Journal* 16 (1).

Otoshi, T.Y. (1968). The effect of mismatched components on microwave noise-temperature calibrations. *IEEE Transactions on Microwave Theory and Techniques* MTT-16: 675–686.

Penfeld, P. (1962). Wave representations of amplifier noise. *IRE Transactions on Circuit Theory*: 84–86.

Rohde, U.L., Whitaker, J.C., and Bucher, T.T.N. (1997). Chapter 3: Receiver system planning. In *Communications Receivers: Principles and Design*, 2e. New York: McGraw-Hill, ISBN 0-07-053608-2.

Rorabaugh, C. (1992). *Circuit Design and Analysis–Featuring C Routines.* McGraw-Hill.

Sannino, M. (1979). On the determination of device noise and gain parameters. *Proceedings of the IEEE* 67: 1364–1366.

Radom, Inc. SYNMAT matching synthesis program, developed by Douglas Mellor.

Van der Ziel. (1986). *Noise in Solid State Devices and Circuits.* New York: Wiley.

Vondran, D. (1999). Noise figure measurement: corrections related to match and gain. *Microwave Journal* 22–38.

Wait, D.F. (1968). Thermal noise from a passive linear multiport. *IEEE Transactions on Microwave Theory and Techniques* MTT-16: 687–691.

Yarman, S. and Carlin, H.J. (1982). A simplified real frequency technique applied to broad-band multistage microwave amplifiers. *IEEE Transactions on Microwave Theory and Techniques* MTT-30 (12): 2216–2222.

Young, P. and Scanlan, S.O. (1981). Matching network design studies for microwave transistor amplifiers. *IEEE Transactions on Microwave Theory and Techniques* MTT-29 (10): 1027–1035.

Niclas, K.B. (1980). GaAs MESFET feedback amplifier: design consideration and characteristics. *Microwave Journal* 39–40.

Niclas, K.B. (1980). The matched feedback amplifier: ultrawide-band microwave amplification with GaAs MESFET's. *IEEE Transactions on Microwave Theory and Techniques* MTT-28 (4): 285–294.

Perez, F. and Ortega, V. (1981). A graphical method for the design of feedback networks for microwave transistor amplifiers: theory and applications. *IEEE Transactions on Microwave Theory and Techniques* MTT-29 (10): 1019–1026.

Bor, S.S. et al. (1996). Using feedback techniques to design a stable and matching condition for microwave transistor amplifiers. *International Journal of Electronics* 81 (6): 713–721.

Schwaz. (1999). EE217 class MESFET Transistor, Spring.

Pozar, D.M. (1998). *Microwave Engineering*, 2e, 252–254. New York: Wiley.

9

Mixer Design

9.1 Introduction

Radiocommunication requires that we shift a baseband information signal to a frequency or frequencies suitable for electromagnetic propagation to the desired destination. At the destination, we reverse this process, shifting the received radio frequency (RF) signal back to the baseband to allow the recovery of the information it contains. This frequency-shifting function is traditionally known as *mixing* and the stages that perform it as *mixers*. Any device that exhibits amplitude-nonlinear behavior can serve as a mixer because nonlinear distortion results in the production of signals at new frequencies from the signals present at the input of a device. Even a rusty screw or bolt on an antenna element can act as a mixer, producing unwanted intermodulation distortion (IMD) products that appear at the receiver input.

Although mixers are equally important in wireless transmission and reception, traditional mixer terminology favors the receiving case because mixing, as such, was first applied in receiving applications. Thus, the signal to be frequency-shifted is applied to the mixer's RF port, and the frequency-shifting power or voltage from a *local oscillator* (LO) is applied to the mixer's LO port, resulting in two outputs at the mixer's *intermediate frequency* (IF) port. If the wanted IF is lower in frequency than the RF signal, the mixer is a *downconverter*; if the wanted IF is higher than the RF, the mixer is an *upconverter*. *Converter* may also be used as a term for a single stage that simultaneously acts as a mixer and LO.

For a given RF signal, an ideal mixer with a perfect LO (i.e., an LO with no harmonics and no noise sidebands) would produce only two IF outputs: one at the frequency sum of the RF and LO, and another at the frequency difference between the RF and LO. Filtering can be used to select the desired IF product and reject the unwanted one, which is sometimes referred to as the *IF image* [1–4].

The simultaneous generation of LO + RF and LO − RF outputs results not from a departure of mixer performance from the ideal, but from the mathematics of mixing itself. Another unavoidable mixing artifact, the *RF image* response, also results from the mathematics of mixing rather than mixer nonideality. Just as a given RF/LO combination produces two IF outputs (LO + RF and LO − RF, the IF and IF image), the mixer will produce output at the desired IF (LO + RF or LO − RF) in response to *two* possible RF inputs: one at LO + IF and another at LO − IF (Figure 9.1). The undesired response, the *RF image* (traditionally referred to merely as the *image*), is $2f_{\mathrm{IF}}$ removed from the desired response. Even if no man-made signals exist at the RF image frequency, reducing a mixer's RF image response can be important because noise at that frequency, including that produced by circuitry between the mixer and antenna, will still be mixed to the desired IF,

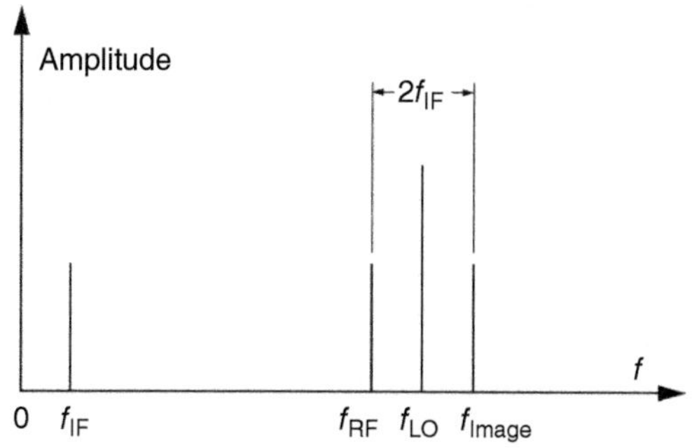

Figure 9.1 Relationship between a mixer's image and desired signal responses. The image is $2f_{\mathrm{IF}}$ away from the desired signal [4] / John Wiley & Sons.

degrading the signal-to-noise ratio. Filtering and phasing techniques can be used to reduce the RF or IF image responses: filtering if the image is sufficiently removed from the desired response, for filtering to provide the necessary rejection, and phasing if the desired and image responses are insufficiently spaced for filtering to work, as in the case of a double-conversion receiver in which signals at a high first IF (e.g., 50–70 MHz), must be converted to a very low first IF, such as 25 kHz.

The output of every real mixer includes a vast number of additional unwanted products, including noise, the fundamentals of the mixer's RF and LO signals and their harmonics, and the sums and differences of the RF and LO and their harmonics. IMD between multiple signals present at the RF port and IF output, resulting from the mixing of LO noise-sideband energy to IF by strong adjacent signals (*reciprocal mixing*), further complicate a mixer's output spectrum and may compromise system performance [4].

9.2 Simulation Examples of Different Types of Downconverters

9.2.1 Simulation Example of Downconverter to Explain Its Noise Analysis

The following discussion about system noise and large signal analysis was in part triggered by the Maxim Integrated Tutorial 5594, "System Noise Figure Analysis for Modern Radio Receivers" [1], and has been enhanced to show nonlinearities.

Let us get started with the block diagram as shown in Figure 9.2a. Here the purpose is to show that the image frequency band of the mixer can add noise to the overall system and for wideband active mixers, where the Friis formula is not applicable. The following will be a set of system analyses using a system simulator.

At the output, there is no selectivity, so all the noise contributions are added. In addition to the noise analysis, we will also look at the budget and distortion analysis.

At the center frequency of 2 GHz of Figure 9.2b, the result assumes a perfect match between the stages. In the simulator, realistic matching conditions are considered; see the noise figure (NF) and gain figure (Figure 9.2b), so there is a minor difference between [1] and the simulated result shown in Figure 9.2b.

Note that the applicable noise figure (NF) is neither the double-sideband (DSB) nor the single-sideband (SSB) noise figure of the mixer. Instead, it is an effective noise figure that lies somewhere between these two values. In this case, with a DSB noise figure of 3 dB, the equivalent noise figure of the two-port network can be calculated to be 4.757 dB, as already noted in Ref. [1]. Using this value in the overall cascade calculation (as per [1]) results in a system noise figure of 7.281 dB.

Manual calculations as per [1] show that this result is consistent with the standard Friis equation, using 4.757 dB for the mixer noise figure.

The gain distribution is shown in Figure 9.2c and the two-tone analysis is shown in Figure 9.2d.

In general, when substituting an equivalent two-port network for the mixer and its adjacent components, the input port should be the latest node in the signal flow for which the image response is rejected. The output port should be the earliest node where the image and wanted responses are combined (usually the output port of the mixer).

If the image response of the mixer is not effectively suppressed by the architecture, then the Friis equation cannot be used without modification.

9.2.2 Simulation Example: Proposed Downconverter Architecture for a Frequency Band at 10.5 GHz

For radio amateurs, Figure 9.3a shows a proposed downconverter architecture for a frequency band at 10.5 GHz. The IF of 435 MHz is really too low to get a good image rejection, but overall the performance of this downconverter is acceptable. For the input filter, an interdigital filter design is used. This is shown in Figure 9.3a.

And by the way, a good systems simulator usually has a filter synthesis program incorporated.

Because of the high resolution, the NF in Figure 9.3b seems to vary a lot, but the variation is really insignificant. In the pass band, the noise figure changes from 2.1 to 1.9 dB (average NF is 2 dB).

The gain distribution and the intermodulation characteristics are shown in Figure 9.3c,d.

Because of the high numerical resolution of the simulator used, we can look as low as −300 dBm, but in practice and for this drive level, this would not matter, as any value of the spurious below −100 dBm would not be noticeable, as the mathematical resolution is beyond the capabilities of the test equipment.

9.2.3 Simulation Example: Dual-conversion Receiver

Finally, for the same noise and gain budget analysis, a dual-conversion receiver is shown in Figure 9.4a.

The aforementioned examples, shown in Figures 9.2–9.4, are generated using the Ansoft Symphony Program.

Figure 9.5 shows the noise formula for different structures, their applications, and the relevant equations.

9.3 Additive or Multiplicative Mixing

All mixers are *multipliers* in the sense that the various new outputs they produce can be described mathematically as the multiplicative products of their inputs. From an implementation standpoint, however, a given mixer circuit can be characterized as additive or multiplicative depending on how RF and LO signals are applied to it. Additive mixing occurs when the RF and LO signals are applied to the same nonlinear device, as in all diode- or transistor-based mixers, as shown in

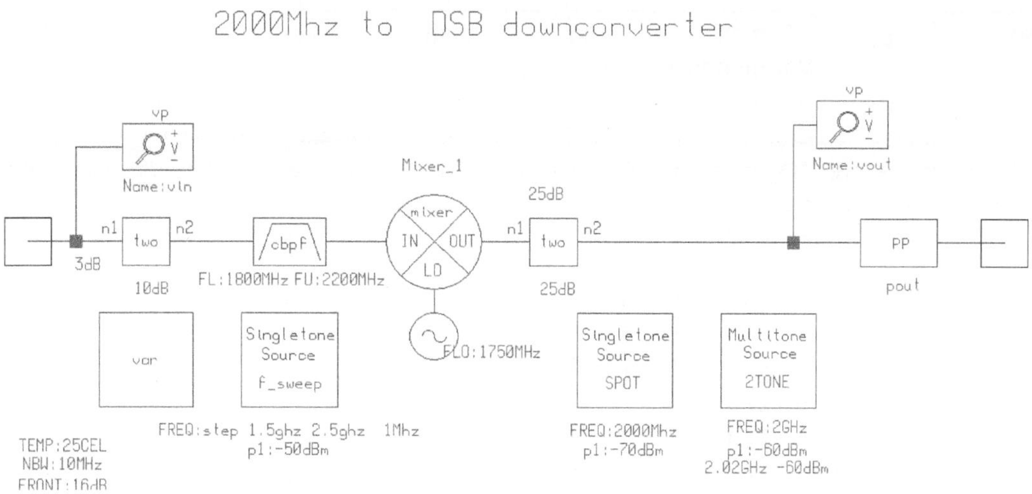

(a)

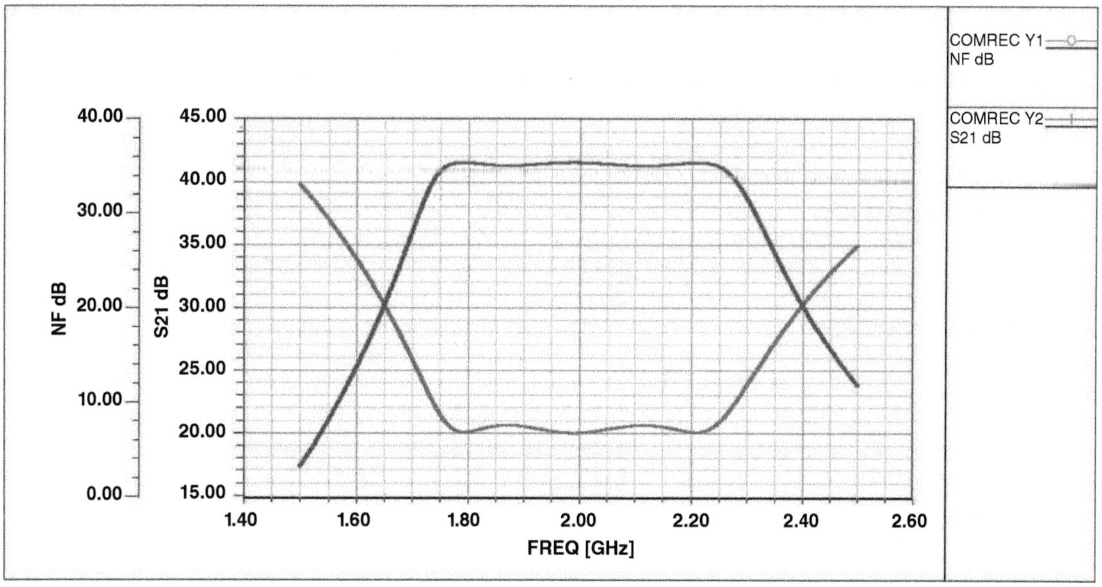

(b)

Figure 9.2 (a) Block diagram of a downconverter that will be used to explain the noise analysis. (b) Simulated gain and noise performance of the downconverter of (a).

(Continued)

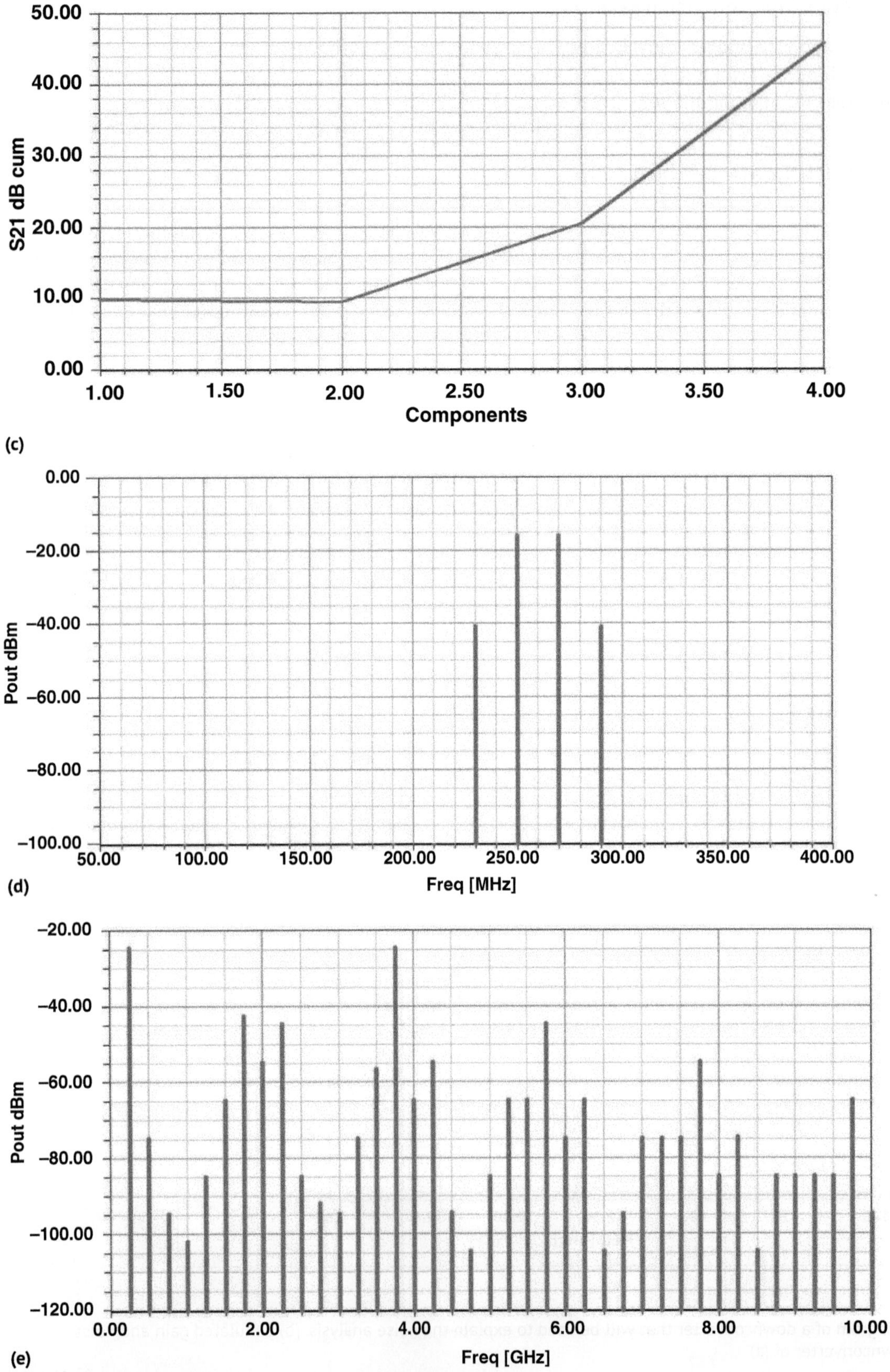

Figure 9.2 (Cont'd) (c) Gain distribution of the downconverter of (a) for stages 1–4. (d) Two-tone analysis of the downconverter of (a). (e) Two-tone (wideband) analysis of the downconverter of (a).

Figure 9.6a,b. Multiplicative mixing occurs when the RF and LO signals are applied to separate semiconductors, as in Figure 9.6c. As a rule, multiplicative mixers afford better isolation between their LO and RF ports than additive mixers, and this enhanced interport isolation is their principal merit. Multiplicative mixing does not in itself suppress unwanted products; the spurious response of a basic multiplicative mixer cell is poor unless it is used in a push–pull or quad configuration [4, 25].

The lower transistor in the dual-gate arrangement is a constant current source, so the second transistor does not add to the nonlinearities. The top transistor is essentially an ON-OFF switch, so the RF and LO currents are separated, and the output has the LO power and the sum and difference powers of the input frequency with the LO.

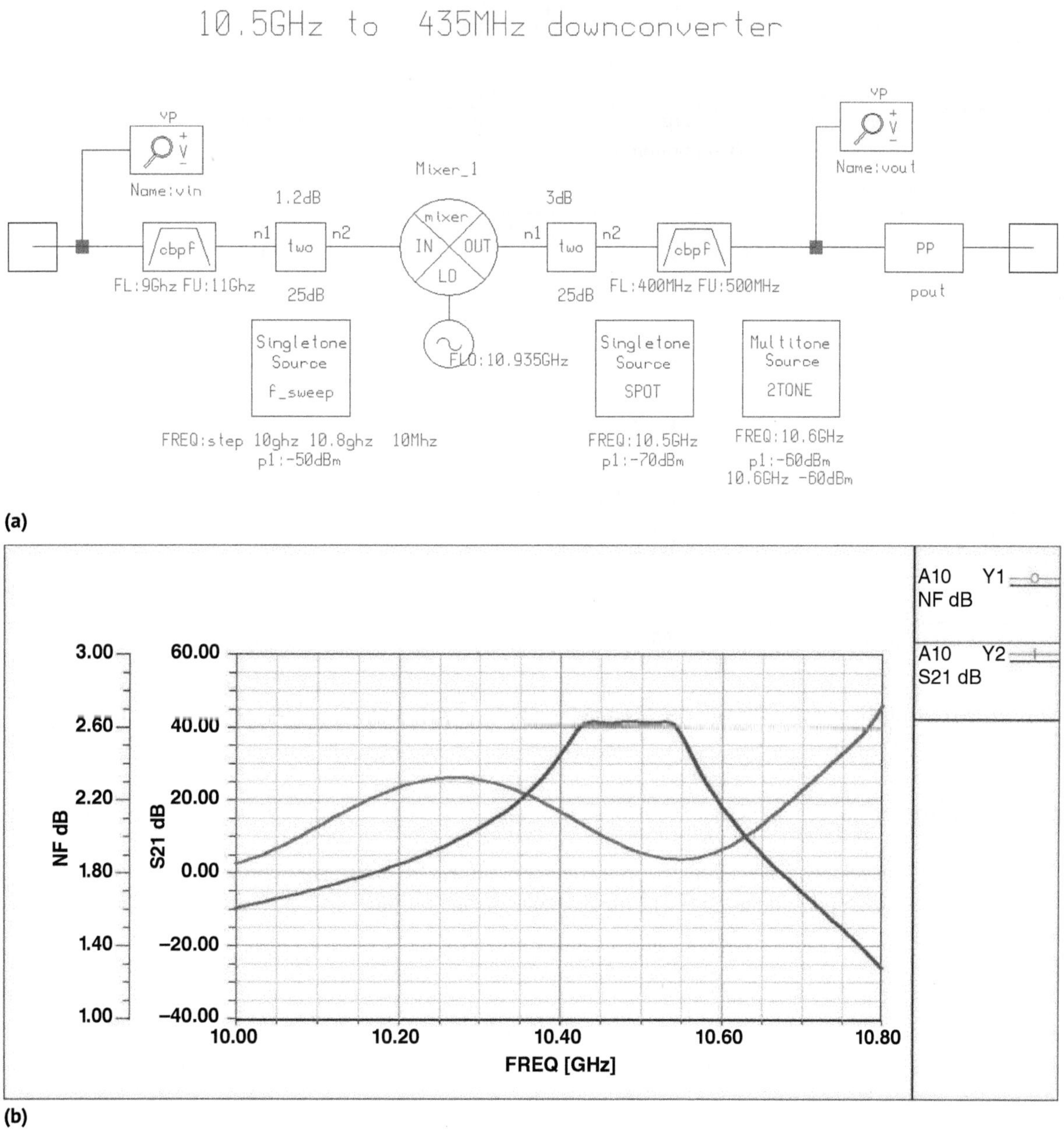

Figure 9.3 (a) Block diagram of the 10.5 GHz–435 MHz downconverter. (b) Simulation of the gain and noise figure of the 10.5 GHz converter of (a).

(Continued)

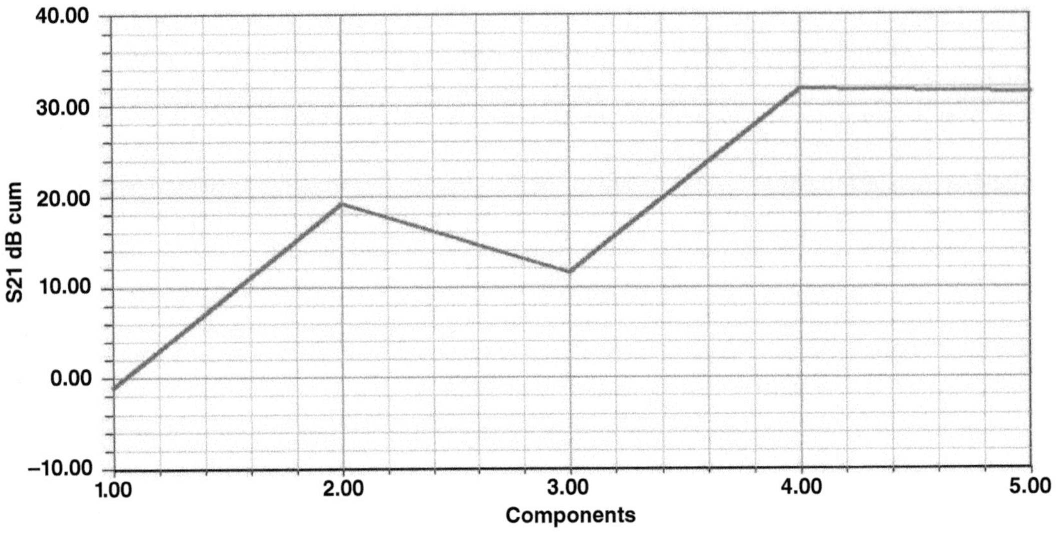

(c)

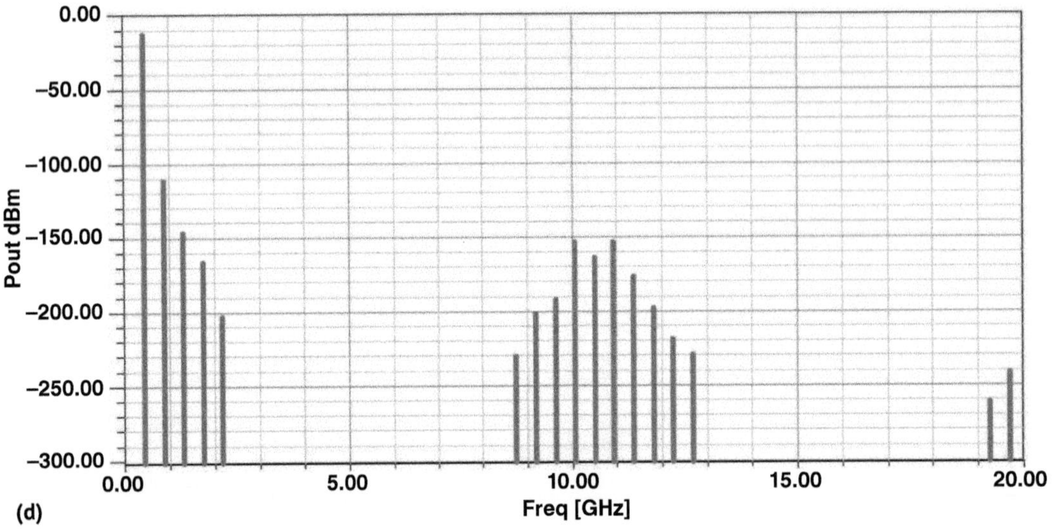

(d)

Figure 9.3 (Cont'd) (c) Gain distribution of the 10.5 GHz converter of (a) between stages 1 and 5. (d) Intermodulation product simulation of the 10.5 GHz converter of (a).

9.4 Basic Mixer Theory

Having seen the examples earlier, let us now consider the basic theory of mixers [4]. Mixing is achieved by the application of two signals to a nonlinear device. Depending upon the particular device, the nonlinear characteristic may differ. However, it can be generally expressed in the form:

$$I = K(V + \upsilon_1 + \upsilon_2)^n \tag{9.1}$$

The exponent n is not necessarily an integer, V may be a DC offset voltage, and the signal voltages υ_1 and υ_2 may be expressed as $\upsilon_1 = V_1 \sin(\omega_1 t)$ and $\upsilon_2 = V_2 \sin(\omega_2 t)$.

When $n = 2$ (9.1) may then be written as:

$$I = K[V + V_1 \sin(\omega_1 t) + V_2 \sin(\omega_2 t)]^2 \tag{9.2}$$

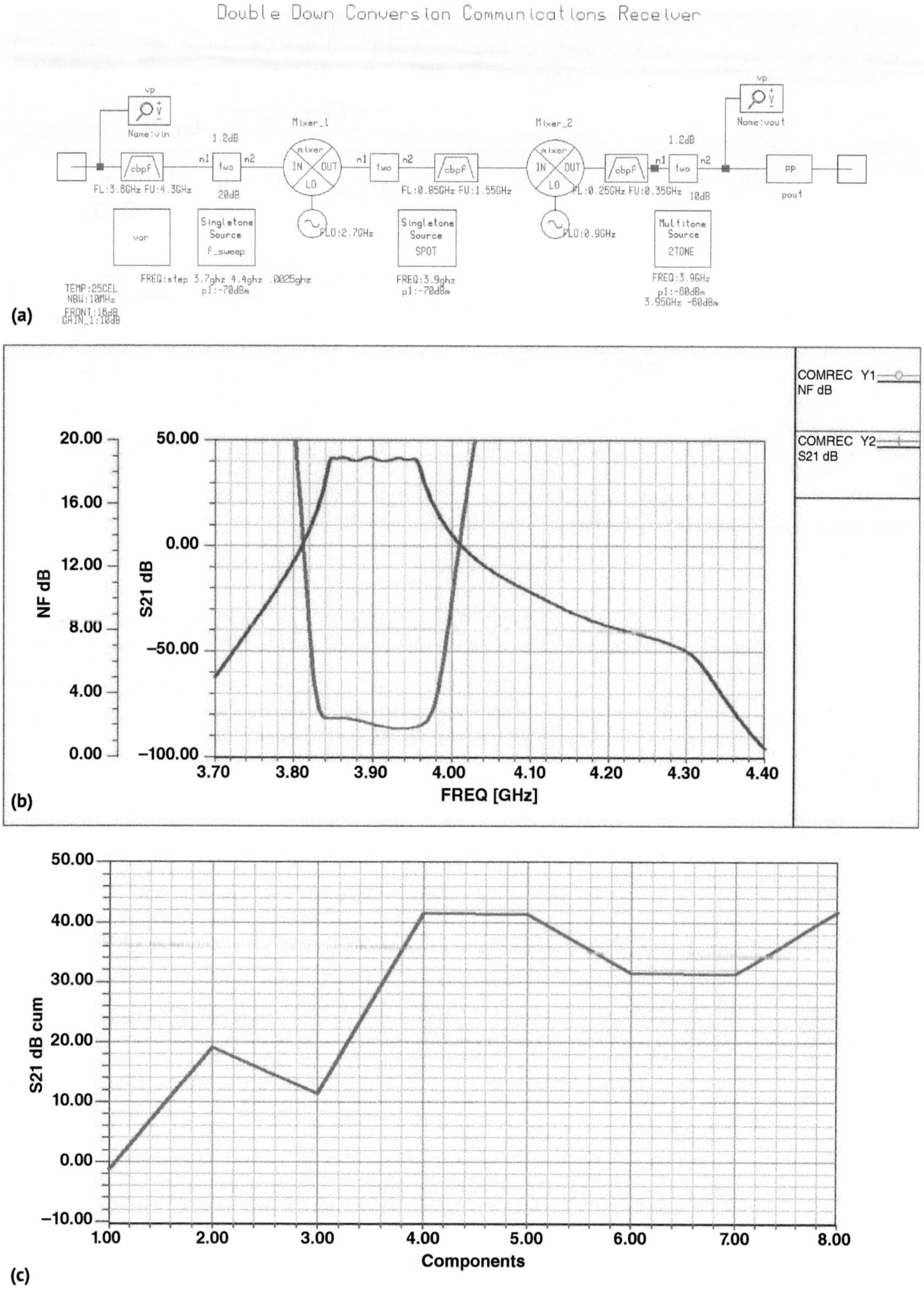

Figure 9.4 (a) Block diagram of a dual-conversion receiver. (b) Simulated gain S21 dB and Noise Figure (NFdB) of the dual-conversion receiver of (a). (c) Gain distribution of the dual-conversion receiver of (a).

Structure	Application	Cascaded F Equation
Three gain blocks	Any	$F = F_1 + \frac{F_2 - 1}{G_1} + \frac{F_3 - 1}{G_1 G_2}$
Heterodyne mixer	SSB, ideal image filtering	$F = F_1 + \frac{2(F_{2DSB} - 1)}{G_1} + \frac{F_3 - 1}{G_1 G_2}$
Complex downconverter	ZIF	$F = F_1 + \frac{F_{2DSB} - 1}{G_1} + \frac{F_3 - 1}{2G_1 G_2}$
Complex downconverter	LIF, no image suppression	$F = 2F_1 + \frac{2(F_{2DSB} - 1)}{G_1} + \frac{F_3 - 1}{G_1 G_2}$
Complex downconverter	LIF, image reject combining	$F = F_1 + \frac{F_{2DSB} - 1}{G_1} + \frac{F_3 - 1}{2G_1 G_2}$

Figure 9.5 Noise formula for different structures [1] / Maxim Integrated Products, Inc.

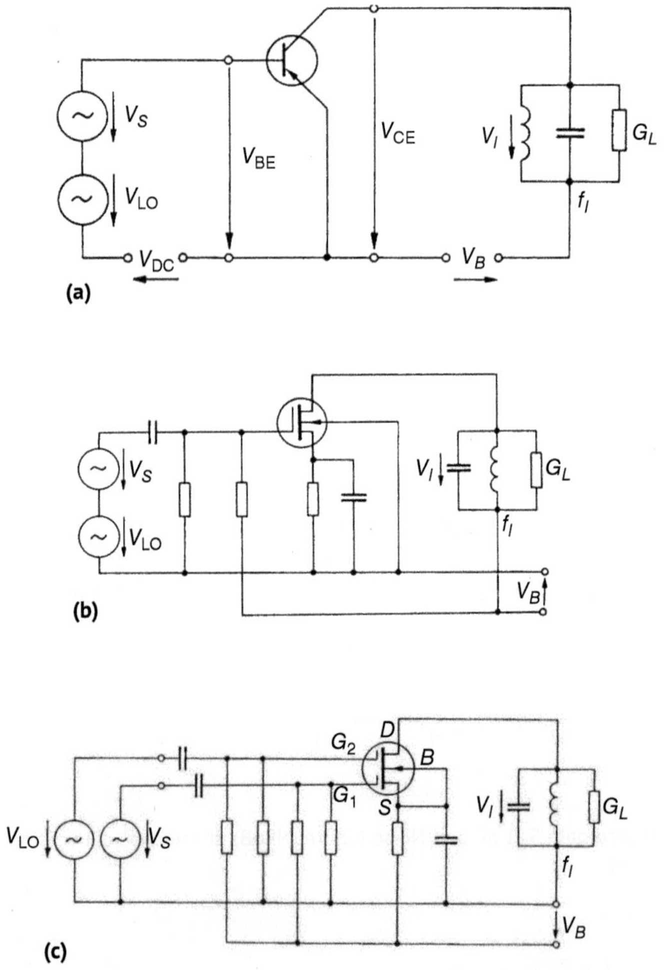

Figure 9.6 (a) Additive mixing in a BJT with two sources (V_s) and (V_{LO}) in series. *Source*: [4] / John Wiley & Sons. (b) Additive mixing in a single-gate MOSFET. *Source*: [4] / John Wiley & Sons. (c) Multiplicative mixing in a dual-gate MOSFET. A dual-gate device is actually two single-gate devices in series. *Source*: [4] / John Wiley & Sons.

This assumes the use of a device with a square-law characteristic. A different exponent will result in the generation of other mixing products, but this is not relevant for a basic understanding of the process. Expanding (9.2),

$$I = K\Big[V^2 + V_1^2\sin^2(\omega_1 t) + V_2^2\sin^2(\omega_2 t) + 2VV_1\sin(\omega_1 t) + 2VV_2\sin(\omega_2 t) + 2V_2V_1\sin(\omega_2 t)\sin(\omega_1 t)\Big] \tag{9.3}$$

The output comprises a direct current and a number of AC contributions. We are interested only in that portion of the current that generates the IF; so, if we neglect those terms that do not include both V_1 and V_2, we may write:

$$\begin{aligned} I_{\mathrm{IF}} &= 2KV_1V_2\sin(\omega_1 t)\sin(\omega_2 t) \\ I_{\mathrm{IF}} &= KV_2V_1\{\cos[(\omega_2-\omega_1)t] - \cos[(\omega_2+\omega_1)t]\} \end{aligned} \tag{9.4}$$

This means that at the output, we have the sum and difference signals available, and the one of interest may be selected by the IF filter [4].

9.5 Properties of Mixers

9.5.1 Conversion Gain/Loss

Even though a mixer works by means of amplitude-nonlinear behavior in its device(s), we generally want (and expect) it to act as a linear frequency shifter. The degree to which the frequency-shifted signal is attenuated or amplified is an important mixer property. *Conversion gain* can be positive or negative; by convention, negative conversion gains are often stated as *conversion loss*.

In the case of a diode (passive) mixer, the insertion loss is calculated from the various loss components:

$$\text{Loss (dB)} = \text{Conversion loss} + \text{transformer loss} + \text{losses due to harmonic generation} + \text{diode loss} \tag{9.5}$$

In the case of a double-balanced mixer (DBM), we must add the transformer losses (on both sides) and the diode losses as well as the mixer sideband conversion, which accounts, by definition, for 3 dB. Ideally, the mixer produces only one upper sideband (USB) and one lower sideband (LSB), which results in the 3 dB loss compared to the input signal. Also, the input and output transformers add about 0.75 dB on each side, and of course, there are the diode losses because of the series resistances of the diodes [4].

Figure 9.7 shows the equivalent circuit of a diode. It consists of a series (loss) resistor R_S and a time-variable electronic resistor, typically called the *diffusion resistance*, R_D, which equals 26 mV$/I_D$, and a capacitance C_D shunting R_D. C_D can be found from

$$C_D = \frac{W^2}{2D}\frac{I_D}{V_T} \tag{9.6}$$

where D is the diffusion constant, a material-dependent value, and W is the physical width. The average value for R_D is somewhere between the calculated value of 26 mV$/I_D$ and some leakage current, simply because it is generated by a rectification mechanism, which turns the LO power into an RF current and then into a combination of DC and RF currents.

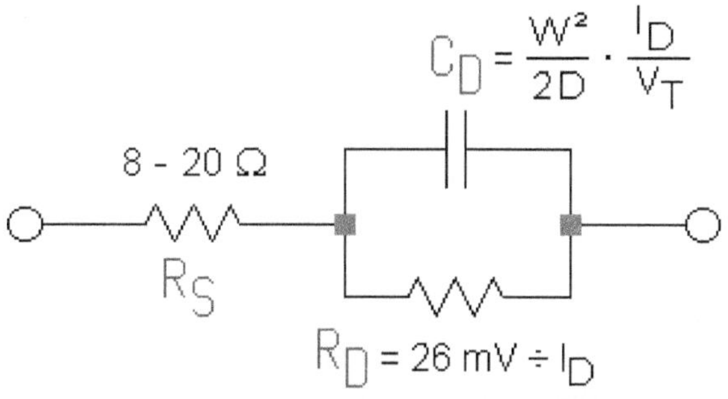

Figure 9.7 Equivalent circuit of a mixer diode. *Source*: [4] / John Wiley & Sons.

We can calculate the diode loss according to:

$$\text{Diode loss (dB)} = \log 10\left(\frac{50 + (2 \times R_S)}{50}\right) \tag{9.7}$$

Assuming that $R_S = 8\ \Omega$, the diode loss for a diode-ring mixer

$$\text{Diode loss (dB)} = \log 10\left(\frac{50 + (2 \times 8)}{50}\right) = 0.5\ \text{dB} \tag{9.8}$$

From (9.5), the insertion loss for this mixer is therefore:

$$\begin{aligned}\text{Loss (dB)} &= 3\ \text{dB (conversion loss)} + 1.5\ \text{dB (transformer loss)}\\ &\quad + 1\ \text{dB (losses from harmonic generation)} + 0.5\ \text{dB (diode loss)}\\ &= 6\ \text{dB}\end{aligned} \tag{9.9}$$

This assumes mixing at the fundamental frequency. Sometimes the diode loss resistor R_S is as high as 25 Ω per arm (as in a MOSFET switch), or 50 Ω total. This now results in

$$\text{Loss (dB)} = 3\ \text{dB} + 1.5\ \text{dB} + 1\ \text{dB} + 3\ \text{dB} = 8.5\ \text{dB (insertion loss)} \tag{9.10}$$

Since the value of R_S is partially determined by the threshold voltage of the diode and the diode diffusion resistance, R_D, a wide range of values can be noticed for different drive levels and mixer topologies. Figure 9.5 shows a shunt capacitance C_D, the so-called *diode diffusion capacitance*. When the diode is conducting, the influence of this nonlinearity is frequency-dependent, which adds to the insertion loss. In this discussion, we have not considered its frequency dependency. At wireless frequencies, modern Schottky diodes, also frequently called *hot carrier* diodes, are operated far from their cutoff frequency, resulting in less than 1 dB of additional losses. There are also mixers with special circuitry to terminate the IF image. This is done with a diplexer circuit or equivalent circuitry. This makes insertion loss values as low as 4 dB possible; however, the large-signal condition or intercept point suffers [4].

9.5.2 Noise Figure

Like any network, a mixer contributes noise to the signals it frequency shifts. The degree to which a mixer's noise degrades the signal-to-noise ratio of the signals it frequency-shifts is evaluated in terms of noise factor (F) and noise figure (NF). A part of the noise in a mixer has to do with the many noise sources. Figure 9.8 is a sketch that shows the various noise contributions.

For a long time, the literature has stated that the noise figure of a passive mixer, which is pretty much independent of its circuit arrangement, is equal to the mixer's insertion loss. But this neglects the influence of the white noise contribution of the mixer's diode(s). This is ironic, considering that RF noise generators were long based on thermionic diodes operated in saturation (the 5722 noise diode was a popular type, as in the Rohde & Schwarz SKTU). Such a diode's noise power output can be readily determined from its saturation current. On the other hand, all Schottky diodes, while conducting, generate white noise that follows the same principle as mentioned earlier. This fact has been practically recognized only by a few companies that make modern noise measurement equipment. Modern noise measurement devices measure the noise figure of a system by "hot and cold" technique, an approach based on knowledge of the absolute noise energy emitted under hot conditions (conductance). This method has the advantage that it can be used up to several tens of gigahertz, while the old vacuum tube-based noise generators ran out of steam at around 1 GHz due to the inability to match the tube to the 50 Ω termination. This was typically accomplished by connecting a 50 Ω resistor between anode and ground (without DC connection), followed by a low-pass filter, which would match the tube capacitance and other parasitics to the required termination of 50 Ω, purely resistive.

In reality, we can take the loss calculation from above and add the Schottky noise generated by the diodes as they are driven by the LO.

If a Schottky diode is used as a noise generator in the conductive mode, it generates a continuous frequency spectrum, possibly up to several gigahertz. There is a mathematical relationship between the noise power spectrum emitted by the diode and the time-averaged current of this diode, which generates the noise. If the noise source impedance is set (typically 50 Ω), the available noise power can be calculated according to:

$$I_R = \sqrt{2e \times I_s \times \Delta f} \tag{9.11}$$

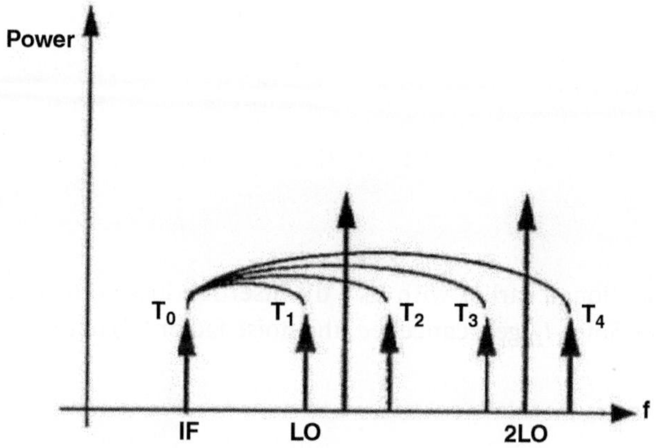

Noise sources where the noise at each sideband contributes to the output noise at the IF through frequency conversion.

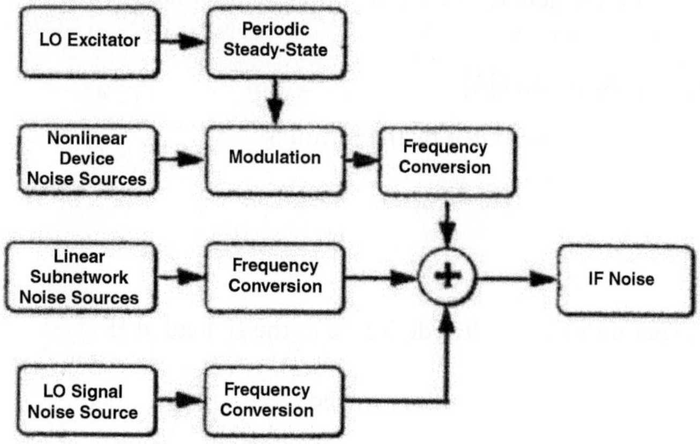

Noise mechanisms.

Figure 9.8 The noise mechanism inside a mixer. *Source*: [4] / John Wiley & Sons.

where

$e = 1.6 \times 10^{-19}$ coulombs
I_s = saturation current of the diode
Δf = effective noise bandwidth

For $S_{11} = 0$ or proper termination of this circuit ($R_G = R_{\text{term}}$):

$$\begin{aligned} P_R &= \left(\frac{I_R}{2}\right)^2 \times R_i \\ &= \frac{e}{2} \times I_s \times \Delta f \times R_i \end{aligned} \tag{9.12}$$

Calculated at a bandwidth of 1 Hz:

$$\frac{P_R}{\Delta f} = \frac{e}{2} \times I_s \times R_i \tag{9.13}$$

If

$$\frac{P_R}{\Delta f} = kT_0 \times F \tag{9.14}$$

the noise factor becomes

$$F = \frac{e \times I_s \times R_i}{2kT_0} \tag{9.15}$$

If the values for e and kT_0 are inserted,

$$F = 20 \times I_D \times \frac{26\text{ mV}}{I_D} + \frac{R_S + R_G}{2R_G} \tag{9.16}$$

Example: Assume the passive mixer mentioned earlier with its 6 dB insertion loss is considered and a DC current of 15 mA results as a function of the LO drive. Since I_D gets cancelled, the noise factor (F) of the diode portion equals

$$\begin{aligned} F &= 20 \times I_D \times \frac{26\text{ mV}}{I_D} + \frac{R_S + R_G}{2R_G} \\ &= 0.52 + 0.58 \\ &= 1.1 \end{aligned} \tag{9.17}$$

The NF is 10 log F, or 0.413 dB. Now this number and the insertion loss must be added. The resulting noise figure would be 6.413 dB. This is consistent with the general published measurement data [4].

9.5.3 Mathematical Nonlinear Approach [4]

The exact noise factor of a real mixer is computed by the formula

$$F = \frac{N_0(\omega_{\text{IF}}) + kT_0}{K_B T_0 G_{\text{Tc}}(\omega_{\text{RF}})} \tag{9.18}$$

where

$N_0(\omega_{\text{IF}}) =$ total noise power (per unit bandwidth) delivered to the IF load at IF
$K_B =$ Boltzmann's constant
$T_0 =$ Reference temperature (290 or 300 K is commonly used)
$G_{\text{Tc}}(\omega_{\text{RF}}) =$ Transducer conversion gain from ω_{RF} to ω_{RF}.

Let us now further elaborate on (9.18). We may write

$$N_0(\omega_{\text{IF}}) = N_S(\omega_{\text{IF}}) + N_{\text{INT}}(\omega_{\text{IF}}) + N_L(\omega_{\text{IF}}) + kT_0 \tag{9.19}$$

where N_S is noise generated by the RF source resistance and transferred to the IF load through frequency conversion, N_{INT} is noise generated internally to the mixer, and N_L is noise generated by the IF termination. If the source resistance is held at temperature T_0, N_S will basically originate from noise generated at the RF and image frequencies, which are transferred to the IF with approximately the same conversion gain, plus a relatively small contribution transferred from other sidebands with a smaller conversion gain. We may then write synthetically

$$N_S(\omega_{\text{IF}}) = 2aK_B T_0 G_{\text{Tc}}(\omega_{\text{RF}}) \tag{9.20}$$

where a is a coefficient slightly larger than 1. $N_{\text{INT}}(\omega_{\text{RF}})$ is generated by transformer losses, by the diode Schottky noise, and by the diode resistive parasitics, and in principle may take on any value; in particular, it may be zero if both the transformers and the diodes are ideal (i.e., if the latter are pure nonlinear resistors). As for $N_L(\omega_{\text{IF}})$, by Nyquist's theorem, the IF load resistor R_L may be described as a noiseless resistor in series with a noise voltage source whose mean-square voltage (per unit bandwidth) is

$$|V_L|^2 = 4K_B T_L R_L \tag{9.21}$$

where T_L is the IF termination temperature. If the IF load is driven by a source with an output impedance $Z_{\text{out}}\,(\omega_{\text{IF}})$, the noise power actually delivered to the load will obviously be

$$N_{\text{out}} = \frac{4K_B T_L R_L^2}{|Z_{\text{out}}(\omega_{\text{IF}}) + R_L|^2} \tag{9.22}$$

In addition to N_{out}, the thermal noise originating from the IF termination delivered to the IF load at ω_{IF} will also include contributions from other sidebands that are back-converted by the mixer nonlinearities with a relatively small conversion gain. Thus we may write

$$N_L(\omega_{\text{IF}}) = \frac{4bK_B T_L R_L^2}{|Z_{\text{out}}(\omega_{\text{IF}}) + R_L|^2} \tag{9.23}$$

where b is slightly larger than 1. If we now introduce the mixer conversion loss, namely

$$L_C = \frac{1}{G_{\text{Tc}}(\omega_{\text{RF}})} \tag{9.24}$$

and combine (9.18) with (9.19), (9.20), and (9.23), we finally get the noise factor expression

$$F = 2a + \frac{N_{\text{INT}}(\omega_{\text{IF}}) + kT}{K_B T_0} L_C + \frac{4bT_L R_L^2}{T_0|Z_{\text{out}}(\omega_{\text{IF}}) + R_L|^2} L_C \tag{9.25}$$

In the normal region of operation of the mixer (sufficient LO drive), we may assume

$$Z_{\text{out}}(\omega_{\text{IF}})R_L \tag{9.26}$$

so that (9.25) becomes

$$F = 2a + \frac{N_{\text{INT}}(\omega_{\text{IF}}) + kT}{K_B T_0} L_C + \frac{bT_L}{T_0} L_C \tag{9.27}$$

Multiplying the $\log_{10}$ of F by 10 gives us the exact mixer noise figure in dB.

The method just discussed is the basis for the far noise calculation for oscillators as mentioned in the oscillator chapter, in which it is referred to as conversion noise. This includes the AM-to-PM conversion noise [4].

Table 9.1 shows how the noise figure and conversion gain vary with LO power for a generic diode DBM. "Starving" a diode mixer by decreasing its LO drive rapidly degrades its performance in all respects [4].

Table 9.1 Noise figure and conversion gain versus LO power for a diode DBM [4].

LO power (dBm)	NF (dB)	Conversion gain (dB)
−10.0	45.3486	−45.1993
−8.0	32.7714	−32.5264
−6.0	19.8529	−19.2862
−4.0	12.1154	−11.3228
−2.0	8.85188	−8.05585
0.0	7.26969	−6.51561
2.0	6.42344	−5.69211
4.0	5.85357	−5.15404
6.0	5.50914	−4.84439
8.0	5.31796	−4.66871
10.0	5.19081	−4.54960
12.0	5.08660	−4.45887
14.0	4.99530	−4.38806
16.0	4.91716	−4.33322
18.0	4.85920	−4.29407
20.0	4.82031	−4.26763

9.5.4 SSB Versus DSB Noise Figure

In radiocommunication, we typically use mixers as "SSB" devices; that is, we are interested in only one of the two possible RF inputs $(f_{LO} + f_{IF})$ or $(f_{LO} - f_{IF})$ that produce output at the desired IF. The unused input is the *RF image* or simply *image*. A mixer's RF image response can be important even if no man-made signals are present at f_{image} because noise there, including that produced by the mixer's RF-port termination, will be converted to the IF, possibly compromising the system's signal-to-noise ratio. As a result of this and the IEEE definition of noise figure [4, 6], which, for a mixer, considers only the noise associated with a system's principal frequency transformation, some controversy exists about the definition and measurement of mixer noise figure. IEEE's NF definition for mixers assumes *no* noise contribution at f_{image}, not even from the mixer's RF-port termination at that frequency; yet, by convention (and out of necessity, since an absolutely noiseless, $T = 0$ f_{image} termination is unavailable on the average test bench), the noise figures we measure always include *some* noise at f_{image}, even if, as a result of filtering, it arises only from the mixer's RF-port termination.

The NF measurements and simulations presented in this book reflect this conventional SSB NF scenario. Stephen Maas provides in-depth material on these issues, including the importance of DSB NF for some systems, and the limitations of NF as a figure of merit, in [5].

9.5.5 Linearity [4, p. 650]

In the RF system, the linearity of the RF mixer is more important than the noise figure.

9.5.5.1 1 dB Compression Point

Like other networks, a mixer is amplitude-nonlinear above a certain input level; above this point, the output level fails to track input-level changes proportionally. This figure of merit, $P_{-1\,dB}$, identifies the single-tone input-signal level at which the output of the mixer has fallen 1 dB below the expected output level. The 1 dB compression point in a conventional double-balanced diode mixer is approximately 6 dB below the LO power. For lower distortion mixers, it is usually 3 dB below the LO power [4].

9.5.5.2 1 dB Desensitization Point

This specification is another figure of merit similar to the 1 dB compression point. However, the 1 dB desensitization point refers to the level of an interfering (undesired) input signal that causes a 1 dB decrease in nominal conversion gain for the desired signal. For a diode-ring DBM, the 1 dB desensitization point is usually 2–3 dB below the 1 dB compression point [4].

9.5.5.3 Dynamic Range

The dynamic range of any RF/wireless system can be defined as the difference between the 1 dB compression point and the minimum discernible signal (MDS). These two points are specified in units of power (dBm), giving dynamic range in dB. When the RF input level approaches the 1 dB compression point, harmonic and intermodulation products begin to interfere with the system performance. High dynamic range is obviously desirable, but cost, power consumption, system complexity, and reliability must also be considered [4].

9.5.5.4 Harmonic Intermodulation Products

Harmonic intermodulation products (HIP) are spurious products that are harmonically related to the f_{LO} and f_{RF} input signals.

$$\text{HIP} = Mf_{LO} + Nf_{RF} \tag{9.28}$$

Table 9.2 shows the relative harmonic intermodulation product levels for a high-level diode DBM.

9.5.5.5 Intermodulation Distortion

Nonlinearities in the mixer devices give rise to IMD products whenever two or more signals are applied to the mixer's RF port. Testing this behavior with two (usually closely spaced) input signals of equal magnitude can return several figures of merit depending on how the results are interpreted. A mixer's third-order output intercept point ($IP_{3,out}$) is defined as the output power level where the spurious signals generated by $(2f_{RF1} \pm f_{RF2}) \pm f_{LO}$ and $(f_{RF1} \pm 2f_{RF2}) \pm f_{LO}$ are equal in amplitude to the desired output signal as shown in Figure 9.9 [4].

Table 9.2 Typical spurious responses of high-level DBM (decibels below $f_{LO} \pm f_{RF}$ response) [4].

RF input signal harmonics		f_{LO}	$2f_{LO}$	$3f_{LO}$	$4f_{LO}$	$5f_{LO}$	$6f_{LO}$	$7f_{LO}$	$8f_{LO}$
$8f_{RF}$	100	100	100	100	100	100	100	100	100
$7f_{RF}$	100	97	102	95	100	100	100	90	100
$6f_{RF}$	100	92	97	95	100	100	95	100	100
$5f_{RF}$	90	84	86	72	92	70	95	70	92
$4f_{RF}$	90	84	97	86	97	90	100	90	92
$3f_{RF}$	75	63	66	72	72	58	86	58	80
$2f_{RF}$	70	72	72	70	82	62	75	75	100
f_{RF}	60	0	35	15	37	37	45	40	50
		60	60	70	72	72	62	70	70

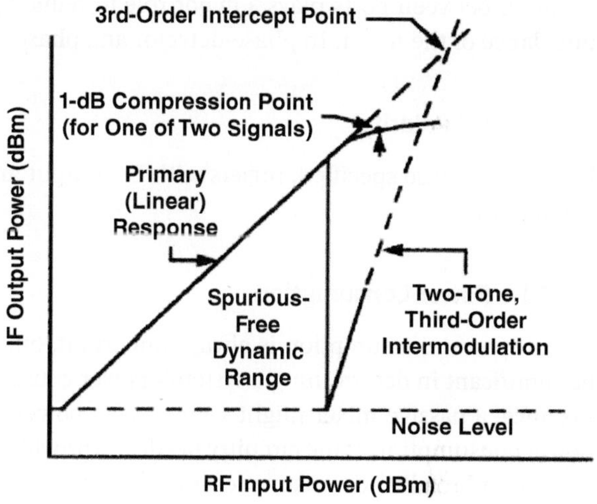

Figure 9.9 Mixer linearity evaluation, including compression and two-tone IMD dynamic range. $P_{-1\,dB}$ for a single-tone cannot be read directly from this graph because the values shown are the result of two-equal-tone drive. *Source:* [4] / John Wiley & Sons.

The third-order input intercept point $IP_{3,in}$ – IP_3 referred to the input level is particularly of useful value and is the most commonly used mixer IMD figure of merit.

IP_{3in} can be calculated according to:

$$IP_{n,in} = IMR \div (n-1) + \text{input power (dBm)} \tag{9.29}$$

where IMR is the intermodulation ratio (the difference in dB between the desired output and the spurious signal, and n is the IM order—in this case, 3). In a conventional diode DBM, IP_{3in} is approximately 14 dB above the single-tone 1 dB compression point ($P_{-1\,dB}$)—approximately 8 dB greater than the LO power. As will be seen later, this does not apply to feedback active mixers. They have their own agenda: above a particular input level, their IMD products increase almost exponentially.

Although designers are usually more concerned with odd-order IM performance, second-order IM can be important in wideband systems (systems that operate over a 2 : 1 or greater bandwidth) as discussed in Ref. [4, Section 1.6.2].

9.5.6 LO Drive Level

A mixer's specifications are usually guaranteed at a particular LO drive level, usually specified as a dBm value that may be qualified with a tolerance. Insufficient LO drive degrades mixer performance; excessive LO drive degrades performance and may damage mixer devices. Commercially available diode mixers are often classified by LO drive level; for example, a "Level 17" mixer requires 17 dBm of LO drive [4].

9.5.7 Interport Isolation

In a mixer, isolation is defined as the attenuation in dB between a signal input at any port and its level as measured at any other port. High isolation numbers are desirable. Figure 9.8 shows LO-to-IF and LO-to-RF isolation versus frequency for a triple-balanced diode DBM. Isolation is dependent mainly on transformer and physical symmetry and device balance, but the level of signals applied to the mixer also plays a role, as shown in Figure 9.9 [4].

9.5.8 Port VSWR

The load presented by a mixer's ports to the outside world can be of critical importance to a designer. For example, high LO-port voltage standing wave ratio (VSWR) may result in inefficient use of available LO power, resulting in LO starvation (underdrive) that degrades the mixer's performance. Figure 9.10 shows LO-port VSWR versus frequency for a high-level diode DBM with two values of LO power. Like interport isolation, port VSWR can vary with the level of the signal applied [4].

9.5.9 DC Offset

Isolation between ports plays a major role in reducing DC offset in a mixer. Like isolation, DC offset is a measure of the unbalance of the mixer. In phase-detector and phase-modulator applications, DC offset is a critical parameter [4].

9.5.10 DC Polarity

Unless otherwise specified, mixers with DC output are designed to have negative polarity when RF and LO signals are of equal phase.

9.5.11 Power Consumption

Circuit power consumption is always important, but in battery-powered wireless designs, it is *critical*. Mixer choice may be significant in determining a system's power consumption, sometimes in ways that seem paradoxical at first glance. For instance, a passive mixer might seem to be a power-smart choice because it consumes *no* power—until we factor in the power consumption of the circuitry needed to provide the (often considerable) LO power a passive mixer requires. If a mixer requires a broadband resistive termination that will be provided by a post-mixer amplifier operating at a high standing current, the power consumption of the amplifier stage must be considered as well. Evaluating the suitability of a given mixer type to a task therefore requires a grasp of its ecology as well as its specifications [4].

LO-IF and LO-RF isolation versus frequency for a high-level triple-balanced diode mixer is shown in Figure 9.10a. The periodic roughness of the traces is a measurement-system artifact, proving that even sophisticated FFT-based instruments are not perfect. Figure 9.10b shows the LO-IF isolation versus frequency and LO drive level for a high-level diode DBM. Figure 9.10c shows the LO-port VSWR versus frequency for a high-level diode DBM [4].

9.6 Diode Mixers

Passive mixers based on diode switches are common in base-station applications, where their high dynamic range and 50 Ω port impedances overcome objections to their inherent conversion loss, unsuitability to integration, and relatively high LO-power requirement. As we will discuss later, FET-based passive mixers are overcoming some of these limitations in portable/mobile wireless applications [4].

9.6.1 Single-diode Mixer

Figure 9.11 shows the schematic of a simplistic single-diode mixer. The LO and RF signals are applied in series to the diode, with no effort made to match the sources to the diode or isolate the sources from each other. The LO and RF signals are both present in the diode simultaneously, so the mixing performed is additive. The LO signal switches the diode on and off, gating the RF signal to the circuit's output at the LO frequency. In this simple circuit, the output port's 50 Ω resistance serves as the diode's AC load and DC return. Because we want the diode to operate as a linear switch with respect to the RF signal, the diode should turn on hard and turn off as completely as possible, and the LO, not the RF source, should switch the diode on and off. The upper frequency limit of the circuit will depend on how rapidly the diode can switch between the on and off states. The LO can be only so strong before the diode's dissipation limits are exceeded; the RF source can be only so strong relative to the LO before it begins to play a role in switching the diode on and off. The LO must therefore be considerably stronger than the RF source: 20 dB or more for low-distortion applications [4].

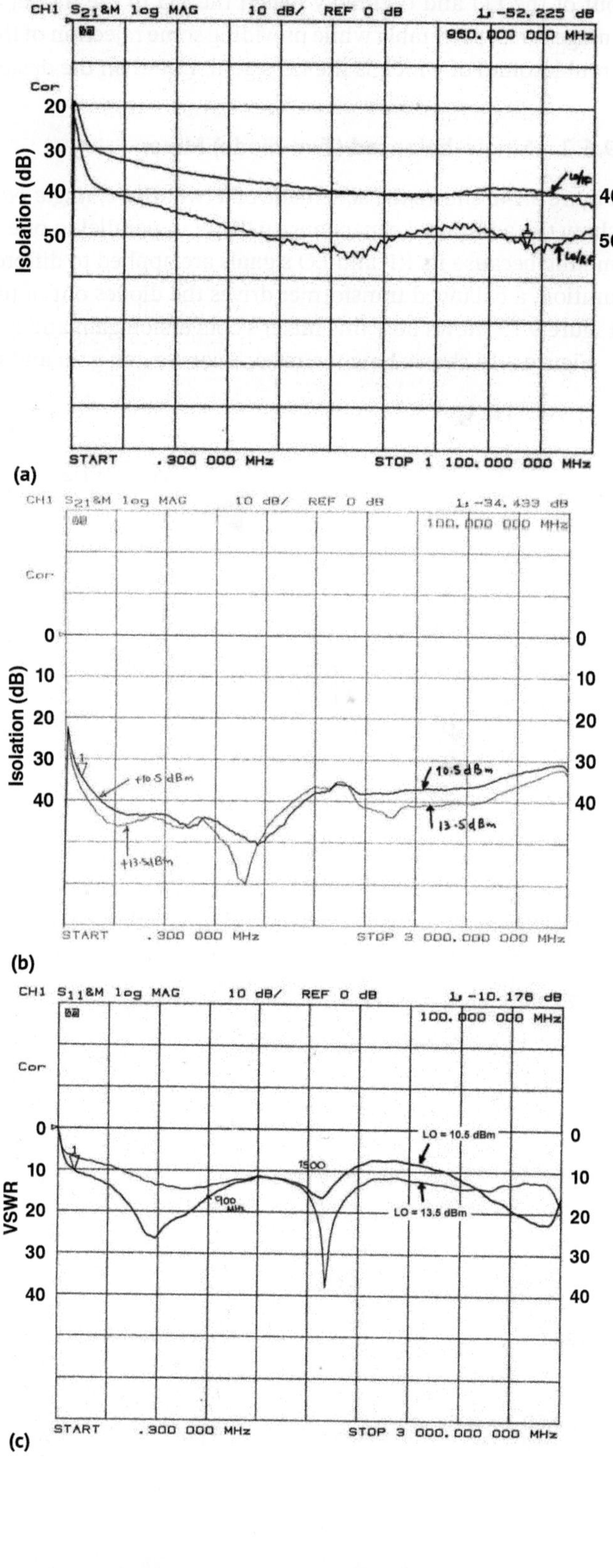

Figure 9.10 (a) LO-IF and LO-RF isolation versus frequency for a high-level triple-balanced diode mixer. The periodic roughness of the traces is a measurement-system artifact, proving that even sophisticated FFT-based instruments are not perfect. *Source*: [4] / John Wiley & Sons. (b) LO-IF isolation versus frequency and LO drive level for a high-level diode DBM. *Source*: [4] / John Wiley & Sons. (c) LO-port VSWR versus frequency for a high-level diode DBM. *Source*: [4] / John Wiley & Sons.

Figure 9.12 shows how the single-diode mixer's conversion gain and noise figure vary with applied LO power. Figure 9.13 shows the mixer's output spectrum.

The following several pages present sample data on diodes suitable for mixer service in three tables (Tables 9.3–9.5) and the some parts of manufacturer's data sheet on the Infineon BAT14-099.

The simple single-diode mixer circuit shown in Figure 9.11 is intended only as an illustration of the basic behavior of diode mixer behavior. A practical single-diode mixer would include filtering at its RF, LO, and IF ports—RF filtering for image rejection, reduced LO radiation, and optimum matching of the RF source to the diode; LO filtering to keep RF out of the LO and optimally match the LO to the diode; and IF filtering to optimally match the diode to its IF (and IF image) load, preferably while providing some rejection of the mixer's unwanted outputs, the strongest (and most potentially troublesome) of which is the LO signal. Detail on the design of single-diode mixers can be found in [26].

9.6.2 Single-balanced (Two-diode) Mixer

Figure 9.14a shows the schematic of a two-diode, *single-balanced* mixer. For the purpose of simulation in Figure 9.14b, we show the additional capacitance 0.01 pF in parallel to one diode. Unlike the single-diode mixer, it performs multiplicative mixing because its RF and LO signals are applied to different ports. In this more commonly seen two-diode mixer configuration, a balanced transformer drives the diodes out of phase for the LO and in phase for signals present at the RF port. Figure 9.15 shows how this mixer's conversion gain and noise figure vary with applied LO power.

Figure 9.16 shows how the mixer's conversion gain and noise figure vary with frequency for a constant LO power.

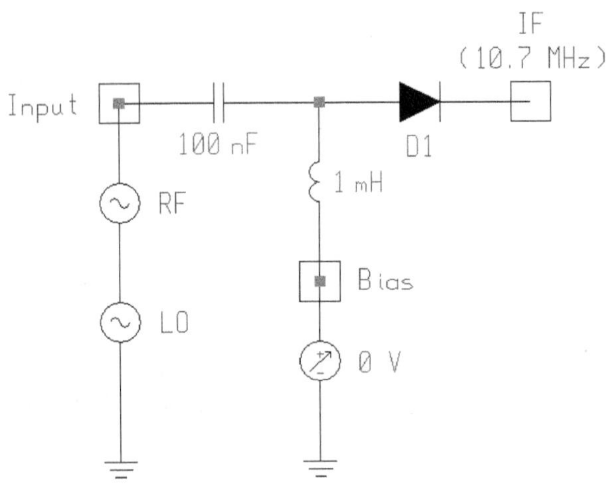

Figure 9.11 Schematic of the single-diode mixer. The circuit's 50 Ω output termination serves as the diode's DC return, as well as completing the circuit for RF, LO, and IF. *Source*: [4] / John Wiley & Sons.

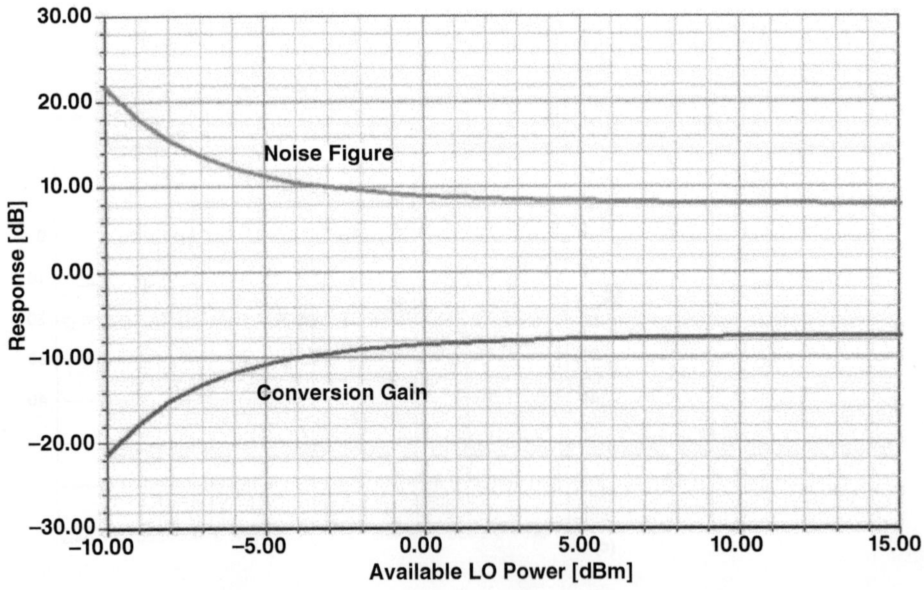

Figure 9.12 Conversion gain and noise figure versus LO power for the single-diode mixer. The values reported are worse than those predicted by theory because RF, LO, and IF matching has not been attempted. In this analysis, LO = 310.7 MHz (−10 to 25 dBm), RF = 300 MHz (−50 dBm), and IF = 10.7 MHz. *Source*: [4] / John Wiley & Sons.

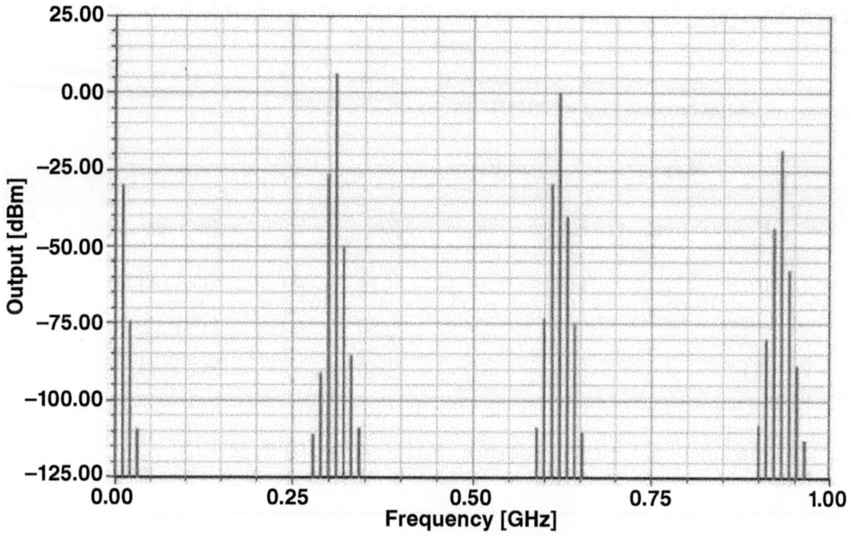

Figure 9.13 Output spectrum of the single-diode mixer. Three LO harmonics and three LO sidebands were used in this analysis. In this analysis, LO = 310.7 MHz (13 dBm), RF = 300 MHz (−20 dBm), and IF = 10.7 MHz. *Source:* [4] / John Wiley & Sons.

Table 9.3 Example of RF Schottky diodes (Infineon) [4].

	Maximum ratings		Characteristics ($T_A = 25\,°C$)					
				V_F at I_F		V_F at I_F		
Type	V_R V	I_F mA	C_T pF	mV	mA	mV	mA	Package
BAT14-03W	4	90	0.22	430	1	550	10	SOD-323
BAT14-099 (Dual)	4	90	0.22	430	1	550	10	SOT-143
BAT14-099R (Quad)	–	90	0.38	400	1	480	10	SOT-143
BAT15-03W	4	110	0.21	230	1	320	10	SOD-323
BAT15-099 (Dual)	4	110	0.21	230	1	320	10	SOT-143
BAT15-099R (Quad)	–	110	0.37	230	1	320	10	SOT-143
BAT17	4	130	0.55	340	1	425	10	SOT-23
BAT17-04 (Dual)	4	130	0.55	340	1	425	10	SOT-23
BAT68	8	130	0.75	320	1	395	10	SOT-23
BAT68-04 (Dual)	8	130	0.75	320	1	395	10	SOT-23
BAT68-05 (Dual)	8	130	0.75	320	1	395	10	SOT-23
BAT68-06 (Dual)	8	130	0.75	320	1	395	10	SOT-23
BAT68-07 (Dual)	8	130	0.75	320	1	395	10	SOT-143
BAT68-03W	8	130	0.75	320	1	395	10	SOD-323
BAT114-099 (Dual)	4	90	0.22	580	1	680	10	SOT-143
BAT114-099R (Quad)	–	90	0.22	580	1	680	10	SOT-143

Table 9.4 Example of Schottky detector diodes (Infineon) [4].

	Maximum ratings		Characteristics ($T_A = 25\,°C$)								
			C_T at V_R		V_F at I_F		R_O at V_F		I_R at V_R		
	V_R	I_F									
Type	V	mA	pF	V	V	mA	kΩ	V	µA	V	Package
BAT62	40	20	0.4	0	0.53	2	160	0	≤ 10	40	SOT-143
BAT62-03W	40	20	0.4	0	0.53	2	160	0	≤ 10	40	SOD-323
BAT63	3	100	0.65	0.2	0.19	1	30	0	≤ 10	3	SOT-143

Table 9.5 Part of datasheet BAT 14-099 (Infineon). S_{11}-parameters. Typical impedance characteristics (with external bias I and $Z_0 = \Omega$).

f	$I = 0.02$ mA		$I = 0.05$ mA		$I = 0.1$ mA		$I = 0.2$ mA		$I = 0.5$ mA	
GHz	MAG	ANG	MAG	ANG	MAG	ANG	MAG	ANG	MAG	ANG
1	0.99	−15.89	0.91	−16.40	0.79	−16.40	0.57	−16.60	0.13	−17.30
2	0.96	−30.40	0.88	−30.80	0.76	−31.09	0.56	−30.70	0.13	−28.40
3	0.95	−45.30	0.87	−46.20	0.75	−47.30	0.55	−47.00	0.11	−43.99
4	0.93	−59.60	0.86	−61.60	0.73	−62.40	0.53	−62.40	0.10	−54.40
5	0.93	−74.80	0.85	−77.10	0.72	−78.70	0.51	−78.70	0.07	−80.70
6	0.91	−89.50	0.83	−93.10	0.69	−95.70	0.48	−95.70	0.04	−102.30
7	0.89	−106.60	0.80	−110.50	0.66	−112.70	0.45	−114.00	0.02	158.01
8	0.88	−123.40	0.79	−129.40	0.64	−132.40	0.43	−135.40	0.06	118.40
9	0.86	−143.20	0.76	−150.20	0.62	−154.20	0.40	−161.20	0.12	96.20
10	0.83	−166.10	0.72	−174.10	0.58	−179.10	0.37	171.10	0.19	72.10
11	0.82	168.10	0.71	158.10	0.59	153.80	0.39	140.80	0.25	62.60
12	0.80	138.20	0.72	127.20	0.60	121.20	0.44	108.20	0.33	49.20

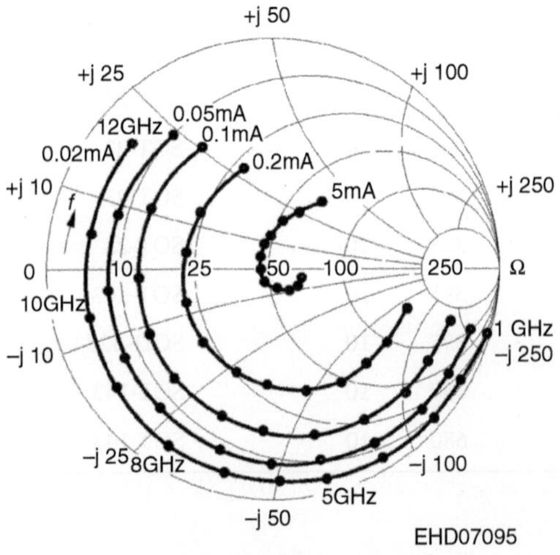

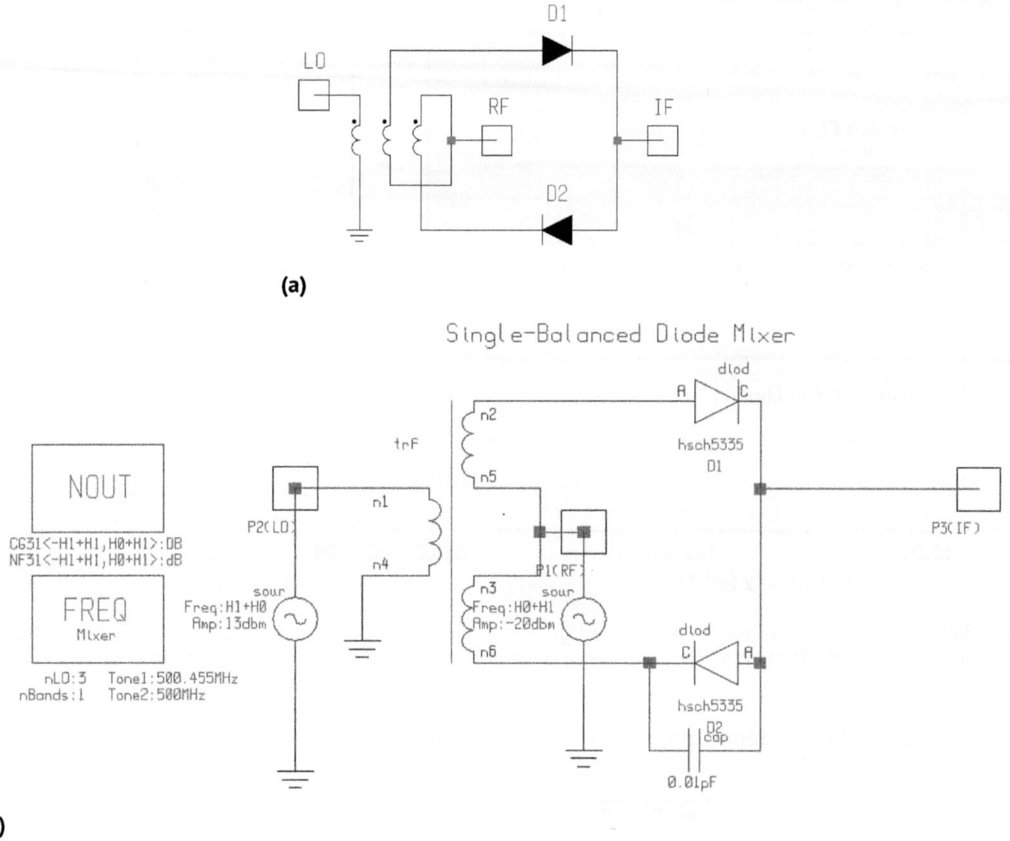

Figure 9.14 (a) Schematic of the two-diode (also known as *single-balanced*) mixer. *Source*: [4] / John Wiley & Sons. (b) Simulation of the circuit in (a) with an additional capacitance 0.01 pF in parallel to one diode. *Source*: [4] / John Wiley & Sons.

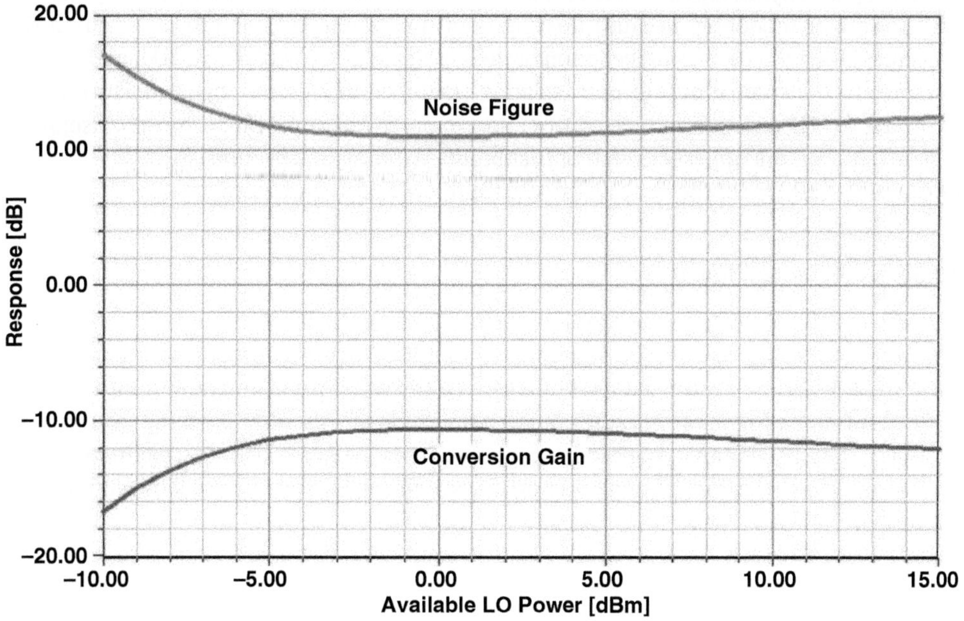

Figure 9.15 The nonideal mixer's conversion gain and noise figure vary with available LO power. In this analysis, LO = 500 MHz (13 dBm), RF = 500.455 MHz (−20 dBm), and IF = 455 kHz.

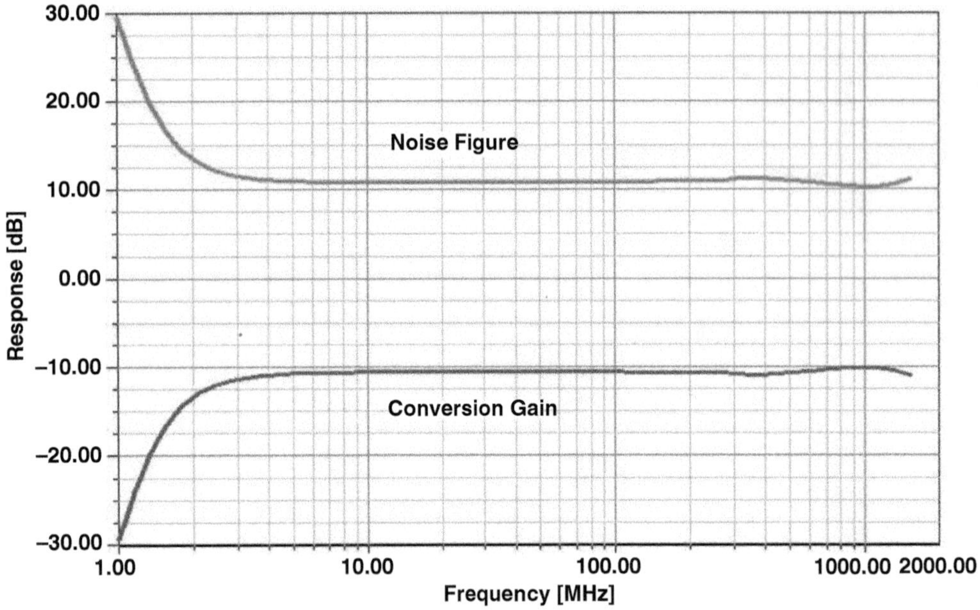

Figure 9.16 The nonideal two-diode mixer's conversion gain and noise figure vary with frequency for a constant LO power. In this analysis, LO = 1–1500 MHz (2 dBm), RF = 1.455–1500.455 MHz (−40 dBm), and IF = 455 kHz. *Source*: [4] / John Wiley & Sons.

The two-diode mixer is used mostly in the frequency range above 1 GHz in a manner akin to a phase discriminator, using step-recovery diodes in the LO feed for enhanced harmonic mixing. Such mixers are mainly used in medium-cost spectrum analyzers or microwave receivers up to several tens of gigahertz, with the necessary transformers and Baluns printed on the circuit board.

With perfectly matched diodes and perfect transformer and constructional symmetry, no LO energy arrives at the IF and RF ports, and there is only slight attenuation between the RF and IF ports. Both building and computer modeling such a mixer are impossible: building, because perfectly matched diodes and perfect transformer and constructional symmetry cannot be achieved in practice; and computer modeling, because floating-point mathematics runs out of handling the infinite amplitude spread involved in calculating the perfect cancellation of the LO signal as it travels to the RF and IF ports. That said, Figure 9.17 compares the mixer's *quasi*-ideal port-to-port isolation (matched diodes, a perfect transformer, no stray inductances and capacitances, and a 10 MΩ resistors connected from port to port) and nonideal port-to-port isolation (slightly mismatched diodes and 0.5 pF between the upper terminal of the middle winding and ground) [4].

Figure 9.18a,b shows the mixer's output spectrum for the quasi-ideal and nonideal cases, respectively.

9.6.3 Subharmonically Pumped Single-balanced Mixer

Figure 9.19 shows a single-balanced mixer with a difference: antiparallel diode pairs take the place of single diodes, the RF and IF are buffered from each other only by filtering, and the LO is applied at half the frequency necessary to provide the desired frequency conversion. The RF-to-IF isolation is limited to that provided by the series input and output filtering, but the LO-to-IF isolation is higher at f_{LO}, and much higher at $2f_{LO}$, than that achievable with a DBM with the LO signal at $2f_{LO}$ (Figure 9.20). Although the example shown is for an upconverting high frequency (HF) receiver, this technique finds application well into the microwave range as the basis for (I/Q) modulators, in which carrier leakage must be reduced to a level difficult to achieve with conventional DBMs [4, 7–12].

9.6.4 Diode-ring Mixer

Adding two more diodes and another transformer to the singly balanced mixer results in a DBM as shown in Figure 9.21. A DBM's frequency response is largely determined by the frequency response of its transformers, which act as transmission lines. The low-frequency limit is determined by the inductance of the transformer windings, the reactance of which, at the lowest frequency of interest, should be at least four times the impedance at which the transformer operates. The upper frequency limit is determined mainly by the degradation of the transformers' transmission line behavior at higher frequencies, although the increasing importance of diode capacitance also plays a role [4].

A DBM's interport isolation is determined by the symmetry of its transformers, diodes, and physical construction. In practice, the effects of diode mismatch can be minimized by using a dual diode, such as—in the case of the BAT15—the BAT15-099, the maximum V_F spread between the diodes, which is specified as 20 mV.

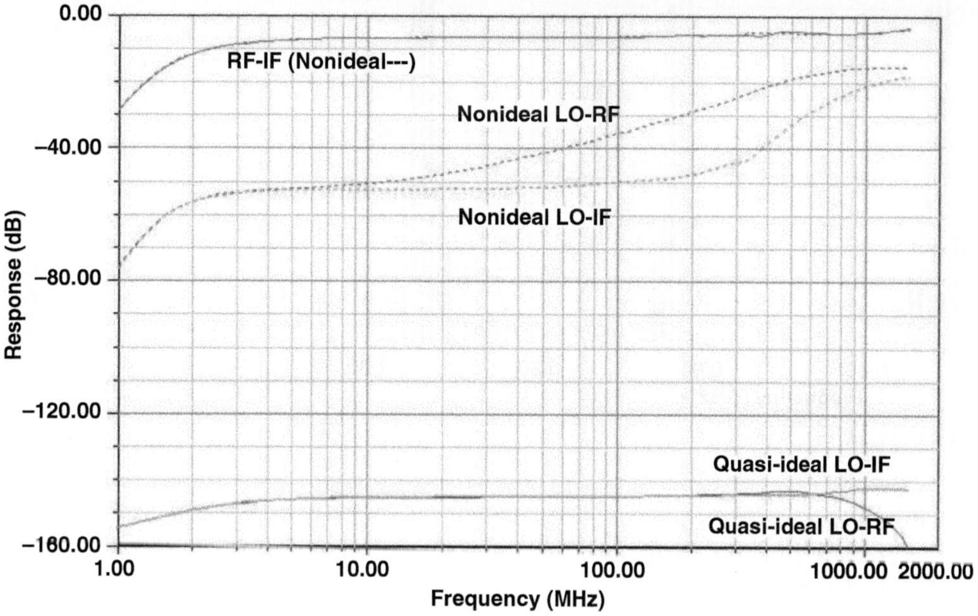

Figure 9.17 Port-to-port isolation of the quasi-ideal (identical diodes and no stray capacitance) and nonideal (slightly mismatched diodes and 0.5 pF of stray capacitance between the upper terminal of the middle transformer winding and ground) two-diode mixer. In this analysis, LO = 1–1500 MHz (2 dBm), RF = 1.455–1500.455 MHz (−40 dBm), and IF = 455 kHz. *Source*: [4] / John Wiley & Sons.

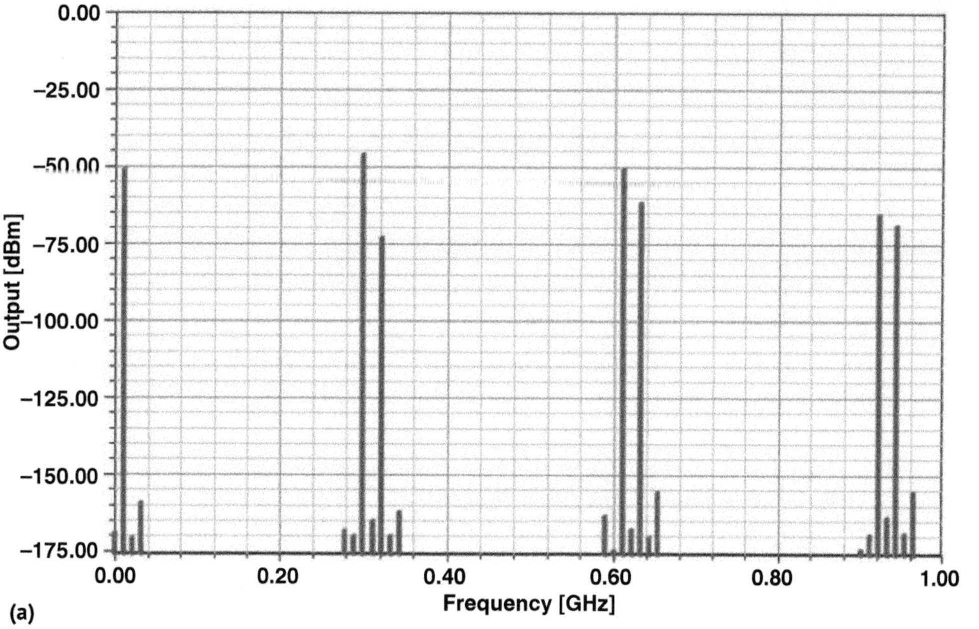

(a)

Figure 9.18 The nonideal mixer's output spectrum with (a) identical diodes and no stray capacitance and (b) slightly mismatched diodes and 0.5 pF of stray capacitance between the upper terminal of the middle transformer winding and ground. In these analyses, LO = 310.7 MHz (2 dBm), RF = 300.0 MHz (−40 dBm), and IF = 10.7 MHz. Four LO harmonics and three LO sidebands were used. *Source*: [4] / John Wiley & Sons.

(Continued)

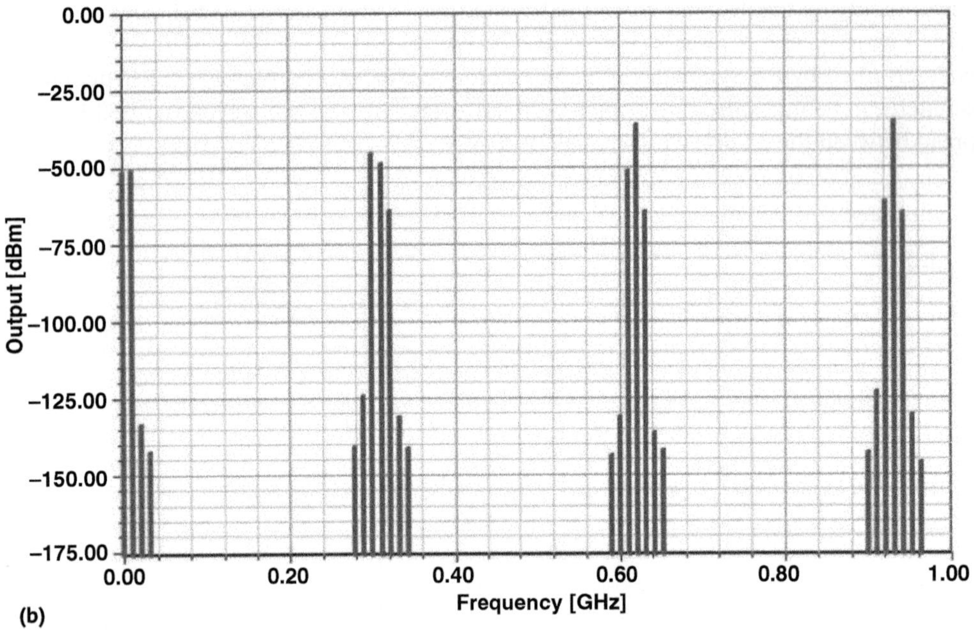

(b)

Figure 9.18 (Cont'd)

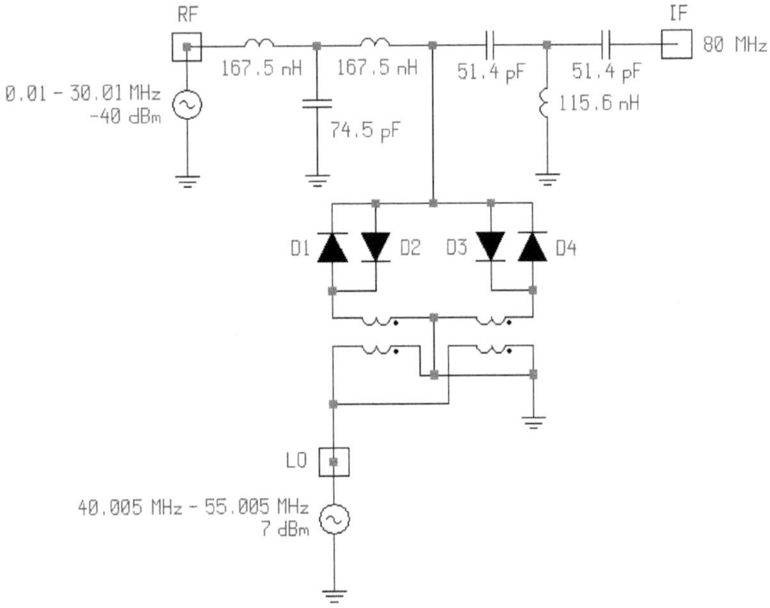

Figure 9.19 A subharmonically pumped single-balanced mixer using antiparallel diode pairs. The LO operates from 40.005–55.005 MHz to mix 0.01–30.01 MHz RF to an IF of 80 MHz. *Source*: [4] / John Wiley & Sons.

Figure 9.22 shows how the DBM's conversion gain and noise figure vary with applied LO power. Figure 9.23 shows how the DBM's conversion gain and noise figure vary with frequency for an LO power of 7 dBm, with quasi-ideal and nonideal balance. Figure 9.24 shows how the DBM's port-to-port isolation differs with quasi-ideal and nonideal balance for an LO power of 7 dBm.

Figure 9.25 shows how the DBM's RF- and LO-port return loss varies with frequency; the sharp peak corresponds to a resonance caused by one of the stray capacitances added to simulate less-than-ideal balance in the modeled mixer.

Figure 9.26 shows the DBM's output spectrum. Figure 9.27 shows the DBM's output waveform over 50 cycles of the LO signal.

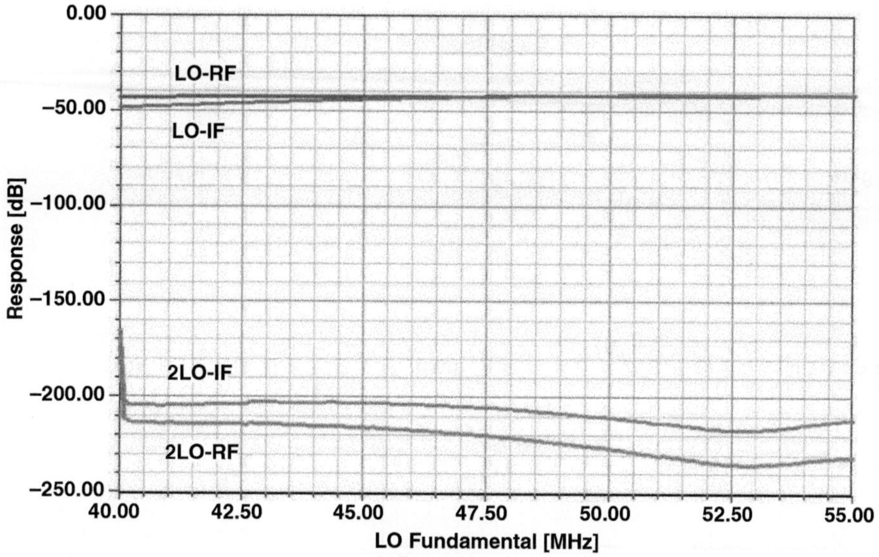

Figure 9.20 Simulated interport isolation of the subharmonic SBM. For realism, the diodes and transformers are slightly mismatched. *Source*: [4] / John Wiley & Sons.

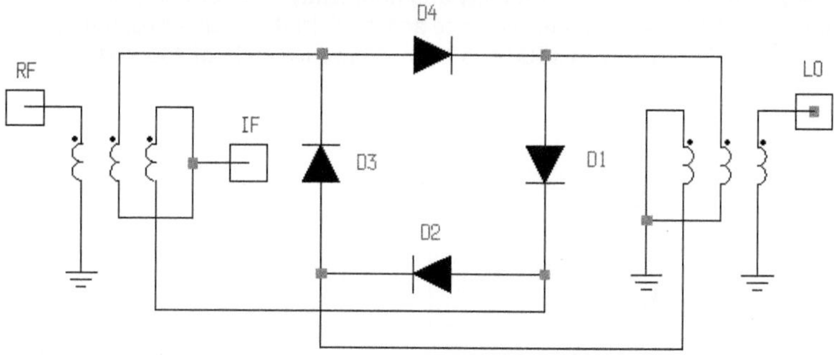

Figure 9.21 Schematic of the diode-ring DBM. *Source*: [4] / John Wiley & Sons.

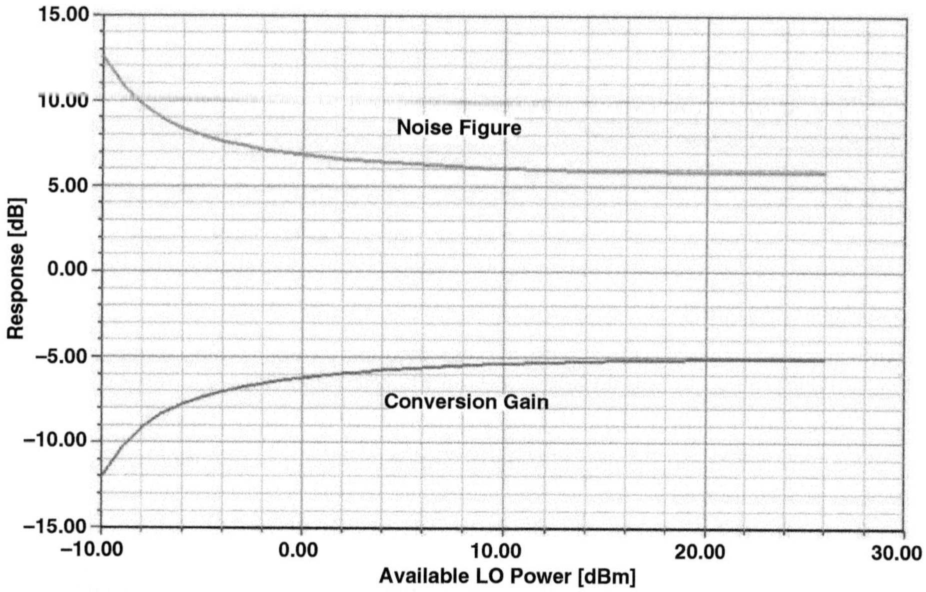

Figure 9.22 DBM conversion gain and noise figure versus LO power. In this analysis, LO = 310.7 MHz (−10 to 26 dBm), RF = 300 MHz (−40 dBm), and IF = 10.7 MHz. *Source*: [4] / John Wiley & Sons.

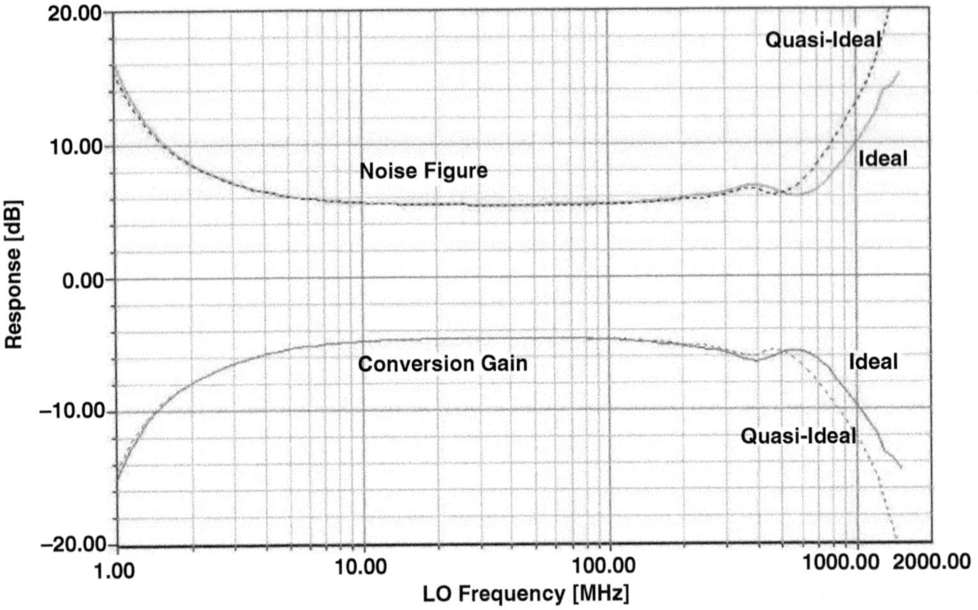

Figure 9.23 This plot of conversion gain and noise figure versus frequency for quasi-ideally and nonideally balanced versions of the same DBM reveals that balance plays a relatively minor role in the CG and NF performance achieved. In these analyses, the LO (−7 dBm) sweeps from 1 to 1500 MHz and the RF (−40 dBm) sweeps from 1.455 to 1500.455 MHz to produce an IF of 455 kHz. *Source*: [4] / John Wiley & Sons.

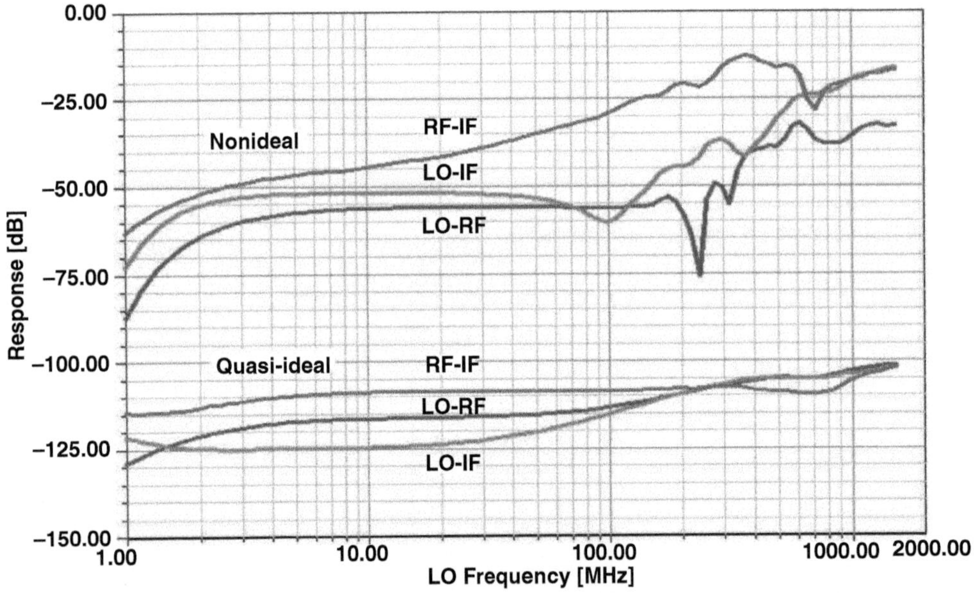

Figure 9.24 Interport isolation for DBMs with quasi-ideal and nonideal balance. In these analyses, the LO (−7 dBm) sweeps from 1 to 1500 MHz and the RF (−40 dBm) sweeps from 1.455 to 1500.455 MHz to produce an IF of 455 kHz. *Source*: [4] / John Wiley & Sons.

Two-tone testing of the DBM allows us to characterize its IP_3 figures of merit. Figure 9.28 shows the nonideal DBM's IF and IM_3 responses for LO powers of −5, 1, 7, and 13 dBm [4].

Figure 9.29 details how the DBM's IP_3 increases with LO drive, and Figure 9.30 shows the desired IF outputs and close third-order spurs near 10.7 MHz.

Although these curves indicate that the simulated mixer's IP_3 generally increases with LO drive, the improvement in IP_3 is not as great as we might expect. The reason for this is that these four analyses, as well as the other diode-mixer analyses in this chapter, were done using diode models with a threshold voltage (V_J) of 0.23. If high-level diodes with a V_J of about 0.8 V had been used, IP_{3out} for the 13 dBm LO case shown here would increase to +13 dBm. IP_{3in} for the 13 dBm LO case would turn out to be 13 dBm + insertion loss = 13 dBm + 7 dB = 20 dBm [4].

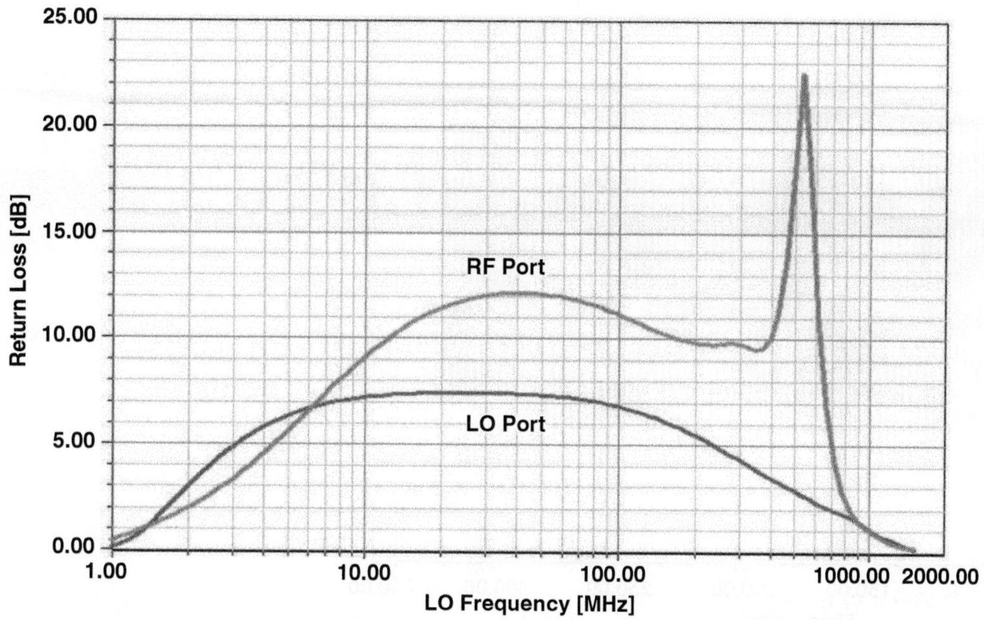

Figure 9.25 Return loss versus frequency for the DBM's RF and LO ports. The sharp peak results from a stray resonance. *Source*: [4] / John Wiley & Sons.

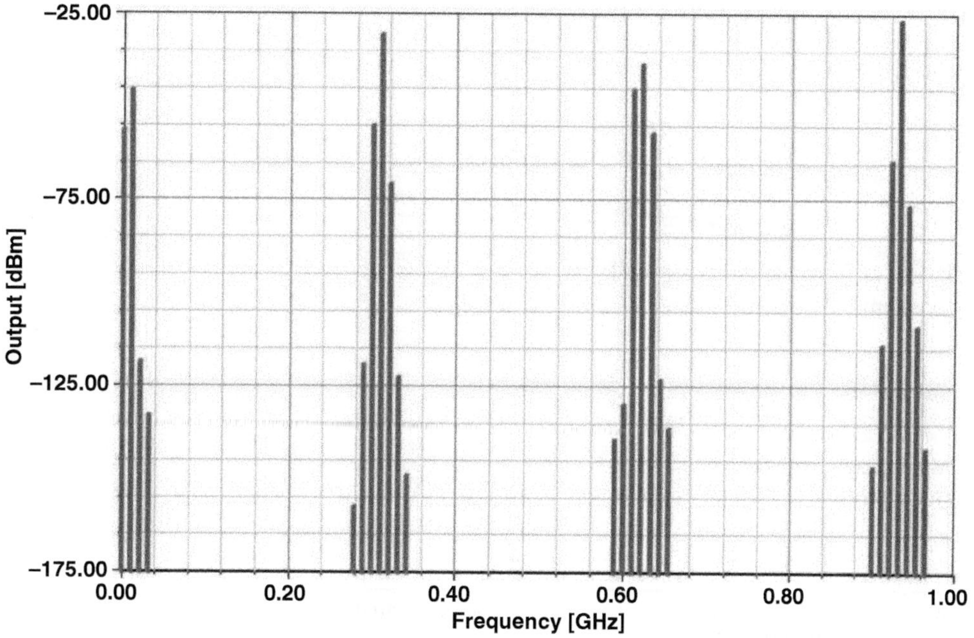

Figure 9.26 Output spectrum of a nonideally balanced DBM. In this analysis, LO = 310.7 MHz (−7 dBm) and RF = 300 MHz (−40 dBm) for an IF of 10.7 MHz. Four LO harmonics and three LO sidebands were used. *Source*: [4] / John Wiley & Sons.

The issue of diode damage aside, attempting to increase IP_3 merely by driving a low- or medium-barrier diode harder eventually results in diminishing returns. High-barrier diodes are essential in getting the best IP_3 performance with high LO drive [4].

Figure 9.31 shows the DBM's output voltage over 50 cycles of the LO signal, and Figure 9.32 shows the anode–cathode voltage of one of the ring's diodes, also over 50 LO cycles, both under two-tone IMD test conditions.

A DBM, unless it is termination-insensitive, is extremely sensitive to nonresistive termination. This is because the transmission line transformers do not operate properly when they are not properly terminated, and the reflected power generates high voltage across the diodes. This effect results in much higher distortion levels than in a properly terminated transformer.

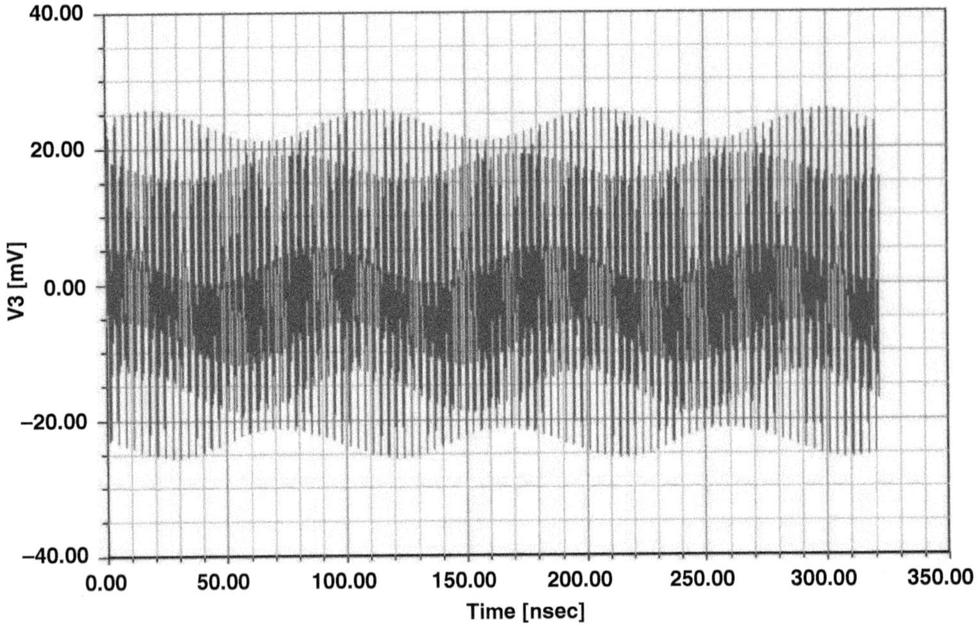

Figure 9.27 IF-port voltage waveform of the DBM over 100 cycles of the LO signal. The 310.7 MHz LO and 10.7 MHz IF components are clearly evident. *Source*: [4] / John Wiley & Sons.

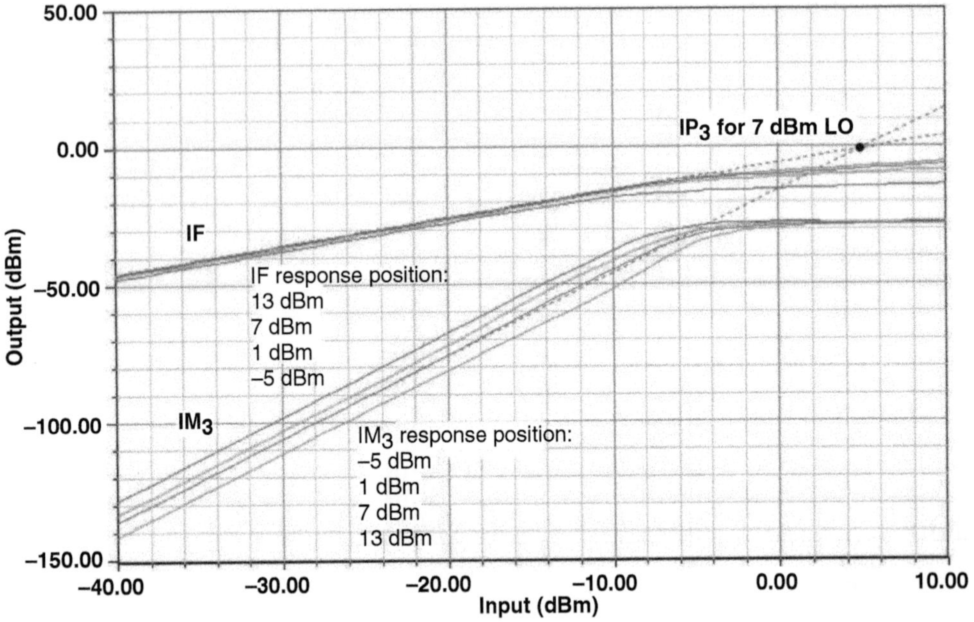

Figure 9.28 Diode DBM IF and IM_3 output versus RF power for four LO-drive levels. The responses for LO = 7 dBm have been extrapolated to show IP_3. Within limits, varying a mixer's LO drive affects its linear IF output relatively little while significantly affecting IMD. See Figure 9.29. For all four analyses, LO = 310.7 MHz, RF1 = 300.0 MHz (−40 to 10 dBm), and RF2 = 300.3 MHz (−40 to 10 dBm); four LO harmonics and three LO sidebands were used. *Source*: [4] / John Wiley & Sons.

9.6.5 Termination-insensitive Mixer

Figure 9.33 shows a mixer circuit that tolerates a fairly high VSWR at its output without significant degradation of its third-order IM performance [4, 13].

Because of the difficulties in manufacturing and cost, it appears that this mixer (Figure 9.33) fell out of grace, as we have not seen any recent advertising. A major problem is the printed transformer indicated in the dashed rectangular box.

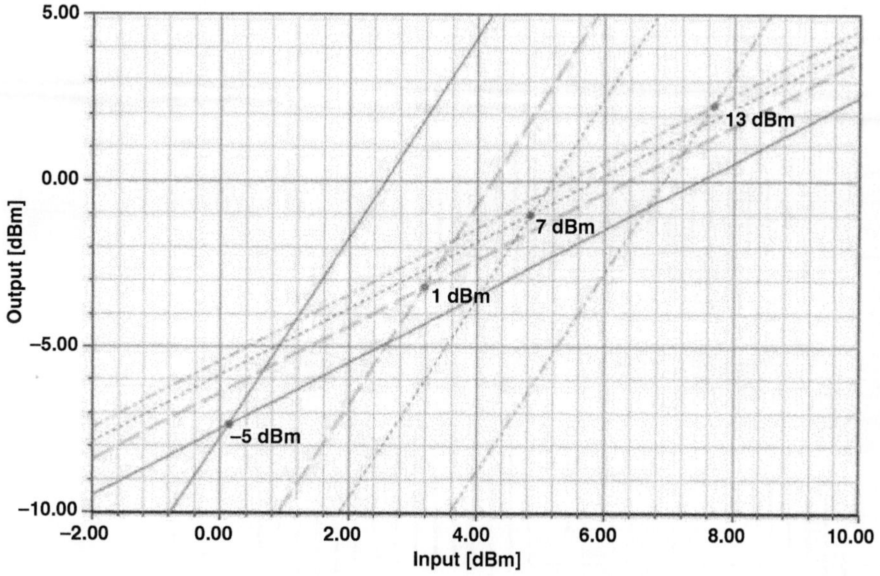

Figure 9.29 Extrapolating the responses for all four LO levels represented in Figure 9.28 shows how varying a diode DBM's LO drive shifts its third-order intercept point.

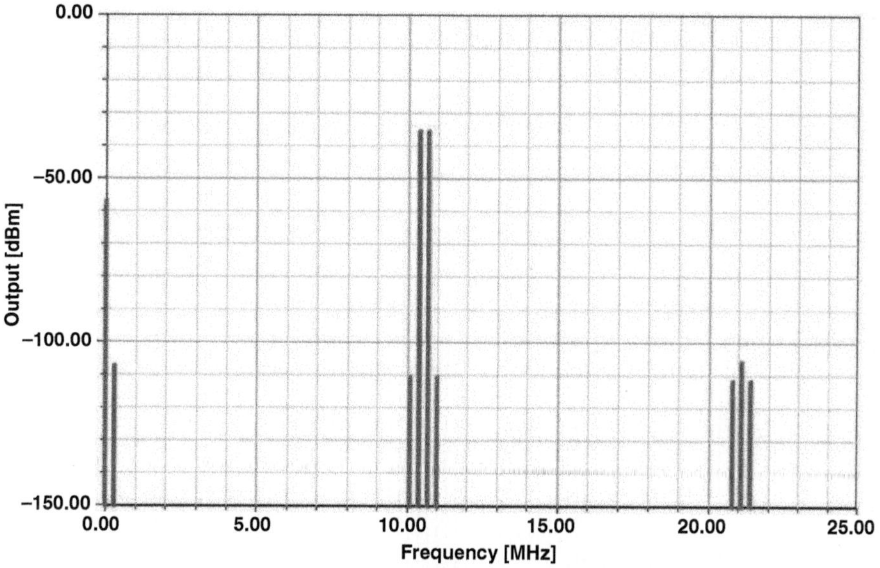

Figure 9.30 The DBM's output in the 11 MHz region during two-tone testing. The third-order products are clearly visible above and below the desired output signals. The test conditions for this analysis are those for Figure 9.28 with LO = 13 dBm. *Source*: [4] / John Wiley & Sons.

9.6.6 Image-reject Mixer

The image-reject mixer consists of a basic quadrature IF mixer with an additional 90° hybrid at the IF ports, as shown in Figure 9.34. The primary function is to differentiate between the real signal and the image signal. This type of device is especially useful in applications where the desired RF signal and image are so close in frequency that rejecting the image with filtering is not practical. This type of mixer is now heavily used for downconversion of the frequency spectrum, going into an analog-to-digital converter [4].

9.6.7 Quadrature IF Mixer

A quadrature IF mixer produces two IF outputs in phase quadrature. Its basic structure consists of two DBMs: a 90° splitter and 0° splitter. The basic block diagram is shown in Figure 9.35 [4].

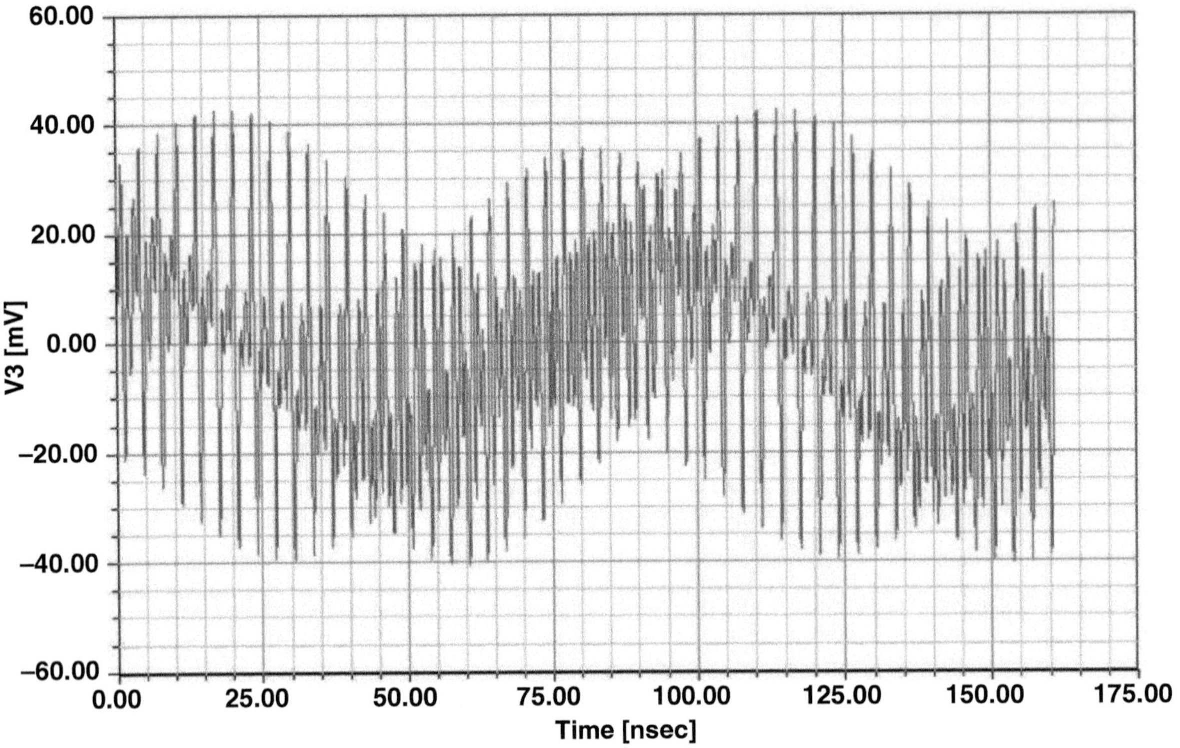

Figure 9.31 The DBM's IF-output voltage over 50 cycles of the LO signal. The test conditions for this analysis are those for Figure 9.28 with LO = 13 dBm. *Source*: [4] / John Wiley & Sons.

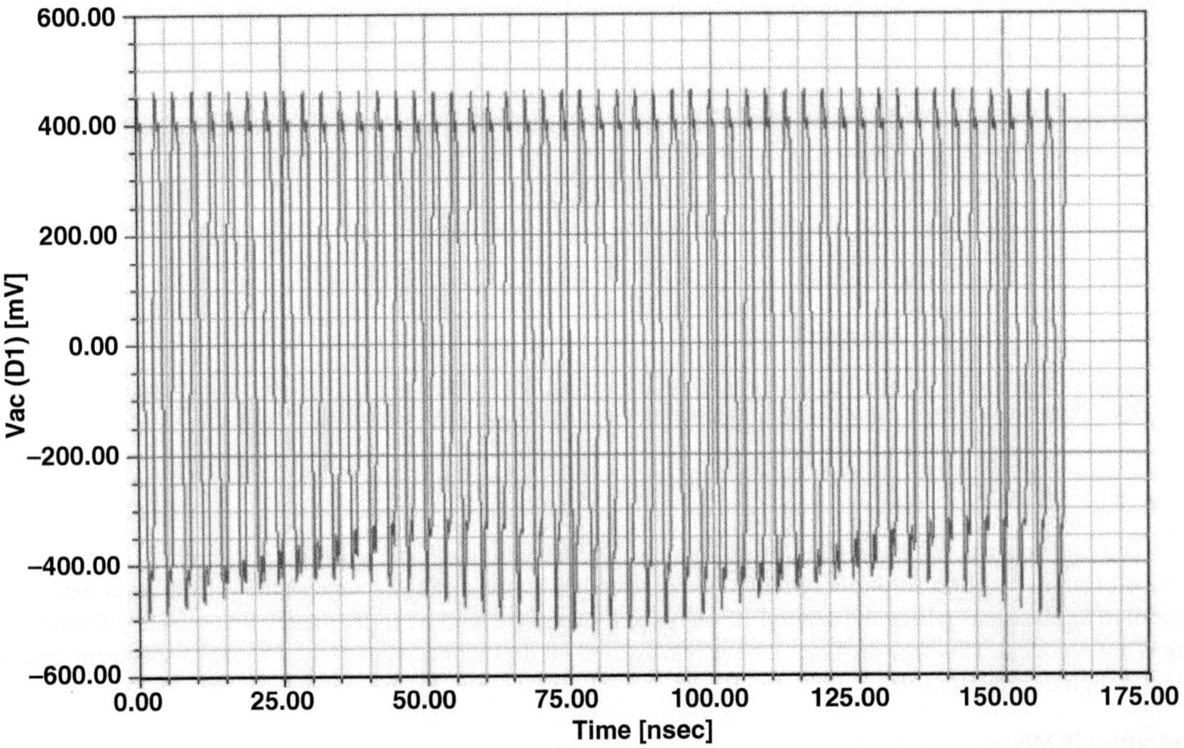

Figure 9.32 The anode–cathode voltage of one of the DBM's diodes, also over 50 LO cycles under two-tone IMD test conditions. The test conditions for this analysis are those for Figure 9.28 with LO = 13 dBm. *Source*: [4] / John Wiley & Sons.

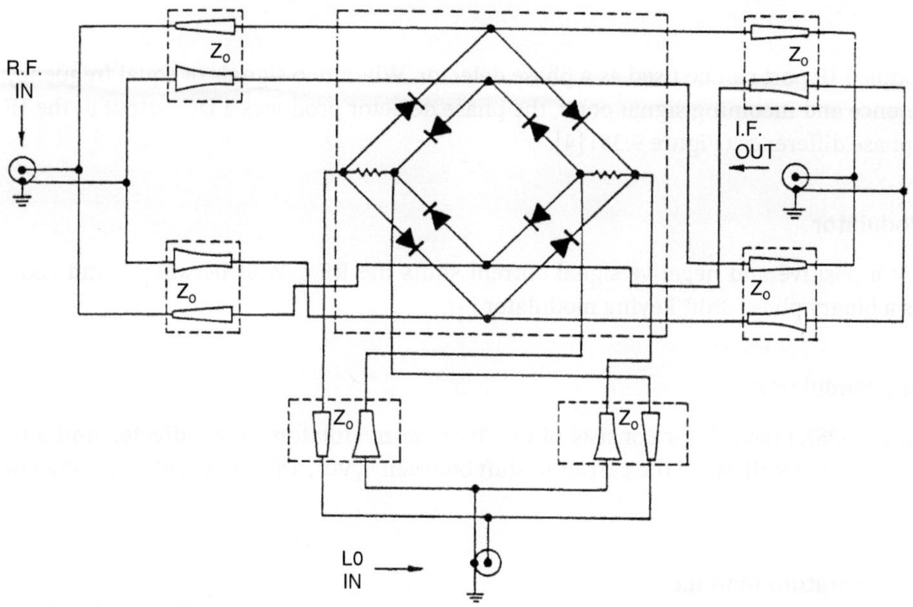

Figure 9.33 Example of a termination-insensitive mixer from Ref. [4, p 675, 13].

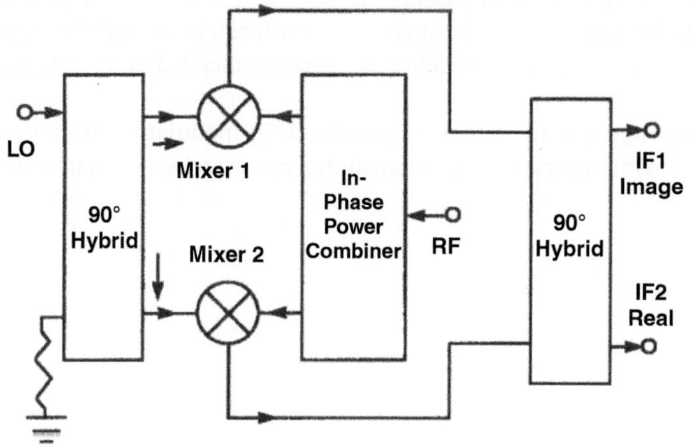

Figure 9.34 An image-reject mixer uses phasing to differentiate between its (LO + IF) and (LO − RF) IF outputs. *Source*: [4] / John Wiley & Sons.

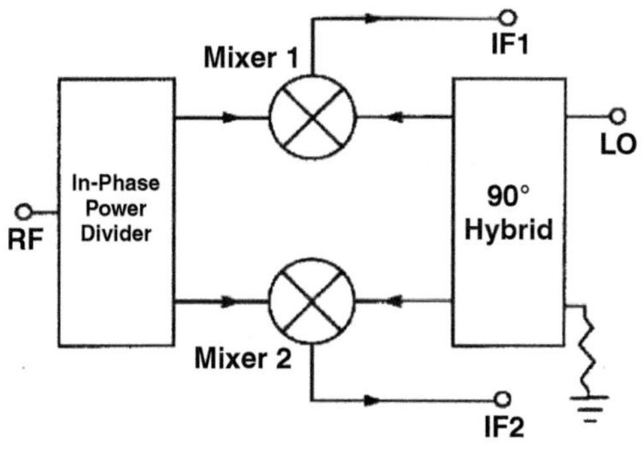

Figure 9.35 Quadrature mixer. *Source*: [4] / John Wiley & Sons.

9.7 Phase Detector

Theoretically, any mixer with a DC-coupled IF port can be used as a phase detector. When two signals of equal frequency are applied simultaneously to the reference and incoming signal ports, the phase detector produces a DC output at the IF port proportional to the cosine of the phase difference (Figure 9.36) [4].

9.7.1 Binary Phase-shift Keying Modulator

Binary phase modulation occurs when a positive and negative signal current-shifts the RF carrier between 0 and 180°. Figure 9.37 shows a DBM operating as a binary phase-shift keying modulator [4].

9.7.2 Quadrature Phase-shift Keying Modulator

A typical quadrature phase-shift keying (QPSK) modulator consists of two biphase modulators, a 90° divider, and a 0° power combiner. Data inputs at the control ports will cause the carrier to shift between 0, 90°, 180°, and 270°, as shown in Figure 9.38 [4].

9.7.3 Single-sideband or In-phase/Quadrature Modulator

SSB or in-phase/quadrature (I/Q) modulators are useful in discriminating and removing the LSB or USB generated during frequency conversion, especially when the sidebands are very close in frequency and attenuation of one of the sidebands cannot be achieved with filtering. This is the case with audio and video modulation, where signals from DC to 10 MHz must be converted to a higher frequency that is appropriate for transmission. In such cases, both sidebands will be very close in frequency to the carrier frequency. With the I/Q modulator, one of the sidebands is easily cancelled or attenuated along with its carrier [4].

Attenuation of the carrier has been the most troublesome aspect in the design of passive (I/Q) modulators. Isolation between the LO port and the RF port of the mixers, which is the main parameter in determining carrier rejection, is usually insufficient at frequencies above 200 MHz.

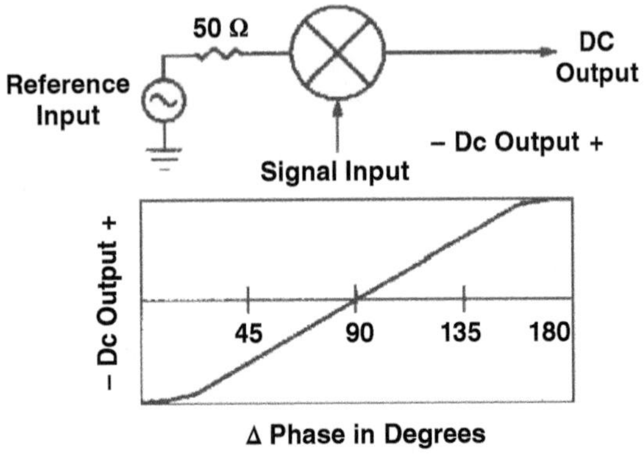

Figure 9.36 A mixer with a DC-coupled IF port can be used as a phase detector. *Source*: [4] / John Wiley & Sons.

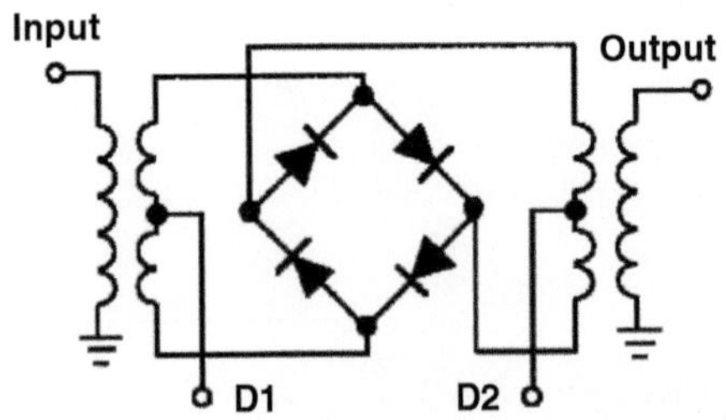

Figure 9.37 A diode-ring mixer as phase modulator. *Source*: [4] / John Wiley & Sons.

A ring of PIN diodes can be used as electronic attenuators by applying variable forward bias to the diodes (Figure 9.39). Maximum attenuation is achieved when the current at the control port is zero. The maximum attenuation is the isolation between the input and output ports. Minimum attenuation (insertion loss) is achieved when the IF port current is 20 mA [4].

I/Q modulator designs have basically comprised two DBMs (Figure 9.40). The mixers are fed at the LO ports by a carrier phase-shifted through a 90° hybrid. Thus, the carrier signal's relative phase is 0° to one mixer and 90° to the other mixer. Modulation signals are fed externally in phase quadrature to the two mixers' IF ports. The mixers' modulated output signals are combined through a two-way, in-phase power divider/combiner.

The circuit forms a phase-cancellation network to one of the sidebands and a phase-addition network to the other sideband. The carrier is somewhat attenuated and is directly dependent on the inherent LO-to-RF isolation of the mixers and the modulating signal level. In industry-standard (I/Q) modulators, USB suppression results when the first modulation port (MOD 1) is fed with a signal that is 90° in advance of the signal feeding the second modulation port (MOD 2). Opposite

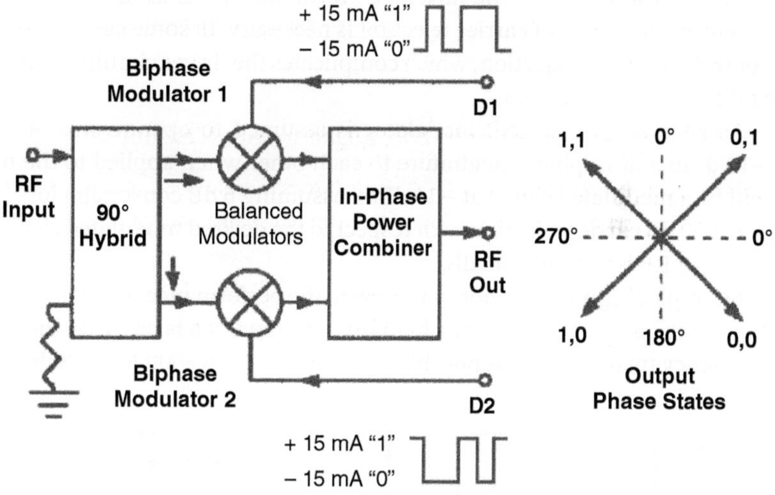

Figure 9.38 Two biphase modulators form the basis for a QPSK modulator. *Source*: [4] / John Wiley & Sons.

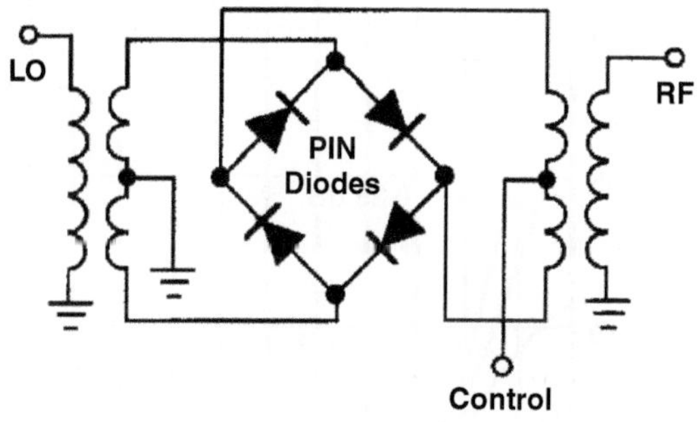

Figure 9.39 A diode DBM can be used as a DC-controlled attenuator if PIN diodes are used instead of Schottky devices in its ring. *Source*: [4] / John Wiley & Sons.

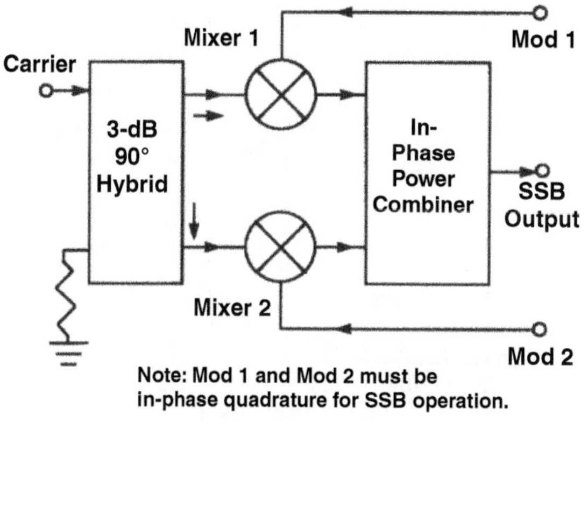

Figure 9.40 An SSB modulator matches two high-frequency mixers, a 90° hybrid, and an in-phase power combiner to produce an SSB output signal. *Source*: [4] / John Wiley & Sons.

phasing can be arranged by changing the internal phase polarity of the mixers or by interchanging the 90° hybrid output ports to the LO ports of the mixers.

The phase and amplitude imbalances between the various components used in the manufacturing of the I/Q modulators must be tightly maintained for optimum SSB rejection. Matching of the two mixers for conversion loss and insertion phase is extremely critical, since differences in these parameters will add to amplitude- and phase-imbalance errors. The 90° hybrid in the LO port must be in nearly perfect phase quadrature.

Phase- and amplitude-imbalance errors adversely affect sideband suppression (Figure 9.41). In most cases, a typical passive (I/Q) modulator operates with a carrier input level of +10 dBm, which is required to drive the diodes in the mixers to operate in the linear range [4].

The dynamic range of these mixers can be significantly improved by using diodes with a higher barrier height. The LO signal in this case must be increased in order to drive these diodes into conduction in their linear range.

Carrier rejection is also a problem when designing an SSB modulator, since only a few decibels of suppression can be achieved in standard high-frequency models. In the past, the major contributor to carrier suppression was the inherent LO-to-RF isolation through the mixers. Unfortunately, this isolation is usually poor at cellular frequencies (800–1000 MHz), where at least 25 dB of carrier rejection is necessary. In some cases, designers feed a small amount of DC into the IF ports to control the carrier rejection, which complicates the driver circuitry and calls for temperature compensation when operating at different temperatures.

As an example, an SSB modulator is assumed to operate with +10 dBm LO drive with each modulating signal at −10 dBm and in phase quadrature to each other when applied to the modulating ports (MOD 1 and MOD 2). The result will be a modulated signal at −16 dBm, assuming 6 dB conversion loss [4].

For 20 dB carrier rejection with respect to the desired modulated signal, the carrier must be at −36 dBm, which translates to LO-to-RF isolation of 46 dB.

By employing a subharmonic approach, the performance of SSB modulators can be extended beyond the limits of conventional designs as reported by Joshi in [9]. The approach is based on the use of subharmonic mixers in place of fundamental-frequency mixers and is applicable from about 140 to 3000 MHz. Subharmonic mixers use antiparallel diode pairs in their

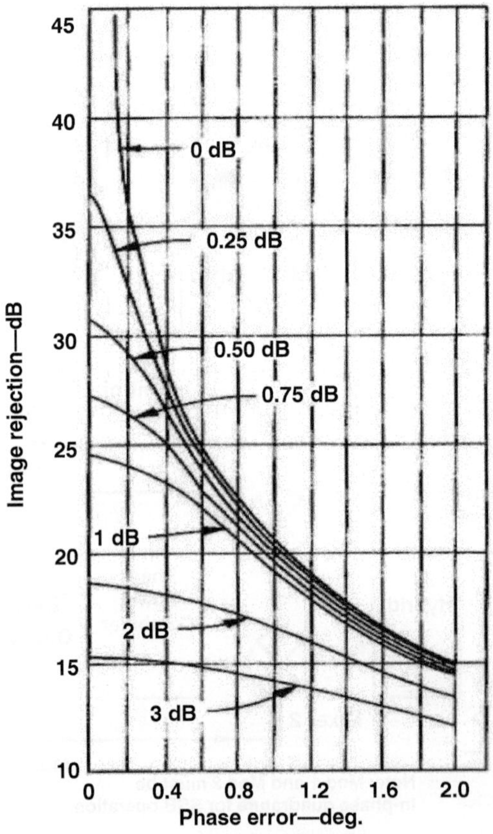

Figure 9.41 The level of SSB rejection improves as the phase- and amplitude-imbalance performance of an SSB modulator improves.

construction [4, 14–16]. Matched antiparallel diode pairs used in single-ended or single-balanced mixer configurations cancel even-order intermodulation products (such as $2f_{LO} \times 2f_{RF}$, $3f_{LO} \times 3f_{RF}$, etc.) at all ports.

Single-ended mixers lack the port-to-port isolation needed for SSB modulator applications. Odd-order products of the RF and LO frequencies (even $f_{LO} \times$ odd f_{RF}) and (odd $f_{LO} \times$ even f_{RF}) appear on all ports, requiring extensive filtering for satisfactory performance.

For a single-balanced mixer, even harmonics of the LO combining with odd harmonics of the RF appear at the IF port, whereas odd harmonics of the LO combining with even harmonics of the RF appear at the RF and IF ports. This assumes that a balanced transformer is placed at the LO port, which is a logical choice due to the fact that the highest level signal appears at the LO port. Since the desired odd-order IF products appear at both the RF and IF ports, a diplexing network is needed to isolate the RF and IF signals.

The subharmonic modulator design provides a unique way to isolate the RF and IF signals. A single-balanced harmonic mixer offers good LO-to-RF and LO-to-IF isolation but poor RF-to-IF isolation. Fortunately, harmonically related signals are spaced well apart in the frequency spectrum, simplifying the filtering of harmonically related signals.

Harmonic mixing also works well with low LO power levels, with somewhat lower 1 dB compression on the RF port than with fundamental-frequency mixing. The ability to operate with LO frequencies that are a fraction of the carrier frequency (1/2, 1/4, 1/6, etc.) significantly reduces the cost of an LO source, especially at higher frequencies. Also, using lower-frequency LO sources helps avoid the signal-leakage problems inherent with higher-frequency LO sources. Minimizing signal leakage, especially at higher frequencies, becomes expensive and bulky. Subharmonic mixing offers several advantages:

- The technique offers the ability to operate at LO frequencies that are 1/2, 1/4, or 1/6 of the carrier frequency. For example, for an IF of 100 MHz at an RF of 2 GHz, the LO can be $(2000 \pm 100) \div 2 = 950$ or 1050 MHz.
- The LO's even harmonics are strongly attenuated.
- The filtering requirements for fundamental frequency and odd harmonic signals of the LO are not critical.
- The cost of generating the LO is reduced due to the fact that the LO frequency needs to be only a fraction of the carrier frequency.

As an example of the performance improvements possible with the subharmonic mixers, units were evaluated at both cellular (935 to 960 MHz) and PCN/PCS (1.8 to 1.9 GHz) bands. For a conventional SSB modulator at 1.9 GHz fed with +10 dBm modulation signals, carrier rejection is barely 10 dB (Figure 9.42).

Sideband rejection can be improved by tuning, but the carrier rejection is controlled by the LO-to-RF isolation of the DBMs. Conventional DBMs with high isolation at cellular and PCN bands are very expensive and large when special techniques are used to improve LO-to-RF isolation. In contrast, the subharmonic nature of the new approach allows the use of lower-frequency, less-expensive components in the modulators' construction.

The subharmonic modulators offer an improvement of more than 15 dB in carrier suppression compared to the conventional approach. The measured VSWR (return loss) at the LO and RF ports is better than 1.50:1 (Figures 9.43 and 9.44). Measurements made on a cellular-band SSB modulator reveal carrier rejection on the order of 40 dB. Typical insertion loss is 7 dB while sideband rejection is 30 dB (Figure 9.45).

By virtue of harmonic mixing, even-order mixing products are attenuated by about 30 dB with respect to the desired modulated output signal. The fundamental-frequency feedthrough into the output port is approximately 5 dB lower than

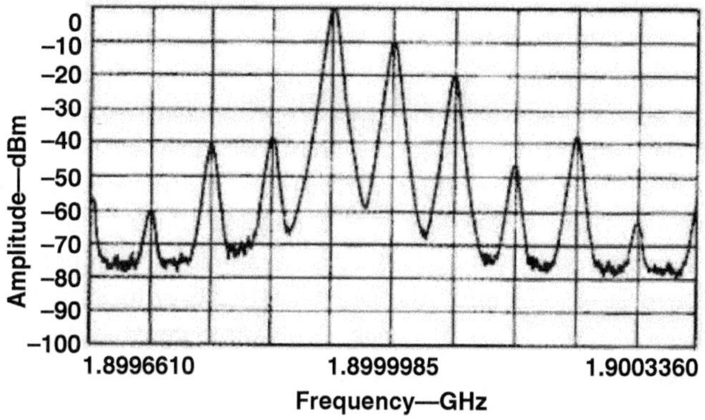

Figure 9.42 This plot of carrier and sideband rejection was measured for a conventional SSB modulator operating at 1.9 GHz. *Source*: [4] / John Wiley & Sons.

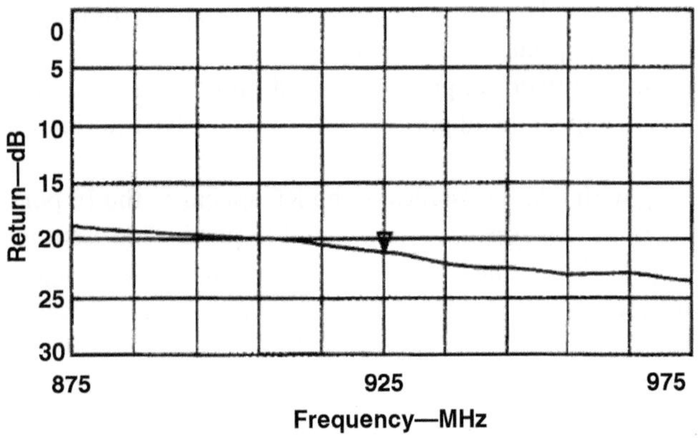

Figure 9.43 The SSB modulator's return loss as measured at the LO port. *Source*: [4] / John Wiley & Sons.

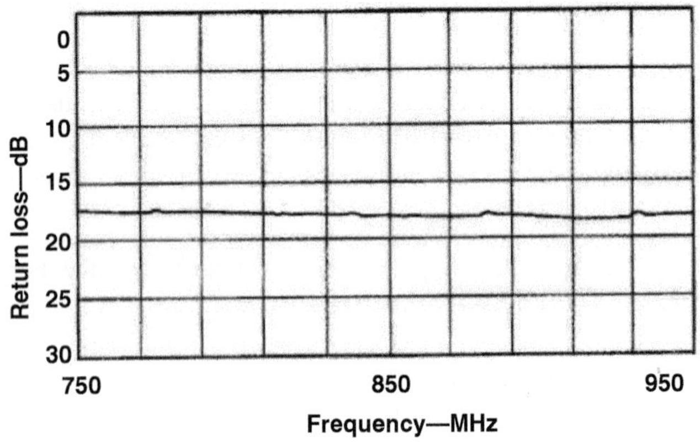

Figure 9.44 The novel harmonic SSB modulator's return loss as measured at the RF port. *Source*: [4] / John Wiley & Sons.

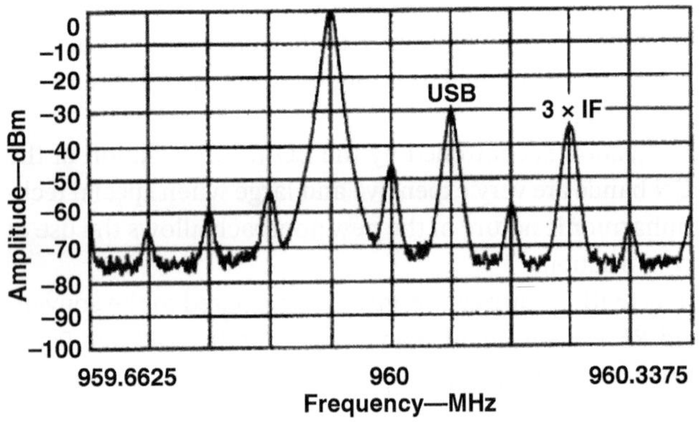

Figure 9.45 This plot of carrier and sideband rejection as measured for the novel harmonic SSB modulator operating at cellular frequencies. *Source*: [4] / John Wiley & Sons.

the desired modulated signal, whereas the fourth harmonic mixing with the modulating signal is approximately 10 dB lower. Typical loss for fourth harmonic mixing is 17–19 dB while maintaining 30 dB of carrier rejection.

Since harmonically related products are well spaced in frequency, filtering undesired signals is relatively inexpensive using standard octave-bandwidth filters. Low-cost commercial bandpass filters typically offer better than 40–50 dB attenuation of unwanted harmonic signals. Constant-impedance bandpass filters offering a good impedance match at the desired stopbands can also be used in cases where harmonically related products require impedance termination within a system.

The subharmonic modulator design is easily applied at custom frequencies. Conversion of an SSB modulator with output frequency corresponding to twice the LO frequency to one with output corresponding to four times the LO frequency requires only one component change, in the form of a signal-combining network at the modulator's output. Although the conversion loss of the fourth-harmonic LO component mixing with the modulating signal is in the vicinity of 18 dB, the cost of generating the LO is drastically reduced with the subharmonic modulator. In spite of higher signal loss, the carrier rejection is still at least 30 dB at the fourth harmonic, and harmonically related products can be eliminated with an inexpensive filter [4].

9.8 Triple-balanced Mixer

Rings can be combined to form a doubly DBM, or triple-balanced mixer (TBM), as shown in Figure 9.46. TBMs achieve a higher dynamic range and interport isolation than double-balanced designs at the expense of LO power and increased complexity and size.

Figure 9.47 shows the Figure 9.46 circuit's interport isolation with the circuit configured in a less than ideally balanced form, with small variations in transformer-winding inductance and diode parameters introduced for more realistic modeling. Note that the mixer's interport isolation generally increases with frequency, rather than decreasing with frequency as with the DBM (Figure 9.24) [4].

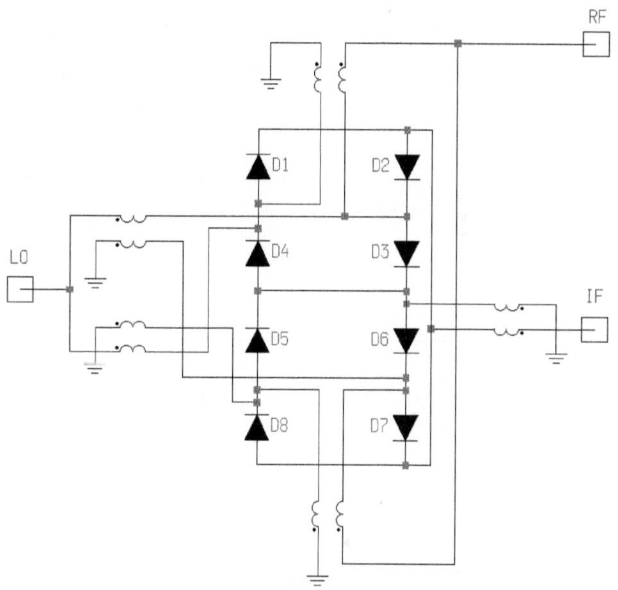

Figure 9.46 A triple-balanced diode mixer. A limitation of this configuration is that the internal DC common connections associated with its RF and LO transformers disallow usable IF response down to DC. *Source*: [4] / John Wiley & Sons.

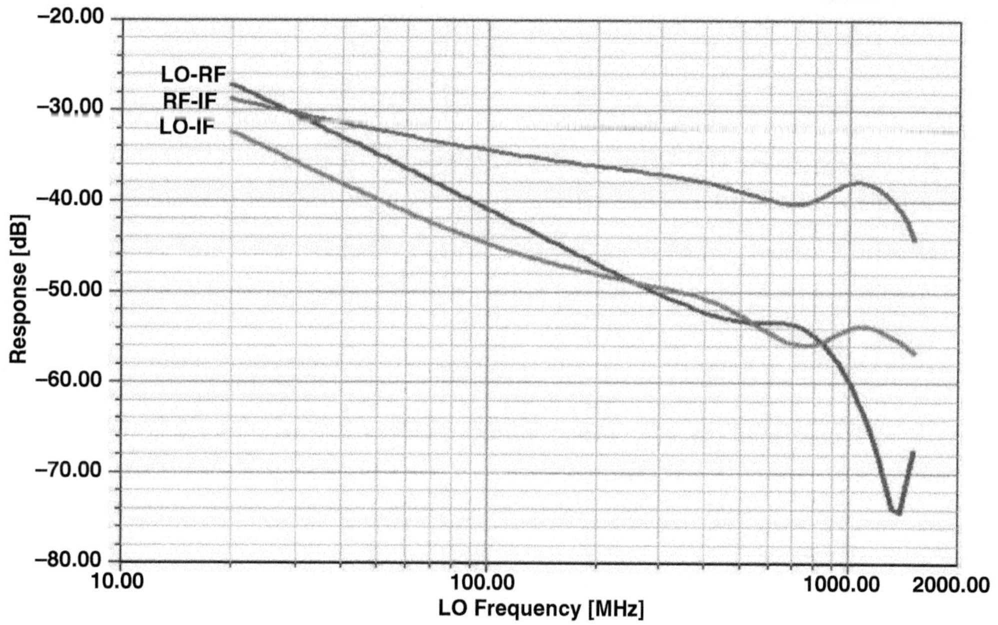

Figure 9.47 The TBM offers improved high-frequency isolation over a standard DBM. *Source*: [4] / John Wiley & Sons.

9.9 Rohde & Schwarz Subharmonically Pumped DBM

Figure 9.48 shows an example of a complete microwave diode mixer used as the input stage of a spectrum analyzer. Its LO is applied at 13.2 GHz, one-third the frequency necessary to mix the 40.1 GHz input signal to the 500 MHz IF, with distortion in the diodes providing the 3× frequency multiplication. The circuit's conversion gain is −18.8 dB; its noise figure is 20.5 dB.

The subharmonic-drive technique exemplified by this mixer is important for another reason in addition to LO isolation. Depending on the application, directly generating a sufficiently phase-quiet LO signal at higher wireless frequencies and

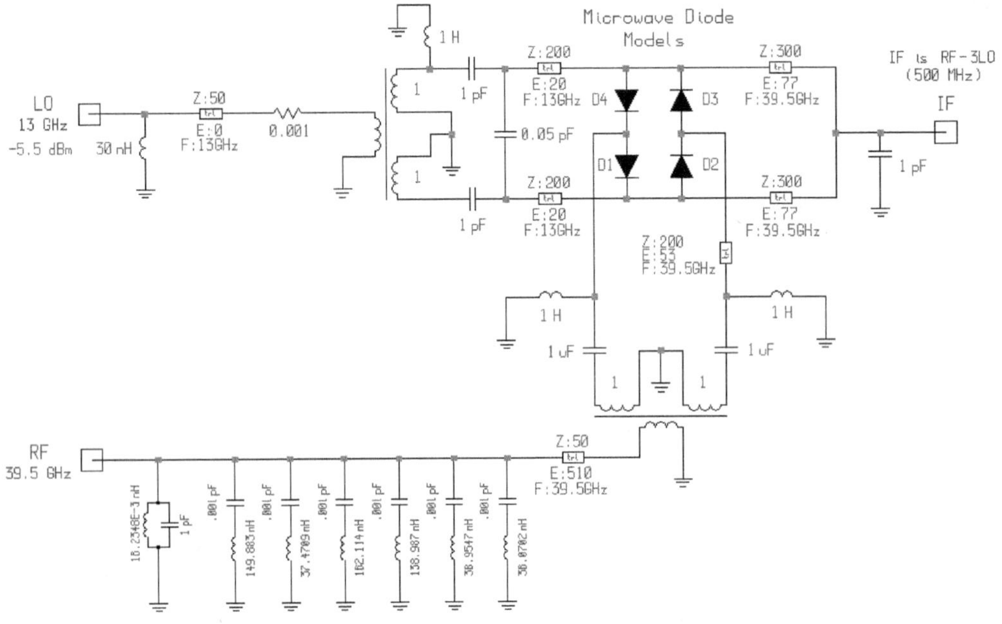

(a)

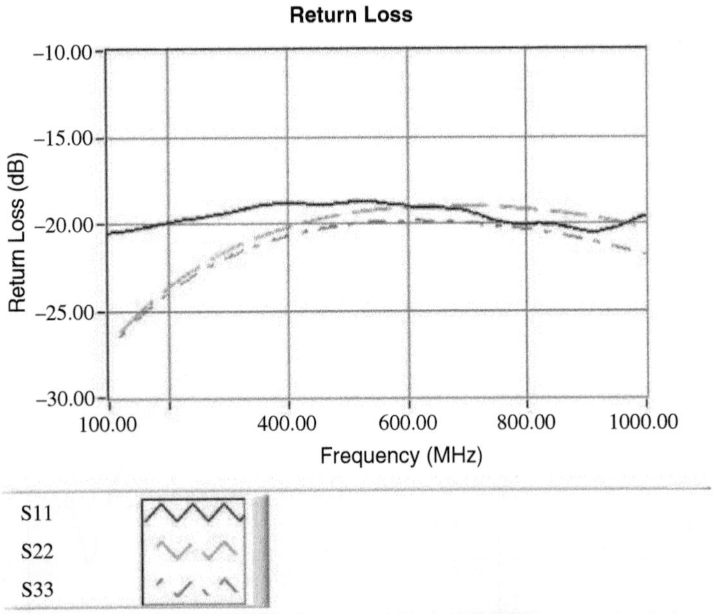

(b)

Figure 9.48 (a) Schematic of the spectrum-analyzer mixer. *Source:* [4] / John Wiley & Sons. (b) Measured data at Synergy Microwave of the return loss as a function of frequency of the circuit in (a).

(Continued)

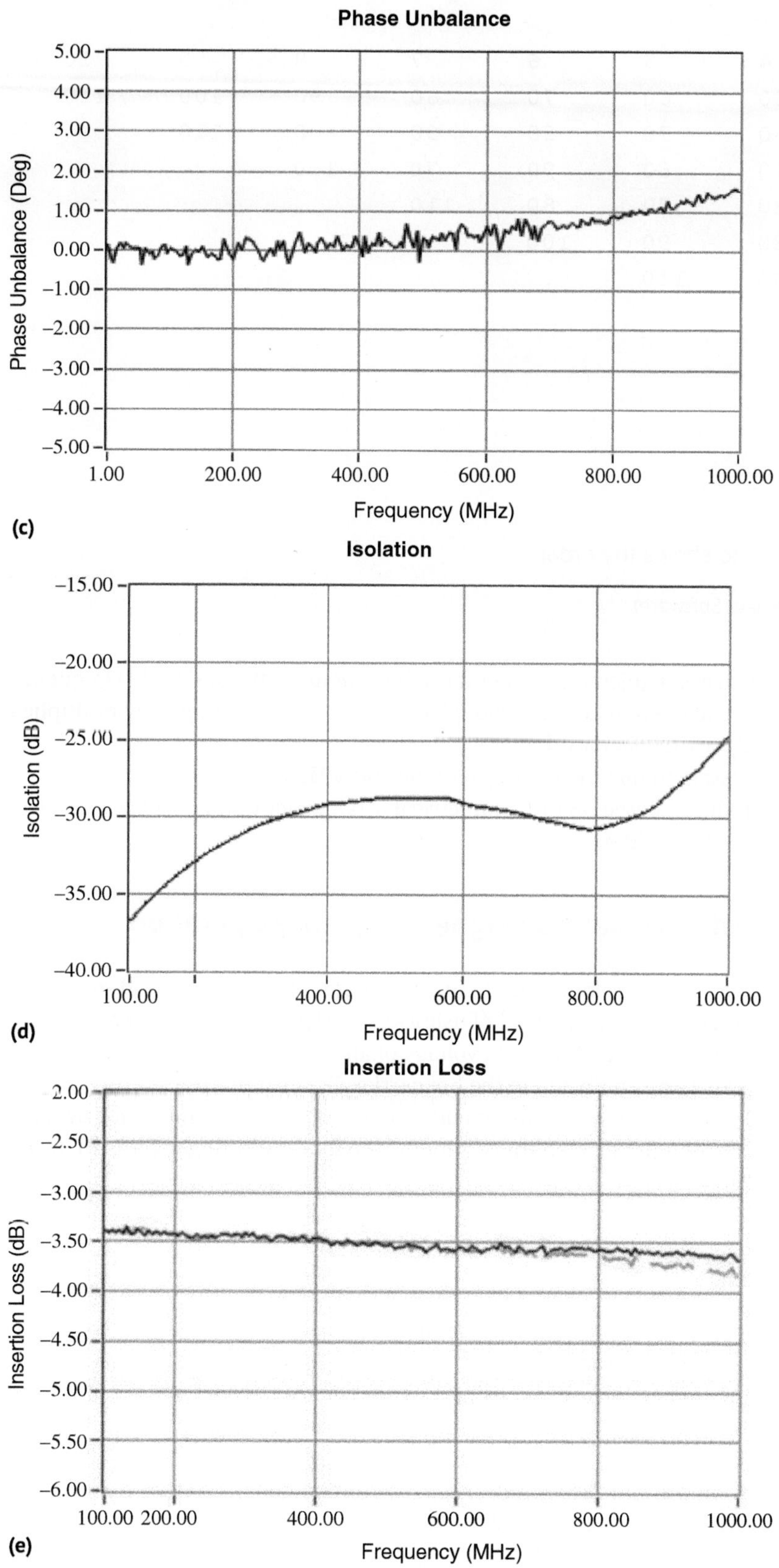

Figure 9.48 (Cont'd) (c) Measured data at Synergy Microwave of the phase unbalance as a function of frequency of the circuit in (a). (d) Measured data at Synergy Microwave of the isolation as a function of frequency of the circuit in (a). (e) Measured data at Synergy Microwave of the insertion loss as a function of frequency of the circuit in (a).

MIXERSPURS MAXORDER = 10

*0	1	2	3	4	5	6	7	8	9	10
20	18	32	40	50	60	70	80	90	100	110
30	0	40	50	60	70	80	90	100	110	
40	20	50	60	70	80	90	70	110		
50	30	60	70	80	70	80	110			
60	40	70	70	80	90	100				
70	50	80	90	100	110					
80	60	90	100	110						
90	70	100	100							
100	80	110								
110	100									
120										
↓										

Vertical lines are the attenuation in dB, the first line shows the order.

Figure 9.49 A typical spurious table for a DBM (Compact Software).

into the microwave range may not be feasible. In such cases, injecting the LO at a subharmonic of the desired LO frequency may provide better phase noise performance than a fundamental LO, even though the phase noise of a frequency-multiplied source increases by $\log_{10} n$ dB, where n is the multiplication factor [4].

The LO is applied at one-third the frequency necessary to mix the RF signal to the 500 MHz IF.

The DBM generates internally many spurious products. Based on the symmetrical arrangement, some will be canceled. The following is a typical spurious table (Figure 9.49) for a DBM.

9.10 DBM Using GaAs FETS Connected as Diodes for Higher-frequency Operation

Figure 9.50 shows a DBM using GaAs FETs connected as diodes for higher-frequency operation, as the gate and source are connected. Each (FET) transistor is connected as a diode (passive). Instead of using hot carrier diodes, for millimeter-wave applications, GaAs FETs can be used. The drawback is the diode's threshold voltage of about 1 V, and they are not selected for symmetry. On the other hand, they can be less expensive compared to the hot carrier diodes.

An experimental mixer is shown in Figure 9.51. This mixer was built by Hittite under a DARPA contract for Synergy Microwave as a test vehicle. The mixer consists of rectangular printed transformers. The four multi-finger GaAs FETs are shown in 4 rows of dark gray squares (Synergy Microwave).

9.11 Analysis of a Balanced Microwave Mixer

The following is an analysis/simulation of a very popular balanced microwave mixer using a Lange wideband coupler at the input. The necessary bond wires limit the upper frequency range.

Figure 9.52 shows the schematic of a typical balanced microwave mixer using a Lange wideband coupler at the input and Schottky barrier diodes prepared for simulation.

Figure 9.53 shows the voltage waveform at the IF output. Again, notice that the filter provides rejection of the input signals. The IF component is clearly seen in the waveform, and the RF and LO harmonic components are small.

Figure 9.54 shows the output spectrum of the mixer. This project includes a low-pass filter in the IF output, so there is excellent RF and LO rejection at the IF port.

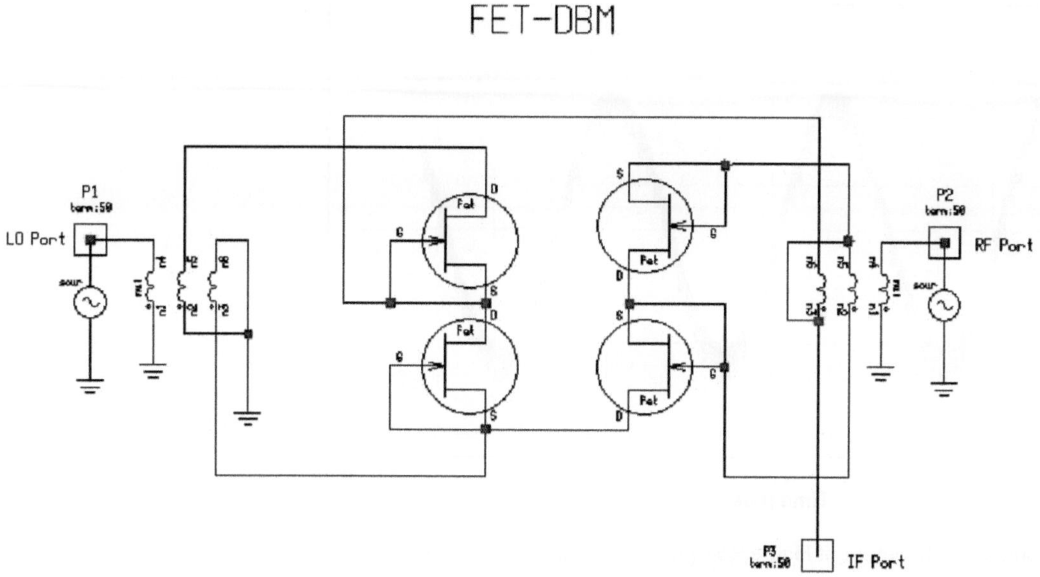

Figure 9.50 DBM using GaAs FETs connected as diodes for higher-frequency operation—as the gate and source are connected.

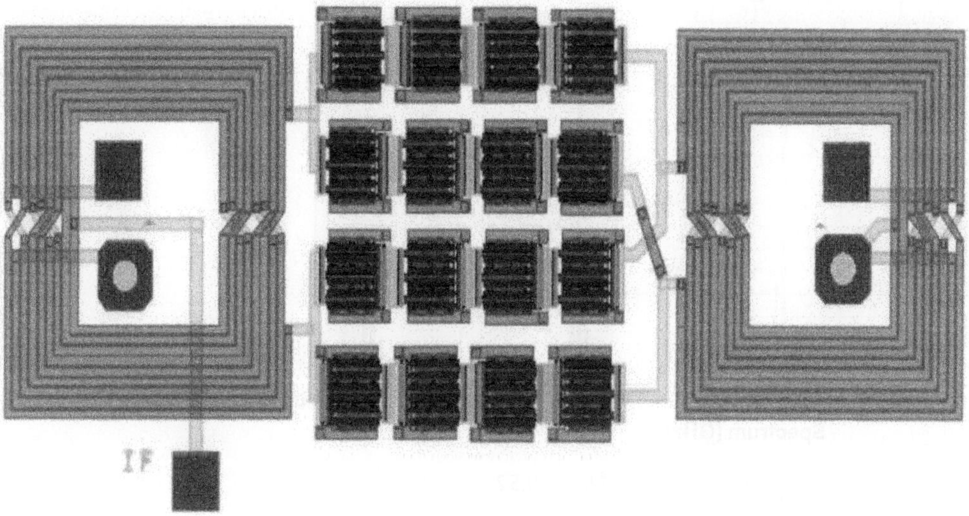

Figure 9.51 Mixer layout by Hittite for Synergy Microwave.

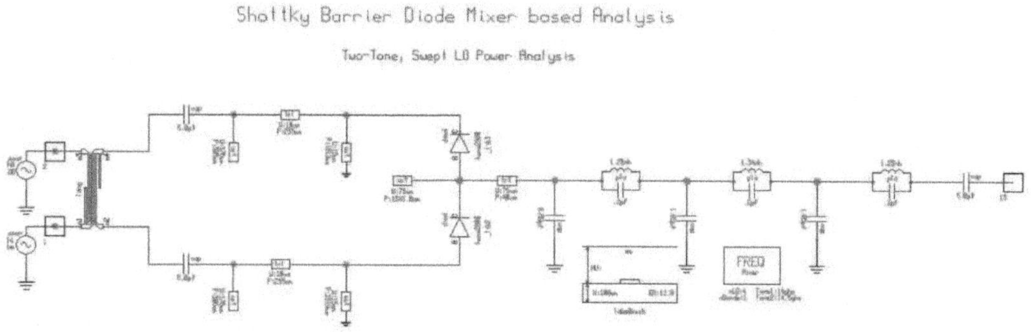

Figure 9.52 Schematic of a typical balanced microwave mixer using a Lange wideband coupler at the input and Schottky barrier diodes prepared for simulation [50, pp 3–93].

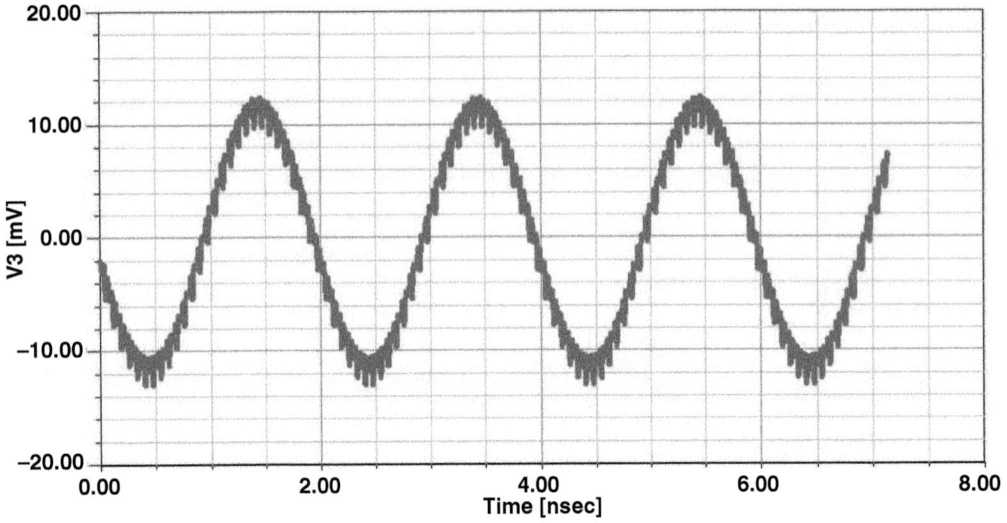

Figure 9.53 Voltage waveform at the IF output of Figure 9.52 [50, pp 3–95].

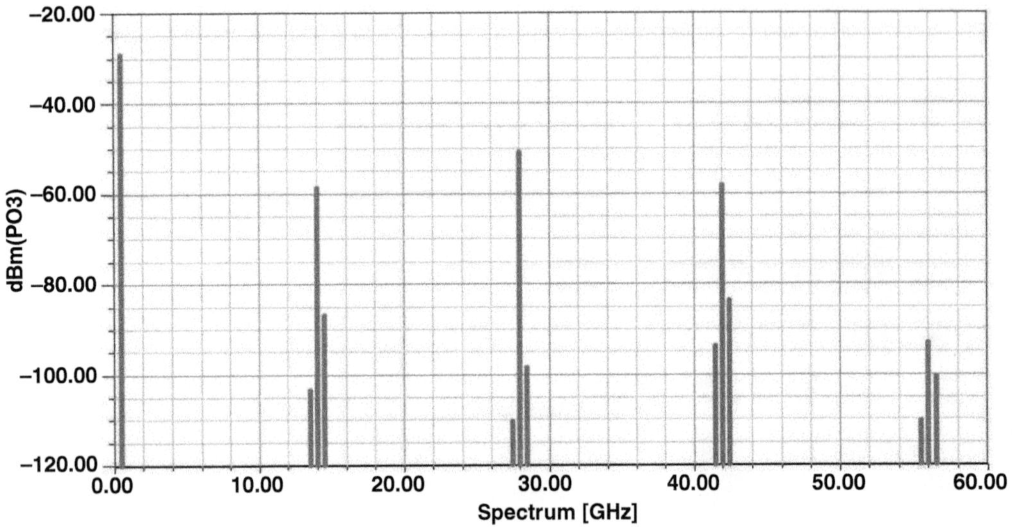

Figure 9.54 Simulated output power spectrum in dBm of the mixer of Figure 9.52.

Next, we modify the project so that both LO and RF frequencies are swept together, thus producing a constant IF frequency.

Figure 9.55 shows conversion gain, RF return loss, and IF spectral purity, all as a function of input frequencies.

Figure 9.56 shows the mixer conversion gain as a function of LO input power. From this graph, we can determine that the optimal operating point is about +6 dBm, which provides a good balance between the cost of providing LO power and diminishing returns for conversion gain.

9.11.1 Three-tone Mixer Analysis with Swept Power

Now we perform a three-tone mixer analysis with swept power:

The typical demand placed on mixers today requires arbitrary three-tone analysis capabilities. Basically, this means that the mixer is being driven by a standard LO, but there are two RF signals being input simultaneously, and the RF signals are very close to each other in frequency (Figure 9.57). The goal of the mixer designer is to minimize the intermodulation products at the IF. We will analyze our mixer with the LO at its nominal 14 GHz and the RF signals at 14.49 and 14.51 GHz.

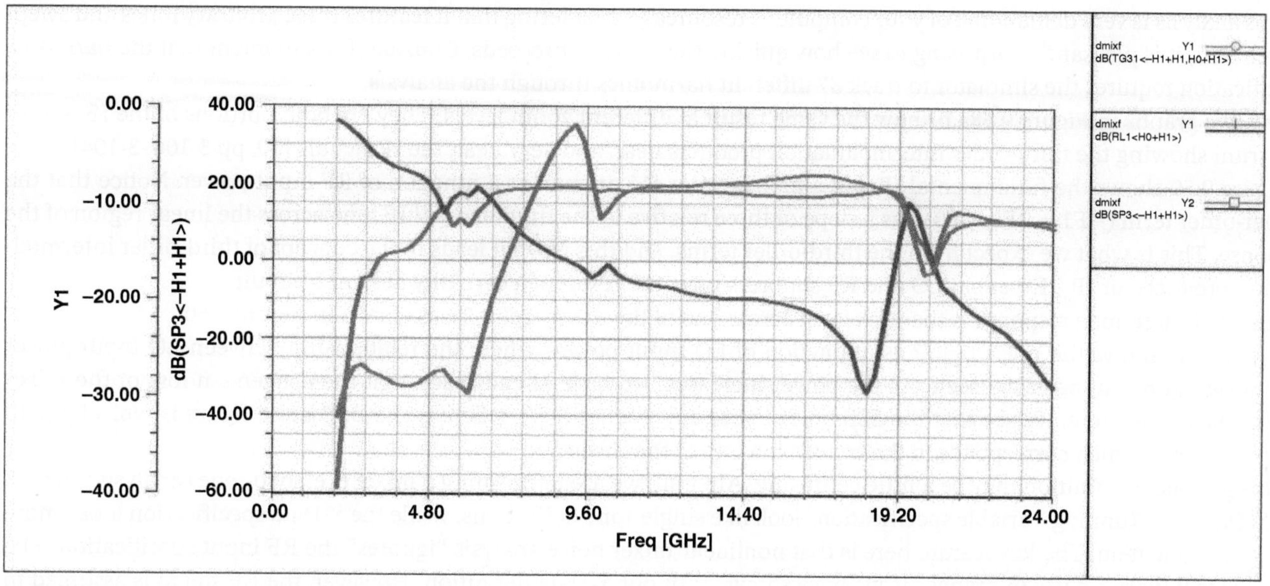

Figure 9.55 Conversion gain, RF return loss, and IF spectral purity, all as a function of input frequencies (of Figure 9.52) [50, pp 3-97].

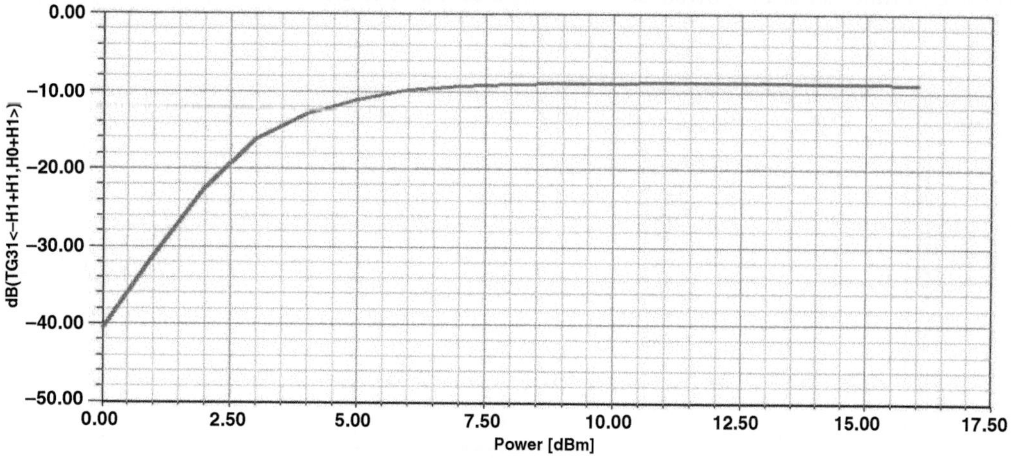

Figure 9.56 Mixer conversion gain as a function of LO input power (of Figure 9.52) [50, pp 3-99].

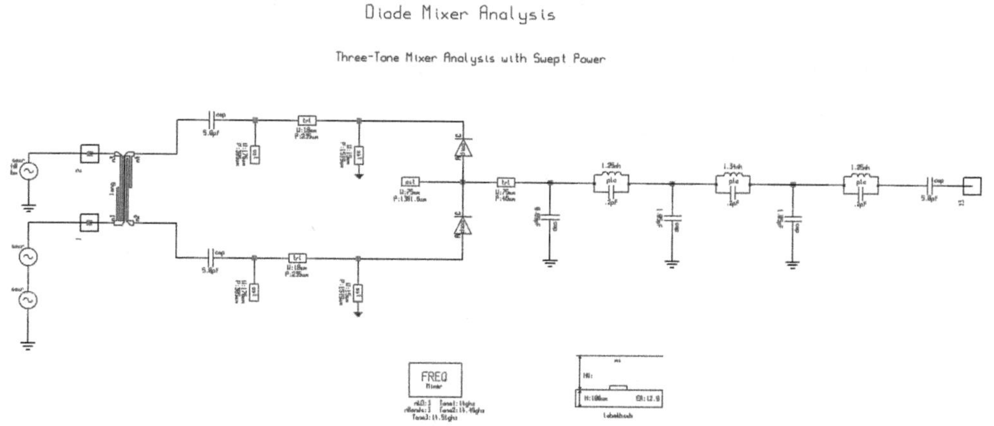

Figure 9.57 Schematic of a three-tone (LO and two RF signals) mixer analysis with swept power [50, pp 3-100-3-104].

This analysis is very demanding of your computer's resources, considering that it features three arbitrary tones and swept power. Still, it is pleasantly surprising to see how quickly the analysis proceeds. Consider for a moment that the harmonic specification requires the simulator to track 87 different harmonics through the analysis.

The two graphs in Figure 9.58a,b show the same result at different zoom levels. They are both portions of the IF output spectrum showing the third-order intermodulation products near, and very near, the IF signals [50, pp 3-100–3-104].

Figure 9.59 shows the fundamental IF and a higher-order IM product as a function of RF input power. Notice that the higher-order term (−F1 + 2F2 − F3) has a slope of three relative to the fundamental IF tone across the linear region of the response. This is what we expect from the third-order terms, and this is what leads to the concept of third-order intermodulation products, or IP_3. Excellent LO and RF suppression is the result of a carefully designed circuit.

The final three-tone response is the waveform across one of the diodes (chip2), as shown in Figure 9.60.

Figure 9.61 shows the noise figure as a function of LO input power, where the relationship between LO input power and noise figure can be easily seen. As the power increases, we have already seen that the conversion loss of the mixer approaches 9 dB. Here, we see that the SSB noise figure approaches 6 dB, whereas the DSB noise figure is typically 3 dB higher, or 9 dB, which corresponds to the conversion loss of the mixer.

This graphical definition may be a little confusing. After all, we are performing a mixer (i.e., two-tone) analysis, and yet the X-Data and Running Variable specifications look like single-tone definitions, while the Y-Data specification looks more like a two-tone item. The key feature here is that nonlinear mixer noise analysis "ignores" the RF input specification. The signal is still there for use in graphs, as we have shown with our Y-Data definition. However, the RF signal is assumed to be low enough that it qualifies as a small signal input, so its absolute level is irrelevant. The definition shown is the proper way to display mixer noise figure as a function of LO power. Based on the discussion earlier, it is not possible to analyze

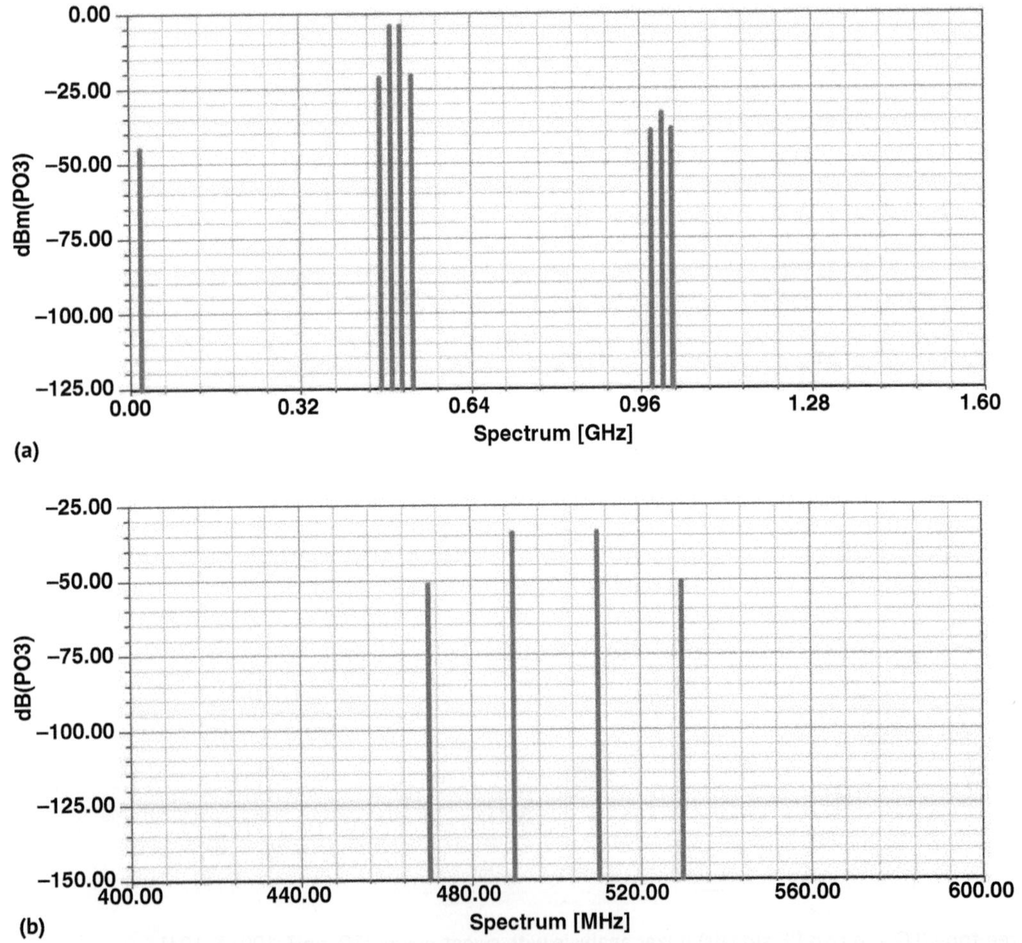

Figure 9.58 (a) IF output spectrum showing the third-order intermodulation products near the IF signals [50, pp 3-100–3-104]. (b) IF output spectrum showing the third-order intermodulation products very near the IF signals [50, pp 3-100–3-104].

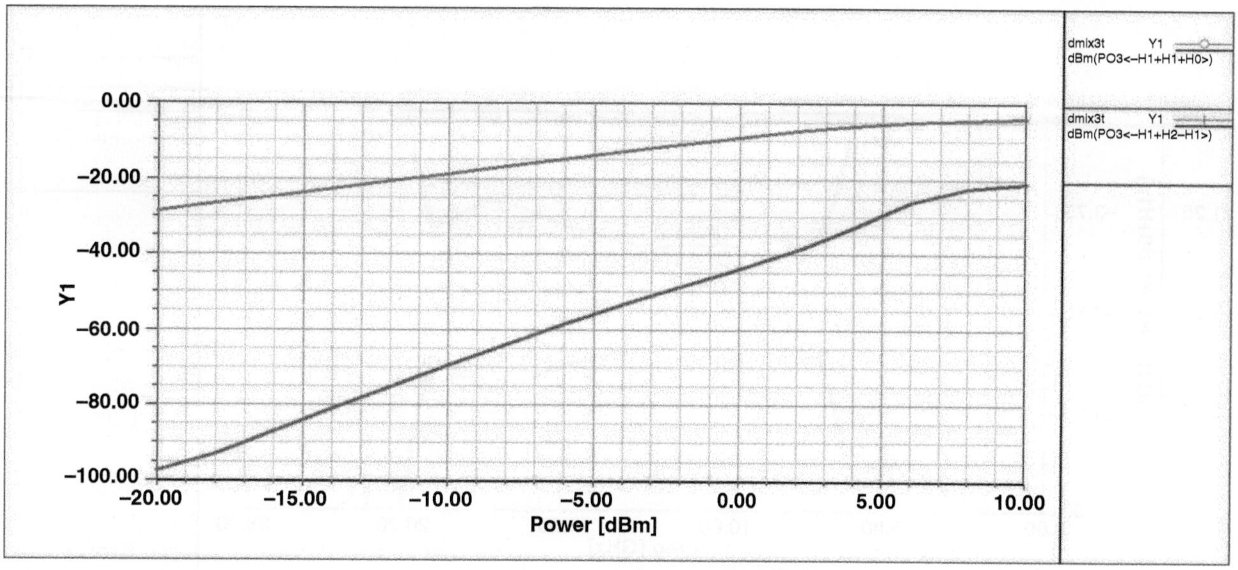

Figure 9.59 Fundamental IF and a higher-order IM product as a function of RF input power [50, pp 3-100–3-104].

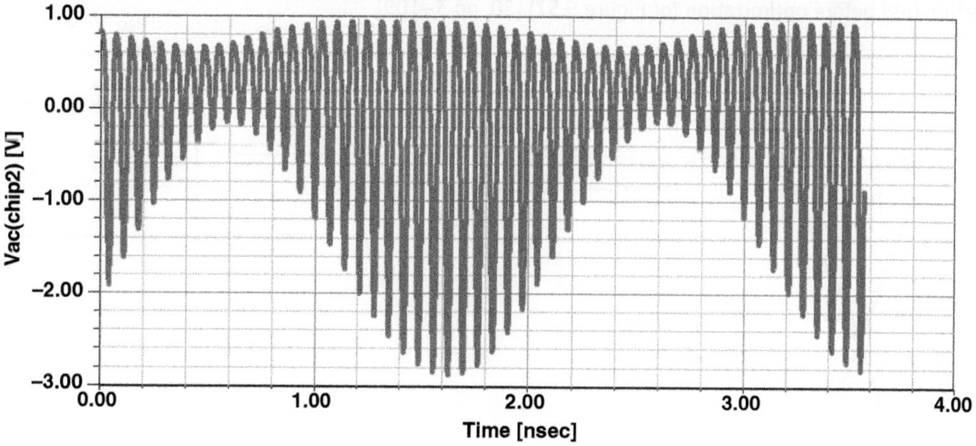

Figure 9.60 Three-tone response for Figure 9.57 [50, pp 3-100–3-104].

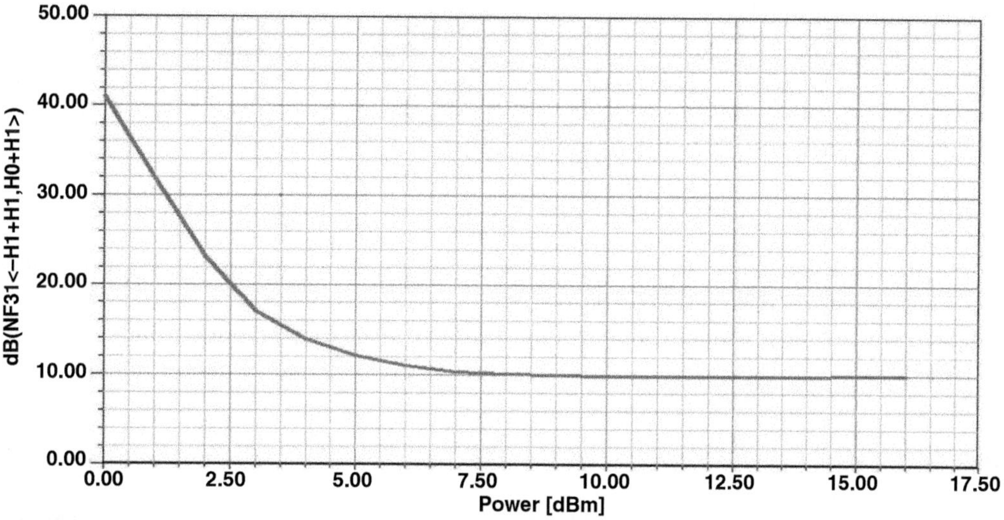

Figure 9.61 Noise figure as a function of LO input power of Figure 9.57 [50, pp 3-106].

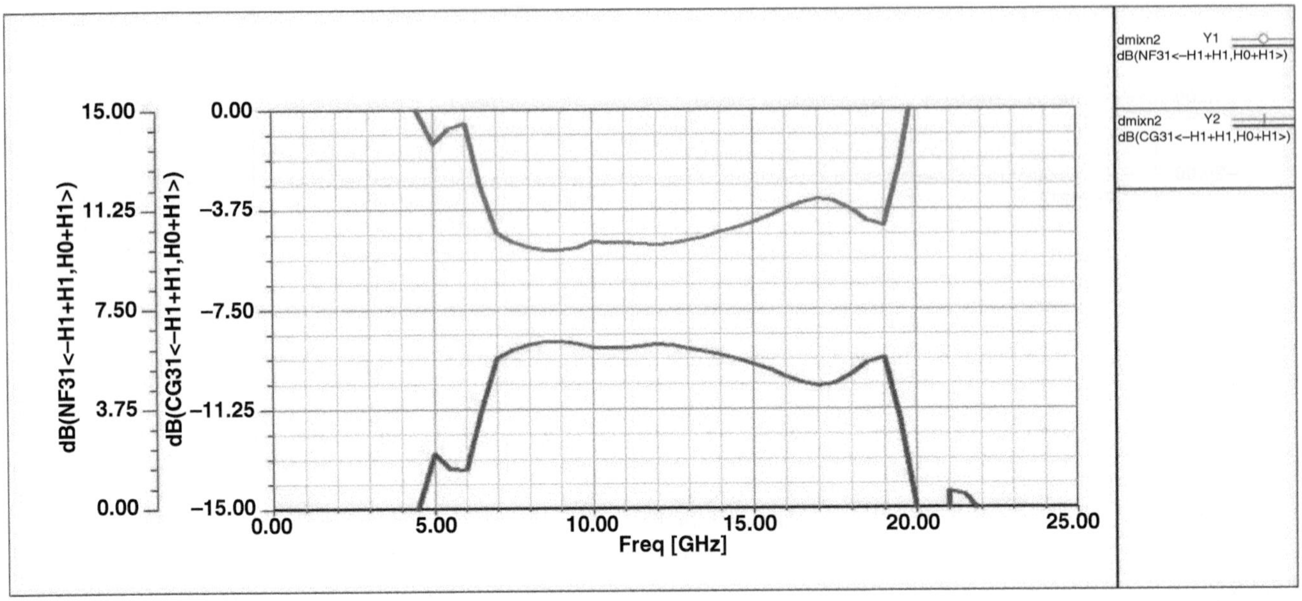

Figure 9.62 Noise figure and insertion loss before optimization (of Figure 9.57) [50, pp 3–109].

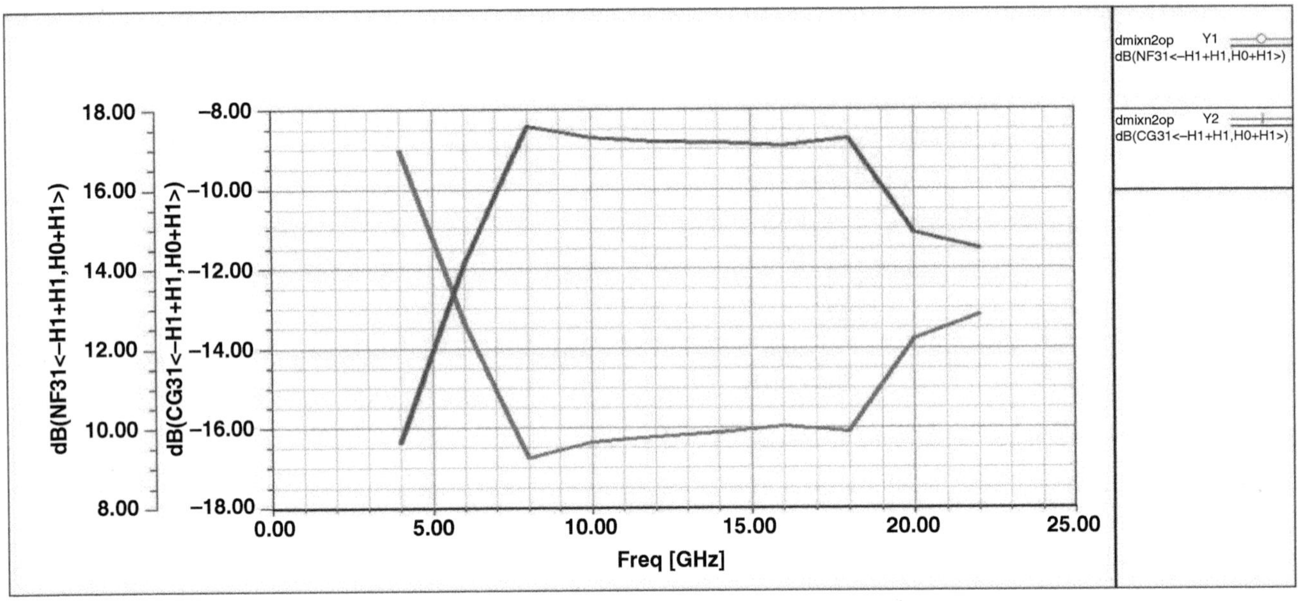

Figure 9.63 Simulation: optimized for low oscillator drive (of Figure 9.57).

mixer noise figure as a function of RF drive. The RF specification is ignored, so the simulator will simply not see the swept variable. The analysis will still run, however, and you will be able to see the noise figure at a single drive level. The results here would be valid for any LO "small signal" RF drive level [50, pp 3–107].

Figure 9.62 shows the noise figure and insertion loss before optimization, and Figure 9.63 shows the optimized conversion loss and the noise figure for a low LO Drive.

9.12 Transistor Mixers

Diode mixers are lossy, termination-sensitive, and require considerable LO power. Their attractiveness to designers of highly integrated wireless products is further reduced by their dependence on transformers for balance and port-to-port isolation. Transistor mixers are therefore used where high integration and reduced current drain are paramount—that is, in most non-base station wireless applications. All transistor-based mixers, unless used in a switching mode, are active mixers.

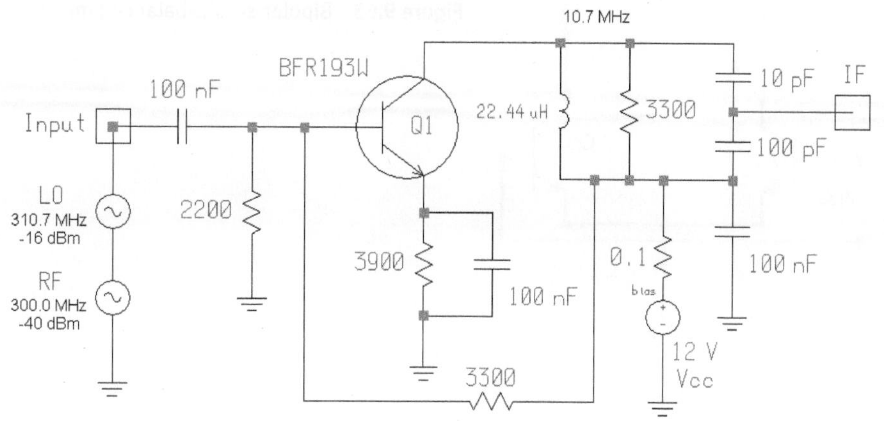

Figure 9.64 A single-BJT mixer. *Source*: [4] / John Wiley & Sons.

Until 10 years ago, single BJTs were commonly used in a simple additive-mixing arrangement, an example of which appears in Figure 9.64. Such a circuit behaves like a combination of a preamplifier and a single-diode mixer. Single-BJT mixers were used in early AM-FM radios until 1980, and unfortunately in some handheld 2-m and 70-cm ham equipment, to achieve the lowest possible power consumption.

By definition, however, the presence of such a mixer also destroys any possibility of achieving a high intercept point. In addition to this, many combination frequencies occur despite elaborate input filtering. Although single-BJT mixers can exhibit considerable conversion gain (>10 dB), their dynamic range is restricted, and their port-to-port isolation is poor. BJT mixers in today's competitive wireless designs use multiple transistors and are based almost exclusively on the Gilbert multiplier cell. Due to the poor performance of the single-transistor mixers, they are no longer used [4].

9.13 The Differential Configuration–Differential LO Frequency Mixer

If two transistors are connected in a differential configuration, their total emitter current, I_k, is supplied by a constant current source (normally another transistor). The basic differential structure is shown in Figure 9.65. This circuit has an exponential nonlinear characteristic.

For the differential structure, to obtain an expression for the nonlinear characteristics, we note that

$$i_1 = I_{S1}e^{v_{BE1}q/kT},\ i_2 = I_{S2}e^{v_{BE2}q/kT}$$

and

$$v_{BE1} - v_{BE2} = v_1 - v_2 \tag{9.30}$$

Therefore, if $I_{S1} = I_{S2}$ (which should be true if both transistors are integrated on the same "chip" and are identical in size and construction), then

$$\frac{i_1}{i_2} = e^{(v_1-v_2)q/kT} \tag{9.31}$$

If we now set $i_1 + i_2 = I_k$, we obtain

$$i_2 = \frac{I_k}{1+e^z} \text{ and } i_1 = \frac{I_k}{1+e^{-z}} \tag{9.32}$$

where $z = (v_1 - v_2)q/kT$.

It is of interest to note that both i_1 and i_2 possess odd symmetry of their average value $I_k/2$. Specifically

$$i_2 - \frac{I_k}{2} \equiv -i(z) = \frac{I_k}{2}\left(\frac{1-e^z}{1+e^z}\right) = -\frac{I_k}{2}\tanh\frac{z}{2}, \tag{9.33}$$

from which we observe that $i(z) = -i(-z)$. With this, we can write i_2 in the alternative form

$$\frac{i_2}{I_k} = \frac{1}{2}\left(1 - \tanh\frac{z}{2}\right) \tag{9.34}$$

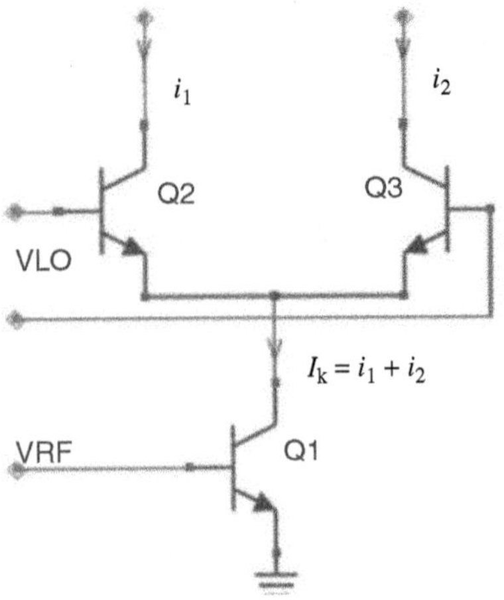

Figure 9.65 Bipolar single-balanced mixer.

and, in a similar fashion, we write

$$\frac{i_1}{I_k} = \frac{1}{2}\left(1 + \tanh\frac{z}{2}\right) \tag{9.35}$$

If this circuit is operated about its point of symmetry ($z = 0$), as it normally is, then the DC component in either current remains at $I_k/2$ for all symmetrical driving signals ($v_1 - v_2$). If, in addition, the driving signals are periodic, then even-order harmonics will not exist. In particular, if $v_1 - v_2$ is a square wave of zero average value and peak amplitude V_1, then i_1 is a square wave in phase with the input and i_2 is a square wave 180° out of phase with the input. Both i_1 and i_2 are symmetric and have peak values.

Both i_1 and i_2 are symmetric in magnitude and are 180° out of phase. For small values of $z(z \ll 1)$ or equivalently ($|v_1 - v_2| \ll 26$ mV), we may approximate $\tanh(z/2)$ by $z/2$ to obtain

$$i_2 = \frac{I_k}{2}\left(1 - \frac{z}{2}\right) = \frac{I_k}{2} - \frac{g_{\text{in}}}{2}(v_1 - v_2) \tag{9.36}$$

and

$$i_1 = \frac{I_k}{2}\left(1 + \frac{z}{2}\right) = \frac{I_k}{2} + \frac{g_{\text{in}}}{2}(v_1 - v_2) \tag{9.37}$$

where $g_{\text{in}} = qI_k/2kT$. Since g_{in} is just the DC small-signal conductance "seen," if looking into the emitter of transistor 1 or 2 with the base grounded and with an emitter current of $I_k/2$, it is clear that the single current represents simply the expected small-signal transfer for a differential pair of transistors. In addition, the small-signal DC transconductance is given by

$$g_m = \frac{\alpha i_{2\text{ac}}}{v_1 - v_2} = \frac{-\alpha i_{1\text{ac}}}{v_1 - v_2}$$

$$g_m = \frac{\alpha g_{\text{in}}}{2} \tag{9.38}$$

where α is the ratio of collector to emitter current.

If now $v_1 - v_2 = V_1 \cos \omega t$ and we again define $x = qV_1/kT$, then

$$i_1 = \frac{I_k}{2} + i \text{ and } i_2 = \frac{I_k}{2} - i \tag{9.39}$$

$$i = \frac{I_k}{2} \tanh\left(\frac{x}{2} \cos \omega t\right) \tag{9.40}$$

As it is quite apparent that the Fourier series of i contains no even harmonics, i may be expanded in the form

$$\begin{aligned} i &= I_1 \cos \omega t + I_3 \cos 3\omega t + I_s \cos 5\omega t + \dots \\ &= I_k \sum_{n=2}^{\infty} a_{2n-1}(x) \cos(2n-1)\omega t \end{aligned} \tag{9.41}$$

As (Vs.ωt) the waveform varies with x. It is quite apparent that i may be expanded in the Fourier series of i, which contains no even harmonics; $a_n(x) = \frac{1}{\pi}\int_{-\pi}^{\pi}\left[\frac{1}{2}\tanh\left(\frac{x}{2}\cos\theta\right)\right]\cos n\theta d\theta = \frac{I_n}{I_k}$

The coefficients $a_n(x)$ do not appear to be expandable in terms of tabulated functions.

The values for $x = \infty$ are obtained by noting that, for very large drive voltages, the currents i_1 and i_2 must have the form of square waves. In addition, for small values of x, we may expand $\frac{1}{2}$ tanh $[(x/2)\cos\omega t]$ in a MacLauren expansion of the form

$$\begin{aligned}\frac{1}{2}\tanh\left(\frac{x}{2}\cos\omega t\right) &= \frac{1}{2}\left(\frac{x}{2}\cos\omega t - \frac{x^3}{24}\cos^3\omega t\right)\\ &= \frac{1}{2}\left[\left(\frac{x}{2} - \frac{x^3}{32}\right)\cos\omega t - \frac{x^3}{96}\cos 3\omega t\right]\end{aligned} \tag{9.42}$$

The limits and linearity are determined not by the presence of the third-harmonic current component but rather by the presence of the nonlinear term in $a_1(x)$. This increase in input drive by a factor of 6.3 over the single-transistor broadband amplifier is due primarily to the symmetry of the differential characteristic and the resultant absence of even (principally second) harmonics from the output. Intuitively, what we are doing is compensating for the nonlinear input characteristic of a single transistor by using a second transistor as a nonlinear emitter impedance. When a tuned circuit that extracts the fundamental component of current is placed in the collector of either transistor, it is convenient to again define a large signal average fundamental transconductance (not frequency-dependent) as

$$G_m(x) = \frac{\alpha I_1}{V_1} = \frac{\alpha q I_k}{kT}\frac{I_1}{xI_k} = 2\alpha g_{\text{in}}\frac{a_1(x)}{x} = g_m\frac{4a_1(x)}{x}, \tag{9.43}$$

where g_m is the small-signal transconductance. With this definition of $G_m(x)$, if a high Q_T tuned circuit with a resistance R_L at resonance is placed in the collector of transistor 2, the tuned-circuit voltage v_T is given by

$$v_T(t) = G_m(x)V_1R_L\cos\omega t \tag{9.44}$$

Finally, it should be noted that the differential characteristic has many of the same properties as the exponential characteristic. For small input signals, the differentially connected circuit operates as a linear amplifier; for large input signals, I_1/I_k approaches a constant and the circuit functions as a limiter. In addition, i_1 and i_2 are directly proportional to I_k; therefore, linear gain control or amplitude modulation can be achieved by controlling I_k. In practice, it turns out to be more effective to control I_b for the differential pair than I_{de} for the single transistor. There the RF input should be placed at the constant current transistor [28].

In real life, it would be appropriate to replace g_m with Y_{21*}, which represents the operating point and frequency-dependent transconductance. Following the noise correlation *ABCD* matrix, the noise contribution from the semiconductor will be $\text{Re}(Y_{21*})$, which is needed later to correctly calculate the mixer noise!

9.14 Mixer Gain

Due to nonideal switching and frequency conversion, the process changes the concept of gain in mixer circuit. For a sinusoidal LO signal, the detailed evaluation of gain in a single-balance mixer is shown here [49, 52]. Evaluating only the differential output current due to small signal current, $i_{\text{rf}} = I_{\text{RF}}\sin(\omega_{\text{RF}}t)$, yields

$$i_{\text{out, if}} = I_{\text{RF}}\sin(\omega_{\text{RF}}t)\tanh\left(\frac{v_{\text{LO}}}{2V_{\text{Th}}}\right) \tag{9.45}$$

This nonlinear operation captures the saturating behavior of current-steering pair Q2 and Q3. And $I_{\text{RF}} = g_{m,\text{RF}}V_{\text{RF}}$.

9.14.1 Low LO Levels [52]

When $V_{\text{LO}} \ll 2V_{\text{Th}}$, the tanh function is approximately linear and the conversion calculation for $v_{\text{LO}} = V_{\text{LO}}\sin(\omega_{\text{LO}}t)$ is simply

$$i_{\text{out, if}} = I_{\text{RF}}\sin(\omega_{\text{RF}}t)\tanh\left(\frac{V_{\text{LO}}\sin(\omega_{\text{LO}}t)}{2V_{\text{Th}}}\right) \tag{9.46}$$

$$\approx I_{RF}\sin(\omega_{RF}t)\left(\frac{V_{LO}}{2V_{Th}}\right)\sin(\omega_{LO}t) \tag{9.47}$$

$$= \frac{I_{RF}\,V_{LO}}{4V_{Th}}(\cos((\omega_{RF}-\omega_{LO})t)) - (\cos((\omega_{RF}+\omega_{LO})t)) \tag{9.48}$$

$$= \frac{I_{RF}\,V_{LO}}{4V_{Th}}(\cos(\omega_{IF})t) - (\cos((\omega_{RF}+\omega_{LO})t)) \tag{9.49}$$

Ignoring the high-frequency component, the overall conversion transconductance is

$$G_{mC} = \frac{g_{m,RF}\,V_{LO}}{4V_{Th}} \ll \frac{g_{m,RF}}{2} \tag{9.50}$$

9.14.2 Strong LO Levels [52]

For $V_{LO} \geq 2V_{Th}$, tanh begins to distort the LO sinusoid, producing harmonics of LO signal. It affects gain. The tanh nonlinearity is an odd-order distortion, i.e., only odd harmonies are produced. From Eq. (6.8)

$$i_{out,\,if} = I_{RF}\sin(\omega_{RF}t)\sum_{i=1}^{\infty} C_{2i-1}\sin((2i-1)\omega_{LO}t) \tag{9.51}$$

where C_{2i-1} is the amplitude of the $(2i-1)$ harmonic. Evaluating Fourier coefficients

$$C_{2i-1} = \frac{1}{2\pi}\int_0^{2\pi} \tanh\left(\frac{V_{LO}\sin(\omega_{LO}t)}{2V_{Th}}\right)\sin((2i-1)\omega_{LO}t)(d\omega_{LO}t) \tag{9.52}$$

At ω_{IF} frequency

$$i_{out,\,if} = I_{RF}C_1\sin(\omega_{RF}t)\sin(\omega_{LO}t) = \frac{I_{RF}C_1}{2}\sin(\omega_{RF}-\omega_{LO})t + \ldots \tag{9.53}$$

Gain is about $\frac{C_1}{2}$.

Figure 9.66 shows the amplitude of harmonic components from applying single-frequency sinusoids to tanh nonlinearity and the numerical evaluations for first, third, fifth, and seventh harmonics for various input peak amplitudes. Under strong LO drive, C_1 approaches a value of $\frac{4}{\pi}$.

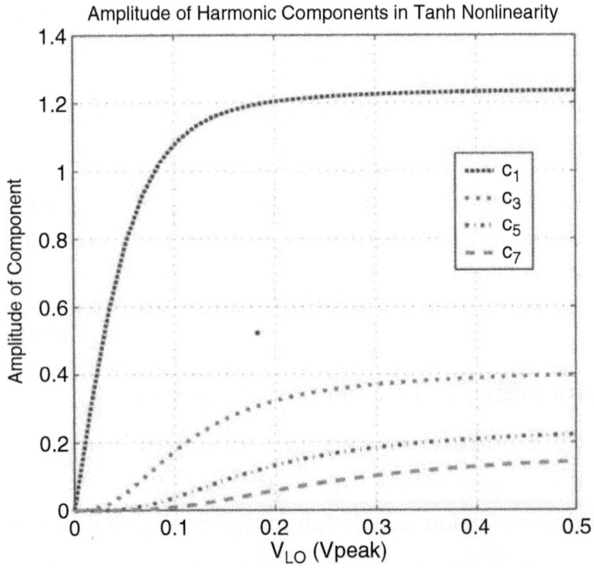

Figure 9.66 Amplitude of harmonic components from applying single-frequency sinusoids to tanh nonlinearity. *Source*: [52] / Massachusetts Institute of Technology (MIT).

We can calculate an effective gain when LO has strong switching using C_1 values. Considering the mixer as a cascade of stages and the frequency conversion gain as $\frac{C_1}{2}$, conversion transconductance is

$$G_{mC} = \frac{C_1 g_{m,\mathrm{RF}}}{2} \rightarrow \lim_{V_{\mathrm{LO}} \geq 2V_{\mathrm{Th}}} \left(\frac{2}{\pi}\right) g_{m,\mathrm{RF}} \tag{9.54}$$

Strong LO drive levels improve both gain and noise performance. To some extent, conversion linearity can be improved by using a strong LO drive [52].

9.15 Single-balanced Gilbert Cell Mixer

The circuit Figure 9.67a is a Gilbert-type single-balanced mixer; input is at the current source, lower transistor (q3) and the LO at the differential input, and the collector has an impedance down-transformer. Output impedances are typically a few hundred ohms.

Figure 9.67b shows the transconductance of the device q1.

Note the phase shift due to the nonideal transistor in Figure 9.67(c). This fact is frequently overlooked in the literature.

From Figure 9.67d, LO drive of 80 mV RMS gives a noise figure of about 6.5 dB and a gain of about 30 dB. We will get to the dynamic range later. This circuit is optimized for voltage gain and not for power gain. For power gain and low distortion (IP_3, 1 dB compression point), the transformer impedance needs to be reduced.

9.16 Calculation of Single-balanced Mixer Gain

The single-balanced mixer is shown in Figure 9.68. The LO signal (310 MHz at −10 dBm) is connected between the transistor Q2 and Q3 differential pair. Its primary purpose is switching too fast to provide square-wave characteristics. This condition has been ensured by operating Q2 and Q3 into the saturation region. The RF (455 MHz at −50 dBm) signal coupled through a DC blocking capacitor connected to Q1 is the transconductance (g_m) stage, and it is biased at 10 mA of DC bias current. This should be calculated theoretically through a mathematical equation on how g_m can affect the conversion gain [56].

Since we are interested in seeing the circuit's conversion gain and noise figure, the transistor at the RF stage does not have emitter degeneration. Literature shows that a degenerating emitter helps to improve linearity at the expense of less conversion gain and may have a higher noise figure (if a resistive element is used in the design). We are performing downconversion at the IF port (145 MHz). The output V_{out} is differential. At the top of the differential pairs, a tuned circuit connects differential outputs and serves as load. These loads are significant for the mixer, and they play a major role in the conversion gain of the mixer.

To achieve an optimum gain and noise figure, it is necessary to do conjugate matching at the input. I have tried to match the RF and IF ports shown in Figure 9.69. The spectrum of the downconversion mixer is shown in Figure 9.70.

To calculate a mixer conversion gain, consider the small-signal equivalent circuit of the given mixer. The RF stage is represented by the hybrid-π model, and the LO differential pair is represented by the T model. The effect of C_π (0.5 pF) and C_μ (2.6 fF) was found by performing a small-signal analysis under the harmonic balance simulation in ADS at RF frequency and hence ignored; they are very small at 455 MHz.

The load impedance of the tuned circuit is computed by determining the real part of Z_L at resonance. So, $Z_L = \left(SL||\frac{1}{SC}\right)$. The total real part of load $Re\{Z_L\} = \sqrt{\frac{L}{C}}$ is connected between two differential outputs.

Figures 9.71 and 9.72 show the simulated transconductance of the RF stage at the device bias level using ADS software, and it is roughly about 299 mS.

The total voltage conversion gain of a given mixer shown in Figure 9.68 is as follows:

$$\text{Conversion gain} = \frac{2}{\pi} g_{m,\mathrm{RF}} R_L$$

$$= \frac{2}{\pi} \cdot \frac{V_{\mathrm{IF}}}{V_{\pi,\mathrm{RF}}} \cdot \frac{V_{\pi,\mathrm{RF}}}{V_{\mathrm{RF}}} (R_L||R_{B1}) = \frac{2}{\pi} \cdot A_{vB} \cdot \frac{V_{\pi\mathrm{RF}}}{V_{\mathrm{RF}}} \tag{9.55}$$

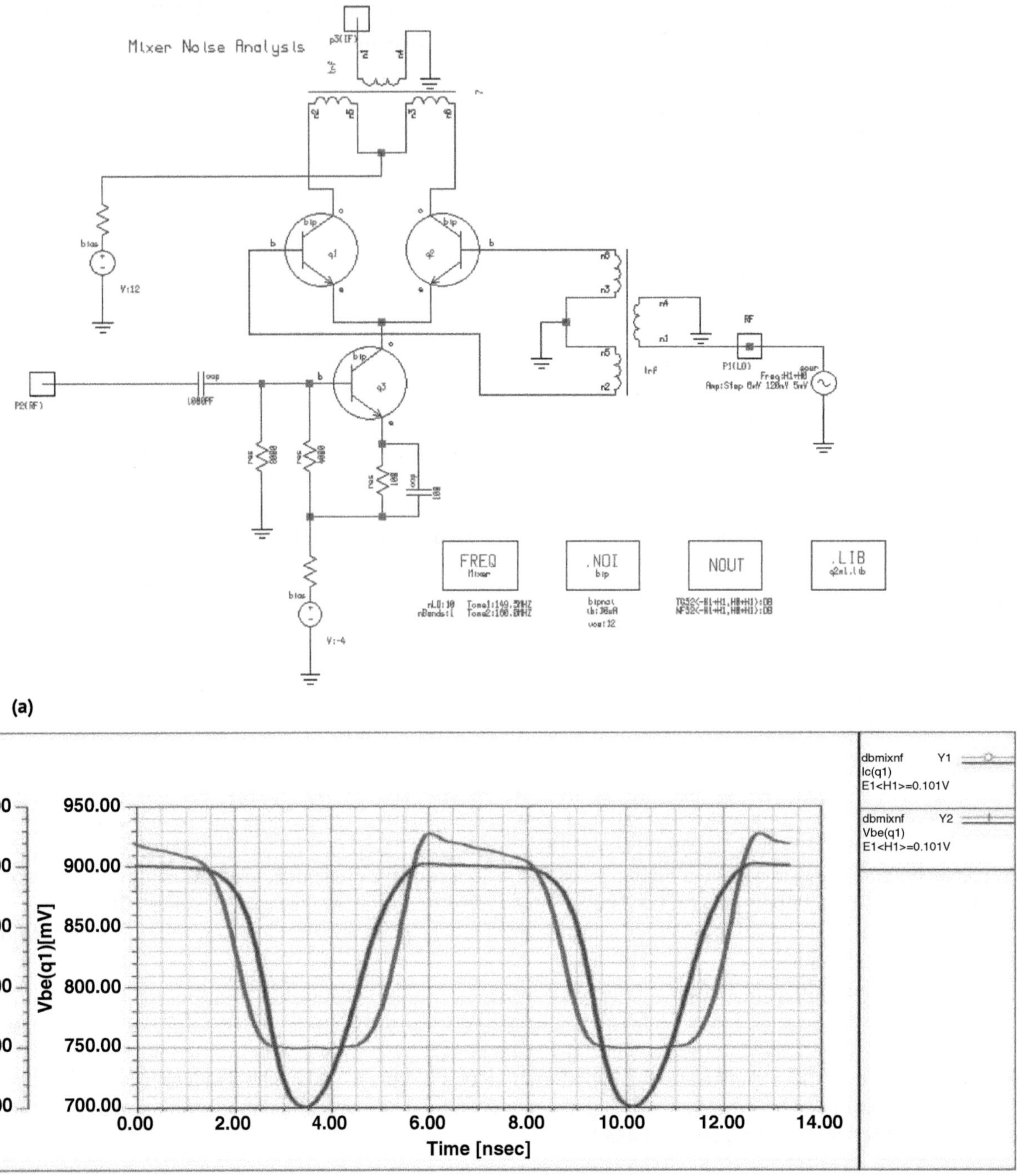

Figure 9.67 (a) Single-balanced Gilbert cell mixer (input at the current source, LO at the differential stage) [50, pp 3–52]. (b) Differential collector I_{cq1} current as a function V_{be} ($Y_{21} = 1$ mA/50 mV > 20 mA/V).

(Continued)

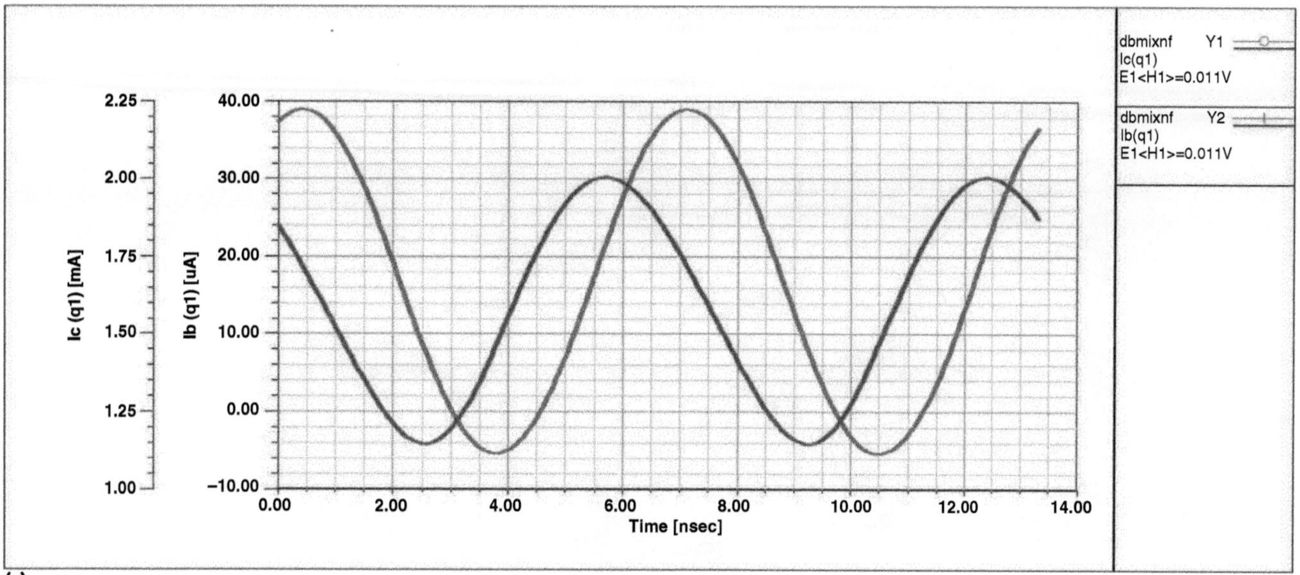

(c)

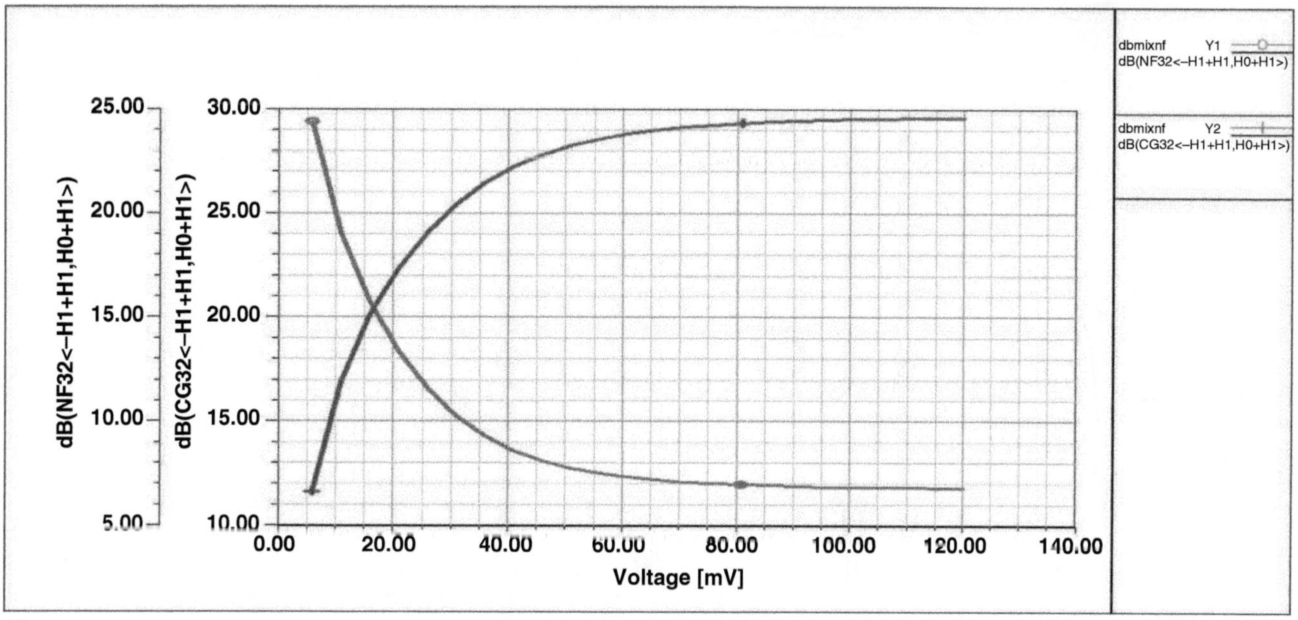

(d)

Figure 9.67 (Cont'd) (c) $\mathrm{H}_{31} = 250\frac{\text{uA}}{10}\text{uA} = 25$, the RF current gain ($I_c/I_b$). (d) Simulation of the gain and noise figure as a function of the LO drive.

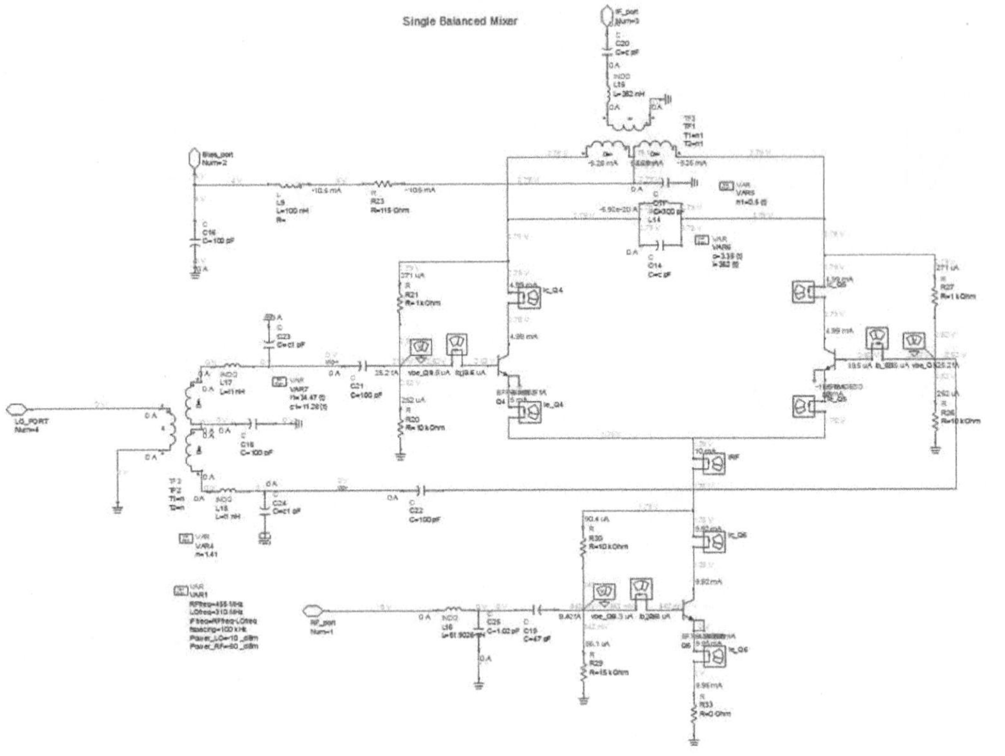

Figure 9.68 Single-balanced mixer. *Source*: Modified from [56].

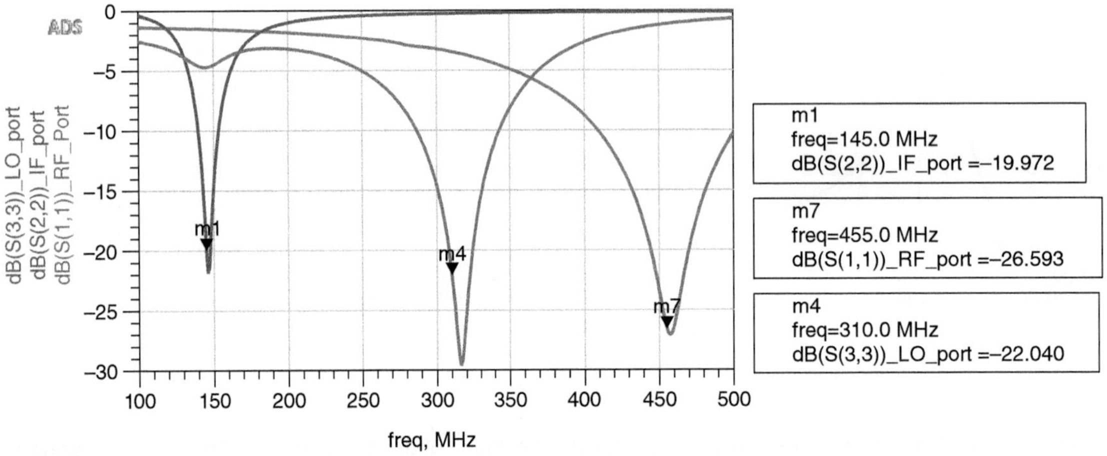

Figure 9.69 The simulated return loss of RF and IF ports matched to the system impedance. *Source*: Modified from [56].

Since R_{B1} is connected between the base and collector of Q1, the Miller equivalent resistance representation is shown in a small-signal model. The value of R_{B1_1} and R_{B1_2} is given by

$$R_{B1\_1} = \frac{R_{B1}}{1 - A_{vB}} \tag{9.56}$$

and

$$R_{B1\_2} = \frac{A_{vB}R_{B1}}{A_{vB} - 1} \approx R_{B1} \tag{9.57}$$

$$A_{vB} = \frac{I_{\mathrm{RF}}}{2V_{th}}(R_L||R_{B1}) = 39.\ 5538 \tag{9.58}$$

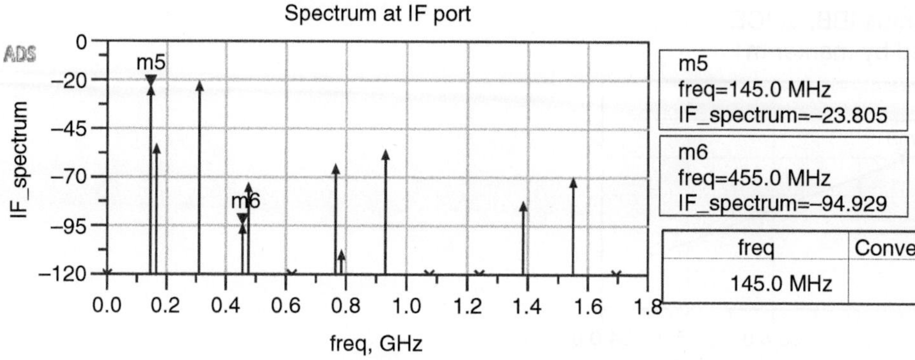

freq	ConversionGain
145.0 MHz	26.20

Figure 9.70 Simulated gain at IF frequency and IF spectrum is shown at 145 MHz. *Source*: Adapted from [56].

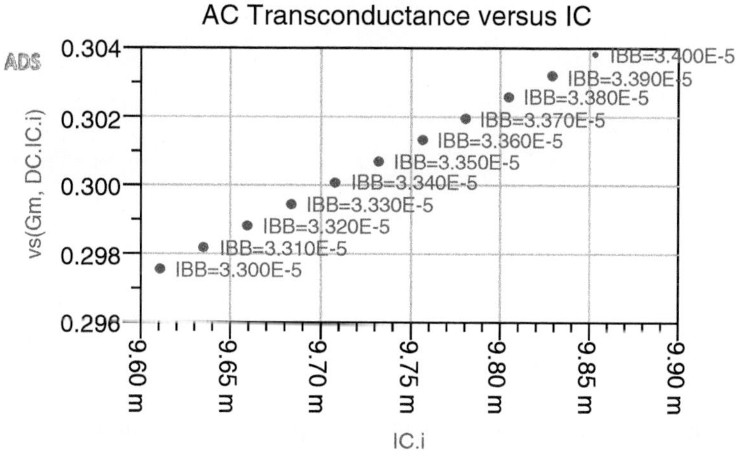

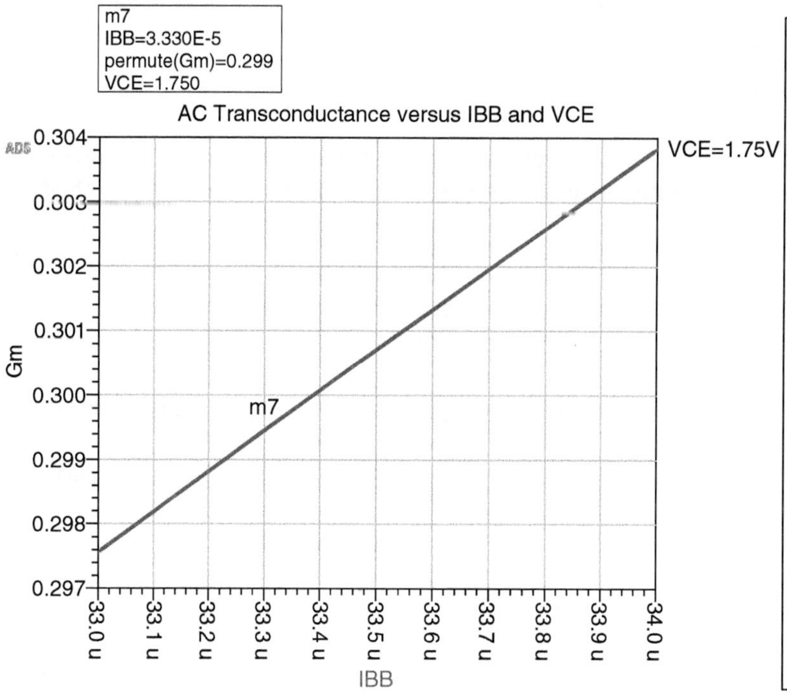

	VCE	Gm
IBB=3.300E-5	1.750	0.298
IBB=3.310E-5	1.750	0.298
IBB=3.320E-5	1.750	0.299
IBB=3.330E-5	1.750	0.299
IBB=3.340E-5	1.750	0.300
IBB=3.350E-5	1.750	0.301
IBB=3.360E-5	1.750	0.301
IBB=3.370E-5	1.750	0.302
IBB=3.380E-5	1.750	0.303
IBB=3.390E-5	1.750	0.303
IBB=3.400E-5	1.750	0.304

Figure 9.71 AC transconductance of RF stage versus collector current. *Source*: Adapted from [56].

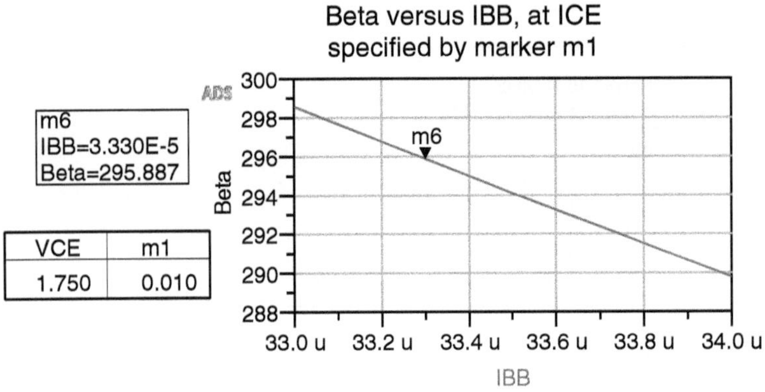

Figure 9.72 The graph shows a value of beta at a bias current of 10 mA. *Source*: Adapted from [56].

From small-signal model,

$$\frac{V_{\pi \mathrm{RF}}}{V_{\mathrm{RF}}} = \frac{(r_\pi || R_{B1\_1} || R_{B2})}{R_S + (r_\pi || R_{B1\_1} || R_{B2})} \tag{9.59}$$

Here, $r_\pi = \dfrac{\beta_0}{g_{m,\mathrm{RF}}}$

Solving for $\dfrac{V_{\pi \mathrm{RF}}}{V_{\mathrm{RF}}} \approx 0.7976$

$$\text{Conversion gain CG} = 20 \log\left(\frac{2}{\pi} \cdot A_{vB} \cdot \frac{V_{\pi \mathrm{RF}}}{V_{\mathrm{RF}}}\right) = 26.0614\ \mathrm{dB} \tag{9.60}$$

The simulated gain is about 26.20 dB. The detail calculations are explained in "optimizing switching mixer" dissertation, a work in progress by Trusha Kared, PhD student of Prof. Ulrich Rohde at the Brandenburg Technical University, Cottbus, Germany.

9.17 Noise Mixing in Single-balanced Mixer

The noise of single-balanced mixer topology, along with analysis and design tools, has been presented in the literature [27, 28, 31, 40], [31–34], [36–39], [41–45], [47], [48]. I have summarized the primary source of noise and a basic method for calculating these noise spectral densities in a single-balanced mixer. The most significant noise contribution mechanisms are listed as follows:

- When the switching transistors (Q2 and Q3) are entirely in the switched state, noise in the signal path, including noise from the RF input stage, appears at the output as differential noise since only one of the outputs is active at any given time, while another switching transistor has less contribution to the noise.
- When Q2 and Q3 are changing switch states, noise from the RF input stage Q1 appears as common mode in the output currents, i_2 and i_3, adding nothing to the differential output current, i_{out}. However, the switch devices share the DC current and contribute differential noise to the output. In the current commuting mixer, the length of this period is defined by the magnitude of the LO and reduces with a stronger LO drive.
- Additionally, noise present in the LO signal may be considerable during the switch phase. This can impact the output similarly to the noise generated by Q2 and Q3.
- The output I-to-V stage consistently introduces differential noise. Often, a resistive load is used to implement this stage, and in that case, it adds thermal noise.
- Summing linear sources and linear sources shaped by the circuit's periodic and time-varying nature yields the noise figure [51].

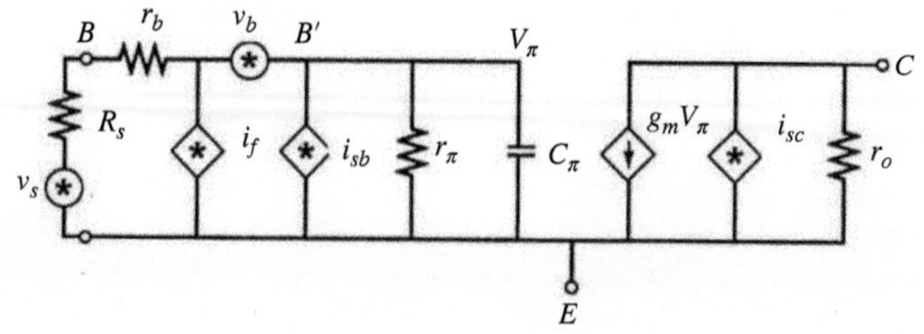

Figure 9.73 Noise model of the transconductance stage. *Source*: [53] / MDPI / CC BY 4.0.

9.17.1 Noise Due to RF Stage or Transconductance Amplifier [46, 53, 54]

Figure 9.73 shows the noise model of the transconductance stage [53].

Base resistance r_b provides a thermal noise, whose power spectrum density is given by $v_b^2 = 4KTr_b$.

Currents i_b and i_c provide shot noise, which is given by $i_{sb}^2 = 2qI_b$ and $i_{sc}^2 = 2qI_c$, respectively.

The current passing through the B-E junction will have some flicker noise $i_f^2 = \frac{KI_B^\gamma}{f\alpha}$; α is set to 1. K can be replaced by $2qf_L$, where f_L is the lower corner frequency and ranges from 3 KHz to 7 MHz. So the flicker noise $i_f^2 = \frac{2qf_L I_B^\gamma}{f}$.

Noise voltage can be found by multiplying flicker noise current with base resistance r_b

$$v_f^2 = \frac{2qf_L i_B^\gamma r_b'^2}{f} \tag{9.61}$$

Experimental result shows effective $r_b' = \frac{r_b}{2}$, and c_μ and r_μ have been neglected.

Considering frequencies below $\frac{f_t}{\sqrt{\beta}}$, above this frequency, some of the noise will be correlated and have more noise than the model indicated.

Common emitter device assuming the source impedance R_s; noise due to source resistance is given by $\overline{v_s^2} = 4KTR_s$.

i. **Noise at the output**

$$\overline{i_{no}^2} = i_{sc}^2 + (g_m v_\pi)^2 \tag{9.62}$$

v_π is the noise voltage between B' and E

$$\overline{v_\pi^2} = \frac{(v_b^2 + v_s^2)z_\pi^2}{(r_b + R_s + z_\pi)^2} + \frac{\left(i_{sb}^2 + i_f^2\right)z_\pi^2(R_s + r_b)^2}{(r_b + R_s + z_\pi)^2} \tag{9.63}$$

where $z_\pi = \frac{r_\pi}{1 + j2\pi f r_\pi C_\pi}$

$$i_o = g_m v_\pi = g_m\left[\frac{z_\pi}{r_b + R_s + z_\pi}\right]v_s \tag{9.64}$$

$$\text{Gain (A)} = \frac{i_o}{v_s} = \frac{g_m V_\pi}{r_b + R_s + Z_\pi} \tag{9.65}$$

ii. Equivalent noise at the input [51]

From Eqs. (9.62) and (9.64), we can write

$$\overline{v_{ni}^2} = \frac{1}{(\text{Gain})^2} \cdot \overline{i_{no}^2} = \frac{1}{\frac{g_m^2 z_\pi^2}{(r_b+R_s+z_\pi)^2}} \cdot \left[i_{sc}^2 + g_m^2 v_\pi^2\right] \tag{9.66}$$

$$\overline{v_{ni}^2} = v_s^2 + v_b^2 + i_{sb}^2(R_s + r_b)^2 + i_f^2(R_s + r_b)^2 + \frac{i_{sc}^2(r_b + R_s + z_\pi)^2}{g_m^2 z_\pi^2}$$

$$\overline{v_{ni}^2} = 4KT\,(r_b + R_s) + 2qI_B(r_b + R_s)^2 + \frac{2qf_L i_B^\gamma(R_s + r_b)^2}{f} + \frac{2qI_C(R_s + r_b + z_\pi)^2}{g_m^2 z_\pi^2}$$

At a low frequency, $\frac{2qI_c(R_s + r_b + z_\pi)^2}{g_m^2 z_\pi^2}$, the effect of C_π will be neglected and can be written as $\frac{2qI_c(R_s + r_b + r_\pi)^2}{g_m^2 r_\pi^2}$ and we know that $\beta_0 = g_m r_\pi$; so, $\frac{2qI_C(R_s + r_b + r_\pi)^2}{\beta_0^2}$.

At a higher frequency up toward $\frac{f_T}{\sqrt{\beta_0}}$

$$\frac{2qI_c\left(R_s + r_b + \frac{1}{\omega C_\pi}\right)^2}{\frac{g_m^2}{\omega^2 C_\pi^2}} \cong 2qI_c(r_b + R_s)^2\left(\frac{f}{f_T}\right)^2$$

Divided into low-frequency and higher-frequency components, we get

$$\overline{v_{ni}^2} = 4KT\,(r_b + R_s) + 2qI_B(r_b + R_s)^2 + \frac{2qI_c(R_s + r_b + r_\pi)^2}{\beta_0^2} + \frac{2qf_L i_B^\gamma(R_s + r_b')^2}{f} + 2qI_c(r_b + R_s)^2\left(\frac{f}{f_T}\right)^2 \tag{9.67}$$

Equivalent noise voltage source v_n can be found by setting $R_s = 0$; in the expression for v_{ni}^2 in (9.67)

$$\overline{v_n^2} = 4KTr_b + 2qI_B r_b^2 + \frac{2qI_c r_\pi^2}{\beta_0^2} + \frac{2qf_L I_\beta^\gamma r_b'^2}{f} + 2qI_c r_b^2\left(\frac{f}{f_T}\right)^2 \tag{9.68}$$

Since $r_\pi = \beta_0 r_e$ and $r_b^2 \ll \beta_0 r_e^2$, $r_e = 1/g_m$ can be simplified to

$$\overline{V_n^2} = 4KTr_b + \frac{2qI_c}{g_m^2} + \frac{2qf_L I_\beta^\gamma r_b'^2}{f} + 2qI_c r_b^2\left(\frac{f}{f_T}\right)^2 \tag{9.69}$$

Equivalent noise current source I_n can be found by letting R_s be very high, dividing all terms in the expression (6.24) by R_s^2, and letting R_s go against ∞.

$$\overline{I_n^2} = 2qI_B + \frac{2qI_C}{\beta_0^2} + \frac{2qf_L I_\beta^\gamma}{f} + 2qI_c\left(\frac{f}{f_T}\right)^2 \tag{9.70}$$

Since $I_c/\beta_0^2 << I_B$, the first term will dominate the second term, which thus can be removed, so

$$\overline{I_n^2} = 2qI_B + \frac{2qf_L I_\beta^\gamma}{f} + 2qI_c\left(\frac{f}{f_T}\right)^2. \tag{9.71}$$

iii. Mid-band Noise or Minimum noise [51]

From Eqs. (9.69) and (9.71), v_n and I_n are frequency-dependent. At low-frequency flicker, the noise of BJT will be a dominant source of noise, while at a higher frequency, we get an additional frequency-dependent part of shot noise in the collector. We can say that mid-band, where noise is not strongly frequency-dependent but parasitic, will be a dominant source of noise. If we remove the frequency-dependent part in (9.69) and (9.71)

$$\overline{V_n^2} = 4KTr_b + \frac{2qI_c}{g_m^2} = 4KT\left(r_b + \frac{1}{2g_m} + \frac{g_m}{2\beta}\right) \tag{9.72}$$

$$\overline{I_n^2} = 2qI_B = 4KT\left(\frac{g_m}{2\beta}\right) \tag{9.73}$$

From Eqs. (9.72) and (9.73), total input referred noise voltage for the RF stage amplifier for a finite source impedance of R_s

$$\overline{V_{n,i}^2} = 4KT\left(R_s + r_{b1} + \frac{1}{2g_{m1}} + \frac{g_{m1R_s}{}^2}{2\beta}\right) \tag{9.74}$$

9.17.1.1 Flicker Noise ($1/f$) of Transconductor

$1/f$ noise exists in the baseband, which is not at the same frequency as the RF signal. Therefore, it is upconverted to the LO frequency and its odd harmonics. In an ideal situation, the transconductor transistor's $1/f$ noise is moved out of the baseband, which should not affect the SNR at the mixing output [51].

9.17.1.2 White Noise of Transconductor

Figure 9.74 shows the spectral content of several signals in the single-balanced mixer. The signal resulting from LO mixing comprises odd harmonics of the fundamental frequency f_{LO}. The output current of the RF stage at small-signal levels exhibits noise. We assume the power spectral density of noise from the transconductance stage is flat with frequency, i.e., white noise associated in the transconductor stage [4, 35, 51, 56].

Therefore, mixer communication is assumed as square wave-like, and LO frequency and its odd harmonics downconverted the respective component of the white noise to the IF and is written as

$$\overline{V_{n,o\_RF}^2} = n.\left(\frac{2g_{m1}R_C}{\pi}\right)^2\left(4KT\left(r_{b1} + \frac{1}{2g_{m1}}\right)\right) \tag{9.75}$$

Any periodic LO waveform, sine wave or otherwise, switches the mixer results in square-wave communication of transconductance stages' output current, and the factor n is given as

$$n = 2\left(1 + \frac{1}{3^2} + \frac{1}{5^2} + \ldots\right) = \frac{\pi^2}{4} \tag{9.76}$$

Due to the lack of an image sideband, a double-sideband mixer would contribute half of the value shown in (9.76).

The analysis of frequency dependence in RF stage noise necessitates the examination of sidebands of as many harmonics as are required in a mixer circuit. This power spectral density is difficult to calculate due to poor parasitic modeling at very high frequencies [35, 51].

9.17.2 Noise Due to LO Switching Pair

The mixer noise due to the LO switching is characterized as direct switching noise and indirect switching noise [35, 51, 56].

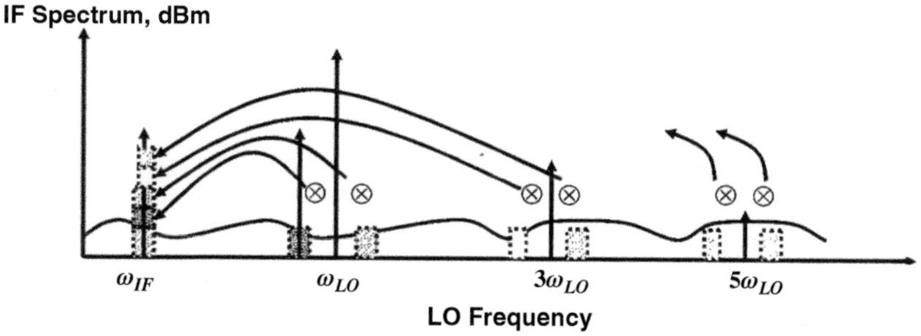

Figure 9.74 The noise spectral content of the LO frequency and its harmonics downconverted to IF port in the single-balanced mixer. *Source*: [55] / MDPI/ CC BY 4.0.

9.17.2.1 Direct Switching Noise

Additional $1/f$ noise, a weak signal overlapped on the LO switching signal, is experienced by the base. Significant changes in the LO signal cause the switching events to occur either forward or backward in time. As a result, the noise causes jitter. Understanding the difference between the switched current with and without jitter is necessary in order to quantify the noise contribution, as illustrated in Figures 9.110 and 9.111. Subtracting the two signals results in a sequence of pulses with the following properties:

- The pulses are located at the edges of switching times.
- The amplitude can be positive or negative, equal to twice the switched current.
- The pulse width $\Delta t = \frac{V_n(t)}{S}$, where $V_n(t)$ is the magnitude of the base noise voltage at the time of switching and S is the slope of the LO voltage applied to the base. Figure 9.111 demonstrates this relationship.

Consequently, jitter may be considered a sampled representation of the original $1/f$ noise, which is further scaled according to the slope S of the signal. The sampling frequency is double the LO frequency due to jitter happening at each switching moment or twice during each LO period (pulse rate of pulse train $2f_{\mathrm{LO}}$). Thus, the normalized jitter noise current for one period is:

$$I_{o,n} = \frac{2}{T_{\mathrm{LO}}}.2I_{ee}.\frac{V_n}{S} \tag{9.77}$$

The slope (S) of the sinusoidal LO signal can be given as follows:

For a differential LO signal, $V_{\mathrm{LO}}d = 2\sin\omega_{\mathrm{LO}}t$; therefore $S = \frac{dV_{\mathrm{LO}}d}{dt} = 2V_{\mathrm{LO}}\omega_{\mathrm{LO}}$ (9.78)

So, ST_{LO} will be $4\pi V_{\mathrm{LO}}$.

The baseband noise spectrum becomes

$$I_{o,n}(f) = \frac{1}{\pi V_{\mathrm{LO}}}.I_{ee}.V_n(f) \tag{9.79}$$

The noise $V_n(t)$ from the LO switch is only transferred to the output during the zero crossing. Both switches contribute noise to the mixer output when they are on; no noise is contributed when one switch is off, and when the other switch is on, it also contributes no noise due to its functioning as a cascade structure with a fixed tail current of I_{RF} by the RF input transconductance stage. The noise current at the mixer output is a series of pulses occurring at twice the LO frequency rate, with a height equal to $2I_{\mathrm{RF}}/S$ and a width that is randomly modulated by noise. The details of the periodic sampling function and autocorrelation of the output noise are given in [57]. The input noise is stationary and white, and its power spectral density is given by:

$$\overline{V_{n,s}^2} = 4KT\left(r_{b2} + \frac{1}{2g_{m2}}\right) \tag{9.80}$$

$$[g_{m2}]_{\mathrm{zerocrossing}} = \frac{2I_{\mathrm{RF}}}{\Delta V} \tag{9.81}$$

g_{m2} is the switch transconductance at zero crossing.

The power spectral density of the output noise current is given by [11, 35]

$$\overline{I_{n0,s}^2} = \frac{2}{T_{\mathrm{LO}}}\int_0^{T_{\mathrm{LO}}} p^2(t)\overline{V_{n,s}^2} = \left(\frac{2}{T_{\mathrm{LO}}}\right)\left(\frac{2I_{\mathrm{RF}}}{S}\right)^2\left(\frac{1}{T_S}\right)\overline{V_{n,s}^2} \tag{9.82a}$$

$$= \left(\frac{4I_{\mathrm{RF}}}{S\cdot T_{\mathrm{LO}}}\right)\left(\frac{2I_{\mathrm{RF}}}{S\cdot T_S}\right)4KT\left(r_{b2} + \frac{1}{2g_{m2}}\right) \tag{9.82b}$$

Substitute the value of $S \cdot T_{\mathrm{LO}} = 4\pi V_{\mathrm{LO}}$ solve it to get:

$$\overline{I_{n0,s}^2} = \left(\frac{I_{\mathrm{RF}}}{\pi V_{\mathrm{LO}}}\right)2KT(2r_{b2}g_{m2} + 1) \tag{9.82c}$$

The output power spectral density depends on LO magnitude (V_{LO}) and bias current (I_{RF}). Detailed calculations can be found in [56].

9.17.2.2 Indirect LO Switching

Indirect switching noise depends on the LO frequency, and BJT capacitances existed at the virtual ground. The noise voltage v_n << LO signal so that it can be treated as a switched voltage of slowly varying amplitude. The switching edges will be a time constant, enabling a current through the capacitance C_p associated with transconductance stage Q1.

$$C_p = C_{be2} + C_{be3} + C_{js1} + C_{bc1} \tag{9.83}$$

$$I_{n,o} = \frac{2}{T_{\text{LO}}} \int_0^{T_{\text{LO}}/2} C_p \left[\frac{d}{dt} v_p(t) \right] \text{dt} \tag{9.84a}$$

which leads to the noise contribution

$$i_{n,o} = \frac{2}{T_{\text{LO}}} C_p v_n = 2 f_{\text{LO}} C_p v_n \tag{9.84b}$$

The spectrum analysis reveals that indirect switching noise behaves similarly to direct switching noise. Current bursts at the virtual ground change polarity as the transistor switches on and off, but the differential pair compensates for this shift. As a result, $1/f$ noise is sampled at twice the LO frequency and appears at the output. In any case, indirect switching noise is significantly lower than direct switching noise. Although faster switching can minimize direct conversion noise, it does not entirely eliminate switching noise [35, 51, 56].

9.17.3 Noise Due to Load [51, 56]

Noise due to load R_L can be written as $4KTR_L$ since the output is differential; total thermal noise downconverted to IF will be

$$\overline{V^2_{n,R_L}} = 8KTR_L \tag{9.85}$$

9.18 Total Mixer Output Noise (BJT) [51, 56]

From (6.32), (6.47), and (6.48), total noise at mixer output can be written as

$$\overline{V^2_{n0,\text{total}}} = \overline{V^2_{n,R_L}} + 2(\overline{I^2_{n0,s}})\, R_L^2 + \overline{V^2_{n,o\_RF}} \tag{9.86}$$

$$\overline{V^2_{n0,\text{total}}} = 8KTR_L + \left(\frac{I_{\text{RF}}}{\pi V_{\text{LO}}} \right) 4KT(2r_{b2}g_{m2} + 1){R_L}^2 + n.\left(\frac{2g_{m1}R_L}{\pi} \right)^2 \left(4KT\left(r_{b1} + \frac{1}{2g_{m1}} \right) \right) \tag{9.87}$$

The simplified expression of total mixer noise downconverted to IF is given by

$$\overline{V^2_{n0,\text{total}}} = 4KT{R_L}^2 \left[\frac{2}{R_L} + (2r_{b2}g_{m2} + 1)\left(\frac{I_{\text{RF}}}{\pi V_{\text{LO}}} \right) + {g_{m1}}^2 \left(r_{b1} + \frac{1}{2g_{m1}} \right) \right] \tag{9.88}$$

The input referred noise of the given mixer is

$$\overline{V^2_{ni,\text{total}}} = \frac{1}{(\text{gain})^2} \overline{V^2_{n0,\text{total}}} \tag{9.89}$$

$$\overline{V^2_{ni,\text{total}}} = \frac{4KT{R_L}^2 \left[\frac{2}{R_L} + (2r_{b2}g_{m2} + 1)\left(\frac{I_{\text{RF}}}{\pi V_{\text{LO}}} \right) + {g_{m1}}^2 \left(r_{b1} + \frac{1}{2g_{m1}} \right) \right]}{\left(\frac{2g_{m1}R_L}{\pi} \right)^2} \tag{9.90}$$

Transconductance seen at the RF stage is twice of the switching stages, so $g_{m1} = \frac{I_{\text{RF}}}{2V_{Th}} \approx g_{m2} \approx g_m$

$$\overline{V^2_{ni,\text{total}}} = \pi^2 KT \left[\frac{2}{{g_m}^2 R_L} + \left(r_{b1} + \frac{1}{2g_m} \right) + \left(\frac{1}{{g_m}^2} (2r_{b2}g_m + 1)\left(\frac{I_{\text{RF}}}{\pi V_{\text{LO}}} \right) \right) \right] \tag{9.91}$$

where $K =$ Boltzmann constant
$T =$ Temperature in Kelvin (290K)
$I_{\text{RF}} =$ DC bias current or tail current
$r_{b1} =$ Internal base resistance of RF stage
$r_{b2} =$ Internal base resistance of switching stage
$g_m =$ Transconductance of mixer
$V_{\text{LO}} =$ Amplitude of the LO signal

Note: At given IF $\neq$ 0, this noise results from both the signal and image bands, yielding the SSB noise figure [54].

$$NF_{\text{SSB}} = 1 + \frac{\overline{V^2_{ni,\text{total}}}}{4KTR_S} \tag{9.92}$$

9.18.1 Calculation of Noise Figure of a Single-balanced Mixer [56]

For the single-balanced mixer shown in Figure 9.68, the SSB noise can be calculated using (9.96). The internal base resistance of BFP840 and BFP640 can be found in the spice parameter file; values are 16.87 Ω and 4.406 Ω, respectively. The simulated transconductance g_m is about 0.175S, respectively. Power at the LO port is −10 dBm; converting into mV to the 50 Ω system will be roughly 70.7 mV.

Solving for individual terms listed in Eq. (9.91)

Terms	Calculated values
$\frac{2}{{g_m}^2 R_L}$	0.345535
$r_{b1} + \frac{1}{2g_m}$	19.72714
$\frac{1}{{g_m}^2}$	32.65606
$(2r_{b2}g_m + 1)$	2.5421
$\frac{I_{\text{RF}}}{\pi V_{\text{LO}}}$	0.04446

Substitute all the values in Eq. (6.54), solving for $\frac{\overline{V^2_{ni,\text{total}}}}{4KTR_S}$ will be roughly 1.171477.

$$NF_{\text{SSB}} = 10\log(1 + 1.171477) = 10\log(2.171477) = 3.367552 \text{ dB}.$$

Simulation shows total noise, which is downconverted to the IF port at 145 MHz.

noisefreq	NF dsb	NF ssb
145.0 MHz	2.159	3.343

The simulated and calculated SSB noise figure results are close. The figure shows simulated conversion gain and noise figure as a function of LO power. The optimum power level for the LO drive is about −12 to −6 dBm, where we can achieve a minimum noise figure [56].

Figure 9.75 shows conversion gain and noise figure as a function of RF frequency, and Figure 9.76 shows the SSB noise figure and conversion gain versus RF input frequency for fixed IF at 145 MHz. Due to the narrowband impedance match only designed at RF frequency, we can get maximum power transfer to the IF frequency (Figure 9.76).

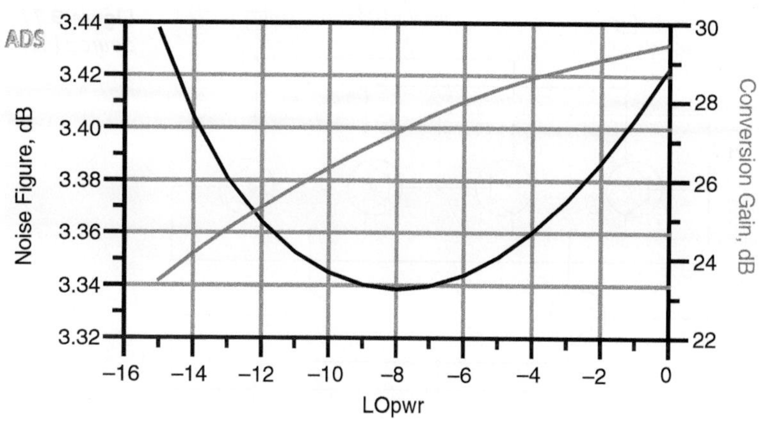

Figure 9.75 Noise figure and conversion gain at IF frequency, 145 MHz. *Source*: Adapted from [56].

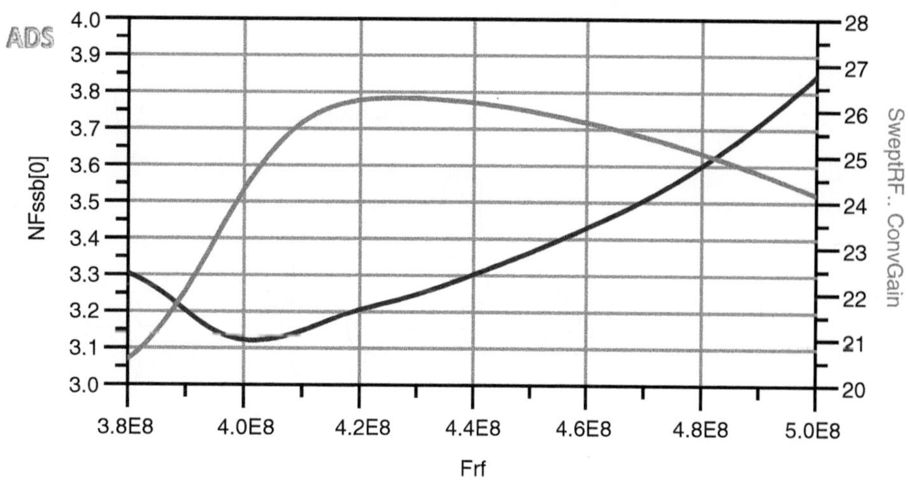

Figure 9.76 SSB noise figure and conversion gain versus RF input frequency for fixed IF at 145 MHz. *Source*: Adapted from [56].

9.19 BJT Gilbert Cell

The classic active mixer (Figure 9.77), conceived by Barrie Gilbert in 1967, is the basis for most active mixers used in wireless products today. Figure 9.78 shows a Gilbert cell mixer implementation for analysis. Figure 9.79 shows its conversion gain and noise figure versus LO power. Figure 9.80 shows its IM_3 and IP_3 responses.

As Gilbert wrote in a 1994 monograph [4, 17]: This circuit is attractive because (a) it can be monolithically integrated with other signal-processing circuitry; (b) it can provide conversion gain, whereas a diode-ring mixer always has conversion loss; (c) it requires very low power to drive the LO port; (d) it provides excellent isolation between the signal ports; and (e) it is far less fussy about load-matching.

The two major advantages afforded by the Gilbert cell are port isolation (it should really be called a multiplicative mixer) and its significant reduction of even-order frequency combinations. As with the single-transistor mixer, the Gilbert cell does not particularly shine with high intercept point and low current consumption. Another major drawback is that it requires a high-impedance output, which gives the user headaches because most of the better filters are in 50 Ω technology, and only very simple monolithic filters offer input and output impedances around 1 kΩ. Impedances of this order can result in significant crosstalk in high-density circuit boards.

The reason the Gilbert cell can be so linear for small input signal levels is twofold: It is a differential amplifier and it makes use of the tanh function, which is linear over a wide range around the zero-crossing point [4, 18]. Attempts have been made to increase the intercept point.

The Plessey SL6440 IC was a special version of the Gilbert cell that achieved high IP_{3in} (+30 dBm) at the expense of DC current.

It also appears that the Gilbert cell is a derivative of the Jones Cell patent US 3.241.078 March 15, 1966, though Jones was not properly given credit in the literature!

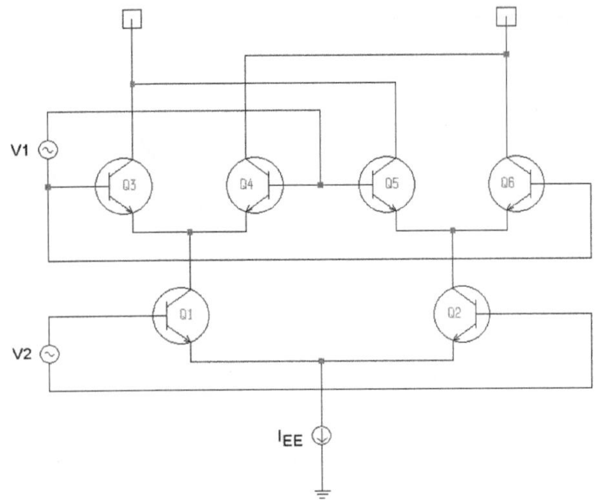

Figure 9.77 The basic double-balanced Gilbert multiplier cell. *Source*: [4] / John Wiley & Sons.

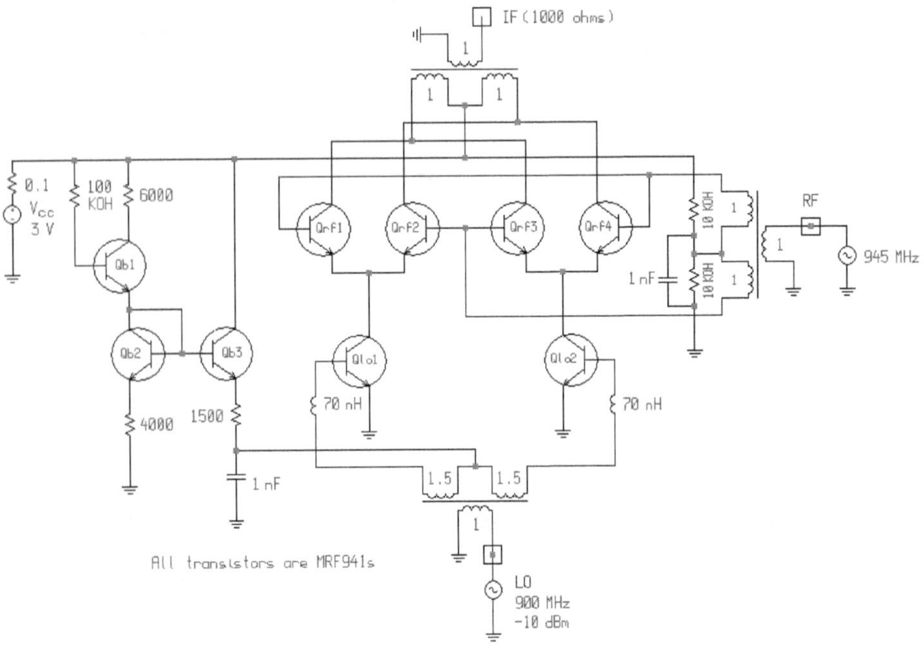

Figure 9.78 Double-balanced Gilbert cell mixer validation circuit. *Source*: [4] / John Wiley & Sons.

The doubly balanced Gilbert cell in Figure 9.78, having four cells, based on symmetry, has the ability to cancel unwanted spurious. This is analogous to the doubly balanced diode mixer.

Our suspicion, however, is that a combination of a modern double-balanced passive mixer using medium- to high-level diodes followed by a feedback FET amplifier, or even one of the late Community Antenna Television (CATV) transistors with f_Ts of more than 25 GHz, would allow the achievement of a significantly higher intercept point. As pointed out in our amplifier chapter, CATV transistors are now available that combine a noise figure of less than 1 dB with a IP_{3in} of more than 30 dBm. Built-in push–pull structures using such transistors can achieve an IP_{2in} of 70 dBm, matching the numbers for the diode DBM. In the amplifier chapter, we learned that feedback always improves the dynamic range, mostly at the expense of gain, since the gain-bandwidth product gets reduced. Motorola has recently come out with a clever extension of its high-level preamplifier and modified it to become a mixer.

The two tones were 10 and 30 GHz, and the resulting difference IF is either 20 GHz with gain or 40 GHz with much less gain, as the cutoff frequency of the semiconductors shows off.

We now look for the optimum drive level for the maximum conversion gain of an NF. The collector load is 2350 Ω (T 7:1).

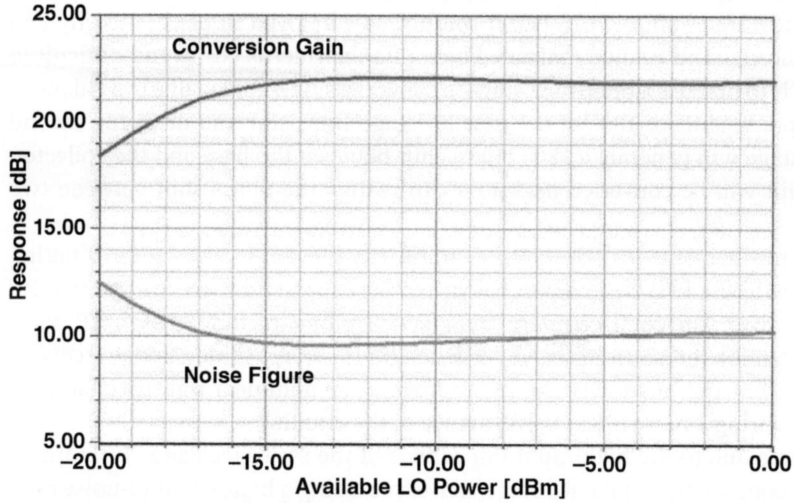

Figure 9.79 Conversion gain and noise figure versus LO power for the Gilbert cell mixer. In this analysis, LO = 900 MHz (−20 to 0 dBm) and RF = 945 MHz (−50 dBm) for an IF of 45 MHz. *Source*: [4] / John Wiley & Sons.

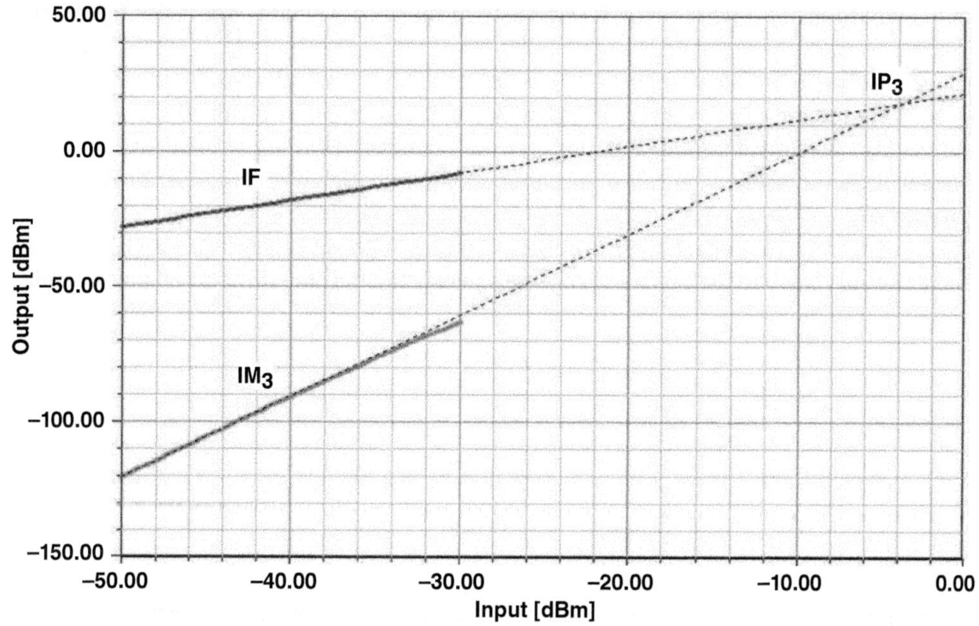

Figure 9.80 IM_3 and IP_3 responses for the BJT Gilbert cell mixer with RF signals at 945 and 946 MHz. In this analysis, LO = 900 MHz (−10 dBm), RF1 = 945 MHz (−50 to −30 dBm), and RF2 = 946 MHz (−50 to −30 dBm). Four LO harmonics and three LO sidebands were used. *Source*: [4] / John Wiley & Sons.

9.20 Gain and Noise Simulation of a Double-balanced Gilbert Cell Differential Mixer (Gilbert Cell) [51, 56]

Figure 9.81 shows a novel doubly balanced mixer using bipolar transistors. This circuit and the following explanation are based on the dissertation in progress by Trusha Kared, PhD student of Prof. Ulrich Rohde at the Brandenburg Technical University, Cottbus, Germany.

The part of the circuit diagram shown shows the two differential mixers called "Gilbert cells." The following is a simulation of an advanced dual-cell mixer. The enhancement is shown in the source of the two differential stages.

Implementing a Gilbert cell mixer presents a challenge in using differential amplifiers within the cell itself. The optimal differential driver of the RF port is the most critical trade-off. Here, a better solution is employed in place of costly and potentially asymmetrical transformers, such as those found in diode mixers. Such a transformer is costly and difficult to implement at the RF stage. The proposed solution is to first use a pre-driver stage to convert a single-ended line to a balanced line [51]. As shown in Figure 9.81, we use the phase shift of 180° of the first RF stage transistor and drive the second transistor. Ideally, the first RF common emitter stage will generate a 180° phase shift between the base and the collector. At the second stage, the base–collector phase shift will be converted back to 0°. In reality, the phase shift between two transistors is never ideal.

Each RF stage incorporates multiple feedback mechanisms to address the nonlinearity present in transistors. Simultaneously, the series feedback from the second stage to the first stage serves to correct the phase shift between the two RF currents and enhances the dynamic range of the mixer. The circuit employs dual feedback linearization at each RF stage, combining series inductive feedback (at the input side) and shunt resistive feedback (at the output side). Series feedback stabilizes input current, while shunt feedback controls output voltage, hence improving the nonlinear transfer characteristics of the transistor, reducing distortion, and ensuring a more linear performance of the circuit.

The mixer circuit is less influenced by the source due to the high input impedance of the series feedback. It improves impedance matching at the input, reducing noise contribution at the RF stage, hence providing a high signal-to-noise ratio at the input. Shunt feedback limits noise propagation at the output as it decreases the output impedance, making it easier for the circuit to drive a load. This improves signal transfer to the next stage with minimal loss. The combined effect results in enhanced linearity, a wide dynamic range, and a low noise figure. A 1000 MHz RF frequency and a fixed IF frequency of 145 MHz are used in the design to achieve the best possible performance in terms of conversion gain, noise figure, and linearity.

Thus, this portion will discuss the analysis and results of Figure 9.81, as developed by Trusha Kared. Figure 9.82 shows the conversion gain and noise figure as a function of LO power to derive the optimal condition for the double-balanced downconversion mixer circuit. She has also developed a set of equations for the prediction of noise figure calculation and linearity analysis. The details of her work are presented in her dissertation, "A Novel High Dynamic Range Wideband Differential Mixer Using SiGe Transistors."

In the ideal case, we assume the LO differential pair operates in a saturation region with square-wave characteristics. At the collector current output, we force the sine wave to look like a square wave, ensuring that the differential pair operates near a saturation region.

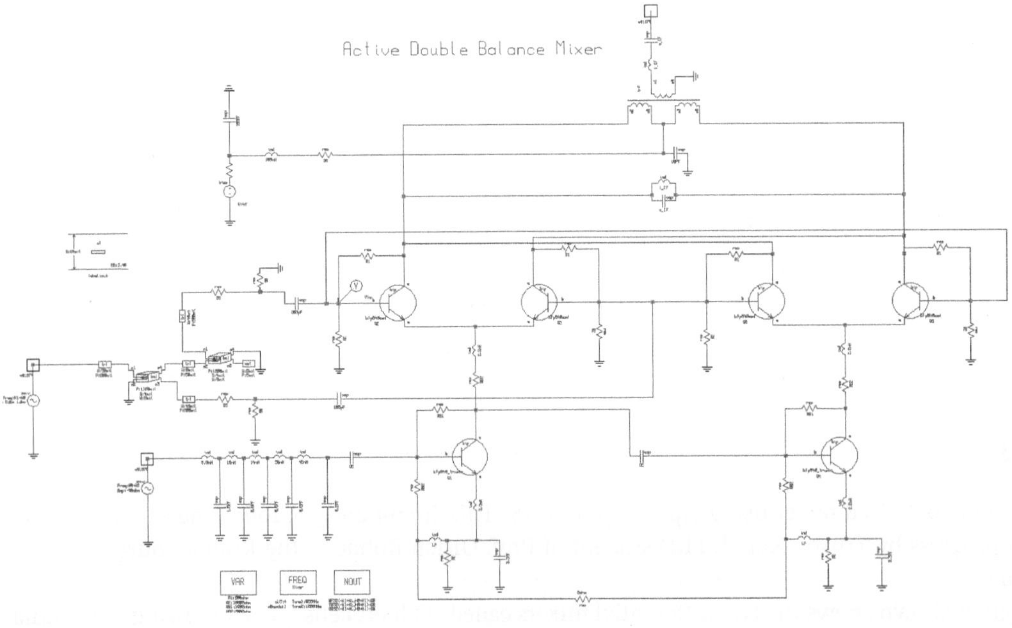

Figure 9.81 Active double-balanced downconversion mixer [51, 56].

Figure 9.83 shows the collector current waveforms of upper transistors (Q2 and Q3, and Q5 and Q6), and Figure 9.84 shows the input base voltage and current of LO differential pairs. Due to the phase shift of the transconductance Y_{21}, at higher frequencies, we lose the device symmetry in the collector current; the collector current waveforms at the lower side show ringing and limited voltage, while the upper side shows saturation.

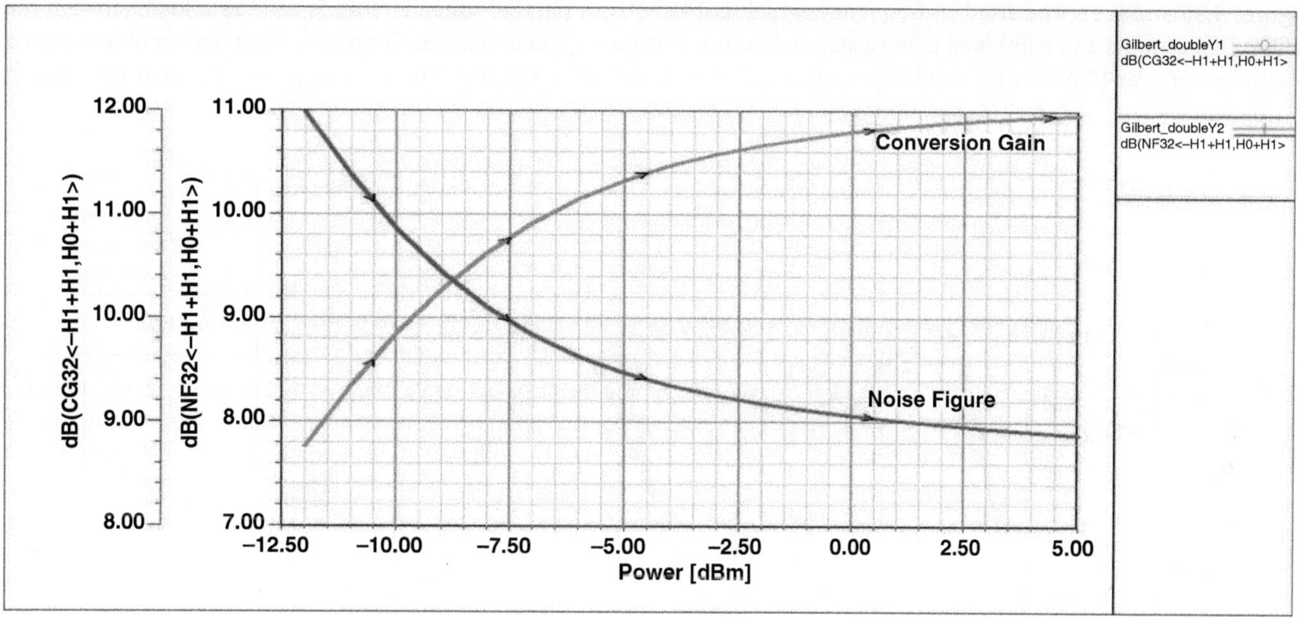

Figure 9.82 Conversion gain and noise figure as a function of LO power (test condition RF: 1000 MHz at −30 dBm; LO: 855 MHz at −15 dBm to 5 dBm) [51, 56].

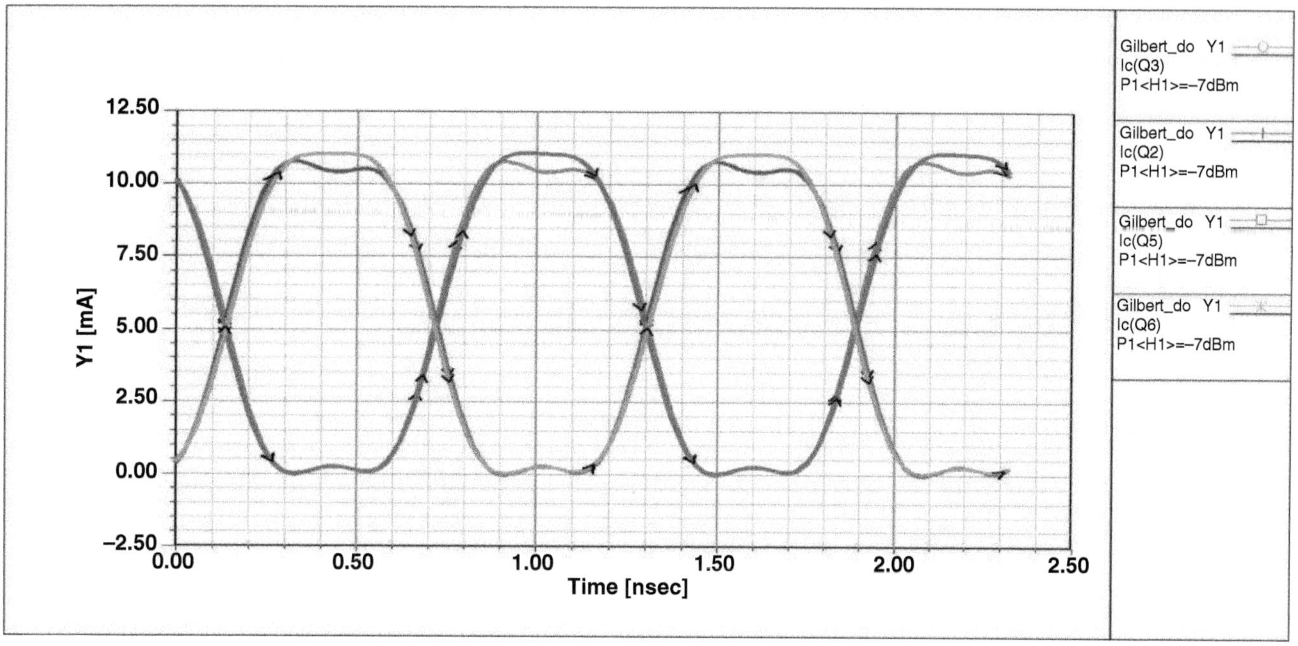

Figure 9.83 Collector current waveform of Q2 and Q3, and Q5 and Q6 LO differential pair showing nonideal square-wave characteristics [51, 56].

9.20.1 Mixer LO and RF Bandwidth

Figure 9.85 shows a frequency sweep of the RF frequency from 400 MHz to 2.5 GHz with constant IF to find the operating bandwidth of the downconversion mixer. Figure 9.85 shows the simulated conversion gain and noise figure as a function of RF frequency.

The return loss of the double-balanced downconversion mixer (Figure 9.81) matched to the 50 Ω system is shown in Figures 9.86 and 9.87. The fixed IF frequency, which is 145 MHz, a parallel-tuned circuit, is used as a load between the differential outputs and combined using a step-down transformer matched to the 50 Ω system. The mixer achieves a good return loss of over 10 dB for RF and LO ports in the desired band and a simulated return loss of over 25 dB at 145 MHz of IF frequency.

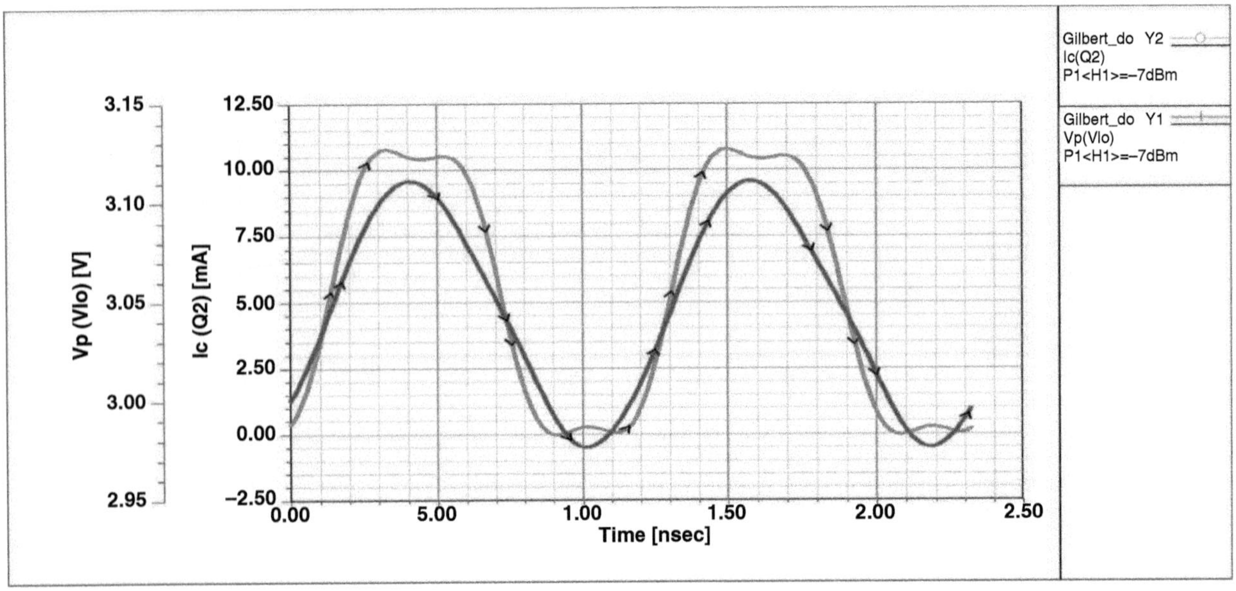

Figure 9.84 Transconductance: input base voltage and current of LO differential pairs at −7 dBm of LO power level [51, 56].

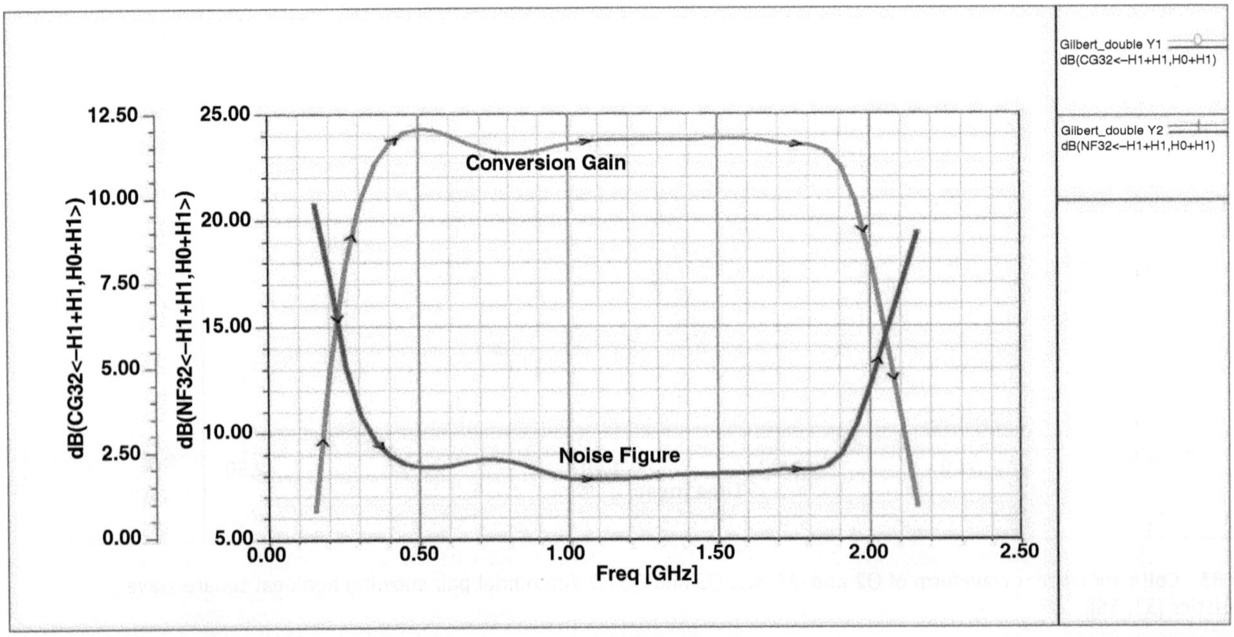

Figure 9.85 Conversion gain and noise figure as a function of RF frequency for constant IF difference of 145 MHz—test condition: RF frequency: 300–2300 MHz at −30 dBm; LO frequency: 155–2155 MHz at −5 dBm [51, 56].

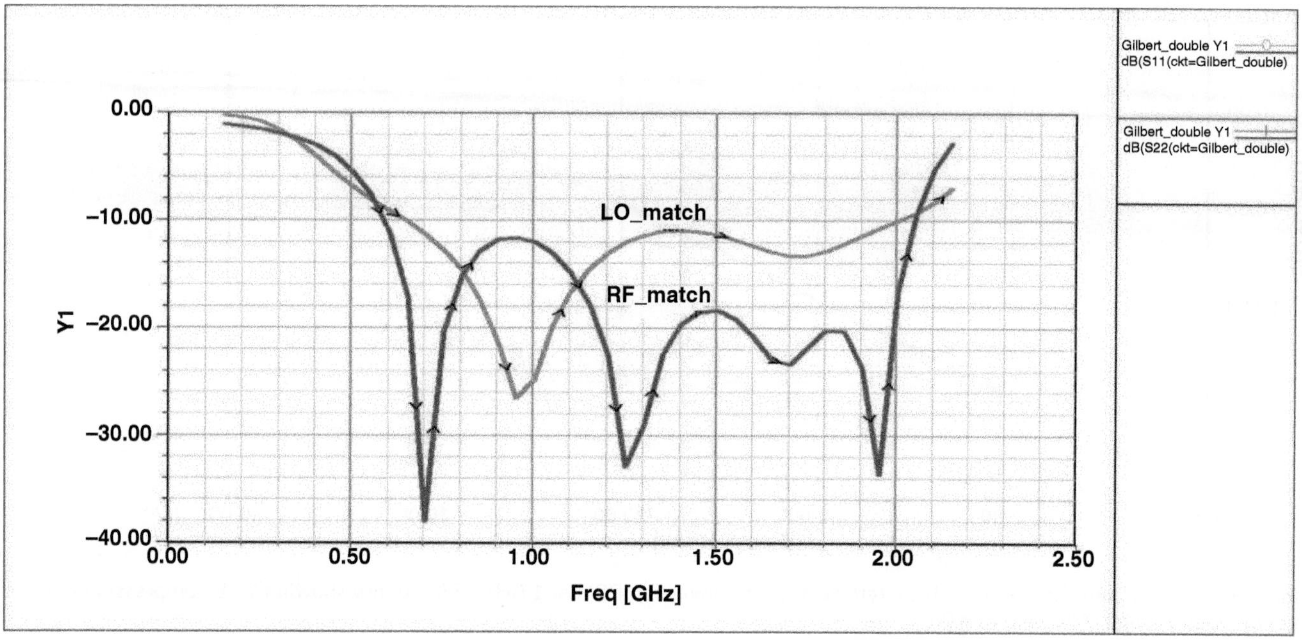

Figure 9.86 LO and RF port return loss of the DBM in a 50 Ω system. *Source*: Adapted from [56].

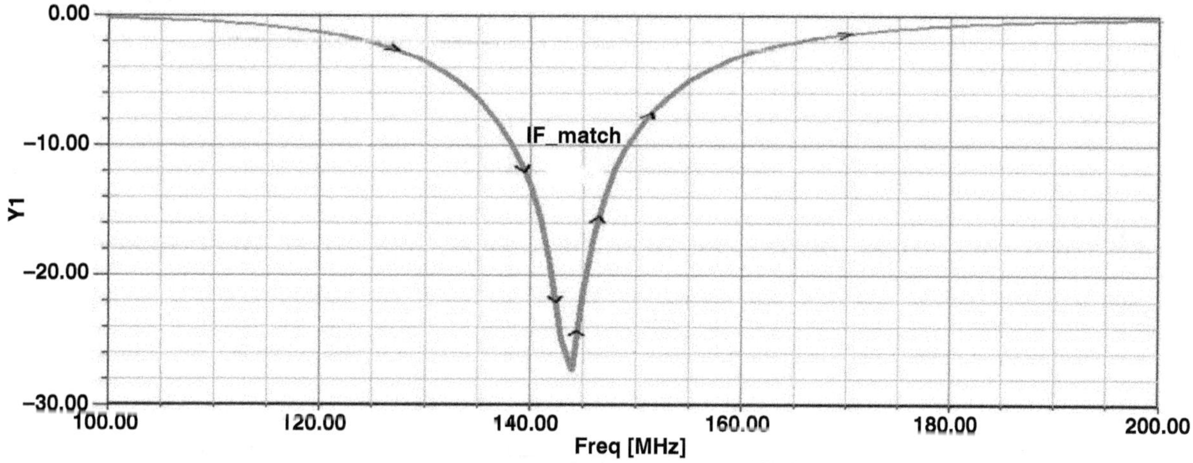

Figure 9.87 IF port return loss of the DBM in a 50 Ω system. *Source*: Adapted from [56].

9.20.2 Linearity of DBM: IP_3 and 1 dB Compression Point

Linearity in a mixer is vital because it provides a proportional relationship between the input and output signals, minimizing distortion and undesirable harmonics, which are critical for sustaining signal integrity. The simulated 1 dB compression point, fundamental and the third harmonic distortion, and IP_3 point are shown in Figures 9.88, 9.89, and 9.90, respectively.

The circuit shown in Figure 9.81a employs dual feedback linearization at each RF stage, combining series inductive feedback (at the input side) and shunt resistive feedback (at the output side). Series feedback stabilizes input current, while shunt feedback controls output voltage, hence improving the nonlinear transfer characteristics of the transistor, reducing distortion, and ensuring a more linear performance of the circuit.

Figure 9.88 shows the compression characteristic of the downconversion mixer as a function of RF input power. Here, $f_{LO} = 855$ MHz, $f_{RF} = 1000$ MHz, and RF power increases in each simulation. The uncompressed gain is about 11.4 dB, about 1 dB less than our estimate, decreasing by 1 dB at −6.9 dBm (≈ 143 mV). LO feedthrough and signal distortion

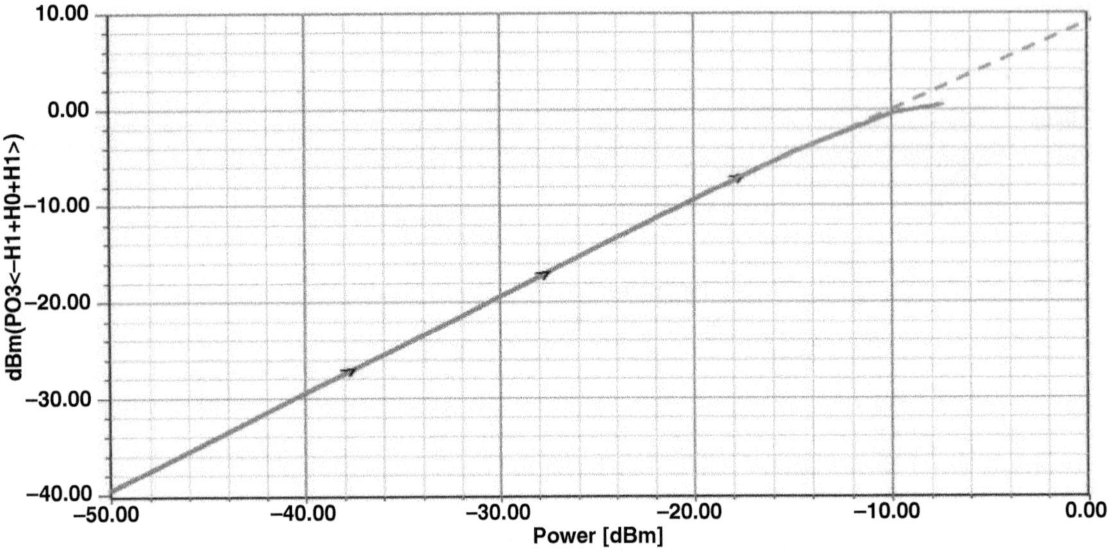

Figure 9.88 Simulated compression characteristic of downconversion DBM at 1 GHz RF frequency showing 1 dB compression point is about −7 dBm. *Source*: Adapted from [56].

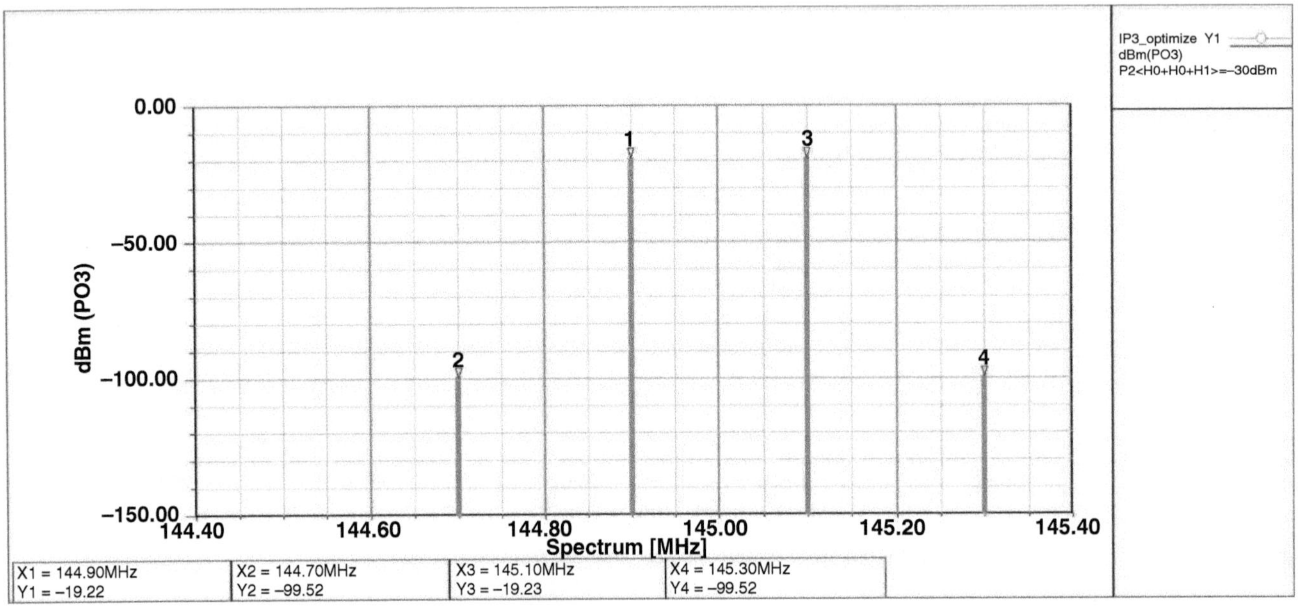

Figure 9.89 Two-tone test of 1 GHz downconversion DBM test condition: RF frequency: 1599.9 MHz and 1600.1 MHz at −30 dBm; LO frequency: 1455 MHz at −5 dBm. *Source*: Modified from [51].

make it challenging to measure 145 MHz IF amplitude in the time domain. Therefore, the fast Fourier transform (FFT) of the input and output is investigated.

In order to measure the input IP_3 of the mixer, we apply two sinusoidal tones to the input with frequencies equal to 999.95 MHz and 1000.05 MHz. The peak amplitude of each tone is −30 dBm. If the tone amplitude is too small, the IM_3 components are corrupted by the FFT noise floor, and if it is too large, the circuit may experience higher-order nonlinearities. Figure 9.89 shows the mixer's downconverted spectrum, revealing a difference of $\Delta P = P_{out} - P_{IM3} = 80.30$ dB between the fundamental and IM_3 tones. We divide this value by 2 and add the input tone power, obtaining $IP_3 = +10.15$ dBm in a 50 Ω system, while Figure 9.90 shows a geometric interpretation of the above relationship.

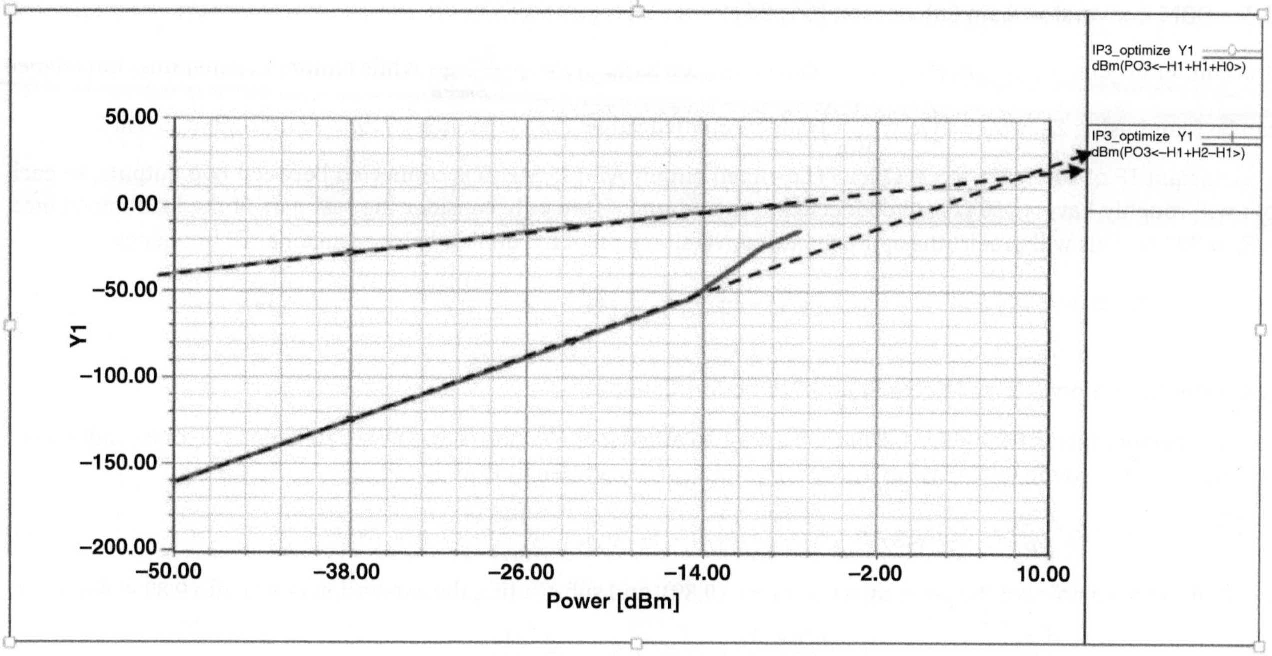

Figure 9.90 Extrapolation of IP_3 from fundamental and third harmonic tone power. *Source*: Modified from [51].

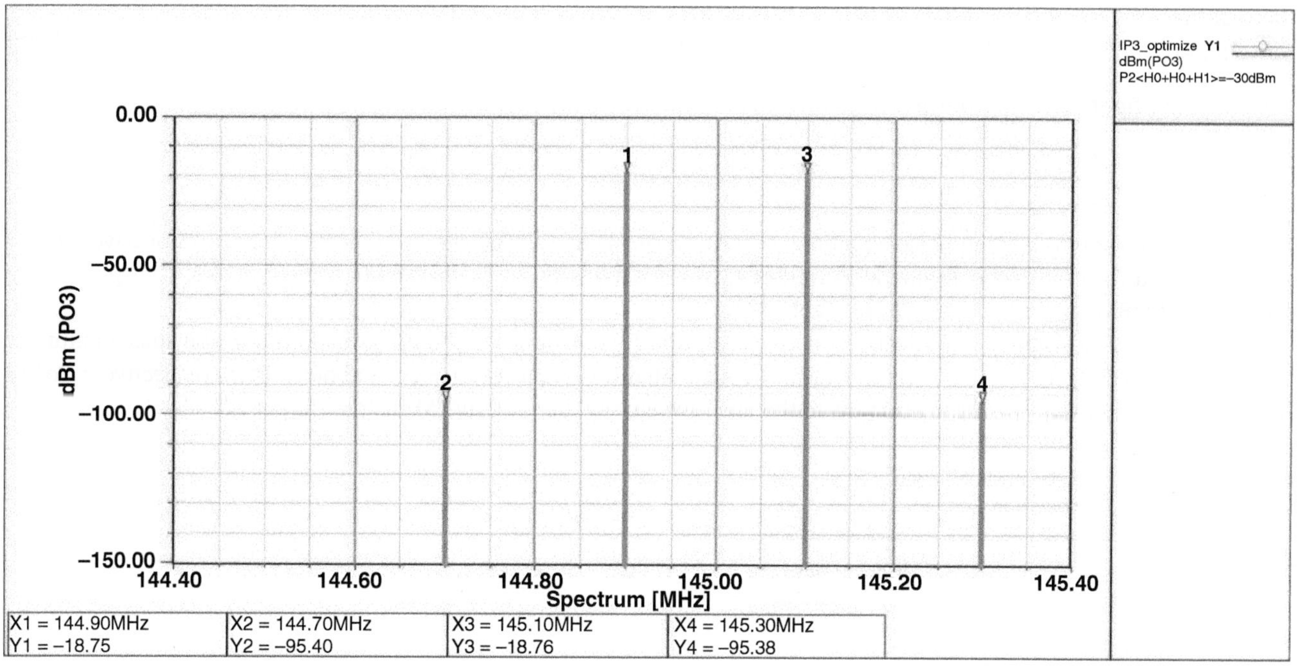

Figure 9.91 Two-tone test of 1.6 GHz downconversion DBM; test condition: RF frequency: 1599.9 MHz and 1600.1 MHz at −30 dBm; LO frequency: 1455 MHz at −5 dBm. *Source*: Modified from [51].

The actual IP_3 point is the calculated value and cannot be measured because the required drive level will destroy the transistor.

Figure 9.91 shows the two-tone test of the DBM (Figure 9.81) at an RF frequency of 1600 MHz and the IP3 point is calculated from the marker readouts as of $\Delta P = P_{\text{out}} - P_{\text{IM3}} = (95.40) - (18.75) = 76.65$ dB. We divide this value by 2 and add the input tone power (−30 dBm), obtaining $IP_3 = +8$ dBm in a 50 Ω system.

9.20.3 DBM Conversion Gain Calculation [51, 56]

For the emitter degeneration case, effective g_m can be written as $g_{m\_eff} = \frac{g_m}{1 + g_m Z_E}$ while emitter degenerating impedance Z_E can be found using $Z_E \approx SL_E + \left(R_E || \frac{1}{SC_E}\right)$, and calculated value of $Z_E \approx 21.983 + j19.577$ and $|Z_E| \approx 29.436\ \Omega$.

At a constant IF of 145 MHz, a 500 Ω load (LC-tuned circuit AT 145 MHz) is connected between two outputs, so each output will roughly have a 250 Ω load connected to the ground. If we only consider the real part of the load impedance, then $R_L \approx 227\ \Omega$. This will predict the approximate conversion gain of the double-balance mixer at IF frequency.

$$\text{So, voltage conversion gain } A_v = \frac{2}{\pi} g_{m\_eff} R_L = \frac{2}{\pi} \frac{g_m}{1 + g_m R_E} R_L = \frac{2}{\pi} \frac{0.182 * 227}{1 + (0.182 * 29.436)} \approx 12.338 \text{ dB} \tag{9.93}$$

9.20.4 Double-balance Mixer Noise Figure Prediction Calculation

In a double-balance mixer, the noise contribution downconverted to the IF port is double because of twice transconductance stages and switches, so the total noise at the mixer output can be written as from Eq. (9.88)

$$\overline{V^2_{n0,\text{total}}} = 4KTR_L^{\,2}\left[\frac{2}{R_L} + 2\left((2r_{b2}g_m + 1)\left(\frac{I_{\text{RF}}}{\pi V_{\text{LO}}}\right)\right) + 2\left(g_m^{\,2}\left(r_{b1} + \frac{1}{2g_m}\right)\right)\right] \tag{9.94}$$

The input referred noise of the given mixer from Eq. (9.89), and substituting the expression (9.94) into 9.89 yields

$$\overline{V^2_{ni,\text{total}}} = \frac{4KTR_L^{\,2}\left[\frac{2}{R_L} + 2\left((2r_{b2}g_m + 1)\left(\frac{I_{\text{RF}}}{\pi V_{\text{LO}}}\right)\right) + 2\left(g_m^{\,2}\left(r_{b1} + \frac{1}{2g_m}\right)\right)\right]}{\left(\frac{2g_{m1}R_L}{\pi}\right)^2} \tag{9.95}$$

$$\overline{V^2_{ni,\text{total}}} = \pi^2 KT\left[\frac{2}{g_m^{\,2} R_L} + 2\left(r_{b1} + \frac{1}{2g_m}\right) + \left(\frac{2}{g_m^{\,2}}(2r_{b2}g_m + 1)\left(\frac{I_{\text{RF}}}{\pi V_{\text{LO}}}\right)\right)\right] \tag{9.96}$$

So, SSB noise figure can be given as

$$NF_{SSB} = 1 + \frac{\overline{V^2_{ni,\text{total}}}}{4KTR_S} \tag{9.97}$$

The output and input-referred noise equations derived in (9.94) and (9.96) assume that RF stages do not have emitter degeneration. In the case of the degenerated emitter g_m is replaced with g_{m_eff} derived in Section 9.20.3. Let's predict the noise figure of DBM shown in Figure 9.93a (Table 9.6).

The internal base resistance of BFP840 (r_{b1}at RF stage) and BFP640 (r_{b2} at LO stage) was found in the spice parameter file; values are 16.87 Ω and 4.406 Ω, respectively. The simulated transconductance g_m is about 0.182S, respectively. Power at the LO port is −5 dBm; converting into mv to the 50 Ω system will be roughly 355 mVp-p (as per the simulation).

Table 9.6 Prediction of the noise figure of double-balanced mixer shown in Figure 9.81.

Terms	Calculated values
$g_{m\_eff} = \frac{g_m}{1 + g_m z_E}$	0.028628
$\frac{2}{g_m^{\,2} R_L}$	10.75013
$2 * \left(r_{b1} + \frac{1}{2g_m}\right)$	68.67051
$\frac{2}{g_m^{\,2}}$	2440.28
$(2r_{b2}g_m + 1)$	1.252272
$\frac{I_{\text{RF}}}{\pi V_{\text{LO}}}$	0.008954

Solving for individual terms listed in Eq. (9.96)

Substitute all the values in Eq. (7.7) and then into (7.8), solving for $\frac{\overline{V^2_{ni,\text{total}}}}{4KTR_S}$ will be roughly 5.264251.

$$NF_{\text{SSB}} = 10\log(1 + 5.264251) = 10\log(6.264251) = 7.968692\text{ dB}.$$

Figure 9.82 shows the optimum condition for gain and noise figure as a function of LO drive. At −5 dBm of LO power simulated noise figure is about 8.31 dB, while the predicated approximation is about 7.96 dB.

9.21 BJT Gilbert Cell with Feedback

Motorola's MC13143 low-power 2.4 GHz mixer IC (now discontinued) uses a patented topology consisting of a class AB-biased Gilbert cell augmented by feedback. Its linearity can be programmed via an external current source to achieve an $IP_{3\text{in}}$ of 20 dBm at the expense of additional supply current.

The MC13143 contains 29 active transistors. Figure 9.92 shows the core of the MC13143 circuit reduced to its essentials and configured for computer analysis per Motorola's MC13143 test circuit. Figure 9.93 shows the circuit's conversion gain and noise figure versus LO drive, and Figure 9.94 shows the circuit's typical IF, IM_3 and IP_3 responses, both as calculated by Ansoft Corporation's Serenade 8.0 circuit simulator. A reproduction of the MC13143 datasheet, used by permission, follows Figure 9.94 [4].

This Motorola's MC13143 mixer IC uses DC feedback for improved linearity that can be programmed by applying a control current (0–2.3 mA) to its MIXER LINEARITY CONTROL pin. Per Motorola, an $IP_{3\text{in}}$ of 20 dBm may be achieved with a control current of 2.3 mA, at the expense of approximately 7 mA of additional supply current. In this validation circuit, R_1 (the resistor in Q10's emitter) has been adjusted for a total current drain (supply + linearity control) of 8 mA [4].

In considering the approach used in the MC13143, one has to remember that feedback around many stages gives an IMD ripple, which means that the higher-order IPs are no longer a straight-line calculation. Above a certain level, the 3-dB/dB law fails, and as a result of this, the IMD characteristic can be as bad as 10 or 20 dB/dB—almost like a breakdown point.

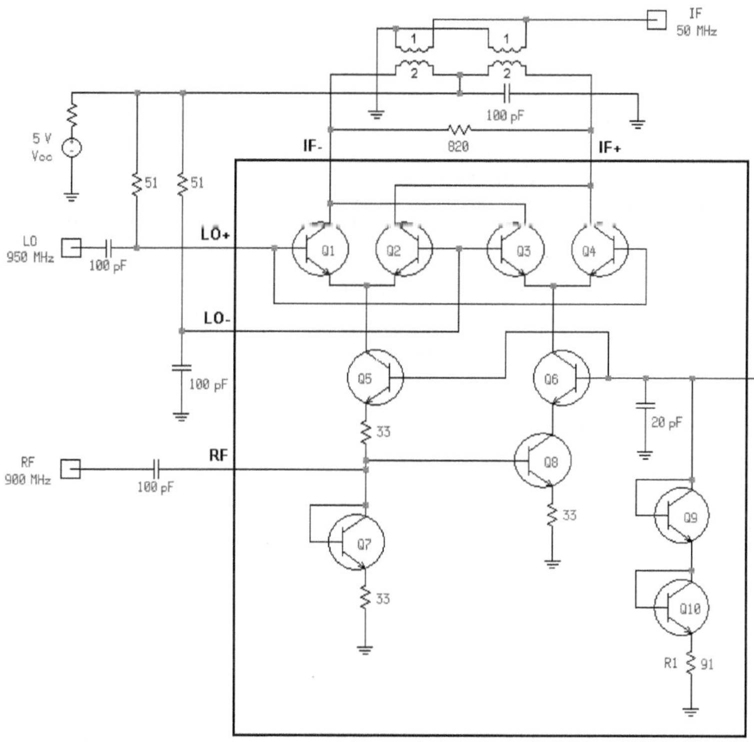

Figure 9.92 Motorola's MC13143 mixer IC. *Source:* [4] / John Wiley & Sons.

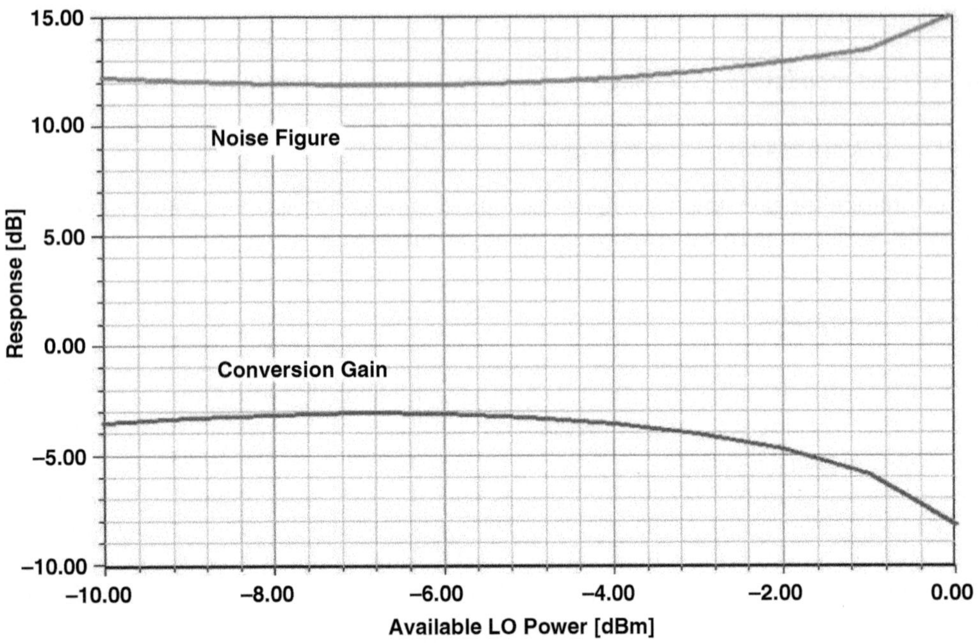

Figure 9.93 Calculated conversion gain and noise figure versus LO drive for the Motorola MC13143 IC operating at $V_{CC} = 2$. Motorola's specifications for the typical values of these characteristics are −5.0 dB and 12 dB, respectively, at an LO drive level of 0 dBm. The RF signal is at 900 MHz and the LO is at 950 MHz, for an IF of 50 MHz. In this analysis, LO = 950 MHz (−10 to 5 dBm) and RF = 901 MHz (−50 dBm) for an IF of 49 MHz. *Source*: [4] / John Wiley & Sons.

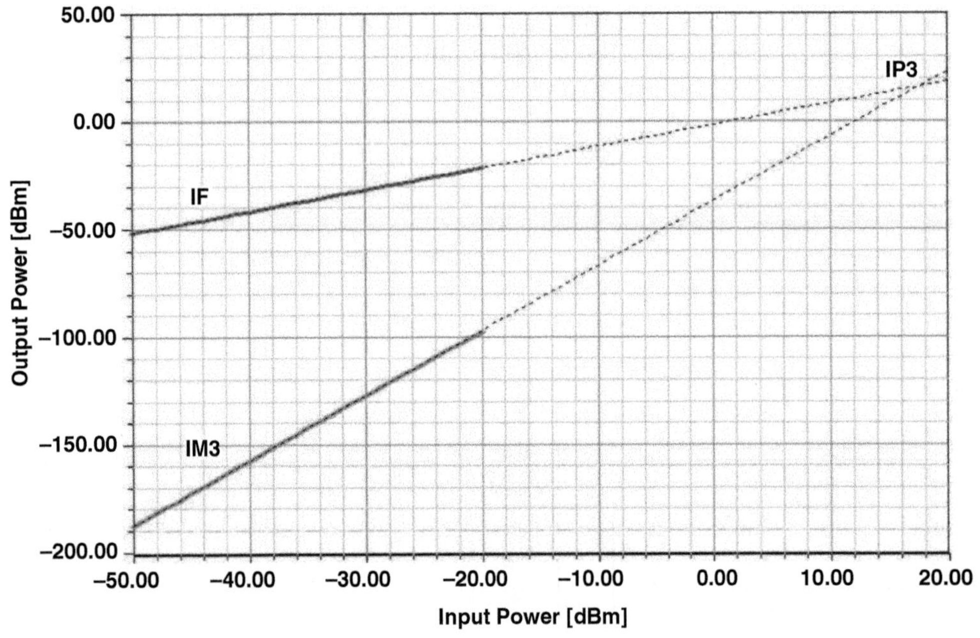

Figure 9.94 Calculated typical IF, IM_3 and IP_3 responses for the Motorola MC13143 mixer configured as shown in Figure 9.104. In this analysis, LO = 950 MHz (−5 dBm), RF1 = 900 MHz (−50 to −20 dBm), RF2 = 901 MHz (−50 to −20 dBm), and $V_{CC} = 5$. Four LO harmonics and three LO sidebands were used. *Source*: [4] / John Wiley & Sons.

High-frequency capability is an issue with silicon BJTs; the upper frequency limit of the technique is in the range of 3 GHz. Bipolar technology will be extended to 10 GHz by SiGe, but when these parts will be commercially available at competitive prices is currently unclear. There has not been enough published material on the actual performance of Gilbert cell-type mixers using SiGe transistors, but we assume that because of coupling on the silicon wafer, things are no longer as well behaved as we are used to at much lower frequencies.

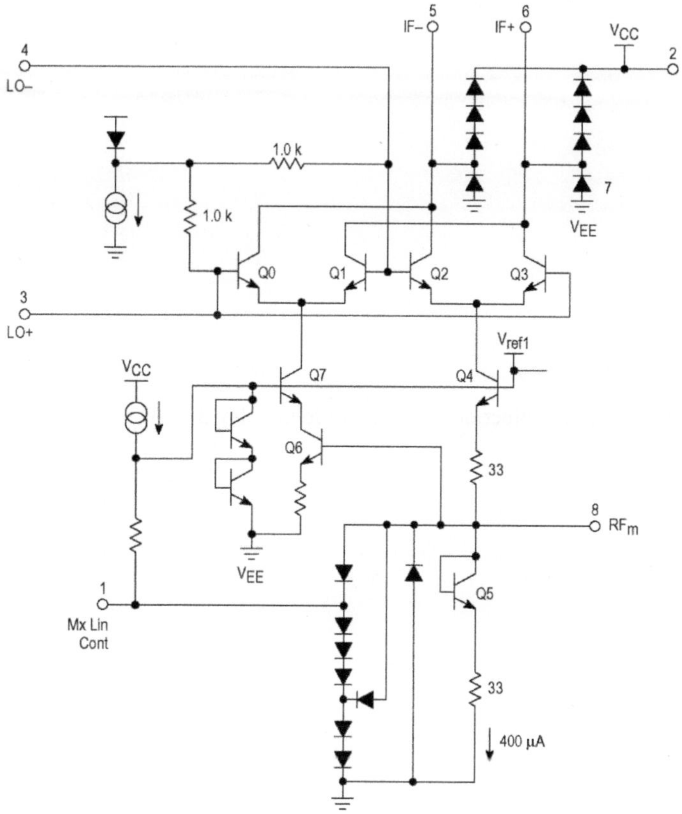

Figure 9.95 Internal circuit of the MC13143. *Source*: [4] / John Wiley & Sons. *Note*: The MC13143 uses a unique and patented circuit technology.

The actual internal circuit of the MC13143 is shown in Figure 9.95.

Final comment: in order to get the high dynamic range, the DC current has to be much higher than the quoted 400 μA in the circuit of Figure 9.95.

9.22 J-FETs/GaAs FET Mixers

FETs, used in active and passive circuits, are still a popular solution for low-power integrated mixers up to 100 MHz (passive quad), 1 GHz (dual-gate MOSFET). Cost-sensitive applications may require the use of a single-gate FET. These types of single-gate microwave mixers are frequently used to validate the quality of software but the user keeps forgetting that this is a quality of the model in question and not of the simulator. Unfortunately we have not seen the same circuit analyzed with modern tools, nor do any publications provide sufficient details of the circuit and the transistor to do this ourselves [4, 19].

Figure 9.96 [4, Figure 4.81] shows a basic, active FET mixer. Like the single-diode and single-BJT mixers, it is additive, although it's common to lessen RF-LO interaction by injecting one signal at the gate and the other at the source as shown. Single-gate GaAs and LDMOS FETs are also used in this arrangement with gate biasing as appropriate to the device type. Figure 9.97 as a reminder shows a BJT in an additive mixing configuration.

The linearity of FETs is based on the fact that a FET follows a square law and therefore the first derivative, its transconductance, is supposed to be constant. This is valid within a wide amplitude range. FET mixers have a noise figure similar to those found in bipolar mixers: typically around 6–9 dB, depending on the configuration. A FET mixer's intercept point is subject to load-impedance variations, with a purely resistive termination providing the best case. Terminating the mixer with a filter is problematic in that a filter looks purely resistive only within its 3 dB passband; in the transition band and beyond, the filter impedance rises rapidly and the mixer intercept point goes down.

The best means devised so far to minimize this effect is to configure the FET output as a high-pass filter, using capacitive coupling from the output tuned circuit to a 50 Ω bandpass filter, as shown in Figure 9.98. Outside its passband, this

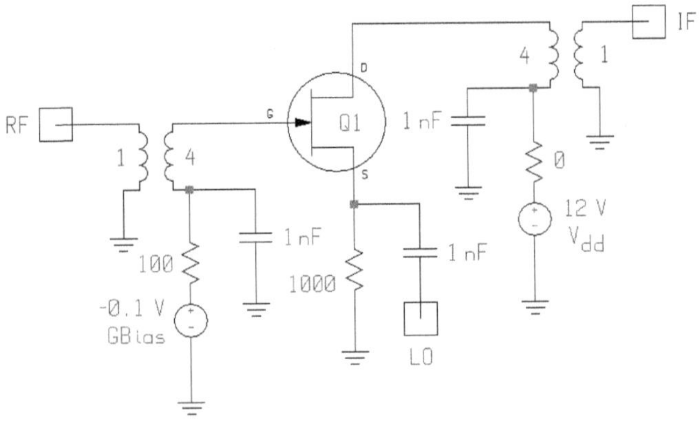

Figure 9.96 An additive JFET mixer cell using gate RF injection and source LO injection. *Source*: [4] / John Wiley & Sons.

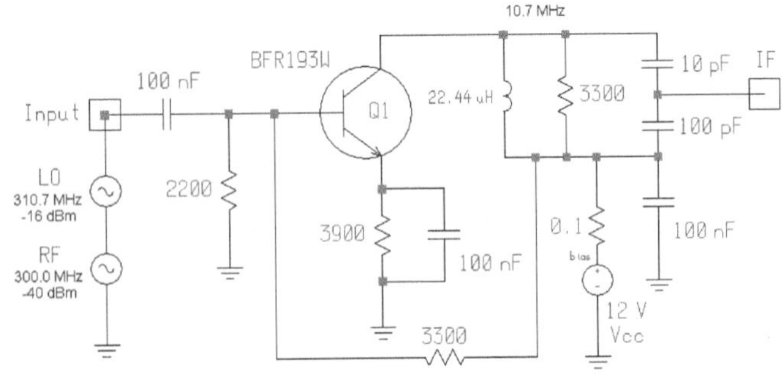

Figure 9.97 A reminder of building an additive mixer with a BJT rather than a multiplicative mixer. *Source*: [4, p 656] / John Wiley & Sons.

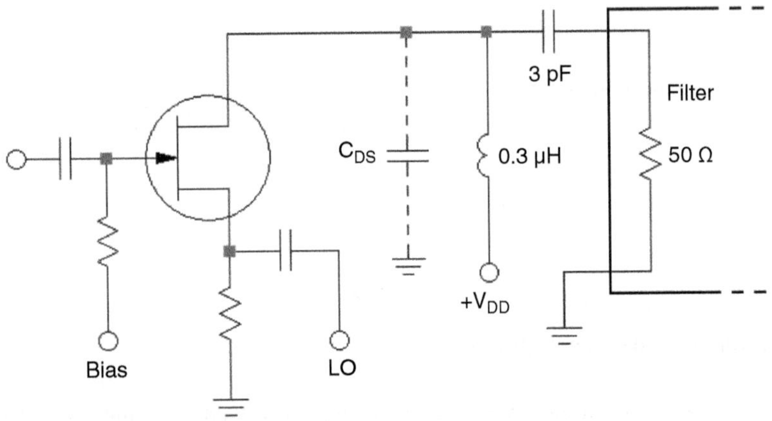

Figure 9.98 FET mixer output coupling that minimizes the effects of filter reactance outside the filter passband. *Source*: [4] / John Wiley & Sons.

impedance inverter acts more like a short circuit and maintains IMD products at a reasonable level. The alternative to this is the popular diplexer, which requires more components and has more insertion loss. The high-pass configuration has been barely mentioned in the literature.

Technically, this dual-gate multiplicative mixer can be described the same way as its bipolar transistor-based cousin. Since the NPN transistor has a much higher transconductance, it will have a higher gain. The FET version was popular because of the higher input impedance, but ultimately gave way to the bipolar or MOS-based differential mixer configuration.

The active FET mixer achieves gain at the expense of intercept point; the difference can be as much as 20 dB. As an example, the passive MOSFET quad mixer, which is to be referenced, achieves an intercept point of greater than 40 dBm while similar active silicon FETs barely make 30 dBm. This 30 dBm performance is possible only at fairly low frequencies

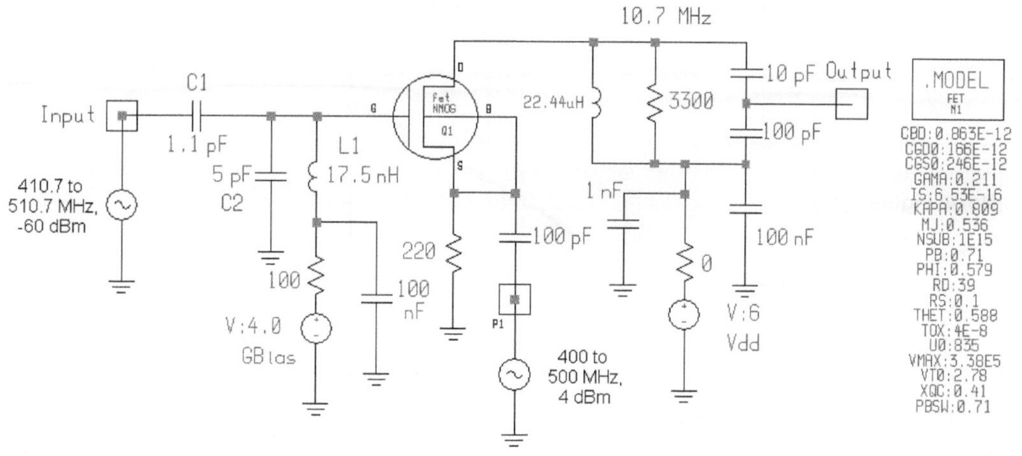

Figure 9.99 The N-JFET circuit of Figure 9.96 modified to work with an LDMOS FET. *Source*: [4] / John Wiley & Sons.

and requires careful selection and adjustment of the DC bias point of all transistors of the circuit. This performance also requires that the FETs be operated grounded-gate, which limits their gain. The other drawback of these active arrangements is that higher impedances at the output reduce the third-order intercept point over-proportionally; as an example, a factor of two in impedance change can cost as much as 6 dB intercept point. In these discussions, we assume that the IMD products are generated solely by the active device and not by any passive devices, such as transformers. Actually, there are also some capacitors whose linearity varies with RF current—another unpleasant effect that is frequently overlooked.

On the other hand, one can use any FET as a passive device similar to a diode mixer, in which the source-drain channel gets switched on and off. This impedance modulation is somewhat similar to a diode mixer, but the gate electrode is isolated from both source and drain. It nonetheless falls in the category of additive mixers because there is sufficient interaction between gate and source, although the impedance at the gate changes significantly less than in an additive diode mixer. Implementation is a challenge in that building a high-performance passive FET mixer requires a pair or a quad of mixer cells that are sufficiently matched to suppress even-order IMD products. Intercept points, depending on the LO drive, vary between +20 and +45 dBm. The lower number is more applicable for microwave and RF frequencies, while the +45 dBm level is easily obtainable between 5 and 30 MHz. It is necessary to remember that the FET is a voltage-driven switch, and LO matching becomes an issue: How do you generate 30 V peak-to-peak across a few picofarads over a wide frequency range? Physical layout can also be critical: Symmetry is not only important for the active devices, but also for the input, output and LO circuit. We have seen cases where fractions of an inch in different lengths leading to the gate electrodes have cost up to 10 dB of IP_3—just because of the resulting lack of symmetry.

9.23 The N-JFET Circuit Modified to Work with an LDMOS FET

Figure 9.99 shows the circuit of Figure 9.96 modified to work with an LDMOS FET, the harmonic-balance version of the SPICE level 3 model. Figure 9.100 shows the LDMOS circuit's conversion gain and noise figure versus frequency. The "noisy" traces reflect the model's numerical problems. This model, which works up to about 0.8 micron and has been validated against the Motorola data and SPICE results, has already been made mathematically continuous but still shows the inability to properly model the real RF transistor. This is why more research needs to be done on RF and microwave applications for MOS in the modeling area. As can be seen from the references, several attempts have been made to improve the model, but the major headache remains with parameter extraction, and as the models get more complex, the modeling quality does not seem to improve as a linear function. The reader should be reminded of the load-pull measurement as outlined earlier [4].

9.24 Dual-gate MOS/GaAs Mixers

Significantly better LO-RF isolation can be obtained by applying the RF to LO signal at separate gates of a dual-gate FET (Figure 9.101). The result is multiplicative mixing with fewer constraints on LO and RF filtering.

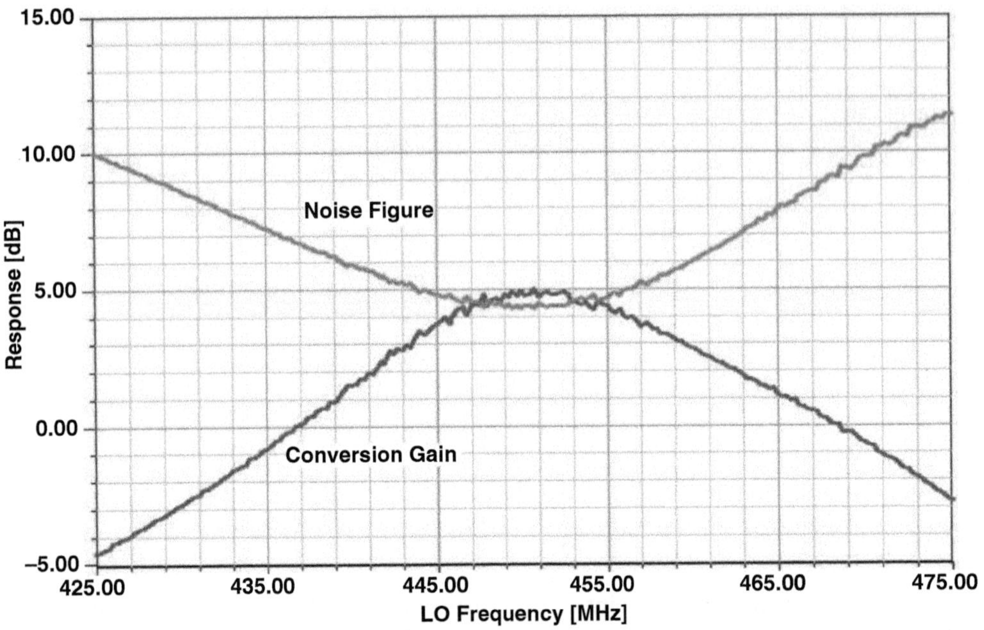

Figure 9.100 The LDMOS FET mixer's calculated conversion gain and noise figure. Although the "noisy" traces reflect numerical problems in the device model, the results can be put to good use in circuit design. *Source*: [4] / John Wiley & Sons.

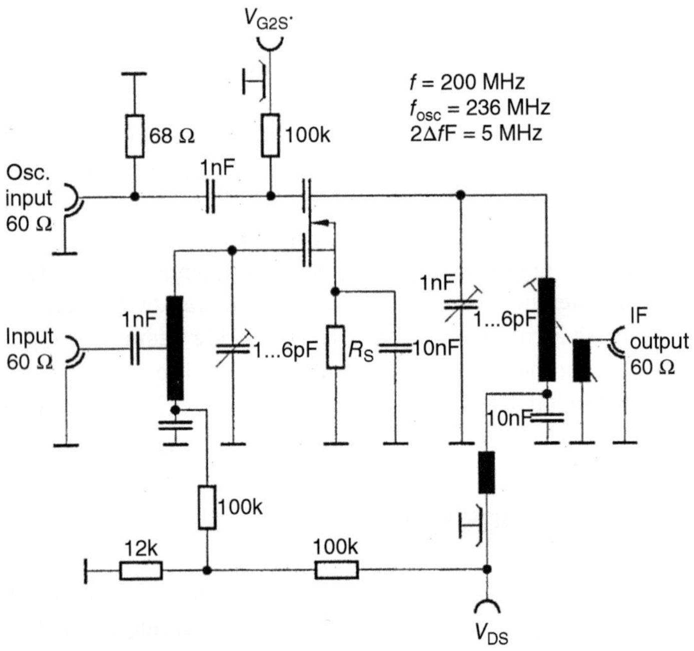

Figure 9.101 Test circuit for multiplicative mixing using a dual-gate N-MOSFET BF900. *Source*: [4] / John Wiley & Sons.

The following is a single-gate FET mixer, using a GaAs FET, and the transistor model was the modified Materka model. The circuit itself is self-explanatory. These types of circuits are used in some very simple applications, lack high dynamic range, and show frequency pulling.

9.24.1 Nonlinear Noise Analysis of a Single-gate FET Mixer

Figure 9.102a shows a single-gate GaAsFET mixer. It belongs to the group of additive mixers. The two signal sources on the left are in series. The FET requires a negative voltage that is fed via some inductors to the gate. The total input circuit is to match Γ_{opt}. The output network uses a transmission line to match the impedance. The parameters of the substrate are shown in the box below the FET. It is a modified Materka model, developed jointly with Raytheon, by Compact Software. The parameter extraction was nontrivial. In the meantime, some more complex models are being used [19].

The first simulation of the GaAsFET mixer shows the DC–IV curve and the load line. As the load line is a tuned circuit storing energy, we do not get a straight line but an elliptical curve with a dent. It needs to be noted that the RF current also goes below zero, about −8 mA.

Figure 9.102b shows the DC–IV curve of the mixer circuit. The large signal performance can be seen from the elliptical load line.

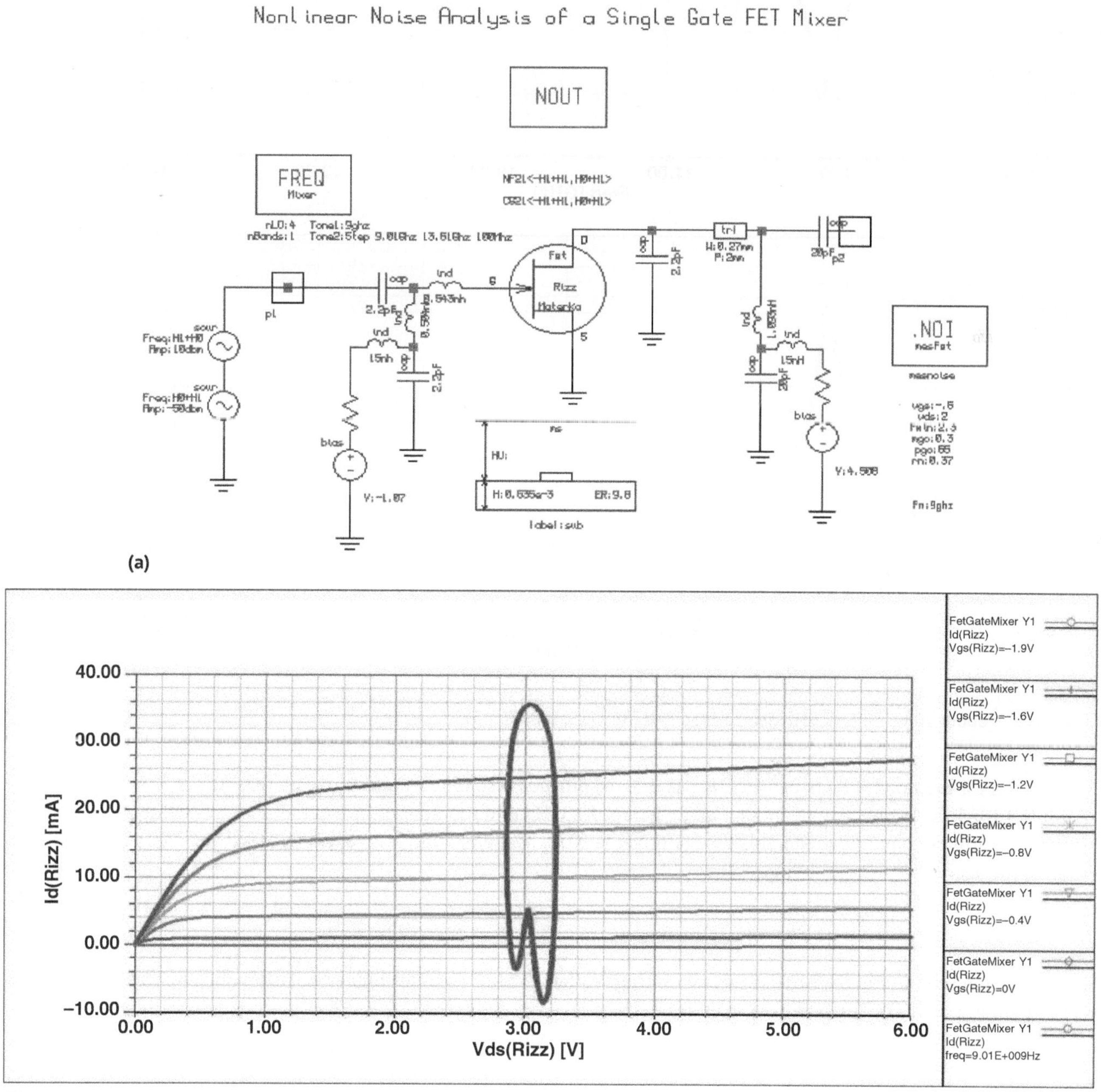

(b)

Figure 9.102 (a) Nonlinear noise analysis of a single-gate GaAsFET mixer. *Source*: Modified from [19, p 862]. (b) DC (RF)–IV curve and elliptical load line of the circuit in (a). *Source*: Modified from [19, p 863].

(Continued)

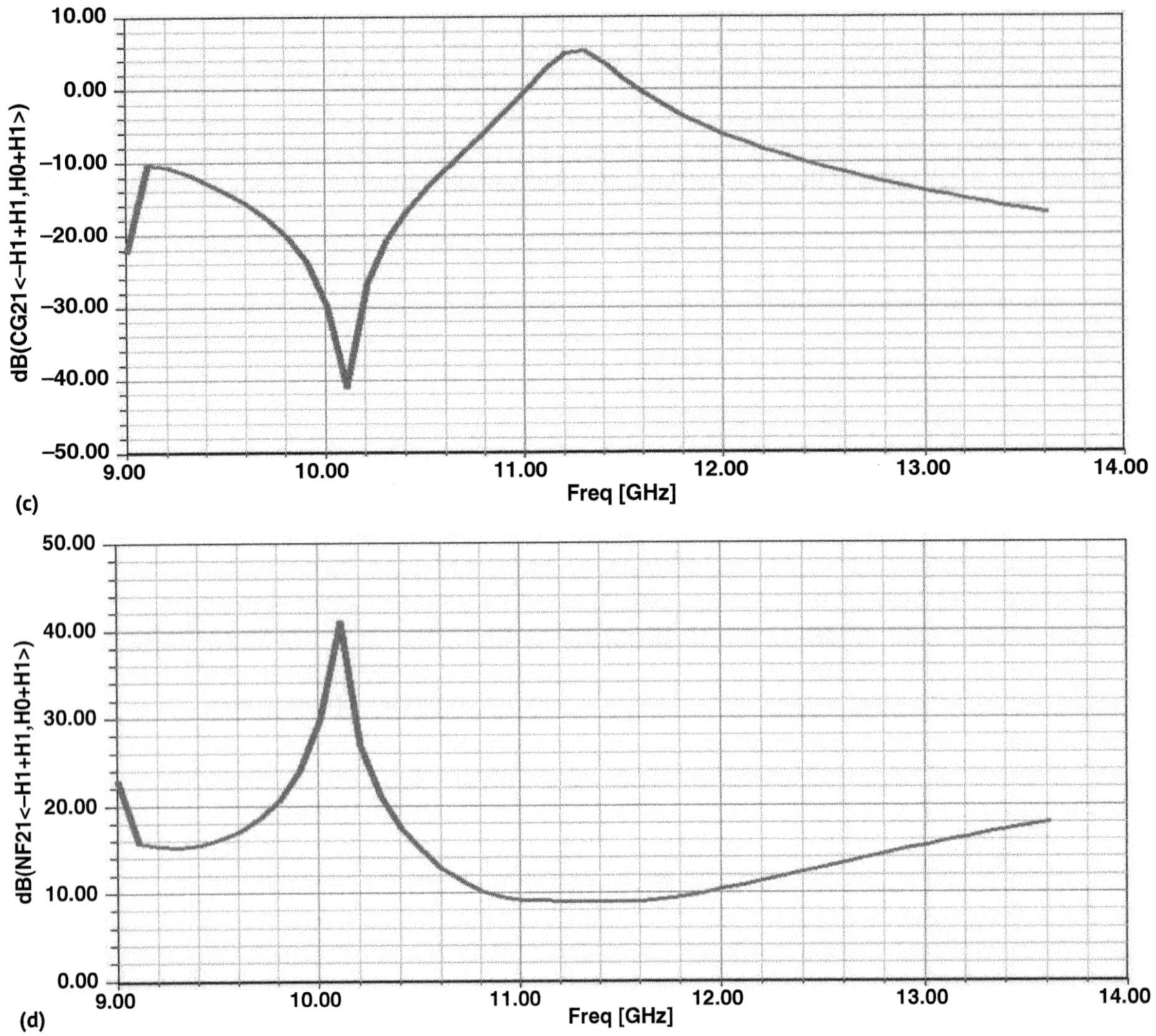

Figure 9.102 (Cont'd) (c) Simulated maximum conversion gain is ~6 dB at 11.3 GHz. *Source*: Modified from [19, p 863]. (d) Noise figure of the mixer is about 10 dB at 11.3 GHz. *Source*: Modified from [19, p 863].

The reason for the notch (at about 10.1 GHz) is due to spurious resonance, as seen in Figure 9.102c.

Figures 9.102d shows the gain of about 6 dB at about 11.3 GHz and the noise figure of the mixer is about 10 dB.

Due to the limited dynamic range, this type of mixer is more academic and less practical. However, it is a good test case to look at the accuracy of the simulator.

9.24.2 FET-based Triple-balanced Mixer

This application seen in Figure 9.103 of two mixer rings called "TBM" provides higher isolation and slightly higher losses. It is built around the FET ring and used to be popular.

The photograph (Figure 9.104) shows the TBM circuit on a chip.

9.25 MOSFET Gilbert Cell

The trend to go to smaller voltages has created some CMOS implementations of the Gilbert cell. While this may allow us to integrate reasonable mixers on the same chip, nobody should expect any miracles from these mixers. They are frequently starved in operating voltage and current and rarely fare better than a single diode mixer with proper drive level.

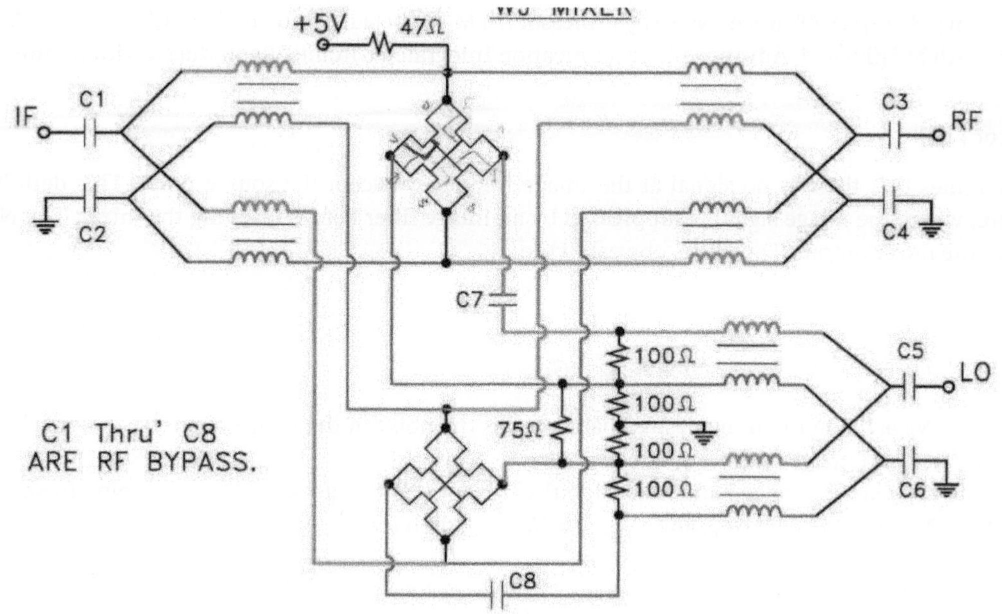

Figure 9.103 An example of a TBM schematic based on FETs (Synergy Microwave).

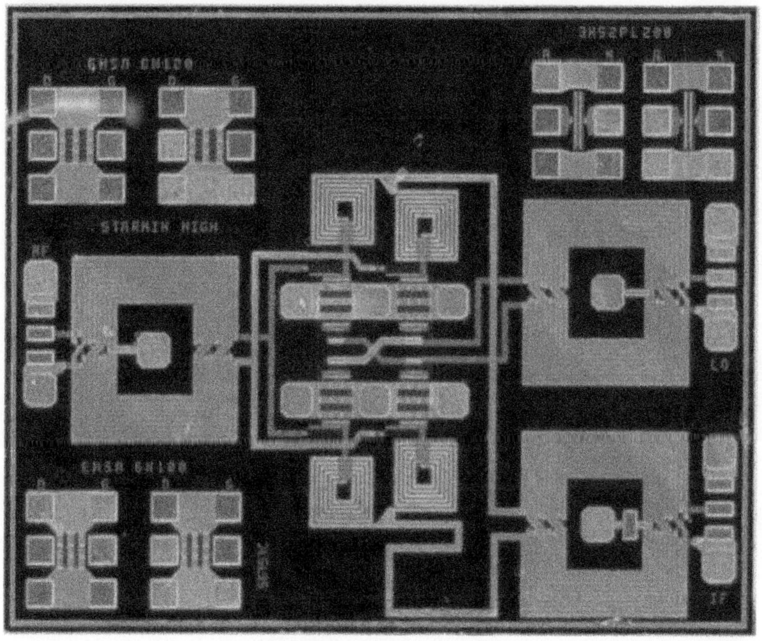

Figure 9.104 Integrated TBM circuit (TriQuint).

On the other hand, the symmetry reduces some of the unwanted spurious frequencies, and because of the high impedance, the actual RF power level is much less than the diode would need. There are so many CMOS transistors available in different processes that we cannot provide information about the general behavior, but this "low-cost" solution, as pointed out, is really only attractive because one can stay in the same technology while building the amplifier and other stages. Because of the flicker corner frequency of MOS, the noise figure of the mixer will be high compared to other devices, but not as bad as its GaAsFET brothers. In any case, to combine reasonable performance and simplicity, one needs to have a preamplifier that reduces the noise figure as well as the third-order intercept point of the mixer by the amount of preamplification [4].

Below are some thoughts on [28–30] CMOS Mixer Noise by Professor Ali M Niknejad, based on his EECS lecture on Mixer Noise and Design, Professor Ali M Niknejad, Advanced Communication Integrated Circuits, Reproduced with permission.

9.25.1 SSB Noise Factor [28]

The SSB NF definition assumes that there is no signal at the image frequency except the source noise. This definition assumes an IF architecture, where the image signal is suppressed by an image filter before reaching the mixer. The NF is the reduction of the S/N at the mixer output. Therefore, one can write:

$$S_{\text{out}} = S_d G_{\text{mix\_d}}$$
$$N_{\text{out}} = N_d G_{\text{mix\_d}} + N_{\text{im}} G_{\text{mix\_im}} + N_{\text{mix\_d}} G_{\text{mix\_d}} + N_{\text{mix\_im}} G_{\text{mix\_im}} \tag{9.98}$$

where S_d is the desired signal, N_d is the noise in the desired band, N_{im} is the noise in the image band, $G_{\text{mix\_d}}$ and $G_{\text{mix\_im}}$ are the mixer gain at both desired and image frequencies, respectively.

$N_{\text{mix\_d}}$ and $N_{\text{mix\_im}}$ are the desired and image band noise due the mixer circuit itself referred to its input. To simplify analysis, we will assume $N_d = N_{\text{im}}$, $N_{\text{mix\_d}} = N_{\text{mix\_im}}$ and $G_{\text{mix\_d}} = G_{\text{mix\_im}}$; therefore,

$$N_{\text{out}} = 2N_d G_{\text{mix\_d}} + 2N_{\text{mix}} G_{\text{mix\_d}}$$
$$\Rightarrow \frac{S_{\text{out}}}{N_{\text{out}}} = \frac{S_d}{2N_d + 2N_{\text{mix}}} = \frac{S_d}{N_d}\left(\frac{1}{2 + \frac{2N_{\text{mix}}}{N_d}}\right) \tag{9.99}$$
$$\Rightarrow F_{\text{SSB}} = 2 + \frac{2N_{\text{mix}}}{N_d}$$

9.25.2 DSB Noise Factor [28]

The DSB NF definition assumes that the image band contains both noise and an image signal identical to the desired band signal. This needs to be applied in the direct conversion receiver where the image is the signal itself. Therefore, one can write:

$$S_{\text{out}} = S_d G_{\text{mix\_d}} + S_{\text{im}} G_{\text{mix\_im}}$$
$$N_{\text{out}} = N_d G_{\text{mix\_d}} + N_{\text{im}} G_{\text{mix\_im}} + N_{\text{mix\_d}} G_{\text{mix\_d}} + N_{\text{mix\_im}} G_{\text{mix\_im}} \tag{9.100}$$

where S_{im} is the image signal.

To simplify analysis, we will assume $S_d = S_{\text{im}}$, $N_d = N_{\text{im}}$, $N_{\text{mix\_d}} = N_{\text{mix\_im}}$ and $G_{\text{mix\_d}} = G_{\text{mix\_im}}$. Therefore,

$$N_{\text{out}} = 2N_d G_{\text{mix\_d}} + 2N_{\text{mix}} G_{\text{mix\_d}} \tag{9.101}$$

Therefore,

$$\Rightarrow \frac{S_{\text{out}}}{N_{\text{out}}} = \frac{2S_d}{2N_d + 2N_{\text{mix}}} = \frac{S_d}{N_d}\left[\frac{1}{1 + \frac{N_{\text{mix}}}{N_d}}\right]$$
$$\Rightarrow F_{\text{DSB}} = 1 + \frac{N_{\text{mix}}}{N_d} \tag{9.102}$$

Core Noise Model

Figure 9.105 shows a MOSFET-based differential mixer circuit showing the resulting waveform of the base voltage and collector currents.

$$I_{o1} = p_0(t) + p_1(t) \cdot i_s$$
$$p_1(t) = \frac{g_{m1}(t) - g_{m2}(t)}{g_{m1}(t) + g_{m2}(t)} \tag{9.103}$$

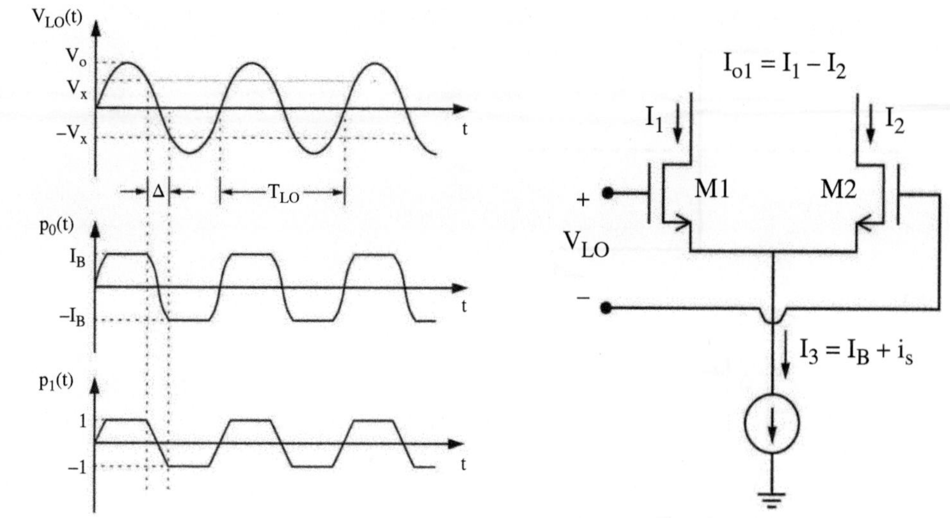

Figure 9.105 MOSFET-based differential mixer circuit showing the resulting waveform of the base voltage and collector currents. *Source:* Modified from [28].

9.25.3 Noise from G_m Stage [28]

$$y_{n3} = n_3(t) \cdot p_1(t) \quad n_3(t) \quad \text{WSS} \qquad \text{PSD} \quad S_{n3}(f) \tag{9.104}$$

A cyclostationary process is a random process whose statistics are periodic functions of time.

The PSD of a cyclostationary process is given by $S(f, t)$. If we measure a cyclostationary process over a bandwidth <1/period, we observe $\overline{S(f, t)}$, a stationary process

Example:

$$y(t) = n(t) \cdot p(t) \leftarrow \text{deter periodic func.}$$
$$S^o_{n3}(f) = \sum_{n=-\infty}^{\infty} \left|p_{1,n}\right|^2 S_{n3}(f - nf_{\mathrm{LO}}). \tag{9.105}$$

If $S_{n3}(t) = N_{n3}$ (white), then

$$S^o_{n3}(f) = N_{n3} \cdot \sum_{n=-\infty}^{\infty} \left|p_{1,n}\right|^2 = \alpha N_{n3}$$
$$\alpha = \sum_{n=-\infty}^{\infty} \left|p_{1,n}\right|^2 = \frac{1}{T_{\mathrm{LO}}} \int_{-\infty}^{\infty} (p_1(t))^2 dt \tag{9.106}$$

For a square wave $p_1(t)$,

$$\alpha = 1$$

For first sidebands ($f_{\mathrm{LO}} \pm f_{\mathrm{F}}$), this noise accounts for $2 \times \left(\frac{2}{\pi}\right)^2 \approx 81\%$ of noise.

($3f_{\mathrm{LO}} \pm f_{\mathrm{IF}}$) accounts for $2 \times \left(\frac{2}{3\pi}\right)^2 \approx 7\%$

The remaining harmonics account for the rest ~ 10%.

Assume $p_1(t)$ is a straight line during period Δ when both devices are on.

Figure 9.106a shows an input square wave used for driving the differential mixer.

Figure 9.106b shows the translation of the driving voltage into the output voltage.

$$\alpha = 1 - \frac{4}{3}\Delta f_{\mathrm{LO}}$$
$$\pi\Delta f_{\mathrm{LO}} = \sin^{-1}\left(\frac{v_x}{v_o}\right) \tag{9.107}$$

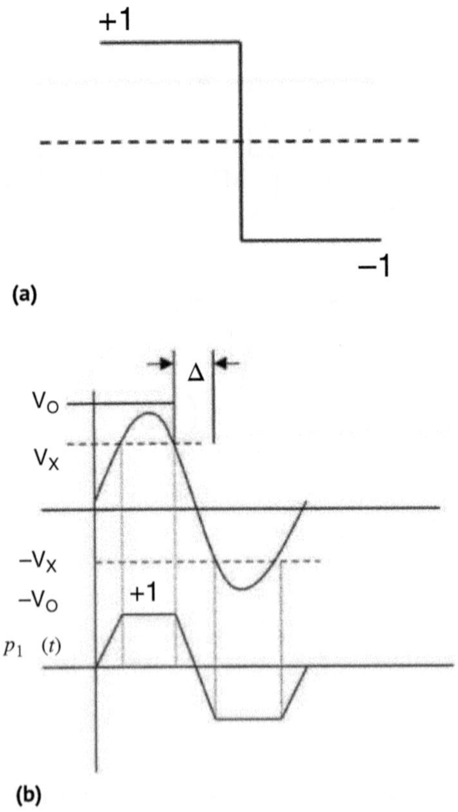

Figure 9.106 (a) Input square wave used for driving the differential mixer. *Source*: Modified from [28]. (b) Translation of the driving voltage into the output voltage. *Source*: Modified from [28].

9.26 Single-balanced Mixer Noise [28]

Gilbert Cell:

$$\underbrace{S_{n3}(f)}_{\text{Noise due to M3 (transconductor)}} = \underbrace{4kT\left(R_s + r_{g3} + \frac{\gamma}{g_{m3}}\right)}_{\text{Input referred noise voltage}} \cdot \underbrace{g_{m3}^{\;2}}_{G_{m3}^{2}\text{ for M3 with degeneration}} \cdot \underbrace{\alpha}_{\text{Time variance of problem}} \tag{9.108}$$

When we switch hard, the noise is due to M3 only (transconductance) (neglecting capacitance and output impedance of M3).

9.26.1 Noise of Switching Pair [28]

Hard switching is good for low noise output. When both M1 and M2 are on, during the switching period Δ, then the noise PSD at output is

$$S_{I_1} = 4kT\gamma\left(\frac{1}{g_{m1}}\left(\frac{g_{m1}}{1+\frac{g_{m1}}{g_{m2}}}\right)^2 + \frac{1}{g_{m2}}\left(\frac{g_{m2}}{1+\frac{g_{m2}}{g_{m1}}}\right)^2\right)^2 = 4kT\gamma\frac{g_{m1}\cdot g_{m2}}{g_{m1}+g_{m2}} \tag{9.109}$$

Figure 9.107 shows the CMOS-based single-cell differential mixer.

Definition: In communications, noise spectral density (NSD), noise power density, noise power spectral density, or simply noise density (N_0) is the power spectral density of noise or the noise power per unit bandwidth. It has the dimension of power over frequency, whose SI unit is watt per hertz (equivalent to watt-second or joule).

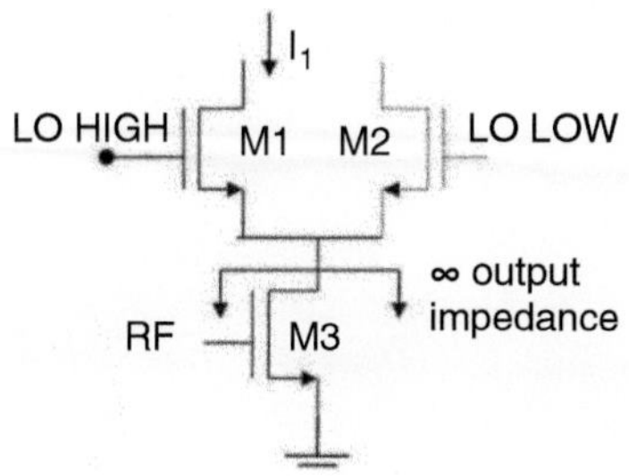

Figure 9.107 CMOS-based single-cell differential mixer. *Source*: Modified from [28].

The current NSD of the current is based on
$I_n^2 = kT/(\text{Re}(R_t))$; ($R_t$ = termination resistance)

$$I_3 = I_1 + I_2 I_{o1,n} = 2I_{1,n} \Rightarrow 4 \times \text{power}$$

$$S_{n1,2}(f,t) = 16kT\gamma\frac{g_{m1}g_{m2}}{g_{m1}+g_{m2}} = 8kT\gamma G(t)$$

where $G(t) = 2g_{m1}g_{m2}$

is the total transconductance of differential pair from the LO port to differential output current.

$$\begin{aligned} S_{n1,2}(f) &= S_{n1,2}(f,t) = 8kT\gamma\left(\frac{1}{T_{LO}}\int_0^{T_{LO}} G(t)dt\right) \\ &= 8kT\gamma\overline{G} \end{aligned} \tag{9.110}$$

To convert the current NSD I_n^2 to power, we need to multiply S by the reference termination.
P (NSP) = $8kTGY21*$; G~1
As the noise floor power is kT, the added NSD (power): kT is the noise factor.
From the RF analysis, we know that $Y21* = 17\text{E}-3$ (mS)
F = $8 \times kT \times 1 \times 17\text{E}-3 \times 50/kT = 136 \times 50/1000 = 6.8$; NF = 10 log(6.8) = 8.32 dB

$$\begin{aligned} v_{LO} &= v_o \sin(2\pi f_{LO} t) \\ \overline{G} &= \frac{1}{\pi v_o}\int_{-v_x}^{v_x} G(v_{LO})\frac{1}{\sqrt{1-\left(\frac{v_{LO}}{v_o}\right)^2}}dv_{LO} \\ &\approx \frac{1}{\pi v_o}\int_{-v_x}^{v_x}\frac{dI_{o1}}{dv_{LO}}\cdot dv_{LO} = \frac{2I_B}{\pi v_o} \\ S_{n1,2}(f) &\approx \frac{16kT\gamma}{\pi}\left(\frac{I_B}{v_o}\right) \end{aligned} \tag{9.111}$$

v_o = LO amplitude
This is roughly $8kT \times Y21*$ the RF time-averaged transconductance, which can be derived via the Bessel function. A numerical example was given earlier [28].

9.27 Differential CMOS Mixer–A More Detailed Analysis

Let us consider the noise generated in a differential switching CMOS mixer. Its topology is shown in Figure 9.108. We will follow the derivation of the individual origins of noise as detailed in [3]. The discussion will focus on $1/f$ noise for the reason that this kind of noise exists only in relative narrow bands at DC or, if upconverted, as noise sidebands of a carrier. It is interesting to see at which frequencies these noise sidebands will arise and why. Also, in direct conversion (zero IF), the baseband $1/f$ noise will be the governing type of noise. But of course, the mechanisms controlling $1/f$ noise affect white

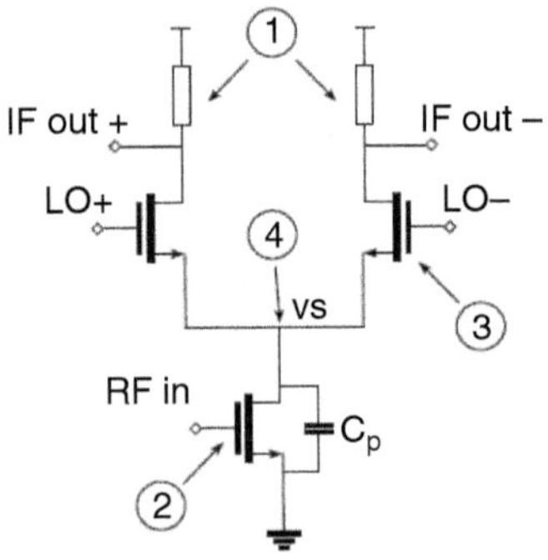

Figure 9.108 Differential CMOS mixer, indicating the origins of noise: (1) load noise, (2) transconductance noise, (3) direct switching noise, (4) indirect switching noise. *Source*: Adapted from [35].

noise in the same manner. The difference is, however, that the white noise is constant over frequency; therefore, it doesn't change anything if it is translated in frequency [35, p 989].

This investigation aims at providing a basic understanding of the noise mechanisms present in an ideal differential switching mixer. Imperfections such as simultaneous conduction of both transistors during switching will cause additional noise. For simplicity, only a single-balanced mixer is discussed. There is, however, no principal difference to DBMs, so that the theory is easily extended to this case. Only switching-type mixers will be discussed, but also a Gilbert cell mixer will show qualitatively the same effects.

9.28 The Four Noise Mechanisms

The following four noise mechanisms are present in this type of mixer: load noise, transconductance noise, and direct and indirect switching noise. These four effects will now be discussed. In Figure 9.108, it is indicated by the numbers (1–4) where these mechanisms are located. Throughout this section, the transistor noise is modeled by a single equivalent voltage noise source at the gate of the respective transistor. Figure 9.109 shows the resulting output $1/f$ noise spectra caused by each of the mechanisms [35].

9.28.1 Load Noise

The load of the differential mixer adds noise to the signal. This noise is simply superimposed to the signal at the output of the mixer. Minimizing this noise is done by selecting appropriate load transistors, or even resistors that provide no $1/f$ noise [35].

9.28.2 Transconductance Noise

The current source transistor is used to modulate the switched current by the RF signal. The switching is then responsible for the mixing of the RF signal. However, any noise generated in this transistor will also be mixed by the differential pair just as any other signal. The good news concerning $1/f$ noise is, however, that it is not at the same frequency as the RF signal, but it is in baseband. It therefore gets upconverted to the LO frequency and to its odd harmonics. Under ideal conditions, the $1/f$ noise caused by the transconductance transistor is shifted out of the baseband and should not degrade SNR at the mixer output [35].

9.28.3 Direct Switching Noise

After discussing the current source and the load, it is time to discuss the $1/f$ noise of the switching transistors. Their noise is not just mixed; it impacts the switching. Let us consider the ideal case. The switching transistors are assumed to perform the transition from on to off instantaneously when the gate voltage crosses a certain value. With no noise, the switching would occur exactly at each half-cycle of the LO frequency.

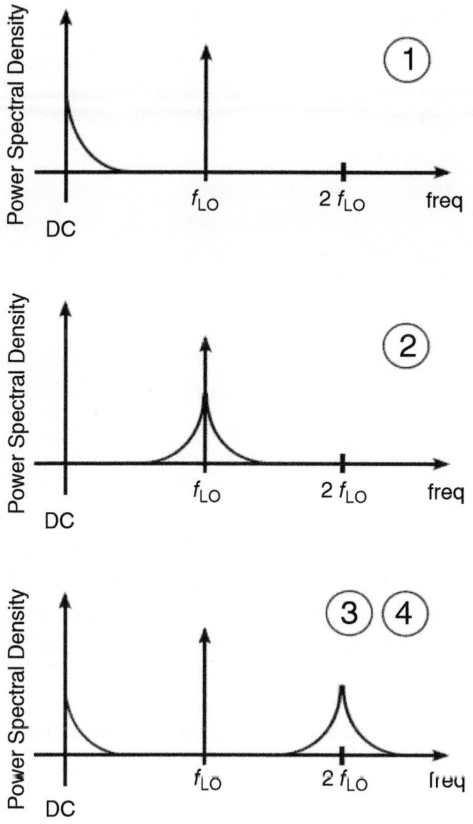

Figure 9.109 Contributions of the four basic noise sources to the output mixer noise: (1) load noise, (2) transconductance noise, (3) direct switching noise, and (4) indirect switching noise. *Source*: Adapted from [35, p 990].

Now consider additional $1/f$ noise at the gate; it is a slow signal that is superimposed to the LO switching signal. The LO signal will be just slightly increased or decreased, and the switching events are consequently shifted in time, a bit earlier or later. Thus, the noise causes jitter.

In order to quantify the noise contribution, we need to describe the difference between the switched current with jitter and without, as shown in Figure 9.110. If the two signals are simply subtracted, one obtains a series of pulses with the following properties:

- The pulses are located at the edges of the switching times.
- The amplitude is either positive or negative and is equal to twice the switched current.
- The width (and the sign) of the pulse depends on the ratio between the magnitude of the gate noise voltage $V_n(t)$ at the switching time and the slope S of the LO voltage applied to the gate. Figure 9.111 highlights this relation.

$$\Delta t = \frac{V_n(t)}{S} \tag{9.112}$$

Thus, the jitter can be understood as a sampled replica of the original $1/f$ noise that is additionally scaled according to the signal's slope S. The sampling frequency equals twice the LO frequency since the jitter occurs at every switching event, thus twice per LO period (pulse rate of pulse train $2f_{LO}$).

The jitter noise current normalized for one period is therefore

$$I_{n,o} = \frac{2}{T_{LO}} \cdot 2I \cdot \frac{v_n}{S} \tag{9.113}$$

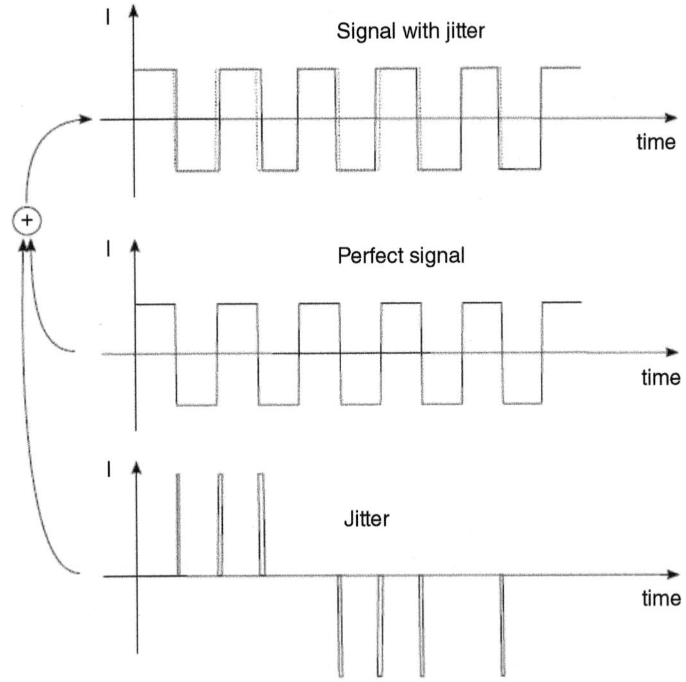

Figure 9.110 Direct switching noise: the LO signal with jitter can be understood as an undisturbed signal superimposed with a pulse train. *Source*: Adapted from [35].

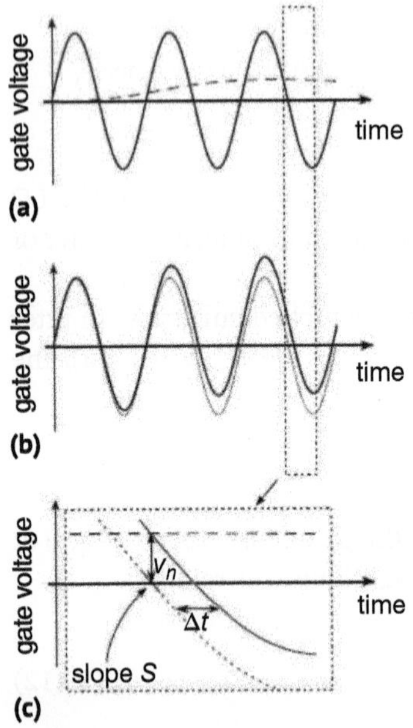

Figure 9.111 (a) Ideal LO gate voltage (solid line), $1/f$ noise voltage (dashed line). (b) Actual (solid line) and ideal (dotted line) LO gate voltage. (c) The time-offset (jitter) at a zero-crossing is calculated from the noise voltage and the slope of the LO signal. *Source*: Adapted from [35].

With the period of the LO signal $T_{LO} = 1/f_{LO}$, the switched current I, the noise voltage at the switching times v_n, the slope (S) of sinusoidal LO signal can be given as shown in Figure 9.111 [35].

For a differential LO signal $V_{LO}d = 2\sin\omega_{LO}t$; therefore, $S = \dfrac{dV_{LO}d}{dt} = 2V_{LO}\omega_{LO}$

So, ST_{LO} will be $4\pi V_{LO}$.

For a sinusoidal signal, $T_{\mathrm{LO}} \cdot S$ is given by $4\pi A$, where a factor of 2 accounts for the fact that the noise is compared to a differential signal, the baseband noise spectrum thereby becomes:

$$i_{n,o}(f) = \frac{1}{\pi A} \cdot \upsilon_n(f) \tag{9.114}$$

Due to the sampling of this signal, this spectrum appears at baseband and as noise sidebands around all even harmonics of the LO frequency. In the ideal case, there is no interaction with the other noise sources. But if the mixer is unbalanced due to an offset voltage, noise of the transconductance transistor can leak through and add to the jitter. It affects the output noise spectrum in the same manner as the switching noise discussed earlier [35].

9.28.4 Indirect Switching Noise

The direct switching noise can be reduced by fast switching. Applying an ideal square-wave voltage to the gates would suppress this kind of noise. However, $1/f$ noise of the switching transistors adds to the overall noise in a second way that only depends on LO frequency and transistor/FET capacitances.

Let us assume ideal switching and investigate the voltage at the virtual ground. Under this assumption, it will be constant. However, if $1/f$ noise exists at the switching transistor's gate, things change. Due to the balanced structure, it is allowed just to regard the unsymmetrical noise. At one transistor (base) gate, the $1/f$ noise will be switched between two states: υ_n if the BJT/FET is conducting and 0 if the device is turned off. However, due to the differential structure, the virtual ground potential will anticipate this imbalanced switching and will also shift by υ_n when the respective transistor is on. The noise signal is stochastic in nature, but much slower than the LO signal. Let us treat it like a switched voltage of slowly varying amplitude. The relevant time constants in our case are the switching edges, as these will enable a current through the capacitance C_p associated with the transconductance of the transistor [35].

$$C_p = C_{be2} + C_{be3} + C_{js1} + C_{bc1} \tag{9.115}$$

$$I_{n,o} = \frac{2}{T_{\mathrm{LO}}} \int_0^{T_{\mathrm{LO}}/2} C_p \cdot \left[\frac{d}{dx}\upsilon_n(f)\right] dt \tag{9.116}$$

which leads to the noise contribution

$$i_{n,o} = \frac{2}{T_{\mathrm{LO}}} \cdot C_p \cdot \upsilon_n = 2 f_{\mathrm{LO}} C_p \upsilon_n \tag{9.117}$$

Concerning the spectrum, the same observations are made as for the direct switching noise. The current bursts at the virtual ground change their polarity depending on whether the transistor is switched on or off. But regarding the load, this change in polarity is compensated by the differential pair that routes one burst through one branch and the next through the other. Thus, at the output, again, $1/f$ noise appears, sampled with twice the LO frequency.

Anyway, the indirect switching noise should usually be much lower than direct switching noise. Direct conversion noise can be reduced by fast switching, but switching noise can never be suppressed completely [35].

9.29 A Practical Example of a Simulated Differential MOS Mixer

Figure 9.112a shows a practical example of a single-balanced differential MOS mixer with its simulated gain and noise analysis seen in Figure 9.112b. The gain of the transistor stage is responsible for converting RF input voltage to the current signal, essentially acting as a current source that is then switched by the LO signal to produce the desired IF. The working principle behind this type of mixer is the square law characteristics of MOS devices. Figure 9.112c shows the collector waveform of the lower transistor depicting the voltage saturation

Simulation of the noise and gain values of the differential mixer is shown in Figure 9.112a,b. Figure 9.112c shows the collector waveform of the lower transistor depicting the voltage saturation.

Figure caption: Figure 9.112(c): Collector waveform of the lower transistor depicting the voltage saturation

Figure 9.112d shows the sweep of the differential mixer with the LO drive changed between −2 V and −5 V.

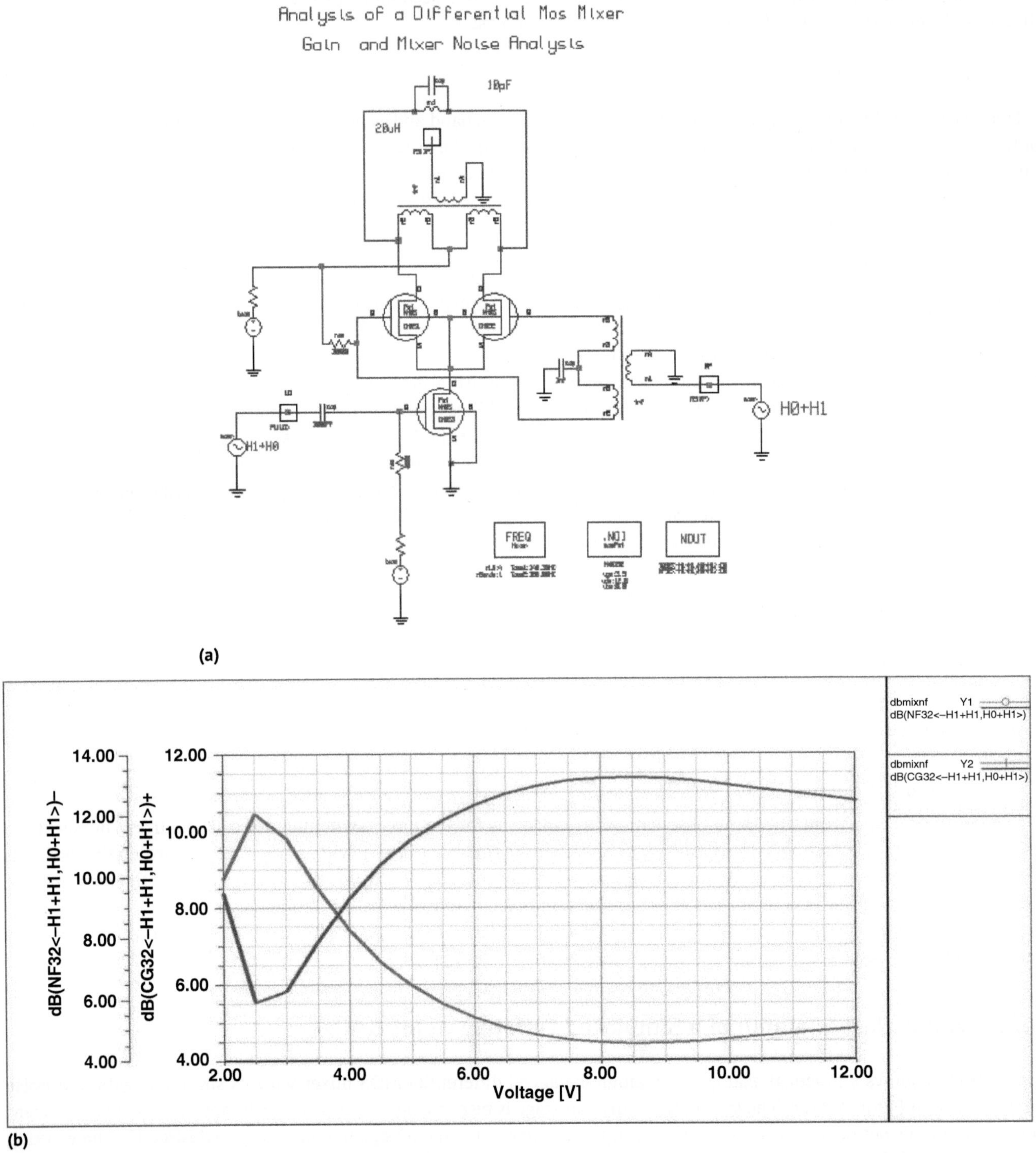

Figure 9.112 (a) A practical example of a differential MOS mixer—gain and noise analysis. (b) Differential MOS mixer—simulation of the gain and noise analysis.

(Continued)

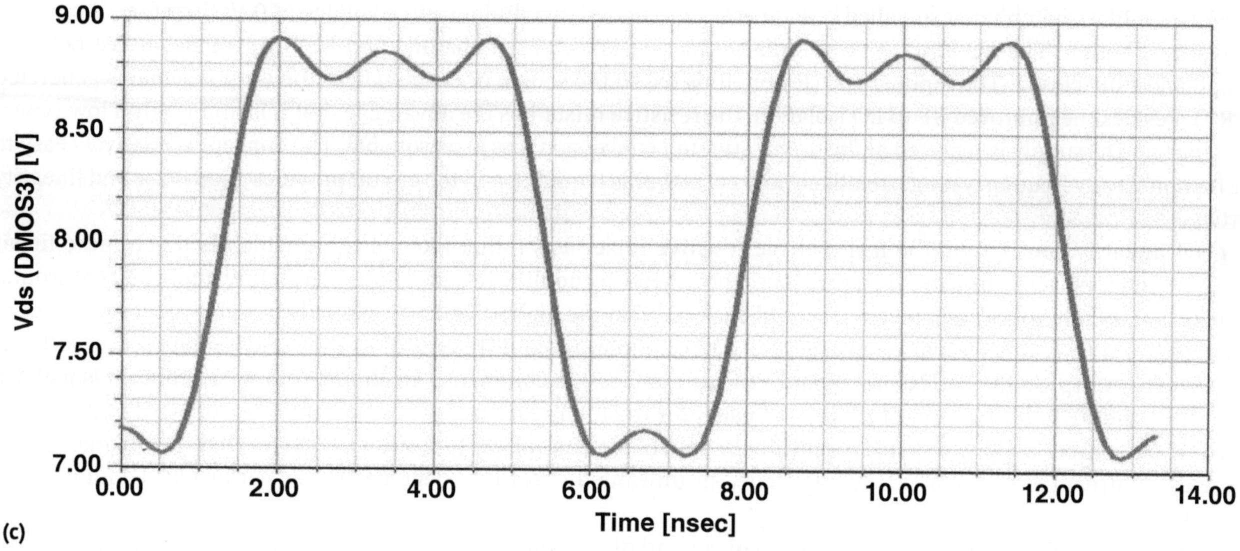

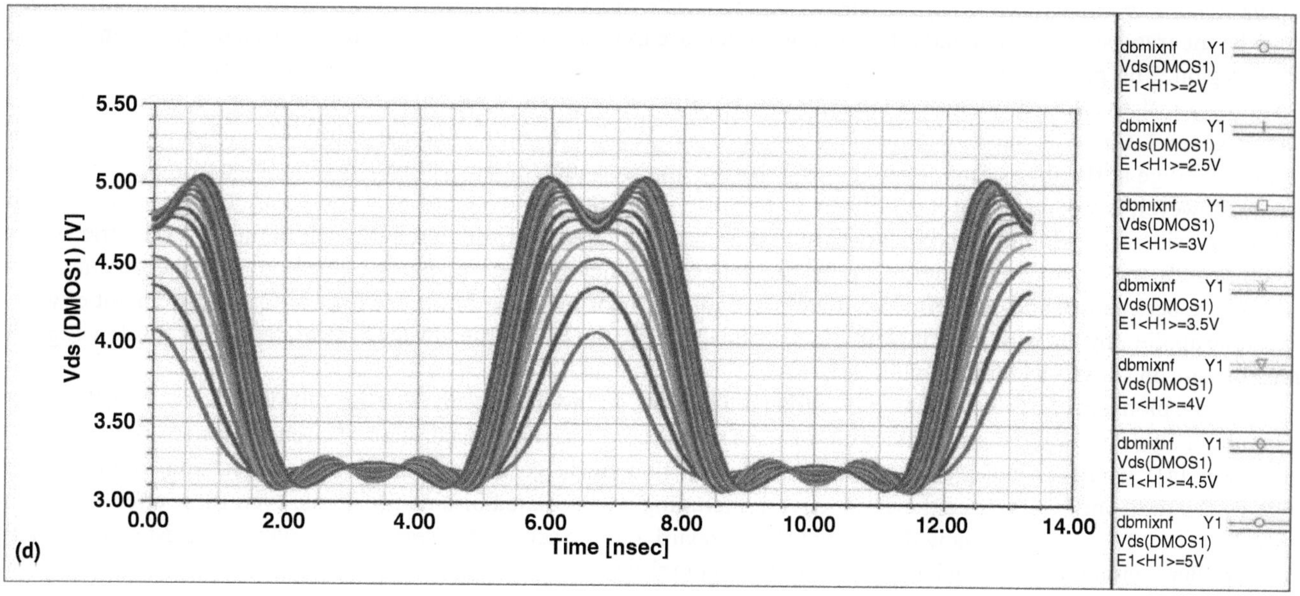

Figure 9.112 (Cont'd) (c) Differential MOS mixer—collector waveform of the lower transistor. (d) Differential MOS mixer—sweep of the differential mixer with the LO drive changed between −2 V and −5 V.

9.30 GaAsFET Single-gate Switch: Resistive Mixer

It is a drawback of the diode mixers that the mixing process is performed by a one-port device. LO, RF, and IF signals are all present at the same port and need to be carefully separated through filters or balanced topologies. It is especially critical to suppress the LO signal on the RF line, since the signals are closely spaced in frequency and the LO is at a much higher power level. Replacing the diode by a FET offers a number of advantages. Good RF to LO isolation is achieved when the LO signal is applied at the gate, while the port for RF and IF signals is at the drain.

The FET in a FET switching or so-called resistive mixer is biased at a drain–source voltage of 0 V; therefore, it operates in the resistive region. The LO at the gate modulates the resistance of the channel and ideally switches it fast between fully open and fully closed. The input signal (RF or IF) is fed to the source or drain, where the output signal is generated (IF or RF). Besides the improved RF to LO isolation, the resistive mixer has the advantage that it is inherently a low-power circuit, since no DC supply is required at the drain. Still, the conversion loss is reasonable. The concept is relatively easy to realize for almost any frequency range, relatively low LO power is required, and the circuit provides good noise and linearity properties.

The good linearity comes from the fact that the resistive mixer acts rather like a switch, which is much more linear regarding RF input power than active mixers that use the inherently nonlinear transconductance. Intercept points of around 40 dBm are common in this type of mixer. It should be noted that the highest linearity can only be achieved if the source of the mixing transistors is directly connected to ground. For example, a record IP_3 of 53 dBm was achieved for such a "vertical" DBM by Synergy Corp. on the basis of a standard TriQuint process. In contrast, if the differential mixer concept is realized by connecting the FETs in the form of a ring—as usual for diode mixers—a typical maximum IP_3 of 30 dBm cannot be exceeded. If the sources are floating, it is practically impossible to control the gate-source voltage. High RF voltage swings therefore easily drive the gate into conducting state, an unwanted effect that degrades mixer performance and possibly device lifetime.

Before this type of mixer is discussed in more detail, a word of caution regarding simulating it. It is often observed that a simulation of a resistive mixer is not sufficiently accurate. The cause is not found in the numerics of the simulation tools; it can be the transistor model. Standard transistor models are most accurate for the common transistor operation, i.e., the active region. The performance of a passive transistor is often not considered during model development. Thus, before designing a resistive mixer, one needs to verify that the models used are specified for switching operation at zero volts bias. In general, GaAsFET technology is the most expensive. While GaAsFET competes with SiGe technology in terms of frequency range at the time of this writing, it is a mature and well-understood technology. The price of gallium arsenide wafers, however, is still such that it is unlikely to compete with other technologies on a cost basis. There is also a hybrid technology, silicon on sapphire, that tries to go after the GaAsFET domain. The cutoff frequency of the GaAsFET transistors is significantly higher than any other FETs, but its corner frequency f_c, also frequently referred to as flicker frequency, is somewhere between 10 and 100 MHz and therefore results in poor mixer performance from a noise figure point of view. The same applies, by the way, to oscillators. Another headache with GaAs is the low Q of the material, but this low Q gives transformers on GaAs a wider bandwidth [4].

One of the pioneers of ICs for the wireless market is TriQuint, and we have been working with Wes Hayward of TriQuint in circuit analysis and validation. In doing so, we used some commercial and experimental circuits. In the amplifier chapter (of Ref. [4]), we have already pointed out one amplifier that shows good agreement between measured and predicted data. By the time one has realized the entire circuit, including its associated biasing problems, even these circuits tend to grow. Figures 9.113 and 9.114 show the printed circuit board and the schematic, respectively, which consist of an oscillator/amplifier, switching mixer, and a differential IF post amplifier.

Since the highest accuracy is needed to model these types of circuits, we found that only very recent simulators, such as Serenade 8 from Ansoft, can handle such a large number of FETs without barking at the user. Initially, we had all kinds of difficulties with these circuits because of the TriQuint TOM model, which had been developed for SPICE, and its IF/ELSE statements did not make it mathematically continuous. The types of GaAsFETs used by TriQuint include both regular FETs and enhancement FETs, which require a positive voltage at the gate. Further details of this should be looked up in the foundry manual by TriQuint. In the modeling chapter (of Ref. [4]), we have shown some comparisons between different models and their prediction accuracy. Such converters need to resort to inductors in GaAs technology, and compared to other parts, they become quite large (Figure 9.115) [4].

9.31 Simplified Switching FET Mixer

In order to reduce some of the complexity in modeling and yet deal with the problems of mixers, here is a simpler mixer that is the combination of a set of switching transistors and a post amplifier for the IF. The reader may notice that the small inductances of 15 nH have an extremely high loss modeled by a series resistor of 18 Ω. The first to recommend this type of switching mixer was probably Stephen Maas. The circuit diagram itself (Figure 9.116) is not self-explanatory, so we need to point out the complexity it ends up with [4].

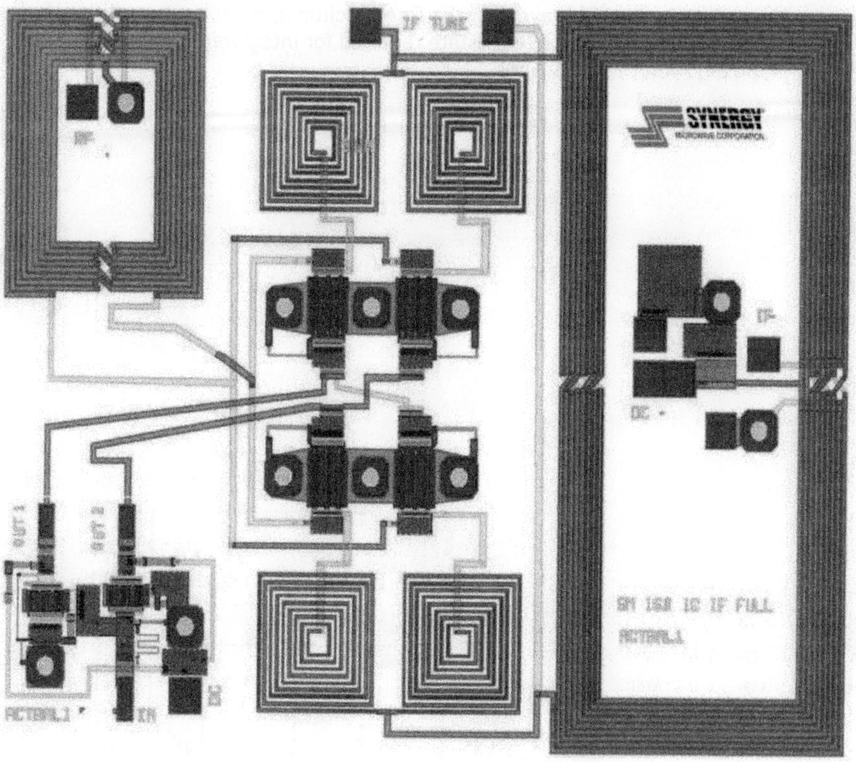

Figure 9.113 Experimental mixer developed by Synergy Microwave (Synergy Microwave).

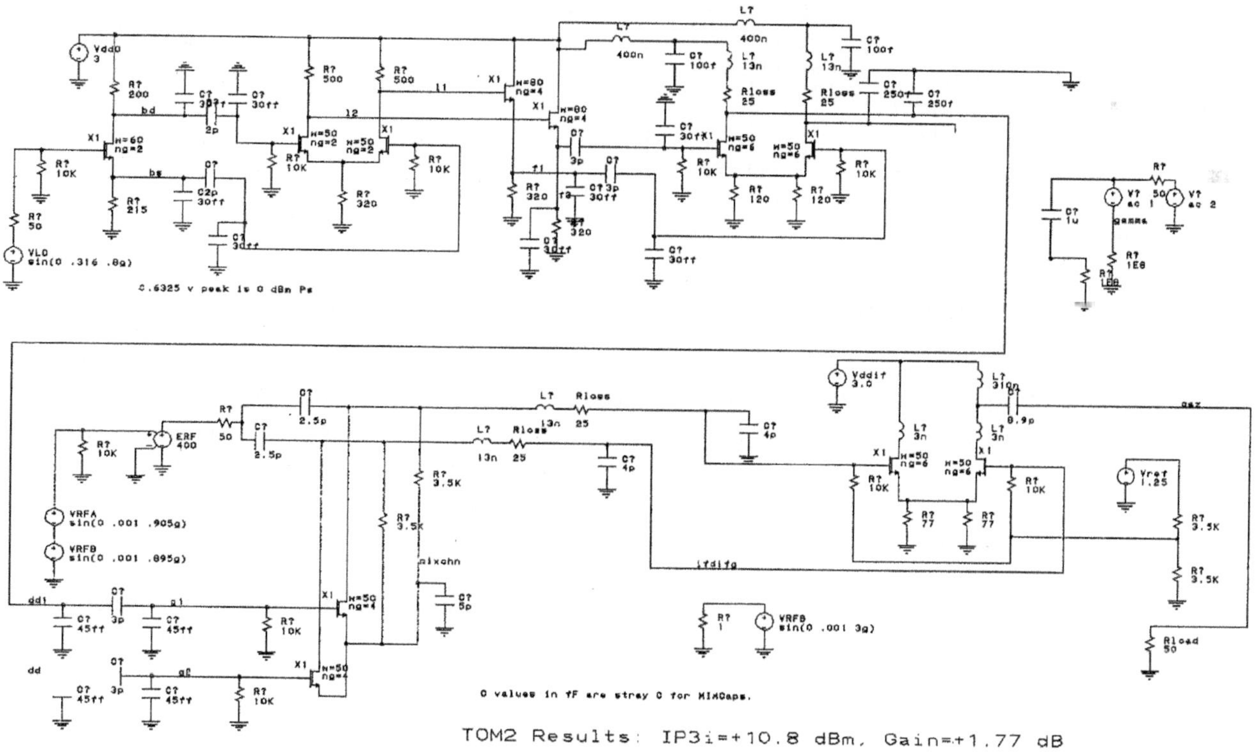

Figure 9.114 The complete switching mixer circuit. *Source*: [4, p 703] / John Wiley & Sons.

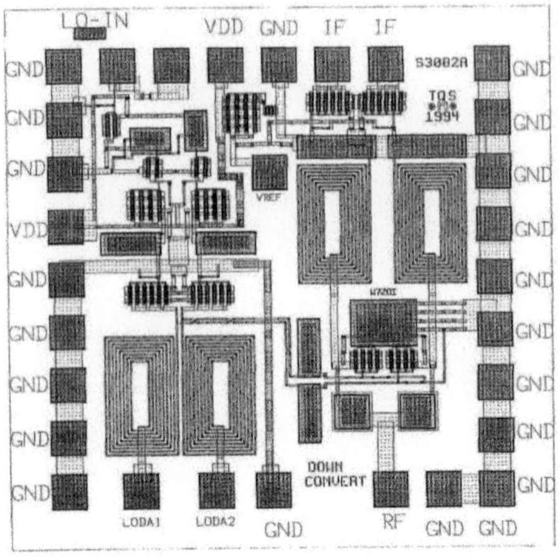

Figure 9.115 Physical layout of the GaAs oscillator/mixer/post-amplifier IC, showing the considerable real estate required for integrated inductors. *Source*: [4, p 704] / John Wiley & Sons.

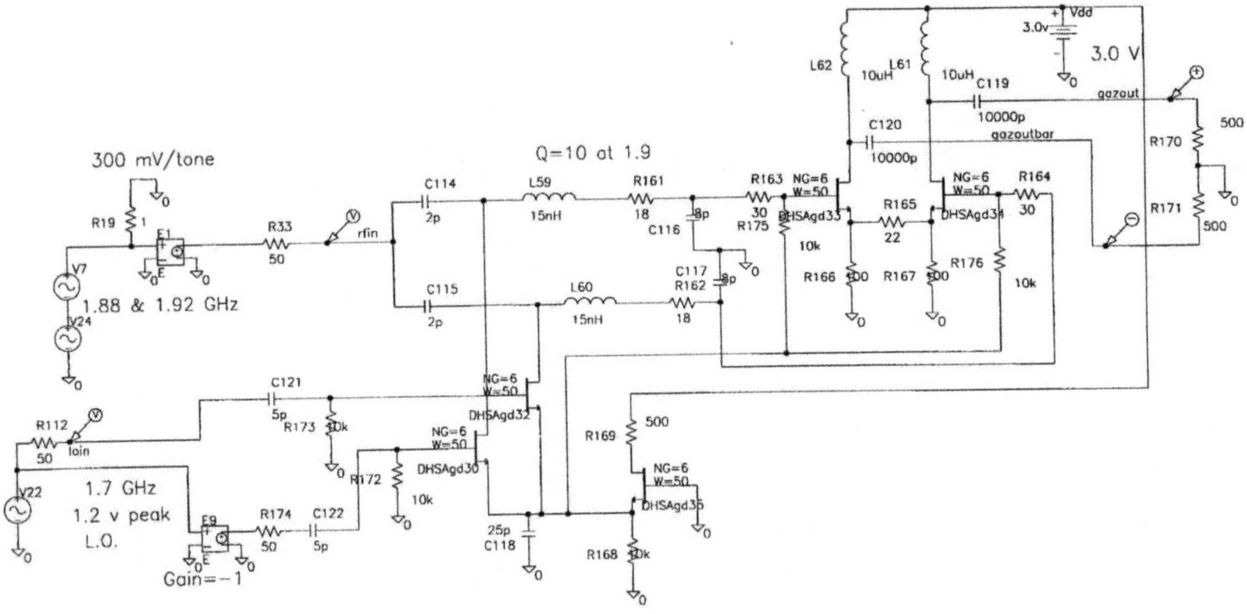

Figure 9.116 Simplified switching FET mixer. *Source*: [4] / John Wiley & Sons.

On the top left, there are two tones being switched ON and OFF by the two FETs labeled DHSAgd30 and 32. The switching is done by applying the LO at 1.7 GHz, 1.2 V peak. The following stages on the top right are like IF amplifiers. The loads are labeled R170 and R171.

9.32 Noise in Resistive Mixers [20]

A mixer's noise performance is not as easy to comprehend as it might be in a linear circuit. The nonlinear operation of the transistor or diode is a prerequisite for the mixer to operate at all. The signal is shifted in frequency; therefore, we need to consider the relation between RF and IF noise. And finally, we don't simply have DC currents driving our noise sources and only small-signal RF. In this case, DC, if applied at all, might be smaller than the LO signal's amplitude.

However, let us have a look at the different mechanisms one after the other. First, consider the nonlinear behavior and the many frequencies involved. This is not a real issue, if we assume for a moment that all noise sources are known. Noise currents and voltages are stochastic signals, but still signals that propagate through an electrical circuit like any other small-signal electrical signal would do. That is a standard task for any circuit simulator.

Starting with an initial noise correlation matrix describing the noise sources, it needs to be calculated how these small-signal quantities are converted to other frequencies. The result is a number of noise correlation matrices, each describing the noise associated with one mixing product. The total mixer noise is obtained by adding all these contributions.

The second issue is the noise sources. If we first consider only white noise, the issue becomes simpler in a resistive mixer. With the transistor acting as a controlled resistor, only thermal noise is observed. There is no excess noise due to *pn*-junctions or hot electrons in the channel. Thus, it can be assumed that the noise figure of a resistive mixer follows the same rule as the noise figure of any passive mixer.

Things get a bit more involved when it comes to $1/f$ noise. The low-frequency noise becomes an issue when the $1/f$ noise corner is higher than the IF frequency, or if the RF is directly converted into baseband. In this case, $1/f$ noise is the dominant noise. This is obviously the case in baseband, but the noise also gets upconverted and forms sidebands at the harmonics of the LO signal. Since the spacing between RF signal sidebands and LO equals the IF frequency, it can be assumed that the upconverted $1/f$ noise is also dominant at all other RF frequencies. And again, the mixing process is straightforward to calculate or simulate [4].

Based on Ref. [21]. Figures reprinted with permission. The mixer noise modeling approach was also published in Refs. [22–24].

The complicated issue in resistive mixer noise calculation is the question of how the currents control the transistor's noise sources. Classical theory considers only DC current. The result is that no $1/f$ noise is expected at all, since the circuit topology fixes V_{DS} at 0 V and no DC current is flowing. Unfortunately, measurement reveals that this optimistic guess is wrong.

In order to predict the $1/f$ noise in a resistive mixer accurately, it is necessary that the transistor model implement cyclostationary noise sources.

Cyclostationary noise sources are controlled by all harmonics of the large-signal current, not only by the DC component. Thereby, even without DC current flowing, noise sidebands appear at all LO frequencies. This noise gets downconverted and shows up as baseband $1/f$ noise, even in the absence of DC current.

The $1/f$ noise performance of resistive mixers was investigated at the circuit setup shown in Figure 9.117. It is basically a Fujitsu HEMT SHC40LG on a circuit board, in a configuration that allows for a detailed investigation of the transistor noise performance. The whole configuration is shielded in order to suppress the coupling of low-frequency noise. An important detail of this setup is the switch S1 that allows the capacitance C1 to be short-circuited. Closing the switch enables DC current to flow through the transistor, which will impact the noise performance, as we will see later on.

In order to minimize the resistive mixer noise, it is necessary to make sure that the noise performance is not additionally degraded, for example, through the following mechanisms:

1. A high LO amplitude potentially drives the gate into forward conduction mode. The result is additional shot noise and excess $1/f$ noise of the Schottky junction. It is, however, advisable to prevent this from happening, as the gate is not

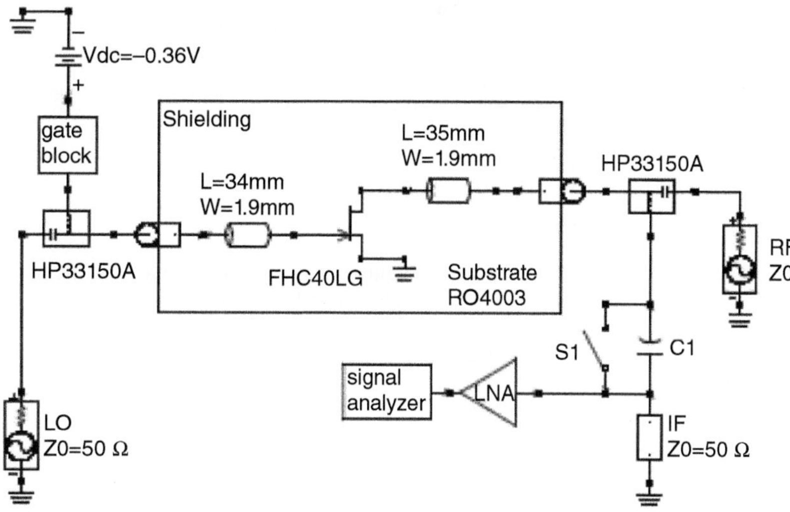

Figure 9.117 FET resistive mixer setup used to investigate the noise performance. From [4, 21].

meant to draw current, and driving it in forward direction could possibly reduce lifetime. Anyway, monitoring that no gate DC current flows prevents these problems.
2. The LO itself will have phase noise. In the investigation, it is necessary to sort out which part of the noise is generated by the mixer and which part is mixed LO noise.
3. If the RF power is significant, it will add to the $1/f$ noise, due to the same mechanisms that are responsible for the noise caused by the LO signal.

Measured and simulated SSB $1/f$ noise at 1 kHz is shown in Figure 9.118 as a function of LO frequency. The LO power is set to 0 dBm. Simulations are shown as straight lines, while the symbols refer to measured points. The higher values were obtained when the switch S1 was open, so that no DC current can flow. Lower $1/f$ noise is observed with S1 closed, providing a DC path through the FET.

When the switch is open, classical noise models would not predict any $1/f$ noise at all, since its power level would be controlled by DC current only—no DC current, no $1/f$ noise. This example therefore is a proof of the cyclostationary noise source concept: The simulations were performed assuming that the $1/f$ noise sources are to be modeled this way. But still, assuming cyclostationary noise sources alone cannot fully explain the measured $1/f$ noise. It is expected that any of the harmonics of the LO signal generates $1/f$ noise sidebands. However, the DC component is nonexistent as long as the switch is open.

An important mechanism responsible for the baseband noise observed in this case is the cross-coupling of the LO signal from the gate to the channel. Figure 9.119 shows a sketch highlighting the coupling paths. The LO leaks through C_{gd}, and in packaged FETs also through C_{gs}, if L_s is significantly large. This cross-coupled LO signal undergoes a mixing process as any other signal in the channel does: it is downconverted to DC. Indeed, a DC current is observed when the switch is closed. Opening it can suppress DC, but it cannot change the fact that the LO fundamental noise sideband is downconverted with the cross-coupled signal. In short, the reason for the $1/f$ noise at baseband is the fact that the LO signal causes $1/f$-type noise sidebands and that the cross-coupled LO signal downconverts these sidebands.

The impact of the LO power level is shown in Figure 9.120 [4, p 715] for an LO frequency of 9 GHz, obtained under the same conditions as the previous graph. The $1/f$ noise first increases with LO power due to decreased conversion loss and increasing cross-coupled power, and finally saturates.

Getting back to Figure 9.118, it seems that the $1/f$ noise can be reduced by providing a DC path to the mixing FET. At first sight, this seems to be a strange behavior, since a DC current causes additional noise. One might expect that the measured noise would increase.

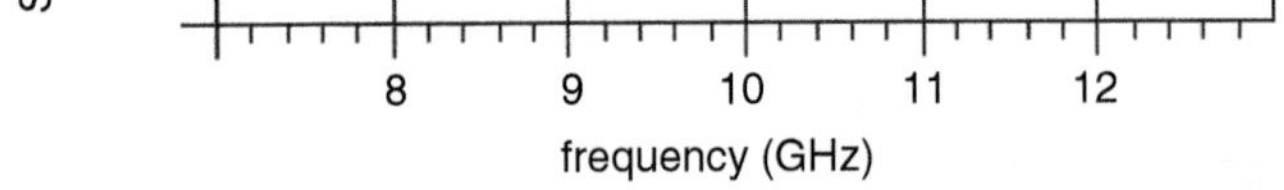

Figure 9.118 SSB noise power density at 1 kHz and LO power of 0 dBm. Simulations are shown as lines and measurements as symbols. Thin lines and crosses refer to the condition without DC current; thick lines and bullets refer to the circuit set up with a DC path. From [4, 21].

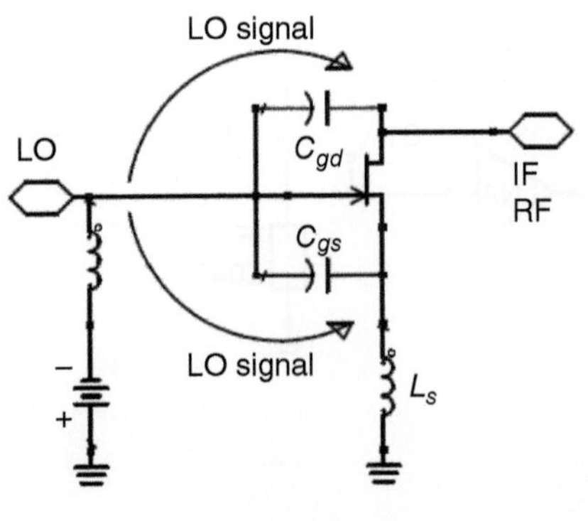

Figure 9.119 The LO signal couples through the capacitances C_{gs}, C_{gd}, and the source inductance L_s into the RF and IF paths. From [4, 21].

In fact, the DC current causes the $1/f$ noise that is superimposed to the downconverted noise. The noise at a certain offset frequency, however, is correlated between the sidebands at all harmonics. The DC current's noise obviously cancels the downconverted noise. Depending on the type of GaAs FET, around 5 dB of improvement is to be expected, which can even be improved to around 10 dB by forcing very quiet DC currents through the device.

The following example of the GaAsFET passive switching mixer, the Siemens model CMY210 (Figure 9.121), was probably the best performing on the market. A reproduction of its datasheet, used by permission, follows the schematic [4, Ch4, p 706].

General Description and Notes:

The CMY 210 is an all-port, single-ended, general-purpose up- and downconverter. It combines small conversion losses and excellent intermodulation characteristics with a low demand of LO- and DC power.

The internal level-controlled LO buffer enables a good performance over a wide LO level range. The internal mixers principle with one port RF and IF requires a frequency separation at pins 1 and 6, respectively.

Parameters	Symbol	min	typ	max	Unit
Drain-source breakdown voltage $I_{IF} = 500\ \mu A$ $V_{LO\text{-}GND} = 0\ V$ $V_{RF\text{-}GND} = 4\ V$ CAP – pin not connected	$V_{(BR)}$ IF-GND	8	–	–	V
Drain current $V_{RF\text{-}GND} = 0\ V$ $V_{LO\text{-}GND} = 0\ V$ $V_{IF\text{-}GND} = 3\ V$ CAP-pin not connected	I_D	0.8	1	1.4	mA
Conversion gain $f_{RF} = 920$ MHz $f_{LO} = 965$ MHz $f_{IF} = 45$ MHz $P_{LO} = -3$ dBm	G_c	–	5.5	–	dB
Single sideband noise figure $f_{RF} = 920$ MHz $f_{LO} = 965$ MHz $f_{IF} = 45$ MHz $P_{LO} = -3$ dBm	F_{SSB}	–	9	–	dB
Third-order intermodulation $f_{RF} = 920$ MHz $f_{LO} = 965$ MHz $f_{IF} = 45$ MHz $P_{LO} = -3$ dBm	IP3	–	−2	–	dBm
LO/RF isolation $f = 965$ MHz	$ISO_{LO/RF}$	–	11	–	dB

$T_A = 25\ °C/V_D = 3\ V$; CAP pin connected to ground by 680-Ω resistor.

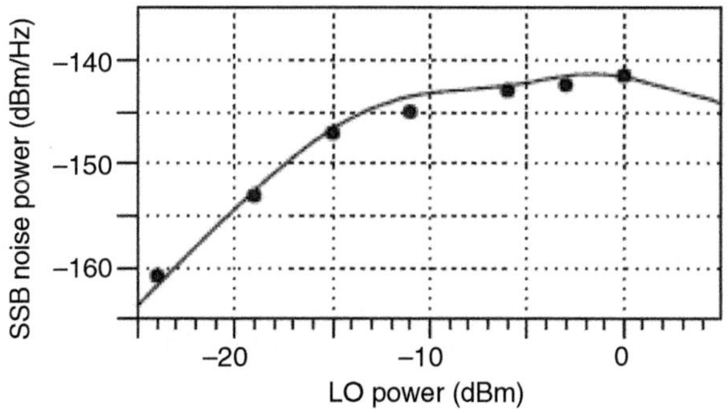

Figure 9.120 Dependence of SSB noise power density at 1 kHz on LO power at an LO frequency of 9 GHz. The line shows simulation, and bullets show the actual measurements. From [4, 21].

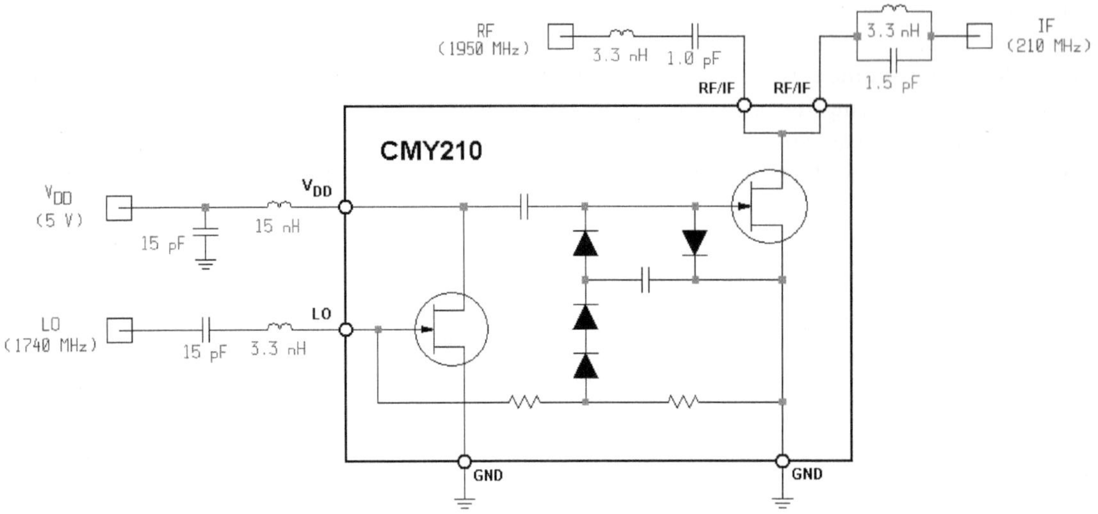

Figure 9.121 The Siemens CMY210 IC includes a shunt switch mixer and AGC-equipped LO amplifier for more constant LO drive. Despite its simplicity, this mixer achieves a typical IP_{3in} of 25 dBm. *Source*: [4, p 706] / John Wiley & Sons.

Note 1: Best performance with the lowest conversion loss is achieved when each circuit or device for the frequency separation meets the following requirements:

Input filter: Throughpass for the signal to be mixed and reflection of the mixed signal and the harmonics of both.
Output filter: Throughpass for the mixed signal and reflection of the signal to be mixed and the harmonics of both.

The impedance for the reflecting frequency range of each filter toward the ports 1 and 6 should be as high as possible.

In the simplest case, a series- and a parallel-resonator circuit will meet these requirements, but also others, as an appropriate drop in filters or microstripline elements can be used.

The two branches with filters should meet immediately at the package leads of the ports 1 and 6.

Parasitic capacitances at these ports must be kept as small as possible.

The mixer can also be driven with a source and a load impedance different to 50 Ω, but performance will degrade at larger deviations.

Note 2: The LO-buffer needs an external inductor L4 at port 4; the value of inductance depends on the LO frequency. It is tuned for minimum l_{op} consumption into port 4.

At lower LO frequencies, it can be reduced by an additional capacitor C5.

Note 3: The LO input impedance at port 3 can be matched with a series inductor. It can also be tuned for a minimum current I_{op} into port 4. C3 is a DC blocking capacitor.

Since the input impedance of port 3 can be slightly negative at lower frequencies, the source reflection coefficient should be kept below 0.8 ($Z_0 = 50\ \Omega$) within this frequency range.

The conversion noise figure F_{SSB} is corresponding with the value of conversion loss *Lc*. The LO signal must be clean of noise and spurious at the frequencies $f_{LO} \pm f_{IF}$.

The CMY210 circuit is a clever arrangement of a switching-type mixer with an automatic level control arrangement that provides the mixer with a reasonably constant drive level following the LO amplifier. Similar to cases we have shown, there is a clear relationship between the LO power (invested DC power) and large signal performance. This mixer externally has a series-tuned circuit at the input, which allows the ultra-high frequency (UHF) to pass, and a parallel-tuned circuit, which keeps the UHF frequency from getting into the IF stage. The resulting conversion loss of 5.5 dB is consistent with the example we presented before of 6 dB, and while the noise figure is also stated as 5.5 dB, that value is deceiving because it was determined using a 120 MHz IF. If an IF of 10 MHz or less was used, the noise figure would be tremendously worse. This is due to the frequently mentioned flicker-noise contribution. As a comparison, Figure 9.123 shows the predecessor to the CMY210, the CMY91, which includes a built-in IF amplifier. Figure 9.122 shows its partial specifications. Since it

Electrical Characteristics

Test conditions: $T_a = 25\,°C$; $V_{DD} = 3$ V, see test circuit; $f_{RF} = 1620$ MHz; $f_{LO} = 1500$ MHz; $P_{LO} = 0$ dBm; $f_{IF} = 120$ MHz, unless otherwise specified:

Parameter, Test Conditions	Symbol	min	typ	max	Unit
Operating current	I_{op}	–	7	–	mA
Conversion loss	L_c	–	5,5	–	dB
SSB noise figure	F_{SSB}	–	5,5	–	dB
2 Tone third-order order IMD $P_{RF1} = P_{RF2} = -3$dBm $f_{RF1} = 1615$MHz; $f_{RF2} = 1620$MHz; $f_{LO} = 1500$MHz	d_{IM3}	–	56	–	dBc
Third-order input intercept point	$IP3_{in}$	–	25	–	dBm
P_{-1dB} input power	$P_{-1\,dB}$	–	20	–	dBm
LO leakage at RF/IF port (1,6)	$P_{LO\,1,6}$	–	–6	–	dBm

Test circuit/application example

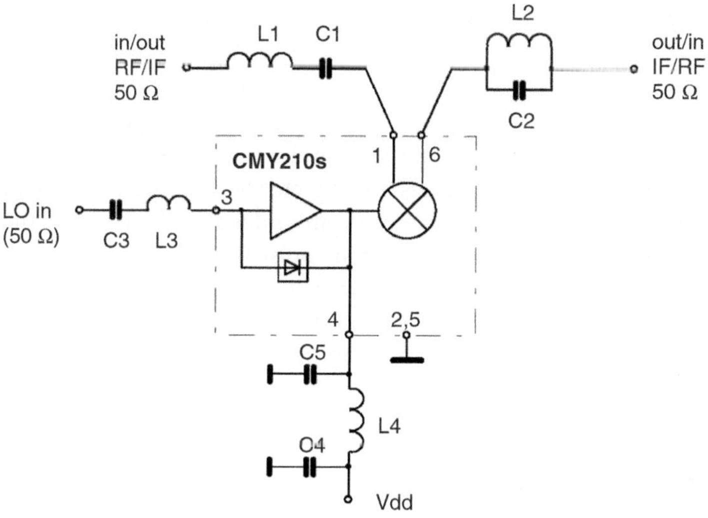

Notes for external elements:

L1, C1: Filter for upper frequency
C2, L2: Filter for lower frequency: each filter is a through path for the desired frequency (RF or IF) and isolates the other frequency (IF or RF) and its harmonics (Figure 9.122). These two filters must be connected to pins 1 and 6 directly.
Parasitic capacitances at ports 1 and 6 must be as small as possible.
L4 and C5 are optimized by indicating lowest I_{op} at used LO frequency: same procedure applies for L3.
The ports 1, 3, and 6 must be DC open.

f_{LO}	L1	C1	L2	C2	L3	C3	L4	C4	C5
MHz	nH	pF	nH	pF	nH	pF	nH	pF	pF
500	11	8.2	19	4.7	6	47	19	47	3.3
1000	7	4.7	9	3.3	6	33	16	33	0
1500	4	2.2	7,5	2.2	6	18	8.5	18	0
2000	*	2.2	*	1.5	4	15	*	15	0
2500	*	2.2	*	1.5	3	15	*	15	0
Notes (discussed as follows)	1	1	1	1	3	3	2	2	2

Figure 9.122 (a) Part of the datasheet of the Siemens CMY210 IC (discontinued product). *Source*: [4] / John Wiley & Sons. (b) Part of the datasheet of the Siemens CMY210 IC (discontinued product). *Source*: [4] / John Wiley & Sons.

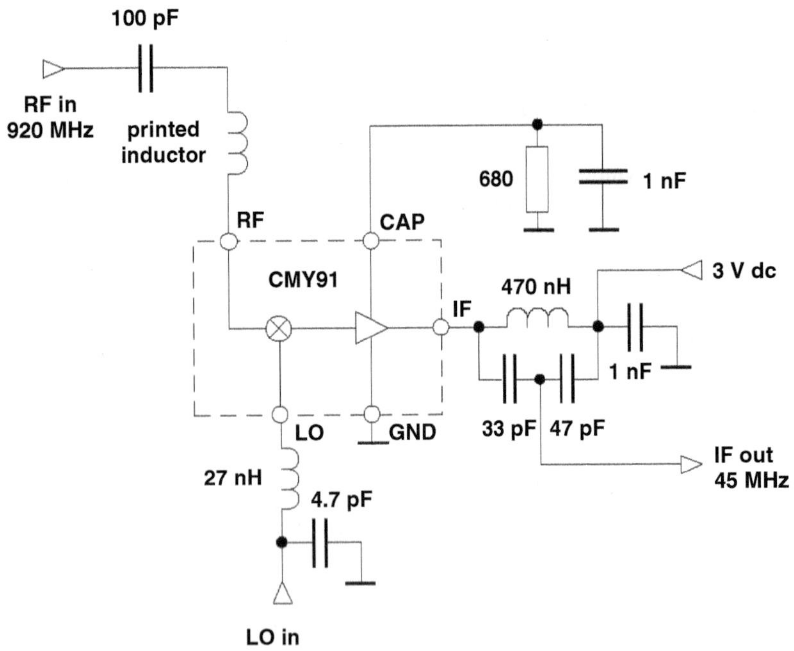

Figure 9.123 The CMY91 IC includes a post-mixer amplifier instead of an AGC LO amplifier. Its $IP_{3,in}$ is typically −2 dBm (drain supply, 3 V at 1 mA) with no connection made to its CAP pin; in the configuration shown earlier, its typical $IP_{3,in}$ is 0 dBm (drain supply, 3 V at 2.5 mA). *Source*: Adapted from [46].

has a 5.5 dB gain, one can assume that the IF amplifier has something like 10 dB of gain, and the mixer portion, 6 dB losses. The resulting noise figure is the sum of the mixer noise figure and the IF amplifier noise figure. The IF transistor is probably not "noise-matched" with the mixer, but the resulting 9 dB noise figure is not uncommon in this arrangement. Going by previous examples, we would say that the noise figure of the mixer is somewhere around 6 dB, and therefore the mismatched (open at the input) IF amplifier has a noise figure of 3 dB, resulting in this total noise figure of 9 dB as stated. The third-order IP of −2 dBm is no match to the CMY210's IP_{3in} performance, which is typically 25 dBm.

For readers who are interested in what mixers look like when pushed well into the microwave region, following is an example we present—the layout for a 20-GHz diode mixer. Figure 9.124 presents its layout, which includes radial stubs, circular stubs, and a Lange coupler. Figure 9.125 shows the layout implemented on a ceramic substrate.

9.33 Notes from TriQuint Semiconductor's GaAs Design Class

As further food for thought and experimentation, Figure 9.126a–e presents mixer and related circuit notes as presented by Wes Hayward in TriQuint Semiconductor's GaAs Design Class.

The complications lie in the input transformer and the LO amplifier. The following are a set of measurements for this type of mixer.

To summarize the mixer applications, Figure 9.126f shows a receiver (downconverter) for data transmission and the right circuit is a pulsed, DRO-based oscillator. The receiver portion uses a microstrip filter at the input, a matching circuit for the FET, and half of the branch-line coupler. The other side of the coupler is fed by the LO, which is stabilized by a DRO. At the output of the branch-line coupler, the IF is generated by rectifying the signal using two hot carrier diodes. This example is important as it combines the critical circuit details covered in this book.

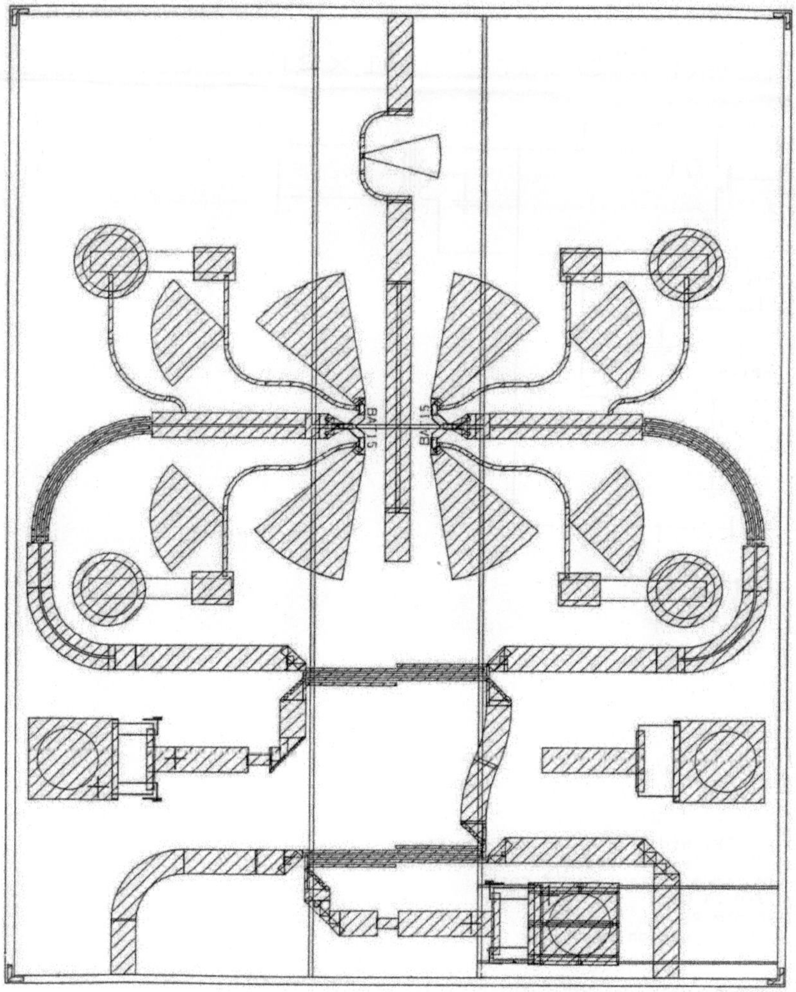

Figure 9.124 Layout of the 20 GHz diode mixer. *Source*: [4, p 716] / John Wiley & Sons.

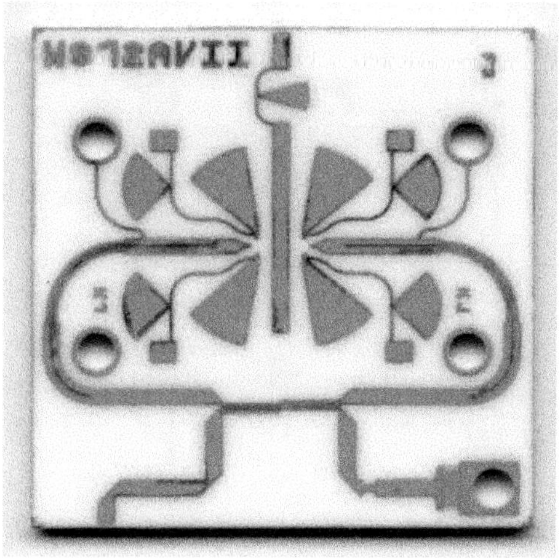

Figure 9.125 The 20 GHz diode mixer circuit board. *Source*: [4, p 717] / John Wiley & Sons.

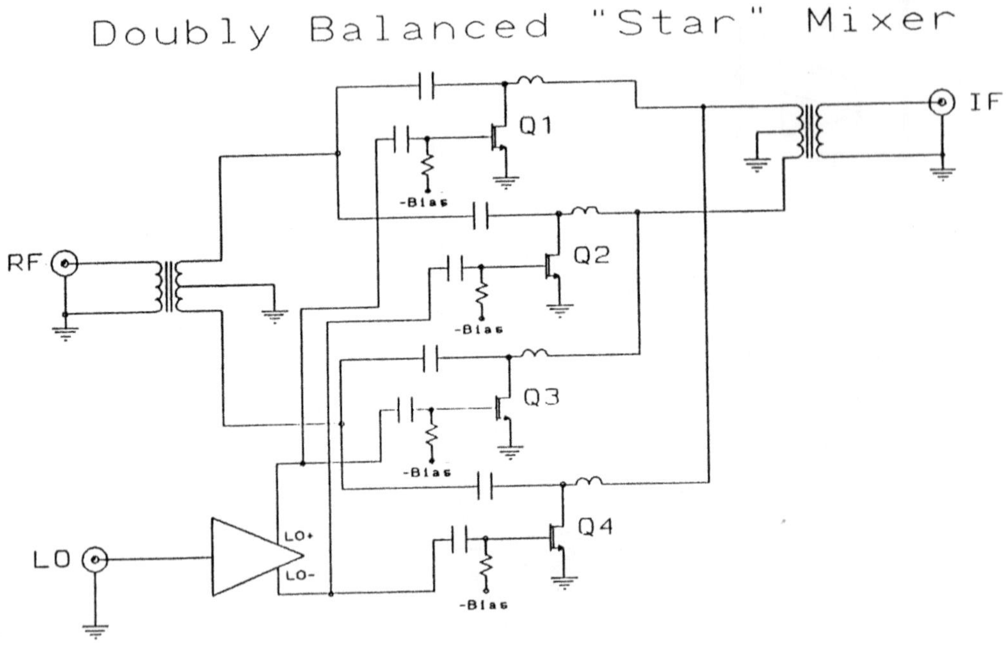

(a)

HA Mixer Family Measurements

The mixers are all identical except for FET width. Mixers used the HA process and are half micron gate length. Bias was adjusted to best gain at +6 LO. The LO baluns were half wave lines of semi-rigid coax.
F-LO=1.65, RF=1.9, and IF=0.25 GHz.
Input Power is -13 or -10 dBm RF.
All measurements show down-conversion results.

FET Width	P-av RF	P-LO dBm	P-out at IF	IMDR dB	Conv Gain dB	IP3-out dBm	IP3in dBm
50	-13	6	-31	44.7	-11.7	-2.35	9.35
50	-13	10	-30.5	53.5	-11.2	2.55	13.75
50	-10	10	-27.8	49	-11.5	3	14.5
100	-10	6	-25.7	41.5	-9.4	1.35	10.75
100	-10	10	-24.8	48	-8.5	5.5	14.0
100	-10	14	-24.5	68.8	-8.2	16.2	24.4
200	-10	6	-24.3	40.5	-8	2.25	10.25
200	-10	10	-23.7	45	-7.4	5.1	12.5
200	-10	14	-23.5	60.8	-7.2	13.2	20.4
200	-10	17	-24	60.7	-7.7	12.65	20.3
400	-10	6	-24	42.8	-7.7	3.7	11.4
400	-10	10	-23.5	52	-7.2	8.8	16.0
400	-10	14	-23.3	67.7	-7	16.85	23.85
400	-10	17	-23.2	57.5	-6.9	11.85	18.75
50	-10	6	-30.8	56	-14.5	3.5	18.0
50	-10	6	-28.6	42.5	-12.3	-1.05	11.25
50	-10	10	-28	49.2	-11.7	2.9	14.6
50	-10	14	-28.2	60.8	-11.9	8.5	20.4
50	-10	17	-28.3	63.5	-12	9.75	21.75
50	-10	17	-28	62.8	-11.7	9.7	21.4
100	-10	6	-29.5	42.5	-13.2	-1.95	11.25
100	-10	10	-25	47.2	-8.7	4.9	13.6
100	-10	14	-24.7	63.7	-8.4	13.45	21.85
100	-10	16	-25.5	62	-9.2	11.8	21.0

(b)

Figure 9.126 (a) Popular doubly balanced resistive "star" mixer. *Source*: [4, p 718] / John Wiley & Sons. (b) Measurements of the mixer in Figure 9.118a. *Source*: [4, p 719] / John Wiley & Sons.

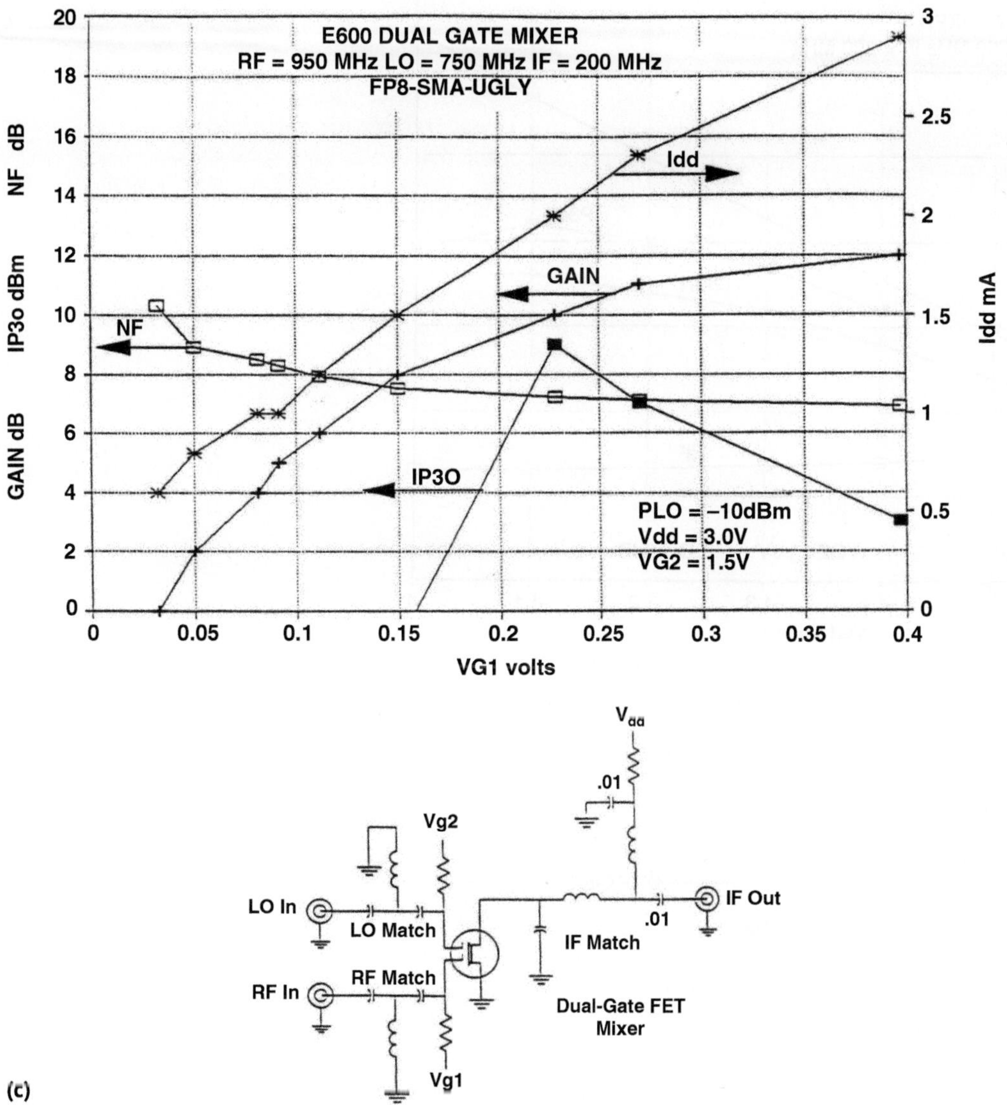

(c)

Figure 9.126 (c) Test circuit for the measurement shown earlier. *Source*: [4, p 721] / John Wiley & Sons.

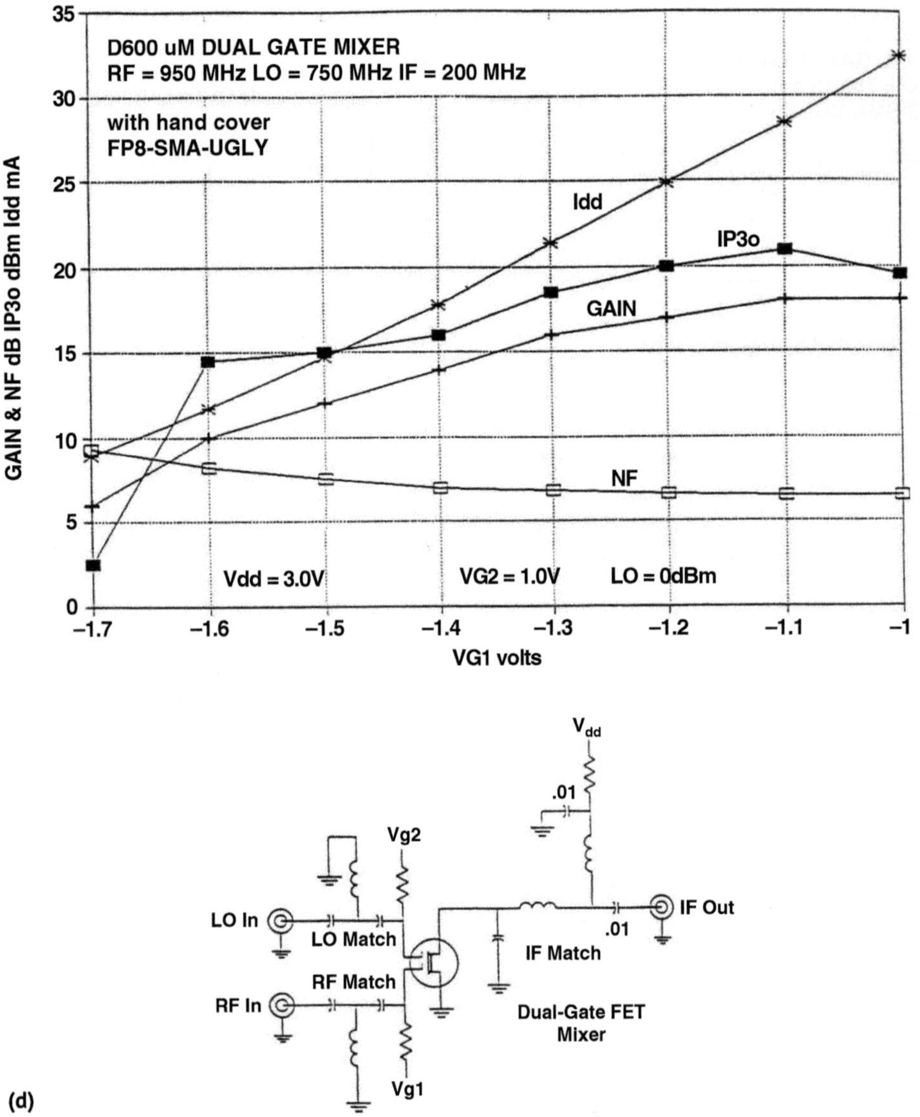

(d)

Figure 9.126 (Cont'd) (d) Measured result of the dual-gate FET mixer circuit shown here. *Source*: [4, p 722] / John Wiley & Sons.

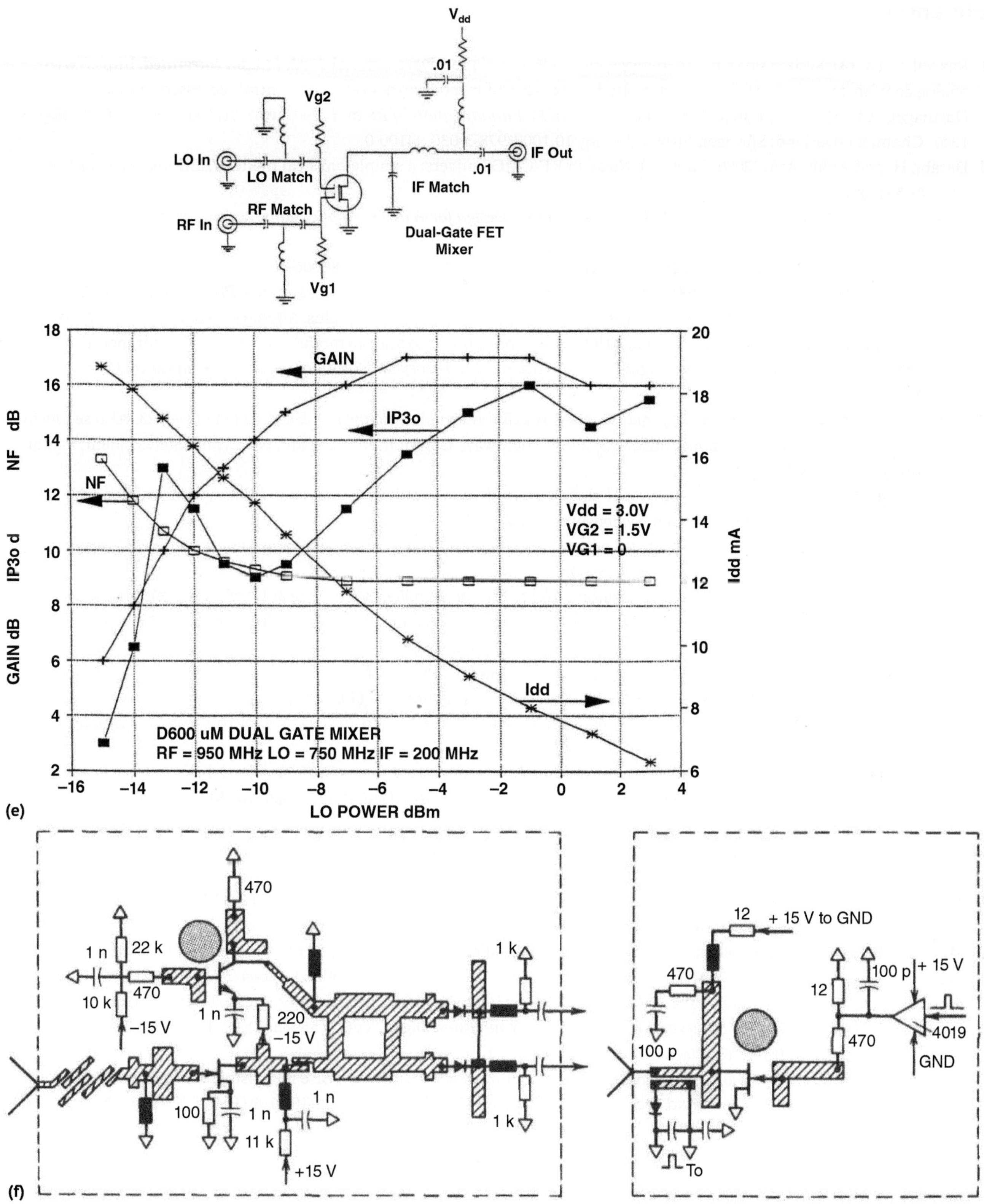

Figure 9.126 (Cont'd) (e) Measured result of the dual-gate FET mixer circuit shown here. *Source*: [4, p 723] / John Wiley & Sons. (f) Microstrip-based receiver (downconverter) and transmitter using a DRO as the frequency source. *Source*: Adapted from [19, p 931].

References

1 Razzell, C. (2013). System noise figure analysis for modern radio receivers, tutorial 5594. *Maxim Integrated*. https://www.analog.com/en/technical-articles/system-noisefigureanalysis-for-modern-radio-receivers.html (accessed 25 April 2023).

2 Hartnagel, H.L., Quay, R., Rohde, U.L. et al. (eds.) (2023). *Fundamentals of RF and Microwave Techniques and Technologies*, 1241. Cham, Switzerland: Springer. https://doi.org/10.1007/978-3-030-94100-0.

3 Darabi, H. and Abidi, A.A. (2000, January). Noise in RF-CMOS mixers: a simple physical model. *IEEE Journal of Solid-State Circuits* 35 (1): 15–25.

4 Rohde. U.L. and Rudolph, M. (2013). *RF/Microwave Circuit Design for Wireless Applications*, 2e. Hoboken, NJ: John Wiley & Sons.

5 Maas, S.A. (1993). *Microwave Mixers*, 2e, 153–171. Boston: Artech House. ISBN 0-89006-601-1.

6 IRE Subcommittee 7.9 on Noise (1963). Description of noise performance in receiving systems. *Proceedings of the IRE* 51: 436.

7 Qian, C. (1998, September). Subharmonic mixers simplify digital-modulation systems. *Microwaves & RF* 37 (9): 62–70.

8 Itoh, K., Shlmozawa, M., Kawakami, K. et al. (1996). Even-harmonic quadrature modulator with low vector modulation error and low distortion for microwave digital radio. In *Microwave Theory and Techniques Society Symposium Digest*, (967–970). IEEE.

9 Joshi, S.R. (1994, November). Novel I/Q modulators mix cellular signals. *Microwaves & RF*, 33 (11) 119–123. Also see Joshi, S.R. (1995). Modulator with harmonic mixers. Synergy Microwave Corporation, Paterson, NJ, United States patent, Patent No. 5,416,449, 16 May 1995.

10 Schneider, M.V., Trambarulo, R., Gnauck, A. et al. (1990). High-speed millimeter-wave modulator/demodulator. In *Microwave Theory and Techniques Society Symposium Digest*, (611–614). IEEE.

11 Charamiec, J. (1978, September). Subharmonically pumped Schottky diode single-sideband modulator. *IEEE Transactions on Microwave Theory and Techniques* 26 (9): 635–638.

12 Doyle, I. (1998, October). A Simplified Subharmonic I/Q Modulator. *Applied Microwave & Wireless*, 10 (8), 34–40.

13 Will, P. (1980). Broadband doubly balanced mixer having improved termination insensitivity characteristics. Adams-Russell Co., Inc., Waltham, MA, United States Patent, Patent No. 4,224,572, 23 September 1980.

14 Lee, J.T. (1983, August). Balanced subharmonic mixers. *Microwave Journal*. 26 (8), 129–135.

15 Neuf, D. (1984). Fundamental versus harmonic mixing. *Microwave Journal* 27 (7), 121–134.

16 Henderson, B. (1982, January). Full-range orthogonal circuit mixers reach 2 to 26 GHz. *Microwave Systems News*, January Issue.

17 Gilbert, B. (1994). *Demystifying the Mixer*, 27. self-published monograph.

18 Clarke, K.K. and Hess, D.T. (1971). *Communication Circuits: Analysis and Design*, 115. Reading, MA: Addison-Wesley. ISBN 0-201-01040-2.

19 Vendelin, G.D., Pavio, A.M., and Rohde, U.L. (2005). *Microwave Circuit Design Using Linear and Nonlinear Techniques*, 2e. Hoboken, NJ: John Wiley & Sons.

20 Tie, G.K. and Aitchison, C.S. (1983). Noise figure and associated conversion gain of a microwave MESFET gate mixer. In *Conference Proceedings of the 13th European Microwave Conference*, (03–08 September 1983).

21 Margraf, M. (2004). Niederfrequenz-Rauschen und Intermodulationen von resistiven FET-Mischern. PhD dissertation. Berlin Institute of Technology, 2004 (in German). http://opus.kobv.de/tuberlin/volltexte/2004/572/pdf/margrafmichael.pdf.

22 Margraf, M. and Boeck, G. (2003, June). A new scalable low frequency noise model for field-effect transistors used in resistive mixers. In *IEEE MTT-S International Microwave Symposium Digest*, Vol. 1 (559–562), (8–13 June 2003). https://doi.org/10.1109/MWSYM.2003.1211000

23 Margraf, M. and Boeck, G. (2004, July). Analysis and modeling of low-frequency noise in resistive FET mixers. *IEEE Transactions on Microwave Theory and Techniques* 52 (7): 1709–1718. https://doi.org/10.1109/TMTT.2004.830486.

24 Margraf, M. and Boeck, G. (2003). 1/f noise optimum for field-effect transistors in single-ended resistive mixers. In *Proceedings of 33rd European Microwave Conference*, Vol. 3 (1015–1017), (7–9 October 2003). https://doi.org/10.1109/EUMC.2003.1262825.

25 Zinke, O. and Hartnagel, H.L. (1987). *Lehrbuch der Hochfrequenztechnik, Band 2: Elektronik und Signalverarbeitung*, 429–430. New York: Springer Verlag.

26 Zinke, O. and Brunswig, H. (1987). *Lehrbuch der Hochfrequenztechnik*, Vol. 2, 3e, 435. New York: Springer Verlag. ISBN 0-387-17042-1.

27 Razavi, B. (1998). *RF Microelectronics*. New Jersey: Prentice Hall Inc.

28 Niknejad, A.M. (2015). *Current/Voltage Communicating Mixer*. U.C. Berkeley. http://rfic.eecs.berkeley.edu/142/pdf/module17.pdf.

29 Havens, J.H. (1998, July). Mixer Noise Analysis Using Advice. *Technical Memorandum ITD-98-34557M*, Lucent Technology-Bell Labs.

30 Hull, C.D. and Meyer, R.G. (1993, December). A systematic approach to the analysis of noise in mixers. *IEEE Transactions on Circuits and Systems* 40 (12): 909–919.

31 Roychowdhury, J., Long, D., and Feldmann, P. (1998, March). Cyclostationary noise analysis of large RF circuits with multitone excitations. *IEEE Journal of Solid –State Circuits* 33 (3): 324–336.

32 Ham, D. and Hajimiri, A. (2000, May). Complete noise analysis for CMOS switching mixers via stochastic differential equations. In *Proceedings of the IEEE 2000 Custom Integrated Circuits Conference (Cat. No.00CH37044)*, (439–442). IEEE.

33 Terrovitis, M.T. and Meyer, R.G. (1999, June). Noise in current-commuting CMOS mixers. *IEEE Journal of Solid–States Circuits* 34 (6): 772–783.

34 Gardner, W.A. (1993). *Cyclostationary in Communication and Signal Processing*. Piscataway, NJ: IEEE.

35 Vendelin, G.D., Pavio, A.M., Rohde, U.L. et al. (2021). *Microwave Circuit Design Using Linear and Nonlinear Techniques*, 3e. Hoboken, NJ: John Wiley & Sons.

36 Meyer, R.G. (1986, August). Intermodulation in high-frequency bipolar transistor integrated-circuits mixers. *IEEE Journal of Solid-State Circuits* SC-21 (4): 534–537.

37 Yuan, F. and Opal. A. (1998, May). Distortion analysis of periodically switched nonlinear circuits. In *IEEE Canadian Conference on Electrical Engineering*, (ed. R. Horsey), Vol. 1 (153–156).

38 Ruhumbika, A.G. and Abdulla, R. (2014). Analytical method for gain analysis of a double balanced Gilbert cell mixer. In *Proceedings of Third International Conference on E-Learning and E-Technologies in Education (ICEEE)*, (18–20 March 2014). Malaysia: SDIWC. (240–245). ISBN: 978-0-9891305-4-7.

39 Niknejad, A.M. (2014). *Mixer Noise*. U.C. Berkeley. http://rfic.eecs.berkeley.edu/ee242/pdf/Module_5_2_MixerNoise.pdf

40 BSIM Research Group (1999). *BSIMS Version 3.2 MOSFET Model: User's Manual*, 2.2e. Berkeley: University of California. Department of Electrical Engineering and Computer Sciences.

41 Johns, D. and Martin, K. (1997). *Analog Integrated Circuit Design*. John Wiley &Sons.

42 Terrovitis, M.T. and Meyer, R.G. (1999, June). Noise in current-commutating CMOS mixers. *IEEE Journal of Solid-State Circuits* 34 (6): 772–783.

43 Lee, T.H. (1998). *The Design of CMOS Radio-Frequency Integrated Circuits*. Cambridge University Press.

44 Hull, C.D. and Meyer, R.G. (1993, December). A systematic approach to the analysis of noise in mixers. *IEEE Transactions on Circuits and Systems-I: Fundamental Theory and Applications* 40: 909–919.

45 Kamen, E.W. and Heck. B.S. (2000). *Fundamentals of Signals and Systems Using the Web and MATLAB*, 2e. Upper Saddle River, NJ: Prentice Hall.

46 Rohde, U.L. and Newkirk, D.P. (2000). *RF Microwave Circuit Design for wireless Applications*, 641–644. John Wiley & Sons.

47 Hayward, W. and Taylor, S. (1991). Compensation method and apparatus for enhancing single ended to differential conversion. TriQuint Semiconductor, Inc., Beaverton, OR, United States Patent, Patent No. 5,068,621, 26 November 1991.

48 W. Hayward, "RFIC Lore and Philosophy" and "Non-Linear Designs," in TriQuint Semiconductor GaAs Design Class Notes, Section 5, TriQuint Semiconductor, Hillsboro, OR, USA; B. Hickey, "Design Example: Gilbert Cell Mixer with Active Baluns," in TriQuint Semiconductor GaAs Design Class Notes, Section 7, TriQuint Semiconductor, Hillsboro, OR, USA.

49 Kundert, K. Intercept Point," *The Designer's Guide Community*, May 2002. [Online]. Available: https://designers-guide.org/analysis/intercept-point.pdf (accessed March 26, 2026).

50 Harmonica User's Manual, Example's Volume, *Non-Linear Designs*, Compact Software, Inc., Paterson, NJ, USA, 1997.

51 Kared, T., Silaghi, H., Rudolph, M. et al. (2025). Design of a fixed IF down-conversion double-balanced mixer for UHF band applications. *Sensors* 25 (3): 608. https://doi.org/10.3390/s25030608.

52 Baker, M.W. (2002). A BiCMOS RF mixer for 5-GHz receivers. Doctoral dissertation. Massachusetts Institute of Technology.

53 J. M. Østby, Electronic noise-estimates and countermeasures, Lecture 8 (Mot 5): Noise in bipolar transistors, INF 5460, Department of Informatics, University of Oslo, Oslo, Norway, 2016. [Online]. Available: https://www.uio.no/studier/emner/matnat/ifi/INF5460/h16/undervisningsmateriale/f8-1p.pdf (accessed March 26, 2026).

54 Razavi, B. (2001). *Design of Analog CMOS Integrated Circuits*. Boston: McGraw Hill.

55 D. Chen, "Mixer Design," slide 94, *EE 507: Mixed-Signal IC Design*, Dept. Elect. Computer. Eng., Iowa State Univ., Ames, IA, USA. [Online]. Available: https://slideplayer.com/slide/6012717 (accessed Mar. 26, 2026).

56 Kared, T. (2025). A novel high dynamic range wideband differential mixer using SiGe transistors. Ph.D. Dissertation. Brandenburg University of Technology, Cottbus, Germany.

57 Darabi, H. and Abidi, A.A. (2000, January). Noise in RF-CMOS miners: a simple physical model. *IEEE Transactions on Solid-State Circuits* 35 (1): 15–25.

Further Reading

D. L. Ash, "High Performance Receiver," *IEEE Trans. Consum. Electron.*, vol. CE-24, no. 1, pp. 39–46, Feb. 1978. [Online]. Available: https://doi.org/10.1109/TCE.1978.266995 (accessed Mar. 26, 2026).

Borremans, M., Steyaert, M., and Yoshitomi, T. (1998). A 1.5 V, wide band 3GHz, CMOS quadrature direct up-converter for multi-mode wireless communications. In *Proceedings of the IEEE 1998 Custom Integrated Circuits Conference (Cat. No.98CH36143)*, (79–82). IEEE.

Babiker, S., Asenov, A., Camerson, N. et al. (1998, August). Complete Monte Carlo RF analysis of 'real' short-channel compound FET's. *IEEE Transactions on Electron Devices* 45 (8): 1644–1643.

Brunel, E., Douglas, K., Gallagher, J. et al. (1996, February). A downconverter for use in a dual-mode AMPS/CDMA chip set. *Microwave Journal*, 39 (2), 20–42.

Chan, P.Y., Rofougaran, A., Ahmed, K.A. et al. (1993, September). A highly linear 1-GHz CMOS downconversion mixer. *Proceedings of ESSCIRC '93: Nineteenth European Solid-State Circuits Conference*, Seville (210–213).

Crols, J. and Stayaert, M.S.J. (1995, July). A 1.5-GHz highly linear CMOS downconversion mixer. *IEEE Journal of Solid-State Circuits* 30 (7): 736–742.

Datta, S., Shi, S., Roenker, K.P. et al. (1998, August). Simulation and design of InAlAs/InGaAs pnp heterojunction bipolar transistors. *IEEE Transactions on Electron Devices* 46 (8): 1634–1641.

de la Fuente, M.L., Pascual, J.P., and Artal, E. (1998). Analyze and design a cascade MESFET mixer. *Microwaves & RF* 37 (5): 129–138.

Goyal, R. (1995). *High-Frequency Analog Integrated Circuit Design*. New York: John Wiley & Sons. ISBN 0-471-53043-3.

Hewlett-Packard (1995). Low-Power Mixing Design Example Using HP Advanced Design System. *Solutions from HP EEsof.*

Kanazawa, K., Kazumura, M., Nambu, S. et al. (1985). A GaAs double-balanced dual-gate FET mixer IC for UHF receiver front-end applications. In *IEEE MTT International Microwave Symposium Digest*, (60–62). IEEE.

Kimura, K. (1994, January). Bipolar four-quadrant analog quarter-square multiplier. *IEEE Journal of Solid-State Circuits* 29 (1): 46–55.

Liu, S.-Y. and Chuang, H.-R. (1998). A 2.4-GHz transceiver RF front-end for ISM-band digital wireless communications. *Applied Microwave & Wireless* 10 (5), 32–48.

Meyer, R.G. (1986). Intermodulation in high-frequency bipolar transistor integrated-circuit mixers. *IEEE Journal of Solid-State Circuits* SC-21 (4): 534–537.

Narayanan, R.M. (1987). Multioctave, double-balanced mixers designed using MIC technology. *MSN & CT*, 17 (1) 108–115.

Rabaey, D. and Sevenhans, J. (1993). The challenges for analog circuit design in mobile radio VLSI chips. In *Proceedings of AACD Workshop*, Leuven, Vol. 2, (225–236).

Titus, W., Holway, L., Pucel, R.A. et al. (1988, October). Broadband Control Components–FET Mixers. Final Technical Report. Contract No. F33615-85-C-1725, PAY/RD/S-4270, Raytheon Company Research Division, Lexington, Massachusetts.

Sevenhans, J., Vanwelsenaers, A., Wenin, J. et al. (1991, May). An integrated Si bipolar transceiver for a zero IF 900 MHz GSM digital mobile radio front-end of a hand portable phone," In *Proceedings of the IEEE 1991 Custom Integrated Circuits Conference*, (7.7.1–7.7.4). IEEE.

Sevenhans, J., Haspeslagh, D., Delarbre, A. et al. (1994, February). An analog radio front-end chip set for a 1.9 GHz mobile radio telephone application. In *Proceedings of IEEE International Solid-State Circuits Conference - ISSCC '94*, San Francisco (44–45).

TQ9203 Mixer Noise Figure Measurement Note. Application Note, TriQuint Semiconductor, Inc., Hillsboro, OR, USA, 1994.TriQuint Semiconductor, Inc.

Abrie, P.L.D. (1985). *The Design of Impedance-Matching Networks for Radio-Frequency and Microwave Amplifiers.* Norwood, MA: Artech House.

10

VHF-to-EHF Oscillators

10.1 Introduction

Practically all modern telecommunication equipment and test equipment use frequency control techniques based on *frequency synthesis*. The production of the many frequencies involved in a radiocommunication system's modulation, transmission, reception, and demodulation functions through the combination and mathematical manipulation of a very few input frequencies. Although the frequency synthesis techniques now ascendant in wireless systems, phase-locked loops—and, less commonly, direct digital synthesizers—are fundamentally different, though all are ultimately based on RF oscillators. This chapter covers oscillator theory, evaluation, and design.

Two types of oscillators are needed in a phase-locked loop system (Figure 10.1). One is a *crystal oscillator*, which generates the synthesizer's reference signal. As Figure 10.1 reflects, the reference oscillator may be built into the synthesizer IC in highly integrated systems. The other oscillator, a *voltage-controlled oscillator* (VCO), is varied in frequency by the system to produce the synthesizer's output signal. Although designing good oscillators remains somewhat like black magic or a special art, we will show that mathematics and CAD tools, applied in conjunction with practical experience, can keep the oscillator design process well under control.

In this case, the designer has control over the VCO and the loop filter; the reference oscillator is part of the chip. In most cases (up to 2.5 GHz), the dual-modulus prescaler is inside the chip. Refer [1, Figure 6.1] or [2, Figure 11.1].

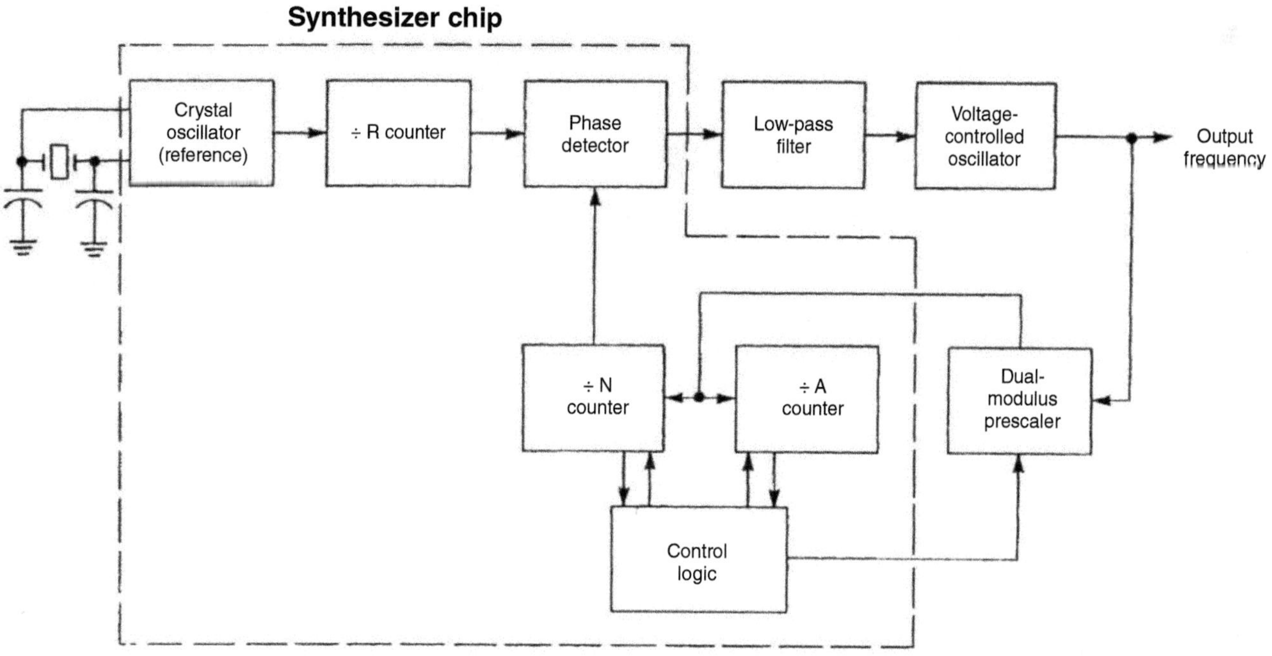

Figure 10.1 Block diagram of a modern, integrated frequency synthesizer [1, 2].

10.2 Background

The oscillator was probably first discovered by people who wanted to build an amplifier back in the vacuum tube days. The modern equivalent schematic diagrams of RF connections for common oscillator circuits are shown in Figure 10.2.

In Figure 10.2, we see an active device, a bipolar transistor (or field-effect transistor can be used), with a tuned input and output circuit. Under ideal circumstances, the voltage at the input is 180° out of phase and the amplifier is "stable." Because of the unavoidable feedback capacitance, a certain amount of energy is transferred from the output to the input at about 90° out of phase. If one of the tuned circuits is detuned to the point where the phase shift is −180°, oscillation will occur. This means that the small energy from the output will be amplified, brought back to the input in phase and further amplified, causing the stage to "take off" and oscillate. Unless there is some mechanism to limit the amplitude of the oscillation at the input or output, its amplitude can theoretically increase to a level sufficient to destroy the device through breakdown effects or thermal runaway.

The criterion for oscillation was first described by Barkhausen (see Eq. (10.6)). The conditions for oscillation exist when the feedback gain is high enough to cancel all losses and the difference between forward gain and reverse gain is less than zero. This implies also that oscillator circuits (Figure 10.2) provide a negative resistance, which is responsible for oscillation.

Here is list of important figures of merit [3].

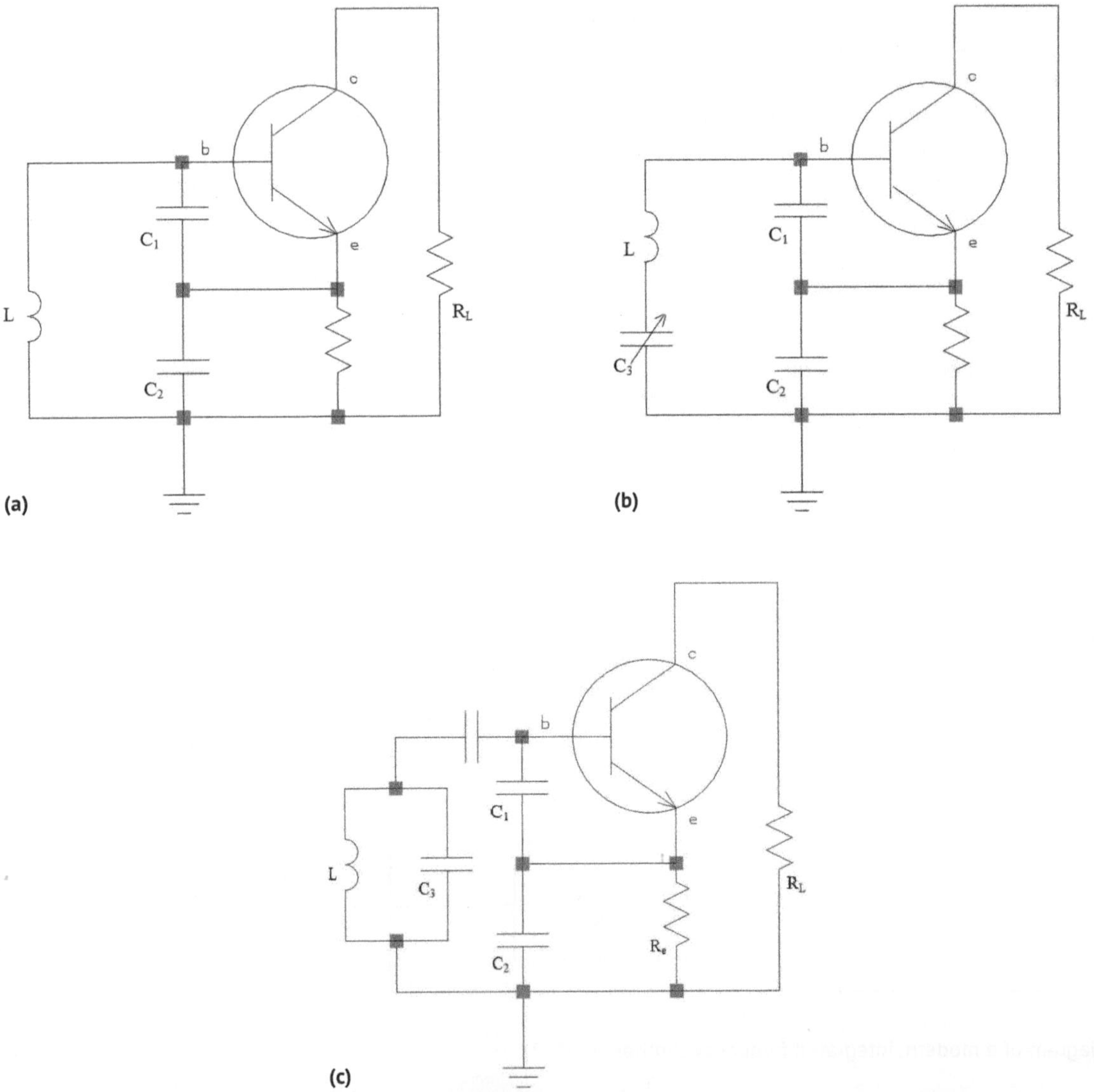

Figure 10.2 Schematic diagrams of RF connections for common oscillator circuits (DC and biasing circuits not shown) [3, 4]: (a) standard Colpitts oscillator, (b) modified Colpitts (Clapp–Gouriet) configuration, and (c) modified Colpitts oscillator.

10.2.1 Oscillator Specifications

Although we have tended to emphasize phase noise as a critical indicator of oscillator quality, phase noise is only one of a number of oscillator performance criteria that must be considered in a wireless system design. In addition to phase noise, the following characteristics are used by commercial companies to describe oscillator performance [3]:

Frequency pushing characterizes the degree to which an oscillator's frequency is affected by its supply voltage. A sudden current surge caused by activating a transceiver's RF power amplifier may produce a spike on the VCO's DC power supply and a consequent frequency jump. Like tuning sensitivity (below), pushing is specified in frequency/voltage form and is tested by varying the VCO's DC supply voltage (typically ±1 V) with its tuning voltage held constant.

Harmonic output power: The harmonic content is measured relative to the output power. Typical values are 20 dB or more suppression relative to the fundamental. This suppression can be improved by additional filtering.

Output power: The output power of the oscillator, typically expressed in dBm, is measured into a 50-Ω load. The output power is always combined with a specification for flatness or variation. A typical spec would be 0 dBm ± 1 dB.

Output power as a function of temperature: All active circuits vary in performance as a function of temperature. An oscillator's output power over a temperature range should vary less than a specified value, such as 1 dB.

Post-tuning drift: After a voltage step is applied to the tuning diode input, the oscillator frequency may continue to change until it settles to a final value. This post-tuning drift is one of the parameters that limits the bandwidth of the VCO input.

Power consumption: This characteristic conveys the DC power, usually specified in milliwatts and sometimes qualified by operating voltage, required by the oscillator.

Sensitivity to load changes: To keep costs down, many wireless applications use a VCO alone, without the buffering action of a high reverse-isolation amplifier stage. In such applications, *frequency pulling*, the change of frequency resulting from partially reactive loads, is an important oscillator characteristic. Pulling is commonly specified in terms of the frequency shift that occurs when the oscillator is connected to a load that exhibits a non-unity VSWR (such as 1.75, usually referenced to 50 Ω), compared to the frequency that results with a unity-VSWR load (usually 50 Ω). Frequency pulling must be minimized, especially in cases where power stages are close to the VCO unit and short pulses may affect the output frequency. Such feedback may make phase-locking impossible.

Spurious outputs: A VCO's spurious output specification, expressed in decibels, enumerates the strength of unwanted and non-harmonically related components relative to the oscillator fundamental. Because a stable, properly designed oscillator is inherently clean, such *spurs* are typically introduced only by external sources in the form of radiated or conducted interference. See *harmonic output power*.

Temperature drift: Although the synthesizer is responsible for locking and maintaining the oscillator's frequency, the VCO's frequency change as a function of temperature is a critical parameter and must be specified. Its value varies between 10 kHz/°C to several hundred kHz/°C depending on the center frequency and tuning range.

Tuning characteristic: This specification shows the relationship, depicted as a graph, between a VCO's operating frequency and the tuning voltage applied. Ideally, the correspondence between operating frequency and tuning voltage is linear. See *tuning linearity*.

Tuning linearity: For stable synthesizers, a constant deviation of frequency versus tuning voltage is desirable. It is also important to make sure that there are no breaks in the tuning range—for example, that the oscillator does not stop operating with a tuning voltage of 0 V. See *tuning characteristic*.

Tuning sensitivity, tuning performance: This datum, typically expressed in megahertz per volt (MHz/V), enumerates how much a VCO's frequency changes per unit of tuning-voltage change.

Tuning speed: This characteristic is defined as the time necessary for the VCO to reach 90% of its final frequency on the application of a tuning-voltage step. Tuning speed depends on the internal components between the input pin and tuning diode—including, among other things, the capacitance present at the input port. The input port's parasitic elements determine the VCO's maximum possible modulation bandwidth.

The following mathematical treatment now explains the theory of oscillator design.

10.3 Oscillator Design

A phase-locked loop generally has two oscillators: the oscillator at the output frequency and the reference oscillator. The reference oscillators of today can be a direct digital synthesizer, which is part of the system, and the VCO is controlled by either the reference or the oscillator loop. The VCO is one of the most important parts in a phase-locked loop system, the performance of which is determined only by the loop filter at offsets inside the loop bandwidth, and only by the quality of the VCO design at offsets outside the loop bandwidth.

To some designers, VCO design appears to be magic. Shortly, we will go through the mathematics of the oscillator and some of its design criteria, but the results have only limited meaning. This is due to component tolerances, stray effects, and most of all, nonlinear performance of the oscillator device, which can be modeled with only a certain degree of accuracy. However, after building oscillators for a while, a certain feeling will be acquired for how to do this, and certain performance behavior will be predicted on a rule-of-thumb basis rather than on precise mathematical effort. For reasons of understanding, we will deal with the necessary mathematical equations, but we consider it essential to explain that these are only approximations.

10.3.1 Basics of Oscillators

An electronic oscillator is a device that converts DC power to a periodic output signal (ac power). If the output waveform is approximately sinusoidal, the oscillator is referred to as *sinusoidal*. There are many other oscillator types normally referred to as *relaxation* oscillators. For application in frequency synthesizers, we will explore only sinusoidal oscillators for reasons of purity and noise sideband performance.

All oscillators are inherently nonlinear. Although the nonlinearity results in some distortion of the signal, linear analysis techniques can normally be used for the analysis and design of oscillators. Figure 10.3 shows, in block diagram form, the necessary components of an oscillator. It contains an amplifier with frequency-dependent forward loop gain $G(j\omega)$ and a frequency-dependent feedback network $H(j\omega)$ [3].

The output voltage is given by

$$V_o = \frac{V_{\text{in}}G(j\omega)}{1 + G(j\omega)H(j\omega)} \tag{10.1}$$

For an oscillator, the output V_o is nonzero even if the input signal $V_i = 0$. This can only be possible if the forward loop gain is infinite (which is not practical), or if the denominator

$$1 + G(j\omega)H(j\omega) = 0 \tag{10.2}$$

at some frequency ω_o. This leads to the well-known condition for oscillation (the *Nyquist criterion*), where at some frequency ω_o

$$G(j\omega_o)H(j\omega_o) = -1 \tag{10.3}$$

That is, the magnitude of the open-loop transfer function is equal to 1:

$$|G(j\omega_o)H(j\omega_o)| = 1 \tag{10.4}$$

and the phase shift is 180°:

$$\arg[G(j\omega_o)H(j\omega_o)] = 180° \tag{10.5}$$

This can be more simply expressed as follows: If in a negative-feedback system, the open-loop gain has a total phase shift of 180° at some frequency ω_o, the system will oscillate at that frequency provided that the open-loop gain is unity. If the

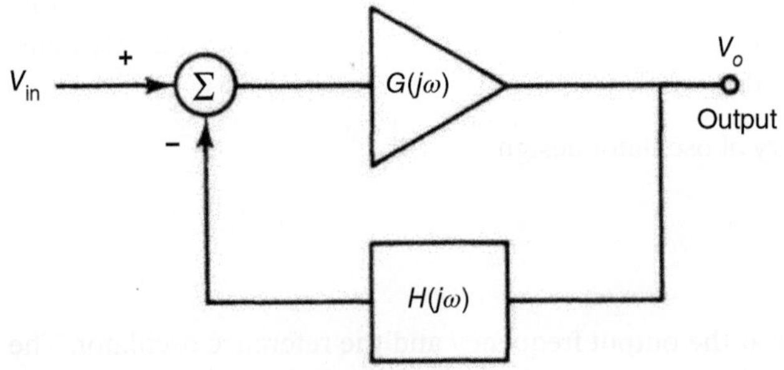

Figure 10.3 Block diagram of an oscillator showing forward and feedback loop components. *Source*: [3]/John Wiley & Sons.

gain is less than unity at the frequency where the phase shift is 180°, the system will be stable, whereas if the gain is greater than unity, the system will be unstable.

This statement is not correct for some complicated systems, but it is correct for those transfer functions normally encountered in oscillator design. The conditions for stability are also known as the *Barkhausen criterion*, which states that if the closed-loop transfer function is

$$\frac{V_o}{V_i} = \frac{\mu}{1 - \mu\beta} \tag{10.6}$$

The system will oscillate provided that $\mu\beta = 1$. This is equivalent to the Nyquist criterion, the difference being that the transfer function is written for a loop with positive feedback. Both versions state that the total phase shift around the loop must be 360° at the frequency of oscillation and the magnitude of the open-loop gain must be unity at that frequency [3].

The following analysis of the relatively simple oscillator shown in Figure 10.4 illustrates the design method. The linearized (and simplified) equivalent circuit of Figure 10.4 is given in Figure 10.5, where, h_{rb} has been neglected, and $1/h_{\mathrm{ob}}$ has been assumed to be much greater than the load resistance R_L and is also ignored.

Note that the transistor is connected in the common-base configuration, which has no voltage phase inversion (the feedback is positive), so the conditions for oscillation are

$$|G(j\omega_o)H(j\omega_o) = 1| \tag{10.7}$$

and

$$\arg[G(j\omega_o)H(j\omega_o)] = 0° \tag{10.8}$$

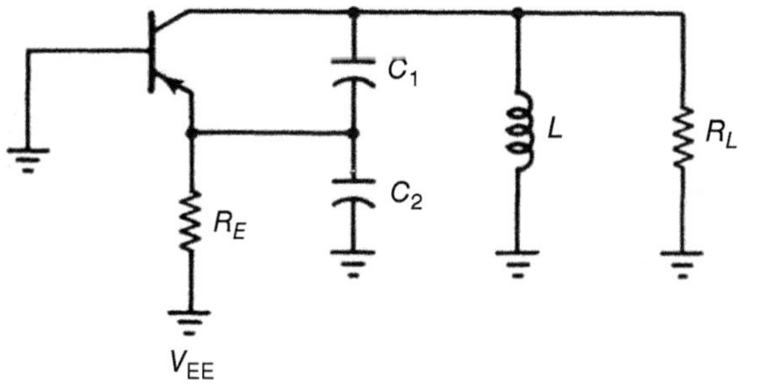

Figure 10.4 Oscillator with capacitive voltage divider. *Source*: [1]/John Wiley & Sons.

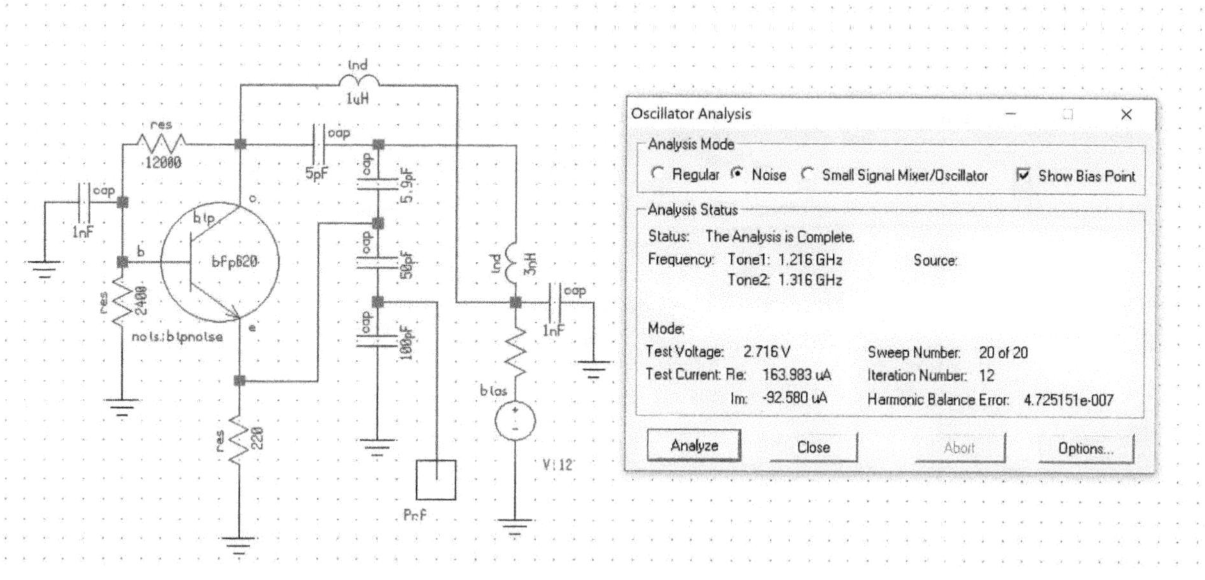

Figure 10.5 Detailed circuit of Figure 10.4 tuned to about 1296 MHz.

The circuit analysis can be greatly simplified by assuming that C_1 is 6.8 pF and $C_2 = 50\text{ pF}/2 = 25\text{ pF}$.

$$\frac{1}{\omega(C_2 + C_1)} << \frac{h_{\text{ib}} R_E}{h_{\text{ib}} + R_E} \tag{10.9}$$

and also that the Q of the load impedance is high. In this case the circuit reduces to that of Figure 10.6, where

$$V = \frac{V_o C_1}{C_1 + C_2} \tag{10.10}$$

and

$$R_{\text{eq}} = \frac{h_{\text{ib}} R_E}{h_{ib} + R_E} \left(\frac{C_1 + C_2}{C_1} \right)^2 \tag{10.11}$$

Figure 10.7 shows how the input impedance of an oscillator circuit—in this case, a BJT Colpitts oscillator minus its resonator—satisfies this condition for oscillation.

Then the forward gain

$$G(j\omega) = \frac{h_{\text{fb}}}{h_{\text{ib}}} Z_L = \frac{\alpha}{h_{\text{ib}}} Z_L \tag{10.12}$$

and

$$H(j\omega) = \frac{C_1}{C_1 + C_2} \tag{10.13}$$

where

$$Y_L = \frac{1}{Z_L} = \frac{1}{j\omega L} + \frac{1}{R_{\text{eq}}} + \frac{1}{R_L} + \frac{1}{j\omega C} \tag{10.14}$$

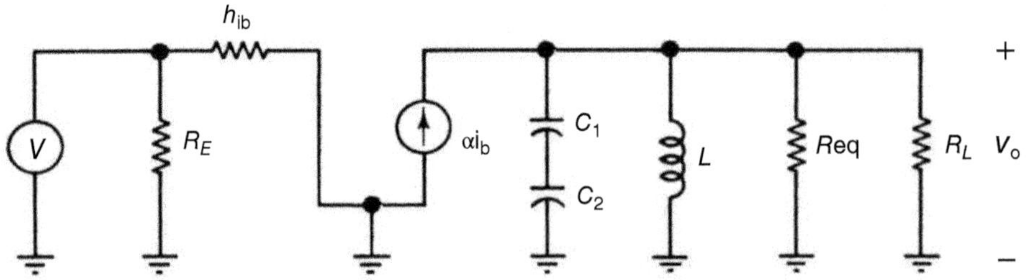

Figure 10.6 Further simplification of Figure 10.4, assuming high-impedance loads. *Source*: [1]/John Wiley & Sons.

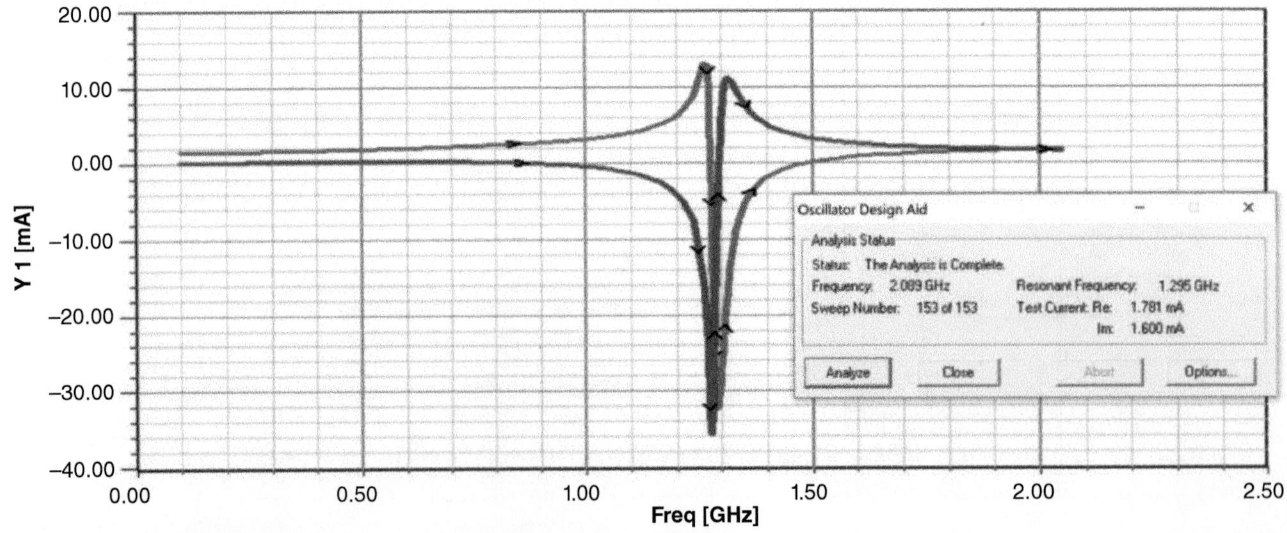

Figure 10.7 Calculated BJT Colpitts oscillator resonance condition.

A necessary condition for oscillation is that

$$\arg[G(j\omega)H(j\omega)] = 0° \tag{10.15}$$

Since H does not depend on frequency in this example, if arg GH is zero, the phase shift of the load impedance Z_L must also be zero. This occurs only at the resonant frequency of the circuit [1]

$$\omega_o = \frac{1}{\sqrt{L[C_1C_2/(C_1 + C_2)]}} \tag{10.16}$$

At this frequency

$$Z_L = \frac{R_{\text{eq}}R_L}{R_{\text{eq}} + R_L} \tag{10.17}$$

and

$$GH = \frac{h_{\text{fb}}}{h_{\text{ib}}}\left(\frac{R_{\text{eq}}R_L}{R_{\text{eq}} + R_L}\right)\frac{C_1}{C_1 + C_2} \tag{10.18}$$

The other condition for oscillation is the magnitude constraint that

$$G(j\omega)H(j\omega) = \frac{\alpha}{h_{\text{ib}}}\left(\frac{R_{\text{eq}}R_L}{R_{\text{eq}} + R_L}\right)\frac{C_1}{C_1 + C_2} = 1 \tag{10.19}$$

As to the output power, Figure 10.8a is the simulated result; a more detailed and mathematically analyzed example will follow.

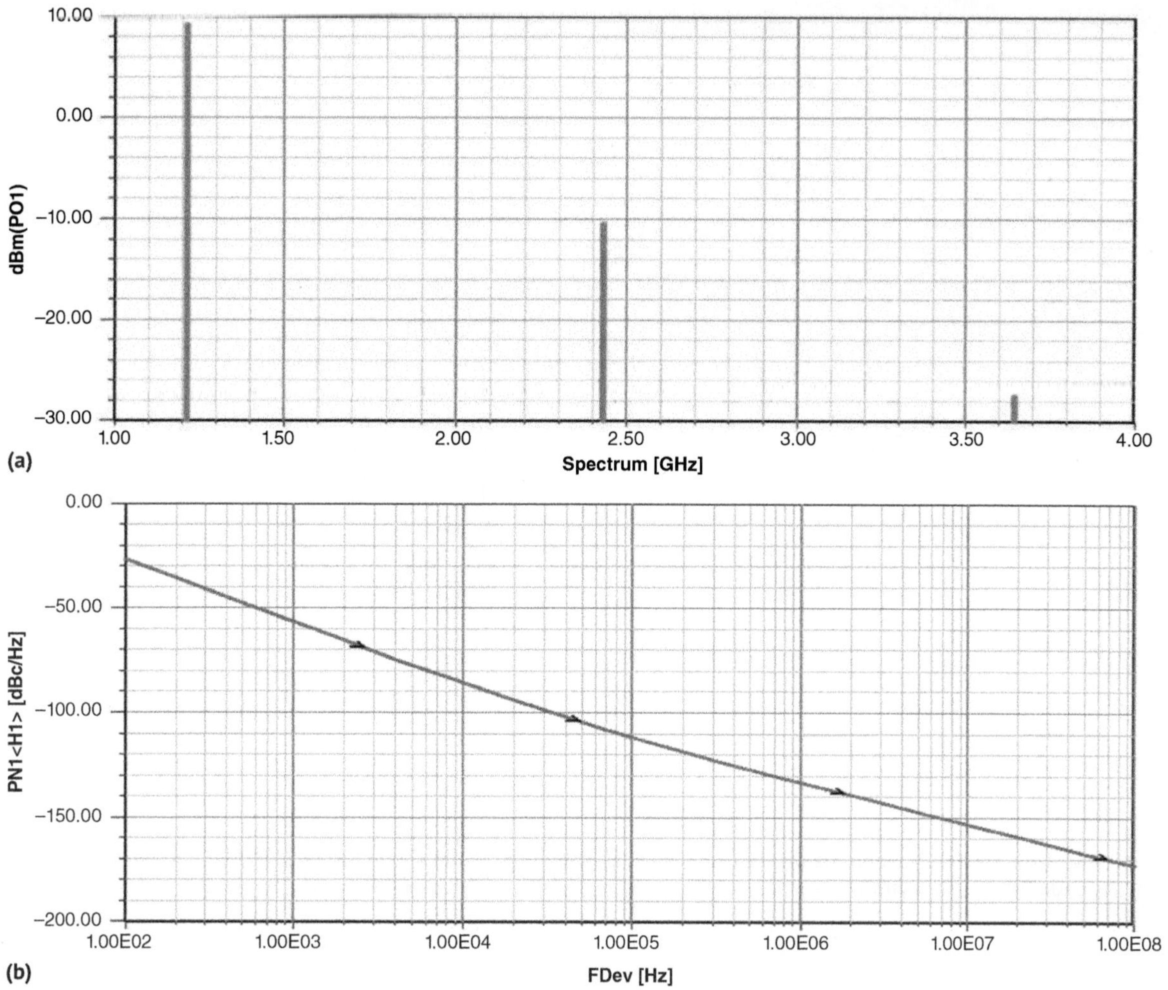

Figure 10.8 (a) Simulated output power of the circuit in Figure 10.5. (b) Simulated phase noise of the circuit in Figure 10.5.

And the simulated phase noise of the circuit in Figure 10.5 is shown in Figure 10.8b.

Figure 10.9a shows a 640 MHz oscillator using a helix resonator and other discrete components. That is about the practical limit of building oscillators with discrete components.

This generator was developed around 1968 and is discontinued. The helical resonator was a novelty of this generator. Each frequency band used such a resonator for obtaining high Q. A simulation of this oscillator section is shown in the example of Figure 10.9b, a 640 MHz FET oscillator.

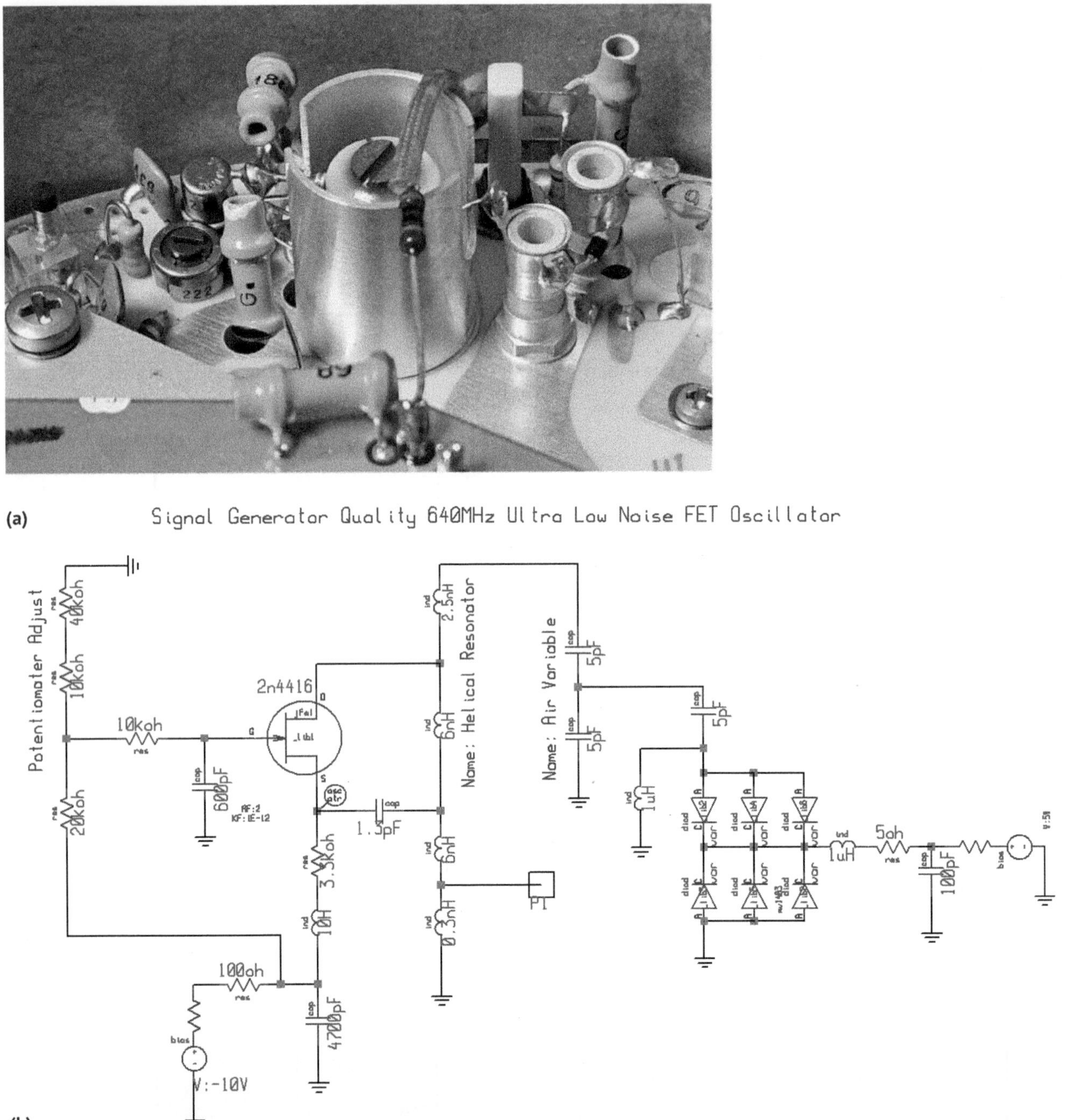

Figure 10.9 (a) 640 MHz helix resonator using the old-style mechanical assembly (Rohde and Schwarz SMDU Signal Generator: Oscillator Section). (b) 640 MHz FET oscillator.

For the 640 MHz FET oscillator shown in Figure 10.9b, a buffer amplifier is required for this operation, and the tuning diodes load the circuit, resulting in higher phase noise, as seen in Figure 10.9d, while Figure 10.9c shows its simulated output power.

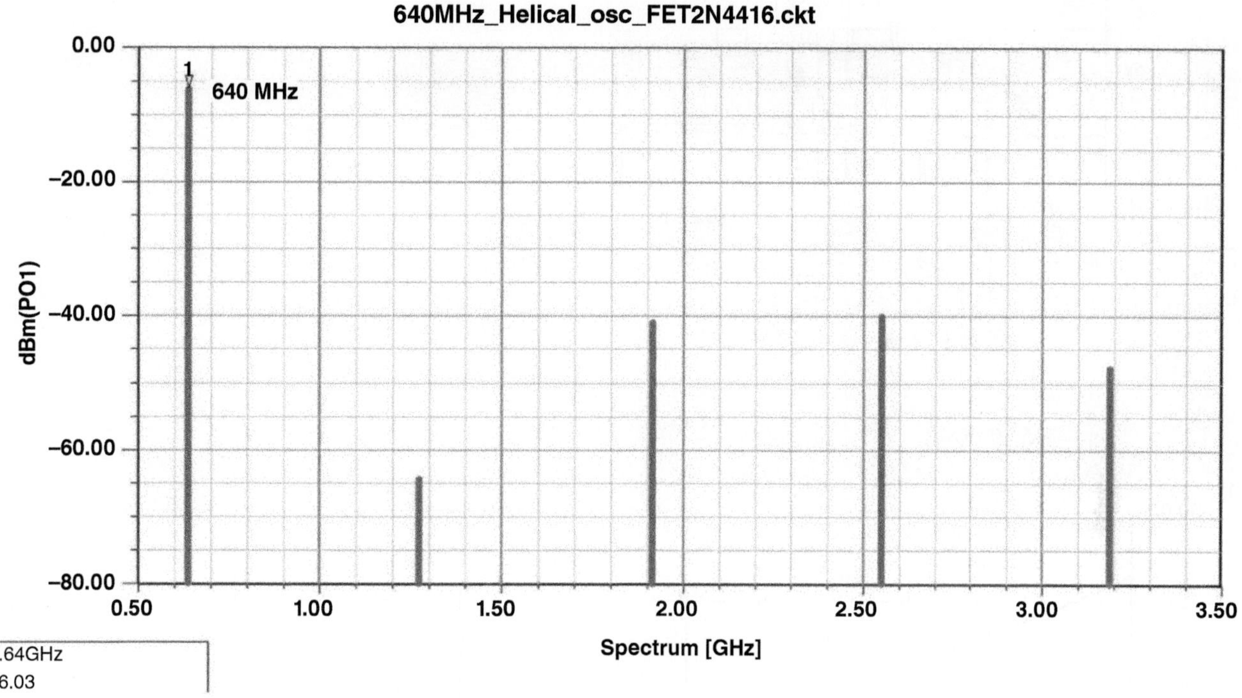

(c)

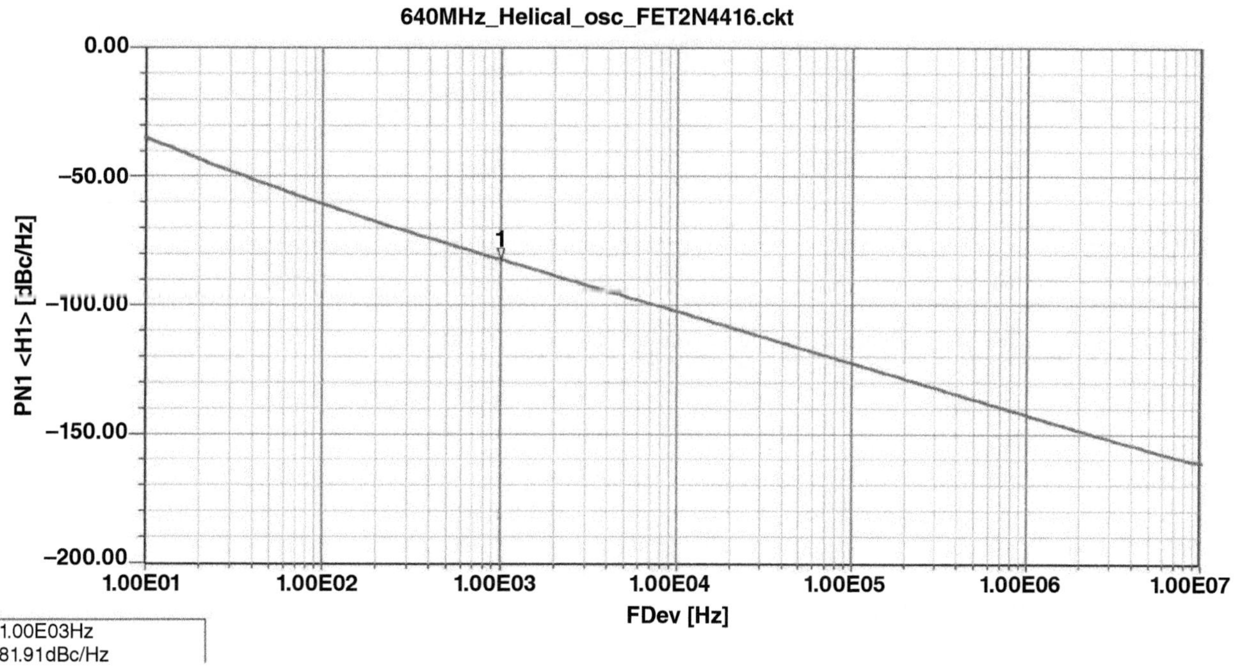

(d)

Figure 10.9 (Cont'd) (c) Simulated output power of the 640 MHz oscillator shown in (b). (d) Simulated phase noise of the 640 MHz oscillator shown in (b).

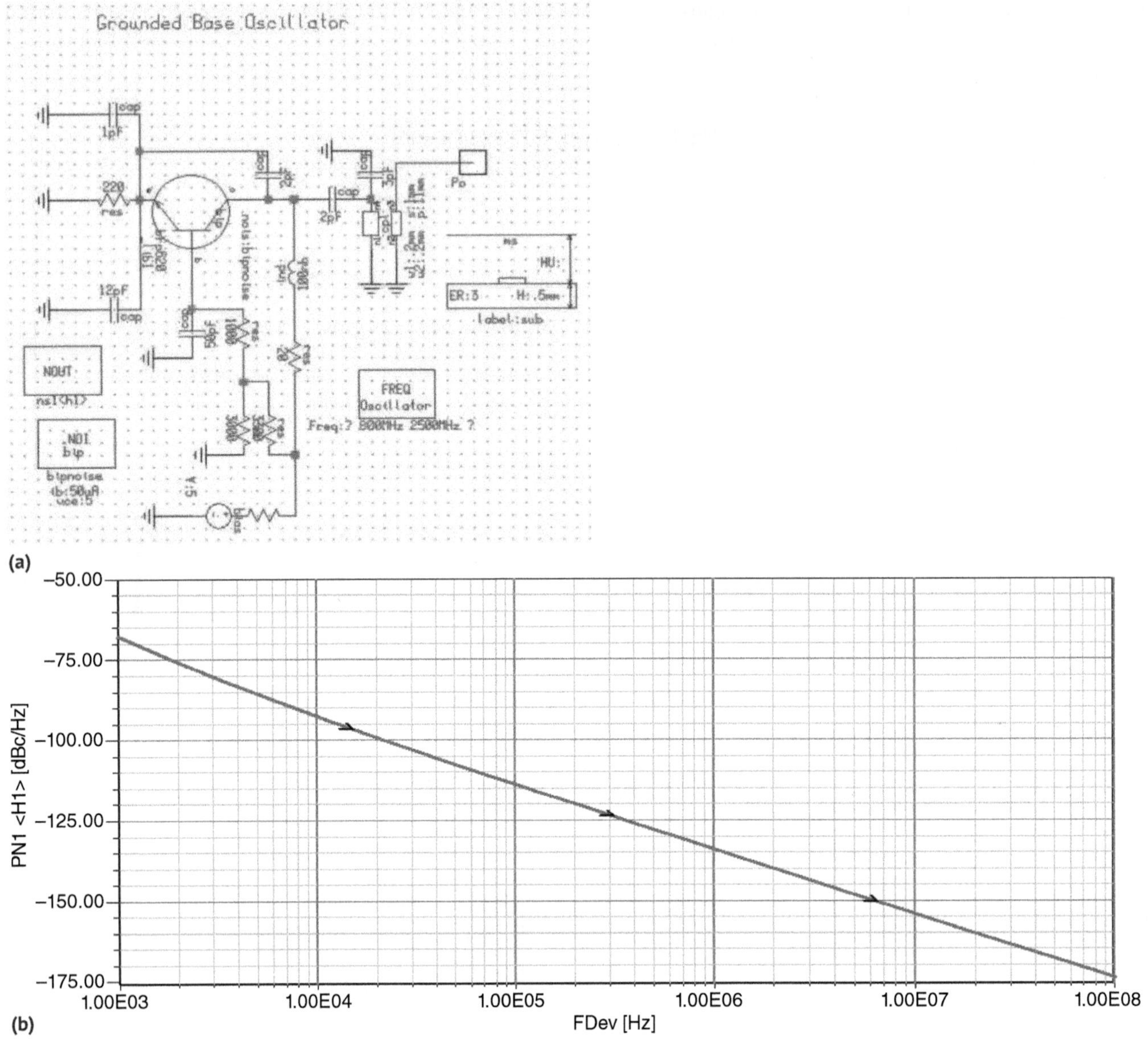

Figure 10.10 (a) Grounded base oscillator using coupled transmission lines as resonator, having 10 dBm output power, harmonic suppression > 30 dB. (b) Simulated phase noise of the circuit in (a).

One more similar example:

Figure 10.10a is a grounded base oscillator with coupled transmission lines used as resonators. Two transmission lines allow good coupling and is mechanically stable. The frequency of oscillation, about 2000 MHz, depends upon the length of the transmission line. Figure 10.10b shows the simulated phase noise of the circuit in Figure 10.10a.

10.3.2 Three-reactance Oscillators

Although the block diagram formulation of the stability criteria is the easiest to express mathematically, it is frequently not the easiest to apply since it is often difficult to identify the forward loop gain $G(j\omega)$ and the feedback ratio $H(j\omega)$ in electronic systems. A direct analysis of the circuit equations is frequently simpler than the block diagram interpretation (particularly for single-stage amplifiers). Figure 10.11 shows a generalized circuit for an electronic amplifier. The Colpitts oscillator is the configuration with Z1 and Z2 being a capacitor and Z3 an inductance, for the resonance condition [3].

By inspecting Figure 10.11, one can see the three necessary reactances and the load. All other known circuits are derivatives of this by rotation. The small-signal equivalent circuit is given in Figure 10.12 (where h_{re} has been neglected).

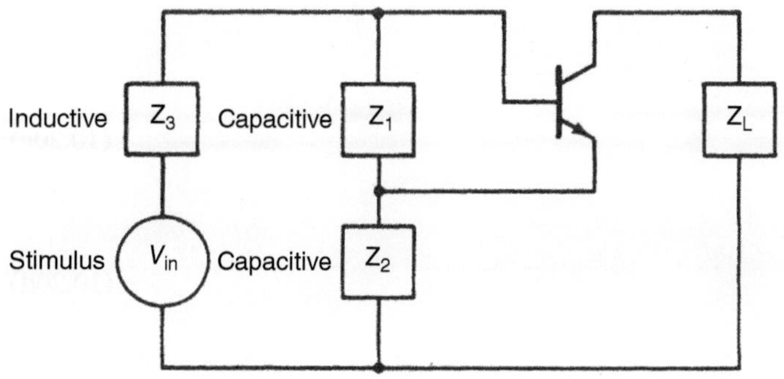

Figure 10.11 Generalized circuit for an oscillator using an amplifier model. Z_3 is inductive even with a capacitor in series with it. *Source*: Modified from [5].

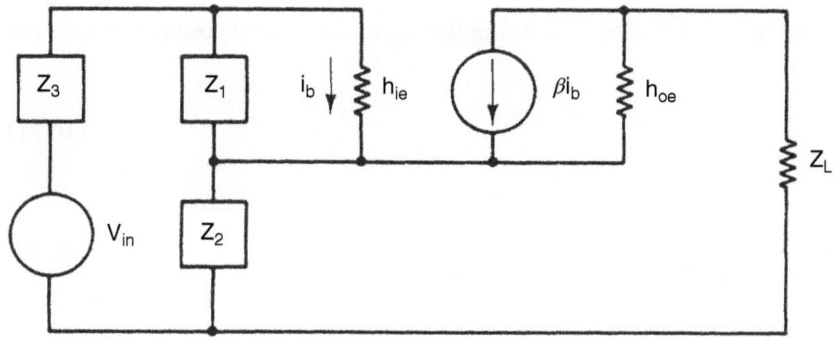

Figure 10.12 Small-signal equivalent circuit of Figure 10.11. *Source*: Modified from [5].

10.4 Linear Approach

For many years, until recently, oscillators were analyzed with a linear approach as will be shown as follows. Figure 10.13a,b illustrates the oscillator subcircuit for the purpose of calculating the negative resistance.

From Figure 10.13b, the circuit equation is given from Kirchhoff's voltage law as

$$V_{\text{in}} = I_{\text{in}}(X_{C_1} + X_{C_2}) - I_b(X_{C_1} - \beta X_{C_2}) \tag{10.20a}$$

$$0 = -I_{\text{in}}(X_{C_1}) + I_b(X_{C_1} + h_{\text{ie}}) \tag{10.20b}$$

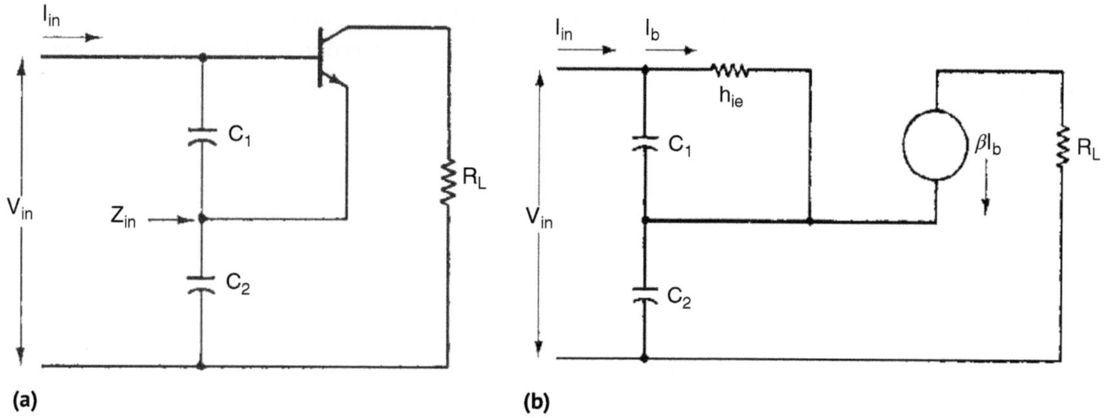

Figure 10.13 (a) Oscillator subcircuit for impedance analysis. (b) Equivalent sub-oscillator circuit for the calculation of the negative resistance. *Source*: [4]/IEEE.

Considering, $\frac{1}{Y_{11}} = h_{\text{ie}}$

$$Z_{\text{in}} = \frac{V_{\text{in}}}{I_{\text{in}}} = \frac{(1+\beta)X_{C_1}X_{C_2} + h_{\text{ie}}(X_{C_1} + X_{C_2})}{X_{C_1} + h_{\text{ie}}} \tag{10.20c}$$

$$Z_{\text{in}} = \frac{\left(-\frac{(1+\beta)}{\omega^2 C_1 C_2} + \frac{(C_1+C_2)}{j\omega C_1 C_2}\frac{1}{Y_{11}}\right)}{\left(\frac{1}{Y_{11}} + \frac{1}{j\omega C_1}\right)} \tag{10.20d}$$

The input impedance (Z_{in}) of this Colpitts oscillator circuit, including the parasitics, is given as [4, 6]:

$$Z_{\text{in}}|_{\text{package}} = -\left[\frac{Y_{21}}{\omega^2(C_1+C_p)C_2}\frac{1}{(1+\omega^2 Y_{21}^2 L_p^2)}\right] - j\left[\frac{(C_1+C_P+C_2)}{\omega(C_1+C_p)C_2} - \frac{\omega Y_{21} L_P}{(1+\omega^2 Y_{21}^2 L_P^2)}\frac{Y_{21}}{\omega(C_1+C_P+)C_2}\right] \tag{10.20e}$$

The resonator losses are expressed by the R_{s1} (see Figure 10.14). Now splitting the Z_{in} of the Colpitts oscillator into real and imaginary parts, including parasitics, we obtain:

$$R_{\text{NEQ}} = \frac{R_N}{(1+\omega^2 Y_{21}^2 L_p^2)} \tag{10.21}$$

$$\frac{1}{C_{\text{EQ}}} = \left\{\left[\frac{1}{\frac{(C_1+C_P)C_2}{(C_1+C_2+C_P)}}\right] - \left[\frac{\omega^2 Y_{21} L_P}{(1+\omega^2 Y_{21}^2 L_P^2)}\right]\left[\frac{Y_{21}}{\omega(C_1+C_P)C_2}\right]\right\} \tag{10.22}$$

$$R_N = -\frac{Y_{21}}{\omega^2 C_1 C_2} \tag{10.23}$$

where

R_N is the negative resistance without lead inductance and package capacitance.

Important: This is a noisy electronically generated negative resistance. It combines all internal noise sources to be modulated on the oscillator carrier frequency.

R_{NEQ} is the negative resistance with base-lead inductance and package capacitance.

C_{EQ} is the equivalent capacitance with base-lead inductance and package capacitance

The method shown above is called the one-port oscillator design [7]. Figure 10.14 shows the general schematic diagram of a one-port negative-resistance model. The negative real part of Z_{in} is used to compensate the losses of the parallel-tuned circuit.

As stated, R_N is noisy; it makes sense to explain noise. Following is the phase noise calculation and its noise contribution based on R_N in more exact details.

The stability or phase noise of an oscillator can be determined in the time or frequency domain. Phase noise is a short-term phenomenon and has various components. Figure 10.15a shows the typical illustration of the stability and phase noise in the time and frequency domain. The major noise contributors are thermal noise, Schottky noise, and the flicker

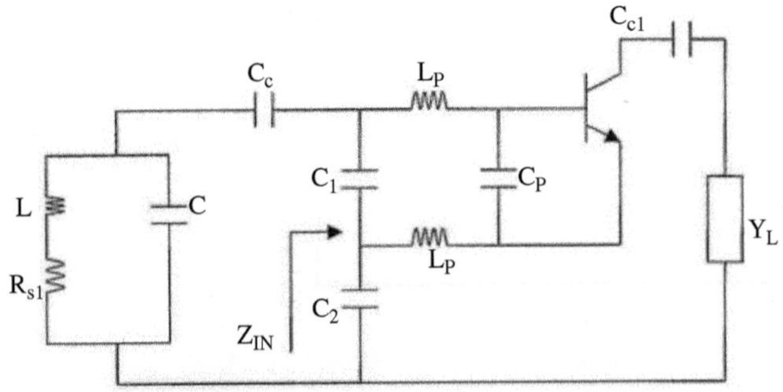

Figure 10.14 Colpitts oscillator with base-lead inductances and package capacitance. *Source*: [4]/IEEE.

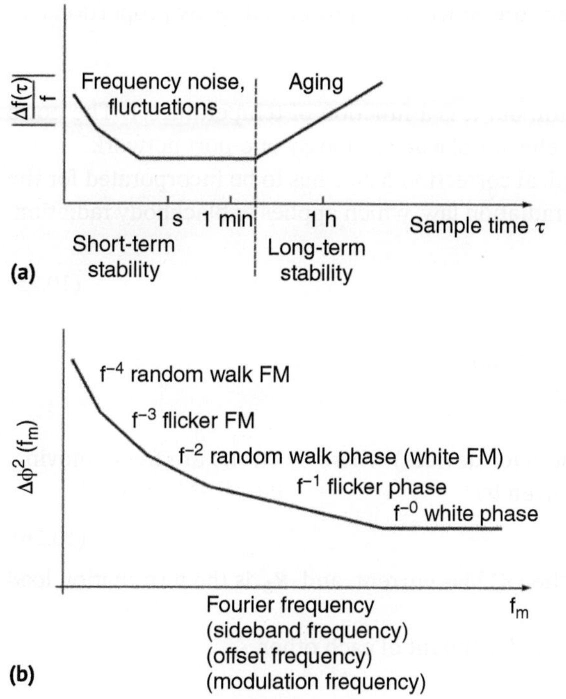

Figure 10.15 Typical characterization of the noise sideband in the time and frequency domain and its contributions: (a) time domain and (b) frequency domain. *Source*: Adapted from [9].

noise from the active device. Flicker noise depends on the transistor type and its biasing. The noise contribution from the resonator is mainly thermal noise. The minimum phase noise is at far offsets from the carrier, the best number being ($-P_{out}$ (dBm) + kT (−174 dBm) + NF [large-signal noise figure of the amplifier transistor of the oscillator in dB]) [6, 8].

Note that two different effects are considered, such as aging in Figure 10.15a and phase noise in Figure 10.15b.

10.5 Noise in Semiconductors and Circuits

Microwave applications generally use transistors, and following are their noise contributions [1–11].

Johnson Noise

The Johnson noise (thermal noise) is due to the movement of molecules in solid-state devices called "Brown's molecular movements."

- It is expressed as $v_n^2 = 4kT_0RB\,(\text{emf})\,(\text{V}^2/\text{Hz})$ (10.24a).
- The power of thermal noise can thus be written as:

$$\text{Noise power} = \frac{v_n^2}{4R} = kT_0B\ (\text{W/Hz}) \tag{10.24a}$$

It is most common to perform noise evaluations using noise power density, in W/Hz. By setting $B = 1$ Hz, we get:

$$\text{For } B = 1\text{ Hz},\ \text{noise power} = kT_0$$

$$\text{T} = 290\text{ K and } k\text{ (Boltzmann's constant)} = 1.38 \times 10^{-23}\ \text{J/K}$$

$$\text{By Thevenin, noise power} = 1.38 \times 10^{-23} \times 290 = 4\ \times 10^{-23}\ \text{W}$$

- Noise floor below the carrier for 0 dBm output is given by:

$$L(\omega) = 10\log\left(\frac{v_n^2/R}{1mW}\right) = -173.97\text{ dBm or about} - 174\text{ dBm} \tag{10.24b}$$

- In order to reduce this noise, the only option is to lower the temperature, since noise power is directly proportional to temperature.
- The Johnson noise sets the theoretical noise floor.
- The available noise power does not depend on the value of resistor, but it is a function of temperature T. The noise temperature can thus be used as a quantity to describe the noise behavior of a general lossy one-port network.
- For high frequencies and/or low temperature, a quantum mechanical correction factor has to be incorporated for the validation of equation. This correction term results from Planck's radiation law, which applies to blackbody radiation.

$$P_{\text{av}} = kT.B \cdot p(f,T), \text{ with } p(f,T) = \left[\frac{hf}{kT} / \left(e^{\left(\frac{hf}{kT}\right)} - 1\right)\right] \tag{10.25}$$

- where $h = 6.626.10^{-34}$ J^{-s}, Planck's constant (**Planck's radiation noise**)

Schottky/Shot Noise

- The Schottky noise occurs in conducting PN junctions (semiconductor devices) where electrons are freely moving. The root mean square (RMS) noise current in 1 Hz bandwidth is given by

  $$\overline{i_n^2} = 2 \times q \times I_{\text{DC}}\ P = i_n^2 \times R_L \tag{10.26}$$

  where q is the charge of the electron, P is the noise power, I_{DC} is the DC bias current, and R_L is the termination load (can be complex).
- Since the origin of this noise generated is totally different, they are independent of each other.

Flicker Noise

The electrical properties of surfaces or boundary layers are influenced energetically by states, which are subject to statistical fluctuations and therefore lead to the flicker noise or $1/f$ noise for the current flow.

- $1/f$ noise is observable at low frequencies and generally decreases with increasing frequency f according to the $1/f$ law until it will be covered by frequency-independent mechanism, like thermal noise or shot noise.

Example: The noise for a conducting diode is bias-dependent and is expressed in terms of AF and KF.

$$\left\langle i_{\text{Dn}}^2 \right\rangle_{\text{AC}} = 2qI_{\text{dc}}B + KF\frac{I_{\text{DC}}^{\text{AF}}}{f}B \tag{10.27}$$

- The AF is generally in range of 1–3 (dimensionless quantity) and is a bias-dependent curve fitting term, typically 2.
- The KF value is ranging from 1E^{-12} to 1E^{-6} and defines the flicker corner frequency.

Transit Time and Recombination Noise

When the transit time of the carriers crossing the potential barrier is comparable to the periodic signal, some carriers diffuse back and this causes noise. This is really seen in the collector area of NPN transistors.

- The electron and hole movements are responsible for this noise. The physics for this noise has not been fully established.

Avalanche Noise

- When a reverse bias is applied to semiconductor junction, the normally small depletion region expands rapidly.
- The free holes and electrons then collide with the atoms in depletion region, thus ionizing them, and produce spiked current called the avalanche current.
- The spectral density of avalanche noise is mostly flat. At higher frequencies, the junction capacitor with lead inductance acts as a low-pass filter.
- Zener diodes are used as voltage reference sources and the avalanche noise needs to be reduced by big bypass capacitors.

10.6 Prediction of Phase Noise

The definition of phase noise was first given by Baghdady et al. [12].

To gain a better understanding of the design criteria from the perspective of noise, the prevalent oscillator noise models are studied, starting with the popular Leeson's noise model, and their advantages and drawbacks are discussed.

The first phase noise plot was suggested by Leeson in 1966 [13]. The key quality factor considered here is the phase noise. Leeson developed an approximation equation in which Dieter Scherer added the flicker component (f_c/f_m). Jeremy Everard showed the best ratio between the unloaded and loaded Q (see definition of m in (1.1)) and Rohde added the VCO noise term—the last term in Equation (10.28) that uses the voltage gain of the oscillator for noise modulation.

In designing an oscillator, it is important to have a concept of how to do this and hopefully validate the same.

The basic equation (Rohde's modified Leeson's equation) needed to calculate the phase noise is found in *The Design of Modern Microwave Oscillators for Wireless Applications: Theory and Optimization* [3].

$$£(f_m) = 10\log\left\{\left[1 + \frac{f_0^2}{[2f_m Q_0 m(1-m)]^2}\right]\left(1 + \frac{f_c}{f_m}\right)\frac{FkT}{2P_0} + \frac{2kTRK_0^2}{f_m^2}\right\} \tag{10.28}$$

where $£(f_m)$, f_m, f_0, f_c, Q_L, Q_0, F, k, T, P_0, R, K_0, and m are the ratio of the sideband power in a 1 Hz bandwidth at f_m to total power in dB, offset frequency, flicker corner frequency, loaded Q, unloaded Q, noise factor, Boltzmann's constant, temperature in Kelvin, average output power, equivalent noise resistance of tuning diode, voltage gain, and ratio of the loaded and unloaded Q, respectively.

The following nonlinear approach, which is the basis of the mathematics used here, is based in part on Ulrich L. Rohde, Chao-Ren Chang and Jason Gerber, "Design and Optimization of Low-Noise Oscillators Using Nonlinear CAD Tools," *IEEE Frequency Control Symp. Proc.*, 1994, pp 548-554.

Many applications are for low-frequency crystal oscillators, so before we jump in to the 100 MHz crystal oscillator topic, let us look at a lower-frequency crystal oscillator first. Modern test equipment uses a 5 or 10 MHz crystal for long-term stability, small drift but not as a reference for phase noise. The phase noise improves if the signal-to-noise ratio is increased by running more power through the crystal with the drawback of higher aging of the crystal resonator. Therefore, the 100 MHz crystal oscillator is synchronized with a lower frequency like 10 MHz, at a fairly narrow bandwidth to correct the drift in aging while the 100 MHz oscillator is running at a much higher power to improve the phase noise performance. The following starts with an example (Figure 10.16a–c) for a 10 MHz crystal oscillator, which can be easily modified to another close frequency between 5 and 20 MHz, even above 100 MHz, using the same approach. Table 10.1 shows the comparison of the measured and simulated data of modern day and the older crystal oscillators [14–19].

Hewlett Packard (HP) became famous at the time, 1980, with the best crystal oscillator model HP10811A. The concept of using the crystal also as a filter came from U. L. Rohde, "Crystal oscillator provides low noise," Electronic Design, 1975, which has been widely used, and the description of these crystal oscillators was again done by Anisha Apte in her PhD dissertation [8].

Table 10.1 Measured and simulated data of modern day and the older crystal oscillators.

Phase noise of various designs	**Axtal 10 MHz XO measured data**	**HP10811A 10 MHz XO (1980) measured data**
Offset frequency	**dBc/Hz**	**dBc/Hz**
@ 1 Hz	−115	−111
@ 10 Hz	−146	−135
@ 100 Hz	−157	−154
@ ≥ 1 kHz	−160	−166

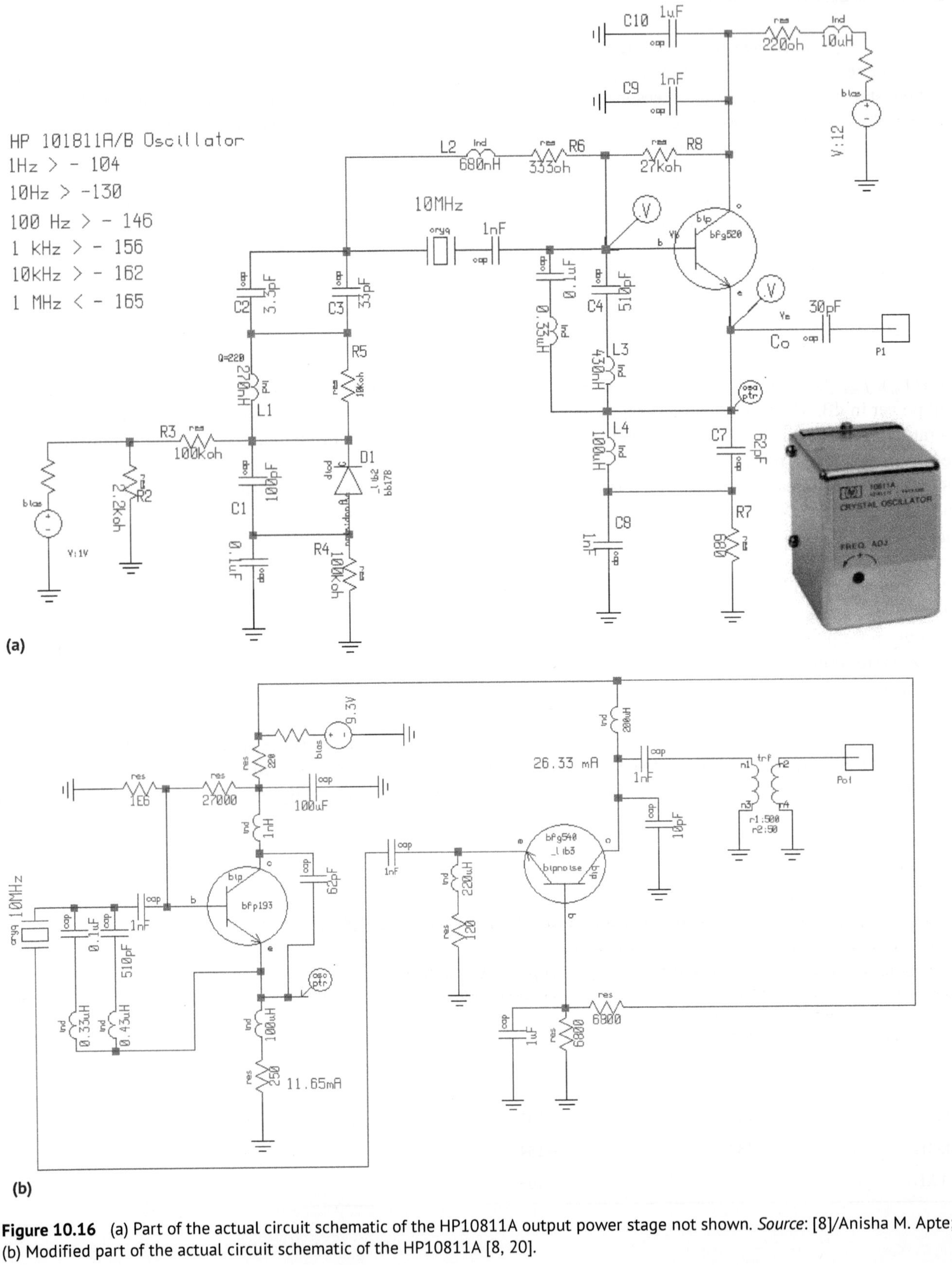

Figure 10.16 (a) Part of the actual circuit schematic of the HP10811A output power stage not shown. *Source*: [8]/Anisha M. Apte; [20]. (b) Modified part of the actual circuit schematic of the HP10811A [8, 20].

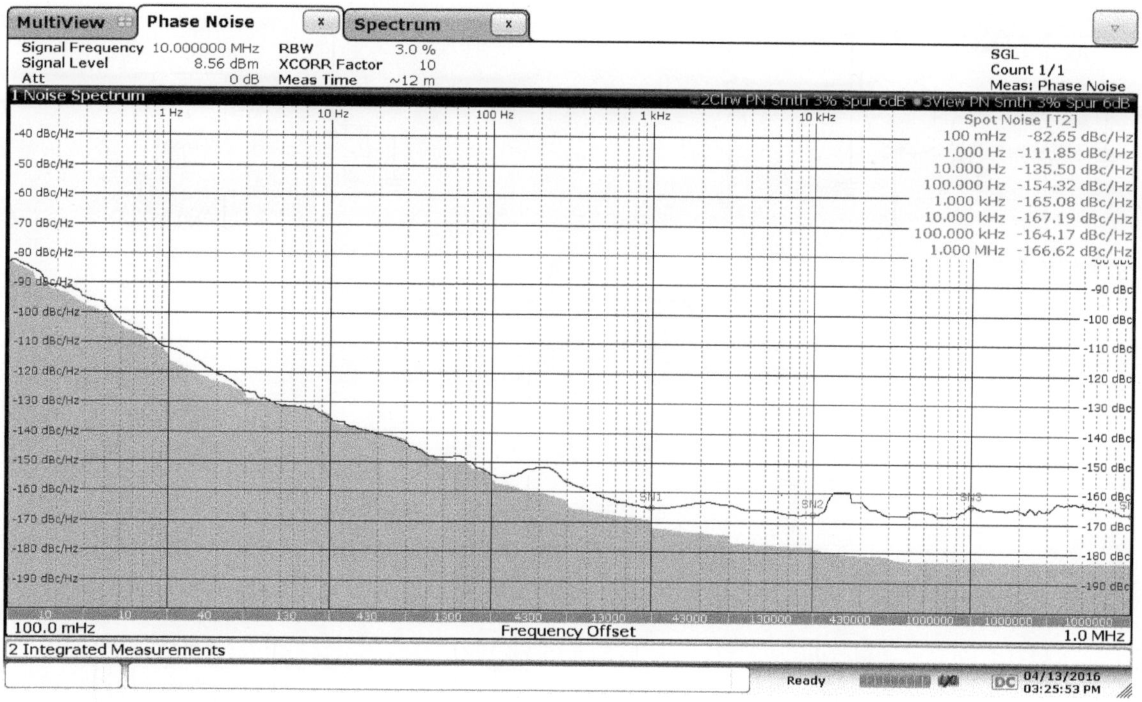

Date: 13.APR.2016 15:25:52

(c)

Figure 10.16 (Cont'd) (c) Phase noise of the actual circuit schematic of the HP10811A.

Example: 15 MHz crystal oscillator—Colpitts type The following 15 MHz crystal oscillator, shown in Figure 10.17a, is a Colpitts type with the crystal operating in series mode, 30 pF in series, tuned to 15 MHz. Rather than grounding the quartz crystal, which has a loss resistance in the vicinity of 70 Ω, it is connected to the emitter of the second stage circuit, a grounded base amplifier. The emitter input impedance is less than 2 Ω, so the resonator Q is not affected. The grounded base circuit has minimal feedback compared to an emitter circuit. To provide a termination into the circuit, the collector is loaded with 5000 Ω and the transformer using a tuned circuit, and a capacitive voltage divider transforms this impedance down to 50 Ω. It is a CAD-simulator-based optimization and uses an SC cut crystal. The resulting phase noise is shown in Figure 10.17b.

The bent shape of the noise curve in Figure 10.17b shows the very close-in flicker corner frequency, the middle part is the filtering effect of the crystal, and the rest is the residual noise of the amplifier.

The next plot, Figure 10.17c, shows the spectral power performance of the oscillator, with the fundamental at about 10.5 dBm and harmonic suppression of about 38 dB.

Now comes an interesting calculation for the large signal oscillator noise figure:

Dynamic range is $175 + 10 = 185$ (the signal-to-noise ratio), since kT is -177, plus 10 is 187, so the oscillator noise figure is 2 dB, at the conditions given.

For the following circuit, we now show the verification of 100 MHz crystal oscillator using the CAD simulation tool (Ansoft Designer), our analytical model, and then the measured values.

Example: Figure 10.18 is the typical simplified Colpitts 100 MHz crystal oscillator circuit, grounded base buffer circuit, showing noise figure plots, phase noise plots, and output power.

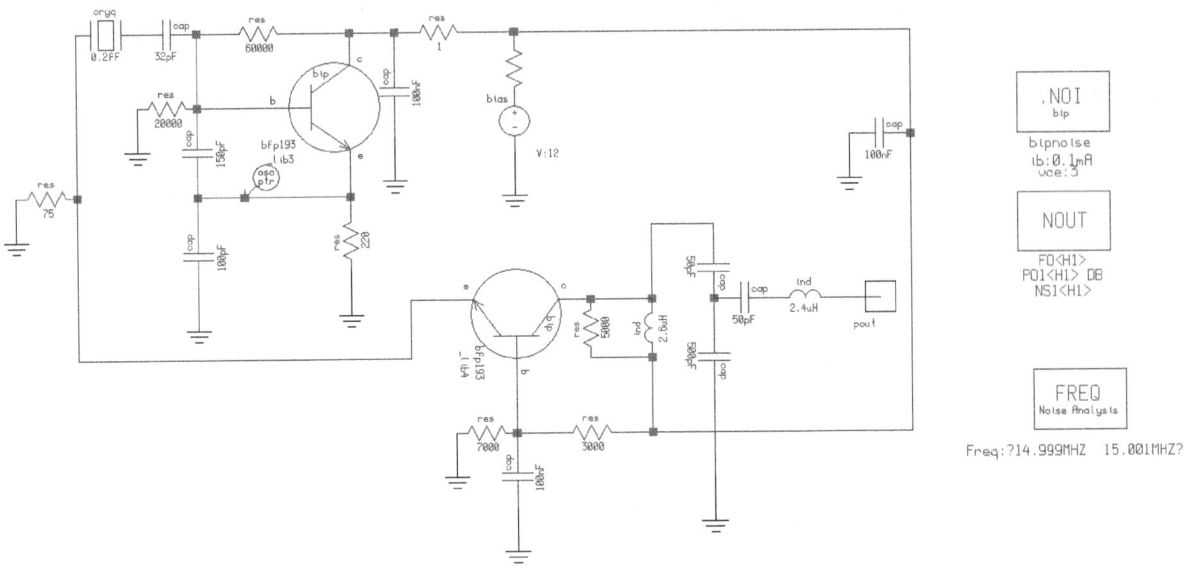

(a)

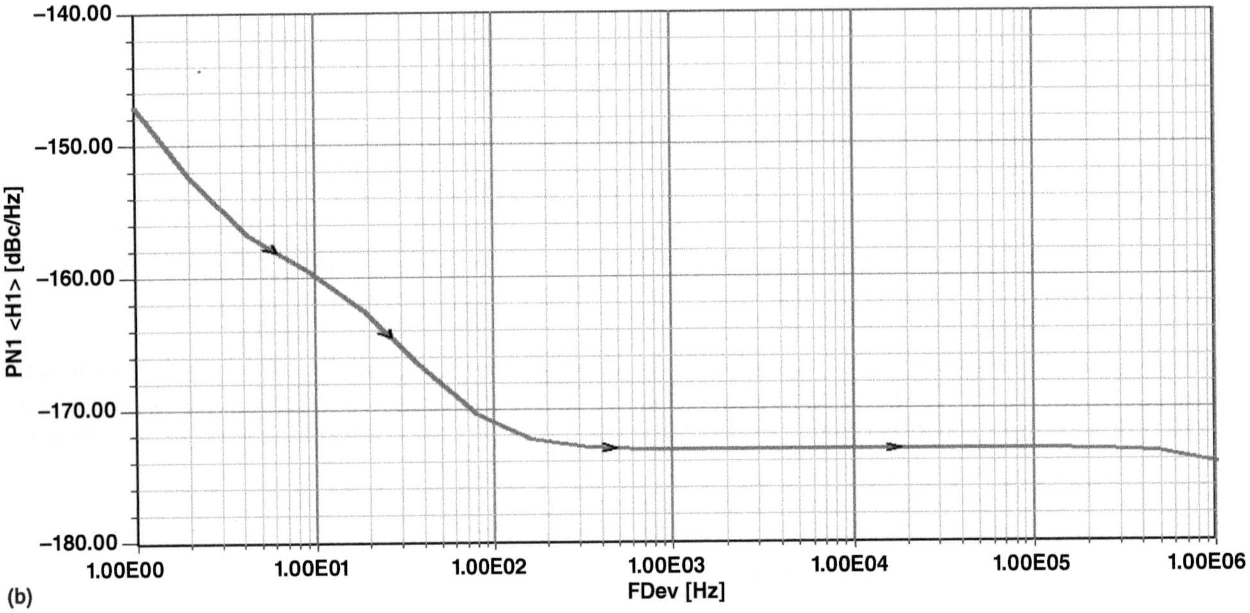

(b)

Figure 10.17 (a) 15 MHz low-noise oscillator. (b) Phase noise of 15 MHz low-noise oscillator of (a).

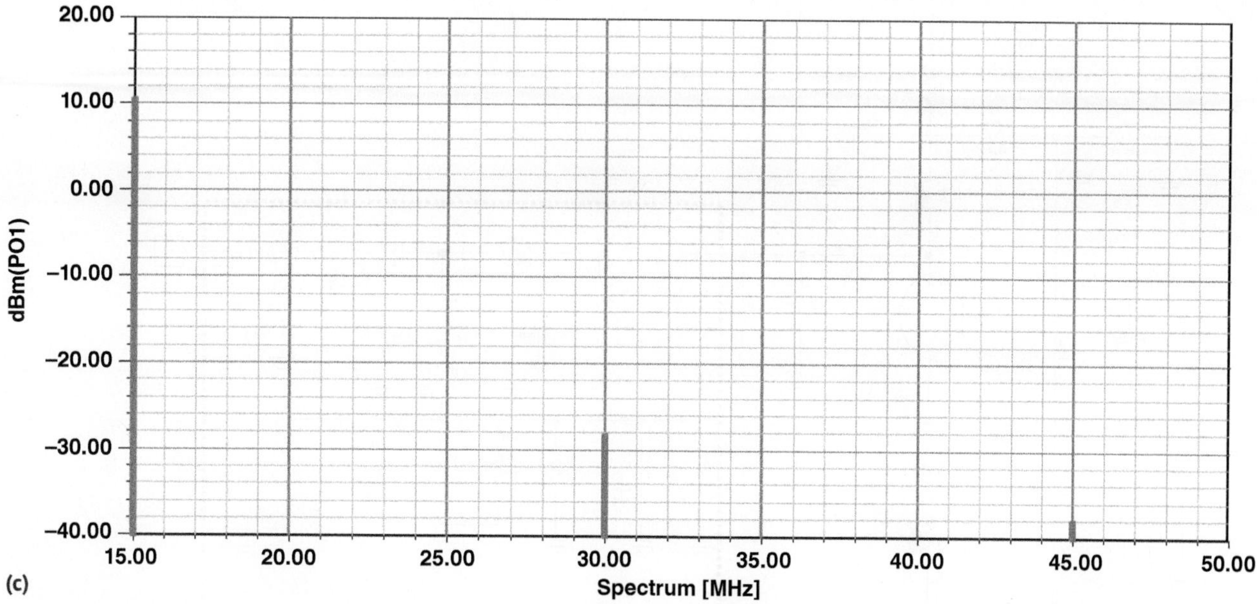

(c)

Figure 10.17 (Cont'd) (c) Power output of 15 MHz low-noise oscillator of (a).

$C1 := 15.6\ \text{pF}$ $C2 := 12.5\ \text{pF}$ $kT := 4.143 \cdot 10^{-21} \cdot J$ $R := 0.2\ \Omega$

$qcharge := 1.602 \cdot 10^{-19}\ \text{coul}$ $Ic := 8.2\ \text{mA}$ $Ib = 220\ \mu A$

$y11 := (0.000884 - 0.0000158j) \cdot \text{mho}$

$y21 := (0.0105 - 0.00084j) \cdot \text{mho}$

$L := 6.4949 \cdot 10^{-3} \cdot \text{henry}$ $af := 2$ $kf := 1 \cdot 10^{-10}$ $y := \dfrac{C1}{C2}$

$Vcc := 10V$ $i := 0.7$ $p := 1.45$ $q := 1.05$

$Q := 25000$ $Q0 := 180000$ $nfdB := 4.747$ $PoutdB := 13$

$fc := 100\ \text{MHz}$ $wc := 2\pi \cdot fc$ $gm := |y21| \cdot y^q$

$fo_i := 10^i \cdot \text{Hz}$ $wo_i := 2\pi \cdot fo_i$ $B1i := (wo_i)^2 \cdot L^2 \cdot Vcc^2$

$kconstant := \dfrac{KT \cdot R}{wc^2 \cdot C2^2}$ $k0_i := \dfrac{kconstant}{B1_i}$ $b := \dfrac{|y21| \cdot y^p}{|y11|}$

$$k1constant_i := qcharge \cdot Ic \cdot gm^2 + \frac{kf \cdot Ib^{af} \cdot gm^2}{wo_i}$$

$k1_i := \dfrac{k1constant_i}{wc^2 \cdot B1_i}$ $k3_i := wc^2 \cdot gm^2$ $k2_i := wc^4 \cdot b^2$ $k_i := \dfrac{k3_i}{k2_i \cdot C2^2}$

$t2_i := k0_i \dfrac{(1+y)^2}{y^2}$ $t1_i := \left[\left(\dfrac{b^2}{gm^3}\right)^2 \cdot \dfrac{k_i^{\,3} \cdot k1_i \cdot (wc)^2}{y^2 + k_i}\right] \cdot \dfrac{(1+y)^2}{y^2}$ $1_i := t1_i + t2_i$

$$m_i := 10 \cdot \log\left[1_i \cdot \left(kg^{-2} \cdot m^{-4} \cdot s^5 \cdot A^2\right) \cdot \frac{Q0^2}{Q^2}\right]$$

$$L_i := \text{if}(m_i < (-177 + Pout(\text{dBm}) - nfdB)),\ (-(177 + PoutdB - nfdB)),\ m_i$$

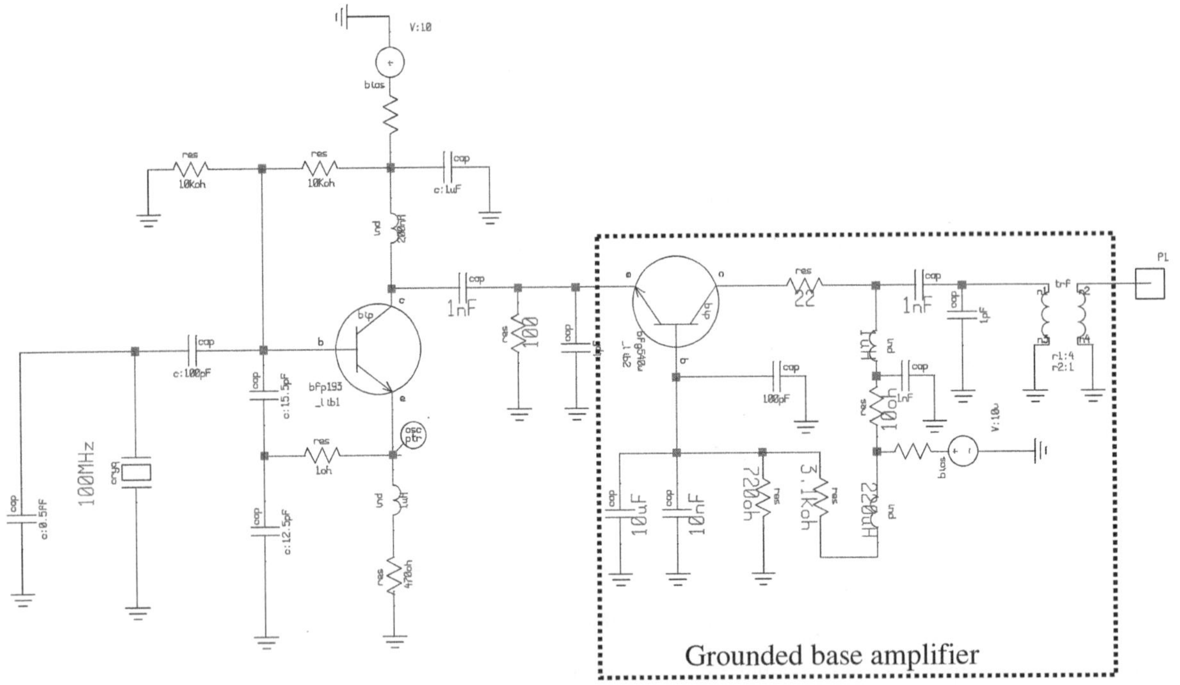

Figure 10.18 100 MHz crystal oscillators with the grounded base low-noise amplifier [8, 10].

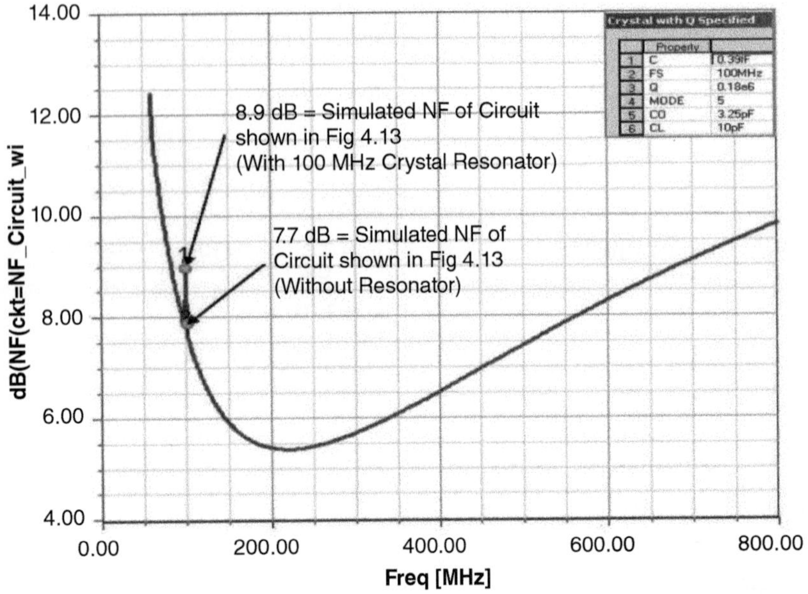

Figure 10.19 Simulated plot showing NF of 100 MHz oscillator shown in Figure 10.18. *Source*: [21]/Brandenburg University of Technology.

The basic equation needed to predict the phase noise using CAD simulation for the circuit shown in Figure 10.18 is found in [8]. The measured phase noise for a 100 MHz fifth overtone OCXO optimized using PI + MC + NF (phase-injection, mode-coupled, noise filtering) is shown in Figure 10.23. Figures 10.19–10.22b all show the various simulation results and theoretical calculations.

$$£(f_m) = 10\log\left\{\left[1 + \frac{f_0^{\,2}}{\left[2 f_m Q_0 m(1-m)\right]^2}\right]\left(1 + \frac{f_c}{f_m}\right)\frac{FkT}{2P_0} + \frac{2kTRK_0^2}{f_m^2}\right\} \tag{10.29}$$

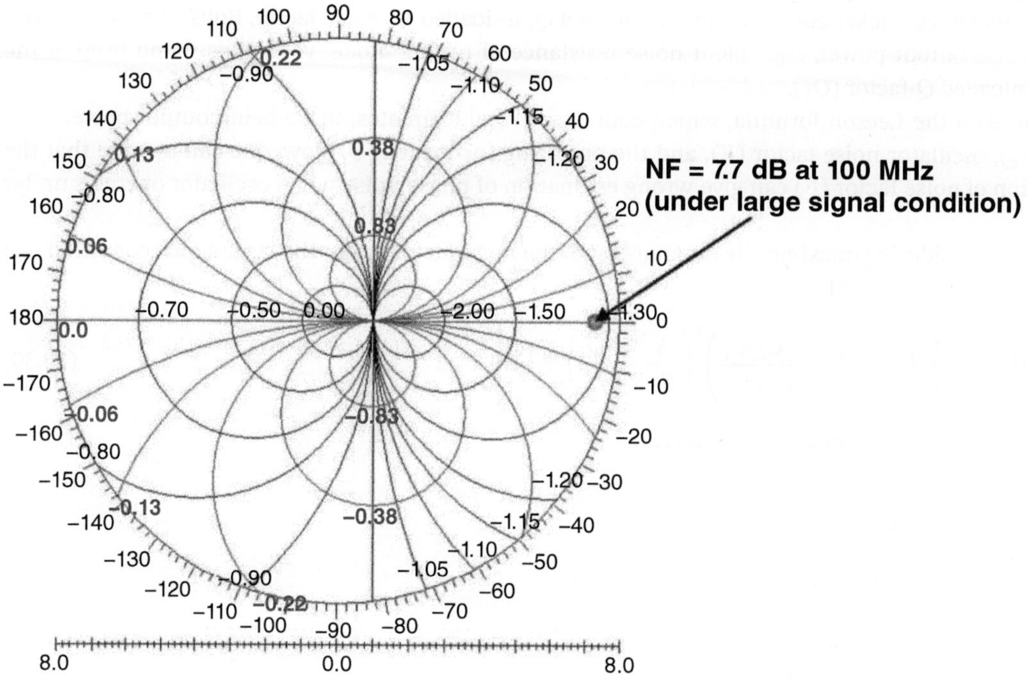

Figure 10.20 CAD simulated polar plot of noise of the oscillator circuit shown in Figure 10.18. *Source*: [21]/Brandenburg University of Technology.

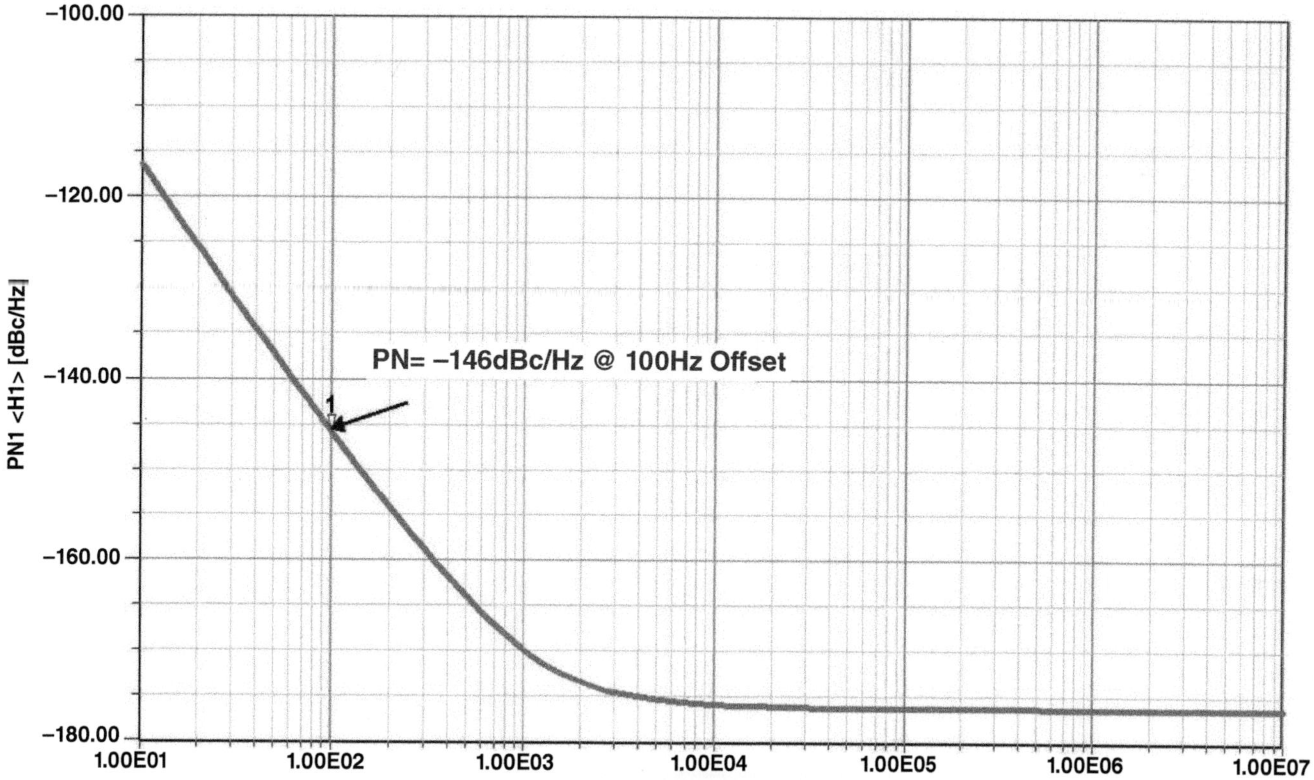

Figure 10.21 CAD simulated phase noise plot of 100 MHz crystal oscillator with buffer stage. *Source*: [21]/Brandenburg University of Technology.

where $£(f_m)$, f_m, f_0, f_c, Q_L, Q_0, F, k, T, P_o, R, K_0 and m are the ratio of the sideband power in a 1 Hz bandwidth at f_m to total power in dB, offset frequency, flicker corner frequency, loaded Q, unloaded Q, noise factor, Boltzmann's constant, temperature in Kelvin, average output power, equivalent noise resistance of tuning diode, voltage gain and ratio of the loaded Q-factor (Q_o) and unloaded Q-factor (Q_L).

In the past this was done with the Leeson formula, which contains several estimates, those being output power (P_o), flicker corner frequency (f_c), oscillator noise factor (F), and the operating (or loaded) Q. Now, one can assume that the small signal linear estimation of noise factor (F) can give wrong estimation of phase noise when oscillator operates under large signal condition.

The approximate formulae (considering quasi-nonlinear analysis) of the noise factor under the large signal condition can be given by the following equation [3, p. 135].

$$F \cong 1 + \frac{Y_{21}^{+} C_2 C_c}{(C_1 + C_2)C_1}\left[r_b + \frac{1}{2r_e}\left(r_b + \frac{(C_1 + C_2)C_1}{Y_{21}^{+} C_2 C_c}\right)^2 \left(\frac{1}{\beta^{+}} + \frac{f^2}{f_T^2}\right) + \frac{r_e}{2}\right] \tag{10.30}$$

where Y_{21}^{+} is the large signal [Y] parameter of the active device.

m_i	fo_i	L_i
−88.581	$1 \cdot s^{-1}$	−88.581
−118.515	$10 \cdot s^{-1}$	−118.515
−147.901	$100 \cdot s^{-1}$	−147.901
−174.251	$1 \cdot 10^3 \cdot s^{-1}$	−174.251
−195.999	$1 \cdot 10^4 \cdot s^{-1}$	−185.253
−216.22	$1 \cdot 10^5 \cdot s^{-1}$	−185.253
−236.243	$1 \cdot 10^6 \cdot s^{-1}$	−185.253
−256.245	$1 \cdot 10^7 \cdot s^{-1}$	−185.253

(a)

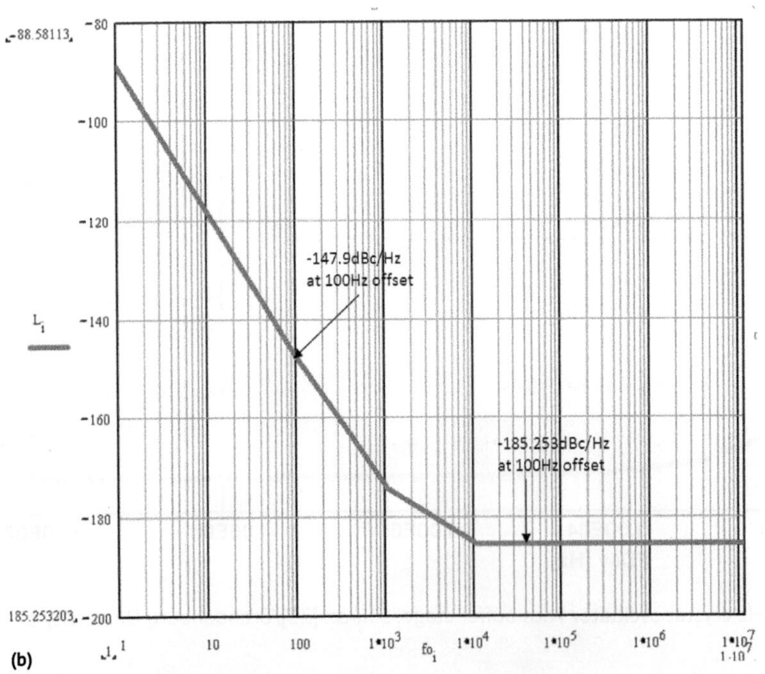

(b)

Figure 10.22 (a) The theoretical phase noise model. (b) Theoretical Calculated phase noise plot of the 100MHz crystal oscillator with buffer stage. *Source:* [8]/Anisha M. Apte.

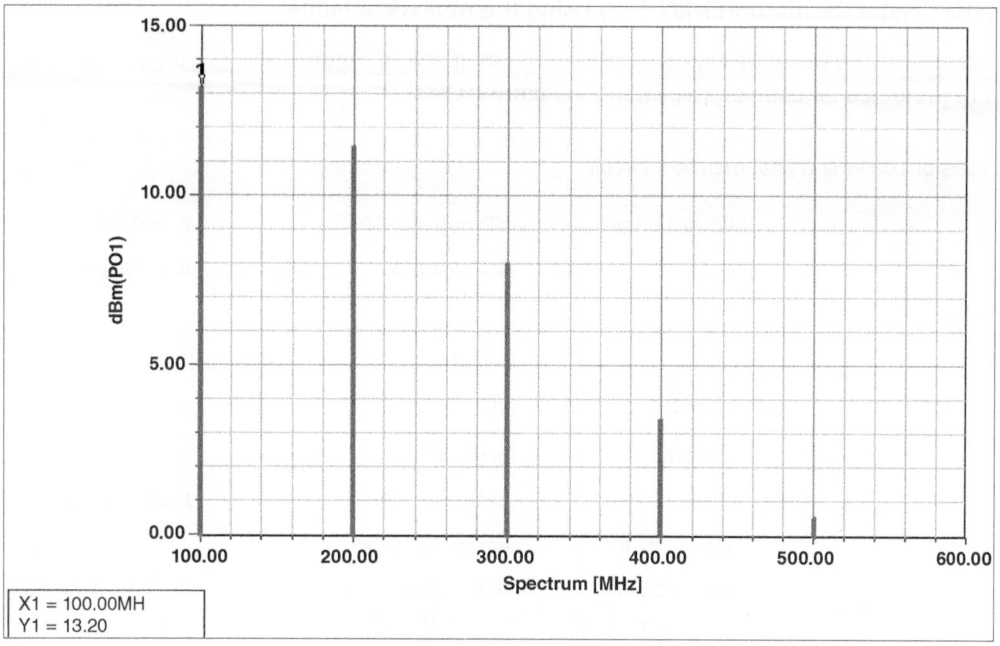

(c)

Figure 10.22 (Cont'd) (c) CAD Simulated power output plot of the 100MHz crystal oscillator with buffer stage.

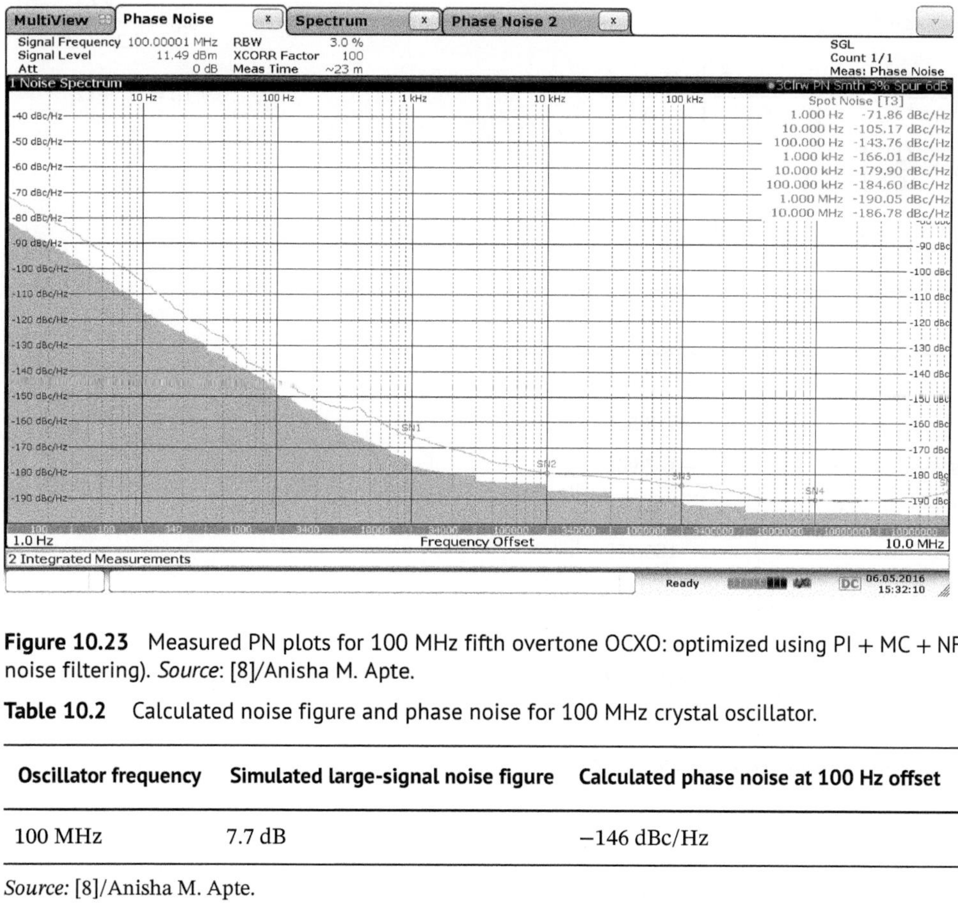

Figure 10.23 Measured PN plots for 100 MHz fifth overtone OCXO: optimized using PI + MC + NF (phase-injection, mode-coupled, noise filtering). *Source*: [8]/Anisha M. Apte.

Table 10.2 Calculated noise figure and phase noise for 100 MHz crystal oscillator.

Oscillator frequency	Simulated large-signal noise figure	Calculated phase noise at 100 Hz offset
100 MHz	7.7 dB	−146 dBc/Hz

Source: [8]/Anisha M. Apte.

Table 10.2 shows the calculated noise figure and phase noise for 100 MHz crystal oscillator using (10.30) for unloaded $Q_o = 180000$, and time average loaded Q ($Q < Q_o/4$) under large signal drive level condition is 25000.

10.6.1 Verification of 100 MHz Crystal Oscillator (LNXO 100) Using the Analytical Model

The theoretical calculated parameters of 100 MHz crystal oscillator circuit are given in Figure 10.23 [8]; after defining all the values, the phase noise can be predicted for comparative analysis (Figure 10.24).

Theoretical calculated parameters of 100 MHz crystal oscillator circuit

$C1 := 15.6\ \mathrm{pF}$ $\quad C2 := 12.5\ \mathrm{pF}$ $\quad kT := 4.143 \cdot 10^{-21}\ \mathrm{J}$ $\quad R := 0.2\ \Omega$

$qcharge := 1.602 \cdot 10^{-19} \cdot coul$ $\quad Ic := 8.2\ \mathrm{mA}$ $\quad Ib = 220\ \mu\mathrm{A}$

$y11 := (0.000884 - 0.0000158j) \cdot mho$

$y21 := (0.0105 - 0.00084j) \cdot mho$

$L := 6.4949 \cdot 10^{-3} \cdot henry$ $\quad af := 2$ $\quad kf := 1 \cdot 10^{-10}$ $\quad y := \dfrac{C1}{C2}$

$Vcc := 10V$ $\quad i := 0.7$ $\quad p := 1.45$ $\quad q := 1.05$

$Q := 25000$ $\quad Q0 := 180000$ $\quad nfdB := 4.747$ $\quad PoutdB := 13$

$fc := 100\ \mathrm{MHz}$ $\quad wc := 2\pi \cdot fc$ $\quad gm := |y21| \cdot y^q$

$fo_i := 10^i \cdot Hz$ $\quad wo_i := 2\pi \cdot fo_i$ $\quad B1i := (wo_i)^2 \cdot L^2 \cdot Vcc^2$

$kconstant := \dfrac{KT \cdot R}{wc^2 \cdot C2^2}$ $\quad k0_i := \dfrac{kconstant}{B1_i}$ $\quad b := \dfrac{|y21| \cdot y^p}{|y11|}$

$$k1constant_i := qcharge \cdot Ic \cdot gm^2 + \frac{kf \cdot Ib^{af} \cdot gm^2}{wo_i}$$

$k1_i := \dfrac{k1constant_i}{wc^2 \cdot B1_i}$ $\quad k3_i := wc^2 \cdot gm^2$ $\quad k2_i := wc^4 \cdot b^2$ $\quad k_i := \dfrac{k3_i}{k2_i \cdot C2^2}$

$t2_i := k0_i \dfrac{(1+y)^2}{y^2}$ $\quad t1_i := \left[\left(\dfrac{b^2}{gm^3}\right)^2 \cdot \dfrac{k_i^{\ 3} \cdot k1_i \cdot (wc)^2}{y^2 + k_i}\right] \cdot \dfrac{(1+y)^2}{y^2}$ $\quad 1_i := t1_i + t2_i$

$$m_i := 10 \cdot \log\left[1_i \cdot \left(kg^{-2} \cdot m^{-4} \cdot s^5 \cdot A^2\right) \cdot \frac{Q0^2}{Q^2}\right]$$

$$L_i := \mathrm{if}(m_i < (-177 + Pout(dBm) - nfdB)),\ (-(177 + PoutdB - nfdB)),\ m_i$$

m_i	fo_i	L_i
−88.581	$1 \cdot s^{-1}$	−88.581
−118.515	$10 \cdot s^{-1}$	−118.515
−147.901	$100 \cdot s^{-1}$	−147.901
−174.251	$1 \cdot 10^3 \cdot s^{-1}$	−174.251
−195.999	$1 \cdot 10^4 \cdot s^{-1}$	−185.253
−216.22	$1 \cdot 10^5 \cdot s^{-1}$	−185.253
−236.243	$1 \cdot 10^6 \cdot s^{-1}$	−185.253
−256.245	$1 \cdot 10^7 \cdot s^{-1}$	−185.253

Figure 10.24 The theoretical phase noise model expressed in Refs. [8, 22].

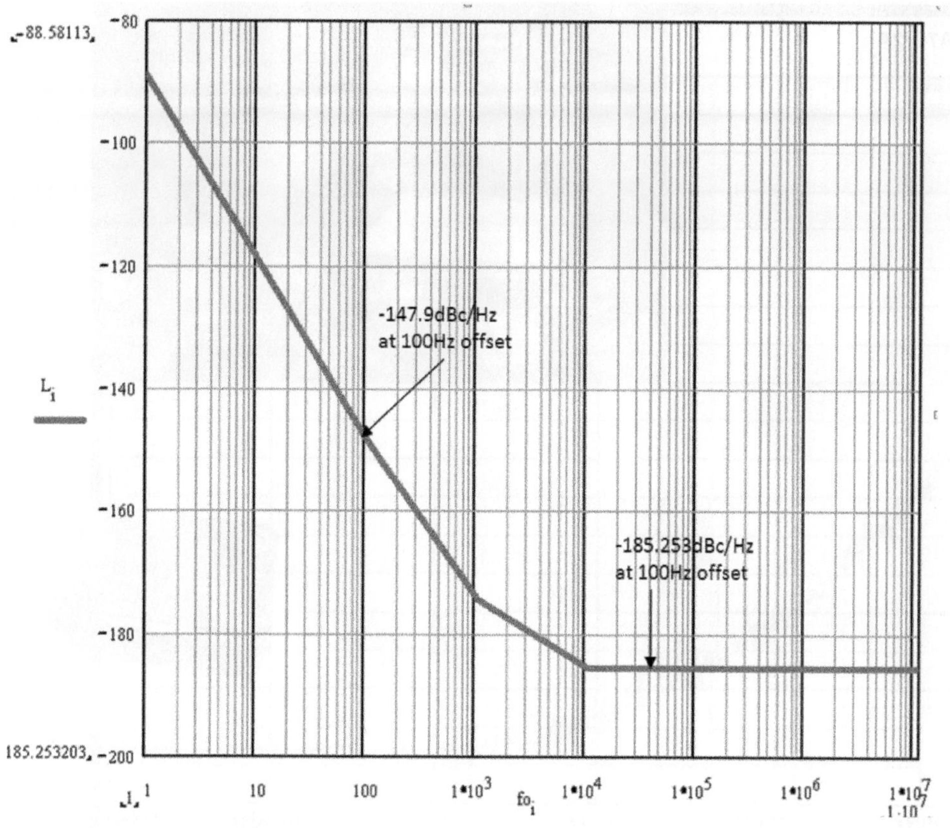

Figure 10.25 Theoretically calculated phase noise plot for 100 MHz crystal oscillator (LNXO 100). *Source*: [8]/Anisha M. Apte.

While the noise at 100 Hz and more offset is very good, the flicker noise values we had to guess and the formulae can be adjusted to get a better match at 1 and 10 Hz. The calculated and measured phase noise of the 100 MHz crystal oscillator is shown in Figure 10.25 and Figure 10.26, respectively.

Example: 432 MHz Capacitive Feedback: Figure 10.27a is an example of a grounded base UHF oscillator for 432 MHz with capacitive feedback:

The oscillator has 18 dBm output power, with a harmonic surppression greater than 30 dB (see Figure 10.27b). Please note that the flicker noise contribution is part of the simulation.

The negative noise resistance that is assumed to modulate/upconvert the semiconductor noise has the following value [3]:

$$R_n = \left[\frac{g_m^2}{\omega^2 C_1^2 \left(\frac{C_2^2}{C_1^2} g_m^2 + \omega^2 \beta^2 C_2^2 \right)} \right] \frac{qI_C}{kT} \tag{10.31}$$

Based on

$$Z_{\text{in}} = \frac{\left(-\frac{(1+\beta)}{\omega^2 C_1 C_2} + \frac{(C_1+C_2)}{j\omega\, C_1 C_2} \frac{1}{Y_{11}} \right)}{\left(\frac{1}{Y_{11}} + \frac{1}{j\omega\, C_1} \right)} \tag{10.32}$$

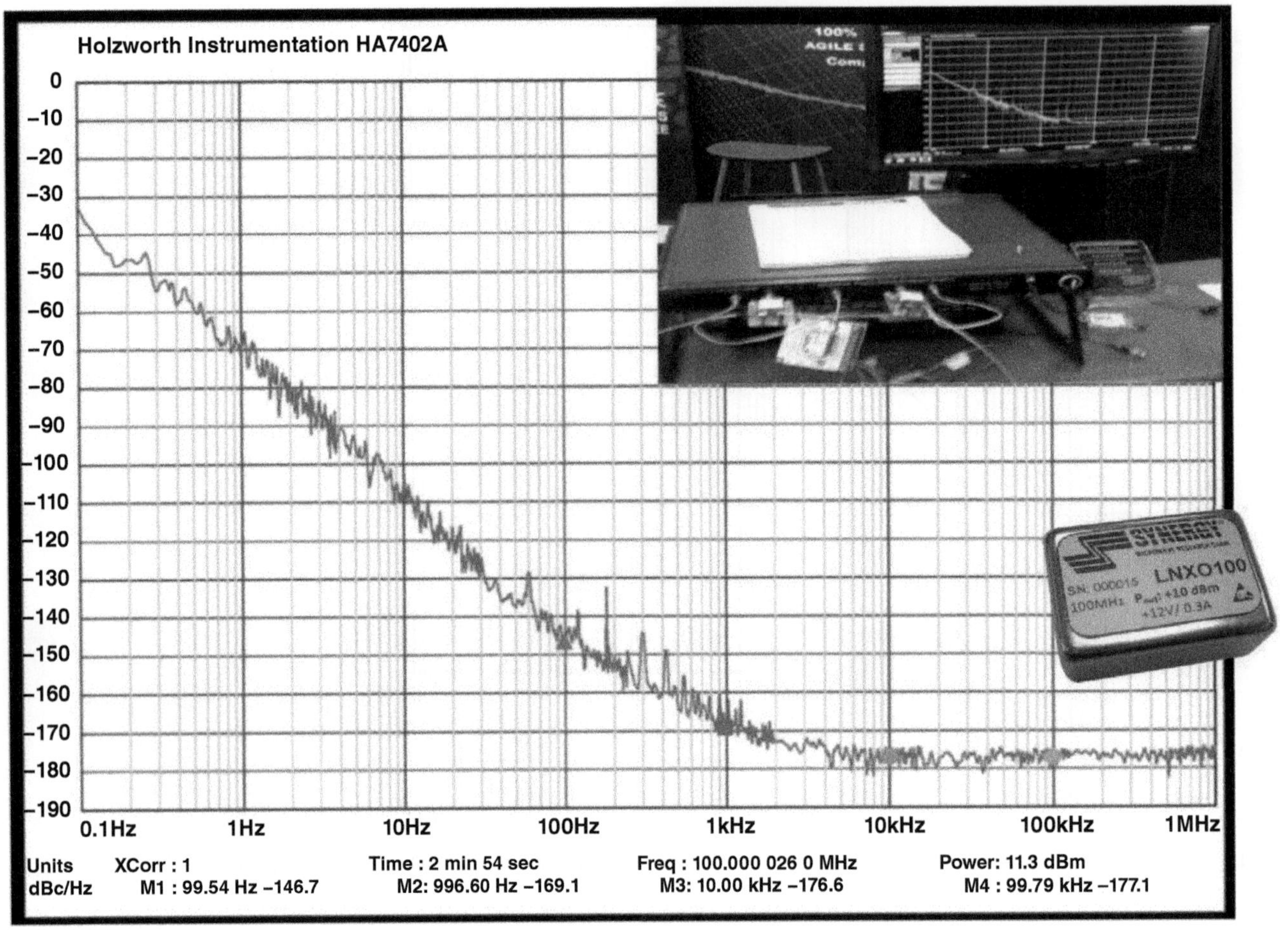

Figure 10.26 Measured phase noise of the 100 MHz crystal oscillator (Synergy, LNXO100) using the Holzworth HA7402A phase noise analyzer. *Source*: [8]/Anisha M. Apte; [22].

And after the introduction of the flicker noise component, the final phase noise in dBc/Hz is

$$£(\omega) = 10 \times \log \left[\left[k_0 + \left(\frac{k^3 k_1 \left[\frac{Y_{21}^+}{Y_{11}^+}\right]^2 [y]^{2p}}{\left[Y_{21}^+\right]^3 [y]^{3q}}\right)\left(\frac{1}{(y^2+k)}\right)\right]\left[\frac{[1+y]^2}{y^2}\right]\right] \tag{10.33}$$

where

$$k_0 = \frac{kTR}{\omega^2 \omega_0^2 L^2 V_{cc}^2} \tag{10.34a}$$

$$k_1 = \frac{qI_c g_m^2 + \frac{K_f I_b^{AF}}{\omega} g_m^2}{\omega^2\, \omega_0^4 L^2 V_{cc}^2}\ k_2 = \omega_0^2 (\beta^+)^2,\ \text{and} \tag{10.34b}$$

$$\beta^+ = \left[\frac{Y_{21}^+}{Y_{11}^+}\right]\left[\frac{C_1}{C_2}\right]^p \tag{10.34c}$$

$$g_m = \left[Y_{21}^+\right]\left[\frac{C_1}{C_2}\right]^q,\ \text{newly defined.} \tag{10.34d}$$

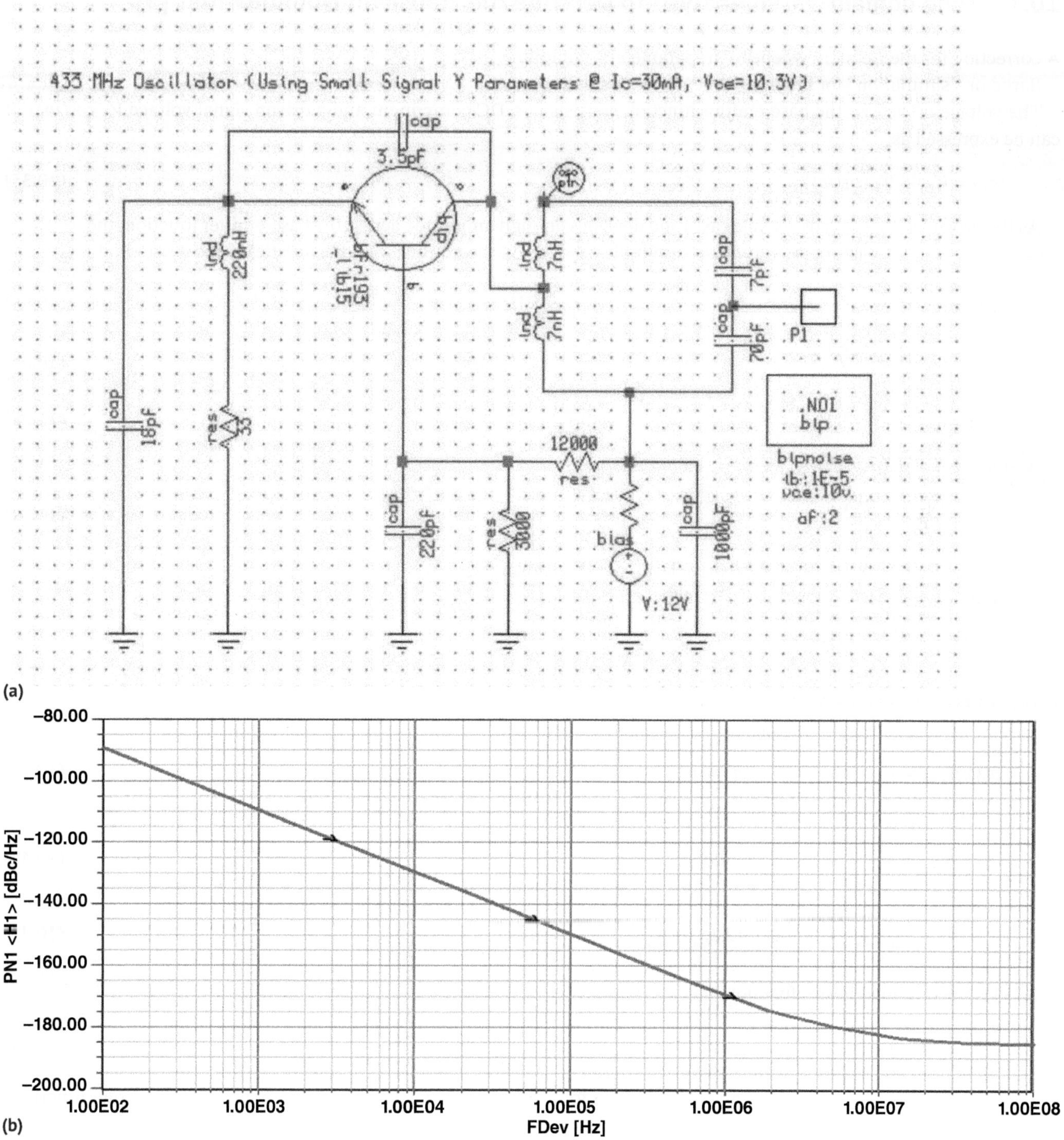

Figure 10.27 (a) Grounded base oscillator example at 433 MHz [23]. (b) Simulated phase noise plot of the grounded base oscillator example at 433 MHz in (a). *Source*: Modified from [23].

10.7 Time domain-based Analysis to Describe the Transistor Nonlinearities [3]

A correction for the frequency-dependent parameters will follow.

Based on "simulation" for larger drive level, we can assume:

The voltage $v(t)$across the base-emitter junction consists of a DC component and a driven signal voltage $V_1 \cos(wt)$. It can be expressed as:

$$v(t) = V_{dc} + V_1 \cos(wt) \tag{10.35}$$

As the driven voltage $V_1 \cos(wt)$increases and develops enough amplitude across the base-emitter junction, the resulting current is a periodic series of pulses whose amplitude depends on the nonlinear characteristics of the device and is given as

$$i_e(t) = I_s e^{\frac{qv(t)}{kT}} \tag{10.36}$$

$$i_e(t) = I_s e^{\frac{qV_{dc}}{kT}} e^{\frac{qV_1 \cos(wt)}{kT}} \tag{10.37}$$

$$i_e(t) = I_s e^{\frac{qV_{dc}}{kT}} e^{x \cos(wt)} \tag{10.38}$$

Assuming $I_c \approx I_e(\beta > 10)$

$$x = \frac{V_1}{(kT/q)} = \frac{qV_1}{kT} \tag{10.39}$$

$i_e(t)$ is the emitter current and x is the drive level, which is normalized to kT/q.

From the Fourier series expansion, $e^{x \cos(wt)}$ is expressed as

$$e^{x \cos(wt)} = \sum_n a_n(x) \cos(nwt) \tag{10.40}$$

$a_n(x)$ is a Fourier coefficient and given as

$$a_0(x)|_{n=0} = \frac{1}{2\pi} \int_0^{2\pi} e^{x \cos(wt)} d(wt) = I_0(x) \tag{10.41}$$

$$a_n(x)|_{n>0} = \frac{1}{2\pi} \int_0^{2\pi} e^{x \cos(wt)} \cos(nwt) d(wt) = I_n(x) \tag{10.42}$$

$$e^{x \cos(wt)} = \sum_n a_n(x) \cos(nwt) = I_0(x) + 2 \sum_1^{\infty} I_n(x) \cos(nwt) \tag{10.43}$$

$I_n(x)$ is the modified Bessel function.

$$\text{As } x \to 0 \Rightarrow I_n(x) \to \frac{(x/2)^n}{n!} \tag{10.44}$$

$I_0(x)$ are monotonic functions having positive values for $x \geq 0$ and $n \geq 0$; $I_0(0)$ is unity, whereas all higher order functions start at zero.

The short current pulses are generated from the growing large-signal drive level across the base-emitter junction, which leads to strong harmonic generation [3]. The advantage of this pulse performance is the reduction of phase noise, due to the smaller duty cycle of the transistor. The emitter current represented earlier can be expressed in terms of harmonics as

$$I_{dc} = I_s e^{\frac{qV_{dc}}{kT}} I_0(x) \tag{10.45}$$

$$V_{dc} = \frac{kT}{q} \ln\left[\frac{I_{dc}}{I_s I_0(x)}\right] \Rightarrow \frac{kT}{q} \ln\left[\frac{I_{dc}}{I_s}\right] + \frac{kT}{q} \ln\left[\frac{1}{I_0(x)}\right] \tag{10.46}$$

I_s is the collector saturation current

$$V_{dc} = V_{dcQ} - \frac{kT}{q} \ln I_0(x) \tag{10.47}$$

$$i_e(t) = I_{dc}\left[1 + 2\sum_{1}^{\infty} \frac{I_n(x)}{I_0(x)} \cos(nwt)\right] \tag{10.48}$$

V_{dcQ} and I_{dc} are the operating DC bias voltage and the DC value of the emitter current. Furthermore, the Fourier transform of $i_e(t)$, a current pulse or series of pulses in the time domain, yields a number of frequency harmonics common in oscillator circuit designs using nonlinear devices.

The peak amplitude of the harmonic content of the output current is defined as $\left[\frac{I_N(x)}{I_1(x)}\right]$, and the DC offset voltage is calculated analytically in terms of the drive level, as shown in Table 10.3. It gives good insight into the nonlinearities involved in the oscillator design.

It may be of interest to see the start-up condition of an oscillator; the transient response is shown in Figure 10.28, while Figure 10.29 shows a schematic diagram of a one-port negative resistance model.

Now we will take a Ge transistor (see Figure 10.30) from the past (PNP TV44 SIEMENS) and try to design an oscillator with it, which actually was done in 1964.

The TV 44 was an early coaxial Germanium "Microwave" transistor, around 1966, by Siemens at the time. With better understanding of microwave microstrip lines, the area of coaxial transistors died. Figure 10.31a shows a simulation of the circuit in Figure 10.30. The original circuit had a Ge PNP transistor, while the simulation uses a Si-Ge NPN transistor. Figure 10.31b shows the simulated phase noise of the circuit in Figure 10.31, 10 dBm output power, with harmonic suppression greater than 50 dB.

In those days the industry switched from coaxial transistors to microstrip based design. A good example would be the coaxial microwave tubes (pencil tubes), like RH7C, which were a nightmare to build a circuit around.

Now we move from bipolar transistor types CSiGe to GaAs FET oscillators using distributed elements.

Table 10.3 For T = 300 K, data are generated at a different drive-level.

Drive level $[x]$	Drive voltage $\left(\left[\frac{kT}{q}\right] * x\right)$ mV	Offset coefficient $\ln[I_0(x)]$	DC offset $\frac{kT}{q}[\ln I_0(x)]$mV	Fundamental current $2[I_1(x)/I_0(x)]$	Second harmonic $[I_2(x)/I_1(x)]$
0.00	0.000	0.000	0.000	0.000	0.000
0.50	13.00	0.062	1.612	0.485	0.124
1.00	26.00	0.236	6.136	0.893	0.240
2.00	52.00	0.823	21.398	1.396	0.433
3.00	78.00	1.585	41.210	1.620	0.568
4.00	104.00	2.425	63.050	1.737	0.658
5.00	130.00	3.305	85.800	1.787	0.719
6.00	156.00	4.208	206.180	1.825	0.762
7.00	182.00	5.127	330.980	1.851	0.794
8.00	208.00	6.058	459.600	1.870	0.819
9.00	234.00	6.997	181.922	1.885	0.835
10.00	260.00	7.943	206.518	1.897	0.854
15.00	390.00	12.736	331.136	1.932	0.902
20.00	520.00	17.590	457.340	1.949	0.926

Source: [3]/John Wiley & Sons.

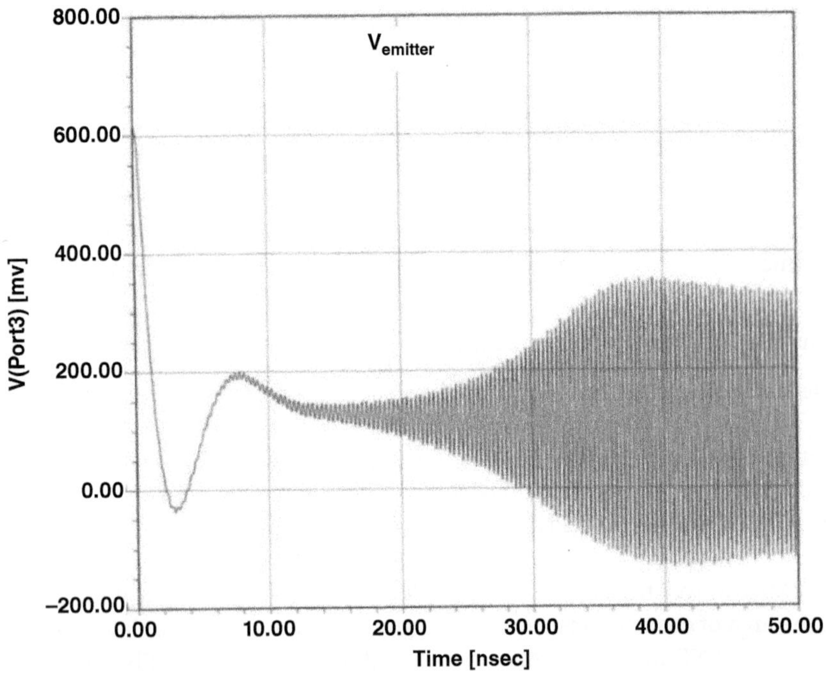

Figure 10.28 Example of the transient simulation of a ceramic resonator based high-Q oscillator showing the DC offset as shown in column 4 (Table 10.3) (the voltage displayed is taken from the emitter). *Source*: [3]/John Wiley & Sons.

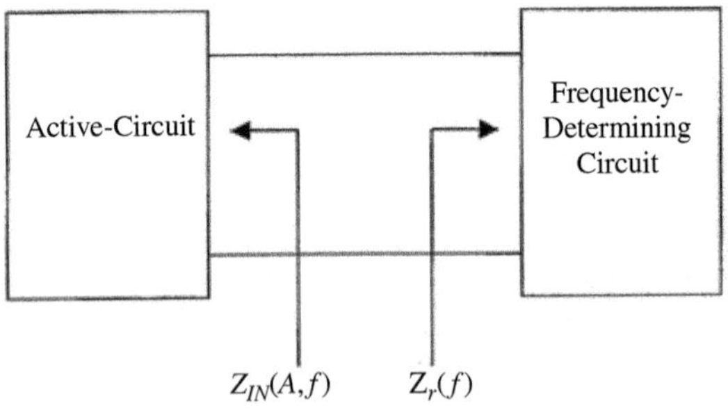

Figure 10.29 Schematic diagram of a one-port negative resistance model. *Source*: [3]/John Wiley & Sons.

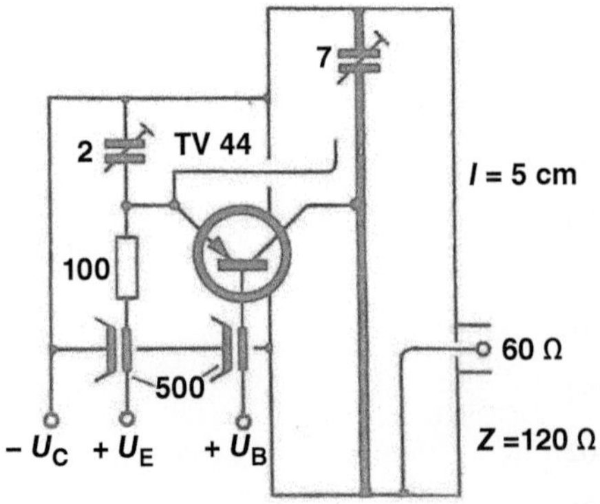

Figure 10.30 Microwave transistor circuit assembly using SIEMENS TV44 Ge transistor. *Source*: [24, pp. 111]/Verlag für Radio-Foto-Kinotechnik.

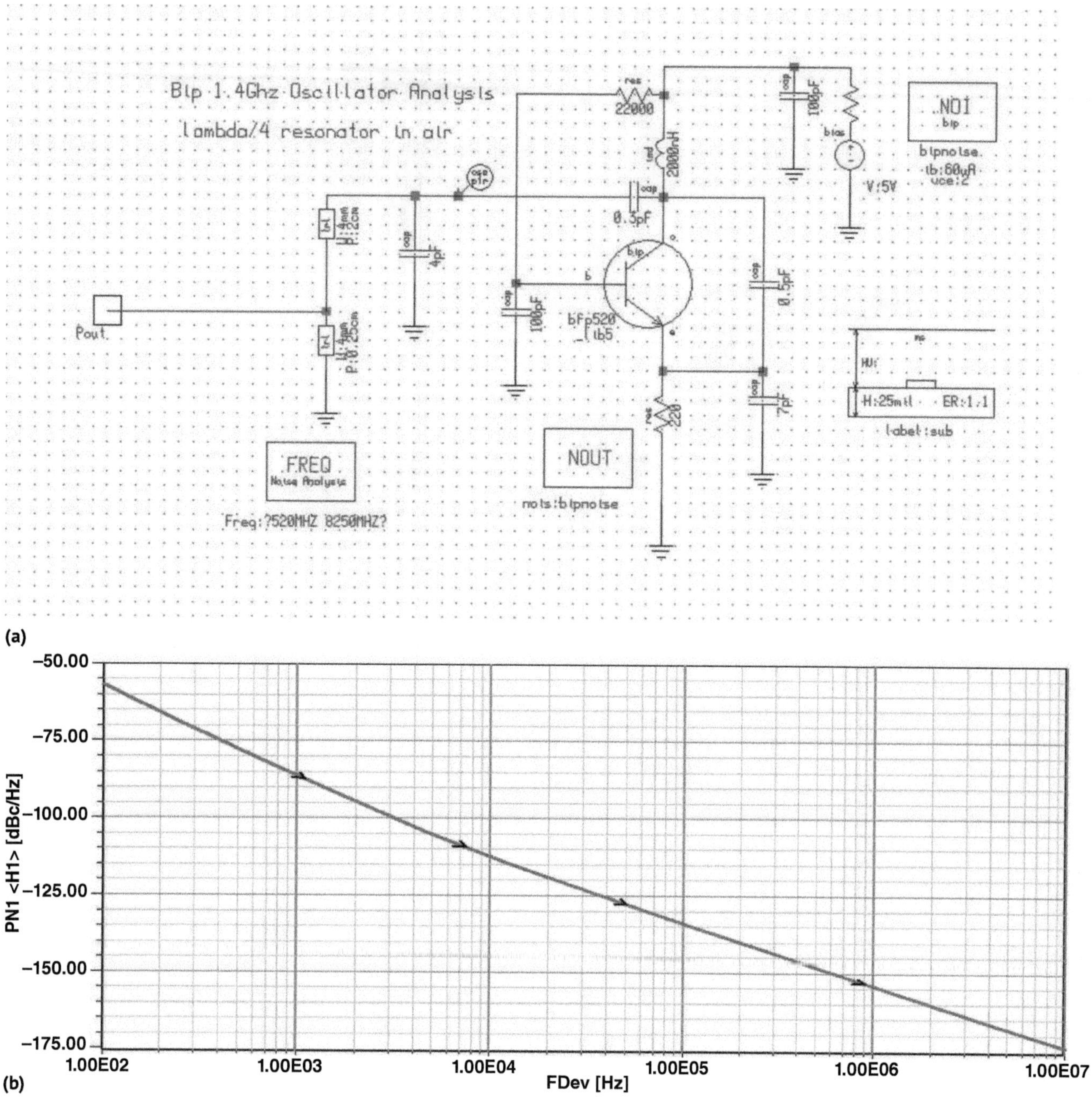

Figure 10.31 (a) CAD simulation of the circuit shown in Figure 10.30 using a Si NPN transistor as a test circuit. (b) Simulated phase noise of the circuit in (a), 10 dBm output power, harmonic suppression > 50 dB.

Example: 2000 MHz GaAs FET-based Oscillator Low-cost applications are frequently implemented as an RFIC. For further validation, a GaAs FET-based 2000 MHz Colpitts oscillator was designed and built. Figure 10.32a shows the circuit diagram of the oscillator. It uses a combination of transmission lines and rectangular inductors as resonators. The inductor in the middle of the schematic in Figure 10.32a, connected to a "via" hole, is needed as a DC return. If a tuning diode is connected to the capacitor on the left of the schematic in Figure 10.32a, then a DC control voltage can be applied, and the center inductor becomes an RF choke. The output is taken from the source. An additional external DC decoupling capacitor will be needed because of the DC coupling. The transistor and the circuit were constructed using the TriQuint GaAs foundry, and the transistor was optimized for DC current. Figure 10.32b shows the predicted phase noise of this oscillator [3, 21].

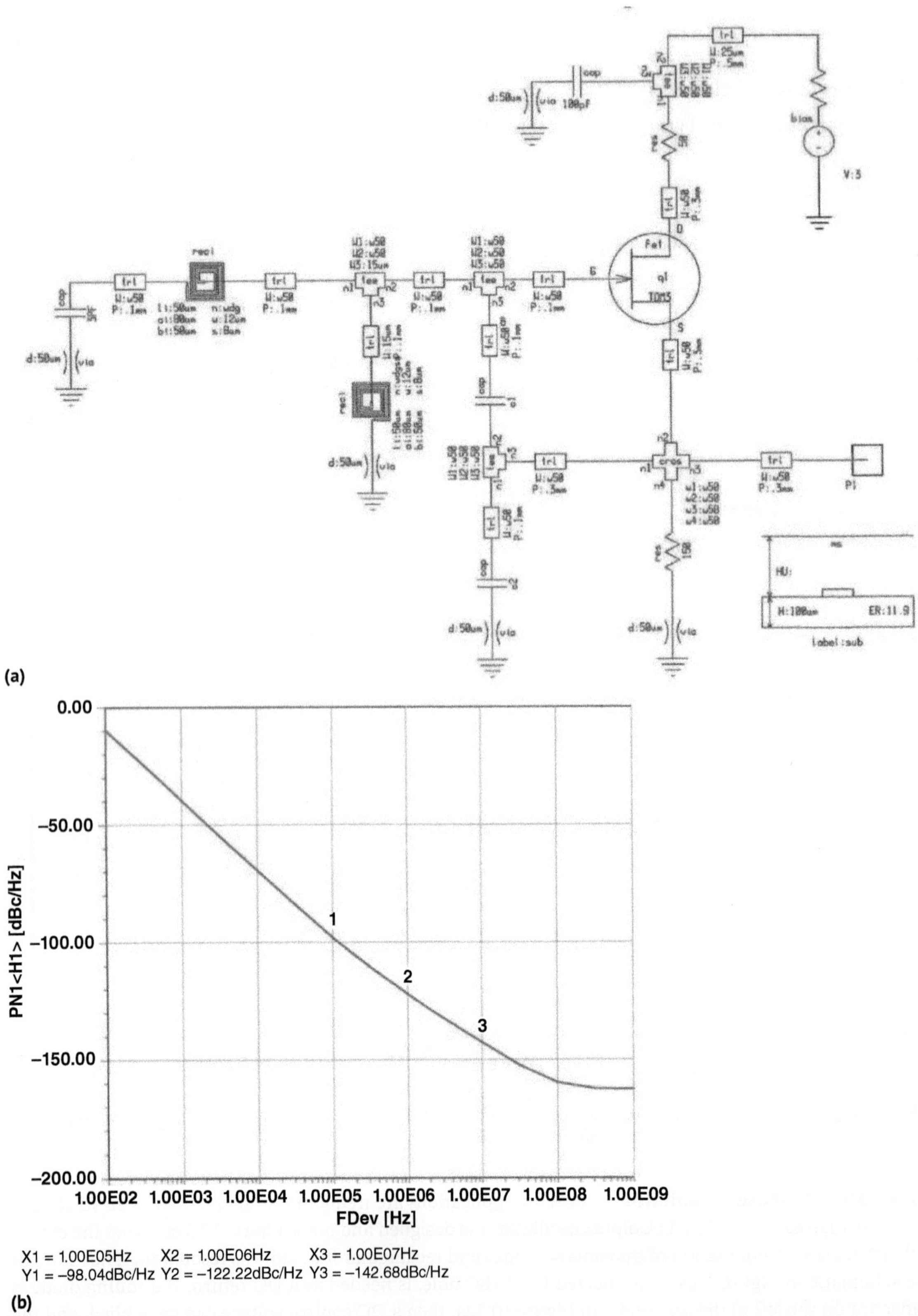

Figure 10.32 (a) Circuit diagram of the 2 GHz distributed elements GaAs FET oscillator [3, 21]. (b) Predicted phase noise of the oscillator shown in (a) [3, 21].

This is how a microwave oscillator on a printed circuit should really be simulated in detail.

The measured values were 100 dBc/Hz at 100 kHz and 120 dBc/Hz at 1 MHz offsets. There is a deviation of about 2 dB compared to simulation.

It is interesting to examine the load line of this oscillator, which is shown in Figure 10.33. This circuit is operated in a fairly linear range.

Figure 10.34 shows the layout of the 2 GHz GaAs FET oscillator. Its output power is 1.8 dBm.

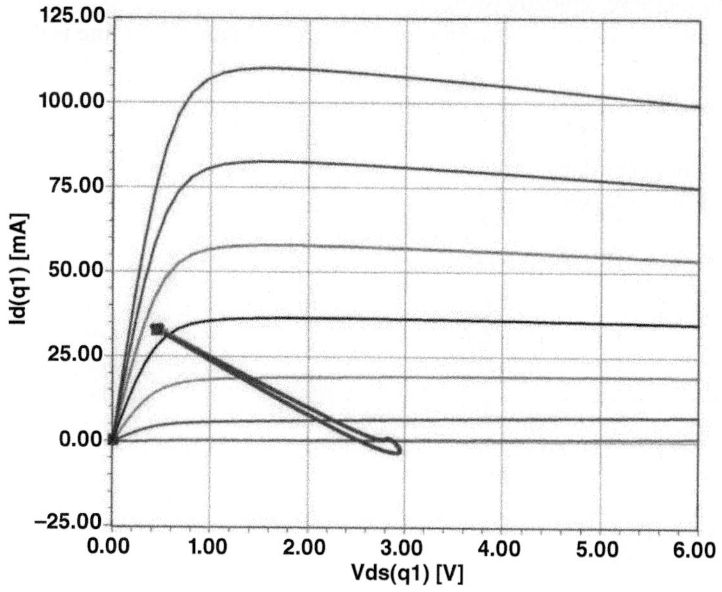

Figure 10.33 Trace of the DC–IV and the load line for the GaAs FET oscillator [3, 21].

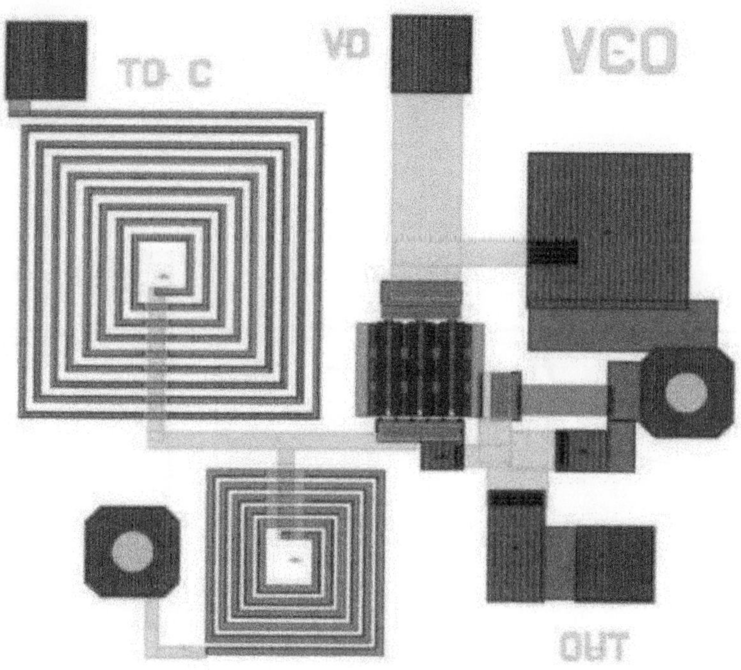

Figure 10.34 Layout of the 2 GHz GaAs FET oscillator (Courtesy Synergy Microwave, NJ, USA) [3, 21].

Now a detailed example of a typical bipolar transistor oscillator simulation at 3500 MHz is shown in Figure 10.35a, and the various simulation results, theoretical calculation of phase noise, etc., are shown in Figures 10.35b–f, 10.36, and 10.37, including the various current, voltage waveforms, and simulated output power.

3500 MHz Oscillator design

It should be noted here that under large signal conditions, this frequency changes. Based on the larger emitter current, the base-emitter diffusion capacitor theoretically is linearly proportional to the emitter current, thus increasing the time average. Likewise the collector depletion capacitance will increase typically twofold.

It is important to observe that the DC parameters are not exceeded. The reason that the load is not a straight line is due to the flywheel effect, which stores energy.

In theory, the collector and the base currents should be in-phase, but due to the inherent transient delay, they are retarded as shown (see antenna theory).

The simulated voltages (Vce and Vbe) also show the similar retardation (delay).

The following is a numerical calculation of the simulation as shown earlier.

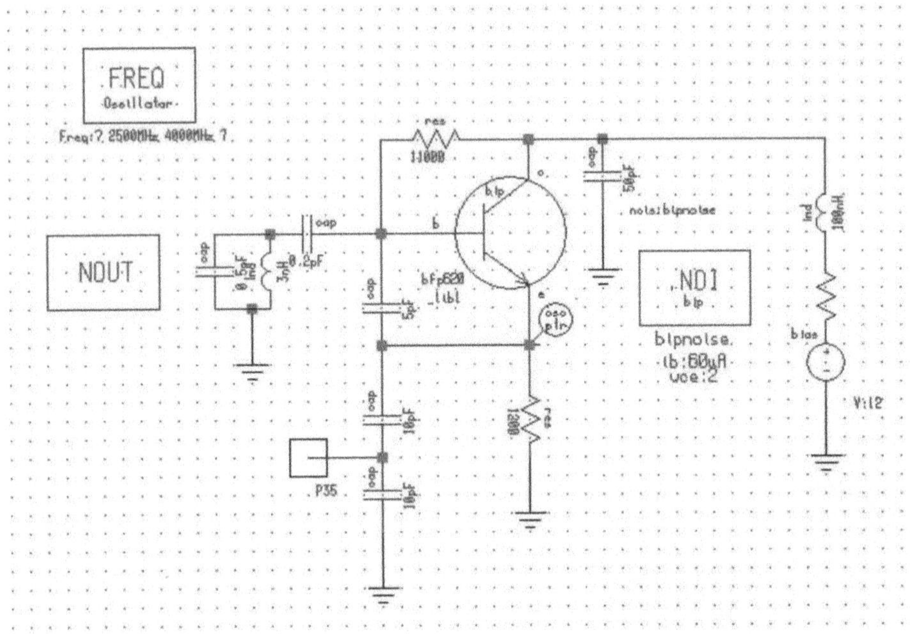

(a)

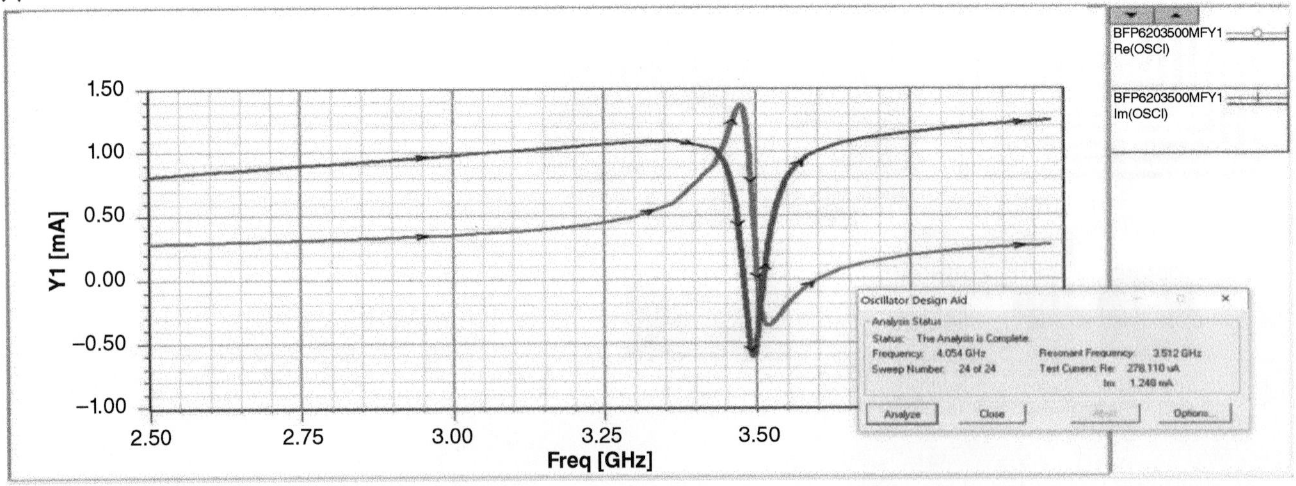

(b)

Figure 10.35 (a) A 3500 MHz Colpitts oscillator design. (b) Linear oscillator analysis to determine the likely resonance frequency.

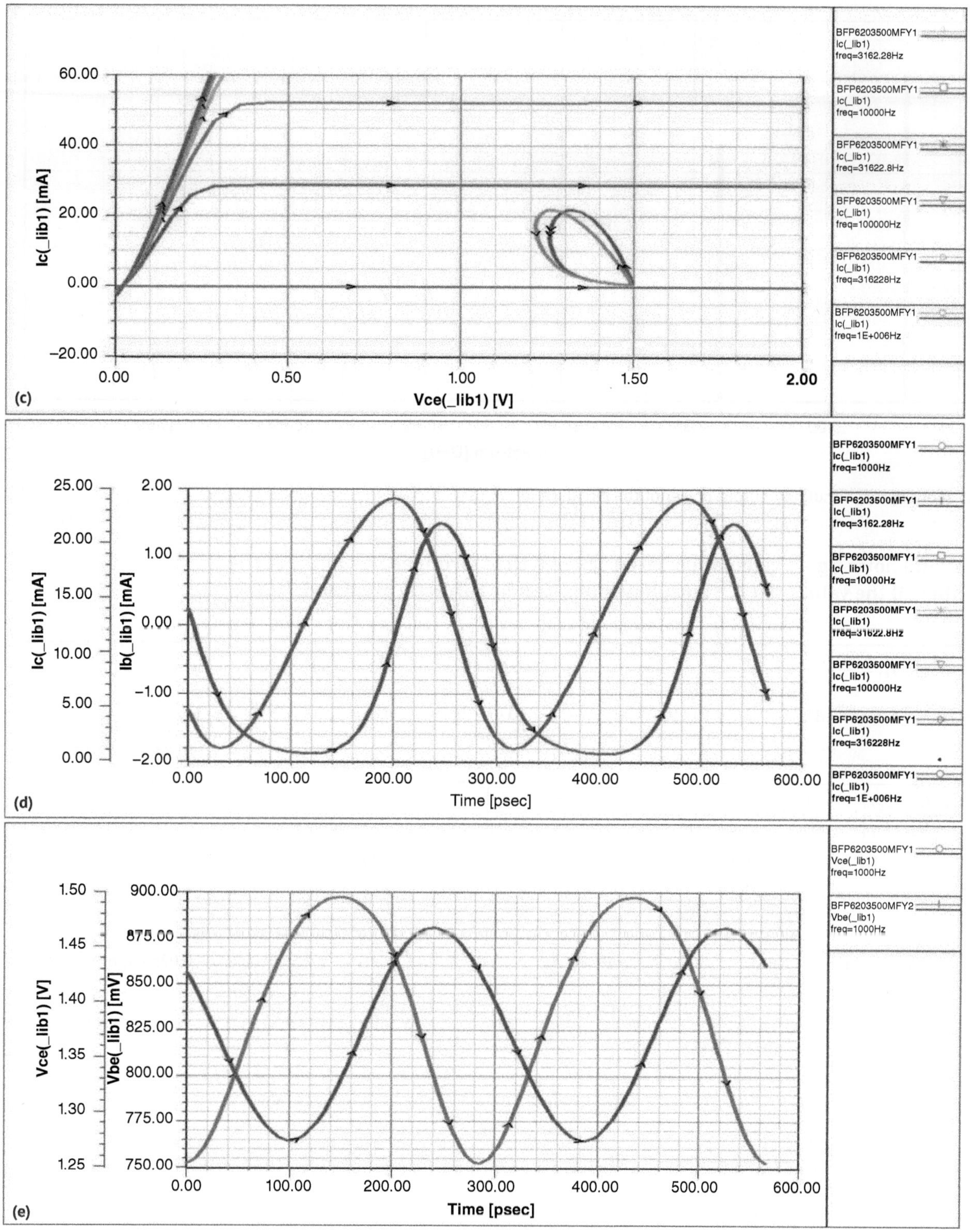

Figure 10.35 (Cont'd) (c) DC–IV curve and load line under large signal conditions for the circuit in (a). (d) Simulated collector current (Ic_lib1) (trace with the higher peak) and base current (Ib_lib1). (e) Simulated voltages Vce (Re [OSC] lower trace) and Vbe (Im [OSC] upper trace).

(Continued)

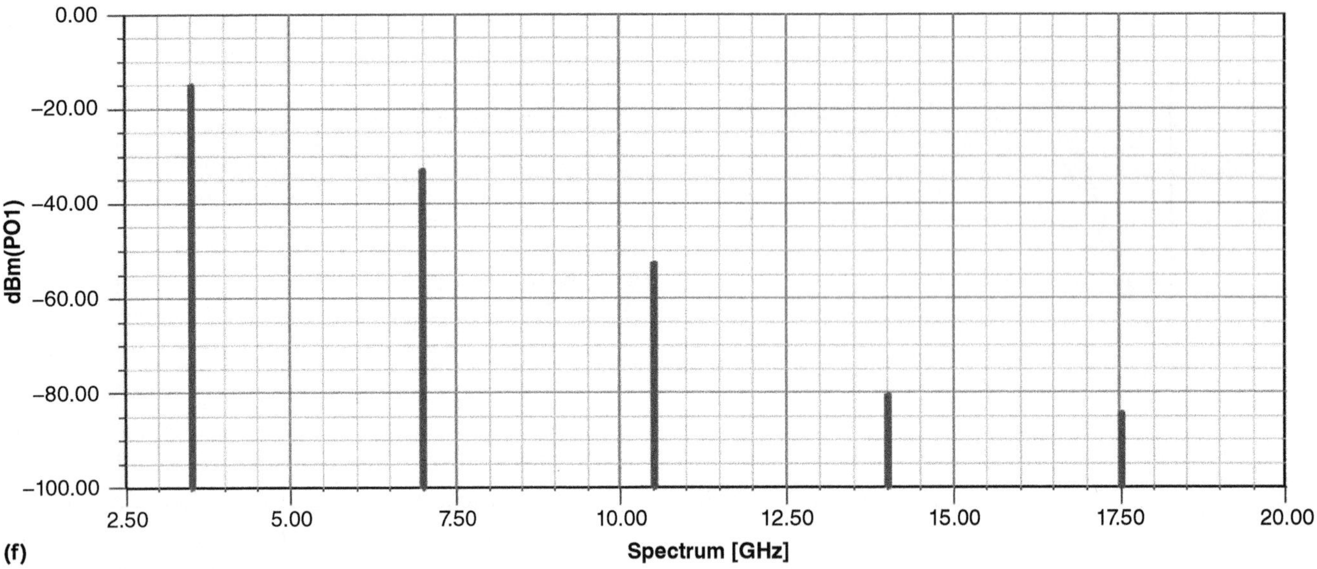

(f)

Figure 10.35 (Cont'd) (f) Simulated output power of the circuit in (a).

Phase noise and noise figure calculation based on analytical formulae
After defining all the values, the NF and phase noise can be correctly predicted.

$Ic := 8.1\ mA$ $\quad r_e := \frac{26\ mV}{Ic}$ $\quad Y_{21} := \frac{1}{2 \cdot r_e \cdot \pi}$ $\quad Y_{21} := 0.05 \cdot S$ $\quad r_b := 10\ \Omega$ $\quad I_b := 0.048\ mA$

$f := 3500\ MHz$ $\quad f_T := 45\ GHz$ $\quad C_1 := 10 \cdot 10^{-12}\ F$ $\quad C_2 := 10 \cdot 10^{-12}\ F$ $\quad C_c := 0.2 \cdot 10^{-12}\ F$ $\quad Icmax := 24\ mA\ +$

$Ibmax := 1.2\ mA$

$$F := 1 + \frac{(Y_{21} \cdot C_2 \cdot C_c)}{(C_1 + C_2) \cdot C_1}\left[r_b + \frac{1}{2r_e}\left[r_b + \frac{(C_1 + C_2) \cdot C_1}{(Y_{21} \cdot C_2 \cdot C_c)}\right]^2 \cdot \left(\frac{1}{\beta} + \frac{f^2}{f_T^2}\right) + \frac{r_e}{2}\right] \qquad \beta := \frac{Icmax}{Ibmax} \qquad \beta := 20$$

$F = 18.789$ $\quad NF\_dB := 10 \cdot \log(F)$ $\quad NF\_dB := 12.739$ $\quad KT := 4.143 \cdot 10^{-21} \cdot J$

$L := 3.0 \cdot 10^{-9}\ henry$ $\quad qcharge := 1.602 \cdot 10^{-19} coul$ $\quad kf := 1 \cdot 10^{-11}$ $\quad Qmac := 200$ $\quad Q := 100$

$Vcc := 12$ $\quad wc := 2 \cdot \pi \cdot f$ $\quad R := wc \cdot L \cdot Q$ $\quad R := 6.597 \cdot 10^3\ \Omega$ $\quad af := 2$ $\quad y_{11} := (0.00852 + 0.02591j)\ mho$

$i := 0..8$ $\quad fo_i := 10^i\ Hz$ $\quad p := 2$ $\quad q := 0.2$ $\quad y_{21} := (0.227 - 0.002239j)\ mho$

$wo_i := 2 \cdot \pi \cdot fo_i$ $\quad PoutdBm := -15$ $\quad B1_i := (wo_i)^2 \cdot L^2 \cdot Vcc^2$

$wo_i := 2 \cdot \pi \cdot fo_i$ $\quad PoutdBm := -15$ $\quad B1_i := (wo_i)^2 \cdot L^2 \cdot Vcc^2$

$y := \frac{C_1}{C_2}$ $\quad kconstant := \frac{KT \cdot R}{wc^2 \cdot C_2^2}$ $\quad b := \frac{|y21|y^p}{|y11|}$ $\quad gm := |y21|y^q$ $\quad k0_i := \frac{kconstant}{B1_i}$

$k1constant_i := q\ charge \cdot Ic \cdot gm^2 + \frac{kf \cdot Ib^{af} \cdot gm^2}{wo_i}$ $\quad k1_i := \frac{k1constant_i}{wc^2 \cdot B1_i}$ $\quad k2_i := wc^4 \cdot b^2$ $\quad k3_i := wc^2 \cdot gm^2$ $\quad k_i := \frac{k3_i}{k2_i \cdot C_2^2}$

$$t1_i := \left[\left(\frac{b^2}{gm^3} \right)^2 \cdot \frac{(k_i)^3 \cdot k1_i (wc)^2}{(y^2 + k_i)} \right] \frac{(1+y)^2}{y^2} \qquad t2_i := k0_i \cdot \frac{(1+y)^2}{y^2} \qquad l_i := t1_i + t2_i$$

$$m_i := 10.\log\left[l_i \left(k_g^{-2} \cdot m^{-4} \cdot s^5 \cdot A^2\right) \cdot \frac{Qmac^2}{Q^2} \right] \qquad L_i := \text{if}[m_i < (-(174 + PoutdBm - NF\_dB)), ((-(174 + PoutdBm - NF\_dB)), m_i]$$

An oscillator circuit known as the *Clapp circuit* or *Clapp–Gouriet circuit* is shown in Figure 10.38. This oscillator is equivalent to the one just discussed, but it has the practical advantage of being able to provide another degree of design freedom by making C_o much smaller than C_1 and C_2.

It is possible to use C_1 and C_2 to satisfy the condition of Eq. (10.47) and then adjust C_o for the desired frequency of oscillation ω_o, which is determined from

$$\omega_o L - \frac{1}{\omega_o C_o} - \frac{1}{\omega_o C_1} - \frac{1}{w_o C_2} = 0 \tag{10.49}$$

This is the basis of all modern crystal oscillators. A good example is, as shown in Figure 10.39, the schematic of a 100 MHz crystal oscillator, theoretically scaled from the 10 MHz HP10811A. The harmonically tuned LC circuit across the base-emitter helps in the reduction of undesired modes [8].

This is the schematic of a high-performance 100 MHz crystal oscillator. The equivalent circuit of the crystal is a L-R-C series combination, the equivalent circuit of the SAW resonator is similar.

Figure 10.40a is an example of a 1 GHz SAW oscillator similar to the Synergy 1 GHz Voltage Controlled SAW Oscillator Surface Mount MODEL HFSO1000-5H. Figure 10.40b shows its DC–IV curve with load line. Figure 10.40c shows the simulated collector and base currents, and the simulated phase noise is shown in Figure 10.40d.

The C-L-R component sequence in series models the actual resonator.

Figure 10.41 shows the measured phase noise of Synergy's low-cost 1 GHz SAW oscillator. These are typical results for this type of product.

m_i
−7.526
−27.526
−47.526
−67.527
−87.527
−107.527
−127.527
−147.527
−167.527

fo_i
$1 \cdot s^{-1}$
$10 \cdot s^{-1}$
$100 \cdot s^{-1}$
$1 \cdot 10^3 \cdot s^{-1}$
$1 \cdot 10^4 \cdot s^{-1}$
$1 \cdot 10^5 \cdot s^{-1}$
$1 \cdot 10^6 \cdot s^{-1}$
$1 \cdot 10^7 \cdot s^{-1}$
$1 \cdot 10^8 \cdot s^{-1}$

L_i
−7.526
−27.526
−47.526
−67.527
−87.527
−107.527
−127.527
−146.261
−146.261

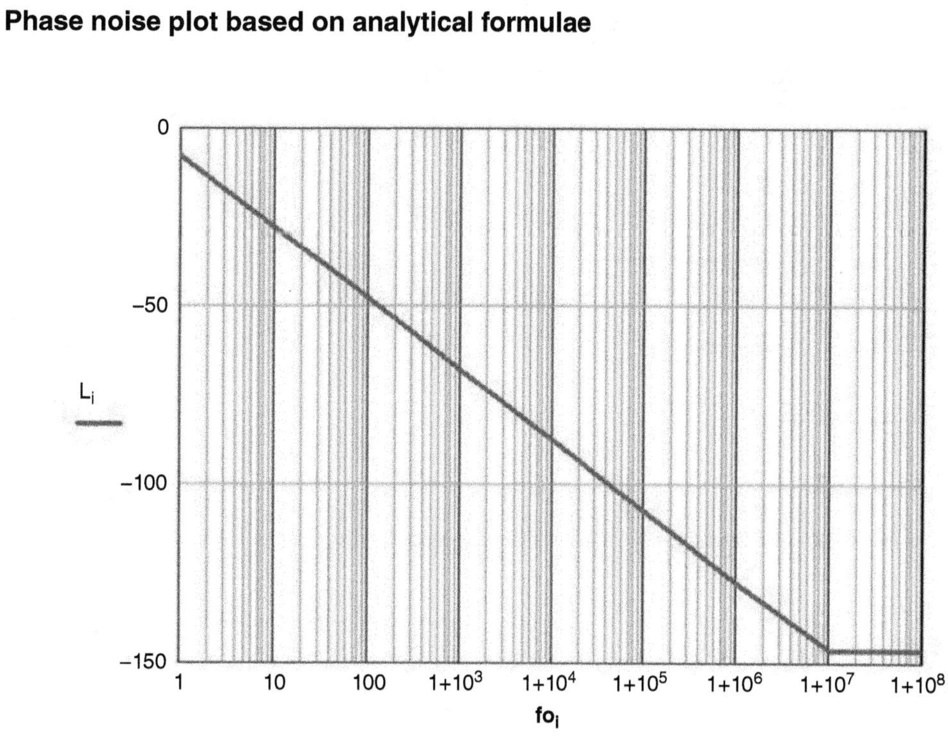

Figure 10.36 Calculated phase noise for the circuit in Figure 10.35a based on the modified Leeson equation.

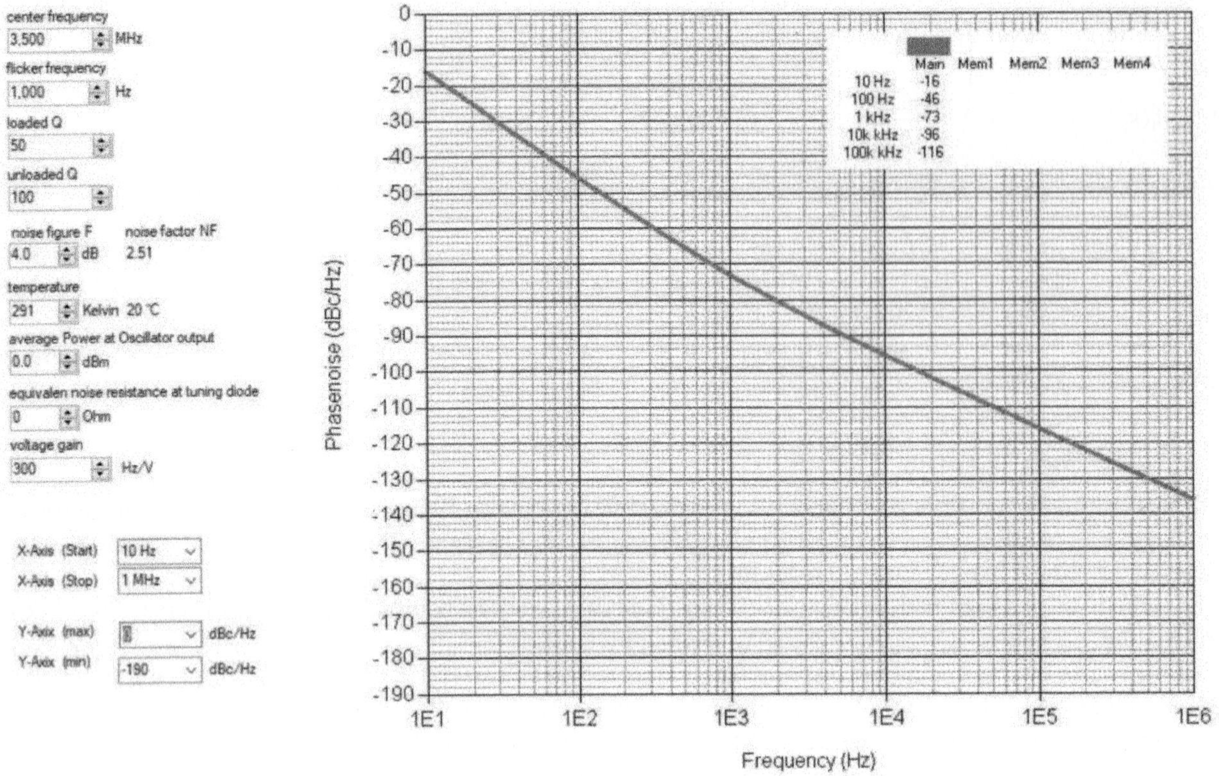

Figure 10.37 A detailed plot of the 3500 MHz circuit of Figure 10.35a on a graphic display using a software program designed by Joerg Bartls.

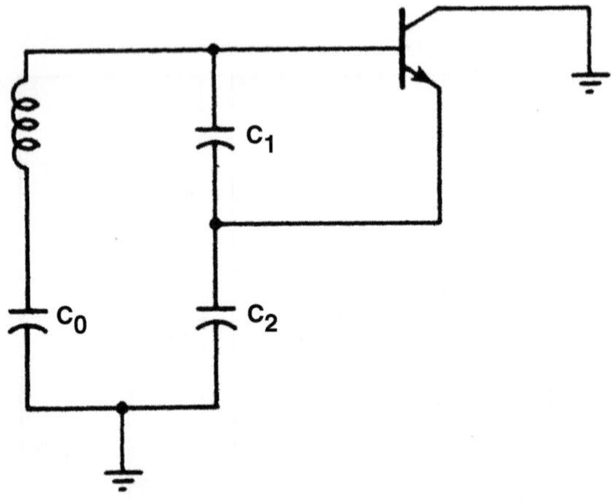

Figure 10.38 Circuit of a Clapp oscillator. *Source*: Modified from [5].

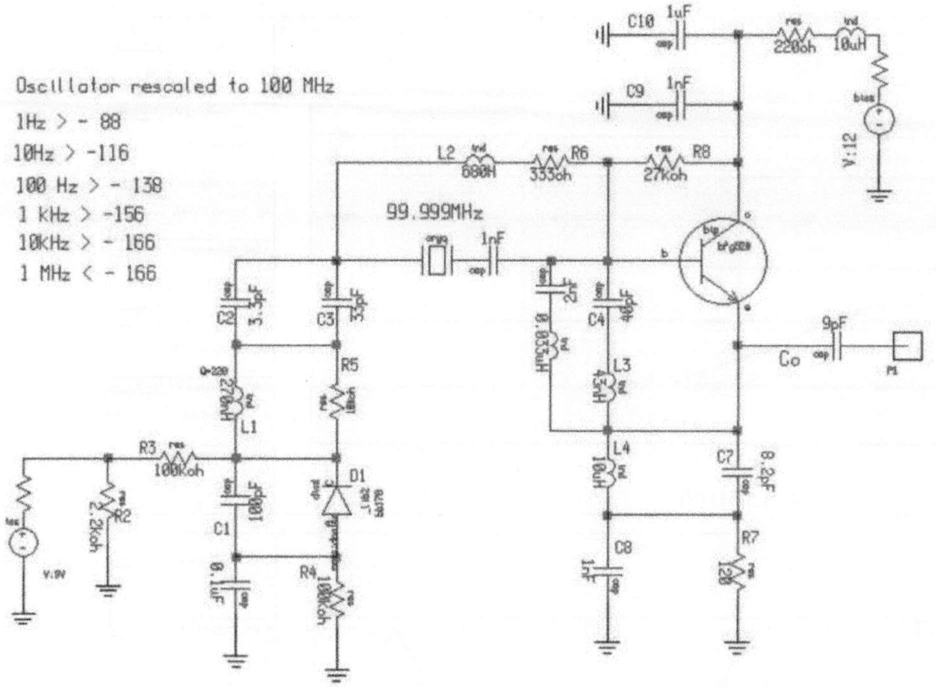

Figure 10.39 HP 10 MHz crystal oscillator scaled to 100 MHz. *Source*: Modified from [8].

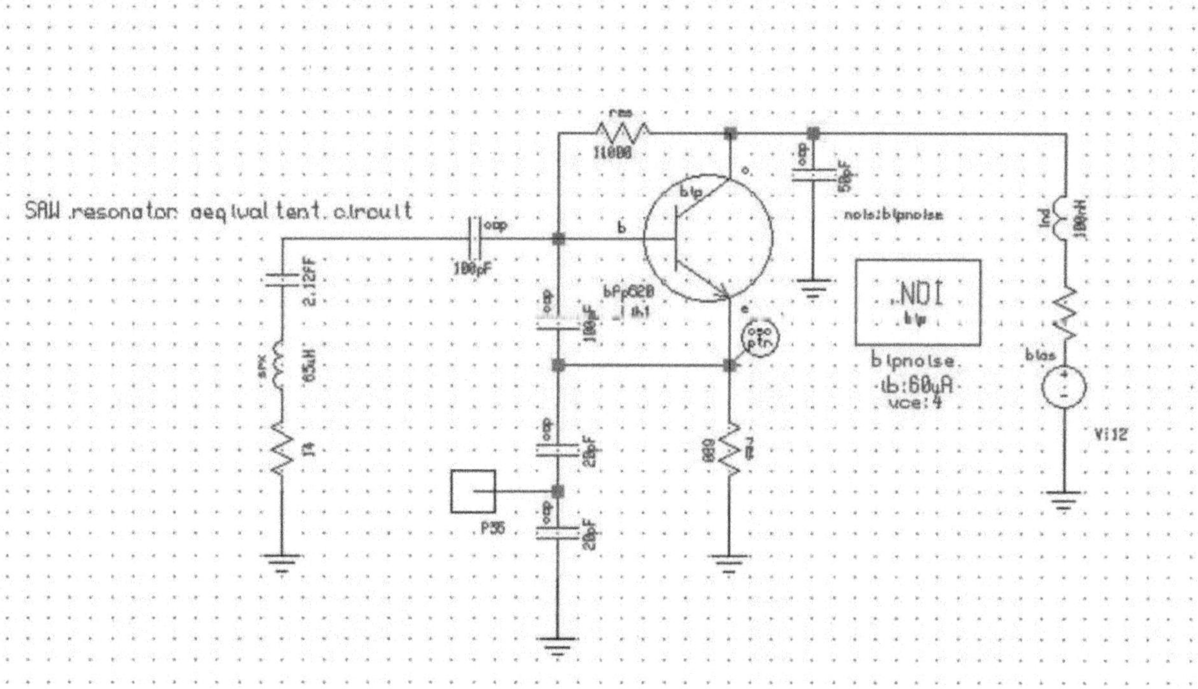

(a)

Figure 10.40 (a) Schematic circuit of 1 GHz SAW resonator-based oscillator circuit.

(Continued)

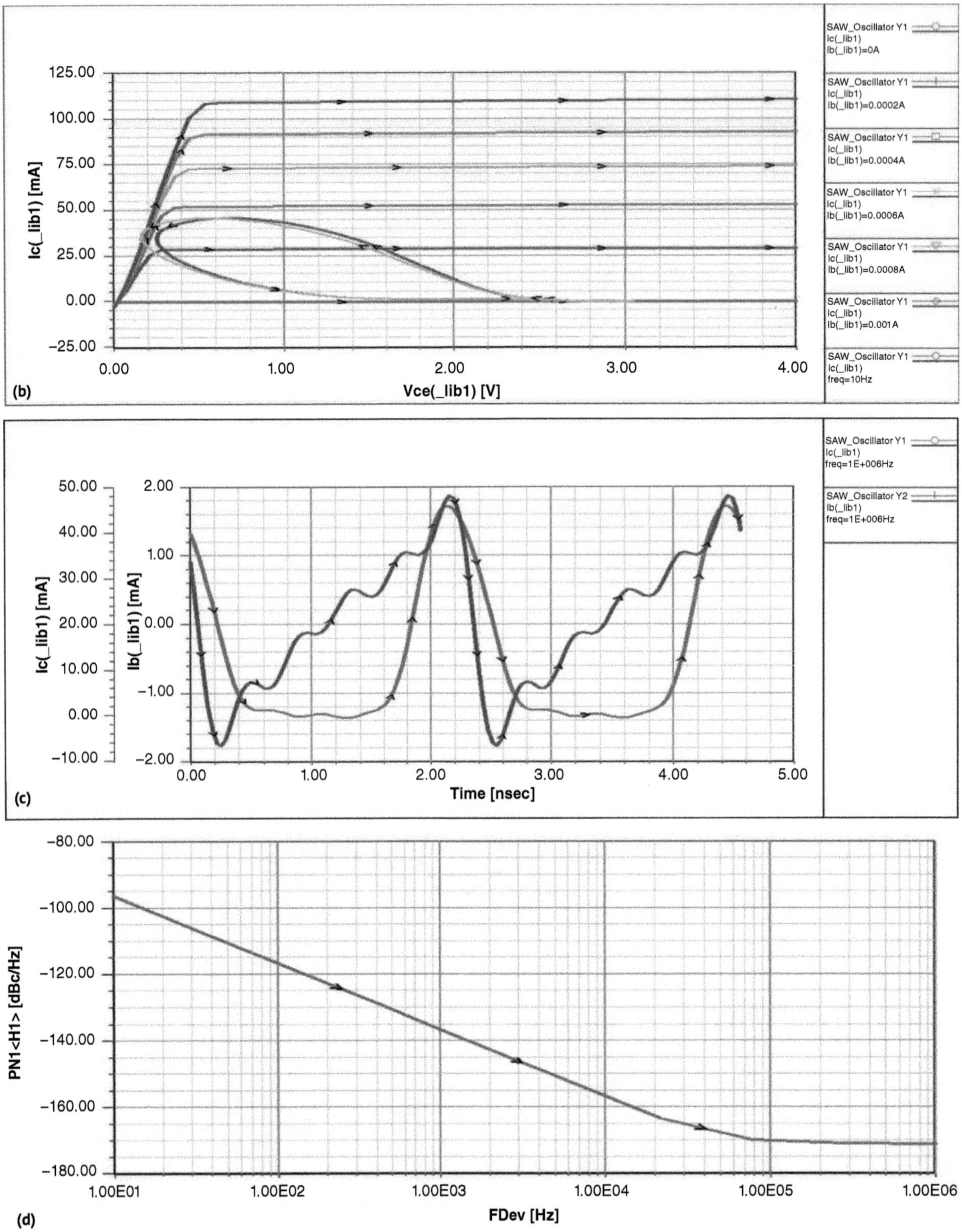

Figure 10.40 (Cont'd) (b) 1 GHz SAW resonator oscillator DC–IV curve with load line. (c) Simulated collector and base currents of the 1 GHz SAW oscillator of (a). (d) Simulated phase noise of the 1 GHz SAW oscillator of (a).

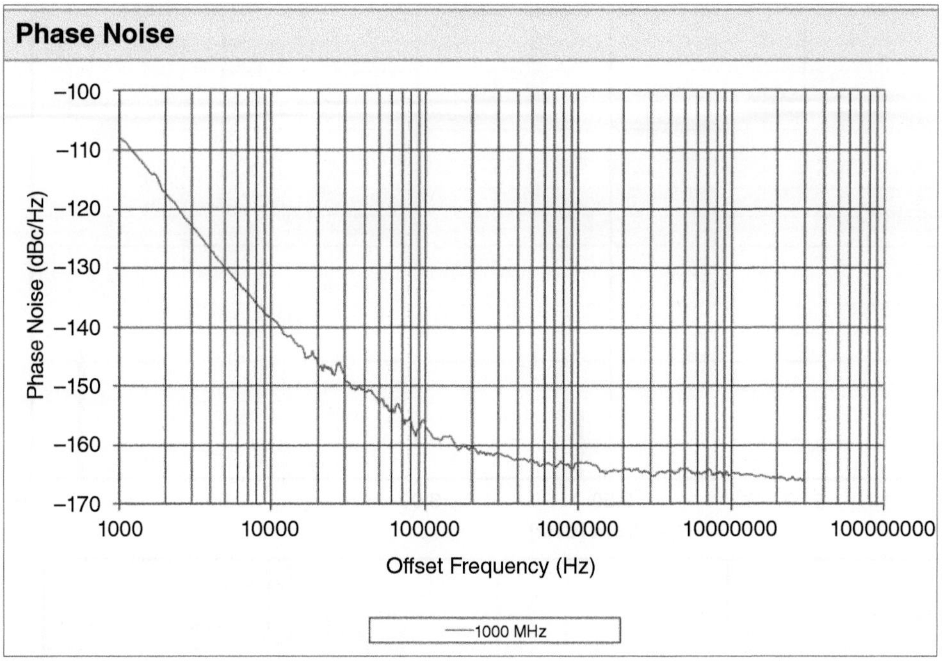

Figure 10.41 Measured phase noise of a low-cost 1 GHz SAW oscillator (Synergy Microwave).

Example of a 1.4 GHz oscillator using quarter-wavelength transmission lines: Now we show an example of an oscillator using quarter-wavelength transmission lines. The output is obtained by a tapped transmission line as can be seen in Figure 10.42a.

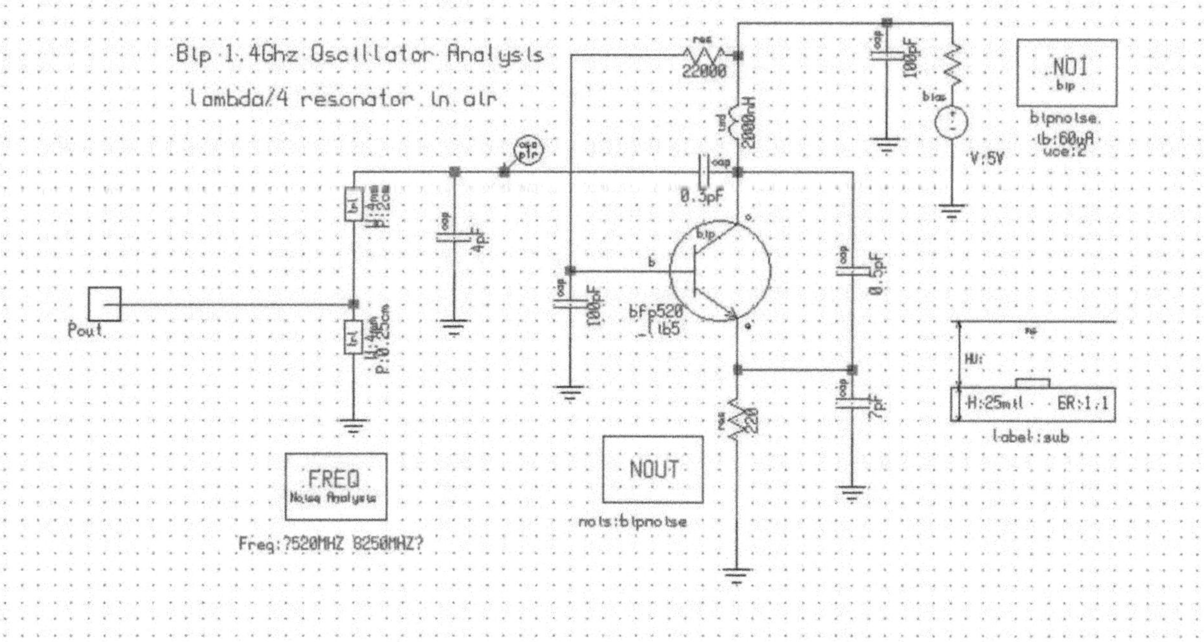

(a)

Figure 10.42 (a) 1.4 GHz oscillator analysis.

(Continued)

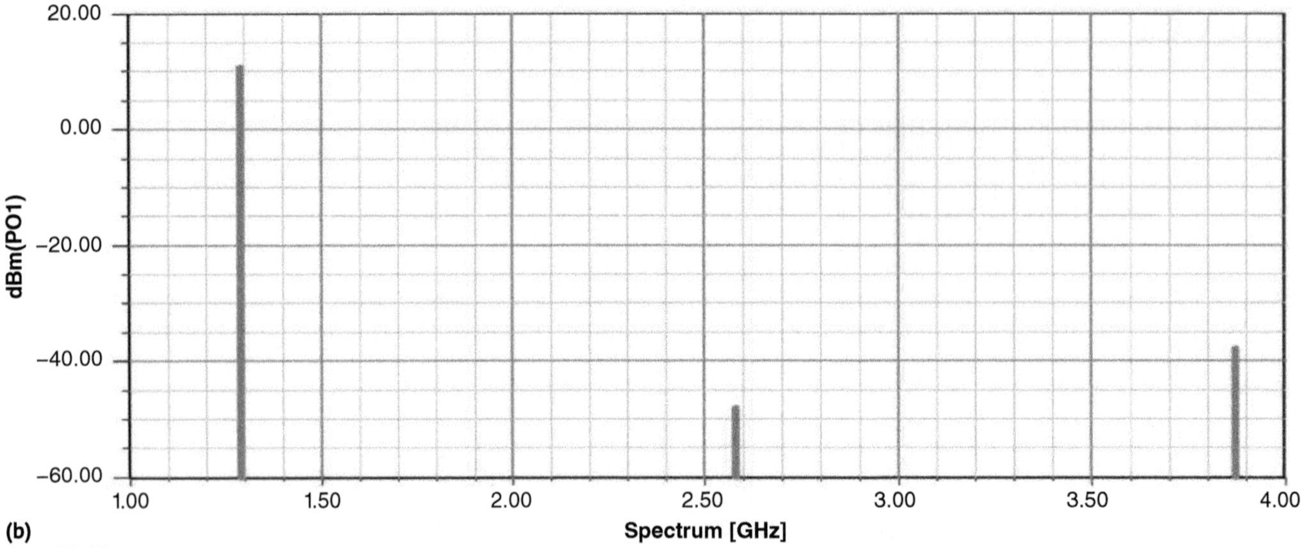

(b)

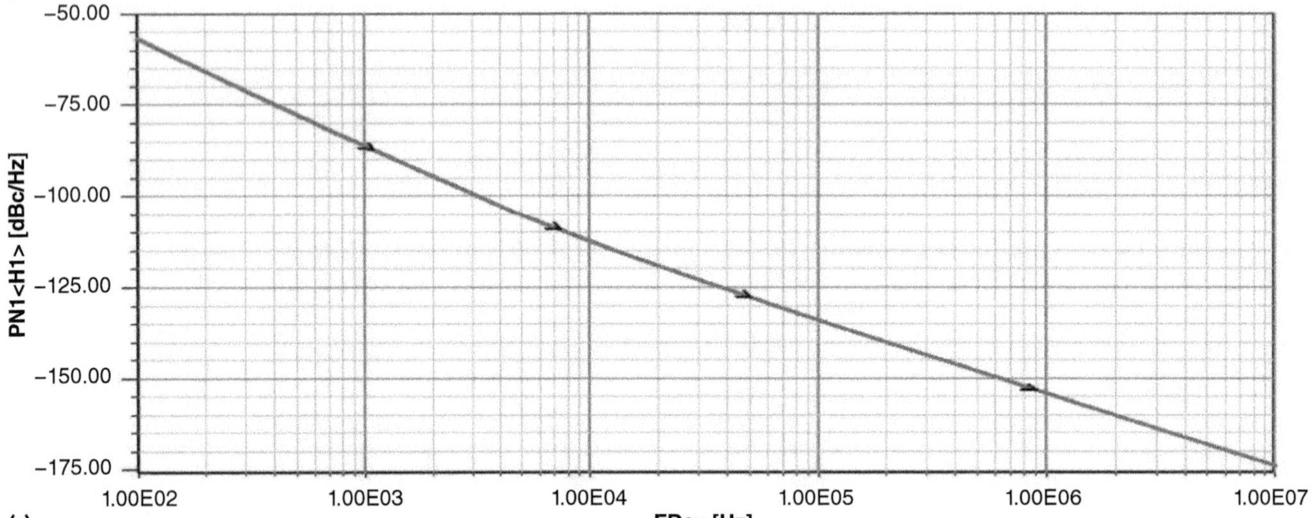

(c)

Figure 10.42 (Cont'd) (b) Simulated output power of the 1.4 GHz TL oscillator of (a). (c) Simulated phase noise of the 1.4 GHz TL oscillator of (a).

This Colpitts type oscillator operates at 1.4 GHz. The results of the simulation showing the power output in dBm are shown in Figure 10.42b and the simulated phase noise in Figure 10.42c.

Now we are going to be courageous. Having access to some of the most advanced semiconductor parameters from foundries, let's try and "build" a sub-terahertz oscillator. Here we go:

Similar to 640 MHz being a standard frequency in many synthesizer applications, 370 GHz seems to be a popular frequency in various publications. Hence, an example is considered, as shown in Figure 10.43a for simulation.

The 370 GHz oscillator consists of a grounded base circuit, using an experimental transistor, called the BFP950F. Unfortunately, not via mail order. The circuit is a Colpitts type oscillator with capacitive feedback, where the resonator is a transmission line, which is tapped. The transmission line is modeled correctly by using an END element, which models the radiation. The output is emitter-follower as decoupling (see the left side of the circuit in Figure 10.43a). In order to evaluate the circuit and whether we are operating in a safe region, we look at the currents and voltages, as seen in the RF voltage analysis in Figure 10.43b.

No breakdown effects are indicated. Vbe_lib1 trace (trace with the highest peak), the input of the oscillator transistor shows some distortion, while the Vbe_lib2 (trace with the flat peak) of the output (amplifier) transistor shows a slight saturation.

The Vce_lib1 trace is well-behaved and sinusoidal. There is also a minimal phase shift (emitter follower), as is evident in Figure 10.43b.

Since there is no low-pass filter, the first harmonic is only suppressed by 3 dB as seen in Figure 10.43c, which indicates not-so-good phase noise, as seen in Figure 10.43d, as the loaded Q is not very high. The purpose here is really only to show that a stable oscillation is possible.

The phase noise at 100 Hz offset appears to be higher than the simulated output power. How is this possible? This is explained in the publications [22, 25].

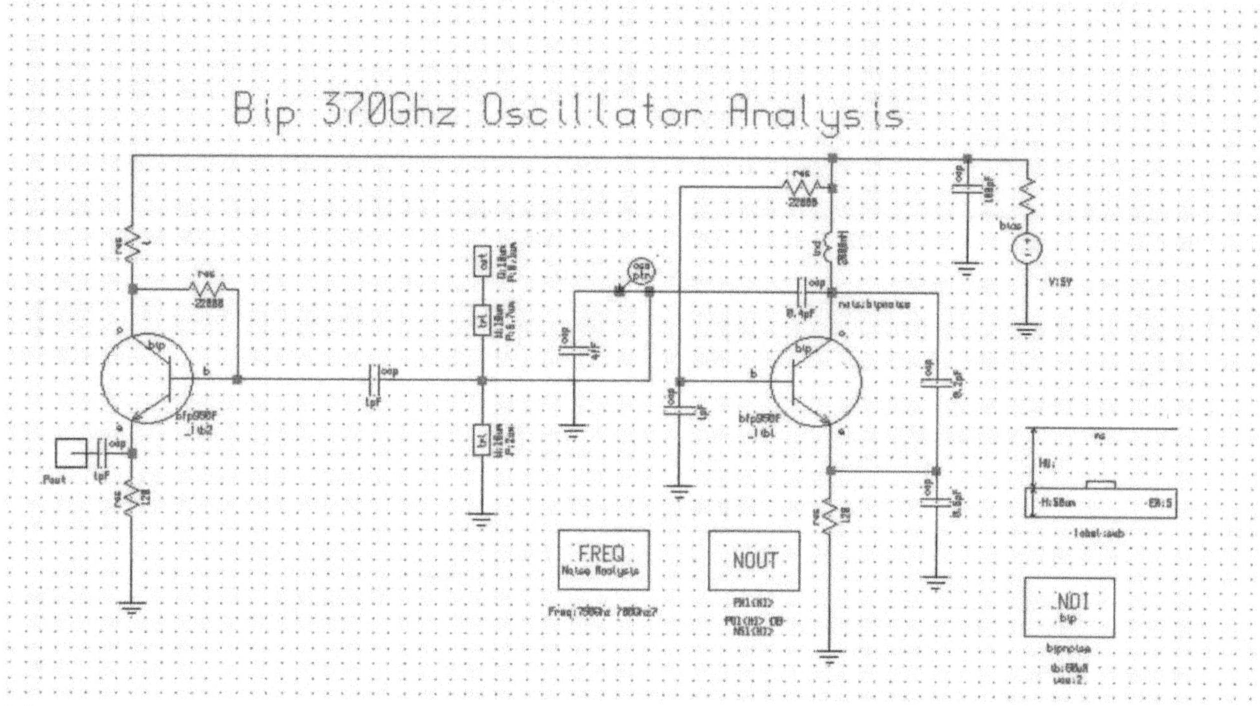

(a)

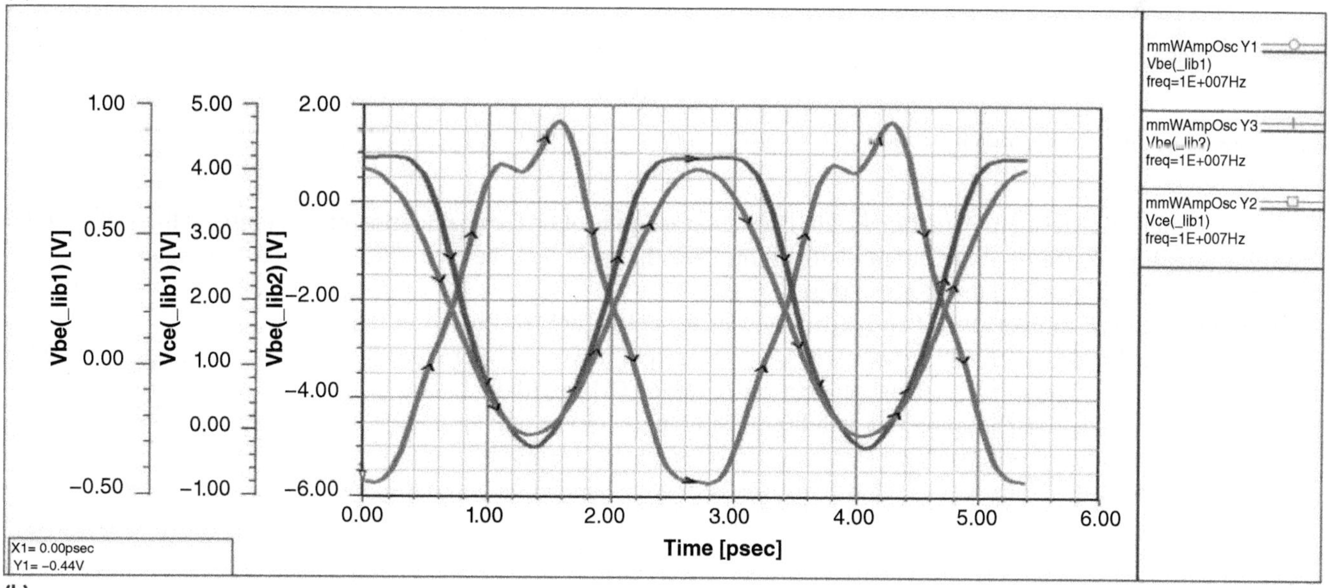

(b)

Figure 10.43 (a) 370 GHz oscillator design and analysis. (b) 370 GHz oscillator–RF voltage analysis (Vce).

(Continued)

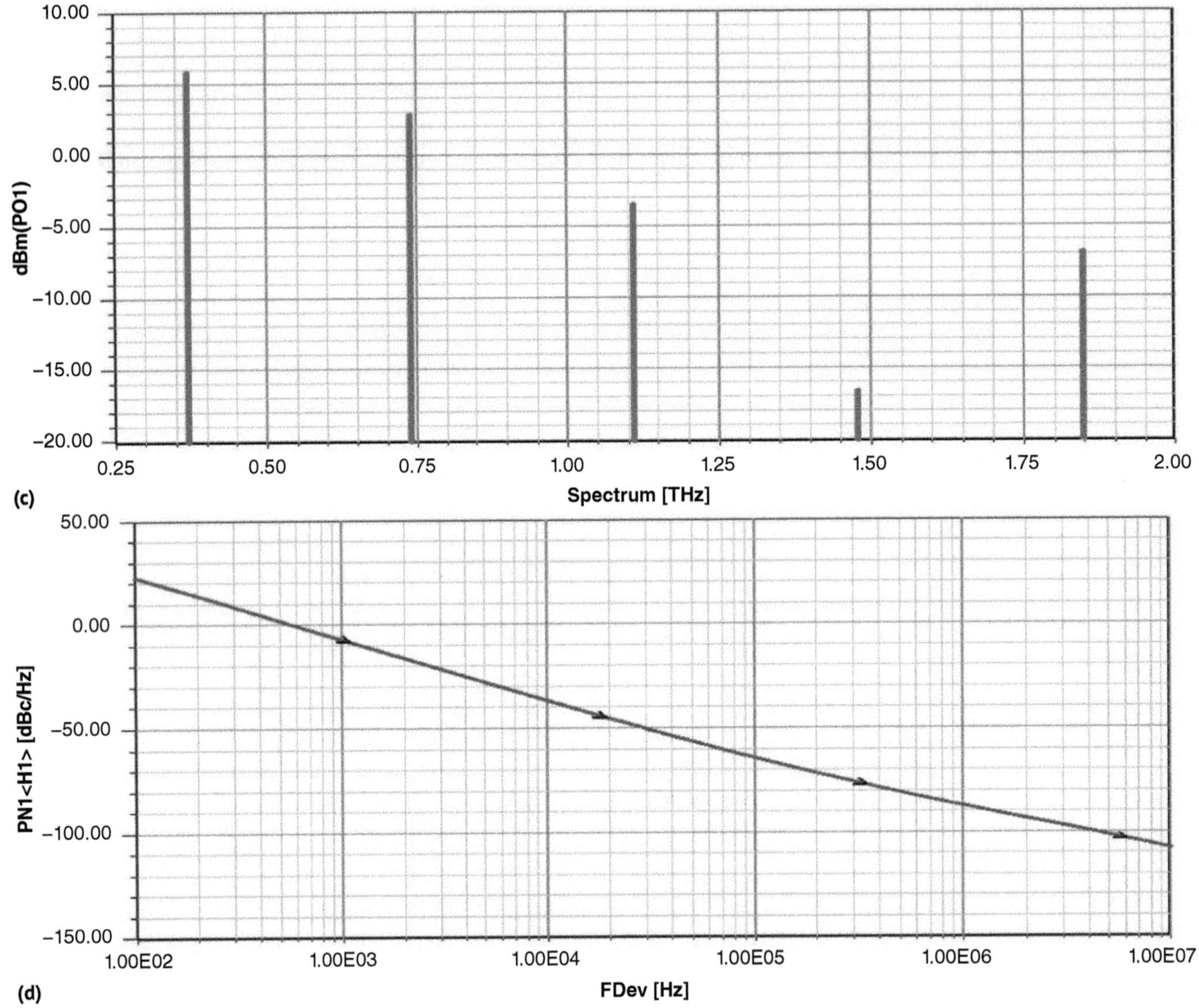

Figure 10.43 (Cont'd) (c) 370 GHz oscillator output power. (d) 370 GHz oscillator phase noise simulation.

Another interesting resonator is a radial stub. They are used quite often for termination of transmission lines. The radial stub has low impedance on one side and is open on the other side, and behaves like a quarter-wavelength resonator. Figure 10.44a shows an oscillator based on this.

A harmonic suppression of about 10 dB is expected, as seen in Figure 10.44b.

This example shows a much higher operating Q, as the output power is larger than the predicted phase noise at 100 Hz offset. Figure 10.44c shows the simulated phase noise of this 68 GHz oscillator.

Another top parameter is the amplitude stability, which is necessary to provide good performance in the sense of a stable oscillation and avoiding limiting effects. Here are some thoughts on amplitude stability.

Amplitude stability: Linearized analysis of the oscillator is convenient for determining the frequency but not the amplitude of the oscillation. The Nyquist stability criterion defines the frequency of oscillation as the frequency at which the loop phase shift is 360°, but it says nothing about the oscillation amplitude. If no provisions are taken to control the amplitude, it is susceptible to appreciable drift. Two frequently used methods for controlled amplitude are; operating the transistor in the nonlinear region or to use a second stage for amplitude limiting. For the single-stage oscillator, amplitude limiting is accomplished by designing an unstable oscillator; that is the loop gain is made greater than 1 at the frequency at which the phase shift is 180°. As the amplitude increases, the β of the transistor decreases, causing the loop gain to decrease until the amplitude stabilizes. This is a self-limiting oscillator. There are nonlinear analysis techniques predicting the amplitude of oscillation, but their results are approximate except in idealized cases, forcing the designer to resort to an empirical approach.

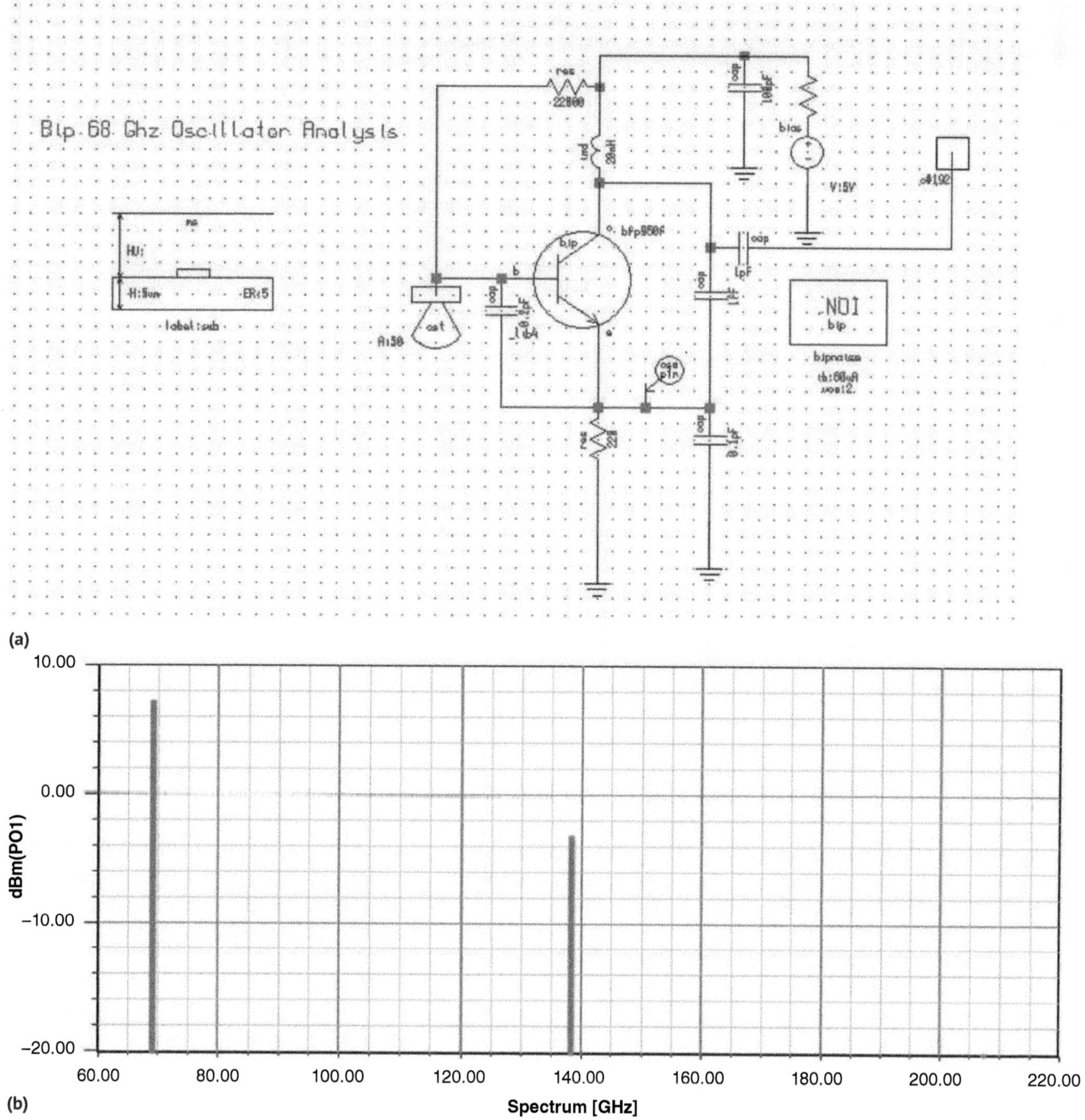

Figure 10.44 (a) 68 GHz oscillator design example. (b) Simulated output power of the 68 GHz oscillator of (a).

(Continued)

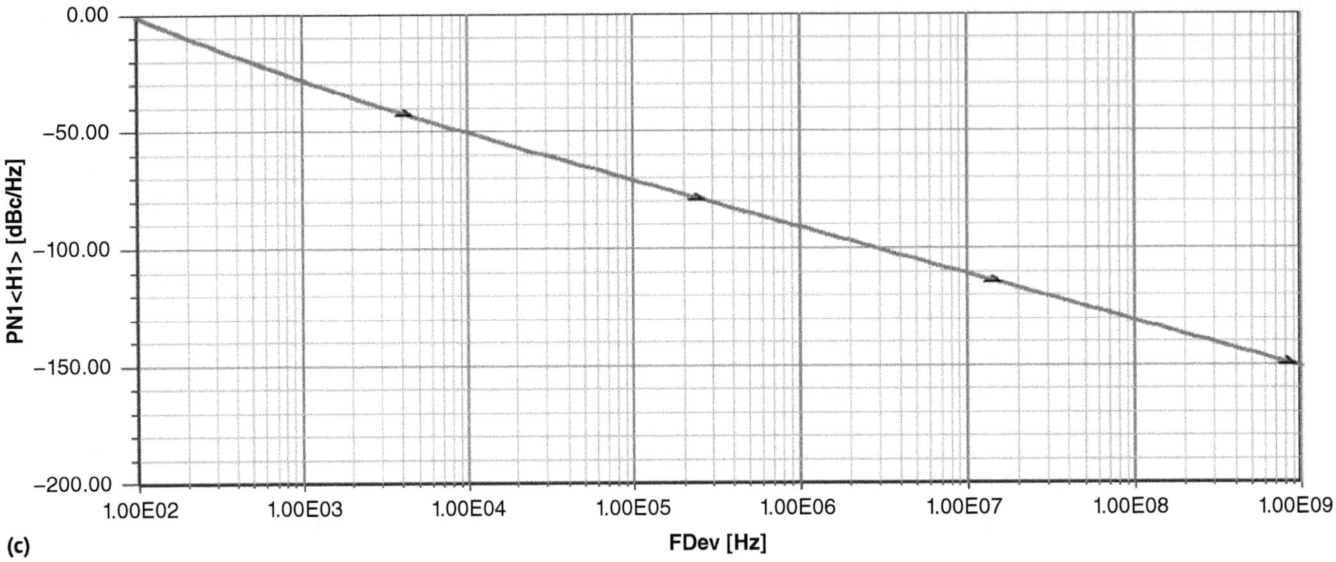

Figure 10.44 (Cont'd) (c) Simulated phase noise of the 68 GHz oscillator of (a).

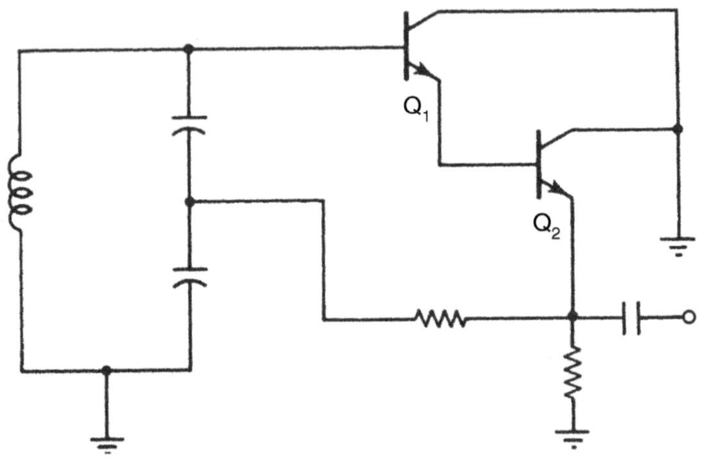

Figure 10.45 Two-stage, emitter-coupled oscillator. *Source*: Modified from [5].

An example of a two-stage emitter-coupled oscillator is shown in Figure 10.45. In this circuit, amplitude stabilization occurs as a result of current limiting in the second stage. This circuit has the additional advantage that it has output terminals that are isolated from the feedback path. The emitter signal of Q_2, having a rich harmonic content, is normally used as output. Harmonics of the fundamental frequency can be extracted at the emitter of Q_2 by using an appropriately tuned circuit. Note that the collector of Q_2 is isolated from the feedback path [5].

Phase stability: An oscillator has a frequency or phase stability that can be considered in two separate parts. First, there is *long-term* stability, in which the frequency changes over a period of minutes, hours, days, weeks, or even years. This frequency stability is normally limited by the circuit components' temperature coefficients and aging rates. The other part, *short-term* frequency stability, is measured in terms of seconds. One form of short-term instability is due to changes in phase of the system; here the term *phase stability* is used synonymously with frequency stability. It refers to how the frequency of oscillation reacts to small changes in phase shift of the open-loop system. It can be assumed that the system with the largest rate of change of phase versus frequency $d\phi/df$ will be the most stable in terms of frequency stability. Figure 10.46 shows the phase plots of two open-loop system used in oscillators. At the system crossover frequency, the phase shift is −180°. If some external influence causes a shift in phase, and for example, it adds 10° of phase lag, then the frequency will change until the total phase shift is again 0°. In this case, the frequency will decrease to the point where the open-loop phase shift is 170°. Figure 10.46 shows that Δf_2, the change in frequency associated with the 10° change in phase of GH_2, is greater than the change in frequency Δf_1, associated with the open-loop system GH_1, whose phase is changing more rapidly near the open-loop crossover frequency [5].

This qualitative discussion illustrates that $d\phi/df$ at $f = f_o$ is a measure of an oscillator's phase stability. It provides a good means of quantitatively comparing the phase stability of two oscillators. Consider the simple parallel-tuned circuit shown in Figure 10.47.

For this circuit, the two-port is

$$\frac{V_o(j\omega)}{I(j\omega)} = \frac{R}{1 + jQ[(\omega/\omega_o) - (\omega_o/\omega)]} \tag{10.50}$$

where

$$\omega_o = \frac{1}{\sqrt{LC}} \text{ and } Q = \frac{R}{\omega_o L} \tag{10.51}$$

The circuit phase shift is

$$\arg \frac{V_o}{I} = \theta = \tan^{-1} Q\left(\frac{\omega}{\omega_o} - \frac{\omega_o}{\omega}\right) \tag{10.52}$$

and

$$\frac{d\theta}{d\omega} = \frac{1/Q}{1/Q^2 + \left[(\omega^2 - \omega_o^2)/\omega_o\omega\right]^2} \frac{\omega^2 + \omega_o^2}{(\omega_o\omega)^2} \tag{10.53}$$

at the resonant frequency ω_o,

$$\left.\frac{dQ}{d\omega}\right|_{\omega=\omega_o} = \frac{2Q}{\omega_o} \tag{10.54}$$

The frequency stability factor is S_F defined as the change in phase $d\phi/d\omega$ divided by the normalized change in frequency $\Delta\omega/\omega_o$. That is,

$$S_F = 2Q \tag{10.55}$$

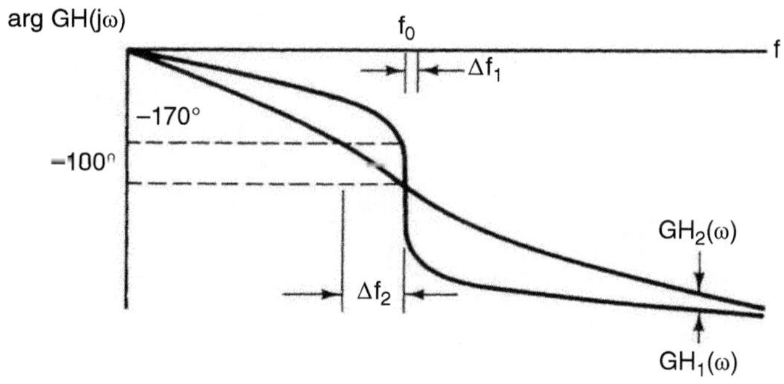

Figure 10.46 Phase plot of two open-loop systems with different resonator Qs. *Source*: Modified from [5].

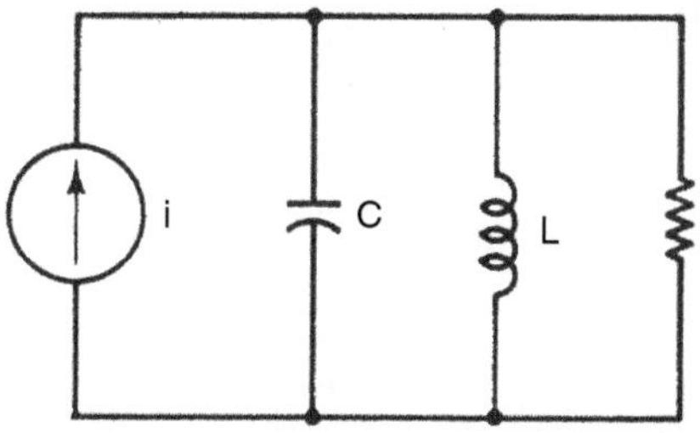

Figure 10.47 Parallel-tuned circuit for phase-shift analysis. *Source*: Modified from [5].

S_F is a measure of the short-term stability of an oscillator. Equation (10.54) indicates that the higher the circuit Q, the higher the stability factor. This is one reason for using high-Q circuits in oscillator circuits. Another reason is the ability of the tuned circuit to filter out undesired harmonics and noise.

The two oscillator types shown so far can be considered two-port oscillators because one port is responsible for oscillation and the other port supplies the output power. This type of output supposedly has better isolation (although this is not true at higher frequencies) but has a higher noise floor. A better way of extracting energy is shown in Figure 10.48.

The collector has no additional selectivity and therefore has more wideband noise. The emitter output is part of the tuned circuit, the crystal in reality is the inductor. Therefore the output is less noisy. This can be shown in Figure 10.49.

These types of oscillators are essentially what is referred to as *two-port* oscillators. The name comes from the fact that the oscillating device has two ports, with the output energy taken from the output port, typically the collector or drain, while the tuned circuit (resonator) is at the input port. For years, it was widely held that this type of design provides better isolation between the tuned circuit and termination; unfortunately, because of the Miller effect, this is not true. At the higher frequencies, changes in output-port loading affect the input port, resulting in *pulling*, a detuning effect that increases with frequency. In addition, taking the output from the collector or the drain results in a poorer signal-to-noise ratio than that achievable by extracting energy from the tuned circuit itself. The only real advantages the collector output provides are, more power and higher efficiency, but definitely at the expense of phase noise, which is often referred to as *SSB phase noise* because it is expressed in terms of the decibel ratio of the noise power in a single (the upper or lower) noise sideband, in a 1-Hz bandwidth centered at a specified frequency offset from the oscillator carrier, to the carrier power. Figure 10.49 compares the phase noise of an oscillator with the energy taken out at the collector and at the tuned circuit; the difference is highly visible.

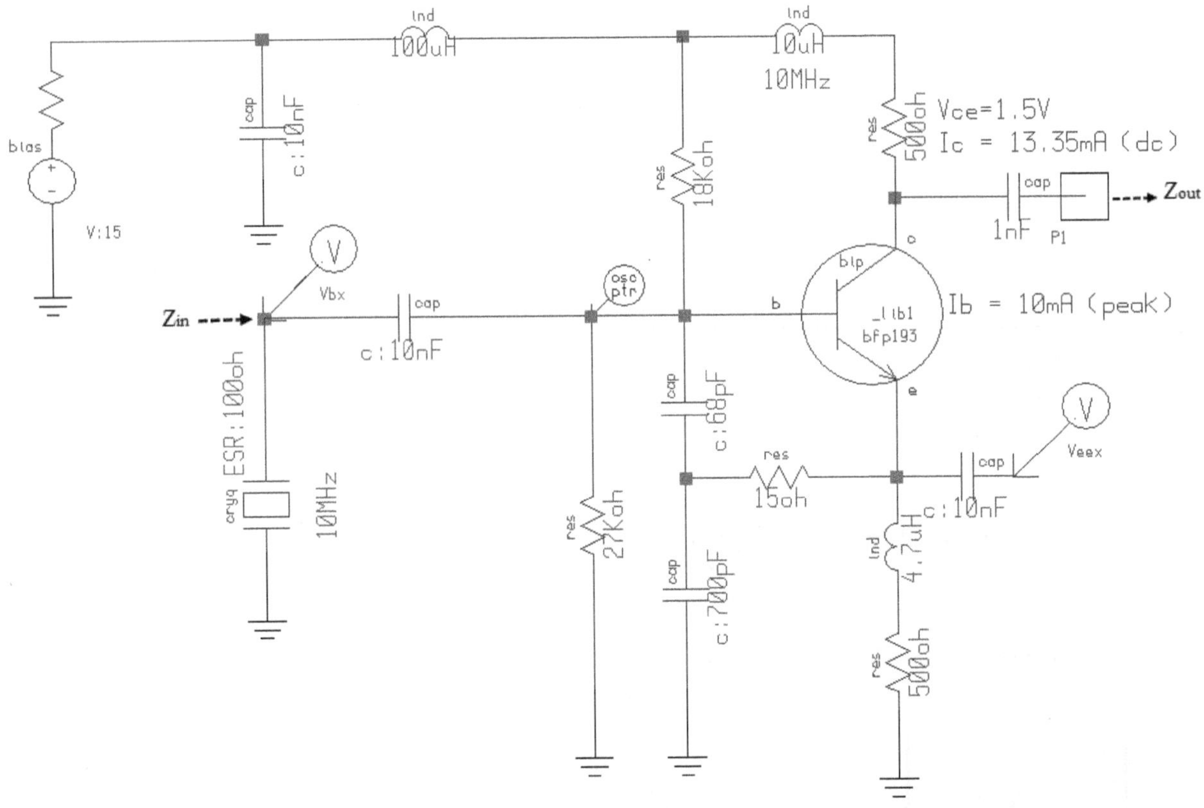

Figure 10.48 A crystal oscillator with two possible outputs: one at the collector and another at the emitter. *Source*: [8, Figure 5.10]/Anisha M. Apte.

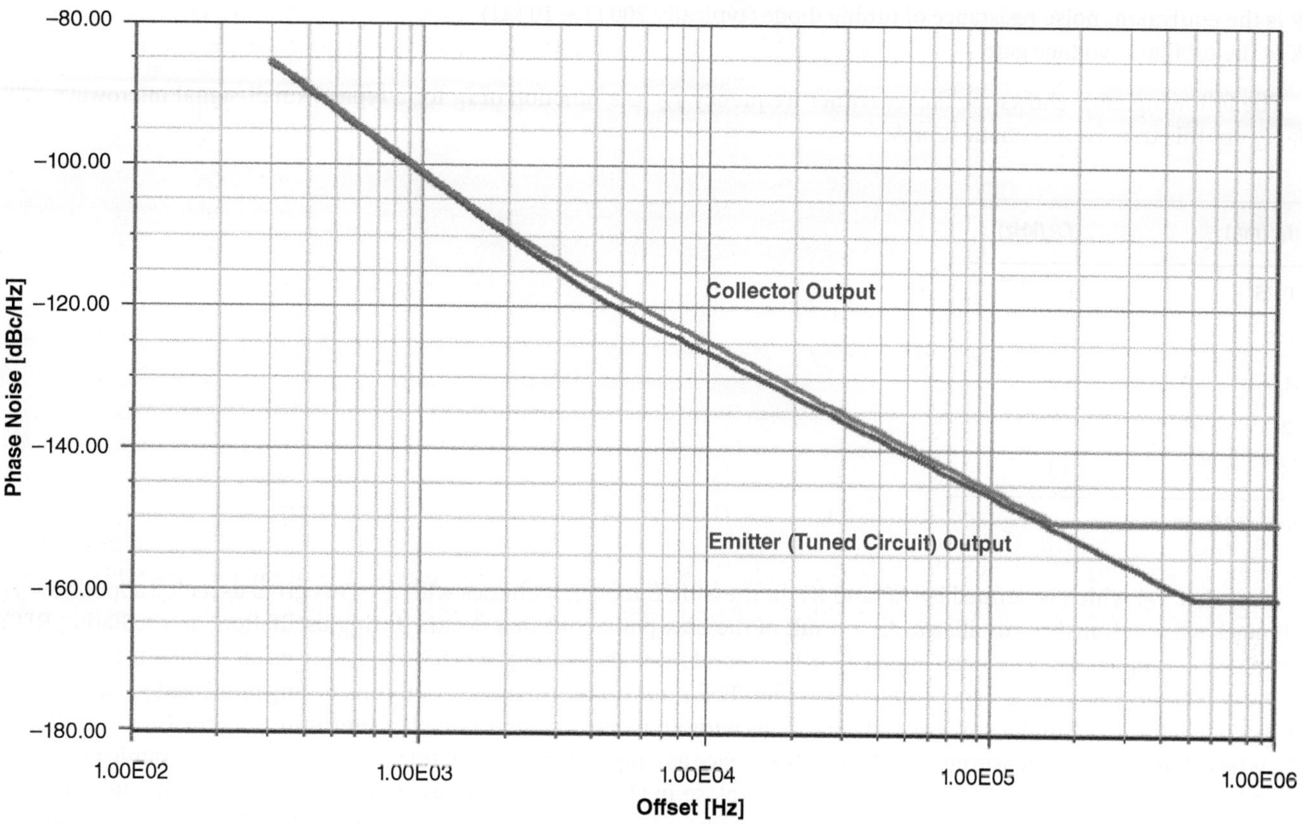

Figure 10.49 Phase noise with oscillator output taken from collector (upper trace) versus emitter (lower trace) of a BJT oscillator. *Source*: [1, Figure 5.20]/John Wiley & Sons.

Flicker Noise

We just state this here without having delved into oscillator noise theory itself. This equation is based on David Leeson's work, and work done by Dieter Scherer of Hewlett-Packard about 1978 and Ulrich L. Rohde in 1985. Scherer was the first to introduce the flicker effect to the Leeson equation and Rohde by adding the AM-to-PM (VCO, last term in the equation) conversion effect, which is caused by the nonlinear capacitance of the active devices. This equation must be further explained [3]:

$$£(f_m) = 10\log\left\{\left[1 + \frac{{f_0}^2}{(2f_m Q_{\text{load}})^2}\right]\left(1 + \frac{f_c}{f_m}\right)\frac{FkT}{2P_{\text{sav}}} + \frac{2kTRK_0{}^2}{{f_m}^2}\right\} \tag{10.56}$$

where

$£(f_m)$ is the ratio of sideband power in 1-Hz bandwidth at f_m to total power in dB
f_m is the frequency offset
f_0 is the center frequency
f_c is the flicker frequency
Q_{load} is the loaded Q of the tuned circuit
F is the noise factor
$kT = 4.1 \times 10^{-21}$ at 300 K_0 (room temperature)
P_{sav} is the average power at oscillator output

R is the equivalent noise resistance of tuning diode (typically 200 Ω − 10 kΩ)
K is the oscillator voltage gain

The following table shows the flicker corner frequency f_C as a function of I_C for a typical small-signal microwave BJT. $I_{C(\max)}$ of this transistor is about 10 mA.

I_C (mA)	f_C (kHz)
0.25	1
0.5	2.74
1	4.3
2	6.27
5	9.3

Source: Motorola

Note that f_C, which is defined by AF and KF in the SPICE model, increases with I_C. This gives us a clue about how f_C changes when a transistor oscillates. As a result of the bias-point shift that occurs during oscillation, an oscillating BJT's average I_C is higher than its small-signal I_C. KF is therefore higher for a given BJT operating as an oscillator than for the same transistor operating as a small-signal amplifier. This must be kept in mind when considering published f_C data, which is usually determined under small-signal conditions without being qualified as such. Determining a transistor's oscillating f_C is best done through measurement: Operate the device as a high-Q UHF oscillator (we suggest using a ceramic resonator-based tank in the vicinity of 1 GHz), and measure its close-in (10 Hz–10 kHz) phase noise versus offset from the carrier. f_C will correspond to a slight decrease in the slope of the phase noise versus offset curve. Generally, f_C varies with device type as follows: silicon JFETs, 50 Hz and higher; microwave RF BJTs, 1–10 kHz (as above); MOSFETs, 10–100 kHz; GaAsFETs, 10–100 MHz. Figure 10.50 shows the phase noise of oscillators using different semiconductors and resonators.

Equation (10.56) is based on Rohde et al. [3]. The additional term introduces a distinction between a conventional oscillator and a VCO. Whether the voltage- or current-dependent capacitance is internal or external makes no difference; it simply affects the frequency.

JFETs have very low Flicker contribution, approx. 50 Hz. We will use the 2N4417 in the simulation example of a 144 MHz JFET-based LC Hartley oscillator shown in Figure 10.51a, where Figure 10.51b shows its simulated power output and Figure 10.51c shows this oscillator's simulated phase noise.

Why 144 MHz? This frequency is the beginning of the 2 m HAM radio frequency band and some individuals may be interested in playing with it.

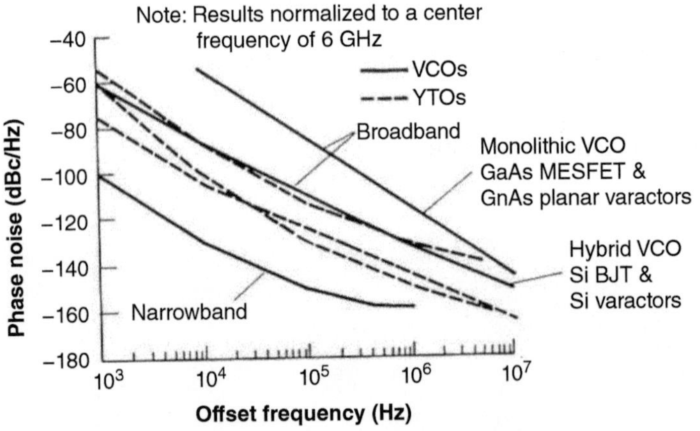

Figure 10.50 Phase noise of oscillators using different semiconductors and resonators. *Source*: [3]/John Wiley & Sons.

Example: JFET 144 MHz Hartley Oscillator It should be noted that the noise contribution of the FET is added in the simulation.

During 1965, the first police and taxi two-way radios came with a channel spacing of 25 kHz. The radios were typically operating at around 150 MHz. In order to be able to determine the synthesizer generated noise-blocking of the receiver,

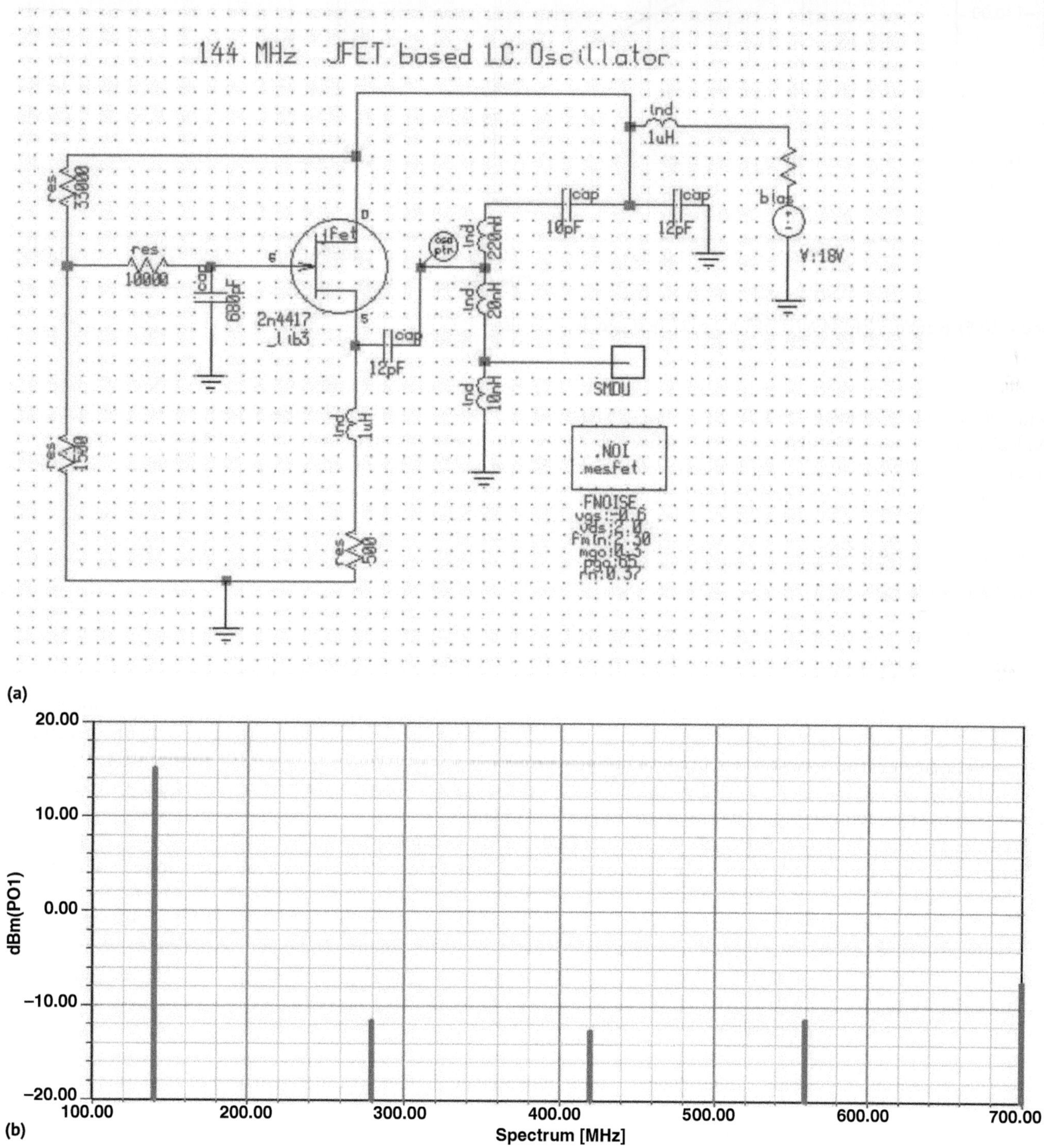

Figure 10.51 (a) 144 MHz JFET-based LC Hartley oscillator. (b) 144 MHz JFET-based LC oscillator simulated power output.

(Continued)

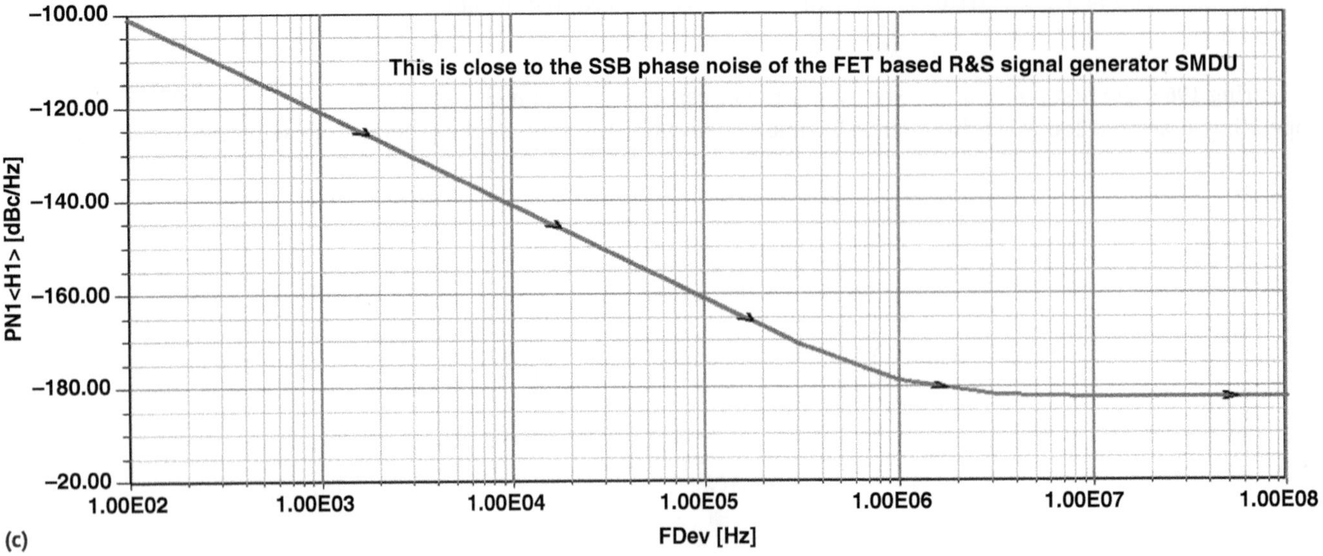

(c)

Figure 10.51 (Cont'd) (c) 144 MHz JFET-based LC oscillator phase noise.

there was need to have a better signal generator. Figure 10.51c shows the phase noise plot, where at 25 kHz offset a predicted value of the phase noise of −150 dBc/Hz is seen. There was no radio synthesizer on the market which was that good. Typical values were −135 dBc/Hz at 25 kHz channel spacing. Modern radios now resort to 6.25 kHz spacing. This circuit produces −135 dBc/Hz at 6.25 kHz. However, modern radios now use DDS-based synthesizers, where the question is how many spurs they generate rather than phase noise.

Next topic is the use of MOS transistors. They are famous for their high flicker corner frequency, but can be found in almost all integrated circuits.

MOS FET Oscillator

The following example is a MOSFET-based oscillator. Typically, 100–120 nm devices are being used. The MOSFET behaves very similar to JFET but has no diode from gate to GND and has a much higher flicker corner frequency. Also, there are sub-THz MOSFETs available. The MOSFETs are typically found only in integrated circuits and cannot be obtained individually. The circuit shown in Figure 10.52a explains a design of such an oscillator. The MOSFET on the left side is used for stabilizing the bias voltage of the oscillator. The general disadvantage in integrated circuits and the active power supply stabilization is that the active power supply adds noise to the circuit. The gate voltage of the oscillator transistor DMOS2 is determined by the 100 Ω resistor divided by the differential resistor of the supply transistor DMOS1. Figure 10.52c shows the various DC bias points for this MOS oscillator in Figure 10.52a.

Because of CMOS, the output power of −20 dBm at 36 GHz is low as expected.

In this example, Figure 10.52a shows a 36 GHz MOSFET oscillator, Figure 10.52b shows the validation of oscillation in the linear mode for a resonant frequency of approx. 36 GHz. Figure 10.52c depicts the various DC bias points for the MOS oscillator in Figure 10.52a, while Figure 10.52d shows the CAD simulated nonlinear analysis of the oscillator showing convergence. Figure 10.52e shows the drain current and voltage as a function of time and Figure 10.52f shows the simulated output power of the circuit. Lastly, the Figure 10.52g shows the CAD Simulated phase noise plot of MOS oscillator of the circuit of Figure 10.52a.

Comparing to the previous example with the JFET at 150 MHz, the resulting value of −75 dBc/Hz compared to −150 dBc is that the noise is much higher.

Now we move from single-transistor oscillators to push–pull oscillators. This is the design of choice of all cell phones and related products.

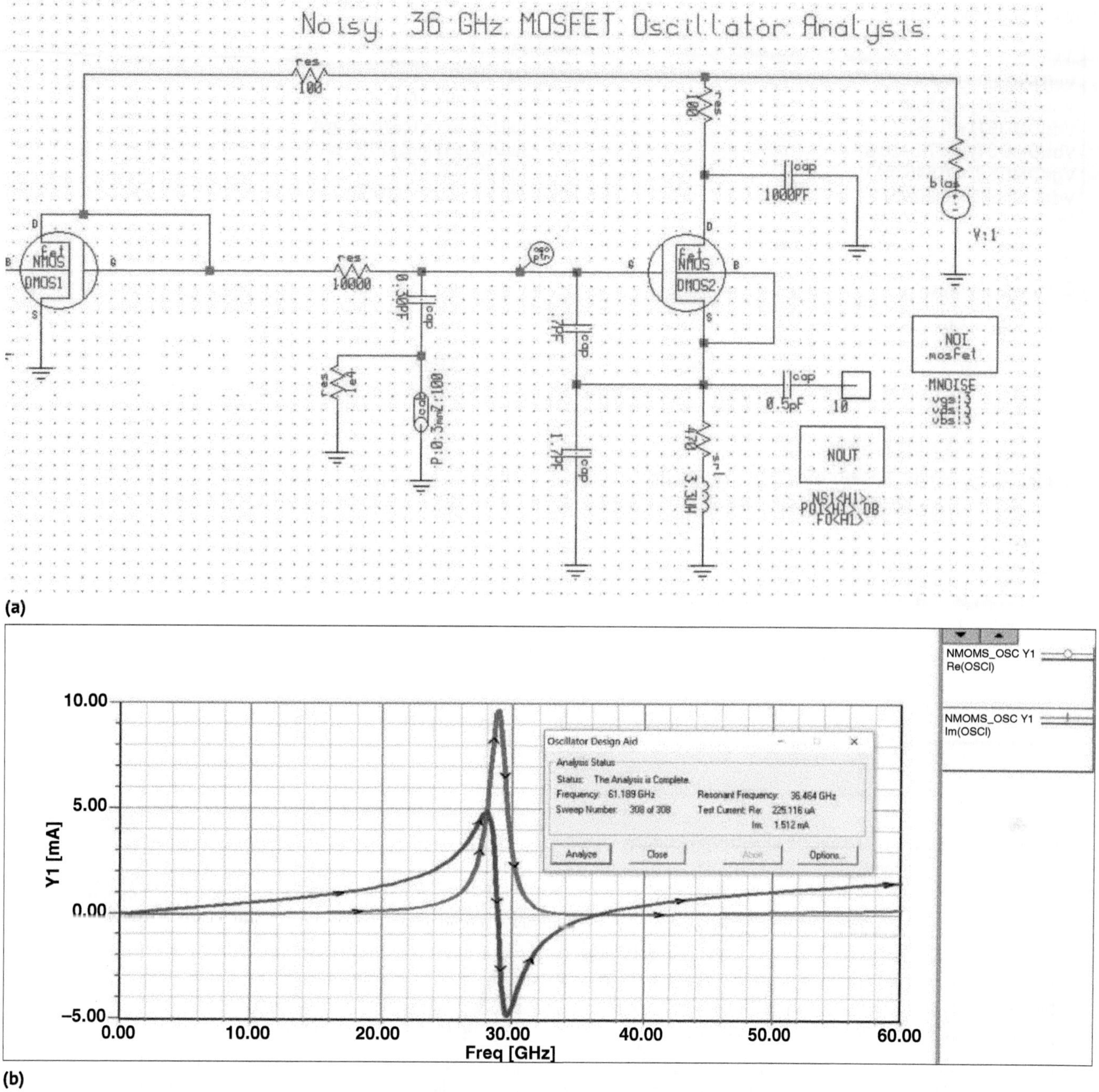

Figure 10.52 (a) 36 GHz MOSFET oscillator analysis. (b) Validation of oscillation in the linear mode for a resonant frequency of approx. 36 GHz.

(Continued)

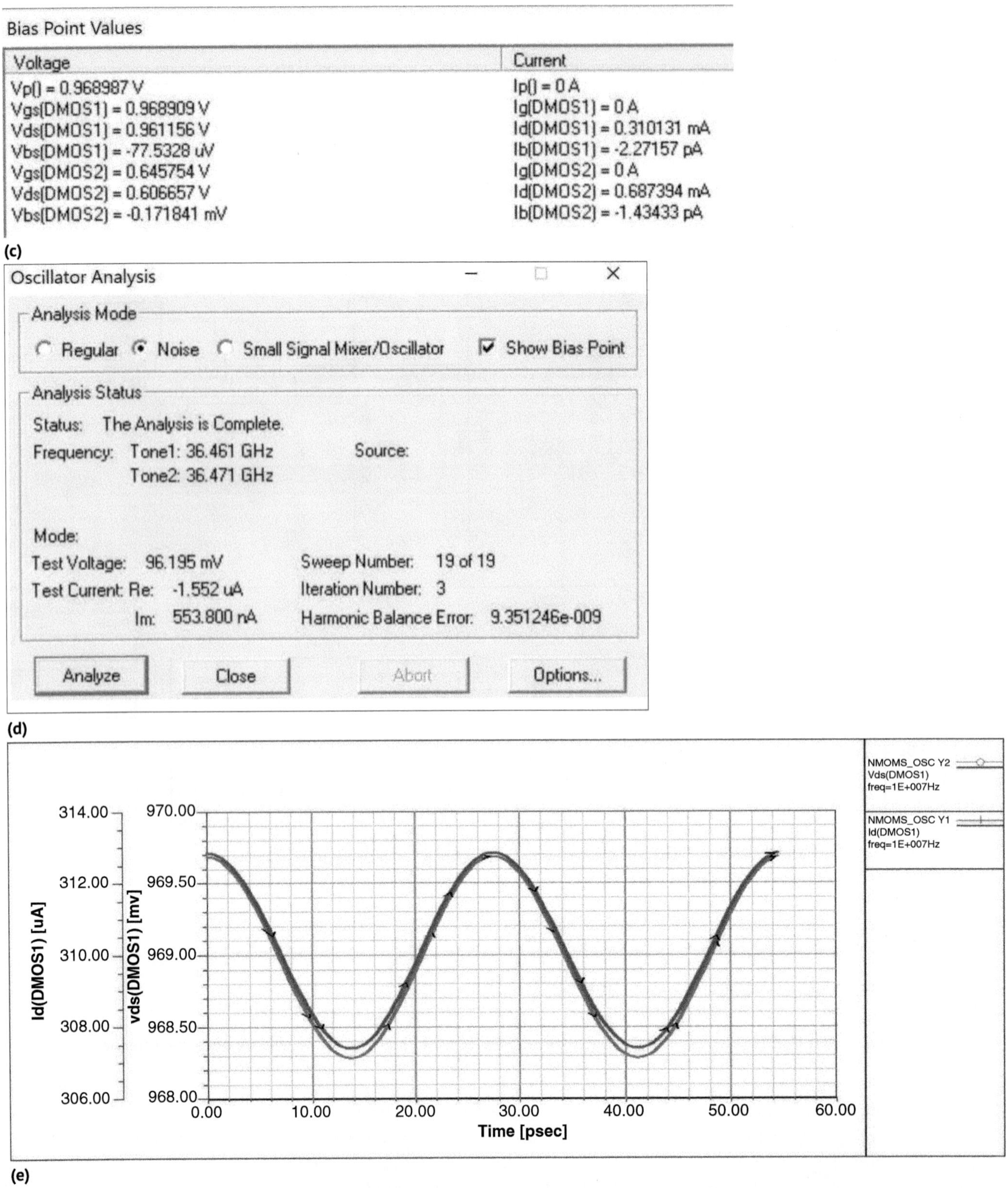

Figure 10.52 (Cont'd) (c) Various DC bias points for the MOS oscillator in (a). (d) Nonlinear analysis of the oscillator of (a) showing convergence. (e) Drain current and voltage as a function of time (for the circuit of (a)).

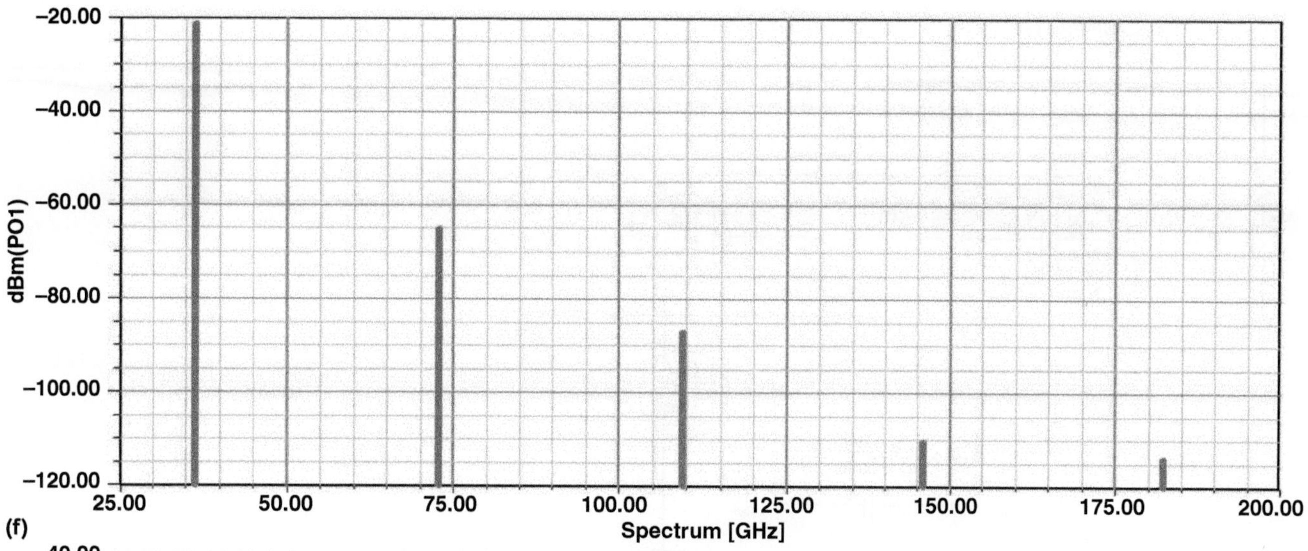

(f)

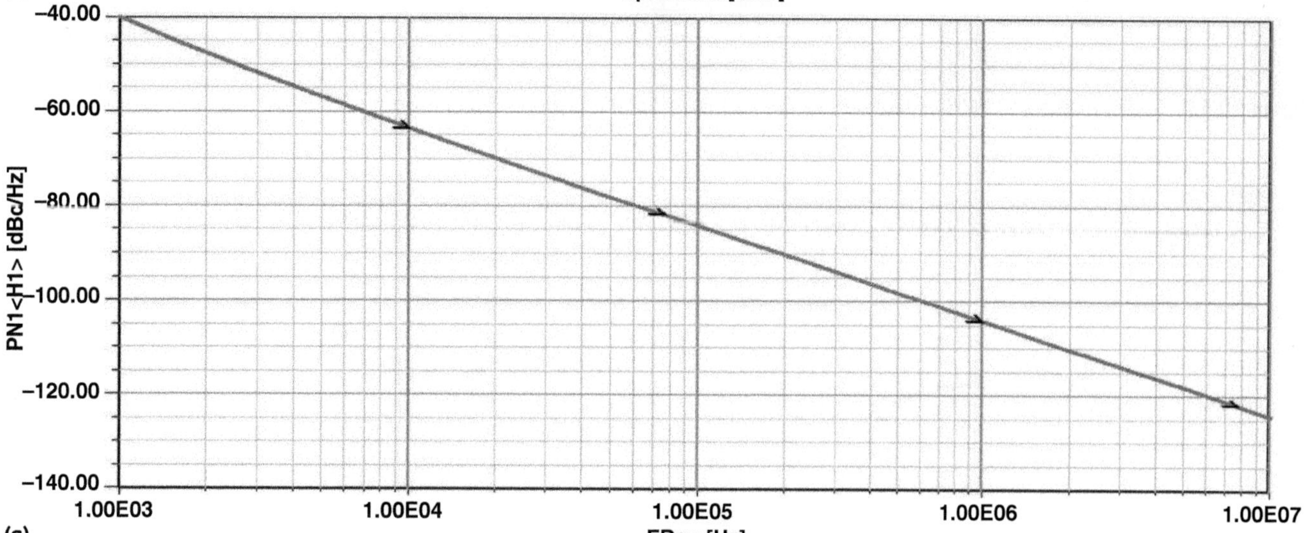

(g)

Figure 10.52 (Cont'd) (f) Simulated output power of the circuit of (a). (g) Simulated phase noise plot of MOS oscillator of the circuit of (a).

10.8 Push–Pull Oscillator

2400 MHz MOSFET-based Push–Pull Oscillator [3, Ch. 8]

Wireless applications are extremely cost-sensitive, and when implemented as an RFIC, they are designed using silicon technology. Most mixers in RFIC's are built on the principle of differential amplifiers (Gilbert cell) and require a phase and out of phase signal (symmetrical drive). For these symmetrical requirements, this is best achieved using a push–pull technology with two outputs. The design choices are SiGe transistors or BiCMOS transistors. The submicron devices in 0.35 micron technology and smaller are ideally suited for this frequency application. The 0.25 and smaller technology is more costly, but does not provide a significant advantage. As will be seen, the critical phase noise is determined by the Q of the inductor and other elements of the resonator and by the flicker noise from the device.

Figure 10.53 shows the circuit of the 2400 MHz integrated CMOS oscillator 0.35 um in cross-coupled (push–pull) configuration [26–44].

The aforementioned circuit (Figure 10.53) [3, Figure 8.23] uses a cross-coupled CMOS-NMOS pair as an oscillator. The advantage compared to an all NMOS structure is that it generates a large symmetrical signal swing and balances out the

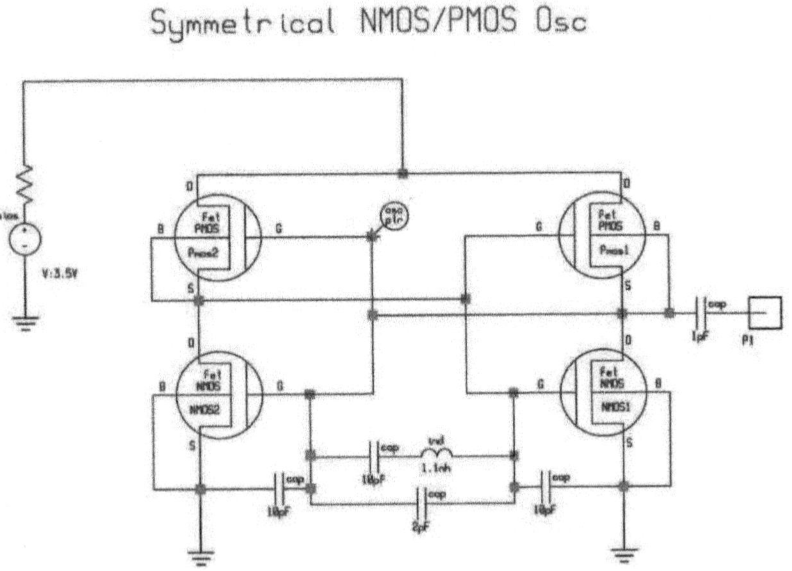

Figure 10.53 Circuit of the 2400 MHz integrated CMOS oscillator. *Source*: [3]/John Wiley & Sons.

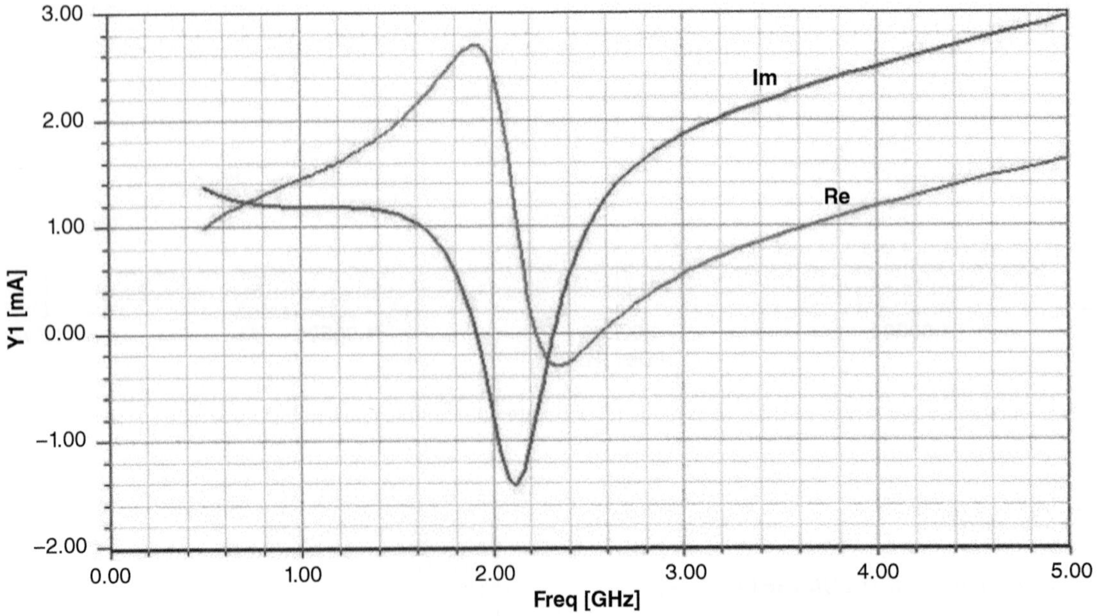

Figure 10.54 The real and imaginary currents, which cause the negative resistance for oscillation. *Source*: [3]/John Wiley & Sons.

pull-up and pull-down signals, resulting in a better noise. This type of topology rejects the common mode noise and substrate noise.

Figure 10.54 shows the starting condition that requires a negative resistance and a cancellation of the reactances at the frequency of oscillation. The currents shown in Figure 10.24 indicates that this condition is met.

It is important to notice that the condition of zero reactance does not quite occur at the point of most negative current. Since the circuit is totally symmetrical, only the condition $C_1 = C_2$ can be met (Section 10.2). C_1 and C_2 refer to the gate source capacitance of the field-effect transistors. As outlined previously, this is not necessarily the best condition for phase noise.

Figures 10.55 and 10.56 show the predicted phase and RF output power, including harmonic contents.

Another example operating at the same frequency, with a different scaled MOS transistor, is considered to show the sensitivity of the circuit, and shown in Figure 10.67a, followed by the simulation results. Figure 10.130a is another example of the NMOS/PMOS oscillator simulated to operate at 57 GHz. The simulation results are also shown.

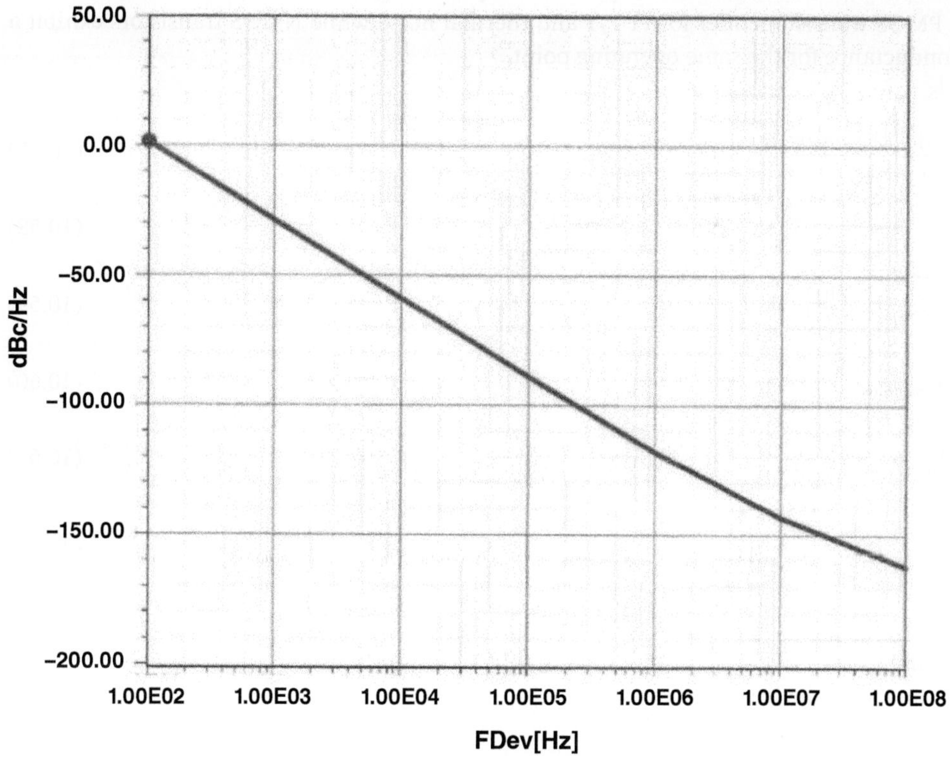

Figure 10.55 Predicted phase noise of the 2400 MHz MOSFET oscillator. *Source*: [3]/John Wiley & Sons.

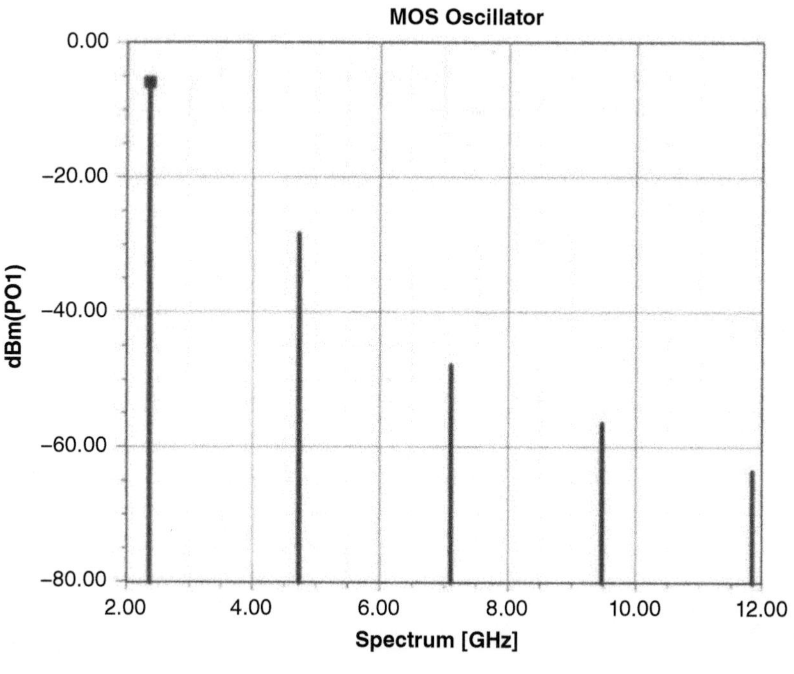

Figure 10.56 Predicted output spectrum of the 2400 MHz MOSFET oscillator. *Source*: [3]/John Wiley & Sons.

10.9 Design Parameters [3]

10.9.1 Design Equations

1. The transconductance.

   Figure 10.57 shows a cross-coupled PMOS and a cross-coupled NMOS pair using CMOS devices.

   According to the literature, PMOS transistors offer lower 1/f and thermal noise while NMOS transistors exhibit a higher f_T and a higher transconductance for the same operating point.

   The total transconductance is

$$[g_m]_{\text{large-signal}} = -\frac{[g_m]_{\text{PMOS}} + [g_m]_{\text{NMOS}}}{2} \tag{10.57}$$

$$[g_m]_{\text{large-signal}} = \frac{\partial I_{\text{ds}}}{\partial V_{\text{gs}}} \tag{10.58}$$

$$[g_m]_{\text{NMOS}} = \sqrt{2 I_{\text{ds}} \mu_{\text{nmos}} C_{\text{ox-nmos}} \left[\frac{w}{L}\right]_{\text{nmos}}} \tag{10.59}$$

$$[g_m]_{\text{PMOS}} = \sqrt{2 I_{\text{ds}} \mu_{\text{pmos}} C_{\text{ox-nmos}} \left[\frac{w}{L}\right]_{\text{pmos}}} \tag{10.60}$$

$$\frac{\partial I_{\text{ds}}}{\partial V_{\text{gs}}} = K_p \frac{w}{L}(V_{\text{gs}} - V_{\text{th}}) \tag{10.61}$$

$$[g_m]_{\text{large-signal}} = \sqrt{\frac{2 w K_p I_{\text{ds}}}{L}} \tag{10.62}$$

where

K_p is the transconductance parameter
μ_{pmos} is the carrier mobility of the PMOS device

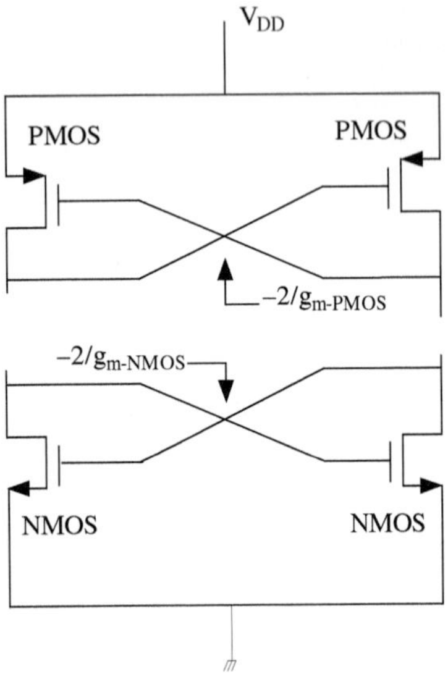

Figure 10.57 Determining the transconductance of the differential circuit of the cross-coupled PMOS and NMOS pair. *Source*: [3]/John Wiley & Sons.

μ_{nmos} is the carrier mobility of the NMOS device
C_{ox} is the unit capacitance of the gate oxide

2. The transconductance parameter.
   The transconductance parameter is defined as

$$K_p = \mu \, C_{\text{ox}} \tag{10.63}$$

where μ is the carrier mobility and C_{ox} is the unit capacitance of the gate oxide.

3. The gate oxide capacitance of the device.
   The unit capacitance of the gate oxide C_{ox} is given as

$$C_{\text{ox}} = \varepsilon_{\text{ox}} \left[\frac{w_i l_i}{t_{\text{ox}}} \right] \tag{10.64}$$

where
ε_{ox} is the permittivity of the oxide
t_{ox} is the thickness of the oxide layer between spiral and substrate
w_i is the width of the spiral line
l_i is the length of the spiral line.

4. The drain current.
   The drain current is

$$I_{\text{ds}} = \frac{1}{2} \left[K_p \frac{w}{L} (V_{\text{gs}} - V_{\text{th}})^2 \right] \tag{10.65}$$

$$I_{\text{ds}} = \frac{g_m}{2} (V_{\text{gs}} - V_{\text{th}}) \tag{10.66}$$

where $(V_{\text{gs}} - V_{\text{th}})$ is defined as

$$(V_{\text{gs}} - V_{\text{th}}) = \sqrt{\frac{2 I_{\text{ds}} L}{K_p w}} \tag{10.67}$$

5. The size of the device.
   The size of the device determines the transconductance of the transistor and the large signal transconductance needs to be large enough to sustain oscillation and compensate for the losses of the resonator.
   The expression of the ratio of the channel width (gate) and channel length (gate) is

$$\frac{w}{L} = \frac{(2SG_P)^2}{2K_p I_{\text{ds}}} \rightarrow \frac{(g_m)^2}{2K_p I_{\text{ds}}} \tag{10.68}$$

where w is the width of the channel (gate) and L is the length of the channel (gate) of the device.

6. The total equivalent resistance at resonant frequency.
   Figure 10.58a,b shows the equivalent cross-coupled oscillator resonant circuit and the corresponding equivalent resistances at resonance condition.

The total equivalent parallel resistor at resonance frequency is

$$R_T = \frac{2R_P}{2 - R_P(g_{m-\text{NMOS}} + g_{m-\text{PMOS}})} \tag{10.69}$$

$$Q_L = \frac{R_P}{\omega L} \tag{10.70}$$

where $g_{m-\text{NMOS}}$ and $g_{m-\text{PMOS}}$ are the corresponding large signal transconductances of the NMOS and PMOS devices.

For a symmetrical output signal, the large signal transconductance of the NMOS and PMOS transistors has to be ideally equal as $g_{m-\text{NMOS}} = g_{m-\text{PMOS}} = g_m$ and the equivalent resistance at resonance condition is

$$R_T = \frac{1}{\left[\frac{1}{R_P} - g_m \right]} \tag{10.71}$$

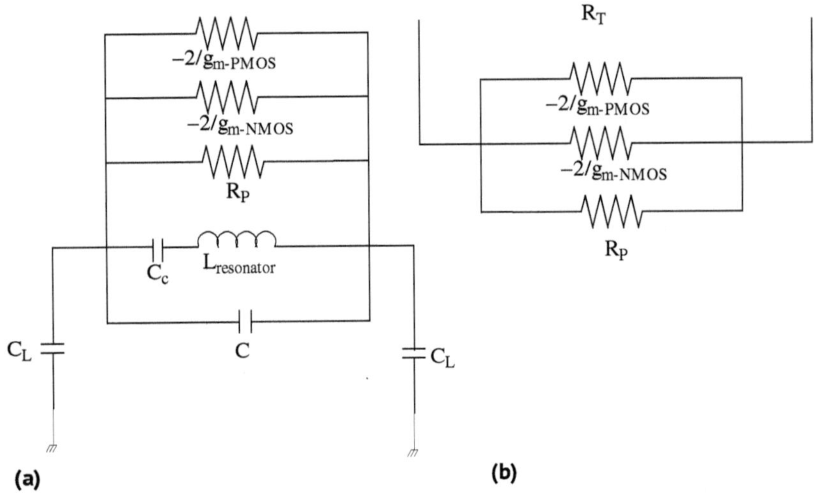

Figure 10.58 (a) The equivalent cross-coupled oscillator resonator circuit and (b) the equivalent resistances at resonance condition. *Source*: [3]/John Wiley & Sons.

The differential negative resistance generated by the cross-coupled NMOS and PMOS transistor pair compensates the parallel loss resistance R_P of the resonator circuit.

7. Start-up Conditions

For the start-up condition and guaranteed sustained oscillation condition, the value of R_T must be negative and

$$R_T = \frac{1}{\left[\frac{1}{R_P} - g_m\right]} < 0 \Rightarrow g_m > \frac{1}{R_P} \tag{10.72}$$

From the loop gain criteria using a stability factor of 2 (loop gain = 2, the gain is adjusted to 1 by self-adjusting the conducting angle of the circuit), the start-up condition is

$$\frac{g_m}{SG_p} \rightarrow 2 \tag{10.73}$$

where

$$G_P = \frac{1}{R_P} \tag{10.74}$$

S is the stability factor
R_P is the equivalent parallel loss resistance of the resonator

10.9.2 Design Calculations [3]

Parallel Loss Resistance of the Resonator

The equivalent parallel loss resistance of the resonator is given as

$$R_P = (1 + Q^2)R_s \Rightarrow 101R_s \quad \text{for} \quad (Q = 10) \tag{10.75}$$

where R_s is the series loss resistance.

$$[R_P]_{f=2400\text{MHz}} = Q\omega L_{\text{ind}} = 190\,\Omega \tag{10.76}$$

where $Q = 10$, $L_{\text{ind}} = 1.1$ nH and $G_P = \frac{1}{R_P} \Rightarrow 6.577$ mS

10.9.3 Large Signal Transconductance

$$[g_m]_{\text{large-signal}} = \sqrt{\frac{2wK_pI_{ds}}{L}} = \sqrt{\frac{2*(250E-6)*(35.6E-6)*(14.8E-3)}{0.35E-6}} = 27.435\,\text{mS} \tag{10.77}$$

10.9.4 Size of the Device

The width of the CMOS is given as

$$\frac{w}{L} = \frac{(2SG_P)^2}{2K_pI_{ds}} \rightarrow \frac{(g_m)^2}{2K_pI_{ds}} \tag{10.78}$$

$$\frac{w}{L} = \frac{(2SG_P)^2}{2K_pI_{ds}} = 714.3\ \mu\text{m} \tag{10.79}$$

For $L = 0.35\ \mu\text{m}$; $w = 250\ \mu\text{m}$

where

$K_p = 35.6\ E-6$

$I_{ds} = 14.8\,\text{mA}$

$G_P = 6.577\,\text{mS}$

$S = 2$

10.9.5 Oscillation Frequency

The frequency of the oscillation is given as

$$f_0 = \frac{1}{2\pi\sqrt{L_{\text{resonator-tan}\,k}C_{\text{resonator-tan}\,k}}} \tag{10.80}$$

$$C_{\text{resonator-tan}\,k} = \frac{1}{2}[C_{\text{NMOS}} + C_{\text{PMOS}} + C_L + C] \tag{10.81}$$

where

$$C_{\text{NMOS}} = 4C_{\text{gd-nmos}} + C_{\text{gs-nmos}} + C_{\text{db-nmos}} \tag{10.82}$$

$$C_{\text{PMOS}} = 4C_{\text{gd-pmos}} + C_{\text{gs-pmos}} + C_{\text{db-pmos}} \tag{10.83}$$

For the cross-coupled configuration, $C_{\text{NMOS-Pair}}$ is the series combination of the two C_{NMOS} and is given as

$$C_{\text{NMOS-Pair}} = 2C_{\text{gd-nmos}} + \frac{1}{2}C_{\text{gs-nmos}} + \frac{1}{2}C_{\text{db-nmos}} \tag{10.84}$$

Similarly, $C_{\text{PMOS-Pair}}$ is the series combination of the two C_{PMOS} and is given as

$$C_{\text{PMOS-Pair}} = 2C_{\text{gd-pmos}} + \frac{1}{2}C_{\text{gs-pmos}} + \frac{1}{2}C_{\text{db-pmos}} \tag{10.85}$$

The capacitance of the resonator is given as

$$C_{\text{resonator- tan}\,k} = \frac{1}{2}[C_{\text{NMOS}} + C_{\text{PMOS}} + C_L + C] \tag{10.86}$$

where

$C_L = 10$ pF (load capacitance)

$C = \frac{1}{2}$-resonator parallel capacitance

$$f_0 = \frac{1}{2\pi\sqrt{L_{\text{resonator- tan }k} C_{\text{resonator- tan }k}}} = \frac{1}{2\pi\sqrt{1.1E-9 * 3.3E-12}} = 2400\,\text{MHz} \tag{10.87}$$

where

$L_{\text{resonator- tan }k} = 1.1\,\text{nH}$

$C_{\text{resonator- tan }k} = 3.3\,\text{pF}$

10.10 Phase Noise

The phase noise of CMOS oscillators has been subject to endless discussions. The main contributors still are the resonant circuit with a low Q and the flicker frequency contribution from the device. From Rohde et al. [3, Ch. 8], we take the following equations and adapt them to the CMOS device.

$$PN_{\text{inr}}(\omega_0+\omega) = \frac{4KT}{R_P}[\text{NFT}_{\text{inr}}(\omega_0)]^2 = \frac{4KT}{R_P}\left\{\frac{1}{2}\left[\frac{1}{2\omega_0 C_{\text{eff}}}\right]\left[\frac{\omega_0}{\omega}\right]\right\}^2 \tag{10.88a}$$

$\longrightarrow$ Phase noise contribution from the resonator

$$PN_{\text{Vgn}}(\omega_0+\omega) = 4KTr_b[\text{NFT}_{\text{Vgn}}(\omega_0)]^2 = 4KTr_b\left\{\frac{1}{2}\left[\frac{C_1+C_2}{C_2}\right]\left[\frac{1}{2Q_0}\right]\left[\frac{\omega_0}{\omega}\right]\right\}^2 \tag{10.88b}$$

$\longrightarrow$ Phase noise contribution from the gate resistance

$$PN_{\text{ign}}(\omega_0+\omega) = 2qI_b[\text{NFT}_{\text{ign}}(\omega_0)]^2 = 2qI_b\left\{\frac{1}{2}\left[\frac{C_2}{C_1+C_2}\right]\left[\frac{1}{\omega_0 C_{\text{eff}}}\right]\left[\frac{\omega_0}{\omega}\right]\right\}^2 \tag{10.88c}$$

$\longrightarrow$ Phase noise contribution from the gate current

$$PN_{\text{ibn}}(\omega_0+\omega) = \frac{K_f I_b^{AF}}{f_m}[\text{NF}_{\text{ibn}}(\omega_0)]^2 = \frac{K_f I_b^{AF}}{\Delta\omega}\left\{\frac{1}{2}\left[\frac{C_2}{C_1+C_2}\right]\left[\frac{1}{2\omega_0 Q_0 C_{\text{eff}}}\right]\left[\frac{\omega_0}{\omega}\right]\right\}^2 \tag{10.88d}$$

$\longrightarrow$ Phase noise contribution from the flicker noise of the transistor

10.88c plus 10.88d gives the total noise contribution from the gate.

$$PN_{\text{idn}}(\omega_0+\omega) = 2qI_d[NFT_{\text{ign}}(\omega_0)]^2 = 2qI_d\left\{\frac{1}{2}\left[\frac{C_1}{C_1+C_2}\right]\left[\frac{1}{2\omega_0 C_{\text{eff}}}\right]\left[\frac{\omega_0}{\Delta\omega}\right]\right\}^2 \tag{10.88e}$$

$\longrightarrow$ Phase noise contribution from the drain current

The following values were used.

$R_p = 190\,\Omega$

$f_0 = 2.4\,\text{GHz}$

$L = 1.1\,\text{nH}$

$C_0 = 2\,\text{pF}$

$C_1 = C_2 = 0.2\,\text{pF}$

$n = 2$

$I_g = 100\,\mu\text{A}$

$I_d = 14\,\text{mA}$

$$AF = 2$$

$$KF = 5E - 5$$

$$q = 1.6E - 19$$

$$T = 290 \text{ K}$$

and the following contributions were obtained at 1 MHz offset

$$\text{PN1} = -117.78 \text{ dBc/Hz}$$

$$\text{PN2} = -146.37 \text{ dBc/Hz}$$

$$\text{PN3} = -123.4 \text{ dBc/Hz}$$

$$\text{PN4} = -140.9 \text{ dBc/Hz}$$

These calculations show that the phase noise contribution from the tuned circuit dominates and sets the value at −117.78 dBc/Hz.

The circuit was then analyzed using Microwave Harmonica/Ansoft Designer, using a lossy circuit with a Q_0 of 10 and using the SPICE-type parameters, which were obtained from the manufacturer.

The output power measured single-ended was −7 dBm. Figure 10.59 shows the simulated output power and harmonic contents. The accuracy of the prediction is within 1 dB.

Figure 10.60 shows the predicted phase noise from Designer and the phase noise prediction from the set of equations shown earlier. It should be pointed out that at close-in frequencies the flicker noise contribution dominates, while in the medium range, the resonator dominates and for high currents, the drain current adds significant noise.

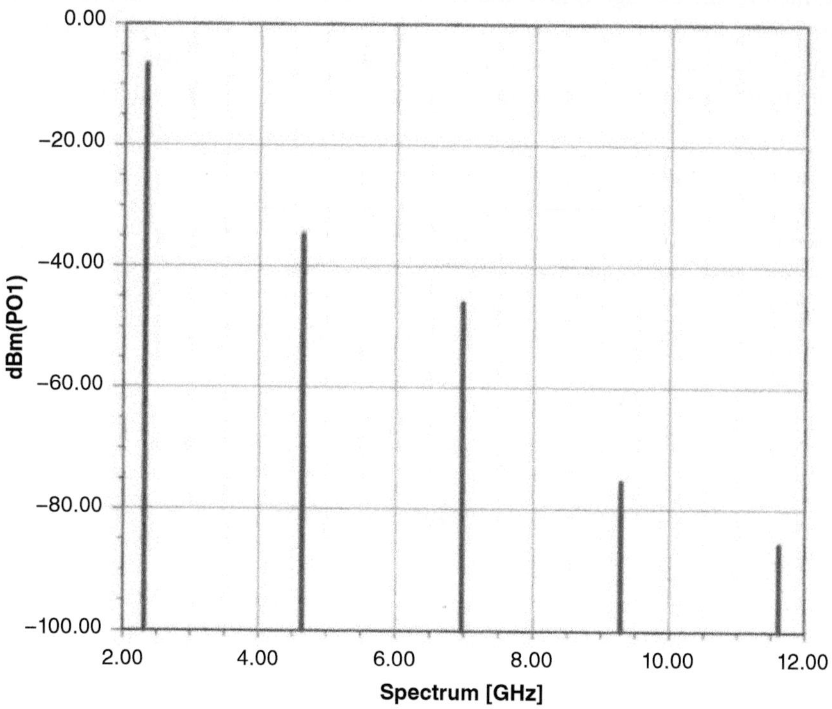

Figure 10.59 The predicted output spectrum of the CMOS oscillator. *Source*: [3]/John Wiley & Sons.

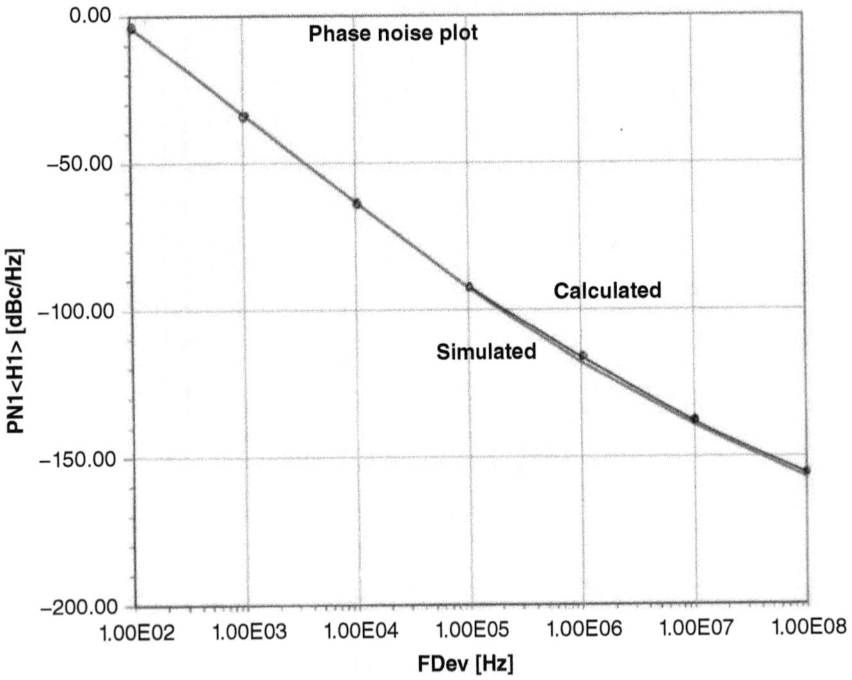

Figure 10.60 The predicted phase noise from Ansoft Designer. *Source*: [3]/John Wiley & Sons.

This approach has shown a very good agreement between the simulations and calculations, as demonstrated. Table 10.4 shows a list of oscillators implemented in various technologies. It is apparent from the list that this design is state-of-the-art. The publications, which cover this topic, have analyzed various other contributions, both from the transistor and the tuning mechanism. When FETs are used as varactors, the average Q is in the vicinity of 30, which means that the low Q inductor is still responsible for the overall phase noise. The three areas of improvement are the power supply voltage, the Q, and the device selection. So far, the power supply voltage has not been addressed; however, latest designs operating at 1.5 V show a poorer noise performance. Their distinct trade-offs and the application dictate if such degradation is allowable.

Table 10.4 A list of oscillators implemented in various technologies.

Technology	Type	Frequency	Bias	Phase noise
0.8 μm BiCMOS	*LC*	200 MHz	40 μA@2 V	−104dBc/Hz@24 kHz
0.35 μm CMOS	Ring	450 MHz	@3.3 V	−99 dBc/Hz@100 kHz
1.2 μm CMOS	Ring	900 MHz	1 mA@5 V	−83 dBc/Hz@100 kHz
0.8 μm CMOS	Ring	800 MHz	@5 V	−80 dBc/Hz@100 kHz
0.6 μm CMOS	Ring	900 MHz	10 mA@3 V	−101 dBc/Hz@100 kHz
0.6 μm CMOS	*LC*	900 MHz	1.2 mA@2.5 V	−102 dBc/Hz100 kHz
0.5 μm CMOS	Ring	900 MHz	6.2 mA@2.5 V	−105 dBc/Hz@600 kHz
25 GHz bipolar	*LC*	800 MHz	1 mA@2.7 V	−106 dBc/Hz@100 kHz
0.8 μm CMOS	*LC*	1.1 GHz	4.7 mA@2.7 V	−112 dBc/Hz@100 kHz
0.8 μm CMOS	Ring	1.27 GHz	2.64 mA@5 V	

Technology	Type	Frequency	Bias	Phase noise
0.35 μm CMOS	LC	1.2 GHz	3.7 mA@2.5 V	−153 dBc/Hz@3 MHz
0.35 μm CMOS	LC	1.3 GHz	6 mA@2 V	−119 dBc/Hz@600 kHz
0.6 μm CMOS	LC	1.6 GHz	12 mA@3 V	−99 dBc/Hz@100 kHz
0.7 μm CMOS	LC	1.8 GHz	8 mA@3 V	−115 dBc/Hz@200 kHz
0.7 μm CMOS	LC	1.8 GHz	4 mA@1.5 V	−116 dBc/Hz@600 kHz
0.6 μm CMOS	LC	1.8 GHz	3.8 mA@2.7 V	−100 dBc/Hz@100 kHz
0.6 μm CMOS	LC	1.8 GHz	2.7 mA@2.7 V	−101 dBc/Hz@100 kHz
0.4 μm CMOS	LC	1.8 GHz	3.7 mA@3 V	−118 dBc/Hz@600 kHz
0.35 μm CMOS	LC	2 GHz	4 mA@2.5 V	−117 dBc/Hz@600 kHz
0.35 μm CMOS	LC	1.9 GHz	9.2 mA@3 V	−122 dBc/Hz@600 kHz
0.35 μm CMOS	LC	2 GHz	7.5 mA@3 V	−87 dBc/Hz@100 kHz
0.35 μm CMOS	LC	2 GHz	7.5 mA@3 V	−87 dBc/Hz@100 kHz
0.25 μm CMOS	LC	1.9 GHz	5 mA@2.5 V	−100 dBc/Hz@100 kHz
0.25 μm CMOS	LC	1.8 GHz	7.8 mA@2.5 V	−143 dBc/Hz@3 MHz
0.2 μm SIMOX	LC	2 GHz	6 mA@0.5 V	−110 dBc/Hz@1 MHz
0.65 μm BiCMOS	LC	1.8 GHz	19 mA@1.8 V	−125 dBc/Hz@600 kHz
18 GHz BiCMOS	Ring	2 GHz	0.3 mA@3 V	−75 dBc/Hz@1 MHz
25 GHz 0.5 μm bipolar	LC	1.9 GHz	6.4 mA@3.3 V	−100 dBc/Hz@100 kHz
20 GHz bipolar	LC	2.5 GHz	7 mA@2 V	−104 dBc/Hz@100 kHz
35 GHz BiCMOS	LC	4.4 GHz	4 mA@2.5 V	−102 dBc/Hz@600 kHz
0.25 μm CMOS	LC	5.3 GHz	4.7 mA@1.5 V	−93 dBc/Hz@100 kHz

Source: [3]/John Wiley & Sons.

10.11 Push–Push Configuration [3, Ch. 10]

Figure 10.61 shows the two identical oscillator circuits coupled through the arbitrary coupling network under push–push configuration.

The evaluation of pushing factor of the push–push configuration is carried out by considering uncorrelated noise voltage perturbations, Δv_{n1} and Δv_{n2}, associated with the two identical oscillator circuits as shown in Figure 10.61. From refs. [45–47], due to the symmetry of the push–push oscillator topology, two modes (common and differential modes) exist, and the corresponding pushing factor is calculated in terms of the common mode (CM) and differential mode (DM) pushing factor.

The frequency noise spectral density for the push–push topology can be given by

$$\left[\overline{\Delta f_n^2}\right]_{\text{push–push}} = \left([\Delta f_n]_{\text{CM}} + [\Delta f_n]_{\text{DM}}\right)^2 \tag{10.89}$$

$$\left[\overline{\Delta f_n^2}\right]_{\text{push–push}} = [K_{\text{PF}}]_{\text{CM}}^2\left[\overline{\Delta v_n^2}\right]_{\text{DM}} + [K_{\text{PF}}]_{\text{DM}}^2\left[\overline{\Delta v_n^2}\right]_{\text{DM}} + 2[K_{\text{PF}}]_{\text{CM}}[K_{\text{PF}}]_{\text{DM}}\left([\Delta v_n]_{\text{CM}} * [\Delta v_n]_{\text{DM}}\right) \tag{10.90}$$

where $[K_{\text{PF}}]_{\text{CM}}$ and $[K_{\text{PF}}]_{\text{DM}}$ are common and differential mode pushing factors, and $[\Delta v_n]_{\text{CM}}$ and $[\Delta v_n]_{\text{DM}}$ are the common and differential mode noise perturbations.

The effect of the differential noise perturbation, due to the symmetry for the push–push topology, gives insignificant variation of the oscillating frequency, so $[K_{\text{PF}}]_{\text{DM}} \to 0$.

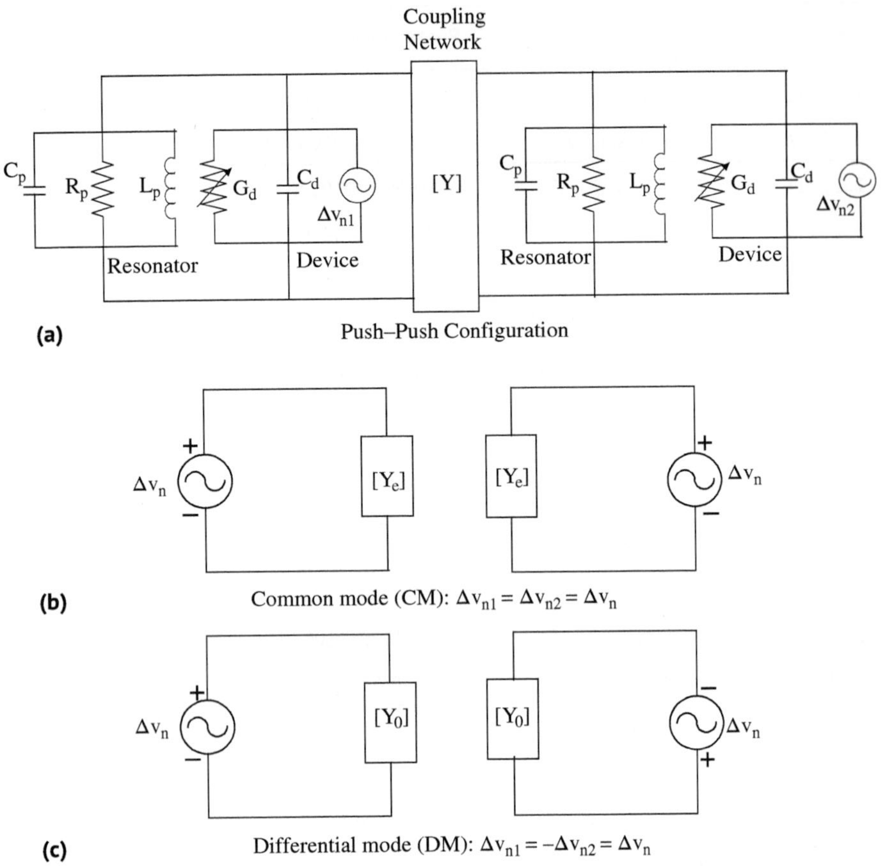

Figure 10.61 Two identical oscillator circuits coupled through the arbitrary coupling network under push–push configuration. *Source:* [3]/John Wiley & Sons.

The common mode input noise perturbation [3, circuit of Figure 10.20] can be given as

$$\overline{[\Delta v_n^2]}_{\text{CM}} = \left[\frac{\Delta v_{n1} + \Delta v_{n2}}{2}\right]^2 = \frac{1}{4}\left(\overline{[\Delta v_{n1}^2]} + \overline{[\Delta v_{n2}^2]} + \overline{[\Delta v_{n1} * \Delta v_{n2}]}\right) \tag{10.91}$$

Since the input noise voltage perturbations Δv_{n1} and Δv_{n2} associated with the two identical active devices are uncorrelated to each other, $\overline{[\Delta v_{n1} * \Delta v_{n2}]} = 0$. Considering the active device (transistor) for the two identical oscillator circuits in push–push topology operates under the same working condition, their input noise voltage perturbation can be described by the same statistic and given as $\overline{[\Delta v_{n1}^2]} = \overline{[\Delta v_{n2}^2]}$.

Equation (10.91) can be rewritten as

$$\overline{[\Delta V_n^2]}_{\text{CM}} = \left[\frac{\Delta v_{n1} + \Delta v_{n2}}{2}\right]^2 = \left[\frac{\overline{[\Delta v_n^2]}}{2}\right] \tag{10.92}$$

From (10.90),

$$\overline{[\Delta f_n^2]}_{\text{push-push}} = [K_{\text{PF}}]_{\text{CM}}^2 \overline{[\Delta v_n^2]}_{\text{CM}} = [K_{\text{PF}}]_{\text{CM}}^2 \left[\frac{\overline{[\Delta v_n^2]}}{2}\right] \tag{10.93}$$

From (10.93),

$$[\mathrm{L}(f_m)]_{\text{push-push}(f=f_0)} = 20\log\left\{\frac{[\Delta f_n]_{\text{push-push}}}{\sqrt{2}f_m}\right\} = -3\text{ dB} + 20\log\left\{\frac{[K_{\text{PF}}]_{\text{CM}}\Delta v_n}{\sqrt{2}f_m}\right\} \tag{10.94}$$

From (10.94), there is a 3 dB improvement in the phase noise with respect to the individual oscillator oscillating at half the frequency of the push–push frequency, and the analysis agrees with the general equation of the N-coupled oscillator described by [3, Eq. (10.206)].

The improvement of the phase noise of the push–push topology, referring to one individual oscillator that oscillates at fundamental frequency f_0, can be expressed as

$$\frac{[\mathrm{L}(f_m)]_{\text{push-push}(f=f_0)}}{[\mathrm{L}(f_m)]_{\text{fundamental}\left(f=2\frac{f_0}{2}\right)}} = \frac{20\log\left\{\frac{[\Delta f n]_{\text{push-push}(f=f_0)}}{\sqrt{2}f_m}\right\}}{20\log\left\{\frac{[\Delta f n]_{\text{fundamental}\left(f=2\frac{f_0}{2}\right)}}{\sqrt{2}f_m}\right\}} \tag{10.95}$$

From Rohde et al. [3, Eqs. (10.208) and (10.217)],

$$[\mathrm{L}(f_m)]_{\text{push-push}(f=f_0)} = -9\,\text{dB} + [\mathrm{L}(f_m)]_{\text{fundamental}\left(f=2\frac{f_0}{2}\right)} \tag{10.96}$$

where $f_0/2$ is the fundamental frequency of the subcircuit of the oscillator in push–push topology.

From (10.96), push–push topology gives 9 dB improvement in the phase noise compared to the fundamental frequency of the individual oscillator oscillating at f_0, twice the designed oscillating frequency of $f_0/2$.

Example: Figure 10.62a shows the schematic of the push–push oscillator that has two identical oscillators. It consists of the two individual oscillator circuits, osc #1 and osc #2, that are oscillating at half of the push–push frequency ($f_0/2$). The individual oscillator circuit corresponding to the half-resonator oscillates at $f_0/2$ (1000 MHz) and is used as a starting point to verify the above noise analysis with respect to the push–push configuration.

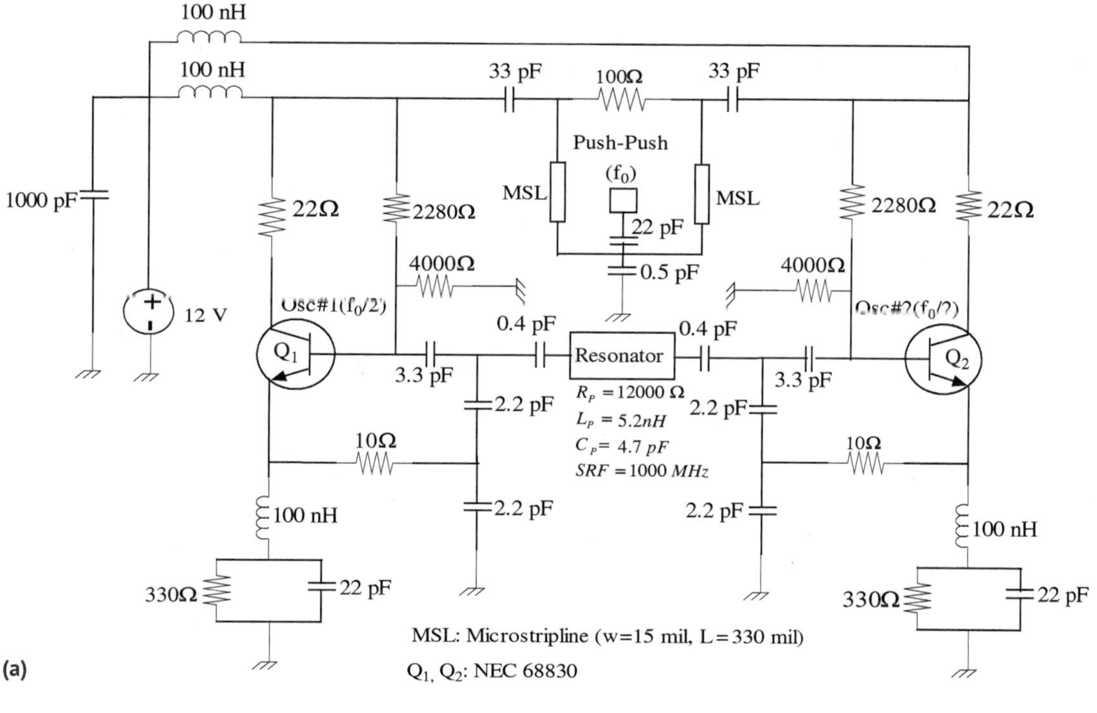

(a)

Figure 10.62 (a) Two identical oscillator circuits coupled through the resonator under the push–push configuration.

(Continued)

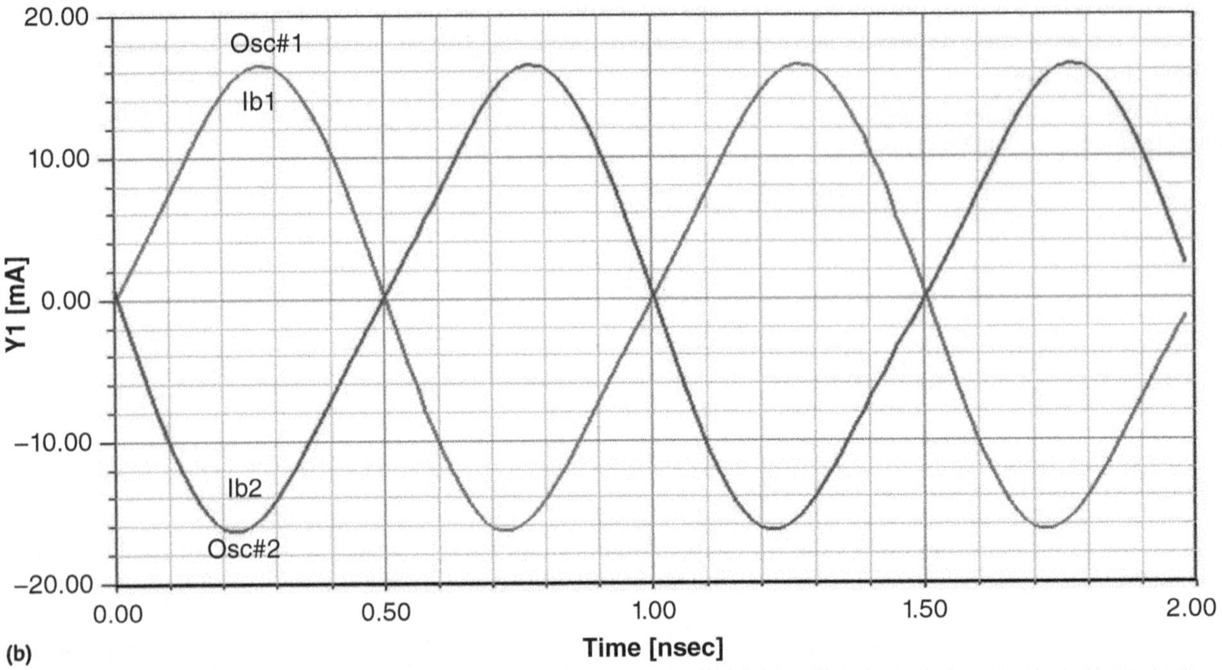

(b)

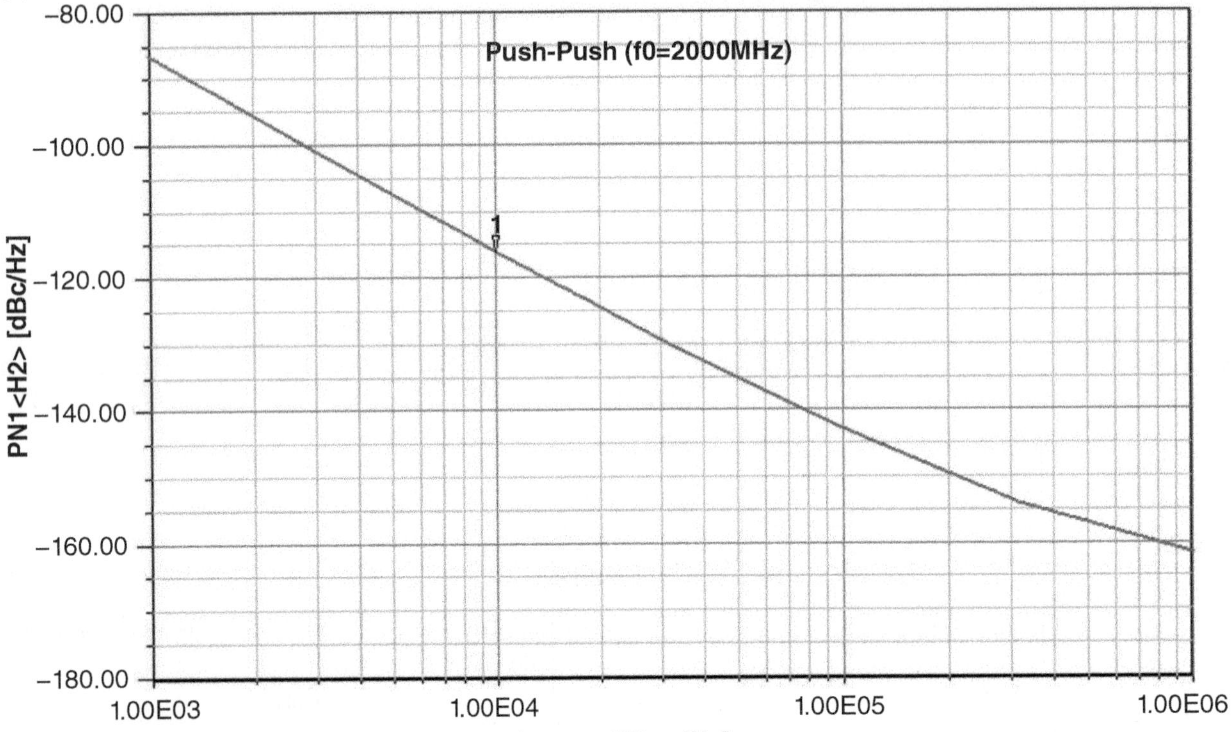

(c)

Figure 10.62 (Cont'd) (b) Simulated plot of the RF base current for the push–push configuration. (c) Simulated phase noise plot for the push–push configuration.

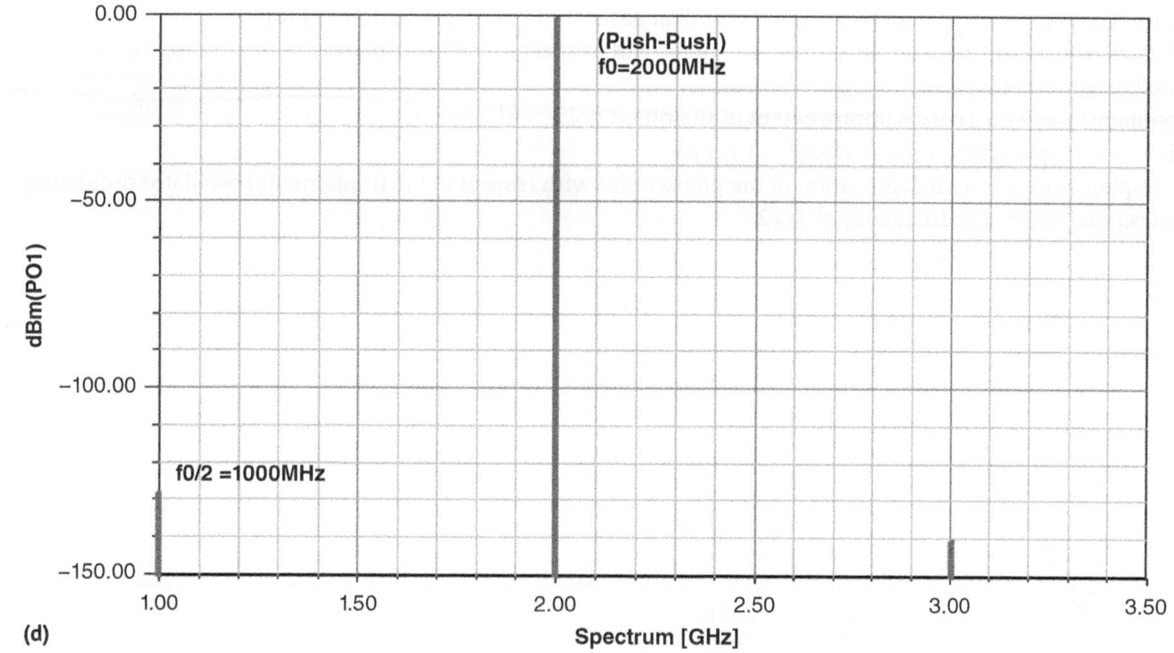

Figure 10.62 (Cont'd) (d) Simulated plot of the output power for the push–push configuration. *Source*: [3]/John Wiley & Sons.

The circuit shown in Figure 10.62a is fabricated on 32-mil thickness Roger substrate of dielectric constant 3.38 and loss tangent 2.7E-4. As shown in Figure 10.62b, the resonator is equivalently represented by the parallel RLC model (measured data at self-resonance frequency of 1000 MHz as $R_P = 12000\ \Omega$, $L_P = 5.2$ nH, $C_P = 4.7$ pF), which acts as a resonator and also provides the required phase shift of 180° for the synchronization.

Figure 10.62b–d shows the simulated (Ansoft Designer) plot of the RF base current, phase noise, and output power for the push–push configuration (which is phase-shifted by 180° in a mutually synchronized condition, $f_0 = 2000$ MHz).

Figure 10.63b shows the schematic of the uncoupled individual oscillator circuit osc #1 for a comparative analysis of the performance. The circuit is fabricated on 32-mil thickness Roger substrate of dielectric constant 3.38 and loss tangent 2.7E-4.

As shown in Figure 10.63b, the resonator is equivalently represented by a parallel RLC model (measured data at a self-resonance frequency of 2000 MHz as $R_P = 12000\ \Omega$, $L_P = 1.24$ nH, $C_P = 4.7$ pF).

The oscillator circuit osc #1, as shown in Figure 10.63c, oscillates at f_0 (2000 MHz) and is used as a reference for the comparative noise analysis with respect to the push–push configuration.

Figure 10.63b,c shows the simulated (Ansoft Designer) plot of the phase noise and the output power for the individual uncoupled oscillator circuit as shown in Figure 10.63a.

From Figure 10.63c, the improvement of the phase noise for the push–push topology at 10 KHz offset from the carrier is 8.46 dB, which agrees with the analytical expression given within 1 dB, Equation (10.96).

Figure 10.63d shows the measured phase noise plot for the push–push ($f_0 = 2000$ MHz) configuration and the uncoupled individual oscillator ($f_0 = 2000$ MHz) circuit as shown in Figure 10.63a,d. The measured phase noise for the push–push and the uncoupled individual oscillator at 10 kHz offset from the carrier is −113 dBc/Hz and −105 dBc/Hz, respectively.

The relative improvement of the measured phase noise of the push–push configuration, with respect to fundamental frequency of the oscillator composing push–push topology, is 8 dB and agrees with the simulated (Ansoft Designer) and analytical values within 1 dB.

The discrepancy of 1 dB of the measured phase noise from the analytical value, Equation (10.96), can be attributed to the package parasitics, dynamic loaded Q, and tolerances of the component values of the two uncoupled individual oscillator circuits.

The aforementioned relative noise analysis gives a theoretical basis of the noise prediction as [48]:

Fundamental oscillator: 12 dB degradation of the phase noise with respect to the fundamental oscillator oscillating at f_0, twice the designed oscillating frequency of $f_0/2$.

Push–push topology ($f = 2f_0/2$): 9 dB improvement of the phase noise with respect to the fundamental oscillator oscillating at f_0, twice the designed oscillating frequency of $f_0/2$.

Frequency multiplier/doubler: 6 dB degradation of the phase noise with respect to the fundamental oscillator oscillating at f_0, twice the designed oscillating frequency of $f_0/2$.

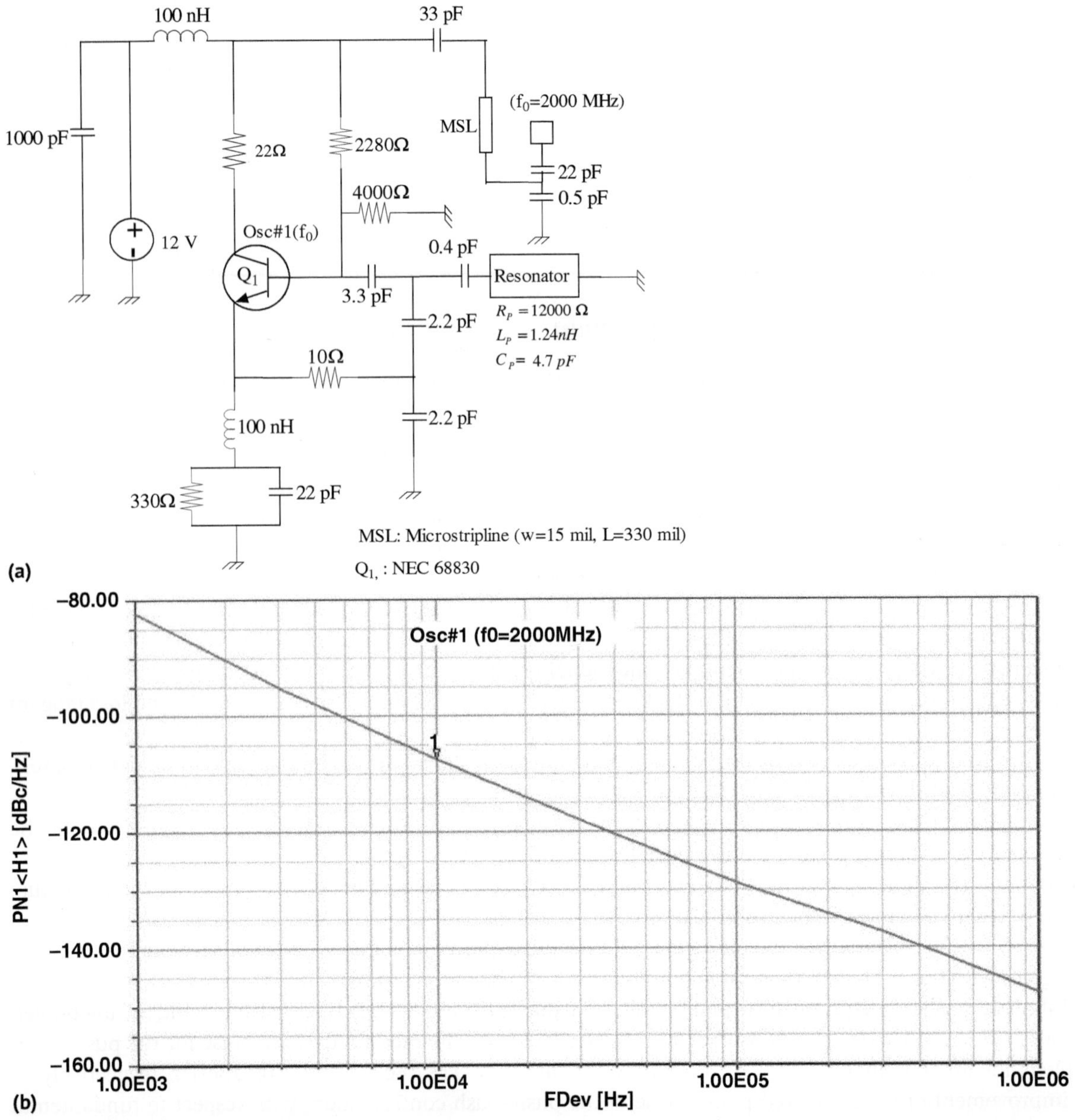

Figure 10.63 (a) Schematic of the uncoupled individual oscillator circuit osc. #1 ($f_0 = 2000$ MHz). (b) Simulated phase noise plot for the oscillator circuit shown in (a).

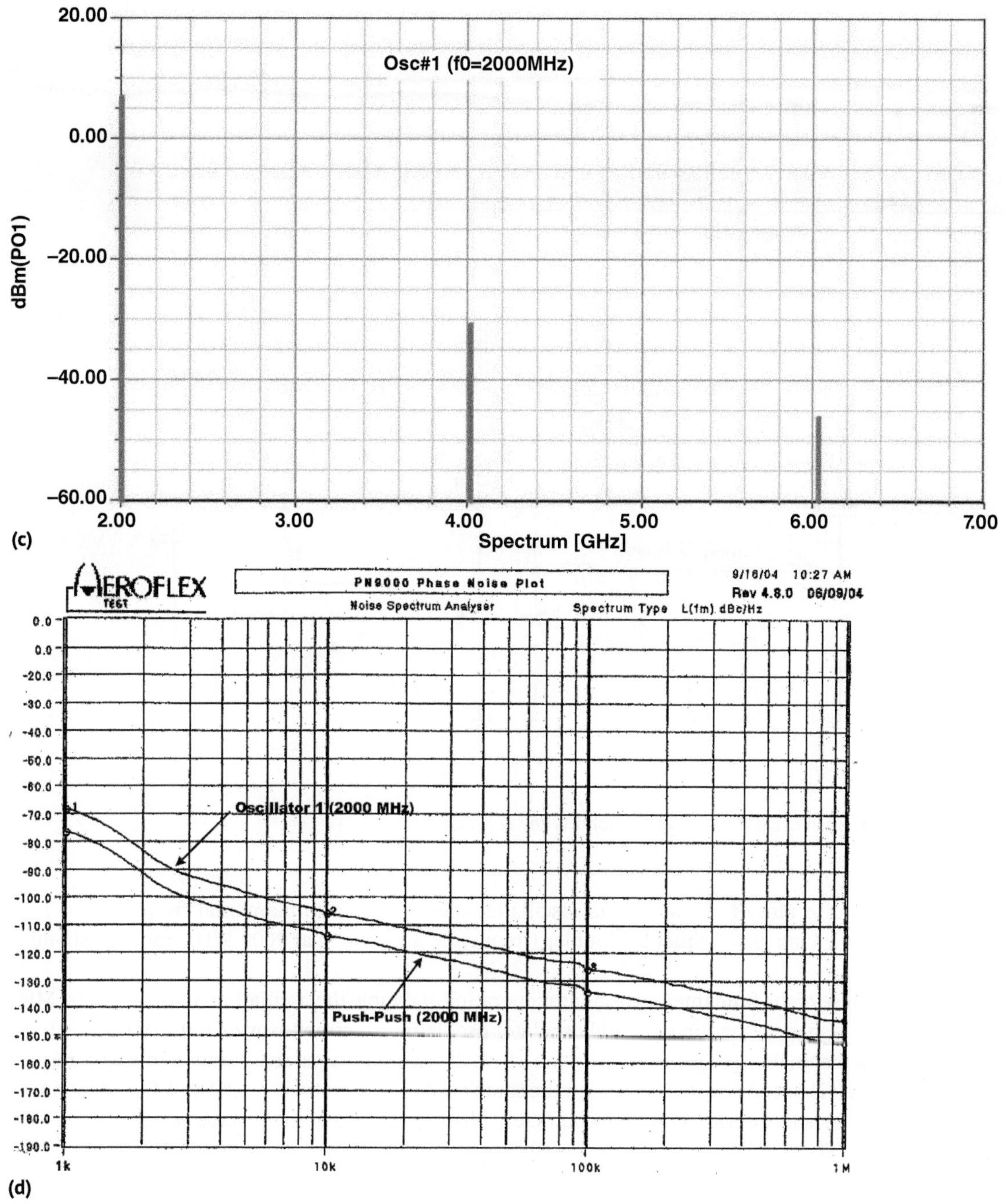

Figure 10.63 (c) Simulated plot of the output power for the oscillator circuit shown in (a). (d) Measured phase noise plot for the push–push ($f_0 = 2000$ MHz) configuration and the uncoupled individual oscillator ($f_0 = 2000$ MHz) circuit as shown in (a). *Source*: [3]/John Wiley & Sons.

Validation

Figure 10.64 shows the schematic of the push–push oscillator that has two identical oscillators. It consists of two individual oscillators that are oscillating at half the push–push frequency ($f_0/2$). The individual oscillator, corresponding to the half-resonator, oscillates at $f_0/2$ (1000 MHz) and is used as a starting point to verify the above noise analysis with respect to the frequency multiplier and push–push configuration.

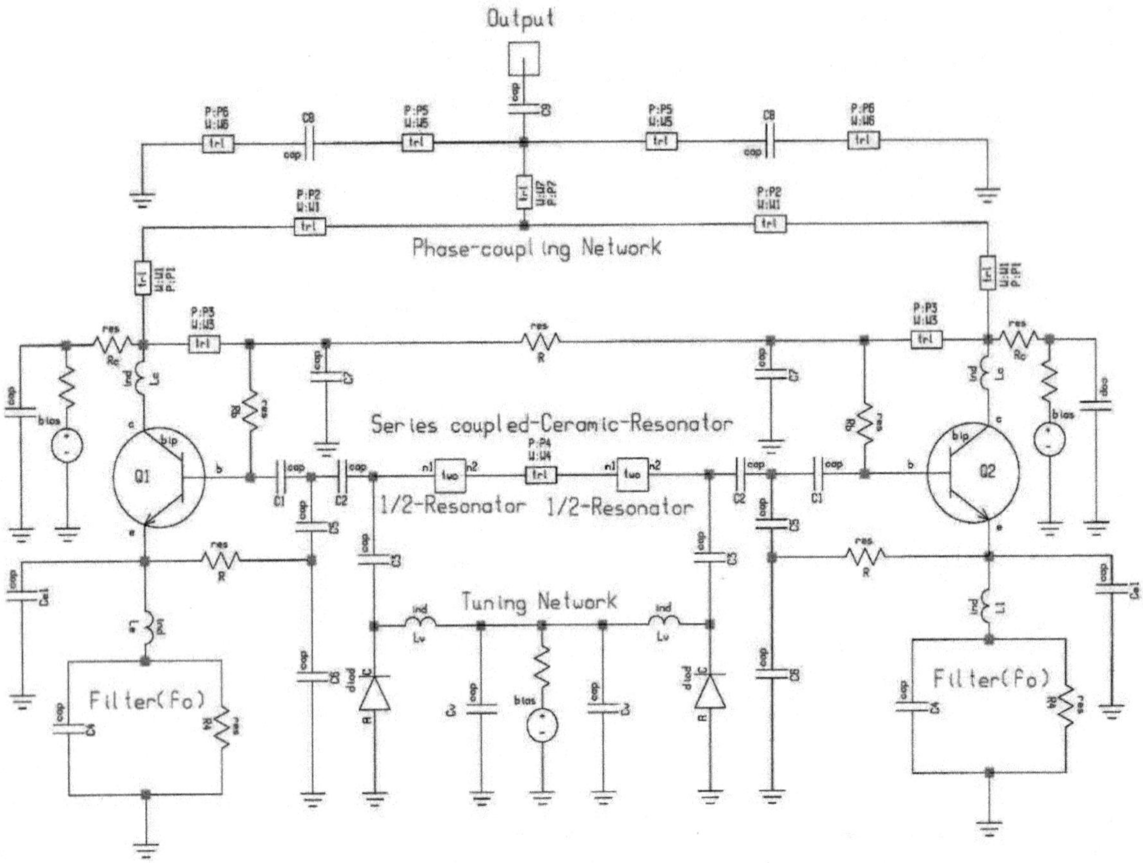

Figure 10.64 Schematic of the push–push oscillator. *Source*: [3]/John Wiley & Sons.

Figure 10.64 shows the phase noise plot of the individual oscillator operating at a fundamental frequency of $f_0/2$ (1000 MHz) and f_0 (2000 MHz), working as a frequency doubler (frequency multiplier) at 2000 MHz and a push–push configuration at 2000 MHz.

As discussed earlier, the phase noise of the fundamental oscillator operating at double the oscillating frequency 2000 MHz is worsening by 12 dB/octave with respect to the fundamental oscillator oscillating at a frequency of 1000 MHz.

The simulated graph is based on the unchanged parameters of the active device and the passive components with respect to the two frequencies $f_0/2$ and f_0. It is not an easy task to design the same oscillator operating at $f_0/2$ and f_0 and maintain the same operating parameters of the active device, coupling coefficient, drive level, and quality factor, etc.

For the case of the frequency doubler, the phase noise is degraded by 6 dB with respect to the fundamental frequency as shown in Figure 10.65. Measured phase noise of a free-running LC oscillator at 7.69/2 GHz is shown in Figure 10.66.

The relative improvement of the phase noise of the push–push configuration, with respect to the fundamental frequency of the oscillator composed of push–push topology, is 9 dB. This is shown in Figure 10.65 and agrees with the theoretically predicted result [48].

As the Q of the components deteriorates at higher frequencies, the push–push arrangement has a big advantage; see 0.2 THz push–push VCO with low power, low phase noise, and high oscillation efficiency in 65 nm CMOS [50].

Example: The following is an example of a symmetrical NMOS/PMOS state-of-the-art oscillator.

Figure 10.67a shows the schematic used for the simulation of a symmetrical NMOS/PMOS oscillator, with the simulation results in the following figures. Figure 10.67b shows the voltages (V_{ds}) of the NMOS and PMOS transistors of the NMOS/PMOS oscillator. Figure 10.67c shows the gate voltage of the NMOS part as a function of time.

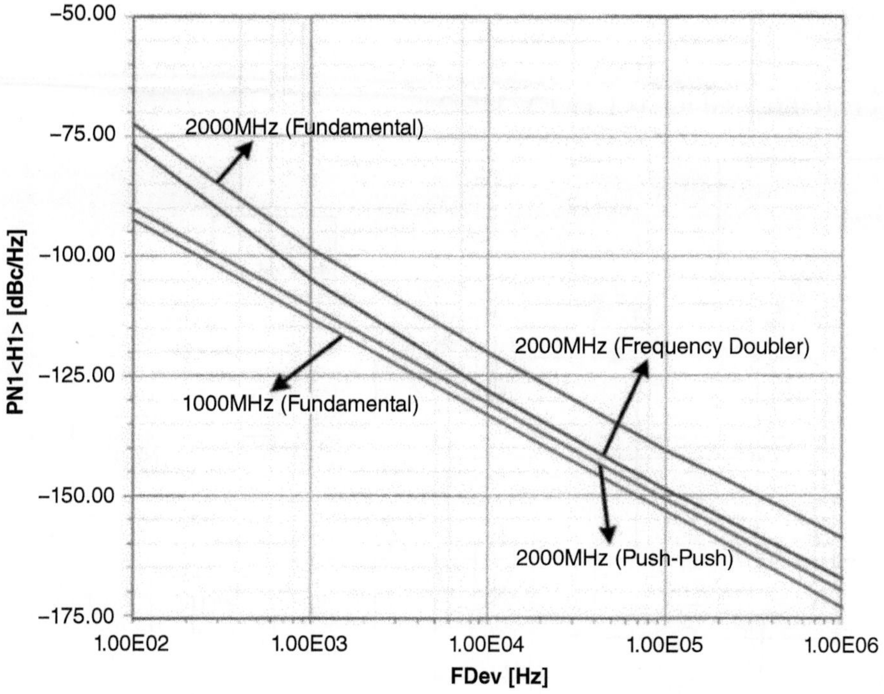

Figure 10.65 Phase noise plot of fundamental oscillator at $f_0/2$ and f_0, frequency doubler at f_0, and push–push oscillator at f_0. *Source*: [3]/John Wiley & Sons.

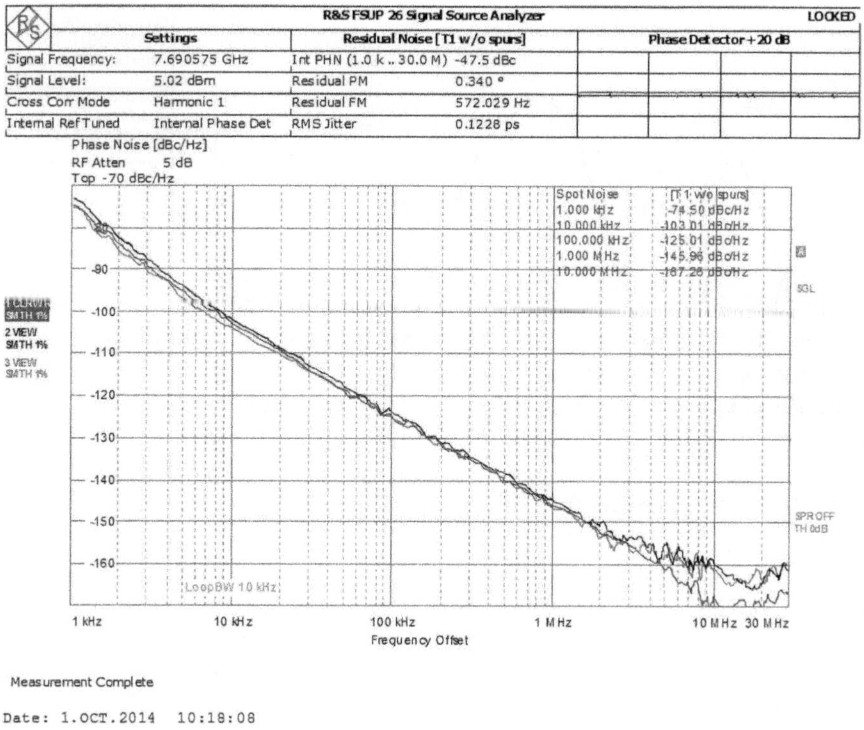

Figure 10.66 Measured phase noise of a free-running LC oscillator at 7.69/2 GHz. *Source*: [49, p. 316]/John Wiley & Sons.

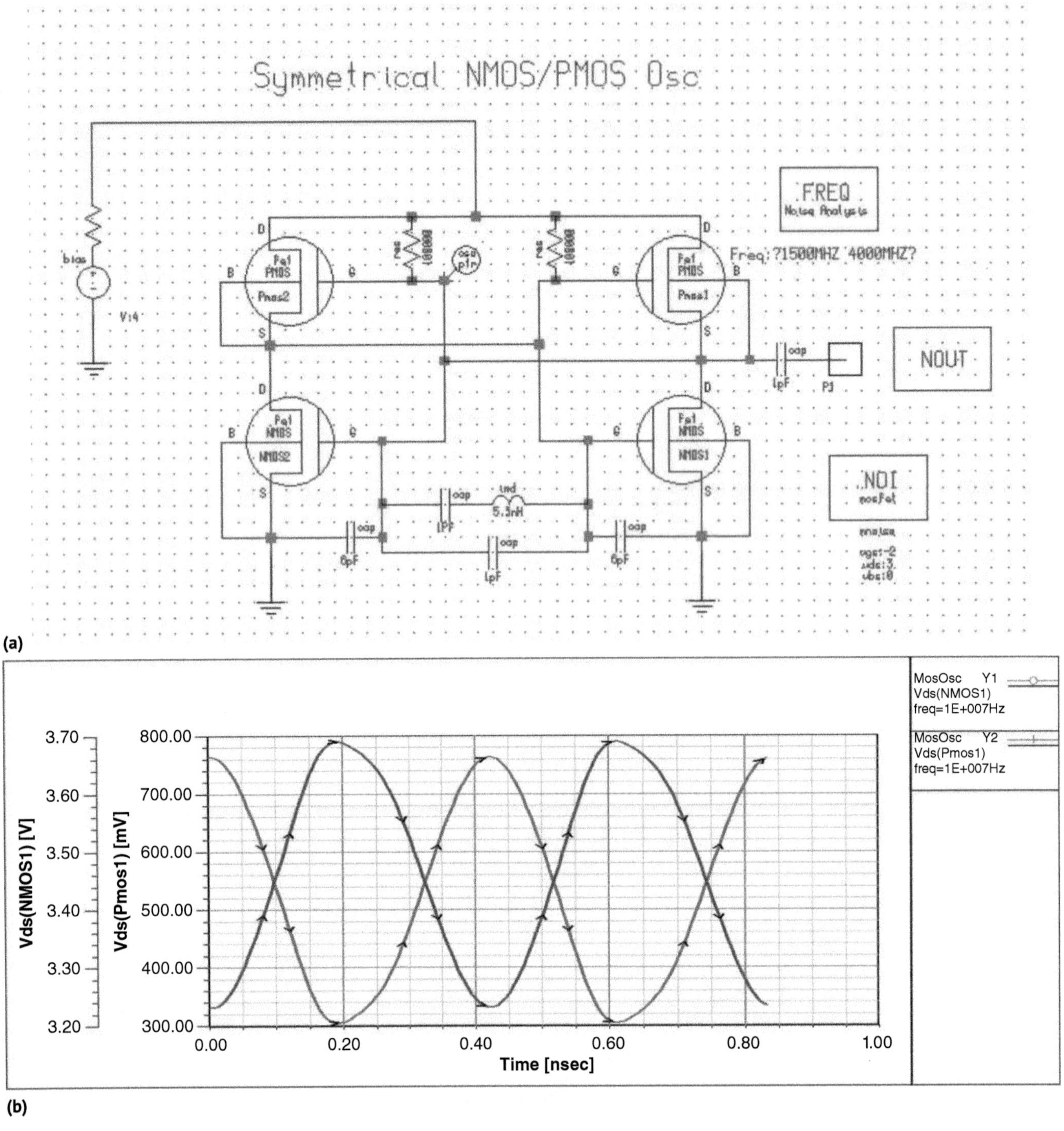

Figure 10.67 (a) Symmetrical NMOS/PMOS oscillator. *Source*: [3, Ch. 8]/John Wiley & Sons. (b) Voltages (V_{ds}) of the NMOS and PMOS transistors of the NMOS/PMOS oscillator (a).

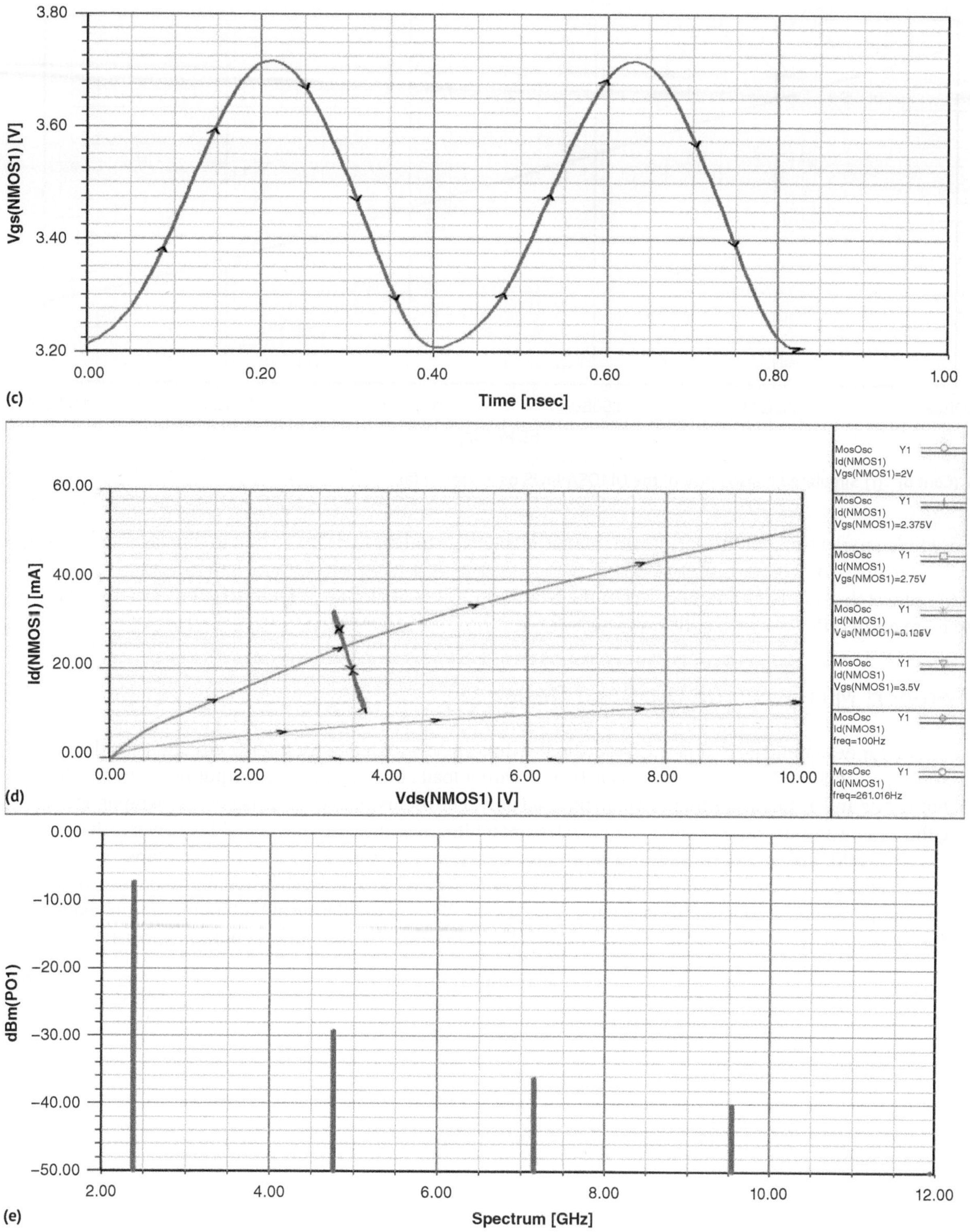

Figure 10.67 (Cont'd) (c) Gate voltage of the NMOS part as a function of time. (d) DC–IV curve and load line of the NMOS/PMOS oscillator of (a). (e) Simulated output power of the NMOS/PMOS oscillator of (a).

(Continued)

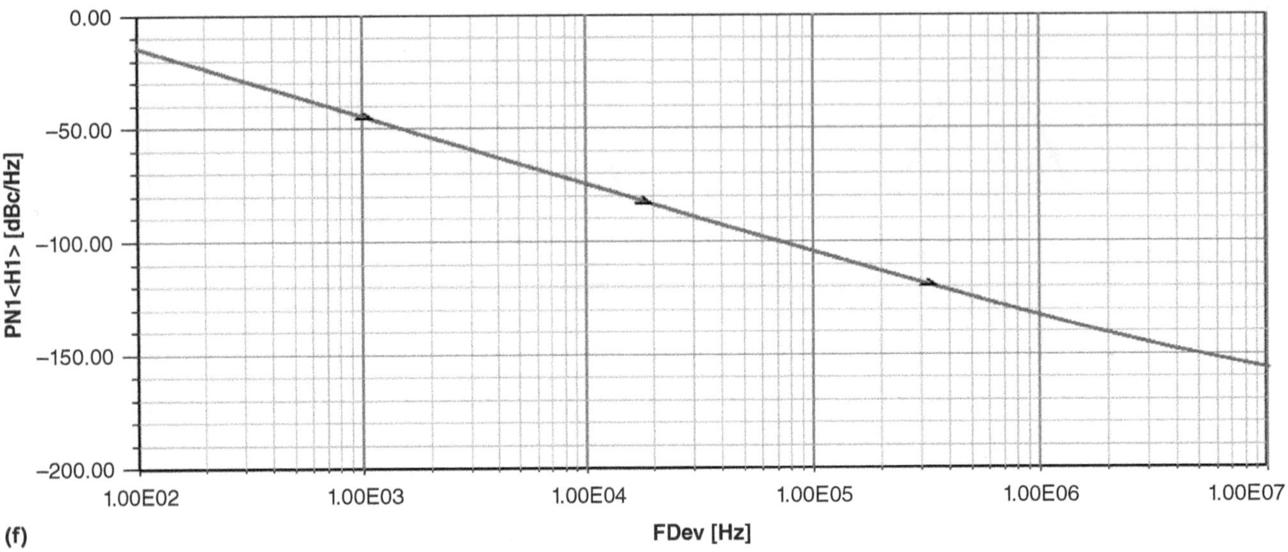

Figure 10.67 (Cont'd) (f) Simulated phase noise of the NMOS/PMOS oscillator of (a).

There is very little energy stored in the load (load line) as it shows a straight line. Compared to bipolar transistors, the DC–IV curves are really curved (see Figure 10.67d).

Figure 10.67e shows the simulated output power of the NMOS/PMOS oscillator and Figure 10.67f shows the CAD simulated phase noise of this NMOS/PMOS oscillator.

10.12 Two-port Microwave/RF Oscillator Design [1]

A common method for designing oscillators is to resonate the input port with a passive high-Q circuit at the desired frequency of resonance. It will be shown that if this is achieved with a load connected on the output port, the transistor is oscillating at both ports and is thus delivering power to the load port. The oscillator may be considered a two-port structure, with M_3 being the lossless resonating port and M_4 the lossless matching such that all the external RF power is delivered to the load (see Figure 10.68). The resonating network has been described. Nominally, only parasitic resistance is present at the resonating port, since a high-Q resonance is desirable for minimizing oscillator noise. It is possible to have loads at both the input and output ports if such an application occurs, since the oscillator is oscillating at both ports simultaneously.

The simultaneous oscillation condition is proved as follows [1]. Assume that the oscillation condition is satisfied at port 1:

$$1/S'_{11} = \Gamma_G \tag{10.97}$$

Thus,

$$S'_{11} = S_{11} + \frac{S_{12}S_{21}\Gamma_L}{1 - S_{22}\Gamma_L} = \frac{S_{11} - D\Gamma_L}{1 - S_{22}\Gamma_L} \tag{10.98}$$

$$\frac{1}{S'_{11}} = \frac{1 - S_{22}\Gamma_L}{S_{11} - D\Gamma_L} = \Gamma_G \tag{10.99}$$

By expanding Eq. (8.61), we find

$$\begin{aligned}\Gamma_G S_{11} - D\Gamma_L\Gamma_G &= 1 - S_{22}\Gamma_L \\ \Gamma_L(S_{22} - D\Gamma_G) &= 1 - S_{11}\Gamma_G \\ \Gamma_L &= \frac{1 - S_{11}\Gamma_G}{S_{22} - D\Gamma_G}\end{aligned} \tag{10.100}$$

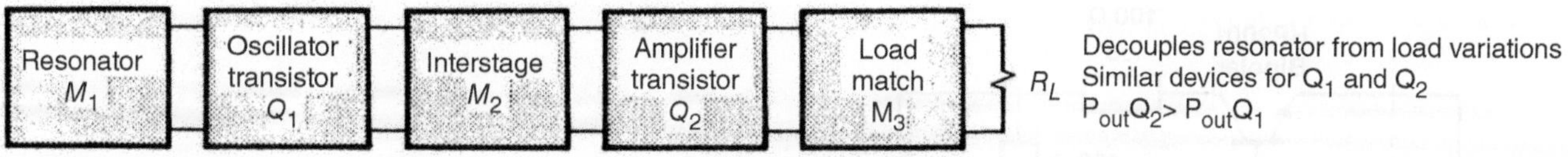

Figure 10.68 Buffered oscillator design. *Source*: [1]/John Wiley & Sons.

Thus,

$$S'_{22} = S_{22} + \frac{S_{12}S_{21}\Gamma_G}{1 - S_{11}\Gamma_G} \tag{10.101}$$

$$\frac{1}{S'_{22}} = \frac{1 - S_{11}\Gamma_G}{S_{22} - D\Gamma_G} \tag{10.102}$$

Comparing Eqs. (10.100) and (10.102), we find

$$1/S'_{22} = \Gamma_L \tag{10.103}$$

which means that the oscillation condition is also satisfied at port 2; this completes the proof. Thus, if either port is oscillating, the other port must be oscillating as well. A load may appear at either or both ports, but normally the load is in Γ_L, the output termination. This result can be generalized to an n-port oscillator by showing that the oscillator is simultaneously oscillating at each port:

$$\Gamma_1 S'_{11} = \Gamma_2 S'_{22} = \Gamma_3 S'_{33} = \cdots = \Gamma_n S'_{nn} \tag{10.104}$$

Before concluding this section on two-port oscillator design, the buffered oscillator shown in Figure 10.68 must be considered. This design approach is used to provide the following:

1. A reduction in loading-pulling, which is the change in oscillator frequency when the load reflection coefficient changes;
2. A load impedance that is more suitable to wideband applications ($k < 1$);
3. A higher output power from a working design, although the higher output power can also be achieved by using a larger oscillator transistor; and
4. This type of oscillator is not optimized for low phase noise. However, buffered oscillator designs are quite common in wideband YIG applications, where changes in the load impedance must not change the generator frequency.

Two-port oscillator design may be summarized as follows [1]:

1. Select a transistor with sufficient gain and output power capability for the frequency of operation. This may be based on oscillator data sheets, amplifier performance, or S-parameter calculation.
2. Select a topology that gives $k < 1$ at the operating frequency. Add feedback if $k < 1$ has not been achieved.
3. Select an output load-matching circuit that gives $|S'_{11}| > 1$ over the desired frequency range. In the simplest case, this could be a 50 Ω load.
4. Resonate the input port with a lossless termination so that $\Gamma_G S'_{11} = 1$. The value of S'_{22} will be greater than unity with the input properly terminated.

In all cases, the transistor delivers power to a load and the input of the transistor. Practical considerations of realizability and DC biasing will determine the best design.

For both bipolar and FET oscillators, a common topology is the common base or common gate, since a common-lead inductance can be used to raise S_{22} to a large value, usually greater than unity even with a 50 Ω generator resistor. However, it is not necessary for the transistor S_{22} to be greater than unity, since the 50 Ω generator is not present in the oscillator design. The requirement for oscillation is $k < 1$; then resonating the input with a lossless termination will ensure that $|S'_{11}| > 1$.

A simple example will clarify the design procedure. A common-base bipolar transistor (HP2001) was selected to design a fixed-tuned oscillator at 2 GHz. The common-base S-parameters and stability factor are given in following table [1].

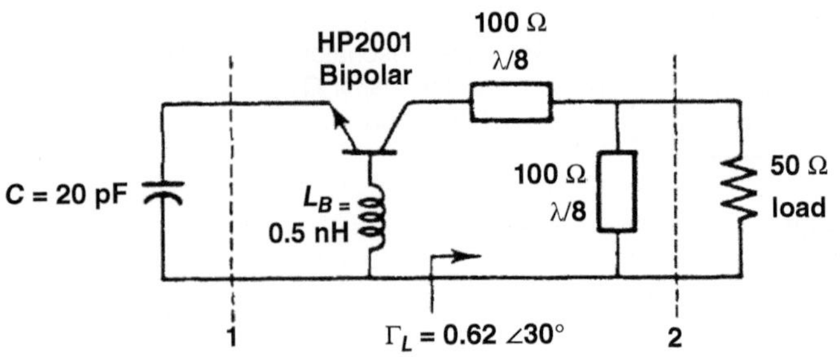

Figure 10.69 Oscillator example at 2 GHz. *Source*: [1]/John Wiley & Sons.

($V_{CE} = 15$ V, $I_c = 25$ mA)	
$L_B = 0$	$L_B = 0.5$ nH
$S_{11} = 0.94\underline{/174°}$	$1.04\underline{/173°}$
$S_{21} = 1.90\underline{/-28°}$	$2.00\underline{/-30°}$
$S_{12} = 0.013\underline{/98°}$	$0.043\underline{/153°}$
$S_{22} = 1.01\underline{/-17°}$	$1.05\underline{/-18°}$
$k = -0.09$	-0.83

Using the load circuit in Figure 10.69, we see that the reflection coefficients are as follows:

$$\Gamma_L = 0.62\underline{/30°}$$

$$S'_{11} = 1.18\underline{/173°}$$

Thus, a resonating capacitance of $G = 20$ pF resonates the input port. In a YIG-tuned oscillator, this reactive element could be provided by the high-Q YIG element. For a dielectric resonator oscillator (DRO), the puck would be placed to give $\Gamma_G \approx 1.0\underline{/-173°}$.

Another two-port design procedure is to resonate the Γ_G port and calculate S'_{22} until $|S'_{22}| > 1$, then design the load port to satisfy. This design procedure is summarized in Figure 10.70.

An example using this procedure at 4 GHz is given in Figure 10.71 using an AT41400 silicon bipolar chip in the common-base configuration with a convenient value of base and emitter inductance of 0.5 nH. The feedback parameter is the base inductance, which can be varied if needed.

Since a lossless capacitor at 4 GHz of 2.06 pF gives $\Gamma_G = 1 - 0\underline{/-137.7°}$, this input termination is used to calculate S'_{22}, giving $S'_{22} = 0.637\underline{/44.5°}$. This circuit will oscillate into any passive load. Varying the emitter capacitor about 20° on the Smith chart to 1.28 pF gives $S'_{22} = 1.16\underline{/-5.5°}$, which will oscillate into a load of $\Gamma_L = 0.861\underline{/5.5°}$. The completed lumped element design is given in Figure 10.72.

These types of oscillators with an inductor or transmission line from base to ground are frequently called as reflection oscillators because of S_{11} or $S_{22} >> 1$.

A good example for a two-port oscillator and actual application is the following oscillator based on the dielectric resonator.

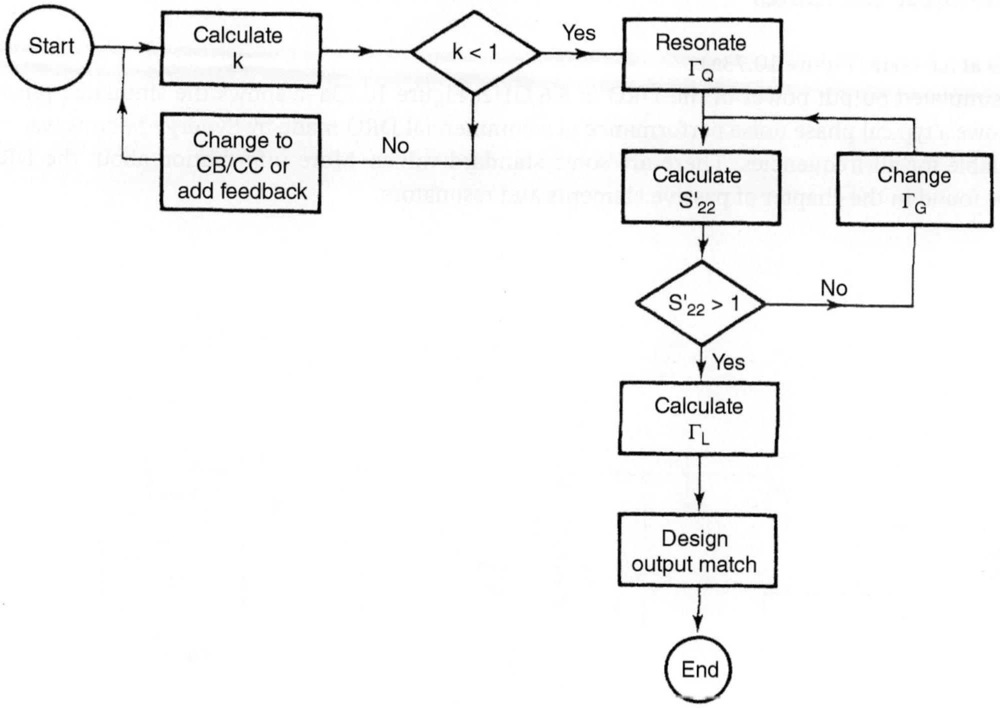

Figure 10.70 Oscillator design flowchart. *Source*: [1]/John Wiley & Sons.

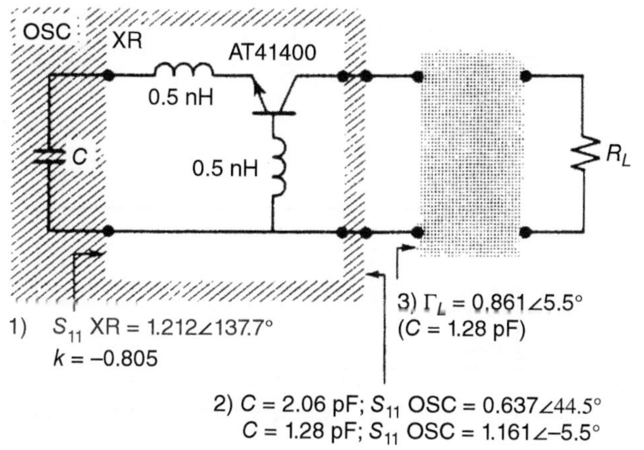

Figure 10.71 A 4 GHz lumped resonator oscillator (LRO) using the AT41400. *Source*: [1]/John Wiley & Sons.

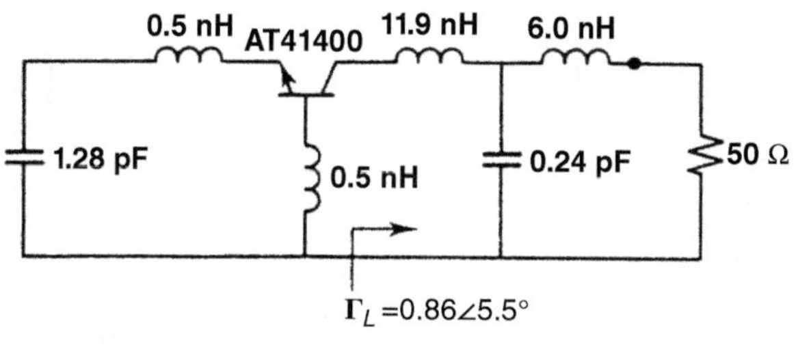

Figure 10.72 Completed lumped resonator oscillator. *Source*: [1]/John Wiley & Sons.

10.13 Dielectric Resonator Oscillator

Here is an example of a DRO at 6.6 GHz (Figure 10.73a).

Figure 10.73b shows the simulated output power of the DRO at 6.6 GHz. Figure 10.73a–c shows the simulated phase noise, while Figure 10.74 shows a typical phase noise performance of a commercial DRO made by Synergy Microwave.

These DROs are not available for all frequencies. There are some standard values. More information about the DRs (dielectric resonators) can be found in the chapter of passive elements and resonators.

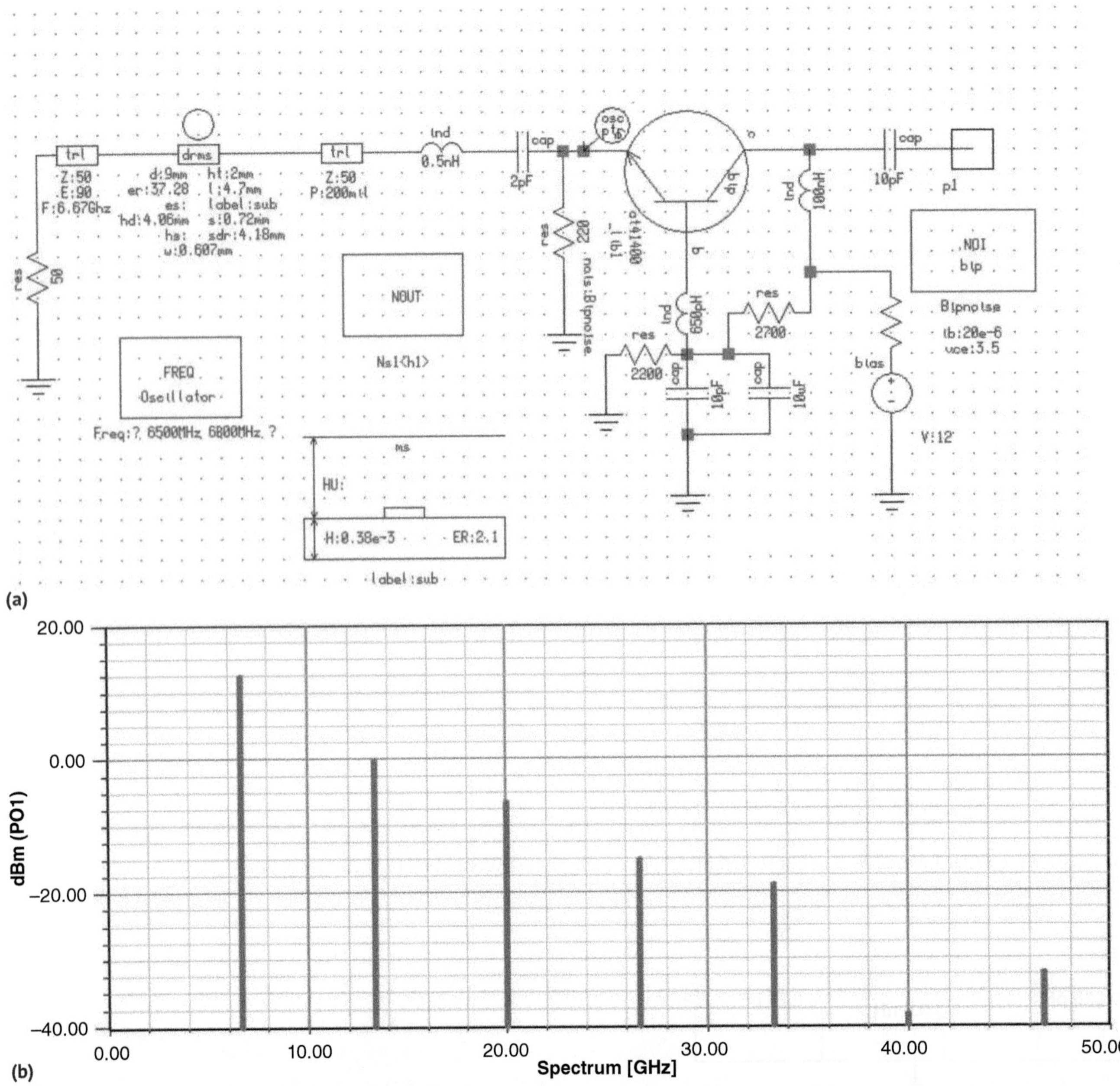

Figure 10.73 (a) CAD simulation example of a DRO at 6.6 GHz [9, Figure 10.9; 2]. (b) Simulated output power of the DRO at 6.6 GHz shown in (a).

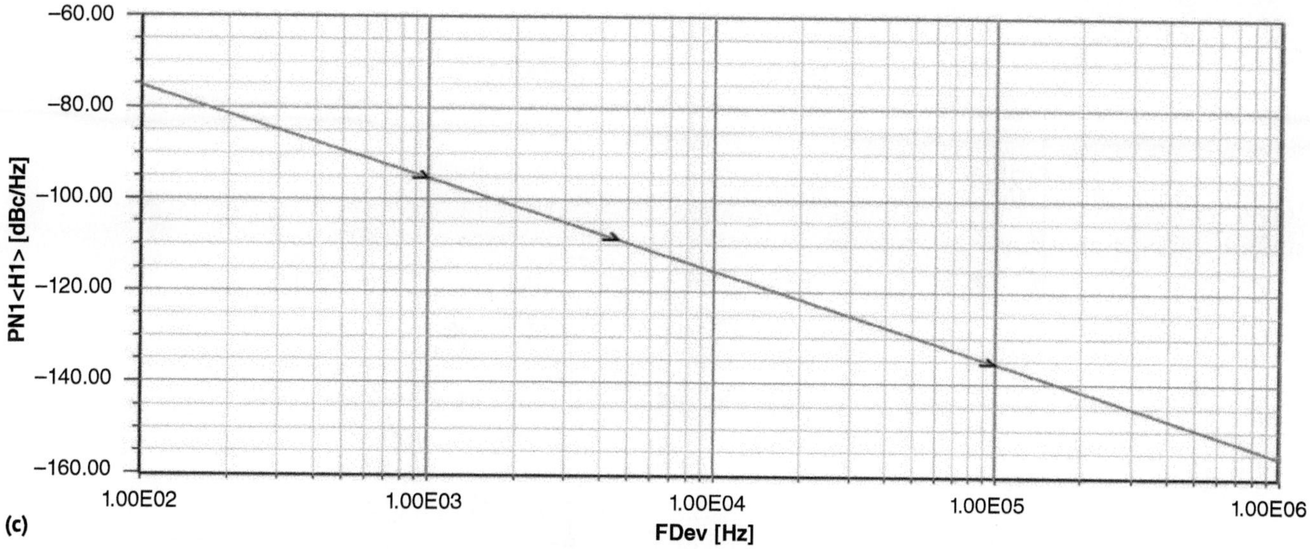

Figure 10.73 (Cont'd) (c) Simulated phase noise of the DRO at 6.6 GHz shown in (a).

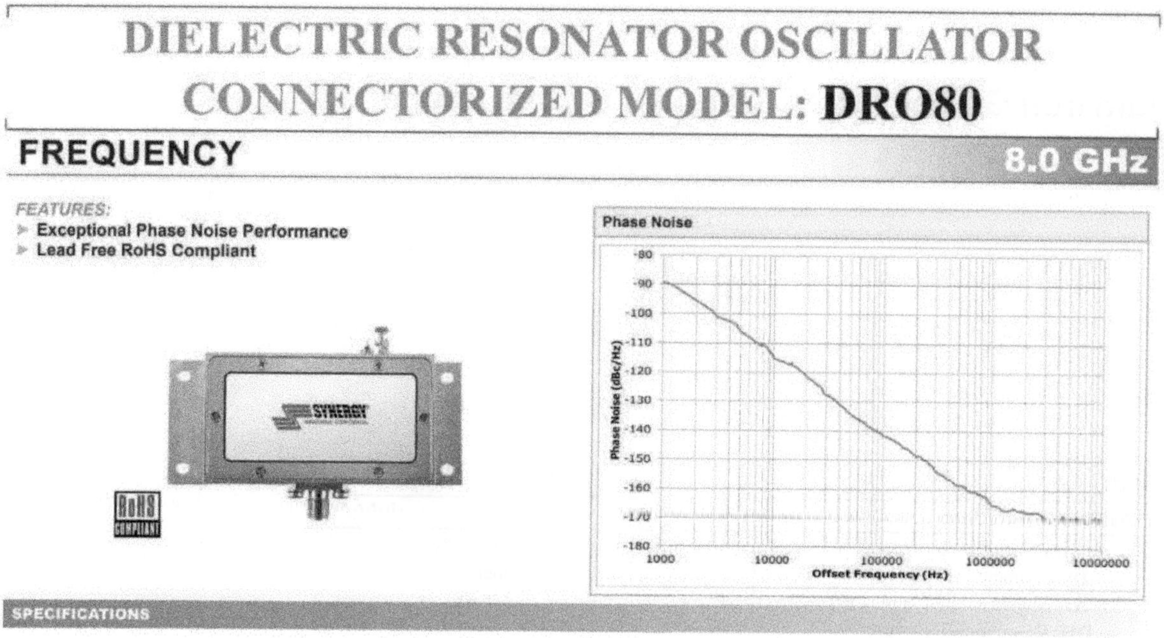

Figure 10.74 A typical phase noise performance of a commercial DRO made by Synergy Microwave. *Source*: Synergy Microwave Corporation.

10.14 10 GHz YIG Oscillator

Another member of the two-port oscillator is the YIG oscillator. YIG resonators (based on yttrium–iron–garnet crystals) are either used for input filter design of microwave receivers or as resonators in wideband signal generators. So far there is no substitute for these types of resonators. Figure 10.75a shows the YIG oscillator schematic; the paralled-tuned circuit is its equivalent circuit.

The top left part shows the typical equivalent circuit of the YIG resonator. The two transmission lines are critical to obtain the resonance at the required frequency.

The YIG resonator always has two loops for a current tuning (coarse and fine).

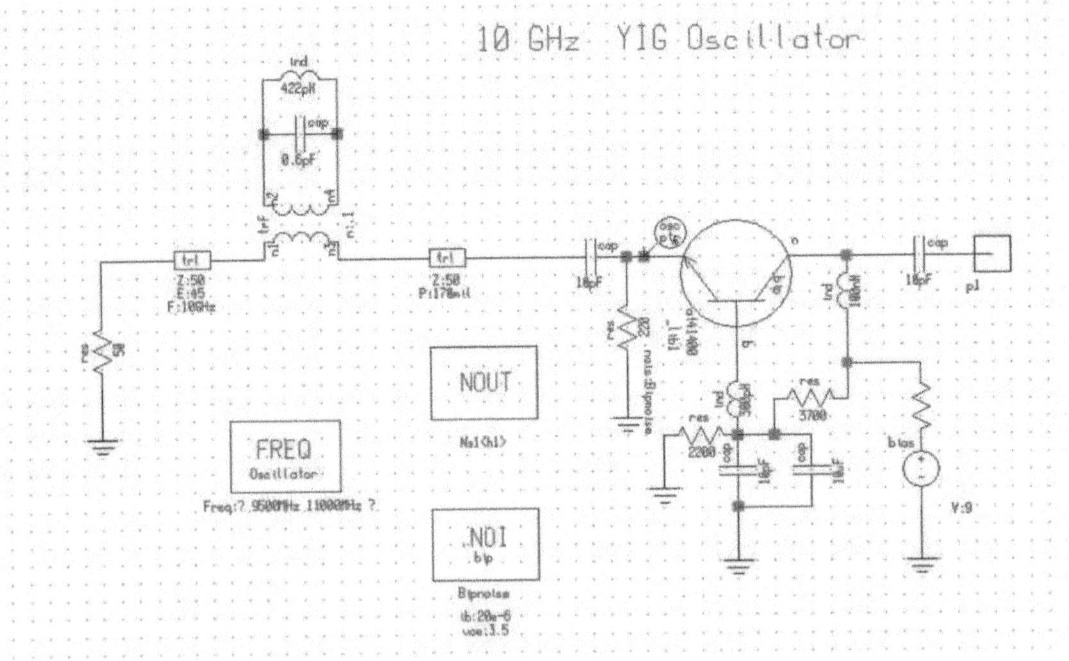

(a)

YIG - Yttrium Iron Garnet Sphere Coupled By Two Loops

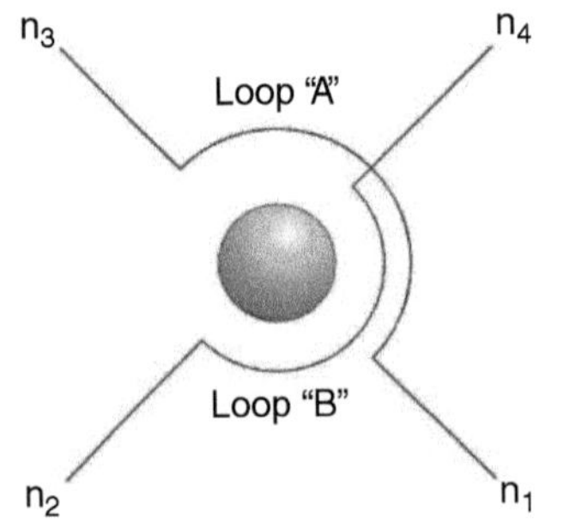

Keywords

keyword	description	unit	default
H	DC bias field	Oersteds	
DS	Diameter of sphere	meter	
MSAT	Sphere saturation magnetization (see notes)	Gauss	
DH	Resonance line width	Oersteds	
AA	Angle loop A	degree	
DLA	Diameter of loop A	meter	
DWA	Wire diameter of loop A	meter	
LA	Inductance of loop A	henry	0.0
AB	Angle loop B	degree	
DLB	Diameter of loop B	meter	
DWB	Wire diameter of loop B	meter	
LB	Inductance of loop B	henry	
AAB	Angle between loop A and loop B	degree	0.0
BC	Spacing between ground planes	meter	

(b)

Figure 10.75 (a) Simulation schematic of a 10 GHz YIG oscillator [2, 9]. (b) YIG sphere coupled by two loops [51].

Figure 10.75c shows the linear analysis of the YIG oscillator to find the approximate resonant frequency, which is 10 GHz in this case, Figure 10.75d shows the simulated output power of the YIG oscillator, and Figure 10.75e shows the simulated phase noise.

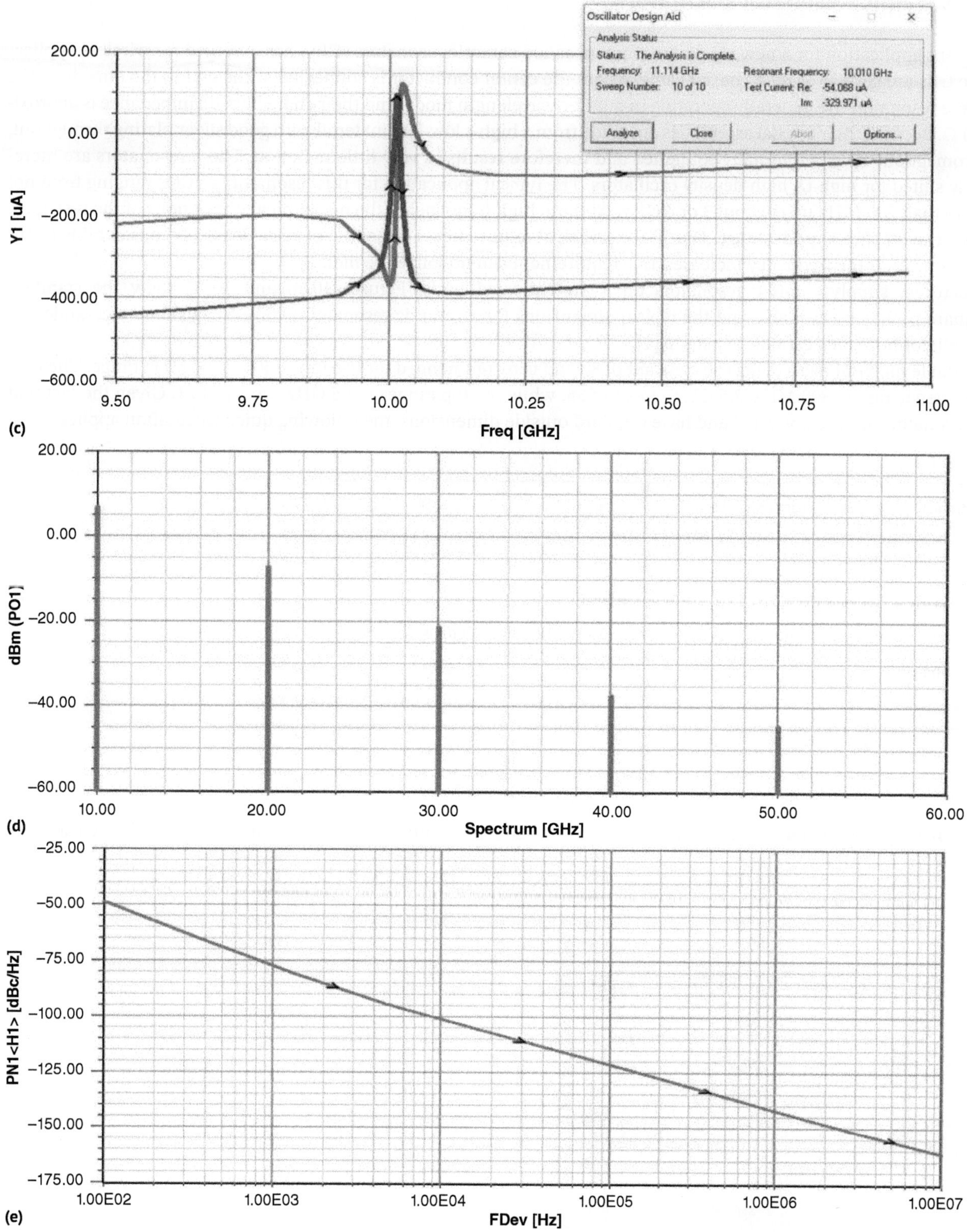

Figure 10.75 (Cont'd) (c) Linear analysis of the YIG oscillator to find the approximate resonant frequency: 10 GHz in this case. (d) Simulated output power (YIG oscillator) of (a). (e) Simulated phase noise of the YIG oscillator of (a).

10.15 Ceramic Resonator Oscillators

An important application for a new class of resonators called ceramic resonators (CRs) has emerged for wireless applications. The CRs are similar to rigid coaxial cable, where the center conductor is connected at the end to the outside of the cable. These resonators are generally operating in quarter-wavelength mode, and their characteristic impedance is approximately 10 Ω. Because their coaxial assemblies are made from a high-ε low-loss material with good silver plating throughout, the electromagnetic field is internally contained and therefore results in very little radiation. These resonators are therefore ideally suited for high-Q, high-density oscillators. The typical application for this resonator is VCOs ranging from not much more than 200 MHz up to about 3 or 4 GHz. At these high frequencies, the mechanical dimensions of the resonator become too tiny to offer any advantage. One of the principal requirements is that the physical length is considerably larger than the diameter. If the frequency increases, this requirement can no longer be met.

Manufacturers supply ceramic resonators on a prefabricated basis. Figures 10.76 and 10.77 show the standard round/square packaging available and the typical dimensions for a ceramic resonator. Figure 10.78 shows a ceramic resonator oscillator built on a ceramic substrate [1].

The available material has a dielectric constant of 88 and is recommended for use in the 400–1500 MHz range. The next higher frequency range (800 MHz–2.5 GHz) uses ε of 38, while the top range (1–4.5 GHz) uses ε of 21. Given the fact that ceramic resonators are prefabricated and have standard outside dimensions, the following quick calculation applies:

Relative dielectric constant of resonator material	$\epsilon_r = 21$	$\epsilon_r = 38$	$\epsilon_r = 88$
Resonator length in millimeters	$l = \frac{16.6}{f}$	$l = \frac{12.6}{f}$	$l = \frac{8.2}{f}$
Temperature coefficient (ppm/°C)	10	6.5	8.5
Available temperature coefficients	−3 to + 12	−3 to + 12	−3 to + 12
Typical resonator Q	800	500	400
Frequency range	1-4.5 GHz	800–2500 MHz	400–1500 MHz

Calculation of equivalent circuit: The equivalent parallel-resonant circuit afforded by a ceramic resonator has a resistance at resonance of

$$R_p = \frac{2(Z_0)^2}{R^* l} \tag{10.105}$$

Figure 10.76 Ceramic resonators. *Source*: [1]/John Wiley & Sons.

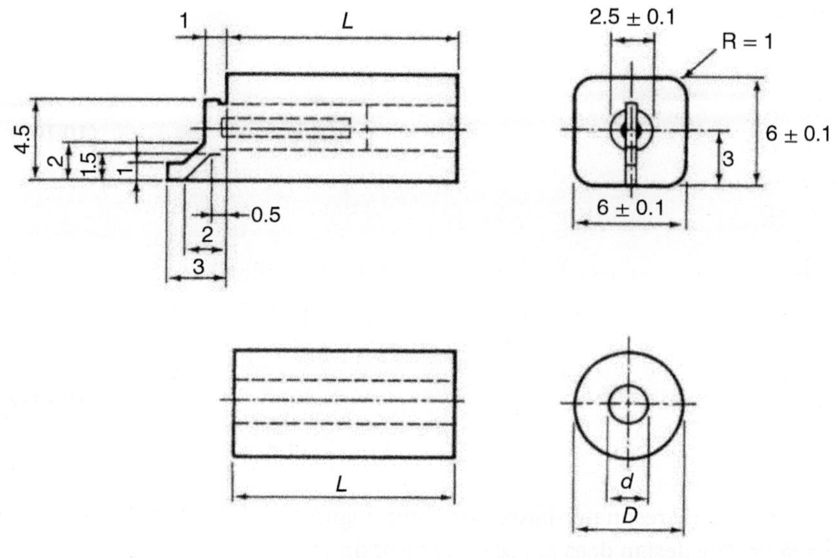

Figure 10.77 Standard round/square ceramic resonator packaging and dimensions. *Source*: [1]/John Wiley & Sons.

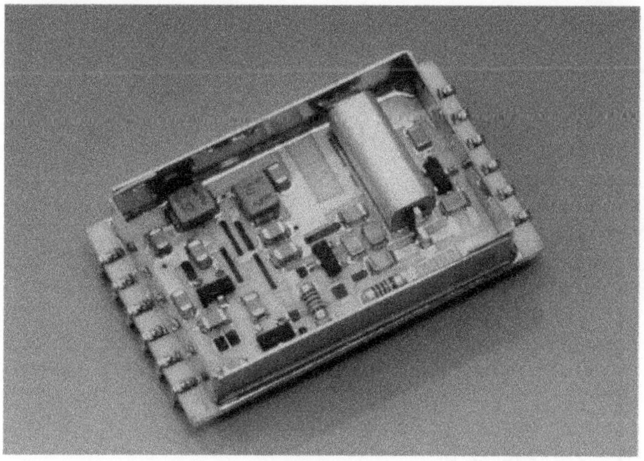

Figure 10.78 Modern ceramic resonator oscillator. *Source*: [1]/John Wiley & Sons.

where

Z_0 is the characteristic impedance of the resonator
l is the mechanical length of the resonator
R^* is the equivalent resistor due to metallization and other losses

As an example, one can calculate

$$C^* = \frac{2\pi\varepsilon_0\varepsilon_r}{\log_e(D/d)} = 55.61 \times 10^{-12}\frac{\varepsilon_r}{\log_e(D/d)} \tag{10.106}$$

and

$$L^* = \frac{\mu_r\mu_0}{2\pi} = \log_e\left(\frac{D}{d}\right) = 2 \times 10^{-7}\log_e\left(\frac{D}{d}\right) \tag{10.107}$$

$$Z_0 = 60\,\Omega \frac{1}{\sqrt{\varepsilon_r}} \log_e\left(\frac{D}{d}\right)$$
$$= 60 \frac{1}{\sqrt{88}} \log_e\left(\frac{6}{2.5}\right) \tag{10.108}$$
$$= 5.6\ \Omega$$

A practical example for $\varepsilon_r = 88$ and 450 MHz is

$$C_p = \frac{C * l}{2} = 49.7\ \text{pF} \tag{10.109}$$

$$L_p = 8L * l = 2.52\ \text{nH} \tag{10.110}$$

$$R_p = 2.5\ \text{k}\Omega \tag{10.111}$$

Figure 10.79 shows the schematic of a published ceramic resonator-based oscillator. Figures 10.80 and 10.81 show the simulated and measured phase noise of the oscillator. The design does not seem to be optimized.

By using ceramic resonator-based oscillators in conjunction with miniature synthesizer chips, it is possible to build extremely small phase-locked loop systems for cellular telephone operation.

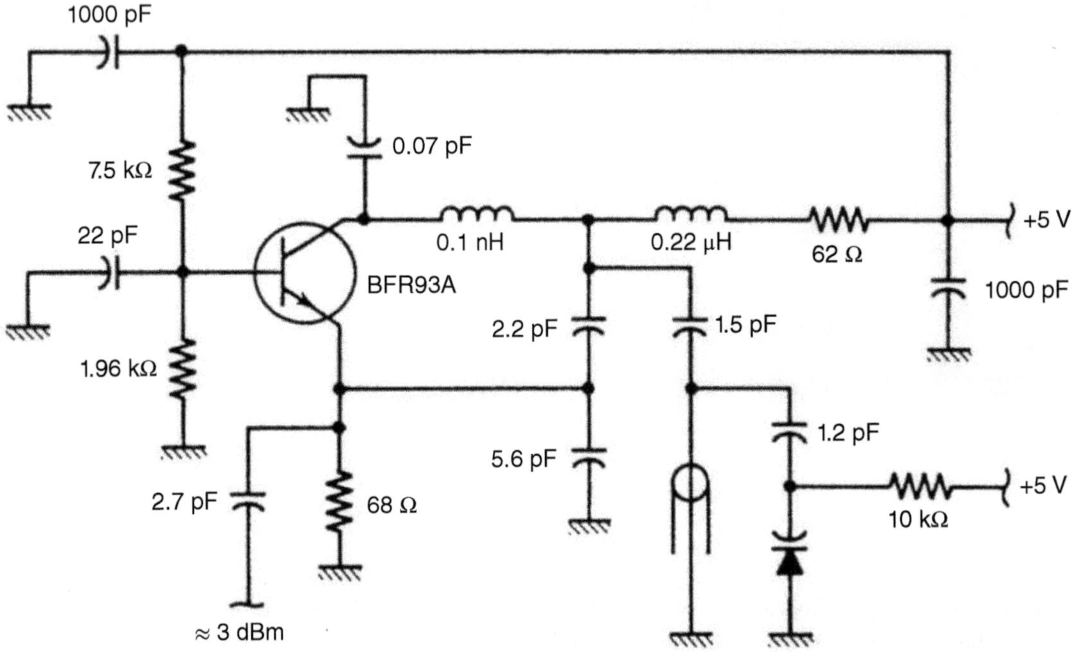

Figure 10.79 Schematic of a high-performance ceramic resonator-based oscillator that can be used from 500 MHz to 2 GHz. *Source:* [1]/John Wiley & Sons.

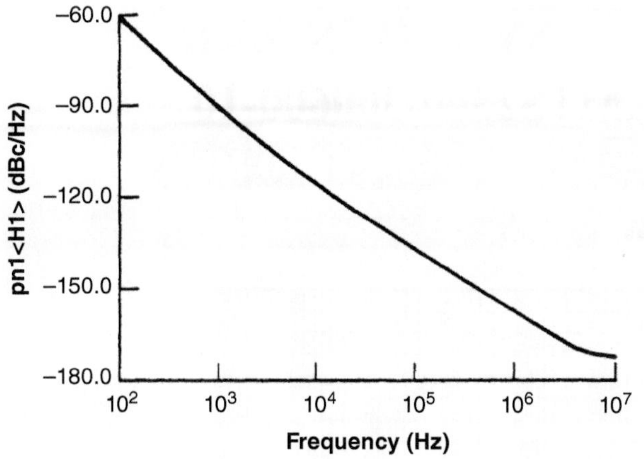

Figure 10.80 Simulated phase noise of an NPN bipolar 1 GHz ceramic resonator-based oscillator. *Source*: [1]/John Wiley & Sons.

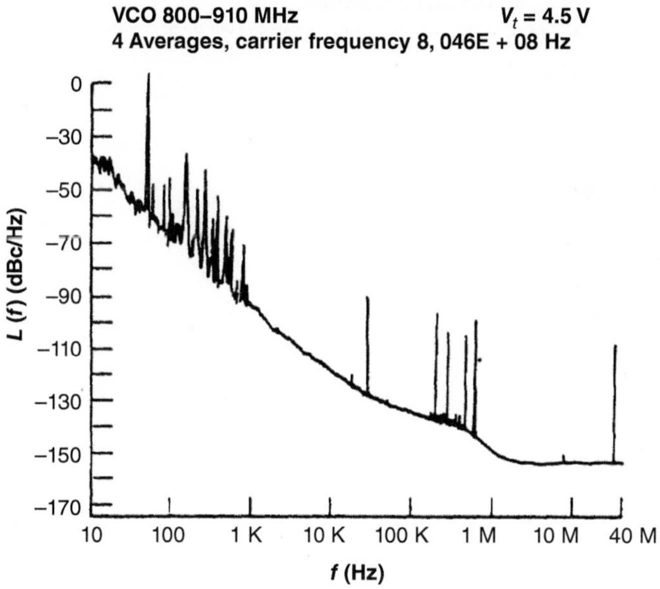

Figure 10.81 Measured phase noise of a ceramic resonator-based oscillator. *Source*: [1]/John Wiley & Sons.

Figure 10.82 shows one of the smallest currently available PLL-based synthesizers manufactured by Synergy Microwave Corporation. Because of the high-Q resonator, these types of oscillators exhibit extremely low phase noise. Values better than −130 dBc/Hz at 1 MHz off the carrier are achievable. The ceramic resonator reduces the oscillator's sensitivity to microphonic effects and proximity effects caused by other components.

A table comparing CRO and SAW oscillators at 2488.32 MHz is shown in Figure 10.83.

The novel oscillator topology results in CROs capable of operating over a temperature range as wide as −40 to +85 °C with minimal frequency drift due to temperature. As an example of the technology, a CRO was developed for use at 2488.32 MHz. A comparison of the CRO with a SAW oscillator at the same frequency (see Figure 10.83) reveals remarkably similar noise performance and tunability. Computer-aided engineering (CAE) modeling tools used in the development of the 2488.32 MHz CRO proved to be accurate in predicting expected performance, when compared to measurements.

A typical example for a ceramic resonator oscillator is shown in Figure 10.84a, a product of Synergy Microwave, and is a part of a synthesizer (PLL). Figure 10.84b shows a CRO free-running phase noise.

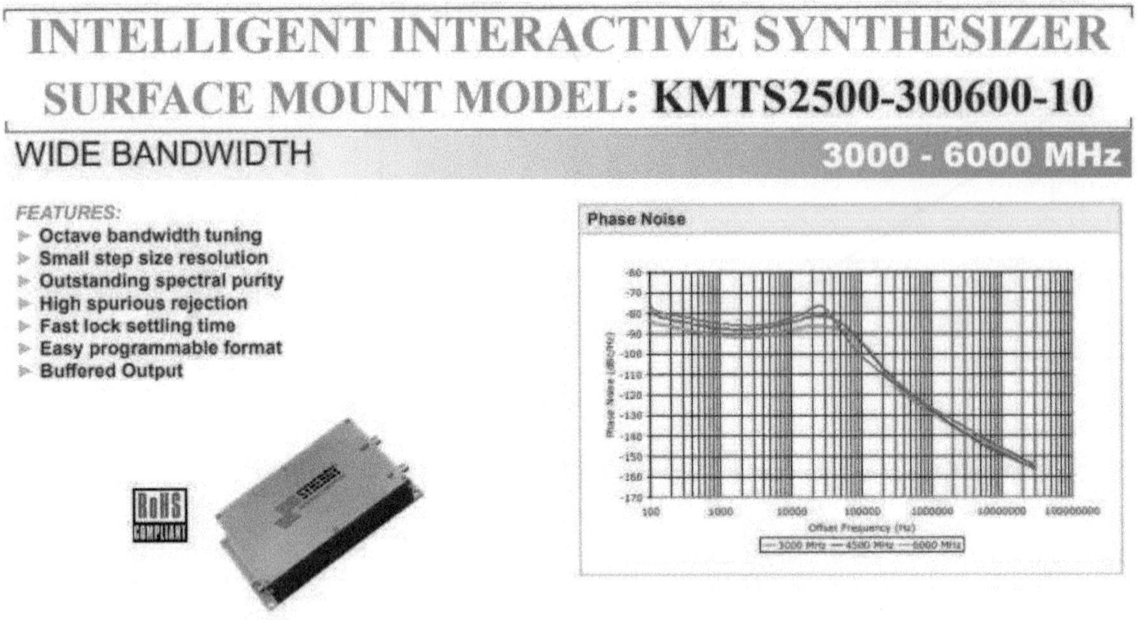

Figure 10.82 PLL-based synthesizer. *Source*: Synergy Microwave Corporation.

Comparing CRO and SAW oscillators at 2488.32 MHz		
Parameter	SAW-VCO	CRO-VCO
Center Frequency	2488.32 MHz	2488.32 MHz (Can be set per requirement)
Tuning Frequency	250 PPM	250 PPM
Tuning Voltage	1 to 5 V	1 to 5 V
Supply Voltage	60 mA @ +5 VDC	30 mA @ +12 VDC
Output Power	+3 dBm	0 dBm (Can be set per requirement)
Harmonic Rejection	20 dB	20 dB
Typical Phase Noise		
Offset 10 kHz	–122 dBc/Hz	–123 dBc/Hz
Offset 100 kHz	–142 dBc/Hz	–143 dBc/Hz
Operating Temperature	–20 to +70°C	–40 to +85°C
Thermal Drift	Thermal Drif	Thermal Drif
–20 to +70°C	±300kHz	±400kHz
–40 to +85°C	N/A	±600kHz

Figure 10.83 A table comparing CRO and SAW oscillators at 2488.32 MHz. *Source*: [1]/John Wiley & Sons.

In this case (Figure 10.85a), the impedance of the cable is typically 50 Ω. If teflon is used as a material, a fairly good Q is obtained. The tricky part of this circuit is that the output port is part of the resonator, which results in a much better phase noise than taking the energy out from the collector. The drawback is that the coupling to the 50 Ω load should not be too tight, not optimized for maximum power. Figure 10.85b shows the simulated power output for the circuit.

The 2.5 GHz oscillator simulated phase noise in Figure 10.85c shows −135 dBc/Hz at 25 kHz offset, which beats the 144 MHz FET oscillator. See Figure 10.51c: 144 MHz JFET-based LC oscillator phase noise.

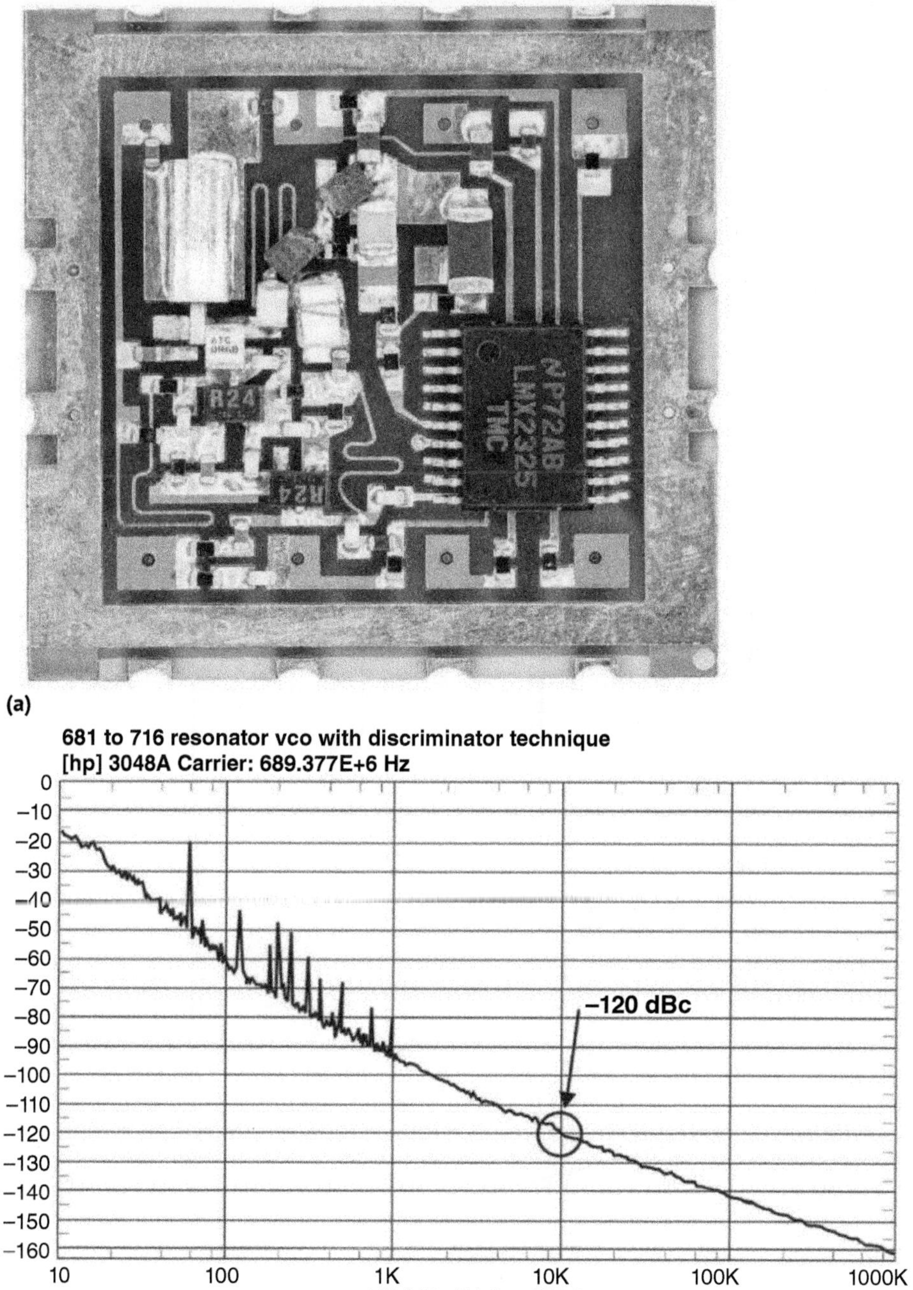

Figure 10.84 (a) Photograph of a typical ceramic coaxial resonator-based VCO in a PLL (Synergy Microwave, NJ, USA). (b) The measured phase noise of a 689.377 MHz free-running CRO. *Source*: [1]/John Wiley & Sons.

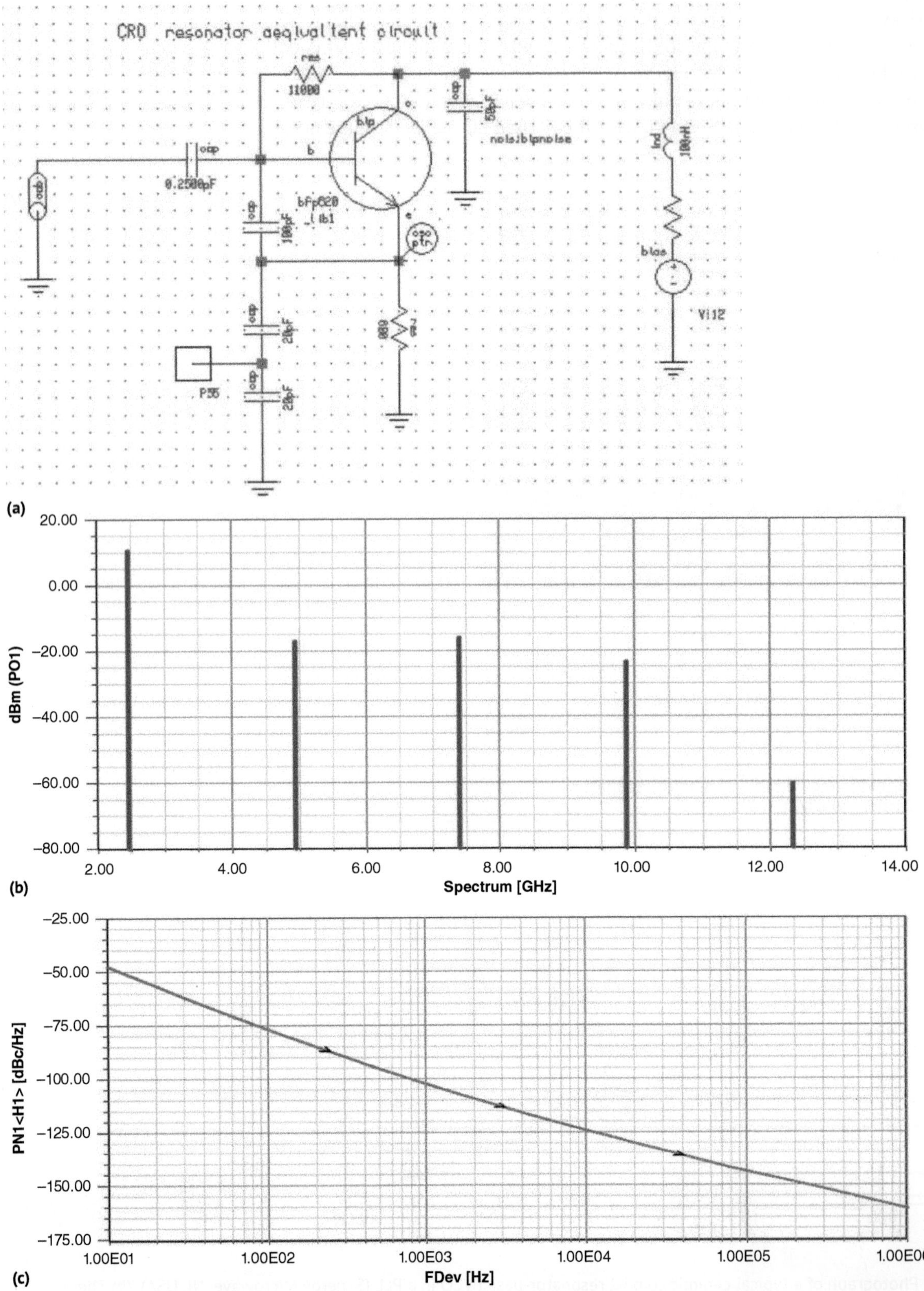

Figure 10.85 (a) Using a coaxial cable as the equivalent circuit of a ceramic resonator. (b) Simulated power output for the circuit of (a). (c) Simulated phase noise for the circuit of (a).

10.16 GaAs FET-based Oscillator

Figure 10.86a is a 5 GHz GaAsFET oscillator schematic, using the Materka model for CAD simulation purposes. The DC–IV curve is rounder than a BJT; the flicker corner frequency is much higher but the input S-parameter S_{11} is less dependent on the input power. As the GaAs FET is a high-impedance device, there is a much higher voltage swing, and the key nonlinearities are the transconductance Y_{21} and the output admittance Y_{22}.

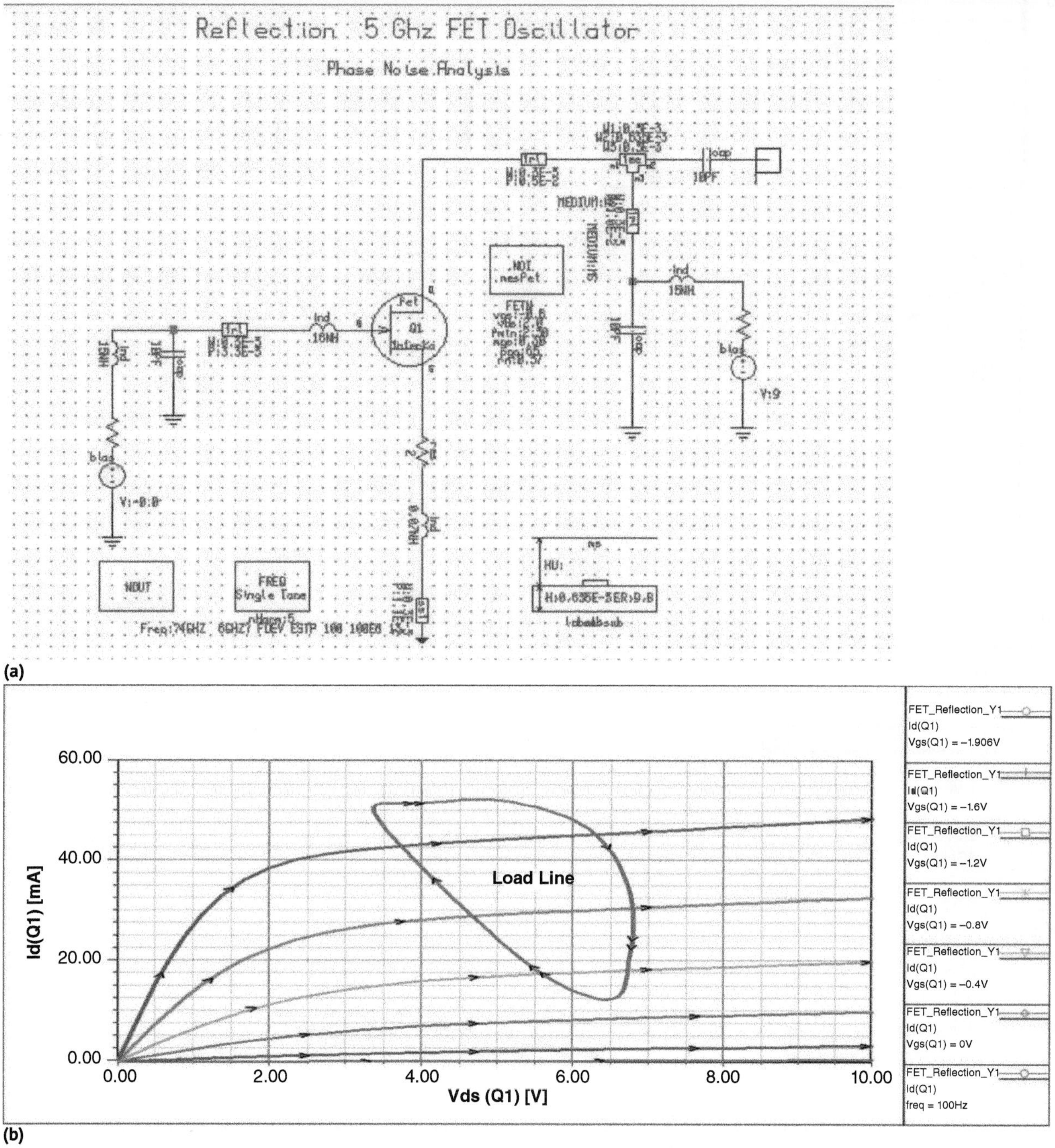

Figure 10.86 (a) 5 GHz FET oscillator schematic for CAD simulation. (b) Load line simulation for the 5 GHz FET oscillator of (a).

(Continued)

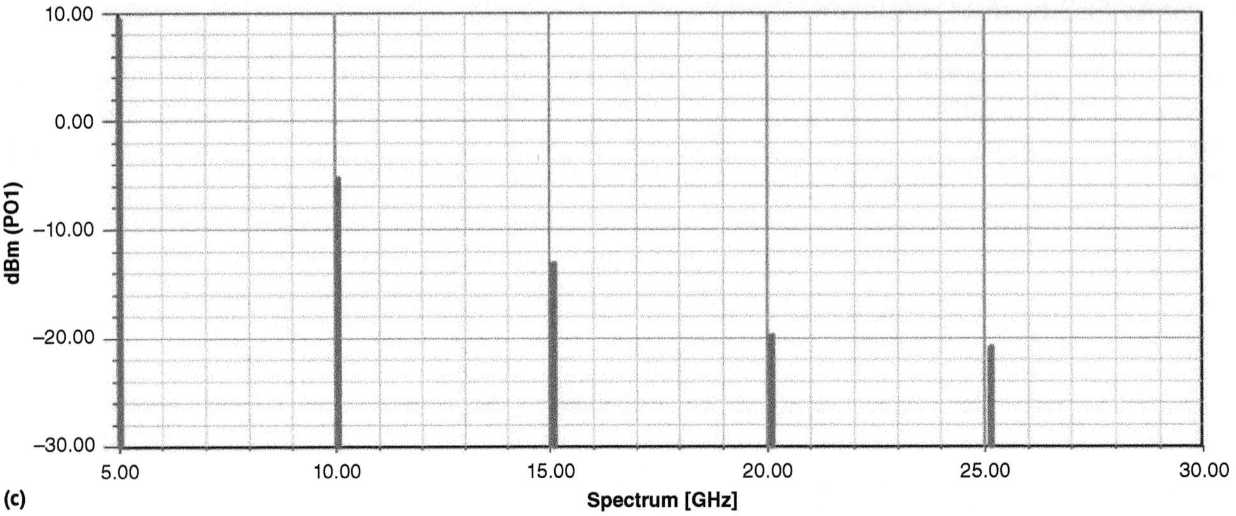

(c)

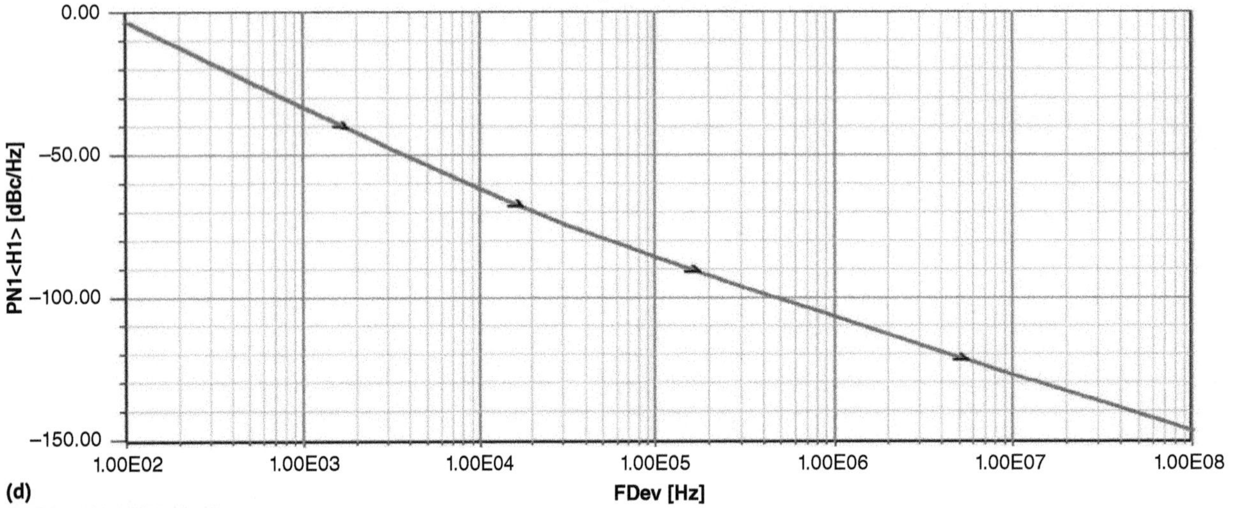

(d)

.Noise model (MESFET)

	Property	Value
1	af	
2	fc	1.0E7
3	fcp	
4	fmin	2.30
5	fn	4GHZ
6	kf	
7	mgo	0.30
8	name	FETN
9	pgo	65
10	rn	0.37
11	vds	2.0
12	vgs	-0.6
13		

(e)

Figure 10.86 (Cont'd) (c) Simulated output power of the 5 GHz FET oscillator of (a). (d) Simulated phase noise of the 5 GHz FET oscillator of (a). (e) Noise model parameters of the GaAs FET used in the simulation of (a).

Example of a 5 GHz FET Oscillator Note the high flywheel effect of the output load line in Figure 10.86b.

Figure 10.86c shows the simulated output power of the 5 GHz FET oscillator, Figure 10.86d shows the simulated phase noise, and Figure 10.86e shows the noise model parameters of the GaAs FET used in the simulation of Figure 10.86a.

Example of a 5 GHz CFY65-based Oscillator with Distributed Elements: Figure 10.87a shows a 5 GHz CFY65-based oscillator, modeled using mostly distributed elements. This design is mostly based on distributed models and turned out to be quite accurate.

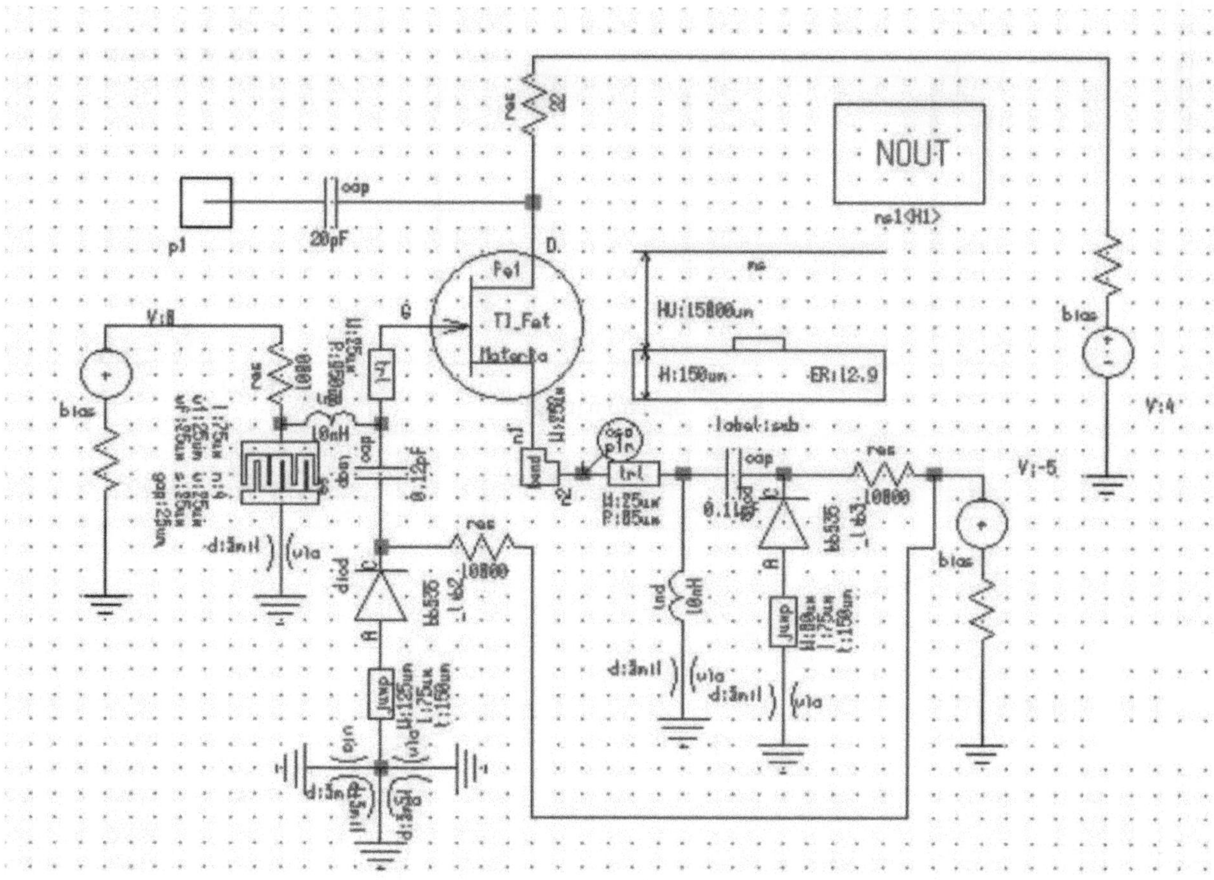

(a)

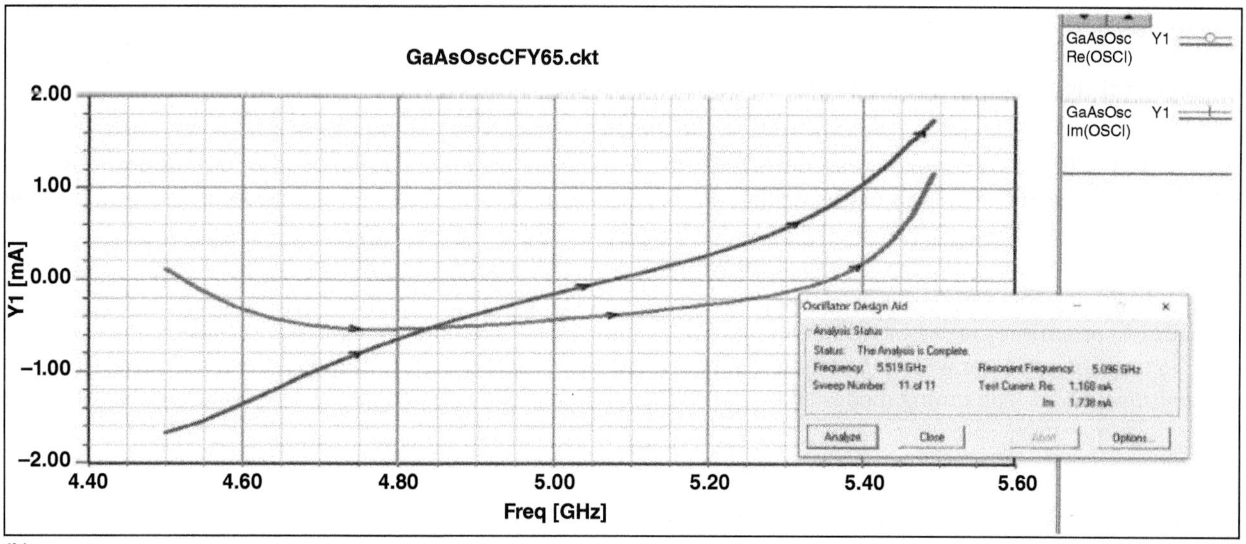

(b)

Figure 10.87 (a) 5 GHz CFY65-based oscillator. *Source*: [49, p. 438]/John Wiley & Sons. (b) Linear analysis of the oscillator frequency of the circuit in (a).

(Continued)

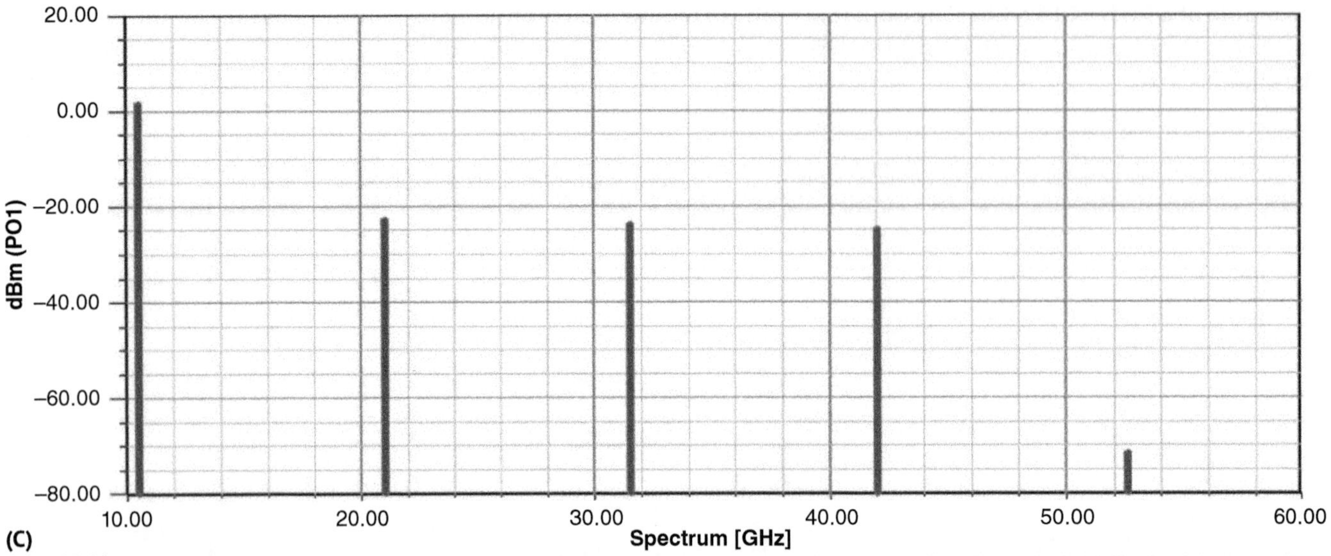

(C)

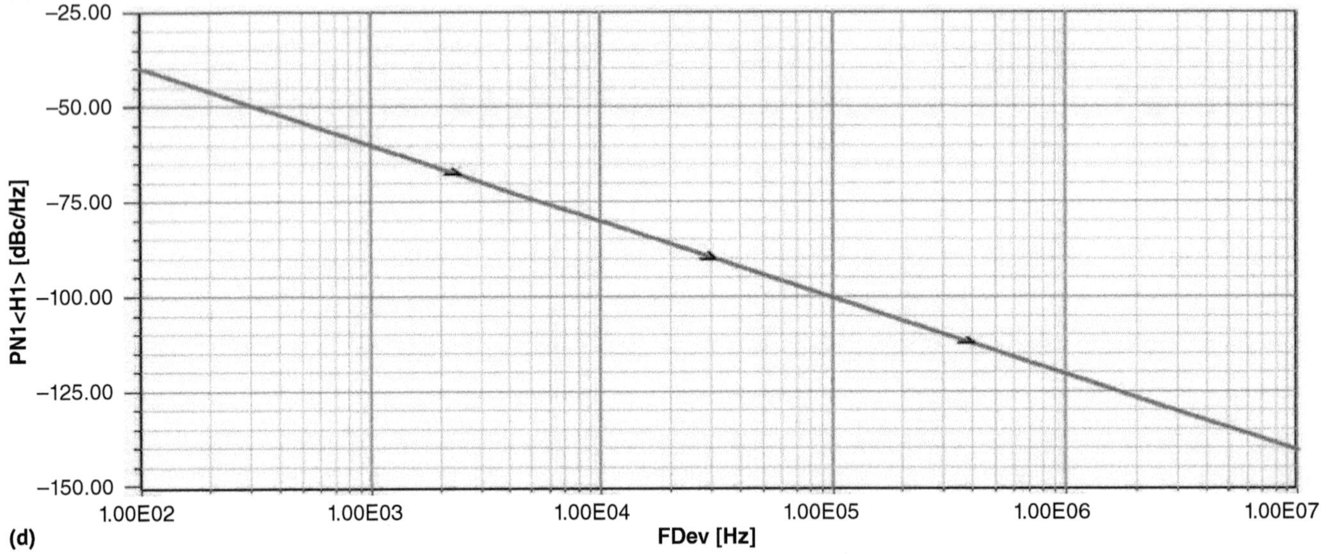

(d)

Figure 10.87 (Cont'd) (c) Simulated output power of the oscillator in (a). (d) Simulated phase noise of the oscillator in (a).

The 5 GHz GaAs FET-based oscillator uses the SIEMENS CFY65 GaAs FET for simulation, though it is now obsolete. Figure 10.87b shows the linear analysis of the oscillator frequency, Figure 10.87c shows the simulated output power of the 5 GHz FET oscillator, and Figure 10.87d shows the simulated phase noise.

This is a typical example of a microwave transistor with distributed elements and detailed modeling. Most designs are now done with foundry libraries, so this level of design is no longer obvious.

A schematic of a ceramic resonator-based VCO (Figure 10.79) has already been shown in the section on ceramic resonator oscillators. To tune the oscillator within the required range, tuning diodes are used. These diodes are often called *varactors* or *voltage-sensitive diodes*. The tuning diodes add noise to the oscillator. How this happens is explained in the following section.

10.17 Tuning Diode-based AM-to-PM Conversion

Now some calculations about influence of tuning diodes in the VCOs. In this section, we show some practical results.

The loading effect of the tuning diode is due to losses, and these losses can be described by a resistor in parallel with the tuned circuit [1].

It is possible to define an equivalent noise R_{aeq} that, inserted in Nyquist's equation,

$$V_n = \sqrt{4kT_oR\Delta f} \tag{10.112}$$

where $kT_o = 4.2 \times 10^{-21}$ at about 300 K, R is the equivalent noise resistor, and Δf is the bandwidth, determines an open-circuit noise voltage across the tuning diode. Practical values of R_{aeq} for carefully selected tuning diodes are in the vicinity of 200 Ω – 50 kΩ. If we now determine the noise voltage $V_n = \sqrt{4 \times 4.2 \times 10^{-21} \times 10,000}$, the resulting voltage value is $1.296 \times 10^{(-8)}\text{V}\sqrt{\text{Hz}}$.

This noise voltage generated from the tuning diode is now multiplied with the VCO gain K_o, resulting in the rms frequency deviation

$$(\Delta f_{rms}) = K_o \times (1.296 \times 10^{-8}\ \text{V})\ \text{in 1-Hz bandwidth} \tag{10.113}$$

To translate this into an equivalent peak phase deviation,

$$\theta_d = \frac{K_o\sqrt{2}}{f_m}(1.296 \times 10^{-8})\ \text{radin 1-Hz bandwidth,} \tag{10.114}$$

for a typical microwave oscillator operating at 36 GHz, having a voltage gain of 1 GHz/V,

$$\theta_d = \frac{0.00183}{f_m}\ \text{radin 1-Hz bandwidth} \tag{10.115}$$

For $f_m = 1$ MHz (typical spacing just outside the PLL loop bandwidth), t $\theta_c = 1.5 \times 10^{-5}$. This can be converted now into the SSB signal-to-noise ratio:

$$\mathcal{L}(f_m) = 20\ \log_{10}\frac{\theta_c}{2} = -105\ \text{dBc/Hz} \tag{10.116}$$

For example, without a tuning diode, a 3.5 GHz oscillator having a QL of 100, and 6 dBm output power, we can expect the phase noise as seen in Figure 10.88.

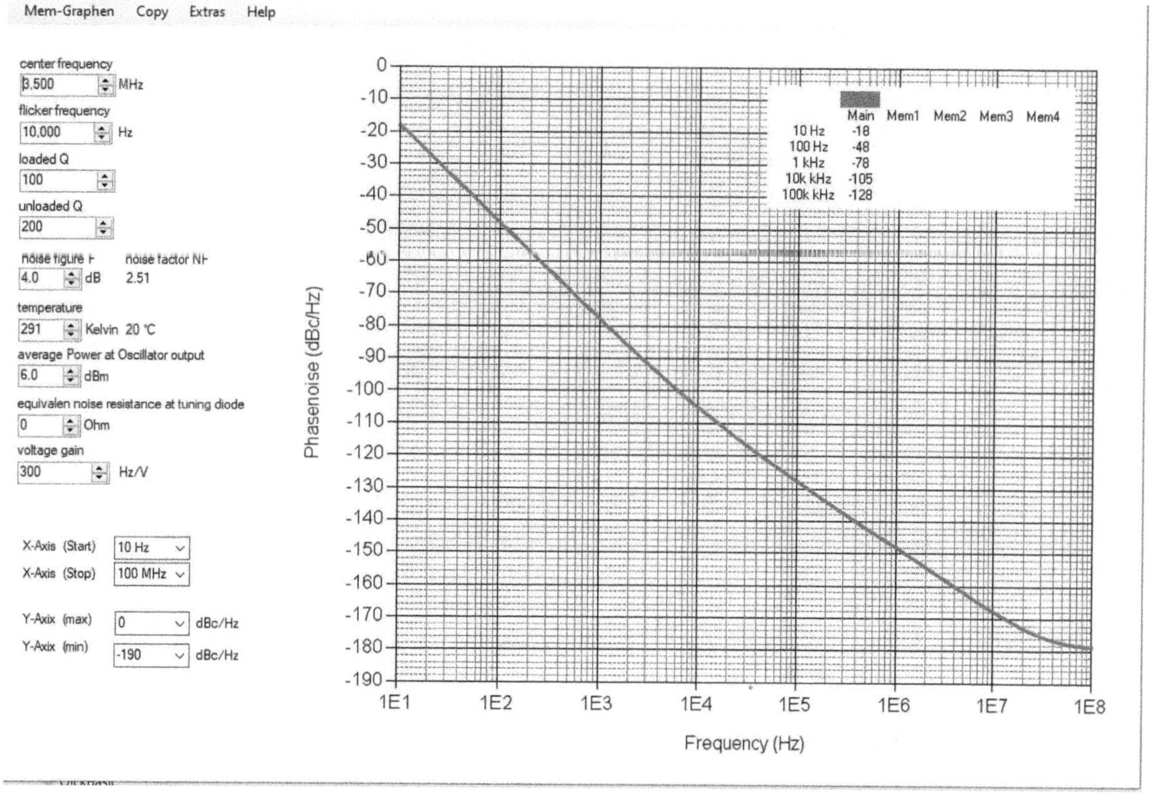

Figure 10.88 Graphic display of (a 3.5 GHz oscillator without a tuning diode) the modified Leeson equation, using a software tool designed by Joerg Bartls.

Adding a tuning diode with a sensitivity of 100 MHz/V (scaled-down example from above at 35 GHz, Ka band), the phase noise changes drastically. Figure 10.88 shows a graphic display of a 3.5 GHz oscillator without a tuning diode, which was simulated using the modified Leeson equation, with a software tool designed by Joerg Bartls, while Figure 10.89 shows the simulated graphic display of the same oscillator with a tuning diode.

For the typical oscillator gain of 10 MHz/V found in wireless applications, the resulting phase noise will be 20 dB better [10 log (1 GHz ÷ 10 MHz)]. However, the best tuning diodes, like the BB104, have an R_n of 200 Ω instead of 10 kΩ, which again changes the picture. With $kT_o = 4.2 \times 10^{-21}$, the resulting noise voltage will be

$$\begin{aligned} V_n &= \sqrt{4 \times 4.2 \times 10^{-21} \times 200} \\ &= 1.833 \times 10^{-9}\,\mathrm{V}\sqrt{\mathrm{Hz}} \end{aligned} \tag{10.117}$$

From (10.114), the equivalent peak phase deviation for a gain of 10 MHz/V in a 1 Hz bandwidth is then

$$\theta_d = \frac{1 \times 10^7 \sqrt{2}}{f_m}(1.833 \times 10^{-9})\,\mathrm{rad} \tag{10.118}$$

or

$$\theta_d = \frac{0.026}{f_m}\,\mathrm{rad\,in\,1\text{-}Hz\,bandwidth} \tag{10.119}$$

With $f_m = 25$ kHz, $\theta_c = 1.04 \times 10^{-6}$. Expressing this as phase noise:

$$\mathcal{L}(f_m) = 20\,\log_{10}\frac{\theta_c}{2} = -126\,\mathrm{dBc/Hz} \tag{10.120}$$

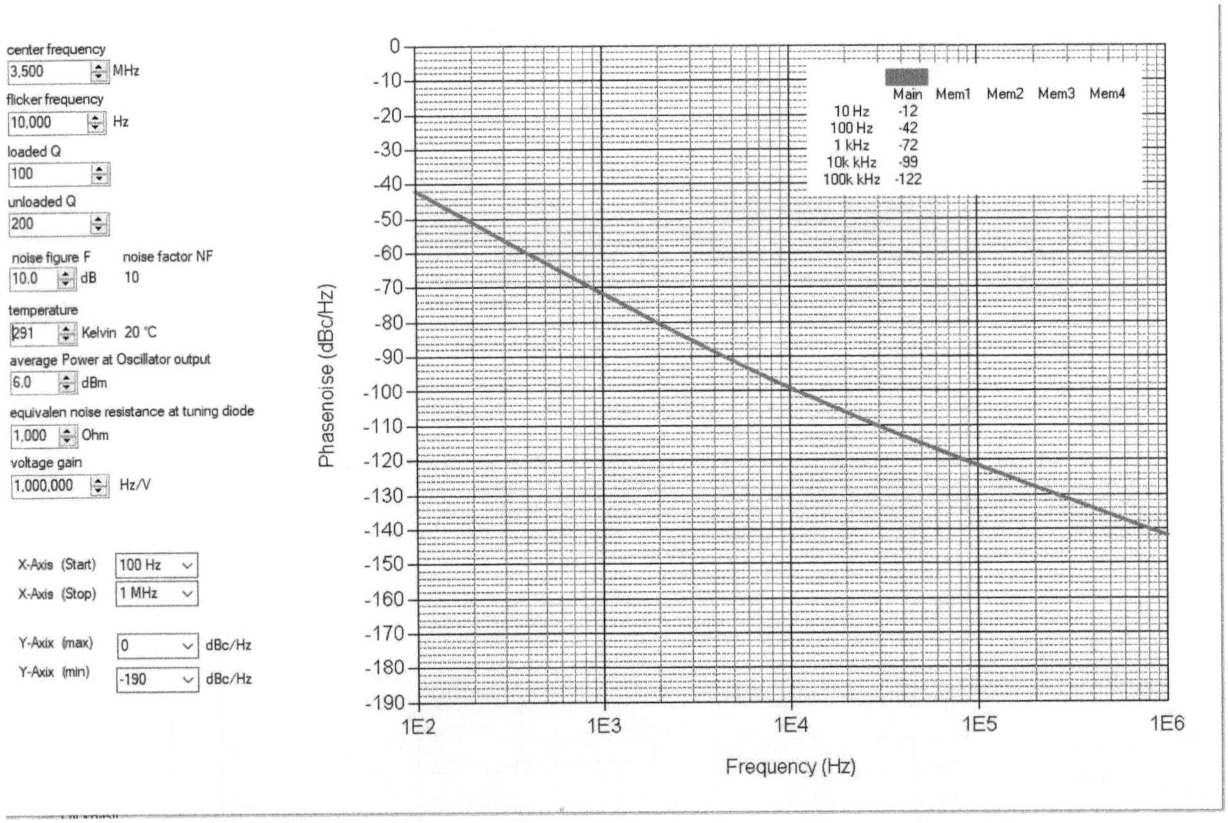

Figure 10.89 Simulated graphic display of (a 3.5 GHz oscillator with a tuning diode) the modified Leeson equation using a software tool designed by Joerg Bartls.

The phase noise value in (10.120) is that typically achieved in the Rohde & Schwarz SMDU or with the Hewlett-Packard 8640 signal generator and considered state-of-the-art for a free-running oscillator. It should be noted that both signal generators use a slightly different tuned circuit; the Rohde & Schwarz generator uses a helical resonator, whereas the Hewlett-Packard generator uses an electrically shortened quarter-wavelength cavity. Both generators are mechanically pre-tuned, and a tuning diode with a gain of about 100 kHz/V is used for AFC and frequency modulation purposes. It is apparent that, because of the nonlinearity of the tuning diode, the gain is different for low DC voltages than for high DC voltages. The impact of this is that the noise varies with the tuning range.

If this oscillator had to be used for a synthesizer, its 1 MHz tuning range would be insufficient; therefore, a way had to be found to segment the band into the necessary ranges. In VCOs, this is typically done with switching diodes that allow the proper frequency bands to be selected. These switching diodes insert in parallel or series, depending on the circuit, with additional inductors or capacitors, depending on the design.

In low-energy-consuming circuits, the VCO frequently is divided into a coarse-tuning section using tuning diodes and a fine-tuning section with a tuning diode. In the coarse-tuning range, this results in very high gains, such as 1–10 MHz/V, for the diodes. The noise contribution of those diodes is therefore very high and can hardly be compensated by the loop. For low-noise applications, which automatically mean higher power consumption, it is unavoidable to use switching diodes.

Let us now examine some test results. Referring back to Eq. (10.106), Figure 10.90 shows the noise sideband performance as a function of Q, whereby the top curve with $Q_L = 100$ represents a somewhat poor oscillator and the lowest curve with $Q_L = 100,000$ probably represents a crystal oscillator where the unloaded Q of the crystal was in the vicinity of 3×10^6. Figure 10.91 shows the influence of flicker noise [1].

Corner frequencies of 10 Hz–10 kHz have been selected, and it becomes apparent that around 1 kHz the influence is fairly dramatic, whereas the influence at 20 kHz off the carrier is not significant.

Figures 10.91 and 10.92 show the influence of the tuning diodes on a high-Q oscillator.

Curve A in Figure 10.92 uses a lightly coupled tuning diode with a K_o of 10 kHz/V; the lower curve is the noise performance without any diode. As a result, the two curves are almost identical, which can be seen from the somewhat smeared form of the graph. Curve B shows the influence of a tuning diode at 100 kHz/V and represents a value of −143 dBc/Hz from −155 dBc/Hz, already some deterioration. Curve C shows the noise if the tuning diode results in a 1 MHz/V VCO gain, and the noise sideband at 25 kHz has now deteriorated to −123 dBc/Hz. These curves speak for themselves.

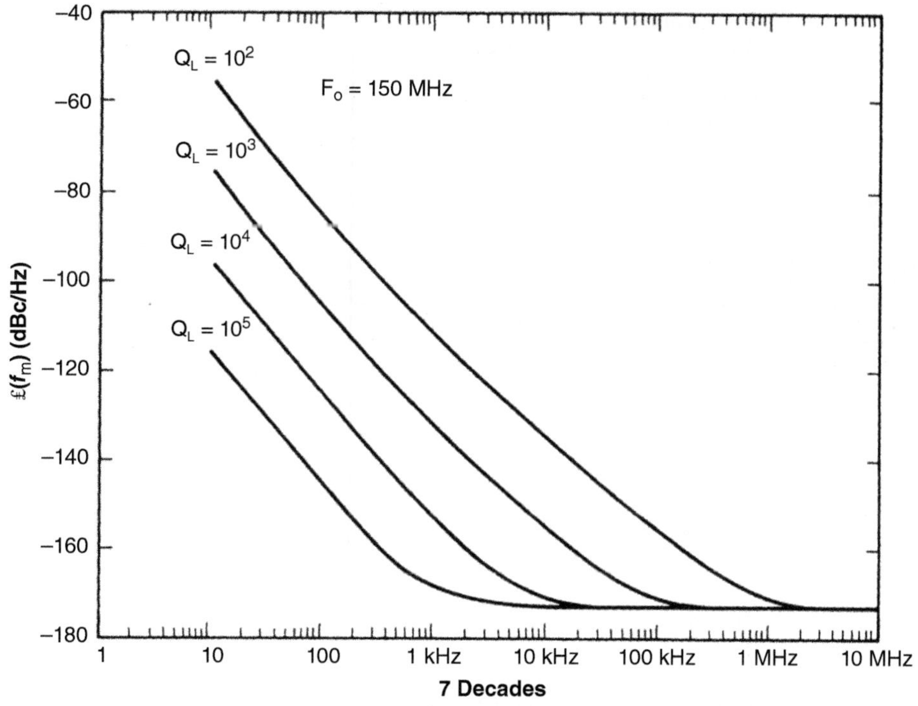

Figure 10.90 Noise sideband of an oscillator at 150 MHz as a function of the loaded Q of the resonator. *Source*: [1]/John Wiley & Sons.

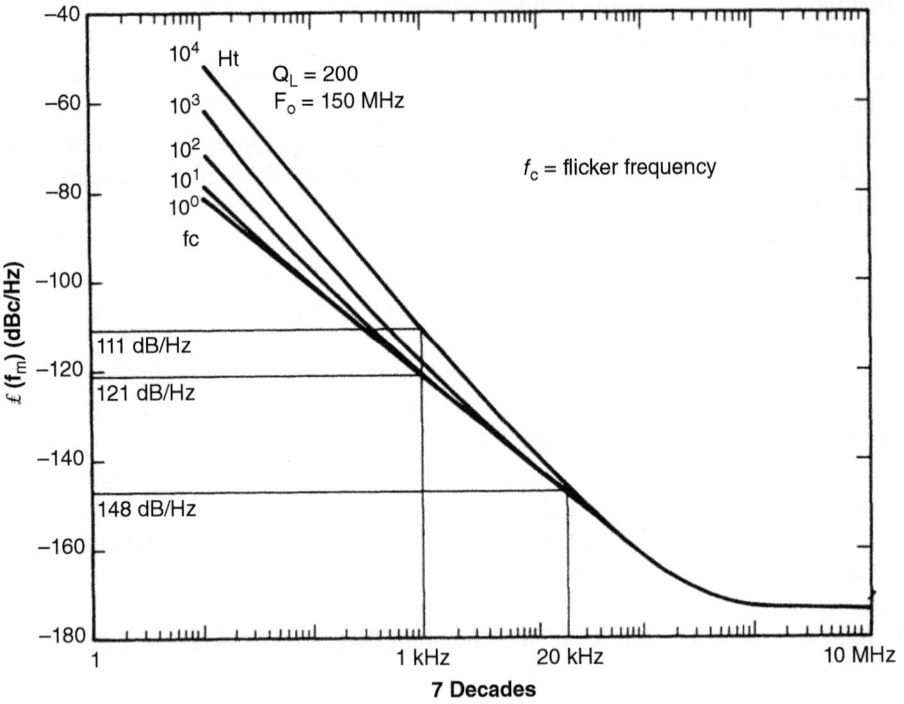

Figure 10.91 Noise sideband performance as a function of the flicker frequency ωC varying from 10 Hz to 10 kHz. *Source*: [1]/John Wiley & Sons.

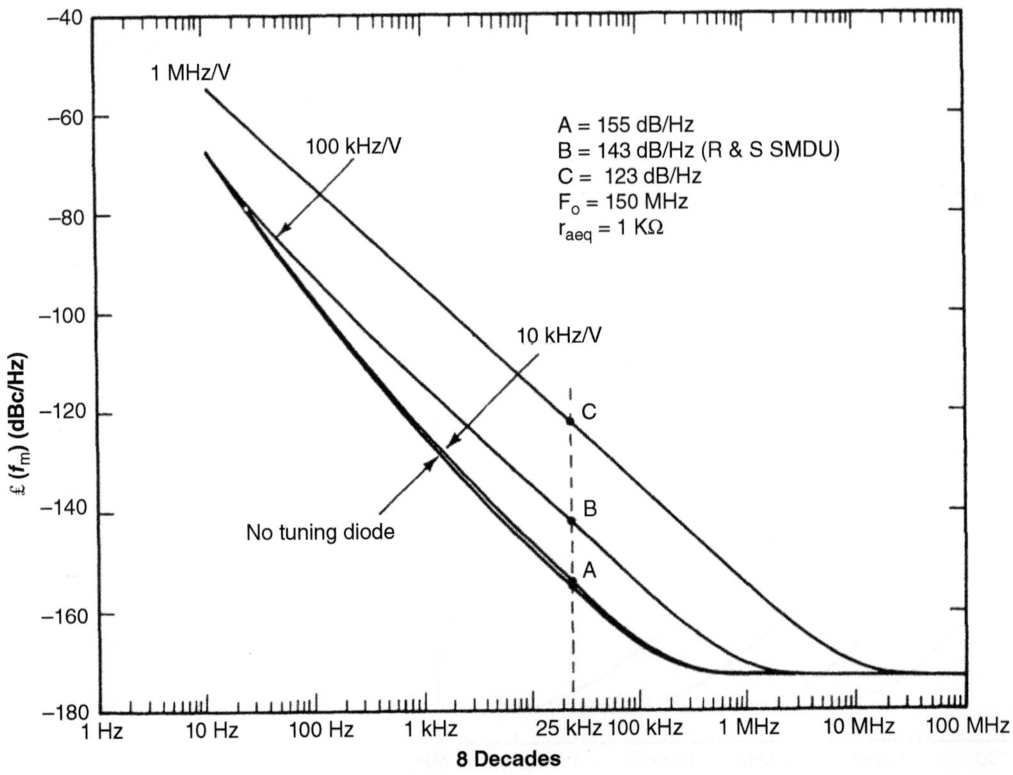

Figure 10.92 Noise sideband performance of an oscillator at 150 MHz, showing the influence of various tuning diodes. *Source*: [1]/John Wiley & Sons.

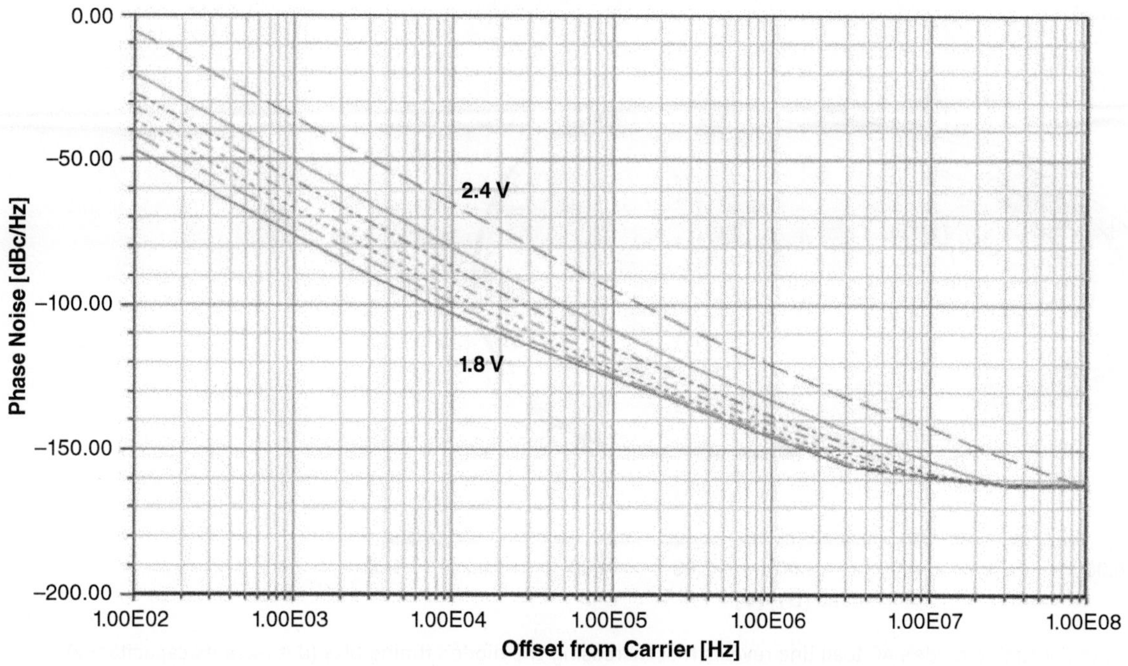

Figure 10.93 Noise sideband performance of an oscillator at 147 MHz, showing the influence of a tuning diode operated at bias voltages from 1.8 to 2.4 V in 0.1 V increments. *Source*: [1]/John Wiley & Sons.

Figure 10.93 shows how the phase noise of a 147 MHz VCO varies with the bias applied to its tuning diode. Because a tuning diode's Q increases with the reverse bias applied to it, we expect a VCO's phase noise performance to improve, even if only slightly, as its tuning voltage is increased. Yet Figure 10.92 illustrates the reverse of this relationship. Inspecting this VCO's circuit (Figure 10.93), we see that the diode is in series with the circuit's 220 nH resonator. *All* of the RF current flowing through the resonator must flow through the relatively low-Q diode, and even small decreases in the diode's capacitance will significantly modify the tank's LC ratio [1].

Figure 10.94 shows the schematic of a voltage controlled oscillator (VCO) for its performance analysis using CAD simulation software.

Graphing the diode's AC load line (Figure 10.95), we see that as the tuning voltage increases (that is, as the tuning diode capacitance decreases), the RF voltage across the diode becomes disproportionately high. We note also that at every value of tuning voltage, the RF voltage across the diode goes positive (relative to the anode) during part of each cycle.

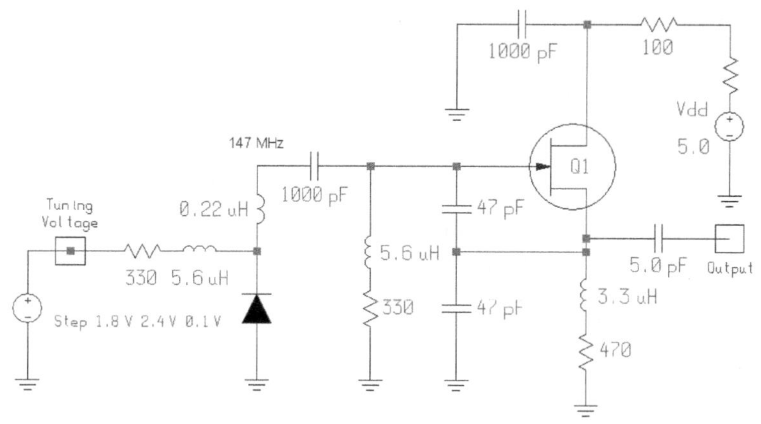

Figure 10.94 The VCO schematic. Because the tuning diode is in series with the resonator, it passes all of the tank's RF current and plays a disproportionately large role in determining the circuit's phase noise performance. *Source*: [1]/John Wiley & Sons.

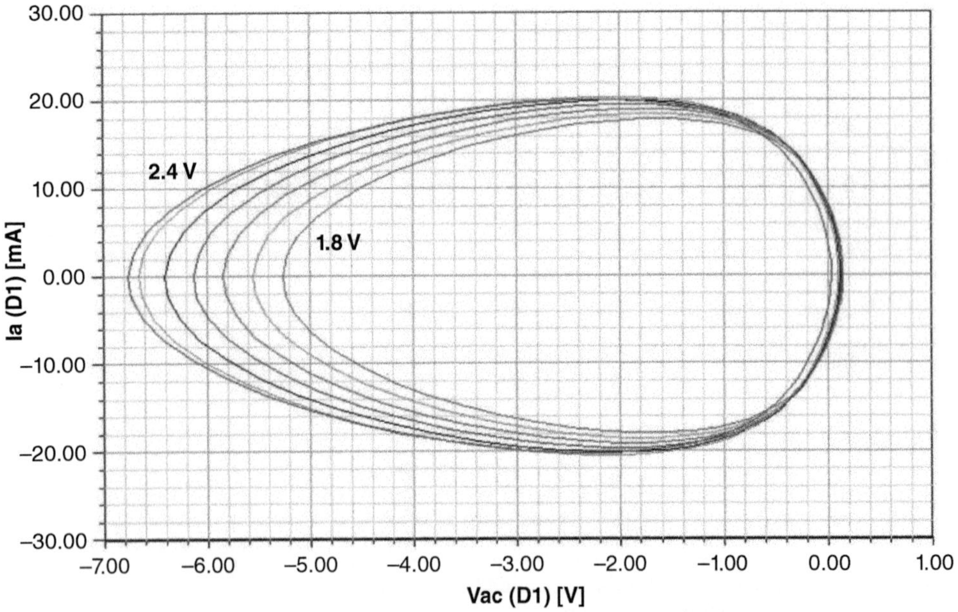

Figure 10.95 Graphing the tuning diode's AC load line reveals that increasing the diode's tuning bias (decrease its capacitance) results in a disproportionally large increase in the RF voltage across the diode. At all tuning voltages, the RF voltage across the diode is slightly positive (relative to the anode) during part of each cycle. *Source*: [1]/John Wiley & Sons.

Modifying the oscillator tank to add capacitance in parallel with the tuning diode (Figure 10.96) reduces the tuning diode's effect on tank Q and LC ratio, and therefore its contribution to phase noise. With this arrangement, we now see the expected relationship between tuning bias and phase noise: The higher the tuning bias, the lower the phase noise (Figure 10.96). Figure 10.97 shows that the diode's AC load line is now virtually unchanged across the tuning-voltage range. The RF voltage across the diode still goes positive (relative to the anode) during part of each cycle [1].

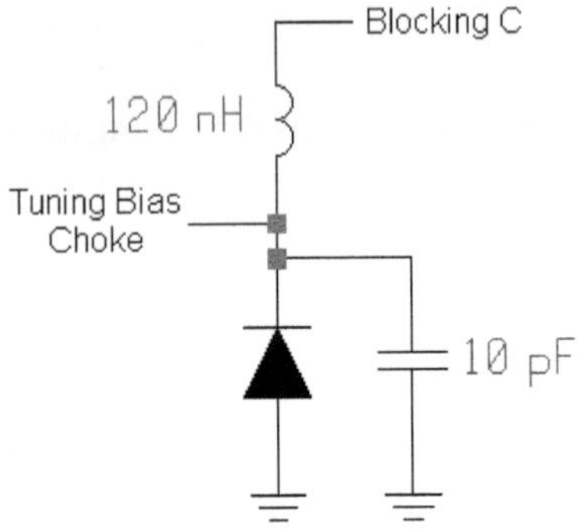

(a)

Figure 10.96 (a) Adding capacitance across the tuning diode reduces the effect of its nonlinearities on phase noise performance. We have also revalued the resonator to 120 nH to maintain oscillation near 147 MHz. (b) Now that the tuning diode passes only part of the tank's RF current, varying its capacitance does not result in the large change in phase noise performance shown in Figure 10.92. Note also that the phase noise versus tuning bias relationship has reversed relative to that shown in Figure 10.92: the highest tuning voltage now corresponds to the lowest phase noise *Source*: [1]/John Wiley & Sons.

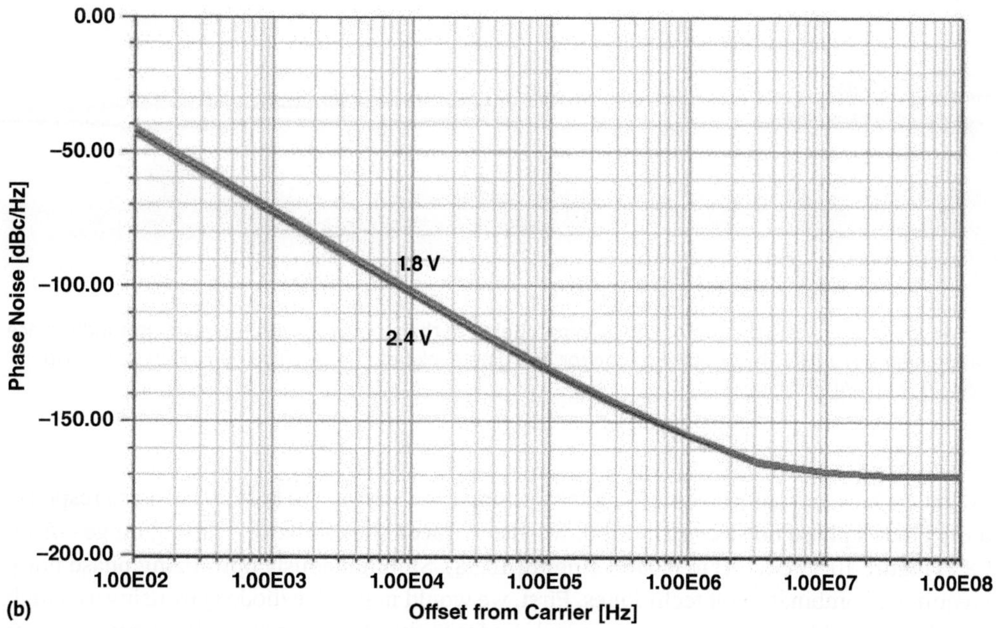

Figure 10.96 (Cont'd)

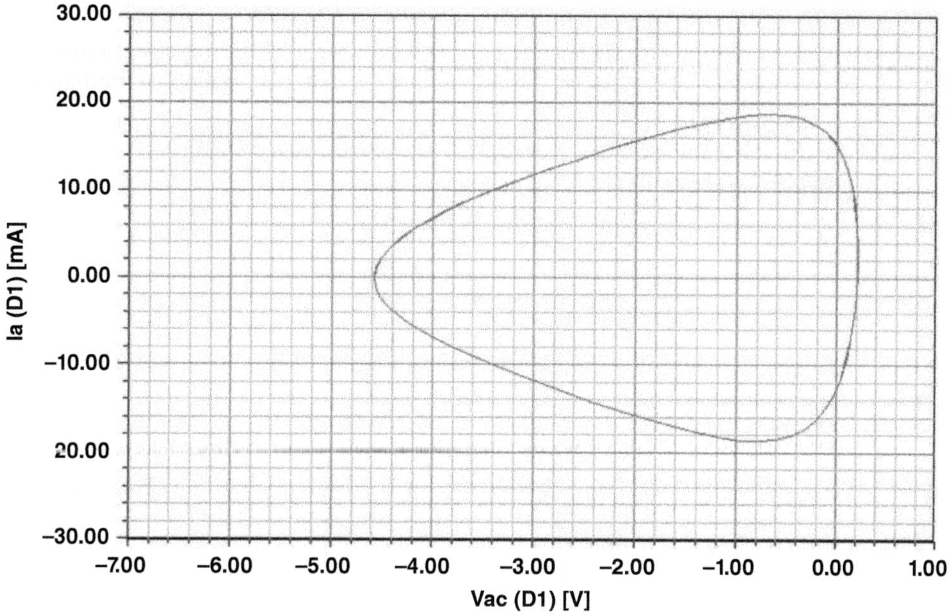

Figure 10.97 Ac load line for the tuning diode in Figure 10.94. Now the voltage swing across the diode is virtually the same at all tuning voltages. The RF voltage across the diode still goes positive (relative to the anode) during part of the cycle. *Source*: [1]/John Wiley & Sons.

Figure 10.98 shows one more modification to the VCO tank. Now, the tuning diode is lightly coupled to the hot end of the tank, in parallel with the resonator. The result is phase noise performance that is essentially unchanged across the 1.8–2.4 V tuning-voltage range. Figure 10.99 compares this circuit's phase noise performance with the worst-case performance of the tank configurations in Figures 10.93 and 10.95. Figure 10.100 compares the diode AC load lines that correspond to the phase noise curves in Figure 10.99. The best phase noise performance corresponds to the diode AC load line that exhibits the narrowest voltage swing and never goes positive relative to the anode.

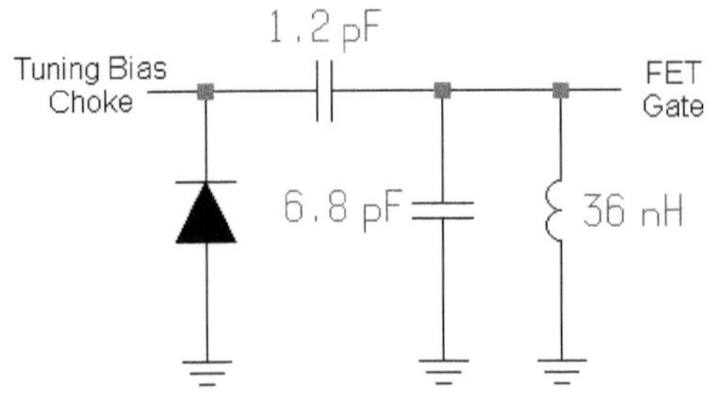

Figure 10.98 Tank circuit reconfigured to lightly couple the tuning diode in parallel with the resonator, with the resonator inductance again decreased to maintain oscillation near 147 MHz. The 1000 pF capacitor and gate choke of Figure 10.93 (5.6 μH in series with 330 Ω) are no longer needed. *Source*: [1]/John Wiley & Sons.

It's important to keep in mind that we have not gained something for nothing in achieving the best phase noise response in Figure 10.99. As we reduced the diode's phase noise contribution, we also reduced the oscillator's tuning range—from 7.1% (Figure 10.93 circuit) to 1.8% (Figure 10.95 circuit) to 0.046% (Figure 10.98). Simultaneously optimizing phase noise performance and tuning range requires a combination of techniques. First, we would maximize diode *Q* by using as a high a tuning voltage as feasible. Second, we would use antiseries-connected diodes to minimize the RF voltage swing across them (Figure 10.99). Both of these techniques work against the achievement of wide tuning ranges by minimizing the capacitance swing the diodes can contribute. As a result of this, and because hyperabrupt diodes are not available for the frequencies at which wireless VCOs operate, we would therefore use multiple antiseries diode pairs in parallel.

It is of interest to compare various oscillators. Figure 10.101 shows the performance of a 10 MHz crystal oscillator, a 40 MHz *LC* oscillator, the HP8640 cavity-tuned oscillator at 500 MHz, the 310–640 MHz switched-reactance oscillator of the HP8662 oscillator, and a 2–6 GHz YIG oscillator at 6 GHz [1].

The phase noise plot in Figure 10.102 shows the phase noise of a modern PLL synthesizer (R&S SMA100B). The phase noise on the far right is outside the loop bandwidth, of about 300 kHz.

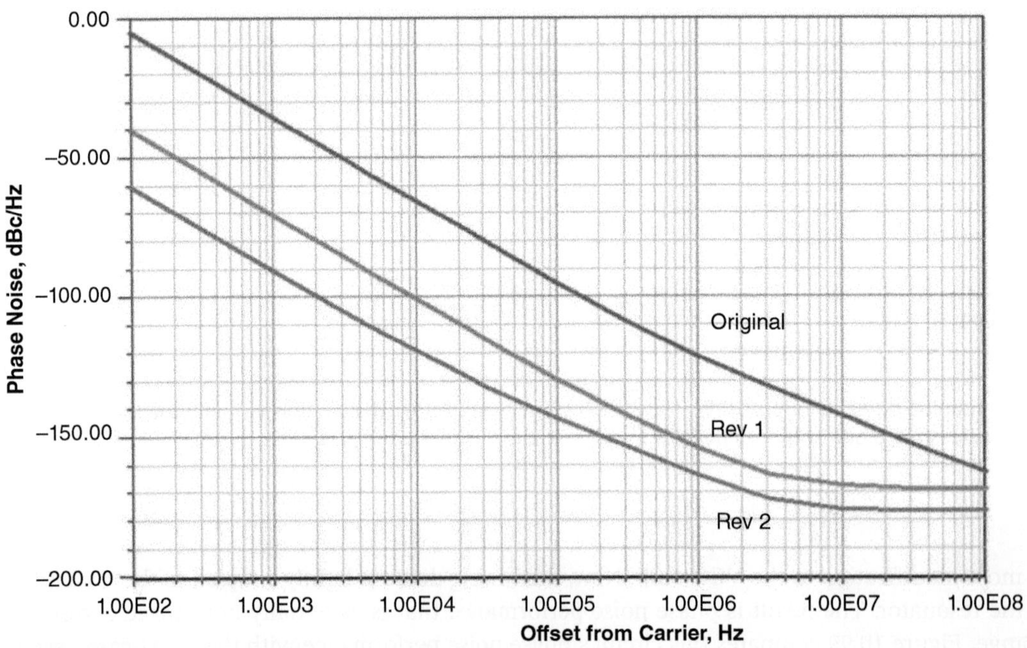

Figure 10.99 Comparison of worst-case phase noise performance obtained with the tank configurations of Figures 10.93 (**Original**), 10.95 (**Rev 1**), and 10.98 (**Rev 2**). *Source*: [1]/John Wiley & Sons.

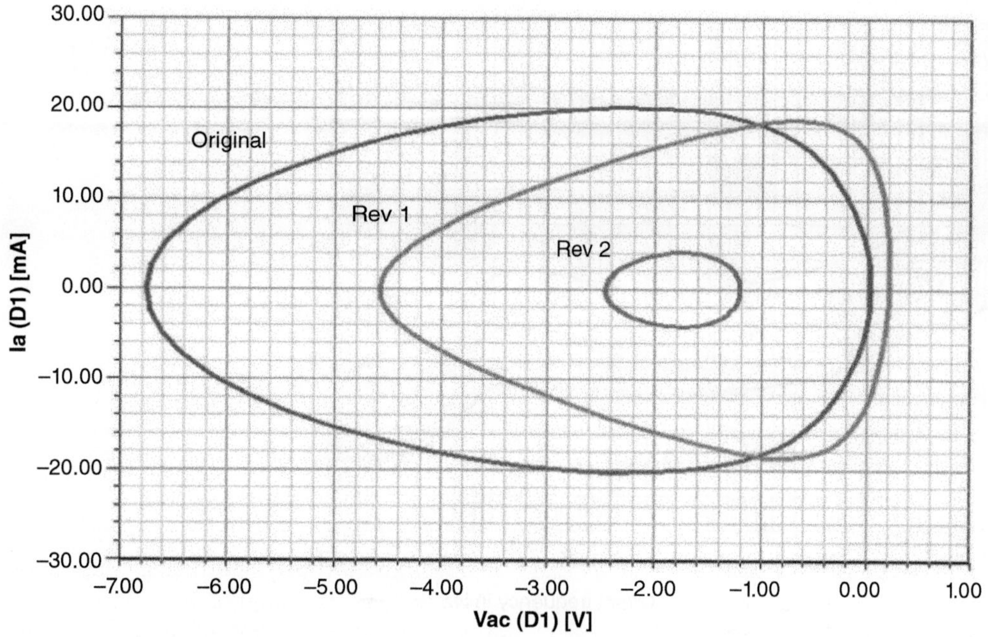

Figure 10.100 Comparison of the tuning diode's AC load line for the worst-case phase noise curves of Figure 10.97. *Source*: [1]/John Wiley & Sons.

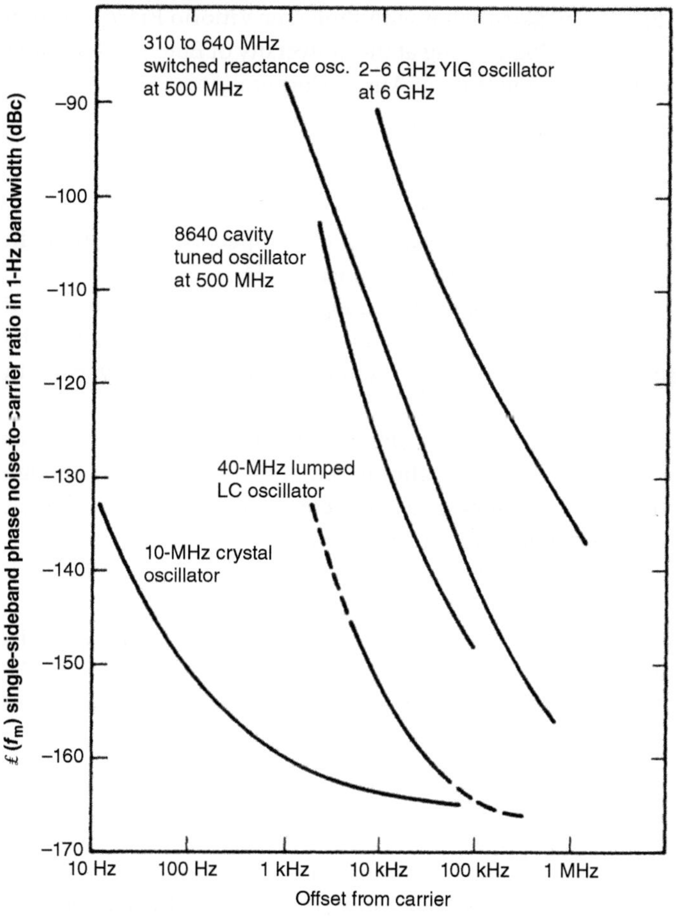

Figure 10.101 Comparison of noise-sideband performances of a crystal oscillator, *LC* oscillator, cavity-tuned oscillator, switched-reactance oscillator, and YIG oscillator. *Source*: [1]/John Wiley & Sons.

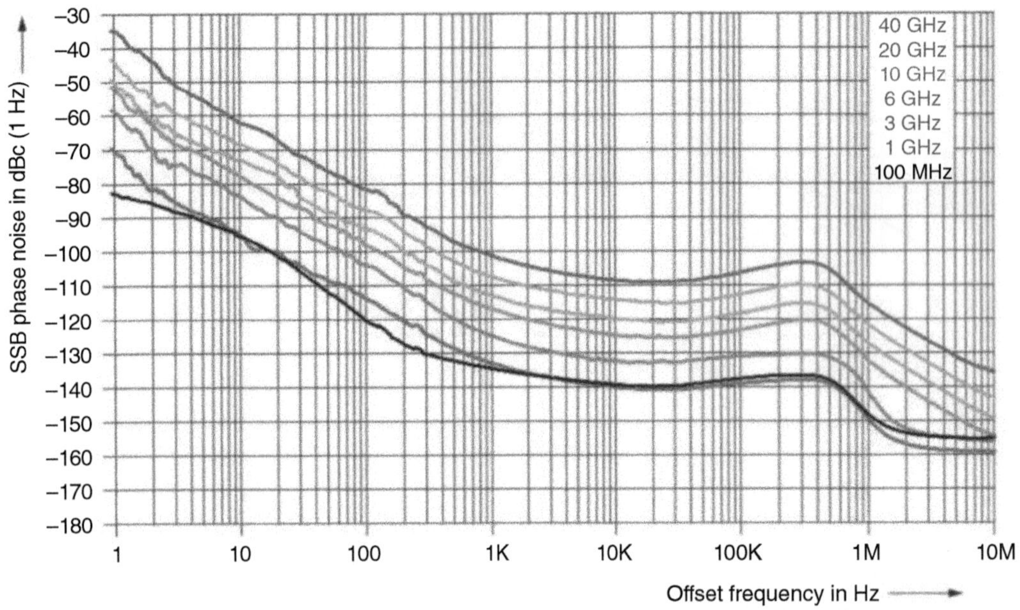

Figure 10.102 Measured phase noise of the Rohde & Schwarz synthesizer model SMA100B.

***Acknowledgment*:** Some of the research work for nonlinear operation in oscillators discussed here was partly sponsored by the US Department of Defense MIMIC program and was performed in cooperation with Professor Vittorio Rizzoli's group of the Department of Electrical Engineering, Information and Systems Engineering at the University of Bologna, Italy, and the Motorola Portable Products Division, Florida, as well as the engineering team of Ansoft Corporation.

10.18 More Practical Circuits

Because of the need for integration, we will present three IC oscillators of increasing complexity. The circuit in Figure 10.103a is found in early Siemens ICs. Figure 10.103b shows its phase noise performance [1].

Figure 10.104a shows the basic push-pull oscillator topology underlying the circuit in Figure 10.103a. The omission of the current source and active bias components improves the simpler circuit's phase noise performance over that of the circuit shown in Figure 10.103. Figure 10.104b compares the phase noise of this basic circuit to the integrated version in Figure 10.104a.

Figure 10.105a is for those who speculate on the possibility of replacing the BJTs in Figure 10.104a with LDMOS FETs or JFETs. Figure 10.105b compares its phase noise to that of the BJT case. The rule of thumb is that FETs do not have enough gain for high-Q oscillation in oscillators above 400 MHz, and bipolar transistors are a better choice. Because of their higher flicker noise contribution, GaAs FETs should be considered only at frequencies above 4 or 5 GHz.

Figure 10.106a shows a circuit based on two differential amplifiers and frequently found in Gilbert cell IC applications. Figure 10.106b shows its phase noise performance.

Figure 10.107 is a validation circuit that demonstrates a comparison of measured phase noise performance versus its prediction by circuit simulation software. The circuit is based on the core of Motorola's MC12148 ECL oscillator IC, the topology of which is derived from the superseded MC1648. Motorola specifies the MC12148's typical phase noise performance as −90 dBc/Hz at an offset of 25 kHz. The close agreement between the measured and simulated results highlights the value of state-of-the-art CAD tools in wireless oscillator design (Figure 10.108).

Figure 10.108 presents the results of *two* simulations: one with ideal L and C components, and another with realistic values specified for these reactances. In using a circuit simulator to validate and compare oscillator topologies, it's important to start with ideal components because we are interested in comparing the inherent phase noise performance of various circuits—particularly the role played by the nonlinear reactances in active devices. Because the Qs of practical components

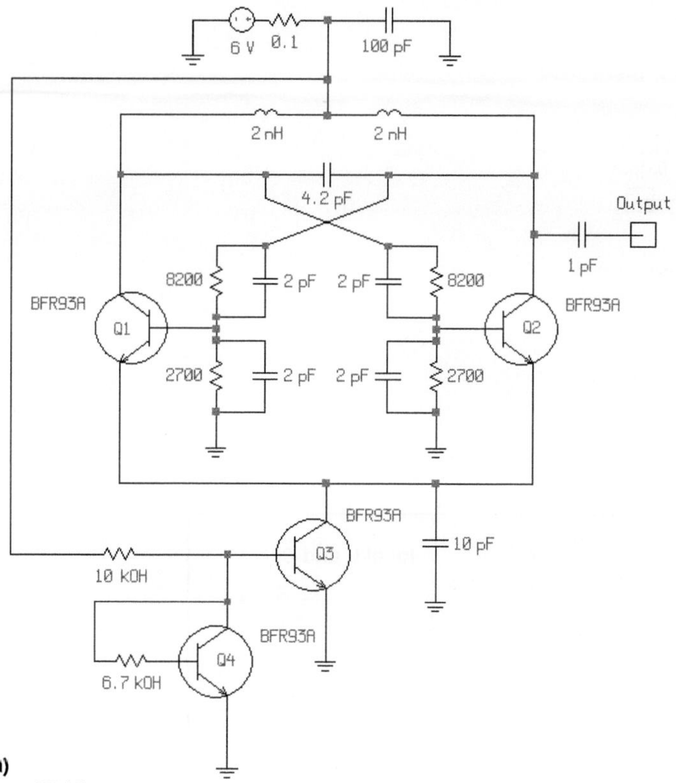

(a)

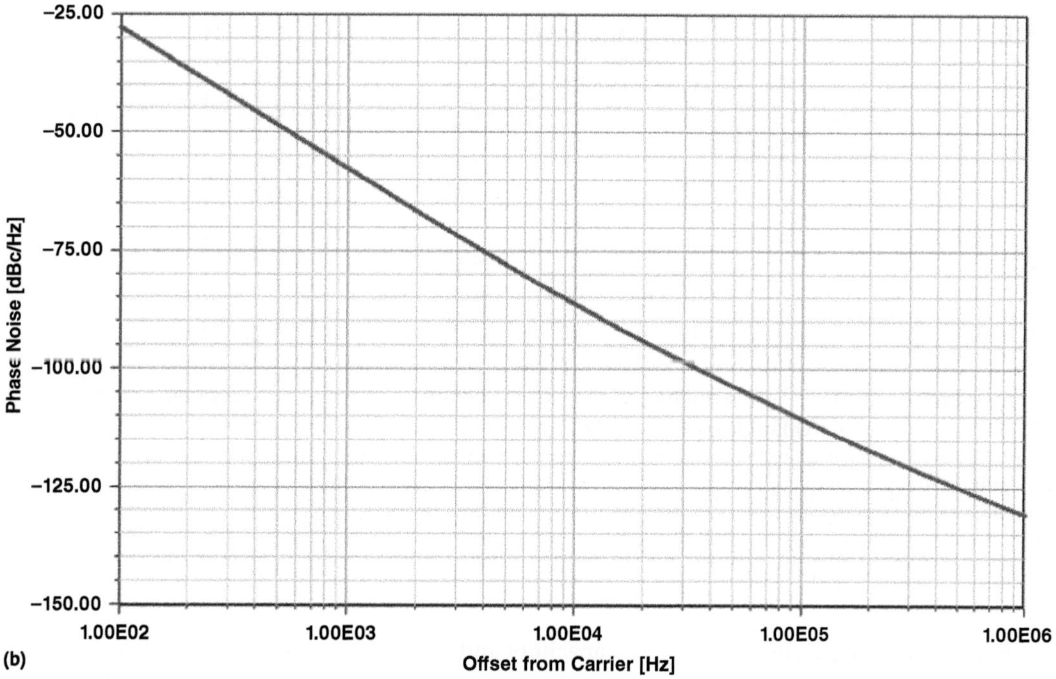

(b)

Figure 10.103 (a) Schematic of a Siemens IC oscillator project. (b) Phase noise of the Siemens IC oscillator. The frequency of oscillation is 1.051 GHz. *Source*: [1]/John Wiley & Sons.

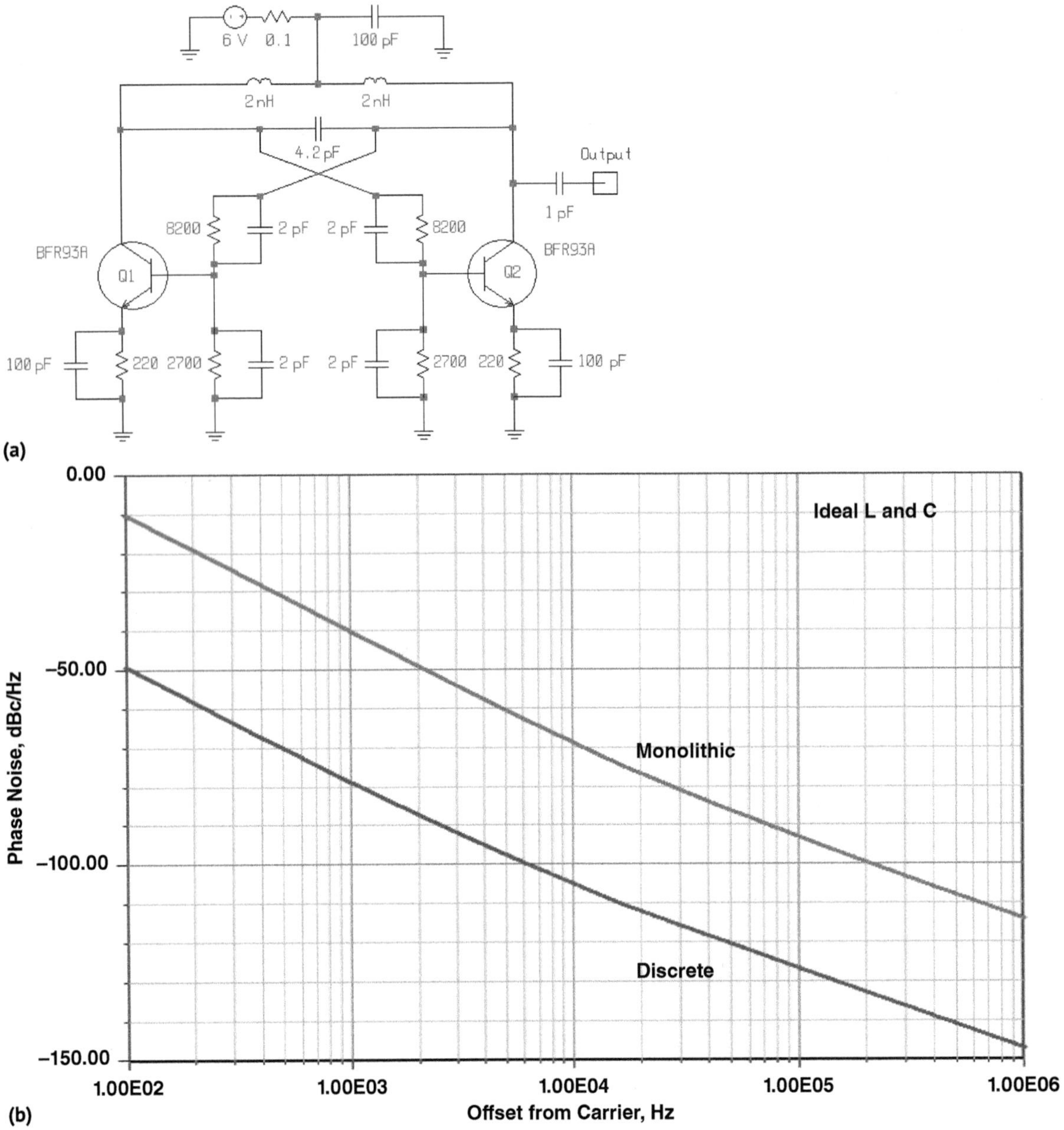

Figure 10.104 (a) Basic push-pull oscillator underlying the circuit shown in Figure 10.99. (b) The basic version of the oscillator (**discrete** curve) is a better phase noise performer than the integrated version (**monolithic** curve). The frequency of oscillation is 1.039 GHz. The main reason for the higher noise is the lower Q of the IC resonator. *Source*: [1]/John Wiley & Sons.

are so low in the microwave range (very generally, 200–300 for capacitors and 60–70 for inductors), the inherent merits of one topology, or one device, over another would be masked by loading effects if we conducted our investigations using nonideal reactances. Once initial evaluation is completed, realistic *Q*s should be specified for increased accuracy in phase noise, output power, and output spectrum simulation.

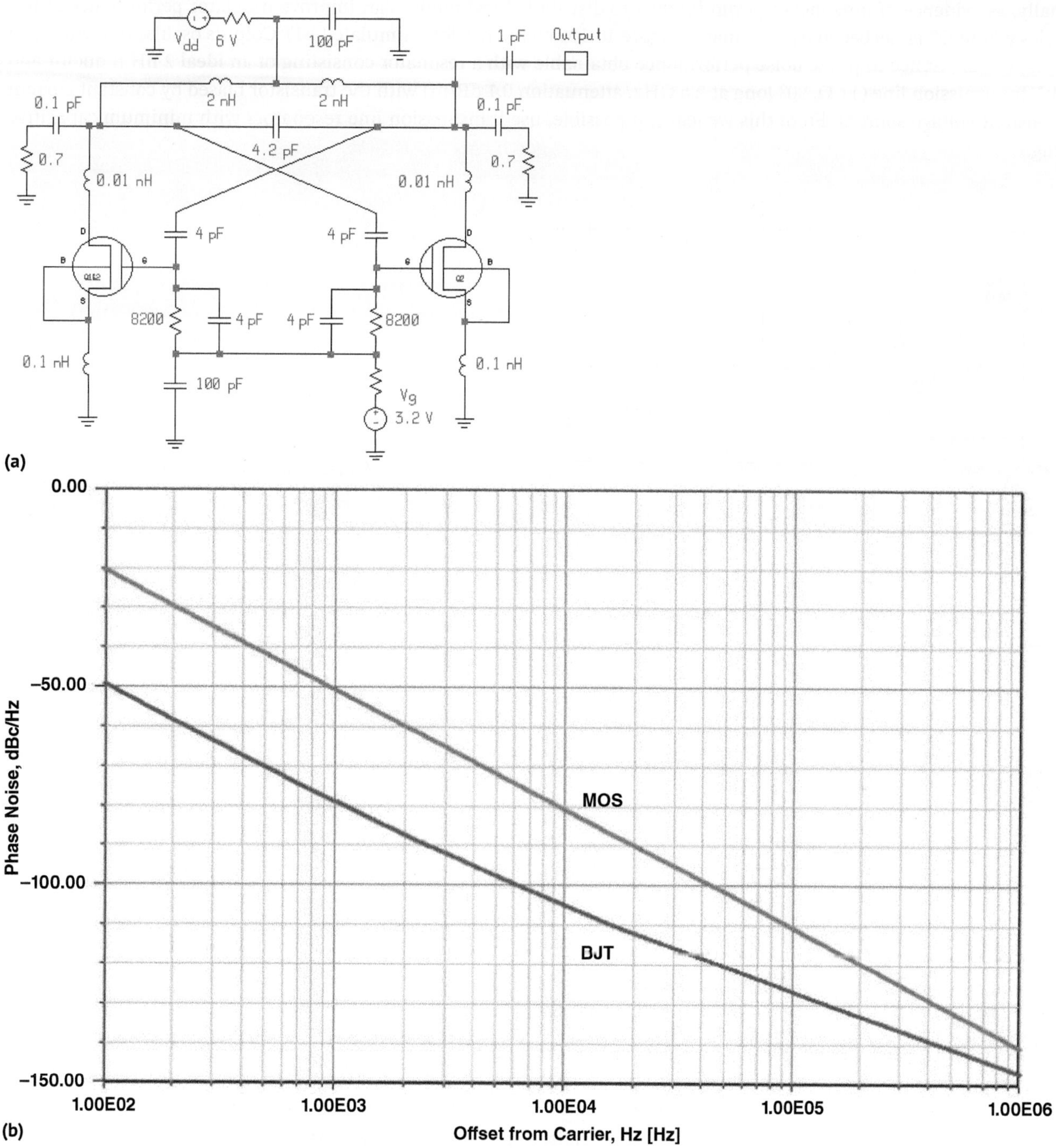

Figure 10.105 (a) 1.04 GHz push-pull oscillator using LDMOS FETs instead of BJTs (LDMOS FET parameters courtesy of David K. Lovelace, Motorola, Inc.). (b) Phase noise comparison of the BJT and MOSFET oscillators. *Source*: [1]/John Wiley & Sons.

Finally, as evidence of how moving from lumped to distributed techniques can improve oscillator performance at frequencies where *LC* tanks become problematic, Figure 10.109 compares, for a simulated BJT Colpitts oscillator operating at 2.3 GHz, the difference in phase noise performance obtainable with a resonator consisting of an ideal 2 nH inductor and a 1/4-λ transmission line (11 Ω, 90° long at 2.6 GHz, attenuation 0.1 dB/m) with the transistor biased by constant current and constant voltage sources. From this we learn, if possible, use transmission line resonators with minimum capacitive loading.

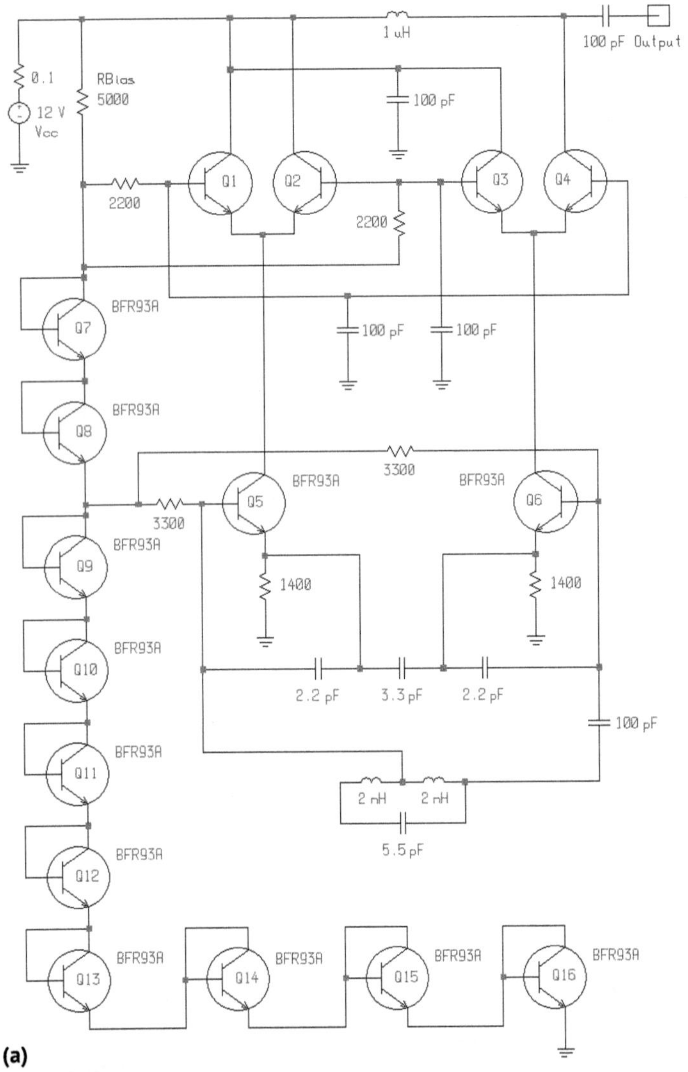

(a)

Figure 10.106 (a) 1.022 GHz IC oscillator based on two differential amplifiers. With an RF signal fed to the bases of $Q_1 - Q_4$ and $Q_2 - Q_3$ in push-pull, and IF output extracted from the collectors of $Q_1 - Q_3$ and $Q_2 - Q_4$ in push-pull, the circuit would act as a Gilbert cell converter. In the circuit shown, we have modified the differential amplifiers' collector circuitry to give single-ended output and avoid cancellation of the oscillator signal. *Source*: [1]/John Wiley & Sons. (b) Phase noise of the two differential amplifier oscillator circuits [1].

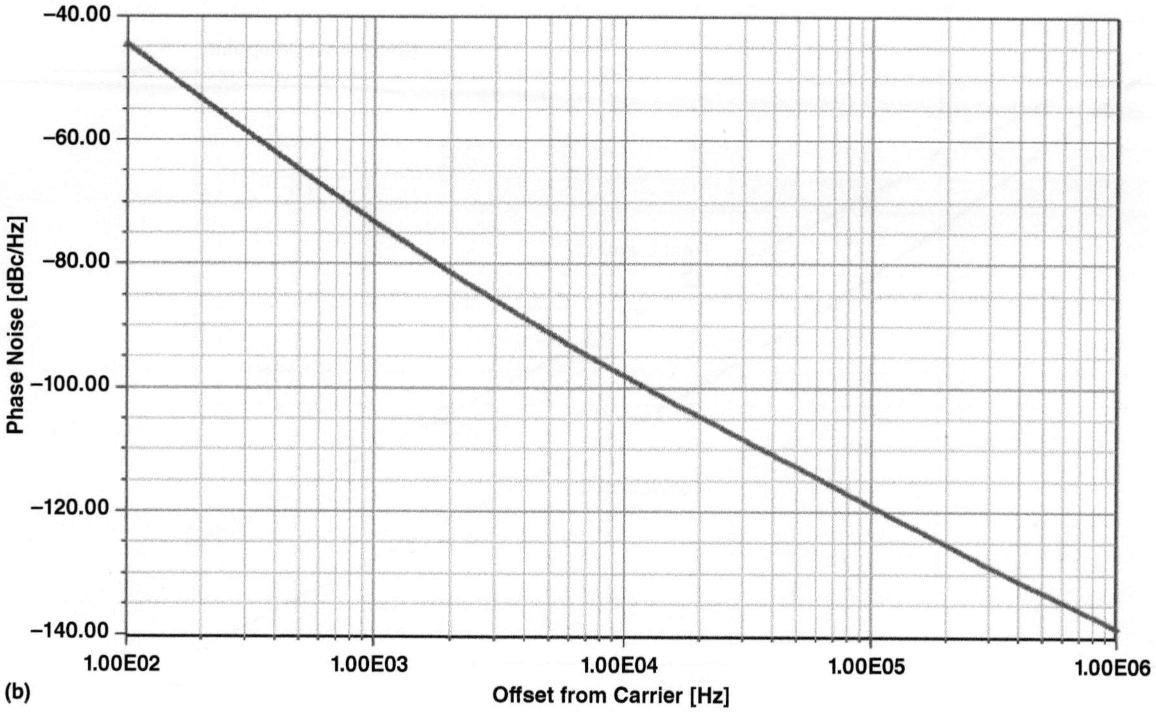

(b)

Figure 10.106 (Cont'd)

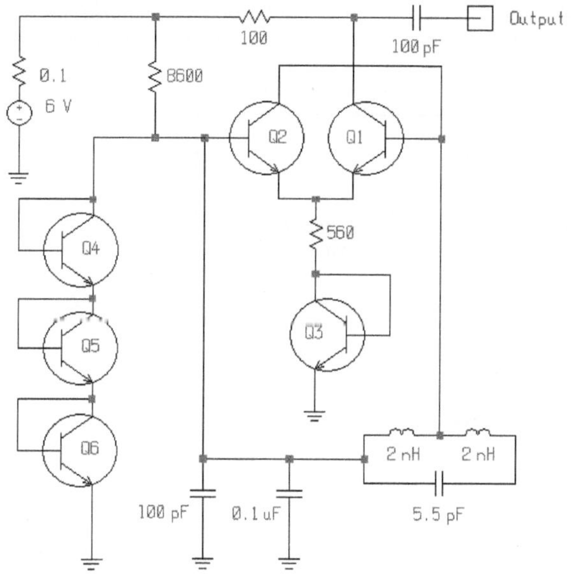

Figure 10.107 This validation circuit models the oscillator topology at the core of Motorola's MC12148 ECL oscillator IC. *Source:* [1]/John Wiley & Sons.

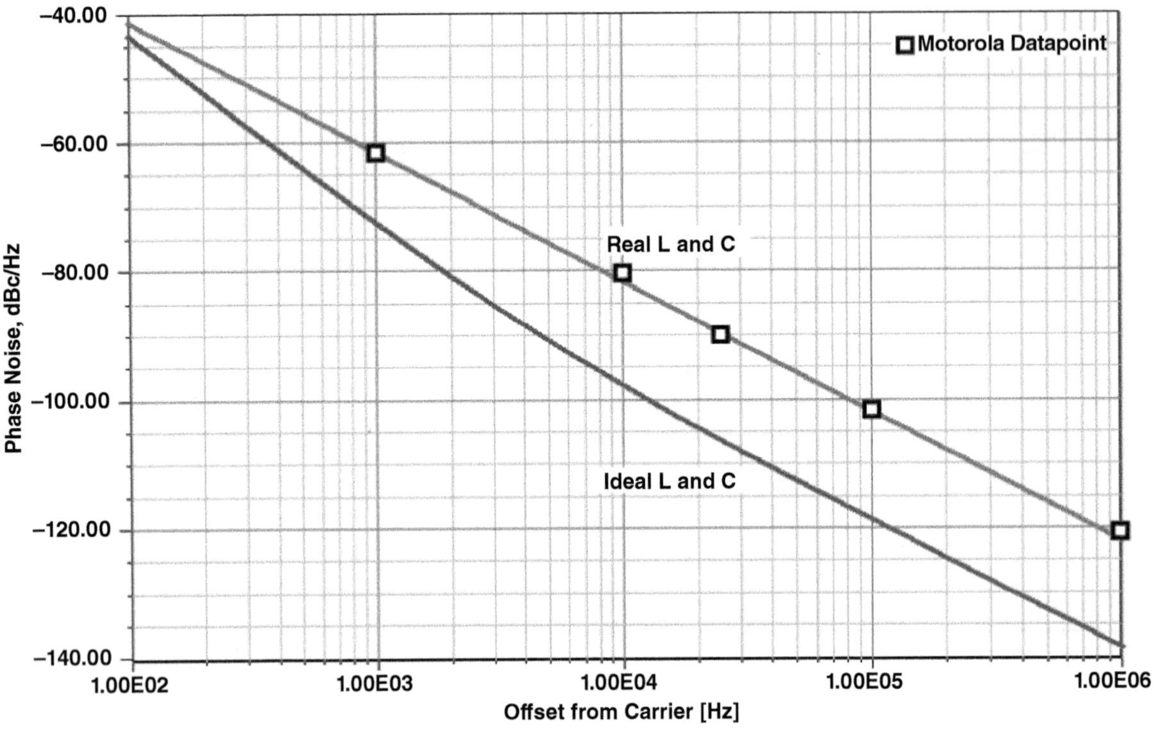

Figure 10.108 Comparison of measured versus simulated phase noise of the MC12148 differential oscillator. The simulation (1.016 GHz) predicts a phase noise of −89.54 dBc/Hz at an offset of 10 kHz (**Real L and C** curve), which agrees well with Motorola's specification of −89.54 dBc/Hz (typical at 930 MHz) at this offset. The **Ideal L and C** curve plots the simulation's phase noise performance with ideal reactances, as opposed to the **Real L and C** curve, which graphs the results with realistic Qs specified for its reactive components ($Q = 250$ for its Cs and $Q = 65$ for its L, all at 1 GHz) [1].

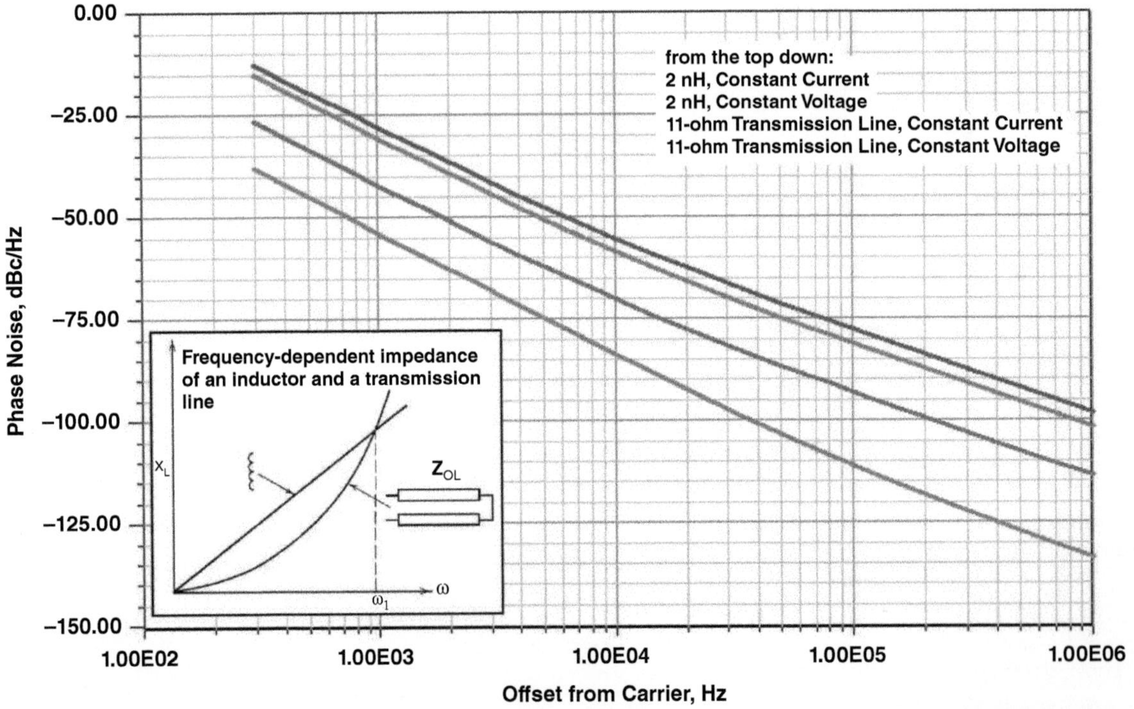

Figure 10.109 Phase noise performance of a 2.3 GHz BJT oscillator with a resonator consisting of an inductor (2 nH) and a $1/4$-λ transmission line (11 Ω, approximating the behavior of a dielectric resonator) with bias from a constant current source and a low-impedance, resistive constant voltage source [1].

10.19 Silicon/GaAs-based Integrated VCOs and Possible Difficulties

Due to process variations, the center frequency of the VCO cannot be repeated without trimming. Trimming can be avoided by making the VCO gain large enough to compensate for tank circuit variations, but this degrades phase noise performance and increases the VCO susceptibility to unwanted spurs. For example, a VCO with a gain of 10 MHz/V has a phase noise of −80 dBc/Hz at 10 kHz while a VCO with a gain of 75 MHz/V has a phase noise of only −60 dBc/Hz at 10 kHz.

All but the tank inductors are integrated.

Tank inductors are generally not integrated for reasons of size and Q. The area of a 1 nH inductor is approximately 42,000 mm^2, which is ≈ 3.5% of a typical IC; the area of a 10 nH inductor is approximately 167,000 mm^2, which is ≈ 14% of a typical IC. For GaAs substrate, the resulting Q would be about 5; for silicon, as we already saw, 20 is already a very high value.

The customer can use PCB traces for the required inductance instead of external surface mount inductors (cost: $0.03/each, most expensive passive component!). It must be pointed out that even for a non-integrated approach, the higher VCO gain is noisier by definition. These problems are not unique to silicon; they apply to GaAs also. Figure 10.110 shows the size relationship of on-chip capacitors and inductors compared to the actual transistor; therefore, on-chip inductors should really be avoided because of reasons of space and Q. Just imagine how the picture would change if the 1.1 nH inductor had to be 11 nH instead! It would dominate the IC area, and therefore the IC cost as well [1].

Using passive distributed elements, the following oscillator with a wide tuning range is a good example. How to do this?

In addition, this circuit has a noise cancelation loop using a PNP transistor in the collector, sampling the collector current using a 38 Ω resistor. This patented circuit at Synergy Microwave is very effective.

The oscillator shown in Figure 10.111 [18, pp. 827–830] uses a Colpitts configuration with four tuning diodes. It can be found in the R&S EB200 receiver model.

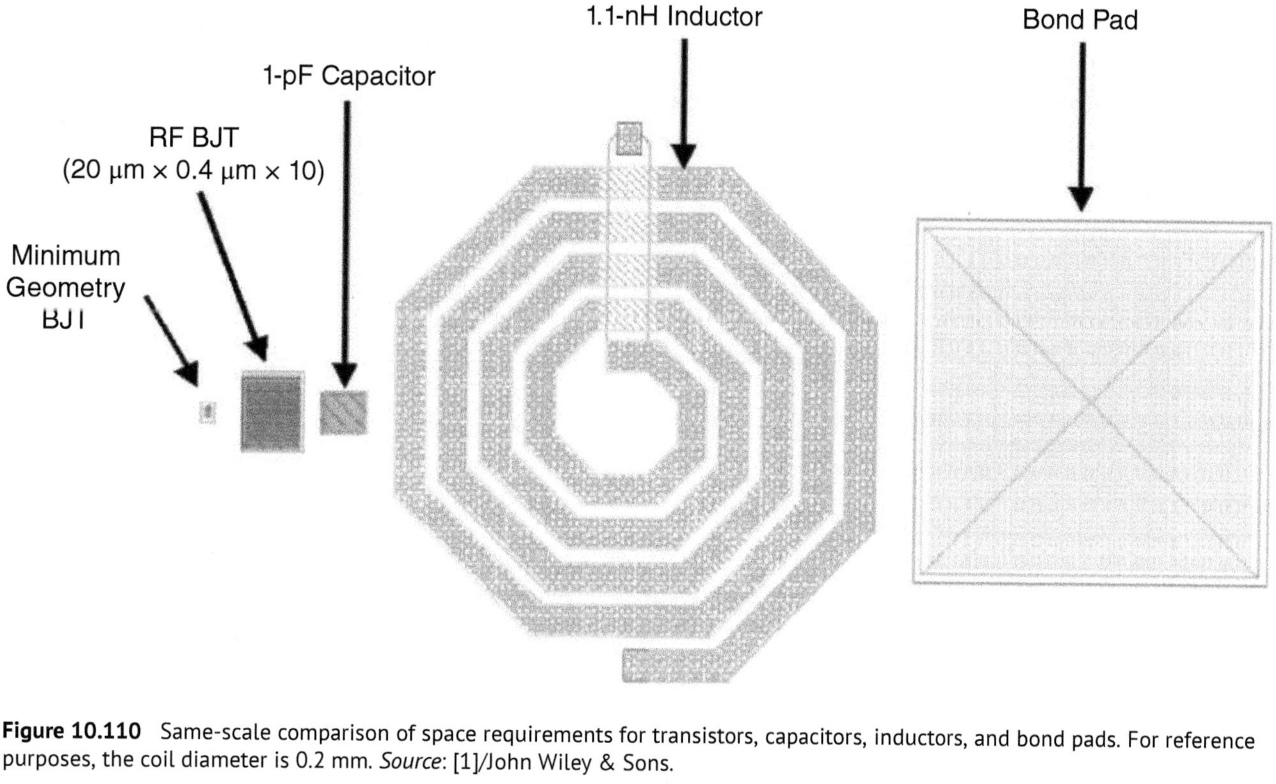

Figure 10.110 Same-scale comparison of space requirements for transistors, capacitors, inductors, and bond pads. For reference purposes, the coil diameter is 0.2 mm. *Source*: [1]/John Wiley & Sons.

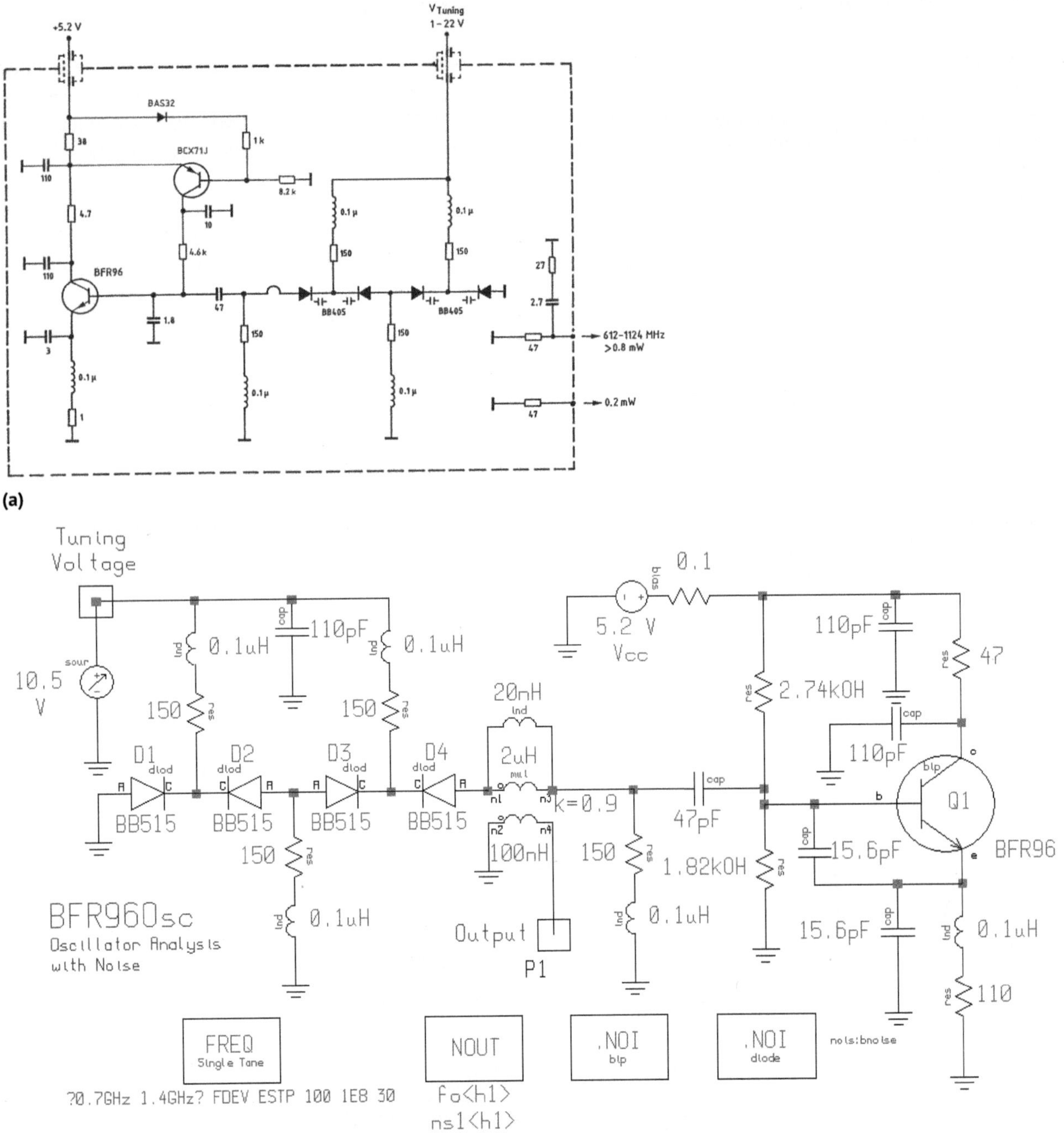

Figure 10.111 (a) Circuit schematic of a 612–1124 MHz BJT microwave wide band oscillator. *Source*: [1]/John Wiley & Sons. (b) Simulation circuit of a BJT microwave wide band oscillator. *Source*: [18]/John Wiley & Sons.

10.20 Time-domain Simulation

SPICE, a general-purpose simulation program capable of nonlinear DC, linear AC, and nonlinear transient analysis, is probably the most common of the time-domain methods used for nonlinear circuit analysis. Its ability to represent arbitrary time-domain waveforms and model time delays, level shifts, and the sinking and sourcing characteristics of such circuits has made it the simulator of choice for digital designers concerned with analog issues. SPICE's application in RF, microwave, and wireless circuit design is limited by its difficulties with simulating multitone excitations, and coupling and transmission line effects—including losses, parasitics, and dispersion for distributed elements—and its inefficiency in simulating steady-state conditions.

One area in which SPICE's transient analysis capabilities can be particularly useful to RF and wireless designers is the simulation of oscillator startup. To illustrate this, we present a 3.8 GHz DRO modeled with Ansoft Corporation's Super-Spice simulator. Figure 10.112a shows how the oscillator's S_{11} response, positive throughout the simulation's frequency span of 3–5 GHz, peaks at 3.77 GHz. Figure 10.112b shows the oscillator's output voltage as the circuit starts oscillating

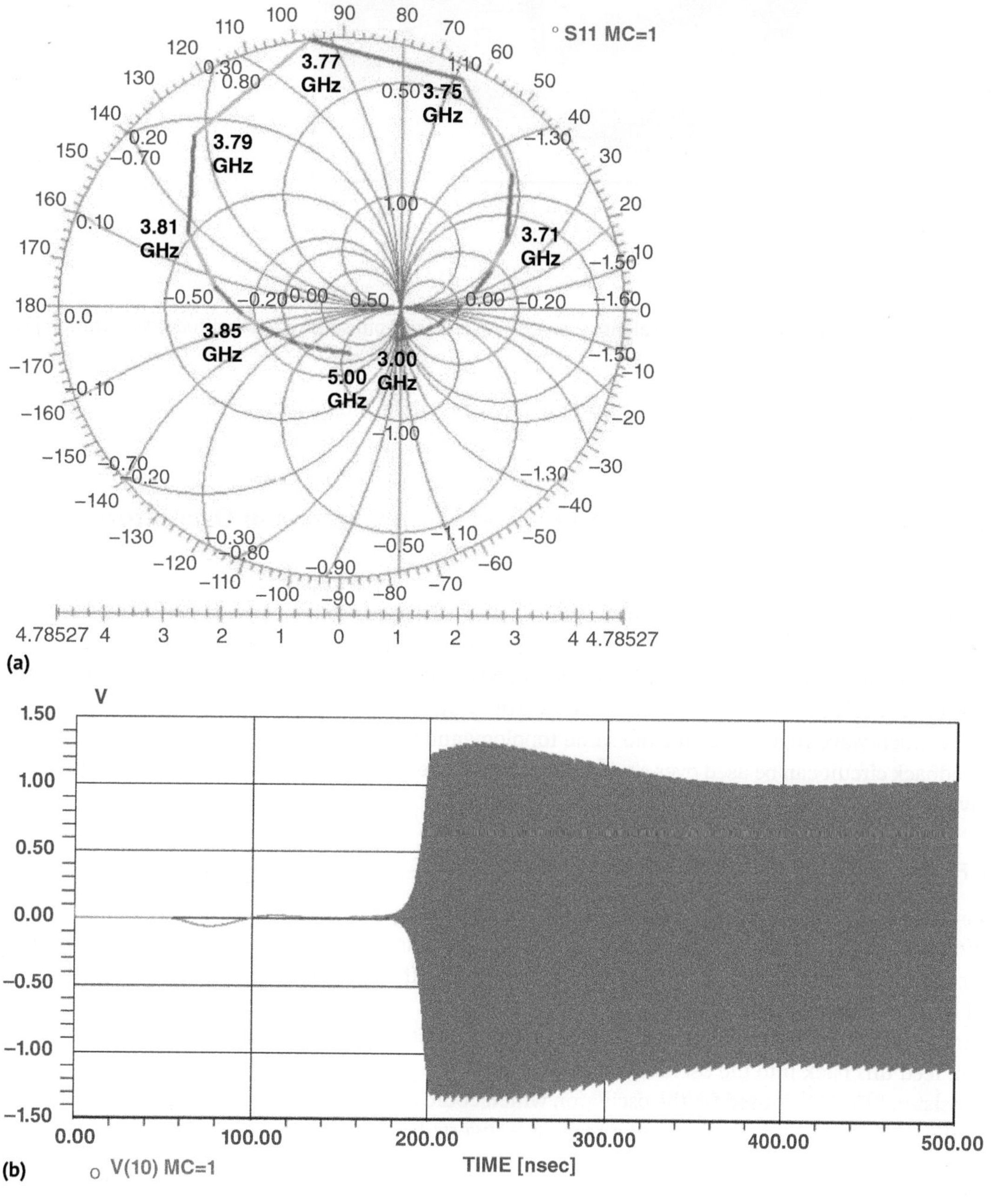

Figure 10.112 (a) S_{11} response of a 3.8 GHz BJT DRO as modeled by Ansoft's Super-Spice time-domain simulator. *Source*: [18]/John Wiley & Sons. (b) Output voltage of the BJT DRO as simulated by Super-Spice [18, 52].

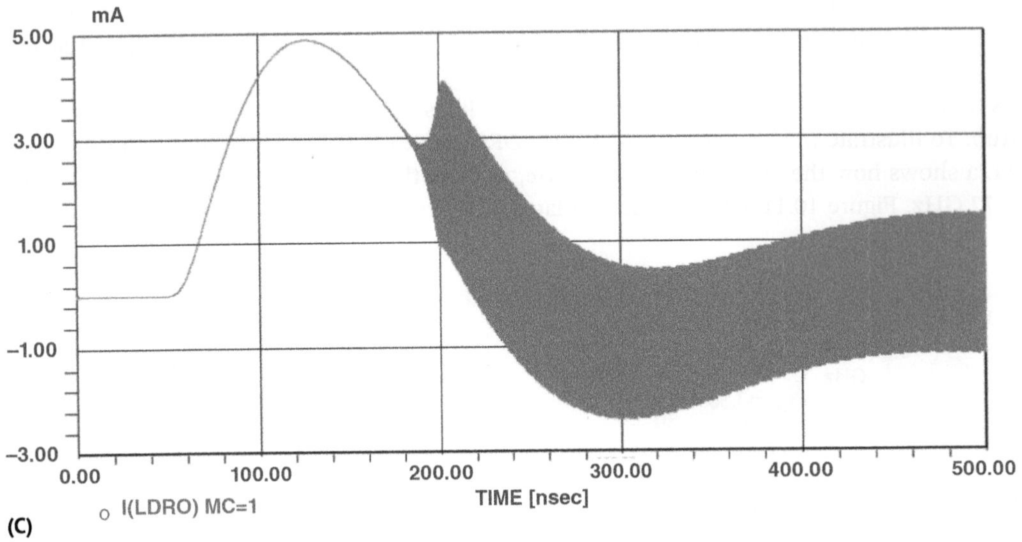

(C)

Figure 10.112 (Cont'd) (c) Current through the BJT DRO's equivalent resonator inductance from startup to amplitude stabilization [18, 49, p. 513, 52].

and amplitude stabilizes; Figure 10.112c shows the current through the equivalent inductance of the circuit's dielectric resonator across the same time span.

10.21 Phase Noise Improvements of Integrated RF and Millimeter-wave Oscillators

The generation of microwave and millimeter-wave frequencies can be done by lower-frequency VCOs multiplied up in frequency, such as that provided by comb generators; by using hybrid GaAs FET-based oscillators with external resonators; or—for the best phase noise obtainable—YIG oscillators whose physical size and cost does not always provide a practical solution. A significant improvement in oscillator noise performance can be obtained with a novel feedback circuit that uses internally generated noise and cancels the close-in noise in a bandwidth of up to 1 MHz. This results in more than 15 dB phase noise improvement at microwave frequencies for the same topology and the feedback system can be made part of the biasing circuit. This feedback circuit can be used over a wide frequency range and works well at VHF/UHF frequencies and in the millimeter-wave area [1].

10.21.1 Reduction of Flicker Noise

From inspecting a device datasheet, one is always pleasantly surprised about very low spot noise figures, while the $1/f$ corner frequency is rarely specified at all. The mechanism that transforms the noise, which is generated by the $1/f$ contribution into what we refer to as modulation noise, is the AM-to-PM conversion. If one could "linearize" the transistor, the flicker noise effect would be drastically reduced.

In the case of the bipolar transistor, you could sample the noise of the 56 Ω resistor shown in Figure 10.113, and consistent with the previous method, feed this back into the oscillator system in this case and modulate the base voltage of Q1 and Q2. The higher-power transistor, BFR93A, is used for the oscillator, which could equally be an HBT gate "stabilizer" via Q2 in such a manner that all the sample noise across the emitter resistor is inverted, amplified, and brought back. This noise-canceling scheme works inside the loop bandwidth determined by the circuit of Q2. Figure 10.114 shows the phase noise performance of the circuits of Figure 10.113 with and without noise feedback [1].

10.21.2 Applications to Integrated Oscillators

Figure 10.115a shows a silicon-based bi-CMOS technology wireless oscillator. The three big inductors can be seen easily. This design by Motorola is a good candidate for using this innovative technology. Applying this feedback technology, we measure the phase noise shown in Figure 10.115b.

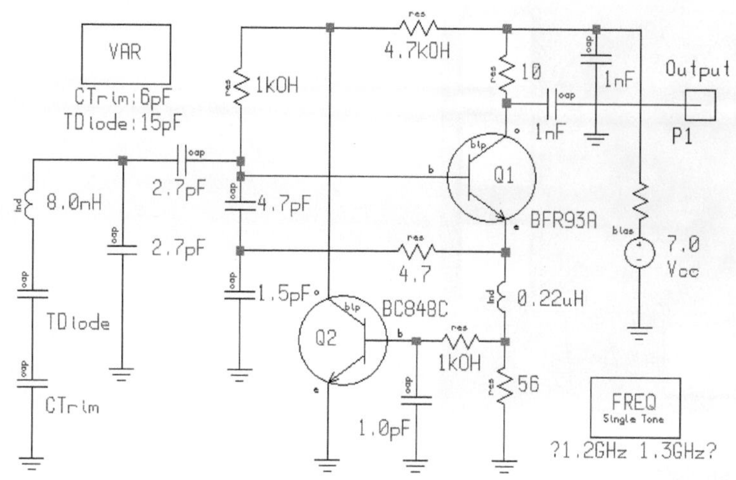

Figure 10.113 BJT-based oscillators with noise feedback. The noise sampling is done in the transistor (Q1) emitter. The noise sampling done in the collector using a variation of the active biasing circuit shown in Figure 10.113. *Source*: [1]/John Wiley & Sons.

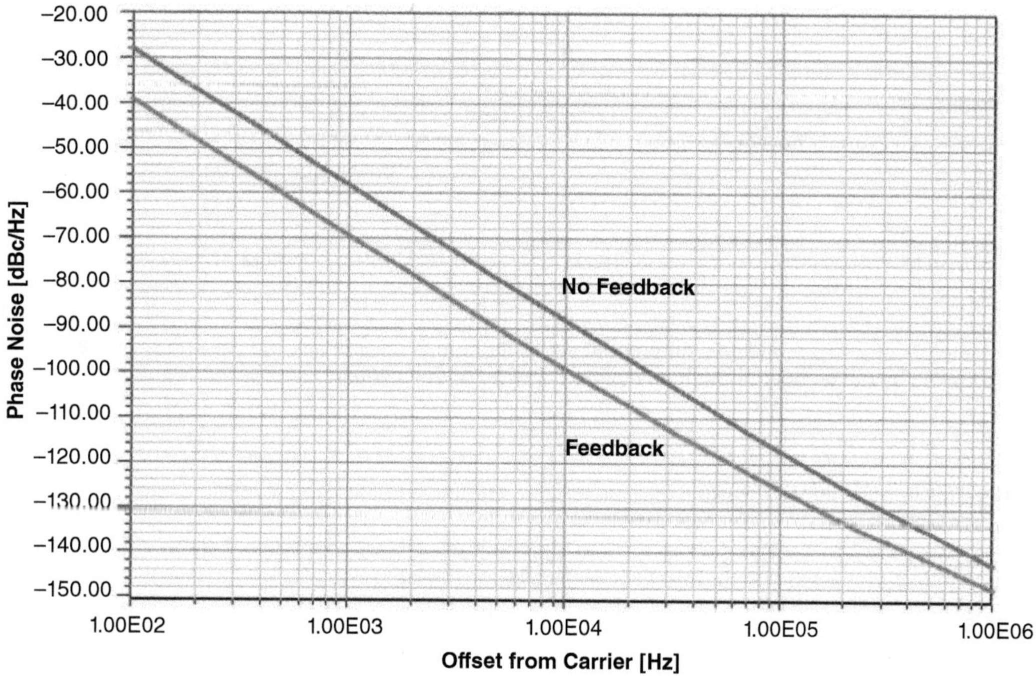

Figure 10.114 Phase noise of Figure 10.113 oscillator with and without noise feedback. *Source*: [1]/John Wiley & Sons.

Moving up into the millimeter-wave area, one of the target oscillators is the Ka-band MMIC VCO designed for Ka-band smart ammunitions applications (Figure 10.116). This VCO MMIC employs 0.25-μm gate length, double-heterojunction, pseudomorphic, high-electron mobility transistor (PHEMT) technology. This is a custom chip developed by Martin Marietta under the US Government MMIC program. Ansoft Serenade tools were used in this design.

Features:

- Fundamental mode, differential-ring VCO
- Electronically tunable
- Output power > 16 dBm
- 13% power-added efficiency (PAE)

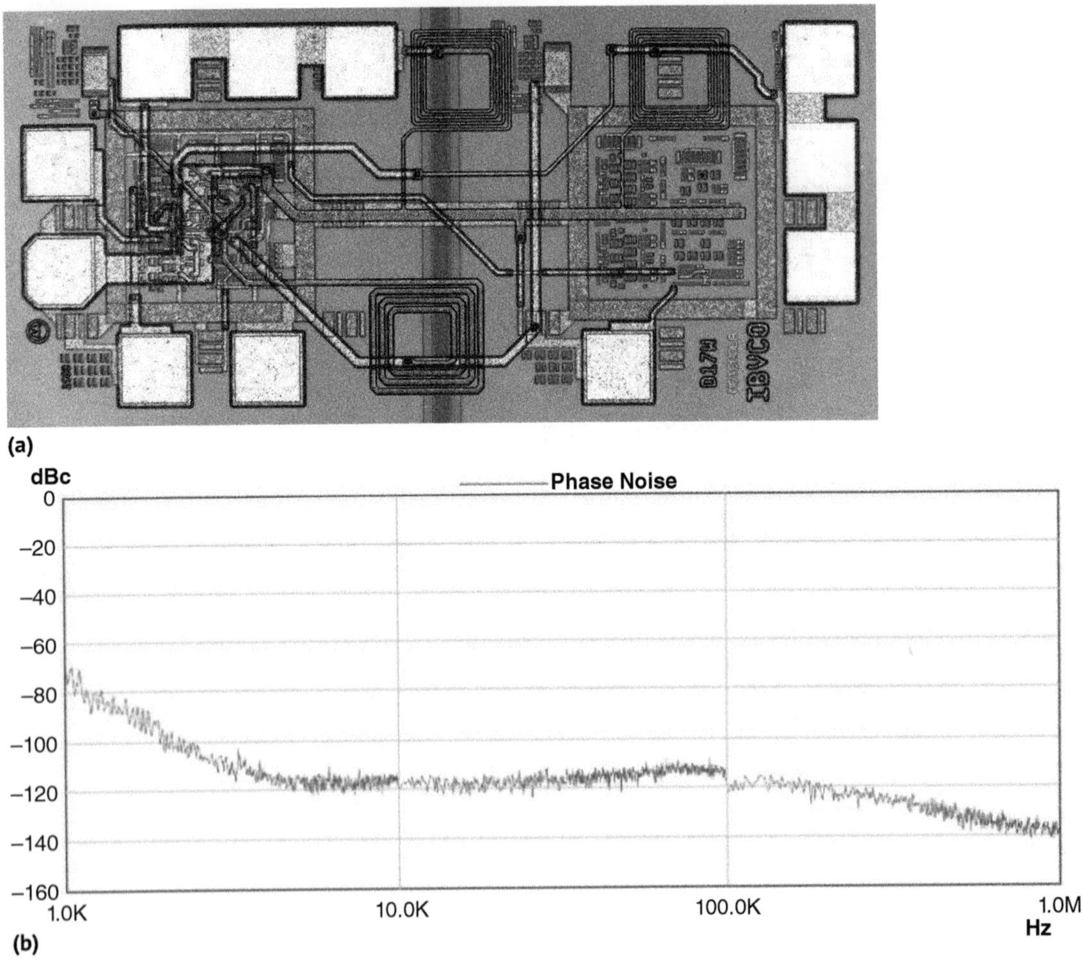

Figure 10.115 (a) Layout of the Motorola 800 MHz monolithic differential oscillator IC in Si technology. (b) Phase noise of the oscillator of (a). The noise floor is limited by the on-board isolation amplifier. *Source*: [1]/John Wiley & Sons.

- Compact size for easy integration with power amplifier
- 0.25 μm pseudomorphic HEMTs

Specifications:

Frequency range: Ka band (26.5 ...40 GHz)
Output power: 16 dBm minimum

When applying the same technology as shown in the previous bipolar example, the phase noise can be significantly reduced. Since field-effect transistors require a negative gate voltage, it is possible to sample the noise in the drain current using a resistor, amplify the noise with an appropriate loop amplifier, and modulate the gate voltage within a 1 MHz bandwidth.

A spectrum analyzer measurement of this oscillator type, as shown in Figure 10.116b, indicates a phase noise of −78.6 dBc/Hz at an offset frequency of 100 kHz and a center frequency of 38 GHz, while the closed-loop phase noise using the compensation scheme indicates a phase noise of about −108 dBc/Hz at 100 kHz. This is consistent with reported results from other researchers who have used complicated phase-locked loops to achieve similar performance as seen in Figure 10.117.

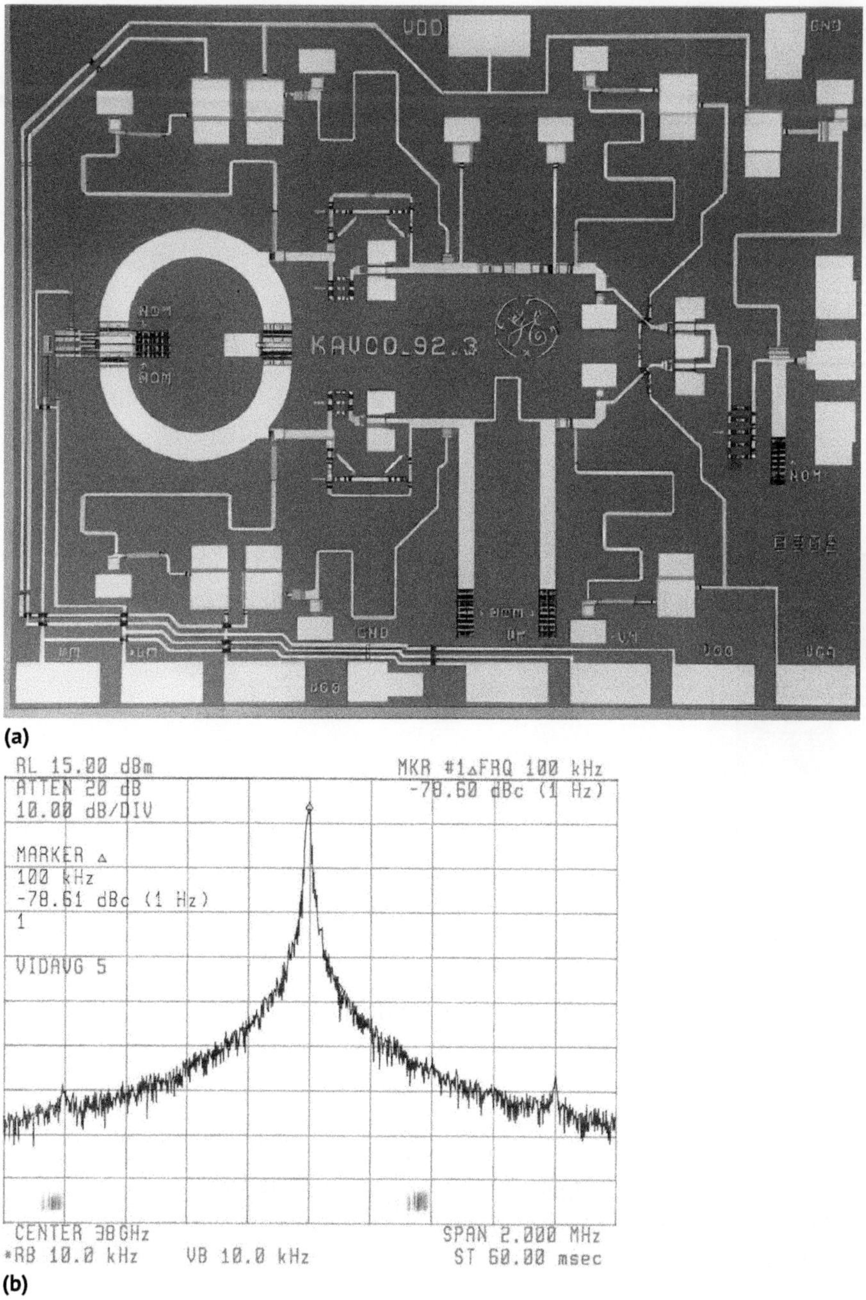

Figure 10.116 (a) Layout of the Ka-band VCO. (b) Spectrogram of the 38 GHz free-running push-pull VCO shown in Figure 10.127. *Source*: [1]/John Wiley & Sons.

To build these type of oscillators before 1990 was very difficult, and in order to be able to select the right resonant frequency, the resonator was split in smaller pieces, which can be connected to get the desired resonance frequency. Needless to say, not a high-Q arrangement.

Figure 10.118 shows an experimental VCO that was a result of collaboration between the Ti-Raytheon and Compact Software Teams.

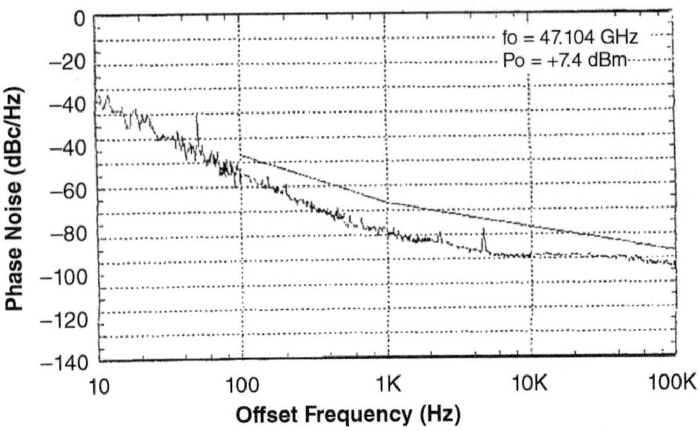

Figure 10.117 A 47 GHz loop-stabilized VCO with phase noise performance. *Source*: [1]/John Wiley & Sons.

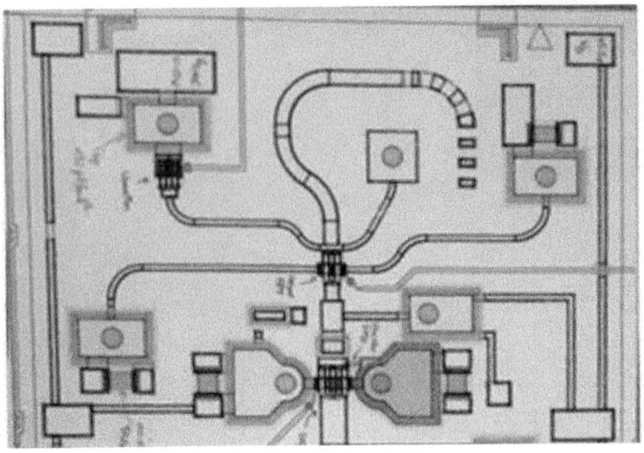

Figure 10.118 Experimental VCO from the Ti-Raytheon Compact Software Team work. *Source*: [53]/IEEE.

10.22 Reverse Engineering

One of the difficulties we face is the fact that most if not all universities and research institutes receive money in support of their work and the companies paying for the effort are most concerned about their intellectual properties.

As a result, the publications or research reports only show a minimum of relevant data, mostly the target performance and the results obtained. For the purpose of this book, we did some reverse engineering of the transistor parameters necessary to obtain the published results. It was done on a best effort level and has become most educational.

So finally we look at some of the circuits and their results from the dissertation [54]. Material (Section 10.22) used from Hossain [54] with permission.

In the first part, it presents 96 and 197 GHz fundamental sources using a 0.8 μm InP TS-DHBT process, which deliver +9 and 0 dBm output power with 25% and 4.6% overall DC-to-RF efficiency, respectively. The first part also comprises a 290 GHz harmonic oscillator, which exhibits −9.5 dBm output power and shows 0.5% overall DC-to-RF efficiency. Furthermore, it demonstrates 162 and 270 GHz push–push sources utilizing InP TS-DHBTs on a BiCMOS process, which achieve −4.5 and −9.5 dBm output power and combined overall DC-to-RF efficiencies of 1.5% and 0.4%, respectively.

Figure 10.119 shows the structure of a W-band fundamental oscillator analyzed post-layout using 2.5D EM ADS momentum. Figure 10.120 shows the equivalent circuit of bipolar transistor (BJT) and Figure 10.121 shows the simplified schematic of the reflection type oscillator.

A transmission line section TL_b is used to boost the negative impedance seen when looking into the emitter and collector of the transistor (T). In addition, the impedances formed by the lines TL_{c1} and TL_{c2} on the collector side cause

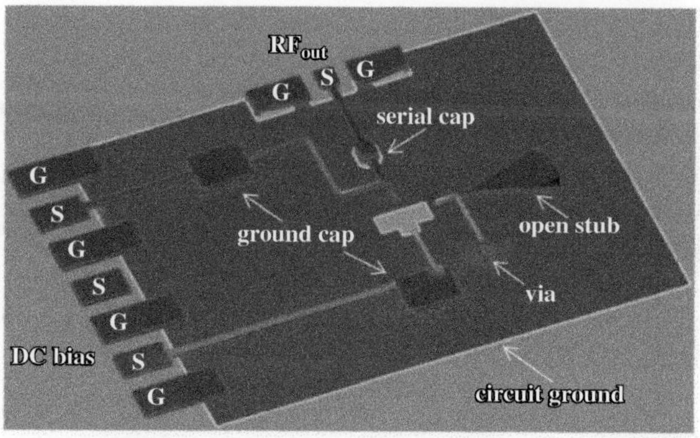

Figure 10.119 Structure of a W-band fundamental oscillator analyzed post-layout using 2.5D EM ADS momentum. *Source*: [1]/John Wiley & Sons.

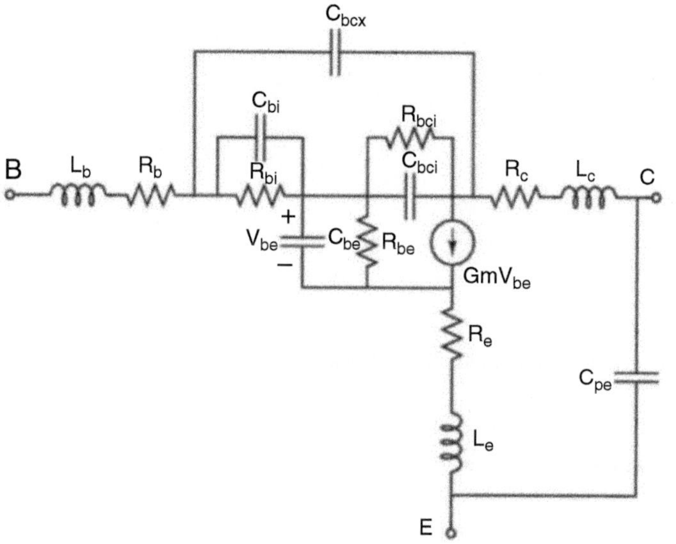

Figure 10.120 Equivalent circuit of bipolar transistor (BJT). *Source*: [54]/Muhammad Maruf Hossain.

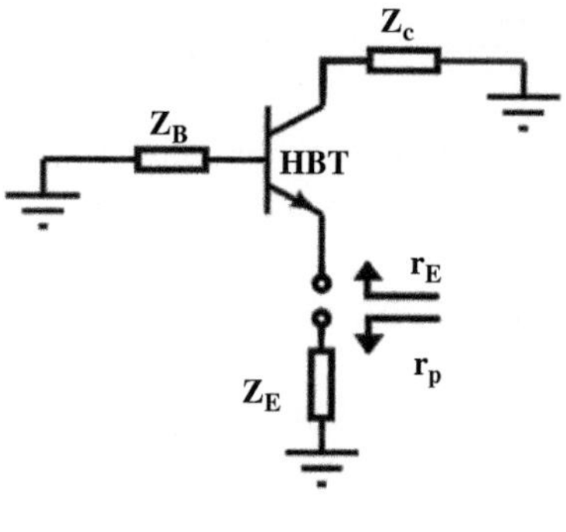

Figure 10.121 Simplified schematic of the reflection type oscillator. *Source*: [54]/Muhammad Maruf Hossain.

the transistor (T) to become more unstable. To achieve the required oscillation condition in the passive part of the oscillator, the transmission line TL_e and the parallel MIM capacitor are used to adjust the phase slope and magnitude. Figure 4.5 shows the passive part of the oscillator circuit. In order to avoid a frequency shift due to a small MIM capacitor value and its parasitic, a radial stub (STB_e) has been introduced. All transmission lines are realized as thin-film microstrip geometries. To fulfil the stable oscillation condition, the phase steepness of the reflection coefficient on the emitter port is set to zero by optimizing TL_{c1}, TL_{c2}, TL_e, and the radial stub (STB_e) with a ground via. The collector lines TL_{c1} and TL_{c2} are also used for impedance matching and to maximize the output power. The circuit is designed so that it can operate without any resistive feedback or particular biasing topology, which would decrease efficiency. A blocking capacitor (C_{blk}) is included at the output to facilitate future integration with frequency multipliers or mixers. Finally, post-layout simulation has been performed and optimized using ADS harmonic balance simulation tool, assisted by a 2.5D planar EM simulator [54].

A schematic of the reflection type oscillator is shown in Figure 10.122. Note the use of the radial stub used as the resonator.

Example: 95 GHz Oscillator Figure 10.123a shows the schematic of the 95 GHz oscillator for analysis, where Figure 10.123b shows the linear analysis for the resonance frequency of the proposed 95 GHz oscillator of Figure 10.123a. The harmonic balance error is small enough with 9e-09, as seen in Figure 10.123c.

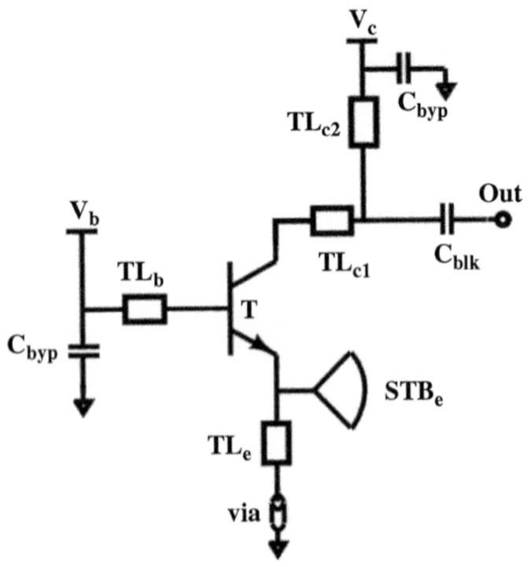

Figure 10.122 Schematic of a 96 GHz fundamental oscillator. *Source*: [54]/Muhammad Maruf Hossain.

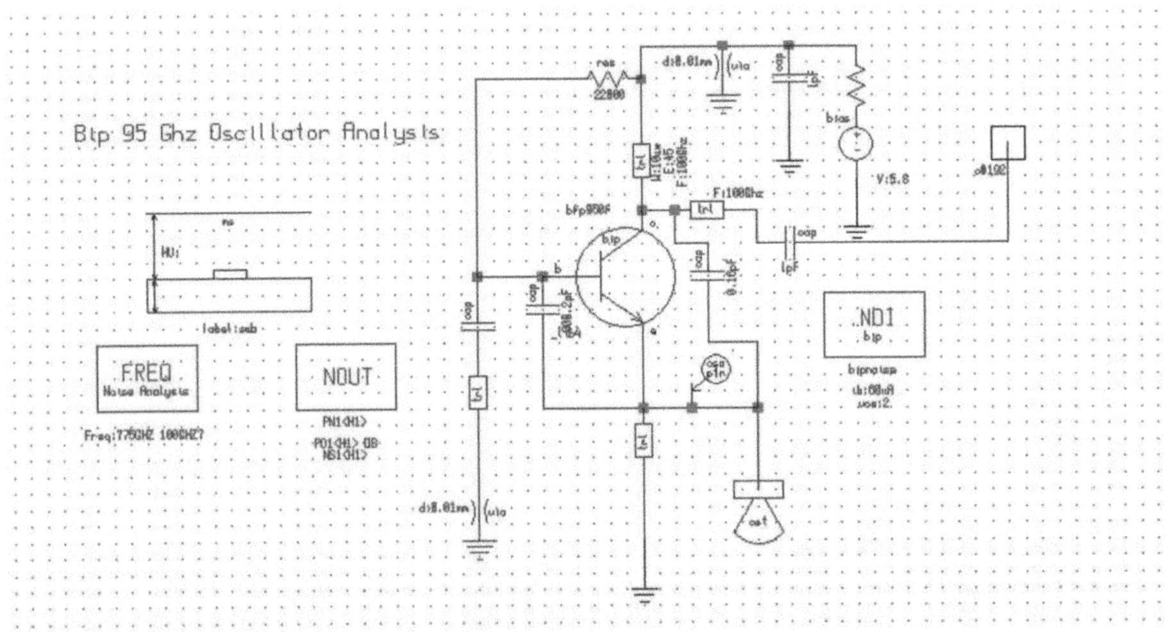

(a)

Figure 10.123 (a) 95 GHz oscillator analysis.

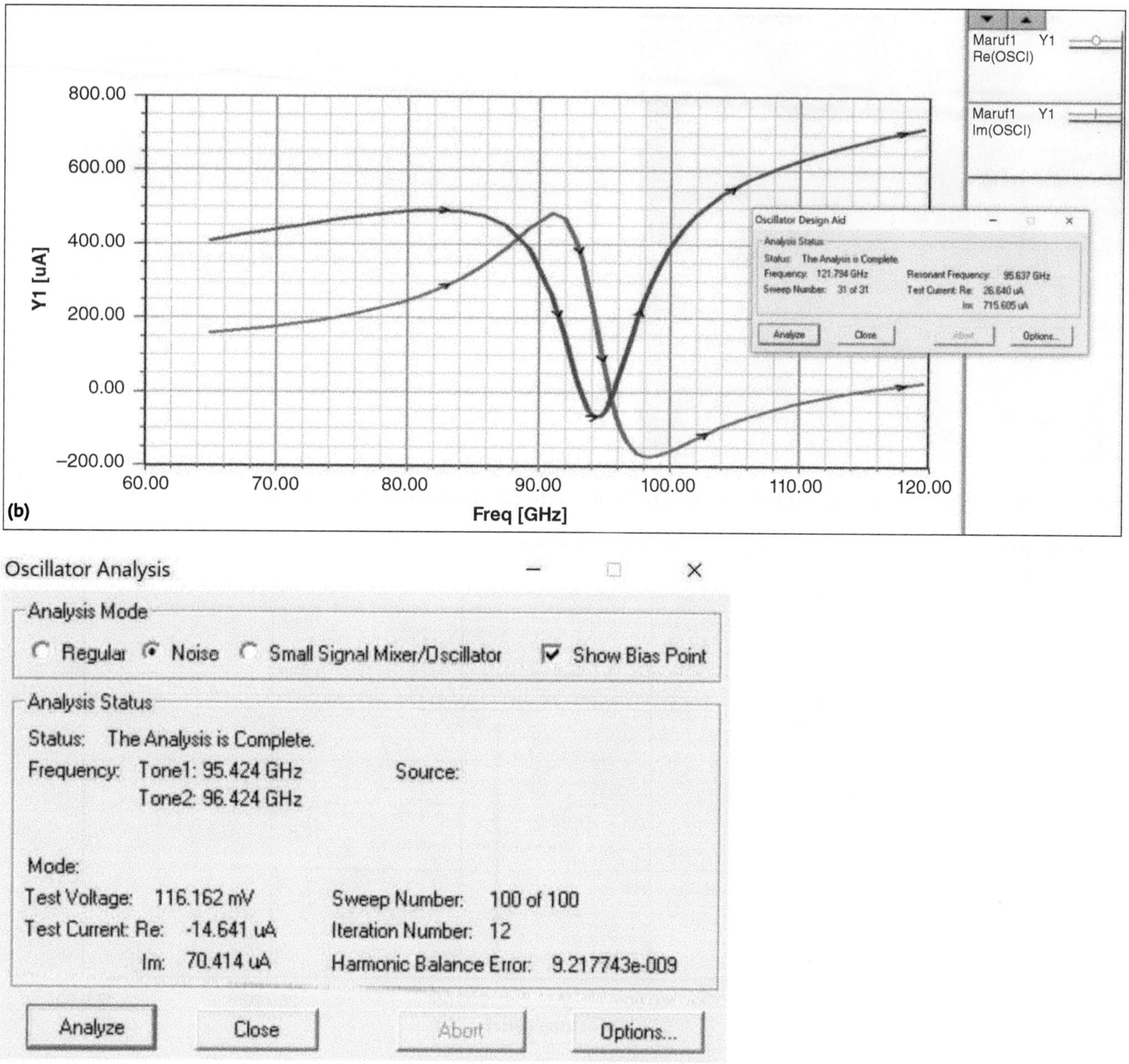

(C)

Figure 10.123 (Cont'd) (b) Linear analysis for the resonance frequency of the proposed 95 GHz oscillator of (a). (c) Large signal validation of the 95 GHz oscillator of (a).

Next, Figure 10.124 is the PCB layout, where the radial stub can be seen in the chip photo of the 96 GHz fundamental oscillator. Figure 10.125 shows the simulated output power of the 96 GHz oscillator and Figure 10.126 shows its measured phase noise performance.

We have been experimenting with these type of oscillators. The following is an example of a 161 GHz oscillator, which is part of a PLL system. The measured phase noise is shown in Figure 10.127.

Figure 10.128 shows a simplified circuit diagram of 200 GHz oscillator and Figure 10.129 shows the chip photo of the 200 GHz fundamental oscillator.

Following is another example of a 57 GHz NMOS/PMOS oscillator shown in Figure 10.130a. It follows the aforementioned explanation and uses interdigital capacitances and a planar inductor. This design is typical for integrated circuits.

Figure 10.130b shows the linear analysis for finding the resonant frequency of the proposed 57 GHz circuit, while Figure 10.130c shows the voltages (V_{ds}) of the NMOS and PMOS of the oscillator: note the symmetrical waveforms. Figure 10.130d shows the drain current of NMOS1 and NMOS2 as a function of time.

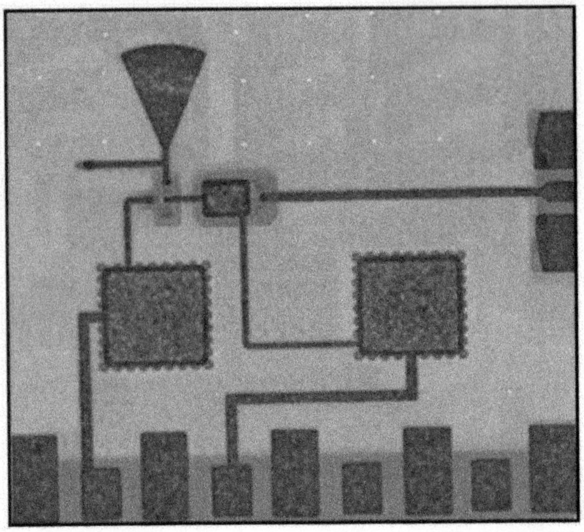

Figure 10.124 Chip photo of the 96 GHz fundamental oscillator. *Source*: [54]/Muhammad Maruf Hossain.

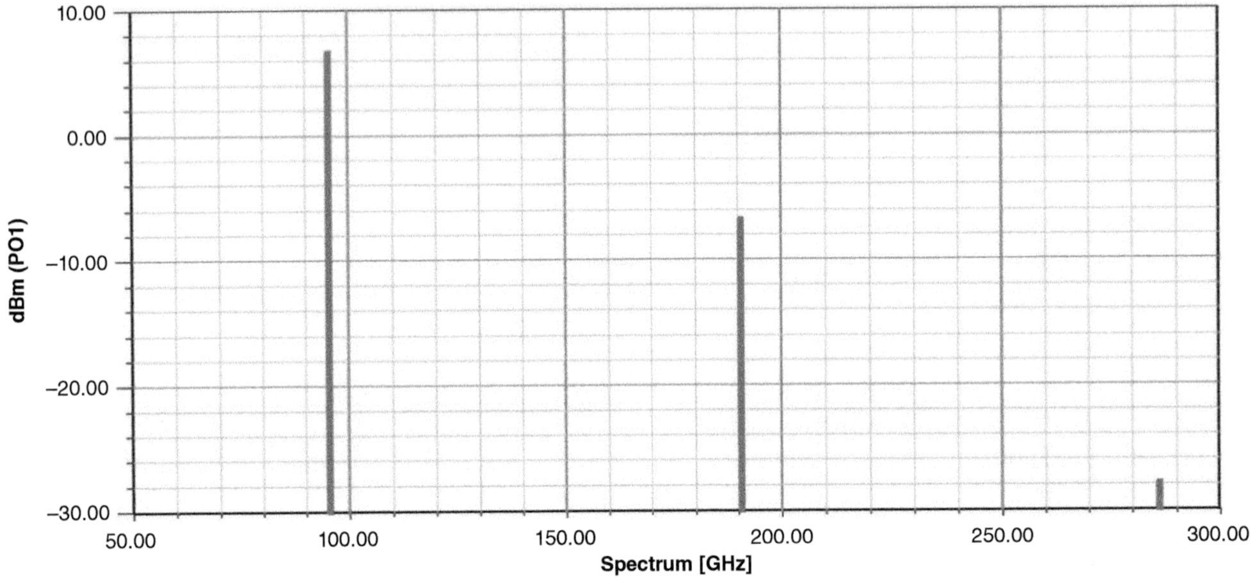

Figure 10.125 Simulated output power of the 96 GHz oscillator of Figure 10.123a.

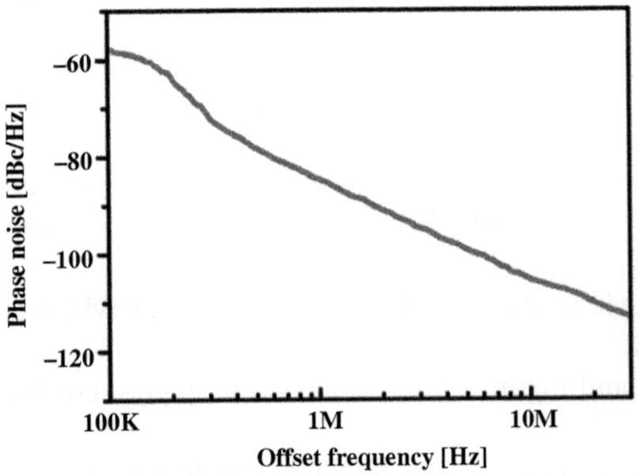

Figure 10.126 Measured phase noise performance of the 96 GHz fundamental oscillator. *Source*: [54]/Muhammad Maruf Hossain.

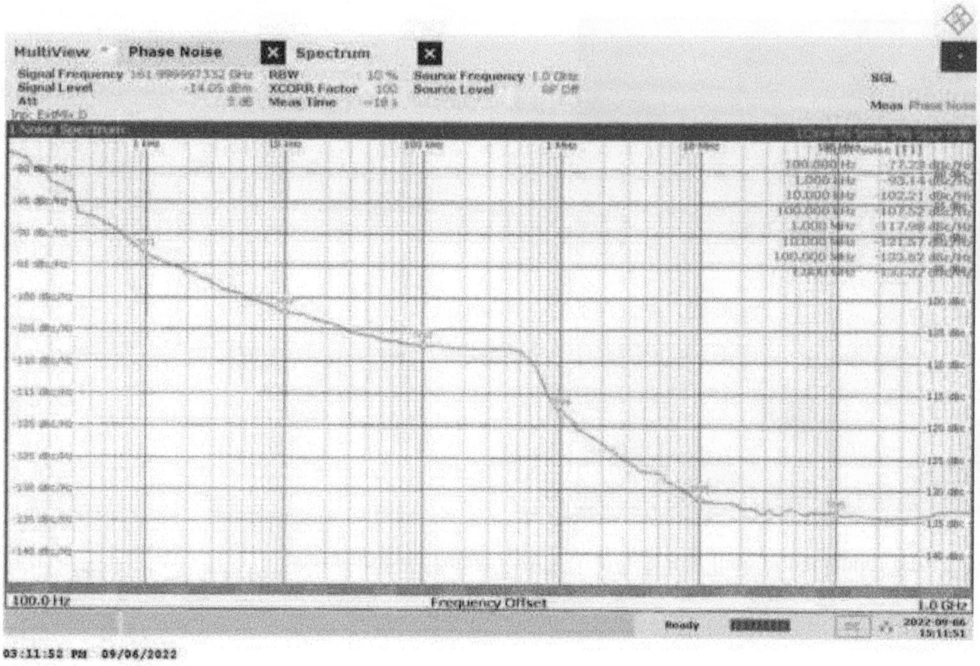

Figure 10.127 Measured phase noise performance of a 162 GHz experimental oscillator in a PLL system (Courtesy Rohde & Schwarz, Munich).

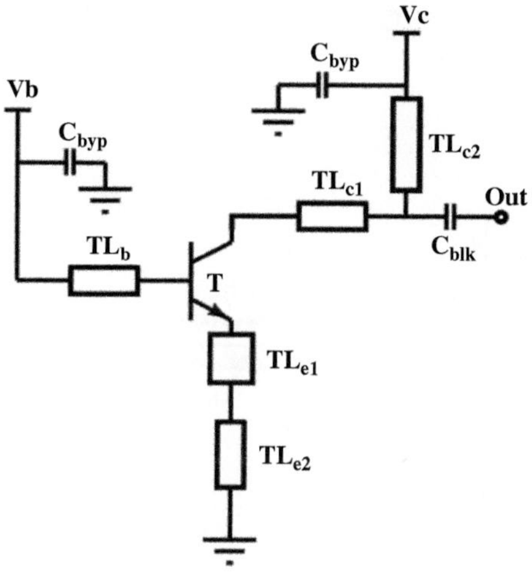

Figure 10.128 Simplified circuit diagram of 200 GHz oscillator. *Source*: [54]/Muhammad Maruf Hossain.

A minimal flywheel effect can be seen in Figure 10.130e.

As shown in Figure 10.130f,g, it can be seen that the 57 GHz oscillator is noisier than its lower frequency colleague, as shown in the example of Figure 10.67a.

Manufacturers frequently supply a foundry manual, and we recommend a sanity check by using the provided parameters of the active devices to simulate circuits to see if the results match the expectation. We also occasionally run harmonic balance simulators against each other to validate the numerical stability.

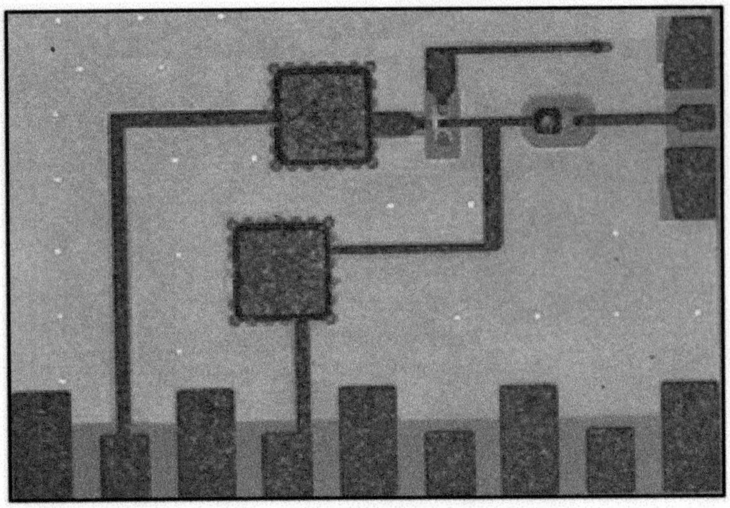

Figure 10.129 Chip photo of the 200 GHz fundamental oscillator. *Source*: [54]/Muhammad Maruf Hossain.

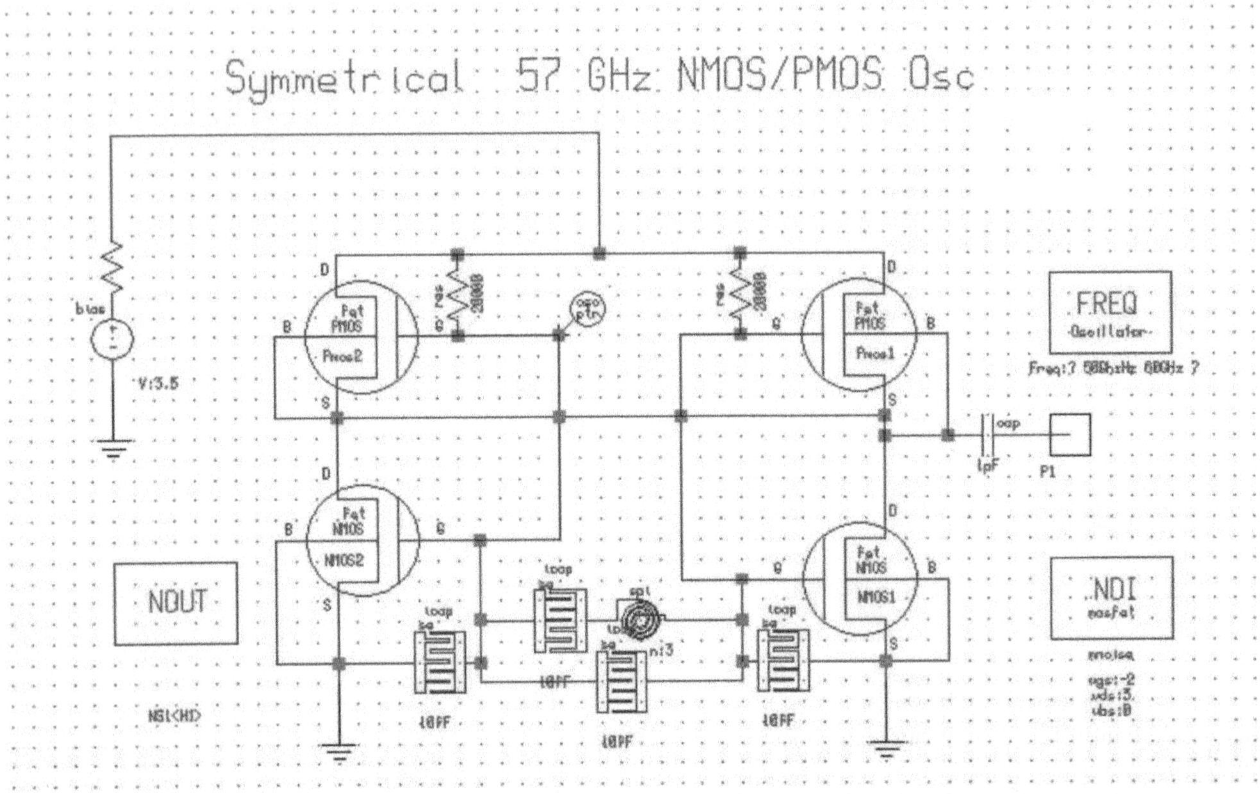

(a)

Figure 10.130 (a) Symmetrical 57 GHz NMOS/PMOS oscillator.

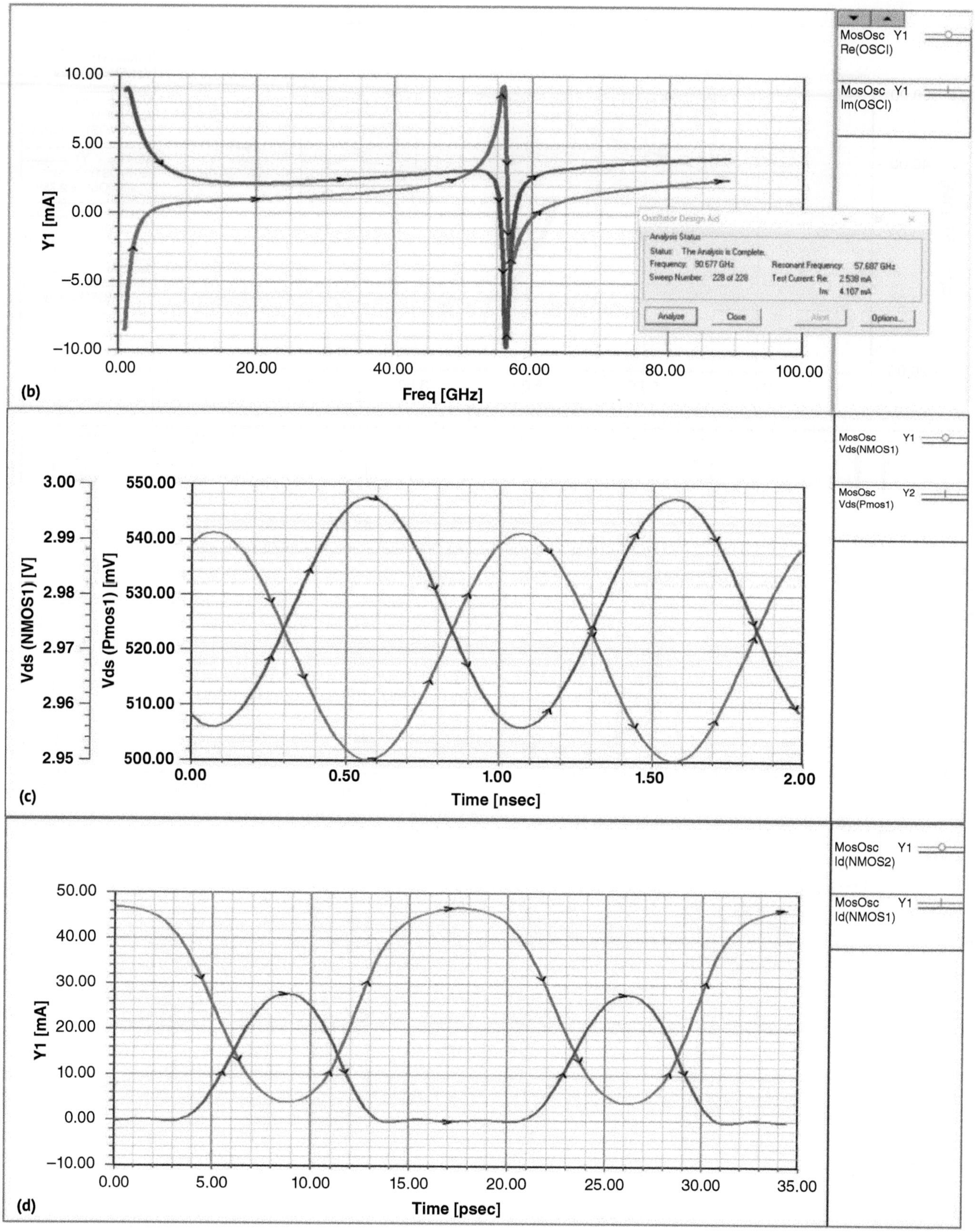

Figure 10.130 (Cont'd) (b) Linear analysis for finding the resonant frequency of the proposed 57 GHz circuit in Figure 10.131a. (c) Voltages (V_{ds}) of the NMOS and PMOS of the oscillator in Figure 10.131a—note the symmetrical waveforms. (d) Drain current of NMOS1 and NMOS2 as a function of time of the circuit of (a).

(Continued)

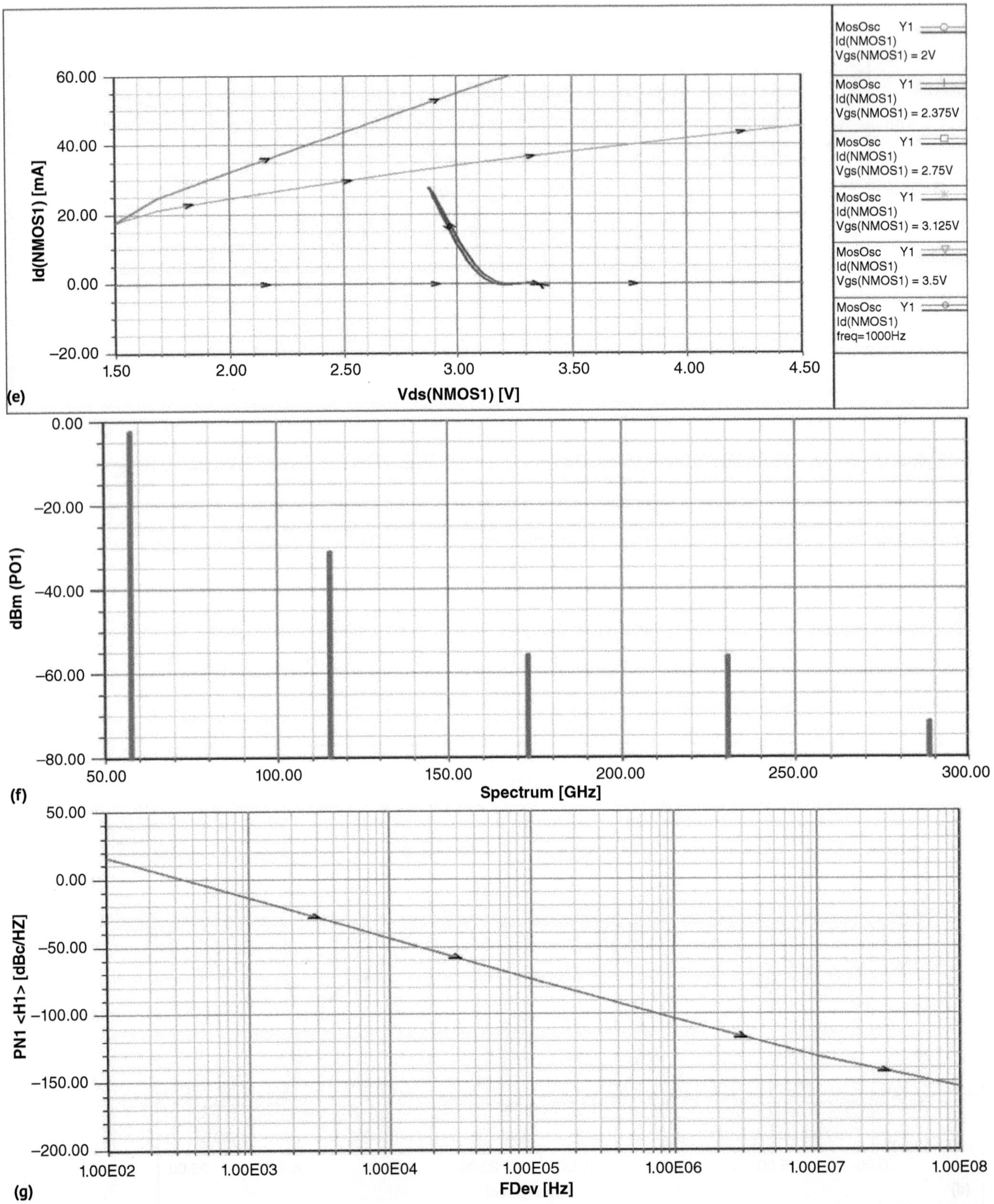

Figure 10.130 (Cont'd) (e) Simulated DC–IV curve including the load line of the circuit of (a). (f) Simulated output power of the circuit of (a). (g) Simulated phase noise of the circuit of (a).

The following 460 MHz oscillator is such an example for validation, and not verification.

The circuit (Figure 10.131a) is a Colpitts type oscillator with a loosely coupled parallel circuit. The oscillator will oscillate at a frequency where the parallel-tuned resonant circuit acts like an inductor. The linear analysis (Figure 10.131b) predicts the resonant frequency close to 463 MHz, a popular police UHF frequency.

In Figure 10.131c, the simulated output power shows a harmonic suppression of 25 dB, and as seen in Figure 10.131d, the flicker corner frequency is about 80 kHz.

An interesting circuit is the self-oscillating mixer. Figure 10.132a shows an oscillator, nonlinear by definition, driven additionally by a second signal, resulting in the combination f_{LO} (local oscillator) plus or minus the applied RF frequency. With very little LO suppression, the two sidebands will appear in the collector output power. To be useful, this circuit needs a tuned circuit at the IF, like LO = 145 MHz, RF = 116 MHz, IF = 29 MHz. At 20 MHz, the tuned circuit will sufficiently suppress the other RF frequencies. This case is called downconversion.

In the self-oscillating mixer circuit shown in Figure 10.132a [2, Figure 10.211], P1 is the input point. Both amplitude and frequency can be varied in the simulation. P2 is the IF output.

The real part (Re[OSC] or the upper trace) is negative and the circuit at resonance crosses zero at 4.4 GHz as seen by the Im (OSC) or the lower trace in Figure 10.132b. The simulated output power of the LO of the circuit is about −6 dBm at 4.5 GHz, as shown in Figure 10.132c.

Figure 10.132d shows a distorted-looking load line, which is not uncommon for a strong oscillation. The simulation in Figure 10.132e shows that the mixer of Figure 10.132a has a strong notch at about 1.6 GHz.

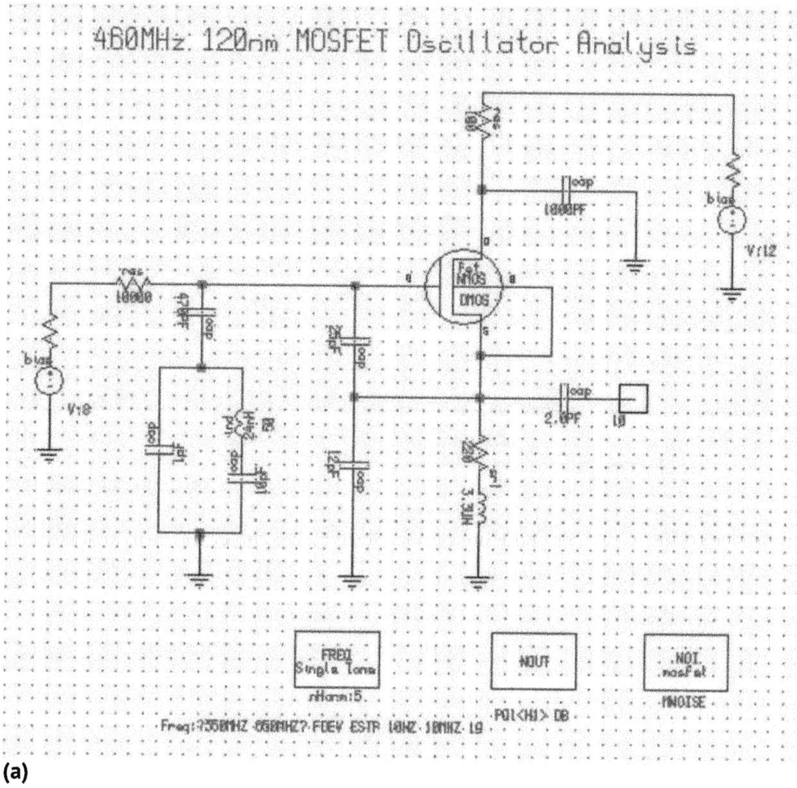

(a)

Figure 10.131 (a) 460 MHz 120 nm MOSFET oscillator analysis.

(Continued)

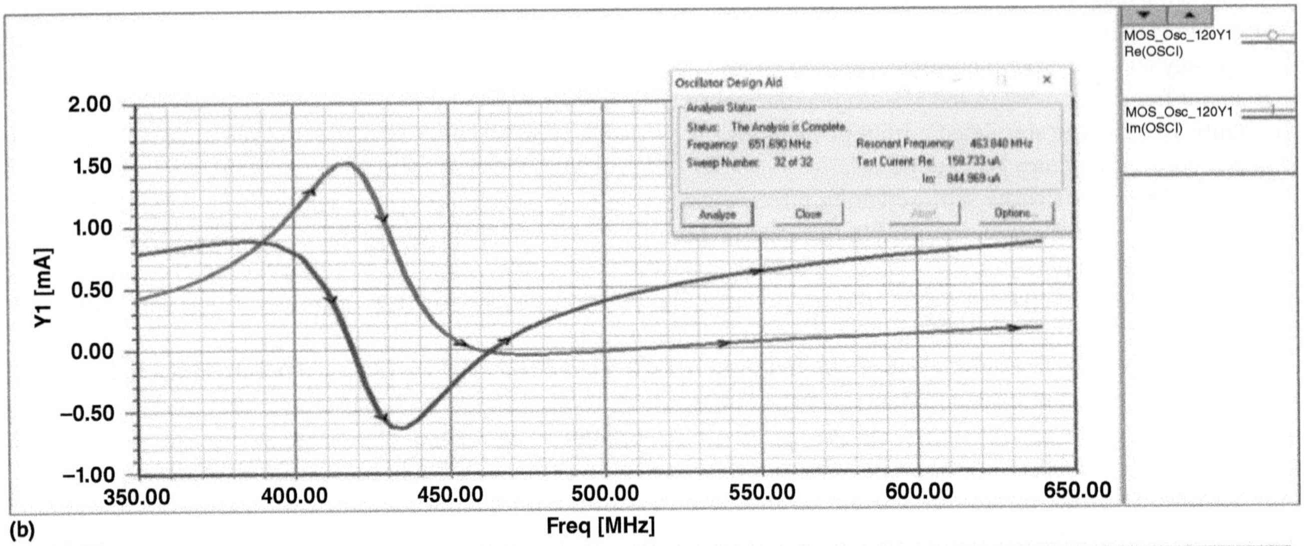

(b)

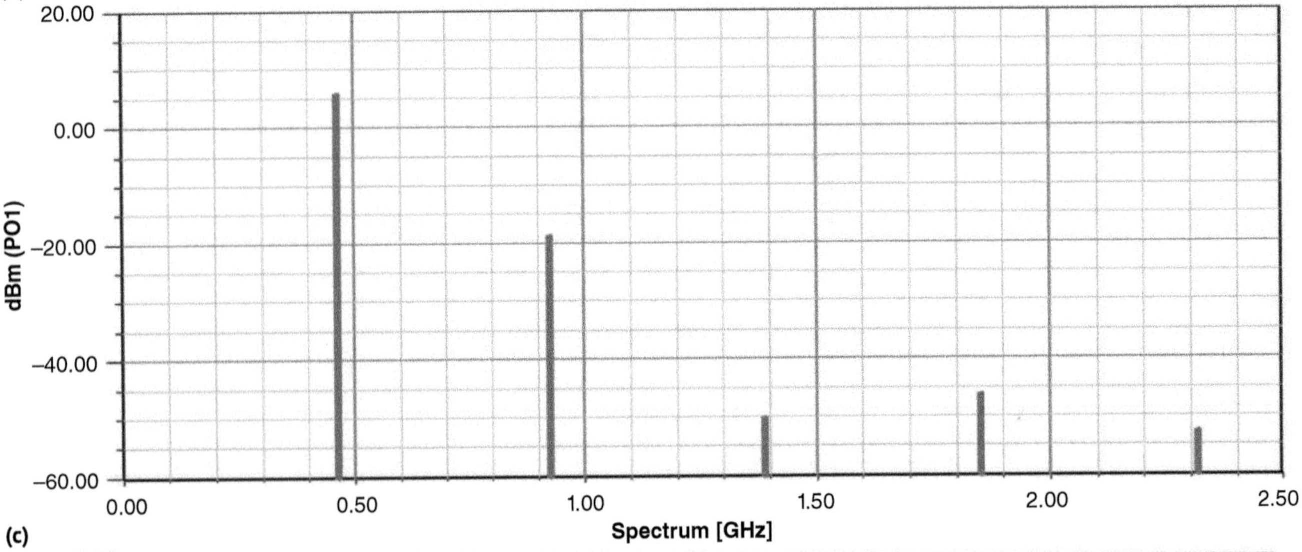

(c)

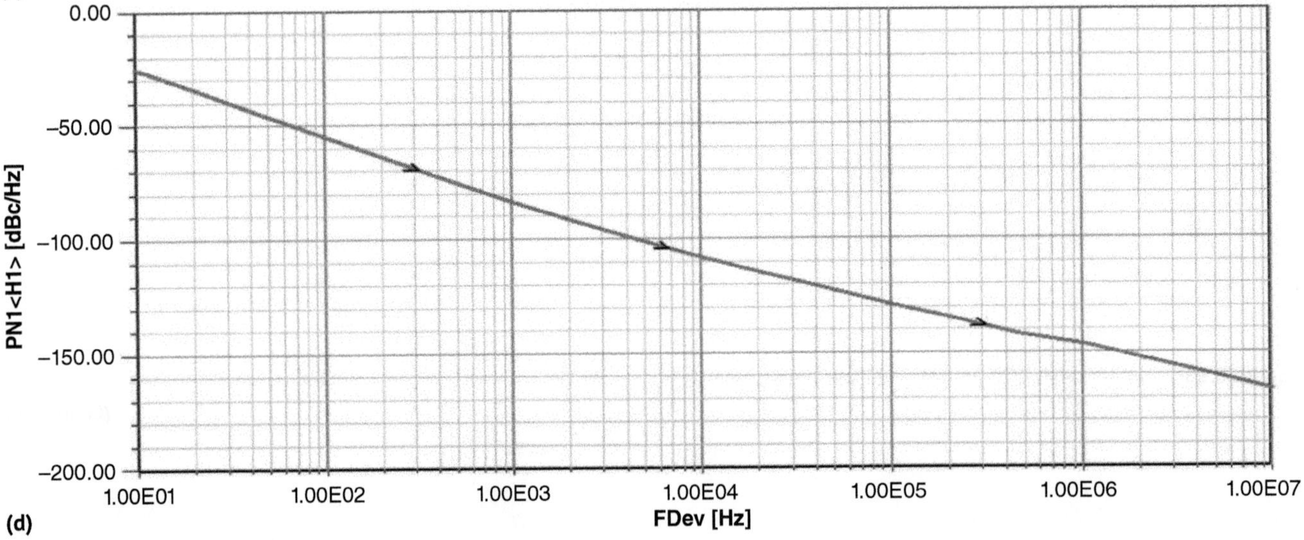

(d)

Figure 10.131 (Cont'd) (b) Linear analysis for finding the resonant frequency of an experimental 460 MHz circuit in (a). (c) Simulated output power of the circuit of (a), showing a harmonic suppression of 25 dB. (d) Simulated phase noise of the circuit of (a).

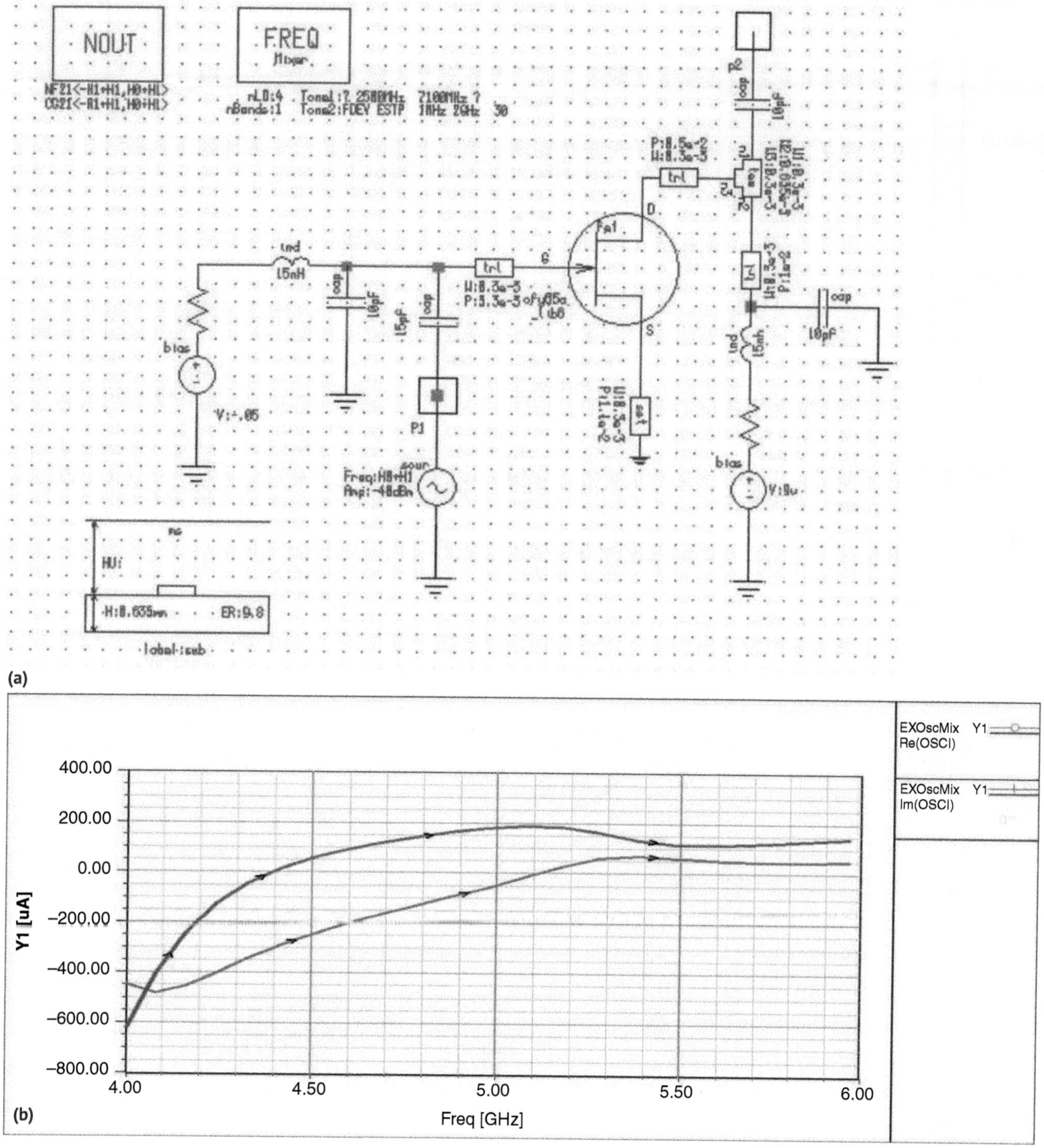

Figure 10.132 (a) Circuit of the self-oscillating mixer. *Source*: [2]/John Wiley & Sons. Linear analysis shows that the likely resonant frequency is 4.4 GHz. (b) Simulated output power of the LO of the circuit is about −6 dBm at 4.5 GHz.

(Continued)

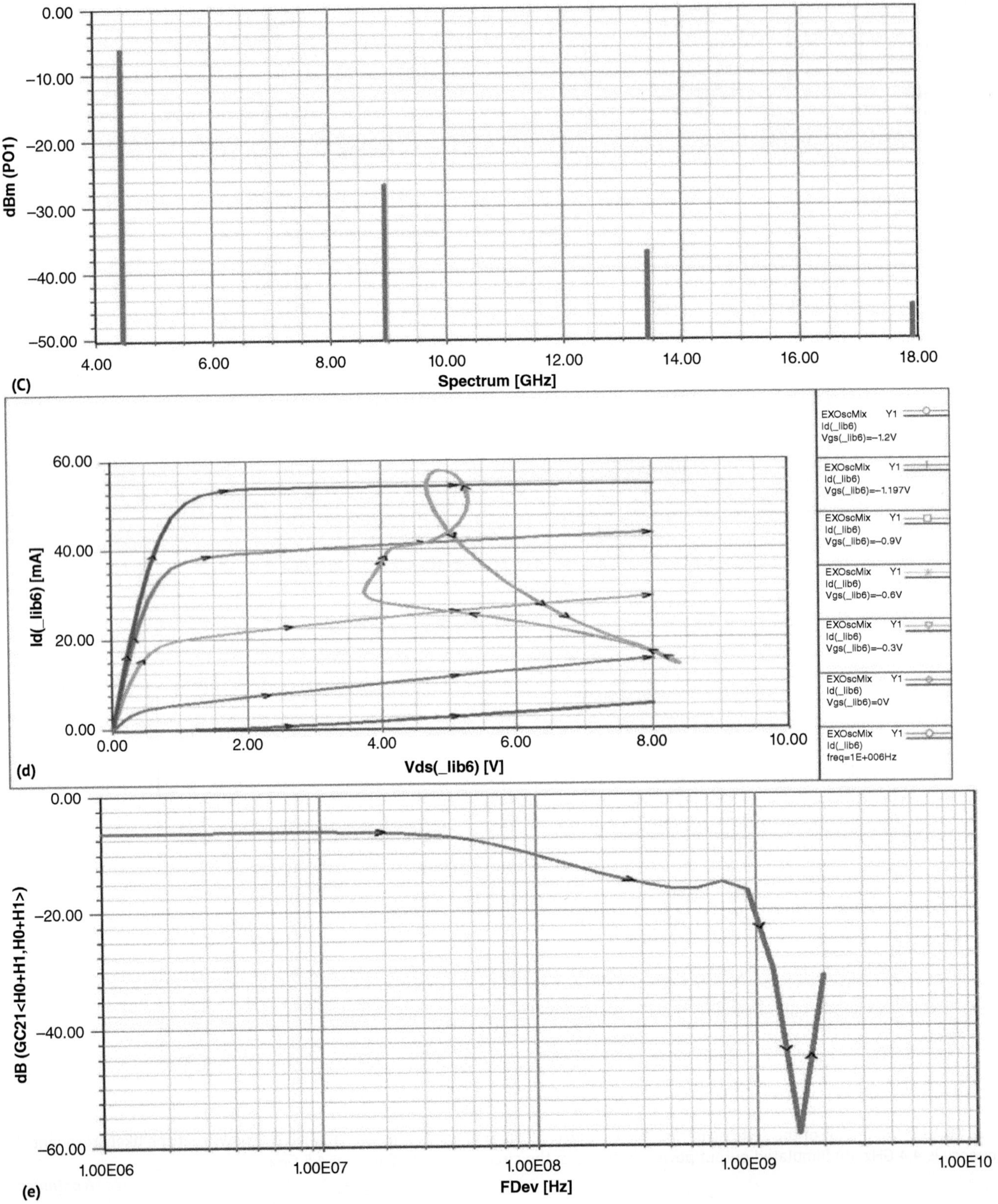

Figure 10.132 (Cont'd) (c) DC–IV curve and a distorted-looking load line simulation of the circuit in (a). (d) The simulation shows that the mixer of (a) has a strong notch at about 1.6 GHz.

Example: Our next example in Figure 10.133 is an FET mixer with both sources at the gate, like a single-diode mixer.

We encourage our readers to simulate the circuit as a good exercise.

Another important application of an oscillator is an injection-locked oscillator. Figure 10.134 is the schematic of a free-running oscillator that gets locked on the frequency applied to the input on the lower left of the schematic. Actually, sub-harmonic locking is also possible based on the nonlinearity of the circuit. Enough harmonic energy will be generated.

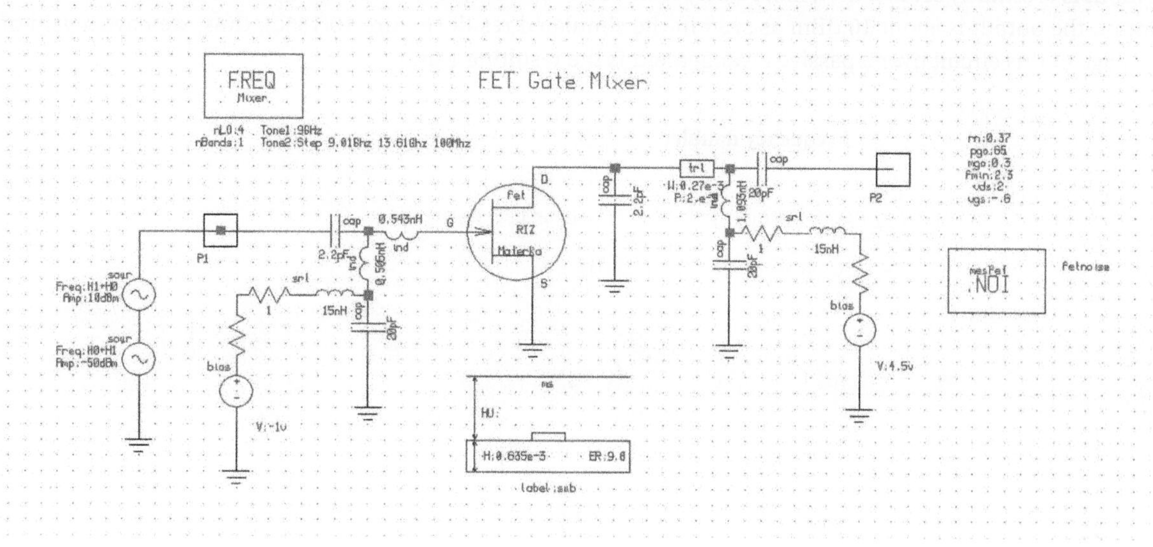

Figure 10.133 FET gate mixer.

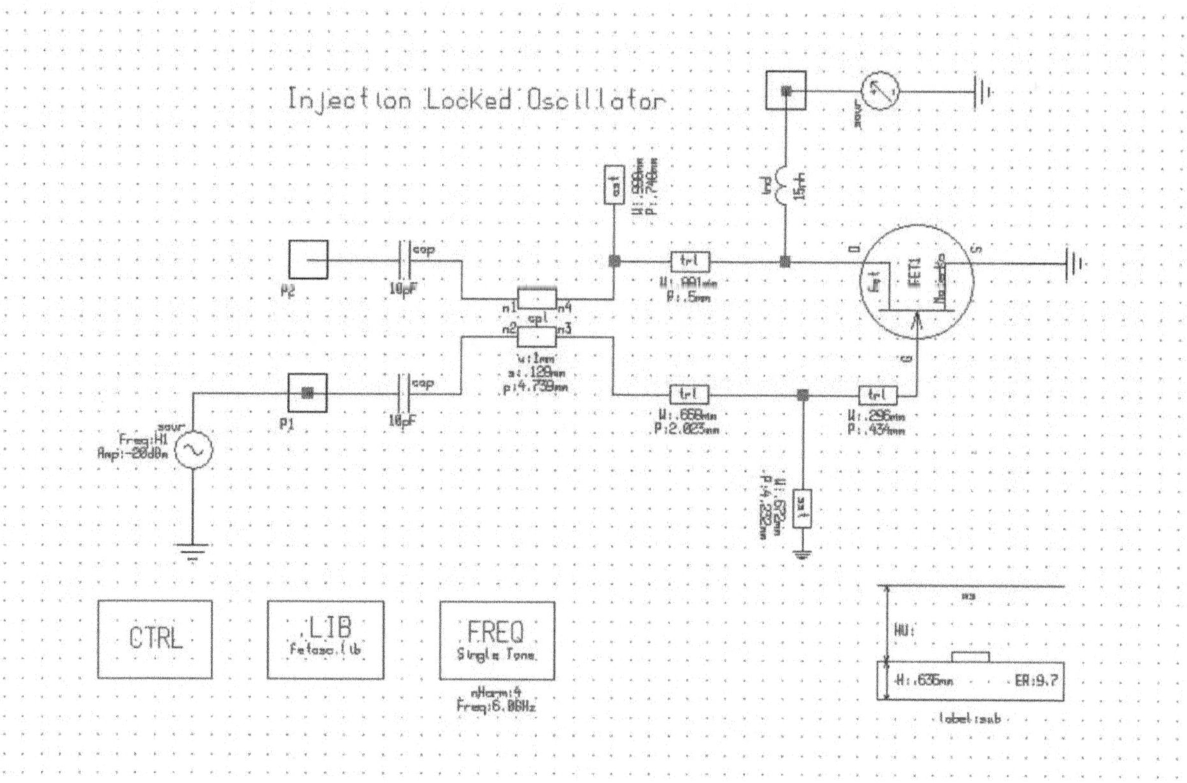

Figure 10.134 Injection-locked oscillator.

Talking about synchronizing with a subharmonic, the example in Figure 10.135a is a frequency doubler design approach [51, pp. 3–18]. The nonlinear FET is operated in the range based on the square-law characteristic of the transconductance; the input frequency will be doubled and can be taken off the drain circuit.

Frequency Doubler

Figure 10.135b shows the input and output matching of the frequency doubler. As seen in Figure 10.135b, the input S_{11} parameter shows a good matching at 5 GHz, the output S_{22} matching is much wider, and there is notch filter for the input frequency. The input is arbitrarily set at 2 mW, more likely it will be 7–10 dBm, so we will run both cases.

At −4 dBm input, the output is about 10 dBm as seen in the simulated result (Figure 10.136c); thus, we have a gain of 14 dB. The advantage of this circuit over a passive arrangement is the resulting gain.

Frequency doubler

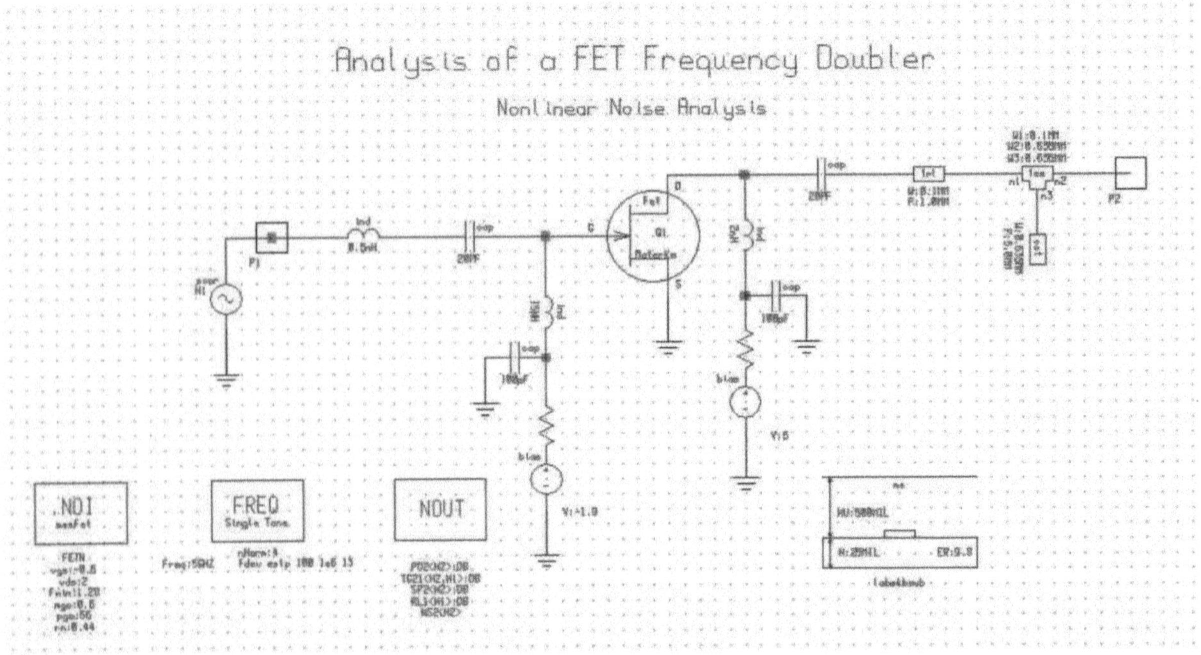

(a)

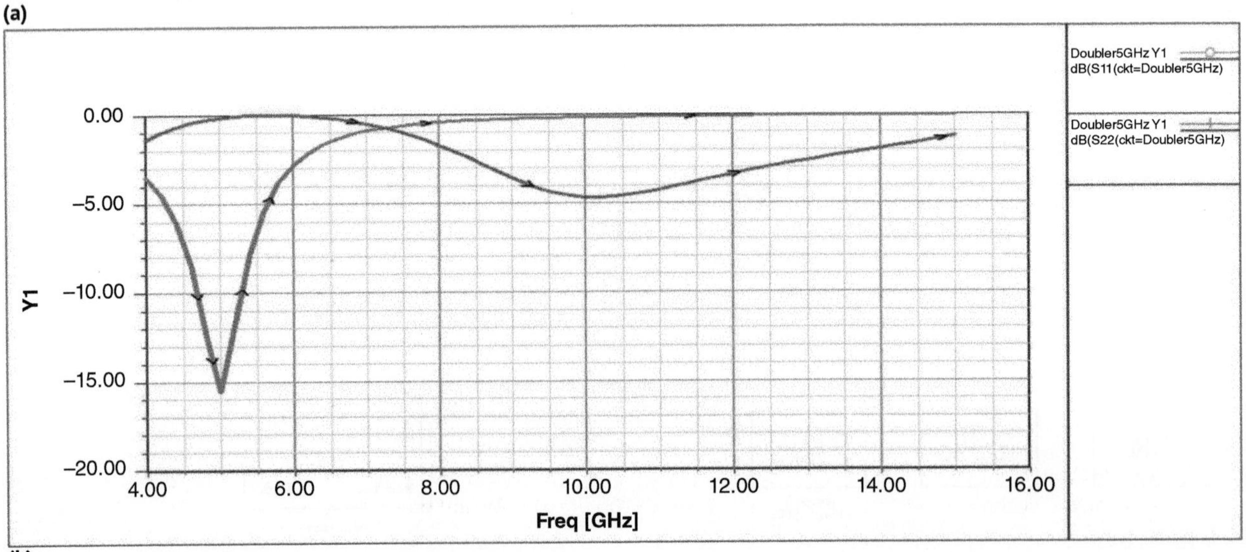

(b)

Figure 10.135 (a) Frequency doubler circuit using an FET. (b) Simulation of the input (S_{11} is -15 dB) at 5 GHz and output (S_{22} is -5 dB) at 10 GHz.

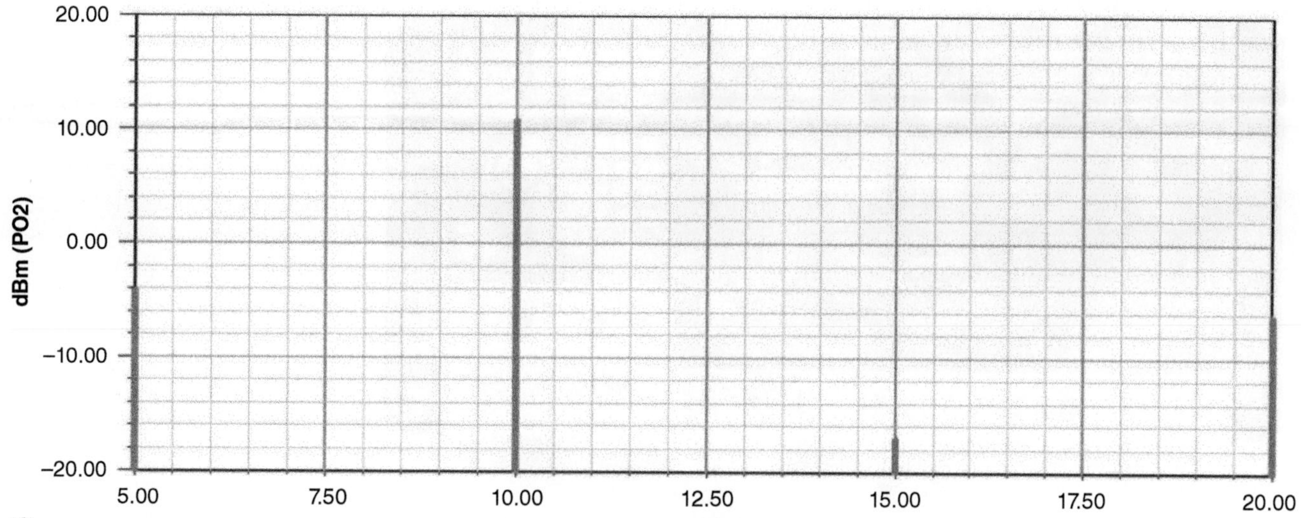

(C)

Reduced Noisy Source (Amplitude)

	Property	Value	Visible
1	Freq	H1	Value
2	Amp	0.002WATT	Nothing
3	option	P	Nothing
4	Phase		Nothing
5	Fdev	100 1E3 10E3 100E3 1E6	Nothing
6	Rfup	-55 -85 -115 -135 -155	Nothing
7	Amn	-130 -145 -155 -165 -170	Nothing
8			

(d)

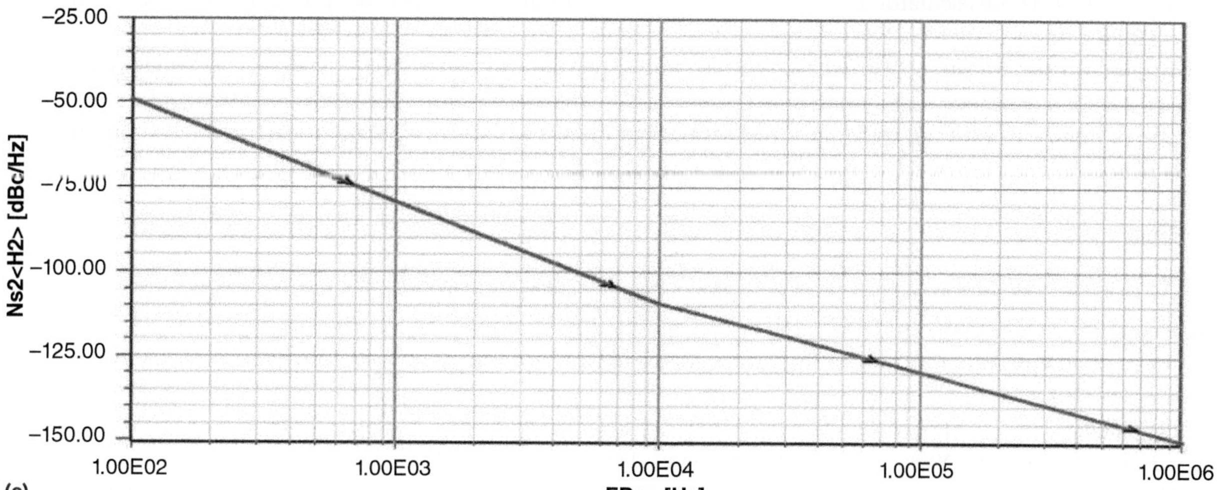

(e)

Figure 10.135 (Cont'd) (c) Simulated gain of the frequency doubler of (a). (d) Noise table for the frequency doubler of (a). (e) Simulated phase noise for the frequency doubler of (a). *Source*: [51]/Verlag für Radio-Foto-Kinotechnik.

In Figure 10.135d showing the noise table, Fdev shows the frequncy offset from the carrier, and the values of the SSB upper side band in dB below the carrier are shown underneath as Rfup. Amn refers to the transistor's minimum noise at the offset, but this is corrected by the noise correlation table applicable for the transistor type.

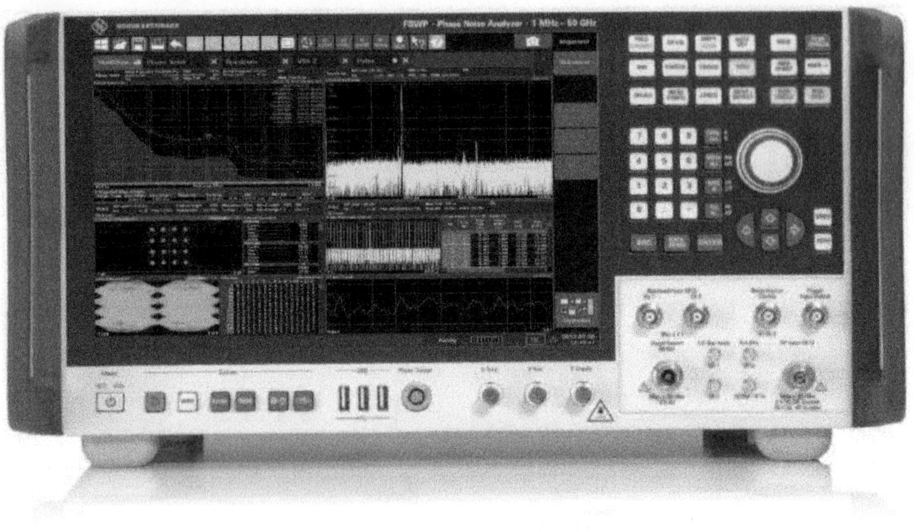

Figure 10.136 Rohde & Schwarz FSWP (50 GHz phase noise analyzer). *Source*: Courtesy of Dr. Ulrich L. Rohde.

As expected, the FET adds very little noise so does the phase noise, at 10 GHz is exactly 6 dB worse. Figure 10.135e shows the simulated phase noise for this frequency doubler.

10.23 Conclusion

This chapter showed an in-depth discussion on oscillators, noise performance, and other parameters. It addresses various resonators, from crystal, to LC oscillator, different transistors, from Si-Ge to JFETs, MESFETs, and finally MOSFETs. It also covers the frequencies from 10 MHz (crystal oscillators) upto PLL systems at 167 GHz (see Figure 10.127). In addition, this chapter has an in-depth look into the noise sources of the transistors as this noise gets modulated on top of the carrier frequency. The 100 MHz crystal oscillator details are from the PhD dissertation of Anisha Apte, Feb 2020 under the guidance of the authors (Rohde and Rudolph).

The test equipment used for the measured phase noise and other data was extremely critical—the Rohde & Schwarz FSWP was used—which is now available upto 85 GHz and much wider offset from the carrier. A fully configured FSWP (see photograph in Figure 10.136) can cost upto US$500,000. Not many companies have such sophisticated equipment.

References

1 Rohde, U.L. and Rudolph, M. (2013). *RF/Microwave Circuit Design for Wireless Applications*, 2e, 745–746. New York: John Wiley & Sons, ISBN 1118431405.

2 Pavio, A.M. and Rudolph, R. (2021). *Microwave Circuit Design: Using Linear and Nonlinear Techniques*. Wiley, ISBN: 10.1002/9781119741725.

3 Rohde, U.L., Poddar, A.K., and Boeck, G. (2005). *The Design of Modern Microwave Oscillators for Wireless Applications: Theory and Optimization*. New York, NY: John Wiley & Sons,, ISBN 0-471-72342-8.

4 Rohde, U.L. and Apte, A. (2016). Everything you always wanted to know about Colpitts oscillators. *IEEE Microwave Magazine* 17 (8): 59–76.

5 Rohde, U.L. (1983). *Digital PLL Frequency Synthesizers, Theory and Design*. New York: Wiley.

6 Rohde, U.L. (1997). *Microwave and Wireless Synthesizers, Theory and Design*. New York: Wiley, ISBN 0-471-52019-5.

7 Vendelin, G.D. (1982). *Design of Amplifiers and Oscillators, By the S-parameter Method*. New York: Wiley.

8 Apte, A.M. (2020). A new analytical design method of ultra-low-noise voltage controlled VHF crystal oscillators and its validation. Thesis.

9 Vendelin, G., Pavio, A.M., and Rohde, U.L. (2005). *Microwave Circuit Design Using Linear and Nonlinear Techniques.* New York: Wiley, 2005.

10 Rohde, U., Poddar, A., and Apte, A. (2013). Getting its measure. *IEEE Microwave Magazine* 14 (6): 73–86.

11 Rohde, U.L., Poddar, A.K., and Apte, A. (2013). Phase noise measurements and system comparisons. *Microwave Journal* 56 (4): 22–46.

12 Baghdady, E.J., Lincoln R.N., and Nelin, B.D. (1965). *Short-term Frequency Stability NASA SP-80*, pp. 65–87.

13 Leeson, D.B. (1966). A simple model of feedback oscillator noise spectrum. *IEEE, Proc.*, 329–332.

14 Driscoll, M.M. (1989). Low noise crystal controlled oscillator. US Patent No. US4797639A.

15 Rohde, U.L. and Poddar, A.K. (2012). Crystal oscillators. In: *Wiley Encyclopedia and Electronics Engineering*, 1–38. John Wiley & Sons, Inc.

16 Rohde, U.L. and Poddar, A.K. (2012). Crystal oscillator design. In: *Wiley Encyclopedia of Electrical and Electronics Engineering*, 1–47. John Wiley & Sons, Inc.

17 Scherer, D. (1983). The "art" of phase noise measurement. *RF & Microwave Measurement Symposium and Exhibition*, Hewlett Packard.

18 Rohde, U.L. and Newkirk, D.P. (2000). *RF/Microwave Circuit Design for Wireless Applications.* John Wiley & Sons, ISBN 0-471-29818-2.

19 Apte, A., Rohde, U.L., Poddar, A. et al. (2017). "Optimizing phase-noise performance: theory and design techniques for a crystal oscillator". *IEEE Microwave Magazine* 18 (4): 108–123. https://doi.org/10.1109/MMM.2017.2679998

20 Packard, H. (1985). *RF and Microwave Phase Noise Measurement Seminar.* http://www.hparchive.com/seminar_notes/HP_PN_seminar.pdf

21 Poddar, A.K. (2014). Slow wave resonator based tunable multi-band multi-mode injection-locked oscillators. Dr.-Ing.-habil Thesis, Chapter 5, 163–177, BTU Cottbus, Germany.

22 Poddar, A., Rohde, U., and Apte, A. (2013). How low can they go, oscillator phase noise model, theoretical, experimental validation, and phase noise measurements. *IEEE Microwave Magazine* 14 (6): 50–72.

23 Rohde, U.L. and Poddar, A.K. (2018). *An Optimized Grounded Base Oscillator Design for VHF/UHF.* QEX.

24 Rohde, U.L. (1965). *Transistoren bei hochsten Frequnzen, Theorie und Schaltungspraxis von Diffusionstransistoren im VHF und UHF-Bereich.* Verlag Fur Radio-Foto-Kinotechnik GMBH.

25 Rohde, U.L. (2017). Noise analysis – then and today. https://www.microwavejournal.com/articles/29151-noise-analysis-then-and-today?v=preview

26 Porret, A.S., Melly, T., Enz, C.C. et al. (2000). Design of high Q varactors for low-power wireless applications using a standard CMOS process. *IEEE Journal of Solid-State Circuits* 35: 337–345.

27 Merrill, R.B., Lee, T.W., You, H. et al. (1995). "Optimization of high Q integrated inductors for multi-level metal CMOS. *IEEE International Electron Devices Meeting*, 983–986.

28 Ham, D. and Hajimiri, A. (2001). Concepts and methods in optimization of integrated LC VCOs. *IEEE Journal of Solid-State Circuits* 36 (6): 896–909.

29 Tiemeijer, L.F., Leenaerts, D., Pavlovic, N. et al. (2001). Record Q spiral inductors in standard CMOS. *IEEE Electron Devices Meeting Technical Digest*, 40.7.1–40.7.3.

30 Song, S.S. and Shin, H. (2003). An RF model of the accumulation-mode MOS varactor valid in both accumulation and depletion regions. *IEEE Transactions on Electron Devices* 50 (9): 1997–1999.

31 Holnar, K., Rappitsch, G., Huszka, Z. et al. (2002). MOS varactor modeling with a subcircuit utilizing the BSIM3v3 model. *IEEE Transactions on Electron Devices* 49 (7): 1206–1211.

32 Fong, N.H.W., Plouchart, J.O. Zamdamer, N. et al. (2003). Design of wide-band CMOS VCO for multiband wireless LAN applications. *IEEE Journal of Solid-State Circuits* 38 (8): 1333–1342.

33 Hung, C.M., Floyd, B.A., Park, N. et al. (2001). Fully integrated 5.35 GHz CMOS VCOs and prescalers. *IEEE Transactions on Microwave Theory and Techniques* 49 (1): 17–22.

34 Maget, J., Tiebout, M., and Kraus, R. (2002). Influence of novel MOS varactors on the performance of a fully integrated UMTS VCO in standard 0.25 μm CMOS technology. *IEEE Journal of Solid-State Circuits* 37 (7): 953–958.

35 Andreani, P. and Mattisson, S. (2000). On the use of MOS varactors in RF VCOs. *IEEE Journal of Solid-State Circuits* 35 (6): 905–910.

36 Zhang, X. and Ding, K. (1993). Capacitance-time transient characteristics of pulsed MOS capacitor application in measurement of semiconductor parameters. *IEEE Proceedings-G* 140 (6): 449–452.

37 Bunch, R.L. and Raman, S. (2003). Large-signal analysis of MOS varactors in CMOS – Gm LC VCOs. *IEEE Journal of Solid-State Circuits* 38 (8): 1325–1332.

38 Kral, A., Behbahani, F., and Abidi, A.A. (1998). RF-CMOS oscillators with switched tuning. *IEEE Custom Integrated Circuits Conference*, 555–558.

39 Maget, J., Tiebout, M., and Kraus, R. (2003). MOS varactors with n- and p-type gates and their influence on an LC-VCO in digital CMOS. *IEEE Journal of Solid-State Circuits* 38 (7): 1139–1147.

40 Magierowski, S.K., Iniewski, K., and Zukotynski, S. (2002). Differentially tunable varactor with built-in common-mode rejection. *IEEE 45th Midwest Symposium on Circuits and Systems* 1: 559–562.

41 Tang, Y., Aktas, A., Ismail, M. et al. (2001). A fully integrated dual-mode frequency synthesizer for GSM and wideband CDMA in 0.5 μm CMOS. *44th IEEE Symposium on Circuits and Systems* 2: 866–869.

42 Levantino, S., Samori, C., Bonfanti, A. et al. (2002). Frequency dependence on bias current in 5-GHz CMOS VCOs: impact on tuning range and flicker noise upconversion. *IEEE Journal of Solid-State Circuits* 37 (8): 1003–1011.

43 Fong, N.H.W., Plouchart, J.O., Zamdamer, N. et al. (2003). A 1-V 3.8-5.7-GHz wide-band VCO with differentially tuned accumulation MOS varactors for common-mode noise rejection in CMOS SOI technology. *IEEE Trans. Microwave Theory and Techniques* 51: (8), 1952–1959.

44 Cheah, J., Kwek, E., Low, E. et al. (2002). Design of a low-cost integrated 0.25 μm CMOS bluetooth SOC in 16.5 mm^2 silicon area. *IEEE International Solid-State Circuits Conference*, 90–91.

45 Freitag, R.G., Lee, S.H., Krafcsik, D.M. et al. (2003). Stability and improved circuit modeling considerations for high power MMIC amplifiers. *IEEE, Microwave and Millimeter-Wave Monolithic Circuits Symposium*, 2169–2172.

46 Heinbockel, J. and Mortazawi, A. (1992). A periodic spatial power combining MESFET oscillator. *IEEE, MTT-S Digest*, 545–548.

47 Naishadham, K. and Durak, T. (2002). Measurement-based closed-form modeling of surface mounted RF components. *IEEE Trans.* 50 (10): 2276–2286.

48 Rohde, U.L. and Poddar, A.K. (2004). Noise analysis of systems of coupled oscillators. *Integrated Nonlinear Microwave and Millimeterwave Circuits (INMMIC) Workshop, Monte Porzo Catone, Italy*, November 15–16, 2004.

49 Rohde, U.L., Rubiola, E., and Whitaker, J.C. (2021). *Microwave and Wireless Synthesizers: Theory and Design, 2e*. Wiley, ISBN: 978-1-119-66600-4.

50 Zhang, H. and Xue, Q. Stage Key Lab of Millimeter Waves (HK), Department of Electronic Engineering City University of Hong Kong, Hong Kong, P. R. China.

51 Ansoft Corporation (2001). Serenade Examples Volume, Version 8.7.

52 User's Manual, Compact Software SUPER SPICE 1.1.

53 Cohen, E.D. (2012). The MIMIC program—a retrospective. *IEEE Microwave Magazine* 13 (4): 77–88. https://doi.org/10.1109/MMM.2012.2189989.

54 Hossain, M.H. (2016). *Signal Generation for Millimeter Wave and THz Applications in InP-DHBT and InP-on-BiCMOS Technologies, Faculty IV – Electrical Engineering und Informatics*. Technischen Universität Berlin.

Further Reading

Suarez, A. (2026). New strategies for simulating nonlinear circuits with complex solution curves. *IEEE Microwave Magazine*, 27 (8): 25.

A

AM and FM Noise in Oscillators

A.1 Introduction

When measuring the single sideband (SSB) noise of oscillators, until recently all specifications were in reference to frequency modulation (FM) noise at specific offsets of the carrier. Modern signal generators and vector signal generators have a baseband in which the various types of modulations are created and then added/mixed to the wanted output frequency. Such modern signal generators are available to 74 GHz and higher.

In principle, an oscillator is an amplifier with feedback. Because of the positive feedback, a signal amplitude of the carrier at a specific frequency builds up until the transistor gain based on the nonlinearities of the active device limits amplitude.

An oscillator has a well-defined amplitude which is controlled by the non-linearity of the circuit. If there is an amplitude perturbation, it is naturally rejected by the oscillator.

- This occurs because the oscillation occurs at a frequency when the loop gain is unity. If the amplitude grows, due to compressive characteristics of the non-linearity, the loop gain decreases and the oscillation amplitude dampens. Likewise, if the amplitude drops, the loop gain goes over unity due to expansive characteristics of the non-linearity, and the amplitude grows back.
- The phase of the oscillator, on the other hand, is "free running." Any phase-shifted solution to the oscillator is a valid solution. So if a perturbation changes the phase of the oscillator, there is no "restoring force" and the phase error persists.

They are frequently used to determine the dynamic range of a receiver, specifically the blocking dynamic range. Blocking is defined as the degradation of receiver sensitivity, usually by 3 dB, in the presence of a much stronger (blocking) signal. Therefore, we must first understand that *Sensitivity* is the measure of what the smallest possible detectable signal is. Sensitivity is measured through signal-to-noise ratio (SNR) at the output of the receiver. Different SNR values can be chosen according to the specific task. Blocking is therefore defined as the reduction in SNR caused by an interfering signal [1–4].

Another more modern test method is the determination of the "Adjacent Carrier Power Ratio" (ACPR). A typical cell phone has 200 kHz channel spacing.

It is the ratio between the total powers in the adjacent channels (intermodulation signal) to the main channel's power (wanted signal). There are two ways of measuring ACPR. The first way is by finding 10*log of the ratio of the total output power to the power in the adjacent channel. The second (and much more popular method) is to find the ratio of the output power in a smaller bandwidth around the center of carrier to the power in the adjacent channel. The smaller bandwidth is equal to the bandwidth of the adjacent channel signal. The second way is more popular because it can be measured easily. **Measurement**: ACPR is usually measured in decibels (dB). Higher ACPR values indicate that the signal in the main channel is cleaner and there is less interference to adjacent channels.

When evaluating receivers by measuring their oscillator's noise there is a distinct difference between amplitude modulation (AM) and FM noise generated in the generator used for determining the ACPR (see Figure A.1).

Figure A.1a shows the R&S SMBV100B synthesized signal generator. As seen in the figure, at the test frequency of 14 MHz, the AM noise dominates, whereas at higher frequencies, FM noise dominates.

Now, being interested in knowing how to design a good low AM noise generator, let us look at Figure A.1b, c.

Using a tunable LC (helical-resonator)-based free-running oscillators (VCOs) and a field-effect transistor-based oscillator, the resulting design will be very good. In this discussion, we want to show a mathematical approach to derive this information. Here, the measured AM (very low) and FM (very high) noises on the carrier are shown [5].

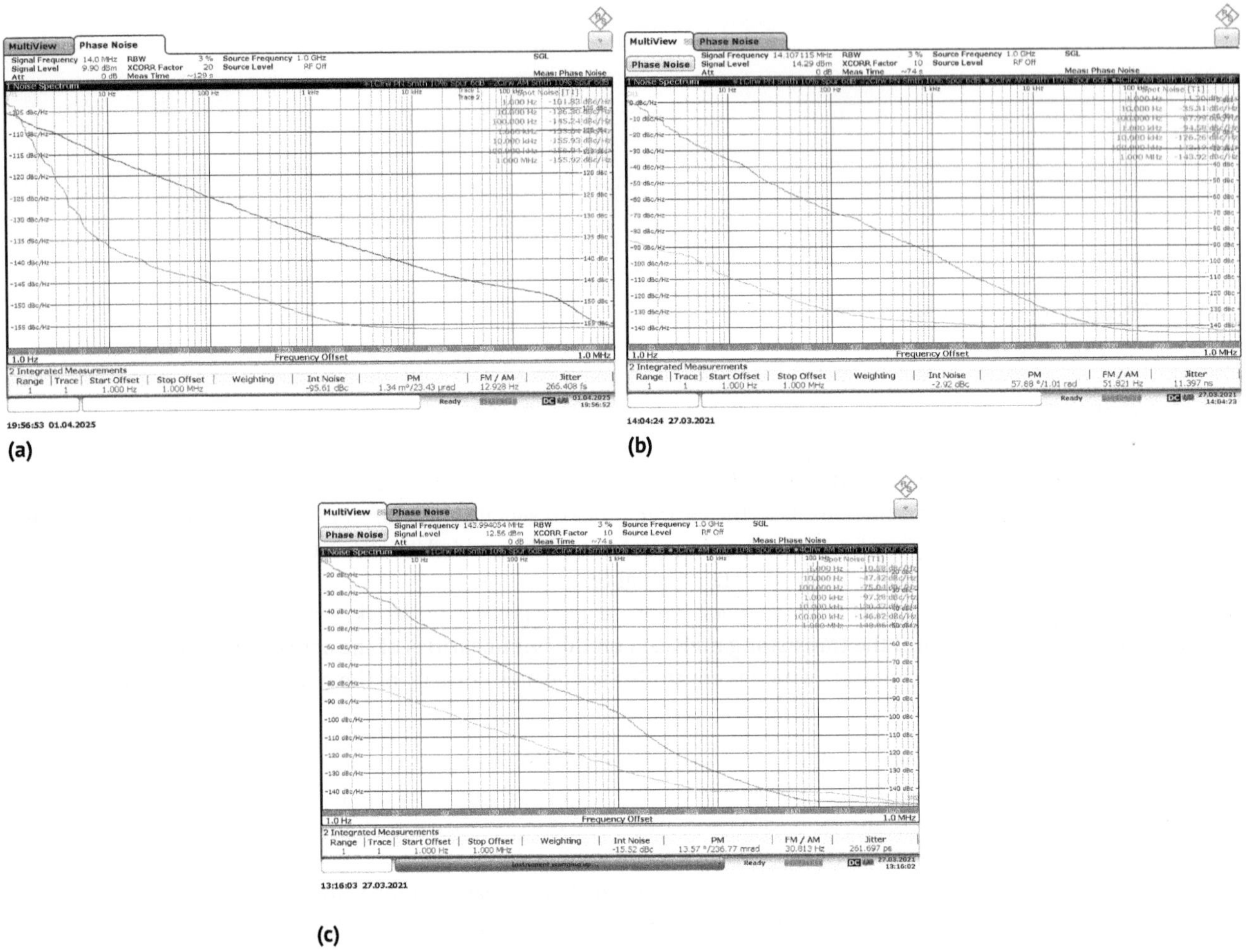

Figure A.1 (a) R&S SMBV100B signal generator at test frequency 145 MHz showing the AM and FM noise; (b) R&S SMBV100B signal generator at test frequency 14 MHz showing the AM and FM noise; (c) measured results of a circuit with helical resonators at the 2 m band.

See the measurement figure (Figure A.1c); it's the circuit with the helical resonator at the 2 m band.

A.2 Phase Noise Analysis Based on the Negative Noisy Resistance Model

The following noise analysis of the oscillator, while based on the approach of Kurokawa [1], is an attempt to introduce the concept of a "noisy" negative resistance, which is time dependent. This was first proposed by Kurokawa [6], addressing the question of synchronized oscillators, provided insight in the general case of a series oscillator. The method introduced here is specific for a real oscillator and real noise sources.

This concept gets started by connecting a parallel-tuned circuit to a transistor in Colpitts configuration. Since the two capacitors C_1 and C_2 are similar in value, not different by more than a factor of 2 or 3, and connected to a parallel-tuned circuit via a small coupling capacitor C_c, the output impedance of the emitter follower circuit gets transformed to the base. The differential output impedance at the emitter is $1/Y_{21}^{+}$ (large signal), while the input impedance itself is $1/Y_{21}^{+} \times \beta$. Because of this, the contribution of Y_{11} can be neglected for the basic analysis. This is only valid for this particular case. Consistent with Equation (6.1), which is based on the same approximation but includes the parasitics, the transistor circuit now provides a negative resistance (or a negative conductance). This negative conductance cancels the losses concentrated in the loss resistorR_P, which for infinite Q would also be infinite.

Figure A.2 is a Colpitts oscillator arrangement, simplified for the purpose of showing the circuit components. On the left side, we see the resonator tank circuit with the loss resistorR_P, and on the right side, we see the negative conductance, which is time dependent. The time dependence comes from the fact that the collector current is a series of pulses and the negative conductance is only present during a small period of time. This explanation is necessary to justify the existence of a loaded Q. If there would be a negative resistance or conductance present all the time, the compensating circuit would reduce the bandwidth to essentially zero or an infinite Q. In reality, however, for most of the time, the transistor "loads" the tuned circuit, and, therefore, the operating Q is less than the unloaded Q.

The following two circuits show the transition from a series tuned circuit connected with the series time-dependent negative resistance as outlined in Equation [1, Ch 6, 6.1] and the resulting input capacitance marked C_{IN}. Translated, the resulting configuration consists of a series circuit with inductance L and the resulting capacitance C'. The noise voltage $e_N(t)$ describes a small perturbation, which is the noise resulting from R_L and $-R_N(t)$.

Figure A.3 shows the equivalent representation of the oscillator circuit in the presence of noise.

The circuit equation of the oscillator circuit of Figure A.3 can be given as

$$L\frac{di(t)}{dt} + (R_L - R_N(t))i(t) + \frac{1}{C'}\int i(t)dt = e_N(t) \tag{A.1}$$

where $i(t)$ is the time varying resultant current. Due to the noise voltage $e_N(t)$, Eq. (A.1) is a nonhomogeneous differential equation. If the noise voltage is zero, it translates into a homogeneous differential equation.

For a noiseless oscillator, the noise signal $e_N(t)$ is zero and the expression of the free-running oscillator current $i(t)$ can be assumed to be a periodic function of time and can be given as

$$i(t) = I_1\cos(\omega t + \phi_1) + I_2\cos(2\omega t + \phi_2) + I_3\cos(3\omega t + \phi_3) + \dots + I_n\cos(n\omega t + \phi_n) \tag{A.2}$$

where $I_1, I_2, \dots, I_n$ are peak harmonic amplitudes of the current and $\varphi_1, \varphi_2, \dots, \varphi_n$ are time invariant phases.

In the presence of the noise perturbation $e_N(t)$, the current $i(t)$ may no longer be a periodic function of time and can be expressed as

$$\begin{aligned} i(t) &= I_1(t)\cos[\omega t + \phi_1(t)] + I_2(t)\cos[2\omega t + \phi_2(t)] + I_3(t)\cos[3\omega t + \phi_3(t)] \\ &+ \dots I_{n-2}(t)\cos[(n-2)\omega t + \phi_{n-2}(t)] + I_{n-1}(t)\cos[(n-1)\omega t + \phi_{n-1}(t)] \\ &+ I_n(t)\cos[n\omega t + \phi_n(t)], \end{aligned} \tag{A.3}$$

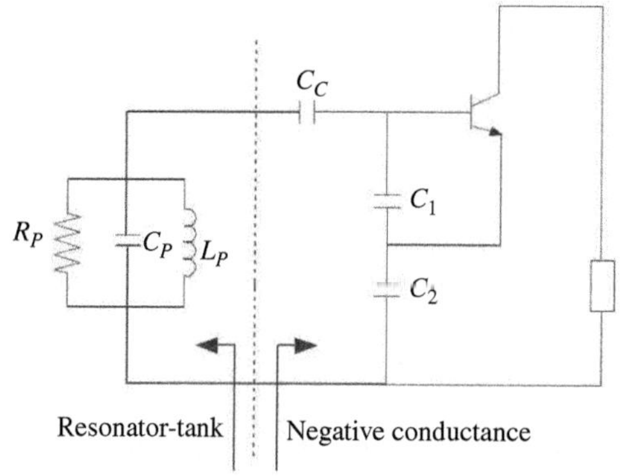

Figure A.2 Colpitts oscillator configuration for the intrinsic case, no parasitics assumed, and an ideal transistor considered.

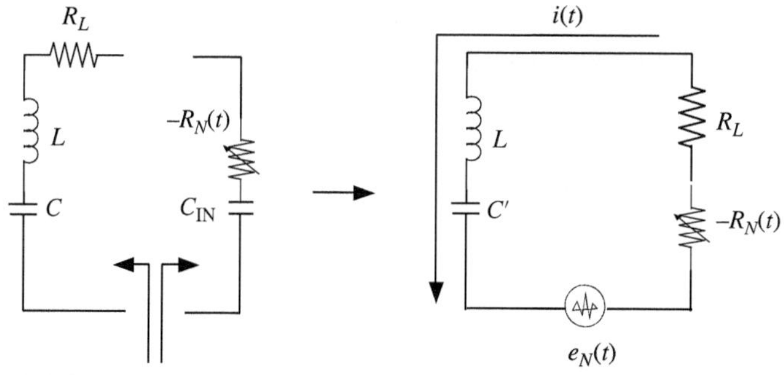

Figure A.3 Equivalent representation of the oscillator circuit in presence of noise.

where $I_1(t),\ I_2(t), \ldots, I_n(t)$ are time variant amplitudes of the current and $\varphi_1(t),\ \varphi_2(t), \ldots, \varphi_n(t)$ are time variant phases.

Considering that $I_n(t)$ and $\varphi_n(t)$ do not change much over the period of $2\pi/n\omega$; each corresponding harmonic over one period of oscillation cycle remains small and more or less variant. The solution of the differential equation becomes easy since the harmonics are suppressed due to a $Q > 10$, which prevents $i(t)$ to flow for the higher terms.

After the substitution of the value of $\frac{di}{dt}$ and $\int i(t)dt$, the complete oscillator circuit equation, as given in Eq. (A.1), can be rewritten as

$$\begin{aligned}
&L\left\{-I_1(t)\left(\omega + \frac{d\varphi_1(t)}{dt}\right)\sin[\omega t + \varphi_1(t)] + \frac{dI_1(t)}{dt}\cos[\omega t + \varphi_1(t)] +\right.\\
&-I_2(t)\left(2\omega + \frac{d\phi_2(t)}{dt}\right)\sin[2\omega t + \phi_2(t)] + \frac{dI_2(t)}{dt}\cos[2\omega t + \phi_2(t)] +\\
&-I_3(t)\left(3\omega + \frac{d\phi_3(t)}{dt}\right)\sin[3\omega t + \phi_3(t)] + \frac{dI_3(t)}{dt}\cos[3\omega t + \phi_3(t)] + \cdots\\
&\left.-I_n(t)\left(n\omega + \frac{d\varphi_n(t)}{dt}\right)\sin[n\omega t + \varphi_n(t)] + \frac{dI_n(t)}{dt}\cos[n\omega t + \varphi_n(t)]\right\} +\\
&[(R_L - R_N(t))i(t)] + \frac{1}{C'}\left\{\left[\frac{I_1(t)}{\omega} - \frac{I_1(t)}{\omega^2}\left(\frac{d\phi_1(t)}{dt}\right)\right]\sin[\omega t + \phi_1(t)]\right.\\
&\left.+\frac{1}{\omega^2}\left(\frac{dI_1(t)}{dt}\right)\cos[\omega t + \phi_1(t)]\right\}\\
&\frac{1}{C'}\left\{\left[\frac{I_2(t)}{2\omega} - \frac{I_2(t)}{4\omega^2}\left(\frac{d\phi_2(t)}{dt}\right)\right]\sin(2\omega t + \phi_2(t)) + \frac{1}{4\omega^2}\left(\frac{dI_2(t)}{dt}\right)\cos(2\omega t + \phi_2(t))\right\} +\\
&\frac{1}{C'}\left\{\left[\frac{I_3(t)}{3\omega} - \frac{I_3(t)}{9\omega^2}\left(\frac{d\phi_3(t)}{dt}\right)\right]\sin[3\omega t + \phi_3(t)] + \frac{1}{9\omega^2}\left(\frac{dI_3(t)}{dt}\right)\cos[3\omega t + \phi_3(t)]\right\} + \cdots\\
&\frac{1}{C'}\left\{\left[\frac{I_n(t)}{n\omega} - \frac{I_n(t)}{n^2\omega^2}\left(\frac{d\phi_n(t)}{dt}\right)\right]\sin[n\omega t + \phi_n(t)] + \frac{1}{n^2\omega^2}\left(\frac{dI_n(t)}{dt}\right)\cos[n\omega t + \phi_n(t)]\right\} = e_N(t)
\end{aligned} \tag{A.4}$$

After solving a set of complex equations, we determine

$\overline{R_N(t)}$ is the average negative resistance under large signal condition of the oscillator

$$\overline{R_N(t)} = \left[\frac{2}{T_0 I}\right]\int_{t-T_0}^{t} R_N(t)I(t)\cos^2[(\omega t + \phi)]dt \tag{A.5}$$

A.3 Calculation of the Region of the Nonlinear Negative Resistance

Under steady-state free-running oscillation condition,

$$\frac{dI(t)}{dt} \to 0$$

implies steady current, and

$$e_N(t) \to 0$$

with I is the fundamental RF current. Solving now the homogeneous differential equation for $R_L - R_N(t)$,we obtain

$$\left[\frac{2}{T_0}\right]\int_{t-T_0}^{t} e_N(t)\cos[\omega t + \phi(t)]dt = \frac{dI}{dt}\left[L + \frac{1}{\omega^2 C'}\right] + \left[R_L - \overline{R_N(t)}\right]I(t); \tag{A.6}$$

now we introduce γ; $\gamma = \frac{\Delta R}{\Delta I}$; for $\Delta \to 0$, $\gamma \to 0$ and

$$\left[R_L - \overline{R_N(t)}\right] = \gamma\Delta I,\ \gamma \to 0 \Rightarrow \left[R_L - \overline{R_N(t)}\right]I(t) \to 0 \tag{A.7}$$

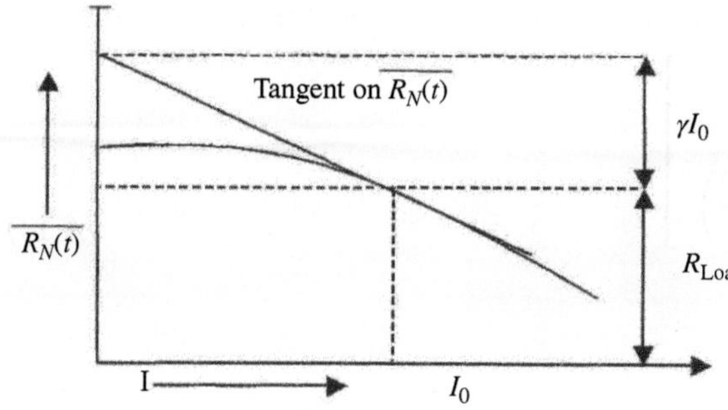

Figure A.4 Plot of negative resistance of $\left[\overline{R_N(t)}\right]$ vs. amplitude of current I.

$$R_L - \overline{R_N(t)} = R_{\text{Load}} - \left[\frac{2}{T_0}\right] \int_{t-T_0}^{t} R_N(t)\cos^2[\omega t + \phi(t)]dt \to 0 \tag{A.8}$$

$\left[R_L - \overline{R_N(t)}\right] I(t) \to 0$ gives the intersection of $\left[\overline{R_N(t)}\right]$ and $[R_L]$. This value is defined as I_0, which is the minimum value of the current needed for the steady-state sustained oscillation condition.

Figure A.4 shows the plot of the nonlinear negative resistance, which is a function of the amplitude of the RF current. As the RF amplitude gets larger, the conducting angle becomes narrower.

For a small variation of the current ΔI from I_0, the relation above is expressed as

$$\left[R_L - \overline{R_N(t)}\right] = \gamma \Delta I \tag{A.9}$$

$\gamma \Delta I$ can be found from the intersection on the vertical axis by drawing the tangential line on $\left[\overline{R_N(t)}\right]$ at $I = I_0$. $|\Delta I|$ decreases exponentially with time for $\gamma > 0$.

Hence, I_0 represents the stable operating point. On the other hand, if $\left[\overline{R_N(t)}\right]$ intersects $[R_L]$ from the other side for $\gamma < 0$ then $|\Delta I|$ grows indefinitely with time. Such an operating point does not support stable operation.

A.4 Calculation of the Noise Signal in Time Domain

From solving the two orthogonal equations, we need to obtain information about current $I(t)$ and $\varphi(t)$.

$$\left[\frac{2}{IT_0}\right] \int_{t-T_0}^{t} e_N(t)\sin[\omega t + \phi(t)]dt = -\frac{d\phi(t)}{dt}\left[L + \frac{1}{\omega^2 C'}\right] + \left[-\omega L + \frac{1}{\omega C'}\right] \tag{A.10}$$

$$\left[\frac{2}{T_0}\right] \int_{t-T_0}^{t} e_N(t)\cos[\omega t + \phi(t)]dt = \frac{dI(t)}{dt}\left[L + \frac{1}{\omega^2 C'}\right] + \left[R_L - \overline{R_N(t)}\right] I(t) \tag{A.11}$$

The analysis of the noise signal can be accomplished by decomposing the noise signal $e_N(t)$ to an infinite number of random noise pulses represented by

$$\varepsilon\delta(t - t_0) \tag{A.12}$$

where ε is the strength of the pulse at the time instant t_0, and both ε and t_0 are independent random variables from one pulse to next pulse!

The time average of the square of the current pulses over a period of time can be shown to be

$$\frac{1}{2T} \int_{-T}^{T} \left[\sum \varepsilon\delta(t - t_0)\right]^2 dt = \overline{e_N^2(t)} \tag{A.13}$$

$$|I(f)|^2 = \frac{|e_N(f)|^2}{8L^2}\left[\frac{1}{(\omega-\omega_0)^2+\left(\frac{|e_N(f)|^2}{4L^2I_0^2}\right)^2}+\frac{1}{(\omega+\omega_0)^2+\left(\frac{|e_N(f)|^2}{4L^2I_0^2}\right)^2}\right] \tag{A.14}$$

$$+\frac{|e_N(f)|^2}{8L^2}\left[\frac{1}{(\omega-\omega_0)^2+\left(\frac{\gamma I_0}{2L}\right)^2}+\frac{1}{(\omega+\omega_0)^2+\left(\frac{\gamma I_0}{2L}\right)^2}\right]$$

with

$$\frac{|e_N(f)|^2}{8L^2}\left[\frac{1}{(\omega-\omega_0)^2+\left(\frac{|e_N(f)|^2}{4L^2I_0^2}\right)^2}+\frac{1}{(\omega+\omega_0)^2+\left(\frac{|e_N(f)|^2}{4L^2I_0^2}\right)^2}\right]\rightarrow \text{FM noise} \tag{A.15}$$

$$\frac{|e_N(f)|^2}{8L^2}\left[\frac{1}{(\omega-\omega_0)^2+\left(\frac{\gamma I_0}{2L}\right)^2}+\frac{1}{(\omega+\omega_0)^2+\left(\frac{\gamma I_0}{2L}\right)^2}\right]\rightarrow \text{AM noise} \tag{A.16}$$

Since $\frac{\gamma I_0}{2L} >> \frac{2|e_N(\tau)|^2}{4L^2I_0^2}$.

For $\omega \rightarrow \omega_0$, FM noise dominates over the AM noise.

For $\omega >> \omega_0$, both the FM noise and AM noise terms give equal contribution.

Now the rest of the FM noise related calculation

Considering $\omega + \omega_0 >> \omega - \omega_0$, then

$$|I(f)|^2 = \frac{|e_N(f)|^2}{8L^2}\left[\frac{1}{(\omega-\omega_0)^2+\left(\frac{|e_N(f)|^2}{4L^2I_0^2}\right)^2}+\frac{1}{(\omega-\omega_0)^2+\left(\frac{\gamma I_0}{2L}\right)^2}\right] \tag{A.17}$$

$$P_{\text{noise}}(f) = 2R_L|I(f)|^2 \tag{A.18}$$

$$P_{\text{noise}}(f) = 2R_L\left(\frac{|e_N(f)|^2}{8L^2}\right)\left[\frac{1}{(\omega-\omega_0)^2+\left(\frac{|e_N(f)|^2}{4L^2I_0^2}\right)^2}+\frac{1}{(\omega-\omega_0)^2+\left(\frac{\gamma I_0}{2L}\right)^2}\right] \tag{A.19}$$

Since $R_{\text{Load}} = R_L + R_o$, the effective dynamic resistance of the free-running oscillator is given by

$$\sum_{\text{effective}} |R_{\text{tot}}| = R_N(t) - R_{\text{Load}}] = R_o \tag{A.20}$$

where R_o is the output resistance; $R_0 - R_{\text{tot}} = 0$.

The Q of the resonator circuit is expressed as

$$Q_L = \frac{\omega L}{R_0} \tag{A.21}$$

The oscillator output noise power in terms of Q is given by

$$P_{\text{noise}}(f) = \frac{\omega_0^2}{2Q_L^2}\frac{|e|^2}{2\overline{R_N(t)}}\left[\frac{1}{(\omega-\omega_0)^2 + \left(\frac{\omega_0^2}{4Q_L^2}\right)^2\left(\frac{|e|^2}{2\overline{R_N(t)}P_{\text{out}}^2}\right)^2} + \frac{1}{(\omega-\omega_0)^2 + \left(\frac{\omega_0^2}{Q_L^2}\right)\left(\frac{\gamma I_0}{2\overline{R_N(t)}}\right)^2}\right] \tag{A.22}$$

Figure A.5 shows the Colpitts oscillator with a series resonator and the small signal AC equivalent circuit. From the analytical expression of the noise analysis above, the influence of the circuit components on the phase noise can be explicitly calculated as

$$|\phi(f)|^2 = \frac{1}{4\omega^2 L^2 I_0^2(f)}|n_1(f)|^2 \tag{A.23}$$

$$\frac{1}{4\omega^2 L^2 I_0^2(f)}|n_1(f)|^2 = \frac{2|e_N(f)|^2}{4\omega^2 L^2 I_0^2(f)} \Rightarrow |\phi(f)|^2 = \frac{2|e_N(f)|^2}{4\omega^2 L^2 I_0^2(f)}, \tag{A.24}$$

where the frequency f varies from $-\infty$ to $+\infty$.

The resulting SSB phase noise is

$$L = \frac{|e_N(f)|^2}{4\omega^2 L^2 I_0^2(f)} \tag{A.25}$$

The unknown variables are $|e_N(f)|^2$ and $I_0^2(f)$, which need to be determined next.

$I_0^2(f)$ will be transformed into $I_{c0}^2(f)$ by multiplying $I_0^2(t)$ with the effective current gain $Y_{21}^+/Y_{11}^+ = \beta^+$.

Calculation of $I_{c0}^2(f)$

From Figure A.5, the LC-series resonant circuit is in shunt between the base and the emitter with the capacitive negative conductance portion of the transistor. We now introduce a collector load R_{Load} at the output, or better yet, an impedance Z.

The oscillator base current $i(t)$ is

$$i(t) = |I_0|\cos(\omega t) = \frac{V_{bc}(t)}{Z} \tag{A.26}$$

and the collector current is

$$|I_{c0}| = \left|\frac{[0.7 - V_{ce}]}{R_{\text{Load}} + j\left(\omega L - \frac{1}{\omega C_{\text{IN}}}\right)}\right| \approx \left|\frac{V_{ce}}{R_{\text{Load}} + j\left(\omega L - \frac{1}{\omega C_{\text{IN}}}\right)}\right| \text{ or} \tag{A.27}$$

$$\overline{I_{c0}^2(f)} \approx \left\{\frac{\overline{V_{ce}^2(f)}}{[R_{\text{Load}}]^2 + \left(\omega L - \frac{1}{\omega C_{\text{IN}}}\right)^2}\right\} = \left\{\frac{\overline{V_{ce}^2(f)}}{\left[\frac{\omega L}{Q}\right]^2 + \left(\omega L - \frac{1}{\omega C_{\text{IN}}}\right)^2}\right\} \tag{A.28}$$

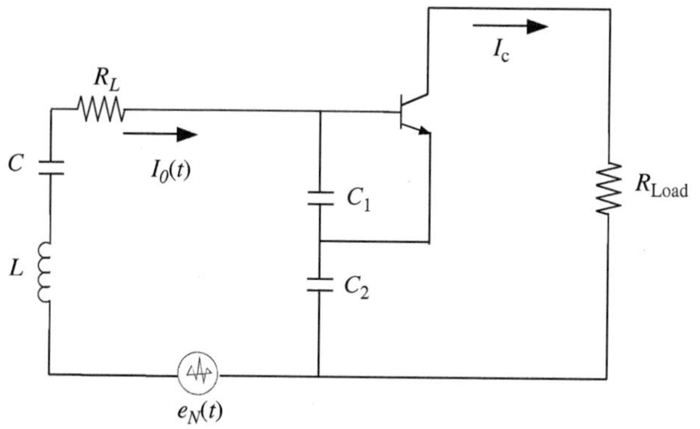

Figure A.5 Colpitts oscillator with series resonator and small signal AC equivalent circuit.

The voltage V_{ce} is the RF voltage across the collector-emitter terminals of the transistor. Considering the steady-state oscillation $\omega \to \omega_0$, the total loss resistance is compensated by the negative resistance of the active device as $R_L = \overline{R_N(t)}$. The expression of $\left|\overline{I_{c0}^2(f)}\right|_{\omega=\omega_0}$ is

$$\left|\overline{I_{c0}^2(f)}\right|_{\omega=\omega_0} = \left|\frac{\overline{V_{ce}^2(f)}}{\left[\frac{\omega_0 L}{Q}\right]^2 + \left(\omega_0 L - \frac{1}{\omega_0 C_{IN}}\right)^2}\right| = \left|\frac{\overline{V_{ce}^2(f)}}{(\omega_0 L)^2\left[\frac{1}{Q^2} + \left(1 - \frac{1}{\omega_0^2 LC_{IN}]}\right)^2\right]}\right| \tag{A.29}$$

$$\left|\overline{I_{c0}^2(f)}\right|_{\omega=\omega_0} = \left|\frac{\overline{V_{ce}^2(f)}}{(\omega_0 L)^2\left[\frac{1}{Q^2} + \left(1 - \frac{1}{\omega_0^2 L}\frac{C_1+C_2}{C_1C_2}\right)^2\right]}\right|, \tag{A.30}$$

where C_{IN} is the equivalent capacitance of the negative resistor portion of the oscillator circuit.

$$C\prime = \frac{CC_{IN}}{C + C_{IN}},\ C_{IN} = \frac{C_1C_2}{C_1 + C_2} \tag{A.31}$$

$$Q = \frac{\omega L}{R_L} \tag{A.32}$$

For a reasonably high Q resonator $\left|\overline{I_{c0}^2(f)}\right|_{\omega=\omega_0} \propto [C_{IN}]_{\omega=\omega_0}$

A.5 Calculation of the Noise Voltage $e_N(f)$

The equivalent noise voltage from the negative resistance portion of the oscillator circuit is given an open-circuit noise voltage (EMF) of the circuit as shown in Figure A.6.

The noise voltage associated with the resonator loss resistance R_s is

$$\left|\overline{e_R^2(f)}\right|_{\omega=\omega_0} = 4k\text{TBR}_s \tag{A.33}$$

R_s denotes the equivalent series loss resistor, which can be calculated from the parallel loading resistor R_{load}, see Figure A.6.

$$\left|\overline{e_R^2(f)}\right|_{\omega=\omega_0} = 4kTR \text{ For B} = 1\text{ Hz bandwidth} \tag{A.34}$$

The total noise voltage power within 1 Hz bandwidth can be given as

$$\left|\overline{e_N^2(f)}\right|_{\omega=\omega_0} = \overline{e_R^2(f)} + \overline{e_{NR}^2(f)} = [4kTR] + \left[\frac{4qI_c g_m^2 + \frac{K_f I_b^{AF}}{\Delta\omega} g_m^2}{\omega_0^2 C_1^2\left(\omega_0^2(\beta^+)^2 C_2^2 + g_m^2 \frac{C_2^2}{C_1^2}\right)}\right] \tag{A.35}$$

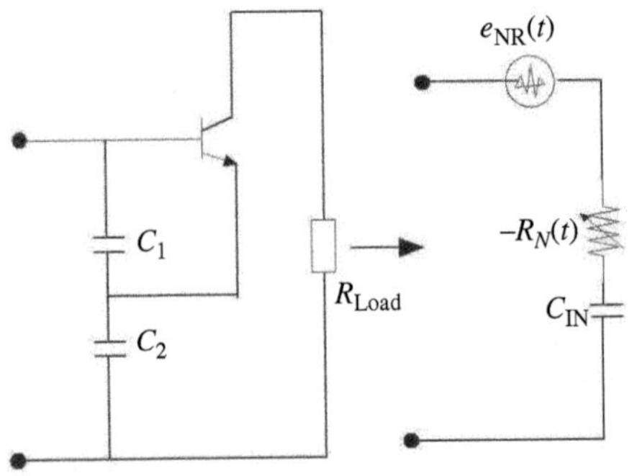

Figure A.6 Equivalent representation of negative resistance portion of the circuit at the input for the open circuit noise voltage.

Derivation of Eq. (A.35):

The total noise voltage power within 1 Hz bandwidth can be given as

$$\left|\overline{e_N^2(f)}\right|_{\omega=\omega_0} = \overline{e_R^2(f)} + \overline{e_{NR}^2(f)} \tag{A.36}$$

The first term in Eq. (A.36) is the noise voltage power due to the loss resistance R, and the second term is associated with the negative resistance of the active device R_N.

Figures A.7 and A.8 illustrate the oscillator circuit for the purpose of the calculation of the negative resistance.

From Figure A.8, the circuit equation is given from Kirchhoff's voltage law (KVL) as

$$V_{\text{in}} = I_{\text{in}}(X_{C_1} + X_{C_2}) - I_b(X_{C_1} - \beta X_{C_2}) \tag{A.37}$$

$$0 = -I_{\text{in}}(X_{C_1}) + I_b(X_{C_1} + h_{ie}) \tag{A.38}$$

Considering, $\dfrac{1}{Y_{11}} = h_{ie}$

$$Z_{\text{in}} = \frac{V_{\text{in}}}{I_{\text{in}}} = \frac{(1+\beta)X_{C_1}X_{C_2} + h_{ie}(X_{C_1} + X_{C_2})}{X_{C_1} + h_{ie}} \tag{A.39}$$

$$Z_{\text{in}} = \frac{\left(-\frac{(1+\beta)}{\omega^2 C_1 C_2} + \frac{(C_1+C_2)}{j\omega C_1 C_2}\frac{1}{Y_{11}}\right)}{\left(\frac{1}{Y_{11}} + \frac{1}{j\omega C_1}\right)} \tag{A.40}$$

$$Z_{\text{in}} = \frac{-jY_{11}(1+\beta) + \omega(C_1 + C_2)}{\omega C_2(Y_{11} + j\omega C_1)} \tag{A.41}$$

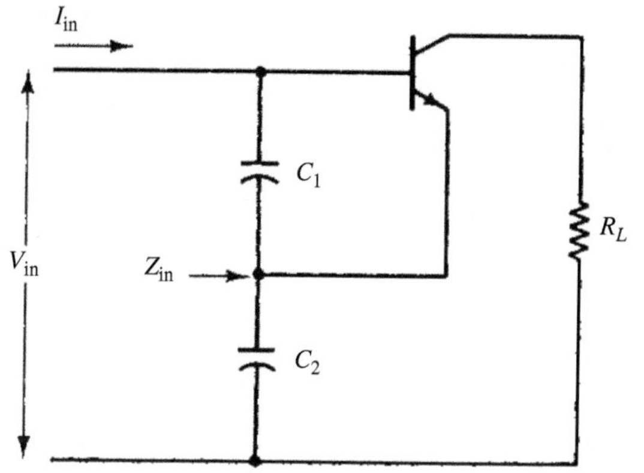

Figure A.7 Oscillator circuit for the calculation of the negative resistance.

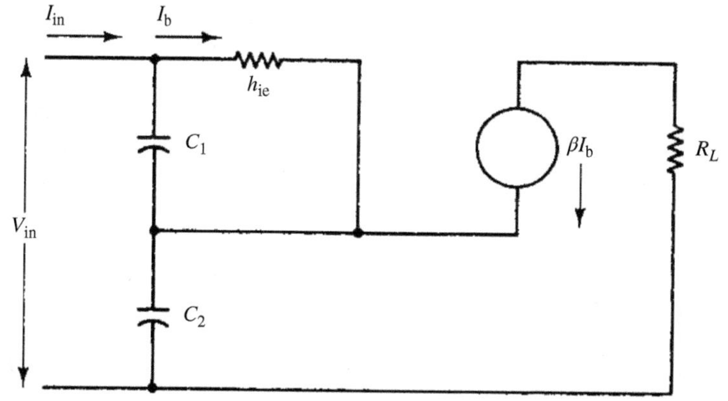

Figure A.8 Equivalent oscillator circuit for the calculation of the negative resistance.

$$Z_{\text{in}} = \frac{[\omega(C_1 + C_2) - jY_{11}(1+\beta)][Y_{11} - j\omega C_1]}{\omega C_2\left(Y_{11}^2 + \omega^2 C_1^2\right)} \tag{A.42}$$

$$Z_{\text{in}} = \left[\frac{\omega Y_{11}(C_1 + C_2) - (1+\beta)\omega C_1 Y_{11}}{\omega C_2\left(Y_{11}^2 + \omega^2 C_1^2\right)}\right] - j\left[\frac{Y_{11}^2(1+\beta) + \omega^2 C_1(C_1 + C_2)}{\omega C_2\left(Y_{11}^2 + \omega^2 C_1^2\right)}\right] \tag{A.43}$$

$$Z_{\text{in}} = -R_n - jX \tag{A.44}$$

$$R_n = \frac{(1+\beta)\omega C_1 Y_{11} - \omega Y_{11}(C_1 + C_2)}{\omega C_2\left(Y_{11}^2 + \omega^2 C_1^2\right)} = \frac{(1+\beta)C_1 Y_{11} - Y_{11}(C_1 + C_2)}{C_2\left(Y_{11}^2 + \omega^2 C_1^2\right)} \tag{A.45}$$

$$R_n = \frac{\beta C_1 Y_{11} - Y_{11} C_2}{C_2\left(Y_{11}^2 + \omega^2 C_1^2\right)} = \frac{\beta Y_{11}}{\frac{C_2}{C_1}\left(Y_{11}^2 + \omega^2 C_1^2\right)} - \frac{Y_{11}}{\left(Y_{11}^2 + \omega^2 C_1^2\right)} \tag{A.46}$$

Considering $\beta = \dfrac{Y_{21}}{Y_{11}} \approx \dfrac{g_m}{Y_{11}}$

$$R_n = \frac{g_m}{\frac{C_2}{C_1}\left(\frac{g_m^2}{\beta^2} + \omega^2 C_1^2\right)} - \frac{g_m/\beta}{\left(\frac{g_m^2}{\beta^2} + \omega^2 C_1^2\right)} \tag{A.47}$$

$$R_n = \frac{g_m \beta^2 C_1}{\left(g_m^2 C_2 + \omega^2 \beta^2 C_1^2 C_2\right)} - \frac{g_m \beta}{\left(g_m^2 + \beta^2 \omega^2 C_1^2\right)} \tag{A.48}$$

$$R_n = \frac{g_m \beta^2 \omega^2 C_1 C_2}{\omega^2 C_1^2\left(\frac{C_2^2}{C_1^2} g_m^2 + \omega^2 \beta^2 C_2^2\right)} - \frac{g_m \beta \omega^2 C_2^2}{\omega^2 C_1^2\left(\frac{C_2^2}{C_1^2} g_m^2 + \beta^2 \omega^2 C_2^2\right)} \tag{A.49}$$

$$R_n = \left[\frac{g_m^2}{\omega^2 C_1^2\left(\frac{C_2^2}{C_1^2} g_m^2 + \omega^2 \beta^2 C_2^2\right)}\right]\left[\frac{\beta^2 \omega^2 C_1 C_2}{g_m} - \frac{\beta \omega^2 C_2^2}{g_m}\right] \tag{A.50}$$

$$R_n = \left[\frac{g_m^2}{\omega^2 C_1^2\left(\frac{C_2^2}{C_1^2} g_m^2 + \omega^2 \beta^2 C_2^2\right)}\right]\left[g_m\left[\left(\frac{\omega C_1}{Y_{11}}\right)\left(\frac{\omega C_2}{Y_{11}}\right) - \frac{\omega^2 C_2^2}{\beta Y_{11}^2}\right]\right] \tag{A.51}$$

$$R_n = \left[\frac{g_m^2}{\omega^2 C_1^2\left(\frac{C_2^2}{C_1^2} g_m^2 + \omega^2 \beta^2 C_2^2\right)}\right]\left[g_m\left[\left(\frac{\omega C_1}{Y_{11}}\right)\left(\frac{\omega C_2}{Y_{11}}\right) - \frac{1}{\beta}\left(\frac{\omega C_2}{Y_{11}}\right)\left(\frac{\omega C_2}{Y_{11}}\right)\right]\right] \tag{A.52}$$

$$\text{Considering}\left(\frac{\omega C_1}{Y_{11}}\right)\left(\frac{\omega C_2}{Y_{11}}\right) >> \frac{1}{\beta}\left(\frac{\omega C_2}{Y_{11}}\right)\left(\frac{\omega C_2}{Y_{11}}\right) \text{ and } \left(\frac{\omega C_1}{Y_{11}}\right)\left(\frac{\omega C_2}{Y_{11}}\right) \approx 1 \tag{A.53}$$

$$\left[g_m\left[\left(\frac{\omega C_1}{Y_{11}}\right)\left(\frac{\omega C_2}{Y_{11}}\right) - \frac{1}{\beta}\left(\frac{\omega C_2}{Y_{11}}\right)\left(\frac{\omega C_2}{Y_{11}}\right)\right]\right] \cong \frac{I_C}{V_T} = \frac{I_C}{kT/q} \Rightarrow \frac{qI_C}{kT} \tag{A.54}$$

From (A.52) and (A.54)

$$R_n = \left[\frac{g_m^2}{\omega^2 C_1^2\left(\frac{C_2^2}{C_1^2} g_m^2 + \omega^2 \beta^2 C_2^2\right)}\right]\frac{qI_C}{kT} \tag{A.55}$$

From (A.36) the total noise voltage power within 1 Hz bandwidth can be given as

$$\left|\overline{e_N^2(f)}\right|_{\omega=\omega_0} = \overline{e_R^2(f)} + \overline{e_{NR}^2(f)} \tag{A.56}$$

$$\left|\overline{e_N^2(f)}\right|_{\omega=\omega_0} = [4kTR] + \left[\frac{4qI_c g_m^2 + \frac{K_f I_b^{AF}}{\omega} g_m^2}{\omega_0^2 C_1^2\left(\omega_0^2 (\beta+)^2 C_2^2 + g_m^2 \frac{C_2^2}{C_1^2}\right)}\right], \tag{A.57}$$

Where

$$\beta^+ = \left[\frac{Y_{21}^+}{Y_{11}^+}\right]\left[\frac{C_1}{C_2}\right]^p, g_m = [Y_{21}^+]\left[\frac{C_1}{C_2}\right]^q, \text{ redefined}$$

$$\omega = 2\pi f$$

The values of p and q depend upon the drive level.

$$\left|\overline{e_N^2(f)}\right|_{\omega=\omega_0} = [4kTR] + \left[\frac{4qI_{c0}g_m^2 + \frac{K_f I_b^{AF}}{\Delta\omega} g_m^2}{\omega_0^2 C_1^2 \left(\omega_0^2(\beta^+)^2 C_2^2 + g_m^2 \frac{C_2^2}{C_1^2}\right)}\right], \tag{A.58}$$

where

$$\beta^+ = \left[\frac{Y_{21}^+}{Y_{11}^+}\right]\left[\frac{C_1}{C_2}\right]^p$$

$$g_m = [Y_{21}^+]\left[\frac{C_1}{C_2}\right]^q, \text{ redefined}$$

The values of p and q depend upon the drive level.

The flicker noise contribution in Eq. (A.36) is introduced by adding term $\frac{K_f I_b^{AF}}{\omega}$ in I_{c0}, where K_f is the flicker noise coefficient and AF is the flicker noise exponent. This is valid only for the bipolar transistor. For an FET, the equivalent currents have to be used.

The first term in the expression above is related to the thermal noise due to the loss resistance of the resonator tank, and the second term is related to the shot noise and flicker noise in the transistor.

Now, the phase noise of the oscillator can be expressed as

$$\left|\overline{\phi^2(\omega)}\right| = \frac{2\left|\overline{e_N^2(\omega)}\right|}{4\omega_0^2 L^2 I_0^2(\omega)} \tag{A.59}$$

$$\left|\overline{\phi^2(\omega)}\right|_{\text{SSB}} = \frac{1}{2}\left|\overline{\phi^2(\omega)}\right| = \frac{\left|\overline{e_N^2(\omega)}\right|}{4\omega_0^2 L^2 I_0^2(\omega)} \tag{A.60}$$

$$\left|\overline{\phi^2(\omega)}\right|_{\text{SSB}} = \left\{[4kTR] + \left[\frac{4qI_c g_m^2 + \frac{K_f I_b^{AF}}{\omega} g_m^2}{\omega_0^2 C_1^2 \left(\omega_0^2(\beta^+)^2 C_2^2 + g_m^2 \frac{C_2^2}{C_1^2}\right)}\right]\right\}\left[\frac{(\omega_0)^2\left[\frac{1}{Q^2} + \left(1 - \frac{1}{\omega_0^2 L}\frac{C_1+C_2}{C_1C_2}\right)^2\right]}{4\omega^2\left|V_{ce}^2(\omega)\right|}\right] \tag{A.61}$$

$$\left|\overline{\phi^2(\omega)}\right|_{\text{SSB}} = \left[4kTR + \left[\frac{4qI_c g_m^2 + \frac{K_f I_b^{AF}}{\omega} g_m^2}{\omega_0^2 C_1^2 \left(\omega_0^2(\beta^+)^2 C_2^2 + g_m^2 \frac{C_2^2}{C_1^2}\right)}\right]\right]\left[\frac{\omega_0^2}{4\omega^2 V_{ce}^2}\right]\left[\frac{1}{Q^2} + \left(1 - \frac{1}{\omega_0^2 L}\frac{C_1+C_2}{C_1C_2}\right)^2\right] \tag{A.62}$$

Considering $\left(\frac{1}{\omega_0^2 L}\frac{C_1+C_2}{C_1C_2}\right) >> 1$; For $\omega_0 = 2\pi f = 6.28 \times 10^9$ Hz, $L = 1 \times 10^{-9}$ H, $C_1 = 1 \times 10^{-12}$ F, and $C_2 = 1 \times 10^{-12}$ F.

$$\left(\frac{1}{\omega_0^2 L}\frac{C_1+C_2}{C_1C_2}\right) = 50.7$$

Since the phase noise is always expressed in dBc/Hz, we now calculate, after simplification of Eq. (A.40),

$$£(\omega) = 10\text{Log}\left\{\left[4kTR + \left[\frac{4qI_c g_m^2 + \frac{K_f I_b^{AF}}{\omega} g_m^2}{\omega_0^2 C_1^2 \left(\omega_0^2(\beta^+)^2 C_2^2 + g_m^2 \frac{C_2^2}{C_1^2}\right)}\right]\right]\left[\frac{\omega_0^2}{4\omega^2 V_{cc}^2}\right]\left[\frac{1}{Q^2} + \frac{[C_1+C_2]^2}{C_1^2 C_2^2 \omega_0^4 L^2}\right]\right\} \tag{A.63}$$

For the bias condition (which is determined from the output power requirement), the loaded quality factor, and the device parameters (transconductance and β^+), the best phase noise can be found by differentiating $\left|\overline{\phi^2(\omega)}\right|_{SSB}$ with respect to $\frac{C_1}{C_2}$.

Considering that all the parameters of $\left|\overline{\phi^2(\omega)}\right|_{SSB}$ are constants for a given operating condition (except the feedback capacitor), the minimum value of the phase noise can be determined for any fixed value of C_1 as

$$\left|\overline{\varphi^2(\omega)}\right| = \left[k_0 + \frac{k_1}{k_2C_1^2C_2^2 + k_3C_2^2}\right]\left[\frac{C_1+C_2}{C_1C_2}\right]^2 \tag{A.64}$$

$$k_0 = \frac{kTR}{\omega^2\omega_0^2L^2V_{ce}^2} \tag{A.65}$$

$$k_1 = \frac{qI_{c0}g_m^2 + \frac{K_fI_b^{AF}}{\omega}g_m^2}{\omega^2\omega_0^2L^2V_{ce}^2} \tag{A.66}$$

$$k_2 = \omega_0^4(\beta^+)^2 \tag{A.67}$$

$$k_3 = g_m^2 \tag{A.68}$$

where k_1, k_2, and k_3, are constant only for a particular drive level, with $y = \frac{C_1}{C_2}$.

Making k_2 and k_3 also dependent on y, as the drive level changes, the final noise equation is

$$£(\omega) = 10 \times \log\left[\left[k_0 + \left(\frac{k^3k_1\left[\frac{Y_{21}^+}{Y_{11}^+}\right]^2[y]^{2p}}{\left[Y_{21}^+\right]^3[y]^{3q}}\right)\left(\frac{1}{(y^2+k)}\right)\right]\left[\frac{[1+y]^2}{y^2}\right]\right] \tag{A.69}$$

where

$$k_0 = \frac{kTR}{\omega^2\omega_0^2L^2V_{cc}^2}$$

$$k_1 = \frac{qI_cg_m^2 + \frac{K_fI_b^{AF}}{\omega}g_m^2}{\omega^2\omega_0^4L^2V_{cc}^2}$$

$$k_2 = \omega_0^2(\beta^+)^2$$

A.6 Practical Examples

The following section is based on the work published in the References [5–8].

To predict the phase noise of the oscillator shown in Figure A.9, the Harmonica Simulator (Ansoft) was used and duplicated the details of the actual circuit as seen in Figure A.10.

Now, we go to 430 MHz (see Figure A.11) we look at three critical values 1, 10, and 100 KHz. They are −120, −140, and −160 dBc/Hz. The output available after analysis will be the noise sideband spectrum, NS1<H1>, the phase noise, PN1<H1>, and the amplitude noise, AN1<H1>.

And finally, Figure A.12a shows the simulated output power of the 430 MHz Oscillator and the simulation results of the three noise spectrums showing the simulated phase noise of a free-running JFET-based 960 MHz oscillator using a helix resonator in Figure A.12b.

Red curve (top trace) shows the standard phase noise, the green (just below the top trace) one shows the spectral density, and the purple curve (lowest trace) shows the amplitude noise, which means at 150 kHz off the carrier, the AM noise dominates. This fact is commonly not attended to, just ignored.

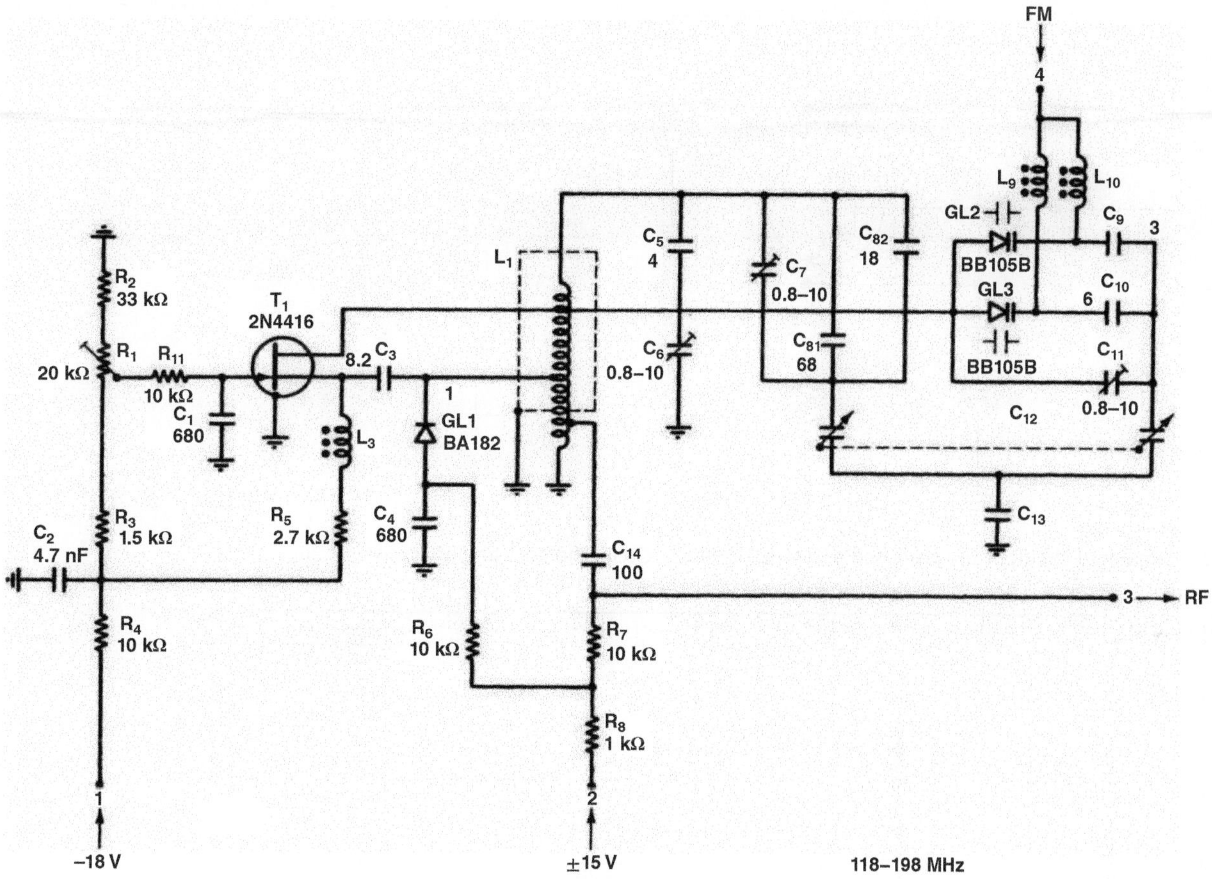

Figure A.9 Schematic diagram of the R&S SMDU (signal generator) oscillator section.

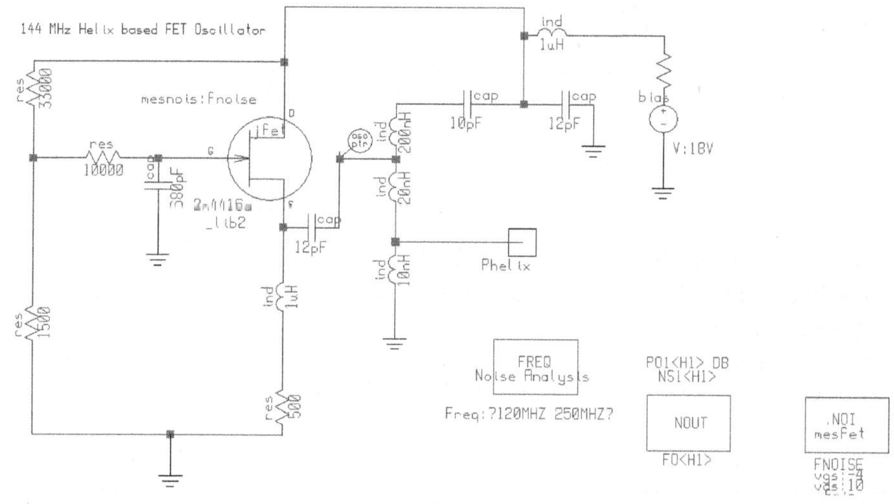

Figure A.10 144 MHz FET oscillator circuit for simulation.

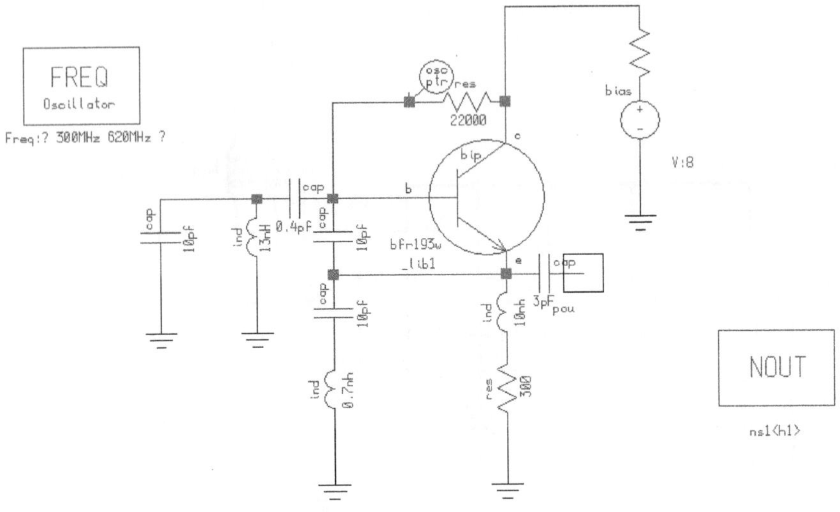

Figure A.11 Simulated circuit of the proposed 430 MHz Oscillator circuit, no post amplifier.

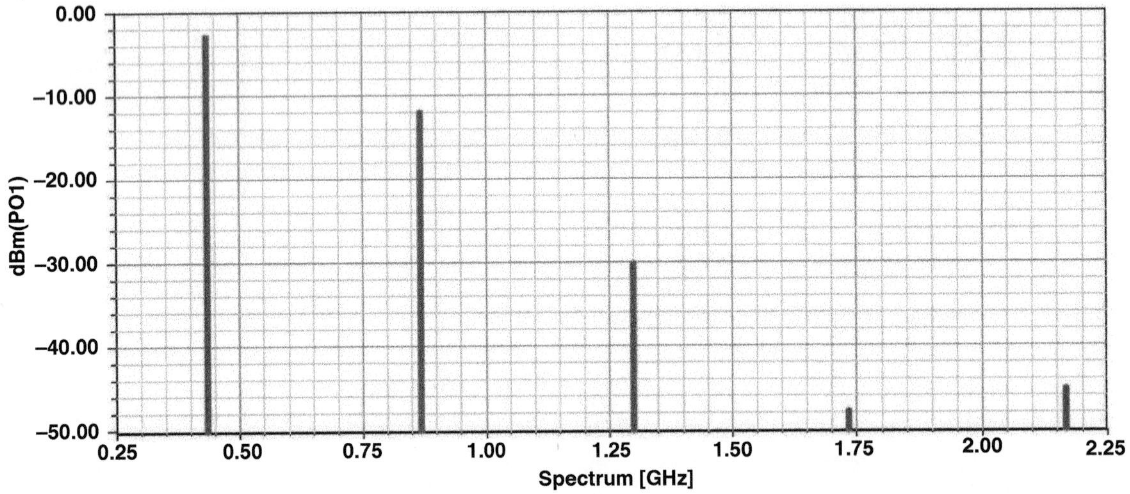

(a)

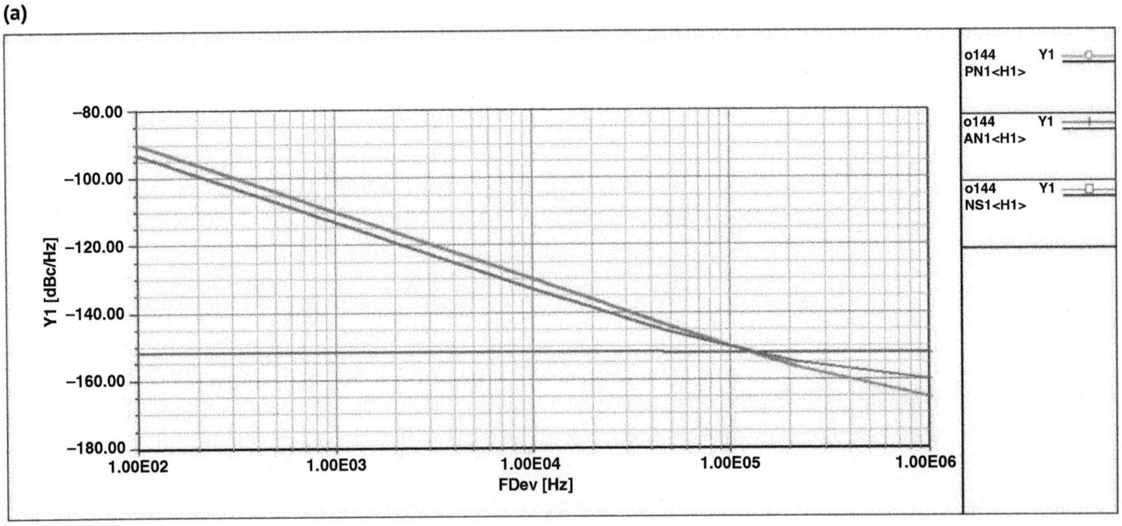

(b)

Figure A.12 (a) Simulated output power of the 430 MHz oscillator; (b) simulated phase noise of a free-running JFET-based 960 MHz oscillator using a helix resonator.

Because of the influence of the tuning diodes the measured noise differs from the prediction. Figure A.13 shows the measured results, where the blue line (top) is the FM noise and the gray (bottom) is the AM noise. The AM noise is caused by the ALC circuit and does not affect the performance significantly

The AM noise contribution of a digitally generated signal has a huge contribution. In Figure A.14, the blue line (top) is the Phase Noise, typically shown in the datasheet, but the almost similar AM noise is not mentioned.

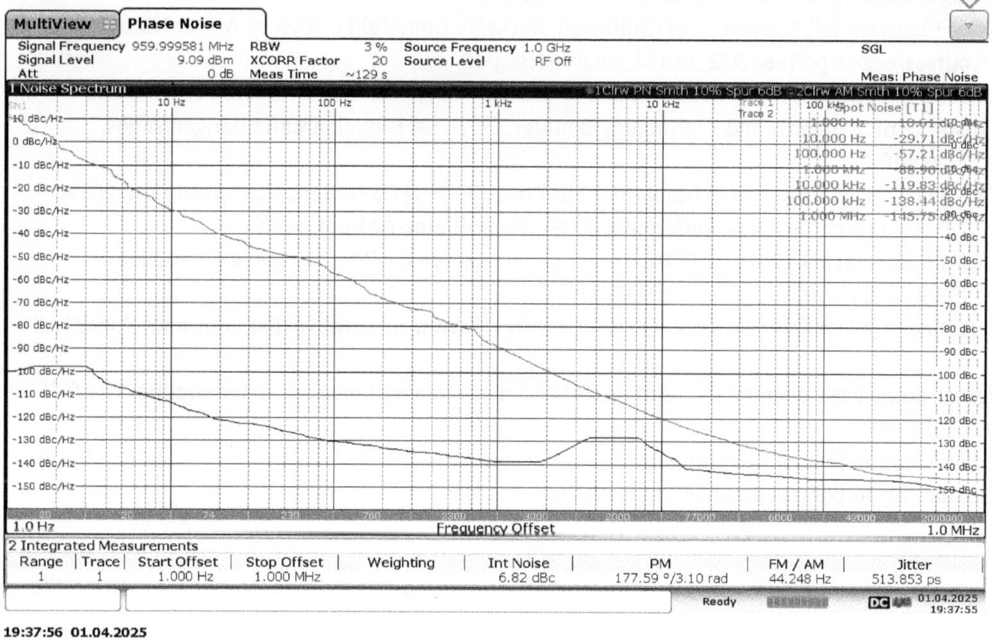

Figure A.13 Measured phase noise of a free-running JFET-based 960 MHz oscillator using a helix resonator show the influence of the tuning diodes—the blue line (top trace) is the FM noise and the gray (bottom trace) is the AM noise

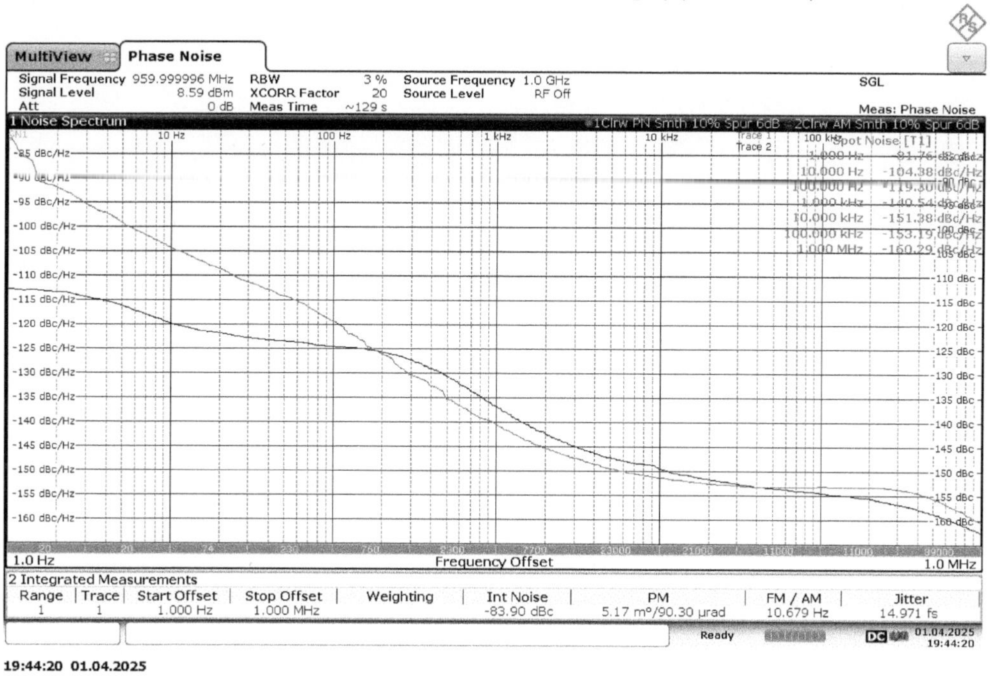

Figure A.14 Measurement of the AM and FM noise contribution of a synthesized 960 MHz oscillator. Top—FM noise, bottom—AM noise.

At the crossover point of the AM and FM noise, it can be seen that the AM noise dominates thereafter.
If using this generator to measure the ACPR, one would get a poor value for a probably much better device under test.

References

1 Rohde, U.L., Poddar, A.K. and Boeck, G. (2005, May). *The Design of Modern Microwave Oscillators for Wireless Applications: Theory and Optimization*. New York: John Wiley & Sons. ISBN 0-471-72342-8.

2 Prof. Ali M. Niknejad, "Oscillator Phase Noise", University of California, Berkeley Copyright c 2009 by Ali M. Niknejad https://rfic.eecs.berkeley.edu/courses/ee242/pdf/eecs242_lect22_phasenoise.pdf

3 Correlation Measurements between PM and AM Noise in Oscillators A. Hati, C. W. Nelson and D. A. Howe, Time and Frequency Division National Institute of Standards and Technology Boulder, CO/USA E-mail: archita.hati@nist.gov, https://tf.nist.gov/general/pdf/2724.pdf

4 E. Rubiola and R. Brendel, "The AM noise mechanism in oscillators," in *Proc. EFTF–IFCS*, Besançon, France, Apr. 2009, pp. 33–39. [hal.science] https://rubiola.org/pdf-articles/conference/2009-ifcs-oscillator-AM-noise.pdf

5 Rohde, U.L. and Rudolph, M. (2012, December). (2013). *RF/Microwave Circuit Design for Wireless Applications*, 2e. New York: John Wiley & Sons. ISBN 10: 0470901810/ISBN 13: 9780470901816

6 Microwaves101 (2011), Receiver Blocking. Available at: Microwaves 101. https://www.microwaves101.com/encyclopedias/receiver-blocking

7 Kurokawa, K. (1969, July–August). Some basic characteristics of broadband negative resistance oscillator circuits. *The Bell System Technical Journal* 48 (6): 1937–1955.

8 Rohde, U.L. (2004). A new and efficient method of designing low noise microwave oscillators. https://depositonce.tu-berlin.de/bitstream/11303/1306/1/Dokument_16.pdf (accessed 16 March 2026).

B

Modeling of a Two-stage Amplifier Operating from 2 to 20 GHz

To show the real complexity of designing an amplifier over such a frequency range, an example of how this is accomplished is given below. Most publications show only a simplified schematic and provide no insight into the parasitics, where the real magic lies. Therefore, although fairly lengthy, the following shows the actual full level of detail required to build such an amplifier. At the time this work was performed, no schematic capture program was available in the CAD tool. We invited interested parties to solve the puzzle and enter it into our modern tool. The amplifier shown in Figure B.1, based on the simulation, covers essentially 2–20 GHz. The simulated noise figure at the time was about 5 dB, and the field-effect transistor (FET) sizes were 150 µm and 250 µm.

```
The netlist of the simulation for the two-stage amplifier is given below:
blk
  trl 1  2          w= 12.5µµm          p= 127µm       gaas
  trl 2  3  4  5  z= (28.914/63.5) p= 209.0166µµm k=6.8
  ind 4  6  l=0.02nh
  ind 5  6  l=0.02nh
gcap:2por 1  6
end
```

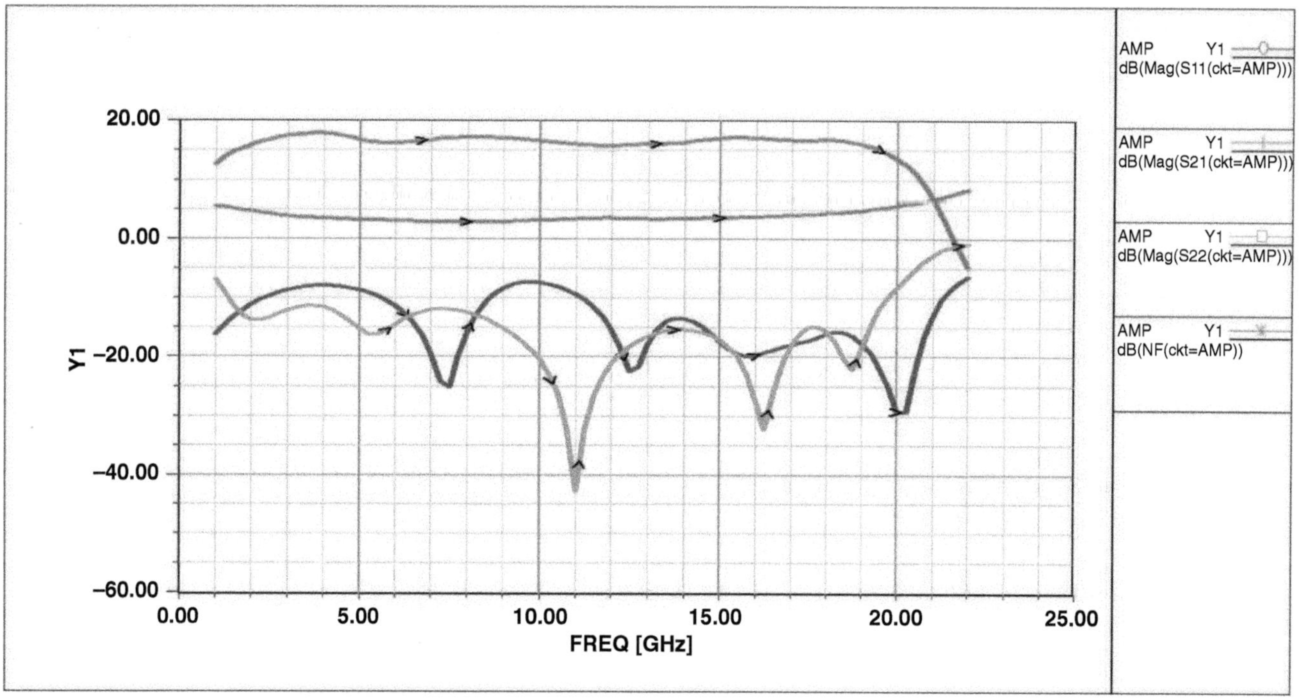

Figure B.1 Frequency response of a two-stage amplifier operating from 2 to 20 GHz.

```
*
*
* ====================== [ 125μm fet ] ====================== *
*
blk
  fet 1  2  3 rg=5.2 cdg=.025pf cgs=.113pf g=.022 ri=5.72
+   gds=.00426 cds=.0136pf rd=0.721 rs=0.396 t=4e-12
  fet 4  5  2 rg=5.2 cdg=.0076pf cgs=.075pf g=.022 ri=5.72
+   gds=.00426 cds=.0136pf rd=1.44 rs=0.2 t=4e-12
* --------cg1d2-------
  cap 1  5 c=.0025pf
* ------------------gate 2 bypass cap (cg2s1)------------------ *
  gcap   4  9
  gcap   4 10
* --- via inductance <shared> with gate2 feedback inclusive --- *
  ind 3  7 l=.015nh
  ind 3  8 l=.015nh
* --------cg1g2-------
  cap 1  4 c=.0062pf
* --------outer cascode drain to source cap (cd2s1)------------ *
  cap 5  3 c=.039pf
* ------------gate 2 bias network from cell library------------ *
  trl 4 40 w= 17.78μm  p= 203.2μm  gaas
  res 40 0 r=988
* -----------------drain neutralizing network------------------ *
  cap 5  6 c=.005pf
  tfr 6  3 w= 20.32μm  p= 76.2μm  rs=400 gaas
* ------------------------------------------------------------- *
  ind 7  9 l=0.015nh
  ind 8 10 l=0.015nh
125μμmmod:4por 1  5  9  10
end
*
*
* ====================== [ 250μm fet ] ====================== *
*
blk
  fet 1  2  3 rg=2.6 cdg=.05pf cgs=.226pf g=.044 ri=2.86
+   gds=.00852 cds=.0272pf rd=0.361 rs=0.20 t=4e-12
  fet 4  5  2 rg=2.6 cdg=.0152pf cgs=.15pf g=.044 ri=2.86
+   gds=.00852 cds=.0272pf rd=0.72 rs=0.10 t=4e-12
* --------cg1d2------- *
  cap 1  5 c=.005pf
* --------gate 2 bypass cap (cg2s1)--------- *
  gcap   4  9
  gcap   4 10
* --- via inductance <shared> with gate2 feedback inclusive --- *
  ind 3  7 l=.015nh
  ind 3  8 l=.015nh
* --------cg1g2------- *
  cap 1  4 c=.0123pf
* -----------outer cascode drain to source cap (cd2s1)--------- *
```

```
  cap 5  3 c=.070pf
* -----------gate 2 bias network from cell library----------- *
  trl 4 40 w= 17.78µm  p= 203.2µm  gaas
  res 40 0 r=988
* ----------------drain neutralizing network------------------ *
  cap  5  6 c=.005pf
  tfr  6  3 w= 33.02µm  p= 76.2µm  rs=400 gaas
* ----------------------------------------------------------- *
  ind 7  9 l=0.015nh
  ind 8 10 l=0.015nh
250µmmod:4por 1  5  9  10
end
*
* ========================================================== *
*
*
blk
  tee     2  2  1  w1=40µm  w2=40µm  w3=30.5µm  gaas
  trl 2  3  5  6  z= (28.914/66.0)  p= 55.6µm  k=6.8
  trl 3  4  6  7  z= (28.914/66.0)  p= 55.6µm  k=6.8
  trl 4  9  7  10 z= (28.914/66.0)  p= 111.1µm k=6.8
  srl 5  8  r=0.1oh  l=0.02nh
  srl 7  8  r=0.1oh  l=0.01nh
  srl 10 8  r=0.1oh  l=0.02nh
  srl 8  0  r=0.02oh l=0.011nh
c4r4: 1por  2
end
*
blk
  tee     2  2  1  w1=57µm  w2=57µm  w3=25.0µm  gaas
  trl 2  3  5  6  z= (28.914/114.0) p= 73.0µm  k=6.8
  trl 3  4  6  7  z= (28.914/114.0) p= 73.0µm  k=6.8
  trl 4  9  7  10 z= (28.914/114.0) p= 146.0µm k=6.8
  srl 5  8  r=0.1oh  l=0.02nh
  srl 7  8  r=0.1oh  l=0.01nh
  srl 10 8  r=0.1oh  l=0.02nh
  srl 8  0  r=0.02oh  l=0.011nh
c10: 1por  2
end
*
blk
  step  1  2  w1= 30.0µm    w2= 30.0µm    gaas
  trl    2  3  w=  30.0µm    p=  25.5µm    gaas
  step  3  4  w1= 30.0µm    w2= 15.0um    gaas
* - - - - - - - - - - - - - - - - - - - - - - - - - - - - - - *
  tfr   4  5  w= 15.0µm     p= 108.4µm    rs=6.0oh gaas
* pyhsical length=(p+30um) due to overlap.
* - - - - - - - - - - - - - - - - - - - - - - - - - - - - - - *
  step  5  6  w1= 15.00µm   w2= 30.0µm    gaas
  trl   6  7  w=  30.0µm    p=  25.5µm    gaas
  step  7  8  w1= 30.0µm    w2= 30.0µm    gaas
* port 2 (node 8) to cap.
```

```
r43: 2por   1  8
end
*
blk
  step  1  2  w1= 12.50µm   w2= 16.0µm    gaas
  trl   2  3  w=  16.0µm    p=  19.0µm    gaas
  step  3  4  w1= 16.0µm    w2= 10.0µm    gaas
* - - - - - - - - - - - - - - - - - - - - - - - - - - - - - - *
  tfr   4  5  w= 10.0µm     p= 25.0µm     rs=6.0oh gaas
* 25.0µm
* physical length=(p+30µm) due to overlap.
* - - - - - - - - - - - - - - - - - - - - - - - - - - - - - - *
  step  5  6  w1= 10.00µm   w2= 16.0µm    gaas
  trl   6  7  w=  16.0µm    p=  19.0µm    gaas
  step  7  8  w1= 16.0µm    w2= 16.0µm    gaas
* port 2 (node 8) to cap.
r15: 2por   1  8
end
*
blk
  step  1  2  w1= 12.5µm    w2= 30.0µm    gaas
  trl   2  3  w=  30.0µm    p=  19.0µm    gaas
  step  3  4  w1= 30.0µm    w2= 15.0µm    gaas
* - - - - - - - - - - - - - - - - - - - - - - - - - - - - - - *
  tfr   4  5  w= 15.0µm     p= 83.0µm     rs=6.0oh gaas
* 83.0µm
* pyhsical length=(p+30µm) due to overlap.
* - - - - - - - - - - - - - - - - - - - - - - - - - - - - - - *
  step  5  6  w1= 15.00µm   w2= 30.0µm    gaas
  trl   6  7  w=  30.0µm    p=  19.0µm    gaas
  step  7  8  w1= 30.0µm    w2= 12.5µm    gaas
* port 2 (node 8) to cap.
r50: 2por   1  8
end
*
blk
  step  1  2  w1= 12.50µm   w2= 16.0µm    gaas
  trl   2  3  w=  16.0µm    p=  19.0µm    gaas
  step  3  4  w1= 16.0µm    w2= 10.0µm    gaas
* - - - - - - - - - - - - - - - - - - - - - - - - - - - - - - *
  tfr   4  5  w= 10.0µm     p= 118.16µm   rs=6.0oh gaas
* physical length=(p+30µm) due to overlap.
* - - - - - - - - - - - - - - - - - - - - - - - - - - - - - - *
  step  5  6  w1= 10.00µm   w2= 16.0µm    gaas
  trl   6  7  w=  16.0µm    p=  19.0µm    gaas
  step  7  8  w1= 16.0µm    w2= 12.5µm    gaas
* port 2 (node 8) to cap.
rg1: 2por   1  8
end
*
blk
  step  1  2  w1= 12.50µm   w2= 16.0µm    gaas
```

```
  trl   2  3  w=  16.0µm    p=  18.0µm    gaas
  step  3  4  w1= 16.0µm    w2= 10.0µm    gaas
* - - - - - - - - - - - - - - - - - - - - - - - - - - - - - - *
  tfr   4  5  w= 10.0µm     p= 94.003µm  rs=6.0oh gaas
* physical length=(p+30µm) due to overlap.
* - - - - - - - - - - - - - - - - - - - - - - - - - - - - - - *
  step  5  6  w1= 10.00µm   w2= 16.0µm    gaas
  trl   6  7  w=  16.0µm    p=  30.5µm    gaas
  step  7  8  w1= 16.0µm    w2= 16.0µm    gaas
* port 2 (node 8) to cap.
rd1: 2por   1  8
end
*
*
blk
  step  1  2  w1= 12.50µm   w2= 16.0µm    gaas
  trl   2  3  w=  16.0µm    p=  19.0µm    gaas
  step  3  4  w1= 16.0µm    w2= 10.0µm    gaas
* - - - - - - - - - - - - - - - - - - - - - - - - - - - - - - *
  tfr   4  5  w= 10.0µm     p= 50.5µm     rs=6.0oh gaas
* physical length=(p+30µm) due to overlap.
* - - - - - - - - - - - - - - - - - - - - - - - - - - - - - - *
  step  5  6  w1= 10.00µm   w2= 16.0µm    gaas
  trl   6  7  w=  16.0µm    p=  19.0µm    gaas
  step  7  8  w1= 16.0µm    w2= 12.5µm    gaas
* port 2 (node 8) to cap.
rg2: 2por   1  8
end
*
blk
  step  1  2  w1= 26.0µm    w2= 26.0µm    gaas
  trl   2  3  w=  26.0µm    p=  5.5µm     gaas
  step  3  4  w1= 26.0µm    w2= 20.0µm    gaas
* - - - - - - - - - - - - - - - - - - - - - - - - - - - - - - *
  tfr   4  5  w= 20.0µm     p= 235.0µm    rs=6.0oh gaas
* physical length=(p+30µm) due to overlap.
* - - - - - - - - - - - - - - - - - - - - - - - - - - - - - - *
  step  5  6  w1= 20.00µm   w2= 26.0µm    gaas
  trl   6  7  w=  26.0µm    p=  33.6µm    gaas
  step  7  8  w1= 26.0µm    w2= 26.0µm    gaas
* port 2 (node 8) to cap.
rd2: 2por   1  8
end
*
blk
  rg1   1  2
  c10   2  0
tg1:1por 1
end
*
blk
  rg2   1  2
```

```
  c10    2  0
  trl    2  3  w= 10.16µm   p= 571.5µm     gaas
  r15    3  4
  trl    4  5  w= 10.16µm   p= 571.5µm     gaas
  c10    5  0
  r50    5  0
  trl    5  6  w= 10.16µm   p= 502.92µm    gaas
  r50    6  7
  trl    7  8  w= 10.16µm   p= 393.7µm     gaas
  ind    8  0  l=.2nh
tg2:1por 1
end
*
blk
  r43    1  2
  c4r4   2  0
rcs: 1por   1
end
*
*  [ dual spiral network-1 ]
blk
  trl    1  2  p= 327.152µm  w= 25.4µm     gaas
  step   2  3  w1= 25.4µm    w2= 83.82µm   gaas
  trl    3  4  p= 22.606µm   w= 83.82µm    gaas
  two    5  4  s70
  trl    5  6  p= 129.54µm   w= 25.4µm     gaas
* - - - - - - - - - - - - - - - - - - - - - - - - - - - - - - *
  rcs    6  0
* ( src 6  0  r=43.3oh   c=4.4pf )
* - - - - - - - - - - - - - - - - - - - - - - - - - - - - - - *
  trl    6  7  p= 66.802µm   w= 25.4µm     gaas
  step   7  8  w1= 25.4µm    w2= 83.82µm   gaas
  trl    8  9  p= 22.606µm   w= 83.82µm    gaas
  two   10  9  s70
  trl   10 11  p= 611.124µm  w= 12.5µm     gaas
dual1:  2por 1  11
end
*
*  [ dual spiral network-2 ]
blk
  trl    1  2  p= 405.13µm  w= 25.4µm     gaas
  step   2  3  w1= 25.4µm   w2= 83.82µm  gaas
  trl    3  4  p= 22.606µm   w= 83.82µm   gaas
  two    5  4  s70
  trl    5  6  p= 129.54µm    w= 25.4µm     gaas
* - - - - - - - - - - - - - - - - - - - - - - - - - - - - - - *
  rcs    6  0
* ( src 6  0  r=43.3oh   c=4.4pf )
* - - - - - - - - - - - - - - - - - - - - - - - - - - - - - - *
  trl    6  7  p= 66.802µm   w= 25.4µm      gaas
  step   7  8  w1= 25.4µm    w2= 83.82µm    gaas
  trl    8  9  p= 26.162µm   w= 83.82µm     gaas
```

```
   two   10  9  s70
   trl   10 11  p= 127µm       w= 25.4µm      gaas
 dual2:  2por 1  11
 end
 *
 blk
   trl     1   3  w= 12.5µm   p= 10µm        gaas
   dual1   3   4
   cap     4   0  c=.1uf
   rd1     1   2
   c10     2   0
 td1:1por 1
 end
 *
 blk
   trl    1  9  w= 25.4µm      p= 10µm         gaas
   dual2 10  9
   cap   10  0    c=.1uf
   rd2    1  2
   c10    2  0
 td2:1por  1
 end
 *
 *  [ dc block-1 ]
 blk
   step  1  2  w1= 71.12µm      w2= 142.748µm gaas
   trl   2  3  w= 142.748µm     p= 71.12µm    gaas
   src   3  4  r=0.1oh          c=3.97pf
   trl   4  5  w= 142.748µm     p= 71.12µm    gaas
   step  5  6  w1= 142.748µm    w2= 12.5µm    gaas
 cblock:  2por  1  6
 end
 *
 *  [ dc block-2 ]
 blk
   step   1   2  w1= 15.24µm    w2= 82.042µm  gaas
   trl    2   3  w= 82.042µm    p= 169.926µm  gaas
   src    3  4   r=0.1oh        c=5.0pf
   trl    4  5   w= 82.042µm    p= 169.926µm  gaas
   step   5  6   w1= 82.042µm   w2= 15.24µm   gaas
 cline:   2por  1  6
 end
 *
 *
 * ====================== [ stage 1 ] ====================== *
 *
 l1:  531.81µm
 l2p: 287.54µm
 l3p: 419.06µm
 l4p: 388.14µm
 l5:  875.72µm
 * - - - - - - - - - - - - - - - *
```

```
l6p: 382.35µm
l7p: 337.36µm
l8p: 315.52µm
l9p: 177.92µm
l10: 730.82µm
* - - - - - - - - - - - - - - *
l11: 600µm
l12: 100.0µm
l13: 100.0µm
l14: 499.25µm
*
*
* -----------------------------------------------------------
* --------------- gate line - first stage -------------------
* -----------------------------------------------------------
blk
  trl 1   204 w= 12.5µm  p=l1   gaas
  tee 203 204 205  w1= 12.5µm   w2= 12.5µm   w3= 12.5µm   gaas
  trl 203 201 w= 12.5µm  p=l2p  gaas
  two 201 202 sse40041
  trl 202 304 w= 12.5µm  p=l2p  gaas
  tee 303 304 305  w1= 12.5µm   w2= 12.5µm   w3= 12.5µm   gaas
  trl 303 301 w= 12.5µm  p=l3p  gaas
  two 302 301 sse40041
  trl 302 404 w= 12.5µm  p=l3p  gaas
gate1a: 4por   1   205 305 404
end
blk
  tee 403 404 405  w1= 12.5µm   w2= 12.5µm   w3= 12.5µm   gaas
  trl 403 401 w= 12.5µm  p=l4p  gaas
  two 401 402 sse40050
  trl 402 504 w= 12.5µm  p=l4p  gaas
  tee 503 504 505  w1= 12.5µm   w2= 12.5µm   w3= 12.5µm   gaas
  trl 503   6  w= 12.5µm  p=l5  gaas
  tg1 6
gate1b: 3por   404 405 505
end
*
* ----------------------------------------------------------- *
* ----------------- drain line - first stage ------------------ *
* ----------------------------------------------------------- *
blk
  td1 11
  trl 11  111 w= 12.5µm  p=l6p gaas
  two 111 121 sse40041
  trl 121 123 w= 12.5µm  p=l6p gaas
  tee 123 124 125  w1= 12.5µm   w2= 12.5µm   w3= 12.5µm   gaas
* ----------------- first fet m derived line ------------------ *
  trl 125 140 w= 12.5µm  p=l11 gaas
  trl 124 122 w= 12.5µm  p=l7p gaas
  two 122 131 sse40050
  trl 131 133 w= 12.5µm  p=l7p gaas
```

```
  tee 133 134 135  w1= 12.5µm   w2= 12.5µm   w3= 12.5µm   gaas
* ----------------- second fet m derived line ----------------- *
  trl 135 160 w= 12.5µm  p=l12 gaas
  trl 134 132 w= 12.5µm  p=l8p gaas
  two 132 141 sse40080
  trl 141 143 w= 12.5µm  p=l8p gaas
drain1a: 3por  140 160 143
end
blk
  tee 143 144 145  w1= 12.5µm   w2= 12.5µm   w3= 12.5µm   gaas
* ----------------- third fet m derived line ----------------- *
  trl 145 180 w= 12.5µm  p=l13 gaas
  trl 144 142 w= 12.5µm  p=l9p gaas
  two 142 151 sse40080
  trl 151 153 w= 12.5µm  p=l9p gaas
  tee 153 154 155  w1= 12.5µm   w2= 12.5µm   w3= 12.5µm   gaas
* ----------------- fourth fet m derived line ----------------- *
  trl 155 200 w= 12.5µm  p=l14 gaas
  trl 154 16  w= 12.5µm  p=l10 gaas
drain1b: 4por  143 145 155 16
end
blk
* ------------------------------------------------------------- *
* ----------------------- first stage ------------------------- *
* ------------------------------------------------------------- *
  cblock  1   2
  gate1a  2   205 305 404
  gate1b  404 405 505
  trl 205 206 w= 12.5µm  p= 19.05µm  gaas
  125µmmod    206 140 207 208
  trl 305 306 w= 12.5µm  p= 19.05µm  gaas
  125µmmod    306 160 307 308
  trl 405 406 w= 12.5µm  p= 19.05µm  gaas
  125µmmod    406 180 407 408
  trl 505 506 w= 12.5µm  p= 19.05µm  gaas
  125µmmod    506 200 507 508
  ind 207 0   l=.0056nh
  ind 508 0   l=.0056nh
  res 208 307 r=0.001oh
  res 308 407 r=0.001oh
  res 408 507 r=0.001oh
  ind 208 0   l=.0056nh
  ind 308 0   l=.0056nh
  ind 408 0   l=.0056nh
  drain1a 140 160 143
  drain1b 143 180 200 16
stage1:2por    1   16
end
*
*
* ======================= [ stage 2 ] ======================= *
*
```

```
p1:  208.23µm
p1a:  11.43µm
p2:  635.56µm
p2a:  49.119µm
p3:  673.12µm
p3a:  80.0µm
p4:  690.04µm
p4a:  76.2µm
p5:  462.23µm
* - - - - - - - - - - - - - - - - *
p6:     805.1546µm
p7pa:  203.2µm
p7pb:  130.0277µm
p8pa:  235.6383µm
p8pb:  203.2µm
p9pa:  267.2588µm
p9pb:  203.2µm
p10:   243.84µm
* - - - - - - - - - - - - - - - - *
p11:  141.9073µm
p12:  115.7478µm
p13:  165.4835µm
p14:  66.74612µm
*
* ------------------------------------------------------------- *
* ---------------- gate line - second stage ------------------- *
* ------------------------------------------------------------- *
blk
  trl 1   204 w= 14.859µm  p=p1 gaas
  tee 203 204 205  w1= 14.859µm   w2= 14.859µm   w3= 14.859µm   gaas
  trl 205 20   w= 14.859µm  p=p1a gaas
  trl 203 304 w= 14.859µm  p=p2 gaas
  tee 303 304 305  w1= 14.859µm   w2= 14.859µm   w3= 14.859µm   gaas
  trl 305 30   w= 14.859µm  p=p2a gaas
  trl 303 404 w= 14.859µm  p=p3 gaas
gate2a: 4por 1  20  30  404
end
blk
  tee 403 404 405  w1= 14.859µm   w2= 14.859µm   w3= 14.859µm   gaas
  trl 405 40   w= 14.859µm  p=p3a gaas
  trl 403 504 w= 14.859µm  p=p4 gaas
  tee 503 504 505  w1= 14.859µm   w2= 14.859µm   w3= 14.859µm   gaas
  trl 505 50   w= 14.859µm  p=p4a gaas
  trl 503 6    w= 14.859µm  p=p5 gaas
  tg2 6
gate2b: 3por 404  40  50
end
* ------------------------------------------------------------- *
* ----------------- drain line - second stage ----------------- *
* ------------------------------------------------------------- *
blk
  td2 11
```

```
  trl 11  123 w= 12.5µm  p=p6 gaas
  tee 124 125 123 w1= 12.5µm   w2= 12.5µm   w3= 12.5µm  gaas
* ----------------- first fet m derived line ----------------- *
  trl 125 140 w= 12.5µm  p=p11 gaas
  trl 124 133 w= 12.5µm  p=p7pa gaas
  two 133 136 s20
  trl 136 137 w= 12.5µm  p=p7pb gaas
  tee 137 134 135 w1= 12.5µm   w2= 12.5µm   w3= 12.5µm  gaas
* ----------------- second fet m derived line ---------------- *
  trl 135 160 w= 12.5µm  p=p12 gaas
  trl 134 143 w= 12.5µm  p=p8pa gaas
  two 143 146 s20
  trl 146 147 w= 12.5µm  p=p8pb gaas
  tee 145 147 144 w1= 12.5µm   w2= 12.5µm   w3= 12.5µm  gaas
* ----------------- third fet m derived line ----------------- *
  trl 145 180 w= 12.5µm  p=p13 gaas
  trl 144 153 w= 12.5µm  p=p9pa gaas
  two 153 156 s20
  trl 156 157 w= 12.5µm  p=p9pb gaas
  tee 157 154 155 w1= 12.5µm   w2= 12.5µm   w3= 12.5µm  gaas
* ----------------- fourth fet m derived line ---------------- *
  trl 155 200 w= 12.5µm  p=p14 gaas
  trl 154 16  w= 12.5µm  p=p10 gaas
* ------------------------------------------------------------ *
  gate2a  2    20  30  404
  gate2b  404  40  50
* ------------------------------------------------------------ *
* ----------------- 250µm dual gate fets --------------------- *
* ------------------------------------------------------------ *
  250µmmod    20  140 207 208
 vvg   20  21  m=1 r1=1e6 r2=1e-6
  250µmmod    30  160 307 308
 vvg   30  31  m=1 r1=1e6 r2=1e-6
  250µmmod    40  180 407 408
 vvg   40  41  m=1 r1=1e6 r2=1e-6
  250µmmod    50  200 507 508
 vvg   50  51  m=1 r1=1e6 r2=1e-6
  ind 207 0   l=.0056nh
  ind 508 0   l=.0056h
  res 208 307 r=0.001oh
  res 308 407 r=0.001oh
  res 408 507 r=0.001oh
  ind 208 0   l=.011nh
  ind 308 0   l=.011nh
  ind 408 0   l=.011nh
 fet1:2por   2  21
 fet2:2por   2  31
 fet3:2por   2  41
 fet4:2por   2  51
 fets1:4por  2  21  31  41
 fets2:4por  2  31  41  51
 cline       1  2
```

```
 cline        16 17
stage2:2por  1  17
end
*
*
* ==================== [ stages 1 & 2 ] ===================== *
*
blk
  wire  1  2  d=17.5µm        l=350µm       h=100µm
  wire  1  2  d=17.5µm        l=350µm       h=100µm
  trl   2  3  w=101.6µm       p=101.6µm     gaas
  open  3     w=101.6µm       gaas
  trl   2  4  w=101.6µm       p=101.6µm     gaas
  step  4  5  w1= 101.6µm     w2=72.0µm     gaas
  trl   5  6  w= 72.0µm       p=2470.0µm    gaas
  stage1   6 7
  stage2   7 8
twostg: 2por 1  8
end
*
* ========================================================= *
*
*
freq
  step 1.5ghz 21.5ghz 0.25ghz
end
*
out
  plo twostg s
* - - - - - - - *
  plo stage1 s
* pri stage1 s
* - - - - - - - *
  plo stage2 s
  plo fets1  s
  plo fets2  s
end
*
opt
  stage1 r1=50 r2=50 f 1.8 ghz 20.5ghz ms21=8.0db w 10
  stage2 r1=50 r2=50 f 1.8 ghz 20.5ghz ms21=8.0db w 10
  twostg r1=50 r2=50 f 1.8ghz 20.5ghz ms21=16.0db w 20
+ ms11 -12db lt w 5 ms22 -10db lt w 3
  fet1 mvg3 0db -2db w 6
  fet2 mvg3 0db -2db w 6
  fet3 mvg3 0db -2db w 6
  fet4 mvg3 0db -2db w 6
end
*
data
  gaas:ms er=12.9 h= 101.6µm  tand=.0004 met1=3.7 5µm
  s20:s20 file=new020.s
```

```
  s70:s70 file=new070.s
  sse40041:sse40041 file=new041.s
  sse40050:sse40050 file=new050.s
  sse40080:sse40080 file=new080.s
end
```

```
2-6 GHz 0.5 W Amplifier Chip
```

Talking about detailed level of modeling, the frequency response of a two-stage 0.5 W amplifier chip operating from 2 to 6 GHz is shown in Figure B.2.

The following netlist includes all the necessary descriptions of the elements, and the frequency response is shown above.

```
* 1uM FET 1.2MM AND 2MM WIDE                      *
* ******************************************
OPTIMIZATION VAR.
INPC1:?10UM 49.761UM 70UM?
INPC2:?40UM 77UM 90UM?
INPLL1:?300UM 593.04UM 900UM?
INPLL2:?300UM 489.24UM 900UM?
INPLL3:?300UM 318.16UM 900UM?
INPLR1:?10UM 12.442UM 90UM?
INTC3:?10UM 30UM 60UM?
INTL4:?220UM 385.04UM 550UM?
INTLR2:?12UM 15.074UM 30UM?
INTL1:?300UM 1306.91UM 2500UM?
INTL2:?300UM 1354.7UM 3000UM?
INMAT:?200UM 413.65UM 900UM?
INMFB:?250UM 305.64UM 450UM?
```

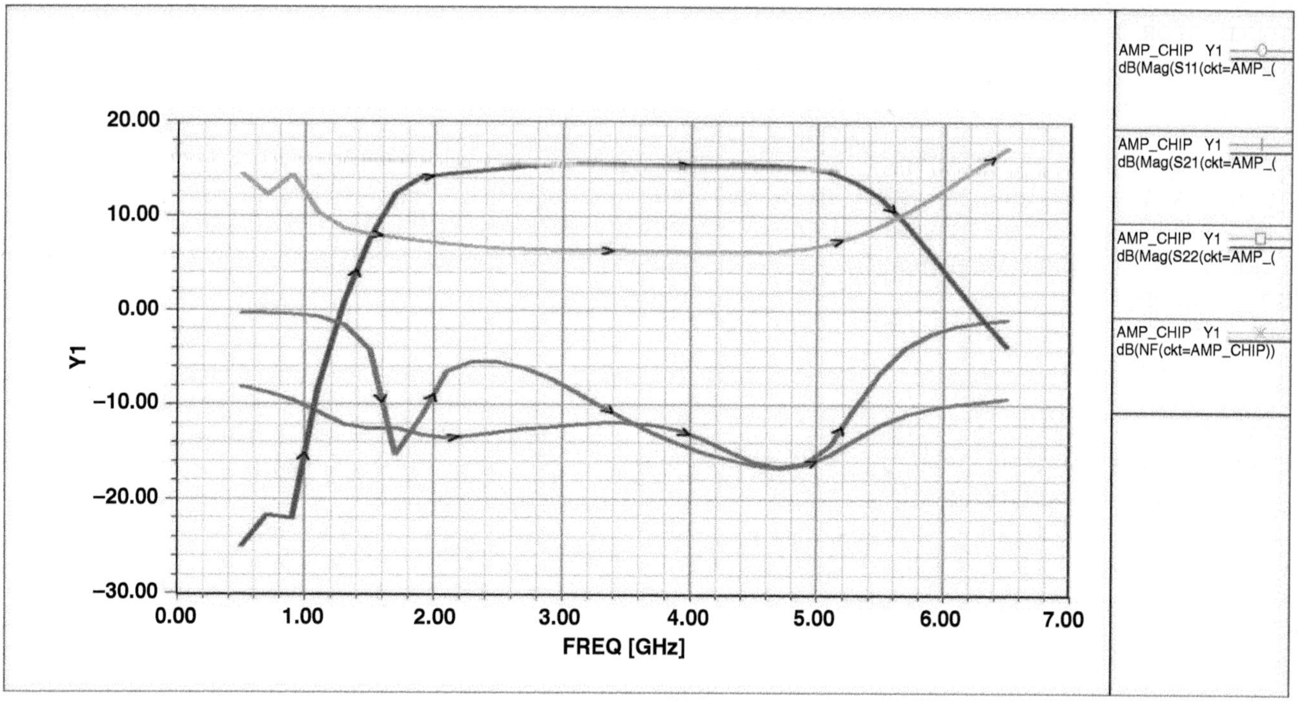

Figure B.2 Frequency response of the 0.5 W amplifier chip (netlist shown below).

```
INMFC:?250UM 257.41UM 450UM?
INTSS:?050UM 280.09UM 400UM?
NETIND:? 0.10NH .42422NH 0.50NH?
* SCALING*
* FET1 Z1=1       1.2mm WIDE 1uM LONG POWER FET
* FET2 Z2=1.66    2mm   WIDE 1uM LONG POWER FET
Z1:1
Z2:1.66
* * * * * * * * * * * * * inductor subcircuits L1 THRU L6
BLK
  TRL 1  2       W=12UM  P=150UM              SUB
  CPL 2 3 4 5  W=12UM  S=10UM  P=INPLL1  SUB
  TRL 5  3       W=12UM  P=22UM               SUB
  TRL 4  6       W=12UM  P=100UM              SUB
INDUC1:2POR 1 6
END
*
BLK
  TRL 1 2       W=12UM P=130UM             SUB
  CPL 2 3 4 5  W=12UM  S=10UM  P=INPLL2  SUB
  TRL 5  3      W=12UM  P=22UM               SUB
  TRL 4  6      W=12UM  P=100UM              SUB
INDUC2:2POR 1 6
END
*
BLK
  TRL 1 2       W=12UM P=100UM             SUB
  CPL 2 3 4 5  W=12UM  S=10UM  P=INPLL3  SUB
TRL 5  3      W=12UM  P=22UM             SUB
TRL 4  6      W=12UM  P=100UM           SUB
INDUC3:2POR 1 6
END
*
BLK
  TRL 1 2       W=12UM P=100UM             SUB
  CPL 2 3 4 5  W=12UM  S=10UM  P=INTL4    SUB
  TRL 5  3      W=12UM  P=22UM               SUB
  TRL 4  6      W=12UM  P=50UM               SUB
INDUC4:2POR 1 6
END
*
BLK
  TRL 1 2       W=15UM P=100UM             SUB
  CPL 2 3 4 5  W=15UM  S=12UM  P=470UM    SUB
  TRL 5  3      W=15UM  P=26UM               SUB
  TRL 4  6      W=15UM  P=100UM              SUB
INDUC5:2POR 1 6
END
*
BLK
  TRL 1 2       W=12UM P=100UM             SUB
  CPL 2 3 4 5  W=15UM  S=12UM  P=550UM    SUB
```

```
  TRL 5   3      W=12UM   P=22UM              SUB
  TRL 4   6      W=12UM   P=100UM             SUB
INDUC6:2POR 1 6
END
* * * * * * * * * * * * * Q1 definition
BLK
 FET 1 2  90  G=(0.10*Z1) CGS=(1.07PF*Z1) T=1E-12 CDG=(.133PF*Z1)
+ CDS=(.18PF*Z1) RG=(.99/Z1) RI=(4.5/Z1) RS=(.46/Z1) RD=(.725/Z1)
+ LS=.043NH GDS=(.00834*Z2)
+ LG=.03NH LD=.01NH CDC=(0.015PF*Z1) CGE=(.1PF*Z1)
AFET12:3POR 1 2 90
END
* * * * * * * * * * * * * Q2 definition
BLK
 FET 1 2  90  G=(0.10*Z2) CGS=(1.07PF*Z2) T=1E-12 CDG=(.133PF*Z2)
+ CDS=(.18PF*Z2) RG=(.99/Z2) RI=(4.5/Z2) RS=(.46/Z2) RD=(.725/Z2)
+ LS=.043NH GDS=(.00834*Z2)
+ LG=.03NH LD=.01NH CDC=(0.015PF*Z2) CGE=(.1PF*Z2)
AFET20:3POR 1 2 90
END
* * * * * * * * * * * * * Input Network * * * * * * * * * * * * * * * * * * *
BLK
  IND 1  2      L=0.15NH
  TRL 2  3      W=70UM  P=400UM                                   SUB
  TFC 3 10      W=INPC1 L=INPC1  T=0.20UM  ER=6.8  TAND=0.01  SUB
  TRL 10 80     W=180UM P=INMAT                                   SUB
  IND 80 0      L=0.10NH
  TFC 82 90     W=300UM L=300UM  T=0.20UM  ER=6.8  TAND=0.01  SUB
  INDUC1 3 4
  TFC 4  81     W=INPC2 L=INPC2  T=0.20UM  ER=6.8  TAND=0.01  SUB
  IND 81 0      L=0.10NH
  INDUC2 4 5
  TFR 5 11      W=50UM  P=INPLR1 RS=100                           SUB
  INDUC3 11 12
  TFC 12 90     W=300UM L=300UM  T=0.20UM  ER=6.8  TAND=0.01  SUB
  RES 12 0      R=1200OH
  AFET12 5 6 90
  TFC 6 90      W=INTC3 L=INTC3  T=0.20UM  ER=6.8  TAND=0.01  SUB
  TRL 90 81     W=300UM P=INMFB                                   SUB
  TRL 90 80     W=300UM P=INMFC                                   SUB
INPUT: 3POR 1 6 82
END
* 1 is Input, 2 is Output, 82 is top of decoupling cap
*
* * * * * * * * Interstage Matching Network * * * * * * * * * * * * * * * * * *
BLK
  TRL 1 2      W=10UM P=INTL1                                   SUB
  INDUC4 2 10
  INDUC4 10 11
  TFC 2  3     W=200UM L=200UM  T=0.20UM  ER=6.8  TAND=0.01  SUB
  TRL 3  4     W=200UM P=200UM                                  SUB
  TFR 4  20    W=80UM  P=INTLR2 RS=100                          SUB
```

```
  TRL 20 21     W=10UM   P=INTL2                                  SUB
  TFC 21 90     W=300UM L=300UM  T=0.20UM   ER=6.8   TAND=0.01   SUB
  SRL 21 0      R=1000OH L=2NH
  AFET20 4 5 90
  SST 90 0      W=300UM P=INTSS                                   SUB
  SST 90 0      W=300UM P=INTSS                                   SUB
INTER: 4POR 1 5 11 90
END
*
* * * * * * * * * * * * * Output Network * * * * * * * * * * * * * * * * * *
BLK
  INDUC5  1  10
  INDUC5 10  11
  TFC   11 12    W=300UM  L=300UM  T=0.20UM  ER=6.8  TAND=0.01  SUB
  TFC   1  2     W=80UM   L=75UM   T=0.20UM  ER=6.8  TAND=0.01  SUB
  INDUC6 2 20
  INDUC6 20 0
TRL 2 3    W=70UM   P=150UM                                SUB
IND 3 4    L=NETIND
OUTPUT: 4POR 1 4 11 12
END
*
* * * * * * * * * * * * * COMPLETE AMPLIFIER * * * * * * * * * * * * * * * * *
BLK
  INPUT 1 2 20
  INTER 2 3 20 30
  OUTPUT 3 4 31 30
  IND 20 21 L=0.20NH
  RES 21 22 R=10OH
  IND  22  31  L=0.20NH
  IND  21   0  L=2.0NH
  IND 22   0 L=2.0NH
AMP_CHIP:2POR 1 4
END
* * * * * * * * * * * * * FREQ/DATA BLOCKS * * * * * * * * * * * * * * * * * *
FREQ
      STEP 0.5GHZ   2.50GHZ    0.200GHZ
*     STEP 1.0GHZ   2.50GHZ    0.100GHZ
      STEP 2.5GHZ   6.50GHZ    0.200GHZ
*     STEP 7.0GHZ   8.00GHZ    0.250GHZ
*     STEP 1.5GHZ   4.50GHZ    0.125GHZ
*     STEP 4.0GHZ   8.00GHZ    0.250GHZ
*     STEP 6.0GHZ   8.0GHZ     0.500GHZ
END
*
OUT
   PRI AMP_CHIP S
END
*
* * * * * * * * * * * * * * * * * *
OPT
  AMP_CHIP
```

```
    F=2.0GHZ 6.0GHZ
    MS21=15DB          W=15
    MS21=AVG           W=50
    F=0.5GHZ 0.8GHZ
    MS21=-25DB -5DB W=5
 * F=6.0GHZ
 * MS21=15DB W=30
  INPUT
    F=2.5GHZ 4.5GHZ
    MS11=-12DB -22DB W=10
  OUTPUT
    F=3.0GHZ 4.0GHZ
    MS22=-8DB -30DB
END
* * * * * * * * * * * * * DATA BLOCKS * * * * * * * * * * * * * * * * * *
DATA
      SUB: MS H=100UM ER=12.9 MET1=AU 3UM RGH=0.0001MM TAND=0.0005
*
END
```

Conclusion

While this is a NETLIST, dating back to earlier design practices, the modern CAD tools would normally use a schematic capture for this purpose. The advantage of the NETLIST is that all the details, including the component values, can be easily identified.

C

Testing True Performance of ADCs Using R&S®SMA100B Signal Generator

C.1 Overview

Analog-to-digital converters (ADCs) have taken giant steps in speed and accuracy over the past years: Such devices allow high-speed video processing and even software-defined radio (SDR) applications, where digitization is being performed at ever-higher intermediate frequencies (IF). This has clearly raised the bar for ADC test equipment, especially regarding the noise performance of the analog input signal and sample clock sources used in the test setup. The R&S®SMA100B offers outputs for the analog input signal of the ADC as well as the clock input signal in one single instrument—both with outstanding signal purity.

C.2 Introduction

Many performance parameters for an ADC are identical to those for analog circuits in general. These include the well-known signal-to-noise ratio (SNR), spurious-free dynamic range (SFDR), signal to noise-and-distortion (SINAD), total harmonic distortion (THD), and third-order input intercept (TOII or IIP3).

With high-speed ADC products, where 16 bits at a few hundred MSPS are quite common and sampling rates can easily reach multiple GSPS, very pure signal sources are necessary. With current high-performance ADCs, SNRs of more than 90 dBFS become possible, requiring an input signal with an SNR of greater than 100 dB. With sampling rates surpassing the 10 GSPS mark, phase noise performance becomes even more important. Typically, the pure phase noise performance and phase noise-induced SNR for a fixed input bandwidth gets worse with increasing carrier frequency. In addition to that, the higher the clock frequency, the harder it gets to realize very narrow bandpass filters. Thus, not only the phase noise performance gets worse, but the ADC also is affected by noise with a larger bandwidth, increasing the relevance of a low-phase noise signal source even more.

Figure C.1 depicts a traditional test setup to determine the performance of an ADC. It consists of two signal sources and, depending on the application, sometimes one or two bandpass filters. To minimize measurement error, the specifications of the signal sources should always exceed those of the ADC itself.

In Figure C.2, a modern test setup is shown. Instead of two signal generators, only one R&S®SMA100B is used to excite the ADC under test. The R&S®SMA100B not only offers an RF output, but also an optional clock output within the same

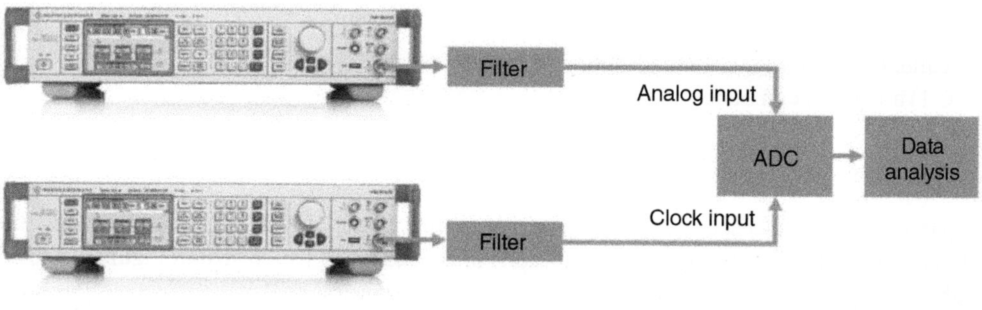

Figure C.1 Traditional ADC performance test setup.

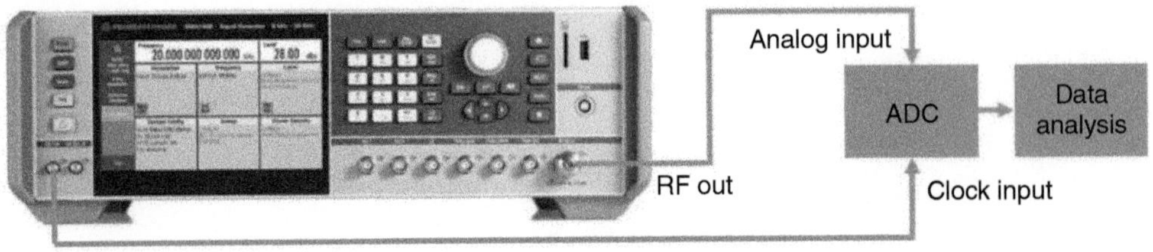

Figure C.2 Modern ADC performance test setup.

instrument. Both, the RF output and the clock output, provide signals with highest purity. Therefore, in many cases neither a second signal generator, nor filters are needed to obtain accurate measurements.

This document outlines the most important aspects concerning ADC noise and distortion performance and presents some typical performance data such as jitter and SNR for the analog high-end signal generator R&S®SMA100B when used as signal or clock source.

C.3 Fundamentals

C.3.1 Ideal Signal-to-noise Ratio (in dB)

The SNR is the ratio of the RMS signal amplitude to the RMS value of the sum of all spectral components except the desired signal, the first six harmonics and DC. As the signal level is decreased, SNR typically decreases decibel-for-decibel in a linear fashion.

For an ideal ADC with a full scale sinusoidal at the input and exact rounding, the SNR is determined by the quantization noise of the converter and can be estimated as:

$$\text{SNR} = 20\log_{10}\left(\sqrt{1.5}\cdot 2^n\right) \approx 6.02\cdot n + 1.76\text{ dB} \tag{C.1}$$

where n is the number of bits.

The ideal SNR for typical ADC resolutions is provided in following table:

Number of bits	8	10	12	14	16
SNR/dB	50	62	74	86	98

For oversampled systems, where the sampling frequency is higher than twice the Nyquist frequency and with adequate filtering after AD conversion, the SNR must be calculated as:

$$\text{SNR} = 20\log_{10}\left(\sqrt{3}\cdot 2^{n-1}\cdot\sqrt{\frac{f_{\text{clk}}}{f_{\text{analog}}}}\right) \tag{C.2}$$

$$\text{SNR} \approx 6.02\cdot n + 1.76\text{ dB} + 10\log_{10}\left(\frac{f_{\text{clk}}}{2\cdot f_{\text{analog}}}\right) \tag{C.3}$$

where f_{clk} is sampling or clock frequency and f_{analog} is analog input frequency.

A detailed derivation for Eqs. (C.1) and (C.2) can be found in [1].

C.3.2 Jitter

Jitter is one of the most critical parameters when discussing the SNR performance of a real ADC [2, 3]. It describes the stability of a signal in the time domain and specifies the signal's fluctuations of zero crossings as shown in Figure C.3 for a 10 MHz sine signal with very high jitter (t_{jitter1}) and lower jitter (t_{jitter2}).

Figure C.4 is a basic block diagram of the track & hold unit of an ADC input stage. At the beginning of the sampling process, the track and hold switch is closed and the voltage at the hold capacitor follows the input voltage. The zero crossing of the ADC's clock signal opens the track and hold switch, and the voltage across the capacitor is held. The conversion process in the quantizer is then performed. Any error in the zero crossing point may therefore translate to a voltage error at the hold capacitor in general. In a first-order approximation, the voltage error is proportional to the jitter and input slope of the signal, as shown in Figure C.5:

The approximate voltage error ΔV of a signal $s(t)$ due to sampling jitter in Figure C.5 can be expressed as:

$$\Delta V = \text{slewrate}(s(t)) \cdot t_{\text{jitter}} = \frac{d}{dt}s(t) \cdot t_{\text{jitter,clk}} \tag{C.4}$$

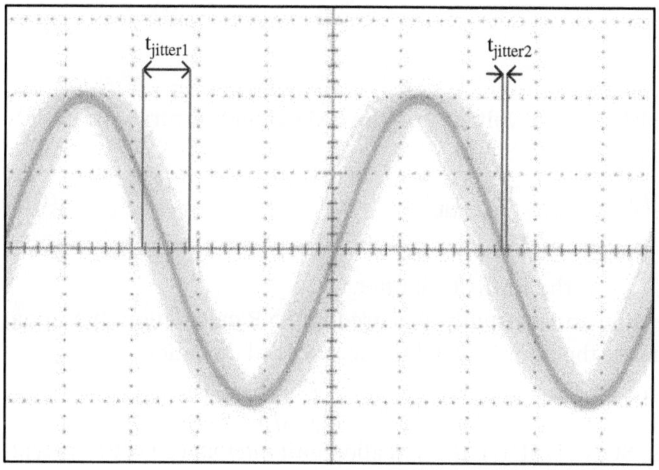

Figure C.3 10 MHz signal with very high jitter (yellow outer thick/highlighted trace) and with lower jitter (blue (inner/thin) trace).

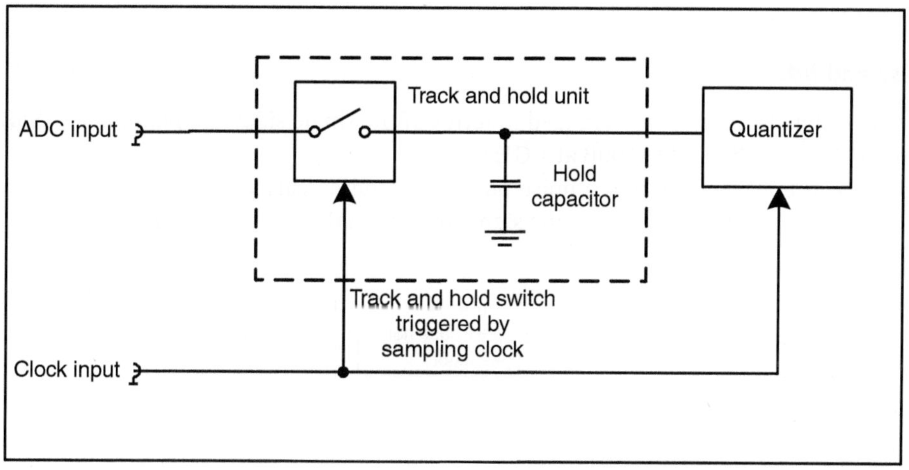

Figure C.4 Basic block diagram of track & hold unit of an ADC input stage.

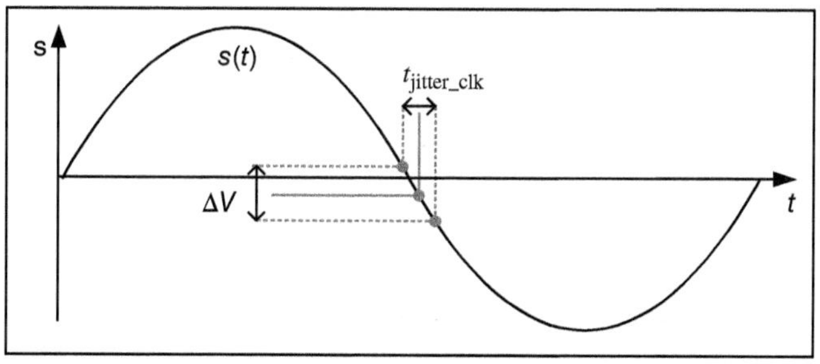

Figure C.5 Voltage error ΔV due to jitter on the sampling clock $t_{\text{jitter,clk}}$.

For a sine wave signal, $s(t)$ is determined as follows:

$$s(t) = A \cdot \sin(2\pi \cdot f_{\text{analog}} \cdot t) \tag{C.5}$$

This yields a voltage error ΔV of:

$$\Delta V = A \cdot 2\pi \cdot f_{\text{analog}} \cdot \cos(2\pi \cdot f_{\text{analog}} \cdot t) \cdot t_{\text{jitter}} \tag{C.6}$$

Its RMS value can be expressed as:

$$\Delta V_{\text{RMS}} = \sqrt{\frac{1}{T} \cdot \int_0^T \Delta V(t)^2 dt} = \frac{A}{\sqrt{2}} \cdot 2\pi \cdot f_{\text{analog}} \cdot t_{\text{jitter,clk}} \tag{C.7}$$

The SNR due to clock jitter is then the relation between the RMS signal voltage and the RMS voltage error:

$$\text{SNR}_{\text{jitter,clk}} = 20 \log\left(\frac{\frac{A}{\sqrt{2}}}{\frac{A}{\sqrt{2}} \cdot 2\pi \cdot \text{f}_{\text{analog}} \cdot t_{\text{jitter,clk}}}\right) = -20 \log(2\pi \cdot f_{\text{analog}} \cdot t_{\text{jitter,clk}}) \tag{C.8}$$

where f_{analog} is the analog input frequency to the ADC and $t_{\text{jitter,clk}}$ is the RMS clock jitter.

The RMS jitter of a signal is the most commonly used expression and is usually also used for SNR calculation. If a signal with a Gaussian-distributed jitter is considered, the RMS value for the jitter is equal to the standard deviation.

In case the jitter is given as peak-to-peak measure, the RMS jitter equates to $t_{\text{jitter,rms}} = \frac{t_{\text{jitter,clk}}}{2 \cdot \sqrt{2}}$. For some digital data transmission standards working with different data rates (e.g., SONET/SDH), the so-called *unit interval* (UI) jitter is typically used instead of the RMS jitter. This normalized jitter allows a consistent comparison of jitter for different data rates. In this case, the RMS jitter can be recovered using $t_{\text{jitter,rms}} = \frac{t_{\text{jitter,UI}}}{f_0}$ with the signal frequency f_0.

C.3.3 Relation Between Phase Noise and Jitter

Generally, the phase noise performance of a signal generator is described in terms of the single sideband phase noise (SSB phase noise) $L(f_{\text{off}})$, as shown in Figure C.6 for the R&S®SMA100B at 1 GHz.

The SSB phase noise $L(f_{\text{off}})$ is defined as the ratio of the phase noise power spectral density in 1 Hz bandwidth at a carrier offset f_{off} to the carrier power. The SSB phase noise describes the signal quality in the frequency domain.

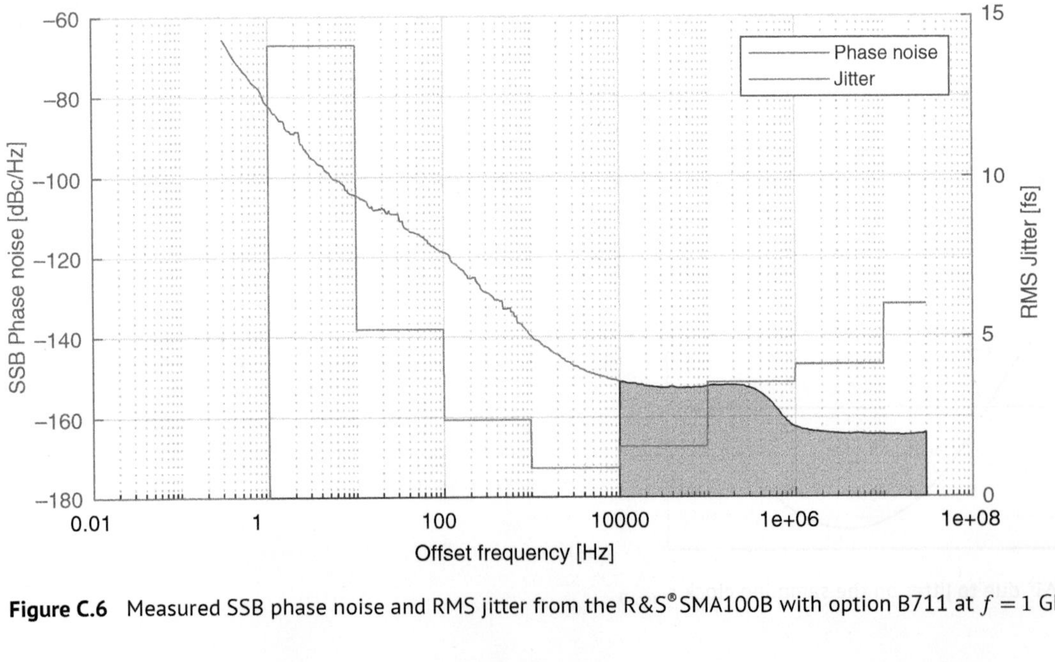

Figure C.6 Measured SSB phase noise and RMS jitter from the R&S®SMA100B with option B711 at $f = 1$ GHz.

There is a direct relationship between phase noise and jitter. RMS jitter is evaluated by integrating the phase noise within a defined bandwidth, as graphically illustrated by the shaded area in Figure C.6 for each decade from 1 Hz to 10 MHz and the last half decade. The RMS jitter can be calculated from the SSB phase noise using the expression:

$$t_{\text{jitter}} = \frac{1}{2\pi f_0}\sqrt{2\int_{f_1}^{f_2} 10^{\frac{L(f_{\text{off}})}{10}}\, df_{\text{off}}} \tag{C.9}$$

where f_0 is the carrier frequency, f_1 is the lower offset frequency for integration, f_2 is the upper offset frequency for integration, and $L(f_{\text{off}})$ is the SSB phase noise at offset f_{off}.

The factor 2 in front of the integral in Eq. (C.9) represents the contributions from both sidebands, as the integrand $L(f_{\text{off}})$ only describes the SSB phase noise.

In Figure C.6, the RMS jitter for the SSB phase noise is calculated for each decade and plotted as the red trace. As the figure clearly shows, the jitter contribution close to the carrier rises since the phase noise increases very rapidly for decreasing offsets in this area. However, a question remains: Why does the jitter per decade increase again for offsets above 1 kHz while the phase noise decreases in this area? The reason is the increasing integration bandwidth used for the jitter calculation. For example, the bandwidth for jitter calculation from 1 to 10 MHz is 9 MHz as compared to 90 kHz for the decade from 10 to 100 kHz. For this example, the integration bandwidth has increased by a factor of 100 (= 20 dB) whereas the phase noise contribution is only approximately 10 dB lower.

Since frequency offsets close to the carrier are of no consequence to high-speed applications (see also the next section), it becomes clear that wideband noise is the most significant parameter affecting jitter performance for high-speed ADCs.

If the total jitter contribution for multiple decades or for jitter sources is needed, it is necessary to take the root sum square of the individual jitter contributors per decade or adapt the integration limits accordingly.

C.3.4 Jitter Bandwidth

One important aspect when calculating jitter is the clock bandwidth (difference in offset frequencies f_1 and f_2 in Eq. (C.9)), which should be considered for integration.

The lower offset frequency f_1 depends on the frequency resolution of the system under consideration. Phase fluctuations, which are slower than the observation time for one FFT shot, do not add to the measurement result when evaluated in the frequency domain. Alternatively, looking upon this fact from the time domain perspective, phase error contributions with sufficiently low frequencies look roughly constant for short time periods. This roughly constant phase error can be ignored, as the noncoherent sampling viewed here is not phase-locked to any signal, so that the constant phase difference between analog input and clock is unknown. With those considerations, the lower frequency for jitter integration can be calculated as:

$$f_1 = \frac{f_{\text{clk}}}{\text{FFT length}} \tag{C.10}$$

where f_{clk} clock or sampling frequency of the ADC and FFT length number of FFT samples, e.g., 65536 for a 64k FFT.

Example: A 64k FFT performed with an ADC running at 1 GSPS encode rate leads to a frequency resolution of

$$f_1 = \frac{1\ \text{GHz}}{65536} = 15.2\ \text{kHz}$$

Thus, for most high-speed ADC applications, the phase noise close to the carrier brings no significant distortion to the sampling process. It becomes more dominant, though, in systems with longer capturing durations and thus higher frequency resolutions or in systems with lower sampling rate.

The upper offset frequency f_2 depends on the maximum bandwidth of the clock signal path. The analog input bandwidth usually equals the Nyquist band edge ($f_{\text{clk}}/2$), but could be much lower for narrowband applications. The bandwidth of the clock path depends on the filtering used and can range from several MHz up to a few GHz. Since Eq. (C.9) performs the integration for both sidebands of SSB phase noise, the upper offset frequency f_2 is always half the clock bandwidth, assuming symmetric phase noise sidebands. This will integrate the noise from the carrier, which is assumed to be in the middle of the clock path passband, to the edge of the clock path passband.

Example: For an ADC running at 125 MSPS encode rate and a clock bandwidth of 200 MHz, the upper offset frequency for jitter calculation is 100 MHz.

Example: The SONET communications standard [4] states its jitter requirements in terms of an integration bandwidth from 12 kHz to 20 MHz.

Although those values are aimed at data transmissions with rates up to 2488.32 Mbps using 155.32 MHz reference clocks, they are still widely used to characterize and specify jitter based on phase noise. For RF sampling applications with input frequencies up to 10 GHz and higher, IC package and technology for the ADC need to allow appropriate bandwidths. This also enables increasing clock bandwidths, in case there is no additional filtering implemented in the clock path.

Regardless of jitter bandwidth, the lowest possible jitter—and therefore the best SNR—can only be achieved by using proper filtering on both the analog input and clock paths.

One important aspect needs to be mentioned here: Since an ADC is a sampled system, the analog input signal is convolved with the clock signal in the frequency domain. This means that the spectral distribution of the clock signal shapes the sampled signal at the ADC output. If the bandwidth of the clock signal is greater than the Nyquist band, the wideband noise of the clock signal will be aliased completely into the Nyquist band [5], as graphically shown in Figure C.7.

As a result, the energy of the clock signal's wideband noise is accumulated many times within the Nyquist band, thus reducing the SNR significantly.

Example: For an ADC running at a clock rate of 80 MSPS and a clock bandwidth of 160 MHz, the wideband noise of the clock signal is aliased 4 times into the Nyquist band (40 MHz), causing a degradation of the SNR contribution due to the clock source of 6 dB.

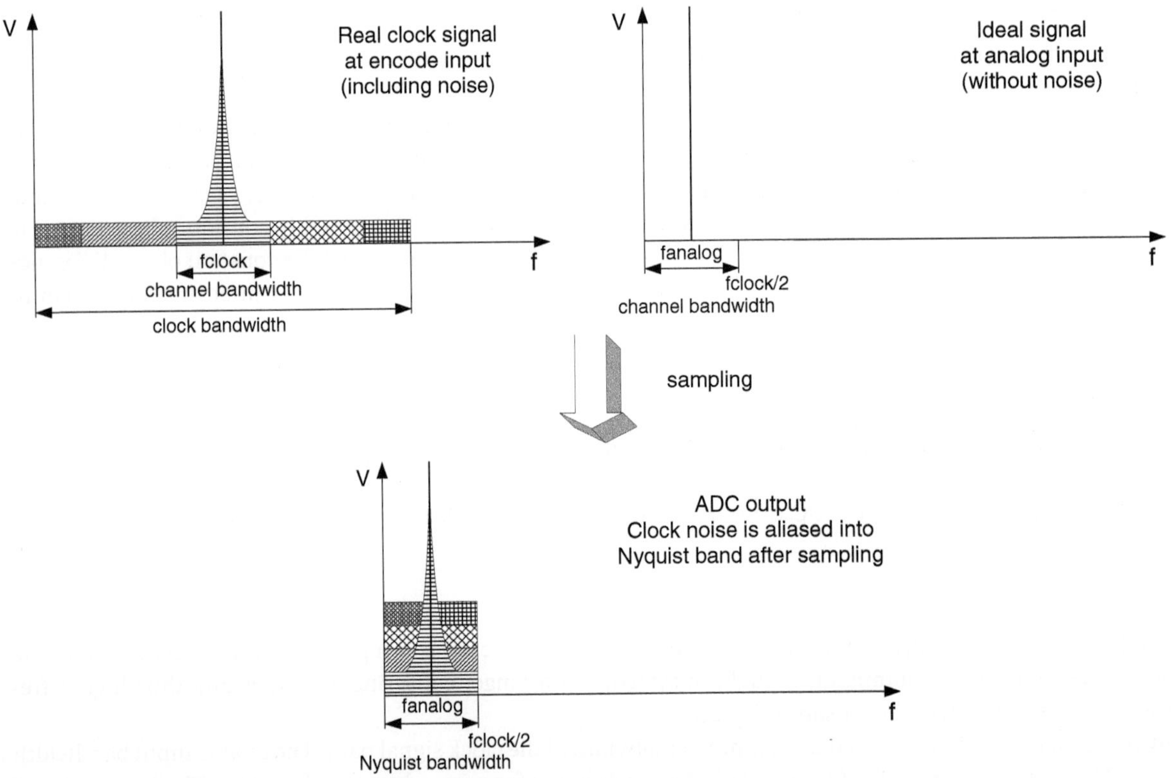

Figure C.7 Graphical illustration of clock noise aliasing into Nyquist band.

C.3.5 Real Signal-to-noise Ratio (in dB)

The SNR of a real ADC is degraded from the imperfections explained above, and it is illustrated for an arbitrarily chosen 14-bit ADC in Figure C.8.

Figure C.8 shows that for a real 14-bit converter, it is not possible to reach the ideal SNR of 86 dB even at very low input frequencies. This is due to factors such as jitter on the clock and analog input signals as well as factors associated with ADC design, including intrinsic aperture jitter, differential nonlinearities (DNL) in the ADC quantizer, and other internal noise such as thermal noise.

The real signal-to-noise ratio due to these effects can be approximated and summed together as:

$$SNR_{ADC,real} = -10 \log_{10}\left(10^{-\frac{SNR_{ADC,Noise}}{10}} + 10^{-\frac{SNR_{jitter}}{10}}\right) \tag{C.11}$$

with

$$SNR_{ADC,Noise} = -20 \log_{10} \sqrt{\left(\frac{U_{thermal}}{2^n}\right)^2 + \left(\frac{1+e}{2^n}\right)^2} \tag{C.12}$$

and

$$SNR_{jitter} = -20 \log_{10} \sqrt{(2\pi f_{analog})^2 \cdot \left(t^2_{jitter,ADC} + t^2_{jitter,clk} + t^2_{jitter,ana}\right)}, \tag{C.13}$$

where $SNR_{ADC,Noise}$ is the signal-to-noise contribution due to ADC noise and nonlinearities, SNR_{jitter} is the signal-to-noise contribution due to jitter, $U_{thermal}$ internal thermal noise of the converter, n number of bits, e average DNL of the converter, f_{analog} analog input frequency, $t_{jitter,ADC}$ intrinsic aperture RMS jitter of the converter, $t_{jitter,clk}$ RMS jitter of the clock source, and $t_{jitter,ana}$ RMS jitter of the analog signal source.

Equation (C.12) describes the device's intrinsic noise floor due to thermal noise and converter nonlinearities. This noise floor is constant over input frequency and can therefore be neglected at high signal frequencies, where SNR degradation due to jitter becomes dominant. However, at low input frequencies, these effects determine the maximum available SNR, e.g., 73 dB for the ADC shown in Figure C.8.

At higher frequencies, the SNR performance of a converter is determined mainly by jitter as seen in the set of curves in Figure C.8. Eq. (C.13) provides the mathematical description for the SNR performance of an ADC due to jitter and shows that the impact of jitter increases with higher signal input frequency.

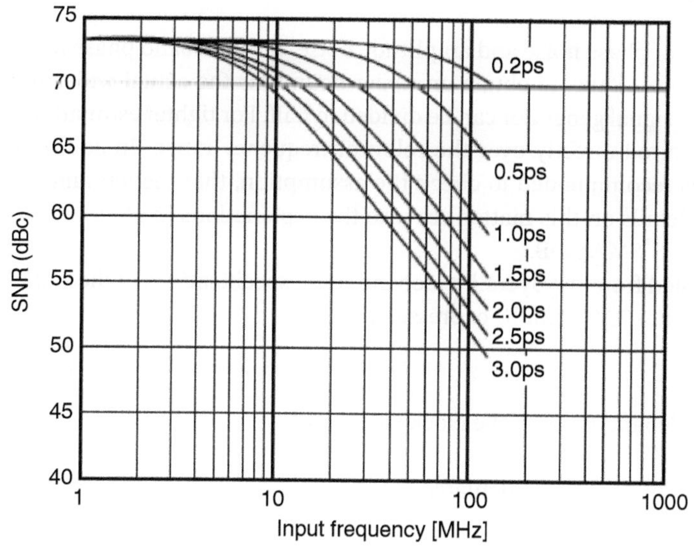

Figure C.8 SNR vs. analog input frequency and clock jitter [6].

Substituting (C.13) and (C.12) into (C.11) yields [2]:

$$\mathrm{SNR}_{\mathrm{ADC,real}} = -20\log_{10}\sqrt{(2\pi f_{\mathrm{analog}} t_{\mathrm{jitter}})^2 + \left(\frac{U_{\mathrm{thermal}}}{2^n}\right)^2 + \left(\frac{1+e}{2^n}\right)^2} \tag{C.14}$$

where t_{jitter} is the RMS sum of all jitter contributions.

$$t_{\mathrm{jitter}} = \sqrt{t^2_{\mathrm{jitter,clk}} + t^2_{\mathrm{jitter,ana}} + t^2_{\mathrm{jitter,ADC}}} \tag{C.15}$$

or in exponential form

$$\mathrm{SNR}_{\mathrm{ADC,real}} = -10\log_{10}\left(10^{-\frac{\mathrm{SNR}_{\mathrm{ADC,Noise}}}{10}} + 10^{-\frac{\mathrm{SNR}_{\mathrm{jitter,ADC}}}{10}} + 10^{-\frac{\mathrm{SNR}_{\mathrm{jitter,clk}}}{10}} + 10^{-\frac{\mathrm{SNR}_{\mathrm{jitter,ana}}}{10}}\right) \tag{C.16}$$

One common way to isolate the SNR contributors of an ADC is to perform an SNR measurement at a specific clock rate with a very low input frequency (yielding $\mathrm{SNR}_{\mathrm{ADC,Noise}}$) and a second SNR measurement with the same clock rate but at a very high input frequency (yielding $\mathrm{SNR}_{\mathrm{ADC,real}}$). Using Eq. (C.13), the SNR due to jitter can be calculated, but it is not possible to determine which part is from the ADC and which part is from the clock or analog source. If the jitter contribution of the analog source and clock source is known, it is possible to isolate the ADC jitter using Eq. (C.16).

Of course, this technique does not provide exact values of the ADC's noise and jitter. All individual considered noise sources and their relation to the SNR inside the ADC as well as from the signal generator are only approximations. As such, an input signal with high purity is desirable for ADC testing. In this case, SNR calculations for the ADC are less affected by modeling inaccuracies (e.g., simplifications like averaging over DNL effects or ignoring non-linearities) between SNR contribution and signal impurity, as the latter becomes smaller.

C.3.6 Relation Between Wideband Phase Noise, Jitter, and SNR at Clock Input

For most ADCs, the amplitude of the clock signal is hard-limited by the use of internal differential input and buffer stages. This minimizes amplitude effects. Therefore, only fluctuations of zero crossings (= phase noise) of the clock signal influence the sampling process.

As already pointed out, the close-in phase noise is negligible for most high-speed applications. Thus, often only the wideband phase noise contributes to the SNR performance of the ADC induced by clock imperfections in a significant way. The spectral density of the wideband phase noise is often assumed to remain constant for offsets >10 MHz to simplify jitter estimations. As this estimation can often turn out to be incorrect, other estimations or upper bounds are preferable for system design and evaluation.

Assuming that the overall wideband noise has equal parts of phase noise and amplitude noise, the wideband phase noise can be estimated to be 3 dB below the overall wideband noise. Thus, to obtain an upper bound for the actual wideband phase noise, the specified overall wideband noise level of the signal generator can be divided in half. For tighter estimations, the signal generator's overall wideband noise can be measured directly over the relevant frequency range. In addition, when considering sufficiently large integration bands, it is recommended to check the assumption, that the left and the right sideband contribute equally to the jitter. For further details on this matter, see also the explanations in section C.4.2 regarding the overall wideband noise performance of the R&S®SMA100B.

In the case that the wideband phase noise shall be estimated from the overall wideband noise specifications of the signal generator, the integrated noise for the jitter calculation (see Eq. (C.9)) can be expressed as:

$$\mathrm{Noise}_{\mathrm{integrated}} \lesssim \mathrm{BBNoise}_{\mathrm{clk}} + 10\log_{10}\mathrm{BW}_{\mathrm{clk}} - 3\ \mathrm{dB} \tag{C.17}$$

where $\mathrm{BBNoise}_{\mathrm{clk}}$ is the specified maximum broadband/wideband noise of the clock signal and $\mathrm{BW}_{\mathrm{clk}}$ is the clock bandwidth.

Replacing the integral in Eq. (C.9) by Eq. (C.17), the clock jitter can be calculated by:

$$t_{\mathrm{jitter}} = \frac{1}{2\pi f_{\mathrm{clk}}}\sqrt{10^{\frac{\mathrm{Noise}_{\mathrm{integrated}}}{10}}} \tag{C.18}$$

The multiplier of 2 for the integral, representing both SSB phase noise sidebands, is already included in $\mathrm{Noise}_{\mathrm{integrated}}$ in Eq. (C.17) as the complete clock bandwidth is used for integration.

Substituting t_{jitter} in Eq. (C.13) by Eq. (C.18) yields:

$$\text{SNR}_{\text{jitter,clk}} = -20\log_{10}\left(2\pi f_{\text{analog}} \frac{1}{2\pi f_{clk}} \sqrt{10^{\frac{\text{Noise}_{\text{integrated}}}{10}}}\right)$$

$$\text{SNR}_{\text{jitter,clk}} = -\text{Noise}_{\text{integrated}} - 20\log_{10}\frac{f_{\text{analog}}}{f_{\text{clk}}}$$

Substituting $\text{Noise}_{\text{integrated}}$ by Eq. (C.17) yields:

$$\text{SNR}_{\text{jitter,clk}} \gtrsim -\text{BBNoise}_{\text{clk}} - 10\log_{10}\text{BW}_{\text{clk}} + 3\text{ dB} - 20\log_{10}\frac{f_{\text{analog}}}{f_{\text{clk}}} \tag{C.19}$$

where $\text{BBNoise}_{\text{clk}}$ is specified maximum broadband/wideband noise of the clock signal, BW_{clk} is the clock bandwidth, f_{analog} is the analog input frequency of the ADC, and f_{clk} clock/sampling frequency of the ADC.

Equation (C.19) estimates the SNR of an ADC in the frequency domain using the wideband phase noise spectral density, whereas Eq. (C.8) calculates the SNR in the time domain using the jitter. Nevertheless, both equations should yield the very close results for operating regions where the approximation of constant wideband phase noise is adequate. For good approximations of accuracies, both results are usually within 1 dB.

C.3.7 Effects in Under- or Oversampled Systems

More and more modern communication systems implement direct conversion of high IF frequencies into the digital domain. To support lower clock frequencies, a technique called undersampling is used. In this case, the sampling rate is lower than the signal frequency, and the aliasing effect is used to downconvert the signal to the baseband. For this, the spectral purity requirements of the clock signal that is used are considerably higher in such systems as shown below.

An important factor here is the last term of the Eq. (C.19): $-20\log\left(\frac{f_{\text{analog}}}{f_{\text{clk}}}\right)$. It causes the SNR of a sampled system to degrade with 6 dB every time the analog input frequency doubles.

The reason for this becomes clear by graphically showing the effect of jitter for both oversampling and undersampling. In the case of undersampling, the slightest delay in sampling generates a huge variation in the sampled value. In contrast, oversampled systems experience only small sample variations for a given clock delay. This is depicted in Figure C.9.

This is an important element for ADCs used in undersampled IF stage systems since the SNR can deteriorate considerably. It is necessary to verify whether the performance of the clock source complies with the required specification. The requirements for the phase noise of the clock source in undersampled systems are much higher than in baseband systems. The performance in many IF-sampling radio architectures is limited by clock phase noise, not data converter performance.

C.3.8 Relation between Wideband Noise, Jitter, and SNR at Analog Input

Any calculation of the SNR contribution of an ADC's analog signal input must take the amplitude and phase noise of the analog signal source into account since no amplitude-limiting input stages are present at the analog input.

As with the clock input, calculating the SNR at the analog input requires to integrate over the signal generator's overall noise spectral density. This is often assumed to remain constant for offsets >10 MHz or to remain below a specified wideband noise spectral density. Alternatively, it can be measured for different frequencies throughout the integration range.

In order to determine a lower bound of SNR_{analog} and therefore an upper bound of its contribution to the total SNR, we can once again use the generator's wideband noise specification. Eq. (C.19) describes an estimate for the SNR due to phase noise at the clock input. For the analog signal input, the relation must be changed to:

$$\text{SNR}_{\text{analog}} \gtrsim -\text{BBNoise}_{\text{analog}} - 10\ \log_{10}\text{BW}_{\text{analog}} \tag{C.20}$$

where $\text{BBNoise}_{\text{analog}}$ is the specified maximum overall wideband noise spectral density of the analog signal and $\text{BW}_{\text{analog}}$ is the bandwidth of the ADC analog input signal (usually $\frac{f_{clk}}{2}$ or below).

As the term $\frac{f_{\text{analog}}}{f_{\text{clk}}}$ was only relevant for jitter calculation, the relation of input and clock frequency has no effect on the SNR contribution of the analog signal source to the ADC's performance.

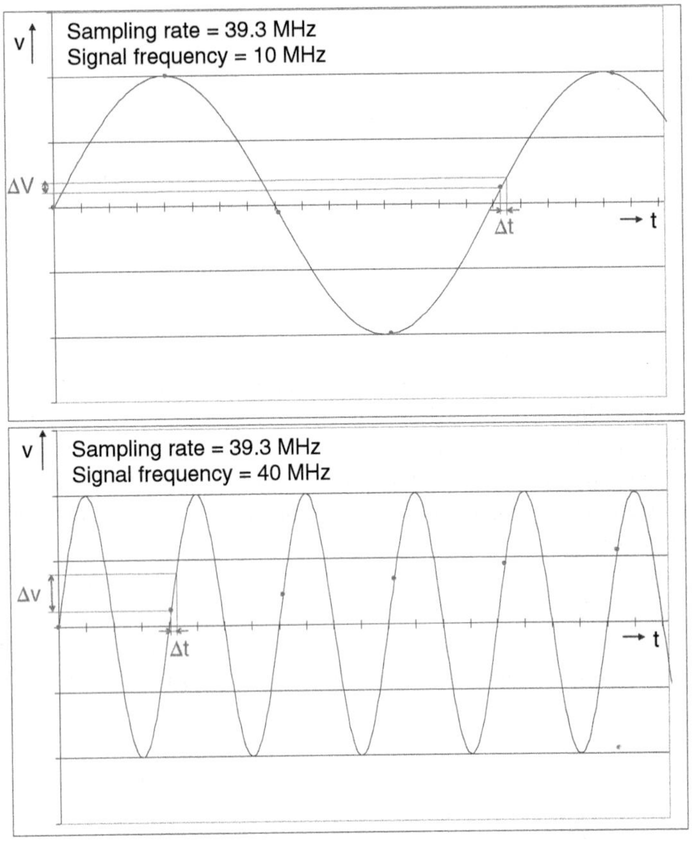

Figure C.9 Effect of jitter on over- and undersampled systems.

C.3.9 Distortion and Spurious

Besides noise performance, the spectral purity due to spurious and signal distortion is the most important parameter of an ADC. Definitions of some commonly used expressions are listed below [7]:

- Signal-to-noise and distortion in dB
  The signal-to-noise and distortion (SINAD) value is the ratio of the RMS signal amplitude to the RMS level of the sum of all spectral components, including harmonics but excluding DC and the desired signal. The difference between SNR and SINAD is the energy contained in the first six harmonics.
- Total harmonic distortion in dBc
  Total harmonic distortion (THD) is the ratio of the RMS signal energy to the RMS value of the sum of the first six harmonics.
- Spurious-free dynamic range
  The spurious-free dynamic range (SFDR) is the ratio of the RMS level of the signal to the RMS level of the peak spurious spectral component at the ADC output.

To measure these parameters, an analog signal source with very low harmonic distortion and nonharmonics is required.

- Third-order input intercept point in dBm
  The third-order input intercept point (IIP3) is determined by applying a two-tone signal to the analog input of the ADC and measuring the emerging intermodulation products.
  The IIP3 is normally far beyond the maximum input level of the analog input, but it is a measure of the linearity of the ADC's input stage.

C.3.9.1 Harmonics

Harmonics are integer multiples of the fundamental signal frequency that might be generated at any active stage within the RF path due to semiconductor nonlinearities. The offset frequency to the carrier can be evaluated exactly, and it makes the use of suitable filters for suppression possible in those cases, where the range of input frequencies is sufficiently small.

Generally, the signal generator's harmonic distortion is less critical than the nonharmonic performance since the frequencies of the harmonic spurs are known (integer multiple of the fundamental signal) and since these spurs are located far from the carrier in the case of most high-speed applications.

Harmonics applied to the ADC's clock input normally do not influence the performance of the converter since the zero crossing of the clock signal is not shifted by harmonic signal components. Furthermore, the ADC's clock input stage normally converts an applied sine signal into a square signal containing strong harmonic signal components in any case.

For the analog input signal, harmonic distortion is still critical to the ADC's performance, as it aliases into the sampled signal. However, for many applications the out-of-the-box harmonics performance of the R&S®SMA100B may be already sufficient. If this is the case, no additional filtering is needed, and there is no restriction on the analog input frequency due to any fixed filter pass or stop bands. As the harmonics of the R&S®SMA100B reach levels around and well below −70 dBc or even lower for higher-order harmonics, this is the preferable route for many applications.

In case the harmonic distortion has to be even less than what the signal generator already provides, additional filtering is required for the input signal. For this, typically lowpass or bandpass filters or combinations of both are used, depending on the specific requirements [7]. With the high performance of the R&S®SMA100B, the requirements for the filters can be lowered.

Lowpass filters suppress the harmonics effectively and have the advantage that the frequency of the source signal can still be varied across a specific range. However, part of the wideband noise is still passed to the ADC, and the noise contribution of the analog signal source can only be improved partially.

Bandpass filters have an identical effect on the harmonics, but they also have the additional advantage that the noise and subharmonics are suppressed more effectively. The disadvantage is that the test can only take place within a narrow frequency range, so several filters are needed in order to perform tests over a wide frequency range.

For both filter types, the high-frequency suppression of the used filter must be examined carefully in order to eliminate high-order harmonics of the input signal. This signal could possibly cross the clock frequency, or even a multiple of it and could alias back to the Nyquist band. If multiple filters are combined in series, a fixed attenuator between the filters should be used to improve the matching between these filters. Otherwise, the filter characteristic may become much worse than expected.

Example: Assume that an ADC is running at a clock rate of 1 GSPS and that the analog input frequency is 428 MHz. Due to insufficient filtering, the 6th harmonic of the input signal at 2568 MHz is only suppressed by 5 dB. Therefore, it is applied with a level of, for example, −90 dBc to the input of the converter. As a result, an alias product might appear in the output spectrum of the converter. This harmonic at 2568 MHz lies between 2.5 and 3 GHz and therefore between two and a half and three times the sample rate:

$$f_{\text{alias}} = (3 \cdot f_{\text{clk}}) - (6 \cdot f_{\text{analog}}) = 432 \text{ MHz}$$

C.3.9.2 Nonharmonics

For signal generators, nonharmonic signal components are more critical than harmonics because adequate filtering is generally not possible. The frequency of a nonharmonic spurious cannot easily be predicted, which means that the offset to the carrier is unknown. A spurious line can appear very close to the carrier (e.g., 20 kHz) for one frequency and very far from the carrier (e.g., 100 MHz) at another frequency.

Nonharmonics are critical to both the clock and the analog signal input of an ADC. In the general case, a single nonharmonic spur is equivalent to amplitude modulation as well as phase modulation of the carrier due to its complex nature in the baseband. Thus, the zero crossings of the clock signal at the ADC input stage fluctuate as well.

For the special case of a pair of spurs that is symmetric around the carrier, a modulation of only the amplitude or only the phase is possible depending on the phase difference between the two spurs as well as between spur and carrier. If the spurs are added in-phase to the carrier, only amplitude modulation occurs. When the spurs are added in quadrature to the carrier instead, the baseband signal has almost unity magnitude but a sinusoidally modulated phase. In the general case, the pair of spurs modulates both the amplitude as well as the phase of the carrier with distorted sinusoidal, though.

For a spurious at the clock input, the level of this nonharmonic at the ADC output signal depends on the relation between the clock and analog signal frequency, just like for the phase noise of the clock signal. Therefore, the spur level of the clock signal must be corrected by the following expression:

$$k = 20\log\left(\frac{f_{\text{analog}}}{f_{\text{clk}}}\right) \tag{C.21}$$

Example: A spur at −80 dBc is present at the clock signal of an ADC running with 20 MSPS clock rate and an analog input frequency of 50 MHz. According to Eq. (C.21), the spurious appears at the converter output at a level of −72 dBc, reducing the SFDR performance of the ADC drastically.

Figure C.10 shows a symmetrical pair of spurs for a 100 MHz carrier frequency at 20 kHz offset generated by a low-performance signal source with a (single sideband) spur level of −70 dBc. Without knowing the relative phase difference between both spurs and to the carrier, it is not possible to predict the exact nature of the modulation. Nevertheless, an upper bound for the generated distortion is possible. Additionally, expressions for SNR and SFDR are valid regardless of the exact modulation, as both only consider the spurious energy. For jitter induced by spurs, it is possible to assume that the symmetric pair is added in quadrature and will cause the maximum possible phase modulation.

When selecting an appropriate signal source for ADC testing, the spurious specification of the source must be carefully considered over the complete frequency range. As a first approach, the performance of the signal source should be approx. 10 dB better than the performance of the tested DUT.

C.3.9.3 Signal deterioration from Nonharmonics

To obtain the RMS jitter of such a symmetrical pair of spurs that cause pure phase modulation, one can evaluate Eq. (C.9) for these spurs and gets:

$$t_{\text{jitter,rms}} = \frac{\Delta\varphi_{\text{rms}}}{\omega_0} = \frac{\frac{1}{\sqrt{2}}\cdot 2\cdot 10^{\frac{a_{\text{spur}}}{20}}}{2\pi\cdot f_0} = \frac{10^{\frac{a_{\text{spur}}}{20}}}{\sqrt{2}\pi f_0} \tag{C.22}$$

where a_{spur} is the (SSB) level of the spur pair in dBc and f_0 is the carrier frequency.

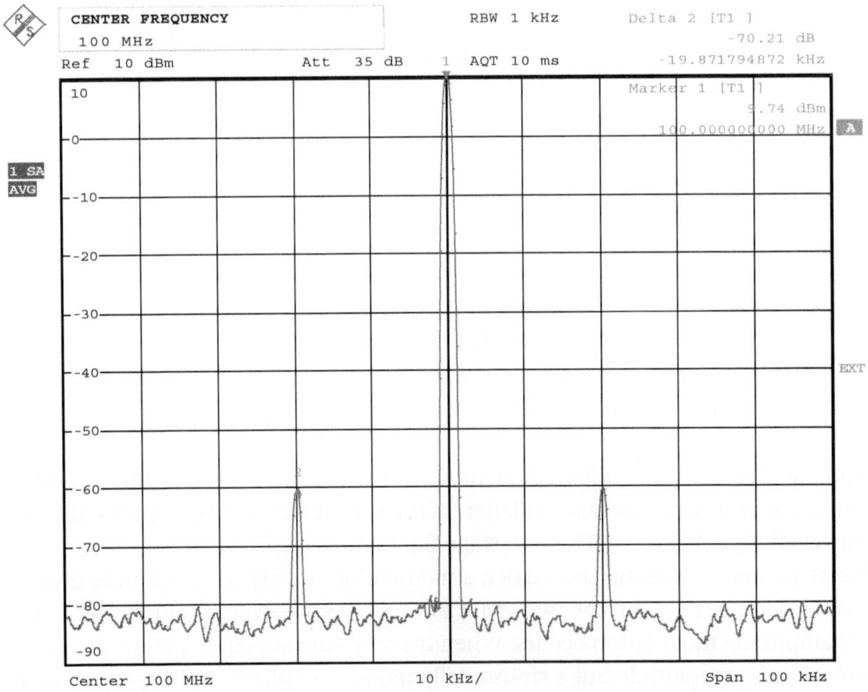

Figure C.10 Symmetrical pair of nonharmonic spurs at 20 kHz carrier offset.

The SNR contribution of a phase modulating pair of nonharmonic spurs at the clock input of an ADC can be calculated by substituting Eq. (C.22) into Eq. (C.8):

$$\mathrm{SNR}_{\mathrm{spur,clk}} = -20\log\left(\frac{10^{\frac{a_{\mathrm{spur}}}{20}} \cdot 2\pi \cdot f_{\mathrm{analog}}}{2\sqrt{2}\pi \cdot f_{\mathrm{clk}}}\right) = -a_{\mathrm{spur}} - 3\ \mathrm{dB} - 20\log\left(\frac{f_{\mathrm{analog}}}{f_{\mathrm{clk}}}\right) \tag{C.23}$$

For a symmetric pair of spurs at the analog input of an ADC, Eq. (C.23) simplifies to:

$$\mathrm{SNR}_{\mathrm{spur,analog}} = -a_{\mathrm{spur}} - 3\ \mathrm{dB} \tag{C.24}$$

The SFDR due to a nonharmonic spurious at the analog signal input of an ADC is equal to the spurious distance of the signal source, which is the relation of spur level to signal level.

For a nonharmonic applied to the clock input, again the relation between analog frequency and clock rate must be considered by applying the correction factor k (see Eq. (C.21)) to the influence of the spurs on the SFDR value.

C.3.9.4 Intermodulation

For testing the third-order intercept point (IIP3) of a DUT, a two-tone signal at the analog ADC input is required. A common way to generate a two-tone signal is to combine the output signals of two signal generators, working at different frequencies, by using a power combiner. However, issues can arise with modern signal generators, which are equipped with an automatic level control (ALC) loop. This can happen due to limited isolation of the power combiner. A small portion of the signal generated by one signal generator travels through the combiner to the output of the other signal generator. Consequently, this part of the signal couples to the signal generator's output detector, resulting in an amplitude modulation of this generator's output signal with the difference frequency of the two generators. Figure C.11 shows an example with the corresponding AM sidebands visible in the power combiner's output spectrum.

This can be problematic in cases where the signal generator's IM3 products are so high that they distort the ADC measurement. Depending on accuracy requirements, one might consider a margin of 10 dB between the generator's IM3 products and the ones from the ADC. Often, the ALC has to be switched off in order to achieve this.

The R&S®SMA100B is able to prevent this problem in many situations. Depending on the hardware options and the configured settings, you can take advantage of the ALC while still getting very low IM3 artifacts at the same time for carrier distances of a few 100 kHz already. The dynamically adapted ALC bandwidth greatly reduces the amount by which the two ALC loops interact with each other. During transient, the ALC bandwidth allows fast settling behavior. After the desired level is reached, the bandwidth is reduced in order to filter out spectral components from other sources. Therefore, the

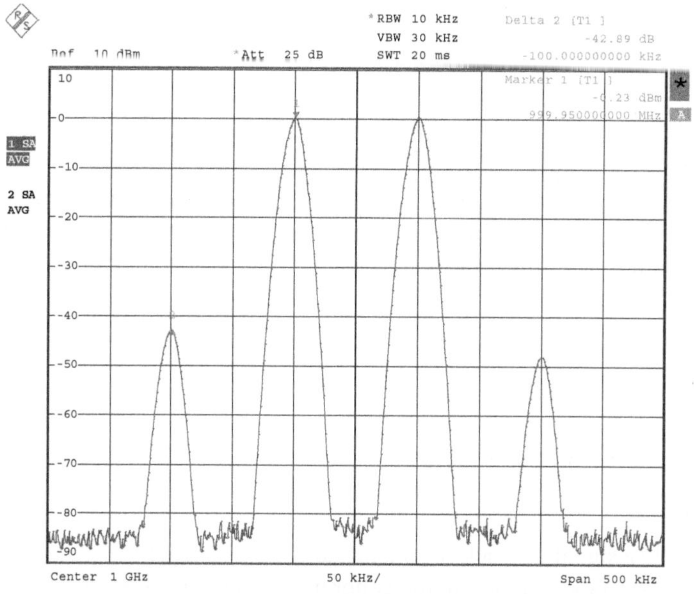

Figure C.11 Example output spectrum showing interacting ALC loops.

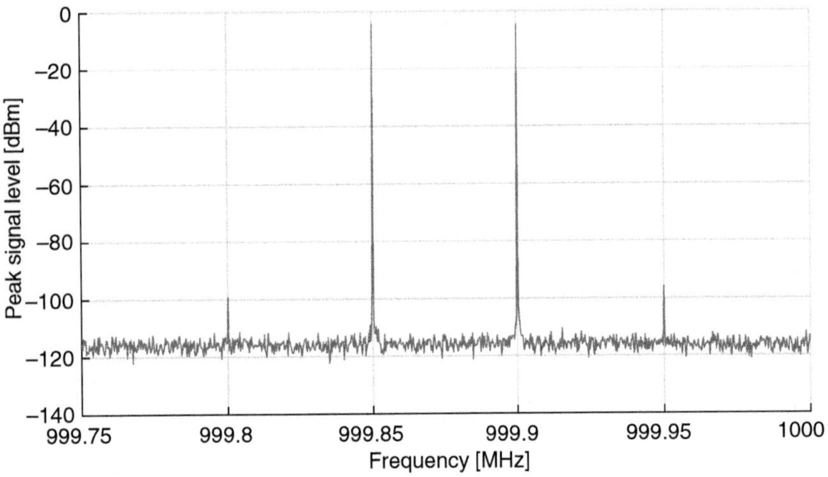

Figure C.12 Output spectrum for two R&S®SMA100B.

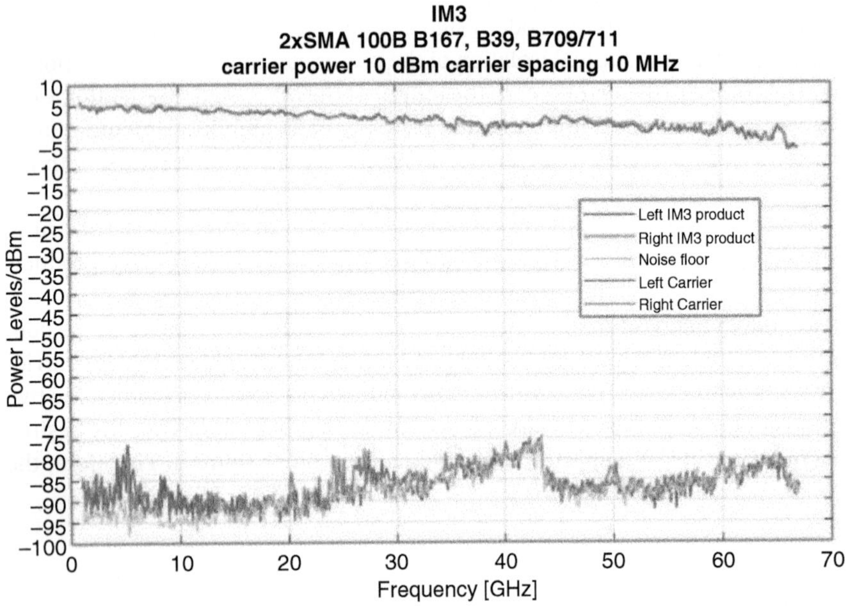

Figure C.13 IM3 products vs. frequency.

ALCs produce much smaller IM3 artifacts that could distort the two-tone test-signal. Figure C.12 shows the output spectrum for two R&S®SMA100B signal generators combined with a power combiner. As showcased in Figure C.13 the used R&S®SMA100B was measured with a minimum IM3 attenuation of 75 dB over its frequency range, using a coupler with 20 dB isolation and a carrier spacing of 10 MHz.

C.4 Performance of the R&S®SMA100B for ADC Testing

Spectral purity is by far the most limiting factor for a potential signal source and is deteriorated mainly by noise, harmonics, and nonharmonic spurs. Thus, this application note focuses predominantly on the generator's performance in each of those metrics.

Ultimately, these factors determine a potential candidate for a signal or clock source for an ADC test setup.

This section provides data for the spectral purity of an R&S®SMA100B analog signal generator with the R&S®SMAB-B711 ultra-low phase noise option installed. The impact on ADC performance measurements is described by means of examples.

C.4.1 Phase Noise Performance

Figure C.14 highlights the outstanding phase noise characteristic of the R&S®SMA100B measured with the R&S®FSWP50.

As explained in Section C.3.3, the jitter is mostly dominated by the wideband phase noise. As noise measurements for large bandwidths can be time consuming and difficult to calculate with in early design stages, Section C.3.6 introduced the approximation of a constant wideband phase noise. Inspecting the R&S®SMA100B's SSB phase noise performance, it becomes clear that the phase noise is not constant for large offsets either. Thus, to obtain accurate phase noise and jitter figures, a phase noise measurement for the full band in question can often be necessary.

Nevertheless, wideband approximations are still possible in many cases. Those approximations estimate the wideband noise to be constant starting at a given frequency offset. To find this offset, a compromise has to be found. If the offset frequency is chosen too low, the approximated noise density is unreasonably high. If, on the other hand, the offset frequency is chosen too high, the approximated noise density might be too low of an estimate. In Figure C.14, the phase noise density at 10 MHz is chosen as a compromise between those two aspects. For most of the carrier frequencies and offset frequencies, the noise density at 10 MHz serves as an upper bound, while still being on the lower side of the slope after 1 MHz. The quality of the approximation always depends on the relevant carrier frequencies and clock bands.

To provide guidance for when the constant wideband phase noise approximation might be feasible, Figure C.15 compares the SSB phase noise density at 10 MHz to the peak phase noise at offsets larger than 10 MHz and to the phase noise power average in the upper band between 10 MHz and 1 GHz away from the carrier.

The figure shows, that for most applications, the phase noise density at 10 MHz is a reliable upper bound for the wideband phase noise. However, the figure also shows that for a few carrier frequencies, the approximation obtained from the phase noise at 10 MHz might be a little low for an upper bound, as it is the case for a carrier frequency of 5.5 GHz for example. For carrier frequencies above 10 GHz, the phase noise density at 100 MHz is a viable candidate for the wideband performance evaluation as well.

For the clock synthesizer, the SSB phase noise power spectral density at 10 MHz can be a conservative approximation without estimating the noise performance too high as well, as visible in Figure C.16.

In case of clock inputs with very large bandwidths, see the next section for a more in-depth depiction of the R&S®SMA100B's wideband noise characteristics.

C.4.2 Wideband Noise Performance

To discuss the wideband noise performance of a signal generator, a distinction must be made between wideband phase noise, wideband amplitude noise and overall wideband noise, which is the sum of both.

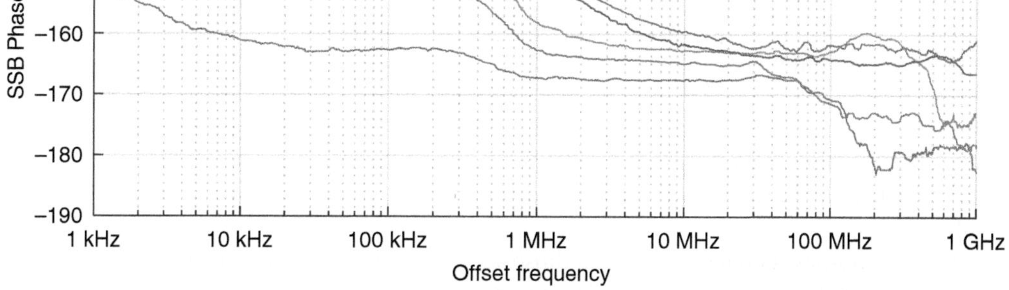

Figure C.14 R&S®SMA100B (with B711) SSB phase noise (meas.).

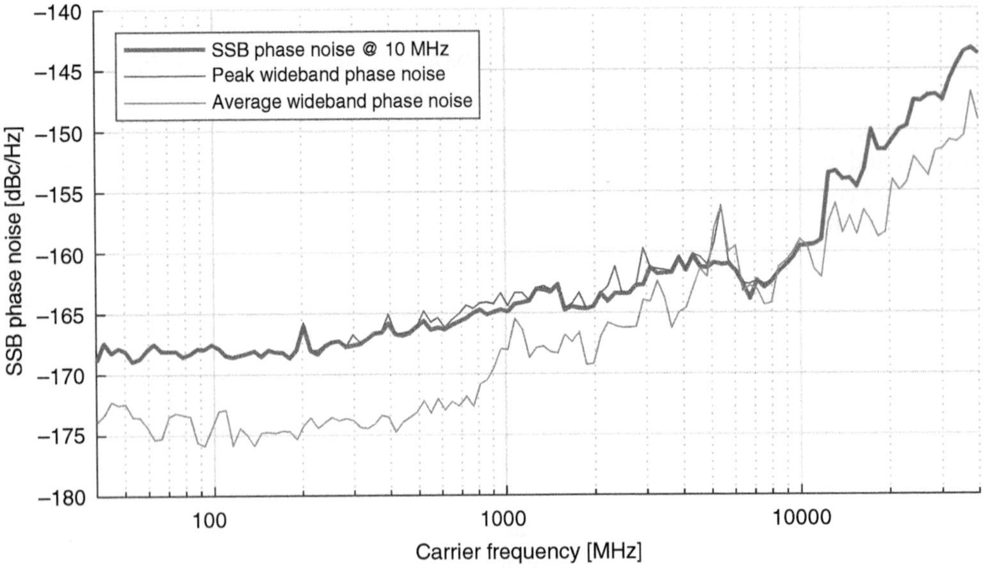

Figure C.15 R&S®SMA100B (with B711) wideband phase noise (meas.)

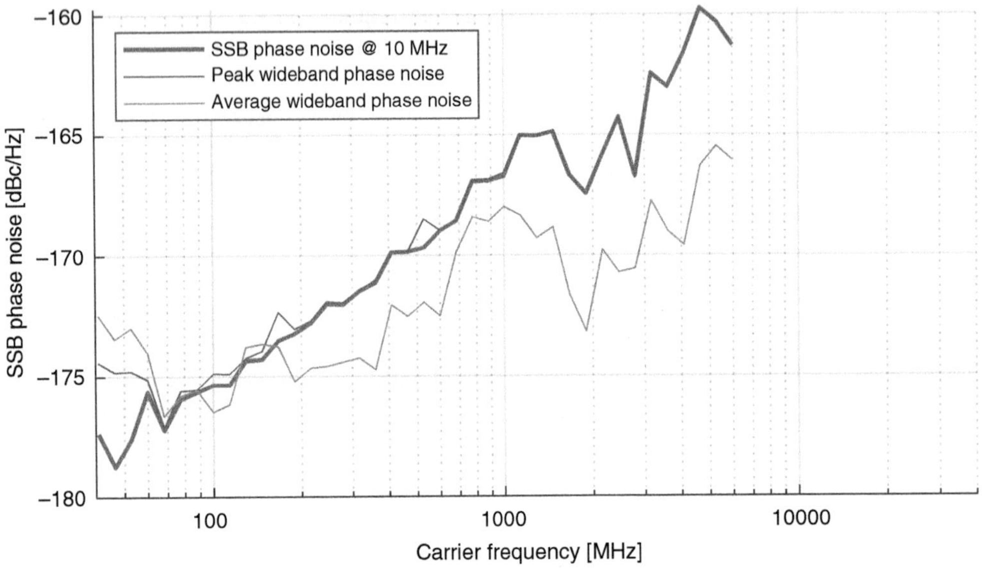

Figure C.16 R&S®SMA100B (with B711) B29 clock synthesizer wideband phase noise (meas.).

Figure C.17 shows measured wideband noise for carrier frequencies up to approximately 100 MHz. The R&S®SMA100B datasheet contains plots showing measured wideband noise for higher carrier frequencies.

As described in the previous section, the noise spectral density for large offsets may deviate from the wideband noise specification at a given offset. This section contains measurements of the noise distribution over large offset frequency ranges.

Usually, for large offsets one assumes that amplitude and phase noise contribute equally to the overall noise. Therefore, the wideband phase noise is normally 3 dB lower than the overall wideband noise.

As the R&S®SMA100B has different subassemblies for the frequency ranges 8 kHz–6 GHz, 6 GHz–20 GHz, and 20 GHz–40 GHz, it is advantageous to evaluate the wideband noise performance for every board each to observe the different behaviors in the respective bands. Due to the band-pass behavior of each of those subassemblies, they do not produce significant noise energy in frequency regions outside of their respective operating ranges.

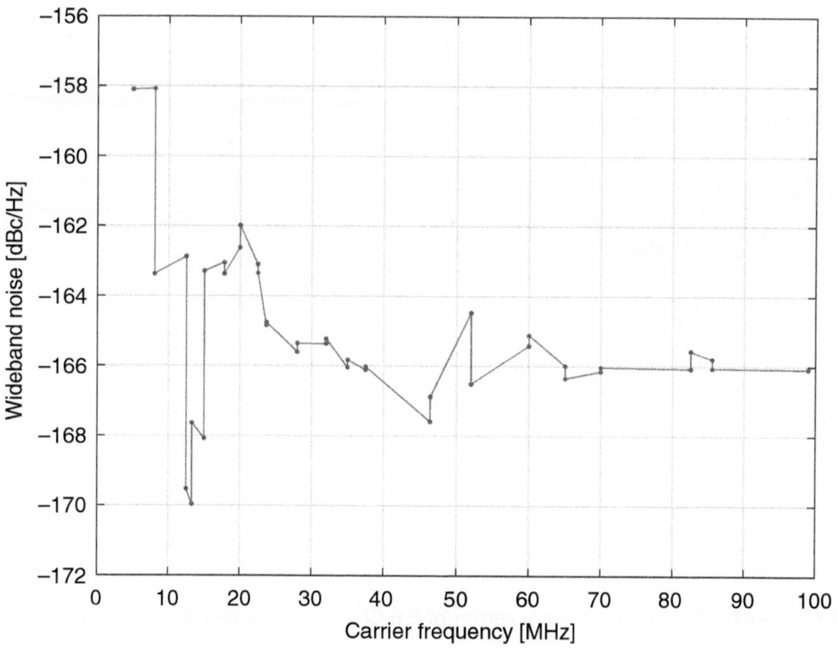

Figure C.17 Measured wideband noise at 10% (of carrier frequency) offset and 10 dBm of an R&S®SMA100B with B711.

For this, we show an R&S®SMA100B's generator noise spectral density for selected carriers as well as the specified wideband noise power spectral density from the datasheet. The appendix will show families of curves for the different frequency ranges for additional carrier frequencies. In the plots below, the carrier frequency as well as both noise spectral densities are shown as in Figure C.18. The vertical line represents the carrier frequency with a small horizontal bar indicating the specified wideband noise at an offset of 10 MHz–40 MHz depending on the carrier frequency. The measured noise power spectral density is plotted in the same color in dBc/Hz vs. the measured frequency.

Thanks to the R&S®SMA100B's excellent noise performance, the measured power spectral densities are far below −150 dBc/Hz, often by large margins. This also means, that the measured noise power spectral densities can and most likely will contain contributions from the signal analyzer's noise and distortion, too. Furthermore, the wideband noise behavior of the R&S®SMA100B also depends on the hardware options built into the signal generator, as different RF modules and chains may have different frequency responses, noise figures, and nonlinearities.

Figure C.19 demonstrates the wideband noise characteristics for carrier frequencies chosen arbitrarily throughout the different sub-paths in the 8 kHz–6 GHz range. For each carrier frequency, the carrier, its measured noise power spectral density as well as the wideband noise specification is visible.

The measurement shows, that typically the actual wideband noise is neither flat nor constant. Often, it is also not symmetric around the carrier and in many cases the noise power spectral density will rise above values actually measured at 10 MHz–40 MHz around the carrier for the phase noise plots. Although the measured noise at 10 MHz or 40 MHz is always below the instrument's specified maximum noise level at this offset, it might be possible that at large offsets for some carrier frequencies, the noise level is above its value at the specified frequency offset. This depends heavily on the frequency response of the RF chain and thus on the carrier frequency. Using the specified wideband noise as an upper bound for the whole bandwidth of interest, one obtains a safe upper bound for the integrated noise, as in many cases the noise power spectral density is 10 dB–20 dB below the specification.

The same characteristic for the 6 GHz–20 GHz range can be seen in Figure C.20. As in the range up to 6 GHz, the noise power spectral density usually is neither constant over the frequency, nor symmetric around the carrier. Again, using the specified wideband noise at the given offsets provides a conservative upper bound for the integrated noise for the same reasons. The increased noise at roughly 4–6 GHz from Figure C.19 does not appear for carrier frequencies above 6 GHz. As those carriers are created by different subassemblies, they also exhibit different noise characteristics. All RF chains have approximately passband behavior. They do not produce a significantly higher noise density in frequency bands of other RF

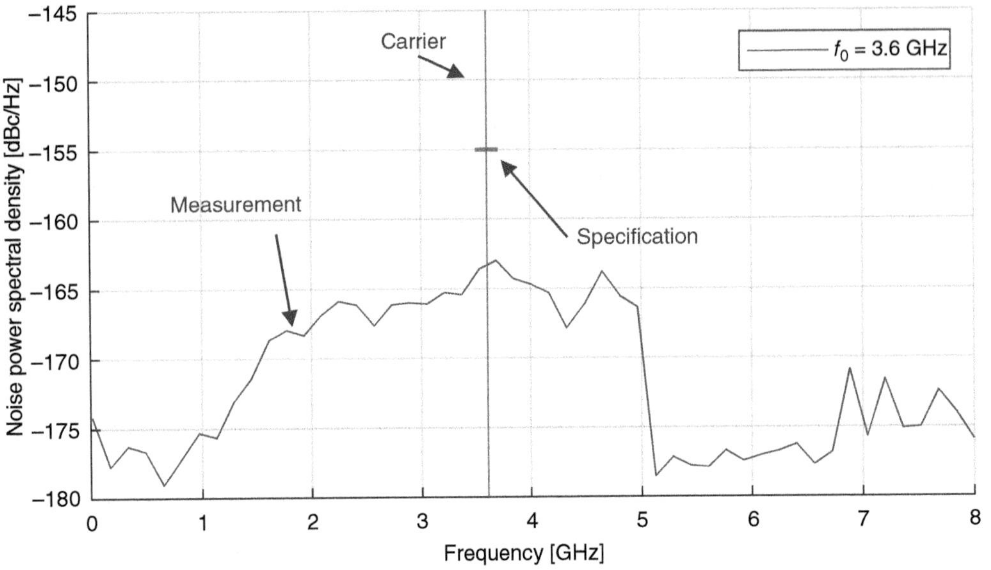

Figure C.18 Example noise characteristic of an R&S®SMA100B (with B711) for a carrier of 3.6 GHz and an output level of 10 dBm (meas.).

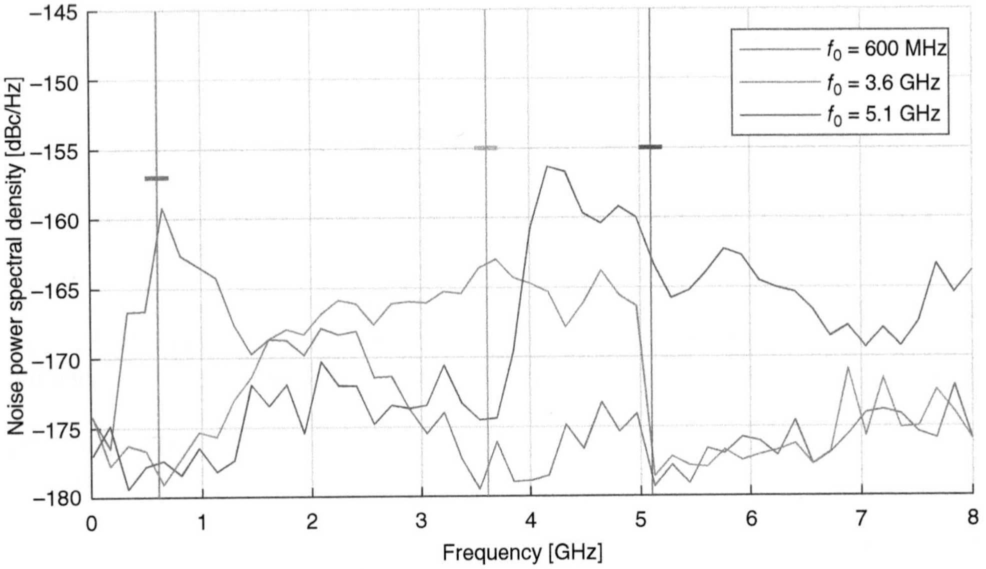

Figure C.19 Wideband noise characteristic of an R&S®SMA100B (with B711) in the 8 kHz–6 GHz range at an output level of 10 dBm (meas.).

chains than shown in the figures at the band edges. For Figure C.20, this means for example that the noise power spectral density of carriers between 6 GHz and 20 GHz measured at frequencies below 4 GHz stay well below −170 dBc/Hz.

For frequencies above 20 GHz, in Figure C.21 a wideband noise behavior as expected can be seen. The noise power spectral density directly around the carrier is at its maximum, although typically still 5–10 dB below the 40 MHz offset specification. At offsets >2 GHz, the noise often reaches its wideband level at approximately 5–10 dB lower again. In this frequency range, the assumption of flat and constant wideband noise power spectral density allows for conservative upper bounds of the integrated noise. Also, for the carrier frequencies where this is not the case, as for example for the 20 GHz curve, the noise characteristic still allows flat and constant wideband noise assumptions.

Finally, Figure C.22 depicts the wideband noise characteristic of the B29 Clock Synthesizer.

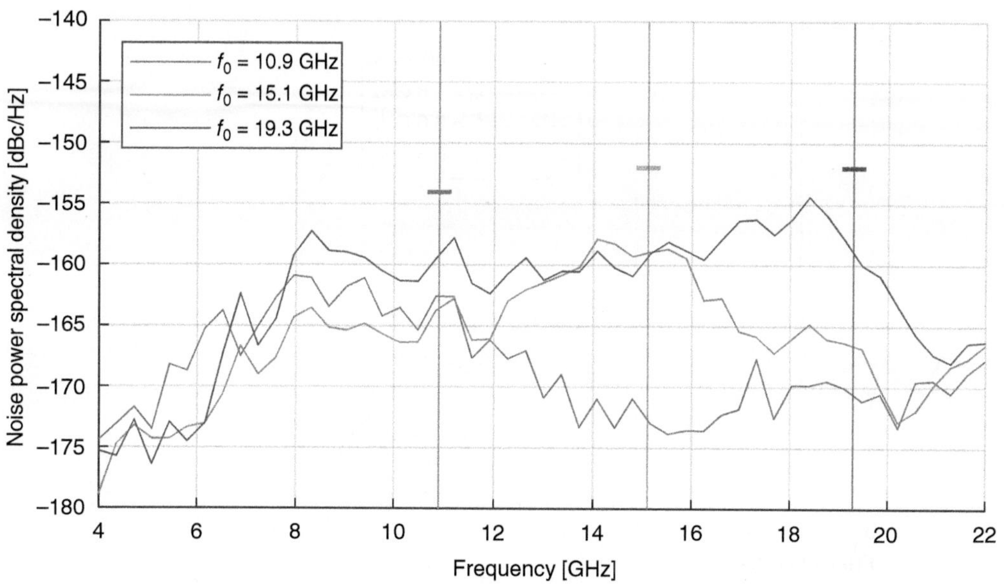

Figure C.20 Wideband noise characteristic of an R&S®SMA100B (with B711) in the 6 GHz to 20 GHz range at an output level of 10 dBm (meas.).

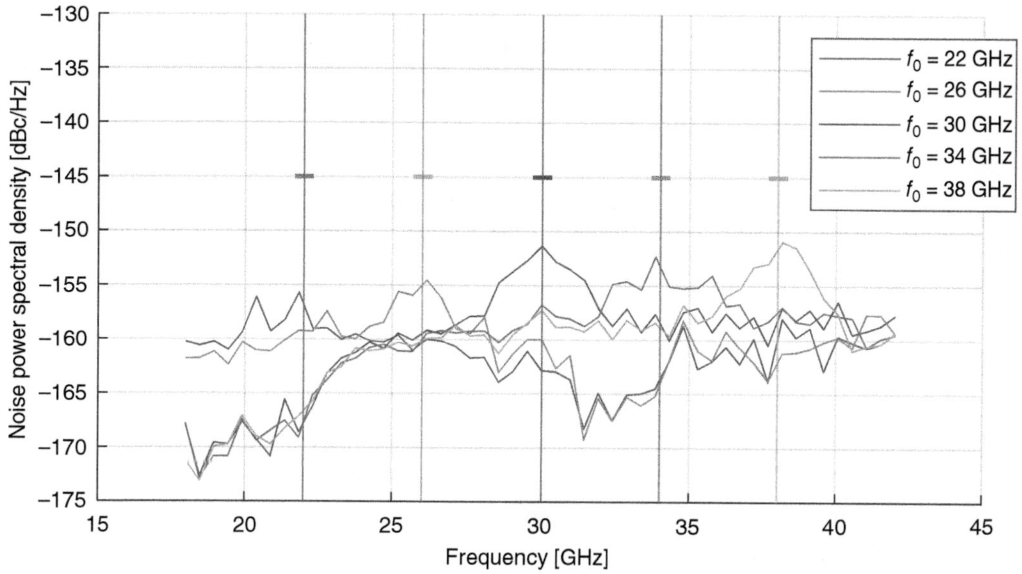

Figure C.21 Wideband noise characteristic of an R&S®SMA100B (with B711) in the 20 GHz to 40 GHz range at an output level of 10 dBm (meas.).

All measurements clearly show different characteristics for different sub-paths. To help a rough overview over the observable characteristics and the variations therein, more measurements are provided in the appendix.

The measurement results presented in this chapter are not only relevant for wideband noise considerations at the analog input. Assuming that for large offsets, the phase noise is 3 dB below the overall noise, the noise characteristics are also relevant when estimating SNR degradations due to jitter at the clock input. For large clock bandwidths, both sidebands are relevant and should be considered individually instead of just assuming symmetry around the carrier.

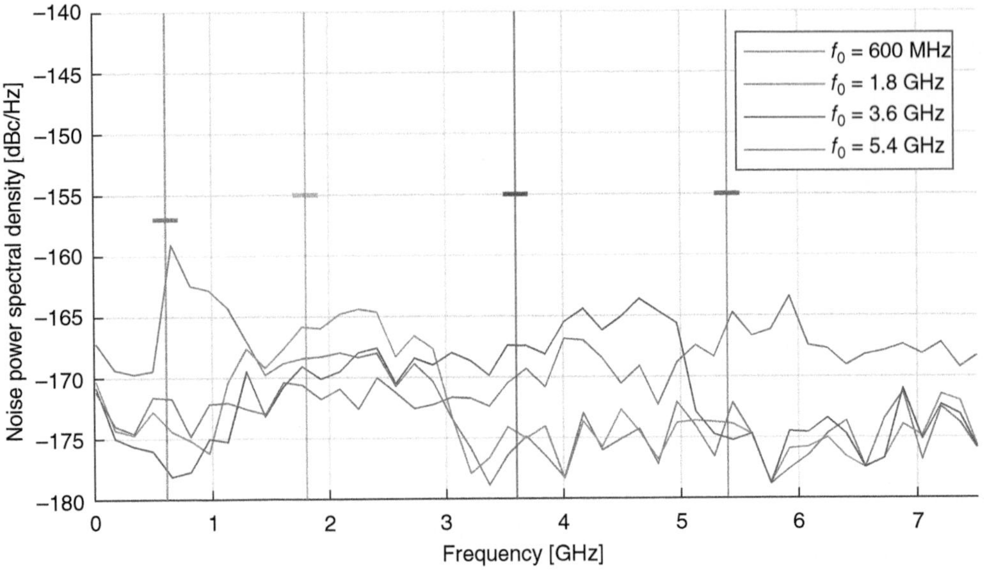

Figure C.22 Wideband noise characteristic of an R&S®SMA100B (with B711) B29 clock synthesizer for a single-ended sine with power 10 dBm (meas.).

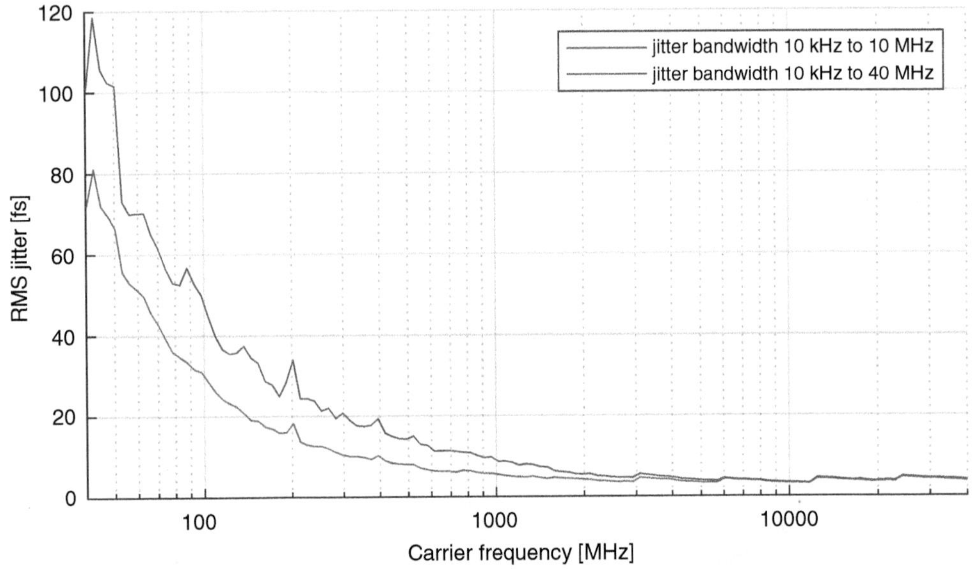

Figure C.23 RMS jitter vs. clock frequency for the RF output of an R&S®SMA100B (with B711) (meas.).

C.4.3 Jitter and SNR Performance of the R&S®SMA100B as a Clock Source

Figures C.23 and C.24 show the RMS jitter derived from phase noise measurements for the two clock bandwidths of 20 and 80 MHz for the R&S®SMA100B and its clock synthesizer, respectively. Although the phase noise decreases for decreasing frequencies, the RMS jitter rises for decreasing frequencies.

While the clock synthesizer's phase noise and jitter are inferior to the main synthesizer's performance for high frequencies, the B29 clock synthesizer is clearly the better choice for narrow-band clock inputs with clock frequencies not rising above a few hundred MHz. In this region, the clock synthesizer shows largely improved jitter performances.

Figure C.25 shows the SNR contribution due to clock jitter derived from the RMS jitter as shown above in Figure C.23 using Eq. (C.8). The SNR is shown vs. the clock frequency for three different analog input frequencies (10 MHz, 100 MHz, and 1 GHz) and for clock bandwidths of 20 and 80 MHz.

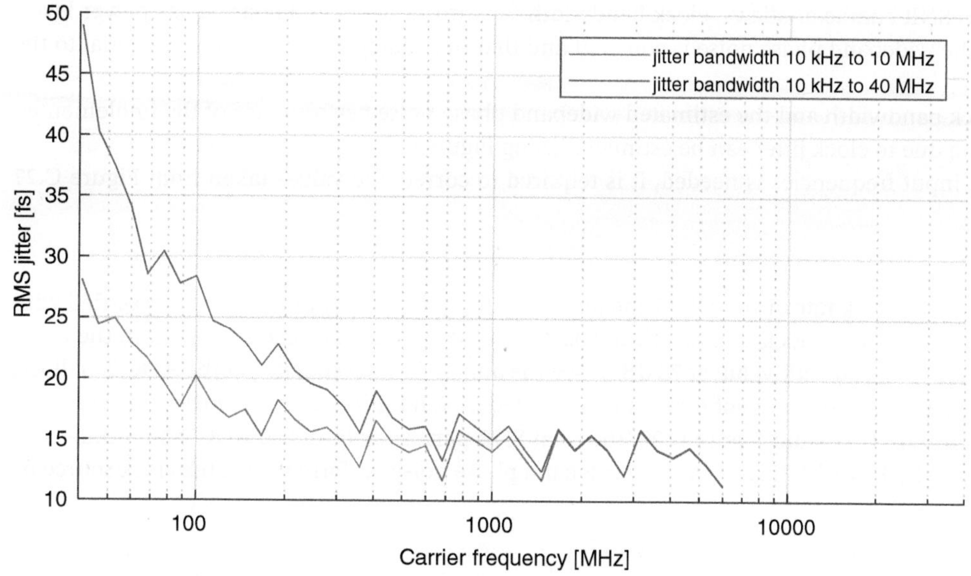

Figure C.24 RMS jitter vs. clock frequency for an R&S®SMA100B (with B711) B29 clock synthesizer (meas.).

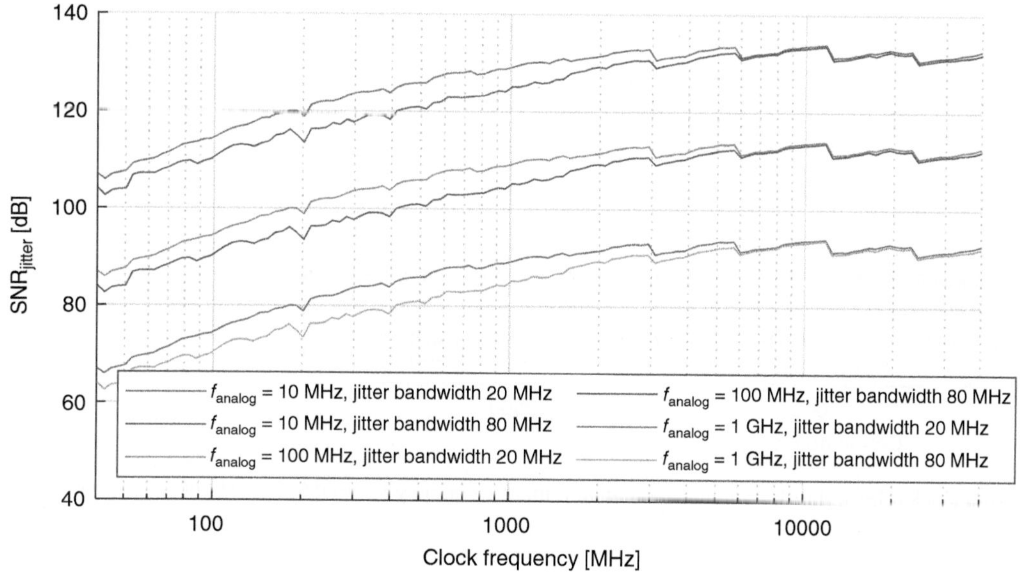

Figure C.25 SNR contribution from clock jitter vs clock frequency and analog input frequency for an R&S®SMA100B (with B711) (meas.).

The B29 clock synthesizer's performance, which is depicted in Figure C.26, again exhibits similar behavior as when comparing the RMS jitter in Figures C.23 and C.24.

C.4.4 SNR Performance from Wideband Approximations of the R&S®SMA100B as a Clock Source

In many applications, the relevant clock bandwidth is much higher than the 20 MHz or 80 MHz used for SNR calculation in the figure above. For such wideband applications, the wideband phase noise of the clock source must be considered for SNR calculation as shown in Section C.3.6. Eq. (C.19), which we derived there, describes the relation between wideband phase noise, clock bandwidth, clock and signal frequency, and SNR. It is repeated here for clarity:

$$\mathrm{SNR}_{\mathrm{jitter,clk}} \gtrsim -\mathrm{BBNoise}_{\mathrm{clk}} - 10\log_{10}(\mathrm{BW}_{\mathrm{clk}}) + 3\ \mathrm{dB} - 20\log_{10}\left(\frac{f_{\mathrm{analog}}}{f_{\mathrm{clk}}}\right)$$

The following diagram provides SNR characteristic vs. clock bandwidth for various wideband phase noise power levels and under the assumptions that the wideband phase noise is constant and that the analog signal frequency is equal to the clock rate.

Knowing the implemented clock bandwidth and the estimated wideband phase noise performance of the implemented clock source, the SNR contribution due to clock jitter can be estimated using Figure C.27.

If the SNR for different analog input frequencies is needed, it is required to correct the values taken from Figure C.27 using the factor $\frac{f_{\text{analog}}}{f_{\text{clk}}}$.

Example: A 14-bit ADC running at a clock rate of 200 MSPS and an analog input frequency of 100 MHz is specified with a max. SNR of 73 dB (see Figure C.8). With a clock bandwidth of 500 MHz and a wideband phase noise from the clock source of −160 dBc/Hz, Figure C.27 yields an SNR value of 73 dB. Since the analog input frequency at the ADC is half the clock rate, according to Eq. (C.21), a correction value of 6 dB must be added, resulting in an SNR contribution from the clock source of 79 dB. According to Eq. (C.16) and Figure C.28, the total SNR performance of the ADC is reduced—as a result of the non-ideal clock source—by 1 −72 dB. Improving the wideband phase noise performance of the clock source by

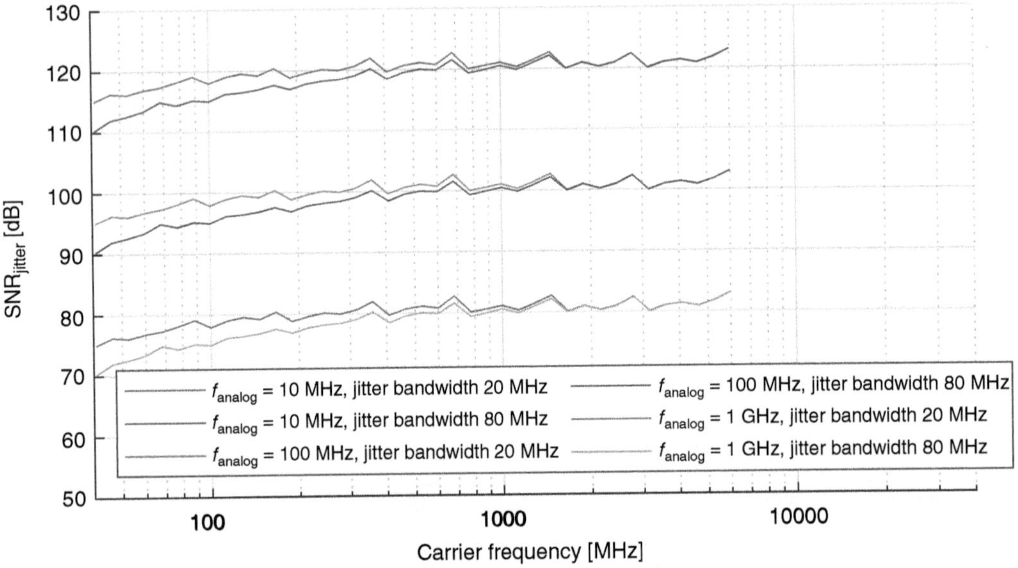

Figure C.26 SNR contribution from clock jitter vs clock frequency and analog input frequency for an R&S®SMA100B (with B711) B29 clock synthesizer (meas.).

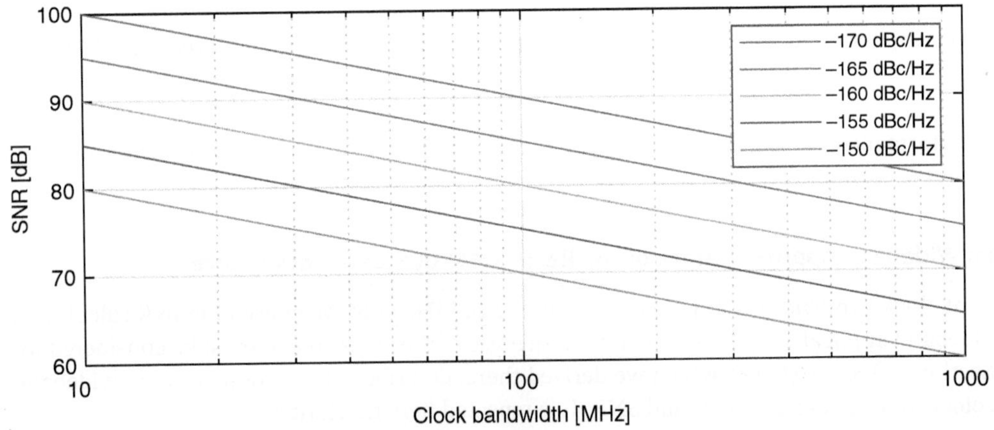

Figure C.27 SNR vs. clock frequency bandwidth and wideband phase noise, $f_{\text{analog}} = f_{\text{clk}}$.

5 dB would reduce this SNR deterioration to 0.33 dB. The same SNR improvement could be achieved by reducing the clock bandwidth from 500 MHz to 155 MHz.

The SNR deterioration due to the addition of two SNR contributors is graphically illustrated in Figure C.28.

The dotted red lines show an SNR reduction of 1 dB if the difference between the two contributors is 6 dB. The dotted green lines show an SNR impairment of 0.33 dB for a difference of 11 dB (see example above).

Figure C.29 depicts approximated as well as calculated estimated SNR contribution from the R&S®SMA100B as a clock source. For each clock rate, the figure shows the SNR calculated from the measured and integrated phase noise as well as the SNR calculated from the wideband phase noise approximation. For the wideband approximation, Eq. (C.19) is used to calculate SNR values again. The analog input frequency is equal to the clock rate for these calculations.

The figure shows the SNR contributions from clock jitter calculated from the measured phase noise power spectral density as well as calculated from the wideband phase noise level and confirms the claim from Section C.3.3, that for large bandwidths the wideband phase noise performance dominates the jitter. For small bandwidths, the phase noise close to the carrier is non-negligible and worsens the actual SNR with respect to the approximation, though. Thus, the viability of using the wideband approximation should be evaluated with care when a small bandwidth of the clock path is expected. With

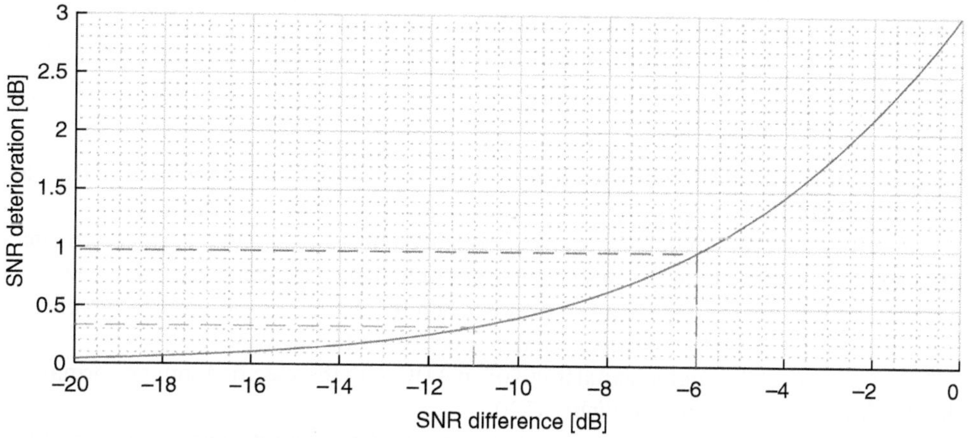

Figure C.28 SNR deterioration vs. SNR difference of two SNR contributions.

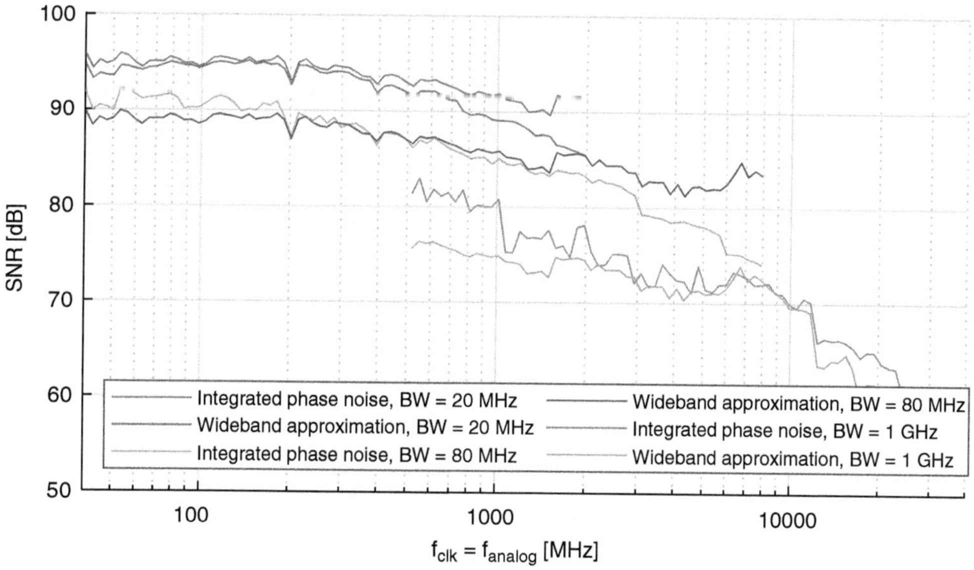

Figure C.29 SNR vs. clock rate and clock bandwidth from constant wideband phase noise for an R&S®SMA100B (with B711) (meas.).

increasing clock bandwidths, the wideband phase noise performance starts to dominate the jitter and the approximation becomes more conservative.

If the SNR for different analog input frequencies is needed, the relation between clock rate and f_{analog} must be considered by adding the correction value given from Eq. (C.21) to the SNR values from Figure C.29. Calculating the SNR value for a different clock bandwidth while still approximating the phase noise as constant is possible by correcting the SNR value from Figure C.29 using the following equation

$$\text{SNR}_{\text{BW2}} = \text{SNR}_{\text{BW1}} - 10\log_{10}\left(\frac{\text{BW2}}{\text{BW1}}\right) \tag{C.25}$$

Example: If the clock bandwidth reduces from 1 GHz to 500 MHz, the SNR will improve by 3 dB. Inversely, if the clock bandwidth increases from 100 to 200 MHz, the SNR will deteriorate by 3 dB.

C.4.5 SNR Performance of the R&S®SMA100B as Analog Source

The SNR contribution from the R&S®SMA100B as an analog signal source is usually dominated by the signal generator's wideband noise in the relevant frequency bands. A detailed depiction of the R&S®SMA100B's wideband noise performance was presented in Section C.4.2.

C.4.6 Nonharmonics

C.4.6.1 SFDR and SNR Due to Nonharmonics

The SFDR performance of a signal generator is normally dominated by its harmonic distortion since the level of the harmonics is much higher than the level of nonharmonic spurious.

However, by using adequate filtering at the generator's output, the harmonic signal components can be suppressed effectively and only nonharmonic spurious will then contribute to the signal generator's SFDR performance. Those spurs are hard to find as usually their frequencies are unknown, and they appear only at discrete frequencies with comparatively low amplitudes. Therefore, we will characterize spurs using the maximum levels specified in the datasheet.

The SFDR contribution for a spur that is present at the analog signal input of an ADC with the specified spur performances of the R&S®SMA100B as well as the R&S®SMA100A is shown in Figure C.30.

Figure C.30 is also valid for the clock input if the analog signal frequency is equal to the clock rate. Otherwise, the SFDR must be corrected just like the phase noise using Eq. (C.21).

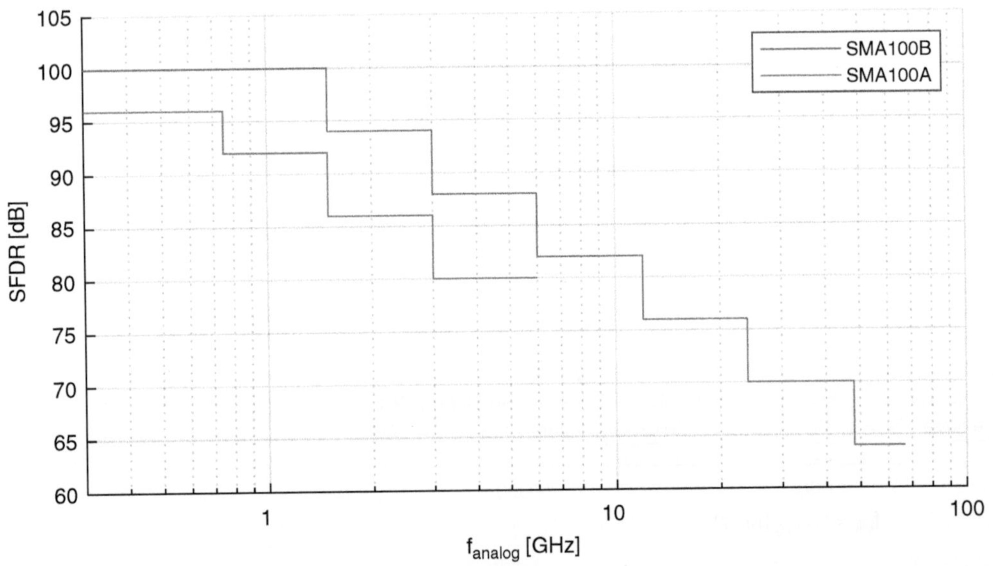

Figure C.30 SFDR performance of the R&S®SMA100 due to nonharmonic spurious versus analog signal frequency.

The specification above is also helpful for determining the SNR due to nonharmonics. The SNR does not distinguish between amplitude modulation and phase modulation. Therefore, the noise energy of each spur is directly visible from its level in the frequency spectrum, regardless if the spur is a single nonharmonic or a pair of spurs. This means, the specified worst-case SFDR also provides the designer with a worst-case spur level for each spur. In the case there are multiple spurs or spur-pairs with the same amplitude, the SNR has to be reduced by 3 dB, 4.7 dB, 6 dB, 7 dB, and so on respectively for each additional spur.

C.4.6.2 Jitter Contribution of Nonharmonics

The following diagram shows the RMS jitter contribution due to a single nonharmonic spur. The typical RMS jitter performance of the R&S®SMA100B, incl. option B711 due to phase noise for 80 MHz bandwidth, is also plotted in Figure C.31 in order to estimate the spur level at which the jitter due to the spur is higher than the jitter due to phase noise. The RMS jitter SMA100B B711 spur spec curve depicts the RMS jitter contribution of a nonharmonic spur equal to the specified performance of the R&S®SMA100B with option B711. The figure also includes the phase noise contributed jitter performance of the R&S®SMA100A for comparison.

Clearly, the jitter performance of the R&S®SMA100B is dominated by its phase noise performance instead of spurious performance over the complete frequency range. The RMS jitter contribution due to phase noise is considerably higher than the maximum possible RMS jitter contribution from nonharmonics based on the spurious specification. Keep in mind though, that for more than one spur, their jitter contributions need to be added as in Eq. (C.15).

C.4.7 Harmonics

The harmonic distortion for the second harmonic of an R&S®SMA100B (for carrier frequencies > 10 MHz) is depicted in Figure C.32 for an output level of +10 dBm. As the R&S®SMA100B produces very low harmonics, the measurement result also includes harmonic contributions and other artifacts from the used R&S®FSW85. The measured intrinsic noise floor of the measuring instrument is shown in the figure as well. This noise floor is measured at two times the set carrier frequency and shown in dBc for comparability. The harmonics with higher orders or higher frequencies are normally far below these values.

The harmonic distortion of a signal generator varies versus the set output level and frequency and is dominated by the harmonic performance and the signal level at the RF output stage, which is followed by an attenuation stage. Contrary to the R&S®SMA100A, the R&S®SMA100B automatically fine-tunes gain and attenuation in order to achieve a good overall performance at the signal output. There is no necessity for the user to adjust attenuation settings when aiming for low harmonic distortion or a high SNR.

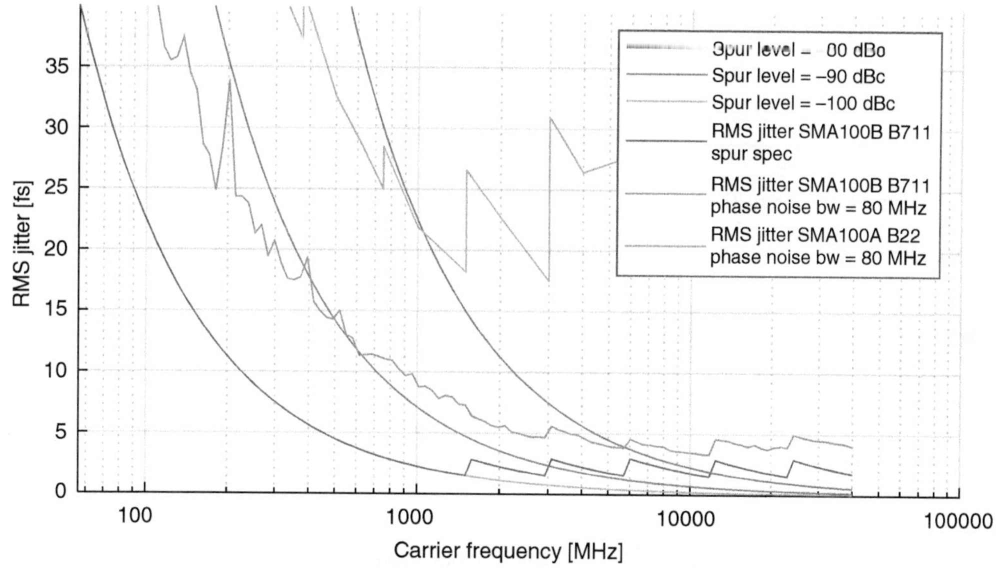

Figure C.31 RMS jitter contribution due to nonharmonic spurs and measured phase noise performance of an R&S®SMA100B.

C.5 Example SNR Calculations for typical ADC Test Setups

This chapter provides some example SNR calculations for the R&S®SMA100B as a clock and analog source using the characteristics shown in the previous chapters.

When sourcing for a signal generator, the following factors must be taken into account:

- The generator's contribution to the wideband noise and close-in noise should be negligible or should have only minimal effect on the SNR measurement procedure of the ADC.
- The generator's nonharmonic distortion products should be far below the distortion products of the ADC.
- Since the demands placed on a signal source may become higher in future challenges, the performance of a purchased generator should optionally already have enough margin to meet these future requirements.

C.5.1 ADC Test Setup with the R&S®SMA-B29 Clock Synthesizer

In general, signal sources for the analog and clock inputs are needed, as well as means for data analysis. For most ADCs, an evaluation board with a PC interface, e.g., USB, and the required software tools are available. Most ADC evaluation boards include some level shifters or signal transformers in order to convert a sine wave signal at the board input to the required signal for the analog or clock input of the DUT, e.g., a differential signal for analog input.

With the R&S®SMA100B, it is also possible to combine analog input signal generator and sample clock generator in one device. Using the B29 option, the test setup is as simple as shown in Figure C.33.

If required, the clock synthesizer provides a differential output signal, thus eliminating the need for a balun at the ADC clock input.

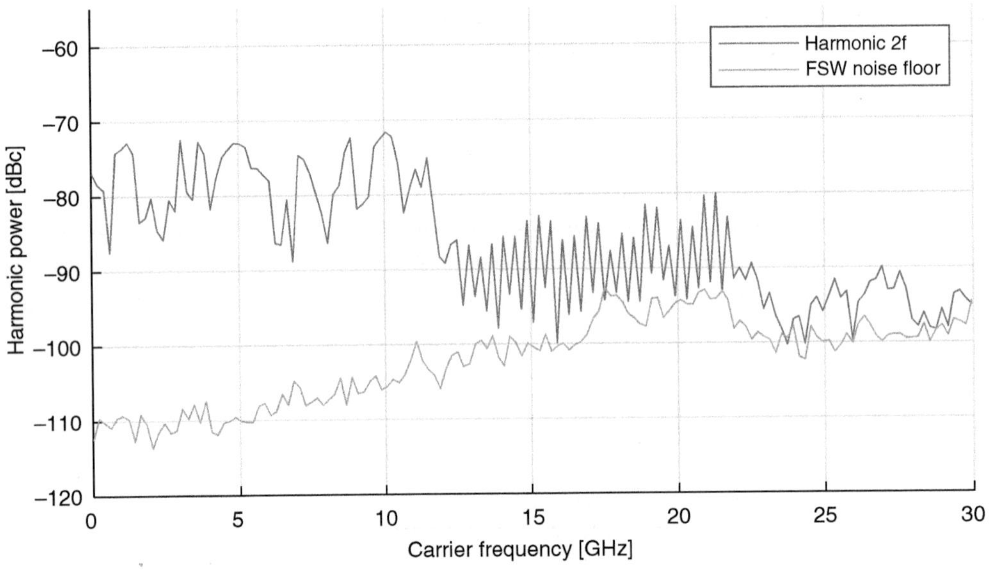

Figure C.32 Second harmonics at +10 dBm vs. carrier frequency for an *R&S®SMA100B (meas.) for carrier frequencies* > *10* MHz.

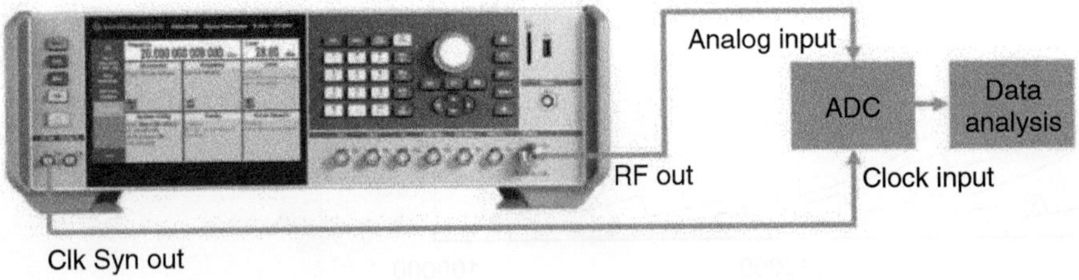

Figure C.33 Simplified ADC test setup using the R&S®SMA-B29 clock synthesizer as clock source.

The clock synthesizer output and the RF output differ from each other in frequency range and properties. As both outputs provide a high-grade signal purity, this setup is the perfect solution for many applications. However, especially applications with both signal and clock frequencies higher than 6 GHz may require using a second R&S®SMA100B.

C.5.2 ADC Test Setup with Two R&S®SMA100B Signal Generators

Figure C.34 depicts a test setup with two R&S®SMA100B signal generators.

The RF output of one R&S®SMA100B serves the analog input of the ADC, and the RF output of the other R&S®SMA100B serves the clock input. If necessary, filters can be added to the signal paths to keep SNR contributions from signal impurities even lower. Both signal generators should be synchronized using a common reference signal. Thereby, the phase noise contributions of the used signal generators are reduced further.

C.5.3 Typical ADC Specifications

The following table lists some typical characteristics of discrete high-speed ADCs.

ADC	Resolution (bits)	Sample rate (MSPS)	SNR (dBFS)	SFDR (dBFS)
Analog Devices AD9213	12	10250	51	65
Texas Instruments ADC14C105	14	105	90	82
Maxim Integrated MAX19588	16	100	79	98
Analog Devices LTC2270	16	20	84	99
Texas Instruments ADC12DJ5200RF	12	10,400	56.7	78

C.5.4 Example SNR Calculation for the R&S®SMA100B

In this example, two R&S®SMA100B generators with B711 are used as a clock source and analog source for an ADC test set, applying a 10 MHz sine signal at the analog input. The ADC under consideration has an SNR of approximately 80 dB. The SNR of the test setup should be far below this value.

The following table lists the frequencies and bandwidths for the clock signal and the analog signal used for the calculations. Note that the given analog and clock bandwidths are a result of bandwidth limitations in the DUT or filters that are inserted between the SMA100B's outputs and the ADC's inputs.

Clock Frequency	Clock Bandwidth	Analog Frequency	Analog Bandwidth
100 MHz	300 MHz	10 MHz	25 MHz

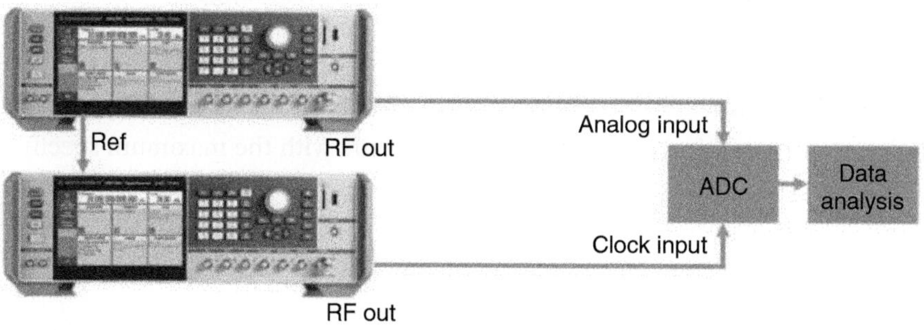

Figure C.34 General ADC performance test setup.

Step 1: Calculation of SNR due to jitter/phase noise of clock source

Using the R&S®SMA100B's wideband noise specification at 100 MHz (−157 dBc/Hz) and Eq. (C.19) (see also Section C.4.4), $\mathrm{SNR_{clk}}$ can be approximated as

$$\begin{aligned}\mathrm{SNR_{clk}} &= -\mathrm{BBNoise_{clk}} - 10\log_{10}(\mathrm{BW_{clk}})\ \mathrm{dB} + 3\ \mathrm{dB} - 20\log_{10}\frac{f_{\mathrm{analog}}}{f_{\mathrm{clk}}}\ \mathrm{dB}\\ &= 157\ \mathrm{dB} - 10\log_{10}(300\ \mathrm{MHz}/1\ \mathrm{Hz})\mathrm{dB} + 3\ \mathrm{dB} - 20\log_{10}\frac{10\ \mathrm{MHz}}{100\ \mathrm{MHz}}\ \mathrm{dB}\\ &= 157\ \mathrm{dB} - 84.77\ \mathrm{dB} + 3\ \mathrm{dB} + 20\ \mathrm{dB} \approx 95\ \mathrm{dB}\end{aligned}$$

This shows that there is still a big margin to the expected performance of the DUT.

Using the measured performance of an R&S®SMA100B instead of its specification indicates that there might be an even bigger margin in reality. Figure C.29 shows the measured SNR performance of an R&S®SMA100B used as a clock source for an ADC test set. For a clock rate of 100 MHz, an analog signal frequency of 100 MHz and a clock bandwidth of 80 MHz, an SNR value of roughly 90 dB can be read from the diagram. Then, the lower analog input frequency of 10 MHz must be considered, and the SNR value must be corrected using Eq. (C.21). Also, the SNR must be corrected due to the higher clock bandwidth of 300 MHz by using Eq. (C.25).

Estimating SNR_{clk} based on measured performance of an R&S®SMA100B yields

$$\begin{aligned}\mathrm{SNR_{clk}} &= \mathrm{SNR_{clk,80\ MHz,BW100\ MHz}} - 20\log_{10}\left(\frac{10\ \mathrm{MHz}}{100\ \mathrm{MHz}}\right)\mathrm{dB} - 10\log_{10}\left(\frac{300\ \mathrm{MHz}}{80\ \mathrm{MHz}}\right)\mathrm{dB}\\ &= 90\ \mathrm{dB} + 20\ \mathrm{dB} - 5.74\ \mathrm{dB} \approx 104\ \mathrm{dB}\end{aligned}$$

and indicates an even bigger margin.

Step 2: Calculation of SNR due to nonharmonics of clock source

It is assumed that the clock signal includes one symmetrical pair of nonharmonic spurs with the maximum specified level at 100 MHz, which is −100 dBc. Evaluating Eq. (C.23) yields

$$\begin{aligned}\mathrm{SNR_{spur,clk}} &= -a_{\mathrm{spur}} - 3\ \mathrm{dB} - 20\log\left(\frac{f_{\mathrm{analog}}}{f_{\mathrm{clk}}}\right)\mathrm{dB}\\ &= 100\ \mathrm{dB} - 3\ \mathrm{dB} - 20\log_{10}\left(\frac{10\ \mathrm{MHz}}{100\ \mathrm{MHz}}\right)\mathrm{dB} = 117\ \mathrm{dB}\end{aligned}$$

Step 3: Calculation of SNR due to wideband noise of analog source

Figure C.17 shows the measured wideband noise performance of an R&S®SMA100B. For an analog signal frequency of 10 MHz, a value of =−163 dBc/Hz can be read from the diagram. Together with the analog bandwidth of 25 MHz, the SNR contribution can be estimated according to Eq. (C.20) as

$$\mathrm{SNR_{analog}} = -\mathrm{BBNoise} - 10\log_{10}(25\ \mathrm{MHz}/1\mathrm{Hz})\mathrm{dB} = 163\ \mathrm{dB} - 74\ \mathrm{dB} = 89\ \mathrm{dB}$$

Using the R&S®SMA100B's specified value of −157 dBc/Hz at a frequency of 100 MHz, the resulting SNR contribution would be 83 dB.

Step 4: Calculation of SNR due to nonharmonics of analog source

It is assumed that the analog signal includes one symmetrical pair of nonharmonic spurs with the maximum specified level at 10 MHz, which is −100 dBc. According to Eq. (C.24), this yields an SNR contribution of 100 dB − 3 dB = 97 dB.

Step 5: Summary

Summing all SNR contributions, the SNR limitation of the test set is

$$\mathrm{SNR_{testset}} = -10\log_{10}\left(10^{-\frac{104}{10}} + 10^{-\frac{117}{10}} + 10^{-\frac{89}{10}} + 10^{-\frac{97}{10}}\right) = 88.2\ \mathrm{dB}$$

when considering the measured performance of an R&S®SMA100B, and 82.6 dB when using specified values.

This SNR estimation shows that a test set using two R&S®SMA100B generators as the clock source and the analog signal source has a margin of approximately 8 dB (estimated from measured values) or approximately 2.6 dB (based on specified values), respectively, from the considered ADC. With these margins, there is an SNR deterioration of roughly 0.6 dB (based on measured values) or roughly 2 dB (based on specified values), respectively, as can be seen in Figure C.28.

If more precise knowledge of the test set performance is required, the relevant parameters can be measured for the particular instruments that are used. In general, further improvement of the test set performance can be achieved by using filters with lower bandwidths if necessary.

C.6 Ordering Information

Designation	Type	Order No.
RF and Microwave Signal Generator	R&S®SMA100B	1419.8888.02
Frequency Option 8 kHz to xx GHz	R&S®SMAB-B1xx	1420.yyyy.02
Ultra Low Phase Noise	R&S®SMAB-B711	1420.8020.02
Differential Clock Synthesizer, 3 GHz	R&S®SMAB-B29	1420.8088.02
Clock synthesizer frequency extension to 6 GHz	R&S®SMAB-K722	1420.9810.02

C.7 Performance Data for the R&S®SMA100B with B711

C.7.1 Wideband Noise

As described previously in Section C.4.2, this appendix shows further noise power spectral density characteristics for different frequency bands to demonstrate variation over the carrier frequency.

For some carrier frequencies and generator configurations, the measured noise power spectral density at very large offsets from the carrier might surpass the wideband noise level specified in the datasheet for an offset of 10 MHz or 40 MHz. In this case, the affected offset frequencies usually cover only a very small portion of the bandwidth and lie far beyond the offsets at which the wideband noise level is specified. The noise power spectral density never surpasses the specified wideband noise level at the specified offset (Figures C.35–C.42)

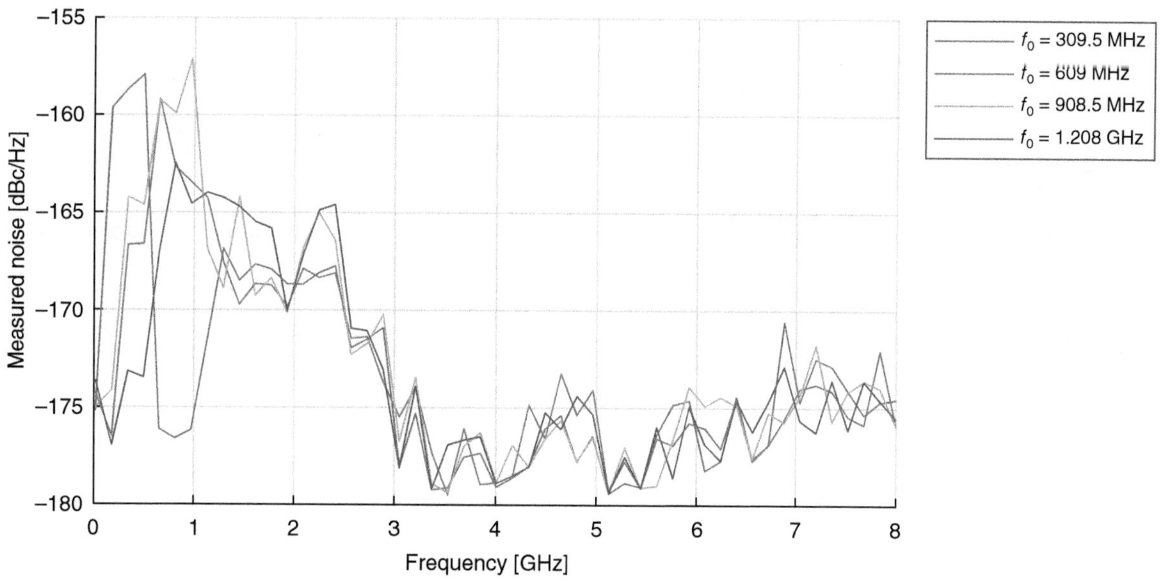

Figure C.35 Noise PSD for multiple carrier frequencies with $f \leq 1.5$ GHz.

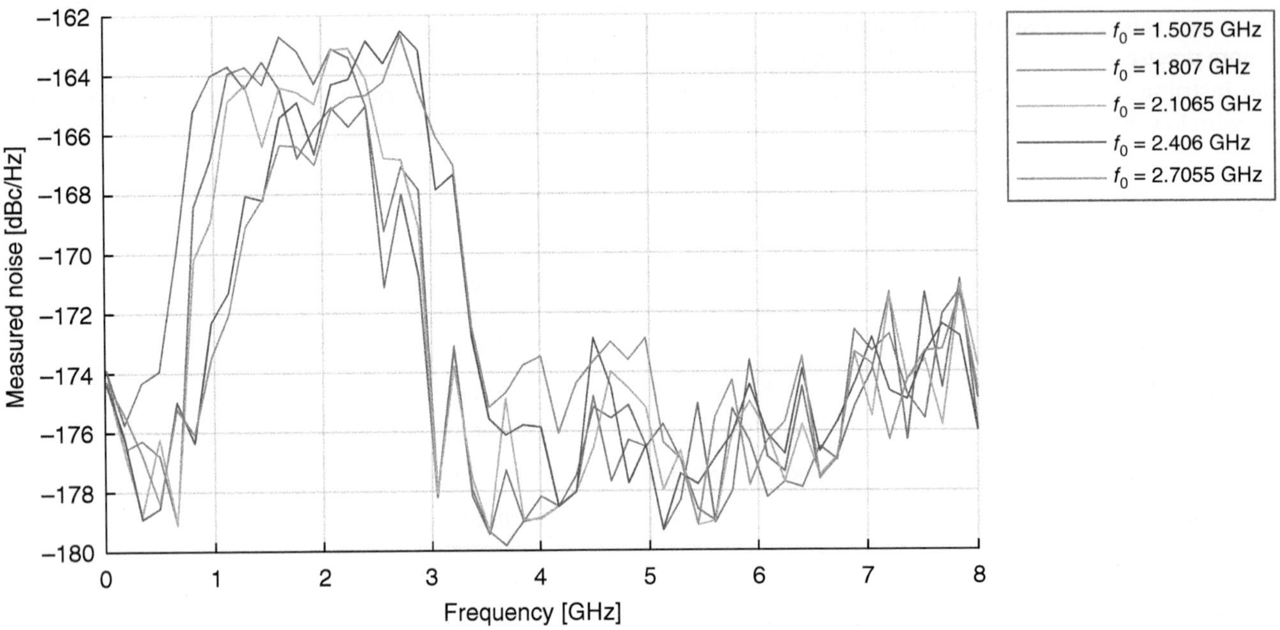

Figure C.36 Noise PSD for multiple carrier frequencies with 1.5 GHz $< f \leq 3$ GHz.

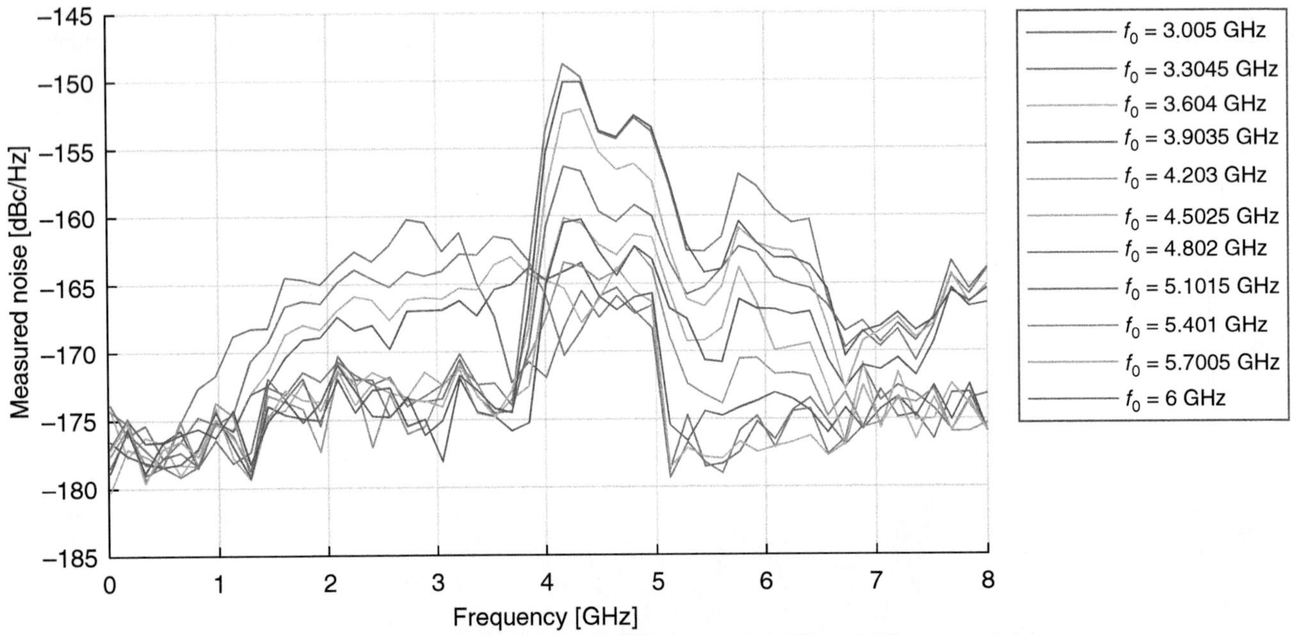

Figure C.37 Noise PSD for multiple carrier frequencies with 3 GHz $< f \leq 6$ GHz.

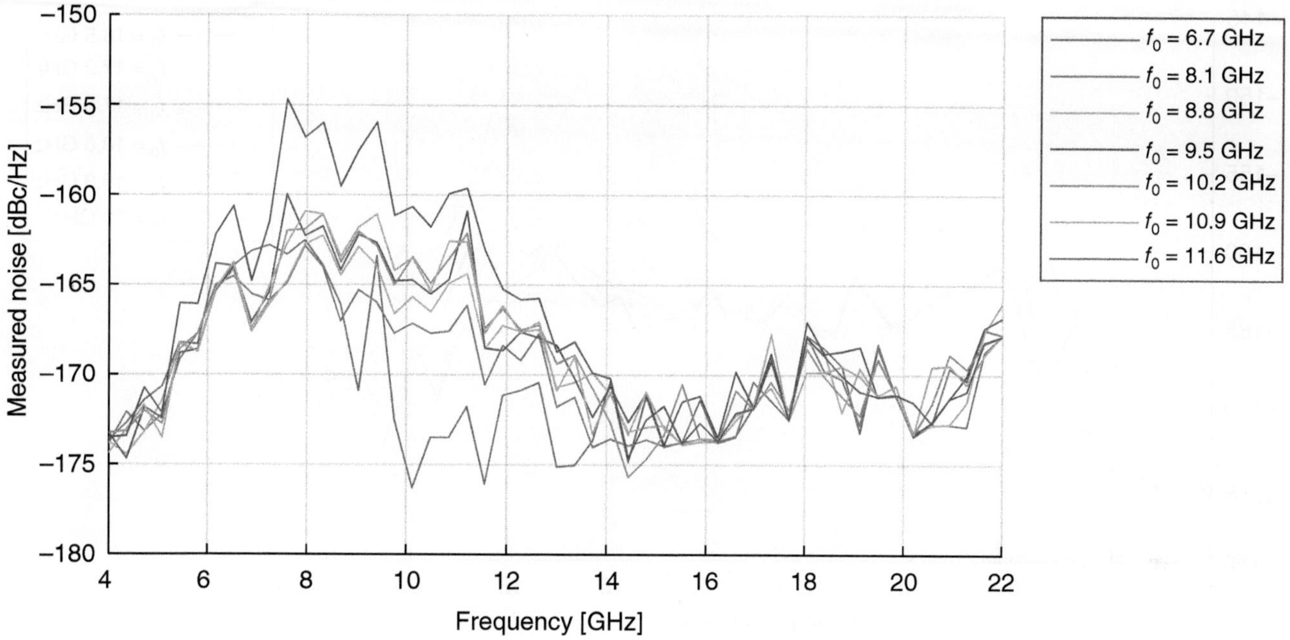

Figure C.38 Noise PSD for multiple carrier frequencies with 6 GHz < $f \leq$ 12 GHz.

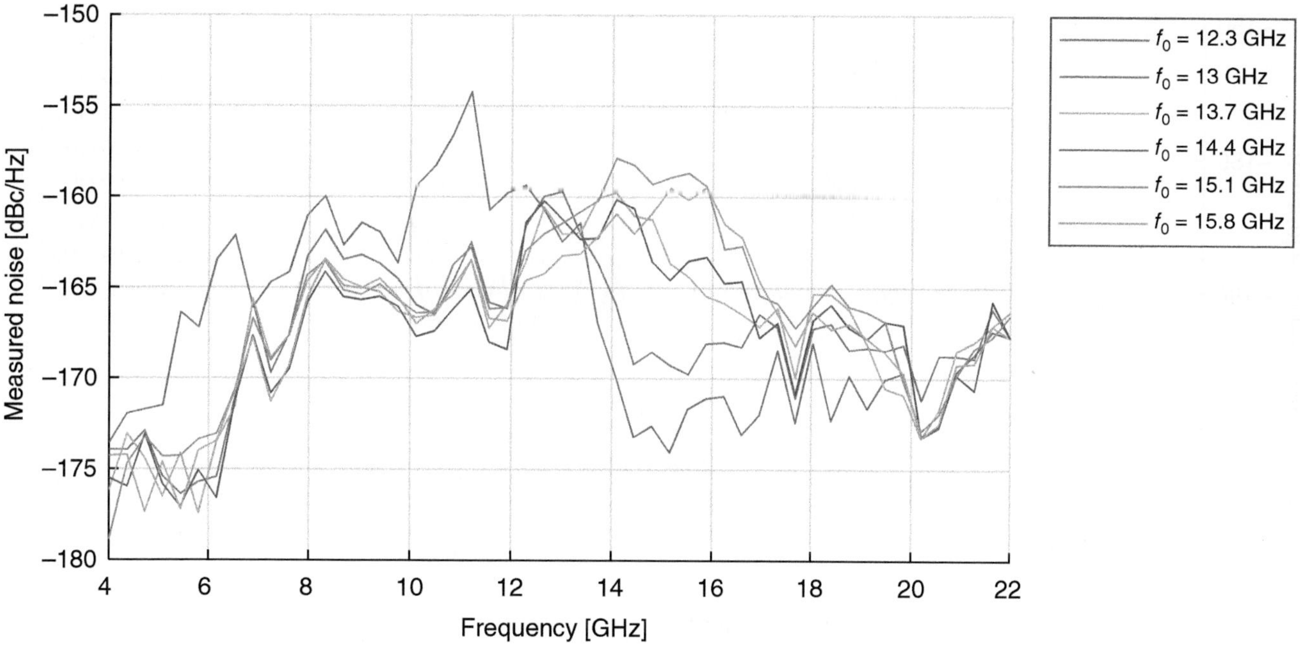

Figure C.39 Noise PSD for multiple carrier frequencies with 12 GHz < $f \leq$ 16 GHz.

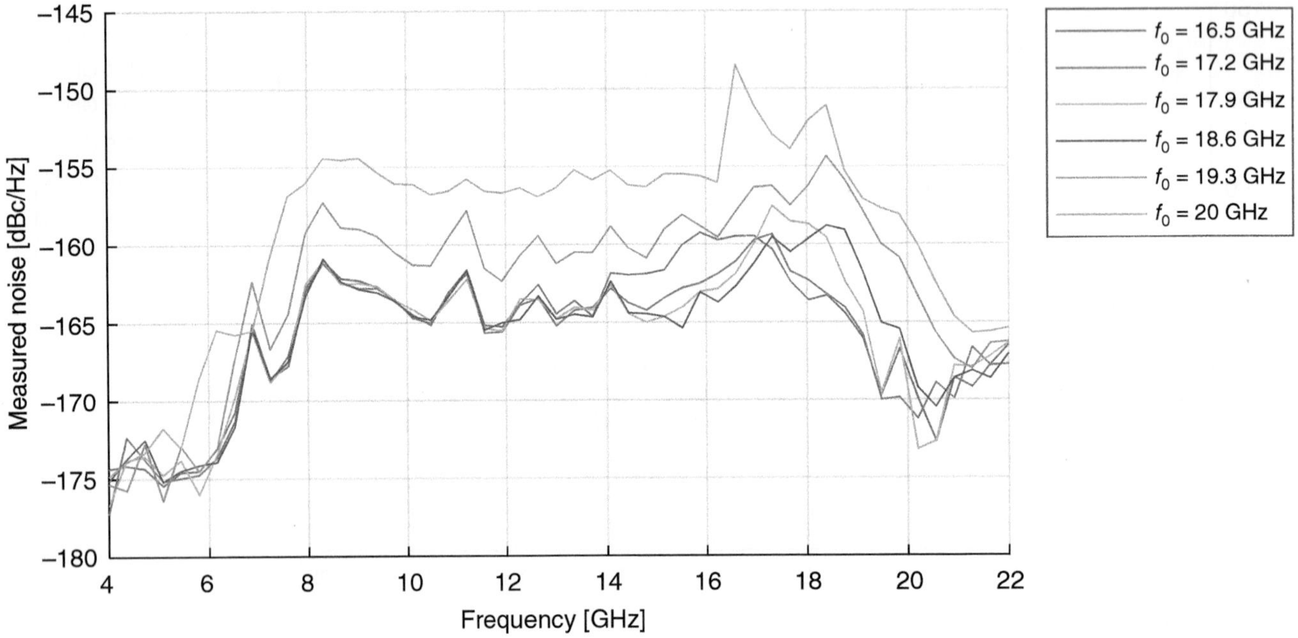

Figure C.40 Noise PSD for multiple carrier frequencies with 16 GHz $< f \leq$ 20 GHz.

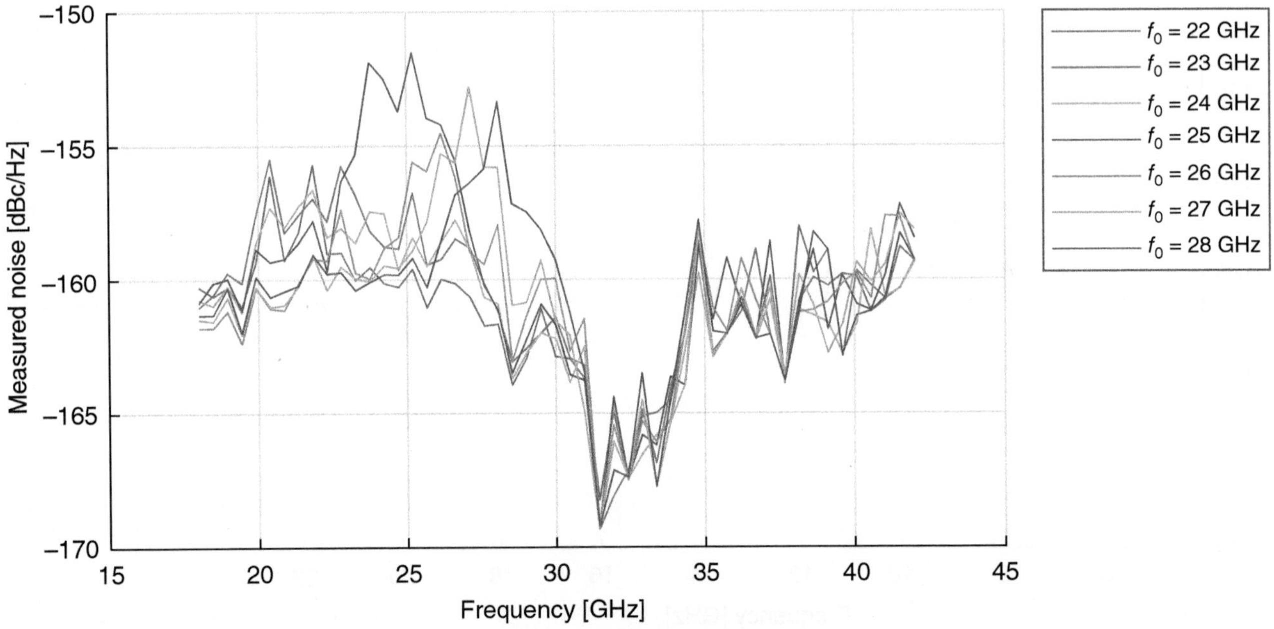

Figure C.41 Noise PSD for multiple carrier frequencies with 20 GHz $< f \leq$ 28 GHz.

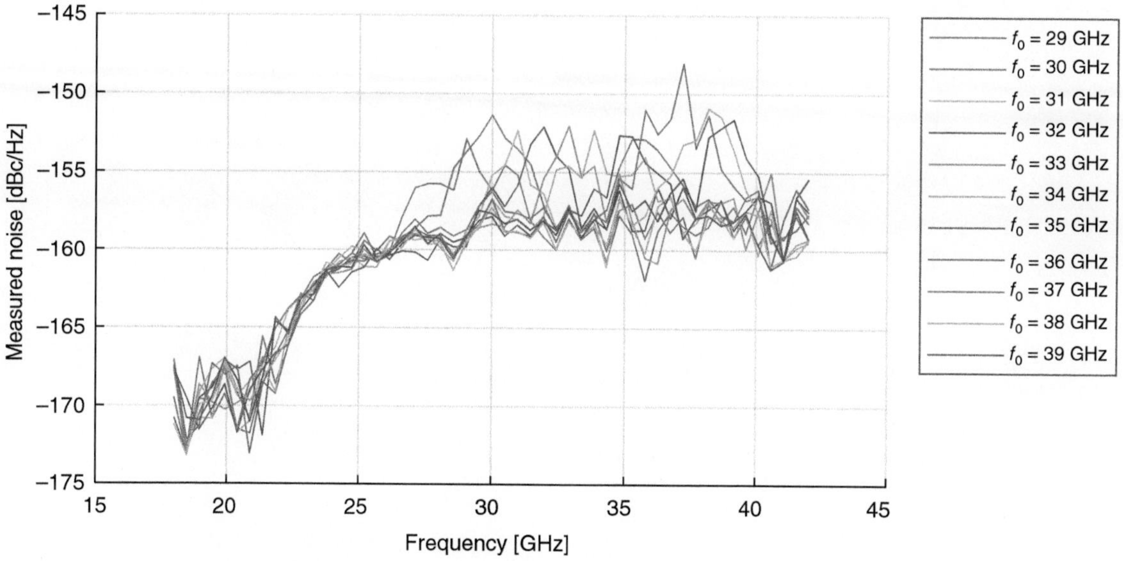

Figure C.42 Noise PSD for multiple carrier frequencies with 28 GHz $< f \leq 40$ GHz.

C.8 Summary

Testing the true performance of ADCs can place high requirements on the input signals. In this application note, background information on the influence of signal quality on the measurement results has been given. It has been shown that the outstanding signal purity of the R&S®SMA100B makes unveiling the true performance of an ADC much easier than it used to be.

References

1 F. Maloberti, Data Converters, Springer US, 2007.
2 B. Brannon, *Aperture Uncertainty and ADC System Performance, Analog Devices application note AN-501*, Analog Devices, 2000.
3 E. Bartolome, V. Mishra, G. Dutta and D. Smith, *Clocking High-Speed Data Converters, Texas Instruments application note*, Texas Instruments, 2005.
4 American National Standards Institute, *T1.105 Synchronous optical networking*.
5 B. Brannon, *Sampled Systems and the Effect of Clock Phase Noise and Jitter, Analog Devices application note AN-756*, Analog Devices, 2004.
6 Analog Devices, Inc., *Data Sheet AD9445, 14-Bit, 125 MSPS ADC*, 2005.
7 B. Brannon and R. Reeder, *Understanding High Speed ADC Testing and Evaluation, Analog Devices application note AN-835*, Analog Devices, 2006.

D

State-of-the-art Cryogenic Low-noise Amplifier

D.1 A Famous Application: The Deep Space Network

Cryogenic low-noise amplifiers are used together with large antennas and systems in systems such as the Deep Space Network (DSN). The DSN is basically the world's largest telecommunications system.

NASA describes the DSN as an "international array of giant radio antennas that supports interplanetary spacecraft missions, plus a few that orbit Earth." The DSN also facilitates radar and radio astronomy observations that are used to study the solar system and the larger universe.

The DSN consists of three complexes strategically located around the world: Goldstone (California), Madrid (Spain), and Canberra (Australia). It uses the following antenna types:

- **70-m antennas:** These are the largest and most sensitive, capable of tracking spacecraft traveling billions of miles from Earth.
- **34-m antennas:** These are also used for tracking and communication, with some being beam waveguide antennas that use a tube to guide signals to below-ground equipment rooms.
- **26-m antennas**

The DSN allows missions to perform the following functions:

- Track spacecraft
- Send commands to spacecraft
- Receive scientific data from spacecraft

The DSN is managed by NASA's Jet Propulsion Laboratory (JPL).

D.1.1 Design Thoughts for a Low-noise Amplifier

The design thoughts for a low-noise amplifier (LNA) that are discussed in this chapter are based on the following three papers: [1–3].

D.2 The GaAs-FET Noise Model and Its Application in Low-noise Amplifiers

D.2.1 Linear Empirical GaAs Model

This linear model is still successfully used as the parameter extraction is easy and it is valid for small-signal operation (Figure D.1).

Noise in a MESFET is produced by sources that are intrinsic to the device. The equivalent noise circuit of an intrinsic FET is represented in Figure D.2.

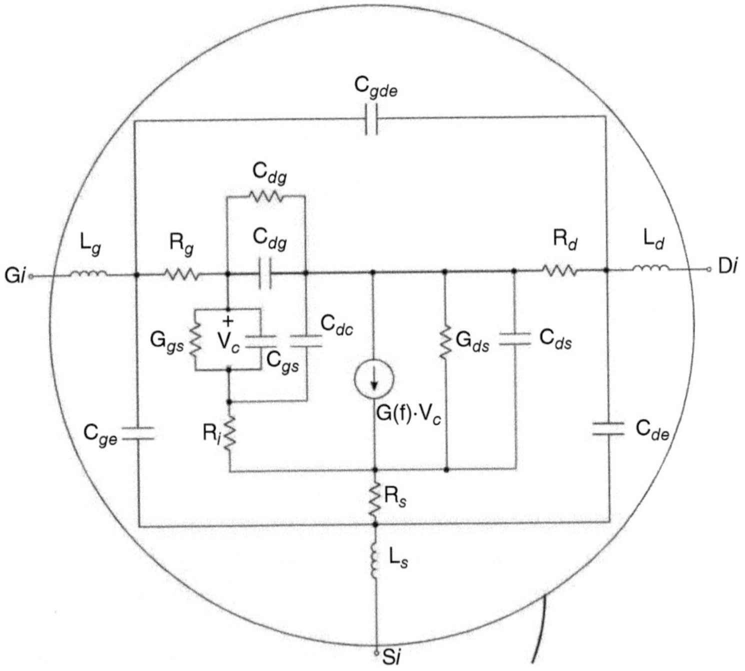

Figure D.1 Linear empirical GaAs model.

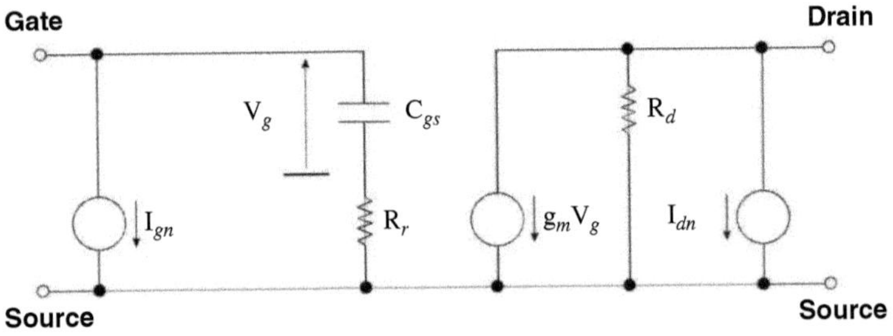

Figure D.2 Equivalent simple noise circuit of an intrinsic FET device.

The intrinsic FET is internally represented as a noiseless nonlinear two-port with one equivalent noise current source connected across the gate-source terminal, and one across the drain-source terminal. The correlations of the gate and drain noise current sources are:

$$\begin{aligned} \left\langle |I_{gn}|^2 \right\rangle &= 4K_B T \Delta f \frac{\omega^2 C_{gs}^2}{g_m} R \\ \left\langle |I_{dn}|^2 \right\rangle &= 4K_B T \Delta f g_m P \\ \left\langle I_{gn} I_{dn}^* \right\rangle &= 4K_B T \Delta f j\omega C_{gs} \sqrt{PRC} \end{aligned} \tag{D.1}$$

The correlation matrix of the noise current sources is

$$C_{dc}(\omega) = \frac{2}{\pi} K_B T d\omega \begin{bmatrix} \frac{\omega^2 C_{gs}^2}{g_m} R & -j\omega C_{gs}\sqrt{PRC} \\ j\omega C_{gs}\sqrt{PRC} & g_m P \end{bmatrix} \tag{D.2}$$

The gate and drain noise parameters R and P and the correlation coefficient C, are related to the physical noise sources acting in the channel and are functions of the device structure and bias point. These noise parameters at a bias point can be explicitly calculated from measured device noise parameters. By defining measured noise parameters, F_{min}, R_n, and G_{opt}, and using a noise deembedding procedure, the parameters R, P, and C and the intrinsic noise correlation matrix of a FET device as functions of device bias are determined by the program that is used.

In addition to the noise sources shown above, the flicker $(1/f)$ noise can also be modeled by means of a noise current source connected in parallel with the intrinsic drain port. The flicker noise component in a narrow band, Δf is expressed in the form

$$\left\langle |I_f|^2 \right\rangle = Q\Delta f \frac{|I_D|^{\mathrm{AF}}}{f^{\mathrm{FCP}}} \tag{D.3}$$

where I_D is the instantaneous value of the channel current, and Q, AF, and FCP are empirical parameters. In most practical cases, AF and FCP are directly obtained from measurements (typically AF = 2, FCP = 1), while Q is not. Q is either provided directly using KF or is computed by providing the flicker corner frequency FC. FC is the frequency for which the flicker noise equals the shot/diffusion noise. The corner frequency is defined by the equation

$$Q\frac{|I_D|^{\mathrm{AF}}}{f_{\mathrm{C}}^{\mathrm{FCP}}} = g_m P \tag{D.4}$$

Given the corner frequency F_{C} and the measurement bias point V_{gs} and V_{ds}, the program automatically computes I_D, g_m, and P, and finally Q.

The following are some of the more popular nonlinear FET-GaAs MESFET and HEMT models from the past:

- Chalmers (Angelov) Model
- Curtice–Ettenberg Cubic Model
- Curtice Quadratic Model
- Modified Materka-Kacprzak Model
- Physics-Based MESFET Model
- Raytheon (Statz) Model

Note that the modified Angelov and advanced physics-based models are now taking over.

D.2.2 General Notes on the Models

The channel current, I_{ds}, is the most significant nonlinearity of the MESFET models when operating in the nominal active region. Through the first-order derivatives, I_{ds} gives rise to the transconductance and output conductance associated with small-signal models. Higher-order derivatives produce the harmonic components and intermodulation products in the simulations.

The capacitances, C_{gs} and C_{gd}, represent the Schottky junction capacitance of the gate-channel interface. C_{ds} represents the stray capacitance through the substrate.

The series resistor–capacitor pair of Rdsd-Cdsd between the intrinsic drain and source serves the purpose of increasing the output conductance of the MESFET at high frequencies. It is known that a MESFET exhibits trapping effects starting at frequencies between 10 and 100 kHz which increase the output conductance. Since the equivalent output conductance of the nonlinear current source I_{ds} is typically calculated from DC I–V data, the dispersion of the output conductance is not included in Ids. The Rdsd–Cdsd pair modifies the output conductance to the first order.

The resistors R_g, R_d, and R_s represent the gate metallization resistance and ohmic resistances of the drain and source contacts.

The extrinsic parameters L_g, L_d, L_s, C_{ge}, C_{de}, and C_{gde} are generally used to model wire-bond and lead inductances and inter-electrode capacitances. These parasitics are often neglected for low frequency work.

The parameters C_{gsb} and C_{dsb} are used to model bond-pad capacitances. The parameters L_{gb}, K_{db}, L_{sb} model wire-bond inductances of the package, while Z_{gt}, L_{gt}, Z_{dt}, L_{dt}, Z_{st}, L_{st} model package transmission-line effects and C_{gsp}, C_{dsp}, C_{gdp} model package capacitances.

The GaAs FET models have been implemented for use in the normal operating region ($V_{ds} < \mathbf{0}$) and in the reverse operating region ($V_{ds} < \mathbf{0}$). The normal operating region uses V_{gs} and V_{ds} as the controlling voltages. The reverse operating region uses V_{gd} and $-V_{ds}$ as the controlling voltages. The equations given for the models are in the normal.

Finally, using foundries, the exact data about the models and their I-parameters are no longer fully disclosed. In building an amplifier here, we make use of the Raytheon–Statz model.

For more information, consult the references on linear operation at the end of this appendix.

D.2.3 GaAs FET Model at Room Temperature

Figures D.3 and D.4 show a noise model of a grounded source FET with noise source at input and the output. See the separate references at the end of this appendix.

The mean square value of the noise sources in the narrow frequency range Δf are given by

$$\overline{i_d^2} = 4kTg_mP\Delta f \tag{D.5}$$

$$\overline{i_g^2} = \frac{4kT(\omega C_{gs})^2 R}{g_m}\Delta f \tag{D.6}$$

$$\overline{i_g i_d^*} = -j\omega C_{gs} 4kTC\sqrt{PR}\Delta f \tag{D.7}$$

$$S(i_d) = \frac{\overline{i_d^2}}{\Delta f} = \left\langle \left|\overline{i_d^2}\right| \right\rangle = 4kTg_mP \tag{D.8}$$

$$S(i_g) = \frac{\overline{i_g^2}}{\Delta f} = \left\langle \left|\overline{i_g^2}\right| \right\rangle = \frac{4kT(\omega C_{gs})^2 R}{g_m} \tag{D.9}$$

$$S(i_g i_d^*) = \left\langle \left|\overline{i_g i_d^*}\right| \right\rangle = -j\omega C_{gs} 4kTC\sqrt{PR} \tag{D.10}$$

where P, R, and C are FET noise coefficients and are defined as

$$P = \left[\frac{1}{4kTg_m}\right] \overline{i_d^2}/\text{Hz};\ P = 0.67 \text{ for JFETs and } 1.2 \text{ for MESFETs} \tag{D.11}$$

$$R = \left[\frac{g_m}{4kT\omega^2 C_{gs}^2}\right] \overline{i_g^2}/\text{Hz};\ R = 0.2 \text{ for JFETs and } 0.4 \text{ for MESFETs} \tag{D.12}$$

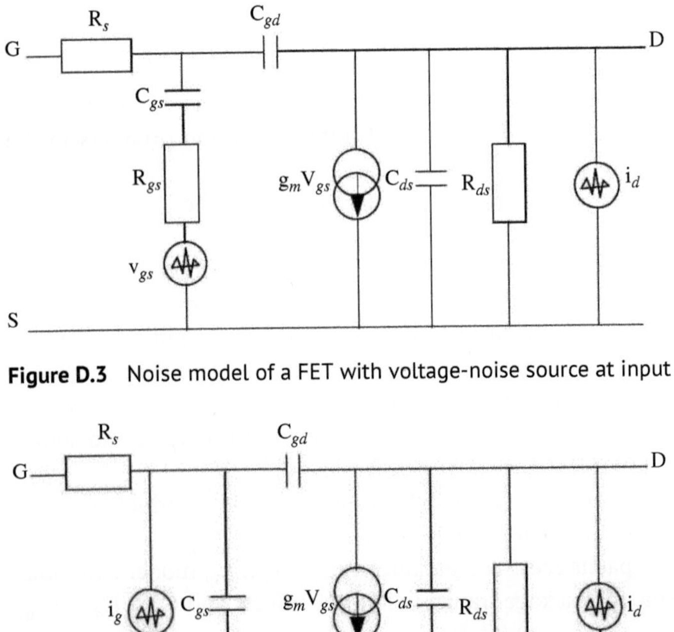

Figure D.3 Noise model of a FET with voltage-noise source at input and the current-noise source at the output.

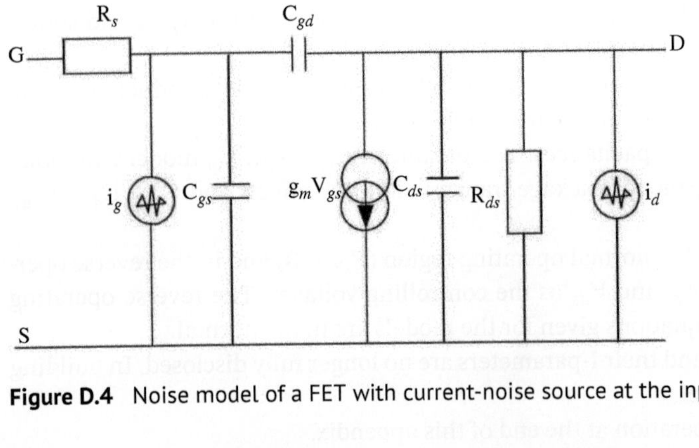

Figure D.4 Noise model of a FET with current-noise source at the input and the output.

$$C = -j\left[\frac{\overline{i_g i_d^*}}{\sqrt{[\overline{i_d^2}\,\overline{i_g^2}]}}\right];\ C = 0.4\ \text{for JFETs and}\ 0.6\text{–}0.9\ \text{for MESFETs} \tag{D.13}$$

D.2.4 Calculation of the Noise Parameters

Figures D.5 and D.6 show the intrinsic FET with noise sources at the input and output.

$$[Y]_{\text{FET-Intrinsic}} = \begin{bmatrix} y_{11} & y_{12} \\ y_{21} & y_{22} \end{bmatrix} \tag{D.14}$$

$$[C_Y] = [N]_{\text{noise-matrix}} = \begin{bmatrix} \overline{i_g i_g^*} & \overline{i_g i_d^*} \\ \overline{i_d i_g^*} & \overline{i_d i_d^*} \end{bmatrix} \tag{D.15a}$$

$$[C_Y]_{\text{FET}} = 4kT\begin{bmatrix} \dfrac{\omega^2 c_{gs}^2 R}{g_m} & -j\omega c_{gs} C\sqrt{PR} \\ j\xi c_{gs} C\sqrt{PR} & g_m P \end{bmatrix} \tag{D.15b}$$

The noise transformation from the output to input can now be done to calculate the noise parameters as follows in great detail.

Figure D.6 shows the equivalent circuit of a FET with the noise source transferred to the input side, and Figure D.7 shows the equivalent circuit representation of FET with noise sources at input.

$$[ABCD]_{\text{FET}} = \begin{bmatrix} \left(\dfrac{sc_{gd}}{sc_{gd}-g_m} + \dfrac{R_s sc_{gd}(g_m + g_{ds} + sc_{gs} + sc_{ds})}{sc_{gd}-g_m}\right) & \left(\dfrac{1}{sc_{gd}-g_m} + \dfrac{R_s(sc_{gd} + g_{ds} + sc_{gs} + sc_{ds})}{sc_{gd}-g_m}\right) \\ \left(\dfrac{sc_{gd}(g_m + g_{ds} + sc_{gs} + sc_{ds})}{sc_{gd}-g_m}\right) & \left(\dfrac{(sc_{gd} + g_{ds} + sc_{gs} + sc_{ds})}{sc_{gd}-g_m}\right) \end{bmatrix} \tag{D.16}$$

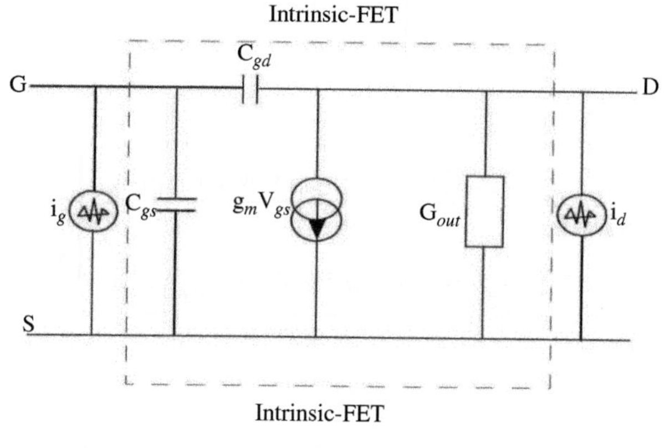

Figure D.5 Intrinsic FET with noise sources at input and output.

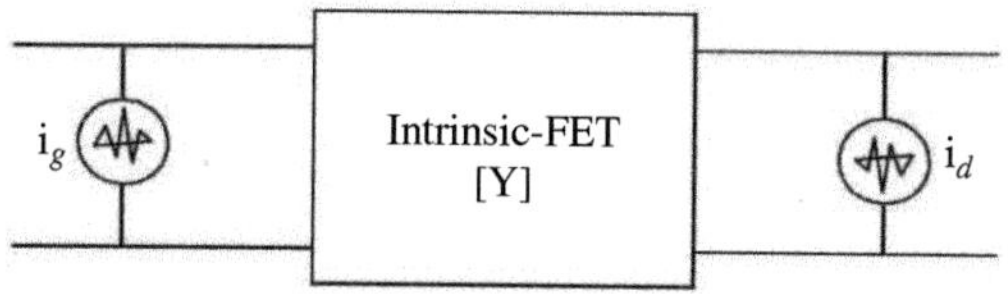

Figure D.6 Intrinsic FET with noise sources at input and output.

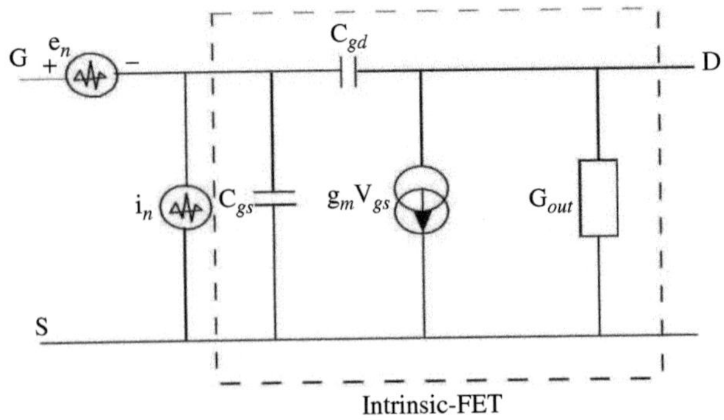

Figure D.7 Equivalent circuit representation of FET with noise sources at input.

$$[C_Y]_{\text{FET}} = 4kT\begin{bmatrix} \dfrac{\omega^2 c_{gs}^2 R}{g_m} & -j\omega c_{gs} C\sqrt{PR} \\ j\omega c_{gs} C\sqrt{PR} & g_m P \end{bmatrix} \tag{D.17}$$

$$[C_a]_{\text{FET}} = [T][C_Y]_{tr}[T]^{+} \tag{D.18}$$

$$[C_a]_{\text{FET}} = \begin{bmatrix} 0 & \left(\dfrac{1}{sc_{gd} - g_m} + \dfrac{R_s(sc_{gd} + g_{ds} + sc_{gs} + sc_{ds})}{sc_{gd} - g_m}\right) \\ 1 & \left(\dfrac{(sc_{gd} + g_{ds} + sc_{gs} + sc_{ds})}{sc_{gd} - g_m}\right) \end{bmatrix} * 4kT\begin{bmatrix} \dfrac{\omega^2 c_{gs}^2 R}{g_m} & -j\omega c_{gs} C\sqrt{PR} \\ j\omega c_{gs} C\sqrt{PR} & g_m P \end{bmatrix} * K1 \tag{D.19}$$

$$K1 = \begin{bmatrix} 0 & 1 \\ \left(\dfrac{1}{sc_{gd} - g_m} + \dfrac{R_s(sc_{gd} + g_{ds} + sc_{gs} + sc_{ds})}{sc_{gd} - g_m}\right)^{\bullet} & \left(\dfrac{(sc_{gd} + g_{ds} + sc_{gs} + sc_{ds})}{sc_{gd} - g_m}\right)^{\bullet} \end{bmatrix} \tag{D.20}$$

$$= 4kT\begin{bmatrix} \left(\dfrac{sc_{gs}C\sqrt{PR}}{sc_{gd} - g_m} + \dfrac{sc_{gs}R_sC\sqrt{PR}(sc_{gd} + g_{ds} + sc_{gs} + sc_{ds})}{sc_{gd} - g_m}\right) & \left(\dfrac{g_mP}{sc_{gd} - g_m} + \dfrac{g_mPR_s(sc_{gd} + g_{ds} + sc_{gs} + sc_{ds})}{sc_{gd} - g_m}\right) \\ \left(\dfrac{\omega^2 c_{gs}^2 R}{g_m} + \dfrac{(sc_{gd} + g_{ds} + sc_{gs} + sc_{ds})sc_{gs}C\sqrt{PR}}{sc_{gd} - g_m}\right) & \left(-j\omega c_{gs}C\sqrt{PR} + \dfrac{(sc_{gd} + g_{ds} + sc_{gs} + sc_{ds})g_mP}{sc_{gd} - g_m}\right) \end{bmatrix} * K2 \tag{D.21}$$

$$K2 = \begin{bmatrix} 0 & 1 \\ \left(\dfrac{1}{sc_{gd} - g_m} + \dfrac{R_s(sc_{gd} + g_{ds} + sc_{gs} + sc_{ds})}{sc_{gd} - g_m}\right)^{\bullet} & \left(\dfrac{(sc_{gd} + g_{ds} + sc_{gs} + sc_{ds})}{sc_{gd} - g_m}\right)^{\bullet} \end{bmatrix} \tag{D.22}$$

$$[C_a]_{\text{FET}} = \begin{bmatrix} C_{uu^\bullet} & C_{ui^\bullet} \\ C_{u^\bullet i} & C_{ii^\bullet} \end{bmatrix} = 4kT\begin{bmatrix} R_n & \dfrac{F_{\min} - 1}{2} - R_n Y_{\text{opt}}^{\bullet} \\ \dfrac{F_{\min} - 1}{2} - R_n Y_{\text{opt}} & R_n\left|Y_{\text{opt}}\right|^2 \end{bmatrix} \tag{D.23}$$

$$C_{uu^\bullet} = 4kT\left[\left(\frac{g_m P}{sc_{gd} - g_m} + \frac{g_m P R_s(sc_{gd} + g_{ds} + sc_{gs} + sc_{ds})}{sc_{gd} - g_m}\right)\left(\frac{1}{sc_{gd} - g_m} + \frac{R_s(sc_{gd} + g_{ds} + sc_{gs} + sc_{ds})}{sc_{gd} - g_m}\right)^\bullet\right] \quad (D.24)$$

$$C_{ui^\bullet} = 4kT\left[\left(\frac{sc_{gs} C\sqrt{PR}}{sc_{gd} - g_m} + \frac{sc_{gs} R_s C\sqrt{PR}(sc_{gd} + g_{ds} + sc_{gs} + sc_{ds})}{sc_{gd} - g_m}\right)\right] + A1 \quad (D.25)$$

$$A1 = \left[\left(\frac{g_m P}{sc_{gd} - g_m} + \frac{g_m P R_s(sc_{gd} + g_{ds} + sc_{gs} + sc_{ds})}{sc_{gd} - g_m}\right)\left(\frac{(sc_{gd} + g_{ds} + sc_{gs} + sc_{ds})}{sc_{gd} - g_m}\right)^\bullet\right] \quad (D.26)$$

$$C_{u^\bullet i} = 4kT\left[\left(-j\omega c_{gs} C\sqrt{PR} + \frac{(sc_{gd} + g_{ds} + sc_{gs} + sc_{ds})g_m P}{sc_{gd} - g_m}\right)\left(\frac{1}{sc_{gd} - g_m} + \frac{R_s(sc_{gd} + g_{ds} + sc_{gs} + sc_{ds})}{sc_{gd} - g_m}\right)^\bullet\right] \quad (D.27)$$

$$C_{ii^\bullet} = 4kT\left[\left(\frac{\omega^2 c_{gs}^2 R}{g_m} + \frac{(sc_{gd} + g_{ds} + sc_{gs} + sc_{ds}) sc_{gs} C\sqrt{PR}}{sc_{gd} - g_m}\right) + B1\right] \quad (D.28)$$

$$B1 = \left(-j\omega c_{gs} C\sqrt{PR} + \frac{(sc_{gd} + g_{ds} + sc_{gs} + sc_{ds})g_m P}{sc_{gd} - g_m}\right)\left(\frac{(sc_{gd} + g_{ds} + sc_{gs} + sc_{ds})}{sc_{gd} - g_m}\right)^\bullet \quad (D.29)$$

and after substituting the values of $C_{uu^\bullet}$, $C_{ui^\bullet}$, $C_{u^\bullet i}$, $C_{ii^\bullet}$, we obtain the noise parameters as follows:

$$R_n = \frac{C_{uu^\bullet}}{2kT} \quad (D.30)$$

$$F_{\min} = \left[1 + \frac{C_{ui^\bullet} + C_{uu^\bullet} Y_{\text{opt}}}{kT}\right] \quad (D.31)$$

$$Y_{\text{opt}} = \sqrt{\frac{C_{ii^\bullet}}{C_{uu^\bullet}} - \left[\text{Im}\left(\frac{C_{ui^\bullet}}{C_{uu^\bullet}}\right)\right]^2} + j\text{Im}\left(\frac{C_{ui^\bullet}}{C_{uu^\bullet}}\right) \quad (D.32)$$

$$Y_{\text{opt}} = G_{\text{opt}} + jB_{\text{opt}} \quad (D.33)$$

$$\Gamma_{\text{opt}} = \frac{Z_{\text{opt}} - Z_o}{Z_{\text{opt}} + Z_o} \Rightarrow \frac{Y_{\text{opt}} - Y_o}{Y_{\text{opt}} + Y_o} \quad (D.34)$$

Finally, we obtain the familiar equation as follows:

$$F = F_{\min} + \frac{R_n}{G_g}\left[(G_{\text{opt}} - G_g)^2 + (B_{\text{opt}} - B_G)^2\right] \quad (D.35)$$

This is the actual resulting noise figure. While amplifiers are traditionally matched for S11 and S22 complex conjugate, here at the input we need to tune for G_{opt} (noise matching), which may result in less power gain.

Neglecting the effect of gate leakage current I_{gd} and the gate to drain capacitance C_{gd}, the models above may be further simplified, as we will demonstrate below.

Figure D.8 shows the equivalent configuration of the FET without gate-drain capacitance.

$$[C_a]_{tr} = \begin{bmatrix} \overline{e_n e_n^\bullet} & \overline{e_n i_n^\bullet} \\ \overline{i_n e_n^\bullet} & \overline{i_n i_n^\bullet} \end{bmatrix} \quad (D.36)$$

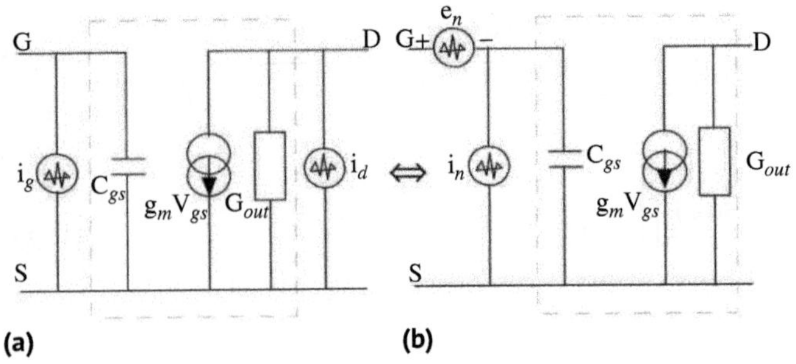

Figure D.8 Equivalent configuration of the FET without gate-drain capacitance.

$$[C_a]_{\text{FET}} = [T][C_Y]_{\text{FET}}[T]^+ \tag{D.37}$$

$$[T] = \begin{bmatrix} 0 & B_{CS} \\ 1 & D_{CS} \end{bmatrix}_{\text{FET}} \tag{D.38}$$

$$[C_a]_{\text{FET}} = \begin{bmatrix} 0 & B_{CS} \\ 1 & D_{CS} \end{bmatrix} [C_Y]_{\text{FET}} \begin{bmatrix} 0 & 1 \\ B^{\bullet}_{CS} & D^{\bullet}_{CS} \end{bmatrix} \tag{D.39}$$

The transformation matrix T comes from the ABCD matrix of the intrinsic FET.

$$[Y]_{\text{FET}} = \begin{bmatrix} y_{11} & y_{12} \\ y_{21} & y_{22} \end{bmatrix} = \begin{bmatrix} sc_{gs} & 0 \\ g_m & G_{\text{out}} \end{bmatrix} \tag{D.40}$$

$$[ABCD]_{\text{FET}} = \begin{bmatrix} A_{CS} & B_{CS} \\ C_{CS} & D_{CS} \end{bmatrix} = \begin{bmatrix} -\dfrac{y_{22}}{y_{11}} & \dfrac{1}{-y_{21}} \\ \dfrac{\Delta}{y_{21}} & \dfrac{-y_{11}}{y_{21}} \end{bmatrix} = \begin{bmatrix} -\dfrac{G_{\text{out}}}{g_m} & \dfrac{-1}{g_m} \\ \dfrac{sc_{gs}G_{\text{out}}}{g_m} & \dfrac{-sc_{gs}}{g_m} \end{bmatrix} \tag{D.41}$$

$$[T] = \begin{bmatrix} 0 & B_{CS} \\ 1 & D_{CS} \end{bmatrix} = \begin{bmatrix} 0 & \dfrac{-1}{g_m} \\ 1 & \dfrac{-sc_{gs}}{g_m} \end{bmatrix} \tag{D.42}$$

$$[T]^+ = \begin{bmatrix} 0 & 1 \\ B^{\bullet}_{CS} & D^{\bullet}_{CS} \end{bmatrix} = \begin{bmatrix} 0 & 1 \\ \dfrac{1}{-g_m} & \dfrac{sc_{gs}}{-g_m} \end{bmatrix} \tag{D.43}$$

$$[C_a]_{\text{FET}} = [T][C_Y]_{\text{FET}}[T]^+ \tag{D.44}$$

$$[C_a]_{\text{FET}} = \begin{bmatrix} 0 & \dfrac{1}{-g_m} \\ 1 & \dfrac{sc_{gs}}{-g_m} \end{bmatrix} * 4kT \begin{bmatrix} \dfrac{\omega^2 c_{gs}^2 R}{g_m} & -j\omega c_{gs} C\sqrt{PR} \\ j\omega c_{gs} C\sqrt{PR} & g_m P \end{bmatrix} * \begin{bmatrix} 0 & 1 \\ \dfrac{1}{-g_m} & \dfrac{sc_{gs}}{-g_m} \end{bmatrix} \tag{D.45}$$

$$[C_a]_{\text{FET}} = \frac{4kT}{g_m} \begin{bmatrix} P & -j\omega c_{gs}\left(P + C\sqrt{PR}\right) \\ j\omega c_{gs}\left(P + C\sqrt{PR}\right) & \omega^2 c_{gs}^2\left(P + R + 2C\sqrt{PR}\right) \end{bmatrix} \tag{D.46}$$

$$[C_a]_{\text{FET}} = \begin{bmatrix} C_{uu^\bullet} & C_{ui^\bullet} \\ C_{u^\bullet i} & C_{ii^\bullet} \end{bmatrix} = 4kT \begin{bmatrix} R_n & \frac{F_{\min}-1}{2} - R_n Y^\bullet_{\text{opt}} \\ \frac{F_{\min}-1}{2} - R_n Y_{\text{opt}} & R_n \left|Y_{\text{opt}}\right|^2 \end{bmatrix} \tag{D.47}$$

and after substituting the values of $C_{uu^\bullet}$, $C_{ui^\bullet}$, $C_{u^\bullet i}$, $C_{ii^\bullet}$, we obtain the noise parameters as follows:

$$R_n = \frac{C_{uu^\bullet}}{4kT} = \frac{P}{g_m} \tag{D.48}$$

$$Y_{\text{opt}} = \sqrt{\frac{C_{ii^\bullet}}{C_{uu^\bullet}} - \left[\text{Im}\left(\frac{C_{ui^\bullet}}{C_{uu^\bullet}}\right)\right]^2} + j\text{Im}\left(\frac{C_{ui^\bullet}}{C_{uu^\bullet}}\right) \tag{D.49}$$

$$Y_{\text{opt}} = G_{\text{opt}} + jB_{\text{opt}} \tag{D.50}$$

$$G_{\text{opt}} = \frac{\omega c_{gs}}{P}\sqrt{PR(1-C^2)} \tag{D.51}$$

$$B_{\text{opt}} = -\omega c_{gs}\left(1 + C\sqrt{\frac{R}{P}}\right) \tag{D.52}$$

$$F_{\min} = 1 + \frac{C_{ui^\cdot} + C_{uu^\cdot} Y_{\text{opt}}}{kT} = 1 + \frac{2\omega c_{gs}}{g_m}\sqrt{PR(1-C^2} \tag{D.53}$$

This equation is valid for the somewhat simpler equivalent circuit. A more detailed noise equivalent model and derivation is found in [4].

D.2.5 Influence of C_{gd}, R_{gs}, and R_s on the Noise Parameters

$$[C_a]_{\text{FET}} = \begin{bmatrix} \overline{e_n e_n^\bullet} & \overline{e_n i_n^\bullet} \\ \overline{i_n e_n^\bullet} & \overline{i_n i_n^\bullet} \end{bmatrix} \tag{D.54}$$

$$[C_a]_{\text{FET}} = \begin{bmatrix} C_{uu^\bullet} & C_{ui^\bullet} \\ C_{u^\bullet i} & C_{ii^\bullet} \end{bmatrix} \tag{D.55}$$

$$C_{uu^\bullet} = \left|\frac{g_m}{g_m - j\omega c_{gd}}\right|^2 \left(\frac{P + R - 2C_r\sqrt{PR}}{g_m}\right) + (R_s + R_{gs}) \tag{D.56}$$

$$C_{ui^\bullet} = \left|\frac{g_m}{g_m - j\omega c_{gd}}\right|^2 \left[\left(\frac{\omega^2 C_{gs}^2 C_{gd}^2}{g_m^2} + \frac{j\omega c_{gd}}{g_m}\right)\left(R - C\sqrt{PR}\right) - \frac{j\omega c_{gd}}{g_m}\left(P - C^\bullet\sqrt{PR}\right)\right] \tag{D.57}$$

$$C_{ii^\bullet} = \left|\frac{g_m}{g_m - j\omega c_{gd}}\right|^2 \left[\left|\frac{\omega^2 C_{gs}^2 C_{gd}^2}{g_m^2} + \frac{j\omega c_{gd}}{g_m}\right|^2 \frac{R}{g_m} + \left|\frac{j\omega c_{gs}}{g_m}\right|^2 P g_m\right] + \tag{D.58}$$

$$\left|\frac{g_m}{g_m - j\omega c_{gd}}\right|^2 \left[2\text{Re}\left\{\left(\frac{\omega^2 C_{gs}^2 C_{gd}^2}{g_m^2} + \frac{j\omega c_{gd}}{g_m}\right)\left(\frac{j\omega c_{gs}}{g_m}\right) C\sqrt{PR}\right\}\right] \tag{D.59}$$

$$C_{u^\bullet i} = \left\{\left|\frac{g_m}{g_m - j\omega c_{gd}}\right|^2 \left[\left(\frac{\omega^2 C_{gs}^2 C_{gd}^2}{g_m^2} + \frac{j\omega c_{gd}}{g_m}\right)\left(R - C\sqrt{PR}\right) - \frac{j\omega c_{gd}}{g_m}\left(P - C^\bullet\sqrt{PR}\right)\right]\right\}^\bullet \tag{D.60}$$

Where

$$\frac{R}{g_m} = \overline{e_n e_n^\bullet} \tag{D.61}$$

$$P g_m = \overline{i_n i_n^\bullet} \tag{D.62}$$

$$C = \frac{\overline{|e_n i_n^{\bullet}|}}{\sqrt{\left(\overline{e_n e_n^{\bullet}}\right)\left(\overline{i_n i_n^{\bullet}}\right)}} = \frac{\overline{|e_n i_n^{\bullet}|}}{\sqrt{\overline{|e_n^2|}\,\overline{|i_n^2|}}} \tag{D.63}$$

$$[C_a]_{FET} = \begin{bmatrix} C_{uu^{\bullet}} & C_{ui^{\bullet}} \\ C_{u^{\bullet}i} & C_{ii^{\bullet}} \end{bmatrix} = \begin{bmatrix} R_n & \frac{F_{\min}-1}{2} - R_n Y_{opt}^{\bullet} \\ \frac{F_{\min}-1}{2} - R_n Y_{opt} & R_n \left|Y_{opt}\right|^2 \end{bmatrix} \tag{D.64}$$

The modified expressions for the noise parameters are now:

$$R_n = C_{uu^{\bullet}} \tag{D.65}$$

$$Y_{opt} = \sqrt{\frac{C_{ii^{\bullet}}}{C_{uu^{\bullet}}} - \left[\mathrm{Im}\left(\frac{C_{ui^{\bullet}}}{C_{ii^{\bullet}}}\right)\right]^2} + j\mathrm{Im}\left(\frac{C_{ui^{\bullet}}}{C_{uu^{\bullet}}}\right) \tag{D.66}$$

$$Z_{opt} = \sqrt{\frac{C_{uu^{\bullet}}}{C_{ii^{\bullet}}} - \left(\frac{\mathrm{Im}C_{ui^{\bullet}}}{C_{ii^{\bullet}}}\right)^2} - j\left(\frac{\mathrm{Im}C_{ui^{\bullet}}}{C_{ii^{\bullet}}}\right) \tag{D.67}$$

$$Y_{opt} = G_{opt} + jB_{opt} \tag{D.68}$$

$$F_{\min} = 1 + 2\left\lfloor Re(C_{ui^{\cdot}}) + C_{ii^{\cdot}}\mathrm{Re}(Z_{opt})\right\rfloor \tag{D.69}$$

$$F_{\min} = 1 + 2\left[\left(\frac{\omega^2 c_{gs}^2}{g_m^2}\right)(R_{gs}+R_s)Pg_m + \sqrt{\frac{\omega^4 c_{gs}^4}{g_m^4}(R_{gs}+R_s)^2 P^2 g_m^2 + \left(\frac{\omega^2 c_{gs}^2}{g_m^2}\right)[PR(1-C^2) - Pg_m R_{gs}]}\right] \tag{D.70}$$

$$R_n = \left|\frac{g_m}{g_m - j\omega c_{gd}}\right|^2 \left(\frac{P + R - 2C_r\sqrt{RP}}{}\right) + (R_{gs} + R_s) \tag{D.71}$$

$$R_{opt} = \frac{1}{\omega c_{gs}}\sqrt{\frac{g_m(R_s + R_{gs}) + R(1-C_r^2)}{P} + \omega^2 c_{gs}^2 (R_s + R_{gs})^2} \tag{D.72}$$

$$X_{opt} = \frac{1}{\omega c_{gs}}\left(1 - C_r\sqrt{\frac{R}{P}}\right) \tag{D.73}$$

$F_{\min}$ is best determined using a suitable CAD tool, which is something we will tackle soon.

D.2.6 Temperature Dependency of the Noise Parameters of an FET

We will now introduce a minimum noise temperature $T_{\min}$ and modify the noise parameters that were previously derived.

As pointed out by the NASA space signal surveillance, the lowest noise factor (figure) is needed. The Nyquist formula states

$$Pn = kT_0o$$

As we reduce k from 290 K (room temperature) to a much lower temperature like 27 K, the sensitivity will increase. This equation will now have temperature dependence factors.

In order to use field-effect transistors in amplifiers in huge dishes for scanning the universe and for evaluating signals coming from outer space, extremely low noise temperatures are needed. These amplifiers are cooled cryogenically to temperatures close to zero degrees kelvin (K). Since there are no measurements available, the following method allows prediction of the noise performance of FETs under such "cooled" conditions. The derivation below shows how to calculate the temperature-dependent noise performance.

Figure D.9 shows the familiar two-port noise representation of the intrinsic FET in admittance (a) and ABCD matrix form (b).

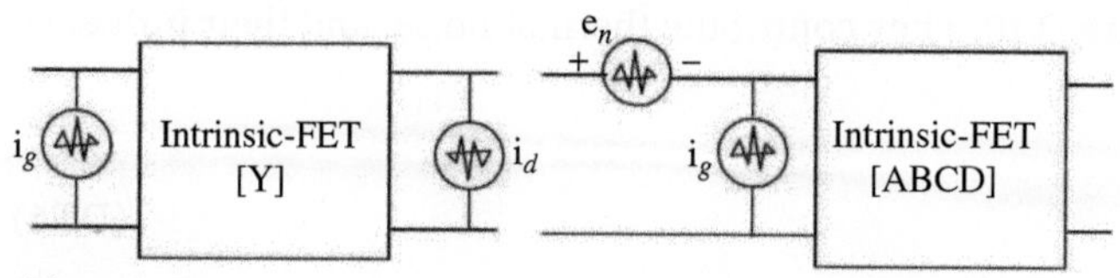

(a) (b)

Figure D.9 Noise representation in linear two-port: (a) current noise source at input and output and (b) current and voltage noise source at input.

The admittance representation of the noise parameters of the intrinsic FET is expressed as follows:

$$G_1 = \frac{\overline{|i_g^2|}}{4kT_0\Delta f} \tag{D.74}$$

$$G_2 = \frac{\overline{|i_d^2|}}{4kT_0\Delta f} \tag{D.75}$$

$$C_r = \frac{\overline{|i_g i_d^{*}|}}{\sqrt{\overline{|i_d^2|}\,\overline{|i_g^2|}}} \tag{D.76}$$

where k is the Boltzmann constant $(k = 1.38 \times 10^{-23})$, T_0 is standard room temperature (290 K), and Δf is the reference bandwidth.

The ABCD matrix representation and the corresponding noise parameters are as follows:

$$R_n = \frac{\overline{|e_n^2|}}{4kT_0\Delta f} \tag{D.77}$$

$$g_n = \frac{\overline{|i_n^2|}}{4kT_0\Delta f} \tag{D.78}$$

$$C_r = \frac{\overline{|e_n i_n^{*}|}}{\sqrt{\overline{|e_n^2|}\,\overline{|i_n^2|}}} \tag{D.79}$$

$$N = R_{\text{opt}} g_n \tag{D.80}$$

where g_n is the noise conductance.

The expression for the noise temperature T_n and a noise measure M of a two-port driven by generator impedance Z_g is expressed as follows (consistent with the previously shown equation for the actual noise factor in case of non-optimal input matching):

$$T_n = T_{\min} + T_0 \frac{g_n}{R_g} \left| Z_g - Z_{\text{opt}} \right|^2 \tag{D.81}$$

$$T_n = T_{\min} + N T_0 \frac{\left| Z_g - Z_{\text{opt}} \right|^2}{R_g R_{\text{opt}}} \tag{D.82}$$

$$T_n = T_{\min} + 4 N T_0 \frac{\left| T_g - T_{\text{opt}} \right|^2}{\left(1 - \left|T_{\text{opt}}\right|^2\right)\left(1 - \left|T_g\right|^2\right)} \tag{D.83}$$

$$T_{\text{opt}} = \frac{Z_{\text{opt}} - Z_o}{Z_{\text{opt}} + Z_o} \tag{D.84}$$

$$M = \frac{T_n}{T_0} \left(\frac{1}{1 - \frac{1}{Ga}} \right) \tag{D.85}$$

where Z_0 is the reference impedance and G_a is the available gain.

An extrinsic FET with parasitic resistances is shown in Figure D.10. They contribute thermal noise, and their influence can be calculated based on the ambient temperature T_a.

$$G_1 = \frac{T_g}{T_o} \frac{R_{gs}(\omega C_{gs})^2}{(1 + \omega^2 C_{gs}^2 R_{gs}^2)} \tag{D.86}$$

$$G_2 = \frac{T_g}{T_o} \frac{g_m^2 R_{gs}}{(1 + \omega^2 C_{gs}^2 R_{gs}^2)} + \frac{T_d}{T_o} g_{gs} \tag{D.87}$$

$$C_{r_c} = C_r \frac{|i_g i_d^*|}{\sqrt{|i_d^2||i_g^2|}} = \frac{-j\omega g_m C_{gs} R_{gs}}{(1 + \omega^2 C_{gs}^2 R_{gs}^2)} \frac{T_g}{T_o} \tag{D.88}$$

The noise properties of the intrinsic FET are treated by assigning an equivalent temperature T_g and T_d to R_{gs} and g_{ds}. No correlation is assumed between the noise sources represented by the equivalent temperature T_g and T_d in Figure D.10. Figure D.11 shows the intrinsic FET with assigned equivalent temperature.
The modified noise parameters are expressed as follows:

$$Z_{opt} = R_{opt} + jX_{opt} \tag{D.89}$$

$$R_{opt} = \sqrt{\left(\frac{f_T}{f}\right)^2 \frac{R_{gs}}{R_{ds}} \frac{T_g}{T_d} + R_{gs}^2} \tag{D.90}$$

$$X_{opt} = \frac{1}{\omega C_{gs}} \tag{D.91}$$

$$T_{min} = 2\frac{f}{f_T} \sqrt{R_{gs} g_{ds} T_g T_d + \left(\frac{f_T}{f}\right)^2 R_{gs}^2 g_{ds}^2 T_d^2} + 2\left(\frac{f_T}{f}\right)^2 R_{gs} g_{ds} T_d \tag{D.92}$$

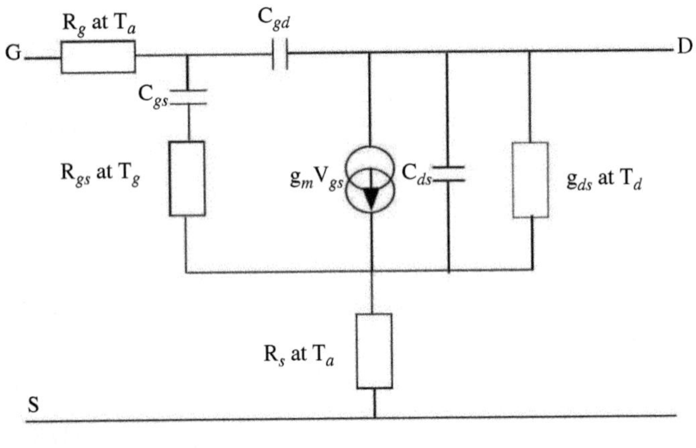

Figure D.10 Extrinsic FET with parasitic resistances.

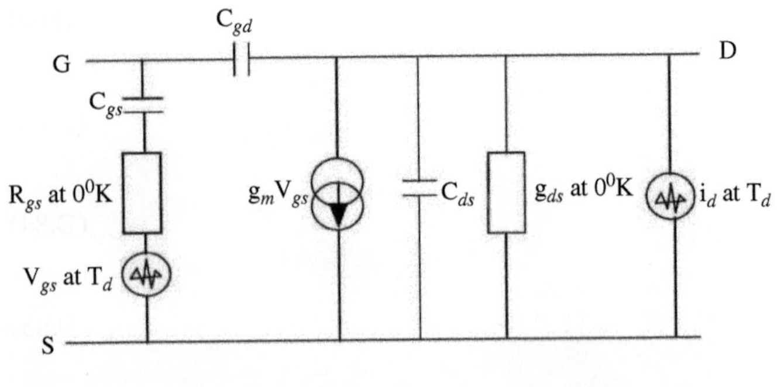

Figure D.11 Intrinsic FET with assigned equivalent temperature.

$$T_{\min} = (F_{\min} - 1)T_0 \tag{D.93}$$

$$g_n = \left(\frac{f_T}{f}\right)^2 \frac{g_{ds}T_d}{T_0} \tag{D.94}$$

$$f_T = \frac{g_m}{2\pi C_{gs}} \tag{D.95}$$

$$\frac{4NT_0}{T_{\min}} = \frac{2}{1 + \frac{R_{gs}}{R_{\text{opt}}}} \tag{D.96}$$

$$R_n = \frac{T_g}{T_o} R_{gs} + \frac{T_d}{T_0} \frac{g_{ds}}{g_m^2} (1 + \omega^2 C_{gs}^2 R_{gs}^2) \tag{D.97}$$

$$C_r = C\sqrt{R_n g_n} = \frac{T_d}{T_0} \frac{g_{ds}}{g_m^2} (\omega^2 C_{gs}^2 R_{gs} + j\omega C_{gs}) \tag{D.98}$$

D.2.7 Approximation and Discussion

With some reasonable approximation, the expression for the noise parameters becomes much simpler (by introducing the following approximation, the obtained values from the calculation typically vary less than 5% from the exact value):

Given the following:

$$if \frac{f}{f_T} \leq \sqrt{\frac{R_{gs}}{R_{ds}} \frac{T_g}{T_d}} \tag{D.99}$$

and $R_{\text{opt}} \geq R_{gs}$, then

$$R_{\text{opt}} \cong \left(\frac{f_T}{f}\right) \sqrt{\frac{r_{gs}}{r_{ds}} \frac{T_g}{T_d}} \tag{D.100}$$

$$X_{\text{opt}} \cong \frac{1}{\omega C_{gs}} \tag{D.101}$$

$$T_{\min} \cong 2\frac{f}{f_T} \sqrt{r_{gs} g_{ds} T_g T_d}$$

$$T_{\min} \cong 2\frac{f}{f_T} \sqrt{r_{gs} g_{ds} T_g T_d} \tag{D.102}$$

$$g_n \cong \left(\frac{f_T}{f}\right)^2 \frac{g_{ds}T_d}{T_0} \tag{D.103}$$

$$f_T = \frac{g_m}{2\pi C_{gs}} \tag{D.104}$$

$$\frac{4NT_0}{T_{\min}} \cong 2 \tag{D.105}$$

Example: A linear FET model with the following intrinsic parameters is assumed:

$R_{gs} = 2.5\ \Omega$	$C_{gd} = 0.042$ pF
$r_{ds} = 400\ \Omega$	$g_m = 57$ mS
$C_{gs} = 0.28$ pF	f = 8.5 GHz
$C_{ds} = 0.067$ pF	

The temperature-dependent noise parameters for the intrinsic FET are now calculated for two cases (room temperature).

Example 1: $T_a = 297$ K, $T_g = 304$ K, $T_d = 5514$ K, $V_{ds} = 2$ V, $I_{ds} = 10$ mA

$$f_T = \frac{g_m}{2\pi C_{gs}} = 32.39\,\text{GHz} \tag{D.106}$$

$$R_{\text{opt}} = \sqrt{\left(\frac{f_T}{f}\right)^2 \frac{r_{gs}}{g_{ds}} \frac{T_g}{T_d} + r_{gs}^2} = 28.42\,\Omega \tag{D.107}$$

$$X_{\text{opt}} = \frac{1}{\omega C_{gs}} = 66.91\,\Omega \tag{D.108}$$

$$T_{\min} = 2\frac{f}{f_T}\sqrt{r_{gs}g_{ds}T_gT_d + \left(\frac{f_T}{f}\right)^2 r_{gs}^2 g_{ds}^2 T_d^2} + 2\left(\frac{f_T}{f}\right)^2 r_{gs}g_{ds}T_d = 58.74K \tag{D.109}$$

$$g_n = \left(\frac{f_T}{f}\right)^2 \frac{g_{ds}T_d}{T_0} = 3.27\,\text{mS} \tag{D.110}$$

$$F_{\min} = \frac{T_{\min}}{T_0} + 1 = \frac{58.7}{290} + 1 = 1.59\,\text{dB} \tag{D.111}$$

$$R_n = \frac{T_g r_{gs}}{T_0} + \frac{g_{ds}T_d}{T_0 g_m^2}(1 + \omega^2 r_{gs}^2 c_{gs}^2) = 17.27\,\Omega \tag{D.112}$$

Example 2: $T_a = 12.5$ K, $T_g = 14.5$ K, $T_d = 1406$ K, $V_{ds} = 2$ V, Ids = 5 mA (cooled down to 14.5 K!)

$$f_T = \frac{g_m}{2\pi C_{gs}} = 32.39\,\text{GHz} \tag{D.113}$$

$$R_{\text{opt}} = \sqrt{\left(\frac{f_T}{f}\right)^2 \frac{r_{gs}}{g_{ds}} \frac{T_g}{T_d} + r_{gs}^2} = 12.34\,\Omega \tag{D.114}$$

$$X_{\text{opt}} = \frac{1}{\omega C_{gs}} = 66.9\,\Omega \tag{D.115}$$

$$T_{\min} = 2\frac{f}{f_T}\sqrt{r_{gs}g_{ds}T_gT_d + \left(\frac{f_T}{f}\right)^2 r_{gs}^2 g_{ds}^2 T_d^2} + 2\left(\frac{f_T}{f}\right)^2 r_{gs}g_{ds}T_d = 7.4K \tag{D.116}$$

$$g_n = \left(\frac{f_T}{f}\right)^2 \frac{g_{ds}T_d}{T_0} = 0.87\,\text{mS} \tag{D.117}$$

$$F_{\min} = \frac{T_{\min}}{T_0} + 1 = \frac{7.4}{290} + 1 = 0.21\,\text{dB} \tag{D.118}$$

$$R_n = \frac{T_g r_{gs}}{T_0} + \frac{g_{ds}T_d}{T_0 g_m^2}(1 + \omega^2 r_{gs}^2 c_{gs}^2) = 3.86\,\Omega \tag{D.119}$$

These results [5] are consistent with those published by Pucel [6] and Pospieszalski [7].

D.2.8 Experimental FET-based Wideband Amplifier Example Simulation with Lumped Lossless Components

Now we will take a closer look at a case using an experimental FET-based wideband amplifier with lumped lossless components. First, we will consider what happens at room temperature.

Figure D.12 shows the schematic of the wideband amplifier with lumped elements, Figure D.13 shows that the amplifier exhibits very flat gain as well as very good input and output return loss, varies from 0.38 to 0.78 dB.

This figure shows that the amplifier has very flat gain (S21) as well as very good return loss (S11 and S22).

Finally, Figure D.14 shows that the noise figure (i.e. 10log noise factor) varies from 0.38 to 0.78 dB. These extremely low values are due to the cooling.

It is not really possible to design an amplifier up to 20 GHz due to the parasitic elements.

Therefore, we must now switch, as expected, to distributed elements,

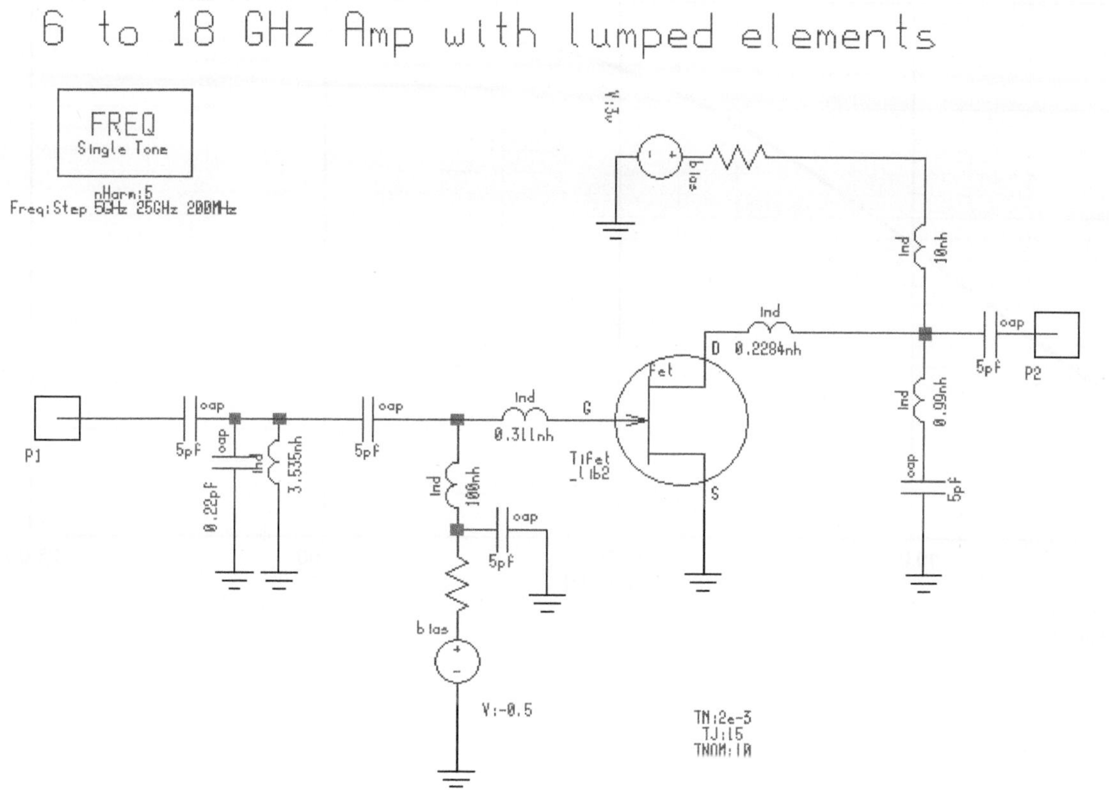

Figure D.12 Schematic of the wideband amplifier with lumped elements.

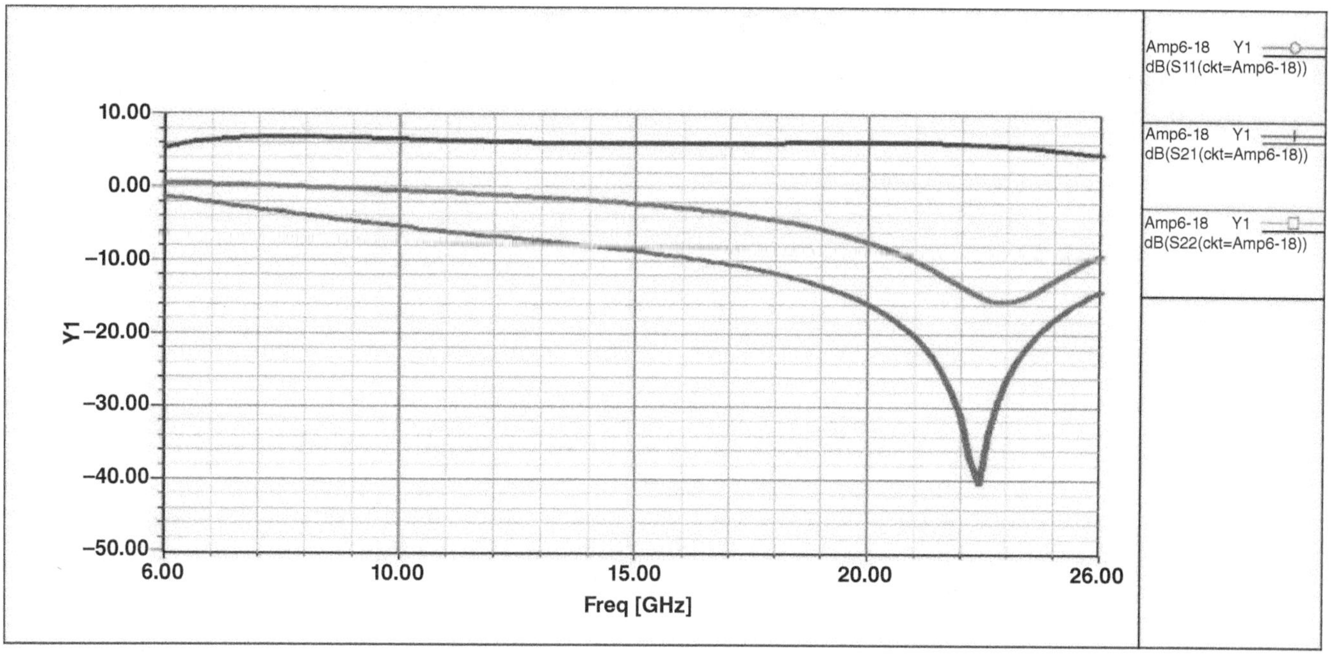

Figure D.13 The amplifier exhibits very flat gain as well as very good input and output return loss.

Figure D.15 shows the wideband amplifier with lossy microwave distributed components, while the Figure D.16 shows the frequency response for the same amplifier of fig. D.15 with distributed elements and the Figure D.17 shows the degraded noise figure due to substrate losses in the distributed elements.

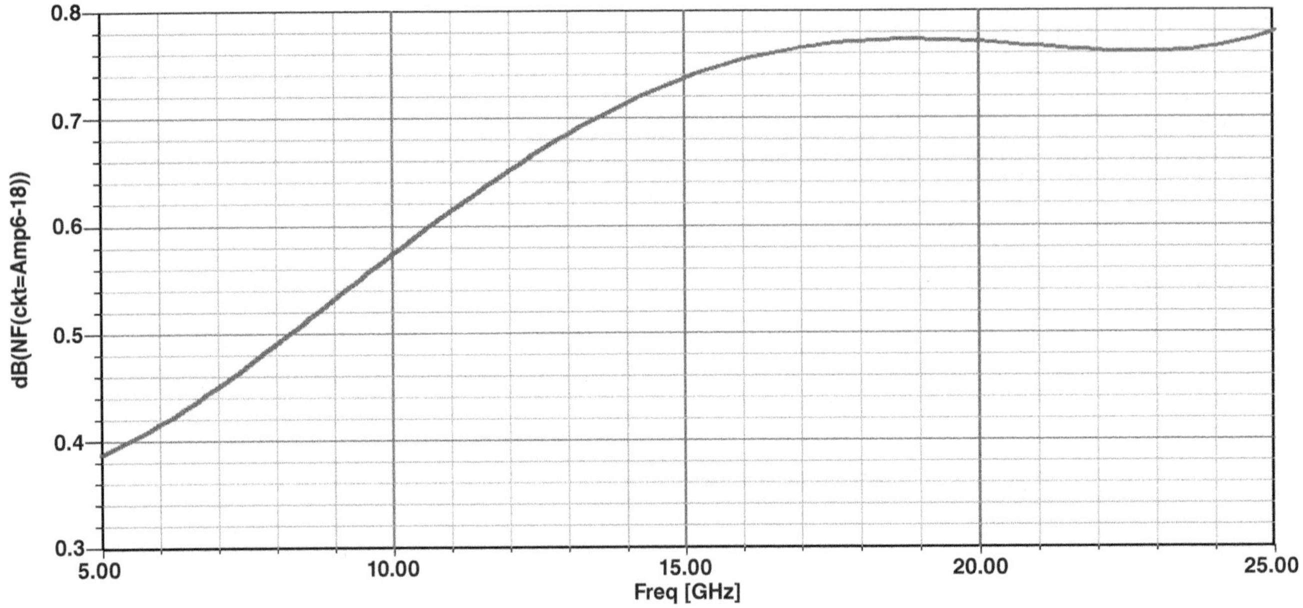

Figure D.14 50 noise figure.

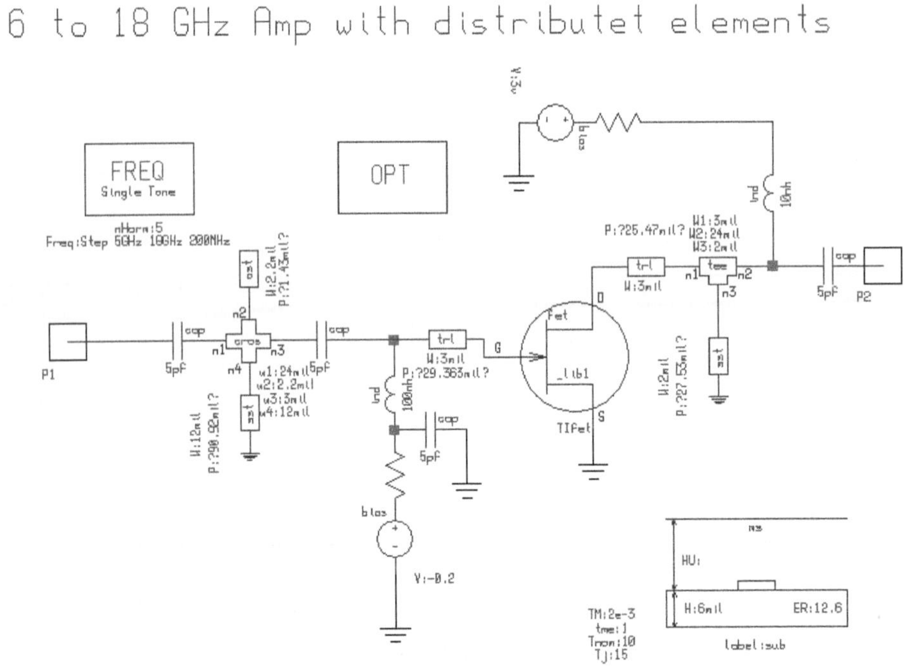

Figure D.15 Wideband amplifier with lossy microwave distributed components.

Figure D.18 is a table showing the conversion between noise figure and noise temperature.

Here, we see the frequency response for the same amplifier but with distributed elements that can actually be built in the real world.

Because of the losses in the distributed elements due to substrate losses, the noise figure for this arrangement suffers a lot.

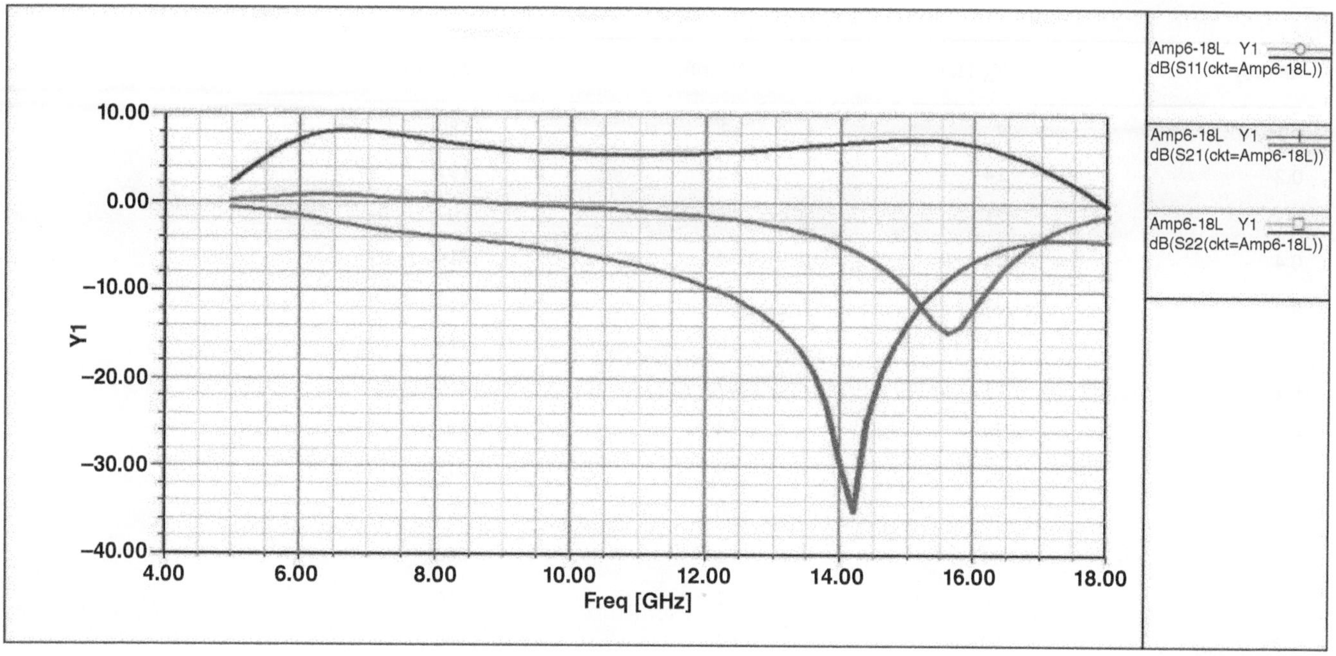

Figure D.16 Frequency response for the same amplifier with distributed elements.

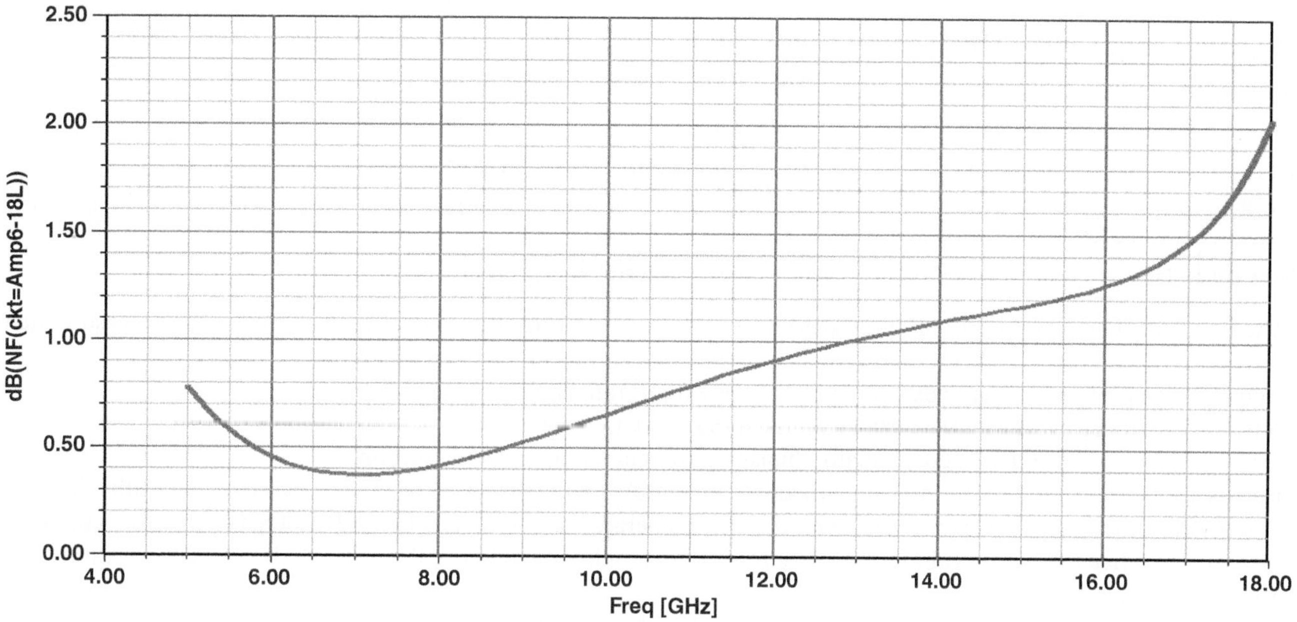

Figure D.17 Degraded noise figure due to substrate losses in the distributed elements.

Most very low temperature transistor amplifiers are quoted in K (degrees Kelvin), rather than in noise figure. We can convert between noise figure and noise temperature as follows:

$$\text{Noise temperature (K)} = T_{\text{REF}} * \left(10^{\frac{\text{NF(dB)}}{10}} - 1\right)$$
$$\text{Noise figure (dB)} = 10 * \log 10\left(\frac{\text{T}_{\text{Noise}}(K)}{\text{T}_{\text{Ref}}(K)} + 1\right) \tag{D.120}$$

Note that unless otherwise specified, $T_{\text{Ref}} = 290$ K.

NF (dB)	T_N (K)	NF (dB)	T_N (K)
0.1	7	2.1	180
0.2	14	2.2	191
0.3	21	2.3	202
0.4	28	2.4	214
0.5	35	2.5	226
0.6	43	2.6	238
0.7	51	2.7	250
0.8	59	2.8	263
0.9	67	2.9	275
1.0	75	3.0	289
1.1	84	3.1	302
1.2	92	3.2	316
1.3	101	3.3	330
1.4	110	3.4	344
1.5	120	3.5	359
1.6	129	3.6	374
1.7	139	3.7	390
1.8	149	3.8	406
1.9	159	3.9	422
2.0	170	4.0	438

Figure D.18 Table–conversion between noise figure and noise temperature.

The frequently quoted value of 17 K translates into a noise figure of 0.265 dB. However, it is very difficult to measure such a low value.

Note that some experts prefer to use the signal-to-noise (S/N) ratio to determine the noise figure.

A useful calculator for this purpose can be found at the following link:
https://www.rfcafe.com/references/calculators/noise-figure-temperature-calculator.htm

Here is a validation circuit using a 14.7 K junction temperature and lossy matching. This is based on the textbook [8].

Figure D.19 shows the experimental design for the low-temperature amplifier and lossy distributed elements. Figure D.20 depicts the significantly higher gain in the low-temperature amplifier, while the Figure D.21 shows the CAD Simulated noise figure for this amplifier.

We can see in the above figure that the non-optimized gain has clearly increased.

Although the noise figure is much improved, it is very tough to measure in the real world.

The Figure D.22 shows the simulated noise figure at different operating temperatures in degrees K.

Since a temperature of 2 K corresponds to a noise figure of 0.03 dB, this is clearly “impossible” to measure!

D.2.9 Dynamic Range of the Amplifier

Now we will look at the three-tone dynamic range test which was popular for designing transistors in the cable TV (CATV) era. Since there are so many signals bombarding the antenna, a test with white noise as a test signal would be better, but it is hard to produce. The CATV test used an FM channel with two sidebands.

The modulation index for HiFi was selected to be 5, the upper audio frequency is 16 kHz. Thus, the bandwidth/sideband = 16 × 50 = 80 kHz, or 160 kHz total bandwidth.

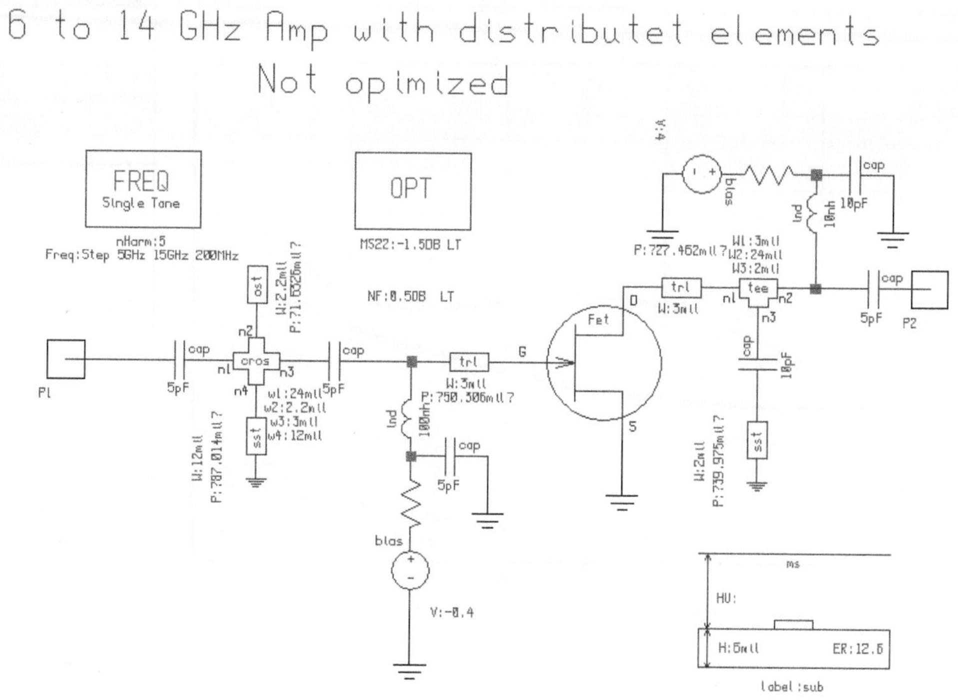

Figure D.19 Experimental design for the low-temperature amplifier and lossy distributed elements.

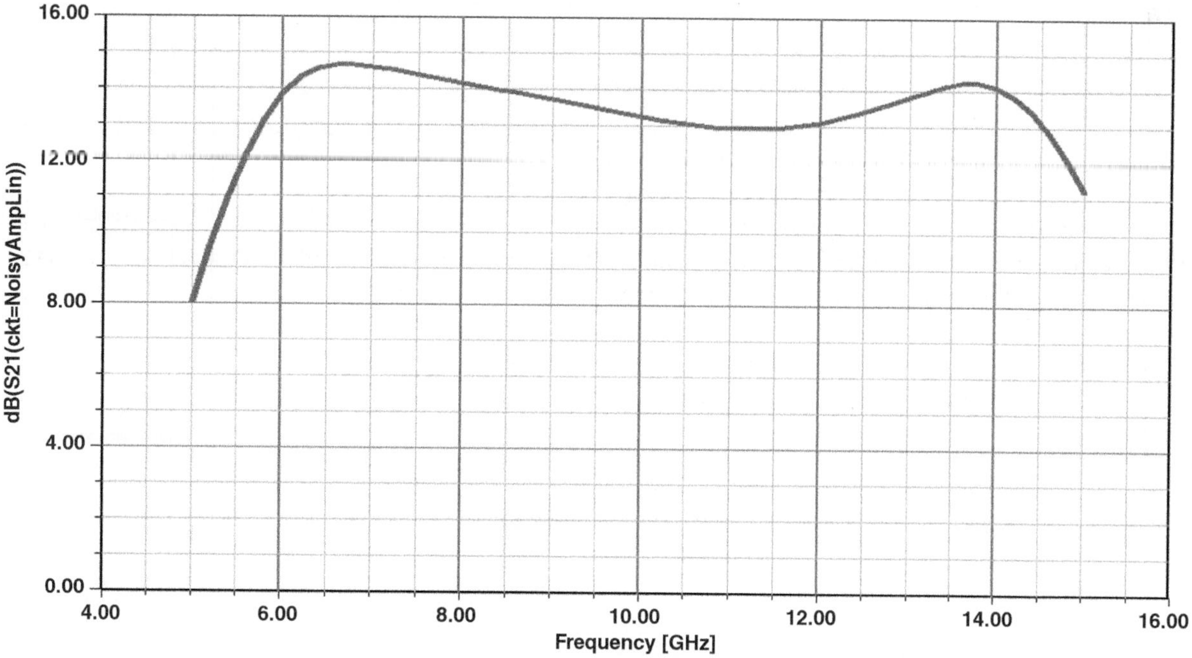

Figure D.20 Significantly higher gain in the low-temperature amplifier.

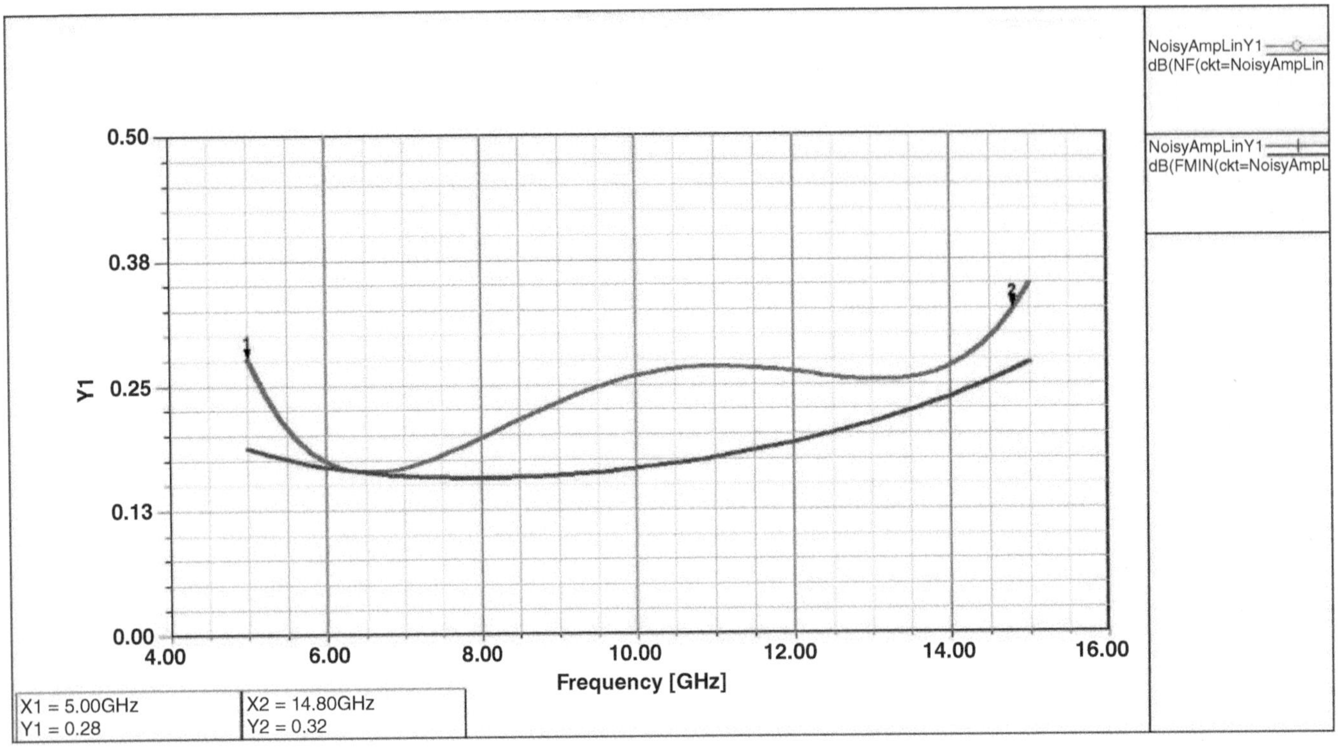

Figure D.21 Simulated noise figure.

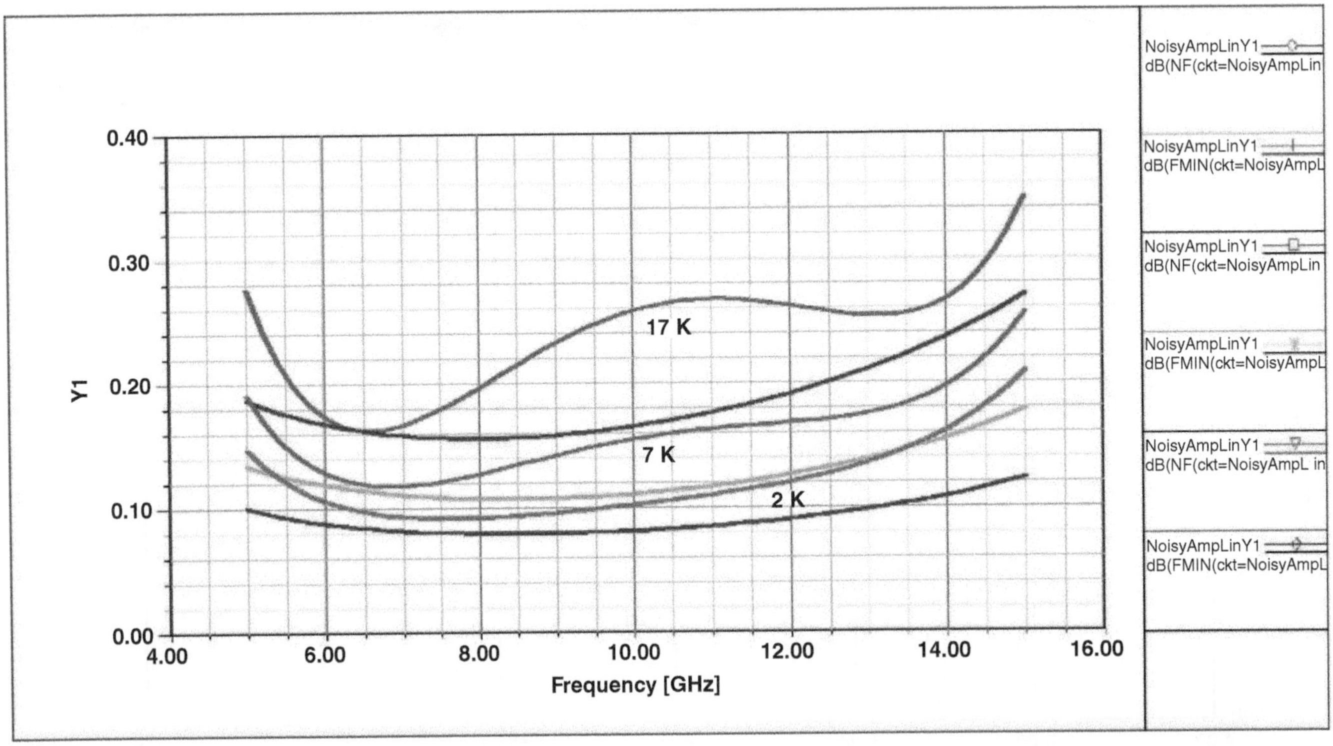

Figure D.22 Simulated noise figure at different operating temperatures.

Analog video bandwidth:

NTSC: The NTSC (National Television System Committee) standard, a common analog TV format, uses a 6 MHz bandwidth for each channel, with the video signal occupying the first 4 MHz.

Luminance (Y): The luminance signal, which carries the black and white information, has a bandwidth of 4.2 MHz.

Chrominance (Color): The chrominance signals (I and Q) carry color information and have lower bandwidths (1.6 MHz for I, 0.6 MHz for Q) because the human eye is less sensitive to color details than to brightness.

Figure D.23 shows an experimental circuit for large-signal operation at room temperature Figure D.24 shows the frequency response at the bias level as seen in Figure D.23, Figure D.25 shows the Noise figure at the bias level as seen in Figure D.23 and Figure D.26 shows its CAD simulated DC current–voltage (I–V) curve for this amplifier.

For further details, see [4, p. 705 (AlGaN/GaN HEMT)].

From our Raytheon–TI joint venture, we had some Materka model parameters that we adjusted to fit the values of the GaN transistor that were published in the Springer book. This involved changing the intrinsic capacitance values and some other parameters to emulate this device. The actual frequency response did not change much.

While the two-tone test is usually recommended, as we explained the three-tone test comes closer to the practical multi-toned environment.

Figure D.27 shows an experimental circuit for operation with larger signals, typical example of a 6–18 GHz amplifier with distributed elements and Figure D.28 showing its three tone output (power sweep) and Figure D.29 shows the simulated output power load lines. Figure D.30 is its predicted dynamic range at 10 GHz and three tone test of about 3 × 1 mV rms.

Note especially the three-tone input in the circuit for operation with larger signals.

As a simulated linearity test, it is important to determine the 1 dB compression point. It occurs at about 3 dBm output or –5.2 dBm input. As long as the model has a good fit, this is perfect.

The simulated output power load lines show the power compression very nicely.

The simulated dynamic range of this circuit of 142 dB relative to –45 dBm is impressive. The textbook three-tone test would have one tone smaller by 6 dB.

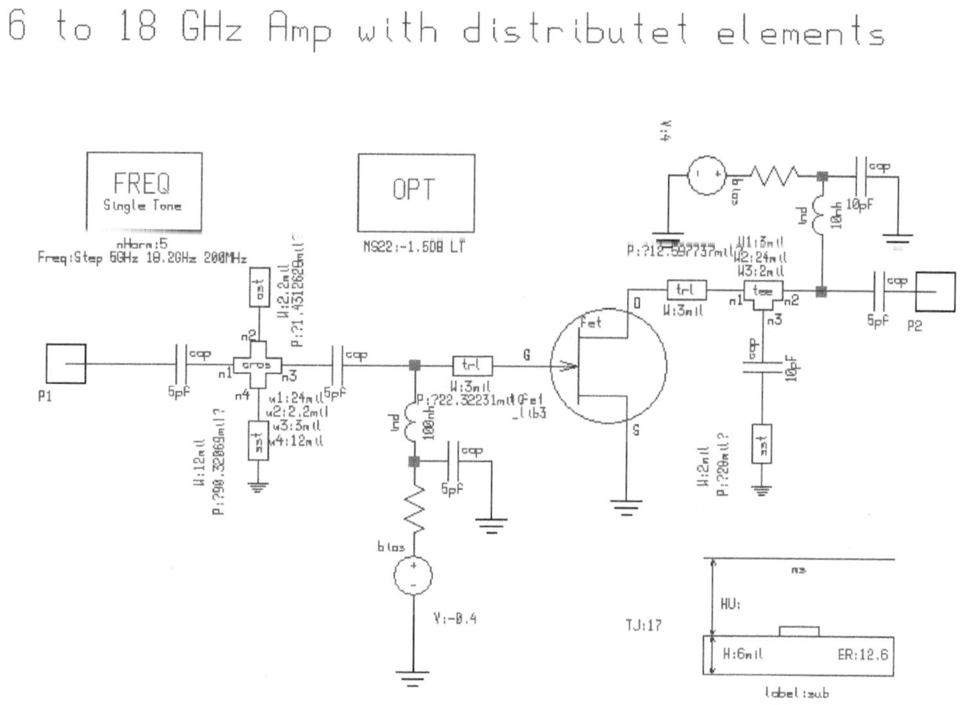

Figure D.23 Experimental circuit for large-signal operation at room temperature.

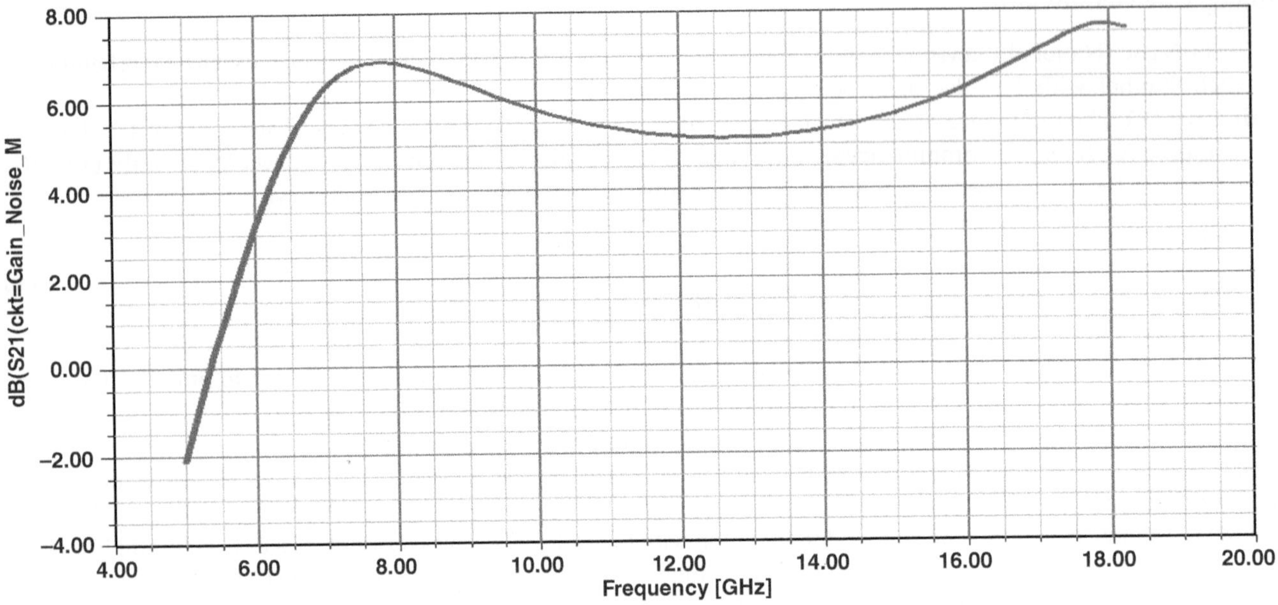

Figure D.24 Frequency response at the bias level as seen in Figure D.23.

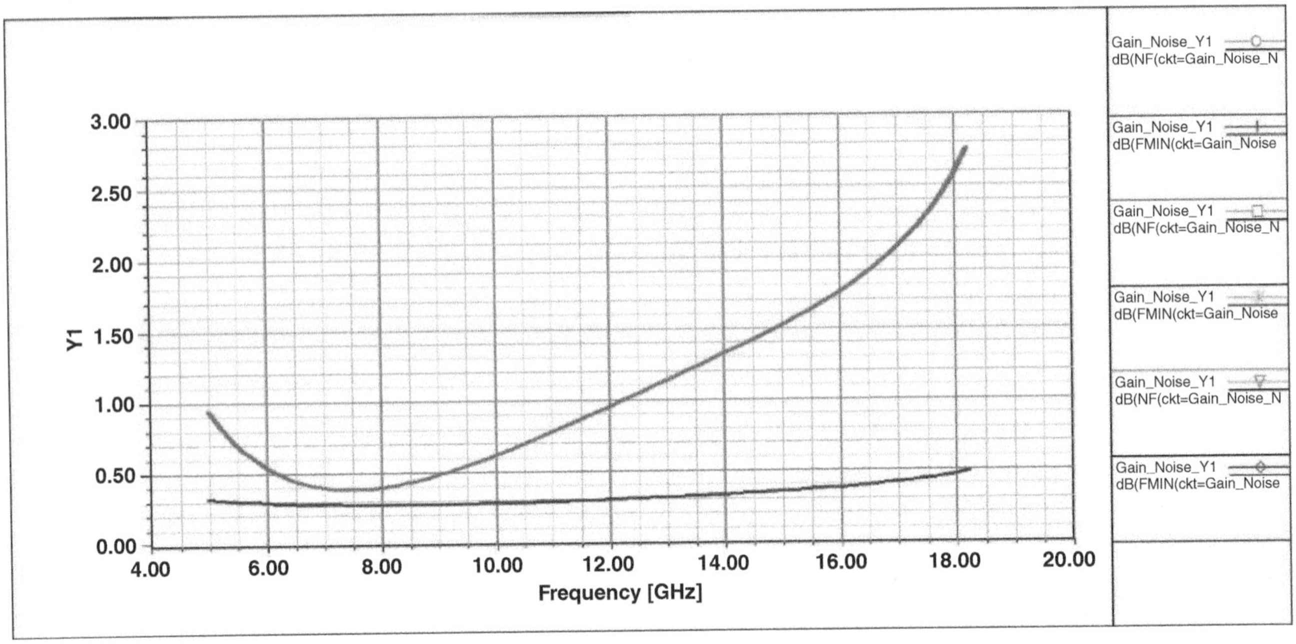

Figure D.25 Noise figure at the bias level as seen in Figure D.23.

D.2.10 Single Stage Low-Noise Amplifier

Now we will look at a 90 GHz amplifier designed for a temperature of 10 K. It consists of CFY99A experimental transistor, uses ideal transmission lines, and shows the simulation results in the following figures.

Figure D.31 shows the example of a cooled amplifier (low-noise 90 GHz amplifier using ideal distributed elements), where Figure D.32 shows its CAD simulated gain (S21) and output return loss (S22) at 10 K, and Figure D.33 shows the simulated noise figure for the same amplifier

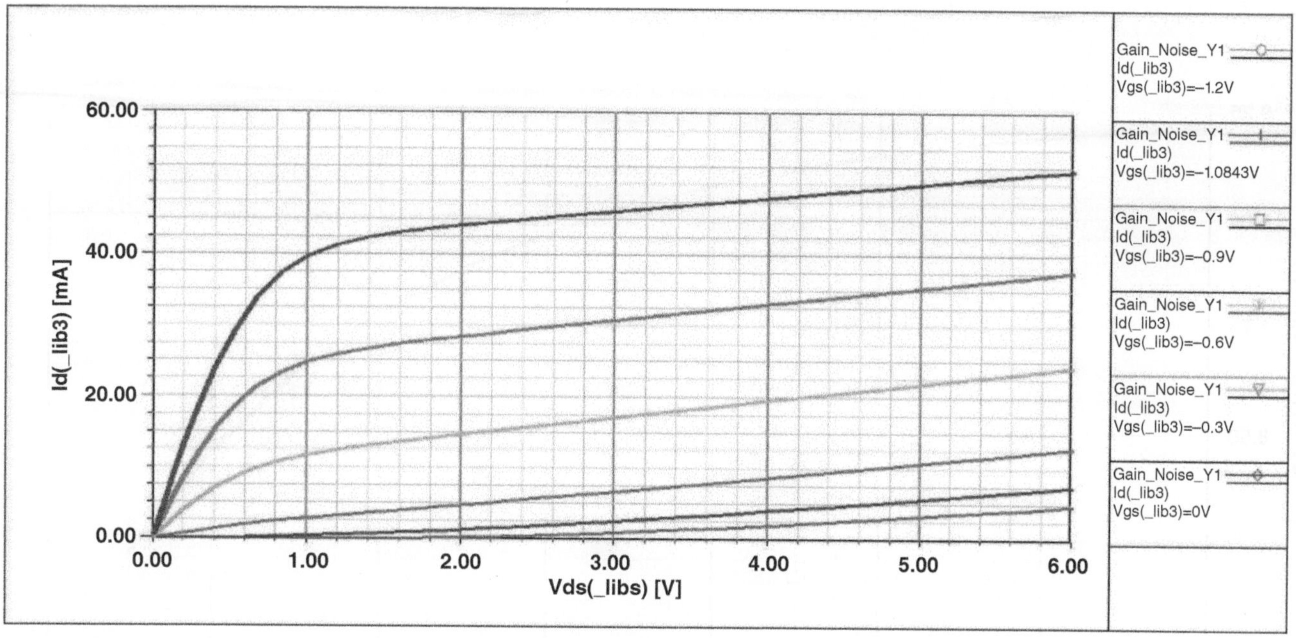

Figure D.26 Simulated DC current–voltage (I–V) curve for the amplifier.

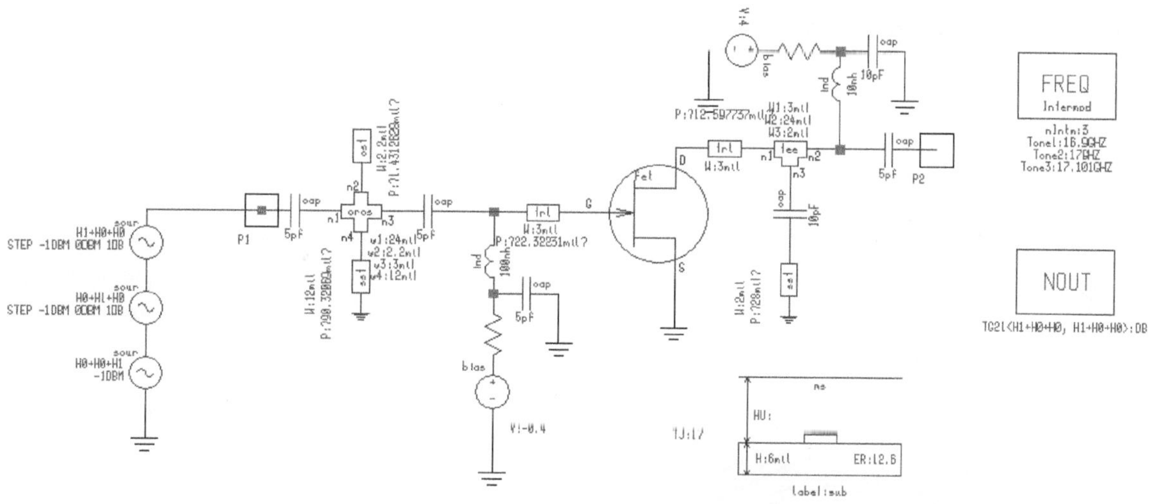

Figure D.27 Experimental circuit for operation with larger signals. 6–18 GHz amplifier with distributed elements.

D.2.11 Reverse-engineered Two-stage Amplifier

Now we will consider an experiment that involves reverse engineering a two-stage amplifier. The amplifier in question is described in the reference [3]:

Figure D.34 shows the circuit diagram of the two-stage amplifier as published in [3]. Figure D.35 shows the CAD Simulation of the gain and return loss of the 100 GHz amplifier described in Fig. D.34. Figure D.36 shows its Noise simulation at 300 K and Figure D.37 shows the Noise simulation at 10 deg K after optimization.

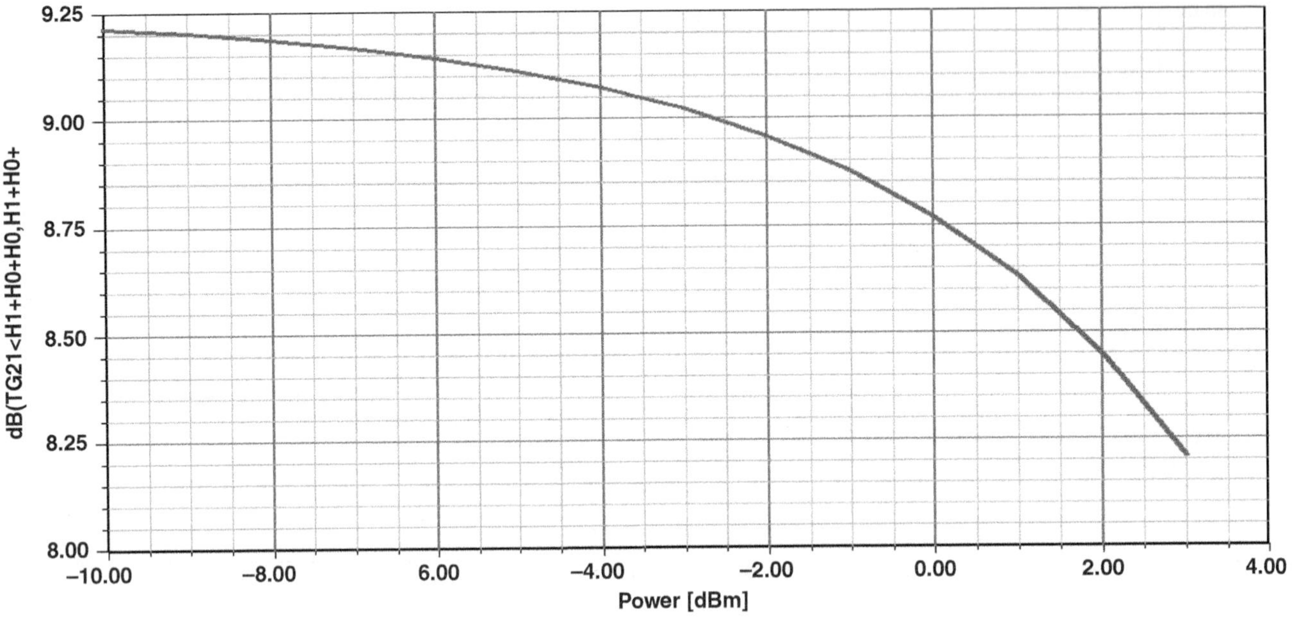

Figure D.28 Three tone output (power sweep).

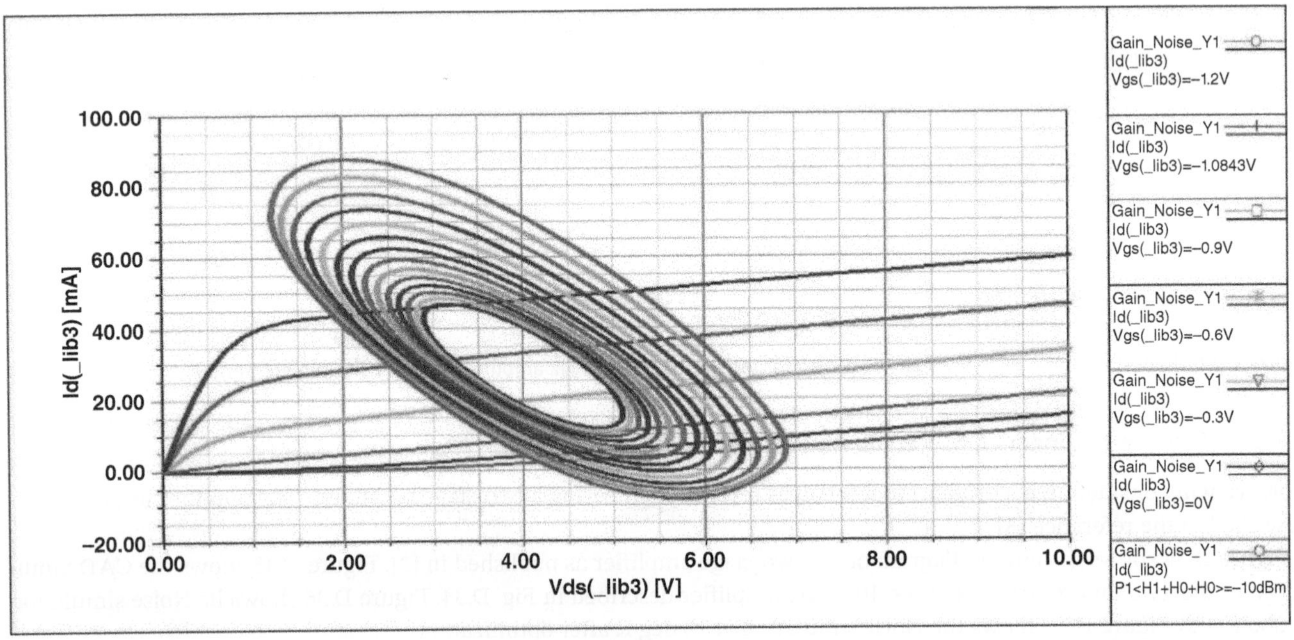

Figure D.29 Simulated output power load lines.

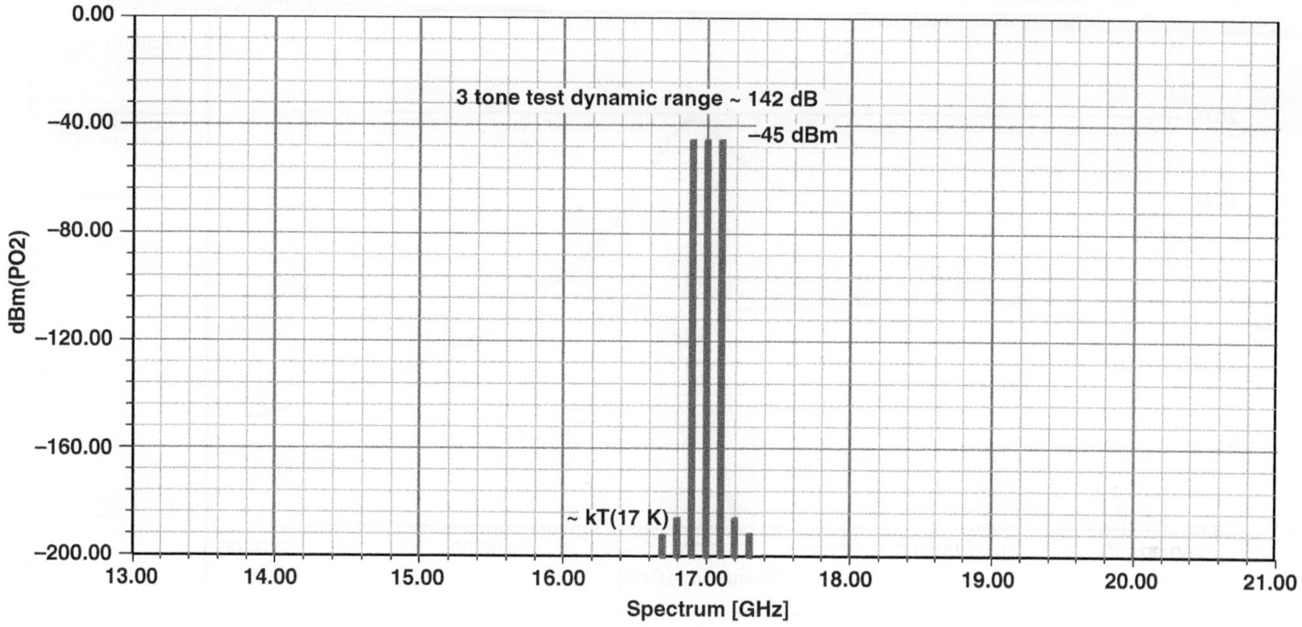

Figure D.30 Predicted dynamic range at 10 GHz and three tone test of about 3 × 1 mV rms.

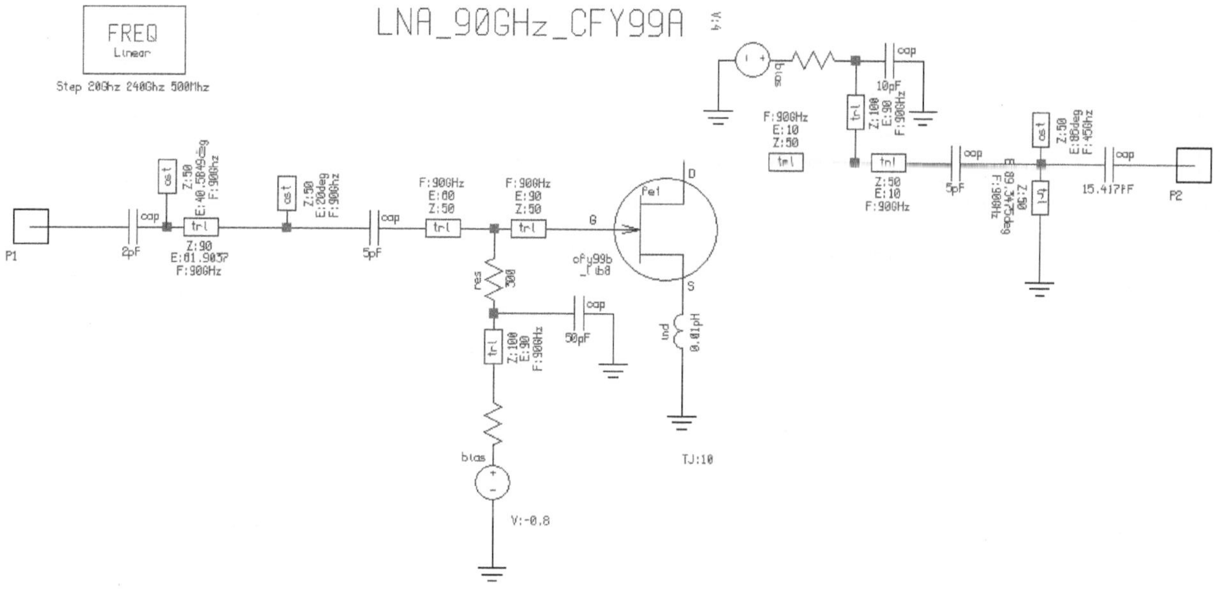

Figure D.31 Cooled amplifier (low-noise 90 GHz amplifier using ideal distributed elements).

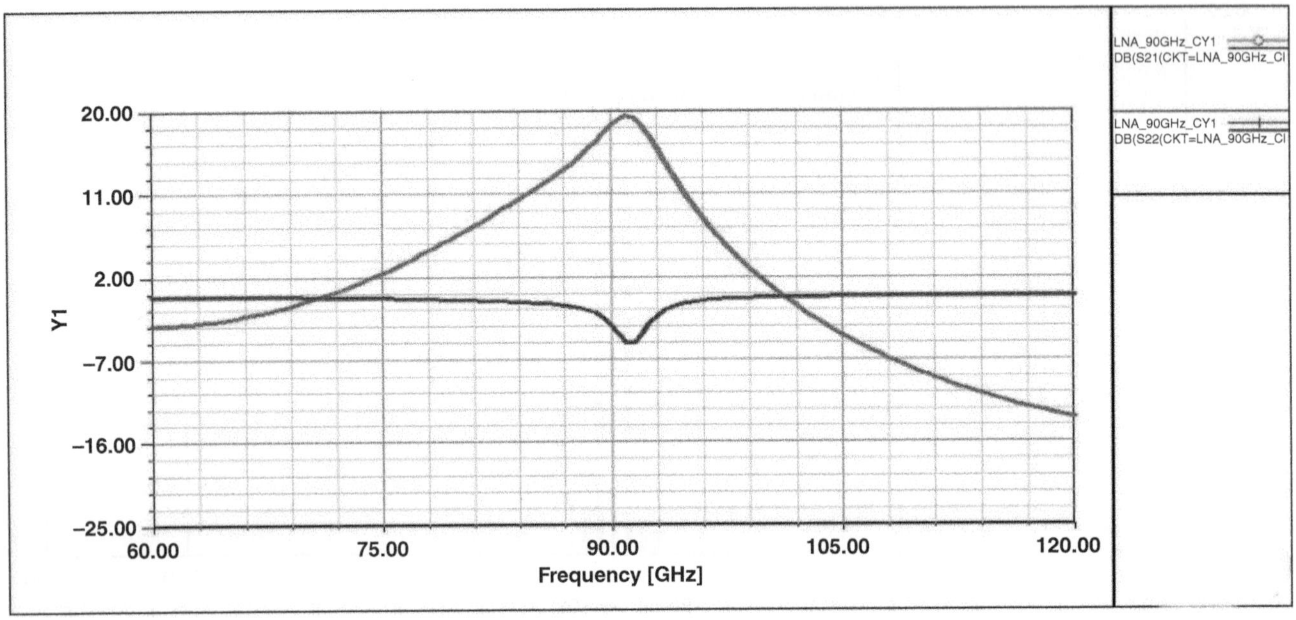

Figure D.32 Simulated gain (S21) and output return loss (S22) at 10 K.

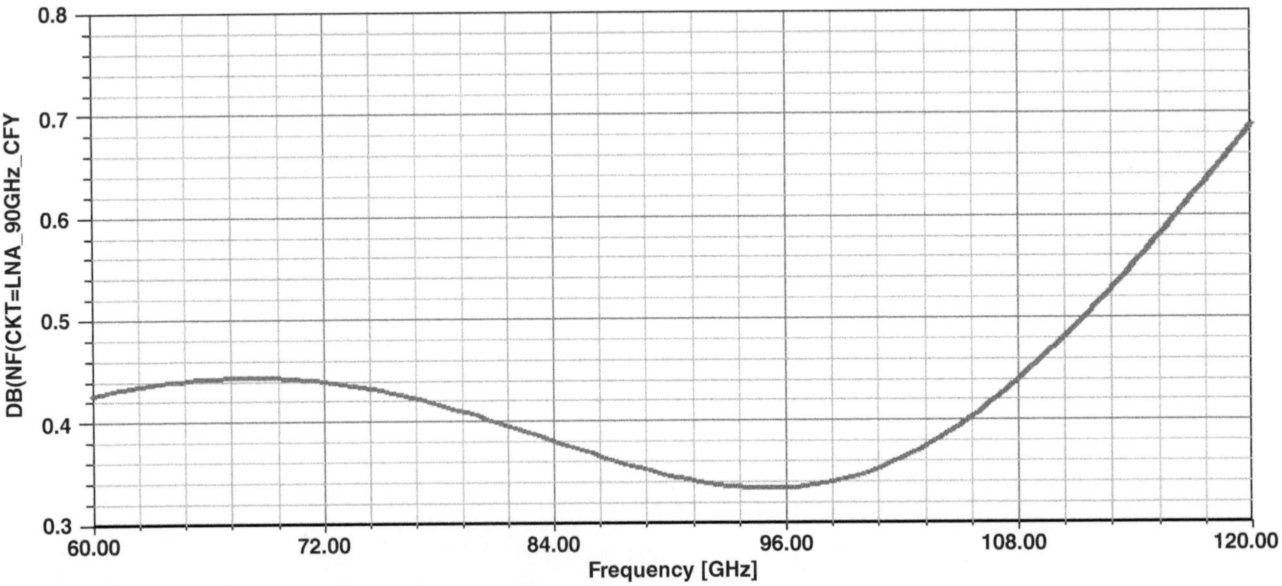

Figure D.33 Simulated noise figure for the same amplifier.

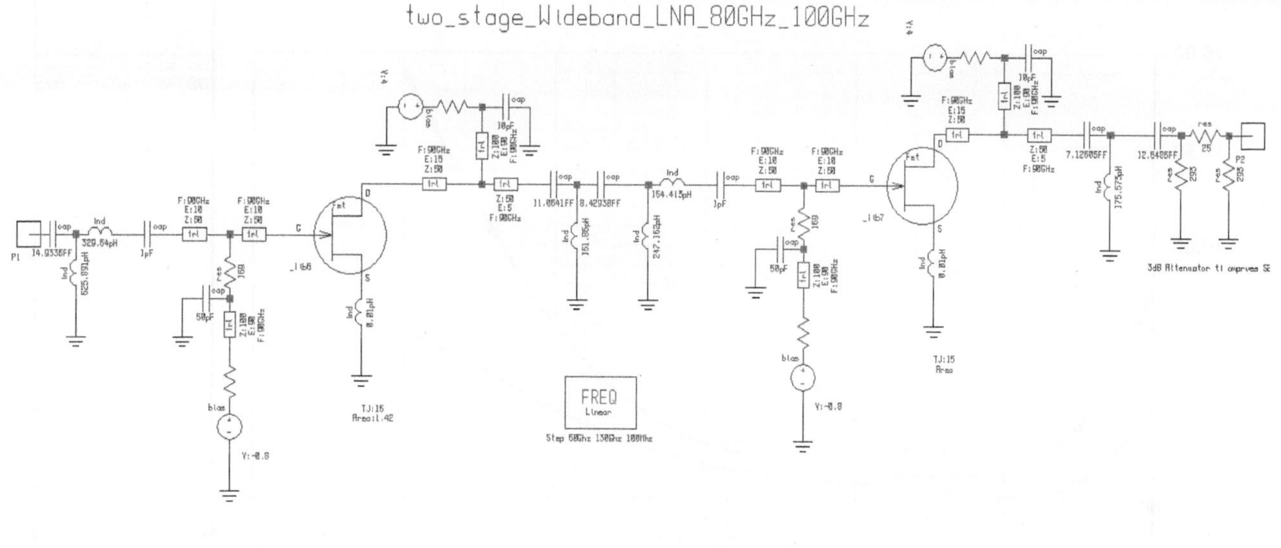

Figure D.34 Circuit diagram of the two-stage amplifier as published in [3].

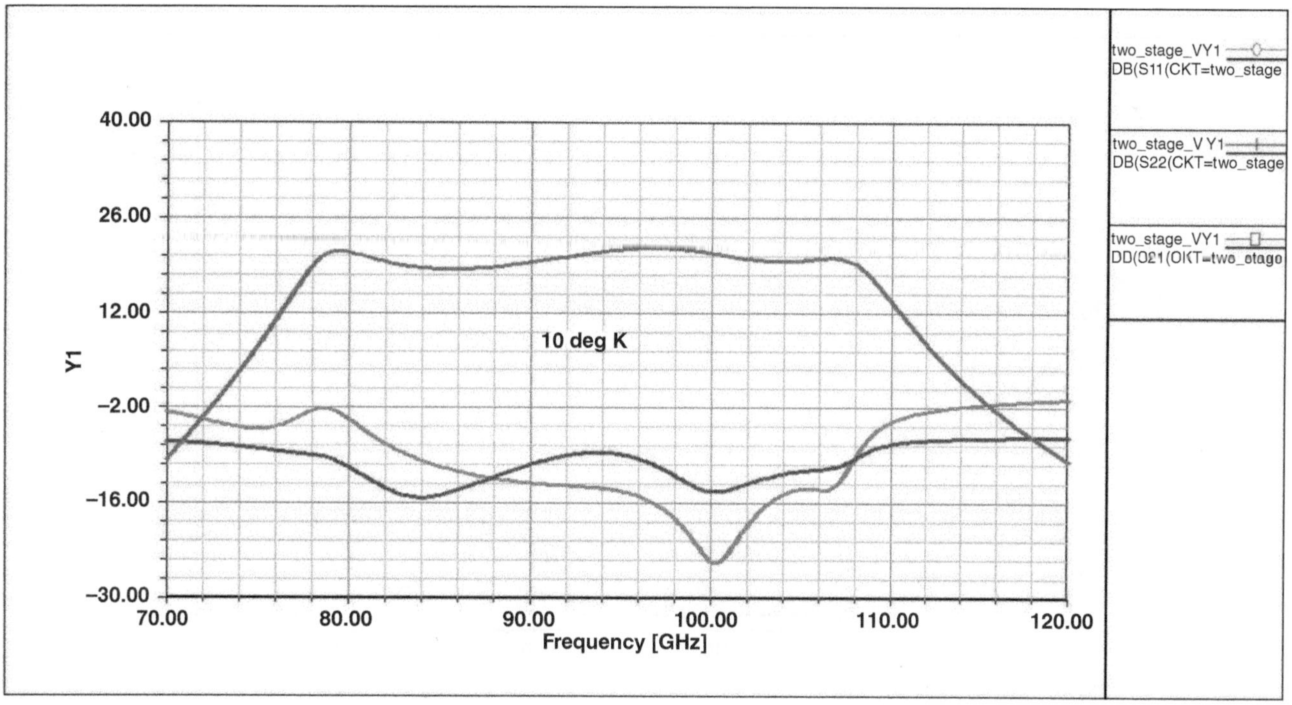

Figure D.35 Simulation of the gain and return loss of the 100 GHz amplifier described above.

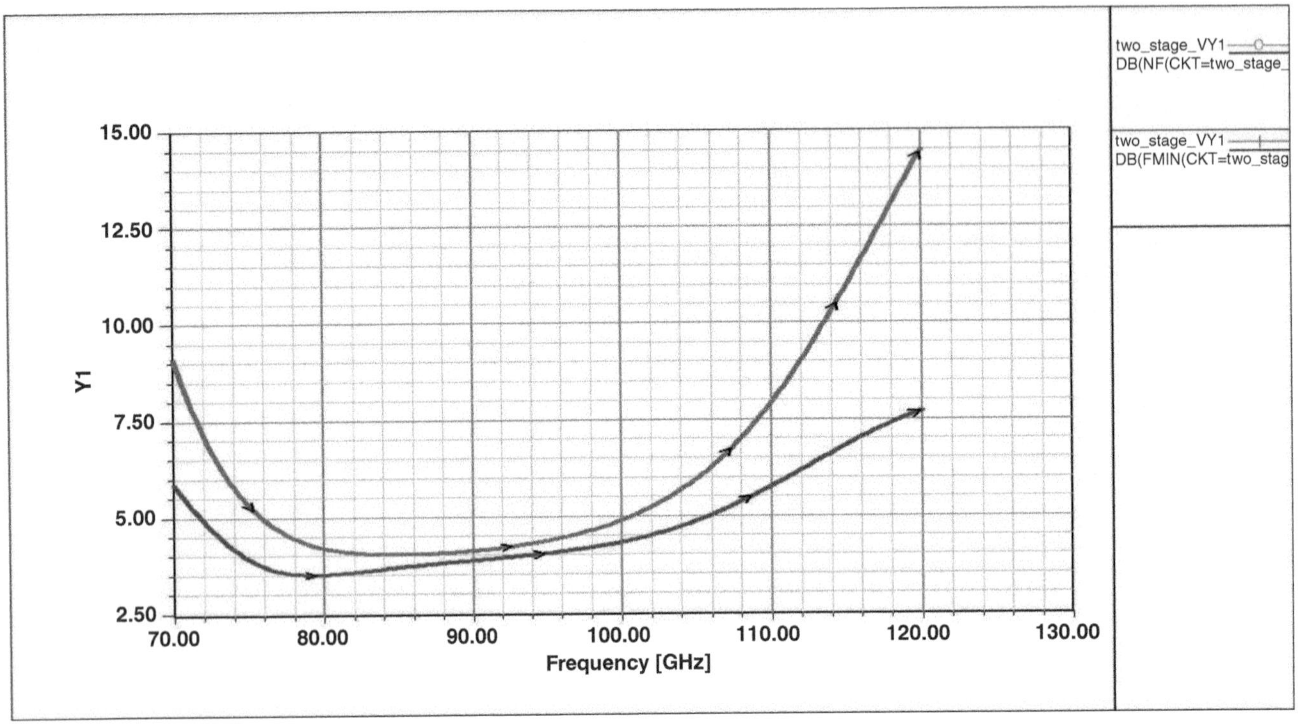

Figure D.36 Noise simulation at 300 K.

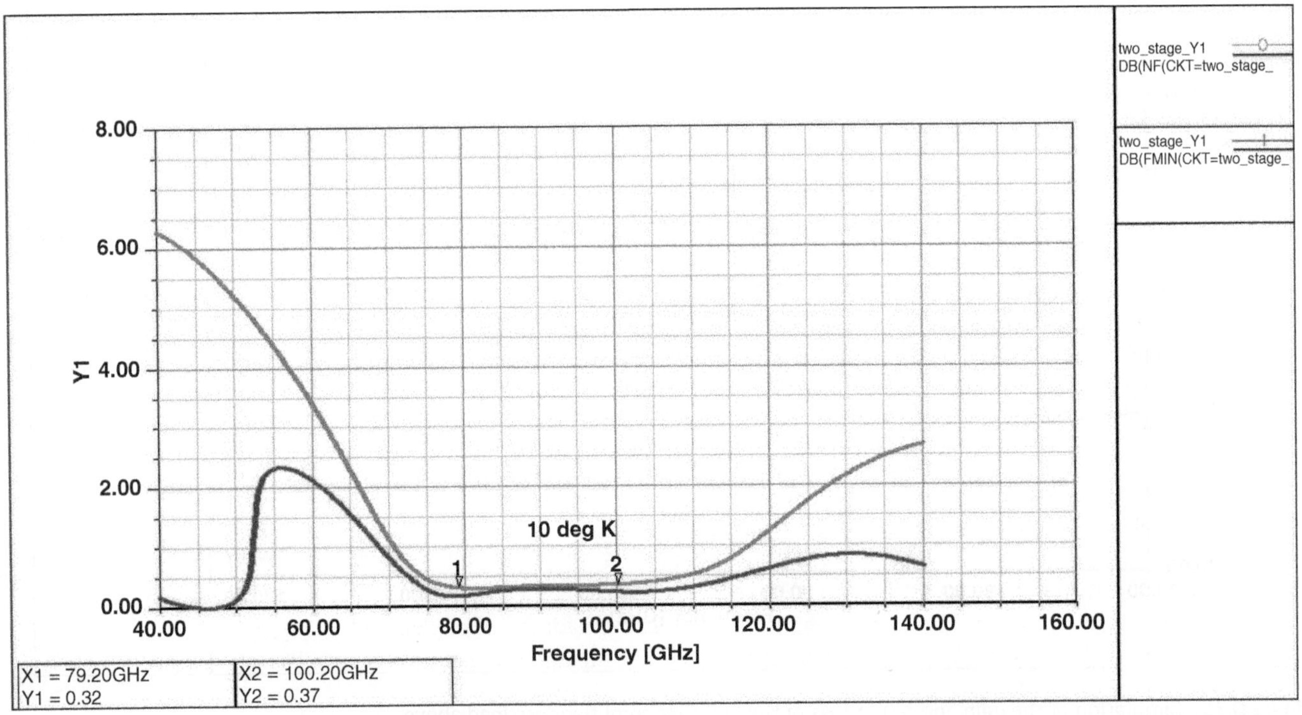

Figure D.37 Noise simulation at 10 deg K after optimization.

References

1 Thome, F., Leuther, A., Massler, H. et al. (2017). Comparison of a 35-nm and a 50-nm gate-length metamorphic HEMT technology for millimeter-wave low-noise amplifier MMICs. In *2017 IEEE MTT-S International Microwave Symposium (IMS)*, Honolulu, HI, USA, 752–755.

2 Chiong, C.-C., Wang, Y., Chang K.-C. et al. (2022). Low-noise amplifier for next-generation radio astronomy telescopes: review of the state-of-the-art cryogenic LNAs in the most challenging applications. *IEEE Microwave Magazine* 23 (1): 31–47.

3 Thome, F., Schäfer, F., Türk, S. et al. (2022). A 67–116-GHz cryogenic low-noise amplifier in a 50-nm InGaAs metamorphic HEMT technology. *IEEE Microwave and Wireless Components Letters* 32 (5): 430–433.

4 Hartnagel, H.L., Quay, R., Rohde, U.L. et al. (2023). *Fundamentals of RF and Microwave Techniques and Technologies.* Springer International Publishing AG, Springer, MOXIC, ISBN 13 9783030940980.

5 Rohde, U.L., Poddar, A.K., and Boeck, G. (2005). *The Design of Modern Microwave Oscillators for Wireless Applications: Theory and Optimization.* New York: John Wiley & Sons. ISBN 0-471-72342-8.

6 Pucel, R.A., Haus, H.A., and Statz, J. (1975). Signal and noise properties of gallium arsenide microwave field effect transistors. *Advances in Electronics and Electron Physics* 38: 195–265. New York: Academic Press.

7 Pospieszalski, M.W. (1989). Modeling of noise parameters of MESFETs and MODFETs and their frequency and temperature dependence. *IEEE Transactions on Microwave Theory and Techniques* 37: 1340–1350.

8 Vendelin, G.D., Pavio, A.M., Rohde, U.L. et al. (2021). *Microwave Circuit Design Using Linear and Nonlinear Techniques*, 3e, 199. John Wiley & Sons.

Further Reading

Linear Operation

Liechti, C.A. (1976, June). Microwave field-effect transistors – 1976. *IEEE Transactions on Microwave Theory and Techniques* MTT-24 (6): 279–300.

Tajima, Y., Wrona, B., and Mishima, K. (1981, February). GaAs FET large-signal model and its application to circuit designs. *IEEE Transactions on Electron Devices* 28 (2): 171–175.

Rohde, U.L., Pavio, A.M., and Pucel, R.A. (1988, December). Accurate noise simulation of microwave amplifiers using CAD. *Microwave Journal* 31 (12): 130–141.

FET Noise Model

Statz, H., Haus, H., and Pucel, R.A. (1974, September). Noise characteristics of gallium arsenide field effect transistors. *IEEE Transactions Electron Devices* ED-21: 549–562.

Rizzoli, V., Mastri, F., and Cecchetti, C. (1989, September). Computer-aided noise analysis of MESFET and HEMT mixers. *IEEE Transactions on Microwave Theory and Techniques* MTT-37: 1401–1410.

Rizzoli, V., Mastri, F., and Masotti, D. (1992). General-purpose noise analysis of forced nonlinear microwave circuits. *Military Microwave* 411–416.

Rohde, U.L. and Chang, C.-R. (1993, May). Parameter extraction for large signal noise models and simulation of noise in large signal circuits like mixers. *Microwave Journal*.36 (5): 24–28.

Heinz, F., Thome, F., Schwantuschke, D. et al. (2022, February). A scalable small-signal and noise model for high-electron-mobility transistors working down to cryogenic temperatures. *IEEE Transactions on Microwave Theory and Techniques* 70 (2): 1097–1110.

Rohde, U.L. (1993). Parameters extraction for large signal noise models and simulation of noise in large signal circuits like mixers and oscillators. In *23rd European Microwave Conference*, Madrid, Spain (September 6–9).

Rohde, U.L., Chang, C.-R., and Gerber, J. (1994). Design and optimization of low-noise oscillators using non-linear CAD tools. In *Presented at the Frequency Control Symposium*, Boston (June 1–2).

Transistors and Noise

Tucker, J.R. (2009). The birth of SIS mixers. In *Proceedings of the Astronomical Society of the Pacific Conference Series*, (Vol. 417, 355–363).

Weinreb, S. (1980, October). Low-noise cooled GASFET amplifiers. *IEEE Transactions on Microwave Theory and Techniques* 28 (10): 1041–1054. https://doi.org/10.1109/TMTT.1980.1130223.

Aja, B., Schuster, K., Schafer, F. et al. (2011, November). Cryogenic low-noise mHEMT-based MMIC amplifiers for 4-12 GHz band. In *IEEE Microwave and Wireless Components Letters* 21 (11): 613–615. https://doi.org/10.1109/LMWC.2011.2167502.

McCulloch, M.A., Melhuish, S.J., and Piccirillo, L. (2013, October). A cryogenic ultra-low-noise MMIC-based LNA with a discrete first stage transistor suitable for radio astronomy applications. *arXiv preprint arXiv: 1310.3088*.

McCulloch, M.A., Grahn, J., Melhuish, S.J. et al. (2017, January–March). Dependence of noise temperature on physical temperature for cryogenic low-noise amplifiers. *Journal of Astronomical Telescopes, Instruments, and Systems* 3 (1): 014003. https://doi.org/10.1117/1.JATIS.3.1.014003.

C.-C. Chiong, "Cryogenic MMIC low-noise-amplifier performance survey." [Online]. Available: https://idv.sinica.edu.tw/ccchiong

Pospieszalski, M.W. (1992, June). Cryogenic-cooled HFET amplifiers and receivers: state-of-the-art and future trends. In *Proceedings of IEEE MTT-S International Microwave Symposium Digest*, (1369–1372). https://doi.org/10.1109/MWSYM.1992.188260.

Pospieszalski, M.W. (2017, June). On the limits of noise performance of field effect transistors. In *Proceedings of IEEE MTT-S International Microwave Symposium (IMS)*. https://doi.org/10.1109/MWSYM.2017.8059045.

Blanter, Y.M. and Büttiker, M. (2000). Shot noise in mesoscopic conductors. *Physics Reports* 336 (1-2): 1–166. https://doi.org/10.1016/ S0370-1573(99)00123-4.

Shell, J. (2007). The cryogenic DC behavior of Cryo3/AZ1 InP 0.1-by-80-micrometer-gate high electron mobility transistor devices. IPN Progress Report. 42–169. Pasadena: NASA-JPL, California Institute of Technology.

Varonen, M., Reeves, R., Kangaslahti, P. et al. (2016, March). An MMIC low-noise amplifier design technique. *IEEE Transactions on Microwave Theory and Techniques* 64 (3): 826–835. https://doi.org/10.1109/TMTT.2016.2521650.

Moschetti, G., Thome, F., Ohlrogge, M. et al. (2016). Stability investigation of large gate-width metamorphic high electron-mobility transistors at cryogenic temperature. *IEEE Transactions on Microwave Theory and Techniques* 64 (10): 3139–3150. https://doi.org/10.1109/TMTT.2016.2598168.

Cha, E., Moschetti, G., Wadefalk, N. et al. (2017, December). Two-finger InP HEMT design for stable cryogenic operation of ultra-low-noise Ka-and Q-band LNAs. *IEEE Transactions on Microwave Theory and Techniques* 65 (12): 5171–5180. https://doi.org/10.1109/TMTT.2017.2765318.

Carpenter, J., Iono, D., Kemper, F. et al. (2020, January). The ALMA development program: roadmap to 2030. International union of radio science, Rome, Italy. Available: https://www.ursi.org/files/CommissionReports/J_news_2020_01.pdf

Kojima, T., Kiuchi, H., Uemizu, K. et al. (2020, August). Demonstration of a wideband submillimeter wave low-noise receiver with 4-21 GHz IF output digitized by a high-speed 32 GSPs ADC. *Astronomy & Astrophysics* 640: L9. https://doi.org/10.1051/0004-6361/202038713.

Pospieszalski, M.W. (2010, October). Interpreting transistor noise. *IEEE Microwave Magazine* 11 (6): 61–69. https://doi.org/10.1109/MMM.2010.937733.

Liu, J., Shan, W., and Shi, S. (2012). Temperature-dependent performances of Nb SIS mixers at millimeter and submillimeter wavelength. In *2012 International Conference on Microwave and Millimeter Wave Technology (ICMMT)*, (1–4).

Rudolph, M., Doerner, R., Heymann, P. et al. Direct extraction of FET noise models from noise figure measurements. *IEEE Transactions on Microwave Theory and Techniques* 50 (2): 461–464. https://doi.org/10.1109/22.982224.

Heymann, P., Rudolph, M., Prinzler, H. et al. Experimental evaluation of microwave field-effect-transistor noise models. *IEEE Transactions on Microwave Theory and Techniques* 47 (2): 156–163. https://doi.org/10.1109/22.744290.

Pucel, R.A., Haus, H.A., and Statz, H. (1975). Signal and noise properties of gallium arsenide microwave field-effect transistors. In: *Advances in Electronics and Electron Physics*, (ed. L. Marton), Vol. 38 195–265. Academic Press. ISSN 0065-2539, ISBN 9780120145386, https://doi.org/10.1016/S0065-2539(08)61205-6.

Acknowledgement

We are grateful for the input from Philip Erickson Ph.D. Director MIT Haystack Observatory, a more detailed paper with him will be on a MIT website.
https://dspace.mit.edu/entities/publication/67a8ff2c-443c-4a58-8a8d-818424fee457

Copyright Note:

E

Fundamentals of Terahertz: Technology for 6G[1]

Dr. Taro Eichler, Robert Ziegler

E.1 Introduction

Terahertz (THz) waves have frequencies extending from 0.1 to 10 THz or wavelengths between 3 mm and 30 μm. They fall in the spectral region between microwaves and optical waves. The prospect of offering large contiguous frequency bands to meet demand for extremely high data transfer rates in the Tbit/s range is making this region a key research area for next-generation wireless communications (6G). Commercial deployment is expected around 2030.

Efforts to explore and "unlock" this frequency region require an interdisciplinary approach with close interaction with high-frequency semiconductor technology for RF electronics but also encompassing alternative approaches using photonic technologies. The THz region also shows great promise for many application areas ranging from imaging to spectroscopy and sensing.

The objective of this white paper is to provide a concise overview of the fundamentals of THz waves and their properties for various applications with a focus on 6G communications.

Section E.2 introduces the key 6G performance requirements and research areas.

Section E.3 discusses potential applications such as THz-based communications and sensing. These applications will require extension of the frequency spectrum to frequencies above 100 GHz. There, the interaction of the THz waves with matter has a strong impact on application scenarios.

Section E.4 highlights various ways to generate THz radiation. Besides using electronic microwave monolithic integrated circuits (MMICs), alternative methods based on photonic technologies will play a key role in the future. Especially with the prospect of miniaturizing today's lab setups into photonic integrated circuits (PICs), these approaches could become mainstream.

With its established production processes and high level of integration, THz generation via electronics is currently still the dominant technology. Frequency limits are being further extended on a continuous basis.

Section E.5 summarizes the current state of the art in high-frequency semiconductor technologies.

At the 2023 ITU World Radio Conference (WRC23), additional frequency bands beyond 100 GHz are expected to be discussed and considered for allocation at the subsequent WRC27. Today's activities are now being pursued to demonstrate the full potential of THz communications in feasibility studies.

In order to fully exploit the THz potential for the development of future communications standards such as 6G, it is crucial to understand the propagation characteristics by performing channel measurements. Subsequent development of new channel models for these new communications frequency bands represents one of the first steps in the standardization process.

Section E.6 describes channel propagation measurement concepts and presents some initial results from channel sounding campaigns at 158 and 300 GHz at Rohde & Schwarz headquarters in Munich.

[1] White Paper Version 01.02

E.2 Toward the Next Wireless Communications Standard 6G

E.2.1 From 5G to 6G—Vision and Key Technologies

Setting up nationwide 5G networks with successive provision of new communications capabilities and services will keep the industry busy for many years. 5G has opened the door for new application scenarios such as Industry 4.0 with the representative service categories **enhanced mobile broadband (eMBB)**, **ultra-reliable low latency communications (URLLC)**, and **massive machine type communications (mMTC)**. While 5G services will continue to improve toward the 3GPP Release 18 evolution step, academia and industry have initiated research on the fundamentals of next-generation wireless communications (6G).

Commercial deployment is expected around 2030, and a pre-commercial launch could possibly occur one year earlier. Excellent overviews illustrating different perspectives on 6G visions and technologies are given, for example, in [1–10]. Here, we highlight some of the key aspects.

The ITU radiocommunication (ITU-R) Working Party 5D (WP 5D) began to develop a new draft recommendation document "IMT Vision for 2030 and beyond" in 2021 to define the framework and overall objectives for the future development of IMT2030+, which will eventually be called "6G." The first 3GPP standardization efforts are anticipated for 2023/2024, starting with the definition of technical performance requirements. The definition of the standard is to be commenced from 2026 to 2027 onwards.

It is difficult to predict which applications and use cases envisioned by the industry will eventually be the main driver for the next-generation wireless standard and which in turn will shape the system requirements (key performance indicators, KPIs) and technologies for 6G. Among the potential application scenarios are, for example, holographic applications, extended realities (XR) (augmented reality [AR], virtual reality [VR], mixed reality [MR]), or digital twins, which pose extremely stringent requirements with respect to data transmission rate and latency. The vision is that in the 6G era, the digital, physical, and human world will seamlessly fuse (cyber-physical fusion), leading to a connected society in which communications connect people, machines, and virtual services with all of the necessary components of our daily life. The deployed sensors and actuators need to securely transfer and process data at extremely high rates to realize "immersive communications" of this kind.

Table E.1 summarizes the estimated KPI requirements. For 6G, the targeted values are more stringent by a factor of approximately 10–100 times compared to 5G and will pose new challenges for wireless communications and transport networks.

In certain time-sensitive control applications for synchronization (e.g., in industrial application environments such as factory automation), instead of the net latency, it is important to have consistent and deterministic end-to-end latency with low fluctuations, i.e., low jitter. This is introduced as a new KPI [11, 12]. Space–time synchronization will allow time synchronization and mutual positioning by wireless technology for remote devices to work cooperatively [9].

Table E.1 Expected enhancements of KPIs from 5G to 6G. Jitter is a new KPI defining a limit for latency fluctuations for time-sensitive operations [11, 12].

Key performance indicators (KPI)	5G	6G	Improvement factors
Peak data rate (in Gbps)	10	100–1000	10–100
User experienced data rate (in Gbps)	0.1	1–10	10–100
User plane latency (in ms)	1	0.1	10
Connection density (devices/km^2)	10^6	10^7 to 10^8	10–100
Reliability	99.999%	99.99999%	100%
Energy efficiency	1×	5× to 100×	5 to 100
Spectral efficiency	1×	2×	2×
Positioning (in cm)	20–100 in 2D	1 in 3D	20 to 100
Jitter, i.e., latency variations (in μs)	—	0.1–1000	—

The new 6G communications standard should also be an important foundation for a human-centered, sustainable, and inclusive society [9]. Living space will be expanded, and communications will also need to cover those areas. For example, the United Nations Sustainable Development Goals (SDG) offer a guideline in the form of a footprint for energy and resources, as well as equal, affordable, and scalable provision of services to remote areas. The current network energy footprint is already a concern—not solely due to the environmental impact but also because of its major impact on network operator OPEX. In order to cope with the exponential increase in data communications, the energy per transferred bit must be reduced.

E.2.2 6G Research Areas

Figure E.1 provides an overview (albeit not comprehensive) of the main 6G research areas that have been identified to realize the vision and enable use cases for future wireless communications.

While some of the technologies represent evolutionary changes already used in previous generations, other technologies are disruptive and could unlock potential beyond Shannon's limits.

E.2.2.1 Evolution of Technologies Already Used in Previous Generations

Some evolutionary technology steps, such as new waveforms as alternatives to orthogonal frequency division multiplexing (OFDM) or full-duplex operation, were previously discussed for 5G but were not included in the standard.

E.2.2.2 Ultra-high-speed Channel Coding

Channel coding is a fundamental component in wireless communications for correcting transmission errors and is thus a key to reliability. At the same time, it is a complex part of baseband processing, and the new KPIs in terms of very high throughput, reliability, low power consumption, etc., require enhancements of modern channel coding schemes such as Turbo, low density parity check (LDPC), and polar codes.

E.2.2.3 New Waveforms and Multiple Access

Efficient operation of wireless systems is highly dependent on adequate waveform design. While OFDM remains a strong candidate for 6G waveforms, alternative application-specific waveforms or unified scalable waveforms need to be explored, for example, for joint communications and radar sensing. Moreover, the different spectrum ranges, device characteristics (phase noise, PA power efficiency), and system characteristics (signal bandwidth) must be taken into account. A number of new waveforms are under consideration, including orthogonal time frequency space modulation (OTFS) for high mobility scenarios with large Doppler spreads. Furthermore, single carrier waveforms may play a more dominant role to fulfill the requirements for power efficiency in future devices. To allow more flexible use of wireless channels, it could be beneficial to relax the orthogonality constraint, as is realized for non-orthogonal multiple access (NOMA).

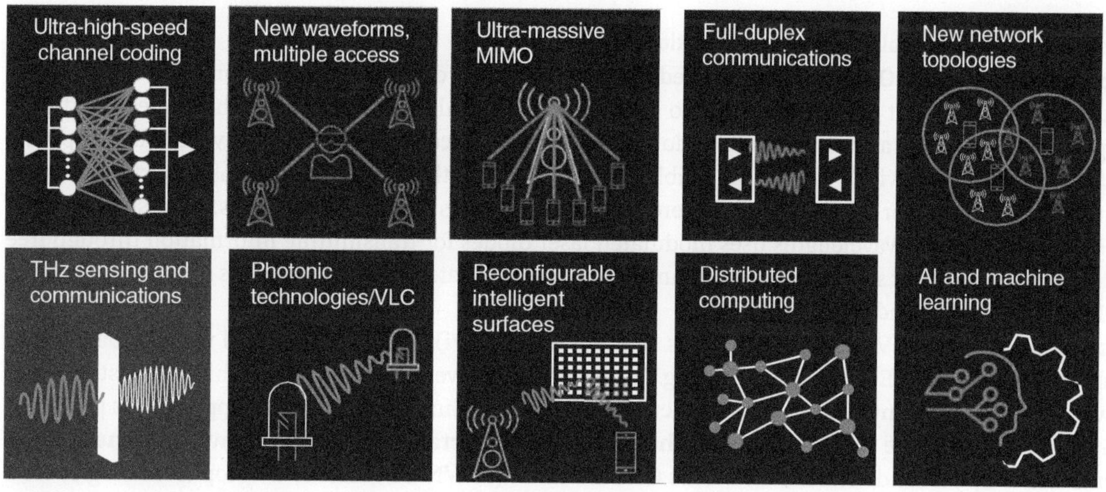

Figure E.1 6G research areas.

E.2.2.4 Ultra-massive Multiple Input Multiple Output (MIMO)

Massive MIMO was introduced in 5G. The trend toward even higher frequencies and shorter wavelengths makes it possible to increase the number of antennas in a small area to provide more directed beams. This is also needed to compensate for higher losses at high frequencies.

Another means to increase spectral efficiency is to apply self-interference cancelation techniques enabling **full-duplex communications** for in-band bidirectional communications and relaying.

E.2.2.5 New Network Topologies

The cellular layout used in the current network architecture is designed to minimize interference at the cell borders between cells. In order to achieve ultra-high-speed, high-capacity (with needed improvement in particular on **the uplink to connect to the "cloud AI brain"**), and high-reliability communications, however, it is ideal to communicate at short distances via a low-loss path and increase the redundancy over multiple communications paths. One possibility for such a spatially distributed topology involves **cell-free networks** where base stations—distributed over a large area—coordinate coherent joint transmission to provide service to each user. This approach will lead to a higher signal-to-noise ratio, higher gain, and a more consistent quality of experience for users at different locations. However, the implementation entails high computing complexity and tight synchronization requirements between the base station locations along with the exchange of large amounts of data between the sites.

In order to provide new services to drones, aircraft, ships, and space stations/satellites and thus bring coverage to remote areas, sea, and space, it is necessary to extend the network coverage three-dimensionally. Thus, the vertical direction is also included in addition to horizontal deployment. Such ubiquitous communications could be realized, for example, with non-terrestrial networks (NTNs) that utilize drones (high-altitude platform stations or HAPS in the stratosphere at 20 km altitude) and low earth orbit (LEO) satellite constellations as mobile base stations in the sky.

E.2.2.6 THz Communications and Sensing

One of the key technology concepts for 6G—albeit just one of the 6G enablers—is the use of THz waves for communications and sensing. These THz applications and technologies will be discussed in detail in the following sections. Unlocking the potential of the sub-THz and THz frequency regions (100 GHz–3 THz) with the available extremely high bandwidths of several GHz represents a technological way forward. Besides ultra-high data rates in wireless communications, this would also benefit sensing and imaging applications as well as possible future medical diagnostic procedures.

The concept of **joint communications and sensing (JCS)**—sometimes called integrated communications and sensing (ICAS)—comprises a system definition to natively support both applications as part of the physical layer design, i.e., the jointly used waveforms and the network architecture. It is not limited to THz frequencies but also encompasses the mmWave (mmWave) range. The wide bandwidths will also benefit high-precision sensing applications. These could include not only positioning, object detection, or high-resolution radar but also spectroscopy-type analysis—of particular interest for environmental sensing.

E.2.2.7 Photonic Technologies and Visible Light Communications

Optical wireless communications (OWC) is already used in some areas as a complementary technology to satisfy the requirements for additional capacity. It has the potential to become more prevalent as it combines high speed and high fidelity with low deployment costs. Key advantages over radio frequency-based access networks is the availability of approximately 300 THz of license-free bandwidth carried on visible and IR wavelengths, robustness against interference, and secure communications, e.g., in indoor environments where the radiation does not penetrate walls. Free-space optical communications (FSO) with infrared wavelengths uses modulated laser diodes for transmitting information through free space, e.g., as a backhaul solution or for space-based communications between satellites. However, it is affected by weather conditions, atmospheric turbulence, and especially fog.

In **visible light communications (VLC)**, also known as "light fidelity" (LiFi), data is transmitted via high bandwidth intensity modulation of commercial LEDs used for lighting. A photodiode serves as the receiver. This is a cost-efficient approach that allows easy integration into existing infrastructure, primarily for line-of-sight indoor applications [13].

The stringent 6G KPIs will also drive the **evolution of the future transportation network in wireless networks**. For example, the Innovative Optical and Wireless Networks Global Forum (IOWN GF) aims to develop technologies for a computing and communications network architecture to achieve scalability, elasticity (regarding sporadic workload fluctuations), energy efficiency, and latency manageability at the same time [11]. **Photonic technologies** can help to cope with these challenges: the proposed open all-photonic network (APN) could help to streamline data transfer and processing and realize a large-capacity, low-latency, and low-energy consumption infrastructure. For example, it enables a direct end-to-end optical path connection across domains/hierarchies between any user terminals or points in the network with minimal photoelectric conversion to provide low-latency service.

Integrated optical devices (PICs) could provide the routing and termination functions to implement such an end-to-end all-optical connection.

Furthermore, extrapolating the trend for the capacity demand increase in the 2030s, Pbit/s class communications capacity will be required for long-haul transmission. The limit of current single-mode fibers (SMFs) is around 100 Tbit/s. The limit can be further increased with ultra-wideband optical transmission using multicore fiber transmission technology, where multiple cores are embedded into a single glass fiber, and by further increasing the number of wavelengths through multiplexing and adding new optical bands.

Another photonic technology that was first demonstrated conceptually more than 25 years ago but has recently attracted increasing interest is **quantum communications** (and quantum networks). This could play a complementary role for 6G as one possibility to ensure **trustworthiness for ultra-secure and reliable communications**. Quantum communications provides an inherently secure means of **quantum key distribution (QKD)** through the exchange of entangled photons. These digital keys would subsequently be used for conventional encrypted transmission. Such technologies could be integrated and transmitted via optical fiber. However, free-space QKD between a satellite and Earth has also been demonstrated [14], which could serve as the basis for realizing a future quantum communications network [15].

E.2.2.8 Reconfigurable Intelligent Surfaces and Metamaterials

The use of **reconfigurable intelligent surfaces** (RIS) when installed on building facades or in indoor environments could direct the energy of a wireless signal toward a certain point, thus providing better coverage in non-line-of-sight environments and reducing energy consumption.

Since the early days of wireless communications, the propagation channel has been considered as a randomly and—depending on the scenario—dynamically changing entity between the transmitter and receiver. It has the effect of degrading the quality of the received signal in an uncontrollable fashion owing to the surrounding environment and objects, as well as the mobility of the receiver. One of the basic assumptions in communications theory (from which, for example, the Shannon theorem is derived) is that the propagation channel is a given within the communications system and cannot be influenced. The use of RISs could provide a tool to change the propagation channel and thus introduce a new programmable entity into the wireless communications network with the prospect of extending the boundaries beyond the Shannon limit.

A **RIS** (also referred to as an **intelligent reflecting surface, IRS**) is a planar structure that is designed to have properties that enable dynamic control of electromagnetic waves. An IRS consists of a large number of low-cost, passive (and thus low-energy) elements, each reflecting the incident signal with a certain phase shift to collaboratively achieve beamforming and suppress interference at one or more designated receivers. The building blocks of such an RIS are so-called metamaterials that are—in contrast to a material with properties based on its atomic constituents—engineered structures designed to interact with electromagnetic radiation in a desired fashion and with special properties. They usually comprise an array of structures smaller than the wavelength of interest.

Targeted RIS scenarios are ultra-dense network deployments predominantly in indoor environments (malls, airports, stadiums, or industrial/factory environments).

In October 2021, the European Telecommunications Standards Institute (ETSI) launched a new Industry Specification Group (ISG) on reconfigurable intelligent surfaces (ISG RIS) for reviewing and establishing global standardization for RIS technology [16]. Smart repeaters, as they are currently discussed in 3GPP for Release 18, may be the stepping stone toward reflective intelligent surfaces.

E.2.2.9 Distributed Computing and Artificial Intelligence Communications Systems

Although future 6G application scenarios still have to be defined, it is evident that performance requirements will be even more demanding compared with 5G in terms of data rate, latency, spectral efficiency, security, reliability, and energy consumption. This will also impact on the **processing architecture: information technologies and communications technologies will further merge**, i.e., the processing of large amounts of data will take place in distributed systems in the network and not necessarily in the end-user device. This leads to challenging data rate and latency requirements. Computing power could be offloaded to the cloud or edge networks separate from devices (thus allowing cheaper terminals).

In the traditional "von Neumann" computing architecture, the computing unit (CPU) and the memory are separated. Moving large amounts of data between the CPU and memory becomes a bottleneck in terms of energy consumption and latency. New approaches in computing architecture, such as neuromorphic computing, take inspiration from the human brain for energy-efficient processing with low latencies. There, the computational tasks are performed in the memory itself. The human brain, for example, has very good computing power versus power consumption (a total of 20–30 W).

Artificial intelligence (AI)-related tasks rely heavily on computational processing power. Further upscaling of the performance of graphical processing units (GPUs) will eventually be limited by the available energy. For such tasks, analog AI cores for in-memory computing are particularly suitable. New concepts such as an analog 3D cross-point array of resistors, used as nonvolatile memory to store the weights (e.g., for convolutional neural networks), can be used to calculate matrix–vector calculations (the prevalent type of calculation for neural networks) in parallel [17, 18].

E.2.2.10 AI and Machine Learning

Already today, AI and machine learning (ML) are being used for self-management and control operations for a number of network applications. These include the initial network planning phase, network control and optimization such as traffic management, dynamic spectrum management, as well as predictive and adaptive resource allocation.

It is expected that in the future, AI will become an integral part of all areas of the wireless communications system. This could include a physical layer design that adapts to the specific propagation channel and environmental conditions with the possibility of end-to-end optimization—instead of optimization of each component of the radio transmission chain [19]. To cope with the increase in complexity of future 6G networks regarding device types, spectrum ranges, and flexible network topologies such as cooperative mesh type networks, AI and ML will play an important role for 6G deployment and operation. This will make it possible to optimize system performance in terms of maximizing user experience and cost efficiency and minimizing energy consumption.

Recently, ETSI has also begun to focus on AI in terms of AI-based testing and testing AI-based systems from the regulatory perspective [20]. Relevant working groups are the ETSI Technical Committee (TC) Core Network and Interoperability Testing (INT) and the ETSI Technical Committee (TC) Methods for Testing and Specification (MTS).

E.3 Properties and Applications of THz Waves

E.3.1 New 6G Spectrum at mmWave and THz Frequencies

E.3.1.1 On the Way to the THz Spectrum

Here, 5G pioneered the use of mmWave frequencies with large bandwidths up to 400 MHz per component carrier to enable the transmission rates necessary for demanding real-time applications such as wireless factory automation, the nascent 6G technology is aiming at significantly higher transmission rates and lower latencies.

However, large contiguous frequency ranges for ultra-high data rates with bandwidths of several GHz are only available in the sub-THz and THz range, i.e., above 100 GHz (Figure E.2).

The Shannon–Hartley theorem, which gives a boundary for the maximum achievable channel capacity (under the influence of noise) $C = B \cdot \log_2(1 + S/N)$, where C is the capacity in bit/s, B the bandwidth in Hz, and S/N the signal-to-noise ratio, states that the maximum achievable data rate scales proportionally to the bandwidth B [21]. This was the motivation for 5G to extend the frequency region into the mmWave range, as frequencies below 6 GHz are already heavily in use. In order to increase data rates even further for future wireless standards such as 6G, even higher frequency bands up into the THz spectrum—where large contiguous bandwidths of several GHz are available—are under consideration. Figure E.2 shows the different frequency ranges allocated or considered for future wireless communications.

Up to 4G (LTE), only the frequency region below 6 GHz was used for wireless communications. This frequency range still plays a key role because of its favorable propagation conditions and the large cell sizes that can be achieved as a result.

In the 3GPP 5G NR nomenclature, this frequency region is called 5G FR1 (410 MHz–7.125 GHz). This region also includes LTE and Wi-Fi such as Wi-Fi 6 (IEEE 802.11ax) and Wi-Fi 7 (IEEE 802.11be).

In 3GPP Release 17, 5G FR2 (see Figure E.2) has been assigned up to 71 GHz (24.25–71 GHz).

For the evolution of 5G ("Beyond 5G" in Figure E.2), even higher frequencies above 71–100 GHz are under consideration: bands are frequency regions considered for future wireless communications by the International Telecommunication Union (ITU) World Radio Conference 2019 (e.g., 81–86 GHz) (see Figure E.2). Although considered initially, the 80 GHz

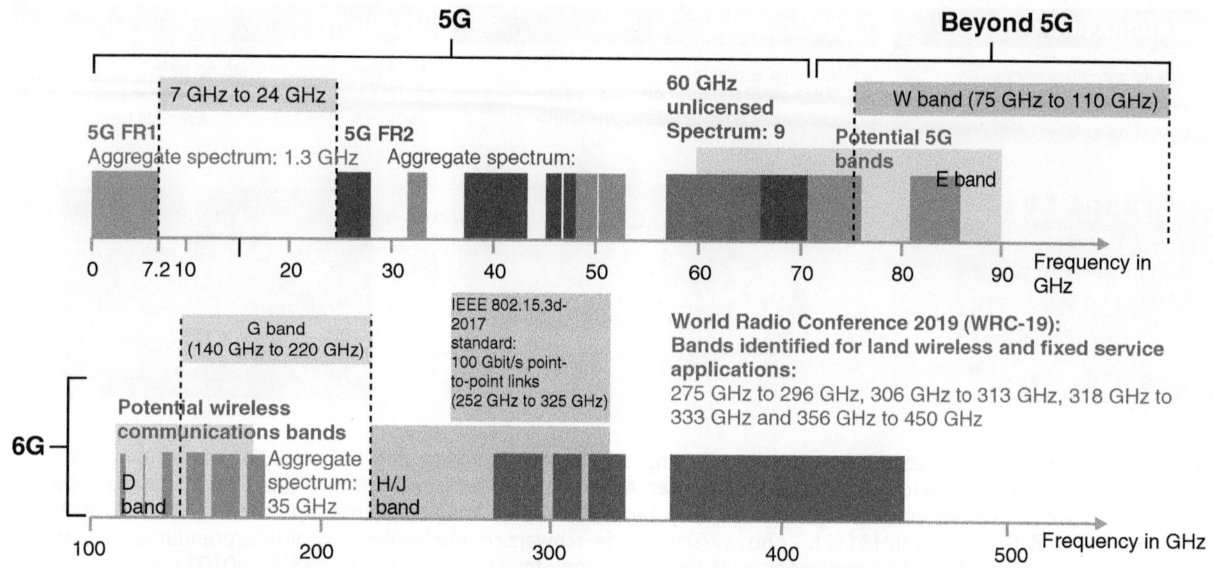

Figure E.2 New spectrum for 5G and 6G. A huge spectrum will be available in the mmWave and THz frequency range. Spectrum of frequency bands available for wireless communications. The 5G bands in FR1 and FR2 are distributed over the range below 71 GHz. 6G research is concentrating on the D and H bands.

frequency region was not approved for IMT-Advanced (5G). This frequency region also contains the 60 GHz unlicensed band (the E band extends from 60 to 90 GHz).

E.3.1.2 Sub-THz Bands

In order to enable terabit-class data rates for 6G, even higher bandwidths are necessary. However, they are only available in the range beyond 100 GHz. Potential new frequency bands have been identified in the D band (110–170 GHz), G band (140–220 GHz), and H/J band (220–330 GHz).

The ITU [22] radio regulations document summarizes the conclusions of the radio regulations adopted by the World Radiocommunication Conference of 1995 (WRC-95) and reviewed by subsequent World Radiocommunication Conferences (WRC-97, WRC-2000, WRC-03, WRC-07, WRC-12, WRC-15, and WRC-19). As stated on pp. 185 in [22], the region 275 GHz–3 THz is not officially allocated. However, the footnote 5.564A comments on the operation of fixed and land wireless service applications in frequency bands in the range 275–450 GHz, which were added at ITU World Radio Conference 2019 (WRC19). From the final acts of the WRC19 [23]: “For the operation of fixed and land wireless service applications in frequency bands in the range 275 GHz-450 GHz: The frequency bands 275–296, 306–313, 318–333, and 356–450 GHz are identified for use by administrations for the implementation of land wireless and fixed service applications, where no specific conditions are necessary to protect Earth exploration-satellite service (passive) applications.”

Standardization activities are currently also underway at the IEEE, where an interest group is introducing a standard for 100 Gbps at these frequencies [24]. This standard specifically targets the frequency range from 252 to 325 GHz with signal bandwidths up to 69.12 GHz.

Development of THz-range bands for wireless backhaul/access is already happening in the W band spectrum (75–110 GHz) around 95 GHz. The next extremely high candidate band, the D band at 110–170 GHz, is being eyed by base station infrastructure vendors. 6G research efforts are now concentrating on the D band and the H/J band around 300 GHz. Since the agenda for WRC27 will be fixed at WRC23, activities are now in full swing to demonstrate the full potential of THz communications through feasibility studies and channel propagation measurements (Section E.6) before WRC23 in 2023.

E.3.1.3 Beyond 7 GHz (7–24 GHz)

Despite strong interest in the sub-THz band, the frequency range 7–24 GHz that lies in the gap between FR1 and FR2 still offers potentially large blocks of available spectrum (Figure E.2). It could provide a “sweet spot,” offering the possibility for higher-order MIMO in a smaller package (compared to FR1) but more favorable propagation characteristics in comparison to the bands operating in FR2. Therefore, this region could be a complementary 6G candidate.

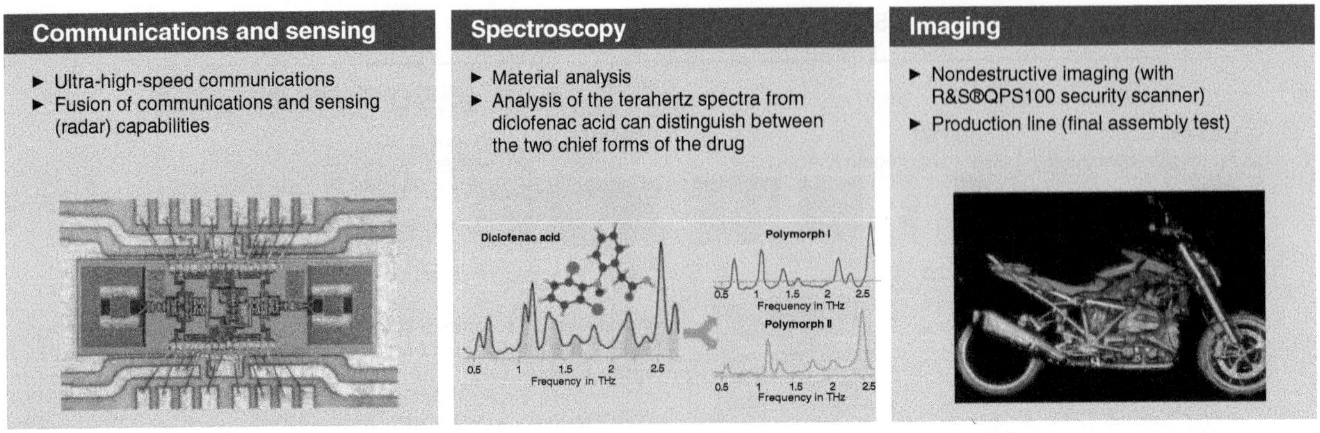

Figure E.3 Application of THz waves in communications and sensing, spectroscopy, and imaging. Left: D-band radar chip from the Leibnitz Institute for High Performance Microelectronics (IHP). Center: Analysis of the THz spectra from a sample of diclofenac acid can distinguish between the two chief forms, or polymorphs, of the drug. Right: Microwave imaging picture captured with the R&S®QPS100 (70–80 GHz). *Source*: Rohde & Schwarz GmbH&Co. KG / https://scdn.rohde-schwarz.com/ur/pws/dl_downloads/premiumdownloads/premium_dl_pdm_downloads/3683_6158_52/Fundamentals-of-THz-technology-for-6G_wp_en_3683-6158-52_v0102.pdf / last accessed on Nov 10, 2025.

E.3.2 THz Applications

E.3.2.1 A Plethora of Applications Yet to Be Explored

THz waves lie in the portion of the frequency spectrum between the realms of optics and microwaves. One of the early drivers of THz spectroscopy was the need for high-sensitivity instrumentation for astrophysical observations and environmental monitoring. However, recent technological innovation in photonics and nanotechnology is now enabling THz research to be applied in many more sectors.

E.3.2.1.1 Spectroscopy and imaging

THz waves show great promise and are already used for many applications in spectroscopy and high-resolution imaging since many forms of condensed matter, molecular compounds, vapors, and gases possess different physical features resonant with THz waves [25, 26]. Many substances and complex molecules can be easily identified using THz spectroscopy techniques such as pharmaceuticals, biomolecules, and proteins. Figure E.3 shows an example of a common drug molecule where the characteristic absorption lines at THz frequencies can be used to distinguish between the two main molecule variants. Security could also benefit from such a sensor, as many common explosives or narcotics could be identified by their spectral "fingerprint" [26]. THz waves allow nondestructive product inspection and can "look inside" plastics, textiles, paper, etc. Unlike X-rays, THz waves do not have any ionizing effect and are generally considered biologically innocuous. Thus, they do not require any additional safety measures when used, for example, for security scanning.

E.3.2.1.2 Communications

Data traffic is increasing exponentially, and Internet Protocol traffic surpassed 100 exabytes $(1 \text{ Ebyte} = 10^{18} \text{ byte})$ per month already in 2017. This is expected to have tripled in 2022. Since the fastest-growing part of data traffic is related to wireless channels, such an increase in network capacity requires much higher wireless transmission data rate links. The prospect of offering large contiguous frequency bands to meet the demand for extremely high data transfer rates in the Tbit/s range with low latency makes it a key research area for 6G wireless communications—also in the context of joint communications and radar sensing.

E.3.2.1.3 Sensing and positioning

There is also a concept emerging that uses wireless frequencies for both communications and sensing capabilities, namely joint communications and sensing (JCS, Section E.2.2). THz frequencies would not simply carry communications data but could work much like a highly precise radar system to detect objects. In this way, THz sensing can provide high-resolution maps of the environment and positioning information at centimeter and sub-centimeter level accuracy. This could be applied, for example, in factories for industrial control and monitoring applications such as robotics or VR.

E.3.2.2 THz Waves for Atmospheric Remote Sensing and Astrophysics Applications

According to the Planck radiation law, the energy of a THz photon corresponds to a temperature of a few Kelvin up to few tens of Kelvin. THz astronomy research; therefore, explores the cold universe such as molecular clouds of interstellar matter. Of special interest are regions where new stars are born. In 1970, carbon monoxide (CO) was detected as the first molecule in an interstellar cloud with a transition in the THz range at 115 GHz [27], commencing the age of THz astronomy.

Radiometer Physics GmbH (RPG, a Rohde & Schwarz Company) originated from the Max Planck Institute for Radio Astronomy and the NASA Jet Propulsion Laboratory and has been involved in this scientific research from early on. The company has a long history of providing cutting-edge instruments for THz space R&D projects for remote sensing, mmWave, sub-mm, and THz instruments and components up to 2 THz.

The absorption and emission properties of molecular species (see Figure E.7 for an example for the Earth's atmosphere) are of particular interest for scientists and meteorologists who are investigating the dynamics of the atmosphere of planets (Earth, Jupiter, Venus, Mars, etc.) using mmWave and submillimeter wave instruments (SWIs). Depending on the applications and science objectives, various types of **passive instruments** operating typically in the 18–664 GHz range for **radiometers** and 0.3–1.2 THz range for **spectrometers** are used, as well as **active instruments** operating in the 35–94 GHz range for **FMCW radars** and at 166 GHz for **scintillation radiometers**. These instruments can be operated from ground-based stations or from space onboard satellites. Figure E.4 shows examples of a passive radiometer, active sources, and radar instruments for atmospheric remote sensing, planetary science, and astrophysics purposes.

Over the past 10 years, RPG has developed THz heterodyne receivers in the 183–664 GHz range for the Ice Cloud Imager (ICI) [28], a conical scanning multichannel radiometer developed by Airbus- automated stall entry ASE for the European

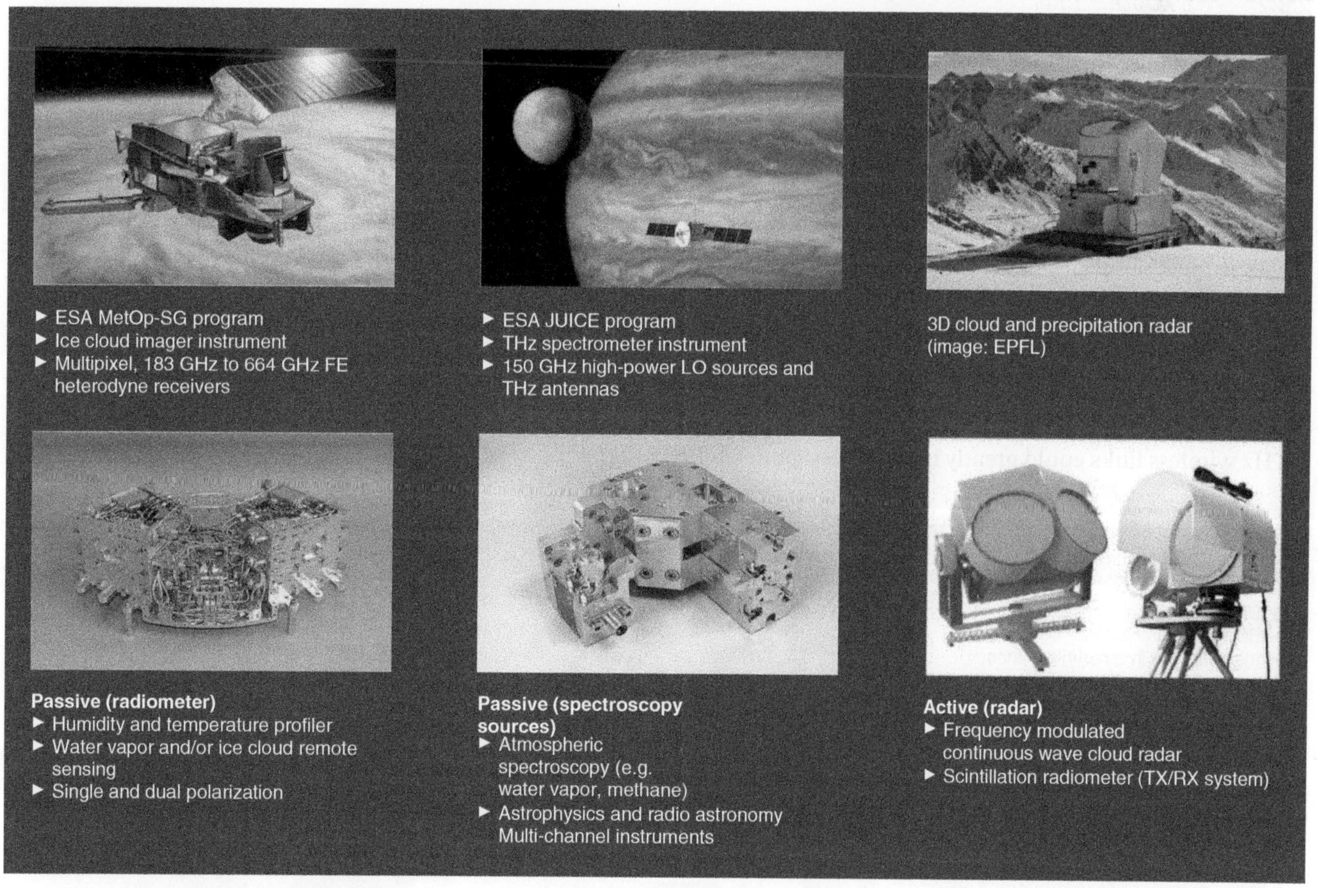

Figure E.4 Examples of THz wave applications in the field of atmospheric remote sensing and astrophysics. Left: Space-based passive radiometer ICI multichannel receivers (RX) operating in the 183–664 GHz range (delivered flight model) [28] (photo courtesy of RPG/Airbus/ESA/EUMETSAT). Center: Example of 132–158 GHz LO sources (TX) developed for the SWI spectrometer for planetary science onboard the ESA JUICE mission. *Source*: [30] / The National Radio Astronomy Observatory. Right: 94 GHz FMCW radar instrument for cloud sensing and 166 GHz scintillation radiometer instrument for field evaporation sensing. *Source*: Rohde & Schwarz GmbH&Co. KG / https://scdn.rohde-schwarz.com/ur/pws/dl_downloads/premiumdownloads/premium_dl_pdm_downloads/3683_6158_52/Fundamentals-of-THz-technology-for-6G_wp_en_3683-6158-52_v0102.pdf / last accessed on Nov 10, 2025.

Space Agency ESA/EUMETSAT Meteorological Operational Satellite Second Generation program (MetOp-SG, Figure E.4, left). It is a European undertaking providing weather data services (including, for example, ocean surface wind vectors and land surface soil moisture) to monitor the climate and improve weather forecasts for the 2023–2043 timeframe. Water vapor, carrying energy as latent heat, is playing a pivotal role in fueling destructive storms and accelerating climate change. With the increase in the oceanic and atmospheric temperature, the amount of vapor in the atmosphere globally has increased by about 4% since the mid-1990s [29]. Since this trend acts as an amplifier for global warming and gives rise to more rapidly intensifying storms, close monitoring of temperature, vapor, and winds is extremely valuable for predicting weather patterns.

For the ESA Jupiter Icy Moon Explorer (JUICE) program (Figure E.4, center), RPG developed high-power sources in the 132–158 GHz range for the SWI [30], a dual-channel 0.6–1.2 THz spectrometer developed by the Max Planck Institute for Solar System Research (MPS). The JUICE mission is planned for launch in 2023. It will spend at least 3 years making detailed observations of the giant gaseous planet Jupiter, as well as three of its largest and water-rich moons (Ganymede, Callisto, and Europa). Also of interest is the characterization of Jupiter's atmosphere and magnetosphere.

The technology underlying both instruments (for the MetOp-SG and JUICE mission) relies heavily on GaAs semiconductor devices (amplifiers, multipliers, mixers) and high-precision manufacturing and assembly techniques (horn antennas, RF modules) that were qualified in accordance with the most stringent requirements required for space missions.

E.3.3 THz Waves for Communications: What Are Possible Use Cases?

E.3.3.1 Use Absorption Windows, Power, and Antenna Arrays for Directivity

The technical challenges of millimeter waves are even more pronounced in the THz range. This includes higher path loss and shorter range. However, similar to millimeter waves in 5G, these problems can be mitigated by focusing the waves through beamforming. The shorter wavelengths also have the advantage that a larger number of antennas can be fitted into a small package to produce highly pinpointed beams.

One of the main use cases for ultra-high-speed communications at THz frequencies involves high-capacity front- and backhaul point-to-point (quasi-static) communications links, also for infrastructure in remote locations (using high-gain antennas or antenna arrays) or "last-mile" connections. The expectation is that capacities up to 100 Gbps can be supported over a distance of a few kilometers [31]. Furthermore, the narrow beams could also enable very dense deployment of communications devices. Other applications include short-range communications such as ultrafast kiosk downloads (e.g., at vending machines) or chip-to-chip communications (Figure E.5).

Today, data centers are an integral part of enterprise computing infrastructures. Contemporary data center networks mostly rely on copper or optical fiber cables for intra-rack or inter-rack network connections. This is associated with high cabling complexity.

THz wireless links could greatly increase design flexibility and reduce costs [32].

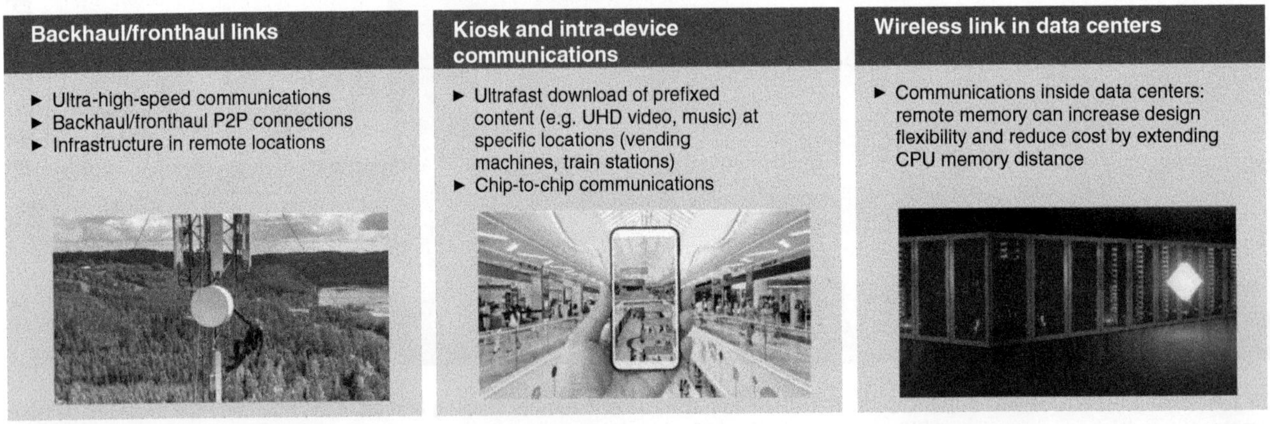

Figure E.5 THz communications use cases. *Source*: Rohde & Schwarz GmbH&Co. KG / https://scdn.rohde-schwarz.com/ur/pws/dl_downloads/premiumdownloads/premium_dl_pdm_downloads/3683_6158_52/Fundamentals-of-THz-technology-for-6G_wp_en_3683-6158-52_v0102.pdf / last accessed on Nov 10, 2025.

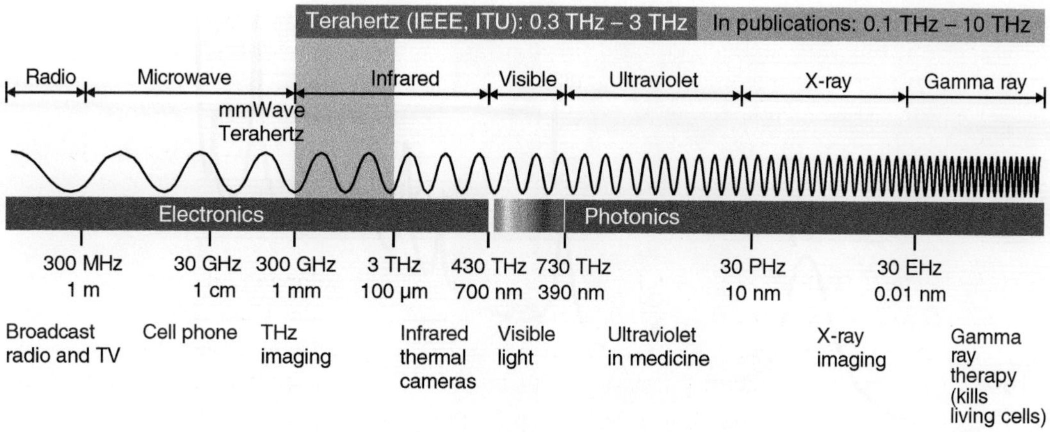

Figure E.6 The electromagnetic spectrum and applications with the THz spectrum on the borderline between electronics and photonics.

While both outdoor and indoor THz use cases are feasible, indoor use cases are likely to dominate THz applications. In indoor settings, the short range and weak penetration capabilities of the THz spectrum are an advantage. For example, THz communications can be highly secure since it can be limited to a specific indoor space.

E.3.4 The Electromagnetic Spectrum and Its Applications

Figure E.6 shows the electromagnetic spectrum and various applications ranging from broadcast and cellular communications at the lower frequency end to X-ray imaging and gamma ray therapy at the highest energies (and thus frequencies). The THz spectrum lies on the borderline between electronics and photonics. The definition of the THz band varies in literature, but the IEEE and ITU definitions are 300 GHz–3 THz (THF or tremendously high frequency), just above the mmWave frequency range (30–300 GHz; EHF or extremely high frequency) and the microwave range (3–30 GHz; SHF or super high frequency). However, in publications, the range above 100 GHz to 10 THz is often denoted as THz frequencies. The frequency range from 100 to 300 GHz is also often called the sub-THz region.

E.3.5 Interaction of THz Waves with Matter

The energy of THz waves is comparably low (by a factor of 100× to 1000× smaller than optical frequencies) and does not initiate any chemical structure changes. It lies in the region of molecular rotational transitions of gas molecules and vibrational transitions of weak bonds. THz waves can thus propagate through dry, nonmetallic, and nonpolar materials with low attenuation. This ability, combined with its small wavelength (versus microwaves) and photon energy (versus X-rays), also makes THz waves an ideal option for nonionizing medical imaging, such as burn injury assessment or skin cancer diagnosis.

Figure E.7 shows the specific atmospheric attenuation within the mmWave and THz spectrum. The resonance frequencies reflect the rotational excitation spectra of various molecules present in the atmosphere. The atmospheric attenuation increases steadily with increasing operating frequency. Between these resonance frequencies, there exists a number of "atmospheric windows," which can be used for various applications. Furthermore, the characteristic water vapor and ice absorption line at 183 GHz (in the G band; 140–220 GHz) is exploited in earth and planetary exploration for remote humidity sensing (detection of ice). An example of radio-astronomical observations in the sub-THz band is the first ever direct picture of a black hole in the center of the galaxy M87. It was observed at 230 GHz by the Event Horizon Telescope (EHT) array by coherently combining the detected signals of radio telescopes distributed over different locations around the earth [33].

E.3.5.1 mmWaves and THz Waves for Wireless Communications Links

As shown in Figure E.2, the frequency bands at 24–30 GHz and 37–40 GHz establish the mmWave region for 5G cellular networks. Furthermore, frequencies from 43 to 50 GHz and around 60 GHz are used for satellite communications links.

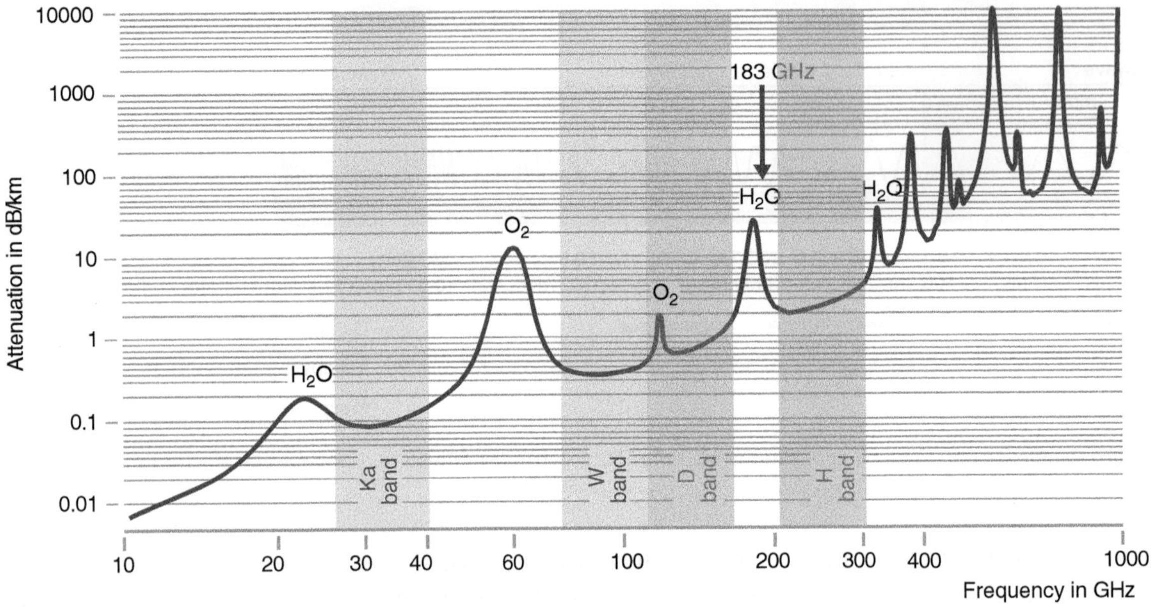

Figure E.7 Specific atmospheric attenuation within the mmWave and THz spectrum. At air pressure of 1013 hPa, temperature of +15 °C, and water vapor density of 7.5 g/m^3. The rotational excitations of different molecules present in the atmosphere (i.e., water, oxygen) are reflected in the absorption spectra.

The increased propagation loss at 60 GHz owing to the resonance of oxygen molecules offers the possibility of establishing intrinsically secure wireless links with high data throughput for indoor applications (WLAN systems).

E.3.5.2 THz Challenges: Path Loss for Frequencies Above 100 GHz

The next-generation communications standard 6G is expected to also rely heavily on components that operate in the frequency range beyond 100 GHz such as the D band (110–170 GHz) or the H/J band (220–330 GHz).

Free-space path loss (FSPL) is proportional to the square of the signal frequency ν. A link at 280 GHz, for example, has 20 dB additional path loss compared to a link at 28 GHz.

However, there is a common misconception that increasing the carrier frequency of a wireless link will inevitably lead to a significantly higher channel loss. This is true only under the assumption that both the transmitter (TX) and receiver (RX) antennas are omnidirectional. The path loss in the THz band can be overcome by using very large antenna arrays at the base station, i.e., ultra-massive MIMO (Figure E.1).

The **FSPL** of a wireless link is described by the **Friis equation**

$$\text{FSPL} = \frac{P_{\text{TX}}}{P_{\text{RX}}} = \frac{1}{G_{\text{RX}} \bullet G_{\text{TX}}} \left(\frac{4\pi d}{\lambda}\right)^2 \propto \upsilon^2$$

where d is the distance between the TX and RX antenna, λ is the operating wavelength, and G_{RX} and G_{TX} are the antenna gains of the transmitter and receiver, respectively. Given that the distance between TX and RX as well as their antenna gains are fixed, the FSPL value increases quadratically with increasing frequency $\nu = c/\lambda$. Figure E.8 (left) illustrates this relationship, Figure E.8 (right) depicts the scenario where both TX and RX have directional antennas with a fixed effective aperture size of 1 cm^2, i.e., the FSPL versus the distance between TX and RX at three different frequencies for the case in which the antennas are omnidirectional ($G_{\text{RX}} = G_{\text{TX}} = 1$).

The antenna gain G_{ant} depends on the effective aperture A_{eff} of the antenna, which is proportional to its physical size, as well as on the wavelength:

$$G_{\text{ant}} = \frac{4\pi A_{\text{eff}}}{\lambda^2}$$

A directional antenna with a fixed effective aperture of 1 cm^2 has theoretically an antenna gain of 12.3, 18.4, and 24.4 dBi when operated at frequencies of 35, 70, and 140 GHz, respectively. With constant physical dimensions of the antennas, the antenna gain increases quadratically with increasing frequency.

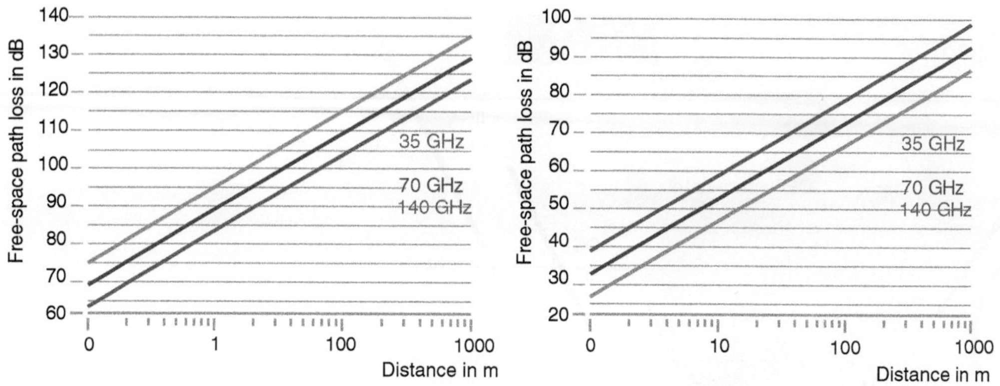

Figure E.8 Theoretical Friis free-space path loss for mmWave wireless links. Left: Scenario where both TX and RX are omnidirectional. Right: Scenario where both TX and RX have directional antennas with a fixed effective aperture size of 1 cm^2.

Combining the last two equations, we can conclude that by using a directional antenna with a fixed effective aperture at either the transceiver or receiver, the FSPL is independent of the carrier frequency.

Therefore, employing highly directional and steerable antennas should enable wireless communications links operating at higher frequencies (with higher bandwidth and throughput) without a significantly reduced signal-to-noise ratio due to atmosphere-induced attenuation when compared with systems at lower carrier frequencies.

E.4 THz Generation and Analysis by Electronic and Photonic Technologies

E.4.1 From Electronics to Photonics

During the past two decades, THz science—occupying the frequency gap between the worlds of microwave electronics and photonics—has attracted increasing interest due to the exciting opportunities that exist in the fields of sensing, imaging, and data communications. Since the pioneering work nearly 100 years ago to create a link between the electrical and optical/infrared regions, development of efficient, stable, and compact THz sources and receivers has been actively pursued and has made THz science a practical reality.

E.4.2 Closing the "THz Gap"

In communications, the frequency range 0.1–10 THz is still a rather untapped region. However, research in this area has recently attracted increasing interest because these high carrier frequencies are associated with the promise of unprecedented channel capacities.

For a long time, it was difficult to generate intensive, directional THz radiation, and the THz range was considered the final frontier in the electromagnetic spectrum. This main technical constraint, the so-called "THz gap," is shown in Figure E.9, where THz emission power is plotted as a function of frequency for various electronic and photonic devices. Certain of these examples will be discussed in greater detail later. Clearly, there is a power drop visible toward the THz frequency spectrum in the region of 0.1–10 THz.

THz frequencies are too high for electronic devices, mainly due to excessive loss and limited carrier velocity. On the other hand, they are too low for photonic devices due to the lack of materials with a sufficiently small bandgap (see Section E.4.4).

Despite the significant efforts that have been made in electronics (e.g., high-mobility semiconductor materials) and optics (e.g., a quantum-cascade laser with cryogenic cooling), the available power around the THz region is still much lower than in other spectral regions. A similar trend also occurs in signal detection, where such a gap keeps this two-decade spectrum isolated from our spectrum-congested world.

E.4.3 Sources for THz Radiation on the Borderline Between Electronics and Photonics

There are three major approaches for generating THz radiation (Figure E.10).

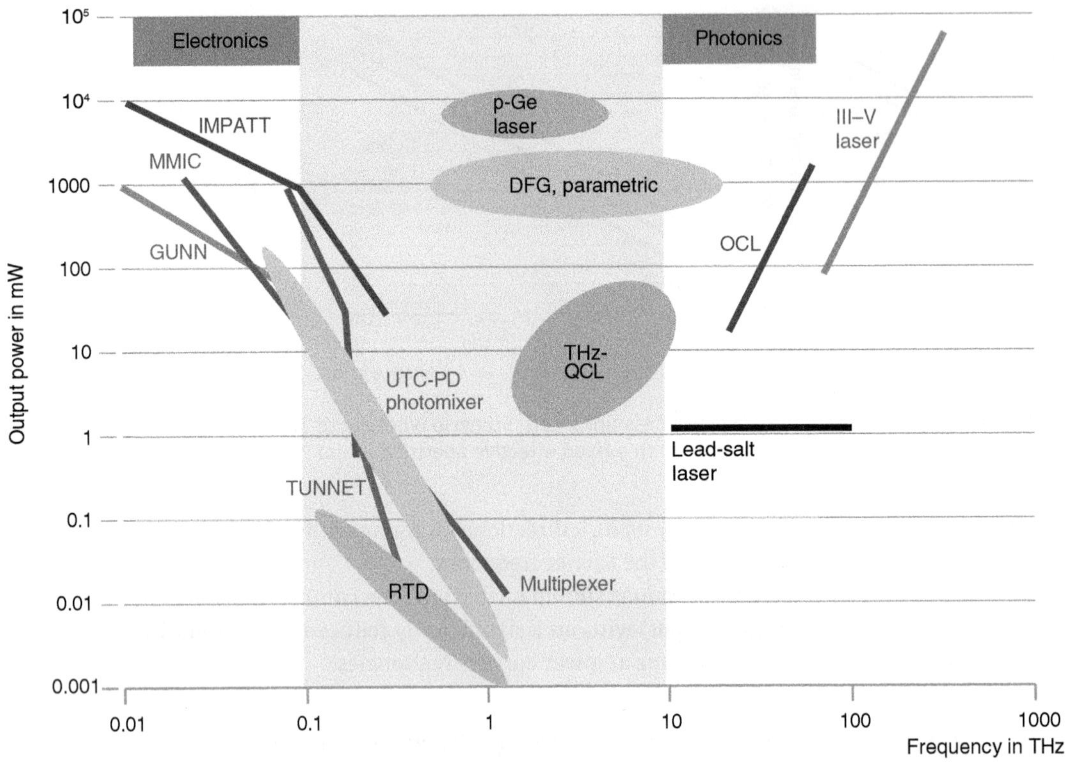

Figure E.9 THz emission power as a function of frequency [34]. Solid lines are for conventional THz sources; IMPATT diode stands for impact ionization avalanche transit-time diode, MMIC stands for monolithic microwave integrated circuit, and TUNNET stands for tunnel injection transit time. Ovals denote recently developed THz sources. THz-QCL: quantum cascade laser, RTD: resonant tunneling diode, and UTC-PD: uni-traveling carrier photo diode.

Some examples will be described in more detail in the following sections.

E.4.3.1 Electronic Sources

The "classic" approach using electronic sources has evolved tremendously in the past years, especially with the refinement of various semiconductor material components (see Section E.5).

These sources include frequency multiplier chains ("upconversion"), resonant tunneling diodes (RTD), transistors, and diodes. The advantage is that the components are very compact and can be operated at room temperature, but they have certain limitations in bandwidth and efficiency. However, electronic sources become comparably inefficient at THz frequencies and provide rather limited frequency tuning.

E.4.3.2 Direct Generation: QCL

Direct THz generation with optical sources includes the very elegant quantum cascade laser (QCL) approach, nonlinear optics (parametric optical processes), and molecular lasers (inefficient and bulky). Quite reasonable power levels could already be reached with QCL, but the efficiency is still limited, and often they must operate at cryogenic temperatures.

E.4.3.3 Indirect Generation: Optoelectronics

In recent years, frequencies between 0.5 and 10 THz have also become the domain of laser-based techniques. Optoelectronic (photonic) approaches use either tunable diode lasers or femtosecond lasers. Photomixers, photoconductive switches, or nonlinear crystals convert the near-infrared laser light into THz waves, either spectrally resolved or broadband.

One particular approach that has gained increasing interest over the past years is "downconversion" from the optical frequency regime using ultrafast photodiodes and photoconductors. The main advantage is the tunability over a large range, operation at room temperature, and the possibility to reuse mature technologies developed for fiber-optic communications. The power envelope has been pushed considerably, but there are still limitations in efficiency.

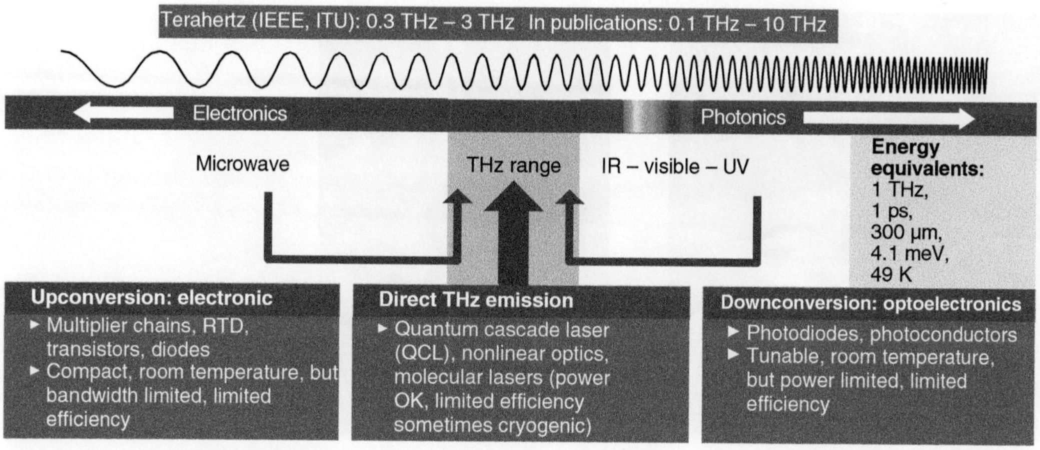

Figure E.10 The three major approaches for THz radiation generation. Note that the energy equivalent of 1 THz corresponds to 49 K, necessitating the use of cryogenic cooling for certain technologies (e.g., QCL) due to relaxation processes at higher temperatures.

E.4.4 Upconversion: Electronic THz Wave Generation and Analysis

Since the founding of Rohde and Schwarz in 1933, the company's core competency has been in the development and manufacturing of advanced RF test and measurement equipment for all relevant electronic and microelectronic areas.

In order to support the multitude of 6G research activities, including semiconductor industry research for device and circuit characterization in the mmWave and THz region, Rohde & Schwarz offers a broad range of development and analysis measurement tools. Some examples are given in the following sections:

- Vector network analyzer (VNA) device characterization with frequency converters up to 1.1 THz
- Extended frequency range support for signal and spectrum analyzers via external harmonic mixers, e.g., the D band (110–170 GHz) and other frequency bands up to 500 GHz (R&S®FS-Zxx series)
- Extended frequency range support for signal generators using frequency multipliers up to 170 GHz (R&S®SMZ series)
- Signal generation and analysis with transmit and receive converters for the D-band spectrum (110–170 GHz)
- Antenna radiation performance measurements in the D band (110–170 GHz) using anechoic chambers

E.4.4.1 VNA THz Measurements with Metrology-level Accuracy

VNAs that perform measurements in the mmWave and THz ranges are more in demand than ever. In the E band (60–90 GHz), they are used, among other applications, to characterize active and passive components in wafer prober systems and waveguide assemblies. Even at higher frequencies, network analyzers can be used in a broad range of applications for testing integrated circuits, sensors and antennas, imaging systems, radio astronomy systems, and materials. The 5G wireless communications standard and the production of automotive radar sensors have also created significant testing demand.

While measurements at frequencies of several GHz up to 67 GHz are part of the standard repertoire of network analyzers, tests in the mmWave and THz ranges are significantly more demanding as they require external frequency converters (which can cover up to 1.1 THz with the R&S®ZC1100). These frequency extenders upconvert stimulus signals and downconvert response signals to characterize devices that operate at frequencies in the THz range.

E.4.4.1.1 On-wafer Characterization of Active Components

Characterizing active components in linear and nonlinear ranges requires defined input power at the probe tip. Since on-wafer power calibration is not possible, the power at the waveguide output is calibrated, and losses in additional waveguides, 1 mm cables, and the probe tip are taken into account in the calibration process. For power sweeps and compression point measurements, the R&S®ZNA has an integrated calibration routine that is able to compensate for mmWave converter nonlinearities for maximum measurement dynamic range and reproducibility. The measurements commonly performed on active components at lower frequencies can be carried out with the R&S®ZNA (see Figure E.11) just as conveniently at high frequencies using system-integrated mmWave converters at metrology-level precision [35].

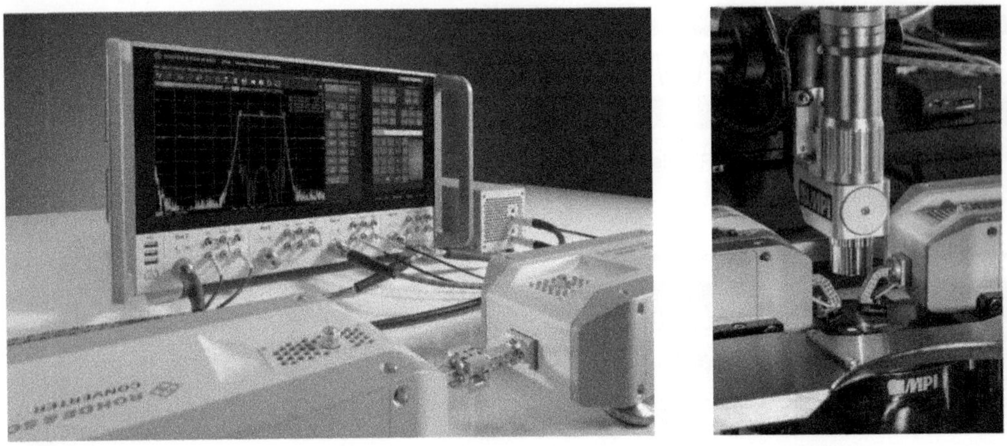

Figure E.11 Measurements in the THz range. With the R&S®ZNA VNA, measurements in the THz range using mmWave converters are configured just as conveniently as in lower frequency ranges (left). The MPI TS150-THZ integrated probe system with microscope configured with the R&S®ZNA for wafer-level measurements up to 330 GHz with GGB Industries Inc. waveguide probes directly mounted at the output of the mmWave R&S®ZC330 converters (right). *Source*: MPI Corporation.

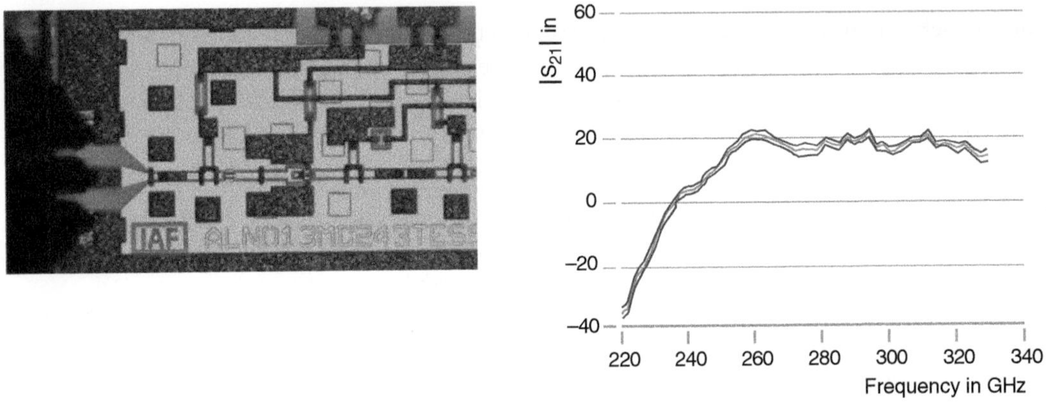

Figure E.12 Fraunhofer IAF four-stage, 325 GHz MMIC LNA (left) and measured scalar gain |S21| with the ±95% confidence intervals (right). *Source*: [35] / EBSCO Industries, Inc.

To demonstrate system capabilities, a four-stage 325 GHz MMIC low-noise amplifier (LNA) developed by the Fraunhofer Institute for Applied Solid State Physics IAF was measured (see Figure E.12). The scalar gain $|S_{21}|$ is shown with the ±95% confidence intervals.

E.4.4.2 Wideband Signal Generation and Analysis in the D Band

Extending the carrier frequencies beyond 100 GHz, one of the focus frequency bands of 6G research lies in the D band (110–170 GHz) due to the prospect of using much wider bandwidths of several GHz for the highest data rates. Figure E.13 shows a test and measurement setup to support research for components and transceivers in this frequency region.

Shown on the left is the R&S®FE170ST transmitter (TX) frontend, which upconverts the modulated signals (e.g., potential 6G waveforms) from the R&S®SMW200A vector signal generator to the frequency range of 110–170 GHz. On the right is the counterpart: the R&S®FE170SR receiver (RX) frontend, which downconverts the signals and transmits the intermediate frequency (IF) to the R&S®FSW signal and spectrum analyzer. The demodulated signal exhibits excellent EVM performance, underlining the extremely low phase noise of the generated signal.

E.4.4.2.1 MMIC design

Figure E.14 shows two sample MMICs designed by Rohde & Schwarz and cooperation partners that are used in T&M equipment.

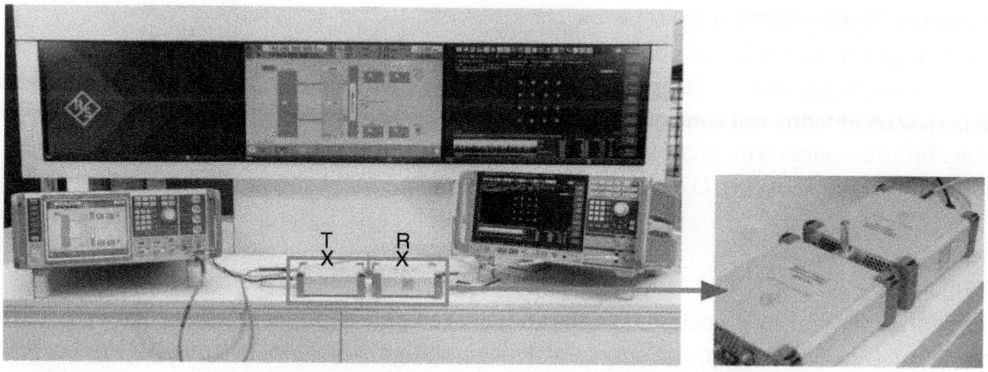

Figure E.13 D-band (110–170 GHz) signal generation and analysis test setup with R&S®FE170ST frontend (transmitter, TX) and R&S®FE170SR frontend (receiver, RX). In this sample setup, the TX and RX are directly connected with an attenuator.

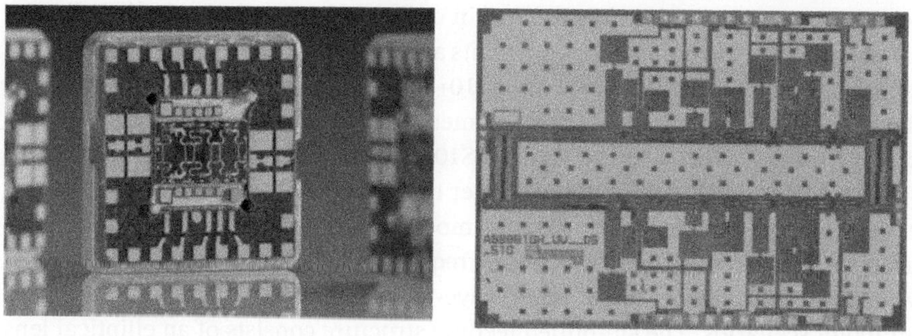

Figure E.14 27–41 GHz GaN/SiC MMIC amplifier in proprietary SMT compatible package (left); 40–70 GHz GaN/SiC MMIC amplifier (right). *Source*: Rohde & Schwarz GmbH&Co. KG / https://scdn.rohde-schwarz.com/ur/pws/dl_downloads/premiumdownloads/premium_dl_pdm_downloads/3683_6158_52/Fundamentals-of-THz-technology-for-6G_wp_en_3683-6158-52_v0102.pdf / last accessed on Nov 10, 2025.

Figure E.15 RPG local oscillators for heterodyne receivers in the HIFI instrument. Onboard the Herschel space observatory (left) [36] and for the SWI onboard the ESA JUICE planetary mission [30] (left: RPG, Herschel/HIFI local oscillators, 480–1100 GHz; right: RPG, JUICE/SWI local oscillator, 136–158 GHz).

The 40–70 GHz amplifier that is shown with dimensions of approximately 4 mm × 3 mm reflects the typical size of such circuits as used in the latest R&S®SMA100B analog signal generator from Rohde & Schwarz. This instrument offers the highest commercially available output power in the microwave range.

As another example, THz sources have also been developed by Radiometer Physics GmbH (RPG) with a combination of high-power amplifier GaAs MMICs up to the W band and high-power, high-frequency multipliers based on GaAs Schottky diode MMICs. Figure E.15 shows an example of such a frequency-tunable source up to 1100 GHz for the heterodyne instrument for the far infrared (HIFI) onboard the Herschel space telescope (left) [36], covering the spectral range from far-infrared to submillimeter wavelengths. The picture on the right shows a frequency- and power-agile source up to 158 GHz for the SWI onboard the JUICE satellite from ESA (European Space Agency) [30].

E.4.4.3 Antenna Radiation Performance Measurements in the D Band

5G pioneered the use of mmWave frequencies for wireless communications in connection with the development of over-the-air (OTA) testing concepts. This is due to the fact that large-scale and highly miniaturized antenna arrays are no longer accessible for conducted testing [37]. OTA antenna test concepts can be extended into the D band and beyond for exploring THz communications and sensing. Future devices will incorporate even more highly integrated active antenna systems for ultra-massive MIMO and sensing applications. However, OTA testing will also be of interest to investigate the characteristics of RISs.

E.4.4.3.1 Over-the-air testing (OTA) with the R&S®ATS1000

Research toward 6G is now putting the focus on frequencies above 100 GHz. There, the significant available spectrum is the key to enable much higher data rates. In this context, not only new wideband high-gain antenna concepts are needed but also advances in applied antenna measurement procedures.

Moving from legacy sub-6 GHz cellular services to 5G NR frequency range 2 (FR2) was already a major technological leap (Figure E.2). Since path loss increases with the square of frequency, higher gain antennas with electronic Beamsteering capabilities were introduced in the user equipment and network infrastructure in order to ensure appropriate quality for the radio links. Due to the dramatic increase in the complexity of integrated circuits as a function of frequency, a majority of developments are now targeting a new incremental step at waveguide D band (110–170 GHz) and G band (140–220 GHz).

In the following discussion, we consider a spherical scanning solution for measuring radiation performance in the D band with the unprecedented dynamic range that is realized in the R&S®ATS1000 chamber. The solution uses a new probe design with direct downconversion that provides a dynamic range of greater than 50 dB at 170 GHz.

The R&S®ATS1000 simplifies the test requirements because no mechanical modification or additional RF cabling is needed to measure the amplitude and phase-coherent response of a DUT in the frequency range from 110 to 170 GHz.

The device under test (DUT) is a newly designed D-band lens-based leaky-wave-fed antenna by IMST (Figure E.16). It could be used in 6G fronthaul point-to-multipoint scenarios. The simplified feeding structure consists of an elliptical lens made of low permittivity $\varepsilon r = 2.34$ and low-loss high-density polyethylene (HPDE) with 35 mm diameter (20λ at 170 GHz), which enables a cost-effective design. The feed consists of a $\lambda/2$ leaky-wave air cavity, excited by a WR6 waveguide. The radiation pattern can be steered by displacing the feeder along the lens focal plane.

This antenna was both modeled numerically and evaluated experimentally with a compact system using spherical near-field scanning. The radiation pattern measurements were carried out in the R&S®ATS1000 mobile spherical scanning range (Figure E.17). This fully anechoic chamber includes a distributed axis positioner.

A DUT feeding assembly is used to perform phase-coherent and time-stable measurements. It is shown below the lens antenna in Figure E.17. This chain consists of a D-band subharmonic mixer, identical to the one that is used at the probe, and a D-band isolator that is attached to the WR6 split-block of the DUT. Measurements are performed with the R&S®ZNA43 4-port VNA, where one port at the front feeds the IF signal to the DUT.

Figure E.17 (inset) shows the measurement probe concept. An orthomode transducer (OMT) is connected to a 20 dBi squared horn antenna with a 3 dB beamwidth of 16° and a cross-polarization isolation of 25 dB over the complete D-band frequency range. The assembly works reciprocally and is able to transmit or receive two orthogonally polarized fields when

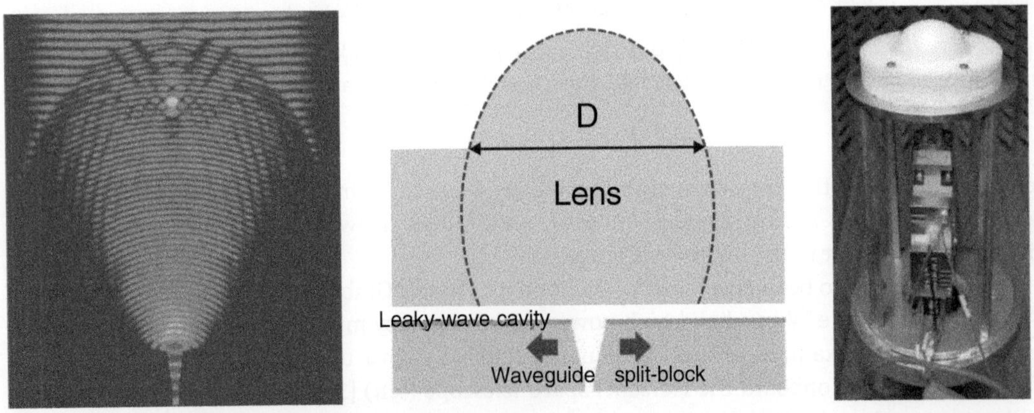

Figure E.16 D-band leaky-wave-fed lens antenna by IMST. *Source*: [38] / IEEE. Simulated E-field distribution inside and outside the lens at 140 GHz (left). Schematic of the lens (center) and photo of the setup in the R&S®ATS1000 chamber (right).

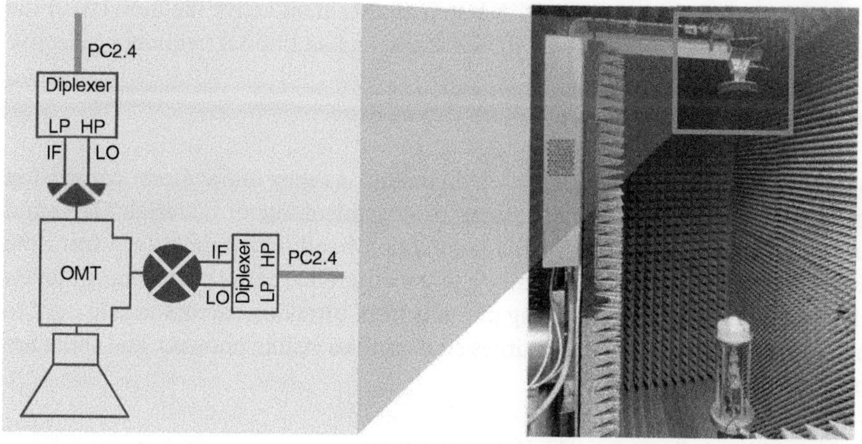

Figure E.17 Measurement setup using a compact spherical scanning system with a frequency converted probe and DUT at the azimuth pole. Block diagram of the frequency converted probe concept (inset). *Source*: Rohde & Schwarz GmbH&Co. KG / https://scdn.rohde-schwarz.com/ur/pws/dl_downloads/premiumdownloads/premium_dl_pdm_downloads/3683_6158_52/Fundamentals-of-THz-technology-for-6G_wp_en_3683-6158-52_v0102.pdf / last accessed on Nov 10, 2025.

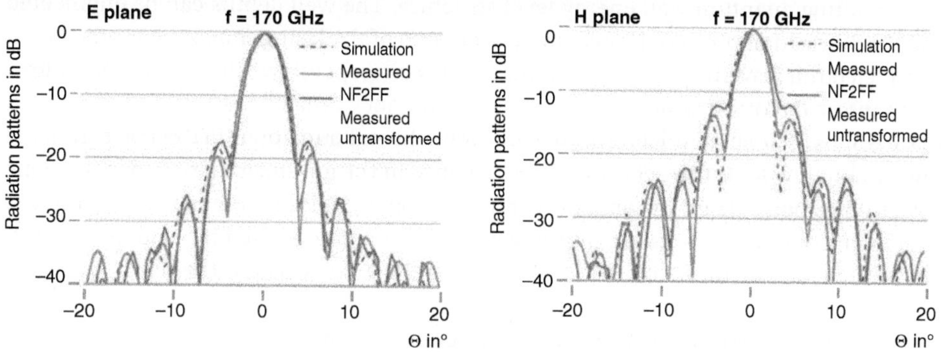

Figure E.18 E- and H-planes. Simulated, measured, and near-field to far-field (NF2FF) transformed normalized directivity patterns as a function of frequency for boresight (direction of maximum gain) radiation.

the DUT is set to receive (RX) or transmit (TX). Downconversion or upconversion is realized directly at the probe, thereby removing any cable loss at RF frequencies. Both polarizations can be measured simultaneously.

The results in Figure E.18 reveal excellent agreement between the DUT full-wave simulations and the measurements. This confirms the high accuracy of the measurement system and technique involving the new probe design. Phase-coherent data acquisition, such as near-field to far-field (NF2FF) transformation can be successfully realized for passive antenna measurements. Interestingly, the untransformed measurement results (NF2FF-measured untransformed trace in Figure E.18) show that the main beam of the radiation pattern is already close to far-field asymptotic behavior.

In conclusion, a high-efficiency D-band lens antenna design was presented with realized gain greater than 30 dB over 42% bandwidth. Accurate characterization of this antenna was performed with a spherical scanning test system, capable of stable phase-coherent measurements, with direct frequency conversion at the DUT input and at the test probe outputs. Phase coherence is a must to support precise application of the near-field to far-field transformation algorithms that are essential for accurate determination of radiation pattern nulls and sidelobe levels.

E.4.5 Direct THz Photon Generation: QCL

E.4.5.1 Interband Diode Laser

E.4.5.1.1 Optical Versus THz Frequencies: Interband versus Intraband Transition

Generation of phase-coherent radiation in a laser (light amplification by stimulated emission of radiation) has been an established concept for many years. It has led to many applications, among them optical communications in conjunction with optical fiber technology, the technical foundation of today's internet. A laser consists of an active medium (with the laser transition between energy levels with population inversion), a cavity for feedback, and additional frequency-selective components.

E.4.5.1.2 Interband Diode Laser

Semiconductor laser diodes are among the most important optoelectronic components, as they allow direct conversion of electrical current into coherent light. For optoelectronics, the direct bandgap III–V semiconductor materials GaAs and GaN are most important. The recombination of an electron (e^-) from the conduction band with a hole (e^+) from the valence band leads to the emission of a photon with a frequency corresponding to the difference between the energy levels (Figure E.19). Interband diode lasers are cheap and efficient for generating photons from ultraviolet across visible light to the IR frequency region. However, THz photons have energies 100–1000 times smaller than visible photons, and there are no materials with such a small bandgap and population inversion.

E.4.5.2 THz QCL: Lasing Transition Between Intersubbands Engineered by Heterostructures

In contrast to the interband diode laser, laser emission is achieved in a QCL through the use of intersubband transitions in a periodic stack of semiconductor multiple quantum well heterostructures ("superlattice"; Figure E.19). The concept was first demonstrated in 1994 by a research group at Bell Labs by Jérôme Faist et al. [39].

Figure E.20 shows an illustration of the periodic quantum well heterostructure (epitaxially grown GaAs or InP structures with a thickness of a few nm) and the resulting quantum well energy level structure. The well depths can be engineered by controlling the layer depths during the fabrication process. Hence, the wavelength of the lasing transition is dependent on the physical structure of the device ("electron wavefunction engineering"). This concept allows the generation of low-energy THz photons not accessible with interband diode lasers.

Operating principle (Figure E.20): Photons of frequency ν are emitted via intersubband transitions of electrons from the excited state $|e\rangle$ to the ground state $|g\rangle$, where $E = h\nu$ is the energy difference between the ground and excited states. One advantage of the concept is that the electron responsible for the emission of the photon tunnels into the next quantum well (active region) with subsequent emission of a photon. As a result, multiple photons can be generated by a single electron, thereby making the process extremely efficient. The tunneling from one well to the next is where the term "quantum cascade" originates from.

For reference, the optical wavelength of 850 nm corresponds to an energy difference of 1.4 eV.

Successful operation of a QCL at THz frequencies was first demonstrated in 2002 [40]. Since then, QCLs have quickly progressed in terms of frequency coverage, increased power output, and increased operating temperature. By carefully designing the quantum wells, lasing has been achieved at wavelengths as short as 2.75 μm (109 THz) and as long as 161 μm (1.9 THz). The longer wavelength devices still require cryogenic cooling, but room temperature operation has been observed to at least 16 μm. Interest has been concentrated in the mid-infrared ($3.5 - 13$ μm) and THz spectrum ($2 - 5$ THz $\approx 60 - 150$ μm). An approach using long-wavelength THz QCL sources with intra-cavity nonlinear frequency mixing has even made frequencies below 1 THz accessible [41].

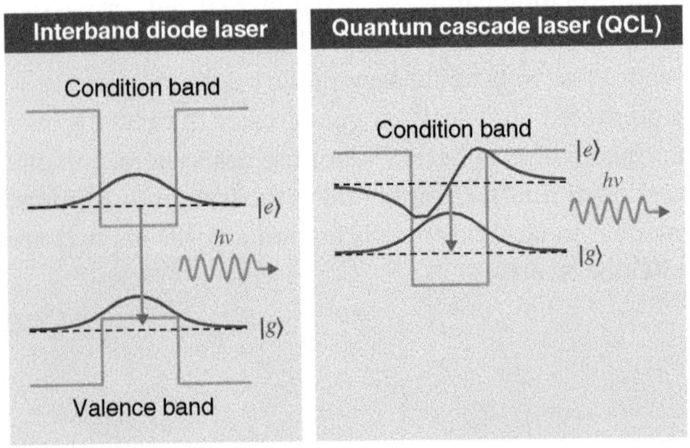

Figure E.19 Interband diode laser versus QCL. Whereas in an interband diode laser, the laser transition between the excited state energy level $|e\rangle$ and the ground state $|g\rangle$ occurs between the conduction band and the valence band, in a QCL, the photon transition occurs between designed quantum well energy levels ("wave function engineering") within the conduction band $h\upsilon$.

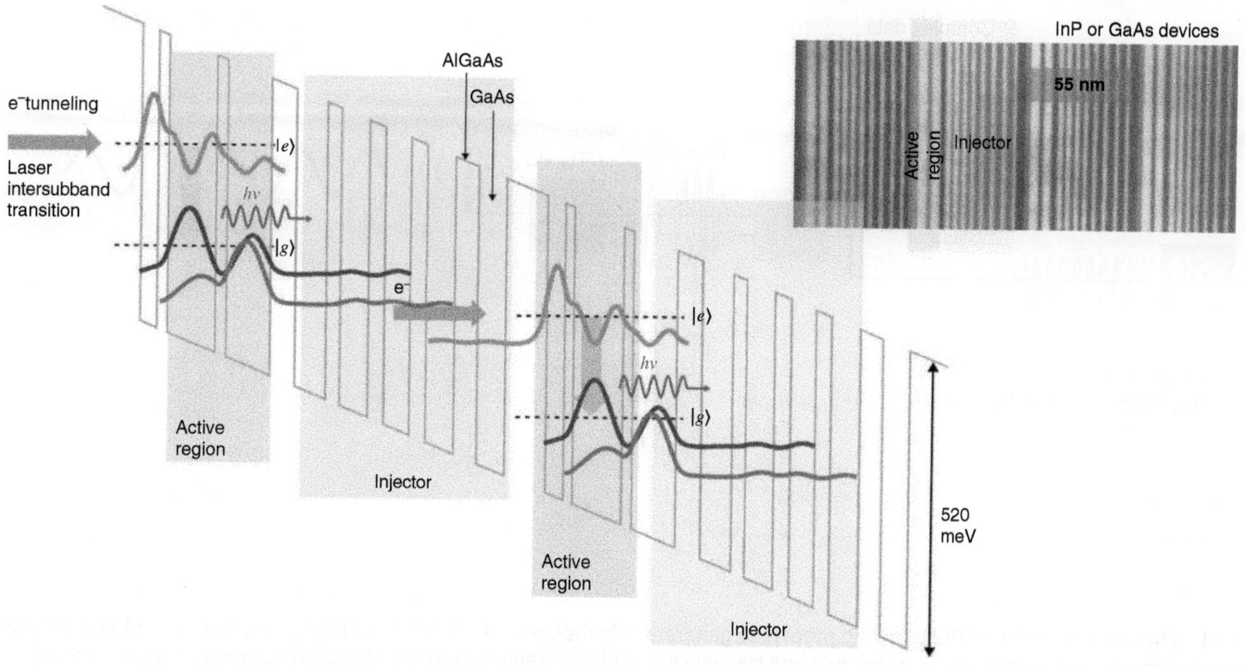

Figure E.20 QCL: Lasing transition between intersubbands engineered by semiconductor heterostructures. "Band structure and electron waveform engineering": Light is emitted as electrons "cascade" through multiple quantum wells, forming a "superlattice." *Source*: Rohde & Schwarz GmbH&Co. KG / https://scdn.rohde-schwarz.com/ur/pws/dl_downloads/premiumdownloads/premium_dl_pdm_downloads/3683 6158 52/Fundamentals-of-THz-technology-for-6G_wp_en_3683-6158-52_v0102.pdf / last accessed on Nov 10, 2025.

The journey toward realizing a THz QCL that operates at room temperature has taken a step forward with the recent publication of a device that operates at −23 °C, within the reach of Peltier coolers [42].

E.4.6 Downconversion Photonic Approach: From Optical to THz via Photomixing

E.4.6.1 Optoelectronic Frequency Domain THz Generation: Uni-traveling Carrier Photodiode (UTC-PD) and PIN Photodiode as Photomixer

An indirect approach for continuous wave (CW) THz generation has recently attracted strong interest: a photodiode can efficiently convert an optical signal into an electrical signal by optical heterodyning in high bandwidth photoconductors via a "photomixing" process. It involves optical/infrared laser light generating free charge carriers in a semiconductor or organic crystal. The charge carriers are accelerated by internal or external electric fields, and the resulting photocurrent becomes the source of the THz wave.

As depicted in Figure E.21, in this approach, the output of two CW single-mode lasers (often at the 1.55 μm "fiber telecom wavelength") with closely spaced emission frequencies at ν_1 and ν_2, respectively, are "mixed" in an ultrafast III–V compound semiconductor photodetector (InGaAs/InP), inducing a photocurrent modulated at the optical beat frequency $\nu_{THz} = \nu_1 - \nu_2$ in the THz region. An antenna structure surrounding the photomixer translates the oscillating photocurrent into the THz wave. State-of-the-art photomixers are based on either GaAs or InGaAs/InP and require laser wavelengths below the semiconductor bandgap (i.e., around 0.8 or 1.5 μm, respectively).

The photomixing technique has the advantage that by tuning the lasers, the difference frequency of the beat note can be varied over a broad spectral range, which translates directly into widely tunable THz radiation [43, 44].

Two types of photodiodes are commonly used: PIN photodiodes (PIN-PD) and uni-traveling carrier photodiodes (UTC-PD) [45]. Both types were originally developed as detectors for fiber-optical communications networks, but later modified to match the requirements of THz emission. UTC-PDs were originally developed by NTT in Japan for multi-stage optical reception at 40 Gbps, but these photodiodes have been enhanced to enable the generation of signals up to 4.5 THz [46, 47].

In the frequency bands suitable for communications applications such as the 300 GHz band, power levels on the order of mW have been achieved [48, 49].

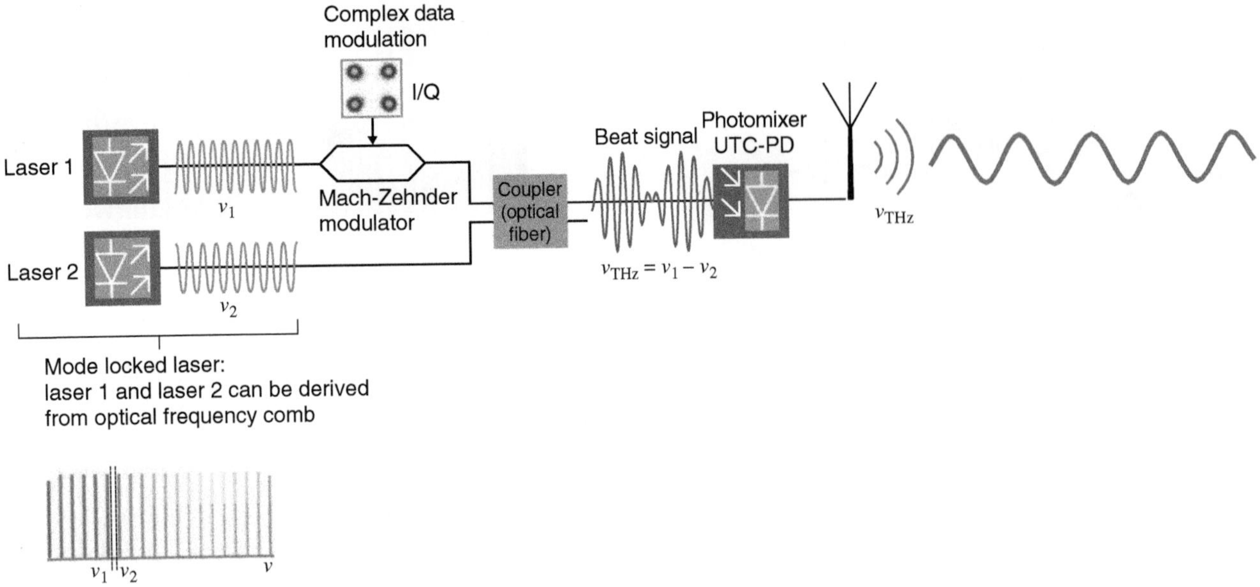

Figure E.21 Photomixing process. Photomixing process to generate THz radiation at the beat frequency $\nu_{THz} = \nu_1 - \nu_2$ of two slightly detuned single-mode lasers. One way to attain extreme frequency and phase stability is to derive both frequencies from an optical frequency comb. For data transmission, one of the lasers is modulated by a Mach–Zehnder modulator (MZM): It consists of an interferometer where the beam is split into two. In one of the interferometer arms, the phase of the laser light is shifted relative to the other path by an electro-optic modulator (EOM), resulting in a constructively or destructively modulated laser beam after recombination of both beams. The beat signal impinges the photomixer uni-traveling carrier photodiode (UTC-PD), and the integrated antenna emits THz radiation.

The combination of the tunable feature of the laser beams with the photomixing technique allows transfer of the techniques for generating optical vector fields developed for optical communications (e.g., with Mach–Zehnder modulators with modulation bandwidths >100 Gbps) and thus technological advancement into the THz range.

Furthermore, if extra frequencies are added, these techniques make it relatively easy to implement multifrequency communications (Figure E.21). The combination also allows easy integration of these wireless links into fiber-optic infrastructure. Furthermore, on-chip communications and future high-speed inter-device communications will also require THz waveguides. Such waveguides have recently been realized utilizing topological valley photonic crystals exhibiting near-zero bending loss and zero backscattering [50]. As shown in Figure E.21, referencing both frequencies to the same frequency comb generator allows transfer of the unique phase and frequency stability of optical combs in a broadband and tunable manner into the THz range [51, 52]. The receiver side, corresponding to the transmitter shown in Figure E.21, could be a Schottky diode or a setup that is symmetrical to the transmitter side. This technology also holds promise for T&M instruments since it can be scaled up to extend the spectrum analysis and vector network analysis frequency range via optoelectronics into the THz region [53, 54].

E.4.6.2 THz Waves for Communications: 300 GHz Point-to-point Transmission

Figure E.22 shows examples of THz data transmission trials (SISO, i.e., with one transmit antenna and one receive antenna) performed in the lab and outdoors. Between 200 GHz and 300 GHz, there is a transmission window with low atmospheric losses (Figure E.7).

In contrast to FSO links, mmWave or THz transmission is much less affected by adverse weather conditions like rain and fog. While the left side of Figure E.22 shows examples of 100 Gbps in the lab, the right photo shows a transmission link at 300 GHz in the harbor of Dunkerque over a distance of 850 m (with a very focused beam).

On the left side of Figure E.22, an UTC-PD InGaAs/InP chip fabricated at IEMN is shown with the response curve photodiode output versus photocurrent (credit: IEMN UMR CNRS 8520/University of Lille-Renatech Network, France) [55]. This photodiode is using a semitransparent top contact with subwavelength apertures for enhanced optical transmission and front-side illumination of the photodiode (1.55 μm wavelength). The responsivity of the device is further increased by using a metallic mirror below the diode mesa through wafer bonding.

Further demonstrations of 100 Gbps transmission in the THz window between 200 and 300 GHz have been realized [56–59].

E.4.7 Time Domain THz Generation: Spectroscopy and Imaging Based on a Femtosecond Laser

Another alternative approach to cover very large frequency ranges (i.e., 0.1–6 THz) involves time domain (TD) systems, which are used for spectroscopy (TDS) and imaging. In such a setup, pulsed THz radiation is generated via femtosecond lasers. The laser pulse is split into two parts with a beam splitter: one part travels to the THz transmitter and the other part to the detector. The ultrashort laser pulse produces a transient current in the emitter, a photoconductive antenna, resulting in an electromagnetic wave packet with a broad spectrum in the THz range [60, 61].

The THz pulse interacts with the spectroscopy sample and reaches the receiver, which works in a "pump and probe" fashion: the incident THz pulse changes certain characteristic properties of the material (e.g., conductivity or birefringence), and the separated laser pulse probes this effect. A variable delay stage scans the THz wave packet with the much shorter "probe" pulse. A Fourier transform of the THz amplitude then reproduces the spectrum.

Time domain spectroscopy offers the advantage of a very large spectrum range and high measurement speed. Commercial systems generate spectra up to 6 THz. THz time domain systems are used in a variety of industrial applications, such as nondestructive testing in quality assurance [25], in particular for synthetic materials such as polymers and ceramics. From the THz measurements, the complex refraction index (permittivity) and geometrical structures can be derived. One application example is thickness gauging via time of flight measurements, which is used, for example, in paper production: the broad spectrum translates into micrometer-level thickness resolution.

Figure E.23 shows a THz image of a prepaid card recorded by a time domain spectrometer (TeraFlash pro, TOPTICA Photonics AG). The sample was probed with frequencies from 100 GHz to 6 THz.

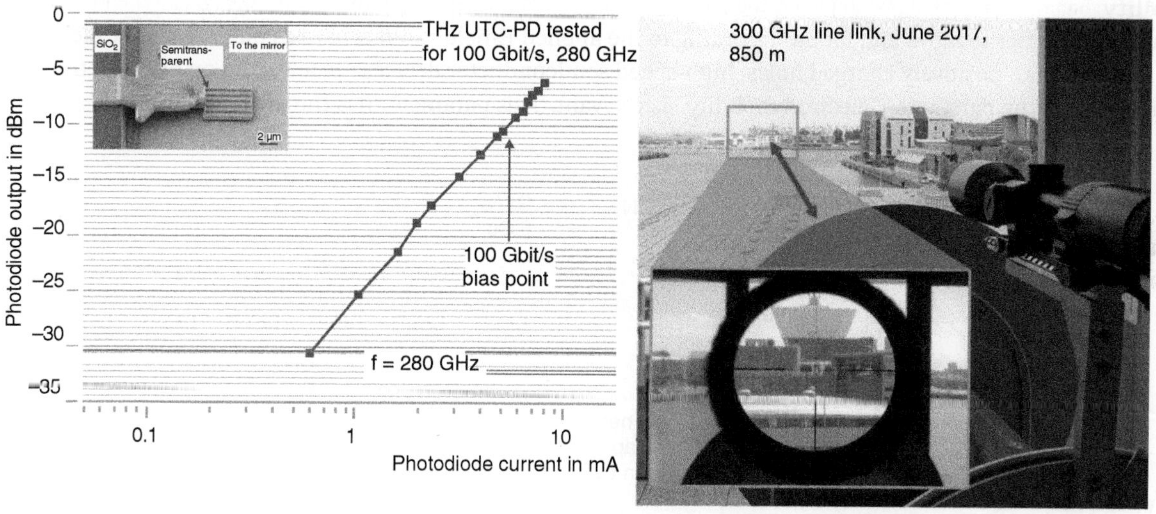

Figure E.22 Examples of data transmission around 300 GHz in the lab (left) and an outdoor trial (right). *Source*: [55] / John Wiley & Sons. A 300 GHz link with live HD video transmission was established in the harbor of Dunkerque over a distance of 850 m where very good directional alignment is crucial (courtesy of Prof. G. Ducournau, IEMN, CNRS-Université de Lille). *Source*: Rohde & Schwarz GmbH&Co. KG / https://scdn.rohde-schwarz.com/ur/pws/dl_downloads/premiumdownloads/premium_dl_pdm_downloads/3683_6158_52/Fundamentals-of-THz-technology-for-6G_wp_en_3683-6158-52_v0102.pdf / last accessed on Nov 10, 2025.

Figure E.23 Photograph (left) and processed THz image (right) of a Japanese prepaid public transport card, revealing an inside view of the underlying electronics (courtesy of TOPTICA Photonics AG).

E.5 Semiconductor Technologies for mmWave and THz Electronics

Modern society's growing demand for wireless connectivity with higher data rates, as well as radar imaging systems with higher resolution, is driving a continuous increase in the frequencies at which these microwave systems operate. While 5G has set the stage for utilizing mmWave frequencies for wireless communications networks, research into 6G as well as satellite communications links is further propelling the development of mmWave systems into the range beyond 100 GHz.

This trend poses an increasing challenge for the semiconductor industry to provide components to deliver high output power at extremely high frequencies. Power amplifiers, often realized as MMICs, are among the most essential components of wireless communications and imaging systems. They have a decisive influence on the overall system's performance in terms of output power, efficiency, bandwidth, linearity, and noise.

Here, we highlight some key properties of the relevant semiconductor materials for high-frequency applications along with the development status of state-of-the-art power amplifiers.

E.5.1 Material Properties of Semiconductors for High-frequency Applications

High power at high frequencies: III–V semiconductors (GaAs, GaN, InP) versus silicon (SiGe): While III–V semiconductors with their direct bandgap are the ideal material for optoelectronics (lasers) and high-frequency applications, silicon technology has the most advanced large-scale manufacturing processes, enabling cost-efficient solutions for commercial applications.

The different high-speed semiconductor material properties are summarized and compared in the radar plot below (Figure E.24).

We provide a brief description of the key parameters as follows:

Electron mobility

The ability of a charge carrier to move and follow according to the applied E-field is referred to as mobility. Since electrons exhibit higher mobility than positively charged holes, high-speed semiconductor devices are all based on electron transfer currents. For this reason, the parameter "electron mobility" is relevant. This parameter has the unit (m^2/[Vs]).

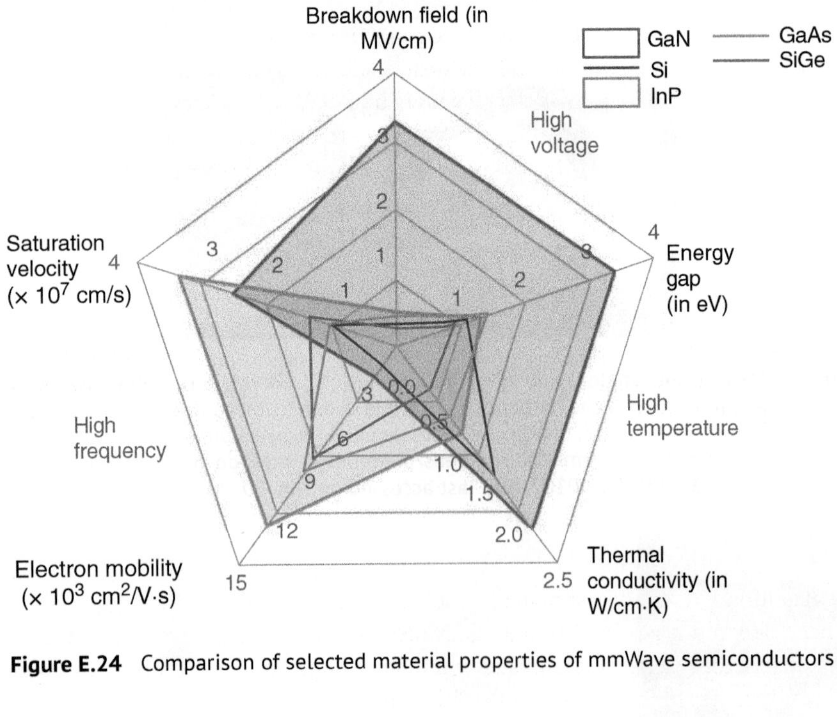

Figure E.24 Comparison of selected material properties of mmWave semiconductors [62].

Saturation velocity
This parameter describes the maximum speed of an electron in a semiconductor device when the charge is driven by high E-field strengths. The unit of the parameter is simply that of velocity (m/s). The operating state of a device operated at high field strength is also called "velocity saturation."

Breakdown field strength
The parameter with the unit (V/m) describes the ability of a material to withstand high voltages before it becomes conductive (which is usually destructive for the device in the absence of current limitation). The maximum operating voltage of the device is strongly defined by this parameter. There are techniques, like the application of field plates, to improve and equalize the field strength distribution in a semiconductor device to increase its breakdown voltage.

Energy gap
The energy gap or bandgap is measured in (eV) and is a quantum physics-based property describing the energy difference between the conduction and valence band, where no energy states are available in an undoped semiconductor. For heterojunction devices, the bandgap structure is shaped by adding additional materials to the semiconductor alloy to improve device performance (e.g., SiGe HBTs or III/V-based HEMT devices).

Thermal conductivity
This parameter describes the ability of a substrate material to conduct thermal energy and is measured in (W/[mK]). Especially for power applications like power amplifiers, this is an essential parameter since very often the device output power is thermally limited. This is especially true for GaN HEMTs, which need to be operated far below their breakdown voltage for thermal reasons. Note that for this reason, high-end GaN devices are manufactured on lattice-matched substrates exhibiting high thermal conductivity such as SiC or even diamond.

The degree to which the above parameters are more or less pronounced in a given semiconductor material strongly impacts its application fields.

Gallium arsenide (GaAs)
GaAs has a relatively high electron mobility and saturation velocity (Figure E.24) depending on the doping concentration. Thus, extremely high frequencies can be achieved. The material cost is high, but there exists a great deal of industry heritage.

GaAs technologies are available nowadays for Schottky diodes, pseudomorphic high electron mobility transistors (pHEMT), and heterojunction bipolar transistors (HBT). GaAs pHEMT technology exhibits good broadband noise up to high frequencies. HBTs are especially well suited in cases where low 1/f noise is desirable in combination with high breakdown voltages. GaAs Schottky diodes can be used to produce very low-noise receivers up to 4.7 THz. It can also be used for high-efficiency multiplication up to several THz, making it a very versatile mmWave technology.

GaAs mixer and multiplier devices can either be integrated as an MMIC into a module, or flip-chip transferred onto host substrates to improve dielectric loading and/or thermal dissipation.

Indium phosphide
Indium phosphide (InP) has the highest electron mobility and saturation velocity. Thus, extremely high frequencies can be achieved. However, the material cost is high, and wafer/chip handling is difficult.

InP technologies are available in the form of heterojunction bipolar transistors (HBT) and high electron mobility transistors (HEMT). The bipolar version is usually applied for analog integrated circuits where the speed and/or breakdown of state-of-the-art SiGe HBTs are insufficient. InP HEMTs have their domain in applications beyond 400 GHz and are used predominately for ultra-LNAs at frequencies up to 700 GHz.

From an application point of view, indium gallium arsenide (InGaAs) metamorphic HEMT (mHEMT) technologies are comparable to InP HEMT technologies except that they use GaAs substrates.

Gallium nitride
Gallium nitride (GaN) received considerable attention when it was shown to enable realization of blue LEDs (InGaN) and subsequently, blue lasers. This leads to a Nobel Prize in physics in 2014 [63]. GaN also provides certain favorable material features for high-power applications, namely a high energy gap, a high breakdown field strength, and, in combination with silicon carbide (SiC) substrates, good thermal conductivity for heat dissipation. This allows GaN transistors to operate under high bias voltages and thus deliver high output power. The lack of mature bulk GaN source material and the insufficient thermal conductivity of GaN led to the need for growing GaN heterostructures on carrier substrates such as silicon carbide

(SiC), Si, and lately even diamond. SiC is the mainstream substrate material for high-power, high-frequency applications owing to its excellent thermal conductivity and low lattice mismatch to GaN. GaN on Si is a more cost-effective solution that makes perfect sense for monolithic integration with CMOS. Si substrates exhibit significantly higher dielectric losses compared to SiC, which makes a significant difference for mmWave MMICs. Furthermore, the thermal conductivity at room temperature differs by a factor of three in favor of SiC (400 W/(m · K) versus 130 W/(m · K)). Since the safe operating area of GaN HEMT transistors is limited by thermal boundary conditions, the substrate's thermal conductivity is of utmost importance in order to achieve high output power combined with a reasonable mean time to failure (MTTF).

Due to efficiency advantages and the possibility to operate at much higher channel temperatures, GaN has almost completely replaced silicon-based LDMOS technologies for sub-10 GHz base stations. It has also recently been used for microwave backhaul applications.

The above-mentioned high bandgap of GaN in combination with maximum channel temperatures up to +200 °C makes the technology an excellent candidate for harsh environments. Especially space-borne applications benefit from the radiation hardness of the technology due to the high bandgap and the ability to operate in a huge temperature range ("day and night"). Compared to other semiconductor technologies, GaN amplifier systems exhibit a very low mass-to-output-power and volume-to-output-power ratio.

Silicon germanium (SiGe) HBT and CMOS

SiGe heterojunction bipolar transistors (HBT) also offer comparably good electron mobility, allowing their use for high-speed analog applications. The full advantage of this technology can be exploited if combined with complementary metal oxide semiconductor (CMOS) technology to realize mixed signal ASICs containing high-speed analog and digital functionality at a moderate cost. The combination of CMOS and SiGe HBTs is often referred to as "BiCMOS" processes. The prospect of utilizing mature CMOS manufacturing processes and combining high-frequency components with analog/digital signal processing circuitry on the same die enables unprecedented levels of integration and flexibility.

SiGe HBTs on their own represent a widely used material for automotive radar applications or short-range communications. Here, the breakdown voltage of the technology is sufficient to provide the necessary output power. Recently, there has also been interest in this material for microwave backhauls in order to obtain lower production costs per chip with high volumes and high integration density.

Hetero-integration as a 6G research topic

Since there is not just one single semiconductor technology that is relevant for mmWave applications, a recent research area for 6G is hetero-integration of various semiconductor technologies to optimize performance and energy consumption, e.g., combinations of InP (speed), CMOS (integration density), and GaN (output power). Table E.2 shows some standard Semiconductor materials with their corresponding properties and applications.

Table E.2 Semiconductor materials with their corresponding properties and applications.

Technology	Properties	Typical transistor types	Application
• Si FET	• Material cheap, 8″/12″ wafers • Very high integration capability • Low output power, medium speed for Si transistors • Si substrate	• MOS/CMOS (FinFET configuration for small node sizes) • Laterally diffused MOS (LDMOS)	• Digital (CPU, DSP) • RF-CMOS (e.g., low power IoT devices) • Sub-10 GHz power applications such as base station amplifiers
SiGe HBT	• Material cheap, 8″/12″ wafers • High integration capability • Low output power, very high speed for SiGe transistors • Si substrate	• Heterojunction bipolar transistors (HBT) • CMOS and HBT	• Analog high-speed integrated circuits (amplifier, mixer, VCO, multiplier, divider, transceiver, etc.)

Technology	Properties	Typical transistor types	Application
BiCMOS	• Material cheap, 8″/12″ wafers • High integration capability • Mixed signal capability for most complex on-chip systems • Si substrate	• CMOS and HBT	• Mixed signal devices (ADC, DAC, monolithic RF/analog/digital sub-systems in general, etc.)
GaAs	• Material expensive, 4″/6″ wafers • Low integration capability • Medium output power, high speed • GaAs substrate	• pHEMT • HBT (e.g., InGaP) • mHEMT	• RF blocks (amplifier, mixer, switches, etc.) • Analog RF with good 1/f noise performance where breakdown voltage of SiGe is insufficient (e.g., VCO, mixer) • Ultra-high-speed and low-noise LNA, comparable to InP HEMT
GaN	• Material expensive, needs different substrate since there are no GaN wafers, 4″/6″ wafers • Low integration capability • Very high output power, medium speed • SiC, Si, or diamond substrate	• HEMT	• High-power applications (amplifier, switches)
• InP	• Material expensive, 3″/4″ wafers • Low integration capability • Medium output power, very high speed • InP substrate	• HEMT • HBT	• Ultra-high-speed and low-noise LNA, comparable to mHEMT • Analog RF functions (amplifier, mixer, multiplier, divider, etc.) • Alternative to SiGe HBT if higher speed and breakdown voltage can be traded against integration density

E.5.2 Overview of State-of-the-art mmWave Power Amplifiers

Applications that go beyond 5G and 6G are pushing the frequency boundaries. The relevant data points are "bunched" around the driving applications.

An overview of state-of-the-art mmWave amplifiers based on data collected in [64, 65] is shown in Figure E.25. Each data point represents a measurement published in a paper over the years. The left side includes measurements up to 2016, whereas the right graph summarizes works until 2020.

It is apparent that the data points (in different colors for different semiconductor technologies) are not equally distributed over the frequency range but rather are bunched around certain frequency bands. This is obviously due to the fact that the applications driving development are grouped around certain frequency allocations. For example, a large group of results was published around 28 GHz that is associated with the development of circuits for infrastructure and terminals for 5G New Radio (NR) mmWave bands such as n258 (24–28 GHz), n257 (26–30 GHz), and n260 (37–40 GHz). Another group of results is concentrated around 60 GHz for WiGig with a large contribution of CMOS-based circuits.

The last conglomeration of data points between 70 and 95 GHz comprises a variety of applications. 77 GHz is used for automotive radar, while some imaging radar systems operate around 94 GHz. Moreover, E band backhaul and point-to-point high data rate communications links are also found here.

It is interesting to note that before 2016, only a few results were published above 100 GHz. However, tremendous progress has been made in recent years among the higher mmWave bands. This is thanks to designs based on InP and to a lesser extent on SiGe and InGaAs mHEMT systems. The driver behind this progress is clearly the next generation of 6G communications.

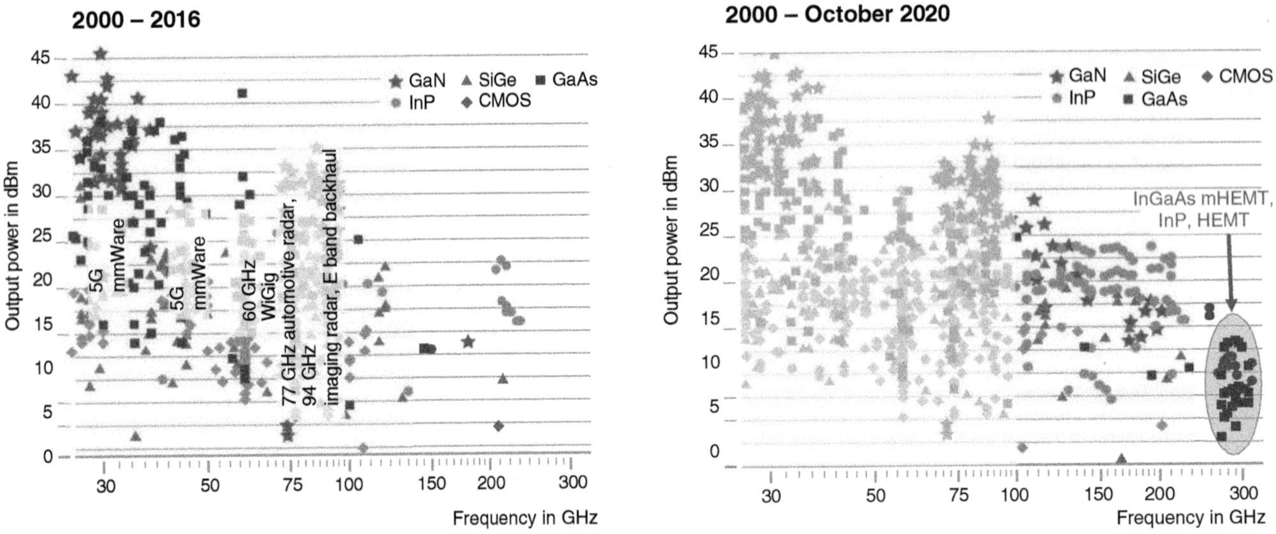

Figure E.25 Overview of state-of-the-art high frequency power amplifiers as of September 2016 (left) and October 2020 (right) [64, 65].

The frequency envelope for power amplifiers has even been pushed beyond 300 GHz—albeit at lower power levels—to 500 GHz with a few milliwatts of power [65, 66] and up to 1 THz with sub-milliwatts power levels [67].

E.6 Channel Propagation Measurements above 100 GHz

E.6.1 From Channel Sounding to Channel Models

E.6.1.1 MmWave and THz Frequency Propagation Characteristics as the Foundation for a New Physical Layer

The development of sub-THz communications as envisaged for 6G is only possible on the basis of a solid understanding of the properties of electromagnetic wave propagation in this as yet insufficiently researched frequency range.

This section explains the time domain channel sounding concept and presents exemplary results for a channel measurement campaign in the H/J band around 300 GHz and the D band (110–170 GHz) in representative outdoor and indoor scenarios. This research activity aims to obtain a better understanding of the mmWave and submillimeter wave (sub-THz and THz) mobile radio channel also with respect to "the technical feasibility of IMT in-bands above 100 GHz" as currently discussed in the ITU-R WP5D [68].

E.6.1.2 Channel Models as the Basis for Utilizing New Frequency Ranges

Before a new communications standard can be developed, the propagation properties in the designated frequency band must be understood and characterized. Then, channel models can be derived to enable system-level simulations of the new standard. Physical layer parameters include the distribution of pilot signals in the time-frequency domain, which allow the receiver to equalize and compensate channel effects on the transmitted signal. Such characterization is needed to develop and verify coding and error correction schemes.

The underlying measurement data must correctly reproduce the investigated environment. Geometry-based stochastic channel models (GSCM) such as 3GPP TR 38.901 [69], which is valid up to 100 GHz, are based on a large number of channel measurements in different environment scenarios.

The 3GPP channel model development and specification up to 4G was limited to the frequency range below 6 GHz and quasi-static environments. Dynamic scenarios and other types of environmental scenarios corresponding to new use cases, such as automotive, high-speed trains, and industrial environments, became relevant with 5G as the frequency range has been extended to the mmWave region. The resulting channel models, however, cannot simply be extended to the range above 100 GHz, where 6G is intended to operate. They have to be verified and fine-tuned to correctly reflect the impact of the environment. Even more than in the mmWave range, propagation in this frequency range is strongly influenced by human bodies, vehicles, and environmental conditions such as rain.

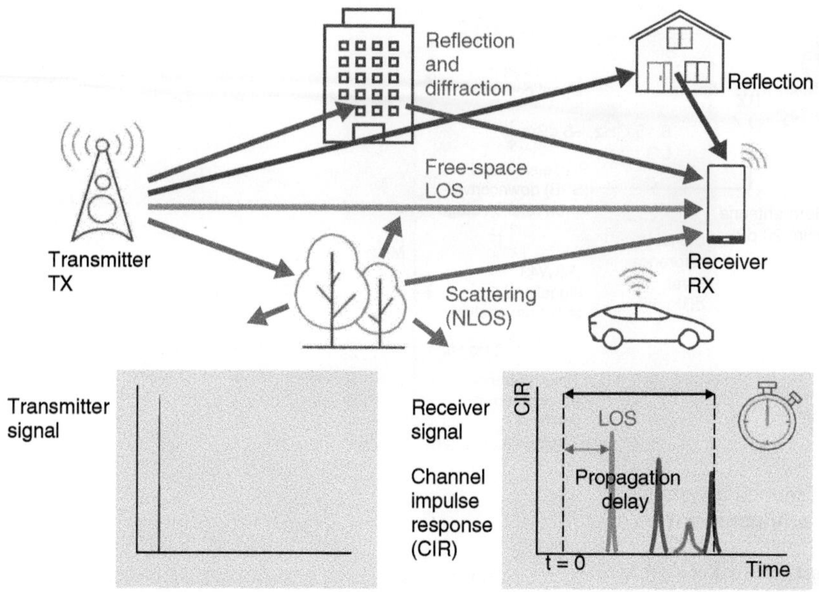

Figure E.26 Operating principle of time domain channel sounding. The channel impulse response (CIR) is measured by emitting an electromagnetic "ping" at the frequency of interest and capturing all returning signal components.

E.6.1.3 From Channel Sounding to Channel Models

Channel measurements by channel sounding deliver an image of the propagation properties of electromagnetic waves at a particular frequency. The term "channel sounding" comes from sonar technology, where short acoustic pulses are sent out from a ship or submarine and the reflections are recorded in the time domain [70]. This provides a viable image of the surroundings. With sonar, the transmitter and receiver are in the same place; with channel sounding for electromagnetic waves, the transmitter and receiver are spatially separated. In time domain channel sounding, a modulated pulse signal with excellent autocorrelation properties such as a Frank–Zadoff–Chu (FZC) sequence serves as a "ping" whose channel impulse response (CIR) is recorded. This propagation time measurement is very similar to the time delay measurements performed in a GPS receiver in reference to the GPS satellites (and subsequently inferring the position information), where each satellite transmits its individual correlation sequence. The CIR includes both the direct propagation components (line of sight, LOS) and all reflection and scattering components (non-line-of-sight, NLOS) from objects in the environment (Figure E.26). The channel model parameters and their values can be derived and determined from the results.

As a rule, objects are only "physically visible" to electromagnetic waves and act as reflectors or scatterers when they are at least as large as the wavelength of the incident wave. This means that at higher frequencies such as 30 GHz, objects with dimensions in the centimeter range already act as reflectors.

E.6.1.4 Collaborative Research

Rohde & Schwarz has many years of experience in channel sounding projects, including a high-resolution channel measurement campaign at 67 GHz in a street canyon in Tokyo, Japan [71]. In 2019, the focal point of a 3GPP research initiative was the development of new 5G channel models in industrial scenarios such as production environments. To support the 3GPP efforts, Rohde & Schwarz in collaboration with the Fraunhofer Heinrich Hertz Institute (HHI) carried out measurements in its Memmingen and Teisnach plants, not only in the 28 GHz and 66 GHz mmWave frequency bands but also in the 3.7–3.8 GHz band designated for private campus networks in Germany [72].

E.6.2 Time Domain Channel Sounding at 300 GHz

For 6G, the focus is on channels at significantly higher frequencies. In a collaborative effort with the Fraunhofer HHI and the Fraunhofer Institute for Applied Solid State Physics (IAF), a research setup was developed that enables signal generation and analysis in the range from 275 to 325 GHz with a bandwidth of 2 GHz [73]. The signal can be used to perform channel measurements and also modulated with novel waveforms for transmission experiments. Figures E.27 and E.28 show the

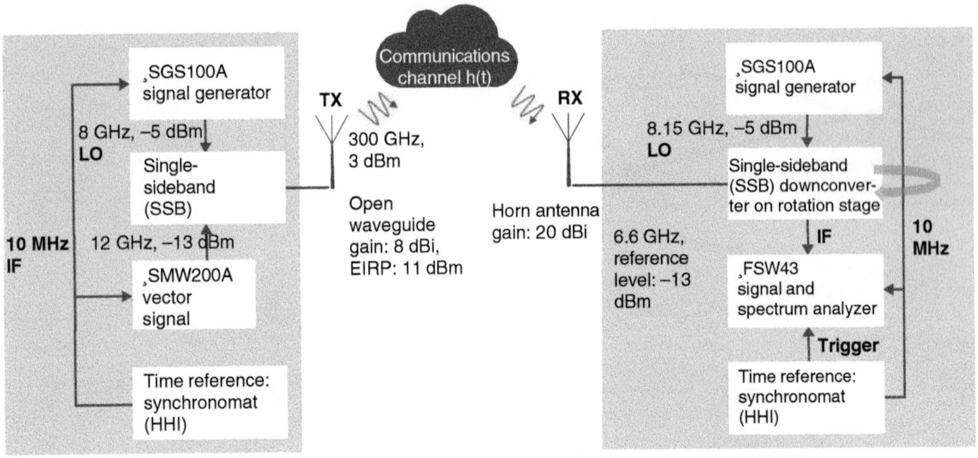

Figure E.27 Block diagram of the 300 GHz channel sounding system.

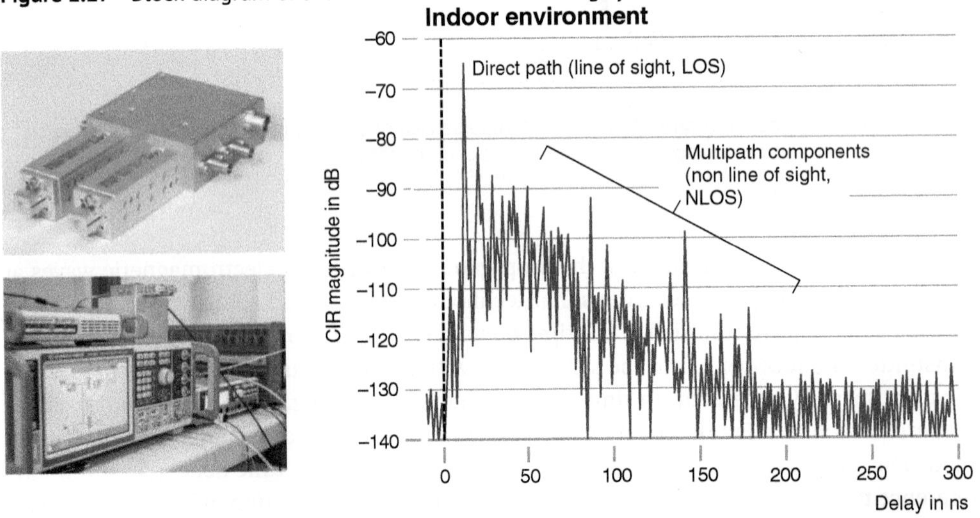

Figure E.28 Channel impulse response (CIR) of indoor environment at 300 GHz. Test setup for channel measurements at 300 GHz with the R&S®SMW200A vector signal generator, R&S®SGS100A signal generator, and R&S®FSW43 signal and spectrum analyzer. A transceiver with an integrated horn antenna is shown at the top of the picture. The setup can be used in channel sounding for channel characterization as well as for transmission experiments with new waveforms (left). CIR at 300 GHz in an indoor environment with multiple reflections (right). The distance between the transmitter and receiver was approximately 4 m. The pronounced shoulder with many reflections is typical for a very confined (small volume) indoor environment. An electromagnetic wave travels approximately 30 cm in 1 ns. *Source*: Rohde & Schwarz GmbH&Co. KG / https://scdn.rohde-schwarz.com/ur/pws/dl_downloads/premiumdownloads/premium_dl_pdm_downloads/3683_6158_52/Fundamentals-of-THz-technology-for-6G_wp_en_3683-6158-52_v0102.pdf / last accessed on Nov 10, 2025.

test setup. One of the key technologies from the IAF is InGaAs mHEMT MMICs (see Section E.5.1) for extremely low-noise and broadband applications at room temperature.

E.6.2.1 Test Setup for Propagation Delay Measurements Between Transmitter and Receiver

The measurements were captured using an instrument-based time domain channel sounder as shown in Figure E.27. On the transmitter side (TX), an R&S®SMW200A broadband vector signal generator produces a digital baseband "ping sequence" with a bandwidth of 2 GHz (suitable for autocorrelation) at an intermediate frequency (IF). The single-sideband upconverter (incorporating the THz transceiver from Fraunhofer IAF) raises this to the wanted RF transmit frequency, with an R&S®SGS100A acting as a local oscillator (LO) source. On the receiver side (RX), the antenna signal is amplified and mixed into the IF domain with a downconverter and LO generator. The IF signal is sampled by the R&S®FSW signal and spectrum analyzer, and the I/Q samples are stored for further processing. The transmitter and receiver are synchronized with two rubidium-based reference clocks and trigger units (synchronomat), enabling coherent measurements and allowing evaluation of phase and coherent averaging as well as determination of the absolute time of flight [73] shows further setup details.

Initial measurements at 300 GHz with this test system demonstrated its wide dynamic range, which is unrivaled in this frequency range (Figure E.28). The following sections discuss joint measurement campaigns that were carried out with the Fraunhofer HHI in various environmental scenarios for systematically characterizing the sub-THz frequency range.

E.6.3 THz Channel Measurements at Rohde & Schwarz Headquarters in Munich

The characterization of the channel sounding system at 300 GHz [73] described above and the initial measurements using the system were carried out on the premises of the Fraunhofer Heinrich Hertz Institute (HHI). The results were published in [74] for the outdoor scenario and in [75] for the indoor conference room scenario.

Recently, a more systematic study of the channel propagation characteristics at sub-THz frequencies (158 and 300 GHz) was conducted at Rohde and Schwarz headquarters in Munich. The frequencies were chosen in accordance with ongoing future 6G network discussions. The study focused on two representative scenarios: a street canyon-like (urban micro-cellular UMi) outdoor scenario in a corridor between two R&D buildings and an indoor measurement campaign in an atrium similar to a shopping mall or airport scenario.

Figure E.29 shows a block diagram of the correlative channel sounder for 158 GHz. This is similar to the setup at 300 GHz (Figure E.27). With the applied time domain channel sounder, the measurement duration is short. This allows multiple measurement iterations to be performed, covering the complete spatial angle of 360° at numerous positions in the measurement campaign.

At the **transmitter (TX)**, the setup comprises a single-sideband upconverter and an amplifier connected to an open waveguide with a gain of 7 dBi. The R&S®SMW200A vector signal generator provides a precomputed sounding sequence at an intermediate frequency (IF) of 14 GHz. The used sounding sequence is a Frank–Zadoff–Chu sequence with a length of 100 μs and bandwidth of $B = 2$ GHz (and a corresponding time resolution of $\tau = 1/B = 0.5$ ns).

The **receiver (RX)** consists of an R&S®FSW43 signal and spectrum analyzer and a downconverter that is fed by an LO generator at 12.24 GHz. A horizontally polarized E-plane horn antenna with 20 dBi antenna gain and an opening angle of approx. 15° in the E-plane (azimuth plane) serves as the receive antenna. The D-band frontend downconverts the received signal to an IF at 11.12 GHz. The signal analyzer samples the IF signal with a frequency of 2.5 GHz, where one measurement covers a total number of 250 periods of the sequences. The receiving antenna and downconverter are mounted on a precision rotation stage to allow angle-resolved measurements with respect to the azimuth angle.

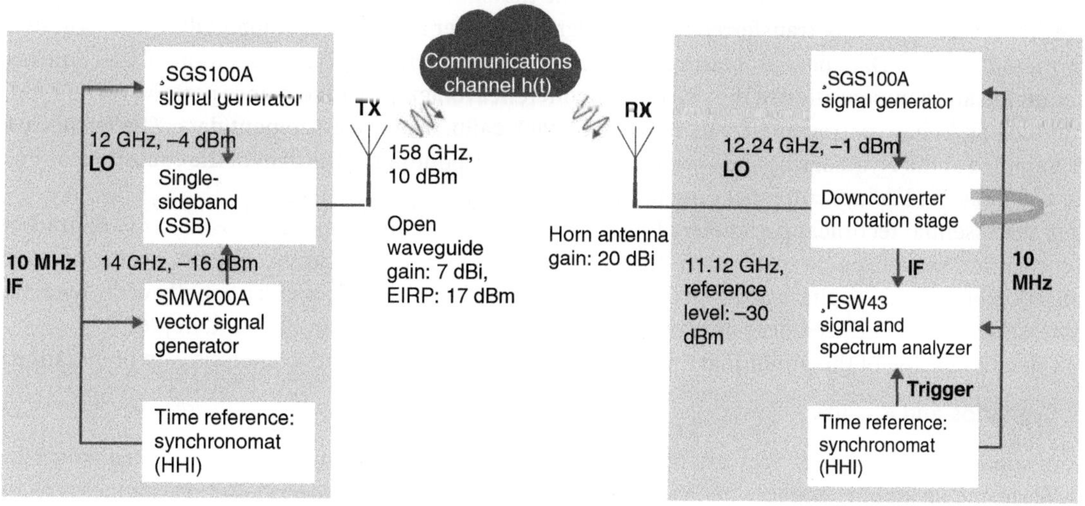

Figure E.29 Block diagram of the channel sounding measurement setup for 158 GHz.

Table E.3 Channel sounding measurement parameters at 158 and 300 GHz.

Measurement parameters	Value at 158 GHz	Value at 300 GHz
Frequency	158 GHz	300 GHz
Bandwidth	2 GHz	2 GHz
Temporal resolution	0.5 ns	0.5 ns
Autocorrelation sequence	Frank–Zadoff–Chu	Frank–Zadoff–Chu
Period length	200,000 samples	200,000 samples
Period duration (maximum measurable channel length)	100 μs	100 μs
Number of measured periods	250	250
Measurement duration	25 ms	25 ms
Theoretical processing gain	77 dB	77 dB
Dynamic range	>70 dB	>70 dB
Transmit power	10 dBm	3 dBm
TX antenna gain	8 dBi	8 dBi
Effective isotropic radiated power (EIRP)	18 dBm	11 dBm
RX antenna gain	20 dBi	20 dBi
Angular resolution	15° (24 steps)	15° (24 steps)

The receiver was mounted on a cart (camera dolly) for easy and precise movement to the different receiver positions (Figure E.31). Since the used antenna has a beamwidth of approximately 15°, the radio channel is sampled in steps of 15° in the azimuth angle domain at the receiver.

All instruments are connected to a time reference (Fraunhofer HHI synchronomat) to ensure coherent sampling between the transmitter and receiver. Besides a 10 MHz reference signal derived from a high-precision rubidium atomic clock, the synchronomat also enables simultaneous and coherent triggering at the transmitter and receiver.

The received I/Q time domain samples were transferred from the signal analyzer as raw measurement data to a computer after completion of the measurements. Data post-processing includes resampling and filtering, estimation of the common phase drift per sequence period and compensation of the phase drift, coherent averaging of all sequence periods, correlation, and application of amplitude and phase corrections by using back-to-back calibration measurement data. The results are generic calibrated CIR, sampled at the Nyquist rate and with magnitudes corresponding to the gain of the communications channel (including the antennas) and delays corresponding to the time of flight.

Table E.3 summarizes the essential technical parameters for the time domain channel sounder. Using a configuration with perfect complex correlation sequences (Frank–Zadoff–Chu sequences) and with additional coherent averaging after estimation and compensation of drift due to phase noise enables a large processing gain and an exceptionally wide dynamic range. Fundamental assessments of the utilized time domain channel sounding principle can be found in [76].

The following section describes some of the preliminary results. A more detailed quantitative analysis is in preparation.

E.6.4 Measurement Scenarios and Results

The measurements were conducted at Rohde & Schwarz headquarters in Munich, Germany. They represented an urban micro-street canyon scenario and an indoor shopping mall/airport scenario (Figure E.30). The outdoor measurements were performed between the two shaded buildings on the left. The width of the street is 15.5 m, and the height of the surrounding buildings is approximately 20 m.

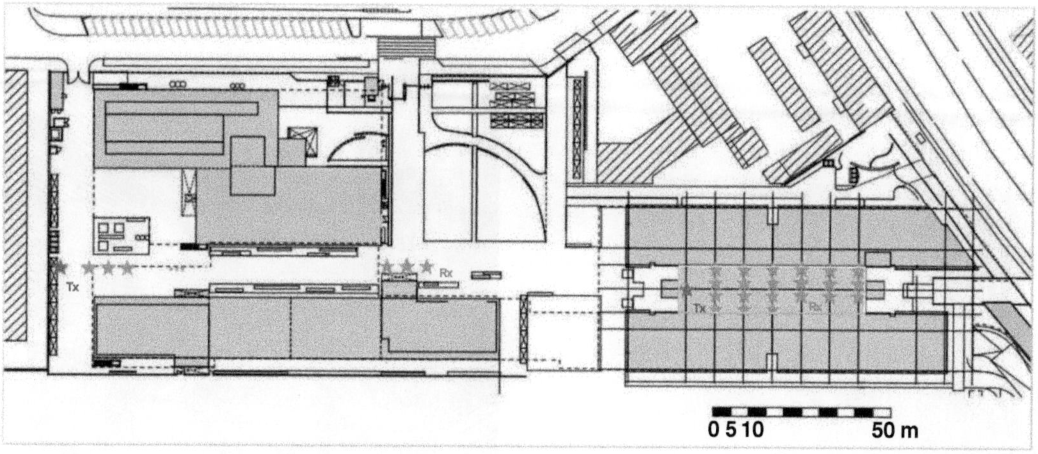

Figure E.30 Map of the investigated scenarios. Street canyon (left), indoor shopping mall/airport scenario (right). The positions of the fixed transmitter (TX) are marked with stars and the different positions of the receiver (RX) with stars.

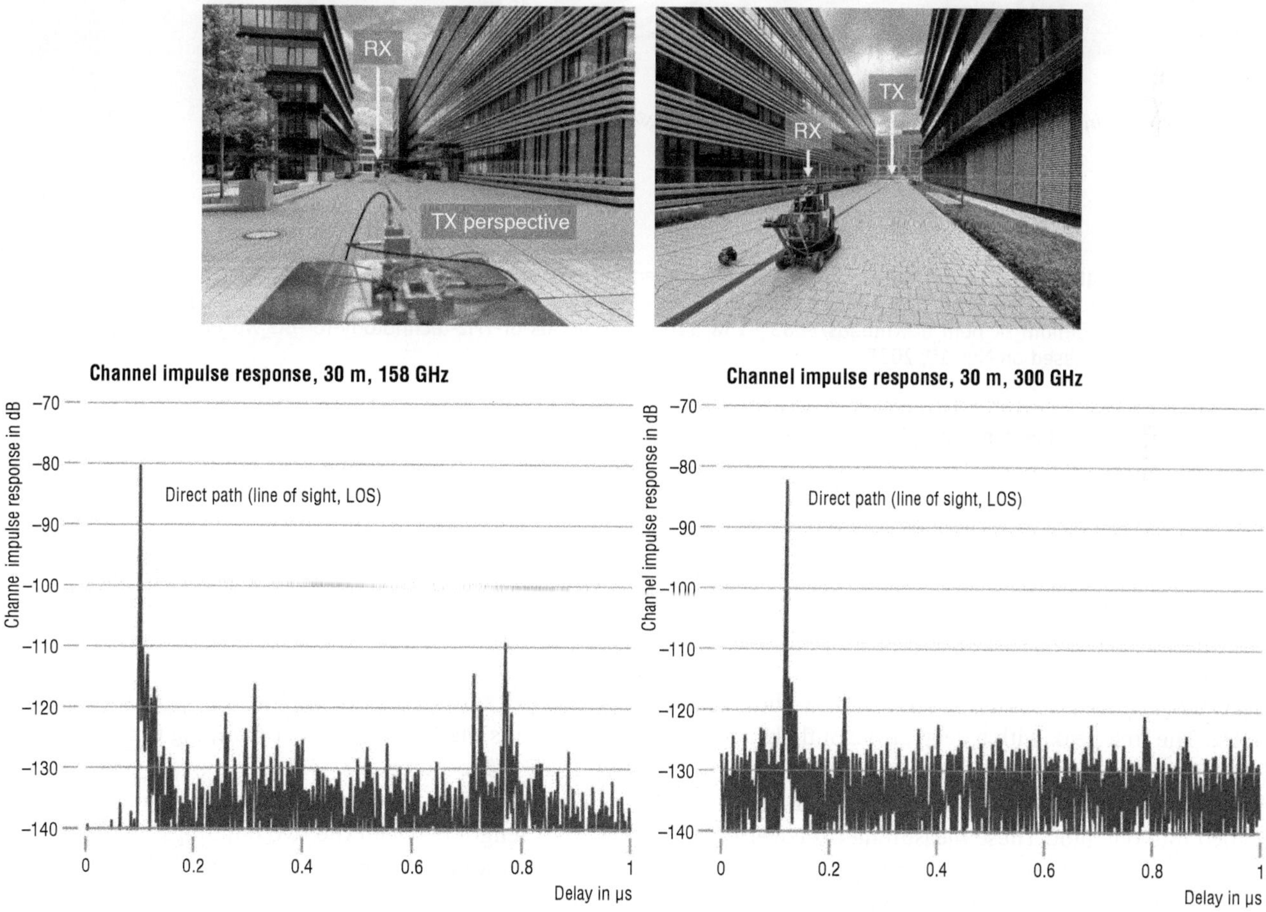

Figure E.31 Angle-resolved CIR THz channel measurements at 158 and 300 GHz (D band) in an outdoor street canyon environment at Rohde & Schwarz headquarters in Munich. The left photo shows the transmitter (TX) perspective at the end of the street canyon (see also Figure E.30). The right photo shows the RX setup mounted on a wireless platform (camera dolly) for measurements at variable positions. Measurement examples are shown below with channel impulse responses at 30 m distance from the transmitter at 158 GHz (left) and at 300 GHz at the same position (right). Note that 1 μs delay corresponds to 300 m distance. *Source*: Rohde & Schwarz GmbH&Co. KG / https://scdn.rohde-schwarz.com/ur/pws/dl_downloads/premiumdownloads/premium_dl_pdm_downloads/3683_6158_52/Fundamentals-of-THz-technology-for-6G_wp_en_3683-6158-52_v0102.pdf / last accessed on Nov 10, 2025.

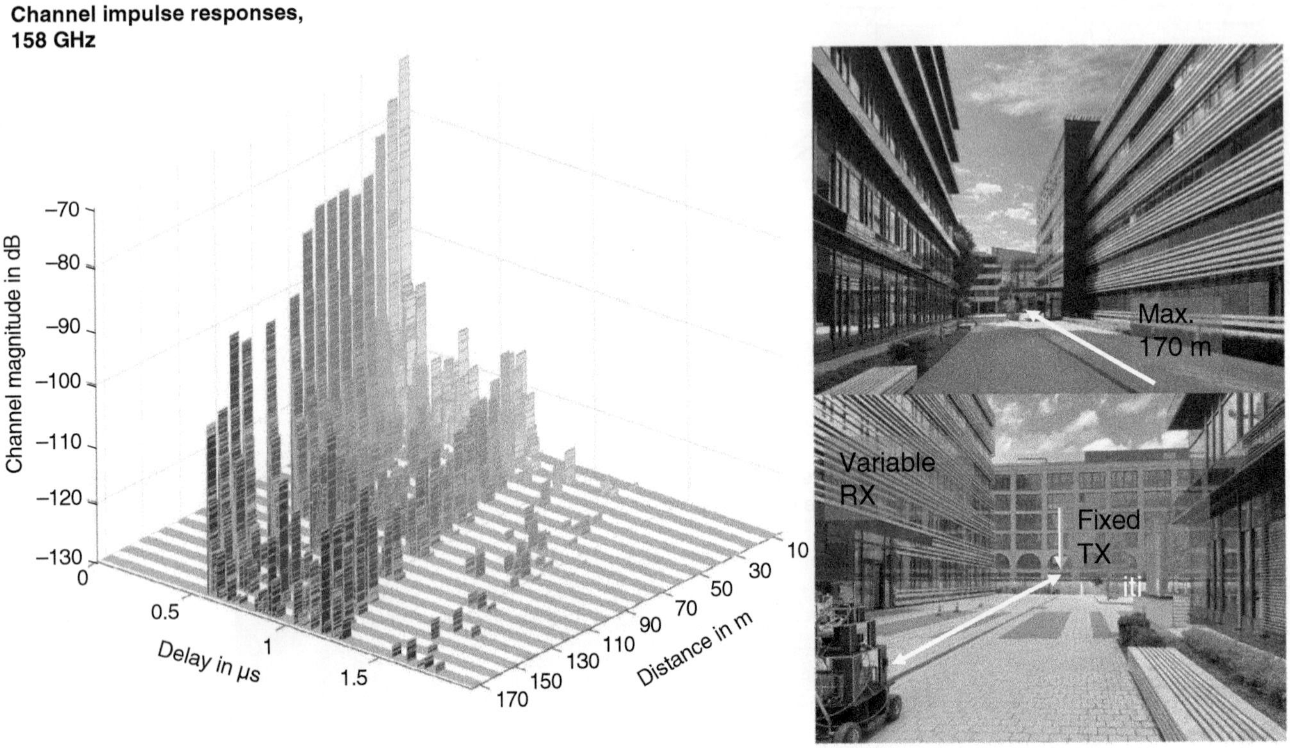

Figure E.32 Large-scale outdoor street canyon scenario measurements. The graph shows the CIRs at 158 GHz with aligned antennas over the distance from 10 to 170 m. *Source*: Rohde & Schwarz GmbH&Co. KG / https://scdn.rohde-schwarz.com/ur/pws/dl_downloads/premiumdownloads/premium_dl_pdm_downloads/3683_6158_52/Fundamentals-of-THz-technology-for-6G_wp_en_3683-6158-52_v0102.pdf / last accessed on Nov 10, 2025.

E.6.4.1 Outdoor Street Canyon Scenario (Urban Micro-cellular UMi)

The first measurement scenario was located in a corridor between two R&D buildings resembling a street canyon scenario (urban micro-cellular UMi), as shown in Figures E.30 and E.31. The stationary transmitter (TX) was placed at a height of 1.5 m at a bicycle stand (star on left side of Figure E.30) at the end of the corridor. There, the scenery leads into a more open space with a small plaza and isolated trees (Figure E.31). The receiver (RX) including the T&M equipment was mounted on a wireless platform (camera dolly) at a height of 1.5 m and positioned at measurement positions at various distances from the transmitter up to a maximum distance of 170 m (Figure E.32). Most of the measurements were in a line of sight (LOS) situation.

The example measurements in Figure E.31 show the CIR for aligned antennas at a distance of 30 m for 158 and 300 GHz. The first peak with a delay (time of flight) of 0.1 μs rep, represents the LOS path, corresponding to a distance of 30 m. Multipath components are also visible and more pronounced at 158 GHz than at 300 GHz.

Figure E.32 shows the instantaneous CIRs in an outdoor scenario at 158 GHz at various distances from 10 to 170 m combined into one plot. These measurements cover the complete street length between the two buildings. The antennas were always aligned. From this set of large-scale measurements, the path loss exponent can be derived. The multipath components at larger delays are present almost over the complete measurement setup.

E.6.4.1.1 Angle-resolved Measurements

Further evaluation of the measured datasets involved analysis of the angular information. At each measurement point, the receiver was rotated to 24 equally spaced angular positions, resulting in spatial scanning of the radio channel in the azimuth plane with 15° resolution.

Figure E.33 shows the result of a path estimation for both frequencies (158 and 300 GHz) at the same measurement point (30 m, outdoor). The angle axis represents a cyclic dimension, meaning the first and last rows in these plots are identical. This is also reflected by the colors. For these evaluations, proper control of the noise threshold is necessary in order to

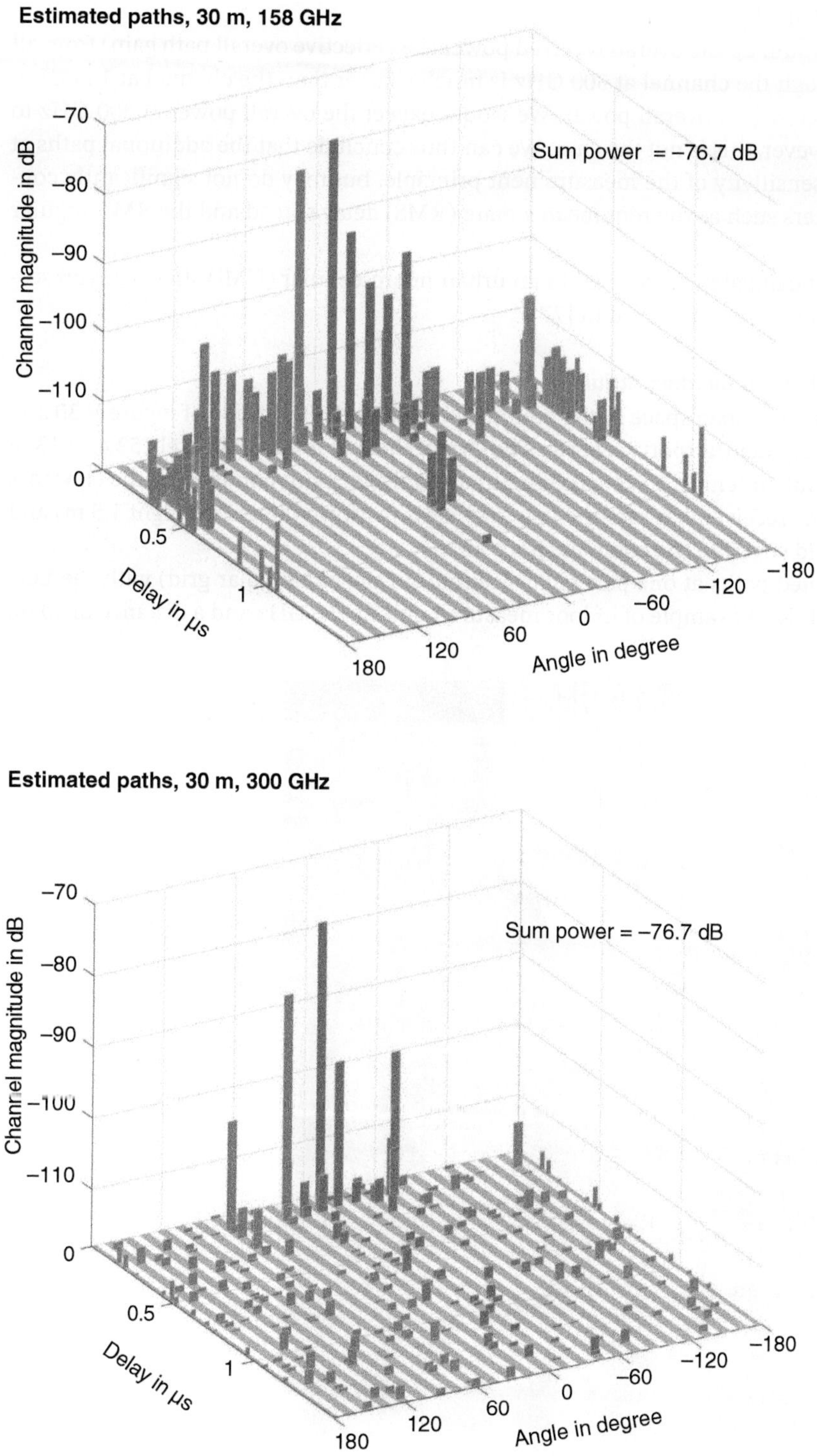

Figure E.33 Estimated paths in the delay angle domain at 30 m distance (outdoor) for 158 GHz (top) and 300 GHz (bottom).

clearly discriminate the signal paths from noise. For the example given in Figure E.33, the absolute noise thresholds were set to −120 dB for 158 GHz and to −118 dB for 300 GHz.

Based on this path estimation, it is possible to sum up the overall received power (i.e., effective overall path gain) from all paths. This is also indicated in the plots. Although the channel at 300 GHz is much sparser than the channel at 158 GHz, we can clearly see that there is not a large difference in overall power. We would expect the overall power at 300 GHz to be around 6 dB less compared to 158 GHz. However, this is not the case. We can thus conclude that the additional paths at 158 GHz could be resolved owing to the high sensitivity of the measurement principle, but they do not significantly contribute to the overall power. Statistical parameters such as the root mean square (RMS) delay spread and the RMS angular spread can also be evaluated from these results.

Further results for the channel measurement campaign performed in an urban micro-cellular (UMi) street canyon scenario at 158 and 300 GHz as described above are also summarized in [77].

E.6.4.2 Indoor Shopping Mall/Airport Scenario in R&D Building Atrium

The indoor measurements were performed in a large open space in the building shown on the right side of Figure E.30 and in Figure E.34. The scenario resembles a shopping mall/airport scenario. The size of the hall is approximately 52 m × 13 m with a ceiling height of around 20 m. The transmit antenna (TX) was configured at a fixed position in front of the elevators at the building entrance at a height of 1.5 m. The receiver was mounted on a wireless platform (antenna height 1.5 m) and moved to distinct positions on a rectangular grid covering the complete floor area.

Figure E.35 shows the angle-resolved estimated paths at one particular position (in the rectangular grid) with the LOS peak and a number of multipath components. It is an example of indoor measurements at 158 GHz and a distance of 40 m.

Figure E.34 Photograph of the indoor atrium scenario (shopping mall/airport) with the receiver on a rotation stage in the front facing toward the transmitter position at the elevators in the back (see also Figure E.30).

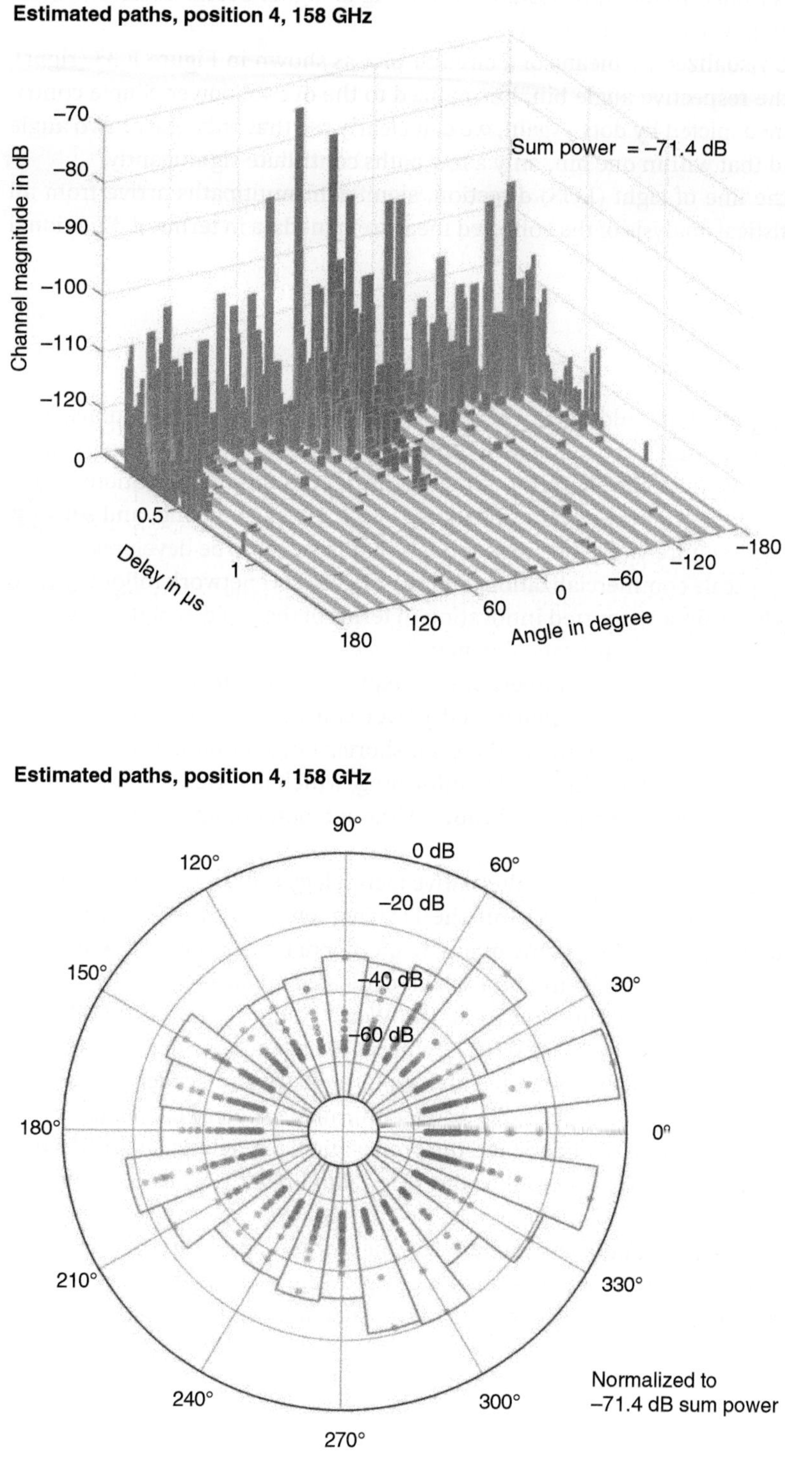

Figure E.35 Estimated paths of one particular indoor position at 158 GHz for the atrium scenario (shopping mall/airport) with an angular range covering −180° to 180°. In the circular plot on the bottom, the same data set is shown in a polar plot where each dot corresponds to one peak and the triangle ("pie") represents the overall power in the respective angle bin. A number of multipath components can be observed.

As expected, there are more multipath components from different directions contributing to higher overall received power compared to outdoor measurements (−71.4 dB at 40 m versus −75.1 dB at 30 m).

Another interesting result representation can be visualized by means of a circular plot as shown in Figure E.35 (right). Here, each "pie" represents the overall power in the respective angle bin, normalized to the overall power. Single contributions from distinct paths within one angle bin are depicted by dots. Again, we can clearly see that only one or two angle bins account for almost all of the overall power and that within one bin, only a few paths contribute significantly.

Although most of the power is received from the line of sight (LOS) direction, significant multipaths arrive from all measured azimuth directions. A more detailed statistical analysis of the collected measurement data in terms of directional channel gain can be found in [78].

E.7 Conclusion

THz technology and applications represent just one potential building block for future 6G wireless communications. Nevertheless, the technology promises to be indispensable—not only to achieve target requirements in terms of maximum throughput on the Tbit/s level as well as extremely low latencies, but also to realize intriguing novel applications. Envisioned 6G use cases span a plethora of applications ranging from communications, spectroscopy, imaging, and sensing. However, successful commercial implementation will require viable business models that have yet to be developed.

THz technology may need some time to reach large-scale commercialization after the targeted 6G network rollout around 2030. Nevertheless, the initiated research activities have already spurred innovation in terms of the rapid evolution of high-frequency semiconductor technologies for RF electronics as well as promising new devices.

While semiconductor technology continues to evolve toward extremely high frequencies, the technical challenges are known and inherited from mmWaves. Namely, they involve efficiency and power consumption. In comparison to mmWaves, the challenges only intensify in the THz spectral region due to the even shorter range of radio wave propagation. However, such problems can be mitigated by focusing signals through beamforming, which involves accommodating an even larger number of antennas in a small volume to produce pinpoint beams. Although outdoor and indoor THz use cases are feasible, indoor use cases are likely to dominate.

Using photonic technologies to generate THz waves could represent an alternative technology evolution branch with the potential to transfer the existing power efficiency of optoelectronic devices into the THz frequency region. Furthermore, by miniaturizing today's lab setups into PICs, these approaches could become mainstream. It will be fascinating to see which technology will eventually prevail or which of the approaches will be used for which dedicated application.

The next milestone for initial 6G technology trials and feasibility studies is the World Radio Conference (WRC) in 2023. While decisions about the 6G spectrum will not be made in 2023, WRC23 will set the agenda for the next WRC in 2027—when 6G standardization will be in full swing. That is why research activities in industry and academia are striving to have proof in hand by 2023 that THz communications are both usable and technically feasible. This includes THz channel measurement campaigns as described in the last section to understand the propagation characteristics and develop channel models for this novel communications frequency region—as well as the THz radio links that have already been demonstrated by various companies and research institutions.

Rohde & Schwarz has been a close partner and leading test and measurement supplier for industry and academia since the beginning of the digital wireless communications era. Today, the company is already providing solutions and expertise to accompany 6G research projects, including THz research, and pave the way toward commercialization of the next wireless communications standard.

References

1 NTT DOCOMO, Inc. (2022, January). 5G Evolution and 6G. White Paper. (Version 4.0).

2 Latva-aho, M. and Leppänen, K. (Eds.) (2019). *Key Drivers and Research Challenges for 6G Ubiquitous Wireless Intelligence*. 6G Flagship Research Program. Finland: University of Oulu.

3 Samsung (2021). 6G the Next Hyper-Connected Experience for All. White Paper.

4 The 5G Infrastructure Association (2021). European Vision for the 6G Network Ecosystem. *White Paper*.

5 Nokia Bell Labs (2021). Communications in the 6G Era. White Paper.

6 Pärssinen, A., Alouini, M., Berg, M. et al. (eds.) (2021, April). *White Paper on RF Enabling 6G - Opportunities and Challenges from Technology to Spectrum*. 6G Research Visions, No. 13, 6G Flagship, University of Oulu.
7 MediaTek Inc. (2022, January). 6G Vision Whitepaper.
8 IMT-2030 (6G) Promotion Group (2021, June). 6G Vision and Candidate Technologies. China: CAICT, MIIT.
9 National Institute of Information and Communications Technology (NICT) (2021, August). Beyond 5G/6G White Paper. Tokyo: NICT.
10 KDDI Research, Inc. (2021, October). Beyond 5G/6G White Paper ver.2.0.1.
11 Innovative Optical Wireless Network Global Forum (IOWN GF) (2021). IOWN GF System and Technology Outlook. IOWN GF.
12 5G Americas (2021, November). Mobile Communications towards 2030 (White Paper).
13 Chun, H., Gomez, A., Quintana, C. et al. (2019). A wide-area coverage 35 Gb/s visible light communications link for indoor wireless applications. *Nature Scientific Reports* 9: 4952.
14 Yin, J., Cao, Y., Li, Y.H. et al. (2017). Satellite-based entanglement distribution over 1200 kilometers. *Science* 356 (6343): 1140–1144.
15 National Institute of Information and Communications Technology (NICT) (2021, August). Quantum Network White Paper. Tokyo: NICT.
16 ETSI (2021, October). Industry Specification Group (ISG) on Reconfigurable Intelligent Surfaces (RIS). https://www.etsi.org/committee/ris (accessed 10 July 2026).
17 Joshi, V., Le Gallo, M., Haefeli, S. et al. (2020). Accurate deep neural network inference using computational phase- change memory. *Nature Communications* 11: 2473.
18 Sebastian, A., Le Gallo, M., Khaddam-Aljameh, R. et al. (2020). Memory devices and applications for in-memory computing. *Nature Nanotechnology* 15: 529–544.
19 Luo, F.-L. (2020). *Machine Learning for Future Wireless Communications*. Wiley-IEEE Press.
20 ETSI White Paper #34 (2020, June). Artificial Intelligence and Future Directions for ETSI. ETSI.
21 Shannon, C.E. (1948). A mathematical theory of communication. *The Bell System Technical Journal* 27: 379–423, 623–656.
22 ITU Radiocommunication (ITU-R) (2020). Radio Regulations, Articles, Edition of 2020. ITU.
23 ITU (2019, October-Novomber). World Radiocommunication Conference 2019 (WRC19) Final Acts. ITU.
24 IEEE 802.15 WPAN™ Terahertz Interest Group (IGthz). https://grouper.ieee.org/groups/802/15/pub/SCTHz.html (accessed 10 July 2026).
25 Naftaly, M., Vieweg, N., and Deninger, A. (2019). Industrial applications of terahertz sensing: state of play. *Sensors* 19: 4203.
26 VDI / VDE (March 2020). Terahertzsysteme und Anwendungsfelder, VDI-Statusreport. VDI / VDE-Gesellschaft Mess- und Automatisierungstechnik.
27 Wilson, R.W., Jefferts, K.B., and Penzias, A.A. (1970, July). Carbon monoxide in the Orion Nebula. *Astrophysical Journal* 161: L43–L44.
28 Thomas, B., Brandt, M., Goliasch, A. et al. (2017). The ice cloud imager front end receivers onboard MetOp-SG satellite – preliminary design and results. In *28th International Symposium on Space Terahertz Technology (ISSTT 2017)*, Cologne, Germany.
29 Francis, J.A. (2021, November). Vapor storms. *Scientific American* 325 (5): 26–33.
30 Thomas, B., Philipp, M., Lichius, K. et al. (2019). Digitally tunable 150 GHz local oscillator chain for the submillimeter wave instrument onboard the ESA JUICE mission. In *Proceedings of the 30th International Symposium on Space Terahertz Technology (ISSTT2019)*, Gothenburg, Sweden.
31 Ericsson (2017, February). Microwave backhaul beyond 100 GHz. Ericsson Technology Review.
32 Hamza, A.S., Deogun, J.S., and Alexander, D.R. (2016). Wireless communication in data centers: a survey. *IEEE Communications Surveys & Tutorials* 18 (3): 1572–1595.
33 The Event Horizon Telescope Collaboration, Akiyama, K., Alberdi, A. et al. (2019). First M87 Event Horizon Telescope Results. IV. Imaging the central supermassive black hole. *The Astrophysical Journal Letters* 875 (1): L4.
34 Tonouchi, M. (2007, February). Cutting-edge terahertz technology. *Nature Photonics* 1: 97–105.
35 Rumiantsev, A., Swe, T.N., and Henkel, A. (2021, December). Achieving metrology-level accuracy when making THz measurements. *Microwave Journal*. The Latest in mmWave and THz Test and Measurement Technology. https://www.microwavejournal.com/articles/27082.
36 de Graauw, Th., Helmich, F.P., Phillips, T.G. et al. (2010, July). The Herschel-Heterodyne Instrument for the Far-Infrared (HIFI). *Astronomy and Astrophysics* 518: L6.

37 Stuhlfauth, R. and Mellein, H. (2021). Over-the-Air RF Conformance Measurements on 5G NR Devices. Rohde & Schwarz white paper.

38 Derat, B., Schmitz, S., Lachner, S. et al. (2020). Numerical modeling and experimental validation of a D-band lens-based antenna design for beyond 5G communications. In *2020 Antenna Measurement Techniques Association Symposium (AMTA)*, (1–5).

39 Faist, J., Capasso, F., Sivco, D.L. et al. (1994). Quantum cascade laser. *Science* 264 (5158): 553–556.

40 Köhler, R., Tredicucci, A., Beltram, F. et al. (2002). Terahertz semiconductor-heterostructure laser. *Nature* 417: 156–159.

41 Fujita, K., Hayashi, S., Ito, A. et al. (2019). Sub-terahertz and terahertz generation in long-wavelength quantum cascade lasers. *Nanophotonics* 8 (12): 2235–2241.

42 Khalatpour, A., Paulsen, A.K., Deimert, C. et al. (2021). High-power portable terahertz laser systems. *Nature Photonics* 15: 16–20.

43 Nagatsuma, T., Ducournau, G., and Renaud, C. (2016). Advances in terahertz communications accelerated by photonics. *Nature Photonics* 10: 371–379.

44 Liebermeister, L., Nellen, S., Globisch, B. et al. (2021). Optoelectronic frequency-modulated continuous-wave terahertz spectroscopy with 4 THz bandwidth. *Nature Communications* 12: 1071.

45 Nellen, S., Ishibashi, T., Deninger, A. et al. (2020). Experimental comparison of UTC- and PIN-photodiodes for continuous-wave terahertz generation. *Journal of Infrared, Millimeter, and Terahertz Waves* 41: 343–354.

46 Ishibashi, T., Muramoto, Y., Yoshimatsu, T., and Ito, H. (2014). Unitraveling-carrier photodiodes for terahertz applications. *IEEE Journal of Selected Topics in Quantum Electronics* 20 (6): 79–88.

47 Ishibashi, T. and Ito, H. (2022, March-April). Uni-traveling carrier photodiodes: development and prospects. *IEEE Journal of Selected Topics in Quantum Electronics* 28 (2): 1–6.

48 Song, H., Ajito, K., Muramoto, Y. et al. (2012, July). Uni-travelling-carrier photodiode module generating 300 GHz power greater than 1 mW. *IEEE Microwave and Wireless Components Letters* 22 (7): 363–365.

49 Latzel, P., Pavanello, F., Billet, M. et al. (2017, November). Generation of mW level in the 300-GHz band using resonant-cavity-enhanced unitraveling carrier photodiodes. *IEEE Transactions on Terahertz Science and Technology* 7 (6): 800–807.

50 Webber, J., Yamagami, Y., Ducournau, G. et al. (2021, December). Terahertz band communications with topological valley photonic crystal waveguide. *Journal of Lightwave Technology* 39 (24): 7609–7620.

51 Gotti, R., Puppe, T.A., Mayzlin, Y. et al. (2020). Comb-locked frequency-swept synthesizer for high precision broadband spectroscopy. *Nature Scientific Reports* 10: 2523.

52 Fortier, T. and Baumann, E. (2019). 20 years of developments in optical frequency comb technology and applications. *Communication Physics* 2: 153.

53 Fernandez Olvera, A.D.J., Krause, B.L., and Preu, S. (2021). A true optoelectronic spectrum analyzer for millimeter waves with Hz resolution. *IEEE Access* 9: 114339–114347.

54 Fernandez Olvera, A.D.J., Mukherjee, A.K., and Preu, S. (2021, October). A fully optoelectronic continuous-wave 2-port vector network analyzer operating from 0.1 THz to 1 THz. *IEEE Journal of Microwaves* 1 (4): 1015–1022.

55 Chinni, V.K., Latzel, P., Zegaoui, M. et al. (2018). Single-channel 100 Gbit/s transmission using III–V UTC-PDs for future IEEE 802.15.3d wireless links in the 300 GHz band. *Electronics Letters* 54: 638–640.

56 Nagatsuma, T., Fujita, Y., Yasuda, Y. et al. (2016). Real-time 100-Gbit/s QPSK transmission using photonics-based 300-GHz-band wireless link. In *2016 IEEE International Topical Meeting on Microwave Photonics (MWP)*, (27–30). IEEE.

57 Castro, C., Nellen, S., Elschner, R. et al. (2019). 32 GBd 16QAM wireless transmission in the 300 GHz band using a PIN diode for THz upconversion. In *Optical Fiber Communications Conference and Exhibition (OFC)*, (1–3). IEEE.

58 Dan, I., Ducournau, G., Kalfass, I. et al. (2020, May). A 300-GHz wireless link employing a photonic transmitter and an active electronic receiver with a transmission bandwidth of 54 GHz. *IEEE Transactions on Terahertz Science and Technology* 10 (3): 271–281.

59 Koenig, S., Lopez-Diaz, D., Antes, J. et al. (2013). Wireless sub-THz communication system with high data rate. *Nature Photonics* 7: 977–981.

60 Stanze, D., Deninger, A., Roggenbuck, A. et al. (2011). Compact CW terahertz spectrometer pumped at 1.5 μm wavelength. *Journal of Infrared, Millimeter, and Terahertz Waves* 32: 225–232.

61 Vieweg, N., Rettich, F., Deninger, A. et al. (2014). Terahertz-time domain spectrometer with 90 dB peak dynamic range. *Journal of Infrared, Millimeter, and Terahertz Waves* 35: 823–832.

62 Roccaforte, F. and Leszczynski, M. (2020). *Nitride Semiconductor Technology: Power Electronics and Optoelectronic Devices*. Wiley-VCH Verlag GmbH & Co. KGaA.

63 Nakamura, S. (2015). Nobel lecture: Background story of the invention of efficient blue InGaN light emitting diodes. *Reviews of Modern Physics* 87: 1139.

64 Cwiklinski, M. (2021). Design of millimeter-wave power amplifiers in GaN HEMT technology. Ph.D. Dissertation. Fraunhofer Verlag.

65 Wang, H., Eleraky, M., Abdelaziz, B. et al. Power Amplifiers Performance Survey 2000-Present. ETH Zurich & Georgia Institute of Technology, 2025. https://gems.ece.gatech.edu/PA_survey.html (accessed 10 July 2026).

66 Gashi, B., John, L., Meier, D. et al. (2021, November). Broadband 400-GHz InGaAs mHEMT transmitter and receiver S-MMICs. *IEEE Transactions on Terahertz Science and Technology* 11 (6): 660–675.

67 Mei, X., Yoshida, W., Lange, M. et al. (2015, April). First demonstration of amplification at 1 THz using 25-nm InP high electron mobility transistor process. *IEEE Electron Device Letters* 36 (4): 327–329.

68 ITU-R (2022). Draft working document towards a preliminary draft new Report ITU-R M. 2541-0. Technical Feasibility of IMT in Bands Above 100 GHz. ITU.

69 3rd Generation Partnership Project (3GPP) (2019, October). 3GPP TR 38.901 (V16.0.0): Study on channel model for frequencies from 0.5–100 GHz.

70 Salous, S. (2013). *Radio Propagation Measurement and Channel Modelling*. Wiley.

71 Peter, M., Keusgen, W., Eichler, T. et al. (2018). High-resolution directional channel measurements at 67 GHz and advanced analysis of interactions using geometric information. In *2018 IEEE International Symposium on Antennas and Propagation & USNC/URSI National Radio Science Meeting*, (77–78).

72 Schmieder, M., Eichler, T., Wittig, S. et al. (2020). Measurement and characterization of an indoor industrial environment at 3.7 and 28 GHz. In *2020 14th European Conference on Antennas and Propagation (EuCAP)*, (1–5).

73 Schmieder, M., Keusgen, W., Peter, M. et al. (2020). THz channel sounding: Design and validation of a high performance channel sounder at 300 GHz. In *2020 IEEE Wireless Communications and Networking Conference Workshops (WCNCW)*, (1–6). IEEE.

74 Undi, F., Schultze, A., Keusgen, W. et al. (2021). Angle-resolved THz channel measurements at 300 GHz in an outdoor environment. In *2021 IEEE International Conference on Communications Workshops (ICC Workshops)*, (1–7).

75 Schultze, A., Undi, F., Peter, M. et al. (2021). Angle-resolved THz channel measurements at 300 GHz in a conference room environment. In *2021 XXXIVth General Assembly and Scientific Symposium of the International Union of Radio Science (URSI GASS)*, (1–4).

76 Wittig, S., Keusgen, W., Peter, M. et al. (2020). On the dynamic range of digital correlative time domain radio channel measurements. *arXiv:2008.07805*.

77 Keusgen, W., Schultze, A., Peter, M. et al. (2022). Sub-THz channel measurements at 158 GHz and 300 GHz in a street canyon environment. In *Submitted to 2022 Joint European Conference on Networks and Communications & 6G Summit (EuCNC/6G Summit)*, Grenoble. https://arxiv.org/abs/2203.04404.

78 Schultze, A., Keusgen, W., Peter, M. et al. (2022). Observations on the angular statistics of the indoor sub-THz radio channel at 158 GHz. In *2022 IEEE USNC-URSI Radio Science Meeting (Joint with AP-S Symposium)*, (9–10). https://doi.org/10.23919/USNC-URSI52669.2022.9887443. https://ieeexplore.ieee.org/document/9887443.

Note: All links have been checked and were functional when this document was created. However, we cannot rule out subsequent changes to the links in the reference list.

Rohde & Schwarz

The Rohde & Schwarz technology group is among the trailblazers when it comes to paving the way for a safer and connected world with its leading solutions in test and measurement, technology systems and networks and cybersecurity.

Founded more than 85 years ago, the group is a reliable partner for industry and government customers around the globe. The independent company is headquartered in Munich, Germany, and has an extensive sales and service network with locations in more than 70 countries.

www.rohde-schwarz.com

Rohde & Schwarz customer support

www.rohde-schwarz.com/support

F

Noise Figure Test Under Pulsed Conditions and Compression, Utilizing a Vector Network Analyzer

Through reciprocal exchange in response to evolving technological requirements, vector network analyzers (VNA) have mutated from purely S-parameter test units into versatile device test solutions. Especially, the so-called "high-end" class VNAs provide a comprehensive feature range for an extensive characterization of active devices. In this connection, the noise figure (NF) testing plays a particularly important role, due to its complexity and importance for a multitude of applications, such as testing of low noise amplifiers (LNAs), T/R-modules, satellite receivers, and OnWafer chip testing in general. Moreover, the capability to combine measurement features is crucial for an extensive device characterization.

In this regard, the present chapter reveals how the VNA noise figure test on an LNA under pulsed RF conditions can be configured and performed.

Exemplary measurements have been carried out utilizing the high-end vector network analyzer R&S®ZNA from Rohde & Schwarz. This VNA stands out for an excellent RF performance and can be tailored to the specific task by means of a wide range of configurable test personalities/options. All functionalities to optimize such a VNA for pulsed NF testing and the resulting performance are outlined in the following (Figure F.1).

Measurement results are presented, captured on an LNA with ~10 dB gain and NF $\lesssim$ 1 dB, in $f = 2$–7 GHz.

F.1 Theoretical Introduction

F.1.1 NF of Cascaded Networks

In a system with many circuits connected in cascade, we must consider the contributions of the various circuits (Figure F.2). Considering the equivalent noise resistor R_n in series with the input circuit, we have

$$F = \frac{R_G + R_n}{R_G}$$

$$F = 1 + \frac{R_n}{R_G}$$

The excess noise added by the circuit is R_n/R_G.

In considering two cascaded circuits a and b, by definition, the available noise at the output of b is

$$N_{ab} = F_{ab}G_{ab}kTB$$

where B is the equivalent noise bandwidth in which the noise is measured.

The total available gain G is the product of the individual available gains, so

$$N_{ab} = F_{ab}G_aG_bkTB$$

The available noise from network a at the output of network b is

$$N_{a/b} = N_aG_b = F_aG_aG_bkTB$$

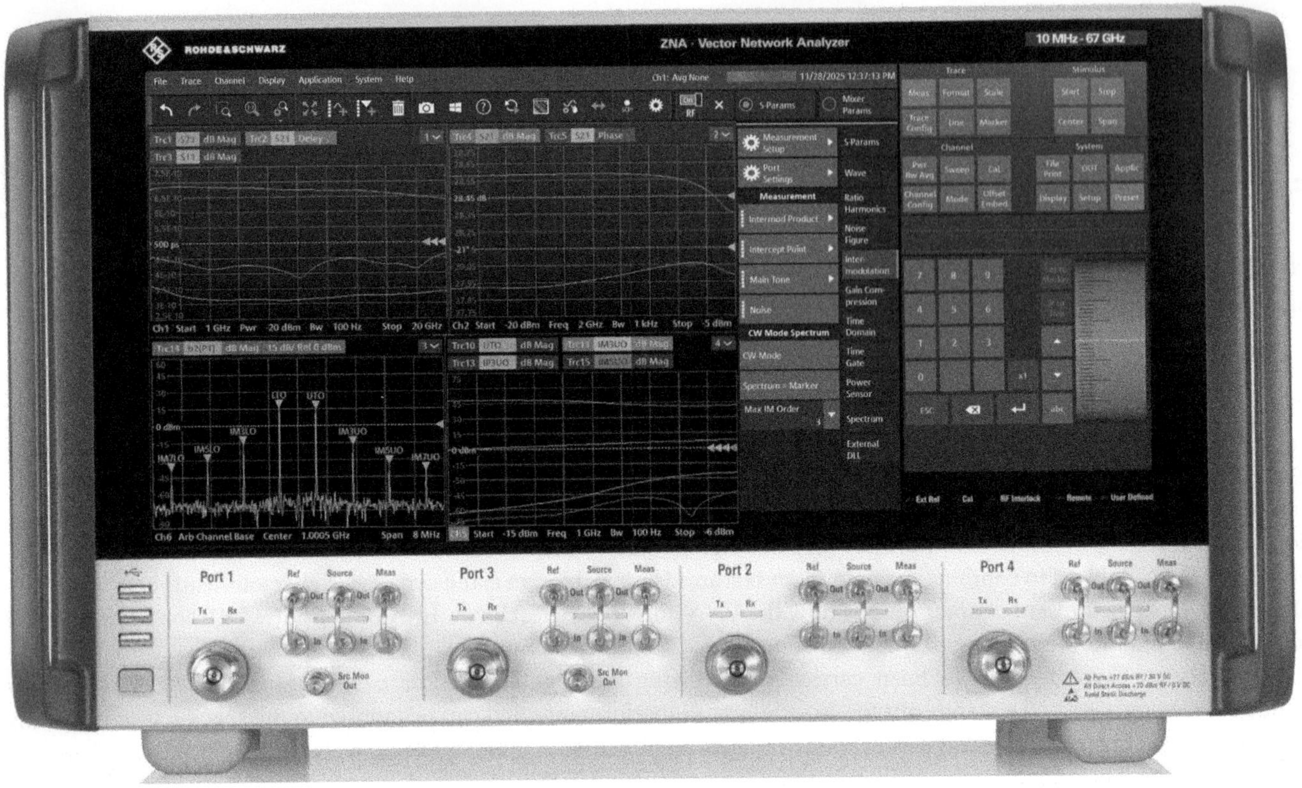

Figure F.1 Vector network analyzer R&S®ZNA, high-end class, with 2 or 4 test ports.

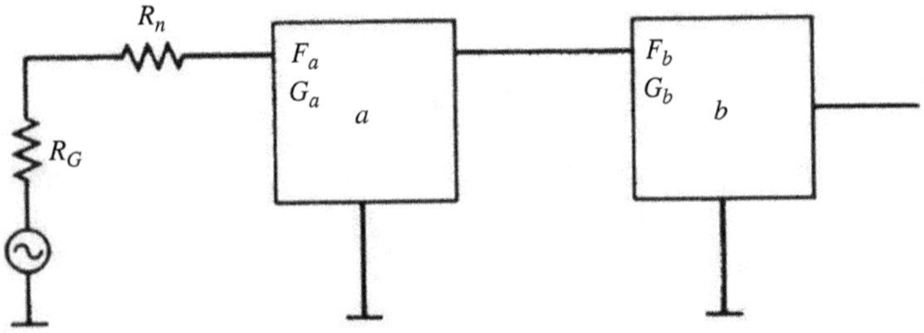

Figure F.2 Cascaded noisy two-ports with noise figures F_a and F_b and gain figures G_a and G_b.

The available noise added by network b (its excess noise) is

$$N_{b/b} = (F_b - 1)G_b kTB$$

The total available noise N_{ab} is the sum of the available noise contributed by the two networks:

$$\begin{aligned} N_{ab} &= N_{a/b} + N_{b/b} = F_a G_a G_b kTB + (F_b - 1)G_b kTB \\ &= \left(F_a + \frac{F_b - 1}{G_a}\right) G_a G_b kTB \\ &\text{let: } F_{ab} = F_a + \frac{F_b - 1}{G_a} \end{aligned}$$

For any number of circuits, this can be extended as follows:
Careful, this calculation refers to "Noise Factor F." The conventional term noise figure NF means

$$\mathrm{NF} = 10 * \log(F)$$

$$F = F_1 + \frac{F_2 - 1}{G_1} + \frac{F_3 - 1}{G_1 G_2} + \frac{F_4 - 1}{G_1 G_2 G_3} + \cdots$$

When considering a long chain of cascaded amplifiers, there will be a minimum NF achievable for this chain. This is a figure of merit and was proposed by Haus and Adler. It is calculated by rearranging the above equation:

$$(F_{\mathrm{tot}})_{\min} = (F_{\min} - 1) + \frac{F_{\min} - 1}{G_A} + \frac{F_{\min} - 1}{G_A^2} + \cdots + 1$$

where $F_{\min}$ is the minimum NF for each stage and G_A is the available power gain of the identical stages. Using

$$\frac{1}{1 - X} = 1 + X + X^2 + \cdots$$

We find a quantity $(F_{\mathrm{tot}} - 1)$, which is defined as the noise measure M. The minimum noise measure

$$(F_{\mathrm{tot}})_{\min} - 1 = \frac{F_{\min} - 1}{1 - 1/G_A} = M_{\min}$$

refers to the noise of an infinite chain of tuned optimally, low-noise stages. This represents a lower limit on the noise of an amplifier.

The minimum noise measure $M_{\min}$ is an invariant parameter and is not affected by feedback. It is somewhat similar to a gain-bandwidth product in its use as a system invariant. The minimum noise measure is achieved when the amplifier is tuned for the available power gain and $\Gamma_G = \Gamma_{\mathrm{on}}$.

From

$$\text{Power gain} = \frac{\text{Power delivered to load}}{\text{Power input to network}}:$$

$$G = \frac{|S_{21}|^2\left(1 - |\Gamma_L|^2\right)}{\left(1 - |S_{11}|^2\right) + |\Gamma_L|^2\left(|S_{22}|^2 - |D|^2\right) - 2\mathrm{Re}(\Gamma_L N)}$$

$$\text{Available power gain} = \frac{\text{Power available from network}}{\text{Power available from source}}:$$

$$G_A = \frac{|S_{21}|^2\left(1 - |\Gamma_S|^2\right)}{\left(1 - |S_{22}|^2\right) + |\Gamma_S|^2\left(|S_{11}|^2 - |D|^2\right) - 2\mathrm{Re}(\Gamma_S M)}$$

$$\text{Transducer power gain} = \frac{\text{Power delivered to load}}{\text{Power available from source}}:$$

$$G_T = \frac{|S_{21}|^2\left(1 - |\Gamma_S|^2\right)\left(1 - |\Gamma_L|^2\right)}{|(1 - S_{11}\Gamma_S)(1 - S_{22}\Gamma_L) - S_{12}S_{21}\Gamma_L\Gamma_S|^2}$$

Unilateral transducer power gain ($S_{12} = 0$):

$$G_{Tu} = \frac{|S_{21}|^2\left(1 - |\Gamma_S|^2\right)\left(1 - |\Gamma_L|^2\right)}{|1 - S_{11}\Gamma_S|^2|1 - S_{22}\Gamma_L|^2}$$

$$= G_0 G_1 G_2$$

$$G_0 = |S_{21}|^2$$

$$G_1 = \frac{1 - |\Gamma_S|^2}{|1 - S_{11}\Gamma_S|^2}$$

$$G_2 = \frac{1 - |\Gamma_L|^2}{|1 - S_{22}\Gamma_L|^2}$$

Maximum unilateral transducer power gain when $|S_{11}| < 1$ and $|S_{22}| < 1$:

$$G_u = \frac{|S_{21}|^2}{\left|\left(1 - |S_{11}|^2\right)(1 - |S_{22}|)^2\right|}$$

$$= G_0 G_{1\max} G_{2\max}$$

$$G_{i\max} = \frac{1}{1 - |S_{ii}|^2} \quad \text{where } i = 1, 2$$

This set of equations is very important as a non-purely resistive generator and load for this measurement, in reality, must be taken into account to be able to produce a valid result.

F.1.2 Why Pulsed NF Measurements?

Let us assume a simple, non-optimized feedback amplifier, which operates initially simply as a medium-gain/medium-noise amplifier and at room temperature (Figure F.3).

Now, if we sweep the NF as a function of a particular RF interfering signal, then the gain and noise will change. Even before the 1 dB compression point, the frequency-dependent gain of this non-optimized amplifier changes, but the frequency-dependent NF changes much more. Because of the higher input/output power, the transistor temperature will increase, but the gain compression will reduce the output power.

As is predictable, the gain will go down, and the NF will go up. This, in part, has to do with the optimum noise match Γopt changes too.

The following graphs were obtained from a CAD simulation using a harmonic-balance-based optimized stimulator. The test setup to follow is to validate this phenomenon (Figures F.4–F.6).

Summing up the previous sections, it can be stated that the NF of an amplifier increases already at RF levels, where no significant reduction of the gain has yet been observed. Furthermore, already in this overdrive range, the NF increases with increased frequency, and the matching is reduced. To evaluate these intricate correlations, measurements under pulsed RF—as described in the following—reveal valuable information for the amplifier optimization.

F.2 VNA Noise Figure Test Implementation—General

Typically used with spectrum analyzers, the Y-factor technique utilizes a noise source connected to the input of the device under test (DUT) and measures the output noise of the DUT under the ON and OFF states of the noise source. In contrast to this, the NF personality of the utilized VNA is based upon the cold source technique, which terminates the input of the DUT with a resistance equal to the system impedance and measures the output noise power of the DUT. In addition, the gain of the DUT is measured, and for calibration (w/o DUT), the terminated VNA receiver noise power is determined. Based on these results, the NF is calculated. This VNA solution stands out due to the following aspects:

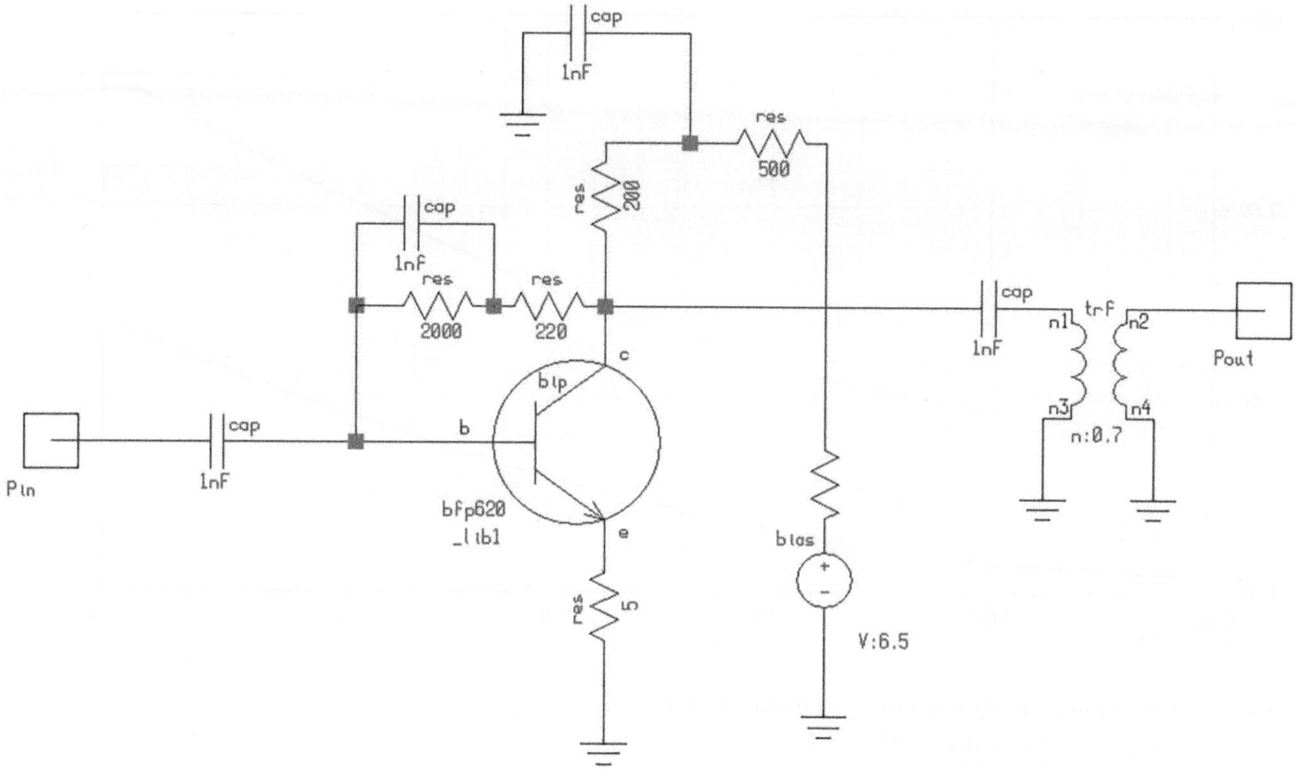

Figure F.3 Circuit (exemplary) of a not optimized feedback amplifier.

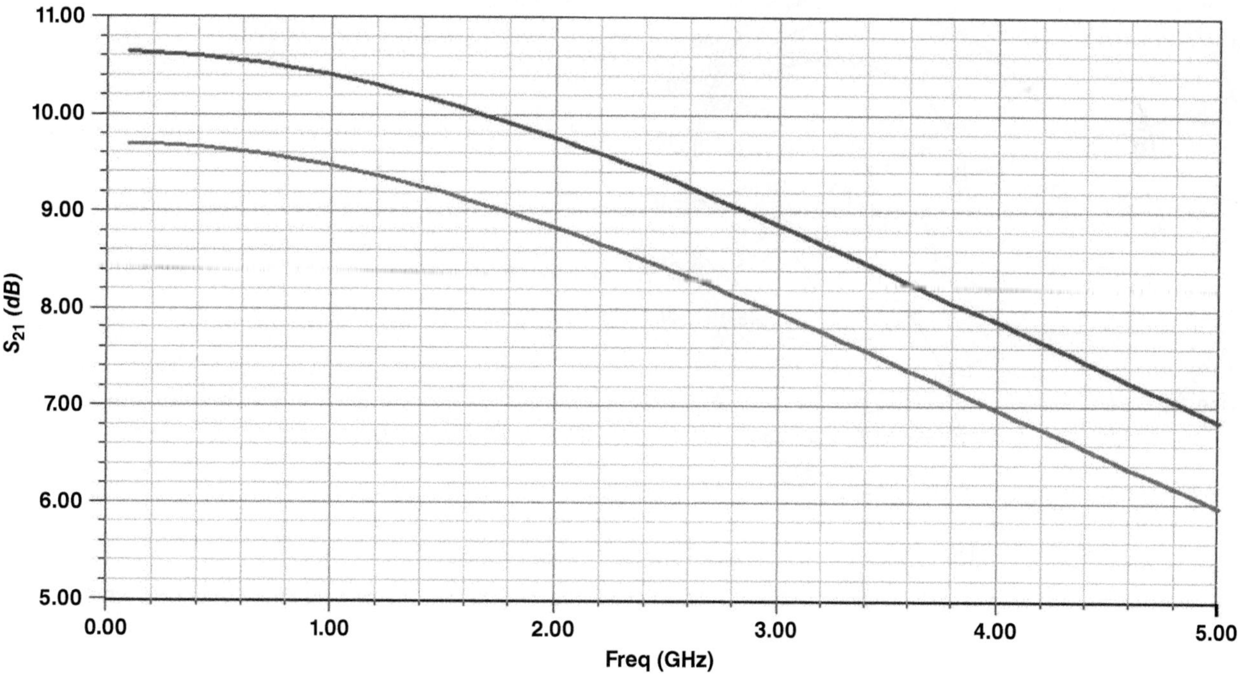

Figure F.4 Gain reduction as the power dissipation and the operation temperature increase, particularly at the 1 dB compression point. As we will see, the frequency dependency is also affected.

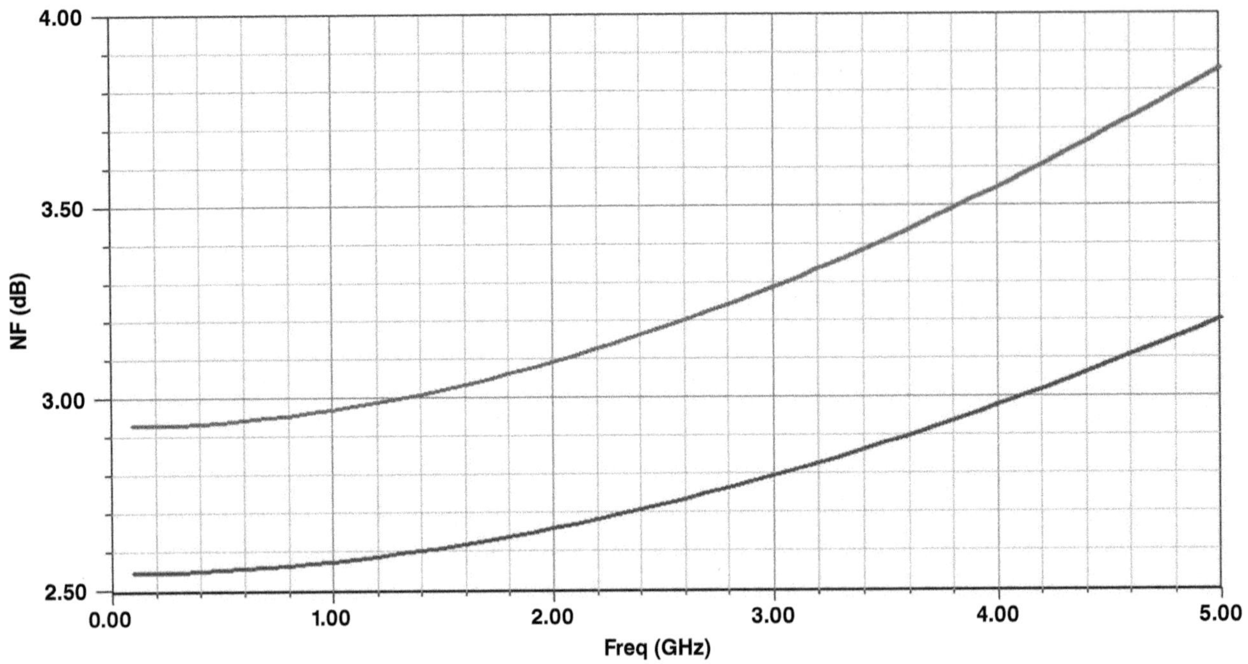

Figure F.5 Frequency dependence of the NF in the overdrive range.

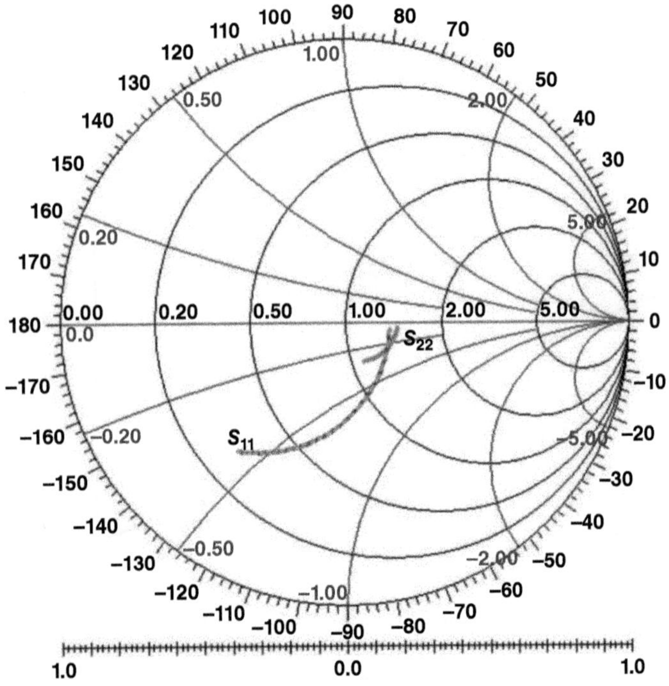

Figure F.6 Frequency dependence of the matching of an amplifier, when driven in the over-range (amplifier not optimized). The frequency range is 100 MHz–5 GHz.

- No noise source is required, i.e., the calibration and measurement do not require any additional component compared to a system error correction w/o NF calibration
- The NF test can be carried out with the DUT permanently connected to the test cables; no additional (dis)connection of the noise source is required

- The VNA result—especially the required gain measurement—includes the system error correction, which overcomes accuracy degradations caused by finite (source) port matching
- Because the gain measurement is a basic VNA function, the cold source technique is a technique preferred for VNAs
- The flexible combination of dedicated HW features (such as pre-amplification, flexible signal path routing, etc.) allows for an optimized configuration of the test instrument corresponding to the type and characteristics of the DUT

F.3 VNA Noise Figure Test Setup for Testing an LNA

F.3.1 External Test Setup Components

In the following, a common external setup for making an NF measurement with a VNA.

(Exemplary: R&S®ZNA. The NF test by itself just requires a 2-port model. However, the depicted VNA with four ports and up to four RF sources plus high power source and eight receivers supports further test scenarios, e.g., intermodulation test on a converter with an external LO signal).

The LNA setup requires two test cables, partially a 10 dB attenuator, a power sensor, and the calibration standards, which are most conveniently provided with a USB-based automatic calibration unit. An alternative to the automatic calibration unit is a manual calibration kit. Further, a 50 Ω termination will be required for the receiver noise calibration (e.g., from a manual calibration kit) (Figure F.7).

F.3.2 Additional HW VNA Features for NF Testing

As mentioned, the NF test presupposes hard- and software personalities, and further optional hardware add-ons help to optimize the source and receive port configuration for handling low stimulus power, minimizing the receiver NF, etc. At which Port 1 and Port 2 are designated to act as source and receive ports, respectively.

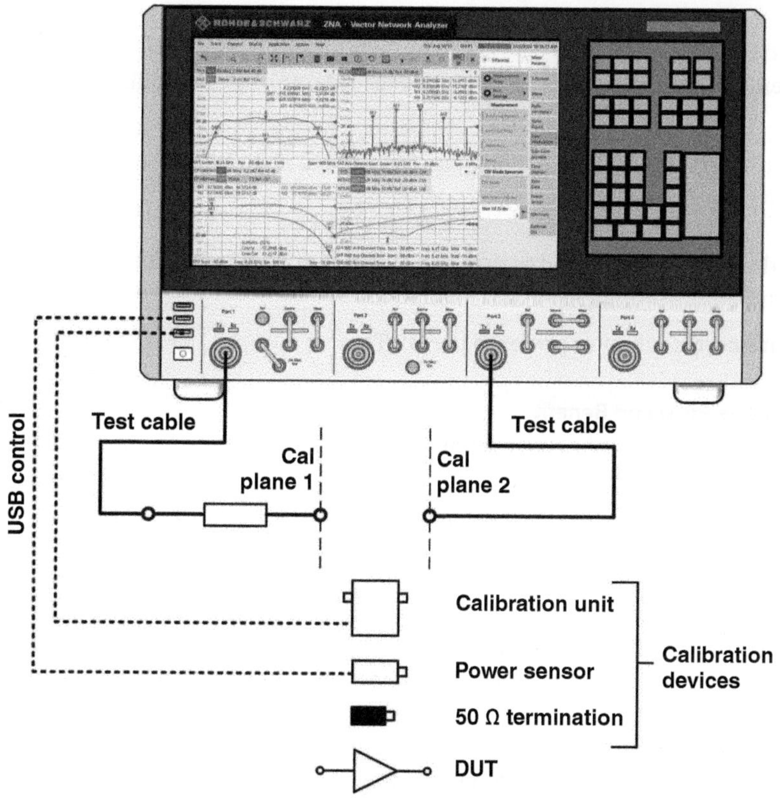

Figure F.7 Schematic setup for a VNA-based NF test and required components (because of installed HW add-ons and to optimize the signal paths for high or low power levels, the front panel jumpers are arranged as required).

General Presupposed Features

NF software personality: The VNA firmware (FW) supports NF testing with an additional dedicated personality. It enables the NF calculation algorithms, adds menus for the entry of essential test parameters, and helps to find the optimum settings derived from the DUT characteristics. The HW configuration is supported by a graphical user interface (details are described further down).

Path access functionality: Depending on the characteristics of the DUT (e.g., gain, NF), the VNA can be optimized by fitting additional HW components for Port 1, Port 2, and partially Port 3. The functionality and requirements of these add-ons are discussed in the following. However, most of them are combined with rerouting of the signal paths. Thus, the access to the source, measurement, and reference paths by means of the front panel jumpers is prerequisite ("direct channel access" functionality; see Figure F.18).

F.3.2.1 Details: HW Add-ons for Port 1 (Typically Source Port) and Benefit

The following options increase the sensitivity of the measurement and reference receivers of Port 1, and ensure spur-free and accurate stimulation signals:

1. **Direct Source Monitor Path Access (Adds Selectable Reference Signal Access)**
   Function: Enables the reference signal access ahead of the source attenuator, to ensure an adequate reference power level and thus an accurate reference signal with low trace noise (Figure F.9).

   Benefit: Increase of the reference receiver sensitivity, for accurate results despite of very low input power.
2. **Measurement Receiver Step Attenuator**
   Function: With this attenuator installed, an otherwise fitted 10 dB attenuator in the receive path is omitted ex-factory (instead, the step attenuator is set to 10 dB after PRESET). Setting the step attenuator to 0 dB increases the receiver sensitivity by 10 dB (Figure F.8).

   Benefit: Increase of the measurement receiver sensitivity.
3. **Measurement Receiver (Isolation) Amplifier**
   Function: Adds 30 dB gain in the measurement path and provides isolation versus internal LO crosstalk and spurs (Figure F.8).

   Benefit: Increase of the measurement receiver sensitivity by 30 dB, and spur-free source signal despite of very low power levels for high-gain amplifiers and receivers.
4. **Source Step Attenuator**
   Function: Adding mechanical attenuation to the source signal, and acting as a thermal noise source (set typically to 40 dB with NF test configuration, including temperature compensation) (Figure F.8).

   Benefit: Offering a precisely known temperature-corrected noise power to the DUT.

F.3.2.2 Details: HW Add-ons for Port 2 (Typically Receive Port) and Benefit

During the NF calibration and the NF measurement, the measurement receiver of the receiving port is used to measure the b-wave at the receiving port to determine the following parameters:

- Noise power when the receiver is terminated with a 50 Ω load (receiver calibration)
- Noise power of the DUT measurement
- Transmitted power during the gain measurement of the DUT

The b-wave during the gain measurement of the DUT typically measures a high-level signal at the output of the DUT. On the other hand, the measurement of the noise power at the output of the DUT is a more challenging measurement, since it requires a very sensitive receiver. Thus, most settings on Port 2 are selected with the goal of improving the sensitivity of the receiver—thus reducing the VNA receiver NF—during the output noise power measurement.

The following options increase the sensitivity of the measurement and reference receivers of Port 2 [1–4]:

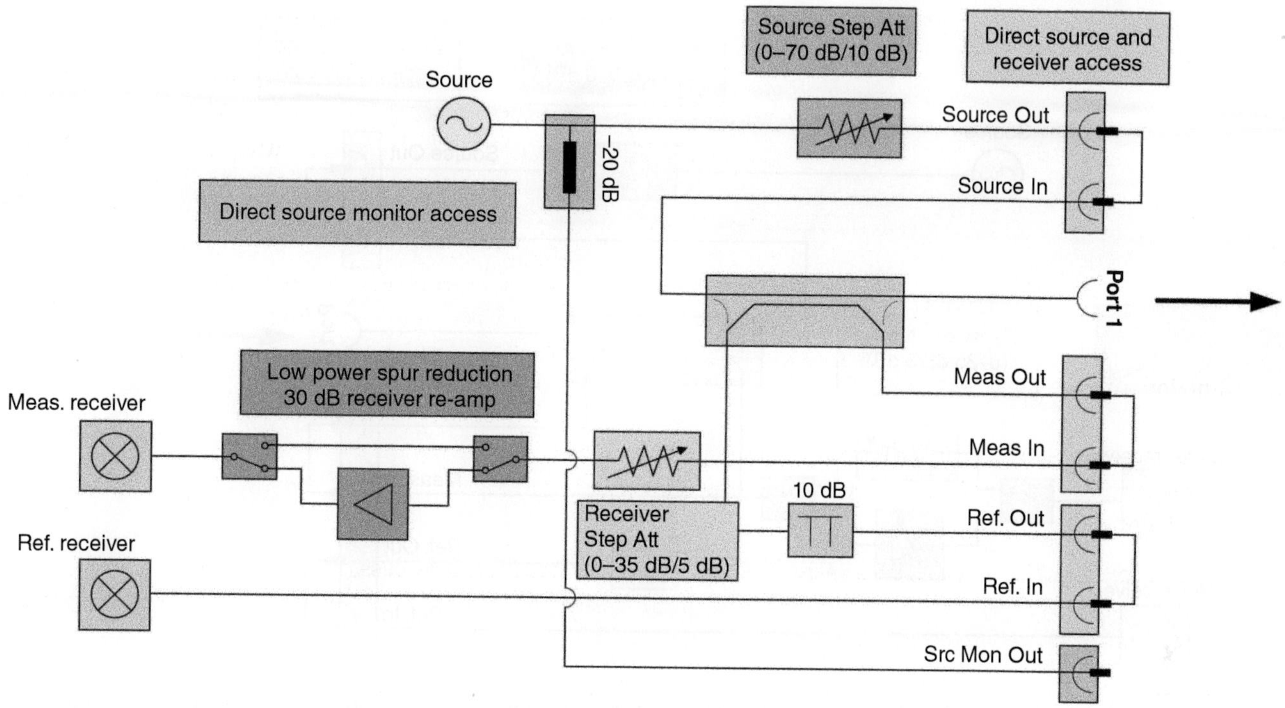

Figure F.8 Simplified depiction of Port 1 HW add-ons for optimized NF test capability: direct path access loops, selectable reference-signal access (details see Figure F.9), source and measurement receiver step attenuators, measurement receiver amplifier for both, isolation (signal purity), and increased sensitivity.

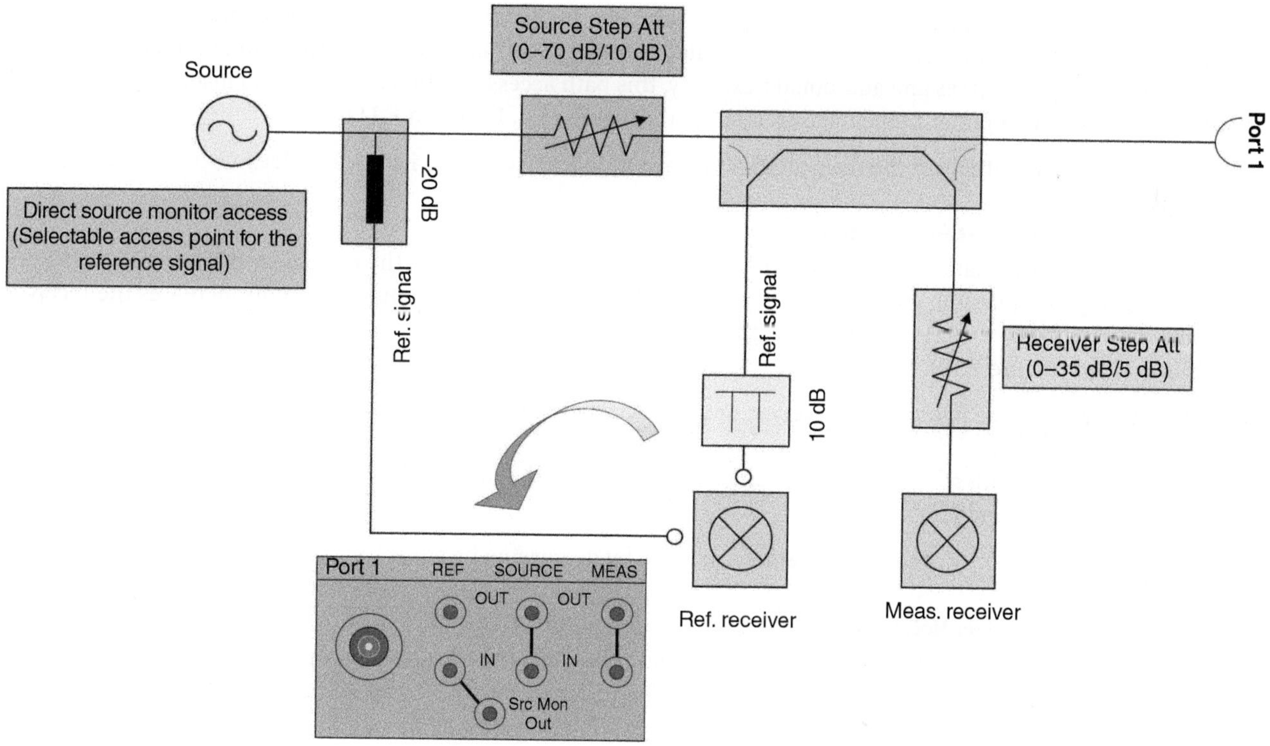

Figure F.9 Schematic depiction of the "Source Monitor Connection" functionality at Port 1 (cutout of Figure F.8): with the REF front panel jumper turned to the "Src Mon Out" connector, the reference signal from the source is not picked up by the coupler, but ahead of the source step attenuator. Hence, the reference signal is not lowered by the source step attenuator, and is high enough for a low trace noise reference signal.

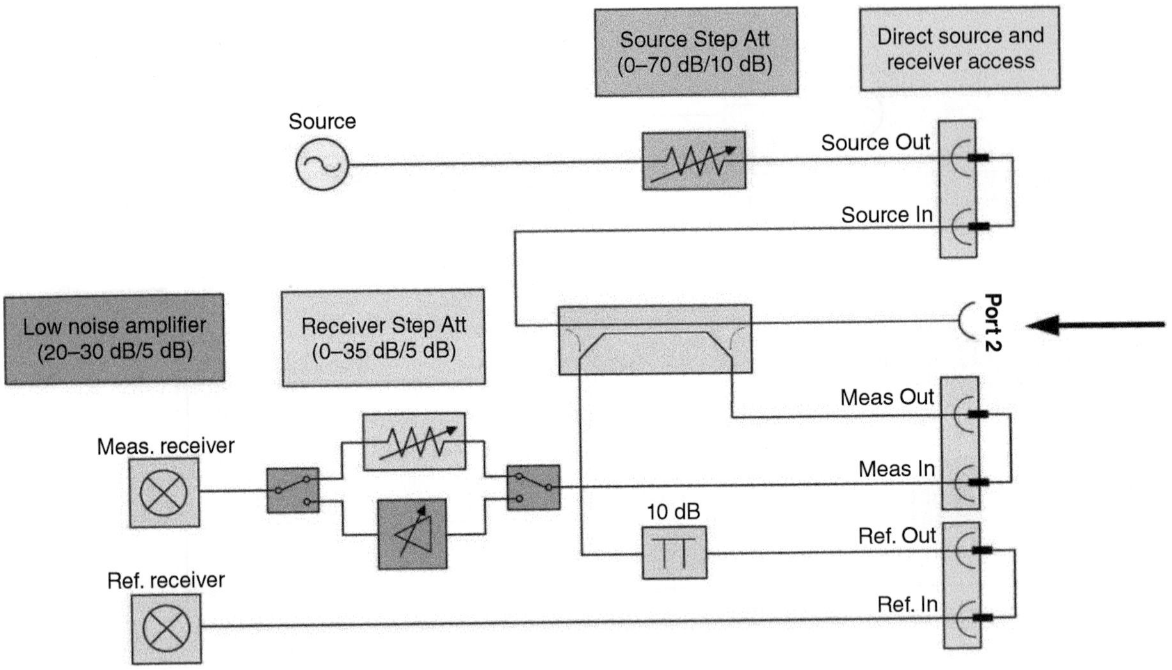

Figure F.10 Simplified depiction of Port 2 HW add-ons for optimized NF test capability: Direct path access loops, measurement receiver step attenuator combined with the low-noise preamplifier (the source step attenuator is useful for amplifier testing).

1. **Direct Source and Receiver Access (Used for Reversed Coupler Operation)**
   Function: By changing the position of the front panel jumpers, the paths of the measurement and output signals are swapped when passing the coupler. The b-wave is routed through the straight section of the coupler, thus avoiding the coupling loss. For special cases and additional flexibility, this path access functionality allows to insert an (additional) external low noise preamplifier in the measurement path (Figures F.8, F.10, and F.11).

   Benefit: Increase of sensitivity of the measurement receiver, hence decreasing its NF. Provides flexibility for external test setups.
2. **Measurement Receiver Step Attenuator**
   Function: With this attenuator installed, an otherwise fitted 10 dB attenuator in the receive path is omitted ex-factory (Instead, the step attenuator is set to 10 dB after PRESET). Setting the step attenuator to 0 dB increases the receiver sensitivity by 10 dB. On the other hand, with up to 35 dB of mechanical attenuation, compression with high-power signals is avoided (Figure F.10).

   Benefit: Improve the noise figure of the receiver, or avoid receiver compression.
3. **Low Noise Preamplifier**
   Function: Adds up to 30 dB gain to the measurement receiver, with the result of improving the receiver noise figure by 10–20 dB (Figures F.10 and F.11).

   Benefit: Increased sensitivity of the measurement receiver by decreasing its NF.

F.3.2.3 Resulting Port 2 Receiver Noise Figure

When utilizing the described add-ons and functionalities, the receivers of the utilized network analyzer can achieve the receiver NF as depicted in the following evaluation result. The minimal receiver NF is achieved with the Port 2 coupler in reversed operation, the receiver attenuator set to 0 dB, and 30 dB gain added by the measurement receiver preamplifier (bottom trace [5]) (Figure F.12).

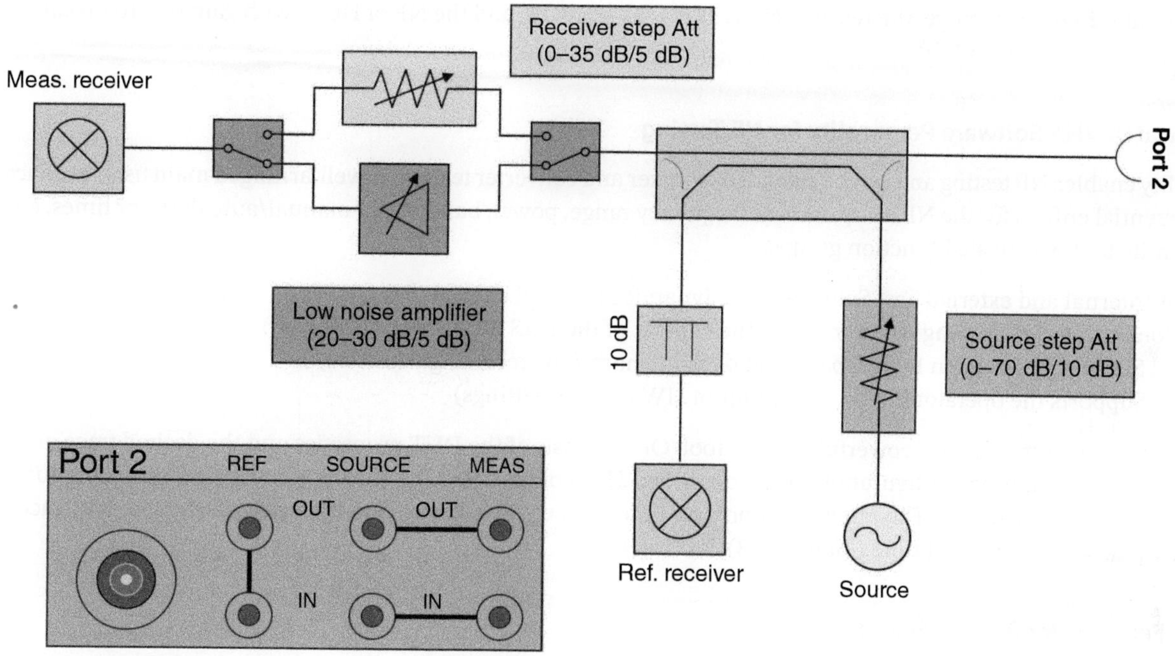

Figure F.11 Schematic depiction of the signal flow using the "Reversed Coupler" operation: with the SOURCE and MEAS front panel jumpers in horizontal position, the paths of the measurement and output signal are swapped. The received signal passes the coupler straight ahead, without being lowered by the coupling loss (at the cost of the maximum output power, which is reduced by the coupling loss).

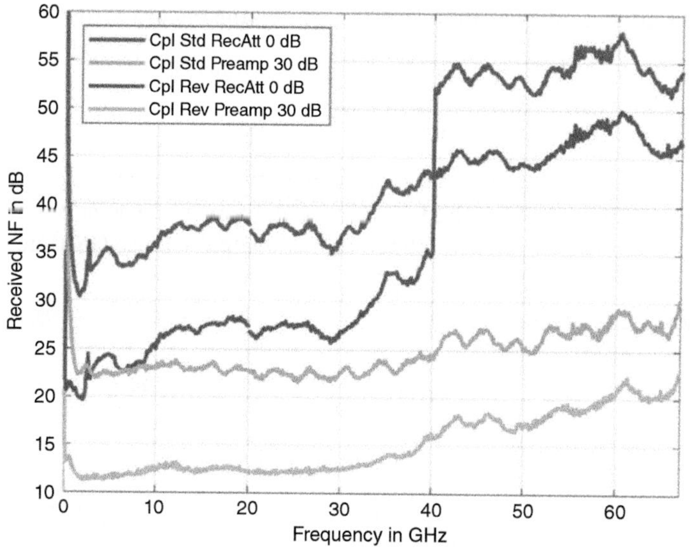

Figure F.12 Effect of the HW add-ons to the receiver NF of the utilized VNA series (up to 67 GHz) [5–8].
[5]: *The setup with the highest receiver NF is the standard coupler configuration with receiver step attenuator set to 0 dB.*
[6]: *Reversing the coupler will decrease the NF by about 10 dB.*
[7]: *Further decrease of the receiver NF can be achieved over a broad frequency range by using the standard coupler configuration, but activating the LNA.*
[8]: *The minimal receiver NF result can be reached with the reversed coupler configuration in combination with the receiver LNA at 30 dB gain.*

In addition to the optimization by means of HW add-ons, detector times and averaging can be increased as required. This way, in the medium frequency range, the receiver NF is as low as $\gtrsim 10$ dB, and the NF of DUTs with gain of ≤10 dB can be evaluated even down to an NF <1 dB.

F.3.3 Dedicated VNA Software Personality for NF Testing

This personality enables NF testing and can be extended to mixer and converter testing. A well-arranged main user interface accepts the essential entries for the NF setup, such as frequency range, power, bandwidth, manual/auto detector times, etc. Submenus ramify to the dedicated function groups:

- The VNA internal and external Test Setup setting (dynamic graphical display of the HW add-ons)
- Calibration settings (NF testing is calibrated in the context of the R&S SMARTerCal concept)
- Advanced Settings (Correction for sideband and parasitic harmonic receiving windows)
- Quickset (Supports the operator to find the optimum HW and SW settings)

Especially the “Quickset” GUI is a powerful support tool: On the basis of the DUT properties and the desired trace noise of the result trace, the required configuration for all described HW components is indicated (Isolation and preamplifier, attenuators, reversed coupler, etc.). The test and calibration time can be optimized to find the tradeoff between low uncertainty and acceptable measurement time (Figure F.13).

F.3.4 VNA Features for Pulsed Measurements

Correspondingly equipped, the utilized VNA is a one-box solution for measurements under pulsed conditions:

For pulse generation, RF pulse modulators are available per port. Control occurs by four of eight available pulse generators, which can be configured individually. Pulse parameters, such as pulse length, pulse repetition rate, time delay offsets, double pulses, or arbitrary sequences of pulses, can be defined as required. A comprehensive trigger architecture supports output of the trigger and timing signals, and vice versa, accepts external trigger signals. In addition to the standard trigger function, the “Trigger board” interface allows for enhanced synchronization, e.g., for sophisticated test execution timing for OnWafer chip testing (pulsed RF combined with pulsed supply voltage). Measurements can be carried out in various operating modes: Point-in-Pulse sweep, via (multiple) pulse profile analysis time sweeps, and by means of narrowband pulse-average sampling (Figure F.14).

F.3.5 Operation Principle: VNA Noise Figure Testing Under Pulsed Conditions

A completed NF measurement setup can be combined with the pulsed source and receive operation mode: Thereby, the trigger and pulse sequencing persist during all measurements. With the source attenuator permanently set, the noise power of the DUT is captured during the pulse OFF state, whereas the required gain sampling occurs during the pulse ON time.

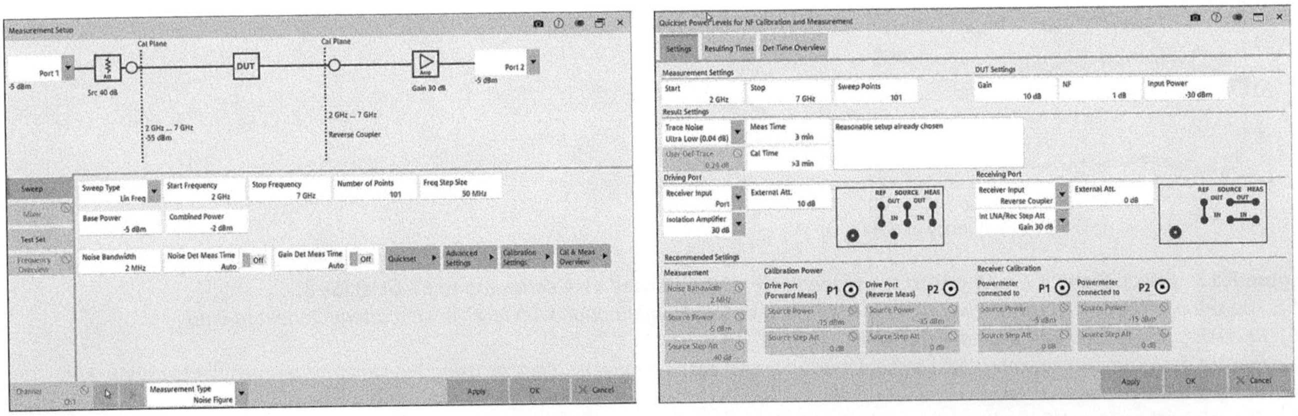

Figure F.13 Main GUI of the NF personality (left), and “Quickset” dialog (right). Starting with the DUT characteristics, the latter guides the operator to the optimum HW settings, including a dynamic indication of the selected front panel jumper setting.

However, the measurement of the noise power requires extended detector acquisition times per point, typically in the range of milliseconds (ms). Hence, the measurement time per point conflicts with the typical pulse length, often being in the range of microseconds (μs). An unsophisticated measurement of one data point during each pulse-ON/OFF state is not possible. The utilized VNA overcomes this challenge by splitting the overall sampling time: The total detector time is automatically distributed to multiple pulses, and capturing samples for the gain and the DUT noise power occurs just during pulse-ON and -OFF state resp.

All four detector times (calibration and measurement time separate for both, NF and gain measurements) are calculated in real time for the actual test parameters and presented in a dedicated GUI window.

Thus, a detailed insight into the test execution is ensured: The number of pulses per point gets evident, reliable synchronization is facilitated, and time-consuming test execution steps can be identified and optimized (Figure F.15).

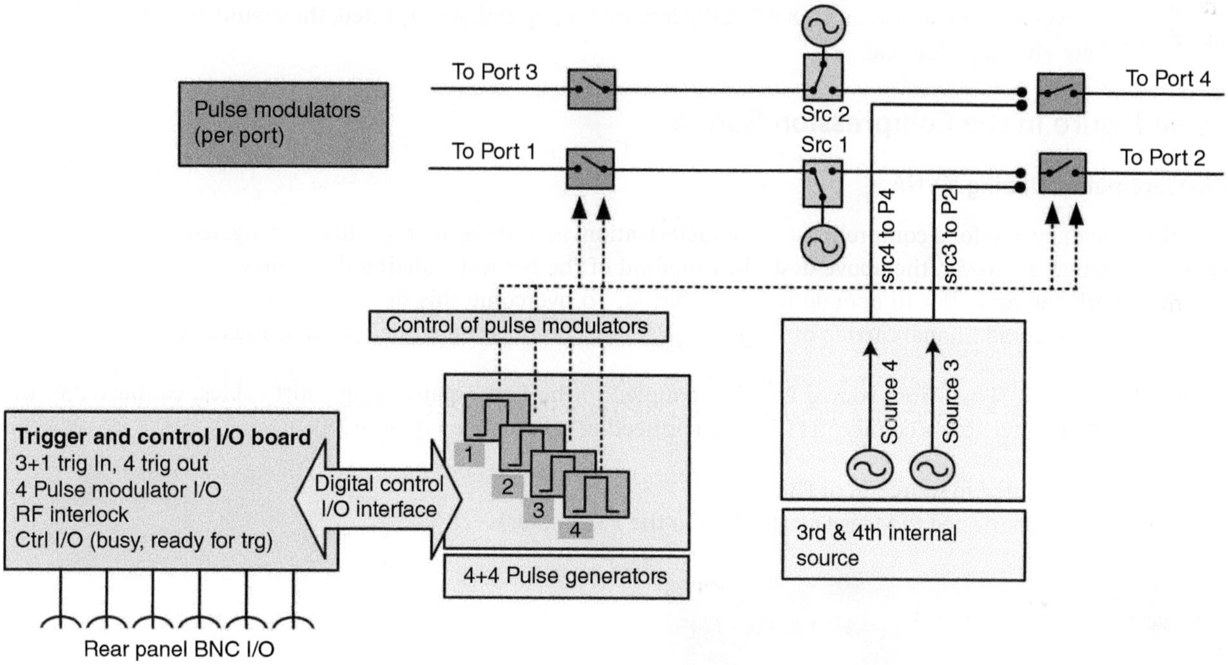

Figure F.14 Architecture for measurements under pulsed conditions: 4+4 pulse generators, one pulse modulator per port, 4 independent pulse streams using up to 4 RF sources, I/O interface for synchronization.

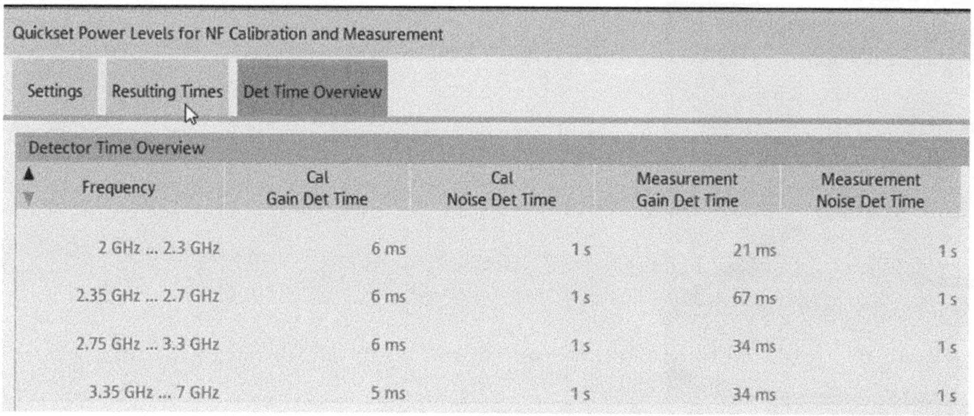

Frequency	Cal Gain Det Time	Cal Noise Det Time	Measurement Gain Det Time	Measurement Noise Det Time
2 GHz ... 2.3 GHz	6 ms	1 s	21 ms	1 s
2.35 GHz ... 2.7 GHz	6 ms	1 s	67 ms	1 s
2.75 GHz ... 3.3 GHz	6 ms	1 s	34 ms	1 s
3.35 GHz ... 7 GHz	5 ms	1 s	34 ms	1 s

Figure F.15 The dynamic indication of the detector times allows for a clear insight into the test time, pulse timing.

F.4 Noise Figure Testing Under Pulsed Conditions: Exemplary Amplifier Measurement and Test Results

As a concrete example in this outline, results from a pulsed RF noise figure test are presented.

As DUT, a commercial amplifier was used with the following key parameters:

Frequency range 2–7 GHz
Gain 10–16 dB
NF ~1 dB.

The amplifier has a convenient power supply over USB (Figures F.16 and F.17).

The pulse parameters are exemplary, set to 4 μs pulse length and 20 μs pulse period.

Noise figure and gain are measured in both swept CW mode and pulsed mode.

The measurement reveals an NF as low as <0.8 dB at the low frequency end. As expected, the results in swept CW mode and with pulsed RF are virtually identical.

F.5 Noise Figure in the Compression Range

F.5.1 Measurement Utilizing a VNA

To optimize the efficiency and for a comprehensive characterization, amplifiers are typically investigated close to or in the gain compression range. However, the above-described method of the NF test utilizing the source attenuator as a noise resistor conflicts with the amplifier to operate in compression. To overcome this demand, a second internal source of the R&S®ZNA is used as an additional carrier, driving the LNA into compression while the NF is measured.

VNA Setup: The driving signal from source Port 3 is coupled to the Port 1 path using short cables, connected to the corresponding direct channel access connectors (For the connection scheme see Figure F.18):

- source Out of P1 is routed to source In of P3
- At P3, source Out is connected to Meas Out (leading to the P3 coupler)

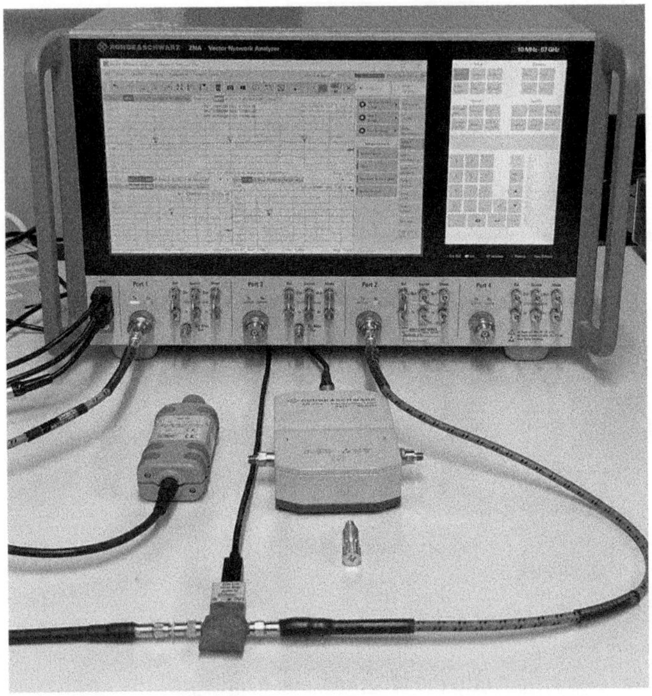

Figure F.16 Test setup for the pulse NF test (VNA, DUT/LNA, calibration unit, power test head, termination).

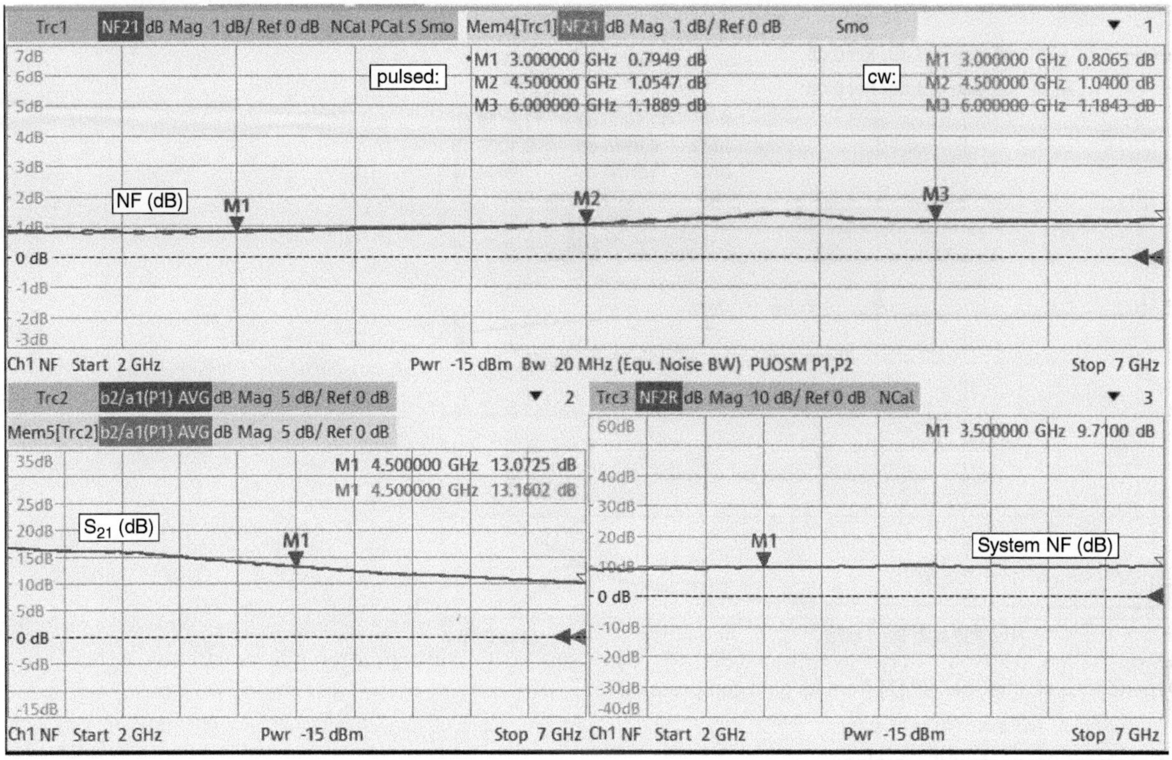

Figure F.17 Measurement result traces from the LNA. Comparison of NF and gain (1) under CW and pulsed conditions: *CW frequency sweep and pulsed RF sweep match very well; differences are only visible in the values read by the markers. The lower right trace shows the receiver noise figure.*

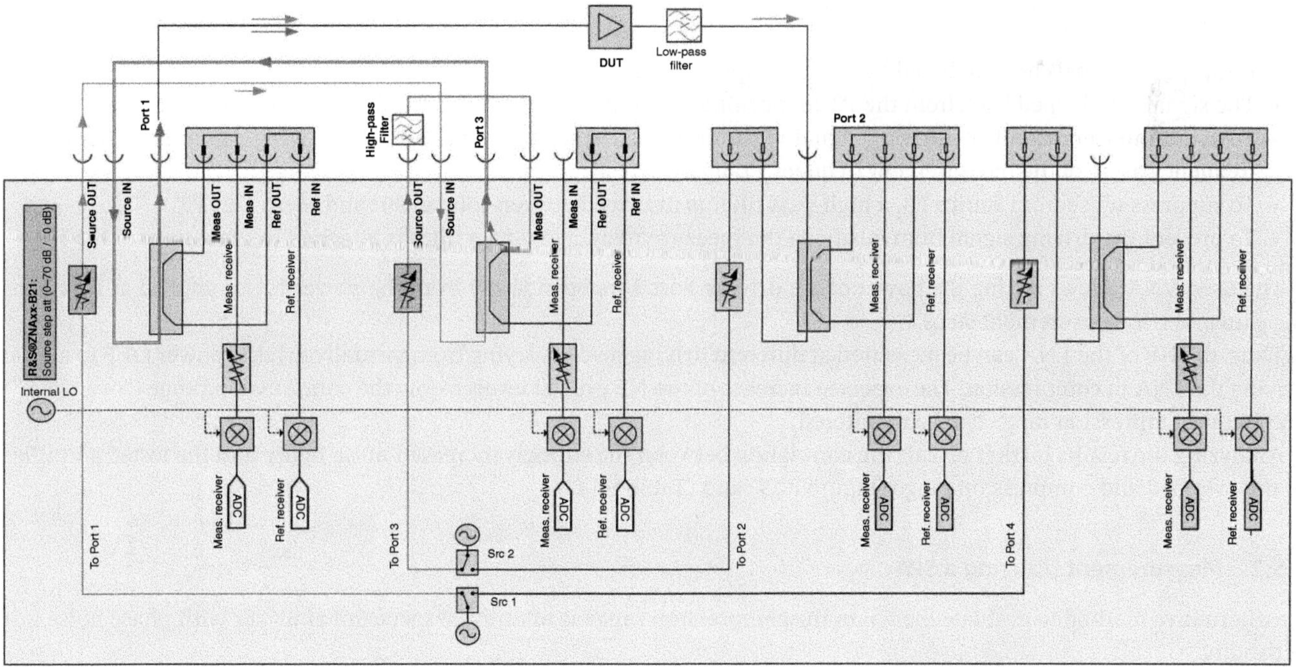

Figure F.18 Signal routing for combining the sources from Port 1 and Port 3, utilizing the Port 3 coupler.

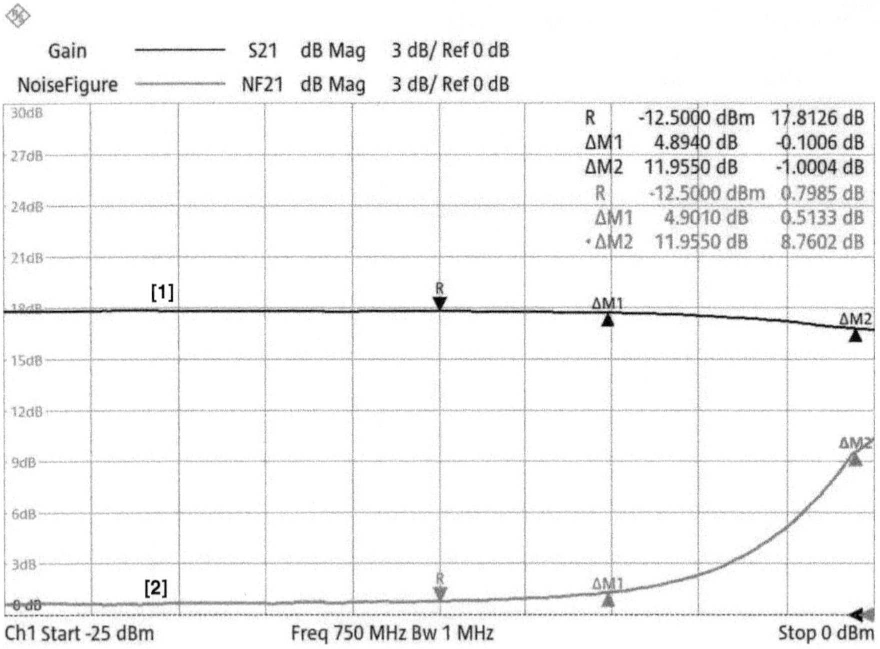

Figure F.19 Dependence of amplifier NF [7] from its gain compression [6].

Table F.1 Measurement Parameters (Exemplary):

NF measurement frequency	750 MHz (CW)
Driving signal frequency	800 MHz
Driving signal level	−25 to 0 dBm

- Hence, both signals are combined by the P3 coupler
- The signals are looped back from the P3 connector to P1 source In, making both signals available at P1
- To avoid interference of the driving signal with the measurement signal range, the driving signal is set to a fixed frequency aside of the measurement frequency range.
- To suppress noise from source P3, a high-pass filter is inserted between source Out and Meas Out P3
- To prevent the driving signal from falsifying the measurement, a low-pass filter is attached to the output of the LNA

The used VNA allows setting the power of the driving Port 3 independently from the power levels applied at Port 1 for the gain and NF measurement steps.

Thus, the NF of the LNA can be measured at different driving levels, varying from virtually driving power OFF to a level driving the LNA in compression. The expected increase of the NF even at levels below the compression range ("overdrive") and in the compression range can be monitored.

Analyzing the results further reveals the correlation between the strongly increased noise figure and the tested amplifier in the overload and compression range (Figure F.19) and (Table F.1).

F.5.2 Measurement Utilizing a SPA

An alternative method to evaluate the NF in the compression range is to utilize a spectrum analyzer with phase noise test capability.

Basic idea behind these methods is that the phase noise at an offset larger than 1 MHz (far away from flicker noise) is dominated by thermal noise. In combination with a high-performance signal source, it is possible to stimulate the amplifier with different input power levels and measure wideband phase noise.

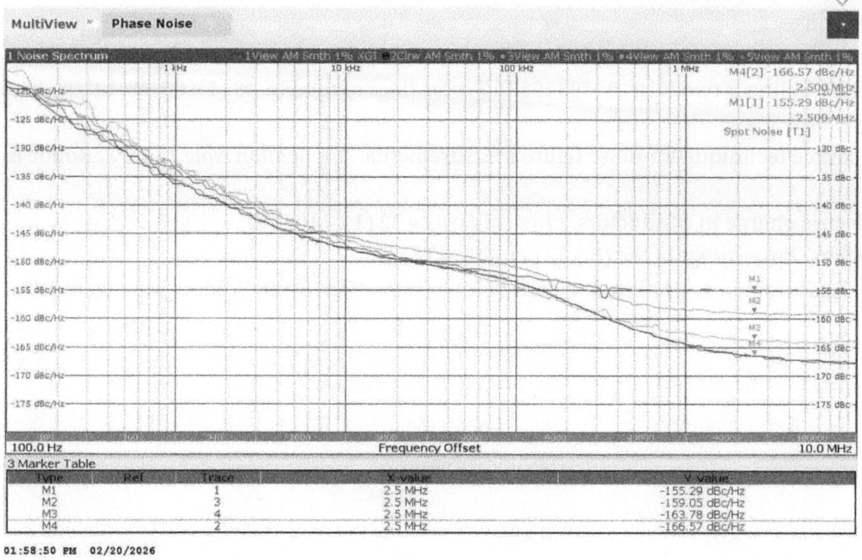

Figure F.20 Additive phase noise of a device with varying drive level.

Calculation of the amplifier's NF is given by the following equation:

$$\mathbf{NF} = \boldsymbol{L(f)} - \boldsymbol{N}_{\mathbf{th}} + \boldsymbol{P}_{\mathbf{in}}$$

$L(f)$: phase noise measured at wanted frequency offset (dBm/Hz)
N_{th}: thermal phase noise (−177 dBm/Hz SSB @ 27 °C)
P_{in}: signal level applied to DUT (dBm)

In the present case, the R&S®FSWP signal source analyzer was used. Since an NF indicates the wideband noise generated by a device, a wideband phase noise measurement can be used to calculate the NF. Plus, the integrated high-performance signal source for additive phase noise measurements establishes operation conditions close to the final environment.

The result screen Figure F.20 presents results from wide-band phase noise measurements, taken at different drive levels. The results shown in Figure F.20 are summarized in the following table.

DUT drive level [dBm]	Phase noise [dBc/Hz]	NF [dB]
−6	−166,57	4,43
−11	−163,78	2,22
−16	−159,05	1,95
−21	−155,29	0,71

Another application example and further details can be found in [9].

References

1 Hartnagel, H., Quay, R., Rohde, U.L., and Rudolph, M. (eds.) (2023). *Fundamentals of RF and Microwave Techniques and Technologies*. Springer.

2 Vendelin, G., Pavio, A.M., Rohde, U.L. et al. (2021). *Microwave Circuit Design Using Linear and Nonlinear Techniques*, 3e, p. 199. John Wiley & Sons.

3 Rohde, U.L., Poddar, A.K., Boeck, G. (2005). *The Design of Modern Microwave Oscillators for Wireless Applications: Theory and Optimization*. New York, NY: John Wiley & Sons.
4 Rohde, U.L., Rudolph, M. (2013). *RF / Microwave Circuit Design for Wireless Applications*, 2e. New York, NY: John Wiley & Sons.
5 IRE standards on methods of measuring noise in linear twoports, 1959 (1960). *Proceedings of the IRE* 48 (1): 60–68. https://ieeexplore.ieee.org/document/4065915.
6 Leffel, M. and Stumpf, C. (2021). The cold source technique for noise figure measurements. *Application Note 1SL378*. Rohde & Schwarz.
7 Nyquist, H. (1928). Thermal agitation of electric charge in conductors. *Physical Review* 32 (1): 110–113.
8 Friis, H. (1944). Noise figures of radio receivers. *Proceedings of the IRE* 32 (7): 419–422.
9 Gleißner, A. (2025). Dynamic-noise-figure-measurements. *Application Card 5215.4998.92*, Ver. 02.00, Product Management Network Analyzers, Rohde & Schwarz.

Index

Note: Page numbers in *italics* and **bold** refers to figures and tables respectively.

m

u

v

w

y